AutoCAD and Its Applications

COMPREHENSIVE

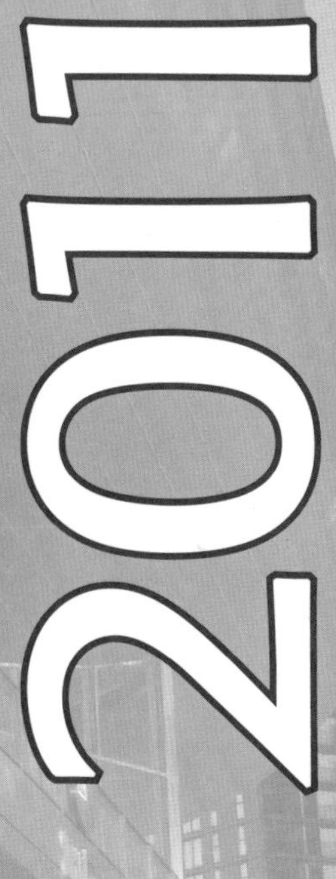

by

Terence M. Shumaker
Faculty Emeritus
Former Chairperson
Drafting Technology
Autodesk Premier Training Center
Clackamas Community College, Oregon City, Oregon

David A. Madsen
President, Madsen Designs Inc.
Faculty Emeritus, Former Department Chairperson Drafting Technology
Autodesk Premier Training Center
Clackamas Community College, Oregon City, Oregon
Director Emeritus, American Design Drafting Association

David P. Madsen
Vice President, Madsen Designs Inc.
Computer-Aided Design and Drafting Consultant and Educator
Autodesk Developer Network Member
American Design Drafting Association Member

Jeffrey A. Laurich
Instructor, Mechanical Design Technology
Fox Valley Technical College
Appleton, WI

Publisher
The Goodheart-Willcox Company, Inc.
Tinley Park, Illinois
www.g-w.com

Manufactured in the United States of America.

Library of Congress Catalog Card Number 2010009068

ISBN 978-1-60525-329-9

1 2 3 4 5 6 7 8 9 – 11 – 15 14 13 12 11 10

The Goodheart-Willcox Company, Inc. Brand Disclaimer: Brand names, company names, and illustrations for products and services included in this text are provided for educational purposes only and do not represent or imply endorsement or recommendation by the author or the publisher.

The Goodheart-Willcox Company, Inc., Safety Notice: The reader is expressly advised to carefully read, understand, and apply all safety precautions and warnings described in this book or that might also be indicated in undertaking the activities and exercises described herein to minimize risk of personal injury or injury to others. Common sense and good judgment should also be exercised and applied to help avoid all potential hazards. The reader should always refer to the appropriate manufacturer's technical information, directions, and recommendations; then proceed with care to follow specific equipment operating instructions. The reader should understand these notices and cautions are not exhaustive.

The publisher makes no warranty or representation whatsoever, either expressed or implied, including, but not limited to, equipment, procedures, and applications described or referred to herein, their quality, performance, merchantability, or fitness for a particular purpose. The publisher assumes no responsibility for any changes, errors, or omissions in this book. The publisher specifically disclaims any liability whatsoever, including any direct, indirect, incidental, consequential, special, or exemplary damages resulting, in whole or in part, from the reader's use or reliance upon the information, instructions, procedures, warnings, cautions, applications or other matter contained in this book. The publisher assumes no responsibility for the activities of the reader.

Library of Congress Cataloging-in-Publication Data

Shumaker, Terence M.
AutoCAD and its applications: advanced, 2011/ by Terence M. Shumaker, David A. Madsen and Jeffrey A. Laurich—18th edition
p. cm.

Includes index
ISBN 978-1-60525-329-9
1. Computer graphics. 2. AutoCAD. I. Madsen, David A. II. Laurich, Jeffrey A. III. Title.

T385.S46123 2011
620'.00420285536—dc22 2010009068

AutoCAD and Its Applications—Advanced provides complete instruction in mastering three-dimensional design and modeling using AutoCAD. This text also provides complete instruction in customizing AutoCAD and introduces programming AutoCAD. These topics are covered in an easy-to-understand sequence and progress in a way that allows you to become comfortable with the commands as your knowledge builds from one chapter to the next. In addition, *AutoCAD and Its Applications—Advanced* offers:

- Examples and discussions of industrial practices and standards.
- Professional tips explaining how to effectively and efficiently use AutoCAD.
- Exercises to reinforce the chapter topics. These exercises should be completed where indicated in the text as they build on previously learned material.
- Chapter tests for review of commands and key AutoCAD concepts.
- A large selection of modeling and customizing problems supplement each chapter. Problems are presented as 3D illustrations, actual plotted drawings, and engineering sketches.

Fonts Used in This Text

Different typefaces are used throughout each chapter to define terms and identify AutoCAD commands. Important terms appear in ***bold-italic face, serif*** type. AutoCAD menus, commands, variables, dialog box names, and toolbar button names are printed in **bold-face, sans serif** type. File names, folder names, and paths appear in the body of the text in Roman, sans serif type. Keyboard keys are shown inside of square brackets [ ] and appear in Roman, sans serif type. For example, [Enter] means to press the enter (return) key.

Other Text References

This text focuses on advanced AutoCAD applications. Basic AutoCAD applications are covered in *AutoCAD and Its Applications—Basics,* which is available from Goodheart-Willcox Publisher. *AutoCAD and Its Applications* texts are also available for previous releases of AutoCAD.

Introducing the AutoCAD Commands

There are several ways to select AutoCAD drawing and editing commands. Selecting commands from the ribbon is slightly different than entering them from the keyboard. When a command is introduced, the command-entry methods are illustrated in the margin next to the text reference.

CONE

Ribbon
Home > Modeling Cone
Type
CONE

The example in the margin next to this paragraph illustrates the various methods of initiating the **CONE** command to draw a solid cone primitive while the 3D environment is active. The 3D environment consists of a drawing file based on the acad3D.dwt template and the 3D Modeling workspace current, as described in Chapter 1. This book assumes the 3D environment is current for all procedures and discussions.

Flexibility in Design

Flexibility is the keyword when using *AutoCAD and Its Applications—Advanced.* This text is an excellent training aid for both individual and classroom instruction. It is also an invaluable resource for any professional using AutoCAD. *AutoCAD and Its Applications—Advanced* teaches you how to apply AutoCAD to common modeling and customizing tasks.

When working through the text, you will see a variety of notices. These include Professional Tips, Notes, and Cautions that help you develop your AutoCAD skills.

PROFESSIONAL TIP

These ideas and suggestions are aimed at increasing your productivity and enhancing your use of AutoCAD commands and techniques.

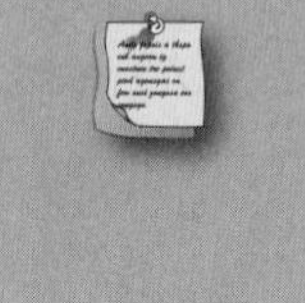

NOTE

A note alerts you to important aspects of a command, function, or activity that is being discussed. These aspects should be kept in mind while you are working through the text.

CAUTION

A caution alerts you to potential problems if instructions or commands are incorrectly used or if an action can corrupt or alter files, folders, or storage media. If you are in doubt after reading a caution, always consult your instructor or supervisor.

AutoCAD and Its Applications—Advanced provides several ways for you to evaluate your performance. Included are:

- **Exercises.** The student website contains exercises for each chapter. These exercises allow you to perform tasks that reinforce the material just presented. You can work through the exercises at your own pace. However, the exercises are intended to be completed when called out in the text.
- **Chapter test.** Each chapter includes a written test at the end of the chapter. Questions require you to give the proper definition, command, option, or response to perform a certain task. You may also be asked to explain a topic or list appropriate procedures. An electronic version of the test is available on the student website.

- **Drawing problems.** There are a variety of drawing, design, and customizing problems at the end of each chapter. These are presented as real-world CAD drawings, 3D illustrations, and engineering sketches. The problems are designed to make you think, solve problems, use design techniques, research and use proper drawing standards, and correct errors in the drawings or engineering sketches. Graphics are used to represent the discipline to which a drawing problem applies.

These problems address mechanical drafting and design applications, such as manufactured part designs.

These problems address architectural and structural drafting and design applications, such as floor plans, furniture, and presentation drawings.

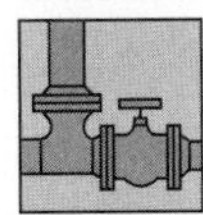

These problems address piping drafting and design applications, such as tank drawings and pipe layout.

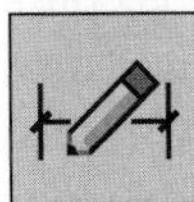

These problems address a variety of general drafting, design, and customization applications. These problems should be attempted by everyone learning advanced AutoCAD techniques for the first time.

NOTE

Some problems presented in this text are given as engineering sketches. These sketches are intended to represent the kind of material from which a drafter is expected to work in a real-world situation. As such, engineering sketches often contain errors or slight inaccuracies and are most often not drawn according to proper drafting conventions and applicable standards. Additionally, other drawings may contain errors or inaccuracies. Errors in these problems are *intentional* to encourage you to apply appropriate techniques and standards in order to solve the problem. As in real-world applications, sketches should be considered preliminary layouts. Always question inaccuracies in sketches and designs and consult the applicable standards or other resources.

Student Website

The student website is located at www.g-wlearning.com/CAD. On this home page, select the entry for *AutoCAD and Its Applications—Advanced* to access the material for this book. The student website contains the exercises and chapter test for each chapter. Additionally, the programming chapters, Chapters 27 through 29, are provided on the student website, not in the printed book. The appendix material is also presented on the student website. The icon shown in the margin here appears throughout the text to indicate a reference to the student website.

As you work through each chapter, exercises on the student website are referenced. The exercises are intended to be completed as the references are encountered in the text. The NURBS modeling tutorial in Appendix A should be completed after Chapter 3. The solid modeling tutorial in Appendix B should be completed after Chapter 14. The remaining appendix material is intended as reference material.

Also included on the student website is the student software supplement. This contains a variety of student activities that are intended to supplement the exercises on the student website. These activities are referenced within the appropriate exercises and can be completed as additional practice. For each chapter that includes associated student software supplemental activities, the student website offers a link to the DWG files for the activities.

About the Authors

Terence M. Shumaker is Faculty Emeritus, the former Chairperson of the Drafting Technology Department and former Director of the Autodesk Premier Training Center at Clackamas Community College. Terence taught at the community college level for over 28 years. He has professional experience in surveying, civil drafting, industrial piping, and technical illustration. He is the author of Goodheart-Willcox's *Process Pipe Drafting* and coauthor of the *AutoCAD and Its Applications* series (Releases 10 through 2010 editions) and *AutoCAD Essentials.*

David A. Madsen is the president of Madsen Designs Inc. (www.madsendesigns.com). David is Faculty Emeritus and the former Chairperson of Drafting Technology and the Autodesk Premier Training Center at Clackamas Community College in Oregon City, Oregon. David was an instructor and a department chairperson at Clackamas Community College for nearly 30 years. In addition to teaching at the community college level, David was a Drafting Technology instructor at Centennial High School in Gresham, Oregon. David is a former member of the American Design Drafting Association (ADDA) Board of Directors. He was honored with Director Emeritus status by the ADDA in 2005. David has extensive experience in mechanical drafting, architectural design and drafting, and building construction. He holds a Master of Education degree in Vocational Administration and a Bachelor of Science degree in Industrial Education. David is coauthor of the *AutoCAD and Its Applications* series (Releases 10 through 2010 editions), *Architectural Drafting Using AutoCAD, Geometric Dimensioning and Tolerancing*, and other drafting and design textbooks.

Jeffrey A. Laurich has been an instructor in Mechanical Design Technology at Fox Valley Technical College in Appleton, WI, since 1991. He has also taught business and industry professionals in the Autodesk Premier Training Center at FVTC. Jeff teaches drafting, AutoCAD, Autodesk MAP, 3ds max, and other programs. He created a certificate program at FVTC entitled Computer Rendering and Animation that integrates the 3D capabilities of AutoCAD and 3ds max. Jeff has professional experience in furniture design, surveying, and cartography and holds a degree in Natural Resources Technology. In his consulting business, Jeff uses 3ds max to create renderings and animations for manufacturers and architects. He was a contributing author on *AutoCAD and Its Applications—Advanced* for releases 2007, 2008, and 2009.

Acknowledgments

The authors and publisher would like to thank the following individuals and companies for their assistance and contributions.

Contributing Authors

The authors wish to acknowledge the following contributors for their professional expertise in providing in-depth research and testing, technical assistance, reviews, and development of new materials.

J.C. Malitzke for Chapters 3 and 11 through 14. J.C. is the department chair of Computer Integrated Technologies and professor of Mechanical Design and Drafting/CAD at Moraine Valley Community College in Palos Hills, IL. He also manages and teaches for the Authorized Autodesk Training Center at Moraine Valley. J.C. has been teaching for over 30 years and actively using and teaching Autodesk products for nearly 25 years. He is a founding member and past chair of the Autodesk Training Center Executive Committee and currently is chair of the Autodesk Advisory Board. J.C. has been the coauthor and principal investigator on two National Science Foundation grants. He has won numerous awards including: Educator of the Year by the Illinois Drafting Educators Association; the Instructor Quality Award by Autodesk; Autodesk University Instructor Award; Professor of the Year, Co-Innovator of the Year, and Co-Master Teacher awards at Moraine Valley Community College; and the Illinois Outstanding Faculty Member of the Year awarded by the Illinois Community College Trustees Association. J.C. holds a Bachelor's degree in Education and a Master's degree in Industrial Technology from Illinois State University.

Craig Black for Chapters 21, 23, 24, and 27 through 29. Craig is an instructor of Mechanical Design and the former manager of the Autodesk Premier Training Center at Fox Valley Technical College in Appleton, WI. He has been teaching at Fox Valley Technical College since 1990. Craig has served two terms on the Autodesk Training Center Executive Committee (now known as the Advisory Board) and chaired the committee in 2001. He has presented various topics at a number of Autodesk University annual training sessions and has been contracted to teach training sessions on Autodesk products across the United States. Craig not only teaches, but also does AutoCAD customization and AutoLISP and DCL programming for area businesses and industries. Prior to his current position, Craig worked in the civil, architectural, electrical, and mechanical drafting and design disciplines.

Adam Ferris for Chapters 22, 25, and 26. Adam is the Director of Digital Technology at Wilson Architects in Boston, MA. Wilson Architects is a medium-size architecture firm specializing in the science and technology field. Adam is an award-winning Autodesk Certified Instructor with over 16 years of experience on training Autodesk products in the Boston area. He has trained over 4000 students in New England. Prior to joining Wilson Architects, Adam owned an Autodesk products/CAD consulting firm aiding businesses in efficient AutoCAD configurations and management, network management, and 3D design with Autodesk 3ds max and Autodesk VIZ. Adam attends Autodesk University every year and has lectured there several times.

Contribution of Materials

Autodesk, Inc.
Bill Fane
CADENCE magazine
EPCM Services Ltd.
Fitzgerald, Hagan, & Hackathorn
Kunz Associates

ADDA

The content of this text is considered a fundamental component to the design drafting profession by ADDA International. This publication covers topics and related material, as stated in the ADDA Curriculum Certification Standards and the ADDA Professional Certification Examination Review Guides. Although this publication is not conclusive, with respect to ADDA standards, it should be considered a key reference tool in pursuit of a professional career.

Brief Contents

Three-Dimensional Design and Modeling

Model Visualization and Presentation

Customizing AutoCAD

Programming AutoCAD

www.g-wlearning.com/CAD

Expanded Contents

Three-Dimensional Design and Modeling

Chapter 9

Chapter 10

Chapter 11

Chapter 12

Chapter 13

Model Visualization and Presentation

Customizing AutoCAD

Programming AutoCAD

Student Website Contents

www.g-wlearning.com/CAD

Using the Student Website

Chapters 27 through 29

Chapter Exercises

Chapter Tests

Student Software Supplement

Appendices

Reference Materials

Related Web Links

CHAPTER 1

Introduction to Three-Dimensional Modeling

Learning Objectives

After completing this chapter, you will be able to:

✓ Describe how to locate points in 3D space.
✓ Describe the right-hand rule of 3D visualization.
✓ Explain the function of the ribbon.
✓ Display 3D objects from preset isometric viewpoints.
✓ Display 3D objects from any desired viewpoint.
✓ Set a visual style current.

Three-dimensional (3D) design and modeling is a powerful tool for use in design, visualization, testing, analysis, manufacturing, assembly, and marketing. Three-dimensional models also form the basis of computer animations, architectural walkthroughs, and virtual worlds used in the entertainment industry and for gaming platforms. Drafters who can design objects, buildings, and "worlds" in 3D are in demand for a wide variety of positions, both inside and outside of the traditional drafting and design disciplines.

The first twelve chapters of this book present a variety of techniques for drawing and designing in 3D. The skills you learn will provide you with the ability to construct any object in 3D and prepare you for entry into an exciting aspect of graphic communication.

To be effective in creating and using 3D objects, you must first have good 3D visualization skills. These skills include the ability to see an object in three dimensions and to visualize it rotating in space. Visualization skills can be obtained by using 3D techniques to construct objects and by trying to see two-dimensional sketches and drawings as 3D models. This chapter provides an introduction to several aspects of 3D drawing and visualization. Subsequent chapters expand on these aspects and provide a detailed examination of 3D drawing, editing, visualization, and display techniques.

Using Rectangular 3D Coordinates

In two-dimensional drawing, you see one plane defined by two dimensions. These dimensions are usually located on the X and Y axes and what you see is the XY plane. However, in 3D drawing, another coordinate axis—the Z axis—is added. This results in two additional planes—the XZ plane and the YZ plane. If you are looking at a standard AutoCAD screen after AutoCAD is launched using the acad.dwt template, the positive Z axis comes directly out of the screen toward you. AutoCAD can only draw lines in 3D if it knows the X, Y, and Z coordinate values of each point on the object. For 2D drawing, only two of the three coordinates (X and Y) are needed.

Compare the 2D and 3D coordinate systems shown in **Figure 1-1.** Notice that the positive values of Z in the 3D coordinate system come up from the XY plane. In a new drawing based on the acad.dwt template, consider the surface of your computer screen as the XY plane. Anything behind the screen is negative Z and anything in front of the screen is positive Z.

The object in **Figure 1-2A** is a 2D drawing showing the top view of an object. The XY coordinate values of the origin and each point are shown. Think of the object as being drawn directly on the surface of your computer screen. However, this is actually a 3D object. When displayed in a pictorial view, the Z coordinates can be seen. Notice in **Figure 1-2B** that the first two values of each coordinate match the X and Y values of the 2D view. Three-dimensional coordinates are always expressed as (X,Y,Z). The 3D object was drawn using positive Z coordinates. Therefore, the object comes out of your computer screen when it is viewed from directly above. The object can also be drawn using negative Z coordinates. In this case, the object would extend behind, or into, the screen.

Study the nature of the rectangular 3D coordinate system. Be sure you understand Z values before you begin constructing 3D objects. It is especially important that you carefully visualize and plan your design when working with 3D constructions.

Figure 1-1.
A comparison of 2D and 3D coordinate systems.

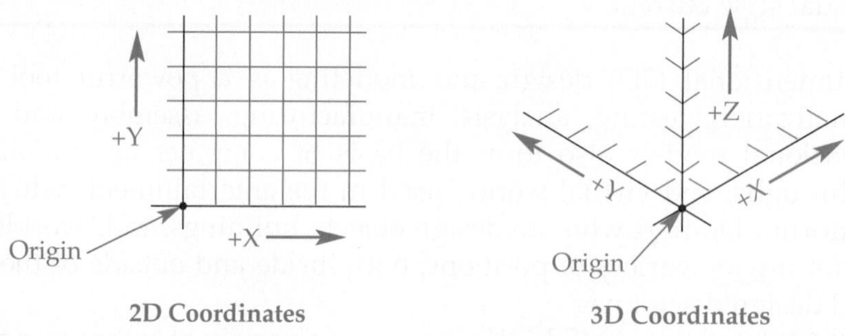

Figure 1-2.
A—The points making up a 2D object require only two coordinates. B—Each point of a 3D object must have an X, Y, and Z value. Notice that the first two coordinates (X and Y) are the same for each endpoint of a vertical line.

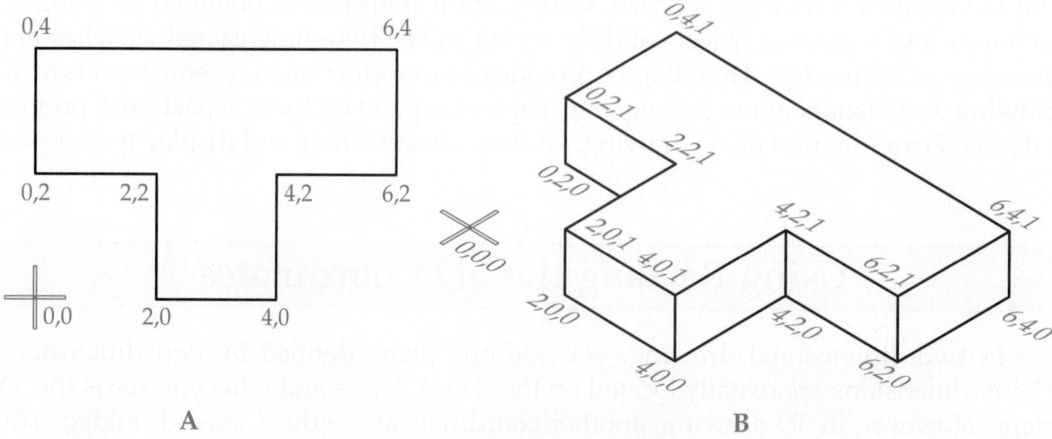

PROFESSIONAL TIP

All points in three-dimensional space can be drawn using one of three coordinate entry methods—rectangular, spherical, or cylindrical. This chapter uses the rectangular coordinate entry method. Complete discussions on the spherical and cylindrical coordinate entry methods are provided in Chapter 6.

Exercise 1-1

Complete the exercise on the student website.
www.g-wlearning.com/CAD

Right-Hand Rule of 3D Drawing

In order to effectively draw in 3D, you must be able to visualize objects in 3D space. The ***right-hand rule*** is a simple method for visualizing the 3D coordinate system. It is a representation of the positive coordinate values in the three axis directions. AutoCAD's world coordinate system (WCS) and various user coordinate systems (UCS) are based on this concept of visualization.

To use the right-hand rule, position the thumb, index finger, and middle finger of your right hand as shown in Figure 1-3. Although this may seem a bit unusual, it can do wonders for your understanding of the three axes. Imagine that your thumb is the X axis, your index finger is the Y axis, and your middle finger is the Z axis. Hold your hand in front of you so that your middle finger is pointing directly at you, as shown in Figure 1-3. This is the plan view of the XY plane. The positive X axis is pointing to the right and the positive Y axis is pointing up. The positive Z axis comes toward you and the origin of this system is the palm of your hand.

Figure 1-3.
Positioning your hand to use the right-hand rule to understand the relationship of the X, Y, and Z axes.

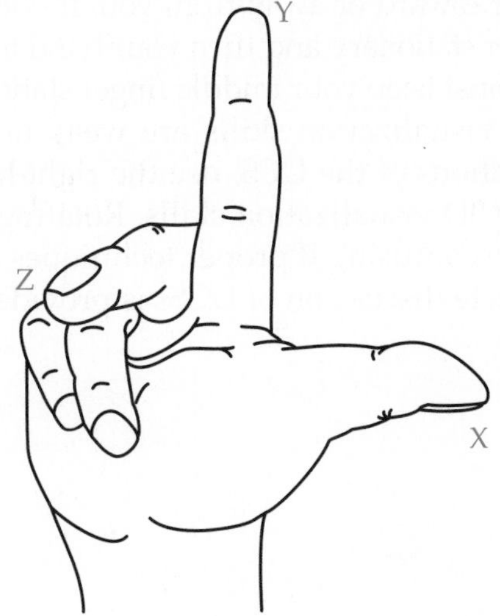

Figure 1-4.
A comparison of the UCS icon and the right-hand rule.

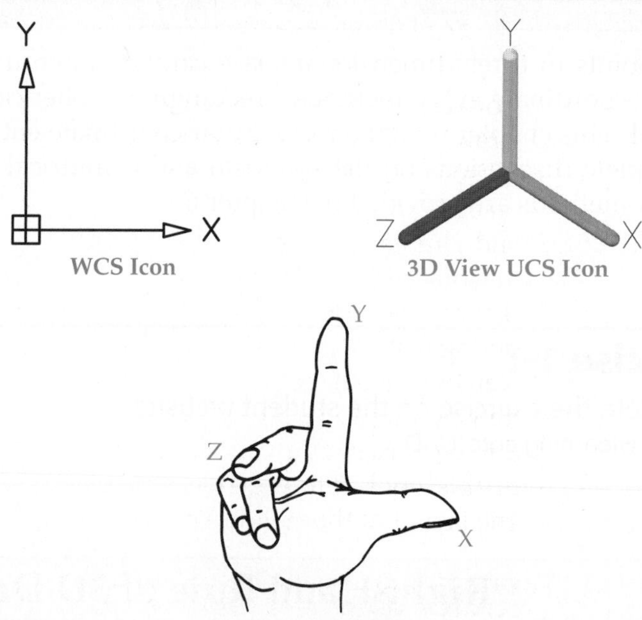

UCSICON

Ribbon
View > Coordinates
Show UCS Icon
Type
UCSICON

The concept behind the right-hand rule can be visualized even better if you are sitting at a computer and the AutoCAD graphics window is displayed. Make sure the current drawing is based on the acad.dwt template. If the UCS icon is not displayed in the lower-left corner of the screen, turn it on by using the **UCSICON** command. Now, orient your right hand as shown in Figure 1-3 and position it next to the UCS (or WCS) icon. Your index finger and thumb should point in the same directions as the Y and X axes, respectively, on the UCS icon. Your middle finger will be pointing out of the screen directly at you, which is the Z axis. See Figure 1-4. Notice the illustration on the right in the figure. This is the shaded UCS icon. It is displayed when the visual style is not 2D Wireframe. Visual styles are introduced later in this chapter.

The right-hand rule can be used to eliminate confusion when rotating the UCS. The UCS can rotate on any of the three axes, just like a wheel rotates on an axle. Therefore, if you want to visualize how to rotate about the X axis, keep your thumb stationary and turn your hand either toward or away from you. If you wish to rotate about the Y axis, keep your index finger stationary and turn your hand to the left or right. When rotating about the Z axis, you must keep your middle finger stationary and rotate your entire arm.

If your 3D visualization skills are weak or you are having trouble visualizing different orientations of the UCS, use the right-hand rule. It is a useful technique for improving your 3D visualization skills. Rotating the UCS around one or more of the axes can become confusing if proper techniques are not used to visualize the rotation angles. A complete discussion of UCSs is provided in Chapter 6.

Basic Overview of the Interface

AutoCAD provides four working environments tailored to either 2D or 3D drawing or annotating a drawing. These environments are called *workspaces* and can be quickly restored. The workspace for 2D development based on the traditional AutoCAD screen layout is called AutoCAD Classic. The 2D Drafting & Annotation workspace is designed for drawing in 2D and annotating a drawing. It is similar to a streamlined version of the AutoCAD Classic layout with the ribbon displayed in place of toolbars and pull-down menus. The workspace for basic 3D modeling and editing is called 3D Basics. The full range of 3D modeling features is found in the 3D Modeling workspace.

Workspaces can be created, customized, and saved to allow a variety of graphical user interface configurations. The 3D Modeling workspace provides quick access to the full range of tools required to construct, edit, view, and visualize 3D models. It is the workspace used throughout this book. This section provides an overview of the 3D Modeling workspace and the layout of the ribbon and its panels.

NOTE

In order to use the default 3D "environment," you must start a new drawing file based on the acad3D.dwt template and set the 3D Modeling workspace current. All discussions in the remainder of this book assume that AutoCAD is in this default 3D environment.

Workspaces

A ***workspace*** is a drawing environment in which menus, toolbars, palettes, and ribbon panels are displayed for a specific task. A workspace stores not only which of these tools are visible, but also their on-screen locations. You can quickly change workspaces using the **Workspace Switching** tool on either the status bar or **Quick Access** toolbar, **WSCURRENT** command, or **Workspaces** toolbar, as shown in **Figure 1-5.** By default, toolbars are only displayed in the AutoCAD Classic workspace. The default 3D Modeling workspace is shown in **Figure 1-6.**

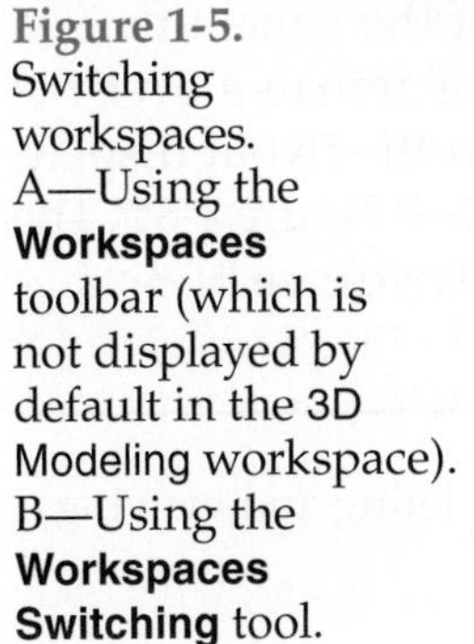

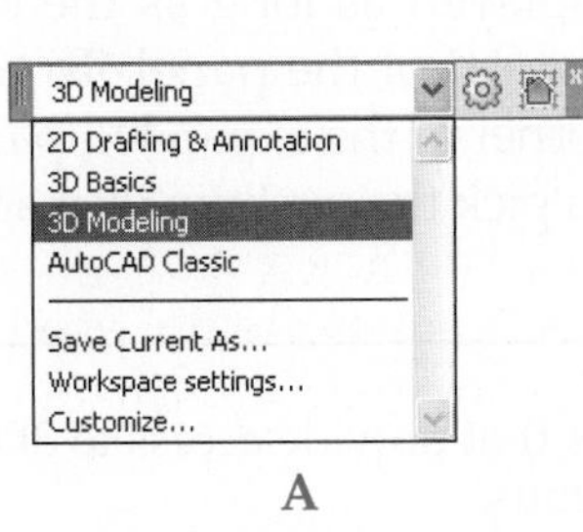

A

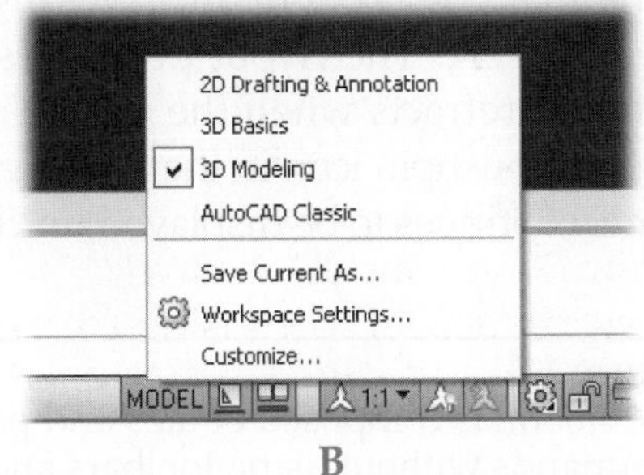

B

Figure 1-5. Switching workspaces. A—Using the **Workspaces** toolbar (which is not displayed by default in the 3D Modeling workspace). B—Using the **Workspaces Switching** tool.

Figure 1-6.
The 3D Modeling workspace with a drawing file based on the acad3D.dwt template.

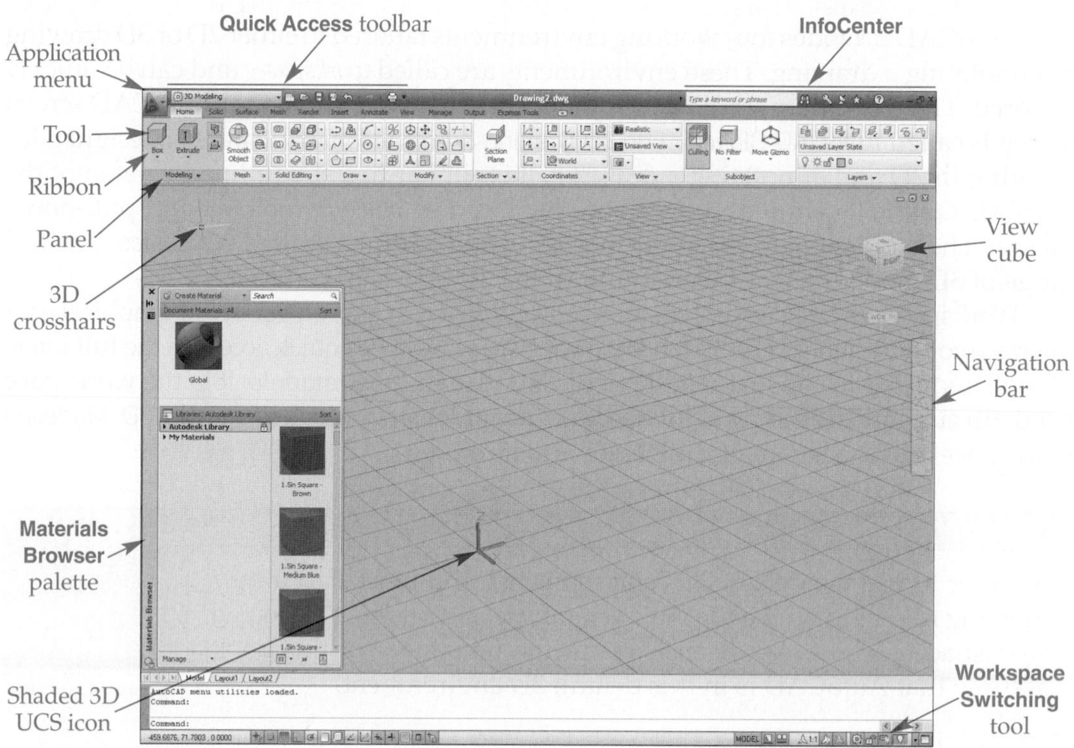

Ribbon Panels

Within the 3D Modeling workspace, the **Home** tab on the ribbon displays ten panels. The tools in these panels provide all of the functions needed to design and view your 3D model. See **Figure 1-7.** The title appears at the bottom of the panel. The button at the right-hand end of the tab names controls the display of the ribbon tabs, panels, and panel titles. Picking it cycles through the display of the full ribbon, tab and panel titles only, or tab titles only.

Many panels contain more tools than are displayed in the default view. These panels can be expanded by picking the flyout arrow to the right of the panel title. See **Figure 1-8A.** The flyout portion is displayed as long as the cursor remains over the panel. It retracts when the cursor moves off of the panel. To retain the flyout display, pick the pushpin icon in the lower-left corner of the expanded panel. See **Figure 1-8B.** The panel continues to be displayed until you pick the pushpin icon again to release it.

Figure 1-7.
The ribbon is composed of tabs and panels that provide access to 3D modeling and viewing commands without using toolbars and menus.

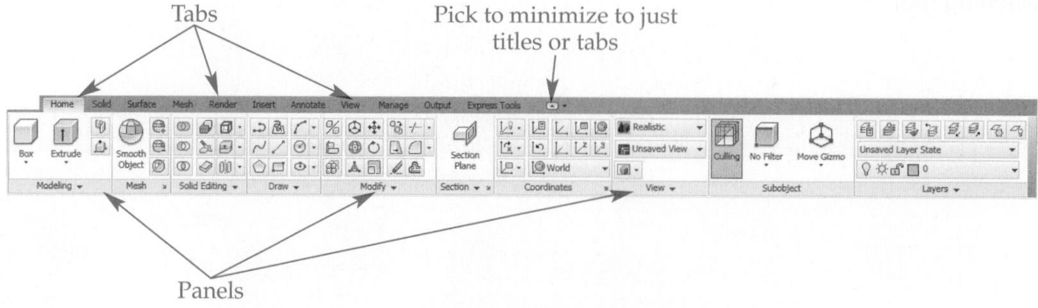

Figure 1-8.
A—Picking the flyout arrow expands the panel. The flyout is displayed as long as the cursor remains over the panel. It retracts when the cursor moves off of the panel. B—Multiple panel flyouts can be displayed using the pushpin, but may overlap in the process. Pick the panel you wish to use to display its tools in full.

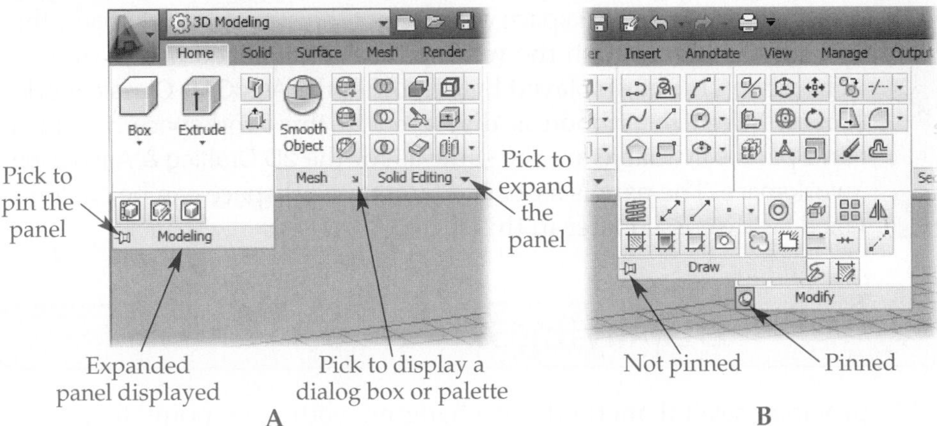

Multiple panel flyouts can be displayed using the pushpin, but may overlap in the process. This may obscure menu tools in adjacent panels. Simply pick on the panel you wish to use and its tools are displayed in full.

In some cases, a dialog box or palette may be related to the tools in a panel. This is indicated by an arrow in the lower-right corner of the panel title. See **Figure 1-8A.** Picking on the arrow displays the related dialog box.

You can display only those control panels that you need. Right-click anywhere on the ribbon tab to display the shortcut menu. Select **Panels** to display a cascading menu that contains a list of the control panels available for that tab. See **Figure 1-9.** The panels that are currently displayed in the ribbon have a check mark next to their name. Select any of the checked control panels that you wish to remove from the current display. Unchecked panels can be displayed by selecting their name. Each tab has its own set of available control panels.

Each panel contains command tools. These are discussed in detail throughout the Modeling and Presentation sections of this book.

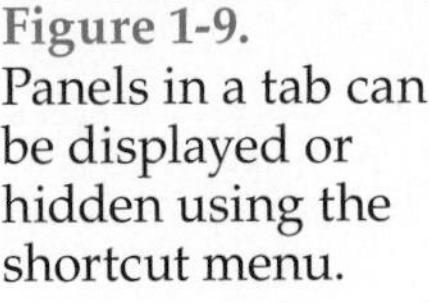
Figure 1-9.
Panels in a tab can be displayed or hidden using the shortcut menu.

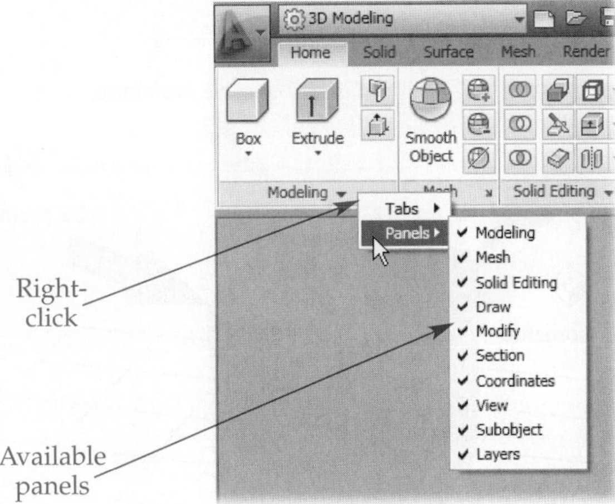

NOTE

The 2D Drafting & Annotation, 3D Basics, and 3D Modeling workspaces display the ribbon by default. In each workspace, the ribbon displays a set of panels specifically related to that workspace. As soon as you select a different workspace or re-select the current workspace, the panels associated with the workspace are displayed in the ribbon. The ribbon is not displayed by default in the AutoCAD Classic workspace. When the ribbon is displayed in this workspace, the tabs and panels available are the same as for the 2D Drafting & Annotation workspace. The panels associated with a workspace can be customized, as discussed later in this book.

Displaying 3D Views

AutoCAD provides several methods of changing your viewpoint to produce different pictorial views. The default view in the 2D environment based on the acad.dwt template is a plan, or top, view of the XY plane. The default view in the 3D environment based on the acad3D.dwt template is a pictorial, or 3D, view. The ***viewpoint*** is the location in space from which the object is viewed. The methods for changing your viewpoint include preset isometric and orthographic viewpoints, the view cube, and camera lens settings. Camera settings are discussed in detail in Chapter 19.

Isometric and Orthographic Viewpoint Presets

A 2D isometric drawing is based on angles of 120° between the three axes. AutoCAD provides preset viewpoints that allow you to view a 3D object from one of four isometric locations. See **Figure 1-10.** Each of these viewpoints produces an isometric view of the object. In addition, AutoCAD has presets for the six standard orthographic views of an object. The isometric and orthographic viewpoint presets are based on the world coordinate system (WCS).

Figure 1-10.
There are four preset isometric viewpoints in AutoCAD. This illustration shows the direction from which the cube will be viewed for each of the presets. The grid represents the XY plane of the WCS.

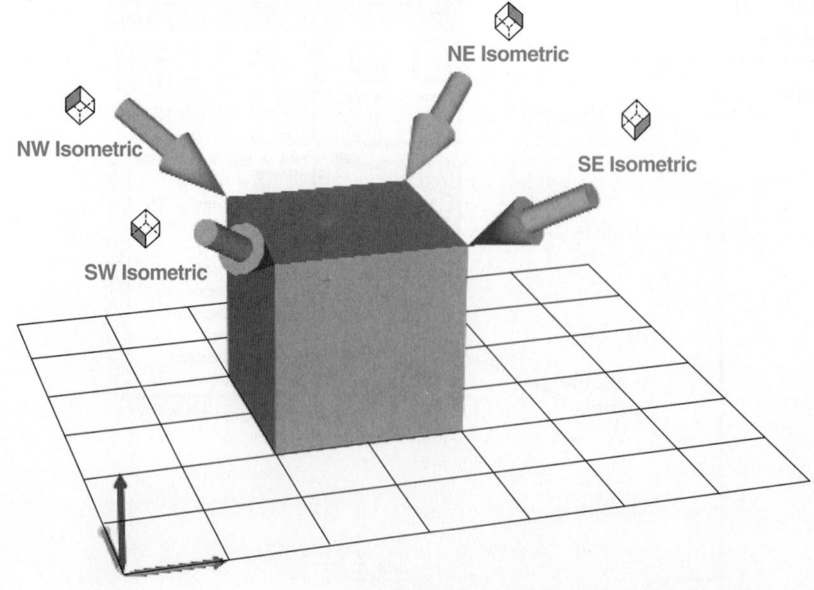

Figure 1-11.
Selecting preset views using the **View** panel in the **Home** tab of the ribbon.

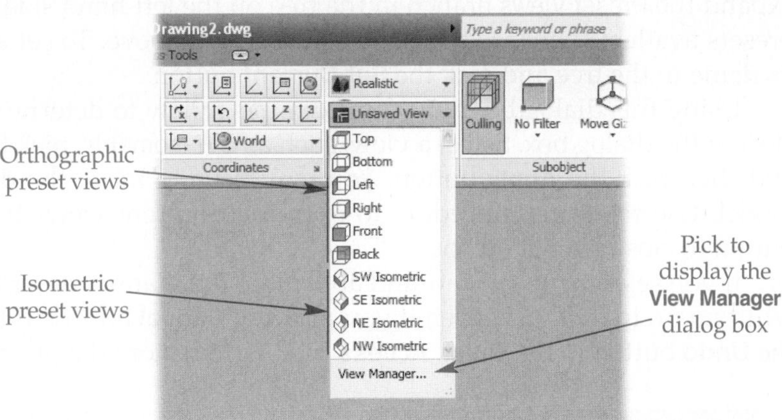

> **Ribbon**
> View
> > Views
> > SW Isometric
> > SE Isometric
> > NW Isometric
> > NE Isometric
>
> VIEW

> **Ribbon**
> View
> > Views
> > Top
> > Bottom
> > Left
> > Right
> > Front
> > Back
>
> VIEW

The four preset isometric views are southwest, southeast, northeast, and northwest. The six orthographic presets are top, bottom, left, right, front, and back. To switch your viewpoint to one of these presets, pick the view in the list in the **Views** panel in the **View** tab of the ribbon and select the view name. See Figure 1-11. There is also a drop-down list in the **View** panel on the **Home** tab.

Once you select a view, the viewpoint in the current viewport is automatically changed to display an appropriate isometric or orthographic view. Since these presets are based on the WCS, selecting a preset produces the same view of the object regardless of the current user coordinate system (UCS).

A view that looks straight down on the current drawing plane is called a ***plan view.*** An important aspect of the orthographic presets is that selecting one not only changes the viewpoint, but, by default, it also changes the UCS to be plan to the orthographic view. All new objects are created on that UCS instead of the WCS (or previous UCS). Working with UCSs is explained in detail in Chapter 6. However, to change the UCS to the WCS type UCS to access the **UCS** command and then type W for the **World** option.

When an isometric or other 3D view is displayed, you can easily switch to a plan view of the current UCS by typing the **PLAN** command and selecting the **Current** option. The **PLAN** command is discussed in more detail in Chapter 4.

> **Ribbon**
> View
> > Views
>
> 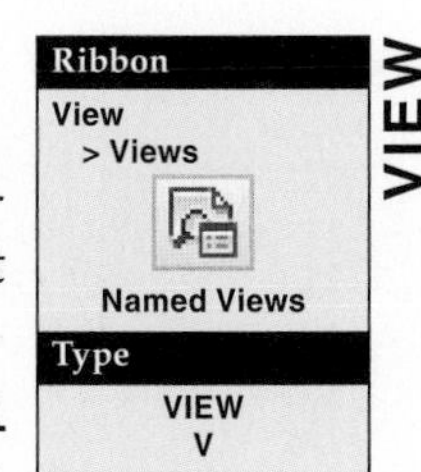
>
> Named Views
>
> **Type**
> VIEW
> V
>
> VIEW

View Manager Dialog Box

The **View Manager** dialog box allows you to work with any named view, orthographic preset, or isometric preset. See Figure 1-12. To select a preset viewpoint, first

Figure 1-12.
The **View Manager** dialog box allows you to work with any named view and orthographic and isometric preset views.

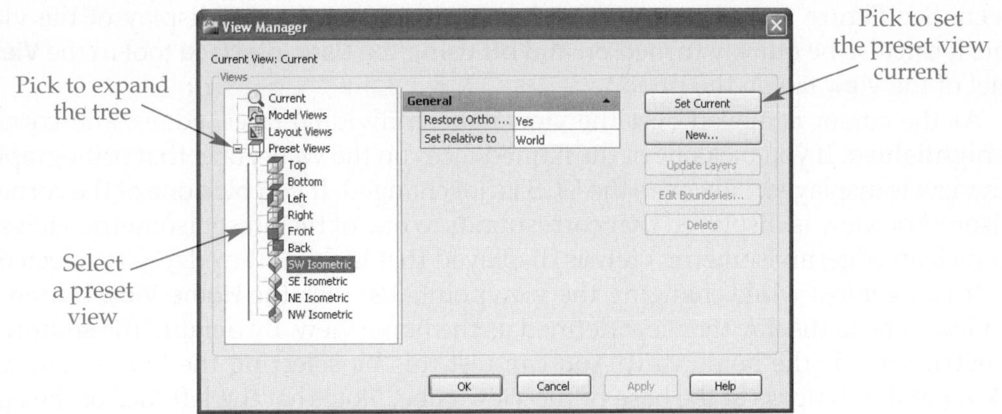

expand the Preset Views branch in the tree on the left-hand side of the dialog box. The presets available here are the same as described above. To set a preset current, select its name in the tree and pick the **Set Current** button.

Using this dialog box, you can examine a view to determine if you like it before closing the dialog box. Select a view, such as SW Isometric, pick the **Set Current** button, and then pick the **Apply** button. You may have to move the dialog box to view the model. Use the same procedure to examine different views before you pick the **OK** button to close the dialog box.

If you select a preset view and apply it in the manner described above, the display will remain in that view even if you pick the **Cancel** button. If this happens, just pick the **Undo** button in the **Quick Access** toolbar to restore the previous view.

NOTE

Selecting an orthographic view of a model using one of the methods described above produces a plan view, but it may not achieve the results you desire. Three-dimensional models can be displayed in AutoCAD using either parallel or perspective projection. Displaying a plan view in either projection is possible. However, a true plan view, as used in 2D orthographic projections, can only be created when the model is displayed as a parallel projection. You can quickly change the display from perspective to parallel, or vice versa, by right-clicking on the view cube, then picking either **Parallel** or **Perspective** in the shortcut menu. The current projection is checked in the shortcut menu.

Exercise 1-2

Complete the exercise on the student website.
www.g-wlearning.com/CAD

Introduction to the View Cube

You are not limited to the preset isometric viewpoints. In fact, you can view a 3D object from an unlimited number of viewpoints. The ***view cube*** allows you to display all of the preset isometric and orthographic views without making panel or menu selections. It also provides quick access to additional pictorial views and easy dynamic manipulation of all orthographic views. It allows you to dynamically rotate the view of the objects to create a new viewpoint.

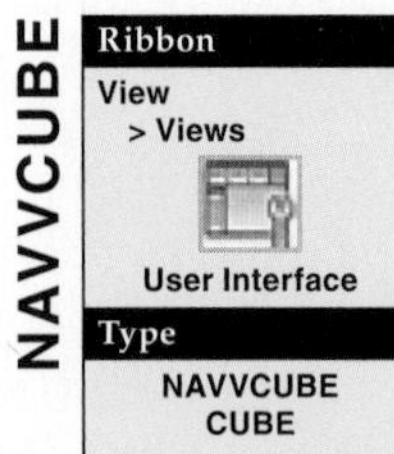

The view cube is displayed by default. It appears in the upper-right corner of the screen. See Figure 1-13. The **NAVVCUBE** command controls the display of the view cube. It can also be quickly turned on and off using the **User Interface** tool in the **Views** panel of the **View** tab on the ribbon.

As the cursor is moved over the view cube, individual faces, edges, and corners are highlighted. If you pick one of the named faces on the view cube, that orthographic plan view is displayed. However, the UCS is not changed. If you pick one of the corners, an isometric view is displayed that corresponds to one of the preset isometric views. If you pick an edge, an isometric view is displayed that looks at the edge you selected.

If you get lost while changing the viewpoint, just pick the **Home View** button on the view cube to display the view defined as the home view. By default, the southwest isometric view is the home view. You can test this by selecting the home view and looking at the compass at the base of the view cube. Note that the left face of the cube

Figure 1-13.
Using the view cube to change the viewpoint.

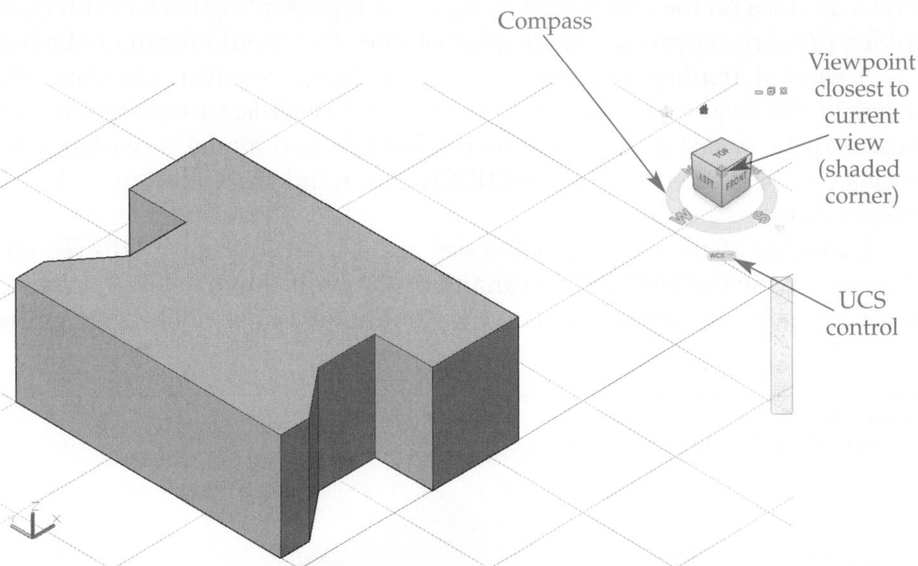

aligns with the west point on the compass. The front surface aligns with the south point on the compass. Therefore, the top-front corner points to the southwest.

One edge, corner, or face of the view cube is always shaded or highlighted. Refer to Figure 1-13. This shading indicates the viewpoint on the cube that is closest to the current view.

The **NAVVCUBE** command has many options. This discussion is merely an introduction to the command. The command options and view cube features are covered in detail in Chapter 4.

PROFESSIONAL TIP

The **UNDO** command reverses the effects of the **NAVVCUBE** command.

Exercise 1-3

Complete the exercise on the student website.
www.g-wlearning.com/CAD

Introduction to 3D Model Display Using Visual Styles

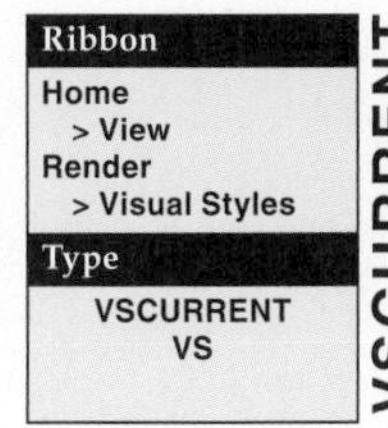

The ***display*** of a 3D model is how the model is presented. This does not refer to the viewing angle, but rather colors, edge display, and shading or rendering. An object can be shaded from any viewpoint. A model can be edited while still keeping the object shaded. This can make it easier to see how the model is developing without having to reshade the drawing. However, when editing a shaded object, it may also be more difficult to select features.

A 3D model can be displayed in a variety of visual styles. A ***visual style*** is a combination of settings that control the display of edges and shading in a viewport.

There are ten basic visual styles—2D wireframe, wireframe, hidden, conceptual, realistic, shaded, shaded with edges, shades of gray, and sketchy. A ***wireframe display*** shows all lines on the object, including those representing back or internal features. A ***hidden display*** suppresses the display of lines that would normally be hidden.

A ***shaded display*** of the model can be created by setting the Conceptual, Realistic, or one of the shaded visual styles current. The Realistic visual style is considered the most realistic *shaded* display. A more detailed shaded model, a ***rendered display*** of the model, can be created with the **RENDER** command. A rendering is the most realistic presentation.

VSCURRENT

Ribbon
Home >View View >Visual Styles
Type
VSCURRENT **VS**

Examples of the basic visual styles are shown in **Figure 1-14.** To change styles, you can use the **VSCURRENT** command or the drop-down list in the **View** panel on the **Home** tab or **Visual Styles** panel on the **Render** tab in the ribbon. See **Figure 1-15.**

Figure 1-14.
The default AutoCAD visual styles.

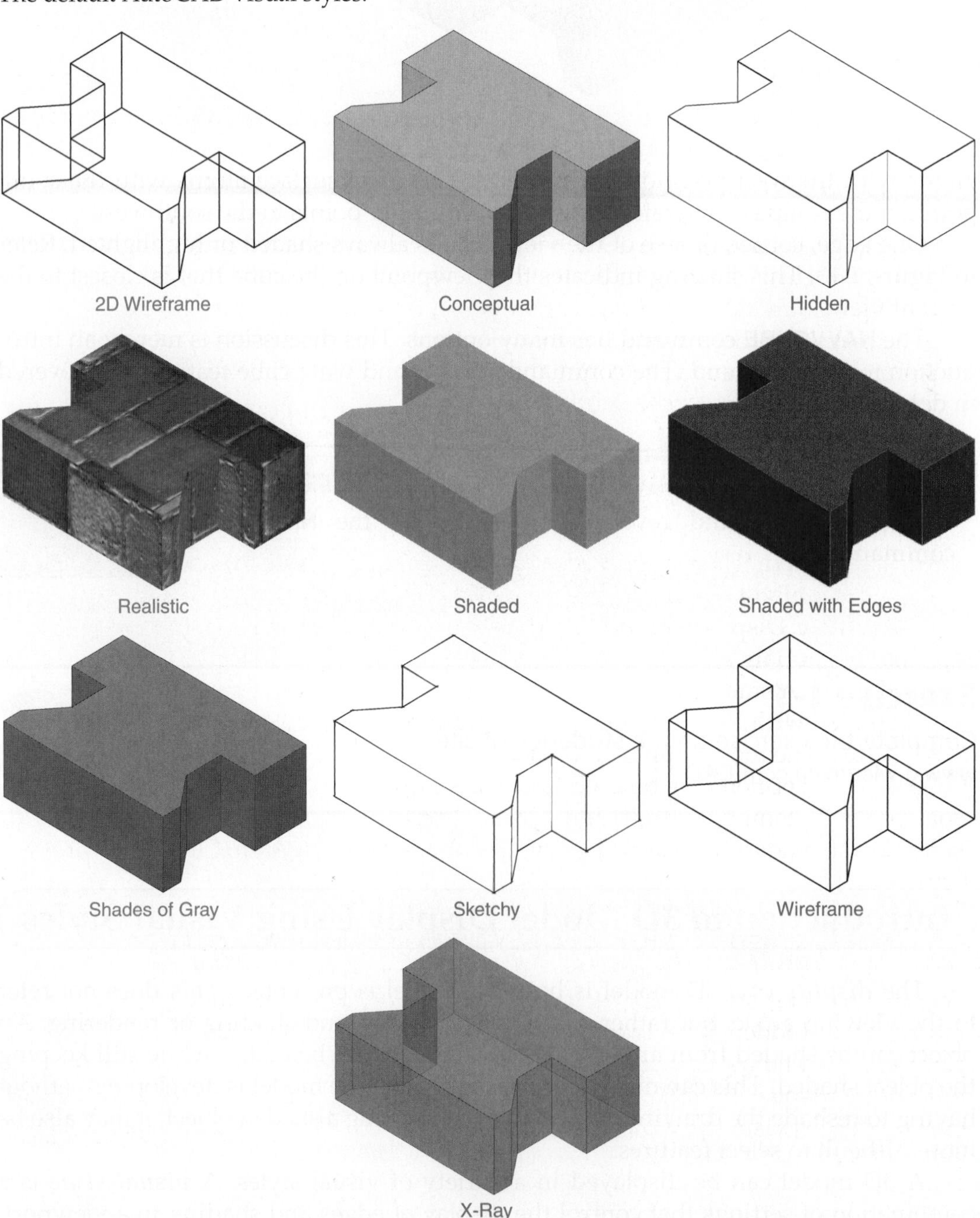

Figure 1-15.
A visual style can be set current using the drop-down list in the **View** panel in the **Home** tab or the **Visual Styles** panel in the **Render** tab of the ribbon.

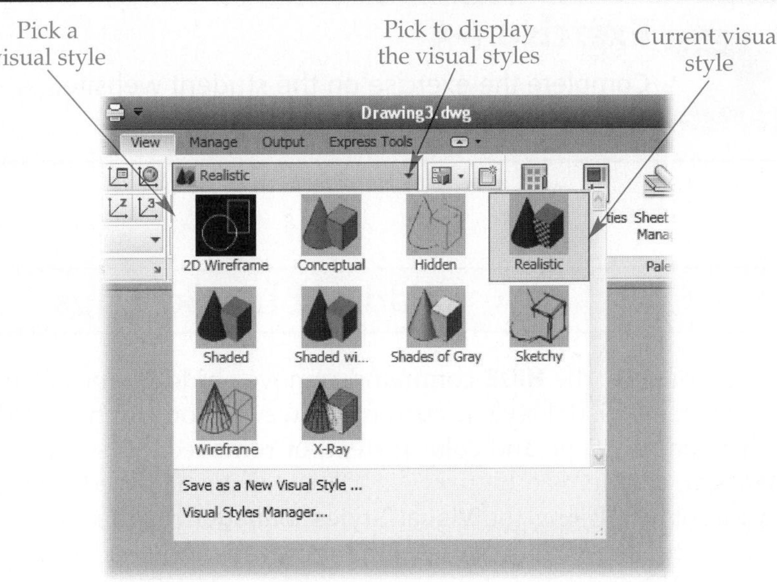

In the default 3D environment based on the acad3D.dwt template, the default display mode, or visual style, is Realistic. In this visual style, all objects appear as solids and are displayed in their assigned layer colors. Other display options are available. These options are discussed in detail in Chapter 15, but are given here as an introduction.

- **2D Wireframe.** Displays all lines of the model using assigned linetypes and lineweights. The 2D UCS icon and 2D grid are displayed, if turned on. If the **HIDE** command is used to display a hidden-line view, use the **REGEN** command to redisplay the wireframe view.
- **Wireframe.** Displays all lines of the model. The 3D grid and the 3D UCS icon are displayed, if turned on.
- **Hidden.** Displays all visible lines of the model from the current viewpoint and hides all lines not visible. Objects are not shaded or colored.
- **Sketchy.** Edges appear hand-sketched.
- **Shades of Gray.** Gray shades are shown with highlighted edges.
- **Conceptual.** The object is smoothed and shaded with transitional colors to help highlight details.
- **Realistic.** Displays the shaded and smoothed model using assigned layer colors and materials.
- **Shaded.** A smooth-shaded model is displayed, but edges are not shown.
- **Shaded with Edges.** Edges are displayed on the smooth-shaded model.
- **X-Ray.** The model appears transparent.

A variety of options can be used to change individual components of each of these visual styles. A complete discussion of visual styles is given in Chapter 15. Detailed discussions on rendering, materials, lights, and animations appear in Chapter 15 through Chapter 18.

NOTE

When the visual style is 2D Wireframe, you can quickly view the model with hidden lines removed by typing HIDE. The **HIDE** command can be used at any time to remove hidden lines from a wireframe display. If **HIDE** is used when any other visual style is current, Hidden visual style is set current.

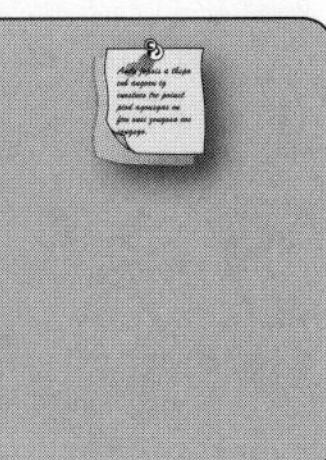

Exercise 1-4

Complete the exercise on the student website.
www.g-wlearning.com/CAD

Hidden Line Settings

VISUALSTYLES

Ribbon
View
> Visual Styles
> Visual Style Manager...

Type
VISUALSTYLES
VSM

By default, the **HIDE** command removes hidden lines from the display when the 2D Wireframe visual style is current. However, you can have hidden lines displayed in a different linetype and color instead of removed. To set this, open the **Visual Styles Manager** palette. See Figure 1-16. This can be accessed by typing the command or on the ribbon. Using the **Visual Styles Manager** palette, you can control all available settings for wireframe, hidden-line removed, and shaded displays in AutoCAD. See Chapter 15 for a detailed discussion of this palette.

To change the hidden line style in the 2D Wireframe visual style, select the corresponding image tile at the top of the **Visual Styles Manager**. In the **2D Hide - Occluded Lines** category of the palette are drop-down lists from which you can select a linetype and color.

When the **Linetype** drop-down list is set to Off, the display of hidden lines is suppressed by the **HIDE** command. This is the default setting. When a linetype is selected from the drop-down list, hidden lines are displayed in that linetype after the **HIDE** command is used. The linetypes available in the drop-down list are not the same as the linetypes loaded into your drawing.

You can also change the display color of the hidden lines. Simply pick a color in the **Color** drop-down list. To have the hidden lines displayed in the same color as the object, select ByEntity, which is the default. The color setting has no effect when the **Linetype** drop-down list is set to Off.

Figure 1-16.
The manner in which hidden lines appear in hidden displays when the 2D Wireframe visual style is current is controlled in the **Visual Styles Manager** palette.

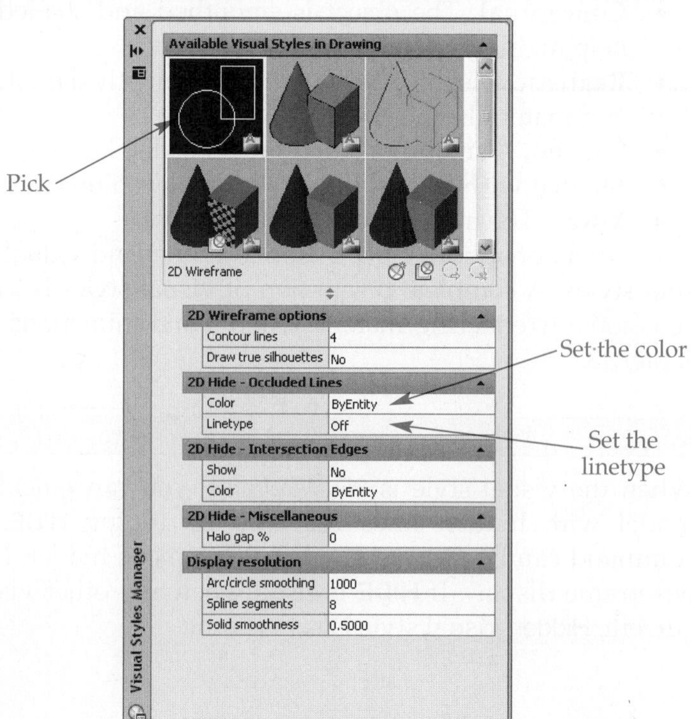

NOTE

Making changes in the **Visual Styles Manager** redefines the selected visual style.

Rendering a Model

The **RENDER** command creates a realistic image of a model, Figure 1-17. However, rendering an image takes longer than shading an image. There are a variety of settings that you can change with the **RENDER** command that allow you to fine-tune renderings. These include lights, materials, backgrounds, fog, and preferences. Render settings are discussed in detail in Chapter 15 through Chapter 18.

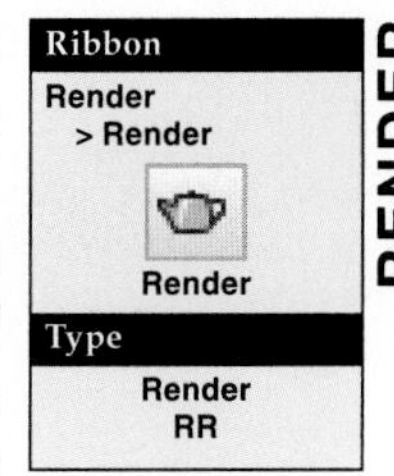

When the command is initiated, the render window is displayed and the image is rendered. See Figure 1-18. The rendering that is produced is based on a variety of advanced render settings that are discussed in Chapter 15 through Chapter 18. The default render settings create an image using a single light source located behind the viewer. The light intensity is set to 1 and, if no materials are applied, the objects are rendered with a matte material that is the same color as the object display color.

Figure 1-17.
Rendering produces the most realistic display and can show shadows and materials.

Figure 1-18.
The rendered model is displayed in the **Render** window.

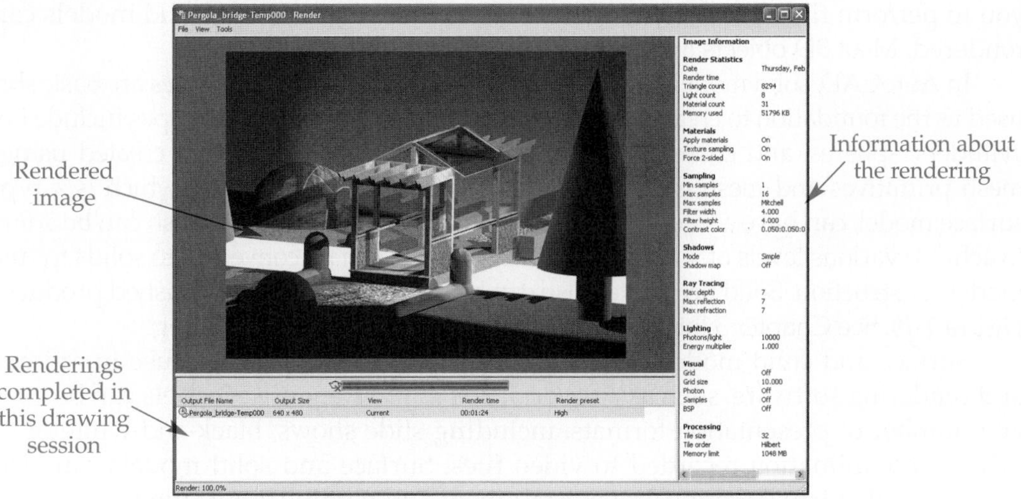

NOTE

If the image is rendered in the viewport, clean the screen using the **ZOOM**, **PAN**, **REGEN**, or **REDRAW** command. Setting the rendering destination as the viewport is discussed in Chapter 15.

Exercise 1-5

Complete the exercise on the student website.
www.g-wlearning.com/CAD

3D Construction Techniques

Before constructing a 3D model, you should determine the purpose of your design. What will the model be used for—manufacturing, analysis, or presentation? This helps you determine which tools you should use to construct and display the model. Three-dimensional objects can be drawn as solids, meshes, or surfaces and displayed in wireframe, hidden-line removed, and shaded views.

A ***wireframe object,*** or model, is an object constructed of lines in 3D space. Wireframe models are hard to visualize because it is difficult to determine the angle of view and the nature of the surfaces represented by the lines. The **HIDE** command has no effect on a wireframe model because there is nothing to hide. All lines are always visible because there are no surfaces or faces between the lines. Wireframe models have very limited applications.

Surface modeling represents solid objects by creating a skin in the shape of the object. However, there is nothing inside of the object. Think of a surface model as a balloon filled with air. A surface model looks more like the real object than a wireframe and can be used for rendering. Surface models are often constructed for applications such as civil engineering terrain modeling, automobile body design, sheet metal design and fabrication, and animation.

Like surface modeling, ***solid modeling*** represents the shape of objects, but it also provides data related to the physical properties of the objects. Solid models can be analyzed to determine mass, volume, moments of inertia, and centroid. A solid model is not just a skin, it represents a solid object. Some third-party programs allow you to perform finite element analysis on the model. In addition, solid models can be rendered. Most 3D objects are created as solid models.

In AutoCAD, solid models can be created from primitives. ***Primitives*** are basic shapes used as the foundation to create complex shapes. Some of these basic shapes include boxes, cylinders, spheres, and cones. Detailed shapes and primitives can be created using 3D mesh primitives and mesh modeling techniques. A 3D ***mesh object,*** which is a type of surface model, can have a free-flowing shape because the size of the mesh can be adjusted to achieve various levels of smoothness. Mesh objects can be converted to solids for use in model construction. Solid primitives also can be modified to create a finished product. See Figure 1-19. See Chapter 3 for a detailed discussion of 3D mesh modeling.

Surface and solid models can be exported from AutoCAD for use in animation and rendering software, such as Autodesk 3ds max®. Rendered models can be used in any number of presentation formats, including slide shows, black-and-white or color prints, and animation recorded to video files. Surface and solid models can also be used to create virtual worlds for entertainment and gaming applications.

Figure 1-19.
A—These two cylinders and the box are solid primitives. B—With a couple of quick modifications, the large cylinder becomes a shaft with a machined keyway.

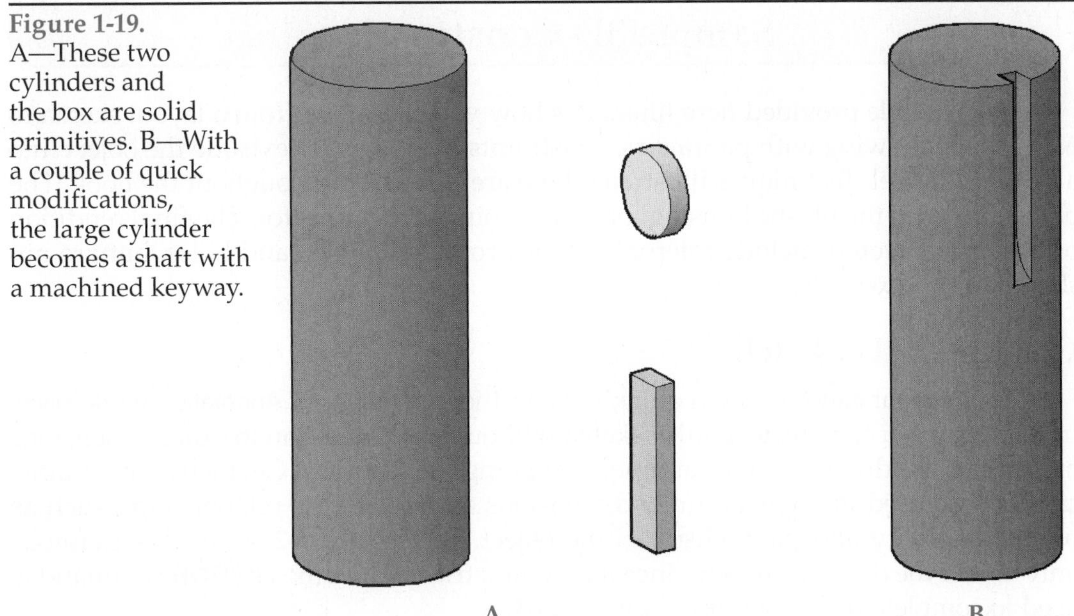

3D Object Snaps

The construction and editing of a 3D model can be more efficient with the use of 3D object snaps. These work in the same manner as the standard 2D object snaps and can be set using the **3D Object Snap** tab of the **Drafting Settings** dialog box. If you use 3D object snaps, turn on only those options you need to construct the object. See Figure 1-20.

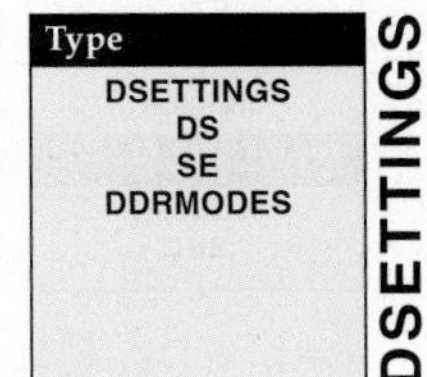

DSETTINGS

There are six 3D object snaps. With the exception of the knot snap, these should be familiar to you from your work in 2D. The knot 3D object snap option refers to a fit curve spline and the point at which one curve ends and the next curve begins.

Figure 1-20.
The 3D object snaps can be set in the **Drafting Settings** dialog box.

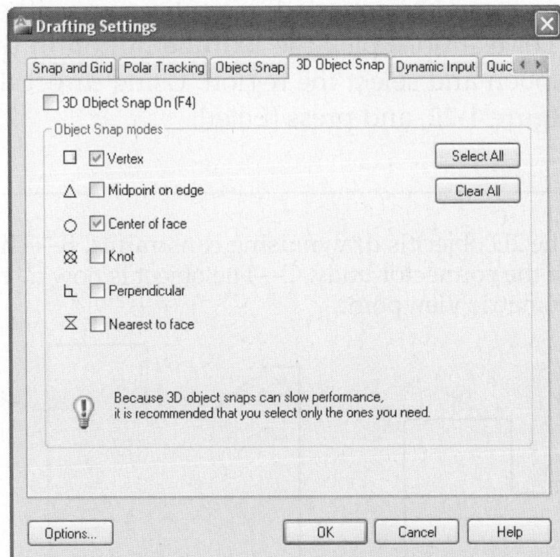

Sample 3D Construction

The example provided here illustrates how you can move from a layout sketch to an exact 2D drawing with parametric constraints, then quickly extrude the object into a 3D solid model. Techniques illustrated here are discussed throughout the book. The object shown is the plastic housing for a 24/15-pin cable connector. The final rendition of this object would include internal cutouts, rounded edges, and holes, but are not shown in this example.

Creating a 2D Sketch

A 2D layout can be drawn quickly using the 2D Drafting & Annotation workspace. In doing so, you can create a quick sketch without close attention to exact dimensions or you can use direct distance and polar tracking for accuracy. Geometric constraints can then be used to apply accurate dimensions and geometric relationships, such as perpendicularity and parallelism, to the object. See Figure 1-21A. In this example, only half of the object is drawn, since it is symmetrical. Then the **MIRROR** command is used to complete the layout. See Figure 1-21B.

Converting to a Region

REGION

Ribbon
Home
> Draw
Region
Type
REGION
REG

The object is still composed of just 2D lines. Therefore, it must be changed to an object that can be converted into a solid model. The **REGION** command is used for this purpose. After selecting this command you are prompted to select objects. Be sure to pick all of the lines of the object. After pressing [Enter], you are prompted that a loop is extracted, a region is created, and, if constraints were applied, that they are removed. The appearance of the object will not change.

Converting to a Solid

The object is now a "region" which can be quickly converted to a solid. But first change to the 3D Modeling workspace, then use the **View** tab or the view cube to display the object from the southwest isometric viewpoint. Working with models is more intuitive when a 3D display is used. See Figure 1-21C.

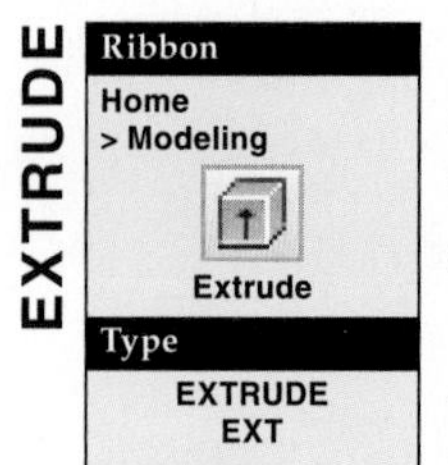

The region can be converted to a solid using either the **EXTRUDE** or **PRESSPULL** command. For example, pick the **Extrude** button in the **Modeling** panel on the **Home** tab of the ribbon and select the region. Using direct distance, enter a height of .55, as shown in Figure 1-22, and press [Enter].

Figure 1-21.
A—Half of the 2D object is drawn using constraints. B—The object is mirrored to create a full profile of the connector body. C—The object is now a "region" and is displayed in the southwest isometric viewpoint.

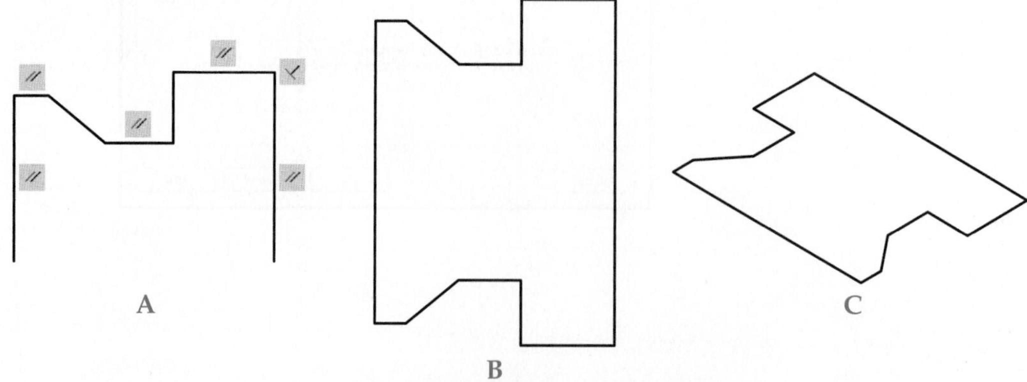

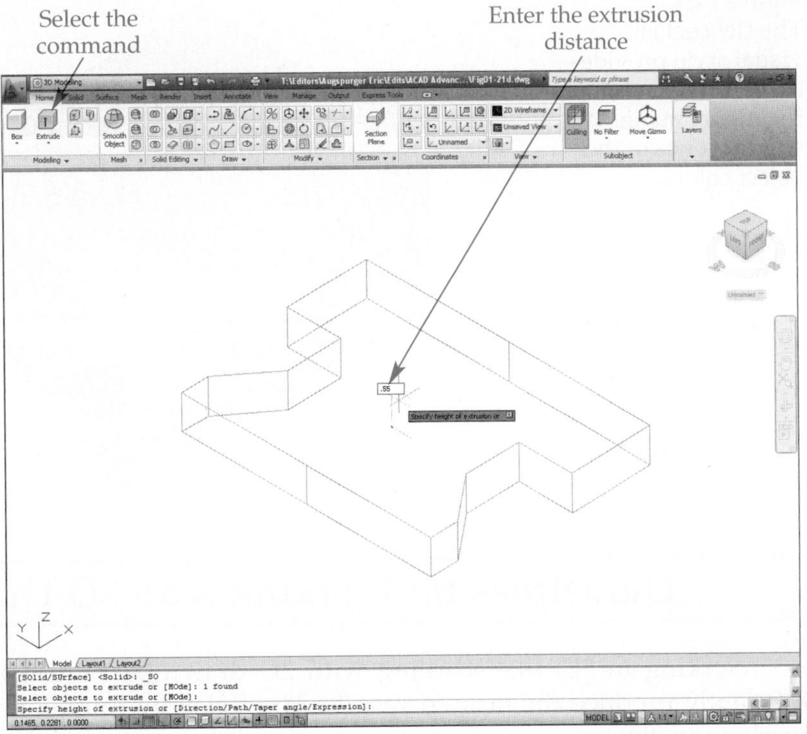

Figure 1-22.
The object is extruded into a solid and displayed as a wireframe.

Using a Visual Style

The object you have created is now a 3D solid. If the drawing was based on the acad3D.dwt template, the model is displayed in the Realistic visual style. If the drawing was based on the acad.dwt template, the model is displayed in the 2D Wireframe visual style.

As mentioned previously, there are ten different visual styles from which to choose. They can be selected from the **View** panel on the **Home** tab or the **Visual Styles** panel on the **View** tab of the ribbon. See Figure 1-23. The Conceptual visual style provides a quick display of the model using shaded tones of the object color, as shown in Figure 1-24.

The techniques used in this example are just a brief introduction to the creation of 3D models. Detailed descriptions of modeling, display, and editing techniques are included in the following chapters.

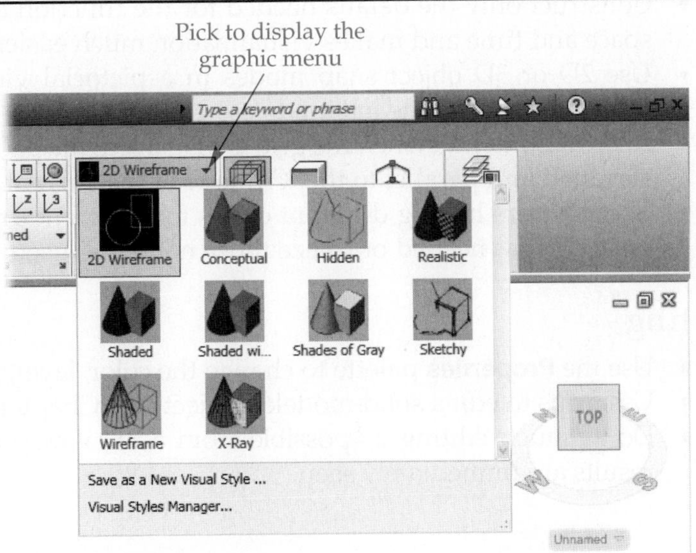

Figure 1-23.
One of ten different visual styles can be selected on the ribbon.

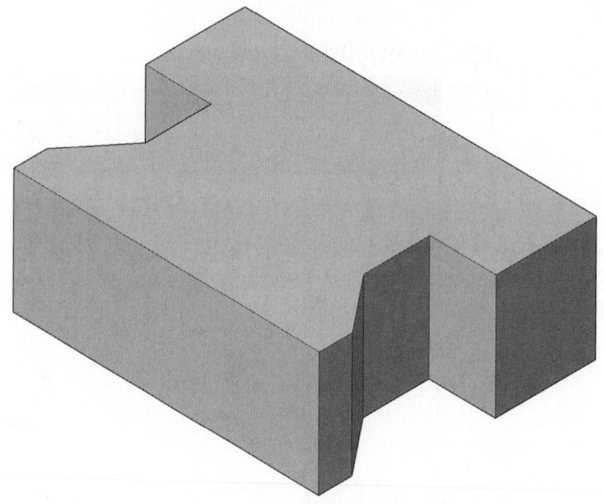

Figure 1-24.
The **Conceptual** visual style provides a quick display of the model using shaded tones of the object colors.

Guidelines for Working with 3D Drawings

Working in 3D, like working with 2D drawings, requires careful planning to efficiently produce the desired results. The following guidelines can be used when working in 3D.

Planning

- Determine the type of final drawing you need and the manner in which it will be displayed. Then, choose the method of 3D construction that best suits your needs—wireframe, surface, mesh, or solid.
- If appropriate for the project, use 2D geometric constraints to create a 2D layout or sketch.
- For an object requiring only one pictorial view, it actually may be quicker to draw an object in 3D rather than in AutoCAD's isometric mode. AutoCAD's 3D solid modeling tools enable you to quickly create an accurate model, then display it in the required isometric format using preset views. The **FLATSHOT** command can then be used to create a 2D drawing of the model.
- It is best to use AutoCAD's 3D commands to construct objects and layouts that need to be viewed from different angles for design purposes.
- Construct only the details needed for the function of the drawing. This saves space and time and makes visualization much easier.
- Use 2D or 3D object snap modes in a pictorial view in conjunction with a dynamic UCS to save having to create new UCSs.
- Keep in mind that when the grid is displayed, the pattern appears at the current elevation and parallel to the XY plane of the current UCS.
- Create layers having different colors for different drawing objects. Turn them on and off as needed or freeze those not being used.

Editing

- Use the **Properties** palette to change the color, layer, or linetype of 3D objects.
- Use grips to edit a solid-modeled object (see Chapter 12).
- Do as much editing as possible from a 3D viewpoint. It is quicker and the results are immediately seen.

Displaying

- Use the **HIDE** command and visual styles to help visualize complex drawings.
- To quickly change views, use the preset isometric views, view cube, and **PLAN** command.
- Use the **VIEW** command to create and save 3D views for quicker pictorial displays. This avoids having to repeatedly use the **NAVVCUBE** command.
- Freeze unwanted layers before displaying objects in 3D and especially before using **HIDE**. AutoCAD regenerates layers that are turned off, which may cause an inaccurate hidden display to be created. Frozen layers are not regenerated.
- You may have to slightly move objects that touch or intersect if the display removes a line you need to see or plot. However, be sure to move the objects back to maintain accuracy in the model.

Chapter Test

Answer the following questions. Write your answers on a separate sheet of paper or complete the electronic chapter test on the student website.
www.g-wlearning.com/CAD

1. What are the three coordinates needed to locate any point in 3D space?
2. In a 2D drawing, what is the value for the Z coordinate?
3. What purpose does the right-hand rule serve?
4. Which three fingers are used in the right-hand rule?
5. What is the definition of a *viewpoint?*
6. What is the function of the *ribbon* and its panels?
7. How do you turn the display of individual panels on or off in the ribbon?
8. How can you quickly change the display from perspective projection to parallel projection, or vice versa?
9. How many preset isometric viewpoints does AutoCAD have? List them.
10. How does changing the UCS impact using one of the preset isometric viewpoints?
11. List the six preset orthographic viewpoints.
12. When selecting a preset orthographic viewpoint, what happens to the UCS?
13. Which command allows you to dynamically change your viewpoint using an on-screen cube icon?
14. Define *wireframe display.*
15. Define *hidden display.*
16. Define *surface model.*
17. Define *solid model.*
18. Define *primitive.*
19. Define *mesh object.*
20. What is a *visual style?*

CHAPTER 2

Creating Primitives and Composites

Learning Objectives

After completing this chapter, you will be able to:

- ✓ Construct 3D solid primitives.
- ✓ Explain the dynamic feedback presented when constructing solid primitives.
- ✓ Create complex solids using the **UNION** command.
- ✓ Remove portions of a solid using the **SUBTRACT** command.
- ✓ Create a new solid from the interference volume between two solids.
- ✓ Create regions.

Overview of Solid Modeling

In Chapter 1 you were introduced to the three basic forms of 3D modeling—wireframe objects, solid models, and surface models. Solid models are probably the most useful and, hence, most common type of 3D modeling. A solid model accurately and realistically represents the shape and form of a final object. In addition, a solid model contains data related to the object's volume, mass, and centroid.

Solid modeling is very flexible. A model can start with solid primitives, such as a box, cone, or cylinder, and a variety of editing functions can then be performed. Think of creating a solid model as working with modeling clay. Starting with a basic block of clay, you can add more clay, remove clay, cut holes, round edges, etc., until you have arrived at the final shape and form of the object.

PROFESSIONAL TIP

Snaps can be used on solid objects. For example, using 2D object snaps you can snap to the center of a solid sphere using the **Center** object snap. The **Endpoint** object snap can be used to select the corners of a box, apex of a cone, corners of a wedge, etc. Using 3D object snaps you can snap to the vertex or midpoint of an edge or to the center of a face.

Constructing Solid Primitives

As you learned in Chapter 1, a primitive is a basic building block. The eight ***solid primitives*** in AutoCAD are a box, cone, cylinder, polysolid, pyramid, sphere, torus, and wedge. These primitives can also be used as building blocks for complex solid models. This section provides detailed information on drawing all of the solid primitives. All of the 3D modeling primitive commands can be accessed using the **Modeling** panel in **Home** tab of the ribbon, the **Modeling** toolbar, or by typing the name of the 3D modeling primitive. See Figure 2-1.

The information required to construct a solid primitive depends on the type of primitive being drawn. For example, to draw a solid cylinder you must provide a center point for the base, a radius or diameter of the base, and the height of the cylinder. A variety of command options are available when creating primitives, but each primitive is constructed using just a few basic dimensions. These are shown in Figure 2-2.

Certain familiar editing commands can be used on solid primitives. For example, you can fillet or chamfer the edges of a solid primitive. In addition, there are other editing commands that are specifically for use on solids. You can also perform Boolean operations on solids. These operations allow you to add one solid to another, subtract one solid from another, or create a new solid based on how two solids overlap.

PROFESSIONAL TIP

Solid objects of a more free-form nature can be created by first constructing a mesh primitive, editing it, and then converting it to a solid. This process is covered in detail in Chapter 3.

Using Dynamic Input and Dynamic Feedback

Dynamic input allows you to construct models in a "heads up" fashion with minimal eye movement around the screen. When a command is initiated, the command prompts are then displayed in the dynamic input area, which is at the lower-right corner of the crosshairs. As the pointer is moved, the dynamic input area follows it. The dynamic input area displays values of the cursor location, dimensions, command prompts, and command options (in a drop-down list). Coordinates and dimensions are displayed in boxes called *input fields*. When command options are available, a

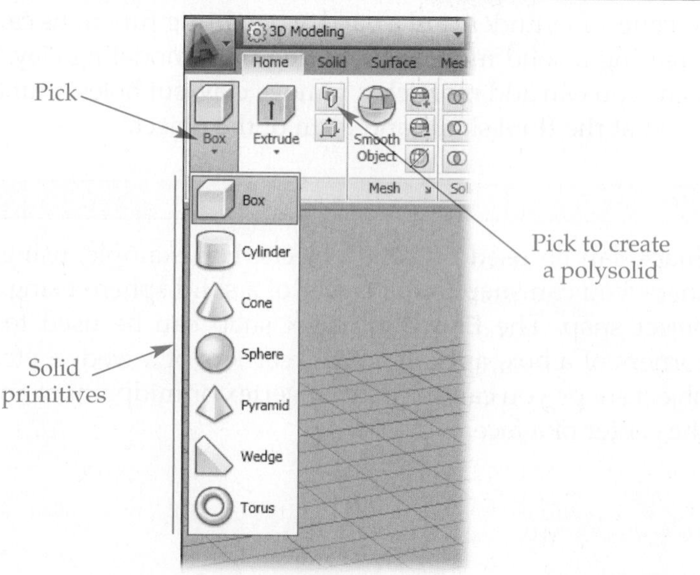

Figure 2-1.
The **Modeling** panel in the **Home** tab of the ribbon.

Figure 2-2.
An overview of AutoCAD's solid primitives and the dimensions required to draw them.

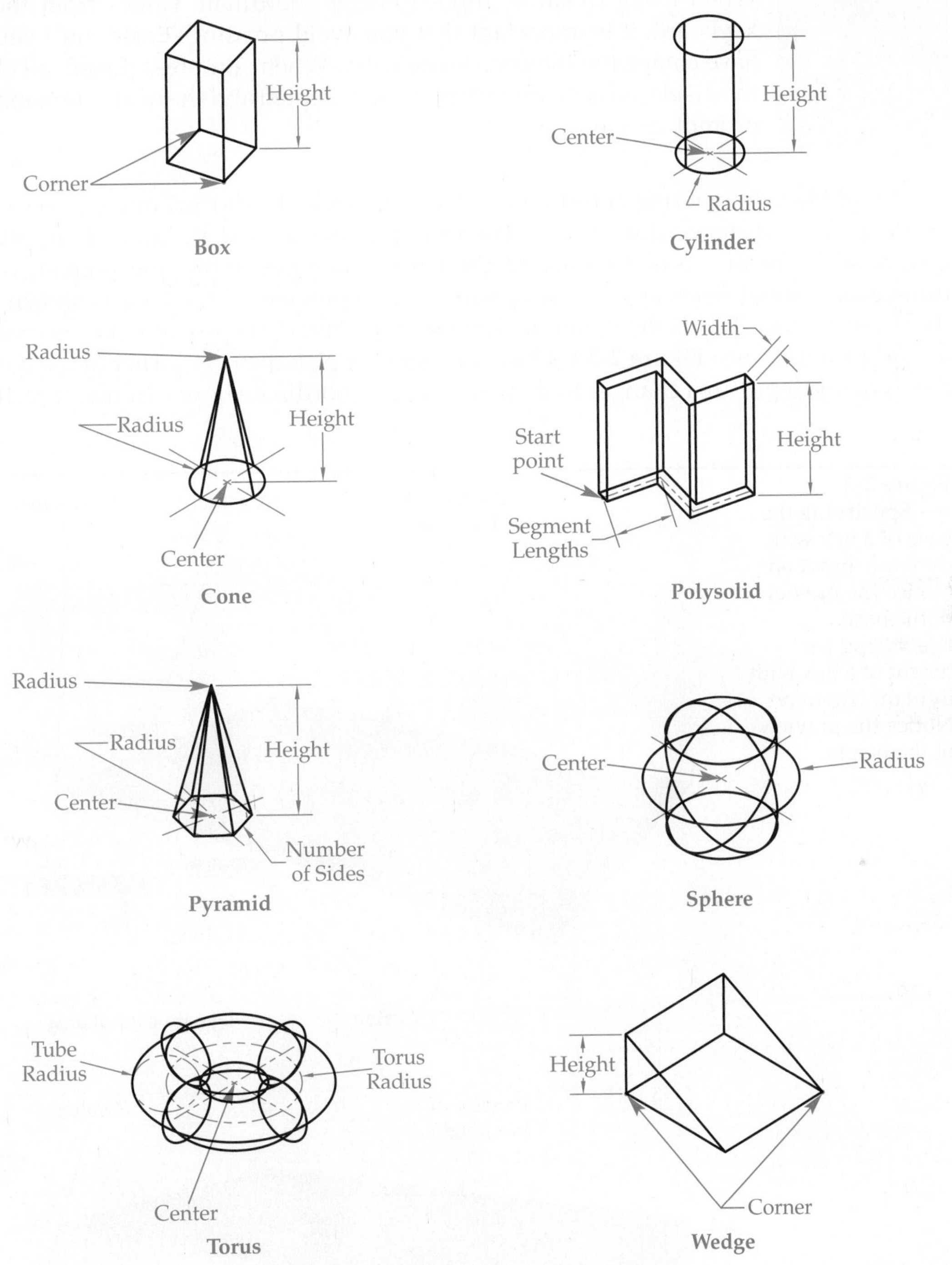

drop-down list arrow appears. Press the down arrow key on the keyboard to display the list. You can use your pointer to select the option or press the down arrow key until a dot appears by the desired option and press [Enter].

For example, after selecting the modeling command **BOX**, the first item that appears in the dynamic input area is the prompt to specify the first corner and a display of the X and Y coordinate values of the crosshairs. At this point you can use the pointer to specify the first corner or type coordinate values. Type the X value and then a comma or the [Tab] key to move to the Y value input box. This locks the typed value and any movement of the pointer will not change it.

CAUTION

When using dynamic input to enter coordinate values from the keyboard, it is important that you avoid pressing [Enter] until you have completed the coordinate entry. When you press [Enter], all of the displayed coordinate values are accepted and the next command prompt appears.

In addition to entering coordinate values for sizes of solid primitives, you can provide direct distance dimensions. For example, the second prompt of the **BOX** command is for the second corner of the base. When you move the pointer, two dimensional input fields appear. Also, notice that a preview of the base is shown in the drawing area. This is the ***dynamic feedback*** that AutoCAD provides as you create a solid primitive. See **Figure 2-3A.** Once you enter or pick the first corner of the base, the dynamic input area changes to display X and Y coordinate boxes. In this case, the

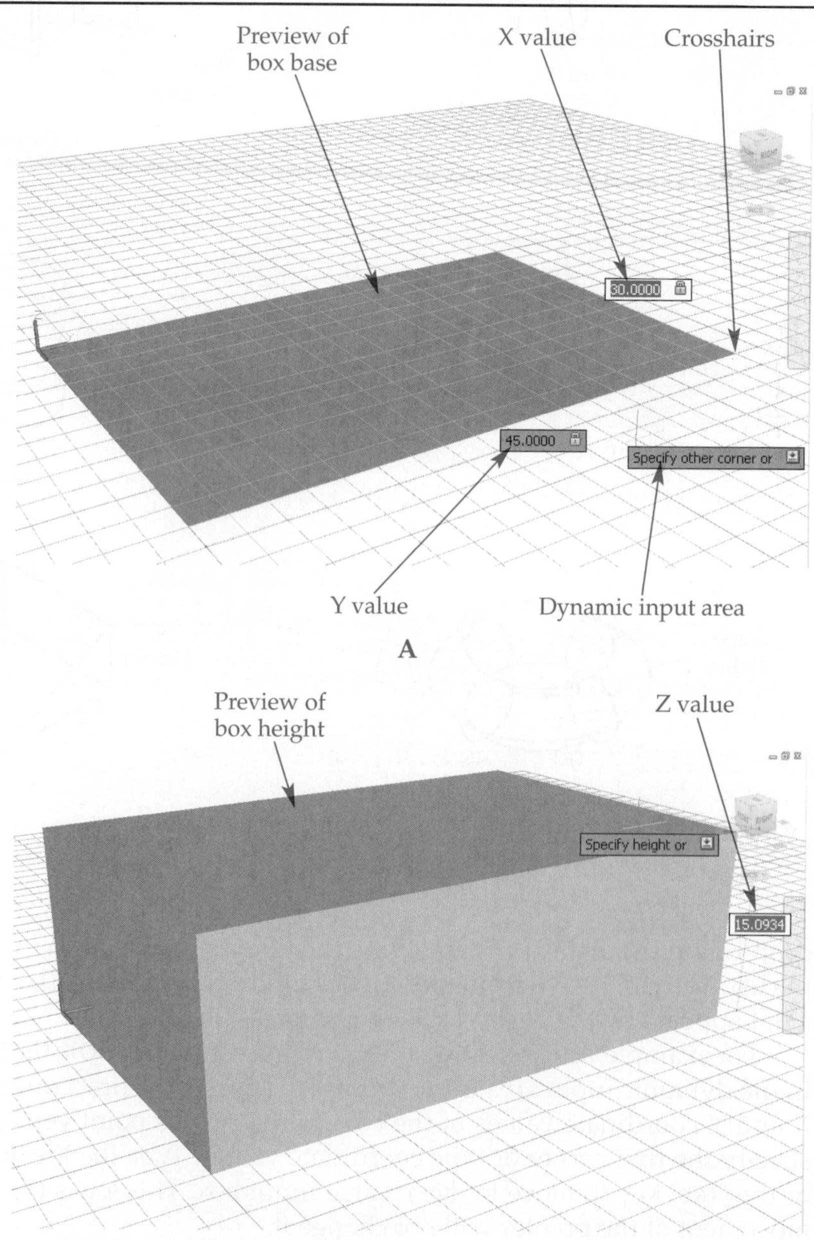

Figure 2-3.
A—Specifying the base of a box with dynamic input on. Notice the preview of the base.
B—Setting the height of a box with dynamic input on. Notice the preview of the height.

values entered are the X and Y coordinates of the opposite corner of the box base. If X, Y, *and* Z coordinates are entered, the point is the opposite corner of the box.

After establishing the location and size of the box base, the next prompt asks you to specify the height. Again, you can either enter a direct dimension value and press [Enter] or select the height with the pointer. See Figure 2-3B. AutoCAD provides dynamic feedback on the height of the box as the pointer is moved.

NOTE

The techniques described above can be used with any form of dynamic input. The current input field is always highlighted. You can always enter a value and use the [Tab] key to lock the input and move to the next field.

Box

Ribbon
Home
> Modeling
Solid
> Primitive
Box
Type
BOX

BOX

A ***box*** has six flat sides forming square corners. It can be constructed starting from an initial corner or the center. See Figure 2-4. A cube can be constructed, as well as a box with unequal-length sides.

When the command is initiated, you are prompted to select the first corner or enter the **Center** option. The first corner is one corner on the base of the box. The center is the geometric center of the box, as shown in Figure 2-4. If you select the **Center** option, you are next prompted to select the center point.

After selecting the first corner or center, you are prompted to select the other corner or enter the **Cube** or **Length** option. The "other" corner is the opposite corner of the box base if you enter an XY coordinate or the opposite corner of the box if you enter an XYZ coordinate. If the **Length** option is entered, you are first prompted for the length of one side. If dynamic input is on, you can also specify a rotation angle. After entering the length, you are prompted for the width of the box base. If the **Cube** option is selected, the length value is applied to all sides of the box.

Once the length and width of the base are established, you are prompted for the height, unless the **Cube** option was selected. Either enter the height or select the **2point** option. This option allows you to pick two points on screen to set the height. The box is created.

PROFESSIONAL TIP

If dynamic UCS is on, you can select a surface that is not parallel to the current UCS on which to locate the object. This feature is called a *dynamic UCS* and discussed in detail in Chapter 5.

Figure 2-4.
A—A box created using the **Cube** option. B—A box created by selecting the center point.

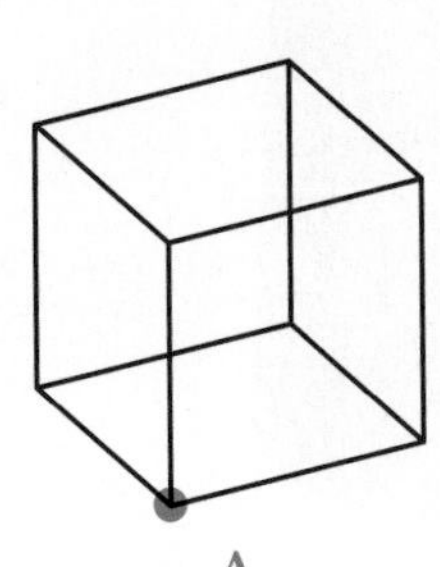

A

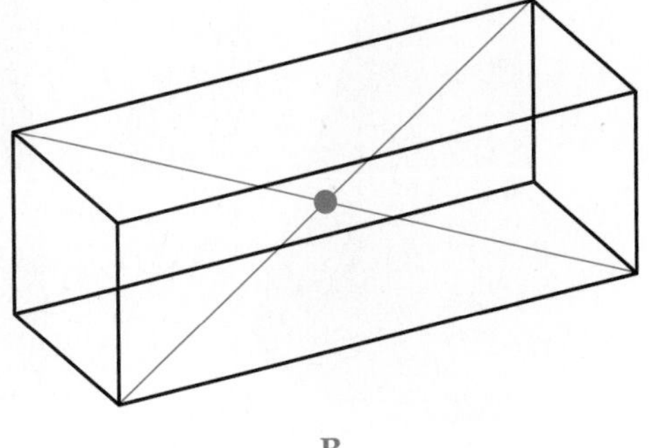

B

Exercise 2-1

Complete the exercise on the student website.
www.g-wlearning.com/CAD

Cone

CONE

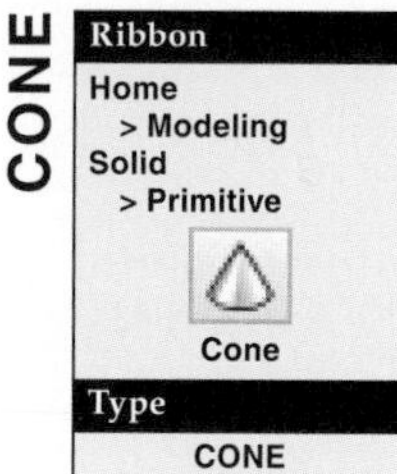

A *cone* has a circular or elliptical base with edges that converge at a single point. The cone may be *truncated* so the top is flat and the cone does not have an apex. See Figure 2-5. When the command is initiated, you are prompted for the center point of the cone base or to enter an option. If you pick the center, you must then set the radius of the base. To specify a diameter, enter the **Diameter** option after specifying the center.

The **3P**, **2P**, and **Ttr** options are used to define a circular base using either three points on the circle, two points on the circle, or two points of tangency on the circle and a radius. The **Elliptical** option is used to create an elliptical base.

If the **Elliptical** option is entered, you are prompted to pick both endpoints of one axis and then one endpoint of the other axis of an ellipse that defines the base. If the **Center** option is entered after the **Ellipse** option, you are asked to select the center of the ellipse and then pick an endpoint on each of the axes.

After the base is defined, you are asked to specify a height. You can enter a height or enter the **2point**, **Axis endpoint**, or **Top radius** option. The **2point** option is used to set the height by picking two points on screen. The distance between the points is the height. The height is always applied perpendicular to the base.

The **Axis endpoint** option allows you to orient the cone at any angle, regardless of the current UCS. For example, to place a tapered cutout in the end of a block, first create a construction line. Refer to Figure 2-6. Then, locate the cone base and give a coordinate location of the apex, or axis endpoint. You can then use editing commands to subtract the cone from the box to create the tapered hole. See Chapters 11 and 12 for details on model editing.

Figure 2-5.
A—A circular cone. B—A frustum cone. C—An elliptical cone.

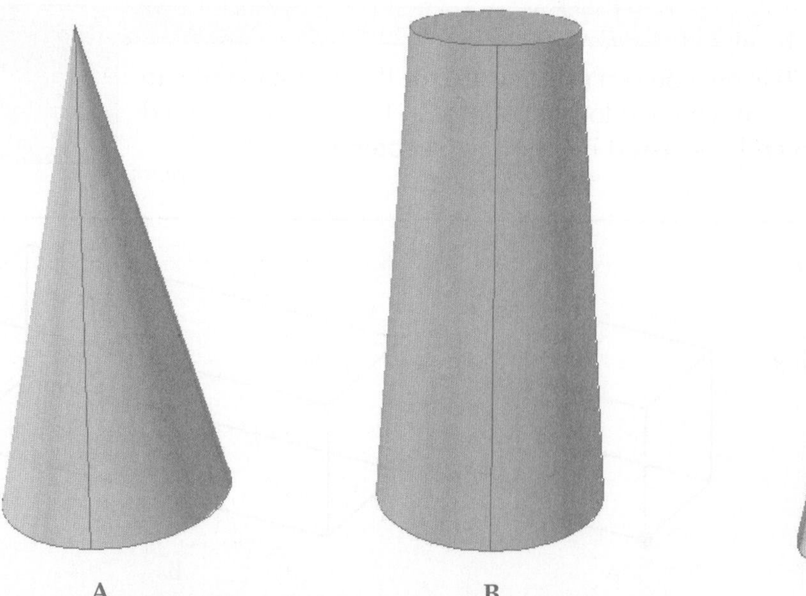

Figure 2-6.
A—Cones can be positioned relative to other objects using the **Axis endpoint** option.
B—The cone is subtracted from the box.

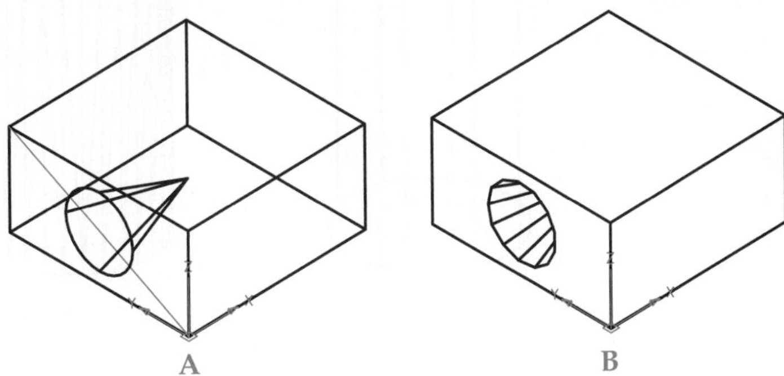

The **Top radius** option allows you to specify the radius of the top of the cone. If this option is not used, the radius is zero, which creates a pointed cone. Setting the radius to a value other than zero produces a ***frustum cone,*** or a cone where the top is truncated and does not come to a point.

Cylinder

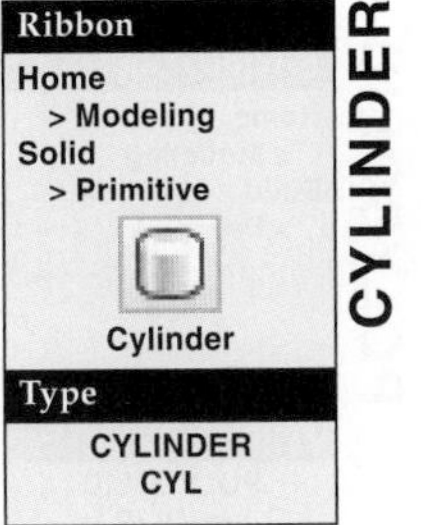

CYLINDER

A ***cylinder*** has a circular or elliptical base and edges that extend perpendicular to the base. See Figure 2-7. When the command is initiated, you are prompted for the center point of the cylinder base or to enter an option. If you pick the center, you must then set the radius of the base. To specify a diameter, enter the **Diameter** option after specifying the center.

The **3P**, **2P**, and **Ttr** options are used to define a circular base using either three points on the circle, two points on the circle, or two points of tangency on the circle and a radius. The **Elliptical** option is used to create an elliptical base.

If the **Elliptical** option is entered, you are prompted to pick both endpoints of one axis and then one endpoint of the other axis of an ellipse defining the base. If the **Center** option is entered, you are asked to select the center of the ellipse and then pick an endpoint on each of the axes.

After the base is defined, you are asked to specify a height or to enter the **2point** or **Axis endpoint** option. The **2point** option is used to set the height by picking two points on screen. The distance between the points is the height. The **Axis endpoint** option allows you to orient the cylinder at any angle, regardless of the current UCS, just as with a cone.

The **Axis endpoint** option is useful for placing a cylinder inside of another object to create a hole. The cylinder can then be subtracted from the other object to create a hole.

Figure 2-7.
A—A circular cylinder. B—An elliptical cylinder.

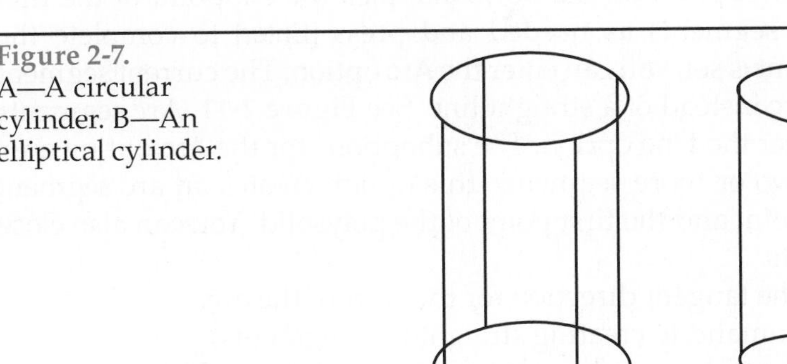

Figure 2-8.
A—A cylinder is drawn inside of another cylinder using the **Axis endpoint** option.
B—The large cylinder has a hole after **SUBTRACT** is used to remove the small cylinder.

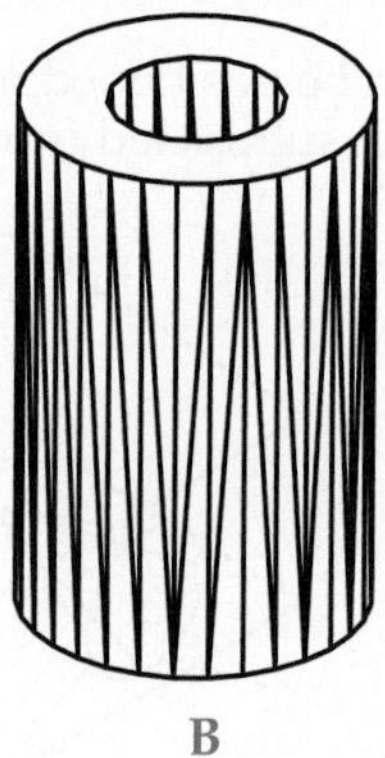

Refer to Figure 2-8. If the axis endpoint does not have the same X and Y coordinates as the center of the base, the cylinder is tilted from the XY plane.

If polar tracking is on when using the **Axis endpoint** option, you can rotate the cylinder axis 90° from the current UCS Z axis and then turn the cylinder to any preset polar increment. See Figure 2-9A. If the polar tracking vector is parallel to the Z axis of the current UCS, the tooltip displays a positive or negative Z value. See Figure 2-9B.

Polysolid

POLYSOLID

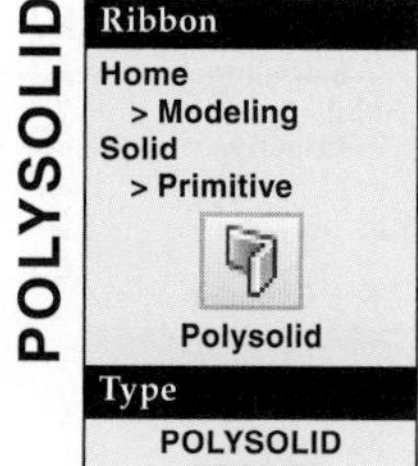

The *polysolid* primitive is simply a polyline constructed as a solid object by applying a width and height to the polyline. Many of the options used to create polylines are used with the **POLYSOLID** command. The principal difference is that a solid object is constructed using **POLYSOLID**.

When the command is initiated, you are prompted to select the first point or enter an option. By default, the width of the polysolid is equally applied to each side of the line you draw. This is center justification. Using the **Justify** option, you can set the justification to center, left, or right. The justification applies to all segments created in this command session. See Figure 2-10. If you select the wrong justification option, you must exit the command and begin again.

The default width is .25 units and height is four units. These values can be changed using the **Height** and **Width** options of the command. The height value is saved in the **PSOLHEIGHT** system variable. The width value is saved in the **PSOLWIDTH** system variable. Using these system variables, the default width and height can be set outside of the command.

The **Object** option allows you to convert an existing 2D object into a polysolid. AutoCAD entities such as lines, circles, arcs, polylines, polygons, and rectangles can be converted. The 2D object cannot be self intersecting. Some objects, such as 3D polylines and revision clouds, cannot be converted.

Once you have set the first point on the polysolid, pick the endpoint of the first segment. Continue adding segments as needed and press [Enter] to complete the command. After the first point is set, you can enter the **Arc** option. The current segment will then be created as an arc instead of a straight line. See Figure 2-11. Arc segments will be created until you enter the **Line** option. The suboptions for the **Arc** option are:

- **Close.** If there are two or more segments, this option creates an arc segment between the active point and the first point of the polysolid. You can also close straight-line segments.
- **Direction.** Specifies the tangent direction for the start of the arc.
- **Line.** Returns the command to creating straight-line segments.
- **Second point.** Locates the second point of a two-point arc. This is not the endpoint of the segment.

Figure 2-9.
A—If polar tracking is on, you can rotate the cylinder axis 90° from the current UCS Z axis and then move the cylinder to any angle in the XY plane. B—If the polar tracking vector is moved parallel to the current Z axis of the UCS, the tooltip displays a positive or negative Z dimension.

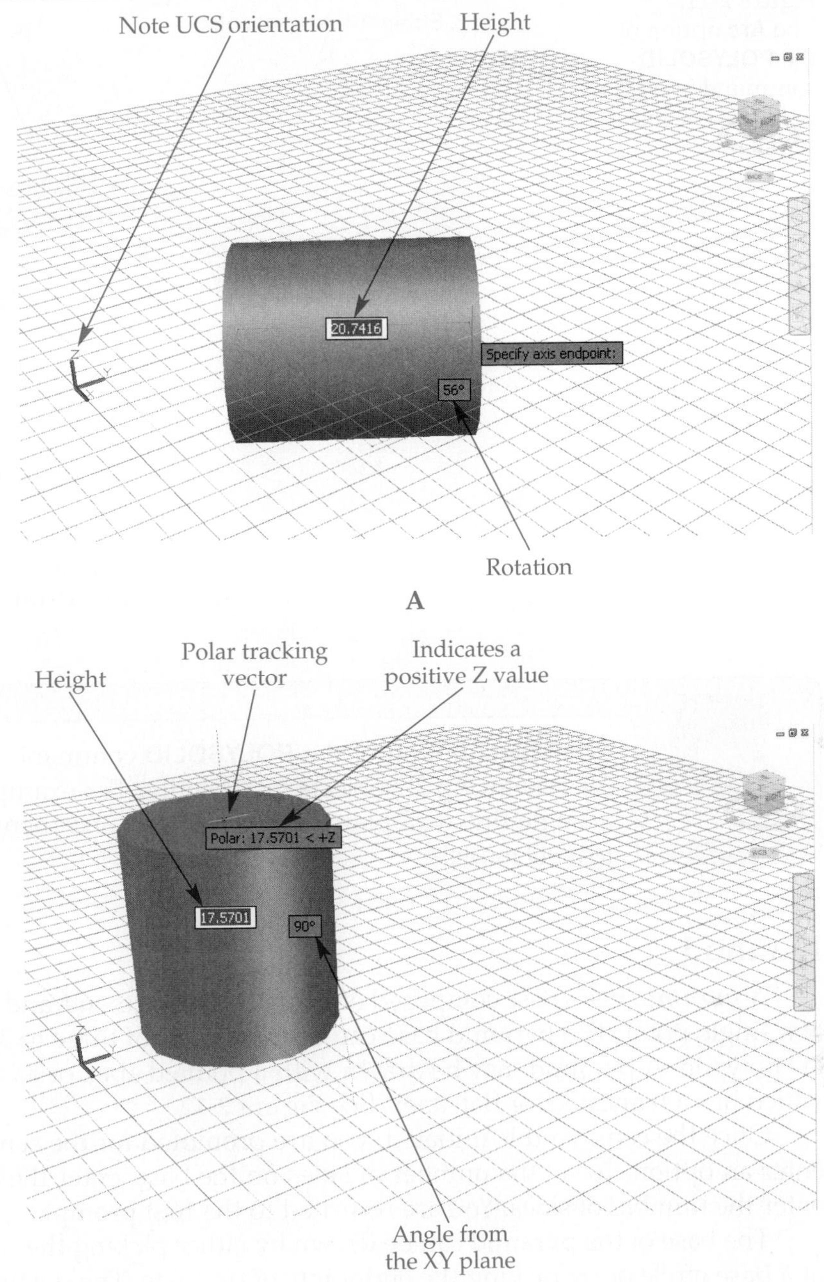

Figure 2-10.
When you begin the **POLYSOLID** command, use the **Justify** option to select the alignment.

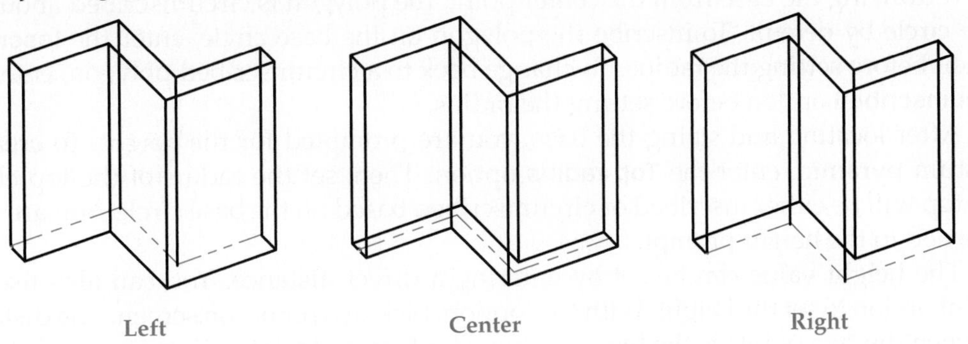

Figure 2-11.
The **Arc** option of the **POLYSOLID** command is used to create curved segments.

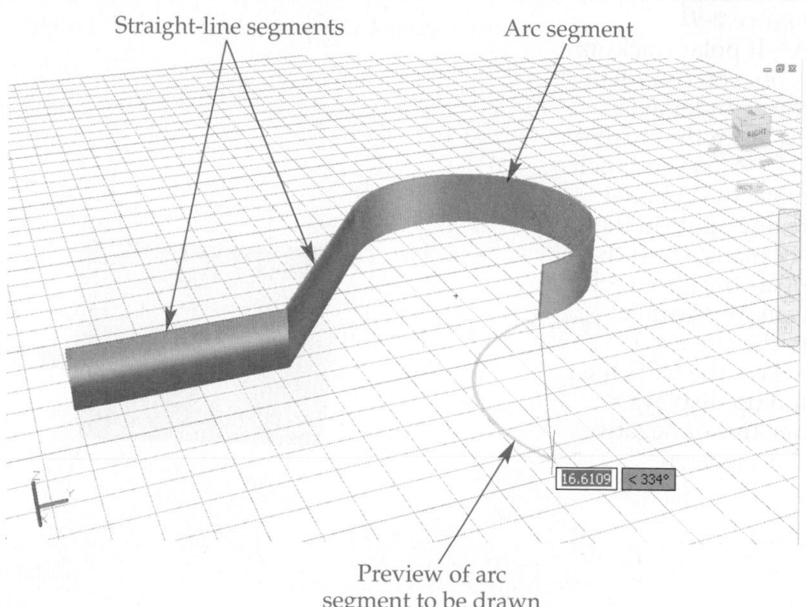

PROFESSIONAL TIP

The **Object** option of the **POLYSOLID** command is a powerful tool for converting 2D objects to 3D solids. For example, you can create a single-line wall plan using a polyline and then quickly convert it to a 3D model.

Pyramid

PYRAMID

Ribbon
Home
> Modeling
Solid
> Primitive

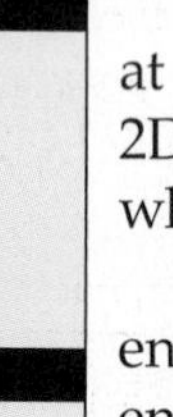

Pyramid

Type
PYRAMID
PYR

A ***pyramid*** has a base composed of straight-line segments and edges that converge at a single point. The pyramid base can be composed of three to 32 sides, much like a 2D polygon. A pyramid may be drawn with a pointed apex or as a ***frustum pyramid,*** which has a truncated, or flat, apex. See **Figure 2-12.**

Once the command is initiated, you are prompted for the center of the base or to enter an option. To set the number of sides on the base, enter the **Sides** option. Then, enter the number of sides. You are returned to the first prompt.

The base of the pyramid can be drawn by either picking the center and the radius of a base circle or by picking the endpoints of one side. The default method is to pick the center. Simply specify the center and then set the radius. To pick the endpoints of one side, enter the **Edge** option. Then, pick the first endpoint of one side followed by the second endpoint. If dynamic input is on, you can also set a rotation angle for the pyramid.

If drawing the base from the center point, the polygon is circumscribed about the base circle by default. To inscribe the polygon on the base circle, enter the **Inscribed** option before setting the radius. To change back to a circumscribed polygon, enter the **Circumscribed** option before setting the radius.

After locating and sizing the base, you are prompted for the height. To create a frustum pyramid, enter the **Top radius** option. Then, set the radius of the top circle. The top will be either inscribed or circumscribed based on the base circle. You are then returned to the height prompt.

The height value can be set by entering a direct distance. You can also use the **2point** option to set the height. With this option, pick two points on screen. The distance between the two points is the height value. The **Axis endpoint** option can also be used to specify the center of the top in the same manner as for a cone or cylinder.

Figure 2-12.
A sampling of pyramids that can be constructed with the **PYRAMID** command

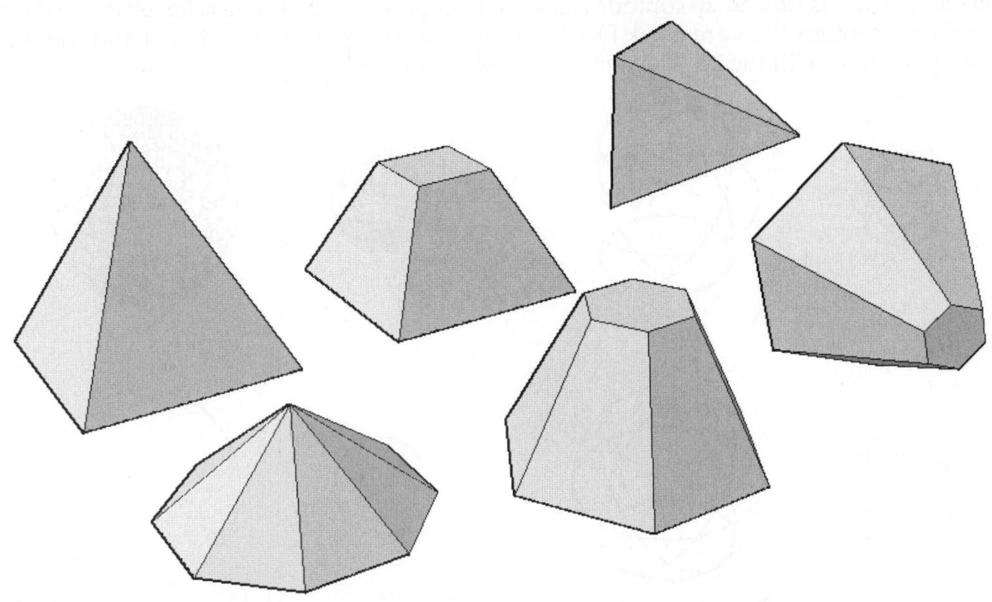

PROFESSIONAL TIP

Six-sided frustum pyramids can be used for bolt heads and as "blanks" for creating nuts.

Sphere

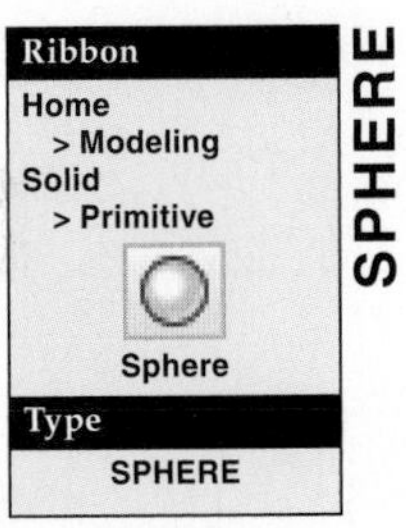

SPHERE

A *sphere* is a round, smooth object like a baseball or globe. Once the command is initiated, you are prompted for the center of the sphere or to enter an option. If you pick the center, you must then set the radius of the sphere. To specify a diameter, enter the **Diameter** option after specifying the center. The **3P**, **2P**, and **Ttr** options are used to define the sphere using either three points on the surface of the sphere, two points on the surface of the sphere, or two points of tangency on the surface of the sphere and a radius.

Spheres and other curved objects can be displayed in a number of different ways. The manner in which you choose to display these objects should be governed by the display requirements of your work. Notice in **Figure 2-13A** the lines that define the shape of the spheres in a wireframe display. These lines are called ***contour lines,*** also known as *tessellation lines* or *isolines*. The **Visual Styles Manager** can be used to set the display of contour lines and silhouettes on spheres and other curved 3D surfaces for a given visual style. See **Figure 2-14.**

With the **Visual Styles Manager** displayed, select the 2D Wireframe image tile. The Contour lines setting in the **2D Wireframe options** area establishes the number of lines used to show the shape of curved objects. A similar setting appears in all of the other visual styles if their Show property in the **Edge Settings** area is set to Isolines. The default value is four, but can be set to a value from zero to 2047. **Figure 2-13B** displays spheres with 20 contour lines. It is best to use a lower number during construction and preliminary displays of the model and, if needed, higher settings for more realistic visualization. The contour lines setting is also available in the **Display** tab of the **Options** dialog box or by typing ISOLINES.

The Draw true silhouettes setting in the **2D Wireframe options** area controls the display of silhouettes on 3D solid curved surfaces. The setting is either Yes or No. Notice the sphere silhouette in **Figures 2-13C** and **2-13D.** The Draw true silhouettes setting is stored in the **DISPSILH** system variable.

Figure 2-13.
A—The Draw true silhouettes setting is No and four contour lines are used. B—The Draw true silhouettes setting is No and 20 contour lines are used. C—The Draw true silhouettes setting is Yes and four contour lines are used. D—The Draw true silhouettes setting is Yes and the **HIDE** command is used with the 2D Wireframe visual style set current.

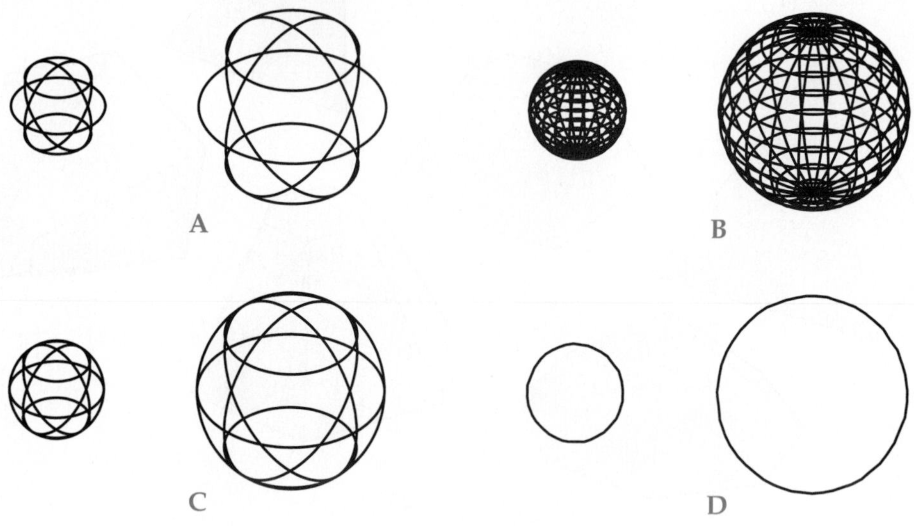

Figure 2-14.
The **2D Wireframe options** area of the **Visual Styles Manager** is used to control the display of contour lines and silhouettes on spheres and other curved 3D surfaces in a given visual style.

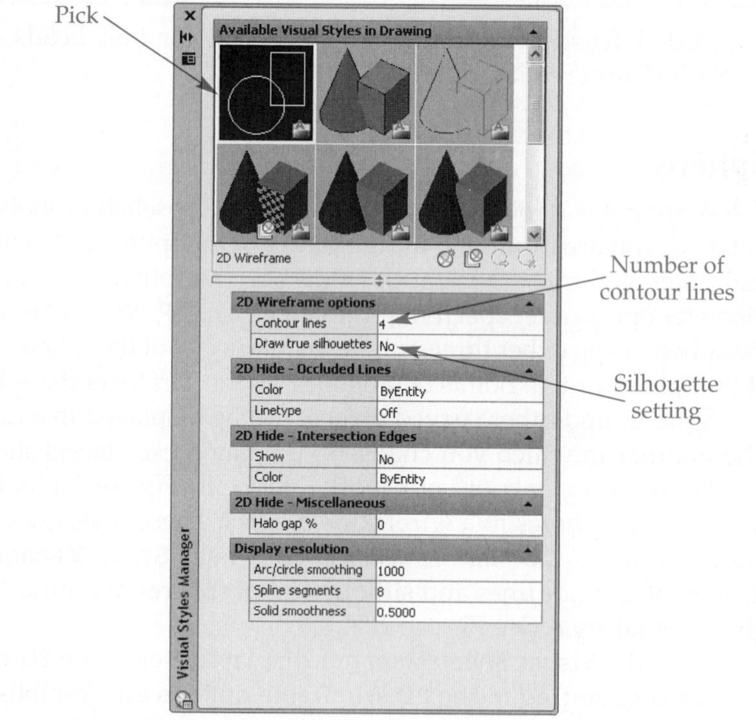

Torus

TORUS

Ribbon
Home
> Modeling Solid
> Primitive
Torus
Type
TORUS
TOR

A basic ***torus*** is a cylinder bent into a circle, similar to a doughnut or inner tube. There are three types of tori. See Figure 2-15. A torus with a tube diameter that touches itself is called ***self intersecting*** and has no center hole. To create a self-intersecting torus, the tube radius must be greater than the torus radius. The third type of torus looks like a football. It is drawn by entering a negative torus radius and a positive tube radius of greater absolute value, i.e. –1 and 1.1.

Once the command is initiated, you are prompted for the center of the torus or to enter an option. If you pick the center, you must then set the radius of the torus. To

Figure 2-15.
The three types of tori are shown as wireframes and with hidden lines removed.

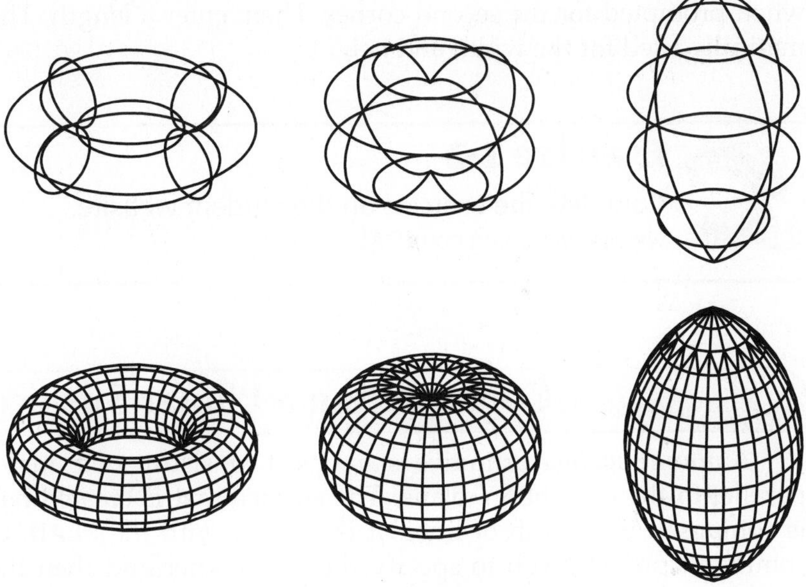

specify a diameter, enter the **Diameter** option after specifying the center. This defines a base circle that is the centerline of the tube. The **3P**, **2P**, and **Ttr** options are used to define the base circle of the torus using either three points, two points, or two points of tangency and a radius.

Once the base circle of the torus is defined, you are prompted for the tube radius or to enter an option. The tube radius defines the cross-sectional circle of the tube. To specify a diameter of the cross-sectional circle, enter the **Diameter** option. You can also use the **2point** option to pick two points on screen that define the diameter of the cross-sectional circle.

Wedge

Ribbon
Home
> Modeling
Solid
> Primitive
Wedge
Type
WEDGE
WE

WEDGE

A ***wedge*** has five sides, four of which are at right angles and the fifth at an angle other than 90°. See Figure 2-16. Once the command is initiated, you are prompted to select the first corner of the base or to enter an option. By default, a wedge is constructed by picking diagonal corners of the base and setting a height. To pick the center point, enter the **Center** option. The center point of a wedge is the middle of the angled surface. You must then pick a point to set the width and length before entering a height.

After specifying the first corner or the center, you can enter the length, width, and height instead of picking a second corner. When prompted for the second corner, enter the **Length** option and specify the length. You are then prompted for the width. After the width is entered, you are prompted for the height.

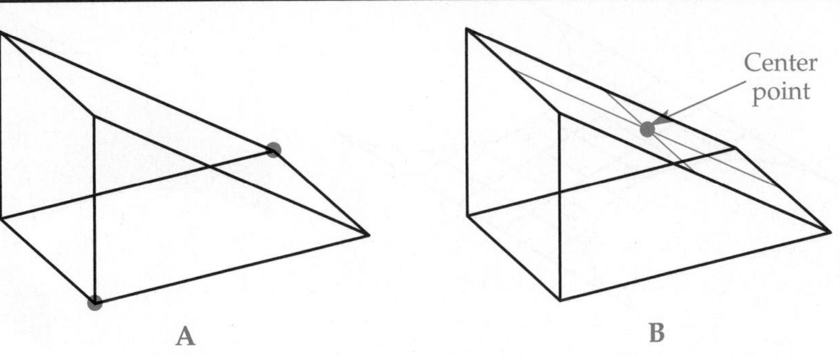

Figure 2-16.
A—A wedge drawn by picking corners and specifying a height. B—A wedge drawn using the **Center** option. Notice the location of the center.

To create a wedge with equal length, width, and height, enter the **Cube** option when prompted for the second corner. Then, enter a length. The same value is automatically used for the width and height.

Exercise 2-2

Complete the exercise on the student website.
www.g-wlearning.com/CAD

Constructing a Planar Surface

PLANSURF

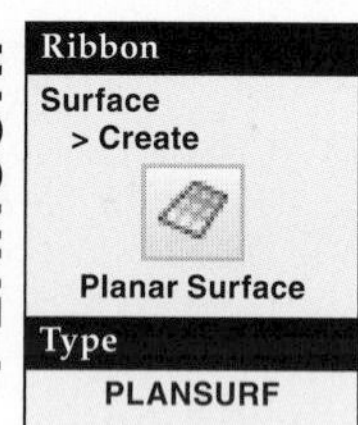

A ***planar surface*** primitive is an object consisting of a single plane and is created parallel to the current XY plane. The surface that is created has zero thickness and is composed of a mesh of lines. It is created with the **PLANESURF** command. The command prompts you to specify the first corner and then the second corner of a rectangle. Once drawn, the surface is displayed as a mesh with lines in the X and Y directions. See Figure 2-17A. These lines are called *isolines* and do not include the object's boundary. The **SURFU** (Y axis) and **SURFV** (X axis) system variables determine how many isolines are created when the planar surface is drawn. The isoline values can be changed later using the **Properties** palette. The maximum number of isolines in either direction is 200.

Figure 2-17.
A—A rectangular planar surface with four isolines in the Y direction and eight isolines in the X direction. B—These two arcs and two lines form a closed area and lie on a single plane. C—The arcs and curves are converted into a planar surface. D—The planar surface is converted into a solid.

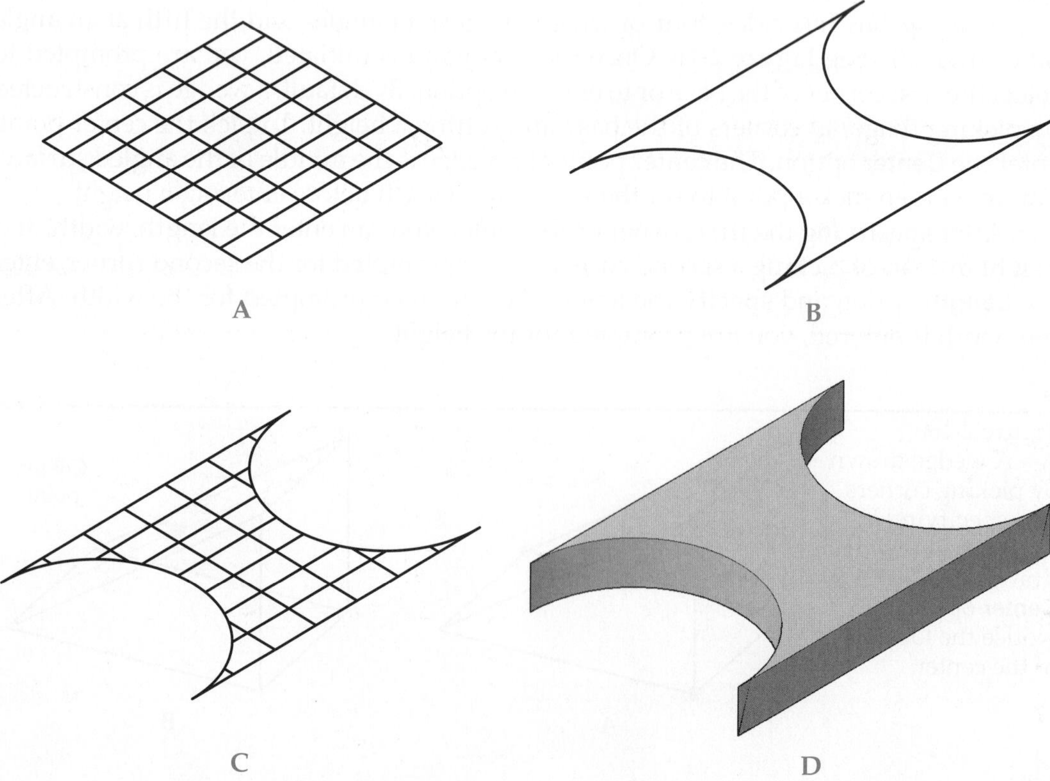

The **Object** option of the **PLANESURF** command allows you to convert a 2D object into a planar surface. Any existing object or objects lying in a single plane and forming a closed area can be converted to a planar surface. The objects in Figure 2-17B are two arcs and two lines connected at their endpoints. The resulting planar surface is shown in Figure 2-17C.

Although a planar surface is not a solid, it can be converted into a solid in a single step. For example, the object in Figure 2-17C is converted into a solid using the **THICKEN** command. See Figure 2-17D. The object that started as two arcs and two lines is now a solid model and can be manipulated and edited like any other solid. This capability allows you to create intricate planar shapes and quickly convert them to a solid for use in advanced modeling applications. Model editing procedures are discussed in detail in Chapters 11 through 13.

Creating Composite Solids

A ***composite solid*** is a solid model constructed of two or more solids, often primitives. Solids can be subtracted from each other, joined to form a new solid, or overlapped to create an intersection or interference. The commands used to create composite solids are found in the **Solid Editing** panel of the **Home** tab or the **Boolean** panel of the **Solid** tab in the ribbon. See Figure 2-18.

Introduction to Booleans

There are three operations that form the basis of constructing many complex solid models. Joining two or more solids is called a ***union*** operation. Subtracting one solid from another is called a ***subtraction*** operation. Forming a solid based on the volume of overlapping solids is called an ***intersection*** operation. Unions, subtractions, and intersections as a group are called ***Boolean operations.*** George Boole (1815–1864) was an English mathematician who developed a system of mathematical logic where all variables have the value of either one or zero. Boole's two-value logic, or *binary algebra,* is the basis for the mathematical calculations used by computers and, specifically, for those required in the construction of composite solids.

NOTE

Boolean operations used to create composite solids can also be used on meshes that have been converted to solids. See Chapter 3 for a complete discussion of meshes.

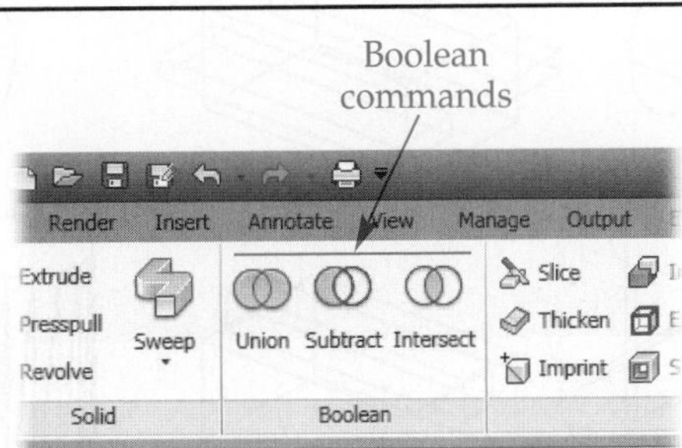

Figure 2-18.
Selecting a Boolean command in the **Boolean** panel on the **Solid** tab of the ribbon. These commands are also located in the **Solid Editing** panel on the **Home** tab.

Joining Two or More Solid Objects

UNION

Ribbon
Home
> Solid Editing
Solid
> Boolean
Solid, Union
Type
UNION
UNI

The **UNION** command is used to combine solid objects, Figure 2-19. The solids do not need to touch or intersect to form a union. Therefore, accurately locate the primitives when drawing them. After selecting the objects to join, press [Enter] and the action is completed.

In the examples shown in Figure 2-19B, notice that lines, or edges, are shown at the new intersection points of the joined objects. This is an indication that the features are one object, not separate objects.

Subtracting Solids

SUBTRACT

Ribbon
Home
> Solid Editing
Solid
> Boolean
Solid, Subtract
Type
SUBTRACT
SU

The **SUBTRACT** command allows you to remove the volume of one or more solids from another solid. Several examples are shown in Figure 2-20. The first object selected in the subtraction operation is the object *from* which volume is to be subtracted. The next object is the object to be subtracted from the first. The completed object will be a new solid. If the result is the opposite of what you intended, you may have selected the objects in the wrong order. Just undo the operation and try again.

Figure 2-19.
A—The solid primitives shown here have areas of intersection and overlap. B—Composite solids after using the **UNION** command. Notice the lines displayed where the previous objects intersected.

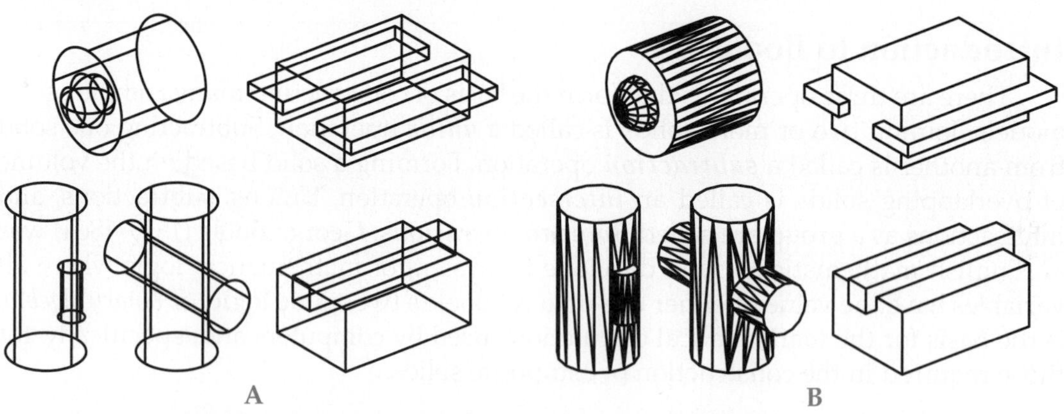

Figure 2-20.
A—The solid primitives shown here have areas of intersection and overlap. B—Composite solids after using the **SUBTRACT** command.

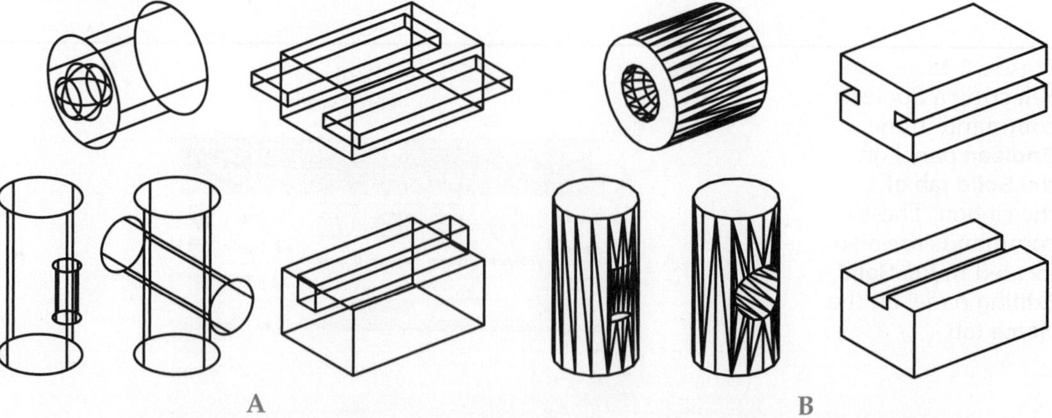

Creating New Solids from the Intersection of Solids

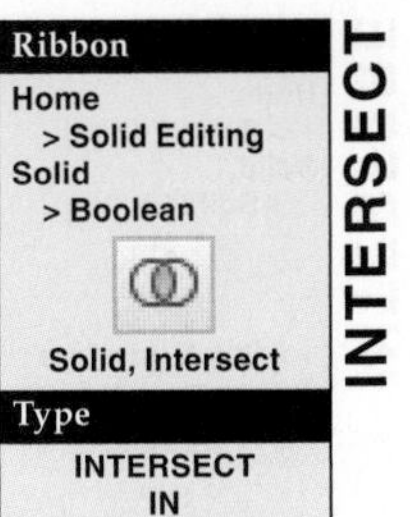

When solid objects intersect, the overlap forms a common volume, a space that both objects share. This shared space is called an ***intersection.*** An intersection (common volume) can be made into a composite solid using the **INTERSECT** command. Figure 2-21 shows several examples. A solid is formed from the common volume. The original objects are removed.

The **INTERSECT** command is also useful in 2D drawing. For example, if you need to create a complex shape that must later be used for inquiry calculations or hatching, draw the main object first. Then, draw all intersecting or overlapping objects. Next, create regions of the shapes. Finally, use **INTERSECT** to create the final shape. The resulting shape is a region and has solid properties. Regions are discussed later in this chapter.

Exercise 2-3

Complete the exercise on the student website.
www.g-wlearning.com/CAD

Joined first using the **UNION** command

A

B

Figure 2-21.
A—The solid primitives shown here have areas of intersection and overlap.
B—Composite solids after using the **INTERSECT** command.

Creating New Solids Using the Interfere Command

INTERFERE

Ribbon
Home
> Solid Editing
Solid
> Solid Editing

> Solid, Interfere
Type
INTERFERE
INF

When you use the **SUBTRACT**, **UNION**, and **INTERSECT** commands, the original solids are deleted. They are replaced by the new composite solid. The **INTERFERE** command does not do this. A new solid is created from the interference (common volume) as if the **INTERSECT** command is used, but the original objects can be either deleted or retained.

Once the command is initiated, you are prompted to select the first set of solids or to enter an option. The **Settings** option opens the **Interference Settings** dialog box, which is used to change the visual style and color of the interference solid and the visual style of the viewport. The **Nested selection** option allows you to check the interference of separate solid objects within a nested block. A ***nested block*** is one that is composed of other blocks. When any needed options are set, select the first set of solids and press [Enter].

You are prompted to select the second set of solids or to enter an option. Entering the **Check first set** option tells AutoCAD to check the objects in the first set for interference. There is no second set when this option is used. Otherwise, select the second set of solids and press [Enter].

AutoCAD zooms in on the highlighted interference solid and displays the **Interference Checking** dialog box. See Figure 2-22. The visual style is set to a wireframe display by default and the interference solid is shaded in a color, which is red by default.

In the **Interfering objects** area of the **Interference Checking** dialog box, the number of objects selected in the first and second sets is displayed. The number of interfering pairs found in the selected objects is also displayed.

The buttons in the **Highlight** area of the dialog box are used to highlight the previous or next interference object. If the **Zoom to pair** check box is checked, AutoCAD zooms to the interference objects when the **Previous** or **Next** button is selected.

To the right of the **Highlight** area are three navigation buttons—**Zoom Realtime**, **Pan Realtime**, and **3D Orbit**. Selecting one of these display options temporarily hides

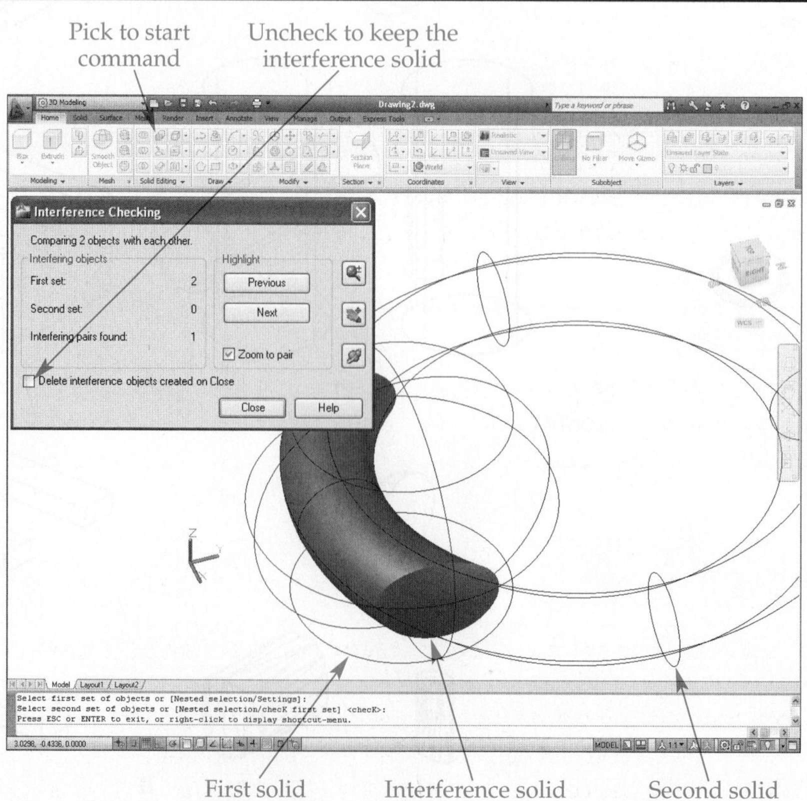

Figure 2-22.
The **Interference Checking** dialog box is used to check for interference between solids. To retain the interference solid, uncheck the **Delete interference objects created on Close** check box.

Figure 2-23.
A—Two solids form an area of intersection. B—After using **INTERFERE**, a new solid is defined (shown here in color) and the original solids remain. C—The new solid can be moved or copied.

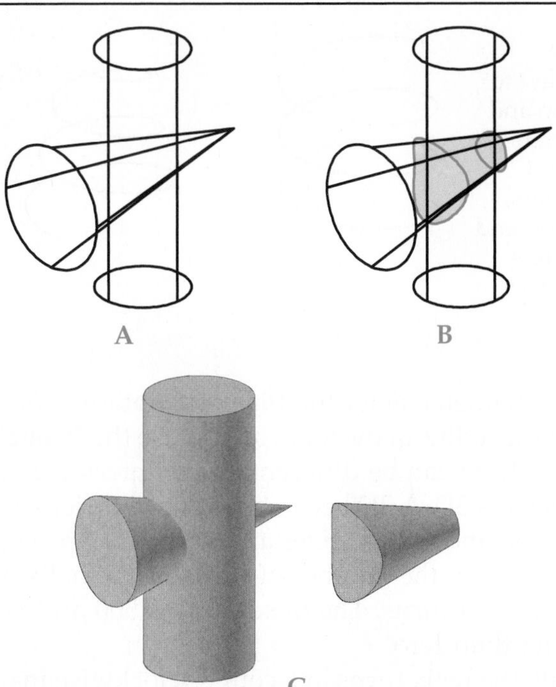

the dialog box and activates the selected command. This allows you to navigate in the viewport. When the navigation mode is exited, the dialog box is redisplayed.

By default, the **Delete interference objects created on Close** check box is checked. This means that the object(s) created by interference is deleted. In order to retain the new solid(s), uncheck this box.

An example of interference checking and the result is shown in Figure 2-23. Notice that the original solids are intact, but new lines indicate the new solid. The new solid is retained as a separate object because the **Delete interference objects created on Close** check box was unchecked. The new solid can be moved, copied, and manipulated just like any other object. Figure 2-23C shows the new object after it has been moved and a conceptual display generated.

When the **INTERFERE** command is used, AutoCAD compares the first set of solids to the second set. Any solids that are selected for both the first and second sets are automatically included as part of the first selection set and eliminated from the second. If you do not select a second set of objects or the **Check first set** option is used, AutoCAD calculates the interference between the objects in the first selection set.

Exercise 2-4

Complete the exercise on the student website.
www.g-wlearning.com/CAD

Creating a Helix

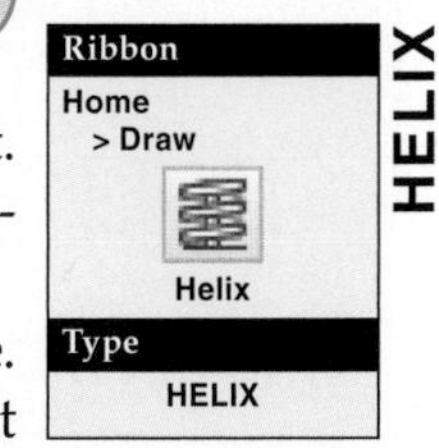

A *helix* is a spline in the form of a spiral and can be created as a 2D or 3D object. See Figure 2-24. It is not a solid object. However, it can be used as the path or framework for creating solid objects such as springs and spiral staircases.

When the command is initiated, you are prompted for the center of the helix base. After picking the center, you are prompted to enter the radius of the base. If you want

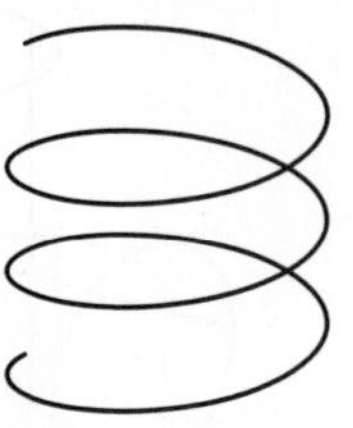
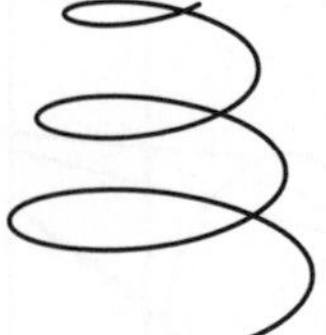
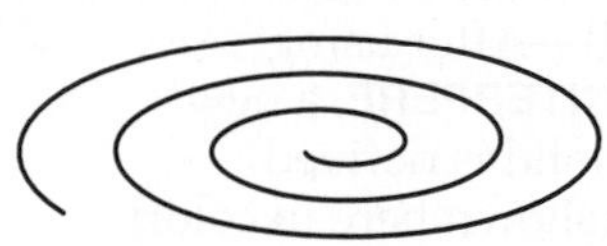

Figure 2-24.
Three types of helices. From left to right, equal top and bottom diameters, unequal top and bottom diameters, and unequal top and bottom diameters with the height set to zero.

to specify the diameter, enter the **Diameter** option. After the base is defined, you are prompted for the radius of the top. You can use the **Diameter** option to enter a diameter. The top and bottom can be different sizes. Entering different sizes creates a tapered helix, if the helix is 3D. A 2D helix should have different sizes for the top and bottom.

After the top and bottom sizes are set, you are prompted to set the height or enter an option. To specify the number of turns in the helix, enter the **Turns** option. Then, enter the number of turns. The maximum is 500 and you can enter values less than one, but greater than zero.

By default, the helix turns in a counterclockwise manner. To change the direction in which the helix turns, enter the **Twist** option. Then, enter CW for clockwise or CCW for counterclockwise.

The height of the helix can be set in one of three ways. First, you can enter a direct distance. To do this, type the height value or pick with the mouse to set the height. To create a 2D helix, enter a height of zero.

You can also set the height for one turn of the helix using the **Turn height** option. In this case, the total height is the number of turns multiplied by the turn height. If you provide a value for the turn height and then specify the helix height, the number of turns is automatically calculated and the helix is drawn. Conversely, if you provide values for both the turn height and number of turns, the helix height is calculated by AutoCAD.

Finally, you can pick a location for the axis endpoint using the **Axis endpoint** option. This is the same option available with a cone, cylinder, or pyramid.

As an example, a solid model of a spring can be created by constructing a helix and a circle and then using the **SWEEP** command to sweep the circle along the helix path. See Figure 2-25. The **SWEEP** command is discussed in detail in Chapter 10.

First, determine the diameter of the spring wire and then draw a circle using that value. For this example, you will create two springs each with a wire diameter of .125 units, so draw two circles of that diameter, Figure 2-26. Their locations are not important. Next, determine the diameter of the spring and draw a corresponding

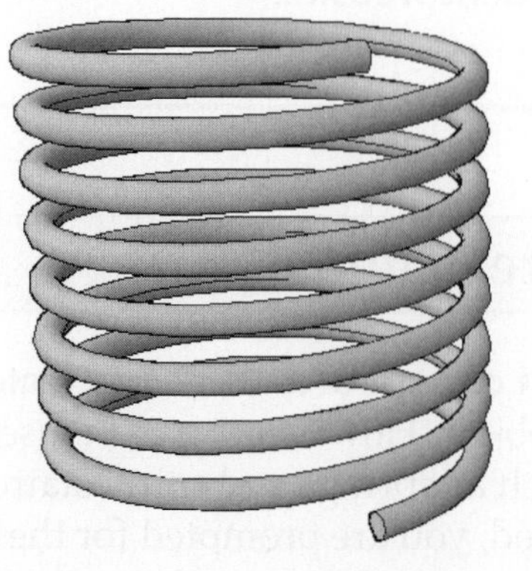
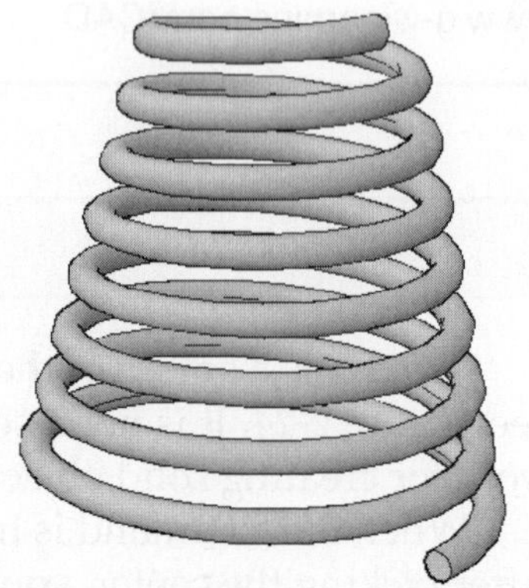

Figure 2-25.
A helix can be used as a path to create a spring.

Figure 2-26.
To create a spring, first draw a circle the same diameter as the spring wire. Then, draw the helix and sweep the circle along the helix. Shown here are the two helices used to create the springs in Figure 2-25.

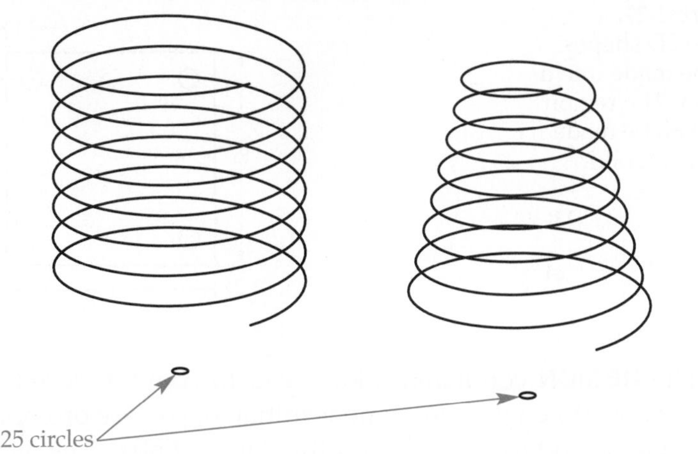

helix. For this example, draw a helix anywhere on screen with a bottom diameter of one unit and a top diameter of one unit. Set the number of turns to eight and specify a height of two units. Draw another helix with the same settings, except make the top diameter .5 units.

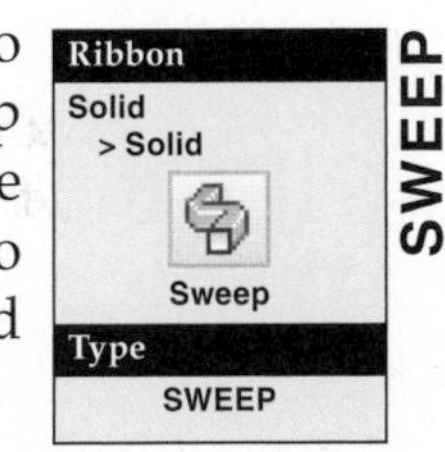

Initiate the **SWEEP** command. You are first prompted to select the objects to sweep; pick one circle and press [Enter]. Next, you are prompted to select the sweep path. Select one of the helices. The first sweep, or spring, is completed. Repeat the procedure for the other circle and helix. The drawing is now composed of the two original, single-line helices and the two new swept solids. The circles are consumed by the **SWEEP** command.

Exercise 2-5

Complete the exercise on the student website.
www.g-wlearning.com/CAD

Working with Regions

A ***region*** is a closed, two-dimensional solid. It is a solid model without thickness (Z value). A region can be analyzed for its mass properties. Therefore, regions are useful for 2D applications where area and boundary calculations must be quickly obtained from a drawing.

Boolean operations can be performed on regions. When regions are unioned, subtracted, or intersected, a ***composite region*** is created. A composite region is also called a ***region model.***

A region can be quickly and easily given a thickness, or ***extruded*** into a 3D solid object. This means that you can convert a 2D shape into a 3D solid model in just a few steps. An application is drawing a 2D section view, converting it into a region, and extruding the region into a 3D solid model. Extruding is covered in Chapter 9.

Constructing a 2D Region Model

The following example creates, as a region, the plan view of a base for a support bracket. In Chapter 9, you will learn how to extrude the region into a solid. First, start a new drawing. Next, create the profile geometry in Figure 2-27 using the **RECTANGLE** and **CIRCLE** commands. These commands create 2D objects that can be converted into regions. The **PLINE** and **LINE** commands can also be used to create closed 2D objects.

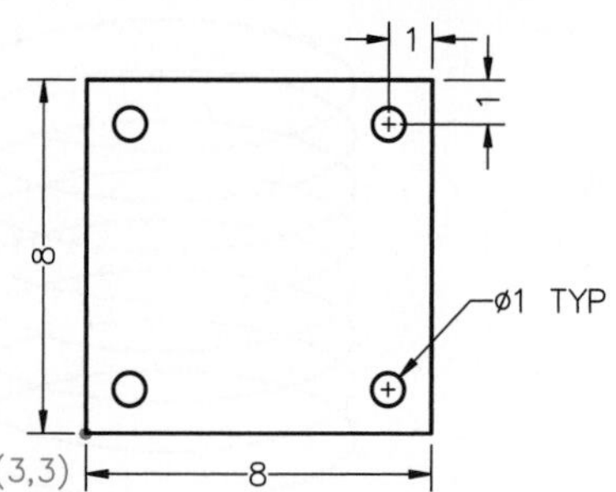

Figure 2-27.
These 2D shapes can be made into a region. The region can then be made into a 3D solid.

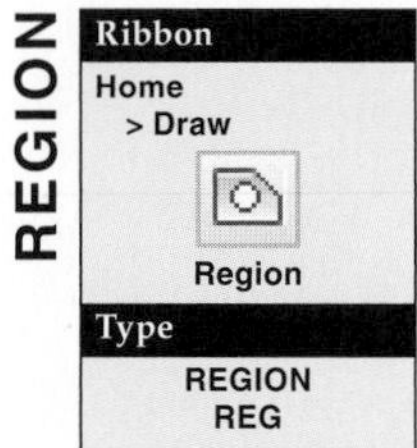

The **REGION** command allows you to convert closed, two-dimensional objects into regions. When the command is initiated, you are prompted to select objects. Select the rectangle and four circles and then press [Enter]. The rectangle and each circle are now separate regions and the original objects are deleted. You may need to switch to a wireframe visual style in order to see the circles. You can individually pick the regions. If you pick a circle, notice that a grip is displayed in the center, but not at the four quadrants. This is because the object is not a circle anymore. However, you can still snap to the quadrants.

In order to create the proper solid, the circular regions must be subtracted from the rectangular region. Using the **SUBTRACT** command, select the rectangle as the object to be subtracted *from* and then all of the circles as the objects to subtract. Now, if you select the rectangle or any of the circles, you can see that a single region has been created from the five separate regions. If you set the Conceptual or Realistic visual style current, you can see that the circles are now holes in the region. See Figure 2-28.

Using the Boundary Command to Create a Region

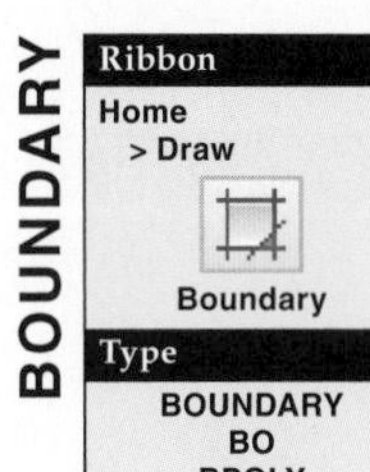

The **BOUNDARY** command is often used to create a polyline for hatching or an inquiry. In addition, this command can be used to create a region. When the command is initiated, the **Boundary Creation** dialog box is displayed. See Figure 2-29.

Next, select **Region** from the **Object type:** drop-down list in the **Boundary retention** area of the dialog box. Also, you can refine the boundary selection method by turning island detection on or off. When the **Island detection** check box above the **Boundary retention** area is checked, island detection is on.

- **On.** When an internal point is selected in the object, AutoCAD creates separate regions from any islands that reside within the object.
- **Off.** When an internal point is selected in the object, AutoCAD ignores islands that reside within the object when creating the region.

Finally, select the **Pick Points** button. The dialog box is closed and you are prompted to select an internal point. Pick a point inside of the object that you wish to convert to a region. Press [Enter] when you are finished and the region is created. You can always check to see if an object is a polyline or region by selecting the object, right-clicking, and selecting **Properties...** from the shortcut menu.

Figure 2-28.
Once the circular regions are subtracted from the rectangular region, they appear as holes. This is clear when a shaded visual style is set current.

Figure 2-29.
Regions can be created using the **Boundary Creation** dialog box.

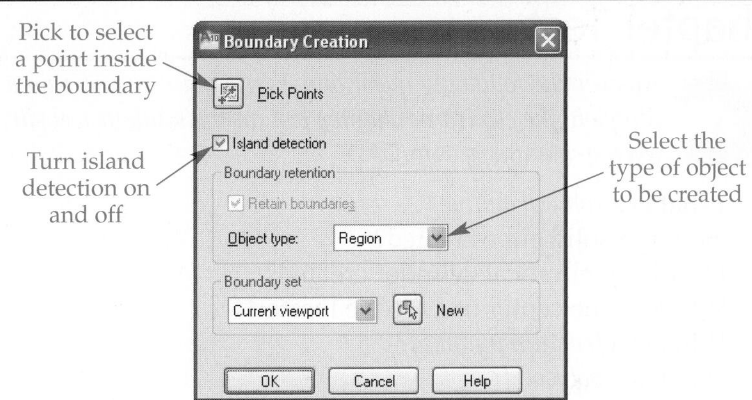

Calculating the Area of a Region

A region is not a polyline. It is an enclosed area called a ***loop.*** Certain properties of the region, such as area, are stored as a value of the region. The **AREA** command can be used to determine the length of all sides and the area of the loop. This can be a useful advantage of a region.

For example, suppose a parking lot is being repaved. You need to calculate the surface area of the parking lot to determine the amount of material needed. This total surface area excludes the space taken up by planting dividers, sidewalks, and lampposts because you will not be paving under these items. If the parking lot and all objects inside of it are drawn as a region, the **AREA** command can give you this figure in one step using the **Object** option. If a polyline is used to draw the parking lot, all internal features must be subtracted each time the **AREA** command is used.

PROFESSIONAL TIP

Regions can prove valuable when working with many items:

- Roof areas excluding chimneys, vents, and fans.
- Bodies of water, such as lakes, excluding islands.
- Lawns and areas of grass excluding flower beds, trees, and shrubs.
- Landscaping areas excluding lawns, sidewalks, and parking lots.
- Concrete surfaces, such as sidewalks, excluding openings for landscaping, drains, and utility covers.

You can find many other applications for regions that can help in your daily tasks.

Exercise 2-6

Complete the exercise on the student website.
www.g-wlearning.com/CAD

Chapter Test

Answer the following questions. Write your answers on a separate sheet of paper or complete the electronic chapter test on the student website.
www.g-wlearning.com/CAD

1. What is a *solid primitive?*
2. How is a solid cube created?
3. How is an elliptical cylinder created?
4. Where is the center of a wedge located?
5. What is a *frustum pyramid?*
6. What is a *polysolid?*
7. Name at least four AutoCAD 2D entities that can be converted to a polysolid.
8. What type of entity does the **HELIX** command create and how can it be converted into a solid model?
9. What is a *composite solid?*
10. Which type of mathematical calculations are used in the construction of solid models?
11. How are two or more solids combined to make a composite solid?
12. What is the function of the **INTERSECT** command?
13. How does the **INTERFERE** command differ from **INTERSECT** and **UNION**?
14. What is a *region?*
15. How can a 2D section view be converted to a 3D solid model?

Drawing Problems

Draw the objects in the following problems using the appropriate solid primitive commands and Boolean operations. Apply geometric constraints to the base 2D drawing as appropriate. Do not add dimensions to the models. Save the drawings as P2-(problem number). Display and plot the problems as indicated by your instructor.

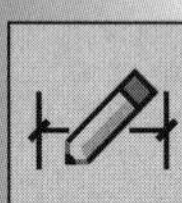

1.

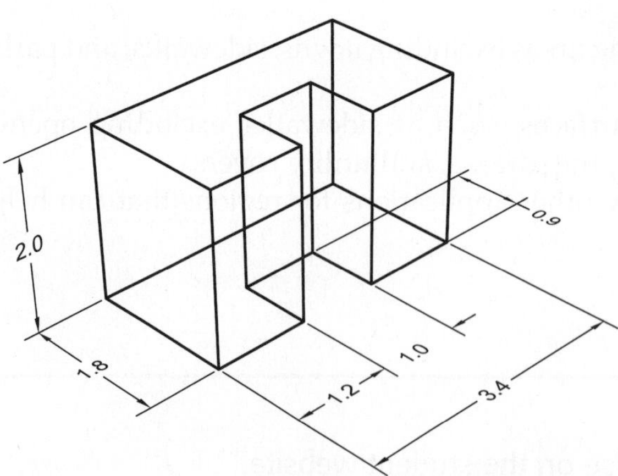

Drawing Problems - Chapter 2

2.

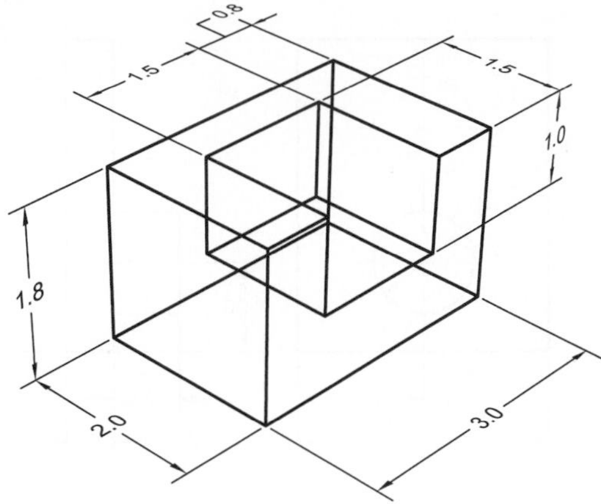
0.8
1.5
1.5
1.0
1.8
2.0
3.0

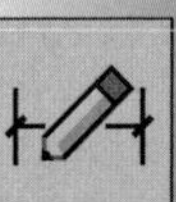

3.

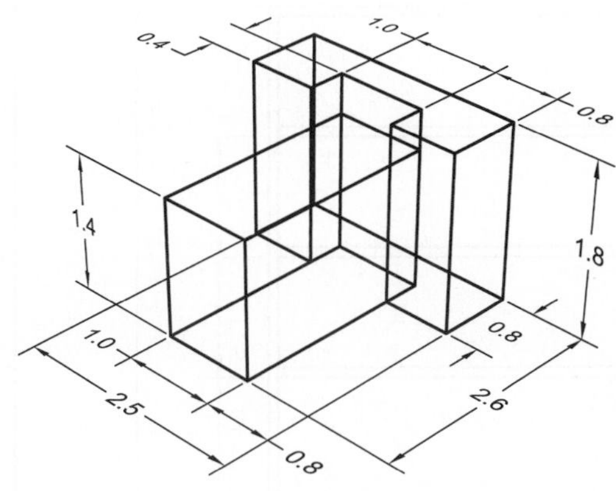
0.4
1.0
0.8
1.4
1.8
0.8
1.0
2.5
2.6
0.8

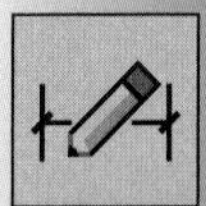

4.

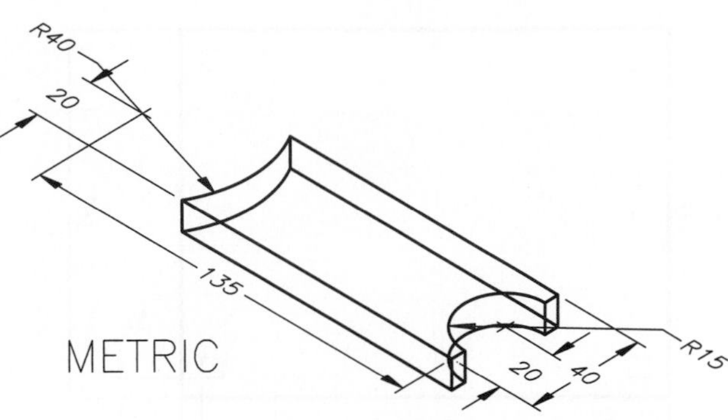
R40
20
135
METRIC
R15
20
40

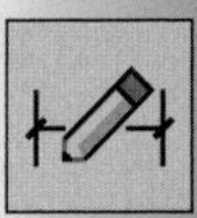

5.

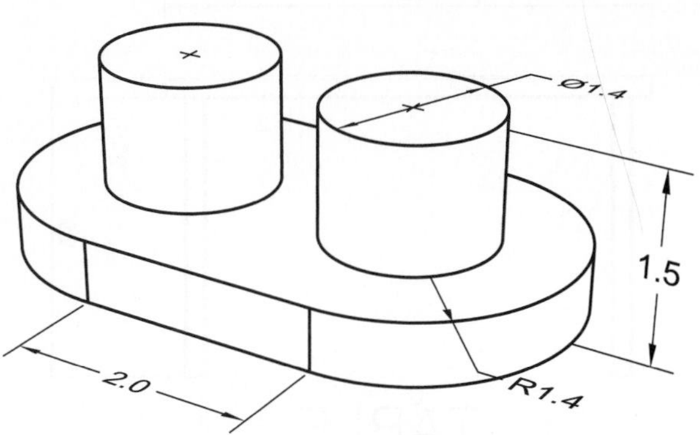
Ø1.4
1.5
2.0
R1.4

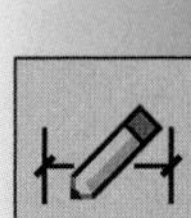

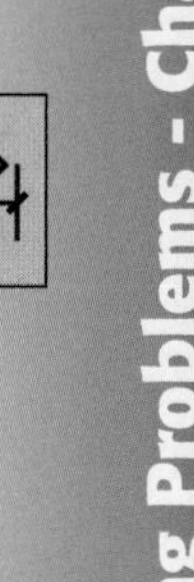
Drawing Problems - Chapter 2

6.

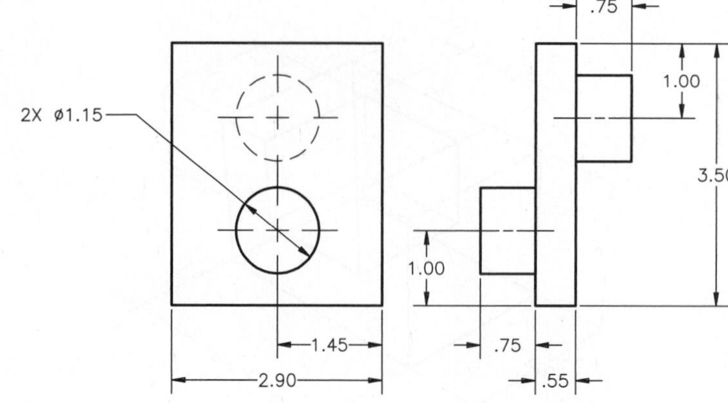

7.

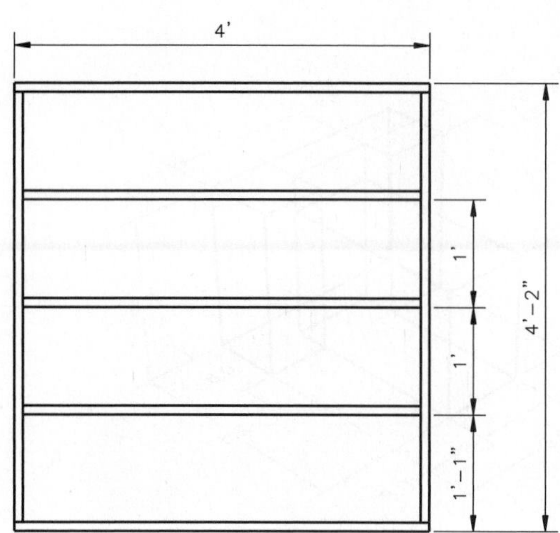

8.

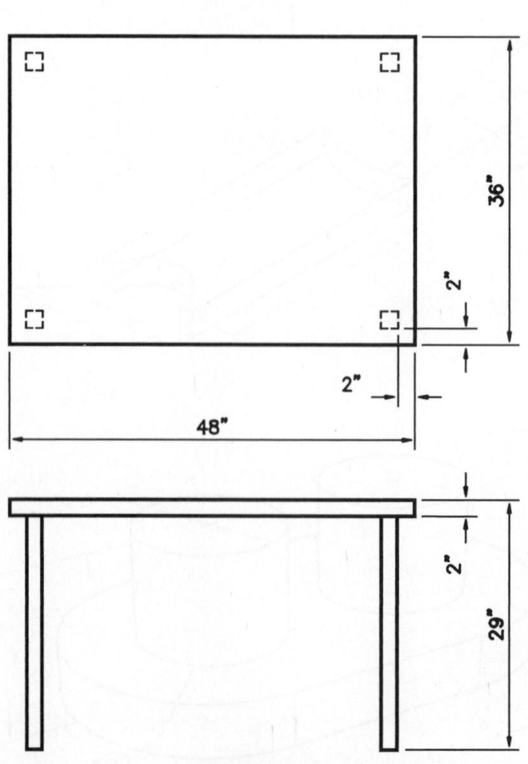

9.

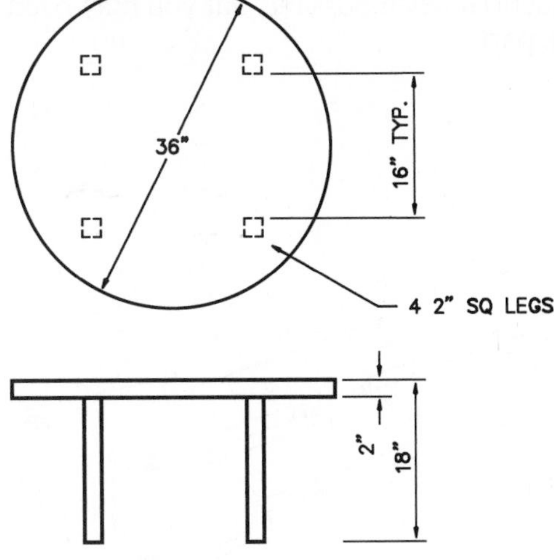

10.

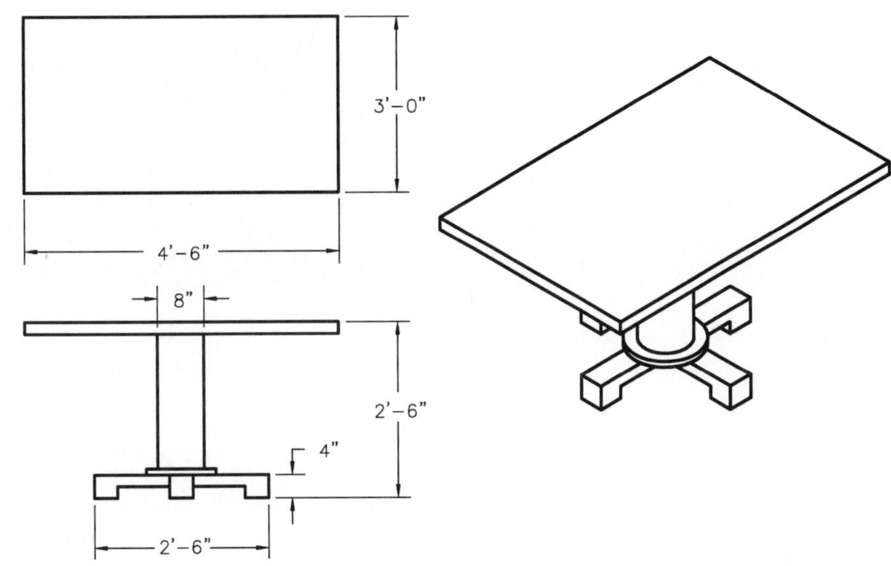

11. Draw the seat for a kitchen chair shown below. In later chapters, you will complete this chair. The seat is 1″ thick.

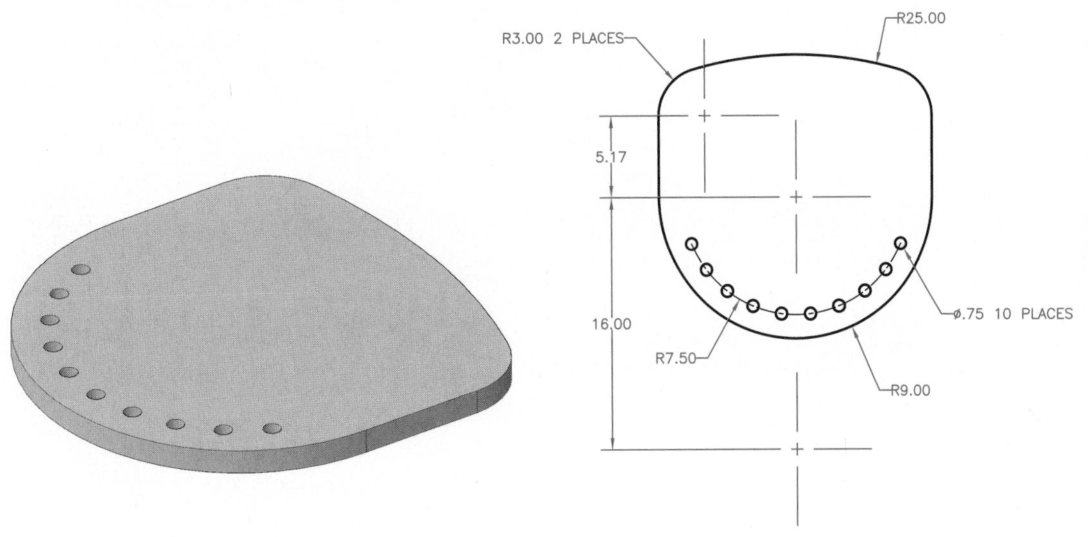

Three-dimensional models are not typically dimensioned, but there are cases where dimensions may be applied to assist workers on the shop floor. Whenever a 3D model is dimensioned, follow accepted drafting practices when possible, but you may need to violate certain rules in order to clearly describe the part.

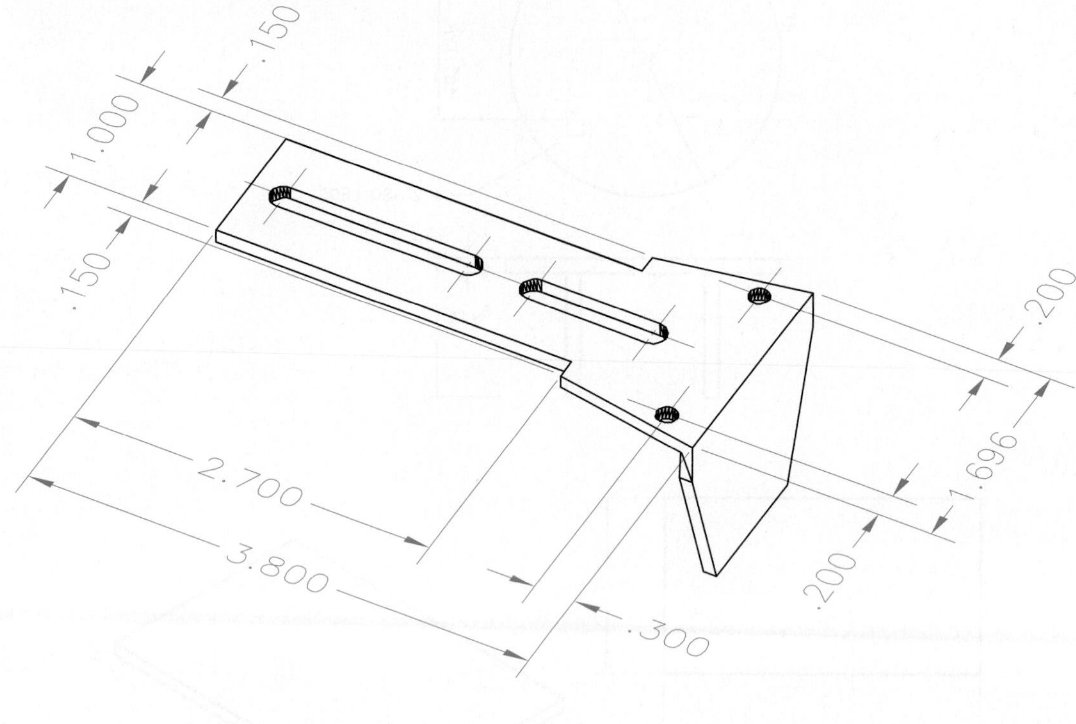

CHAPTER 3

Mesh Modeling

Learning Objectives

After completing this chapter, you will be able to:

✓ Explain tessellation division and values.
✓ Create mesh primitives.
✓ Create a smoothed mesh object.
✓ Create a refined mesh object.
✓ Generate a mesh by converting a solid or surface.
✓ Generate a solid or surface by converting a mesh.
✓ Execute editing techniques on mesh objects.
✓ Create a split face on a mesh.
✓ Produce an extruded mesh face.
✓ Apply a crease to mesh subobjects.
✓ Create and close mesh object gaps.
✓ Create a new mesh face by collapsing a mesh face or edge.
✓ Merge mesh faces to form a single mesh face.
✓ Construct a new mesh face by spinning a triangular mesh face.
✓ Construct mesh forms.

Overview of Mesh Modeling

Mesh primitives and mesh forms can be used to create freeform designs. See Figure 3-1. The tools for creating and editing meshes extend the capability of AutoCAD's 3D modeling tools. There are two key workflows that the designer considers:

- The creation of 3D models, which can be solids, surfaces, or meshes.
- Editing the 3D models to create unique shapes.

Mesh models can be created as mesh primitives, mesh forms, or freeform mesh shapes. A ***mesh model*** consists of vertices, edges, and faces. You can modify or refine a mesh by adding smoothness, creases, extrusions, and splits. You can also distort a mesh to create unique freeform shapes.

Mesh models are a type of surface model. ***Subdivision surfaces*** is another term for mesh models. They do *not* have volume or mass. Rather, mesh models only define the shape of the design.

Figure 3-1.
Constructing an ergonomic mouse as a mesh model. A—The basic mesh primitive. B—Using editing tools, the mesh is reformed. C—The completed mesh model.

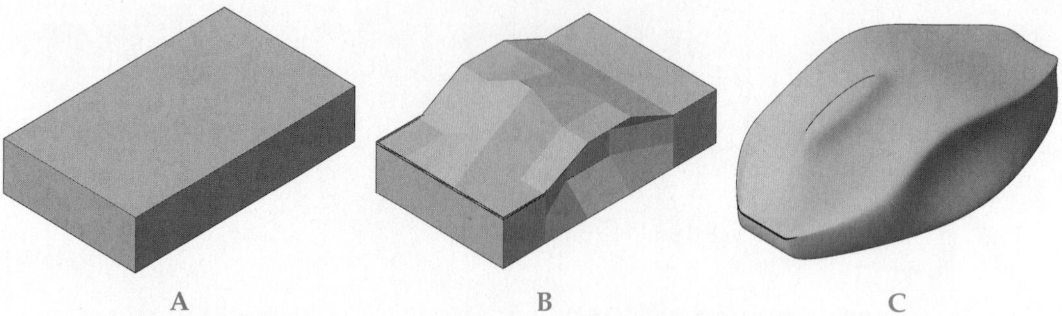

Mesh objects can be created using one of these methods:

- Construct mesh primitives (**MESH** command).
- Convert an existing solid or surface into a mesh object (**MESHSMOOTH** command).
- Construct mesh forms that are ruled, revolved, tabulated, or edge-defined objects (**RULESURF**, **REVSURF**, **TABSURF** and **EDGESURF** commands).
- Convert legacy surface objects into mesh objects using commands such as **3DFACE**, **3DMESH** and **PFACE** commands.

The tools used to create and modify meshes are found on the **Mesh** tab in the ribbon. The commands used to create mesh primitives are similar to those used to create solid primitives, which are discussed in Chapter 2. However, mesh primitives have face mesh objects that are divided into smaller faces. These divisions are based on tessellation division values (smoothness), as discussed in this chapter.

Tessellation Division Values

Tessellation divisions are the basic foundation for the smoothness of a mesh object. Tessellation divisions on a mesh object consist of planar shapes that fit together to form the surface. See Figure 3-2. These divisions display the edges of a mesh face that can then be edited.

When creating mesh primitives, set the mesh tessellation divisions *before* creating a mesh primitive shape. Setting the proper mesh tessellation division value ensures the model has enough faces, edges, and vertices for editing. The default tessellation divisions are listed in **Mesh Primitive Options** dialog box, which is discussed in the next section.

You can change the default smoothness in the **Mesh Primitive Options** dialog box or, if the primitive is being drawn using the command line, by entering the **Settings** option of the **MESH** command. Figure 3-3 shows an example of a box mesh primitive created using the default tessellation divisions. The default settings create a box with

Figure 3-2.
Tessellation divisions are key to mesh modeling as they define the smoothness of the mesh model.

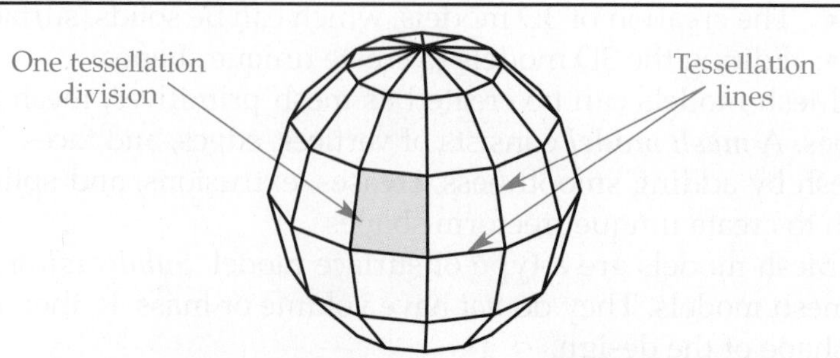

Figure 3-3.
This box mesh primitive is drawn with the default settings. A—Displayed with the 2D Wireframe visual style current. B—After the **HIDE** command is used.

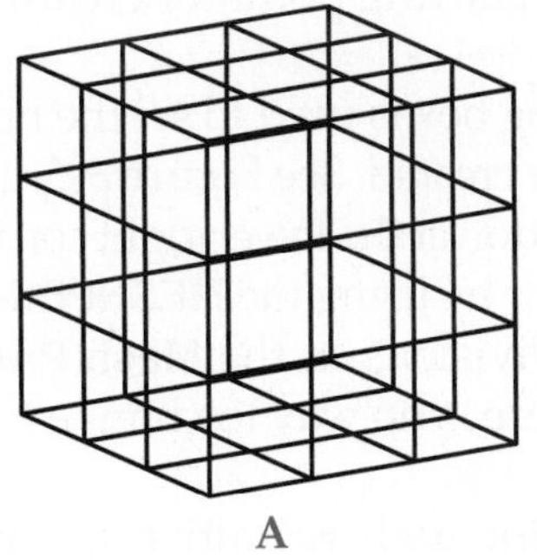

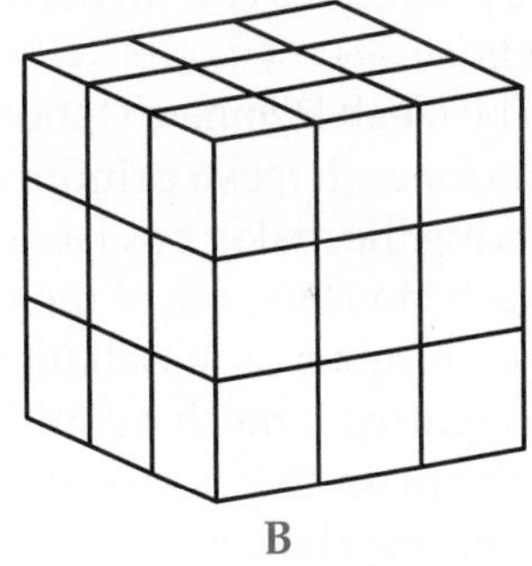

no smoothness, length divisions of three, width divisions of three, and height divisions of three.

Drawing Mesh Primitives

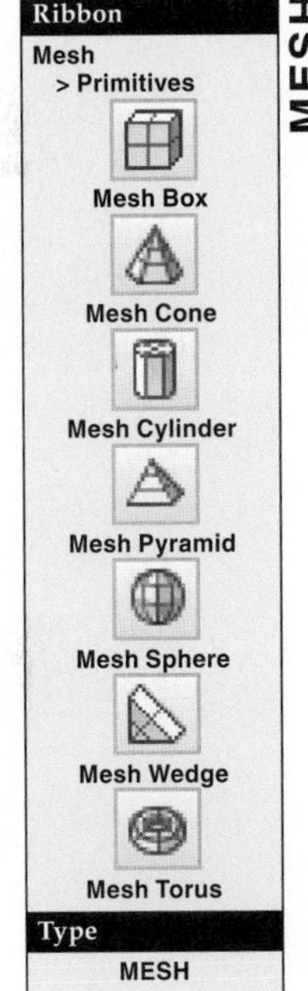

MESH

A primitive is a basic building block. Just as there are solid primitives in AutoCAD, there are mesh primitives. The seven ***mesh primitives*** are mesh box, mesh cone, mesh cylinder, mesh pyramid, mesh sphere, mesh wedge, and mesh torus. See **Figure 3-4.**

Figure 3-4.
An overview of AutoCAD's mesh primitives and the dimensions required to draw them.

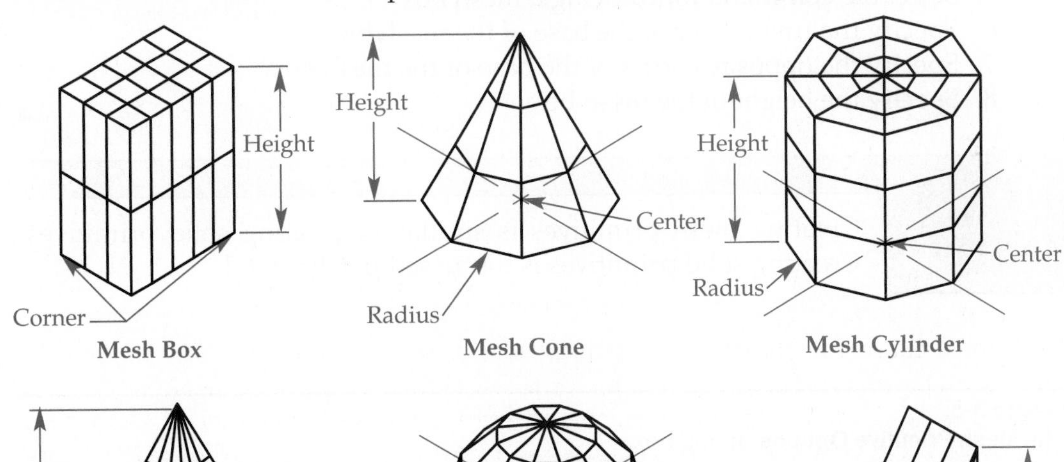

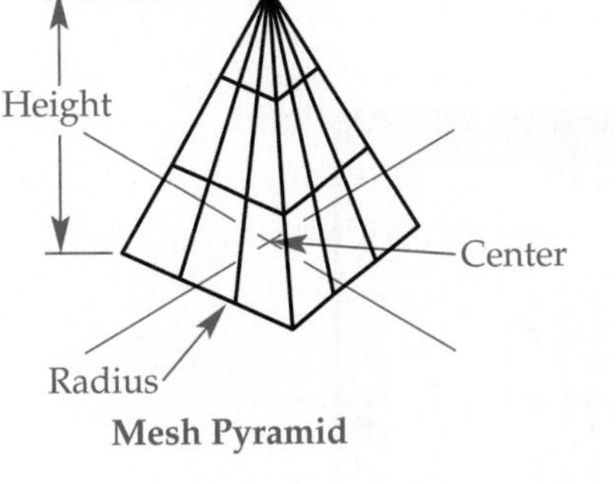

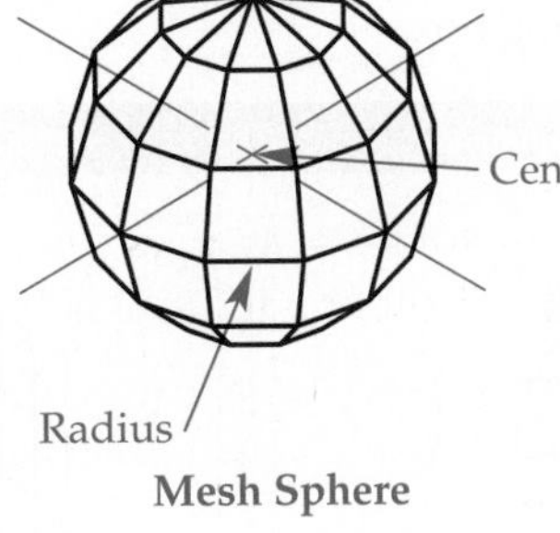

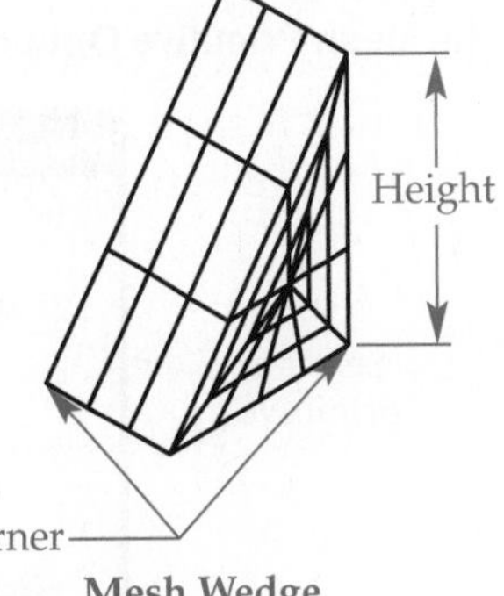

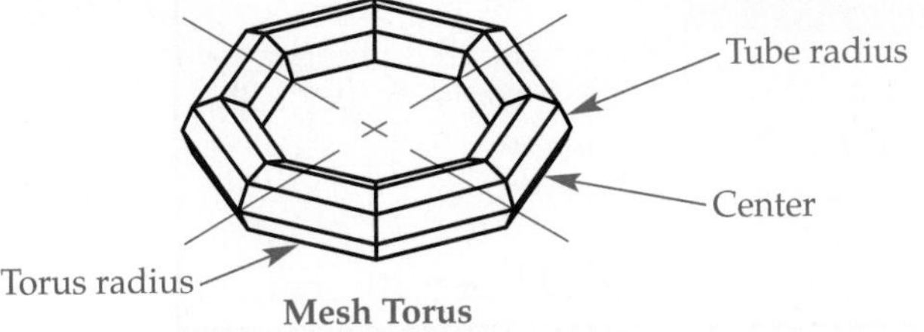

These primitives can be used as the starting point for creating complex freeform mesh models.

The **Mesh Primitive Options** dialog box is used to set the number of tessellation subdivisions for each mesh primitive object created. See Figure 3-5. This dialog box is displayed by picking the dialog box launcher button at the lower-right corner of the **Primitives** panel in the **Mesh Modeling** tab of the ribbon or by using the **MESHPRIMITIVEOPTIONS** command. Set the number of tessellation subdivisions in the **Mesh Primitive Options** dialog box *before* creating a mesh primitive. There is no way to change the number of subdivisions after the primitive is created.

The tessellation subdivisions for each primitive are based on the dimensions required to create the primitive. For example, a box has length, width, and height subdivisions. On the other hand, a mesh cylinder has axis, height, and base subdivisions. To set the subdivisions, select the primitive in the tree on the left-hand side of the **Mesh Primitive Options** dialog box. The subdivision properties are then displayed below the tree. Enter the number of subdivisions as required and then close the dialog box. All new primitives of that type will have this number of subdivisions until the setting is changed in the dialog box. Existing primitives are *not* affected.

The following is an example of how to create a mesh box primitive. The mesh box is the mesh primitive used for most base shapes. Refer to Figure 3-4 for the information required to draw mesh primitives.

1. Open the **Mesh Primitive Options** dialog box.
2. Select the box primitive in the tree.
3. Enter 3 in each of the Length, Width, and Height property boxes to set the subdivisions.
4. Close the dialog box.
5. Select the command for drawing a mesh box.
6. Specify the first corner of the base of the mesh box.
7. Specify the opposite corner of the base of the mesh box.
8. Specify the height of the mesh box.

PROFESSIONAL TIP

Creating mesh primitives is similar to creating solid primitives. Creating solid primitives is discussed in Chapter 2.

Figure 3-5.
The **Mesh Primitive Options** dialog box.

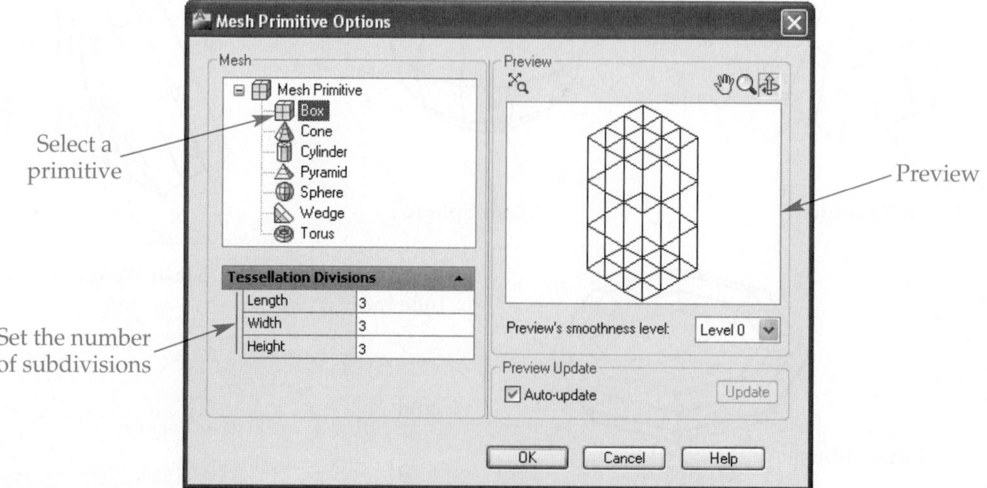

Exercise 3-1

Complete the exercise on the student website.
www.g-wlearning.com/CAD

Converting between Mesh and Surface or Solid Objects

AutoCAD offers the flexibility of converting between solid, surface, and mesh objects. This allows you to select the type of modeling that offers the best tools for the task at hand, then convert the model into a form appropriate for the end result. The next sections discuss converting between solid, surface, and mesh objects and the settings that control the conversion.

PROFESSIONAL TIP

The **DELOBJ** system variable plays an important role when working with meshes, solids, and surfaces. Set the **DELOBJ** system variable to 0 to retain the geometry used to create the mesh, solid, or surface. When the variable is set to 1, the geometry used to create the mesh, solid, or surface is deleted.

Mesh Tessellation Options

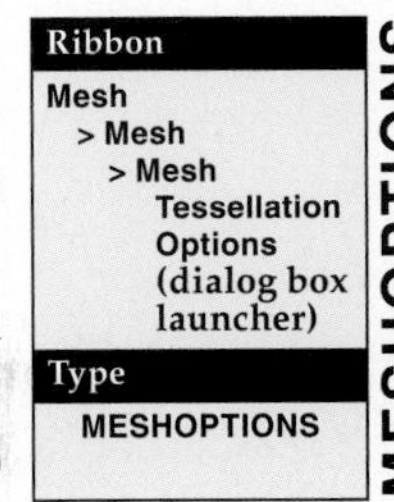

The **Mesh Tessellation Options** dialog box contains many options, Figure 3-6. This dialog box is displayed by picking the dialog box launcher button at the lower-right corner of the **Mesh** panel in the **Mesh Modeling** tab of the ribbon or by using the **MESHOPTIONS** command.

When converting to mesh objects, the resulting mesh is one of three different mesh types. The **FACETERMESHTYPE** system variable controls which type of mesh is created. Selecting Smooth Mesh Optimized in the **Mesh type:** drop-down list converts objects to the optimized mesh type (**FACETERMESHTYPE** = 0). This is the default and recommended

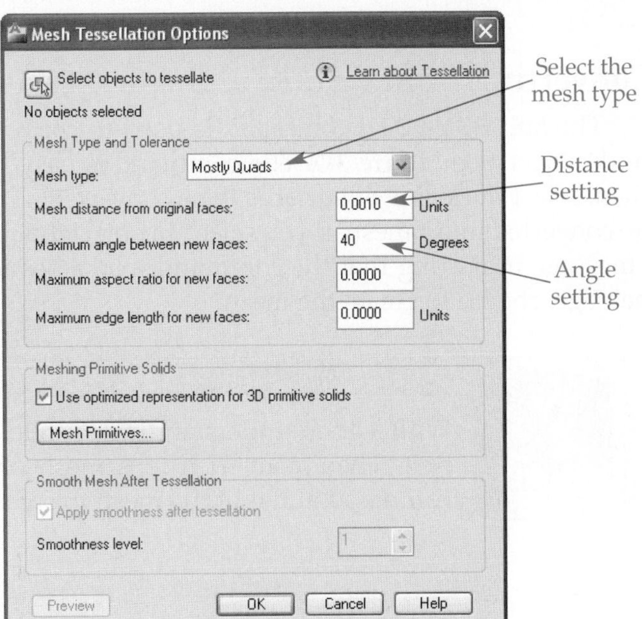

Figure 3-6.
The **Mesh Tessellation Options** dialog box.

setting. Selecting Mostly Quads in the **Mesh type:** drop-down list creates faces that are mostly quadrilateral (**FACETERMESHTYPE** = 1). Selecting Triangle in the **Mesh type:** drop-down list creates faces that are mostly triangular (**FACETERMESHTYPE** = 2).

The **Mesh distance from original faces** setting is the maximum deviation of the mesh faces (**FACETERDEVSURFACE**). Simply put, this setting determines how closely a converted mesh shape matches the original solid or surface shape. No smoothness values are set for this system variable.

The **Maximum angle between new faces** setting is the maximum angle of a surface normal of two adjoining faces (**FACETERDEVNORMAL** system variable). The higher the value, the more faces are created in very curved areas and the less faces are created in flat areas. Increasing the value is good for objects that have curved areas, such as fillets, rounds, holes, or other tightly curved areas. No smoothness values are set for this system variable.

The **Maximum aspect ratio for new faces** setting is the upper limit for the ratio of height to width for new faces (**FACETERGRIDRATIO** system variable). By adjusting this value, long faces can be avoided, such as would be created from a cylindrical object during the conversion process. A low value setting value will create a cleaner look in the formed faces. No smoothness values are set for this system variable.

- 0 = no limitation is applied.
- 1 = height must be equal to width.
- >1 = height may exceed width.
- <1 = width may exceed height.

The **Maximum edge length for new faces** setting is the maximum length any edge can be (**FACETERMAXEDGELENGTH** system variable). The default setting is 0, which allows the size of the mesh to be determined by the size of the 3D model. A setting of 1 or higher results in a reduced number of and less accurate faces compared to the original model. No smoothness values are set for this system variable.

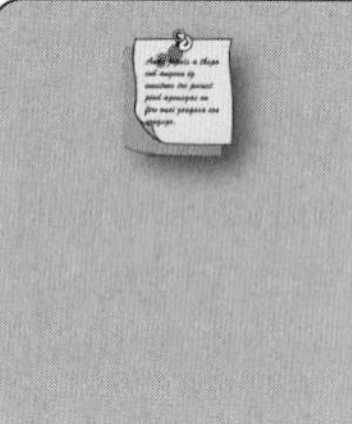

NOTE

Converting swept solids and surfaces, regions, closed polylines, 3D face objects, and legacy polygon and polyface mesh objects may produce unexpected results. If this happens, undo the operation and try making setting adjustments in the **Mesh Tessellation Options** dialog box for better results.

Converting from a Solid or Surface to a Mesh

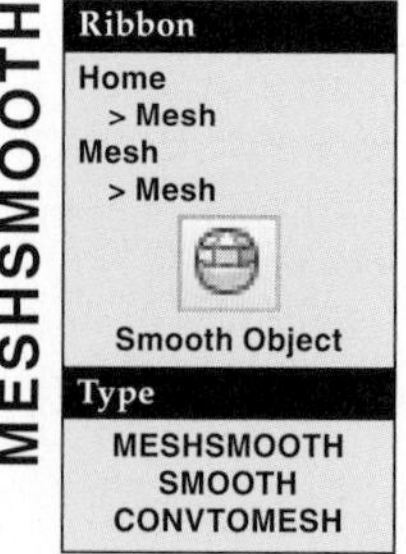

The **MESHSMOOTH** command is used to convert a solid or surface object into a mesh object. See **Figure 3-7**. The command is easy to use. First, enter the command. Then, select the solids or surfaces to be converted. Finally, press [Enter] and the objects are converted into a mesh. If you select an object that is not a primitive, you may receive a message indicating that the command works best on primitives. If you receive this message, choose to create the mesh.

PROFESSIONAL TIP

With a good understanding of solid modeling, create a solid model first. Then, convert it to a mesh object using the **MESHSMOOTH** command and edit the mesh to create a freeform design.

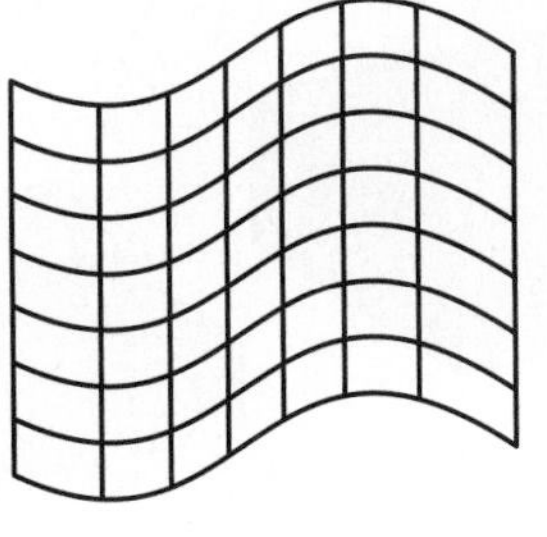

A

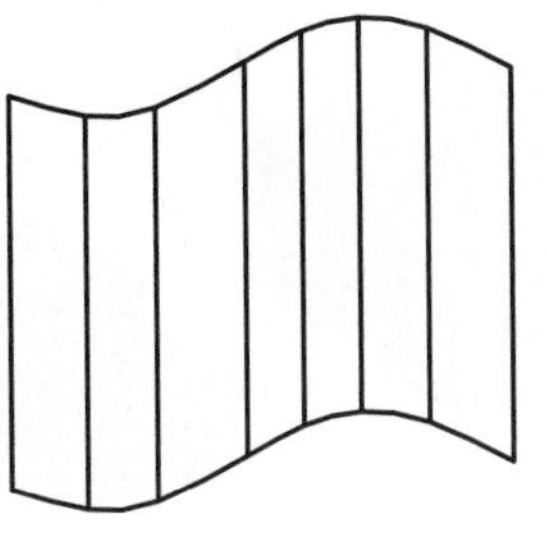

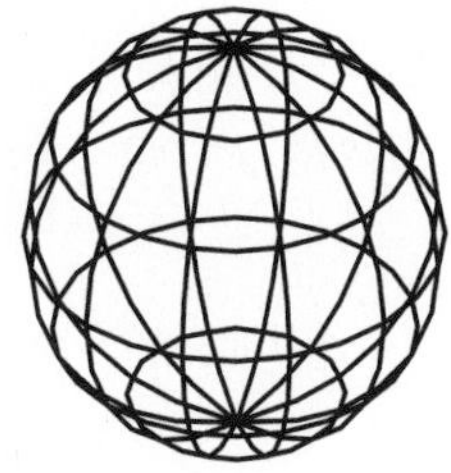

B

Figure 3-7. Converting an existing solid or surface into a mesh object. A—The existing object is a surface (left) or solid (right). B—After converting it into a mesh using the **MESHSMOOTH** command.

Exercise 3-2

Complete the exercise on the student website.
www.g-wlearning.com/CAD

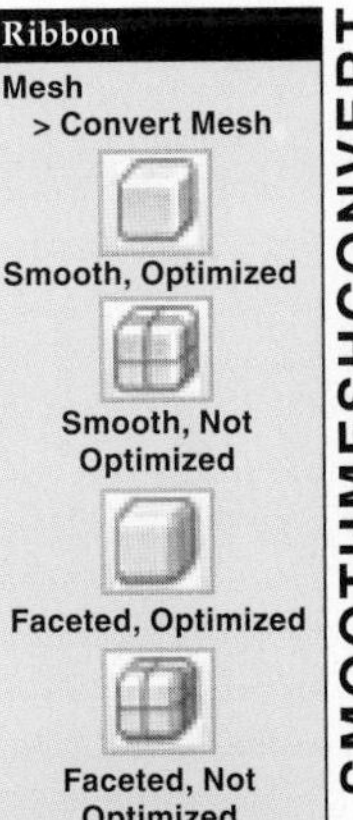

Converting from a Mesh to a Solid or Surface

Mesh objects can be converted into solids or surfaces using the **CONVTOSOLID** and **CONVTOSURFACE** commands. The faces on the resulting solid or surface can be smoothed or faceted and optimized or not. This is controlled by the **SMOOTHMESHCONVERT** system variable. Select one of the four possible settings *before* converting a mesh to a solid or surface.

- **Smoothed and optimized.** Coplanar faces are merged into a single face. The overall shape of some faces can change. Edges of faces that are not coplanar are rounded. (**SMOOTHMESHCONVERT** = 0)
- **Smoothed and not optimized.** Each original mesh face is retained in the converted object. Edges of faces that are not coplanar are rounded. (**SMOOTHMESHCONVERT** = 1)
- **Faceted and optimized.** Coplanar faces are merged into a single, flat face. The overall shape of some faces can change. Edges of faces that are not coplanar are creased or angular. (**SMOOTHMESHCONVERT** = 2)
- **Faceted and not optimized.** Each original mesh face is converted to a flat face. Edges of faces that are not coplanar are creased or angular. (**SMOOTHMESHCONVERT** = 3)

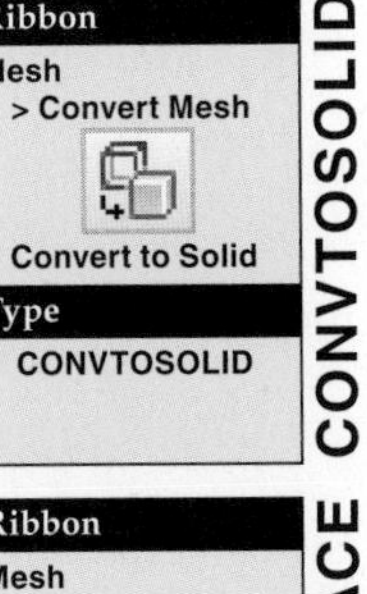

To convert a mesh to a solid, first set the smoothing option, as described above. Then, select the **CONVTOSOLID** command. Next, select the mesh objects to convert and press the [Enter] key. The objects are converted from mesh objects to solid objects based on the selected smoothing option.

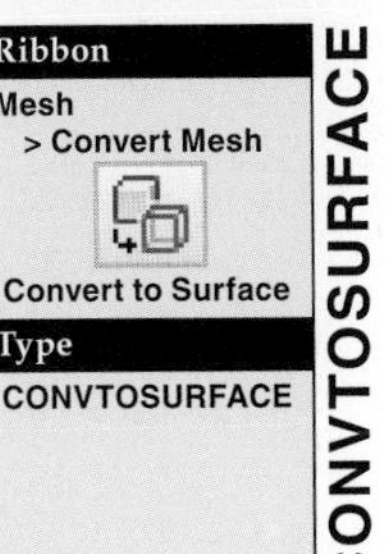

To convert a mesh to a surface, first set the smoothing option, as described above. Then, select the **CONVTOSURFACE** command. Next, select the mesh objects to convert and press the [Enter] key. The objects are converted from mesh objects to surface objects based on the selected smoothing option.

Figure 3-8.
A—Three basic mesh objects.
B—The mesh objects converted into faceted solids.
C—The mesh objects converted into smoothed surfaces.

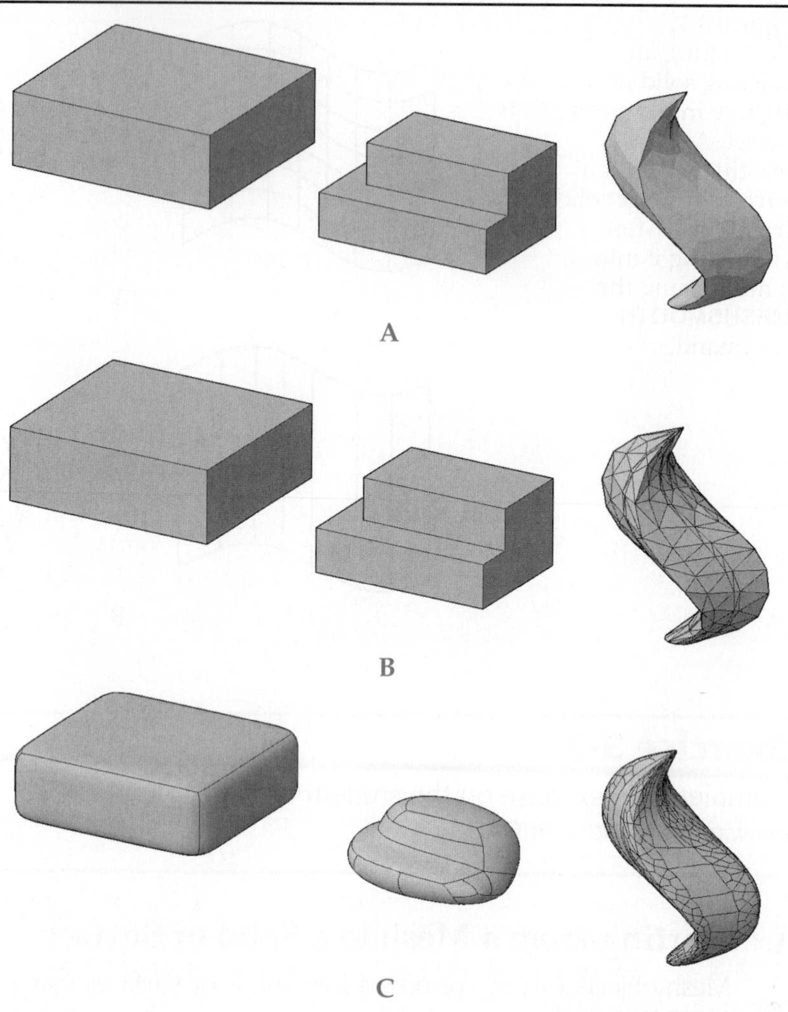

The examples shown in Figure 3-8A are simple mesh objects. In Figure 3-8B, the mesh objects have been converted into solid objects with the **Faceted, Optimized** button selected in the **Convert Mesh** panel on the **Mesh Modeling** tab in the ribbon. In Figure 3-8C, the mesh objects have been converted into surface objects with the **Smooth, Optimized** button selected.

PROFESSIONAL TIP

There are some mesh shapes that cannot be converted to a 3D solid. If using grips to edit the mesh, gaps or holes between the faces may be created. Smooth the mesh object to close the gaps or holes. Also, during the mesh editing process, mesh faces may be created that intersect each other and cannot be converted to a 3D solid. Converting this type of a mesh into a 3D solid will result in the following error message:

Mesh not converted because it is not closed or it self-intersects.
Object cannot be converted.

In some cases, there may be a mesh shape that cannot be converted to a solid object, but can be converted to a surface.

Exercise 3-3

Complete the exercise on the student website.
www.g-wlearning.com/CAD

Smoothing and Refining a Mesh Object

The roundness of a mesh object is increased by increasing the smoothness level. The smoothness can be set before creating the mesh, as described earlier in this chapter. The mesh object can also be refined or have its smoothness increased after it is created. This is described in the following sections. When smoothing or refining a mesh object, the number of tessellation subdivisions either increased or decreased. The lowest smoothness level is 0 and the highest smoothness level is 4. The default smoothness level is 0.

PROFESSIONAL TIP

When creating primitives, begin with the least amount of faces as possible. You can always refine the mesh model to create more faces.

Adjusting the Smoothness of a Mesh

When a mesh is smoothed, it changes form to more closely represent a rounded shape. The **MESHSMOOTHMORE** command is used to increase the level smoothness on a mesh object. There are five levels of smoothness, ranging from level 0 to level 4. Once the command is selected, pick the mesh objects for which to increase the smoothness. Then, press the [Enter] key and the level is increased by one. See Figure 3-9. If the selection set includes objects that are not meshes, you have the opportunity to either filter out the non-mesh objects or convert them to meshes.

When a mesh is desmoothed, it changes form to more closely represent a boxed shape. The **MESHSMOOTHLESS** command is used to decrease the level smoothness on a mesh object. Once the command is selected, pick the mesh objects for which to decrease the smoothness. Then, press the [Enter] key and the level is decreased by one. If the selection set includes objects that are not meshes, you have the opportunity to either filter out the non-mesh objects or convert them to meshes.

Ribbon
Home
> Mesh
> Smooth More
Mesh
> Mesh

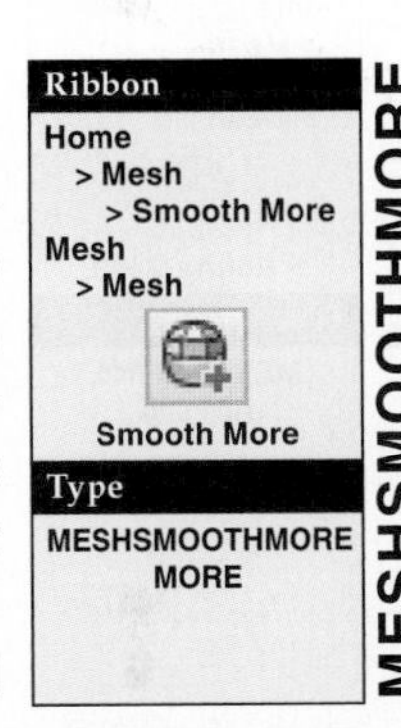

Type
MESHSMOOTHMORE
MORE

MESHSMOOTHMORE

Ribbon
Home
> Mesh
> Smooth Less
Mesh
> Mesh

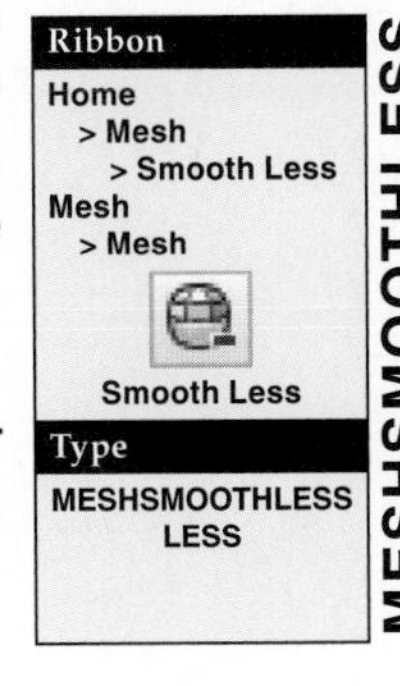

Type
MESHSMOOTHLESS
LESS

MESHSMOOTHLESS

Figure 3-9.
Use the **MESHSMOOTHMORE** command to increase the level smoothness from level 0 to level 4. From left to right, smoothness levels 0, 1, 2, 3, 4.

NOTE

The maximun number of levels is controlled by the **SMOOTHMESHMAXLEV** system variable.

PROFESSIONAL TIP

The **MESHSMOOTHMORE** and **MESHSMOOTHLESS** commands change the smoothing one level at a time. You can use the **Properties** palette to change the smoothness to any level in one step.

Exercise 3-4

Complete the exercise on the student website.
www.g-wlearning.com/CAD

Refining a Mesh

MESHREFINE

Ribbon
Home
> Mesh
Mesh
> Mesh
> Refine Mesh
Type
MESHREFINE
REFINE

Refining a mesh increases the number of subdivisions in a mesh object. Refining adds detail to the mesh to make it more realistic with smooth flowing lines. This also gives a greater selection for editing the mesh faces, vertices, or edges. The **MESHREFINE** command is used to refine a mesh. When the command is used, the number of subdivisions is quadrupled. See **Figure 3-10.** The smoothness level on the original object must be level 1 or higher.

The entire mesh can be refined (all faces at once) or a selected face can be refined. When you refine a mesh *model,* the increased amount of face subdivisions becomes a new smoothness level 0. However, if you refine an individual *face,* the level of smoothness is not reset.

Be careful not to create too dense of a mesh. Creating a mesh that is too dense may result in subobjects (the mesh shapes) that are very small. This may make it difficult to select subobjects and edit the mesh.

Exercise 3-5

Complete the exercise on the student website.
www.g-wlearning.com/CAD

Figure 3-10.
Refining a mesh.
A—The original mesh with a smoothness level of 1.
B—The refined mesh, which now has more faces and a smoothness level of 0.

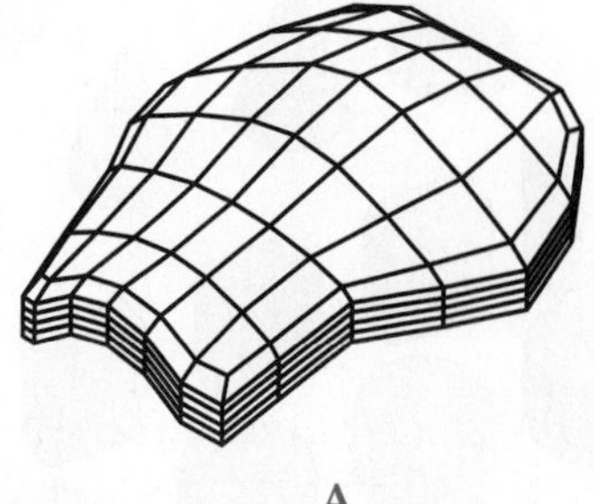
A

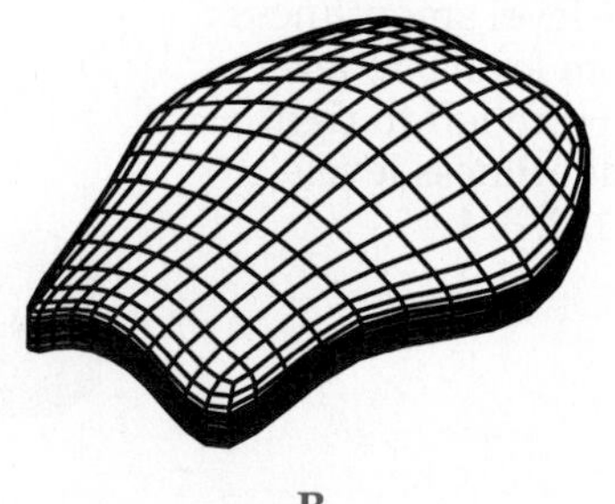
B

Editing Meshes

As discussed earlier, the second key workflow of mesh modeling is the ability to edit the mesh. The tools for editing a mesh are found in the **Mesh Edit** and **Subobject** panels on the **Mesh** tab of the ribbon. See Figure 3-11. The face, edge, and vertex subobjects can be edited to change the shape of the mesh. These subobjects can be moved, rotated, or scaled. A face on the mesh can be split or extruded.

Subobject filters are used to assist in the selection of a mesh face, edge, or vertex *before* it is moved, rotated, or scaled. This is especially true for a very dense mesh. First, right-click in the drawing window and select **Subobject Selection Filter** to display the cascading menu, Figure 3-12. Then, select the filter you wish to use. You can also use the filter in the **Subobject** panel of the **Mesh** tab in the ribbon.

Subobject editing is discussed in detail in Chapter 12. The same procedures discussed in that chapter for solids can be applied to mesh models.

NOTE

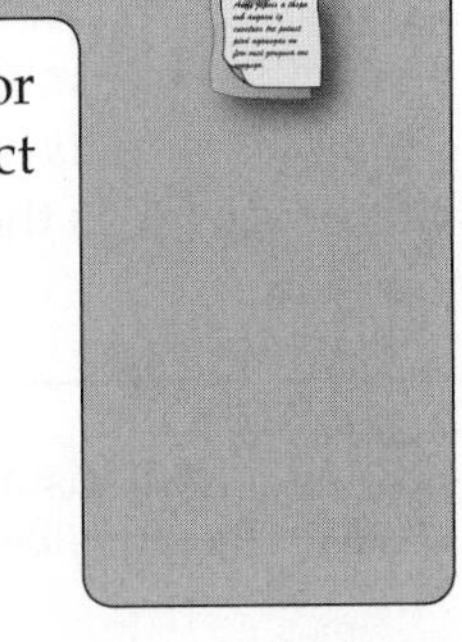

Use the [Ctrl] key and left mouse button to select the subobjects or use the **SUBOBJSELECTIONMODE** system variable for subobject filtering. These settings apply:

- 0 = off
- 1 = vertices
- 2 = edges
- 3 = faces
- 4 = solid history

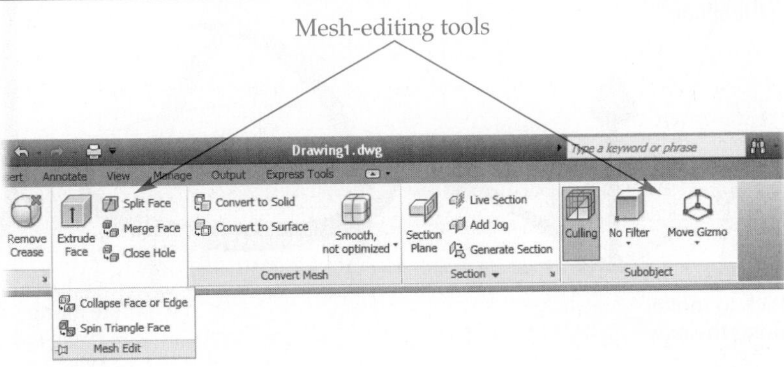

Figure 3-11.
The tools for editing a mesh are found in the **Mesh Edit** and **Subobject** panels of the **Mesh** tab in the ribbon.

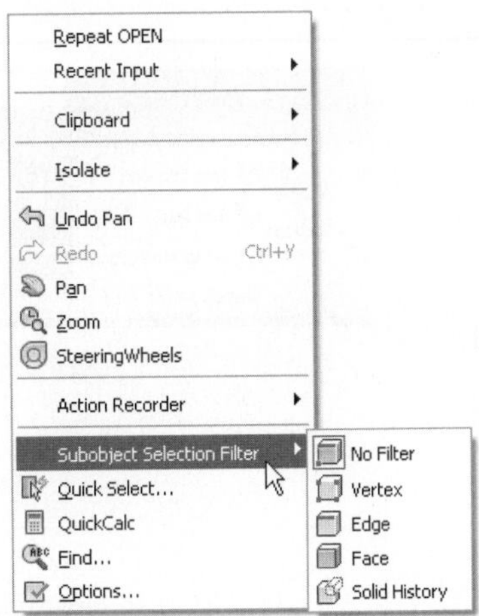

Figure 3-12.
Selecting a subobject filter in the shortcut menu.

Gizmos

A *gizmo,* also called a grip tool, appears when a subobject is selected. This tool is used to specify how the transformation (movement, rotation, or scaling) is applied. A visual style other than 2D Wireframe must be current in order for the gizmo to appear. Using gizmos (grip tools) is discussed in detail in Chapter 11.

The move gizmo allows movement of a subobject along the X, Y, or Z axis or on the XY, XZ, or YZ plane. The rotate gizmo allows rotation about the X, Y, or Z axis. The scale gizmo allows scaling along the X, Y, or Z axis or XY, XZ, or YZ plane.

You can switch between the three gizmos by picking the button in the drop-down list in the **Subobject** panel on the **Mesh** tab of the ribbon. You can also cycle between the gizmos by pressing the [Enter] key.

For example, to use the move gizmo, pick any axis to move along that axis. See Figure 3-13. To move along a plane, pick the rectangular area at the intersection of the axes. To use the rotate gizmo, pick the circle with the center about which you wish to rotate the selection. To use the scale gizmo, pick an axis to nonuniformly scale along that axis. Pick the triangular area at the intersection of the axes to uniformly scale the selection. Pick the area between the inner and outer triangular areas to nonuniformly scale along that plane.

When you make a subobject selection on a mesh model, the ribbon displays ***context-sensitive panels*** based on the selection. For example, if a face subobject is selected, the ribbon displays the **Crease** and **Edit Face** panels, Figure 3-14. Context-sensitive panels

Figure 3-13.
Editing a mesh using one of the three gizmos is an important part of mesh modeling. A—Move gizmo. B—Rotate gizmo. C—Scale gizmo.

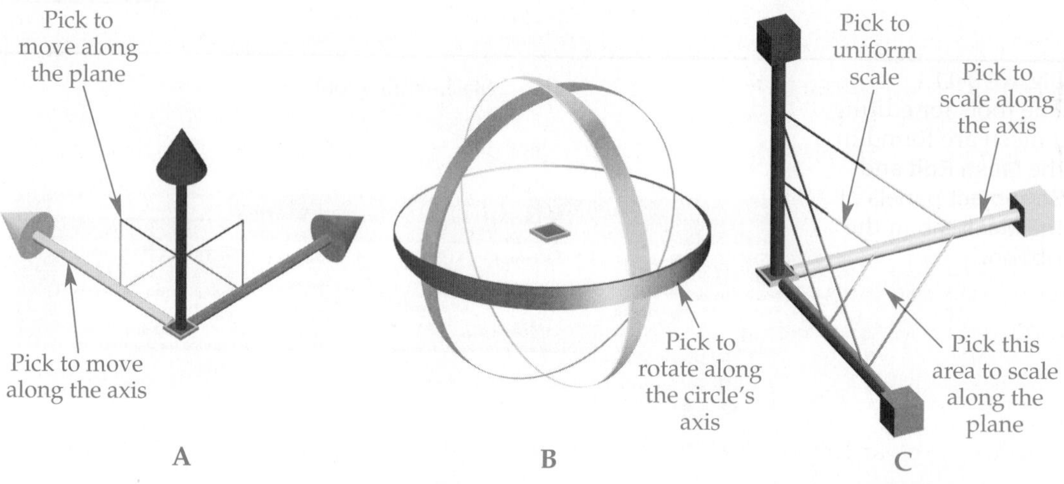

Figure 3-14.
A context-sensitive panel is only displayed when certain objects or subobjects are selected. In this case, the **Crease** and **Edit Face** panels are displayed when a face subobject is selected.

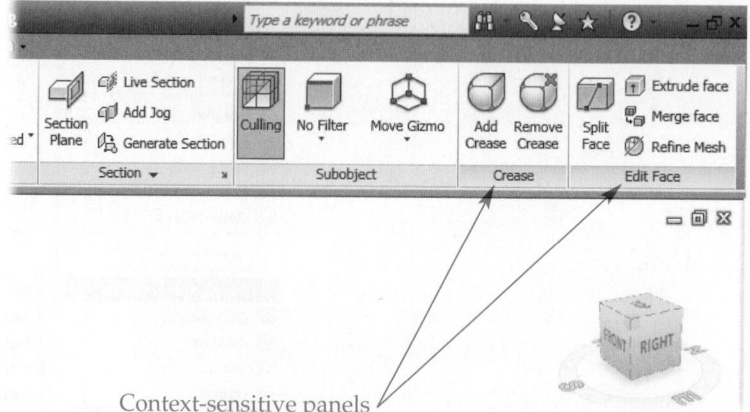

allow improved editing efficiency by displaying tools commonly used for the selected subobject. These panels are indicated by the green bar behind the panel name. Once the selection is canceled, the context-sensitive panels are no longer displayed.

Extrude a Face

You can select a face subobject on a mesh and extrude it. Extruding a mesh face adds new features to the mesh. This creates new faces that can be edited. The **MESHEXTRUDE** command is used to extrude mesh faces. The command sequence to extrude a mesh face is similar to extruding to create a solid shape. This command is covered in detail in Chapter 9.

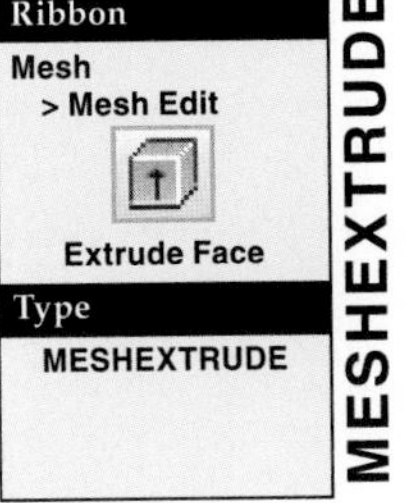

MESHEXTRUDE

To extrude a face, enter the command. The command automatically turns on the face subobject filter. This is indicated by the **Face** button on the **Subobject** panel in the **Mesh** tab on the ribbon. Next, pick the face(s) subobject to extrude. Finally, enter an extrusion height. In Figure 3-15, a camera is being developed as a mesh model. The **MESHEXTRUDE** command is used to extrude an individual face as the first step in creating the lens tube.

PROFESSIONAL TIP

Extrude a face using the **MESHEXTRUDE** command instead of moving a face. This will give greater editing control over an individual face.

Split a Face

Splitting a mesh face is used to increase the number of faces on the model without refining a mesh. Splitting one face creates two faces. This is easier than refining the mesh. Think of these new split faces as subdivisions of an existing face.

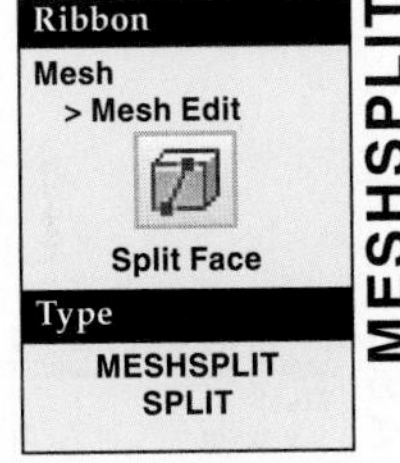

MESHSPLIT

The **MESHSPLIT** command is used to split a face. Once the command is entered, select the face to split. Then, specify the starting point and the ending point of a line defining the split. You can specify any two points on the mesh face. Figure 3-16 shows a mesh model with three faces that have been split.

Once the faces are split, mesh editing options, such as extruding, moving, rotating, or scaling, can be used on the new faces. In Figure 3-17, the split faces from Figure 3-16 have been extruded and then the model smoothed to level 3.

Figure 3-15.
Extruding a mesh face. A—Select the face to extrude. B—The process adds faces to the model.

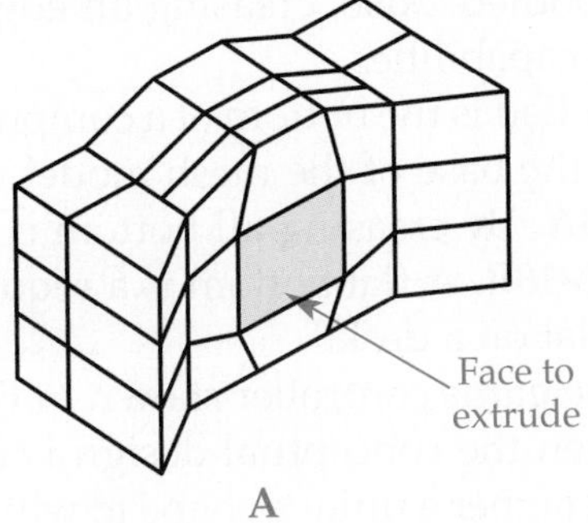

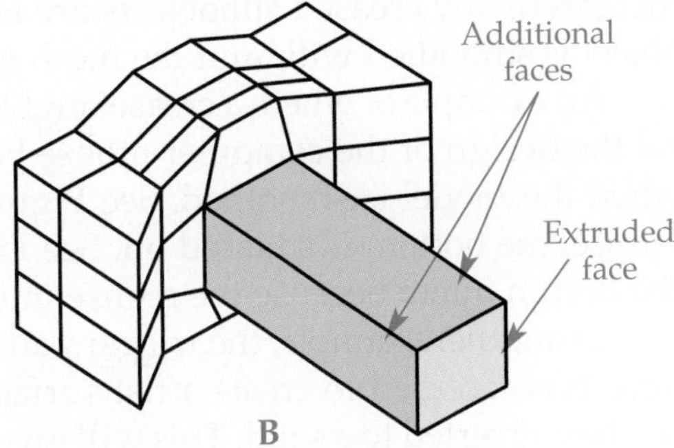

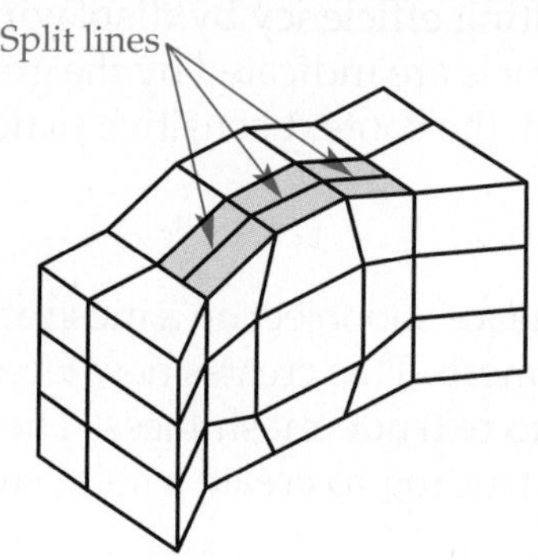

Figure 3-16.
Three faces have been split on this mesh model. Each face has been split into two faces.

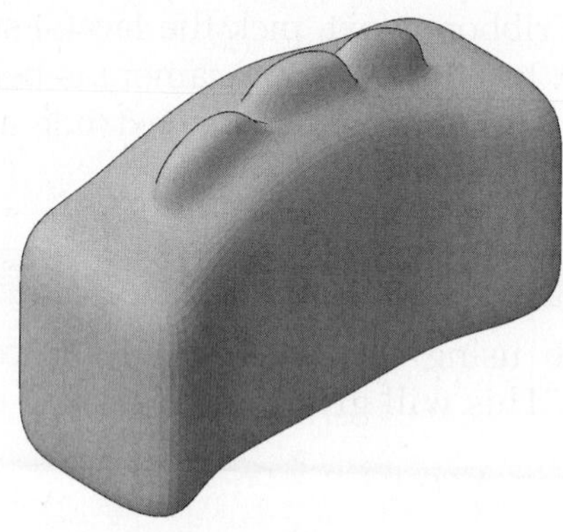

Figure 3-17.
Smoothing is applied to this model. Notice how its edges are rounded.

Exercise 3-6

Complete the exercise on the student website.
www.g-wlearning.com/CAD

Applying a Crease to a Mesh Model

MESHCREASE

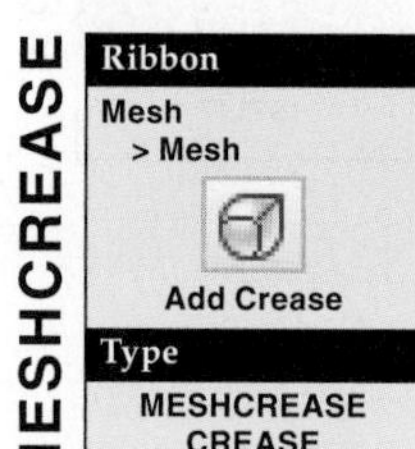

A ***crease*** is a sharpening of a mesh subobject, much like a crease in folded paper. A crease sharpens or squares off an edge or flattens a face. This prevents the subobject from being smoothed. Creases can be applied to faces, edges, and vertices. A smoothness level of 1 or higher must be assigned to the mesh object for the creases to have an effect, but they can be applied at any smoothness level. Once a crease is applied, any existing smoothing is removed from the subobject. If the mesh smoothness is increased, any creased subobjects are not smoothed. Also, creasing an edge *before* an object is smoothed will limit the mesh editing capabilities.

An example of where a crease might be applied is the bottom of a computer mouse. As the design of the computer mouse begins, the base of the mesh model is rounded when the model is smoothed. See **Figure 3-18A.** By creasing all bottom edges of the mouse, the bottom is squared off. See **Figure 3-18B.** A flat bottom is a requirement of the design intent because the mouse must sit flat on a desk.

In another example, the trigger pads on the game controller shown in **Figure 3-19** have been creased to create a flat surface. When the conceptual design is finished, it can be converted to a solid. This will give the designer a unique shape to which buttons and joysticks can be added. This unique shape would be difficult, if not impossible, to create from scratch as a solid model.

Figure 3-18.
Adding creases to a mesh. A—Before the creases are applied, the bottom of the mouse is rounded. This is obvious if the object is rotated (right). B—After the creases are added, the bottom is flat.

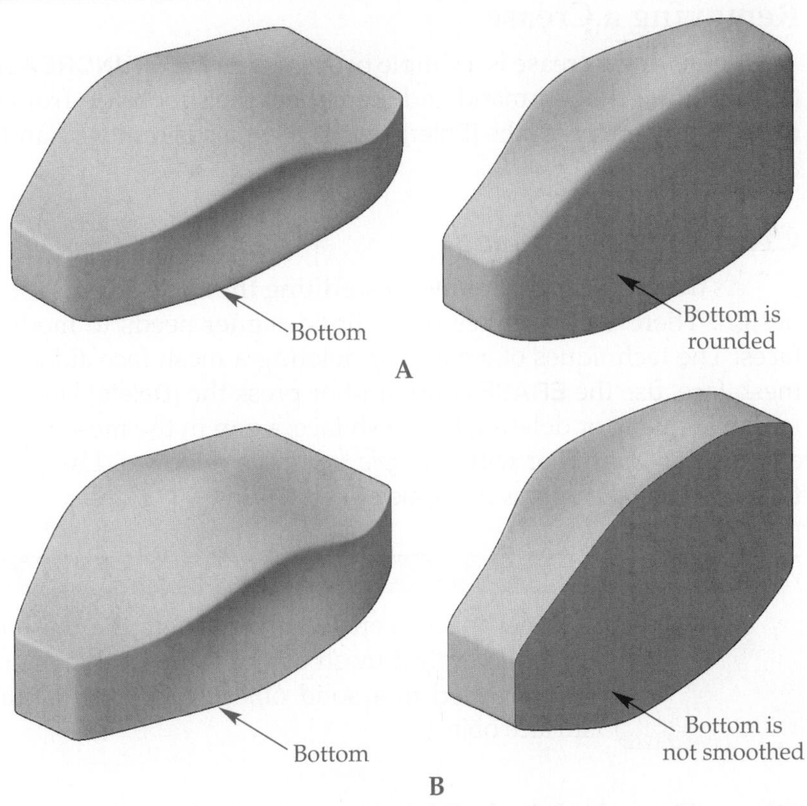

Figure 3-19.
In this example, creases have been added to the mesh to create flat areas on the game pad for buttons.

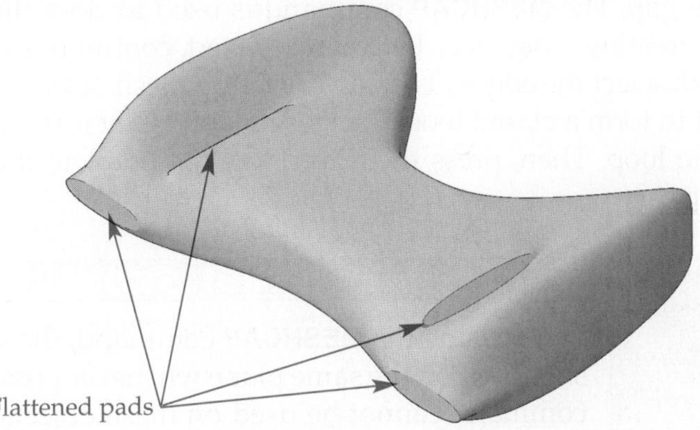

Once the command is entered, select the subobjects to crease. You do not need to press the [Ctrl] key to select subobjects, but filters should be used to make selection easier. Once the subobjects are selected, press the [Enter] key. This prompt appears:

Specify crease value or [Always] <Always>:

If you enter a crease value, this is the highest smoothing level for which the crease is retained. If the smoothing level is set higher than the crease value, the creased subobject is smoothed. The Always option forces the crease to be retained for all smoothness levels. In most cases, this is the recommended option.

Exercise 3-7

Complete the exercise on the student website.
www.g-wlearning.com/CAD

Removing a Crease

MESHUNCREASE
Ribbon
Mesh
> Mesh
Remove Crease
Type
MESHUNCREASE
UNCREASE

Removing a crease is a simple process. The **MESHUNCREASE** command is used to do this. Enter the command and then select the subobjects from which a crease is to be removed. Then, press the [Enter] key. The crease is removed and smoothing is applied as appropriate.

Creating a Mesh Gap

As the designer creates meshes, editing the mesh faces is the next step toward final design. There will be times when the designer needs to modify a mesh by deleting faces. The techniques of erasing or deleting a mesh face aid in the design. To delete a mesh face, use the **ERASE** command or press the [Delete] key after selecting the mesh face. By erasing or deleting the mesh face, a gap in the mesh occurs. Figure 3-20 shows a game pad controller with *multiple* top faces removed. Use the [Ctrl] key to select the top faces and then press the [Delete] key.

PROFESSIONAL TIP

If a mesh face is erased or removed, the mesh object is not considered a *watertight* mesh object. A mesh that is not watertight cannot be converted to a solid object. However, it can be converted to a surface object.

Close Gaps in Mesh Objects

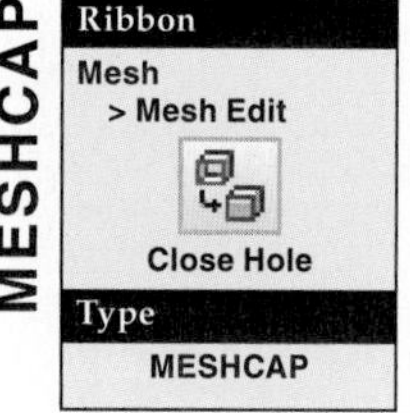

After a mesh face has been erased or deleted, the designer has the ability to close the mesh gap. The **MESHCAP** command is used to close the gap, Figure 3-21. This is done by creating a new face between selected, continuous edges. Once the command is entered, select the edges of the surrounding mesh object faces. The selected edges do not need to form a closed loop. If a closed loop is not formed, AutoCAD automatically closes the loop. Then, press the [Enter] key to complete the command and close the mesh gap.

PROFESSIONAL TIP

When using the **MESHCAP** command, the selected mesh face edges should be on the same plane whenever possible. Also, the **MESHCAP** command cannot be used on mesh objects that use adjacent edges or closed and bounded geometry, such as a circle.

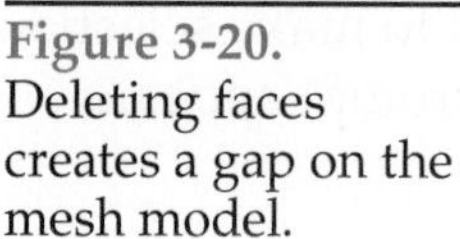

Figure 3-20.
Deleting faces creates a gap on the mesh model.

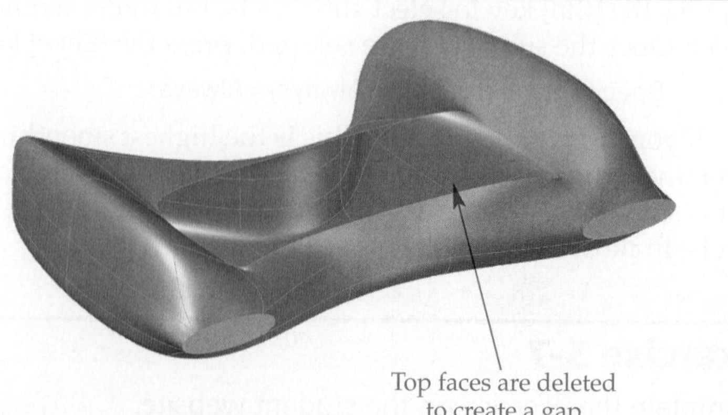

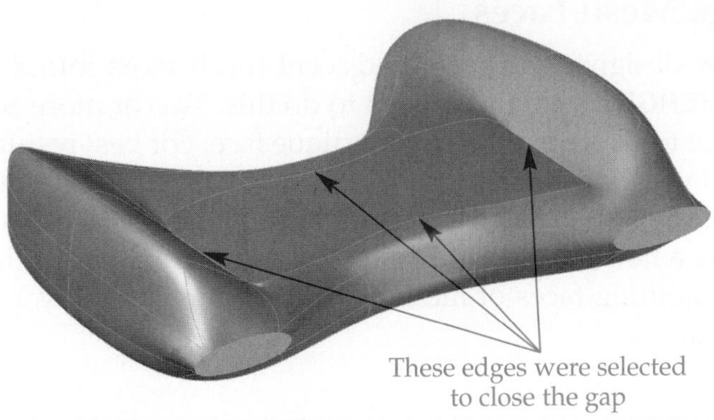

Figure 3-21.
A gap on a mesh model can be closed with the **MESHCAP** command.

Exercise 3-8

Complete the exercise on the student website.
www.g-wlearning.com/CAD

Collapse a Mesh Face or Edge

After a mesh is created the surrounding mesh faces may be converged to create a different and unique mesh face shape. This is called collapsing a mesh. The designer can collapse surrounding mesh faces at the *center* of a selected mesh edge or face. New mesh faces are created. This helps the designer create new mesh face shapes for further editing.

The **MESHCOLLAPSE** command is used to collapse a mesh. Once the command is selected, pick the face or edge to collapse. The command is immediately applied and then ends.

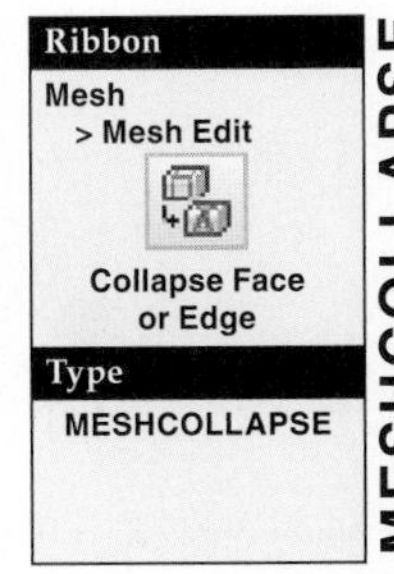

Figure 3-22 shows an in-progress conceptual design for a computer mouse. The side must be collapsed to a point and then moved inward. Enter the command and select the middle edge. Be sure to use filters to help select the edge. Once the edge is collapsed, the new vertex can be selected for subobject editing and moved inward.

Exercise 3-9

Complete the exercise on the student website.
www.g-wlearning.com/CAD

Figure 3-22.
Collapsing a mesh. A—The edge to be selected. B—The mesh is collapsed and the new vertex is edited.

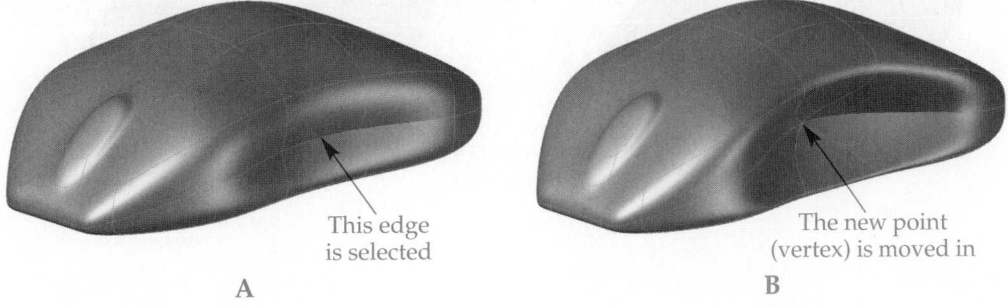

Merge Mesh Faces

MESHMERGE

Ribbon
Mesh
> Mesh Edit

Merge Face
Type
MESHMERGE

The designer can merge adjacent mesh faces into a new, single mesh face. The **MESHMERGE** command is used to do this. Two or more adjacent faces can be selected to merge to create a different or unique face. For best results, the faces should be on the same plane. Do *not* try to merge faces that are not adjacent or are on corners.

Figure 3-23 shows a tape dispenser. The three faces on the front need to be merged to create a new mesh shape. Enter the command and set the face filter. Then, select the three adjoining faces of mesh objects to merge. Finally, press the [Enter] key to apply the command.

Exercise 3-10

Complete the exercise on the student website.
www.g-wlearning.com/CAD

Spin Mesh Faces

MESHSPIN

Ribbon
Mesh
> Mesh Edit

Spin Triangle Face
Type
MESHSPIN

Mesh faces have different and unique shapes when created. Additional unique shapes can be created by spinning a triangular mesh face. Spinning a mesh face spins adjoining edges of *two* triangle mesh faces. Modify a mesh face by spinning rotates the newly shared mesh edges. The **MESHSPIN** command is used to spin a face.

PROFESSIONAL TIP

Use the **Vertex** option of the **MESHSPLIT** command to create triangular mesh faces.

Figure 3-24 shows a pocket camera. The camera's top button has been split to create two triangular faces. Before the button top can be extruded and smoothed, the top adjoining faces must be spun.

Select the command, then pick the two triangular faces. Press the [Enter] key to complete the command. Notice how the diagonal line of the triangular faces is now in a different orientation. You can then extrude one of the faces up or down to create a new button. Smooth the camera to create its final design shape.

Figure 3-23.
Merging faces. A—The faces to be merged. B—The resulting model.

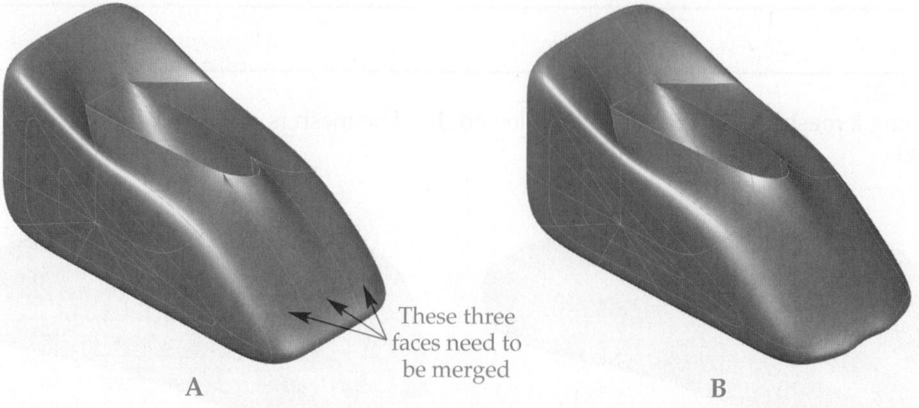

Figure 3-24.
Spinning a face. A—The face has been split, but the orientation is wrong. B—The face is spun and the orientation is correct.

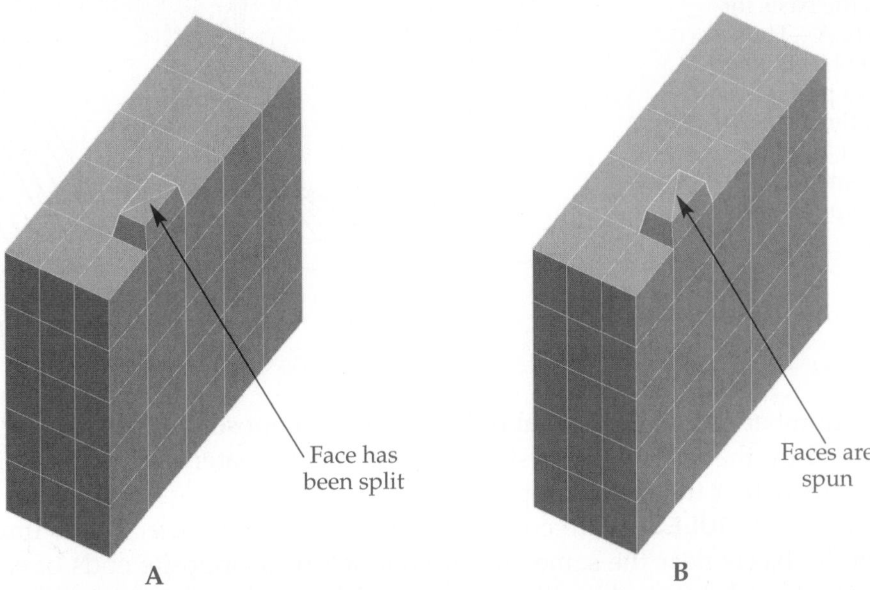

Drawing Mesh Forms

You are not limited to the mesh primitives discussed earlier. You can also create ***mesh forms*** based on arcs, lines, and polylines. Mesh forms can be created by four basic methods:

- Using the **RULESURF** command to add a mesh between two objects such as arcs, lines, or polylines. This is called a ruled-surface mesh.
- Revolving an open shape with the **REVSURF** command. This is called a revolved-surface mesh.
- Using the **TABSURF** command to extrude an open shape along a directional path. This is called a tabulated-surface mesh.
- Using the **EDGESURF** command to create the mesh based on four adjoining edges, such as lines, arcs, polylines or splines. This is called an edged-surface mesh.

These four commands can create either meshes or legacy surfaces. The **MESHTYPE** system variable controls which type of object is created. The default setting of 1 results in the commands creating mesh objects. A setting of 0 results in the commands creating legacy surfaces (polygon or polyface objects).

NOTE

It is strongly recommended the **MESHTYPE** system variable be set to 1 so the **RULESURF**, **REVSURF**, **TABSURF**, and **EDGESURF** commands will create mesh objects. Legacy surfaces are not widely used.

Ruled-Surface Mesh

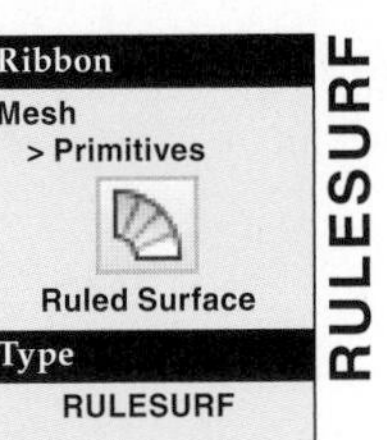

RULESURF

A mesh can be constructed between two objects using the **RULESURF** command, Figure 3-25. This is called a ***ruled-surface mesh.*** The two objects can be points, lines, arcs, circles, polylines, splines, or enclosed objects. A ruled-surface mesh can be created between a point and any of the objects listed. However, AutoCAD cannot generate a ruled-surface mesh between an open and a closed object, such as an arc and a circle.

Figure 3-25.
Lines, arcs, polylines, or splines can be used as the basis for the mesh. A—The original objects. B—If you pick incorrectly, the mesh is twisted. C—The mesh created by correctly picking the objects.

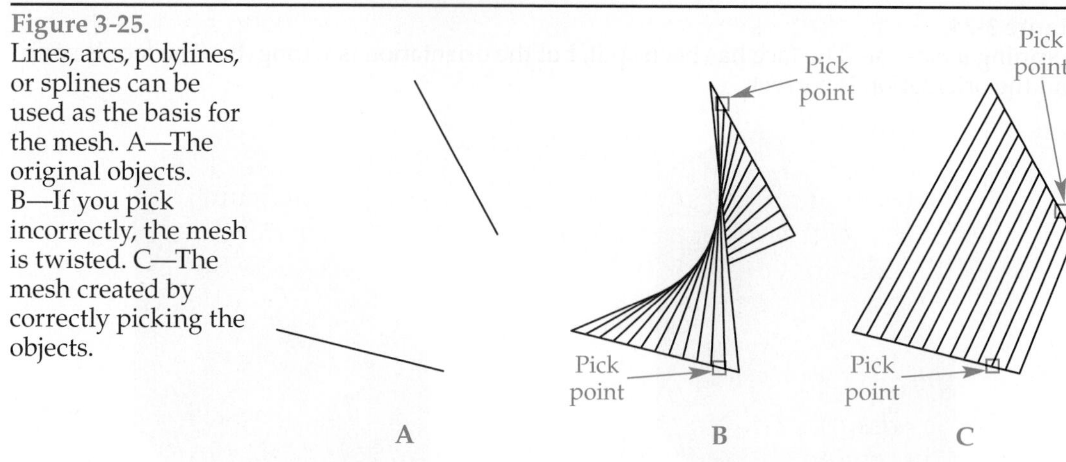

The number of tessellation subdivisions that compose the ruled-surface mesh is determined by the **SURFTAB1** system variable. The greater the number of subdivisions, the smoother the surface appears.

When using **RULESURF** to create a mesh between two objects, it is important to select both objects near the same end. If you pick near opposite ends of each object, the resulting mesh may not be what you want. The mesh may be twisted or reversed, somewhat like a butterfly shape. Refer to Figure 3-25.

Revolved-Surface Mesh

REVSURF

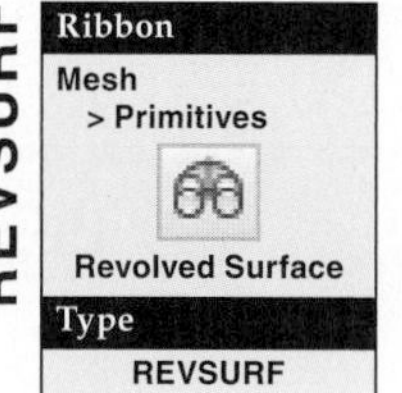

With the **REVSURF** command, you can draw a profile and then rotate that profile around an axis to create a symmetrical object. This is called a ***revolved-surface mesh.*** The **REVSURF** command is a powerful tool and will greatly assist anyone who needs to draw a symmetrical mesh. The profile, or path curve, can be drawn using lines, arcs, circles, ellipses, elliptical arcs, polylines, or donuts. The rotation axis can be a line or an open polyline. Notice the initial layout path curves shown in Figure 3-26 and the resulting meshes. The **REVSURF** command is powerful because it can create a symmetrical surface from any profile.

Figure 3-26.
Creating revolved-surface mesh objects. The profile shape and axis can be polylines or splines.

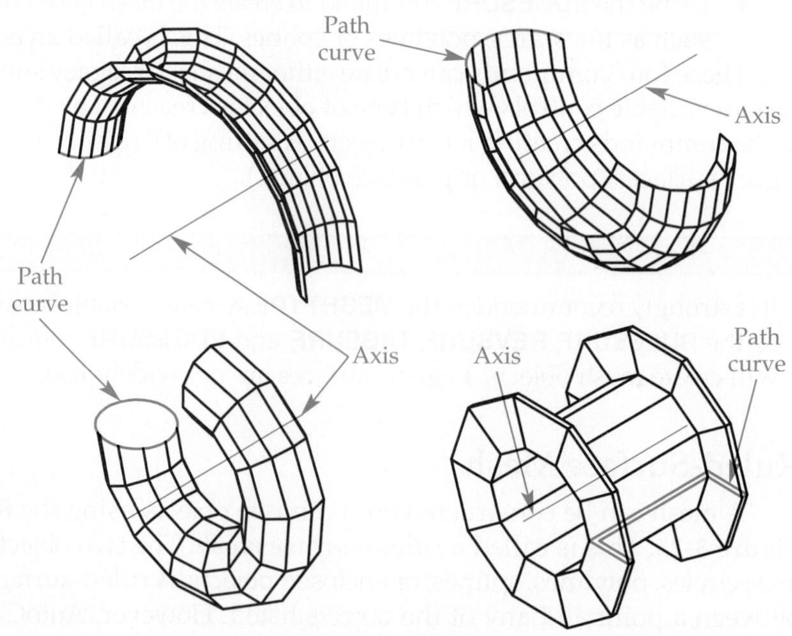

After selecting the path curve and the axis, you are prompted to specify the start angle. This allows you to specify an offset angle at which to start the surface revolution. This is useful when the profile is not revolved through 360°. The next prompt requests the included angle. This lets you create the object through 360° of rotation or just a portion of that, as shown in Figure 3-26.

The **SURFTAB1** and **SURFTAB2** system variables control the number of tessellation subdivisions on a revolved-surface mesh. The **SURFTAB1** value determines the number of divisions in the direction of rotation around the axis. The **SURFTAB2** value divides the path curve into the specified number of subdivisions of equal size.

Tabulated-Surface Mesh

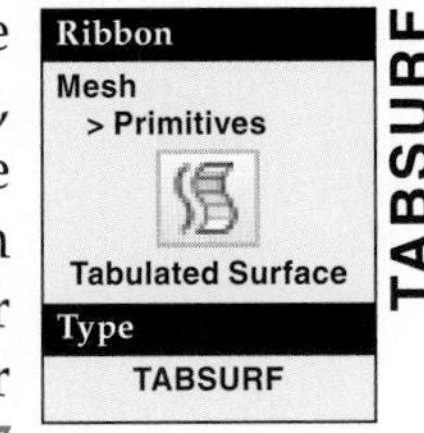

A ***tabulated-surface mesh*** is similar to a ruled-surface mesh. However, the shape is based on only one entity, Figure 3-27. This entity is called the path curve. Lines, arcs, circles, ellipses, 2D polylines, and 3D polylines can all be used as the path curve. A line called the direction vector is also required. This line indicates the direction and length of the tabulated-surface mesh. AutoCAD finds the endpoint of the direction vector closest to your pick point. It sets the direction toward the *opposite* end of the vector line. The mesh follows the direction and length of the direction vector. Figure 3-27 shows the difference the location of the pick point makes when selecting the vector. The **SURFTAB1** system variable controls the number of "steps" that are constructed.

Edged-Surface Mesh

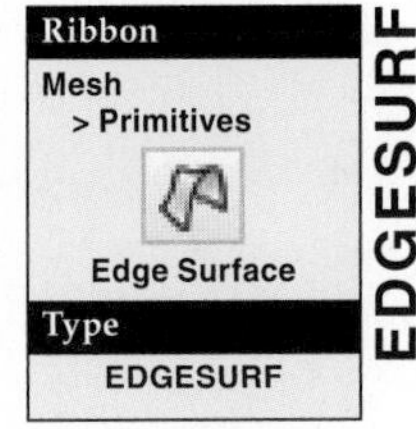

The **EDGESURF** command allows you to construct a 3D mesh between four edges, Figure 3-28. The edges can be lines, polylines, splines, or arcs. The endpoints of the objects must precisely meet. However, a closed polyline cannot be used. The four objects can be selected in any order. The resulting mesh is smooth and called an ***edged-surface mesh***.

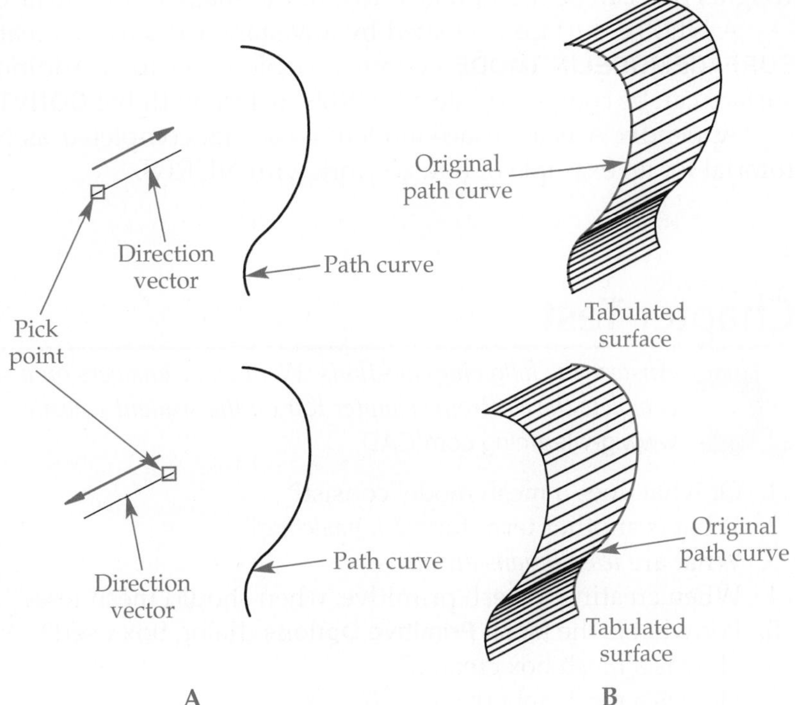

Figure 3-27.
Creating a tabulated-surface mesh object. Lines, arcs, polylines, or splines can be used as the shape and vector. A—The objects on which the mesh will be based. B—The completed tabulated-surface mesh.

Figure 3-28.
Creating an edged-surface mesh object. Four lines, arcs, polylines, or splines can be used as the four edges. A—The objects on which the mesh will be based. B—The completed edged-surface mesh.

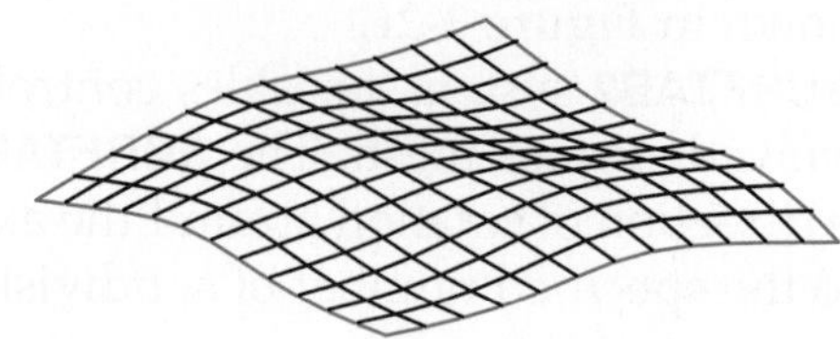

The number of tessellation subdivisions is determined by the variables **SURFTAB1** and **SURFTAB2**. The default value for each of these variables is 6. Higher values increase the smoothness of the mesh. The current values are always displayed on the command line when one of the mesh surfacing commands is executed.

Exercise 3-11

Complete the exercise on the student website.
www.g-wlearning.com/CAD

NURBS

Non-uniform, rational B-spline (NURBS) surfaces can be created in AutoCAD. These surfaces are based on Bézier splines or curves. NURBS surfaces are used to create organic, freeform models. NURBS surfaces have fit points, control vertices, and weights that can be manipulated to edit the shape of the surface.

A NURBS surface is created by any standard surface creation method when the **SURFACEMODELINGMODE** system variable is set to 1. Additionally, any "standard" surface can be converted into a NURBS surface with the **CONVTONURBS** command.

Appendix A is a surface modeling tutorial completed as NURBS. Complete this tutorial for an example of how to work with NURBS.

Chapter Test

Answer the following questions. Write your answers on a separate sheet of paper or complete the electronic chapter test on the student website.
www.g-wlearning.com/CAD

1. Of what does a mesh model consist?
2. What is another term for *mesh modeling?*
3. What are *tessellations divisions?*
4. When creating a mesh primitive, when should mesh tessellation divisions be set?
5. For what is the **Mesh Primitive Options** dialog box used?
6. How is a mesh box created?
7. How is a mesh sphere created?
8. How is a mesh torus created?
9. List the four commands used to create mesh forms.
10. What is the purpose of the **DELOBJ** system variable?

11. Which command converts a mesh object to a surface object?
12. Which command converts a mesh object to a solid object?
13. Describe the smoothness of a mesh object.
14. How many smoothness levels are there?
15. Which command is used to convert an existing solid or surface to a mesh object?
16. List two ways to decrease the smoothness of a mesh.
17. What happens to the mesh when you refine it?
18. How many types of subobjects does a mesh have? List them.
19. Which keyboard key is used to select subobjects for editing?
20. What is a *context-sensitive panel?*
21. Name the three operations that can be performed with a gizmo.
22. How do you cycle through the three different gizmos?
23. Which command is used to extrude a mesh face?
24. Briefly describe the process for extruding a mesh face.
25. What is the process for splitting a mesh face?
26. Why would you crease a mesh model?
27. Which command is used to remove a crease?
28. Explain why you would erase or delete a mesh face during the design process.
29. Which command is used to close gaps in mesh object?
30. What is the purpose of collapsing a mesh face or edge?

Drawing Problems

1. In this problem, you will create an ergonomic computer mouse as a mesh shape. Change visual styles as needed throughout your work.
   A. Set the tessellation divisions for the box primitive to length = 5, width = 3, and height = 3.
   B. Create a 5 × 3 × 1.04 mesh box primitive.
   C. Move the middle faces inward .25 units on both sides of the mesh box, as shown below in A.
   D. Move edges on the mesh model to form the mouse shape shown below in B.
   E. Smooth the model.
   F. Continue editing faces, edges, and vertices to create a mouse shape similar to the one shown below in C.
   G. Crease the model to create a flat bottom.
   H. Save the model.

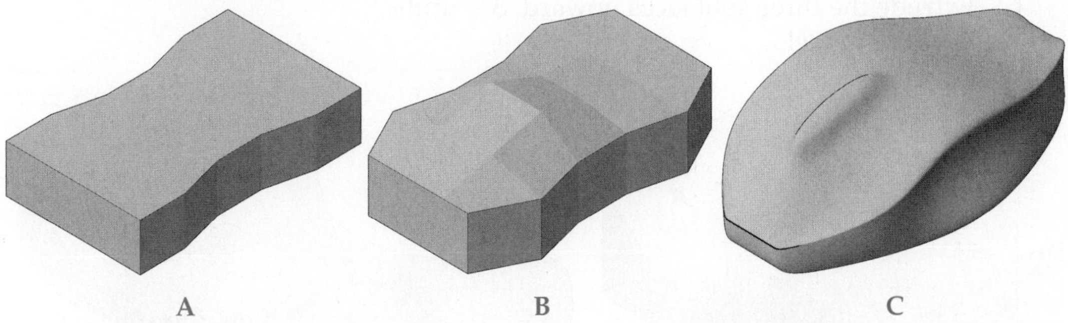

2. In this problem, you will create a wireless phone charger cradle as a mesh shape. Change visual styles as needed throughout your work.
   A. Set the tessellation divisions for the box primitive to length = 5, width = 3, and height = 2.
   B. Create a 3 × 2 × 1.5 mesh box primitive.

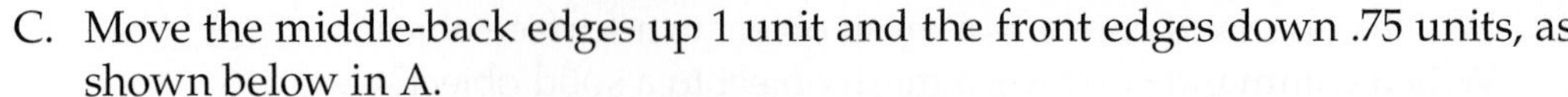

C. Move the middle-back edges up 1 unit and the front edges down .75 units, as shown below in A.
D. Crease the middle face, then move it down 1 unit.
E. Crease all of the faces on the bottom.
F. Smooth the mesh to a level 3 mesh.
G. On the top of the model, nonuniformly scale the middle face (which is now round) to create an elliptical shape. Then, move the face toward the back of the model.
H. Move the middle three edges in the front of the model .5 units toward the back of the model, as shown below in B.
I. Save the model.

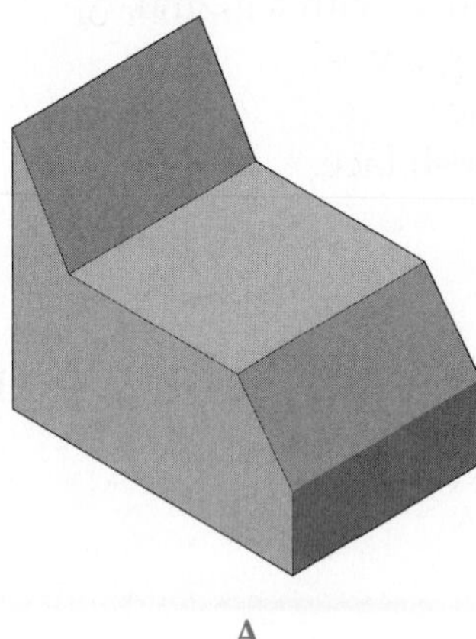

A

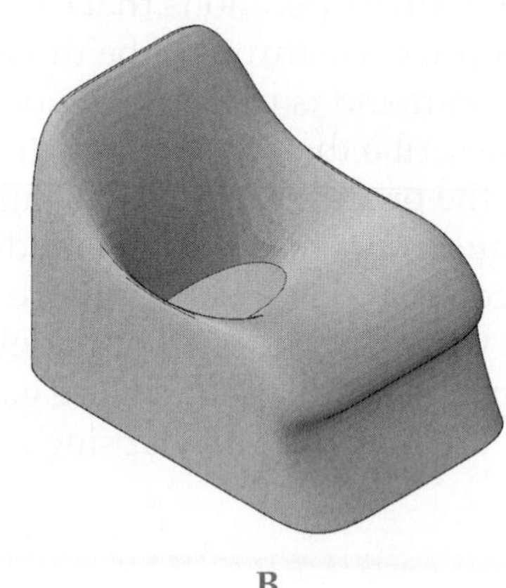

B

3. In this problem, you will create a game pad controller as a mesh shape. Change visual styles as needed throughout your work.
   A. Set the tessellation divisions for the box primitive to length = 5, width = 3, and height = 2.
   B. Create a 6 × 4 × 1 mesh box primitive.
   C. Move the middle faces on the back side (long side) inward 1.25 units.
   D. Move top edges on the right and left side up 1 unit, as shown below in A.
   E. Move the front middle face inward .5 units.
   F. Crease the front two corner faces. The top corner of these faces is formed by the edges moved up earlier.
   G. Smooth the model to level 3.
   H. Move the two creased areas outward .25 units.
   I. Split the top face to the left of the indention into three unique faces, as shown below in B.
   J. Extrude the three split faces upward .375 units.
   K. Save the model.

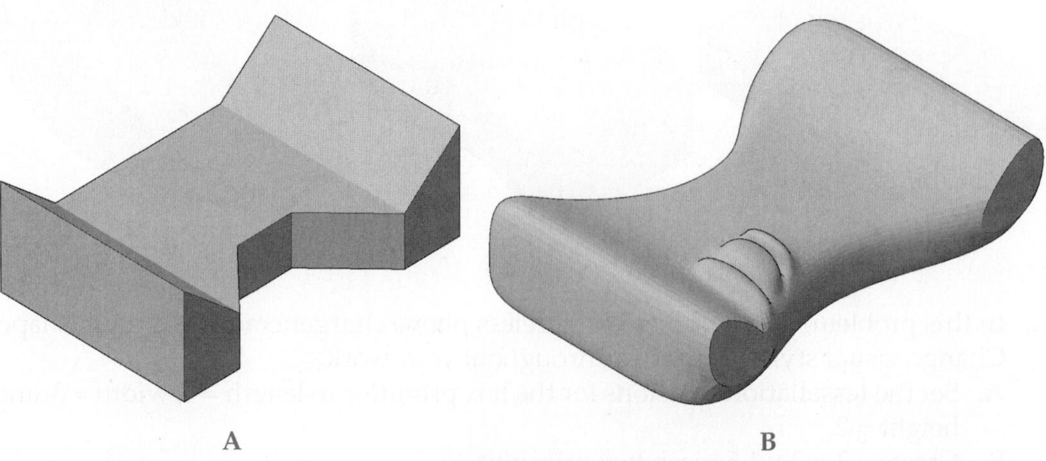

A B

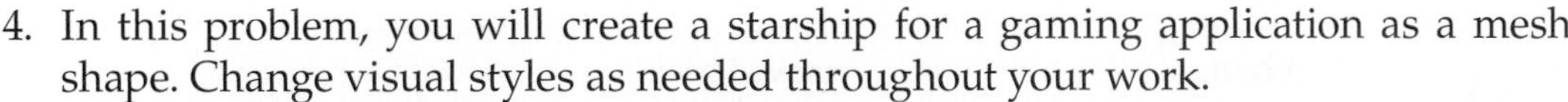

4. In this problem, you will create a starship for a gaming application as a mesh shape. Change visual styles as needed throughout your work.
   A. Set the tessellation divisions for the sphere primitive to axis = 12 and height = 6.
   B. Create a ∅4 mesh sphere primitive.
   C. Move the vertex on the top of the sphere up 4 units.
   D. Move the vertex on the right and left sides outward 2 units, as shown below in A.
   E. Move the vertex on the bottom of the sphere into the sphere by 1 unit.
   F. Smooth the model to level 2.
   G. On the bottom of the ship, as indicated in A, crease the four middle faces, then move them into the sphere by 1 unit.
   H. Extrude the four middle faces a distance of .75 with a taper of 30°. This forms the landing gear.
   I. Select two faces on the front of the ship and crease them to create windows, as shown below in B.
   J. Increase the smoothness to level 4.
   K. Save the model.

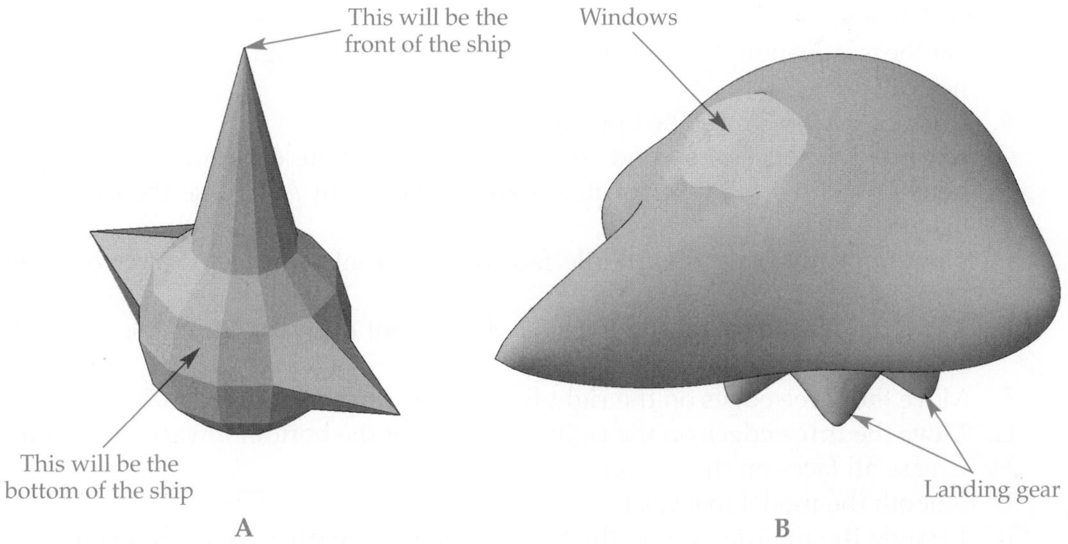

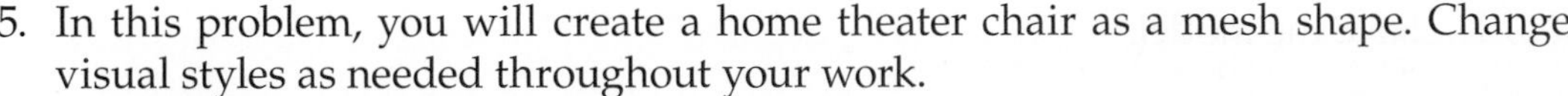

5. In this problem, you will create a home theater chair as a mesh shape. Change visual styles as needed throughout your work.
   A. Set the tessellation divisions for the box primitive to length = 5, width = 3, and height = 2.
   B. Create a 3 × 2 × 1.5 mesh box primitive.
   C. Move the top-back edges (short side) up 1 unit and the top-front edges down .75 units.
   D. Crease the one face in the middle of the top, then move it down 1 unit, as shown on the next page in A.
   E. Crease all faces on the bottom.
   F. Smooth the model to level 3.
   G. Nonuniformly scale the middle face on the top, which is now circular, to create an elliptical shape. Then, move the face .5 units toward the chairback.
   H. Move the three top edges on the front .5 units toward the chairback.
   I. Move the middle face in the front top down 1 unit, as shown on the next page in B.
   J. Rotate the top three back edges 45° to create a sloped chairback.
   K. Move the single edge in the middle of the back inward .5 units, as shown on the next page in C.
   L. Increase the smoothness to level 4.
   M. Save the model.

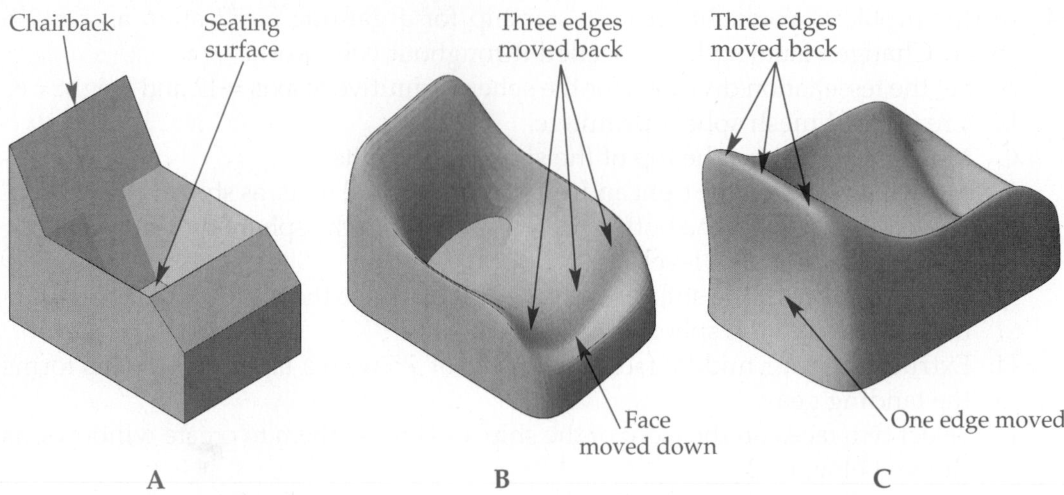

6. In this problem, you will create a conceptual design for a pocket camera as a mesh shape. Change visual styles as needed throughout your work.
   A. Set the tessellation divisions for the box primitive to length = 5, width = 3, and height = 5.
   B. Create a 3 × 1 × 3 mesh box primitive.
   C. Extrude the top face second from the edge to create a button. Extrude to a height of .25 with a taper of 10°, as shown below in A. Crease the top of the extruded face.
   D. Nonuniformly scale the middle face on the front by a factor of two. Then, crease the face.
   E. Move the five edges on the left side of the front outward .25 units, as shown below in B.
   F. Move the three edges on the right-hand side of the top inward .375 units.
   G. Move the three edges on the right-hand side of the bottom inward .375 units.
   H. Crease all faces on the bottom.
   I. Smooth the model to level 4.
   J. Extrude the middle face on the front (which was earlier creased) inward .125 units, as shown below in C.
   K. Save the model.

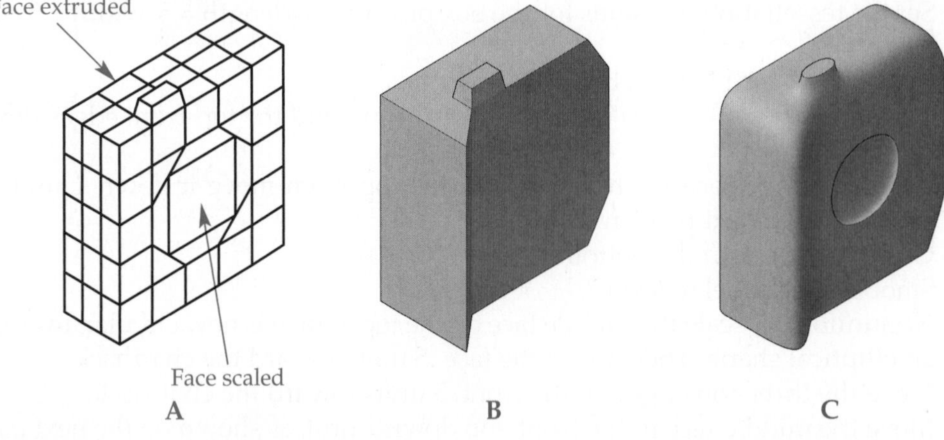

CHAPTER 4

Viewing and Displaying Three-Dimensional Models

Learning Objectives

After completing this chapter, you will be able to:

- ✓ Use the navigation bar to perform a variety of display manipulation functions.
- ✓ Use the view cube to dynamically rotate the view of the model in 3D space.
- ✓ Use the view cube to display orthographic plan views of all sides on the model.
- ✓ Use steering wheels to display a 3D model from any angle.
- ✓ Render a 3D model.

AutoCAD provides several tools with which you can display and present 3D models in pictorial and orthographic views:

- Preset isometric viewpoints, discussed in Chapter 1.
- Dynamic model display using the view cube. This on-screen tool provides access to preset and dynamic display options.
- Complete 3D model display using steering wheels.
- The **3DORBIT**, **3DFORBIT**, and **3DCORBIT** commands provide dynamic display and continuous orbiting tools for demonstrations and presentations.

Once a viewpoint has been selected, you can enhance the display by applying visual styles. The **View** panel in the **Home** tab of the ribbon provides a variety of ways to display a model, including wireframe, hidden line removal, and simple rendering. An introduction to visual styles is provided in Chapter 1 and additional details are provided in Chapter 15.

A more advanced rendering can be created with the **RENDER** command. It produces the most realistic image with highlights, shading, and materials, if applied. Figure 4-1 shows a 3D model of a cast iron plumbing cleanout after using **HIDE**, setting the Conceptual visual style current, and using **RENDER**. Notice the difference in the three displays.

Using the Navigation Bar to Display Models

By default, the navigation bar appears below the view cube on the right side of the screen. See Figure 4-2. It allows you to quickly use the navigation tools described in this chapter, in addition to the **ZOOM** and **PAN** commands and three orbit options. The **NAVBAR** command is used to turn on or off the display of the navigation bar. The navigation bar

Figure 4-1.
A—Hidden display (hidden lines removed). B—The Conceptual visual style set current. C—Rendered with lights and materials.

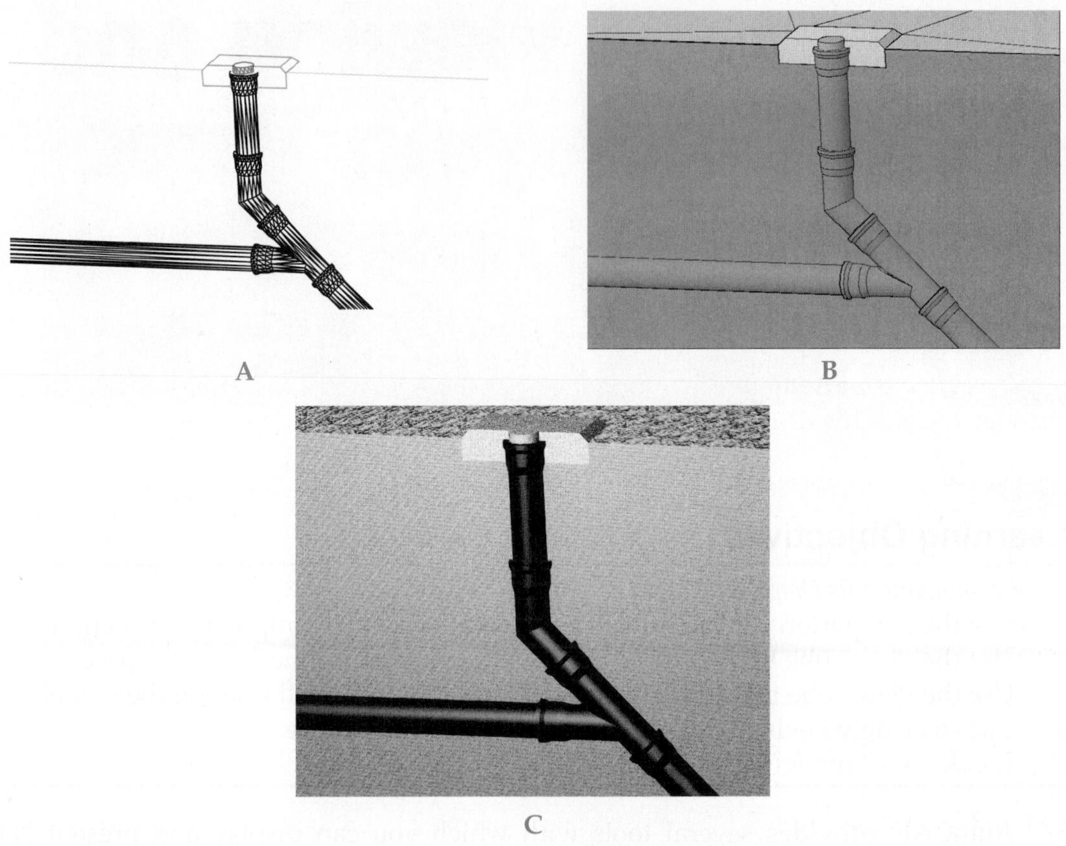

Figure 4-2.
The navigation bar appears below the view cube by default. This bar allows you to quickly use navigation tools for drawing-display purposes.

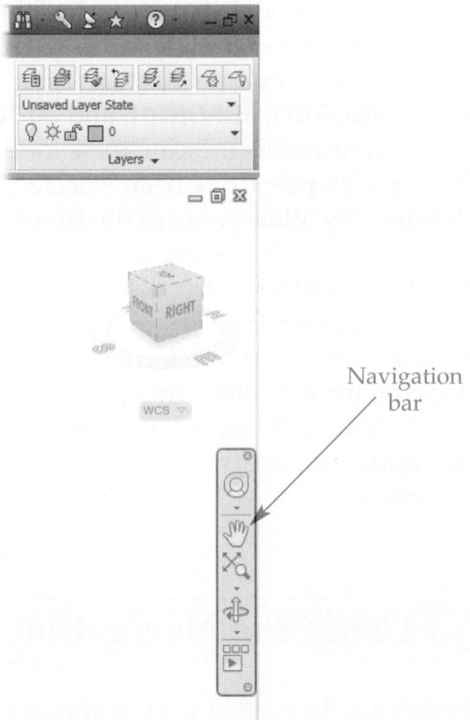

Navigation bar

can be controlled using the ribbon by selecting **View**>**Windows**>**User Interface** and then checking **Navigation Bar** in the drop-down menu. The appearance and location of the navigation bar can be customized, as discussed later in this chapter.

Dynamically Displaying Models with the View Cube

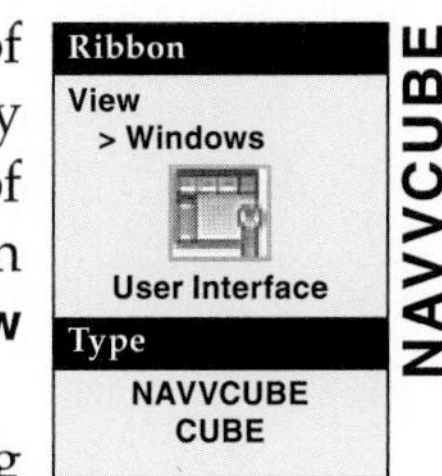

The ***view cube*** navigation tool allows you to quickly change the current view of the model to a preset pictorial or orthographic view or any number of dynamically user-defined 3D views. The tool is displayed, by default, in the upper-right corner of the drawing area. If the view cube is not displayed, pick **View Cube** in the drop-down menu displayed by picking the **User Interface** button on the **Windows** panel in the **View** tab of the ribbon.

The **View Cube Settings** dialog box offers many settings for expanding and enhancing the view cube. This dialog box is discussed in detail later in this chapter.

PROFESSIONAL TIP

The location of the view cube can be quickly changed by picking the customize button on the navigation bar. Then, select **Docking positions** in the shortcut menu and select the location of your choice in the cascading menu.

Understanding the View Cube

The view cube tool is composed of a cube labeled with the names of all six orthographic faces. See Figure 4-3. A compass rests at the base of the cube and is labeled with the four compass points (N, S, E, and W). The compass can be used to change the view. It also provides a visual cue to the orientation of the model in relation to the current user coordinate system (UCS). The ***user coordinate system*** describes the orientation of the X, Y, and Z axes. A complete discussion of user coordinate systems is given in Chapter 6.

Below the cube and compass is a button labeled WCS, which displays a shortcut menu. The label on this button displays the name of the current user coordinate system, which is, by default, the world coordinate system (WCS). This shortcut menu gives you the ability to switch from one user coordinate system to another. A user coordinate system (UCS) is any coordinate system that is not the world coordinate system. The benefit of creating new user coordinate systems is the increase in efficiency and productivity when working on 3D models.

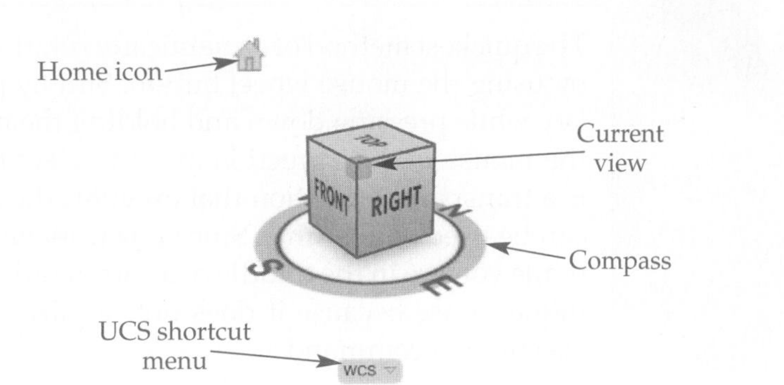

Figure 4-3.
The cube in the view cube tool is labeled with the names of all six orthographic faces. The view cube also contains a **Home** icon, compass, and UCS shortcut menu.

When the cursor is not over the view cube, the tool is displayed in its dimmed or ***inactive state.*** When the cursor is moved onto the view cube, the tool is displayed in its ***active state*** and a little house appears to the upper-left of the cube. This house is the **Home** icon. Picking this icon always restores the same view, called the ***home view.*** Later in this chapter, you will learn how to change the home view.

Notice that a corner, edge, or face on the cube is highlighted. In the default view of a drawing based on the acad3D.dwt template, the corner between the top, right, and front faces is shaded. The shaded part of the cube represents the standard view that is closest to the current view. Move the cursor over the cube and notice that edges, corners, and faces are highlighted as you move the pointer. If you pick a highlighted edge, corner, or face, a view is displayed that looks directly at the selected feature. By selecting standard views on the view cube, you can quickly move between pictorial and orthographic views.

PROFESSIONAL TIP

Picking the top face of the view cube displays a plan view of the current UCS XY plane. This is often quicker than using the **PLAN** command, which is discussed later in this chapter.

Dynamic displays

The easiest way to change the current view using the view cube is to pick and drag the cube. Simply move the cursor over the cube, press and hold the left mouse button, and drag the mouse to change the view. The view cube and model view dynamically change with the mouse movement. When you have found the view you want, release the mouse button. This is similar to the **Orbit** tool in steering wheels, which are discussed later in the chapter.

When dragging the view cube, you are not restricted to the current XY plane, as you are when using the compass. The compass is discussed later.

Pictorial displays

The view cube provides immediate access to 20 different preset pictorial views. When one of the corners of the cube is selected, a standard isometric pictorial view is displayed. See **Figure 4-4A.** Eight corners can be selected. Picking one of the four corners on the top face of the cube restores one of the preset isometric views. You can select a preset isometric view using the **Home** or **View** tab of the ribbon. However, using the view cube may be a quicker method.

Selecting one of the edges of the cube sets the view perpendicular to that edge. See **Figure 4-4B.** There are 12 edges that can be selected. These standard views are not available on the ribbon.

PROFESSIONAL TIP

The quickest method of dynamically rotating a 3D model is achieved by using the mouse wheel button. Simply press and hold the [Shift] key while pressing down and holding the mouse wheel. Now move the mouse in any direction and the view rotates accordingly. This is a transparent function that executes the **3DORBIT** command and can be used at any time. Since it is transparent, it can even be used while you are in the middle of a command. This is an excellent technique to use because it does not require selecting another tool or executing a command.

Figure 4-4.
A—When one of the corners of the cube is selected, a standard isometric view is displayed. B—Selecting one of the edges of the cube produces the same rotation in the XY plane as an isometric view, but a zero elevation view in the Z plane.

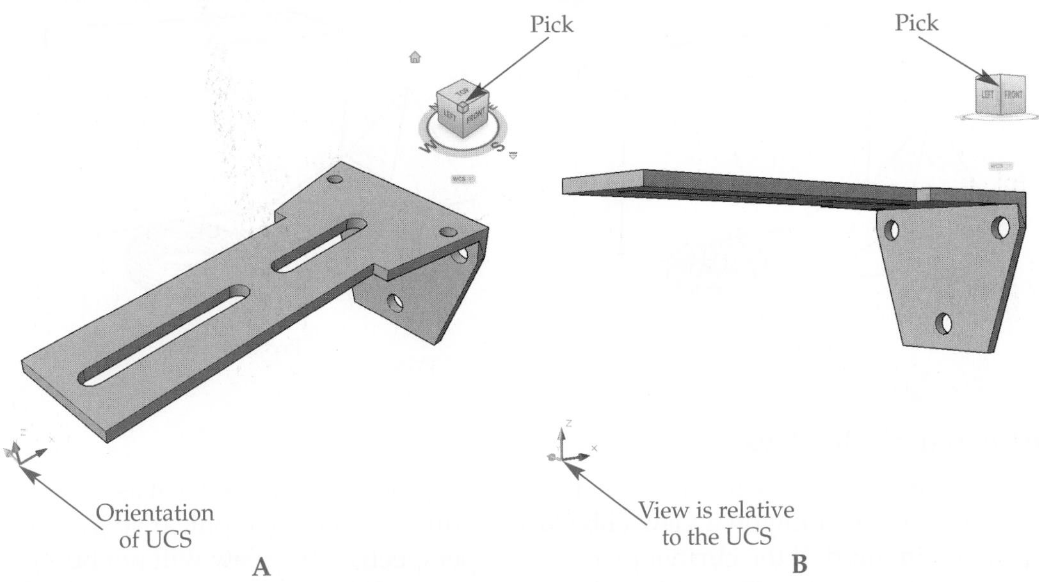

Projection

The view displayed in the graphics window can be in one of two projections. The ***projection*** refers to how lines are applied to the viewing plane. In a pictorial view, lines in a ***perspective projection*** appear to converge as they recede into the background. The points at which the lines converge are called ***vanishing points.*** In 2D drafting, it is common to represent an object in pictorial as a one- or two-point perspective, especially in architectural drafting. In a ***parallel projection,*** lines remain parallel as they recede. This is how an orthographic or axonometric (isometric, dimetric, and trimetric) view is created.

To change the projection, right-click on the view cube to display the shortcut menu. Three display options are given:

- **Parallel.** Displays the model as a parallel projection. This creates an orthographic or axonometric view.
- **Perspective.** Displays the model in the more realistic, perspective projection. Lines recede into the background toward invisible vanishing points.
- **Perspective with Ortho Faces.** Displays the model in perspective projection when a pictorial view is displayed and parallel projection when an orthographic view is displayed. The parallel projection is only set current if a face on the view cube is selected to display the orthographic view. It is not set current if the **PLAN** command is used.

Figure 4-5 illustrates the difference between parallel and perspective projection.

Exercise 4-1

Complete the exercise on the student website.
www.g-wlearning.com/CAD

Figure 4-5.
In a parallel projection, parallel lines remain parallel. In a perspective projection, parallel lines converge to a vanishing point. Notice the three receding lines on the boxes. If these lines are extended, they will intersect.

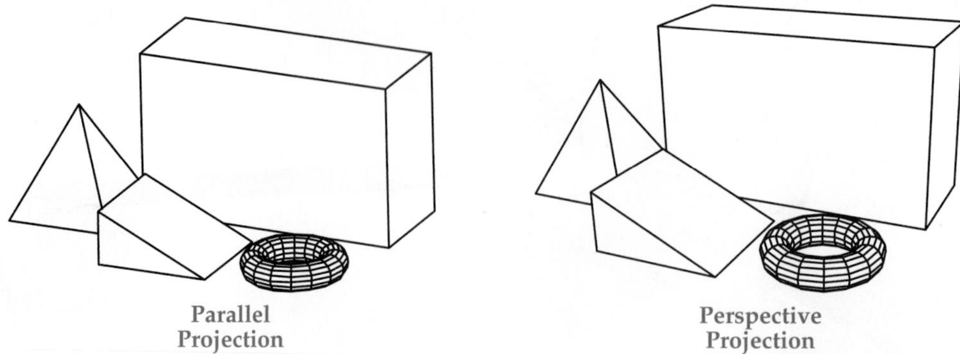

Orthographic displays

The view cube faces are labeled with orthographic view names, such as Top, Front, Left, and so on. Picking on a view cube face produces an orthographic display of that face. Keep in mind, if the current projection is perspective, the view will not be a true orthographic view. See Figure 4-6. If you plan to work in perspective projection, but also want to view proper orthographic faces, turn on **Perspective with Ortho Faces** in the view cube shortcut menu. Another way to achieve a proper orthographic view is to turn on parallel projection.

When a face is selected on the view cube, the cube rotates to orthographically display the named face. The view rotates accordingly. In addition, notice that a series of triangles point to the four sides of the cube (when the cursor is over the tool). See Figure 4-7A. Picking one of these triangles displays the orthographic view corresponding to the face to which the triangle is pointing, Figure 4-7B. This is a quick and efficient method to precisely rotate the display between orthographic views.

Figure 4-6.
A—To properly view orthographic faces using the view cube, turn on **Perspective with Ortho Faces** in the shortcut menu. B—When an orthographic view is set current with perspective projection on, the view is not a true orthographic view. Notice how you can see the receding surfaces.

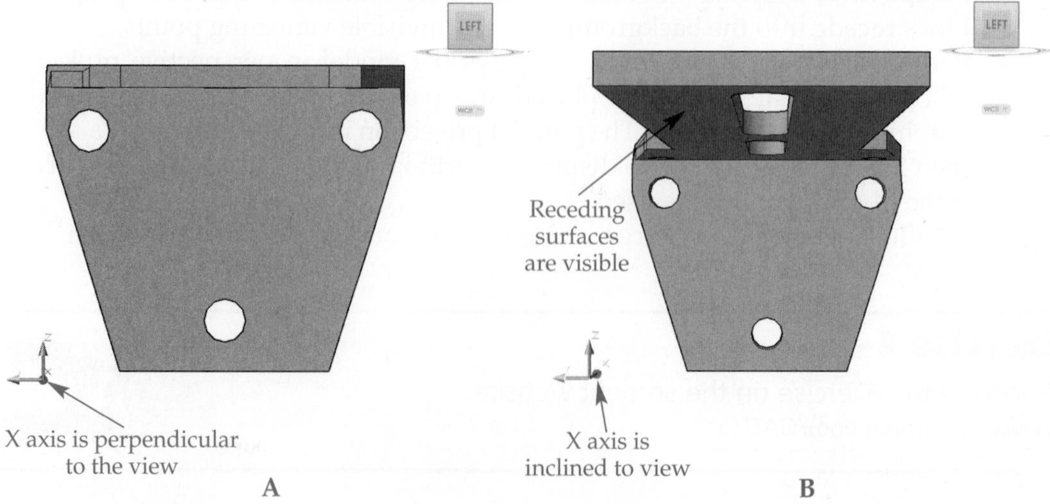

Figure 4-7.
A—The selected orthographic face is surrounded by triangles. Pick one of the triangles to display that orthographic face. B—The orthographic face corresponding to the picked triangle is displayed. Picking a roll arrow rotates the current view 90° in the selected direction. C—The rotated view.

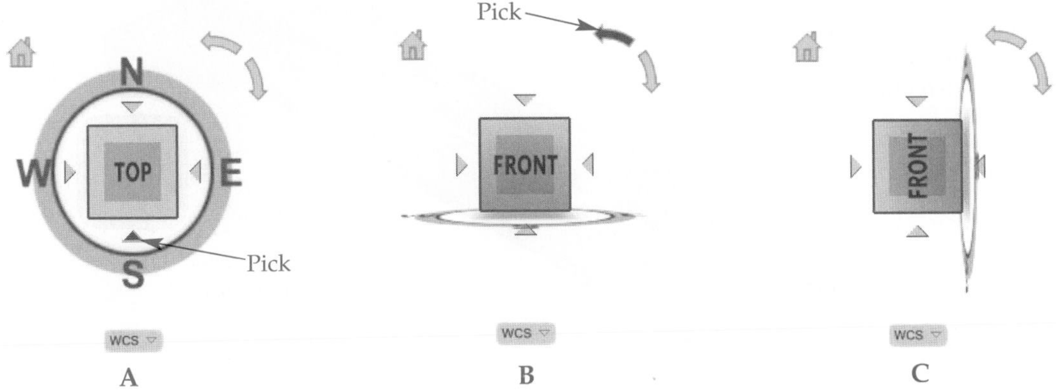

When an orthographic view is displayed, two ***roll arrows*** appear on the view cube. Picking either of these arrows rotates the current view 90° in the selected direction and about an axis perpendicular to the view. See **Figure 4-7C.** Using the triangles and roll arrows on the view cube provides the greatest flexibility in manipulating the model between orthographic views.

Setting views with the compass

The compass allows you to dynamically rotate the model in the XY plane. To do so, pick and hold on one of the four labels (N, S, E, or W). Then, drag the mouse to rotate the view. The view pivots about the Z axis of the current UCS. Try this a few times and notice that you can completely rotate the model by continuously moving the cursor off of the screen. For example, pick the letter W and move the pointer either right or left (or up or down). Notice as you continue to move the mouse in one direction the model continues to rotate in that direction.

You can use the compass to display the model in a view plan to the right, left, front, or back face of the cube. When the pointer is moved to one of the four compass directions, the letter is highlighted, **Figure 4-8A.** Simply single pick on the compass label that is next to the face you wish to view. See **Figure 4-8B.**

Home view

The default home view is the southwest isometric view of the WCS. Picking the **Home** icon in the view cube always displays the view defined as the home view, regardless of the current UCS. You can easily set the home view to any display you wish. First, using any navigation method, display the model as required. Next, right-click on the view cube and pick **Set Current View as Home** from the shortcut menu. Now, when you pick the **Home** icon in the view cube, this view is set current. Remember, the **Home** icon does not appear until the cursor is over the view cube.

PROFESSIONAL TIP

Should you become disoriented after repeated use of the view cube, it is far more efficient to pick the **Home** icon than it is to use **UNDO** or try to select an appropriate location on the view cube.

Figure 4-8.
A—When the pointer is moved over one of the four compass directions, the letter is highlighted.
B—If you pick the letter, the orthographic view from that compass direction is displayed.

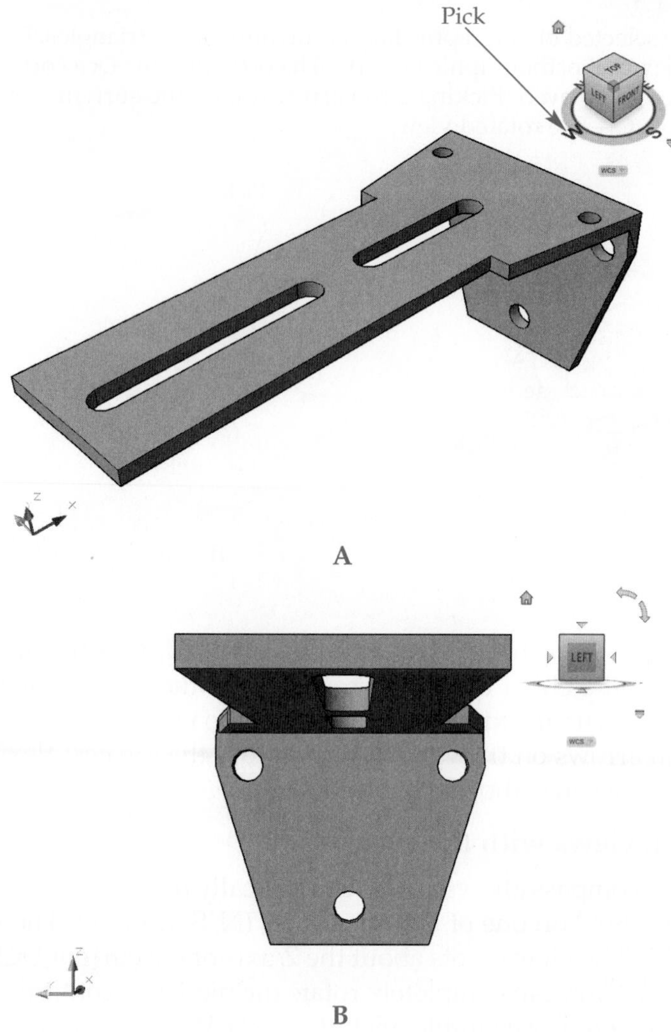

UCS settings

The UCS shortcut menu in the view cube lists the WCS and the names of all named UCSs in the current drawing. If there are no named UCSs, the listing is **WCS** and **New UCS**. See Figure 4-9A. To create a new UCS, pick **New UCS** from the shortcut menu. Next, use the appropriate UCS command options to create the new UCS. See Chapter 6 for complete coverage of the **UCS** command.

As new UCSs are created and saved, their names are added to the UCS shortcut menu. See Figure 4-9B. Now, if you wish to work on the model using a specific UCS, simply select it from the list. The UCS is then restored.

Exercise 4-2

Complete the exercise on the student website.
www.g-wlearning.com/CAD

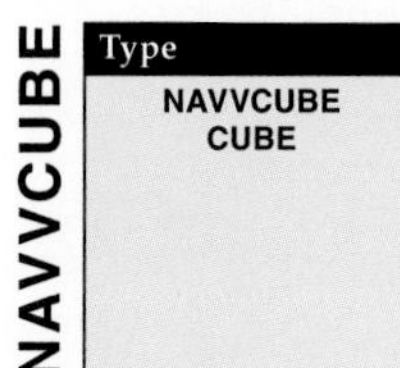

View Cube Settings Dialog Box

The appearance and function of the view cube can be changed using the **View Cube Settings** dialog box, Figure 4-10. This dialog box is displayed by selecting **View Cube Settings...** from the view cube shortcut menu or by using the **Settings** option of the **NAVVCUBE** command. The next sections discuss the options found in the dialog box.

Figure 4-9.
The UCS shortcut menu below the view cube lists all named UCSs in the current drawing. A—The menu entries are **WCS** and **New UCS**. In this example, there are no saved UCSs in the drawing. B—The names of new UCSs are added to the UCS shortcut menu. The current UCS is indicated with a check mark.

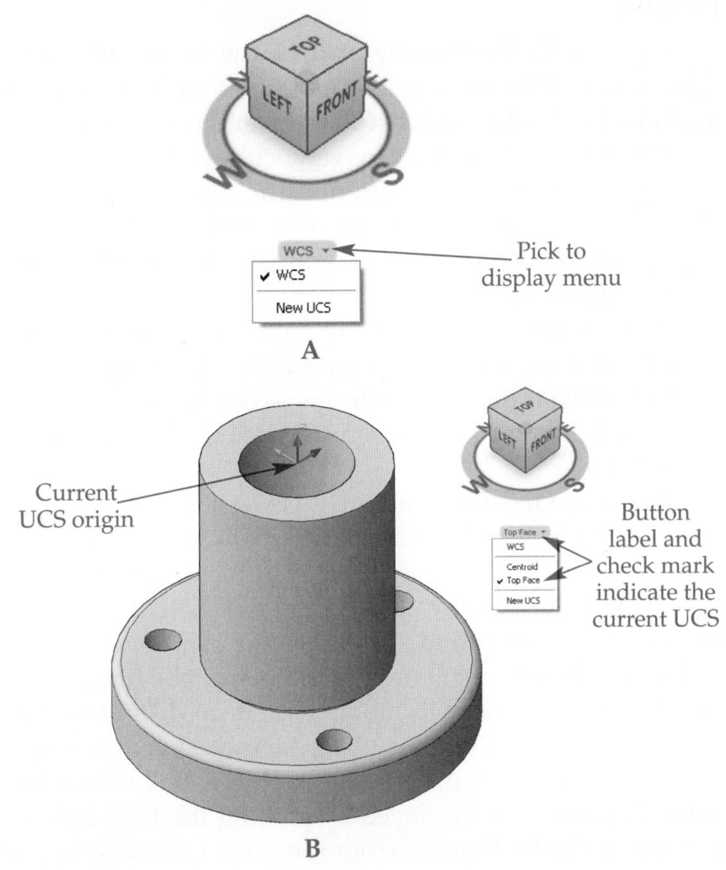

Figure 4-10.
The **View Cube Settings** dialog box.

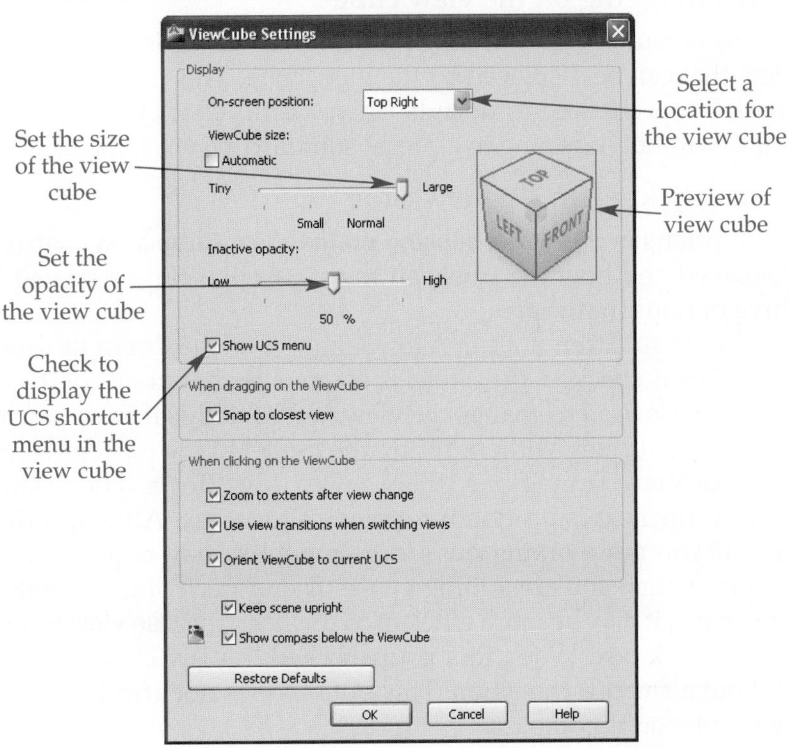

Display

Options in the **Display** area of the **View Cube Settings** dialog box control the appearance of the view cube tool. The thumbnail dynamically previews any changes made to the display options. There are four options in this area of the dialog box.

On-Screen Position. The view cube can be placed in one of four locations in the drawing area: top-right, bottom-right, top-left, or bottom-left corner. By default, it is located in the top-right corner of the screen. To change the location, select it in the **On-screen position:** drop-down list. The **NAVVCUBELOCATION** system variable controls this setting.

View Cube Size. The view cube can be displayed in one of four sizes. Use the **View Cube Size:** slider to set the size to either tiny, small, normal, or large. The slider is unavailable if the **Automatic** check box is checked, in which case AutoCAD sets the size based on the available screen area. The **NAVVCUBESIZE** system variable controls this setting.

Inactive Opacity. When the view cube is inactive, it is displayed in a semitransparent state. Remember, the view cube is in the inactive state whenever the pointer is not over it. When the view cube is in the active state, it is displayed at 100% opacity. Use the **Inactive opacity:** slider to set the level of opacity (transparency). The value can be from 0% to 100% with a default value of 50%. The percentage is displayed below the slider. A value of zero results in the view cube being hidden until the cursor is moved over it. If opacity is set to 100%, there is no difference between the inactive and active states. The **NAVVCUBEOPACITY** system variable controls this setting.

Show UCS menu. By default, the UCS shortcut menu is displayed in the view cube. This menu is displayed by picking the UCS button below the cube. If you wish to remove the UCS menu from the view cube display, uncheck the **Show UCS menu** check box.

When dragging on the view cube

By default, when you drag the view cube, the view "snaps" to the closest standard view that can be displayed by the view cube. This is because the **Snap to closest view** check box is checked by default. Uncheck this check box if you want the view to freely rotate without snapping to a preset standard view as you drag the view cube.

When clicking on the view cube

Options in the **When clicking on the View Cube** area control how the final view is displayed and how the labels on the view cube can be related to the UCS. There are three options in this area.

Zoom to Extents After View Change. If the **Zoom to extents after view change** check box is checked, the model is zoomed to the extents of the drawing whenever the view cube is used to change the view. Uncheck this if you want to use the view cube to change the display without fitting the model to the current viewport.

Use View Transitions When Switching Views. The default transition from one view to the next is a smooth rotation of the view. Although this transition may look nice, if you are working on a large model it may require more time and computer resources than you are willing to use. Therefore, if you are switching views a lot using view cube, it may be more efficient to uncheck the **Use view transitions when switching views** check box. When this is unchecked, a view change just cuts to the new view without a smooth transition. This option does not affect the view when dragging the view cube or its compass.

Orient View Cube to Current UCS. As you have seen, the view cube is aligned to the current UCS by default. In other words, the top face of the cube is always perpendicular to the Z axis of the UCS. However, this can be turned off by unchecking the **Orient View Cube to current UCS** check box. The **NAVVCUBEORIENT** system variable controls this setting. When unchecked, the faces of the view cube are not reoriented

Figure 4-11.
When the **Orient View Cube to current UCS** option is off in the **View Cube Settings** dialog box, the UCS shortcut has WCS displayed above it.

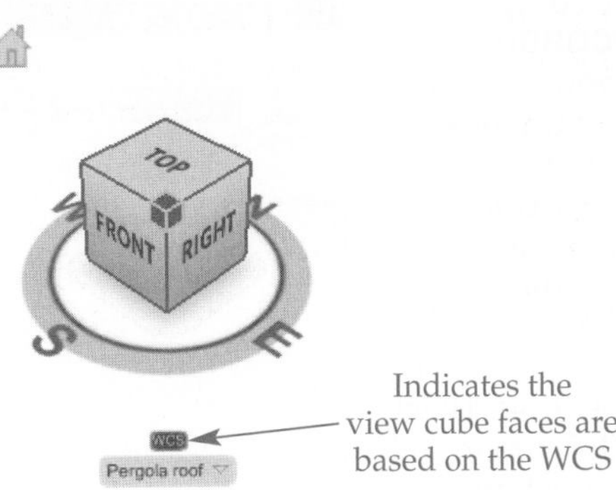

when the UCS is changed. It may be easier to visualize view changes if the view cube faces are oriented to the current UCS.

When the **Orient View Cube to current UCS** check box is unchecked, WCS is displayed above the UCS shortcut menu in the view cube (unless the WCS is current). See Figure 4-11. This is a reminder that the labels on the view cube relate to the WCS and not to the current UCS.

Keep Scene Upright. If the **Keep scene upright** check box is checked, the view of the model cannot be turned upside down. When unchecked, you may accidentally rotate the view so it is upside down. This can be confusing, so it is best to leave this box checked.

Show Compass below the View Cube. The compass is displayed by default in the view cube. However, if you do not find the compass useful, you can turn it off. To hide the compass, uncheck the **Show compass below the View Cube** check box.

Restore Defaults. After making several changes in the **View Cube Settings** dialog box you may get confused about the effect of different option settings on the model display. In this case, it is best to pick the **Restore Defaults** button to return all settings to their original values. Then change one setting at a time and test it to be sure the view cube functions as you intended.

Exercise 4-3

Complete the exercise on the student website.
www.g-wlearning.com/CAD

Creating a Continuous 3D Orbit

The **3DCORBIT** command provides the ability to create a continuous orbit of a model. By moving your pointing device, you can set the model in motion in any direction and at any speed, depending on the power of your computer. An impressive display can be achieved using this command. The command is located in the **Orbit** drop-down list in the **Navigate** panel of the **View** tab on the ribbon. See Figure 4-12A. Additionally, it can be selected on the navigation bar.

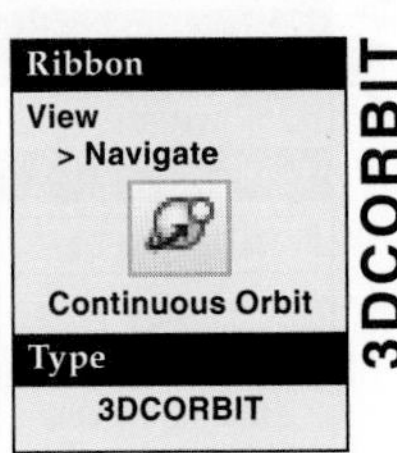

Once the command is initiated, the continuous orbit cursor is displayed. See Figure 4-12B. Press and hold the pick button and move the pointer in the direction that you want the model to rotate and at the desired speed of rotation. Release the button when the pointer is moving at the appropriate speed. The model will continue to rotate until you pick the left mouse button, press [Enter] or [Esc], or right-click and pick **Exit**

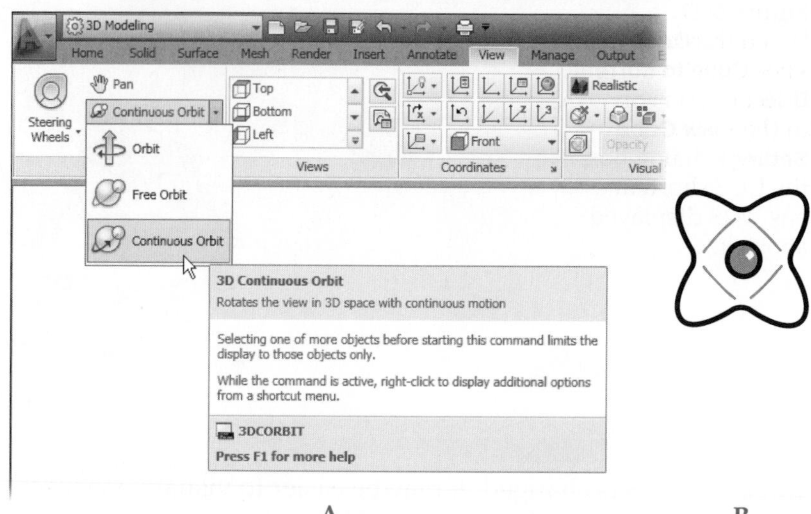

Figure 4-12.
A—Selecting the **3DCORBIT** command. B—This is the continuous orbit cursor in the **3DCORBIT** command (or **Continuous** option of the **3DORBIT** command). Pick and hold the left mouse button. Then, move the cursor in the direction in which you want the view to rotate and release the mouse button.

or another option in the shortcut menu. At any time while the model is orbiting, you can left-click and adjust the rotation angle and speed by repeating the process for starting a continuous orbit.

Plan Command Options

PLAN

Type
PLAN

The **PLAN** command, introduced in Chapter 1, allows you to create a plan view of any user coordinate system (UCS) or the world coordinate system (WCS). The **PLAN** command automatically performs a **ZOOM Extents**. This fills the graphics window with the plan view. The command options are:

- **Current UCS.** This creates a view of the object that is plan to the current UCS.
- **UCS.** This displays a view plan to a named UCS. The preset UCSs are not considered named UCSs.
- **World.** This creates a view of the object that is plan to the WCS. If the WCS is the current UCS, this option and the **Current UCS** option produce the same results.

This command may have limited usefulness when working with 3D models. The dynamic capabilities of the view cube are much more intuitive and may be quicker to use. However, you may find instances where the **PLAN** command is easier to use, such as when the view cube is not currently displayed.

Displaying Models with Steering Wheels

NAVSWHEEL

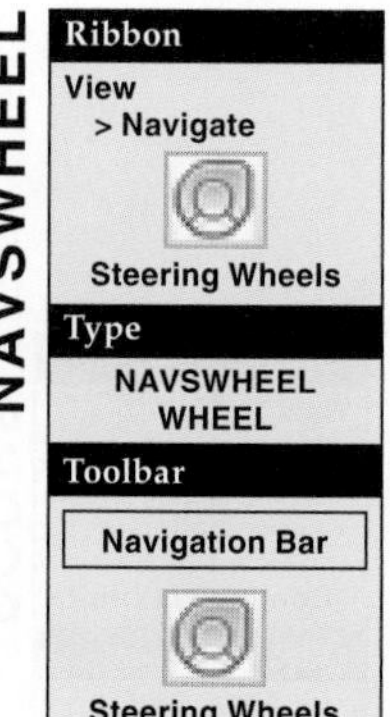

Steering wheels, or *wheels,* are dynamic menus that provide quick access to view-navigation tools. A steering wheel follows the cursor as it is moved around the drawing. Each wheel is divided into wedges and each wedge contains a tool. See Figure 4-13. The **NAVSWHEEL** command is used to display a steering wheel, but the quickest method is to use the steering wheel drop-down menu on the navigation bar. All of the steering wheel options discussed in this chapter can be selected in the drop-down list on the navigation bar.

AutoCAD provides two basic types of wheels: View Object and Tour Building. The options contained in these two wheels are combined to form the Full Navigation wheel. Each of these three types can be used in a full-wheel display or a minimized format.

Figure 4-13.
A—Full Navigation wheel. B—View Object wheel. C—Tour Building wheel.

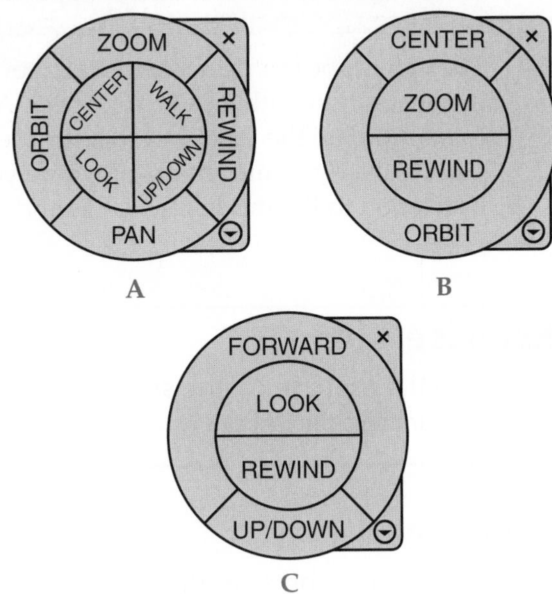

This section discusses the Full Navigation wheel and the two basic wheels in their full and mini formats. There is also a 2D Wheel that is displayed in paper (layout) space or when the **NAVSWHEELMODE** system variable is set to 3. The functions of this wheel are covered in detail in *AutoCAD and Its Applications—Basics.*

Using a Steering Wheel

Once a steering wheel is displayed, move the cursor around the screen. Notice as you move the cursor, the wheel follows it. When you stop the cursor, the wheel stops. If you move the cursor anywhere inside of the wheel, the wheel remains stationary. Note also that as you move the cursor inside of the wheel, a wedge (tool) is highlighted. If you pause the cursor over a tool, a tooltip is displayed that describes the tool.

To use a specific tool, simply pick and hold on the highlighted wedge, then move the cursor as needed to change the view of the model. Once you release the pick button, the tool ends and the wheel is redisplayed. Some options display a *center point* about which the display will move. The **Center** tool is used to set the center point. These features are all discussed in the next sections.

To change between wheels, right-click to display the shortcut menu. Then, select **Full Navigation Wheel** from the menu to display that wheel. Or, select **Basic Wheels** to display a cascading menu. Select either **View Object Wheel** or **Tour Building Wheel** from the cascading menu to display that wheel. Refer to **Figure 4-14.**

Figure 4-14.
The shortcut menu allows you to switch between steering wheels.

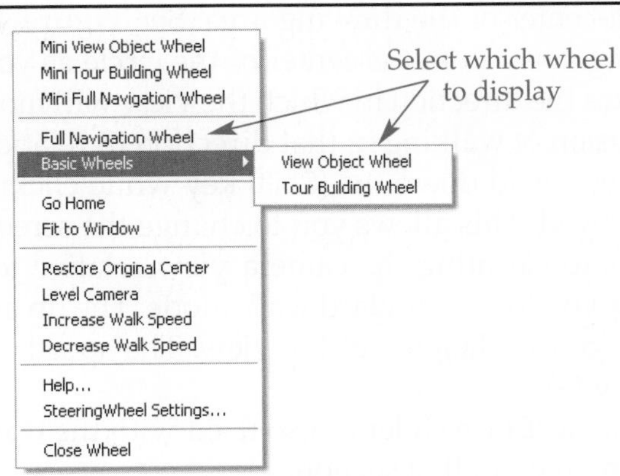

PROFESSIONAL TIP

The view cube is a valuable visualization tool, especially when used in conjunction with a steering wheel. As you use the wheel options to manipulate the model, watch the view cube move. This provides you with a dynamic "map" of where your viewpoint is at all times in relation to the model, the UCS, and the view cube compass.

Exercise 4-4

Complete the exercise on the student website.
www.g-wlearning.com/CAD

Full Navigation Wheel

The Full Navigation wheel contains all of the tools available in the View Object and Tour Building wheels. Refer to Figure 4-13A. These tools are discussed in the following sections. It also contains the **Pan** and **Walk** tools, which are not available in either of the other two wheels. These tools are discussed next. The tools shared with the View Object and Tour Building tools are discussed in the sections corresponding to those tools.

In addition, a number of settings are available to change the appearance of the steering wheel and the manner in which some of the tools function. Refer to the Steering Wheel Settings section later in this chapter for a complete discussion of these settings.

Pan

The **Pan** tool allows you to move the model in the direction that you drag the cursor. This tool functions exactly the same as the AutoCAD **RTPAN** command. When you pick and hold on the tool, the cursor changes to four arrows with the label Pan Tool below the cursor.

If you use the **Pan** tool with the perspective projection current, it may appear that the model is slowly rotating about a point. This is not the case. What you are seeing is merely the effect of the vanishing points. As you pan, the relationship between the viewpoint and the vanishing points changes. You can quickly test this by closing the wheel, right-clicking on the view cube, and picking **Parallel** from the shortcut menu. Now display the wheel again and use the **Pan** tool. Notice the difference. The model pans without appearing to rotate.

Walk

The **Walk** tool is used to simulate walking toward, through, or away from the model. When you pick and hold on the tool, the center circle icon is displayed at the bottom-center of the drawing area. See Figure 4-15. The cursor changes to an arrow pointing away from the center of the circle as you move it off of the circle. The arrow indicates the direction in which the view will move as you move the mouse. This gives the illusion of walking in that direction in relation to the model.

If you hold down the [Shift] key while clicking the **Walk** tool, the **Up/Down** slider is displayed. This allows you to change the screen Y axis orientation of the view. This equates to elevating the camera view relative to the object. Releasing the [Shift] key returns you to the standard walk mode. The up and down arrow keys can also be used to change the "height" of the view. The speed of walking can be increased with the plus key (+).

The **Up/Down** slider is also used with the **Up/Down** tool on the Tour Building wheel. It is explained in that section.

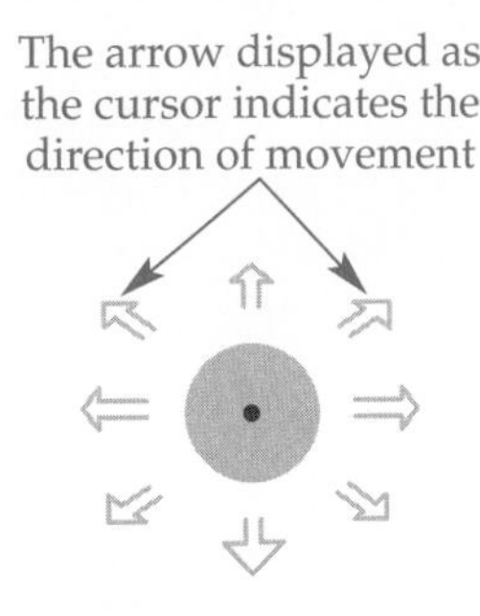

Figure 4-15.
The **Walk** tool displays the center circle icon. As you move the cursor around the icon, one of the arrows shown here is displayed to indicate the direction in which the view is being moved.

View Object Wheel

If you think of your model as a building, the tools in the View Object wheel are used to view the outside of the building. This wheel contains four navigation tools: **Center**, **Zoom**, **Rewind**, and **Orbit**. Refer to Figure 4-13B.

Center

The **Center** tool is used to set the center point for the current view. Many tools, such as **Zoom** and **Orbit**, are applied in relation to the center point. Pick and hold the **Center** tool, then move the cursor to a point on the model and release. The display immediately changes to center the model on that point. See Figure 4-16. The selected point must be on an object, but it does not need to be on a solid. Notice that the center point icon resembles a globe with three orbital axes. These axes relate to the three axes of the model shown on the UCS icon.

Zoom

The **Zoom** tool is used to dynamically zoom the view in and out, just as with the AutoCAD **RTZOOM** command. The tool uses the center point set with the **Center** tool. When you pick and hold the **Zoom** tool in the wheel, the center point is displayed at its current location. The cursor changes to a magnifying glass with the label Zoom Tool displayed below it. See Figure 4-17. There are three different ways to use the **Zoom** tool, as discussed in the next sections.

When the **Zoom** tool is accessed in the Full Navigation wheel, the center point is relocated to the position of the steering wheel. If you wish to zoom on the existing center point when using the Full Navigation wheel, first press the [Ctrl] key and then access the **Zoom** tool. This prevents the tool from relocating the center point. You can also move the center point using the **Center** tool, then switch to the View Object wheel

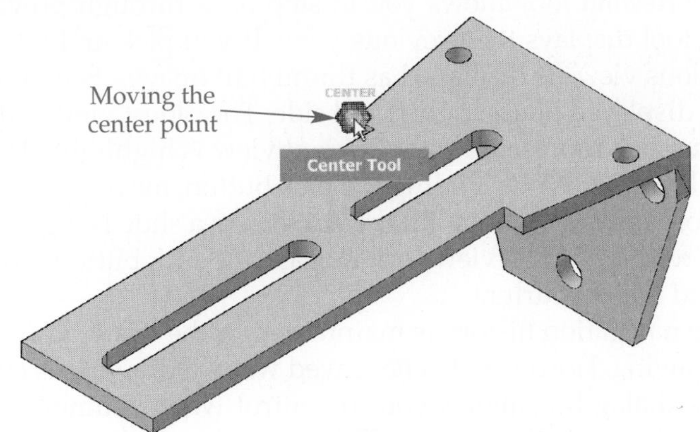

Figure 4-16.
The **Center** tool allows you to select a new center point for the current view. This becomes the point about which many steering wheel tools operate.

Figure 4-17.
The **Zoom** tool operates in relation to the center point. If the tool is selected from the Full Navigation wheel, the center point is automatically relocated to the cursor location.

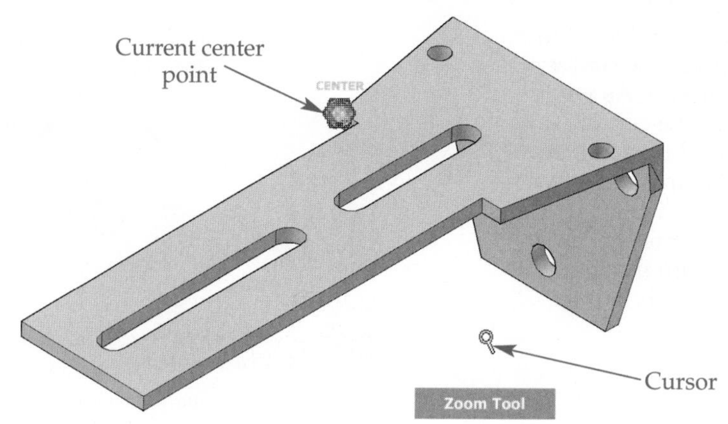

and access the **Zoom** tool from that wheel. In either case, the zoom is relative to the location of the center point.

NOTE

Once you close the steering wheel, the center point is reset. The next time a steering wheel is displayed, the center point will be in the middle of the current view.

Pick and drag. To dynamically zoom, similar to realtime zoom, pick and drag the cursor. As the cursor is moved up or to the right, the viewpoint moves closer (zoom in). Move the pointer to the left or down and the viewpoint moves farther away (zoom out). The zooming is based on the current center point. When you have achieved the appropriate zoom location, release the pointer button.

Single click. If you select the **Zoom** tool with a single click (view object wheel), the view of the model zooms in by an incremental percentage. Each time you single click on the tool, the view is zoomed by 25%. The zoom is in relation to the center point.

In order for this function to work when the **Zoom** tool is accessed from the Full Navigation wheel, you must check the **Enable single click incremental zoom** check box in the **Steering Wheel Settings** dialog box. This is discussed in detail later.

Shift and click. If you press and hold the [Shift] key and then pick the **Zoom** tool (view object wheel), the view is zoomed out by 25%. As with the single-click method, the **Enable single click incremental zoom** check box must be checked in order to use this with the Full Navigation wheel.

Rewind

The **Rewind** tool allows you to step back through previous views. A single pick on this tool displays the previous view. If you pick and hold the tool, a "slide show" of previous views is displayed as thumbnail images. See Figure 4-18. The most recent view is displayed on the right-hand side. The oldest view is displayed on the left-hand side. The slide representing the current view is highlighted with an orange frame and a set of brackets. While holding the pick button, move the cursor to the left. Notice that the set of brackets moves with the cursor. As a slide is highlighted, the corresponding view is restored in the viewport. Release the pick button when you find the view you want and it is set current.

The navigation history is maintained in the drawing file and is different for each open drawing. However, it is not saved when a drawing is closed. The **Steering Wheel Settings** dialog box allows you to control when thumbnail images are created and saved in the navigation history. This dialog box is discussed later in the chapter.

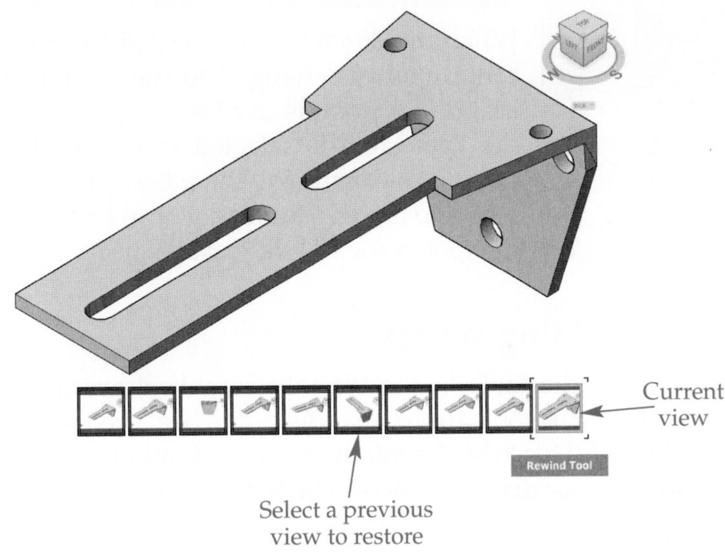

Figure 4-18.
A single pick on the **Rewind** tool displays the previous view. A "slide show" of previous views is displayed as thumbnail images if you press and hold the pick button.

Orbit

The **Orbit** tool allows you to completely rotate your point of view around the model in any direction. The view pivots about the center point set with the **Center** tool. When using the Full Navigation wheel, the center point can be quickly set by pressing and holding the [Ctrl] key, then picking and holding the **Orbit** tool. Next, drag the center point to the desired pivot point on the model and release. Now you can use the **Orbit** tool.

To use the **Orbit** tool, pick and hold on the tool in the steering wheel. The current center point is displayed with the label Pivot. Also, the cursor changes to a point surrounded by two circular arrows. See Figure 4-19. Move the cursor around the screen and the view of the model pivots about the center point. If this is not the result you wanted, just reset the pivot point.

PROFESSIONAL TIP

Since the **Orbit** tool is most often used to quickly move your viewpoint to another side of the model, it is more intuitive to locate the pivot point somewhere on the model. First use the **Center** tool to establish the pivot point. Then, when you use the **Orbit** tool, the view pivots about that part of the model. If the center point is not set, it defaults to the center of the screen.

Figure 4-19.
The **Orbit** tool allows you to rotate your point of view completely around the model in any direction. This is the cursor displayed for the tool.

CAUTION

When the **Zoom** tool is selected from the Full Navigation wheel, the center point is changed to the steering wheel location. Therefore, that center point is used as the pivot point for the **Orbit** tool. When the **Zoom** tool is selected from the View Object wheel, the center point is not changed. Therefore, it is best to switch to the View Object wheel to use the **Zoom** tool or to press the [Ctrl] key before accessing the tool in the Full Navigation wheel.

Tour Building Wheel

Where the tools in the View Object wheel are used to view the outside of the "building," the tools in the Tour Building wheel are used to move around inside of the "building." This wheel contains four navigation tools. The **Forward**, **Look**, and **Up/Down** tools are discussed here. The **Rewind** tool is discussed earlier.

Forward

The principal tool in the Tour Building wheel is the **Forward** tool. It is similar to the **Walk** tool in the Full Navigation wheel. However, it only allows forward movement from the current viewpoint. This tool requires a center point to be set on the model from within the tool. The existing center point cannot be used.

First, move the cursor and steering wheel to the point on the model that will be the target (center point). Next, pick and hold the **Forward** tool. The pick point becomes the center point and a drag distance indicator is displayed. See Figure 4-20. This indicator shows the starting viewpoint, the center point, and the surface of the model that you selected. Hold the mouse button down while moving the pointer up. The orange location slider moves to show the current viewpoint relative to the center point. As you move closer to the model, the green center point icon increases in size, which also provides a visual cue to the zoom level.

Look

The **Look** tool is used to rotate the view about the center of the view. When the tool is activated, the cursor appears as a half circle with arrows. See Figure 4-21. As you move the cursor down, the model moves up in the view as if you are actually tilting your head down to see the top of the model. Similarly, as you "look" away from the model to the right or left, the model appears to move away from your line of sight. If this does not seem intuitive to you, it is easy to change. Open the **Steering Wheel Settings** dialog box and check the **Invert vertical axis for Look tool** check box. Now when using the **Look** tool, moving the cursor in one direction also moves the object in the same direction.

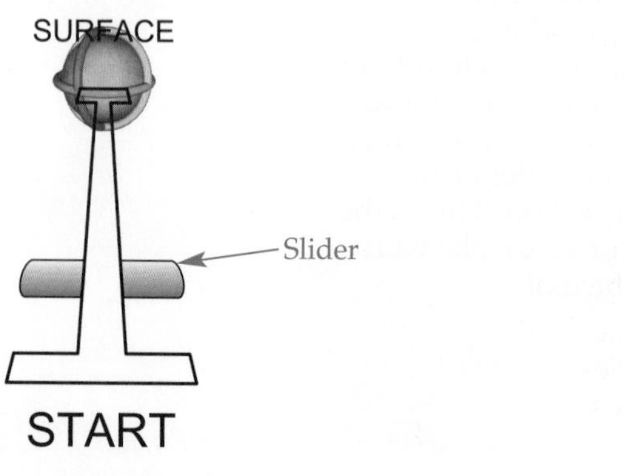

Figure 4-20.
The drag distance indicator is displayed when using the **Forward** tool. This indicator shows the start point of the view and the selected surface of the model. The slider indicates the current view position relative to the starting point.

Figure 4-21.
The **Look** tool cursor appears as a half circle with arrows.

The distance between you and the model remains the same and the orientation of the model does not change. Therefore, you would not want to use this tool if you wanted to see another side of the model.

Up/Down

As the name indicates, the **Up/Down** tool moves the view up or down along the Y axis of the screen, regardless of the orientation of the current UCS. Pick and hold on the tool and the vertical distance indicator appears. See Figure 4-22. Two marks on this indicator show the upper and lower limits within which the view can be moved. The orange slider shows the position of the view as you move the cursor. When the tool is first activated, the view is at the top position. The **Up/Down** tool has limited value. The **Pan** tool is far more versatile.

PROFESSIONAL TIP

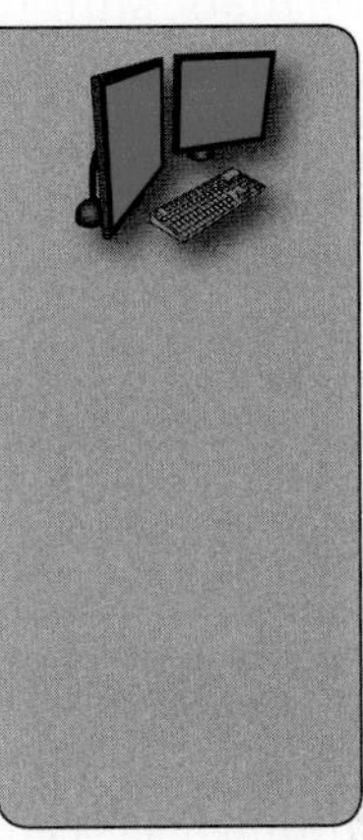

Using steering wheel tools such as **Forward**, **Look**, **Orbit**, and **Walk**, you can manipulate your view to fully explore the model. The **Rewind** tool can then be used to replay all of the previous views saved in the navigation history. This process allows you to find views of the model that you may want to save as named views for later use in model construction or for shots created with show motion (discussed in Chapter 5). Remember, views created using steering wheels are not saved with the drawing file. Therefore, it may be a time-saver to create named views in this manner if there is a possibility they will be needed later.

Exercise 4-5

Complete the exercise on the student website.
www.g-wlearning.com/CAD

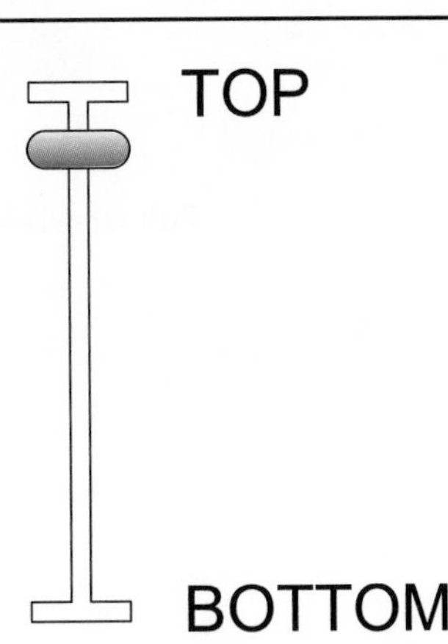

Figure 4-22.
The indicator displayed when using the **Up/Down** tool shows the upper and lower limits within which the view can be moved. The top position is the location of the view when the tool is selected.

Mini Wheels

The three wheels discussed above were presented in their *full wheel* formats. As you gain familiarity with the use and function of each wheel and its tools, you may wish to begin using the abbreviated formats. The abbreviated formats are called ***mini wheels.*** See **Figure 4-23.** The mini wheels can be selected in the steering wheel shortcut menu. When you select a mini wheel, it replaces the cursor pointer.

As you move the mouse, the mini wheel follows. Slowly move the mouse in a small circle and notice that each wedge of the wheel is highlighted. The name of the currently highlighted tool appears below the mini wheel. When the tool you need is highlighted, simply click and hold to activate the tool. All tools in the mini wheels function the same as those in the full-size wheels. The only difference is the appearance of the wheels.

Steering Wheel Shortcut Menu

A quick method for switching between the different wheel formats is to use the steering wheel shortcut menu. Pick the menu arrow at the lower-right corner of a full wheel to display the menu. See **Figure 4-24.** Additionally, you can right-click when any full or mini wheel is displayed to access the menu. To select a wheel or change wheel formats, simply select the appropriate entry in the menu. Note that a check mark is *not* placed by the current wheel.

In addition to selecting a wheel, the shortcut menu provides options for viewing the model. These additional options are:

- **Go Home.** Returns the display to the home view. This is the same as picking the **Home** icon in the view cube.
- **Fit to Window.** Resizes the current view to fit all objects in the drawing inside of the window. This is essentially a zoom extents operation.

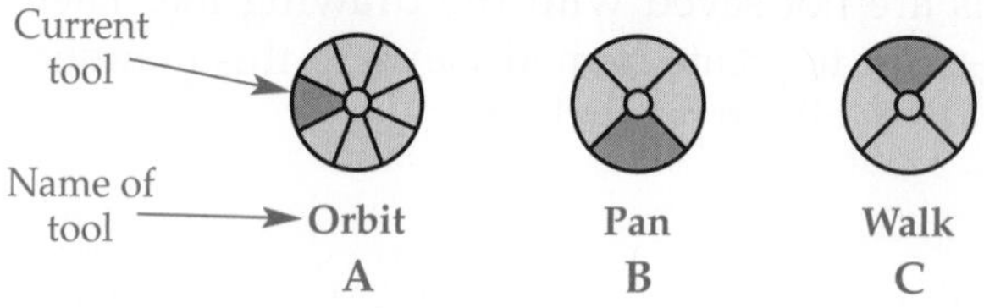

Figure 4-23. In addition to full-size steering wheels, mini wheels can be used. A—Mini Full Navigation wheel. B—Mini View Object wheel. C—Mini Tour Building wheel.

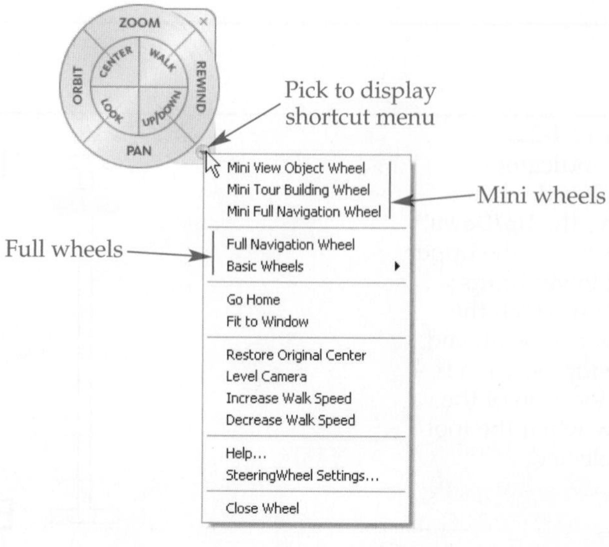

Figure 4-24. Select the down arrow to display the steering wheel shortcut menu. Both full-size wheels and mini wheels can be displayed using this menu.

- **Restore Original Center.** Restores the original center point of the drawing using the current drawing extents. This does not change the current zoom factor. Therefore, if you are zoomed close into an object and pick this option, the object may disappear from view by moving off of the screen.
- **Level Camera.** The camera (your viewpoint) is rotated to be level with the XY ground plane.
- **Increase Walk Speed.** The speed used by the **Walk** tool is increased by 100%.
- **Decrease Walk Speed.** The speed used by the **Walk** tool is decreased by 50%.
- **Help.** Displays the online documentation (help file) for steering wheels.
- **Steering Wheel Settings.** Displays the **Steering Wheel Settings** dialog box, which is discussed in the next section.
- **Close Wheel.** Closes the steering wheel. This is the same as pressing [Esc] to close the wheel.

Exercise 4-6

Complete the exercise on the student website.
www.g-wlearning.com/CAD

Steering Wheel Settings

The **Steering Wheels Settings** dialog box provides options for wheel appearance. See Figure 4-25. It also contains settings for the display and operation of several tools. It is displayed by picking **Steering Wheels Settings** in the shortcut menu.

Changing wheel appearance

The two areas at the top of the **Steering Wheels Settings** dialog box allow you to change the size and opacity of all wheels. The settings in the **Big Wheels** area are for the full-size wheels. The settings in the **Mini Wheels** area are for the mini wheels. The **Wheel size:** slider in each area is used to display the wheels in small, normal, or large

Figure 4-25.
The **Steering Wheels Settings** dialog box provides options for wheel appearance and the display and operation of several tools.

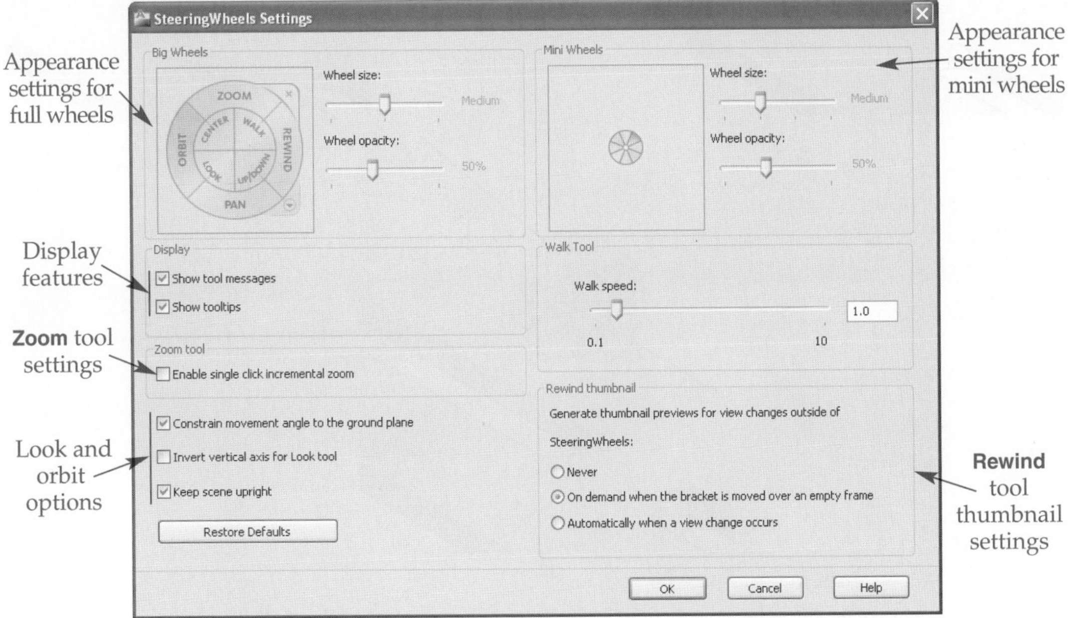

Figure 4-26.
The size of a steering wheel can be set in the **Steering Wheels Settings** dialog box or by using a system variable.

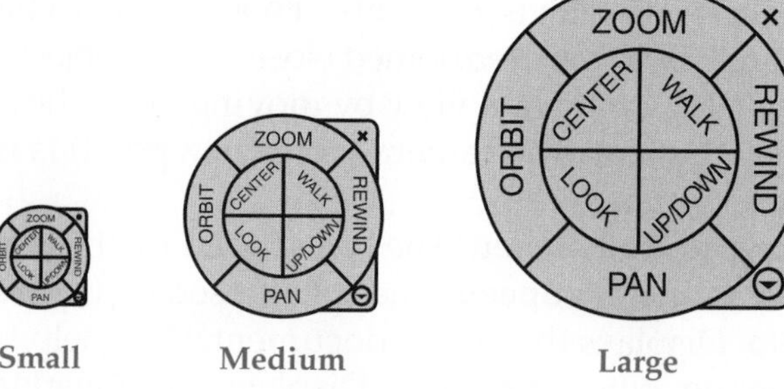

size. See Figure 4-26. The mini wheel has a fourth, extra large size. These sliders set the **NAVSWHEELSIZEBIG** and **NAVSWHEELSIZEMINI** system variables.

The **Wheel opacity:** slider in each area controls the transparency of the wheels. These sliders can be set to a value from 25% to 90% opacity. The sliders control the **NAVSWHEELOPACITYBIG** and the **NAVSWHEELOPACITYMINI** system variables. The appearance of the full wheel in three different opacity settings is shown in Figure 4-27.

Display features

The **Display** area of the **Steering Wheels Settings** dialog box controls three features of the wheel display. These settings determine whether messages and tooltips are displayed.

Figure 4-27.
The opacity of a steering wheel can be changed. A—Opacity of 25%. B—Opacity of 50%. C—Opacity of 90%.

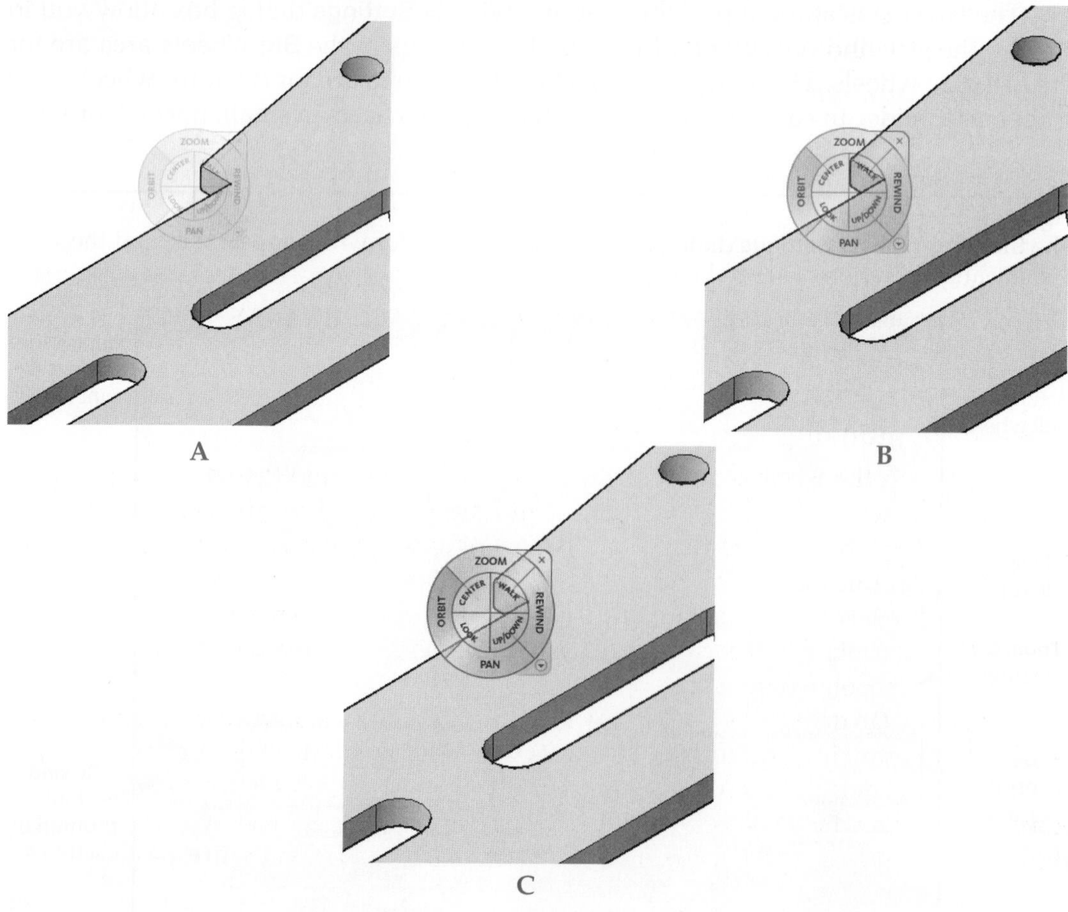

When a tool is selected, its name is displayed below the cursor. In addition, some tools have features or restrictions that can be indicated in a tool message. To see these messages, the **Show tool messages** check box must be checked. Otherwise, the messages are not displayed, but the restrictions remain in effect.

A tooltip is a short message that appears below the wheel when the cursor is hovered over the wheel. When the **Show tooltips** check box is checked, you can hold the cursor stationary over a tool for approximately three seconds and the tooltip will appear. Then, as you move the cursor over tools in the wheel, the appropriate tool tip is immediately displayed.

Walk tool

By default, when the **Walk** tool is used, you move parallel to the ground plane. This is because the **Constrain movement angle to ground plane** check box is checked. Test this by selecting the **Walk** tool and then move the cursor toward the top of the screen. You appear to be "walking" over the top of the model. Now, open the **Steering Wheels Settings** dialog box and uncheck the **Constrain movement angle to ground plane** box. This allows you to "fly" in the direction the cursor is moved when using the **Walk** tool. Exit the dialog box and again select the **Walk** tool. Move the cursor toward the top of the screen. This time it appears that you are flying directly toward or into the model.

The speed at which you walk through or around the model is controlled by the **Walk speed:** slider. The value can also be changed by typing in the text box at the right-hand end of the slider. The greater the value, the faster you will move as the cursor is moved away from the center circle icon.

Zoom tool

As discussed earlier, a single click on the **Zoom** tool zooms in on the current view by a factor of 25%. This is controlled by the **Enable single click incremental zoom** check box. When this check box is not checked, a single click on the tool has no effect. Some users find the single-click zoom confusing.

Look, up/down, and orbit tool options

By default, when the **Look** tool is used and the cursor is moved downward, the view of the model moves up, just as if you were moving your eyes down. If you check the **Invert vertical axis for Look tool** option, this movement is reversed. In this case, the model moves in the same direction as the cursor.

If the **Keep scene upright** check box is not checked, it is possible to turn the model upside down while using the **Orbit** tool. This may not be desirable because it can be disorienting. To prevent this, be sure to leave the option checked. Uncheck this option only when you want to use **Orbit** in a "free-floating" mode.

Rewind thumbnail

Options in the **Rewind thumbnail** section of the **Steering Wheels Settings** dialog box control when and how thumbnail images are generated for use with the **Rewind** tool when view changes are made without using a wheel. There are three options in this area. Only one option can be on.

When the **Never** radio button is on, thumbnail images are never generated for view changes made outside of a wheel. Thumbnail images are created for view changes made with a wheel, which is true for all three options.

When the **On demand when the bracket is moved over an empty frame** radio button is on, thumbnail images are not automatically generated for view changes made outside of a wheel. The frames for these views display a double arrow icon when the **Rewind** tool in a wheel is used. However, as the brackets are moved over these frames, thumbnail images are generated. This is the default setting. See Figure 4-28.

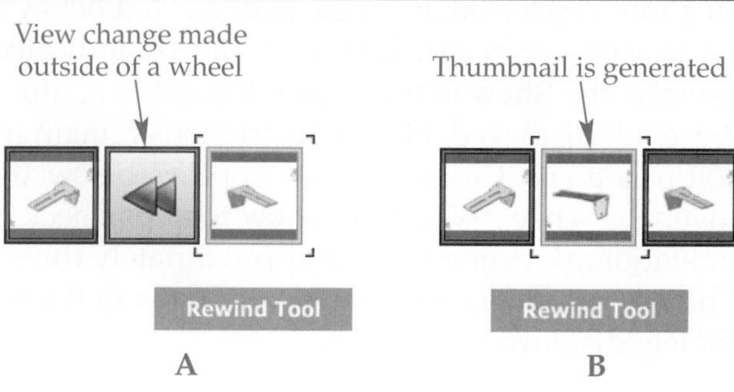

Figure 4-28.
A—By default, when the **Rewind** tool is selected, frames representing view changes made outside of a steering wheel display double-arrow icons.
B—As the brackets are moved over the blank frames, thumbnail images are generated of the views created outside of a wheel.

When the **Automatically when a view change occurs** option is selected, a thumbnail image is generated any time a view change is made outside of a wheel. When the **Rewind** tool in a wheel is used, the frames for these views automatically display thumbnail images.

Restore defaults

Picking the **Restore Defaults** button in the **Steering Wheels Settings** dialog box returns all of the settings in the dialog box to their default values. Select this when at any time you are not sure how the settings are affecting the appearance and function of the steering wheel tools. Then, make changes one at a time as needed.

Exercise 4-7

Complete the exercise on the student website.
www.g-wlearning.com/CAD

Controlling the Display of the Navigation Bar

The navigation bar is a flexible tool that can be customized or positioned to suit the needs of your drawing project. There are five possible tools that can be displayed in the bar. The three principal navigation tools are for the view cube, steering wheels, and show motion. These tools can be removed or added as needed. The bar can be placed at any location around the edge of the screen. In addition, the bar can be linked to the location of the view cube.

Repositioning the Navigation Bar

The navigation bar can be moved around the screen by picking the customize button at the lower-right corner of the bar, then select **Docking positions** to display the cascading menu. See Figure 4-29. The default setting is for the bar to be linked to the location of the view cube. This means if the position of the view cube is changed, the navigation bar will follow. Test this by first confirming that **Link to View Cube** is checked in the menu. Next, move the view cube to one of the other locations, such as top left, using the **View Cube Settings** dialog box. Notice that the view cube and navigation bar move to the new location. The **Undo** button on the **Quick Access** toolbar can quickly restore the previous location.

Figure 4-29.
The navigation bar can be moved around the screen by selecting the customize button at the lower-right corner of the bar, then selecting **Docking positions** in the menu.

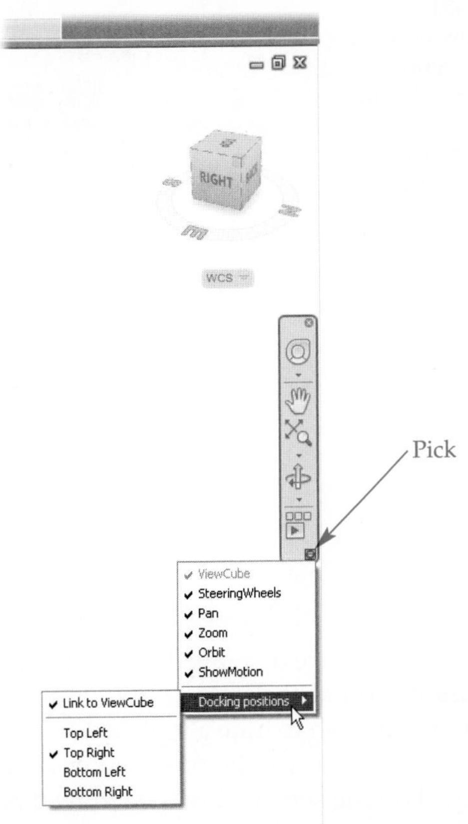

You can also move the navigation bar to a position of your choosing without moving the view cube. Simply uncheck the **Link to View Cube** option. Notice that a band appears at the top of the bar. When the pointer is moved into this band and you click, it becomes a move cursor. See Figure 4-30. Drag the bar to any edge location on the screen. The bar will dock at the closest screen edge. You cannot place it in a floating position in the middle of the screen.

Figure 4-30.
Click and hold to move the navigation bar to any position along the edge of the screen.

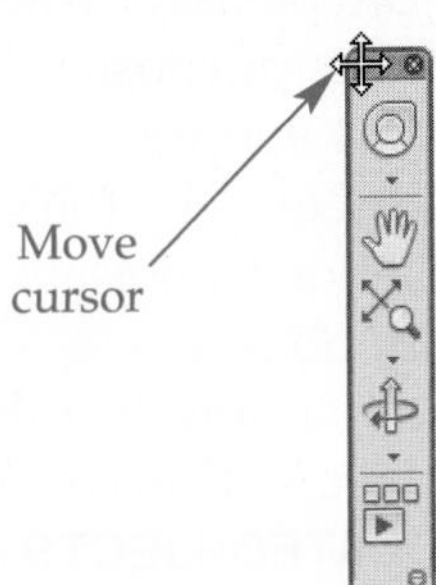

Figure 4-31.
If the view cube is not currently displayed, a button for displaying it appears on the navigation bar.

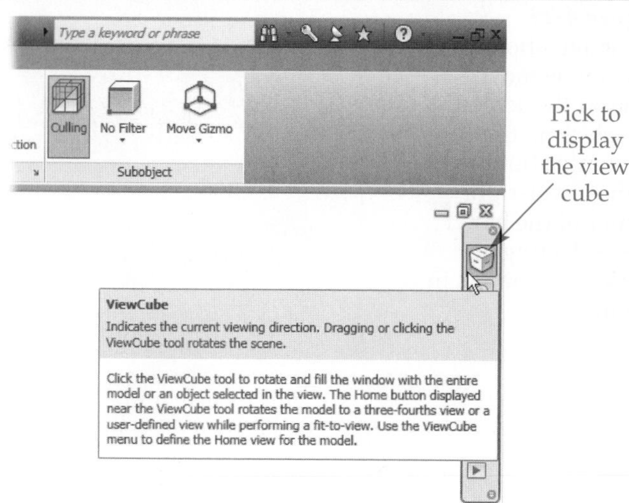

Customizing the Navigation Bar

The tools that are displayed in the navigation bar can be changed by selecting the **Customize** button at the lower-right corner of the bar. By default, all of the tools are checked. If you wish to hide a tool, just select it to uncheck it in the list. All of the tools in the navigation bar can be removed or redisplayed in this manner. If the view cube is displayed on the screen, it will be grayed out in the list. If the view cube is currently not displayed, an option for it appears in the menu. A button for the view cube also appears on the navigation bar. See Figure 4-31. The view cube can then be turned on using the navigation bar.

HIDEOBJECTS

Type
HIDEOBJECTS
Toolbar
Status Bar

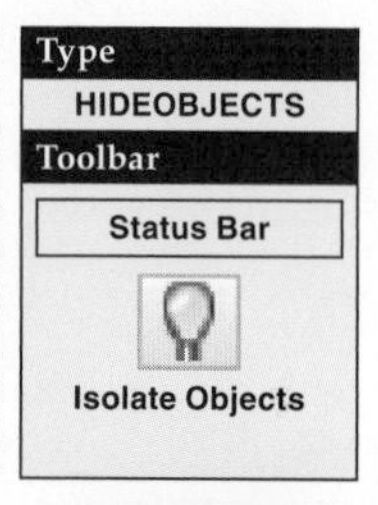

ISOLATEOBJECTS

Type
ISOLATEOBJECTS
ISOLATE
Toolbar
Status Bar
Isolate Objects

UNISOLATEOBJECTS

Type
UNISOLATEOBJECTS
UNISOLATE
UNHIDE
Toolbar
Status Bar

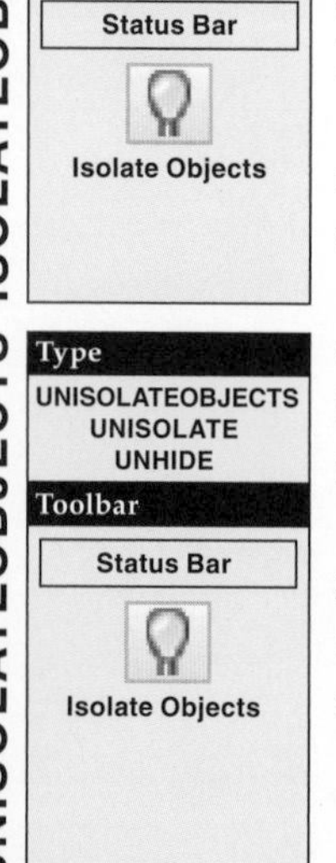

Hiding and Isolating Objects

Complex 3D models require the manipulation of viewpoints and even freezing layers in order to work on details. Instead of freezing layers, the **HIDEOBJECTS** and **ISOLATEOBJECTS** commands can be used to selectively view any aspect of the model for ease of construction and editing purposes.

The **HIDEOBJECTS** command suppresses *selected* objects from view. This allows you to work on objects that are obscured by the selected objects. The **ISOLATEOBJECTS** command suppresses the display of all objects except those objects that are selected. In other words, it hides *unselected* objects.

Once the work is completed, simply use the **UNISOLATEOBJECTS** command to restore the view of all hidden objects. This command displays all objects hidden with either the **HIDEOBJECTS** or **ISOLATEOBJECTS** command.

The commands can also be selected from the **Isolate** flyout of the shortcut menu displayed by right-clicking in the drawing window. See Figure 4-32. In this menu, the **End Object Isolation** selection issues the **UNISOLATEOBJECTS** command. A similar menu is displayed when the **Isolate Objects** button is picked on the status bar.

Consider an example of how to use these commands. The roof panel in Figure 4-33A needs to be hidden from view to construct framing details. Simply select **Hide Objects** from the **Isolate** flyout in the shortcut menu and the select the object(s) to hide. Figure 4-33B shows one roof panel hidden from view.

Figure 4-32.
Using the shortcut menu to hide and show objects.

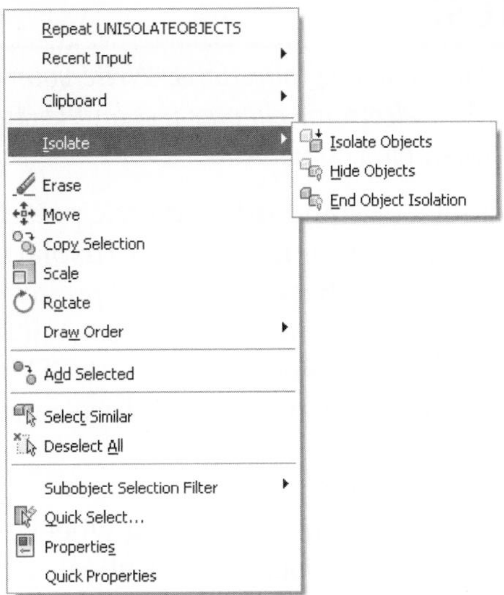

Conversely, suppose the floor framing joists need to be exposed for work. For example, you may need to design or determine hardware for the joist connections. In this case, select **Isolate Objects** from the **Isolate** flyout in the shortcut menu and the select the object(s) you want to view. After pressing [Enter], all of the *unselected* objects in the model are removed from the view, thus isolating the selected objects. The result is shown in Figure 4-33C.

PROFESSIONAL TIP

If you are confused about what parts of the model you have hidden or isolated, just pick **End Object Isolation** from the **Isolate** flyout of the cursor menu to restore the entire model.

Figure 4-33.
A—The roof panel is in the way and needs to be hidden. B—The roof panel is hidden. C—Only the floor joists were selected. The unselected objects are hidden.

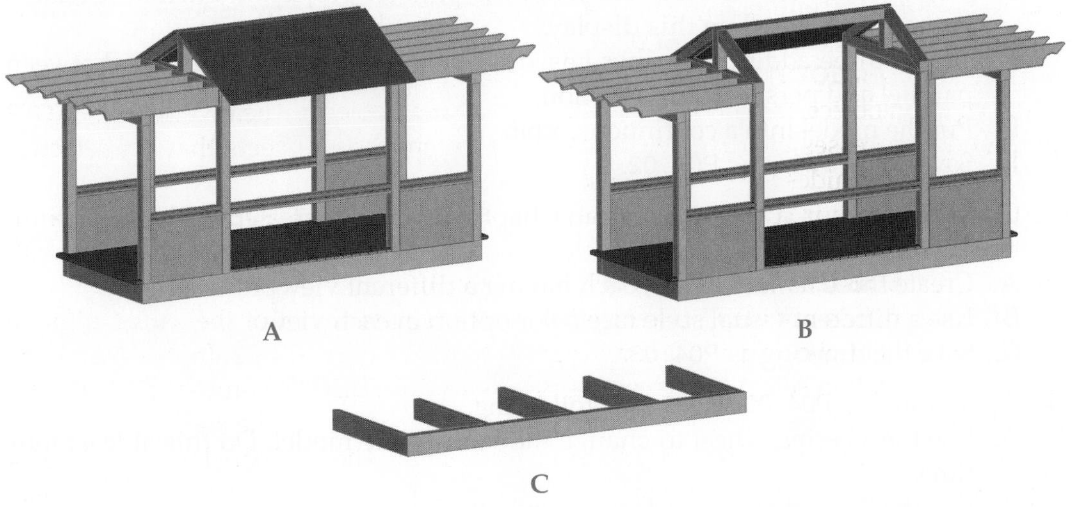

Chapter Test

Answer the following questions. Write your answers on a separate sheet of paper or complete the electronic chapter test on the student website.
www.g-wlearning.com/CAD

1. How do you select a standard isometric preset view using the view cube?
2. How is a standard orthographic view displayed using the view cube?
3. What is the difference between *parallel projection* and *perspective projection?*
4. What happens when one of the four view cube compass letters is picked?
5. Which command generates a continuous 3D orbit?
6. Which command can be used to produce a view that is parallel to the XY plane of the current UCS?
7. What is a *steering wheel?*
8. Briefly describe how to use a steering wheel.
9. The principal tool in the Tour Building wheel is the **Forward** tool. What is the purpose of this tool?
10. What are the three principal navigation tools found in the navigation bar?

Drawing Problems

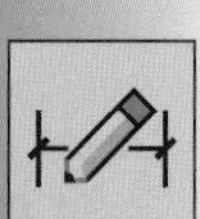

1. Open one of your 3D drawings from Chapter 2 or 3. Do the following.
   A. Use the view cube to create a pictorial view of the drawing.
   B. Set the Conceptual visual style current.
   C. Using the steering wheel, display different views of the model.
   D. Use the **Rewind** tool in the steering wheel to restore a previous view.
   E. Save the restored view as a named view.
   F. Save the drawing as P04_01.
2. Open one of your 3D drawings from Chapter 2 or 3. Do the following.
   A. Toggle the projection from parallel to perspective and turn off the view cube compass.
   B. Pick an edge on the view cube. Set the parallel projection current. Create a named view based on this display.
   C. Create three additional views based on view cube edges. Alternate between parallel and perspective projection.
   D. Put the model into a continuous orbit.
   E. Save the drawing as P04_02.
3. Open one of your 3D drawings from Chapter 2 that was created with solid primitives. Do the following.
   A. Create three named views, each having a different viewpoint.
   B. Use a different visual style face color option in each view.
   C. Save the drawing as P04_03.
4. Open drawing P04_03 and do the following.
   A. Use the steering wheel to change the view of the model. Do this at least four times.
   B. Use the **Rewind** tool to restore a previous view.
   C. Save the view as a named view.
   D. Repeat this for each of the previous views recorded with the steering wheel.
   E. Save the drawing as P04_04.

CHAPTER 5

Using Show Motion to View a Model

Learning Objectives

After completing this chapter, you will be able to:

✓ Explain the show motion tool.
✓ Create still shots of 3D models.
✓ Create walk shots of 3D models.
✓ Create cinematic shots of 3D models.
✓ Replay single shots and a sequence of shots.
✓ Change the properties of a shot.

AutoCAD's ***show motion*** tool is a powerful function that allows you to create named views, animated shots, and basic walkthroughs. It can quickly display a variety of named shots. It is also used to create basic animated presentations and displays. This capability is especially useful for animating 3D models that do not require the complexity and detail of fully textured, rendered animations and walkthroughs. Advanced rendering and walkthroughs are presented later in this book.

A ***view*** is a single-frame display of a model or drawing from any viewpoint. A ***shot*** is the manner in which the model is put in motion and the way the camera *moves* to that view. Therefore a single, named view can be modified to create several different shots using camera motion and movement techniques. Using show motion, you can create shots in one of three different formats:

- Still.
- Walk.
- Cinematic.

A ***still shot*** is exactly the same as a named view, but show motion allows you to add transition effects to display it. A ***walk shot*** requires that you use the **Walk** tool to define a camera motion path to create an animated shot. A ***cinematic shot*** is a single view to which you can add camera motion and movement effects to display the view. These shots are all created using the **New View/Shot Properties** dialog box. This is the same dialog box used to create named views.

Understanding Show Motion

Show motion is simply a means for creating, manipulating, and displaying named views. The **NAVSMOTION** command is used for show motion. The process involves the creation of shots using the **ShowMotion** toolbar and the **New View/Shot Properties** dialog box. Once a shot is created, you can give it properties that allow it to be displayed in many different ways. If a saved shot does not display in the manner you desire, it is easily modified.

A ***view category*** is a heading under which different views are filed. It is not necessary to create categories, especially if you will be making just a few views. On the other hand, if you are working on a complex model and need to create a number of views with a variety of cinematic and motion characteristics, it may be wise to create view categories.

The process of using show motion to create, modify, and display shots begins by first using the **ShowMotion** toolbar. All of your work with shots will be performed using the **New View/Shot Properties** dialog box. Options in this dialog box change based on the type of shot that is selected. These options are discussed in the sections that follow relating to each kind of shot.

PROFESSIONAL TIP

As you first start working with show motion, you may want to create categories named Still, Cinematic, and Walk. As you create shots, file each under its appropriate category name. This will assist you in seeing how the different shot-creation techniques work. Later, you can apply the view categories to specific components of a complex model. In that case, for example, a subassembly of a 3D model may have a view category that contains all three types of shots for use in different types of modeling work or presentations. Creating and using view categories is discussed in detail at the end of this chapter.

Show Motion Toolbar

NAVSMOTION

Type
NAVSMOTION
MOTION

Toolbar
Navigation Bar
ShowMotion

The **ShowMotion** toolbar is displayed at the bottom of the screen when the **NAVSMOTION** command is entered. See Figure 5-1. The quickest way to display the **NAVSMOTION** command is to pick the **ShowMotion** button on the navigation bar. The **ShowMotion** toolbar is displayed at the bottom-center of the screen and provides controls for creating and manipulating views:

- **Unpin ShowMotion/Pin ShowMotion**
- **Play All**
- **Stop**
- **Turn on Looping**
- **New Shot...**
- **Close ShowMotion**

When the toolbar is pinned, it remains displayed if you execute other commands, minimize the drawing, change ribbon panels, or switch to another software application. The **Unpin ShowMotion** button is used to unpin the toolbar. The **Pin ShowMotion** button is then displayed in its place. If you unpin the toolbar you must execute the **NAVSMOTION** command each time you wish to use show motion.

Press the **Play All** button to play all of the views and categories displayed as thumbnails above the toolbar. The playback of these views will loop (repeat) if the **Turn on**

Figure 5-1.
The **ShowMotion** toolbar is displayed at the bottom of the screen and provides controls for creating and manipulating shots.

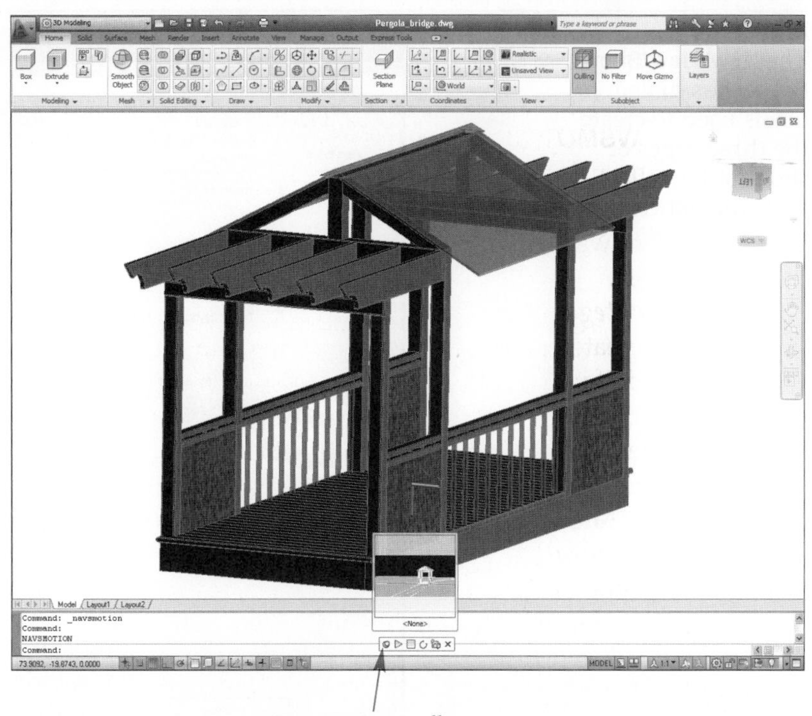

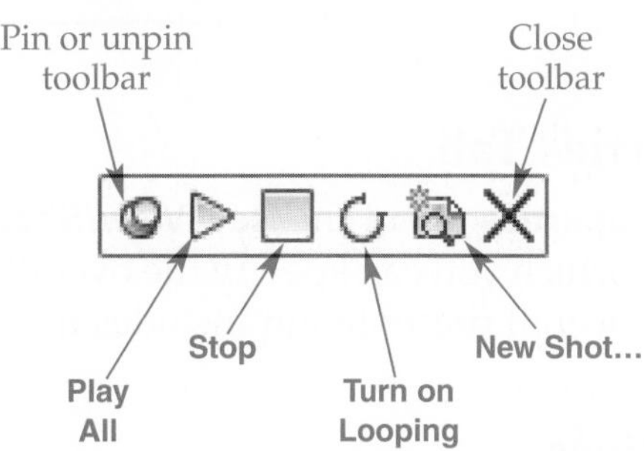

Looping button is selected. If looping is turned on, a view, category, or all categories are displayed in a loop whenever the **Play All** button is selected. The **Turn on Looping** button is a toggle. The image changes to indicate whether or not looping is turned on.

The **New Shot...** button opens the **New View/Shot Properties** dialog box. This dialog box is discussed in the next section. Picking the **Close ShowMotion** button closes the **ShowMotion** toolbar.

New View/Shot Properties Dialog Box

NEWSHOT

All shot creation takes place inside of the **New View/Shot Properties** dialog box. It is accessed by using the **ShowMotion** toolbar described above or using the **NEWSHOT** command. The dialog box can also be displayed from within the **View Manager** dialog box by picking the **New...** button. The **New View/Shot Properties** dialog box is shown in Figure 5-2.

You must supply a shot name and a shot type. A view category is not required, but can help organize shots. This feature is discussed later in the chapter. The dialog box provides two tabs containing options that allow you to define the overall view and then specify the types of movement and motion desired in the shot. These tabs are discussed in the next sections.

Figure 5-2.
The **New View/Shot Properties** dialog box is used to create the three types of shots: still, walk, and cinematic.

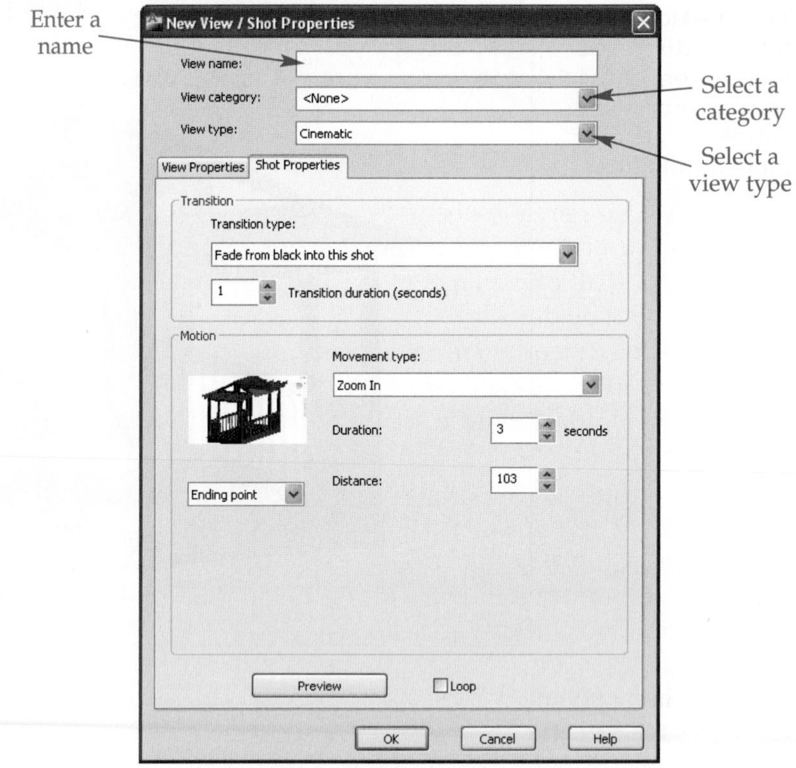

View Properties Tab

The **View Properties** tab of the **New View/Shot Properties** dialog box is composed of three areas in which you can specify the overall presentation of the model view. See Figure 5-3. The overall presentation includes the view boundary, visual settings, and background.

Boundary settings

The setting in the **Boundary** area of the **View Properties** tab determines what is displayed in the shot. The **Current display** button is on by default. This means the shot will be composed of what is currently shown in the drawing area.

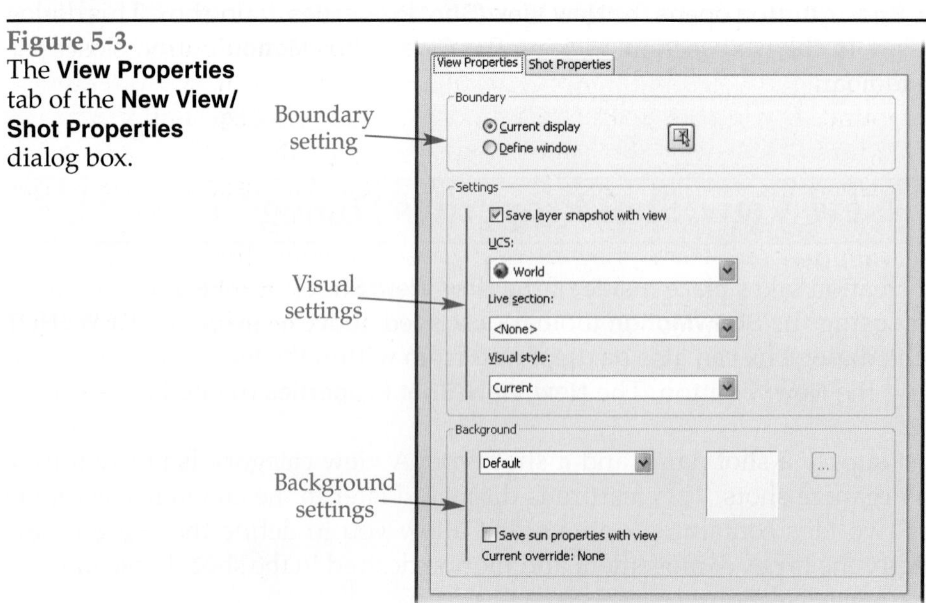

Figure 5-3.
The **View Properties** tab of the **New View/Shot Properties** dialog box.

You can adjust the view by picking the **Define window** radio button. This temporarily closes the dialog box so you can draw a rectangular window to define the view. After picking the second corner, you can adjust the view by picking the first and second corners of the window again. Press [Enter] to accept the window and return to the dialog box.

Visual settings

The **Settings** area of the **View Properties** tab contains options that apply to the overall display of the model in the shot. When the **Save layer snapshot with view** check box is checked, all of the current layer visibility settings are saved with the new shot. This is checked by default.

Any UCS currently defined in the drawing can be selected for use with the new shot. Use the **UCS:** drop-down list to select the UCS. When the shot is restored, that UCS is restored, too. If you select <None> in the drop-down list, there is no UCS associated with the shot.

Live sectioning is a tool that allows you to view the internal features of 3D solids that are cut by a section plane object. This feature is covered in detail in Chapter 14. When a section plane object is created, it is given a name. Therefore, if a model contains one or more section plane objects, their names appear in the **Live section:** drop-down list. If a section plane object is selected in this list, the new shot shows the live sectioning for that plane.

The **Visual style:** drop-down list contains all visual styles in the drawing plus the options of Current and <None>. Selecting a visual style from the list results in that style being set current when the shot is played. Selecting Current or <None> will cause the shot to be displayed in the visual style currently displayed on the screen. A detailed discussion of visual styles is provided in Chapter 15.

NOTE

A shot created with live sectioning on will be displayed in that manner even though live sectioning may be currently turned off in the drawing. However, subsequent displays of the model that were created with live sectioning off are shown with it on. Keep this in mind as you develop shots for show motion.

Background settings

The **Background** area of the **View Properties** tab provides options for changing the background of the new shot. The drop-down list in this area allows you to select a solid, gradient, image, or sun and sky background. You can also choose to retain the default background. If you pick Solid, Gradient, or Image, the **Background** dialog box appears. See Figure 5-4. If you select Sun & Sky, the **Adjust Sun & Sky Background** dialog box is displayed. See Figure 5-5. These dialog boxes are used to set the background. They are discussed in detail in Chapter 17.

The **Save sun properties with view** setting is used to apply the sunlight data to the view. If you are displaying an architectural model using sunlight, you will likely want this check box checked for show motion. Sunlight and geographic location are discussed in detail in Chapter 17.

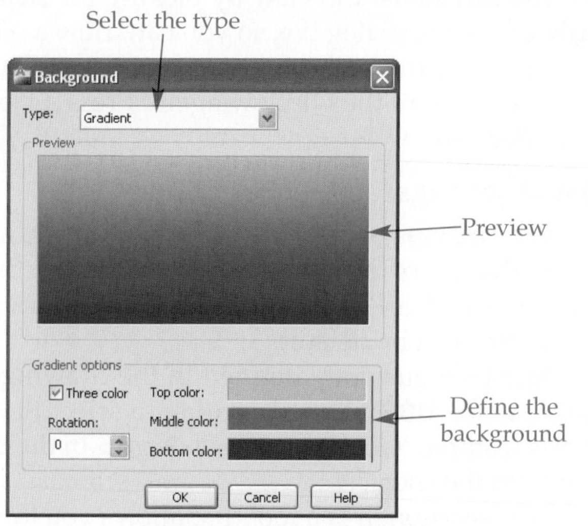

Figure 5-4.
The **Background** dialog box is used to add a background to the shot. Here, a gradient background is being created.

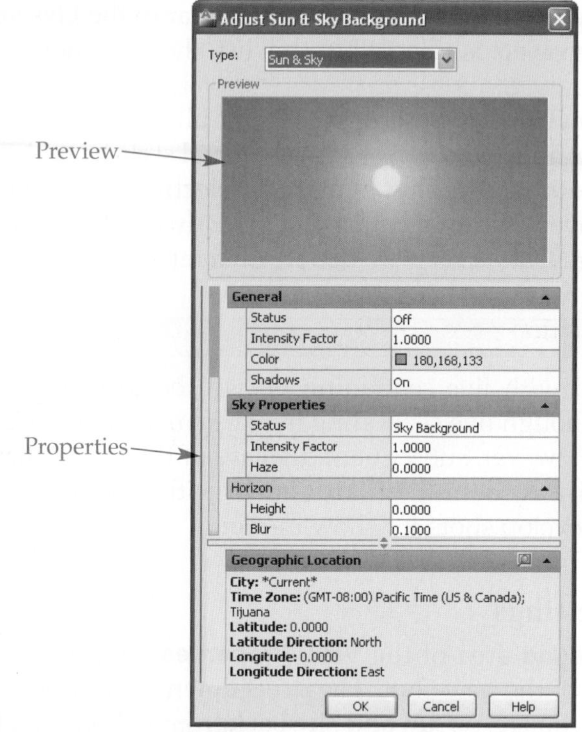

Figure 5-5.
When the background is set to Sun & Sky, the **Adjust Sun & Sky Background** dialog box is used to change settings for the background.

Shot Properties Tab

Settings in the **Shot Properties** tab of the **New View/Shot Properties** dialog box provide options for controlling the transition and motion of shots. There are numerous movement and motion options in this tab. The specific options available are based on the type of shot selected: still, walk, or cinematic. In addition, the cinematic shot contains a variety of motions that can be applied to the shot and each type of motion contains a number of variables. The options in the **Shot Properties** tab are discussed later in this chapter as they apply to different shots, movements, and motions.

Creating a Still Shot

A still shot is the same as a named view, but with a transition. Open the **New View/Shot Properties** dialog box and enter a name in the **View name:** text box. Next, pick in the text box for the **View category:** drop-down list and type a category name or select an existing category. Remember, it is not necessary to create or select a category at this time. Now, select Still from the **View type:** drop-down list.

In the **Shot Properties** tab, select a transition from the **Transition type:** drop-down list. You can select from one of three transitions:

- **Fade from black into this shot.** The screen begins totally black and fades into the current background color.
- **Fade from white into this shot.** The screen begins totally white and fades into the current background color.
- **Cut to shot.** The view is immediately displayed without a transition and the shot movements are applied.

The fade transitions will not function unless hardware acceleration is enabled. Turn on hardware acceleration in the following manner. Refer to Figure 5-6.

1. Enter the **3DCONFIG** command to display the **Adaptive Degradation and Performance Tuning** dialog box.
2. Pick the **Manual Tune** button to display the **Manual Performance Tuning** dialog box.
3. Check the **Enable hardware acceleration** check box.
4. Close both dialog boxes.

After hardware acceleration has been enabled, display the **New View/Shot Properties** dialog box, enter a name, and select a transition. Next, in the **Transition duration (seconds)** text box, enter a length of time over which the transition will occur. If you want a fade transition to be complete, be sure to enter a value in the **Duration:** text box

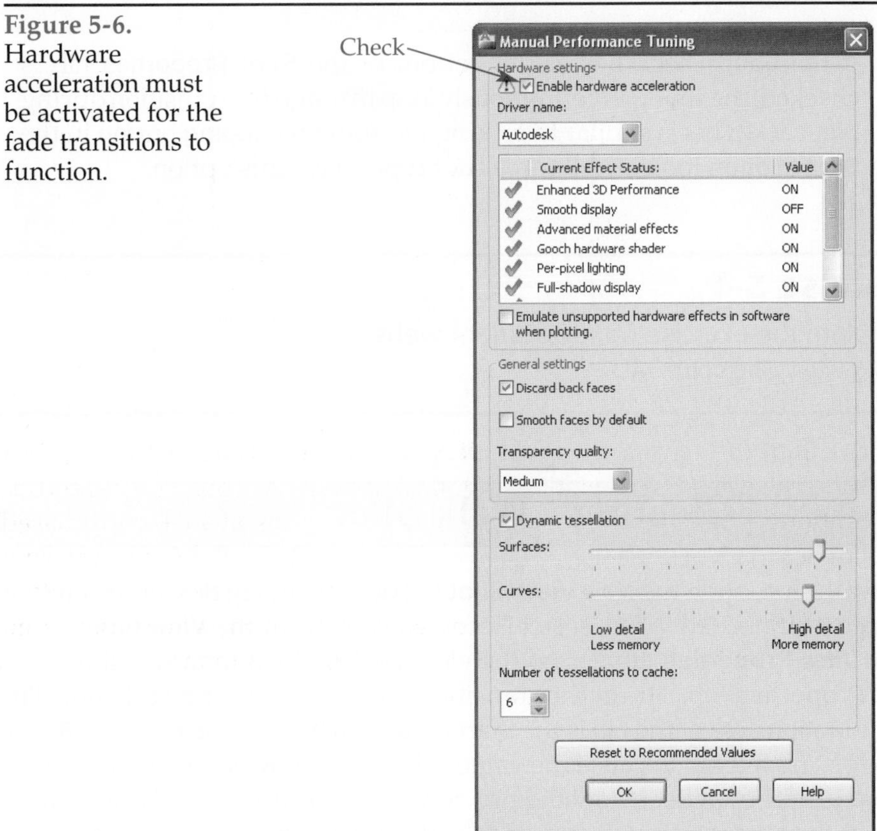

Figure 5-6.
Hardware acceleration must be activated for the fade transitions to function.

Figure 5-7.
A—The large thumbnail image on the bottom is for the category. The small thumbnail image is for the shot just created. Its name may be truncated. B—Move your cursor into the shot thumbnail image and it converts to a large image and the category thumbnail image is reduced to a small image.

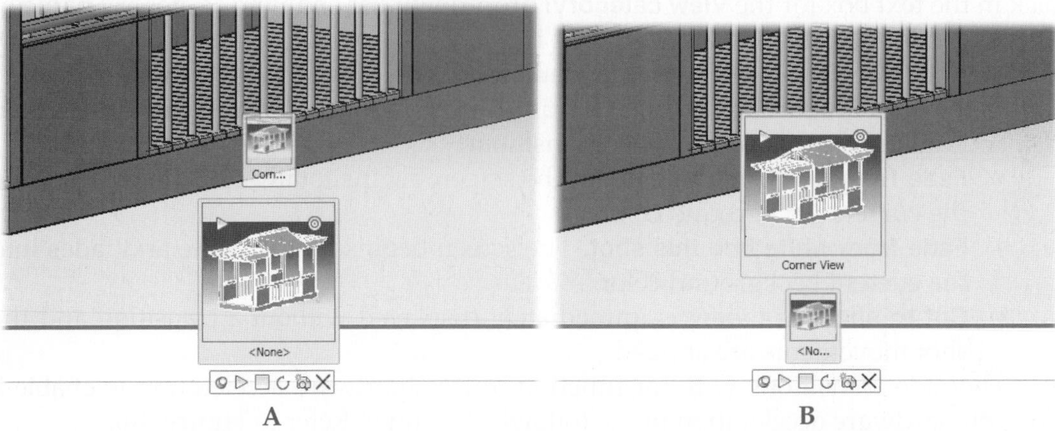

that is equal to or greater than the transition duration. Pick the **Preview** button to view the shot, then edit the transition and motion values as needed.

When finished, pick the **OK** button to close the **New View/Shot Properties** dialog box. The thumbnail image for the new shot is displayed above the **ShowMotion** toolbar. See **Figure 5-7A.** The large thumbnail image represents the category. Since there was no view category selected for the view, the name <None> is displayed. The small thumbnail image represents the shot just created. Its name may be truncated. Move the cursor into the shot thumbnail image and a large image is displayed. The name of the shot should appear in its entirety. The category thumbnail image is reduced to a small image. See **Figure 5-7B.**

NOTE

If the **Loop** check box at the bottom of the **Shot Properties** tab is checked, the shot will continuously loop through the transition during playback. This is similar to picking the **Turn on Looping** button in the **ShowMotion** toolbar. All three shot types have this option.

Exercise 5-1

Complete the exercise on the student website.
www.g-wlearning.com/CAD

Creating a Walk Shot

To create a walk shot, open the **New View/Shot Properties** dialog box, enter a name, and select a category, if needed. Next, select Recorded Walk from the **View type:** drop-down list. In the **View Properties** tab, set up the view as described for a still shot.

In the **Shot Properties** tab, set up the transition as described for a still shot. The only option in the **Motion** area of the tab is the **Start recording** button. See **Figure 5-8.** The **Duration:** text box is grayed out because the value is based on how long you record the walk. The camera drop-down list below the preview image is also grayed out because the walk begins at the current display and ends at the point where you terminate it.

Figure 5-8.
Creating a walk shot.

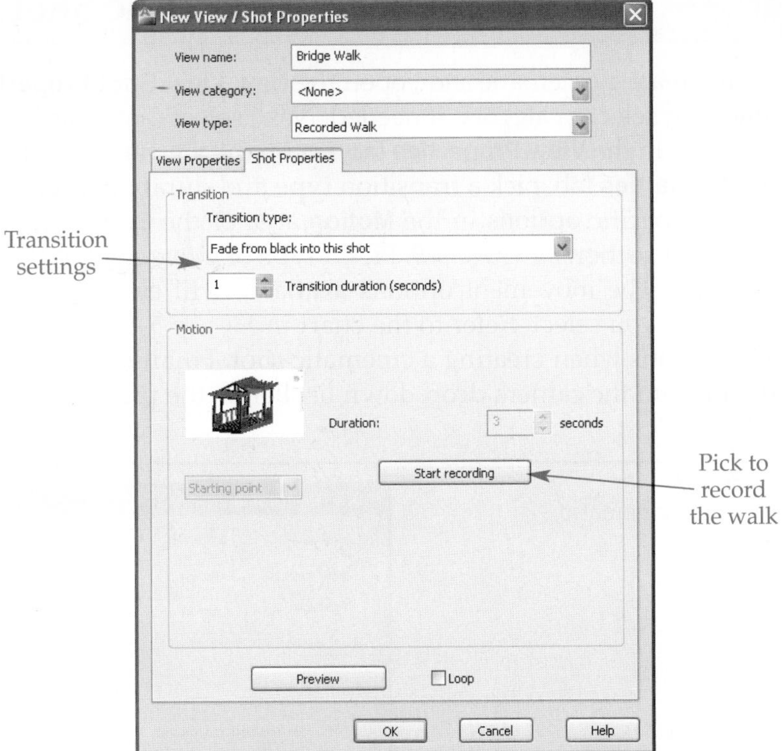

This type of shot uses the same **Walk** tool found in the steering wheels, discussed in Chapter 4. After picking the **Start recording** button, the **New View/Shot Properties** dialog box is hidden and the **Walk** tool message is displayed. Pick and drag to activate the **Walk** tool. As soon as you pick, the center circle icon is displayed. If needed, you can hold down the [Shift] key to move the view up or down. When the [Shift] key is released (with the mouse button still held down), you can resume walking. The mouse button must be depressed the entire time to record all movements. As soon as you release the button, the recording ends and the dialog box is redisplayed.

Finally, preview the shot. If you need to re-record it, pick the **Start recording** button and begin again. You cannot add to the shot; you must start over. Pick the **OK** button to save the shot and a new thumbnail is displayed in the **ShowMotion** toolbar.

NOTE

The walk shot capability of show motion is limited in its ability to produce a true "walkthrough." The path of your "walk" is inexact and it may take several times to create the effect you need. Should you wish to create a genuine walkthrough of an architectural or structural model, it is better to use tools such as **3DWALK** (walkthrough), **3DFLY** (flyby), or a **ANIPATH** (motion path animation). These powerful commands are used to create professional walkthroughs and animations and are covered in Chapter 19.

Exercise 5-2

Complete the exercise on the student website.
www.g-wlearning.com/CAD

Creating a Cinematic Shot

To create a cinematic shot, open the **New View/Shot Properties** dialog box, enter a name, and select a category, if needed. Next, select Cinematic from the **View type:** drop-down list. In the **View Properties** tab, set up the view as described for a still shot. In the **Shot Properties** tab, pick a transition type and duration as described for a still shot.

The specific options in the **Motion** area of the **Shot Properties** tab are discussed in the next sections, Figure 5-9. Pick a type of movement in the **Movement type:** drop-down list. The movement options available will change depending on the type of movement you select. Refer to the chart in Figure 5-10 as a quick reference for movement options when creating a cinematic shot. Finally, pick the current position of the camera from the camera drop-down list below the preview image.

Figure 5-9.
Creating a cinematic shot.

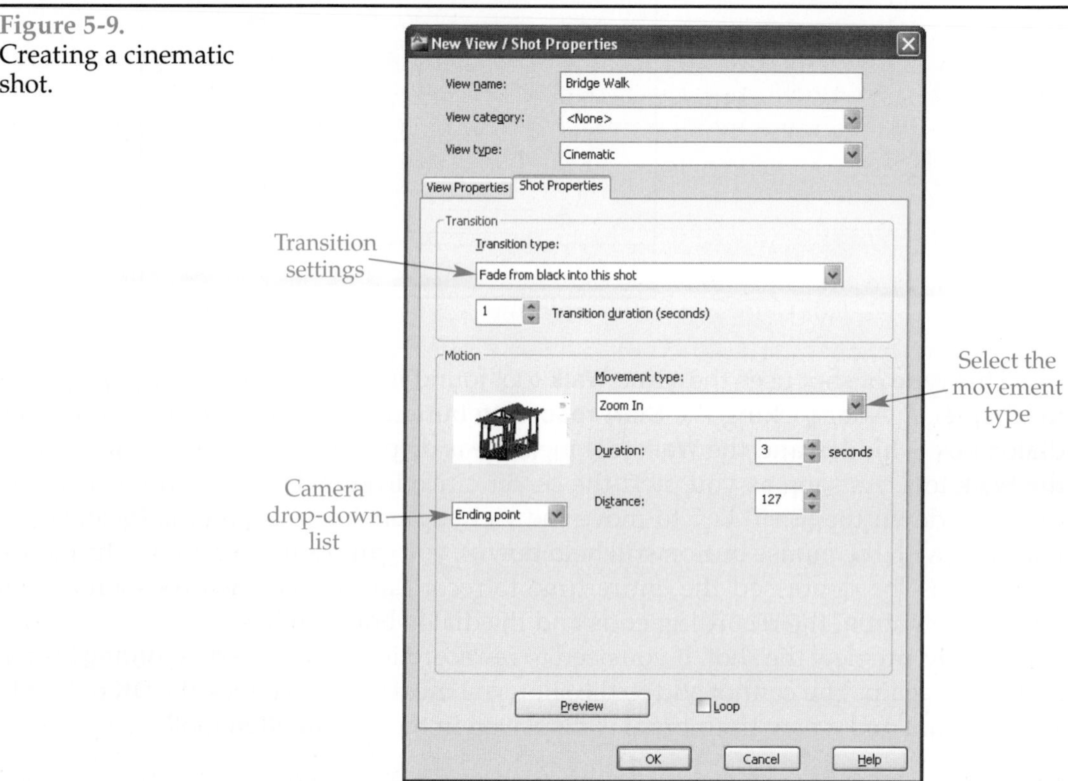

Figure 5-10.
This chart shows the different movement options based on the movement type selected in the **Movement type:** drop-down list in the **New View/Shot Properties** dialog box.

	Duration	Distance	Look at Camera Point	Distance Up	Distance Back	Distance Down
Zoom in	X	X				
Zoom out	X	X				
Track left	X	X	X			
Track right	X	X	X			
Crane up	X		X	X	X	
Crane down	X		X			X
Look	X					
Orbit	X					

Figure 5-11.
The thumbnail image for the new cinematic shot is displayed above the category thumbnail image.

Pick the **Preview** button to view the shot. Make any changes required before picking the **OK** button to save the shot. Once the **OK** button is picked, the new view is displayed as a small thumbnail image above the large category thumbnail image. See Figure 5-11.

Cinematic Basics

The motion and movement options available for a cinematic shot allow you to create a final display that appears to move into position as if the camera is traveling in a path toward the object. The **Motion** area of the **Shot Properties** tab contains a variety of options for creating an array of cinematic shots. Refer to Figure 5-9. The options can be confusing unless you understand some basics about essential components of a cinematic shot. The most important aspect is the preview of the current view. All of the motion actions revolve around this view. This image tile is the current position of the camera and is also referred to as the ***key position*** of a shot. This is the position that is displayed when you pick the **Go** button on a thumbnail image above the **ShowMotion** toolbar.

The two elements of a cinematic shot are motion and movement. ***Motion*** relates to the behavior of the object and how it appears to be in motion during the cinematic shot. In addition, it refers to the position of the model at a specified point in the animation. ***Movement*** in a cinematic shot is the manner in which the camera moves in relation to the object.

PROFESSIONAL TIP

If your goal is to create a series of shots that blend together, it is a good idea to first develop a storyboard of the entire sequence. This could be as simple as a few notes indicating how you want the shots to move or even a few sketches noting the required movements and motion values. Planning your shots will save time when you begin creating them in AutoCAD.

Camera Drop-Down List

The camera drop-down list is located below the preview image in the **Motion** area of the **Shot Properties** tab. The preview image represents the position of the camera based on the option selected in the camera drop-down list. The following three options are available in the drop-down list, Figure 5-12.

- **Ending point.** The view in the preview image is the display that will be shown at the end of the cinematic shot. All movement options take the shot to this point.
- **Starting point.** The view in the preview image is the display that will be shown at the start of the cinematic shot. All movement options begin at this point.

Figure 5-12.
Selecting what the key position represents.

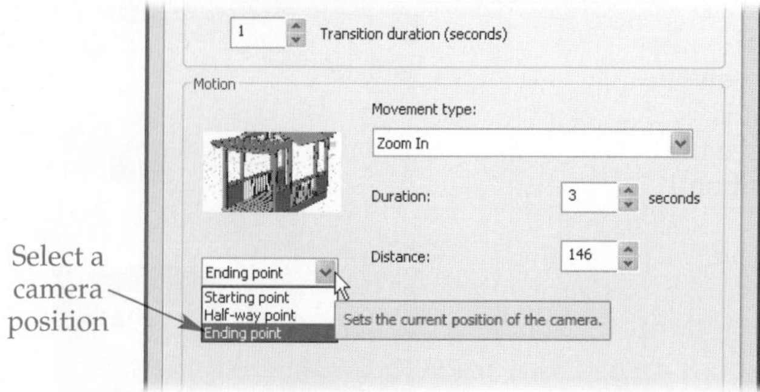

- **Half-way point.** The view in the preview image is the display that will be shown at the half-way point of the cinematic shot.

These options, in part, determine how the cinematic shot is created. If Starting point is selected, the cinematic shot begins with the current view. During the animation, the model may move off of the screen. If this happens, you may want to select Ending point so the model appears to move into position and stay there.

PROFESSIONAL TIP

The **Preview** button in the **New View/Shot Properties** dialog box is the best way to test any changes you make to the motion options. Each time you make a change, pick the button to see if the effect is what you want. Previewing each change is far more efficient than making several changes before you examine the results.

Movement Type

The **Movement type:** drop-down list in the **Shot Properties** tab of the **New View/Shot Properties** dialog box is used to select the motion for the cinematic shot. There are eight types of camera movement that can be used with a cinematic shot:

- Zoom in.
- Zoom out.
- Track left.
- Track right.
- Crane up.
- Crane down.
- Look.
- Orbit.

The options available in the **Motion** area of the tab are based on which movement type is selected. The movement types and their options are discussed in the next sections.

Zoom in

When Zoom in is selected in the **Movement type:** drop-down list, the camera appears to zoom into the model in the shot. This movement type has two options, Figure 5-13.

Figure 5-13.
The settings for the Zoom in movement type.

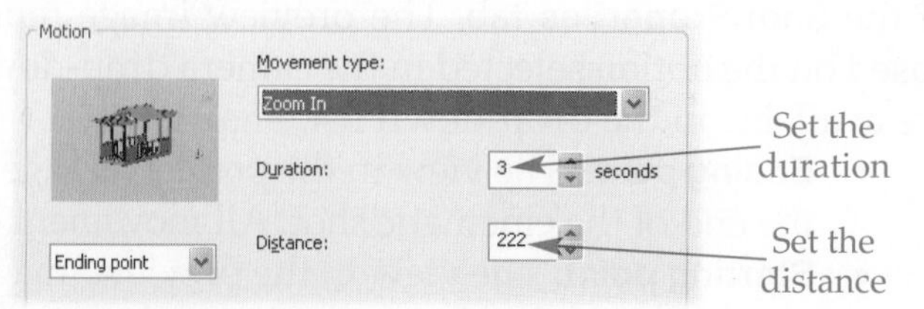

The value in the **Duration:** text box is the length of time over which the animation is recorded. The value in the **Distance:** text box is the distance the camera travels during the animation. The camera zooms in to cover the distance in the specified duration of time. When Zoom in is selected, the camera drop-down list is automatically set to Ending point. Keep in mind that with this option the current position of the camera represents the final display after the cinematic shot is complete.

Zoom out

When Zoom out is selected in the **Movement type:** drop-down list, the camera appears to zoom away from the model in the shot. This movement type has **Duration:** and **Distance:** text options, as described for the Zoom in movement type.

When Zoom out is selected, the camera drop-down list is automatically set to Starting point. With this setting, the camera will zoom out the specified distance and the final image in the shot will be smaller than the preview image. If this is not the effect you want, it may be better to pick Ending point for the current position of the camera. Then, the model will appear to move from behind your view to stop at the current view.

Track left

When Track left is selected in the **Movement type:** drop-down list, the camera will move from right to left. This results in the view moving from left to right. This movement occurs over the specified distance and duration. The Track left movement type has **Duration:** and **Distance:** text boxes, as described above, and the **Always look at camera pivot point** check box, **Figure 5-14.**

If the **Always look at camera pivot point** check box is checked, the center of the view remains stationary. As a result, the view in the shot appears to rotate about this point instead of sliding across the screen. The best way to visualize this motion is to use the **Preview** button with the option checked and then unchecked. This option is available for all "track" and "crane" movement types.

When Track left is selected, the camera drop-down list is automatically set to Half-way point. This means the preview image is the middle point of the animation. It will be displayed at the midpoint of the **Duration:** value.

PROFESSIONAL TIP

If the **Distance:** value is large, the screen may be blank for a few moments until the camera moves enough to bring the model into view. If this is not what you want, decrease the **Distance:** value or check the **Always look at camera pivot point** check box.

Track right

When Track right is selected in the **Movement type:** drop-down list, the camera will move from left to right. This results in the view moving from right to left. This movement type has **Distance:** and **Duration:** text boxes and the **Always look at camera pivot**

Figure 5-14.
The settings for the Track left movement type.

Set the duration
Set the distance
Check to rotate the view about the camera pivot point

point check box. These options function the same as described for Track left. When Track right is selected, the camera drop-down list is automatically set to Half-way point.

Crane up

Where the "track" movement types move the view left and right, the "crane" movement types move the view up and down. When Crane up is selected in the **Movement type:** drop-down list, the camera will move from bottom to top and then backward. This results in the view moving from top to bottom and zooming out. This movement type has **Distance:** and **Duration:** text boxes and the **Always look at camera pivot point** check box described above, which is checked by default.

This movement type has three additional options, Figure 5-15. The value in the **Distance Up:** text box is the distance the camera is moved upward. The value in the **Distance Back:** text box is the distance the camera is moved backward. The backward movement is typically short compared to the upward movement.

You can add more interest to the motion in the shot by shifting the view left or right. First, enable the shift option by checking the check box below the **Distance Back:** setting. Refer to Figure 5-15. Then, select either Shift left or Shift right from the drop-down list. Finally, enter a distance in the text box next to the drop-down list. The camera will be shifted left or right for the distance specified in this text box, resulting in the view shifting in the opposite direction. This shifting option is available for both "crane" motion types.

When Crane up is selected, the camera drop-down list is automatically set to Starting point. This means the preview image shows the beginning of the shot before the movement is applied. If the **Always look at camera pivot point** check box is not checked, and depending on the movement settings, the model may move off of the screen in the shot.

Crane down

When Crane down is selected in the **Movement type:** drop-down list, the camera will move from top to bottom and then forward. This results in the view moving from bottom to top and zooming in. This movement type has **Distance:** and **Duration:** text boxes and the **Always look at camera pivot point** check box described above, which is checked by default. It also has the left/right shifting option described in the previous section.

Instead of **Distance Up:** and **Distance Back:** settings, this movement type has **Distance Down:** and **Distance Forward:** settings. The value in the **Distance Down:** text box is the distance the camera cranes down in the shot. The value in the **Distance Forward:** text box is the distance the camera is moved forward in the shot.

When Crane down is selected, the camera drop-down list is automatically set to Ending point. This means the preview image shows the end of the shot before the movement is applied. If the **Always look at camera pivot point** check box is not checked, and depending on the movement settings, the model may start off of the screen in the shot.

Figure 5-15.
The settings for the Crane up movement type.

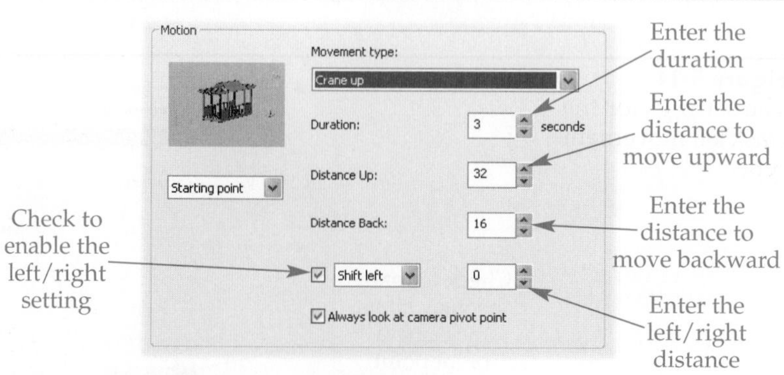

NOTE

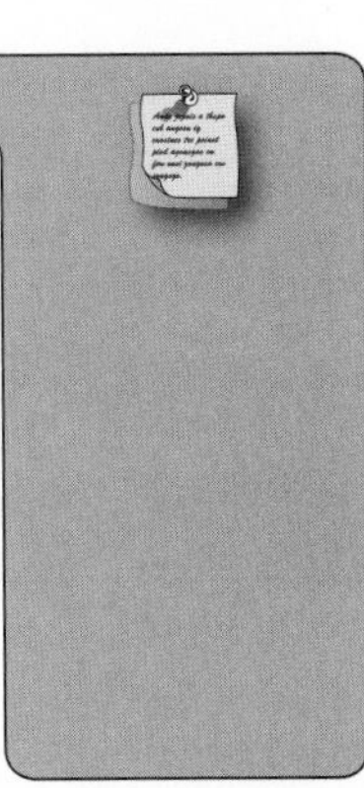

All distance values related to the Crane up and Crane down movement types represent how far away from the view the camera must begin before the cinematic shot is started. For example, a large **Distance Back:** value means that the camera may have to begin at a point beyond or even around the view in order to travel the distance back to display the current view (when Ending point is selected in the camera drop-down list). It is always good practice to make a single change to the movements, then preview the shot. This is especially true for the Crane up and Crane down movement types.

Look

When Look is selected in the **Movement type:** drop-down list, the camera pans based on the values for left/right and up/down to display the view, Figure 5-16. For example, if the movement is set to look up 45° and Ending point is selected in the camera drop-down list, then the camera begins at a 45° angle below the view and looks up to display it.

The value in the **Duration:** text box is the length of time over which the animation is recorded. This option is the same as described for the other movement types.

The first drop-down list below the **Duration:** text box is used to specify either left or right movement. Select Degrees left in the drop-down list to have the camera move from right to left, resulting in the view moving from left to right. Select Degrees right to have the camera move from left to right and the view right to left. Next, specify the angular value for this movement in the degrees text box to the right of the drop-down list. For example, suppose you select Degrees left and enter an angle of 15°. In this case, the camera will start 15° to the *right* of the model and rotate to the left in the shot.

The second drop-down list below the **Duration:** text box is used to specify either up or down movement. Select Degrees up in the drop-down list to have the camera move from bottom to top, resulting in the view moving from top to bottom. Select Degrees down to have the camera and view move in the opposite direction. Specify the angular value for this movement in the degrees text box to the right of the drop-down list.

When Look is selected in the **Movement type:** drop-down list, the camera drop-down list is automatically set to Starting point. This means the shot will start with the current view and then apply the movement settings.

Orbit

When Orbit is selected in the **Movement type:** drop-down list, the camera rotates in place based on the values set for left/right and up/down movements. The Orbit movement type has the same options as the Look movement type, as described in the previous section.

When Orbit is selected in the **Movement type:** drop-down list, the camera drop-down list is automatically set to Starting point. This means the shot will start with the current view and then apply the movement settings. Unlike the Look movement type,

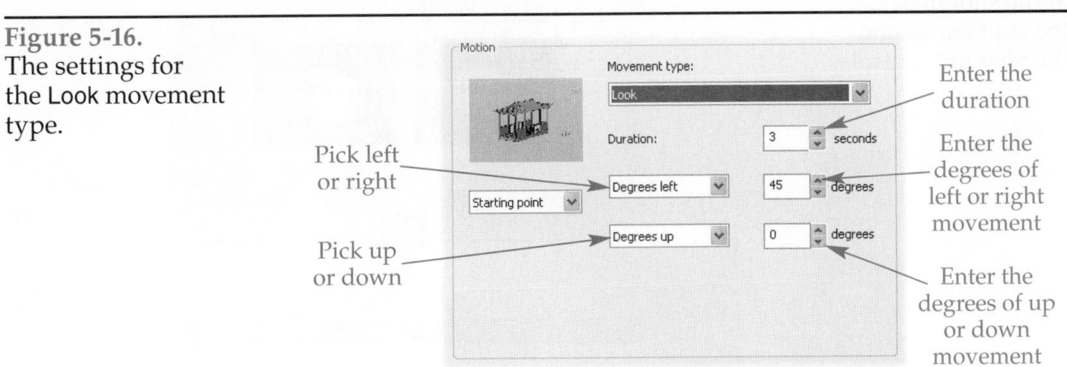

Figure 5-16.
The settings for the Look movement type.

the view center remains stationary in the shot. As a result, you do not need to worry about the model moving off of the screen.

Exercise 5-3

Complete the exercise on the student website.
www.g-wlearning.com/CAD

Displaying or Replaying a Shot

As each shot is created, its thumbnail image is placed above the **ShowMotion** toolbar. The shot name is displayed below the thumbnail image. By default, the category thumbnail image is large and each shot thumbnail image is small. If the cursor is moved over one of the shot thumbnail images, all of the shot thumbnail images are enlarged and the category thumbnail image is reduced in size.

Each thumbnail image is composed of an image of the view in the shot, the shot name, and viewing controls. See Figure 5-17. The viewing controls are only displayed when the cursor is moved over the thumbnail image. The three viewing controls are:

VIEWGO VIEWPLAY

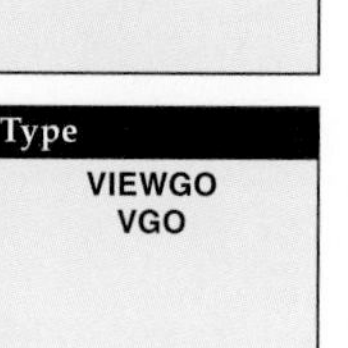

Type
VIEWPLAY
VPLAY

Type
VIEWGO
VGO

- **Play.** Plays the shot. If looping is enabled, the shot repeats. Otherwise, the shot is played once. When this button is picked it changes to **Pause**. You can also play the shot by picking anywhere on its thumbnail image. The **VIEWPLAY** command can be used to replay a shot.
- **Pause.** Pauses the shot. When this button is picked it changes to **Play**. Picking anywhere inside the image or on the **Play** button restarts playing of the shot.
- **Go.** Displays the key position of the view without playing the shot. The **VIEWGO** command also restores a named view.

You can play all of the shots in sequence by picking the **Play** button in the category thumbnail image or by picking anywhere inside of the category thumbnail image. Additionally, you can move to the key position of the first shot in a view category by picking the **Go** button on the view category thumbnail image.

Exercise 5-4

Complete the exercise on the student website.
www.g-wlearning.com/CAD

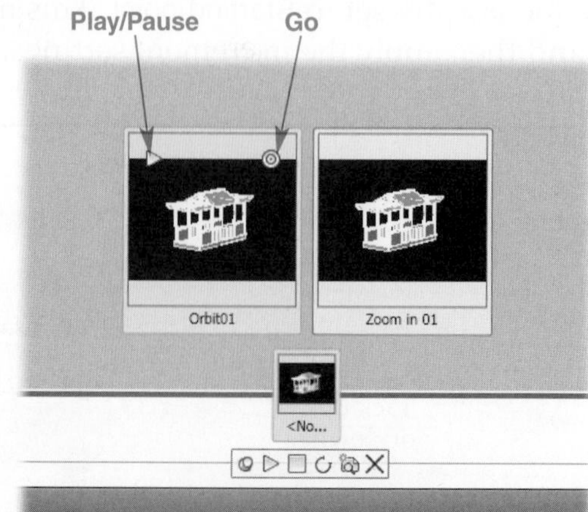

Figure 5-17.
Each thumbnail image is composed of the shot image, the shot name, and the viewing controls.

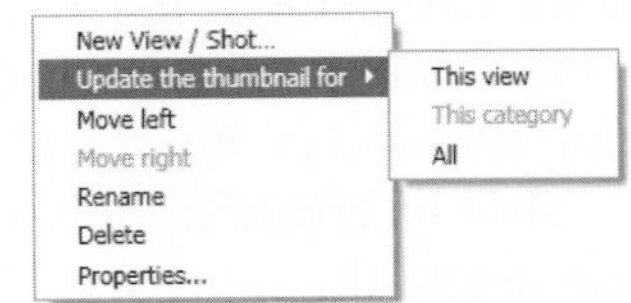

Figure 5-18.
Right-click on a shot or view category thumbnail image in the **ShowMotion** toolbar to display this shortcut menu.

Thumbnail Shortcut Menu

Right-clicking on a shot or view category thumbnail image displays the shortcut menu shown in Figure 5-18. This menu provides quick access for modifying and manipulating shots and their thumbnail images.

Picking the **New View/Shot...** entry displays the **New View/Shot Properties** dialog box. Picking the **Properties...** entry displays the **View/Shot Properties** dialog box. This dialog box is the same as the **New View/Shot Properties** dialog box. However, all of the settings of the shot are displayed and can be changed. This option is only available to change the properties of a shot, not a category.

To rename a shot, pick the **Rename** entry in the shortcut menu. The name is highlighted below the thumbnail image. Type the new name and press [Enter]. To delete a shot, select **Delete** from the shortcut menu. There is no warning; the shot is simply deleted. The **UNDO** command can reverse this action.

If there is more than one shot in a category, you can rearrange the order. Right-click on a thumbnail image and pick either **Move left** or **Move right** in the shortcut menu. This moves the shot one step in the selected direction.

When a change is made to the model, it is not automatically reflected in the thumbnail images. If you do not update thumbnail images, they will remain in their original format, regardless of how many changes are made to the model. Selecting **Update the thumbnail for** in the shortcut menu displays a cascading menu with options for updating the thumbnail image:

- **This view.** Updates only the thumbnail image for the shot on which you right-clicked.
- **This category.** Updates all thumbnail images in the category. This option is only enabled when you right-click on a category thumbnail image.
- **All.** This option updates all thumbnail images in all categories and shots.

Creating and Using View Categories

A ***view category*** is a grouping that can be created in order to separate different types of shots. A view category can also contain shots arranged in a sequence that appear connected as they are played together. This is an optional feature, but an efficient method of separating different types of shots or grouping shots to use for a specific purpose. When a new category is created, it is represented by a category thumbnail image displayed above the **ShowMotion** toolbar.

To create a new view category, simply pick in the **View category:** drop-down list text box in the **New View/Shot Properties** dialog box. Then, enter a name. See Figure 5-19.

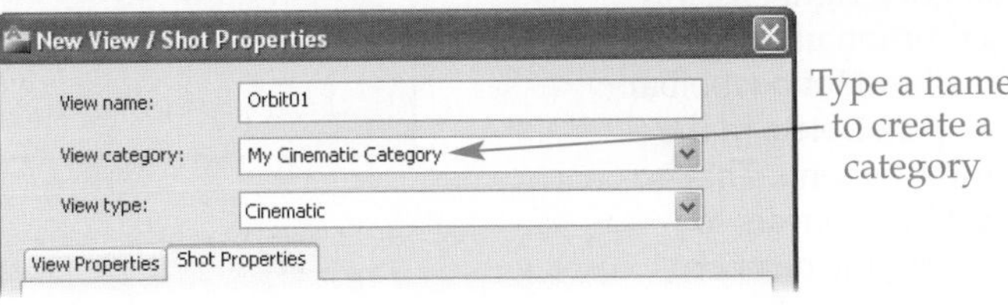

Figure 5-19.
To create a new view category, type its name in the **View category:** drop-down list text box.

The name is added to the drop-down list. If you wish to organize shots by categories, be sure to select the view category from the drop-down list before picking the **OK** button to exit the dialog box and create the shot.

You must create a shot to create a view category. If you delete the only shot in a view category, you also delete the category. The view category will no longer be available in the **New View/Shot Properties** dialog box.

View Category Basics

After a new view category is created, it is represented by a thumbnail image above the **ShowMotion** toolbar. The category name is shown below the thumbnail image. As you create more shots within a category, the thumbnail images for the new shots are placed to the right of existing shots above the category thumbnail image. The thumbnail image for the first shot in a view category is displayed as the view category thumbnail image.

An entire view category and all of the views included in it can be quickly deleted. Simply right-click on the view category thumbnail image and pick **Delete** in the shortcut menu. An alert box appears asking you to confirm the deletion, Figure 5-20. The shots in a deleted view category cannot be recovered, so be sure there are no shots in the category you wish to save before deleting. However, you can use the **UNDO** command to reverse the deletion.

PROFESSIONAL TIP

Remember, you can change the order of the shots in a view category using the shortcut menu. Picking **Move left** or **Move right** moves the shot one step. Continue moving shots until the order is appropriate.

Playing and Looping Shots in a View Category

TYPE
SEQUENCEPLAY
SPLAY

A view category is a useful tool for grouping shots to be played together in a sequence. To play the shots in a view category, move the cursor over the view category thumbnail image above the **ShowMotion** toolbar. Then, pick the **Play** button in the thumbnail image. All shots in the category will be played. You can also use the **SEQUENCEPLAY** command to play the shots in a view category.

If you want the shots in a category to run on a continuous loop, pick the **Turn on Looping** button in the **ShowMotion** toolbar. Then, when the **Play** button is picked, the shots in the view category will display until the **Pause** button is picked either on the **ShowMotion** toolbar or in the view category thumbnail image. Pressing the [Esc] key also stops playback. To return to single-play mode, pick the **Turn off Looping** button on the **ShowMotion** toolbar. This button replaces the **Turn on Looping** button.

Changing a Shot's View Category and Properties

If you put a shot in the wrong view category, it is simple to move the shot to a different category. Right-click on the shot thumbnail image and select **Properties...** in the shortcut menu. This displays the **New View/Shot Properties** dialog box. Select the proper category

Figure 5-20.
To remove a view category, right-click on its thumbnail image in the **ShowMotion** toolbar and pick **Delete** in the shortcut menu. This alert box appears to confirm the deletion.

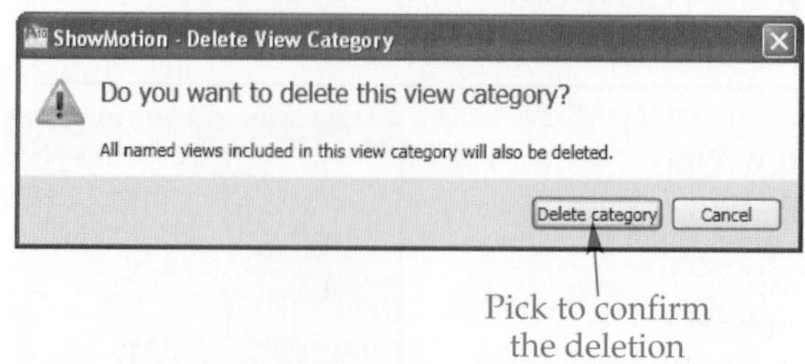

name in the **View category:** drop-down list and pick the **OK** button. The shot thumbnail image will move into position above its new view category thumbnail image.

To change shot properties, first right-click on the shot thumbnail image and pick **Properties...** in the shortcut menu. The **New View/Shot Properties** dialog box is displayed. All properties of the shot can be changed. Change the settings as needed. Always preview the shot before you pick the **OK** button to save it.

PROFESSIONAL TIP

Try this procedure for creating a series of shots in a view category that are to be played in sequence.

1. Play each shot to determine where they should be located in the sequence.
2. Move shots left or right in the category as needed to create the proper sequence.
3. Play the category to determine if the shots properly transition from one to the next.
4. If a subsequent shot does not begin where the previous shot ended, right-click on that shot, pick **Properties...**, and make the necessary adjustments in the **View/Shot Properties** dialog box.
5. Play the category again.
6. Repeat the editing process with each shot until the category plays smoothly.

Exercise 5-5

Complete the exercise on the student website.
www.g-wlearning.com/CAD

Chapter Test

Answer the following questions. Write your answers on a separate sheet of paper or complete the electronic chapter test on the student website.
www.g-wlearning.com/CAD

1. For what is the *show motion tool* used?
2. Define *view* as it relates to the show motion tool.
3. Define *shot* as it relates to the show motion tool.
4. List the formats in which a shot can be created.
5. List the six buttons on the **ShowMotion** toolbar.
6. What is *live sectioning* and how can it be included in a shot?
7. Which type of shot is the same as a named view, but with a transition?
8. Which type of shot requires you to navigate through the view as you record the motion?
9. What is the *key position* of a cinematic shot?
10. Define *motion* and *movement* as they relate to a cinematic shot.
11. What is the purpose of the camera drop-down list in the **Shot Properties** tab of the **New View/Shot Properties** or **View/Shot Properties** dialog box?
12. List the eight types of camera movement for a cinematic shot.
13. Briefly describe two ways to play a single shot.
14. What is a *view category?*
15. How is an entire category of shots replayed?

Drawing Problems

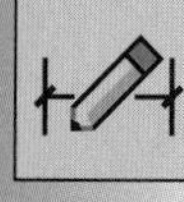

1. Open one of your 3D drawings from Chapter 2 or 3. Do the following.
   A. Display a pictorial view of the drawing.
   B. Create a still shot. Use settings of your choice. Create a view category and place the shot in it.
   C. Display a different view of the drawing and create another still shot. Place the shot in the view category you created.
   D. Display a third view of the drawing and create a third still shot. Place the shot in the view category you created.
   E. Play each shot. If necessary, rearrange the shots. Then, play all shots in the category.
   F. Save the drawing as P05_01.

2. Open one of your 3D drawings from Chapter 2 or 3. Do the following.
   A. Display the objects in the Conceptual or Realistic visual style.
   B. Toggle the projection from parallel to perspective.
   C. Create two still shots and four cinematic shots of the model. Use a different motion type with each of the cinematic shots.
   D. Create two new categories and place one still and two cinematic shots in each category.
   E. Edit the shots in each category to create a smooth motion sequence.
   F. Save the drawing as P05_02.

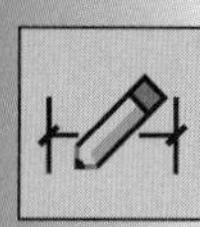

3. Open one of your 3D drawings from Chapter 2 that was created with solid primitives. Do the following.
   A. Display a pictorial view.
   B. Create a walk shot. Place it in a view category named Walk Shots.
   C. Play the shot. How does this compare to a cinematic shot as far as ease of creation?
   D. Save the drawing as P05_03.

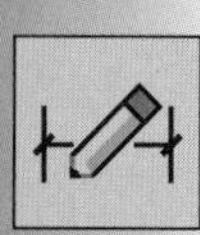

4. Open drawing P05_03 and do the following.
   A. Create a new cinematic shot using the Orbit motion type. Place it in a view category named Cinematic Shots.
   B. Create another cinematic shot using the Track left or Track right movement type. Check the **Always look at camera pivot** check box. Place the view in the Cinematic Shots category.
   C. Create a third cinematic shot using the Crane up or Crane down movement type. Make sure the **Always look at camera pivot** check box is checked. Place the view in the Cinematic Shots category.
   D. Play the Cinematic Shots category. Edit the shots as needed to create a smooth display.
   E. Create three still shots. Place them in a view category named Still Shots. Try creating three different gradient backgrounds to simulate dawn, noon, and dusk.
   F. Play the Still Shots category. Edit the shots as needed to create a smooth display.
   G. Save the drawing as P05_04.

CHAPTER 6

Understanding Three-Dimensional Coordinates and User Coordinate Systems

Learning Objectives

After completing this chapter, you will be able to:

- ✓ Describe rectangular, spherical, and cylindrical methods of coordinate entry.
- ✓ Draw 3D polylines.
- ✓ Describe the function of the world and user coordinate systems.
- ✓ Move the user coordinate system to any surface.
- ✓ Rotate the user coordinate system to any angle.
- ✓ Change the user coordinate system to match the plane of a geometric object.
- ✓ Use a dynamic UCS.
- ✓ Save and manage user coordinate systems.
- ✓ Restore and use named user coordinate systems.
- ✓ Control user coordinate system icon visibility in viewports.

As you learned in Chapter 1, any point in space can be located using X, Y, and Z coordinates. This type of coordinate entry is called *rectangular coordinates.* Rectangular coordinates are most commonly used for coordinate entry. However, there are actually three ways in which to locate a point in space. The other two methods of coordinate entry are spherical coordinates and cylindrical coordinates. These two coordinate entry methods are discussed in the following sections. In addition, this chapter introduces working with user coordinate systems (UCSs).

Introduction to Spherical Coordinates

Locating a point in 3D space with ***spherical coordinates*** is similar to locating a point on Earth using longitudinal and latitudinal values, with the center of Earth representing the origin. Lines of longitude connect the North and South Poles and provide an east-west measurement on Earth's surface. Lines of latitude horizontally extend around Earth and provide a north-south measurement. The origin (Earth's center) can be that of the default world coordinate system (WCS) or the current user coordinate system (UCS). See Figure 6-1A.

When entering spherical coordinates, the longitude measurement is expressed as the angle *in* the XY plane and the latitude measurement is expressed as the angle *from* the XY plane. See Figure 6-1B. A distance from the origin is also provided. The coordinates

Figure 6-1.
A—Lines of longitude, representing the highlighted latitudinal segments in the illustration, run from north to south. Lines of latitude, representing the highlighted longitudinal segments, run from east to west. B—Spherical coordinates require a distance, an angle in the XY plane, and an angle from the XY plane.

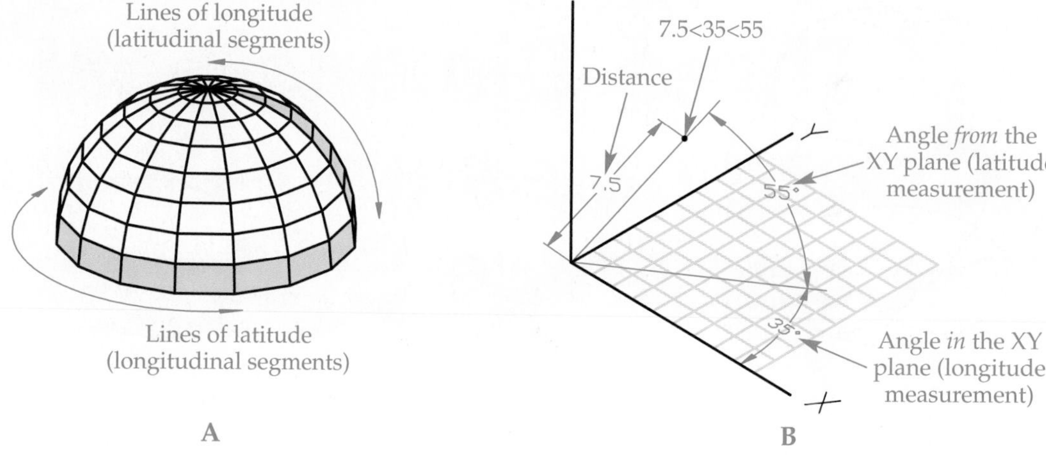

represent a measurement from the equator toward either the North or South Pole on Earth's surface. The following spherical coordinate entry is shown in Figure 6-1B.

7.5<35<55

This coordinate represents an ***absolute*** spherical coordinate, which is measured from the origin of the current UCS. Spherical coordinates can also be entered as ***relative*** coordinates. For example, a point drawn with the relative spherical coordinate @2<35<45 is located two units from the last point, at an angle of 35° *in* the XY plane, and at a 45° angle *from* the XY plane.

If dynamic input is turned on, the "second" or "next" coordinate entry is automatically a *relative* entry (by default). The @ symbol is not entered. To enter *absolute* coordinates with dynamic input turned on, enter an asterisk (*) before the first coordinate.

PROFESSIONAL TIP

Spherical coordinates are useful for locating features on a spherical surface. For example, they can be used to specify the location of a hole drilled into a sphere or a feature located from a specific point on a sphere. If you are working on such a spherical object, you might consider locating a UCS at the center of the sphere, then creating several different user coordinate systems rotated at different angles on the surface of the sphere. Any time a location is required, spherical coordinates can be used. Working with UCSs is introduced later in this chapter.

Using Spherical Coordinates

Spherical coordinates are well-suited for locating points on the surface of a sphere. In this section, you will draw a solid sphere and then locate a second solid sphere with its center on the surface of the first sphere.

To draw the first sphere, select the **SPHERE** command. Specify the center point as 7,5 and a radius of 1.5 units. Display a southeast isometric pictorial view of the sphere.

Figure 6-2.
A—A three-unit diameter sphere shown from the southeast isometric viewpoint.
B—A .8-unit diameter sphere is drawn with its center located on the surface of the original sphere. Also, lines have been drawn between the poles of the spheres. Notice how the polar axes are parallel. C—The objects after the Conceptual visual style is set current.

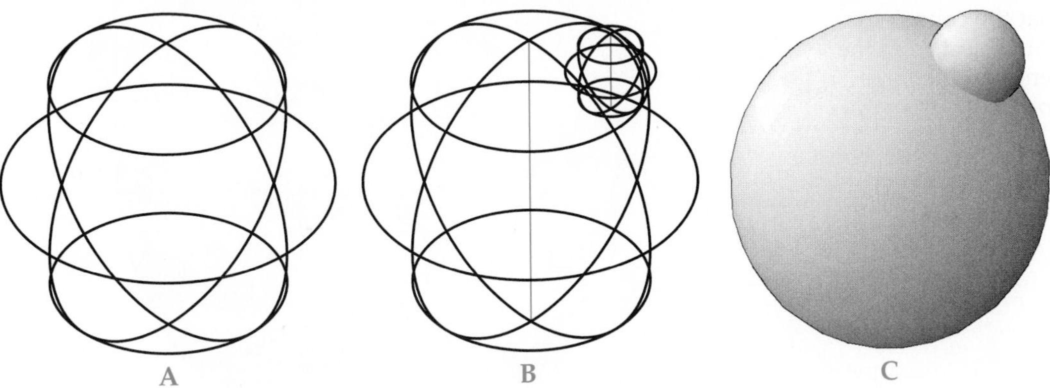

Alternately, you can use the view cube to create a different pictorial view. Also, set the Wireframe visual style current and switch to a parallel projection. Your drawing should look similar to Figure 6-2A.

Since you know the radius of the sphere, but the center of the sphere is not at the origin of the current UCS (the WCS), a relative spherical coordinate will be used to draw the second sphere. The sphere you drew is a solid and, as such, you can snap to its center using object snap. Set the center running object snap and then enter the **SPHERE** command again to draw the second sphere:

Specify center point or [3P/2P/Ttr]: **FROM**↵
Base point: *(use the* ***Center*** *object snap to select the center of the existing sphere)*
<Offset>: **@1.5<30<60**↵ *(1.5 is the radius of the first sphere)*
Specify radius or [Diameter]: **.4**↵

The objects should now appear as shown in Figure 6-2B. The center of the new sphere is located on the surface of the original sphere. This is clear after setting the Conceptual visual style current, Figure 6-2C. If you want the surfaces of the spheres to be tangent, add the radius value of each sphere (1.5 + .4) and enter this value when prompted for the offset from the center of the first sphere:

<Offset>: **@1.9<30<60**↵

Notice in Figure 6-2B that the polar axes of the two spheres are parallel. This is because both objects were drawn using the same UCS, which can be misleading unless you understand how objects are constructed based on the current UCS. Test this by locating a cone on the surface of the large sphere, just below the small sphere. First, display a 3D wireframe view of the objects. Then, select the **CONE** command and continue as follows.

Specify center point of base or [3P/2P/Ttr/Elliptical]: **FROM**↵
Base point: **CEN**↵
of *(pick the large sphere)*
<Offset>: **@1.5<30<30**↵
Specify base radius or [Diameter]: **.25**↵
Specify height or [2Point/Axis endpoint/Top radius]: **1**↵

The result of this construction with the Conceptual visual style set current is shown in Figure 6-3. Notice how the axis of the cone is parallel to the polar axis of the sphere. To draw the cone so that its axis projects from the center of the sphere, you will need to change the UCS. This is discussed later in the chapter.

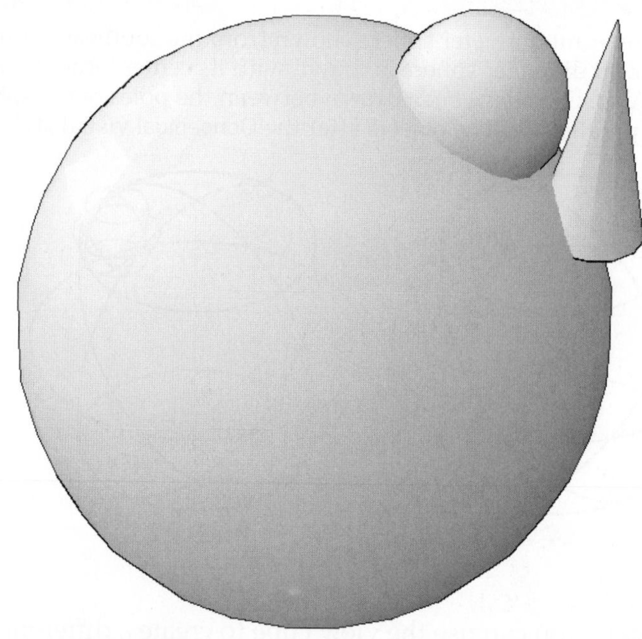

Figure 6-3.
The axis lines of objects drawn in the same user coordinate system are parallel. Notice that the cone does not project from the center of the large sphere.

Introduction to Cylindrical Coordinates

Locating a point in space with ***cylindrical coordinates*** is similar to locating a point on an imaginary cylinder. Cylindrical coordinates have three values. The first value represents the horizontal distance from the origin, which can be thought of as the radius of a cylinder. The second value represents the angle in the XY plane, or the rotation of the cylinder. The third value represents a vertical dimension measured up from the polar coordinate in the XY plane, or the height of the cylinder. See **Figure 6-4.** The absolute cylindrical coordinate shown in the figure is:

7.5<35,6

Like spherical coordinates, cylindrical coordinates can also be entered as relative coordinates. For example, a point drawn with the relative cylindrical coordinate @1.5<30,4 is located 1.5 units from the last point, at an angle of 30° in the XY plane of the previous point, and at a distance of four units up from the XY plane of the previous point.

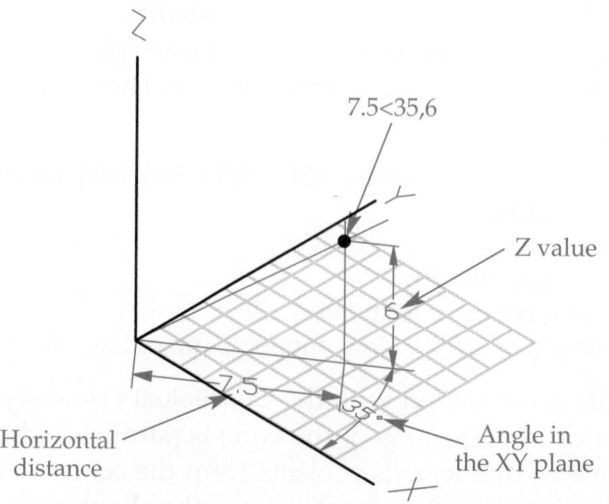

Figure 6-4.
Cylindrical coordinates require a horizontal distance from the origin, an angle in the XY plane, and a Z dimension.

NOTE

Turn off dynamic input before entering cylindrical coordinates. Dynamic input does not correctly interpret the entry when cylindrical coordinates are typed.

Using Cylindrical Coordinates

Cylindrical coordinates work well for attaching new objects to a cylindrical shape. An example of this is specifying coordinates for a pipe that must be attached to another pipe, tank, or vessel. In **Figure 6-5,** a pipe must be attached to a 12′ diameter tank at a 30° angle from horizontal and 2′-6″ above the floor. In order to properly draw the pipe as a cylinder, you will have to change the UCS, which you will learn how to do later in this chapter. An attachment point for the pipe can be drawn using the **POINT** command and cylindrical coordinates. First, set the **PDMODE** system variable to 3. Then, enter the **POINT** command and continue:

> Current point modes: PDMODE=3 PDSIZE=0.0000
> Specify a point: **FROM**↵
> Base point: **CEN**↵
> of *(pick the base of the cylinder)*
> <Offset>: **@6′<30,2′6″**↵ *(The radius of the tank is 6′.)*

The point can now be used as the center of the pipe (cylinder), **Figure 6-5B.** However, if you draw the pipe now, it will be parallel to the tank (large cylinder). By changing the UCS, as shown in **Figure 6-5C,** the pipe can be correctly drawn. Working with the UCS is introduced later in this chapter.

Exercise 6-1

Complete the exercise on the student website.
www.g-wlearning.com/CAD

Figure 6-5.
A—A plan view of a tank shows the angle of the pipe attachment. B—A 3D view from the southeast quadrant shows the pipe attachment point located with cylindrical coordinates. C—By creating a new UCS, the pipe can be drawn as a cylinder and correctly located without editing.

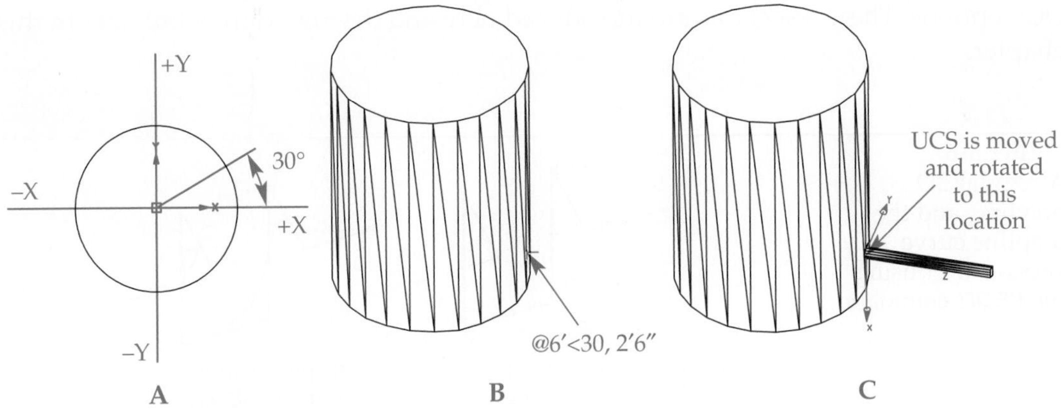

3D Polylines

A polyline drawn with the **PLINE** command is a 2D object. All segments of the polyline must be drawn parallel to the XY plane of the current UCS. A ***3D polyline,*** on the other hand, can be drawn in 3D space. The Z coordinate value can vary from point to point in the polyline.

3DPOLY

Ribbon
Home
> Draw
3D Polyline

Type
3DPOLY
3P

The **3DPOLY** command is used to draw 3D polylines. Any form of coordinate entry is valid for drawing 3D polylines. If polar tracking is on when using the **3DPOLY** command, you can pick points in the Z direction if the polar tracking alignment path is parallel to the Z axis.

The **Close** option can be used to draw the final segment and create a closed shape. There must be at least two segments in the polyline to use the **Close** option. The **Undo** option removes the last segment without canceling the command.

The **PEDIT** command can be used to edit 3D polylines. The **PEDIT Spline** option is used to turn the 3D polyline into a B-spline curve based on the vertices of the polyline. A regular 3D polyline and the same polyline turned into a B-spline curve are shown in Figure 6-6. The **SPLFRAME** system variable controls the display of the original polyline frame and is either turned on (1) or off (0).

Exercise 6-2

Complete the exercise on the student website.
www.g-wlearning.com/CAD

Introduction to Working with User Coordinate Systems

All points in a drawing or on an object are defined with XYZ coordinate values (rectangular coordinates) measured from the 0,0,0 origin. Since this system of coordinates is fixed and universal, AutoCAD refers to it as the ***world coordinate system (WCS).*** A ***user coordinate system (UCS),*** on the other hand, can be defined with its origin at any location and with its three axes in any orientation desired, while remaining at 90° to each other. The **UCS** command is used to change the origin, position, and rotation of the coordinate system to match the surfaces and features of an object under construction. When set up to do so, the UCS icon reflects the changes in the orientation of the UCS and placement of the origin.

UCS

Ribbon
Home
> Coordinates
View
> Coordinates
UCS

Type
UCS

The **UCS** command is used to create and manage UCSs. This command and its options can be accessed on the **Coordinates** panel of the **Home** and **View** tabs on the ribbon or by typing the command. Three selections in the ribbon provide access to all UCS options. These selections are introduced here and discussed in detail later in this chapter.

Figure 6-6.
A regular 3D polyline and the B-spline curve version after using the **PEDIT** command.

Regular 3D Polyline

B-spline Curve (**SPLFRAME** =1)

- **UCS.** This button executes the **UCS** command, which allows you to create a named UCS, and provides access to all of the command options. The **UCS** command can also be selected by picking on the UCS drop-down list below the view cube. All of these options are covered later in this chapter.
- **Named UCS Combo Control.** This drop-down list contains the six orthographic UCS options, which are covered later in this chapter. Any saved UCSs will be listed here, too.
- **UCS, Named UCS.** This button displays the **UCS** dialog box. The three tabs in the dialog box contain a variety of UCS and UCS icon options and settings. These options and settings are described as you progress through the chapter.

Earlier in this chapter, you used spherical coordinates to locate a small sphere on the surface of a larger sphere. You also drew a cone with the center of its base on the surface of the large sphere. However, the axis of the cone, which is a line from the center of the base to the tip of the cone, is not pointing to the center of the sphere. Refer to Figure 6-3. This is because the Z axes of the large sphere and cone are parallel to the world coordinate system (WCS) Z axis. The WCS is the default coordinate system of AutoCAD.

In order for the axis of the cone to project from the sphere's center point, the UCS must be changed using the **UCS** command. Working with different UCSs is discussed in the next section. However, the following is a quick overview and describes how to draw a cone with its axis projecting from the center of the sphere.

First, draw a three-unit diameter sphere with its center at 7,5. Display the drawing from the southeast isometric preset. To help see how the UCS is changing, make sure the UCS icon is displayed at the origin of the current UCS. The UCS icon drop-down list is found in the **Coordinates** panel of the **Home** and **View** tabs of the ribbon and provides access to three options of the **UCSICON** command. See Figure 6-7. Pick the **Show UCS Icon at Origin** button to ensure the icon is displayed at the origin. Also, set the Wireframe visual style current.

Now, the sphere is drawn and the UCS icon is displayed at the origin of the current UCS (or at the lower-left corner of the screen, depending on the zoom level). However, the WCS is still the current user coordinate system. You are ready to start changing the UCS to meet your needs.

Begin by moving the UCS origin to the center of the sphere using the **Origin** option of the **UCS** command and the center object snap. Notice that the UCS icon is now displayed at the center of the sphere, Figure 6-8. Also, if the grid is displayed, the red X and green Y axes of the grid intersect at the center of the sphere.

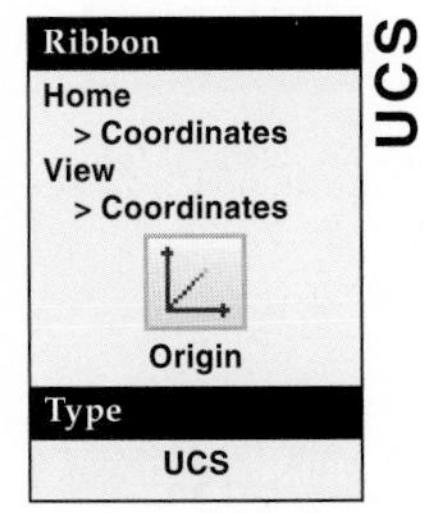

UCS

Study Figure 6-9 and continue as follows. Keep in mind that the point you are locating—the center of the cone on the sphere's surface—is 30° from the X axis and 30° from the XY plane. For ease of visualization, zoom in so the sphere fills the screen.

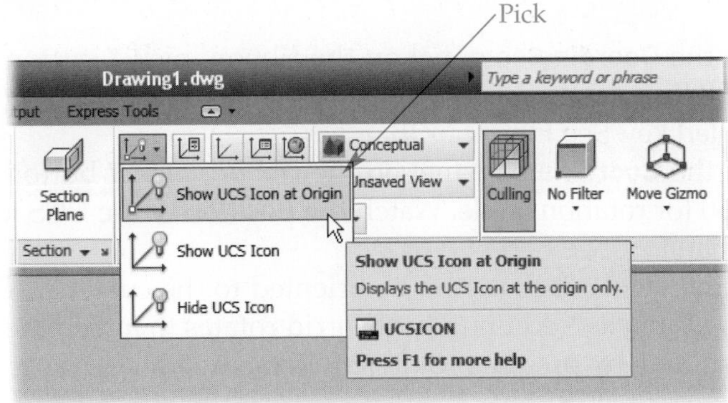

Figure 6-7.
The UCS icon drop-down list provides access to the **Origin**, **Off**, and **On** options of the **UCSICON** command.

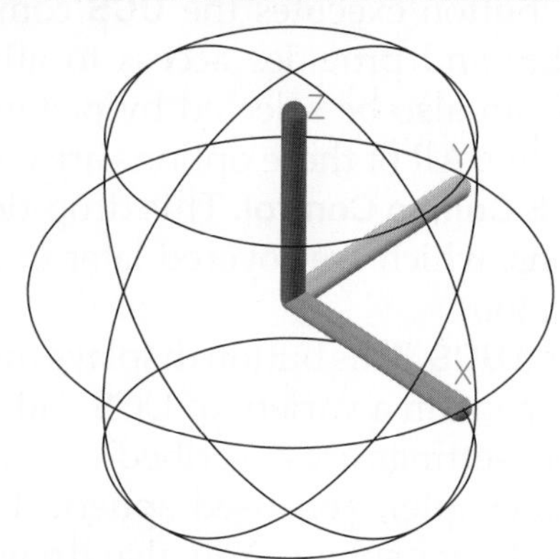

Figure 6-8.
The UCS origin is moved to the center of the sphere.

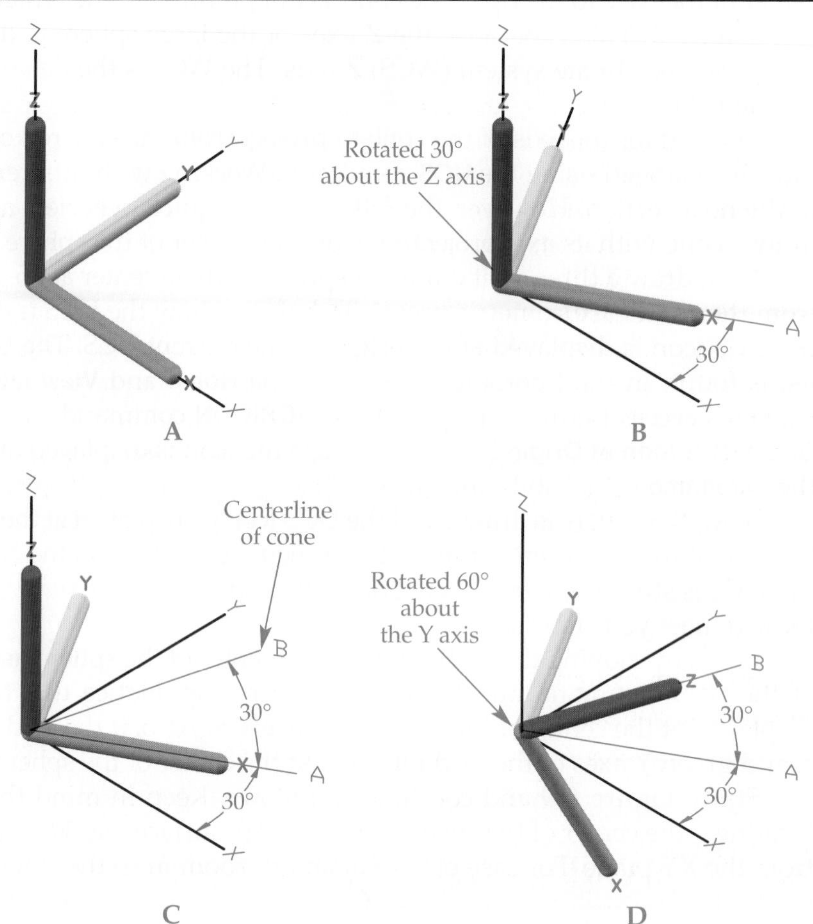

Figure 6-9.
A—The world coordinate system.
B—The new UCS is rotated 30° in the XY plane about the Z axis.
C—A line rotated up 30° from the XY plane represents the axis of the cone.
D—The UCS is rotated 60° about the Y axis. The centerline of the cone coincides with the Z axis of this UCS.

UCS

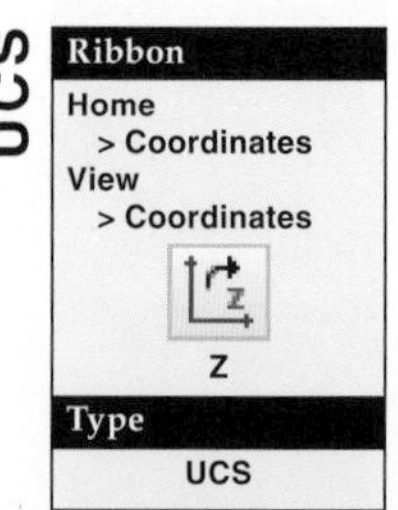

In the **Coordinates** panel on the ribbon, pick **Z** button in the axis rotation flyout. Enter 30 for rotation angle. Watch the position of the UCS icon change when you press the [Enter] key. See Figure 6-9B.

UCS

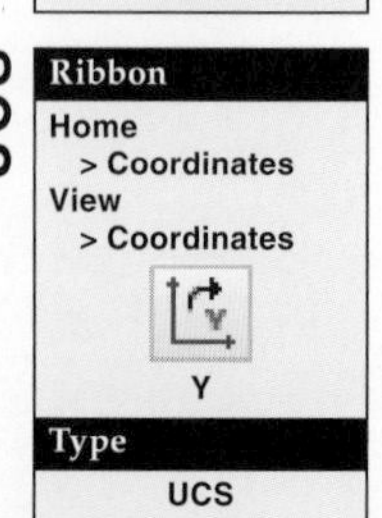

In the **Coordinates** panel on the ribbon, pick **Y** button in the axis rotation flyout. Enter 60 for rotation angle. Watch the position of the UCS icon change when you press the [Enter] key. See Figure 6-9D.

If the view cube is set to be oriented to the current UCS, then it rotates when the UCS is changed. Additionally, the grid rotates to match the new UCS. If the grid is not on, turn it on by pressing [Ctrl]+[G]. Remember, the grid is displayed on the XY plane of the current UCS.

The new UCS can be used to construct a cone with its axis projecting from the center of the sphere. Figure 6-10A shows the new UCS located at the center of the

Figure 6-10.
A—A new UCS is created with the Z axis projecting from the center of the sphere. B—A cone is drawn using the new UCS. The axis of the cone projects from the center of the sphere. C—The objects after the Conceptual visual style is set current.

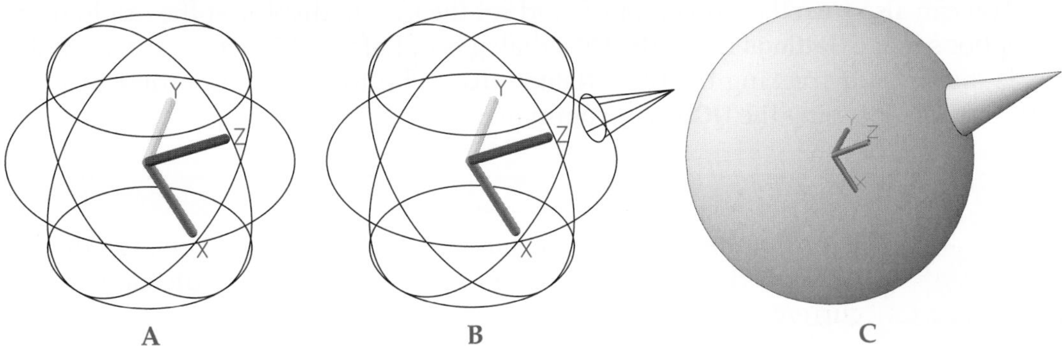

sphere. With the UCS rotated, rectangular coordinates can be used to draw the cone. Enter the **CONE** command and specify the center as 0,0,1.5 (the radius of the sphere is 1.5 units). Enter a radius of .25 and a height of 1. The completed cone is shown in **Figure 6-10B.** You can see that the axis projects from the center of the sphere. **Figure 6-10C** shows the objects after setting the Conceptual visual style current.

This same basic procedure can be used in the tank and pipe example presented earlier in this chapter. To correctly locate the pipe (cylinder), first rotate the UCS 30° about the Z axis. Then, rotate the UCS 90° about the Y axis. The Z axis of this new UCS aligns with the long axis of the pipe. Finally, use rectangular coordinates to draw the cylinder with its center at the point drawn in **Figure 6-5B.**

Once you have changed to a new UCS, you can quickly return to the WCS by picking WCS in the view cube drop-down list or by using the **World** option of the **UCS** command. The WCS provides a common "starting place" for creating new UCSs.

Exercise 6-3

Complete the exercise on the student website.
www.g-wlearning.com/CAD

Working with User Coordinate Systems

Once you understand a few of the basic options of user coordinate systems, creating 3D models becomes an easy and quick process. The following sections show how to display the UCS icon, change the UCS in order to work on different surfaces of a model, and name and save a UCS. As you saw in the previous section, working with UCSs is easy.

Displaying the UCS Icon

The symbol that identifies the orientation of the coordinate system is called the ***UCS icon.*** When AutoCAD is first launched based on the acad3D.dwt template, the UCS icon is located at the WCS origin in the middle of the viewport. The display of this symbol is controlled by the **UCSICON** command. Turn the UCS icon on and off using the **Hide UCS Icon** and **Show UCS Icon** buttons on the **Coordinates** panel in the ribbon. Refer to **Figure 6-7.**

If your drawing does not require viewports and altered coordinate systems, you may want to turn the icon off. The icon disappears until you turn it on again. If you redisplay the UCS icon and it does not appear at the origin, simply pick the **Show UCS Icon at Origin** button on the **Coordinates** panel in the ribbon.

You can also turn the icon on or off and set the icon to display at the origin using the options in the **Settings** tab of the **UCS** dialog box. Refer to Figure 6-11. This dialog box is displayed by picking the **UCS, Named UCS...** button on the **Coordinates** panel in the ribbon or typing the **UCSMAN** command.

PROFESSIONAL TIP

It is recommended that you have the UCS icon turned on at all times when working in 3D drawings. It provides a quick indication of the current UCS.

Changing the Coordinate System

To construct a three-dimensional object, you must draw shapes at many different angles. Different planes are needed to draw features on angled surfaces. To construct these features, it is easiest to rotate the UCS to match any surface on an object. The following example illustrates this process.

The object in Figure 6-12 has a cylinder on the angled surface. A solid modeling command called **EXTRUDE**, which is discussed in Chapter 9, is used to create the base of the object. The cylinder is then drawn on the angled feature. In Chapter 8, you will learn how to dimension the object as shown in Figure 6-12.

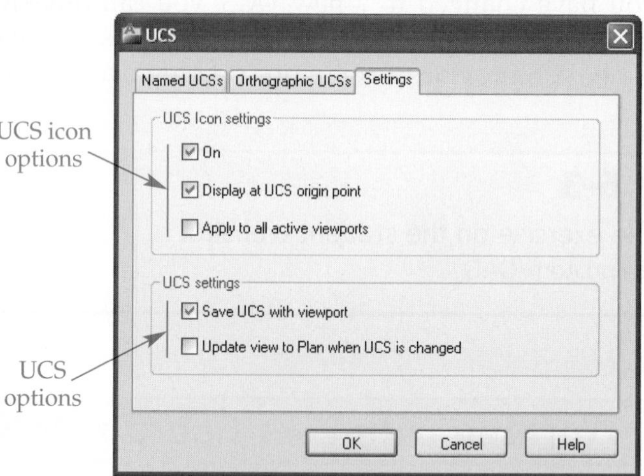

Figure 6-11.
Setting UCS and UCS icon options in the **UCS** dialog box.

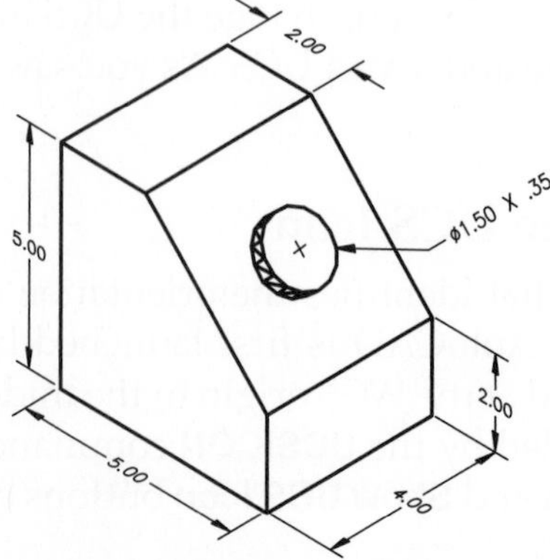

Figure 6-12.
This object can be constructed by changing the orientation of the coordinate system.

The first step in creating this model is to draw the side view of the base as a wireframe. You could determine the X, Y, and Z coordinates of each point on the side view and enter the coordinates. However, a lot of typing can be saved if all points share a Z value of 0. By rotating the UCS, you can draw the side view entering only X and Y coordinates. Start a new drawing and display the southeast isometric view. If the UCS icon is off, turn it on and display it at the origin.

Now, rotate the UCS 90° about the X axis. The new UCS is parallel to the side of the object. The UCS icon is displayed at the origin of the UCS. If needed, pan the view so the UCS icon is near the center of the view.

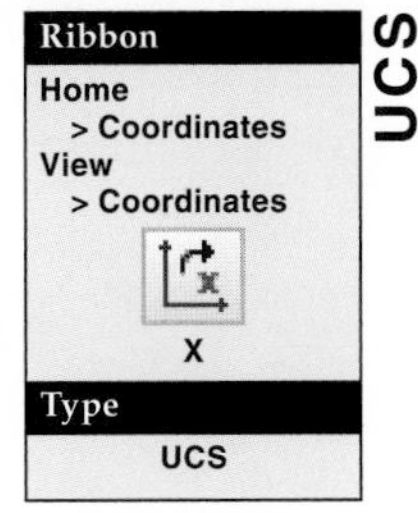

UCS

Next, use the **PLINE** command to draw the outline of the side view. Refer to the coordinates shown in Figure 6-13. When entering coordinates, you may want to turn off dynamic input so direct-distance entry is disabled. The **PLINE** command is used instead of the **LINE** command because a closed polyline can be extruded into a solid. Be sure to use the **Close** option to draw the final segment. A wireframe of one side of the object is created. Notice the orientation of the UCS icon.

Now, the **EXTRUDE** command is used to create the base as a solid. This command is covered in detail in Chapter 9. Make sure the same UCS used to create the wireframe side is current. Then, select the **EXTRUDE** command by picking the **Extrude** button on the **Modeling** panel of the **Home** tab in the ribbon. When prompted to select objects, pick the polyline and then press the [Enter] key. Next, move the mouse so the preview extends to the right and enter an extrusion height of 4. If dynamic input is off, you must enter –4.

The base is created as a solid. See Figure 6-14. You may want to switch to a parallel projection, as shown in the figure.

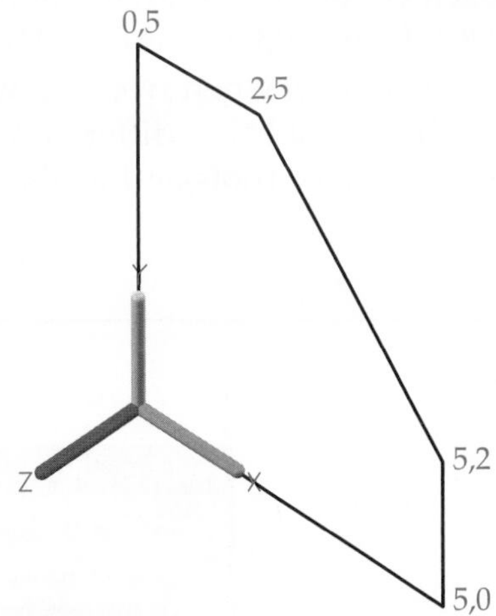

Figure 6-13.
A wireframe of one side of the base is created. Notice the orientation of the UCS.

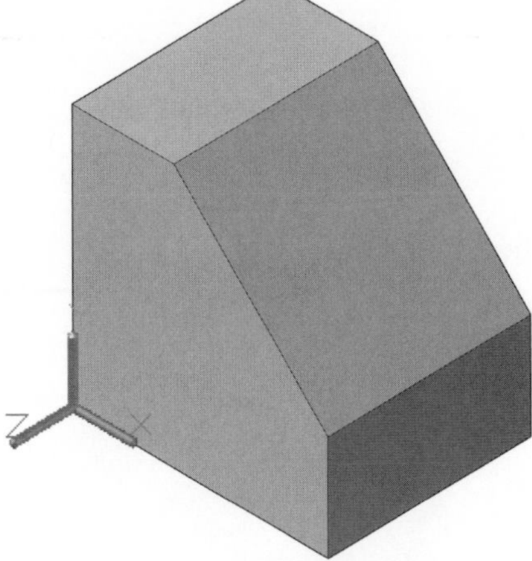

Figure 6-14.
The wireframe is extruded to create the base as a solid.

UCS

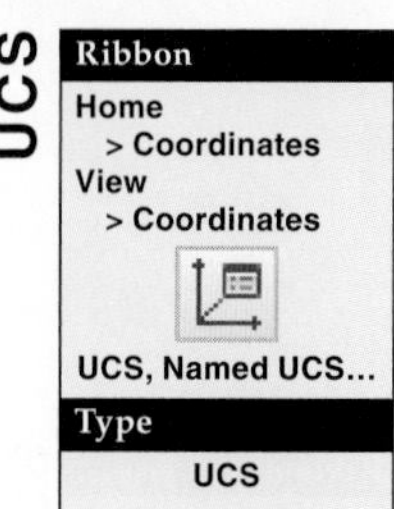

Saving a Named UCS

Once you have created a new UCS that may be used again, it is best to save it for future use. For example, you just created a UCS used to draw the wireframe of one side of the object. You can save this UCS using the **Named**>**Save** option of the **UCS** command or the **UCS** dialog box.

If using the dialog box, right-click on the entry Unnamed and pick **Rename** in the shortcut menu. See Figure 6-15. You can also pick once or double-click on the highlighted name. Then, type the new name in place of Unnamed and press [Enter]. A name can have up to 255 characters. Numbers, letters, spaces, dollar signs ($), hyphens (–), and underscores (_) are valid. Use this method to save a new UCS or to rename an existing one. Now, the coordinate system is saved and can be easily recalled for future use.

NOTE

If Unnamed does not appear in the **UCS** dialog box, AutoCAD is having a problem. Exit the dialog box and enter the **UCS** command on the command line. Then, use the **Named**>**Save** option to save the unnamed UCS. The saved UCS will then appear in the **UCS** dialog box.

PROFESSIONAL TIP

Most drawings can be created by rotating the UCS as needed without saving it. If the drawing is complex with several planes, each containing a large amount of detail, you may wish to save a UCS for each detailed face. Then, restore the proper UCS as needed. For example, when working with architectural drawings, you may wish to establish a different UCS for each floor plan and elevation view and for roofs and walls that require detail work.

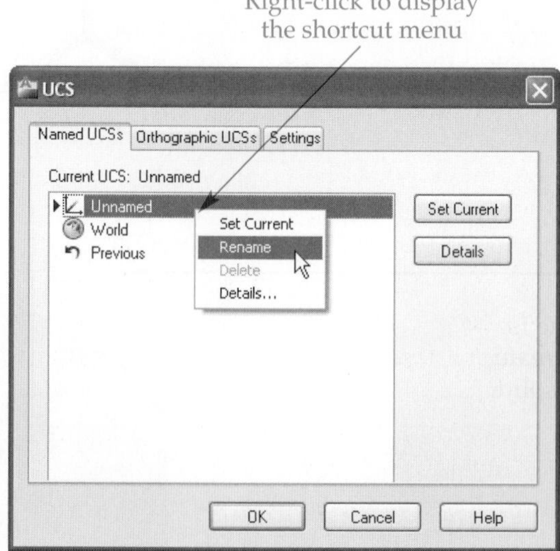

Figure 6-15.
Saving a new UCS. Select **Rename** to enter a name and save the Unnamed UCS.

Dynamic UCS

A powerful tool for 3D modeling is the ***dynamic UCS function.*** A dynamic UCS is a UCS temporarily located on any existing face of a 3D model. The function is activated by picking the **Allow/Disallow Dynamic UCS** button in the status bar, pressing the [Ctrl]+[D] key combination, or setting the **UCSDETECT** system variable to 1. When the pointer is moved over a model surface, the XY plane of the UCS is aligned with that surface. This is especially useful when adding primitives or shapes to model surfaces. In addition, dynamic UCSs are useful when inserting blocks and xrefs, locating text, editing 3D geometry, editing with grips, and area calculations.

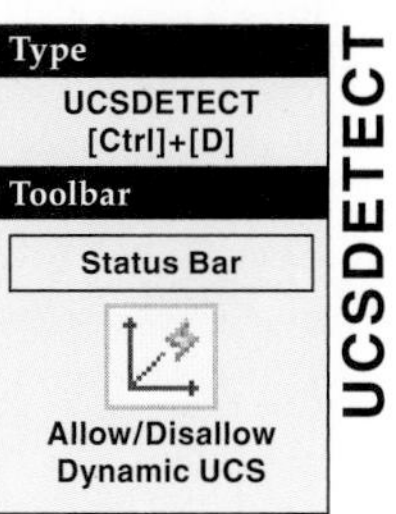

UCSDETECT

An example of using a dynamic UCS is to draw the cylinder on the angled face of the object shown in Figure 6-12. First, select the **CYLINDER** command. Make sure the dynamic UCS function is on. Then, move the pointer over one of the surfaces of the object. Notice that the 3D crosshairs change when they are moved over a new surface. The red (X) and green (Y) crosshairs are flat on the face. For ease of visualizing the 3D crosshairs as they are moved across different surfaces, right-click on the **Allow/ Disallow Dynamic UCS** button in the status bar and select **Display crosshair labels** to turn on the XYZ labels on the crosshairs.

As you move the pointer over the object faces, note that hidden faces are not highlighted, therefore you cannot work on those faces. If you wish to work on a hidden face you must first change the viewpoint to make that face visible. The view cube or [Shift]+ mouse wheel button can be used to dynamically change the viewpoint without interrupting the current command.

The **CYLINDER** command is currently prompting to select a center point of the base. If you pick a point, this sets the center of the cylinder base *and* temporarily relocates the UCS so its XY plane lies on the selected face. Once the point is selected and the dynamic UCS created, the UCS icon moves to the temporary UCS. When the command is ended, the UCS reverts to its previous orientation. To locate a 1.5″ diameter cylinder in the center of the angled face, use the following procedure.

1. At the "specify center point of base" prompt, [Shift] + right-click in the drawing area and pick **3D Osnap** from the shortcut menu, then pick **Center of face** from the cascading menu. See Figure 6-16A.
2. Move the pointer over the angled face. Notice the crosshairs are aligned with the angled face. A circular 3D snap pick point appears at the center of the face. Select this point. See Figure 6-16B.
3. Specify the 1.5 unit diameter for the base.
4. Specify a cylinder height of .35 units. See Figure 6-16C.
5. The cylinder is properly located on the angled face. See Figure 6-16D.

When setting a dynamic UCS, experiment with the behavior of the crosshairs as they are moved over different surfaces. The orientation of the crosshairs is related to the edge of the face that they are moved over. Can you determine the pattern by which the crosshairs are turned? The X axis of the crosshairs is always aligned with the edge that is crossed.

If you want to temporarily turn off the dynamic UCS function while working in a command, press and hold the [Shift]+[Z] key combination while moving the pointer over a face. As soon as you release the keys, dynamic UCS function is reinstated.

Figure 6-16.
Using a dynamic UCS allows you to draw a cylinder on the angled face of the object shown here without creating a new UCS. A—To set the center point of the base and select the angled face for the dynamic UCS, use the **Center of face** 3D object snap and select the center of the face. B—Set the radius or diameter of the base. C—Set the height of the cylinder. D—When the command is ended, the previous UCS is restored.

Pick

Specify the radius or diameter

A

B

Previous UCS is restored

C

D

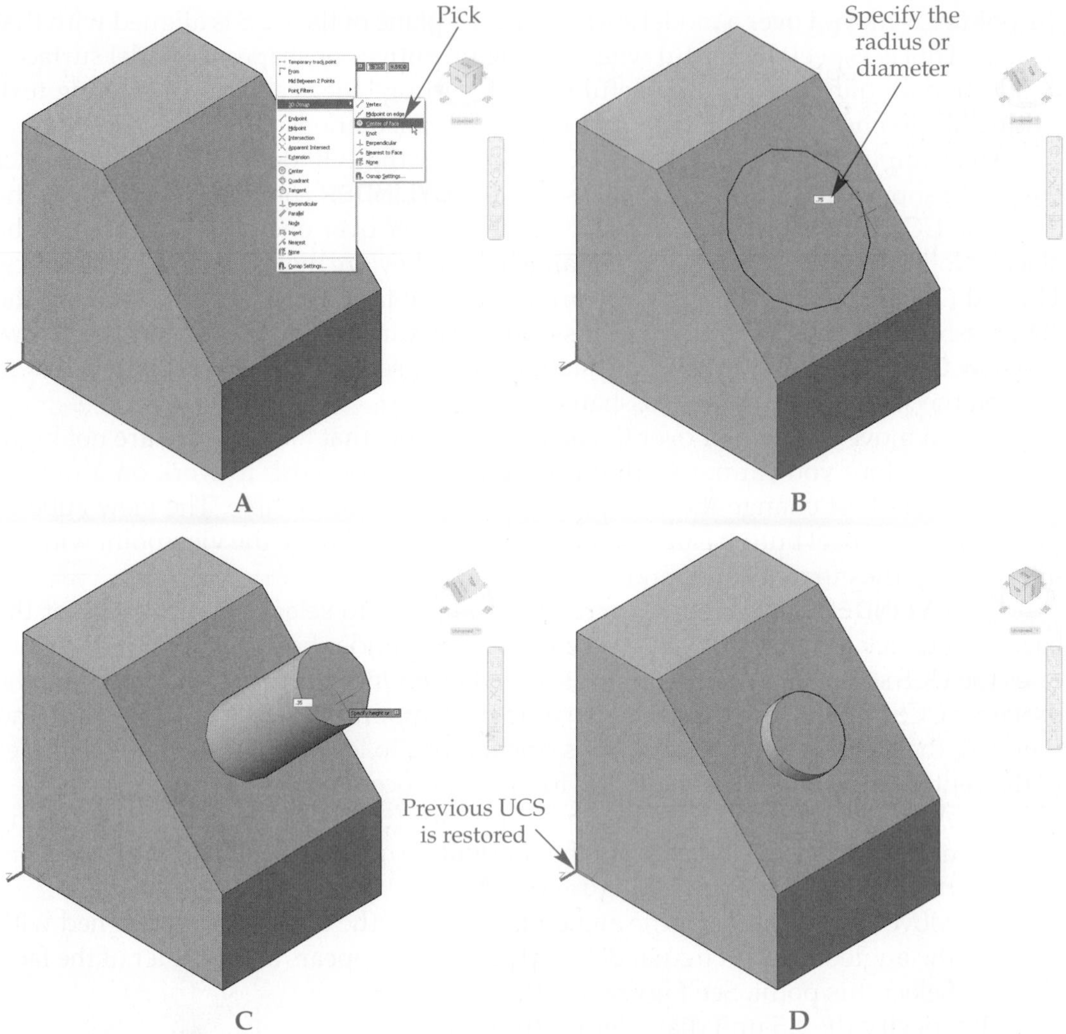

PROFESSIONAL TIP

Selecting 3D object snaps can be automated for solid model construction. Simply right-click on the **3D Object Snap** button on the status bar and select the snap mode you wish to use. To set multiple snap modes, pick the **Settings...** option, and then select the appropriate modes in the **Drafting Settings** dialog box. Be sure to turn on the **3D Object Snap** button before using a modeling command.

Exercise 6-4

Complete the exercise on the student website.
www.g-wlearning.com/CAD

Additional Ways to Change the UCS

There are other ways to change the UCS. These options include picking three points, selecting a new Z axis, and setting the UCS to an existing object. The next sections cover these options.

Selecting Three Points to Create a New UCS

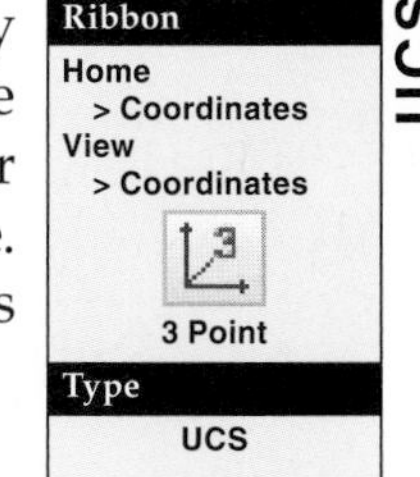

UCS

The **3 Point** option of the **UCS** command can be used to change the UCS to any flat surface. This option requires that you first locate a new origin, then a point on the positive X axis, and finally a point on the XY plane that has a positive Y value. Refer to Figure 6-17. Use object snaps to select points that are not on the current XY plane. After you pick the third point—the point on the XY plane—the UCS icon changes its orientation to align with the plane defined by the three points.

PROFESSIONAL TIP

When typing the **UCS** command, enter 3 at the Specify origin of UCS or [Face/NAmed/OBject/Previous/View/World/X/Y/Z/ZAxis] <World>: prompt. Notice that the option is not listed in the prompt.

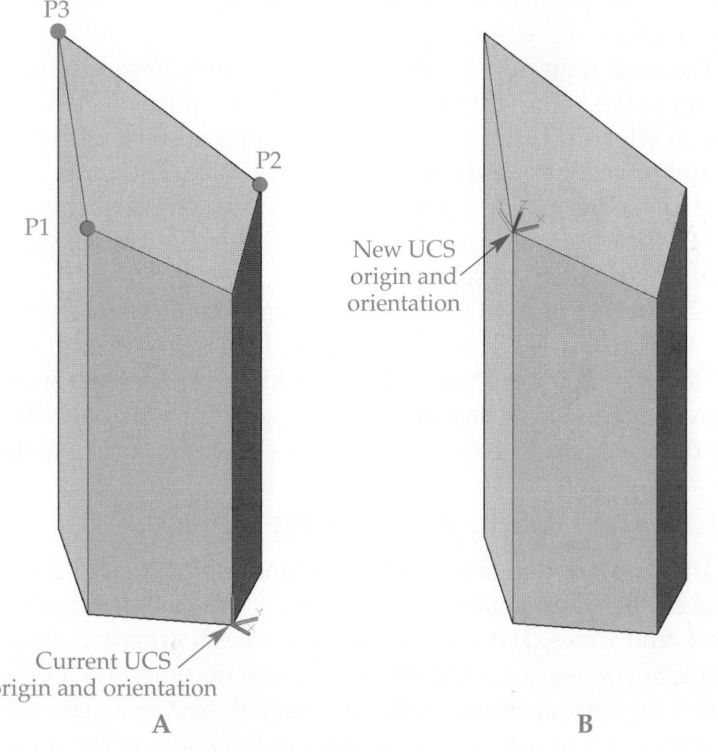

Figure 6-17.
A—A new UCS can be established by picking three points. P1 is the origin, P2 is on the positive X axis, and P3 is on the XY plane and has a positive Y value. B—The new UCS is created.

Figure 6-18.
A—Using the **ZAxis** option to establish a new UCS. B—The new UCS is used to create a cylinder, which is then subtracted from the base to create a hole.

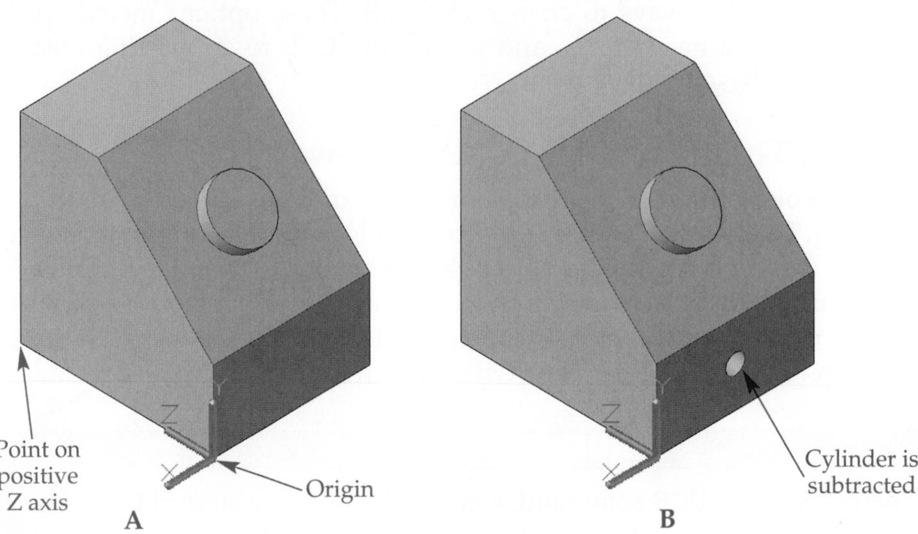

Selecting a New Z Axis

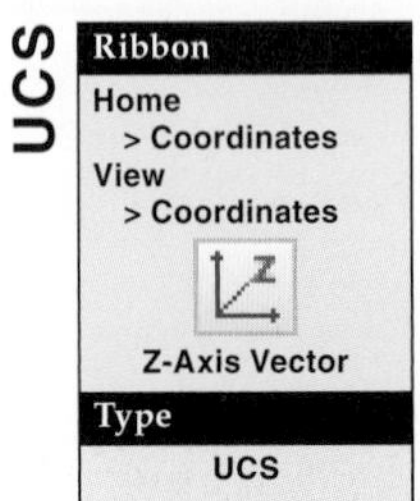

The **ZAxis** option of the **UCS** command allows you to select the origin point and a point on the positive Z axis. Once the new Z axis is defined, AutoCAD sets the new X and Y axes.

You will now add a cylinder to the lower face of the base created earlier. The cylinder extends into the base. Refer to the location of the UCS in **Figure 6-16D.** This is the UCS after adding the cylinder to the angled face with a dynamic UCS. The Z axis does not project perpendicular to the lower face. Therefore, a new UCS must be created on the lower-right face. Change the UCS after entering the **ZAxis** option as follows.

1. Pick the origin of the new UCS. See **Figure 6-18A.** You may have to use an object snap to select the origin.
2. Pick a point on the positive portion of the new Z axis.
3. The new UCS is established and it can be saved if necessary.

Now, use 3D object snaps, auto-tracking, or 2D object snaps to draw a ∅.5″ cylinder centered on the lower face and extending 3″ into the base. Then, subtract the cylinder from the base part to create the hole, as shown in **Figure 6-18B.**

Setting the UCS to an Existing Object

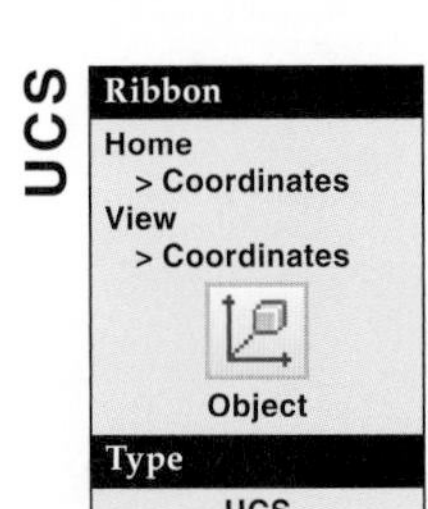

The **Object** option of the **UCS** command can be used to define a new UCS on an object. However, there are some objects on which this option cannot be used: 3D polylines, 3D meshes, and xlines. There are also certain rules that control the orientation of the UCS. For example, if you select a circle, the center point becomes the origin of the new UCS. The pick point on the circle determines the direction of the X axis. The Y axis is relative to X and the UCS Z axis may or may not be the same as the Z axis of the selected object.

Look at **Figure 6-19A.** The circle is rotated an unknown number of degrees from the XY plane of the WCS. However, you need to create a UCS in which the circle is lying on the XY plane. Select the **Object** option of the **UCS** command and then pick the circle. The UCS icon may look like the one shown in **Figure 6-19B.** Notice how the X and Y axes are not aligned with the quadrants of the circle, as indicated by the grip locations. This may not be what you expected. The X axis orientation is determined by the pick point on the circle. Notice how the X axis is pointing at the pick point.

Figure 6-19.
A—This circle is rotated off of the WCS XY plane by an unknown number of degrees. It will be used to establish a new UCS. B—The circle is on the XY plane of the new UCS. However, the X and Y axes do not align with the circle's quadrants. C—The **ZAxis** option of the **UCS** command is used to align the UCS with the quadrants of the circle.

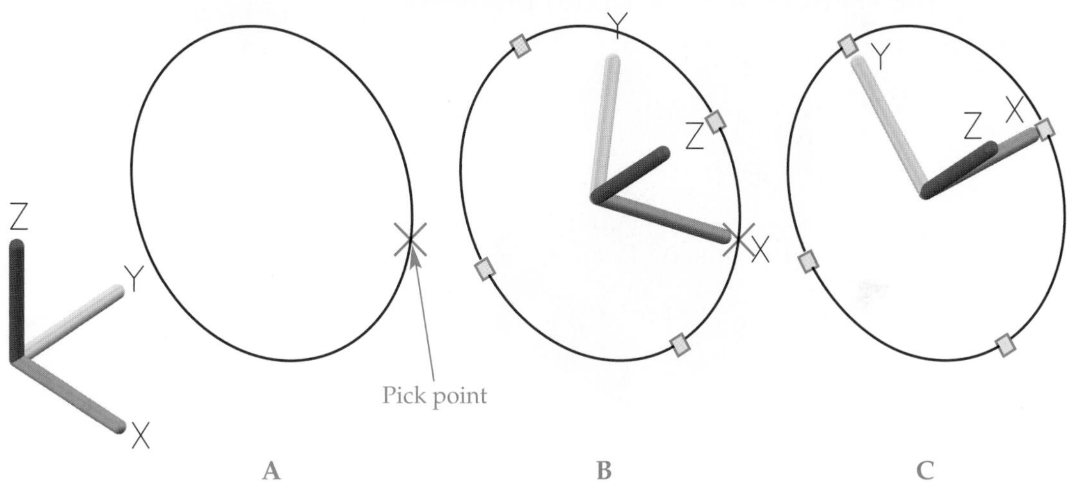

To rotate the UCS in the current plane so the X and Y axes of the UCS are aligned with the quadrants of the circle, use the **ZAxis** option of the **UCS** command. Select the center of the circle as the origin and then enter the absolute coordinate 0,0,1. This uses the current Z axis location, which also forces the X and Y axes to align with the object. Refer to Figure 6-19C. This method may not work with all objects.

Setting the UCS to the Face of a 3D Solid

The **Face** option of the **UCS** command allows you to orient the UCS to any face on a 3D solid object. This option does not work on surface objects. Select the command and then pick a face on the solid. After you have selected a face on a 3D solid, you have the options of moving the UCS to the adjacent face or flipping the UCS 180° on the X, Y, or both axes. Use the **Next**, **Xflip**, or **Yflip** options to move or rotate the UCS as needed. Once you achieve the UCS orientation you want, press [Enter] to accept. Notice in Figure 6-20 how many different UCS orientations can be selected for a single face.

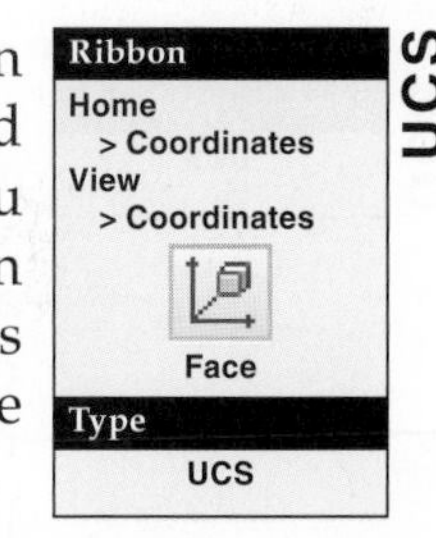

UCS

Figure 6-20.
Several different UCSs can be selected from a single pick point using the **Face** option of the **UCS** command. Given the pick point, five of the eight possibilities are shown here.

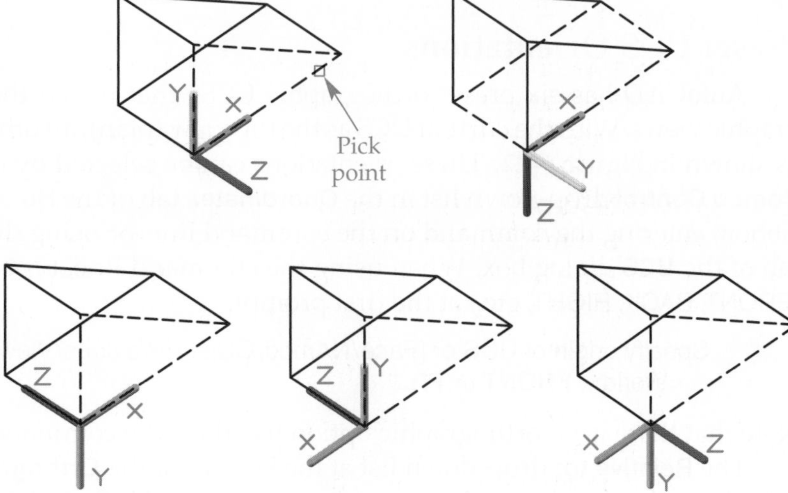

Figure 6-21.
The **View** option of the **UCS** command allows you to place text plan to the current view.

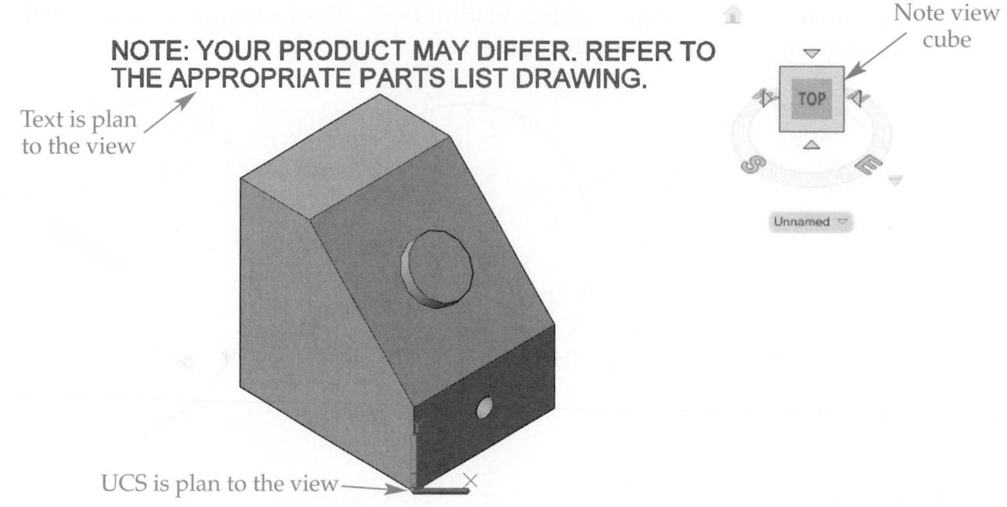

Setting the UCS Perpendicular to the Current View

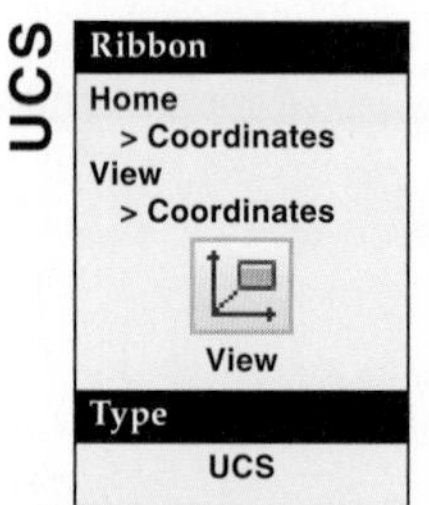

You may need to add notes or labels to a 3D drawing that are plan to the current view, such as that shown in Figure 6-21. The **View** option of the **UCS** command makes this easy to do. Immediately after selecting the **View** option, the UCS rotates to a position so the new XY plane is perpendicular to the current line of sight (parallel to the screen). Now, anything added to the drawing is plan to the current view. The command works on the current viewport only; other viewports are unaffected.

Applying the Current UCS to a Viewport

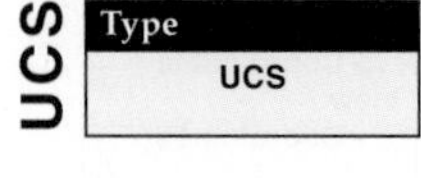

The **Apply** option of the **UCS** command allows you to apply the UCS in the current viewport to any or all model space or paper space viewports. Using the **Apply** option, you can have a different UCS displayed in every viewport or you can apply one UCS to all viewports. With the viewport that contains the UCS to apply active, enter the **Apply** option. This option is only available on the command line. However, the option does not appear in the command prompt. Enter either A or APPLY to select the option. Then, pick a viewport to which the current UCS will be applied and press [Enter]. To apply the current UCS to all viewports, enter the **All** option. See Chapter 7 for a complete discussion of model space viewports.

Preset UCS Orientations

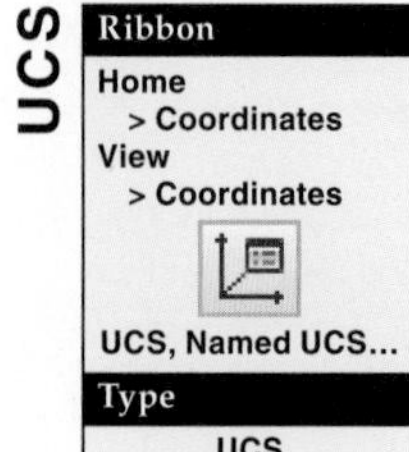

AutoCAD has six preset orthographic UCSs that match the six standard orthographic views. With the current UCS as the top view (plan), all other views are arranged as shown in Figure 6-22. These orientations can be selected by using the **Named UCS Combo Control** drop-down list in the **Coordinates** tab of the **Home** or **View** panel in the ribbon, entering the command on the command line, or using the **Orthographic UCSs** tab of the **UCS** dialog box. When using the command line, type the name of the UCS (FRONT, BACK, RIGHT, etc.) at the first prompt:

Specify origin of UCS or [Face/NAmed/OBject/Previous/View/World/X/Y/Z/ZAxis] <World>: **FRONT** *or* **FR**↵

Note that there is no orthographic option listed for the command.

The **Relative to:** drop-down list at the bottom of the **Orthographic UCSs** tab of the **UCS** dialog box specifies whether each orthographic UCS is relative to a named UCS or absolute to the WCS. For example, suppose you have a saved UCS named Front Corner

Figure 6-22.
The standard orthographic UCSs coincide with the six basic orthographic views.

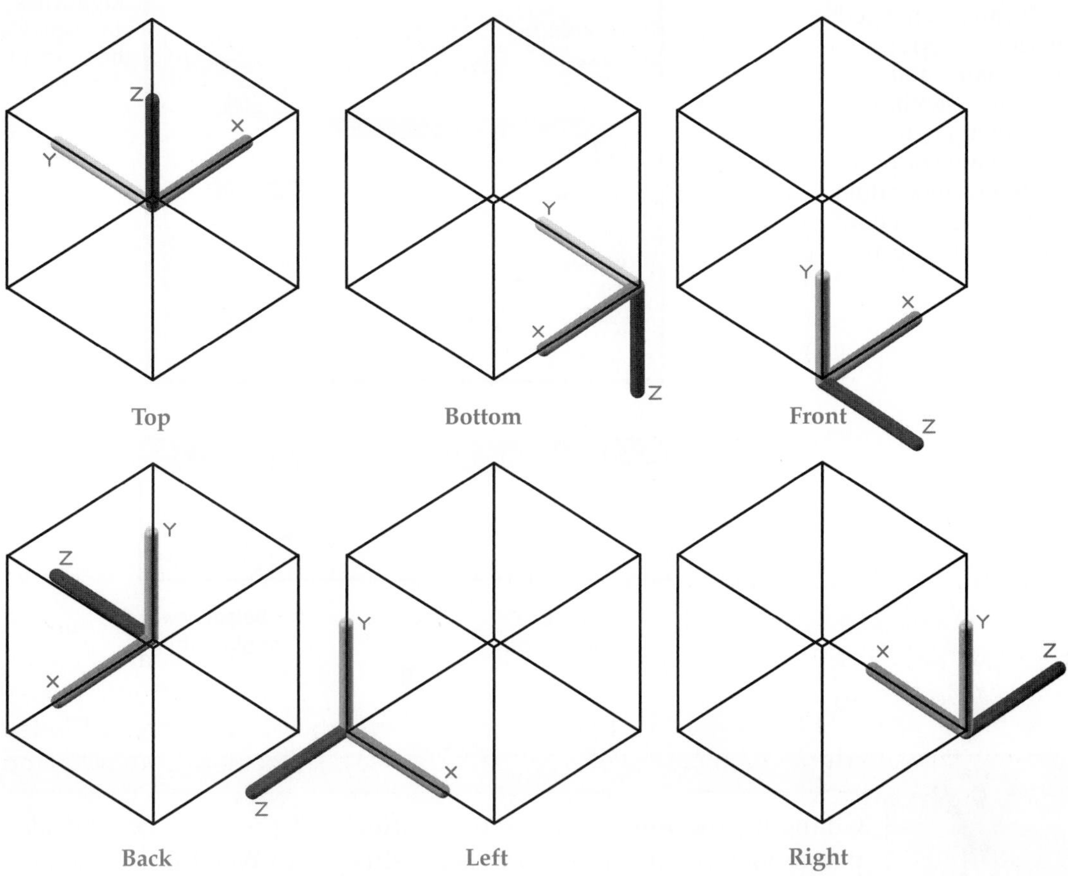

that is rotated 30° about the Y axis of the WCS. If you set current the top UCS relative to the WCS, the new UCS is perpendicular to the WCS, Figure 6-23A. However, if the top UCS is set current relative to the named UCS Front Corner, the new UCS is also rotated from the WCS, Figure 6-23B.

The Z value, or depth, of a preset UCS can be changed in the **Orthographic UCSs** tab of the **UCS** dialog box. First, right-click on the name of the UCS you wish to change. Then, pick **Depth** from the shortcut menu, Figure 6-24A. This displays the **Orthographic UCS depth** dialog box. See Figure 6-24B. You can either enter a new depth value or specify the new location on screen by picking the **Select new origin** button. Once the new depth has been selected, it is reflected in the preset UCS list.

Figure 6-23.
The **Relative to:** drop-down list entry in the **Orthographic UCSs** tab of the **UCS** dialog box determines whether the orthographic UCS is based on a named UCS or the WCS. The UCS icon here represents the named UCS. A—Relative to the WCS. B—Relative to the named UCS.

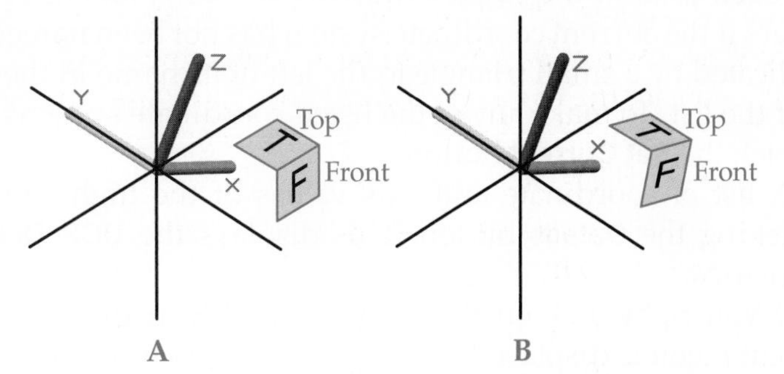

Figure 6-24.
A—The Z value, or depth, of a preset UCS can be changed by right-clicking on its name and selecting **Depth**.
B—Enter a new depth value or pick the **Select new origin** button to pick a new location on screen.

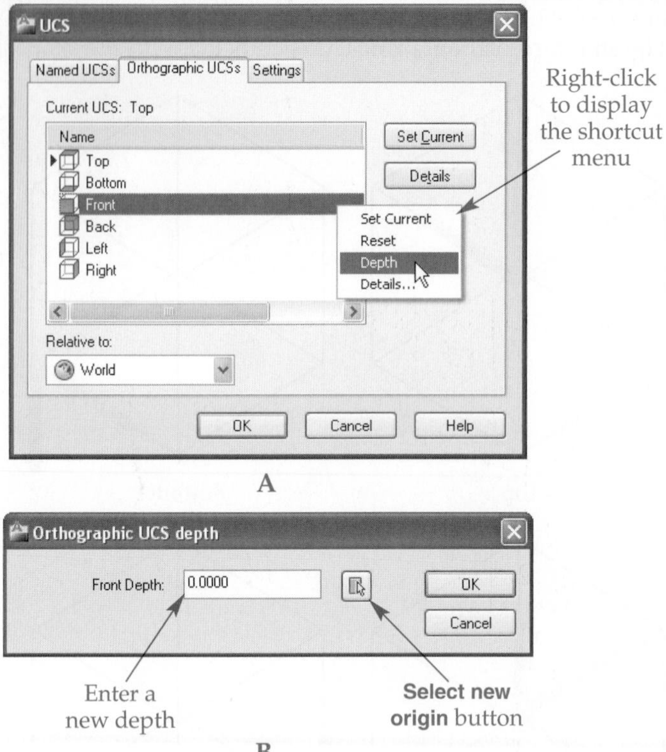

CAUTION

Changing the **Relative to:** setting affects *all* preset UCSs and *all* preset viewpoints! Therefore, leave this set to World unless absolutely necessary to change it.

Managing User Coordinate Systems and Displays

You can create, name, and use as many user coordinate systems as needed to construct your model or drawing. As you saw earlier, AutoCAD allows you to name (save) coordinate systems for future use. User coordinate systems can be created, renamed, set current, and deleted using the **Named UCSs** tab of the **UCS** dialog box, Figure 6-25.

The **Named UCSs** tab contains the **Current UCS:** list box. This list box contains the names of all saved coordinate systems plus World. If other coordinate systems have been used in the current drawing session, Previous appears in the list. Unnamed appears if the current coordinate system has not been named (saved). The current UCS is indicated by a small triangle to the left of its name in the list and by the label at the top of the list. To make any of the listed coordinate systems active, highlight the name and pick the **Set Current** button.

A list of coordinate and axis values of the highlighted UCS can be displayed by picking the **Details** button. This displays the **UCS Details** dialog box shown in Figure 6-26.

If you right-click on the name of a UCS in the list in the **Named UCSs** tab, a shortcut menu is displayed. Using this menu, you can rename the UCS. Saving a UCS is discussed earlier in this chapter. You can also set the UCS current or delete it using the shortcut menu. The Unnamed UCS cannot be deleted, nor can World be deleted.

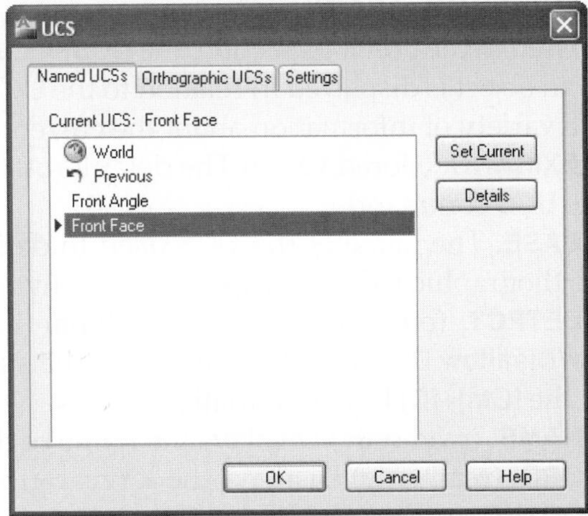

Figure 6-25.
The **UCS** dialog box allows you to rename, list, delete, and set current an existing UCS.

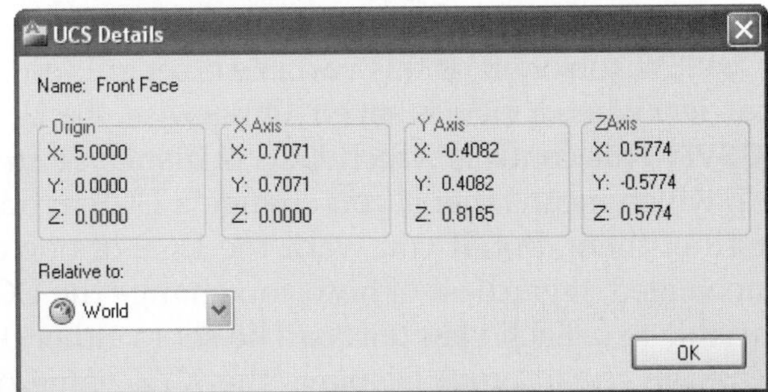

Figure 6-26.
The **UCS Details** dialog box displays the coordinate values of the selected UCS.

PROFESSIONAL TIP

You can also manage UCSs on the command line using the **UCS** command.

Setting an Automatic Plan Display

After changing the UCS, a plan view is often needed to give you a better feel for the XYZ directions. While you should try to draw in a pictorial view when possible as you construct a 3D object, some constructions may be much easier in a plan view. AutoCAD can be set to automatically make your view of the drawing plan to the current UCS. This is especially useful if you will be changing the UCS often, but want to work in a plan view.

The **UCSFOLLOW** system variable is used to automatically display a plan view of the current UCS. When it is set to 1, a plan view is automatically created in the current viewport when the UCS is changed. Viewports are discussed in Chapter 7. The default setting of **UCSFOLLOW** is 0 (off). After setting the variable to 1, a plan view will be automatically generated the next time the UCS is changed. The **UCSFOLLOW** variable generates the plan view only after the UCS is changed, not immediately after the variable is changed. However, if you select a different viewport, the previous viewport is set plan to the UCS if **UCSFOLLOW** has been set to 1 in that viewport. The **UCSFOLLOW** variable can be individually set for each viewport.

PROFESSIONAL TIP

To get the plan view displayed without changing the UCS, use the **PLAN** command, which is discussed in Chapter 4.

UCS Settings and Variables

As discussed in the previous section, the **UCSFOLLOW** system variable allows you to change how an object is displayed in relation to the UCS. There are also system variables that display a variety of information about the current UCS. These variables include:

- **UCSAXISANG.** (stored value) The default rotation angle for the **X**, **Y**, or **Z** option of the **UCS** command.
- **UCSBASE.** The name of the UCS used to define the origin and orientation of the orthographic UCS settings. It can be any named UCS.
- **UCSDETECT.** (on or off) Turns the dynamic UCS function on and off. The **Allow/Disallow Dynamic UCS** button on the status bar controls this variable, as does the [Ctrl]+[D] key combination.
- **UCSNAME.** (read only) Displays the name of the current UCS.
- **UCSORG.** (read only) Displays the XYZ origin value of the current UCS.
- **UCSORTHO.** (on or off) If set to 1 (on), the related orthographic UCS setting is automatically restored when an orthographic view is restored. If turned off, the current UCS is retained when an orthographic view is restored. Depending on your modeling preferences, you may wish to set this variable to 0.
- **UCSVIEW.** (on or off) If this variable is set to 1 (on), the current UCS is saved with the view when a view is saved. Otherwise, the UCS is not saved with the view.
- **UCSVP.** This controls which UCS is displayed in viewports. The default value is 1, which means that if you change a UCS in one viewport, the UCS changes in all of them. But, if you want the UCS of one or more viewports to remain unchanged, regardless of how you change the UCS in other viewports, set this variable to 0. Each viewport can be set to either 0 or 1.
- **UCSXDIR.** (read only) Displays the XYZ value of the X axis direction of the current UCS.
- **UCSYDIR.** (read only) Displays the XYZ value of the Y axis direction of the current UCS.

UCS options and variables can also be managed in the **Settings** tab of the **UCS** dialog box. See Figure 6-11. The settings in this tab are:

- **Save UCS with viewport.** If checked, the current UCS settings are saved with the viewport and the **UCSVP** system variable is set to 1. This variable can be set for each viewport in the drawing. Viewports in which this setting is turned off, or unchecked, will always display the UCS settings of the current active viewport.
- **Update view to Plan when UCS is changed.** This setting controls the **UCSFOLLOW** system variable. When checked, the variable is set to 1. When unchecked, the variable is set to 0.

Chapter Test

Answer the following questions. Write your answers on a separate sheet of paper or complete the electronic chapter test on the student website.
www.g-wlearning.com/CAD

1. Explain *spherical coordinate entry.*
2. Explain *cylindrical coordinate entry.*
3. A new point is to be drawn 4.5″ from the last point. It is to be located at a 63° angle in the XY plane, and at a 35° angle from the XY plane. Write the proper spherical coordinate notation.
4. Write the proper cylindrical coordinate notation for locating a point 4.5″ in the horizontal direction from the origin, 3.6″ along the Z axis, and at a 63° angle in the XY plane.
5. Name the command that is used to draw 3D polylines.
6. Why is the command in question 5 needed?
7. Which command is used to change a 3D polyline into a B-spline curve?
8. How does the **SPLFRAME** system variable affect the B-spline curve created with the command in question 7?
9. What is the *WCS?*
10. What is a *user coordinate system (UCS)?*
11. What effect does the **Origin** option of the **UCSICON** command have on the UCS icon display?
12. Describe how to rotate the UCS so that the Z axis is tilted 30° toward the WCS X axis.
13. How do you return to the WCS from any UCS?
14. Which command controls the display of the user coordinate system icon?
15. What is a *dynamic UCS* and how is one activated?
16. What is the function of the **3 Point** option of the **UCS** command?
17. How do you automatically create a display that is plan to a new UCS?
18. What do you do so that the UCS icon is displayed at the origin of the current user coordinate system?
19. How do you move the UCS along the current Z axis?
20. What is the function of the **Object** option of the **UCS** command?
21. The **Face** option of the **UCS** command can be used on which types of objects?
22. What is the function of the **Apply** option of the **UCS** command?
23. In which dialog box is the **Orthographic UCSs** tab located?
24. Which command displays the **UCS** dialog box?
25. What appears in the **Named UCSs** tab of the **UCS** dialog box if the current UCS has not been saved?

Drawing Problems

For Problems 1–4, draw each object using solid primitives and Boolean commands to create composite solids. Measure the objects directly to obtain the necessary dimensions. Apply geometric constraints to the base 2D drawing as appropriate. Plot the drawings at a 3:1 scale using display methods specified by your instructor. Save the drawings as **P06_***(problem number).*

1.

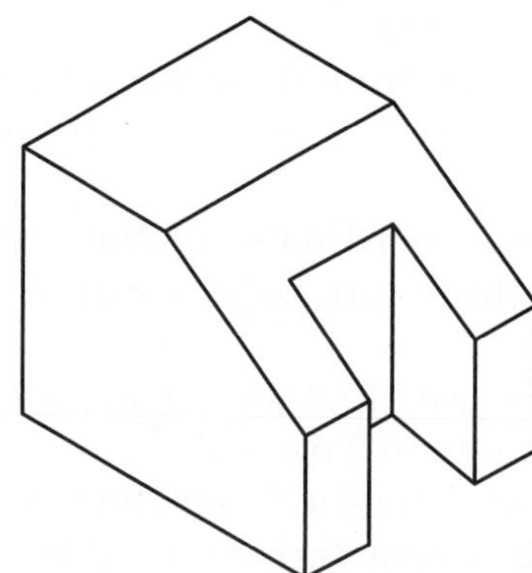

2.

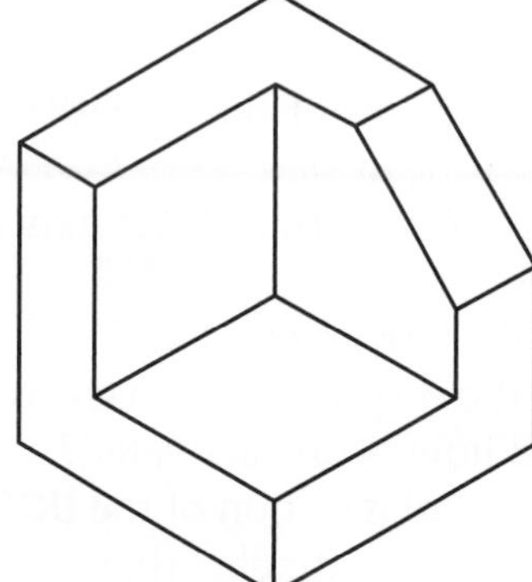

3.

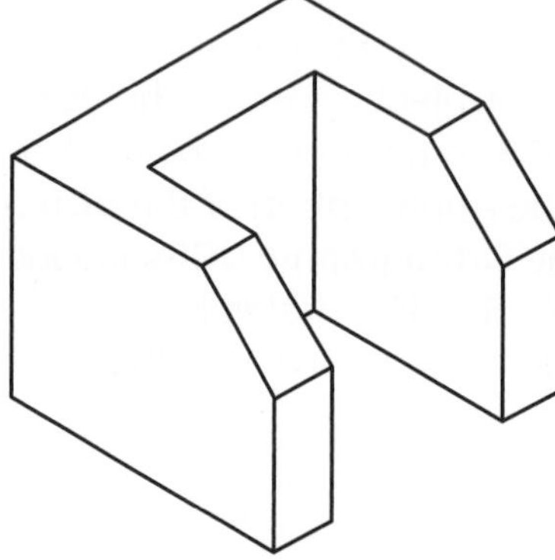

4.

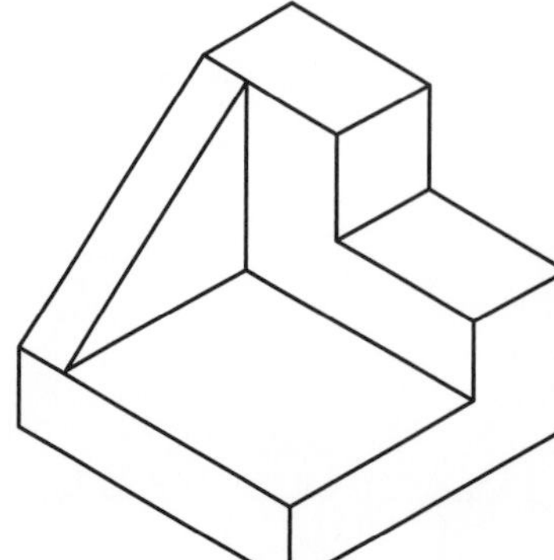

5. Create the mounting bracket shown below. Save the file as P06_05.

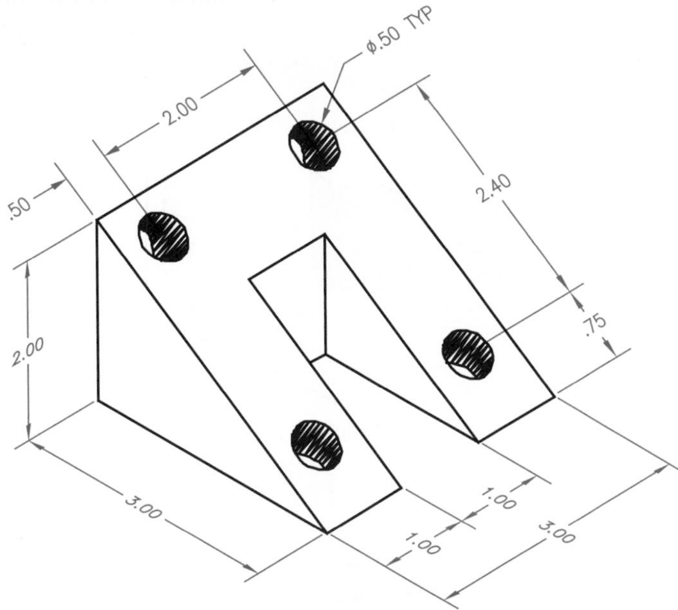

6. Create the computer speaker as shown below. The large-radius, arched surface is created by drawing a three-point arc. The second point of the arc passes through the point located by the .26 and 2.30 dimensions. Save the file as P06_06.

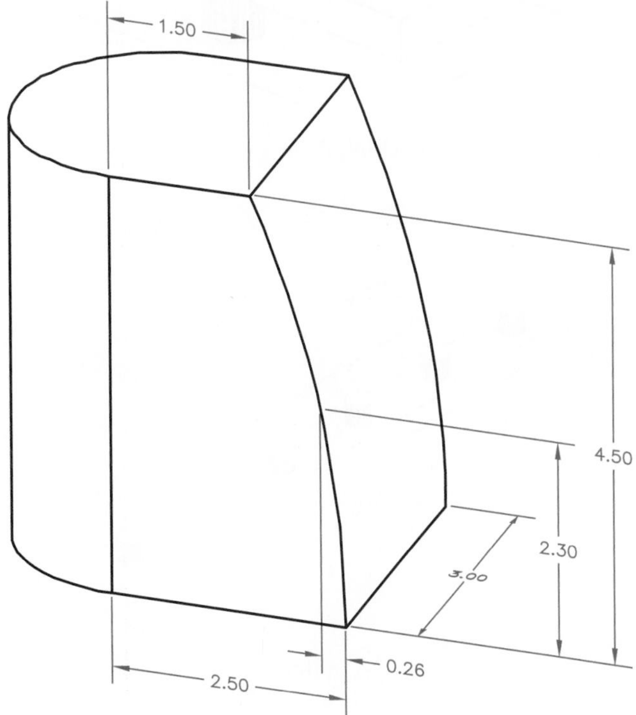

Drawing Problems - Chapter 6

For Problems 7–9, draw each object using solid primitives and Boolean commands to create composite solids. Use the dimensions provided. Save the drawings as **P06_***(problem number).*

7.

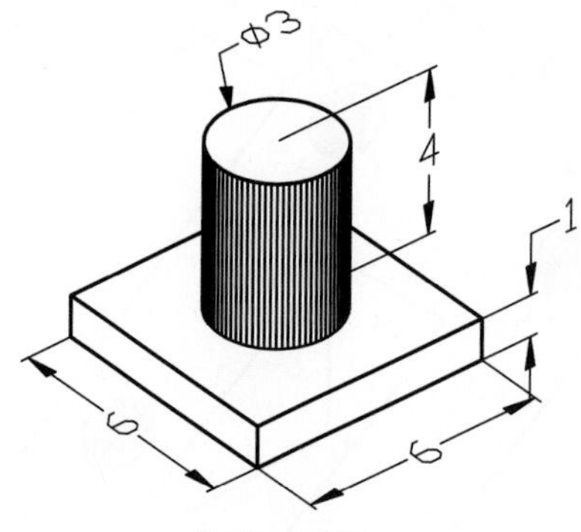

Pedestal #1

8.

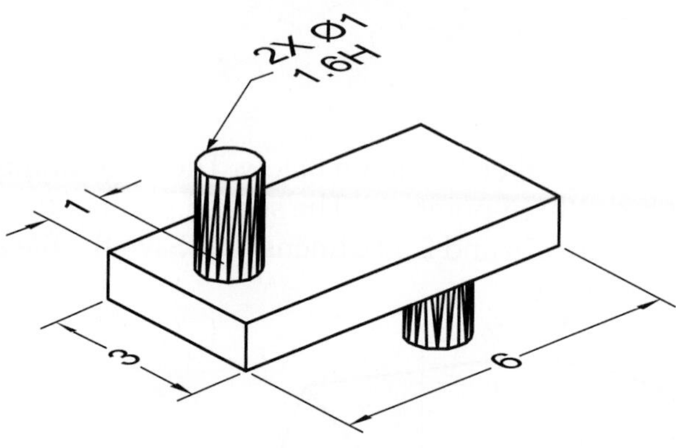

Locking Plate

9.

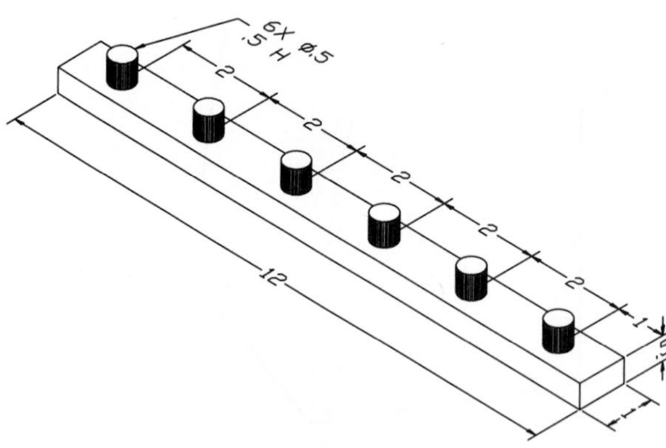

Pin Bar

Drawing Problems - Chapter 6

10. Draw the ∅8″ pedestal shown. It is .5″ thick. The four feet are centered on a ∅7″ circle and are .5″ high. Save the drawing as P06_10.

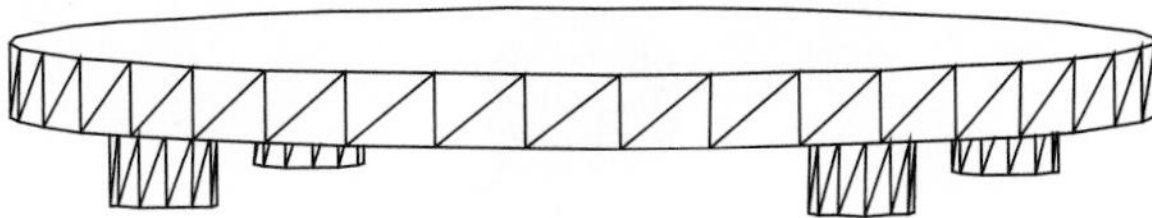

Pedestal #2

11. Four legs (cones), each 3″ high with a ∅1″ base, support this ∅10″ globe. Each leg tilts at an angle of 15° from vertical. The base is ∅12″ and .5″ thick. The bottom surface of the base is 8″ below the center of the globe. Save the drawing as P06_11.

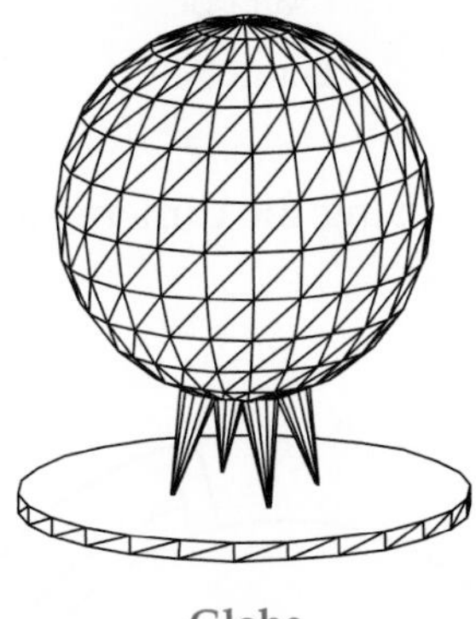

Globe

12. The table legs (A) are 2″ square and 17″ tall. They are 2″ in from each edge. The tabletop (B) is 24″ × 36″ × 1″. Save the drawing as P06_12.

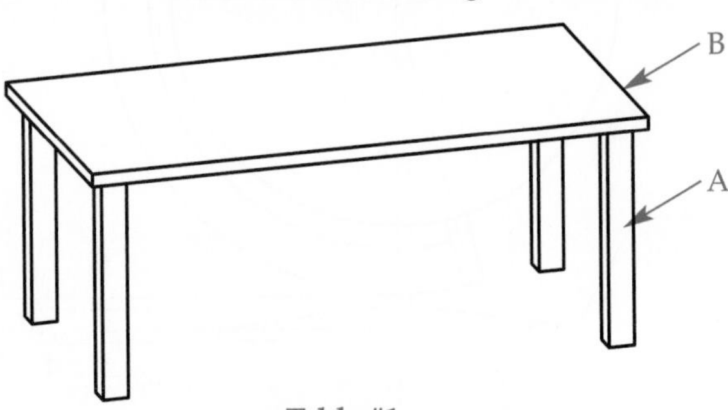

Table #1

13. The table legs (A) for the large table are ∅2″ and 17″ tall. The tabletop (B) is 24″ × 36″ × 1″. The table legs (C) for the small table are ∅2″ and 11″ tall. The tabletop (D) is 24″ × 14″ × 1″. All legs are 1″ in from the edges of the table. Save the drawing as P06_13.

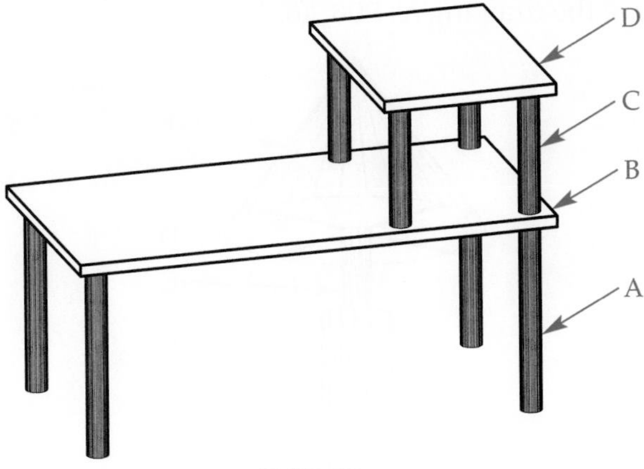

Table #2

Drawing Problems - Chapter 6

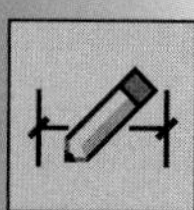

14. The spherical objects (A) are Ø4″. Object B is 6″ long and Ø1.5″. Save the drawing as P06_14.

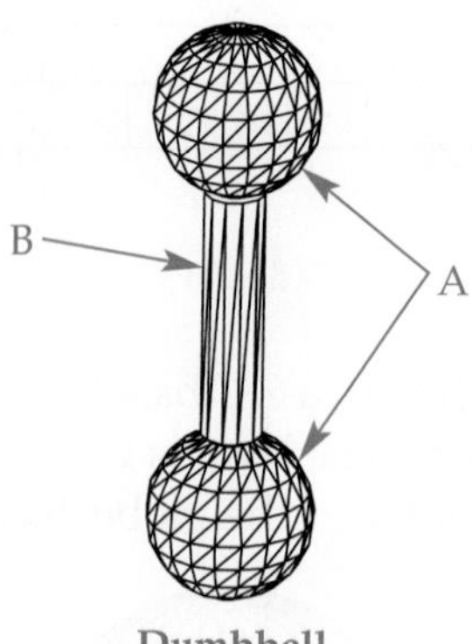

Dumbbell

15. Create the model of the globe using the dimensions shown. Save the file as P06_15.

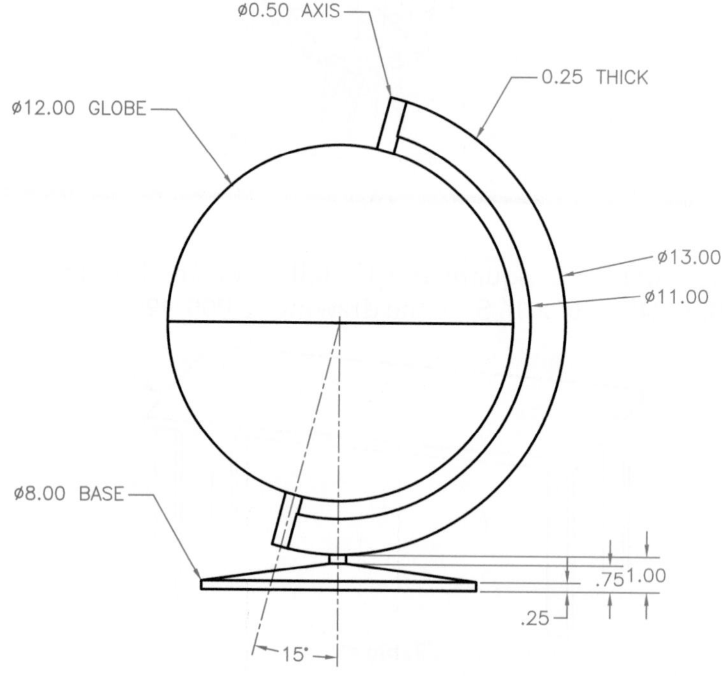

16. Object A is a Ø8″ cylinder that is 1″ tall. Object B is a Ø5″ cylinder that is 7″ tall. Object C is a Ø2″ cylinder that is 6″ tall. Object D is a .5″ × 8″ × .125″ box, and there are four pieces. The top surface of each piece is flush with the top surface of Object C. Object E is a Ø18″ cone that is 12″ tall. Create a smaller cone and hollow out Object E. Save the drawing as P06_16.

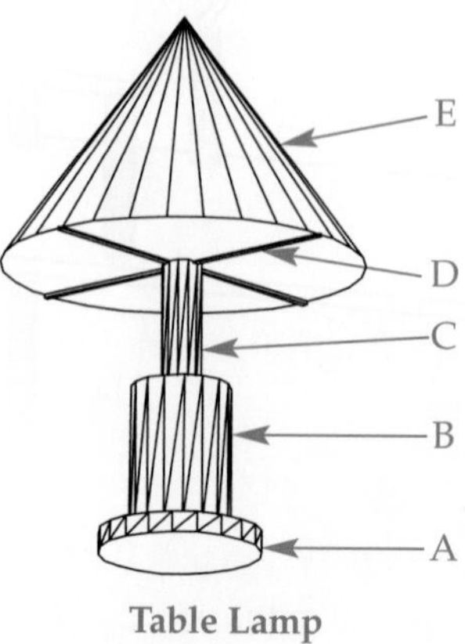

Table Lamp

17. Objects A and B are brick walls that are 5′ high. The walls are two bricks thick. Research the dimensions of standard brick and draw accordingly. Wall B is 7′ long and Wall A is 5′ long. Lamps are placed at each end of the walls. Object C is ∅2″ and 8″ tall. The center is offset from the end of the wall by a distance equal to the width of one brick. Object D is ∅10″. Save the drawing as P06_17.

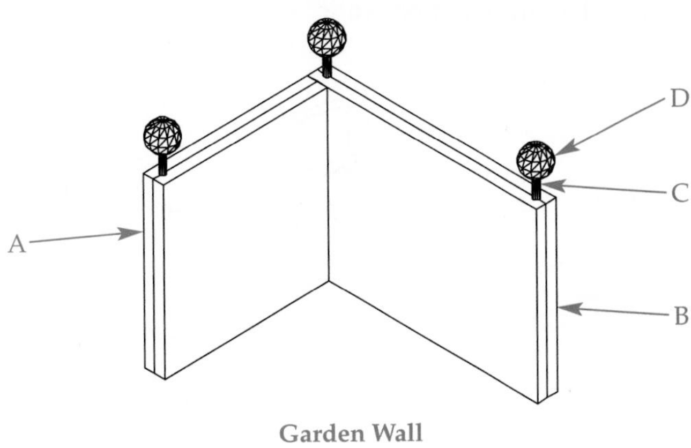

Garden Wall

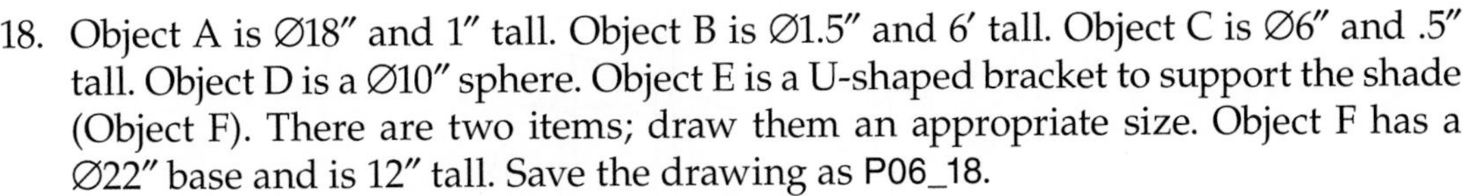

18. Object A is ∅18″ and 1″ tall. Object B is ∅1.5″ and 6′ tall. Object C is ∅6″ and .5″ tall. Object D is a ∅10″ sphere. Object E is a U-shaped bracket to support the shade (Object F). There are two items; draw them an appropriate size. Object F has a ∅22″ base and is 12″ tall. Save the drawing as P06_18.

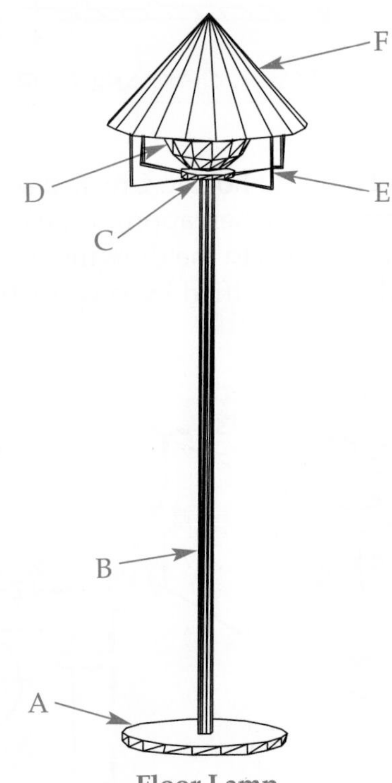

Floor Lamp

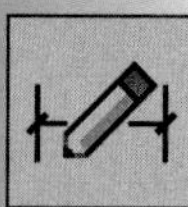

19. This is a concept sketch of a desk organizer. Create a solid model using the dimensions given. Use a dynamic UCS when appropriate or create and save new UCSs as needed. Inside dimensions of compartments can vary, but the thickness between compartments should be consistent. Do not add dimensions to the drawing. Plot your drawing on a B-size sheet of paper in a visual style specified by your instructor. Save the drawing as P06_19.

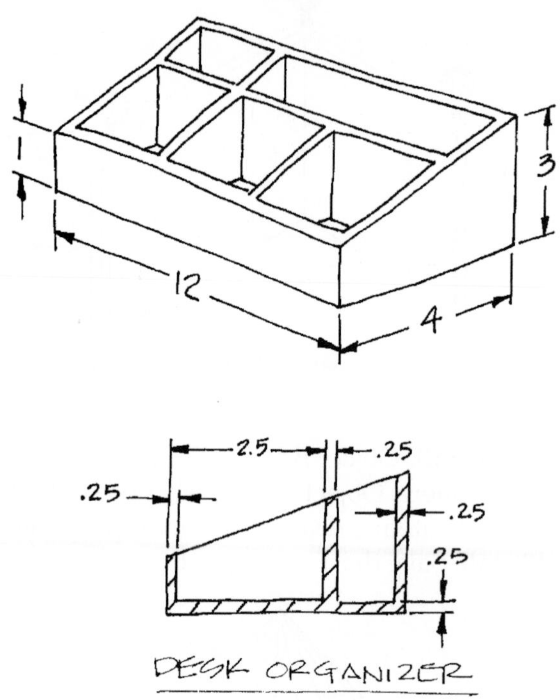

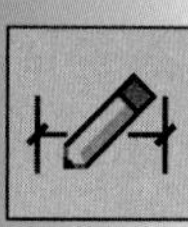

20. This is a concept sketch of a pencil holder. Create a solid model using the dimensions given. Use a dynamic UCS when appropriate or create and save new UCSs as needed. Do not add dimensions to the drawing. Plot your drawing on a B-size sheet of paper in a visual style specified by your instructor. Save the drawing as P06_20.

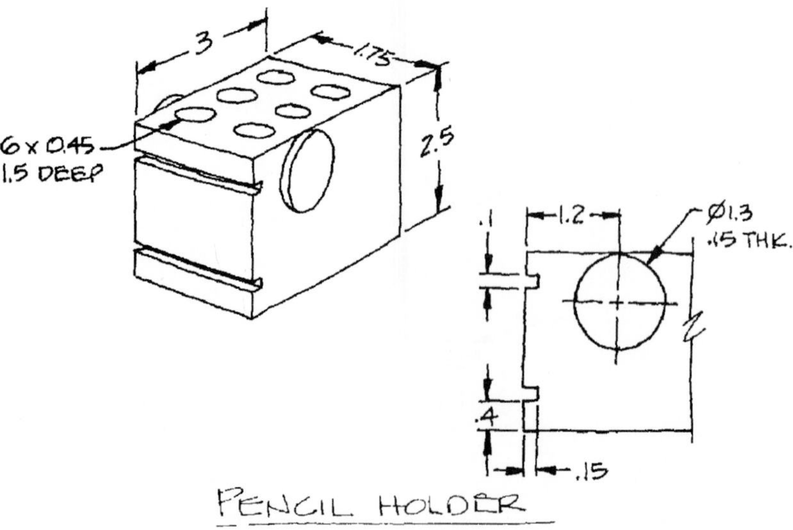

Drawing Problems - Chapter 6

21. This is an engineering sketch of a window blind mounting bracket. Create a solid model using the dimensions given. Use a dynamic UCS when appropriate or create and save new UCSs as needed. Do not add dimensions to the drawing. Create two plots, each of a different view, on B-size paper in the visual styles specified by your instructor. Save the drawing as P06_21.

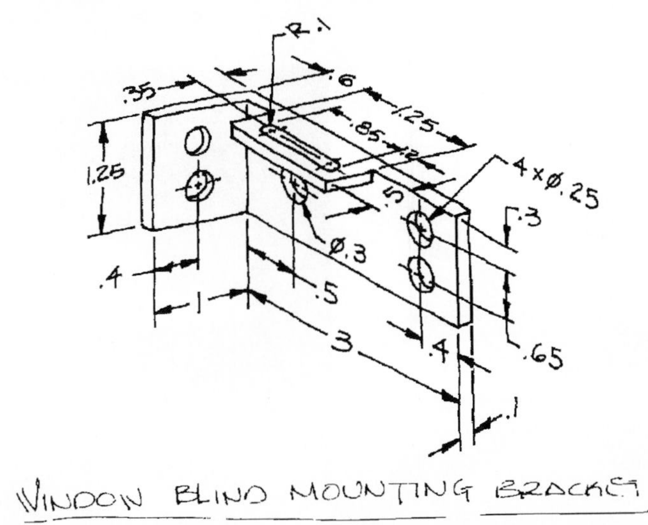

Drawing Problems - Chapter 6

Viewports can be used to increase your modeling efficiency. However, too many viewports can actually decrease your efficiency. Generally, the maximum number of viewports you should use is three or four.

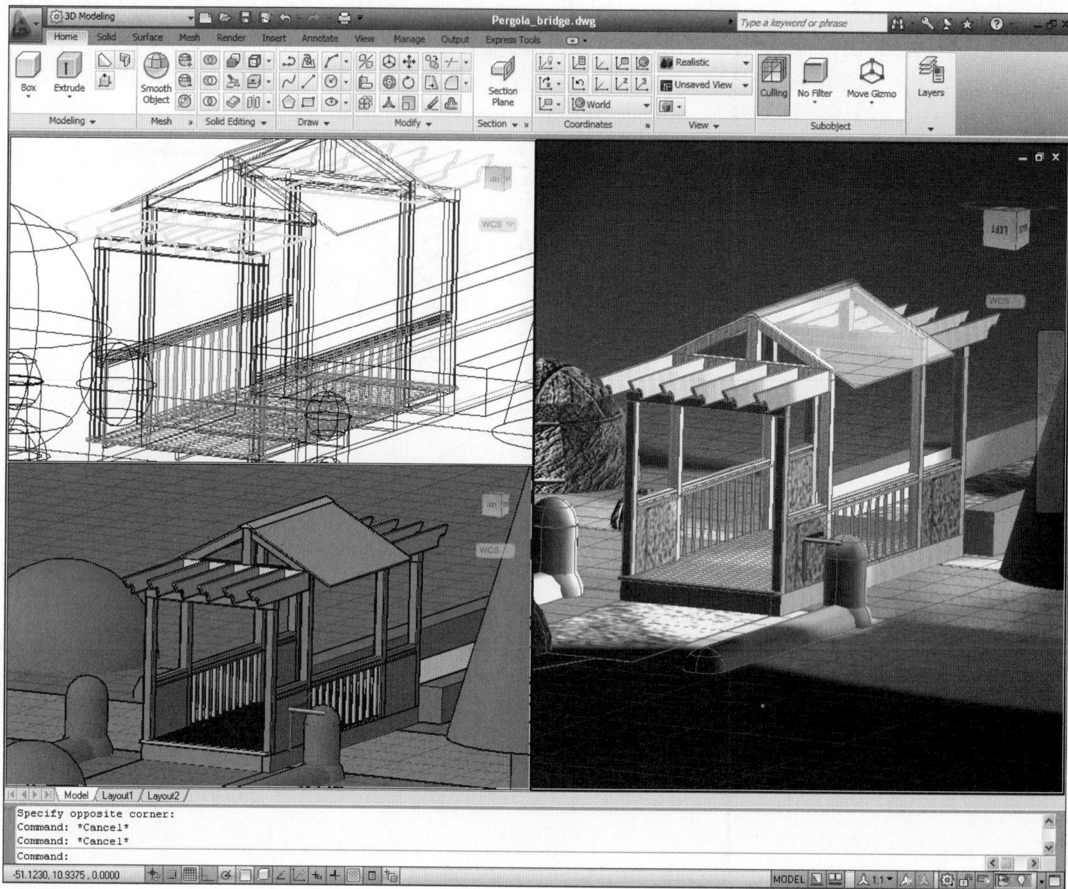

CHAPTER 7

Using Model Space Viewports

Learning Objectives

After completing this chapter, you will be able to:

- ✓ Describe the function of model space viewports.
- ✓ Create and save viewport configurations.
- ✓ Alter the current viewport configuration.
- ✓ Use multiple viewports to construct a drawing.

A variety of views can be displayed in a drawing at one time using model space viewports. This is especially useful when constructing 3D models. Using the **VPORTS** command, you can divide the drawing area into two or more smaller areas. These areas are called ***viewports.*** Each viewport can be configured to display a different 2D or 3D view of the model.

The ***active viewport*** is the viewport in which a command will be applied. Any viewport can be made active, but only one can be active at a time. As objects are added or edited, the results are shown in all viewports. A variety of viewport configurations can be saved and recalled as needed. This chapter discusses the use of viewports and shows how they can be used for 3D constructions.

Understanding Viewports

The AutoCAD drawing area can be divided into a maximum of 64 viewports. However, this is impractical due to the small size of each viewport. Usually, the maximum number viewports practical to display at one time is four. The number of viewports you need depends on the model you are drawing. Each viewport can show a different view of an object. This makes it easier to construct 3D objects.

NOTE

The **MAXACTVP** (maximum active viewports) system variable sets the number of viewports that can be used at one time. The initial value is 64, which is the highest setting.

There are two types of viewports used in AutoCAD. The type of viewport created depends on whether it is defined in model space or paper space. ***Model space*** is the space, or mode, where the model or drawing is constructed. ***Paper space,*** or layout space, is the space where a drawing is laid out to be plotted. Viewports created in model space are called ***tiled viewports.*** Viewports created in paper space are called ***floating viewports.***

Model space is active by default when you start AutoCAD. Model space viewports are created with the **VPORTS** command. These viewport configurations cannot be plotted because they are for display purposes only. If you plot from model space, the content of the active viewport is plotted. Tiled viewports are not AutoCAD objects. They are referred to as *tiled* because the edges of each viewport are placed side to side, as with floor tile, and they cannot overlap.

Floating (paper space) viewports are used to lay out the views of a drawing before plotting. They are described as *floating* because they can be moved around and overlapped. Paper space viewports are objects and they can be edited. These viewports can be thought of as "windows" cut into a sheet of paper to "see into" model space. You can then insert, or *reference,* different scaled drawings (views) into these windows. For example, architectural details or sections and details of complex mechanical parts may be referenced. Detailed discussions of paper space viewports are provided in *AutoCAD and Its Applications—Basics.*

The **VPORTS** command can be used to create viewports in a paper space layout. The process is very similar to that used to create model space viewports, which is discussed next. You can also use the **MVIEW** command to create paper space viewports.

Creating Viewports

Creating model space viewports allows you to work with multiple views of the same model. To work on a different view, simply pick with your pointing device in the viewport in which you wish to work. The picked viewport becomes active. Using viewports is a good way to construct 3D models because all views are updated as you draw. However, viewports are also useful when creating 2D drawings.

The project on which you are working determines the number of viewports needed. Keep in mind that the more viewports you display on your screen, the smaller the view in each viewport. Small viewports may not be useful to you. Four different viewport configurations are shown in Figure 7-1. As you can see, when 16 viewports are displayed, the viewports are very small. Normally, two to four viewports are used.

Quick Viewport Layout

VPORTS
Ribbon
Home
> View
Multiple Viewports

A four-view layout can be instantly displayed by picking the **Multiple viewports** button on the **View** panel in the **Home** tab of the ribbon. See Figure 7-2. This selection automatically creates a top, front, side, and pictorial view based on the current UCS. See Figure 7-3. This layout may be a good one to start from when working on models that require two or more views for construction purposes.

VPORTS

The screen display can be quickly returned to a single view by picking **Single Viewport** on the **View** panel in the **Home** tab of the ribbon. Keep in mind, the resulting view will be the viewport that is current, which is the one surrounded by a highlighted frame.

Figure 7-1.
A—Two vertical viewports. B—Two horizontal viewports. C—Three viewports, with the largest viewport positioned at the right. D—Sixteen viewports.

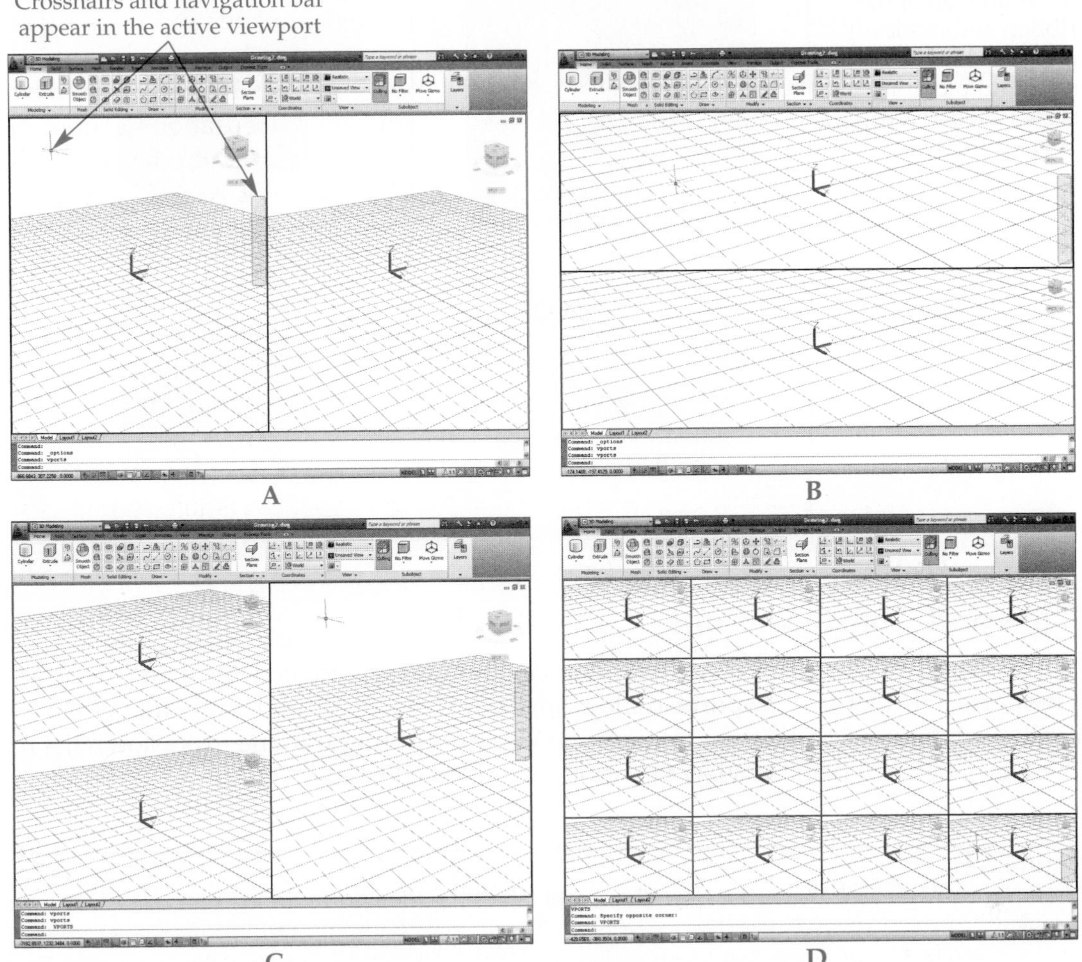

Figure 7-2.
A four-viewport layout can be instantly displayed by picking the **Multiple Viewports** button in the **View** panel on the **Home** tab of the ribbon.

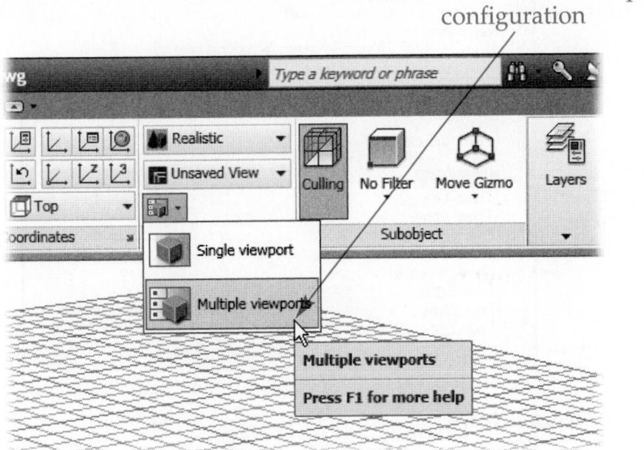

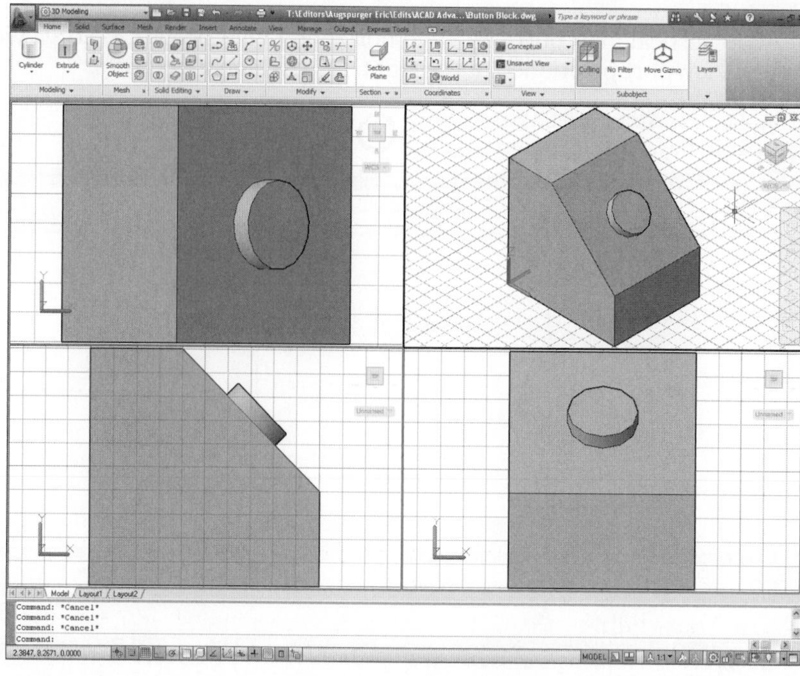

Figure 7-3.
The **Multiple Viewports** button in the **View** panel on the **Home** tab of the ribbon automatically creates a top, front, side, and pictorial view based on the current UCS.

Viewport Configurations

VPORTS

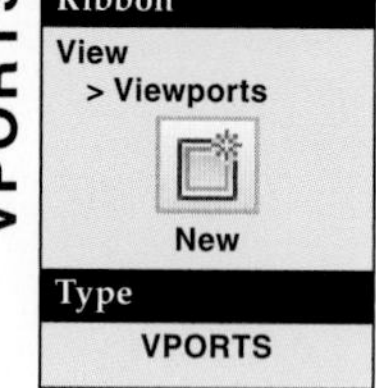

A layout of viewport configurations can be quickly created by using the **New Viewports** tab of the **Viewports** dialog box, which is displayed with the **VPORTS** command, **Figure 7-4.** You can also use the **Viewport Configurations** drop-down list in the **Viewports** panel on the **View** tab of the ribbon. See **Figure 7-5.**

Figure 7-4.
Viewports are created using the **New Viewports** tab of the **Viewports** dialog box.

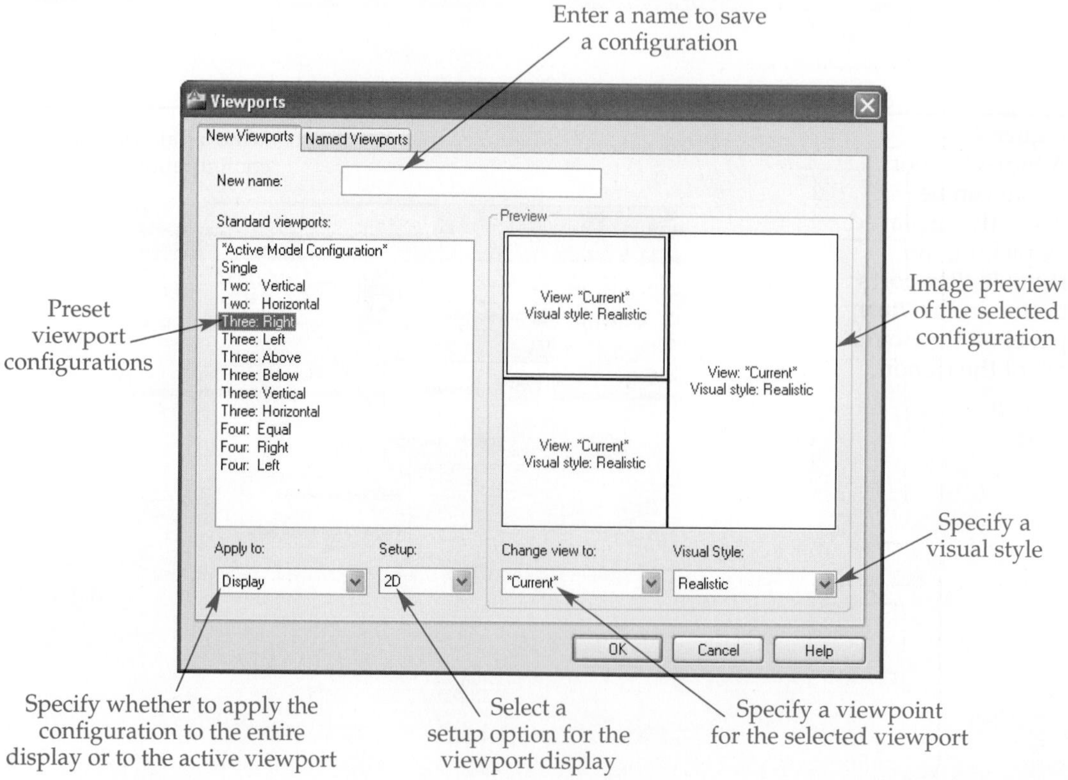

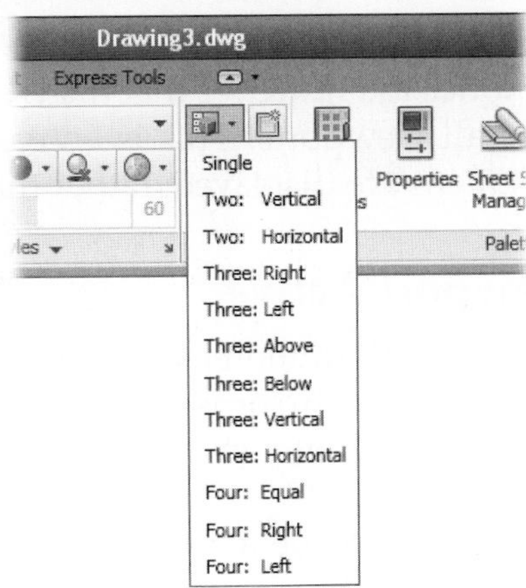

Figure 7-5.
The **Viewport Configurations** drop-down list in the **Viewports** panel on the **View** tab of the ribbon offers a quick way to recall a preset viewport configuration.

There are 12 preset viewport configurations from which to choose in the **New Viewports** tab, including six different options for three-viewport configurations. See Figure 7-6. When you pick the name of a configuration in the **Standard viewports:** list, the viewport arrangement is displayed in the **Preview** area. After you have made a selection, you can save the configuration by entering a name in the **New name:** text box and then picking **OK** to close the dialog box. When the **Viewports** dialog box closes, the configuration is displayed on screen.

Figure 7-6.
Twelve preset tiled viewport configurations are provided in the **Viewports** dialog box.

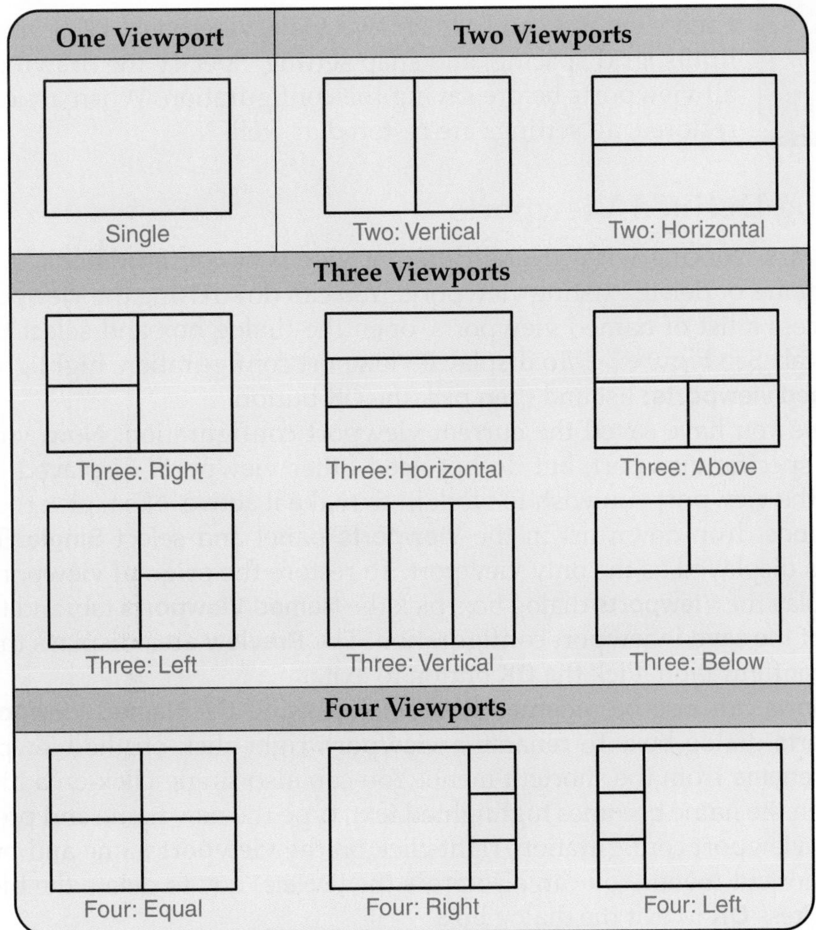

PROFESSIONAL TIP

Notice in Figure 7-1 that the UCS icon and view cube are displayed in all viewports. This is an easy way to tell that several separate screens are displayed, rather than different views of the drawing.

Making a Viewport Active

After a viewport configuration has been created, a thick line surrounds the active viewport. When the screen cursor is moved inside of the active viewport, it appears as crosshairs. When moved into an inactive viewport, the standard Windows cursor appears.

Any viewport can be made active by moving the cursor into the desired viewport and pressing the pick button. You can also press the [Ctrl]+[R] key combination to switch viewports, or use the **CVPORT** (current viewport) system variable. Only one viewport can be active at a time.

Command: **CVPORT**↵
Enter new value for CVPORT <*current*>: **3**↵

The current value given is the ID number of the active viewport. The ID number is automatically assigned by AutoCAD, starting with 2. To change viewports with the **CVPORT** system variable, simply enter a different ID number. This technique may be used in custom programming for AutoCAD. Using the **CVPORT** system variable is also a good way to determine the ID number of a viewport. The number 1 is not a valid viewport ID number.

PROFESSIONAL TIP

Each viewport can have its own view, viewpoint, UCS, zoom scale, limits, grid spacing, and snap setting. Specify the drawing aids in all viewports before saving the configuration. When a viewport is restored, all settings are restored as well.

Managing Defined Viewports

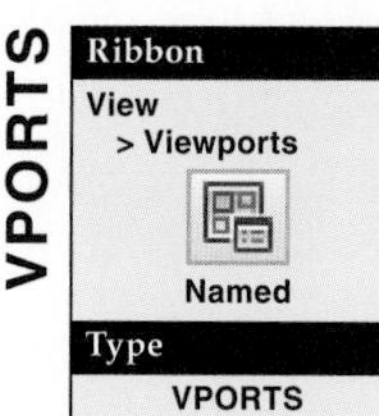

If you are working with several different viewport configurations, it is easy to restore, rename, or delete existing viewports. You can do so using the **Viewports** dialog box. To access a list of named viewports, open the dialog box and select the **Named Viewports** tab. See Figure 7-7. To display a viewport configuration, highlight its name in the **Named viewports:** list and then pick the **OK** button.

Assume you have saved the current viewport configuration. Now, you want to work in a specific viewport, but do not need other viewports displayed on screen. First, pick the viewport you wish to work in to make it active. Next, pick the **Viewport Configurations** drop-down list in the **Viewports** panel and select **Single**. The active viewport is displayed as the only viewport. To restore the original viewport configuration, display the **Viewports** dialog box, pick the **Named Viewports** tab, and then select the name of the saved viewport configuration. The **Preview** area displays the selected viewport configuration. Pick the **OK** button to exit.

Viewports can also be renamed and deleted using the **Named Viewports** tab of the **Viewports** dialog box. To rename a viewport, right-click on the viewport name and pick **Rename** from the shortcut menu. You can also single-click on a highlighted name. When the name becomes highlighted text, type the new name and press [Enter]. To delete a viewport configuration, right-click on the viewport name and pick **Delete** from the shortcut menu. You can also press the [Delete] key to delete the highlighted viewport. Press **OK** to exit the dialog box.

Figure 7-7.
The **Named Viewports** tab of the **Viewport** dialog box lists all named viewports and displays the selected configuration in the **Preview** area.

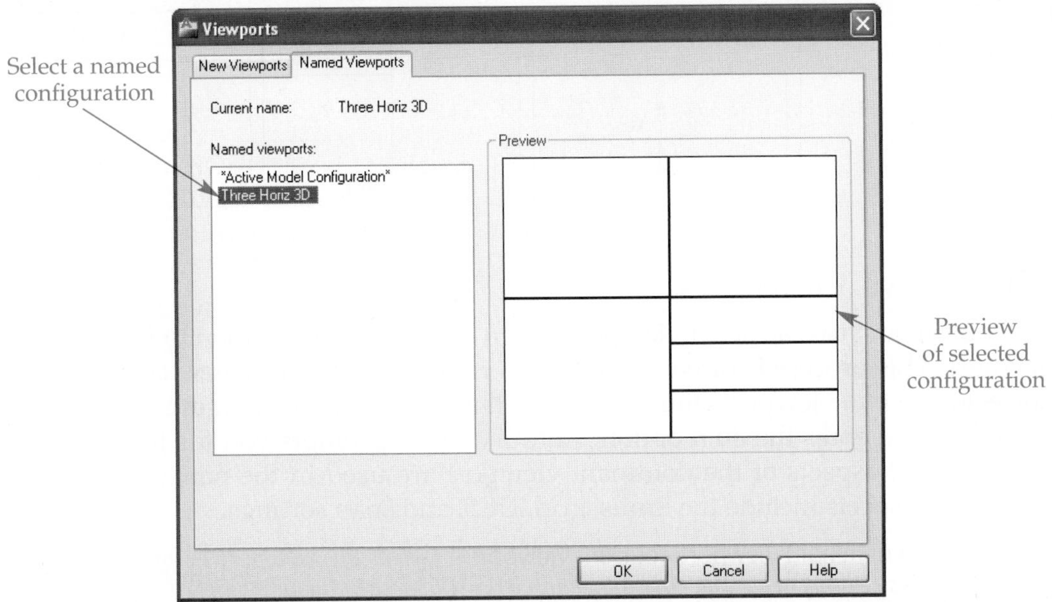

PROFESSIONAL TIP

The **-VPORTS** command can be used to manage viewports on the command line. This may be required for some LISP programs where the dialog box cannot be used. AutoLISP is covered in Chapters 27 and 28, which are provided on the student website.
www.g-wlearning.com/CAD

Using the Viewports Panel

The **Viewports** panel in the **View** tab of the ribbon is shown in Figure 7-8. It is used with both model space and paper space viewports. The **Named** and **New** buttons on the panel display the **Viewports** dialog box. The **Join** button is used to combine two viewports into a single viewport, as described in the next section. The **Viewport Configurations** drop-down list contains the twelve preset viewport configurations. Selecting a configuration in the list sets it current. The remaining buttons apply to paper space viewports in a layout. A complete discussion of paper space viewports is given in *AutoCAD and Its Applications—Basics.*

Figure 7-8.
The **Viewports** panel in the **View** tab of the ribbon.

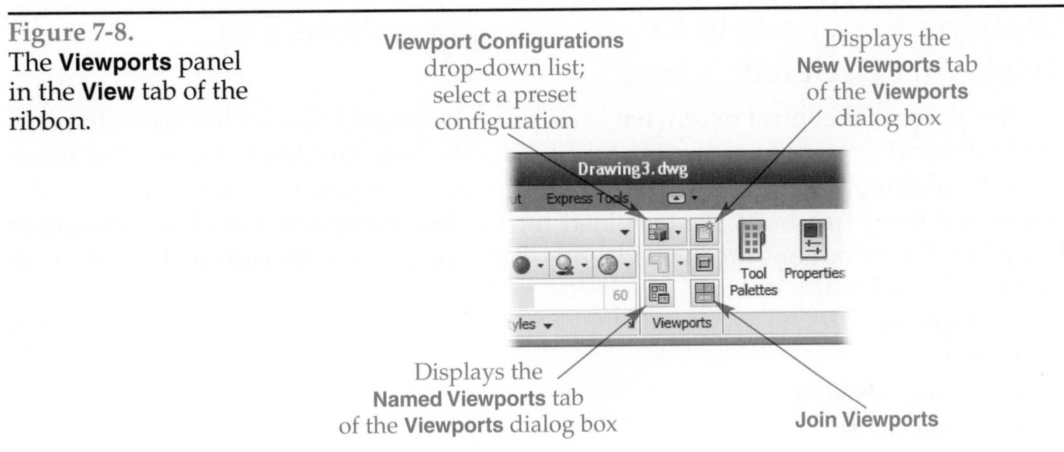

Figure 7-9.
Two viewports can be joined if they will form a rectangle. If the two viewports will not form a rectangle, they cannot be joined.

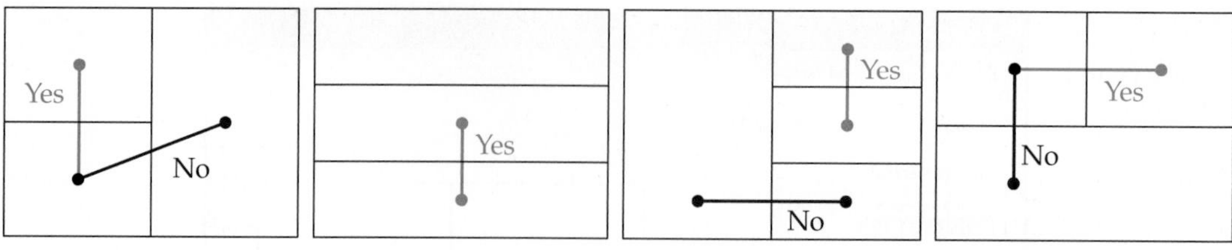

Joining Two Viewports

You can join two adjacent viewports in an existing configuration to form a single viewport. This process is often quicker than creating an entirely new configuration. However, the two viewports must form a rectangle when joined, Figure 7-9.

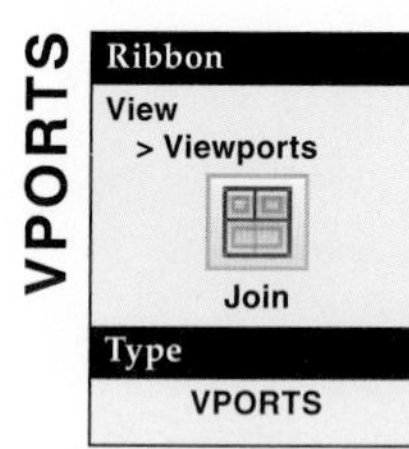

When you enter the **Join** option, AutoCAD first prompts you for the ***dominant viewport.*** All aspects of the dominant viewport are used in the new (joined) viewport. These aspects include the limits, grid, UCS, and snap settings.

Select dominant viewport <current viewport>: *(select the viewport by picking in it or press [Enter] to set the current viewport as the dominant viewport)*
Select viewport to join: *(select the other viewport)*

The two selected viewports are joined into a single viewport. If you select two viewports that do not form a rectangle, AutoCAD returns the message:

The selected viewports do not form a rectangle.

PROFESSIONAL TIP

Create only the number of viewports and viewport configurations needed to construct your drawing. Using too many viewports reduces the size of the image in each viewport and may confuse you. Also, it helps to zoom each view so that the objects fill the viewport.

Exercise 7-1

Complete the exercise on the student website.
www.g-wlearning.com/CAD

Applying Viewports to Existing Configurations and Displaying Different Views

You have total control over what is displayed in model space viewports. In addition to displaying various viewport configurations, you can divide an existing viewport into additional viewports or assign a different viewpoint to each viewport. The options for these functions are provided in the **New Viewports** tab of the **Viewports** dialog box. These options are located along the bottom of the **Viewports** dialog box, as shown in Figure 7-4:

- **Apply to.**
- **Setup.**
- **Change view to.**
- **Visual style.**

Apply to

When a preset viewport configuration is selected from the **Standard viewports:** list, it can be applied to either the entire display or the current viewport. The previous examples have shown how to create viewports that replace the entire display. Applying a configuration to the active viewport rather than the entire display can be useful when you need to display additional viewports.

For example, first create a configuration of three viewports using the Three: Right configuration option. Then, with the right (large) viewport active, open the **Viewports** dialog box again. Notice that the drop-down list under **Apply to:** is grayed out. Now, pick one of the standard configurations. This enables the **Apply to:** drop-down list. The default option is Display, which means the selected viewport configuration will replace the current display. Pick the drop-down list arrow to reveal the second option, Current Viewport. Pick this option and then pick the **OK** button. Notice that the selected viewport configuration has been applied to only the active (right) viewport. See **Figure 7-10.**

Setup

Viewports can be set up to display views in 2D or 3D. The 2D and 3D options are provided in the **Setup:** drop-down list. Displaying different views while working on a drawing allows you to see the results of your work on each view, since changes are reflected in each viewport as you draw. The selected viewport **Setup:** option controls the types of views available in the **Change view to:** drop-down list.

Change view to

The views that can be displayed in a selected viewport are listed in the **Change view to:** drop-down list. If the **Setup:** drop-down list is set to 2D, the views available to be displayed are limited to the current view and any named views. If 3D is active, the options include all of the standard orthographic and isometric views along with named views. When an orthographic or isometric view is selected for a viewport, the resulting orientation is shown in the **Preview** area. To assign a different viewpoint to a viewport, simply pick within a viewport in the **Preview** area to make it active and

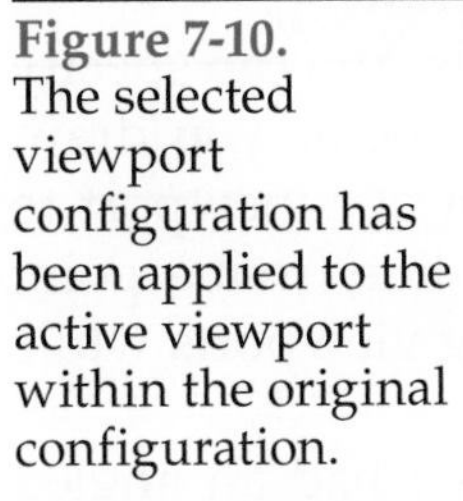
Figure 7-10.
The selected viewport configuration has been applied to the active viewport within the original configuration.

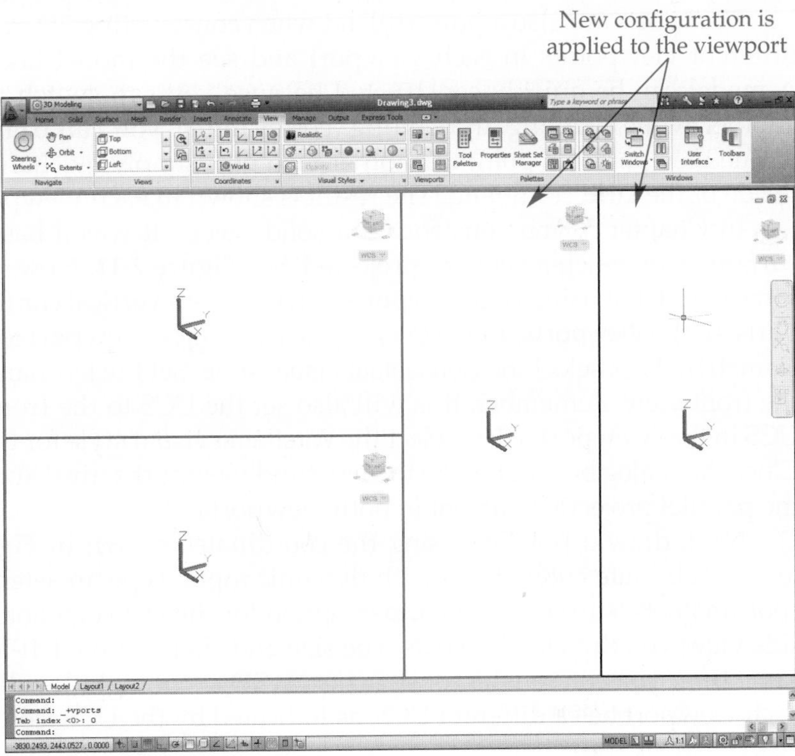

then pick a viewpoint from the **Change view to:** drop-down list. Important: if you set a viewport to one of the orthographic preset views, the UCS is also (by default) changed in that viewport to the corresponding preset.

Visual style

A visual style can be specified for a viewport. Pick within a viewport in the **Preview** area to make it active and then select a visual style from the **Visual Style:** drop-down list. All preset and saved visual styles are available in the drop-down list.

PROFESSIONAL TIP

Displaying saved viewport configurations can be automated by using custom menus. Custom menus are easy to create. If a standard naming convention is used, these named configurations can be saved with template drawings. This creates a consistent platform that all students or employees can use.

Exercise 7-2

Complete the exercise on the student website.
www.g-wlearning.com/CAD

Drawing in Multiple Viewports

When used with 2D drawings, viewports allow you to display a view of the entire drawing, plus views showing portions of the drawing. This is similar to using the **VIEW** command, except you can have several views on screen at once. You can also adjust the zoom magnification in each viewport to suit different areas of the drawing.

Viewports are also a powerful aid when constructing 3D models. You can specify different viewpoints in each viewport and see the model take shape as you draw. A model can be quickly constructed because you can switch from one viewport to another while drawing and editing. For example, you can draw a line from a point in one viewport to a point in another viewport simply by changing viewports while inside of the **LINE** command. The result is shown in each viewport.

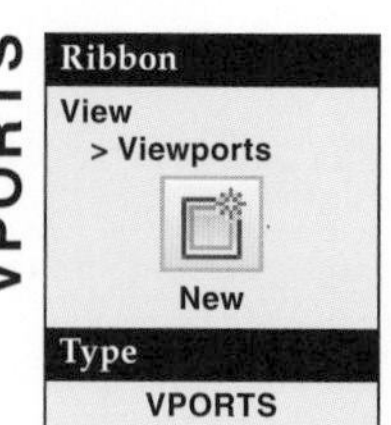

In Chapter 5, you constructed a solid object. It was a base that had an angled surface from which a cylinder projected. See Figure 7-11. Now, you will construct the same object, but using two viewports. First, create a vertical configuration of two viewports. In the **Viewports** dialog box, set the right-hand viewport to display the southeast isometric. Also, select the Conceptual visual style. Set the left-hand viewport to display the front view. Remember, this will also set the UCS to the front preset orthographic UCS in that viewport. Also, select the Wireframe visual style for the left-hand viewport. Close the dialog box and make the left-hand viewport active. You may also want to set the parallel projection current in both viewports.

Next, draw a polyline using the coordinates shown in Figure 7-12. Remember, to enter absolute coordinates with dynamic input, type an asterisk (*) before the first coordinate. Be sure to use the **Close** option for the last segment. As you construct the side view, you can clearly see its true size and shape in the left-hand viewport. At the same time, you can see the construction in 3D in the right-hand viewport. Notice that each viewport has a different UCS, as indicated by the UCS icon.

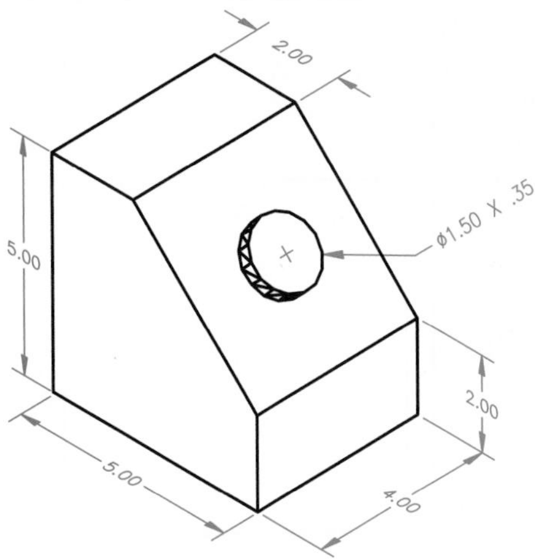

Figure 7-11.
You will construct the object from Chapter 6 using multiple viewports.

The next step is to extrude the shape to create the base. The **EXTRUDE** command is used to do so, as was the case in Chapter 6. With the left-hand viewport current, enter the **EXTRUDE** command, pick the polyline, and enter an extrusion height of –4 units. The front face of the object is now complete, Figure 7-13.

Now, the cylinder needs to be created on the angled face. First, split the left-hand viewport into two horizontal viewports (top and bottom) using the **New Viewports** tab of the **Viewports** dialog box. Set both of the new viewports to display the current view. Pick the **OK** button to close the dialog box. Then, make the upper-left viewport current and set it up to always display a plan view of the current UCS by setting the **UCSFOLLOW** system variable to 1.

Next, create a new UCS on the angled face. Use the **3 Point** option of the **UCS** command, which is described in Chapter 6. The pick points are shown in Figure 7-13; pick them in the right-hand viewport. Notice how the view in the upper-left viewport automatically changes to a plan view of the new current UCS.

With the right-hand viewport current, turn on 3D object snap. Turn off dynamic UCS, if it is on. Then, enter the **CYLINDER** command. When locating the center of the cylinder base, move the pointer to the angled face until the center of face object snap is displayed and then pick the point. Next, enter a diameter of 1.5 units and a height of .35 units.

The object is now complete, Figure 7-14. Notice how the lower-left and right-hand viewports have different UCSs. Each viewport can have its own UCS. The view in the upper-left viewport is the plan view of the current UCS. If the lower-left viewport is

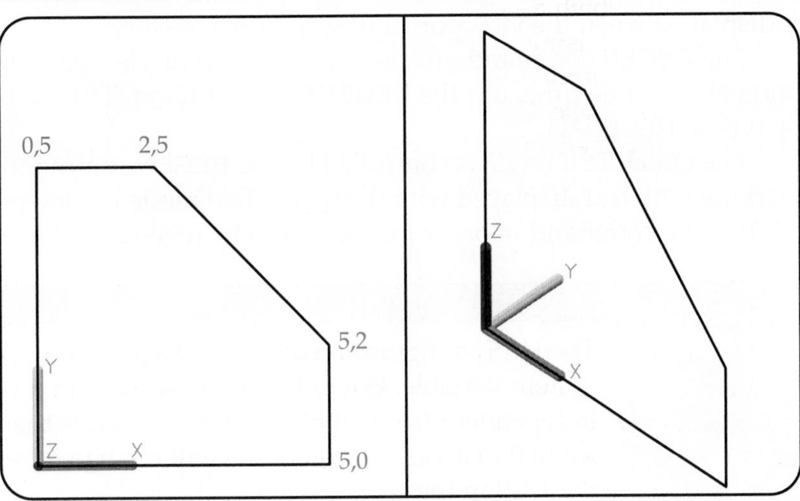

Figure 7-12.
The screen is divided into two viewports. A side view of the object appears in the left-hand viewport and a 3D view appears in the right-hand viewport. Notice the UCS icons.

Figure 7-13.
The base of the object is now complete. A new UCS will be created based on the pick points shown here. Alternately, a dynamic UCS can be used, as shown in Chapter 6.

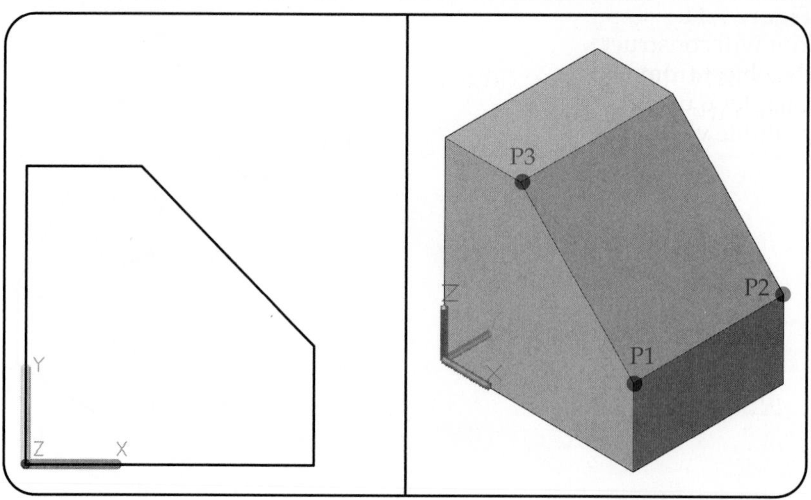

Figure 7-14.
The cylinder is drawn to complete the object. Notice the plan view in the upper-left viewport.

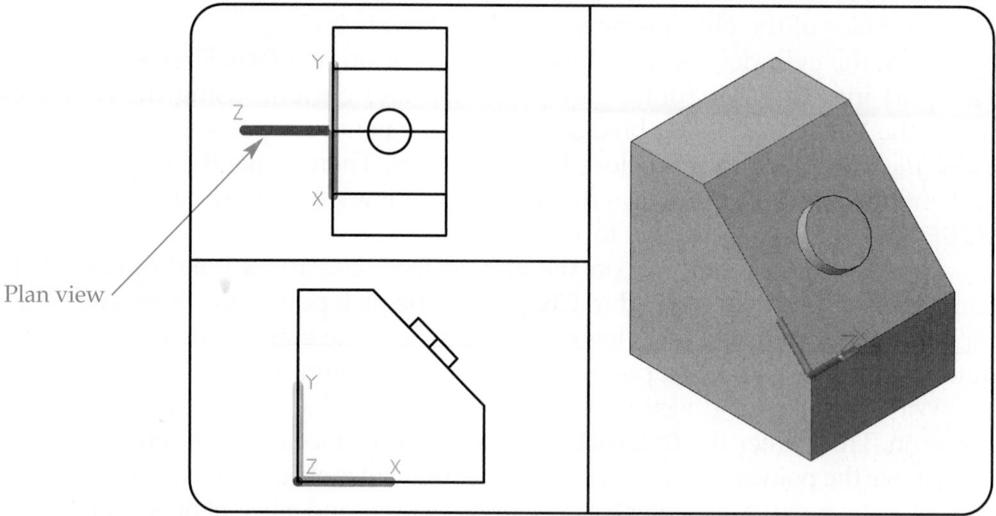

made active, the plan view will be of the UCS in that viewport. The UCS orientation in one viewport is not affected by a change to the UCS in another viewport unless the **UCSFOLLOW** system variable is set to 1 in a viewport. If a viewport arrangement is saved with several different UCS configurations, every named UCS remains intact and is displayed when the viewport configuration is restored.

The **REGEN** command affects only the current viewport. To regenerate all viewports at the same time, use the **REGENALL** command. This command can be entered by typing REGENALL.

The Quick Text mode is controlled by the **REGEN** command. Therefore, if you are working with text displayed with the Quick Text mode in viewports, be sure to use the **REGENALL** command in order for the text to be regenerated in all viewports.

NOTE

The UCS configuration in each viewport is controlled by the **UCSVP** system variable. When **UCSVP** is set to 1 in a viewport, the UCS is independent from all other UCSs, which is the default. If **UCSVP** is set to 0 in a viewport, its UCS will change to reflect any changes to the UCS in the current viewport.

Exercise 7-3

Complete the exercise on the student website.
www.g-wlearning.com/CAD

Chapter Test

Answer the following questions. Write your answers on a separate sheet of paper or complete the electronic chapter test on the student website.
www.g-wlearning.com/CAD

1. What is the purpose of *viewports?*
2. How do you name a configuration of viewports?
3. What is the purpose of saving a configuration of viewports?
4. Explain the difference between *tiled* and *floating* viewports.
5. Name the system variable controlling the maximum number of viewports that can be displayed at one time.
6. How can a named viewport configuration be redisplayed on screen?
7. How can a list of named viewport configurations be displayed?
8. What relationship must two viewports have before they can be joined?
9. What is the significance of the dominant viewport when two viewports are joined?
10. When creating a new viewport configuration, how can you set a visual style in a viewport?

Drawing Problems

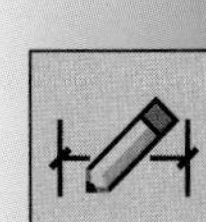

1. Construct seven template drawings, each with a preset viewport configuration. Use the following configurations and names. Save the templates under the same name as the viewport configuration.

Number of Viewports	Configuration	Name
2	Horizontal	TWO-H
2	Vertical	TWO-V
3	Right	THREE-R
3	Left	THREE-L
3	Above	THREE-A
3	Below	THREE-B
3	Vertical	THREE-V

2. Construct one of the problems from Chapter 4 using viewports. Use one of your template drawings from problem 7-1. Save the drawing as P07_02.

3. This is an orthographic drawing of a light fixture bracket. Create it as a solid model, using geometric constraints on the base 2D shapes whenever possible. Use solid primitives and Boolean commands as needed. Use the dimensions given. Similar holes have the same offset dimensions. Use multiple viewports to construct the

drawing. Begin with a four-viewport layout, then switch viewports as needed to work on specific areas of the model. Create new UCSs as needed. Display an appropriate pictorial view of the drawing in the upper-right viewport. Plot the 3D view of the drawing to scale on a B- or C-size sheet of paper. Save the drawing as P07_03.

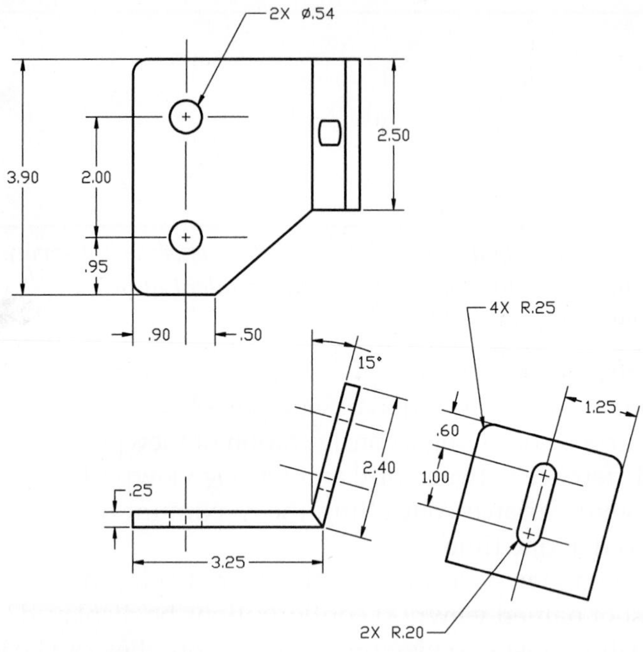

4. This is an orthographic drawing of an angle bracket. Create it as a solid model, using geometric constraints on the base 2D shapes whenever possible. Use solid primitives and Boolean commands as needed. Use the dimensions given. Similar holes have the same offset dimensions. Use multiple viewports to construct the drawing. Begin with a four-viewport layout, then switch viewports as needed to work on specific areas of the model. Create new UCSs as needed. Display an appropriate pictorial view of the drawing in the upper-right viewport. Plot the 3D view of the drawing to scale on a B- or C-size sheet of paper. Save the drawing as P07_04.

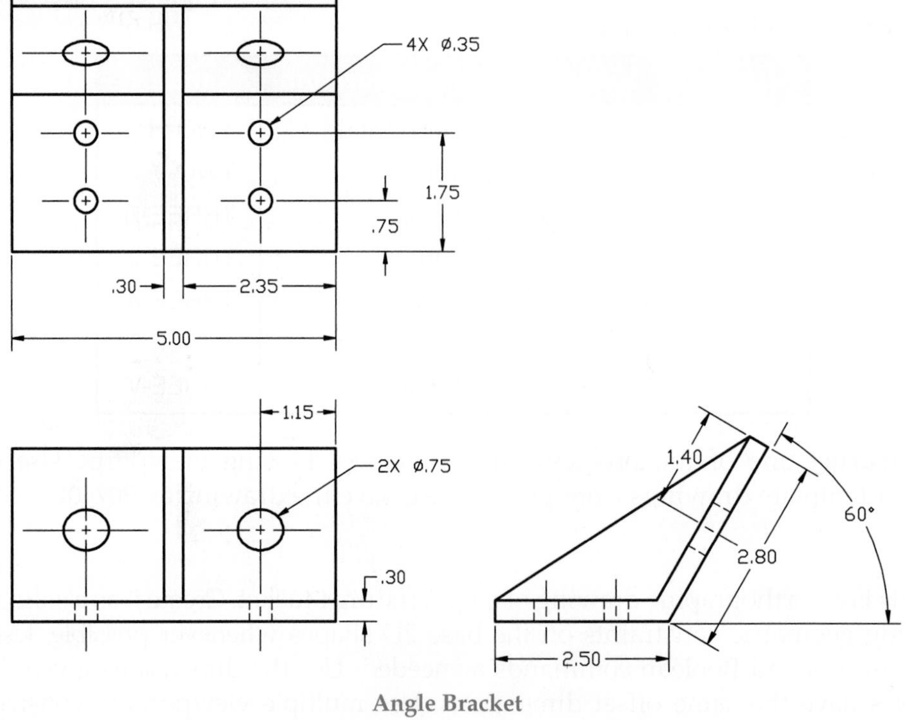

Angle Bracket

CHAPTER 8

Text and Dimensions in 3D

Learning Objectives

After completing this chapter, you will be able to:

✓ Create text with a thickness.
✓ Draw text that is plan to the current view.
✓ Dimension a 3D drawing.

Creating Text with Thickness

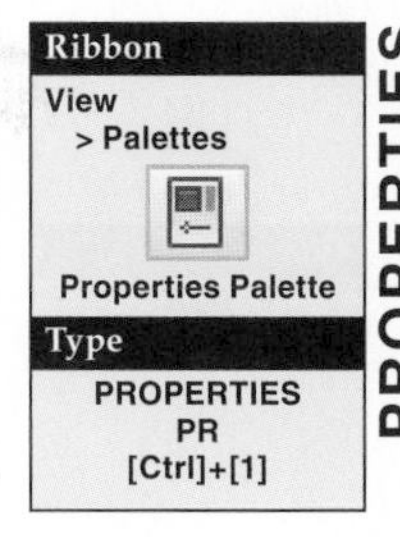

A thickness can be applied to text after it is created. This is done using the **Properties** palette. The thickness setting is located in the **General** section. Once a thickness is applied, the hidden lines can be removed using the **HIDE** command or a shaded visual style. Figure 8-1 shows six different fonts as they appear after being given a thickness and with the Conceptual visual style set current.

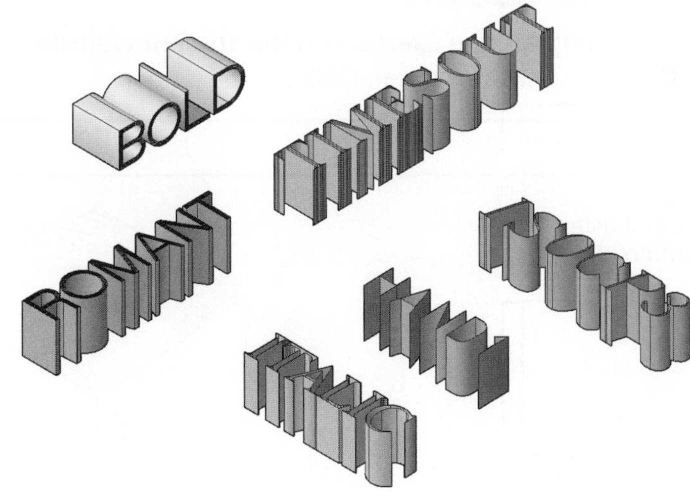

Figure 8-1.
Six different fonts with thickness after the Conceptual visual style is set current.

Only text created using the **TEXT** and **DTEXT** commands (text object) can be assigned thickness. Text created with the **MTEXT** command (mtext object) cannot have thickness assigned to it. In addition, only AutoCAD SHX fonts can be given thickness. AutoCAD SHP shape fonts can be compiled into SHX fonts with the **COMPILE** command. The compiled fonts can then be used to create text with thickness. Windows TrueType fonts *cannot* be used to create text with thickness.

Text and the UCS

Text is created parallel to the XY plane of the UCS in which it is drawn. Therefore, if you wish to show text appearing on a specific plane, establish a new UCS on that plane before placing the text. You can make use of a dynamic UCS. Figure 8-2 shows several examples of text on different UCS XY planes.

Changing the Orientation of a Text Object

If text is improperly placed or created using the wrong UCS, it can be edited using grips or editing commands. Editing commands and grips are relative to the current UCS. For example, if text is drawn with the WCS current, you can use the **ROTATE** command to change the orientation of the text in the XY plane of the WCS. However, to rotate the text so it tilts up from the XY plane of the WCS, you will need to change the UCS. Rotate the UCS as needed so the Z axis of the new UCS aligns with the axis about which you want to rotate. Then, the **ROTATE** command can be used to rotate the text. The **3DROTATE** command can also be used to avoid rotating the UCS. This command is discussed in Chapter 11.

Using the UCS View Option to Create a Title

UCS

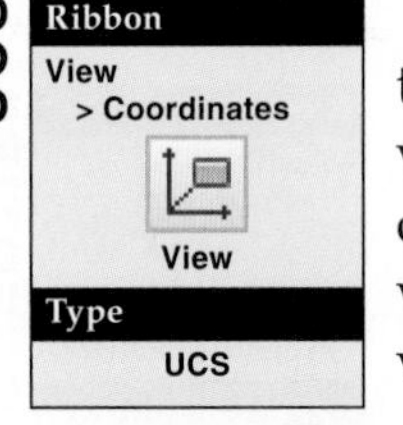

It is often necessary to create a pictorial view of an object, but with a note or title that is plan to your point of view. For example, you may need to insert the title of a 3D view. See Figure 8-3. This is done with the **View** option of the **UCS** command, which is discussed in Chapter 6. With this option, a new UCS is created perpendicular to your viewpoint. However, the view remains unchanged. Inserted text will be horizontal (or vertical) in the current view. Name and save the UCS if you will use it again.

Exercise 8-1

Complete the exercise on the student website.
www.g-wlearning.com/CAD

Figure 8-2.
Text located using three different UCSs.

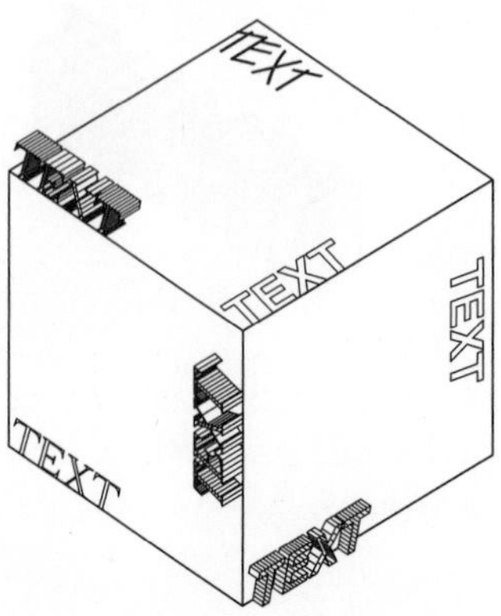

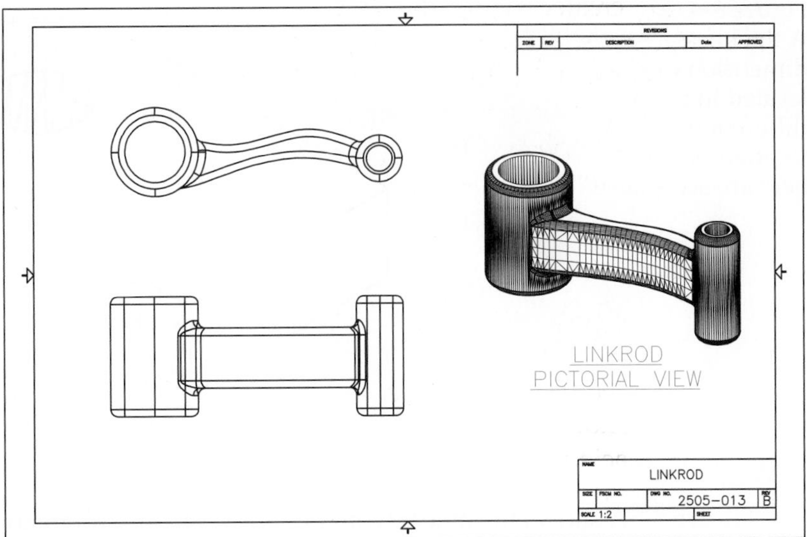

Figure 8-3.
This title (shown in color) has been correctly placed using the **View** option of the **UCS** command.

Dimensioning in 3D

Three-dimensional objects are seldom dimensioned for manufacturing, but may be used for assembly. Dimensioned 3D drawings are most often used for some sort of presentation, such as displays, illustrations, parts manuals, or training manuals. All dimensions, including those shown in 3D, must be clear and easy to read. The most important aspect of applying dimensions to a 3D object is planning. That means following a few basic guidelines.

Creating a 3D Dimensioning Template Drawing

If you often create dimensioned 3D drawings, make a template drawing containing a few 3D settings. Starting a drawing based on one of these templates will speed up the dimensioning process because the settings will already be made for you.

- Create named dimension styles with appropriate text heights. See *AutoCAD and Its Applications—Basics* for detailed information on dimensioning and dimension styles.
- Establish several named user coordinate systems that match the planes on which dimensions will be placed.
- If the preset isometric viewpoints will not serve your needs, establish and save several 3D viewpoints that can be used for different objects. These viewpoints will allow you to select the display that is best for reading dimensions.

Placing Dimensions in the Proper Plane

The location of dimensions and the plane on which they are placed are often a matter of choice. For example, Figure 8-4 shows several options for placing a thickness dimension on an object. All of these are correct. However, several of the options can be eliminated when other dimensions are added. This illustrates the importance of planning.

The key to good dimensioning in 3D is to avoid overlapping dimension and extension lines in different planes. A freehand sketch can help you plan this. As you lay out the 3D sketch, try to group information items together. Dimensions, notes, and item tags should be grouped so that they are easy to read and understand. This technique is called ***information grouping.***

Figure 8-4.
A thickness dimension can be located in many different places. All locations shown here are acceptable.

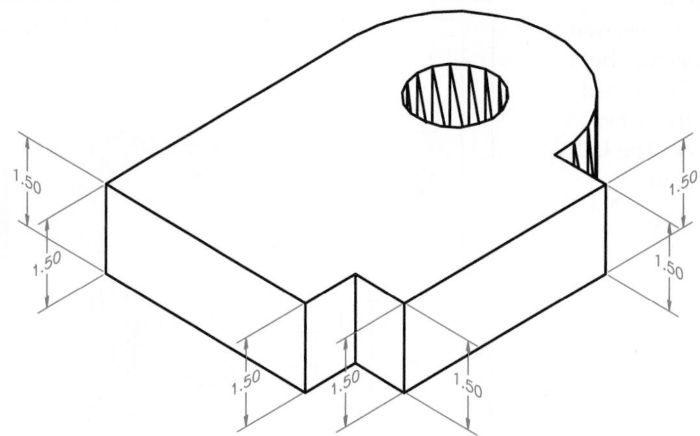

Figure 8-5A shows the object from Figure 8-4 fully dimensioned using the aligned technique. Notice that the location dimension for the hole is placed on the top surface. This avoids dimensioning to hidden points. Figure 8-5B shows the same object dimensioned using the unilateral technique.

To create dimensions that properly display, it may be necessary to modify the dimension text rotation. The dimension shown in Figure 8-6A is inverted because the positive X and Y axes are incorrectly oriented. Using the **Properties** palette, change the text rotation value to 180. The dimension text is then properly displayed, Figure 8-6B. Alternately, you can rotate the UCS before drawing the dimension, but this may be more time-consuming.

PROFESSIONAL TIP

Prior to placing dimensions on a 3D drawing, you should determine the purpose of the drawing. For what will it be used? Just as dimensioning a drawing for manufacturing purposes is based on the function of the part, 3D dimensioning is based on the function of the drawing. This determines whether you use chain, datum, arrowless, architectural, or some other style of dimensioning. It also determines how completely the object is dimensioned.

Figure 8-5.
A—An example of a 3D object dimensioned using the aligned technique. B—The object dimensioned with unilateral dimensions.

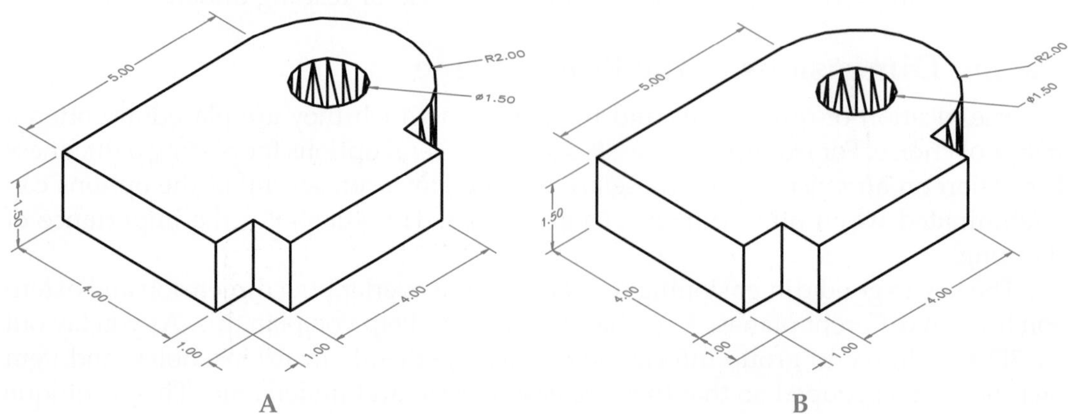

Figure 8-6.
A—This dimension text is inverted. B—The rotation value of the text is changed and the text reads correctly.

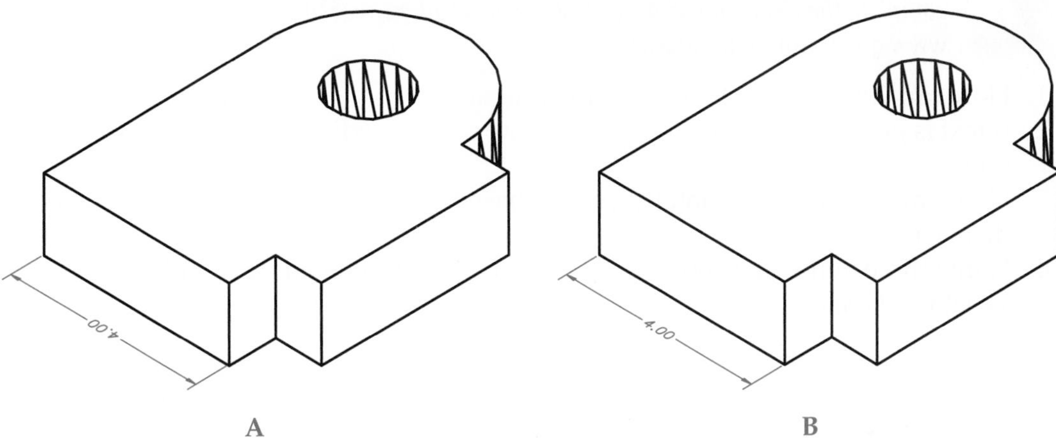

Placing Leaders and Radial Dimensions in 3D

Although standards such as ASME Y14.5 should be followed when possible, the nature of 3D drawing and the requirements of the project may determine how dimensions and leaders are placed. Remember, the most important aspect of dimensioning a 3D drawing is its presentation. Is it easy to read and interpret?

Leaders and radial dimensions can be placed on or perpendicular to the plane of the feature. Figure 8-7A shows the placement of leaders on the plane of the top surface. Figure 8-7B illustrates the placement of leaders and radial dimensions on two planes that are perpendicular to the top surface of the object. Remember that text, dimensions, and leaders are always created on the XY plane of the current UCS. Therefore, to create the layout in Figure 8-7B you must use more than one UCS.

Exercise 8-2

Complete the exercise on the student website.
www.g-wlearning.com/CAD

Figure 8-7.
A—Leaders placed in the plane of the top surface. B—Leaders placed using two UCSs that are perpendicular to the top face.

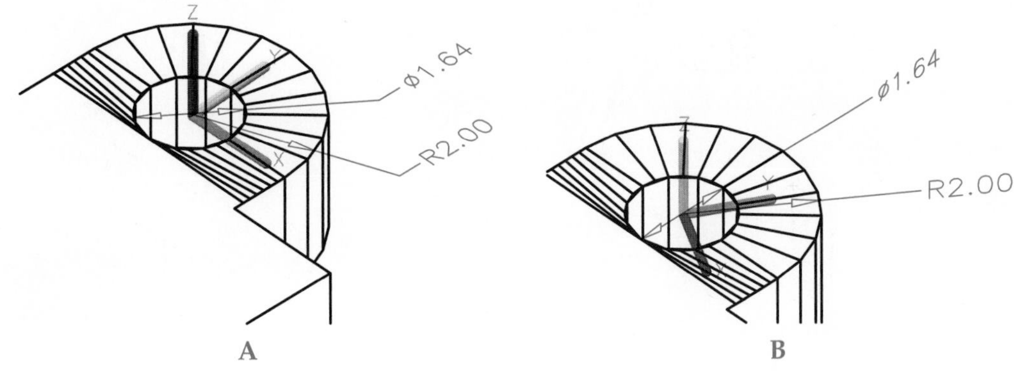

Chapter Test

Answer the following questions. Write your answers on a separate sheet of paper or complete the electronic chapter test on the student website.
www.g-wlearning.com/CAD

1. How can you create 3D text with thickness?
2. If text is placed using the wrong UCS, how can it be edited to appear on the correct one?
3. How can text be horizontally placed based on your viewpoint if the object is displayed in 3D?
4. Name three items that should be a part of a 3D dimensioning template drawing.
5. What is *information grouping?*

Drawing Problems

Apply geometric constraints to the base 2D drawing as appropriate. If you create profiles of a surface that can be converted into regions for solid modeling, apply geometric constraints and save the drawing with a different name than the final solid model.

1. This is a two-view orthographic drawing of a window valance mounting bracket. Create it as a solid model. Use solid primitives and Boolean commands as needed. Use the dimensions given. Similar holes have the same offset dimensions. Create new UCSs as needed. Display an appropriate pictorial view of the drawing. Then, add dimensions. Finally, add the material note so it is plan to the 3D view. Plot the drawing to scale on a C-size sheet of paper. Save the drawing as P08_01.

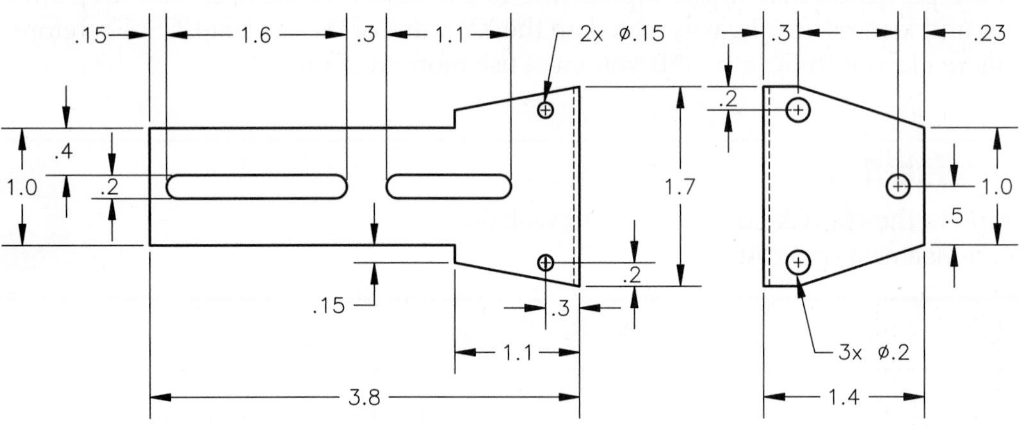

2. Open P07_03. If you have not completed this problem, construct it using the directions for the problem in Chapter 7. Display the 3D view in a single viewport. Then, add dimensions to the 3D view, as shown in Chapter 7. Plot the drawing on a C-size sheet of paper. Save the drawing as P08_02.

3. Create the end table as a solid model using solid primitives and Boolean commands as needed. The end result should be a single object. As a test of your object-editing skills, try drawing the entire model by starting with only a single rectangle. You can copy, resize, extrude, and move objects as you create them from the single rectangle. Use the dimensions given and the following information to construct the model.
   A. Table height is 24″.
   B. Top of bottom shelf is 5″ off of the floor.
   C. Table legs must be located no less than 1/2″ from the tabletop edge.
   D. Shelf must be no closer than .75″ from the outside of table legs.
   E. Dimension the table as shown.
   F. Save the drawing as P08_03.

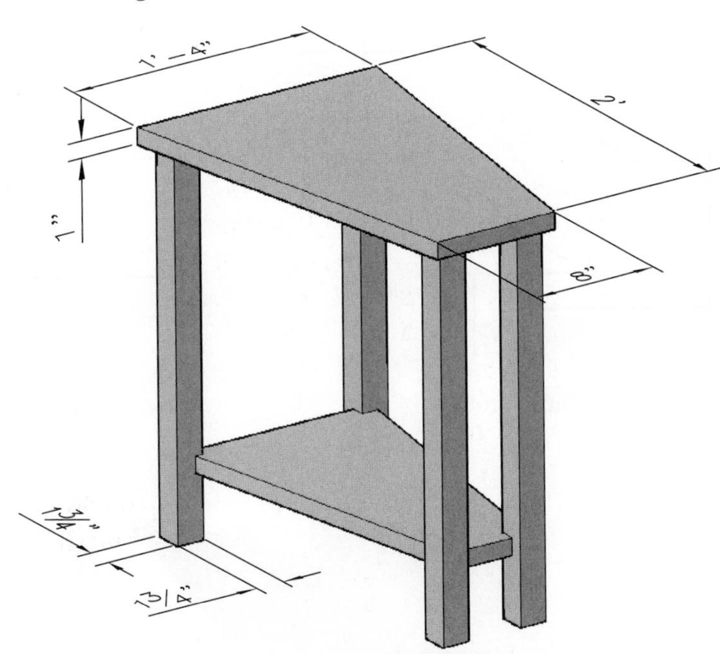

Drawing Problems - Chapter 8

4. Shown below are the profiles of a roof gutter (for the collection of rainwater) and a gutter downspout. Draw the profiles in 3D using the dimensions shown. Use the following additional information to construct a 3D model like the one shown in the shaded view.
   A. Offset the gutter profile to create a material thickness of .025″. Be sure to close the ends to create a closed polyline so a 3D solid is created when it is extruded.
   B. Extrude the gutter profile 12″ to create a one-foot section.
   C. Relocate the downspout profile on the underside of the gutter.
   D. Construct an extrusion path for the downspout. Refer to the shaded view shown below, but use your own design.
   E. Extrude the downspout profile along the path.
   F. Dimension the end of the gutter profile in 3D.
   G. Save the drawing as P08_04.

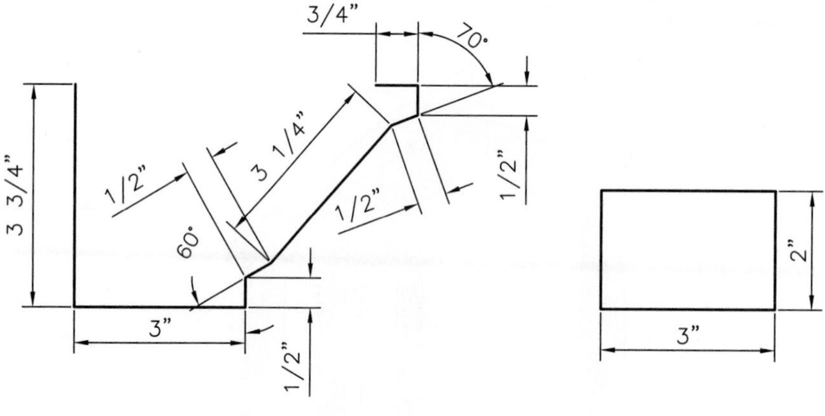

A

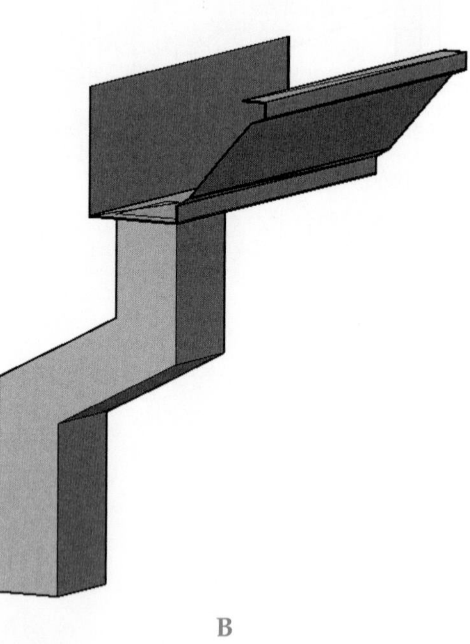

B

5. Open P07_04. If you have not completed this problem, construct it using the directions for the problem in Chapter 7. Display the 3D view in a single viewport. Then, add dimensions to the 3D view. Plot the drawing on a C-size sheet of paper. Save the drawing as P08_05.

Problems 6 and 7 are mechanical parts. Create a solid model of each part. Dimension each model. Place the title of each model so it is plan to the pictorial view. Plot the finished drawings on B-size paper. Save each drawing as ***P08_****(problem number).*

6.

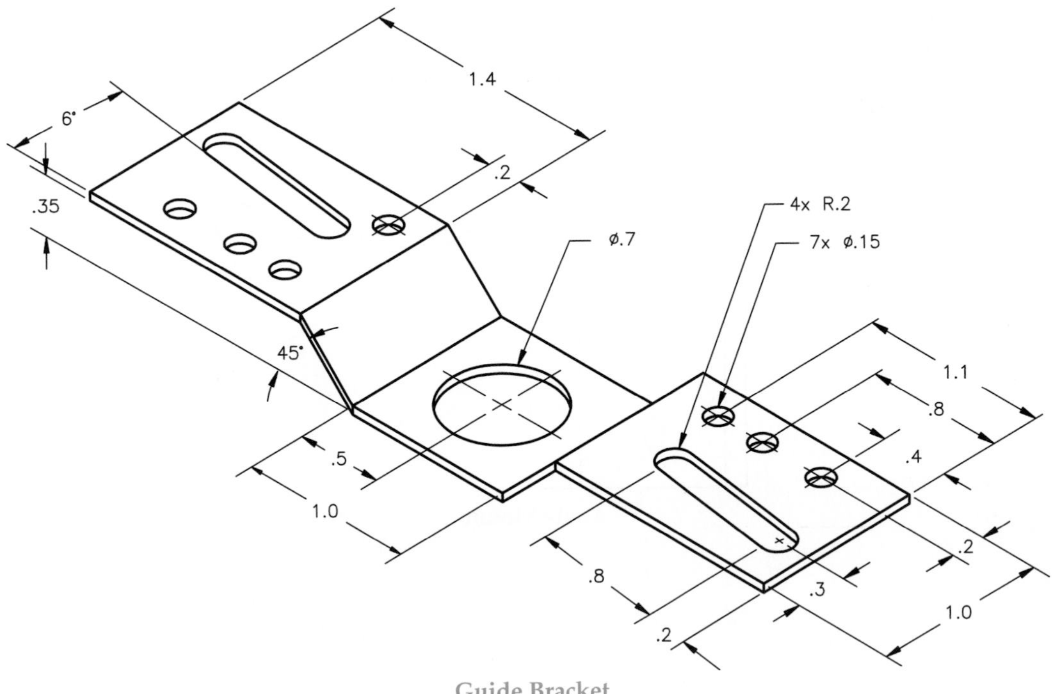

Guide Bracket

Drawing Problems - Chapter 8

7.

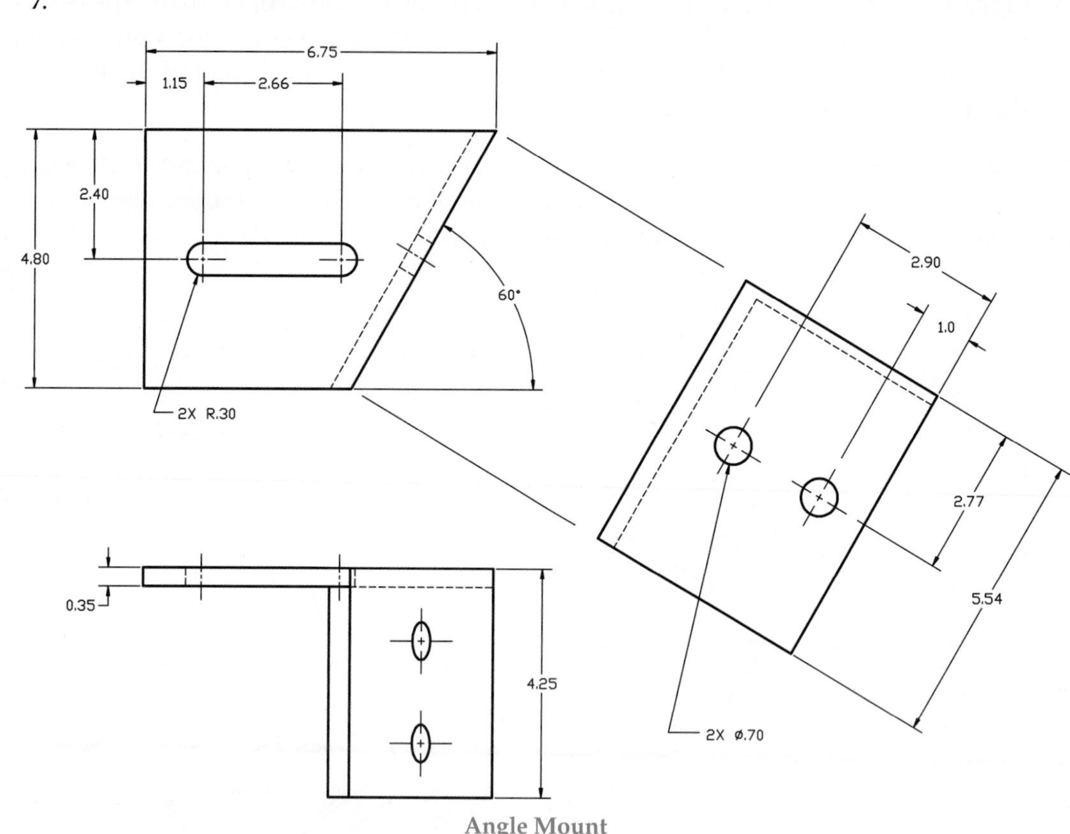

Angle Mount

Drawing Problems - Chapter 8

CHAPTER 9

Model Extrusions and Revolutions

Learning Objectives

After completing this chapter, you will be able to:

- ✓ Create solids and surfaces by extruding 2D profiles.
- ✓ Extrude planar surfaces.
- ✓ Create symmetrical 3D solids and surfaces by revolving 2D profiles.
- ✓ Revolve planar surfaces.
- ✓ Extrude and revolve objects using mathematical expressions and constraints.
- ✓ Use solid extrusions and revolutions as construction tools.

Complex shapes can be created by applying a thickness to a two-dimensional profile. This is called *extruding* the shape. You have been introduced to the operation in previous chapters. Two or more profiles can be extruded to intersect. The resulting union can form a new shape by performing a Boolean operation. Symmetrical objects can be created by revolving a 2D profile about an axis to create a new solid.

Creating Extruded Models

A ***solid extrusion*** is a closed, two-dimensional shape that has been given thickness. The **EXTRUDE** command allows you to create extrusions from lines, arcs, elliptical arcs, 2D polylines, 2D splines, circles, ellipses, 2D solids, regions, planar surfaces, and donuts. Objects in a block cannot be extruded. By default, closed objects, such as circles, polygons, closed polylines, and donuts, are converted to solids when they are extruded. Open-ended objects, such as lines, arcs, polylines, elliptical arcs, and splines, are converted to a ***surface extrusion*** when they are extruded. Surface extrusions have no mass properties.

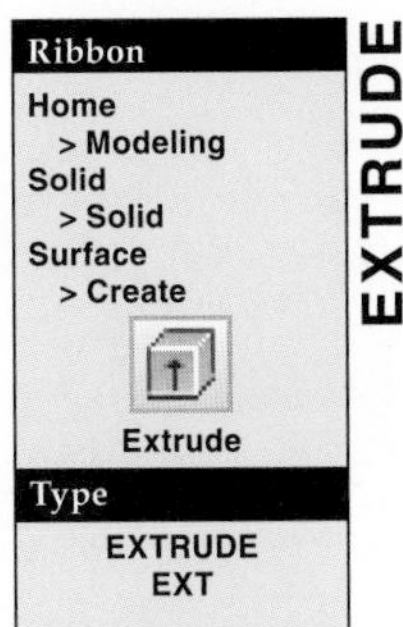

Extrusions can be created along a straight line or along a path curve. A taper angle can also be applied as you extrude an object. Figure 9-1 illustrates a polygon extruded into a solid.

Figure 9-1.
The **EXTRUDE** command creates a solid or surface by adding thickness to a 2D profile. A—The initial, closed 2D profile. B—The extruded solid object shown with hidden lines removed.

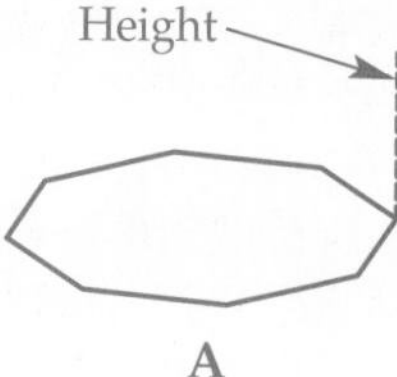

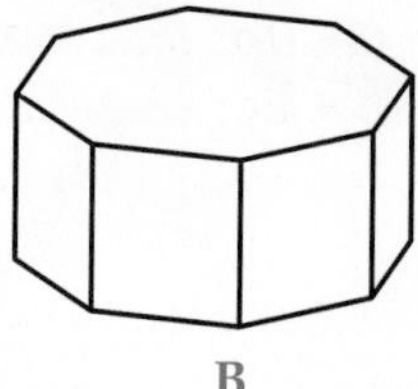

When the **EXTRUDE** command is selected, you are prompted to select the objects to extrude. Select the objects and press [Enter]. You are then prompted for the extrusion height. The height is always applied along the Z axis of the selected object, not the current UCS. A positive value extrudes above the XY plane of the object. A negative height value extrudes below the XY plane. If a pictorial view is displayed, you can drag the mouse to set the extrusion above or below the XY plane and then enter the height value.

Before entering a height, you can specify a taper angle. The taper angle can be any value *between* +90° and –90°. A positive angle tapers to the inside of the object from the base. A negative angle tapers to the outside of the object from the base. See Figure 9-2. However, the taper angle cannot result in edges that "fold into" the extruded object.

If the command is selected from the **Solid** tab in the ribbon, the mode is automatically set to solid. However, if an open profile is selected, a surface is still created. If the command is selected from the **Surface** tab in the ribbon, the mode is automatically set to surface. In either case, you can use the **Mode** option to change the output type.

PROFESSIONAL TIP

Objects such as polylines, lines, and arcs that have a thickness can be converted to surfaces using the **CONVTOSURFACE** command. Circles and closed polylines with a thickness can be converted to solids using the **CONVTOSOLID** command.

Figure 9-2.
A—A positive angle tapers to the inside of the object from the base.
B—A negative angle tapers to the outside of the object.

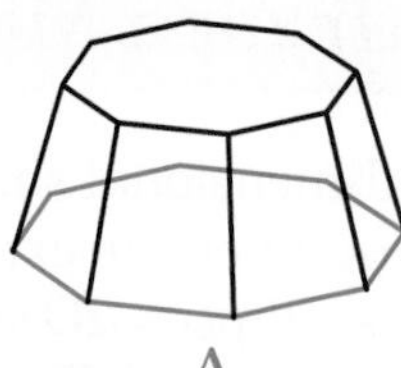

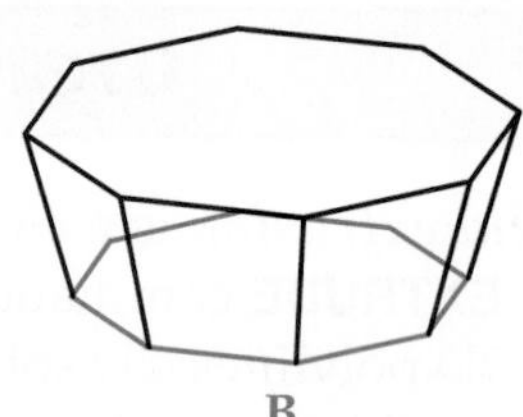

Extrusions along a Path

A 2D shape can be extruded along a path to create a 3D solid or surface. The path can be a line, circle, arc, ellipse, polygon, polyline, or spline. Line segments and other objects can be first joined to form a polyline path. The corners of angled segments on the extruded object are mitered, while curved segments are smooth. See **Figure 9-3.**

When open objects, such as lines, arcs, polylines, elliptical arcs, and splines, are used as the profile, they are converted to a swept surface when extruded along a path. A *sweep* is a solid or surface that is created when an open or closed curve is pulled, or swept, along a 2D or 3D path. An extrusion is really a form of a sweep. Sweeps are discussed in detail in Chapter 10.

The **Mode** option of the **EXTRUDE** command allows you to select the type of object that is created: surface or solid. Keep in mind, objects that are not closed, such as lines and polylines, *always* result in extruded surfaces. A solid can only be created by the **EXTRUDE** command if the original object is a closed curve.

To extrude along a path, enter the **EXTRUDE** command and select the objects to extrude. When prompted for the height of the extrusion, enter the **Path** option. If needed, first enter a taper angle. Then, pick the object to be used as the extrusion path.

Objects can also be extruded along a line at an angle to the base object, **Figure 9-4.** Notice that the plane at the end of the extruded object is parallel to the original object. Also notice that the length of the extrusion is the same as that of the path. The path does not need to be perpendicular to the object.

If the path begins perpendicular to the profile, the cross section of the resulting extrusion is perpendicular to the path, regardless if the path is a straight line, curve, or spline. See **Figure 9-5.** If the path is a spline or curve that does not begin perpendicular to the profile, the profile may not remain perpendicular to the path as it is extruded.

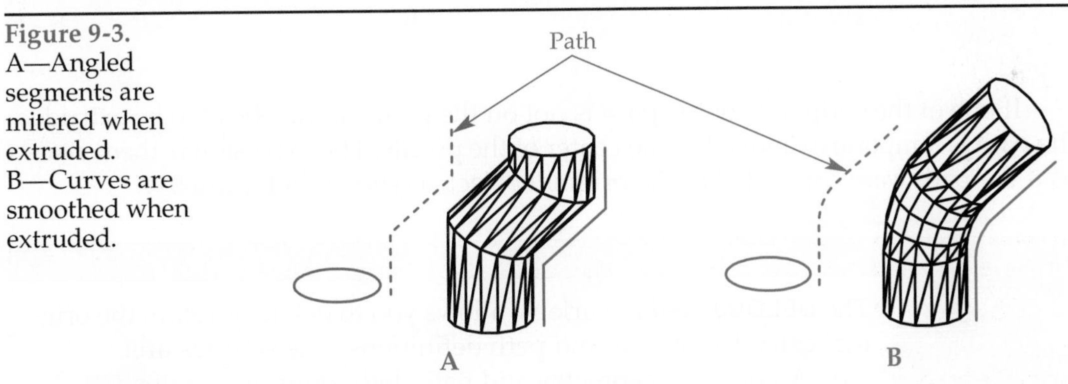

Figure 9-3.
A—Angled segments are mitered when extruded.
B—Curves are smoothed when extruded.

Path

A

B

Figure 9-4.
A—An object extruded along a path. B—The end of an object extruded along an angled path is parallel to the original object.

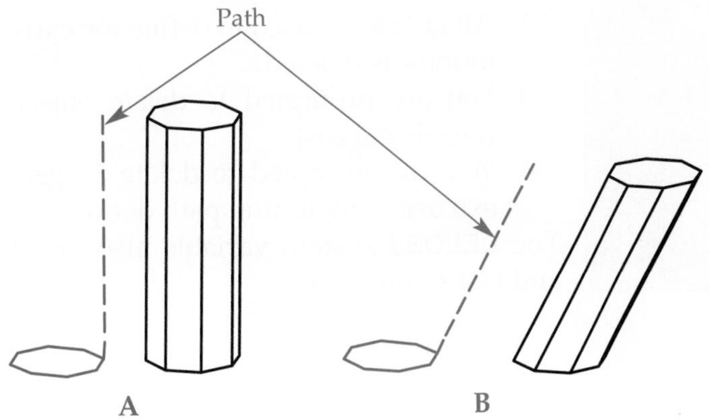

Figure 9-5.
A—Splines can be used as extrusion paths. Notice that the profile on the right is not perpendicular to the start of the path. B—The resulting extrusions.

If one of the endpoints of the path is not on the plane of the object to be extruded, the path is temporarily moved to the center of the profile. The extrusion is then created as if the path were connected to the original object, as shown in Figure 9-4.

NOTE

The **DELOBJ** system variable allows you to delete or retain the original extruded objects and path definitions. The settings are:

- 0 All original geometry and path definitions are retained.
- 1 Objects used for extrusion (profile curves) are deleted. This is the default.
- 2 All geometry used to define the extrusion, including path definitions, is deleted.
- –1 You are prompted to delete objects used for the extrusion (profile curves).
- –2 You are prompted to delete all geometry used to define the extrusion, including path definitions.

The **DELOBJ** system variable also affects the **REVOLVE**, **SWEEP**, and **LOFT** commands.

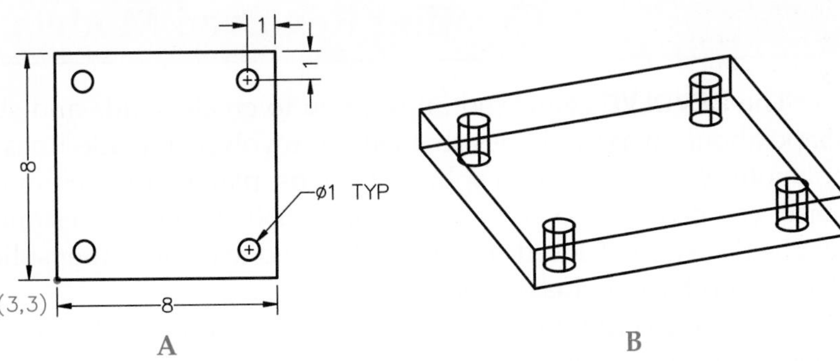

Figure 9-6.
A—The 2D region that will be extruded. B—The solid object created by extruding the region, shown in a 3D wireframe display.

Extruding Regions

In Chapter 2, you learned how to create 2D regions. As an example, you created the top view of the base shown in Figure 9-6A as a region. Regions can be extruded to create 3D solids. The base you created in Chapter 2 can be extruded to create the final solid shown in Figure 9-6B. Any features of the region, such as holes, are extruded the same thickness as the rest of the object. If the profile was created as polylines, the holes must be separately extruded and then subtracted from the solid. Using this method, you can construct a fairly complex 2D region that includes curved profiles, holes, slots, etc. Then, a complex 3D solid can be quickly created. Additional details can be added using editing commands or Boolean operations.

Exercise 9-1

Complete the exercise on the student website.
www.g-wlearning.com/CAD

Extruding a Planar Surface

A planar surface can be extruded into a solid object in the same manner as a region. Nonplanar (curved) surfaces cannot be extruded. Whereas both surfaces and regions have no thickness, the surface is an object composed of a mesh and the region is actually a solid that possesses mass properties. A surface can be quickly converted to a solid using the **EXTRUDE** command. Simply select the surface when prompted to select objects. The surface can be extruded in a specific direction, along a path, or at a taper angle.

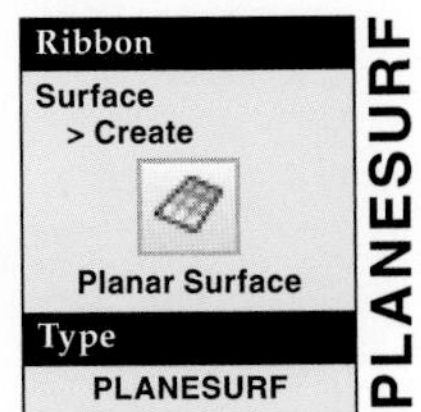

Any closed object, such as a circle, rectangle, polygon, or polyline, can be converted into a surface with the **Object** option of the **PLANESURF** command. Pick the **Planar Surface** button on the **Create** panel of the **Surface** tab in the ribbon. Then, enter the **Object** option and select the closed object. The resulting surface can then be extruded into a 3D solid.

PROFESSIONAL TIP

You can also extrude a face on an existing solid into a new solid. When prompted to select objects, press the [Ctrl] key and pick the face to extrude. A face is a *subobject* of a solid. Subobject editing is covered in detail in Chapter 12.

Creating Revolved Models

REVOLVE

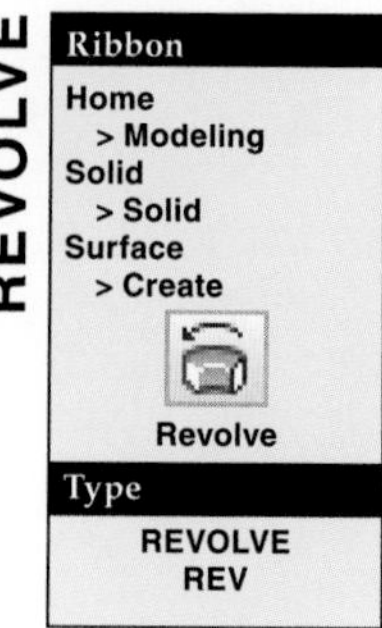

The **REVOLVE** command allows you to create solids and surfaces by revolving a shape about an axis. Shapes that can be revolved include lines, arcs, circles, ellipses, polygons, polylines, closed splines, regions, planar surfaces, and donuts. The selected object can be revolved at any angle up to 360°. Open curves and line segments create revolved surfaces. Closed curves can be used to create either solids or surfaces. Surface revolutions have no mass properties.

When the command is selected, you are prompted to pick the objects to revolve. Then, you must define the axis of revolution. The default option is to pick the two endpoints of an axis of revolution. This is shown in Figure 9-7. You can also revolve about an object or the X, Y, or Z axis of the current UCS. Once the axis is defined, you are prompted to enter the angle through which the profile will be revolved. When the angle is specified, the revolution is created.

If the command is selected from the **Solid** tab in the ribbon, the mode is automatically set to solid. However, if an open profile is selected, a surface is still created. If the command is selected from the **Surface** tab in the ribbon, the mode is automatically set to surface. In either case, you can use the **Mode** option to change the output type.

PROFESSIONAL TIP

When creating solid models, keep in mind that the final part will most likely need to be manufactured. Be aware of manufacturing processes and methods as you design parts. It is easy to create a part in AutoCAD with internal features that may be impossible to manufacture, especially when revolving a profile.

Revolving about an Axis-Line Object

You can select an object, such as a line, as the axis of revolution. Figure 9-8 shows a solid created using the **Object** option of the **REVOLVE** command. Both a full-circle (360°) revolution and a 270° revolution are shown. Enter the **Object** option when prompted for the axis of revolution. Then, pick the axis object and enter the angle through which the profile will be rotated. You can use the **Start Angle** option before entering an angle of revolution. This allows you to specify the point at which the revolution starts and then the angle of revolution.

Figure 9-7.
Points P1 and P2 are selected as the axis of revolution for the profile.

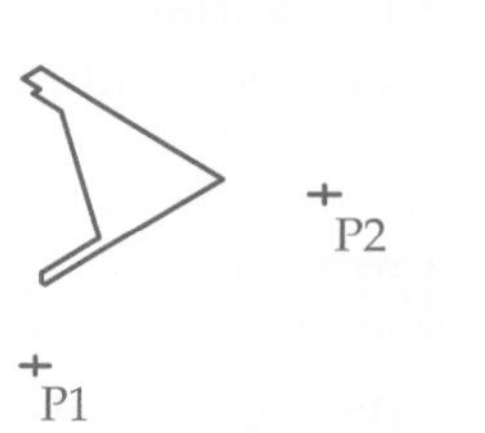

Figure 9-8.
An axis of revolution can be selected using the **Object** option of the **REVOLVE** command. Here, the line is selected as the axis.

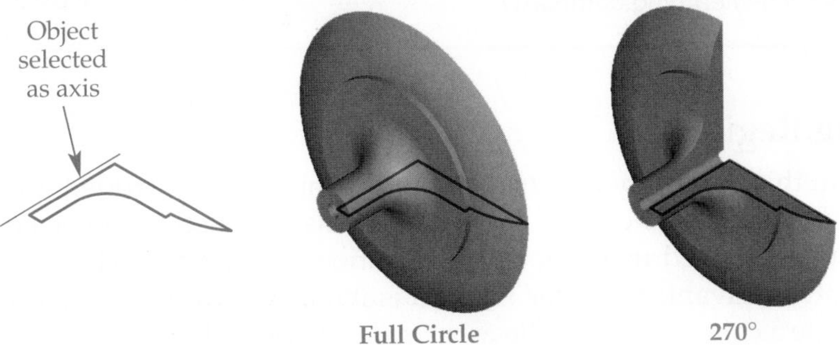

Revolving about the X, Y, or Z Axis

The X axis of the current UCS can be used as the axis of revolution by selecting the **X** option of the **REVOLVE** command. The origin of the current UCS is used as one end of the X axis line. Notice in Figure 9-9 that two different shapes can be created from the same 2D profile by changing the UCS origin. No hole appears in the object in Figure 9-9B because the profile was revolved about an edge that coincides with the X axis. The Y or Z axis can also be used as the axis of revolution. See Figure 9-10.

Figure 9-9.
A—A solid is created using the X axis as the axis of revolution. B—A different object is created with the same profile by changing the UCS origin.

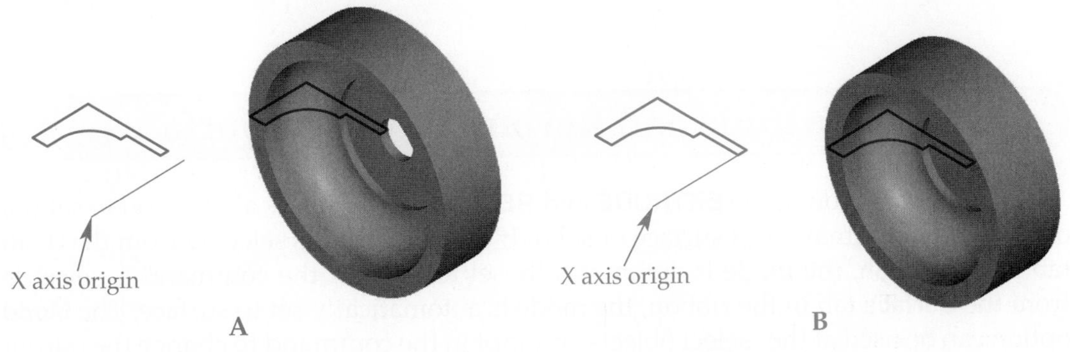

Figure 9-10.
A—A solid is created using the Y axis as the axis of revolution. B—A different object is created by changing the UCS origin.

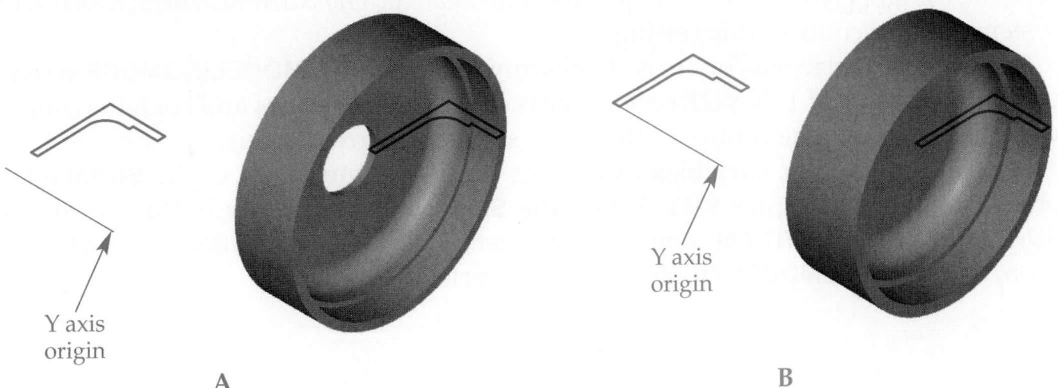

Exercise 9-2

Complete the exercise on the student website.
www.g-wlearning.com/CAD

Revolving Regions

Earlier in this chapter, you learned that regions can be extruded. In this manner, holes, slots, keyways, etc., can be created. Regions can also be revolved. A complex 2D shape can be created using Boolean operations on regions. Then, the region can be revolved. One advantage of this method is it may be easier to create a region than trying to create a complex 2D profile as a single, closed polyline.

Revolving a Planar Surface

Just as a planar surface can be extruded into a solid object, it can also be revolved into a solid object. Nonplanar (curved) surfaces cannot be revolved. When the **REVOLVE** command is selected, simply pick the surface when prompted to select objects. The surface can be revolved about an axis defined by two pick points, an object, or the X, Y, or Z axis of the current UCS.

PROFESSIONAL TIP

You can also revolve a face on an existing solid into a new solid. When prompted to select objects, turn on selection cycling or press the [Ctrl] key and pick the face to revolve. Subobject editing is covered in detail in Chapter 12.

Using Extrude and Revolve to Create Surfaces

The **Mode** option of the **EXTRUDE** and **REVOLVE** commands allows you to set the output of the command to a surface or solid. If the command is selected from the **Solid** tab in the ribbon, the mode is automatically set to solid. If the command is selected from the **Surface** tab in the ribbon, the mode is automatically set to surface. The **Mode** option can be used at the "select objects" prompt in the command to change the output type.

The **SURFACEMODELINGMODE** system variable controls the type of surface created. The default setting of 0 creates a ***procedural surface.*** This is a standard surface composed of multiple flat polygons, but it has no control vertices. By default, this type of surface is associated with the object used to create it. The **SURFACEASSOCIATIVITY** system variable controls this setting.

A NURBS surface can be created by setting the **SURFACEMODELINGMODE** system variable to a value of 1. A ***NURBS surface*** is composed of splines and contains control vertices that enable the control of the curve shape with great precision.

Surface modeling variables can be set on the **Create** panel in the **Surface** tab of the ribbon. See **Figure 9-11.** When the **Surface Associativity** button is on, the **SURFACEASSOCIATIVITY** system variable is set to 1. When the **NURBS Creation** button is on, the **SURFACEMODELINGMODE** system variable is set to 1.

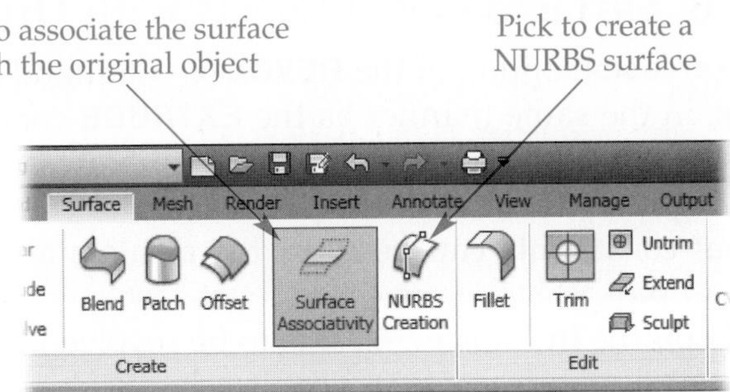

Figure 9-11. Surface modeling variables can be set on the **Create** panel in the **Surface** tab of the ribbon.

Extruding Surfaces from Objects with Dimensional Constraints

When the **Mode** option is set to create a surface, you can use the **Expression** option to create a parametric relationship between a dimensional constraint and the extruded surface. For example, the circle in Figure 9-12 has been given a dimensional constraint named dia1. To create an extruded surface cylinder with a height that is always one-half of diameter, entering the **Expression** option. Then, at the **Enter expression:** prompt, enter:

```
dia1/2
```

In this manner, using appropriate formulaic expressions, you can create an extruded surface that has a mathematical relationship to a 2D object that has a dimensional constraint.

CAUTION

You can enter an expression when creating a solid. The end result is calculated, but the dimensional constraint is removed. Therefore, the parametric aspect of using an expression is lost.

dia1=3.0000

A

B

Figure 9-12. A—The circle has been given a dimensional constraint named dia1. B—The circle is extruded using an expression to create a cylinder surface that is one half the height of dia1. When the diameter is changed, the height changes to maintain the relationship.

Revolving Surfaces from Objects with Dimensional Constraints

The **Expression** option of the **REVOLVE** command can be used with dimensional constraints, in the same manner as the **EXTRUDE** command, in order to construct a revolved surface. Examples of mathematical expressions used with constraints are shown in Figures 9-13 and 9-14. Both examples illustrate how linear and angular dimensional constraints can be used to calculate an angular value for a revolved surface.

The 2D profile in Figure 9-13A is to be revolved at an angle that is one-half that of the angular dimension (ang2). This is achieved by selecting the **Expression** option and then entering:

ang2/2

The result is shown in Figure 9-13B.

In Figure 9-14, the same 2D geometry is used. However, the aligned dimension d2 is used to create a negative value to revolve the object below the XY plane. The following formula is entered:

d2–31

The result of subtracting 31 from the value of d2 is processed as the angular value for the revolution. The result is shown in Figure 9-14.

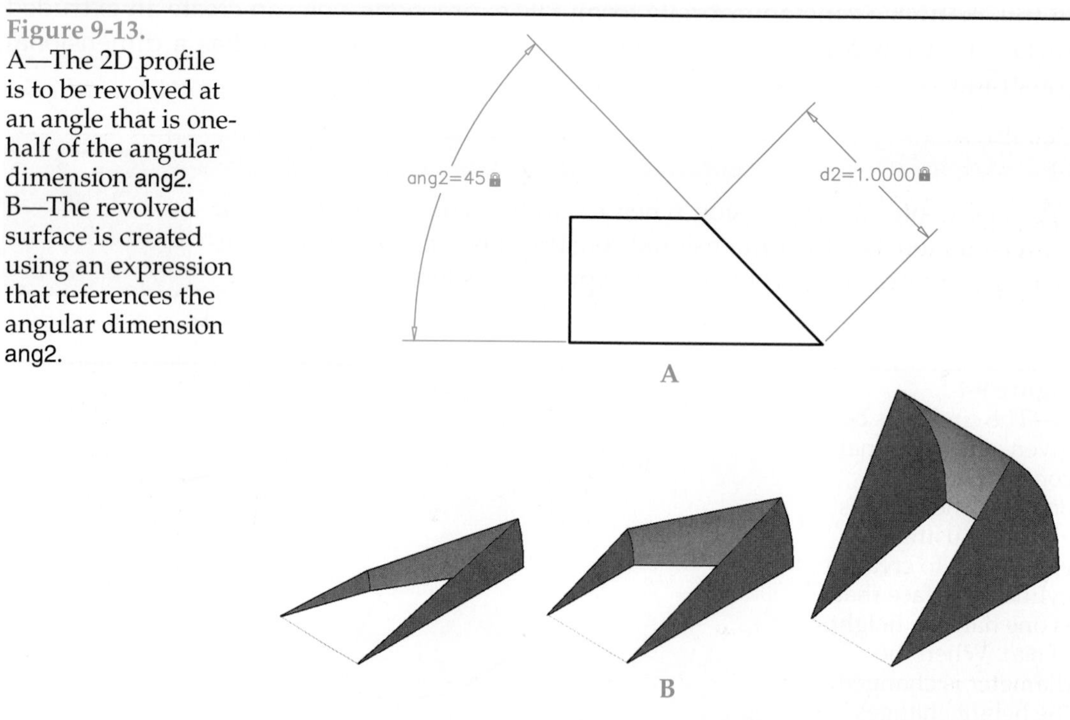

Figure 9-13.
A—The 2D profile is to be revolved at an angle that is one-half of the angular dimension ang2.
B—The revolved surface is created using an expression that references the angular dimension ang2.

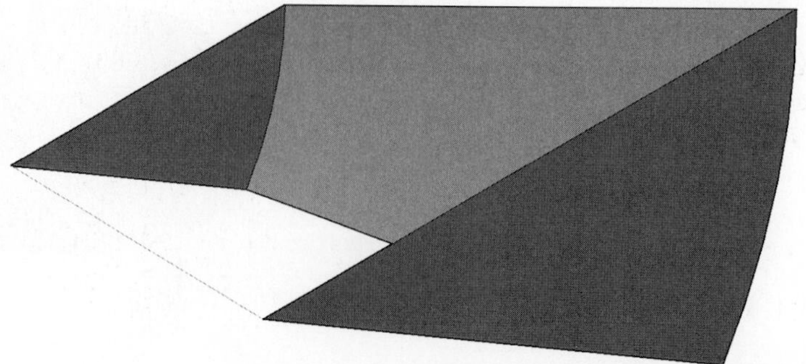

Figure 9-14.
The surface is created using a negative value based on the dimensional constraint d2 to revolve the object below the XY plane.

Using Extrude and Revolve as Construction Tools

It is unlikely that an extrusion or revolution will result in a finished object. Rather, these operations will be used with other solid model construction methods, such as Boolean operations, to create the final object. The next sections discuss how to use **EXTRUDE** and **REVOLVE** with other construction methods to create a finished solid object.

Creating Features with Extrude

You can create a wide variety of features with the **EXTRUDE** command. Study the shapes shown in Figure 9-15. These detailed solid objects were created by drawing a profile and then using the **EXTRUDE** command. The objects in Figures 9-15C and 9-15D must be constructed as regions before they are extruded. For example, the five holes (circles) in Figure 9-15D must be removed from the base region using the **SUBTRACT** command.

Look at Figure 9-16. This is part of a clamping device used to hold parts on a mill table. There is a T-slot milled through the block to receive a T-bolt and one side is stair-stepped, under which parts are clamped. If you look closely at the end of the object, most of the detail can be drawn as a 2D region and then extruded. However, there are also two holes in the top of the block to allow for bolting the clamp to the mill table. These features must be added to the extruded solid.

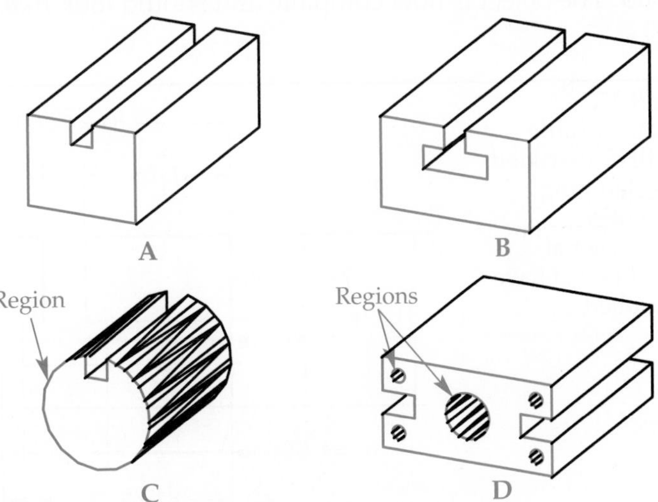

Figure 9-15. Detailed solids can be created by extruding the profile of an object. The profiles are shown here in color.

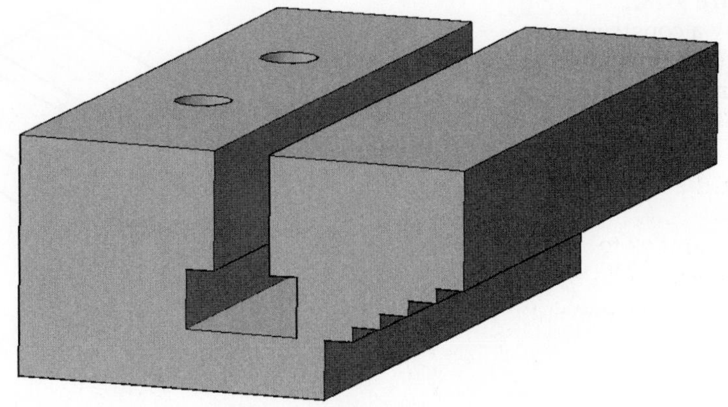

Figure 9-16. Most of this object can be created by extruding a profile. However, the holes must be added after the extruded solid is created.

First, change the UCS to the front preset orthographic UCS. Display a plan view of the UCS. Then, draw the profile shown in Figure 9-17 using the **PLINE** command. You can draw it in stages, if you like, and then use the **PEDIT** command to join all segments into a single polyline.

EXTRUDE

Ribbon
Home > Modeling Solid > Solid
Extrude
Type
EXTRUDE EXT

Next, use the **EXTRUDE** command to create the 3D solid. Extrude the profile a distance of –6 units with a 0° taper. This will extrude the object away from you. Display the object from the southeast isometric preset viewpoint or use the view cube to display a pictorial view. The object should look similar to Figure 9-16 without the holes in the top. Set the Conceptual visual style current, if you like.

The two holes are ∅.5 units and evenly spaced on the surface through which they pass. Change to the WCS and draw a construction line from midpoint to midpoint, as shown in Figure 9-18. Then, set **PDMODE** to an appropriate value, such as 3, and use the **DIVIDE** command to divide the construction line into three parts. The two points created by the **DIVIDE** command are equally spaced on the surface and can be used to locate the two holes.

There are two ways to create a hole. You can draw a circle and extrude it to create a cylinder or you can draw a solid cylinder. Either way, you need to subtract the cylinder to create the hole. Drawing a solid cylinder is probably easiest. When prompted for a center, use the **Node** object snap to select the point. Then, enter the diameter. Finally, enter a negative height so that the cylinder extends into the solid or drag the cylinder down in the 3D view so it extends all of the way through the block. The actual height is not critical, as long as it extends through the block.

You can either copy the first cylinder to the second point or draw another cylinder. When both cylinders are located, use the **SUBTRACT** command to remove them from the solid. The object is now complete and should look like Figure 9-16.

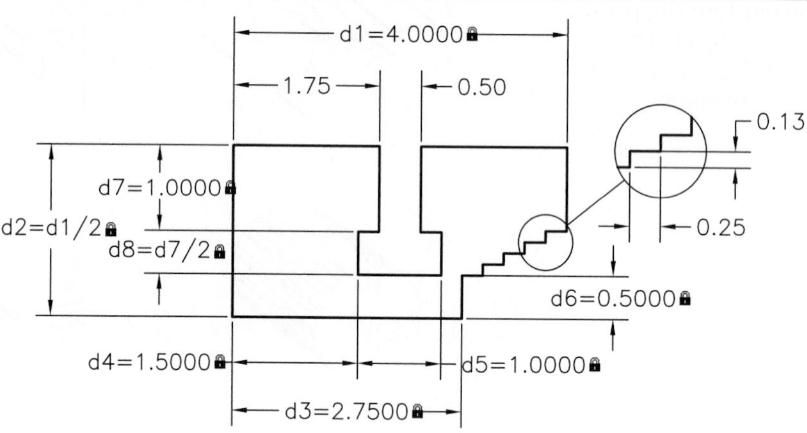

Figure 9-17.
This is the profile that will be extruded for the clamping block. Notice the dimensional constraints that have been applied.

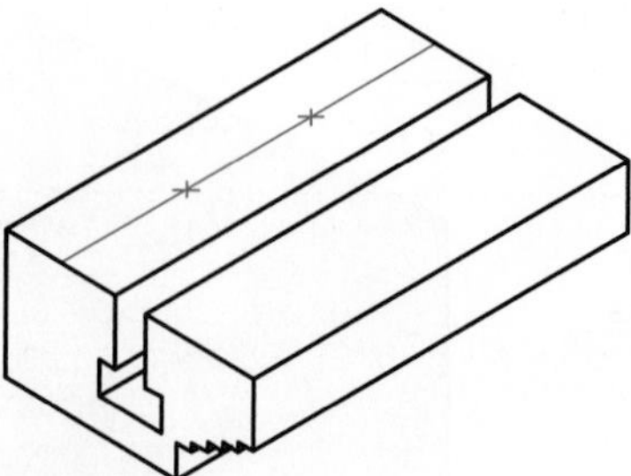

Figure 9-18.
Draw a construction line (shown here in color) and divide it into three parts.

Creating Features with Revolve

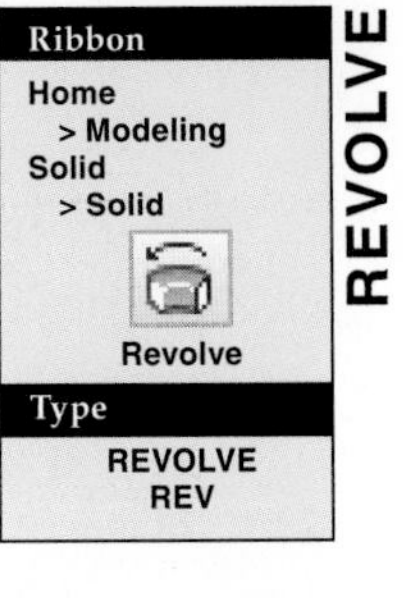

The **REVOLVE** command is very useful for creating symmetrical, round objects. However, many times the object you are creating is not completely symmetrical. For example, look at the camshaft in **Figure 9-19.** For the most part, this is a symmetrical, round object. However, the cam lobes are not symmetrical in relation to the shaft and bearings. The **REVOLVE** command can be used to create the shaft and bearings. Then, the cam lobes can be created and added.

Start a new drawing and make sure the WCS is the current UCS. Using the **PLINE** command, draw the profile shown in **Figure 9-20A.** This profile will be revolved through 360°, so you only need to draw half of the true plan view of the cam profile. The profile represents the shaft and three bearings.

Next, display the drawing from the southwest isometric preset viewpoint. Then, use the **REVOLVE** command to create the base camshaft as a 3D solid. Pick the endpoints shown in **Figure 9-20A** as the axis of revolution. Revolve the profile through 360°. Zoom extents and set the Conceptual visual style current to clearly see the object.

Now, you need to create one cam lobe. Change the UCS to the left orthographic preset. Then, draw a construction point in the center of the left end of the camshaft. Use the **Center** object snap and an appropriate **PDMODE** setting. Next, draw the profile

Figure 9-19.
For the most part, this object is symmetrical about its center axis. However, the cam lobes are not symmetrical about the axis.

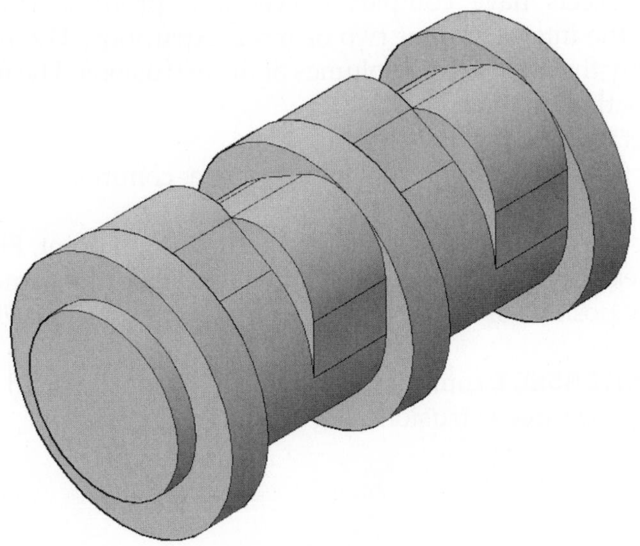

Figure 9-20.
A—This profile will be revolved to create the shaft and bearings. Notice the parallel constraints that have been applied. B—This is the profile of one cam lobe, which will be extruded.

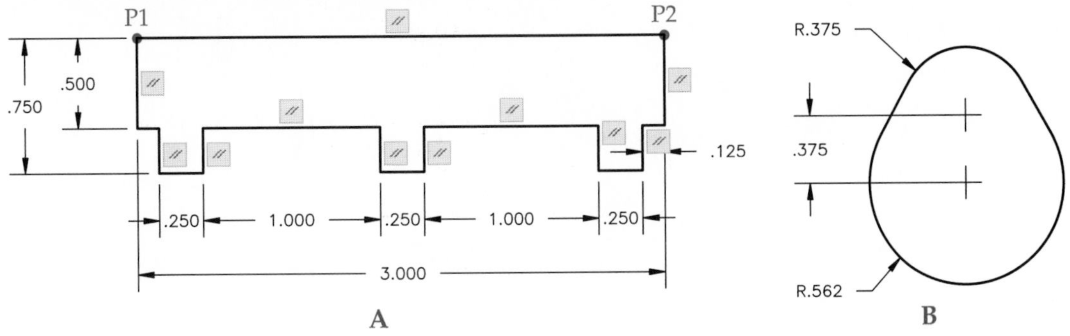

shown in Figure 9-20B. Use the construction point as the center of the large radius. You may want to create a new layer and turn off the display of the base camshaft.

Once the cam lobe profile is created, use the **REGION** command to create a region. Then, use the **EXTRUDE** command to extrude the region a height of –.5 units (into the camshaft). The extrusion should have a 0° taper. If you turned off the display of the base camshaft, turn it back on now.

One cam lobe is created, but it is not in the proper position. With the left UCS current, move the cam lobe –.375 units on the Z axis. If a different UCS is current, the axis of movement will be different. This places the front surface of the cam lobe on the back surface of the first bearing. Now, make a copy of the lobe that is located –.5 units on the Z axis. Finally, copy the first two cam lobes –1.25 units on the Z axis.

You now need to rotate the four cam lobes to their correct orientations. Make sure the left UCS is still current. Then, rotate the first and third cam lobes 30°. If a different UCS is current , you can use the **3DROTATE** command. The center of rotation should be the center of the shaft. There are many points on the shaft to which the **Center** object snap can snap; they are all acceptable. You can also use the construction point as the center of rotation. Rotate the second and fourth cam lobes –30° about the same center.

Finally, use the **UNION** command to join all objects. The final object should appear as shown in Figure 9-19. Use the view cube to see all sides of the object. You can also create a rotating display using the **3DORBIT** command.

Multiple Intersecting Extrusions

Many solid objects have complex curves and profiles. These can often be constructed from the intersection of two or more extrusions. The resulting solid is a combination of only the intersecting volumes of the extrusions. The following example shows the construction of a coat hook.

INTERSECT

Ribbon
Home
> Solid Editing
Solid
> Boolean
Solid, Intersect
Type
INTERSECT
IN

1. Construct the first profile, Figure 9-21A.
2. Construct the second profile located on a common point with the first, Figure 9-21B.
3. Construct the third profile located on the common point, Figure 9-21C.
4. Extrude each profile the required dimension into the same area. Be careful to specify positive or negative heights for each extrusion, Figures 9-21D and 9-21E.
5. Use the **INTERSECT** command to create a composite solid from the volume shared by the three extrusions, Figure 9-21F.

Figure 9-21.
Constructing a coat hook. A—Draw the first profile. B—Draw the second profile. C—Draw the third profile. All three profiles should have a common origin. D—Extrude each profile so that the extruded objects intersect. E—The extruded objects after the Conceptual visual style is set current. F—Use the **INTERSECT** command to create the composite solid. The final solid is shown here with the Conceptual visual style set current.

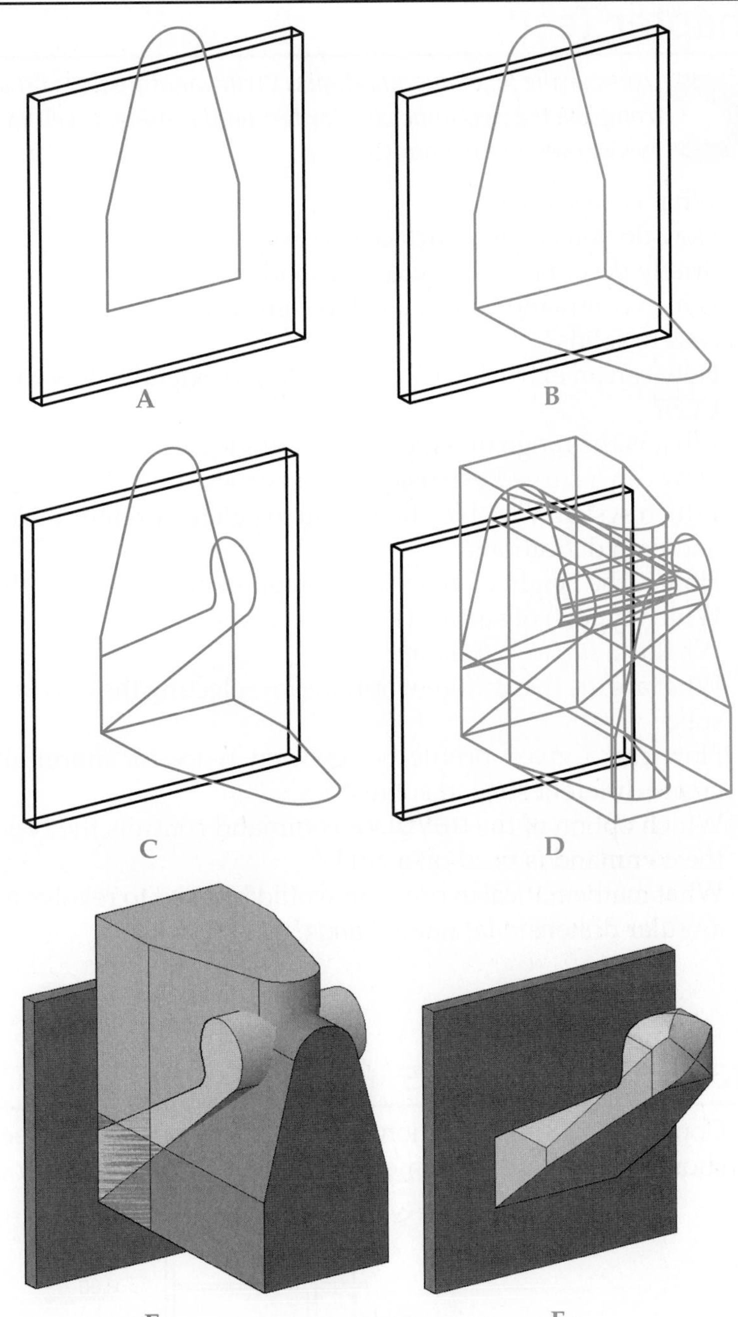

Chapter Test

Answer the following questions. Write your answers on a separate sheet of paper or complete the electronic chapter test on the student website.
www.g-wlearning.com/CAD

1. What is an *extrusion?*
2. How do you create a surface extrusion?
3. Briefly describe how to create a solid extrusion.
4. Which command can be used to convert circles and closed polylines with a thickness to solids?
5. How can an extrusion be constructed to extend below the XY plane of the current UCS?
6. What is the range in which a taper angle can vary?
7. How can a curved extrusion be constructed?
8. Which system variable allows you to delete or retain the original extruded objects and path definitions?
9. How is the height of an extrusion applied in relation to the original object?
10. Which type(s) of surface(s) can be extruded?
11. What is a *surface revolution?*
12. What are the five different options for selecting the axis of revolution for a revolved solid?
13. How can a given profile be revolved twice (or more) about the same axis and create different shaped solids?
14. Which option of the **REVOLVE** command controls the type of object created when the command is used on a circle?
15. What mathematical expression would be used to revolve a profile 90° less than the angular dimensional named ang4?

Drawing Problems

1. Construct a 12′ long section of wide flange structural steel with the cross section shown below. Use the dimensions given. Save the drawing as P09_01.

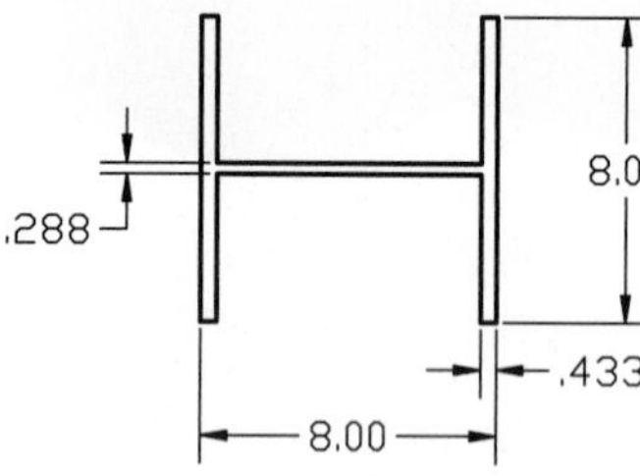

Problems 2–7. These problems require you to use a variety of solid modeling methods to construct the objects. Use **EXTRUDE**, **REVOLVE**, *solid primitives, new UCSs, Boolean commands, and editing tools such as sweeps, extrusions, and revolutions to assist in construction. Apply geometric constraints to the base 2D drawing as appropriate. Do not create section views. Save each as* **P09_***(problem number).*

2.

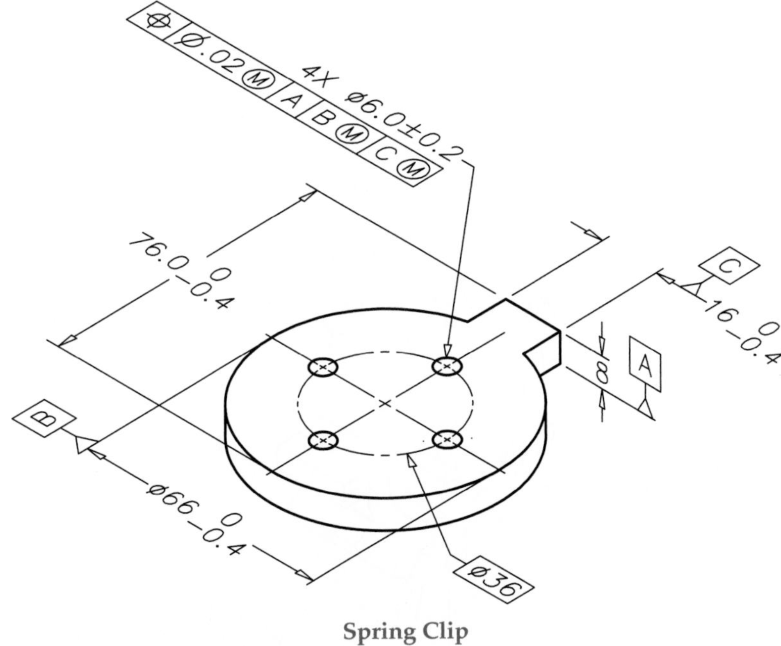

Spring Clip

3.

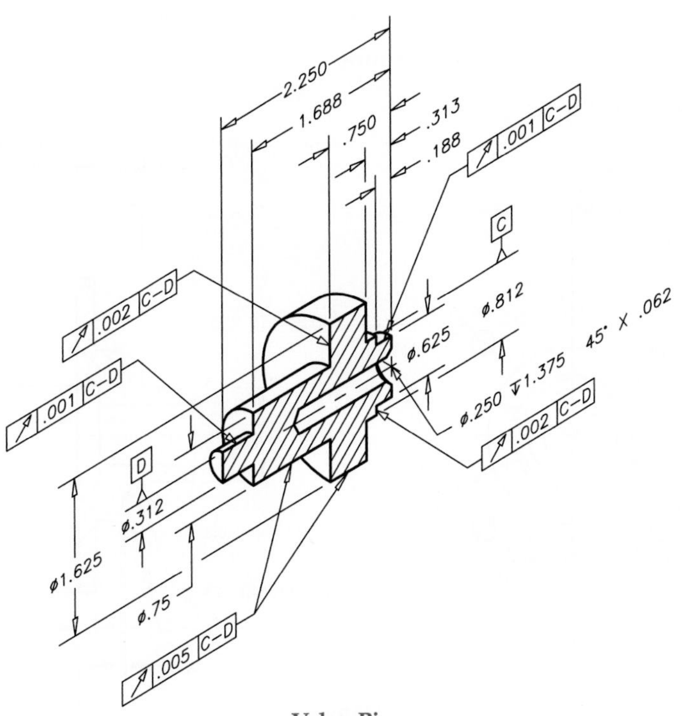

Valve Pin

4.

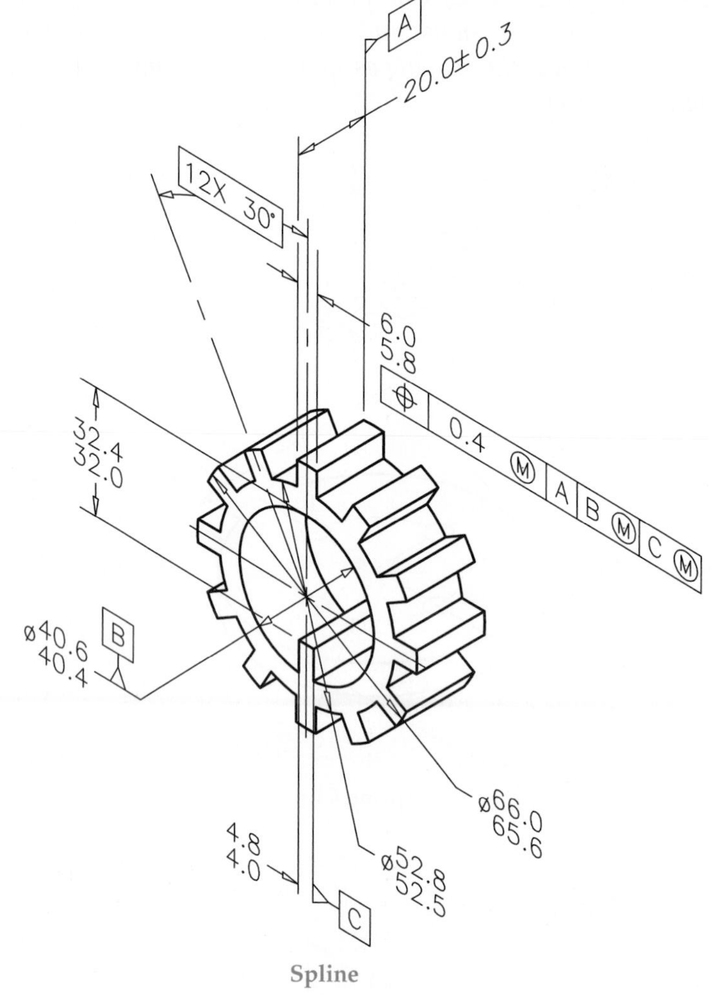

Spline

5.

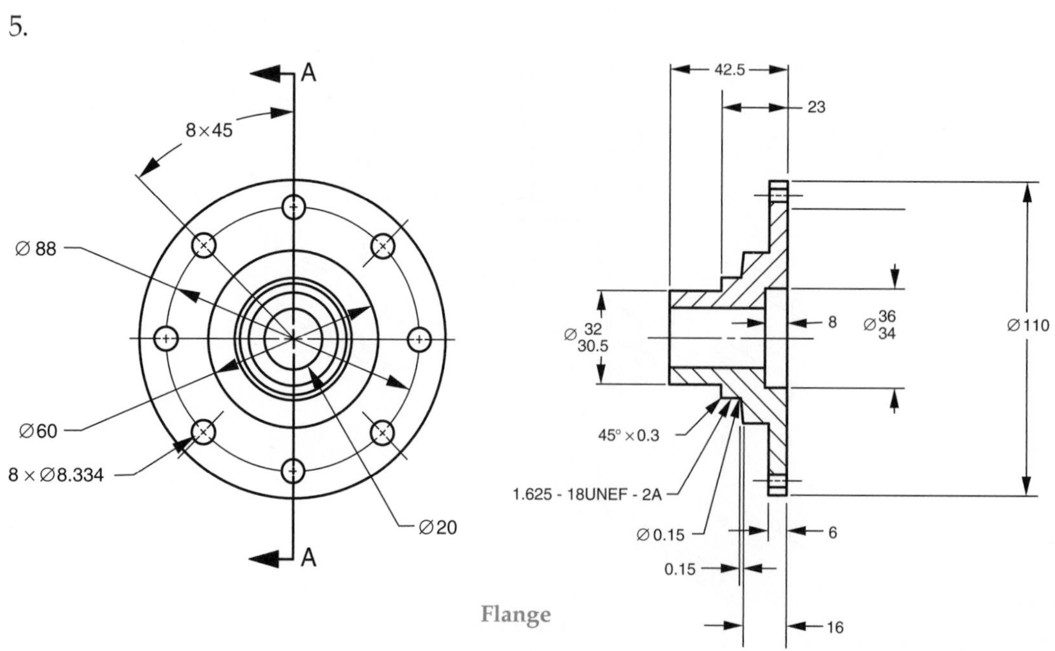

Flange

6.

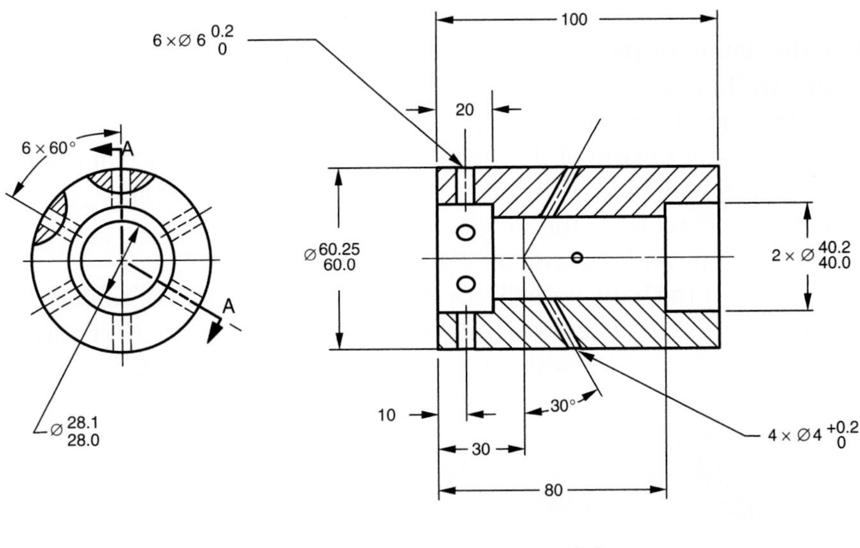

Nozzle

7.

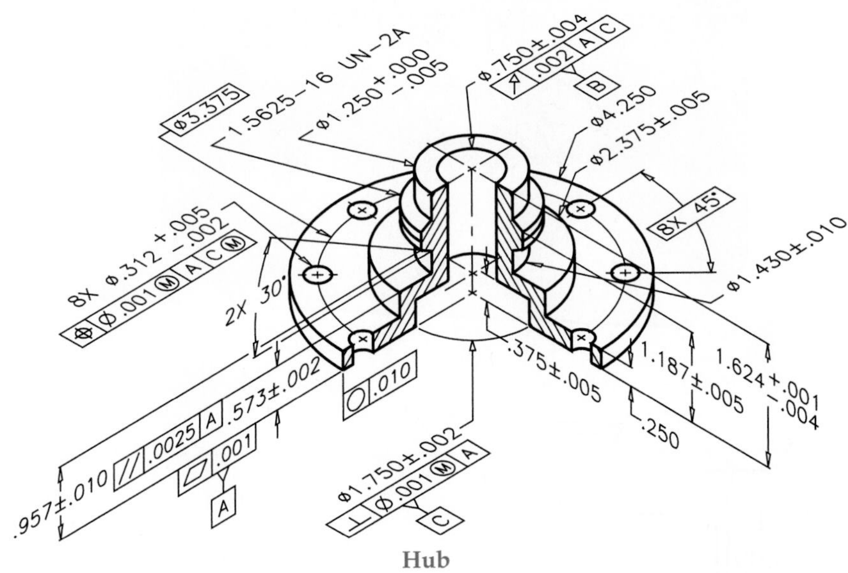

Hub

8. Create the stairway shown below using the following parameters. Save the file as P09_08.
   A. Use the detail for the riser and tread dimensions.
   B. There are 13 risers.
   C. The stairs are 42″ wide.
   D. The landing at the top of the stairs is 48″ long from the face of the last riser.
   E. The vertical wall is 15′-1″ high on the inside and 15′-7″ long.
   F. The floor is 8′ wide on the inside and 15′-7″ long.
   G. Draw the floor and the wall as 1″ thick.
   H. The center of the banister is 3″ away from the wall and 34″ above the steps.
   I. The ends of the banister are directly above the face of the first and last riser.
   J. Use the detail for the profile of the banister. Use **EXTRUDE** as needed.

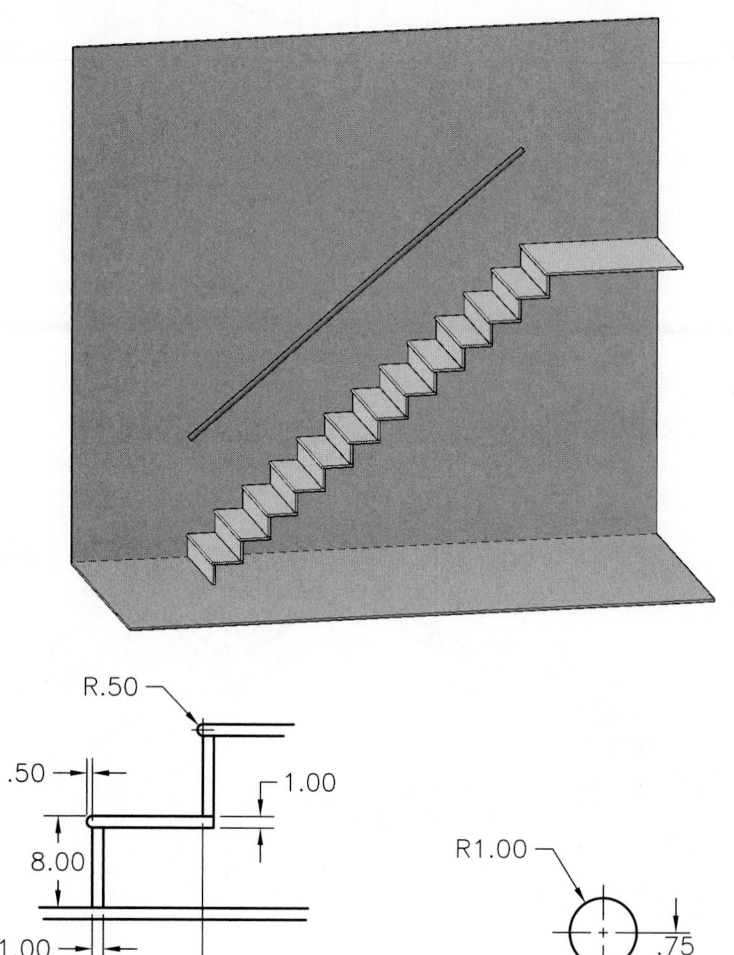

9. Construct picture frame moldings using the profiles shown below.
   A. Draw each of the closed profiles shown. Use your own dimensions for the details of the moldings.
   B. The length and width of A and B should be no larger than 1.5″ × 1″.
   C. The length and width of C and D should be no larger than 3″ × 1.5″.
   D. Construct an 8″ × 12″ picture frame using moldings A and B.
   E. Construct a 12″ × 24″ picture frame using moldings C and D.
   F. Save the drawing as P09_09.

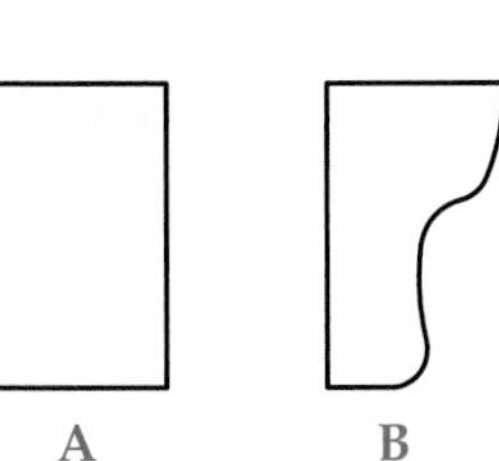

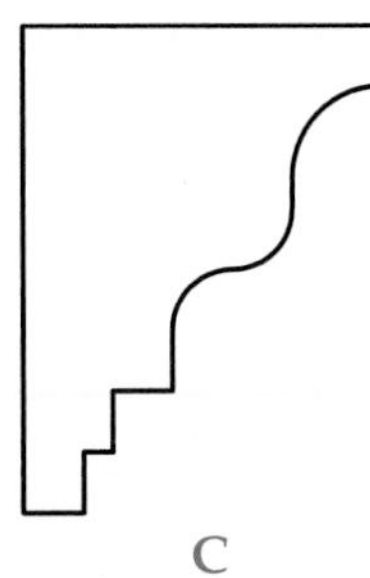

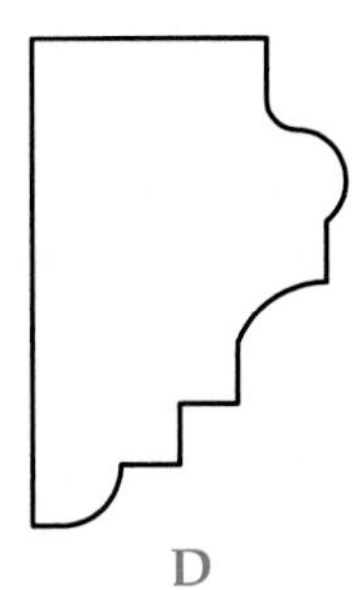

10. In this problem, you will refine the seat of the kitchen chair that you started in Chapter 2. You will use an extrusion following a path to create a curved, receding edge under the seat.
    A. Open P02_11 from Chapter 2. If you have not yet completed this model, do so now.
    B. Create a path for the extrusion by drawing a polyline that exactly matches either the upper or lower edge of the seat. Refer to the drawing shown below.
    C. Change the view and UCS as needed to display a plan view of the edge of the seat. Draw the profile shown below.
    D. Extrude the profile along the path and then subtract the extrusion from the seat.
    E. Save the drawing as P09_10.

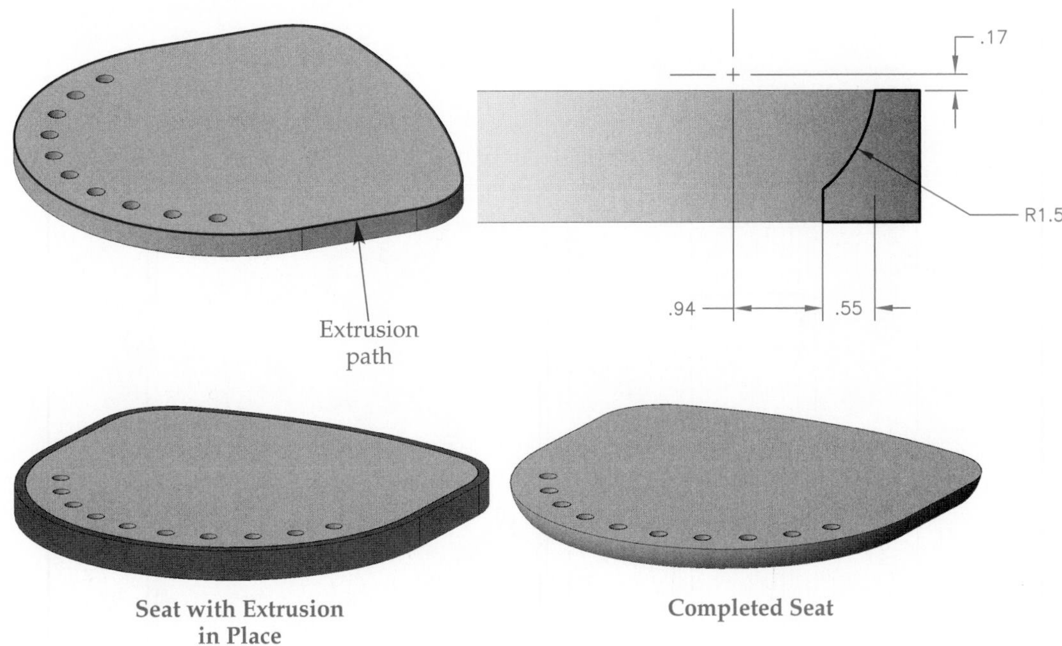

Seat with Extrusion in Place

Completed Seat

11. In this problem, you will be taking a manually drawn layout from an archive. You are to create a 3D model of the garage to update the archive.
    A. Review the manually drawn layout. Make note of the construction details shown.
    B. Research any additional details needed to construct the model. For example, the thickness of the doors is not listed as these are purchased items. However, you need to know these dimensions to draw the 3D model.
    C. Using what you have learned, create the garage as a solid model. Be sure to create all components, including the studs in the walls, the footings, and the anchor bolts.
    D. Create layers as needed. For example, you may wish to place the wall sheathing on a layer so it can be hidden to show the studs.
    E. Save the drawing as P09_11.

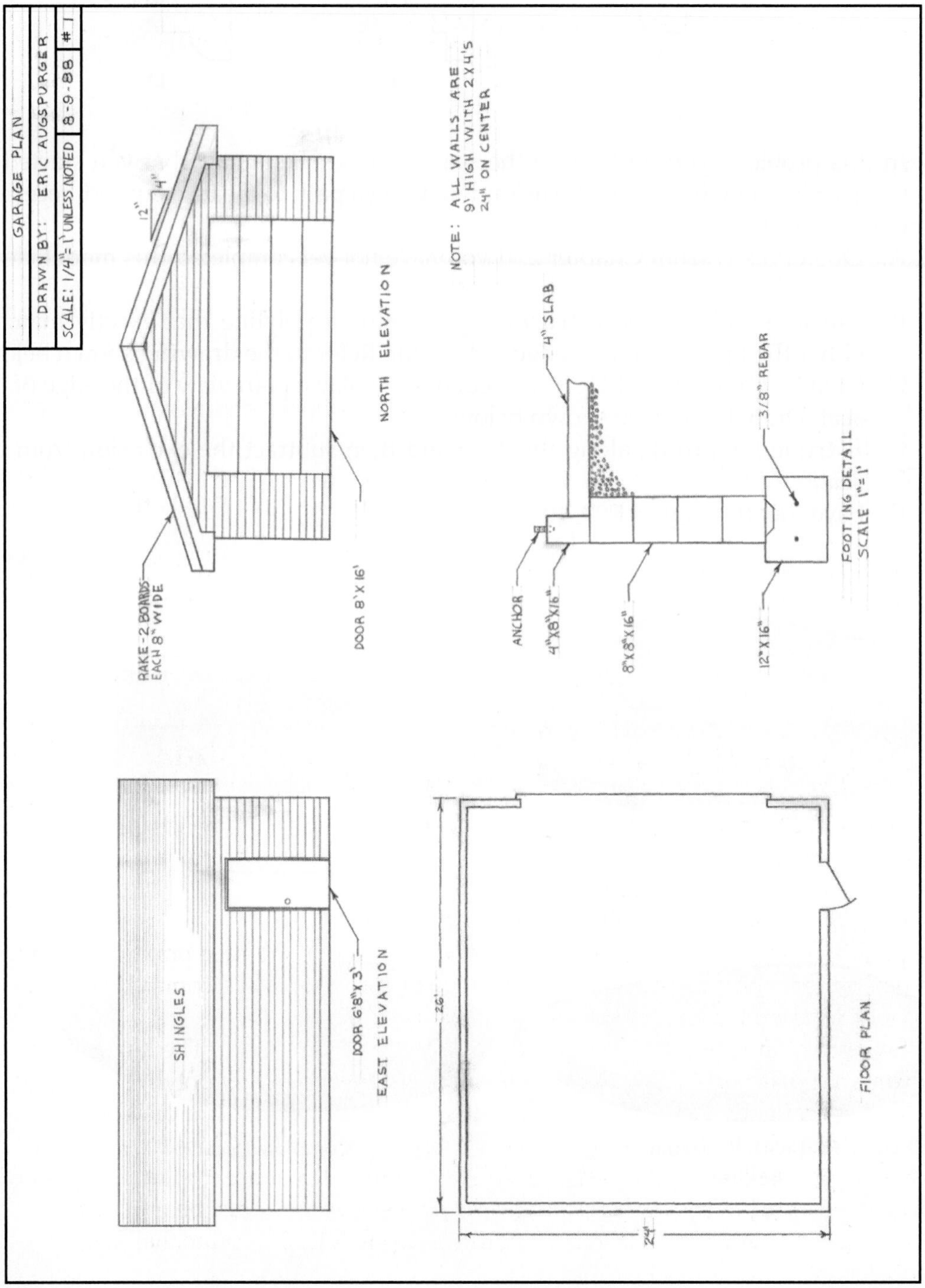

GARAGE PLAN
DRAWN BY: ERIC AUGSPURGER
SCALE: 1/4"=1' UNLESS NOTED | 8-9-88 | # 1

NOTE: ALL WALLS ARE 9' HIGH WITH 2X4'S 24" ON CENTER

12"
4"
RAKE-2 BOARDS EACH 8" WIDE
DOOR 8' X 16'
NORTH ELEVATION

4" SLAB
3/8" REBAR
ANCHOR
4"X8"X16"
8"X8"X16"
12"X16"
FOOTING DETAIL SCALE 1"=1'

SHINGLES
DOOR 6'8"X3'
EAST ELEVATION

26'
24'
FLOOR PLAN

CHAPTER 10
Sweeps and Lofts

Learning Objectives

After completing this chapter, you will be able to:

✓ Sweep 2D shapes along a 2D or 3D path to create a solid or surface object.
✓ Create 3D solid or surface objects by lofting a series of cross sections.

In the previous chapter, you learned about extruded solids and surfaces. Sweeps and lofts are similar to extrusions. In fact, an extrusion is really just a type of sweep. A ***sweep*** is an object created by extruding a single 2D profile along a path object. Sweeping an open shape along the path results in a surface object. If a closed shape is swept, a solid or surface object can be created. A ***loft*** is an object created by extruding between two or more 2D profiles. The shape of the loft object blends from one cross-sectional profile to the next. The profiles can control the loft or it can be controlled by one path or multiple guide curves. As with a sweep, open shapes result in surfaces and closed shapes give you solids. Open and closed shapes cannot be used together in the same loft.

Creating Swept Surfaces and Solids

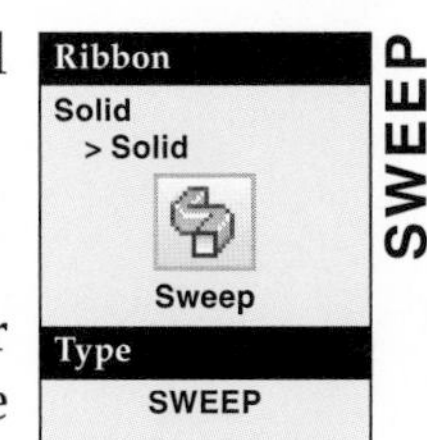

The **SWEEP** command is used to create swept surfaces and solids. The command requires at least two objects:

- 2D shape to be swept.
- 2D or 3D shape to be used as the sweep path.

The profile can be aligned with the path, you can specify the base point, a scale factor can be applied, and the profile can be twisted as it is swept. The command procedure and options are the same for both swept solids and surfaces.

Sweeping an open shape creates a surface. See Figure 10-1. The objects that can be swept to create surfaces include lines, arcs, elliptical arcs, 2D polylines, 2D and 3D splines, traces, and 3D edge subobjects. Sweeping a closed shape creates a solid by default, but surfaces may be created as well. See Figure 10-2. Closed shapes that can be swept include circles, ellipses, closed 2D polylines, closed 2D splines, regions, 2D solids, traces, and 3D solid face subobjects. The sweep path can be a line, arc, circle, ellipse, elliptical arc, 2D polyline, 2D spline, 3D polyline, 3D spline, helix, or 3D edge subobject.

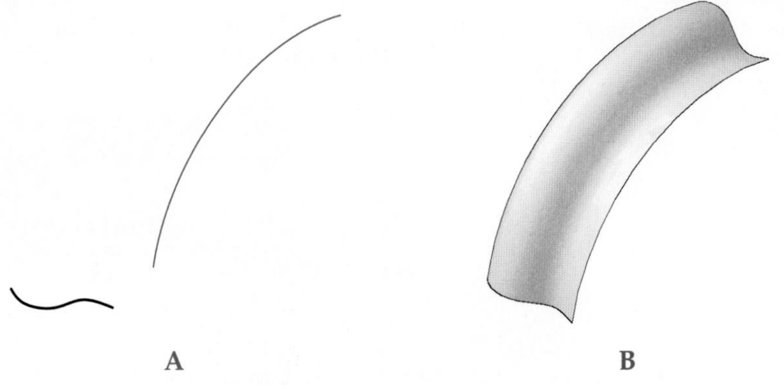

Figure 10-1.
A—This open shape will be swept along the path (shown in color). B—The resulting surface.

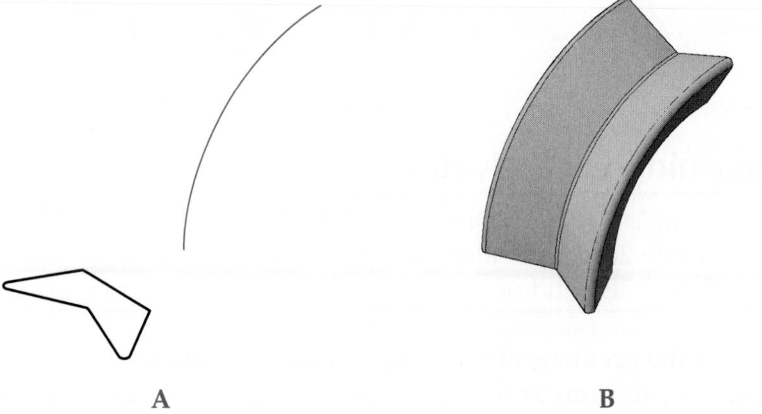

Figure 10-2.
A—This closed shape will be swept along the path (shown in color). B—The resulting solid.

When the command is initiated, you are prompted to select the objects to sweep. Select the profile(s) and press [Enter]. Planar faces of solids may be selected by holding the [Ctrl] key as you select. Multiple profiles can be selected. They are swept along the same path, but separate objects are created.

Next, you are prompted to select the path. The path and profile can lie on the same plane. Select the object to be used as the sweep path and press [Enter]. To select the edge of a surface or solid as the path, press the [Ctrl] key and then select the edge. The profile is then moved to be perpendicular to the path and extruded along the path. The sweep starts at the endpoint of the path nearest to where you selected it.

The **Mode** option that is available when the **SWEEP** command is initiated controls the closed profiles creation mode. This allows you to change the way in which closed shapes are handled. Normally, the **SWEEP** command produces a solid object when a closed shape is swept, but surfaces may be created by changing this mode. For example, a circle swept with a line as the path will normally create a solid cylinder. Setting the **Mode** option to **Surface** will result in a tube created as a surface model.

Exercise 10-1

Complete the exercise on the student website.
www.g-wlearning.com/CAD

Changing the Alignment of the Profile

By default, the profile is aligned perpendicular to the sweep path. However, you can create a sweep where the profile is not perpendicular to the path. See **Figure 10-3.**

Figure 10-3.
A—The profile and path for the sweep. B—By default, the profile is aligned perpendicular to the path when swept. C—Using the **Alignment** option, the profile can be swept so it is not perpendicular to the path.

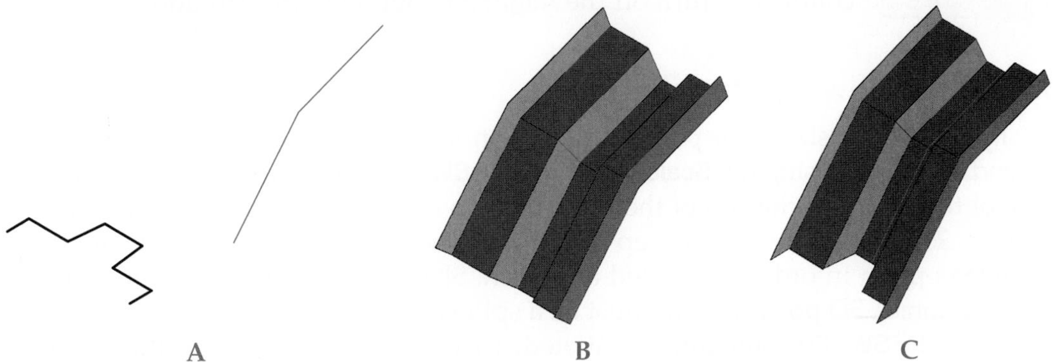

After the **SWEEP** command is initiated, select the profile and press [Enter]. Before selecting the path, enter the **Alignment** option. The default setting of **Yes** means that profile will be moved so it is perpendicular to the path. If you select **No**, the profile is kept in the same position relative to the path as it is swept. The position of the 2D shape determines the alignment.

Changing the Base Point

The base point is the location on the shape that will be moved along the path to create the sweep. By default, if the 2D shape intersects the path, the profile is swept along the path at the point of intersection. If the 2D shape does not intersect the path, the default base point depends on the type of object being swept. When lines and arcs are swept, the default base point is their midpoint. Open polylines have a default base point at the midpoint of their total length.

The base point can be any point on the 2D shape or anywhere in the drawing. See Figure 10-4. To change the base point, use the **Base point** option of the **SWEEP** command. When the command is initiated, select the profile and press [Enter]. Before selecting the path, enter the **Base point** option. Next, pick the new base point. It does not have to be on an existing object. Once the new base point is selected, pick the path to create the sweep.

Figure 10-4.
A—The profile and path for the sweep. B—The sweep is created with the default base point. C—The end of the path is selected as the base point. Notice the difference in this sweep and the one shown in B.

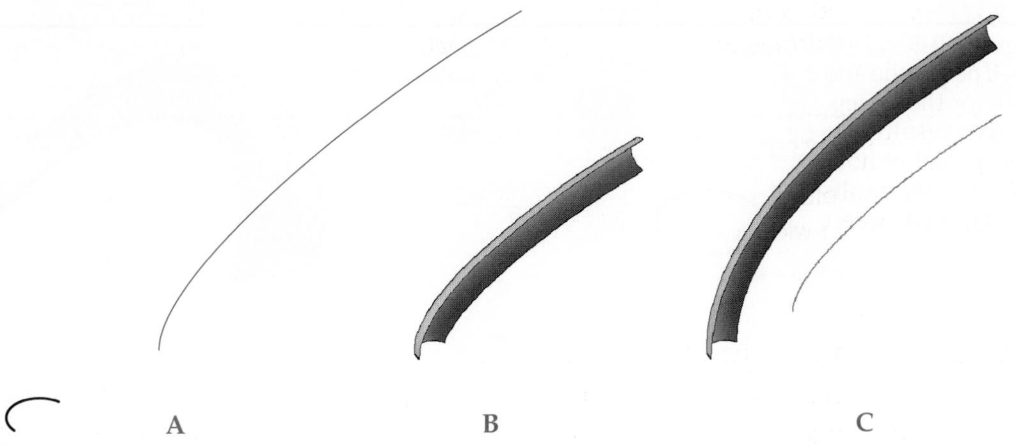

PROFESSIONAL TIP

If the location of the base point in relationship to the path is important, line up the shape with the path before starting the **SWEEP** command. Turn off the **Alignment** option in this situation.

Scaling the Sweep Profile

By default, the size of the profile remains uniform from the beginning of the path to the end. However, using the **Scale** option of the **SWEEP** command, you can change the scale of the profile at the end of the path. This, in effect, tapers the sweep. Figure 10-5 shows a .25 scale applied to a sweep object. A 2D polyline must be edited using the **Fit** or **Spline** option in order to be used as a path. Sharp corners will not work with the **Scale** option. A 3D polyline path must be a spline.

Once the **SWEEP** command is initiated, select the profile and press [Enter]. Before selecting the path, enter the **Scale** option. You are prompted for the scale. Enter the scale value and press [Enter]. The scale value must be greater than zero. You can also enter the **Reference** option. With this option, pick two points for the first reference line and then two points for the second reference line. The difference in scale between the two distances is the scale value. Once the scale is set, pick the path to create the sweep. The **Expression** option allows you enter a mathematical expression to constrain the scaling of the sweep. This option only works when creating a surface sweep.

Twisting the Sweep

The profile can be rotated as it is swept along the length of the path by using the **Twist** option of the **SWEEP** command. The angle that you enter indicates the rotation of the shape along the path of the sweep. The higher the number, the more twists in the sweep. Figure 10-6 shows how a simple, closed profile and a straight line can be used to create a milling tool. The profile was swept with a 270° twist.

Once the **SWEEP** command is initiated, select the profile and press [Enter]. Then, before selecting the path, enter the **Twist** option. You are prompted for the twist angle or to enter the **Bank** option.

Banking is the natural rotation of the profile on a 3D sweep path, similar to a banked curve on a racetrack. See Figure 10-7. The path must be 3D (nonplanar) to set banking. The banking option is disabled for a 2D path, although you can go through the motions of turning it on when creating the sweep. Once you use the **Bank** option to turn banking on, it is on by default the next time the **SWEEP** command is used. To turn it off, enter a twist angle of zero (or the twist angle you wish to use). The **Expression** option allows you to enter a mathematical expression to constrain the number of rotations in the sweep. This option only works when creating a surface sweep.

Figure 10-5.
A—The profile and path for the sweep.
B—The resulting sweep. Notice how the .25 scale results in a tapered sweep.

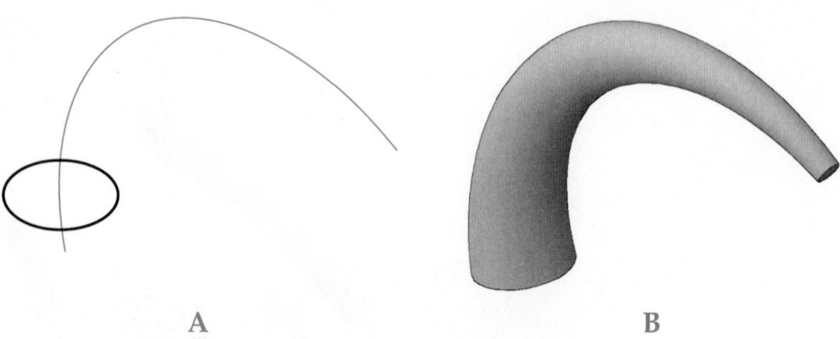

PROFESSIONAL TIP

The sweep options can be changed after the sweep is created using the **Properties** palette. In the **Geometry** section, you will find Profile rotation (alignment), Bank (banking), Twist along path (twist angle), and Scale along path (scale) settings.

Exercise 10-2

Complete the exercise on the student website.
www.g-wlearning.com/CAD

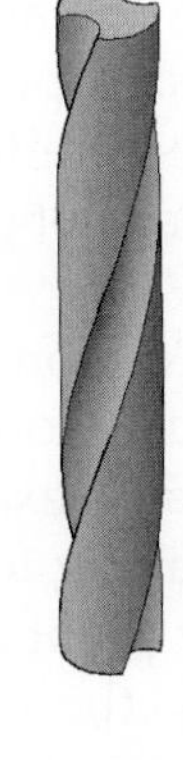

A B

Figure 10-6.
A—The profile and path for creating the end mill. B—The resulting end mill model. Notice how the profile is twisted (rotated) as it is swept.

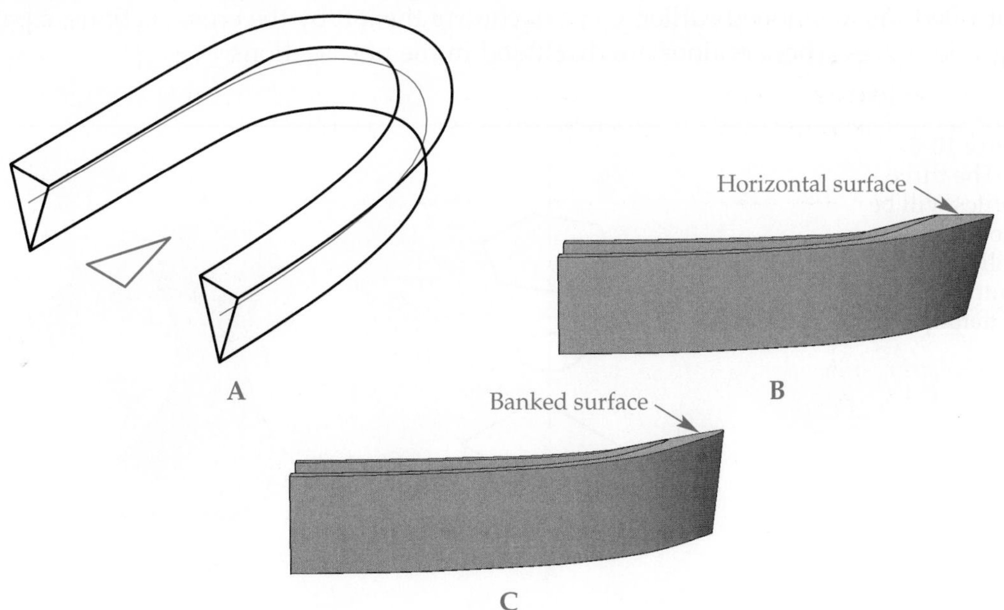

Figure 10-7.
A—The profile and path for the sweep are shown in red. B—Banking is off for this sweep. When viewed from the side, you can see that the profile does not bank through the curve. Look at the upper-right corner. C—Banking is on for this sweep. Notice how the profile banks, or leans, through the curve. Compare this to B.

Creating Lofted Objects

LOFT

Ribbon
Solid > Solid
Loft
Type
LOFT

The **LOFT** command is used to create lofted surfaces and solids based on a series of cross-sectional profiles. Figure 10-8 shows an example of a loft formed from a rectangle, circle, and polygon. The loft may be guided by only the cross sections, as shown in the figure, by a path, or by guide curves. Lofting open shapes results in a surface object, while lofting closed shapes creates a solid. Open and closed shapes cannot be combined in the same loft.

Objects that can be used as cross sections include lines, circles, arcs, points, ellipses, elliptical arcs, 2D polylines, 2D splines, regions, edge subobjects, face subobjects, 2D solids, helices, and traces. Points may be used for the first and last cross sections only. The loft path may be a line, circle, arc, ellipse, elliptical arc, spline, helix, 2D or 3D polyline, or edge subobject. Guide curves may be composed of lines, arcs, elliptical arcs, 2D or 3D splines, 2D or 3D polylines, and edge subobjects. However, 2D polylines are limited to only one segment.

Once the command is initiated, you are prompted to select the cross-sectional profiles. Pick each profile in the order in which it should appear in the loft and press [Enter]. Be sure to individually select the cross sections in the order of the loft creation. You may not get the desired loft if you randomly select them or use a window selection.

The **Mode** option that is available when you start the command controls the closed profiles creation mode and allows you to change the way that closed shapes are handled. It behaves exactly as it does in the **SWEEP** command.

To avoid having to draw point objects for cross sections, the **Point** option allows you to pick any point as either the start point or the end point of the loft. If you pick a point first, it is the start point and you can pick as many shapes as you want as the other cross section. If you pick the other cross sections first, then the point must be the endpoint of the loft. The other cross sections must be closed shapes. Open shapes will not work with the **Point** option.

If you need to use the edges of existing 3D objects as cross sections, the **Join multiple edges** option works well. The edges must be touching at their end points and form a cross section. See Figure 10-9. This option is used to define a single cross section. When you press [Enter], you are again prompted to select cross sections. The option can be used again to select additional cross sections.

After selecting cross sections, you are prompted to select how the loft is to be controlled. As mentioned earlier, you can control the loft by the cross sections, a path, or guide curves. These options are discussed in the next sections.

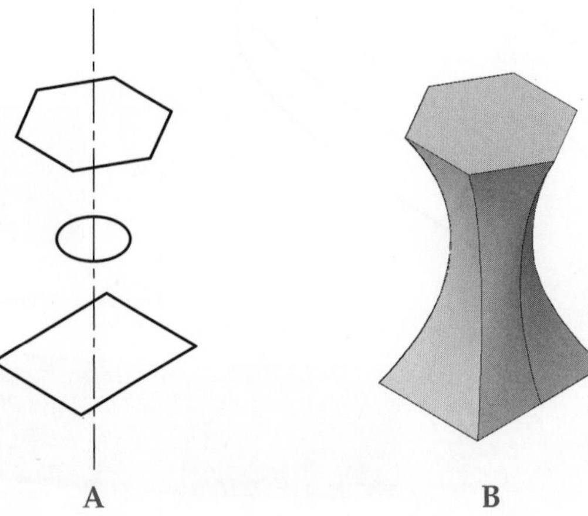

Figure 10-8.
A—The three profiles will be lofted to create a solid. B—The resulting loft with the default settings.

Figure 10-9.
The **Join multiple edges** option allows you to make a quick transition from the edges of one 3D object to another. A—Use the option twice to select two cross sections. B—The resulting loft is a transition between the two objects.

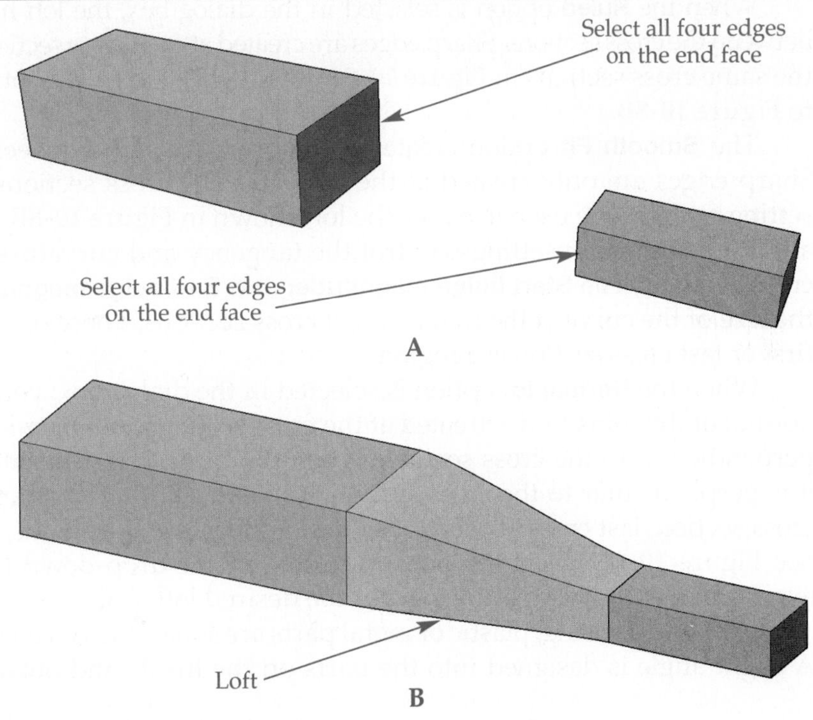

Controlling the Loft with Cross Sections

The **Cross-sections only** option of the **LOFT** command is useful when the 2D cross sections are drawn in their proper locations in space. The command determines the transition from one cross section to the next. The cross sections are not moved by the command.

When you select the **Settings** option, the **Loft Settings** dialog box appears, **Figure 10-10.** The settings in this dialog box control the transition or contour between cross sections. As settings are changed, the preview is updated. When all settings have been made, pick the **OK** button to close the dialog box and create the loft.

Figure 10-10.
The **Loft Settings** dialog box is used to control the transition between profiles.

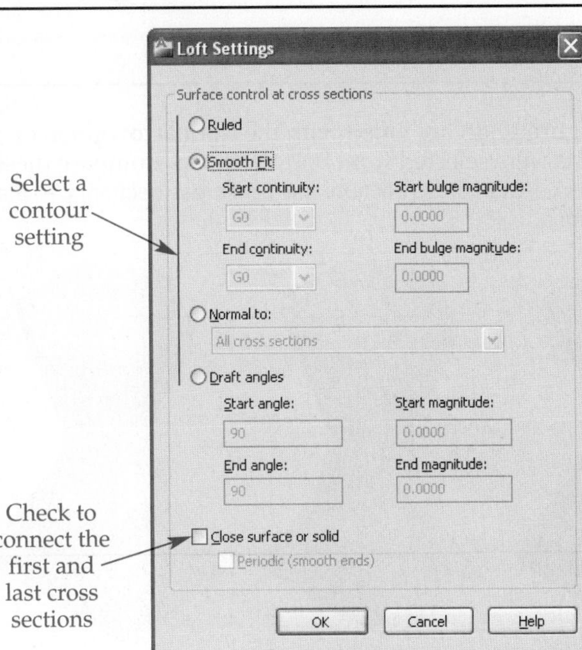

When the **Ruled** option is selected in the dialog box, the loft has straight transitions between the cross sections. Sharp edges are created at each cross section. Figure 10-11 shows the same cross sections in Figure 10-8A lofted with the **Ruled** option on. Compare this to Figure 10-8B.

The **Smooth Fit** option creates a smooth transition between the cross sections. Sharp edges are only created at the first and last cross sections. This is the default setting and the one used to create the loft shown in Figure 10-8B. The **Start continuity:** and **End continuity:** settings control the tangency and curvature of the first and last cross sections. The **Start bulge magnitude:** and **End bulge magnitude:** settings control the size of the curve at the first and last cross sections. These options only apply if the first or last cross section is a region.

When the **Normal to:** option is selected in the dialog box, you can choose how the normal of the transition is treated at the cross sections. A ***normal*** is a vector extending perpendicular to the cross section. When the transition is normal to a cross section, it is perpendicular to the cross section. You can set the transition normal to the first cross section, last cross section, both first and last cross sections, or all cross sections. See Figure 10-12. Select the normal setting in the drop-down list. You will have to experiment with these settings to get the desired loft shape.

In manufacturing, plastic or metal parts are sometimes formed in a two-part mold. A slight angle is designed into the parts on the inside and outside surfaces to make

Figure 10-11.
The profiles in Figure 10-8A are lofted with the **Ruled** option selected in the **Loft Settings** dialog box. Compare this to Figure 10-8B.

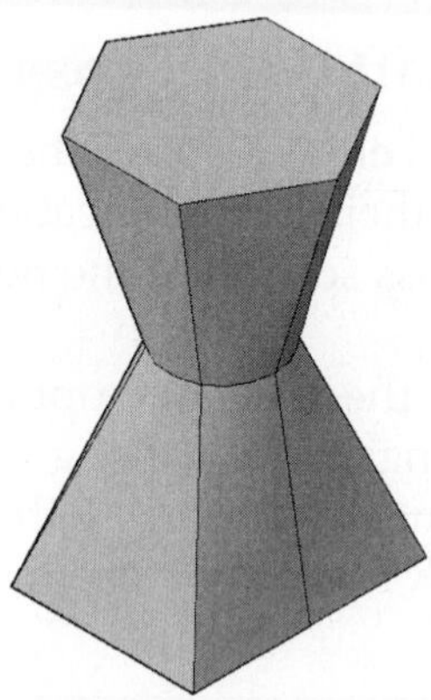

Figure 10-12.
The profiles in Figure 10-8A are lofted with the **Normal to:** option on in the **Loft Settings** dialog box. Cross sections were selected from bottom to top. Compare these results with Figure 10-8B and Figure 10-11. A—Start cross section. B—End cross section. C—Start and End cross sections. D—All cross sections.

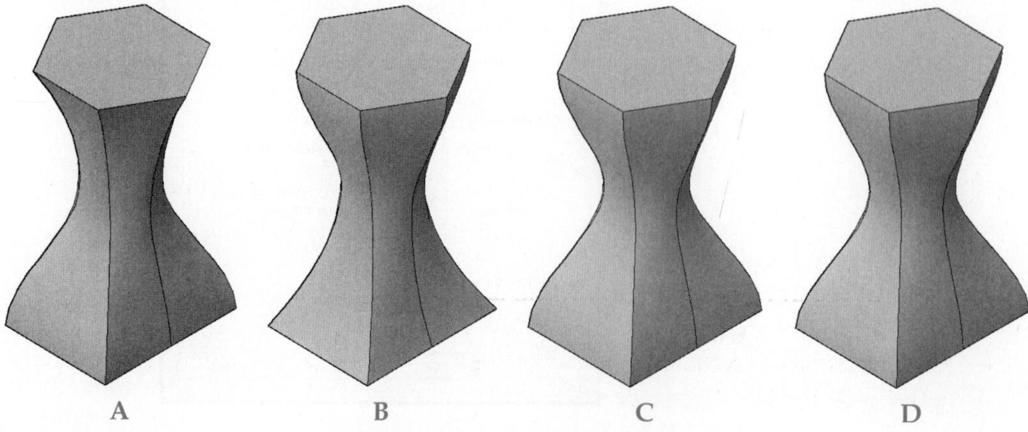

Figure 10-13.
When setting the draft angle, you can set the angle and the magnitude. A—Draft angle of 90° and a magnitude of zero. B—Draft angle of 30° and a magnitude of 180. C—Draft angle of 60° and a magnitude of 180.

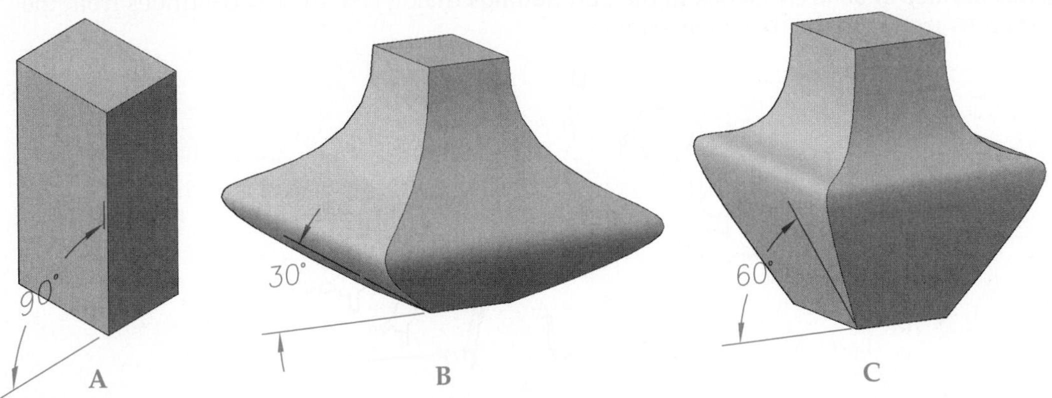

removing the part from the mold easier. This taper is called a ***draft angle.*** The **Draft angles** option allows you to add a taper to the beginning and end of the loft.

When setting the draft angle, you can set the angle and magnitude. See Figure 10-13. The default draft angle is 90°, which means the transition is perpendicular to the cross section. The magnitude represents the relative distance from the cross section, in the same direction as the draft angle, before the transition starts to curve toward the next cross section. Magnitude settings depend on the size of the cross sections, the draft angle values, and the distance between the cross sections. You may have to experiment with different magnitude and angle settings to get the desired loft shape.

The **Close surface or solid** option is used to connect the last cross section to the first cross section. See Figure 10-14. This option "closes" the loft, similar to the **Close** option of the **LINE** or **PLINE** command. The shapes from Figure 10-8 are shown in Figure 10-15 with the **Close** option on. This option is only available when the **Ruled** or **Smooth Fit** radio button is selected.

If the **Smooth Fit** radio button is on and the **Close surface or solid** check box is checked, the **Periodic (smooth ends)** check box is available. If a closed-loop loft is created similar to the loft in Figure 10-14, the seam may kink if the loft is reshaped in some way. Checking the **Periodic (smooth ends)** check box will help alleviate this problem.

Exercise 10-3

Complete the exercise on the student website.
www.g-wlearning.com/CAD

Controlling the Loft with Guide Curves

Guide curves are lines that control the shape of the transition between cross sections. They do not have to be *curves.* They can be lines, arcs, elliptical arcs, splines (2D or 3D), or polylines (2D or 3D). There are four rules to follow when using guide curves:

- The guide curve should start on the first cross section.
- The guide curve should end on the last cross section.
- The guide curve should intersect all other cross sections.
- The surface control in the **Loft Settings** dialog box must be set to **Smooth Fit** (**LOFTNORMALS** = 1).

Figure 10-14.
A—These profiles will be used to create a sealing ring. They should be selected in a counterclockwise direction starting with the first cross section. B—The resulting loft with the default settings. Notice the gap between the first and last cross sections. C—By checking the **Close surface or solid** check box in the **Loft Settings** dialog box, the loft continues from the last cross section to the first cross section.

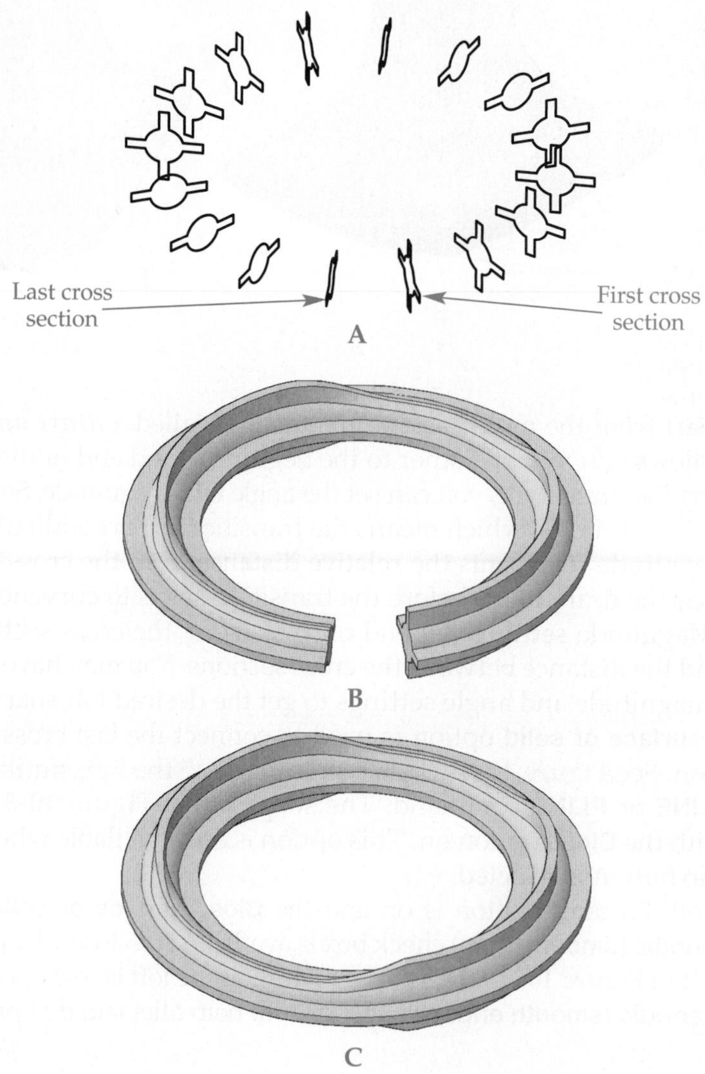

Figure 10-15.
The same cross sectional shapes from Figure 10-8 are used in this loft, however the **Close** option has been applied. Notice how the loft is inside out.

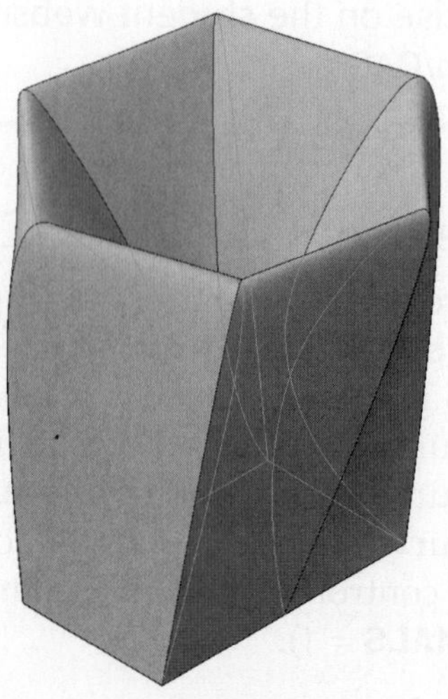

When the **Guides** option is entered, you are prompted to select the guide curves. Select all of the guide curves and press [Enter]. The loft is created. The order in which guide curves are selected is not important.

For example, Figure 10-16A shows two circles that will be lofted. If the **Cross-sections only** option is used, a cylinder is created, Figure 10-16B. However, if the **Guides** option is used and the two guide curves shown in Figure 10-16A are selected, one side of the cylinder is deformed similar to a handle or grip. See Figure 10-16C.

Lofting is used to create open-contour shapes such as fenders, automobile interior parts, fabrics, and other ergonomic consumer products. Figure 10-17 shows the use of open 2D splines in the construction of a fabric covering. Notice how each cross section

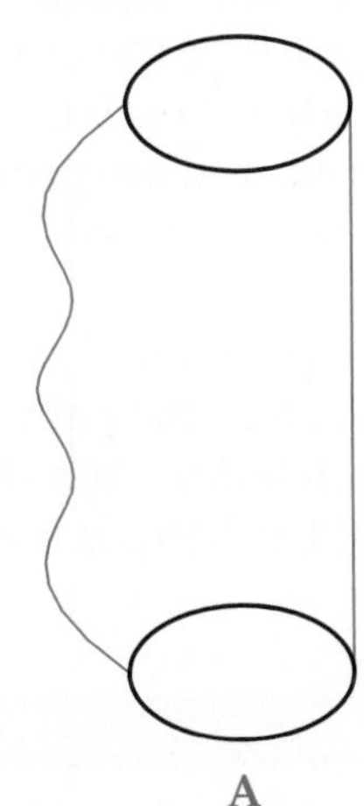

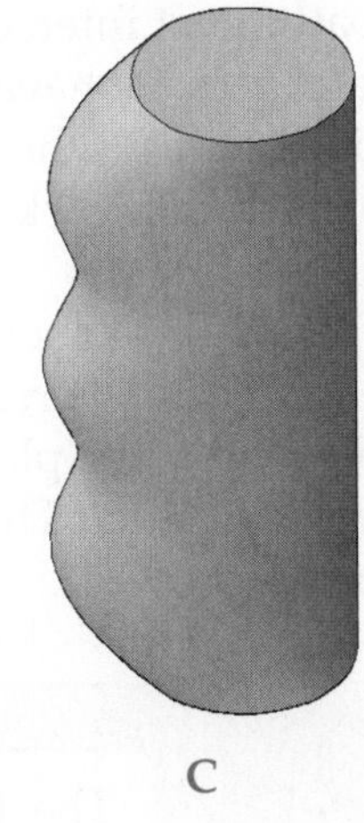

Figure 10-16.
A—These two circles will be lofted. The lines shown in color will be used as guide curves. B—When the circles are lofted using the **Cross-sections only** option, a cylinder is created. C—When the **Guides** option is used and the guide curves shown in A are selected, the resulting loft is shaped like a handle or grip.

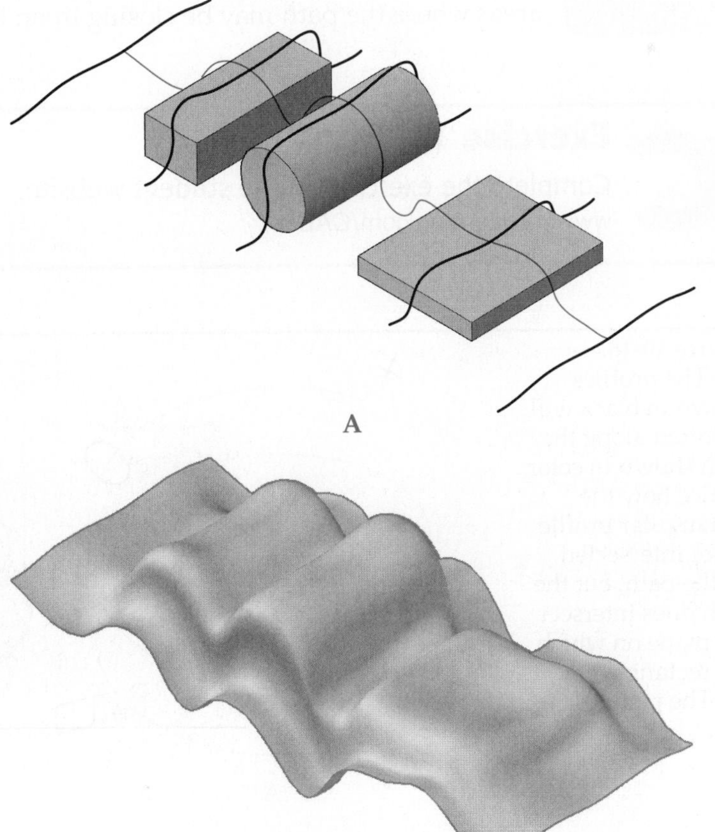

Figure 10-17.
A—The open profiles shown in black and the guide curve shown in color will be used to create a fabric covering for the three solid objects. B—The resulting fabric covering. This is a surface because the profiles were open.

is intersected by the guide curve. There is a cross section at the beginning of the guide curve and one at the end. These conditions fulfill the rules outlined earlier.

CAUTION

Guide curves only work well when the surface control is set to **Smooth Fit** (**LOFTNORMALS** = 1). If you get an error message when using guide curves or the curves are not reflected in the end result, make sure **LOFTNORMALS** is set to 1 and try it again.

Controlling the Loft with a Path

The **Path** option of the **LOFT** command places the cross sections along a single path. The path must intersect the planes on which each of the cross sections lie. However, the path does *not* have to physically touch the edge of each cross section, as is required of guide curves. When the **Path** option is entered, you are prompted to select the path. Once the path is picked, the loft is created. The cross sections remain in their original positions.

Figure 10-18 shows how 2D shapes can be positioned at various points on a path to create a loft. The rectangular shape does not cross the path. However, as long as the path intersects the plane of the rectangle, which it does, the shape will be included in the loft definition. The last shape at the top of the helix is a point object, causing the loft to taper.

PROFESSIONAL TIP

The **LOFT** command does not allow self-intersecting lofts to be created. Unfortunately, the error message you receive only states: The selected entities are not valid. If you see this message, look for areas where the path may be closing in on itself.

Exercise 10-4

Complete the exercise on the student website.
www.g-wlearning.com/CAD

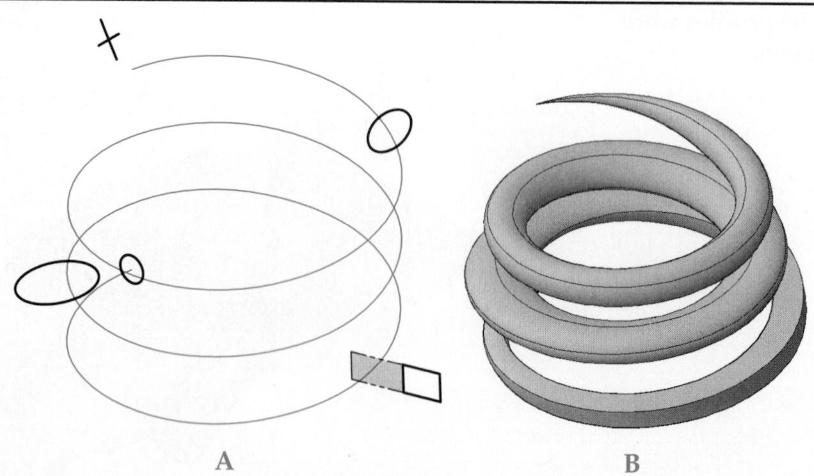

Figure 10-18.
A—The profiles shown in black will be lofted along the path shown in color. Notice how the rectangular profile is not intersected by the path, but the path does intersect the plane on which the rectangle lies.
B—The resulting loft.

Chapter Test

Answer the following questions. Write your answers on a separate sheet of paper or complete the electronic chapter test on the student website.
www.g-wlearning.com/CAD

1. What is a *loft?*
2. What option of the **SWEEP** command determines if it will be a solid or a surface?
3. When using the **SWEEP** command, on which endpoint of the path does the sweep start?
4. What is the purpose of the **Base Point** option of the **SWEEP** command?
5. After the sweep or loft is created, how can the creation options be changed?
6. Which objects may be used as a sweep path?
7. How is the alignment of a sweep set to be perpendicular to the start of the path?
8. Which **SWEEP** command option is used to taper the sweep?
9. What is the difference between the **Ruled** and **Smooth Fit** options in the **LOFT** command?
10. What does the **Bank** option of the **LOFT** command do?
11. Where is the check box that will close the loft, similar to a polyline, and what is its name?
12. List five objects that may be used as guide curves in a loft.
13. What are the four rules that must be followed when using guide curves?
14. When using the **Path** option of the **LOFT** command, what must the path intersect?
15. How can a loft be created so it tapers to a point at its end?

Drawing Problems

1. Create the lamp shade shown below. Create two separate loft objects for the top and the bottom. Then, union the two pieces. Finally, scale a copy and hollow out the lamp shade. Save the drawing as P10_01.

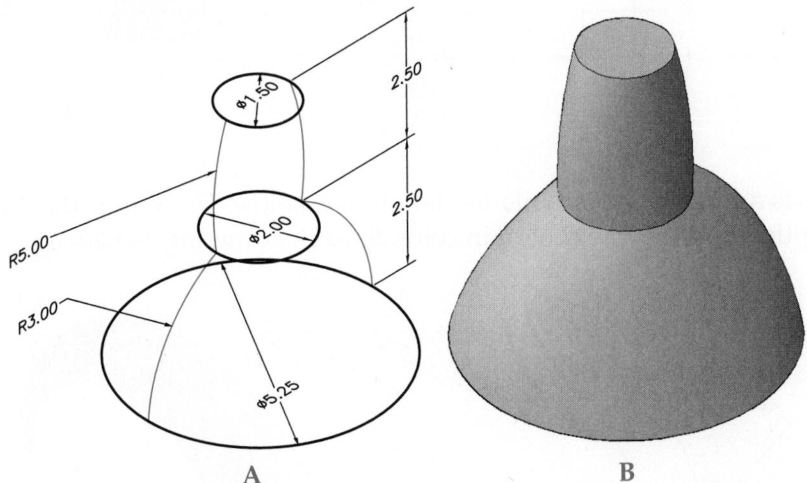

2. Create the two shampoo bottles shown below. One design uses cross sections only and the other uses a guide curve. Each bottle is made up of two loft objects. Join the pieces so each bottle is one solid. Save the drawing as P10_02.

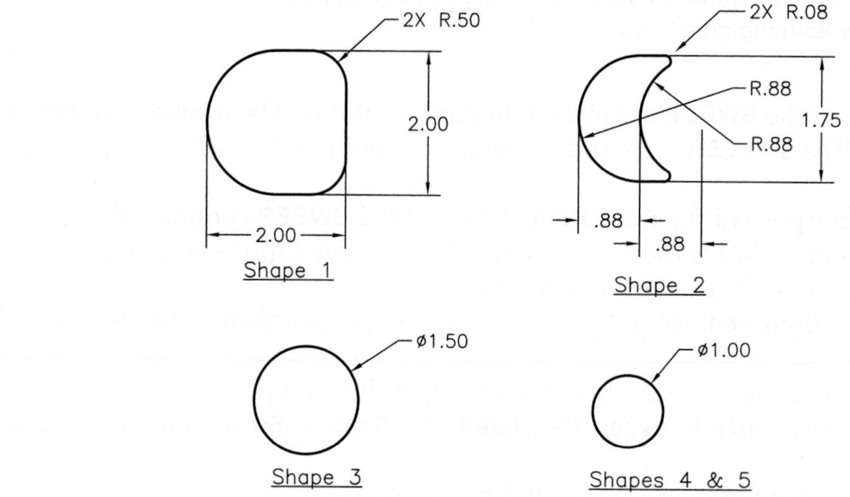

2X R.50
2.00
2.00
Shape 1

2X R.08
R.88
1.75
R.88
.88
.88
Shape 2

Ø1.50
Shape 3

Ø1.00
Shapes 4 & 5

A

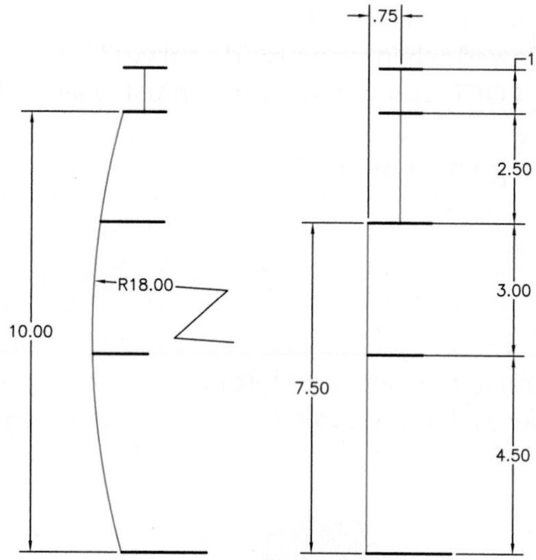

.75
1.00
2.50
R18.00
10.00
3.00
7.50
4.50

B

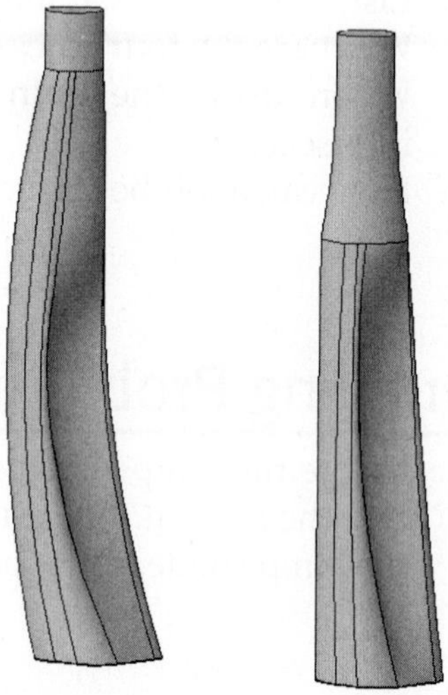

C

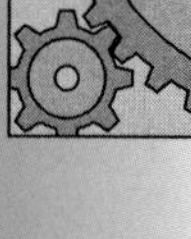

3. Create as a loft the automobile fender shown below. Use either the **Guide** or the **Path** option and the line shown in color. Save the drawing as P10_03.

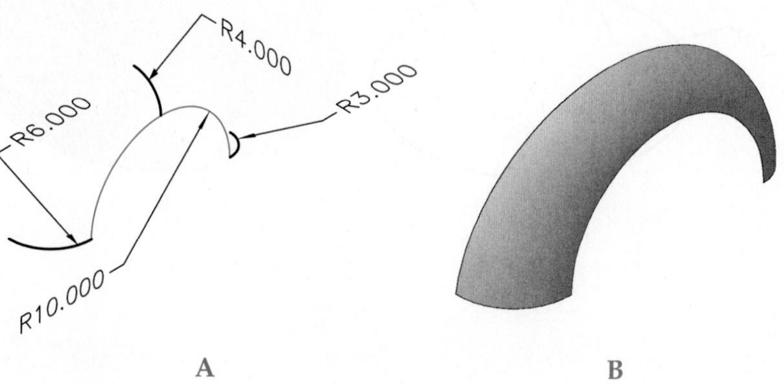

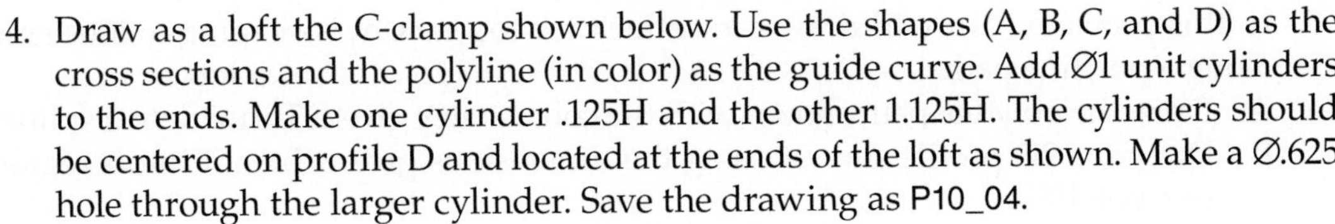

4. Draw as a loft the C-clamp shown below. Use the shapes (A, B, C, and D) as the cross sections and the polyline (in color) as the guide curve. Add Ø1 unit cylinders to the ends. Make one cylinder .125H and the other 1.125H. The cylinders should be centered on profile D and located at the ends of the loft as shown. Make a Ø.625 hole through the larger cylinder. Save the drawing as P10_04.

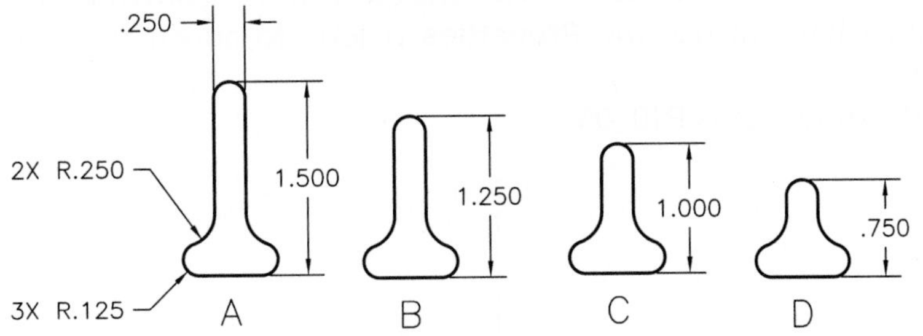

.250
2X R.250
3X R.125
1.500
1.250
1.000
.750
A B C D

Cross Sections

A

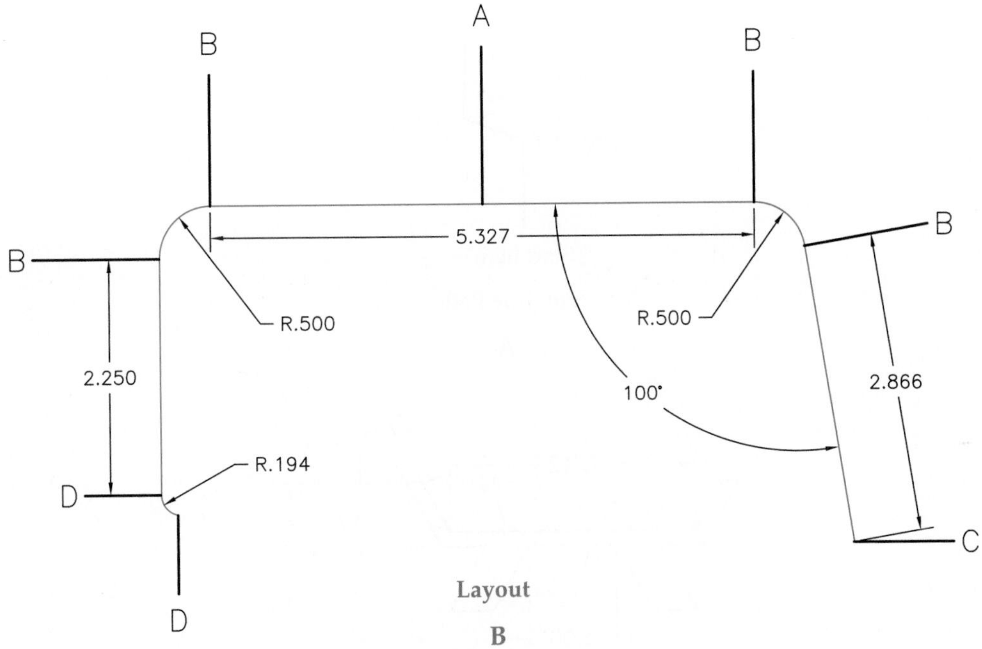

A
B B
B B
5.327
R.500 R.500
2.250
100°
2.866
R.194
D C
D

Layout

B

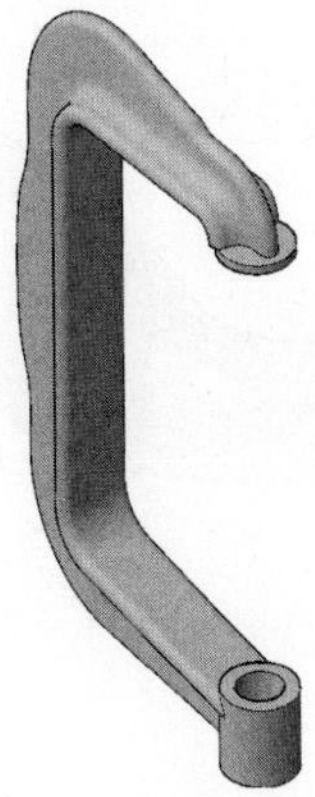

C

5. In this problem, you will draw a racetrack for toy cars by sweeping a 2D shape along a polyline path.
   A. Draw the polyline path shown with the coordinates given. Turn it into a spline.
   B. Draw the 2D profile shown using the dimensions given. Turn it into a region or a polyline.
   C. Use the **SWEEP** command to create the racetrack, as shown in the shaded view.
   D. You may have to use the **Properties** palette to adjust the sweep after it is drawn.
   E. Save the drawing as P10_05.

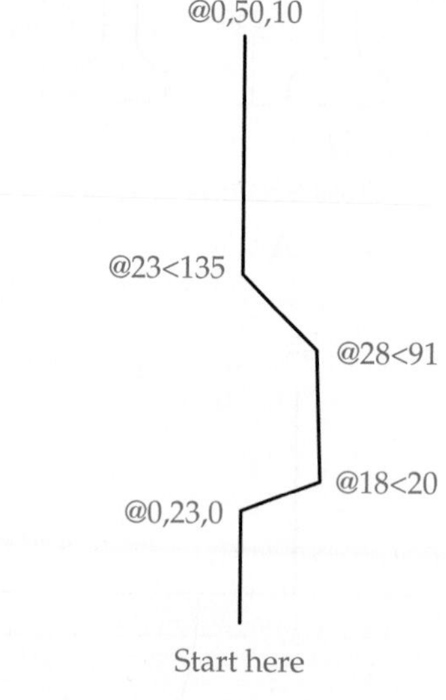

Polyline Path

A

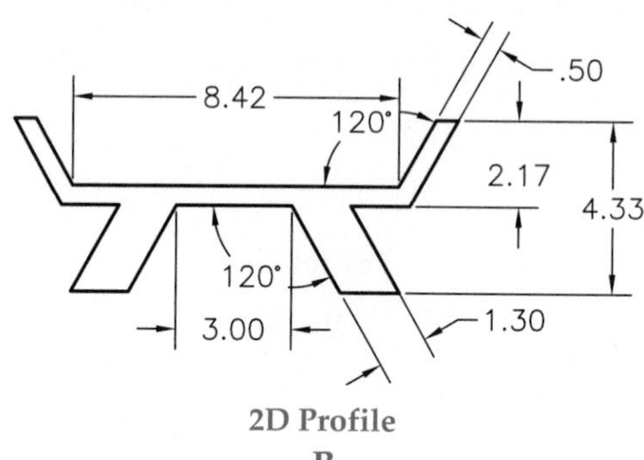

2D Profile

B

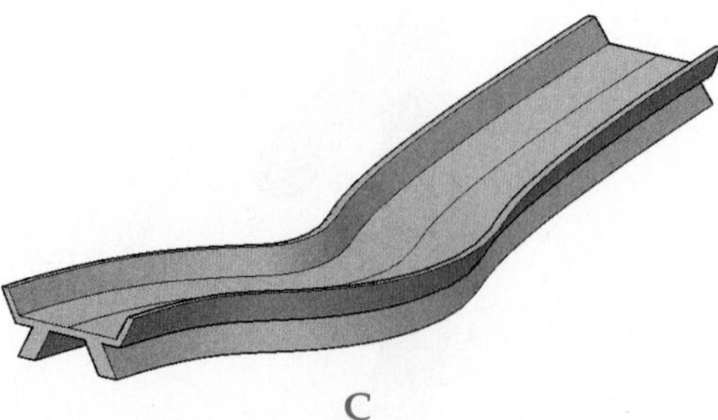

C

Drawing Problems - Chapter 10

6. In this exercise, you will cut a UNC thread in a cylinder by sweeping a 2D shape around a helix and subtracting it.
   A. Draw a Ø.25 cylinder that is 1.00 in height.
   B. Draw the thread cutter profile shown below. The long edge of the cutter should be aligned with the vertical edge of the cylinder.
   C. Draw a helix centered on the cylinder with base and top radii of .125, turn height of .050, and a total height of 1.000.
   D. Sweep the 2D shape along the helix. Then, subtract the resulting solid from the cylinder. Refer to the shaded view shown below.
   E. If time allows, create another cutter profile to cut a .0313 × 45° chamfer on the end of the thread. Use a circle as a sweep path or revolve the profile about the center of the cylinder.
   F. Save the drawing as P10_06.

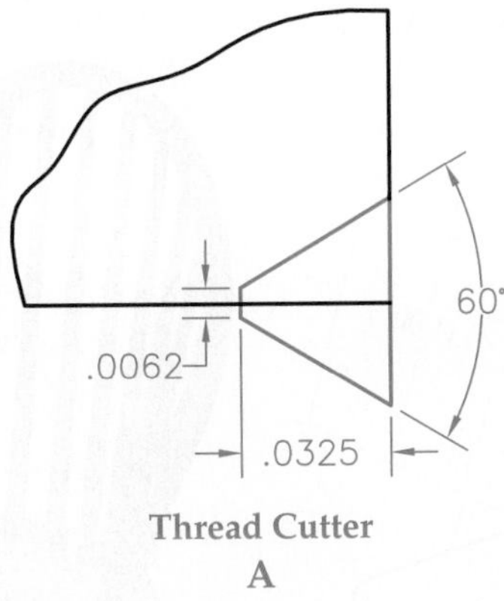

Thread Cutter
A

B

Drawing Problems - Chapter 10

7. In this problem, you will add a seatback to the kitchen chair you started modeling in Chapter 2. In Chapter 9, you refined the seat.
   A. Open P09_10 from Chapter 9.
   B. Draw an arc for the top of the bow. Using a 14.25″ length of the arc, divide it into seven equal segments.
   C. Position eight ∅.50 circles at the division points.
   D. Draw circles at the top of each hole in the seat.
   E. Create a loft between each lower circle and each upper circle.
   F. Using the information in the drawings, create the outer bow for the seatback.
   G. Save the drawing as P10_07.

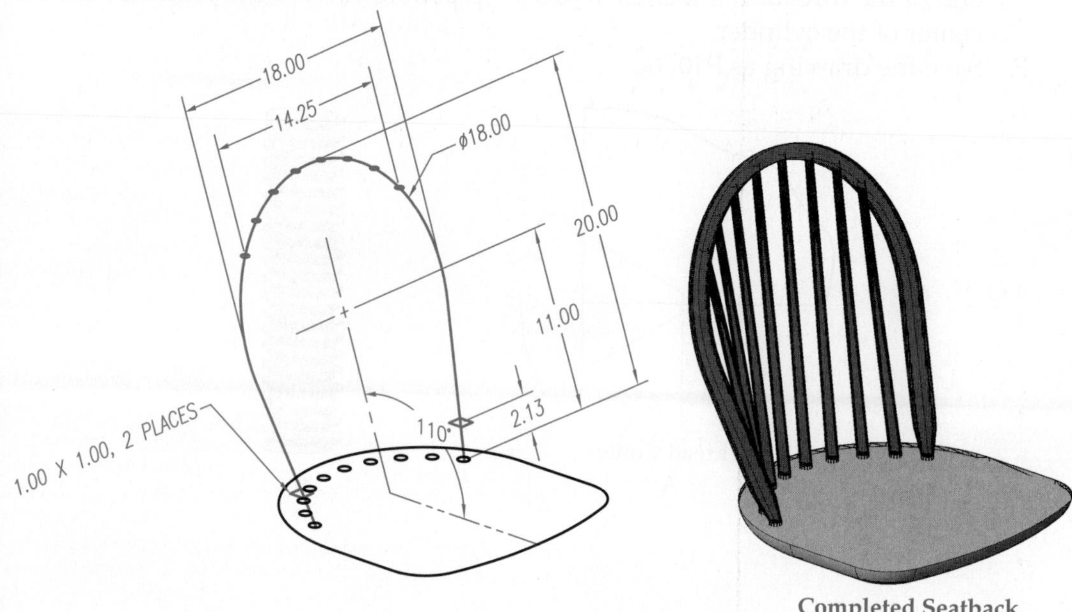

Completed Seatback

CHAPTER 11

Creating and Working with Solid Model Features

Learning Objectives

After completing this chapter, you will be able to:

✓ Change properties on solids.
✓ Align objects.
✓ Rotate objects in three dimensions.
✓ Mirror objects in three dimensions.
✓ Create 3D arrays.
✓ Fillet solid objects.
✓ Chamfer solid objects.
✓ Thicken a surface into a solid.
✓ Convert planar objects into surfaces.
✓ Sculpt watertight surfaces into solids.
✓ Slice a solid using various methods.
✓ Construct details on solid models.
✓ Remove features from solid models.

Changing Properties Using the Properties Palette

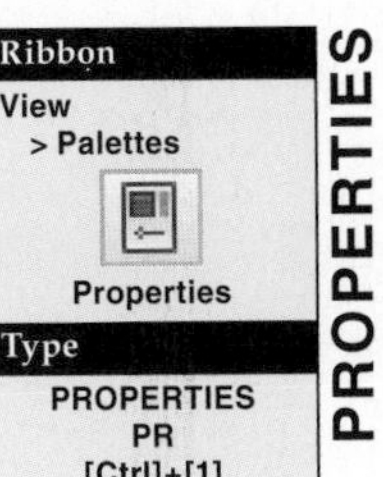

Properties of 3D objects can be modified using the **Properties** palette, which is thoroughly discussed in *AutoCAD and Its Applications—Basics.* This palette is displayed using the **PROPERTIES** command. You can also double-click on a solid object or select the solid, right-click, and pick **Properties** from the shortcut menu.

The **Properties** palette lists the properties of the currently selected object. For example, Figure 11-1 lists the properties of a selected solid sphere. Some of its properties are parameters, which is why solid modeling in AutoCAD can be considered parametric. You can change the sphere's radius; diameter; X, Y, Z coordinates; linetype; linetype scale; color; layer; lineweight; and visual settings. The categories and properties available in the **Properties** palette depend on the selected object.

To modify an object property, select the property. Then, enter a new value in the right-hand column. The drawing is updated to reflect the changes. You can leave the **Properties** palette open as you continue with your work.

Figure 11-1.
The **Properties** palette can be used to change many of the properties of a solid.

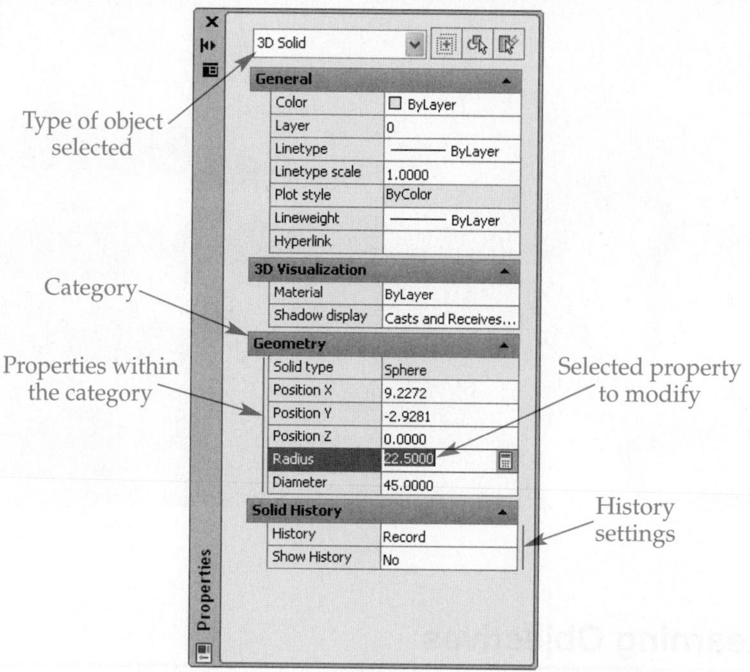

Solid Model History

AutoCAD can automatically record a history of a composite solid model's construction. Solid ***composites*** are created by a Boolean operation or by using the **SOLIDEDIT** command. The control of the history setting is found in the **Solid History** category of the **Properties** palette. By default, the History property in the **Properties** palette is set to Record, which means that the history will be saved. See Figure 11-1. It is generally a good idea to have the history recorded. Then, at any time, you can graphically display all of the geometry that was used to create the model.

If the **SOLIDHIST** system variable is set to a value of 1, all new solids have their History property set to Record. This is the default. If the system variable is set to 0, all new solids have their History property set to None (no recording). With either setting of the system variable, the **Properties** palette can be used to change the setting for individual solids.

To view the graphic history of the composite solid, set the Show History property in the **Properties** palette to Yes. All of the geometry used to construct the model is displayed. If the **SHOWHIST** system variable is set to 0, the Show History property is set to No for all solids and cannot be changed. If this system variable is set to 2, the Show History property is set to Yes for all solids and cannot be changed. A **SHOWHIST** setting of 1 allows the Show History property to be individually set for each solid.

An example of showing the history on a composite solid is provided in Figure 11-2. The object appears in its current edited format in Figure 11-2A. The Conceptual visual style is set current and the Show History property is set to No. In Figure 11-2B, the Show History property is set to Yes. Isolines have also been turned on. You can see the geometry that was used in the Boolean subtraction operations. Using subobject editing techniques, the individual geometry can be selected and edited. Subobject editing is discussed in detail in Chapter 12.

Figure 11-2.
A—The object appears in its current edited format with Show History property turned off.
B—The Show History property is set to Yes and the display of isolines has been turned on.

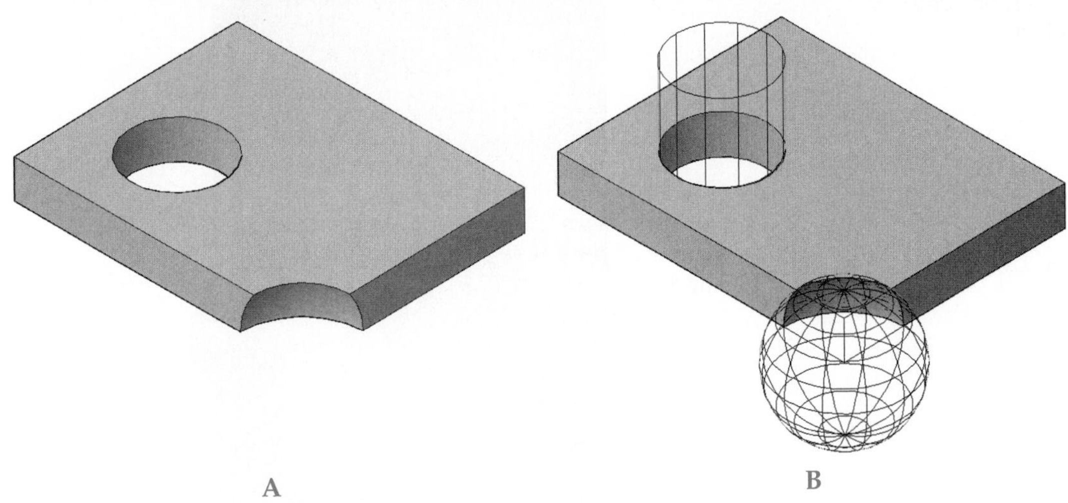

NOTE

If the Solid History property is set to Yes to display the components of the composite solid, as seen in Figure 11-2, the components will appear when the drawing is plotted. Be sure to set the Show History property to No before you print or plot.

Object Cycling

When selecting objects that are on top of another or occupy the same space, ***object cycling*** is the preferred method to select one of the objects. Without object cycling, you may need to erase, move, or copy one of the objects that overlap in order to select the correct object. The [Shift] key and space bar can be pressed at the same time to cycle through objects at a pick point. When you need to cycle through objects:

1. At the "select objects" prompt, hold down the [Shift] key and space bar, then click to select the object you want.
2. Keep clicking until the object you want to select is highlighted.
3. Press the [Enter] key.

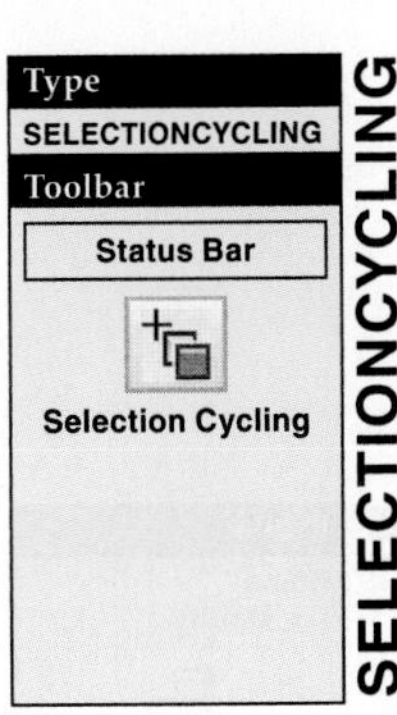

You use the **SELECTIONCYCLING** system variable to turn on object cycling instead of using the [Shift] key and space bar. There is a button on the status bar for toggling object cycling. There are three settings for the system variable:

- Off (0).
- On, but the list dialog box does not display (1).
- On and the list dialog box displays the selected objects that can be cycled through (2).

It is recommended that you turn on selection cycling. With selection cycling turned on, you can cycle and select the many faces that may overlap one another.

In Figure 11-3, the two 3D objects occupy the same space. The tapered 3D object needs to be moved up using the **3DMOVE** command. Turn on selection cycling. When you select the tapered object, the **Selection** dialog box lists the objects overlapping at the pick point. Select which object you want to work with by picking it in the dialog box. As your cursor is over an object in the list, the object is highlighted in the drawing area.

Figure 11-3.
Using selection cycling to choose which object to select.

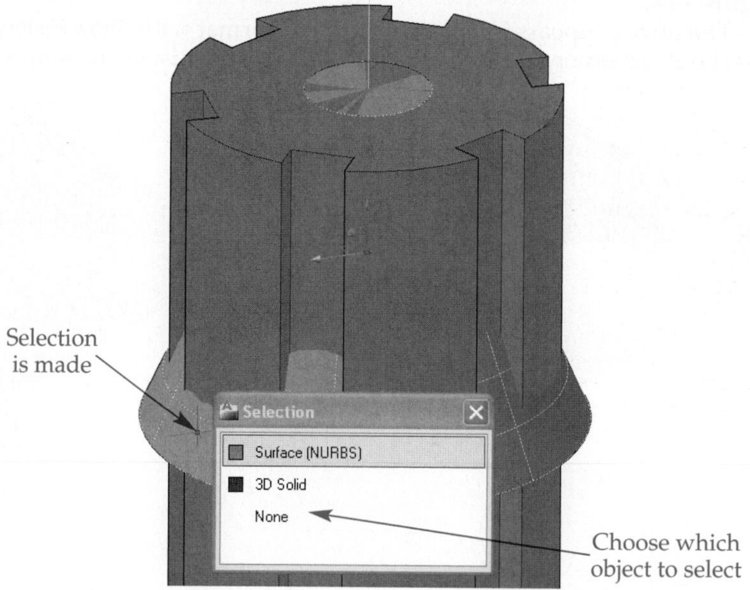

PROFESSIONAL TIP

If multiple 3D objects occupy the same space, selection cycling gives you the ability to select the correct object needed to complete the edit.

Aligning Objects in 3D

AutoCAD provides two different methods with which to move and rotate objects in a single command. This is called ***aligning*** objects. The simplest method is to align 3D objects by picking source points on the first object and then picking destination points on the object to which the first one is to be aligned. This is accomplished with the **3DALIGN** command, which allows you to both relocate and rotate the object. The second method is possible with the **ALIGN** command. The **ALIGN** command aligns 2D or 3D objects by selecting three sets of alignment pairs. You first select the source object, then the first source point, and finally the first destination point on the destination object. This technique is repeated two more times to align one object to another. This method of aligning allows you to not only move and rotate an object, but scale the object being aligned.

PROFESSIONAL TIP

Use the **3DALIGN** command to create assemblies.

Move and Rotate Objects in 3D Space

3DALIGN

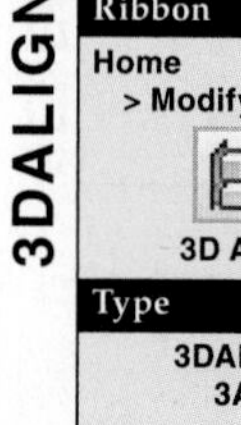

Ribbon
Home
> Modify
3D Align
Type
3DALIGN
3AL

The basic function of moving and rotating an object relative to a second object or set of points is done with the **3DALIGN** command. It allows you to reorient an object in 3D space. Using this command, you can correct errors of 3D construction and quickly manipulate 3D objects. The **3DALIGN** command requires existing points (source) and the new location of those existing points (destination).

For example, refer to **Figure 11-4.** The wedge in **Figure 11-4A** is aligned in its new position in **Figure 11-4B** as follows. Set the **Intersection** or **Endpoint** running object snap to make point selection easier. Refer to the figure for the pick points.

Figure 11-4.
The **3DALIGN** command can be used to properly orient 3D objects. A—Before aligning. Note the pick points. B—After aligning.

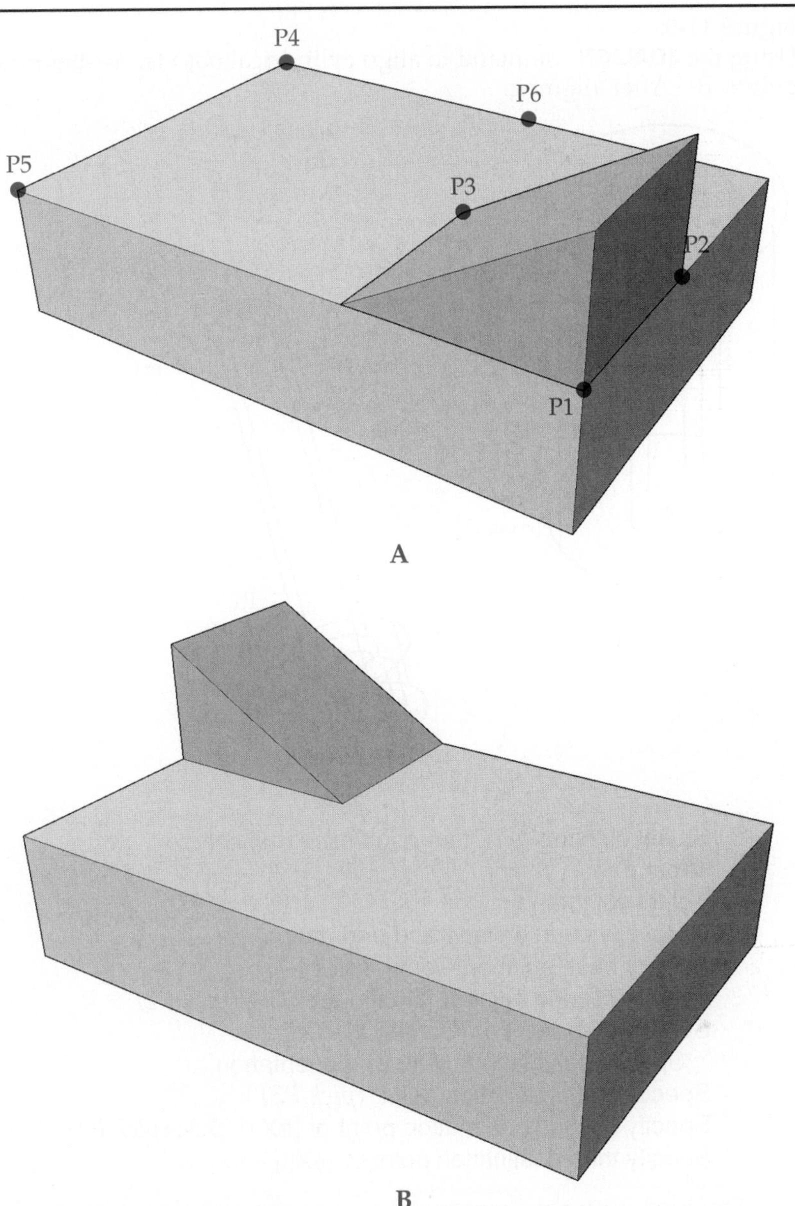


```
Select objects: (pick the wedge)
1 found
Select objects: ↵
  Specify source plane and orientation...
Specify base point or [Copy]: (pick P1)
Specify second point or [Continue] <C>: (pick P2)
Specify third point or [Continue] <C>: (pick P3)
  Specify destination plane and orientation...
Specify first destination point: (pick P4)
Specify second destination point or [eXit] <X>: (pick P5)
Specify third destination point or [eXit] <X>: (pick P6)
```

You can also use the **3DALIGN** command to align cylindrical 3D objects. The procedure is similar to aligning planar objects, which require three pick points. However, you only need two pick points per object.

For example, the socket head cap screw in **Figure 11-5** is aligned to a new position using the **3DALIGN** command. Set the center running object snap to make point selection easier. Refer to the figure for the pick points.

Figure 11-5.
Using the **3DALIGN** command to align cylindrical objects. A—Before aligning. Note the pick points. B—After aligning.

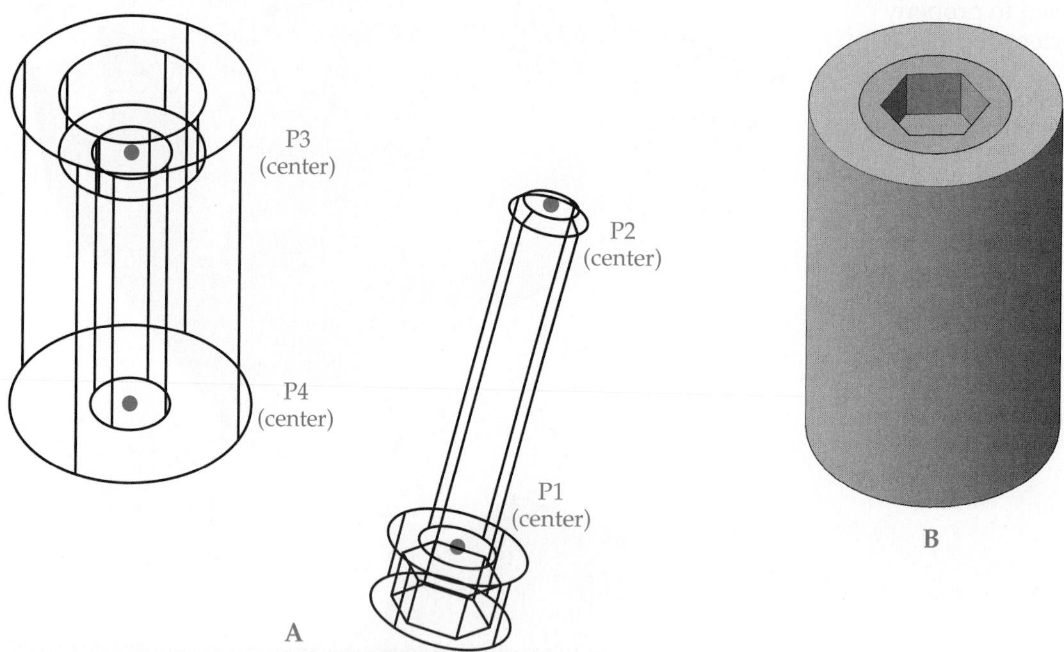

Select objects: *(pick the socket head cap screw)*
1 found
Select objects: ↵
Specify source plane and orientation...
Specify base point or [Copy]: *(pick P1)*
Specify second point or [Continue] <C>: *(pick P2)*
Specify third point or [Continue] <C>: ↵
Specify destination plane and orientation...
Specify first destination point: *(pick P3)*
Specify second destination point or [eXit] <X>: *(pick P4)*
Specify third destination point or [eXit] <X>: ↵

The idea with this command is you are aligning a plane defined by three points with another plane defined by three points. The planes do not need to correspond to actual planar faces. In the example shown in **Figure 11-4,** the alignment planes coincide with planar faces. However, in the example shown in **Figure 11-5,** the alignment planes define planes on which axes lie.

PROFESSIONAL TIP

You can also use the **3DALIGN** command to copy an object, instead of move it, and realign it at the same time. Just select the **Copy** option at the Specify base point or [Copy]: prompt. Then, continue selecting the points as shown above.

Exercise 11-1

Complete the exercise on the student website.
www.g-wlearning.com/CAD

Move, Rotate, and Scale Objects in 3D Space

The **ALIGN** command has the same functions of the **3DALIGN** command, but adds the ability to scale an object. Refer to Figure 11-6. The 90° bend must be rotated and scaled to fit onto the end of the HVAC assembly. Two source points and two destination points are required, Figure 11-6A. Then, you can choose to scale the object.

Ribbon
Home
> Modify
Align
Type
ALIGN
AL

ALIGN

Select objects: *(pick the 90° bend)*
1 found
Select objects: ↵
Specify first source point: *(pick P1)*
Specify first destination point: *(pick P2; a line is drawn between the two points)*
Specify second source point: *(pick P3)*
Specify second destination point: *(pick P4; a line is drawn between the two points)*
Specify third source point or <continue>: ↵
Scale objects based on alignment points? [Yes/No] <N>: **Y**↵

The 90° bend is aligned and uniformly scaled to meet the existing ductwork object. See Figure 11-6B. You can also align using three source and three destination points. However, when doing so, you cannot scale the object.

PROFESSIONAL TIP

Before using 3D editing commands, set running object snaps to enhance your accuracy and speed.

Exercise 11-2

Complete the exercise on the student website.
www.g-wlearning.com/CAD

Figure 11-6.
Using the **ALIGN** command. A—Two source points and two destination points are required. Notice how the bend is not at the proper scale. B—You can choose to scale the object during the operation. Notice how the aligned bend is also properly scaled.

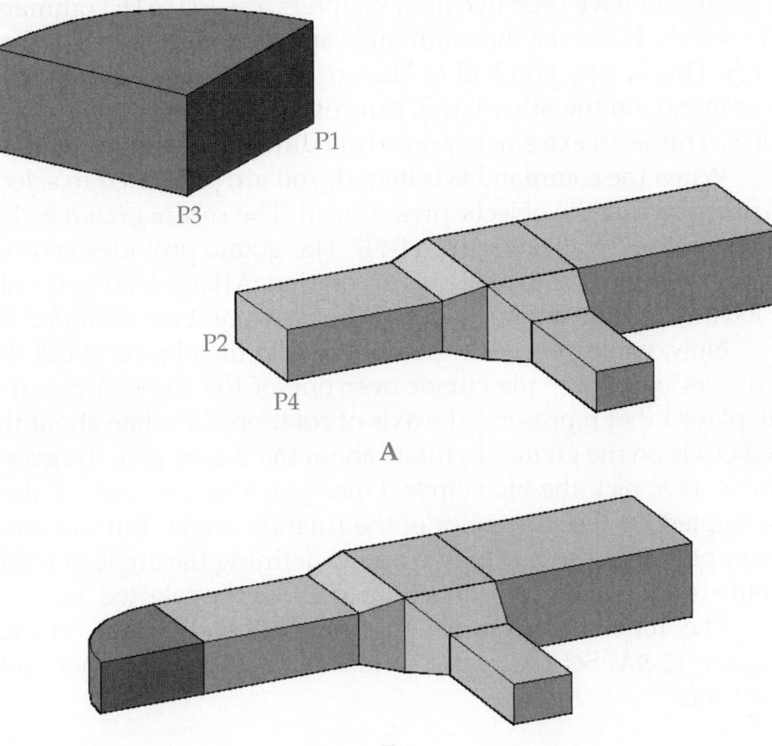

3D Moving

3DMOVE

Ribbon
Home
> Modify
3D Move
Type
3DMOVE
3M

The **3DMOVE** command allows you to quickly move an object along any axis or plane of the current UCS. When the command is initiated, you are prompted to select the objects to move. If the 2D Wireframe visual style is current, the visual style is temporarily changed to the Wireframe visual style because the gizmo is not displayed in 2D mode. After selecting the objects, press [Enter]. The ***move gizmo,*** also called a *grip tool,* is displayed in the center of the selection set. By default, the move gizmo is also displayed when a solid is selected with no command active.

The move gizmo is a tripod that appears similar to the shaded UCS icon. See Figure 11-7A. You can relocate the tool by right-clicking on the gizmo and selecting **Relocate Gizmo** from the shortcut menu. Then, move the gizmo to a new location and pick. You can also realign the gizmo using the shortcut menu.

If you move the pointer over the X, Y, or Z axis of the gizmo, the axis changes to yellow. To restrict movement along that axis, pick the axis. If you move the pointer over one of the right angles at the origin of the gizmo, the corresponding two axes turn yellow. Pick to restrict the movement to that plane. You can complete the movement by either picking a new point or by direct distance entry.

NOTE

If the **GTAUTO** system variable is set to 1, the move gizmo is displayed when a solid is selected with no command active. If the **GTLOCATION** system variable is set to 0, the gizmo is placed on the UCS icon (not necessarily the UCS *origin*). Both variables are set to 1 by default.

3D Rotating

3DROTATE

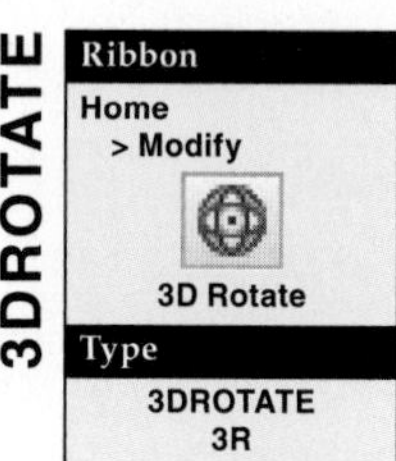

Ribbon
Home
> Modify
3D Rotate
Type
3DROTATE
3R

As you have seen in earlier chapters, the **ROTATE** command can be used to rotate 3D objects. However, the command can only rotate objects in the XY plane of the current UCS. This is why you had to change UCSs to properly rotate objects. The **3DROTATE** command, on the other hand, can rotate objects on any axis regardless of the current UCS. This is an extremely powerful editing and design tool.

When the command is initiated, you are prompted to select the object(s) to rotate. After selecting the objects, press [Enter]. The ***rotate gizmo*** is displayed in the center of the selection set. See Figure 11-7B. The gizmo provides you with a dynamic, graphic representation of the three axes of rotation. After selecting the objects, you must specify a location for the gizmo, which is the base point for rotation.

Now, you can use the gizmo to rotate the objects about the tool's local X, Y, or Z axis. As you hover the cursor over one of the three circles in the gizmo, a vector is displayed that represents the axis of rotation. To rotate about the tool's X axis, pick the red circle on the gizmo. To rotate about the Y axis, pick the green circle. To rotate about the Z axis, pick the blue circle. Once you select a circle, it turns yellow and you are prompted for the start point of the rotation angle. You can enter a direct angle at this prompt or pick the first of two points defining the angle of rotation. When the rotation angle is defined, the object is rotated about the selected axis.

The following example rotates the bend in the HVAC assembly shown in Figure 11-8A. Set the center of face 3D object snap. Then, select the command and continue:

Figure 11-7.
A—The move grip tool is a tripod that appears similar to the shaded UCS icon. B—This is the rotate grip tool. The three axes of rotation are represented by the circles. The origin of the rotation is where you place the center grip.

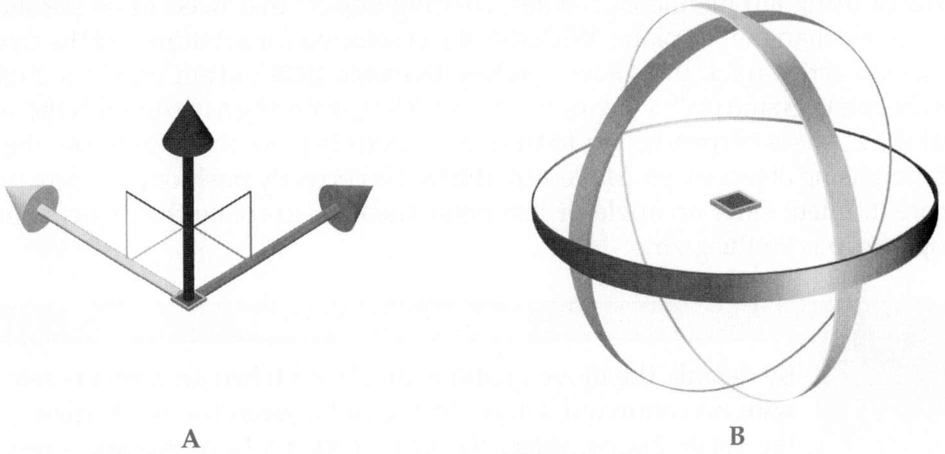

Figure 11-8.
A—Use object tracking or object snaps to place the grip tool in the middle of the rectangular face. Then, select the axis of rotation. B—The completed rotation.

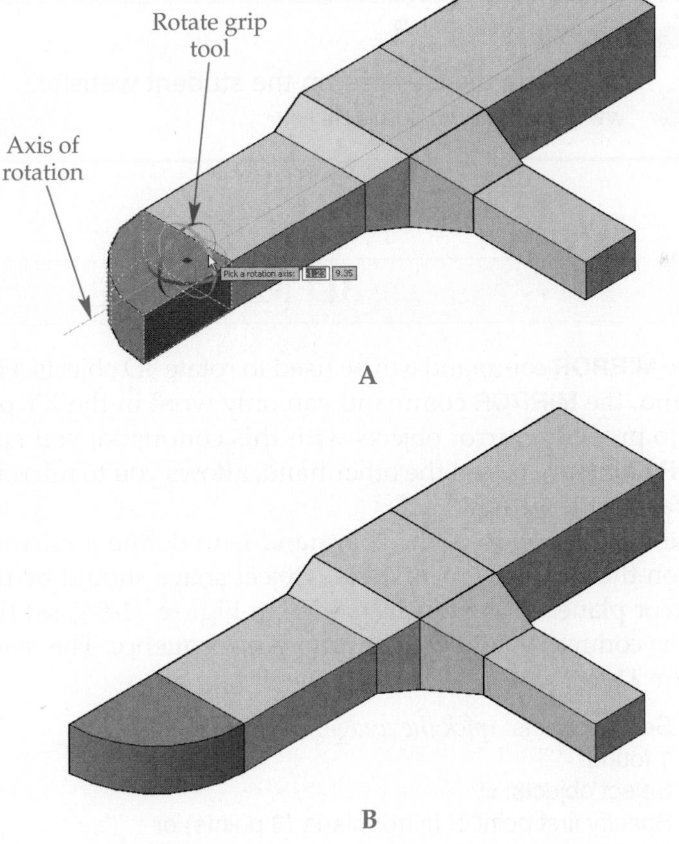

Current positive angle in UCS: ANGDIR=*(current)* ANGBASE=*(current)*
Select objects: *(pick the bend)*
1 found
Select objects: ↵
Specify base point: *(acquire the center of the face and pick)*
Pick a rotation axis: *(pick the green circle)*
Specify angle start point or type an angle: **180**↵

Note that the rotate gizmo remains visible through the base point and the angle of rotation selections. The rotated object is shown in Figure 11-8B.

If you need to rotate an object on an axis that is not parallel to the current X, Y, or Z axes, use a dynamic UCS with the **3DROTATE** command. Chapter 5 discussed the benefits of using a dynamic UCS when creating objects that need to be parallel to a surface other than the XY plane. With the object selected for rotation and the dynamic UCS option active (pick the **Allow/Disallow Dynamic UCS** button on the status bar), move the rotate gizmo over a face of the object. The gizmo aligns itself with the surface so that the Z axis is perpendicular to the face. Carefully place the gizmo over the point of rotation using object snaps. Make sure the tool is correctly positioned before picking to locate it. Then, enter an angle or use polar tracking to rotate the object about the appropriate axis on the gizmo.

PROFESSIONAL TIP

By default, the move gizmo is displayed when an object is selected with no command active. To toggle between the move gizmo and the rotate gizmo, select the grip at the tool's origin and press the space bar. Then, pick a location for the tool's origin. You can toggle back to the move gizmo using the same procedure.

Exercise 11-3

Complete the exercise on the student website.
www.g-wlearning.com/CAD

3D Mirroring

3DROTATE

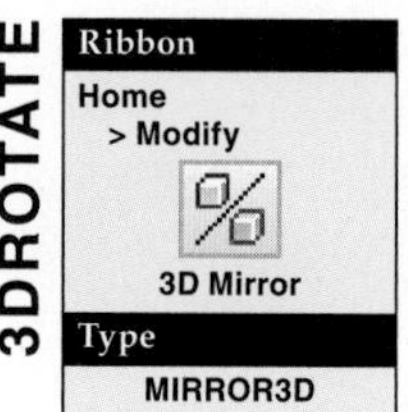

The **MIRROR** command can be used to rotate 3D objects. However, like the **ROTATE** command, the **MIRROR** command can only work in the XY plane of the current UCS. Often, to properly mirror objects with this command, you have to change UCSs. The **MIRROR3D** command, on the other hand, allows you to mirror objects about any plane regardless of the current UCS.

The default option of the command is to define a mirror plane by picking three points on that plane, Figure 11-9A. Object snaps should be used to accurately define the mirror plane. To mirror the wedge in Figure 11-9A, set the midpoint object snap, select the command, and use the following sequence. The resulting drawing is shown in Figure 11-9B.

Select objects: *(pick the wedge)*
1 found
Select objects: ↵
Specify first point of mirror plane (3 points) or
[Object/Last/Zaxis/View/XY/YZ/ZX/3points] <3points>: *(pick P1, which is the midpoint of the box's top edge)*
Specify second point on mirror plane: *(pick P2)*
Specify third point on mirror plane: *(pick P3)*
Delete source objects? [Yes/No] <N>: ↵

There are several different ways to define a mirror plane with the **MIRROR3D** command. These are:

Figure 11-9.
The **MIRROR3D** command allows you to mirror objects about any plane regardless of the current UCS. A—The mirror plane defined by the three pick points is shown here in color. Point P1 is the midpoint of the top edge of the base. B—A copy of the original is mirrored.

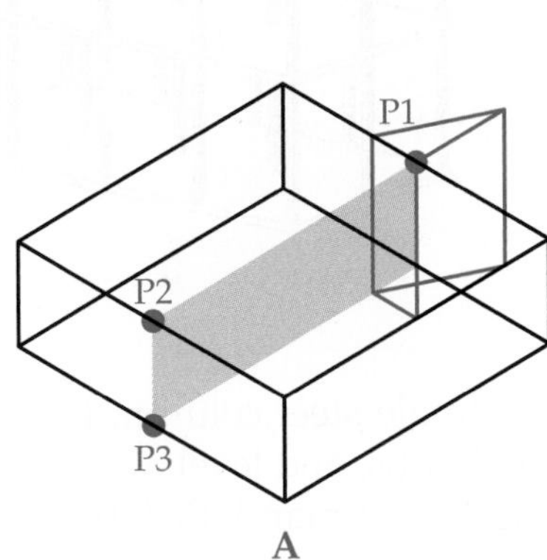

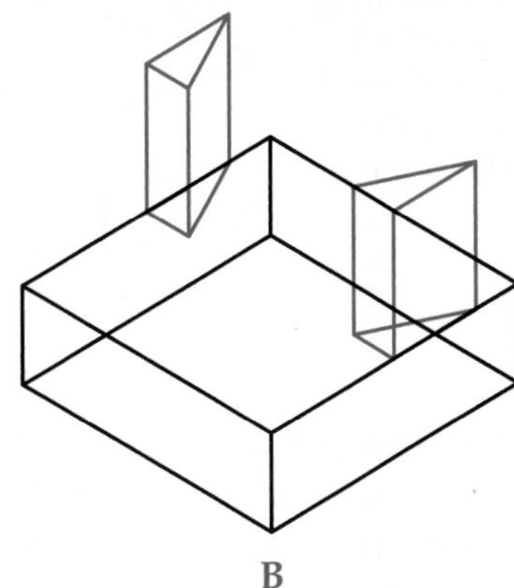

- **Object.** The plane of the selected circle, arc, or 2D polyline segment is used as the mirror plane.
- **Last.** Uses the last mirror plane defined.
- **Zaxis.** Defines the plane with a pick point on the mirror plane and a point on the Z axis of the mirror plane.
- **View.** The viewing direction of the current viewpoint is aligned with a selected point to define the plane.
- **XY, YZ, ZX.** The mirror plane is placed parallel to one of the three basic planes of the current UCS and passes through a selected point.
- **3points.** Allows you to pick three points to define the mirror plane, as shown in the above example.

Exercise 11-4

Complete the exercise on the student website.
www.g-wlearning.com/CAD

Creating 3D Arrays

The **ARRAY** command can be used to create either a rectangular or polar array of a 3D object on the XY plane of the current UCS. You probably used this command to complete some of the problems in previous chapters. The **3DARRAY** command allows you to array an object in 3D space. There are two types of 3D arrays—rectangular and polar.

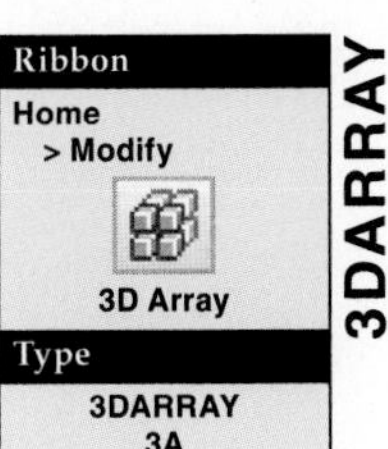

Rectangular 3D Arrays

In a ***rectangular 3D array,*** as with a rectangular 2D array, you must enter the number of rows and columns. However, you must also specify the number of ***levels,*** which represents the third (Z) dimension. The command sequence is similar to that used with the 2D array command, with two additional prompts.

An example of where a rectangular 3D array may be created is the layout of structural steel columns on multiple floors of a commercial building. In Figure 11-10A, you

Figure 11-10.
A—Two floors and one steel column are drawn. B—A rectangular 3D array is used to place all of the required steel columns on both floors at the same time.

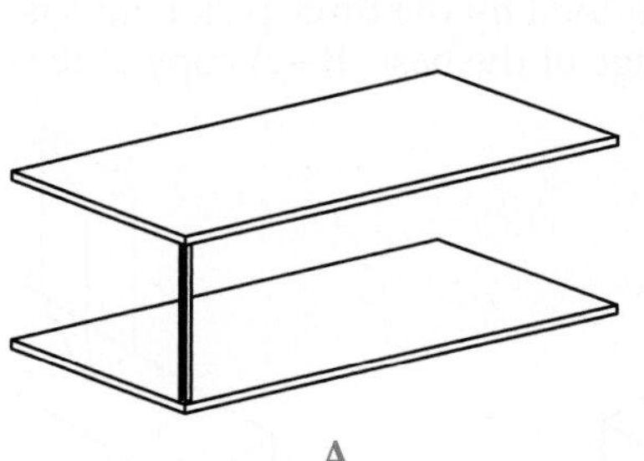

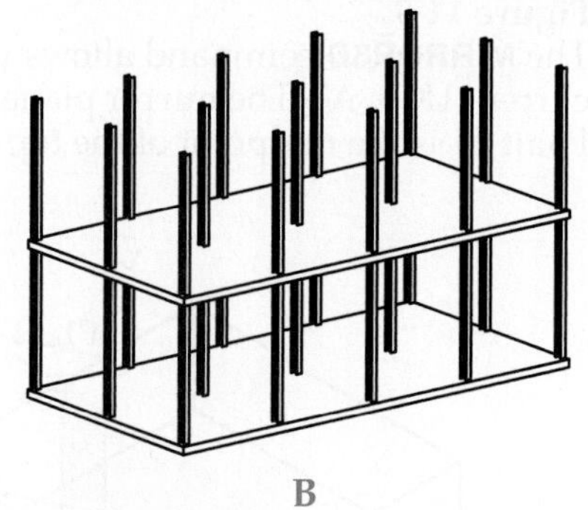

can see two concrete floor slabs of a building and a single steel column. It is now a simple matter of arraying the steel column in rows, columns, and levels.

To draw a rectangular 3D array, select the **3DARRAY** command. Pick the object to array and press [Enter]. Then, specify the **Rectangular** option:

Enter the type of array [Rectangular/Polar] <R>: **R**↵
Enter the number of rows (- - -) <1>: **3**↵
Enter the number of columns (|||) <1>: **5**↵
Enter the number of levels (...) <1>: **2**↵
Specify the distance between rows (- - -): **10'**↵
Specify the distance between columns (|||): **10'**↵
Specify the distance between levels (...): **12'8**↵

The result is shown in Figure 11-10B. Constructions like this can be quickly assembled for multiple levels using the **3DARRAY** command only once.

Polar 3D Arrays

Ribbon
Home
> Modify
3D Array
Type
3DARRAY
3A

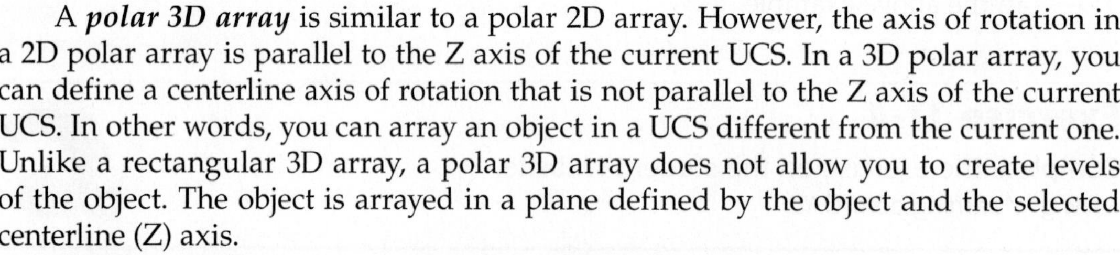

A *polar 3D array* is similar to a polar 2D array. However, the axis of rotation in a 2D polar array is parallel to the Z axis of the current UCS. In a 3D polar array, you can define a centerline axis of rotation that is not parallel to the Z axis of the current UCS. In other words, you can array an object in a UCS different from the current one. Unlike a rectangular 3D array, a polar 3D array does not allow you to create levels of the object. The object is arrayed in a plane defined by the object and the selected centerline (Z) axis.

To draw a polar 3D array, select the **3DARRAY** command. Pick the object to array and press [Enter]. Then, specify the **Polar** option:

Enter the type of array [Rectangular/Polar] <R>: **P**↵

For example, the four mounting flanges on the lower part of the duct in Figure 11-11A must be placed on the opposite end. However, notice the orientation of the UCS. First, copy one flange and rotate it to the proper orientation. Then, use the **3DARRAY** command as follows. Make sure polar tracking is on.

Select objects: *(select the copied flange)*
1 found
Select objects: ↵
Enter the type of array [Rectangular/Polar] <R>: **P**↵
Enter the number of items in the array: **4**↵
Specify the angle to fill (+=ccw, −=cw) <360>: ↵
Rotate arrayed objects? [Yes/No] <Y>: ↵
Specify center point of array: **CEN**↵
of: *(pick the center of the upper duct opening)*
Specify second point on axis of rotation: *(move the cursor so the ortho line projects out of the center of the duct opening and pick)*

Figure 11-11.
A—A ductwork elbow with four flanges in place. Copies of these flanges need to be located on the opposite end. Start by creating one copy as shown. B—By creating a 3D polar array, the flanges are properly oriented without changing the UCS.

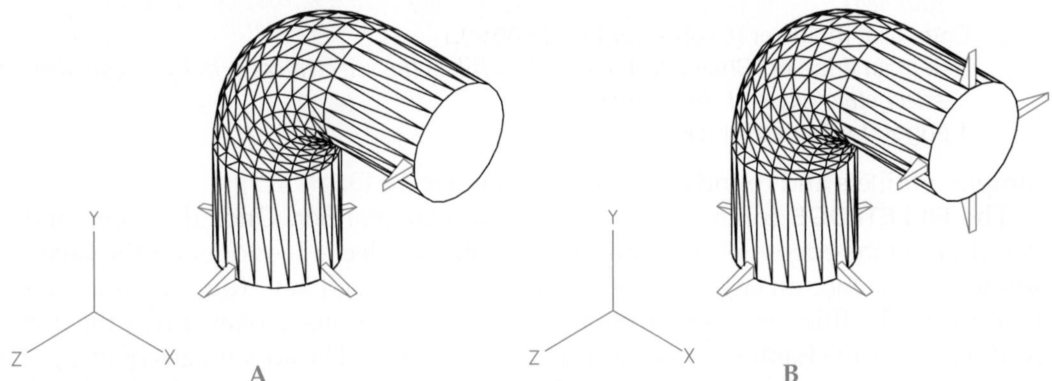

The completed 3D polar array is shown in Figure 11-11B. If additional levels of a polar array are needed, they can be created by copying the array just created.

Exercise 11-5

Complete the exercise on the student website.
www.g-wlearning.com/CAD

Filleting Solid Objects

A ***fillet*** is a rounded interior edge on an object, such as a box. A ***round*** is a rounded exterior edge. The **FILLET** command is used to create both fillets and rounds in 2D and 3D. Additionally, the **FILLETEDGE** command can be used to fillet 3D solids.

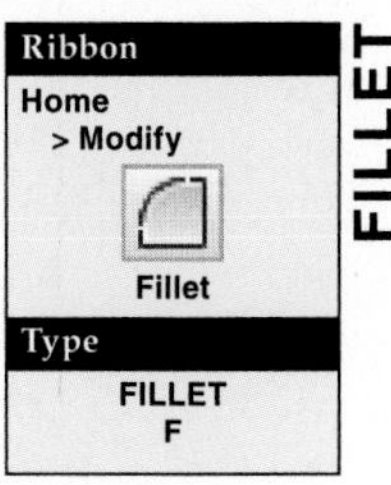

FILLET

Before a fillet or round is created at an intersection, the solid objects that intersect need to be joined using the **UNION** command. Then, use the **FILLET** command. See Figure 11-12. Since the object being filleted is actually a single solid and not two objects, only one edge is selected. In the following sequence, the fillet radius is set at .25, then the fillet is created. First, select the **FILLET** command and then continue as follows.

Figure 11-12.
A—Pick the edge where two unioned solids intersect to create a fillet. B—The fillet after rendering.

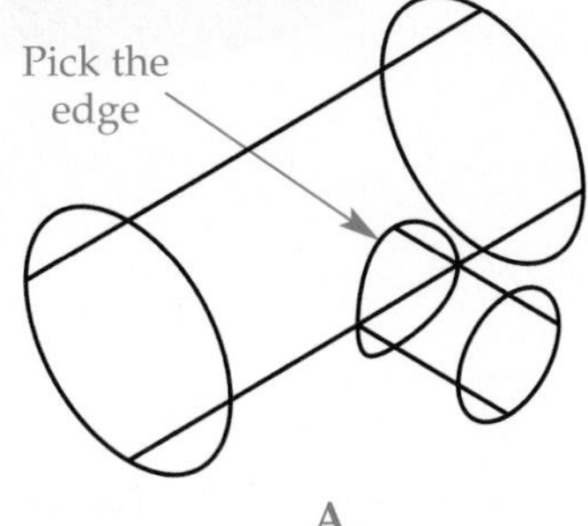

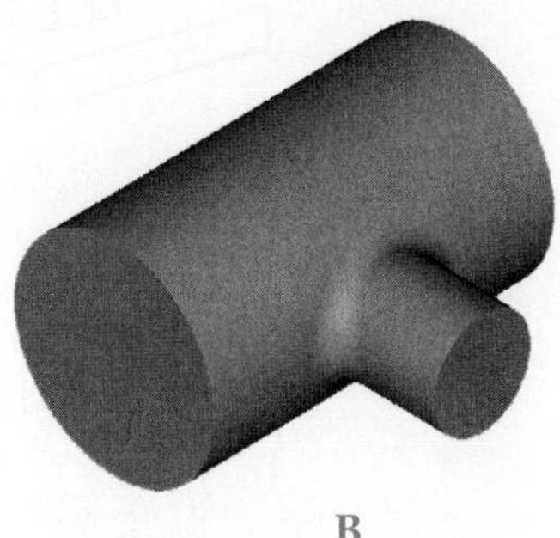

Current settings: Mode = *current*, Radius = *current*
Select first object or [Undo/Polyline/Radius/Trim/Multiple]: **R**↵
Specify fillet radius <*current*>: **.25**↵
Select first object or [Undo/Polyline/Radius/Trim/Multiple]: *(pick edge to be filleted or rounded)*
Enter fillet radius or [Expression] <0.2500>: ↵
Select an edge or [Chain/Radius]: ↵ *(this fillets the selected edge, but you can also select other edges at this point)*
1 edge(s) selected for fillet.

Examples of fillets and rounds are shown in **Figure 11-13.**

The **FILLETEDGE** command works in a similar manner. Once the command is entered, select the edges to fillet. You can continue to select edges or enter the **Chain** or **Radius** option. Once all edges are selected, press the [Enter] key. You are prompted to either accept the fillet or enter a radius. If the current radius is okay, press the [Enter] key. If not, enter the **Radius** option and set the new value. The advantages of using this command are 1) a preview is shown and 2) a parameter grip is associated with the fillet. The parameter grip allows for subobject editing, which is discussed in Chapter 12.

PROFESSIONAL TIP

You can construct and edit solid models while the object is displayed in a shaded view. If your computer has sufficient speed and power, it is often much easier to visualize the model in a 3D view with a shaded visual style set current. This allows you to realistically view the model. If an edit or construction does not look right, just undo and try again.

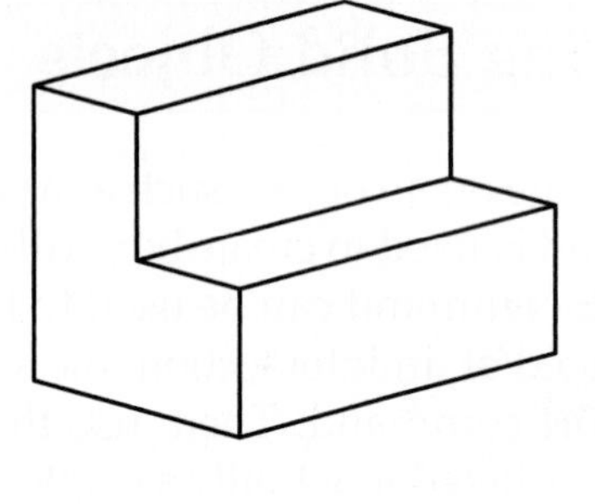

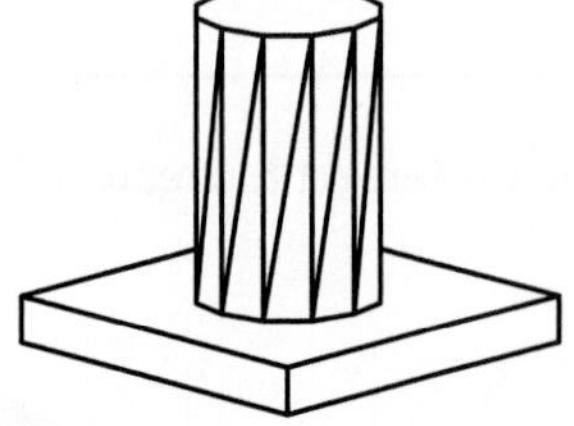

Figure 11-13.
Examples of fillets and rounds. The wireframe displays show the objects before the **FILLET** command is used.

Chamfering Solid Objects

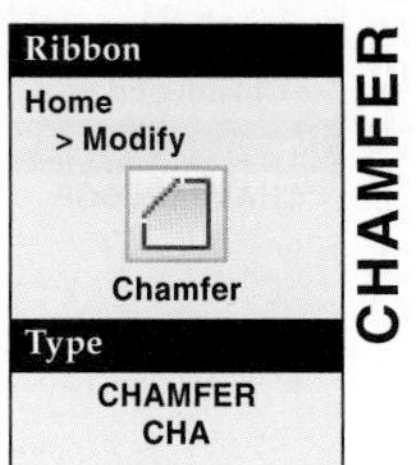

A ***chamfer*** is a small square edge on the edges of an object. The **CHAMFER** command can be used to create a chamfer on a 2D object or 3D solid. Just as when chamfering a 2D line, there are two chamfer distances. Therefore, you must specify which surfaces correspond to the first and second distances. The detail to which the chamfer is applied must be constructed before chamfering. For example, if you are chamfering a hole, the object (cylinder) must first be subtracted to create the hole. If you are chamfering an intersection, the two objects must first be unioned.

After you enter the command, you must pick the edge you want to chamfer. The edge is actually the intersection of two surfaces of the solid. One of the two surfaces is highlighted when you select the edge. The highlighted surface is associated with the first chamfer distance. This surface is called the ***base surface.*** If the highlighted surface is not the one you want as the base surface, enter N at the [Next/OK] prompt and press [Enter]. This highlights the next surface. An edge is created by two surfaces. Therefore, when you enter N for the next surface, AutoCAD cycles through only two surfaces. When the proper base surface is highlighted, press [Enter].

Chamfering a hole is shown in Figure 11-14A. The end of the cylinder in Figure 11-14B is chamfered by first picking one of the vertical isolines, then picking the top edge. The following command sequence is illustrated in Figure 11-14A.

(TRIM mode) Current chamfer Dist1 = *current*, Dist2 = *current*
Select first line or [Undo/Polyline/Distance/Angle/Trim/mEthod/Multiple]: *(pick edge 1)*
Base surface selection...
(if the side surface is highlighted, change to the top surface as follows)
Enter surface selection option [Next/OK (current)] <OK>: **N**↵
(the top surface should be highlighted)
Enter surface selection option [Next/OK (current)] <OK>: ↵
Specify base surface chamfer distance or Expression <*current*>: **.125**↵
Specify other surface chamfer distance or Expression <*current*>: **.125**↵
Select an edge or [Loop]: *(pick edge 2, the edge of the hole)*
Select an edge or [Loop]: ↵

Figure 11-14.
A—A hole is chamfered by picking the top surface, then the edge of the hole. B—The end of a cylinder is chamfered by first picking the side, then the end. Both ends can be chamfered at the same time, as shown here.

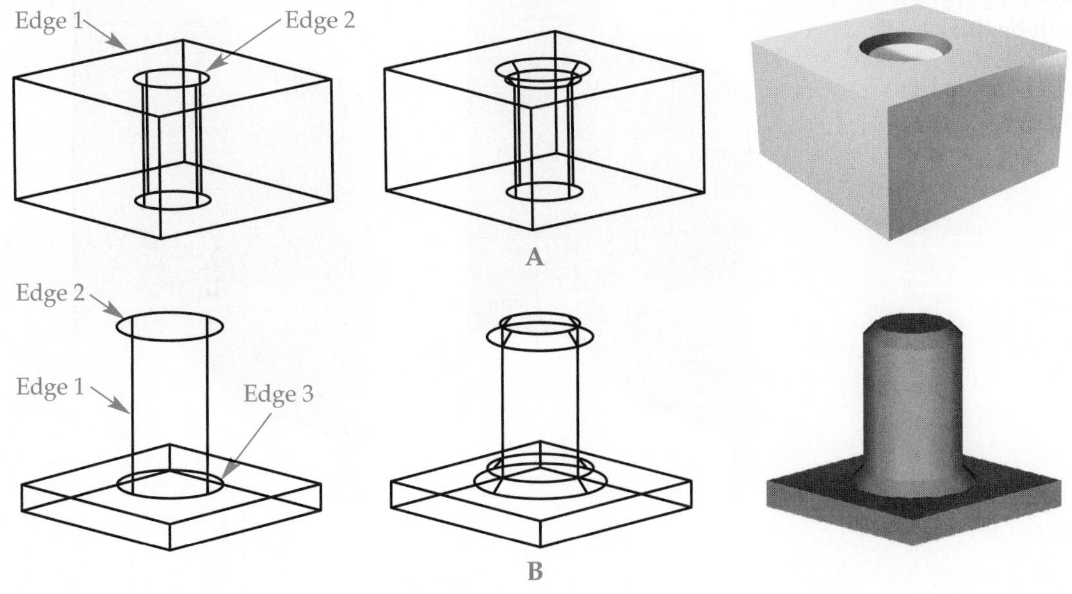

CHAMFEREDGE

Ribbon
Solid
> Solid Editing
Chamfer Edge
Type
CHAMFEREDGE

The **CHAMFEREDGE** command can also be used to create a chamfer on a 3D solid. This command is similar to the **CHAMFER** command. Once the command is entered, select the edge to chamfer. Next, you can select additional edges on the same face. Once the edges are selected, press [Enter]. Then, accept the chamfer or enter the **Distance** option to change the chamfer distance. The advantage of using this command is a preview is shown.

NOTE

Editing a fillet or chamfer is discussed in Chapter 12. The **SOLIDEDIT** command discussed in Chapter 13 can also be used to edit a fillet or chamfer.

Exercise 11-6

Complete the exercise on the student website.
www.g-wlearning.com/CAD

Thickening a Surface into a Solid

THICKEN

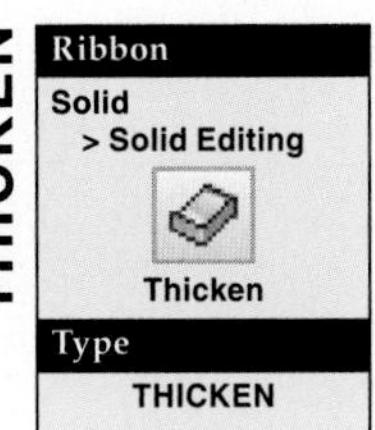

Ribbon
Solid
> Solid Editing
Thicken
Type
THICKEN

A surface has no thickness. The value of the **THICKNESS** command does not affect the thickness of a planar surface, unlike for entities such as lines, polylines, polygons, and circles. But, a surface can be quickly converted to a 3D solid using the **THICKEN** command.

To add thickness to a surface, enter the **THICKEN** command. Then, pick the surface(s) to thicken and press [Enter]. Next, you are prompted for the thickness. Enter a thickness value or pick two points on screen to specify the thickness. See **Figure 11-15.**

By default, the original surface object is deleted when the 3D solid is created with **THICKEN**. This is controlled by the **DELOBJ** system variable. To preserve the original surface, change the **DELOBJ** value to 0.

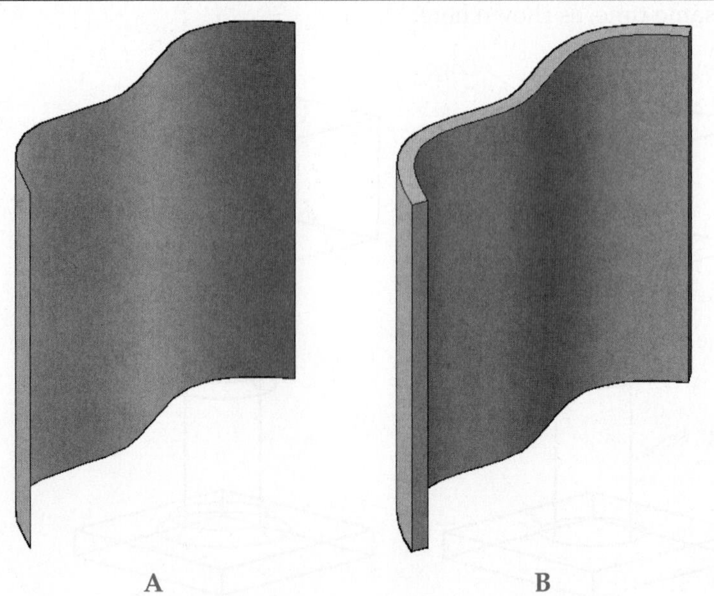

Figure 11-15. A—This surface will be thickened into a solid. B—The thickened surface is a 3D solid.

Converting to Surfaces

AutoCAD provides a great deal of flexibility in converting and transforming objects. For example, a simple line can be quickly turned into a surface, then into a 3D solid in just a few steps. Refer to Figure 11-16.

1. Use the **Properties** palette to give the line a thickness. Notice that the object is still a line object, as indicated in the drop-down list at the top of the **Properties** palette.
2. Select the **CONVTOSURFACE** command.
3. Pick the line. Its property type is now listed in the **Properties** palette as a surface extrusion.
4. Use the **THICKEN** command to give the surface a thickness. Its property type is now a 3D solid.

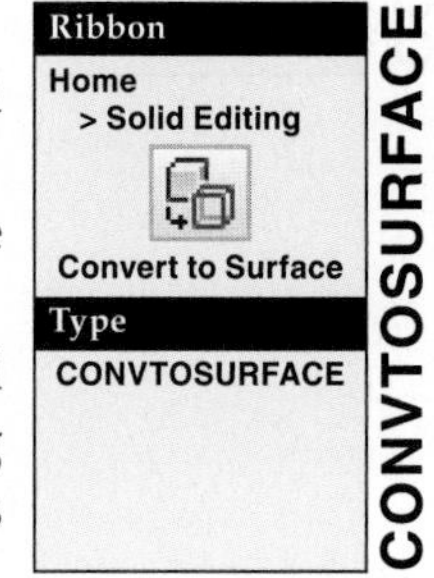

In this process, the **CONVTOSURFACE** and **THICKEN** commands were instrumental in creating a 3D solid from a line. Other objects that can be converted to surfaces using the **CONVTOSURFACE** command are 2D solids, arcs with thickness, open polylines with a thickness and no width, regions, and planar 3D faces.

Exercise 11-7

Complete the exercise on the student website.
www.g-wlearning.com/CAD

Figure 11-16.
The stages of converting a line into a solid. First, draw the line. Next, give the line a thickness using the **Properties** palette. Then, convert the line to a surface using the **CONVTOSURFACE** command. Finally, use the **THICKEN** command to give the surface a thickness.

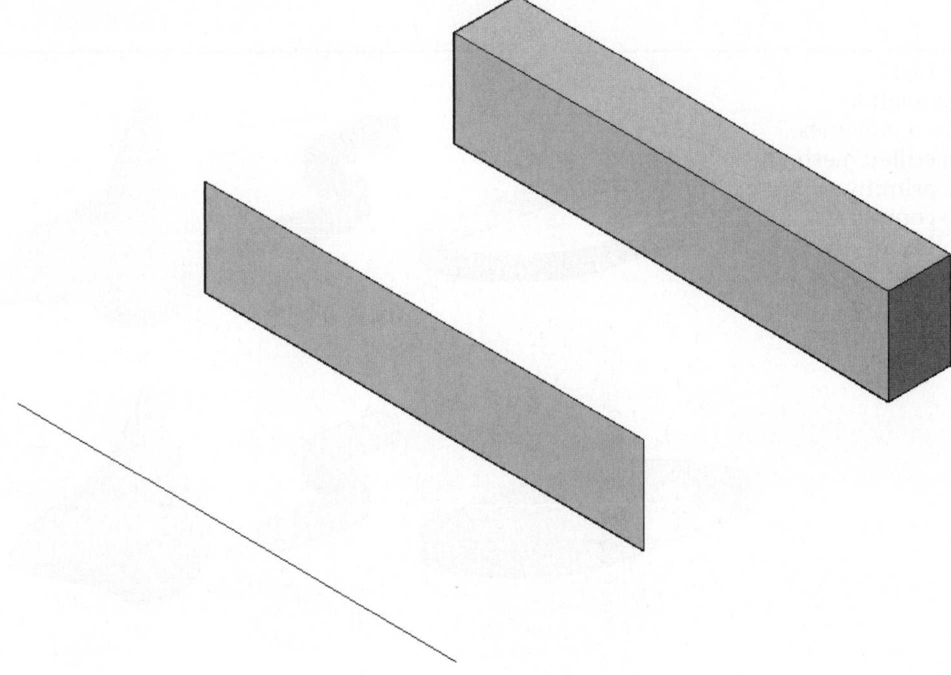

Converting to Solids

CONVTOSOLID

Ribbon
Home
> Solid Editing
Convert to Solid
Type
CONVTOSOLID

Additional flexibility in creating solids is provided by the **CONVTOSOLID** command. This command allows you to directly convert certain closed objects into solids. You can convert:

- Circles with thickness.
- Wide, uniform-width polylines with thickness. This includes polygons and rectangles.
- Closed, zero-width polylines with thickness. This includes polygons, rectangles, and closed revision clouds.
- Mesh primitives and other watertight mesh objects. Keep in mind that the smoothness level applied to the mesh primitives appears on the object when it is converted to a solid.
- Watertight surface models.

First, select the command. Then, select the objects to convert and press [Enter]. The objects are instantly converted with no additional input required. Figure 11-17 shows the three different objects before and after conversion to a solid.

If an object that appears to be a closed polyline with a thickness does not convert to a solid and the command line displays the message Cannot convert an open curve, the polyline was not closed using the **Close** option of the **PLINE** command. Use the **PEDIT** or **PROPERTIES** command to close the polyline and use the **CONVTOSOLID** command again.

PROFESSIONAL TIP

To quickly create a straight section of pipe, draw a donut with the correct ID and OD of the pipe. Then, use the **Properties** palette to give the donut a thickness equal to the length of the section you are creating. Finally, use the **CONVTOSOLID** command to turn the donut into a solid.

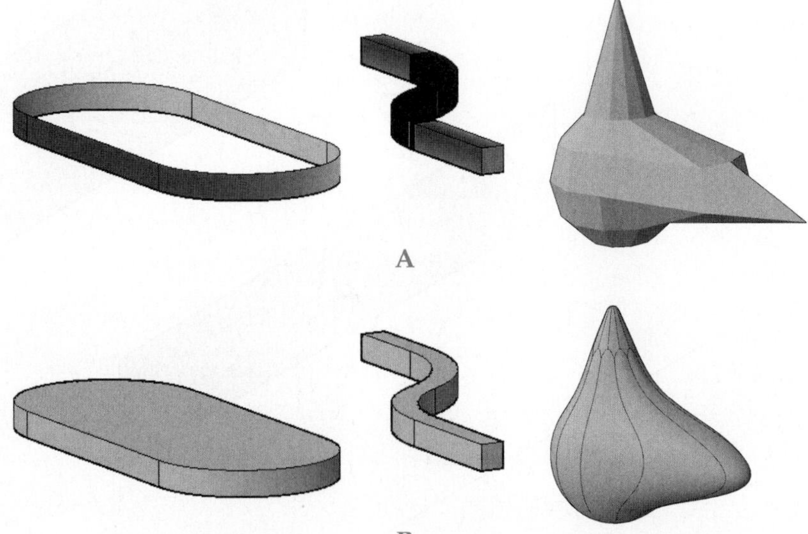

Figure 11-17.
A—From left to right, two polylines and an edited mesh sphere primitive that will be converted into solids. B—The resulting solids.

Sculpting: Surface to Solid

Ribbon
Surface
> Edit
Sculpt
Type
SURFSCULPT
CREATESOLID

SURFSCULPT

When designing using surfaces, you can sculpt a surface into a solid using the **SURFSCULPT** command. This is similar to using the **CONVTOSOLID** command. The main use for sculpting is to create a solid from a watertight area by trimming and combining multiple surfaces. The command can also be used on solid and mesh objects.

Notice the surfaces in Figure 11-18 create a *watertight* volume where trimming of the surface can occur. A single-surface enclosed area or multiple surfaces that create an enclosed area are considered to be watertight objects. Watertight objects can be converted to a solid.

Which command should you use, **CONVTOSOLID** or **SURFSCULPT**? The difference is very subtle.

- **CONVTOSOLID** works the best when you have a watertight mesh and want to convert it to a solid. Also, it works well on polylines and circles with thickness.
- **SURFSCULPT** works the best when you have watertight surfaces or solids that completely enclose a space (no gaps). This is shown in Figure 11-18.

As a best practice, use **CONVTOSOLID** for converting watertight meshes to solids. Use **SURFSCULPT** for converting surfaces or solids.

NOTE

The watertight surface area must have a G0 continuity (position continuity) for the **SURFSCULPT** command to work properly.

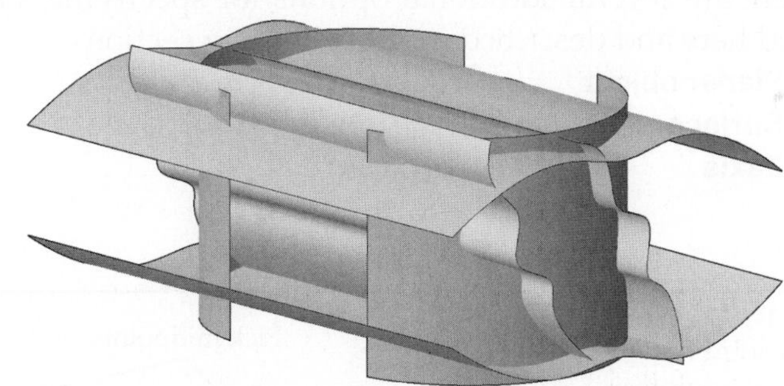

A

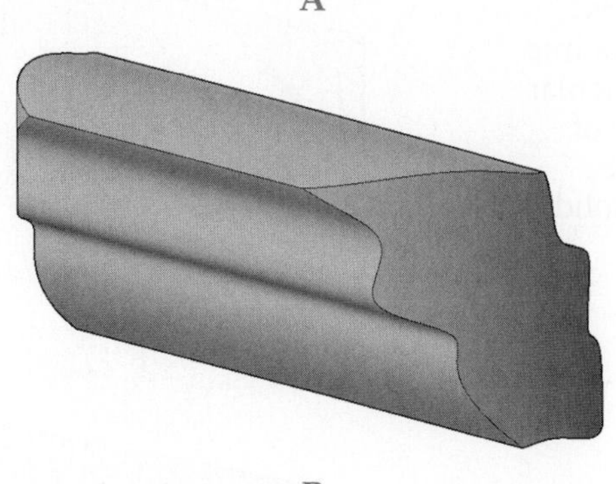

B

Figure 11-18.
A—These six surfaces form a watertight volume around a solid box. B—The solid is sculpted using the surfaces and the **SURFSCULPT** command.

Slicing a Solid

A 3D solid can be sliced at any location by using existing objects such as circles, arcs, ellipses, 2D polylines, 2D splines, or surfaces. Why slice a solid? To create complex, angular, contoured, or organic shapes that traditionally cannot be created by just using solids. Good modeling techniques incorporate slicing of solids. For example, think of a computer mouse. This would be difficult to create with solids. But, you can create a basic solid shape, then slice the solid using contoured surfaces as slicing tools. After slicing the solid, you can choose to retain either or both sides of the model. The slices can then be used for model construction or display and presentation purposes.

SLICE

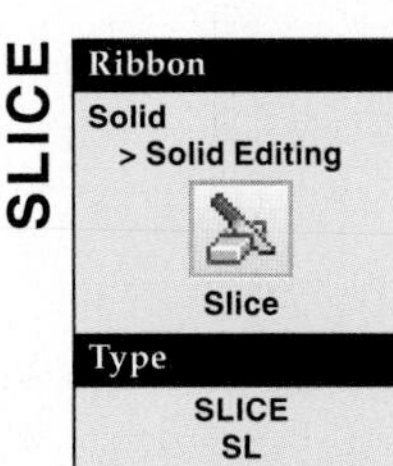

The **SLICE** command is used to slice solids. When the command is initiated, you are asked to select the solids to be sliced. Select the objects and press [Enter]. Next, you must define the slicing path. The default method of defining a path requires you to specify two points on a slicing plane. The plane passes through the two points and is perpendicular to the XY plane of the current UCS. Refer to **Figure 11-19** as you follow this sequence:

1. Select the command and pick the object to be sliced.
2. Pick the start point of the slicing plane. See **Figure 11-19A.**
3. Pick the second point on the slicing plane.
4. You are prompted to specify a point on the desired side to keep. Select anywhere on the back half of the object. The point does not have to be *on* the object. It must simply be on the side of the cutting plane that you want to keep.
5. The object is sliced and the front half is deleted. See **Figure 11-19B.**

When prompted to select the side to keep, you can press [Enter] to keep both sides. If both sides are retained, two separate 3D solids are created. Each solid can then be manipulated for construction, design, presentation, or animation purposes.

There are several additional options for specifying a slicing path. These options are listed here and described in the following sections.

- **Planar object**
- **Surface**
- **Zaxis**

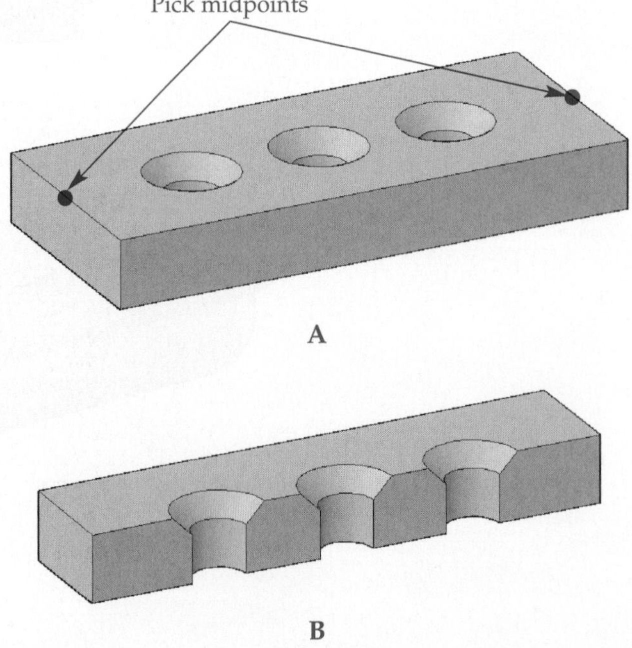

Figure 11-19.
Slicing a solid by picking two points. A—Select two points on the cutting plane. The plane passes through these points and is perpendicular to the XY plane of the current UCS. B—The sliced solid.

- **View**
- **XY**
- **YZ**
- **ZX**
- **3points**

NOTE

Once the **SLICE** command has been used, the history of the solid to that point is removed. If a history of the work is important, then save a copy of the file or place a copy of the object on a frozen layer prior to performing the slice.

Planar Object

A second method to create a slice through a 3D solid is to use an existing planar object. Planar objects include circles, arcs, ellipses, 2D polylines, and 2D splines. See **Figure 11-20A.** The plane on which the planar object lies must intersect the object to be sliced. The current UCS has no effect on this option.

Be sure that the planar object has been moved to the location of the slice. Then, select the **SLICE** command, pick the object to slice, and press [Enter]. Next, enter the **Planar Object** option and select the slicing path object (the circle, in this case). Finally, specify which side is to be retained. See **Figure 11-20B.** Again, if both sides are kept, they are separate objects and can be individually manipulated.

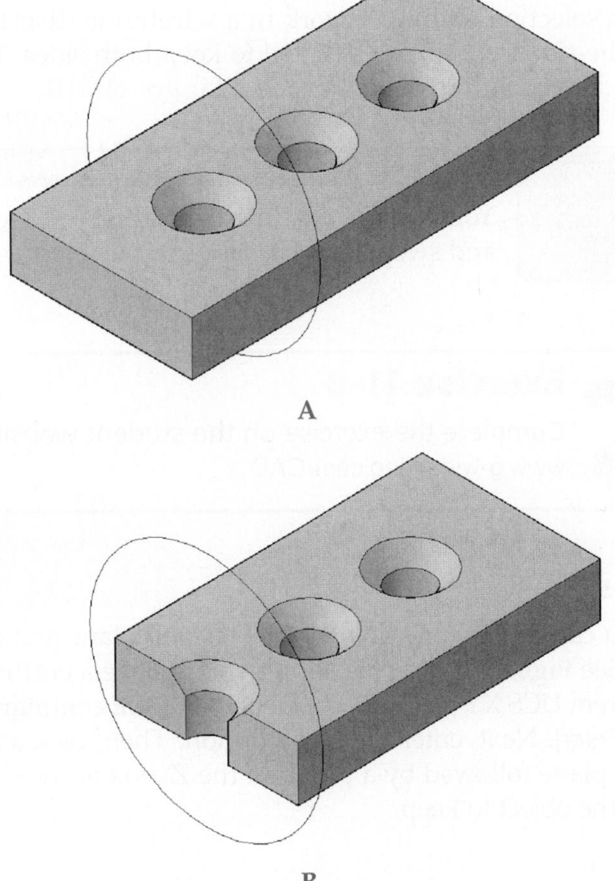

Figure 11-20. Slicing a solid with a planar object. A—The circle is drawn at the proper orientation and in the correct location. B—The completed slice.

Figure 11-21.
Slicing a solid with a surface. A—Draw the surface and locate it within the solid to be sliced. The solid is represented here by the wireframe. B—The completed slice with both sides retained. The top can now be moved and rotated as shown here.

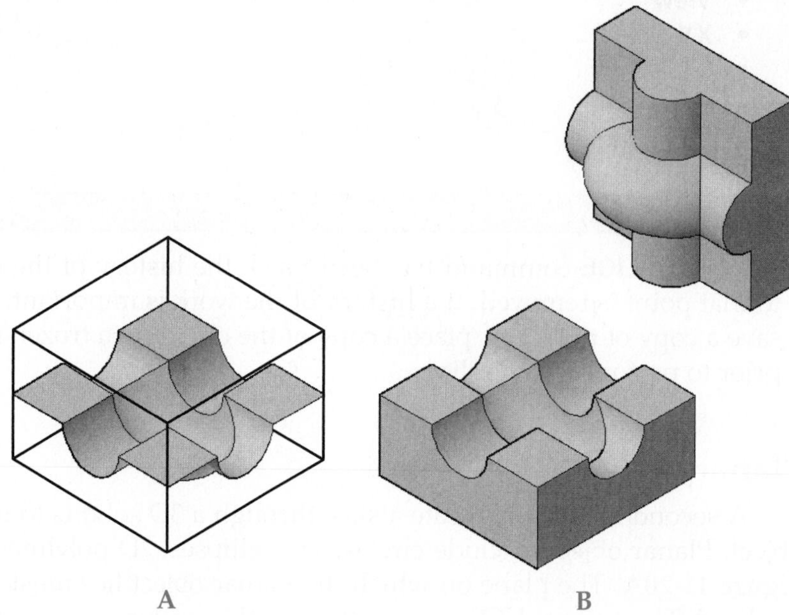

Surface

A surface object can be used as the slicing path. The surface can be planar or nonplanar (curved). This method can be used to quickly create a mating die. For example, refer to Figure 11-21. First, draw the required surface. The surface should exactly match the stamped part that will be manufactured, Figure 11-21A. Then, draw a box that encompasses the surface. Next, select the **SLICE** command, pick the box, and press [Enter]. Then, enter the **Surface** option and select the surface. You may need to turn on selection cycling or work in a wireframe display. Finally, when prompted to select the side to keep, press [Enter] to keep both sides. The two halves of the die can now be moved and rotated as needed, Figure 11-21B.

PROFESSIONAL TIP

You can use extruded surfaces, revolved surfaces, lofted surfaces, and swept surfaces as slicing objects.

Exercise 11-8

Complete the exercise on the student website.
www.g-wlearning.com/CAD

Z Axis

You can specify one point on the cutting plane and one point on the Z axis of the plane. See Figure 11-22. This allows you to have a cutting plane that is not parallel to the current UCS XY plane. First, select the **SLICE** command, pick the object to slice, and press [Enter]. Next, enter the **Zaxis** option. Then, pick a point on the XY plane of the cutting plane followed by a point on the Z axis of the cutting plane. Finally, pick the side of the object to keep.

Figure 11-22.
Slicing a solid using the **Zaxis** option. A—Pick one point on the cutting plane and a second point on the Z axis of the cutting plane. B—The resulting slice.

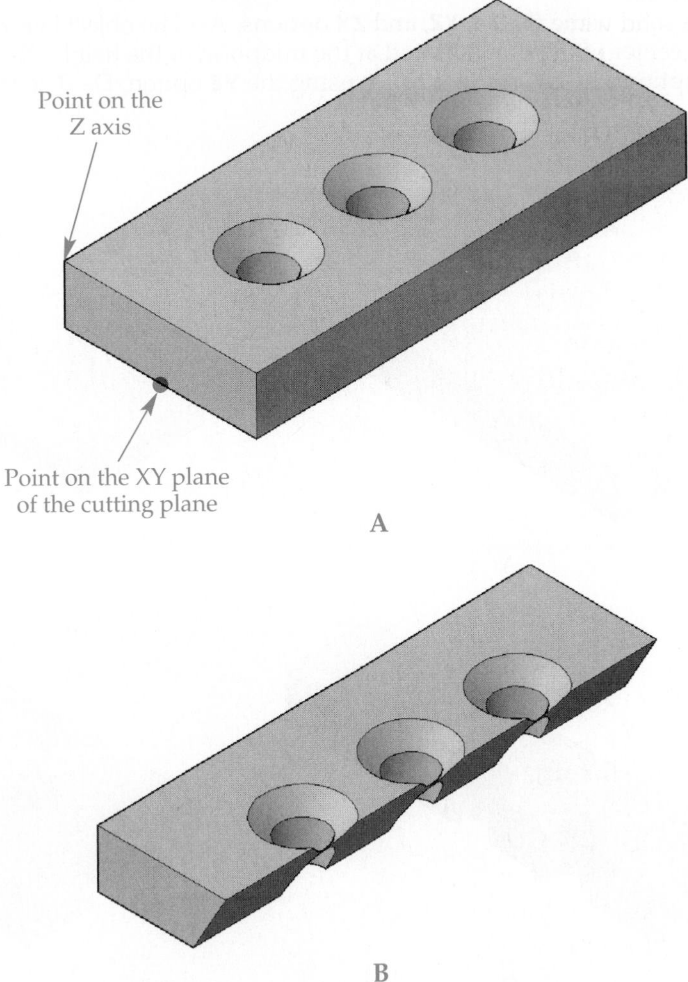

View

A cutting plane can be established that is aligned with the viewing plane of the current viewport. The cutting plane passes through a point you select, which sets the depth along the Z axis of the current viewing plane. First, select the **SLICE** command, pick the object to slice, and press [Enter]. Next, enter the **View** option. Then, pick a point in the viewport to define the location of the cutting plane on the Z axis of the viewing plane. Use object snaps to select a point on an object. The cutting plane passes through this point and is parallel to the viewing plane. Finally, pick the side of the object to keep.

XY, YZ, and ZX

You can slice an object using a cutting plane that is parallel to any of the three primary planes of the current UCS. See Figure 11-23. The cutting plane passes through the point you select and is aligned with the primary plane of the current UCS that you specify. First, select the **SLICE** command, pick the object to slice, and press [Enter]. Next, enter the **XY**, **YZ**, or **ZX** option, depending on the primary plane to which the cutting plane will be parallel. Then, pick a point on the cutting plane. Finally, pick the side of the object to keep.

Figure 11-23.
Slicing a solid using the **XY**, **YZ**, and **ZX** options. A—The object before slicing. The UCS origin is in the center of the first hole and at the midpoint of the height. B—The resulting slice using the **XY** option. C—The resulting slice using the **YZ** option. D—The resulting slice using the **ZX** option.

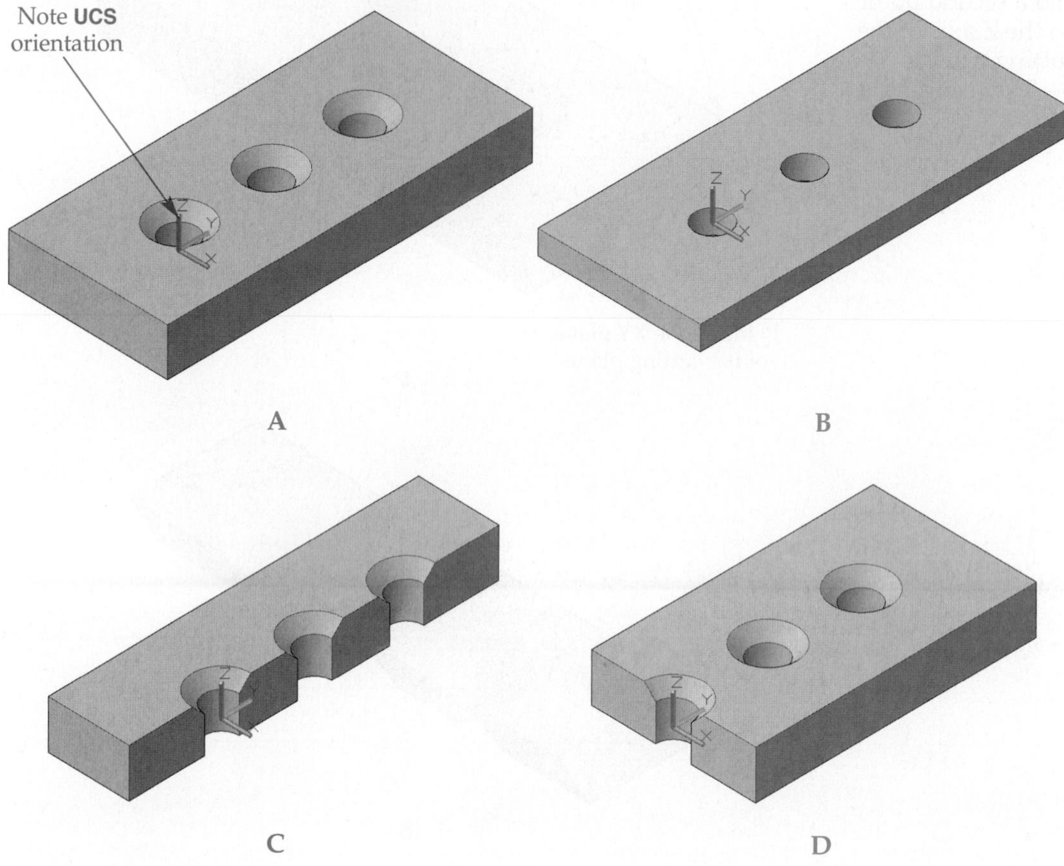

Three Points

Three points can be used to define the cutting plane. This allows the cutting plane to be aligned at any angle, similar to the **Zaxis** option. See Figure 11-24. First, select the **SLICE** command, pick the object to be sliced, and press [Enter]. Then, enter the **3points** option. Pick three points on the cutting plane and then select the side of the object to keep.

Exercise 11-9

Complete the exercise on the student website.
www.g-wlearning.com/CAD

Figure 11-24.
Slicing a solid using the **3points** option. A—Specify three points to define the cutting plane. B—The resulting slice.

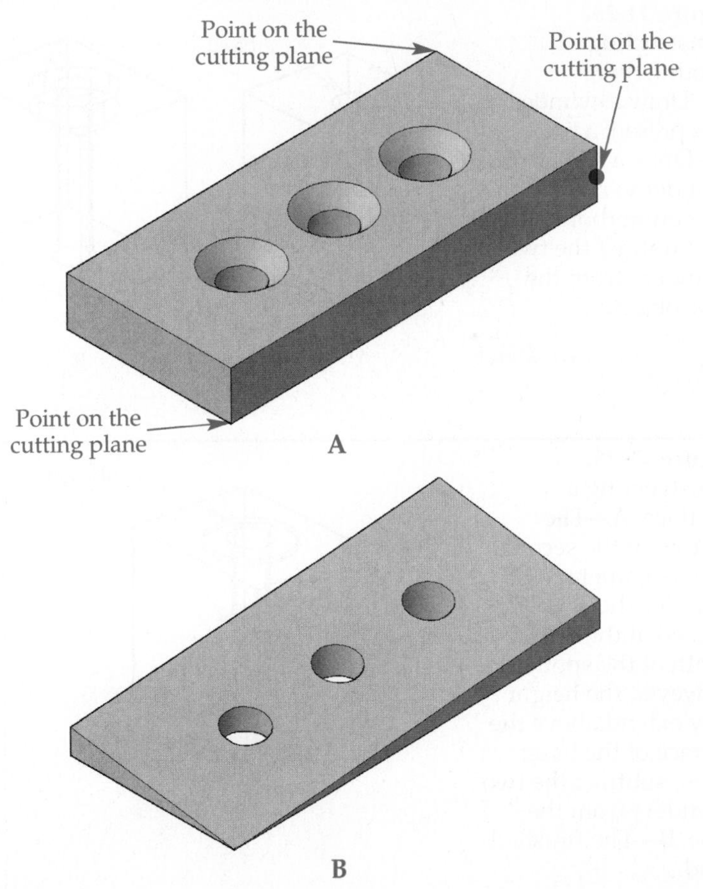

Removing Features

Sometimes, it may be necessary to remove a feature that has been constructed. For example, suppose you placed a R.5 fillet on an object based on an engineering sketch. Then, the design is changed to a R.25 fillet. Subobject editing is the best technique to accomplish this. Subobject editing is covered in Chapter 12.

Constructing Features on Solid Models

A variety of machining, structural, and architectural details can be created using some basic solid modeling techniques. The features discussed in the next sections are just a few of the possibilities.

Counterbore and Spotface

A ***counterbore*** is a recess machined into a part, centered on a hole, that allows the head of a fastener to rest below the surface. Create a counterbore as follows.

1. Draw a cylinder representing the diameter of the hole, **Figure 11-25A.**
2. Draw a second cylinder that is the diameter of the counterbore and center it at the top of the first cylinder. Move the second cylinder so it extends below the surface of the object to the depth of the counterbore, **Figure 11-25B.**
3. Subtract the two cylinders from the base object, **Figure 11-25C.**

A ***spotface*** is similar to a counterbore, but is not as deep. See **Figure 11-26.** It provides a flat surface for full contact of a washer or underside of a bolt head. Construct it in the same way as a counterbore.

Figure 11-25.
Constructing a counterbore.
A—Draw a cylinder to represent a hole.
B—Draw a second cylinder to represent the counterbore.
C—Subtract the two cylinders from the base object.

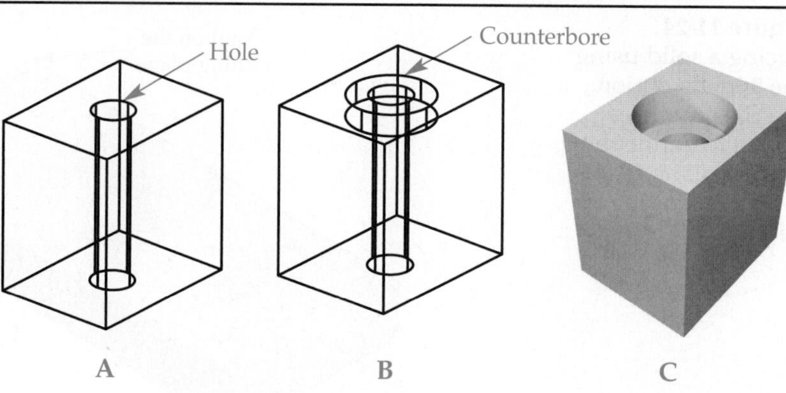

Figure 11-26.
Constructing a spotface. A—The bottom of the second, larger-diameter cylinder should be located at the exact depth of the spotface. However, the height may extend above the surface of the base. Then, subtract the two cylinders from the base. B—The finished solid.

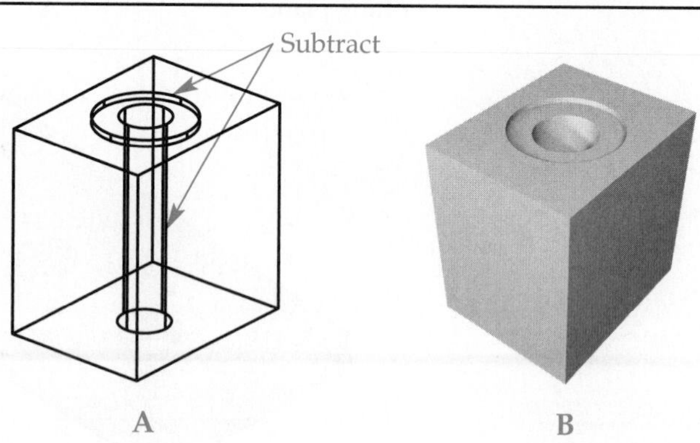

Countersink

A ***countersink*** is like a counterbore with angled sides. The sides allow a flat-head machine screw or wood screw to sit flush with the surface of an object. A countersink can be drawn in one of two ways. You can draw an inverted cone centered on a hole and subtract it from the base or you can chamfer the top edge of a hole. Chamfering is the quickest method.

1. Draw a cylinder representing the diameter of the hole, Figure 11-27A.
2. Subtract the cylinder from the base object.
3. Select the **CHAMFER** or **CHAMFEREDGE** command.
4. Select the top edge of the base object.
5. Enter the chamfer distance(s).
6. Pick the top edge of the hole, Figure 11-27B.

Figure 11-27.
Constructing a countersink.
A—Subtract the cylinder from the base to create the hole.
B—Chamfer the top of the hole to create a countersink.

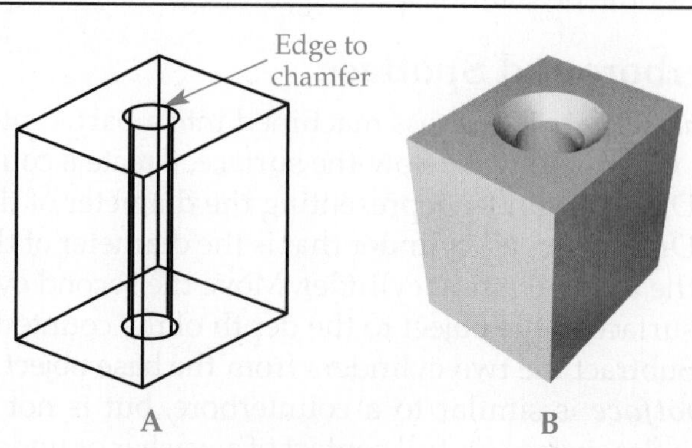

Figure 11-28.
Constructing a boss. A—Draw a cylinder for the hole so it extends above the surface of the object. B—Draw a cylinder the height of the boss on the top surface of the object. C—Union the large cylinder to the base. Then, subtract the small cylinder (hole) from the unioned objects. D—Fillet the edge to form the boss.

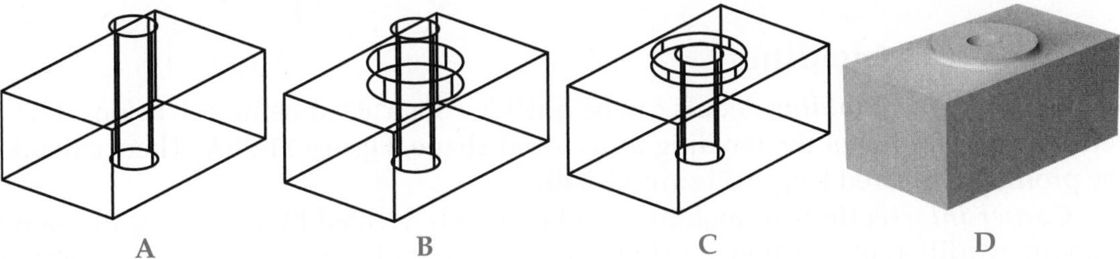

Boss

A ***boss*** serves the same function as a spotface. However, it is an area raised above the surface of an object. Draw a boss as follows.

1. Draw a cylinder representing the diameter of the hole. Extend it above the base object higher than the boss is to be, **Figure 11-28A.**
2. Draw a second cylinder the diameter of the boss. Place the base of this cylinder above the top surface of the base object a distance equal to the height of the boss. Give the cylinder a negative height value so that it extends inside of the base object, **Figure 11-28B.**
3. Union the base object and the second cylinder (boss). Subtract the hole from the unioned object, **Figure 11-28C.**
4. Fillet the intersection of the boss with the base object, **Figure 11-28D.**

O-Ring Groove

An ***O-ring*** is a circular seal that resembles a torus. It sits inside of a groove constructed so that part of the O-ring is above the surface. An ***O-ring groove*** can be constructed by placing the center of a circle on the outside surface of a cylinder. Then, revolve the circle around the cylinder. Finally, subtract the revolved solid from the cylinder.

1. Construct the cylinder to the required dimensions, **Figure 11-29A.**
2. Rotate the UCS on the X axis (or appropriate axis).
3. Draw a circle with a center point on the surface of the cylinder, **Figure 11-29B.**
4. Revolve the circle 360° about the center of the cylinder, **Figure 11-29C.**
5. Subtract the revolved object from the cylinder, **Figure 11-29D.**

Figure 11-29.
Constructing an O-ring groove. A—Construct a cylinder; this one has a round placed on one end. B—Draw a circle centered on the surface of the cylinder. C—Revolve the circle 360° about the center of the cylinder. D—Subtract the revolved object from the cylinder. E—The completed O-ring groove.

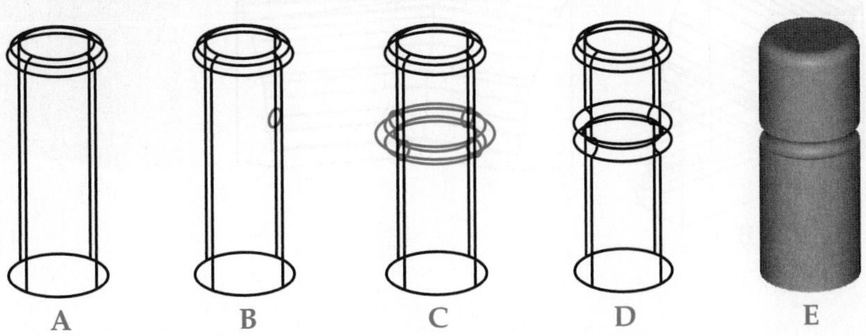

PROFESSIONAL TIP

In many cases, you will draw the O-ring as a torus. A copy of the torus can be used to create the O-ring groove instead of revolving a circle as described above.

Architectural Molding

Architectural molding details can be quickly constructed using extrusions. First, construct the profile of the molding as a closed shape, **Figure 11-30A.** Then, extrude the profile the desired length, **Figure 11-30B.**

Corner intersections of molding can be quickly created by extruding the same shape in two different directions and then joining the two objects. First, draw the molding profile. Then, copy and rotate the profile to orient the local Z axis in the desired direction, **Figure 11-31A.** Next, extrude the two profiles the desired lengths, **Figure 11-31B.** Finally, union the two extrusions to create the mitered corner molding, **Figure 11-31C.**

Exercise 11-10

Complete the exercise on the student website.
www.g-wlearning.com/CAD

Figure 11-30.
A—The molding profile. B—The profile extruded to the desired length.

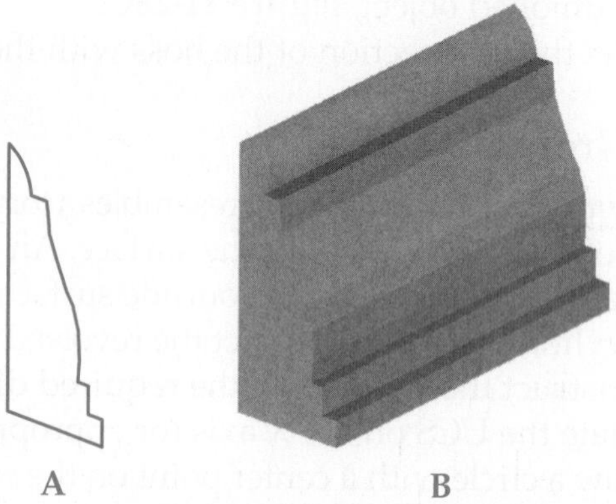

Figure 11-31.
Constructing corner molding. A—Copy and rotate the molding profile. B—Extrude the profiles to the desired lengths. C—Union the two extrusions to create the mitered corner. Note: The view has been rotated. D—The completed corner.

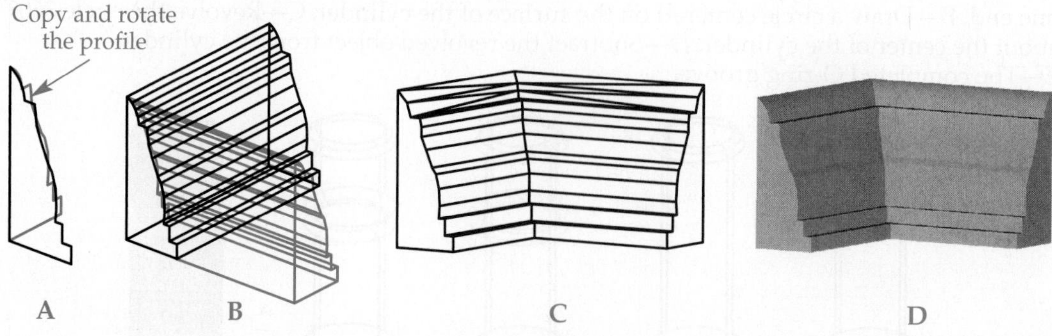

Chapter Test

Answer the following questions. Write your answers on a separate sheet of paper or complete the electronic chapter test on the student website.
www.g-wlearning.com/CAD

1. Which properties of a solid can be changed in the **Properties** palette?
2. What does the History property control?
3. What is the preferred command for aligning objects to create an assembly of parts?
4. How does the **3DALIGN** command differ from the **ALIGN** command?
5. How does the **3DROTATE** command differ from the **ROTATE** command?
6. How does the **MIRROR3D** command differ from the **MIRROR** command?
7. Which command allows you to create a rectangular array by defining rows, columns, and levels?
8. How does a 3D polar array differ from a 2D polar array?
9. How many levels can a 3D polar array have?
10. Which command is used to fillet a solid object?
11. Which command is used to chamfer a solid object?
12. What is the purpose of the **THICKEN** command and which type of object does it create?
13. Which system variable allows you to preserve the original object when the **THICKEN** command is used?
14. List four objects that can be converted to surfaces using the **CONVTOSURFACE** command.
15. What is the preferred command to convert a mesh into a solid?
16. What is the preferred command to convert a watertight series of surfaces into a solid?
17. Name four types of surfaces that can be used to slice objects.
18. Briefly describe the function of the **SLICE** command.
19. Which **SLICE** command option would be used to create a contoured solid object like a computer mouse?
20. Which two commands can be used to create a countersink?

Drawing Problems

1. Construct an 8″ diameter tee pipe fitting using the dimensions shown below. Hint: Extrude and union two solid cylinders before subtracting the cylinders for the inside diameters.
   A. Use **EXTRUDE** to create two sections of pipe at 90° to each other, then **UNION** the two pieces together.
   B. Fillet and chamfer the object to finish it. The chamfer distance is .25″ × .25″.
   C. The outside diameter of all three openings is 8.63″ and the pipe wall thickness is .322″.
   D. Save the drawing as P11_01.

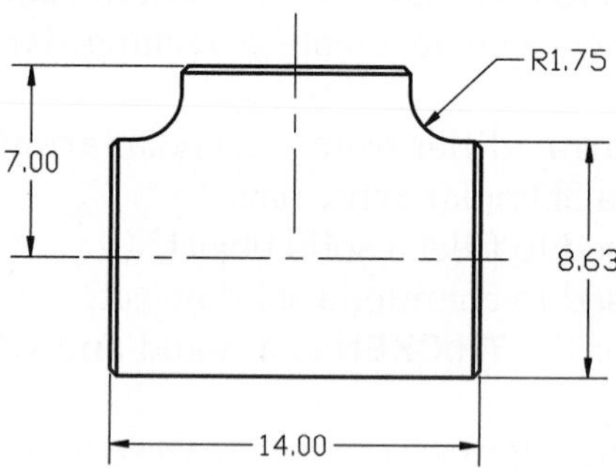

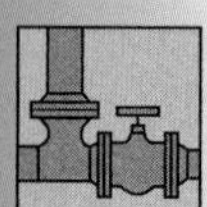

2. Construct an 8″ diameter, 90° elbow pipe fitting using the dimensions shown below.
   A. Use **EXTRUDE** or **SWEEP** to create the elbow.
   B. Chamfer the object. The chamfer distance is .25″ × .25″. Note: You cannot use the **CHAMFER** command.
   C. The outside diameter is 8.63″ and the pipe wall thickness is .322″.
   D. Save the drawing as P11_02.

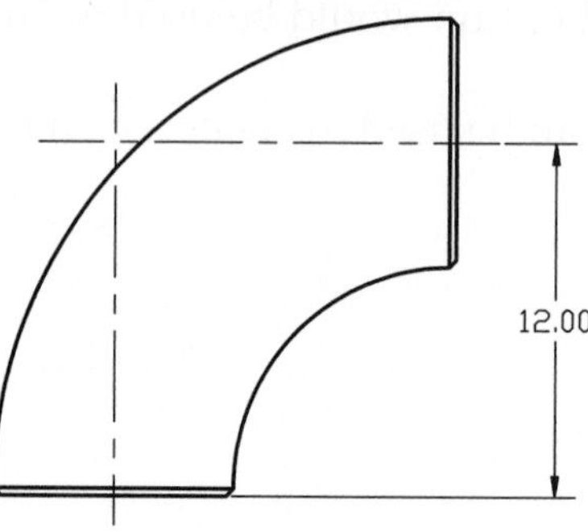

Drawing Problems - Chapter 11

Problems 3–6. These problems require you to use a variety of solid modeling functions to construct the objects. Use all of the solid modeling and editing commands you have learned so far to assist in construction. Use a dynamic UCS when practical and create new UCSs as needed. Use **SOLIDHIST** *and* **SHOWHIST** *to record and view the steps used to create the solid models. Create copies of the completed models and split them as required to show the internal features visible in the section views. Save each drawing as* P11_*(problem number).*

3.

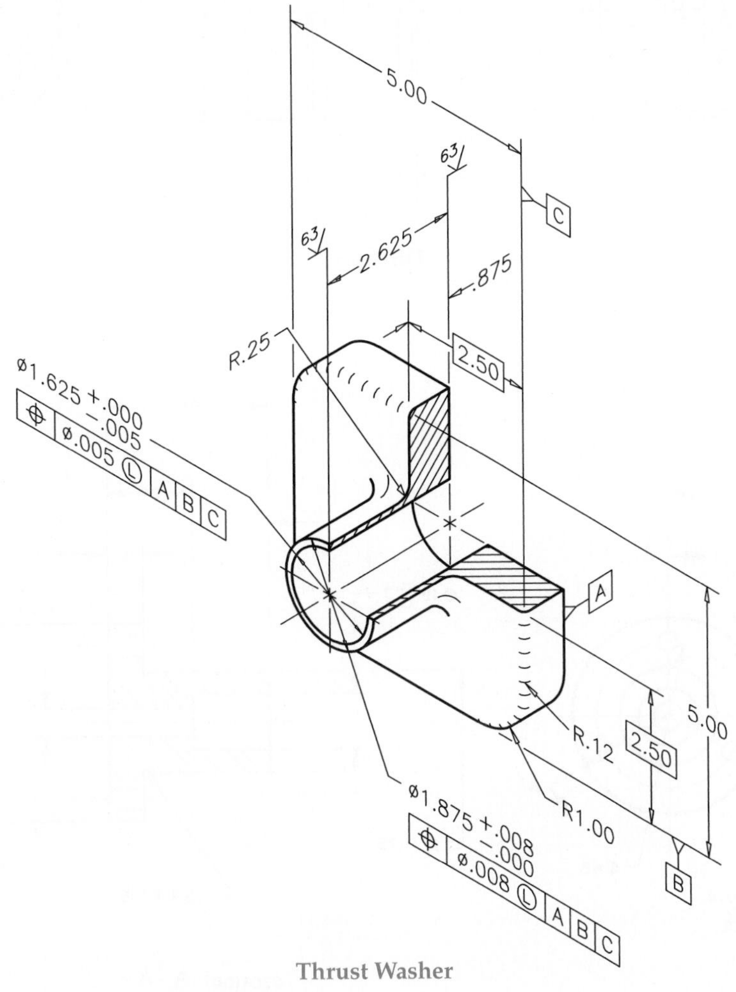

Thrust Washer

4.

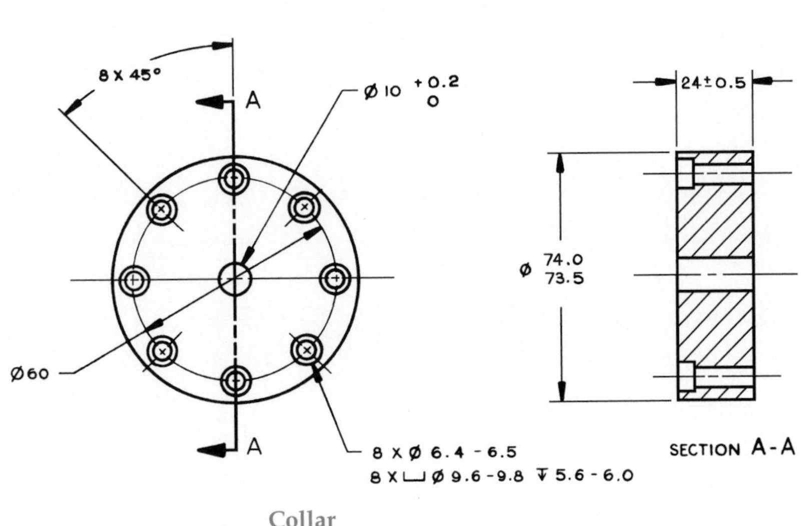

Collar

Drawing Problems - Chapter 11

5.

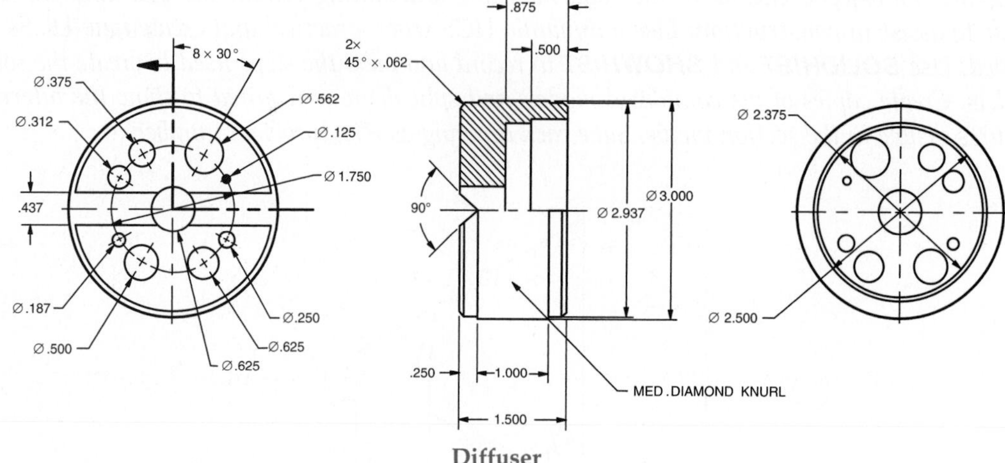

Diffuser

6.

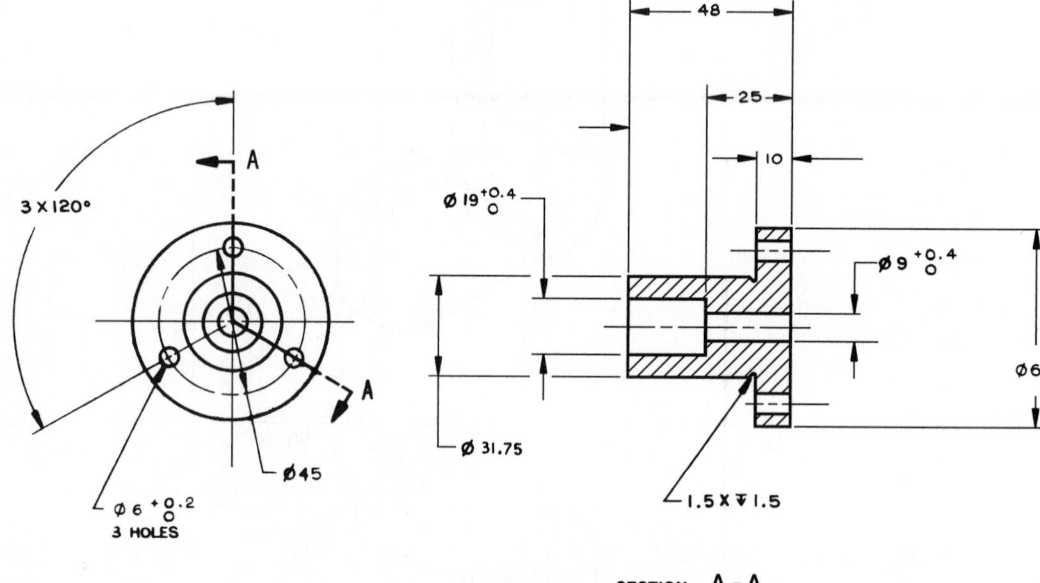

Bushing

7. In this problem, you will add legs to the kitchen chair you started in Chapter 2. In Chapter 9 you refined the seat and in Chapter 10 you added the seatback. In this chapter, you complete the model. In Chapter 17, you will add materials to the model and render it.
   A. Open P10_07 from Chapter 10.
   B. Using the **LINE** command, create the framework for lofting the legs and crossbars as shown below. The crossbars are at the midpoints of the legs. Position the lines for the double crossbar about 4.5″ apart.
   C. Using a combination of the **3DROTATE** and **3DMOVE** commands in conjunction with new UCSs, draw and position circles as shapes for lofting the legs and crossbars. The legs transition from Ø1.00 at the ends to Ø1.25 at the midpoints. The crossbars transition from Ø.75 at the ends to Ø1.00 at a position 2″ from each end.
   D. Create the legs and crossbars. The completed model is shown below.
   E. Save the drawing as P11_07.

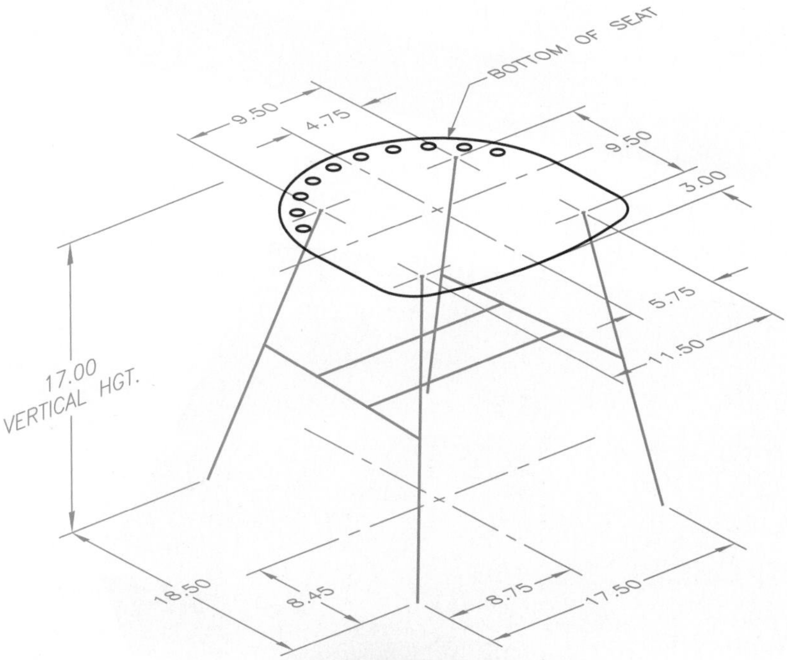

BOTTOM OF SEAT
9.50
4.75
9.50
3.00
5.75
11.50
17.00
VERTICAL HGT.
18.50
8.45
8.75
17.50

Completed Model

Subobject editing is very useful in 3D modeling. As shown here, the size of a hole can be quickly changed. The original hole is shown in the top illustration. The size of the hole is decreased in the middle illustration and increased in the bottom illustration.

CHAPTER 12

Subobject Editing

Learning Objectives

After completing this chapter, you will be able to:

- ✓ Select subobjects (faces, edges, and vertices).
- ✓ Edit solids using grips.
- ✓ Edit fillet and chamfer subobjects.
- ✓ Edit face subobjects.
- ✓ Edit edge subobjects.
- ✓ Edit vertex subobjects.
- ✓ Extrude a closed boundary using the **PRESSPULL** command.

Grip Editing

There are three basic types of 3D solids in AutoCAD. The commands **BOX**, **WEDGE**, **PYRAMID**, **CYLINDER**, **CONE**, **SPHERE**, and **TORUS** create 3D solid ***primitives***. ***Swept objects*** are 2D open or closed profiles given thickness by the **EXTRUDE**, **REVOLVE**, **SWEEP**, and **LOFT** commands to create a 3D solid or surface. Finally, 3D solid ***composites*** are created by a Boolean operation (**UNION**, **SUBTRACT** or **INTERSECT**) or by using the **SOLIDEDIT** command. The **SOLIDEDIT** command is discussed in Chapter 13. Smooth-edged solid primitives are achieved by converting mesh objects to solids. These are also composite solids.

There are two types of grips—base and parameter—that may be associated with a solid or surface object. These grips provide an intuitive means of modifying solids. Base grips are square and parameter grips are typically arrows. The editing that can be performed with these grips are discussed in the next sections.

3D Solid Primitives

The 3D solid primitives all have basically the same grips. However, not all grips are available on all primitives. All primitives have a base grip at the centroid of the base. This grip functions like a standard grip in 2D work. It can be used to stretch, move, rotate, scale, or mirror the solid.

Boxes, wedges, and pyramids have square base grips at the corners that allow the size of the base to be changed. See **Figure 12-1.** The object dynamically changes in the viewport as you select and move the grip or after you type the new coordinate location for the grip and press [Enter]. If ortho is off, the length and width can be changed at the same time by dragging the grip, except in the case of a pyramid. The triangular parameter grips on the base allow the length or width to be changed. The height can be changed using the top parameter grip(s). Each object has one parameter grip for changing the height of the apex and one for changing the height of the plane on which the base sits. A pyramid also has a parameter grip at the apex for changing the radius of the top.

Cylinders, cones, and spheres have four parameter grips for changing the radius of the base, or the cross section in the case of a sphere. See **Figure 12-2.** Cylinders and cones also have parameter grips for changing the height of the apex and the height of the plane on which the base sits. Additionally, a cone has a parameter grip at the apex for changing the radius of the top.

A torus has a parameter grip located at the center of the tube. See **Figure 12-3.** This grip is used to change the radius of the torus. Parameter grips at each quadrant of the tube are used to change the radius of the tube.

A polysolid does not have parameter grips. Instead, a base grip appears at each corner of the starting face of the solid. See **Figure 12-4.** Use these grips to change the

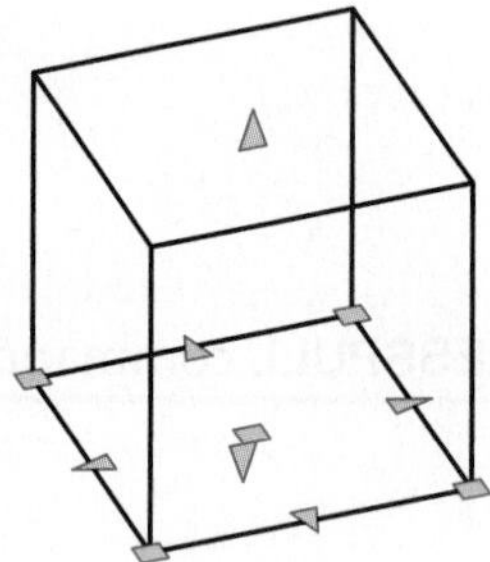
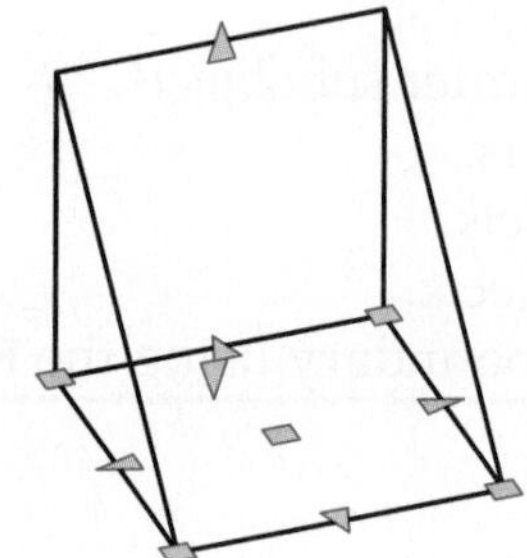
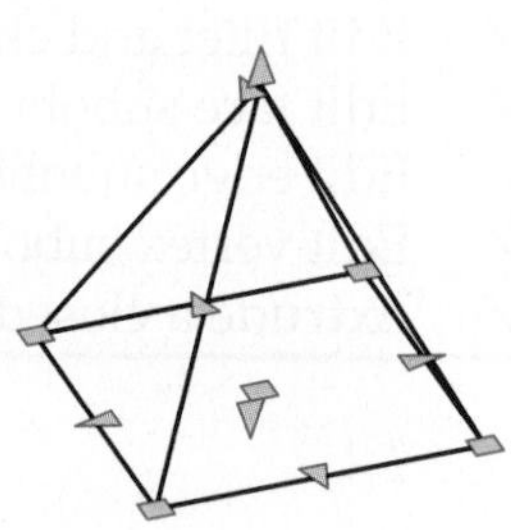

Figure 12-1.
Boxes, wedges, and pyramids have square base grips at the corners and parameter grips on the sides of the base and center of the top face, edge, or vertex.

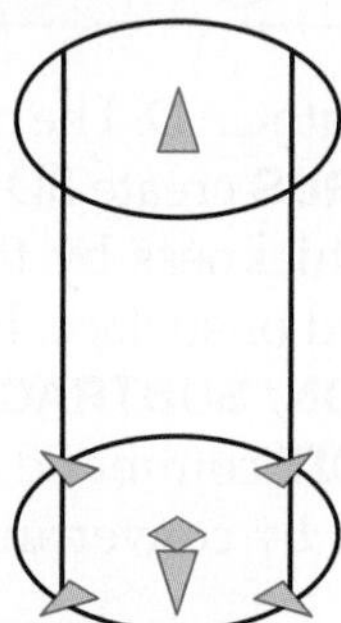
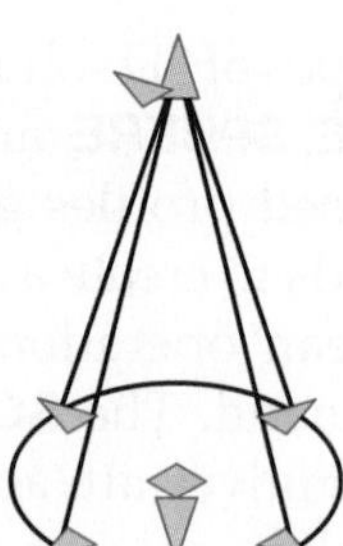
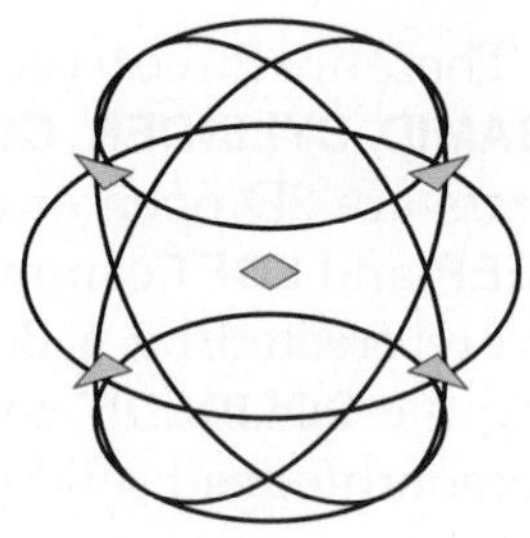

Figure 12-2.
Cylinders, cones, and spheres have four parameter grips for changing their radius. Cylinders and cones have parameter grips for changing their height. Cones also have a parameter grip for changing the radius of the top.

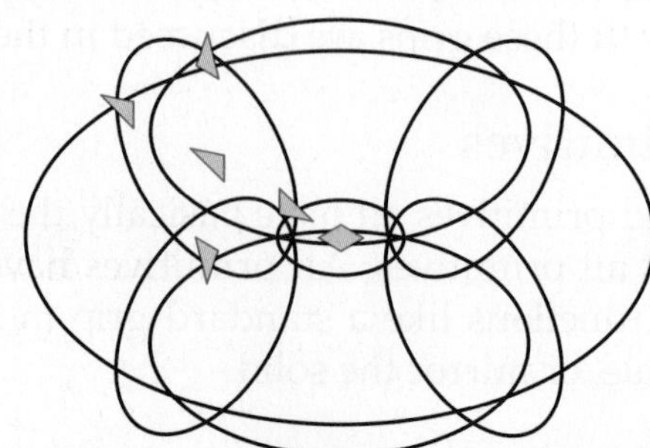

Figure 12-3.
A torus has a parameter grip located at the center of the tube for changing the radius of the torus. There are also parameter grips for changing the radius of the tube.

Figure 12-4.
A polysolid has a base grip at each corner of the starting face of the solid and one at the endpoint of each segment.

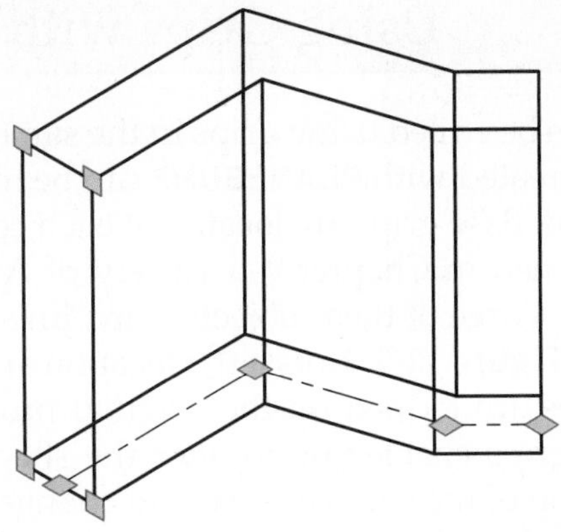

cross-sectional shape of the polysolid. The corners do not need to remain square. Base grips also appear at the endpoint of each segment centerline. Use these to change the location of each segment's endpoints.

PROFESSIONAL TIP

With dynamic input turned on, you can subobject edit a primitive by selecting a parameter grip and typing a delta value for adding geometry to the primitive. For example, suppose a cylinder has a height of 10 units. The new height requirement is 15. Select the cylinder, pick the parameter grip, and drag it so the dynamic input changes to delta entry. In the dynamic input box, type 5 for the new height and press [Enter].

Swept Solids

Extrusions, revolutions, sweeps, and lofts are considered swept solids and typically have base grips located at the vertices of the 2D profiles. These can be used to change the size of the profile and, thus, the solid. Other grips that appear include:

- A parameter grip appears on the upper face of extrusions for changing the height.
- A base grip appears on the axis of revolved solids for changing the location of the axis in relation to the profile.
- Base grips appear on the vertices of the path of sweeps for changing the shape of the path.

Composite Solids

The Boolean commands create composite solids. Solids that have been modified using any of the options of the **SOLIDEDIT** command also become composite solids, as do meshes converted into solids. The solid may still look like a primitive, sweep, loft, etc., but it is a composite. The grips available with the previous objects are no longer available, unless performing subobject editing on a composite created with a Boolean command (discussed later in this chapter). Composite solids have a base grip located at the centroid of the base surface. This grip can be used to stretch, move, rotate, scale, or mirror the solid. Composite solids converted from meshes have several grips, but these all act as base grips.

NOTE

When performing grip editing on a solid, AutoCAD must be able to "solve" the end result. If it cannot, the edit is not applied.

Using Grips with Surfaces

Surfaces can be edited using grips in the same manner as discussed with solids. A planar surface created with **PLANESURF** can be moved, rotated, scaled, and mirrored, but not stretched. Base grips are located at each corner.

As you learned in Chapter 9, a variety of AutoCAD objects can be extruded to create a surface. Three of these objects—arc, line, and polyline—are shown extruded into surfaces in **Figure 12-5.** Notice the location and type of grips on the surface extrusions. Base grips are located on the original profile that was extruded to make the surface. These grips enable you to alter the shape of the surface. A parameter grip located on the top of the surface is used to change the height of the extrusion.

Surfaces that have been extruded, or swept, along a path can be edited with grips. Also, the grips located on the path allow you to change the shape of the surface extrusion. See **Figure 12-6.**

Figure 12-5.
Surfaces extruded from an arc, line, and polyline. Notice the grips.

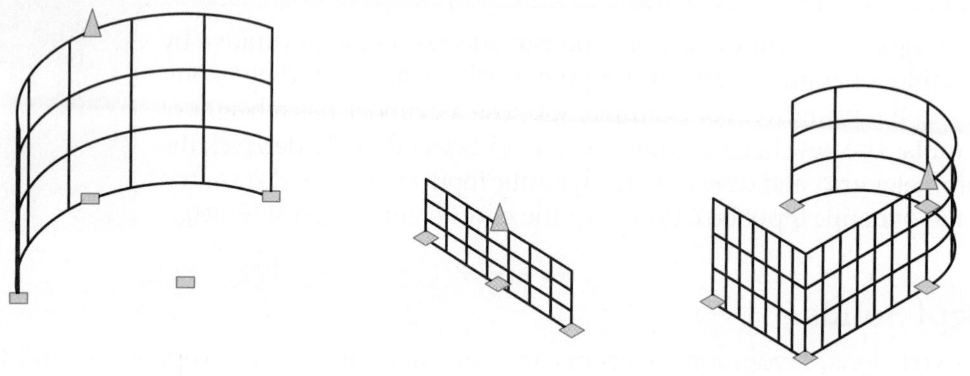

Figure 12-6.
A—Grips can be used to modify the path on this swept surface. B—The swept surface after grip editing.

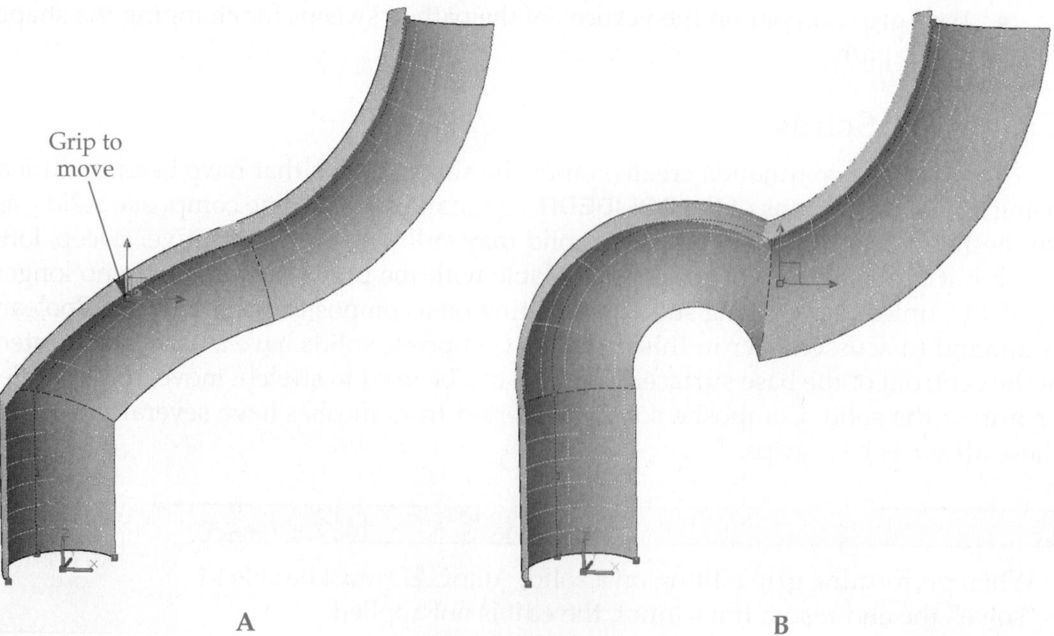

Exercise 12-1

Complete the exercise on the student website.
www.g-wlearning.com/CAD

Overview of Subobject Editing

AutoCAD solid primitives, such as cylinders, wedges, and boxes, and mesh objects, are composed of three types of subobjects: faces, edges, and vertices. In addition, the objects that are used with Boolean commands to create a composite solid are considered subobjects, if the history is recorded. The primitive subobjects can be edited. See Figure 12-7. Once selected, the primitive subobjects can even be deleted or modified as needed from the composite solid. Figure 12-8 illustrates the difference between a composite solid model, the solid primitives used to construct it, and an individual subobject of one of the primitives.

Subobjects can be easily edited using grips, which provide an intuitive and flexible method of solid model design. For example, suppose you need to rotate a face subobject in the current XY plane. You can select the subobject, pick its base grip, and then cycle through the editing functions to the rotate function. You can also use the **ROTATE** command on the selected subobject.

To select a subobject, press the [Ctrl] key and pick the subobject. You can select multiple subobjects and subobjects on multiple objects. To select a subobject that is hidden in the current view, first display the model as a wireframe. After creating a selection set, select a grip and edit the subobject as needed. Multiple objects can be selected in this manner. To deselect objects, press the [Shift]+[Ctrl] key combination and pick the objects to be removed from the selection set.

If objects or subobjects are overlapping, you have two options for cycling through the subobjects. The first and preferred method is to set the **SELECTIONCYCLING** system variable to 2. This is the same as picking the **Selection Cycling** button on the status bar to turn it on. Selection cycling is discussed in Chapter 11. The second method comes into play when selection cycling is turned off. Press the [Ctrl] key and the spacebar to turn on cycling and pick the subobject. Then, release the spacebar, continue holding the [Ctrl] key, and pick until the subobject you need is highlighted. Press [Enter] or the spacebar to select the highlighted subobject.

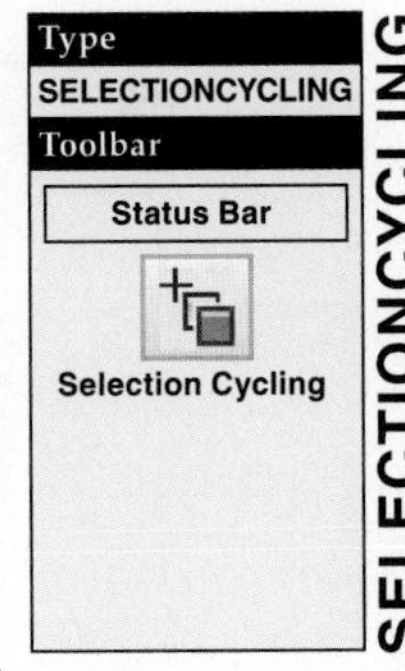

Figure 12-7.
A—Selecting a subobject solid primitive within a composite solid displays its grips.
B—The grips on the primitive can be used to edit the primitive.

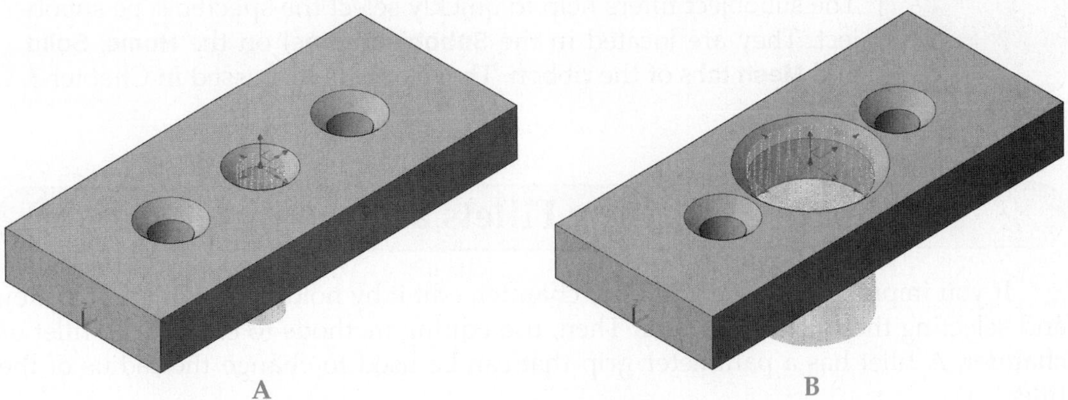

Figure 12-8.
A—The composite solid model is selected. Notice the single base grip. B—The wedge primitive subobject has been selected. Notice the grips associated with the primitive. C—An edge subobject within the primitive subobject is selected for editing.

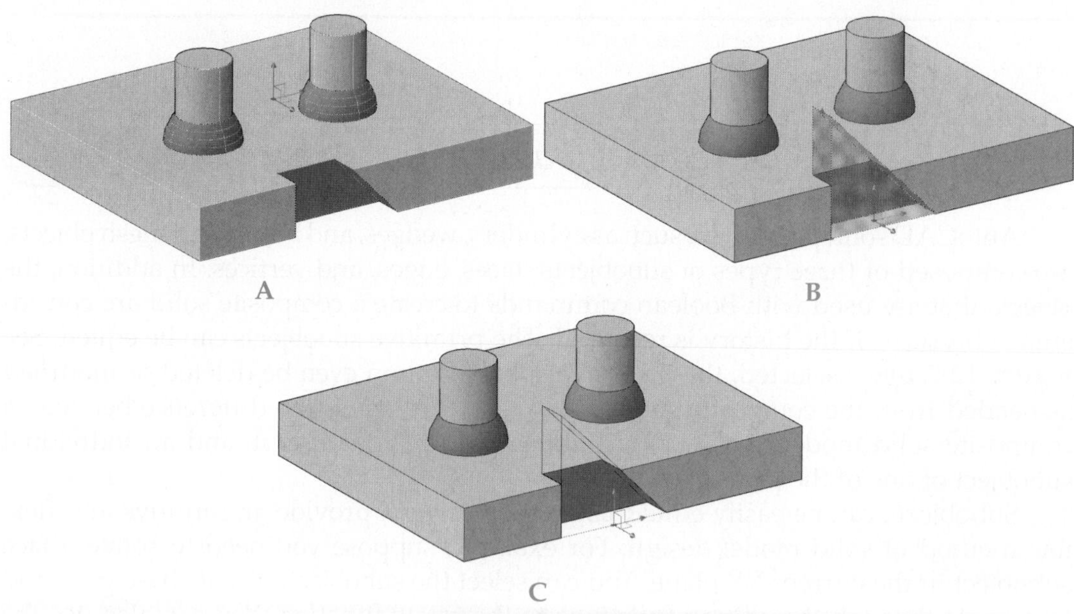

The [Ctrl] key method can be used to select subobjects for use with editing commands such as **MOVE**, **COPY**, **ROTATE**, **SCALE**, **ARRAY**, and **ERASE**. Some commands, like **ARRAY**, **STRETCH**, and **MIRROR**, are applied to the entire solid. Other operations may not be applied at all, depending on which type of subobject is selected. You can also use the **Properties** palette to change the color of edge and face subobjects or the material assigned to a face. The color of a subobject primitive can also be changed, but not the color of its subobjects.

CAUTION

Using the **SOLIDEDIT** command (discussed in Chapter 13) removes the history from a composite solid. Therefore, the original objects—the subobjects—are no longer available for subobject editing. However, you may still be able to perform some subobject edits, such as moving the original objects.

PROFESSIONAL TIP

The subobject filters help to quickly select the specific type subobject. They are located in the **Subobject** panel on the **Home**, **Solid**, and **Mesh** tabs of the ribbon. The filters are discussed in Chapter 3.

Subobject Editing Fillets and Chamfers

If you improperly create a fillet or chamfer, edit it by holding down the [Ctrl] key and selecting the fillet or chamfer. Then, use editing methods to change the fillet or chamfer. A fillet has a parameter grip that can be used to change the radius of the fillet.

On some occasions during the editing process of fillets or chamfers, the fillet radius or the chamfer distances will not appear in the **Properties** palette. If this is the case and you only see a single grip on the surface of the fillet or chamfer, erase the fillet or chamfer and reapply it.

PROFESSIONAL TIP

The **Delete** option of the **SOLIDEDIT** command deletes selected faces. This is a quick way to remove features such as chamfers, fillets, holes, and slots. This technique is discussed in Chapter 13.

Exercise 12-2

Complete the exercise on the student website.
www.g-wlearning.com/CAD

Face Subobject Editing

Faces of 3D solids can be modified using commands such as **MOVE**, **ROTATE**, and **SCALE** or by using grips and gizmos (grip tools). To select a face on a 3D solid, press the [Ctrl] key and pick within the boundary of the face. Do not pick the edge of the face. The face subobject filter can also be used to limit the selection to a face.

Face grips are circular and located in the center of the face, as shown in **Figure 12-9.** In the case of a sphere, the grip is located in the center of the sphere since there is only one face. The same is true of the curved face on a cylinder or cone.

By default, the history is recorded for all solid primitives and composites. If you select a primitive, all of the grips associated with that primitive are displayed. See **Figure 12-10A.** If you edit a 3D solid primitive face, the history of the primitive is deleted and the object becomes a composite solid. Then, when the object is selected, a single base grip is displayed. See **Figure 12-10B.**

The same holds true for composite solids. If you select a primitive subobject in a composite solid, all of the grips associated with the primitive are displayed. If you edit a 3D solid primitive face of the composite solid, the history of the primitive is deleted and the object becomes a new composite solid. Then, when the subobject is selected, a single base grip is displayed.

While pressing the [Ctrl] key and selecting a face, it may be difficult to select the face you want or to deselect faces you do not need. Use the view cube to transparently change the viewpoint or use [Shift]+mouse wheel button to activate the transparent **3DORBIT** command and change your viewpoint.

Figure 12-9.
Face grips are located in the center of face subobjects.

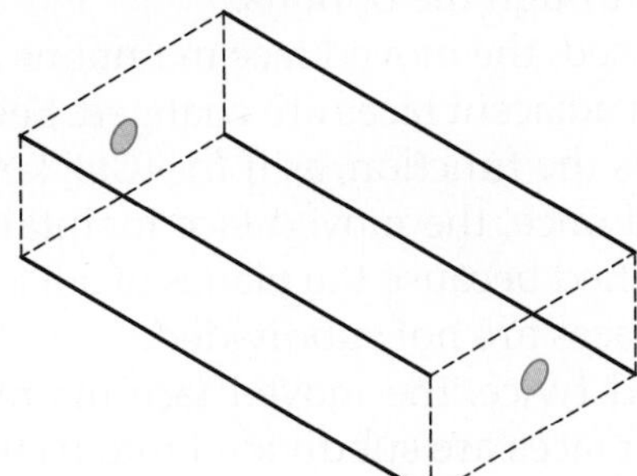

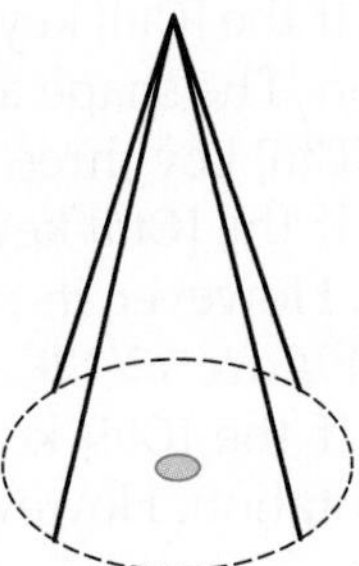

Figure 12-10.
A—This primitive is selected for editing. Notice the grips associated with the primitive.
B—If the primitive is edited, the history of that primitive is deleted and a single base grip is displayed.

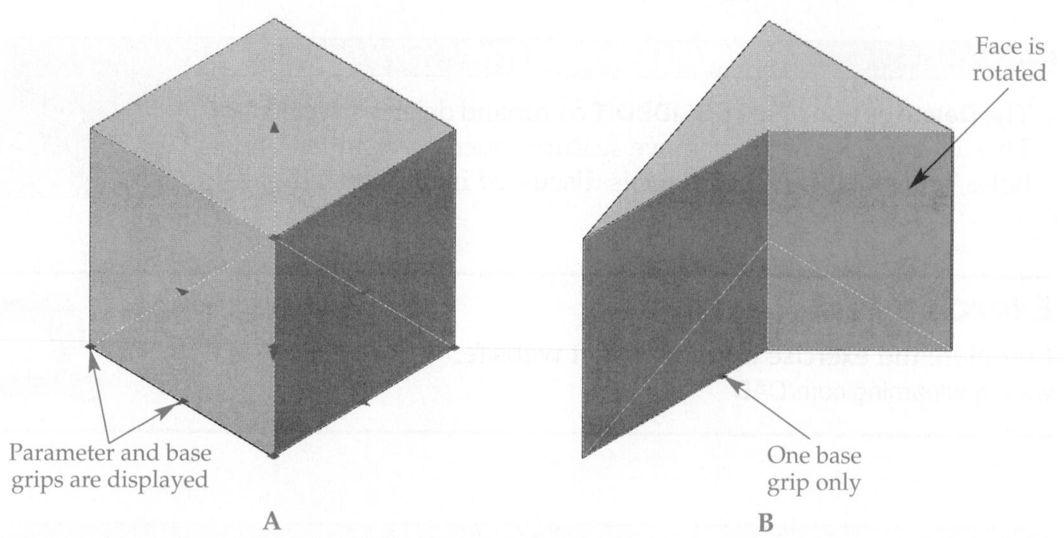

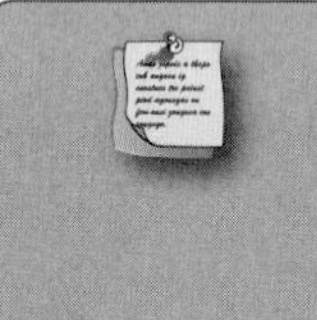

NOTE

The recorded history of a composite solid can be displayed by selecting the solid, opening the **Properties** palette, and changing the Show History property in the **Solid History** category to Yes.

Moving Faces

When a face of a 3D solid is moved, all adjacent faces are dragged and stretched with it. The shape of the original 3D primitive or solid determines the manner in which the face can be moved and how adjacent faces react. A face can be moved using the **MOVE** command, **3DMOVE** command, move gizmo, or by dragging the face's base grip. When moving a face, use the gizmo, polar tracking, or direct distance entry. Otherwise, the results may appear correct in the view in which the edit is made, but, when the view is changed, the actual result may not be what you wanted. See Figure 12-11.

The move gizmo, discussed in Chapter 11, is displayed by default (**GTAUTO** = 1) when the face is selected. To use this gizmo, move the pointer over the X, Y, or Z axis of the gizmo; the axis changes to yellow. To restrict movement along that axis, pick the axis. If you move the pointer over one of the right angles at the origin of the gizmo, the corresponding two axes turn yellow. Pick to restrict the movement to that plane. You can complete the movement by either picking a new point or by direct distance entry.

The [Ctrl] key is used to access options when dynamically moving a face. First, select the face. See Figure 12-12A. Then, pick the face grip or gizmo and press and release the [Ctrl] key to cycle through the options.

If the [Ctrl] key is not pressed, the moved face maintains its size, shape, and orientation. The shape and plane of adjacent faces are changed. See Figure 12-12B. Pressing the [Ctrl] key three times resets the function, as if the [Ctrl] key had not been pressed.

If the [Ctrl] key is pressed once, the moved face maintains its shape and orientation. However, its size is modified because the planes of adjacent faces are maintained. See Figure 12-12C. Adjacent faces are not subdivided.

If the [Ctrl] key is pressed twice, the moved face maintains its size, shape, and orientation. However, adjacent faces are subdivided into triangular faces, if needed.

Figure 12-11.
A—The original solid primitives. B—The box and wedge are dynamically edited without using exact coordinates or distances. C—When the viewpoint is changed, you can see that dynamic editing has produced unexpected results.

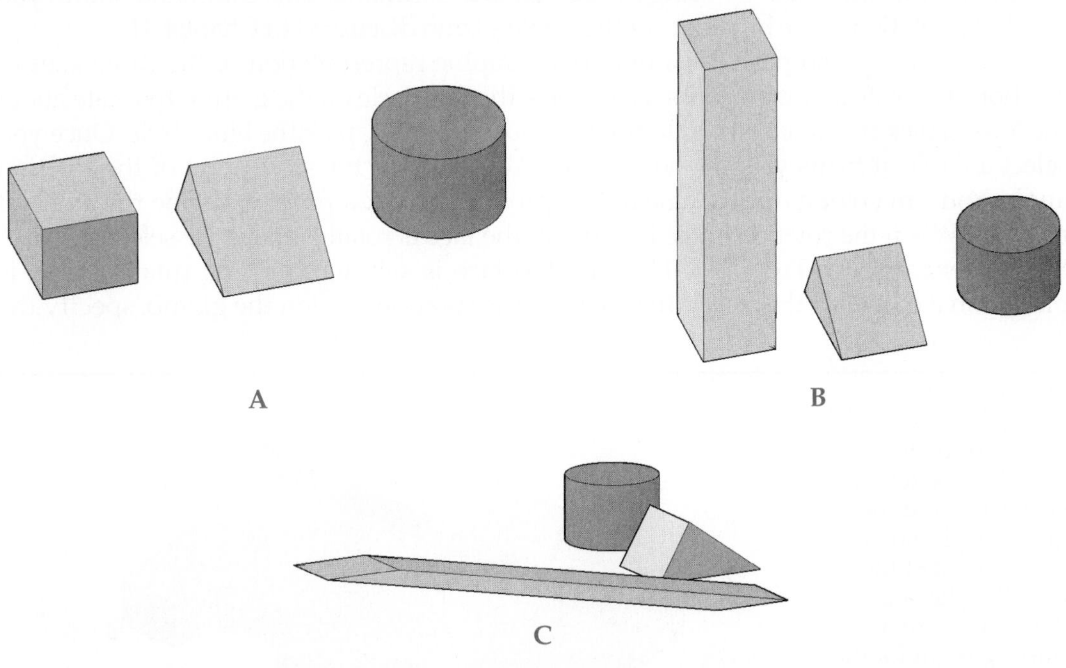

Figure 12-12.
A—The original solid primitive. B—Without pressing the [Ctrl] key, the face maintains its shape and orientation. C—Pressing the [Ctrl] key once keeps the adjacent faces in their original planes, but alters the modified face.

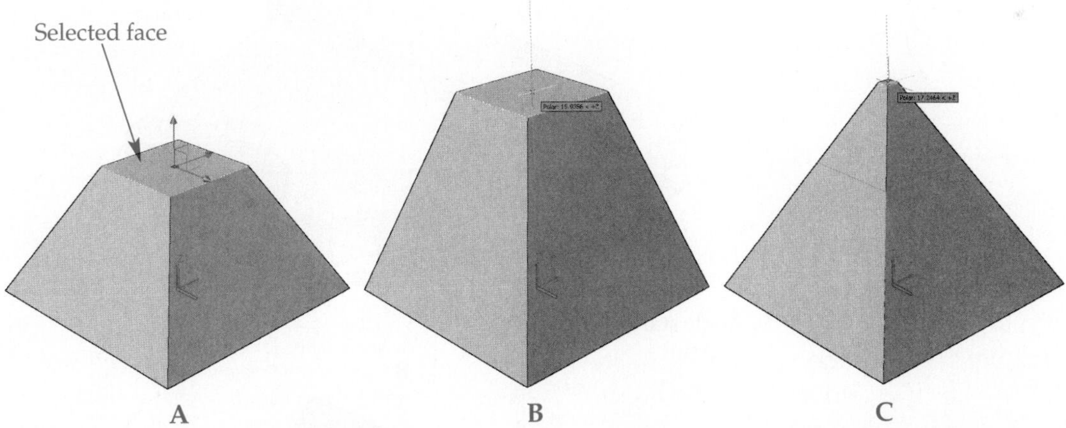

PROFESSIONAL TIP

It is always important to keep the design intent of your solid model in mind. If you are creating a conceptual design, you may be able to use subobject grip editing and gizmos without entering precise coordinates. But, if you are working on a design for manufacturing or production, it is usually critical to use tools such as direct distance entry, gizmos with exact values, and polar tracking for greater accuracy.

Rotating Faces

Before rotating any primitive or subobject, you must know in which plane the rotation is to occur. The **ROTATE** command permits a rotation in the current XY plane. But, you can get around this limitation by using the **3DROTATE** command. This command allows you to select a rotation plane by means of the rotate gizmo discussed in Chapter 11.

The rotate gizmo provides a dynamic, graphic representation of the three axes of rotation. To rotate about the tool's X axis, pick the red circle on the gizmo. To rotate about the Y axis, pick the green circle. To rotate about the Z axis, pick the blue circle. Once you select a circle, it turns yellow and you are prompted for the start point of the rotation angle. You can enter a direct angle at this prompt or pick a point to define the angle of rotation. When the rotation angle is defined, the face is rotated about the selected axis.

For example, in Figure 12-13A, the top face is selected and the rotate gizmo is placed on a corner of the face. After picking the rotation axis on the gizmo, specify the

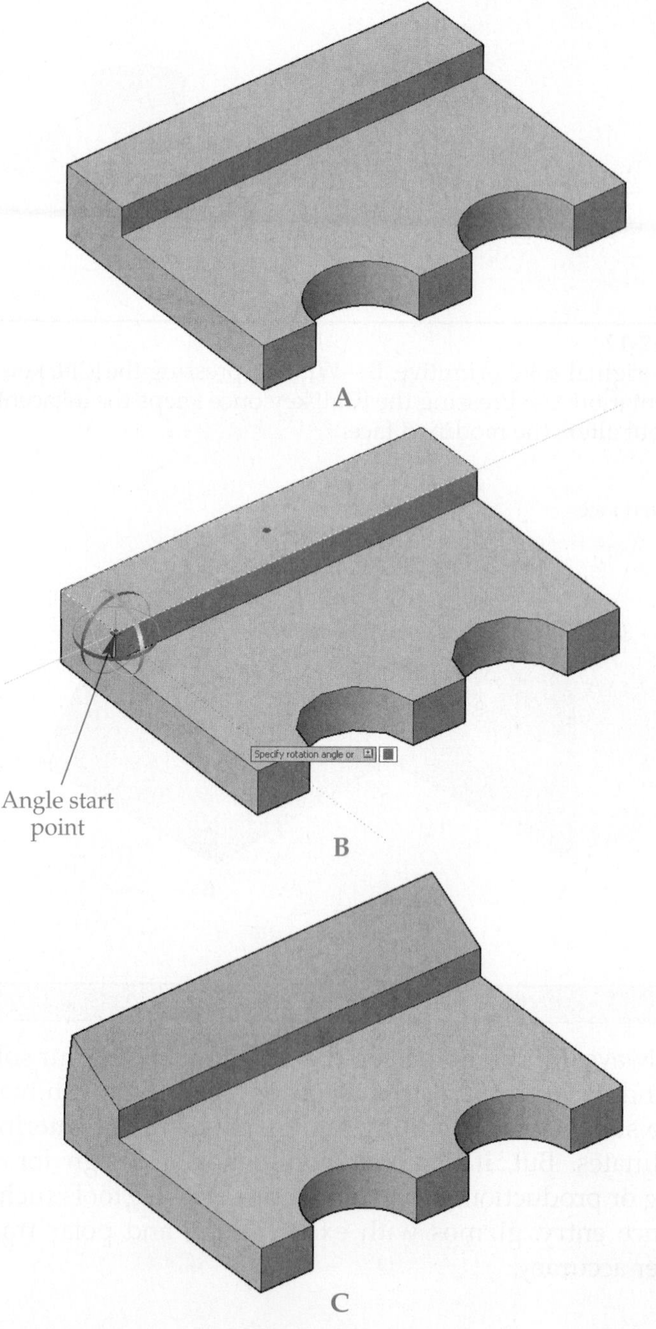

Figure 12-13.
Using the rotate gizmo to rotate a face. A—The top face is selected and the gizmo is placed on a base point of the face. B—The axis of rotation and a starting point for the angle are selected. C—The completed rotation.

angle start point and then the angle end point. Notice in Figure 12-13B that dynamic input can be used to enter an exact angle value. The result is shown in Figure 12-13C.

The [Ctrl] key is used to access options when dynamically rotating a face. Pressing the [Ctrl] key while rotating a face affects adjacent faces in the same manner as discussed for moving faces. Figure 12-14A shows a rotation without pressing [Ctrl]. The shape and size of the face being rotated is maintained, while the adjacent faces change. Figure 12-14B shows a rotation after pressing [Ctrl] once. The shape and size of the face being rotated changes, while the plane and shape of adjacent faces are maintained. Pressing the [Ctrl] key a second time maintains the shape and orientation of the selected face, but triangular faces may be created on adjacent faces. Pressing the [Ctrl] key a third time resets the function.

PROFESSIONAL TIP

When you are working in a 3D view, you may want to use the **3DMOVE** and **3DROTATE** commands exclusively during editing sessions. If this is the case, set the **GTDEFAULT** variable to 1 (the default is 0). This automatically executes the **3DMOVE** and **3DROTATE** commands in a 3D view when you select the **MOVE** or **ROTATE** commands. If the 2D Wireframe visual style is current, the Wireframe visual style is set current for the duration of the command.

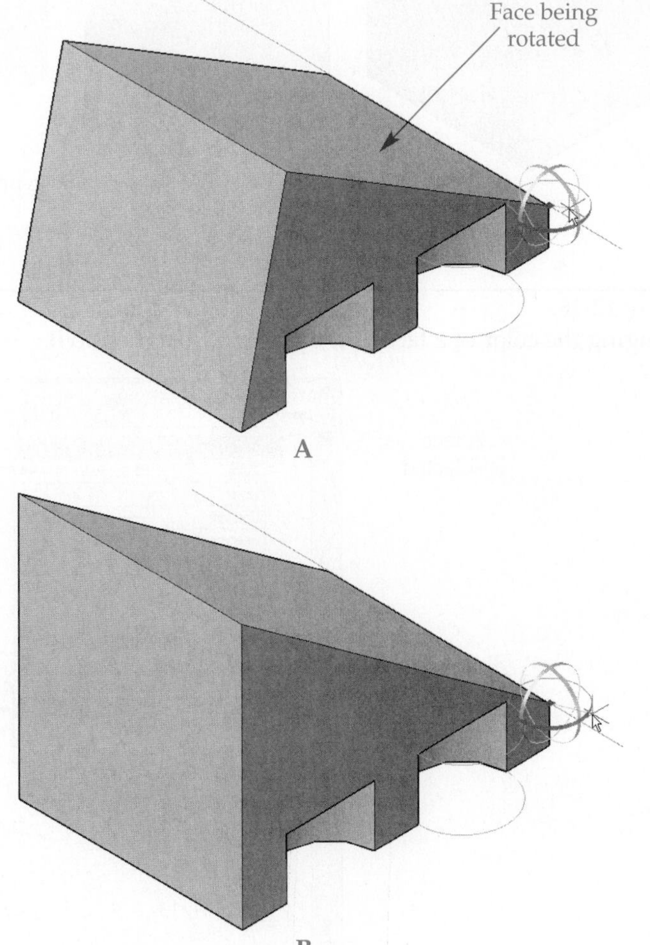

Figure 12-14.
A—Rotating a face without pressing the [Ctrl] key. The large, top face on the object shown in Figure 12-13A has been selected for rotation.
B—Pressing the [Ctrl] key once keeps the adjacent faces in their original planes. The shape and size of the face being rotated changes.

Scaling Faces

Scaling a face is a simple procedure. First, select the face to be scaled. Then, select a base point and dynamically pick to change the scale or use a scale factor. See **Figure 12-15.** Pressing the [Ctrl] key has no effect on the scaling process, except to turn it off or on, if the base point is on the same plane as the face. However, if the base point is not on the same plane as the selected face, then pressing the [Ctrl] key has the same effect as for the other face-editing operations.

Coloring Faces

To change the color of a face, use the [Ctrl] key selection method to select the face. Next, open the **Properties** palette. See **Figure 12-16.** In the **General** category, pick the

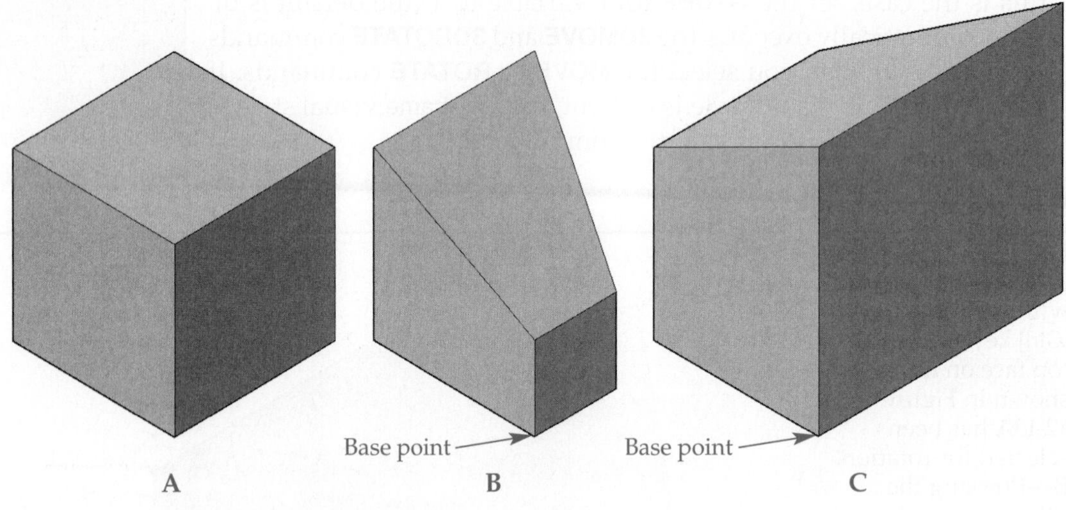

Figure 12-15.
Scaling a face. A—The original solid. B—The dark face is scaled down. C—The dark face is scaled up.

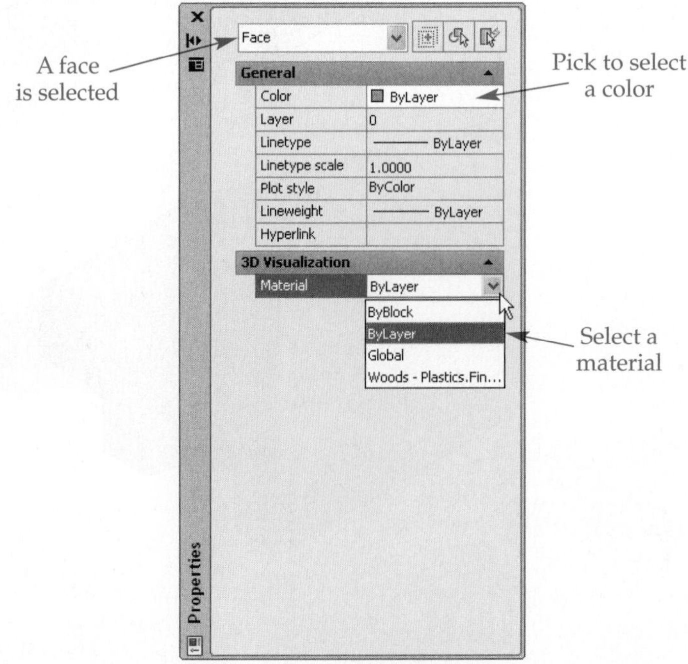

Figure 12-16.
Changing the color of a face or the material assigned to it.

drop-down list for the Color property. Select the desired color or pick Select Color... and choose a color from the **Select Color** dialog box. To change the material applied to the face, pick the drop-down list for the Material property in the **3D Visualization** category. Select a material from the list. A material must be loaded into the drawing to be available in this drop-down list. Materials are discussed in detail in Chapter 16.

Extruding a Solid Face

Planar faces on 3D solids can be extruded into new solids. Refer to the HVAC duct assembly shown in Figure 12-17A. A new, reduced trunk needs to be created on the left end of the assembly. This requires two pieces: a reducer and the trunk.

First, select the **EXTRUDE** command. At the "select objects" prompt, press the [Ctrl] key and pick the face subobject to be extruded. Next, since this is a reduced trunk, specify a taper angle. Enter the **Taper angle** option and specify the angle. In this case, a 15° angle is used. Finally, specify the extrusion height. The height of the reducer is 12". See Figure 12-17B.

Now, the new trunk needs to be created. Select the **EXTRUDE** command. Press the [Ctrl] key and pick the face to extrude. Since this piece is not tapered, enter the extrusion height, which in this case is 44". See Figure 12-17C. The two new pieces are separate solid objects. If the assembly is to be one solid, use the **UNION** command and join the two new solids to the assembly.

Figure 12-17.
A—A new, reduced trunk needs to be created on the left end of the HVAC assembly. The face shown in color will be extruded. B—The **Taper angle** option of the **EXTRUDE** command is used to create the reducer. The face shown in color will be extruded to create the extension. C—The **EXTRUDE** command is used to create an extension from the reducer.

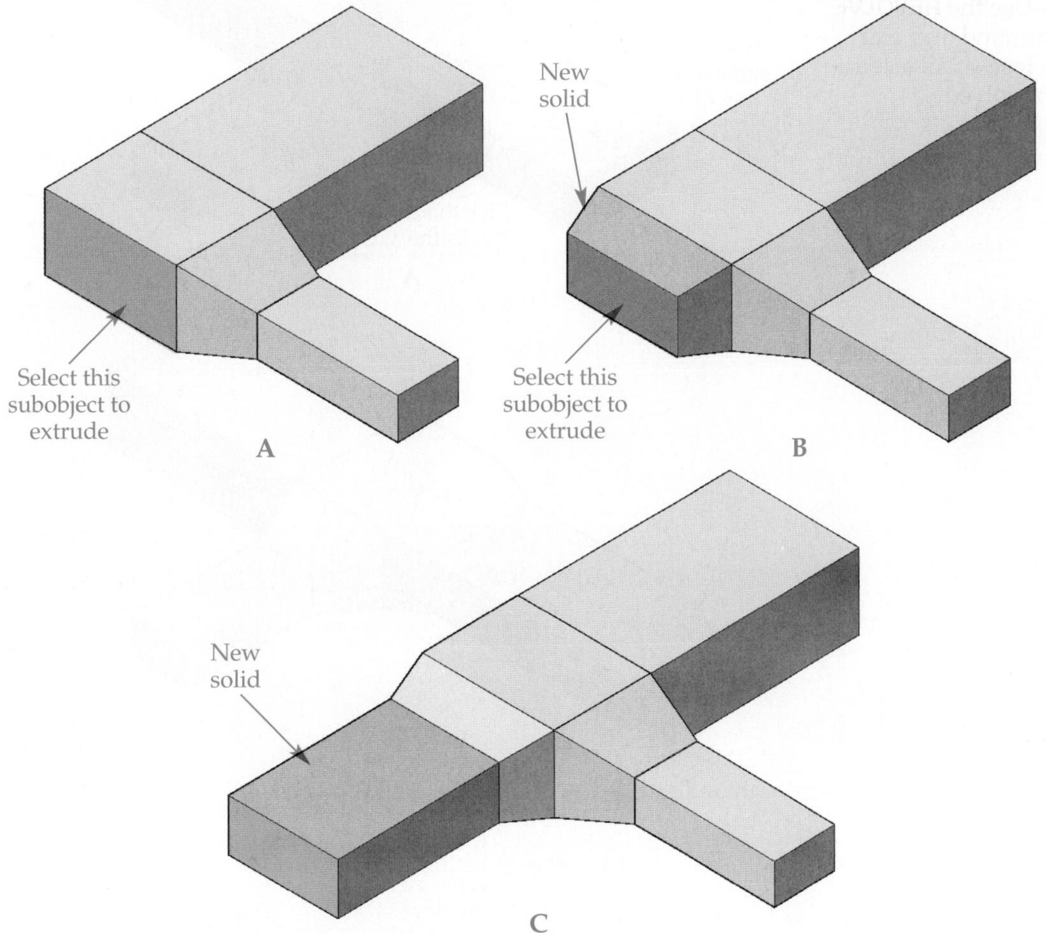

Revolving a Solid Face

Planar faces on 3D solids can be revolved in the same manner as other AutoCAD objects to create new solids. Refer to Figure 12-18A. The face on the left end of the HVAC duct created in the last section needs to be revolved to create a 90° bend. First, select the **REVOLVE** command. At the "select objects" prompt, press the [Ctrl] key and pick the face subobject to be revolved.

Next, the axis of revolution needs to be specified. You can pick the two endpoints of the vertical edge, but you can also pick the edge subobject. Enter the **Object** option of the command, press the [Ctrl] key, and select the edge subobject.

Finally, the 90° angle of revolution needs to be specified. Figure 12-18B shows the face revolved into a new solid. The bend is a new, separate solid. If necessary, use the **UNION** command to join the bend to the assembly.

Exercise 12-3

Complete the exercise on the student website.
www.g-wlearning.com/CAD

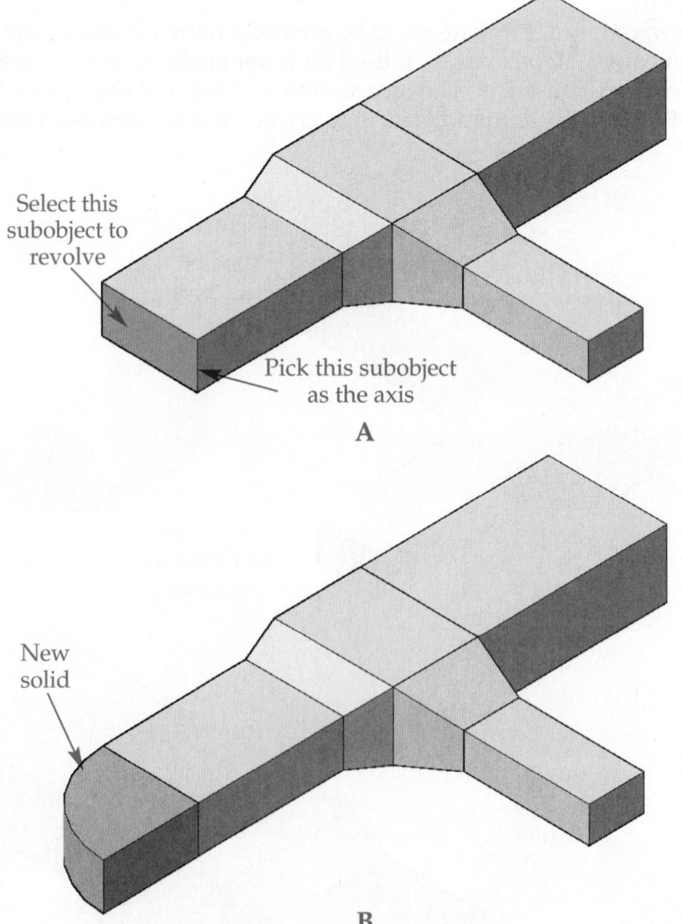

Figure 12-18.
A—The face on the left end of the HVAC duct (shown in color) needs to be revolved to create a 90° bend. B—Use the **REVOLVE** command and pick the face subobject to be revolved.

Edge Subobject Editing

Individual edges of a solid can be edited using grips and gizmos in the same manner as faces. To select an edge subobject, press the [Ctrl] key and pick the edge. The edge subobject filter can also be used to limit the selection to an edge.

Grips on linear edges are rectangular and appear in the middle of the edge, Figure 12-19. In addition to solid edges, the edges of regions can be altered using **MOVE**, **ROTATE**, and **SCALE**, but grips are not displayed on regions as they are on solid subobjects.

Remember, editing subobjects of a primitive removes the primitive's history. This should always be a consideration if it is important to preserve the solid primitives that were used to construct a 3D solid model. Instead of editing the primitive subobjects at their subobject level, it may be better to add or remove material with a Boolean operation, thus preserving the solid's history.

NOTE

There are several ways to perform a 3D edit. Keep the following options in mind when working with subobject editing.

- Selecting the **MOVE**, **ROTATE**, or **SCALE** commands and picking a subobject will not display a gizmo, regardless of the current gizmo button displayed in the **Subobject** panel.
- Selecting the **3DMOVE**, **3DROTATE**, or **3DSCALE** commands and picking a subobject will display the appropriate gizmo, regardless of the current gizmo button displayed in the **Subobject** panel.
- Picking a subobject without having selected a command first will display the gizmo corresponding to the current gizmo button in the **Subobject** panel.

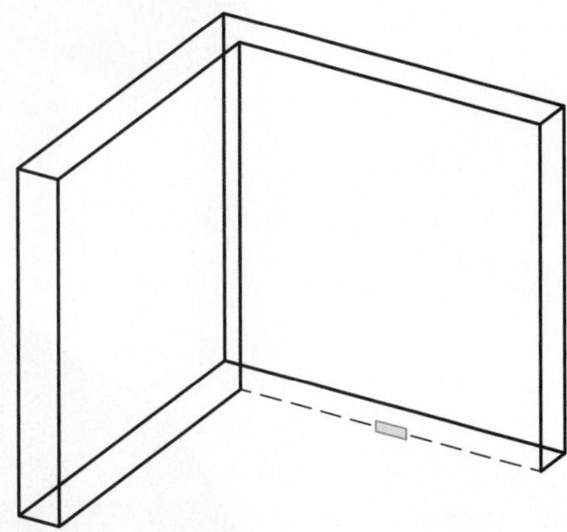

Figure 12-19. Edge grips are rectangular and displayed in the middle of the edge.

Moving Edges

To move an edge, select it using the [Ctrl] key, as previously discussed. See Figure 12-20A. By default, the gizmo corresponding to the gizmo button in the **Subobject** panel appears. If the move gizmo is not current, select the **Move Gizmo** button in the **Subobject** panel. Select the appropriate axis handle and dynamically move the edge or use direct distance entry, Figure 12-20B. The move gizmo remains active until the [Esc] key is pressed to deselect the edge.

If you pick the edge grip to turn it hot, the gizmo is bypassed. This places you in the standard grip editing mode. You can stretch, move, rotate, scale, and mirror the edge. In this case, the stretch function works the same as the move gizmo, but less reliably. You must be careful to use either ortho, polar tracking, or direct distance entry, but the possibility for error still exists.

There are a few options to achieve different results when dynamically moving an edge. The [Ctrl] key is used to access these options. First, select the edge. Then, pick the edge grip or gizmo and press and release the [Ctrl] key to cycle through the options.

If the [Ctrl] key is not pressed, the moved edge maintains its length and orientation. However, the shape and planes of adjacent faces are changed. See Figure 12-21A. If the [Ctrl] key is pressed three times, the function is reset, as if the [Ctrl] key had not been pressed.

If the [Ctrl] key is pressed once, the moved edge maintains its orientation, but its length is modified. This is because the planes and orientation of adjacent faces are maintained. See Figure 12-21B.

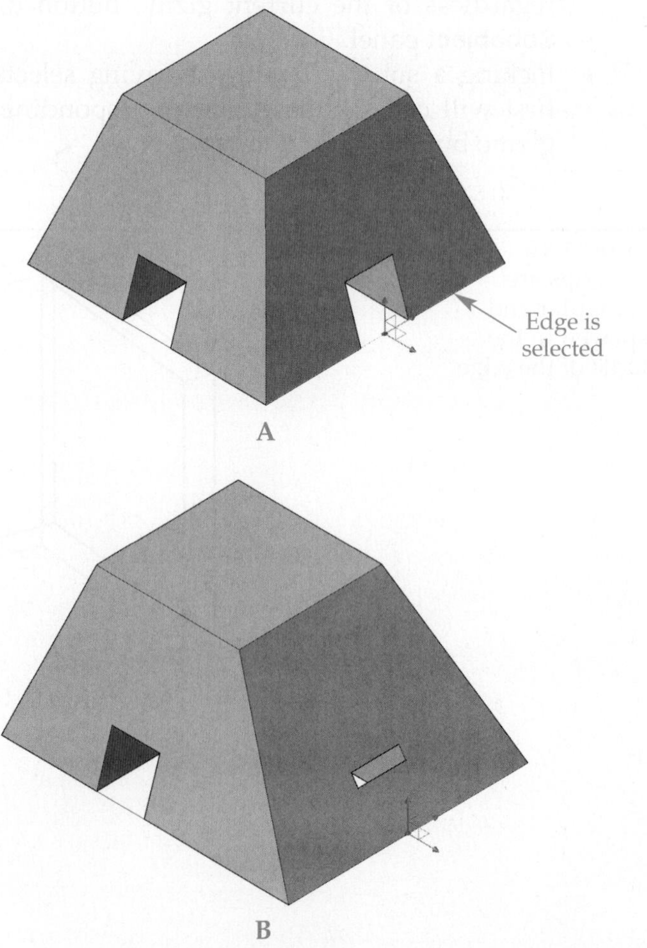

Figure 12-20.
A—Select the edge subobject to be moved. B—The edge is moved. Notice how the size of the primitive used to subtract the cutout is not affected.

If the [Ctrl] key is pressed twice, the moved edge maintains its length and orientation. But, if the move alters the planes of adjacent faces, those faces may become *nonplanar.* In other words, the face may now be located on two or more planes. If this happens, adjacent faces are divided into triangles, **Figure 12-21C.** This is visible when the object in **Figure 12-21C** is displayed in two orthographic views. See **Figure 12-22.**

PROFESSIONAL TIP

You can quickly enter the **MOVE**, **ROTATE**, or **SCALE** command for a subobject by selecting the subobject, right-clicking, and picking the command from the shortcut menu.

Figure 12-21.
Moving an edge. A—If the [Ctrl] key is not pressed, the edge maintains its length and orientation, but the shape and planes of adjacent faces are changed. B—If the [Ctrl] key is pressed once, the moved edge maintains its orientation, but its length is modified because the planes of adjacent faces are maintained. C—If the [Ctrl] key is pressed twice, the adjacent faces may be triangulated.

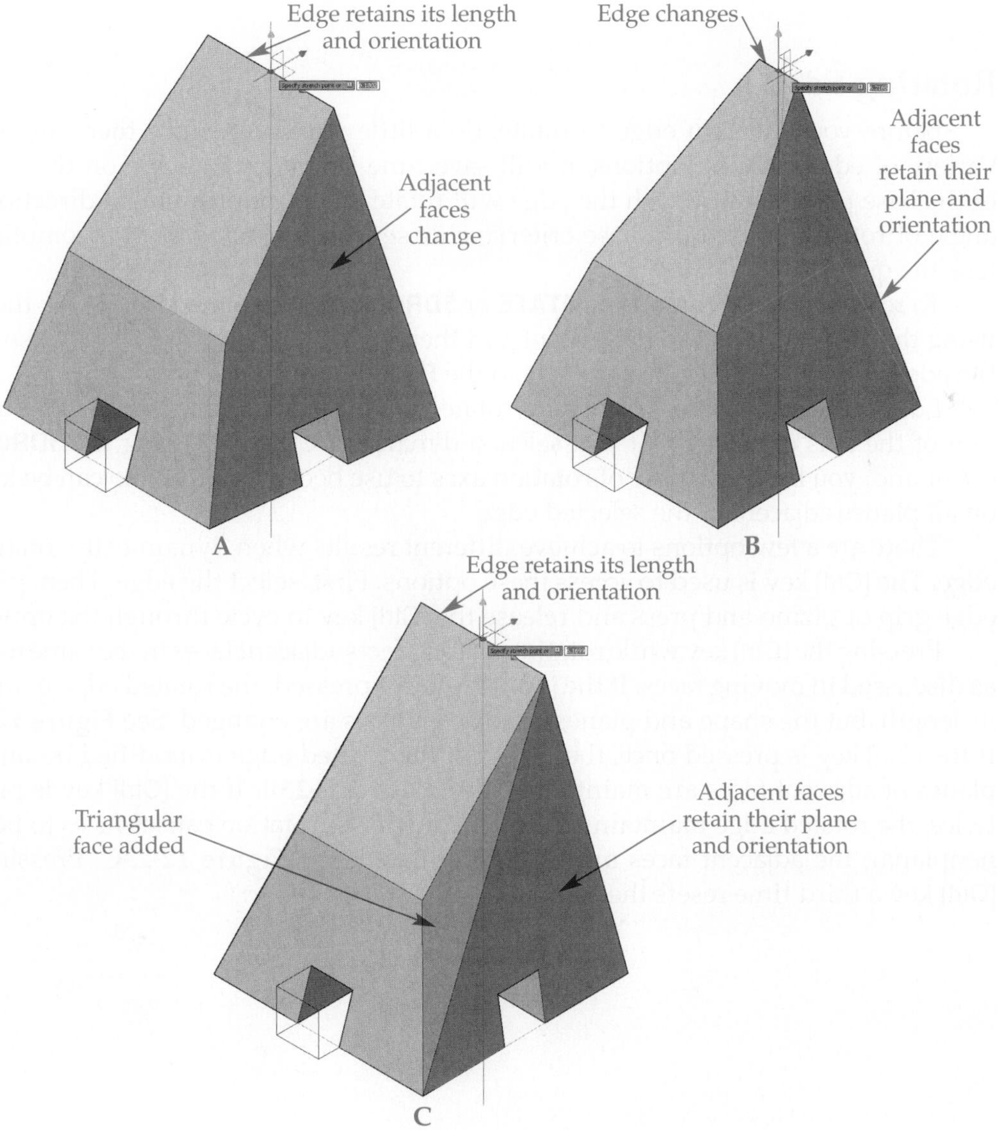

Figure 12-22.
Triangulated faces are clear in plan views. A—Front plan view. B—Side plan view.

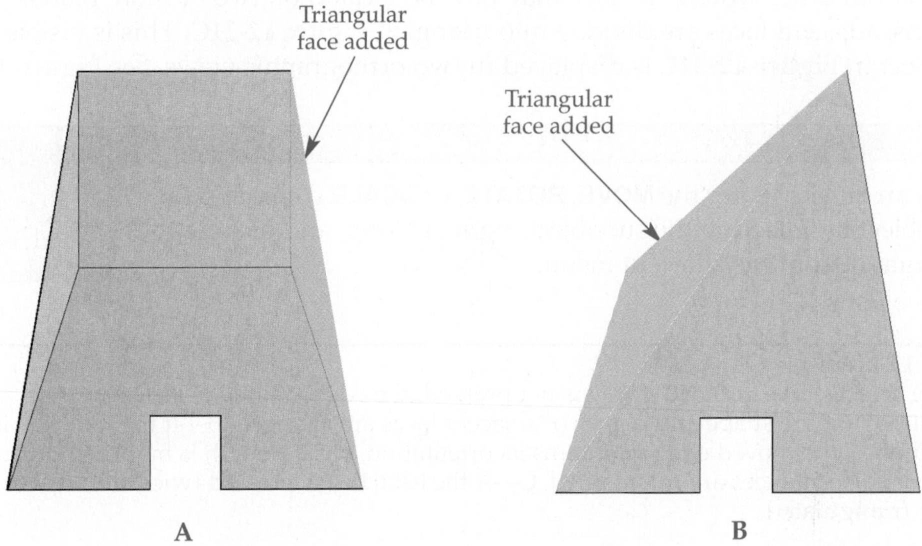

Rotating Edges

Before you select an edge to rotate, do a little planning. Since there are a wide variety of edge rotation options, it will save time if you first decide on the location of the base point about which the edge will rotate. Next, determine the direction and angle of rotation. Based on these criteria, choose the option that will accomplish the task the quickest.

To rotate an edge, enter the **ROTATE** or **3DROTATE** command. Then, pick the edge using the [Ctrl] key. Select a base point and then enter the rotation. You can also select the edge, pick the edge grip, and cycle to the **ROTATE** mode.

Edges are best rotated using the rotate gizmo. It provides a graphic visualization of the axis of rotation. If you select a dynamic UCS while using the **3DROTATE** command, you have a variety of rotation axes to use because the gizmo can be located on all planes adjacent to the selected edge.

There are a few options to achieve different results when dynamically rotating an edge. The [Ctrl] key is used to access these options. First, select the edge. Then, pick the edge grip or gizmo and press and release the [Ctrl] key to cycle through the options.

Pressing the [Ctrl] key while rotating a face affects adjacent faces in the same manner as discussed in moving faces. If the [Ctrl] key is not pressed, the rotated edge maintains its length, but the shape and planes of adjacent faces are changed. See **Figure 12-23A.** If the [Ctrl] key is pressed once, the length of the rotated edge is modified because the planes of adjacent faces are maintained. See **Figure 12-23B.** If the [Ctrl] key is pressed twice, the rotated edge maintains its length, but if the rotation causes faces to become nonplanar, the adjacent faces may be triangulated. See **Figure 12-23C.** Pressing the [Ctrl] key a third time resets the function.

Figure 12-23.
Rotating an edge. A—The [Ctrl] key is not pressed. B—The [Ctrl] key is pressed once. Notice the top edge of the dark face. C—The [Ctrl] key is pressed twice.

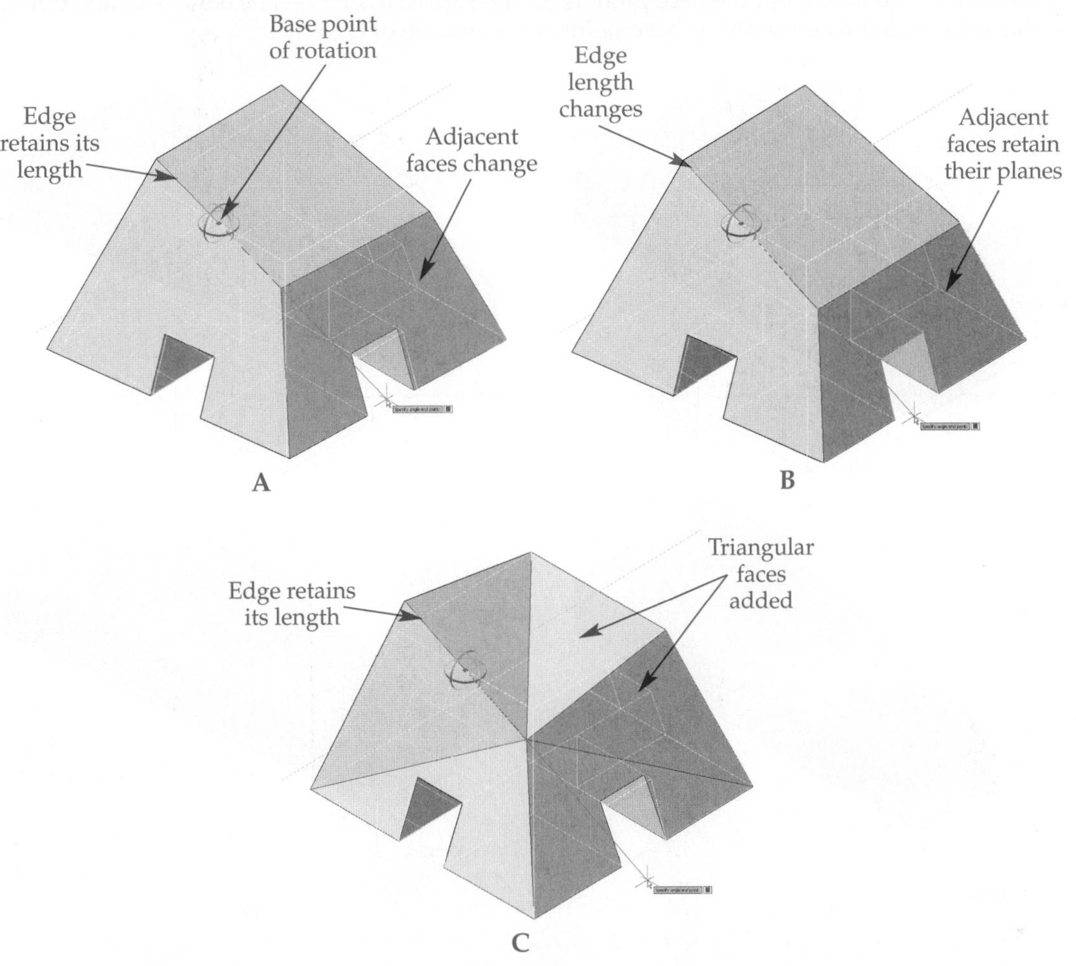

Scaling Edges

Only linear (straight-line) edges can be scaled. Circular edges, such as the ends of cylinders, can be modified using grips or the **SOLIDEDIT** command. These tools can be used to change the diameter or establish taper angles. See Chapter 13 for a complete discussion of the **SOLIDEDIT** command.

To scale a linear edge, enter the **SCALE** command. Select the edge using the [Ctrl] key. Pick a base point for the operation and enter a scale factor. You can also select the edge, pick the edge grip, and cycle to the **SCALE** mode. However, using the scale gizmo may be the best option.

The direction of the scaled edge is related to the base point you select. The base point remains stationary, while the vertices in either direction are scaled. If you enter the **SCALE** command, you are prompted for the base point. If you select the edge grip, the grip becomes the base point. The differences in opposite end and midpoint scaling of an edge are shown in **Figure 12-24.**

There are a few options to achieve different results when dynamically scaling an edge. The [Ctrl] key is used to access these options. First, select the edge. Then, pick the edge grip or enter the **SCALE** command and press and release the [Ctrl] key to cycle through the options.

If the [Ctrl] key is not pressed, the edge is scaled. The shape and planes of adjacent faces are changed to match the scaled edge. See **Figure 12-25A.**

Figure 12-24.
The differences in opposite end and midpoint scaling of an edge. A—The original object. B—The edge is scaled down with a base point on the left corner. C—The edge is scaled down to the same scale factor, but the base point is on the right corner. D—The edge is scaled down to the same scaled factor with the base point at the middle of the edge.

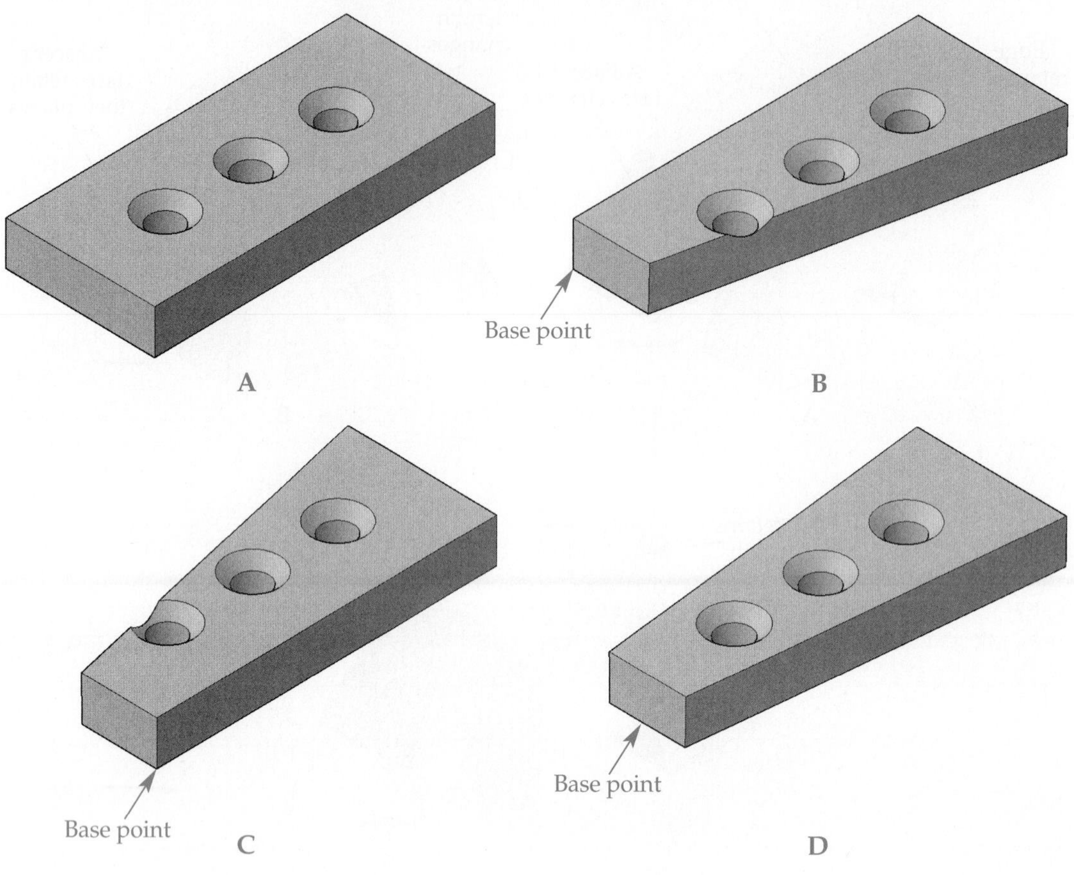

Figure 12-25.
Scaling the front edge with the base point in the middle of the edge. The original object is shown in Figure 12-21A. A—If the [Ctrl] key is not pressed, the edge is scaled and the shape and planes of adjacent faces are changed. B—If the [Ctrl] key is pressed twice, the edge is scaled, as are edges attached to it. If the scaling causes faces to become nonplanar, the adjacent faces may be triangulated.

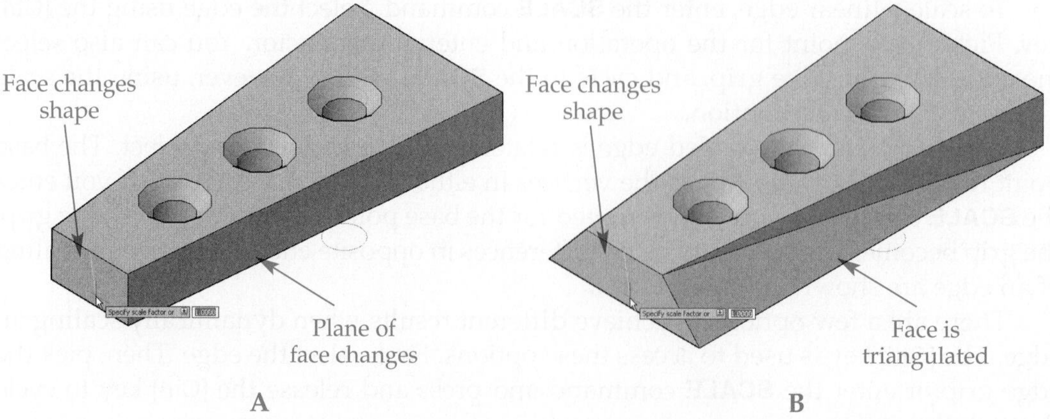

If the [Ctrl] key is pressed once, the edge is, in effect, not scaled. This is because the planes of adjacent faces are maintained.

If the [Ctrl] key is pressed twice, the edge is scaled, as are edges attached to the modified edge. However, if the scaling causes faces to become nonplanar, they may be triangulated. See **Figure 12-25B.**

Coloring Edges

To change the color of an edge, use the [Ctrl] key selection method to select the edge. Next, open the **Properties** palette. In the **General** category, pick the drop-down list for the Color property. Select the desired color or pick Select Color... and choose a color from the **Select Color** dialog box. Edges cannot have materials assigned to them.

Deleting Edges

Edges can be deleted in certain situations. In order for an edge to be deleted, it must completely divide two faces that lie on the same plane. If this condition is met, the **ERASE** command or the [Delete] key can be used to remove the edge. The two faces become a single face.

Exercise 12-4

Complete the exercise on the student website.
www.g-wlearning.com/CAD

Vertex Subobject Editing

The modification of a single vertex involves moving the vertex and stretching all edges and planar faces attached to it. Vertex grips are circular and located on the vertex, as shown in **Figure 12-26.** Use the vertex subobject filter to assist in selecting vertices. A single vertex cannot be rotated or scaled, but you can select multiple vertices and perform rotating and scaling edits. When editing multiple vertices in this manner you are, in effect, editing edges.

As with other subobject editing functions performed on a 3D solid primitive, the solid's history is removed when a vertex is modified. The solid can no longer be edited using the primitive grips; only a single base grip is displayed. Further editing of the solid must be with the **SOLIDEDIT** command, discussed in Chapter 13, or through subobject editing.

Figure 12-26.
Vertex grips are circular and placed on the vertex.

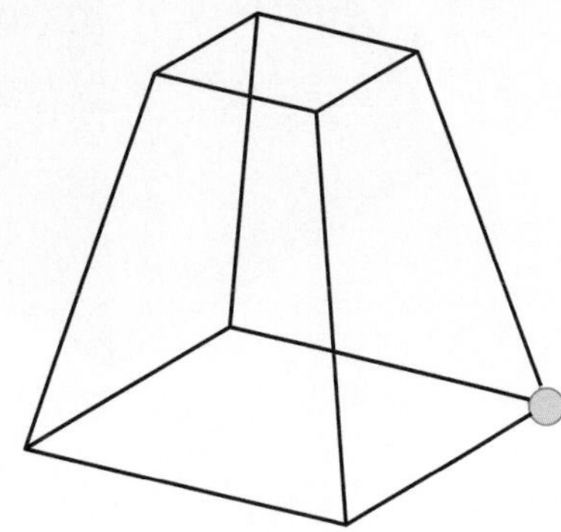

Moving Vertices

To move a vertex, select it using the [Ctrl] key method. If the move gizmo is not displayed, select it from the gizmo drop-down list in the **Subobject** panel. You can use the move gizmo, the **MOVE** command, or standard grip editing modes to move the vertex. If the [Ctrl] key is not pressed while dynamically moving a vertex, adjacent faces are triangulated by the move. Pressing the [Ctrl] key once allows the vertex to be moved without triangulating adjacent faces, but the faces may change shape. In some cases, AutoCAD may deem it necessary to triangulate faces. See **Figure 12-27.**

PROFESSIONAL TIP

If you are dragging a vertex and faces become triangulated, you can transparently change your viewpoint to see the effect of the triangulation. Press and hold the [Shift] key. At the same time, press and hold the mouse wheel button. Now, move the mouse to change the viewpoint. This is a transparent instance of the **3DORBIT** command.

Figure 12-27.
Moving a vertex. A—The original object. B—Without pressing the [Ctrl] key, adjacent faces are triangulated. C—Pressing the [Ctrl] key once moves the vertex and changes some of the adjacent faces.

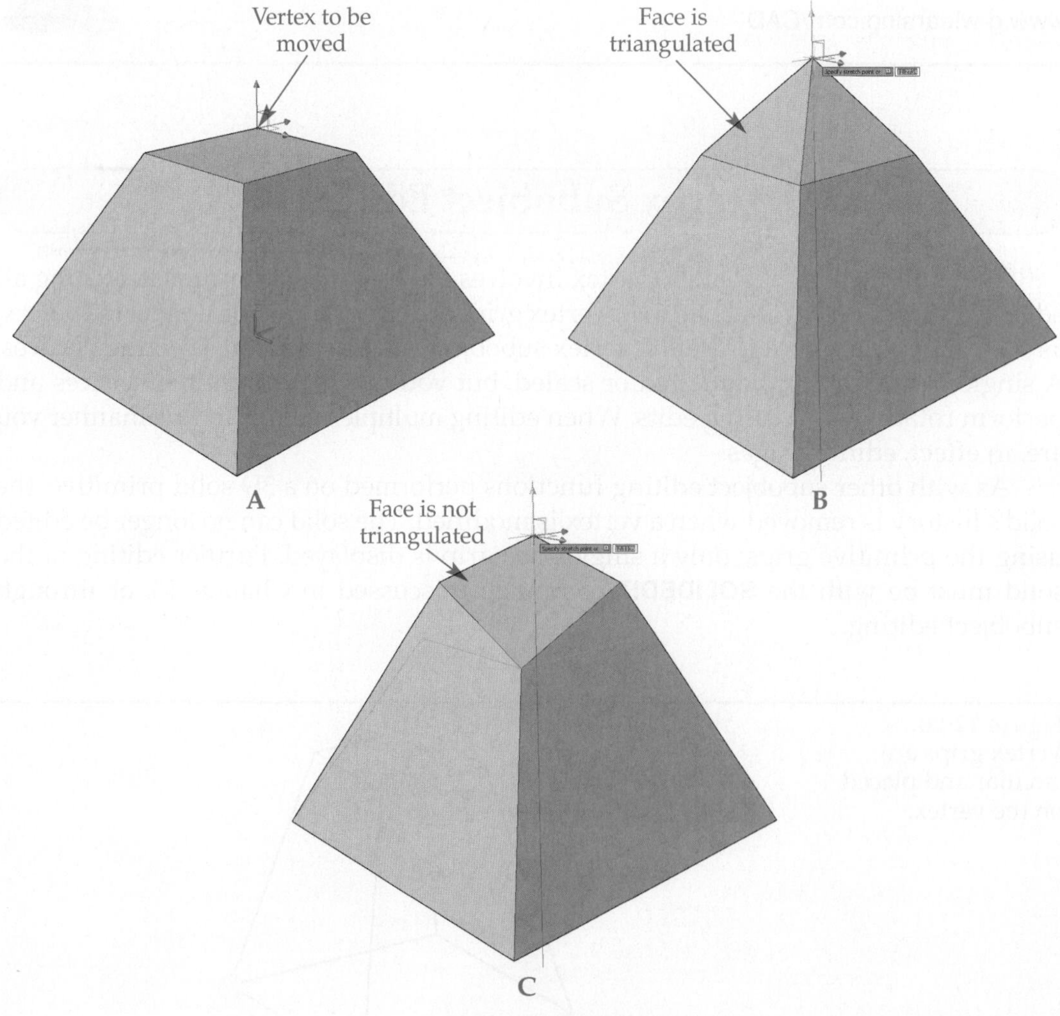

Rotating Vertices

As previously stated, a single vertex cannot be rotated or scaled, but two or more vertices can be. Since two vertices define a line, or edge, any edit is an edge modification. However, the process is slightly different than the edge modifications described earlier in this chapter.

To rotate an edge by selecting its endpoints, press the [Ctrl] key and select each vertex. See **Figure 12-28A.** You may need to use the [Ctrl]+space bar option to turn on cycling. Notice that grips appear at each selected vertex, but the edges between the vertices are not highlighted.

The **ROTATE** command can now be used to rotate the vertices (if **PICKFIRST** is set to 1). However, a more efficient method for rotating vertices is to use the **3DROTATE** command. The combination of the rotate gizmo and the UCS icon enable you to graphically view the rotation plane. Once the command is initiated, move the base point of the rotate gizmo if needed. See **Figure 12-28B.** Then, select the axis of revolution. Finally, pick the angle start point and enter the rotation. See **Figure 12-28C.**

Figure 12-28.
To rotate or scale vertices, multiple vertices must be selected. In effect, the edges are modified. A—Vertices are selected to be rotated. B—The rotate gizmo is placed at the base of rotation and the rotation axis is selected. C—The vertices are rotated.

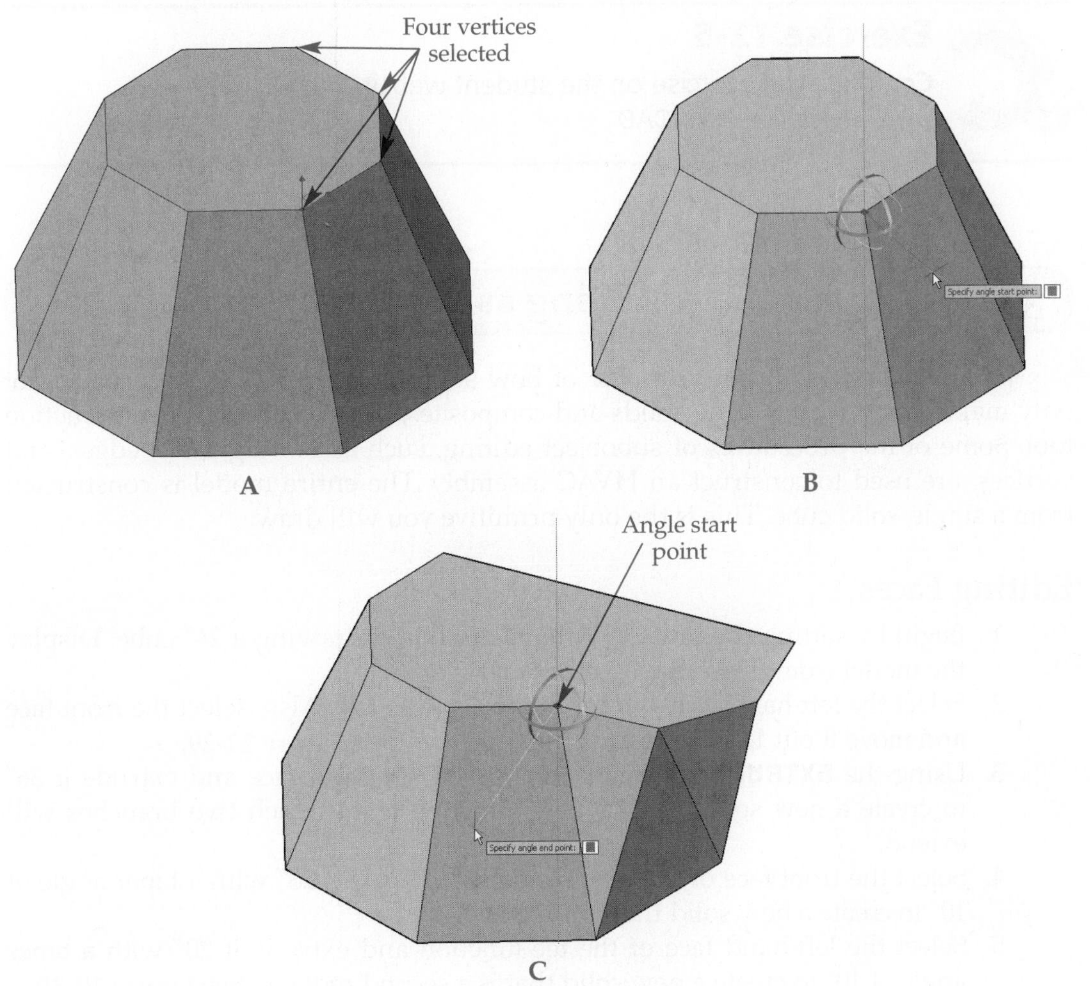

If the [Ctrl] key is not pressed while dynamically rotating the vertices, the area of the selected vertices does not change and adjacent faces are triangulated. This is because the edges of the adjacent faces are attached to the selected vertices, so their edge length changes as the selected edge is rotated. If the [Ctrl] key is pressed once, the adjacent faces are not triangulated unless necessary, but the faces may change shape.

NOTE

If the selected edge does not dynamically rotate at the "angle end point" prompt, then the desired rotation is not possible.

Scaling Vertices

As mentioned earlier, it is not possible to scale a single vertex. However, two or more vertices can be selected for scaling. This, in effect, scales edges. The selection methods are the same as discussed for rotating vertices and the use of the [Ctrl] key while dragging produces the same effects. As the pointer is dragged, the dynamic display of scaled edges may be difficult to visualize. Therefore, it is best to use a scale factor or the **Reference** option to achieve properly scaled edges.

Exercise 12-5

Complete the exercise on the student website.
www.g-wlearning.com/CAD

Using Subobject Editing as a Construction Tool

This section provides an example of how subobject editing can be used to not only make changes to existing solids and composites, but as a powerful construction tool. Some of the procedures of subobject editing, such as editing faces, edges, and vertices, are used to construct an HVAC assembly. The entire model is constructed from a single, solid cube. This is the only primitive you will draw.

Editing Faces

1. Begin by setting the units to Architectural and drawing a 24″ cube. Display the model from the southeast isometric viewpoint.
2. Select the left-hand face and move it 60″ to the left. Also, select the front face and move it out 12″. This forms the first duct. See Figure 12-29.
3. Using the **EXTRUDE** command, select the left-hand face and extrude it 36″ to create a new solid. This is a tee junction from which two branches will extend.
4. Select the front face of the new solid and extrude it 28″ with a taper angle of 10° to create a new solid that is a reducer.
5. Select the left-hand face of the tee junction and extrude it 20″ with a taper angle of 10° to create a new solid that is a second reducer. See Figure 12-30.

Figure 12-29.
The left-hand face of the cube is moved 60″. The front face is then moved 12″.

Figure 12-30.
Two reducers are created by extruding faces from the tee junction.

6. Select the left-hand face of the 20″ reducer and move it 3 17/32″ along the positive Z axis. This places the top surface of the reducer level with the trunk of the duct. Next, extrude the left-hand face of this reducer 60″ into a new solid.
7. Use the **REVOLVE** command to turn the left-hand face of the 60″ extension into a new solid that is a 90° bend. Your drawing should now look like Figure 12-31.

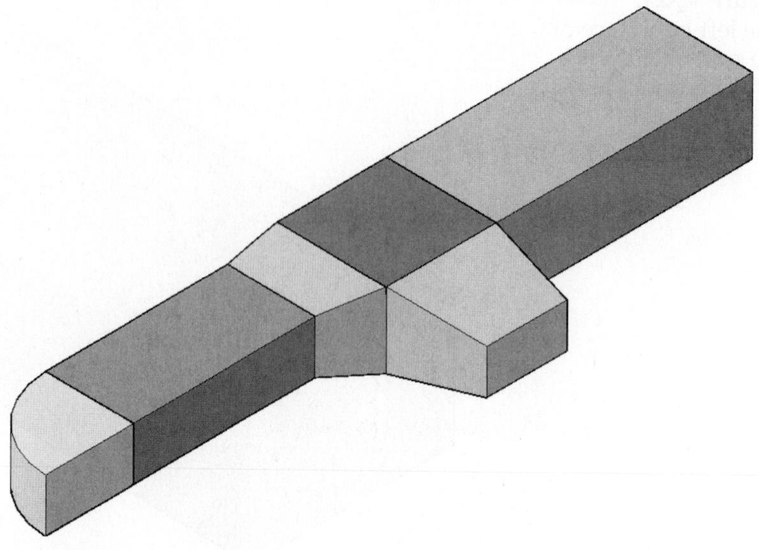

Figure 12-31.
The left end of the 60″ extrusion is revolved 90° to create an elbow.

Editing Edges and Vertices

1. The bottom surface of the 28″ reducer must be level with the bottom of the tee junction and main trunk. Select the bottom edge of the reducer's front face and move it down 4 15/16″.
2. Select the two top vertices on the 28″ reducer's front face and move them down (negative Z) 6″.
3. Select the front, rectangular face of the 90° bend and extrude it 72″ into a new solid.
4. Select the front face of the 28″ reducer and extrude it 108″ into a new solid.
5. Select the front face of the new solid created in step 4 and extrude it 26″ to create a new solid that will be a tee junction.
6. Extrude the left-hand face of the tee junction 20″. See Figure 12-32.
7. Move the top edge of the left-hand face on the 20″ extrusion created in step 6 down 4″.
8. Move each vertical edge of the 20″ extrusion 6″ toward the center of the duct.
9. Mirror a copy of the 20″ extrusion to the opposite side of the tee junction. The completed drawing should look like Figure 12-33.

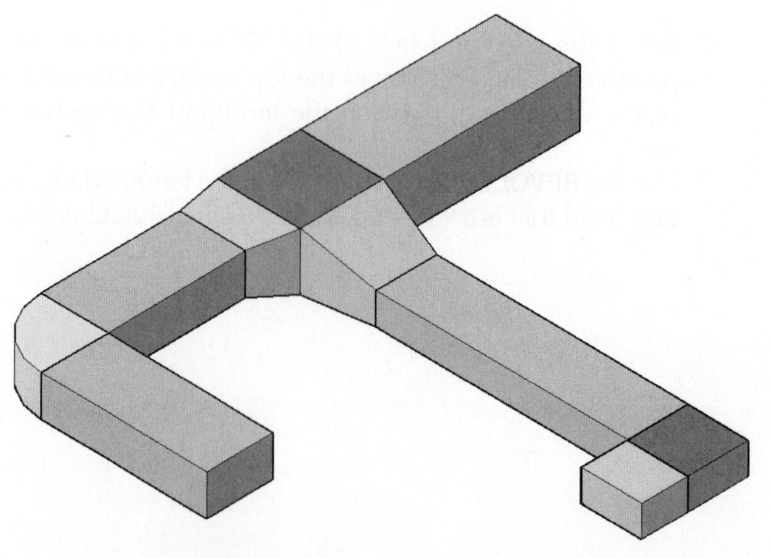

Figure 12-32.
The face of the revolved elbow is extruded 72″. The face of the right branch is extruded 108″. The right duct is then extruded 26″ and the left face of that extrusion is extruded by 20″.

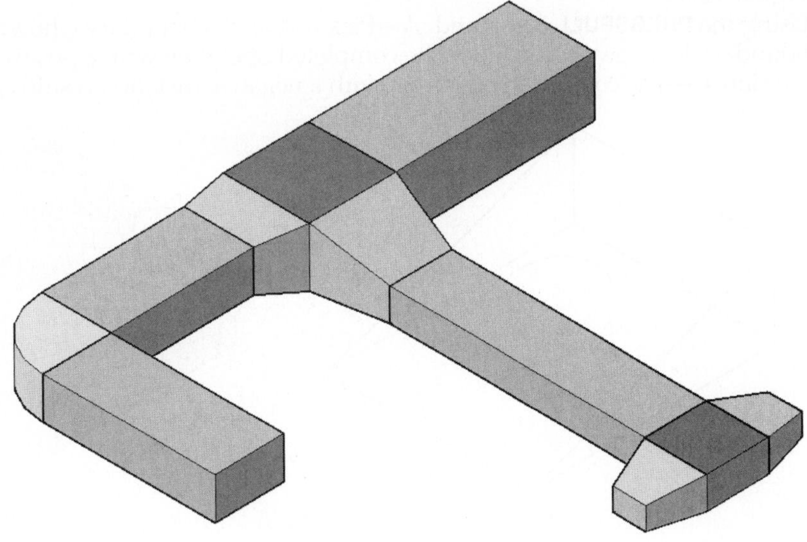

Figure 12-33.
The reducer is mirrored to create the final assembly.

NOTE

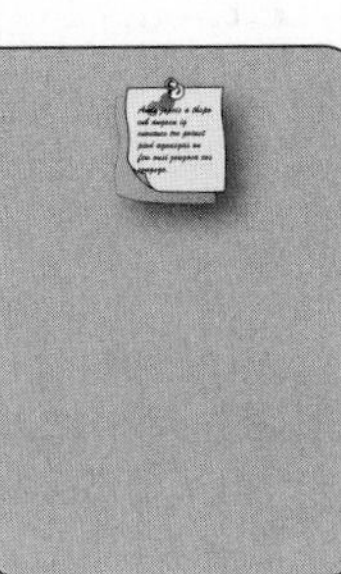

For many of the above subobject editing techniques, you may find it difficult to edit an object after a complex composite solid is created, especially if it has multiple fillets. You may find it necessary to select the subobject, delete it, and start the process over. Or, you may find it necessary to add geometry to the existing solid using the **UNION** command and then edit the new object as needed.

Other Solid Editing Tools

There are other tools that can be used in solid model editing. As you will learn in Chapter 13, the **SOLIDEDIT** tool can be used to edit faces, edges, and vertices much like subobject editing. In addition, you can extrude a closed boundary with the **PRESSPULL** command and explode a solid. The **PRESSPULL** command and exploding a solid are discussed in the next sections.

Presspull

The **PRESSPULL** command allows any closed boundary to be extruded. The boundary can be a flat surface, closed polyline, circle, or region. The extrusion is always applied perpendicular to the plane of the boundary, but can be in the positive or negative direction. When applied to the face of a solid, it is very similar to the **Extrude Face** option of the **SOLIDEDIT** command, though dynamic feedback is provided for the extrusion with **PRESSPULL**.

Ribbon
Home
 > Modeling
Solid
 > Solid

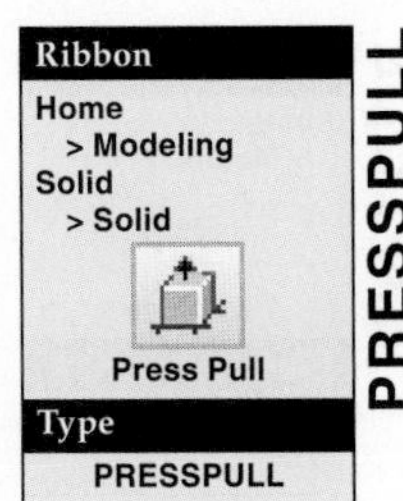

Press Pull
Type
PRESSPULL

PRESSPULL

Once the command is initiated, you are prompted to pick inside of the bounded areas to extrude. Move the pointer inside of a boundary and pick. Then, drag the boundary to a new location and pick or, if dynamic input is on, enter the distance to extrude the face. See Figure 12-34. By extruding the cylinder in a positive direction, the new cylinder is automatically unioned to the existing part Figure 12-34B. Extruding in a negative direction subtracts the cylinder from the wedge, Figure 12-34C.

Figure 12-34.
Using the **PRESSPULL** command. A—Pick inside of a boundary (shown in color) and drag the boundary to a new location. B—The completed operation with a positive distance, resulting in a union. C—The completed operation with a negative distance, resulting in a subtraction.

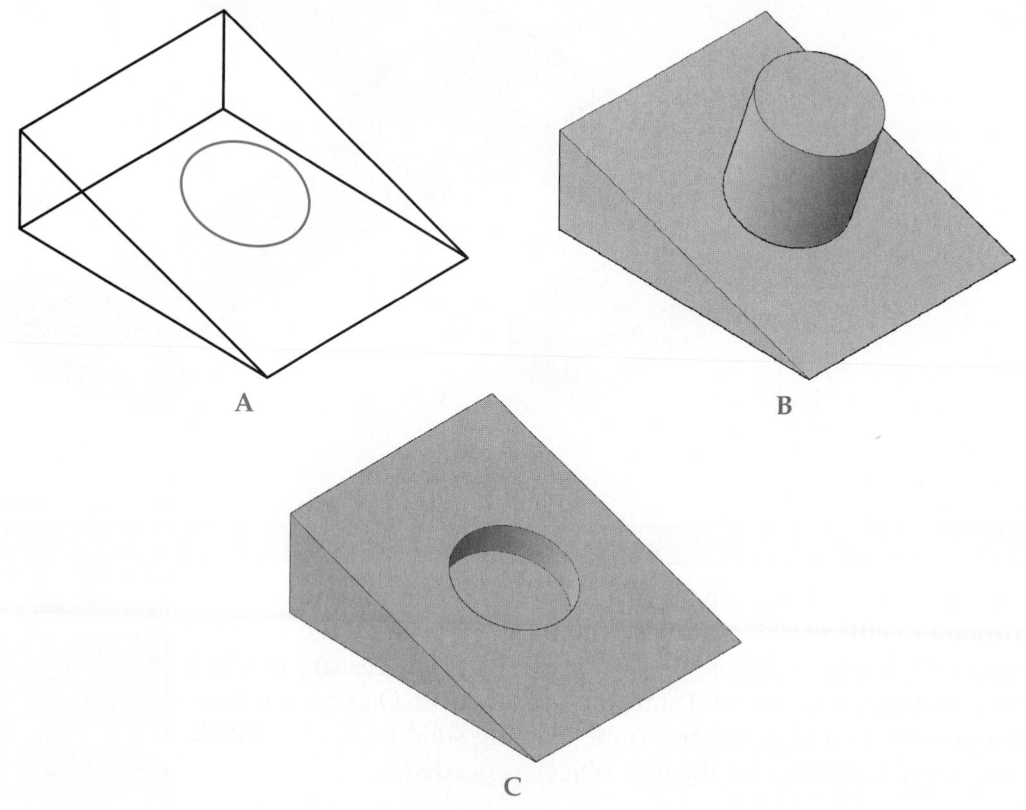

NOTE

The entire boundary must be visible on the screen or the loop will not be found. Also, the **When a command is active** check box must be checked in the **Selection preview** area of the **Selection** tab in the **Options** dialog box.

Exploding a Solid

EXPLODE

Ribbon
Home
> Modify
Explode
Type
EXPLODE
X

A solid can be exploded. This turns the solid into surfaces and/or regions. Flat surfaces on the solid are turned into regions. Curved surfaces on the solid are turned into surfaces. To explode a solid, select the **EXPLODE** command. Then, pick the solid(s) to explode and press [Enter].

Exercise 12-6

Complete the exercise on the student website.
www.g-wlearning.com/CAD

Chapter Test

Answer the following questions. Write your answers on a separate sheet of paper or complete the electronic chapter test on the student website.
www.g-wlearning.com/CAD

1. How is a subobject selected?
2. How is a subobject deselected?
3. When grip editing, two types of grips appear in the part. Name the two types of grips.
4. If you have a cylinder with a height of 10 units, but the height requirement has changed to 15 units, explain the procedure for adding 5 units to the cylinder height.
5. How can you change the radius of a fillet or the distances of a chamfer?
6. When moving a face on a solid primitive, how can you accurately control the axis of movement?
7. How can you change the results of moving a face while dragging it?
8. Describe a major difference of function between the **ROTATE** and **3DROTATE** commands.
9. Which variable enables you to use the **3DROTATE** command in a 3D view even if you select the **ROTATE** command?
10. How does the location and shape of an edge grip differ from a face grip?
11. What is the most efficient tool to use when rotating an edge and how is it displayed?
12. What is the only type of edge that can be scaled?
13. What is the only editing function that can be done to a single vertex?
14. How are two or more vertices selected for editing?
15. What is the function of the **PRESSPULL** command?
16. On which objects can the **PRESSPULL** command be used?
17. When using the **PRESSPULL** command, how do you create a union of two objects?
18. When using the **PRESSPULL** command, how do you create a subtraction of two objects?
19. When a solid object is exploded, which type of object is created from the flat surfaces?
20. When a solid object is exploded, which type of object is created from the curved surfaces?

Drawing Problems

1. Draw the bookcase shown below using the dimensions given. The final result should be a single solid object. Then, use grip and subobject editing procedures to edit the object as follows.
   A. Change the width of the bookcase to 3′.
   B. Change the height of the bookcase by eliminating the top section. The resulting height should be 3′-2″.
   C. Save the drawing as P12_01.

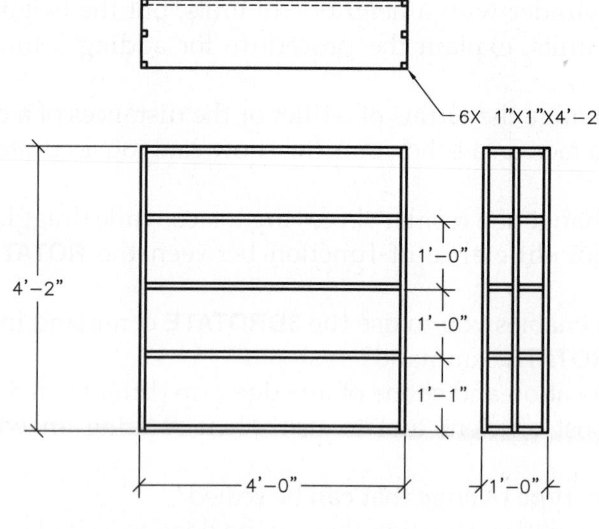

2. Open problem P12_01. Save it as P12_02. Use primitive and subobject editing procedures to create the following edits.
   A. Change the depth of the top of the bookcase to 6-1/2″.
   B. Change the depth of the bottom of the bookcase to 24″.
   C. Reduce the height of the front uprights so they are flush with the top surface of the next lower shelf.
   D. Extend the front of the second lowest shelf to match the front of the bottom. Add two uprights at the front corners between the bottom and this shelf.
   E. Save the drawing.

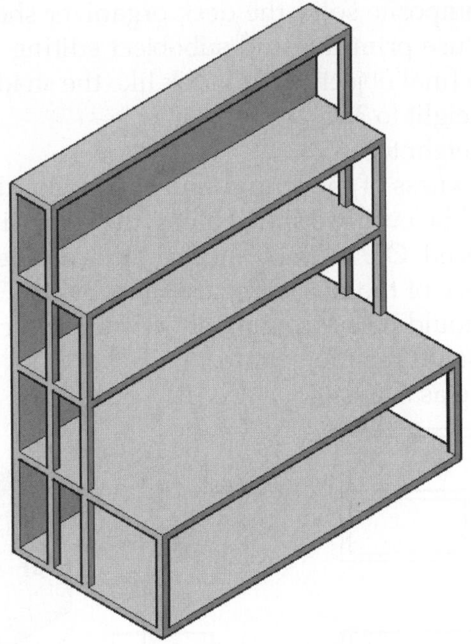

3. Draw the mounting bracket shown below. Then, use primitive and subobject editing procedures to create the following edits.
   A. Change the 3.00″ dimension to 3.50″.
   B. Change the 2.50″ dimension in the front view to 2.75″.
   C. Change the location of the slot in the auxiliary view from .60″ to .70″ and change the length of the slot to 1.15″.
   D. Change the width of each foot in the top view from 2.00″ to 1.50″. The overall dimension (5.00″) should not change.
   E. Change the angle of the bend from 15° to 45°.
   F. Save the drawing as P12_03.

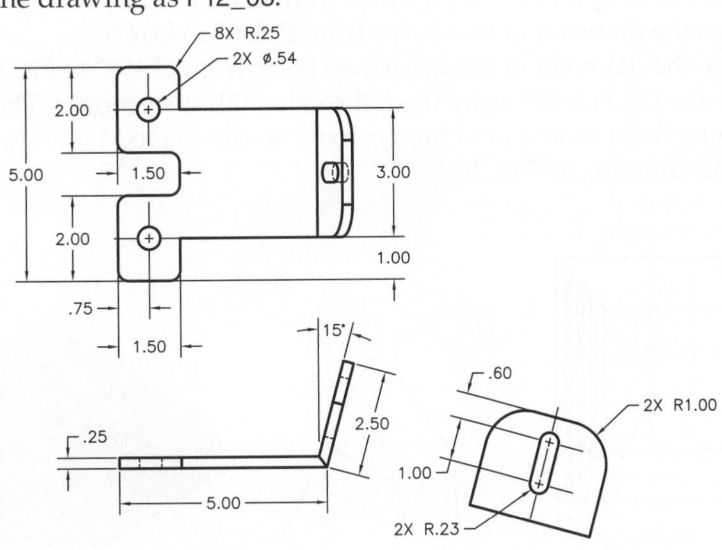

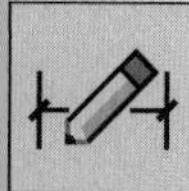

4. Draw as a single composite solid the desk organizer shown in the orthographic views below. Then, use primitive and subobject editing procedures to create the following edits. The final object should look like the shaded view below.
   A. Change the 3″ height to 3.25″.
   B. Change the 2″ height to 1.85″.
   C. Increase the thickness of the long compartment divider to .5″. The increase in thickness should be evenly applied along the centerline of the divider. Locate three evenly spaced, Ø5/16″ × 1.5″ holes in this divider.
   D. Angle the top face of the rear compartments by 30°. The height of the rear of the organizer should be approximately 4.5″ and all corners on the bottom of the organizer should remain square.
   E. Save the drawing as P12_04.

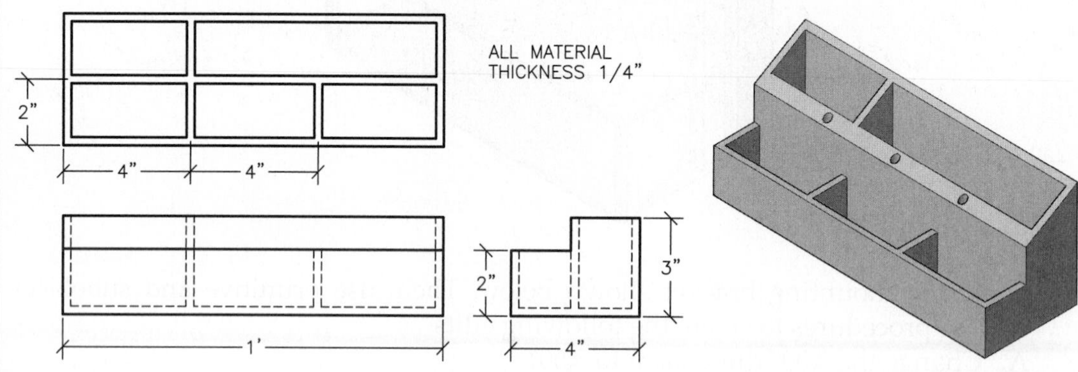

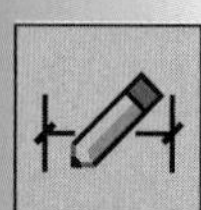

5. Draw the pencil holder shown below. Then, use primitive and subobject editing procedures to create the following edits.
   A. Change the depth of the base to 4.000″. The base should be rectangular, not square, and the grooves should become shorter.
   B. Change the height of the top groove from .250″ to .125″.
   C. Change the diameter of two holes from Ø.450″ to Ø.625″.
   D. Change the diameter of the other two holes from Ø.450″ to Ø1.000″.
   E. Rotate the top face 15° away from the side with the grooves. The planes of the adjoining faces should not change. Refer to the shaded view shown below.
   F. Save the drawing as P12_05.

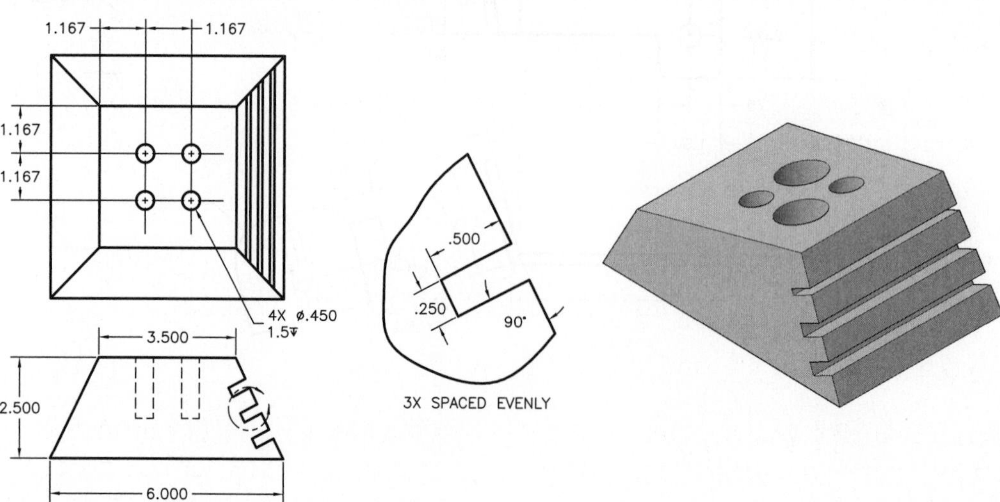

CHAPTER 13

Solid Model Editing

Learning Objectives

After completing this chapter, you will be able to:

✓ Change the shape and configuration of solid object faces.
✓ Copy and change the color of solid object edges and faces.
✓ Break apart a composite solid composed of physically separate entities.
✓ Extract a wireframe to project edges from a 3D solid using the **XEDGES** command.
✓ Use the **SOLIDEDIT** command to construct and edit a solid model.

AutoCAD provides expanded capabilities for editing solid models. As you saw in the previous chapter, grips can be used to edit a solid model. Also, the subobjects that make up a solid, such as faces, edges, and endpoints, can be edited. A single command, **SOLIDEDIT**, allows you to edit faces, edges, or the entire body of the solid.

NOTE

Mesh objects cannot be modified using the **SOLIDEDIT** command. The mesh object must be converted to a solid first. If you select a mesh for editing with the **SOLIDEDIT** command, you are given the option of converting it to a solid, as long as the display of the dialog box has not been turned off.

Overview of the SOLIDEDIT Command

The **SOLIDEDIT** command allows you to edit the faces, edges, and body of a solid. Many of the subobject editing functions discussed in Chapter 12 can also be performed with the **SOLIDEDIT** command. The features of the **SOLIDEDIT** command can be accessed in the **Solid Editing** panel on the **Home** or **Solid** tab of the ribbon or by typing SOLIDEDIT. See Figure 13-1. The quickest method of entering the command is by using the ribbon. For example, directly select a face-editing option from the drop-down list in the **Solid Editing** panel, as shown in Figure 13-1.

Figure 13-1.
Accessing the **SOLIDEDIT** command options.

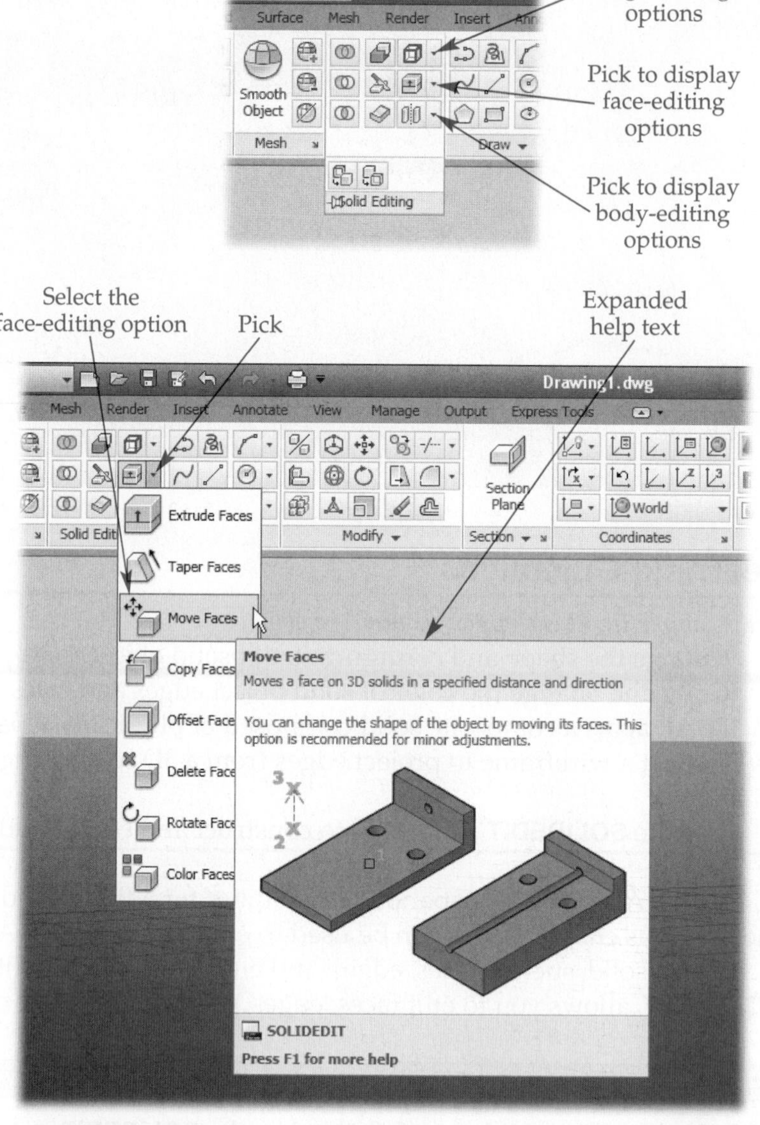

When the **SOLIDEDIT** command is typed, you are first asked to select the component of the solid with which you wish to work. Specify either **Face**, **Edge**, or **Body**. The editing options for the selected component are then displayed and are the same as those seen in **Figure 13-1**.

The editing function is directly entered when the option is selected from the ribbon. This is why using the ribbon is the most efficient method of entering the **SOLIDEDIT** command options.

The following sections provide an overview of the solid model editing features of the **SOLIDEDIT** command. Each option is explained and the results of each are shown. A tutorial later in the chapter illustrates how these options can be used to construct a model.

NOTE

AutoCAD displays a variety of error messages when illegal solid editing operations are attempted. Rather than trying to interpret the wording of these messages, just realize that what you tried to do will not work. Actions that may cause errors include trying to rotate a face into other faces or extruding and tapering an object at too great of an angle. When an error occurs, try the operation again with different parameters or determine a different approach to solving the problem in order to maintain the design intent.

Face Editing

The basic components of a solid are its faces and the greatest number of **SOLIDEDIT** options are for editing faces. All eight face editing options ask you to select faces. It is important to make sure you select the correct part of the model for editing. Remember the following three steps when using any of the face editing options.

1. First, select a face to edit. If you pick an edge, AutoCAD selects the two faces that share the edge. If this happens, use the **Remove** option to deselect the unwanted face. A more intuitive approach is to select the open space of the face as if you were touching the side of a part. AutoCAD highlights only that face.
2. Adjust the selection set at the Select faces or [Undo/Remove/ALL]: prompt. The following options are available.
   - **Undo.** Removes the previous selected face(s) from the selection set.
   - **Remove.** Allows you to select faces to remove from the selection set. This is only available when **Add** is current.
   - **All.** Adds all faces on the model to the selection set. This is only available after selecting at least one face. It can also be used to remove all faces if **Remove** is current.
   - **Add.** Allows you to add faces to the selection set. This is only available when **Remove** is current.
3. Press [Enter] to continue with face editing.

Extruding Faces

SOLIDEDIT

An extruded face is moved, or stretched, in a selected direction. The extrusion can be straight or have a taper. To extrude a face, select the command and pick the **Face>Extrude** option. Remember, the option is directly entered when picking the button in the ribbon. You are then prompted to select the face(s) to extrude. Nonplanar (curved) faces cannot be extruded. As you pick faces, the prompt verifies the number of faces selected. For example, when an edge is selected, the prompt reads 2 faces found. When done selecting faces, press [Enter] to continue.

Next, the height of the extrusion needs to be specified. A positive value adds material to the solid, while a negative value subtracts material from the solid. A taper can also be given.

Specify height of extrusion or [Path]: *(enter height)*
Specify angle of taper for extrusion <0>: *(enter an angle or accept the default)*
Solid validation started.
Solid validation completed.
Enter a face editing option
[Extrude/Move/Rotate/Offset/Taper/Delete/Copy/coLor/mAterial/Undo/eXit] <eXit>: **X**↵
Solids editing automatic checking: SOLIDCHECK=1
Enter a solids editing option [Face/Edge/Body/Undo/eXit] <eXit>: **X**↵

Figure 13-2.
Extruding faces on an object. A—The original object. B—The top face is extruded with a 0° taper angle. C—The top face of the original is extruded with a 30° taper angle. D—The top and right-hand faces of the original are extruded with 15° taper angles.

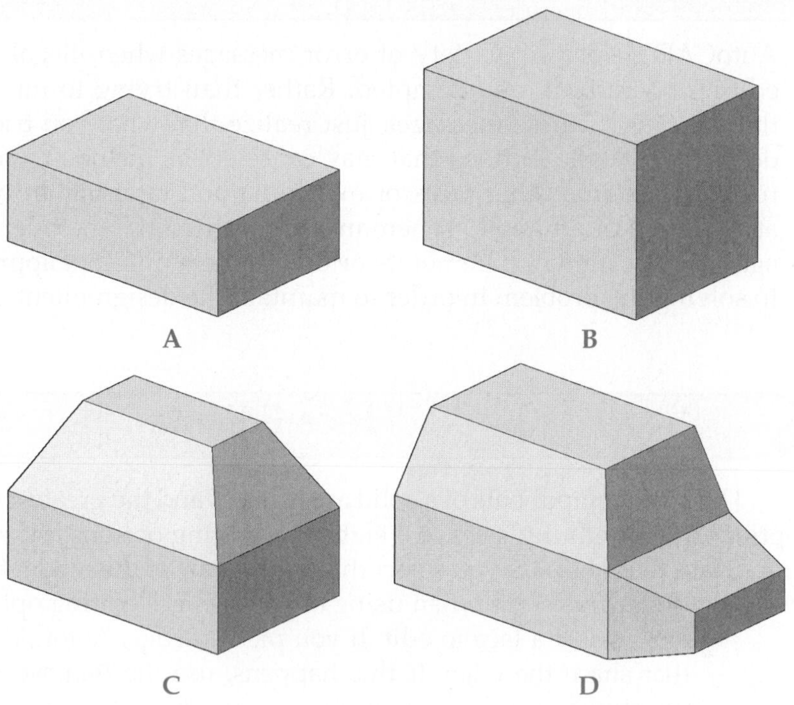

Figure 13-2 shows an original solid object and the result of extruding the top face with a 0° taper angle and a 30° taper angle. It also shows the original solid object with two adjacent faces extruded with 15° taper angles.

In addition to extruding a face perpendicular to itself, the extruded face can follow a path. Select the **Path** option at the Specify height of extrusion or [Path]: prompt. The path of extrusion can be a line, circle, arc, ellipse, elliptical arc, polyline, or spline. The extrusion height is the exact length of the path. See Figure 13-3.

Exercise 13-1

Complete the exercise on the student website.
www.g-wlearning.com/CAD

Figure 13-3.
The path of extrusion can be a line, circle, arc, ellipse, elliptical arc, polyline, or spline. Here, the paths are shown in color.

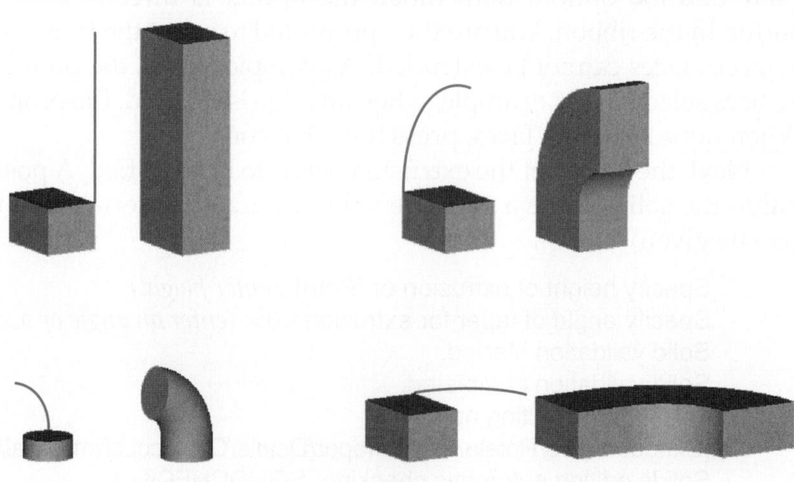

Moving Faces

Ribbon
Home
> Solid Editing
Move Faces
Type
SOLIDEDIT

SOLIDEDIT

The **Move Faces** option moves a face in the specified direction and lengthens or shortens the solid object. In another application, a solid model feature (such as a hole) that has been subtracted from an object to create a composite solid can be moved with this option. Object snaps may interfere with the function of this option, so they may need to be toggled off during the operation.

To move a face, select the command and pick the **Face**>**Move** option. If the button is picked in the ribbon, the option is directly entered. You are then prompted to select the face(s) to move. When done selecting faces, press [Enter] to continue. Next, you are prompted to select a base point of the operation:

Specify a base point or displacement: *(pick a base point)*
Specify a second point of displacement: *(pick a second point or enter coordinates)*
Solid validation started.
Solid validation completed.
Enter a face editing option
[Extrude/Move/Rotate/Offset/Taper/Delete/Copy/coLor/mAterial/Undo/eXit] <eXit>: **X**↵
Solids editing automatic checking: SOLIDCHECK=1
Enter a solids editing option [Face/Edge/Body/Undo/eXit] <eXit>: **X**↵

When adjacent faces are perpendicular, the edited face is moved in a direction so the new position keeps the face parallel to the original. See **Figures 12-4A** and **12-4B.** Faces that are normal to the current UCS can be moved by picking a new location or entering a direct distance. If you are moving a face that is not normal to the current UCS, you can enter coordinates for the second point of displacement, but it may be easier to first use the **Face** option of the **UCS** command to align the UCS with the face to be moved.

When adjacent faces join at angles other than 90°, the moved face will be relocated as stated above, but only if the movement is less than the dimensional offset of the two faces. For example, in **Figure 13-4B** the top edge of the angled face is in .5″ from the vertical face. If the angled face is moved outward a distance of less than .5″, it is altered as shown in **Figure 13-4C.** A portion of the angled face becomes coplanar with the vertical face. If the angled face is moved outward a distance greater than .5″, it is altered so that it forms a single plane with the adjacent face. What has actually happened is that the angled face is moved beyond the adjacent face, while remaining parallel to its original position. Thus, in effect, it has disappeared because the adjacent, vertical face cannot be altered. See **Figure 13-4D.** In this example, the angled face was moved .75″. The new vertical face that is created can now be moved.

Exercise 13-2

Complete the exercise on the student website.
www.g-wlearning.com/CAD

Figure 13-4.
A—The hole will be moved using the **Move** option of the **SOLIDEDIT** command. B—The hole is moved. C—When the angled face is moved, a portion of it is altered to be coplanar with the vertical face. D—If the angled face is moved more, it becomes completely coplanar to the vertical face. This is a new, single face.

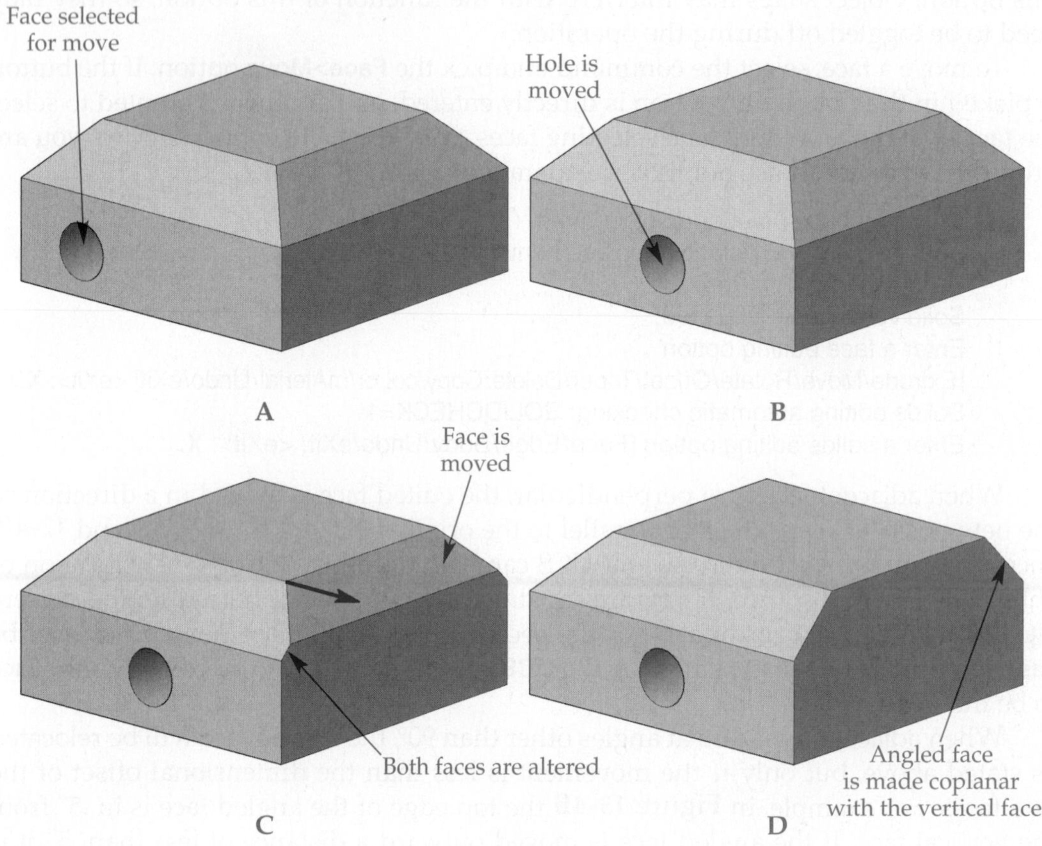

Offsetting Faces

SOLIDEDIT
Ribbon
Home
> Solid Editing
Solid
> Solid Editing
Offset Faces
Type
SOLIDEDIT

The **Offset** option may seem the same as the **Extrude** option because it moves faces by a specified distance or through a specified point. Unlike the **OFFSET** command in AutoCAD, this option moves all selected faces a specified distance. It is most useful when you wish to change the size of features such as slots, holes, grooves, and notches in solid parts. A positive offset distance increases the size or volume of the solid (adds material), a negative distance decreases the size or volume of the solid (removes material). Therefore, if you wish to make the width of a slot wider, provide a negative offset distance to decrease the size of the solid. Picking points to set the offset distance and direct distance entry are always taken as a positive value, so negative values must be entered using the keyboard.

To offset a face, select the command and pick the **Face**>**Offset** option. Remember, the option is directly entered when picking the button in the ribbon. You are then prompted to select the face(s) to offset. When done selecting faces, press [Enter] to continue. Next, enter the offset distance and press [Enter] and then exit the command. See Figure 13-5 for examples of features edited with the **Offset** option.

PROFESSIONAL TIP

Nonplanar (curved) faces cannot be extruded, but can be offset. Using the **Offset** option, you can, in effect, "extrude" a nonplanar face.

Figure 13-5.
Offsetting faces. A—The original objects. The hole is selected to offset. The interior of the L is also selected to offset. B—A positive offset distance increases the size or volume of the solid. C—A negative offset distance decreases the size or volume of the solid.

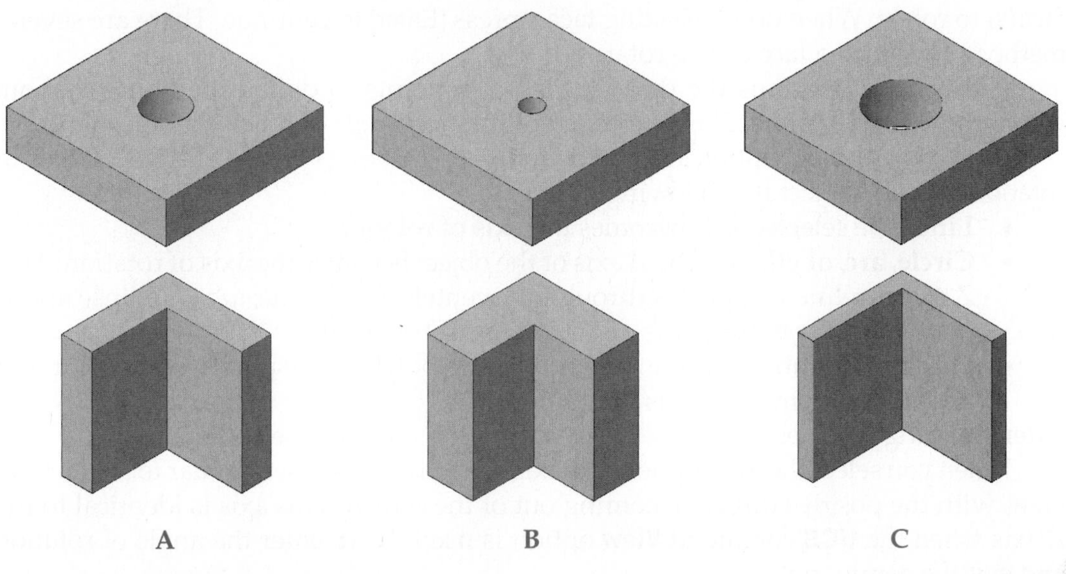

Exercise 13-3

Complete the exercise on the student website.
www.g-wlearning.com/CAD

Deleting Faces

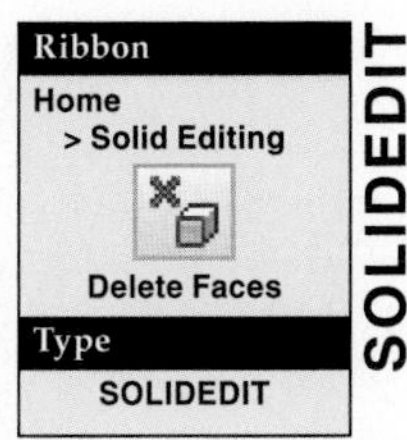

The **Delete** option deletes selected faces. This is a quick way to remove features such as chamfers, fillets, holes, and slots. To delete a solid face, select the command and pick the **Face**>**Delete** option. If the button is picked in the ribbon, the option is directly entered. You are then prompted to select the face(s) to delete. When done selecting faces, press [Enter] to continue and then exit the command. When a face is deleted, existing faces extend to fill the gap. No additional faces are created. For instance, the inclined surface of a wedge cannot be deleted as there are no existing faces that can be extended to fill the gap. When deleting the face that is a chamfered or filleted edge, the adjacent edges are extended to fill the gap. See **Figure 13-6.**

Figure 13-6.
Deleting faces. A—The original objects with three rounds. B—The faces of two rounds have been deleted.

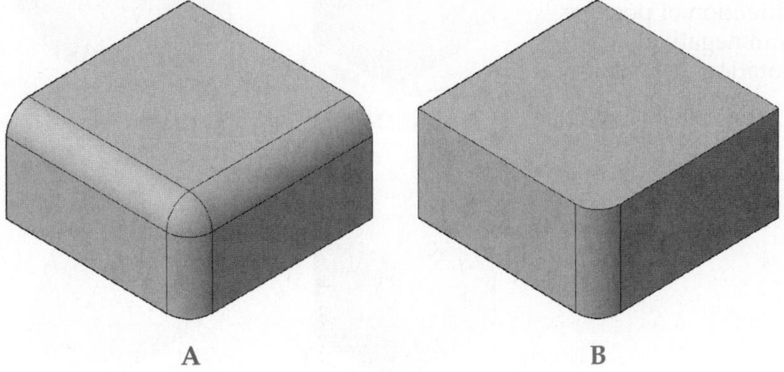

Rotating Faces

SOLIDEDIT

Ribbon
Home
> Solid Editing
Rotate Faces
Type
SOLIDEDIT

The **Rotate** option rotates a face about a selected axis. To rotate a solid face, select the command and pick the **Face>Rotate** option. Remember, the option is directly entered when picking the button in the ribbon. You are then prompted to select the face(s) to rotate. When done selecting faces, press [Enter] to continue. There are several methods by which a face can be rotated.

The **2points** option is the default. Pick two points to define the "hinge" about which the face will rotate. Then, provide the rotation angle and exit the command.

The **Axis by object** option allows you to use an existing object to define the axis of rotation. You can select the following objects.

- **Line.** The selected line becomes the axis of rotation.
- **Circle, arc, or ellipse.** The Z axis of the object becomes the axis of rotation. This Z axis is a line that passes through the center of the circle, arc, or ellipse and is perpendicular to the plane on which the 2D object lies.
- **Polyline or spline.** A line connecting the polyline or spline's start point and endpoint becomes the axis of rotation.

After selecting an object, enter the angle of rotation and exit the command.

When you select the **View** option, the axis of rotation is perpendicular to the current view, with the positive direction coming out of the screen. This axis is identical to the Z axis when the **UCS** command **View** option is used. Next, enter the angle of rotation and exit the command.

The **Xaxis**, **Yaxis**, and **Zaxis** options prompt you to select a point. Either the X, Y, or Z axis passing through that point is used as the axis of rotation. Then, enter the angle of rotation and exit the command.

Figure 13-7 provides several examples of rotated faces. Notice how the first and second pick points determine the direction of positive and negative rotation angles.

NOTE

A positive rotation angle moves the face in a clockwise direction looking from the first pick point to the second. Conversely, a negative angle rotates the face counterclockwise. If the rotated face will intersect or otherwise interfere with other faces, an error message indicates that the operation failed or no solution was calculated. In this case, you may wish to try a negative angle if you previously entered a positive one. In addition, you can try selecting the opposite edge of the face as the axis of rotation.

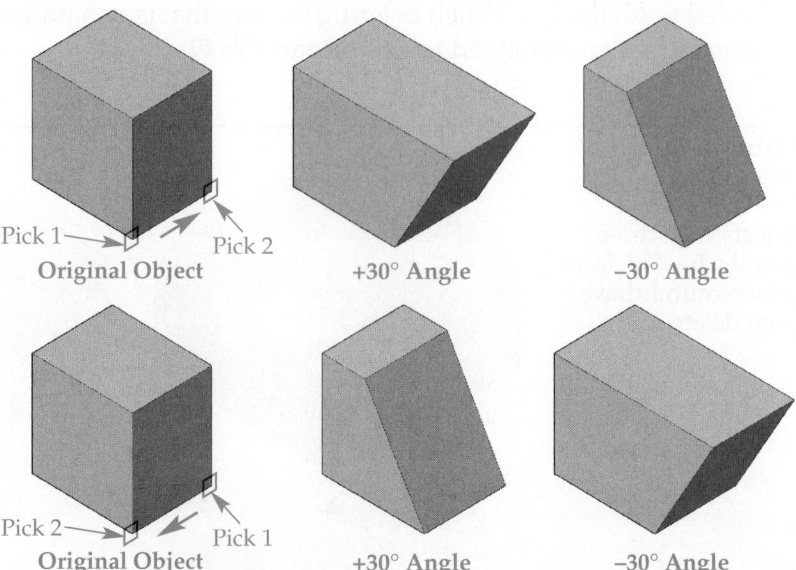

Figure 13-7. When rotating faces, the first and second pick points determine the direction of positive and negative rotation angles.

Exercise 13-4

Complete the exercise on the student website.
www.g-wlearning.com/CAD

Tapering Faces

The **Taper** option tapers a face at the specified angle, from the first pick point to the second. To taper a solid face, select the command and pick the **Face>Taper** option. If the button is picked in the ribbon, the option is directly entered. You are then prompted to select the face(s) to taper. When done selecting faces, press [Enter] to continue:

> Specify the base point: *(pick the base point)*
> Specify another point along the axis of tapering: *(pick a point along the taper axis)*
> Specify the taper angle: *(enter a taper value)*

Tapers work differently depending on whether the faces being tapered describe the outer boundaries of the solid, a cavity, or a removed portion of the solid. A positive taper angle always removes material. A negative taper angle always adds material. For example, if a positive taper angle is entered for a solid cylinder, the selected object is tapered in on itself from the base point along the axis of tapering, thus removing material. A negative angle tapers the object out away from itself to increase its size along the axis of tapering, thus adding material. See **Figure 13-8.**

On the other hand, if the faces of a feature such as a slot or hole are tapered, a positive taper angle increases the size of the feature along the axis of tapering. For

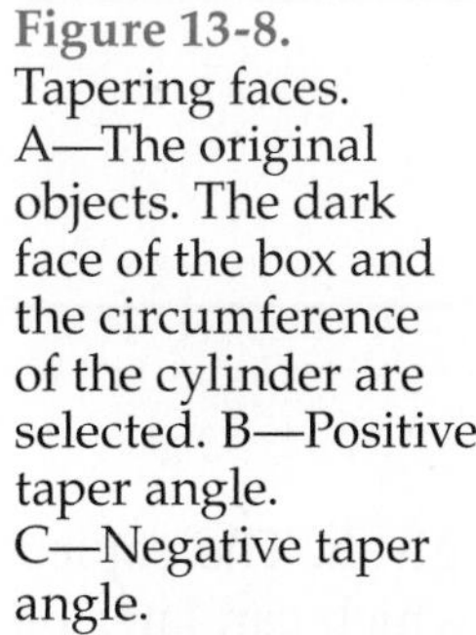

Figure 13-8.
Tapering faces. A—The original objects. The dark face of the box and the circumference of the cylinder are selected. B—Positive taper angle. C—Negative taper angle.

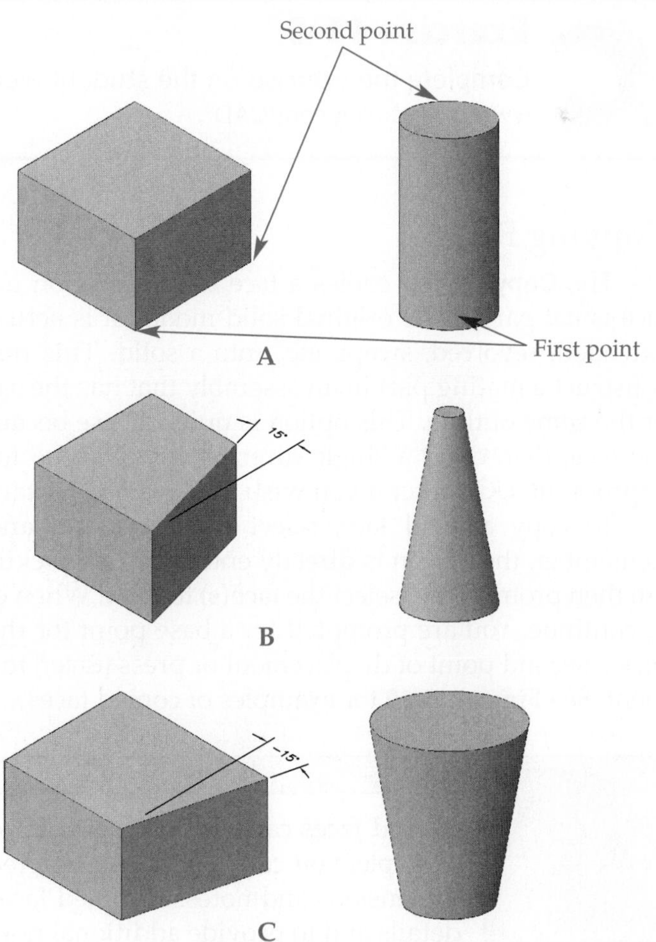

Figure 13-9.
If a hole or slot is tapered using a positive taper angle, its diameter or width increases from the base point along the axis of tapering, thus removing material from the solid. A negative taper angle increases the volume of the solid.

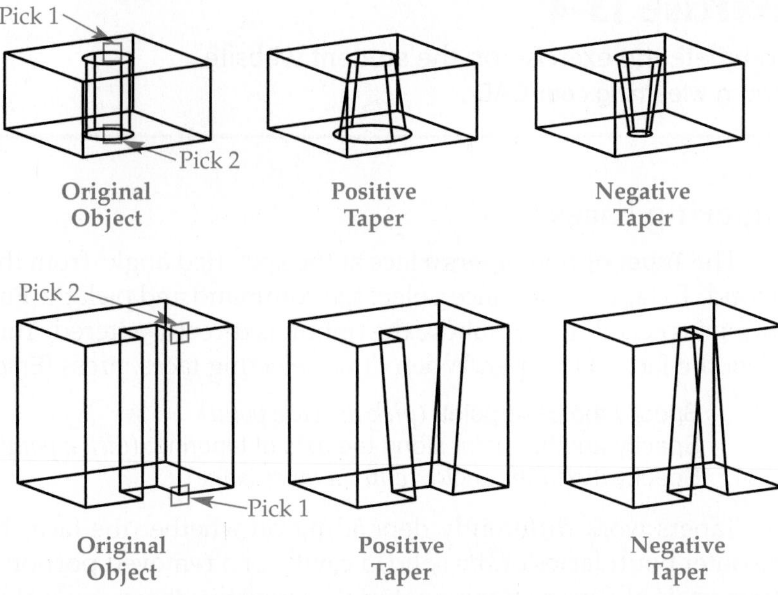

example, if a round hole is tapered using a positive taper angle, its diameter increases from the base point along the axis of tapering, thus removing material from the solid. Conversely, if the same round hole is tapered using a negative taper angle, its diameter decreases from the base point along the axis of tapering, thus adding material to the solid. Figure 13-9 shows some examples of this operation.

Exercise 13-5

Complete the exercise on the student website.
www.g-wlearning.com/CAD

Copying Faces

SOLIDEDIT

The **Copy** option copies a face to the location or coordinates given. The copied face is *not* part of the original solid model. It is actually a region, which can later be extruded, revolved, swept, etc., into a solid. This may be useful when you wish to construct a mating part in an assembly that has the same features on the mating faces or the same outline. This option is quick to use because you can pick a base point on the face, then enter a single direct distance value for the displacement. Be sure an appropriate UCS is set if you wish to use direct distance entry.

To copy a solid face, select the command and pick the **Face>Copy** option. Remember, the option is directly entered when picking the button in the ribbon. You are then prompted to select the face(s) to copy. When done selecting faces, press [Enter] to continue. You are prompted for a base point for the copy. Pick this point and then pick a second point of displacement or press [Enter] to use the first point as a displacement. See Figure 13-10 for examples of copied faces.

PROFESSIONAL TIP

Copied faces can also be useful for creating additional views. For example, you can copy a face to create a separate plan view with dimensions and notes. A copied face can also be enlarged to show details and to provide additional notation for design or assembly.

Figure 13-10.
A face can be quickly copied by picking a base point on the face and then entering a direct distance value for the displacement.

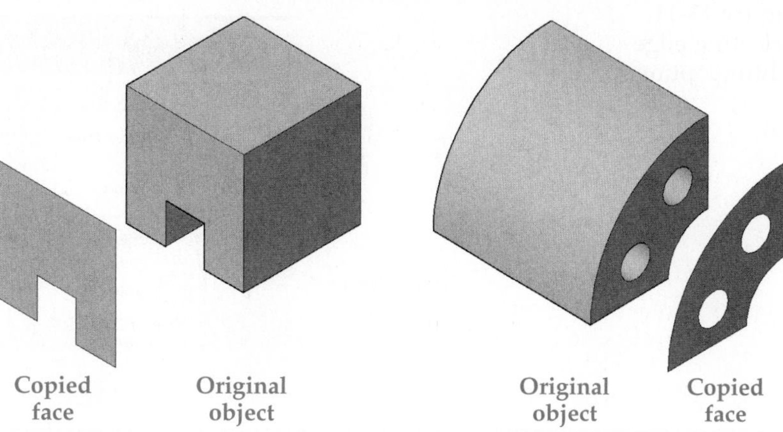

Coloring Faces

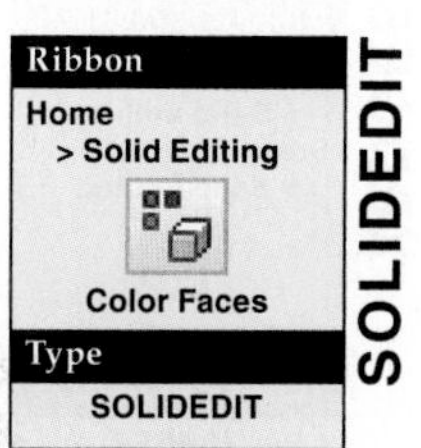

You can quickly change a selected face to a different color using the **Color** option. Select the command and pick the **Face**>**Color** option. If the button is picked in the ribbon, the option is directly entered. You are then prompted to select the face(s) to color. When done selecting faces, press [Enter] to continue. Next, choose the desired color from the **Select Color** dialog box that is displayed. Remember, the color of the object (or face) determines the shaded color.

PROFESSIONAL TIP

Use the **Color** option to enhance features on a solid. For example, if you have a model that has holes or slots, enhance just the color of these features.

Exercise 13-6

Complete the exercise on the student website.
www.g-wlearning.com/CAD

Edge Editing

Edges can be edited in only two ways. They can be copied from the solid. Also, the color of an edge can be changed.

Copying an edge is similar to copying a face. To copy a solid edge, select the command and pick the **Edge**>**Copy** option. Remember, the option is directly entered when picking the button in the ribbon. See Figure 13-11. You are then prompted to select the edge(s) to copy. When done selecting edges, press [Enter] to continue. You are prompted for a base point for the copy. Pick this point and then pick a second point of displacement or press [Enter] to use the first point as a displacement. The edge is copied as a line, arc, circle, ellipse, or spline.

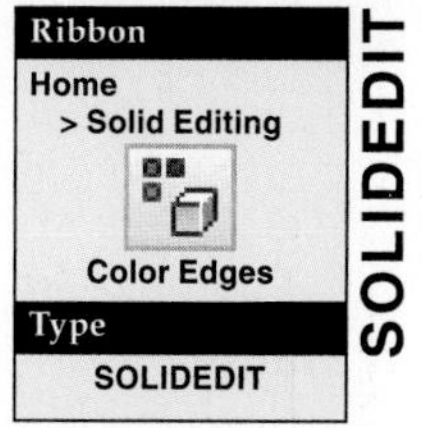

To color a solid edge, select the command and pick the **Edge**>**Color** option. You are then prompted to select the edge(s) to color. When done selecting edges, press [Enter] to continue. Next, choose the desired color from the **Select Color** dialog box that is displayed and pick the **OK** button. The edges are now displayed with the new color. You may need to set a wireframe or hidden visual style current to see the change.

Figure 13-11.
Selecting edge-editing options.

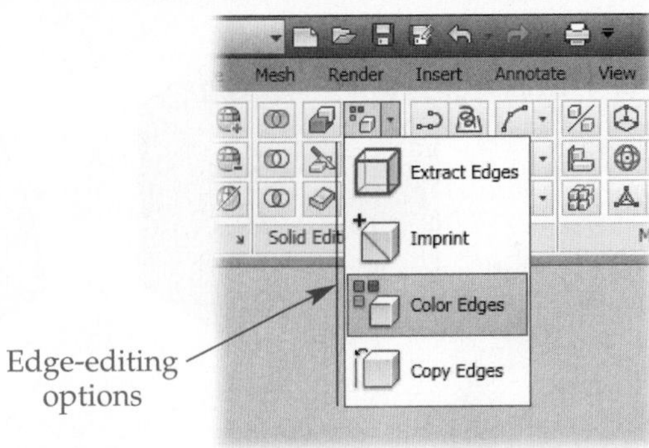

Extracting a Wireframe

XEDGES

Ribbon
Home
> Solid Editing
Solid
> Solid Editing
Extract Edges
Type
XEDGES

The **XEDGES** command creates copies of, or extracts, all of the edges on a selected solid. You can also select edges of a surface, mesh, region, or subobject. Once the command is initiated, you are prompted to select objects. Select one or more solids and press [Enter]. The edges are extracted and placed on top of the existing edges. See **Figure 13-12.** The new objects are created on the current layer.

Straight edges and the curved edges where cylindrical surfaces intersect with flat or other cylindrical surfaces are the only edges extracted. Spheres and tori have no edges that can be extracted. The round bases of cylinders and cones are the only edges of those objects that will be extracted.

Figure 13-12.
Extracting edges with the **XEDGES** command. A—The original object. B—The extracted wireframe (edges).

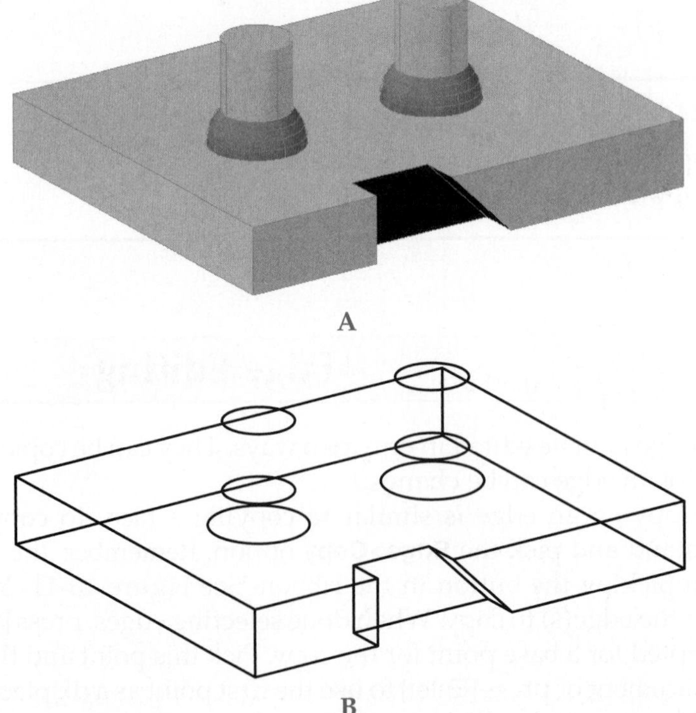

Exercise 13-7

Complete the exercise on the student website.
www.g-wlearning.com/CAD

Body Editing

The body editing options of the **SOLIDEDIT** command perform editing operations on the entire body of the solid model. The body options are **Imprint**, **Separate**, **Shell**, **Clean**, and **Check**. The next sections cover these body editing options.

The **Imprint** option is a body-editing function. However, since it modifies a body by adding edges, the option is located in the edges drop-down list in the **Solid Editing** panel on the **Home** tab. On the **Solid** tab, it is a separate button in the **Solid Editing** panel. In both places, the **Separate**, **Shell**, **Clean**, and **Check** options are found in body drop-down list.

Imprint

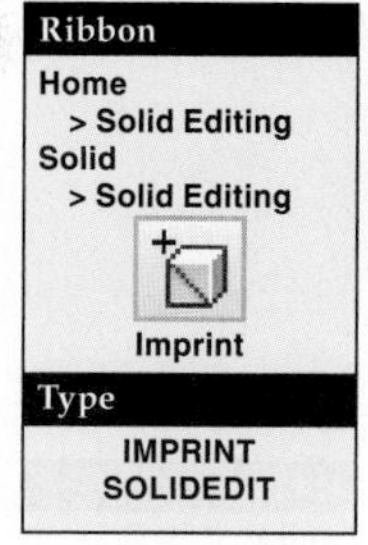

Arcs, circles, lines, 2D and 3D polylines, ellipses, splines, regions, bodies, and 3D solids can be imprinted onto a solid, if the object intersects the solid. The imprint becomes a face on the surface based on the overlap between the two intersecting objects. Once the imprint has been made, the new face can be modified.

To imprint an object on a solid, select the command and pick the **Body>Imprint** option. If IMPRINT is typed or the button is picked on the ribbon, the option is directly entered. Once the option is activated, you are prompted to select the solid. This is the object on which the other objects will be imprinted. Then, select the objects to be imprinted. You have the option of deleting the source objects.

The imprinted face can be modified using face-editing options. Figure 13-13 illustrates objects imprinted onto a solid model. Two of these are then extruded into the solid to create holes. The **PRESSPULL** command is used on the third object to create a cylindrical feature.

Figure 13-13.
Imprinted objects form new faces that can be extruded into the solid. A—A solid box with three objects on the plane of the top face. B—The objects are imprinted, then two of the new faces are extruded through the solid. The **PRESSPULL** command is used on the third new face to create the cylindrical feature.

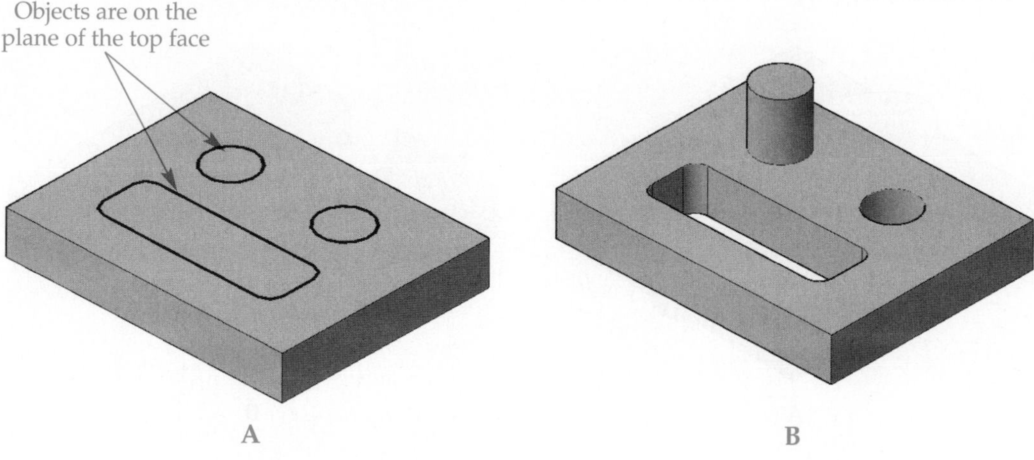

NOTE

Remember that objects are drawn on the XY plane of the current UCS unless you enter a specific Z value. Therefore, before you draw an object to be imprinted onto a solid model, be sure you have set an appropriate UCS for proper placement of the object by using a dynamic UCS or the **UCS** command. Alternately, you can draw the object on the XY plane and then move the object onto the solid object.

Separate

SOLIDEDIT
Ribbon
Home
> Solid Editing
Solid
> Solid Editing
Separate
Type
SOLIDEDIT

The **Separate** option separates two objects that are both a part of a single solid composite, but appear as separate physical entities. This can happen when modifying solids using the Boolean commands. The **Separate** option may be seldom used, but it has a specific purpose. If you select a solid model and an object physically separate from the model is highlighted, the two objects are parts of the same composite solid. If you wish to work with them as individual solids, they must first be separated.

To separate a solid body, select the command and pick the **Body>Separate** option. If the button is picked in the ribbon, the option is directly entered. You are then prompted to select a 3D solid. After you pick the solid, it is automatically separated. No other actions are required and you can exit the command. However, if you select a solid in which the parts are physically joined, AutoCAD indicates this by prompting The selected solid does not have multiple lumps. A "lump" is a physically separate solid entity. In order to separate a solid, it must be composed of multiple lumps. See **Figure 13-14.**

Shell

A ***shell*** is a solid that has been "hollowed out." The **Shell** option creates a shell of the selected object using a specified offset distance, or thickness. To create a shell of a solid body, select the command and pick the **Body>Shell** option. Remember, the option is directly entered when picking the button in the ribbon. You are prompted to select the solid. Only one solid can be selected.

After selecting the solid, you have the opportunity to remove faces. If you do not remove any faces, the new solid object will appear identical to the old solid object when shaded or rendered. The thickness of the shell will not be visible. If you wish to create a hollow object with an opening, select the face to be removed (the opening).

Figure 13-14.
A—After the cylinder is subtracted from the box, the remaining solid is considered one solid.
B—Use the **Separate** option to turn this single solid into two solids.

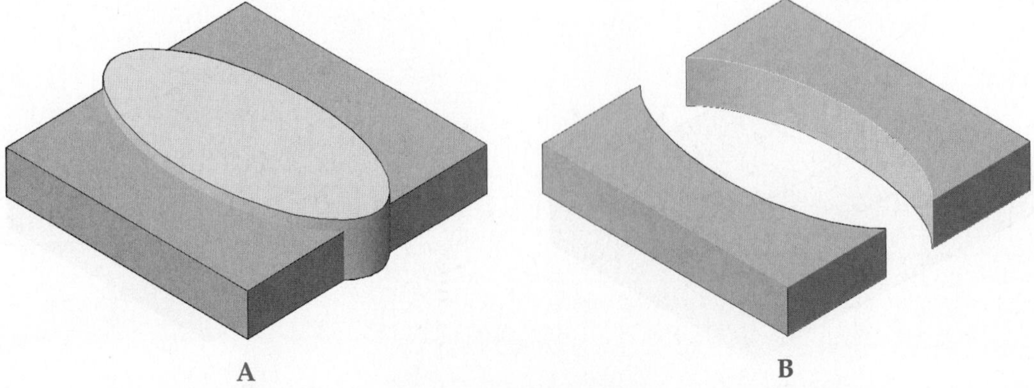

After selecting the object and specifying any faces to be removed, you are prompted to enter the shell offset distance. This is the thickness of the shell. A positive shell offset distance creates a shell on the inside of the solid body. A negative shell offset distance creates a shell on the outside of the solid body. See **Figure 13-15.** If you shell a solid that contains internal features, such as holes, grooves, and slots, a shell of the specified thickness is placed around those features. This is shown in **Figure 13-16.**

If the shell operation is not readily visible in the current view, you can rotate the view by using the [Shift] + the mouse wheel button to enter the transparent **3DORBIT** command or by using the view cube. You can also see the results by picking the **X-Ray Effect** button in the **Visual Styles** panel on the **View** tab of the ribbon.

Figure 13-15.
A—The right-front, bottom, and left-back faces (marked here by gray lines) are removed from the shell operation. B—The resulting object after the shell operation.

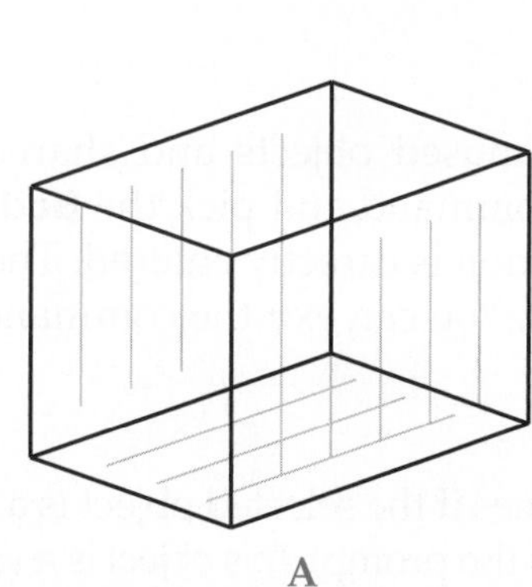

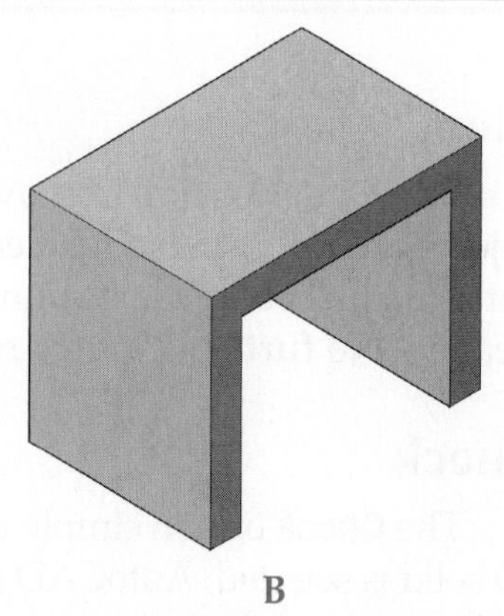

Figure 13-16.
If you shell a solid that contains internal features, such as holes, grooves, and slots, a shell of the specified thickness is also placed around those features. A—Solid object with holes subtracted. B—Wireframe display after shelling with a negative offset. C—The Conceptual visual style is set current.

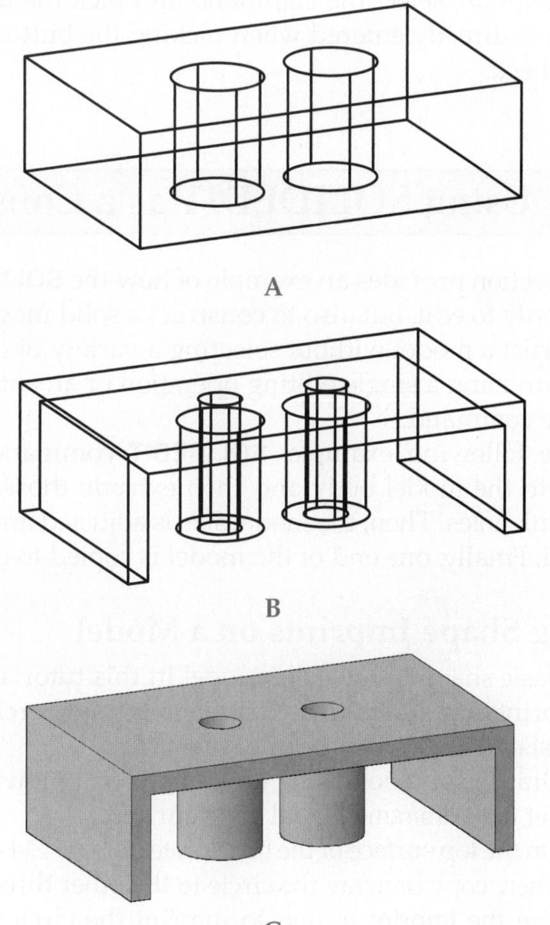

PROFESSIONAL TIP

The **Shell** option of the **SOLIDEDIT** command is very useful in applications such as solid modeling of metal castings or injection-molded plastic parts.

Exercise 13-8

Complete the exercise on the student website.
www.g-wlearning.com/CAD

SOLIDEDIT

Ribbon
Home
> Solid Editing
Solid
> Solid Editing
Clean
Type
SOLIDEDIT

Clean

The **Clean** option removes all unused objects and shared surfaces. Imprinted objects are not removed. Select the command and pick the **Body>Clean** option. If the button is picked in the ribbon, the option is directly entered. Then, pick the solid to be cleaned. No further input is required. You can exit the command.

SOLIDEDIT

Ribbon
Home
> Solid Editing
Solid
> Solid Editing
Check
Type
SOLIDEDIT

Check

The **Check** option simply determines if the selected object is a valid 3D solid. If a true 3D solid is selected, AutoCAD displays the prompt This object is a valid ShapeManager solid. and you can exit the command. If the object selected is not a 3D solid, the prompt reads A 3D solid must be selected. and you are prompted to select a 3D solid. To access the **Check** option, select the command and pick the **Body>Check** option. Remember, the option is directly entered when picking the button in the ribbon. Then, select the object to check.

Using SOLIDEDIT as a Construction Tool

This section provides an example of how the **SOLIDEDIT** command options can be used not only to edit, but also to construct a solid model. This makes it easy to design and construct a model without selecting a variety of commands. It also gives you the option of undoing a single editing operation or an entire editing session without ever exiting the command.

In the following example, **SOLIDEDIT** command options are used to imprint shapes onto the model body and then extrude those shapes into the body to create countersunk holes. Then, the model size is adjusted and an angle and taper are applied to one end. Finally, one end of the model is copied to construct a mating part.

Creating Shape Imprints on a Model

The basic shape of the solid model in this tutorial is drawn as a solid box. Then, shape imprints are added to it. Throughout this exercise, you may wish to change the UCS to assist in the construction of the part.

1. Draw a solid box using the dimensions shown in Figure 13-17.
2. Set the Wireframe visual style current.
3. On the top surface of the box, locate a single Ø.4 circle using the dimensions given. Then, copy or array the circle to the other three corners as shown in the figure.
4. Use the **Imprint** option to imprint the circles onto the solid box. Delete the source objects.

Figure 13-17.
The initial setup for the tutorial model.

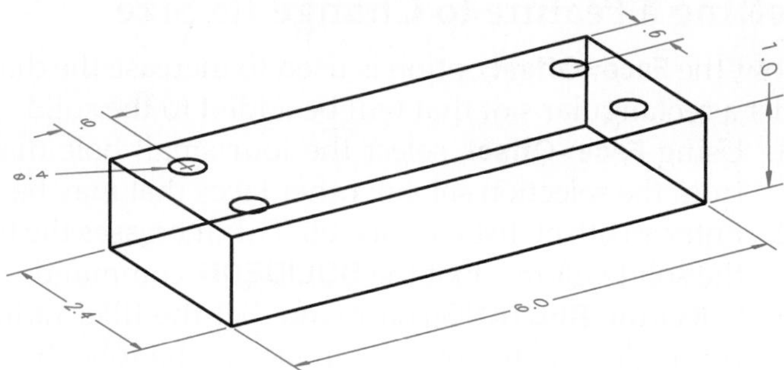

Extruding Imprints to Create Features

The imprinted 2D shapes can now be extruded to create new 3D solid features on the model. Use the **Face>Extrude** option to extrude all four imprinted circles.

1. When you select the edge of the first circle, all features on that face are highlighted, but only the circle you picked and the top face have actually been selected. If you pick inside the circle, only the circle is selected and highlighted. In either case, be sure to also pick the remaining three circles.
2. Remove the top face of the box from the selection set, if needed.
3. The depth of the extrusion is .16 units. Remember to enter –.16 for the extrusion height since the holes remove material. The angle of taper for extrusion should be 35°. Your model should look like **Figure 13-18A.**
4. Extrude the small diameter of the four tapered holes so they intersect the bottom of the solid body. Select the holes by picking the small diameter circles. Instead of calculating the distance from the bottom of the chamfer to the bottom surface, you can simply enter a value that is greater than this distance, such as the original thickness of the object. Again, since the goal is to remove material, use a negative value for the height of the extrusion. There is no taper angle. Your model should now look like **Figure 13-18B.**

Moving Faces to Change Model Size

The next step is to use the **Face>Move** option to decrease the length and thickness of the solid body.

1. Select either end face and the two holes nearest to it. Be sure to select the holes *and* the countersinks. Move the two holes and end face two units toward the other end, thus changing the object length to four units.
2. Select the bottom face and move it .5 units up toward the top face, thus changing the thickness to .5 units. See **Figure 13-19.**

Figure 13-18.
A—The imprinted circles are extruded with a taper angle of 35°. B—Holes are created by further extrusion with a taper angle of 0°.

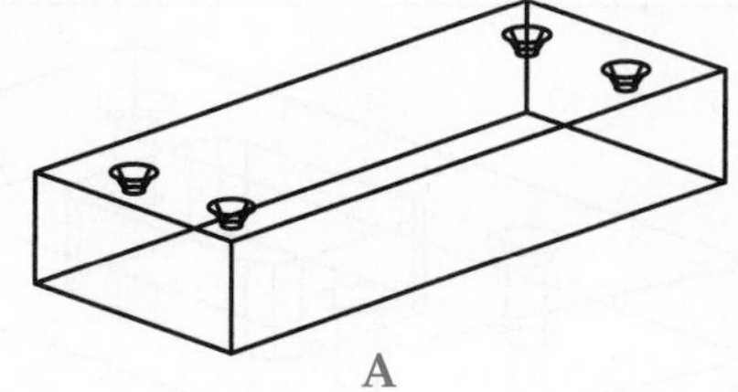
A

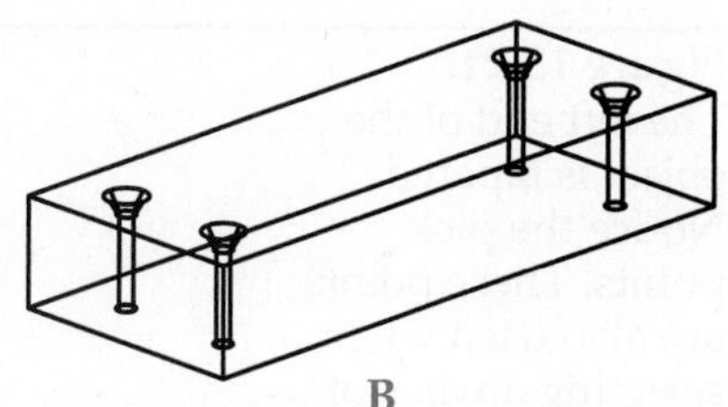
B

Figure 13-19.
The length of the object is shortened and the height is reduced.

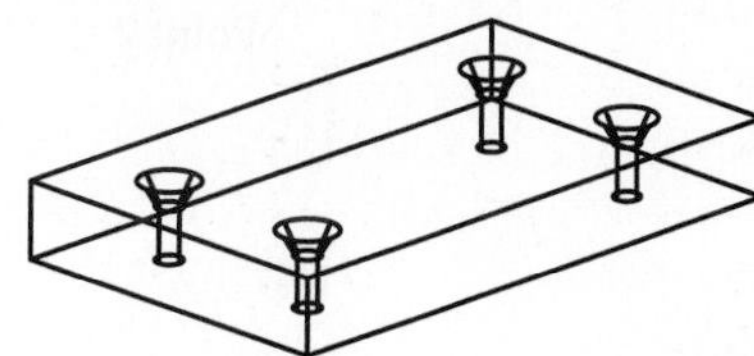

Offsetting a Feature to Change Its Size

Now, the **Face>Offset** option is used to increase the diameter of the four holes and to adjust a rectangular slot that will be added to the solid.

1. Using **Face>Offset**, select the four small hole diameters. Be sure to remove from the selection set any other faces that may be selected.
2. Enter an offset distance of –.05. This increases the hole diameter and decreases the solid volume. Exit the **SOLIDEDIT** command.
3. Select the **RECTANG** command. Set the fillet radius to .4 and draw a 2 × 1.6 rectangle centered on the top face of the solid. See **Figure 13-20A.**
4. Imprint the rectangle on the solid. Delete the source object.
5. Extrude the rectangle completely through the solid (.5 units). Remember to remove from the selection set any other faces that may be selected.
6. Offset the rectangular feature using an offset distance of .2 units. You will need to select all faces of the feature. This decreases the size of the rectangular opening and increases the solid volume. Your drawing should appear as shown in **Figure 13-20B.**

Tapering Faces

One side of the part is to be angled. The **Face>Taper** option is used to taper the left end of the solid.

1. Using **Face>Taper**, pick the face at the left end of the solid.
2. Pick point 1 in **Figure 13-21** as the base point and point 2 as the second point along the axis of tapering.
3. Enter a value of –10 for the taper angle. This moves the upper-left end away from the solid, creating a tapered end.

Figure 13-20.
A—The diameter of the holes is increased and a rectangle is imprinted on the top surface.
B—The rectangle is extruded to create a slot.

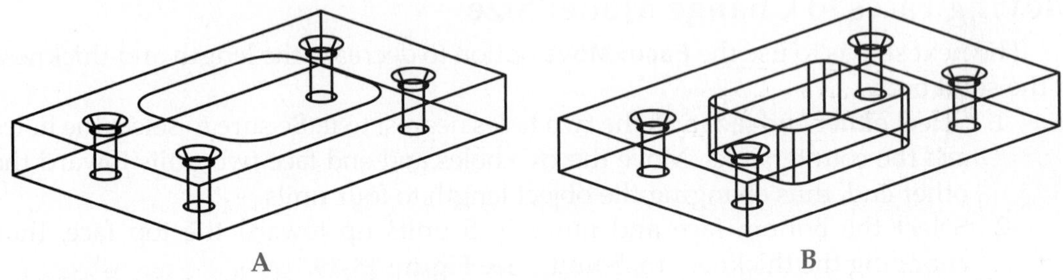

Figure 13-21.
The left end of the object is tapered. Notice the pick points. These points are also used when selecting an axis of rotation.

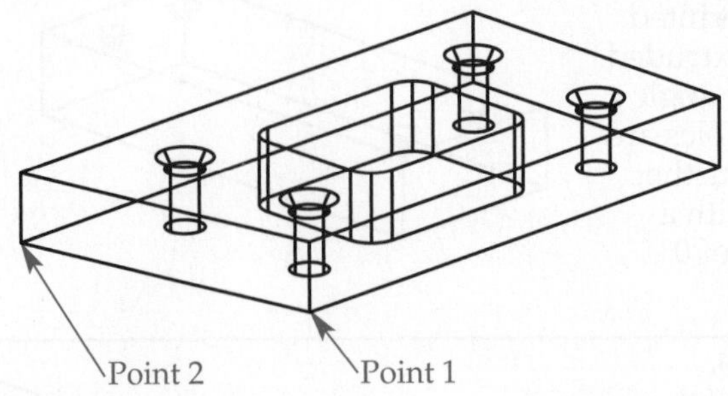

Rotating Faces

Next, use the **Face>Rotate** option to rotate the tapered end of the object. The top edge of the face will be rotated away from the holes, adding volume to the solid.

1. Using **Face>Rotate**, pick the face at the left end of the solid.
2. Pick point 1 in Figure 13-21 as the first axis point and point 2 as the second point.
3. Enter a value of –30 for the rotation angle. This rotates the top edge of the tapered end away from the solid. See Figure 13-22.

Copying Faces

A mating part will now be created. This is done by first copying the face on the tapered end of the part.

1. Using the **Face>Copy** option, pick the angled face on the left end of the solid.
2. Pick one of the corners as a base point and copy the face one unit to the left. This face can now be used to create a new solid. See Figure 13-23A.
3. Draw a line four units in length on the negative X axis from the lower-right corner of the copied face. Use the **EXTRUDE** command on the copied face to create a new solid. Select the **Path** option and use the line as the extrusion path. See Figure 13-23B. If you do not use the **Path** option, the extrusion is projected perpendicular to the face.

NOTE

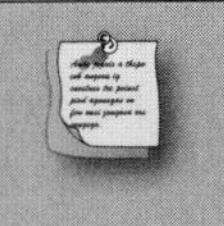

The **Face>Extrude** option of the **SOLIDEDIT** command cannot be used to turn a copied face into a solid body.

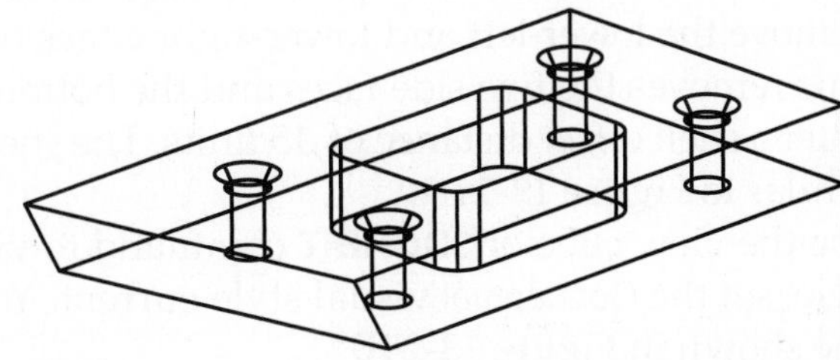

Figure 13-22.
The tapered end of the object is modified by rotating the face.

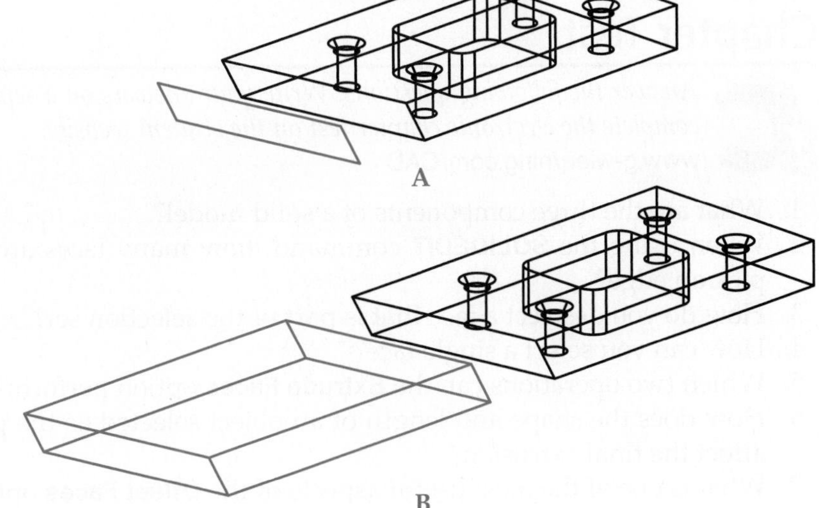

Figure 13-23.
Creating a mating part. A—The angled face is copied. B—The copied face is extruded into a solid.

Figure 13-24.
A—The shelled object. B—The viewpoint is changed and the Conceptual visual style is set current.

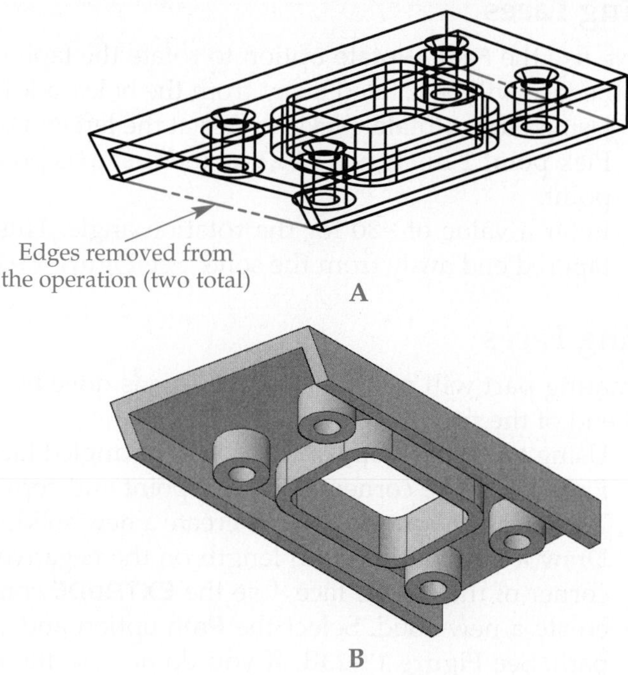

Creating a Shell

The bottom surface of the original solid will now be shelled out. Keep in mind that features such as the four holes and the rectangular slot will not be cut off by the shell. Instead, a shell will be placed around these features. This becomes clear when the operation is performed.

1. Select the **Shell** option and pick the original solid.
2. Remove the lower-left and lower-right edges of the solid. See **Figure 13-24A.** This removes the two side faces and the bottom face.
3. Enter a shell offset distance of .15 units. The shell is created and should appear similar to **Figure 13-24A.**
4. Use the view cube or **3DORBIT** command to view the solid from the bottom. Also, set the Conceptual visual style current. Your model should look like the one shown in **Figure 13-24B.**

Chapter Test

Answer the following questions. Write your answers on a separate sheet of paper or complete the electronic chapter test on the student website.
www.g-wlearning.com/CAD

1. What are the three components of a solid model?
2. When using the **SOLIDEDIT** command, how many faces are highlighted if you pick an edge?
3. How do you deselect a face that is part of the selection set?
4. How can you select a single face?
5. Which two operations can the **Extrude Faces** option perform?
6. How does the shape and length of an object selected as the path of an extrusion affect the final extrusion?
7. What is one of the most useful aspects of the **Offset Faces** option?

8. How do positive and negative offset distance values affect the volume of the solid?
9. How is a single object, such as a cylinder, affected by entering a positive taper angle when using the **Taper Faces** option?
10. When a shape is imprinted onto a solid body, which component of the solid does the imprinted object become and how can it be used?
11. What does the **XEDGES** command do?
12. How does the **Shell** option affect a solid that contains internal features such as holes, grooves, and slots?
13. How can you determine if an object is a valid 3D solid?
14. Describe two ways to change the view of your model while you are inside of a command.
15. How can you extrude a face in a straight line, but not perpendicular to the face?

Drawing Problems

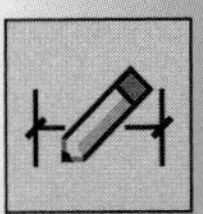

1. Complete the tutorial presented in this chapter. Then, perform the following additional edits to the original solid.
   A. Lengthen the right end of the solid by .5 units.
   B. Taper the right end of the solid with the same taper angle used on the left end, but taper it in the opposite direction.
   C. Fillet the two long, top edges of the solid using a fillet radius of .2 units.
   D. Rotate the face at the right end of the solid with the same rotation angle used on the left end, but rotate it in the opposite direction.
   E. Save the drawing as P13_01.

2. Construct the solid part shown below using as many **SOLIDEDIT** options as possible. After completing the object, make the following modifications.
   A. Lengthen the 1.250″ diameter feature by .250″.
   B. Change the .750″ diameter hole to .625″ diameter.
   C. Change the thickness of the .250″ thick flange to .375″ (toward the bottom).
   D. Extrude the end of the 1.250″ diameter feature .250″ with a 15° taper inward.
   E. Save the drawing as P13_02.

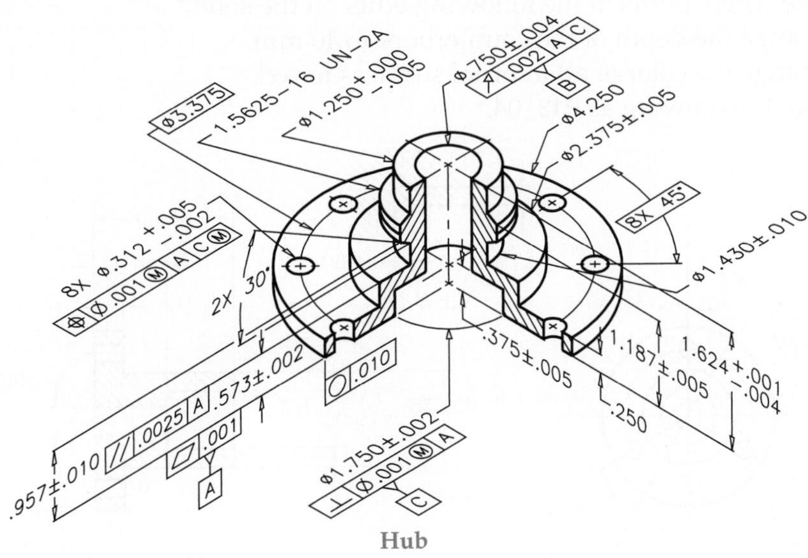

Hub

3. Construct the solid part shown below. Then, perform the following edits on the solid using the **SOLIDEDIT** command.
   A. Change the diameter of the hole to 35.6/35.4.
   B. Add a 5° taper to each inner side of each tooth (the bottom of each tooth should be wider while the top remains the same).
   C. Change the width of the 4.8/4.0 key to 5.8/5.0.
   D. Save the drawing as P13_03.

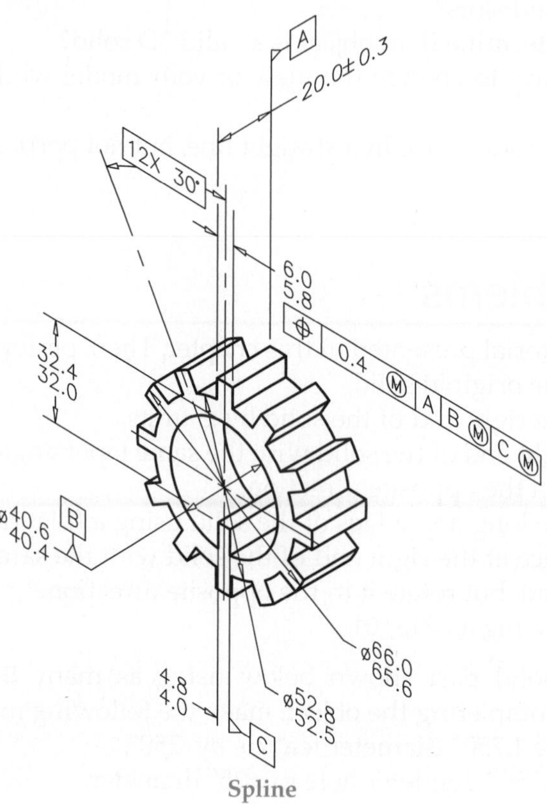

Spline

4. Construct the solid part shown below using as many **SOLIDEDIT** options as possible. Then, perform the following edits on the solid.
   A. Change the depth of the counterbore to 10 mm.
   B. Change the color of all internal surfaces to red.
   C. Save the drawing as P13_04.

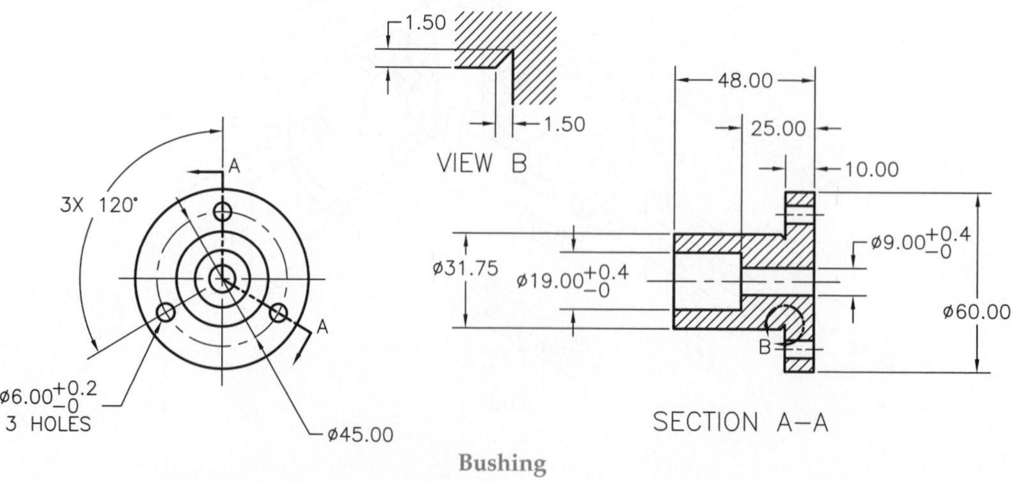

Bushing

Drawing Problems - Chapter 13

5. Construct the solid part shown below using as many **SOLIDEDIT** options as possible. Then, perform the following edits on the solid.
   A. Change the 2.625″ height to 2.325″.
   B. Change the 1.625″ internal diameter to 1.425″.
   C. Taper the outside faces of the .875″ high base at a 5° angle away from the part. Hint: The base cannot be directly tapered.
   D. Save the drawing as P13_05.

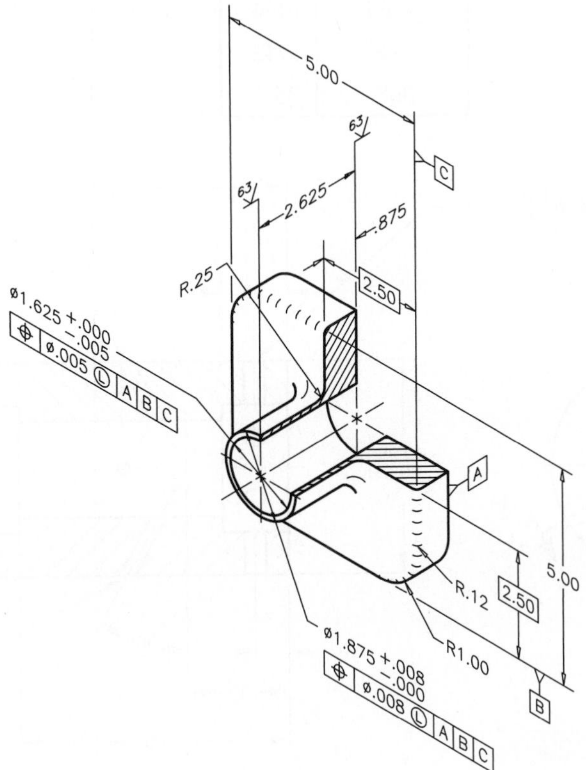

Thrust Washer

6. Construct the solid part shown below using as many **SOLIDEDIT** options as possible. Then, change the dimensions on the model as follows. Save the drawing as P13_06.

Existing	New
100	106
80	82
Ø60	Ø94
Ø40	Ø42
30°	35°

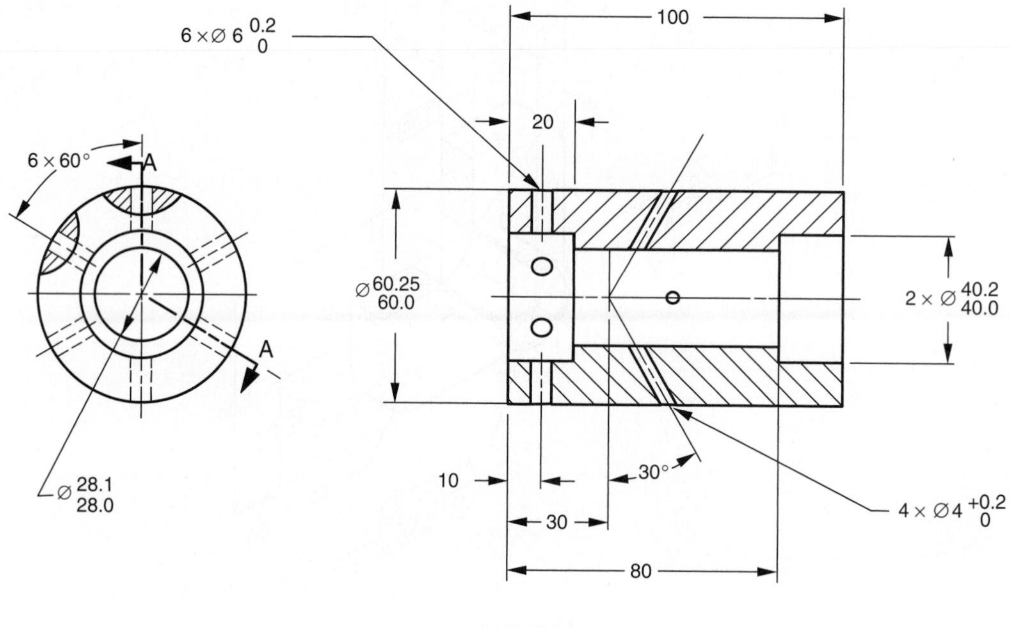

Nozzle

CHAPTER 14

Solid Model Display and Analysis

Learning Objectives

After completing this chapter, you will be able to:

- ✓ Control the display of solid models.
- ✓ Construct a 3D section plane through a solid model.
- ✓ Adjust the size and location of section planes.
- ✓ Create a dynamic section of a 3D solid model.
- ✓ Construct 2D and 3D section blocks.
- ✓ Create a flat, 2D projection of a 3D solid model.
- ✓ Create a multiview layout of a solid model using **FLATSHOT** or **SOLVIEW** and **SOLDRAW**.
- ✓ Construct a profile of a solid using **SOLPROF**.
- ✓ Perform an analysis of a solid model.
- ✓ Export and import solid model data.

Certain aspects of a solid model's appearance are controlled by the **ISOLINES**, **DISPSILH**, and **FACETRES** system variables. The **ISOLINES** system variable controls the number of lines used to define solids in wireframe displays. The **FACETRES** system variable controls the number of lines (smoothness) used to define solids in hidden and shaded displays. The **DISPSILH** system variable is used to display a silhouette edge.

Internal features of the model can be shown using the **SECTIONPLANE** command. This command can create 2D and 3D section views on an object. The **FLATSHOT** command creates a 2D projection of the current view. This chapter also looks at how sections can be combined with 2D projections created with the **SOLVIEW** and **SOLDRAW** commands to create a drawing layout for plotting. In addition, this chapter covers how a profile of a solid can be created using the **SOLPROF** command.

Controlling Solid Model Display

AutoCAD solid models can be displayed as wireframes, with hidden lines removed, shaded, or rendered. A wireframe is the default display when a drawing is started based on the acad.dwt template and is the quickest to display. The hidden, shaded, and rendered displays require a longer regeneration time. When a drawing is started based on the acad3D.dwt or acadiso3D.dwt template, the default display is the Realistic visual style, which is a shaded display.

Isolines

ISOLINES

Type
ISOLINES

The appearance of a solid model in a wireframe display is controlled by the **ISOLINES** system variable. ***Isolines*** represent the edges and curved surfaces of a solid model. This setting does *not* affect the final shaded or rendered object. However, if the Show property in the **Edge Setting** category in the **Visual Styles Manager** palette is set to Isolines, isolines are displayed when the visual style is set current. The default **ISOLINES** value is four. It can have a value from zero to 2047. All solid objects in the drawing are affected by changes to the **ISOLINES** value, as are all visual styles set to display isolines. Figure 14-1 illustrates the difference between **ISOLINES** settings of four and 12.

The setting of the **ISOLINES** system variable can be changed in the **Visual Styles Manager** palette. For the 2D Wireframe visual style, change the Contour lines property in the **2D Wireframe Options** category. For all other visual styles, change the Number of lines property in the **Edge Settings** category. See Figure 14-2A. The Show property in the **Edge Setting** category must be set to Isolines to display the Number of lines property. The **ISOLINES** setting can also be changed in the **Contour lines per surface** text box found in the **Display resolution** area of the **Display** tab in the **Options** dialog box. See Figure 14-2B. Or, type ISOLINES at the command prompt and then enter a new value.

Creating a Display Silhouette

DISPSILH

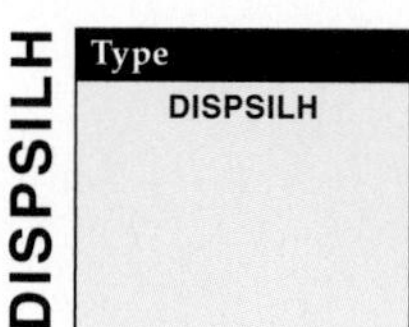

When the Show property in the **Edge Setting** category of a 3D visual style is set to Facet Edges, objects are defined by ***tessellation lines.*** The **Facet Edges** and **Isolines** buttons in the **Visual Styles** panel on the **View** tab of the ribbon also control the Show setting. When the **HIDE** command is used with the 2D Wireframe visual style current, tessellation lines are also displayed. The number of tessellation lines is controlled by the **FACETRES** system variable, which is discussed in the next section.

In the 2D Wireframe visual style, a model can also appear smooth with only a silhouette displayed, similar to the Hidden visual style. This is controlled by the **DISPSILH** (display silhouette) system variable. The **DISPSILH** system variable has two values, 0 (off) and 1 (on). Figure 14-3 shows solids with **DISPSILH** set to 0 and 1 after setting the 2D Wireframe visual style current and then using **HIDE**.

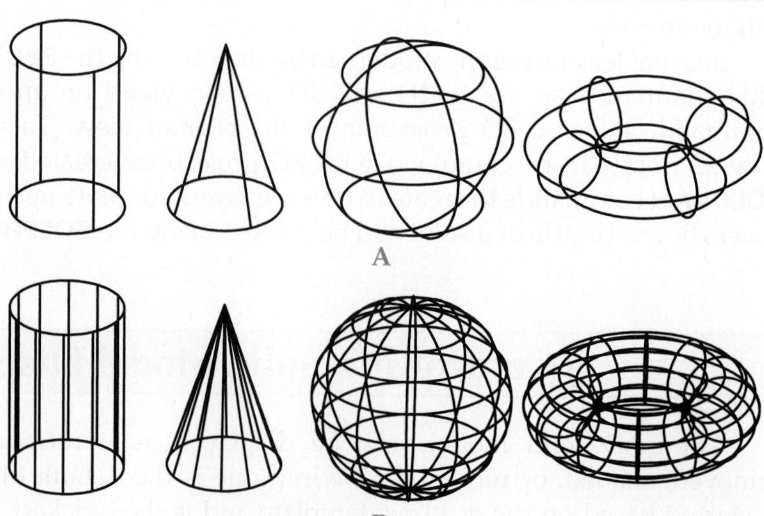

Figure 14-1.
Isolines define curved surfaces.
A—**ISOLINES** = 4.
B—**ISOLINES** = 12.

Figure 14-2.
A—The **ISOLINES** and **DISPSILH** values can be set in the **Visual Styles Manager**.
B—The **ISOLINES**, **FACETRES**, and **DISPSILH** values can be set in the **Options** dialog box.

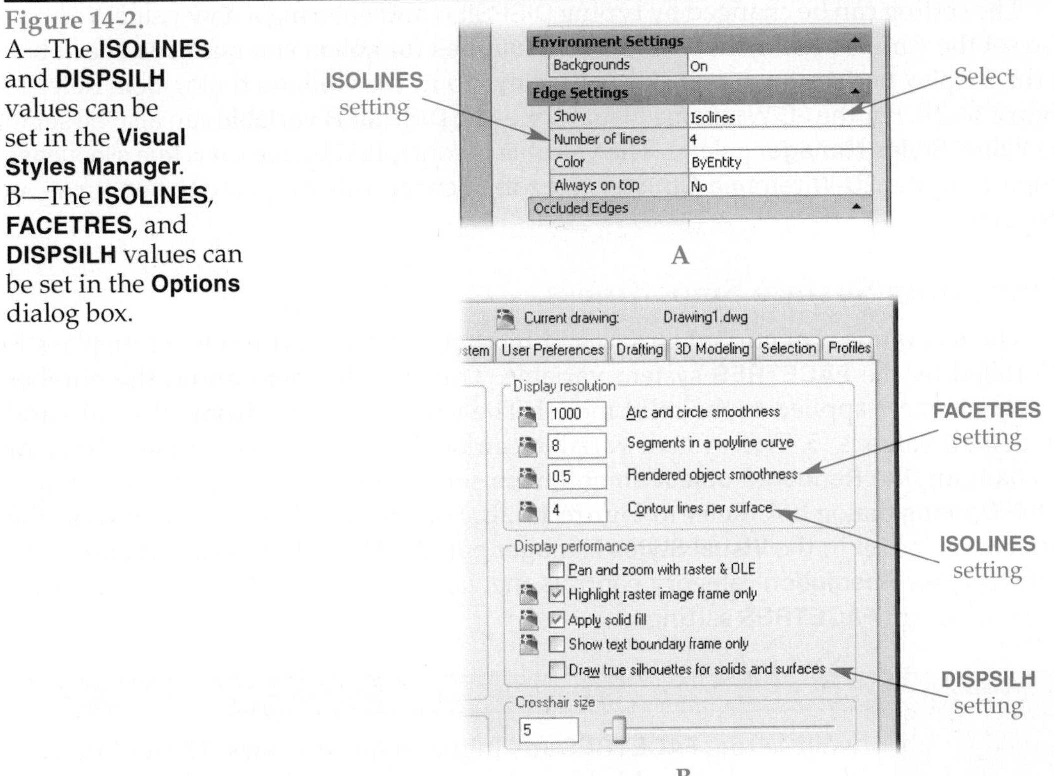

Figure 14-3.
A—The **HIDE** command used when **DISPSILH** is set to 0. Objects are displayed faceted.
B—The **HIDE** command used when **DISPSILH** is set to 1. Facets are eliminated and only the silhouette is displayed.

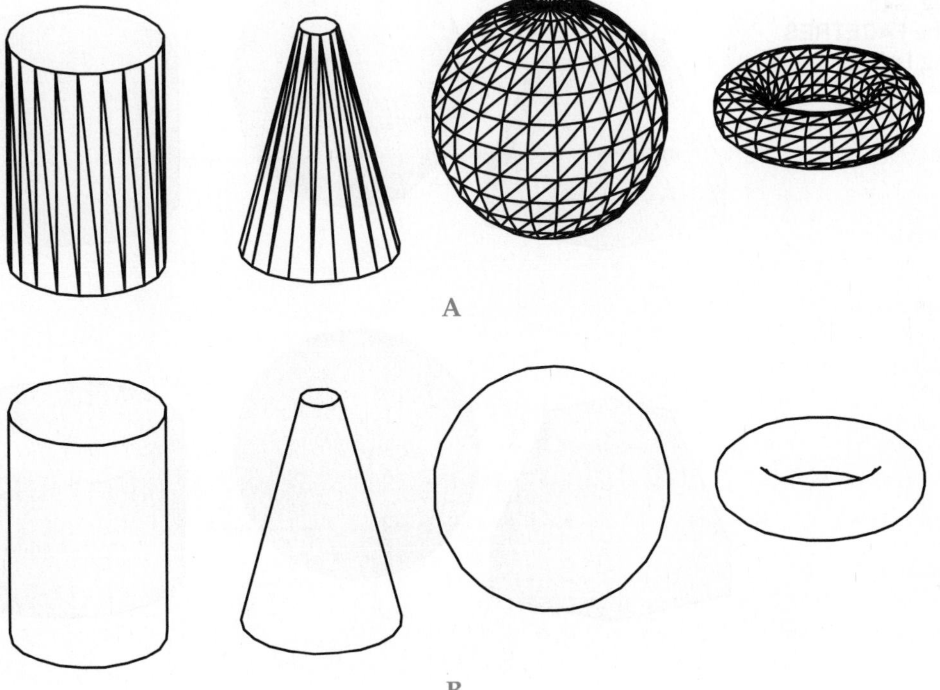

The setting can be changed by typing DISPSILH and entering a new value. You can also set the variable using the **Draw true silhouettes for solids and surfaces** check box in the **Display performance** area of the **Display** tab in the **Options** dialog box. Refer to Figure 14-2B. For the 2D Wireframe visual style, the **DISPSILH** variable can also be set in the **Visual Styles Manager** palette. The variable is controlled by the Draw true silhouettes property in the **2D Wireframe Options** category. Setting this property to Yes turns on silhouettes.

Controlling Surface Smoothness

The smoothness of curved surfaces in hidden, shaded, and rendered displays is controlled by the **FACETRES** system variable. This variable determines the number of polygon faces applied to the solid model. The value can range from .01 to 10.0 and the default value is .5. This system variable can be changed by typing FACETRES or by changing the **Rendered object smoothness** setting in the **Display resolution** area in the **Options** dialog box. Refer to Figure 14-2B. For the 2D Wireframe visual style, the variable can be set in the **Visual Styles Manager** palette. The Solid Smoothness property in the **Display Resolution** category controls the variable. Figure 14-4 shows the effect of two different **FACETRES** settings.

CAUTION

Avoid setting **FACETRES** any higher than necessary. The additional edges and faces added by a higher setting will slow down system performance. Always use the lowest setting that will produce the results required by the project.

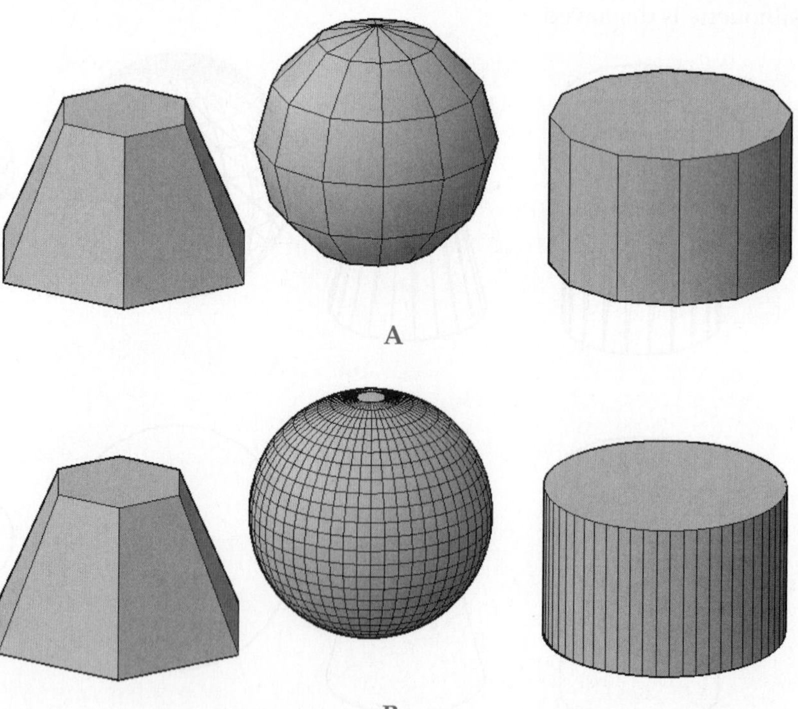

Figure 14-4.
A—The **FACETRES** setting is .5.
B—The **FACETRES** setting is 5.0.

Exercise 14-1

Complete the exercise on the student website.
www.g-wlearning.com/CAD

Creating Section Planes

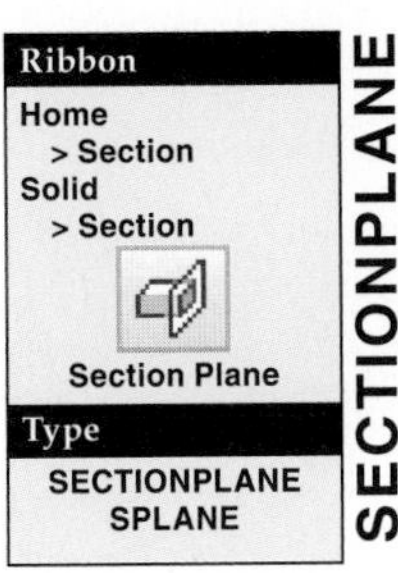

The **SECTIONPLANE** command offers a powerful visualization and display tool. It allows you to construct a section plane, known as an AutoCAD ***section object,*** that can then be used as a plane to cut through a 3D model. Once the section object is drawn, it can be moved to any location, jogs can be added to it, and it can be rendered "live" so that internal features and sectioned material are dynamically visible as the cutting plane is moved. A variety of section settings allow you to customize the appearance of section features. Additionally, you can generate 2D sections/elevations or 3D sections that can be inserted into the drawing as a block. Once the command is initiated, you are prompted to select a face, the first point on the section object, or to enter an option. Sectioning options are located in the **Section** panel of the **Home** or **Solid** tab on the ribbon.

Pick a Face to Construct a Section Plane

The simplest way to create a section plane is to pick a flat face on the 3D object. Once the command is initiated, move the pointer until the face you wish to select is highlighted, then pick it. A transparent section object is placed on the face you selected and the model is cut at the plane. See Figure 14-5. The section plane can now be moved to create a section anywhere along the 3D model.

Pick Two Points to Construct a Section Plane

A second method for defining a section plane is to pick two points through which the section object passes. The section object is perpendicular to the XY plane of the current UCS. When the command is initiated, pick the first point, which cannot be on a face. See P1 in Figure 14-6. It may be best to turn off dynamic UCSs or you could end up picking a face as the first point instead of a point. After picking the first point, move the pointer and notice that the section plane rotates about the first point. Next, pick the second point (P2) to define a line that cuts through the model. After the second point

Figure 14-5.
Creating a section object on a face. A—The object before the face is selected. B—The section object is created.

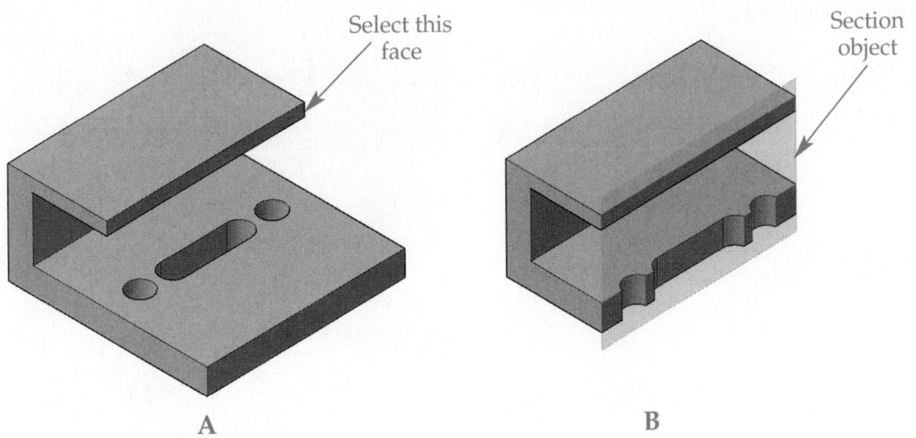

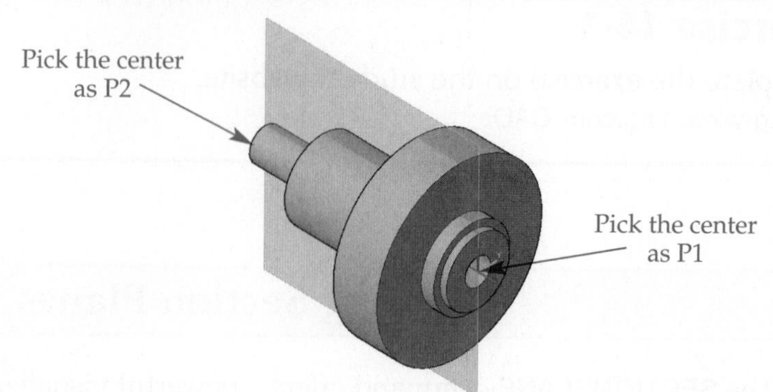

Figure 14-6.
Creating a section object by selecting two points.

is picked, the section object is created. The section plane extends just beyond the edges of the model.

NOTE

When the section object is created by picking two points, notice that the model is not automatically cut as it is when a face is selected. This is because *live sectioning* is not turned on when picking two points, but is turned on when picking a face. To turn live sectioning on or off, select the section object, right-click, and select **Activate live sectioning** from the shortcut menu. Live sectioning is discussed later in this chapter.

Pick Multiple Points to Construct a Section Plane

The previous method accepts only two points to construct a single section plane. Using the **Draw section** option, you can specify multiple points in order to create section plane *jogs*. In engineering drawing terminology, a section object drawn in this manner can represent an ***offset*** or ***aligned*** section plane.

Once the command is initiated, select the **Draw section** option. Pick the start point, using object snaps if necessary. See **Figure 14-7**. Continue picking points as needed. After picking the last point to define the section plane, press [Enter]. You are then prompted to specify a point in the direction of the section view. This point is on the

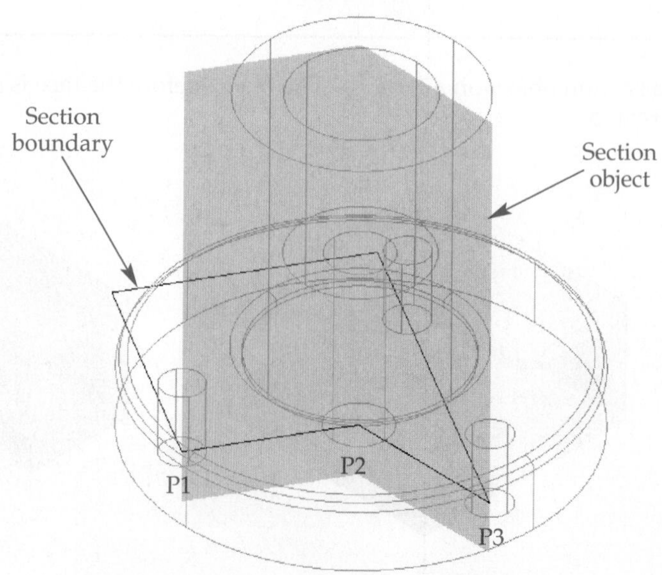

Figure 14-7.
Using the **Draw** option of the **SECTIONPLANE** command to create a section object with multiple segments.

opposite side of the section object as the viewer. Pick a point on the model using object snaps if necessary. The section plane is created.

Notice in Figure 14-7 that the pick points created a section object that does not extend beyond the boundary of the model because the hole centers were selected. Also, the command "squares up" the section boundary to create a closed profile. Using section object grips, the section plane can be easily edited to include the entire solid. This is discussed in detail later in the chapter.

Create Orthographic Section Planes

The **Orthographic** option enables you to quickly place a section plane through the front, back, top, bottom, left, or right side of the object. See Figure 14-8. The origin is the center point of all objects in the model. Once the command is initiated, select the **Orthographic** option. Then, specify which orthographic plane you want to use as the section plane. The section object is created and all objects in the drawing are affected by it.

You may encounter a situation in which there is more than one solid on the screen and you want to use the **Orthographic** option to create a section object based on just one object. In this case, create a new layer, move objects you do not want to section to this layer, and then freeze the layer. The section object will be created based on the object that is visible. However, if the section plane passes through the objects on the frozen layers, those objects will be sectioned when the layers are thawed. A section plane affects all visible objects.

Figure 14-8. Examples of orthographic **SECTIONPLANE** options. A—Top. B—Bottom. C—Left. D—Right.

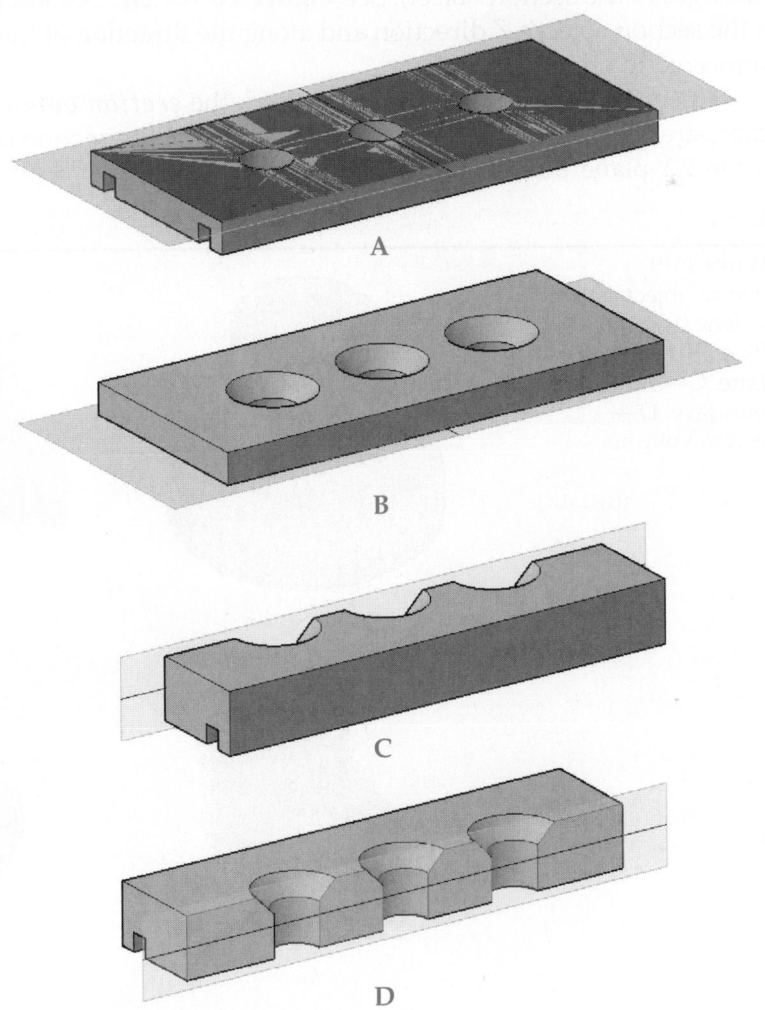

Exercise 14-2

Complete the exercise on the student website.
www.g-wlearning.com/CAD

Editing and Using Section Planes

A wide range of section object editing and display options are available. However, there is not ribbon or command line access to these procedures. Instead, you must first select the section object, then right-click to display the shortcut menu. From this menu, you can access all of the display and editing functions that apply to the section object.

Section Object States

There are three possible states for the section object created by the **SECTIONPLANE** command—section plane, section boundary, and section volume. See Figure 14-9. The section object can be changed from one state to another. Depending on which state is active, the section object produces different results on the solid(s).

When the section object is created by picking a face, picking two points, or using the **Orthographic** option, the object is in the ***section plane state.*** A transparent plane is displayed on each segment of the section object and a line connects the pick points (or the edges of the section object). See Figure 14-9B. The section plane extends infinitely in the section object's Z direction and along the direction of the object segment (unless connected to other segments).

When the **Draw section** option is used, the ***section boundary state*** is applied. A transparent plane is displayed on each segment of the section object. A 2D box extends to the XY-plane boundaries of the section object. See Figure 14-9C. The sectioned

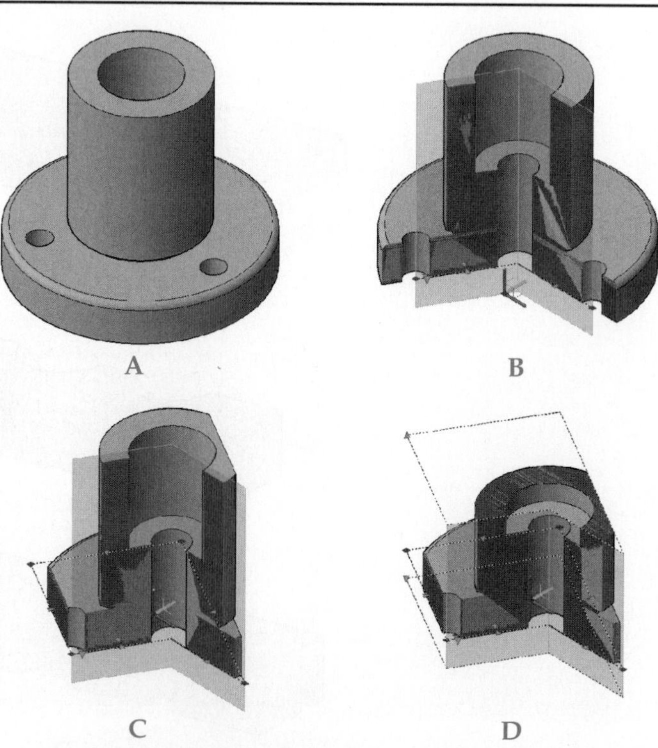

Figure 14-9.
Section object states. A—The original object. B—Section plane. C—Section boundary. D—Section volume.

object fits inside of this footprint. The section plane extends infinitely in the section object's Z direction.

The ***section volume state*** is not applied when the section object is created. The section object must be switched to this state once the object is created, as described in the next section. A transparent plane is displayed on each segment of the section object. In addition, a 3D box extends to the XYZ boundaries of the section object. The sectioned object fits inside of this box. See **Figure 14-9D.**

Section Object Properties

Once created, the properties of the section object can be changed. The **Section Object** category in the **Properties** palette contains properties specific to the section object. See **Figure 14-10.** These properties are described in the next sections.

Name

The default name of the first section object is Section Plane(1). Subsequent section planes are sequentially numbered, such as Section Plane(2), Section Plane(3), and so on. It may be beneficial to rename section objects so the name is representative of the section. For example, Front Half Section is much more descriptive than Section Plane(1). To rename a section object, select the Name property. Then, type a new name in the text box.

Type

As discussed earlier, the section object is in one of three states. The three states are section plane, section boundary, and section volume. To change the state of the section object, select the Type property. Then, pick the state in the drop-down list. The state can also be changed using the menu grip on the section object.

Live section

Live sectioning is a tool that allows you to dynamically view the internal features of a solid, surface, or region as the section object is moved. This tool is discussed later in the chapter. To turn live sectioning on or off, select the Live Section property. Then, pick either Yes (on) or No (off) in the drop-down list. This is the same as turning live sectioning on or off using the shortcut menu or the ribbon.

Figure 14-10.
The properties of a section object can be changed in the **Properties** palette.

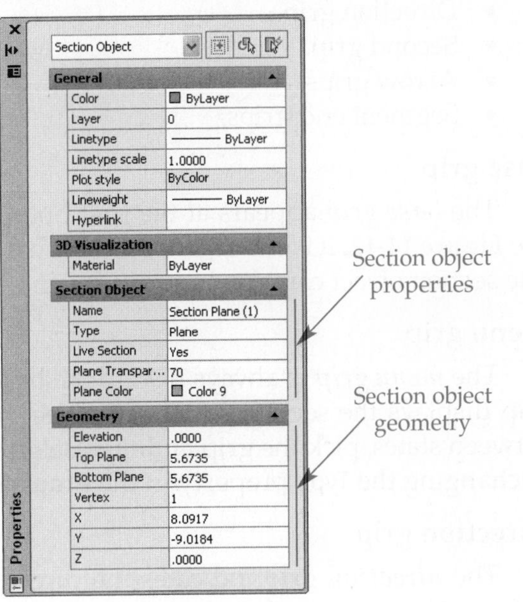

Figure 14-11.
A—The Plane Transparency property of the section plane object is set to 1 (or 1% transparent). B—The Plane Transparency property of the section plane object is set to 85 (or 85% transparent).

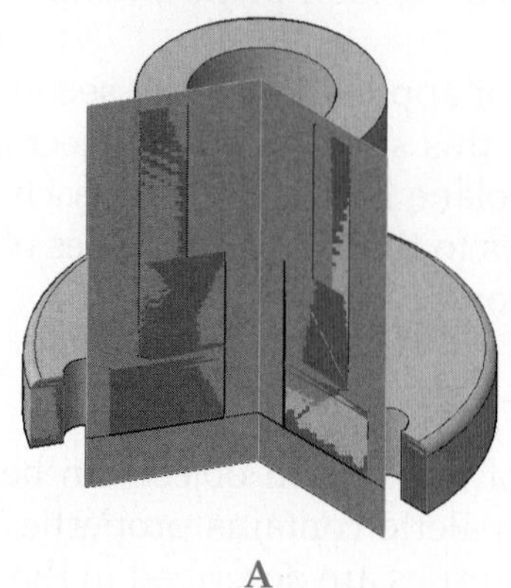

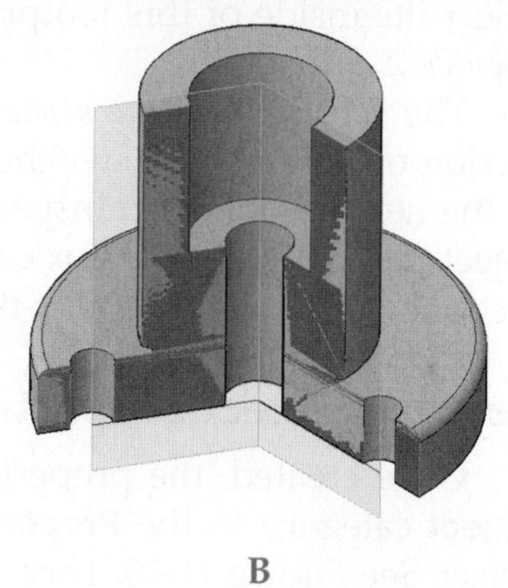

Plane transparency

The Plane Transparency property determines the opacity of the plane for the section object. The property value can range from 1 to 100. The lower the value, the more opaque the section plane object. See **Figure 14-11.**

Plane color

The plane of the section object can be set to any color available in the **Select Color** dialog box. To change the color, select the Plane Color property and then select a color from the drop-down list. To choose a color in the **Select Color** dialog box, pick the Select Color... entry in the drop-down list. This property only affects the plane of the section object, not the lines defining the boundary, volume, or section line. The color of these lines is controlled by the Color property in the **General** category.

Editing the Section Object

When the translucent planes of a section object or the lines representing the section object state are picked, grips are displayed. The specific grips displayed are related to the current section object state. Refer to the grips shown in **Figure 14-9.** The types of grips are:

- Base grip.
- Menu grip.
- Direction grip.
- Second grip.
- Arrow grips.
- Segment end grips.

Base grip

The ***base grip*** appears at the first point picked when defining the section object. See **Figure 14-12.** It is the grip about which the section object can be rotated and scaled. The section object can also be moved using this grip.

Menu grip

The ***menu grip*** is always next to the base grip. Refer to **Figure 14-12.** Picking this grip displays the section state menu. See **Figure 14-13.** To switch the section object between states, pick the grip and then select the state from the menu. This is the same as changing the Type property in the **Properties** palette.

Direction grip

The ***direction grip*** indicates the direction in which the section will be viewed. Refer to **Figure 14-12.** Pick the grip to flip the view 180°. The direction grip also shows the direction of the live section. Live sectioning is discussed later in this chapter.

Figure 14-12
The types of grips displayed on a section object.

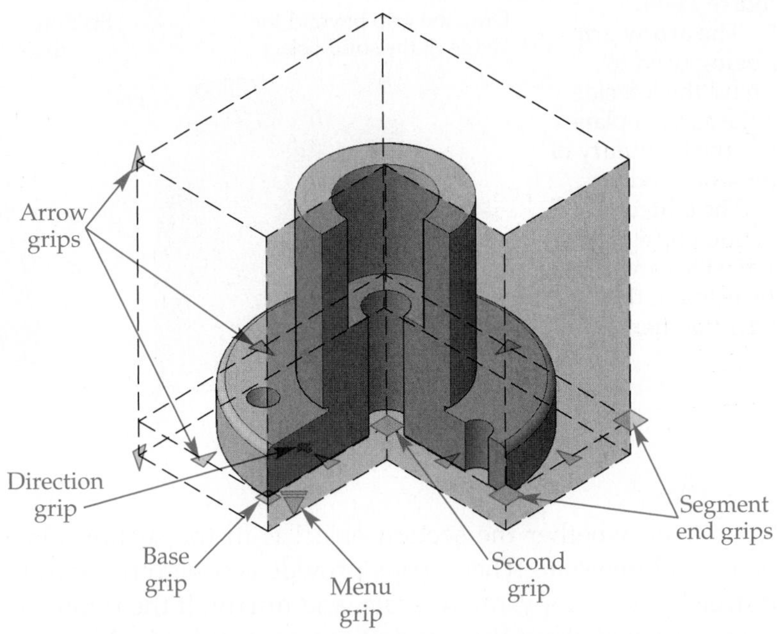

Figure 14-13.
Changing section object states.

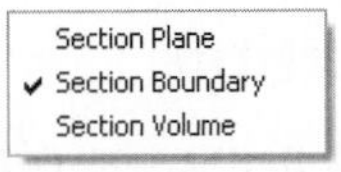

Section Plane
✔ Section Boundary
Section Volume

Second grip

The *second grip* appears at the second point picked when defining the section object. See Figure 14-12. The section object can be rotated and stretched about the base grip using the second grip.

Arrow grips

Arrow grips are located on all of the lines that represent the section plane, boundary, and volume. Refer to Figure 14-12. These grips are used to lengthen or shorten the section plane object segments or adjust the height of the section volume. The arrow grips at the top and bottom of the boundary box are used to change the height. Regardless of where the pointer is moved, the section object only extends in the segment's current plane. Changing the length of one segment of the section plane does not affect other segments.

In Figure 14-7, you saw an example of using the **Draw section** option to create a section object. The way in which the section object was created resulted in the section plane not extending beyond the solid object. This can quickly be corrected using the arrow grips. Notice in Figure 14-14 that the arrow grip is being used to extend the left side of the section plane past the boundary of the solid model. This allows any subsequent section views to display the entire object rather than just a portion of it. The right side of the section plane can be extended in the same manner using the opposite arrow grip.

The arrow grips located on the line segments of the section plane move the position of the section plane. As a segment of the section plane is moved, it maintains its angular relationship and connection to any adjacent section plane segment.

Segment end grips

The *segment end grips* are located at the end of each line segment defining the section object state. Refer to Figure 14-12. The number of displayed segment end grips

Figure 14-14.
A—The arrow grip is being used to extend the left side of the section plane past the boundary of the solid model.
B—The edited section object. The right side can be corrected in the same manner.

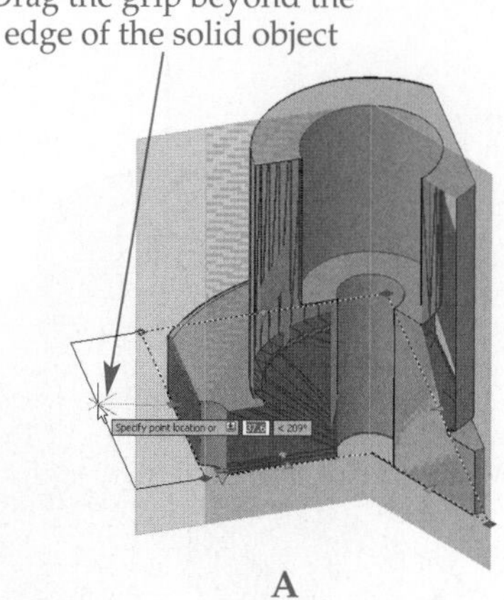

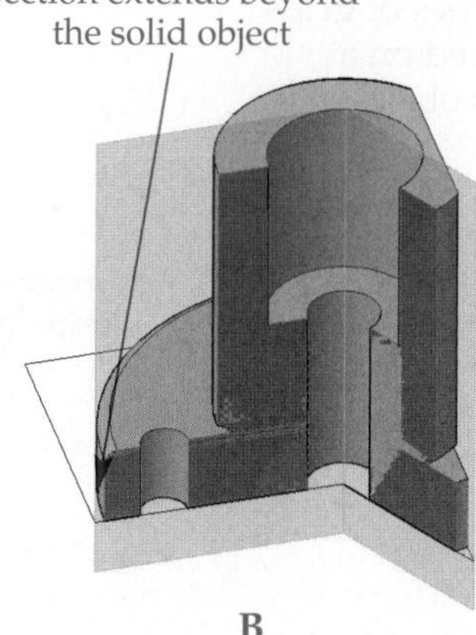

depends on whether the section object is in the section plane, section boundary, or section volume state. These grips provide access to the standard grip editing options of stretch, move, copy, rotate, scale, and mirror. If the rotate option is used, the section plane is rotated about the selected segment end grip. Moving a segment end grip can change the angle between section plane segments.

Adding Jogs to a Section

SECTIONPLANEJOG

Ribbon
Home
> Section
Solid
> Section
Add Jog
Type
SECTIONPLANEJOG
JOGSECTION

You can quickly add a jog, or offset, to an existing section object. First, select the section object. Then, right-click to display the shortcut menu and select **Add jog to section**. You can also enter the **SECTIONPLANEJOG** command. You are then prompted:

Specify a point on the section line to add jog:

Select a point directly on the section line. If any object snap is active, the **Nearest** object snap is temporarily turned on to ensure you pick the line. Once you pick, the jog is automatically added perpendicular to the line segment. See **Figure 14-15A.**

It is not critical that you pick the exact location on the line where you want the jog to occur. Remember, grips allow you to easily adjust the section plane location. Notice in **Figure 14-15B** that the second jog barely cuts through the first hole. The intention is to run the section plane through the middle of the hole. To fix this, drag the arrow grip so the section plane segment is in the desired location, **Figure 14-15C.**

PROFESSIONAL TIP

If the section plane is not properly located, you can quickly change it. Simply pick the section object and right-click to display the shortcut menu. Then, select **Move**, **Scale**, or **Rotate** from the shortcut menu. Finally, adjust the section object location as needed.

Exercise 14-3

Complete the exercise on the student website.
www.g-wlearning.com/CAD

Figure 14-15.
Adding a jog to a section object. A—Pick a point on the section line to add a jog. B—The jog is added, but it is not in the proper location. C—Using the arrow grip, the jog is moved to the proper location.

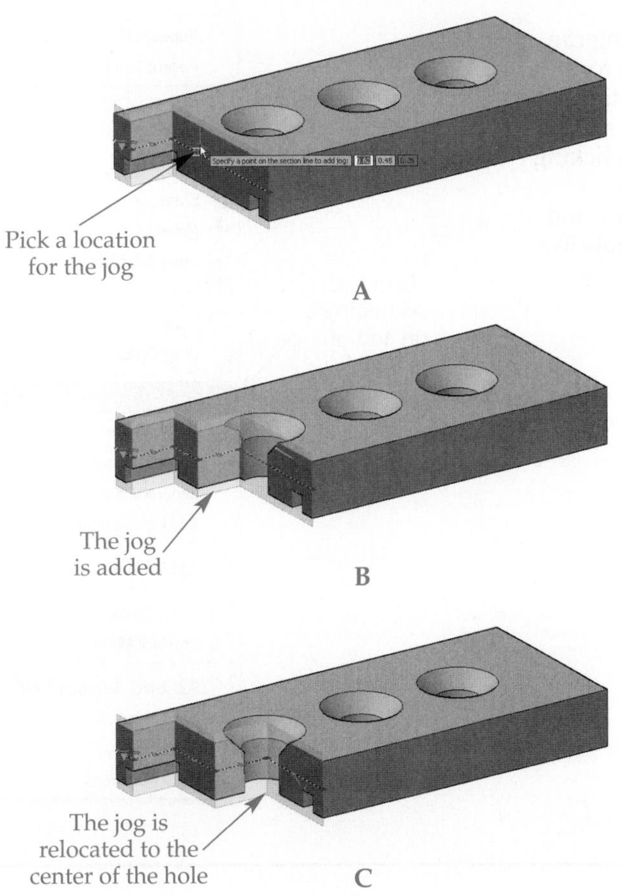

Live Sectioning

Live sectioning is a tool that allows you to view the internal features of 3D solids, surfaces, and regions that are cut by the section plane of the section object. The view is dynamically updated as the section object is moved. This tool is used to visualize internal features and for establishing section locations from which 2D and 3D section views can be created. Live sectioning is either on or off.

As you have seen, if the section plane is created by selecting a face, live sectioning is automatically turned on. However, when picking two points or using the **Draw** option of the **SECTIONPLANE** command, live sectioning is off. Live sectioning can be turned on and off for individual section objects, but only one section object can be "live" at any given time.

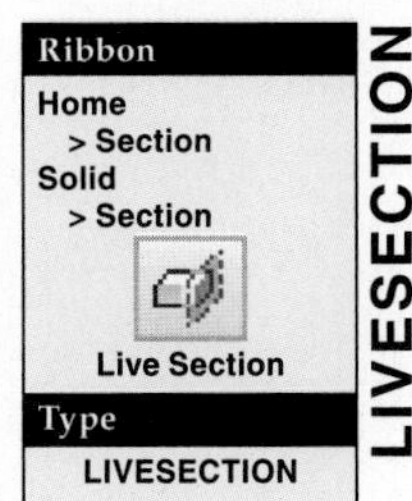

LIVESECTION

To turn live sectioning on or off, select the section object. Then, right-click to display the shortcut menu and pick **Activate live sectioning**. See Figure 14-16. A check mark appears next to the menu item when live sectioning is on. You can also use the ribbon or enter the **LIVESECTION** command and select the section object to toggle the on/off setting. When live sectioning is turned on, the material behind the viewing direction of the section plane is removed. The cross section of the 3D object is shown in gray (by default) and the internal shape of the 3D object is visible.

A wide variety of options allow you to change the appearance of not only the live sectioning display, but also of 2D and 3D section blocks that can be created from the sectioned display. These settings are found in the **Section Settings** dialog box. See Figure 14-17. To open this dialog box, select the section object, right-click, and pick **Live section settings...** in the shortcut menu. You can also pick the dialog box launcher button in the lower-right corner of the **Section** panel in the **Home** or **Solid** tab of the ribbon.

Figure 14-16.
Live sectioning can be turned on for any section state by selecting the section object, right-clicking to display the shortcut menu, and picking **Activate live sectioning**.

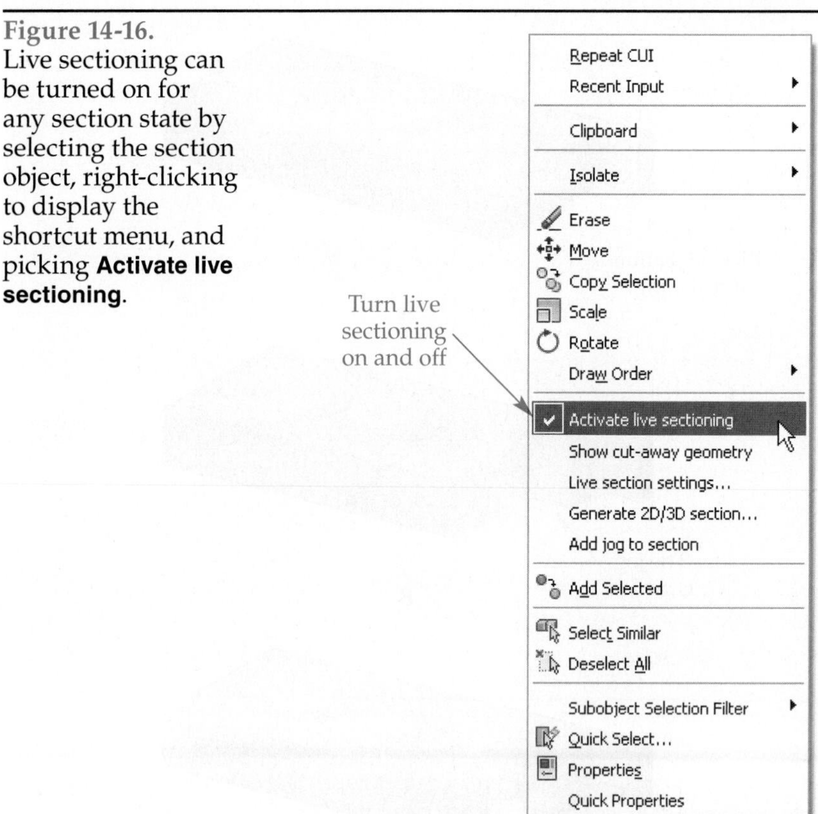

Figure 14-17.
Section settings. A—For a 2D block. B—For a 3D block. C—For live sectioning.

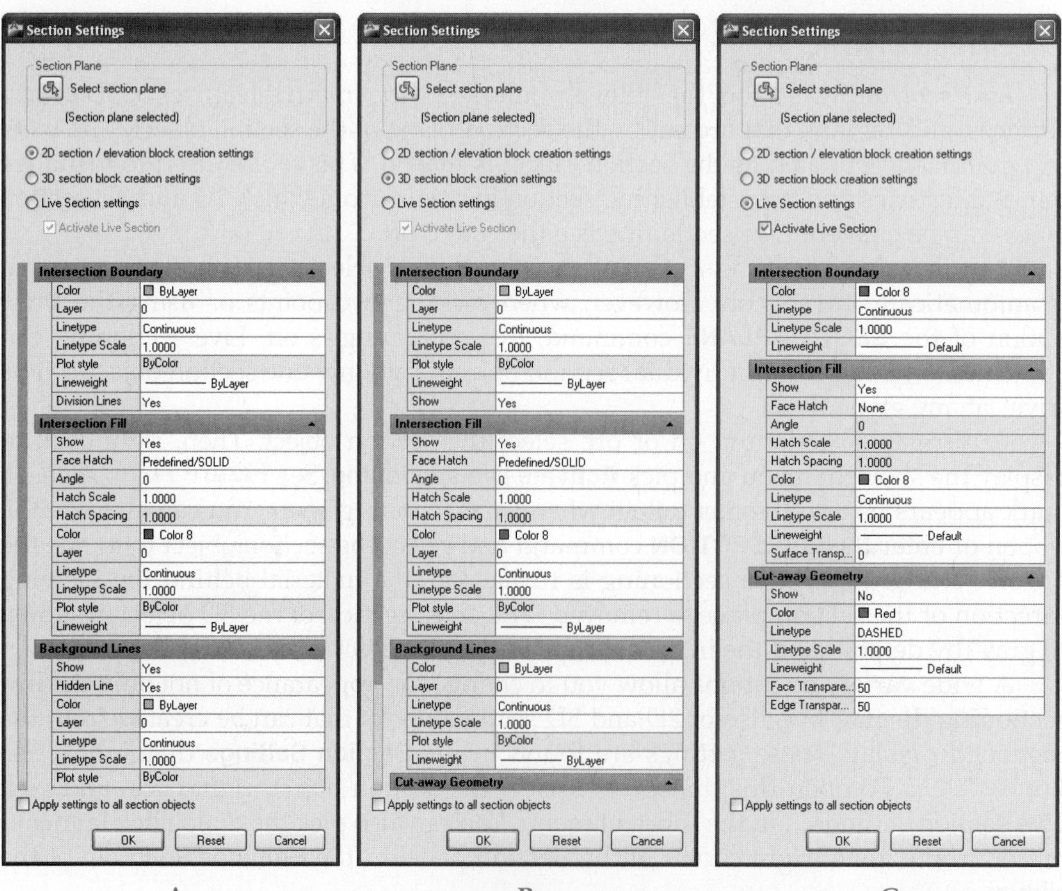

To change the settings for live sectioning, pick the **Live section settings** radio button at the top of the **Section Settings** dialog box. The categories displayed in the dialog box contain properties related to live sectioning. Settings for 2D and 3D sections and elevations are discussed later in this chapter.

The three categories for live sectioning are **Intersection Boundary**, **Intersection Fill**, and **Cut-Away Geometry**. To display a brief description of any property, hover the cursor over the property in the **Section Settings** dialog box. The description is displayed in a tooltip. A check box at the bottom of the **Section Settings** dialog box allows you to apply the properties to all section objects or to just the selected section object.

PROFESSIONAL TIP

Live sectioning can be quickly turned on or off by double-clicking on the section plane object.

Intersection boundary

The intersection boundary is where the model is intersected by the section object. It is represented by line segments. You can set the color, linetype, linetype scale, and line weight of the intersection boundary lines. Any linetype loaded into AutoCAD can be used.

Intersection fill

The intersection fill is the material visible on the model surface where the section object cuts. It is displayed as a solid fill, by default. Any hatch pattern available in AutoCAD can be used as the intersection fill. The angle, hatch scale, and hatch spacing, and color can be set. In addition, the linetype, linetype scale, and line weight can be changed. The fill pattern can even be set to be transparent.

Cutaway geometry

The cutaway geometry is the part of the model removed by the live sectioning. By default, this geometry is not displayed. Changing the Show property to Yes displays the geometry. See **Figure 14-18.** You can set the color, linetype, linetype scale, and line weight of the lines representing the cutaway geometry. In addition, the Face Transparency and Edge Transparency properties allow you to create a see-through effect, as seen in **Figure 14-19.** Each of these two properties is set to 50 by default.

Figure 14-18.
A—The intersection fill can be displayed as a hatch pattern in any specified color. B—The cutaway geometry is displayed.

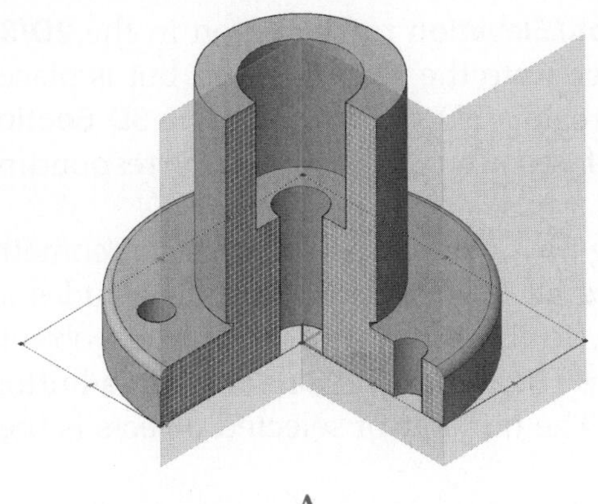

A

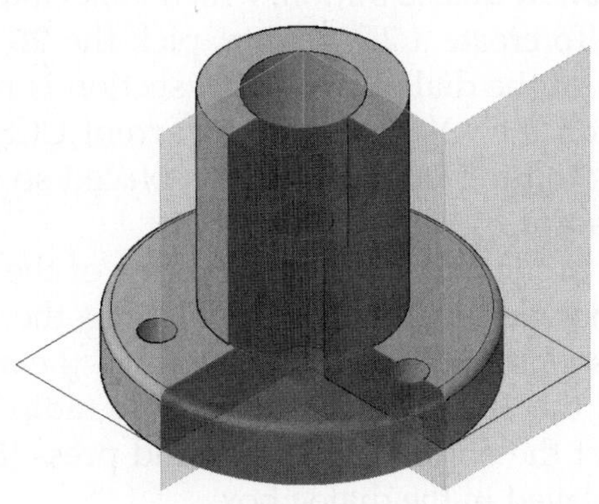

B

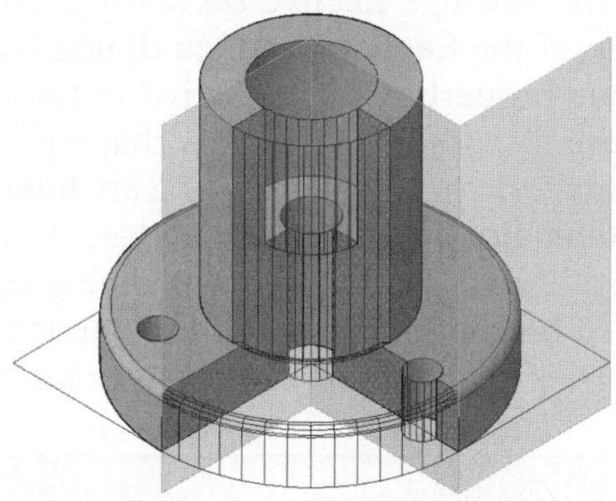

Figure 14-19.
The cutaway geometry is displayed with 100% transparent faces and solid black lines.

PROFESSIONAL TIP

You can also display the cutaway geometry without using the **Section Settings** dialog box. Select the section object, right-click, and pick **Show cut-away geometry** in the shortcut menu. Live sectioning must be on.

Exercise 14-4

Complete the exercise on the student website.
www.g-wlearning.com/CAD

Generating 2D and 3D Sections and Elevations

SECTIONPLANETOBLOCK

Ribbon
Home
> Section
Solid
> Section
Generate Section
Type
SECTIONPLANE-TOBLOCK
GENERATESECTION

The **SECTIONPLANE** command provides a fast and efficient method of creating sections. The sections can be either 2D or 3D. Not only can the sections be displayed on the current drawing, they can also be exported as a file that can then be used in any other drawing or document for display, technical drawing, or manufacturing purposes.

Creating sections

To create a section, enter the **SECTIONPLANETOBLOCK** command. You can also select the section object, right-click, and pick **Generate 2D/3D section...** from the shortcut menu. The **Generate Section/Elevation** dialog box is displayed. See Figure 14-20. In this dialog box, you can specify whether the section will be 2D or 3D, select what is included in the section, and specify a destination for the section. To expand the dialog box, pick the **Show details** button, which looks like a down arrow.

To create a 2D section, pick the **2D Section/Elevation** radio button in the **2D/3D** area of the dialog box. A 2D section is projected onto the section plane, but is placed flat on the XY plane of the current UCS. To create a 3D section, pick the **3D Section** radio button. A 3D section is placed so its surfaces are parallel to the corresponding cut surfaces on the 3D object.

In the **Source Geometry** area of the dialog box, you can specify which geometry is included in the section. Picking the **Include all objects** radio button includes all 3D solids, surfaces, and regions in the section. To limit the section to certain objects, pick the **Select objects to include** radio button. Then, pick the **Select objects** button, select the objects on-screen, and press [Enter]. The number of selected objects is then displayed in the dialog box.

The **Destination** area of the dialog box is where you specify how the section will be placed. To place the section into the current drawing, pick the **Insert as new block** radio button. To update an existing section block, pick the **Replace existing block** radio

Figure 14-20.
The expanded **Generate Section/ Elevation** dialog box.

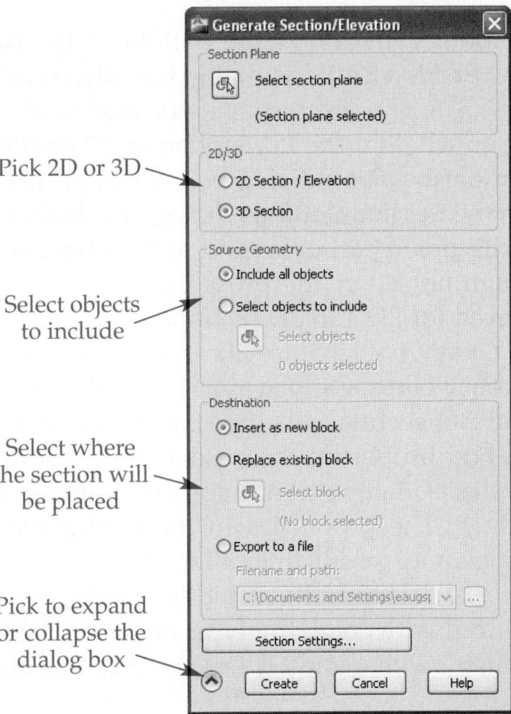

button. Then, pick the **Select block** button, select the block on-screen, and press [Enter]. You will need to do this if the section object is changed. To save the section to a file for use in other drawings, pick the **Export to a file** radio button. Then, enter a path and file name in the text box.

Once all settings have been made, pick the **Create** button. The section is attached to the cursor and can be placed like a regular block. See Figure 14-21. Additionally, the options available are the same as if a regular block is being inserted. Once the block is inserted it can be moved, rotated, and scaled as needed.

Figure 14-21.
Inserting a 2D section block.

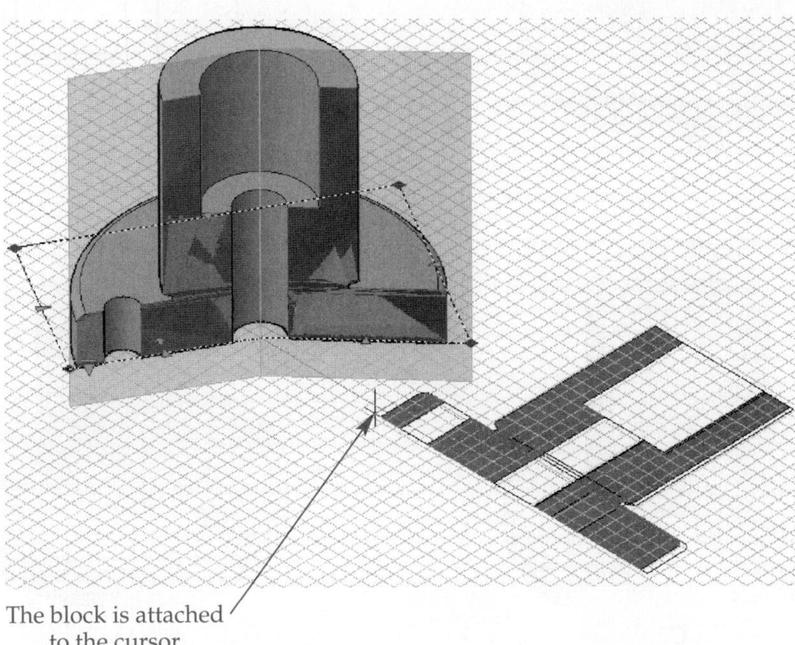

Section settings

The **Section Settings...** button at the bottom of the **Generate Section/Elevation** dialog box opens the **Section Settings** dialog box discussed earlier. Using this dialog box, you can adjust all of the properties associated with the type of section being created. Depending on whether the **2D Section** or **3D Section** radio button is selected in the **Generate Section/Elevation** dialog box, the appropriate categories and properties are displayed in the **Section Settings** dialog box. Refer to Figure 14-17.

The categories discussed earlier related to the **Live Section Settings** radio button are available, although not all of the properties are displayed. Also, two additional categories are displayed for 2D and 3D sections:

- **Background Lines.** Available for 2D and 3D sections.
- **Curved Tangency Lines.** Only available for 2D sections.

Examples of 2D and 3D sections inserted as blocks in the drawing are shown in Figure 14-22. Notice how properties can be set to show cutaway geometry in a different color and to change the section pattern, color, and linetype scale.

Background Lines. The properties in the **Background Lines** category provide control over the appearance of all lines that are not on the section plane. You can choose to have visible background lines, hidden background lines, or both displayed. They can be emphasized with color, linetype, or line weight. The layer, linetype scale, and plot style can also be changed. These settings are applied to both visible and hidden background lines.

Figure 14-22.
A—The section object is created. B—A 2D section block is inserted into the drawing and the view is made plan to the block. C—A 3D section block is inserted into the drawing. Notice how the hatch pattern is displayed. D—The 3D section block is updated and now the cutaway geometry is displayed.

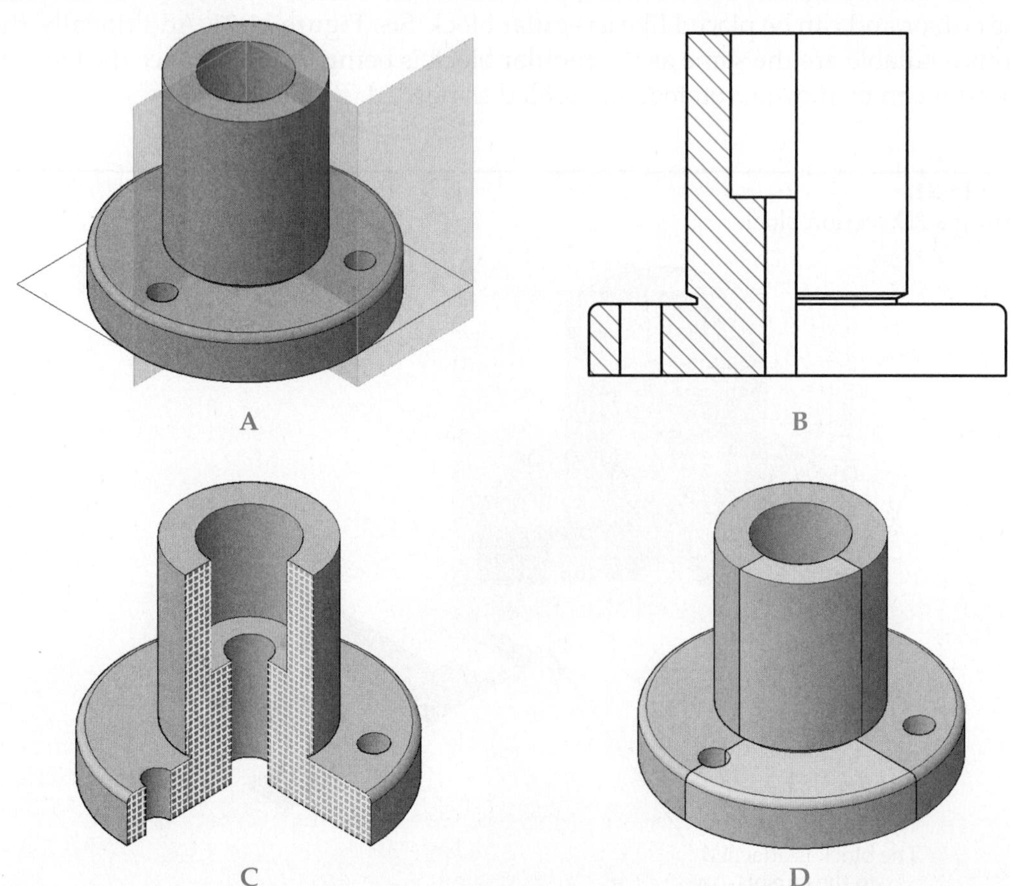

Curved Tangency Lines. The properties in the **Curved Tangency Lines** category apply to lines of tangency behind the section plane. For example, the object shown in Figure 14-22 has a round on the top of the base. This results in a line of tangency behind the section plane where the round meets the vertical edge. You can have these lines displayed or suppressed. In general, lines of tangency are not shown in a section view. If you choose to display these lines, you can set the color, layer, linetype, linetype scale, plot style, and line weight of the lines.

NOTE

When a 3D section is created, you must turn off live sectioning to see the complete sectioned object in the block. With live sectioning on, only the cut surfaces appear in the block.

Updating the section view

Once the section view is created, it is not automatically updated if the section object is changed. To update the section view, select the section object (not the block), right-click, and pick **Generate 2D/3D section...** from the shortcut menu. Then, in the **Destination** area of the **Generate Section/Elevation** dialog box, pick the **Replace existing block** radio button. If necessary, pick the **Select block** button and select the section block in the drawing. If you want to change the appearance of the section view, pick the **Section Settings...** button and adjust the properties as needed. Finally, pick the **Create** button in the **Generate Section/Elevation** dialog box to update the section block.

Exercise 14-5

Complete the exercise on the student website.
www.g-wlearning.com/CAD

Creating a Flat Display of a 3D Model

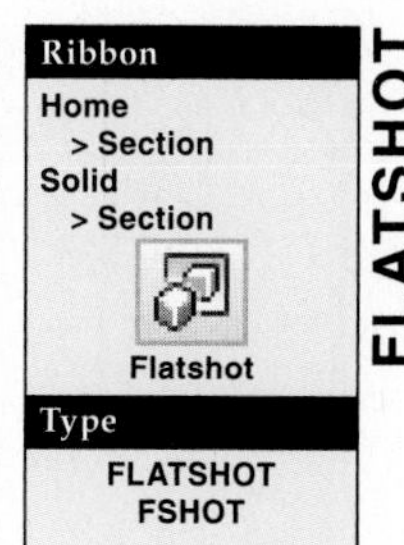

The **FLATSHOT** command creates a flat projection of the 3D objects in the drawing from the current viewpoint. The view that is created is composed of 2D geometry and is projected onto the XY plane of the current UCS. This capability is useful for creating multiview orthographic drawings from a 3D model or technical documents in which pictorial views of 3D objects are required.

Once the command is initiated, the **Flatshot** dialog box is displayed. See Figure 14-23. The options in this dialog box are similar to those found in the **Section Settings** dialog box. However, the display properties of foreground and obscured lines are limited to color and linetype. You can choose whether or not obscured lines are displayed in the flat view. You can also choose whether or not tangential edges are included. From a drafting point of view, think of obscured lines as hidden lines in an orthographic multiview drawing.

Select a destination for the flat view. Then, change the foreground and obscured lines settings as needed. Finally, pick the **Create** button. The flat view is inserted into the drawing as a block. Therefore, all of the ensuing prompts are those of a block insertion.

Use the **PLAN** command to obtain a plan view of the current UCS. Since the flat view is a block, it can be edited using the **BEDIT** command. Figure 14-24 shows a pictorial view of a 3D object and a plan view of the resulting flat view. To update the flat view to one from a different viewpoint, repeat the command and select the **Replace existing block** radio button in the **Destination** area of the dialog box.

Figure 14-23.
The **Flatshot** dialog box.

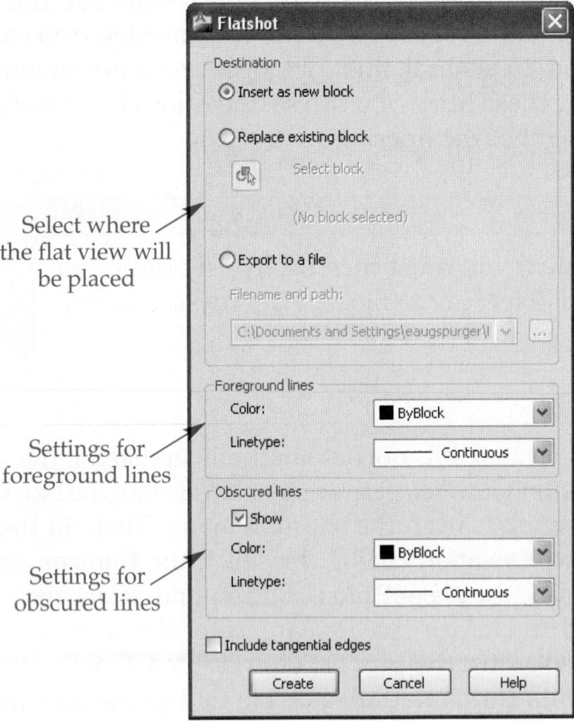

Figure 14-24.
A—The 3D view from which a flat view will be generated. B—The inserted flat view. The viewpoint is plan to the block.

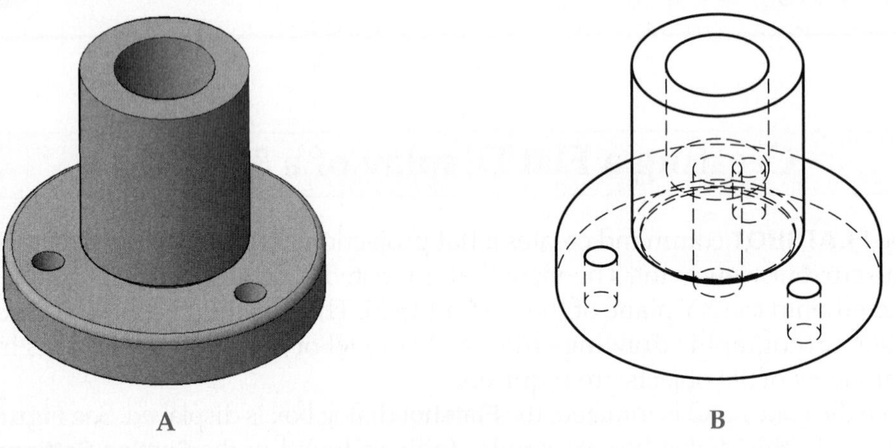

Exercise 14-6

Complete the exercise on the student website.
www.g-wlearning.com/CAD

Creating and Using Multiview Layouts with FLATSHOT

To create a multiview drawing from a 3D model, position the UCS so the XY plane is parallel to a face on the 3D object. You will eventually need to do this once for each orthographic view (front, left side, and top, for example). Next, use the **PLAN** command to display the plan view. Initiate the **FLATSHOT** command and create a 2D view of the 3D model in the current viewing plane. See Figure 14-25. Repeat this for the other two views and you will have three blocks in the drawing, one for each orthographic view. Remember, the XY plane of the UCS must be parallel to the face of the 3D object represented in the orthographic view when creating the block.

Next, switch to layout (paper) space. Create an arrangement of four floating viewports. Scale each to 1:1. Then, display the appropriate view in each viewport. See Figure 14-26. Finally, use the **MVSETUP** command to align the objects, as described in the next section.

Figure 14-25.
Creating an orthographic view using the **FLATSHOT** command. A—Set the UCS parallel to the intended view. B—Set the view plan to the UCS. C—The resulting 2D block.

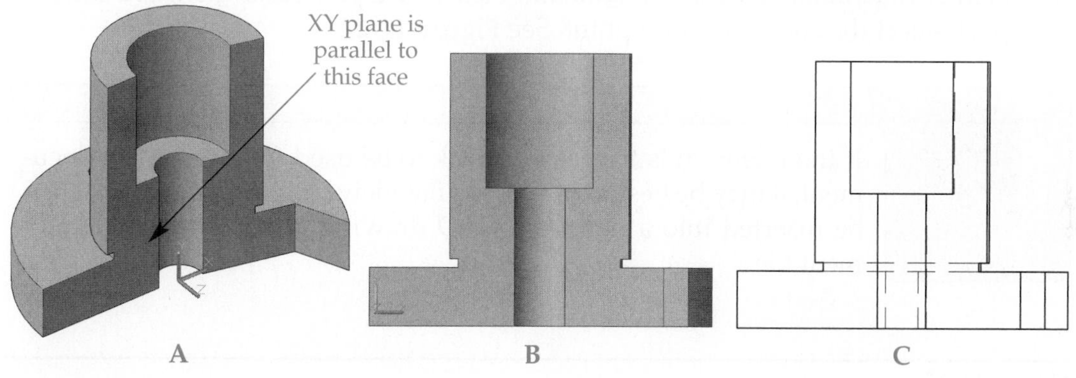

Figure 14-26.
The completed multiview drawing created using the **FLATSHOT** command.

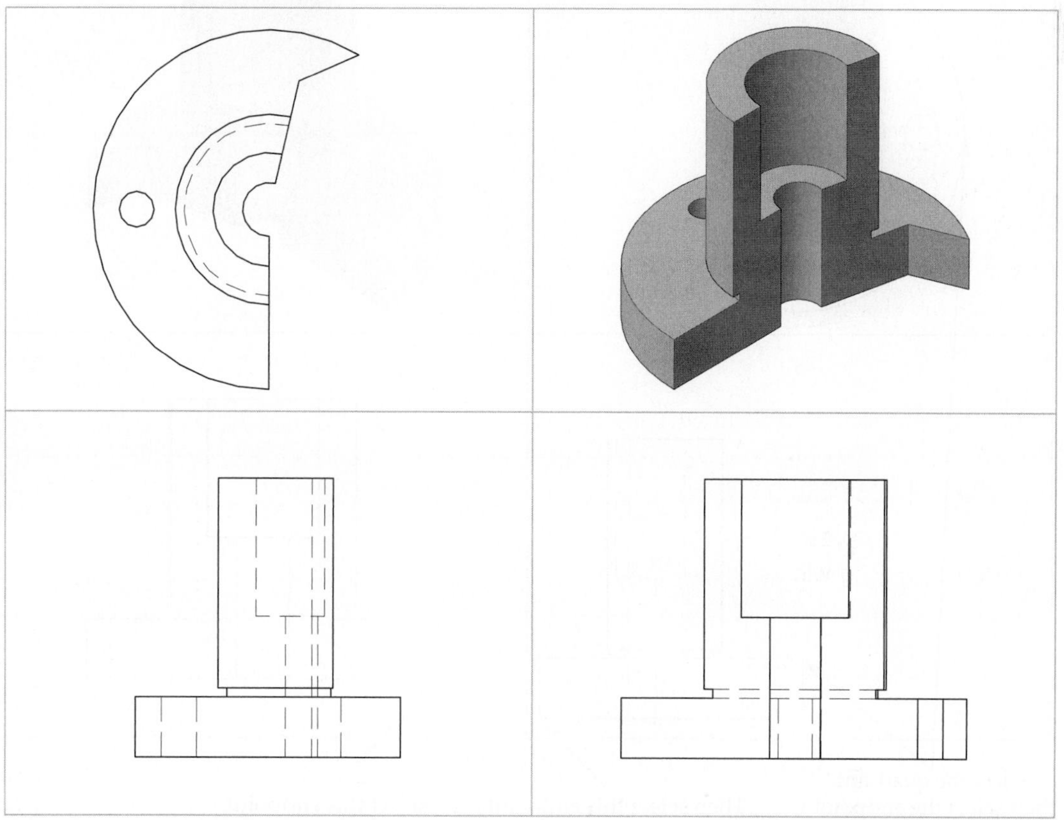

Using MVSETUP to Align Objects

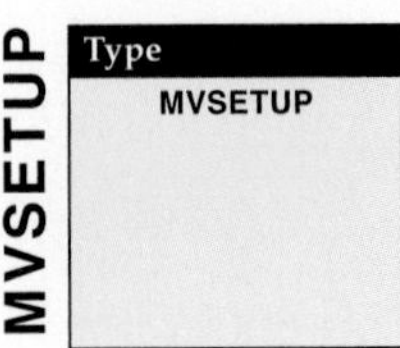

The **MVSETUP** command can be used to set up your model space drawing environment or a layout. The **MVSETUP** command can be used in either model or layout (paper) space.

If you are in model space when entering the command, the prompt will ask you to enable paper (layout) space. You can enter either Yes or No. If you stay in model space, you are prompted for: unit type, drawing scale factor, paper width, and paper height. Once these settings are enabled, a rectangle is drawn in model space showing the width and height. If you turn on a grid, the grid is displayed within the rectangular boundary.

A second use for the **MVSETUP** command is to modify the position of model geometry in each layout. For example, aligning views in a multi-layout drawing, such as created in the last section with the **FLATSHOT** command, is critical. The **MVSETUP** command is used to align views. Enter the command and select the **Align** option. Then, select either **Horizontal** or **Vertical alignment**. Pick a base point and then, in a different viewport, select the corresponding point. See Figure 14-27.

PROFESSIONAL TIP

If the intention is to create a block to be used for a technical document, it may be best to export the flat view block to a file. It can then be inserted into a new AutoCAD drawing or copied into a document file.

Figure 14-27.
The **MVSETUP** command is used to align the views in layout (paper) space.

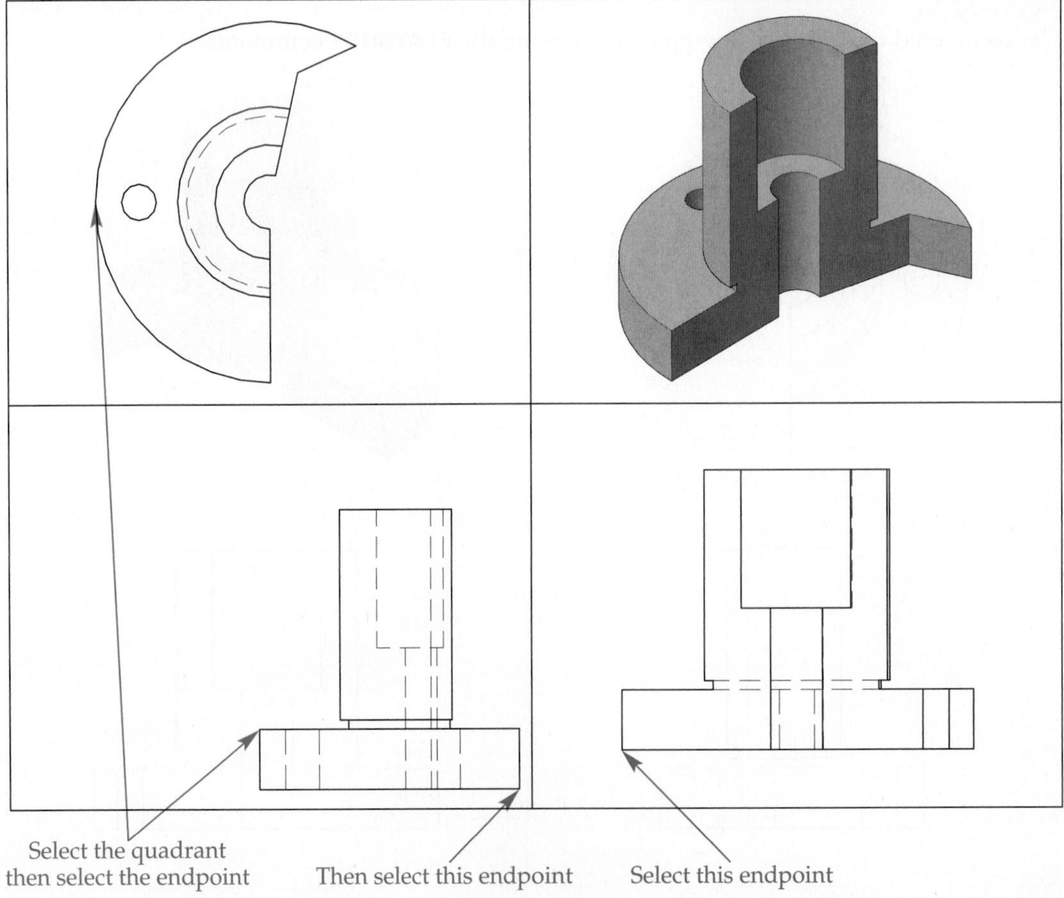

Exercise 14-7

Complete the exercise on the student website.
www.g-wlearning.com/CAD

Creating and Using Multiview Layouts with SOLVIEW

Once a solid model has been constructed, it is easy to create a multiview layout using the **SOLVIEW** command. This command allows you to create a layout containing orthographic, section, and auxiliary views. The **SOLDRAW** command can then be used to complete profile and section views. **SOLDRAW** must be used after **SOLVIEW**. The **SOLPROF** command can be used to create a profile of the solid in the current view. These commands may be typed or selected in the **Modeling** panel of the **Home** tab on the ribbon. See Figure 14-28.

Creating Views with SOLVIEW

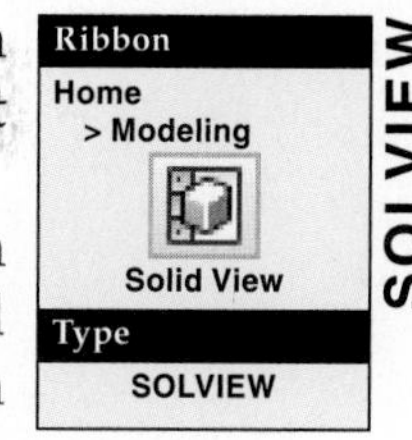

SOLVIEW

The **SOLVIEW** command is used to create new floating viewports and to establish the display within those viewports. Therefore, you may want to delete the default viewport in the layout (paper space) tab before using the **SOLVIEW** command.

First restore the WCS. This will help avoid any confusion. Then, display a plan view. See Figure 14-29. It helps to have additional user coordinate systems created prior to using **SOLVIEW**. This allows you to construct orthographic views based on a specific named UCS.

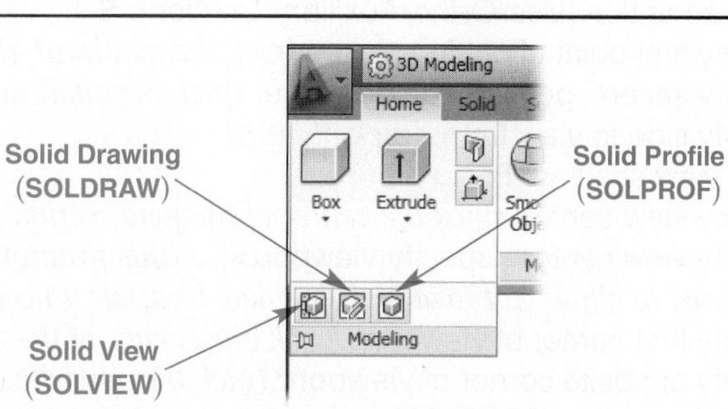

Figure 14-28.
The **SOLVIEW**, **SOLDRAW**, and **SOLPROF** commands are found in the **Modeling** panel of the **Home** tab on the ribbon.

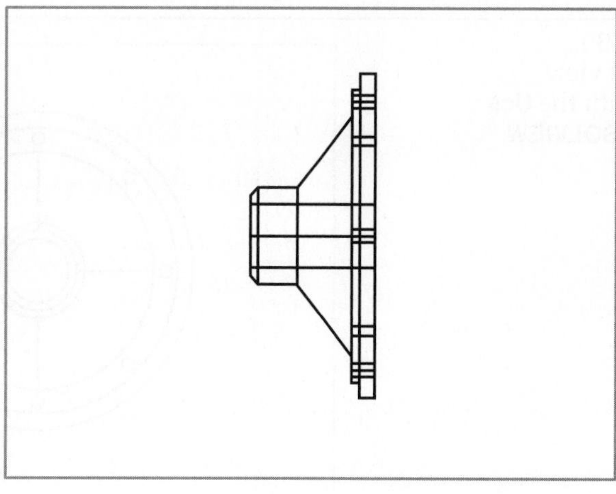

Figure 14-29.
Before using **SOLVIEW**, display a plan view of the WCS.

Before using the **SOLVIEW** command, visualize which view is going to be the top view (or plan view) and how you would like the model rotated in relationship to the layout. With this in mind, look at the current UCS icon and make sure that the X axis is pointing to the "right" and the Y axis is pointing "up" in your imagined layout. If this is not the case, then you must restore the WCS, rotate the current UCS, or restore a saved UCS to correctly align the axes. Then, when you enter the **SOLVIEW** command in the layout, you can simply select the current UCS and you will be creating the top or plan view of your model.

When you enter the **SOLVIEW** command while in model space, AutoCAD automatically switches to layout space (paper space). Next, create an initial view from which other views can project. This is normally the top or front. In the following example, the top view is constructed first by using the plan view of a UCS named Leftside.

Enter an option [Ucs/Ortho/Auxiliary/Section]: **U**↵
Enter an option [Named/World/?/Current] <Current>: **N**↵
Enter name of UCS to restore: **LEFTSIDE**↵
Enter view scale <1.0>: **.5**↵
Specify view center: *(pick a location in the layout for the center of the view)*
Specify view center <specify viewport>: ↵
Specify first corner of viewport: *(pick the first corner of a paper space viewport outside of the object)*
Specify opposite corner of viewport: *(pick the opposite corner of the viewport)*
Enter view name: **TOPVIEW**↵ *(the left of the object in AutoCAD is the top of the part)*
Enter an option [Ucs/Ortho/Auxiliary/Section]: *(leave the command active at this time)*

You must provide a name for the view. The result is shown in Figure 14-30.

The **SOLVIEW** command remains active until you press the [Enter] or [Esc] key. If you exit **SOLVIEW**, you can still return to the drawing and create additional orthographic viewports. With the command active, continue and create a section view to the right of the top view:

Enter an option [Ucs/Ortho/Auxiliary/Section]: **S**↵
Specify first point of cutting plane: *(pick the quadrant at point 1 in Figure 14-31)*
Specify second point of cutting plane: *(pick the quadrant at point 2)*
Specify side to view from: *(pick point 3)*
Enter view scale <0.5>: ↵
Specify view center: *(pick the center of the new section view)*
Specify view center <specify viewport>: ↵ *(this prompt remains active until [Enter] is pressed to allow you to adjust the view location if necessary)*
Specify first corner of viewport: *(pick one corner of the viewport)*
Specify opposite corner of viewport: *(pick the opposite corner of the viewport)*
Enter view name: **SECTION**↵
Enter an option [Ucs/Ortho/Auxiliary/Section]: ↵

Figure 14-30.
The initial view created with the **Ucs** option of **SOLVIEW**.

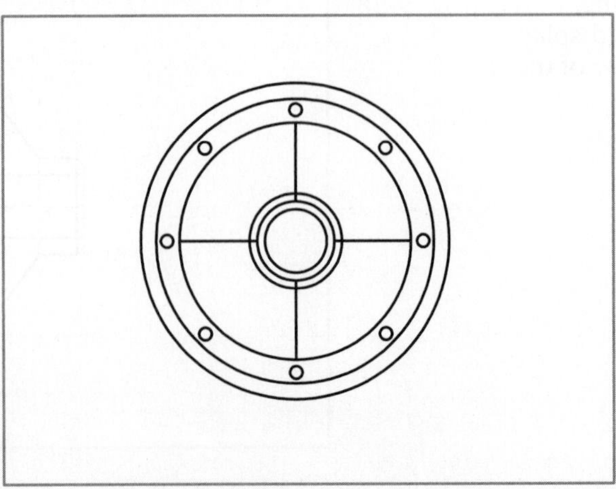

Figure 14-31.
The section view created with **SOLVIEW** (shown on the right) does not show projection lines. The pick points are shown on the left.

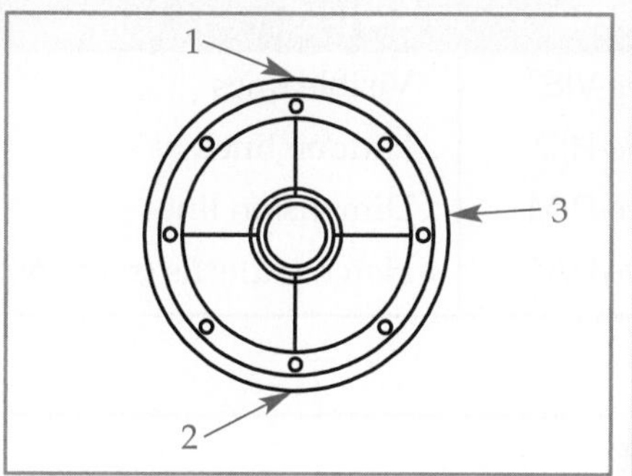

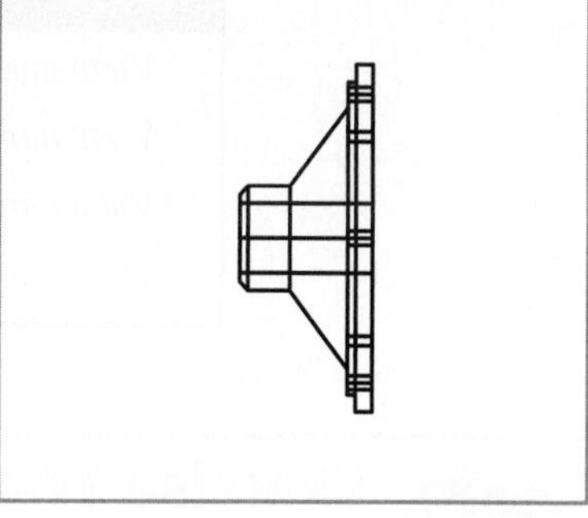

Notice in Figure 14-31 that the new view is shown in the current visual style and not as a section. This is normal. **SOLVIEW** is used to create the views. The **SOLDRAW** command draws the section lines. **SOLDRAW** is discussed later in this chapter.

A standard orthographic view can be created using the **Ortho** option of **SOLVIEW**. This is illustrated in the following example. The new orthographic view is shown in Figure 14-32.

Enter an option [Ucs/Ortho/Auxiliary/Section]: **O**↵
Specify side of viewport to project: *(pick the bottom edge of the left viewport)*
Specify view center: *(pick the center of the new view)*
Specify view center <specify viewport>: ↵
Specify first corner of viewport: *(pick one corner of the viewport)*
Specify opposite corner of viewport: *(pick the opposite corner of the viewport)*
Enter view name: **FRONTVIEW**↵

The **SOLVIEW** command creates new layers that are used by **SOLDRAW** when profiles and sections are created. The layers are used for the placement of visible, hidden, dimension, and section lines. Each layer is named as the name of the view with a three letter tag, as shown in the following table. The use of these layers is discussed later in this chapter.

Figure 14-32.
An orthographic front view is created with the **Ortho** option of **SOLVIEW**. This is the view shown at the lower left.

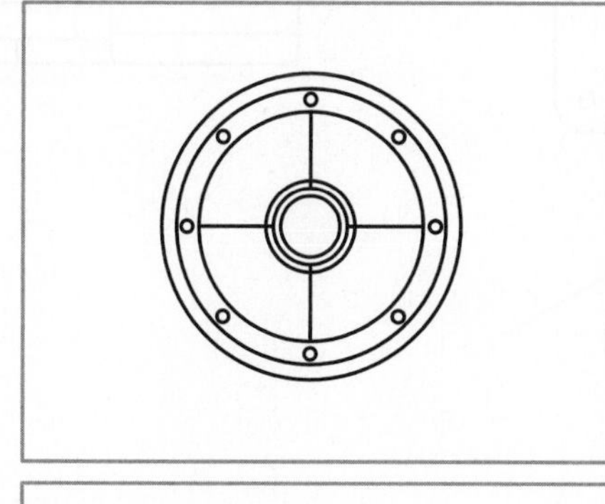

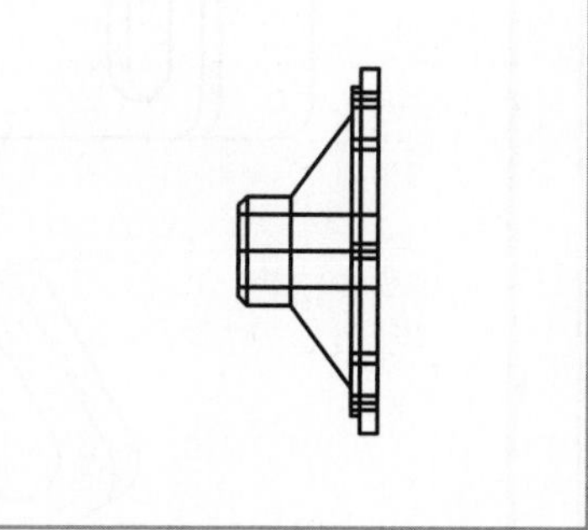

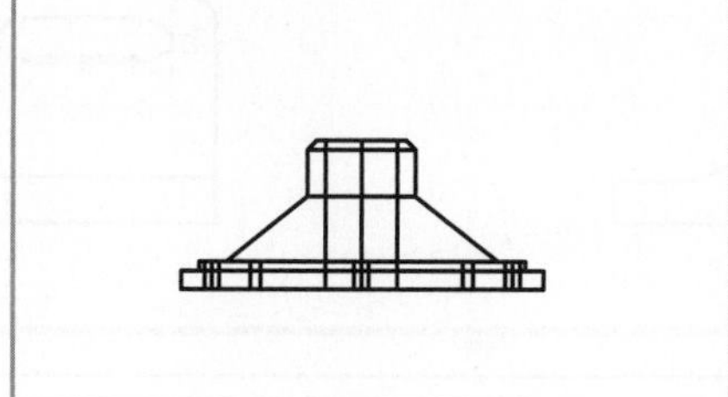

Layer Name	Object
View name-VIS	Visible lines
View name-HID	Hidden lines
View name-DIM	Dimension lines
View name-HAT	Hatch patterns (sections)

Exercise 14-8

Complete the exercise on the student website.
www.g-wlearning.com/CAD

Creating Auxiliary Views with SOLVIEW

Auxiliary views are used to display a surface of an object that is not parallel to any of the standard views. It may be an inclined or oblique surface. Refer to **Figure 14-33.** Sometimes these views are necessary to show or dimension a feature that is not being displayed in true size in any other view. The slot in the inclined surface in **Figure 14-33** is not shown in true size in any of the standard views.

The auxiliary view is taken from one of the other views where the inclined surface is shown as an edge. The auxiliary view will be projected perpendicular to this surface. The auxiliary view is created by picking two points on the surface in the front view and another point to indicate the line of sight.

Figure 14-33.
An auxiliary view (shown in color) is created from the inclined plane in the front view.

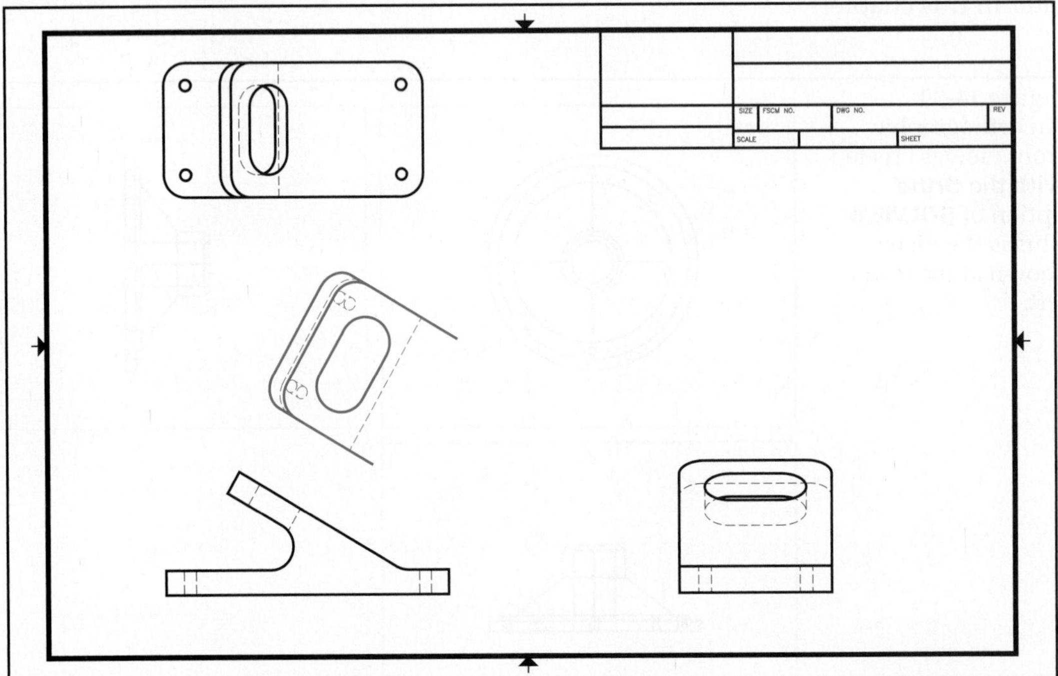

Enter an option [Ucs/Ortho/Auxiliary/Section]: **A**↵
Specify first point of inclined plane: *(using object snaps, pick a point on one end of the inclined surface)*
Specify second point of inclined plane: *(pick a point on the other end of the inclined surface)*
Specify side to view from: *(pick a point on the side of the surface from which you want to view it)*
Specify the view center: *(pick the center)*
Specify the view center: <specify viewport>↵
Specify first corner of viewport: *(pick one corner of the viewport)*
Specify opposite corner of viewport: *(pick the opposite corner of the viewport)*
Enter view name: **AUXILIARYVIEW**↵

Auxiliary views are often incomplete views, so it is acceptable to cut off portions of the view that are not necessary when you specify the corners of the viewport.

PROFESSIONAL TIP

When creating an auxiliary view, you may want to move other viewports that may be in the way to make room for the view.

Creating Finished Views with SOLDRAW

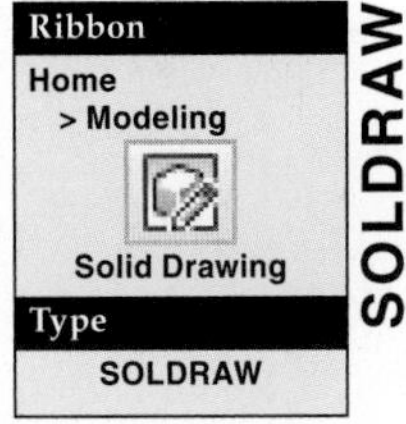

The **SOLVIEW** command saves information specific to each viewport when a new view is created. This information is used by the **SOLDRAW** command to construct a finished profile or section view. **SOLDRAW** first deletes any information currently on the *view name*-VIS, *view name*-HID, and *view name*-HAT layers for the selected view. Visible, hidden, and section lines are automatically placed on the appropriate layer. Therefore, you should avoid placing objects on any of the automatically generated layers other than the *view name*-DIM layer.

The **SOLDRAW** command automatically creates a profile or section in the selected viewport. If you select a viewport that was created using the **Section** option of **SOLVIEW**, the **SOLDRAW** command uses the current values of the **HPNAME**, **HPSCALE**, and **HPANG** system variables to construct the section. These three variables control the angle, scale factor, and name of the hatch pattern.

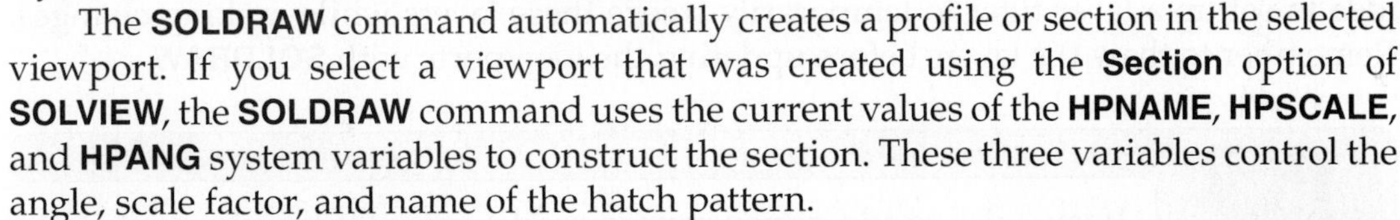

If a view is selected that was not created as a section in **SOLVIEW**, the **SOLDRAW** command constructs a profile view. All new visible and hidden lines are placed on the *view name*-VIS or *view name*-HID layer. All existing objects on those layers are deleted.

Once the command is initiated, you are prompted to select objects. Pick the border of the viewport(s) for which you want the profile or section generated. When all viewports are selected, press [Enter] and the profiles and sections are created.

After the profile construction is completed, lines that should be a hidden linetype are still visible (solid). This is because the linetype set for the *view name*-HID layer is Continuous. Change the linetype for the layer to Hidden and the drawing should appear as shown in **Figure 14-34,** depending on the current visual style and hatch settings. You may also want to change other layer properties such as color, line weight, and plot style.

Figure 14-34.
The new front profile view shows hidden lines after the linetype is set to Hidden for the FRONTVIEW-HID layer.

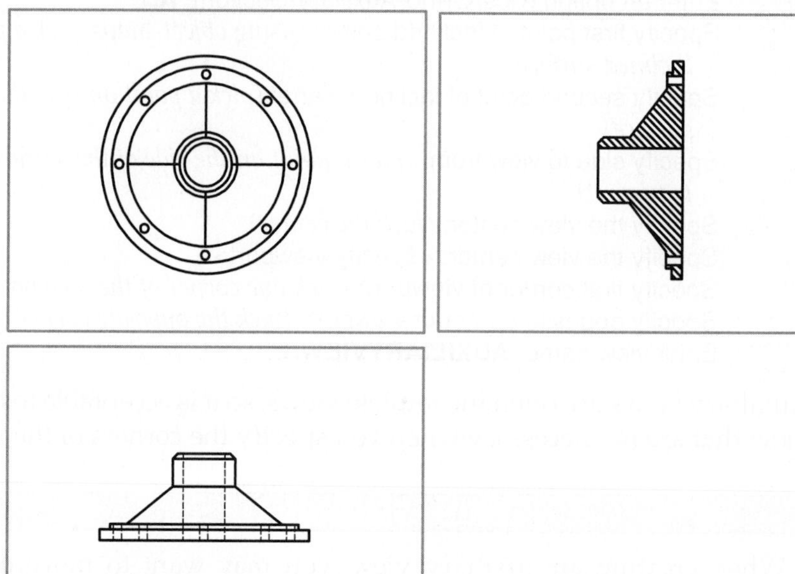

Revising the 3D Model

If changes are needed after these views are created, the best practice is to modify the original 3D solid. However, the views created with **SOLVIEW** and **SOLDRAW** will not immediately reflect changes. To update the views, simply start the **SOLDRAW** command, select the viewports, and press [Enter]. The views are then updated with the changes.

When you go to model space to edit the solid, you may find it difficult to work on the original model. The 2D views created with **SOLDRAW** are projected on the top, bottom, left, and right, and sometimes within the model itself. It may be a good idea to set up a layer filter to temporarily freeze these layers while making changes. Remember to thaw the layers before updating the viewports with **SOLDRAW**.

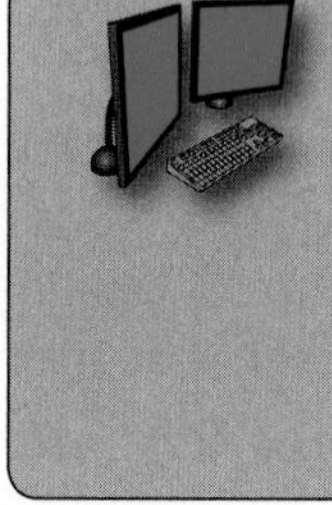

PROFESSIONAL TIP

If you wish to add dimensions in model space for the views created with **SOLVIEW** and **SOLDRAW**, use the view-specific DIM layers. These layers are created for that purpose and are only visible in one view. **SOLDRAW** does not delete information on the DIM layers when it constructs a view. If you prefer to dimension in paper space, use a layer other than the DIM layers created by **SOLVIEW**.

Adding a 3D View in Paper Space to the Drawing Layout

If you want to add a paper space viewport that contains a 3D (pictorial) view of the solid, use the **MVIEW** command. Create a single viewport by picking the corners. The object will appear in the viewport. Next, activate the viewport and use any of the orbit commands or a preset isometric viewpoint to achieve the desired 3D view. Pan and zoom as necessary. Change to the parallel or perspective projection if needed. You can also use the **Visual Styles** panel in the ribbon to adjust the display of the 3D viewport. The visual style set current for this viewport does not affect the displays in the other viewports. See Figure 14-35.

In order to have the hidden display correctly plotted, use the **MVIEW Shadeplot** option on the 3D viewport. Enter the command and select the **Shadeplot** option. Then, set the option to **Hidden**. If you have a hidden display shown in the viewport, you can also select **As displayed**. Then, pick the viewport when prompted to select objects.

Figure 14-35.
Create a 3D viewport with the **MVIEW** command. You can hide the lines in the viewport, as shown at the lower right. To plot the viewport as a hidden display, use the **MVIEW Shadeplot** option and set it to Hidden.

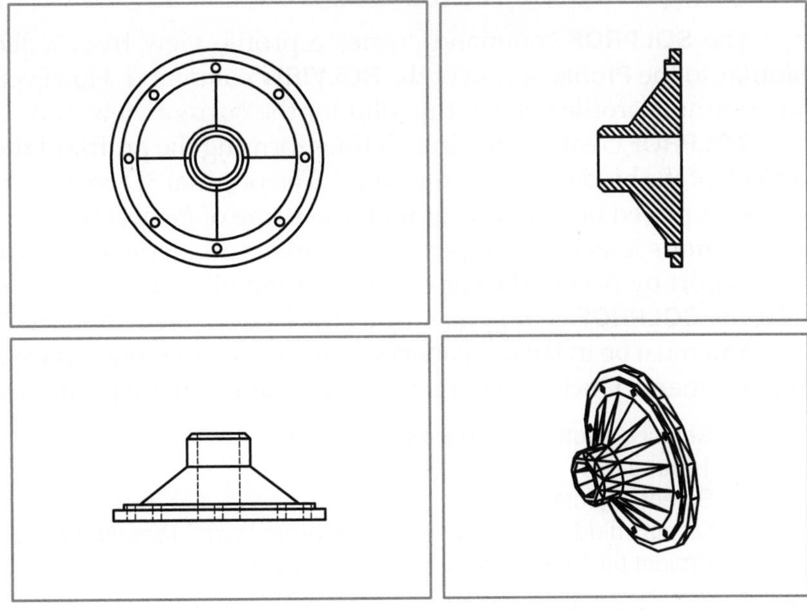

Alternately, you can select the viewport and use the **Properties** palette to set the Shade plot property to As Displayed or Hidden. Any visual style display of the viewport can also be plotted in this manner by setting **MVIEW Shadeplot** to **As Displayed** (when the view is shaded) or **Rendered**.

Tips

Remember the following points when working with **SOLVIEW** and **SOLDRAW**.

- Use **SOLVIEW** first and then **SOLDRAW**.
- Do not draw on the *view name*-HID and *view name*-VIS layers.
- Place model space dimensions for each view on the *view name*-DIM layer for that specific view or simply dimension in paper space on a layer not created by **SOLVIEW**.
- After using **SOLVIEW**, use **SOLDRAW** on all viewports in order to create hidden lines or section views.
- Change the linetype on the *view name*-HID layer to Hidden and adjust other layer properties as needed.
- Create 3D viewports with the **MVIEW** or **VPORTS** command and an orbit command or a preset isometric view. Remove hidden lines when plotting with the **MVIEW Shadeplot** option set to **Hidden**.
- Plot the drawing in layout (paper) space at the scale of 1:1.

Exercise 14-9

Complete the exercise on the student website.
www.g-wlearning.com/CAD

Creating a Profile with SOLPROF

SOLPROF

Ribbon
Home > Modeling
Solid Profile
Type
SOLPROF

The **SOLPROF** command creates a profile view from a 3D solid model. This is similar to the **Profile** option of the **SOLVIEW** command. However, **SOLPROF** is limited to creating a profile view of the solid for the current view only.

SOLPROF creates a block of all lines forming the profile of the object. It also creates a block of the hidden lines of the object. The original 3D object is retained. Each of these blocks is placed on a new layer with the name of PH-*view handle* and PV-*view handle.* A ***view handle*** is a name composed of numbers and letters that is automatically given to a viewport by AutoCAD. For example, if the view handle for the current viewport is 2C9, the **SOLPROF** command creates the layers PH-2C9 and PV-2C9.

You must be in layout (paper) space and have a model space viewport active to use the command. Once the command is initiated, you are prompted to select objects:

Select objects: *(pick the solid)*
1 found
Select objects: ↵
Display hidden profile lines on separate layer? [Yes/No] <Y>: ↵
Project profile lines onto a plane? [Yes/No] <Y>: ↵

If you answer yes to this prompt, the 3D profile lines are projected to a 2D plane and converted to 2D objects. This produces a cleaner profile.

Delete tangential edges? [Yes/No] <Y>: ↵

Answering yes to this prompt produces a proper 2D view by eliminating lines that would normally appear at tangent points of arcs and lines. If you wish to display the profile lines in a 3D view, do not delete the tangential edges. Once the profile is created, freeze the layer of the original object in the viewport. The original object and the profile created with **SOLPROF** are shown in **Figure 14-36.**

NOTE

When plotting views created with **SOLPROF**, hidden lines may not be displayed unless you freeze the layer that contains the original 3D object.

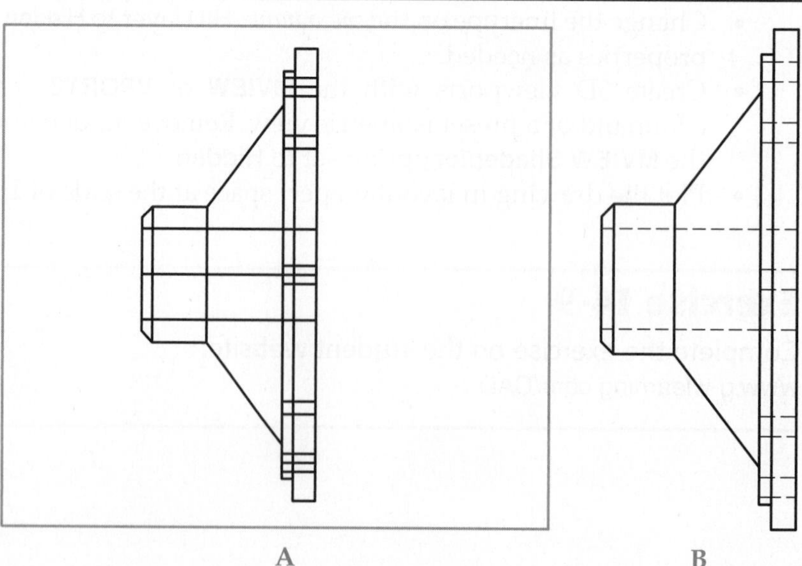

Figure 14-36. A—The original solid. B—A profile created with **SOLPROF**.

Solid Model Analysis

Type
MASSPROP

MASSPROP

The **MASSPROP** command allows you to analyze a solid model for its physical properties. The data obtained from **MASSPROP** can be retained for reference by saving the data to a file. The default file name is the drawing name. The file is an ASCII text file with a .mpr (mass properties) extension. The analysis can be used for third party applications to produce finite element analysis, material lists, or other testing studies.

Once the command is initiated, you are prompted to select objects. Pick the objects for which you want the mass properties displayed and press [Enter]. AutoCAD analyzes the model and displays the results in the AutoCAD text window. See **Figure 14-37.** The following properties are listed.

- **Mass.** A measure of the inertia of a solid. In other words, the more mass an object has, the more inertia it has. Note: Mass is *not* a unit of measurement of inertia.
- **Volume.** The amount of 3D space the solid occupies.
- **Bounding box.** The dimensions of a 3D box that fully encloses the solid.
- **Centroid.** A point in 3D space that represents the geometric center of the mass.
- **Moments of inertia.** A solid's resistance when rotating about a given axis.
- **Products of inertia.** A solid's resistance when rotating about two axes at a time.
- **Radii of gyration.** Similar to moments of inertia. Specified as a radius about an axis.
- **Principal moments and X-Y-Z directions about a centroid.** The axes about which the moments of inertia are the highest and lowest.

PROFESSIONAL TIP

Advanced applications of solid model design and analysis are possible with Autodesk Inventor® software. This product allows you to create parametric designs and assign a wide variety of physical materials to the solid model.

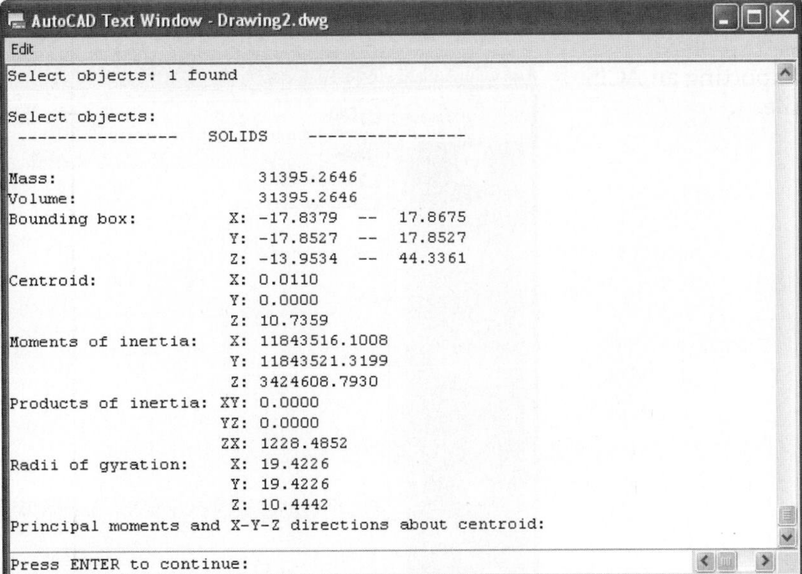

Figure 14-37.
The **MASSPROP** command displays a list of solid properties in the AutoCAD text window.

Solid Model File Exchange

AutoCAD drawing files can be converted to files that can be used for testing and analysis. Use the **ACISOUT** command or **Export Data** dialog box to create a file with a .sat extension. These files can be imported into AutoCAD with the **ACISIN** or **IMPORT** command.

Solids can also be exported for use with stereolithography software. These files have a .stl extension. Use the **STLOUT** command or the **Export Data** dialog box to create STL files.

Importing and Exporting Solid Model Files

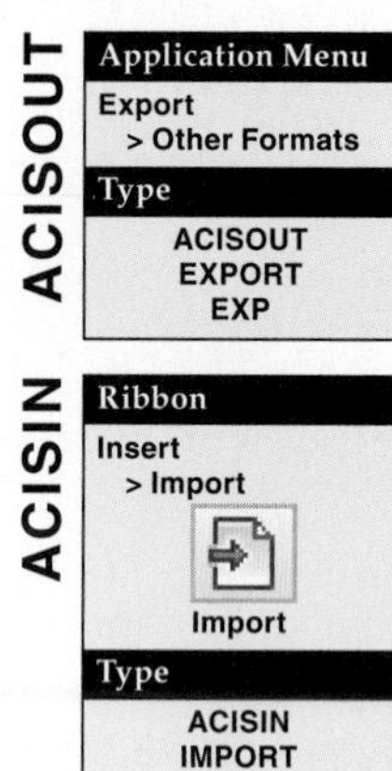

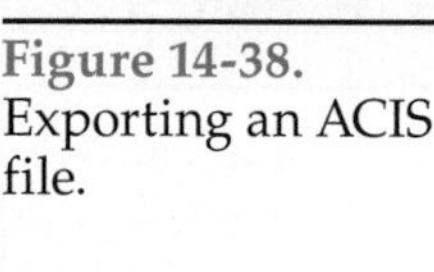

A solid model is frequently used with analysis and testing software or in the manufacture of a part. The **ACISOUT** and **EXPORT** commands allow you to create a type of file that can be used for these purposes. Once the **ACISOUT** command is initiated, you are prompted to select objects. After selecting objects and pressing [Enter], a standard save dialog box is displayed. See **Figure 14-38.** When using the **EXPORT** command, the standard save dialog box appears first. After entering a file name and selecting a file type (SAT), you are then prompted to select objects.

An SAT file can be imported into AutoCAD and automatically converted into a drawing file using the **ACISIN** and **IMPORT** commands. Once either command is initiated, a standard open dialog box appears. Change the file type to SAT, locate the file, and pick the **Open** button.

Stereolithography Files

Stereolithography (SLA) is an additive manufacturing-technology process that creates various plastic 3D model prototypes using a computer-generated solid model, a laser, and a vat of liquid polymer. This technology is also referred to as ***rapid prototyping*** or ***3D printing.*** Some additive manufacturing processes, such as fused-deposition modeling (FDM), add material in "layers" from a filament that is extruded through a heated nozzle. Other additive manufacturing processes are: selective laser sintering (SLS), 3D printing (3DP), multi-jet modeling (MJM), and electron beam melting (EBM). Using one of these processes, a prototype 3D model can be designed and formed in a short amount of time without using standard subtractive manufacturing processes.

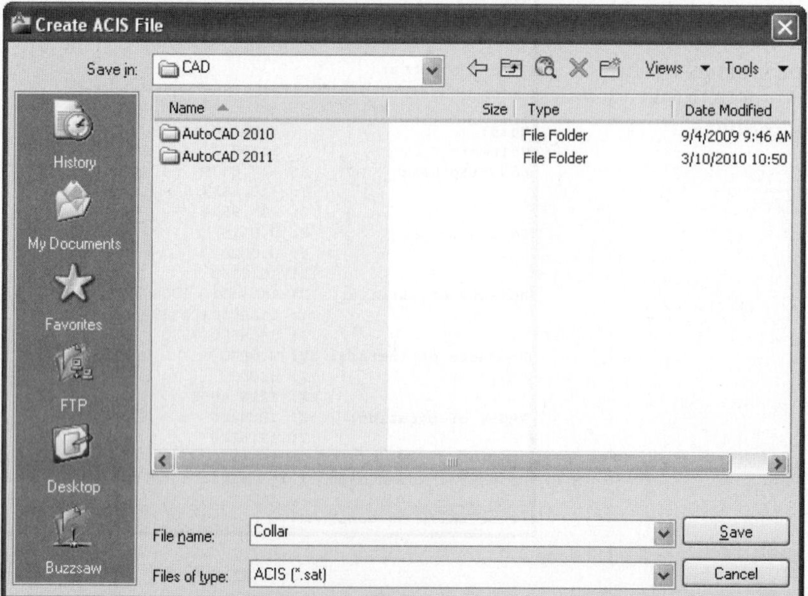

Figure 14-38.
Exporting an ACIS file.

Most CAD software today can create a stereolithograph file (SLT file). AutoCAD can export a drawing file to the STL format, but *cannot* import STL files.

Using the 3DPRINT command to create STL files

The preferred method for creating SLT files is to use the **3DPRINT** command. This command streamlines the process of creating an STL file. It can also be used to send a solid model to a 3D printer service provider. Solids and watertight meshes can be selected for use with the command. A ***watertight mesh*** is completely closed and contains no openings. Watertight meshes are converted to 3D solids when using the **3DPRINT** command.

When the **3DPRINT** command is initiated, a dialog box appears with two options: **Learn about preparing a 3D model for printing** and **Continue**. Select the **Continue** option. Then, select the solids or watertight meshes and press [Enter]. The **Send to 3D Print Service** dialog box is displayed, Figure 14-39. This dialog box allows you to change the scale or select other objects. When done, pick the **OK** button to display a standard save dialog box. Type the file name in the **File name:** text box and pick the **Save** button. Next, a web page is displayed (if you are connected to the Internet). This page offers options for 3D printer service providers. Unless you are sending the file to a service provider, close this window. The STL file is ready to be sent to your additive manufacturing machine to create the model.

NOTE

When using the **3DPRINT** command, the **FACETRES** system variable is automatically set to 10 for the creation of an STL file. The system variable is reset after the command is completed.

Using the STLOUT and EXPORT commands to create STL Files

Another method to create an SLT files is to use the **STLOUT** and **EXPORT** commands. Once the **STLOUT** command is initiated, you are prompted to select solids or watertight meshes. You are then asked if you want to create a binary STL file. If you answer no to the prompt, an ASCII file is created. Keep in mind that a binary STL file may be as much as five times smaller than the same file in ASCII format. Use the file type required by your stereolithography machine or service provider. After you choose the type of file, a standard save dialog box is displayed. Type the file name in the **File name:** edit box and pick **Save** or press [Enter].

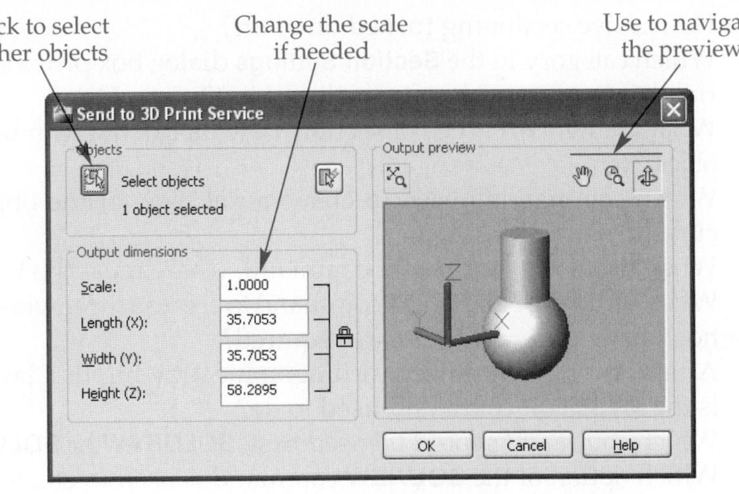

Figure 14-39. This dialog box is displayed when creating an STL file with the **3DPRINT** command. Using this command is the preferred method of creating an STL file.

Once the **EXPORT** command is initiated, the standard save dialog box appears. Name the file and select the STL file type. Once the dialog box is closed, you are prompted to select an object. After you select the object and press [Enter], the file is created. You are not given the option of selecting a binary or ASCII format for the file. A binary file is automatically created.

PROFESSIONAL TIP

For the **STLOUT** command to work properly, the solid model must be positioned in the current UCS in such a way so the entire object has positive XYZ coordinates.

Exercise 14-10

Complete the exercise on the student website.
www.g-wlearning.com/CAD

Chapter Test

Answer the following questions. Write your answers on a separate sheet of paper or complete the electronic chapter test on the student website.
www.g-wlearning.com/CAD

1. What does the **SECTIONPLANE** command create?
2. How is the **Face** option of the **SECTIONPLANE** command used?
3. Which option of the **SECTIONPLANE** command is used to create sections with jogs?
4. When a section object is created by picking a face or two points or using the **Orthographic** option of the **SECTIONPLANE** command, which section object state is established?
5. Which section object grips are used to accomplish the following tasks?
   A. Change the section object state.
   B. Lengthen or shorten the section object segment.
   C. Rotate the section view 180°.
6. How is live sectioning turned on or off?
7. Which category in the **Section Settings** dialog box provides control over the material that is removed by the section object?
8. What are the two types of section view blocks that can be created from a section object?
9. Which command is used to create a flat view of the objects projected from the current viewpoint?
10. When using the **FLATSHOT** command, where must the UCS XY plane be located?
11. When using the **FLATSHOT** command to create a three-view orthographic drawing, how many different UCSs are required?
12. After creating a multiview orthographic drawing in a layout (paper space), what is the **MVSETUP** command used to do?
13. Which command should be used first, **SOLDRAW** or **SOLVIEW**?
14. Which option of the **SOLVIEW** command is used to create an orthographic view?
15. Name the layer(s) that the **SOLVIEW** command automatically create(s).

16. Which layer(s) in question 15 should you avoid drawing on?
17. Which command can automatically complete a section view using the current settings of **HPNAME**, **HPSCALE**, and **HPANG**?
18. Which command creates a profile view from a 3D model?
19. What is the function of the **MASSPROP** command?
20. What is the extension of the ASCII file that can be created by **MASSPROP**?
21. What is a *centroid?*
22. Which commands export and import solid models?
23. Which type of file has an .stl extension?
24. Which three commands can be used to create an STL file?
25. The **3DPRINT** command is used to accomplish two different operations. What are they?

Drawing Problems

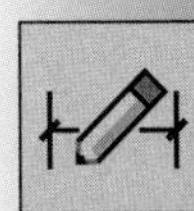

1. Open one of your solid model problems from a previous chapter and do the following.
   A. Use the **Face** option of the **SECTIONPLANE** command to create a section object.
   B. Alter the section so that the section plane object cuts through features of the model.
   C. Change the section settings to display an ANSI hatch pattern.
   D. Save the drawing as P14_01.

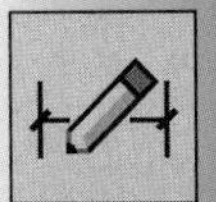

2. Open one of your solid model problems from a previous chapter and do the following.
   A. Construct a section through the model using the **Draw** option of the **SECTIONPLANE** command. Cut through as many features as possible.
   B. Display cutaway geometry with a 50% transparency.
   C. Display section lines using an appropriate hatch pattern.
   D. Generate a 3D section block that displays the cutaway geometry in a color of your choice.
   E. Create a layout with a viewport for the 3D block displayed at half the size of the original model.
   F. Save the drawing as P14_02.

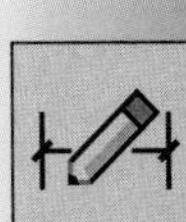

3. Open one of your solid model problems from a previous chapter and do the following.
   A. Create a multiview layout of the model. One of the views should be a section view. Use a total of three 2D views.
   B. Use **SOLVIEW** and **SOLDRAW** to create the views. Be sure that section lines and hidden lines are properly displayed.
   C. Create a fourth viewport that contains a 3D view of the solid. Place the label PICTORIAL VIEW within the viewport.
   D. Plot the drawing so the 3D view is displayed with hidden lines removed.
   E. Save the drawing as P14_03.

4. Open one of your solid model problems from a previous chapter and do the following.
   A. Display the model in a plan view.
   B. Use **SOLPROF** to create a profile view. Save the profile view as a block in a file named P14_04PLN.
   C. Display the original model in a 3D view.
   D. Use **SECTIONPLANE** to construct a 2D front-view section of the model.
   E. Display the section as a plan view.
   F. Insert the block P14_04PLN above the section view. Adjust the views so they are properly aligned.
   G. Save the drawing as P14_04.

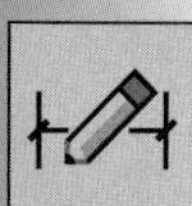

5. Choose five solid model problems from previous chapters and copy them to a new folder. Then, do the following.
   A. Open the first drawing. Export it as an SAT file.
   B. Do the same for the remaining four files.
   C. Compare the sizes of the SAT files with the DWG files. Compare the combined sizes of both types of files.
   D. Begin a new drawing and import one of the SAT files.

6. Draw the object shown below as a solid model. Do not dimension the object. Then, do the following.
   A. Construct a section object that creates a full section along the centerline of the hole.
   B. Generate a 2D section and display it on the drawing at half the size of the original. Specify section settings as desired.
   C. Generate a 3D section and display it on the drawing at half the size of the original. Do not display cutaway geometry. Specify section settings as desired.
   D. Activate live sectioning. Do not display the cutaway geometry.
   E. On the original solid model, display the intersection fill as an ANSI hatch pattern.
   F. Save the drawing as P14_06.

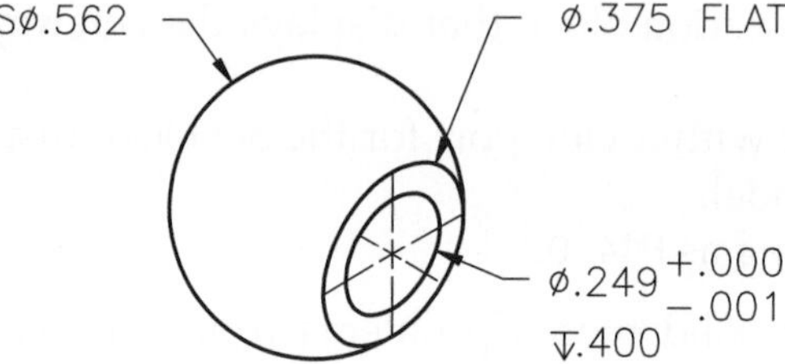

Drawing Problems - Chapter 14

7. Draw the object shown below as a solid model. Only half of the object is shown; draw the complete object. Do not dimension the object. Then, do the following.
   A. Construct a section plane that creates a full section, as shown.
   B. Display the intersection fill as an ANSI hatch pattern.
   C. Activate live sectioning and view the cutaway geometry with a high level of transparency.
   D. Save the drawing as P14_07.

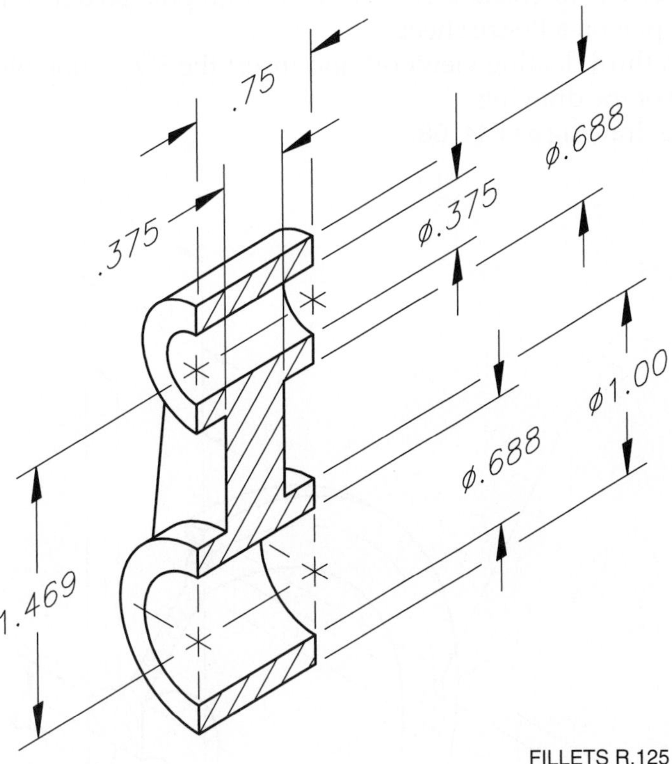

Drawing Problems - Chapter 14

8. Draw the object shown below as a solid model. Only half of the object is shown; draw the complete object. Do not dimension the object. Then, do the following.
   A. Construct a section plane that creates a half section.
   B. Display the intersection fill as an ANSI hatch pattern.
   C. Activate live sectioning and view the cutaway geometry with a low level of transparency.
   D. Generate a 3D section and save it as a block.
   E. Use **SOLVIEW** to create a two-view orthographic layout. Use an appropriate scale to plot on a B-size sheet.
   F. Create a third floating viewport and insert the 3D section block scaled to half the size of the drawing.
   G. Save the drawing as P14_08.

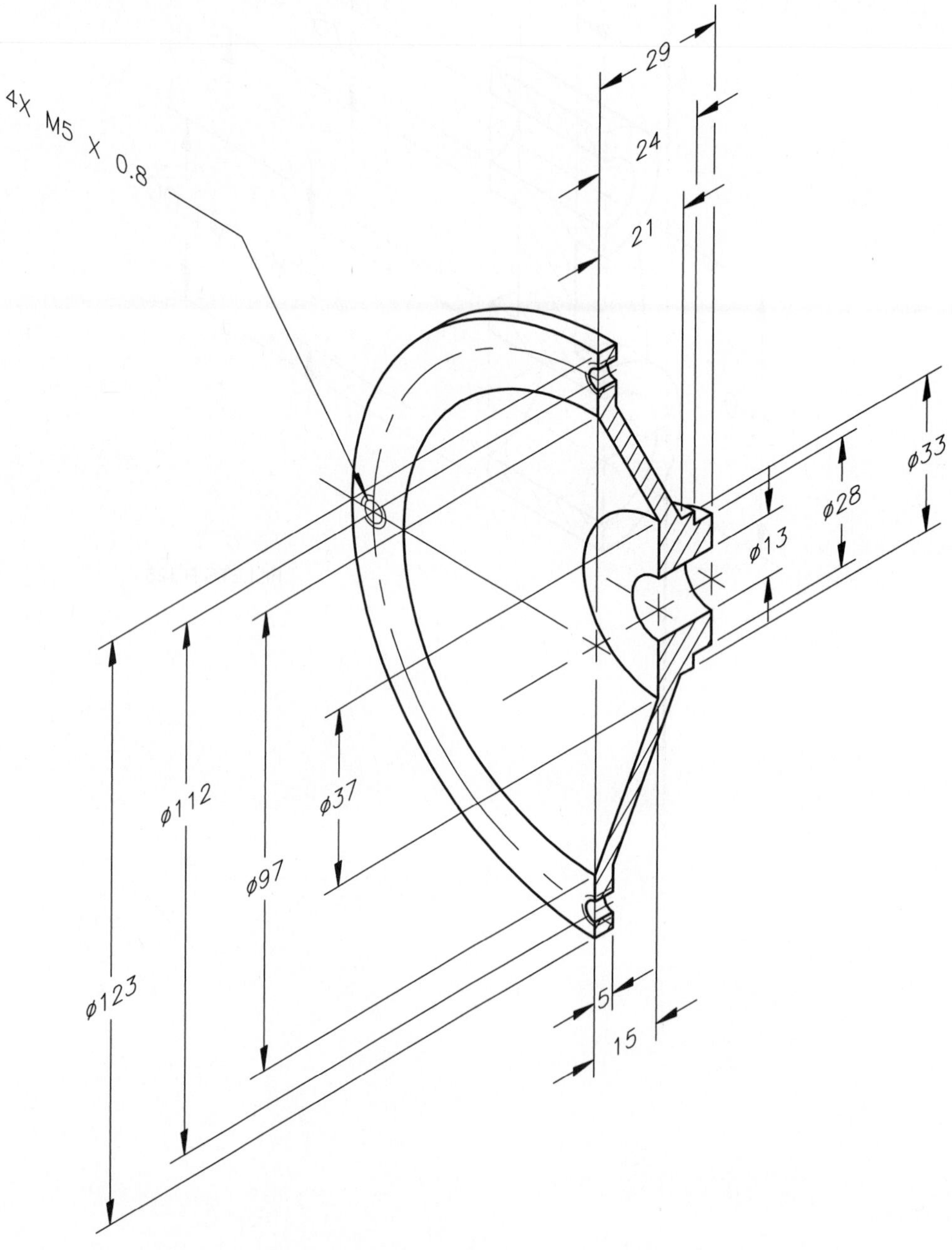

9. Draw the object shown below as a solid model. Use your own dimensions. Then, do the following.
   A. Construct a section plane that creates an offset section. The section should pass through the center of two holes in the base and through the large central hole.
   B. Display the intersection fill as an ANSI hatch pattern.
   C. Activate live sectioning and view the cutaway geometry with a low level of transparency in the color red.
   D. Generate a 3D section of the sectioned solid model and save it as a block.
   E. Use **SOLVIEW** to create a two-view orthographic layout. One view should be a half section. Use an appropriate scale to plot on a B-size sheet.
   F. Create a third floating viewport and insert the 3D section block scaled to half the size of the drawing.
   G. Save the drawing as P14_09.

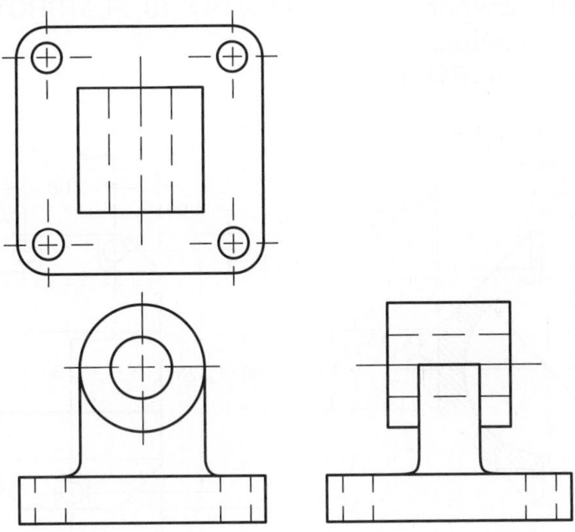

Drawing Problems - Chapter 14

10. Draw the object shown below as a solid model. Do not dimension the object. Then, do the following.
    A. Display a 3D view of the model, generate a flat view, and save it as a block named P14_10_ FLATSHOT.
    B. Construct a section plane that creates a half section.
    C. Display the intersection fill as an ANSI hatch pattern.
    D. Activate live sectioning and view the cutaway geometry with a low level of transparency in the color red.
    E. Alter the section plane to create the section shown below.
    F. Generate a 3D section and save it as a block.
    G. Use **SOLVIEW** to create a two-view orthographic layout. One view should be a full section. Use an appropriate scale to plot on an A-size sheet.
    H. Create a third floating viewport and insert the 3D section block scaled to half the size of the drawing.
    I. Create a fourth viewport and insert the P14_10_FLATSHOT block scaled to half the size of the drawing.
    J. Save the drawing as P14_10.

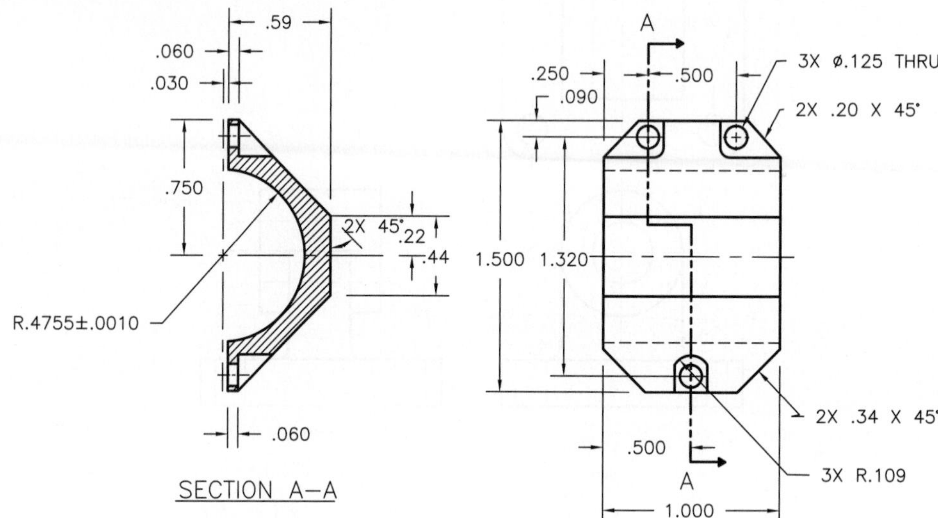

Drawing Problems - Chapter 14

CHAPTER 15

Visual Style Settings and Basic Rendering

Learning Objectives

After completing this chapter, you will be able to:

✓ Describe the **Visual Style Manager** palette.
✓ Change the settings for visual styles.
✓ Create custom visual styles.
✓ Export visual styles to a tool palette.
✓ Render a scene using sunlight.
✓ Save a rendered image from the **Render** window.

In Chapter 1, you were introduced to the default visual styles. In this chapter, you will learn about all visual style settings and how to redefine the visual style. You will also learn how to create your own visual style. Finally, this chapter provides an introduction to lights and rendering.

Overview of the Visual Styles Manager

The **Visual Styles Manager** palette provides access to all of the visual style settings. This palette is a floating window similar to the **Properties** palette. See Figure 15-1. Changes made in the **Visual Styles Manager** redefine the visual style.

At the top of the **Visual Styles Manager** are image tiles for the defined visual styles. See Figure 15-2. The default visual styles are 2D Wireframe, Conceptual, Hidden, Realistic, Shaded, Shaded with Edges, Shades of Gray, Sketchy, Wireframe, and X-Ray. User-defined visual styles also appear as image tiles. The image on the tile is a preview of the visual style settings. Selecting an image tile provides access to the properties of the visual style in the palette below. The name of the currently selected visual style appears below the image tiles and the corresponding image tile is surrounded by a yellow border.

To set a different visual style current using the **Visual Styles Manager**, double-click on the image tile. You can also select the image tile and pick the **Apply Selected Visual Style to Current Viewport** button immediately below the image tiles. An icon consisting of a small image of the button is displayed in the image tile of the visual style that is current in the active viewport, as shown in Figure 15-2. A drawing icon appears in the image tile if the visual style is current in a viewport that is not active. The AutoCAD icon appears in the image tiles of the default visual styles.

Figure 15-1.
The **Visual Styles Manager** palette.

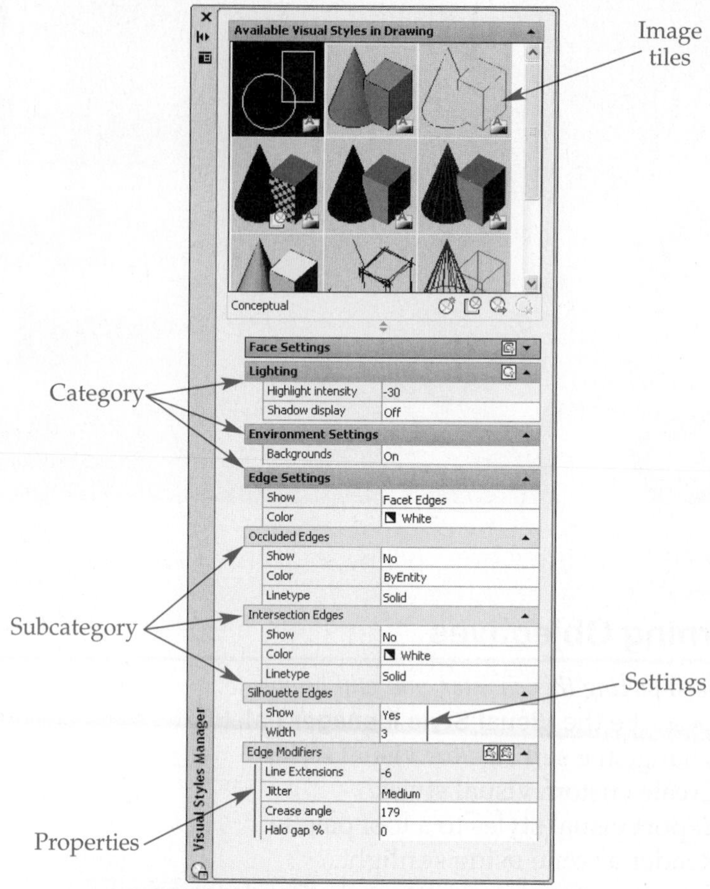

Image tiles

Category

Subcategory

Settings

Properties

Figure 15-2.
The image tiles correspond to the visual styles. The image on the tile is a preview of the visual style's settings.

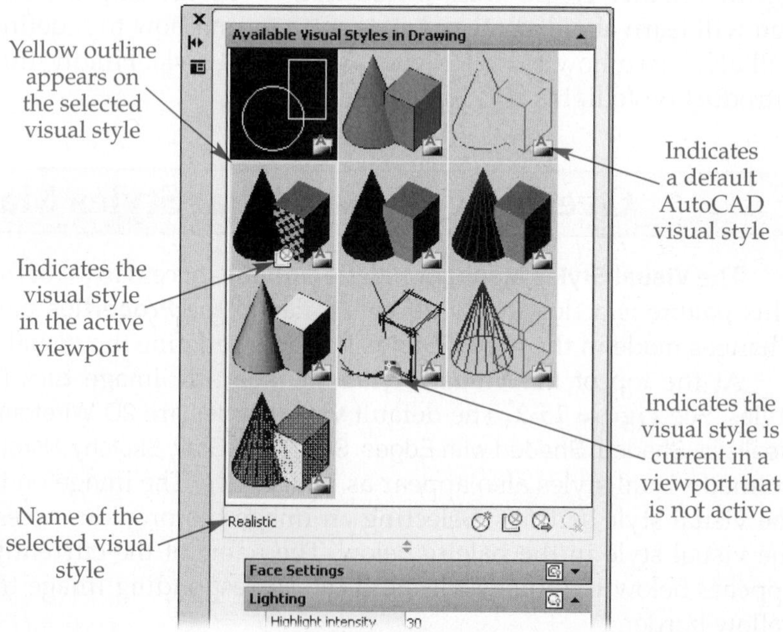

Yellow outline appears on the selected visual style

Indicates a default AutoCAD visual style

Indicates the visual style in the active viewport

Indicates the visual style is current in a viewport that is not active

Name of the selected visual style

Exercise 15-1

Complete the exercise on the student website.
www.g-wlearning.com/CAD

Visual Style Settings

The **Visual Styles** panel on the **View** tab of the ribbon provides several settings for altering the visual style. These settings are also available in the **Visual Styles Manager**. In addition, there are settings in the **Visual Styles Manager** that are not available on the ribbon. The next sections discuss settings available in the **Visual Styles Manager** for the default visual styles.

Changing any setting in the **Visual Styles Manager** *redefines* the visual style. Changes made using the ribbon are temporary. This is important to remember.

2D Wireframe

When the 2D Wireframe visual style is set current, lines and curves are used to show the edges of 3D objects. Assigned linetypes and lineweights are displayed. All edges are visible as if the object is constructed of pieces of wire soldered together at the intersections (thus, the name *wireframe*). Either the 2D or 3D wireframe UCS icon is displayed. OLE objects will display normally. In addition, the drawing window display changes to the 2D Model Space context and parallel projection. For the 2D Wireframe visual style, the **Visual Styles Manager** displays the following categories. See **Figure 15-3.**

- **2D Wireframe Options**
- **2D Hide—Occluded Lines**
- **2D Hide—Intersection Edges**
- **2D Hide—Miscellaneous**
- **Display Resolution**

2D wireframe options

The Contour lines property controls the **ISOLINES** system variable. Isolines are the lines used to define curved surfaces on solid objects when displayed in a wireframe view. The setting is 4 by default and can range from 0 to 2047. Isolines are suppressed when the **HIDE** command is used with the 2D Wireframe visual style set current.

The Draw true silhouettes property controls the **DISPSILH** system variable. This determines whether or not silhouette edges are shown on curved surfaces. It is set to Off by default, which is equivalent to a **DISPSILH** setting of 0.

Figure 15-3.
The categories and properties available for the 2D Wireframe visual style.

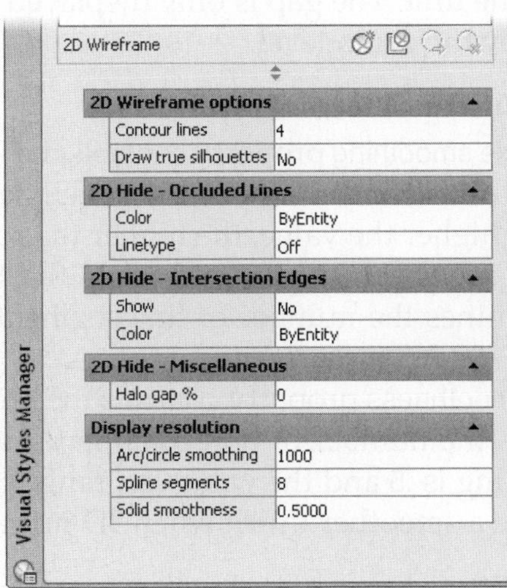

2D hide—occluded lines

The Color property in this category controls the **OBSCUREDCOLOR** system variable. This property determines the color of obscured (occluded) lines when the **HIDE** command is used. The default setting is ByEntity. This means that, when displayed, obscured lines are shown in the same color as the object.

The Linetype property controls the **OBSCUREDLTYPE** system variable. This property determines whether or not obscured lines are displayed and in which linetype they are displayed. The default setting is Off, which means that obscured lines are not displayed when the **HIDE** command is used. The available linetypes are: Solid, Dashed, Dotted, Short Dash, Medium Dash, Long Dash, Double Short Dash, Double Medium Dash, Double Long Dash, Medium Long Dash, and Sparse Dot.

NOTE

The above linetypes are not the same as the linetypes loaded into the **Linetype Manager** dialog box. They are independent of zoom levels, which means the dash size will stay the same when zooming in and out.

2D hide—intersection edges

This category is used to toggle the display of polylines at the intersection of 3D surfaces and set the color of the lines. The Show property controls the **INTERSECTIONDISPLAY** system variable. This property determines whether or not polylines are displayed at the intersection of non-unioned 3D surfaces. The default setting is Off, which means that polylines are not displayed when the **HIDE** command is used.

The Color property in this category controls the **INTERSECTIONCOLOR** system variable. This property determines the color of the polylines displayed at intersection edges. By default, the setting is ByEntity. This means that, when displayed, the polylines at intersection edges are shown in the same color as the object.

2D hide—miscellaneous

The Halo Gap % property controls the **HALOGAP** system variable. This property determines the gap that is displayed where one object partially obscures another (between the foreground edge and where the background edge starts to show). The default setting is 0 and the value can range from 0 to 100. The value refers to a percentage of one unit. The gap is only displayed when the **HIDE** command is used. It is not affected by the zoom level.

Display resolution

The Arc/circle smoothing property controls the zoom percentage set by the **VIEWRES** command. This determines the resolution of circles and arcs. The value can range from 1 to 20,000. The higher the value, the higher the resolution of circles and arcs.

The Spline segments property controls the **SPLINESEGS** system variable. This property determines the number of line segments in a spline-fit polyline. The value can range from –32,768 to 32,767.

The Solid smoothness property controls the **FACETRES** system variable. This property determines the number of polygonal faces applied to curved surfaces on solids. The default setting is .5 and the value can range from .01 to 10.0. A higher **FACETRES** value will create a smoother finish when 3D printing.

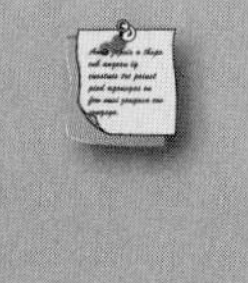

NOTE

Polygon faces will not be visible if the Draw true silhouettes property is set to On. However, a higher setting for the Solid smoothness property will make curved edges smoother.

Exercise 15-2

Complete the exercise on the student website.
www.g-wlearning.com/CAD

Conceptual

When the Conceptual visual style is set current, objects are smoothed and shaded. The shading is a transition from cool to warm colors. The transitional colors help highlight details. The previous projection is retained and that context is set current.

NOTE

The categories and properties for the shaded visual styles (those other than 2D Wireframe) are the same. These settings are discussed in the section Common Visual Style Settings later in this chapter.

Hidden

The Hidden visual style removes obscured lines from your view and makes 3D objects appear solid. The previous projection is retained and that context is set current. The benefit of using Hidden is that you get sufficient 3D display, but it does not push the graphics system too hard. Objects are not shaded or colored. This is very useful when working on complex drawings and/or using a slow computer.

Realistic

As with the Conceptual visual style, the objects have smoothing and shading applied to them when the Realistic visual style is set current. In addition, if materials are applied to the objects, the materials are displayed. The previous projection is retained and that context is set current. This visual style is good for a final look at the scene before rendering.

Shaded

This style is very similar to Realistic, except the lighting is smooth instead of smoothest. Also, textures are turned off.

Shaded with Edges

This style is the same as Shaded, except isolines are displayed with the default number of 4. Intersection edges are turned on and displayed as solid and white. Silhouette edges are also displayed with a width setting of 3.

Shades of Gray

This style is similar to the other shaded styles, except object colors are displayed as monochrome. Materials and textures are turned off. Facet edges are turned on, as are silhouette edges. This style is good for adjusting lights and shadows. It is much easier to see what is happening with lights when all objects are the same color.

Sketchy

The Sketchy visual style has most of the same settings as the Hidden visual style. However, objects look like they are hand-sketched. Line extensions and jitter are turned on, as are silhouette edges.

Wireframe

The Wireframe visual style is almost identical in appearance to the 2D Wireframe visual style. The main difference is that the 3D-mode UCS icon is displayed. The previous projection is retained and that context is set current. When working in 3D, a wireframe view is sometimes necessary to select objects normally hidden from your view.

X-Ray

This visual style is very similar to the Shaded with Edges visual style, except the Opacity property is set to 50%. This makes objects appear somewhat transparent.

Common Visual Style Settings

With the exception of 2D Wireframe, all of the visual styles share similar categories and settings in the **Visual Styles Manager**. The visual styles have **Face Settings**, **Lighting**, **Environmental Settings**, and **Edge Settings** categories. See Figure 15-4. These categories and the properties available in them are discussed in the next sections.

Face settings

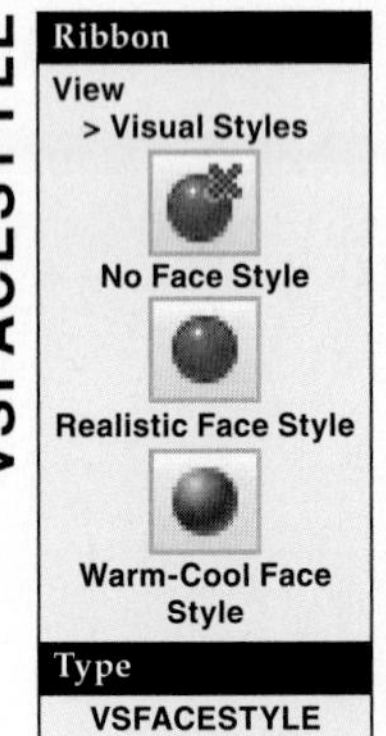

The Face style property controls the **VSFACESTYLE** system variable. This is also the same as selecting the **No Face Style**, **Realistic Face Style**, or **Warm-Cool Face Style** button in the **Visual Styles** panel on the **View** tab of the ribbon. Remember, though, selecting a button on the ribbon does not redefine the visual style. Rather, it is a temporary change to the viewport display.

The Lighting quality property controls the **VSLIGHTINGQUALITY** system variable. This property determines whether curved surfaces are displayed smooth or as a series of flat faces. Lighting quality is unavailable if the Face style property is set to None.

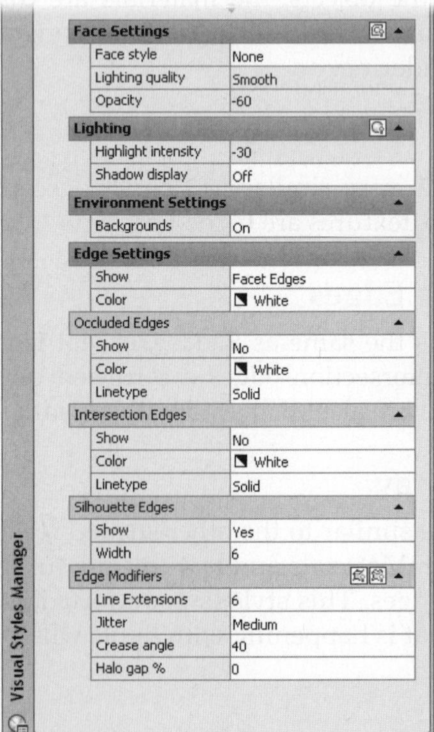

Figure 15-4.
These categories and properties are similar for all of the visual styles except 2D Wireframe.

The Color property controls the **VSFACECOLORMODE** system variable. This property determines how color is applied to the faces of an object. The effects can be temporarily applied by picking a button in the face colors flyout in the **Visual Styles** panel on the **View** tab of the ribbon. The choices are:

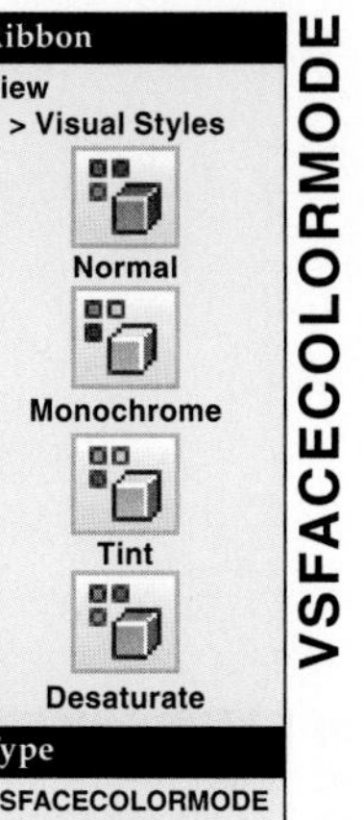

VSFACECOLORMODE

- Normal. The object color is applied to faces.
- Monochrome. One color is applied to all faces. This also displays and enables the Monochrome color property.
- Tint. A combination of the object color and a specified color is applied to faces. This also displays and enables the Tint color property. The Tint property only works when the Material display property is set to Materials.
- Desaturate. The object color is applied to faces, but the saturation of the color is reduced by 30%.

The Monochrome color and Tint color properties control the **VSMONOCOLOR** system variable. This system variable determines the color that is applied when the Face color mode property is set to Monochrome or Tint.

The Opacity property controls the **VSFACEOPACITY** system variable. This property determines how transparent or opaque faces are in the viewport. The value can range from –100 to 100. When the setting is 0, the faces are completely transparent. When the setting is 100, the faces are completely opaque. Settings below 0 set the value, but turn off the effect. To quickly turn the effect on or off, pick the **Opacity** button on the **Face Settings** category title bar. See **Figure 15-5.** This changes the value from negative to positive, or vice versa. This property cannot be changed if the Face style property is set to None.

Opacity may also be controlled by picking the **X-Ray Effect** button in the **Visual Styles** panel on the **View** tab of the ribbon. The adjacent **Opacity** slider sets the value. However, remember, picking a button or making settings on the ribbon is only temporary. The change is not saved to the visual style.

The Material display property controls the **VSMATERIALMODE** system variable. When set to Off, objects display in their assigned color. When the setting is changed to Materials, the objects display the color of the material, but not the textures. When the setting is changed to Materials and textures, full materials are displayed.

CAUTION

Displaying materials and textures on 3D objects in a complex drawing will slow system performance. Set the Face color mode property to Materials and textures only when it is absolutely necessary.

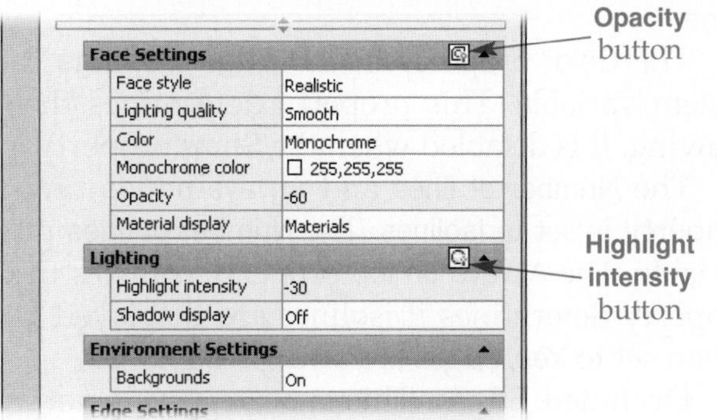

Figure 15-5.
The **Face Settings** category.

Lighting

The Highlight intensity property controls the **VSFACEHIGHLIGHT** system variable. This property determines the size of the highlight on faces to which no material is assigned. A small highlight on an object makes it look smooth and hard. A large highlight on an object makes it look rough or soft. The initial value is 30 or –30, the value can range from –100 to 100. The higher the setting is above 0, the larger the highlight. Settings below 0 set the value, but turn off the effect. To quickly turn the effect on or off, pick the **Highlight intensity** button on the **Lighting** category title bar. See Figure 15-5. This changes the value from negative to positive or vice versa. This property cannot be changed if the Face style property is set to None.

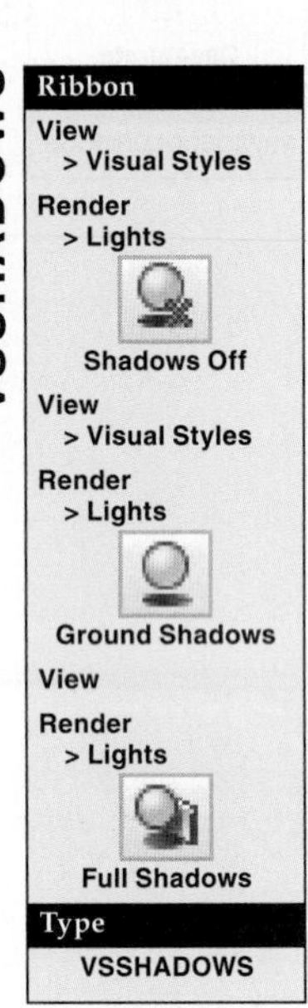

The Shadow display property controls the **VSSHADOWS** system variable. This property controls if and how shadows are cast when the visual style is set current. It is the same as picking a button in the shadows flyout in the **Visual Styles** panel on the **View** tab or in the **Lights** panel on the **Render** tab of the ribbon. Remember, however, a setting made on the ribbon is temporary and does not redefine the visual style.

If the property is set to Ground shadow, objects cast shadows on the ground, but not onto other objects. The "ground" is the XY plane of the WCS. The Mapped Object Shadows setting, which corresponds to the **Full Shadows** button on the ribbon, only works if lights have been placed in the scene and hardware acceleration is enabled.

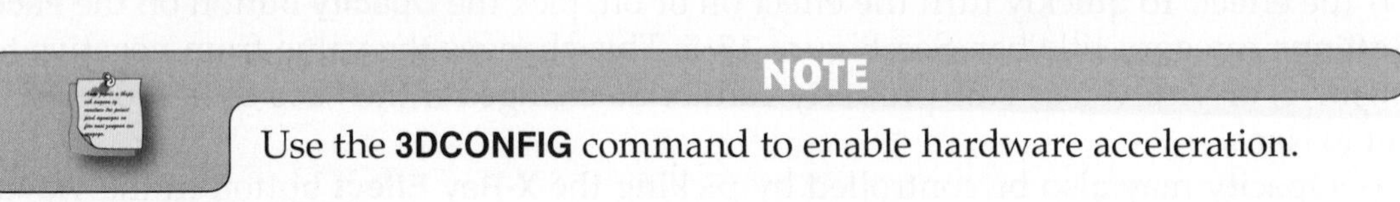

Use the **3DCONFIG** command to enable hardware acceleration.

Environment settings

The Backgrounds property controls the **VSBACKGROUNDS** system variable. This property determines whether or not the preselected background is displayed in the viewport. Backgrounds can only be assigned to a view when a named view is created or edited. After the view is created, restore the view to display the background.

Edge settings

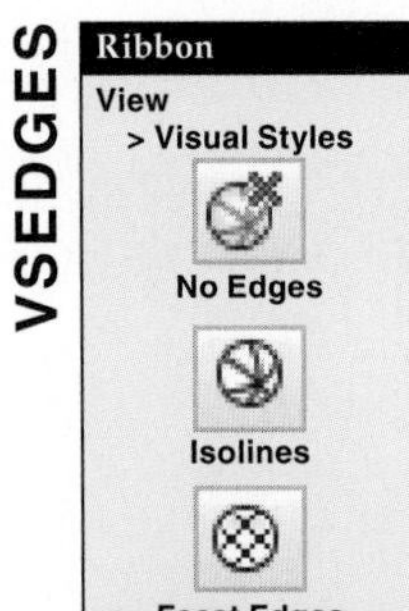

The Show property controls the **VSEDGES** system variable. This property determines how edges on solid objects are represented when the visual style is set current. This is the same as picking a button in the edge flyout in the **Edge Effects** panel on the **View** tab of the ribbon. The settings of Isolines and Facet Edges determine how edges and curved surfaces are displayed on a 3D model. This is discussed in detail in Chapter 14. Setting this property to None turns off isolines and facets and displays no edges. If the Face style property is set to None, the Show property cannot be set to None.

The Color property in the **Edge Settings** category controls the **VSEDGECOLOR** system variable. This property determines the color of all edges on objects in the drawing. It is disabled when the Show property is set to None.

The Number of lines and Always on top properties are displayed when the Show property is set to Isolines. The Number of lines property controls the **ISOLINES** system variable. The Always on top property controls the **VSISOONTOP** system variable. This property determines if isolines are displayed when the model is shaded or hidden. When set to Yes, edges are always displayed.

Occluded Edges. This subcategory is not available if the Show property in the **Edge Settings** category is set to None. The Show property in this subcategory controls the **VSOCCLUDEDEDGES** system variable. This property determines whether or not occluded edges are displayed in a hidden or shaded view. See Figure 15-6.

Figure 15-6.
A—Occluded lines are not shown. B—The Show property is set to Yes and occluded lines are shown.

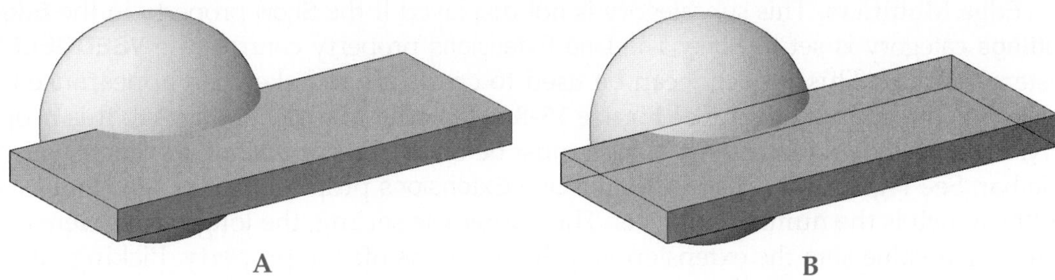

The Color property in this subcategory controls the **VSOCCLUDEDCOLOR** system variable. The Linetype property controls the **VSOCCLUDEDLTYPE** system variable. These properties function similar to those discussed earlier in this chapter in the 2D Hide—Occluded Lines section.

Intersection Edges. This subcategory is not available if the Show property in the **Edge Settings** category is set to None. The Show property in this subcategory controls the **VSINTERSECTIONEDGES** system variable. This property determines whether or not lines are displayed where one 3D object intersects another 3D object. See Figure 15-7.

The Color property in this subcategory controls the **VSINTERSECTIONCOLOR** system variable. The Linetype property controls the **VSINTERSECTIONLTYPE** system variable. These properties function similar to those discussed earlier in this chapter in the 2D Hide—Intersection Edges section.

PROFESSIONAL TIP

Setting the intersection edges Color property to a color that contrasts with the objects in your model is a good way to quickly check for interference between 3D objects.

Silhouette Edges. This subcategory is available for each setting for the Show property in the **Edge Settings** category. The Show property in this subcategory controls the **VSSILHEDGES** system variable. It determines whether or not silhouette edges are displayed around the outside edges of all objects.

Figure 15-7.
A—A line does not appear where these two objects intersect. B—The Show property is set to Yes and a line appears at the intersection.

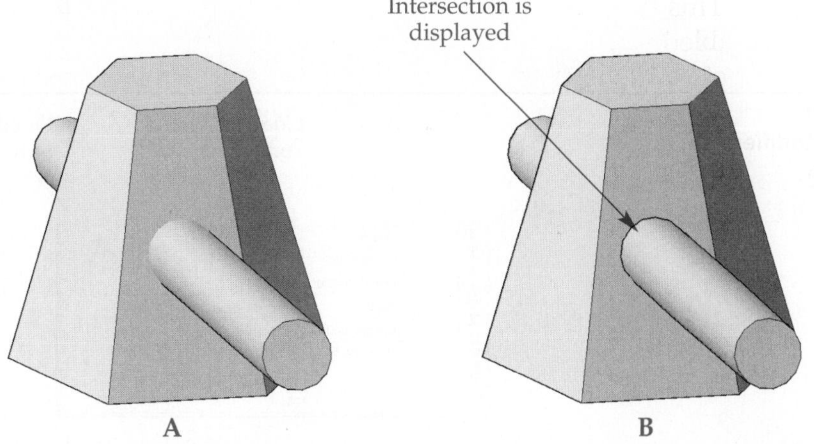

The Width property controls the **VSSILHWIDTH** system variable. This property determines the width of silhouette lines. It is measured in pixels and the value can range from 1 to 25.

Edge Modifiers. This subcategory is not displayed if the Show property in the **Edge Settings** category is set to None. The Line Extensions property controls the **VSEDGELEX** system variable. This property can be used to create a hand-sketched appearance by extending the ends of edges. See **Figure 15-8A.** In order to make changes to this property, the **Line Extensions edges** button must be on in the **Edge Modifiers** subcategory title bar. See **Figure 15-9.** The value for Line Extensions property can range from –100 to 100, which is the number of pixels. The higher the setting, the longer the extension. A negative value sets the extension length, but turns off the property. Picking either button makes the value positive and applies the effect (or makes the value negative and turns off the effect).

The Jitter property controls the **VSEDGEJITTER** system variable. Jitter makes edges of objects look as if they were sketched with a pencil. See **Figure 15-8B.** In order to make changes to this property, the **Jitter edges** button must be on in the **Edge Modifiers** subcategory title bar. See **Figure 15-9.** There are four settings from which to choose: Off, Low, Medium, and High. The number of sketched lines increases at each higher setting.

When the Show property in the **Edge Settings** category is set to Facet Edges, the Crease angle and Halo gap % properties are displayed in the **Edge Modifiers** subcategory. The Crease angle property controls the **VSEDGESMOOTH** system variable. This property determines how facet edges within a face are displayed based on the angle between adjacent faces. It does not affect edges between faces. See **Figure 15-10.** The value can range from 0 to 180. This is the number of degrees between edges below which a line is displayed. The Halo gap % property is similar to the setting discussed earlier in the 2D Hide—Miscellaneous section; however, this property controls the **VSHALOGAP** system variable.

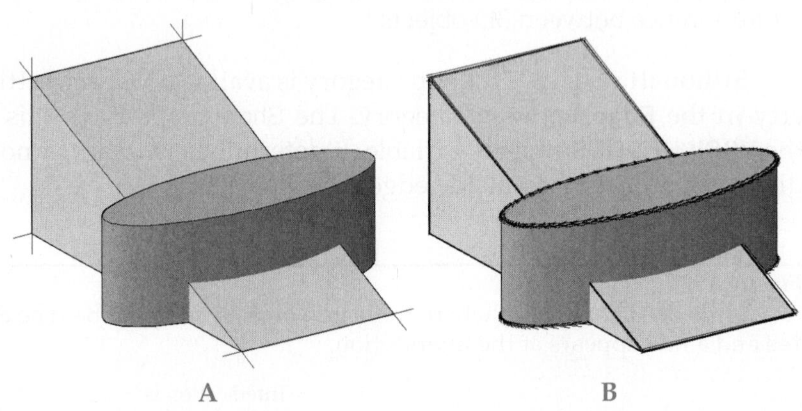

Figure 15-8.
A—Line extensions have been turned on for this visual style.
B—Jitter has been turned on for this visual style.

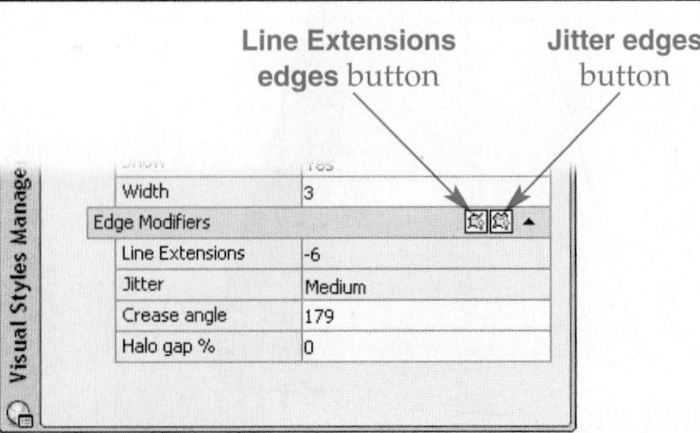

Figure 15-9.
The **Edge Modifiers** subcategory.

Figure 15-10.
A—The **Crease angle** property is set to 0. Notice the edges between facets within each face.
B—The **Crease angle** property is set to 10. The edges are no longer displayed.

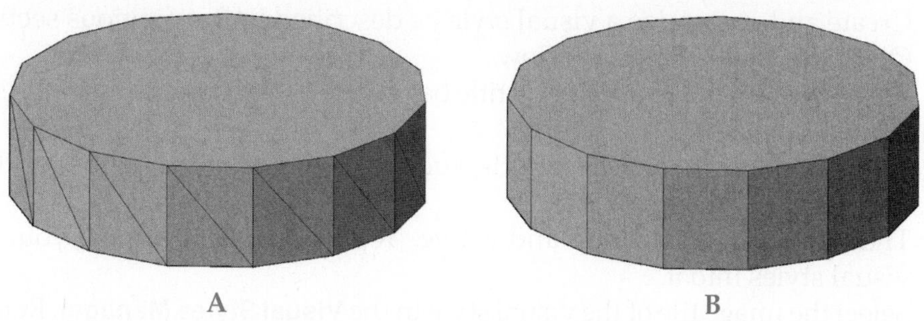

Creating Your Own Visual Style

As you saw in the previous sections, you can customize the default AutoCAD visual styles. However, you may also want to create a number of different visual styles to quickly change the display of the scene. Custom visual styles are easy to create.

To create a custom visual style, open the **Visual Style Manager**. Then, pick the **Create New Visual Style** button below the image tiles. You can also right-click in the image tile area and select **Create New Visual Style...** from the shortcut menu. In the **Create New Visual Style** dialog box that appears, type a name for the new style and give it a description. See Figure 15-11. Then, pick the **OK** button to create the new visual style.

An image tile is created for the new visual style. The name and description of the visual style appear as help text when the cursor is over the image tile. Select the image tile to display the default properties for the new visual style. Then, change the settings as needed to meet your requirements.

Custom visual styles are only saved in the current drawing. They are not automatically available in other drawings. To use the new visual styles in any drawing, they must be exported to a tool palette. This is discussed in the next section.

PROFESSIONAL TIP

To return one of AutoCAD's visual styles to its default settings, right-click on the image tile in the **Visual Styles Manager** and select **Reset to default** from the shortcut menu.

Exercise 15-3

Complete the exercise on the student website.
www.g-wlearning.com/CAD

Figure 15-11.
Creating a new visual style.

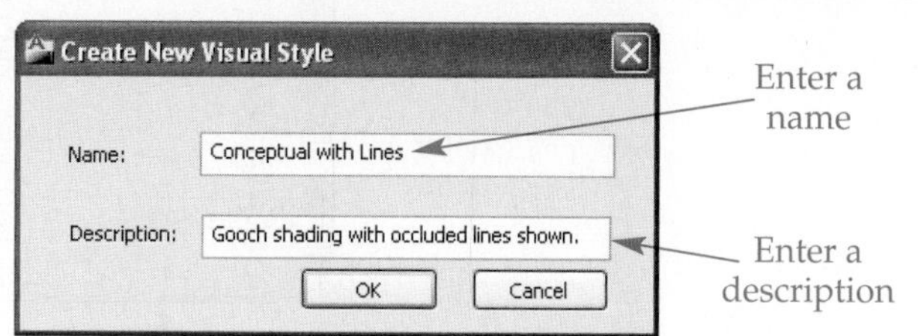

Steps for Exporting Visual Styles to a Tool Palette

To have custom visual styles available in other drawings, export them to a tool palette. Use the following procedure.

1. Create and customize a visual style as described in the previous section.
2. Open the **Tool Palettes** window.
3. Right-click on the **Tool Palettes** title bar and pick **New Palette** from the shortcut menu.
4. Type the name of the new palette, such as My Visual Styles, in the text box that appears. See Figure 15-12A.
5. The new palette is added and active. You are ready to export your custom visual styles into it.
6. Select the image tile of the visual style in the **Visual Styles Manager**. Remember, a yellow border appears around the selected image tile.
7. Pick the **Export the Selected Visual Style to the Tool Palette** button below the image tiles in the **Visual Styles Manager**. You can also right-click on the image tile and select **Export to Active Tool Palette** from the shortcut menu.

TOOLPALETTES

Ribbon
View
> Palettes
Tool Palettes
Type
TOOLPALETTES
TP
[Ctrl]+[3]

A new tool now appears in the palette with the same image, name, and description as the visual style in the **Visual Styles Manager**. See Figure 15-12B. Selecting the tool applies the visual style to the current viewport. You can also right-click on the tool to display a shortcut menu. Using this menu, you can apply the visual style to the current viewport, all viewports, or add the visual style to the current drawing. The shortcut menu also allows you to rename the tool, access the properties of the visual style, and delete the visual style from the palette.

PROFESSIONAL TIP

A visual style can be added as a tool on a tool palette by dragging its image tile from the **Visual Styles Manager** and dropping it onto the tool palette.

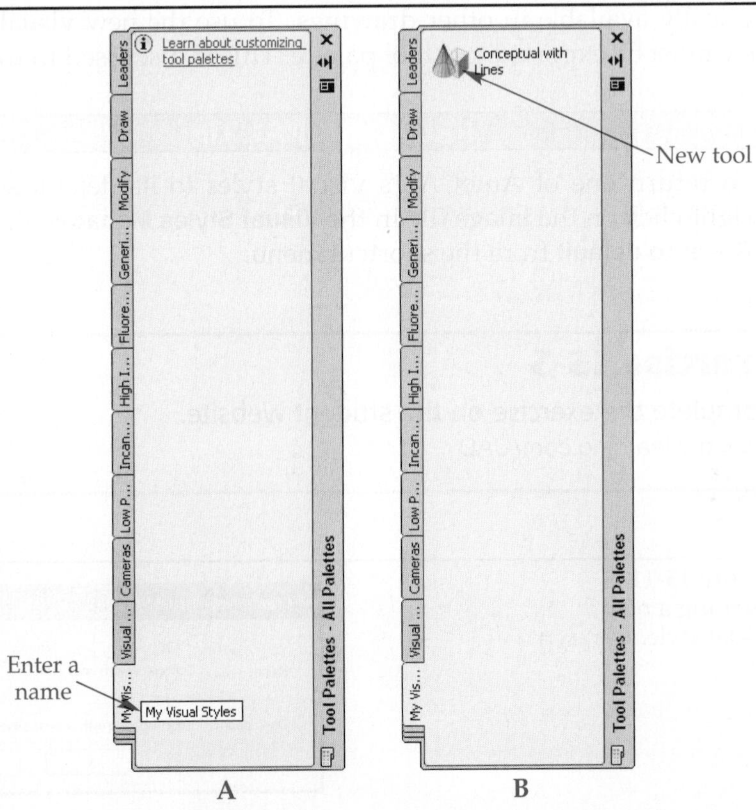

Figure 15-12.
A—Creating a new tool palette on which to place visual style tools.
B—A visual style has been copied to the tool palette as a tool.

Exercise 15-4

Complete the exercise on the student website.
www.g-wlearning.com/CAD

Deleting Visual Styles from the Visual Styles Manager

Custom visual styles can be deleted from the **Visual Styles Manager**. Pick the image tile of the visual style you want to delete. Then, pick the **Delete the Selected Visual Style** button below the image tiles. You can also right-click on the image tile and select **Delete** from the shortcut menu. You are *not* warned about the deletion. The default AutoCAD visual styles cannot be deleted, nor can a visual style that is currently in use.

Plotting Visual Styles

A visual style not only affects the on-screen display, it also affects plots. To plot objects with a specific visual style, use the following guidelines.

Plotting a Visual Style from Model Space

Open the **Plot** dialog box and expand it by picking the **More Options** (>) button. Then, select the desired display from the **Shade plot** drop-down list in the **Shaded viewport options** area. If the desired visual style is current in the viewport, you can also select As displayed from the drop-down list. Finally, plot the drawing.

Plotting a Visual Style from Layout (Paper) Space

When plotting from layout (paper) space, the shade plot properties of the viewports govern how the viewport is plotted. The viewports can be set to plot visual styles in three different ways.

In the first method, select the viewport in layout space and right-click to display the shortcut menu. Pick **Shade plot** to display the cascading menu. Then, select the appropriate visual style.

In the second method, use the **Properties** palette to set the Shade plot property of the viewport. To do this, select the viewport in layout space and open the **Properties** palette. Pick the Shade plot property in the **Misc** category and change the setting to the desired option. The Shade plot property is also available in the **Quick Properties** palette.

In the third method, use the **Visual Styles** suboption of the **Shadeplot** option of the **MVIEW** command. When prompted to select objects, pick the border of the viewport. Do not pick the objects in the viewport.

NOTE

The visual style of the viewport may also be selected when you create a viewport configuration in the **Viewports** dialog box (**VPORTS** command). Select the viewport in the **Preview** area of the dialog box. Then, pick the visual style desired from the **Visual Style:** drop-down list at the bottom of the dialog box. The **VPORTS** command can be used in model space or layout (paper) space.

Introduction to Rendering

RENDER

Ribbon
Render
> Render
Render

Type
RENDER
RR

Visual styles provide a way to plot your 3D scene to paper or a file, but control over the appearance is limited to the visual style settings. In Chapter 1, you were briefly introduced to the **RENDER** command. The **RENDER** command offers complete control over the scene and, with its features, you can create photorealistic images. In this chapter, you will be introduced to AutoCAD's rendering and lighting tools. Materials are discussed in later chapters along with more advanced rendering and lighting features.

When you render a scene, you are making a realistic image of your design that can be printed, displayed on a web page, or used in a presentation. To create an attractive rendering, you have to figure out what view you want to display, where the lights should be placed, what types of materials need to be applied to the 3D objects, and the kind of output that is needed. This section shows you how to create a quick rendering of your scene.

Introduction to Lights

Lights provide the illumination to a scene and are essential for rendering. There are three types of lighting in AutoCAD—default lighting, sunlight, and user-created lighting. AutoCAD automatically creates two default light sources in every scene. These lights ensure that all surfaces on the model are illuminated and visible. The types of lighting are discussed in more detail in Chapter 17.

A scene can be rendered with the default lights, but the results are usually not adequate to produce a photorealistic image. See **Figure 15-13.** The appearance is very artificial and no shadows are created. Shadows anchor objects to the scene and make them look real. See **Figure 15-14.** Without shadows, objects appear to float in space. Because the default lights do not cast shadows, other lights must be added to the scene and set to cast shadows. When a light is added to a scene, the default lights must be turned off. The first time you add a light, you receive a warning to this effect (unless the warning has been disabled).

Figure 15-13.
A scene rendered with the default AutoCAD lighting.

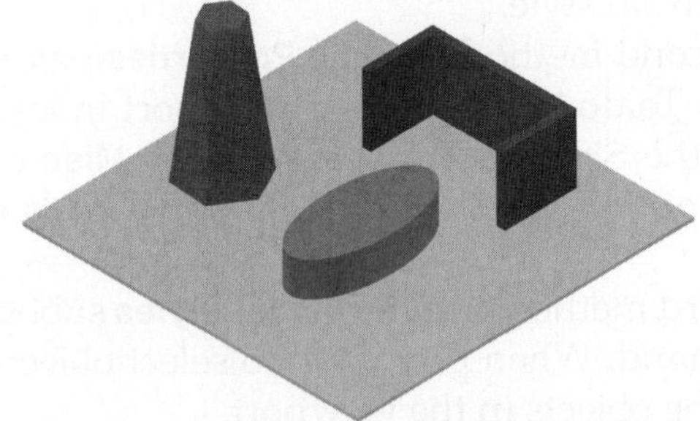

Figure 15-14.
A light has been added and set to cast shadows. Compare this rendering with Figure 15-13.

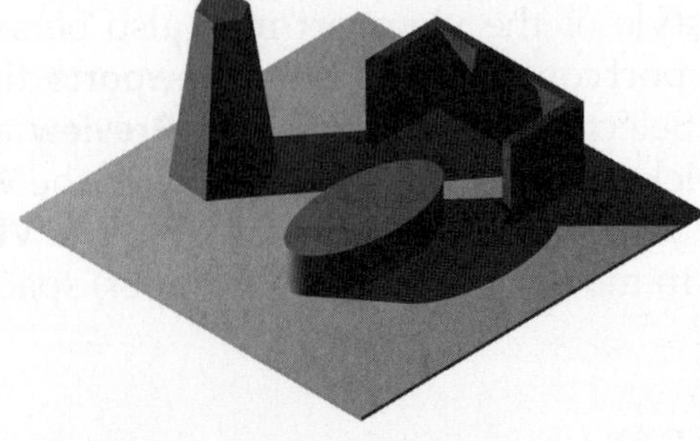

In this section, you will learn how to add sunlight to the scene. Chapter 17 provides detailed information on lighting. Sunlight is produced by an automated distant light. Sunlight can be turned on by picking the **Sun Status** button in the **Sun & Location** panel on the **Render** tab of the ribbon. The button background is blue when sunlight is on. See Figure 15-15.

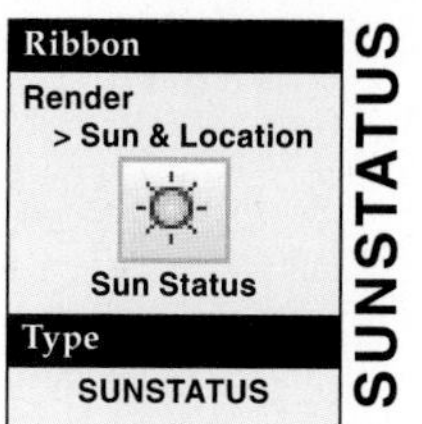

If the **Default Lighting** button is on in the **Lights** panel when the **Sun Status** button is turned on, a warning dialog box is displayed. This dialog box gives you choices to either turn off default lighting or keep it on. You cannot see the effects of sunlight with default lighting turned on, so it is recommended to turn it off.

If the current visual style is set to display full shadows, you should now see shadows in the scene, provided there are areas to receive shadows. Remember, hardware acceleration must be enabled to display full shadows.

The **Date** and **Time** sliders in the **Sun & Location** panel on the **Render** tab of the ribbon are active when sunlight is turned on. You can drag the sliders to adjust the date and time. The current date and time are displayed on the right-hand end of the sliders. As you drag the sliders, the shadows in the scene change to reflect the settings.

Rendering the Scene

The **Render** panel on the **Render** tab of the ribbon is shown in Figure 15-16. There are two buttons located in this control panel to initiate a rendering. If you pick the **Render** button, the **Render** window appears (by default) and AutoCAD immediately

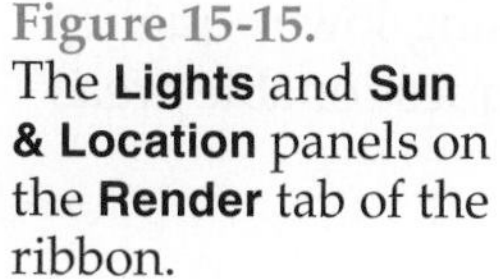

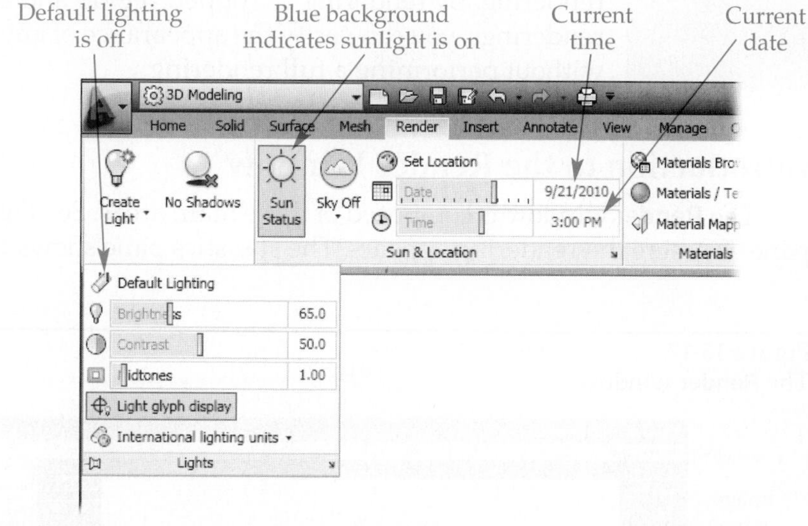

Figure 15-15.
The **Lights** and **Sun & Location** panels on the **Render** tab of the ribbon.

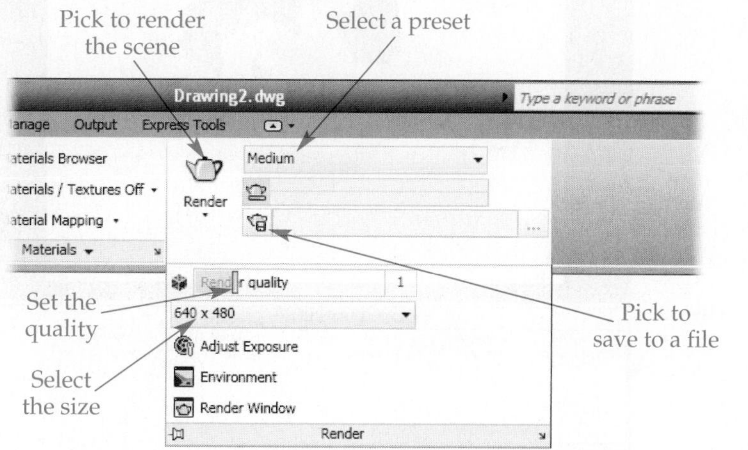

Figure 15-16.
The **Render** panel on the **Render** tab of the ribbon.

starts rendering the viewport. You will see the rendered tiles appear in the image pane as they are calculated. The **Render** window is explained more in the next section.

If you pick the **Render Region** button, which is located in the flyout, you are prompted to pick two points in the viewport, similar to performing a window selection. The selected area is rendered in the viewport. Rendering a cropped area is often used to test areas of the scene for possible problems before performing the final rendering.

Also in the **Render** panel you will find the render preset drop-down list. This is located in the upper-right corner of the panel. This list gives you a selection of rendering presets based on image quality. The choices are:

- Draft
- Low
- Medium
- High
- Presentation

The Draft entry produces the lowest-quality rendering. The Presentation entry produces the highest-quality rendering. The better the quality, the longer it takes to complete the rendering process.

PROFESSIONAL TIP

Rendering a complex drawing may take a very long time and you do not want to repeat it because of some small error. It is important to make sure that everything in the scene is perfect before the final rendering. By rendering a cropped region and using lower-quality renderings, you can verify the appearance of any questionable areas without performing a full rendering.

Introduction to the Render Window

The **Render** window is composed of three main areas. See Figure 15-17. The image pane is where the rendering appears. The statistics pane shows the current rendering

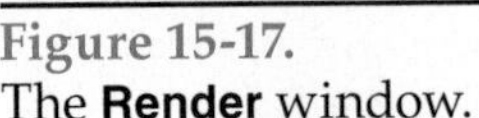

Figure 15-17.
The **Render** window.

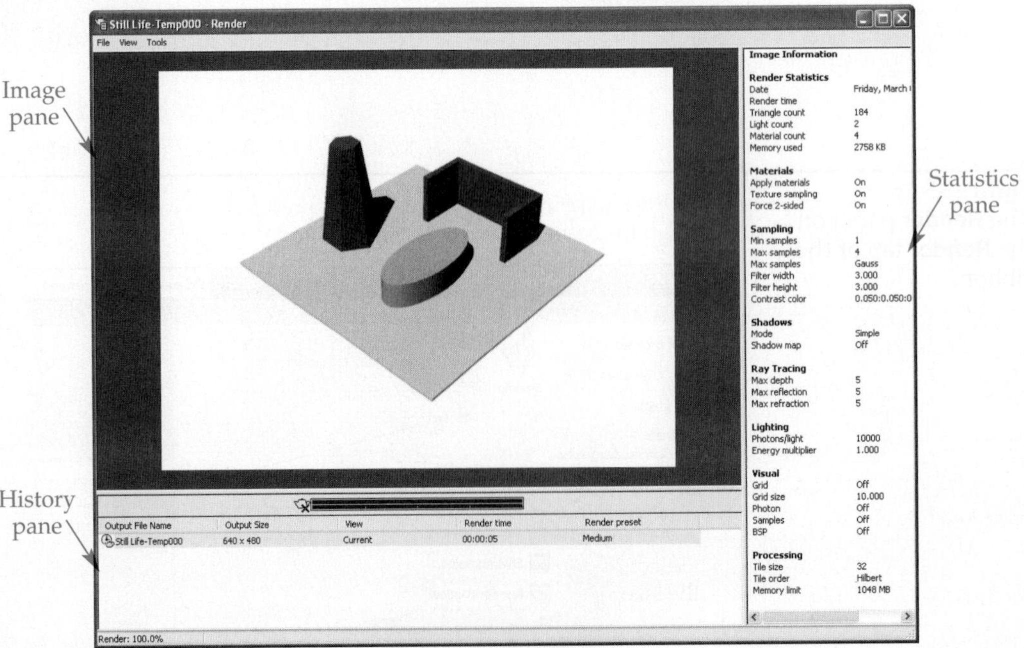

settings. The history pane shows a list of all of the images rendered from the drawing, with the most recent at the top.

You can zoom into the image in the image pane for detailed inspection. Use the mouse scroll wheel or the **Zoom +** and **Zoom –** entries in the **Tools** pull-down menu of the **Render** window. In the **File** pull-down menu of the **Render** window, select **Save...** to save the image selected in the history pane to an image file. The symbol in front of the image in the history pane changes to a folder with a green check mark on it. The **Save Copy...** option in the **File** pull-down menu of the **Render** window creates a copy of the image without modifying the original in the history pane.

NOTE

Advanced rendering is discussed in Chapter 18.

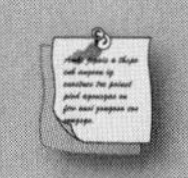

Exercise 15-5

Complete the exercise on the student website.
www.g-wlearning.com/CAD

Chapter Test

Answer the following questions. Write your answers on a separate sheet of paper or complete the electronic chapter test on the student website.
www.g-wlearning.com/CAD

1. What is the **Visual Styles Manager**?
2. Name the ten default AutoCAD visual styles that can be edited in the **Visual Styles Manager**.
3. Describe the difference between setting the Lighting quality property to Smooth and Faceted.
4. What does the Desaturate setting of the Color property in the Face Settings category do?
5. What has to be added to a scene before full shadows are displayed?
6. If you want to make your scene look hand sketched, but the Line Extensions and Jitter properties are not available, what other setting(s) do you have to change?
7. How do you set a visual style to display silhouette edges?
8. List the four settings for the Jitter property.
9. What is an *intersection edge?*
10. How do you make your own visual styles available in other drawings?
11. Which visual styles cannot be deleted?
12. How can you turn on sunlight?
13. Explain the function of the **Render Region** button in the **Render** panel on the ribbon.
14. Name the three main areas of the **Render** window.
15. How can you save a rendered image in the **Render** window?

Drawing Problems

1. This problem demonstrates the differences in rendering time and image quality of the five different rendering presets.
   A. Open any 3D drawing from a previous chapter and display an appropriate isometric view. Change the projection to perspective, if it is not already current.
   B. Turn on sunlight and set the **Date** and **Time** sliders to place the shadows where you want them. Tip: Turning on full shadows allows you to locate the shadows without rendering.
   C. Render the scene once for each rendering preset: Draft, Low, Medium, High, and Presentation.
   D. In the history pane of the **Render** window, note the differences between the rendering time for each rendering.
   E. Save each image with a corresponding name: P15_01_Draft.jpg, P15_01_Low.jpg, P15_01_Medium.jpg, P15_01_High.jpg, and P15_01_Presentation.jpg.
   F. Save the drawing as P15_01.

2. In this problem, you will construct a living room scene using some simple shapes and blocks available through **DesignCenter**.
   A. Draw a 12′ × 12′ × 1″ box.
   B. Draw two boxes to represent walls, 12′ × 4″ × 8′. Position them as shown on the next page.
   C. Open **DesignCenter** and select the **Autodesk Seek Design Content** button (you must be connected to the Internet).
   D. Search for 3D drawing files for the following objects, download the files, open them, and then drag and drop them into the scene: sofa, table, end table, lamp, plant, entertainment center, and chair. The blocks do not have to be exactly the same as shown here and may need to be scaled up or down. Position them as shown.
   E. Apply each of the five default visual styles to the viewport and plot each. Use the As displayed option in the **Plot** dialog box. Note the differences in each one.
   F. Save the drawing as P15_02.

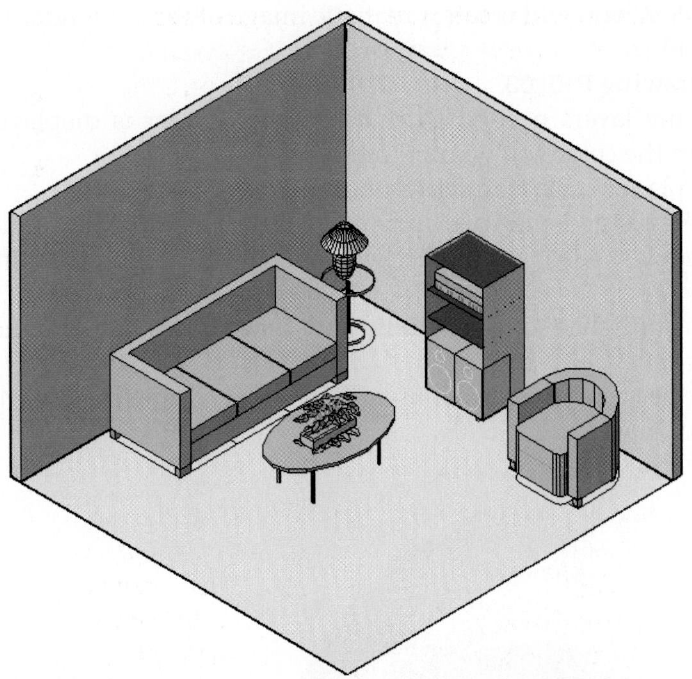

3. In this problem, you will create a new visual style to display the scene as if it is hand sketched.
   A. Open drawing P15_02.
   B. Create a new visual style named Hand Sketched with a description of Displays objects as sketched.
   C. Change the Line Extensions and Jitter settings to make the scene look as shown below.
   D. Change any other settings you like.
   E. Plot the scene. Select the new visual style in the **Shade plot** drop-down list.
   F. Export the new visual style to a tool palette so that it can be used in other drawings.
   G. Save the drawing as P15_03.

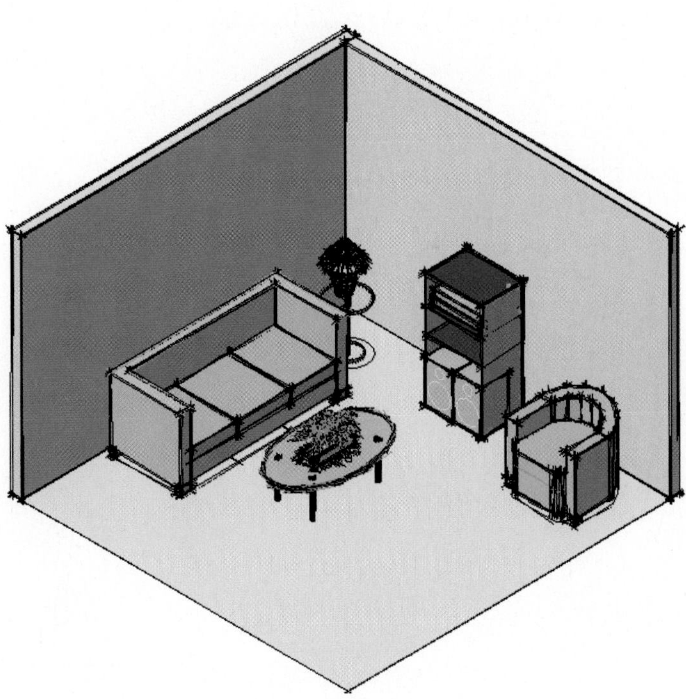

4. In this problem, you will create a realistic image of the car fender that you created in Chapter 10.
   A. Open drawing P10_03.
   B. Freeze any layers needed so that only the fender is displayed. Display the fender in the color you want it to be.
   C. Draw a planar surface to represent the ground.
   D. Set the Realistic visual style current. Then, turn on the highlight intensity and full shadows. Also, set the Show property in the **Edge Settings** category to None.
   E. Turn on sunlight. Adjust the **Date** and **Time** sliders to make the shadows look as shown.
   F. Render the scene and save it as a JPEG image. Name the file P15_04.jpg.
   G. Save the drawing as P15_04.

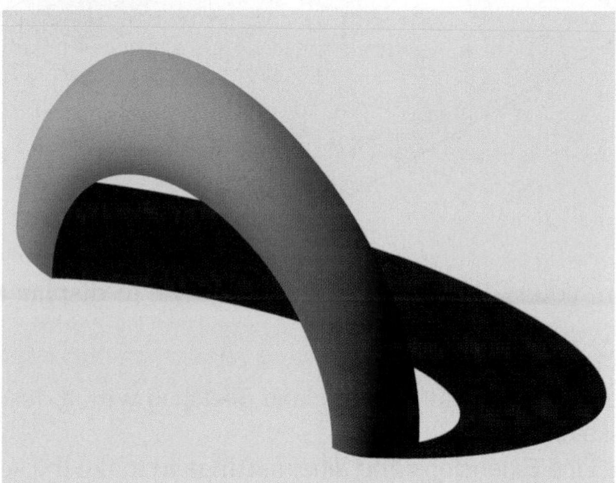

CHAPTER 16

Materials in AutoCAD

Learning Objectives

After completing this chapter, you will be able to:

✓ Attach materials to the objects in a drawing.
✓ Change the properties of existing materials.
✓ Create new materials.

A ***material*** is simply an image stretched over an object to make it appear as though the object is made out of wood, marble, glass, brick, or various other materials. AutoCAD provides an assortment of materials that can be used in your drawings to create a realistic scene. The materials are grouped into categories to make them easier to find.

Materials are easy to attach. They can be dragged and dropped onto the objects, attached to all selected objects, and even attached based on the object's layer. Once the material is attached, you can adjust how the material is *mapped* to the object. If the current visual style is set to display materials in the viewport, you can immediately see the effects on the object. The properties of a material can also be changed to make it look shinier, softer, smoother, rougher, and so on. When you finally render the scene, you will see the full effect of the materials.

The ***materials library*** is the location where all materials are stored. When you install AutoCAD, the Autodesk Material Library is installed. It consists of three sublibraries. The Autodesk Material Library 2011 images are used to show materials in the viewport when the visual style is set to display them. This library is installed by default. The images in the Autodesk Material Library 2011 Base Image Library are low resolution (512 × 512) and are used with AutoCAD materials. This library is also installed by default.

The Autodesk Material Library 2011 Medium Image Library is loaded with medium-resolution images (1024 × 1024) that are good for close-up work or large-scale model rendering. This library is an option that you must select when you install AutoCAD. If you did not choose this option during installation, you can use the installation DVD to add it. For information describing how to do this, pick the **Help** button in the **InfoCenter** or press [F1] to display the online documentation. Under Installation, select Installation Guide. Then, select Stand-Alone Installation>Installing and Running the Product. Finally, select Installing Autodesk Material Libraries.

Materials Browser

MATBROWSEROPEN

Ribbon
Render
> Materials
Materials Browser
Type
MATBROWSEROPEN
RMAT
MAT

In AutoCAD, the **Materials Browser** palette is used to manage the materials library. This is generically called the ***materials browser.*** It provides access to all materials that are available in the Autodesk libraries and from other sources. The **Materials** panel in the **Render** tab of the ribbon contains buttons for accessing the materials browser and the material editor. See **Figure 16-1.**

There are two main areas in the materials browser, **Figure 16-2.** The **Document Materials** area contains materials that have been selected for use in the current drawing. The **Libraries** area shows all available libraries from which materials may be selected. These areas are discussed in more detail in the next sections.

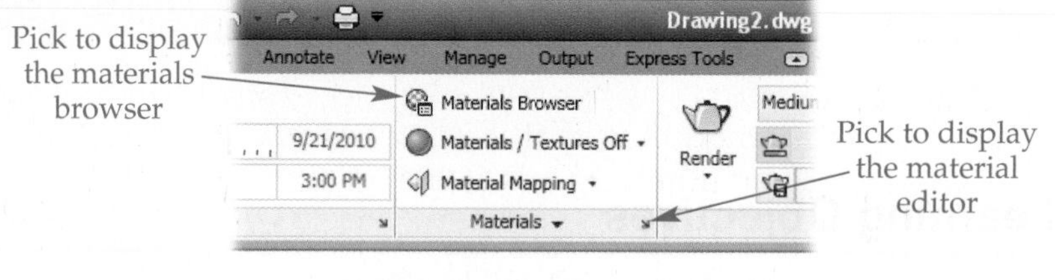

Figure 16-1. The **Materials Browser** button is located on the **Materials** panel in the **Render** tab of the ribbon. The materials editor is displayed by picking the dialog box launcher on the panel.

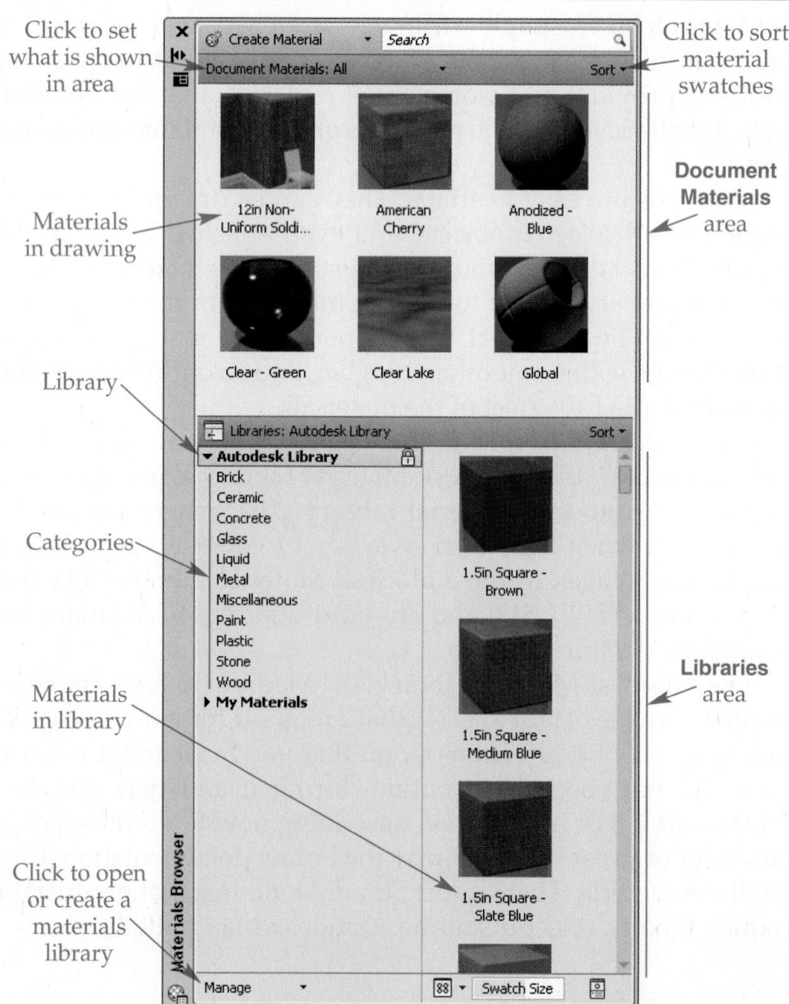

Figure 16-2. The materials browser contains all of the available materials.

Document Materials Area

Materials are added to the **Document Materials** area by selecting them in the **Libraries** area. Single-clicking on a material swatch in the library adds it to the drawing. A swatch for the material then appears in the **Document Materials** area.

Picking the **Documents Materials** title bar opens a drop-down menu. The options in this menu allow you to control which materials are displayed in the **Document Materials** area. You can show all materials, only the materials that are applied to objects in the drawing, selected materials, or unused materials. In addition, you can purge unused materials from the **Document Materials** area.

The options in the **Sort** drop-down menu control how the material swatches are arranged. You can sort materials by name, type, or material color.

Right-clicking on a material swatch displays a shortcut menu. There are several options available in the shortcut menu:

- **Edit.** This displays the material editor for editing the selected material. The material editor is discussed later in this chapter.
- **Duplicate.** A copy of the material is added to the drawing. A new material swatch is added with the same name as the original material, but with a sequential number added to the name.
- **Rename.** This option is used to change the name of the material. It is a good idea to rename all of your materials to meaningful names.
- **Delete.** This option removes the material from the drawing. If it is currently being used in the drawing, a warning alerts you to this fact. Continuing with the deletion removes the material from the objects in the drawing as well as from the **Document Materials** area. It is not, however, removed from the library.
- **Select Objects Applied To.** All objects in the drawing that have the material applied to them are selected in the drawing window.
- **Add to.** This option displays a cascading menu with two choices. The material may be added to the **My Materials** category in the **Libraries** area. This is excellent for organizing the materials that you like to use. You can also add the material to the active tool palette.
- **Purge All Unused.** This option removes from the drawing all materials not currently assigned to an object.

Libraries Area

The **Libraries** area of the materials browser contains the open library files. The libraries are recognizable by the name in bold and a triangle before the name. Pick the triangle to expand or collapse the library.

Libraries may have categories within them. Category names are not bold. Categories may be created by the user or automatically created when a material is added. The material type determines its category.

By default, two libraries exist when AutoCAD is installed: Autodesk Library and My Materials. The Autodesk Library is composed of the materials installed when AutoCAD was installed. My Materials is empty by default, but you may add materials to it. Custom materials are only available in the drawing in which they were created unless saved to a materials library, such as My Materials.

Materials are displayed in the right-hand column as a swatch and name. The materials in this column change when a different category is selected.

The **Manage** bar at the bottom provides options for managing materials libraries and categories:

- **Open Existing Library.** Allows you to select a library file (*.adsklib) so it is displayed in the materials browser.
- **Create New Library.** Gives you the option of saving a library you assembled in the materials browser. Library files are saved with the .adsklib file extension.

- **Remove Library.** Used to delete libraries from the materials browser.
- **Create Category.** Allows you to add categories in the **Library** area. Doing so, you can organize the materials within a materials library you have created.
- **Delete Category.** Removes categories.
- **Rename.** Allows you to rename libraries and categories.

At bottom-center of the materials browser, there is a flyout with three options for the display of material information. **Grid View** is active by default and shows the material swatch and its name. **List View** shows the same information with columns to the right for type and category. **Text View** turns off the swatches and just displays the material name, type, and category. The slider next to this button controls the swatch size in the materials browser. The right-hand button on the bottom of the materials browser opens the material editor.

NOTE

You cannot rename or remove any of the default Autodesk libraries or categories, nor can you add categories to or remove categories from them.

Exercise 16-1

Complete the exercise on the student website.
www.g-wlearning.com/CAD

Applying and Removing Materials

To attach a material to an object in the drawing, you can drag the material from the materials browser and drop it onto an object. To apply a material only to a face on an object, hold the [Ctrl] key and pick the face. To apply a different material to an object, simply select the new material in the materials browser and pick the object again.

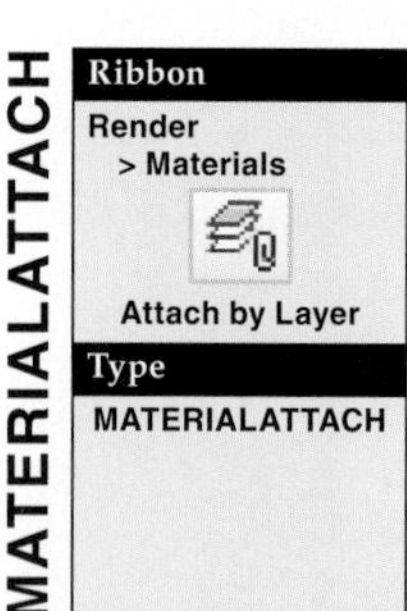

You can use the **MATERIALATTACH** command to assign materials to the layers in your drawing. Once a material is assigned to a layer, any object on that layer is displayed in the material, as long as the object's material property is set to ByLayer. When objects are created in AutoCAD, the default "material" assigned to them is ByLayer. If your objects are organized on layers, this is the easiest way to attach materials. You can override the layer material by applying a material to individual objects.

Remove Materials
Ribbon
Render
> Materials
Remove Materials

Figure 16-3 shows the **Material Attachment Options** dialog box displayed by the **MATERIALATTACH** command. The list on the left side of the dialog box shows the materials loaded into the drawing. The right side of the dialog box shows the layers in the drawing and the material attached to each layer. When no material is attached to a layer, the material is listed as Global. The Global material is a "blank" material in every drawing. To attach a material to a layer, drag the material from the list on the left and drop it onto the layer name on the right. To remove a material from a layer, pick the **X** button next to the material name on the right side of the dialog box.

The easiest way to remove a material from an individual object is with the **Remove Materials** button on the **Materials** panel in the **Render** tab of the ribbon. When the button is picked, a paintbrush selection cursor is displayed. Selecting an object replaces the material assigned to it with the Global material.

A material can also be removed using the **Properties** palette. To remove a material from an object or subobject, simply change its Material property in the **3D Visualization**

Figure 16-3.
Attaching materials to layers.

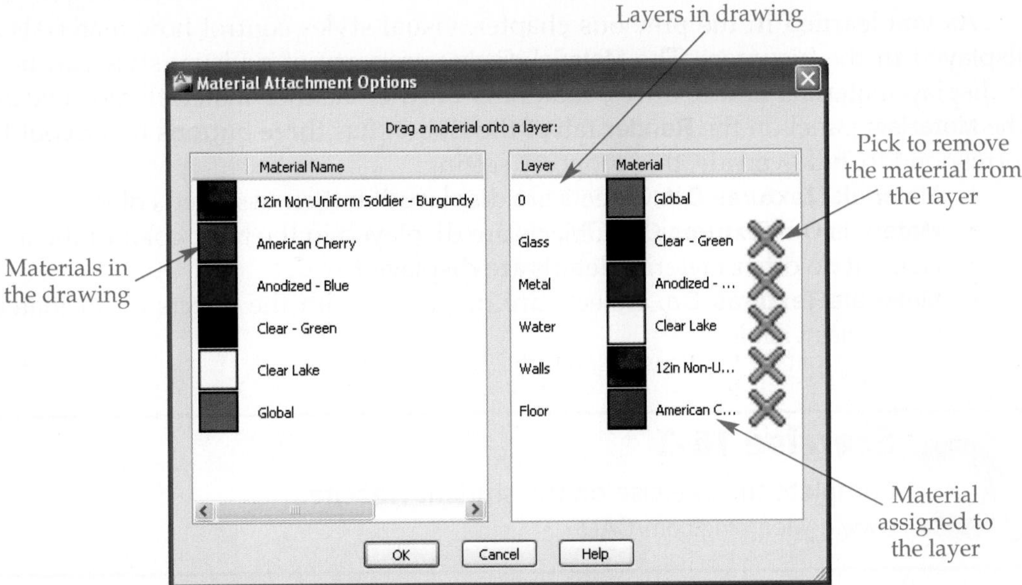

category to Global. If a material has not been assigned to the object's layer, the property can also be set to ByLayer. See **Figure 16-4.**

PROFESSIONAL TIP

A material can be applied to an object by dragging the material from the materials browser and dropping it onto the object. Also, a material can be loaded into the drawing without attaching it to an object by picking the material swatch once in the **Libraries** area of the materials browser or by dragging and dropping it into a blank area of the drawing. This makes the material available in the drawing.

Exercise 16-2

Complete the exercise on the student website.
www.g-wlearning.com/CAD

Figure 16-4.
Removing a material from an object. A—The material is assigned. B—The material is removed.

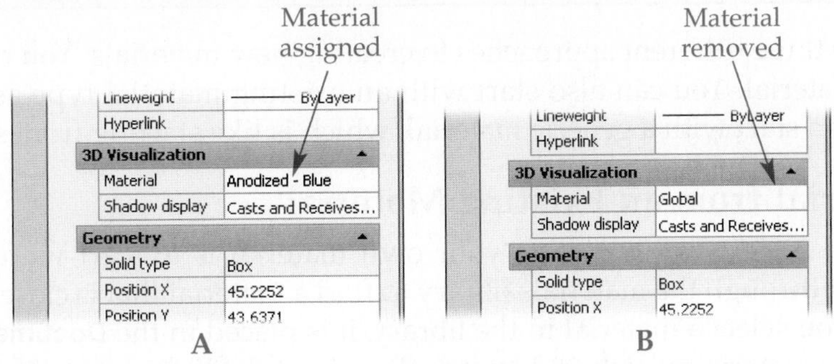

Material Display Options

VSMATERIALMODE

Ribbon
Render
> Materials
Materials/Textures Off
Materials On/Textures Off
Materials/Textures On
Type
VSMATERIALMODE

As you learned in the previous chapter, visual styles control how materials are displayed in the viewport. The Material display property of a visual style can be set to display materials and textures, materials only, or neither materials nor textures. The **Materials** panel on the **Render** tab of the ribbon has three buttons in a flyout that correspond to, but override, this property setting:

- **Materials/Textures Off.** Objects are displayed in their assigned colors.
- **Materials On/Textures Off.** Objects are displayed in the basic color of the material, but no other material details are displayed.
- **Materials/Textures On.** Objects are displayed with the effects of all material properties visible.

Exercise 16-3

Complete the exercise on the student website.
www.g-wlearning.com/CAD

Material Editor

MATEDITOROPEN

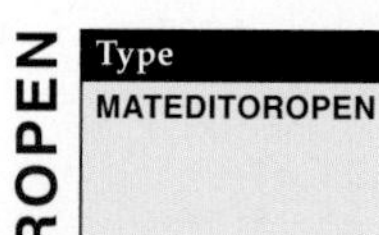

Type
MATEDITOROPEN

The ***material editor*** allows you to create new materials and edit existing materials to your liking. Creating and editing materials are discussed in detail in the next sections.

The **Materials Editor** palette is displayed by picking the dialog box launcher button at the bottom-right corner of the **Materials** panel in the **Render** tab of the ribbon. You can also type the **MATEDITOROPEN** command or double-click on a material swatch in the materials browser.

A preview of the material appears at the top of the material editor. The geometry used for the preview can be changed by picking the button next to the preview and selecting the type of geometry. There are twelve shapes: sphere, cube, cylinder, canvas, plane, object, vase, draped fabric, glass curtain wall, walls, pool of liquid, and utility. The name of the material can be changed by editing it in the **Name** text box.

Materials created or edited in the material editor are then added to the **Document Materials** area of the materials browser. Once the material is in the materials browser, it is a good idea to add it to one of your material libraries. Right-click on the material swatch in the **Document Materials** area, select **Add to**, and then select the library from the shortcut menu.

Creating and Modifying Materials

There are three different approaches to creating new materials. You can duplicate an existing material. You can also start with an existing material type as a template. Finally, you can start with a generic material, which is like starting from scratch.

New Material from an Existing Material

By far the easiest way to create your own material is to start with an existing material. Look through the materials library to find a material that is close to what you want. When you select a material in the library, it is placed in the **Document Materials** area. Select the material swatch and launch the material editor.

Figure 16-5.
This drop-down list can be used to select a material type or to copy the existing material.

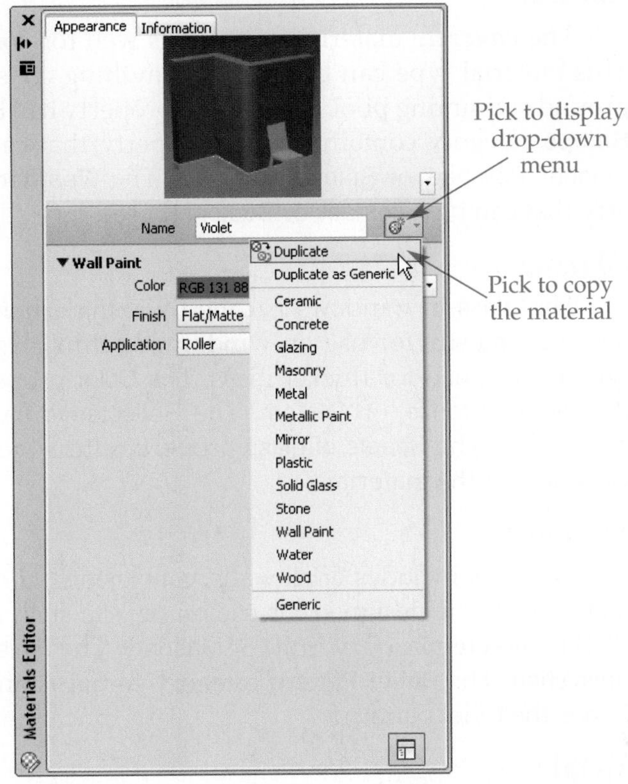

The material selected in the **Document Materials** area of the materials browser is displayed in the material editor. Pick **Duplicate** from the drop-down menu next to the material name in the material editor. See Figure 16-5. This creates a duplicate material with the same name and a sequential number. Change the name and the properties to your liking (discussed later) and close the material editor. The new material is in the **Document Materials** area of the materials browser ready to use.

New Material Using an Existing Material Type

Another way to create a material is by using a material type as a template for your new material. Open the material editor and select one of the material types from the drop-down menu next to the material name. The material type provides certain default settings as a starting point. The following sections discuss the material types, their possible applications, and any special properties the material type may have.

PROFESSIONAL TIP

Often, a material type can be used for a material that is completely unrelated to the name of the material type. For example, the concrete material type may actually serve well for dirt or sand. Be creative and do not limit yourself to what is implied by the name of the material type.

Ceramic

The *ceramic* material type is designed for ceramic floors. However, this material type may serve well for other glossy surfaces, such as countertops, bathtubs, or dinnerware. The Type property can be Ceramic or Porcelain. The Finish property can be High Gloss/Glazed, Satin, or Matte. The **Finish Bumps** category contains a Type property that can be Wavy or Custom. The **Relief Pattern** category contains an Image property. The image determines the relief pattern.

Concrete

The *concrete* material type works well for concrete floors, walls, and sidewalks. This material type can be used for anything constructed of concrete, such as an in-ground swimming pool. The Sealant property can be None, Epoxy, or Acrylic. The **Finish Bumps** category contains a Type property that can be Broom Straight, Broom Curved, Smooth, Polished, or Stamped/Custom. The **Weathering** category contains a Type property that can be Automatic or Custom (based on a selected image).

Glazing

The glass in windows is called glazing. Therefore, the *glazing* material type is designed mostly for use on windows or thin glass objects. The solid glass material type is designed for thicker glass. The Color property can be Clear, Green, Gray, Blue, Blue-green, Bronze, or Custom. The Reflectance value determines how reflective the material is. The Sheets of glass property alters the material effect based on a virtual thickness of the material.

Masonry

Masonry includes brick walls, cobblestone, tile flooring, and so on. The *masonry* material type is designed for use on objects such as these. The Type property can be CMU (concrete masonry unit) or Masonry. The Finish property can be Glossy, Matte, or Unfinished. The **Relief Pattern** category contains an Image property. The image determines the relief pattern.

Metal

The *metal* material type works well for mechanical parts and other objects made from different types of metals. This material type primarily is for raw, or unfinished, metal. The Type property can be Aluminum, Anodized Aluminum, Chrome, Copper, Brass, Bronze, Stainless Steel, or Zinc. The Finish property can be Polished, Semi-polished, Satin, or Brushed. The **Relief Pattern** category contains a Type property that can be Knurl Diamond Plate, Checker Plate, and Custom (based on a selected image). The **Cutouts** category contains a Type property that can be Staggered Circles, Straight Circles, Squares, Grecian, Cloverleaf, Hexagon, or Custom. Depending on which type of metal is selected, additional properties may be available for editing.

Metallic paint

The *metallic paint* material type is for objects made of metal, but with a finish applied. This includes objects like car parts, lawn furniture, and kitchen appliances. The **Flecks** category contains Color and Highlight Spread properties. The **Pearl** category includes a Type property that can be Chromatic or Second Color. The **Top Coat** category includes a Type property that can be Car Paint, Chrome, Matte, or Custom. The **Top Coat** category also includes a Finish property that can be Smooth or Orange Peel.

Mirror

The *mirror* material type is designed for very reflective objects, such as mirrors. It can also be used for water, glass, or any object that should have a high reflectivity. The only property for this material type is Tint Color.

Plastic

The *plastic* material type is designed for use on plastic objects. The plastic can be opaque, translucent, glossy, or textured. The Type property can be Solid, Transparent, Vinyl. The Finish property can be Polished, Glossy, Matte. The **Finish Bumps** category contains an Image property, as does the **Relief Pattern** category. The two images do not have to be the same.

Solid glass

The *solid glass* material type is intended for thick glass objects. The glazing material type is designed for thin glass objects. The Color property can be Clear, Green, Gray, Blue, Blue-green, Bronze, or Custom. The Reflectance property determines the degree of reflectivity of the material. The Refraction property can be Air, Water, Alcohol, Quartz, Glass Diamond, or Custom. The Roughness property determines the polish on the material. The **Relief Pattern** category contains a Type property that can be Rippled, Wavy, or Custom.

Stone

The stone material type works well for stone walls, stone walkways, and marble countertops. The **Stone** category contains an Image property, which is the image applied to the material. This can be a selected image or set to a texture. The **Stone** category also contains a Finish property that can be Polished, Glossy, Matte, or Unfinished. The **Finish Bumps** category contains a Type property that can be Polished Granite, Stone Wall, Glossy Marble, or Custom. The **Relief Pattern** category contains an Image property. This can be a selected image or set to a texture.

Wall paint

The *wall paint* material type works well for interior or exterior painted walls and other objects. The Finish property can be Flat/Matte, Eggshell, Platinum, Pearl, Semi-Gloss, or Gloss. The Application property can be Roller, Brush, or Spray.

Water

The *water* material type is designed for any liquid. Pools, reflecting ponds, rivers, lakes, and oceans are examples of where this material type may be used. The Type property can be Swimming Pool, Generic Reflecting Pool, Stream/River, Pond/Lake, or Sea/Ocean. The Color property (when available) can be Tropical, Algae/Green, Murky/Brown, Generic Reflecting Pool, Generic Pond/Lake, Generic Sea/Ocean, or Custom. The Wave Height property sets the amplitude of ripples in the liquid.

Wood

The *wood* material type works well for various finishes used for flooring, furniture, wood trim, and so on. The **Wood** category contains an Image property, which is the image applied to the material. This can be a selected image or set to a texture. When the **Stain** check box is checked, the color of stain can be set. The **Wood** category also contains a Finish property, which can be Unfinished, Glossy Varnish, Semi-gloss Varnish, or Satin Varnish. The Used For property can be Flooring or Furniture. The **Relief Pattern** category contains a Type property that can be based on Wood Grain or Custom (which is based on a selected image).

Generic

The generic material type is the "blank canvas" material. All properties are available to create any material needed. This material type is used to create a material from scratch, as discussed in the next section.

PROFESSIONAL TIP

AutoCAD provides fantastic-looking materials to dress up the scene and make it look real. However, after you get comfortable with creating materials, start a library of your own materials. If your project is presented to a customer along with projects from competitors, and your competitors are using standard AutoCAD materials, your project will stand out from the crowd.

Exercise 16-4

Complete the exercise on the student website.
www.g-wlearning.com/CAD

New Material from Scratch

Creating a material from scratch gives you complete control over the material. There are many options to consider, but it may be the only way to get exactly the material you are looking for to complete your scene. To start creating a material from scratch, select the generic material type. The properties and settings are discussed in the next sections.

Generic category

The ***diffuse color*** is the color of the object in lighted areas, or the perceived color of the material. See Figure 16-6. It is the predominant color you see when you look at the object. The Color property sets the diffuse color of the material. In AutoCAD, the ambient and specular colors are determined by the diffuse color.

The ***ambient color*** is the color of the object where light does not directly provide illumination. It can be thought of as the color of an object in shadows. In nature, shadows cast by an object typically contain some of the ambient color.

The ***specular color*** is the color of the highlight (the shiny spot). It is typically white or a light color. The amount of specular color shown is determined by the glossiness and reflectivity of the material and the intensity of lighting in the scene.

You have two options for controlling the Color property of the material. Picking the button to the right of the color text box displays a drop-down list. Selecting **Color** in the drop-down list means that you can select whatever color you wish. To set the color, pick in the text box; the **Select Color** dialog box is displayed. See Figure 16-7. If you select **Color by Object** in the drop-down list, the base color of the object is used as the diffuse color.

The Image property allows you to apply an image or texture to control the appearance of the material. Picking in the image preview area displays a standard open

Figure 16-6.
The three colors of a material are illustrated here. In AutoCAD, the Color property sets the diffuse color. The ambient and specular colors are based on the diffuse color.

Figure 16-7.
Setting a color for a material.

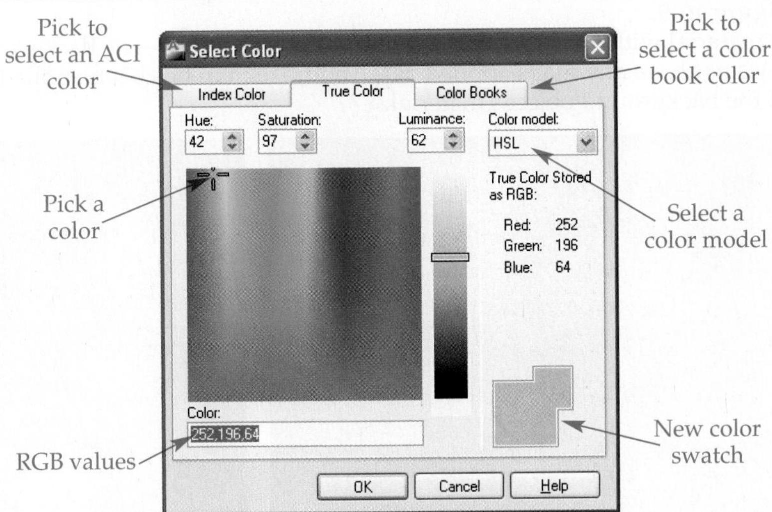

dialog box. When you select an image, it is applied to the material and is displayed in the material preview area, Figure 16-8. Also, the texture editor is displayed with the image loaded and ready for editing, if necessary. The button to the right of the image preview area is used to apply a texture instead of an image. Textures and texture editing are discussed later in this chapter.

When an image, such as a digital photograph, is applied, the material color is replaced with the image. Figure 16-9 shows an object with an image of a forest applied to the material attached to it. The Image Fade property controls the ratio between the image and the object color. It is only available when an image or texture is used. When set to 100, the image completely replaces the object color.

Figure 16-8.
The property settings in the **Generic** category of a material created from scratch.

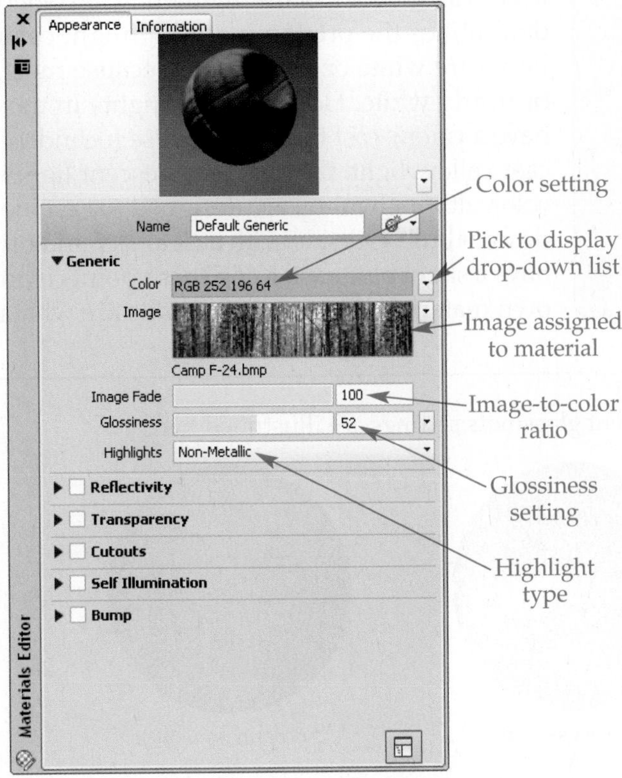

Figure 16-9.
A material with an image applied instead of a color. A—The background object's material does not have any maps applied. B—An image map has been applied to the color component of the background object's material.

A

B

The Glossiness property controls how shiny the material appears, **Figure 16-10.** It is a measure of the surface roughness of a material. A setting of 100 is a very shiny material, such as a smooth surface. A setting of 0 is a matte (dull) material. The button to the right of the setting allows you to add a texture or image to the Glossiness property. The pattern is applied to the glossiness effect. Reflectivity must be turned on for the glossiness effect to be seen. Refer to the next section.

The Highlights property determines how the shiny areas are created. Choices for this property are Metallic or Non-metallic. Highlights are brighter when Metallic is selected.

PROFESSIONAL TIP

Highlights can be seen everywhere. Look around you right now at edges and inclined surfaces. The diffuse color of the surface typically has little to do with the color of the highlight. The color of the light source usually determines the predominant highlight color. The majority of highlights are white or near white because most light sources are white or nearly white. However, highlights in the interior of a home may have a yellow cast to them because incandescent lightbulbs generally cast yellow light. Compact fluorescent lights cast a slightly different color, although many are designed to cast the same color as an incandescent bulb. Outside with a clear sky and bright sun, highlights may have a slight blue cast. Keep these points in mind when creating your own materials. These small details are what make a scene realistic.

Figure 16-10.
Three different glossiness settings are illustrated here.

A

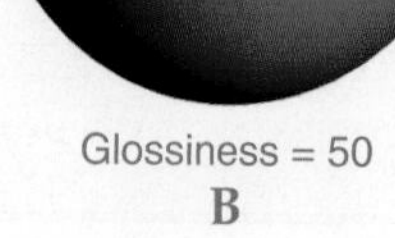

B

C

Figure 16-11.
Setting the reflectivity for a material.

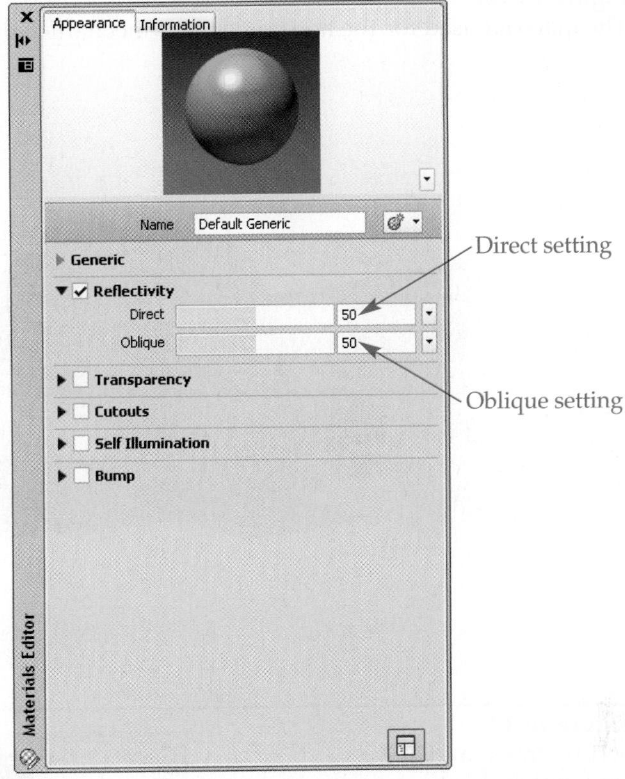

Reflectivity category

Reflectivity is a measure of how much light is bounced off of the surface. There are two basic property settings for reflectivity, Direct and Oblique. See **Figure 16-11.** The Direct property controls how much light is reflected back for surfaces that are more or less facing the camera. The Oblique property controls how much light is reflected back when the surface is at an angle to the camera.

Each property has a slider/text box that controls the amount of reflectivity. No reflections are created with a setting of 0. The maximum reflections are created with a setting of 100. Object color, lighting, surroundings, and other factors also determine just how reflective a material appears in the scene.

The buttons to the right of the text boxes allow you to add a texture or image to control the reflection. The white areas of the texture or image have a reflection. The black areas do not have a reflection. The degree of reflectivity varies for gray areas and the grayscale values of colors.

Transparency category

Transparency is a measure of how much light the material allows to pass through it. Glass, water, crystal, and some plastics, along with other materials, are nearly completely transparent. **Figure 16-12** shows an example of using a transparent material to show the internal workings of a mechanical assembly. The **Transparency** category contains a number of properties that combine to create any transparent or semitransparent material, **Figure 16-13.**

The amount of light that passes through the material is controlled by the Amount property. When this property is set to 0, the material is opaque. A setting of 100 creates a completely transparent material.

The Image property is used to add a transparency map to the material. White areas in the image or texture are transparent. Black areas are opaque. All other colors produce varying degrees of transparency based on their grayscale values. The Amount property is applied to the transparency map. Maps are discussed later in this chapter.

Figure 16-12.
The material used for the housing on this mechanism has a transparency setting of about 50.

Figure 16-13.
Setting transparency for a material.

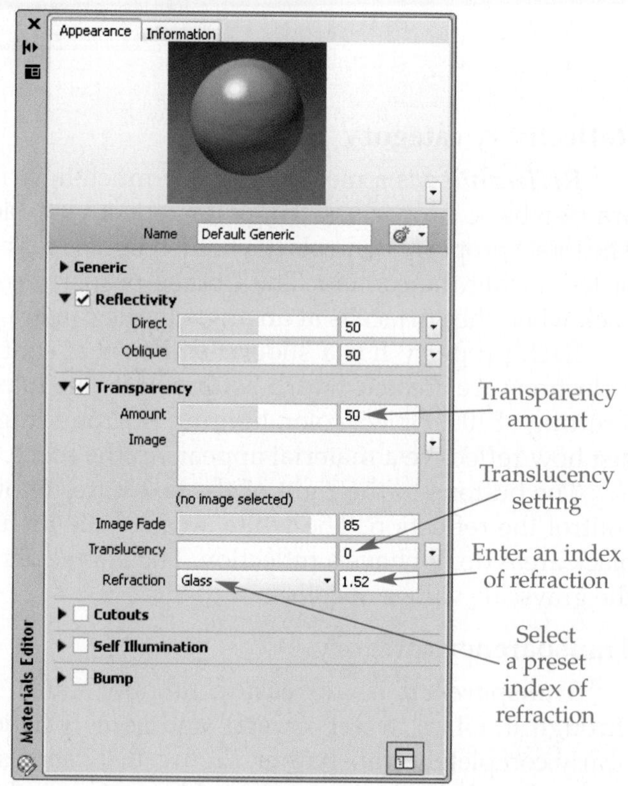

The Image Fade property determines how much of an impact the transparency map has on the transparency of the material. A setting of 100 means the transparency is completely see-through based on the transparency map. As the setting is decreased, a higher percentage of transparency is determined by the Amount property.

Translucency is a quality of transparent and semitransparent materials that causes light to be diffused (scattered) as it passes through the material. See **Figure 16-14.** This makes any object with the material applied to it appear as if it is being illuminated from within, or glowing. The thicker the material, the more pronounced the effect.

Figure 16-14.
The effect of translucency. A—The glass material has a translucency setting of zero. B—When the translucency setting is increased, light is diffused within the material. Notice how the glass appears slightly frosted.

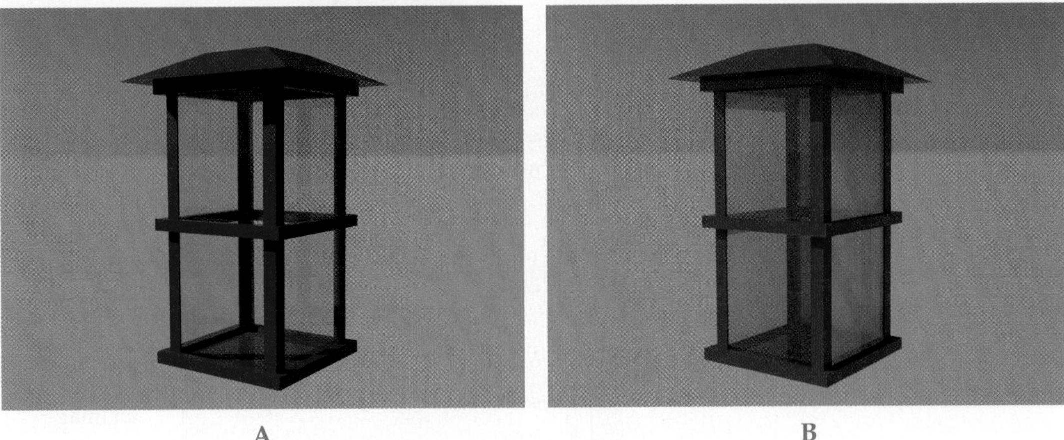

When the Translucency property is 0, light appears to travel through the material, lighting the opposite side. A setting of 100 creates a material similar in appearance to frosted glass.

The ***index of refraction (IOR)*** is a measure of how much light is bent (refracted) as it passes through transparent or semitransparent materials. Refraction is what causes objects to appear distorted when viewed through a bottle or glass of water. See **Figure 16-15.** The Refraction property sets the IOR. The higher the value, the more light is bent as it passes through the material. The IOR of water is 1.3333. You can enter a value in the text box or select a preset IOR by picking the name that is displayed to the left of the text box.

Cutouts category

The Cutouts property allows you to select an image or texture to use for a pattern of cutouts (holes), **Figure 16-16.** Black areas in the image will appear to be see-through, as if there is no object in those areas. White areas in the image have the normal material colors. This is similar to using a transparency map, but without the other transparency settings. **Figure 16-17** shows an example of a cutout map applied to a material.

Figure 16-15.
The effect of refraction. A—The transparent material on the sphere has a refraction setting of zero. B—When the refraction setting is increased, the cylinder behind the sphere is distorted as light is refracted by the material.

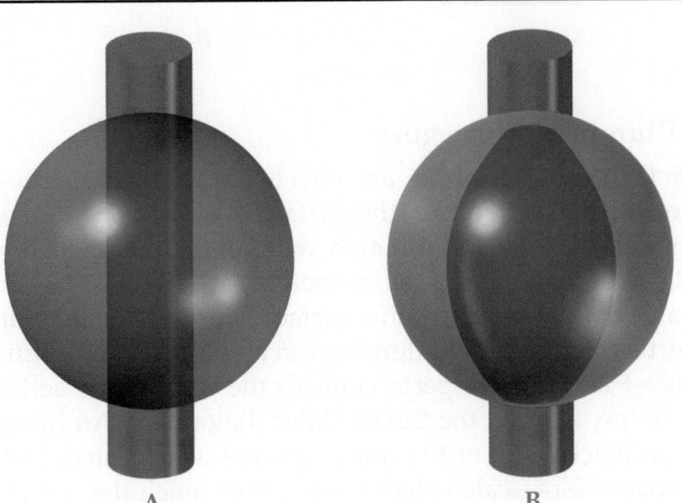

Figure 16-16.
Adding a cutout map to a material.

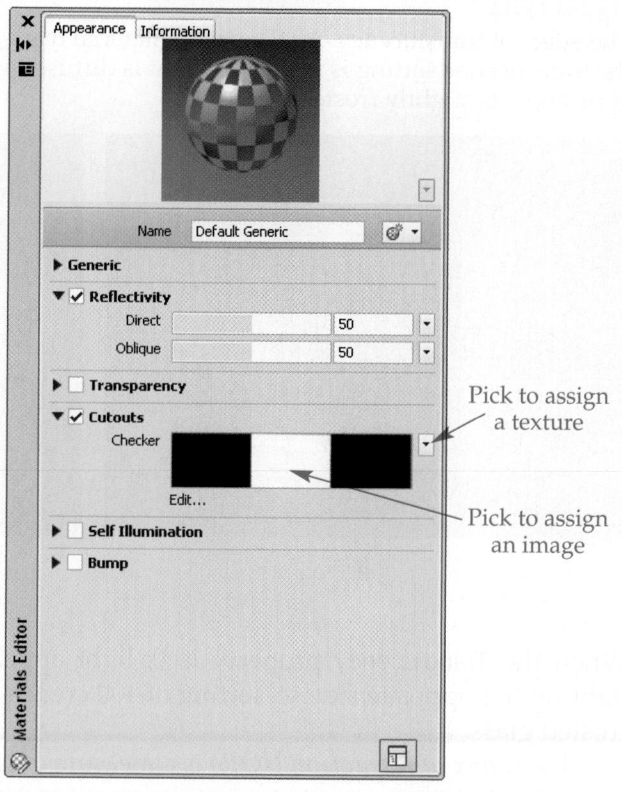

Figure 16-17.
The effect of applying an image to the Cutout property. A—This black and white image will be used as the cutout map. B—The material on the plane is completely opaque. C—When the cutout map is applied to the material, the dark areas of the map produce transparent areas on the object.

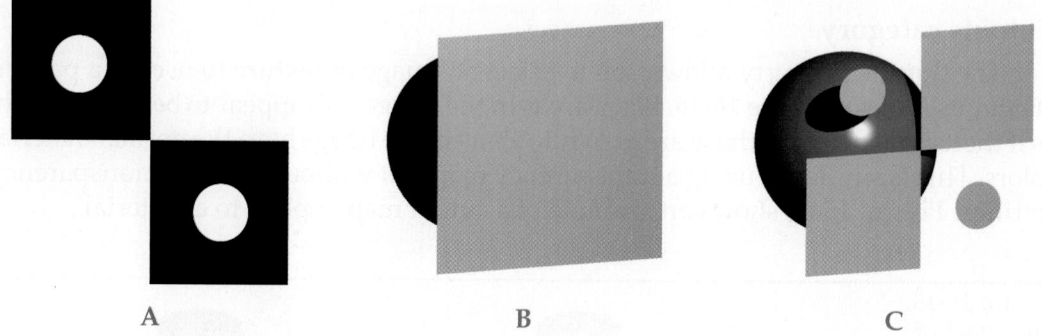

Self illumination category

Self illumination is an effect of a material producing illumination. See **Figure 16-18.** For example, the surface of a neon tube glows. However, in AutoCAD, a material with self illumination will not actually add illumination to a scene. This effect can be simulated with properly placed light sources. ***Luminance*** is defined as the value of light reflected off a surface. The **Self Illumination** category contains several properties related to self illumination and luminance, **Figure 16-19.**

The Filter Color property controls the color of the self illumination effect. Pick in the edit box to open the **Select Color** dialog box. An image may be selected instead of color. Black areas of the image are not illuminated. White areas of the image are illuminated. Grayscale values control how much the rest of the image illuminates the material. To apply a texture to control the illumination, pick the button next to the edit box to display a drop-down list.

Figure 16-18.
The effect of self illumination/luminance. A—The globe of this lightbulb does not have any self illumination. B—Self illumination is applied to the globe material.

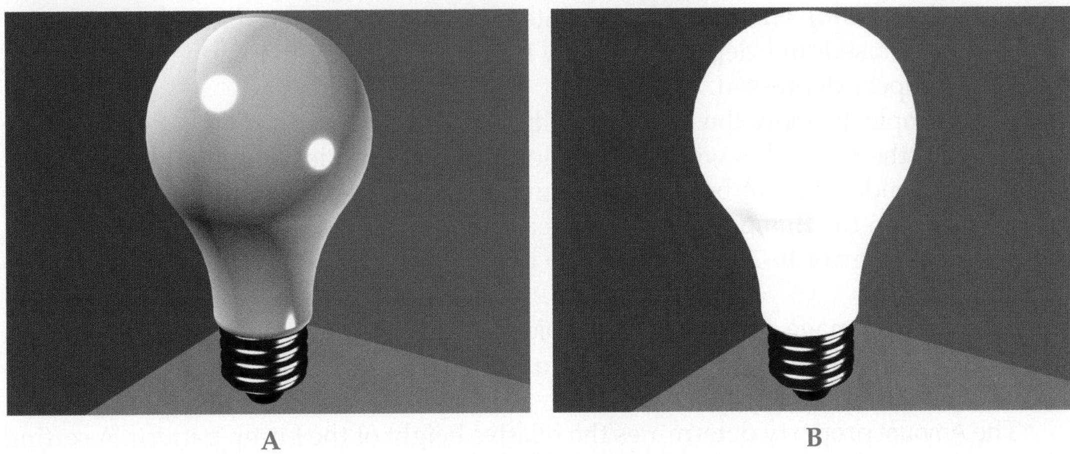

Figure 16-19.
Setting self illumination and luminance for a material.

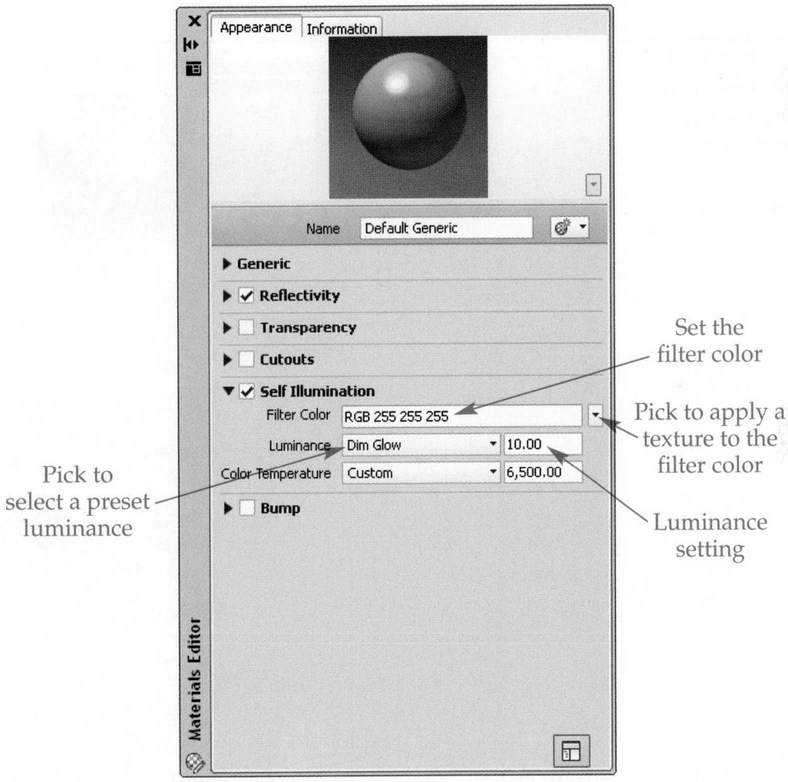

Set the filter color

Pick to apply a texture to the filter color

Pick to select a preset luminance

Luminance setting

Luminance is expressed in candelas per square meter (cd/m^2). For example, 1 cd/m^2 is the equivalent of one candela of light radiating from a surface area that is one square meter. You can enter a specific value in the Luminance property text box. Or, you can pick the name displayed next to the text box to display a drop-down list with choices for typical materials. Some of these choices include Dim Glow, LED Panel, and Cell Phone Screen.

The Color Temperature property determines the warmth or coolness value of the color. The value is expressed as degrees Kelvin. Candles and incandescent bulbs are warm. Fluorescent lights or TV screens are cool. The drop-down gives you choices for typical objects with different color temperatures, but you can enter a value directly in the text box.

Bump category

The **Bump** category contains settings for making some areas of the material appear raised and other areas depressed, Figure 16-20. The image or texture used for this effect is called a ***bump map.*** The black, white, and grayscale values of the map are used to determine raised and depressed areas. Dark areas of the map appear raised and light areas appear depressed.

For example, to show the texture of a brick wall, you could physically model the grooves into the wall. This would take a lot of time to model and would immensely increase the rendering time because of the increased complexity of the geometry. Using the properties in the **Bump** category is an easier and more efficient way to accomplish the same task. Figure 16-21 shows a bump map used to represent an embossed stamp on a metal case.

To apply an image as a bump map, pick the Image property swatch to display a standard open dialog box. To apply a texture as a bump map, pick the button next to the edit box to display a drop-down list.

The Amount property determines the relative height of the bump pattern. A setting of 0 results in a flat material, or no bumps. A setting of 1000 creates the maximum difference between low and high areas of the pattern.

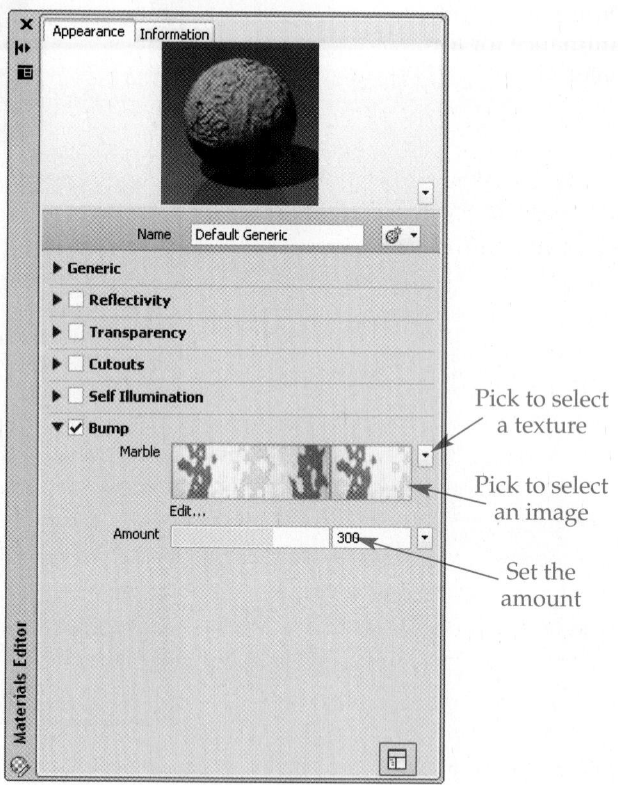

Figure 16-20. Making bump settings for a material.

Figure 16-21.
The effect of a bump map. A—This image will be used as the bump map. B—When applied to the material, the bump map simulates an embossed stamp in the metal case.

A

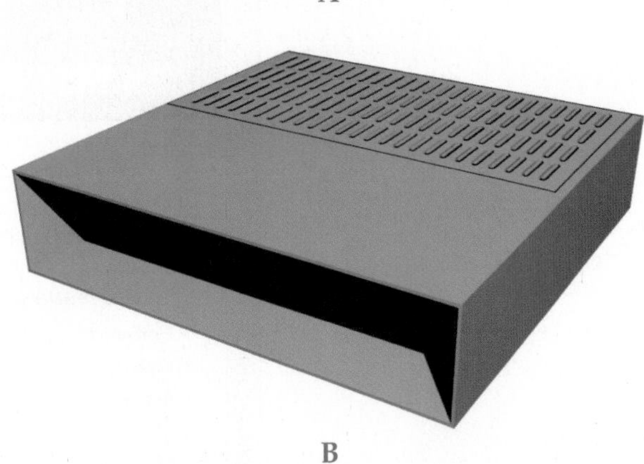

B

Maps

As discussed in previous sections, images and textures may be applied to materials to enhance the color, glossiness, reflectivity, transparency, translucency, cutouts, self illumination, and bumpiness. The applied image or texture is called a ***map.*** A material that has a map applied to at least one of its properties is called a ***mapped material.***

An image applied to a material property is sometimes known as a *2D map* or ***texture map.*** This is because it is composed of a fixed set of pixels, kind of like a mosaic pattern, applied to the object's surfaces.

On the other hand, a ***procedural map*** is mathematically generated based on the colors and values you select. This type of map is sometimes known as a *3D map* because it may extend through the object, depending on its algorithm. For example, if you attach AutoCAD's wood map (procedural map) to a material property and assign the material to an object with a cutout, the grain in the cutout will match the grain on the exterior. An image (texture map) applied to this same object will "reapply" itself to the cutout surface, not necessarily matching the pattern on the adjacent surfaces.

There are nine types of maps in AutoCAD that can be applied to material properties. The image, checker, gradient, and tile maps are texture maps (2D). The marble, noise, speckle, waves, and wood maps are procedural maps (3D). Each map has unique settings, as discussed in the next sections.

PROFESSIONAL TIP

The term *map* may be applied to the type of texture or the property to which the texture is applied. For example, a checker map may be used as a bump map. *Checker* is the type of map and *bump* is the property to which the texture is applied.

Figure 16-22.
The texture editor is used to adjust a map once it has been applied to a material property.

Map preview
12.0000
12.0000
Map type
Image
Source Camp F-24.bmp
Brightness 100
Invert Image
Transforms
Link texture transforms
Position
Scale
Repeat
Settings for map
Texture Editor - Color

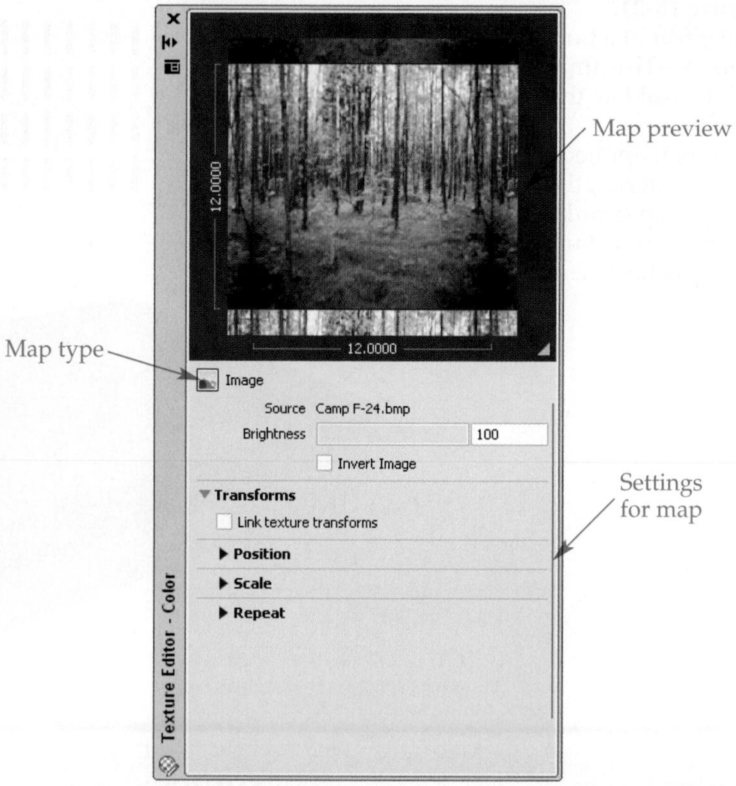

Texture Editor

The ***texture editor*** is used to adjust a map once it has been applied to a material property. It is automatically displayed once a map is assigned. Figure 16-22 shows the texture editor with an image map displayed for editing. A preview of the map appears at the top of the texture editor. The triangle in the lower-right corner of the preview is used to resize the preview. Below the map preview, the type of map is indicated.

Maps rarely appear on the object in the correct position, scale, or angle. For example, if you are using an image map of a logo to be applied to a box for a packaging design, it may not be positioned in the center of the box by default. The map may be rotated in the wrong direction or it may be too large. The texture editor is where adjustments to the map are made.

The bottom of the texture editor contains the settings for the map. The properties that are available depend on the type of map being edited. The next sections discuss the properties for each map type.

Image

Applying an ***image map*** is straightforward. There are several file types that can be applied as a map. When you specify Image as the map for a property, a standard open dialog box is displayed. Navigate to the image file and open it.

In the texture editor, the preview area shows the image with dimensions for the size of an individual tile. If the Sample Size property in the **Scale** category is changed, the preview dimensions reflect the change. Directly below the preview are the image file name and a slider/text box for adjusting the image brightness, Figure 16-23. To select a different image, pick the name to display a standard open dialog box. A brightness value of 100 means the image is at its full brightness. A value of 0 results in the image being all black. The check box below the slider is used to invert the image colors. This produces an effect similar to a photographic negative from an old film camera.

Figure 16-23.
Adjusting the properties of an image map.

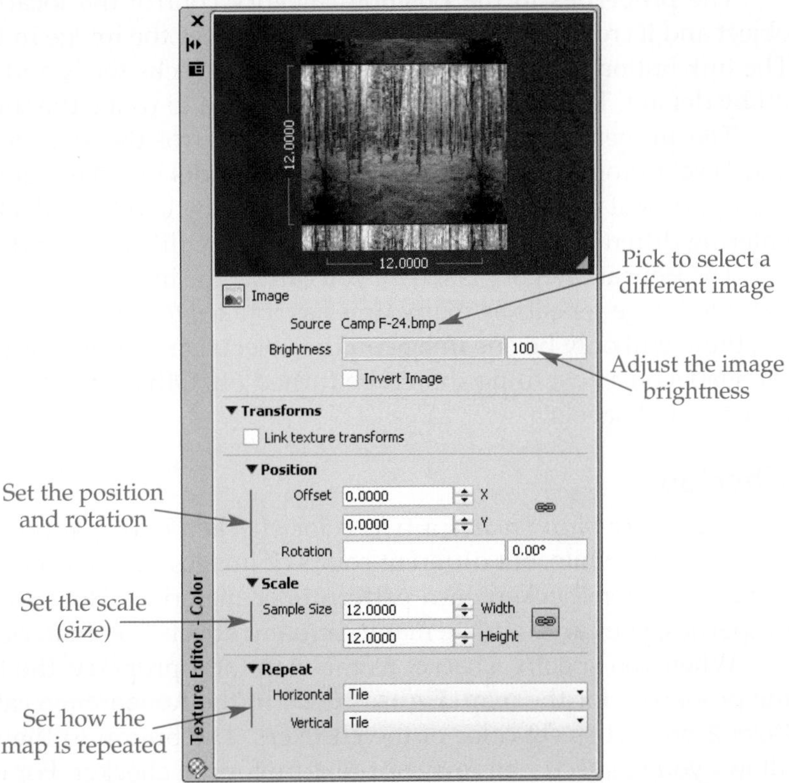

In the **Transforms** category is the **Link texture transforms** check box. It is very important to check this check box if you are using the same map for different properties in the material and need them to be synchronized in appearance. For example, to make a realistic tile floor, an image of tiles is applied to the color property. The same image map is also used as a bump map to make the grout look recessed. If the scaling is changed for the bump map, but the maps are not synchronized, the grout colors and indentations may not match. See **Figure 16-24.** By linking the texture transforms, all transform settings are the same. Transforms are not linked by default and if you turn it on after you have made changes they may not be synchronized. Changing the transform settings forces it to update and synchronize. Linking must be turned on for *each* material map or it will not work.

Figure 16-24.
A—This material has an image map assigned to the color and bump properties. B—If the bump map is not synchronized to transformations, it may not align with the grout lines if the object is transformed.

The properties in the **Position** category control the location of the map on the object and its rotation. The Offset property moves the image in the X and Y directions. The link button to the right of these text boxes locks the X and Y values together. It is off by default. The Rotation property allows you to rotate the image on the material.

The properties in the **Scale** category control the size of the image. The Width and Height properties are locked together by default. This is important to maintain a proportional ***aspect ratio*** for the image. They can be unlocked, but be aware that entering different values for width and height will stretch and distort the image.

The **Repeat** category is where you can set the image to tile or not tile. ***Tiling*** means that the image repeats as many times as it takes to cover the object. If tiling is turned off, there will only be one image on the object. For example, if you are using an image for a label on a box, tiling should be turned off. Otherwise, the box will be completely covered with labels.

Checker

A ***checker map*** creates a two-color checkerboard pattern. By default, the colors are black and white, but different colors or images can be used as well. This map type can be used for checkerboard pattern floor materials. However, by changing various properties, you can simulate many different effects and materials.

When you specify Checker as the map for a property, the texture editor displays the properties for the map, Figure 16-25. In the **Appearance** category, the Color 1 and Color 2 properties set color of the checkers. The button to the right of each property allows you to specify an image or texture for the checker. For example, you may add a texture map to the Color 1 property and a noise map to the Color 2 property. The possibilities are endless.

The Soften property is used to blur the edges between the checkers. To change the setting, enter a value in the text box or use the up and down arrows. A value of 0.00 creates sharp edges between the checkers. The maximum setting of 5.00 produces edges that are very blurred.

Figure 16-25.
Adjusting the properties of a checker map.

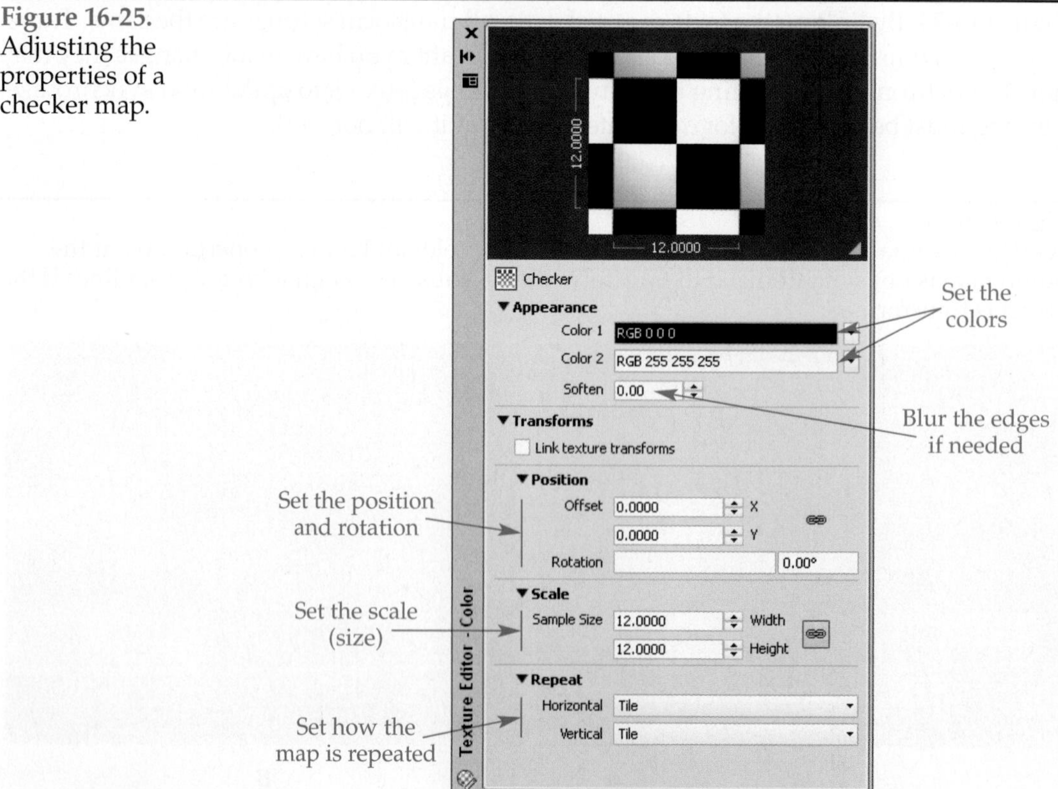

The properties in the **Transforms**, **Position**, **Scale**, and **Repeat** categories are the same as described for an image map. Refer to the Image section for details on these properties.

Gradient

The ***gradient map*** is a texture map that allows you to create a material blending colors in different patterns. It is similar to the **HATCH** command's **Gradient color** option. When you specify Gradient as the map for a property, the texture editor displays the properties for the map, **Figure 16-26.**

The gradient is represented in the **Appearance** category with three nodes at the bottom edge of the ramp. Each node represents a different color in the ramp. By default, the left node (node 1) is black, the middle node (node 3) is gray, and the right node (node 2) is white. The middle node can be moved left or right, as discussed later in this section, to change where the color transitions.

There must be at least three nodes, but you are not limited to three nodes. Picking anywhere in the ramp creates a new node. Selecting a node by picking on it changes the properties directly below the ramp to the settings for that node. The Color property sets the color of the selected node.

Above the ramp is the Gradient Type drop-down list. The setting in the drop-down list controls the pattern of the gradient ramp. See **Figure 16-27.** The default pattern is linear. This results in a typical pattern similar to the ramp display. The other options are:

- **Linear asymmetrical.** Similar to the default linear, but the transition between colors is not symmetrical.
- **Box.** The transition of colors is in the shape of a square.
- **Diagonal.** This transition is linear, but rotated on the surface.

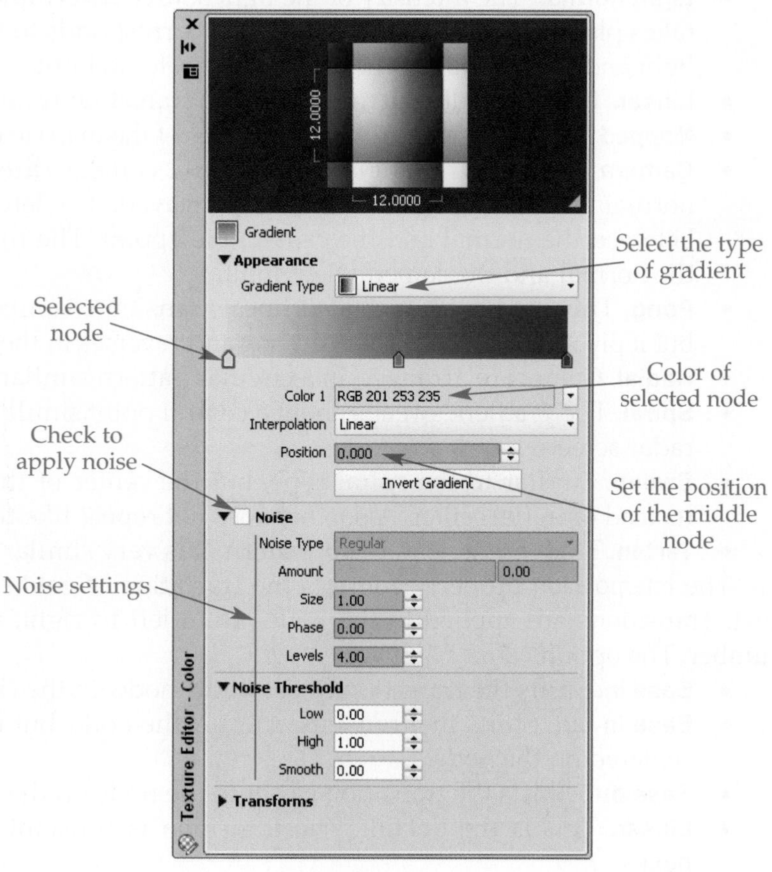

Figure 16-26.
Adjusting the properties of a gradient map.

Figure 16-27.
There are 12 different gradient types available for use in a gradient map. The four shown here are applied to the same object with the same lighting and mapping coordinates. A—Linear asymmetrical. B—Box. C—Diagonal. D—Light normal.

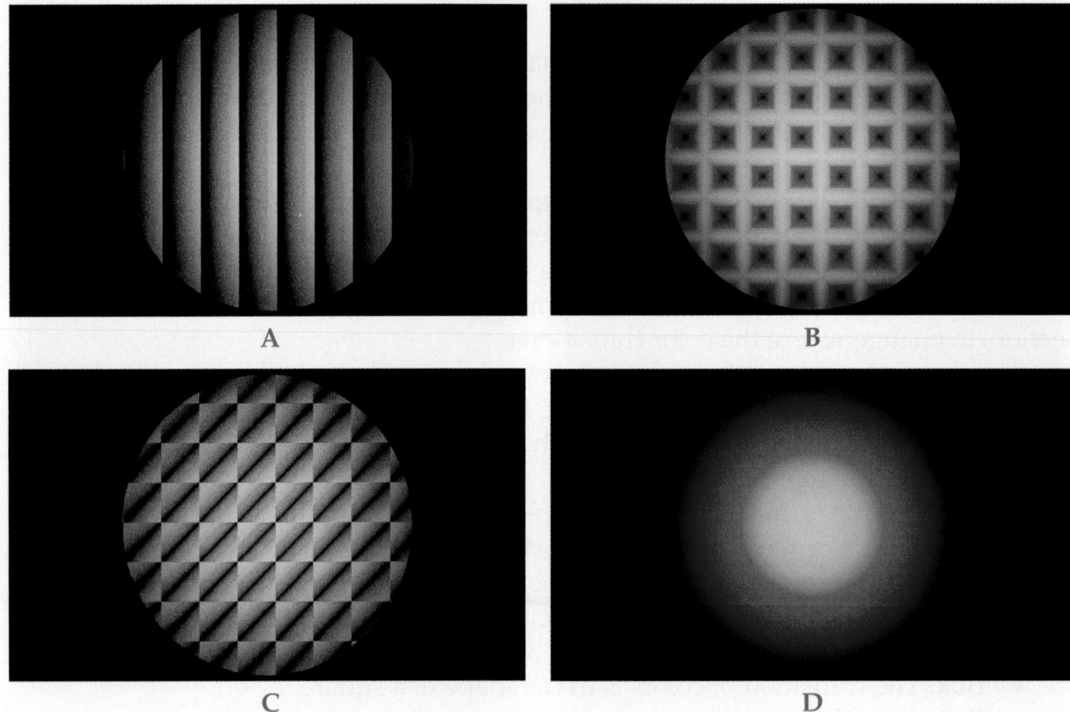

- **Light normal.** The intensity of the light source determines where the transition takes place. The right side of the ramp corresponds to the highest intensity of light and the left side of the ramp is equal to no light.
- **Linear.** This is the default. It is a smooth transition from one node to the next.
- **Mapped.** With this type you can assign one of the maps to determine the gradient.
- **Camera Normal.** The angle between the camera direction and the surface normal controls how the pattern is displayed. The left side of the ramp is 0° between the normal and the camera viewpoint. The right side is 90° between the normal and the camera viewpoint.
- **Pong.** This gradient is a rotated linear transition, similar to the diagonal type, but it pivots about the corner of a box and reverses in the middle of the pattern.
- **Radial.** Colors are arranged in a circular pattern similar to a target.
- **Spiral.** The gradient sweeps about a central point similar to the movement on a radar screen.
- **Sweep.** Similar to the spiral type, but the center of the sweep is at a corner instead of in the center. Also, this does not repeat like the pong type.
- **Tartan.** This resembles a plaid pattern. It is very similar to the box type.

The Interpolation property controls the transition of colors from one node to the next. Transitions are applied to the nodes from left to right, regardless of the node number. The options are:

- **Ease in.** Shifts the transition closer to the node on the right.
- **Ease in out.** Shifts the transition toward the node, but it remains more or less centered on the node.
- **Ease out.** Shifts the transition closer to the node on the left.
- **Linear.** This is the default. The transition is constant from one node to the next.
- **Solid.** No transition between nodes. There is an abrupt change at each node.

The Position property is simply the position of the selected node in the ramp. The node on the left is at the 0 position and the node on the right is at the 1.000 position. Nodes in between will be at varying values between 0 and 1.000.

The **Invert Gradient** button reverses all color values inverting the gradient pattern. In effect, the ramp is flip-flopped from left to right.

Noise may be added to the gradient map to create an uneven appearance. The properties in the **Noise** and **Noise Threshold** categories are similar to those for the noise map. The noise map is described later in this chapter.

The properties in the **Transforms**, **Position**, **Scale**, and **Repeat** categories are the same as described for an image map. Refer to the Image section for details on these properties.

Tiles

A ***tiles map*** is a pattern of rectangular, colored blocks surrounded by colored grout lines. This may be the most versatile map in the whole collection. Tiles are used to simulate tile floors, ceiling grids, hardwood floors, and many different types of brick walls. When you specify Tiles as the map for a property, the texture editor displays the properties for the map, **Figure 16-28.**

You first need to define the pattern for the map. The **Pattern** category contains properties for defining the pattern. For the Type property, select one of the seven predefined tile patterns or Custom Pattern to create your own. The names of the predefined patterns bring to mind brick walls. For example, a mason may use a stack bond to build a brick wall. However, remember these are only *patterns.* You can also use a brick pattern to

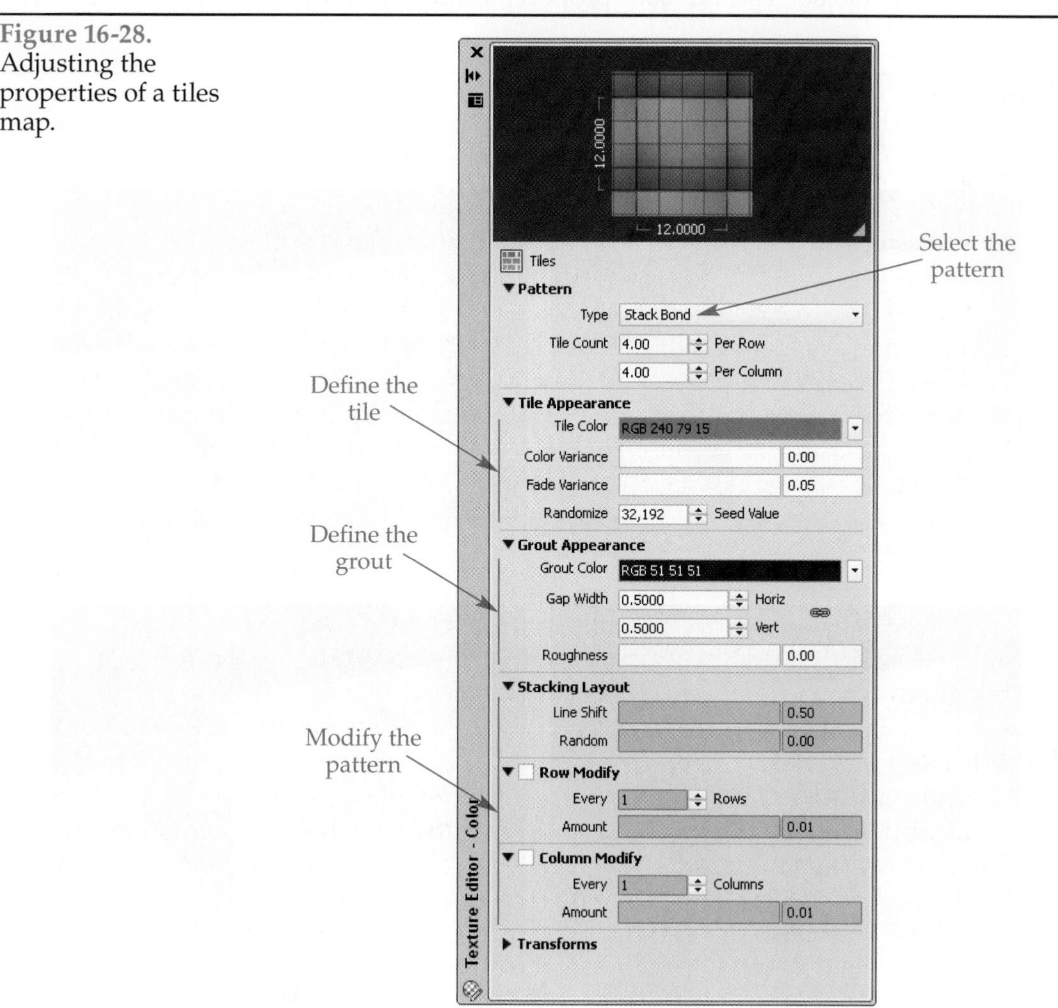

Figure 16-28.
Adjusting the properties of a tiles map.

create tile floors and acoustic ceiling panels. Four of the tile patterns are shown in **Figure 16-29.** The Tile Count property sets how many tiles are in each row and column before the pattern repeats.

As its name implies, the properties in **Tile Appearance** category determine what the tiles look like. You can choose any color you wish or apply a texture or image to the tiles. Ceramic tile floors look more realistic if each tile is slightly different in color. The Color Variance property can be used to alter the color of random tiles to create a more realistic appearance. The Fade Variance property is used to fade the color of random tiles. You will have to experiment with the color variance and fading to create the look you need. Start with very low values. The Randomize property is used to alter the random color variation in the tiles. This variation is automatically applied, but entering a different random seed changes the pattern. If your scene has more than one object with this material applied to it, duplicate the material and change the seed number of the new material, then apply it to the other object.

The properties in the **Grout Appearance** category control what the grout looks like. The grout is the line between the tiles. The Grout Color property is set to dark gray by default, but any color, image, or texture may be used. The Gap Width property determines how wide the grout lines are in relation to the tiles. There are horizontal and vertical settings, which are locked by default so the line widths are equal. In some cases, such as for a hardwood floor material, you will have to scale the pattern differently on the horizontal and vertical axes to make the gap thicker in one direction.

The properties in the **Stacking Layout** category are only available for a custom pattern. The Line shift property changes the location of the vertical grout lines in every other row to create an alternate pattern of tiles. The default value is 0.50 and the range is from 0.00 to 100.00. The Random property randomly moves the same lines. This

Figure 16-29.
A tiles map can have a custom pattern or predefined pattern. Four of the predefined patterns are shown here. A—Running bond. B—English bond. C—Stack bond. D—Fine running bond.

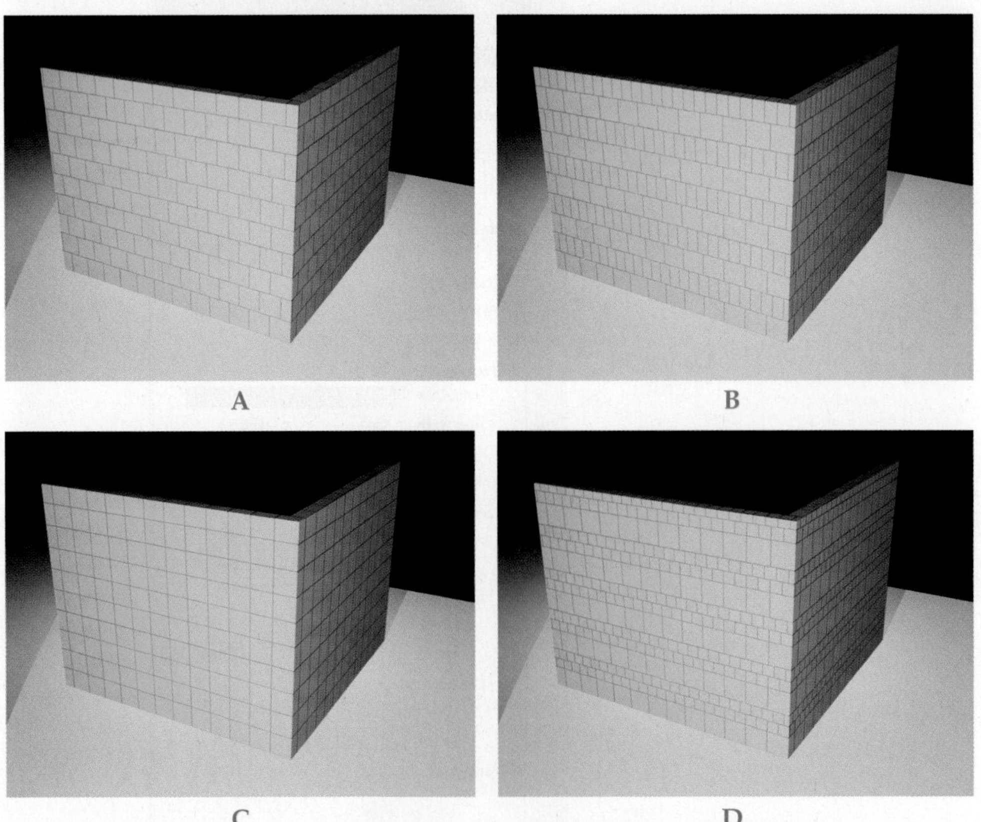

works nicely for hardwood floor materials. Default value is 0.00 and the range is 0.00 to 100.00.

The properties in the **Row Modify** and **Column Modify** categories are available with all tile pattern types, but may be disabled by default. To enable the settings, check the check box by the category name. The settings in these areas allow you to change the number of grout lines in the horizontal and vertical directions to create your own pattern. The two Every properties determine which rows and columns will be changed. When set to 0, no changes take place in the row or column. When set to 1, every row or column will be changed. When set to 2, every other row or column will be changed, and so on. The value must be a whole number. The Amount property controls the size of the tiles in the row or column. A setting of 1 means that the tiles remain their original size. A setting of 0.50 makes the tiles one-half of their original size, a setting of 2 makes the tiles twice their original size, and so on. A setting of 0.00, in effect, completely turns off the row or column and the underlying material color shows through.

The properties in the **Transforms**, **Position**, **Scale**, and **Repeat** categories are the same as described for an image map. Refer to the Image section for details on these properties.

Marble

A ***marble map*** is a procedural map based on the colors and values you set. It is used to simulate natural stone. When you specify Marble as the map for a property, the texture editor displays the properties for the map, Figure 16-30. The viewport may not reflect changes made in the texture editor, even if set to display materials and textures. You may have to render the scene to see the changes.

A marble map is based on two colors—stone and vein. The **Appearance** category contains the Stone Color and Vein Color properties. You can swap the vein and stone colors by picking the button to the right of the color definition and selecting **Swap Colors** from the drop-down list. The Vein Spacing property determines the relative distance between each vein in the marble. The Vein Width property determines the relative width of each vein. Each of these settings can range from 0.00 to 100.00.

Figure 16-30. Adjusting the properties of a marble map.

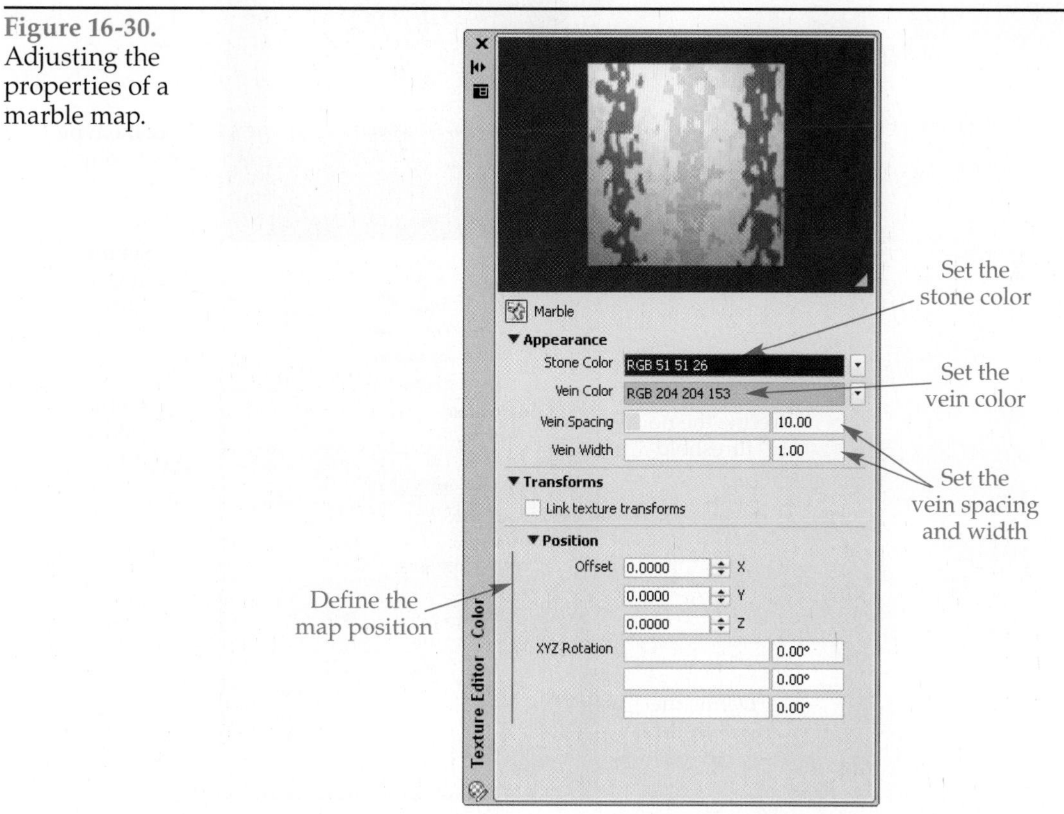

The **Link texture transforms** check box in the **Transforms** category works as described earlier for an image map. Refer to the Image section for details on these properties.

Since this is a procedural (3D) map, the properties in the **Position** category are different from the maps previously discussed. The three Offset properties move the map in the X, Y, and Z directions on the object. Simply enter a value in the text boxes. The XYZ Rotation properties control the rotation of the map around the X, Y, or Z axis. You can move the sliders or enter an angle in the text boxes.

NOTE

The mathematical calculations that create a procedural map are based on the world coordinate system. If you move or rotate the object, a different result is produced.

Noise

A ***noise map*** is a procedural map based on a random pattern of two colors used to create an uneven appearance on the material. It is most often used to simulate materials such as concrete, soil, asphalt, grass, and so on. When you specify Noise as the map for a property, the texture editor displays the properties for the map, **Figure 16-31.**

The properties in the **Appearance** area control how the noise looks. First, you need to select the type of noise. The options for the Noise Type property are:

- **Regular.** This is "plain" noise and useful for most applications.
- **Fractal.** This creates the noise pattern using a fractal algorithm. When this is selected, the Levels property in the **Noise Threshold** category is enabled.
- **Turbulence.** This is similar to fractal, except that it creates fault lines.

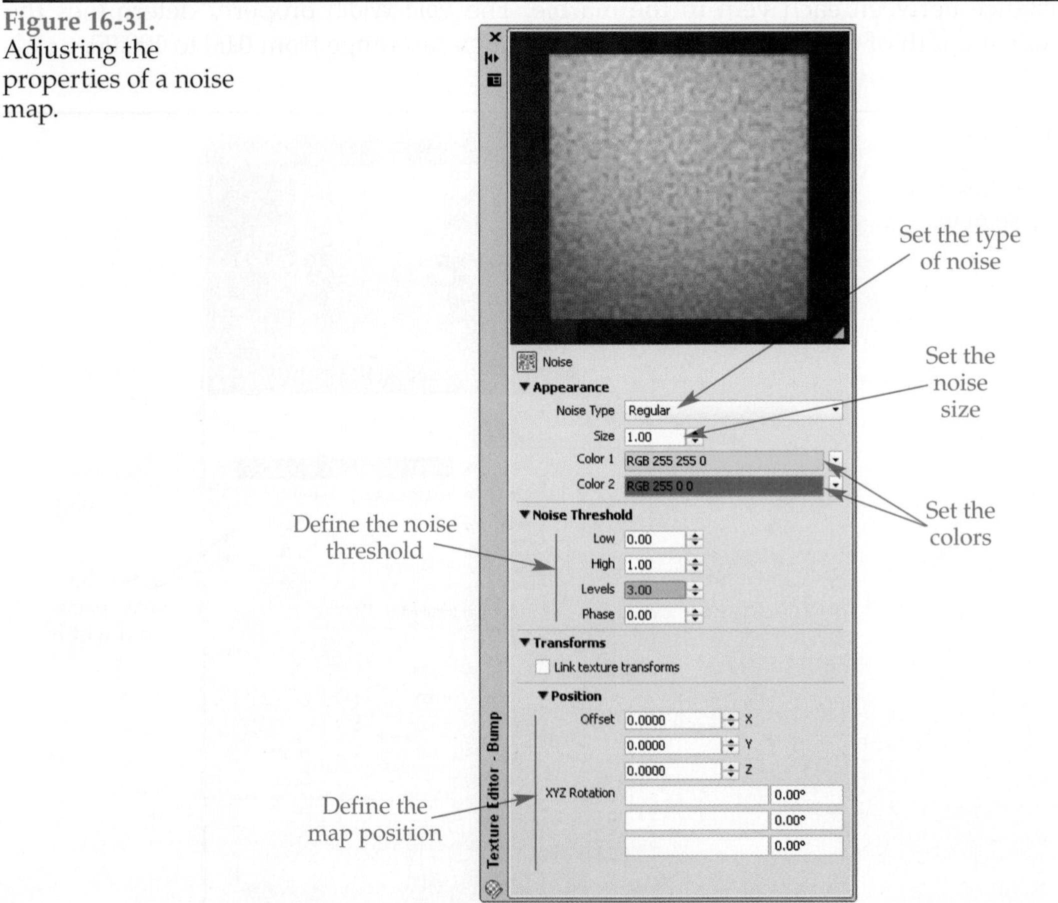

Figure 16-31.
Adjusting the properties of a noise map.

The Size property controls the size scale of the noise. The larger the value, the larger the size of the noise. The default value is 1.00 and the value can range from 0.00 to 1 billion. The Color 1 and Color 2 properties control the color of the pattern of noise. You can assign a color, image, or texture to the property. To swap the color definitions, pick the button next to the properties and select **Swap Colors** from the drop-down list.

The properties in the **Noise Threshold** category are used to fine-tune the noise effect. The properties in this category are:

- **Low.** The closer this setting is to 1.00, the more dominate color 1 is. The default setting is 0.00 and it can range from 0.00 to 1.00.
- **High.** The closer this setting is to 0.00, the more dominate color 2 is. The default setting is 1.00 and it can range from 0.00 to 1.00.
- **Levels.** Sets the energy amount for fractal and turbulence. Lower values make the fractal noise appear blurry and the turbulence lines more defined. The default setting is 3.00 and it can range from 0.00 upward.
- **Phase.** Randomly changes the noise pattern with each value. This allows you to have materials with the same noise map settings look slightly different. You should have different patterns on different materials. This adds a level of realism to your scene.

The properties in the **Transforms** and **Position** categories are the same as described for a marble map. Refer to the Marble section for details on these properties.

Speckle

A ***speckle map*** is a procedural map based on a random pattern of dots created from two colors. This map is great for textured walls, sand, granite, and so on. When you specify Speckle as the map for a property, the texture editor displays the properties for the map, **Figure 16-32.**

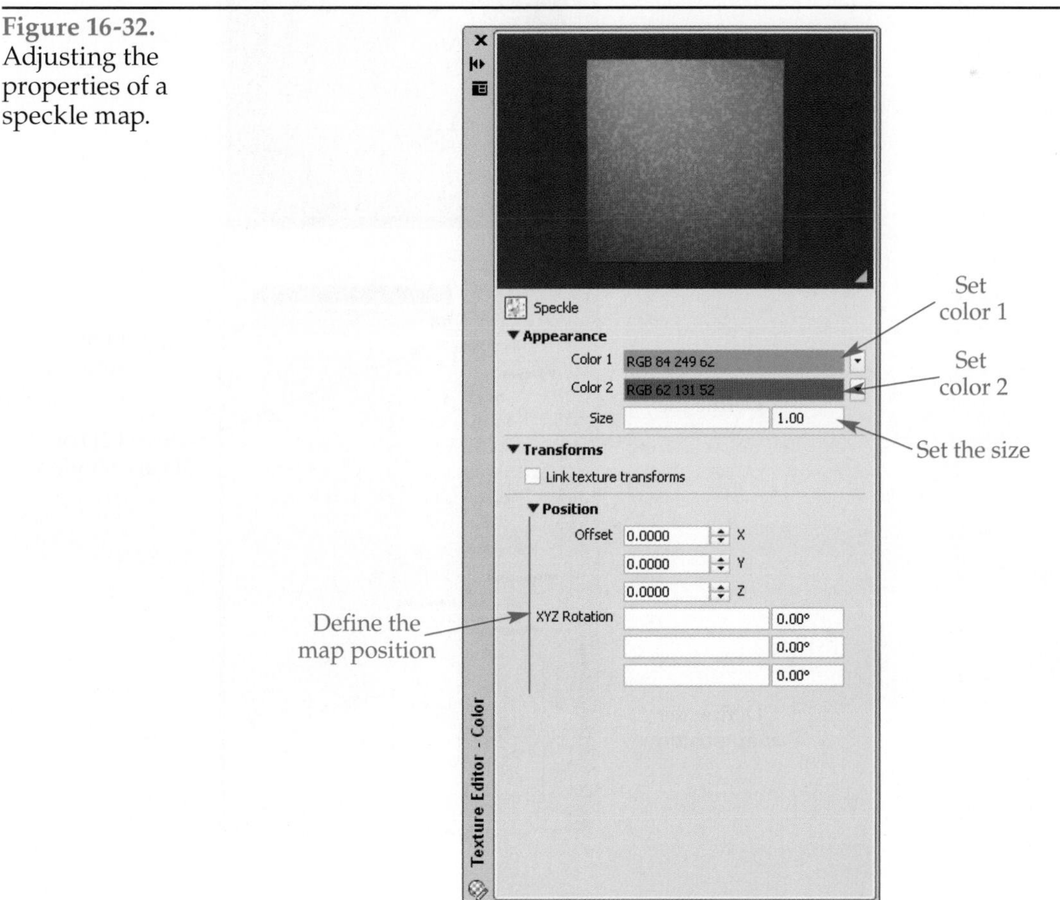

Figure 16-32. Adjusting the properties of a speckle map.

The settings for a speckle map are very simple. In the **Appearance** category, pick colors for the Color 1 and Color 2 properties. You cannot use maps, only colors. To swap the color definitions, pick the button next to the properties and select **Swap Colors** from the drop-down list. The Size property controls the size of the speckles.

The properties in the **Transforms** and **Position** categories are the same as described for a marble map. Refer to the Marble section for details on these properties.

Waves

A ***wave map*** is a procedural map in a pattern of concentric circles. Imagine dropping two or three stones into a pool of water and watching the ripples intersect with each other. A number of wave centers are randomly generated and a pattern created by the overlapping waves is the result. As the name implies, the wave map is usually used to simulate water. When you specify Waves as the map for a property, the texture editor displays the properties for the map, **Figure 16-33.**

In the **Appearance** category, pick colors for the Color 1 and Color 2 properties. You cannot use maps, only colors. To swap the color definitions, pick the button next to the properties and select **Swap Colors** from the drop-down list. The Distribution property can be set to 2D or 3D. This setting determines how the wave centers are distributed on the object. Selecting 3D means that the wave centers are randomly distributed over the surface of an imaginary sphere. This distribution affects all sides of an object. On the other hand, selecting 2D means that the wave centers are distributed on the XY plane. This is much better for nearly flat surfaces, such as the surface of a pond or lake.

Figure 16-33.
Adjusting the properties of a waves map.

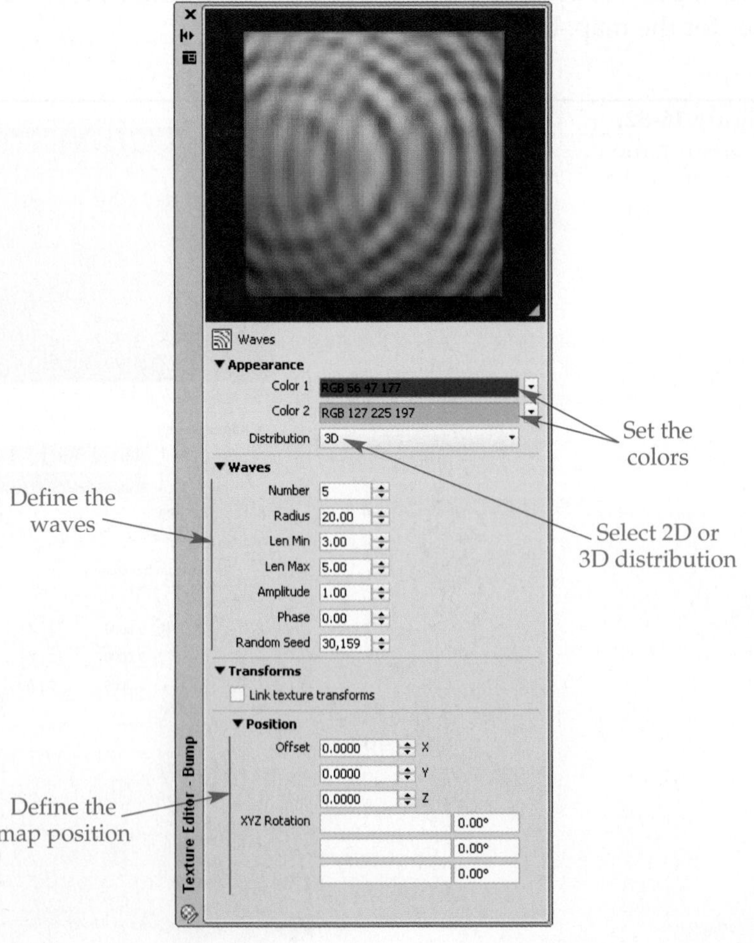

The properties in the **Waves** category define the pattern of waves. The Number property sets the number of wave centers that are generating the waves. The Radius property sets the radius of the circle or sphere from which the waves originate. The Len Min and Len Max properties define the minimum and maximum interval for each wave. The Amplitude property can be thought of as the "power" of the wave. The default value is 1.00, but the value can range from 0.00 to 10000.00. A value less than 1.00 makes color 1 more dominant. For a value greater than 1.00, color 2 is more dominant. The Phase property is used to shift the pattern and the Random Seed property is used to redistribute the wave centers.

The properties in the **Transforms** and **Position** categories are the same as described for a marble map. Refer to the Marble section for details on these properties.

PROFESSIONAL TIP

Remember, you can change the material swatch geometry in the materials editor, such as to a cube, sphere, or cylinder. Some maps, like a wave map, are easier to understand when displayed on a cube.

Wood

A ***wood map*** is a procedural map that generates a wood grain based on the colors and values you select. See Figure 16-34. When you specify Wood as the map for a property, the texture editor displays the properties for the map, Figure 16-35.

A wood map is based on two colors. The Color 1 and Color 2 properties in the **Appearance** category are used to specify these colors, usually one dark and one light color. To swap the color definitions, pick the button next to the properties and select **Swap Colors** from the drop-down list. The Radial Noise property determines the waviness of the wood's rings. The rings are found by cutting a tree crosswise. The Axial Noise property determines the waviness of the length of the tree trunk. The Grain Thickness property determines the relative width of the grain.

The properties in the **Transforms** and **Position** categories are the same as described for a marble map. Refer to the Marble section for details on these properties.

Exercise 16-5

Complete the exercise on the student website.
www.g-wlearning.com/CAD

Figure 16-34.
A wood map is a procedural, or 3D, map. Note how the pattern matches on adjacent surfaces.

Figure 16-35.
Adjusting the properties of a wood map.

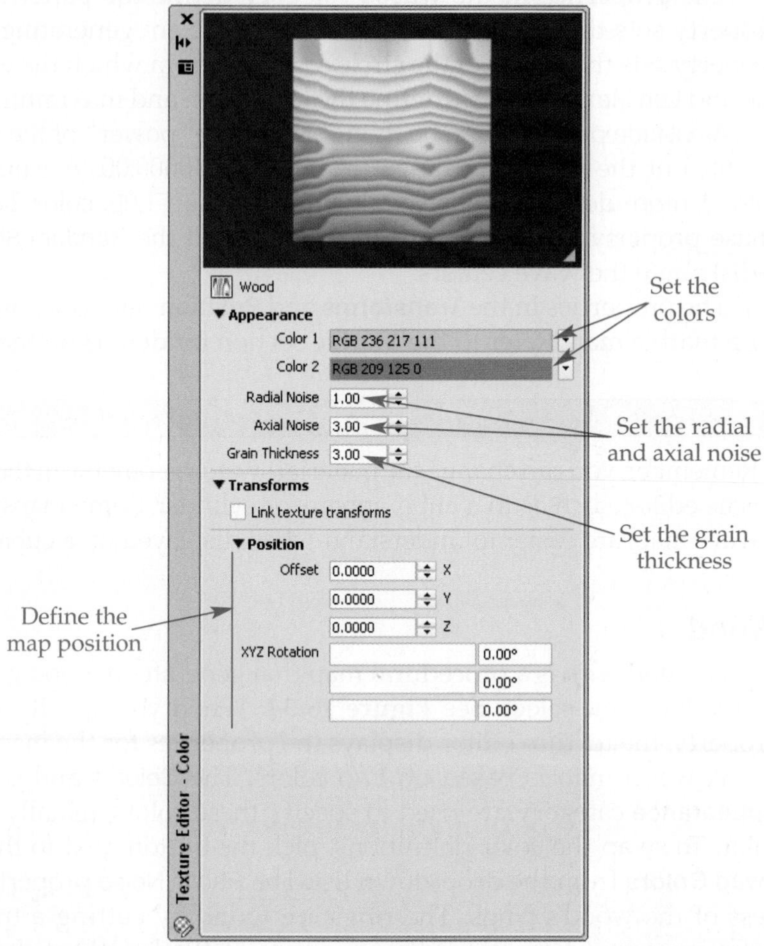

Set the colors

Set the radial and axial noise

Set the grain thickness

Define the map position

Adjusting Material Maps

Simply applying a map to a material property rarely results in a realistic scene when the scene is rendered. The maps usually need to be adjusted to produce the desired results. Maps can be adjusted at the material level or the object level. A combination of these two adjustments is usually required to produce a photorealistic rendering.

Material-Level Adjustments

Material-level adjustments involve changing the properties of the map in the material definition. Map properties are discussed in previous sections. Sometimes, these adjustments may be enough to get the materials looking the way you want them.

Other adjustments become necessary when the same material is applied to more than one object in the same scene. If you make changes at the material level, they affect all objects with that same material. If you need different objects to have different settings, then you will have to make object-level adjustments.

Object-Level Adjustments

Material mapping refers to specifying how a mapped material is applied to an object. When a mapped material is attached to an object, a default set of mapping coordinates, or simply *default mapping*, is used to apply the map to the object.

AutoCAD allows you to adjust mapping at the object level for texture-mapped (2D-mapped) materials. The **MATERIALMAP** command applies a grip tool, or ***gizmo***, based on one of four mapping types: planar, box, spherical, or cylindrical. See Figure 16-36. The colored edge represents the start and end of the map. For best results, select the mapping type based on the general shape of the object to which mapping is applied. Do not be afraid to experiment with other mapping types, however. Any mapping type can be used on any object, regardless of the object's shape. However, only one mapping type can be applied to an object at any given time.

Ribbon
Render > Materials
Planar Mapping
Box Mapping
Cylindrical Mapping
Spherical Mapping
Type
MATERIALMAP SETUV

MATERIALMAP

After one of the mapping types is selected, you are prompted to select the faces or objects. You can select multiple objects or faces. After making a selection, the gizmo is placed on the selection set. The command remains active for you to adjust the mapping or enter an option.

Drag the grips on the gizmo to stretch or scale the material. The effects of editing a color map are dynamically displayed if the current visual style is set to display materials and textures. Otherwise, exit the command and render the scene to see the effect of the edit. To readjust the mapping, select the same mapping type and pick the object again. The gizmo is displayed in the same location as before.

The **Move** and **Rotate** options of the command toggle between the move and rotate gizmos. Using the gizmos, you can move and rotate the map on the object. The **Reset** option of the command restores the default mapping to the object. The **Switch mapping mode** option allows you to change between the four types of mapping.

If the command is typed, there is an additional option. The **Copy mapping to** option is a quick and easy way to apply the changes made to the current object to other objects in the scene. Enter this option, select the face or object to copy from, and then select the faces or objects to copy to. This option is also available if the **Switch mapping mode** option is entered.

For example, look at Figure 16-37A. The grain on the stair risers is running vertically when it should run horizontally. First, apply a planar map to the bottom riser. Next, rotate the mapping 90°, Figure 16-37B. Finally, use the **Copy mapping to** option to copy the mapping to the other risers, Figure 16-37C.

NOTE

Mapping coordinates can be applied to procedural-mapped (3D-mapped) materials, but adjusting the coordinates has no effect. This is because the procedural map is generated from mathematical calculations based on the world coordinate system. Procedural-mapped materials must be adjusted at the material level.

Figure 16-36.
These are the four material map gizmos. From left to right: planar, box, spherical, and cylindrical. The colored edge represents the start and end of the map.

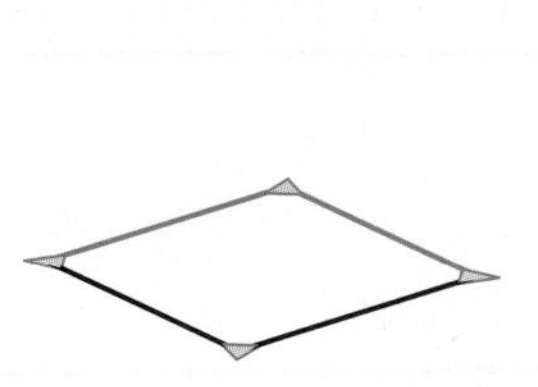

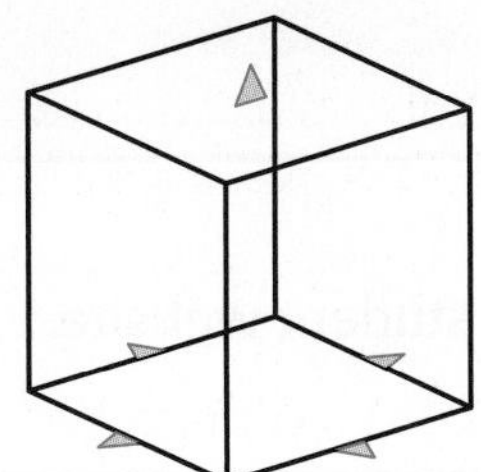

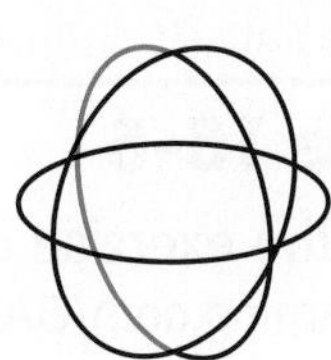

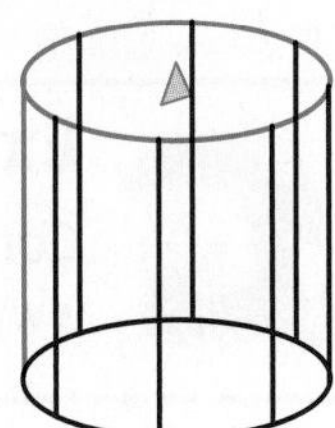

Figure 16-37.
Correcting material mapping. A—The grain on the risers runs vertically instead of horizontally. B—Rotating the map with the rotate gizmo. C—The corrected rendering. (Model courtesy of Arcways, Inc., Neenah, WI)

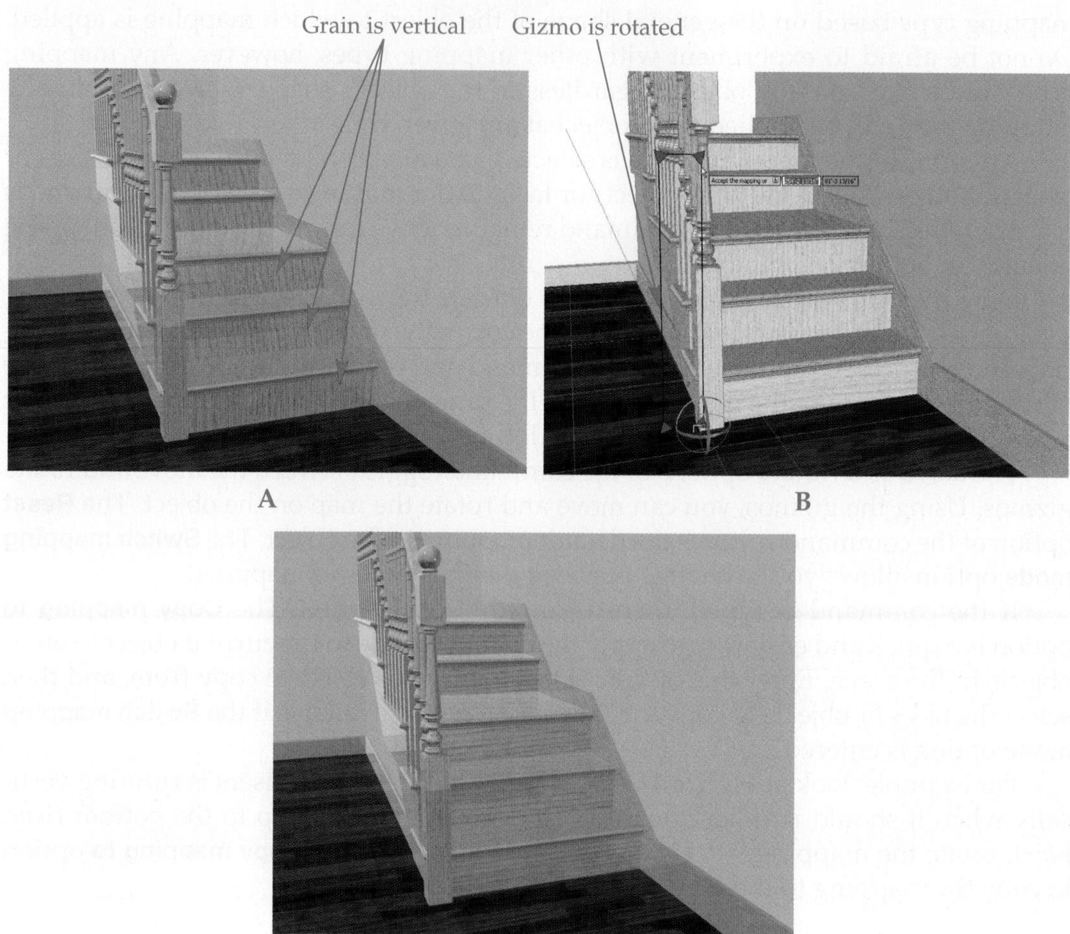

PROFESSIONAL TIP

If you are using a reflectivity, self-illumination, or bump map and need to adjust it at the object level, you cannot see the effects of the mapping change in the viewport. Apply the same map as a color map. Also, set the visual style to display materials and textures. Then, adjust the object mapping as needed. The edits are dynamically displayed in the viewport. When the image is in the correct location, remove the color map from the material.

Exercise 16-6

Complete the exercise on the student website.
www.g-wlearning.com/CAD

Chapter Test

Answer the following questions. Write your answers on a separate sheet of paper or complete the electronic chapter test on the student website.
www.g-wlearning.com/CAD

1. Define *material.*
2. Define *materials library.*
3. What is the My Materials section of the **Libraries** area of the materials browser used for?
4. Describe how to attach a material using the materials browser.
5. How can materials be attached to layers?
6. By default, which material is attached to newly created objects?
7. Which material is used as the base material for creating new materials?
8. Name the twelve shapes that can be used to display the material in the preview in the material editor.
9. How do you know if a material in the materials browser is being used in the drawing?
10. How can the name of an existing material be changed?
11. Name the fourteen basic material types.
12. What is the benefit of using the plastic material type instead of starting from scratch using the generic material?
13. In the **Reflectivity** category of the generic material, there are Direct and Oblique properties. What are these used for?
14. An image mapped to a material property will normally repeat itself to cover the entire object. What is this called and how do you turn it off?
15. Describe the difference between a transparent material and a translucent material.
16. How much illumination does a self-illuminated material add to a scene?
17. How is a marble material created?
18. Explain how black and white areas of a map applied to the Transparency property affect the transparency of a material.
19. Explain what the nodes in the ramp of a gradient map are for.
20. Name the four types of mapping available for adjusting texture maps at the object level.

Drawing Problems

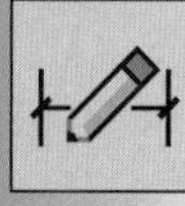

1. In this problem, you will create a scene with basic 3D objects, attach materials to the objects, and adjust the settings of the materials.
   A. Start a new drawing and set the units to architectural.
   B. Draw a 15′ × 15′ planar surface to represent the floor.
   C. Draw two boxes to represent two walls. Make the boxes 15′ × 4″ × 9′. Position them to form a 90° corner. Alternately, you can draw a polysolid of the same dimensions.
   D. Draw a R2′ × 5′H cone in the center of the room.
   E. Open the materials browser and locate a material similar to the one shown on the floor in the figure. Attach this material to the floor.
   F. Locate an appropriate material for the wall and attach it.
   G. Locate an appropriate material for the cone and attach it.
   H. Turn on the sun and adjust the time to create good shadows. Refer to Chapter 15 for an introduction to sun settings.
   I. Render the scene. Save the rendering as P16_01.jpg.
   J. Save the drawing as P16_01.

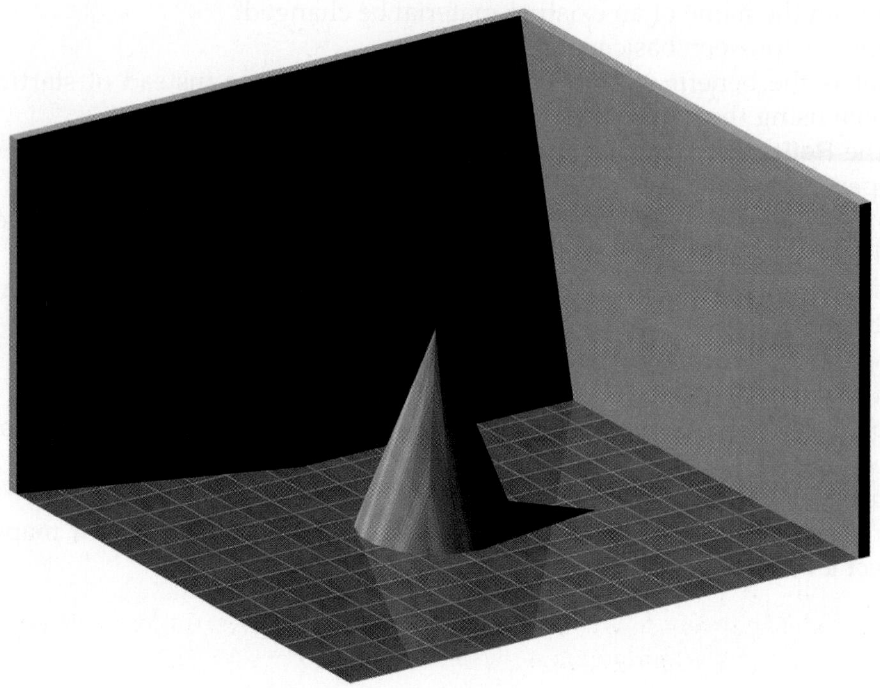

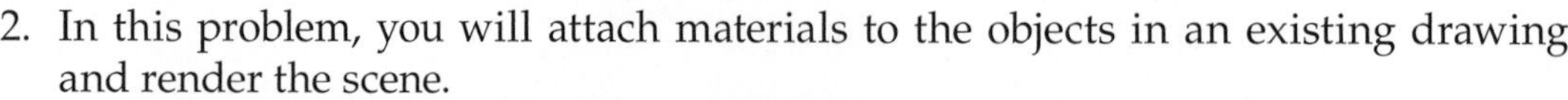

2. In this problem, you will attach materials to the objects in an existing drawing and render the scene.
   A. Open the drawing P15_01 from Chapter 15. If you did not complete this problem, do so now. Save the drawing as P16_02.
   B. Attach the materials of your choice to the objects in the scene. Do not be restricted by the names of the materials. For example, a concrete material may be suitable for foliage or even carpet with a simple color change. Be creative.
   C. If the items in the scene were inserted using the **Autodesk Seek Design Content** button in **DesignCenter**, they may be blocks with nested layers. Instead of exploding the blocks, use the **MATERIALATTACH** command and attach materials to the layers on which the nested objects reside.
   D. Turn on the sun and adjust the time to create good shadows. Refer to Chapter 15 for an introduction to sun settings.
   E. Render the scene. Save the rendering as P16_02.jpg.
   F. Save the drawing.

3. In this problem, you will create custom wood and marble materials.
   A. Start a new drawing and save it as P16_03.
   B. Draw two 5 × 5 × 5 boxes and position them near each other. Using other primitives, cut notches and holes in the boxes. The boxes will be used to test the custom materials.
   C. In the material editor, create two new materials. Name one Wood-*your initials* and the other Marble-*your initials*.
   D. Attach the wood material to one of the boxes and the marble material to the other box.
   E. Render the scene and make note of the wood grain and marble veins.
   F. Use the texture editor to change the properties of the materials.
   G. Render the scene again and make note of the changes. Using the **Render Region** button on the **Render** panel in the **Render** tab of the ribbon can save time when testing material changes.
   H. When you are satisfied with the materials, save the rendering as P16_03.
   I. Save the drawing.

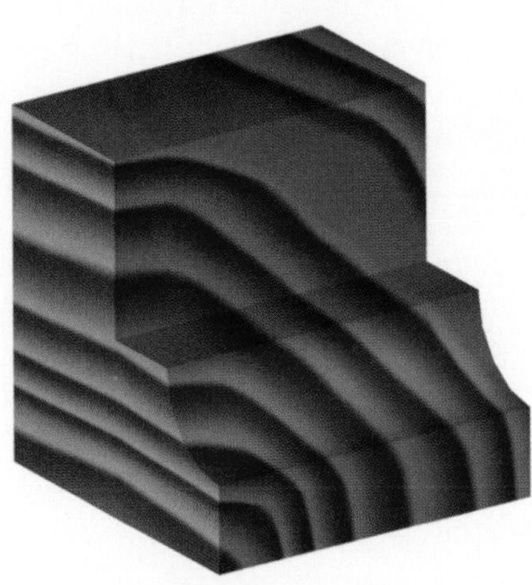

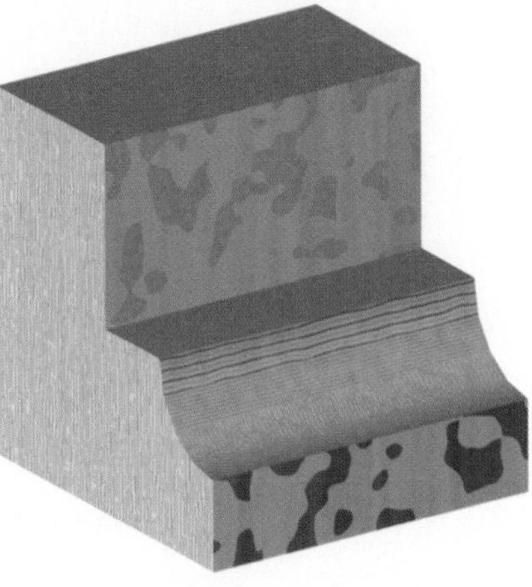

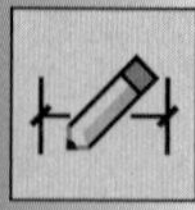

4. In this problem, you will create a bitmap and use it as transparency and bump maps.
   A. Draw a rectangle with an array of smaller rectangles inside of it, as shown below. Sizes are not important and the pattern can be varied if you like.
   B. Display a plan view of the rectangles. Then, copy all of the objects to the Windows clipboard by pressing [Ctrl]+[C] and selecting the objects.
   C. Launch Windows Paint. Then, paste the objects into the blank file. Notice how the AutoCAD background outside of the large rectangle is also included.
   D. Use the select tool (rectangle) in Paint to draw a window around the large rectangle created in AutoCAD and the smaller rectangles within it. Copy this to the Windows clipboard by pressing [Ctrl]+[C].
   E. Start a new Paint file without saving the current one and paste the image from the clipboard into the new blank file. Now, the unwanted AutoCAD background is no longer displayed. If needed, change the small rectangles to black and the lattice to white using the tools in Paint. The colors should be the reverse of what is shown below. Then, save the image file as P16_04.bmp and close Paint.
   F. In AutoCAD, draw a solid box of any size.
   G. Using the material editor, create a new material.
   H. Assign the P16_04.bmp image file you just created as a transparency map. Adjust the map so that it is scaled to fit to the object.
   I. Render the scene and note the effect.
   J. Turn off the transparency property by unchecking the check box in the category name.
   K. Assign the P16_04.bmp image file as a bump map. Adjust the map so that it is scaled to fit to the object.
   L. Render the scene and note the effect.
   M. Save the drawing as P16_04.

CHAPTER 17

Lighting

Learning Objectives

After completing this chapter, you will be able to:

- ✓ Describe the types of lighting in AutoCAD.
- ✓ List the user-created lights available in AutoCAD.
- ✓ Change the properties of lights.
- ✓ Generate and modify shadows.
- ✓ Add a background to your scene and control its appearance.

In Chapter 15, you were introduced to lighting. You learned how to adjust lighting by turning off the default lights and adding sunlight. In this chapter, you will learn all about the lights available in AutoCAD. You will learn lighting tips and tricks to help make the scene look its best.

Types of Lights

Ambient light is like natural light just before sunrise. It is the same intensity everywhere. All faces of the object receive the same amount of ambient light. Ambient light cannot create highlights, nor can it be concentrated in one area. AutoCAD does not have an ambient light setting. Instead, it relies on indirect illumination, which is discussed in Chapter 18.

A ***point light*** is like a lightbulb. Light rays from a point light shine out in all directions. A point light can create highlights. The intensity of a point light falls off, or weakens, over distance. Other programs, such as Autodesk 3ds max®, may call these lights *omni lights.* A ***target point light*** is the same as a standard point light except that a target is specified. The illumination of the target point light is directed toward the target.

A ***distant light*** is a directed light source with parallel light rays. This acts much like the Sun. Rays from a distant light strike all objects in your model on the same side and with the same intensity. The direction and intensity of a distant light can be changed.

A ***spotlight*** is like a distant light, but it projects in a cone shape. Its light rays are not parallel. A spotlight is placed closer to the object than a distant light. Spotlights have a hotspot and a falloff. The light from a standard spotlight is directed toward a target. A ***free spotlight*** is the same as a standard spotlight, but without a target.

A *weblight* is a directed light that represents real-world distribution of light. The illumination is based on photometric data that can be entered for each light. The light from a standard weblight is directed toward a target. A *free weblight* is the same as a standard weblight, but without a target point.

Properties of Lights

There are several factors that affect how a light illuminates an object. These include the angle of incidence, reflectivity of the object's surface, and the distance that the light is from the object. In addition, the ability to cast shadows is a property of light. Shadows are discussed in detail later in this chapter.

Angle of Incidence

AutoCAD renders the faces of a model based on the angle at which light rays strike the faces. This angle is called the ***angle of incidence.*** See **Figure 17-1.** A face that is perpendicular to light rays receives the most light. As the angle of incidence decreases, the amount of light striking the face also decreases.

Reflectivity

The angle at which light rays are reflected off of a surface is called the ***angle of reflection.*** The angle of reflection is always equal to the angle of incidence. Refer to **Figure 17-1.**

The "brightness" of light reflected from an object is actually the number of light rays that reach your eyes. A surface that reflects a bright light, such as a mirror, is reflecting most of the light rays that strike it. The amount of reflection you see is called the ***highlight.*** The highlight is determined by the angle from the viewpoint relative to the angle of incidence. Refer to **Figure 17-1.**

The surface quality of the object affects how light is reflected. A smooth surface has a high specular factor. The ***specular factor*** indicates the number of light rays that have the same angle of reflection. Surfaces that are not smooth have a low specular factor. These surfaces are called ***matte.*** Matte surfaces ***diffuse,*** or "spread out," the light as it strikes the surface. This means that few of the light rays have the same angle of reflection. **Figure 17-2** illustrates the difference between matte and high specular

Figure 17-1.
The amount of reflection, or highlight, you see depends on the angle from which you view the object.

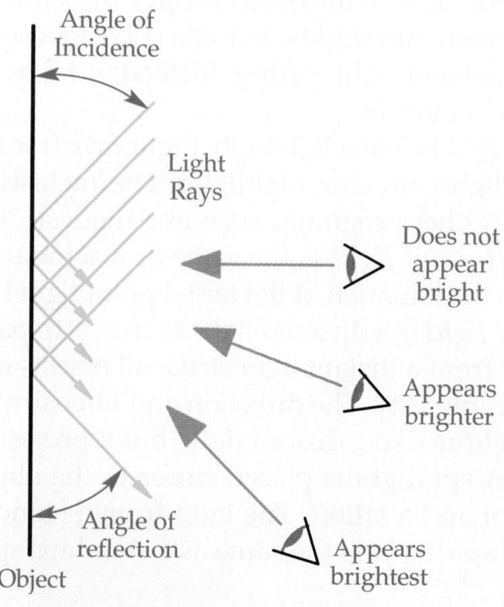

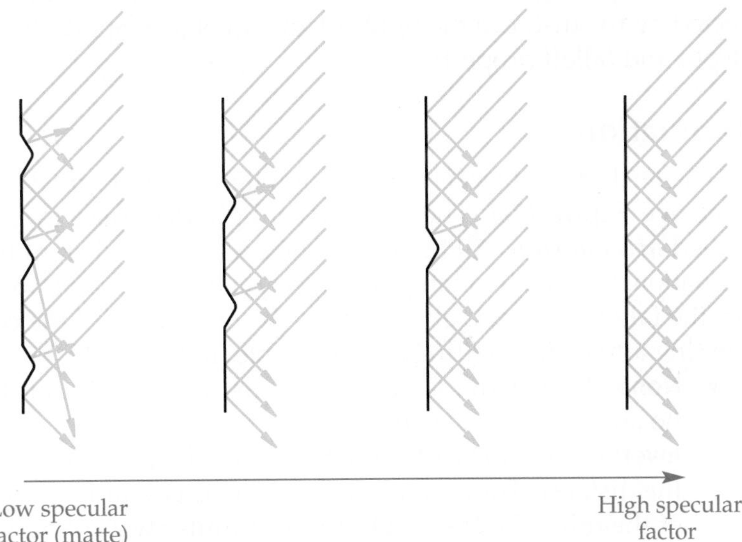

Figure 17-2. Matte surfaces produce diffuse light. This is also referred to as having a low specular factor. Shiny surfaces evenly reflect light and have a high specular factor.

finishes. Surfaces can also vary in ***roughness.*** Roughness is a measure of the polish on a surface. This also affects how diffused the reflected light is.

PROFESSIONAL TIP

When rendering metal and other shiny materials, it helps to have a bright object nearby to create some highlights on the object. Adding a simple box to the left and right of the object (out of the view, however) and applying a self-illuminating material to them will create some nice highlights. Photographers use this trick in the studio.

Hotspot and Falloff

A spotlight produces a cone of light. The ***hotspot*** is the central portion of the cone, where the light is brightest. See Figure 17-3. The ***falloff*** is the outer portion of the cone,

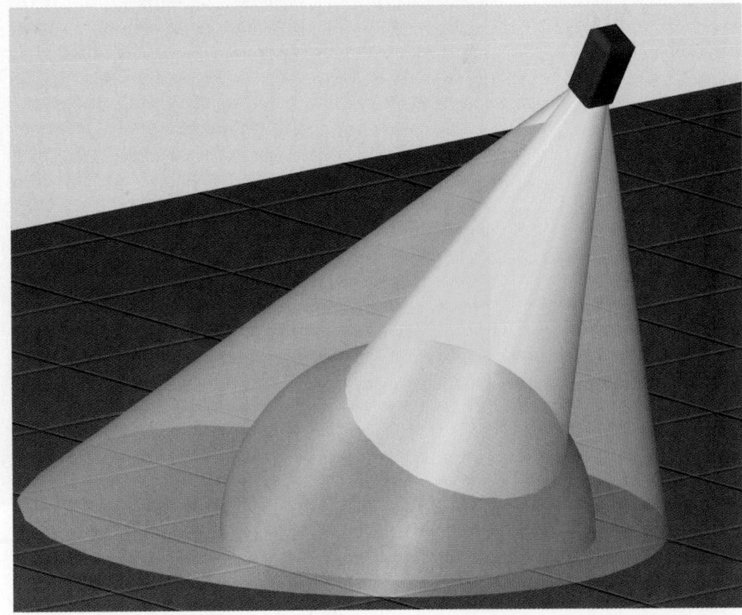

Figure 17-3. The hotspot of a spotlight is the area that receives the most light. The smaller cone is the hotspot. The falloff receives light, but less than the hotspot. The larger cone is the falloff.

where the light begins to blend to shadow. The hotspot and falloff of a spotlight are not affected by the distance the light is from an object. Spotlights are the only lights with hotspot and falloff properties.

Attenuation

The farther an object is from a point light or spotlight, the less light will reach the object. See Figure 17-4. The intensity of light decreases over distance. This decrease is called ***attenuation.*** All lights in AutoCAD, except distant lights, have some kind of attenuation. Often, attenuation is called *falloff* or *decay*. However, do not confuse this with the falloff of a spotlight, which is the outer edge of the cone of illumination. The following attenuation settings are available in AutoCAD.

- **None.** Applies the same light intensity regardless of distance. In other words, no attenuation is calculated.
- **Inverse Linear.** The illumination of an object decreases in inverse proportion to the distance. For example, if an object is two units from the light, it receives 1/2 of the full light. If the object is four units away, it receives 1/4 of the full light.
- **Inverse Squared.** The illumination of an object decreases in inverse proportion to the square of the distance. For example, if an object is two units from the light, it receives $(1/2)^2$, or 1/4, of the full light. If the object is four units away, it receives $(1/4)^2$, or 1/16, of the full light. As you can see, attenuation is greater for each unit of distance with the **Inverse Squared** option than with the **Inverse Linear** option.

PROFESSIONAL TIP

The intensity of the Sun's rays does not diminish from one point on Earth to another. They are weakened by the angle at which they strike Earth. Therefore, since distant lights are similar to the Sun, attenuation is not a factor with distant lights.

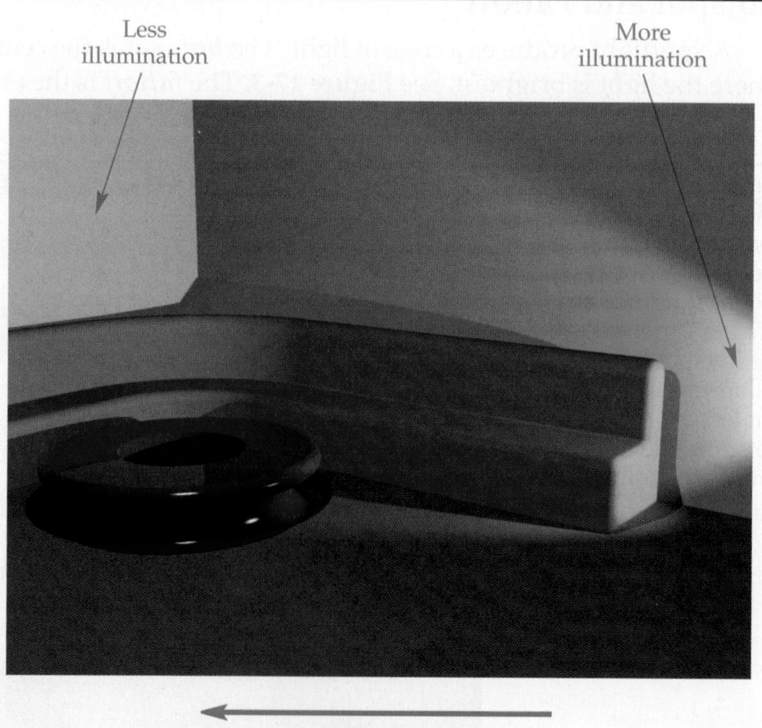

Figure 17-4. Attenuation is the intensity of light decreasing over distance. Attenuation has been turned on in this scene.

AutoCAD Lights

AutoCAD has three types of lighting: default lighting, sunlight with or without sky illumination, and user-created lighting. ***Default lighting*** is the lighting automatically available in the scene. It is composed of two light sources that evenly illuminate all surfaces. As the viewpoint is changed, the light sources follow to maintain an even illumination of the scene. There is no control over default lighting and it must be shut off whenever one of the other types of lighting is used.

As you saw in Chapter 15, ***sunlight*** may be added to any scene. AutoCAD uses a distant light to simulate the parallel rays of the Sun. The date and time of day can be adjusted to create different sunlight illumination. ***Sky illumination*** may also be added with sunlight to simulate light bouncing off of objects in the scene and particles in the atmosphere. This helps create a more-natural feel.

User-created lighting results when you add AutoCAD light objects to the drawing. There are four types of user-created lights: distant light, weblight, point light, and spotlight. See **Figure 17-5.** A distant light is a directed light source with parallel light rays. A weblight is a directional point light containing light intensity (photometric) data. A point light is like a lightbulb with light rays shining out in all directions. A spotlight is like a distant light, but it projects light in a cone shape instead of having parallel light rays.

When created, point lights, weblights, and spotlights are represented by ***light glyphs,*** or icons, in the drawing. To suppress the display of light glyphs, pick the **Light Glyph Display** button in the expanded area of the **Lights** panel on the **Render** tab of the ribbon. The button is blue when light glyphs are displayed. The default lights, sun, and distant lights are not represented by glyphs.

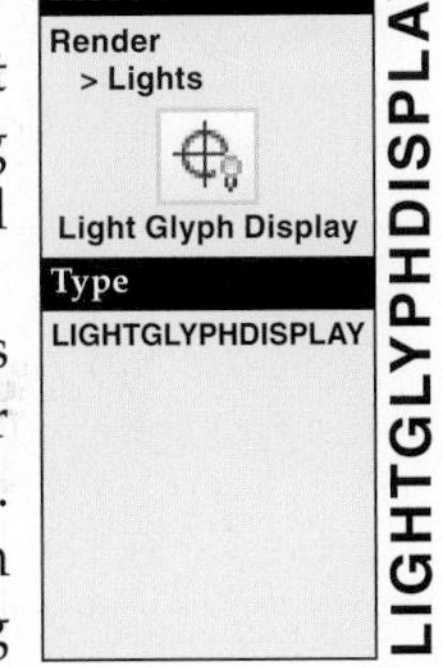

In this section, you will learn how to add lights. You will also learn how to adjust the various properties of sunlight and AutoCAD light objects. The tools for working with lights can be accessed using the command line, the tool palettes, and **Lights** panel on the **Render** tab of the ribbon. See **Figure 17-6.**

So that you will never work with a completely dark scene, default lighting is applied in the viewport and to the rendering if no other lights are added. In order for your lights to be applied, you must switch between default lighting and user lighting. To do this, pick the **Default Lighting** button in the expanded area of the **Lights** panel on the **Render** tab of the ribbon. This button toggles the lighting between default lighting and whatever lights are available in the scene. When default lighting is on, the button is blue. When off, the button is not highlighted. If you elected to do so, AutoCAD will automatically shut off default lighting when sunlight is turned on or a user-created light is added to the scene.

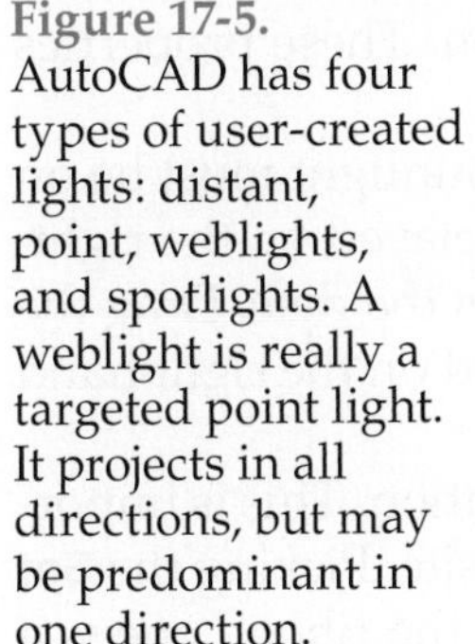

Figure 17-5.
AutoCAD has four types of user-created lights: distant, point, weblights, and spotlights. A weblight is really a targeted point light. It projects in all directions, but may be predominant in one direction.

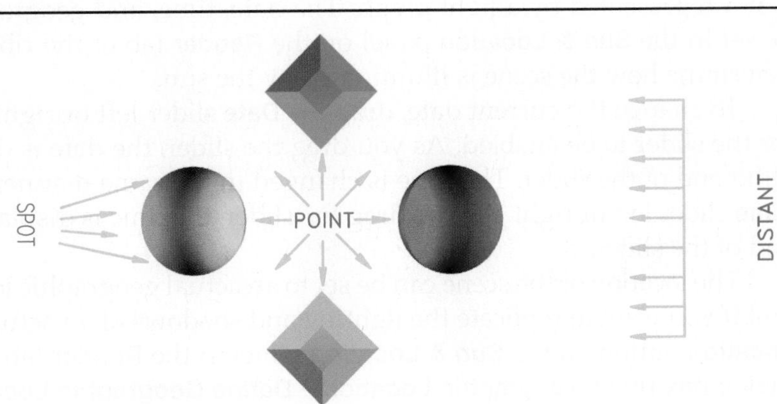

Figure 17-6.
The tools for adding and controlling lights.

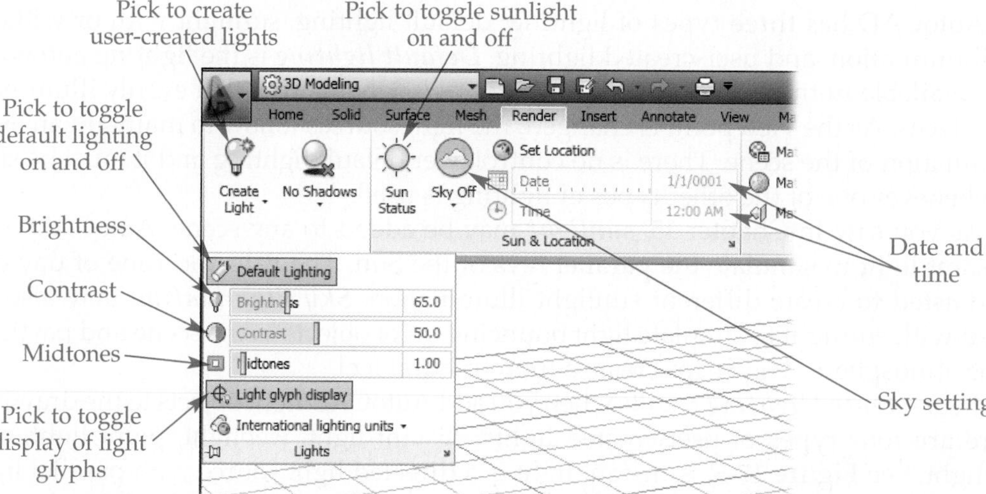

Lighting Units

There are three types of ***lighting units*** available in AutoCAD: standard (generic), international (SI), and US customary (American). Generic lighting is the type of lighting that was used in AutoCAD prior to AutoCAD 2008. This lighting provides very nice results, but the settings are not based on any real measurements. International or US customary lighting is called photometric lighting. ***Photometric lighting*** is physically correct and attenuates at the square of the distance from the source. For more accuracy, photometric data files can be imported from lighting manufacturers.

The **LIGHTINGUNITS** system variable sets which type of lighting is used. A setting of 0 means that standard (generic) lighting is used. However, for more realistic lighting, it is recommended that photometric lighting be used. Enter a value of 1 and US customary (American) lighting units are used. A setting of 2 turns on international lighting units. This is the default setting. A setting of 1 or 2 results in photometric lighting. The only difference between a setting of 1 and 2 is that US customary (American) units are displayed as *candelas* and international units are displayed as *lumens.*

Sunlight

SUNSTATUS

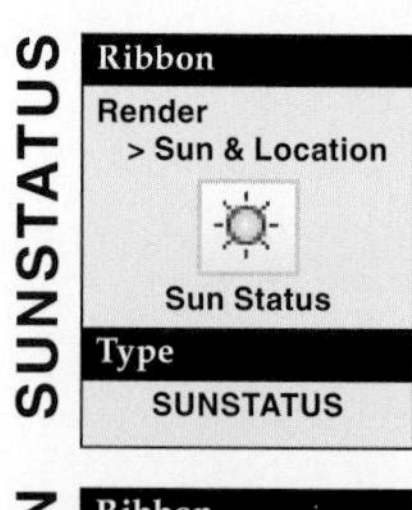

GEOGRAPHICLOCATION

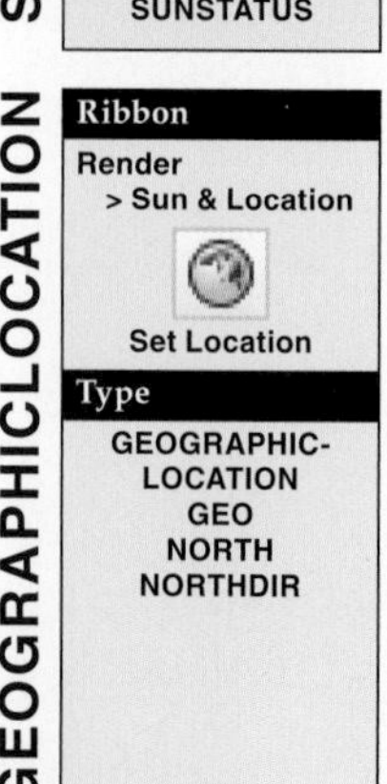

To turn sunlight on or off, pick the **Sun Status** button in the **Sun & Location** panel on the **Render** tab of the ribbon. This button is blue when sunlight is on. Sunlight can also be turned on or off in the **Sun Properties** palette, which is discussed later. Sunlight is not represented by a light glyph. The date, time, and geographic location can also be set in the **Sun & Location** panel on the **Render** tab of the ribbon. These properties determine how the scene is illuminated by the sun.

To change the current date, drag the **Date** slider left or right. Sunlight must be on for the slider to be enabled. As you drag the slider, the date is displayed on the right-hand end of the slider. The time is changed in the same manner as the date. Drag the **Time** slider left or right. As you drag the slider, the time is displayed on the right-hand end of the slider.

The location of the scene can be set to an actual geographic location. This is important if you want to replicate the lighting and shadows of an actual site. Picking the **Set Location** button on the **Sun & Location** panel in the **Render** tab of the ribbon opens a dialog box titled **Geographic Location – Define Geographic Location**. See Figure 17-7. The first option in this dialog box allows you to specify the location by importing a KML or a KMZ file. ***KML*** stands for Keyhole Markup Language. This file contains

Figure 17-7.
This dialog box provides the options for defining the geographic location of the model.

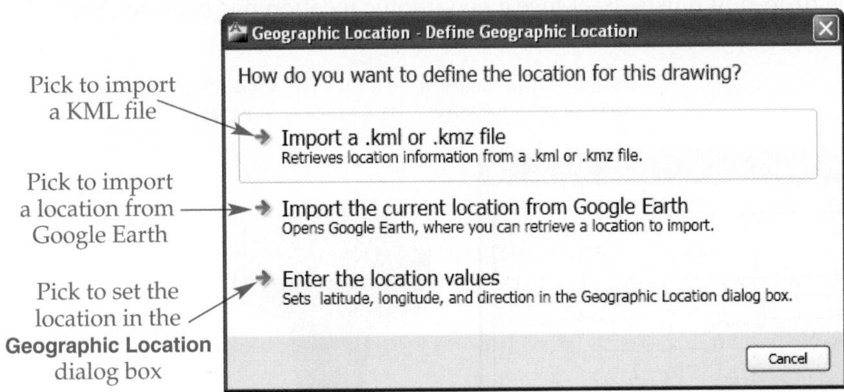

latitude, longitude, and, sometimes, other data to pinpoint a location on Earth. A KMZ file is a zipped KML file. The second option is to import the current location from Google Earth. In order for this to work, Google Earth must be installed and open with the location selected. The third option opens the **Geographic Location** dialog box. See **Figure 17-8.**

When a location is imported from Google Earth, you are asked to select a point in the drawing for the location. After picking a point in the drawing, you are asked for the North vector. Pick a point that should be in the direction of North. Once this is set, a ***geographic marker*** is added at the point selected for the location. This marker looks like a red and white thumbtack. Its visibility is controlled with the **GEOMARKERVISIBILITY** system variable. See **Figure 17-9.**

Figure 17-8.
The **Geographic Location** dialog box is used to input geographic location data.

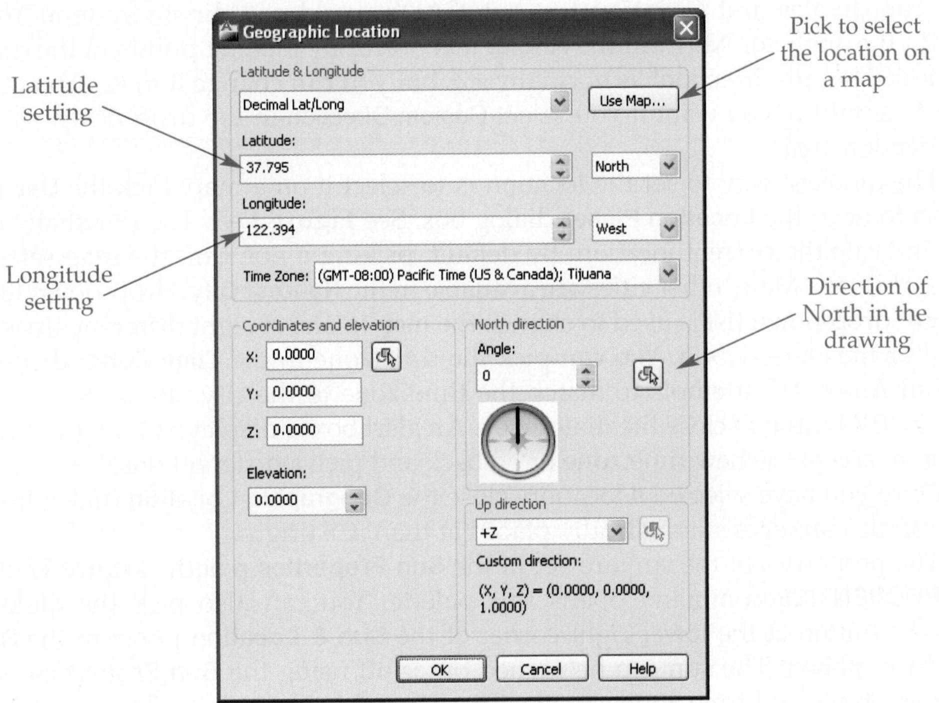

Figure 17-9.
A—The **Location Picker** dialog box is a quick way to set the location. Pick a location on the map or select it in the drop-down lists. B—Once a geographic location has been set, the marker is added to the WCS origin.

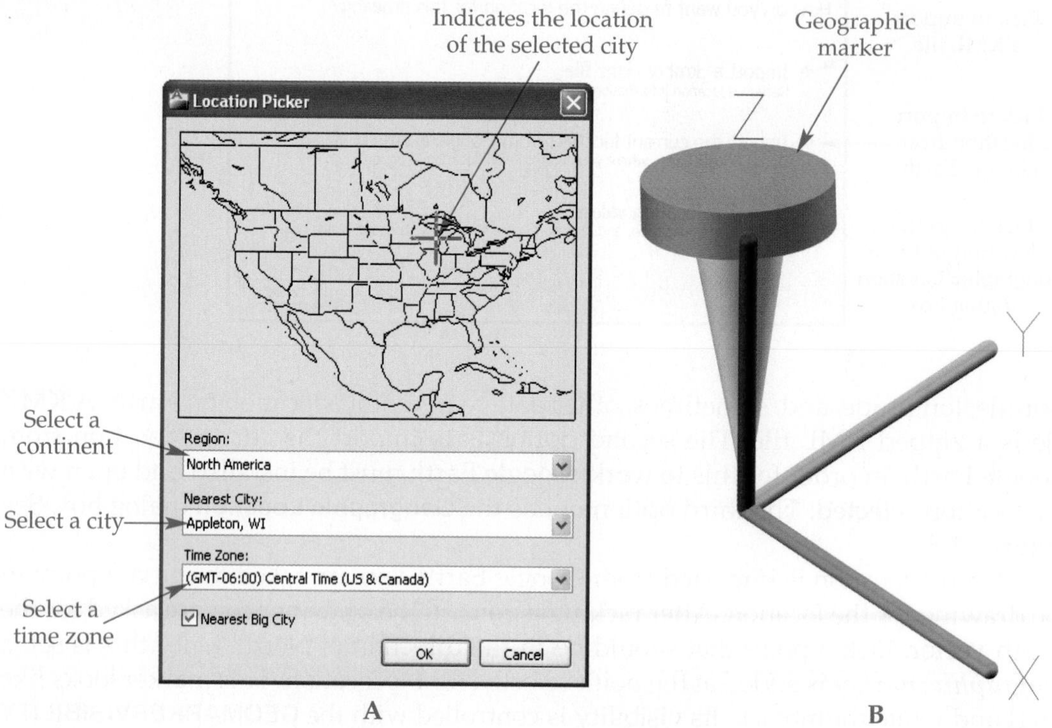

To set the location in the **Geographic Location** dialog box, enter the latitude and longitude. You can select to use either decimal or degrees/minutes/seconds values for these settings. Also, pick a time zone in the **Time Zone:** drop-down list and set XYZ coordinates and elevation for AutoCAD's world coordinate system. You can specify the angle for North in the **Angle:** text box or by picking points in the drawing window. The "up direction" is normally +Z, but you can change it to –Z. You can also set it to anything you want if you select Custom Direction in the drop-down list in the **Up direction** area.

The quickest way to set the location is to select it on a map. Pick the **Use Map...** button to open the **Location Picker** dialog box. See Figure 17-9. The crosshairs on the map indicate the current location. By default, picking a point on the map selects the nearest big city. Many other cities are available in the **Nearest City:** drop-down list. The **Region:** drop-down list is used to change the map to one of eight different areas of the world or the entire world. You can pick the time zone in the **Time Zone:** drop-down list, but AutoCAD attempts to match the time zone to the city you select when you pick the **OK** button to close the dialog box. An alert box is displayed that gives you the option to accept the new time zone or go back and pick a different one.

Once you have selected a location, close the **Geographic Location** dialog box. The geographic marker is automatically placed at the WCS origin.

The properties of the sun are set in the **Sun Properties** palette, Figure 17-10. The **SUNPROPERTIES** command opens this palette. You can also pick the dialog box launcher button at the lower-right corner of the **Sun & Location** panel in the **Render** tab of the ribbon. The sun can be turned on or off using the **Sun Properties** palette. The date, time, and time zone can also be changed in the palette. These settings are the same as previously discussed. There are other properties of the sun that are only available in the **Sun Properties** palette. This palette contains several categories, which are discussed in the next sections.

Figure 17-10.
The **Sun Properties** palette.

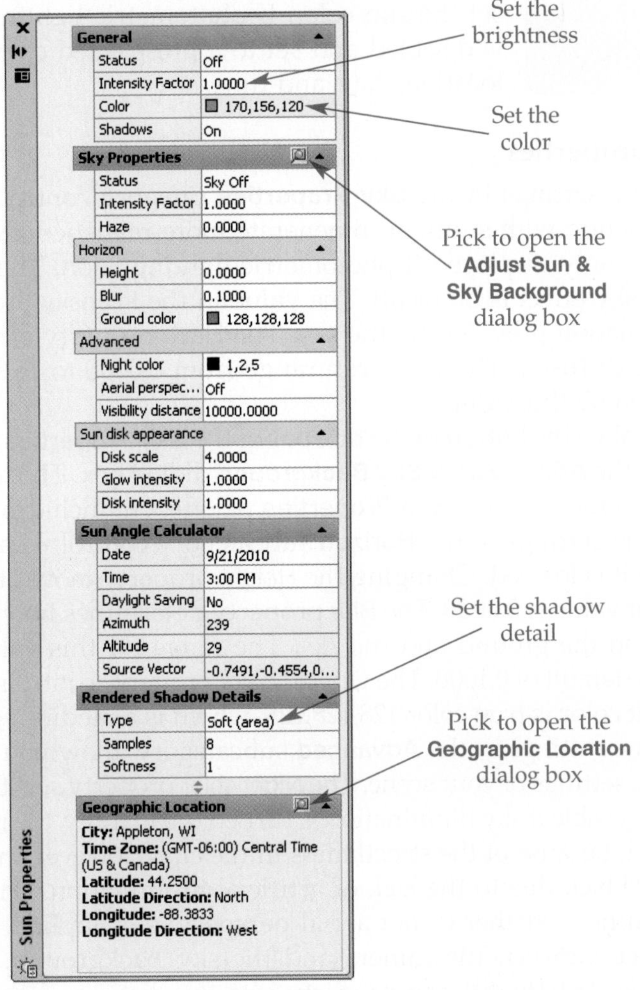

General

The Intensity Factor property in the **General** category determines the brightness of the sun. Setting this property to zero, in effect, turns off sunlight. Increasing the property makes the sunlight brighter. The maximum value for the property is determined by the capabilities of your computer.

The Color property in the **General** category is used to set the color of the sun, if photometric lighting is off. By default, sunlight is white (true color 255, 255, 255). To change the color, pick the drop-down list and choose a new color. If you pick the Select Color... entry, the **Select Color** dialog box is displayed for selecting a color. Sunlight can be changed to any color, but be aware that changing the color of the light may drastically alter the appearance of a scene. This is especially true if materials are attached to objects in the drawing. The color of sunlight is often set to a very light blue for an outdoor scene to help convey a bright blue sky.

The Shadows property in the **General** category determines whether or not the sun casts shadows. The property is either on or off. Shadows are discussed in detail later in this chapter.

NOTE

Photometric lighting must be off (**LIGHTINGUNITS** = 0) in order to set the sun color. If photometric lighting is on, the Color property is disabled and set to a preselected color based on the geographic location, date, and time.

Sky properties

The settings in the **Sky Properties** category control the sky. The sky is used in conjunction with sunlight to generate more-realistic lighting in the scene. This category is only displayed if photometric lighting is on. The Status property determines if the sky effect is on or off. The value in the Intensity property is a multiplier for the illumination provided by the sky. The Haze property controls how the sky illumination is diffused. The value can range from 0.0000 to 15.0000. The preset sun color is affected by this value.

Notice the button in the title bar of the **Sky Properties** category. Picking this button opens the **Adjust Sun & Sky Background** dialog box. This dialog box contains the same settings found in the **Sun Properties** palette, but includes a preview of the sun disk.

The settings in the **Horizon** subcategory control what the horizon looks like and where it is located. Changing the Height property moves the horizon up or down. The default value is 0.0000. The Blur property determines how much the horizon is blurred between the ground and the sky. The range for this value is from 0.0000 to 10.0000 with a default of 0.1000. The Ground color property controls the color of the ground. The default color is true color 128,128,128, which is a medium gray.

The settings in the **Advanced** subcategory allow you to control some of the more artistic settings of your scene. The Night color property sets the color of the night sky. This is only visible if sky illumination is turned on. In a city, the night sky may have an orange tint to it because of the streetlamps in the city. However, in the country, the night sky is nearly black due to the lack of artificial illumination. The Aerial perspective property determines whether or not aerial perspective is applied. This is a way of simulating distance between the camera and the sky/background. The setting is either on or off. The Visibility distance property sets the distance from the camera at which haze obscures 10% of the objects in the background. This is a very useful tool for creating the illusion of depth in a scene. The default value is 10000.0000, but it can range from 0.0000 to whatever is needed.

Normally, lights do not appear in the rendered scene at all, only the illumination provided by the lights. The settings in the **Sun disk appearance** subcategory control what the sun looks like in the sky, or on the background. The Disk scale property sets the size of the sun, or solar disk, as it appears on the background. The default value is 4.0000 with the range of values being from 0.0000 to 25.0000. The Glow intensity value determines the size of the glowing halo around the sun in the sky. Default value is 1.0000 and can range from 0.0000 to 25.0000. The Disk intensity property controls the brightness of the sun on the background. The default value is 1.0000 and the range of values is from 0.0000 to 25.0000.

PROFESSIONAL TIP

The sky at night is always lighter than Earth unless there are other light sources in the scene, such as streetlamps.

Sun angle calculator

The settings in the **Sun Angle Calculator** category determine the angle of the sun in relationship to the XY plane. The Date and Time properties are discussed earlier and can be controlled from the **Sun & Location** panel in the **Render** tab of the ribbon. This category in the **Sun Properties** palette also includes the Daylight Saving property. This property is used to turn daylight saving on or off. The Azimuth, Altitude, and Source Vector properties display the current settings, but are read-only in the **Sun Properties** palette. These values are automatically calculated based on the settings in the **Geographic Location** dialog box.

Rendered shadow details

The Type property in the **Rendered Shadow Details** category determines the type of shadow cast by the sun, if shadows are cast. When the property is set to Sharp, raytraced shadows are cast. These shadows have sharp edges. Raytracing produces accurate shadows, but rendering may take longer. When the property is set to either Soft (mapped) or Soft (area), shadow-mapped shadows are cast. This type of shadow has soft edges. Shadow-mapped shadows may be calculated quicker than raytraced shadows, but the resulting shadows are less precise. In addition, soft shadows do not work with transparent surfaces like windows. When the Type property is set to Soft (mapped), two additional settings are available in the category:

- Map Size. This property determines the number of subdivisions, or samples, used to create the shadow. By default, shadow maps are 256 × 256 pixels in size. If shadows look grainy, increasing this setting will make them look better.
- Softness. This property determines the sharpness of the shadow's edge. The value ranges from 1 to 10. The higher the value, the softer (less sharp) the edge of the shadow.

When the Type property is set to Soft (area), the two additional settings are:

- Samples. This property sets the number of samples used on the solar disk. The value can be from 0.0000 to 1000.0000.
- Softness. This property determines the sharpness of the shadow's edge, as described above.

When the Type property is set to Sharp, the other settings are read only and not applied.

NOTE

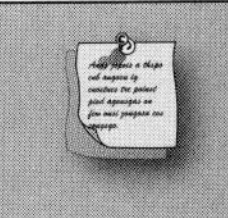

With photometric lighting on (**LIGHTINGUNITS** = 1 or 2), only the Soft (area) selection is available in the Type property drop-down list.

Geographic location

The **Geographic Location** category at the bottom of the **Sun Properties** palette displays the current geographic location settings. Changes cannot be made here, but picking the **Launch Geographic Location** button in the category's title bar opens the **Geographic Location** dialog box, which is described earlier.

Exercise 17-1

Complete the exercise on the student website.
www.g-wlearning.com/CAD

Distant Lights

Distant lights are user-created lights that have parallel light rays. See **Figure 17-11.** When a distant light is created, the location from where the light is originating must be specified along with the direction of the light rays. Distant lights are not represented in the drawing by light glyphs. Distant lights are often used to create even,

Figure 17-11. This example shows the use of a distant light to simulate sunlight shining through a window. Notice how the edges on the shadows of the grill are parallel.

uniform, overhead illumination, such as you would encounter in an office situation. The distance of the objects in the scene to the distant light has no effect on the intensity of the illumination. Distant lights do not attenuate. A distant light can also be used to simulate sunlight without having to set up a time and location. However, the light must be manually moved to change the illumination effect.

DISTANTLIGHT

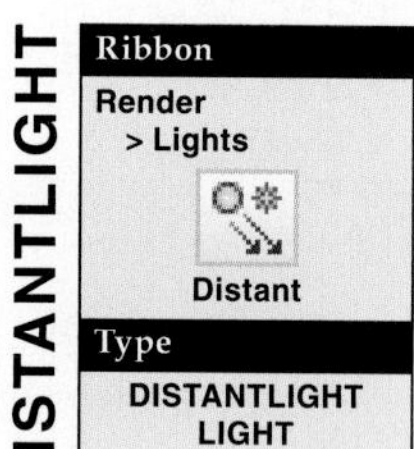

The **DISTANTLIGHT** or **LIGHT** command is used to create a distant light. You are first prompted to specify the direction from which the light is originating or to enter the **Vector** option. If you pick a point, it is the location of the light. Next, you are prompted to specify the point to which the light is pointing. This is simply the location where the light is aimed.

If you enter the **Vector** option instead of picking a "from" point, you must type the endpoint coordinates (in WCS units) of the direction vector. The light will point from the WCS origin to the entered endpoint. This option is not often used.

After the light location and direction are determined, several other options are available. You can name the light, set the intensity, turn the light on or off, determine if and how shadows are cast, and set the light color.

AutoCAD provides a default name for new lights based on the type of light and a sequential number, such as Distantlight1, Distantlight2, and so on. It is a good idea to provide a meaningful name for a light. This is especially true if there are other lights in the scene. To rename a light, enter the **Name** option. Then, type the name of the light and press [Enter].

To set the brightness of the light, enter the **Intensity** option. Then, type a value and press [Enter]. The default value is 1.00. Setting the value to 0.00, in effect, turns off the light. The maximum value depends on the capabilities of your computer. This option is called **Intensity Factor** if photometric lighting is enabled.

When a light is created, it is on. To turn the light off, enter the **Status** option. Then, change the setting to **Off**.

The **Photometry** option is available when photometric lighting is active. It controls the luminous qualities of visible light sources. Once you enter the **Photometry** options, you can select one of three settings:

- **Intensity.** This is the power of the light source and can be entered as candelas (cd), luminous flux (lx), or foot-candles (fc), depending on the current lighting units.
- **Color.** This is the color of the light source and can be changed by typing in a name (to get a list of color names, enter ?) or by the Kelvin temperature value (k).
- **Exit.** Exits the command option.

The **Shadow** option is used to determine if and how shadows are cast by the distant light. To turn off shadow casting, enter the option and select the **Off** setting. The **Sharp** setting creates raytraced shadows. The **Softmapped** setting casts shadow-mapped shadows. Shadows are discussed later in this chapter.

By default, new distant lights cast white light (true color 255, 255, 255). To change the color of the light, enter the **Color** option. This option is called **Filter Color** if photometric lighting is enabled. To specify a new true color, simply specify the RGB values and press [Enter]. To specify a color based on hue, saturation, and luminance (HSL), enter the **Hsl** option and specify the values. To enter an AutoCAD color index (ACI) number, enter the **Index color** option and specify the ACI number. To specify a color book color, enter the **Color Book** option and then specify the name of the color book followed by the name of the color.

Once all settings for the distant light have been made, use the **Exit** option to end the command and create the light. Do not press [Esc] to end the command. Doing so actually cancels the command and the light is not created.

PROFESSIONAL TIP

To provide a visual cue to a distant light's location and direction, draw a line with one endpoint at the "from" coordinate and the other endpoint at the "to" coordinate. Then, when creating the distant light, select the endpoints of the line. Since a distant light is not represented by a light glyph, the line serves the purpose of a glyph.

NOTE

If you attempt to create a distant light with photometric lighting on (**LIGHTINGUNITS** = 1 or 2), you will receive a warning to the effect that photometric distant lights may overexpose the scene. Your choices are to disable or allow distant lights.

Exercise 17-2

Complete the exercise on the student website.
www.g-wlearning.com/CAD

Point Lights

Point lights are user-created lights that have light rays projecting in all directions. See Figure 17-12. When a point light is created, its location must be specified. Since point lights illuminate in all directions, there is no "to" location for a point light. A light glyph represents point lights in the drawing. See Figure 17-13. Point lights can be set to attenuate. In this case, the distance of the objects in the scene to the point light affects the intensity of the illumination.

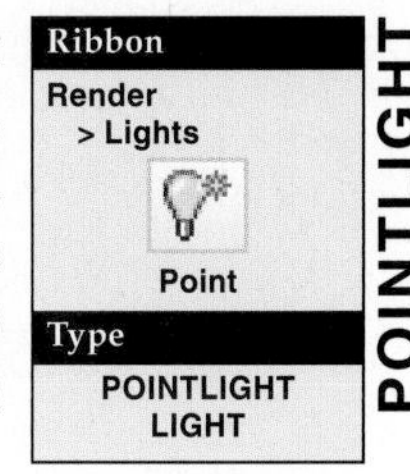

The **POINTLIGHT** or **LIGHT** command is used to create a point light. You are first prompted to specify the location of the point light. Once the location is established, several options for the light are available. You can name the light, set the intensity, turn the light on or off, adjust the photometry settings, determine if and how shadows are cast, set the attenuation, and set the light color. The **Name**, **Intensity** (or **Intensity Factor**), **Status**, **Photometry**, **Shadow**, and **Color** (or **Filter Color**) options work the same as the corresponding options for a distant light.

Figure 17-12.
A—A point light is placed inside of the lamp fixture. Notice how the light projects in all directions. B—When shadow casting is turned on for the light, the lampshade blocks the light from illuminating objects below the shade.

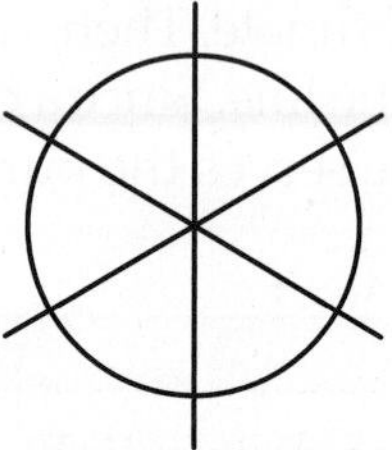

Figure 17-13.
This is the light glyph for a point light.

The **Attenuation** option is used to set attenuation for the point light. When this option is selected, five more options are available:

- **Attenuation Type**
- **Use Limits**
- **Attenuation Start Limit**
- **Attenuation End Limit**
- **Exit**

The **Exit** option returns you to the previous prompt.

The **Attenuation Type** option is used to turn attenuation on and off and to set the type of attenuation. To turn attenuation off, select the option and then enter **None**. To turn attenuation on, select the option and then enter either **Inverse Linear** or **Inverse Squared**. Attenuation is discussed in detail earlier in this chapter.

The **Use Limits** option determines if the attenuation of the light has a beginning and an end. When this option is set to **Off**, attenuation starts at the light and ends when the illumination reaches zero. When set to **On**, attenuation begins at the starting limit and ends at the ending limit.

To set the starting point for attenuation, enter the **Attenuation Start Limit** option. Then, specify the distance from the point light where attenuation will begin. The full intensity of the light provides illumination up to this point. From this point to the attenuation end limit, the light falls off.

To set the point where the illumination attenuates to zero, enter the **Attenuation End Limit** option. Then, specify the distance from the point light where the illumination is zero. Beyond this point, AutoCAD does not calculate the effect of the light.

Target Point Lights

The target point light is like a regular point light except that the command starts by asking for a *source* location and a *target* location. The rest of the options are the same. You can type TARGETPOINT to access this command or pick the **Targetpoint** option in the **LIGHT** command.

CAUTION

It is important to set attenuation limits. If there is no end limit for the light, AutoCAD may calculate the illumination beyond the boundary of the scene, even if there is nothing there to see. To speed processing time, tell AutoCAD where illumination stops for point and spotlights.

PROFESSIONAL TIP

A point light may be used as an incandescent lightbulb, such as in a table lamp. Most of these lightbulbs cast a yellow light. In these cases, you may want to change the color of the light to a light yellow. Compact fluorescent lightbulbs may cast white, light blue, or light yellow light. LED-based lights cast a color based on the color of the LED. Other colors can be used to give the impression of heat or colored lights.

Exercise 17-3

Complete the exercise on the student website.
www.g-wlearning.com/CAD

Spotlights

Spotlights are user-created lights that have light rays projecting in a cone shape in one direction. See **Figure 17-14.** When a spotlight is created, the location from where

Figure 17-14.
Three spotlights are used to simulate recessed ceiling lights. Notice how the light from each spotlight projects in a cone.

the light is originating must be specified along with the direction in which the light rays travel. A light glyph represents spotlights in the drawing. See **Figure 17-15.** Spotlights can be set to attenuate. In this case, the distance of the objects in the scene to the spotlight affects the intensity of the illumination.

SPOTLIGHT

Ribbon
Render
> Lights
Spot

Type
SPOTLIGHT
LIGHT

The **SPOTLIGHT** or **LIGHT** command is used to create a spotlight. You are first prompted to specify the location of the light. This is from where the light rays will originate. Next, you are prompted for the target location. This is simply the location where the light is aimed.

Once the location and target are established, several options for the light are available. The **Name**, **Intensity** (or **Intensity Factor**), **Status**, **Photometry**, **Shadow**, and **Color** (or **Filter Color**) options work the same as the corresponding options for a point light. However, a spotlight also has hotspot and falloff settings.

The ***hotspot*** is the inner cone of illumination for a spotlight. Refer to **Figure 17-16.** This is measured in degrees. To set the hotspot, enter the **Hotspot** option and then specify the number of degrees for the hotspot.

The ***falloff,*** not to be confused with attenuation, is the outer cone of illumination for a spotlight. Like the hotspot, it is measured in degrees. The falloff value must be greater than or equal to the hotspot value. It cannot be less than the hotspot value. In practice, the falloff value is often much greater than the hotspot value. To set the falloff, enter the **Falloff** option and then specify the number of degrees for the falloff.

Once the light is created and you select it in the viewport, grips are displayed. If you hover the cursor over a grip, a tooltip is displayed indicating what the grip will modify. You can use grips to change the location of the spotlight and its target, the hotspot, and the falloff. If you hover over a falloff or hotspot grip, the current angle is displayed in the wireframe cone (if dynamic input is on).

Figure 17-15.
This is the light glyph for a spotlight.

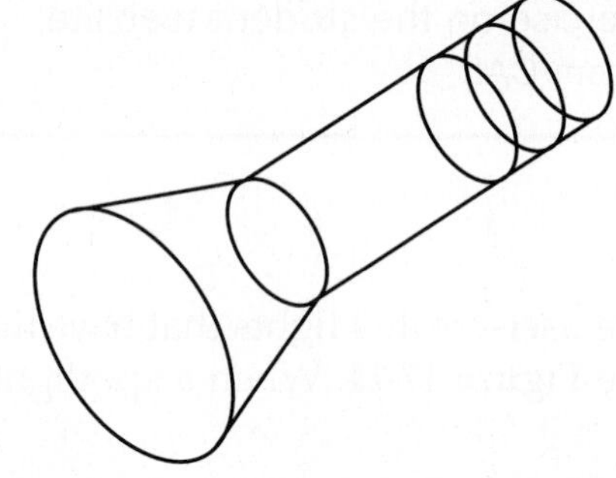

Figure 17-16.
Hotspot and falloff for a spotlight are angular measurements.

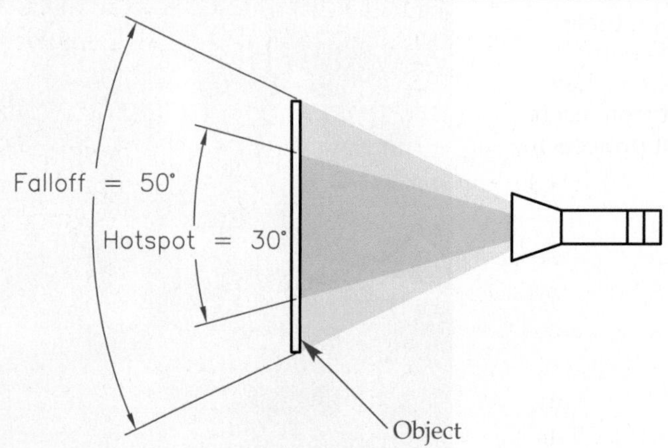

Free Spotlight

A free spotlight is like a standard spotlight except that you do not specify a target, only the light location. The rest of the options are the same. You can type FREESPOT to access this command or pick the **Freespot** option in the **LIGHT** command. When created, a free spotlight points down the Z axis (from positive to negative) of the current UCS. A free spotlight may be easier to control than a standard spotlight because you do not have to worry about the target point. If you want to change the angle or position of the light, use the **3DMOVE** and **ROTATE3D** commands.

PROFESSIONAL TIP

The best way to see how colored lights affect your model is to experiment. Remember, you can render selected areas of the scene. This allows you to see how light intensity and color change objects without performing a full render.

Exercise 17-4

Complete the exercise on the student website.
www.g-wlearning.com/CAD

Weblight

A photometric weblight is really just a targeted point light. The difference is that a weblight provides a more precise representation of the light. Real-world lights appear to evenly illuminate from their source, but, in reality, the shape of the light, the material used in its manufacture, and other factors make all lights distribute their energy in different ways. These data are provided by light manufacturers in the form of light distribution data. Light distribution data can be loaded into the **Photometric Web** subcategory of the **General** category in the **Properties** palette when the light is selected. Select the Web file property, then pick the browse button (**...**) and select an IES file. IES stands for Illuminating Engineering Society. AutoCAD's online documentation has additional information on IES files.

Think of the web of a weblight as a spherical cage surrounding the light source. If the light is evenly distributed from its source, the cage is a true sphere. In actuality, a light may emit more light energy in the X direction than in the Z direction. In this case, the cage bulges out further in the X direction. The position of this bulge may be important to the illumination of the scene and you may need to rotate the web to apply more or less light in one direction or another.

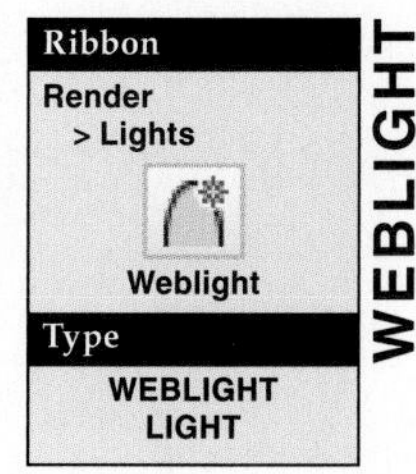

The **WEBLIGHT** or **LIGHT** command is used to create a weblight. You are prompted for source and target locations. The **Name, Intensity Factor, Status, Photometry, Shadow,** and **Filter Color** options work the same as the corresponding options for the previously discussed lights. However, weblights have an additional **Web** option. When this option is activated, these options are presented:

- **File.** Allows you to select an IES file.
- **X.** Rotates the web around the X axis.
- **Y.** Rotates the web around the Y axis.
- **Z.** Rotates the web around the Z axis.
- **Exit.** Exits the **Web** option.

Point and spotlights can be converted to weblights, and vice versa, using the **Properties** palette. Simply select an existing light and open the **Properties** palette. In the **General** category, the Type property determines whether the light is a point light, spotlight, or weblight. Select the type in the drop-down list. Using the **Properties** palette with lights is discussed in detail later in this chapter.

Free Weblight

FREEWEB

Type
FREEWEB LIGHT

A free weblight is the same as a standard weblight except there is no target. Only the source location is specified when placing the light. To change the location and direction of the light, use the **ROTATE3D** and **3DMOVE** commands.

NOTE

To create either standard or free weblights, photometric lighting must be enabled (**LIGHTINGUNITS** = 1 or 2).

Photometric Lights Tool Palette Group

Photometric lights may be easily added to the drawing using the tool palettes in the **Photometric Lights** tool palette group. This palette group contains four palettes: **Fluorescent**, **High Intensity Discharge**, **Incandescent**, and **Low Pressure Sodium**. Lights created with these tools have preset properties for **Intensity Factor**, **Shadow**, and **Filter Color**. The glyph for the long fluorescent lights has a yellow line passing through its center indicating the direction of the light. The high intensity discharge, low-pressure sodium, and regular incandescent lights are point lights. The incandescent halogen lights are free spotlights. The recessed incandescent lights are a special weblight designed for recessed light fixtures.

Lights in Model and Properties Palettes

The **Lights in Model** palette is extremely useful for controlling the lights in your scene, Figure 17-17. Used in conjunction with the **Properties** palette, you can manage and edit all of the lights in a scene.

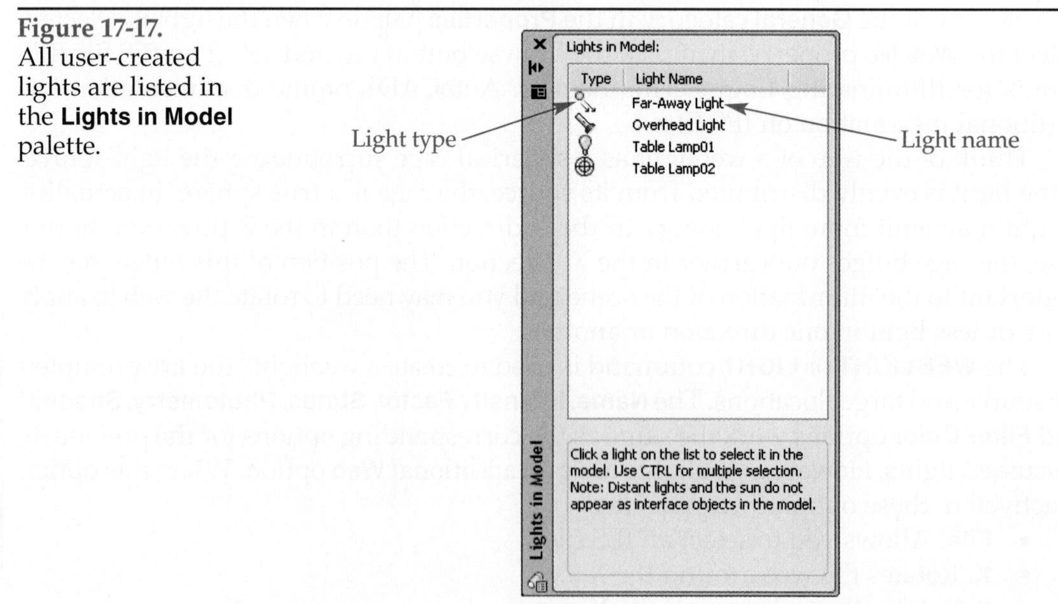

Figure 17-17. All user-created lights are listed in the **Lights in Model** palette.

Light list

Type
LIGHTLIST

The **LIGHTLIST** command displays the **Lights in Model** palette. This palette can also be displayed by picking the dialog box launcher button at the lower-right corner of the **Lights** panel in the **Render** tab of the ribbon. All user-created lights in the scene are displayed in the list. To modify the properties of a light, either double-click on the light name or right-click on it and select **Properties** from the shortcut menu. This opens the **Properties** palette. See Figure 17-18. If the **Properties** palette is already open, you can simply select a light in the **Lights in Model** palette. You can select more than one light by pressing the [Ctrl] key and selecting the names in the **Lights in Model** palette, which allows you to change all of their settings at the same time. This is an excellent way to make the lighting in your scene uniform or to control a series of lights with a single edit.

A light can be deleted from the scene using the **Lights in Model** palette. To do so, simply right-click on the name of the light and select **Delete Light** from the shortcut menu. The light is removed from the drawing. Using the **UNDO** command restores the light.

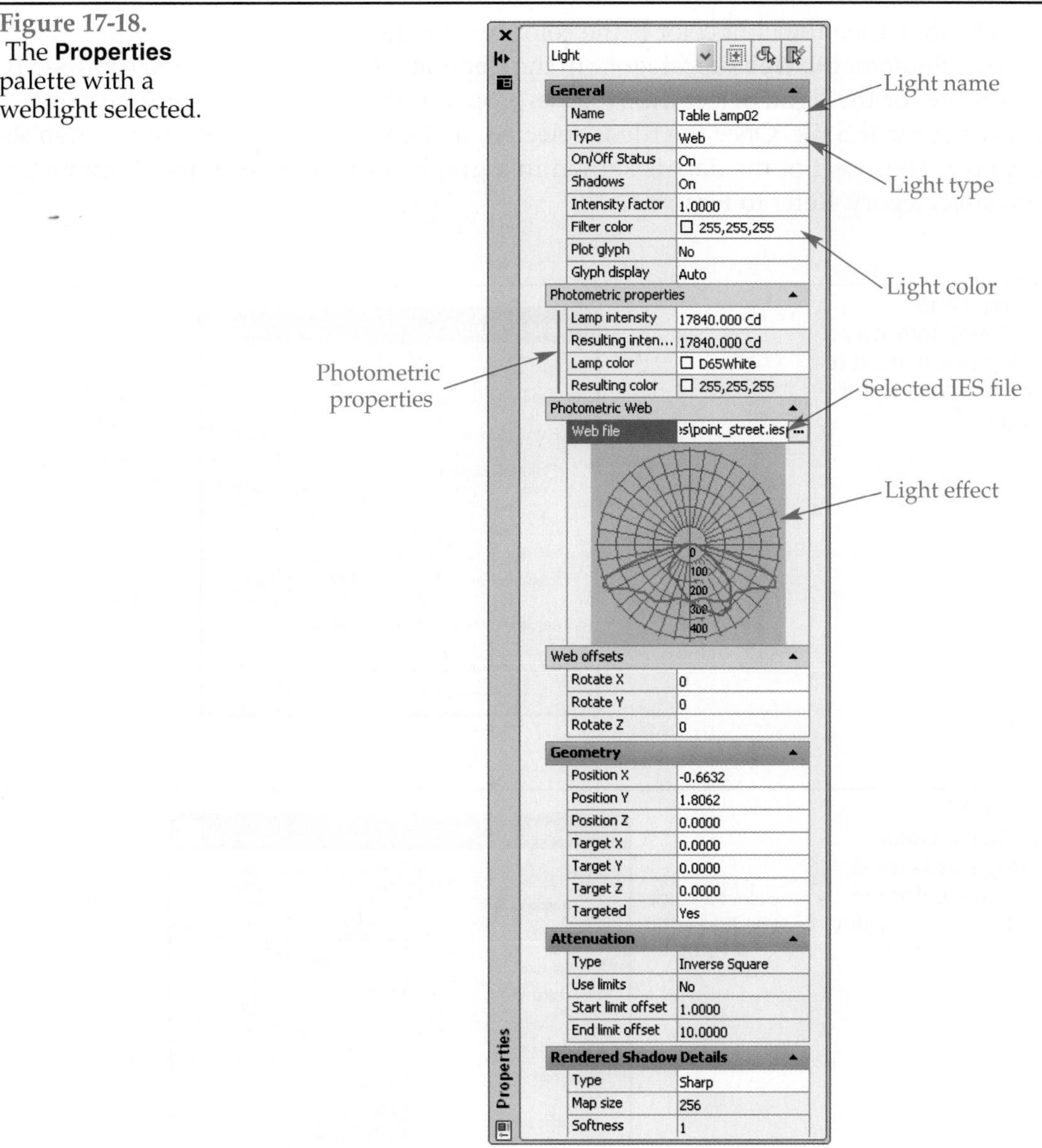

Figure 17-18.
The **Properties** palette with a weblight selected.

Properties palette

The **Photometric Properties** subcategory of the **General** category in the **Properties** palette has special settings for photometric lights. The Lamp intensity property determines the brightness of the light. The value may be expressed in candelas (cd), lumens (lm), or illuminance (lux) values. When you select the Lamp intensity property, a button is displayed to the right of the value. Picking this button opens the **Lamp Intensity** dialog box, Figure 17-19. In this dialog box, you can change the illumination units and set the intensity (Lamp intensity property). You can also set an intensity scale factor. This is multiplied by the Lamp intensity property to obtain the actual illumination supplied by the light. The read-only Resulting intensity property in the **Properties** palette displays the result.

The Lamp color property in the **Photometric Properties** subcategory in the **General** category controls the color of the light. If you select the property, a button is displayed to the right of the value. Picking this button opens the **Lamp Color** dialog box. See Figure 17-20. This dialog box gives you the option to control the color of the light by either standard spectra colors or Kelvin colors. The color selected in the **Filter color:** drop-down list is applied to the color of the light. The **Resulting color:** swatch displays the color cast by the light once the filter color is applied. If the filter color is white (255,255,255), then the light color is the color cast by the light.

The **Photometric Web** subcategory in the **General** category is where you can specify an IES file for the light. Select the Web file property, then pick the browse button (**...**) and select the IES file. Once the file is selected, its location is displayed in the Web file property. The effect of the data is shown in a graph at the bottom of the **Photometric Web** subcategory. Refer to Figure 17-18.

Figure 17-19.
The **Lamp Intensity** dialog box is used to set the intensity for a light.

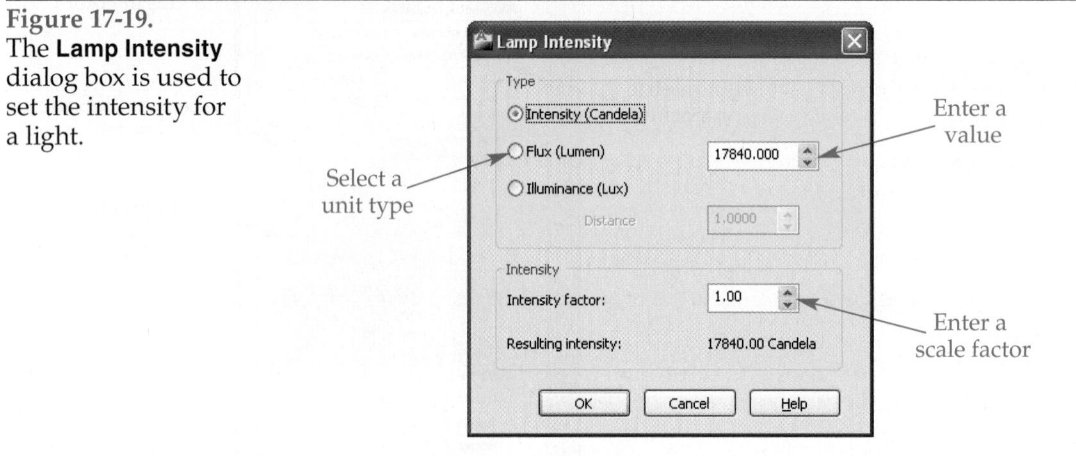

Figure 17-20.
The **Lamp Color** dialog box is used to set the color for the light and a filter color, if needed.

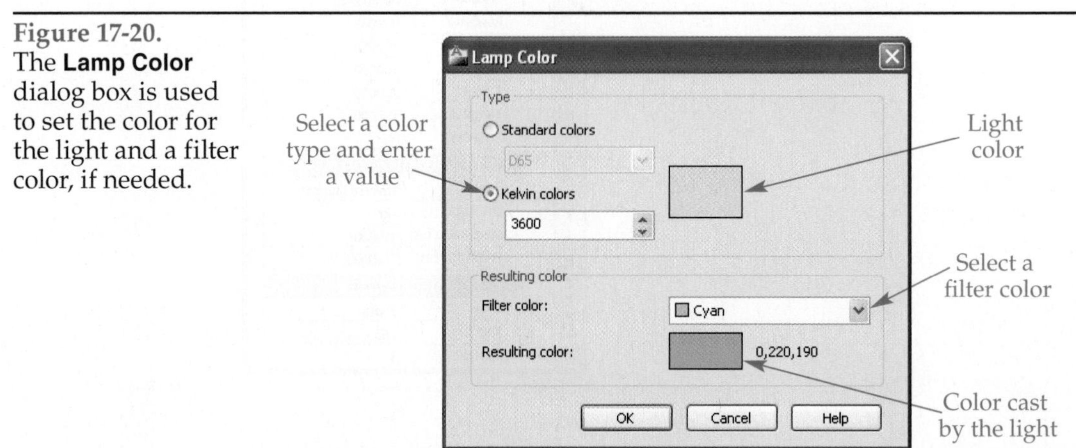

The **Web offsets** subcategory in the **General** category allows you to rotate the web around the X, Y, and Z axes. This is discussed earlier in the Weblight section.

In the **Geometry** category, you can change the X, Y, and Z coordinates of the light. You can also change the X, Y, and Z coordinates of the light's target. If the light is not targeted, the Target X, Target Y, and Target Z properties are not displayed. To change the light from targeted to free, and vice versa, select Yes or No in the Targeted property drop-down list.

The properties in the **Attenuation** category are the same as those discussed earlier in this chapter in the Point Lights section. In order to change these properties in the **Properties** palette, photometric lighting must be off (**LIGHTINGUNITS** = 0).

The last category in the **Properties** palette is **Rendered Shadow Details**. The properties in this category are used to control shadows. Shadows are discussed later in this chapter.

PROFESSIONAL TIP

It is important to give your lights names that make them easy to identify in a list. If you accept the default names for lights, they will be called Pointlight1, Spotlight5, Distantlight7, Weblight2, etc., making them difficult to identify. Use the **Name** option when creating the light or, after the light is created, the **Properties** palette to change the name of the light.

Determining Proper Light Intensity

As a general rule, the object nearest to a point light or spotlight should receive the full illumination, or full intensity, of the light. Full intensity of any light that has an attenuation property is a value of one. Remember, attenuation is calculated using either the inverse linear or inverse square method. Therefore, you must calculate the appropriate intensity.

For example, suppose you have drawn an object and placed a point light and a spotlight. The point light is 55 units from the object. The spotlight is 43 units from the object. Use the following calculations to determine the correct intensity settings for the lights.

- **Inverse linear.** If the point light is 55 units from the object, the object receives 1/55 of the light. Therefore, set the intensity of the point light to 55 so the light intensity striking the object has a value of 1 (55/55 = 1). Since the spotlight is 43 units from the object, set its light intensity to 43 (43/43 = 1).
- **Inverse square.** If the point light is 55 units from the object, the object receives $(1/55)^2$, or 1/3025 ($55^2 = 3025$), of the light. Therefore, set the intensity of the point light to 3025 (3025/3025 = 1). The object receives $(1/43)^2$, or 1/1849 ($43^2 = 1849$), of the spotlight's illumination. Therefore, set the intensity of the spotlight to 1849 (1849/1849 = 1).

However, it should be noted that these settings are merely a starting point. You will likely spend some time adjusting lighting to produce the desired results. In some cases, it may take longer to light the scene than it did to model it.

PROFESSIONAL TIP

If you render a scene and the image appears black, all of the lights may have been turned off or have their intensity set to zero. A scene with no lights placed in it will be rendered with default lighting.

Shadows

Shadows are critical to the realism of a rendered 3D model. A model without shadows appears obviously fake. On the other hand, a model with realistic materials and shadows may be hard to recognize as computer generated. In AutoCAD, the sun, distant lights, point lights, spotlights, and weblights all can cast shadows. AutoCAD's default lighting does not cast shadows. There are two types of shadows that AutoCAD can create: shadow mapped and raytrace. The **Advanced Rendering Settings** palette provides settings for controlling the creation of shadows when rendering. This palette and its options are discussed in the next chapter.

The options for creating shadows are the same for all lights. The options can be set when the light is created or adjusted later using the **Properties** palette. In the case of sunlight, the **Sun Properties** palette is used to set the options.

Shadow-Mapped Shadow Settings

A ***shadow-mapped shadow*** is a bitmap generated by AutoCAD. A shadow map has soft edges that can be adjusted. Creating shadow-mapped shadows is the only way to produce a soft-edge shadow. However, shadow maps do not transmit object color from transparent objects onto the surfaces behind the object. Figure 17-21 shows the difference between shadow-mapped shadows and raytraced shadows.

To specify shadow-mapped shadows and set the quality, or resolution, of the shadow, select the light and open the **Properties** palette (or **Sun Properties** palette). Use the **Lights in Model** palette to select a distant light and open the **Properties** palette for it. At the bottom of the **Properties** palette is the **Rendered Shadow Details** category, Figure 17-22. To specify shadow-mapped shadows, set the Type property to Soft (shadow map).

The Map size property determines the quality of the shadow. The value is the number of samples used to create the shadow. The higher the setting, the better quality of the generated shadow. However, the higher the setting, the longer it will take to render.

The value of the Softness property determines how soft the edge of the shadow is. The higher the value, the softer or blurrier the edge of the shadow. A low value can produce a very hard edge. The value can range from 1 to 5.

Figure 17-21.
The shadow from the object in the foreground is a shadow-mapped shadow. The shadow from the object in the background is a raytraced shadow.

Figure 17-22.
The **Rendered Shadow Details** category in the **Properties** palette.

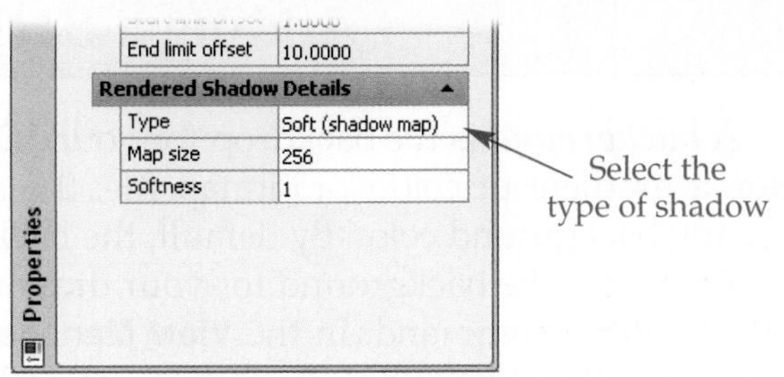

A variation of shadow-mapped shadows is created when the Type property is set to Soft (sampled). This type of shadow map must be used with photometric lighting (**LIGHTINGUNITS** = 1 or 2). In this case, different properties are displayed. The Samples property determines the number of "rays" used to generate the shadows. However, this is not considered raytracing. The Visible in rendering property determines whether the shape of the light is rendered. The Shape property sets the shape of the light. For spotlights, the shape can be either rectangular or circular (disk). For point and weblights, the shape can be linear, rectangular, circular (disk) cylindrical, or spherical. The remaining properties are based on the selected shape and are used to define the size of the shape.

NOTE

For standard shadow-mapped shadows to be created, the Shadow Map property in the **Advanced Render Settings** palette must be set to On. However, for the photometric shadow maps—the Soft (sampled) setting—the Shadow Map property does not affect the shadow generation. Advanced render settings are covered in the next chapter.

Raytrace Shadow Settings

A ***raytrace shadow*** is created by beams, or rays, from the light source. These rays trace the path of light as they strike objects to create a shadow. In addition, rays can pass through transparent objects, such as green glass, and project color onto surfaces behind the object. Raytrace shadows have a well-defined edge. They cannot be adjusted to produce a soft edge. Raytrace shadows can be used with standard and photometric lighting.

All lights set to cast shadows, except those set for shadow-mapped shadows, cast raytraced shadows. To switch from shadow-mapped shadows to raytrace shadows, select the light object and open the **Properties** palette. In the **Rendered Shadow Details** category, set the Type property to Sharp. The other properties are disabled because they only apply to shadow-mapped shadows.

PROFESSIONAL TIP

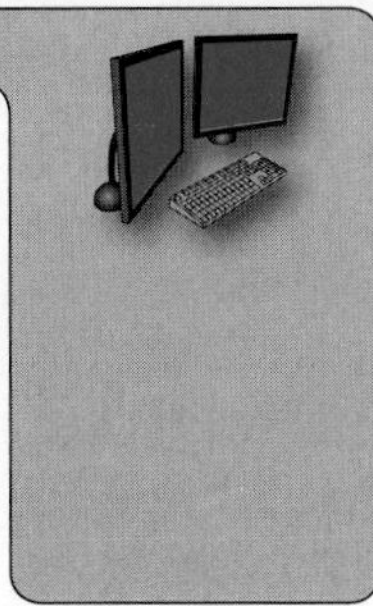

Turning on shadow casting will increase rendering time because of the calculations that AutoCAD has to perform. It is difficult to determine if raytraced or shadow-mapped shadows will be quicker to render because every scene is different and there are many variables that come into play. You will have to experiment with your scene to determine the acceptable level of shadow detail versus rendering time.

Adding a Background

A *background* is the backdrop for your 3D model. The background can be a solid color, a gradient of colors, a bitmap file, the sun and sky, or the current AutoCAD drawing background color. By default, the background is the drawing background.

VIEW

Ribbon
View
> Views

Named Views

Type
VIEW
V

To change the background for your drawing, you must first create a named view with the **VIEW** command. In the **View Manager** dialog box, pick the **New...** button to display the **New View/Shot Properties** dialog box. See **Figure 17-23.** The view name, category, and type are specified at the top of the dialog box. Near the bottom of the **View Properties** tab is the **Background** area. The drop-down list in this area is used to specify the type of background. The choices are: Default, Solid, Gradient, Image, and Sun & Sky. The Default setting uses the current AutoCAD viewport color. Photometric lighting must be on (**LIGHTINGUNITS** = 1 or 2) for Sun & Sky to appear in the drop-down list.

PROFESSIONAL TIP

Before you create the named view, establish the viewpoint from which you want to see the final rendering. Set the perspective projection current, if desired. These settings are saved with the view. The **Views** drop-down list in the **Views** panel on the **View** tab of the ribbon makes it very easy to recall the view.

Solid Backgrounds

If you select Solid in the drop-down list, the **Background** dialog box is displayed. The Type: drop-down list in this dialog box is automatically set to Solid and the default color is displayed in the **Preview** area. See **Figure 17-24.** In the **Solid options** area of the dialog box, pick the horizontal **Color:** bar to open the **Select Color** dialog box. Then, select the background color that you desire. When the **Select Color** dialog box is

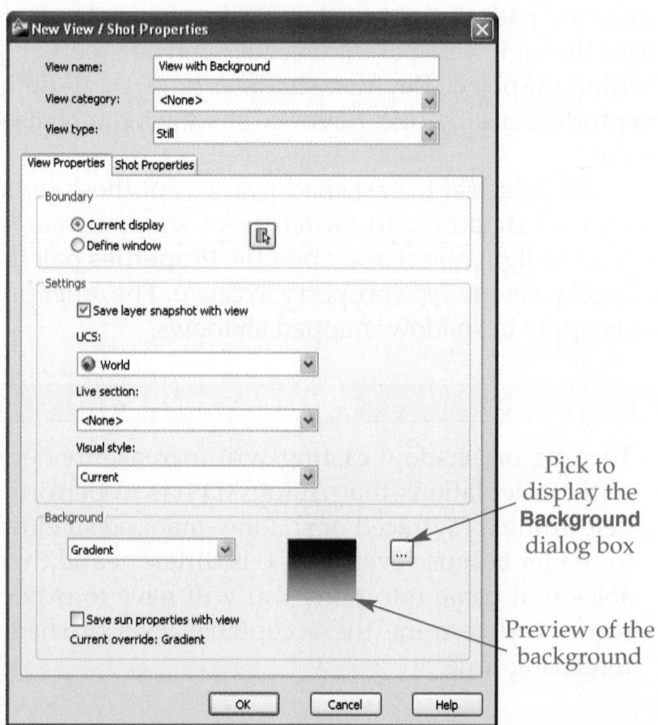

Figure 17-23.
The **View Properties** tab of the **New View/ Shot Properties** dialog box. A named view must be created before you can use a background in your scene.

Figure 17-24.
Creating a solid background.

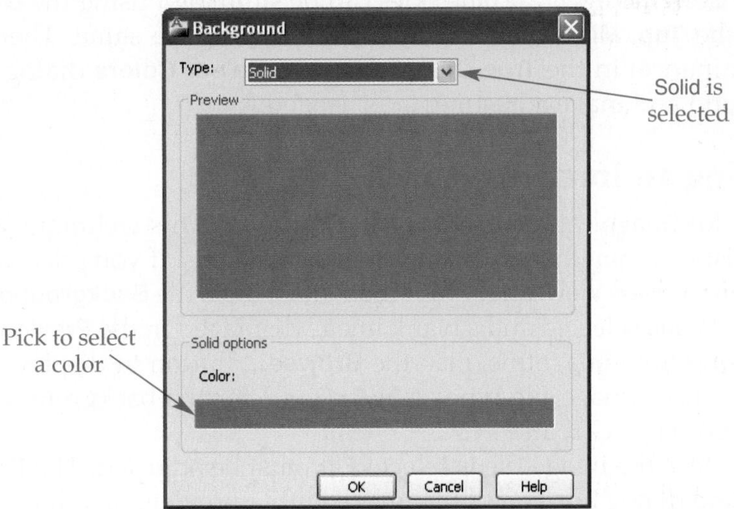

closed, the color you picked is displayed in the **Preview** area of the **Background** dialog box. Close the **Background** dialog box, save the view, set the new view current, and close the **View Manager** dialog box.

NOTE

Make sure to do a test rendering with the background color you selected. The final result may look quite different than your expectations.

Gradient Backgrounds

A gradient background can be composed of two or three colors. If you select Gradient in the drop-down list in the **New View/Shot Properties** dialog box, the **Background** dialog box is displayed with Gradient selected and the default gradient colors displayed in the **Preview** area. See Figure 17-25. The **Top color:**, **Middle color:**, and **Bottom color:** swatches are displayed on the right-hand side of the **Gradient options** area. Selecting a swatch opens the **Select Color** dialog box for changing the color. To create a two-color gradient composed of the top and bottom colors, uncheck the **Three Color** check box. The **Rotation:** text box provides the option of rotating the gradient. Close the **Background** dialog box, save the view, set the view current, and close the **View Manager** dialog box.

Figure 17-25.
Creating a gradient background.

Background
Type: Gradient
Gradient is selected
Preview
Uncheck to create a two-color gradient
Gradient options
Three color
Top color:
Middle color:
Bottom color:
Rotation: 0
Enter a rotation angle
Pick to change the color
OK Cancel Help

Convincing, clear blue skies can be simulated using the **Gradient** option. Initially, set the **Top**, **Middle**, and **Bottom** color values the same. Then, change the lightness (luminance) in the **True Color** tab in the **Select Colors** dialog box. Preview the background and make adjustments as needed.

Using an Image as a Background

An image can be used as a background. This technique can be used to produce realistic or imaginative settings for your models. If you select Image in the drop-down list in the **New View/Shot Properties** dialog box, the **Background** dialog box is displayed with Image selected and a blank image displayed in the **Preview** area. See **Figure 17-26.** To locate the image file, pick the **Browse...** button to display a standard open dialog box. These image file types may be used for the background: TGA, BMP, PNG, JFIF (JPEG), TIFF, GIF, and PCX.

Once the image file is selected, it must be adjusted. The **Preview** area of the **Background** dialog box shows the image with a preview of a drawing sheet. This drawing sheet indicates how the image is going to be positioned in the view. Pick the **Adjust Image...** button to open the **Adjust Background Image** dialog box. See **Figure 17-27.**

In the **Image position:** drop-down list, pick how the image is applied to the viewport. The Center option centers the image in the view without changing its aspect ratio or scale. The Stretch option centers the image and stretches or shrinks it to fill the entire view. This is one way to plot an image file from AutoCAD. The Tile option keeps the image at its original size and shape, but moves it to the upper-left corner and duplicates it, if needed, to fill the view.

After the image is positioned, use the sliders to adjust it further. The sliders are disabled if Stretch is selected in the **Image position:** drop-down list. The slider function is based on which radio button is picked above the image:

- **Offset.** The sliders move the image in the X or Y direction.
- **Scale.** The sliders scale the image in the X or Y direction. This may distort the image if it is scaled too much in one direction. To prevent distortion, check the **Maintain aspect ratio when scaling** check box at the bottom of the dialog box.

The **Reset** button is located at the bottom-right corner of the preview pane. Picking this button returns the scale and offset settings to their original values.

Once the image is adjusted, pick the **OK** button to close the **Adjust Background Image** dialog box. Then, close the **Background** dialog box, save the view, set the view current, and close the **View Manager** dialog box.

Figure 17-26.
Setting an image as the background.

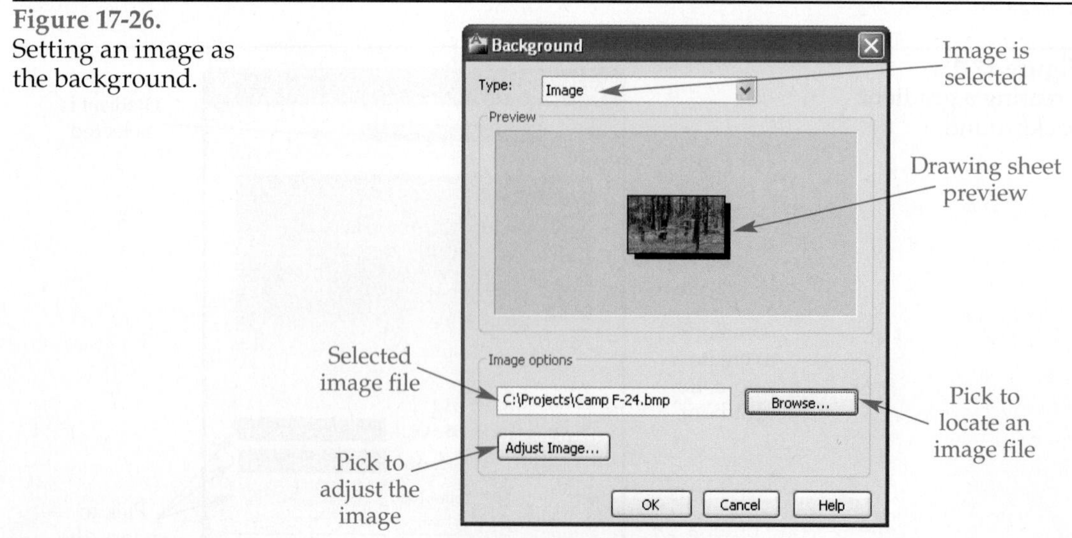

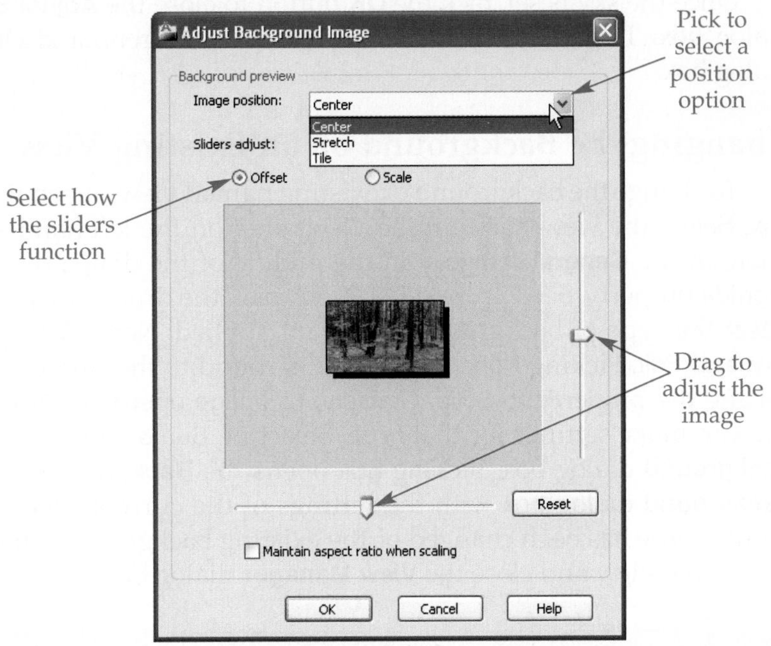

Figure 17-27.
Adjusting the background image.

Sun and Sky

If you select Sun & Sky in the drop-down list in the **New View** dialog box, the **Adjust Sun & Sky Background** dialog box is displayed. See Figure 17-28. AutoCAD uses the settings in this dialog box to simulate the sun in the sky. This dialog box has a preview tile at the top and the **General**, **Sky Properties**, **Sun Angle Calculator**, **Rendered Shadow Details**, and **Geographic Location** categories. The settings in these categories are discussed earlier in this chapter in the Sky Properties section.

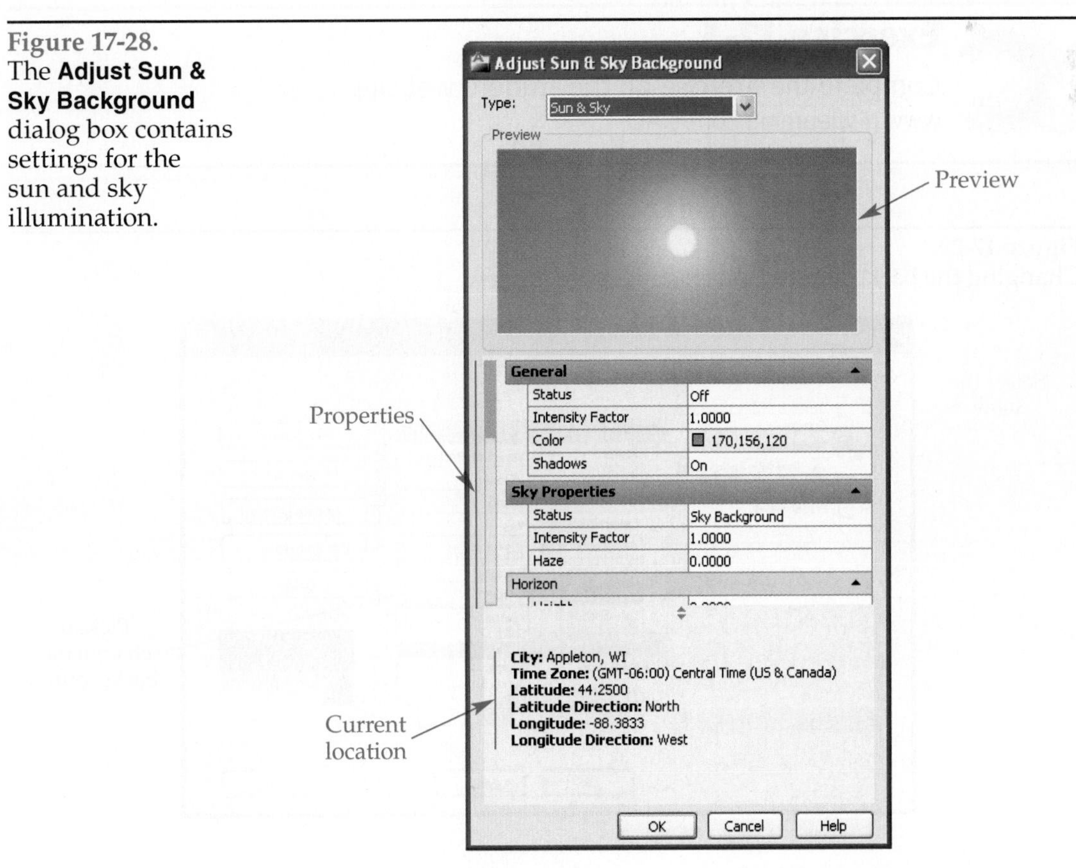

Figure 17-28.
The **Adjust Sun & Sky Background** dialog box contains settings for the sun and sky illumination.

Once the sky is set, pick the **OK** button to close the **Adjust Sun & Sky Background** dialog box. Then, save the view, set the view current, and close the **View Manager** dialog box.

Changing the Background on an Existing View

To change the background of existing named views, open the **View Manager** dialog box. Select the view name in the **Views** tree on the left-hand side of the dialog box. Then, in the **General** category in the middle of the dialog box, select the Background override property. See Figure 17-29. Next, pick the drop-down list for the property and select the type of background you want applied: None, Solid, Gradient, Image, Sun & Sky, and Edit. Picking None sets the background to the AutoCAD default background. Setting the property to Solid, Gradient, or Image opens the **Background** dialog where you can make settings for that type. Selecting Sun & Sky opens the **Adjust Sun & Sky Background** dialog box. Picking Edit opens the **Background** or the **Adjust Sun & Sky Background** dialog box with the settings of the current background. Once the background type has been changed or the existing background edited, pick the **OK** button to save the view and close the **View Manager** dialog box.

NOTE

After exiting the **Background** dialog box, you are returned to the **View Manager** dialog box. Picking the **OK** button to exit the **View Manager** dialog box does not necessarily activate the view that you just created or modified. The view must be set current to see the effects of the changes to the background. A view can be set current using the drop-down list on the **Viewports** panel in the **View** tab of the ribbon or it can be set current in the **View Manager** dialog box.

Exercise 17-5

Complete the exercise on the student website.
www.g-wlearning.com/CAD

Figure 17-29.
Changing the background of an existing, named view.

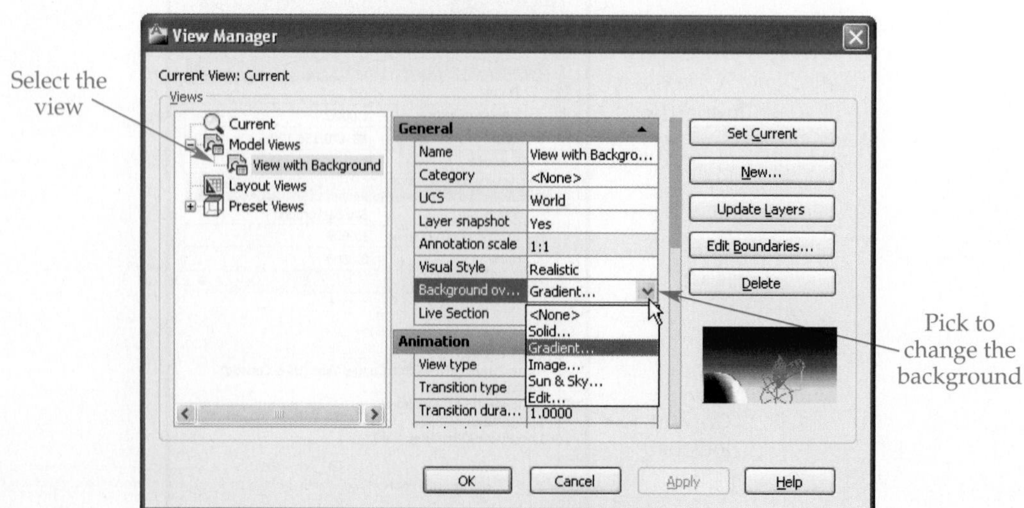

Chapter Test

Answer the following questions. Write your answers on a separate sheet of paper or complete the electronic chapter test on the student website.
www.g-wlearning.com/CAD

1. Compare and contrast *ambient light, distant lights, point lights, spotlights,* and *weblights.*
2. Define *angle of incidence.*
3. Define *angle of reflection.*
4. A smooth surface has a(n) _____ specular factor.
5. Describe *hotspot* and *falloff.* Which lights have these properties?
6. What is *attenuation?*
7. What are the three types of lighting in AutoCAD?
8. What are the four types of light objects in AutoCAD?
9. What are light glyphs and which lights have them?
10. List the types of shadows that can be created in AutoCAD. Which type(s) can have soft edges?
11. Which type of shadow must be created for light to pass through transparent objects?
12. What must be created before a background can be added to a scene?
13. What are the four types of backgrounds in AutoCAD, other than the default background?
14. Why would you draw a line between the "from" point and "to" point of a distant light?
15. Describe how a gradient background can be used to represent a clear blue sky.

Drawing Problems

1. In this problem, you will draw some basic 3D shapes to create a building similar to an ancient structure, place lights in the drawing, and render the scene with shadows.
   A. Begin a new drawing and set the units to architectural. Save the drawing as P17_01.
   B. Draw a 32′ × 22′ planar surface to represent the floor. Using the materials browser, attach a material of your choice to the floor.
   C. Draw cylinders to represent pillars. Make each ∅2′ × 15′ tall. There are ten pillars per side. Attach a suitable material to the pillars.
   D. The roof is 32′ × 22′ and 5′ tall at the ridge. Attach an appropriate material.
   E. Create a perspective viewpoint looking into the building.
   F. Turn on sunlight and turn off the default lighting. Set the geographic location to Athens, Greece. Change the date and time to whatever you wish. Make sure the sun is set to cast shadows.
   G. Render the scene.
   H. Save the drawing.

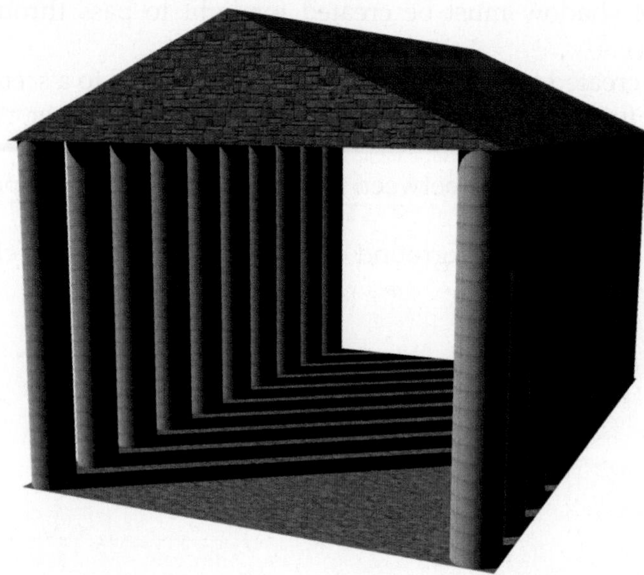

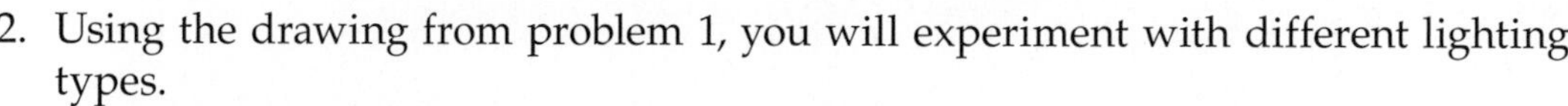

2. Using the drawing from problem 1, you will experiment with different lighting types.
   A. Open drawing P17_01 and save it as P17_02.
   B. Turn off the sun.
   C. Place three point lights inside of the building. Evenly space the lights along the centerline of the ceiling. Adjust the light intensity so that the interior is not washed out. Set the color of the middle light to white. Set the color of the outside lights to red or blue. Render the scene.
   D. Turn off the point lights.
   E. Place two spotlights, one pointing from the front corner to the rear corner and the other pointing between the pillars on the left side of the building. Target them at the floor. Render the scene. Adjust the intensity, hotspot, and falloff as needed.
   F. Turn the point lights back on and render the scene with all six lights active. Adjust the light intensities again if the rendering is too washed out with light.
   G. Save the drawing.

3. Using a previously created mechanical model, you will apply materials and lights to make it ready for presentation.
   A. Open drawing P09_05 created in Chapter 9. Save it as P17_03.
   B. Draw a planar surface below the flange to represent a tabletop.
   C. Using the tool palettes, attach an appropriate material, such as a wood or tile material, to the surface.
   D. Open the materials editor and create a new material based on the Metal material type. Experiment with different finish settings until you find one you like.
   E. Place two spotlights in the drawing and target them at the flange from different angles. Adjust their hotspot, falloff, and intensity to get the proper lighting.
   F. Create a perspective view of the scene. Then, render the scene.
   G. Save the drawing.

4. The building shown below will be used to study passive solar heating at different times of the year. Model the building using the overall dimensions given. Use your own dimensions for everything else. The side with the windows should be facing South (–Y in AutoCAD, by default).
   A. Set the geographical location to a city in the northern hemisphere.
   B. Set the date to midsummer and the time to noon.
   C. Turn on sunlight and the default lighting off.
   D. Render the scene and note the location of the shadows inside the building.
   E. Change the date to late winter, render the scene again, and note the new location of the shadows. You can easily switch between the rendered images in the **Render** window by selecting each rendering in the **History** pane (this is discussed more in the next chapter).
   F. Observing the changes in the shadow locations, what design changes can be made to maximize Sun exposure in the cold winter months? What design changes can be made to minimize Sun exposure in the heat of summer?
   G. Change the geographical location to somewhere closer to the equator. Then, render the scene in summer and follow. How do the shadows compare to those in the previous location?
   H. Save the drawing as P17_04.

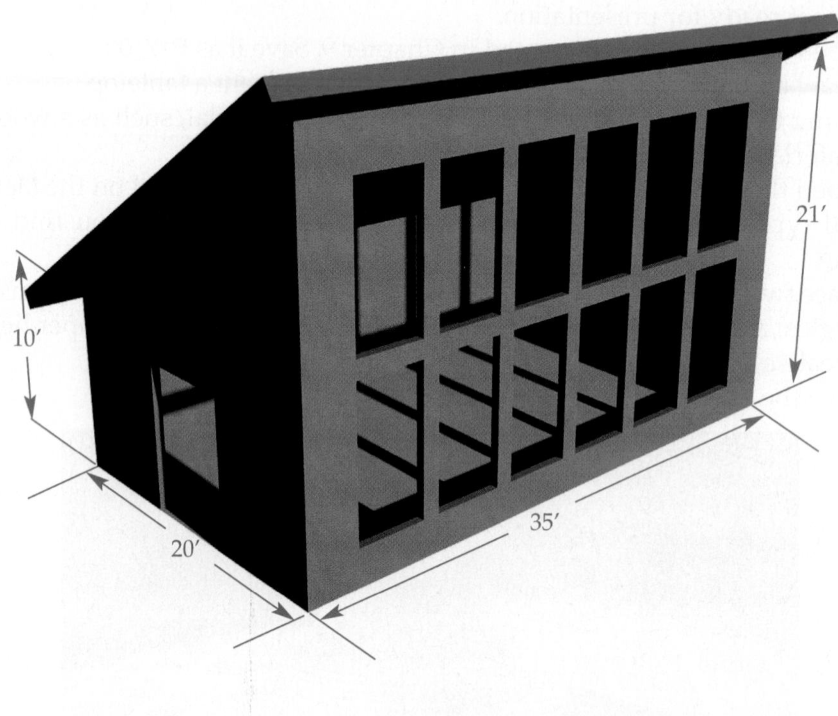

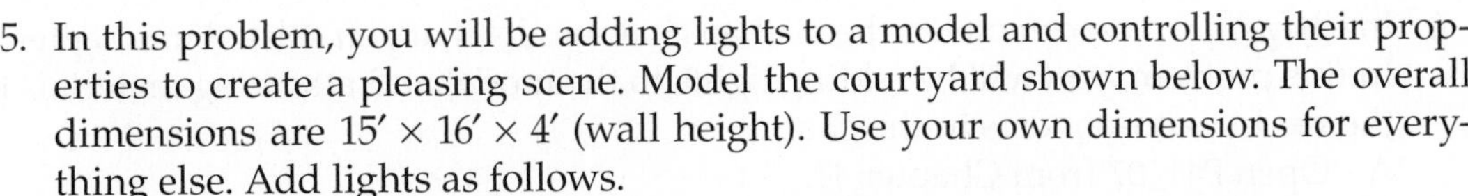

5. In this problem, you will be adding lights to a model and controlling their properties to create a pleasing scene. Model the courtyard shown below. The overall dimensions are 15′ × 16′ × 4′ (wall height). Use your own dimensions for everything else. Add lights as follows.
   A. Turn on the sun and turn off the default lighting. Set the time to late in the day so that the sun is close to the horizon.
   B. In the **Sky Properties** category of the **Sun Properties** palette, select Sky Background and Illumination for the Status property.
   C. Add a point light at the center of each sphere.
   D. Attach a glass material to the spheres that will make them look like they are glowing. There are materials in the Glass category of the Autodesk Library that can be used as a base for creating a new material.
   E. Create a fill light above to illuminate the scene. This can be a point or spotlight.
   F. Render the scene using the low preset to see the lighting effects.
   G. Adjust the sun properties to create the look that you want.
   H. Adjust the properties of the point lights and any other lights in the scene. You may have to increase the intensity of the lights quite a bit to properly illuminate the scene.
   I. When the scene is illuminated the way you want it, render the scene using medium or high preset.
   J. Save the drawing as P17_05.

6. In Chapter 11, you completed the kitchen chair model that you started in Chapter 2. In this problem, you will be adding lights to the model and attaching materials to the various components in the chair.
   A. Open P11_07 from Chapter 11.
   B. Create a layer for each of the chair components: seat, legs/crossbars, seatback bow, and seatback spindles.
   C. Draw a planar surface to represent the floor. Place this on its own layer.
   D. Assign materials to each of the layers. You can use materials from the Autodesk Library or create your own materials.
   E. Set the Realistic visual style current.
   F. Adjust material mapping as needed.
   G. Add lighting to the scene.
   H. Render the scene. If you experience problems with the materials, try attaching by object instead of by layer.
   I. Save the drawing as P17_06.

Drawing Problems - Chapter 17

CHAPTER 18

Advanced Rendering

Learning Objectives

After completing this chapter, you will be able to:

- ✓ Make advanced rendering settings.
- ✓ Set the resolution for a rendering.
- ✓ Save a rendering to an image file.
- ✓ Add fog/depth cueing to a scene.

In Chapter 15, you learned how to create a view of your scene that is more realistic than a visual style. In that chapter, you used AutoCAD's sunlight feature to create a simple rendering with mostly default settings. In this chapter, you will discover how to make a rendering look truly realistic, or *photorealistic*. Some of the advanced rendering features add significantly to the rendering time and you must learn how to balance the quality of the rendering with an acceptable time frame to get the job done.

Render Window

By default, a drawing is rendered in the **Render** window, unless you are rendering a cropped area. This window allows you to inspect the rendering, save it to a file, compare it with previous renderings, and take note of the statistics. See **Figure 18-1.** There are three main areas of the **Render** window—the image, history, and statistics panes.

Image Pane

As AutoCAD processes the scene, the image begins to appear in the image pane in its final form. There may be as many as four phases that the rendering goes through as it is being processed:

- **Translation.** Processes the drawing information and determines light intensity, shadow placement, colors, and so on. This phase is always completed.
- **Photon emission.** ***Photon emission*** is a technique for calculating indirect illumination that traces photons emitted by the light source until they come to rest on a diffuse surface. It determines which areas will be illuminated by indirect, or bounced, light. The photon emission phase may or may not be processed, depending on settings in the **Advanced Render Settings** palette.

Figure 18-1.
The **Render** window.

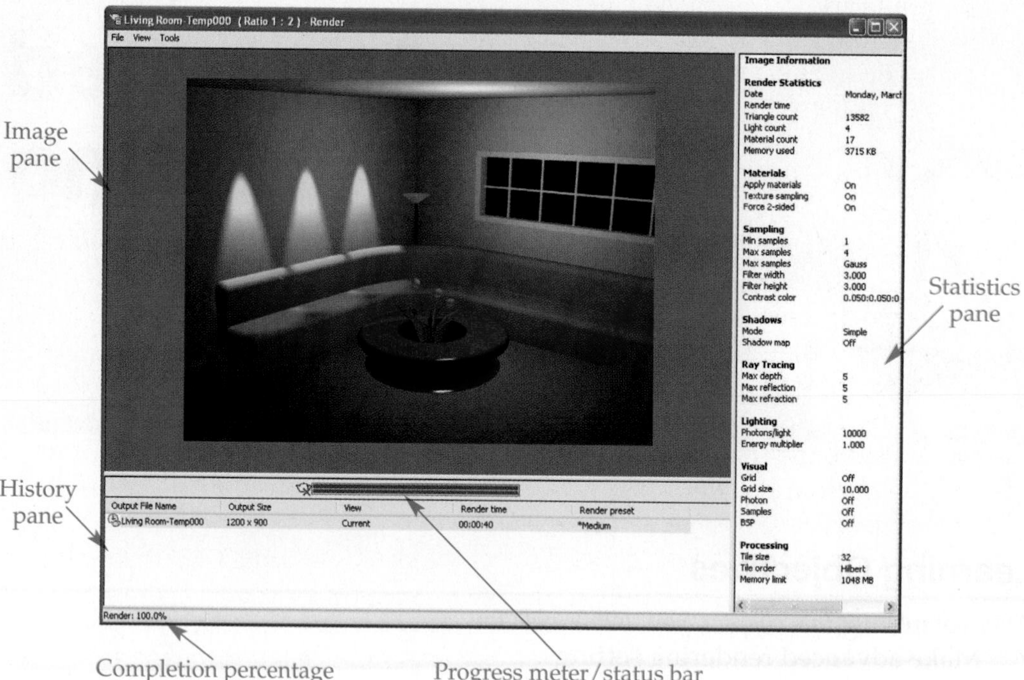

- **Final gather.** ***Final gather*** increases the number of rays used to calculate global illumination (GI). This phase will be processed if it is turned on in the **Advanced Render Settings** palette.
- **Render.** Converts the data into an image. This phase is always completed.

Immediately below the image pane is the progress meter/status bar. The top bar displays the progress of the current phase and the bottom bar indicates the progress of the entire rendering. Also, at the very bottom of the **Render** window, below the history pane, the status of the phase is shown with its percentage complete. The rendering can be cancelled at any time by pressing the [Esc] key or picking the **X** button to the left of the progress meter.

As discussed in Chapter 15, you can zoom the rendering in and out to inspect it. You can also save it to an image file using the **File** pull-down menu in the **Render** window.

History Pane

The history pane contains a list of all of the renderings that were created in this drawing since it was created, not just in this drawing session. The items in this list are called ***history entries.*** There are two types of history entries, which are indicated by icons. See Figure 18-2.

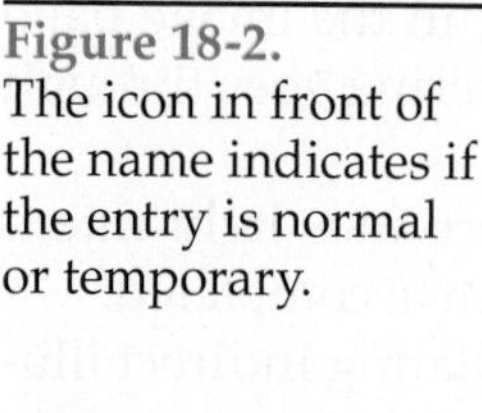

Figure 18-2.
The icon in front of the name indicates if the entry is normal or temporary.

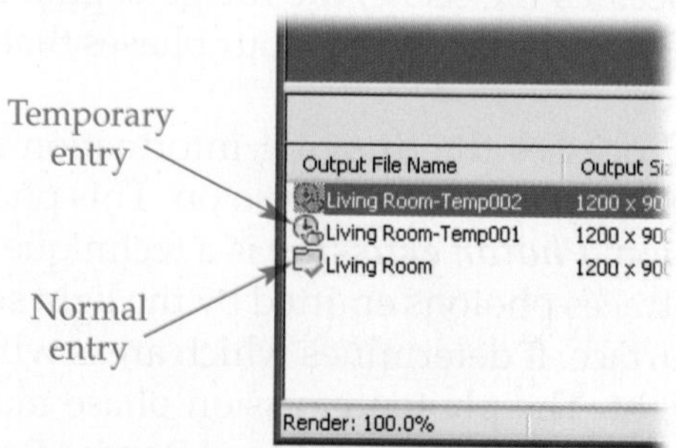

- **Normal.** The entry is saved to file. A link is maintained to that file. If the drawing is saved, closed, and reopened, you can pick the entry to view the rendering in the image pane.
- **Temporary.** The entry is available in the current drawing session, but is not saved to a file. If the drawing is closed, the image is lost. The name of the entry in the Output File Name column ends with -Temp*x*.

Right-clicking on an entry in the history pane displays a shortcut menu. The options in this menu can be used to save the image, render the image again, and manage the entry. The options in the shortcut menu are:

- **Render Again.** Renders the scene again using the same settings. A new entry is not added to the history pane.
- **Save.** Saves the rendered image to a file using a standard save dialog box. This changes the entry from a temporary entry into a normal entry.
- **Save Copy.** Saves the rendered image to a new file without changing the current entry.
- **Make Render Settings Current.** Makes all of the rendering settings of the entry the current rendering settings in the drawing. This allows you to render the current scene using the settings of the entry.
- **Remove From the List.** Deletes the entry from the history pane, but any image files saved from the entry remain.
- **Delete Output File.** Deletes the image file created by saving the entry. The entry remains in the history pane and any image files that were created as copies are retained.

Statistics Pane

The statistics pane shows the details of the rendering that is selected in the history pane. By selecting renderings in the history pane, you can see in the image pane which version provides the best result. Then, you can use the statistics pane to view the settings. The information under the Render Statistics heading (date, render time, etc.) is added when the rendering is completed. The rest of the information reflects the settings in the **Advanced Render Settings** palette and the **Render Presets Manager** dialog box at the time the rendering was created.

Exercise 18-1

Complete the exercise on the student website.
www.g-wlearning.com/CAD

Advanced Render Settings

The quickest and easiest way to control the quality of a rendering is with render presets. AutoCAD provides five standard render presets: Draft, Low, Medium, High, and Presentation. The Draft preset provides the lowest-quality rendering. Each preset above Draft changes the advanced render settings to gradually improve the rendering quality, peaking with the Presentation preset. However, as the quality is improved, the rendering time increases. The presets can be selected in the **Render** panel on the **Render** tab of the ribbon or from the drop-down list at the top of the **Advanced Render Settings** palette. Creating and using your own render presets is covered later in this chapter.

RPREF

Type
RPREF
RPR

The **Advanced Render Settings** palette provides settings that give you complete control over how a rendering is created. The **RPREF** command opens the palette. The palette can also be displayed by picking the dialog box launcher button at the lower-right corner of the **Render** panel in the **Render** tab on the ribbon. There are five main categories in this palette: **General**, **Ray Tracing**, **Indirect Illumination**, **Diagnostic**, and **Processing**. These categories are explained in the next sections.

General

The **General** category provides properties for controlling the rendering destination, materials, sampling, and shadows, Figure 18-3. It contains four subcategories: **Render Context**, **Materials**, **Sampling**, and **Shadows**.

Render context

The **Render Context** subcategory contains general properties that control the rendering. The Procedure property determines what will be rendered. The settings are View, Crop, and Selected. View is the default and renders whatever you see in the drawing window. Crop allows you to specify an area of the scene to render. This is very useful when you want to do a test rendering, but do not want to wait for the whole scene to render. The Selected setting allows you to pick which objects to render.

The Destination property determines where the rendered scene will be displayed. You can choose to have the rendering placed in the viewport or **Render** window.

If the **Determines if File is Written** button in the subcategory title bar is picked (depressed), the Output File Name property is enabled. This property sets the name and location of the file to which the rendering will automatically be saved.

The Output Size property sets the resolution, measured in pixels × pixels, for the rendered image. You can select standard resolutions or pick Specify Output Size... for a custom resolution. In the **Output Size** dialog box that is displayed when Specify Output Size... is selected, you can enter any resolution that you want. See Figure 18-4. If you want to prevent the image from stretching, make sure the **Lock image aspect** button is selected in the dialog box so that the height and width remain proportional. When you change the resolution, it is stored in the drawing.

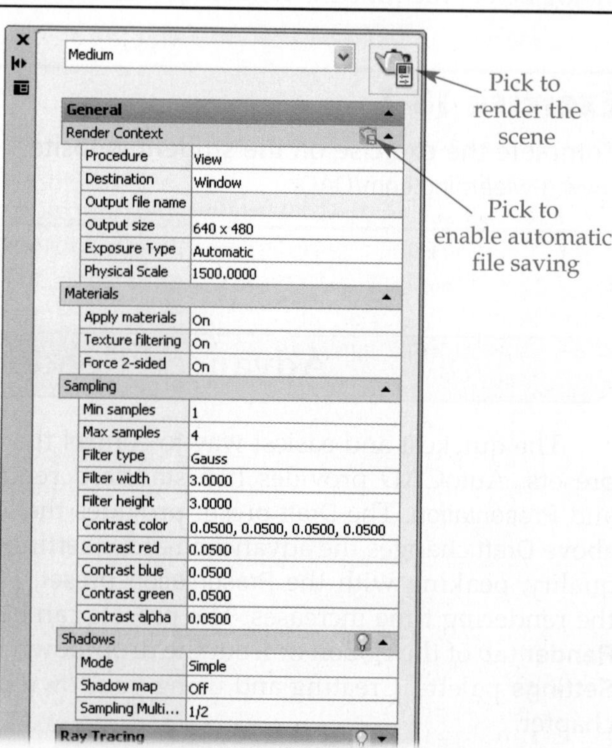

Figure 18-3.
The **General** category of the **Advanced Render Settings** palette.

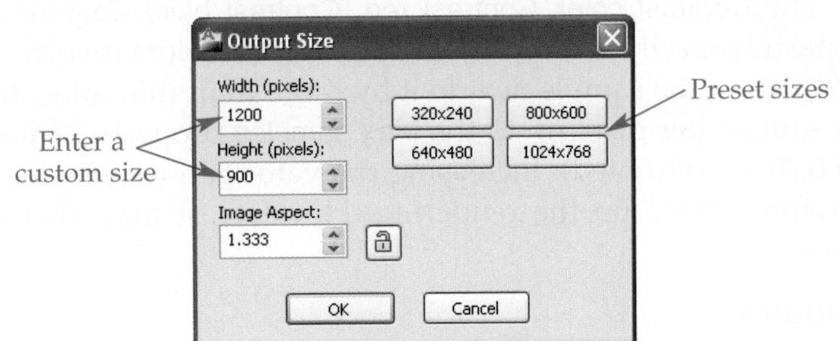

Figure 18-4.
Setting a custom resolution.

The Exposure Type property can be set to Automatic or Logarithmic. When set to Automatic, the entire image is sampled and some of the dim lighting effects are enhanced to make them more visible. When set to Logarithmic, the brightness and contrast are used to map physical values to RGB values. This is better for scenes with high dynamic ranges.

Exposure control needs a scale to work with and, if you are using non-physical lights (**LIGHTINGUNITS** set to 0), the Physical Scale property provides a scale. Standard lights have an Intensity Factor property that is multiplied by the Physical Scale value to determine the actual brightness of the light. The default value is 1500. In other words, a point light with an Intensity Factor value of 2 has an actual lamp intensity of 3000 candelas when the Physical Scale property is set to 1500.

Materials

The properties in the **Materials** subcategory determine how materials are handled in the rendering. The Apply Materials property controls whether or not materials attached to objects are rendered. The property can be set to On or Off. If set to Off, objects are rendered in their own colors. The Texture Filtering property determines whether or not antialiasing is applied to texture maps when rendered. Antialiasing is a way of reducing "jaggies" in the rendered image. The Force 2-sided property determines if AutoCAD renders both sides of all faces. This can fix problems where objects disappear in a rendering, but will increase rendering time.

Sampling

Sampling is a technique that tests the scene color at each pixel and then determines what the final color should be. This is most important in transition areas, such as edges of objects or shadows. Increasing the sampling will smooth out the jagged edges and incorrect coloring, but increase rendering time. You may also notice thicker lines in the final rendering.

The Min samples and Max samples properties set the minimum and maximum number of samples computed per pixel. A value of 1 means one sample per pixel. A value of 1/4 means one sample for every four pixels. The Filter type property determines how the samples are brought together to determine the pixel value:

- Box. Quickest method; evenly combines samples and gives them equal weight.
- Triangle. Weights the samples based on a pyramid with samples in the center of the filter area receiving the most weight.
- Gauss. Weights the samples based on a bell curve with samples in the center of the filter area receiving the most weight.
- Mitchell. Most accurate. Weights samples based on a curve centered on the filter area, like Gauss; however, this curve is steeper.
- Lanczos. Weights samples based on a curve centered on the filter area, like Mitchell, but it diminishes the weight of samples at the edge of the filter area.

The Filter width and Filter height properties determine the size of the filter area. A larger filter area softens the image, but increases rendering time.

The Contrast color, Contrast red, Contrast blue, Contrast green, and Contrast alpha properties specify the threshold value of the colors involved in sampling. If a sample differs from the sample next to it by more than this color, AutoCAD takes more than one sample per pixel up to the Max samples property. Values can be from 0.0 (black) to 1.0 (fully saturated). Increasing the value can reduce the amount of sampling and, therefore, speed up the rendering. However, it may also reduce the quality of the image.

Shadows

The properties in the **Shadows** subcategory control how the renderer handles shadows generated by the lights in the scene. For shadows to be applied, the button in the subcategory title bar must be on (yellow).

The Mode property controls a shader function that calculates light effects. There are three modes that determine how shading is calculated:

- Simple. Shaders are randomly created.
- Sorted. Shaders are called in order from the object to the light.
- Segment. Shaders are called in order from the volume shaders to the segments of the light rays between the object and the light.

The Shadow Map property determines whether shadow-mapped or raytraced shadows are created. When this property is set to On, shadow-mapped shadows are generated. When it is set to Off, raytraced shadows are created.

The Sampling Multiplier property limits shadow sampling for area lights. The values are preset for the rendering presets: Draft = 0, Low = 1/4, Medium = 1/2, High = 1, and Presentation = 1. However, these values can be changed. This is the same principle that is described in the Sampling section, but instead of sampling pixels for object color, it is sampling for shadows.

Raytracing

The **Ray Tracing** category provides properties for controlling how the rendered image is shaded, **Figure 18-5.** ***Raytracing*** is a method of calculating reflections, refractions, and shadows by tracing the path of the light rays from the light sources. This is more accurate at producing shadows than shadow mapping, but it takes more time and the shadow edge is always sharp. To enable raytracing, pick the button in the category's title bar. If this is off (not yellow), there will be no raytracing and the properties are disabled.

The Max reflections property is the maximum number of times that a ray can be reflected. See **Figure 18-6.** The Max refractions property is the maximum number of times that a ray can be refracted. The Max depth property is the maximum number of reflections and refractions. For example, if this property is set to 5 and the Max reflections property is set to 3, then no more than two refractions will occur. A good way to figure out the required maximum depth is to imagine a light ray traveling through transparent objects or bouncing off of reflective objects in your scene. Count how many surfaces the object must contact and that is the maximum depth.

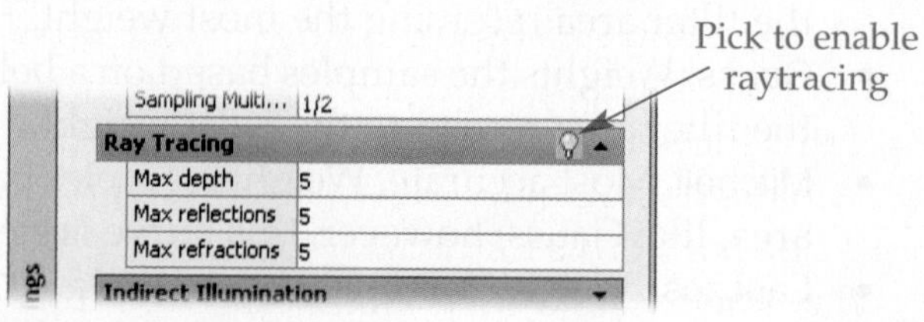

Figure 18-5.
The **Ray Tracing** category of the **Advanced Render Settings** palette.

Figure 18-6.
A—If the **Max reflections** property for raytracing is set to 3, a light ray will bounce three times and then stop at the next surface. B—If transparent objects exist in the scene and the **Max refractions** property is set to 2, the light ray will still make it to the back wall as long as the **Max depth** property is set to 5 or more. C—However, if the **Max depth** property is set to 4, the light ray will not make it to the back wall.

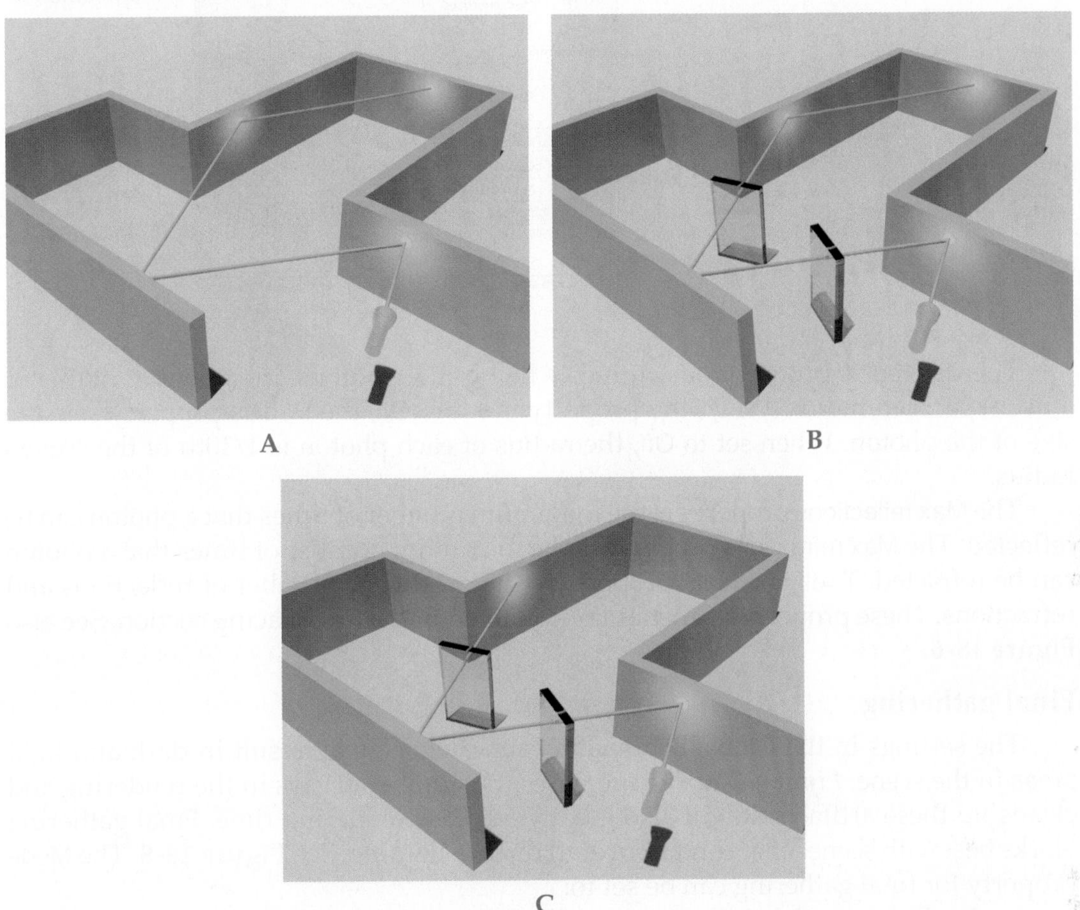

Indirect Illumination

Indirect illumination is a method in AutoCAD that simulates natural, bounced light. If indirect illumination is turned off and light does not directly strike an object, the object is black. Without indirect illumination enabled, other lights must be added to the scene to simulate indirect illumination. The properties in the **Indirect Illumination** category allow you to create a natural-looking scene. There are three subcategories in the **Indirect Illumination** category: **Global Illumination**, **Final Gather**, and **Light Properties**. See Figure 18-7.

Global illumination

Global illumination (GI) is indirect illumination. Bounced light is simulated by generating photon maps on surfaces in the scene. These maps are created by tracing photons from the light source. Photons bounce around the scene from one object to the next until they finally strike a diffuse surface. When a photon strikes a surface, it is stored in the photon map. To enable global illumination, pick the button in the subcategory's title bar. If this button is off (not yellow), there will be no indirect illumination.

The **Photons/sample** property sets the number of photons used to generate the photon map. The higher the value, the less noise global illumination produces. However, rendering time is longer and the image is blurrier.

Figure 18-7.
The **Indirect Illumination** category of the **Advanced Render Settings** palette.

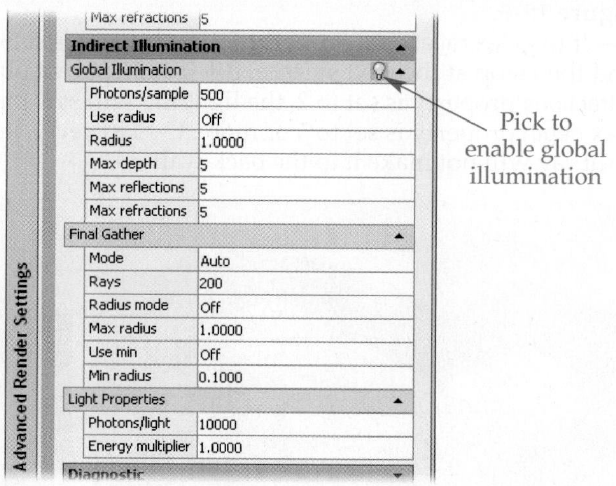

The Use radius property determines whether the photons are a default radius or a user-specified radius. When the property is set to On, the Radius property sets the size of the photon. When set to Off, the radius of each photon is 1/10th of the scene's radius.

The Max reflections property is the maximum number of times that a photon can be reflected. The Max refractions property is the maximum number of times that a photon can be refracted. The Max depth property is the maximum number of reflections and refractions. These properties function as explained in the Ray Tracing section. See also **Figure 18-6.**

Final gathering

The settings in the **Global Illumination** subcategory may result in dark and light areas in the scene. ***Final gathering*** increases the number of rays in the rendering and cleans up these artifacts. It will also greatly increase rendering time. Final gathering works best with scenes that contain overall diffuse lighting. See **Figure 18-8.** The Mode property for final gathering can be set to:

- On. Turns on global illumination for final gathering.
- Off. Turns off global illumination for final gathering.
- Auto. Global illumination is turned on or off based on the sky light status. This is the default setting.

The Rays property sets the number of rays used to calculate indirect illumination. The higher the value, the better the result, but the longer it takes to render the scene.

The Radius mode property determines how the Max radius property is applied during final gathering. There are three possible settings:

- On. The Max radius value is used for final gathering and it is measured in world units.
- Off. The radius of each area processed by final gathering is 10% of the maximum model radius.
- View. The Max radius value is used for final gathering, but it is measured in pixels instead of world units.

The Max radius property determines the maximum radius of each area processed during final gathering. The lower this value, the higher the quality of the rendering because a larger number of smaller areas is processed. However, rendering time is higher.

The Use min property determines whether or not the Min radius property is applied for final gathering. The Min radius property sets the minimum radius of the processed areas. Increasing this improves quality, but increases rendering time.

Figure 18-8.
A—This scene has a single point light. B—Global illumination is turned on. Notice the unevenness of the lighting. This can be seen especially on the sofa and in the corner of the walls. C—Final gathering cleans up artifacts and provides a more even illumination.

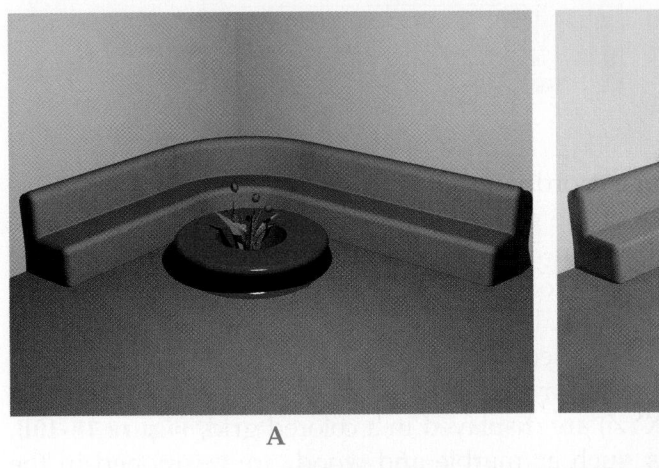

A

B

C

Light properties

The properties in the **Light Properties** subcategory control how the lights in the scene are applied when calculating indirect illumination. The Photons/light property sets the number of photons emitted by each light. Increasing this number makes each light cast more photons and improves the rendering quality. The Energy multiplier property determines how much light energy is used in global illumination. The default value of 1.0000 does not increase or decrease the light energy. Values less than the default decrease the light energy. Values greater than the default increase the light energy.

PROFESSIONAL TIP

If your scene looks washed out (flooded with light) with indirect illumination enabled, experiment with reducing the energy multiplier. This can have a dramatic effect on the scene.

Diagnostic

The properties in the **Diagnostic** category control tools to help you understand why the rendering produced the results it did, Figure 18-9. The scene can be rendered with photon maps, grids, and irradiance shown. These tools can help you diagnose and correct problems.

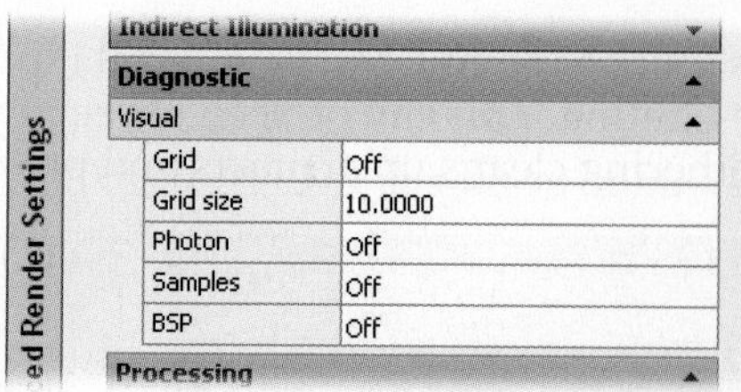

Figure 18-9.
The **Diagnostic** category of the **Advanced Render Settings** palette.

The Grid property determines if a coordinate grid is shown in the rendered image. The Grid size property sets the size of the grid. When the Grid property is set to Off, which is the default, the grid is not shown. There are three other settings:

- Object. A colored grid displays local coordinates (UVW), **Figure 18-10A.** Each object has its own set of local coordinates. UVW can be thought of XYZ on the object. Materials are mapped to objects based on the object's UVW directions. This grid helps to figure out why a material may not be correctly positioned or rotated.
- World. World coordinates (XYZ) are displayed in a colored grid, **Figure 18-10B.** Some procedural materials, such as marble and wood, are positioned in the scene based on the world coordinate system. This grid helps to figure out which way to position the objects in the scene to control the direction of wood grain or marble vein.
- Camera. Coordinates of a UCS corresponding to the camera or current view are displayed in a colored grid, **Figure 18-10C.** Because this grid is based on the camera (viewer) position, it may be used to visualize the horizon location or clipping planes without having to completely render the scene.

The Photon property controls whether or not the effect of a photon map is shown in the rendering. When the property is set to Density or Irradiance and global illumination is on, the scene is rendered and overlaid with an image representing the photon map. See **Figure 18-11.**

Figure 18-10.
Applying a grid to the rendering. A— The Grid property is set to Object. This grid helps to visualize the UVW directions on individual objects: red = U, green = V, blue = W. B— The Grid property is set to World. This helps to visualize the XYZ directions in the entire scene. red = X, green = Y, blue = W. C—The Grid property is set to Camera. This helps to visualize the XYZ directions in the entire scene based on the camera (viewer) direction. red = X, green = Y, blue = W.

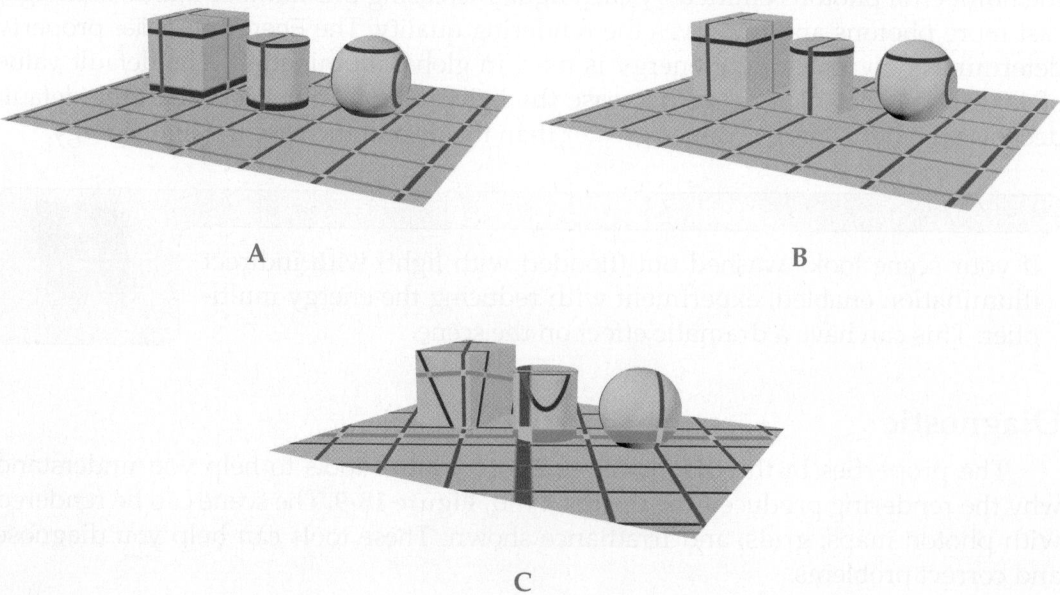

Figure 18-11.
A photon map is applied to the rendering (Photon property set to Density).

- Density. Shows the photon map projected onto the scene. Higher-density areas are red and lower-density areas are the cooler colors.
- Irradiance. Similar to density, but the photons are shaded based on their irradiance value. Maximum irradiance is red and lower irradiance values are shown in the cooler colors.

The Samples property can be set to On or Off. When set to On, a grid is rendered plan to the view and varying shades of gray and white are displayed in the scene. This tool is another way to evaluate the lighting in the scene.

The BSP property determines whether or not the effects of ***binary space partitioning (BSP)*** are shown. BSP is a raytrace-acceleration method. When rendering, if you receive a message about large depth or size values or the rendering is very slow, this tool may help you locate the problem.

- Depth. The depth of the raytrace tree is displayed. Top faces are displayed in bright red. The deeper the faces are in the tree, the cooler the colors in which they are displayed, **Figure 18-12A.**
- Size. The size of the leaves in the raytrace tree are displayed. Different colors are used to identify different leaf sizes, **Figure 18-12B.**

Figure 18-12.
Showing the effects of binary space partitioning. A—The BSP property is set to Depth. B—The BSP property is set to Size.

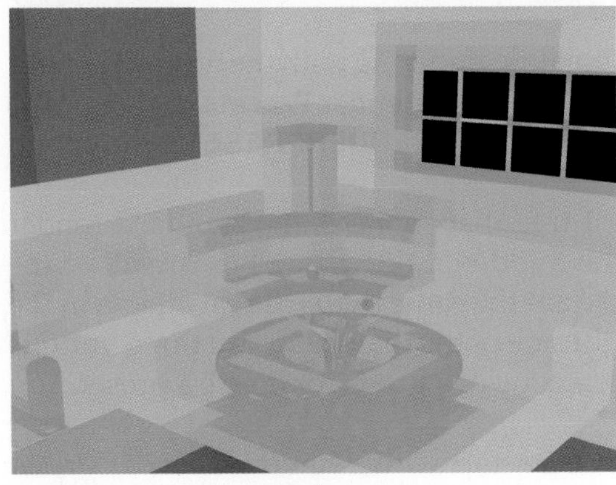

A

B

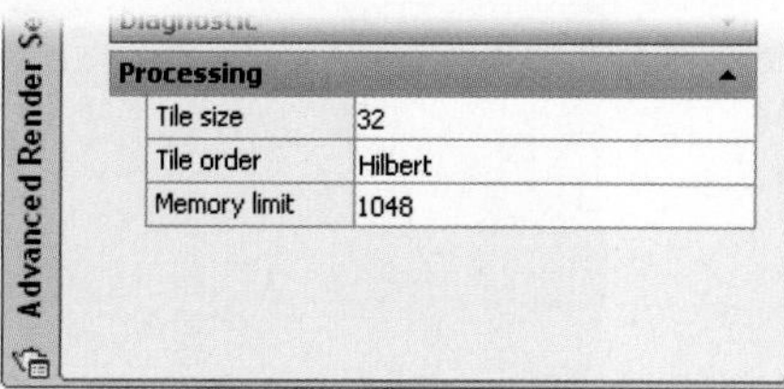

Figure 18-13.
The **Processing** category of the **Advanced Render Settings** palette.

Processing

The properties in the **Processing** category control how the final render processing takes place, Figure 18-13. The Tile size property controls the size of the tiles into which the total image is subdivided. The larger the tile size, the fewer tiles that have to be rendered and the fewer times the image has to update. Larger tiles usually mean a shorter rendering time. The Tile order property controls the order in which the tiles are rendered:

- Hilbert. The "cost" of switching to the next tile determines which tile is rendered next.
- Spiral. The rendering begins with the tiles in the center of the image and then spirals outward.
- Left to Right. The tiles are rendered from bottom to top and left to right in columns.
- Right to Left. The tiles are rendered from bottom to top and right to left in columns.
- Top to Bottom. The tiles are rendered from right to left and top to bottom in rows.
- Bottom to Top. The tiles are rendered from right to left and bottom to top in rows.

The Memory limit property specifies the maximum memory allocated for the rendering process. When this limit is reached, some objects may be removed from rendering.

Exercise 18-2

Complete the exercise on the student website.
www.g-wlearning.com/CAD

Render Presets

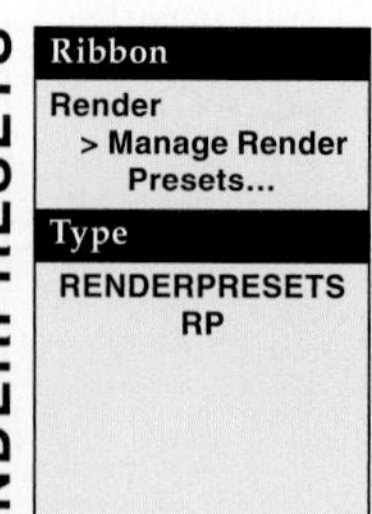

Once settings have been established that create a rendering with the desired results, the settings can be saved to a custom render preset. The **Render Presets** dialog box is used to create custom render presets, Figure 18-14. The **RENDERPRESETS** command opens this dialog box.

The left side of the dialog box displays a tree that contains the standard render presets and any custom render presets. In the middle of the dialog box are all of the properties for the selected render preset. These are the same properties available in the **Advanced Render Settings** palette. On the right side of the dialog box are three buttons that allow you to make a preset current, make a copy of a preset, or delete a preset. You cannot delete one of the default presets.

The easiest way to create a custom preset is to start with a standard render preset and modify the properties until the desired result is produced. This preset will be

Figure 18-14.
The **Render Presets Manager** dialog box.

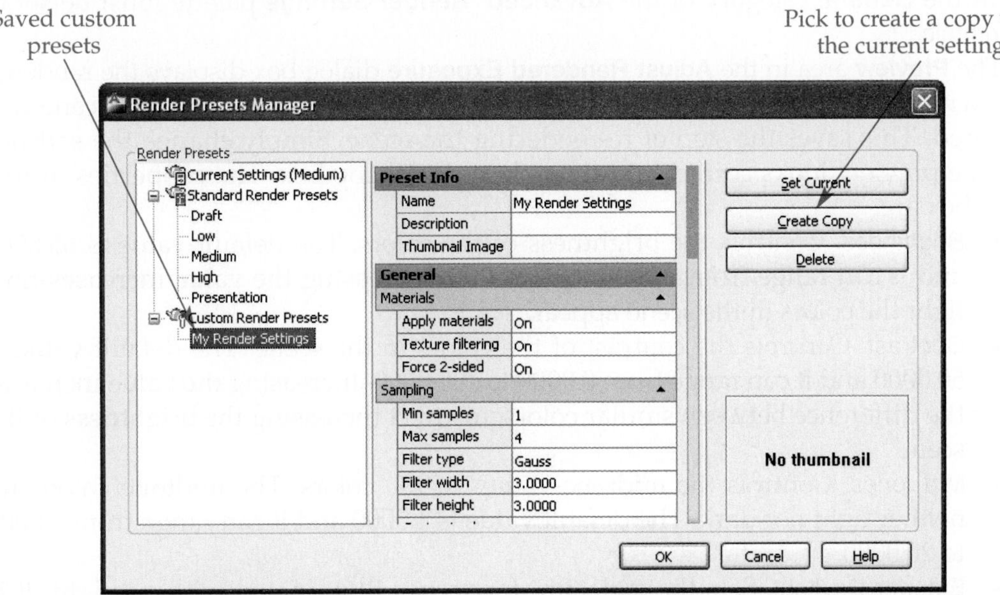

indicated as the current preset in the **Render Presets** dialog box, but there will be an asterisk (*) in front of its name. The asterisk indicates that the preset has been changed from its original settings. Next, pick the **Create Copy** button in the **Render Presets** dialog box to make a copy. In the **Copy Render Preset** dialog box that appears, name the new render preset, provide a description, and pick the **OK** button. The new preset is saved in the Custom Render Presets branch of the tree.

Render Exposure

The **RENDEREXPOSURE** command displays the **Adjust Rendered Exposure** dialog box. See **Figure 18-15.** In this dialog box, you can globally adjust the brightness,

Ribbon
Render
> Render
Adjust Exposure
Type
RENDEREXPOSURE
RENDEREXPOSURE

Figure 18-15.
Using the **RENDEREXPOSURE** command to adjust the rendering.

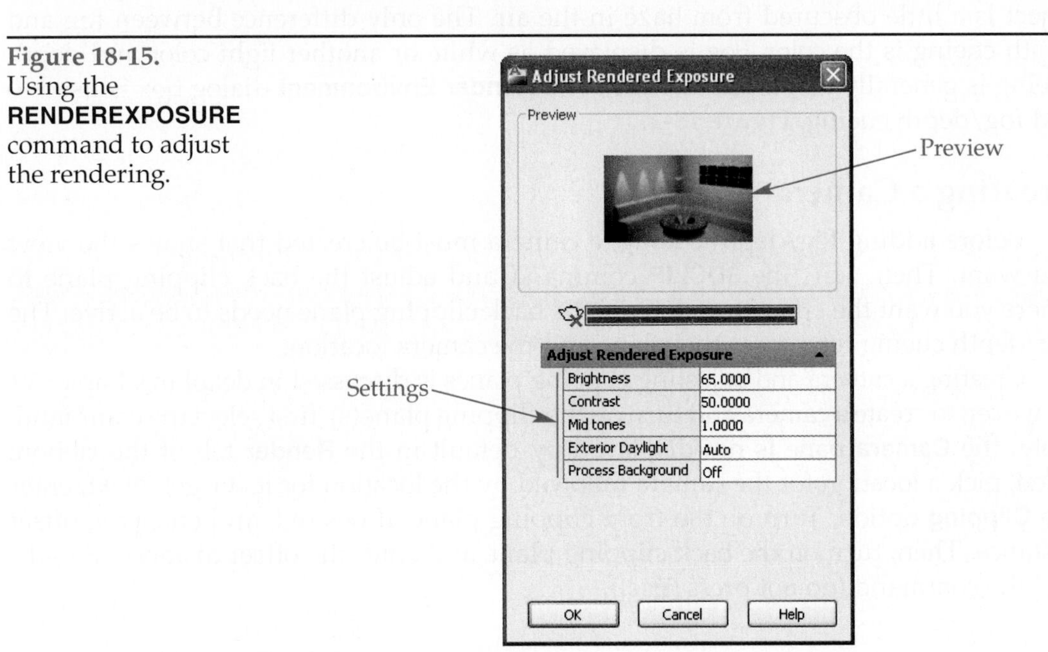

contrast, midtones, and exterior daylight of the scene. In order to use this command, photometric lighting must be on (**LIGHTINGUNITS** = 1 or 2) or the Exposure Type property in the **General** category of the **Advanced Render Settings** palette must be set to Logarithmic.

The **Preview** area in the **Adjust Rendered Exposure** dialog box displays the rendered scene with the changes you make in the dialog box so you can see how the scene will be altered. This saves the step of re-rendering the scene. Simply change the settings until the preview looks correct and then close the dialog box. The properties in this dialog box are:

- Brightness. Controls the brightness of the colors. The default value is 65.0000 and it can range from 0.0000 to 200.0000. Increasing the value increases how light the colors in the scene appear.
- Contrast. Controls the contrast of the colors in the scene. The default value is 50.0000 and it can range from 0.0000 to 100.0000. Increasing the value increases the difference between similar colors, in effect increasing the brightness of the scene.
- Mid tones. Controls the midtone values of the colors. The midtone colors are neither light nor dark. The default value is 1.0000 and it can range from 0.0000 to 20.0000.
- Exterior Daylight. Sets the exposure for scenes illuminated with sunlight. It is either on, off, or automatic. The default setting is Auto.
- Process Background. Specifies whether or not the background is processed by exposure control when the scene is rendered. It is either on or off.

To force the preview to update, pick the button to the left of the rendering progress bars below the preview that looks like a teapot and an X. The preview is updated with the current settings.

Render Environment

The render environment allows for the addition of fog or depth cueing to the scene. ***Fog*** and ***depth cueing*** in AutoCAD are actually ways of using color to visually represent the distance between the camera (viewer) and objects in the model. See **Figure 18-16.** This is similar to looking at an object from a distance and seeing that the object is a little obscured from haze in the air. The only difference between fog and depth cueing is the color. Fog is displayed as white or another light color and depth cueing is generally displayed as black. The **Render Environment** dialog box is used to add fog/depth cueing, **Figure 18-17.**

Creating a Camera

Before adding fog/depth cueing, a camera must be created that shows the view you want. Then, start the **3DCLIP** command and adjust the back clipping plane to where you want the effect to end. Only the back clipping plane needs to be active. The fog/depth cueing references this plane and the camera location.

CAMERA
Ribbon
Render
> Camera
Create Camera
Type
CAMERA
CAM

Creating a camera and adjusting clipping planes is discussed in detail in Chapter 19. However, to create a camera and turn on the clipping plane(s), first select the command. Note: the **Camera** pane is not displayed by default in the **Render** tab of the ribbon. Then, pick a location for the camera followed by the location for its target. Next, enter the **Clipping** option. Turn on the front clipping plane, if desired, and enter the offset distance. Then, turn on the back clipping plane and enter the offset distance. Finally, end the command (do not press [Esc]).

Figure 18-16.
A—This scene has no fog/depth cueing applied. B—The scene has white fog applied (including the background). C—The scene has black depth cueing applied (including the background).

A

B

C

Figure 18-17.
The **Render Environment** dialog box is used to add fog/depth cueing to the scene.

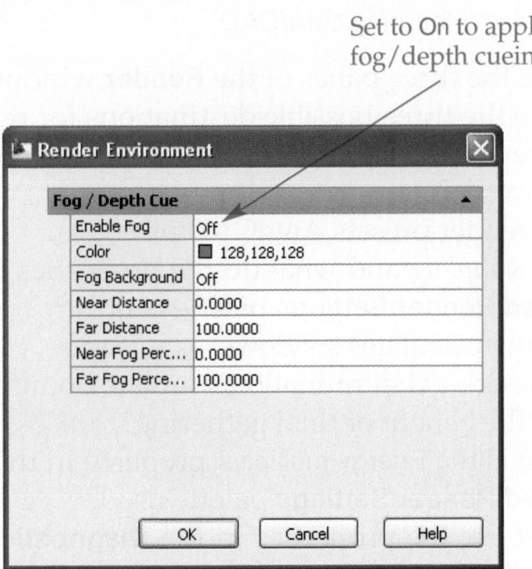

PROFESSIONAL TIP

A camera is automatically created when a view is saved as a named view. This is another good reason to save your views.

Adding Fog

Once a camera is created, open the **Render Environment** dialog box. To turn on fog/depth cueing, set the Enable Fog property to On. To set the color of the effect, select the Color property. Then, choose a color in the drop-down list. The Select Color... entry displays the **Select Color** dialog box. The Fog Background property determines whether or not the background is affected by the fog/depth cueing just like everything else.

The Near Distance property sets where the fog/depth cueing begins. This is a distance from the camera. The value can be from 0.0000 to 100.0000, which is a percentage of the total distance between the camera and the back clipping plane. The back clipping plane is where the target is located. The Far Distance property sets where the fog ends. This is also a distance from the camera. The value is also a percentage of the total distance from the camera to the back clipping plane and can be from 0.0000 to 100.0000. In other words, 100% ends at the back clipping plane.

The Near Fog Percentage property determines the opacity of the fog at its starting location. A value of 100 means the fog is 100% opaque. The near percentage is usually set to 0, or 0% opaque. The Far Fog Percentage property determines the opacity of the fog at its ending location. The fog/depth cueing will increase in opacity from the near distance to the far distance starting with the near fog percentage and ending with the far fog percentage.

Exercise 18-3

Complete the exercise on the student website.
www.g-wlearning.com/CAD

Chapter Test

Answer the following questions. Write your answers on a separate sheet of paper or complete the electronic chapter test on the student website.
www.g-wlearning.com/CAD

1. Describe the three panes of the **Render** window.
2. What are the three possible destinations for render output?
3. Once a rendering is completed and displayed in the **Render** window, how can it be saved to a file?
4. List the render presets AutoCAD provides.
5. What is *sampling* and what do the properties in the **Sampling** subcategory in the **Advanced Render Settings** palette control?
6. Raytracing calculates shadows, _____, and _____.
7. How does global illumination simulate bounced light?
8. What is the benefit of final gathering?
9. For what is the Energy multiplier property in the **List Properties** subcategory in the **Advanced Render Settings** palette used?
10. For what are the properties in the **Diagnostic** category of the **Advanced Render Settings** palette used?
11. Describe how to create a custom render preset.
12. List the properties that can be changed in the **Adjust Render Exposure** dialog box.
13. What is *fog/depth cueing?*
14. Which color is normally used to display depth cueing?
15. What must be created before fog or depth cueing is added to the scene?

Drawing Problems

1. Using the drawing from problem 2 in Chapter 17, you will experiment with advanced render settings.
   A. Open drawing P17_02 and save it as P18_01.
   B. Make sure all of the lights are active and render the scene to the **Render** window using the Medium or High render preset.
   C. In the **Advanced Render Settings** palette, enable global illumination. Then, render the scene again.
   D. Enable final gathering and render the scene again. This time it will probably take much longer to render.
   E. Which rendering has the best quality?
   F. Which setting impacted render time the most?
   G. Save the last image as P18_01.jpg.
   H. Save the drawing.

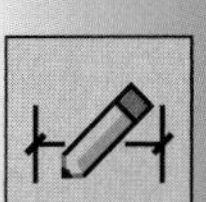

2. In this problem, you will set up fog/depth cueing.
   A. Start a new drawing and save it as P18_02.
   B. Draw a planar surface that is 50 units × 20 units.
   C. Randomly place various objects (cones, boxes, spheres, etc.) on the plane. Assign a different color or material to each object.
   D. Create a viewpoint that is almost at ground level looking down the length of the plane. Try to get as many objects in the view as possible. Save this as a named view and add a background of some type.
   E. Add a distant light source. Position it and adjust its intensity so that interesting shadows are created in the scene, but the objects are sufficiently illuminated.
   F. With the **3DCLIP** command, set up clipping planes with the back clipping plane at the far end of the plane. Make sure the back clipping plane is on.
   G. In the **Render Environment** dialog box, turn on fog and set the color to black. The far distance should be 100 and the percentage should be around 75.
   H. Render the scene.
   I. Change the fog color to white and render the scene again.
   J. Set the fog to affect the background and render the scene again.
   K. Save the image as P18_02.jpg.
   L. Save the drawing.

3. In this problem, you will experiment with the **RENDEREXPOSURE** command and final gathering. Open P17_05 created in Chapter 17 and save it as P18_03. If you did not complete this problem, do so now.
   A. Open the **Advanced Render Settings** palette. In the **General** category, set the Exposure Type property to Logarithmic.
   B. In the **Indirect Illumination** category, change the Mode property in the **Final Gather** subcategory to Off.
   C. Render the scene and note the appearance.
   D. Use the **RENDEREXPOSURE** command to display the **Adjust Rendered Exposure** dialog box. Note the appearance of the preview image.
   E. Change the brightness setting to 80 and note how the preview changes.
   F. Pick the **OK** button to close the **Adjust Rendered Exposure** dialog box and render the scene again. Does the rendered scene match the preview in the **Adjust Rendered Exposure** dialog box?
   G. Turn on final gathering (Mode property = On) and open the **Adjust Rendered Exposure** dialog box. How does the preview look different? Why?
   H. Adjust the brightness setting to get the exposure that you want in the preview. Then, close the dialog box and render the scene again.
   I. Experiment with the other settings in the **Adjust Rendered Exposure** dialog box until you get the scene the way you want it.
   J. Render the scene one last time and save the image as a file called P18_03.jpg.
   K. Save the drawing.

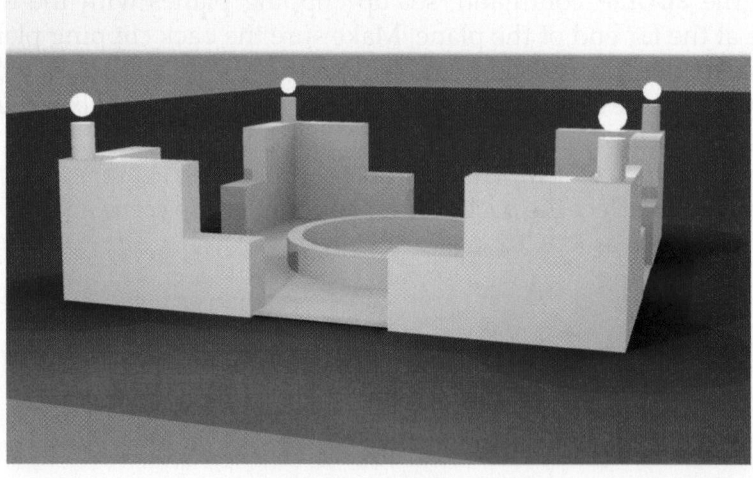

4. Using the same drawing from problem 18-3, you will perform diagnostics to determine the effects of the lights on the final rendering.
   A. Open drawing P18_03 and save it as P18_04.
   B. In the **Advanced Render Settings** palette, select the Medium rendering preset. In the **Indirect Illumination** category, turn off final gathering (Mode property = Off).
   C. In the **Diagnostic** category of the **Advanced Render Settings** palette, set the Grid property to Object. Render the scene.
   D. In the **Diagnostic** category of the **Advanced Render Settings** palette, set the Grid property to World. Render the scene.
   E. In the **Diagnostic** category of the **Advanced Render Settings** palette, set the Grid property to Camera. Render the scene.
   F. Describe the differences and explain why this is helpful in analyzing a scene.
   G. In the **Indirect Illumination** category of the **Advanced Render Settings** palette, turn on global illumination. In the **Diagnostic** category, turn off the grid and set the Photon property to Density.
   H. Render the scene. Describe the effect and what can be learned from it.
   I. Set the Photon property to Irradiance and render the scene. What does this effect tell about the lighting in the scene?
   J. Which diagnostic worked the best and why?
   K. Save the drawing.

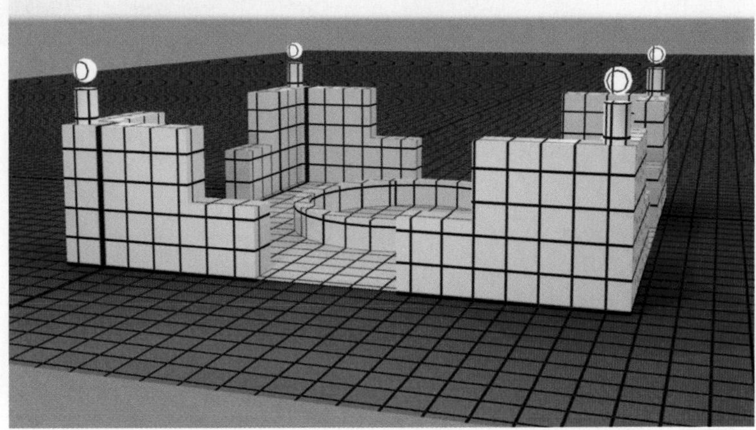

Lighting can have a dramatic effect on the scene. Notice how lighting is used in this model. One light is placed inside of the lamp and casts shadows. Another light is used to create sunlight. This is also set to cast shadows.

CHAPTER 19

Cameras, Walkthroughs, and Flybys

Learning Objectives

After completing this chapter, you will be able to:

✓ Create a camera to define a static 3D view.
✓ Activate and adjust front and back clipping planes.
✓ Record a walkthrough of a 3D model to a movie file.
✓ Record a flyby of a 3D model to a movie file.
✓ Create walkthroughs and flybys by following a path.
✓ Control the viewpoint, speed, and quality of the animation.

Once you have a 3D design complete, or even while still in the conceptual phase of design, you may want to take a stroll through the model and have a look around. You may also want to strap on some wings and fly over and around the model to see it from above. A ***walkthrough animation*** shows a scene as a person would view it walking through the scene. Walkthroughs are typically used to show the interior of a building, but can be created for exterior scenes as well. A ***flyby animation*** is similar to a walkthrough, except that the person is not bound by gravity. In other words, the scene is viewed as a bird flying above would see it. Flybys often show the exterior of a building.

The **3DWALK** command is used to create a walkthrough by recording views as a camera "walks" through the scene. The **3DFLY** command is very similar, but the movement of the camera is not limited to a single Z value. A path can also be drawn and the camera linked to the path. This chapter discusses these commands and other methods needed to create the animation you need. In addition, creating and using cameras are discussed.

Creating Cameras

Cameras are used in AutoCAD to store a viewpoint and easily recall it later when needed for viewing or rendering the scene. After the camera is established, you can zoom, pan, and orbit as needed and then come back to the camera view. It is not necessary to create a camera before using the **3DWALK**, **3DFLY**, and **ANIPATH** commands (discussed later) because these commands create their own cameras.

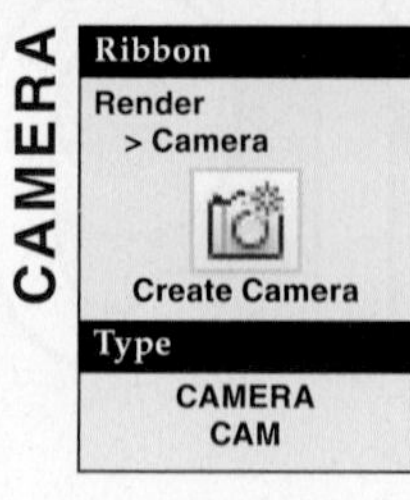

The **CAMERA** command allows you to add a camera to the scene. Note: the **Camera** panel may not be displayed in the **Render** tab of the ribbon. You may have to toggle it on. Cameras are normally placed in the plan view of the scene to make it easy for you to pick where you want to "stand" and where you want to "look." Once the command is selected, you are first prompted to specify the camera location. A camera glyph is placed in the scene at the camera location, Figure 19-1. Next, you must specify the target location. As you move the cursor before picking the target location, a pyramid-shaped field of view indicates what will be seen in the view (if a 3D visual style is current). Once you select the target location, the command remains active for you to select an option:

Enter an option [?/Name/LOcation/Height/Target/LEns/Clipping/View/eXit]<eXit>:

The list, or **?**, option allows you to list the cameras in the drawing. Select this option and type an asterisk (*) to show all of the cameras in the drawing. You can also enter a name or part of a name and an asterisk. For example, entering HOUSE* will list all of the cameras whose name begins with HOUSE, such as HOUSE_SW, HOUSE_SE, and HOUSE_PLAN.

The **Name** option allows you to change the name of the camera as you create it. If you do not rename the camera, it is given a default, sequential name, such as Camera1, Camera2, Camera3, and so on. It is always a good idea to provide meaningful names for cameras. Names such as Living Room_SW, Corner, or Hallway_Looking East leave no doubt as to what the camera shows. If you choose not to rename the camera at this point, it can be renamed later using the **Properties** palette.

The **Location** option allows you to change the placement of the camera. Enter the option and then specify the new location. You can enter coordinates or pick a location in the drawing.

The **Height** option allows you to change the vertical location of the camera. Enter the option and then enter the height of the camera. The value you enter is the number of units from the current XY plane. If you are placing the camera in a plan view, this option is used to tilt the view up or down from the current XY plane.

The **Target** option allows you to change the placement of the camera target. Enter the option and then specify the new location. You can enter coordinates or pick a location in the drawing.

The **Lens** option allows you to change the focal length of the camera lens. If you change the lens focal length, you are really changing the field of view, or the area of the drawing that the camera covers. The lower the lens focal length, the wider the field of view angle. The focal length is measured in millimeters.

The **Clipping** option is used to turn the front and back clipping planes on or off. These planes are used to limit what is shown in the camera view. Clipping planes are discussed later in this chapter.

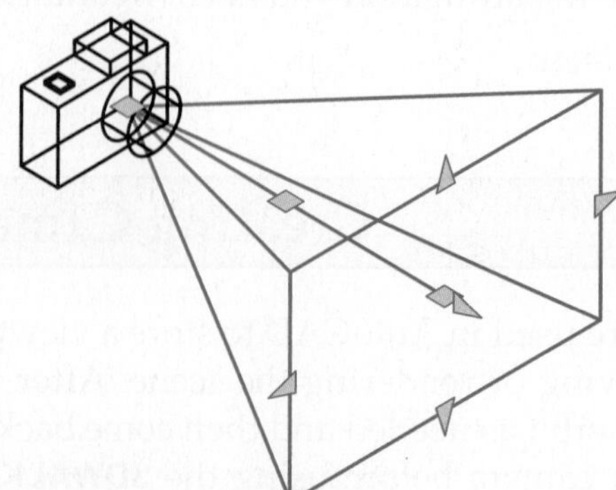

Figure 19-1. A camera is represented by a glyph. When the camera is selected, the field of view (shown in color) and grips are displayed.

The **View** option is used to change the current view to that shown by the camera. This option has two choices—**Yes** or **No**. If you select **Yes**, the active viewport switches to the camera view and the **CAMERA** command ends. If you select **No**, the previous prompt returns.

Once you have made all settings, press [Enter] or select the **Exit** option to end the command. The view (camera) is listed with the other saved views in the drop-down list in the **Views** panel on the **View** tab of the ribbon. It is also listed under the Model Views branch in the **View Manager** dialog box. Selecting the view makes it the current view in the active viewport.

PROFESSIONAL TIP

In addition to the camera name, many other camera properties can be changed in the **Properties** palette. The camera and target locations can be changed, the lens focal length and field of view can be adjusted, and the clipping planes can be modified. Also, you can change the roll angle, which is the rotation about a line from the camera to the target, and set the camera glyph to plot.

Camera System Variables

The **CAMERADISPLAY** system variable controls the visibility of camera glyphs. When set to 1, which is the default, camera glyphs are displayed. When set to 0, camera glyphs are not displayed. Creating a camera automatically sets the variable to 1. The **Show Cameras** button in the **Camera** panel on the **Render** tab of the ribbon toggles the display of camera glyphs off and on. Remember, this panel may not be displayed by default.

When creating a camera, if you pick the camera and target locations without using object snaps, you may assume that the camera and target are located on the XY plane (Z coordinate of 0) of the current UCS. This may or may not be true. The **CAMERAHEIGHT** system variable determines the default height of the camera if a Z coordinate is not provided. It is a good idea to set this variable to a typical eye height before placing cameras. There is no corresponding system variable for the target because the target is usually placed by snapping to an object of interest. If X and Y coordinates are entered for the target location, but a Z coordinate is not provided, the Z value is automatically 0. If a camera was previously created in the drawing session and the **Height** option was used, that height value becomes the default camera height.

Camera Tool Palette

The **Cameras** tool palette provides a quick way to add a camera, but the default tools do not allow for the options described earlier. The **Normal Camera** tool creates a camera with a 50 mm focal length. This camera simulates normal human vision. The **Wide-angle Camera** tool creates a camera with a 35 mm focal length. This type of view is commonly used for scenery or interior views where it is important to show as much as possible with minimal distortion. The **Extreme Wide-angle Camera** tool creates a camera with a 6 mm focal length. This camera produces a fish-eye view, which is very distorted and mainly useful for special effects.

Changing the Camera View

Once the camera is placed, it is easy to manipulate. If you select a camera, the **Camera Preview** window is displayed by default. This window shows the view through the camera, **Figure 19-2.** The view in the window can be displayed in any of the 3D visual styles or any named visual style. Select the visual style in the drop-down list in

Figure 19-2.
The **Camera Preview** window is displayed, by default, when a camera is selected.

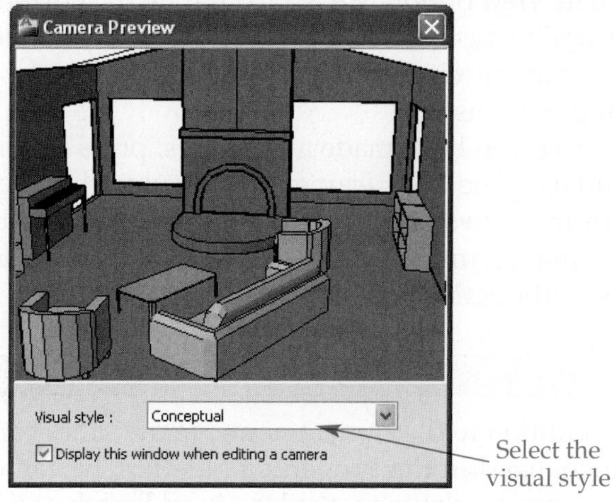

the window. If the **Display this window when editing a camera** check box at the bottom of the window is unchecked, the window is not displayed the next time a camera is selected. The next time the drawing is opened, this setting is restored (checked).

When a camera is selected, grips are displayed. Refer to Figure 19-1. If you hover the cursor over a grip, a tooltip appears indicating what the grip will alter. Picking the base grip on the camera allows you to reposition the camera in the scene. If the **Camera Preview** window is open, watch the preview as you move the camera to help guide you. Selecting the grip on the target allows the target to be repositioned. Again, use the preview in the **Camera Preview** window as a guide. The grip at the midpoint between the camera and target can be used to reposition the camera and target at the same time. If you pick and move one of the arrow grips on the end of the field of view, the lens focal length and field of view are changed.

Camera Clipping Planes

Clipping planes allow you to suppress objects in the foreground or background of your scene. Picture these clipping planes as flat, 2D objects perpendicular to the line of sight that can be moved closer to or farther from the viewer. Only the objects between the front and back clipping planes and within the field of view are seen in the camera view. This is helpful for eliminating walls, roofs, or any other clutter that may take away from the focus of the scene. Also, as mentioned in Chapter 18, the back clipping plane must be enabled when applying fog/depth cueing using the **Render Environment** dialog box. Clipping planes can be set while creating the camera or later using the **Properties** palette and adjusted using grips.

To set the clipping planes while creating the camera, enter the **Clipping** option. You are prompted:

```
Enable the front clipping plane? [Yes/No] <No>:
```

To enable the front clipping plane, enter YES. You are then asked to specify the offset from the target plane. The target plane is described next. Once you enter the offset, or if you answer **No**, you are prompted:

```
Enable the back clipping plane? [Yes/No] <No>:
```

To enable the back clipping plane, enter YES and then specify the offset from the target plane.

The *target plane* is the 2D plane that is perpendicular to the line of sight and passing through the camera's target point. Offsets for both front and back clipping planes are from this plane. Positive values place the clipping planes between the camera and the target plane. Negative values place the planes on the opposite side of the target plane from the camera. You can place the clipping planes anywhere in the scene from the camera location to infinity. You cannot, however, place the back clipping plane between the front clipping plane and the camera.

The best way to adjust clipping planes is using the **Properties** palette or grips. Create the camera and then display a plan view of the camera and target (an approximate plan view is okay). Select the camera and open the **Properties** palette. In the **Clipping** category, select the Clipping property. In the property drop-down list, select Front on, Back on, or Front and back on to turn on the appropriate clipping plane(s). Notice that the clipping planes are visible in the viewport, Figure 19-3. Next, enter offset values for the Front plane and Back plane properties as appropriate or use grips to set the locations of the clipping planes. By displaying a plan view of the camera and target, you can see where the clipping planes are located and visualize their effect on the scene. If the **Camera Preview** window is open, the clipping is reflected in the preview.

Exercise 19-1

Complete the exercise on the student website.
www.g-wlearning.com/CAD

Figure 19-3.
Adjusting the clipping planes for a camera.

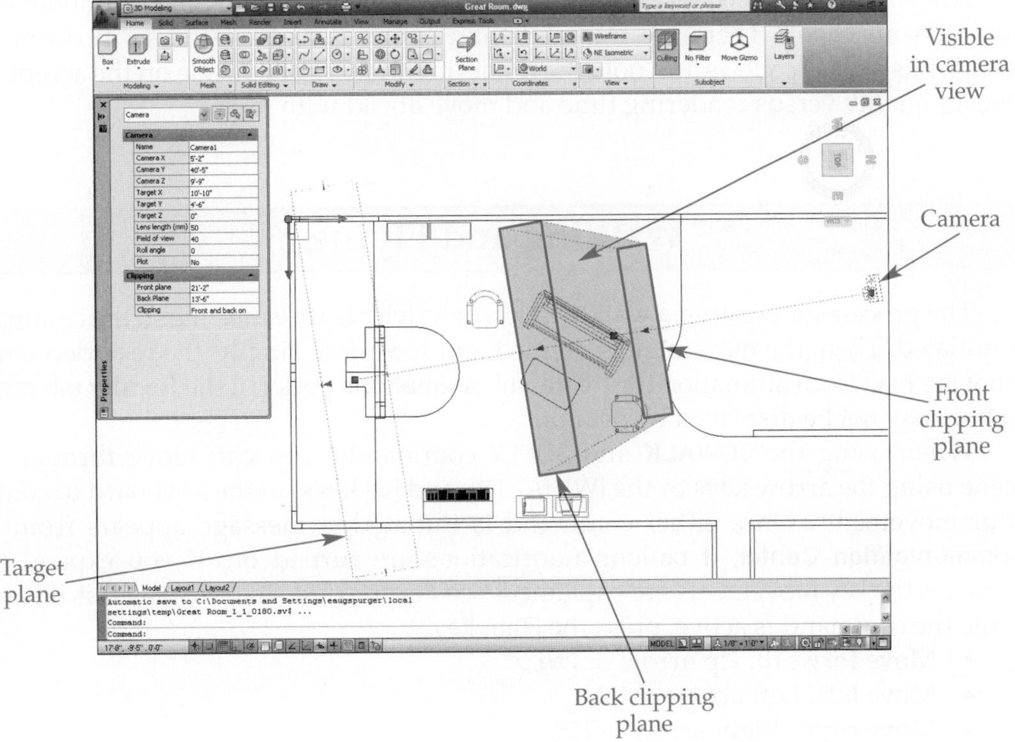

Animation Preparation

The tools presented in this chapter make it easy to lay out a path, plan camera angles, and record the movement of the camera. The resulting animation can be directly output to a number of movie file types that can be shared with others. However, there are some decisions to make first.

It is important to plan out exactly what you want to see in the animation. Think like a movie director and plan the "shots." Ask these questions:

- What will be visible from each camera angle?
- Is there a background in place?
- Is the lighting appropriate?
- Will a simple walkthrough suffice or will a flyby be necessary?
- How close is the viewer (camera) going to be to the objects in the scene?

The answers to these questions will help determine the modeling detail required. Do not model anything that will not be seen. Also, do not place detailed materials on objects that are not the focus of the animation. Processing the animation may take a long time. Unnecessary detail may bog down the computer. In addition, walkthroughs and flybys must be created in views with perspective, not parallel, projection.

The "visual quality" of the scene has the biggest impact on the time involved in rendering the animation. An animation can be rendered in any visual style or using any render preset that is available in the drawing. It is a natural tendency to render at the highest level to make the animation look the best. However, a computer animation has a playback rate of 30 frames per second (fps). If a single frame (view) takes three minutes to render using the Presentation render preset, how long will it take to render a 30 second animation? An animation 30 seconds in length has 900 frames (30 fps × 30 seconds). If each frame takes three minutes to render, the entire animation will take 2700 minutes, or 45 hours, to render.

Are you willing to wait two or three days for a 30 second movie? How about your boss or your client? There are trade-offs and concessions to be made. Perform test renderings on static views and note the rendering time. Then, decide on the acceptable level of quality versus rendering time and move ahead with it.

Walking and Flying

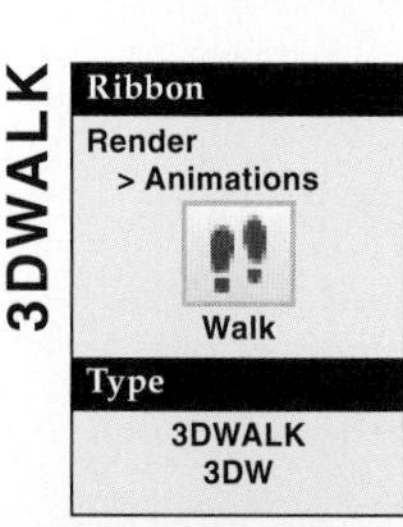

3DFLY
Ribbon
Render > Animations
Fly
Type
3DFLY

The process for creating a walkthrough or a flyby is the same. First, the command is initiated. Then, the movement is defined and recorded. Finally, the recorded movement is saved to an animation file. Note: the **Animations** panel in the **Render** tab on the ribbon may not be displayed by default.

When using the **3DWALK** and **3DFLY** commands, you can move through the scene using the arrow keys or the [W], [A], [S], and [D] keys on the keyboard to control your movements. Once either command is initiated, a message appears from the **Communication Center**, if balloon notifications are turned on. If you expand this message, the key movements are explained. See **Figure 19-4.** To redisplay this message while the command is active, press the [Tab] key.

- **Move forward.** Up arrow or [W].
- **Move left.** Left arrow or [A].
- **Move right.** Right arrow or [D].
- **Back up.** Down arrow or [S].

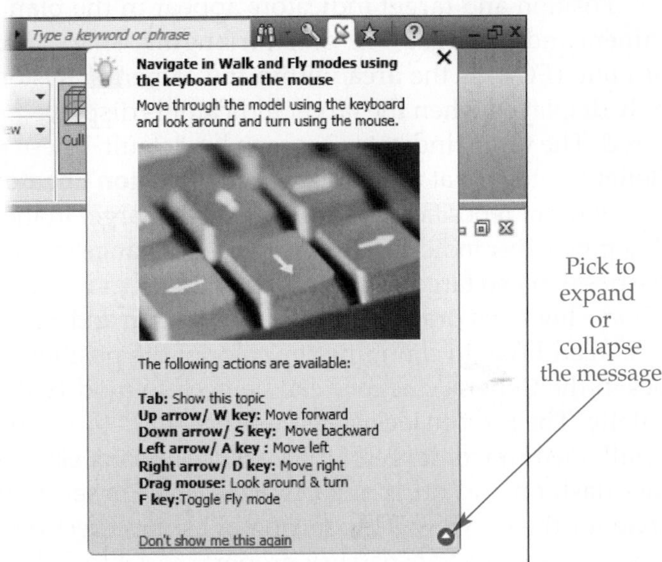

Figure 19-4.
This message from the **Communication Center** shows the keys that can be used to navigate through an animation.

You can also navigate through the scene using the mouse. Press and hold the left mouse button and then drag the mouse in the active viewport to "steer" through the scene. With the **3DWALK** command, the camera remains at the same Z value. With the **3DFLY** command, the Z position of the camera can change. The steps for creating a walkthrough or flyby are provided at the end of this section.

Position Locator

When the **3DWALK** or **3DFLY** command is initiated, the **Position Locator** palette appears. See Figure 19-5. The preview in this palette shows a plan view of the scene. The purpose of this window is to provide an overview of the scene, in plan, while you develop the animation. It does not need to be displayed to create an animation and can be closed if it takes up too much space or slows down the rendering.

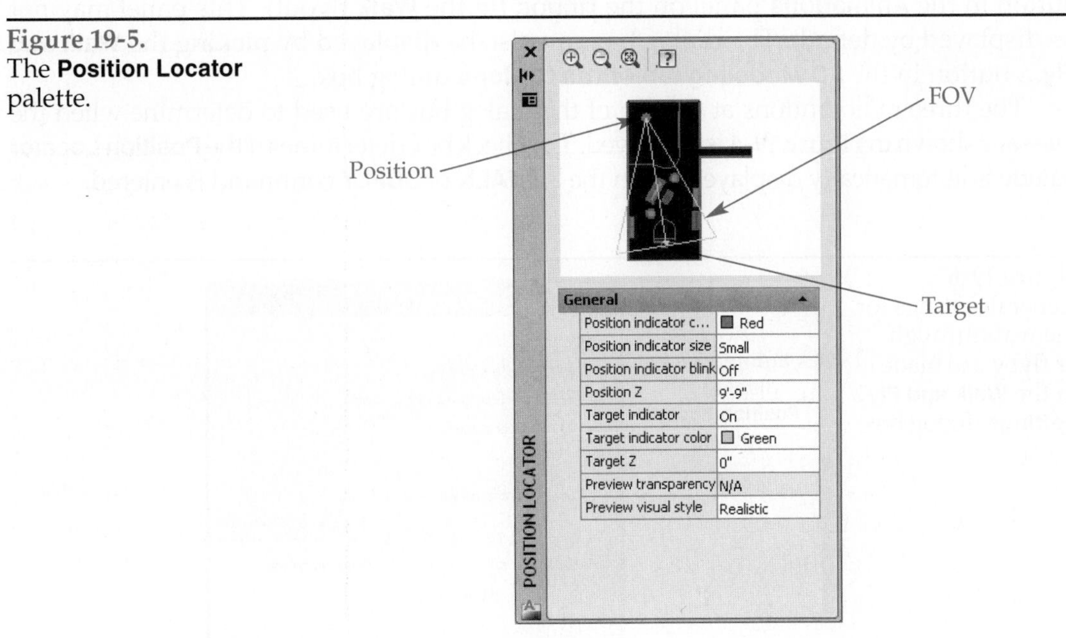

Figure 19-5.
The **Position Locator** palette.

Position and target indicators appear in the plan view to show the location of the camera and its target. The green triangular shape displays the field of view. The ***field of view (FOV)*** is the area within the camera's "vision." The field of view indicator is only displayed when the target indicator is displayed. By default, the position indicator is red. The target indicator is green by default. These properties can be changed in the **General** category at the bottom of the **Position Locator** palette.

You can reposition the camera and the target in the plan view simply by picking and dragging either indicator. The effect of the change is visible in the active viewport. Moving the position and target indicators closer together reduces the field of view. Picking the field of view lines and dragging moves the position and target indicators at the same time.

In addition to changing the color of the position and target indicators, the properties in the **General** category can be used to modify the display in the **Position Locator** palette. The Position indicator size property determines if both indicators are displayed small, medium, or large. If the Position indicator blink property is set to On, both indicators flash on and off in the preview. The Preview visual style property sets the visual style for the preview. This setting does not affect the current viewport or the animation. The Preview transparency property is set to 50% by default, but can be changed to whatever you want. If the view in the **Position Locator** palette is obscured by something (a roof, perhaps), you may want to set the Preview visual style property to Hidden and the Preview transparency property to 80% or 90%. This will make the objects under the roof visible. If hardware acceleration is on (**3DCONFIG** command), then the Preview transparency property is disabled.

Exercise 19-2

Complete the exercise on the student website.
www.g-wlearning.com/CAD

Walk and Fly Settings

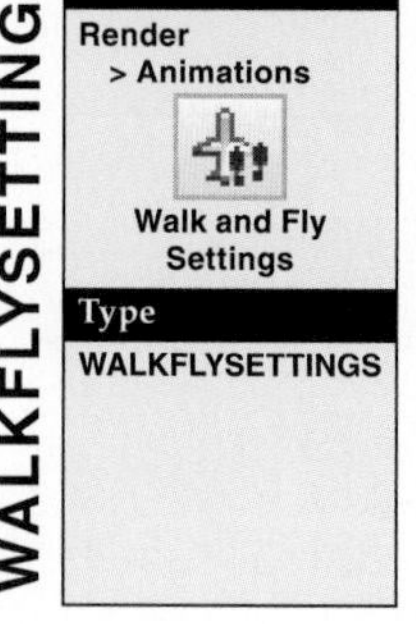

General settings for walkthroughs and flybys are made in the **Walk and Fly Settings** dialog box. See Figure 19-6. Open this dialog box by picking the **Walk and Fly Settings** button in the **Animations** panel on the ribbon (in the **Walk** flyout). This panel may not be displayed by default. The dialog box can also be displayed by picking the **Walk and Fly...** button in the **3D Modeling** tab of the **Options** dialog box.

The three radio buttons at the top of the dialog box are used to determine when the message shown in Figure 19-4 is displayed. The check box determines if the **Position Locator** palette is automatically displayed when the **3DWALK** or **3DFLY** command is entered.

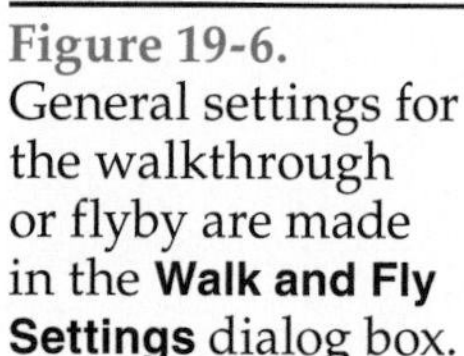

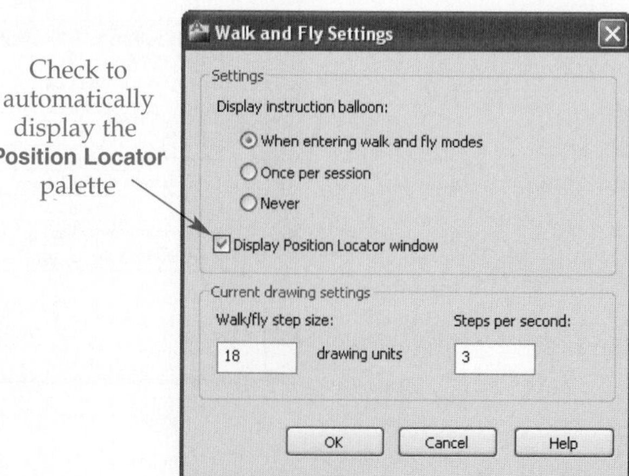

Figure 19-6.
General settings for the walkthrough or flyby are made in the **Walk and Fly Settings** dialog box.

The text boxes in **Current drawing settings** area determine the size of each step and the number of steps per second. The **Walk/fly step size:** setting controls the **STEPSIZE** system variable. This is the number of units that the camera moves in one step. The **Steps per second:** setting controls the **STEPSPERSEC** system variable. This is the number of steps the camera takes each second. Together, these two settings determine how fast the camera moves in the animation.

PROFESSIONAL TIP

You will have to experiment with step size and steps per second values to make an animation that is easy to watch. Start with low numbers (for slow movement) and work your way up. Fast movement is disorienting and makes the viewer feel like they are missing something. The viewer should be able to take their time and get a good look at your design.

To get a feel for the proper speed for a walkthrough, pay attention to the next movie or TV show that you watch. When the director wants you to get a good look at the setting for the scene, the camera very slowly pans around the room. To emphasize distance, the camera slowly zooms in to a target object or person.

Camera tools

The expanded **View** panel on the **Home** tab of the ribbon contains some tools for quickly adjusting the camera before starting the animation. See Figure 19-7. The **Lens length** slider controls how much of the scene is seen by the camera. The ***lens length*** refers to the focal length of the camera lens. The higher the number, the closer you are to the subject. The range is from about 1 to 100,000; 50 is a good starting point. There are stops on the slider for standard lens lengths. You can enter a specific value for the lens length or field of view by selecting the text in the slider, typing the value, and pressing [Enter]. The current view must be a camera view for the slider to be enabled.

Below the **Lens length** slider are text boxes for the camera and target positions. These can be used to change the X, Y, and Z coordinates of the camera or target. It is usually easier to change the X and Y locations in the **Position Locator** palette, but the Z coordinate cannot be set there. The Z coordinate determines eye level.

Animation tools

The **Animations** panel on the **Render** tab of the ribbon contains the tools for controlling the recording and playback of the animation. See Figure 19-8. Remember, this panel may not be displayed by default. The **Record Animation** button is used to initiate recording of camera movement. After the **3DWALK** or **3DFLY** command is activated, pick the button to start recording. Make sure that you are ready to start moving when you pick the button because recording starts as soon as it is picked.

Figure 19-7.
Camera tools are located in the expanded **View** panel on the **Home** tab of the ribbon.

Lens length

Camera position

Target position

Figure 19-8.
The **Animations** panel on the **Render** tab of the ribbon is where you can record and play back the walkthrough or flyby animation. This panel is not displayed by default.

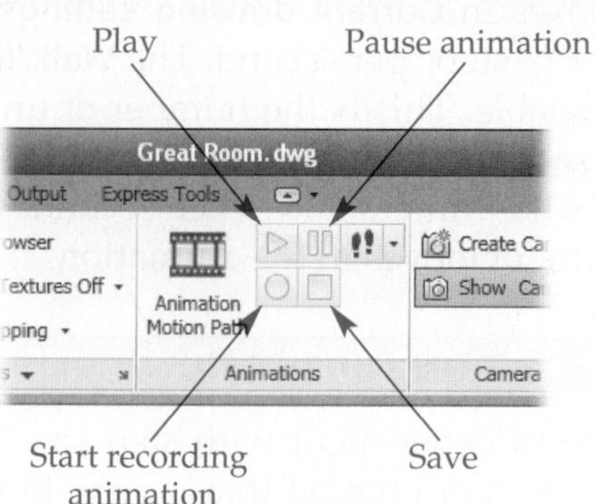

Picking the **Pause Animation** button temporarily stops recording. This allows you to make adjustments to the view without recording the adjustment. When you are ready to begin recording again, pick the record button to resume.

Picking the **Play Animation** button stops the recording and opens the **Animation Preview** dialog box in which the animation is played, Figure 19-9. The controls in this dialog box can be used to rewind, pause, and play the animation. The slider can be dragged to preview part of the animation or move to a specific frame. The visual style can also be set using the drop-down list. If the animation is created using a render preset, the file must be played in Windows Media Player or another media player to view the rendered detail.

Picking the **Save Animation** button in the **Animation** panel stops recording and opens the **Save As** dialog box. Name the animation file, navigate to a location, and pick the **Save** button.

CAUTION

While the **3DWALK** or **3DFLY** command is active and the record button is on, you are creating an animation. If you move the camera in the **Position Locator** palette and start re-recording the animation to correct a problem, but do not first exit the current **3DWALK** or **3DFLY** command session, you are adding another segment to the animation you just previewed. To start over, first exit the current command session.

Figure 19-9.
The animation is played in the **Animation Preview** dialog box.

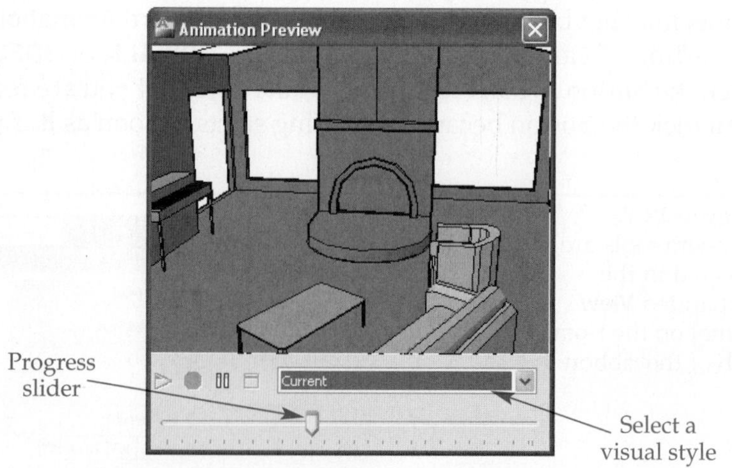

Exercise 19-3

Complete the exercise on the student website.
www.g-wlearning.com/CAD

Animation settings

The **Animation Settings** dialog box may contain the most important settings pertaining to walkthroughs and flybys. See **Figure 19-10.** These animation settings determine how good the animation looks, how long it is going to take to complete, and how big the file will be. The dialog box is displayed by picking the **Animation settings...** button in the **Save As** dialog box displayed when saving an animation.

The **Visual style:** drop-down list is used to set the shading level in the animation. The name of this drop-down list is a little misleading because visual styles and render presets are available. The higher the shading or rendering level selected in this drop-down list, the longer the rendering will take to process and the bigger the file will be. If you have numerous lights casting shadows, detailed materials, and global illumination and final gathering enabled, settle in for a long wait. A simple, straight-ahead walkthrough of 10 or 15 feet can easily result in 300 frames of animation. If each frame takes about five seconds to render, that equals 1500 seconds, or 25 minutes, to create an animation file that is only 10 seconds long.

The **Frame rate (FPS):** text box sets the number of frames per second for the playback. In other words, this sets the speed of the animation playback. The default is 30 fps, which is a common playback rate for computers.

The **Resolution:** drop-down list offers standard choices of resolution, from 160 × 120 to 1024 × 768. These are measured in pixels × pixels. Remember, higher resolutions mean longer processing times and larger file sizes.

The **Format:** drop-down list is used to select the output file type. The file type must be set in this dialog box. It cannot be changed in the **Save As** dialog box. The choices of output file type are:

- **WMV.** The standard movie file format for Windows Media Player.
- **AVI.** Audio-Video Interleaved is the Windows standard for movie files.
- **MOV.** QuickTime® Movie is the standard file format for Apple® movie files.
- **MPG.** Moving Picture Experts Group (MPEG) is another very common movie file format.

Depending on the configuration of your computer, you may not have all of these file type options or you may have additional options not listed here.

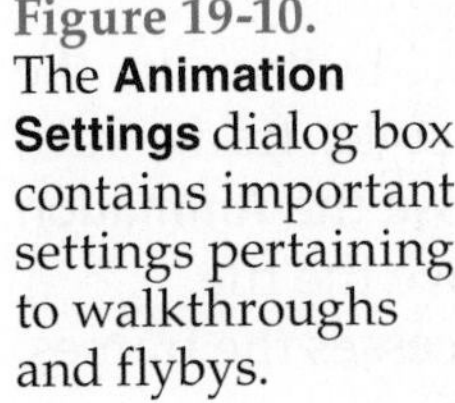

Figure 19-10.
The **Animation Settings** dialog box contains important settings pertaining to walkthroughs and flybys.

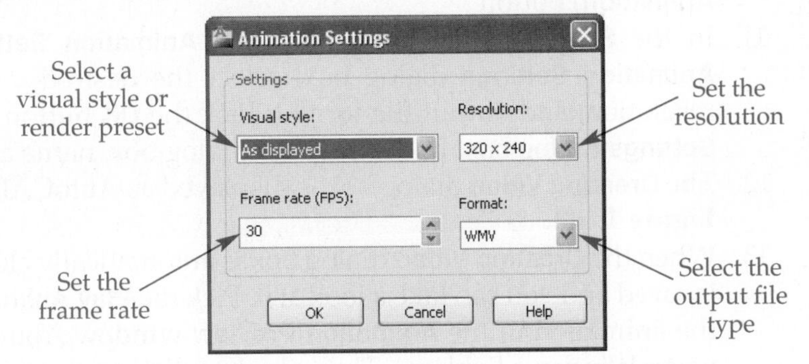

PROFESSIONAL TIP

Other AutoCAD navigation modes may be used to create a walkthrough or flyby. Anytime that you enter constrained, free, or continuous orbit, you can pick the **Record Animation** button to record the movements. After you activate the **3DWALK** or **3DFLY** command, right-click and experiment with some of the other options as you record your animation. You can even combine some of these navigation modes with walking or flying. For example, use **3DFLY** to zoom into a scene, pause the animation, switch to constrained orbit, restart the recording, and slowly circle around your model. Animation settings are also available in this shortcut menu.

Exercise 19-4

Complete the exercise on the student website.
www.g-wlearning.com/CAD

Steps to Create a Walkthrough or Flyby

1. Plan your animation: where you are moving from and to, what you are going to be looking at, and what will be the focal point of the scene.
2. Set up a multiple-viewport configuration of three or four viewports.
3. In one of the viewports, create or restore a named view with the appropriate starting viewpoint. Make sure a background is set up, if desired.
4. Start the **3DWALK** or **3DFLY** command and note in the **Position Locator** palette the location of the camera and target and the field of view. Adjust these in the expanded **View** panel on the ribbon, if necessary.
5. Position your fingers over the navigation keys on the keyboard.
6. Pick the **Record Animation** button.
7. Practice navigating through the view and then pick the **Play Animation** button to preview your animation. When you are done practicing, make sure to cancel the command and reposition the camera at the starting point.
8. Pick the **Record Animation** button to start over.
9. Start navigating through the view. Try to keep the movements as smooth as possible. Any jerks and shakes will be visible in the animation.
10. When you are done, stop moving forward and then pick the stop (**Save Animation**) button.
11. In the **Save As** dialog box pick the **Animation Settings...** button. In the **Animation Settings** dialog box, select the desired visual style, frame rate, resolution, and output file format. Pick the **OK** button to close the **Animation Settings** dialog box. In the **Save As** dialog box, name and save the file.
12. The **Creating Video** dialog box is displayed as AutoCAD processes the frames, Figure 19-11.
13. When the **Creating Video** dialog box is automatically closed, the animation file is saved and you can take a look at it. Pick the **Play Animation** button and watch the animation in the **Animation Preview** window. You can also locate the file using Windows Explorer. Then, double-click on the file to play the animation in Media Player (or whichever program is associated with the file type).
14. Exit the command. If you are not satisfied with the results and want to try it again, make sure to exit the command before you make another attempt at the walkthrough or flyby.

Figure 19-11.
The **Creating Video** dialog box is displayed as AutoCAD generates the animation.

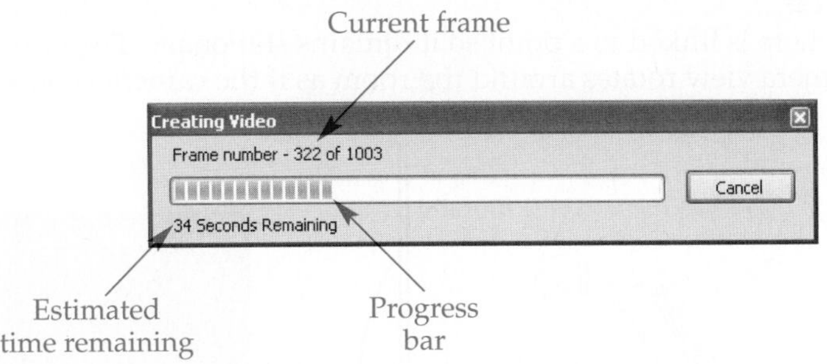

Motion Path Animation

You may have found it hard to create smooth motion using the keyboard and mouse. Fortunately, AutoCAD provides an easy way to create a nice, smooth animated walkthrough or flyby. This is done through the use of a motion path. A ***motion path*** is simply a line along which the camera, target, or both travel during the animation.

One method of using a motion path is to link the camera and target to a single path. The camera and its line of sight then follow the path much like a train follows tracks. See **Figure 19-12.**

Another option when using a motion path is to link the camera to a single point in the scene and the target to a path. For example, the target can be set to follow a circle or arc. The camera swivels on the point and "looks at" the path as if it is being rotated on a tripod. See **Figure 19-13.**

A third way to use a motion path is to have the camera follow a path, but have the target locked onto a stationary point. This is similar to riding in a vehicle and watching an object of interest on the side of the road. As the vehicle moves, your gaze remains fixed on the object. See **Figure 19-14.**

The fourth method of using a motion path is to have both the camera and target follow separate paths. Picture yourself walking into an unfamiliar room. As you walk into the center of the room, your gaze sweeps left and right across the room. In this case, the camera (you) follows a straight line path and the target (your gaze) follows an arc from one side of the room to the other.

Figure 19-12.
A—The camera and target are linked to the same path (shown in color). B—The camera looks straight ahead as it moves along the path.

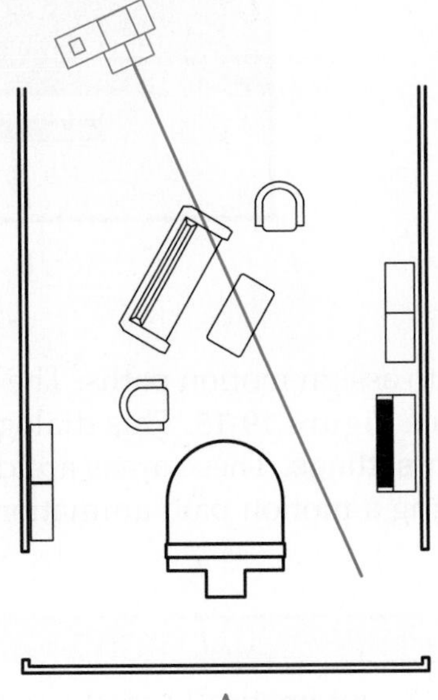

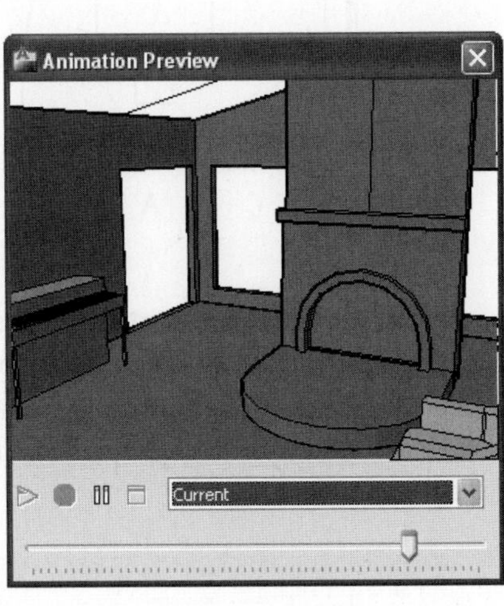

A

B

Figure 19-13.
A—The camera is linked to a point so it remains stationary. The target is linked to the circle.
B—The camera view rotates around the room as if the camera is on a swivel tripod.

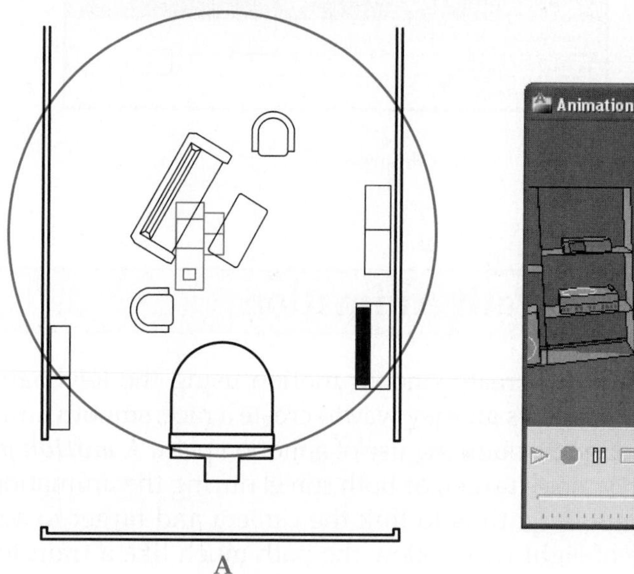

Figure 19-14.
A—The camera is linked to the spline path and the target is linked to the point (shown in color).
B—As the camera moves along the path, it always looks at the point.

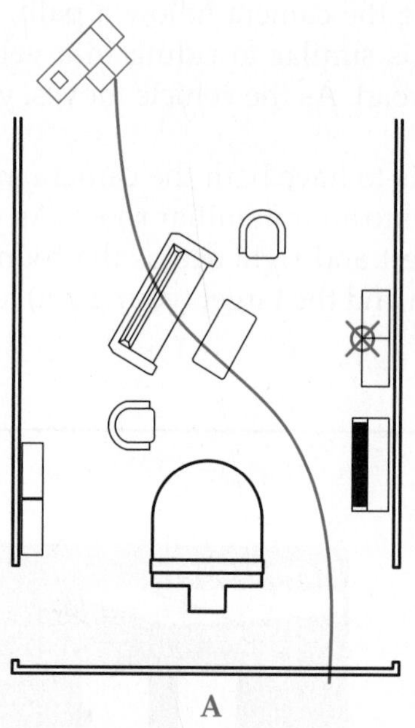

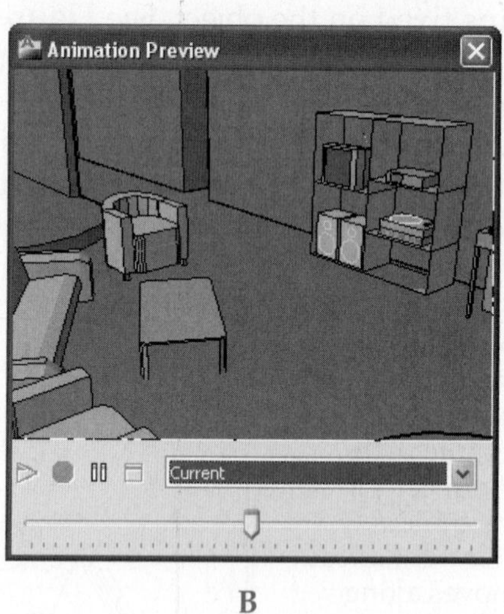

ANIPATH

Ribbon
Render > Animations
Animation Motion Path
Type
ANIPATH

The **ANIPATH** command is used to assign motion paths. The command opens the **Motion Path Animation** dialog box. See Figure 19-15. This dialog box has three main areas: **Camera**, **Target**, and **Animation settings**. These areas are described in detail in the next sections. The steps for creating a motion path animation are provided at the end of this section.

NOTE

Selecting a motion path automatically creates a camera. You cannot add a motion path to an existing camera.

Figure 19-15.
The **Motion Path Animation** dialog box is used to create an animation that follows a path.

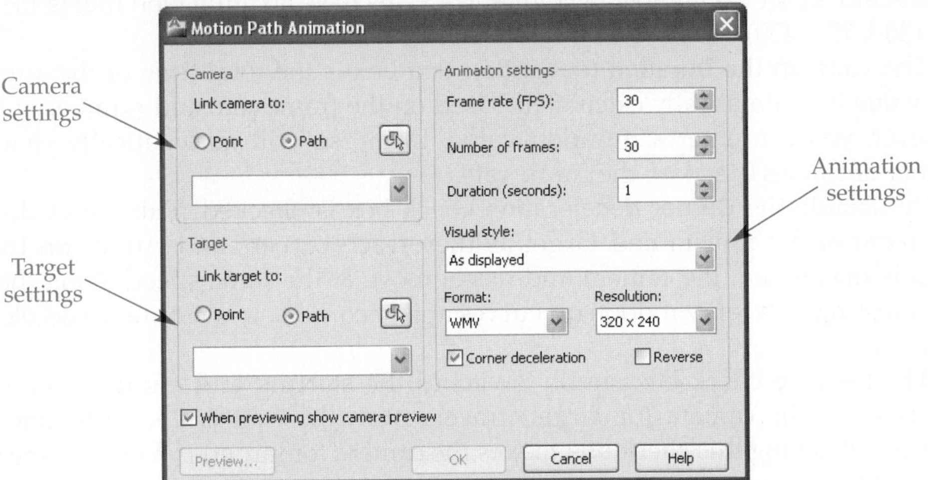

Camera Area

The camera can be linked to a path or a point. To select a path, pick the **Path** radio button and then pick the "select" button next to the radio buttons. The dialog box is temporarily closed for you to select the path in the drawing. The path may be a line, arc, circle, ellipse, elliptical arc, polyline, 3D polyline, spline, or helix, but it must be drawn before the **ANIPATH** command is used. Splines are nice for motion paths because they are smooth and have gradual curves. The camera moves from the first point on the path to the last point on the path, so create paths with this in mind.

To select a stationary point, pick the **Point** radio button. Then, pick the "select" button next to the radio buttons. When the dialog box is hidden, specify the location in the drawing. You can use object snaps or enter coordinates. It may be a good idea to have a point drawn and use object snaps to select the point.

The camera must be linked to either a path or a point. If neither is selected, the command cannot be completed. If you want the camera to remain stationary as the target moves, select the **Point** radio button and then pick the stationary point in the drawing.

Once a point or path has been selected, it is added to the drop-down list. All named motion paths and selected motion points in the drawing appear in this list. Instead of using the "select" button, you can select the path or point in this drop-down list.

Target Area

The target is the location where the camera points. Like the camera, the target can be linked to a point or a path. To link the target to a path, select the **Path** radio button. Then, pick the "select" button and select the path in the drawing. If the camera is linked to a point, the target must be linked to a path. If the camera is set to follow a path, then you actually have three choices for the target. It can be linked to a path, point, or nothing. To link the target to a point, pick the **Point** radio button. Then, pick the "select" button to select the point in the drawing. The None option, which is selected in the drop-down list, means that the camera will look straight-ahead down the path as it moves.

Animation Settings Area

Most of the settings in this area have the same effect as the corresponding settings in the **Animation Settings** dialog box discussed earlier. However, there are four settings unique to the **Motion Path Animation** dialog box.

The **Number of frames:** text box is used to set the total number of frames in the animation. Remember, a computer has a playback rate of 30 fps. Therefore, if the frame rate is set to 30, set the number of frames to 450 to create an animation that is 15 seconds long (30 × 15 = 450).

The value in the **Duration (seconds):** text box is the total time of the animation. This value is automatically calculated based on the frame rate and number of frames. However, you can enter a duration value. Doing so will automatically change the number of frames based on the frame rate.

By default, the **Corner deceleration** check box is checked. This slows down the movement of the camera and target as they reach corners and curves on the path. If this is unchecked, the camera and target move at the same speed along the entire path, creating very jerky motion on curves and at corners. It is natural to decelerate on curves.

The **Reverse** check box simply switches the starting and ending points of the animation. If the camera (or target) travels from the first endpoint to the second endpoint, checking this check box makes the camera (or target) travel from the second endpoint to the first.

Previewing and Completing the Animation

To preview the animation, pick the **Preview...** button at the bottom of the **Motion Path Animation** dialog box. The camera glyph moves along the path in all viewports. If the **When previewing show camera preview** check box is checked, the **Animation Preview** window is also displayed and shows the animation.

To finish the animation, pick the **OK** button in the **Motion Path Animation** dialog box. The **Save As** dialog box is displayed. Name the file and specify the location. If you need to change the file type, pick the **Animation settings...** button to open the **Animation Settings** dialog box. Change the file type, close the dialog box, and continue with the save.

Steps to Create a Motion Path Animation

1. Plan your animation: where you are moving from and to, what you are going to be looking at, and what will be the focal point of the scene.
2. Draw the paths and points to which the camera and target will be linked. Draw the path in the direction the camera should travel (first point to last point). Do not draw any sharp corners on the paths and make sure that the Z value (height) is correct.
3. Start the **ANIPATH** command.
4. Pick the camera path or point.
5. Pick the target path or point (or None).
6. Adjust the frames per second, number of frames, and duration to set the length and speed of the animation.
7. Select a visual style, the file format, and the resolution.
8. Preview the animation. Adjust settings, if needed.
9. Save the animation to a file.

Exercise 19-5

Complete the exercise on the student website.
www.g-wlearning.com/CAD

Chapter Test

Answer the following questions. Write your answers on a separate sheet of paper or complete the electronic chapter test on the student website.
www.g-wlearning.com/CAD

1. Which system variable controls the display of camera glyphs?
2. Name the three camera tools available on the **Camera** tool palette and explain the differences between them.
3. When is the **Camera Preview** window displayed, by default?
4. From where is the offset distance for the camera clipping planes measured?
5. What is the difference between the **3DWALK** and **3DFLY** commands?
6. How do you "steer" your movement when creating a walkthrough or flyby animation?
7. What is the *field of view?*
8. What is the purpose of the **Position Indicator** palette?
9. In the **Walk and Fly Settings** dialog box, which settings combine to control the speed of the animation?
10. How do you start recording a walkthrough or flyby?
11. What must be done before correcting a motion error in a walkthrough or flyby?
12. Motion path animation involves linking a camera or target to _____ or _____.
13. Which types of objects may be used as a motion path?
14. If None is selected as the target "path," what does the camera do in the animation?
15. Explain *corner deceleration.*

Drawing Problems

1. In this problem, you will create and manipulate a camera in a drawing from a previous chapter.
   A. Open drawing P18_02 from Chapter 18 and save it as P19_01.
   B. Create at least two viewports and display a plan view in one of them.
   C. Use the **CAMERA** command to create a camera looking at the objects from the southwest quadrant. Change the camera settings as needed to display a pleasing view of the scene.
   D. Name the camera SW View.
   E. Turn on both the front and back clipping planes. Adjust them to eliminate one object in the front and one object in the back.
   F. Open the **Camera** tool palette and, using the tools in the palette, create three more cameras looking at the scene from various locations. Change their names to Normal, Wide-angle, and Fish-eye to match the type of camera.
   G. Save the drawing.

2. In this problem, you will draw some basic 3D shapes to represent equipment in a small workshop. Then, you will create an animated walkthrough.
   A. Start a new drawing and set the units to architectural. Save it as P19_02.
   B. Draw a planar surface that is 15′ × 30′.
   C. Draw three 9′ tall walls enclosing the two long sides and one short side.
   D. Use boxes and a cylinder to represent equipment and shelves. Refer to the illustration shown below. Use your own dimensions.
   E. Use the **3DWALK** command to create an animation of walking into the workshop. Turn and look at the shelves at the end of the animation.
   F. Set the visual style to Conceptual and the resolution to 640 × 480.
   G. Save the animation to a file named P19_02.avi (or the format of your choice).
   H. Save the drawing.

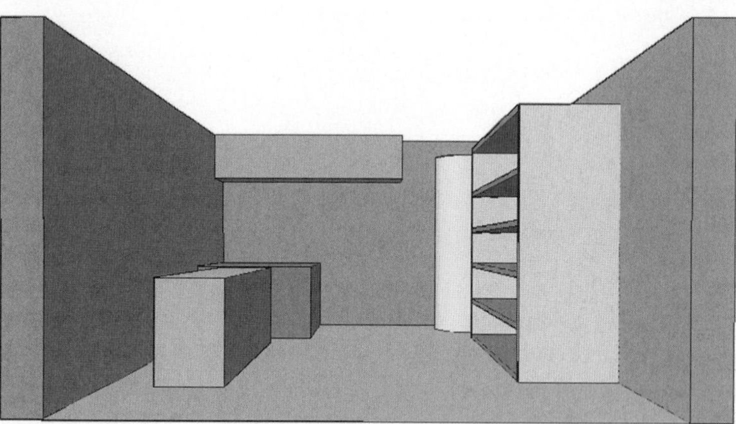

3. In this problem, you will create a motion path animation for the workshop drawn in problem 19-2.
   A. Open drawing P19_02 and save it as P19_03.
   B. Draw a line and an arc similar to those shown in color below. The dimensions are not important.
   C. Move both objects so they are 4′ off of the floor.
   D. Using the **ANIPATH** command, link the camera to the line and the target to the arc. Set the resolution to 640 × 480.
   E. Preview the animation. Adjust the animation settings as necessary. You may need to slow down the animation quite a bit. How do you do this?
   F. Save the animation as a file named P19_03.wmv (or the file format of your choice).
   G. Save the drawing.

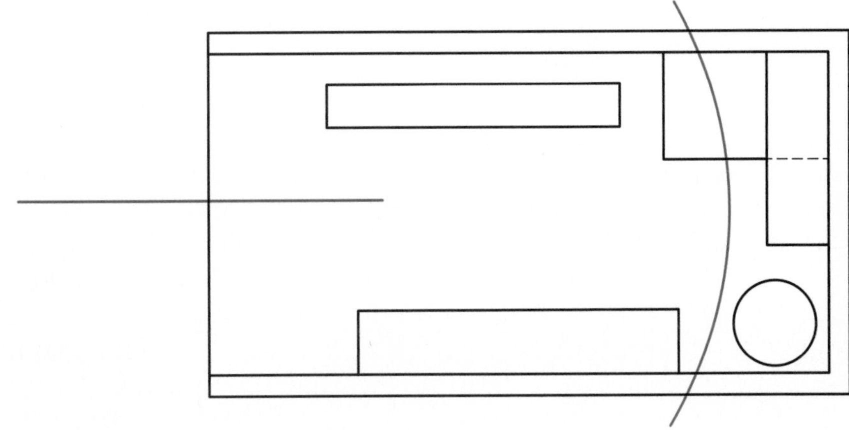

4. In this problem, you will create a motion path animation for presentation of a mechanical drawing from a previous chapter.
   A. Open drawing P17_03 and save it as P19_04.
   B. Draw a circle centered on the flange with a radius of 300.
   C. Move the circle 200 units in the Z direction.
   D. Using the **ANIPATH** command, link the camera to the circle and the target to the center of the flange.
   E. Preview the animation and adjust the animation settings as necessary. Due to the materials in the scene, the preview may play slowly, depending on the capabilities of your computer.
   F. Select a visual style that your computer can handle. Set the resolution to 320 × 240.
   G. Save the animation to a file named P19_04.avi (or the file format of your choice).
   H. Save the drawing.

5. In this problem, you will create a flyby of the building that you created in Chapter 17.
   A. Open drawing P17_01 and save it as P19_05.
   B. Create a perspective view of the scene that shows the building from slightly above it. Save the view.
   C. Start the **3DFLY** command. Practice with the movement keys to make sure you know how to fly around the building. Then, cancel the command.
   D. Restore the starting view and select the **3DFLY** command.
   E. Record the flyby and save the animation as P19_05.wmv (or the file format of your choice).
   F. Save the drawing.

CHAPTER 20

Using Raster, Vector, and Web Graphics

Learning Objectives

After completing this chapter, you will be able to:

✓ Compare raster and vector files.
✓ Import and export raster files using AutoCAD.
✓ Import and export vector files using AutoCAD.
✓ Set image commands to manipulate inserted raster files.
✓ Create DWF, DWFx, and PDF files.

One of the important aspects of drawing in AutoCAD is the ability to share information. Generally, this means sharing drawing data and geometry between CAD software, either other AutoCAD workstations or workstations using different software. AutoCAD creates drawing data files in a format known as a *vector* file. However, you can also share your work, as images, with photo editing and desktop publishing software. In Chapter 15 through Chapter 19, you learned how to create realistic scenes and render them to files. A scene rendered to a file is a *raster* image. However, raster images used in AutoCAD do not have to be created in AutoCAD. They may also come from digital photographs, scanned images, or Internet sources. This chapter introduces using AutoCAD to work with raster and vector graphics files. This includes importing, exporting, and setting various parameters.

Introduction to Raster and Vector Graphics

In the world of electronic imaging, there are two basic types of files—raster and vector. AutoCAD drawings are called vector graphics. A ***vector*** is an object defined by XYZ coordinates. In other words, AutoCAD stores the mathematical definition of an object. ***Pixels*** (picture elements) are the "dots" or "bits" in the monitor that make up the display screen. When drawing vector objects in AutoCAD, your monitor uses pixels to create a representation of the object on the monitor. However, there is no relationship between the physical pixels in your monitor and a vector object. Pixels simply show the object at the current zoom percentage. Some common vector files are DWG, DXF, AI, and EPS.

Many illustrations created with drawing, painting, and presentation software are saved as raster files. A ***raster file*** creates a picture or image file using the location and color of the screen pixels. In other words, a raster file is made up of "dots." Raster files are usually called ***bitmaps.*** There are several types of raster files used for presentation graphics and desktop publishing. Some common raster file types include TIFF, JPEG, and GIF.

Working with Raster Files

Raster images inserted into AutoCAD drawings are treated much like externally referenced drawings (xrefs). Therefore, they are managed in the **External References** palette. See Figure 20-1. This can be displayed with the **EXTERNALREFERENCES** command or by picking the dialog box launcher button at the lower-right corner of the **Reference** pane on the **Insert** tab of the ribbon.

EXTERNALREFERENCES

Type
EXTERNAL-REFERENCES
ER
XREF
IMAGE
IM

Raster images are not added to the drawing database, but are attached and referenced by a path to the file's location. Any changes to the image content must be made to the original file. Settings and commands in AutoCAD can, however, control the portion of the image shown and its appearance. Images can be inserted, removed, and modified using commands found in the **Reference** panel in the **Insert** tab of the ribbon, Figure 20-2. These functions are discussed in detail in this section.

At the top of the **External References** palette is a drop-down list containing buttons for attaching drawings (DWG), image files, DWF files, DGN files, and PDF files. The **File References** area lists all files currently attached to the drawing, whether they

Figure 20-1.
The **External References** palette is used to manage attached images. The tooltip can be configured to display only the file name, a preview, details, or both a preview and details, as shown here.

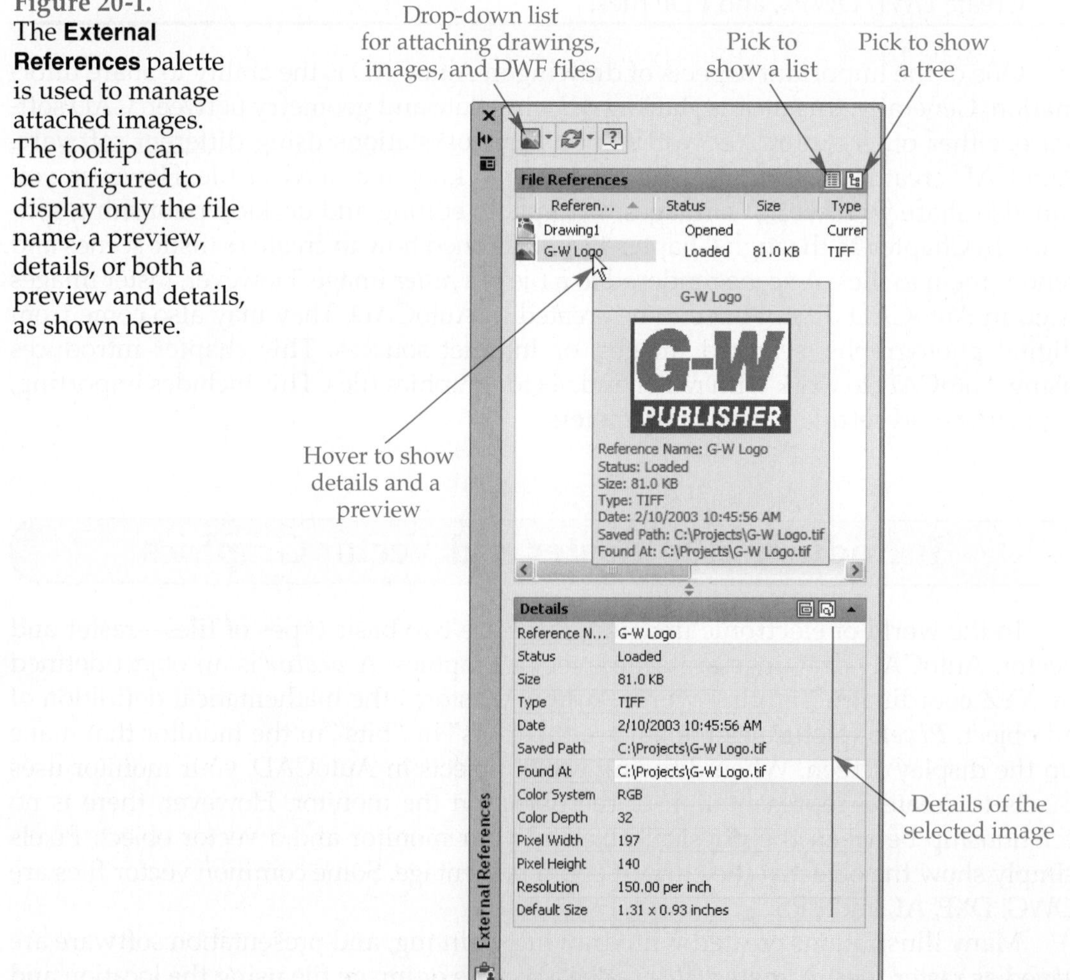

Figure 20-2.
Image commands located on the **Reference** panel in the ribbon.

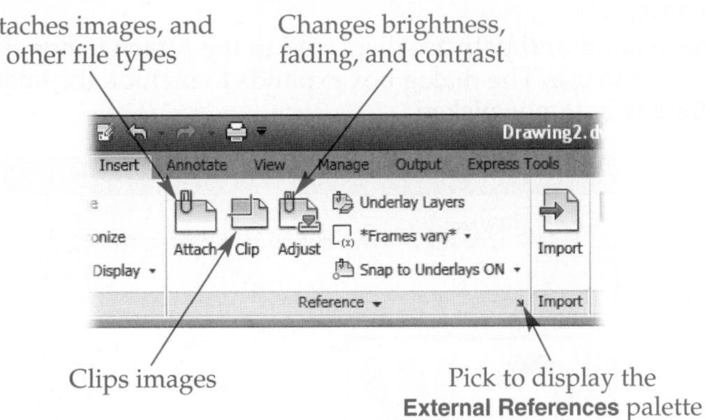

are drawings, images, DWF, DGN, or PDF files. Right-clicking on an entry displays a shortcut menu that allows you to open, attach, unload, reload, and detach the files. If you hover the cursor over an entry in the palette, a preview window is displayed as help text. By default, this preview provides details related to the file and a preview of the file.

There are many different types of raster files. Some raster files used in industry today are:

- **Tagged Image File Format (TIFF).** A file format developed by Aldus Corporation and Microsoft Corporation. This is one of the most commonly used image file types.
- **Joint Photographic Experts Group (JPEG).** A highly compressed graphics image file. This type of file is very common on websites. Also known as a JPG file.
- **Graphics Interchange Format (GIF).** A file format developed to allow the exchange of graphic images over an online computer service, such as the Internet. This type of file is sometimes found on websites, often animated.
- **Portable Network Graphics (PNG).** Developed in the mid 1990s as a replacement for the GIF format. This file type is extensively used for electronic transmission, such as via e-mail or as website graphics.

Other raster file types can also be imported into AutoCAD. If you have a raster image that cannot be directly imported, you will need to first import the file into a paint or draw program. Then, export the image in a format that AutoCAD can read.

Inserting Raster Images

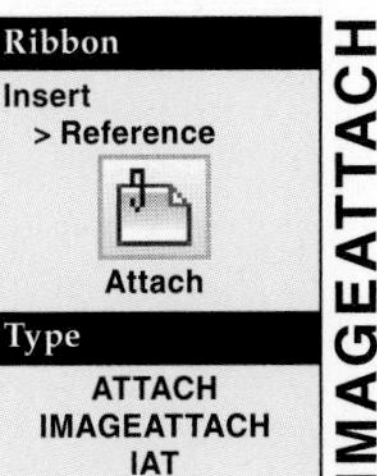

IMAGEATTACH

The **IMAGEATTACH** command is used to attach an image file to a drawing. When the command is selected, the **Select Reference File** dialog box is displayed. Next, navigate to the folder containing the raster file, select the file, and pick **Open**. This displays the **Attach Image** dialog box, Figure 20-3. When the **OK** button is picked and the image placed, it is displayed in the drawing area. See Figure 20-4.

The image name is displayed at the top of the dialog box. You can choose to save the full path, a relative path, or no path. The type of path is selected in the **Path type** drop-down list.

A ***full path*** specifies the complete location of the image file, such as c:\images\building.tif. If the image file is moved from this location, AutoCAD cannot find it.

A ***relative path*** specifies the location of the image file based on the location of the drawing file. For example, the path .\images tells AutoCAD that the image file is located in a subfolder (named images) of the folder where the drawing is located. The entry ..\images tells AutoCAD to look for the file by moving up one folder from where the drawing is stored and then in the subfolder \images. The entry ..\..\images tells AutoCAD to move up two folders and then look in the subfolder \images. The current drawing must be saved in order to specify a relative path.

Figure 20-3.
The image name and path are displayed in the **Attach Image** dialog box. Be sure to select the type of path to use. The dialog box expands to include the **Image Information** area when the **Show Details** button is picked.

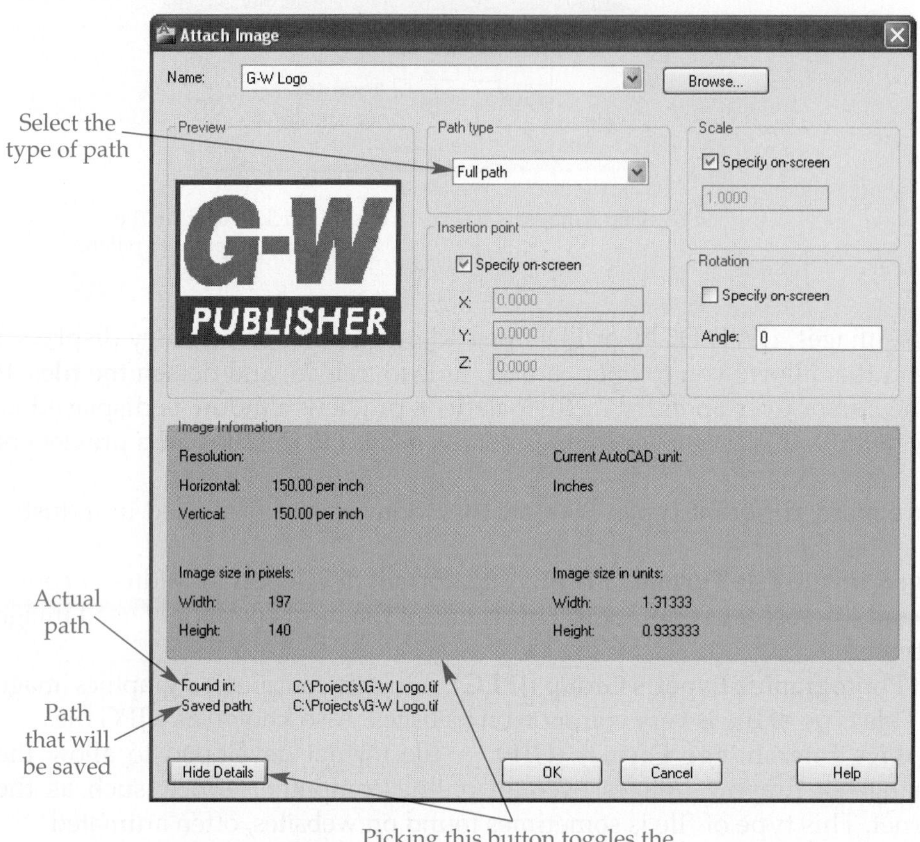

Figure 20-4.
The raster image attached to an AutoCAD drawing.

The *no path* option tells AutoCAD that the image is located in the same folder as the drawing. If the image file is not found in that folder, AutoCAD looks in the path specified by the **PROJECTNAME** system variable, then in the support files search path defined in the **Files** tab of the **Options** dialog box.

The path to the image file is displayed at the bottom of the **Image Information** area. Also displayed is the path that will be saved in the drawing. The **Image Information** area is displayed by picking the **Show Details** button at the bottom of the dialog box.

You can preset image parameters (insertion point, scale, and rotation) or choose to specify them on-screen. Image resolution information is displayed in the **Image Information** area of the expanded dialog box. See **Figure 20-3.**

PROFESSIONAL TIP

If you are working on a project that uses xrefs and attached images, adding a "project subfolder" below the folder where the drawings are stored may be beneficial. Then, use relative paths when inserting images or xrefs. This allows all related files for a project to be found by AutoCAD, even if the folder structure is moved to a different drive or "root" folder.

Managing Attached Images

As stated earlier, the **External References** palette is used to control the raster images inserted into a drawing. When you hover over the file name in the **File References** area, the preview window displays the image name, its status (loaded or unloaded), file size, type, date the image was last saved, and the saved path. Refer to **Figure 20-1.**

Right-clicking on the image name displays a shortcut menu containing options to help you manage the image. The five options are:

- **Open.** This opens the image in the program associated with the file type of the image. For example, if Microsoft Photo Editor is associated with the TIFF file type, the TIFF image is displayed in this program.
- **Attach.** This opens the **Attach Image** dialog box, discussed in the previous section, for attaching an additional image to the drawing.
- **Unload.** Unloads the selected image, but retains its path information. The Status column displays Unloaded if this option is selected. Display the list view to see the columns. An unloaded image is displayed as a frame until reloaded.
- **Reload.** Reloads the selected image file.
- **Detach.** Removes, or detaches, the selected image file from the drawing.

Right-clicking in **File References** area, but not on a file name, displays a different shortcut menu. See **Figure 20-5.** This shortcut menu contains ten options:

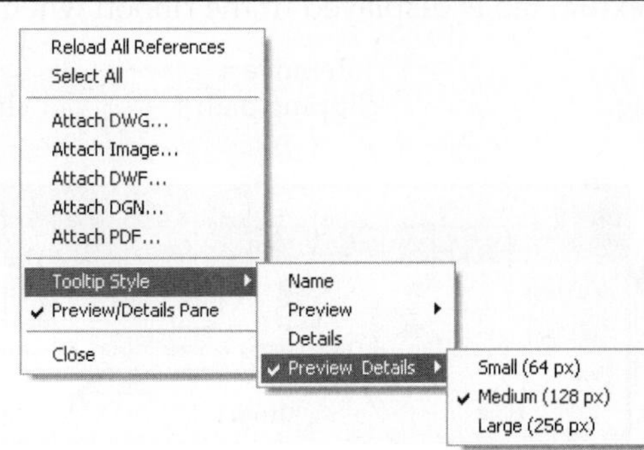

Figure 20-5.
The tooltip in the **External References** palette can be configured to display various information. The settings are located in this shortcut menu.

- **Reload All References.** Reloads any files attached to the current drawing.
- **Select All.** Selects all of the files listed in the **File References** area.
- **Attach DWG.** Allows you to attach other drawings as xrefs.
- **Attach Image.** Allows you to attach additional images.
- **Attach DWF.** Allows you to attach a DWF file as an xref.
- **Attach DGN.** Allows you to attach a DGN file (Microstation drawing) as an xref.
- **Attach PDF.** Allows you to attach a PDF (Adobe Acrobat) file.
- **Tooltip Style.** The tooltip (help text) that appears when you hover over one of the items is customizable. The tooltip can display the file name only, a preview only in the size you select, file details (including the name), or details and a preview.
- **Preview/Details Pane.** Turns on or off the pane at the bottom of the **External References** palette. If the tooltip is set to display details and a preview, you probably will not need to display the preview/details pane.
- **Close.** Closes the **External References** palette.

PROFESSIONAL TIP

If drawing-regeneration time is becoming long, unload attached image files that are not needed for the current drawing session.

Controlling Image File Displays

Once an image is attached to the current drawing, its display can be adjusted if needed. The **IMAGECLIP**, **IMAGEADJUST**, **IMAGEQUALITY**, **TRANSPARENCY**, and **IMAGEFRAME** commands are used to do so. These commands are discussed in this section. All of these commands (except **IMAGEQUALITY**) are available in the **Image** contextual tab. This tab is displayed in the ribbon when the border of the image is selected. See Figure 20-6.

Clipping an image

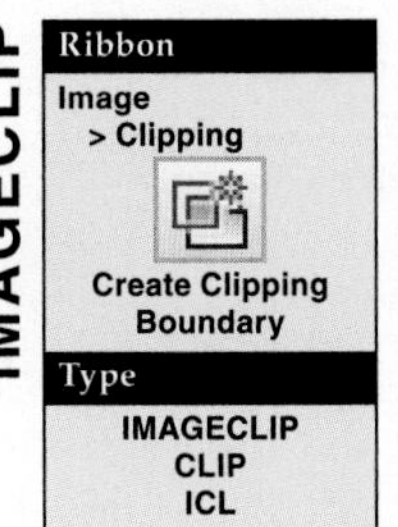

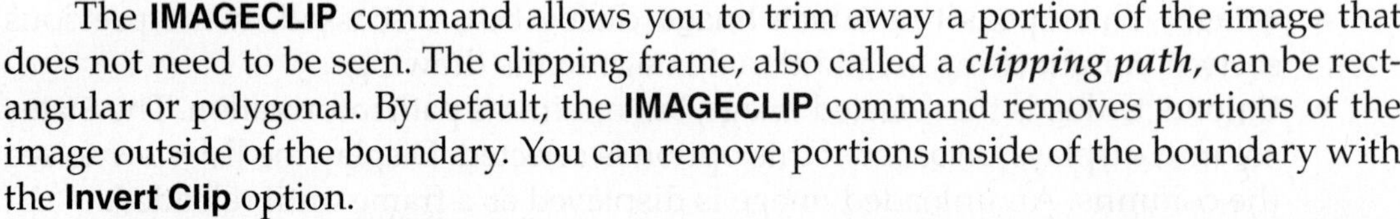

The **IMAGECLIP** command allows you to trim away a portion of the image that does not need to be seen. The clipping frame, also called a ***clipping path,*** can be rectangular or polygonal. By default, the **IMAGECLIP** command removes portions of the image outside of the boundary. You can remove portions inside of the boundary with the **Invert Clip** option.

Once the command is selected, you are prompted to pick the image to clip, unless the image is selected when the command is selected. To create a rectangular clipping frame, continue:

Figure 20-6.
The **Image** contextual tab is displayed in the ribbon when an image is selected.

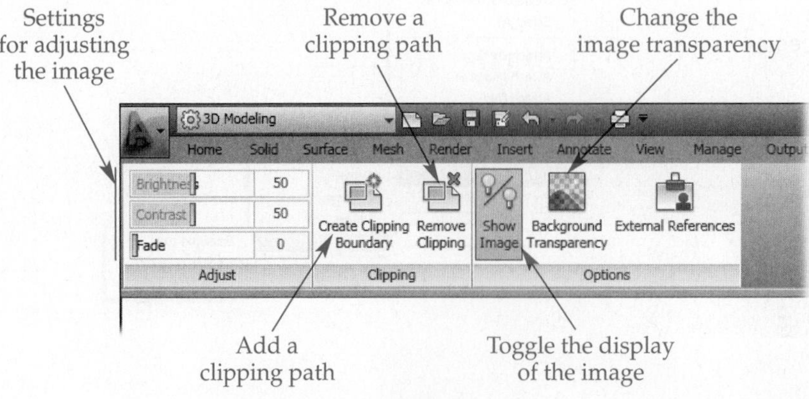

Outside mode – Objects outside boundary will be hidden.
Specify clipping boundary or select invert option:
[Select polyline/Polygonal/Rectangular/Invert clip] <Rectangular>: **R**↵
Specify first corner point: *(pick the first corner of the clipping boundary)*
Specify opposite corner point: *(pick the second corner)*

The image outside of the rectangular frame is hidden. The image is selected and grips are displayed at the corners with an arrow grip on the left-hand edge. Pick and drag the corner grips to change the size of the clip boundary. Picking the arrow grip inverts what is hidden. This is the same as using the **Invert Clip** option.

The **Polygonal** option allows you to construct a clipping frame composed of three or more points. Select the command, pick the image to clip, and continue:

Enter image clipping option [ON/OFF/Delete/New boundary] <New>: **N**↵
Outside mode – Objects outside boundary will be hidden.
Specify clipping boundary or select invert option:
[Select polyline/Polygonal/Rectangular/Invert clip] <Rectangular>: **P**↵
Specify first point: *(pick first point to be used for the clipping boundary)*
Specify next point or [Undo]: *(pick second point)*
Specify next point or [Undo]: *(pick third point)*
Specify next point or [Close/Undo]: *(pick additional points as needed)*
Specify next point or [Close/Undo]: ↵

The image outside of the polygonal frame is hidden. The grips function the same as they do with a rectangular boundary. Figure 20-7 shows the results of using the **Rectangular** and **Polygonal** options of the **IMAGECLIP** command on a raster image. Three additional options of **IMAGECLIP** allow you to work with the display of the clipped image.

- **ON.** Turns the clipping frame on to display only the clipped area.
- **OFF.** Turns off the clipping frame to display the entire original image and frame.
- **Delete.** Deletes the clipping frame and displays the entire original image.

NOTE

You can pick an unclipped image frame to display the grips for editing. The grips are attached to the image itself. If one grip is stretched, it affects the entire image by proportionally enlarging or reducing it. On the other hand, if you select a clipped image for grip editing, the grips are attached to the clipping frame. Stretching the clipping frame does not change the size or shape of the image, but alters the frame and retains the size of the image.

Figure 20-7.
A—A rectangular image clip. B—A polygonal image clip. The path is shown here in color for illustration.

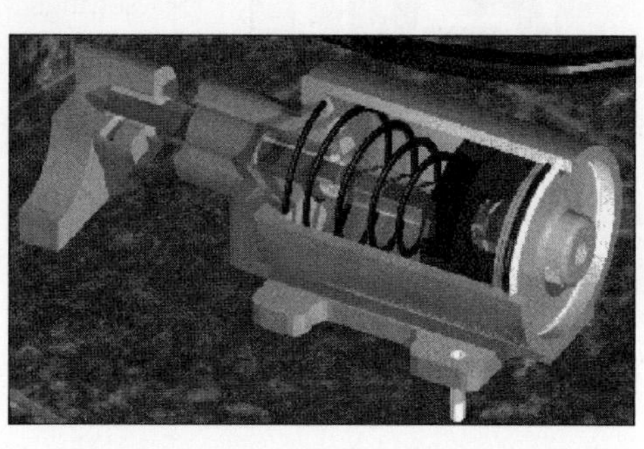

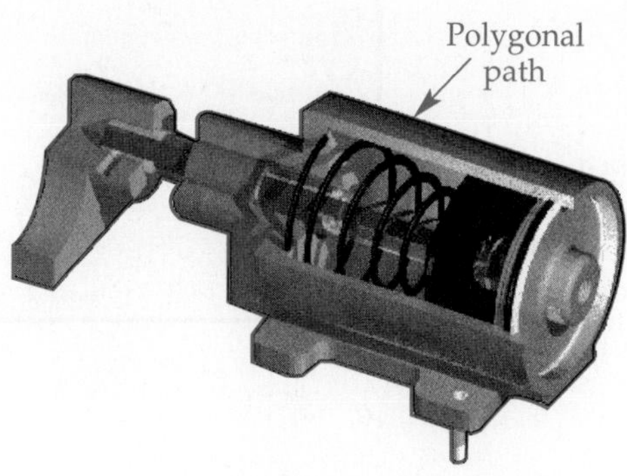

A B

Adjusting an image

IMAGEADJUST

Ribbon
Image
> Adjust
Type
IMAGEADJUST
IAD

The **IMAGEADJUST** command provides control over the brightness, contrast, and fade of the image. These adjustments are made in the **Adjust** panel on the **Image** contextual tab in the ribbon, **Figure 20-6.** The adjustments are applied to all selected images.

If the command is accessed by typing, the **Image Adjust** dialog box is used to make adjustments, **Figure 20-8.** Once the command is selected, you are prompted to pick an image, unless the image is selected before the command is accessed. If you want the same settings applied to multiple images, you can pick them all at the same time. When done picking objects, press [Enter] to display the dialog box.

Values can be changed by typing in the text boxes or by using the slider bars in either the ribbon or dialog box. The preview tile in the dialog box dynamically changes as the sliders are moved. Picking the **Reset** button in the dialog box returns all values to their defaults.

- **Brightness.** Controls pixel whiteness and indirectly affects the contrast. Values can range from 0 to 100, with 50 as the default value. Higher values increase the brightness.
- **Contrast.** Controls the contrast of the image, or how close each pixel is moved toward its primary or secondary color. Values can range from 0 to 100, with 50 as the default value. Higher values increase the contrast.
- **Fade.** Controls the fading of the image, or how close the image is to the background color. Values can range from 0 to 100, with 0 as the default value. Higher values increase the fading.

IMAGEQUALITY

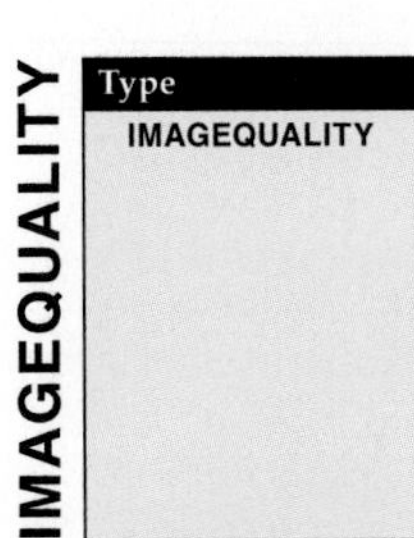

The **IMAGEQUALITY** command provides two options: **High** and **Draft**. The high quality setting produces the best image display, but requires more time to regenerate. If you are working with several images in a drawing, it is best to set the **Draft** option current. The image displayed is lower quality, but requires less time to display. The setting applies to all images in the drawing.

Transparency

TRANSPARENCY

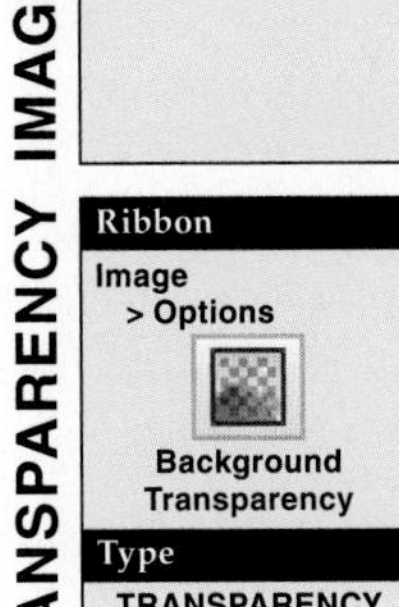

Some raster images have transparent background pixels. The **TRANSPARENCY** command controls the display of these pixels. If **TRANSPARENCY** is on, the drawing will show through the image background. Images are inserted with this feature turned off. The setting applies to individual images. Multiple images can be selected at the same time. Remember, only images containing transparent pixels are affected.

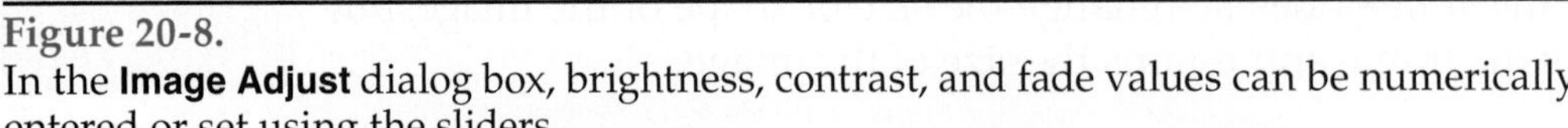

Figure 20-8.
In the **Image Adjust** dialog box, brightness, contrast, and fade values can be numerically entered or set using the sliders.

Image frame

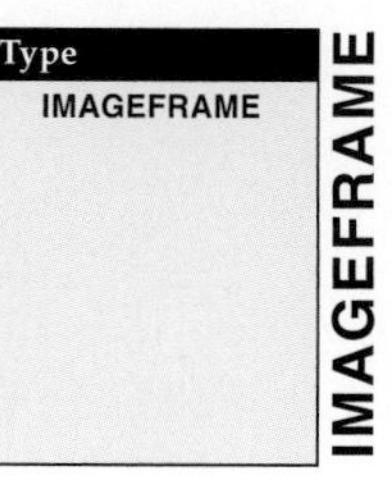

The **IMAGEFRAME** command controls the appearance of frames around all images in the current drawing. When attaching (inserting) images, AutoCAD places a frame around the image in the current layer color and linetype. There are three settings for the **IMAGEFRAME** command. The default setting of 1 turns on the display of the frame and allows the frame to be plotted. A setting of 2 turns on the display of the frame, but the frame is not plotted. A setting of 0 turns off the display of the frame. It is not plotted and cannot be selected for editing. The setting applies to all images in the drawing.

Uses of Raster Files in AutoCAD

One use of raster images is as a background for sketching or tracing. For example, you may need a line drawing of an image that is only available as a continuous tone (print) photograph. The photo can be scanned, which produces a raster image. After importing the raster image with the **IMAGEATTACH** command, use the appropriate drawing commands to sketch or trace the image. After the object is sketched, the original raster image can be deleted, frozen, or unloaded, leaving the tracing. You can then add other elements to the tracing to create a full drawing. See **Figure 20-9.**

Raster files can be combined with AutoCAD drawing and modeling features in many ways to complete or complement the design. For example, company watermarks

Figure 20-9.
Using a raster image as a model for a drawing. A—The imported raster image. B—Use AutoCAD commands to trace the image. Then, either delete the image or freeze its layer. C—The completed drawing plotted on a title block.

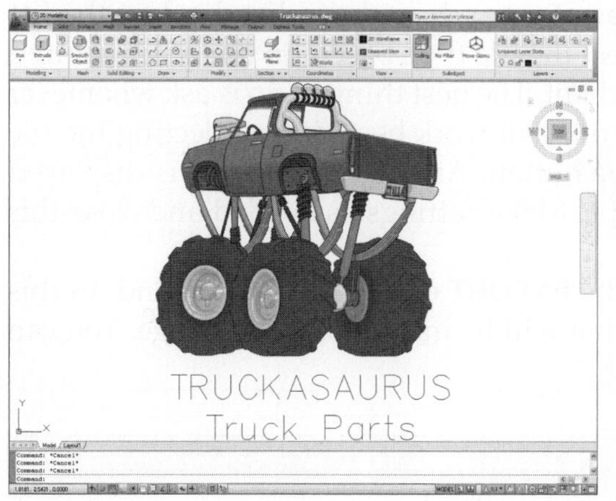

A

B

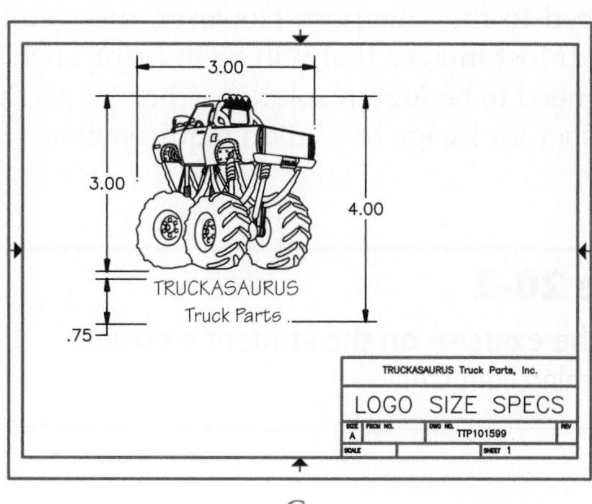

C

or logos can be easily added to title blocks, style sheets, and company drawing standards. Drawings that require designs, labels, and a variety of text fonts can be created using raster files in conjunction with the wide variety of TrueType fonts available with AutoCAD. Archived manual drawings can also be scanned, brought into AutoCAD, and then traced to create a CAD drawing.

You can add features to complement raster files. For example, you can import a raster file, dimension or annotate it, and even add special shapes to it. Then, export it as the same type of file. Now, you can use the revised file in the original software in which it was created. As with any creative process, let your imagination and the job requirements determine how you use this capability of AutoCAD.

Exercise 20-1

Complete the exercise on the student website.
www.g-wlearning.com/CAD

Exporting a Drawing to a Raster File

You can save a rendering to a raster file. This is discussed in Chapter 18. However, 2D objects are not rendered and, therefore, do not appear in the file. If you want what is displayed in the current viewport, including 2D objects, saved as a raster file, you must use the **SAVEIMG** command. This command saves the current AutoCAD viewport as an image file. What you see in the viewport is what you will get in the file, including the effect of the current visual style.

Once the command is selected, the **Render Output File** dialog box is displayed. This is a standard save dialog box. The **Files of type:** drop-down list displays the file types to which the image can be saved. Select the file type based on the type required for a particular process, application, job, or client. The best thing to do is ask whomever you are creating the file for which type of file will work best. After selecting the file type, give the file a name and pick the **Save** button. Another dialog box is displayed that contains settings specific to the file type. Make settings as needed and close this dialog box to save the file.

A BMP file can also be created using the **EXPORT** or **BMPOUT** command. In this manner, you can select individual objects that will be included in the image. You can also save shaded images with this method.

PROFESSIONAL TIP

You can use the **SAVEIMG** command to save a rendering that was rendered to the viewport. However, the result is a low-resolution image. Most images that will be in print, such as in a magazine or book, need to be high resolution. Check with the magazine or book publisher for image resolution requirements.

Exercise 20-2

Complete the exercise on the student website.
www.g-wlearning.com/CAD

Working with Vector Files

A vector file contains objects defined by XYZ coordinates. AutoCAD's native file format (DWG) is a vector file. You can also work with other vector file types. These types include DXF, WMF, SAT, EPS, STL, and DXX. The two most commonly used types, DXF and WMF, are covered in the next sections.

Exporting and Importing DXF Files

DXF is a generic file type that defines AutoCAD geometry in an ASCII text file. Other programs that recognize the DXF format can then "read" this file. The DXF file format retains the mathematical definitions of AutoCAD objects in vector form. The DXF objects imported into other vector-based programs, or opened in AutoCAD, can be edited as needed.

Exporting DXF files

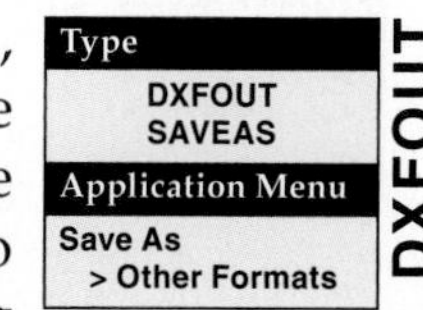

The **DXFOUT** command is used to save a DXF file. Once the command is selected, the **Save Drawing As** dialog box is displayed. See Figure 20-10. Select the DXF file type from the **Files of type:** drop-down list. Name the file and specify a location where you want to save it. Notice that you can select different versions of DXF. This is to ensure that the file you save is "backward compatible." For example, if you are sharing the file with somebody using AutoCAD 2000, save the DXF as that version to ensure AutoCAD 2000 can read the file.

Since this is a "save as" operation, the current drawing is saved as a DXF file. If you continue to work on the drawing, you are working on the DXF version, *not* the DWG version. In order to work on the original drawing, you must open the DWG file. However, if you continue to work on the drawing in DXF form and attempt to save or close the drawing, the **Save Drawing As** dialog box is displayed. You can save the drawing as a DWG or replace the previously saved DXF file. If you save the drawing as a DXF file, you are also informed that the drawing is not a DWG and given the opportunity to save it in the DWG format.

When a DXF file is saved, all geometry in the drawing is saved, regardless of the current zoom percentage or selected objects. The DXF file format saves any surfaced or

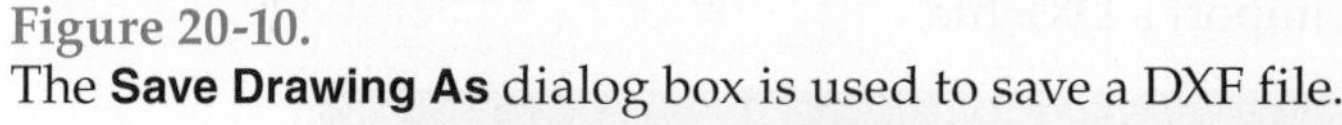

Figure 20-10.
The **Save Drawing As** dialog box is used to save a DXF file.

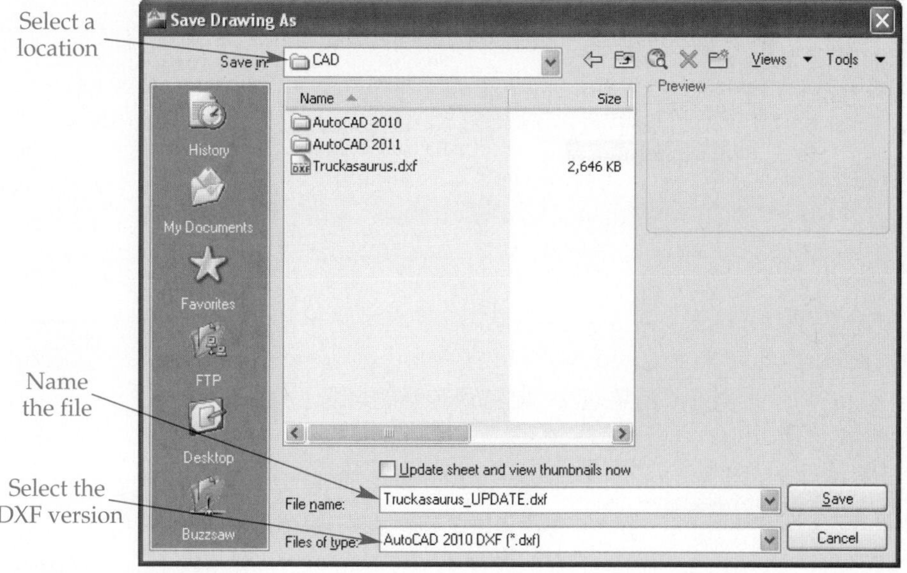

solid 3D objects as 3D geometry. When a DXF file containing 3D geometry is opened, the surfaced or solid 3D geometry remains intact. In addition, the current visual style is saved in the DXF file.

NOTE

Not all programs that can import DXF files are capable of correctly "reading" 3D objects or the visual style.

Importing DXF files

DXFIN

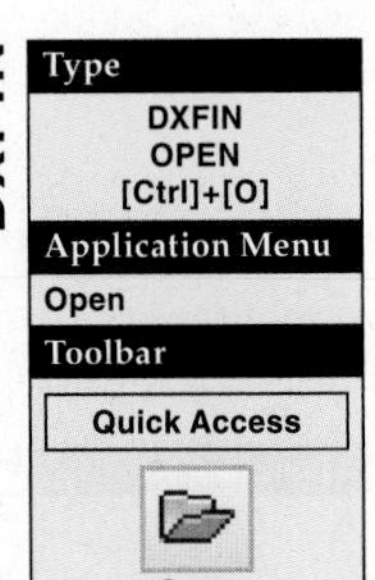

To open a DXF file, use the **DXFIN** command. Once the command is selected, the **Select File** dialog box is displayed, Figure 20-11. Select DXF (*.dxf) from the **Files of type:** drop-down list. Then, select the DXF file you want to open. Notice that there is no preview when the file is selected. AutoCAD does not support previews for the DXF file type. Finally, pick the **Open** button.

The DXF file is opened in a new document window. To place a DXF file into the *current* drawing, insert it as a block. If you do not want it inserted as a block, open the file (**DXFIN**), copy it to the clipboard ([Ctrl]+[C]), and paste ([Ctrl]+[V]) it into the current drawing.

NOTE

If you open a DXF file and try to save it, the **Save Drawing As** dialog box appears. You can save it as DXF, overwriting the existing file, or under a new name or as another file type.

Exporting and Importing Windows Metafiles

The Windows metafile (WMF) format is often used to exchange data with desktop publishing programs. It is a vector format that can save wireframe and hidden displays. Shaded and rendered images cannot be saved. Also, perspective views are saved in parallel projection.

Figure 20-11.
The **Select File** dialog box is used to import a DXF file.

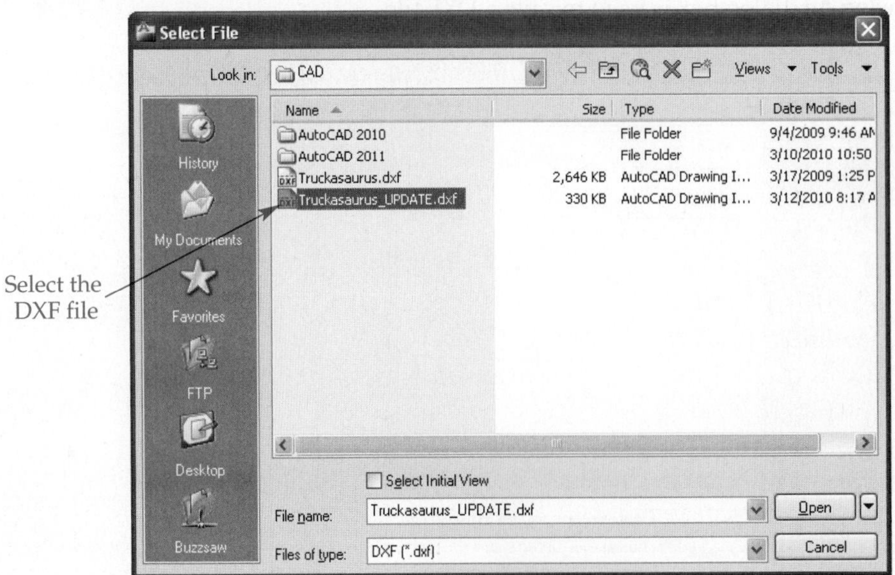

A WMF file cannot retain the definition of all AutoCAD object types. For example, circles are translated to line segments. Also, a WMF file does *not* save three-dimensional data. The view in the current viewport is projected onto the viewing plane and saved as a two-dimensional projection, similar to the result produced by the **FLATSHOT** command.

Exporting WMF files

The **WMFOUT** command is used to create a WMF file. When the command is selected, the **Create WMF File** or **Export Data** dialog box is displayed. These are standard save dialog boxes. Select Metafile (*.wmf) in the **Files of type:** drop-down list. After specifying the file name and folder location and picking the **Save** button, you must select the objects to place in the file. Press [Enter] when all of the objects are selected and the WMF file is saved.

Only the portions of selected objects that are visible on-screen are written into the file. If part of a selected object is not visible on screen, that part is "clipped." Also, the current view resolution affects the appearance of a Windows metafile. For example, when **VIEWRES** is set low, circles in your AutoCAD drawing may look like polygons in the WMF file. When saved to a Windows metafile, curved objects are composed of line segments rather than defined as circles or arcs.

Importing WMF files

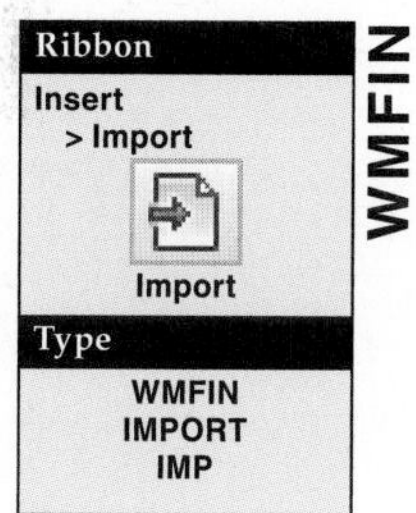

Use the **WMFIN** command to import a Windows metafile into a drawing. When the command is selected, the **Import WMF** or **Import File** dialog box is displayed. Select Metafile (*.wmf) in the **Files of type:** drop-down list, then select a file.

A Windows metafile is imported as a block consisting of all of the objects in the file. You can explode the block if you need to edit the objects within it. If an object is not filled, it is created as a polyline when brought into AutoCAD. This includes arcs and circles. Objects composed of several closed polylines to represent fills are created from solid-fill objects, as if created using the **SOLID** command with the **FILL** system variable off.

There are two settings used to control the appearance of Windows metafiles imported into AutoCAD. Type WMFOPTS to display the **WMF In Options** dialog box, Figure 20-12. You can also pick the **Options…** button in the **Tools** drop-down menu in the "import" dialog box. The **WMF In Options** dialog box contains the following two check boxes.

- **Wire Frame (No Fills).** When checked, filled areas are imported only as outlines. Otherwise, filled areas are imported as filled objects (when **FILL** is on).
- **Wide Lines.** When this option is checked, the relative line widths of lines and borders from the WMF file are maintained. Otherwise, they are imported using a zero width.

PROFESSIONAL TIP

PostScript is a copyrighted page description language developed by Adobe Systems. PostScript files are widely used in desktop publishing. AutoCAD can export PostScript files with the **PSOUT** or **EXPORT** commands. In addition, AutoCAD has several PostScript patterns that can be used as fills for closed polylines. The **PSFILL** command is used to add these patterns. However, professionals rarely use AutoCAD's PostScript functions. AutoCAD cannot import, view, or print PostScript files. In addition, any drawing that contains PostScript patterns (fills) must be sent to a printer or plotter that is PostScript compatible. These printers are not common outside of the graphics industry.

Figure 20-12.
Setting options for imported WMF files.

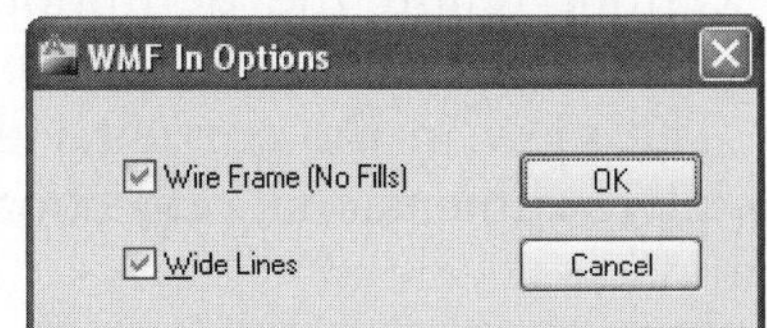

Exercise 20-3

Complete the exercise on the student website.
www.g-wlearning.com/CAD

DWF and PDF Files

Often it is necessary to share AutoCAD drawing data with other "non-AutoCAD" systems. Drawings saved as DWF and PDF files are easily attached to an e-mail or saved to a company server where others can access them. With the associated viewing software, these files can be opened, displayed, and printed as desired without the use of AutoCAD.

You can save an AutoCAD drawing as a ***Design Web Format (DWF)*** file. A DWF file is a highly compressed, vector file that can be viewed using the Autodesk Design Review or Autodesk DWF Viewer. In addition, when one of these programs is installed in conjunction with a supported Internet browser, you can view DWF files on the web. Autodesk Design Review is installed when AutoCAD is installed or it can be downloaded for free from the Autodesk website.

You can also save a drawing in DWFx format. A DWFx file serves the same purpose as a "standard" DWF file, but it is based on Microsoft's XML Paper Specification (XPS) format. DWFx files can be viewed with the Autodesk Design Review or with the Microsoft XPS Viewer, which is supplied with the Microsoft Vista and Windows 7 operating systems. The process for creating and using DWFx files is the same as for DWF files. When a DWF or DWFx file is opened in Autodesk Design Review, the view cube and orbit commands are accessible to view the 3D object from different angles.

You can export an AutoCAD drawing as a ***Portable Document Format (PDF)*** file. A PDF file is a vector-based file, like DWFx and DWF. Anyone can view them using Adobe Reader. This is a free utility that can be downloaded from the Adobe website.

Drawings are saved as DWF, DWFx, or PDF files using either the **3DDWF**, **EXPORT**, or **PUBLISH** command. If you are creating a single-sheet file, use the **3DDWF** or **EXPORT** command. If you are creating a multi-sheet file, use the **PUBLISH** command.

PROFESSIONAL TIP

The **ETRANSMIT** command prepares a transmittal for e-mail that contains font files, plot styles, table files, and xrefs associated with the drawing. This can be a time-saving feature if you share drawings that contain xrefs.

3DDWF
Ribbon
Output
> Export to DWF/PDF
3D DWF
Type
3DDWF

Exporting to DWF and DWFx

The **3DDWF** command must be used to retain the 3D data in the DWF or DWFx file. When the **3DDWF** command is launched, the **Export 3D DWF** dialog box is displayed. See Figure 20-13A. This is a standard save dialog box. You can select either the DWFx

Figure 20-13.
A—The **3DDWF** command displays a standard save dialog box. You can set options using the **Tools** pull-down menu. B—Setting options for 3D DWF files.

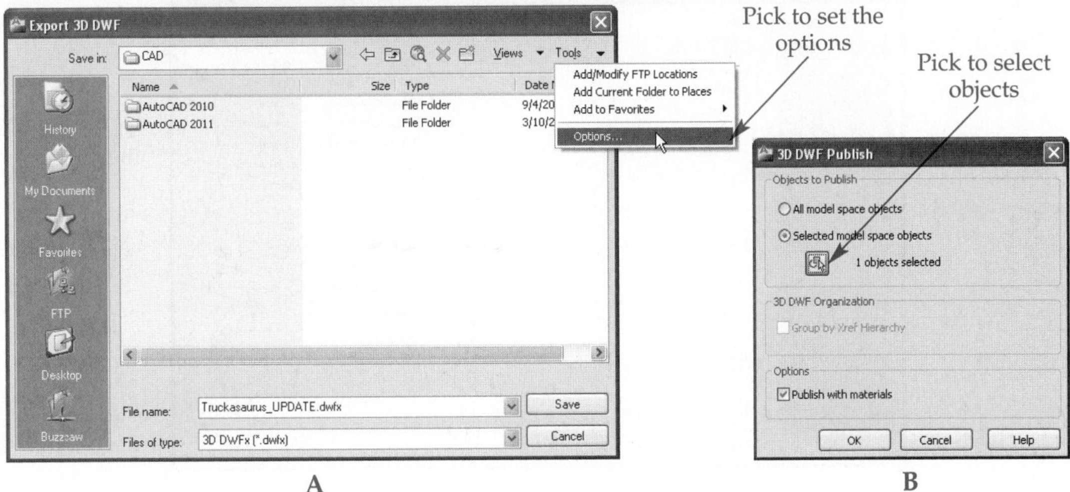

or DWF format from the **Files of type:** drop-down list. Then, enter a file name and select a location.

If you save the file at this point, all of the objects in model space will be exported. Selecting **Options...** in the **Tools** pull-down menu in this dialog box displays the **3D DWF Publish** dialog box, Figure 20-13B. In this dialog box, you can select which objects you wish to export and also whether or not to include materials.

To create the file, pick the **Save** button in the **Export 3D DWF** dialog box. Once the file is saved, a message appears indicating this fact and offering you the opportunity to view the file.

Settings for Exporting to DWF, DWFx, and PDF Files

The **Export to DWF/PDF** panel in the **Output** tab on the ribbon provides different options for exporting to DWF, DWFx, or PDF files, Figure 20-14. The export flyout button contains buttons for exporting to each of these file formats. Each button opens a standard save as dialog box with only the selected file type available. All other options are the same. The **Save As DWFx** dialog box is shown in Figure 20-15.

You can select what to export in either the panel or the dialog box. Your options are display, extents, or window. If you select Current for the page setup, AutoCAD creates the exported file with whatever page setup is current in the drawing. Selecting Override enables up the Page Setup Override... button (in the dialog box), which displays

Figure 20-14.
The **Export to DWF/PDF** panel provides tools for exporting the AutoCAD drawing to DWF, DWFx, and PDF formats.

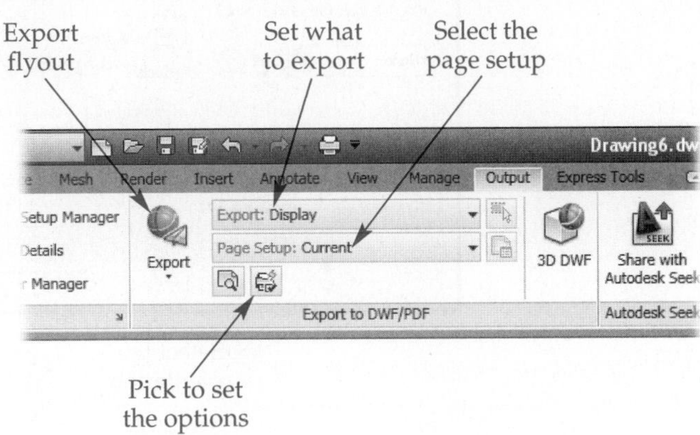

Figure 20-15.
The **Save As DWFx** dialog box is used to save a drawing as a DWFx file.

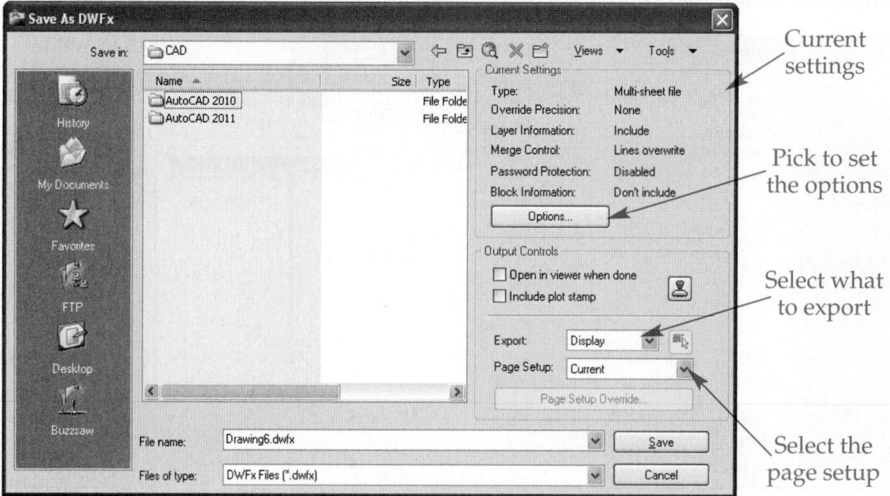

the **Page Setup Override** dialog box. See Figure 20-16. The settings in this dialog box should be familiar as they are the same as the settings in the **Plot** dialog box.

Picking the options button on either the ribbon or in the save as dialog opens the **Export to DWF/PDF Options** dialog box. See Figure 20-17A. This dialog box contains the following settings.

- **Location.** This is where the file will be saved. You can change the location by selecting the property and picking the browse button (...) that is displayed in the property.
- **Type.** Select either a single sheet or multi-sheet file.
- **Override Precision.** The precision is the resolution of the file. Select from a preset precision or select Manage Precision Presets… to open the **Precision Presets Manager** dialog box in which you can create new presets, Figure 20-17B. This is discussed in more detail below.
- **Naming.** Select either to enter a name for the file or to be prompted for one later.

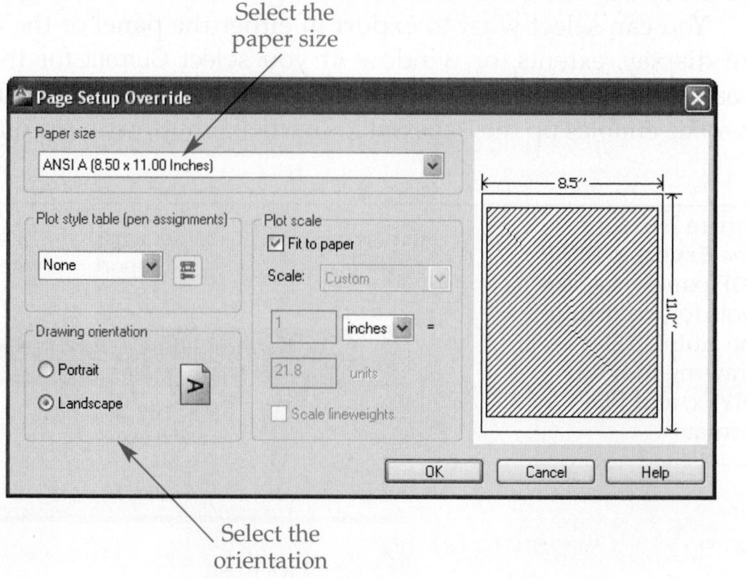

Figure 20-16.
The **Page Setup Override** dialog box has options similar to the **Plot** dialog box.

Figure 20-17.
A—Setting options in the **Export to DWF/PDF Options** dialog box. B—The **Precision Presets Manager** is used to create new presets.

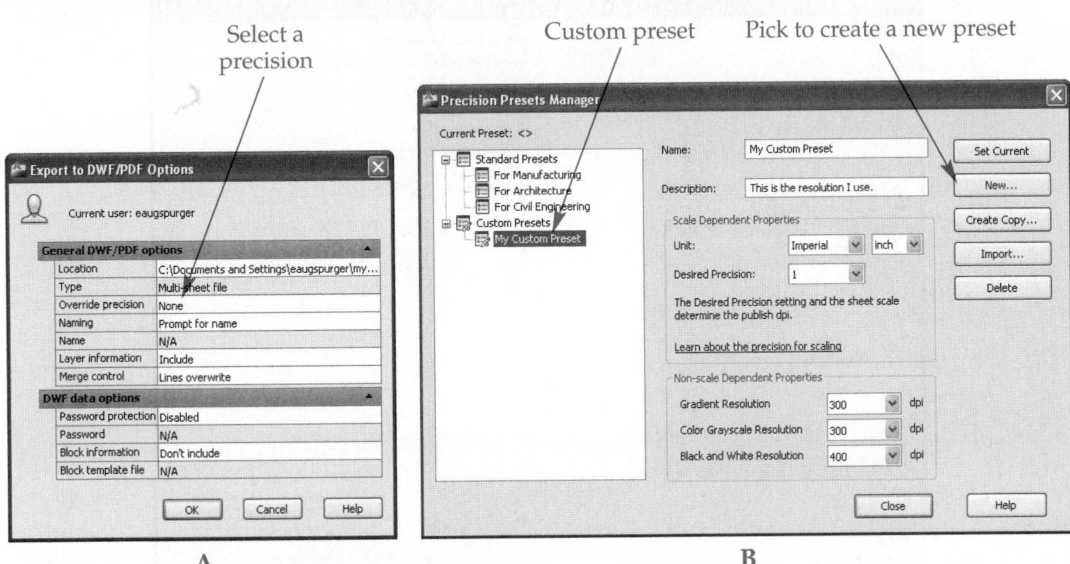

- **Name.** This property is read-only (N/A) unless you have selected to specify a name. In this case, enter the file name here.
- **Layer information.** You can elect to include or not include in the file information about the layers in the drawing.
- **Merge control.** This property determines how the colors of overlapping lines are handled. You can choose to merge (blend) the colors or to have the color of the last plotted line overwrite all other colors.
- **Password protection.** There are three choices for password-protecting the file. The file can be unprotected (password disabled), you can be prompted for a password later, or you can specify a password in this dialog box.
- **Password.** This property is read-only (N/A) unless you have selected to specify a password. In this case, enter the password here.
- **Block information.** You can choose to include or not include in the file attribute information for blocks in the drawing.
- **Block template file.** This property is read-only (N/A) unless you have selected to include block information. If block attribute information is to be included, this property displays a drop-down list and a browse button (...) so that you can either select, create, or edit a block template (DXE) file.

The Override precision property contains standard precision presets for manufacturing, architectural, or civil drafting disciplines. This helps to create scaled prints when the DWF or PDF files are printed. The Manage Precision Presets... option opens the **Precision Presets Manager** dialog box, shown in Figure 20-17B. In this dialog box, you can view the discipline-specific presets or create new custom presets with your own settings. The default presets are read-only and cannot be changed. The properties that do not depend on scale—gradient, color grayscale, and black and white resolutions—determine the quality of the resulting DWF or PDF files.

Exercise 20-4

Complete the exercise on the student website.
www.g-wlearning.com/CAD

Figure 20-18.
Using the **PUBLISH** command to create a DWF, DWFx, or PDF file.

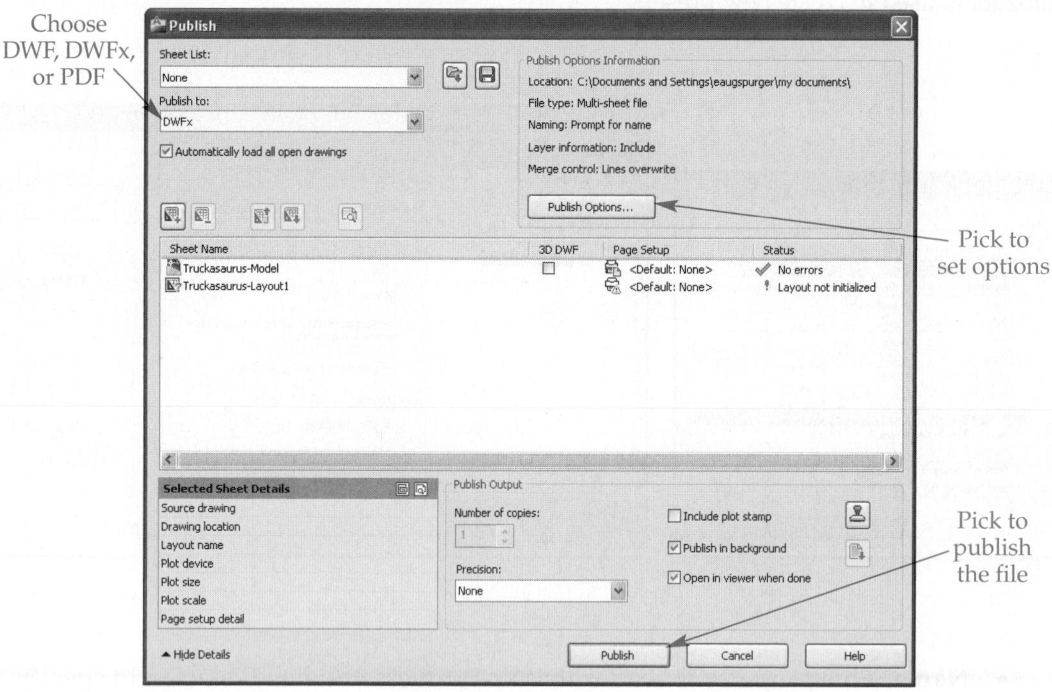

Publishing DWF, DWFx, and PDF Files

To ***publish*** a DWF, DWFx, or PDF file, first enter the **PUBLISH** command. The **Publish** dialog box is displayed, Figure 20-18. Select the file format in the **Publish to:** drop-down list in the upper-left corner of the dialog box. To change settings for the file that will be published, pick the **Publish Options...** button. This displays the **Publish Options** dialog box. The options in this dialog box are the same as those described above for the **Export to DWF/PDF Options** dialog box. Change any settings as necessary and pick the **OK** button to return to the **Publish** dialog box.

When you are ready to create the file, pick the **Publish** button in the **Publish** dialog box. A standard save dialog box appears. Enter a name for the file, select a location, and pick the **Select** button. You cannot change the file type in this dialog box. Next, you are asked if you want to save the current list of sheets. After you make a selection in this message box, the drawing is published to the file. This may take a few seconds to complete.

NOTE

Commands that control the on-screen display of geometry, such as **VIEWRES**, **FACETRES**, and **DISPSILH**, and the current visual style affect the resulting DWFx or DWF file.

PLOT
Ribbon
Output
> Plot
Plot
Type
PLOT
PRINT
Application Menu
Print
> Plot
Toolbar
Quick Access
Plot

Plotting DWF, DWFx, and PDF Files

Instead of publishing DWF, DWFx, and PDF files, you can *plot* them using the **PLOT** command. Use the command to open the **Plot** dialog box. In the **Name:** drop-down list in the **Printer/plotter** area, select the DWF6 ePlot.pc3 (for DWF files), the DWFx ePlot.pc3 (for DWFx files), or DWG To PDF.pc3 (for PDF files).

As discussed in the previous sections, there are settings that control the final output for these files. After the PC3 file is selected, pick the **Properties...** button in the **Printer/plotter** area. In the **Plotter Configuration Editor** dialog box that is displayed, select Custom Properties in the tree on the **Device and Document Settings** tab. Then, pick the **Custom Properties...** button to display the **DWF6 ePlot Properties**, the **DWFx ePlot Properties**, or the **DWG to PDF Properties** dialog box. The latter is shown in Figure 20-19.

The two "resolution" areas in the properties dialog box have settings that control the accuracy of the resulting file. A medium resolution is best in most cases. A file created with high resolution may be too large for practical electronic transmission. A lower resolution will create a smaller file. Small files make for easy electronic transmission. However, the resulting file may not display as accurately as one created at a higher resolution. You can separately set a maximum resolution for vector graphics, gradients, color/grayscale images, and black and white images. See Figure 20-20 for a comparison of resolution settings in a plotted PDF file. The file size increases as the resolution increases.

Once you have made all settings as needed, pick the **OK** button. Then, pick **OK** in the **Plotter Configuration Editor**. If changes were made to the settings, a dialog box appears asking if you want to apply the changes on a one-time basis or save the configuration to a PC3 plotter configuration file.

Use all of the other settings in the **Plot** dialog box just as you would when plotting a hard copy. Refer to *AutoCAD and Its Applications—Basics* for detailed information on plotting. When you pick the **OK** button to "plot", the **Browse for Plot File** dialog box is displayed. This is a standard save dialog box with only the selected file type available. The default filename is the drawing name and current space name separated by a hyphen. Use that name or enter a new name, navigate to the location where you want to save the file, and pick the **Save** button.

PROFESSIONAL TIP

Using the **PLOT** command to produce DWF, DWFx, or PDF files is a better option than exporting when you want to include information in your layout such as borders, title blocks, notes, dimensions, etc. Also, remember that the floating viewports in a layout may be set to any visual style or rendering preset. This gives you the capability of creating a fully rendered DWF, DWFx, or PDF file from your 3D model.

Figure 20-19.
Setting the properties for a PDF file in the **DWG To PDF Properties** dialog box.

Resolution settings

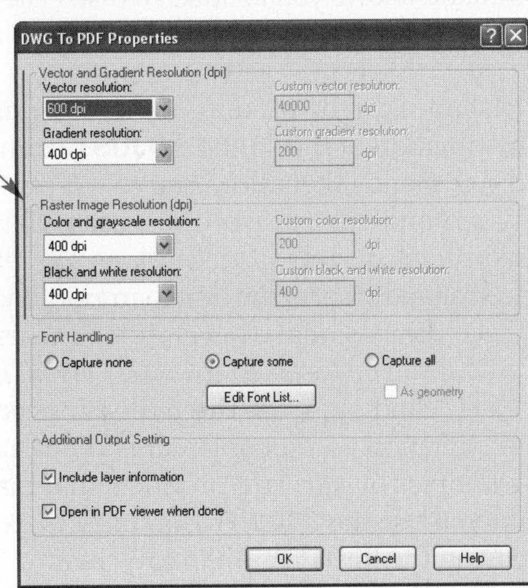

Figure 20-20.
A comparison of low-resolution and high-resolution PDF files. A—The lines in the low-resolution file have jaggies. This is especially apparent on the windows. B—The lines in the high-resolution file are cleaner.

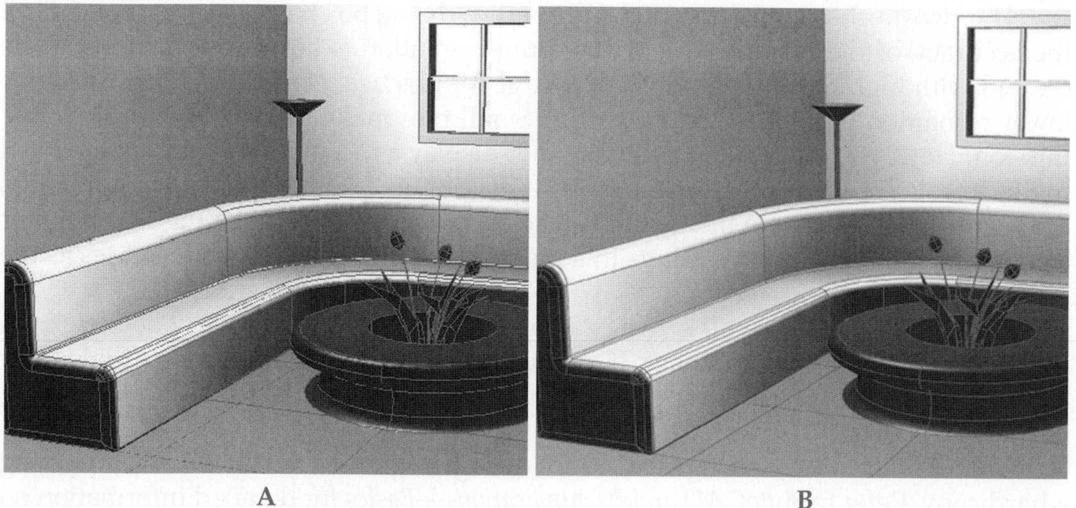

Exercise 20-5

Complete the exercise on the student website.
www.g-wlearning.com/CAD

Chapter Test

Answer the following questions. Write your answers on a separate sheet of paper or complete the electronic chapter test on the student website.
www.g-wlearning.com/CAD

1. Name four common formats of raster images that can be imported into AutoCAD.
2. Which command allows you to attach a raster file to the current AutoCAD drawing?
3. What is the display status in the drawing of an inserted image that has been unloaded?
4. Which two shapes can be used to clip a raster image?
5. What is the function of the **IMAGEADJUST** command?
6. Name two commands that allow you to export bitmap files.
7. Give the name and file type of the vector file that can be exchanged between object-based programs (object definitions are retained), other than DWG.
8. Name the commands that allow you to import and export the file type in question 7.
9. How are three-dimensional objects treated when exported to a WMF file?
10. How can a DXF file be inserted into the *current* drawing?
11. The DWF, _____, and _____ are three file types commonly used to share AutoCAD drawing information with people who do not have AutoCAD.
12. Give two examples of software products that may be used to view DWFx files.
13. Why would you use the **3DDWF** command to export to a DWF or DWFx file?
14. If you are using the **PLOT** command to create a PDF file, what must be selected in the **Name:** drop-down list in the **Printer/plotter** area of the **Plot** dialog box?
15. When plotting to a DWF, DWFx, or PDF file, what is a disadvantage to using high resolution settings? What is a disadvantage of using low resolution settings?

Drawing Problems

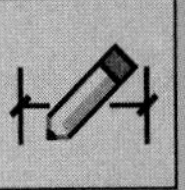

1. Locate some sample raster files with .jpg, .png, or .tif file extensions. These files are often included as samples with software. They can also be downloaded as freeware from the Internet. With the permission of your instructor or supervisor, create a folder on your hard drive and copy the raster files there. Create a new drawing and attach each of the raster files. Place text below each image indicating the source of the image. Save the drawing as P20_01.

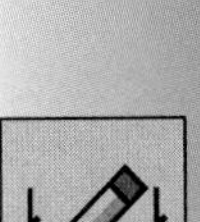

2. Choose a small raster file and attach it to a new AutoCAD drawing.
   A. Insert the image so it fills the entire screen.
   B. Undo and insert the image again using a scale factor that fills half of the screen with the image.
   C. Stretch the original object using grips, then experiment with different clipping boundaries. Stretch the image after it has been clipped and observe the result.
   D. Create a layer named Raster. Create a second layer named Object. Give each layer the color of your choice. Set the current layer to Raster.
   E. Import the same image next to the previous one at the same scale factor.
   F. Set the current layer to Object and use AutoCAD drawing commands to trace the outline of the second raster image.
   G. Unload the raster image or freeze the Raster layer.
   H. Save the drawing as P20_02.

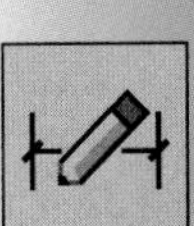

3. For this problem, you will import several raster files into AutoCAD. Then, you will trace the object in each file and save it as a block or wblock to be used on other drawings.
   A. Find several raster files that contain simple objects, shapes, or figures that you might use in other drawings.
   B. Create a template drawing containing Object and Raster layers.
   C. Import each raster file into AutoCAD on the Raster layer using the appropriate command. Set the Object layer current and trace the shape or objects using AutoCAD drawing commands.
   D. Detach the raster information, keeping only the traced lines of the object.
   E. Save the object as a block or wblock using an appropriate file-naming system.
   F. After all blocks have been created, insert each one into a single drawing and label each with its name. Include a path if necessary.
   G. Save the drawing as P20_03.
   H. Print or plot the final drawing.

4. In this problem, you will create a memo outlining your progress on a flange.
   A. Open drawing P17_03 from Chapter 17 and save it as P20_04.
   B. Set the Hidden visual style current.
   C. Set the background color to white.
   D. Use the **SAVEIMG** command and save the scene as a monochrome BMP file.
   E. Render the scene using the Presentation render preset and a resolution of 320 × 480. Save the rendering as a BMP file.
   F. Open a word processor capable of importing BMP files, such as Microsoft Word or Wordpad.
   G. Write a memo related to the project. A sample appears below. The memo should discuss how you created the drawing, the BMP file, and the rendered file. Insert the BMP files as appropriate.
   F. Save the document as P20_04. Print the document.

MEMO

To:	Otto Desque
From:	Ima Drafter
Date:	Thursday, March 14
Subject:	Project Progress

Dear Otto,

I have completed the initial drawing. As you can see from the drawing shown here, the project is complying with design parameters. The drawing is ready for transfer to the engineering department for approval.

I have also included a rendered image of the project. The material spec'ed by the engineering department is represented in the rendering. This may help in evaluation of the design.

Respectfully,

Ima

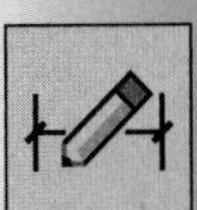

5. Begin a new drawing.
   A. Insert the blocks you created in problem 3. Arrange them in any order.
   B. Add any notes you need to identify this drawing as a sheet of library shapes. Be sure each shape is identified with its file name and location (path).
   C. Create a PDF file of the drawing.
   D. Save the drawing as P20_05.
   E. Open the PDF in Adobe Reader and print it. If Adobe Reader is not installed, obtain permission from your instructor or supervisor to download and install it.

6. Add a raster image to one of your title block template drawings as a design element or a company logo. A sample is shown below. Import an existing raster image or create your own using a program such as Windows Paint. Save the template drawing.

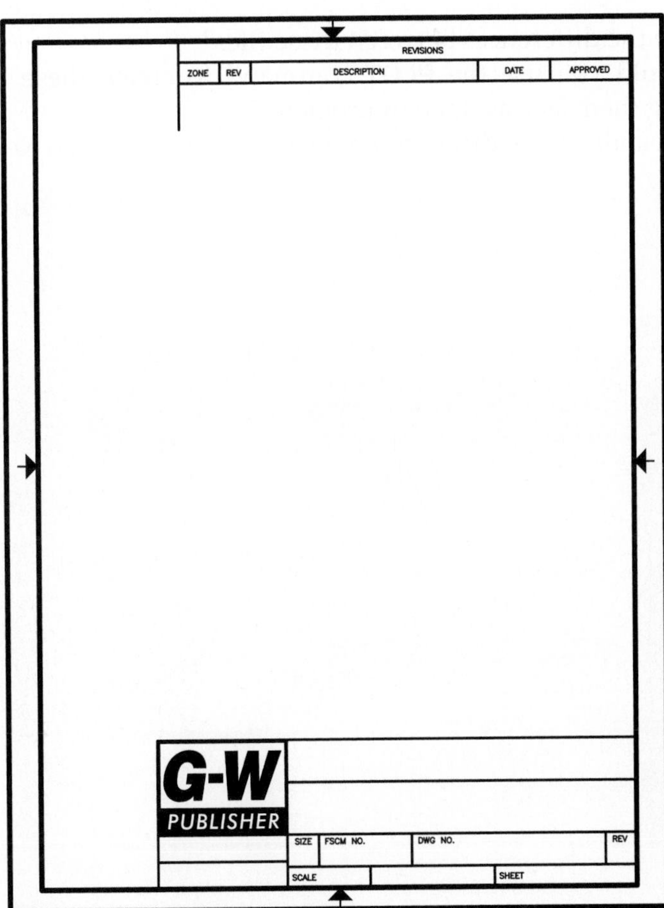

7. Open a 3D drawing from a previous chapter and save it as P20_07.
   A. Export the model to a DWF file. Save it as P20_07.dwf.
   B. Use the **3DDWF** command to export the model as a 3D DWF file. Name this file P20_07.dwf.
   C. Use Autodesk Design Review to open the DWF files and compare them.
   D. Which viewing commands are only available in the 3D file?
   E. Note any other differences between the two files.

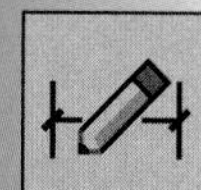

8. Open a 3D drawing from a previous chapter and save it as P20_08.
   A. Use the **PLOT** command to create a DWF plot file. Name it P20_08.dwf.
   B. Use the **PLOT** command to create a DWFx plot file. Name it P20_08.dwfx.
   C. Use the **PLOT** command to create a PDF plot file. Name it P20_08.pdf.
   D. Open these three files with either the Autodesk Design Review or Adobe Reader.
   E. What are the differences between these files?
   F. Why would you use the **PLOT** command to create these files rather than exporting them as was done in problem 7?
   G. What capabilities are different between Autodesk Design Review and Adobe Reader?

CHAPTER 21

Customizing the AutoCAD Environment

Learning Objectives

After completing this chapter, you will be able to:

- ✓ Set system variables.
- ✓ Assign colors and fonts to the text and graphics windows.
- ✓ Control general AutoCAD system variables.
- ✓ Set options that control display quality and AutoCAD performance.
- ✓ Control shortcut menus.

AutoCAD provides a variety of options for customizing the user interface and working environment. These options permit users to configure the software to suit personal preferences. You can define colors for the individual window elements, assign preferred fonts to the command line window, control shortcut menus, and assign properties to program icons.

The options for customizing the AutoCAD user interface and working environment are found in the **Options** dialog box, Figure 21-1. This dialog box is commonly accessed by right-clicking in the drawing area or command line with nothing selected and no command active and picking **Options...** from the shortcut menu. It can also be displayed by picking the **Options** button at the bottom of the application menu (menu browser).

Changes made in the **Options** dialog box do not take effect until either the **Apply** or **OK** button is picked. If you pick the **Cancel** button or the close button (**X**) before picking **Apply**, all changes are discarded. Each time you change the options settings, the system registry is updated and the changes are used in this and subsequent drawing sessions. Settings that are stored within the drawing file have the AutoCAD drawing icon next to them. These settings do not apply to other drawings. Settings without the icon affect all AutoCAD drawing sessions.

Setting AutoCAD System Variables

There are numerous settings that control the manner in which AutoCAD behaves in the Windows environment. These settings are made through the use of system variables. ***System variables*** are used to specify such items as which folders to search for driver and menu files and the location of temporary and support files. The default

Figure 21-1.
The **Options** dialog box is used to customize the AutoCAD working environment. Each tab contains a variety of options and settings.

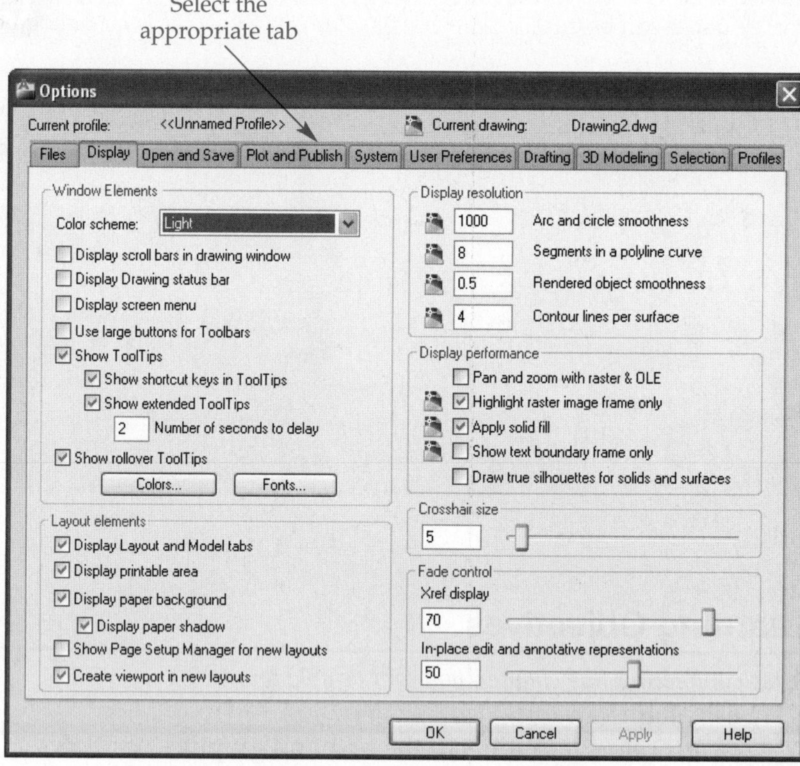

settings created during installation are often adequate, but changing the settings may result in better performance. While several different options exist for setting many of the system variables, the simplest method is to use the **Options** dialog box.

File Locations

When AutoCAD is used in a network environment, some files pertaining to AutoCAD may reside on a network drive so all users can access them. Other files may reside in folders specifically created for a particular AutoCAD user. These files may include external reference files, custom menu files, and drawings containing blocks.

The **Files** tab of the **Options** dialog box is used to specify the path AutoCAD searches to find support files and driver files. It also contains the paths where certain types of files are saved and where AutoCAD looks for specific types of files. Support files include text fonts, menus, AutoLISP files, ObjectARX files, blocks, linetypes, and hatch patterns.

The folder names shown under the Support File Search Path heading in the **Search paths, file names, and file locations:** list are automatically created by AutoCAD during the installation. For example, Figure 21-2 shows that the support files are stored in six different folders. Folders are searched in the order in which they are listed under Support File Search Path. As previously mentioned, some of these paths are created for a specific user. The first path listed is long and ultimately ends with the user-specific \Support folder. In the example shown, this path starts on the C: drive in a folder called \Documents and Settings. The next folders to be searched are, in order, \Support, \Fonts, \Help, \Express, and \Support\Color. These paths are not user specific; they are located in the AutoCAD installation path.

You can add the path of any new folders you create that contain support files. As an example, suppose all of the blocks you typically use are stored in a separate folder named \Blocks. By placing this folder name in the support files search path, it is not necessary to include the entire path to the blocks in any command macros you create to automatically insert your blocks (this is discussed in Chapter 22).

Figure 21-2.
Folder paths can be customized in the **Files** tab of the **Options** dialog box.

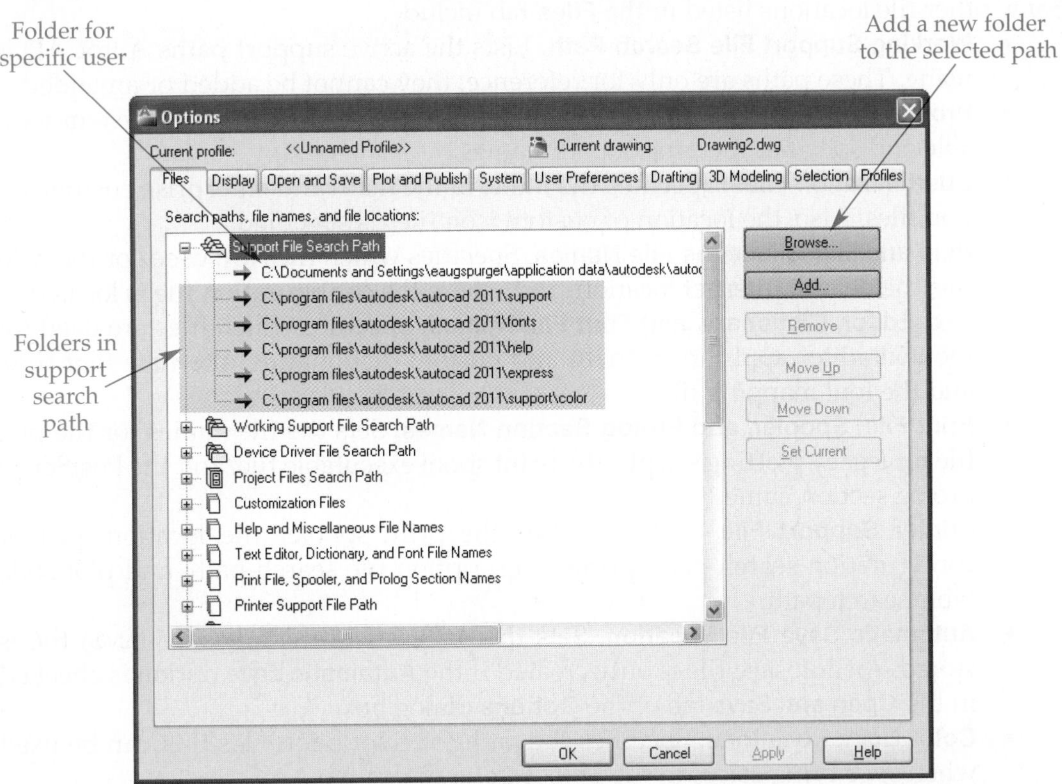

A folder can be added to the existing search path in two ways. The first method is to highlight the Support File Search Path heading and pick the **Add...** button. This places a new, empty listing under the heading. You can now type C:\Blocks to complete the entry. Alternately, instead of typing the path name, after picking **Add...** you can pick the **Browse...** button to display the **Browse for Folder** dialog box. You can then use this dialog box to select the desired folder. The new setting takes effect as soon as you pick **Apply** or the **OK** button and close the **Options** dialog box.

PROFESSIONAL TIP

In addition to the paths listed under Support File Search Path, AutoCAD will search two other folders. The folder that contains the AutoCAD executable file, acad.exe, (typically, C:\Program Files\AutoCAD 2011) is searched, as is the folder that contains the current drawing file. These folders are searched only if the desired file name is not found in any of the listed folders. It is not advisable to store files such as block files in the AutoCAD folder. However, if you store block files in the same folder as the current drawing, you do not need to add that folder to the search path.

Other File Settings

Another setting that can be specified in the **Files** tab is the location of device driver files. ***Device drivers*** are specifications for peripherals that work with AutoCAD and other Autodesk products. By default, the drivers supplied with AutoCAD are placed in the \Drv folder. If you purchase a third-party driver to use with AutoCAD, be sure the driver is loaded into this folder. If the third-party driver must reside in a different folder, you should specify that folder using the Device Driver File Search Path setting. Otherwise, the search for the correct driver is widespread and likely to take longer. Some other file locations listed in the **Files** tab include:

- **Working Support File Search Path.** Lists the active support paths AutoCAD is using. These paths are only for reference; they cannot be added or amended.
- **Project Files Search Path.** Sets the value for the **PROJECTNAME** system variable and specifies the project path names.
- **Customization Files.** Specifies the name of the main and enterprise customization files. Also, the location of custom icon files is specified.
- **Help and Miscellaneous File Names.** Specifies which files are used for the help file, the default Internet location, and where the configuration file is located.
- **Text Editor, Dictionary, and Font File Names.** Specifies which files are used for the text editor application, main and custom dictionaries, alternate font files, and the font mapping file.
- **Print File, Spooler, and Prolog Section Names.** Sets the file names for the plot file for legacy plotting scripts, the print spool executable file, and the PostScript prolog section name.
- **Printer Support File Path.** Specifies the print spooler file location, printer configuration search path, printer description file search path, and plot style table search path.
- **Automatic Save File Location.** Sets the path where the autosave (.sv$) file is stored. An autosave file is only created if the **Automatic save** option is checked in the **Open and Save** tab of the **Options** dialog box.
- **Color Book Locations.** Specifies the path for color book files that can be used when specifying colors in the **Select Color** dialog box.
- **Data Sources Location.** Specifies the path for database source files (.udl).
- **Template Settings.** Specifies the default location for drawing and sheet set template files and the file name for the defaults.
- **Tool Palettes File Locations.** Specifies the path for tool palette support files.
- **Authoring Palette File Locations.** Specifies the location of authoring palette files.
- **Log File Location.** Specifies the path for the AutoCAD log file. A log file is only created if the **Maintain a log file** option is checked in the **Open and Save** tab of the **Options** dialog box.
- **Action Recorder Settings.** Specifies search and storage paths for the action recorder macro files (.actm).
- **Plot and Publish Log File Location.** Specifies the path for the log file for "plot and publish" operations. A log file is only created if the **Automatically save plot and publish log** check box in the **Plot and Publish** tab of the **Options** dialog box is checked.
- **Temporary Drawing File Location.** Sets the folder where AutoCAD stores temporary drawing files.
- **Temporary External Reference File Location.** Indicates where temporary external reference files are placed.
- **Texture Maps Search Path.** Location of texture map files for rendering.
- **Web File Search Path.** Specifies the folders to search for files associated with photometric weblight lighting.
- **i-drop Associated File Location.** Specifies the folder used by default to store downloaded i-drop content.
- **DGN Mapping Setups Location.** Specifies the location of the mapping files for translation to and from Microstation drawings (.dgn). This folder must have read/write file permissions enabled in order for the Microstation commands to properly function.

Customizing the Graphics Window

Numerous options are available to customize the graphics window to your personal liking. Select the **Display** tab in the **Options** dialog box to view the display-control options, Figure 21-3.

The **Window Elements** area has a setting for the color scheme. The color scheme controls the outline color of the ribbon and status bar. Using the drop-down list, you can select between light and dark color schemes.

There are also check boxes for turning the scroll bars, status bar, and screen menu on or off; using large buttons for toolbars; showing tooltips and whether or not to show shortcut keys and extended commands in tooltips; and turning rollover tooltips on or off. The element colors and font settings can be changed using the buttons at the bottom of the **Window Elements** area.

On the right-hand side of the tab near the bottom, the **Crosshair size** setting is a percentage of the drawing screen area. The higher the value, the further the crosshairs extend. The settings in the **Fade control** area determine the display intensity of the unselected objects in reference edit mode and in-place edit and annotative representations. A higher value means the unselected objects are less visible. Other options are discussed in the next sections.

Changing Colors

By customizing colors, you can add your personal touch and make AutoCAD stand out among other active Windows applications. AutoCAD provides this capability with the **Drawing Window Colors** dialog box, Figure 21-4. This dialog box is accessed by picking the **Colors...** button in the **Window Elements** area of the **Display** tab in the **Options** dialog box.

Figure 21-3.
Use the **Display** tab to set up many of the visual elements of the AutoCAD environment.

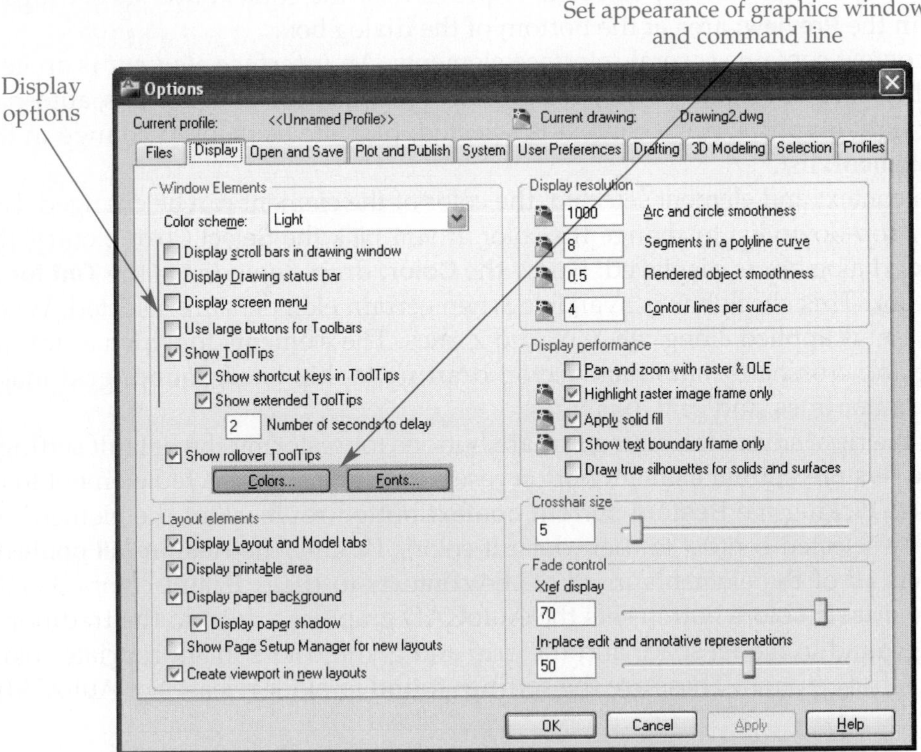

Figure 21-4.
Change AutoCAD color settings using the **Drawing Window Colors** dialog box.

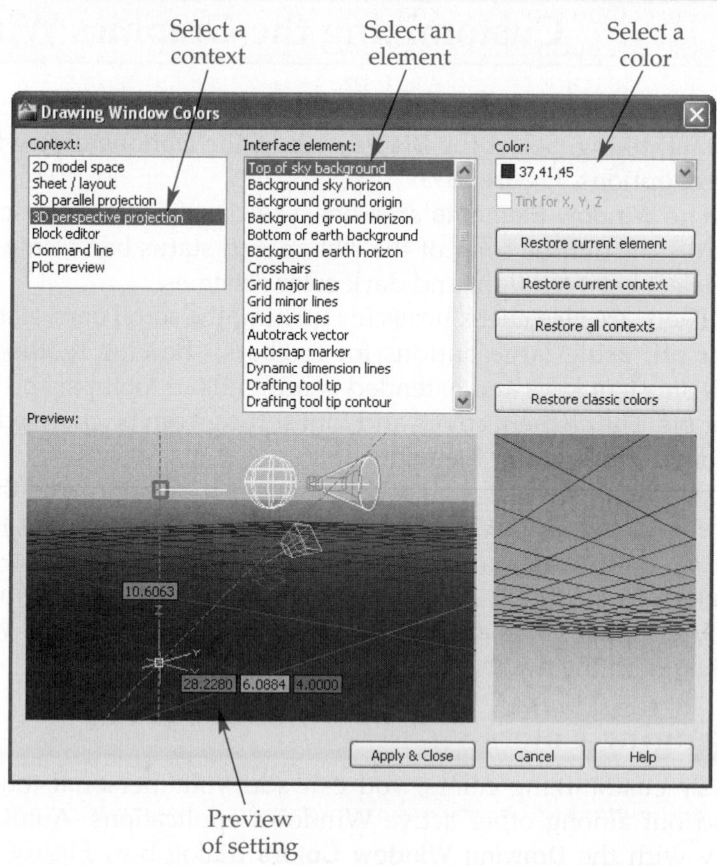

To change a color, first select a context. A ***context*** is one of the environments, or modes, in AutoCAD, such as the 3D perspective projection mode that is set current when a new drawing is started based on the acad3D.dwt template. The **Context:** list box contains the names of all contexts. The context that was current when the **Options** dialog box is opened is initially selected. A preview of the context and its settings is displayed in the **Preview:** area at the bottom of the dialog box.

Each context contains several interface elements. An ***interface element*** is an item that is visible, or can be made visible, in a given context, such as the grid axis, autosnap marker, or light glyphs. Once a context is selected, pick the element to change in the **Interface element:** list.

With a context and element selected, the color of the element can be changed. Use the **Color:** drop-down list to change the color. If you pick the Select Color... entry, the **Select Color** dialog box is displayed. Below the **Color:** drop-down list is the **Tint for X, Y, Z** check box. This check box is available when certain elements are selected. When checked, a tint is applied along the X, Y, and Z axes. The elements to which a tint can be applied are: crosshairs, autotrack vector, drafting tooltip background, grid major lines, grid minor lines, and grid axis lines.

Along the right side of the dialog box are buttons for restoring the default settings. Picking the **Restore current element** button resets the currently selected element to its default color. Picking the **Restore current context** button resets *all* of the elements of the currently selected context to their default colors. Picking the **Restore all contexts** button resets *all* of the elements in *all* of the contexts to their default colors. Lastly, the **Restore classic colors** button sets the AutoCAD graphics screen to the traditional black background in model space and the grip and glyph colors to appropriate colors for the black background. These are the settings found in older releases of AutoCAD.

Once the colors are changed as needed, pick the **Apply & Close** button. Then, pick the **OK** button in the **Options** dialog box. The graphics window regenerates and displays the color changes you made.

PROFESSIONAL TIP

When setting the background color to black, first use the **Restore classic colors** button. Then, adjust the other color settings from that point forward.

Changing Fonts

You can change the fonts used in the **Command Line** window. The font you select has no effect on the text in your drawings, nor is the font used in the AutoCAD dialog boxes, pull-down menus, or screen menus.

To change the font used in the **Command Line** window, pick the **Fonts...** button in the **Display** tab of the **Options** dialog box. The **Command Line Window Font** dialog box appears, **Figure 21-5.**

The default font used by AutoCAD for the graphics window is Courier New. The font style for Courier New is regular (not bold or italic) and the default size is 10 points. Select a new font from the **Font:** list. This list displays the system fonts available for use. Also, set a style and size. The **Sample Command Line Font** area displays a sample of the selected font. Once you have selected the desired font, font style, and font size for the **Command Line** window, pick the **Apply & Close** button to assign the new font.

NOTE

The Windows system "menu" font controls the font style for the text displayed in the screen menu.

CAUTION

The **UNDO** command does not affect changes made to your system using the **Options** dialog box. If you have made changes you do not want to save, pick **Cancel** to dismiss the **Options** dialog box. However, picking **Cancel** does *not* dismiss changes that have been applied using the **Apply** button.

Figure 21-5.
The **Command Line** window can be changed to suit your preference.

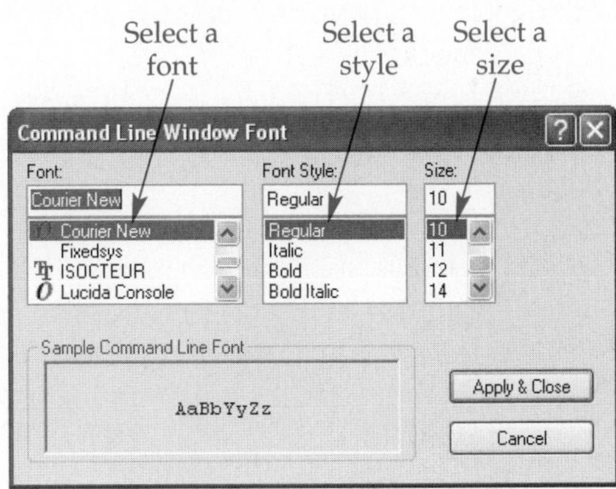

Exercise 21-1

Complete the exercise on the student website.
www.g-wlearning.com/CAD

Layout Display Settings

The appearance of a layout (paper space) tab is different than the appearance of model space. The theory behind the default layout tab settings is to provide a picture of what the drawing will look like when plotted. You can see if the objects will fit on the paper or if some of the objects are outside of the margins. The following options, which are found in the **Layout elements** area of the **Display** tab in the **Options** dialog box, are illustrated in Figure 21-6.

- **Display Layout and Model tabs.** Displays the **Model** and layout tabs at the bottom of the drawing screen area. This is checked by default.
- **Display printable area.** The margins of the printable area are shown as dashed lines on the layout paper. Any portion of an object outside of the margins is not plotted.
- **Display paper background.** Displays the paper size specified in the page setup.
- **Display paper shadow.** Displays a shadow to the right and bottom of the paper. This option is only available if **Display paper background** is checked.
- **Show Page Setup Manager for new layouts.** Determines if the **Page Setup** dialog box is displayed when a new layout is selected or created. By default, this is unchecked.
- **Create viewport in new layouts.** Determines whether a viewport is automatically created when a new layout is selected or created. Many users uncheck this option since they will be creating their own floating viewports.

Figure 21-6.
Customizing the display of layouts. These options are set in the **Options** dialog box.

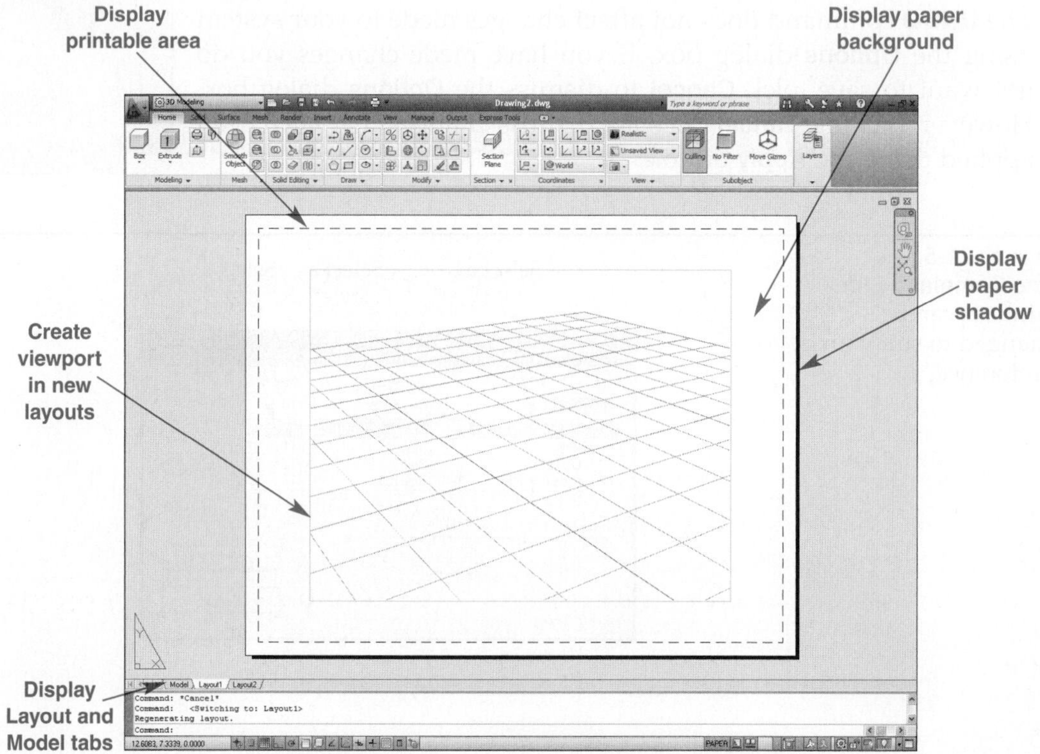

Display Resolution and Performance Settings

The settings in the **Display resolution** and **Display performance** areas of the **Display** tab in the **Options** dialog box affect the performance of AutoCAD. The settings can affect regeneration time and realtime panning and zooming. The following options are available in the **Display resolution** area. If the AutoCAD drawing icon is shown next to the setting, the value is saved in the current drawing, not in the AutoCAD system registry.

- **Arc and circle smoothness.** This setting controls the smoothness of circles, arcs, and ellipses. The default value is 1000; the range is from 1 to 20000. The system variable equivalent is **VIEWRES**.
- **Segments in a polyline curve.** This value determines how many line segments will be generated for each polyline curve. The default value is 8; the range is a nonzero value from –32768 to 32767. The system variable equivalent is **SPLINESEGS**.
- **Rendered object smoothness.** This setting controls the smoothness of curved solids when they are hidden, shaded, or rendered. This value is multiplied by the **Arc and circle smoothness** value. The default value is 0.5; the range is from 0.01 to 10. The system variable equivalent is **FACETRES**.
- **Contour lines per surface.** This value controls the number of contour lines per surface on solid objects. The default value is 4; the range is from 0 to 2047. The system variable equivalent is **ISOLINES**.

The following options are available in the **Display performance** area.

- **Pan and zoom with raster & OLE.** If this is checked, raster images are displayed when panning and zooming. If it is unchecked, only the frame is displayed during the operation. The system variable equivalent is **RTDISPLAY**.
- **Highlight raster image frame only.** If this is checked, only the frame around a raster image is highlighted when the image is selected. If this option is unchecked, the image displays a diagonal checkered pattern to indicate selection. The system variable equivalent is **IMAGEHLT**.
- **Apply solid fill.** Controls the display of solid fills in objects. Affected objects include hatches, wide polylines, solids, multilines, and traces. The system variable equivalent is **FILLMODE**.
- **Show text boundary frame only.** This setting controls the Quick Text mode. When checked, text is replaced by a rectangular frame. The system variable equivalent is **QTEXTMODE**.
- **Draw true silhouettes for solids and surfaces.** Controls whether or not the silhouette curves are displayed for solid objects. The system variable equivalent is **DISPSILH**.

NOTE

After changing display settings, use the **REGEN** or **REGENALL** command to make the settings take effect on the objects in the drawing.

PROFESSIONAL TIP

If you notice performance slowing down, you may want to adjust display settings. For example, if there is a lot of text in the drawing, you can activate Quick Text mode to improve performance. When the drawing is ready for plotting, deactivate Quick Text mode.

File Saving Options

The settings specified in the **Open and Save** tab of the **Options** dialog box deal with how drawing files are saved, safety precautions for files, how file names display in the application menu (menu browser), the behavior of xrefs, and the loading of ObjectARX applications and proxy objects. This tab is shown in Figure 21-7. The options in this tab are discussed in the next sections.

Default Settings for Saving Files

The settings in the **File Save** area determine the defaults for saving files. The setting in the **Save as:** drop-down list determines the default file type. You may want to change this setting if you are saving drawing files as a previous release of AutoCAD or saving drawings as DXF files.

The **Maintain visual fidelity for annotative objects** check box controls how annotative objects are displayed when the drawing is opened in AutoCAD 2007 or earlier versions. If you work primarily in model space, this can be left unchecked. If you use layouts and expect the drawing files to be saved for an older version of AutoCAD, this should be checked. When checked and the drawing is saved and then opened in an older version of AutoCAD, the scaled representations of annotative objects are divided into separate objects. These objects are stored in anonymous blocks saved on separate layers. The block names are based on the layer's original name appended with a number. When the drawing is opened once again in AutoCAD 2008 or later, the annotative objects are restored to normal. The system variable equivalent for this toggle is **SAVEFIDELITY**. Checking the check box sets this variable to 1 (on). Unchecking it sets the variable to 0 (off).

The **Maintain drawing size compatibility** check box determines how drawings with individual objects greater than 256 MB are handled. Files created with versions previous

Figure 21-7.
The **Open and Save** tab settings control default save options, file safety features, how file names are displayed in the application menu, xref options, and ObjectARX application options.

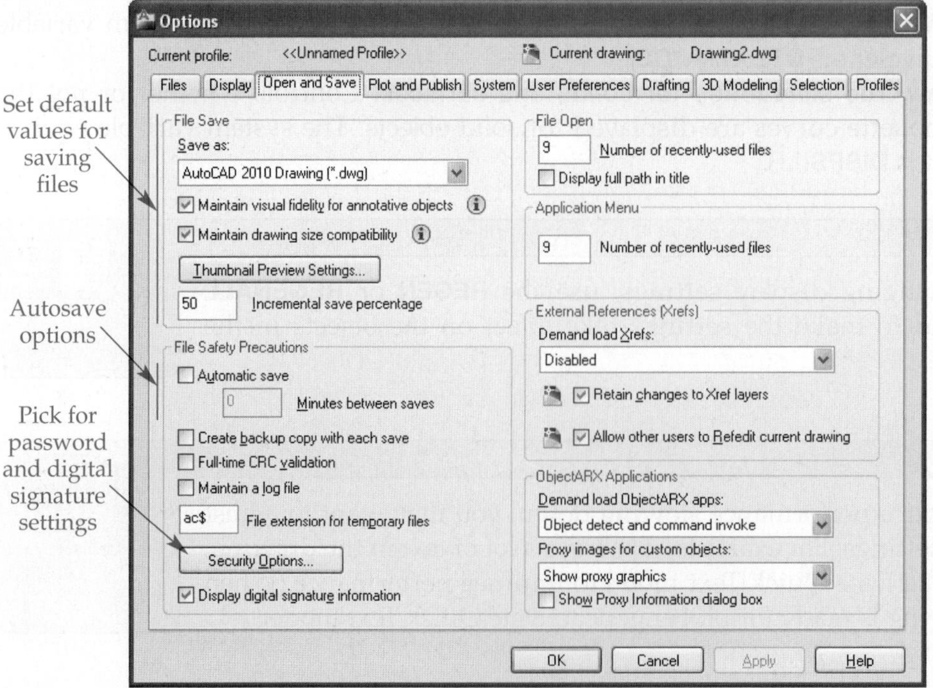

to AutoCAD 2010 are not allowed to have objects larger than 256 MB. When saving a drawing to a legacy format (2009 or earlier), problems may be encountered when attempting to open that drawing due to compatibility issues with these large objects. Checking this option maintains compatibility with earlier versions of AutoCAD by telling AutoCAD to check for objects larger than 256 MB when attempting to save the drawing. An alert box will be displayed noting that the issue needs to be resolved before the file can be saved. If you do not plan on working on your drawings in earlier versions of AutoCAD, this option box can be unchecked. The system variable equivalent for this toggle is **LARGEOBJECTSUPPORT**. Unchecking the **Maintain drawing size compatibility** check box sets the system variable to 1 (on) and allows the ability to create large objects. When the check box is checked, the system variable is set to 0 (off).

The **Incremental save percentage** value determines how much of the drawing is saved when a **SAVE** or **QSAVE** is performed. If the quantity of new data in a drawing file reaches the specified percentage, a full save is performed. To force a full save to be performed each time, set the value to 0.

If you pick the **Thumbnail Preview Settings...** button, the **Thumbnail Preview Settings** dialog box is displayed, Figure 21-8. If the **Save a thumbnail preview image** check box in this dialog box is checked, a preview image of the drawing will be displayed in the **Select File** dialog box when the drawing is selected for opening. The system variable equivalent for this setting is **RASTERPREVIEW**; 1 creates a preview.

The two radio buttons below the **Save a thumbnail preview image** check box set what is used as the basis for the thumbnail. The **Use view when drawing last saved** radio button bases the thumbnail image on the last zoom location of the drawing. The **Use Home View** radio button bases the thumbnail on the view defined as the home view.

The home view can be set to the current view by picking the **Set current View as Home** button in the **Home view** area. This can also be done in the drawing area by using the view cube. See Chapter 4 for a detailed explanation of the view cube. To change the home view to the default setting, pick the **Reset Home to default** button. A preview of the current home view is shown to the left of the buttons.

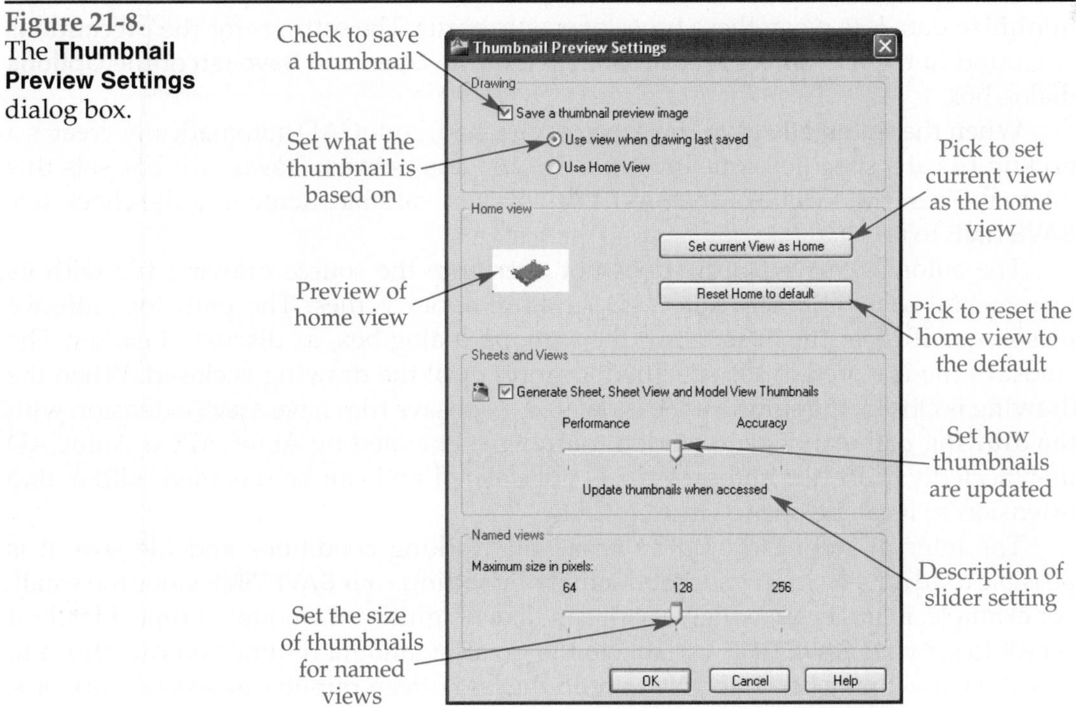

Figure 21-8. The **Thumbnail Preview Settings** dialog box.

When the **Generate Sheet, Sheet View, and Model View Thumbnails** check box is checked in the **Thumbnail Preview Settings** dialog box, the thumbnails in the **Sheet Set Manager** are updated based on the position of the slider below this check box. The slider can be set to one of three positions. A description of the current setting appears below the slider. When the slider is in the middle position (default), thumbnails are updated when they are accessed. When the slider is in the left-hand position, thumbnails must be manually updated. When the slider is in the right-hand position, the thumbnails are updated when the drawing is saved. The system variable equivalent is **UPDATETHUMBNAIL**. The settings are:

- **0.** The **Generate Sheet, Sheet View, and Model View Thumbnails** check box is unchecked.
- **7.** The check box is checked and the slider is in the left-hand position.
- **15.** The check box is checked and the slider is in the middle position.
- **23.** The check box is checked and the slider is in the right-hand position.

The **Maximum size in pixels:** slider in the **Thumbnail Preview Settings** dialog box controls size of thumbnails for named views. The slider can be set to 64, 128, or 256 pixels. This is the square size of the thumbnail. The system variable equivalent is **THUMBSIZE**, where a setting of 0 is 64 pixels, 1 is 128 pixels, and 2 is 256 pixels.

CAUTION

To maintain forward compatibility of drawings, annotative objects should not be edited in older versions of AutoCAD. Doing so may compromise the annotative properties. For example, exploding an annotative block in an older version of AutoCAD then opening that drawing in AutoCAD 2011 results in each of the scaled representations becoming a separate annotative object.

Autosave Settings

When working in AutoCAD, data loss can occur due to a sudden power outage or an unforeseen system error. AutoCAD provides several safety precautions to help minimize data loss when these types of events occur. The settings for the precautions are found in the **File Safety Precautions** area in the **Open and Save** tab of the **Options** dialog box.

When the **Automatic save** check box is checked, AutoCAD automatically creates a backup file at a specified time interval. The **Minutes between saves** edit box sets this interval. This is the value of the **SAVETIME** system variable. Removing the check sets **SAVETIME** to 0.

The automatic save feature does not overwrite the source drawing file with its incremental saves. Rather, AutoCAD saves temporary files. The path for autosave files is specified in the **Files** tab in the **Options** dialog box, as discussed earlier. The autosave file is stored in the specified location until the drawing is closed. When the drawing is closed, the autosave file is deleted. Autosave files have a .sv$ extension with the drawing name and some random numbers generated by AutoCAD. If AutoCAD unexpectedly quits, the autosave file is not deleted and can be renamed with a .dwg extension so it can be opened in AutoCAD.

The interval setting should be based on working conditions and file size. It is possible to adversely affect your productivity by setting your **SAVETIME** value too small. For example, in larger drawings, a save can take a significant amount of time. Ideally, it is best to set your **SAVETIME** variable to the greatest amount of time you can afford to repeat. While it may be acceptable to redo the last fifteen minutes or less of work, it is unlikely that you would feel the same about having to redo the last hour of work.

PROFESSIONAL TIP

Setting and resetting the **SAVETIME** variable according to any given situation is often the best approach. The factors that should influence the current setting include not only file size, but also the working conditions. If your computer system is experiencing frequent lockups or crashes, your automatic saves should occur often. Weather can also be a factor. Wind or electrical storms should be an immediate cue to reduce the value of the **SAVETIME** variable. In addition to automatic saves, most veteran users have developed the positive habit of "save early, save often."

Backup Files

AutoCAD can create a backup of the current drawing file whenever the current drawing is saved. The backup file uses the same name as the drawing, but has a .bak file extension. The backup is not overwritten when a different drawing is opened or saved. When the **Create backup copy with each save** check box in the **Open and Save** tab of the **Options** dialog box is checked, the backup file feature is enabled. If not checked, the file is not backed up when you save. Unless you prefer to take unnecessary risks, it is usually best to have this feature enabled.

CRC Validation

A ***cyclic redundancy check,*** or CRC, verifies that the number of data bits sent is the same as the number received. **Full-time CRC validation** is a feature you can use when drawing files are being corrupted and you suspect a hardware or software problem. When using full-time CRC validation, the CRC is done every time data are read into the drawing. This ensures that all data are correctly received.

Log Files

The log file can serve a variety of purposes. The source of drawing errors can be determined by reviewing the commands that produced the incorrect results. Additionally, log files can be reviewed by a CAD manager to determine the need for staff training or customization of the system.

When the **Maintain a log file** check box is activated in the **Open and Save** tab of the **Options** dialog box, AutoCAD creates a file named with the drawing name, a code, and the .log file extension. The name and location of the log file can be specified using the Log File Location listing in the **Files** tab of the **Options** dialog box. When activated, all prompts, messages, and responses that appear in the **Command Line** window are saved to this file. The log file status can also be set using the **LOGFILEON** and **LOGFILEOFF** commands.

PROFESSIONAL TIP

Toggle the log file open before listing any saved layers, blocks, views, or user coordinate systems. You can then print the log file contents and keep a hard copy at your workstation as a handy reference.

File Opening Settings

The **File Open** area of the **Open and Save** tab in the **Options** dialog box contains two settings. The value in the **Number of recently-used files to list** text box controls the number of drawing files listed in the **File** pull-down menu. The pull-down menus are not displayed, by default, unless the AutoCAD Classic workspace is set current. The value can be from 0 to 9. The **Display full path in title** check box controls whether the entire drawing file path (when checked) or just the file name (when unchecked) is displayed in the title bar of the AutoCAD window.

Application Menu Settings

The **Application Menu** area of the **Open and Save** tab in the **Options** dialog box controls the number of drawing files displayed in the **Recent Documents** entry in the application menu (menu browser). The setting in the **Number of recently used files** text box is the number of files displayed and can range from 0 to 50.

External Reference Settings

The external reference options in the **Open and Save** tab of the **Options** dialog box are important if you are working with xrefs. These options are found in the **External Reference (Xrefs)** area of the tab. The **Demand load Xrefs:** setting can affect system performance and the ability for another user to edit a drawing currently referenced into another drawing. You can select Enabled, Disabled, or Enabled with copy from the drop-down list. This setting is also controlled by the **XLOADCTL** system variable.

If the **Retain changes to Xref layers** option is checked, xref layer settings are saved with the drawing file. The **VISRETAIN** system variable also controls this setting.

The **Allow other users to Refedit current drawing** setting controls whether or not the drawing can be edited in-place when it is referenced by another drawing. This setting is also controlled by the **XEDIT** system variable.

ObjectARX Options

The **ObjectARX Applications** area of the **Open and Save** tab of the **Options** dialog box controls the loading of ObjectARX applications and the displaying of proxy objects. The **Demand load ObjectARX apps:** setting specifies if and when AutoCAD loads third-party applications associated with objects in the drawing. The **Proxy images for custom objects:** setting controls how objects created by a third-party application are displayed. When a drawing with proxy objects is opened, the **Proxy Information** dialog box is displayed. To disable the dialog box, uncheck the **Show Proxy Information dialog box** option. This is unchecked by default.

NOTE

The settings in the **Plot and Publish** tab control how plots are generated. Plotting is covered in detail in *AutoCAD and Its Applications—Basics.*

System Settings

Options for the pointing device, graphic settings, general system options, dbConnect, and web-based help can be found in the **System** tab of the **Options** dialog box, Figure 21-9. These settings affect the interaction between AutoCAD and your operating system.

In the **3D Performance** area is the **Performance Settings** button. Selecting this button displays the **Adaptive Degradation and Performance Tuning** dialog box. The options available in this dialog box are discussed in the next section.

Figure 21-9.
General AutoCAD system options and hardware settings can be controlled in the **System** tab.

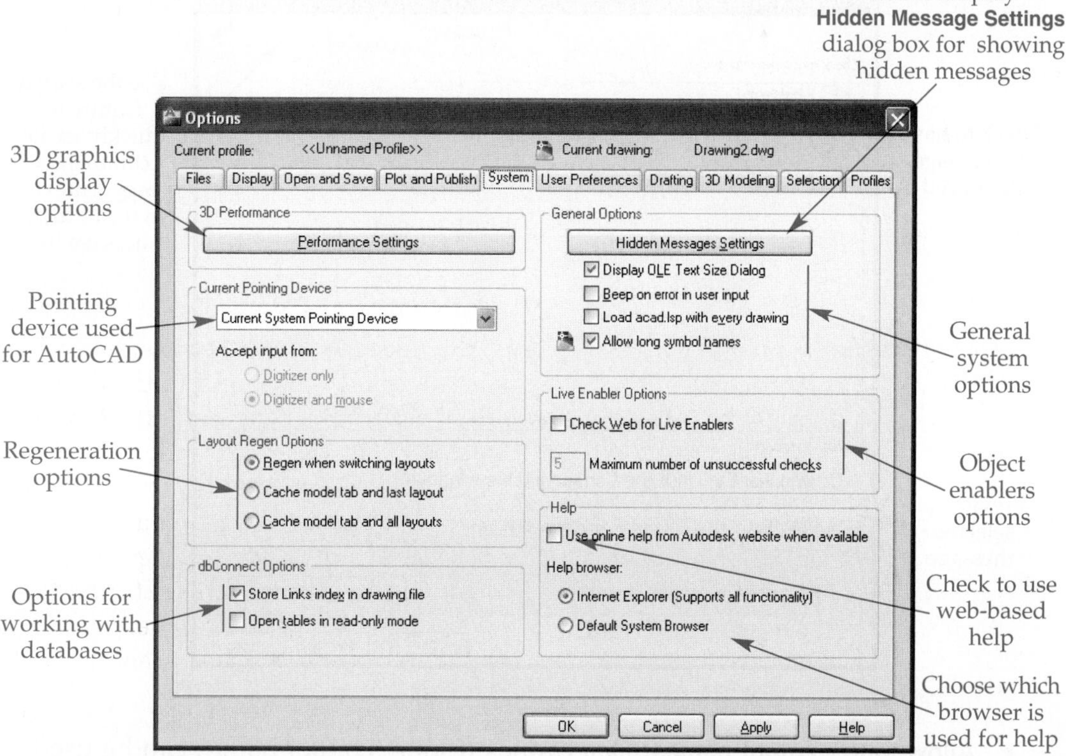

The **Current Pointing Device** area determines the pointing device used with AutoCAD. The default is the current system pointing device (usually your mouse). If you have a digitizer tablet, you will want to select the Wintab Compatible Digitizer option. You must configure your tablet before it can be used.

The **Layout Regen Options** setting determines what is regenerated and when it is regenerated when working with layout tabs. The following options are available in the **dbConnect Options** area.

- **Store Links index in drawing file.** When this option is checked, the database index is saved within the drawing file. This makes the link selection operation quicker, but increases the drawing file size.
- **Open tables in read-only mode.** Determines whether tables are opened in read-only mode.

The settings in the **General Options** area control general system functions. When the **Hidden Messages Settings** button is picked, the **Hidden Message Settings** dialog box is displayed, Figure 21-10. All messages hidden by the user picking the "do not show again" option in a message box are available in this dialog box. For example, if you close the Command Line window, a message is displayed with a warning. If you choose to always hide this message, it appears in the **Hidden Message Settings** dialog box. Check the entry for this message in the dialog box to have it once again displayed when you select a solid. The following additional options are available in the **General Options** area of the **System** tab.

- **Display OLE Text Size Dialog.** When inserting an OLE object, the **OLE Text Size** dialog box is displayed if this option is checked.
- **Beep on error in user input.** Specifies whether AutoCAD alerts you of incorrect user input with an audible beep. By default, this is off.
- **Load acad.lsp with every drawing.** This setting turns the persistent AutoLISP feature on or off. By default, it is off.

Figure 21-10.
Turning on the display of previously hidden messages.

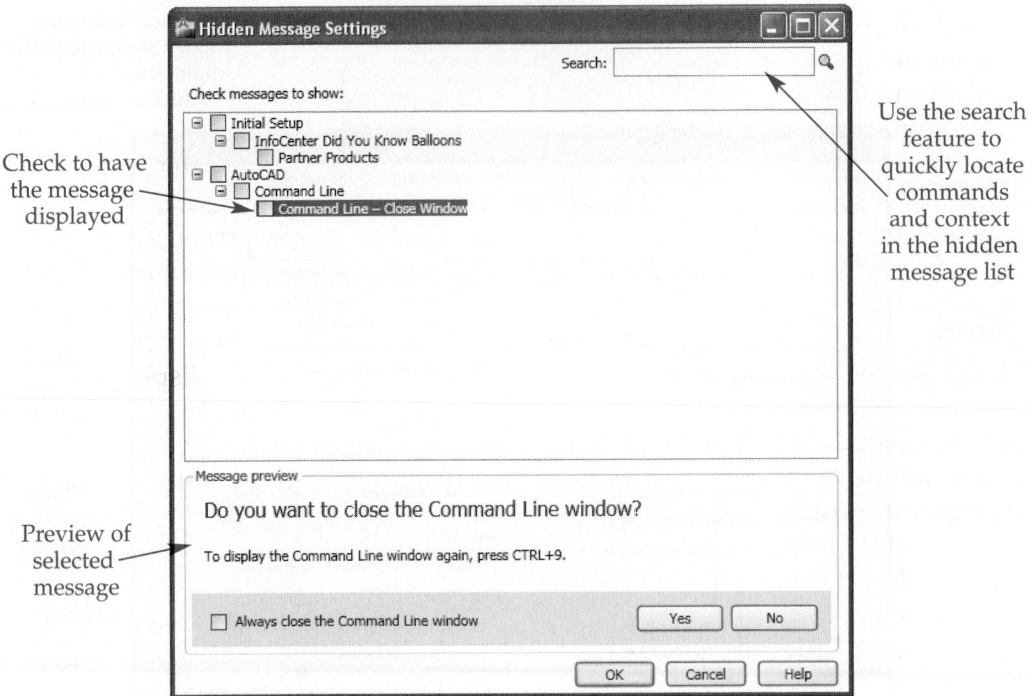

- **Allow long symbol names.** Determines if long symbol names can be used in AutoCAD. If this option is checked, up to 255 characters can be used for layers, dimension styles, blocks, linetypes, text styles, layouts, UCS names, views, and viewport configurations. If unchecked, symbol names are limited to 31 characters. By default, this option is checked. Also, it is saved in the drawing, not the AutoCAD system registry. The system variable is **EXTNAMES**.

The setting in the **Live Enabler Options** area determines whether or not AutoCAD searches the Autodesk website for object enablers. An object enabler allows your version of AutoCAD to open and manipulate drawings created in applications like AutoCAD Architecture without the display of proxy object errors. It also allows the use of the custom-made objects created in those applications. Object enablers are provided for free by Autodesk. The system variable that controls this setting is **PROXYWEBSEARCH**. The **Maximum number of unsuccessful checks** edit box indicates how many times AutoCAD will search the website.

The settings in the **Help** area determine how assistance is obtained when using the **HELP** command. Checking the **Use online help from Autodesk website when available** check box means a web browser will be opened when the **HELP** command is accessed. Online help from Autodesk's website will be displayed. Unchecking the check box allows help to be accessed from the file listed in the Help Location entry under the **Help and Miscellaneous File Names** listing on the **Files** tab of the **Options** dialog box. The radio buttons below the check box allow you to choose Internet Explorer or a default browser that you may be using (i.e., FireFox, Google Chrome). Using Internet Explorer ensures all features of Autodesk's online help pages are supported.

NOTE

All help is provided via a web browser, whether you are using online help from Autodesk or a file on a local computer.

3D Performance Settings

With all of the powerful 3D and solid modeling features that are built into AutoCAD, there are many display-related tasks being handled by AutoCAD, the graphics card, and the computer itself. Materials, lights, shadows, shading, and rendering require a lot of computing power in order to project a quality representation of the model onto the monitor screen. Often there is no reduction in quality to any of the desired effects if the materials are not too complex, few lights are used, or if you have shadows turned off. Sometimes, in order to make one effect look good, fewer resources have to be assigned to other effects. The software and hardware, working together, usually do an adequate job assigning these resources. However, it may be necessary for you to assist in this decision-making process. The settings for this process are made in the **Adaptive Degradation and Performance Tuning** dialog box, Figure 21-11. To display this dialog box, enter the **3DCONFIG** command or pick the **Performance Settings** button in the **Systems** tab of the **Options** dialog box.

The left-hand side of the **Adaptive Degradation and Performance Tuning** dialog box contains settings for controlling adaptive degradation. ***Adaptive degradation*** controls system performance by turning off features or preventing them from using resources. The check box at the top of the area controls whether or not adaptive degradation is being used. When unchecked, adaptive degradation is turned off and all effects are using all resources. This may result in graphics lagging or becoming slow and "choppy" as you zoom and pan around your drawing. The orbiting commands are even more affected by this being turned off. By checking the check box, adaptive degradation is activated.

AutoCAD tracks its graphics performance in terms of frames per second (fps). Just below the **Adaptive degradation** check box is a text box for setting this value. You may enter a new value in the text box or use the arrows to increase or decrease the value. The higher the number, the sooner resources start being reassigned. If performance dips below this level, resources are taken away from the various effects that create the displayed graphics.

The effects that can be controlled while adaptive degradation is turned on are shown in the **Degradation order:** list box. Certain effects that you deem important can be unchecked so they are not degraded and operate using maximum resources. The

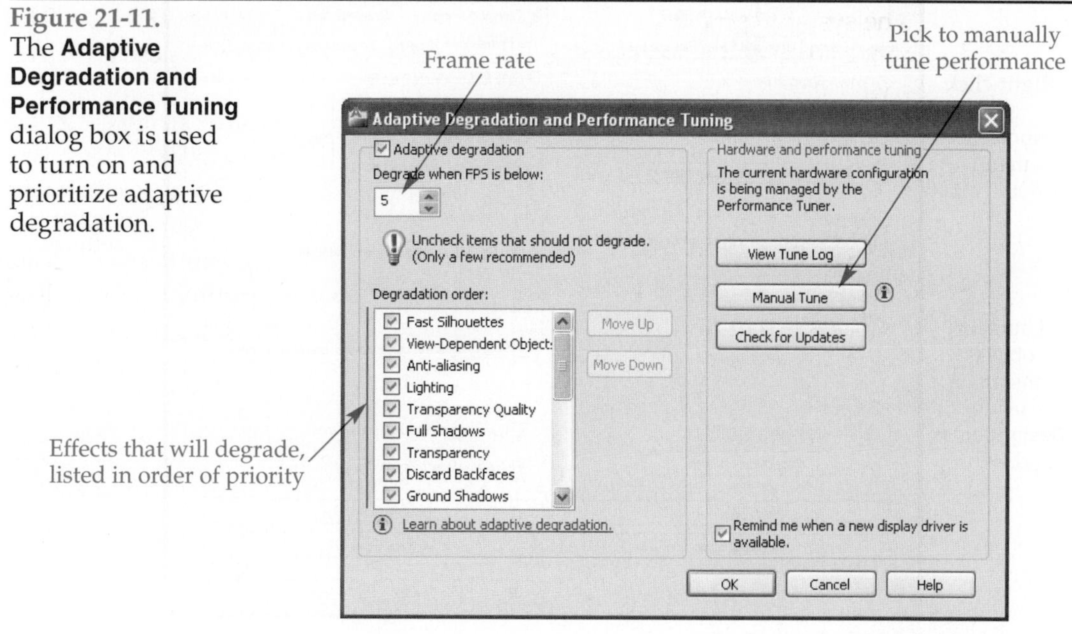

Figure 21-11. The **Adaptive Degradation and Performance Tuning** dialog box is used to turn on and prioritize adaptive degradation.

top-to-bottom order in which the effects are listed determines the priority in which resources are removed. This order can be changed by selecting an effect and picking the **Move Up** or **Move Down** button on the right side of the list.

On the right-hand side of the **Adaptive Degradation and Performance Tuning** dialog box is the **Hardware and performance tuning** area. Picking the **View Tune Log** button displays a log of any features or effects that have been turned off. Information regarding your computer, amount of RAM, and 3D graphics card are also shown. The log can be saved as a file. The **Manual Tune** button displays the **Manual Performance Tuning** dialog box. This dialog box allows control over hardware settings (*hardware acceleration,* graphics card driver name, and the effects the graphics card is capable of), general settings (discard back faces and quality of transparency), and dynamic tessellation settings (surface and curve tessellation settings and number of tessellations to cache). At the bottom of the **Adaptive Degradation and Performance Tuning** dialog box is the **Remind me when a new driver is available** check box. When checked, AutoCAD displays a pop-up message in the graphics area whenever a new graphics card driver is available.

When the settings have been adjusted as desired in the **Adaptive Degradation and Performance Tuning** dialog box, pick the **OK** button to return to the **Options** dialog box. Then, close the **Options** dialog box.

User Preferences

A variety of settings are found in the **User Preferences** tab of the **Options** dialog box. See **Figure 21-12.** AutoCAD allows users to optimize the way they work in AutoCAD by providing options for double-click editing, shortcut menu functions, **DesignCenter** units, working with fields, coordinate data entry, associative dimensions, hyperlink icon display, undo/redo control, default line weight settings, and scale list settings. All of these are controlled in this tab.

Figure 21-12.
The **User Preferences** tab allows you to set up AutoCAD in a manner that works best for you.

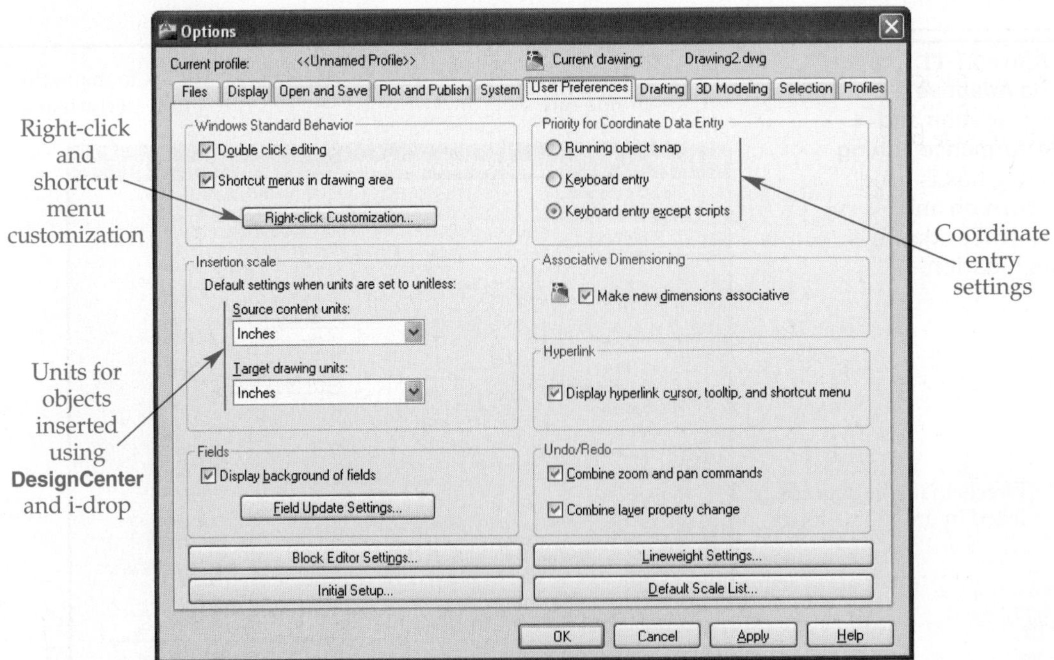

Shortcut Menus and Double-Click Editing

AutoCAD has tools that provide easy access to commonly used editing commands and options. Two of these tools are double-click editing and shortcut menus. Double-clicking on an object calls the most appropriate editing tool for that object type, often the **Properties** palette. Shortcut menus are displayed by right-clicking and are ***context sensitive.*** This means the options available in the shortcut menu are determined by the active command, cursor location, or selected object.

To enable double-click editing, check the **Double click editing** check box in the **Windows Standard Behavior** area in the **User Preferences** tab of the **Options** dialog box. To enable shortcut menus, check the **Shortcut menus in drawing area** check box in the same area. Disabling the shortcut menus makes a right mouse click the equivalent of pressing the [Enter] key. In general, this is not recommended.

You can also customize the setting for the right mouse button. Pick the **Right-click Customization...** button to access the **Right-Click Customization** dialog box. See Figure 21-13. The **Turn on time-sensitive right-click:** check box controls the right-click behavior. A quick click is the same as pressing [Enter]. A longer click displays a shortcut menu. You can set the duration of the longer click in milliseconds. If the check box is checked, the **Default Mode** and **Command Mode** areas of the dialog box are disabled.

Different settings can be used for the three different shortcut menu modes. Each of the three menu modes has a separate area in the **Right-Click Customization** dialog box.

- **Default Mode.** In this mode, no objects are selected and no command is active. The **Repeat Last Command** option activates the last command issued. The **Shortcut Menu** option displays the shortcut menu.
- **Edit Mode.** In this mode, an object is selected, but no command is active. The **Repeat Last Command** option activates the last command issued. The **Shortcut Menu** option displays the shortcut menu.
- **Command Mode.** In this mode, a command is active. The **ENTER** option makes a right-click the same as pressing [Enter]. The **Shortcut Menu: always enabled** option means that the shortcut menu is always displayed in command mode. The **Shortcut Menu: enabled when command options are present** option means the shortcut menu is only displayed when command options are available on the command line. When there are no command options, a right-click is the same as [Enter]. This is the default option.

Figure 21-13.
Use this dialog box to customize the right mouse button.

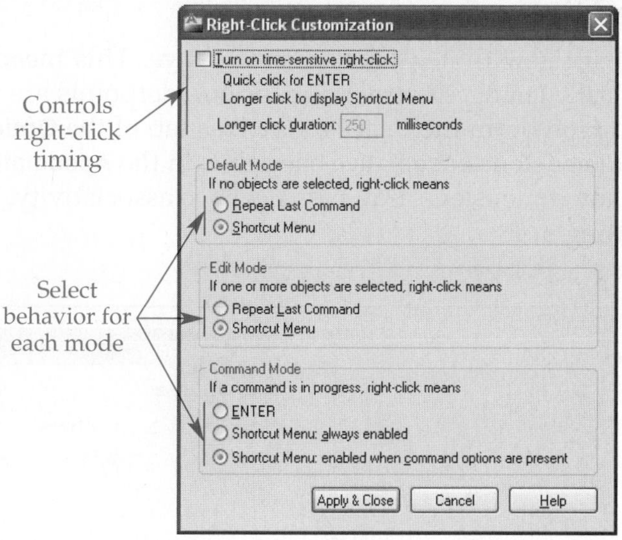

Controls right-click timing

Select behavior for each mode

Right-Click Customization
Turn on time-sensitive right-click:
Quick click for ENTER
Longer click to display Shortcut Menu
Longer click duration: 250 milliseconds
Default Mode
If no objects are selected, right-click means
Repeat Last Command
Shortcut Menu
Edit Mode
If one or more objects are selected, right-click means
Repeat Last Command
Shortcut Menu
Command Mode
If a command is in progress, right-click means
ENTER
Shortcut Menu: always enabled
Shortcut Menu: enabled when command options are present
Apply & Close | Cancel | Help

Insertion Scale

In the **Insertion scale** area of the **User Preferences** tab, unit values can be set for objects when they are inserted into a drawing. This applies to "unitless" objects dragged from **DesignCenter** or inserted using the i-drop method. The **Source content units:** setting specifies the units for objects being inserted into the current drawing. The **Target drawing units:** setting determines the units in the current drawing. These settings are used when there are no units set with the **INSUNITS** system variable.

Fields

A *field* is a special type of text object that displays a specific property value, setting, or characteristic. Fields can display information related to a specific object, general drawing properties, or information related to the current user or computer system. The text displayed in the field can change if the value being displayed changes. Refer to *AutoCAD and Its Applications—Basics* for more information on using fields.

In the **Fields** area of the **User Preferences** tab, you can set whether or not a field is displayed with a nonplotting background. When the **Display background of fields** check box is checked, the field background is displayed in light gray.

Picking the **Field Update Settings...** button in the **Fields** area opens the **Field Update Settings** dialog box, Figure 21-14. In this dialog box, you can set when fields are automatically updated. The five options are **Open**, **Save**, **Plot**, **eTransmit**, and **Regen**. Check as many of the options as appropriate.

Coordinate Data Priority

The **Priority for Coordinate Data Entry** area of the **User Preferences** tab controls how AutoCAD responds to input of coordinate data. The system variable equivalent for this is **OSNAPCOORD**. The three options are:

- **Running object snap.** When this option is selected, object snaps always override coordinate entry. This is equivalent to an **OSNAPCOORD** setting of 0.
- **Keyboard entry.** When this option is selected, coordinate entry always overrides object snaps. This is equivalent to an **OSNAPCOORD** setting of 1.
- **Keyboard entry except scripts.** When this option is selected, coordinate entry will override object snaps except those object snaps contained within scripts. This is the default and equivalent to an **OSNAPCOORD** setting of 2.

Associative Dimensions

By default, all new dimensions are associative. This means that the dimension value automatically changes when a dimension's defpoints are moved. However, you can turn this option off in the **User Preferences** tab of the **Options** dialog box. When the **Make new dimensions associative** check box in the **Associative Dimensioning** area is unchecked, any dimensions drawn do *not* have associativity. This is equivalent to a **DIMASSOC** setting of 1.

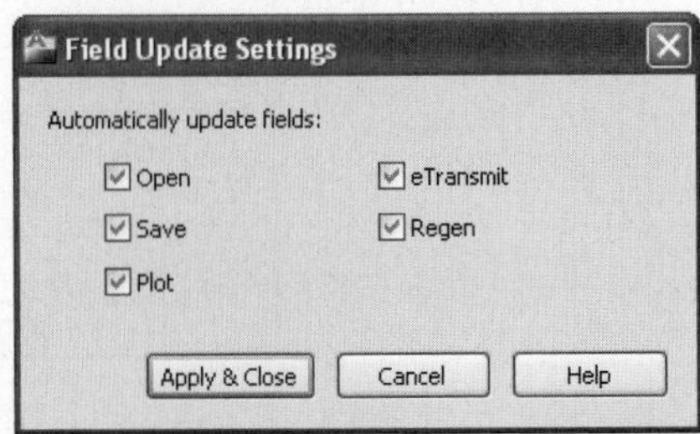

Figure 21-14.
Setting when fields are automatically updated.

Hyperlinks

In the **Hyperlink** area of the **User Preferences** tab, you can set whether or not the hyperlink cursor and tooltip are displayed when the cursor is over a hyperlink. If **Display hyperlink cursor, tooltip, and shortcut menu** is checked, the hyperlink icon appears next to the crosshairs when the cursor is over an object containing a hyperlink. The tooltip is also displayed. Additional hyperlink options are available from the shortcut menu when an object with a hyperlink is selected.

Undo/Redo

The **Undo/Redo** area of the **User Preferences** tab allows you to control how multiple, consecutive zooms and pans are handled within the **UNDO** and **REDO** commands. By checking the **Combine zoom and pan commands** check box, back-to-back zooms and pans are considered a single operation for undo and redo purposes. In other words, performing an undo or redo undoes or redoes the entire zoom/pan sequence. Unchecking the check box allows each zoom or pan to be considered a separate operation.

When the **Combine layer property change** check box is checked, all changes made in the **Layer Properties Manager** palette are considered one operation. If this is unchecked, each change is considered a separate operation when using **UNDO** and **REDO**.

Block Editor Settings, Initial Setup, Line weight Settings, and Edit Scale List

At the bottom of the **User Preferences** tab are the **Block Editor Settings...**, **Initial Setup...**, **Lineweight Settings...**, and **Default Scale List...** buttons. Picking one of these buttons opens a dialog box with corresponding settings.

Picking the **Block Editor Settings...** button opens a dialog box that contains settings for the colors, fonts, and sizes of the various elements that appear when working in the **Block Editor** window, Figure 21-15. The **Block Editor** window is discussed in detail in *AutoCAD and Its Applications—Basics.* At the bottom of the dialog box there are three check boxes that control the highlight of dependent objects during selection, display of tick marks for parameters with value sets, and display of action bars. These check boxes control the **BDEPENDENCYHIGHLIGHT**, **BTMARKDISPLAY**, and **BACTIONBARMODE** system variables.

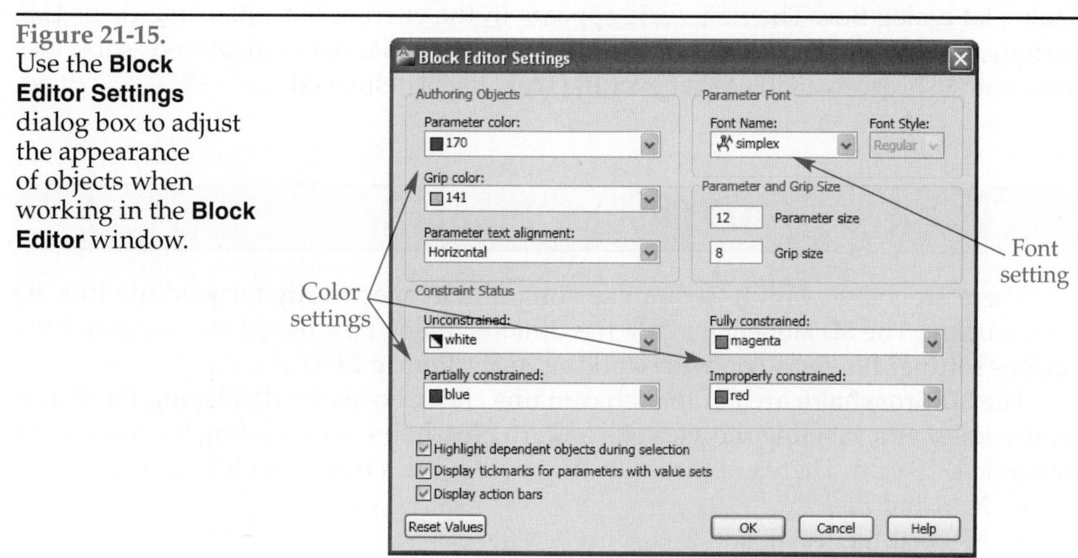

Figure 21-15. Use the **Block Editor Settings** dialog box to adjust the appearance of objects when working in the **Block Editor** window.

Figure 21-16.
The **Edit Scale List** dialog box allows you to change the scale list that appears when using various viewport, page setup, and plot scaling dialog boxes.

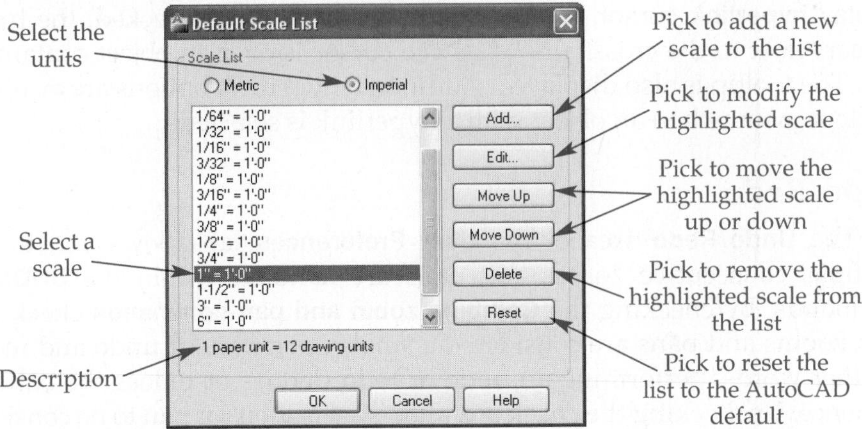

Picking the **Initial Setup...** button on the **User Preferences** tab opens the **Initial Setup** dialog box that was presented when installing the software. This dialog box can be used to adjust any of the settings chosen at installation.

The **Lineweight Settings...** button opens the **Lineweight Settings** dialog box in which you can change default line weight settings. This is discussed in detail in *AutoCAD and Its Applications—Basics.*

A default list of scales appears in various dialog boxes related to viewports, page setups, and plot scaling. You can add custom scales to or remove scales from this list so that it is more appropriate for your application. Picking the **Default Scale List...** button at the bottom of the **User Preferences** tab displays the **Edit Scale List** dialog box. See **Figure 21-16.** The dialog box displays one of the two available lists of scales. The radio buttons at the top of the dialog determine whether a metric scale list or a U.S. customary (Imperial) scale list is displayed. The buttons on the right side of the dialog box allow you to add a new scale, edit an existing scale, move a scale up or down within the list, delete a scale, or reset the list to the default set of scales.

To add a scale, pick the **Add...** button. In the **Add Scale** dialog box that appears, enter a name for the scale in the **Name appearing in scale list:** text box. Then, in the **Scale Properties** area of the dialog box, enter values to indicate how many paper space units equal how many drawing units. Finally, pick the **OK** button to return to the **Edit Scale List** dialog box. The new scale appears in the current list, either metric or U.S. customary (Imperial), and is available wherever the scale list is displayed. All scales from both lists are available whenever the scale list is displayed.

3D Display Properties

There are many ways to customize your system specifically for working in a 3D environment. The **3D Modeling** tab of the **Options** dialog box allows you to control the various settings having to do with working in 3D, **Figure 21-17.**

The **3D Crosshairs** area of the tab contains check boxes for displaying the Z axis on the crosshairs, labeling the axes of standard crosshairs, and labeling the axes of the dynamic UCS icon. There are three labeling possibilities from which to choose:

- X, Y, and Z.
- N (north), E (east), and z.
- Or you can specify custom labels for each axes.

Figure 21-17.
Use the **3D Modeling** tab to customize settings when working in 3D.

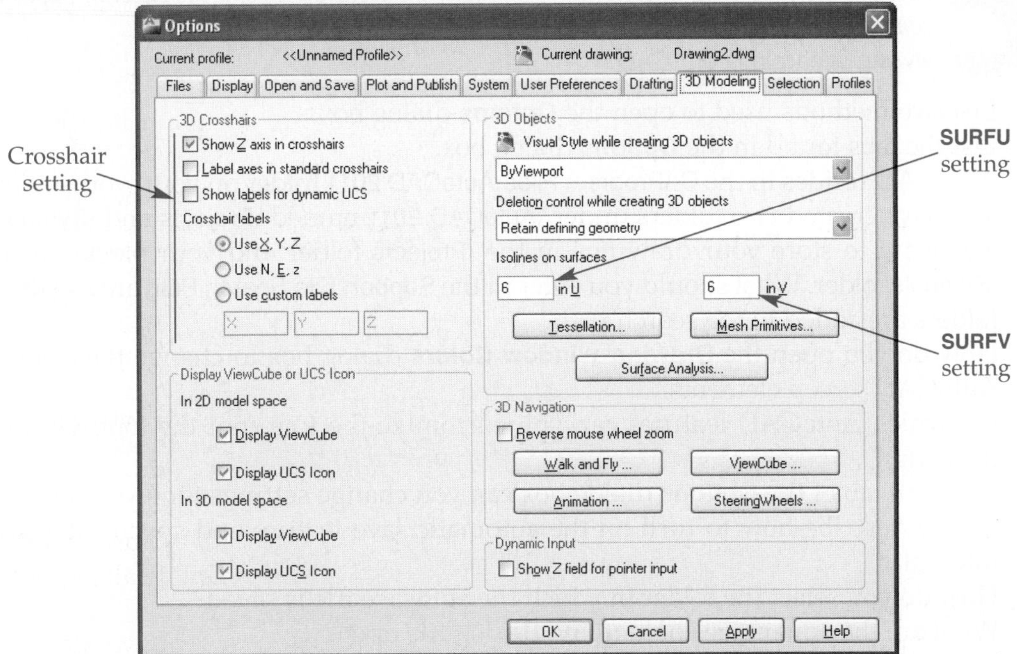

The check boxes in the **Display View Cube or UCS Icon** area determine if the UCS icon and the view cube are displayed in 2D model space. You can choose to have either or both displayed in either 2D or 3D model space. For example, if the view cube is disabled for the 2D model space, it will not be displayed when the 2D Wireframe visual style is set current.

The setting in the **Visual Style while creating 3D objects** drop-down list in the **3D Objects** area determines which visual style is set current when objects are created. The **Deletion control while creating 3D objects** drop-down list determines how geometry is handled when creating 3D objects. For example, when **Delete profile curves** is selected, the profile and path curves are deleted after a sweep is created. The two edit boxes in the **3D Objects** area set the **SURFU** and **SURFV** system variables for mesh objects and old-style surfaces. The **Tessellation...** and **Mesh Primitives...** buttons open dialog boxes that contain settings for controlling the options associated with mesh tessellation and mesh primitives. These settings are discussed thoroughly in Chapter 3. The **Surface Analysis...** button opens a dialog box for changing the display of zebra, curvature, or draft angle analyses.

The **3D Navigation** area has a check box for reversing the zoom direction of the mouse wheel. There are also four buttons in this area that allow access to settings for walkthroughs/flybys, animations, the view cube, and steering wheels. Selecting the **Walk and Fly settings...** button opens the **Walk and Fly Settings** dialog box. This dialog box contains settings used when creating walkthroughs and flybys. Selecting the **Animation settings...** button opens the **Animation Settings** dialog box. This dialog box contains settings that control the actual animation of a walkthrough or flyby. The **Walk and Fly Settings** and **Animation Settings** dialog boxes are discussed in detail in Chapter 19. Selecting the **View Cube...** button opens the **View Cube Settings** dialog box. Selecting the **Steering Wheels...** button opens the **Steering Wheels Settings** dialog box. These dialog boxes are discussed in Chapter 4. The **Dynamic Input** area has a check box for showing the Z field for dynamic input.

Chapter Test

Answer the following questions. Write your answers on a separate sheet of paper or complete the electronic chapter test on the student website.
www.g-wlearning.com/CAD

1. List two methods used to open the **Options** dialog box.
2. List the tabs found in the **Options** dialog box.
3. AutoCAD resides in the C:\Program Files\AutoCAD 2011 folder on your workstation. You have created two folders under \AutoCAD 2011 named \Projects and \Symbols. You want to store your drawings in the \Projects folder and your blocks in the \Symbols folder. What should you enter in the Support File Search Path area so these folders are added to the search path?
4. How do you open the **Drawing Window Colors** dialog box to change the color of AutoCAD screen elements?
5. For which AutoCAD features can you customize the font (not the font within a drawing)?
6. In which tab of the **Options** dialog box can you change settings for layout tabs?
7. Briefly describe how to turn on the automatic save feature and specify the save interval.
8. How do you select the folder in which the autosave file is saved?
9. What are the advantages of toggling the log file open?
10. Name the two commands that toggle the log file on and off.
11. In which tab of the **Options** dialog box can you determine whether the help command accesses Autodesk's website or a local file?
12. How would you set the right mouse button to perform an [Enter], rather than displaying shortcut menus?
13. How do you open the **Edit Scale List** dialog box from within the **Options** dialog box?
14. On which tab of the **Options** dialog box is the **Block Editor Settings...** button located?
15. How do you access the view cube setting from within the **Options** dialog box?

Drawing Problems

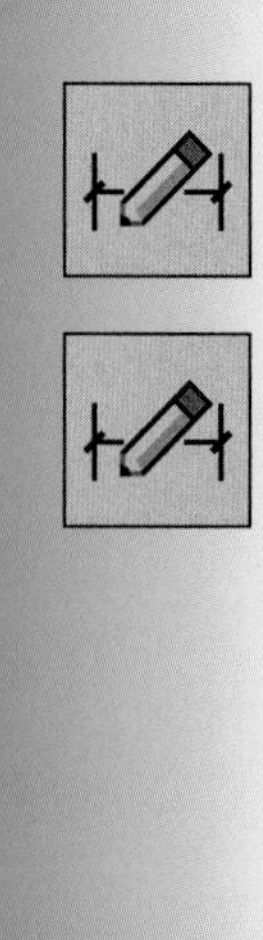

1. Using the methods described in this chapter, change the AutoCAD screen colors to your liking. Customize the 3D parallel and 3D perspective contexts.

2. Set up AutoCAD so that profile curves are retained when creating 3D objects. Also, increase the isolines for meshes. To determine the best default setting, draw various meshes, change the setting, and draw additional meshes.

CHAPTER 22

Customizing Tools and Tool Locations

Learning Objectives

After completing this chapter, you will be able to:

✓ Explain the features of the **Customize User Interface** dialog box.
✓ Describe partial CUIx files.
✓ Create custom commands.
✓ Create new toolbars, ribbon tabs, and ribbon panels.
✓ Customize ribbon tabs and panels with submenus and drop-down lists.
✓ Explain how to customize menus.

One of the easiest ways to alter the AutoCAD environment is by customizing interface elements. ***Interface elements*** are graphic command-entry components of AutoCAD, such as the ribbon, toolbars, and menus. They can be quickly modified by removing and adding commands. New commands can also be created and assigned to an existing ribbon panel, toolbar, or menu or to new interface elements. The most powerful aspect of customizing ribbon panels, toolbars, and menus is the ability to quickly create entirely new functions to help you in your work. The key to good customization can be broken down into four simple rules:

- Always make a backup of the original files, such as the acad.cuix file, *before* customizing.
- Do not over customize your work. Plan your customization in steps to minimize confusion and maximize productivity. Anticipate workflow and where needs exist.
- It is best to locate customized files in folders other than the default AutoCAD folders. This makes it easier to upgrade AutoCAD in the future.
- Thoroughly test your customizations before implementing them. This will save many headaches for you and the end user.

Customize User Interface Dialog Box

The **Customize User Interface** dialog box funnels all major graphical user interface elements of AutoCAD into one central area where they can be tailored for productivity, Figure 22-1. All AutoCAD commands are linked for customization. The interface elements that will be discussed in this chapter are the ribbon tabs and panels, toolbars, and menus located in the menu bar.

Figure 22-1.
The **Customize User Interface** dialog box is used to edit existing interface elements, create new interface elements, create custom commands, and create command icons.

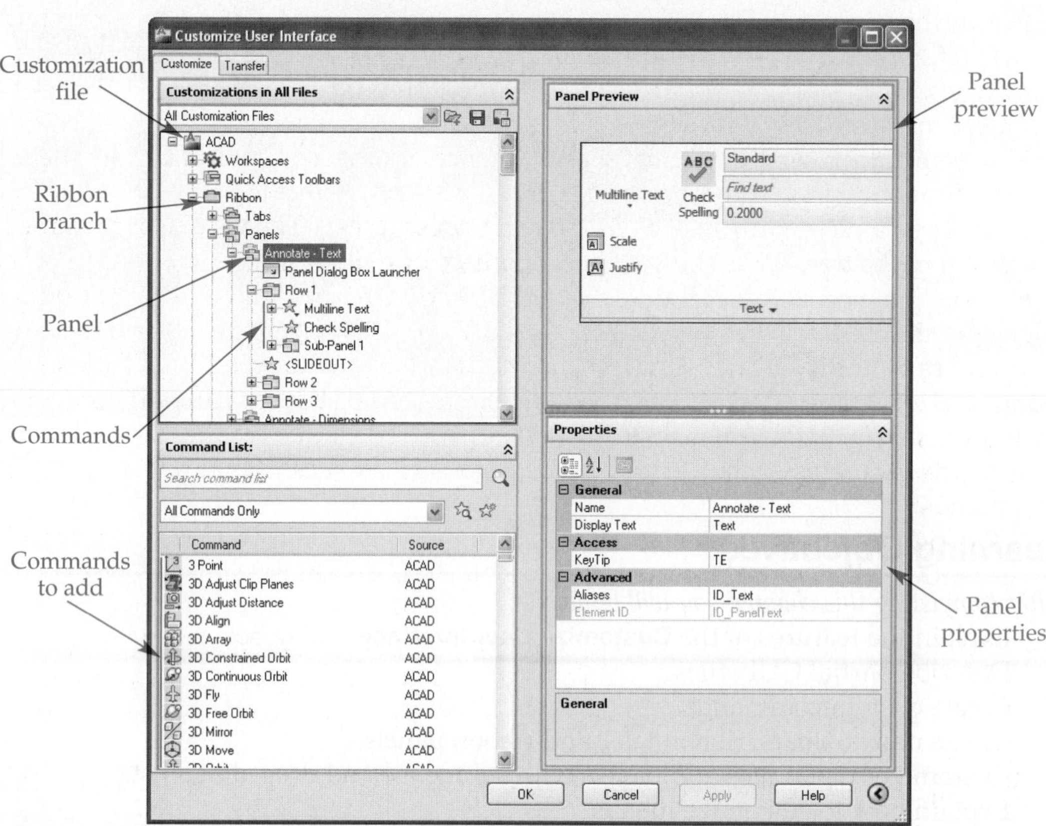

CUI

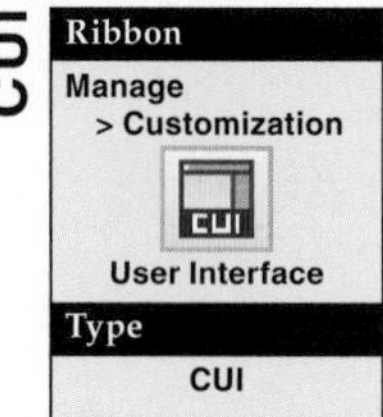

The **CUI** command displays the **Customize User Interface** dialog box. This can also be displayed by right-clicking on a toolbar (when displayed) and picking **Customize...** from the shortcut menu. The changes made in the **Customize User Interface** dialog box are saved in a customization (CUIx) file. By default, this is the acad.cuix file.

The upper-left pane of the **Customize User Interface** dialog box is initially labeled **Customizations in All Files**. The drop-down list located below the pane name contains the name of the main CUIx file and any other currently loaded partial CUIx files. Partial CUIx files are discussed later. By default, the main file acad.cuix and partial CUIx files custom.cuix, autodeskseek.cuix, and acetmain.cuix files are installed. If you select a different entry from the drop-down list, the name of the pane changes to reflect the selection, either **All Customization Files** or **Main CUI File**.

By default, the customization file acad.cuix is the main CUIx file. In the box located below the drop-down list, the selected CUIx file is displayed in a tree. The top level of the tree is the ACAD branch, which is the name of the selected customization file, and the AutoCAD logo icon is shown next to it. The tree under the ACAD branch lists the various customizable interface elements. For example, to see the list of available toolbars, expand the Toolbars branch (node), by picking the plus sign located just to its left. Any other partially loaded CUIx files that have toolbars in them will be listed under the Partial Customization Files branch in the tree and can have their toolbar list similarly expanded.

The shortcut menus displayed in the **Customize User Interface** dialog box provide editing options based on the branch or item selected. Options are available for creating new items; renaming, removing/deleting, copying/pasting, and duplicating items; adding menus or submenus; and inserting separators. For example, to delete a ribbon panel from the interface, expand the Panels branch in the Ribbon branch, locate the desired panel, right-click, and pick **Delete** from the shortcut menu. When prompted, pick the **Yes** button to delete the item. Then, pick the **Apply** or **OK** button in the **Customize User Interface** dialog box to make the deletion permanent. However, a better method is to remove a ribbon panel from the workspace. This way, the ribbon panel is still available to other workspaces. Refer to Chapter 25 for complete details on workspaces.

Content can be renamed. For example, to rename a toolbar, select the toolbar in the Toolbars branch, right-click, and pick **Rename** from the shortcut menu. The existing name of the toolbar in the tree turns into an edit box with the current name highlighted. Type a new name in the edit box and press [Enter]. The new toolbar name is displayed in the tree. You can also rename content by editing the Name property in the **Properties** pane on the right-hand side of the dialog box. Pick the **Apply** or **OK** button in the **Customize User Interface** dialog box to make the change permanent.

You can modify all existing content by deleting and adding commands. The **Command List:** pane of the **Customize User Interface** dialog box contains all commands, including those that are not available by default on an interface element. Custom commands can be assigned to all interface elements, such as ribbon panels, toolbars, and menus.

The right-hand side of the **Customize User Interface** dialog box displays specific information of the highlighted content. Panes that will appear are **Information**, **Preview**, and **Properties**, depending on the selected content. It is easy to read and navigate the **Customize User Interface** dialog box by remembering the general information is stored in the upper-left pane, proceeding down to the **Command List:** pane and over to the panes on the right-hand side of the dialog box for more specific information.

NOTE

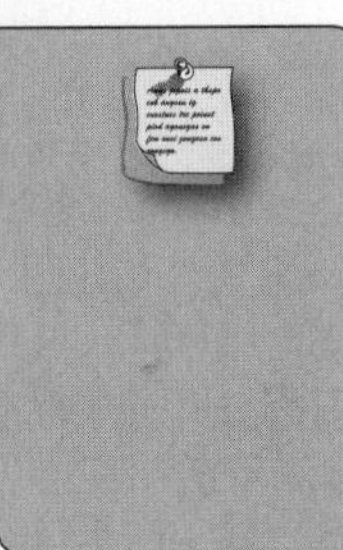

All changes made in the **Customize User Interface** dialog box are saved in the CUIx file, including workspace, ribbon, toolbar, menu, and shortcut key customizations. Customizing the ribbon, toolbars, and menus is discussed later in this chapter. Customizing shortcut keys is discussed in Chapter 23. Customizing workspaces and the **Quick Access** toolbar is covered in Chapter 25.

Adding a Command to an Interface Element

All commands are available in the **Command List:** pane of the **Customize User Interface** dialog box. Included are many commands not found on the default ribbon, toolbars, or menus. Any custom commands you have created are also available in this pane.

To add an existing command to any interface element, first expand the tree for the element in the **Customizations in All Files** pane. For example, to add a command to a toolbar, expand the Toolbars branch so the branch is visible for the toolbar to which you want the command added. Then, select a command from the **Command List:** pane. The list is alphabetized. If you hover the cursor over a command, the macro or command is displayed in a tooltip. See **Figure 22-2.**

You can search the command list by picking the **Find command or text** button at the top of the **Command List:** pane. The drop-down list at the top of the pane can be used to filter the list so that only commands in a certain category appear in the list. You can also filter the list by typing in the text box at the top of the pane. Only those commands containing the characters in this text box are displayed in the list.

Once the command is located, pick and hold on the command in the **Command List:** pane and drag it into the **Customizations in All Files** pane. A horizontal "I-bar" appears in the pane as you drag the command. This represents the location where the

Figure 22-2.
The **Commands List:** pane of the **Customize User Interface** dialog box displays all predefined and custom commands. These commands can be added to interface elements.

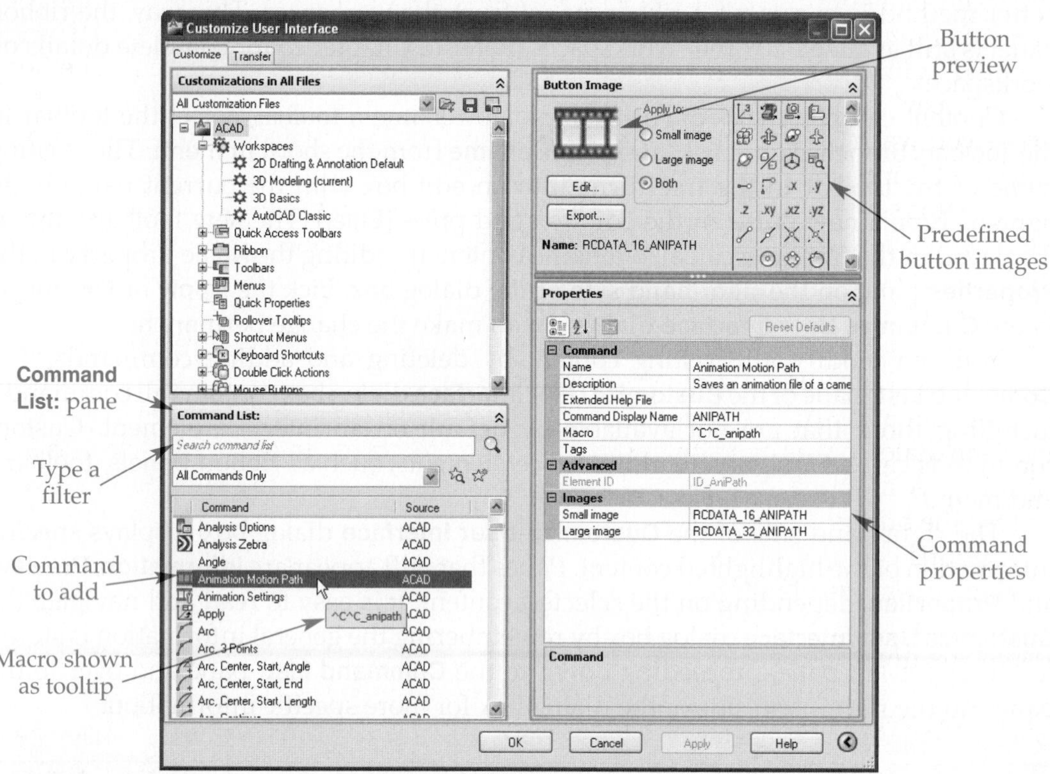

command will be inserted. Position the new command between the commands where you would like it to appear and release the left mouse button. The new command is added to the branch for the interface element. Pick the **Apply** or **OK** button in the **Customize User Interface** dialog box to make the addition permanent.

NOTE

Selecting the **Cancel** button in the **Customize User Interface** dialog box after selecting the **Apply** button does *not* cancel the changes made before the **Apply** button was selected. Also, picking the Windows close button (the X) before picking the **Apply** button *cancels* the changes.

Deleting a Command from an Interface Element

To delete a command from an interface element, expand the tree for the element in the **Customizations in All Files** pane. You may need to expand more than one level to see the command you wish to delete. All of the commands currently on the interface element are displayed as branches below the element name.

Select the command you wish to delete, right-click, and pick **Remove** from the shortcut menu. You can also select the command and press the [Delete] key. Pick the **Apply** or **OK** button in the **Customize User Interface** dialog box to make the deletion permanent. The command is, however, still available in the **Command List:** pane of the **Customize User Interface** dialog box. It can be referenced to another interface element.

Moving and Copying Commands

You can move and copy commands between any interface element. First, in the upper-left pane of the **Customize User Interface** dialog box, expand the tree for both elements that you wish to edit. To move a command from one element to another, pick and hold on the command and drag it to the other interface element. The horizontal "I-bar" cursor appears as you drag. Position the cursor between the commands where you want the new command to appear and release the left mouse button. The command is moved from the first element to the second.

Use this same process to copy a command between elements, but hold the [Ctrl] key before you release the left mouse button. The command remains on the first element and a copy is placed on the second element.

You can also drag commands from the **Customize User Interface** dialog box and drop them onto toolbars and tool palettes that are currently displayed in the AutoCAD window. However, this method cannot be used to add a command tool to the ribbon.

A command can be removed from a displayed toolbar while the **Customize User Interface** dialog box is open by dragging it from the toolbar into the drawing area and releasing. A message appears asking if you want to remove the button. Pick **OK** to remove the button.

Buttons can also be rearranged on displayed toolbars while the **Customize User Interface** dialog box is open by simply dragging a button to a new position. However, it is recommended that you use the **Customize User Interface** dialog box to make edits to toolbars until you are completely comfortable with the drag-and-drop method.

PROFESSIONAL TIP

When dragging a command to a tool palette, if the desired palette is not current (on top), simply pause the cursor over the palette name until the palette is made current. Tool palette customization is discussed in detail in Chapter 24.

Adding a Separator to an Interface Element

A ***separator*** is a vertical or horizontal line that can be used in toolbars and menus to create visual groupings of related commands. On ribbon panels, a separator is a gap between tools. For example, look at the **Draw** panel in the **Home** tab of the ribbon with the 3D Modeling workspace set current. Between the **3D Polyline** button and the **Arc** drop-down list there is a gap between the two indicating a separator. There is also a separator between the **Line** button and the **Circle** drop-down list and between the **Rectangle** button and **Ellipse** drop-down list. These separators are vertical gaps because the ribbon is docked along the top edge. If the ribbon is docked along the left or right side of the screen, the separators will be horizontal gaps. Likewise, if a toolbar is horizontal, the separator is a vertical line. If a toolbar is vertical, the separator is a horizontal line.

To add separators to an interface element, first open the **Customize User Interface** dialog box. Then, in the **Customizations in All Files** pane, expand the branch for the element to which you want separators added.

For a toolbar or menu, right-click on the command in the tree below which you want the separator added. Select **Insert Separator** from the shortcut menu. A separator, represented by two dashes, appears in the tree below the selected command.

For a ribbon panel, right-click on the row in the panel to which you want a separator added. Then, select **Add Separator** from the shortcut menu. A separator is added to the bottom of the row's branch.

Once a separator is added, it can be dragged to a new location in the tree. When done adding and moving separators, pick the **OK** button to close the **Customize User Interface** dialog box.

A ribbon panel can also have a panel separator, called a ***slideout.*** The commands in the tree below the slideout appear in the expanded panel. For example, the **Draw** panel in the **Home** tab of the ribbon (with the 3D Modeling workspace current) has a slideout below the row of buttons containing the **Polygon**, **Rectangle** and **Ellipse** buttons. The slideout is automatically added to all panels.

NOTE

A separator can be removed from an interface element in the same manner as removing a command.

Partial CUIx Files

A ***partial CUIx file*** is any CUIx file that is not the main CUIx file (acad.cuix). To load a partial CUIx file, pick the Open… entry in the drop-down list in the **Customizations in All Files** pane. Remember, the name of this pane may be different, depending on what is currently selected in the drop-down list. You can also pick the **Load partial customization file** button to the right of the drop-down list. Next, in the **Open** dialog box that is displayed, navigate to the folder where the CUIx file is located, select the file, and pick the **Open** button. If the partial CUIx file that has been opened contains any workspaces, the AutoCAD alert shown in **Figure 22-3** is displayed. Any workspace information contained in the CUIx file is not automatically available. Workspaces are covered in Chapter 25.

Once you open the CUIx file, it is automatically selected in the drop-down list. The name of the pane changes to **Customizations in Main CUI**. Now, you can manage the items contained within the partial CUIx.

If you select either the main CUIx or All Customization Files in the drop-down list, the Partial Customization Files branch appears in the tree. Expanding this branch, you can see the partial CUIx files that are loaded. Expanding the branch for a partial CUIx file, you can see the items contained within the CUIx file. These items can be copied from the partial CUI file to the main CUI file as needed.

To unload a partial CUIx file, select All Customization Files in the drop-down list in the "customizations" pane. Then, expand the Partial Customization Files branch, right-click on the name of the CUIx file, and select **Unload** *CUIx_file_name* from the shortcut menu.

PROFESSIONAL TIP

You can also unload a partial CUIx file by typing MENULOAD or MENUUNLOAD at the Command: prompt. Then, in the **Load/Unload Customizations** dialog box, select the CUIx file to unload and pick the **Unload** button.

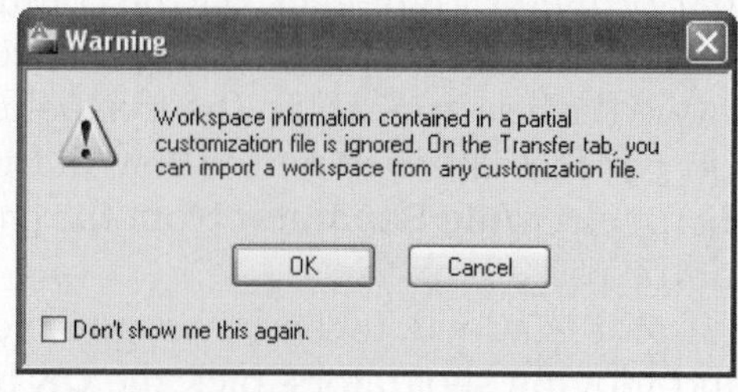

Figure 22-3.
This warning appears when loading a partial CUIx file that contains workspaces. Workspaces are covered in Chapter 25.

Creating New Commands

You are not limited to AutoCAD's predefined commands. ***Custom commands*** can be created and then added to ribbon panels, menus, tool palettes, and toolbars. First, however, you must create the new command. To create a custom command, first pick the **Create a new command** button in the **Command List:** pane of the **Customize User Interface** dialog box. This button is to the right of the drop-down list. A new command is added to the list in the **Command List:** pane. Also, the **Button Image** and **Properties** panes are displayed for the new command. See **Figure 22-4.**

General Command Properties

By default, the new command name is **Command***n*, where *n* is a sequential number. To give the command a descriptive name, highlight the command in the **Command List:** pane. Next, pick in the Name property edit box in the **Command** category of the **Properties** pane. Then, type the new name and press [Enter]. This property is displayed as the command name on the status bar and in the tooltip. The name should be logical and short, such as **Draw Box**. The entry in the Command Display Name property is what appears in the command-line section of the tooltip.

The text that appears in the Description property text box in the **Command** category of the **Properties** pane appears on the AutoCAD status line when the cursor is over the button. This text, called the ***help string,*** should also be logical, but can be longer and more descriptive than the command name.

The Extended Help File property is used to specify an Extensible Application Markup Language (XAML) file to use as extended help. The ***extended help*** is displayed in the tooltip when the cursor is paused over a tool for a longer period of time. By

Figure 22-4.
The first step in adding a custom command to an interface element is to create the custom command.

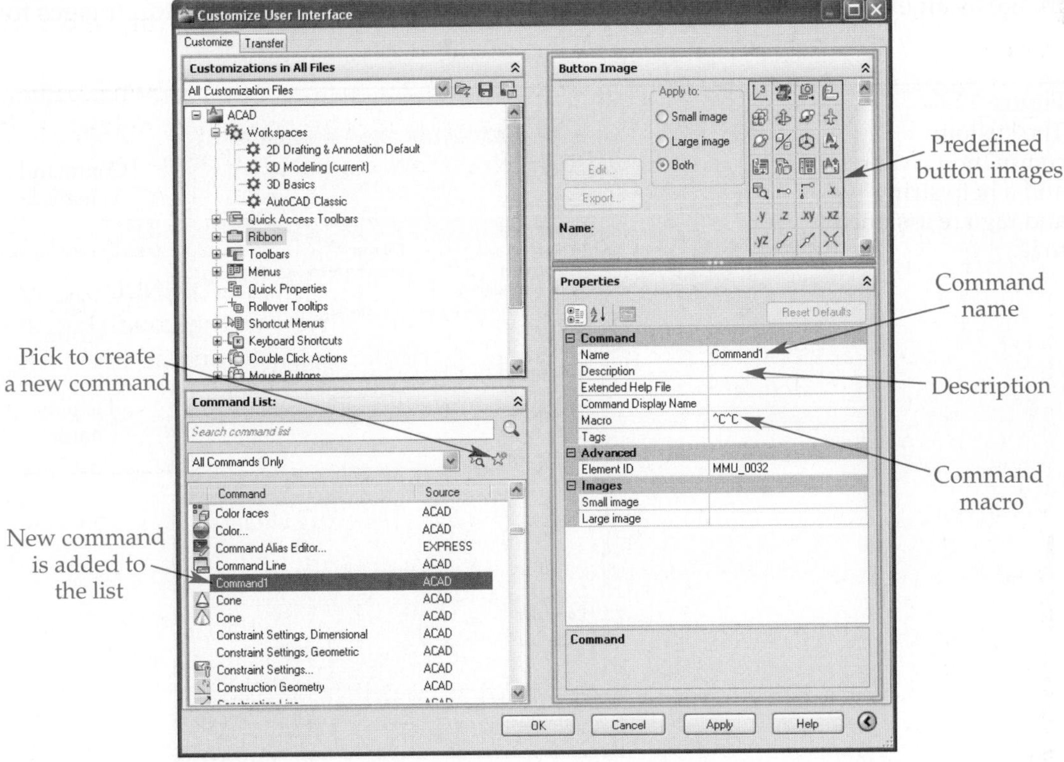

default, if you pause the cursor for two seconds, the extended help is displayed (if the tool contains extended help). To assign an XAML file, select the property and pick the ellipsis button (...) on the right-hand side of the text box. Then, in the standard open dialog box that appears, locate and open the file. For information on creating XAML files, search the Internet for resources. Many resources can be found on the Microsoft website.

As an example, you will create a command that draws a rectangular border for an ANSI E-size sheet (44″ × 34″) using a wide polyline, sets the drawing limits, and finishes with **ZOOM Extents**. To start, create a new command and enter **E-Border** as the name. Also, enter Draws E-size border, sets limits, and zooms extents. for the Description property. In the Command Display Name property, enter Draw Border. See Figure 22-5. In the next sections, you will complete the command and its associated image.

Button Image

The **Button Image** pane in the **Customize User Interface** dialog box is used to define the image that appears on the command button. The image should graphically represent the function of the command. AutoCAD provides several predefined images. One of these can be selected as the button image. You can also right-click on the list of images and select **Import Image...** from the shortcut menu to import an image.

The **CUI Editor—Image Manager** dialog box can be used to control and store custom images in a CUIx file. See Figure 22-6. To display this dialog box, pick the **Image Manager...** button to the right of the drop-down list at the top of the **Customizations in All Files** pane. Any image stored in a loaded CUIx file is available in the list of predefined images.

A different image can be selected for large and small buttons or you can use the same image for both button sizes. It may be a good idea for a button to have a separate image for each of the two button sizes. Pick the appropriate radio button in the **Button Image** pane of the **Customize User Interface** dialog box and select an image. The name of the image appears in the **Images** category in the **Properties** pane. The small image also appears next to the command name in the **Command List:** pane.

However, confusion may arise if your custom command has the same button image as an existing AutoCAD command. It is best to create custom button images for

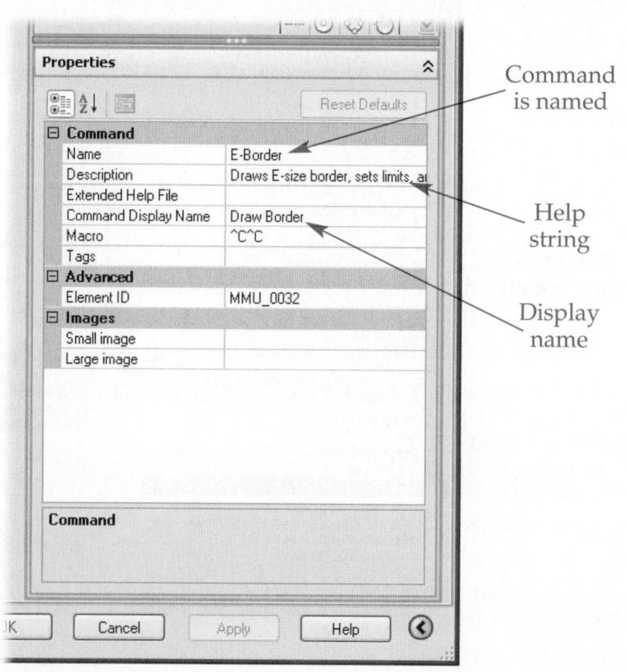

Figure 22-5.
The custom command is named and a help string and tag are assigned to it.

Figure 22-6.
The **CUI Editor—Image Manager** dialog box is used to store button images in a CUIx file.

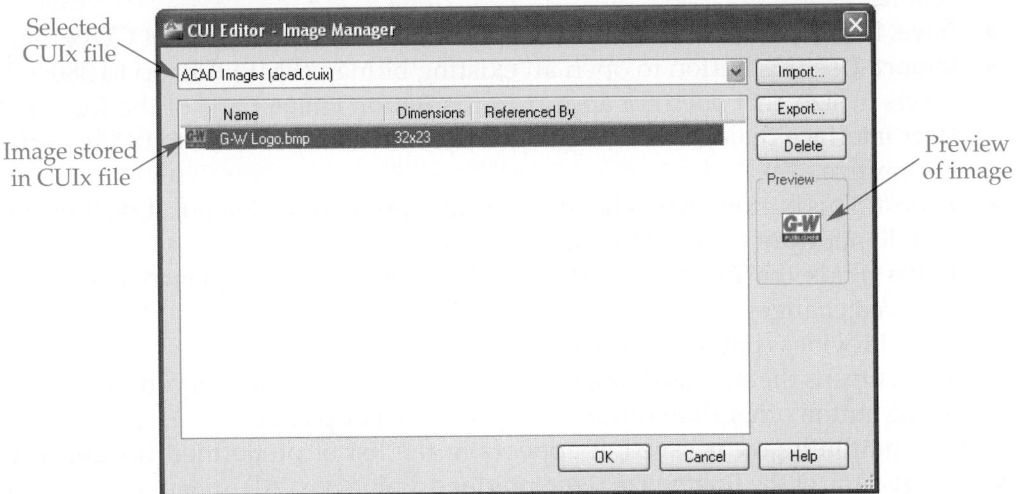

use with your custom commands. The **CUI Editor—Image Manager** dialog box makes it easy to manage the button images. You can either modify an existing button image or create a new image from scratch. In either case, a predefined image must be selected from the list of existing images. Then, pick the **Edit...** button in the **Button Image** pane to open the **Button Editor** dialog box. This is described in the next section.

NOTE

If you use the same image for both small and large buttons, the image is appropriately scaled as needed.

Creating a custom button image

The **Button Editor** dialog box has basic "pixel-painting" tools and several features to simplify the editing process. The four tools are shown as buttons at the top of the dialog box. The pencil paints individual pixels. The line tool allows you to draw a line between two points. The circle tool allows you to draw center/radius style ellipses and circles. The erase tool clears the color from individual pixels. The current color is selected from the color palette on the left-hand side of the dialog box and indicated by a depressed color button. Anything you draw appears in the current color. A preview of the button image appears to the right of the tools.

Drawing a button image is usually much easier with the grid turned on. The grid provides outlines for each pixel in the graphic. Each square represents one pixel. Picking the **Grid** check box toggles the state of the grid.

When the toolbar buttons are set to their default, small size, the button editor provides a drawing area of 16 pixels × 16 pixels. If **Use large buttons for Toolbars** is turned on in the **Display** tab of the **Options** dialog box, then the button image drawing area is 32 pixels × 32 pixels. The images in the **Customize User Interface** dialog box are displayed at the current size setting (small or large).

There are several other tools available in the **Button Editor** dialog box. These include the following.

- **Clear.** If you want to erase everything and start over, pick the **Clear** button to clear the drawing area. This is the button you will use to clear the existing image and start a button image from scratch.

- **Undo.** You can undo the last operation by picking this button. Only the last operation can be undone. An operation that has been undone cannot be redone.
- **Save.** Names the current button image and saves it to the current CUIx file.
- **Import.** Use this button to open an existing bitmap (BMP) file, up to 380 × 380 pixels in size, that does not appear in the **Button Image** pane of the **Customize User Interface** dialog box. The image is automatically resized to fit the current button size.
- **Export.** This button saves a file using a standard save dialog box. Use this when you do not want to alter the original button image.
- **Close.** Ends the **Button Editor** session. A message is displayed if you have unsaved changes.
- **Help.** Provides context-sensitive help.
- **More.** Opens the standard **Select Color** dialog box. This allows you to use colors in the button other than those in the default color palette.

Once a button image is saved, it appears in the list of predefined images in the **Button Image** pane of the **Customize User Interface** dialog box. All images saved for use as button images must be stored where AutoCAD will find them. AutoCAD provides the \Icons folder within the user's support file search path. This is the default folder when using the **Export...** button in the **Button Editor** dialog box. If you choose to use a different folder, it must be added to the support file search path, which is specified in the **Files** tab of the **Options** dialog box.

Rather than using an existing button image for the **E-Border** command, an entirely new button image will be created. With **E-Border** highlighted in the **Command List:** pane, select any one of the images in the **Button Image** pane and pick the **Edit...** button. The **Button Editor** dialog box is displayed. Now, select the **Clear** button to completely remove the existing image.

Figure 22-7A shows a 16 × 16 pixel image created for the **E-Border** button with the **Grid** option activated. Use the pencil and line tools to create this or a similar image. Using the **Save...** button, save your image with a name of E-border; it will be stored in the current CUIx. Pick the **Close** button to return to the **Customize User Interface** dialog box. Your newly created image now appears in the list of existing images in the **Button Image** pane, as shown in **Figure 22-7B.** It is automatically associated with the command.

Figure 22-7.
A—A custom button image is created in the **Button Editor** dialog box. B—The new button image has been saved and appears in the list.

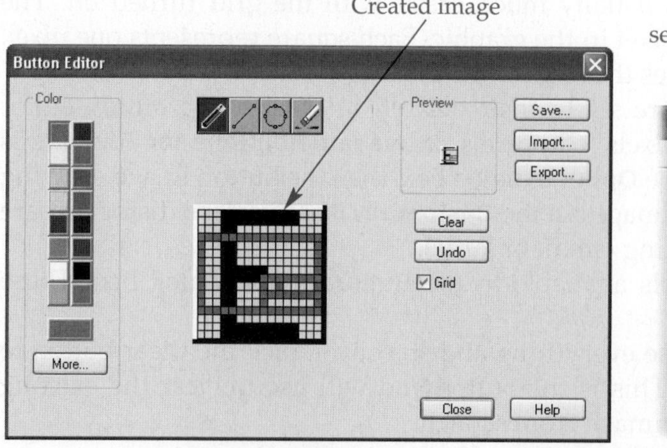

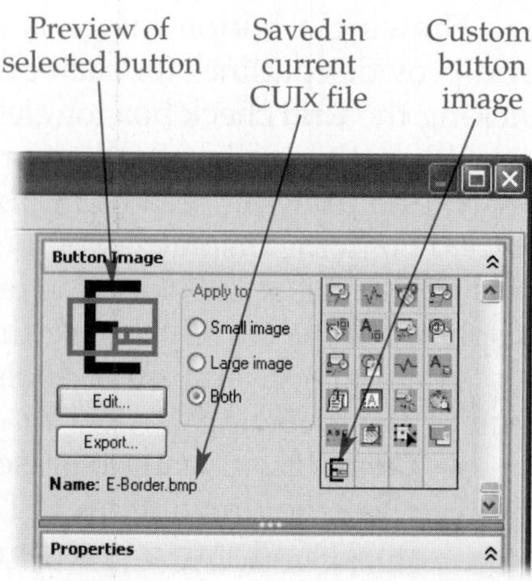

PROFESSIONAL TIP

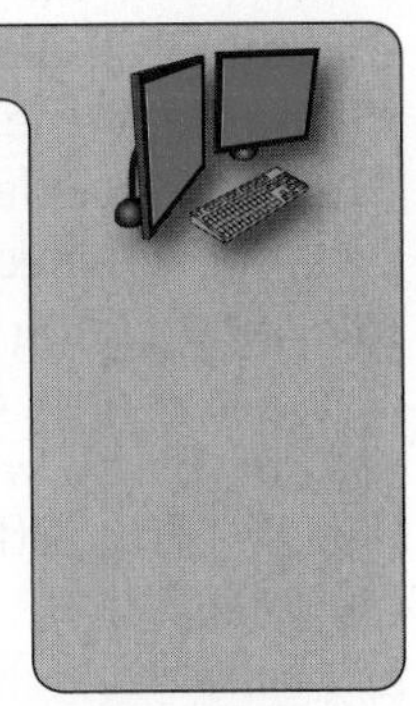

Consider the needs of the persons who will be using your custom commands when you design button images. Simple, abstract designs may be recognizable to you because you created them. However, someone else may not recognize the purpose of the command from the image. For example, the standard buttons in AutoCAD show a graphic that implies something about the command the button executes. A custom command will be most effective if its button image graphically represents the actions the command will perform.

Associating a custom image with a command

There are two ways to associate a new, custom button image with a command. You can use the **Button Image** pane or the **Properties** pane in the **Customize User Interface** dialog box. Once a button image is associated with a command, the image is used for that command on *all* ribbon panels, menus, and toolbars where the command is inserted.

To use the **Button Image** pane to assign an image to a command, first make sure the command is selected in the **Command List:** pane. Then, select the **Large**, **Small**, or **Both** radio button in the **Button Image** pane to determine for which size of button the image will be used. Next, pick the button image in the list of predefined button images. Finally, pick the **Apply** button at the bottom of the **Customize User Interface** dialog box to assign the image to the button.

You can also use the **Properties** pane to associate a saved button image file with the command. Make sure the command is selected in the **Command List:** pane. Then, in the **Properties** pane, expand the **Images** category to display the Small image and Large image properties. If there is an image currently associated with the property, the path to the image is displayed in the text box, Figure 22-8. If there is no path displayed,

Figure 22-8.
The custom button image has been assigned to the custom command.

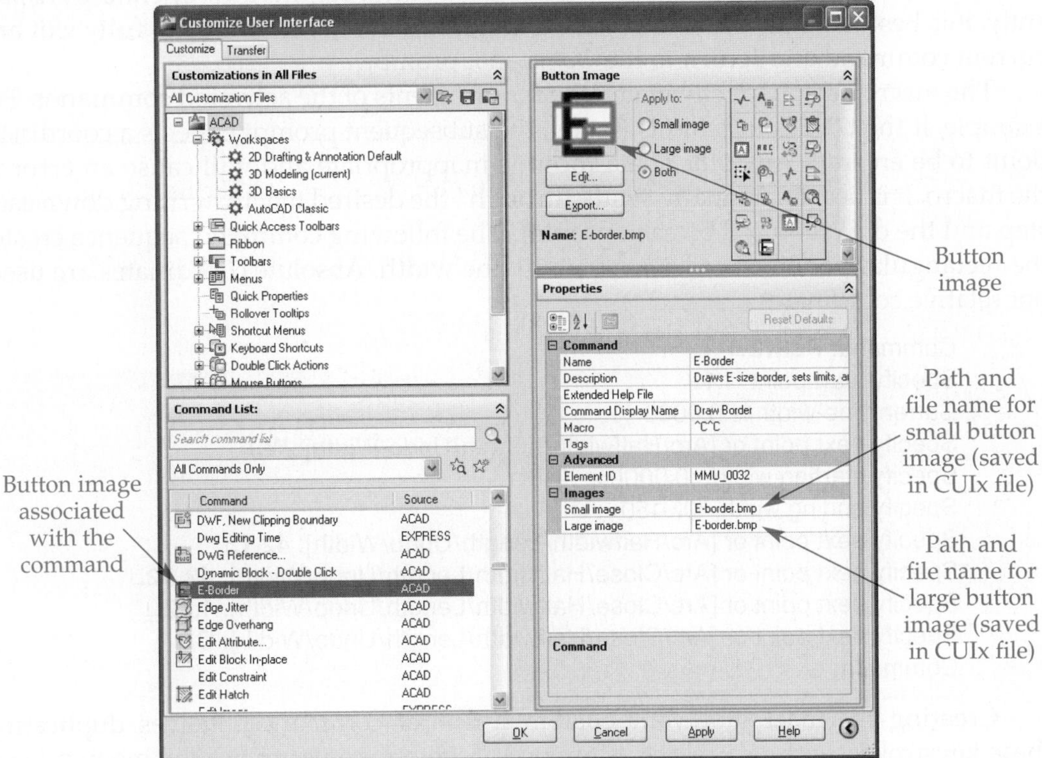

the image is saved in a CUIx file. Pick in each property text box and type the path and file name of the saved image files. Alternately, you can pick the ellipsis button (...) to display a standard open dialog box and locate the file. This button appears when the property is selected. Finally, pick the **Apply** button at the bottom of the **Customize User Interface** dialog box to assign the image(s) to the button.

If you only designate an image file for small buttons, the button for the command will be blank when you switch to large buttons. This is because no image has been designated for that size. Be sure to specify an image for both small and large buttons.

PROFESSIONAL TIP

To open and edit a button image that is not shown in the button image list, select any button image and then pick the **Edit...** button to display the **Button Editor**. Then, use the **Import...** button to open the button image you want to edit. To assign an image that does not appear in the image list, use the **Properties** pane as described above. Additionally, you may use the **CUI Editor—Image Manager** dialog box to import images into a loaded CUIx file, which are then available in the image list.

Defining a Custom Command

Now, you need to define the action that the custom command will perform. A text string called a ***macro*** defines the action performed by the command. This text string appears in the Macro property text box in the **Command** category in the **Properties** pane of the **Customize User Interface** dialog box. In many cases, this "command" is actually a macro that invokes more than one command. By default, the text ^C^C appears in the text box. The text ^C is a cancel command. This is the same as pressing the [Esc] key. The default text, then, represents two cancels.

Two cancels are required to be sure you begin at the Command: prompt. One cancel may not completely exit some commands. In this case, a second cancel is required to fully exit the command. Whenever a command is not required to operate transparently, it is best to begin the macro with two cancel keystrokes (^C^C) to fully exit any current command and return to the Command: prompt.

The macro must perfectly match the requirements of the activated commands. For example, if the **LINE** command is issued, the subsequent prompt expects a coordinate point to be entered. Any other data input is inappropriate and will cause an error in the macro. It is best to manually "walk through" the desired macro, writing down each step and the data required by each prompt. The following command sequence creates the rectangular polyline border with a .015 line width. Absolute coordinates are used, not relative coordinates.

Command: **PLINE**↵
Specify start point: **1,1**↵
Current line-width is 0.0000
Specify next point or [Arc/Halfwidth/Length/Undo/Width]: **W**↵
Specify starting width <0.0000>: **.015**↵
Specify ending width <0.0150>: ↵
Specify next point or [Arc/Halfwidth/Length/Undo/Width]: **42,1**↵
Specify next point or [Arc/Close/Halfwidth/Length/Undo/Width]: **42,32**↵
Specify next point or [Arc/Close/Halfwidth/Length/Undo/Width]: **1,32**↵
Specify next point or [Arc/Close/Halfwidth/Length/Undo/Width]: **C**↵
Command:

Creating the macro for your custom **E-Border** command involves duplicating these keystrokes, with a couple of differences. Some symbols are used in menu macros to represent keystrokes. For example, a cancel (^C) is not entered by pressing [Esc]. Instead, the [Shift]+[6] key combination is used to place the ***caret*** symbol, which is used to represent the [Ctrl] key in combination with the subsequent character (a C in this case). Another keystroke represented by a symbol is the [Enter] key. An [Enter] is placed in a macro as a semicolon (;). A space can also be used to designate [Enter]. However, the semicolon is more commonly used because it is very easy to count to make sure that the correct number of "enters" are supplied.

AutoCAD system variables and control characters can be used in menus. They can be included to increase the speed and usefulness of your menu commands. Become familiar with these variables so you can make use of them in your menus.

- **^B.** Snap mode toggle.
- **^C.** Cancel.
- **^D.** Dynamic UCS toggle.
- **^E.** Isoplane crosshair toggle.
- **^G.** Grid mode toggle.
- **^H.** Issues a backspace.
- **^I.** Issues a tab.
- **^M.** Issues a return.
- **^O.** Ortho mode toggle.
- **^P. MENUECHO** system variable toggle.
- **^Q.** Toggles echoing of prompts, status listings, and input to the printer.
- **^R.** Toggles command versioning. Allows macros written in previous releases to properly function.
- **^T.** Tablet toggle.
- **^V.** Switches current viewport.
- **^Z.** Suppresses the addition of the automatic [Enter] at the end of a command macro.
- **\.** Pauses for user input.

Keeping the above guidelines in mind, the following macro draws the polyline border.

```
^C^CPLINE;1,1;W;.015;;42,1;42,32;1,32;C;
```

Compare this with the command line entry example to identify each part of the macro.

The next steps that the command will perform are to set the limits and zoom to display the entire border. To do this at the command line requires the following entries.

Command: **LIMITS↵**
Reset Model space limits:
Specify lower left corner or [ON/OFF] <0.0000,0.0000>: **0,0↵**
Specify upper right corner <12.0000,9.0000>: **44,34↵**
Command: **ZOOM↵**
Specify corner of window, enter a scale factor (nX or nXP), or
[All/Center/Dynamic/Extents/Previous/Scale/Window/Object] <real time>: **E↵** *(this prompt will differ if the current view is perspective, but the entry is the same)*
Command:

Continue to develop the macro by adding the following text string (shown in color) immediately after the previous one.

```
^C^CPLINE;1,1;W;.015;;42,1;42,32;1,32;C;LIMITS;0,0;44,34;ZOOM;E
```

An "enter" is automatically issued at the end of the macro, so it is not necessary to place a semicolon at the end. The macro for the custom command is now complete.

To assign the macro to your custom **E-Border** command, first make sure the command is selected in the **Command List:** pane of the **Customize User Interface**

Figure 22-9.
The **Long String Editor** dialog box can be used to write longer macros. The text string will automatically "wrap" in this dialog box, which does not affect the macro.

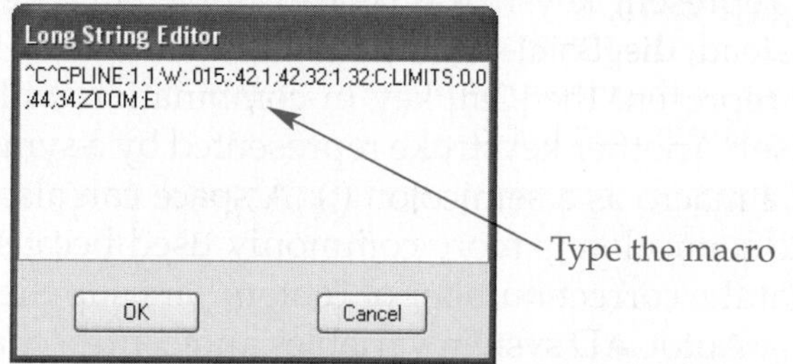

dialog box. Then, pick in the Macro property text box in the **Properties** pane and enter the complete macro shown above. For a long macro such as this one, you can pick the ellipsis button (**...**) at the end of the text box to display the **Long String Editor**. See **Figure 22-9.** Enter the macro in this dialog box and pick the **OK** button to return to the **Customize User Interface** dialog box. Finally, pick the **Apply** button to associate the macro with the custom command.

Exercise 22-1

Complete the exercise on the student website.
www.g-wlearning.com/CAD

Placing a Custom Command on an Interface Element

The custom command is now fully defined. The macro has been written and associated with the command. A custom button image has also been created and associated with the command. Now, you can add the custom command to a ribbon panel, menu, or toolbar just as you would one of the predefined AutoCAD commands. This is introduced earlier and covered in detail later in this chapter in the specific sections on the ribbon, toolbars, and menus.

After adding the custom command to an interface element, it should be fully functional when you exit the **Customize User Interface** dialog box. Once you close the dialog box, test the command to make sure it works. If it does not, edit the macro in the **Customize User Interface** dialog box as needed.

Overview of the Ribbon

By default, the ribbon is docked at the top of the graphics area. The ribbon can be floating or docked to the left or right as well. The ribbon contains commands and tools on ***panels.*** The panels are grouped on ***tabs*** that can be individually displayed. Think of the tabs as the containers that hold the ribbon panels. Together, the panels and tabs make up the ribbon.

You may quickly change from the full ribbon view to one of the minimized options by double-clicking any ribbon tab title. See **Figure 22-10.** By default, the ribbon is set to cycle through all views. To the right of the last tab name is an arrow icon. This button is used to set the appearance of the docked ribbon and its cycling. The three options to set the view of tabs are minimize to tabs, minimize to panel titles, and minimize to panel buttons. When minimized to panel titles or buttons, hover the cursor over the minimized icon and the corresponding panel is displayed. When minimized to tabs, pick the tab name and the tab is displayed.

Figure 22-10.
The four display states of the ribbon when it is docked. Double-click on a tab title to cycle through the views. A—Full. B—Minimized to panel buttons. C—Minimized to panel titles. D—Minimized to tabs.

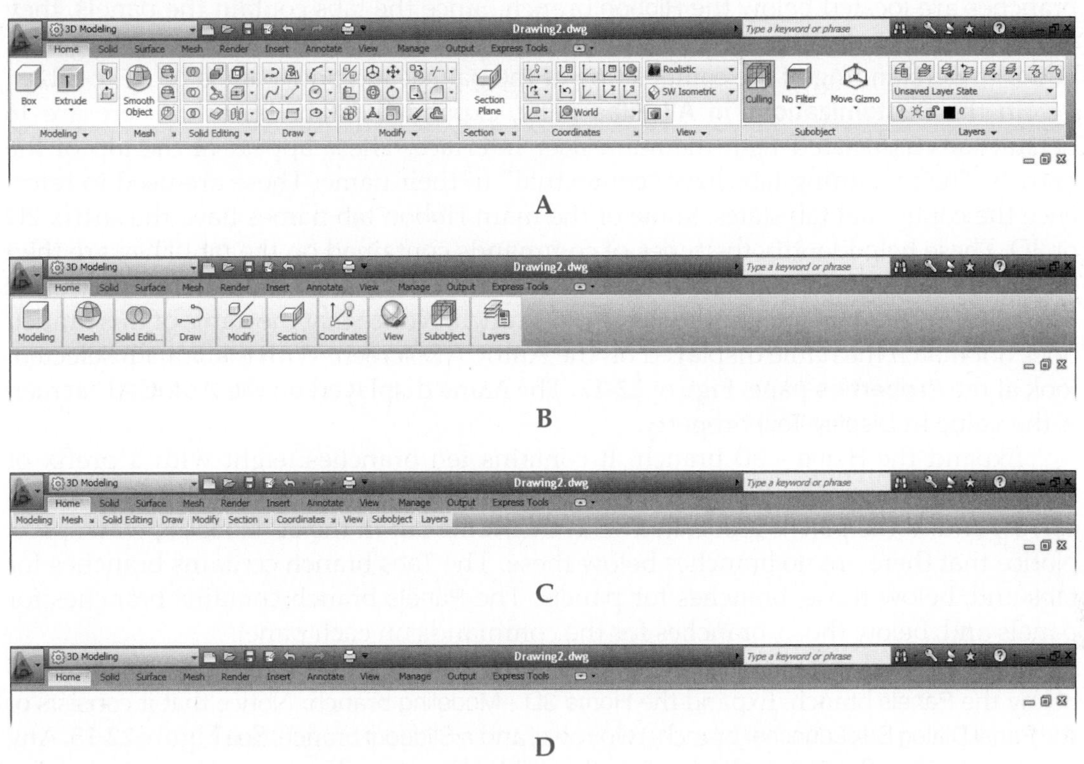

A ribbon panel may contain rows of command buttons, drop-down lists, or sliders. You can choose which panels are visible by right-clicking on the ribbon to display a shortcut menu. See **Figure 22-11.** Select either **Tabs** or **Panels** and choose which content to display. The currently displayed items are checked in the submenus. The items displayed in the **Panels** submenu are based on which tab is current (on top). There are separate panels for each tab.

Workspaces are typically used to set which ribbon components are displayed. Chapter 25 discusses customizing workspaces. Also, the ribbon can dynamically change when commands are accessed. ***Contextual tabs*** may be displayed on the ribbon when a command is active and then hidden when the command is finished. Contextual tabs are discussed further in the next section.

Figure 22-11.
A—The **Show Tabs** submenu is used to choose which tabs are displayed in the ribbon.
B—The **Show Panels** submenu is used to choose which panels are displayed in the current tab.

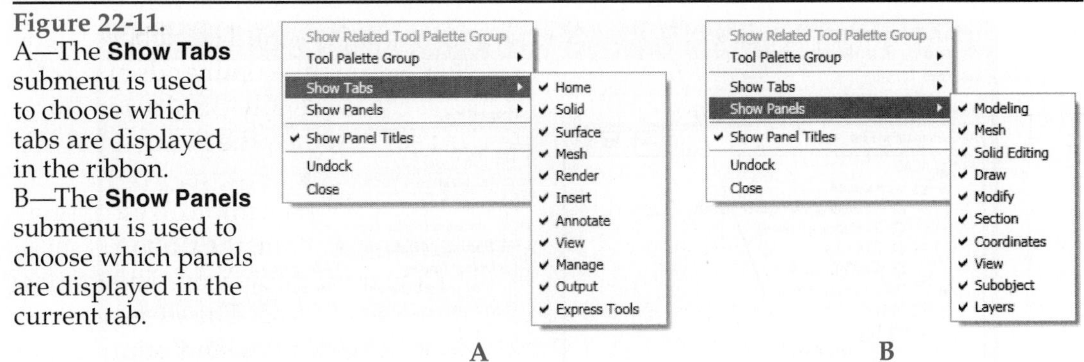

Tabs and Panels

The ribbon has three customization branches in the **Customize User Interface** dialog box: the Tabs branch, Panels branch, and Contextual Tab States branch. These branches are located below the Ribbon branch. Since the tabs contain the panels, they are displayed first in the tree. In this section, you will examine each area to gain a better understanding of the composition of the panels and how panels relate to tabs.

In the **Customizations in All Files** pane, expand the Tabs branch. There are 16 default tabs associated with the main user interface. These appear at the top of the branch. The remaining tabs have "contextual" in their name. These are used to reference the contextual tab states. Some of the main ribbon tab names have the suffix 2D or 3D. These help identify the types of commands contained on the tab. They are then included in the appropriate workspace, either 2D Drafting & Annotation, 3D Modeling, or 3D Basics. For example, select the Home - 3D branch. Notice that the name of this branch does not match the name displayed on the AutoCAD screen. With the branch selected, look at the **Properties** pane, Figure 22-12. The name displayed on the AutoCAD screen is the value in Display Text property.

Expand the Home - 3D branch. It contains ten branches: eight with a prefix of Home 3D and one with a prefix of Home, and one with a prefix of View. These branches correspond to the panels associated with the **Home** tab in the 3D Modeling workspace. Notice that there are no branches below these. The Tabs branch contains branches for tabs and, below those, branches for panels. The Panels branch contains branches for panels and, below those, branches for the commands on each panel.

Now, expand the Panels branch. All of the available panels are shown as branches below the Panels branch. Expand the Home 3D - Modeling branch. Notice that it consists of the Panel Dialog Box Launcher branch, two rows, and a Slideout branch. See Figure 22-13. Any row listed below the Slideout branch is only visible when the ribbon panel is expanded. For the **Modeling** panel in the ribbon, row 2 is located in the expanded portion of the panel.

You can expand the branches for the rows in a tab. Notice row 1 contains two drop-down lists (Solid Primitives Drop-down and Solid Creation Drop-down) and a subpanel. If you expand the Solid Primitives Drop-down branch, you can see the commands associated with the drop-down list. In this case, one drop-down list contains the solid primitive commands and the other one contains commands such as **LOFT** and **EXTRUDE**.

Next, expand the branch for the subpanel (Sub-Panel 1). This subpanel contains three rows. Expanding the branch for each row displays the commands contained in it.

Figure 22-12.
The properties of the Home - 3D tab.

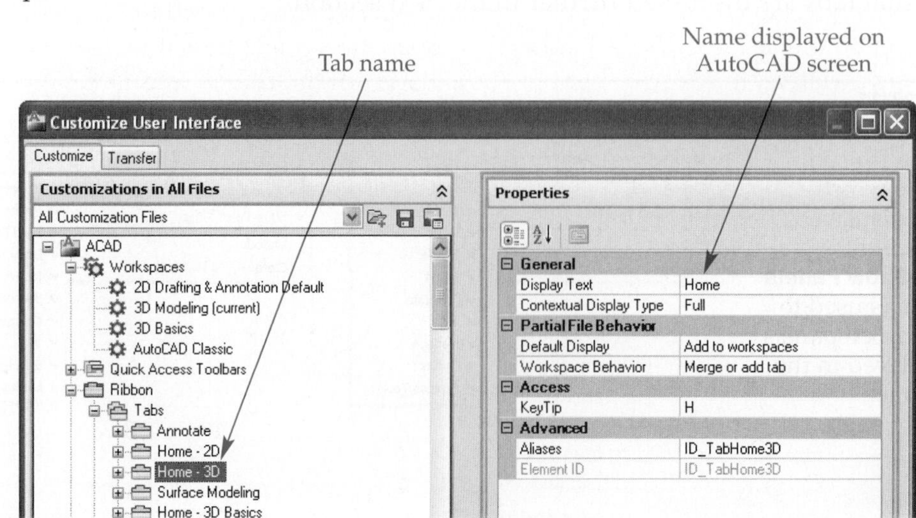

Figure 22-13.
Notice how a ribbon panel is composed in the **Customize User Interface** dialog box.

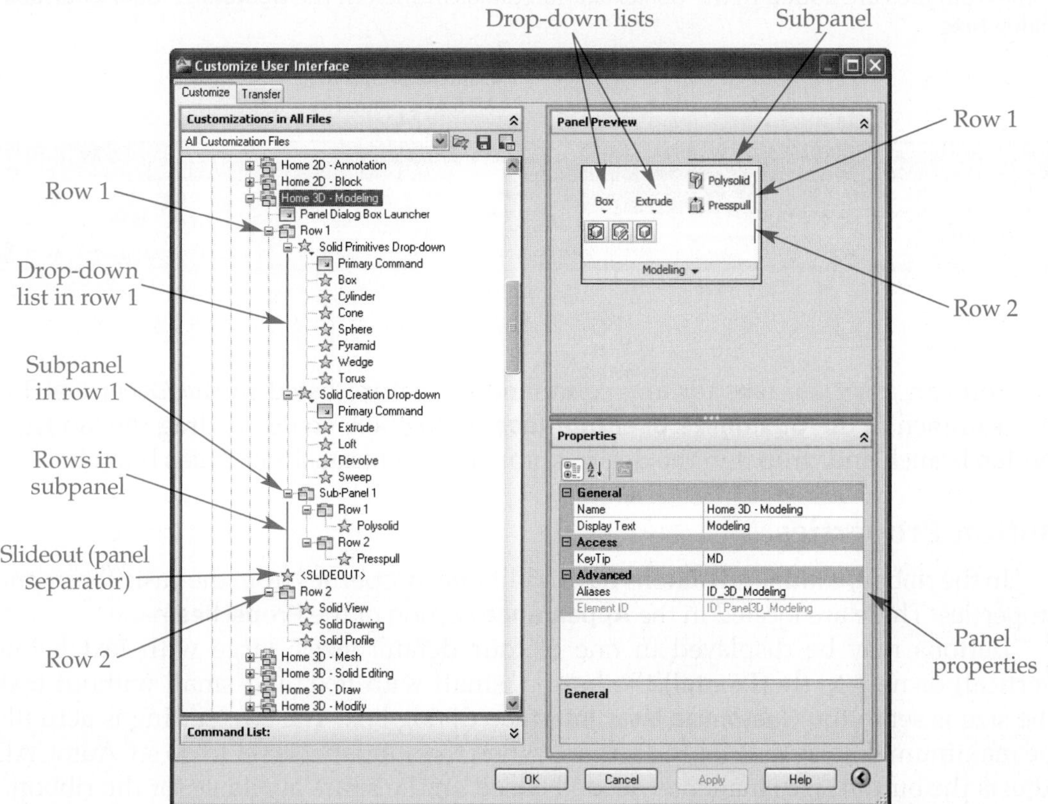

Notice how the drop-down list and subpanel contained in row 1 of the Home 3D Modeling branch are fitted together in the **Modeling** panel in the ribbon. See Figure 22-14.

Expand the branch for row 2 in the Home 3D - Modeling branch. Notice this branch contains commands, but no drop-down lists or subpanels. Also, notice that row 2 is below the Slideout branch, which is the panel separator. This means it is displayed in the expanded panel. Refer to Figure 22-14.

The final branch in the Ribbon branch is Contextual Tab States. Expand this branch and several branches of AutoCAD commands and features are displayed. Some branches have contextual tab panels assigned to them and others do not. For example, expand the Text Editor in progress branch. Below it is the Text Editor Contextual Tab branch. This indicates the **Text Editor** contextual tab will be displayed when text is being created or edited in the drawing.

For example, use the **MTEXT** command to create sample multiline text. While the command is active, the **Text Editor** tab is displayed in the ribbon and made active. See Figure 22-15. When the command is complete, the tab is automatically removed from the ribbon.

Figure 22-14.
Notice how a ribbon panel is composed on the ribbon.

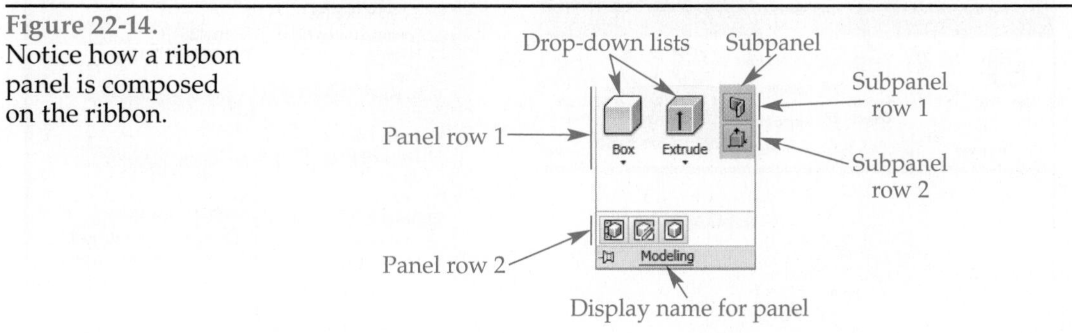

Figure 22-15.
When the **MTEXT** command is active, the **Text Editor** contextual tab is displayed in the ribbon. Contextual tabs are added in the Contextual Tab States branch in the **Customize User Interface** dialog box.

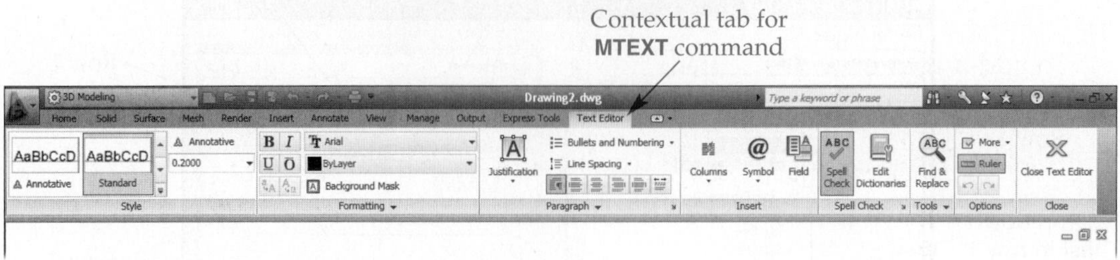

You can associate tabs for any command or feature listed in the Contextual Tab States branch in the **Customize User Interface** dialog box. To do so, drag the tab from the Tab branch and drop it in the desired branch in Contextual Tab States branch.

Button Properties

In the ribbon, command buttons have additional properties from the basic command properties. These are located in the **Appearance** section of the **Properties** pane.

Buttons may be displayed in one of four default sizes: large with text below (vertical) or next to (horizontal) the button, small with text, and small without text. The size is set in the **Customize User Interface** dialog box. The size setting is actually the maximum display size for the button. When the subpanel is set to do so, AutoCAD adjusts the button size smaller as needed based on the space available for the ribbon.

To set the size of a command button, expand the Ribbon branch and then the Panels branch in the **Customizations in All Files** pane. Then, expand the branches for the panel and row that contain the command. Next, select the command in the **Customizations in All Files** pane. If the panel branch is currently selected, you can also pick the button in the **Panel Preview** pane to select the command. Finally, set the Button Style property in the **Appearance** section of the **Properties** pane. See Figure 22-16. Generally, text labels

Figure 22-16.
A—Buttons in a ribbon panel can be displayed in one of four sizes. The small size can be displayed either with or without a label. B—Setting the appearance of a button in a panel.

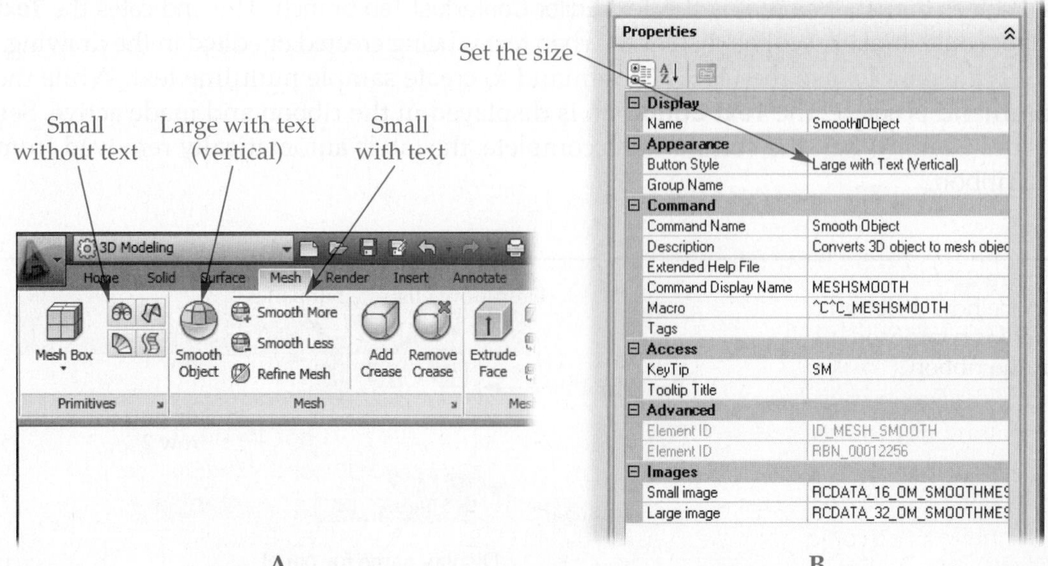

are not shown when small buttons are specified. When large buttons are specified, the vertical orientation is usually selected so the text is below the button. This helps reduce the width of the panel to preserve space on the ribbon.

Customizing a Panel

To add a command to a panel, open the **Customize User Interface** dialog box. In the **Customizations in All Files** pane, expand the Ribbon branch and then the Panels branch. Next, expand the branches for the row to which the command will be added. In the **Commands List:** pane, locate the command to add to the panel. Drag the command from the **Commands List:** pane and drop it into position in the tree in the **Customizations in All Files** pane.

To remove a command from a panel, right-click on the command in the panel's branch in the **Customizations in All Files** pane and select **Remove** from the shortcut menu. You can also select the command and press the [Delete] key.

Row 1 is the top of the panel. In addition, the top of a row branch is the left-hand side of the panel. Commands are displayed in this order on the panel. The commands in a panel can be rearranged. In the **Customizations in All Files** pane, select the command to move and drag it to a new location, either within its current row or in a different row. Rows can also be rearranged by dragging them within the tree in the **Customization in All Files** pane. After you drag a row to a new location, all rows are automatically renumbered. The row at the top of the panel branch is always row 1 and all other rows are sequentially numbered. The panel separator (Slideout branch) can also be dragged to a new location. Remember, rows listed after the panel separator are not displayed until the panel is expanded.

A new row can be added to a panel. In the **Customizations in All Files** pane, right-click on the row *after* which you would like the new row added. To add a new first row, right-click on the panel branch name. Then, select **New Row** from the shortcut menu. The new row is added and all other rows are renumbered. Once a row is added, commands can be added to it.

NOTE

There are certain conditions in which rows in a panel cannot be rearranged. If you attempt to drag a row to a different location and you cannot, just realize you have encountered one of these situations.

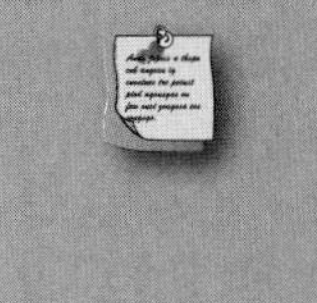

Adding a Drop-Down List to a Ribbon Panel

A drop-down list is added to a row in a ribbon panel using the **Customize User Interface** dialog box. To add a drop-down list, right-click on the row branch in the **Customizations in All Files** pane. Then, select **New Drop-down** from the shortcut menu. A branch for the new drop-down list is added to the bottom of the row's branch. Now you can drag commands from the **Command List:** pane into the drop-down list branch.

Notice the Primary Command branch below the drop-down list branch. The command directly below this is displayed as the button for the drop-down list on the ribbon. For example, the **Modeling** panel on the **Home** tab of the ribbon has the **Box** button displayed for the primitives drop-down list. In the **Customize User Interface** dialog box, the **Box** command is listed directly below the Primary Command branch.

With the branch for the drop-down list selected in the **Customizations in All Files** pane, look at the **Properties** pane. See Figure 22-17. Drop-down lists have appearance properties in addition to the Button Style property: Behavior and Split Button List Style. The Button Style properties are the same as discussed earlier. The other settings determine how the drop-down list on the ribbon is controlled. The Behavior property sets whether the top button in the drop-down list executes a command or displays the

Figure 22-17.
Adding a drop-down list to a ribbon panel.

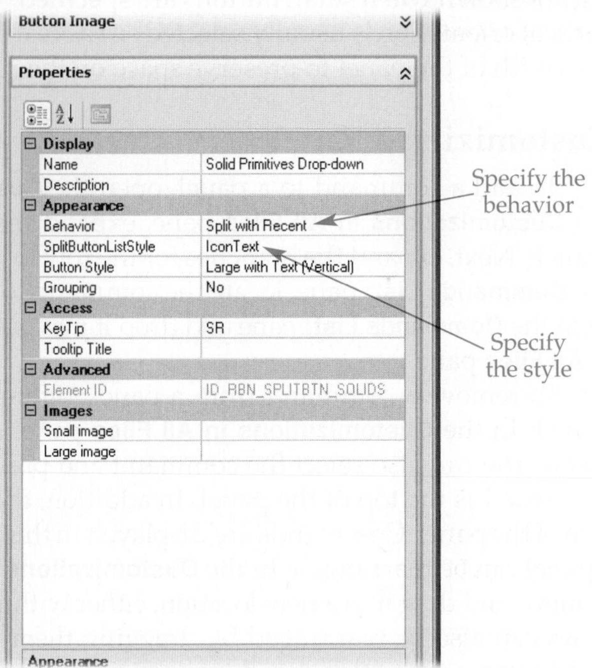

Name property. The Split Button List Style property determines how the buttons and names appear when the list drops down from the panel.

The default Behavior setting is Split with Recent and the default Split Button List Style setting is Icon Text. These settings are typically most desirable for standard AutoCAD workflow. This behavior means that the drop-down list on the ribbon will display in two parts (split). The upper part is the most recent command and the lower part, displayed when the button is pushed, shows additional command icons in the drop-down list.

Creating a New Tab or Panel

To create a new panel, open the **Customize User Interface** dialog box. Then, in the **Customizations in All Files** pane, right-click on the Panels branch and select **New Panel** from the shortcut menu. A new panel with the default name of Panel*x* is added to the bottom of the Panels branch. The name appears in a text box in the tree. Enter a name for the new panel, either in the tree or in the **Properties** pane. Expand the branch for the new panel and notice that the Dialog Box Launcher, Row 1, and Slideout branches are automatically added when the panel is created. Add commands and rows to the new panel as needed. A new tab is similarly created by right-clicking on the Tabs branch.

You will now create a new ribbon panel containing the custom command you created earlier. First, create a new panel and name it My Panel. Then, in the **Command List:** pane, locate the **E-Border** command. Drag the command into the tree in the **Customizations in All Files** panel. The I-bar cursor appears as you drag through the tree. When the I-bar is below the Row 1 branch, drop the command. See Figure 22-18.

Now, create a new tab to hold the new panel. Name the tab My Stuff. In the **Properties** pane for the new tab, change the Display Text property to Border Tools. Next, locate the My Panel branch and drag it into the My Stuff branch. As you drag, the same I-bar appears in the tree. When the I-bar is below the My Stuff branch, release the mouse button.

For the tab to be displayed, it must be added to the current workspace. Workspaces are covered in detail in Chapter 25. To add the tab, select the workspace in the **Customizations in All Files** pane. Then, drag the tab from the **Customizations in All Files** pane and drop it into the Ribbon Tabs branch in the **Workspace Contents** pane.

Figure 22-18.
Adding a command to a custom ribbon panel.

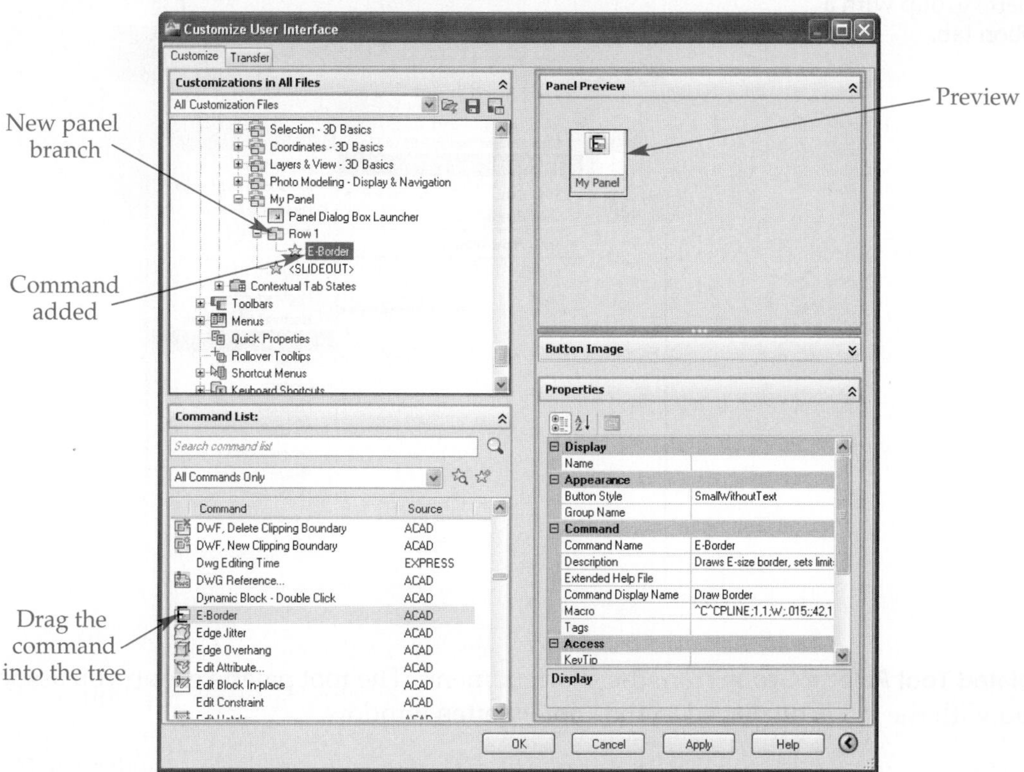

Figure 22-19.
A custom tab, panel, and command have been added to the ribbon. Notice the display name for the tab matches the Display Text property for the tab.

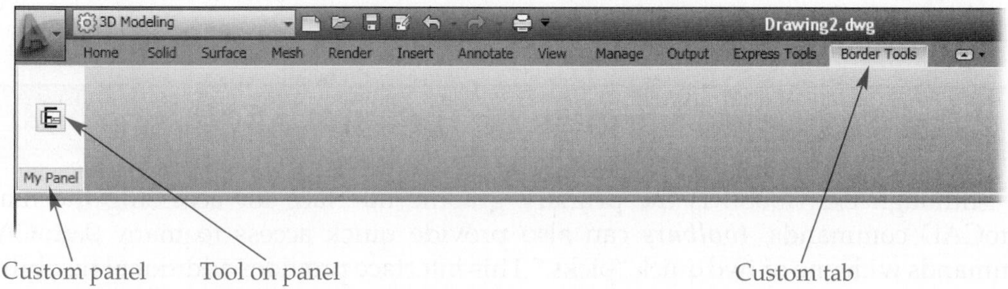

Close the **Customize User Interface** dialog box. After the menu compiles, the new tab is displayed in the ribbon, Figure 22-19. It is located on the right-hand side of the ribbon. Test the **E-Border** button on the panel to make sure the command properly functions.

Associating a Tool Palette Group with a Ribbon Tab

A tool palette group can be associated with a tab in the ribbon. Then, the associated tool palette group is displayed in the **Tool Palettes** window when you right-click on the tab and select **Show Related Tool Palette Group** from the shortcut menu.

To associate a tool palette group with a tab, right-click on the tab in the ribbon (not on a panel). This is done with the **Customize User Interface** dialog box closed. Next, select **Tool Palette Group** in the shortcut menu and then the name of the group in the submenu. See Figure 22-20. You can now right-click on the tab and select **Show**

Figure 22-20.
Associating a tool palette group with a ribbon tab.

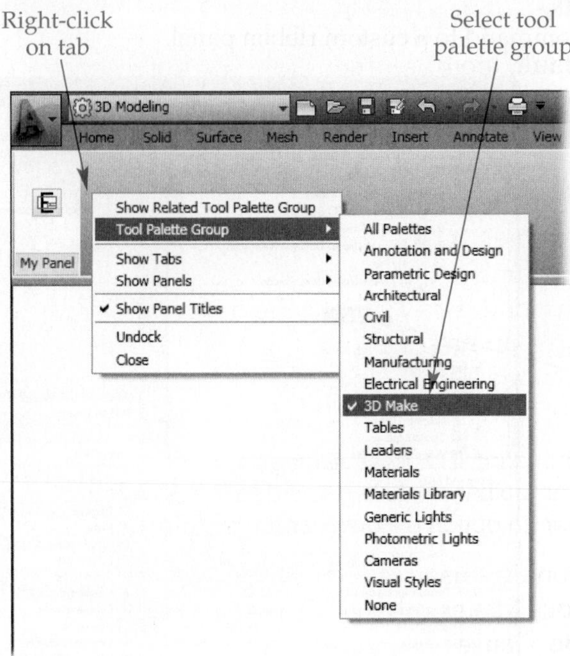

Related Tool Palette Group from the shortcut menu. The tool palette group you associated with the tab is displayed in the **Tool Palettes** window.

Exercise 22-2

Complete the exercise on the student website.
www.g-wlearning.com/CAD

Overview of Toolbars

Although the ribbon is the primary graphic interface for accessing the main AutoCAD commands, ***toolbars*** can also provide quick access to many AutoCAD commands with one or two quick "picks." This interface provides additional flexibility, especially considering toolbars are a fraction of the size of the full ribbon. Toolbars are moved, resized, docked, and floated in the same way as in all Windows-compatible software.

In addition to positioning and sizing toolbars, you can customize the toolbar interface. In the **Customize User Interface** dialog box, all toolbars, except the **Quick Access** toolbar, are listed in the Toolbars branch in the **Customizations in All Files** pane. The **Quick Access** toolbar is customized via workspaces, which are discussed in Chapter 25.

When a command is placed on a toolbar, it is represented by a button. You can add new command buttons or reposition existing command buttons for quicker access. Infrequently used commands can be removed from the toolbar or repositioned to a less prominent location. Entirely new toolbars can be created and filled with predefined or custom commands.

Toolbar Visibility

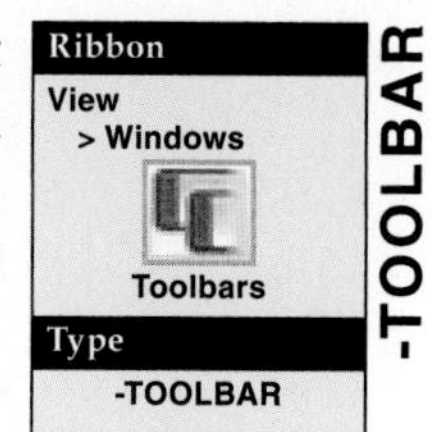

-TOOLBAR

By default, toolbars other than the **Quick Access** toolbar are not displayed except in the AutoCAD Classic workspace. To display toolbars, you may use the **Toolbars** button on the **Windows** panel of the **View** tab in the ribbon or use the **-TOOLBAR** command. Picking the button displays a drop-down list. Select the menu group and then the toolbar name.

When the **–TOOLBAR** command is used, you are first prompted for the toolbar name. The complete toolbar name consists of the menu group and toolbar name, separated by a period. For example, the toolbar name for the **Draw** toolbar is ACAD.DRAW. The menu group name can be omitted when only one menu is currently loaded or if the toolbar name is not duplicated in another menu group. After specifying the toolbar name (or selecting **ALL** for all toolbars), you can select an option.

Command: **-TOOLBAR**↵
Enter toolbar name or [ALL]: **ACAD.DRAW**↵
Enter an option [Show/Hide/Left/Right/Top/Bottom/Float] <Show>:

These options are used to hide, show, or specify a location for the toolbar.

- **Show.** Makes the toolbar visible.
- **Hide.** Causes the toolbar to be invisible.
- **Left.** Places the toolbar in a docked position at the left side of the AutoCAD window.
- **Right.** Places the toolbar in a docked position at the right side of the AutoCAD window.
- **Top.** Places the toolbar in a docked position at the top of the AutoCAD window.
- **Bottom.** Places the toolbar in a docked position at the bottom of the AutoCAD window.
- **Float.** Places the toolbar as a floating toolbar.

For example, to dock the **Zoom** toolbar on the left side of the AutoCAD window, use the following command sequence.

Command: **-TOOLBAR**↵
Enter toolbar name or [ALL]: **ACAD.ZOOM**↵
Enter an option [Show/Hide/Left/Right/Top/Bottom/Float] <Show>: **LEFT**↵
Enter new position (horizontal, vertical) <0,0>: ↵
Command:

Another way to hide a floating toolbar is to pick its menu control button. This is the X in the corner of the toolbar. If you wish to hide a docked toolbar, you can first move it away from the edge to make it a floating toolbar. Then, pick the menu control button. When you hide a previously docked toolbar in this manner, it will appear in the floating position when you again make it visible.

When using floating toolbars, it is also possible to overlap the toolbars to save screen space. To bring a toolbar to the front, simply pick on it. Be sure to leave part of each toolbar showing.

Toolbar Display Options

Located in the **Window Elements** area of the **Display** tab of the **Options** dialog box are four check boxes and a text box relating to toolbars. See Figure 22-21.

When the **Use large buttons for Toolbars** check box is checked, the size of toolbar buttons is increased from 16 × 16 pixels to 32 × 32 pixels. At higher screen resolutions, such as 1280 × 1024, the small buttons may be difficult to see. At lower screen resolutions, such as 800 × 600, the large buttons take up too much of the display area.

When the **Show ToolTips** check box is checked, the name of the button to which you are pointing is displayed next to the cursor. Below this check box is the **Show shortcut**

Figure 22-21.
The **Display** tab of the **Options** dialog box contains settings for toolbars.

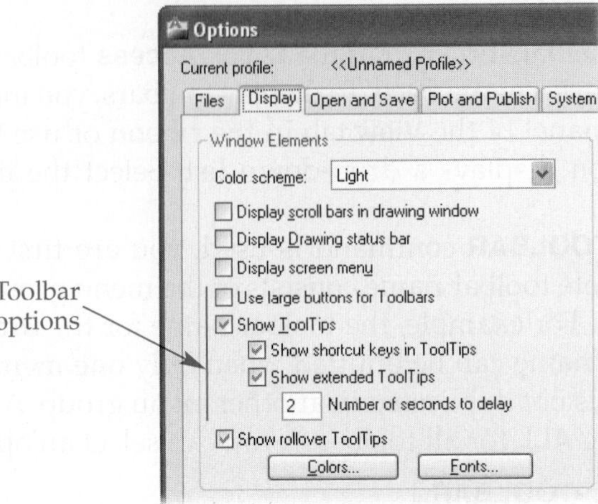

keys in ToolTips check box. When this option is checked, the shortcut key combination for the command is displayed in the tooltip. The **Show extended ToolTips** check box determines whether extended tooltips are displayed. When checked, extended tooltips are displayed when the cursor is hovered over a button for the number of seconds entered in the **Number of seconds to delay** text box. When tooltips are turned off, these two check boxes and the text box are grayed out.

Creating a New Toolbar

To create a new toolbar, open the **Customize User Interface** dialog box. Then, right-click on the Toolbars branch in the upper-left pane and pick **New Toolbar** in the shortcut menu. A new toolbar is added at the bottom of the Toolbars branch. An edit box is displayed in place of the toolbar name with a default name highlighted. Type a descriptive name for the toolbar and press [Enter].

After the new toolbar is named, it is highlighted in the Toolbars branch. The properties for the toolbar are displayed in the **Properties** pane of the **Customize User Interface** dialog box. See Figure 22-22. A preview of the toolbar also appears in the **Toolbar Preview** pane, but since the new toolbar is empty, there is not currently a preview. You can change the name of the toolbar and add a description in the **General** category of the **Properties** pane. The description appears on the AutoCAD status bar when the cursor is over the docked toolbar. In the **Appearance** category, you can specify the default settings for the toolbar, including whether it is included in the current workspace, floating or docked, the location of the toolbar's upper-left corner, and the number of rows for the toolbar. The settings in the **Advanced** category are used for programming applications.

Adding a Command to a Toolbar

To add a command to a toolbar, first expand the Toolbars branch in the **Customizations in All Files** pane in the **Customize User Interface** dialog box. Next, expand the branch for the toolbar to which the command will be added. Then, select a command from the **Command List:** pane and drag it into position in the tree in the **Customizations in All Files** pane. As you drag the command in the tree, an I-bar cursor is displayed. When the I-bar is below the command where you want the new command placed, release the mouse button.

Figure 22-22.
The properties of a toolbar can be changed in the **Properties** pane of the **Customize User Interface** dialog box.

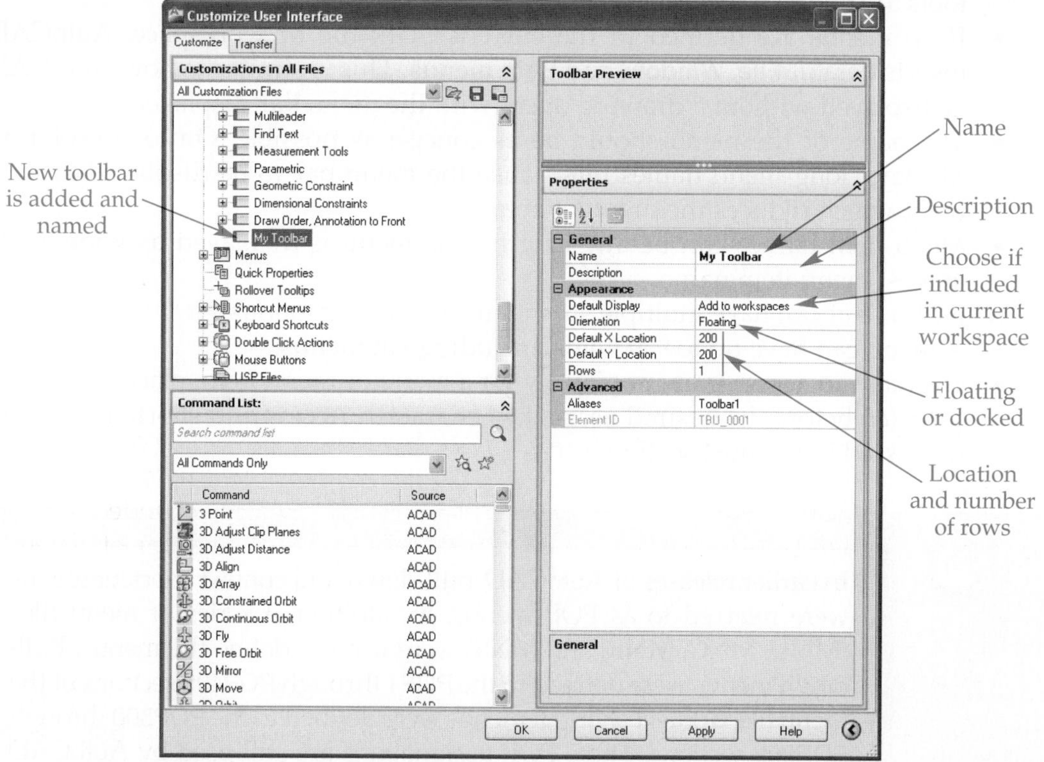

PROFESSIONAL TIP

If the toolbar branch is selected, you can also add a command by dragging and dropping the command into the **Toolbar Preview** pane, rather than into the tree in the **Customizations in All Files** pane. However, the toolbar must contain at least one command in order for the preview to appear.

Exercise 22-3

Complete the exercise on the student website.
www.g-wlearning.com/CAD

Overview of Menus

When the menu bar is displayed, the names of the standard (classic) ***menus*** appear at the top of the AutoCAD graphics window. The menu bar is displayed in the AutoCAD Classic workspace. To display the menu bar in other workspaces, pick the arrow icon at the right-hand end of the **Quick Access** toolbar and select **Show Menu Bar** in the shortcut menu or type the **MENUBAR** command. Menus are selected by placing the cursor over the menu name and picking. You can also use the access (mnemonic) keys to select menus.

Once you understand how menus are designed, you can customize existing menus and create your own. Some basic information about menus includes:

- By default, AutoCAD has 13 menus displayed on the menu bar. If the Express tools are not installed, there are 12 menus.
- If no menus are defined in the current CUIx file or workspace, AutoCAD inserts default **File**, **Window**, and **Help** menus. This is similar to how AutoCAD is displayed without a drawing open (with the menu bar displayed).
- The name of the menu should be as concise as possible. On low-resolution displays, long menu names may cause the menu bar to be displayed on two lines, which reduces the drawing area.
- Menu item names can be any length. The menu is displayed as wide as its longest menu item name.
- Each menu can have multiple submenus (cascading menus).
- A menu can have up to 999 items, including submenus.
- To create an access (mnemonic) key for a menu or menu item, place an ampersand (&) before the desired access key character. Access and shortcut keys are discussed in the next section.

NOTE

In earlier releases of AutoCAD, pull-down and context shortcut menus were referred to as POP menus. In addition, a series of menu files (MNU, MNC, MNR, and MNS) were used to define the menus. Pull-down menus were defined in the POP1 through POP499 sections of the menu file. Context shortcut menus were defined in the POP500 through POP999 sections. These POP designations are still used by AutoCAD for the sake of compatibility with older menus being used in the current version of AutoCAD. In the **Customize User Interface** dialog box, the POP names appear as aliases for these menus.

Shortcut and Access Keys

Before getting started with menu customization, it is important to understand the difference between shortcut keys and access keys. ***Shortcut keys,*** also called ***accelerator keys,*** are key combinations used to initiate a command. For example, [Ctrl]+[1] displays or closes the **Properties** palette. Custom shortcut keys can be created to initiate specific AutoCAD commands or macros. Creating custom shortcut keys is covered in Chapter 23.

Access keys, also called ***mnemonic keys,*** are keys used to access a menu or menu item via the keyboard. Pressing the [Alt] key activates the access keys for the menus. The access keys are shown as underlined (underscored) letters. Most access keys (underscores) are not displayed in the menu bar until the [Alt] key is depressed. For example, notice that the letter M is underlined in the **Modify** menu name. Pressing the [M] key accesses the **Modify** menu.

Any letter in the menu or menu item name can be defined as the access key, but an access key must be unique for a menu or submenu. Notice within the **Modify** pull-down menu that the M is used for **Match Properties**, so **Mirror** and **Move** use the i and v, respectively. The letter T can be used for both **Trim** and **Text** because **Text** is in the **Object** submenu, while **Trim** is in the "main" **Modify** menu. When creating custom menus, you can add custom access keys to the menu.

PROFESSIONAL TIP

Once the access (mnemonic) keys are activated, you can use the arrow keys to navigate through the pull-down menu structure.

Creating a New Menu

A new menu is created within the **Customize User Interface** dialog box. First, a menu is added to the Menus branch. Then, commands are added to the new menu. The process is basically the same as creating a new ribbon panel or toolbar, as described earlier in this chapter. The basic procedure is:

1. Open the **Customize User Interface** dialog box.
2. In the **Customizations in All Files** pane, expand the Menus branch. All of the existing menus are displayed.
3. Right-click on the Menus branch to display the shortcut menu. Pick **New Menu** from the shortcut menu. A new menu is added to the bottom of the list of existing menus. See Figure 22-23A. The name is highlighted in an edit box so the default name can be changed.

Figure 22-23.
A—Adding a new menu. B—Commands have been added to the new menu.

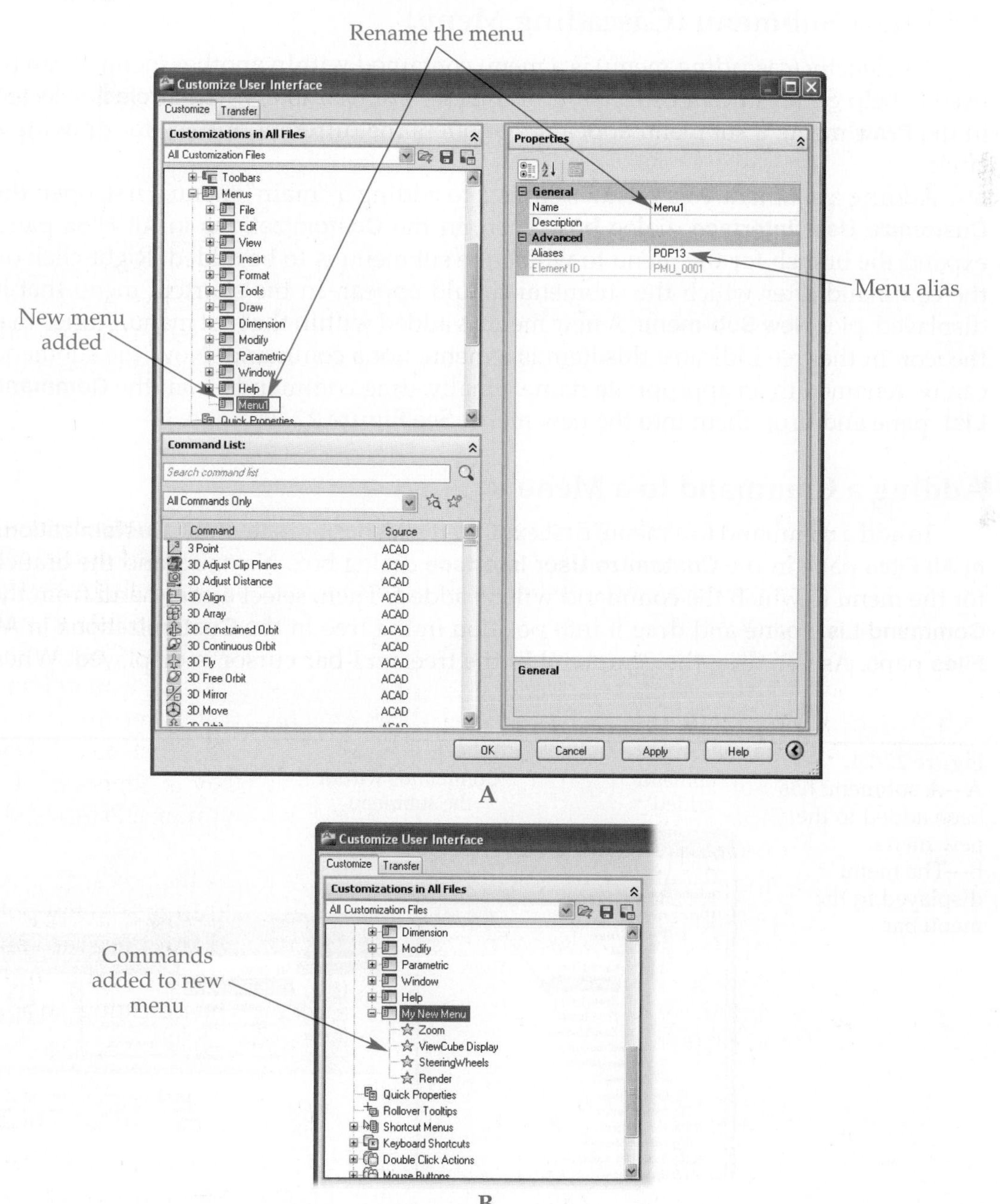

4. Give the menu an appropriate name.
5. Drag the desired commands from the **Commands List:** pane and drop them into the new menu. See Figure 22-23B.
6. To add a separator, right-click on the command below which it should be inserted and select **Insert Separator** from the shortcut menu.

When adding a new menu, it is automatically assigned an alias of POP*n*, where *n* is the next available integer. Also, the menu is automatically available in all workspaces. Workspaces are covered in detail in Chapter 25.

Exercise 22-4

Complete the exercise on the student website.
www.g-wlearning.com/CAD

Adding a Submenu (Cascading Menu)

A ***submenu*** (cascading menu) is a menu contained within another menu. It can be used to help group similar commands or options. For example, when **Circle** is selected in the **Draw** menu, a submenu appears that offers the different options for drawing a circle.

Adding a submenu to a menu is similar to adding a "main" menu. First, open the **Customize User Interface** dialog box. Then, in the **Customizations in All Files** pane, expand the branch for the menu to which the submenu is to be added. Right-click on the command after which the submenu should appear. In the shortcut menu that is displayed, pick **New Sub-menu**. A new menu is added within the first menu. Notice that the icon in the tree indicates this item is a menu, not a command. Now, the submenu can be renamed to an appropriate name. Finally, drag commands from the **Command List:** pane and drop them into the new menu. See Figure 22-24.

Adding a Command to a Menu

To add a command to a menu, first expand the Menus branch in the **Customizations in All Files** pane in the **Customize User Interface** dialog box. Next, expand the branch for the menu to which the command will be added. Then, select a command from the **Command List:** pane and drag it into position in the tree in the **Customizations in All Files** pane. As you drag the command in the tree, an I-bar cursor is displayed. When

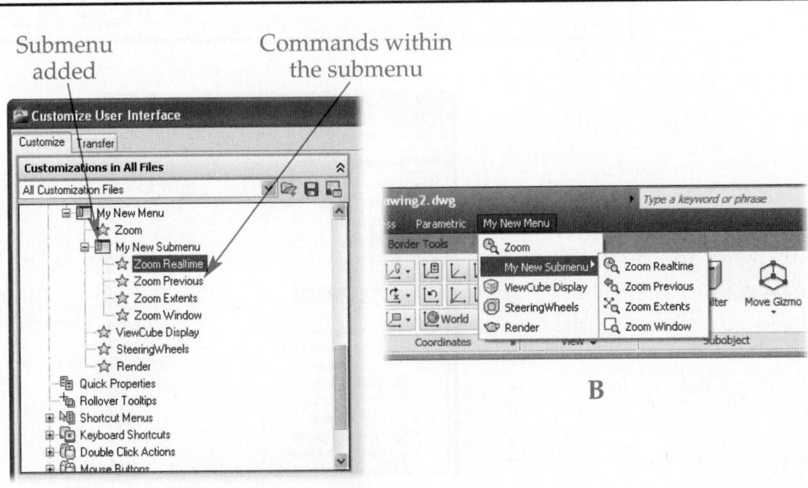

Figure 22-24.
A—A submenu has been added to the new menu.
B—The menu displayed in the menu bar.

the I-bar is below the command where you want the new command placed, release the mouse button.

Removing a Menu

If you want to permanently remove a menu, open the **Customize User Interface** dialog box. Then, expand the tree in the **Customizations in All Files** pane to display the menu to be deleted. Right-click on the menu and pick **Delete** from the shortcut menu. You can also highlight the menu in the tree and press the [Delete] key.

The above procedure is not recommended because the menu is permanently removed. To "restore" the menu in the future, it must be rebuilt. A better way to remove any unwanted menus is by deleting them from the workspace. Managing workspaces is covered in detail in Chapter 25.

Some Notes about Menus

Here are a few more things to keep in mind when developing menus.

- A menu label can be as long as needed, but should be as brief as possible for easy reading. The menu width is automatically created to fit the width of the longest item.
- Menus that are longer than the screen display are truncated to fit on the screen.
- Menus are disabled during **DTEXT** after the rotation angle is entered, while drawing or editing **MTEXT**, and during **SKETCH** after the record increment is set.

Sample Custom Commands

The following examples show how AutoCAD commands and options can be used to create commands. These custom commands can be placed on ribbon panels, toolbars, or menus. Remember, an ampersand (&) preceding a character in a menu or item name defines the keyboard access (mnemonic) key used to enable it. The examples are listed using the following three-step process.

- Step 1. A description of the macro.
- Step 2. The key strokes required for the macro.
- Step 3. The name and macro for the new command as entered in the **Properties** pane of the **Customize User Interface** dialog box.

Example 1

1. This **HEXAGON** command will start the **POLYGON** command and draw a six-sided polygon inscribed in a circle.
2. **POLYGON**↵
   6↵
   (select center)
   I↵
3. Name: &Hexagon
   Macro: *^C^Cpolygon;6;\i

The asterisk in front of the ^C^C repeats the command until it is canceled. The \ in front of i indicates that the macro will wait for user input, in this case the center of the polygon, before continuing.

Example 2

1. This **DOT** command draws a solid dot that is .1 unit in diameter. Use the **DONUT** command. The inside diameter is 0 (zero) and the outside diameter is .1.
2. **DONUT↵**
   0↵
   .1↵
3. Name: &Dot
   Macro: ^C^Cdonut;0;.1

Example 3

1. This **X-POINT** command sets the **PDMODE** system variable to 3 and draws an X at the pick point. The command should repeat.
2. **PDMODE↵**
   3↵
   POINT↵
   (pick the point)
3. Name: &X-Point
   Macro: *^C^Cpdmode;3;point

Example 4

1. This command, named **NOTATION**, could be used by a drawing checker or instructor. It allows the user to circle features on a drawing and then add a leader and text. It first sets the color to red, then draws a circle, snaps a leader to the nearest point that is picked on the circle, and prompts for the text. User input for text is provided, then a cancel [Esc] returns the Command: prompt and the color is set to ByLayer.
2. **-COLOR↵**
   RED↵
   CIRCLE↵
   (pick center point)
   (pick radius)
   LEADER↵
   NEA↵
   (pick a point on the circle)
   (pick end of leader)
   (press [Enter] for automatic shoulder)
   (enter text) ↵
   (press [Enter] to cancel)
   -COLOR↵
   BYLAYER↵
3. Name: &Notation
   Macro: ^C^C-color;red;circle;\\leader;nea;\\;\;-color;bylayer

Example 5

1. This is a repeating command named **MULTISQUARE** that draws one-unit squares oriented at a 0° horizontal angle until the command is canceled.
2. **RECTANG↵**
   (pick lower-left corner)
   @1,1↵
3. Name: &Multisquare
   Macro: *^C^Crectang;\@1,1

PROFESSIONAL TIP

Some commands, such as the **COLOR** command, display a dialog box. Menu macros can provide input to the command line, but cannot control dialog boxes. To access the command-line version of a command, prefix the command name with a hyphen (-), as shown in example 4. However, not all commands that display a dialog box have a command-line equivalent.

Exercise 22-5

Complete the exercise on the student website.
www.g-wlearning.com/CAD

Chapter Test

Answer the following questions. Write your answers on a separate sheet of paper or complete the electronic chapter test on the student website.
www.g-wlearning.com/CAD

1. What is an *interface element* in AutoCAD?
2. Which command is used to access the **Customize User Interface** dialog box?
3. In which pane of the **Customize User Interface** dialog box can you find all predefined commands?
4. How do you add a command to an interface element?
5. How do you remove a command from an interface element?
6. How can you copy a command to a new location on a different interface element?
7. What is a *partial CUI file?*
8. Briefly describe how to create a custom command.
9. What is an *extended help file?*
10. Name the four drawing tools that are provided in the **Button Editor** dialog box.
11. What is the default, small size (in pixels) of the button editor drawing area?
12. Where is the **Use large buttons for Toolbars** check box located? What function does this check box perform?
13. How should you develop and test a new macro before entering it into a custom command definition?
14. Name two ways to specify an [Enter] in a macro. Which of the two methods is recommended?
15. Briefly describe the composition of the ribbon.
16. What determines which commands in a ribbon panel appear in the expanded panel?
17. How do you create a new ribbon tab and add a new panel to it?
18. Briefly describe how to customize a toolbar.
19. How do you create a new toolbar?
20. Explain how to display a toolbar.
21. Briefly describe a contextual tab.
22. What is a *drop-down list?*
23. How do you add a drop-down list to a ribbon panel?
24. How wide is a menu?
25. Interpret the following menu item.

```
^C^Crectang;\@1,1
```

Drawing Problems

Before customizing any toolbars, menus, or the ribbon, check with your instructor or supervisor for specific instructions or guidelines.

1. Create a new toolbar using the following information.
   A. Name the toolbar **Draw/Modify**.
   B. Copy at least three, but no more than six, commonly used drawing commands onto the new toolbar. Use only existing commands; do not create new ones.
   C. Copy at least three, but no more than six, commonly used editing commands onto the new toolbar. Use only existing commands; do not create new ones.
   D. Dock the new **Draw/Modify** toolbar at the upper-left side of the screen.

2. Create a new toolbar using the following information.
   A. Name the toolbar **My 3D Tools**.
   B. Copy the following solid primitive commands onto the new toolbar.
   Box **Pyramid**
   Cone **Sphere**
   Cylinder **Torus**
   C. Copy the following view commands onto the new toolbar.
   Top **Bottom** **Left**
   Right **Front** **Back**
   D. Copy the following UCS commands onto the new toolbar.
   3 Point **Object** **World**
   Face UCS **Origin** **UCS Previous**
   E. Dock the toolbar below the toolbar created in problem 1.

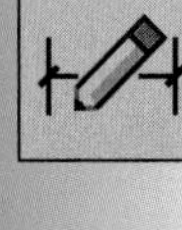

3. Create a new ribbon panel using the following information.
   A. The displayed name of the panel should be **Paper Space Viewports**.
   B. The panel should contain eight custom commands that use the **MVIEW** command to create paper space viewports:
   - **1 Viewport**—allow user to pick location
   - **1 Viewport (Fit)**
   - **2 Viewports (Horizontal)**—allow user to pick location
   - **2 Viewports (Vertical)**
   - **3 Viewports**—allow user to pick orientation and location
   - **3 Viewports (Right)**
   - **4 Viewports**—allow user to pick location
   - **4 Viewports (Fit)**

   C. Construct button graphics for the custom commands. Save the images in the default \Icons folder or create a new folder (be sure to add it to the AutoCAD support environment).
   D. Create a custom command that will switch from one viewport to another.
   E. Place a button on the panel that executes the **PLOT** command.

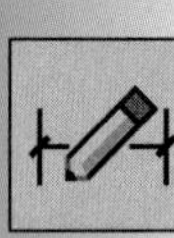

4. Create a new ribbon panel for inserting title block drawings. Name the panel **Title Blocks**.
   A. The panel should contain six custom commands that do the following.
   - Insert the ANSI A title block drawing (plot style of your choice)
   - Insert the ANSI B title block drawing (plot style of your choice)
   - Insert the ANSI C title block drawing (plot style of your choice)
   - Insert the ANSI D title block drawing (plot style of your choice)
   - Insert the ANSI E title block drawing (plot style of your choice)
   - Insert the Architectural title block drawing (plot style of your choice)

   B. Create button graphics for each of the custom commands. Save the images in a new folder and add the folder to the AutoCAD support environment.

5. Add a drop-down list to the **Paper Space Viewports** panel created in problem 3. Add the six custom commands created in problem 4 to this flyout.

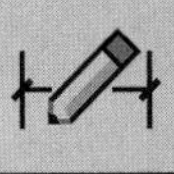

6. Create a new dimensioning menu. Place as many dimensioning commands as you need in the menu. Use submenus if necessary. One or more of the submenus should be dimensioning variables. Include menu access (mnemonic) keys.

7. Create a menu for 3D objects. Include menu access (mnemonic) keys. The menu should include the following items.
   - At least three 3D solid objects
   - **HIDE** command
   - At least three visual style commands
   - **VPORTS** command
   - **NAVSWHEEL** command

8. Create a new menu named **Special**. The menu should include the following drawing and editing commands.

LINE	**MOVE**
ARC	**COPY**
CIRCLE	**STRETCH**
POLYLINE	**TRIM**
POLYGON	**EXTEND**
RECTANGLE	**CHAMFER**
DTEXT	**FILLET**
ERASE	

   Use submenus, if necessary. Include a separator line between the drawing and editing commands and specify appropriate menu access (mnemonic) keys.

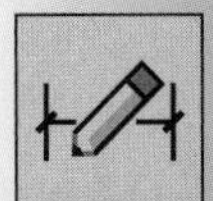

9. Create a menu to insert a variety of blocks or symbols. These symbols can be for any drawing discipline that suits your needs. Use submenus and menu access (mnemonic) keys, if necessary.

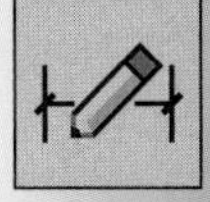

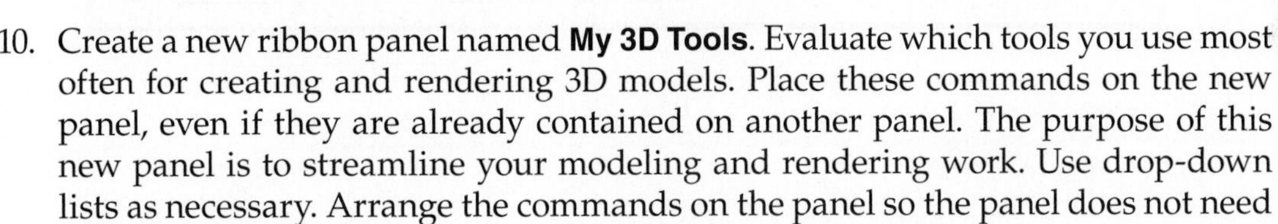

10. Create a new ribbon panel named **My 3D Tools**. Evaluate which tools you use most often for creating and rendering 3D models. Place these commands on the new panel, even if they are already contained on another panel. The purpose of this new panel is to streamline your modeling and rendering work. Use drop-down lists as necessary. Arrange the commands on the panel so the panel does not need to be expanded to access the most frequently used commands.

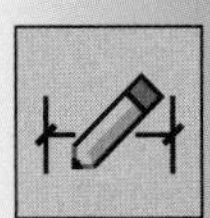

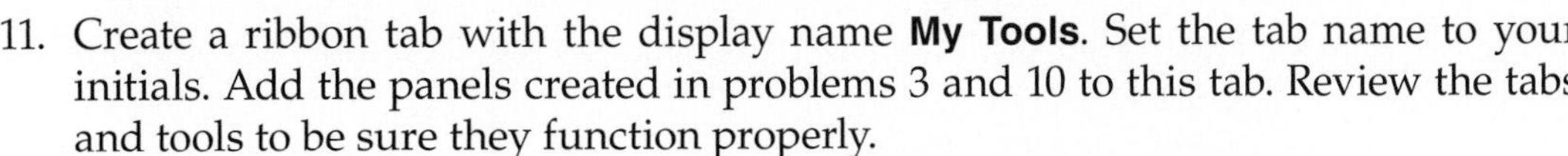

11. Create a ribbon tab with the display name **My Tools**. Set the tab name to your initials. Add the panels created in problems 3 and 10 to this tab. Review the tabs and tools to be sure they function properly.

You can customize the double-click actions in AutoCAD. Here, a custom command has been created that initiates the **SOLIDEDIT** command and it is assigned as the double-click action for a 3D solid. Now, double-clicking on a 3D solid starts the **SOLIDEDIT** command and the user is prompted to select the **Face**, **Edge**, or **Body** option to continue.

CHAPTER 23

Customizing Key and Click Actions

Learning Objectives

After completing this chapter, you will be able to:

- ✓ Assign shortcut keys to commands.
- ✓ Explain how shortcut menus function.
- ✓ Edit existing shortcut menus.
- ✓ Create custom shortcut menus.
- ✓ Customize an object's quick properties.
- ✓ Create custom rollover tooltips.
- ✓ Describe double-click actions.
- ✓ Edit double-click actions.
- ✓ Create custom double-click actions.

AutoCAD has many tools that can be used in "heads-up design." Heads-up design is a concept of working in which your eyes remain focused on the drawing area. For example, when dynamic input is on, you do not need to look at the command line to see the options for the current command. The options are displayed near the cursor in the drawing area. AutoCAD's shortcut menus and double-click actions also contribute to heads-up design. Shortcut menus are displayed by right-clicking. Double-click actions are initiated when an object is double-clicked. Like much of the graphic content in AutoCAD (the ribbon, toolbars, etc.), shortcut menus and double-click actions can be customized.

In order to use the shortcut menu and double-click action customization techniques discussed in this chapter, shortcut menus and double-click editing need to be enabled. To do this, open the **Options** dialog box and select the **User Preferences** tab. Then, check the **Double click editing** and **Shortcut menus in drawing area** check boxes, as shown in Figure 23-1.

The use of shortcut menus can be further refined by picking the **Right-click Customization...** button that appears below the check boxes. This displays the **Right-Click Customization** dialog box. See Figure 23-2. The settings in this dialog box allow you to define what a right-click does when in default mode, edit mode, or command mode. For this chapter, pick the **Shortcut menu** radio buttons in the **Default Mode** and **Edit Mode** areas. Also, pick the **Shortcut Menu: always enabled** radio button in the **Command Mode** area. Then, close the **Right-Click Customization** and **Options** dialog boxes.

Figure 23-1.
The settings for enabling shortcut menus and double-click editing are found in the **Options** dialog box.

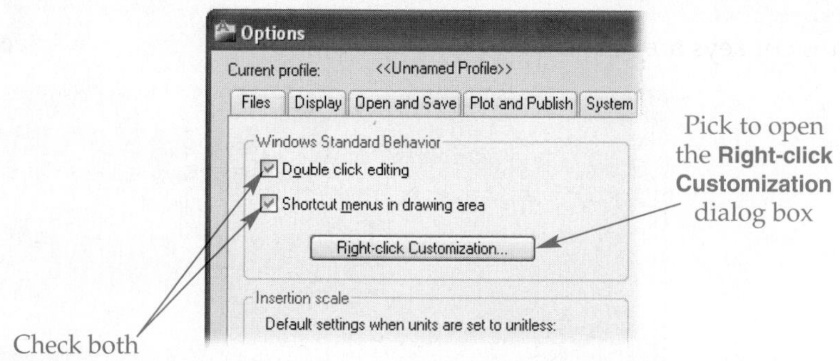

Figure 23-2.
The settings in the **Right-Click Customization** dialog box allow you to define what a right-click does when in default mode, edit mode, or command mode.

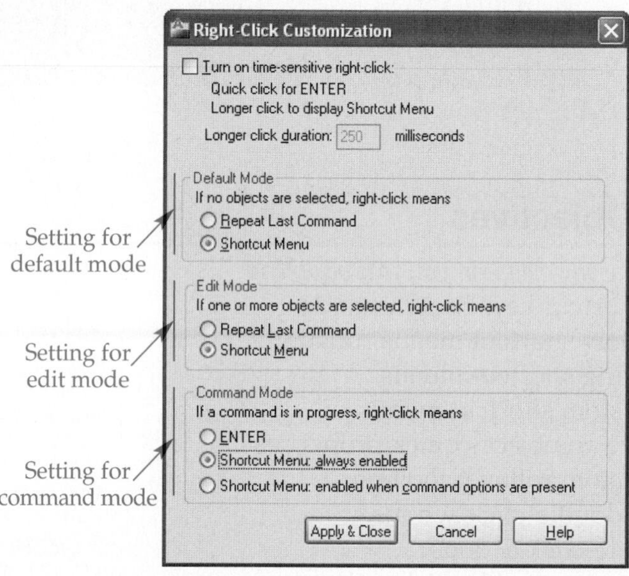

Customizing Shortcut Keys

CUI

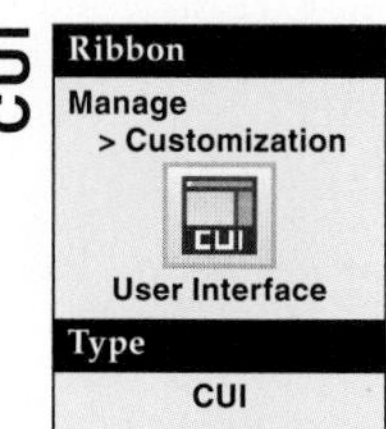

You can define your own custom shortcut keys (accelerator keys) for AutoCAD commands and custom macros. The **Customize User Interface** dialog box is used to define shortcut keys. To see the commands to which shortcut keys are assigned, expand the Keyboard Shortcuts branch in the **Customizations in All Files** pane. Then, expand the Shortcut Keys branch. All commands that have a shortcut key assigned to them appear in this branch. See Figure 23-3.

When the Shortcut Keys branch is selected, the **Shortcuts** pane is displayed in the upper-right corner of the **Customize User Interface** dialog box. A command that has a shortcut key assigned to it can be selected in this pane to display its properties in the **Information** pane in the lower-right corner of the dialog box.

Assigning a Shortcut Key

To assign a shortcut key to a command, first locate the command in the **Command List:** pane of the **Customize User Interface** dialog box. Next, drag the command into the Shortcut Keys branch in the **Customizations in All Files** pane. The command is added to the list of shortcut keys (although it may not be immediately visible) and the **Properties** pane is displayed for the command. See Figure 23-4. In the **Access** category of the **Properties** pane, pick in the Key(s) property text box. Next, pick the ellipsis button (**...**) on the right-hand end of the text box to display the **Shortcut Keys** dialog box. See Figure 23-5.

Figure 23-3.
Shortcut keys are added to commands in the **Customize User Interface** dialog box.

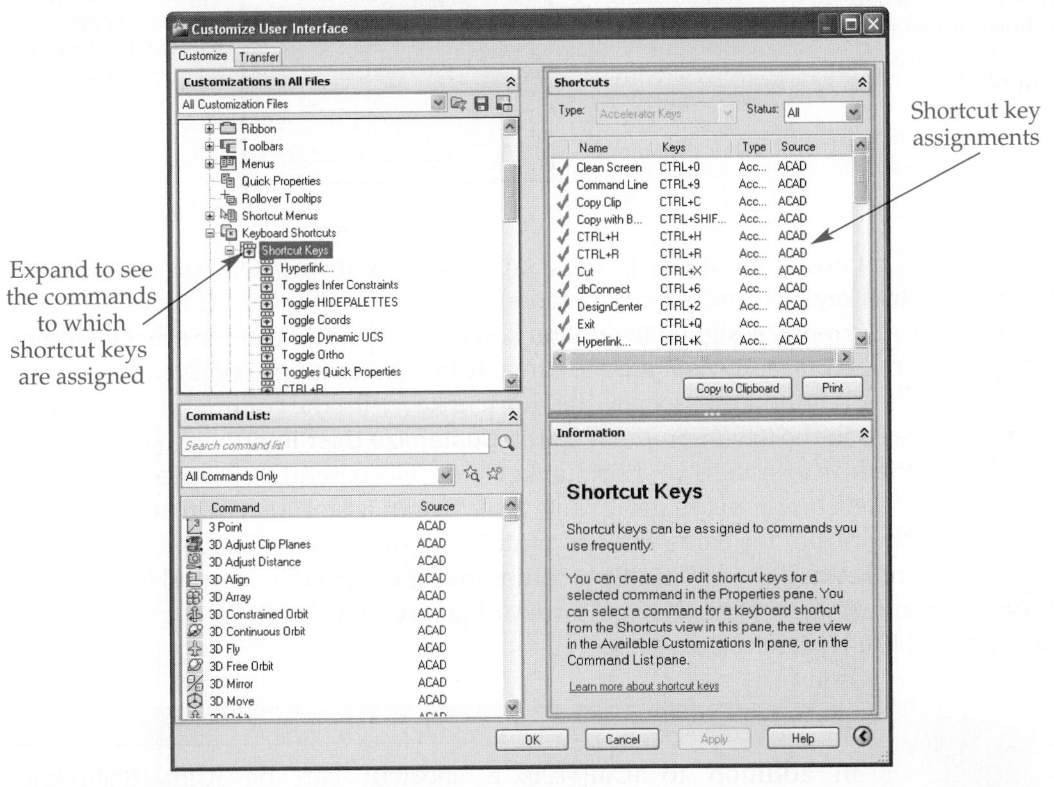

Figure 23-4.
Adding a shortcut key for the **RENDER** command. Drag the **RENDER** command up from the **Command List:** pane to the Shortcut Keys branch.

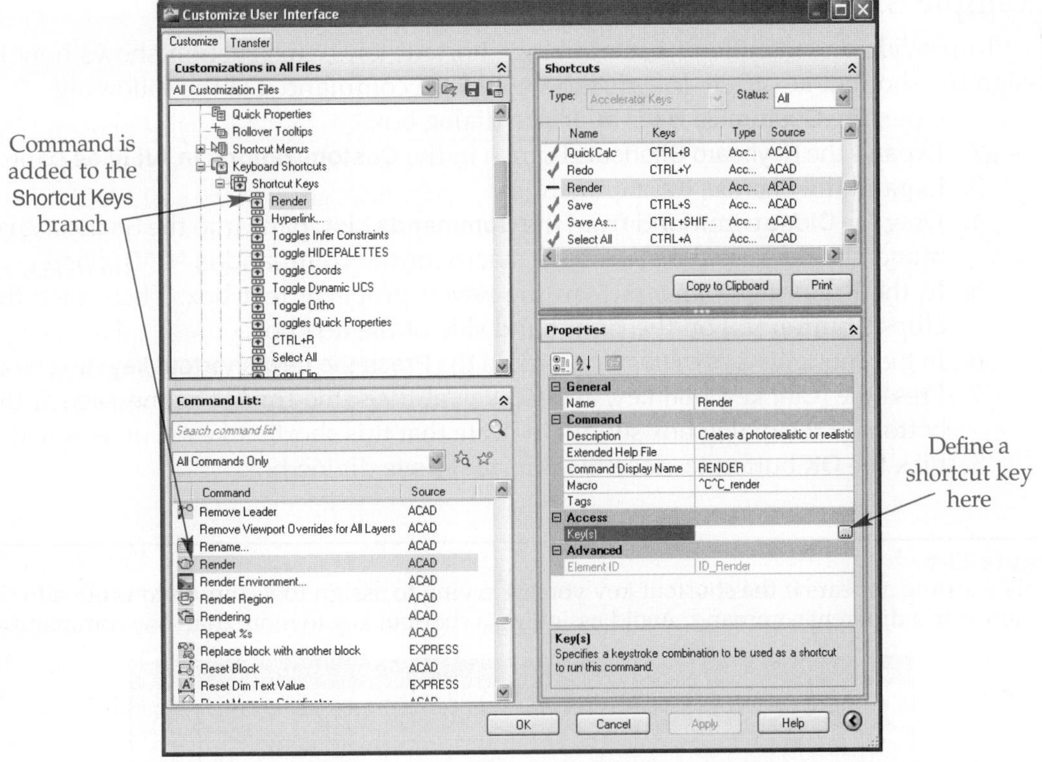

Figure 23-5.
The **Shortcut Keys** dialog box is where a shortcut key is specified for the command.

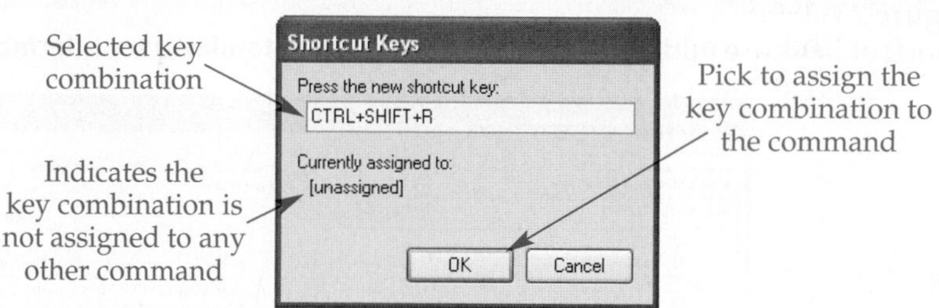

To assign a new shortcut key to the command, pick in the text box labeled **Press the new shortcut key:** and press a combination of [Ctrl] + another key. If the shortcut key combination is currently assigned to another command, the name of the other command is displayed in the **Currently assigned to:** area. If the shortcut key combination is unassigned, pick the **OK** button to associate the shortcut key with the command. The shortcut key then appears in the Key(s) property in the **Customize User Interface** dialog box.

If you attempt to assign a shortcut key that is currently assigned to another command, an alert box appears indicating the shortcut assignment already exists and explaining the priority for using the shortcut. See Figure 23-6. It is not a good idea to have a shortcut key assigned to multiple commands. Be especially careful to ensure the standard Windows keyboard shortcuts are unique, such as [Ctrl]+[X] for cut, [Ctrl]+[C] for copy, and [Ctrl]+[V] for paste.

PROFESSIONAL TIP

In addition to [Ctrl]+*key*, a shortcut can be [Ctrl]+[Shift]+*key*, [Ctrl]+[Alt]+*key*, or [Ctrl]+[Shift]+[Alt]+*key*. The [Caps Lock] key must be off in order to specify the [Shift] key in the **Press new shortcut key:** text box.

Example Shortcut Key Assignment

To provide an example of customizing shortcut keys, this section shows how to assign the shortcut key [Ctrl]+[Alt]+[C] to the **CLOSE** command. Do the following:

1. Open the **Customize User Interface** dialog box.
2. Expand the Keyboard Shortcuts branch in the **Customizations in All Files** pane.
3. Expand the Shortcut Keys branch.
4. Drag the **Close** command from the **Commands List:** pane into the Shortcut Keys branch. Make sure the command macro for the command is ^C^C_close.
5. In the **Properties** pane, pick in the Key(s) property text box. Then, pick the ellipsis button (**...**) on the right-hand side of the text box.
6. In the **Shortcut Keys** dialog box, pick in the **Press the new shortcut key:** text box.
7. Press the [Ctrl] key, [Alt] key, and [C] key at the same time. The message at the bottom of the dialog box should indicate that this shortcut key is unassigned.
8. Pick the **OK** button to close the **Shortcut Keys** dialog box.

Figure 23-6.
This warning appears if the shortcut key you are trying to assign to a command is already assigned to a different command. Avoid assigning a shortcut key to more than one command.

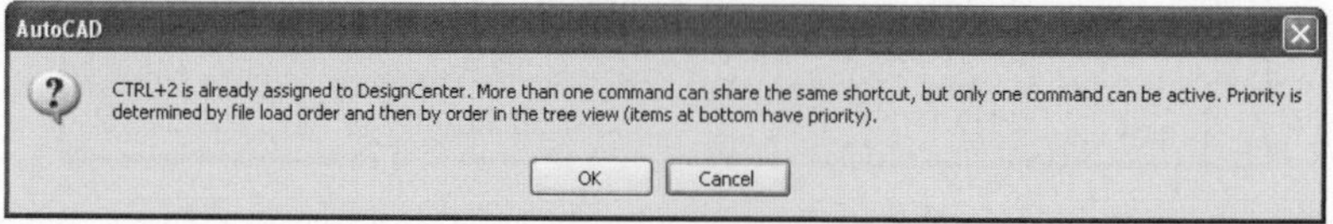

9. Pick the **OK** button to close the **Customize User Interface** dialog box and apply the change.
10. Test the [Ctrl]+[Alt]+[C] shortcut key. When the shortcut key is used, either the current drawing should close or you should be prompted to save the changes to the drawing before closing.

PROFESSIONAL TIP

Shortcut keys (accelerator keys) have some specific limitations. For example, a shortcut cannot pause for user input or use repeating commands. Be aware of this when assigning shortcut keys to custom commands.

Exercise 23-1

Complete the exercise on the student website.
www.g-wlearning.com/CAD

Examining Existing Shortcut Menus

Shortcut menus are context-sensitive menus that appear at the cursor location when using the right-hand button on the mouse (right-clicking). ***Context sensitive*** means that the displayed shortcut menu is dependent on what is occurring at the time of the right-click. For example, if no command is active, there is no object selection, and you right-click in the drawing area, the shortcut menu shown in Figure 23-7A is displayed. If no command is active and you right-click in **Command Line** window, the shortcut menu shown in Figure 23-7B is displayed. If the **CIRCLE** command is active and you right-click in the drawing area before any point is selected, the shortcut menu shown in Figure 23-7C is displayed. Other menus appear when right-clicking in other situations, too, such as when grips are being used or when an object is selected in the drawing window. In the case of a selected object, the shortcut menu is based on the type of object that is selected.

Before learning how to customize shortcut menus, examine the existing shortcut menus. Open the **Customize User Interface** dialog box and look at the **Customizations**

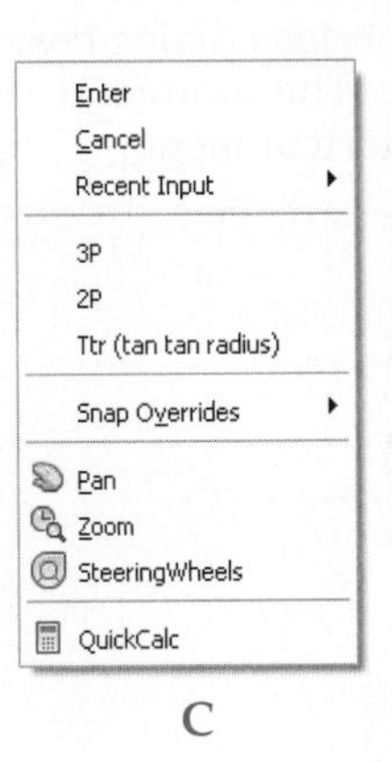

Figure 23-7.
A—Displayed when no command is active and no object is selected.
B—Displayed when no command is active and you right-click in the **Command Line** window.
C—Displayed when the **CIRCLE** command is active and before any point is selected.

in All Files pane. The name of this pane is based on what is selected in the drop-down list. Expand the Shortcut Menus branch in the tree. All of the existing shortcut menu names are displayed as branches. See **Figure 23-8.** There are command-specific, object-specific, and generic shortcut menus. The generic shortcut menus are:

- **Command Menu.** This menu appears when right-clicking in the drawing window while a command is active. Any command options for the active command are inserted into this menu. See **Figure 23-9.**

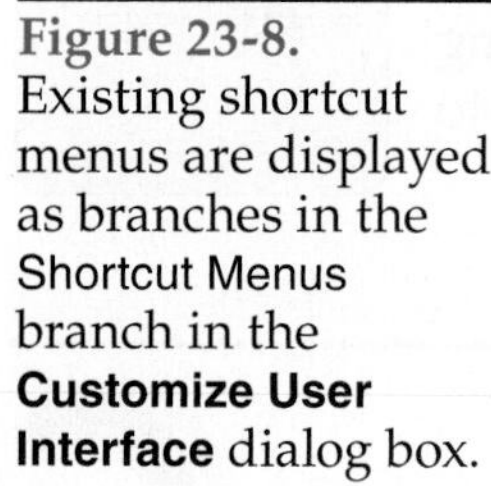

Figure 23-8.
Existing shortcut menus are displayed as branches in the Shortcut Menus branch in the **Customize User Interface** dialog box.

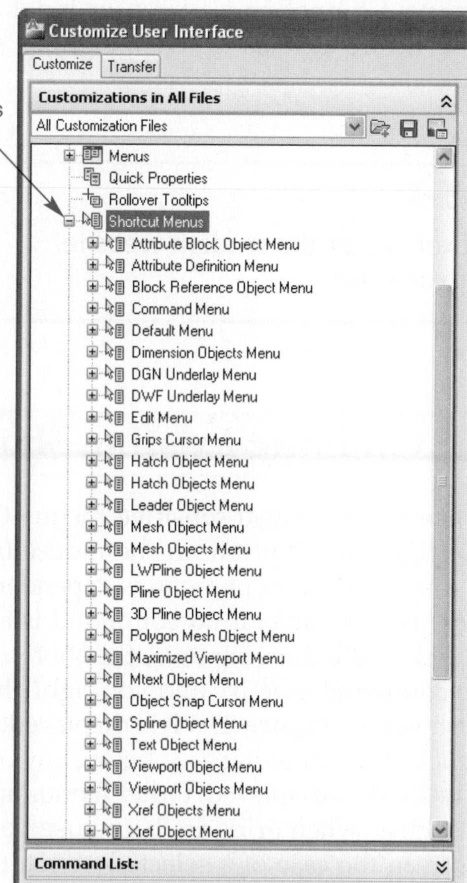

Figure 23-9.
A—The Command Menu branch in the **Customize User Interface** dialog box. B—The command shortcut menu.

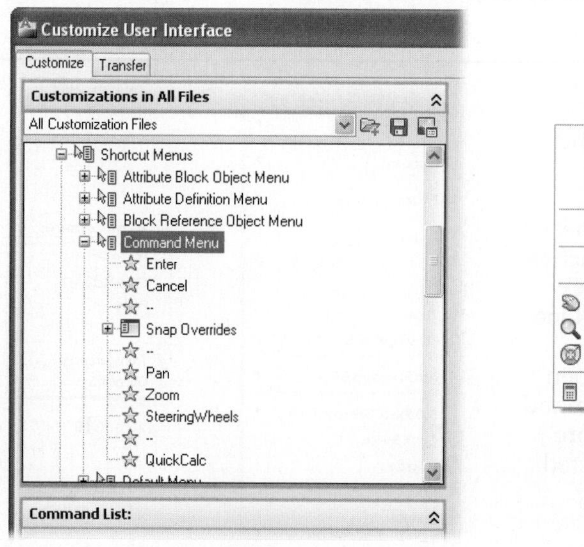

- **Default Menu.** This menu appears when right-clicking in the drawing window while no command is active and no objects are selected. See Figure 23-10.
- **Edit Menu.** This menu appears when right-clicking in the drawing window when no command is active and an object is selected. In order for this menu to be displayed, the **PICKFIRST** system variable must be set to 1. If an object menu is available for the type of object selected, it is inserted into this menu. See Figure 23-11.

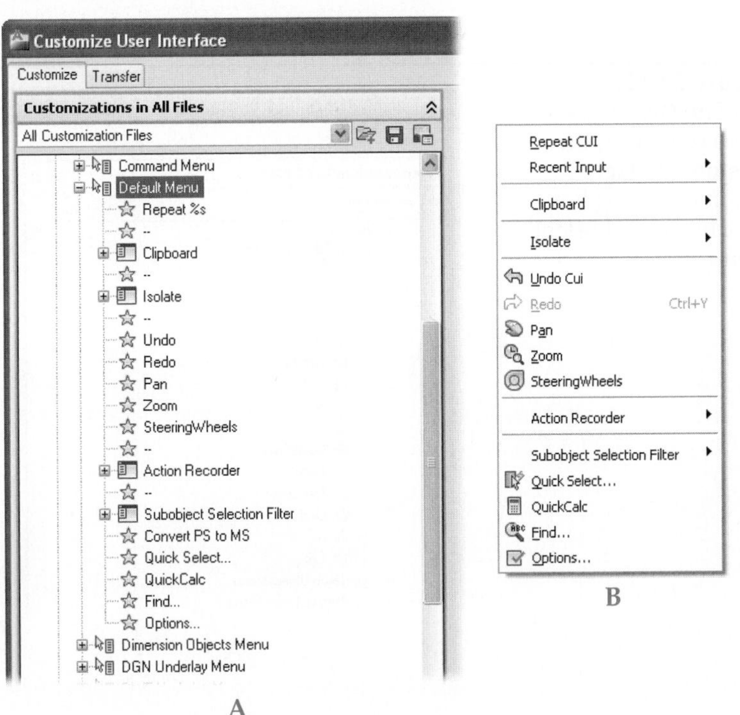

Figure 23-10.
A—The Default Menu branch in the **Customize User Interface** dialog box. B—The default shortcut menu.

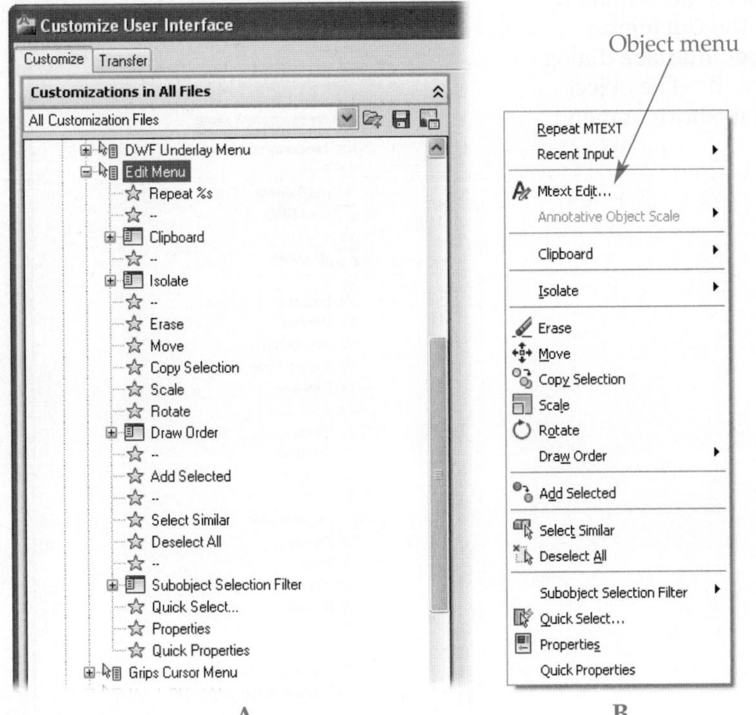

Figure 23-11.
A—The Edit Menu branch in the **Customize User Interface** dialog box. B—The edit shortcut menu.

- **Grips Cursor Menu.** This menu appears when grips are being used. See **Figure 23-12.** An object must be selected and at least one grip must be hot.
- **Object Snap Cursor Menu.** This menu appears when holding the down [Shift] key and right-clicking. See **Figure 23-13.** It also appears as the Snap Overrides branch in the Command Menu branch, meaning it is displayed as a cascading menu.

The remaining menus are object-specific menus that appear when right-clicking while a certain type of object is selected. Notice that there are menu branches named Attribute Block Objects Menu, Block Reference Object Menu, Dimension Objects Menu, Hatch Object Menu, and others. These menus contain items that can be used on the type of

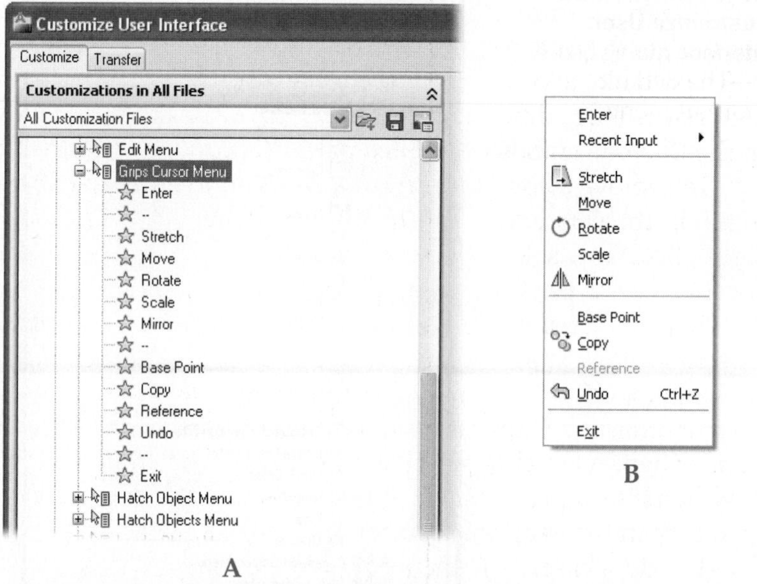

Figure 23-12. A—The Grips Cursor Menu branch in the **Customize User Interface** dialog box. B—The grips shortcut menu.

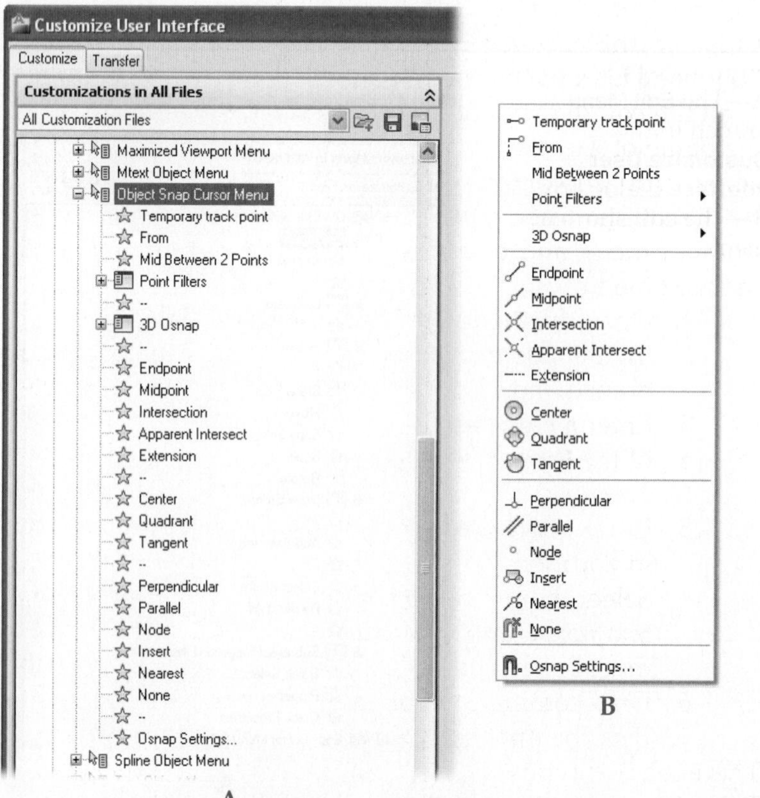

Figure 23-13. A—The Object Snap Cursor Menu branch in the **Customize User Interface** dialog box. B—The object snap shortcut menu.

object selected. For example, in the Shortcut Menus branch, the menu branch Dimension Objects Menu contains commands for editing the dimension text position, dimension text precision, and dimension style.

Customizing Shortcut Menus

The existing shortcut menus can be customized by adding or removing commands. Shortcut menus can also be customized by visually grouping commands using separators. Cascading menus can be added to shortcut menus.

To add a command to a shortcut menu, open the **Customize User Interface** dialog box and, in the **Customizations in All Files** pane, expand the branch for the shortcut menu you would like to customize. Next, locate the command you wish to add in the **Command List:** pane. Then, drag the command into the desired position within the shortcut menu in the **Customizations in All Files** pane and drop it when the bar appears.

To remove a command from a shortcut menu, expand the branch for the shortcut menu in the **Customizations in All Files** pane. Highlight the command to be removed, right-click, and select **Remove** from the shortcut menu. You can also highlight the command and press the [Delete] key.

To add a separator to a shortcut menu, expand the branch for the shortcut menu in the **Customizations in All Files** pane. Highlight the command *after* which you would like the separator to be added. Right-click and select **Insert Separator** from the shortcut menu.

To rename a shortcut menu, highlight the branch in the **Customizations in All Files** pane. Then, right-click and select **Rename** from the shortcut menu. Finally, type the new name and press [Enter]. The shortcut menu can also be renamed using the Name property in the **Properties** pane.

To add a cascading menu to a shortcut menu, expand the branch for the shortcut menu in the **Customizations in All Files** pane. Highlight the command *after* which you would like the cascading menu to appear. Right-click and select **New Sub-menu** from the shortcut menu. A new shortcut menu branch with the default name of Menu*x* is added to the current shortcut menu. The new menu can be renamed. Now, in the **Command List:** pane, locate the commands you wish to add to the new shortcut menu. Drag the commands into the **Customizations in All Files** pane and drop them next to the name of the new shortcut menu. When the arrow appears next to the new menu name, drop the command to add it to the new shortcut menu.

Creating a new, custom shortcut menu is a two-step process. First, make a new shortcut menu and then drag commands into it. Follow these steps to make a new shortcut menu:

1. Open the **Customize User Interface** dialog box.
2. In the **Customizations in All Files** pane, right-click on the Shortcut Menus branch and select **New Shortcut Menu** in the shortcut menu that is displayed.
3. Enter a name for the shortcut menu.
4. In the **Properties** pane, add a description for the shortcut menu in the **General** category.
5. In the **Advanced** category of the **Properties** pane, add an alias. This alias is in addition to the automatic, sequential POP5*xx* alias that AutoCAD creates. Select the property, pick the ellipsis button (**...**) at the right-hand end of the text box, and type the alias in the **Aliases** dialog box that appears. Each alias must be on its own line in this dialog box. Close the **Aliases** dialog box.
6. Drag commands from the **Command List:** pane into the new shortcut menu.
7. Pick the **Apply** or **OK** button to apply the changes.

There are two types of custom shortcut menus: object specific and command oriented. The next sections describe the two types of custom shortcut menus in detail.

Creating Object-Specific Shortcut Menus

When creating an object-specific shortcut menu, there can actually be two menus available. One menu is displayed for instances when just a single object of a given type is selected. The other menu is displayed when more than one object is selected.

The name assigned to the object menu should follow the same syntax used for naming AutoCAD's default object-specific menus: *object_type* **Object Menu** or *object_type* **Objects Menu** (with an S). In this way, when looking at the shortcut menus in the **Customizations in All Files** pane in the **Customize User Interface** dialog box, you will easily recognize which object type that menu applies to and whether it is for multiple selected objects or a single selected object. The use of this syntax is optional. Menus can be named using whatever naming scheme you wish. However, it is recommended to follow the naming syntax described here.

The alias for the shortcut menu has a syntax that *must* be followed. It is this alias that AutoCAD uses in determining to which object or objects the menu applies. The syntax for the alias must take on the form of OBJECT_*type* or OBJECTS_*type* and must be exactly followed in order for AutoCAD to properly display the shortcut menu.

As an example, the following procedure creates a shortcut menu that allows access to the **LENGTHEN** and **BREAK** commands when a single line is selected.

1. Open the **Customize User Interface** dialog box.
2. Right-click on the Shortcut Menus branch in the **Customizations in All Files** pane and select **New Shortcut Menu**.
3. Name the shortcut menu **Line Object Menu**.
4. In the **Properties** pane, select the Aliases property in the **Advanced** category. Then, pick the ellipsis button to open the **Aliases** dialog box. On the second line, enter the alias OBJECT_LINE and then pick the **OK** button to close the **Aliases** dialog box. Since the **LENGTHEN** and **BREAK** commands can only be applied to a single object, be sure to use the OBJECT_*type* syntax (without the S).
5. Drag the **LENGTHEN** and **BREAK** commands from the **Command List:** pane into the Line Object Menu branch in **Customizations in All Files** pane. See Figure 23-14A.
6. Pick the **OK** button to close the **Customize User Interface** dialog box and apply the changes.

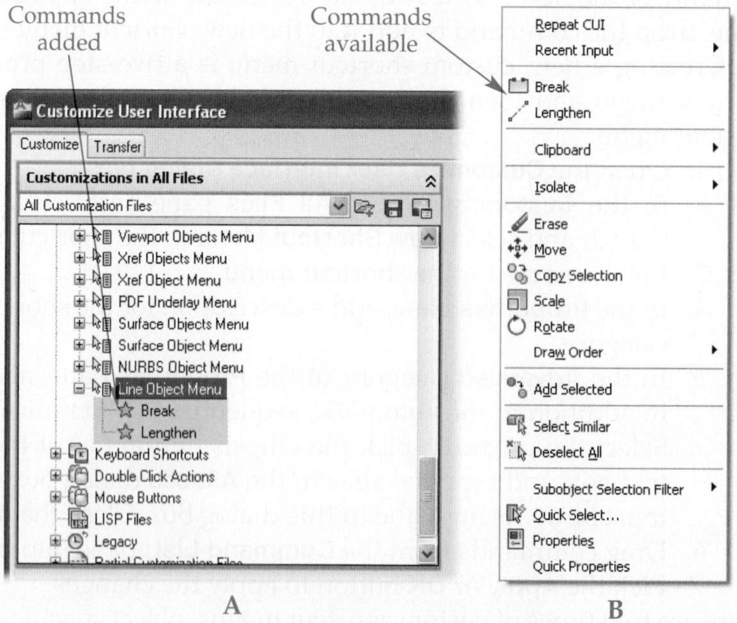

Figure 23-14.
A—The new object-specific shortcut menu is created.
B—The **BREAK** and **LENGTHEN** commands are now available in the shortcut menu.

Now, draw a line, select it, and right-click. Notice that **Lengthen** and **Break** entries appear in the shortcut menu. See Figure 23-14B. Selecting either entry executes the command. If the command accepts a preselected object, it is executed on the selected line. Neither **LENGTHEN** nor **BREAK** accepts preselected objects; you must reselect the line. Having the entries in the shortcut menu provides for quicker access to the command.

Creating Command-Oriented Shortcut Menus

When a command is being executed, any command options appear in the shortcut menu. For example, when the **CIRCLE** command prompts for a radius, you can right-click and select **Diameter** from the shortcut menu. Custom shortcut menus can be created for use when certain commands are active. This allows you to add options to the shortcut menu that is displayed when a command is active. Quicker access to object snaps and object selection methods are just a couple of applications that custom, command-oriented shortcut menus could allow for within commonly used commands.

A command-oriented shortcut menu is created in the same way as an object-oriented shortcut menu, as discussed in the previous section. However, the syntax for the alias is slightly different. The alias must be in the form of COMMAND_*command_name*, where *command_name* is the name of the command with which you want the shortcut menu associated.

Also, if the command step does not have any default options, such as a Select objects: prompt, right-clicking is, by default, interpreted as the [Enter] key. Therefore, in the **Right-Click Customization** dialog box, the **Shortcut Menu: always enabled** radio button must be selected in the **Command Mode** area, as described earlier.

As an example, the following procedure creates a custom shortcut menu that displays **Select Previous**, **Select Last**, and **Select Fence** selection options at the Select objects: prompt for the **MOVE** command.

1. Open the **Customize User Interface** dialog box.
2. Right-click on the Shortcut Menus branch in the **Customizations in All Files** pane and select **New Shortcut Menu**.
3. Name the shortcut menu **Move Command Menu**.
4. In the **Properties** pane, select the Aliases property in the **Advanced** category. Then, pick the ellipsis button to open the **Aliases** dialog box. On the second line, enter the alias COMMAND_MOVE and then pick the **OK** button to close the **Aliases** dialog box. The syntax of COMMAND_*command_name* must be exactly followed in order for AutoCAD to properly display the shortcut menu.
5. Drag the **Select Previous**, **Select Last**, and **Select Fence** commands from the **Command List:** pane into the Move Command Menu branch in **Customizations in All Files** pane. See Figure 23-15A.
6. Pick the **OK** button to close the **Customize User Interface** dialog box and apply the changes.

Now, initiate the **MOVE** command. At the Select objects: prompt, right-click and notice that **Select Previous**, **Select Last**, and **Select Fence** entries are available in the shortcut menu. See Figure 23-15B. Remember, the **Shortcut Menu: always enabled** radio button must be on in the **Right-Click Customization** dialog box for this shortcut menu to appear.

Exercise 23-2

Complete the exercise on the student website.
www.g-wlearning.com/CAD

Figure 23-15.
A—The new command-specific shortcut menu is created. B—The **Previous**, **Last**, and **Fence** commands are available in the shortcut menu.

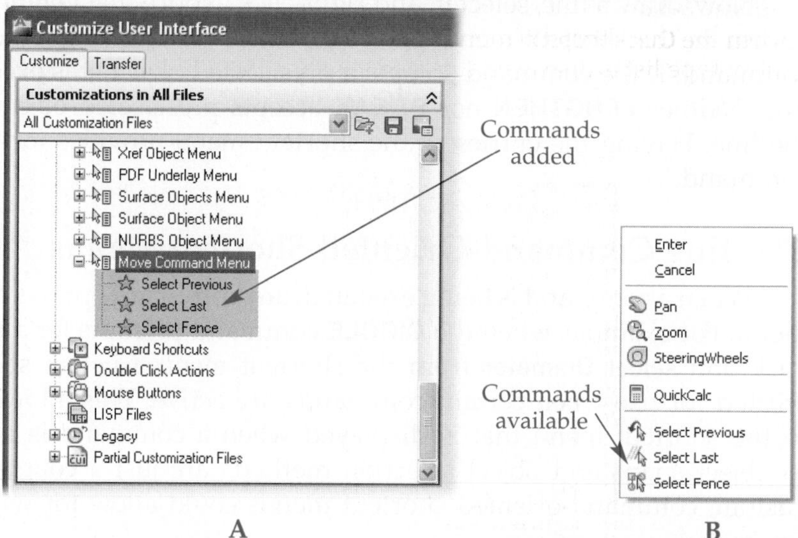

Customizing Quick Properties and Rollover Tooltips

The **Quick Properties** palette appears, by default, when you select objects in the graphics area. This is a streamlined version of the **Properties** palette that displays ***quick properties,*** which provide certain information about the object. In order for the **Quick Properties** palette to be displayed, the **Quick Properties** button on the status bar must be on. The palette can also be displayed by right-clicking with an object selected and selecting **Quick Properties** in the shortcut menu so it is checked.

Rollover tooltips provide quick properties in a graphic tooltip. This tooltip is displayed as the cursor is hovered over the object. A rollover tooltip serves a similar purpose as the **Quick Properties** palette, but it only provides information. Properties cannot be changed in the tooltip.

The quick properties displayed in the **Quick Properties** palette and a rollover tooltip can be customized. Each object type can have different quick properties displayed. Additionally, you can set which quick properties are displayed in the **Quick Properties** palette and rollover tooltip for a given object type. The steps for customizing quick properties are the same for the **Quick Properties** palette and rollover tooltips, as discussed in this section.

Open the **Customize User Interface** dialog box. In the **Customizations in All Files** pane, select the Quick Properties branch. Two columns are displayed on the right-hand side of the dialog box. See **Figure 23-16.** When you select Rollover Tooltips in the **Customizations in All Files** pane, the same two columns are displayed. However, the settings may be different between the Quick Properties branch and the Rollover Tooltips branch.

The left-hand column displays the ***object type list,*** which is a list of AutoCAD object types. You may add or remove object types by picking the **Edit Object Type List** button at the top of the column. This displays the **Edit Object Type List** dialog box, **Figure 23-17.** All available AutoCAD object types are listed in this dialog box. Those that are checked appear in the object type list in the **Customize User Interface** dialog box.

The object type list controls which objects display quick properties. The right-hand column displays a list of quick properties that can be displayed. The properties that are checked appear in the **Quick Properties** palette or rollover tooltip. Remember, the palette and tooltip can display different properties.

All objects have a General category containing similar quick properties. By default, the color and layer properties are turned on for all objects in the list. To display the General category for objects that do not have defined quick properties, pick the **General** button at the bottom of the object type list.

Figure 23-16.
When the Quick Properties branch is selected in the **Customize User Interface** dialog box, the object type list is displayed on the right-hand side of the dialog box.

Object type list
Pick to edit the object type list
Check the properties to display
Pick to list only general properites
Customize User Interface
Customize
Transfer
Customizations in All Files
All Customization Files
Workspaces
2D Drafting & Annotation Default
3D Modeling (current)
3D Basics
AutoCAD Classic
Quick Access Toolbars
Ribbon
Toolbars
Menus
Quick Properties
Rollover Tooltips
Shortcut Menus
Keyboard Shortcuts
Double Click Actions
Mouse Buttons
Command List:
Search command list
All Commands Only
Command
Source
3 Point
3D Adjust Clip Planes
3D Adjust Distance
3D Align
3D Array
3D Constrained Orbit
3D Continuous Orbit
3D Fly
3D Free Orbit
3D Mirror
3D Move
3 Point Angular Dimension
Aligned Dimension
Angular Dimension
Arc
Arc Length Dimension
Block Reference
Circle
Diametric Dimension
Hatch
Jogged Dimension
Leader
Line
Mesh
MText
Multileader
Ordinate Dimension
Polyline
Radial Dimension
Rotated Dimension
Surface
Surface (Extrusion)
Surface (Loft)
Surface (NURBS)
Surface (Planar)
Surface (Revolve)
Surface (Sweep)
Table
Text
Viewport
General
Color
Layer
Linetype
Linetype scale
Plot style
Lineweight
Transparency
Hyperlink
Reset Overrides
The on/off state of general properties can be shared with all defined object types in the current CUI file using the Reset Overrides button.
General
OK
Cancel
Apply
Help

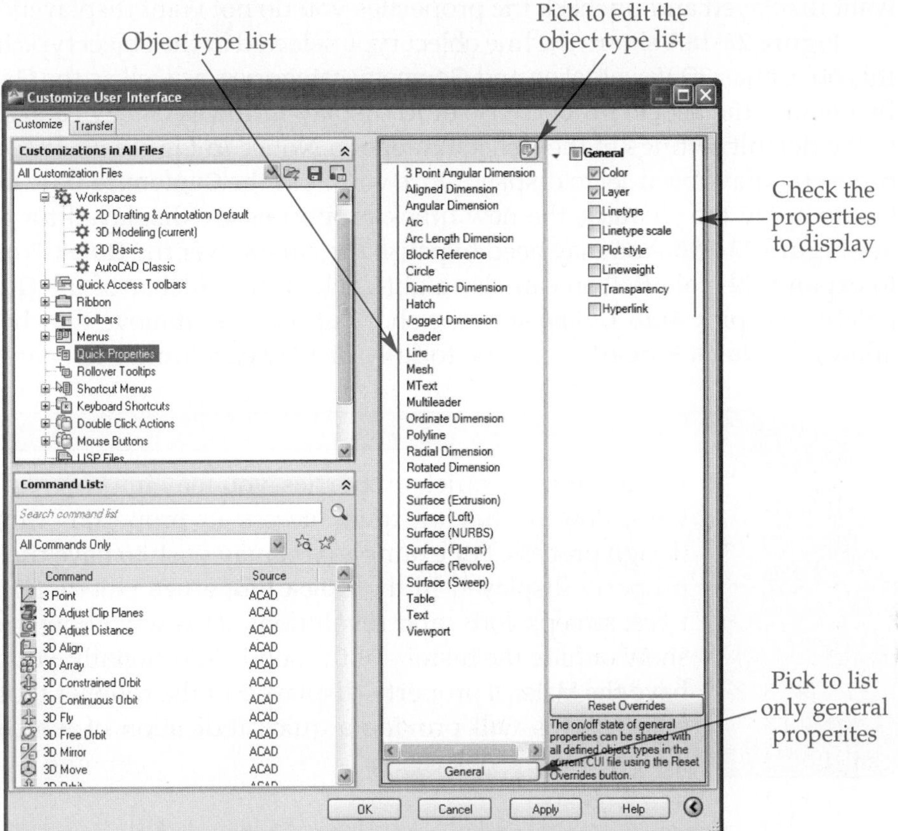

Figure 23-17.
Determining which object types appear in the object type list in the **Customize User Interface** dialog box.

Check the object types to display
Edit Object Type List
2D Polyline
3 Point Angular Dimension
3D Face
3D Polyline
3D Solid
Box
Cone
Cylinder
Pyramid
Sphere
Torus
Wedge
Extrusion
Sweep
Revolve
Loft
None
ACADPROXY_ENTITY
Aligned Constraint Parameter
Aligned Dimension
Alignment Grip
Alignment Parameter
Angular Constraint Parameter
Angular Dimension
Arc
Arc Length Dimension
Array Action
Attribute
Attribute Definition
Base Point Parameter
Block Properties Table
Block Reference
Block Table Grip
Body
Camera
OK
Cancel

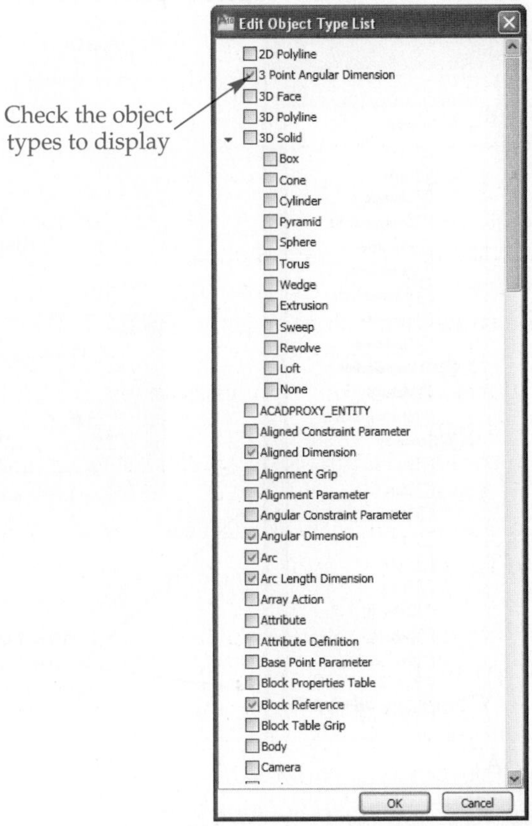

To customize the quick properties displayed for a specific object type, select the object in the object type list. The category list in the right-hand column displays all categories and quick properties available for that object. Check the properties that you want displayed and uncheck the properties you do not want displayed.

Figure 23-18A shows the line object type selected in the object type list. Notice that this object has 3D Visualization and Geometry categories, as well as the General category. By default, the Length property is set to display in the Geometry category in addition to the default settings in the General category. Notice in **Figure 23-18A** that additional properties have been set to display. Once you close the **Customize User Interface** dialog box to apply the changes, the new quick properties are displayed for a selected line. See **Figure 23-18B.** You may need to pause the cursor over the **Quick Properties** palette to expand the palette. You can also right-click on the sidebar of the **Quick Properties** palette and pick **Auto-Collapse** in the shortcut menu to remove the check mark. This allows the **Quick Properties** palette to fully display each time it appears.

PROFESSIONAL TIP

By customizing quick properties, you may improve your AutoCAD work flow by having quick access to properties throughout the design process. For example, you may wish to have the Show History property displayed in the **Quick Properties** palette for solid primitives, sweeps, lofts, and revolutions. This will allow you to quickly show or hide the history of the solid. Additionally, you may wish to have the Material property displayed in the rollover tooltip for these objects. This will provide a quick indication of which material is assigned to the object.

Figure 23-18.
Customizing the quick properties for a line. A—Two properties in addition to the default properties are set to display. B—The **Quick Properties** palette contains the additional properties.

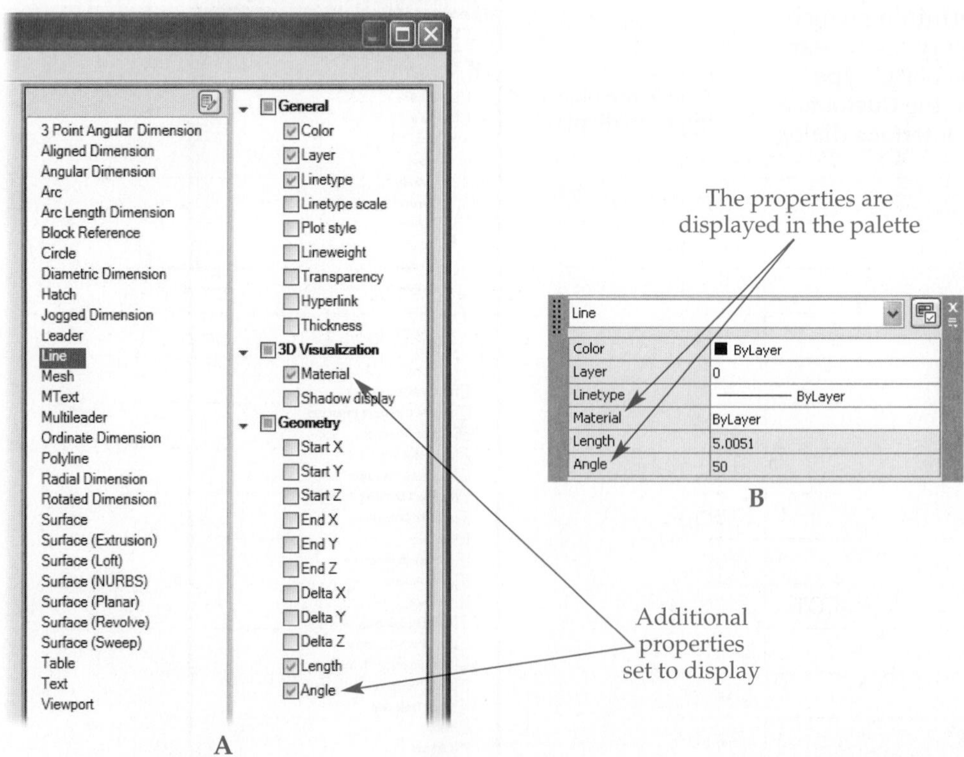

Customizing Double-Click Actions

By double-clicking on certain objects, an appropriate editing command is automatically executed. Which command is initiated is determined by the ***double-click action*** associated with the object type. Some AutoCAD objects have very specific editing tools available. For example, multiline text (mtext) objects are edited with the in-place text editor and polylines are edited with the **PEDIT** command.

A list of AutoCAD objects that have default double-click actions associated with them, other than the **PROPERTIES** command, is shown in Figure 23-19. If you double-click on one of the object types listed in the table, the command or macro listed in the Associated Double-Click Action column is executed. If the object type is not listed in the table, it is likely the **Properties** palette is displayed, by default, when the object is double-clicked. This is the double-click action associated with most objects.

Assigning Double-Click Actions

Double-click actions are assigned to specific object types in the **Customize User Interface** dialog box. In the **Customizations in All Files** pane, expand the Double Click Actions branch. All of the AutoCAD object types are listed. See Figure 23-20. Expand each branch and notice that many double-click actions call the **Properties** palette, while the objects listed in the table in Figure 23-19 have double-click actions that call the object-specific editing command.

Use the following procedure to change the double-click editing action associated with an object type. For this example, the **DDPTYPE** command will be associated with the point object type so the **Point Style** dialog box appears when a point object is double-clicked.

Figure 23-19.
AutoCAD objects to which a default double-click action other than **PROPERTIES** is assigned.

AutoCAD Object Type	Associated Double-Click Action
ATTDEF	**DDEDIT**
ATTBLOCKREF	**EATTEDIT**
ATTDYNBLOCKREF	**EATTEDIT**
ATTRIB	**ATTIPEDIT**
BLOCKREF	$M=$(if,$(and,$(>,$(getvar,blockeditlock),0)),^C^C_properties,^C^C_bedit)
DYNBLOCKREF	$M=$(if,$(and,$(>,$(getvar,blockeditlock),0)),^C^C_properties,^C^C_bedit)
IMAGE	**IMAGEADJUST**
LWPOLYLINE	**PEDIT**
MLINE	**MLEDIT**
MTEXT	**MTEDIT**
POLYLINE	**PEDIT**
SECTIONOBJECT	**LIVESECTION**
SPLINE	**SPLINEDIT**
TEXT	**DDEDIT**
XREF	**REFEDIT**

Figure 23-20.
All of the AutoCAD object types are displayed in the Double Click Actions branch in the **Customize User Interface** dialog box.

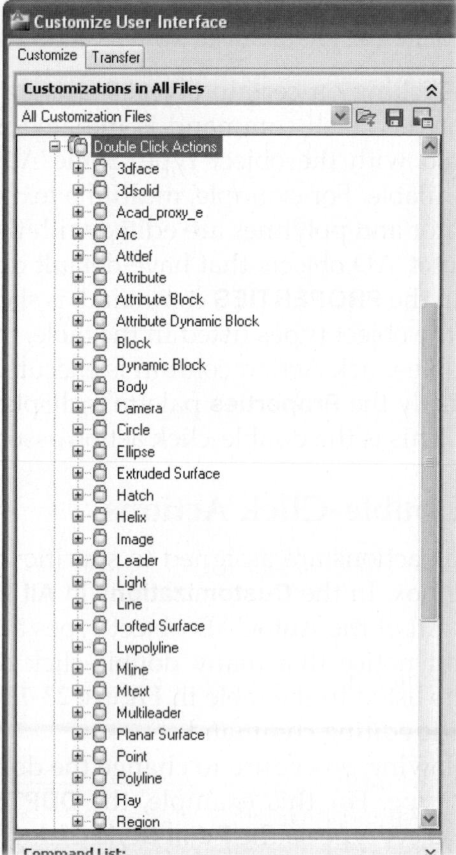

1. Open the **Customize User Interface** dialog box.
2. In the **Customizations in All Files** pane, expand the Double Click Actions branch.
3. Expand the Point branch under the Double Click Actions branch. Notice that the **PROPERTIES** command is associated with the point object type.
4. In the **Command List:** pane, select the **Point Style...** command. This is the **DDPTYPE** command, as indicated in the **Properties** pane when the command is selected.
5. Drag the **Point Style...** command from the **Command List:** pane and drop it into the Point branch in the **Customizations in All Files** pane. See Figure 23-21. The command replaces the existing command as there can only be one double-click action.
6. Pick the **OK** button to close the **Customize User Interface** dialog box and apply the change.

Now, draw a point using the **POINT** command. Double-click on the point and the **Point Style** dialog box appears. Select a new point style in the dialog box and pick the **OK** button. All existing points in the drawing should update to the new style. If not, use the **REGEN** command to update the display.

Custom Double-Click Action

You can also create a custom command and assign it to an object type as a double-click action. In this section, you will create a custom command for editing the radius of a circle to one-half of its current value when the circle double-clicked. Here is a breakdown of what the custom command will do:

- Cancel any commands in progress. (^C^C)
- Execute the **SCALE** command with the circle that was double-clicked being the object to be scaled. (scale;)

Figure 23-21.
The double-click action associated with the point object type is changed.

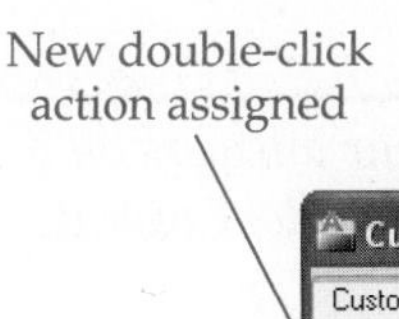

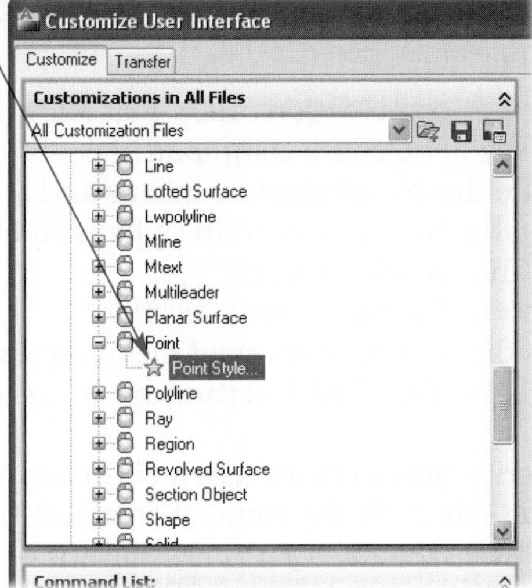

- Set the center of the circle, which is the point last created, as the base point. (@;)
- Scale the circle to half of its original size. (0.5)

This can be a handy tool should you accidentally enter the intended diameter in response to the radius prompt for the **CIRCLE** command. On noticing you just made that mistake, double-click the circle and the problem is fixed.

Follow these steps to create a custom command and assign it as a double-click action for the dimension object type:

1. Open the **Customize User Interface** dialog box.
2. Pick the **Create a new command** button in the **Command List:** pane.
3. Name the custom command **CirRadToDia**.
4. In the **Properties** pane, select the Macro property. Then, enter the macro ^C^Cscale;@;0.5 in the text box.
5. In the **Customizations in All Files** pane, expand the Double Click Actions branch and locate the Circle branch below it. Notice that the **PROPERTIES** command is currently associated with the circle object type.
6. In the **Command List:** pane, select the new **CirRadToDia** command and drag it into the Circle branch in the **Customizations in All Files** pane.
7. Pick the **OK** button to close the **Customize User Interface** dialog box and apply the change.

Draw a circle using the **CIRCLE** command. Now, double-click on the circle. The circle becomes one-half of its original size. In other words, the previous radius value is the new diameter value.

NOTE

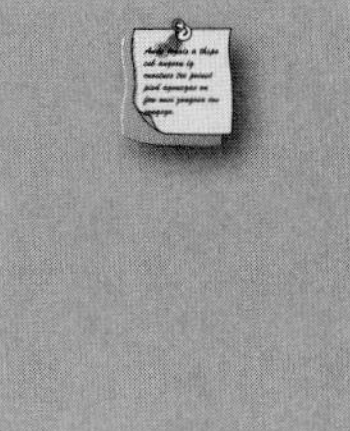

The above macro is intended to be used *immediately* after incorrectly creating the circle. Using the custom double-click action assigned to the circle object type after performing other drawing operations will cause unexpected results as the center of the circle will no longer be the last point created.

Chapter Test

Answer the following questions. Write your answers on a separate sheet of paper or complete the electronic chapter test on the student website.
www.g-wlearning.com/CAD

1. What is the key combination called that allows you to press the [Ctrl] key and an additional key to execute a command?
2. How can you disable all shortcut menus and all double-click actions?
3. In which dialog box can you define what a right-click does when in default mode, edit mode, and command mode?
4. Why are shortcut menus *context sensitive?*
5. When does the command shortcut menu appear?
6. What must the **PICKFIRST** setting be in order for the edit shortcut menu to appear?
7. Briefly describe how to create a new shortcut menu.
8. Describe the syntax for the name of an object-specific shortcut menu.
9. Why is CIRCLE_OBJECT *not* a valid alias for a shortcut menu?
10. Describe the difference between an OBJECT_*type* shortcut menu and an OBJECTS_*type* shortcut menu.
11. What is the syntax for the alias for a command-specific shortcut menu?
12. List the steps to add an alias to a shortcut menu.
13. How do you turn on the display of the **Quick Properties** palette?
14. What is a *quick property?*
15. How does a rollover tooltip differ from the **Quick Properties** palette?
16. In which dialog box do you select the objects that appear in the object type list in the **Customize User Interface** dialog box?
17. What is a *double-click action?*
18. What is the most common double-click action?
19. In which dialog box is a double-click action assigned?
20. List the basic steps for modifying the double-click action associated with an object.

Drawing Problems

Before customizing AutoCAD, check with your instructor or supervisor for specific instructions or guidelines.

1. Create a shortcut key for each of the drawing and editing commands listed below. Be sure not to use any existing shortcut keys.

LINE	**MOVE**
ARC	**COPY**
CIRCLE	**STRETCH**
POLYLINE	**TRIM**
POLYGON	**EXTEND**
RECTANGLE	**CHAMFER**
DTEXT	**FILLET**
ERASE	

2. In this problem, create an object-specific shortcut menu. The shortcut menu should be displayed when an arc object is selected. The shortcut menu should contain the **LENGTHEN** and **BREAK** commands.

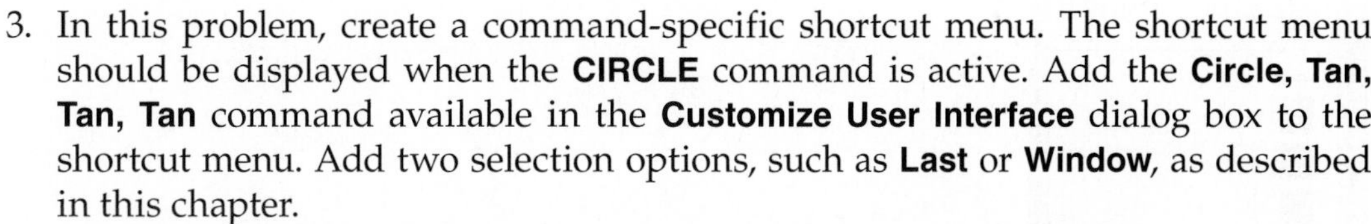

3. In this problem, create a command-specific shortcut menu. The shortcut menu should be displayed when the **CIRCLE** command is active. Add the **Circle, Tan, Tan, Tan** command available in the **Customize User Interface** dialog box to the shortcut menu. Add two selection options, such as **Last** or **Window**, as described in this chapter.

4. By default, double-clicking on a circle displays the **Properties** palette. Take the steps necessary so that a **REGEN** is performed instead.

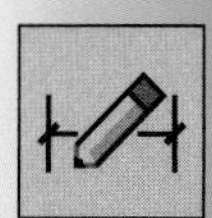

5. Create a custom command that changes the color of an object to blue. Then, assign this command as the double-click action for the hatch object type.

Tool palettes can be easily customized to improve your modeling efficiency. You can modify existing tool palettes or create new palettes populated with your own custom commands. Shown here is a custom tool palette containing a variety of tools.

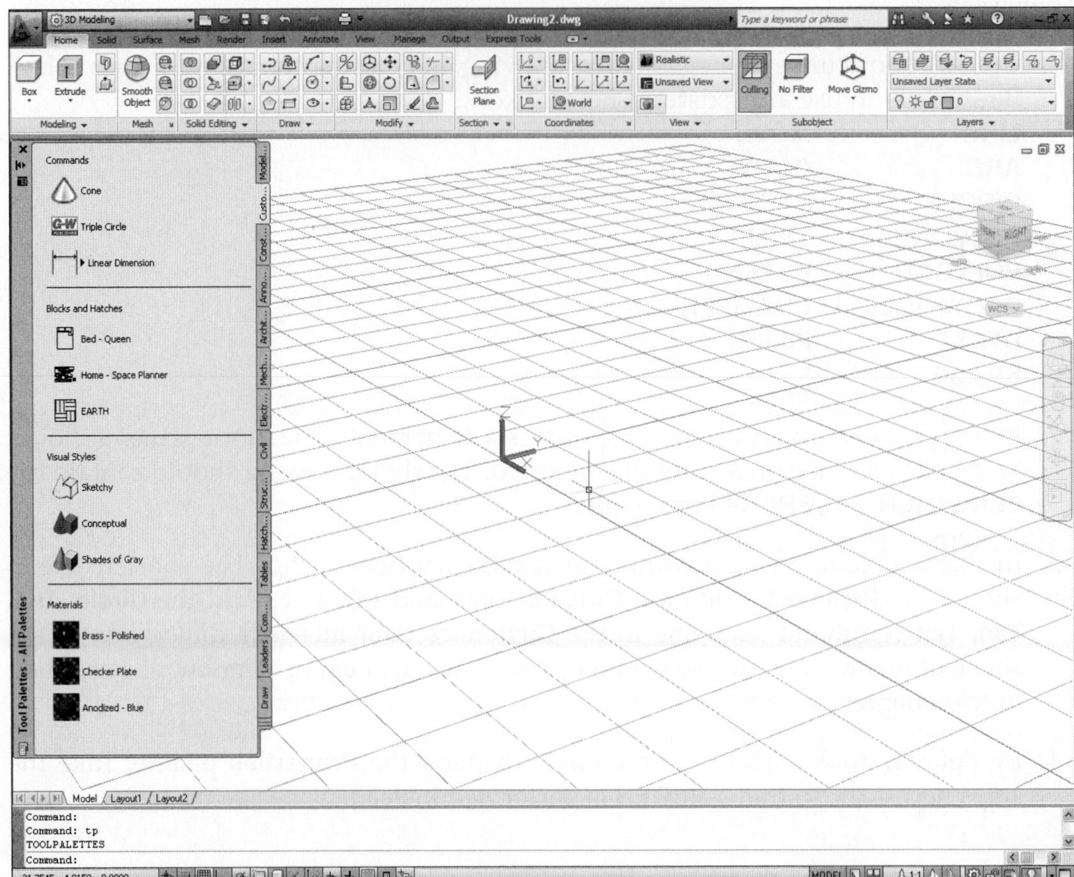

CHAPTER 24

Tool Palette Customization

Learning Objectives

After completing this chapter, you will be able to:

✓ Modify the appearance of the **Tool Palettes** window.
✓ Compare and contrast block insertion, hatch insertion, and command tools.
✓ Create new tool palettes from scratch and using **DesignCenter**.
✓ Add tools to existing tool palettes.
✓ Explain how tool palettes are formatted.
✓ Adjust the properties of tools.
✓ Create a flyout tool.
✓ Organize tool palette tabs into groups.
✓ Export and import tools and tool palettes.

Tool palettes are a user-interface method for the easy insertion of blocks and hatch patterns and for command entry. Blocks and hatch patterns can be simply dragged and dropped from a tool palette directly into a drawing. Tool palettes are contained within the **Tool Palettes** window.

Tool Palette Overview

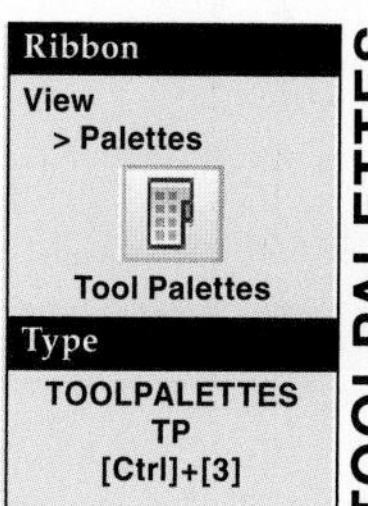

The **TOOLPALETTES** command is used to display the **Tool Palettes** window. Notice that the **Tool Palettes** window has a number of tabs on its edge. Each of these tabs corresponds to a tool palette. To make a tool palette active, pick its tab. If there are more tabs than can be displayed, pick on the "stack" at the bottom of the tabs to display a shortcut menu in which you can select the tool palette to display. To use a tool on any tool palette, drag it from the tool palette and drop it into the drawing.

By default, the **Tool Palettes** window contains the **Modeling**, **Constraints**, **Annotation**, **Architectural**, **Mechanical**, **Electrical**, **Civil**, **Structural**, **Hatches and Fills**, **Tables**, **Command Tool Samples**, **Leaders**, **Draw**, **Modify**, **Cameras**, **Visual Styles**, and five light palettes. Each of these is described below.

- **Modeling palette.** Contains tools for creating specific variations of some solid primitives. Also contains two UCS tools and a 3D align tool.
- **Constraints palette.** Contains tools for inserting geometric and dimensional constraints.

- **Annotation palette.** Contains blocks that are typically inserted in paper space.
- **Architectural, Mechanical, Electrical, Civil, and Structural palettes.** Can be used to insert blocks that are meant to be used within the discipline for which the palette is named. Certain preset properties are already attached to the symbols, such as scale and rotation.
- **Hatches and Fills palette.** Contains some commonly used hatches and sample gradient fills that can be quickly inserted into a drawing.
- **Tables palette.** Allows you to insert sample tables in US customary (Imperial) and metric formats.
- **Command Tool Samples palette.** Allows you to execute certain commands by picking the tool.
- **Leaders palette.** Allows you to place leaders with or without text or with various types of callout balloons in both US customary (Imperial) and metric scales.
- **Draw palette.** Contains many of the same tools found in the **Draw** panel on the **Home** tab of the ribbon and some tools for inserting blocks, attaching images, and attaching xrefs.
- **Modify palette.** Contains many of the same tools found in the **Modify** panel on the **Home** tab of the ribbon.
- **Light palettes.** Contain tools for inserting many types of lights.
- **Cameras palette.** Contains tools for inserting three different cameras into the drawing. The cameras have differing lens lengths and fields of view.
- **Visual Styles palette.** Contains tools for three variations of the default visual styles.

For detailed instruction on how tool palettes can be used to insert blocks and hatch patterns, see *AutoCAD and Its Applications—Basics.* This chapter describes how to customize existing tool palettes and create your own tool palettes.

Tool Palette Appearance

There are a number of methods for altering the appearance of the **Tool Palettes** window, either for productivity or personal preference. For example, the tabs can be renamed. Right-click on the title of the tab and select **Rename Palette** from the shortcut menu. An edit box is displayed near the current name with the name highlighted. Type a new name and press [Enter] to rename the palettes. Other methods of changing the appearance of the **Tool Palettes** window are covered in this section.

Docking

By default, the **Tool Palettes** window is floating on the right side of the screen in the AutoCAD Classic workspace and not displayed in the 2D Drafting & Annotation and 3D Modeling workspaces. In the 3D Basics workspace, it will be displayed if displayed in the previous workspace.

When floating, the **Tool Palettes** window can be moved around the screen just like a floating toolbar. Moving the window to the far side of the drawing window forces the title bar to flip to the other side of the **Tool Palettes** window so the bar is toward the outer edge of the drawing window.

By default, the **Tool Palettes** window can be docked, just like a toolbar. Moving the window outside of the drawing window to the left or right docks it. The **Tool Palettes** window cannot be docked at the top or bottom. To prevent docking, right-click on the **Tool Palettes** window title bar to display the shortcut menu. Select **Allow Docking** to remove the check mark. When a check mark appears next to **Allow Docking,** the window can be docked. If docked, it can be moved to a new, floating location by picking and holding on the title bar, dragging the window to the desired location, and releasing the pick button.

The **Tool Palettes** window can be resized just like standard windows. While floating, the top and bottom edges, the vertical area just below the tabs, and the corners can be used to resize the window. While docked, only the right or left edge can be used for resizing. As with standard windows, move the cursor to one of the edges of the window until a double arrow appears. Then, press and hold the pick button, drag the edge until the window reaches the desired size, and release the pick button.

Transparency

Using the **Tool Palettes** window while it is floating may cause occasional visibility problems because it covers up part of the drawing window. However, the window can be made partially transparent so that the part of the drawing under the window can be seen. Transparency will not be active when the window is in a docked position. Hardware acceleration may need to be turned off to set transparency, depending on your system configuration.

Right-click on the title bar of the **Tool Palettes** window and pick **Transparency...** from the shortcut menu. The window must be floating for this option to appear. The **Transparency** dialog box is then displayed, Figure 24-1. To set the level of transparency, move the slider in the **General** area to a lower value. The slider controls the level of transparency applied to the window. The further to the left that the slider is placed, the more transparent the **Tool Palettes** window, Figure 24-2. Placing the slider all of the way to the right makes the **Tool Palettes** window opaque. The slider in the **Rollover** area sets the transparency level for when the cursor is over the window. This setting must be equal to or greater than the slider setting in the **General** area. To see the transparency level when the cursor is over the window, pick the **Click to Preview** button.

If you find a particular level of transparency that you like, but wish to make the **Tool Palettes** window opaque for a short time while you perform an operation or two, you can just use the toggle to turn off transparency. Open the **Transparency** dialog box and check the **Disable all window transparency (global)** check box. Then, when you want to go back to your previous level of transparency, simply open the **Transparency** dialog box again and uncheck the check box. You will not have to adjust the slider; just toggle the transparency back on.

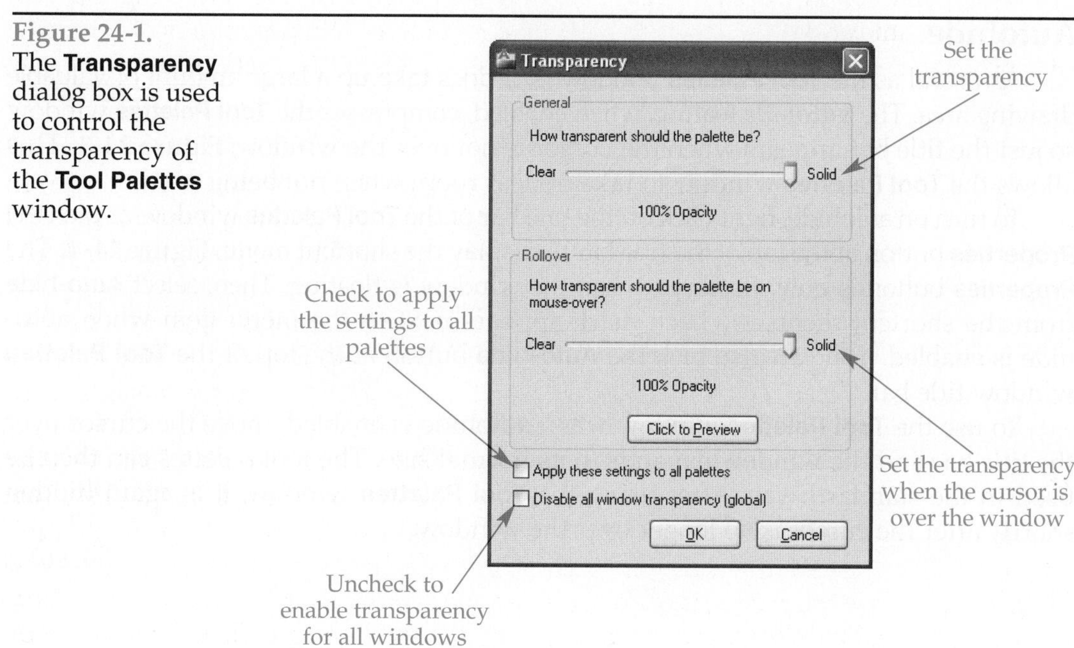

Figure 24-1.
The **Transparency** dialog box is used to control the transparency of the **Tool Palettes** window.

Figure 24-2.
A—The **Tool Palettes** window has a low transparency setting.
B—The **Tool Palettes** window has a high transparency setting.

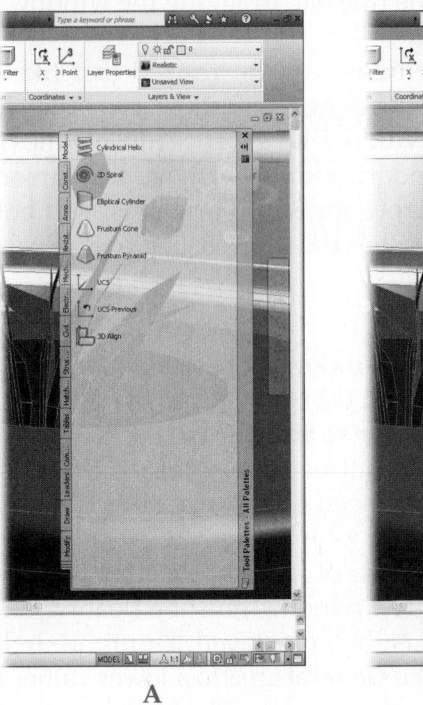

A B

Also notice in the **Transparency** dialog box the **Apply these settings to all palettes** check box. This allows you to apply the changes made in the dialog box to all AutoCAD palettes. Often, a user will prefer the same level of transparency for all palettes.

NOTE

Although transparency allows you to *see* through the **Tool Palettes** window, you cannot *work* through it. You cannot access points behind the window because the cursor is actually on the **Tool Palettes** window, not on the drawing underneath it. This is true of any AutoCAD palette.

Autohide

As useful as the **Tool Palettes** window is, it does take up a large amount of valuable drawing area. The ***autohide*** feature, when enabled, compresses the **Tool Palettes** window so just the title bar appears when the cursor is not over the window, Figure 24-3. This allows the **Tool Palettes** window to take up less room when not being used.

To turn on autohide, right-click on the title bar of the **Tool Palettes** window or pick the **Properties** button at the top of the title bar to display the shortcut menu, Figure 24-4. The **Properties** button is only displayed when the window is floating. Then, select **Auto-hide** from the shortcut menu. A check mark appears next to the menu item when autohide is enabled. You can also pick the **Auto-hide** button at the top of the **Tool Palettes** window title bar.

To use the **Tool Palettes** window when autohide is enabled, move the cursor over the title bar and the window expands to its normal size. The tool palettes can then be used in the standard way. After using the **Tool Palettes** window, it is again hidden shortly after the cursor is no longer over the window.

Figure 24-3.
When the autohide feature is enabled, the **Tool Palettes** window appears as only the title bar when the cursor is not over it.

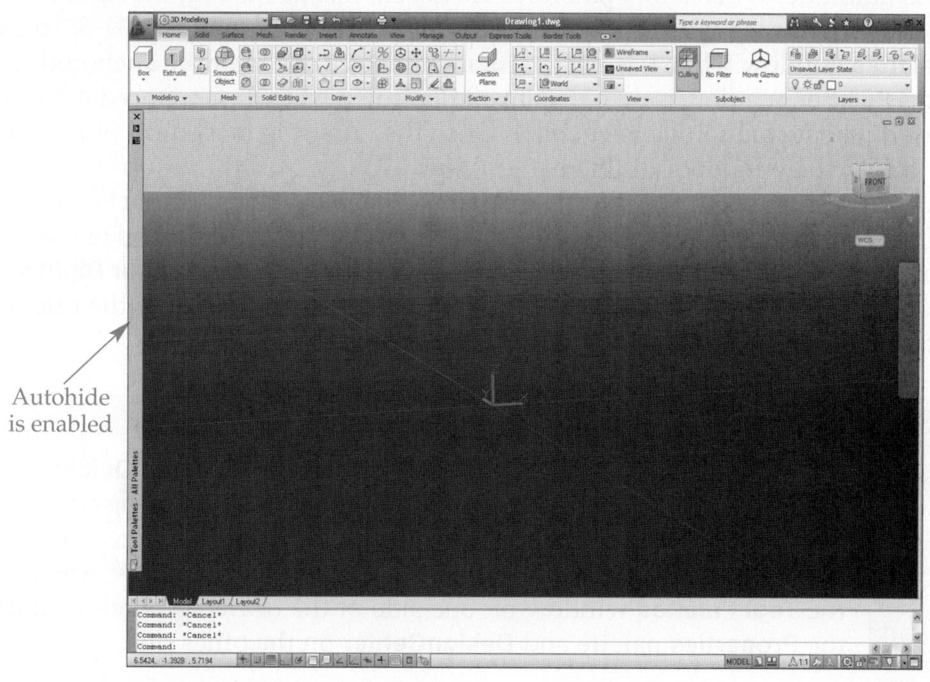

Figure 24-4.
Using the shortcut menu to enable autohide.

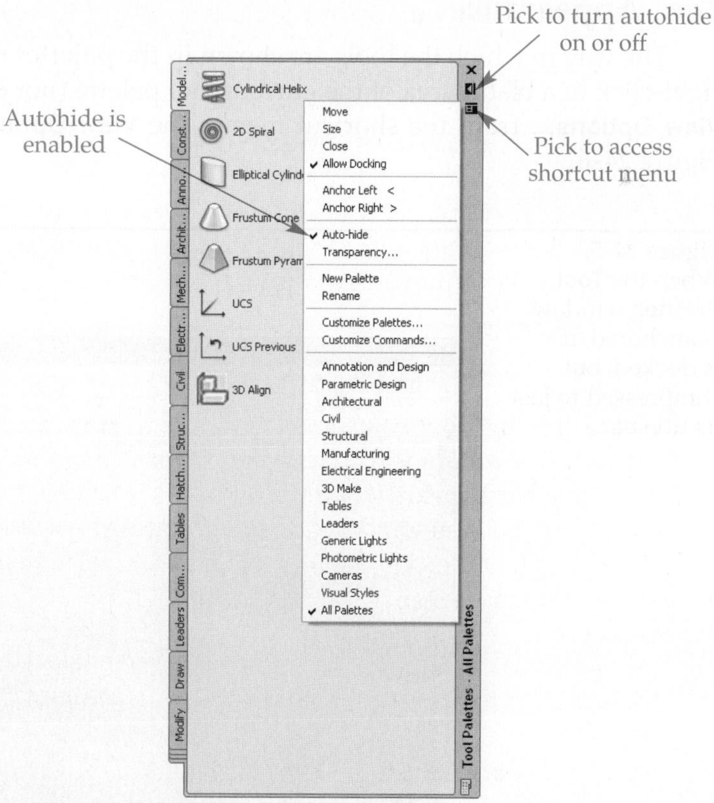

Pick to turn autohide on or off

Autohide is enabled

Pick to access shortcut menu

Anchoring

The **Tool Palettes** window can also be ***anchored.*** Anchoring is a combination of docking and autohide. When a tool palette is anchored, it is docked on the right or left side of the drawing area, but it is compressed to just a title bar. See Figure 24-5. To use the **Tool Palettes** window when it is anchored, move the cursor over the anchored title bar. The **Tool Palettes** window is then displayed floating next to the anchored window. It can be used just as if autohide is enabled. Once the cursor is moved off of the **Tool Palettes** window, the window is hidden.

To anchor the tool palette, docking must first be enabled. Then, right-click on the title bar of the **Tool Palettes** window or pick the **Properties** button at the top of the title bar to display the shortcut menu. Next, select either **Anchor Left <** or **Anchor Right >** in the shortcut menu. To return the window to floating mode, pick and drag the title bar back into the drawing area while the window is displayed.

PROFESSIONAL TIP

The **Properties** palette, **External References** palette, **QuickCalc** palette, **DesignCenter**, and the ribbon also have the anchoring feature. You can create a very productive drawing-window arrangement by anchoring these items along with the **Tool Palettes** window. Anchor the **Tool Palettes** window on one side of the drawing window and the **Properties** palette and **DesignCenter** on the other side.

Tool Appearance

The way in which the tools are shown in the palettes can be customized. To do so, right-click in a blank area of the current tool palette (not on the title bar). Then, select **View Options...** from the shortcut menu. The **View Options** dialog box is displayed, Figure 24-6.

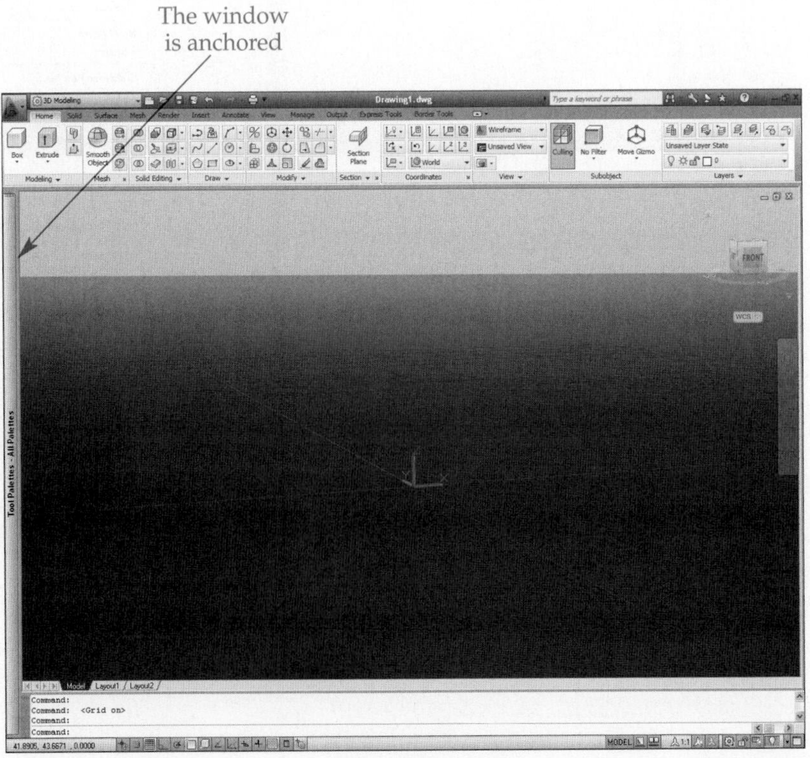

Figure 24-5. When the **Tool Palettes** window is anchored, it is docked, but compressed to just its title bar.

Figure 24-6.
The **View Options** dialog box is used to set how the tools appear in tool palettes.

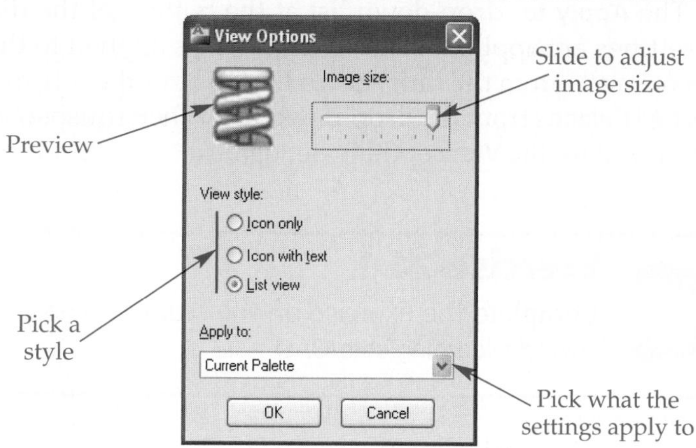

The **Image Size:** area of the dialog box is used to set the size of the tool icons in the palette. Drag the slider to the left or right to change the size. Dragging the slider to the left decreases the size of the icon. Dragging the slider to the right increases the size of the icon. To the left of the slider is a preview that represents the size of the icon.

The **View Style:** area controls how the tools on the tool palettes are displayed. When the **Icon only** radio button is selected, the tools are represented as an image only, Figure 24-7A. Tooltips will be displayed if you hold your cursor over a tool. Selecting the **Icon with text** radio button represents the tools with an image and the tool name below the image, Figure 24-7B. When the **List view** radio button is selected, the tools are represented with an image and the tool name to the side of the image, Figure 24-7C. This is the default view style.

Figure 24-7.
The various ways in which tools can appear in tool palettes. A—Icons only. B—Icons and text. C—As a list.

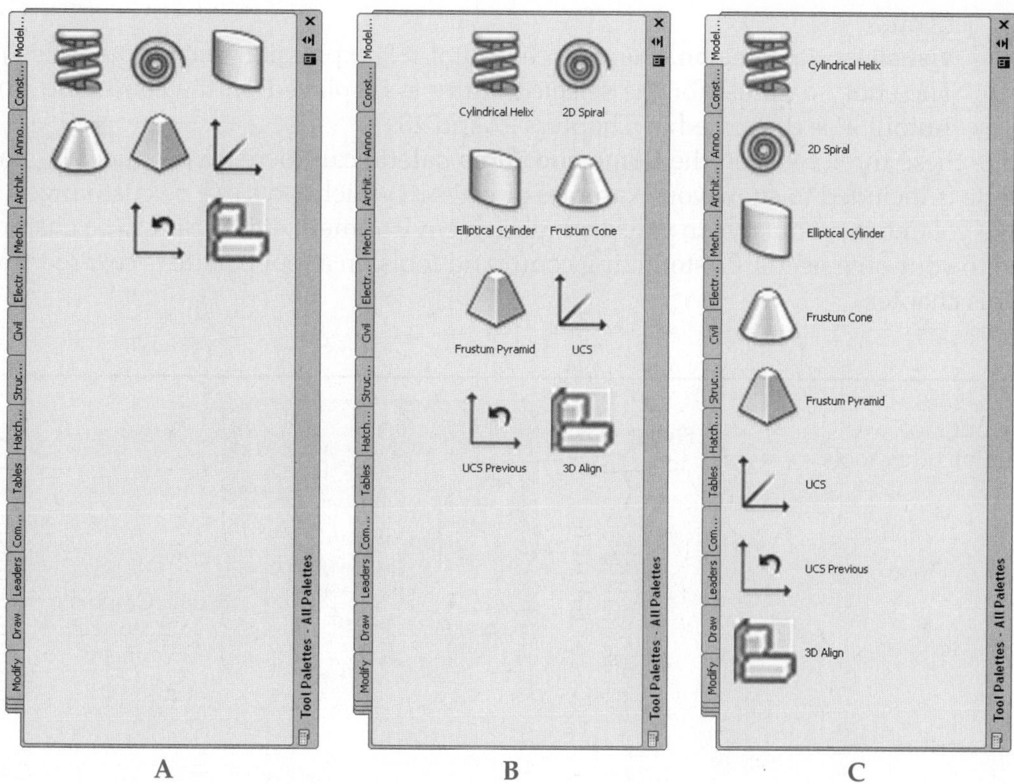

The **Apply to:** drop-down list at the bottom of the dialog box determines where the settings are applied. To have the settings applied to the current tool palette, select Current Palette from the drop-down list. To have the settings applied to all tool palettes, select All Palettes from the drop-down list. When finished making settings, pick the **OK** button to close the **View** options dialog box.

Exercise 24-1

Complete the exercise on the student website.
www.g-wlearning.com/CAD

Commands in a Tool Palette

As indicated earlier, the **Tool Palettes** window not only offers a means to easily insert blocks and hatch patterns, it can be used to execute commands. Make the **Tool Palettes** window active and select the **Command Tool Samples** palette, Figure 24-8. The tools on this palette are provided to demonstrate how tool palettes can be customized by adding commands. This tool palette is provided with the intention that it will be customized by the user. The default tools provided are:

- **Line.** Executes the **LINE** command. Notice that this tool has a small triangle, or arrow, to the right of the icon. The arrow indicates that the tool acts as a ***flyout tool,*** similar to flyout buttons on the ribbon. Picking the arrow displays a graphic shortcut menu containing other command tools attached to this tool, as shown in Figure 24-8. Selecting a command tool from the shortcut menu executes the command and causes that command image to become the default image displayed in the palette.
- **Linear Dimension.** Executes the **DIMLINEAR** command. This tool is also a flyout.
- **VisualLisp Expression.** Executes the AutoLISP expression (entget (car (entsel))). The entity data list for the selected entity is displayed on the command line. AutoLISP is discussed in Chapters 27 and 28.

While these three tools on the **Command Tools** palette can increase productivity, this palette is included to show you examples of ways in which tools can be customized to make you more productive in your own design environment. It is meant to be customized to your own needs. Customizing command tools on a tool palette is covered later in this chapter.

Figure 24-8.
A flyout tool contains other tools.

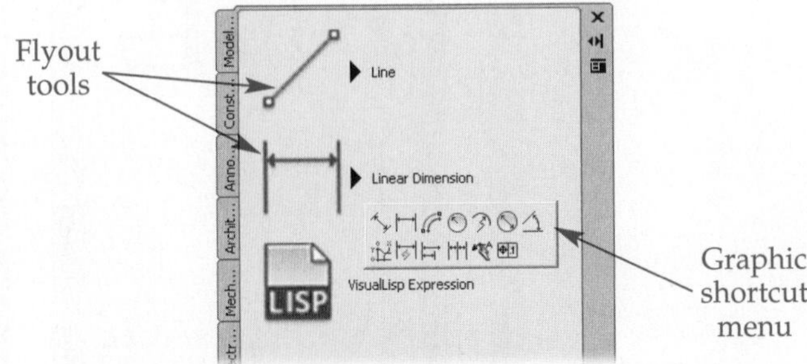

Adding Tool Palettes

As discussed later in this chapter, you can add new tools to tool palettes. You can also create new tool palettes and then add tools to them. To add a new tool palette, right-click on a blank area of an existing tool palette or on the title bar of the **Tool Palettes** window. Then, select **New Palette** from the shortcut menu. A new, blank palette is added and a text box appears next to the name. Type the desired name for the new palette and press [Enter].

Notice the help link at the top of the new tool palette, **Figure 24-9.** This link offers assistance in customizing tool palettes. It will disappear once you add a tool to the tool palette.

The new tool palette can be reordered within the tabs by right-clicking on the tab name to display the shortcut menu. Then, select **Move Up** or **Move Down** to reorder the palettes. You may need to do this several times in order to get the palette in the position you want. The other palettes can be reordered in this same manner. If you need to reorder multiple palettes, you can use the **Customize** dialog box. This dialog box is discussed in detail later in this chapter.

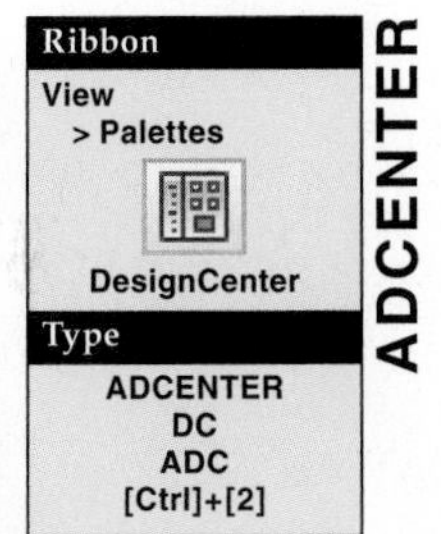

DesignCenter can be used to create a new palette fully populated with all of the blocks contained in a drawing. In the **Folders** tab of the **DesignCenter** palette, navigate to the drawing from which you are making the tool palette, right-click on the drawing name, and select **Create Tool Palette** from the shortcut menu. A new tool palette is added to the **Tool Palettes** window with the same name as the drawing file.

For example, open **DesignCenter** and navigate to the Fasteners-US drawing located in the \Sample\DesignCenter folder. Right-click on the file name and select **Create Tool Palette** from the shortcut menu, **Figure 24-10A.** A tool palette named **Fasteners-US** is added to the **Tool Palettes** window. All of the blocks contained in the Fasteners-US drawing are available on the tool palette, **Figure 24-10B.**

Figure 24-9.
A new, blank tool palette has been created.

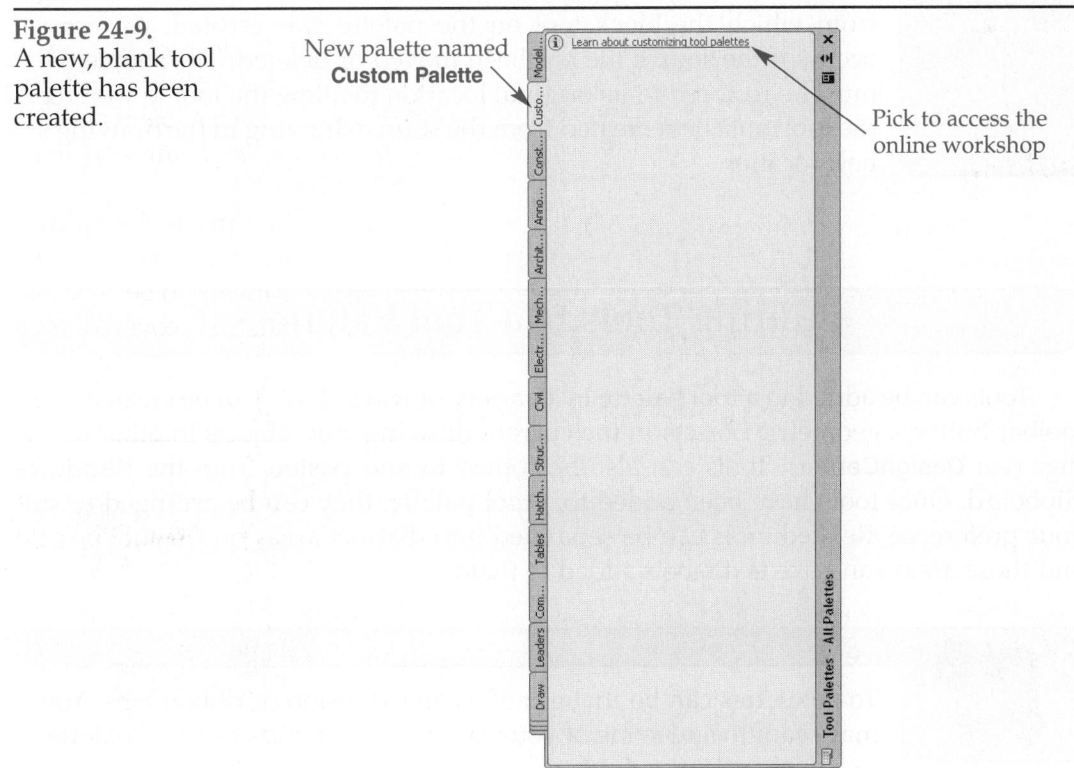

Figure 24-10.
A—Creating a tool palette from the blocks contained within a drawing. B—The new tool palette is added.

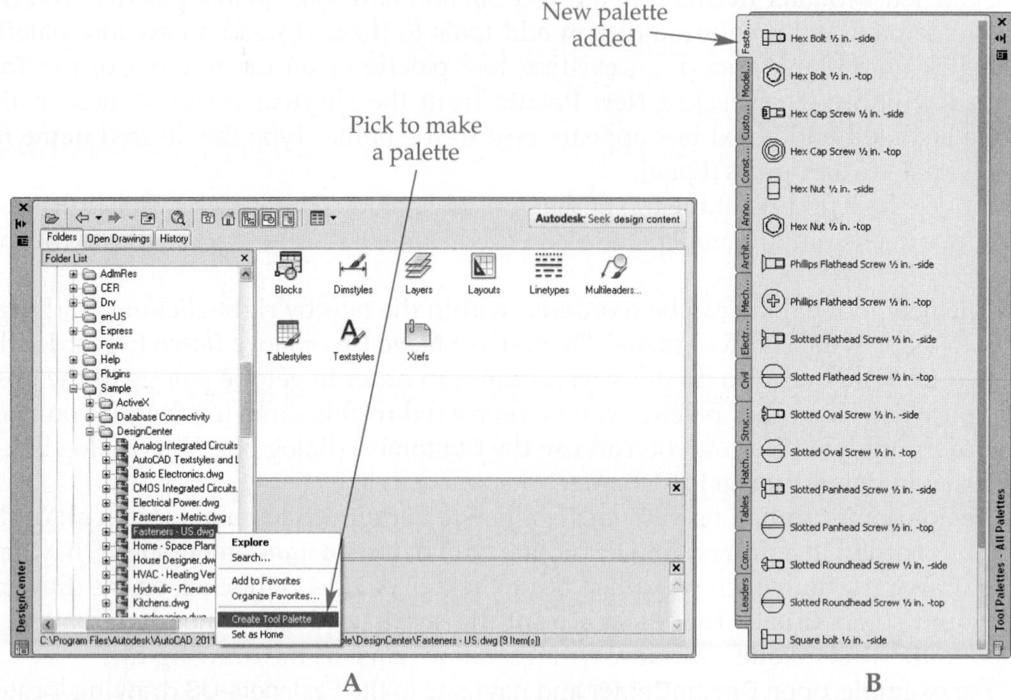

CAUTION

When using any "block insertion" tool from a tool palette, the block is actually being imported from the source drawing—the drawing from which the block tool on the palette was created. An error occurs if the source file has been moved or deleted. The source file must be restored to its original location to allow the tool to work or the tool must be recreated from the source drawing in the drawing's new location.

Adding Tools to a Tool Palette

Tools can be added to a tool palette in a variety of ways. Tools can be created from toolbar buttons, geometric objects in the current drawing, and objects in other drawings (via **DesignCenter**). Tools can also be copied to and pasted from the Windows clipboard. Once tools have been added to a tool palette, they can be arranged to suit your preference. Related tools can be separated into distinct areas on the tool palette and those areas can have text labels added to them.

PROFESSIONAL TIP

Tool palettes can be thought of as an extension of ribbon tabs. You may want to add some of your favorite commands to a tool palette.

Creating a Tool from a Toolbar

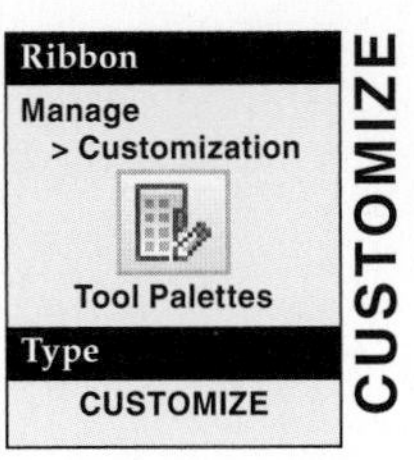

Toolbar buttons can be directly dragged and dropped onto a tool palette. To do this, the **Customize** dialog box for tool palettes must be open. This is *not* the **Customize User Interface** dialog box that is used to customize the ribbon, as described in previous chapters. To display the **Customize** dialog box, use the **CUSTOMIZE** command or right-click on a tool palette and select **Customize Palettes...** from the shortcut menu.

You do not actually use the **Customize** dialog box to copy a toolbar button to a tool palette, but the dialog box must be open. The **Customize** dialog box is discussed in detail later in this chapter.

Make sure the palette you want the button added to is current (on top). With the **Customize** dialog box open, move your cursor to the desired toolbar button. Flyout buttons cannot be added to a tool palette. Pick and hold on the toolbar button and drag it to the desired location in the tool palette, Figure 24-11. A horizontal "I-bar" appears in the tool palette to indicate where the new tool will be inserted. Drop the toolbar button when it is in the desired position. The tool is inserted in the tool palette. Close the **Customize** dialog box.

NOTE

Tools cannot be dragged from the ribbon to a tool palette, only from toolbars. Toolbars are not displayed by default except in the AutoCAD Classic workspace.

Figure 24-11.
Creating a tool on a tool palette from a toolbar button.

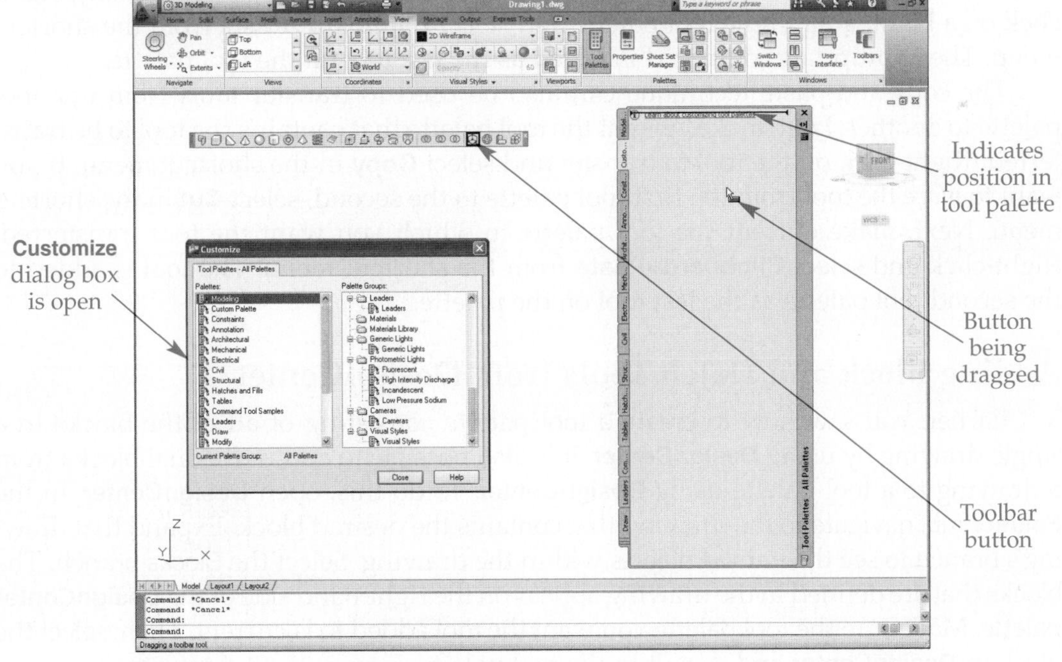

Creating a Tool from an Object in the Current Drawing

Another way to add a drawing command to a tool palette is to drag an object in the current drawing, such as a line, hatch, block, dimension, camera, or light, and drop it onto the tool palette. The appropriate command to create that object is added to the tool palette. Not all objects support this method. For example, solid primitives cannot be dragged onto a tool palette to create a tool.

First, ensure that the **PICKFIRST** system variable is set to 1. Also, make sure the tool palette to which you want the tool added to is current (on top). Next, with no command active, select the desired object. Move the cursor directly onto the selected object (not a grip). Then, press and hold down either mouse button. Finally, drag the object to the desired position on the tool palette and drop it. The appropriate drawing command is inserted into the tool palette.

PROFESSIONAL TIP

You can also drag and drop blocks and hatch patterns from the current drawing onto a tool palette. To drag a block from the drawing to a tool palette, the drawing must be saved (cannot be unnamed) because the tool references the source file of the block.

Using the Windows Clipboard to Create a Tool

The copy-and-paste feature of the Windows operating system is another way to transfer objects in the drawing to a tool palette. First, ensure that the **PICKFIRST** system variable is set to 1. Then, with no command active, right-click on the selected object and select **Clipboard** from the shortcut menu and then **Copy** from the cascading menu. Next, make current the tool palette to which you want the tool added. Finally, right-click on a blank area in the tool palette and select **Clipboard>Paste** from the shortcut menu. The appropriate command is added as the last tool on the tool palette.

The copy-and-paste technique can also be used to transfer tools from one tool palette to another. First, make current the tool palette that contains the tool to be transferred. Right-click on the tool to transfer and select **Copy** in the shortcut menu. If you want to *move* the tool from the first tool palette to the second, select **Cut** in the shortcut menu. Next, make current the tool palette to which you want the tool transferred. Right-click and select **Clipboard>Paste** from the shortcut menu. The tool is added to the second tool palette as the last tool on the palette.

Adding Block and Hatch Tools from DesignCenter

Earlier, you saw how to create a tool palette consisting of all of the blocks in a single drawing by using **DesignCenter**. It is also possible to add individual blocks from a drawing to a tool palette using **DesignCenter**. To do this, open **DesignCenter**. In the **Folders** tab, navigate to the drawing that contains the desired block. Expand that drawing's branch to see the named objects within the drawing. Select the Blocks branch. The blocks that are defined in the drawing appear on the right-hand side of the **DesignCenter** palette. Make sure the tool palette you want the tool added to is current. Then, select the block in **DesignCenter** and drag it to the tool palette, Figure 24-12. Move the cursor to the desired position on the tool palette and drop the block. The new block tool is inserted in the tool palette.

A tool that inserts an entire drawing into the current drawing can also be added to a tool palette using **DesignCenter**. In **DesignCenter**, navigate to the drawing in the **Folders** tab. Select the drawing on the right-hand side of the **DesignCenter** palette, drag it to the tool palette, and drop it in the desired position, Figure 24-13. The new

Figure 24-12.
Adding an individual block contained within a drawing as a tool on a tool palette.

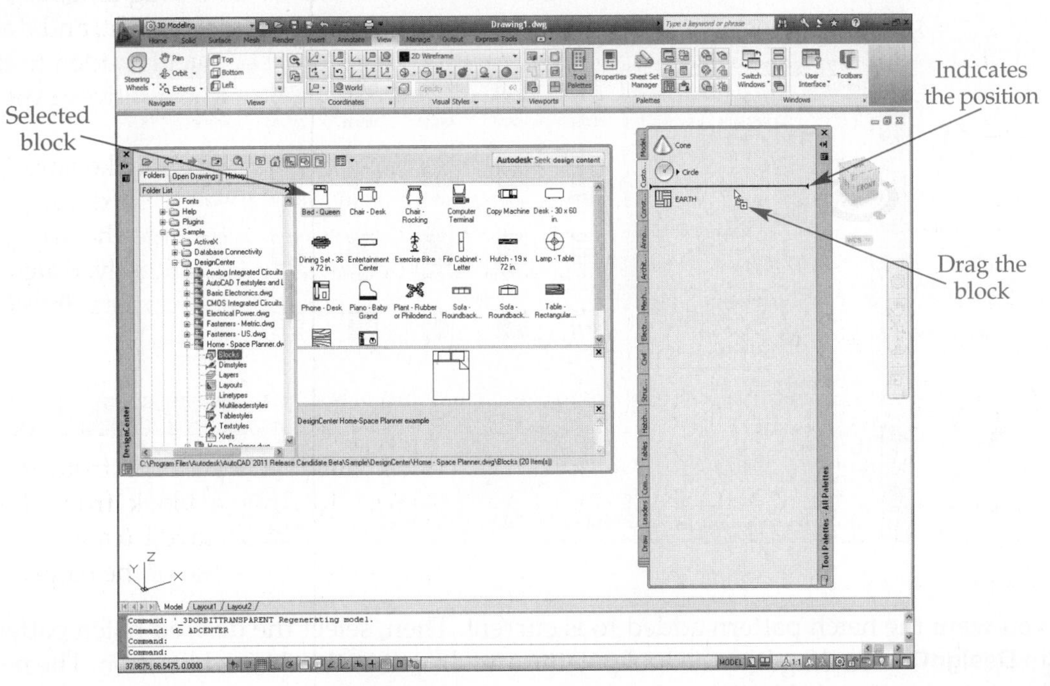

Figure 24-13.
Adding an entire drawing as a tool on a tool palette.

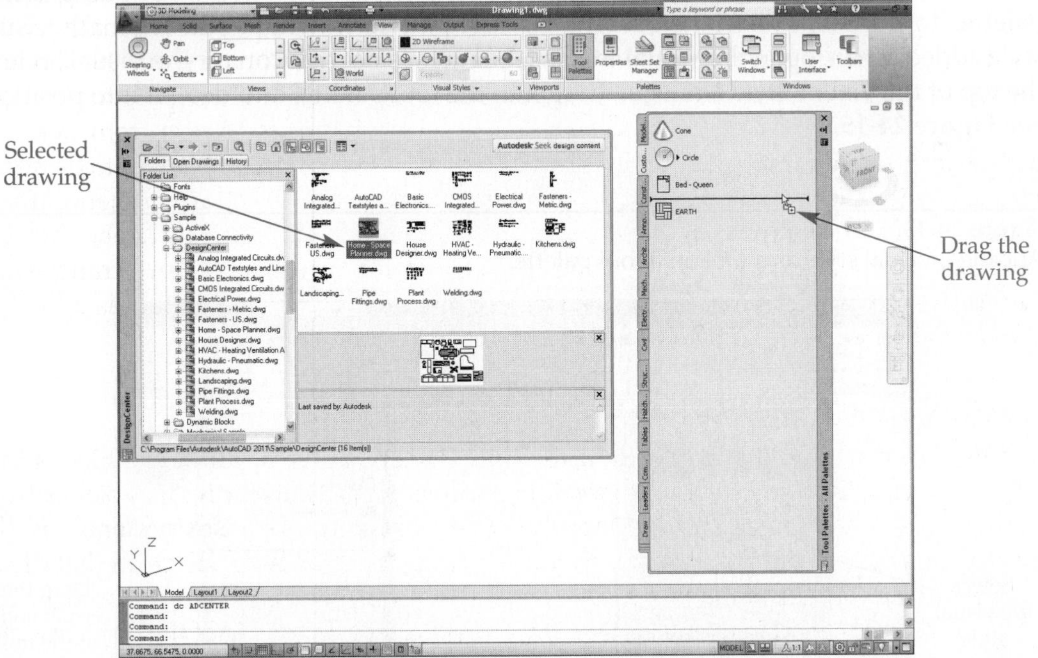

block tool is inserted in the tool palette. When the new tool is used, the entire drawing is inserted into the current drawing as a block.

DesignCenter can also be used to add hatch patterns to a tool palette. Hatch pattern definitions are stored in two files—acad.pat and acadiso.pat. These files are located in the user's \Support folder. In the **Folders** tab of **DesignCenter**, navigate to the acad.pat file and select it. All of the hatch patterns defined within that file are shown on the right-hand side of the **DesignCenter** palette, Figure 24-14. Make sure the tool palette

Figure 24-14.
Hatch patterns can be selected in **DesignCenter** and dragged to a tool palette to create a new tool.

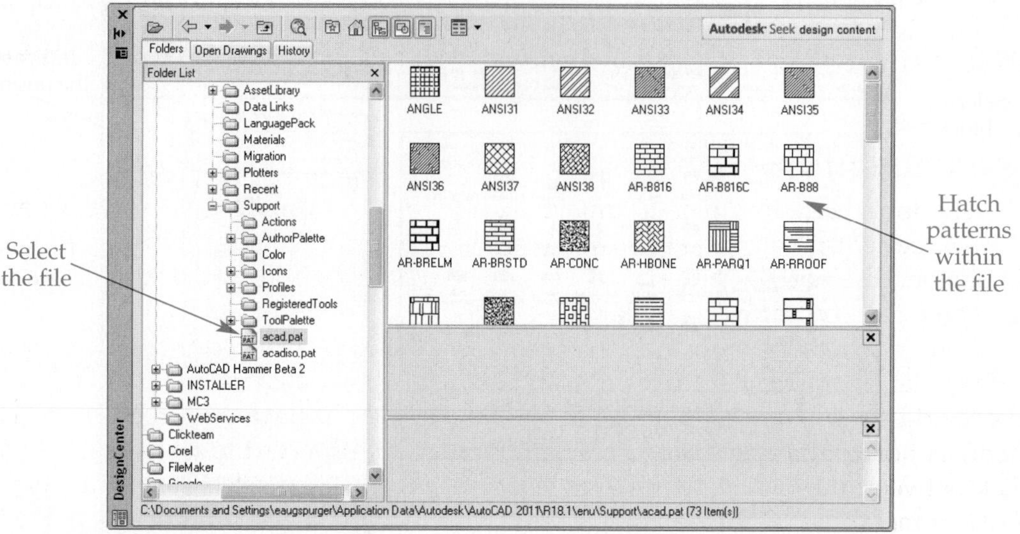

you want the hatch pattern added to is current. Then, select the desired hatch pattern in **DesignCenter**, drag it to the tool palette, and drop it in the desired location. The new hatch tool is inserted in the tool palette.

Adding Visual Style Tools

Visual styles that are available in the drawing can be added as a tool to a tool palette. To do this, first make current the tool palette to which you want the visual style added. Then, open the **Visual Styles Manager**. Select the icon for the visual style at the top of the **Visual Style Manager**, drag it to the tool palette, and drop it into position. See Figure 24-15.

Figure 24-15.
Adding a visual style as a tool on a tool palette.

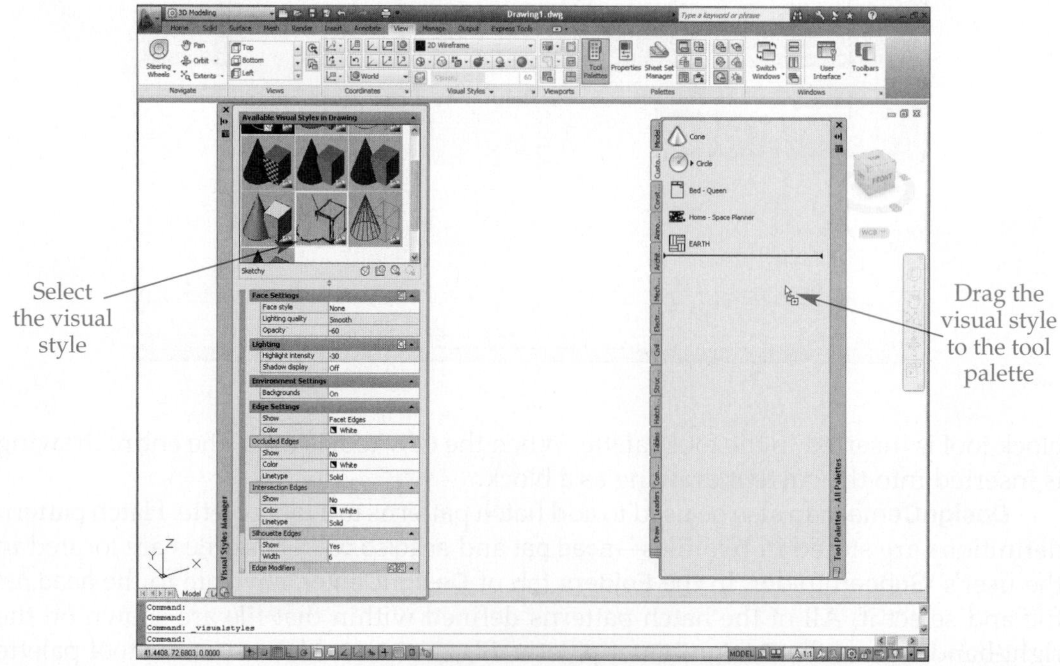

Select
the visual
style

Drag the
visual style
to the tool
palette

PROFESSIONAL TIP

While they can be added to a tool palette, both visual styles and materials are more efficiently and productively managed using the **Visual Styles Manager** and **Materials Browser** palettes.

Rearranging Tools on a Tool Palette

The tools on a tool palette can be rearranged into a more productive order. To move a tool within a tool palette, simply select the tool and drag it to a new location. Remember, the horizontal I-bar indicates where the tool will be moved. In this way, you can place your drawing tools together, your block tools together, and so on.

You can further separate the tools within a tool palette by adding separator bars and text labels. Move the cursor so that it is between the two tools where you would like to add the separator. Then, right-click and select **Add Separator** from the shortcut menu. A horizontal bar is added to the tool palette. To add text to a tool palette, right-click between the two items where the label should be and select **Add Text** from the shortcut menu. A text box is displayed with the default text highlighted. Type the text that you want for the label and press [Enter]. The text label is added to the tool palette. Separators and text labels can be used together to make the visual grouping of tools even more apparent, **Figure 24-16.** You can move separators and text labels to different locations within the palette just as you can tools.

Exercise 24-2

Complete the exercise on the student website.
www.g-wlearning.com/CAD

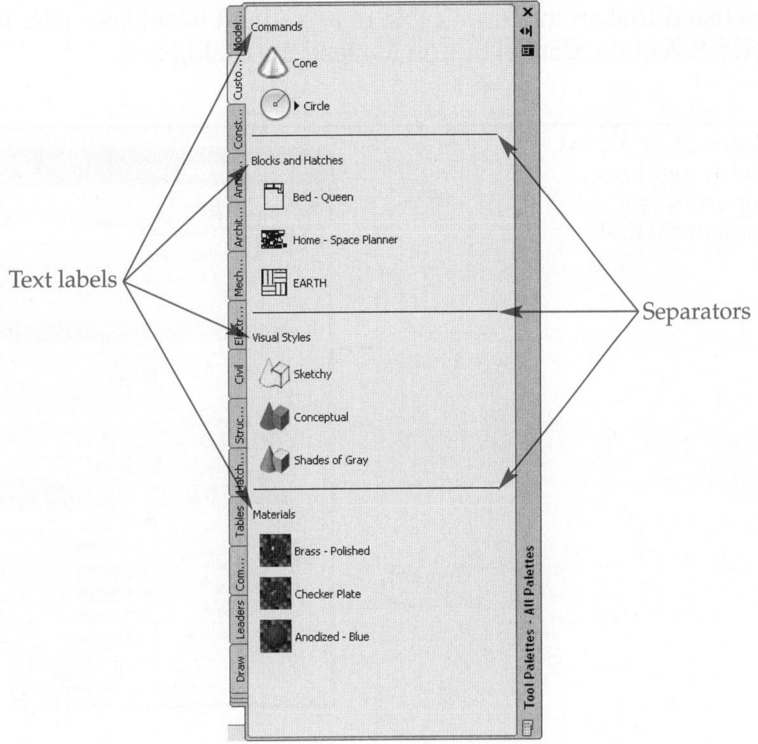

Figure 24-16.
Separators and text labels can be added to tool palettes to help group tools.

Modifying the Properties of a Tool

A tool on a tool palette can basically do one of these operations:

- Insert a block.
- Insert a hatch pattern.
- Insert a gradient fill.
- Initiate a command.
- Insert a light.
- Insert a camera.

Another type of tool called a flyout is discussed later in this chapter. Each of these tools has general properties assigned to it, such as color, layer, or linetype. Each also has some tool-specific properties assigned to it, depending on the operation associated with the tool. Most of these assigned properties can be customized to your own needs.

For example, select the **Hatches and Fills** palette in the **Tool Palettes** window. Right-click on the **Curved** gradient tool and select **Properties...** in the shortcut menu. The **Tool Properties** dialog box is displayed, Figure 24-17. Notice that the lower half of the dialog box is divided into two sections. The **General** category contains settings for properties such as Color, Layer, and Linetype. The **Pattern** category contains settings for properties specific to this tool's operation—inserting a gradient fill. Notice that properties such as Color 1, Color 2, and Gradient angle are shown. Other types of tools will have a different category in place of the **Pattern** category. Select the **Cancel** button to close the dialog box.

Now, select the **Command Tool Samples** palette in the **Tool Palettes** window, right-click on the **VisualLisp Expression** tool in the tool palette, and select **Properties...** from the shortcut menu. The **Tool Properties** dialog box is displayed, Figure 24-18. This is the same dialog box displayed for the **Curved** gradient tool. However, in place of the **Pattern** category is the **Command** category. The settings in the **Command** category are specific to the **VisualLisp Expression** tool. Notice that the **General** category contains the same properties as for the **Curved** gradient tool. There are some other general properties listed that are not applicable to a gradient (Linetype scale, Text style, and Dimension style). Select the **Cancel** button to close the dialog box.

Figure 24-17.
Modifying the properties of a gradient fill tool.

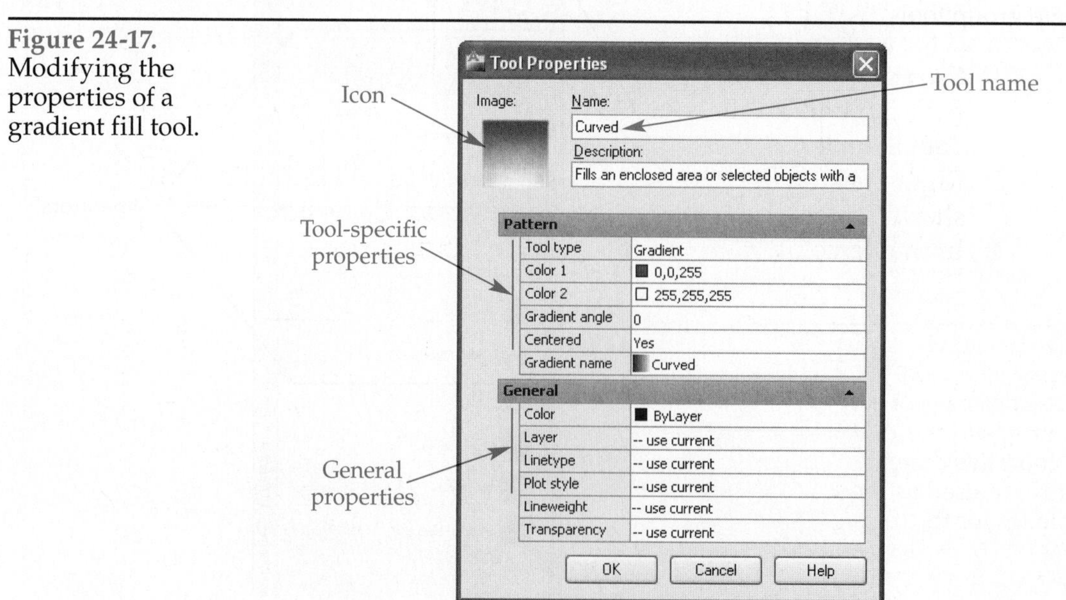

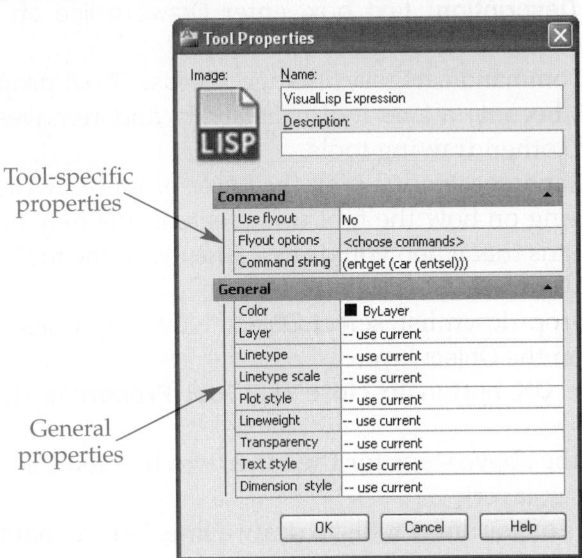

Figure 24-18. Modifying the properties of a command tool.

Customizing General Properties

The three items at the top of the **Tool Properties** dialog box are available for all types of tools. The **Image:** area shows the image that is assigned to the tool. This can be modified within the dialog box for command, camera, light, and visual style tools. The **Name:** text box displays the name of the tool. You can enter a new name for the tool. The **Description:** text box displays the current description of the tool. You can change the existing description or enter a new description. The name and description appear in the tooltip that is displayed when the cursor is held over the tool, Figure 24-19.

The **General** category in the **Tool Properties** dialog box can be used to assign specific values to the general properties that are a part of nearly all AutoCAD objects. The properties that can be customized are Color, Layer, Linetype, Linetype scale, Plot style, Lineweight, Transparency, Text style, and Dimension style. Depending on the type of tool, some of these properties may not be available.

The values assigned in the **Tool Properties** dialog box override the current property settings in the drawing when the tool is used. For instance, create three layers called Object, Hidden, and Center in the current drawing. A separate line tool can now be created for each of these layers:

1. Create a new, blank palette named **Line Tools**.
2. Add three line tools to the tool palette by copying them from the **Command Tool Samples** palette.
3. Right-click on the first of these new line tools and select **Properties** from the shortcut menu to display the **Tool Properties** dialog box.
4. In the **Name:** text box, enter Line-Object as the name.

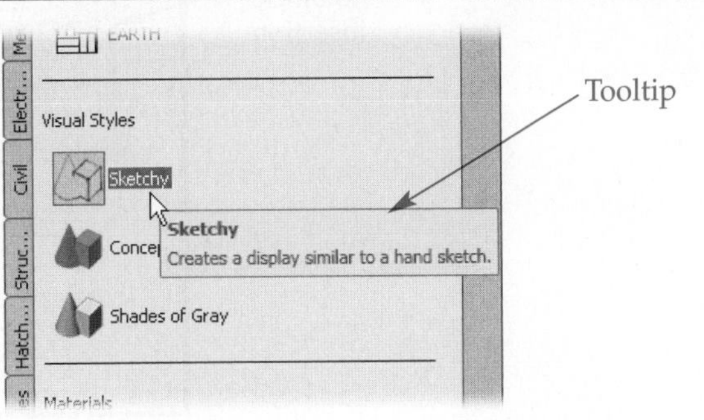

Figure 24-19. The Name: and Description: property settings in the **Tool Properties** dialog box are used as the tooltip for the tool.

5. In the **Description:** text box, enter Draws a line on the Object layer. as the description.
6. In the **Command** category, change the Use flyout property to No. This makes the tool become a Line tool exclusively and removes the flyout that allows access to other drawing tools.
7. In the **General** category, pick the Layer property, which is set to —use current (depending on how the tool was created, this may be set to a specific layer). This means that when you draw a line using the tool, the line is drawn on the current layer.
8. In the drop-down list, select Object. Now, any lines drawn with the tool are placed on the Object layer.
9. Pick the **OK** button to close the **Tool Properties** dialog box and save the changes.
10. Repeat the above steps for the other two line tools setting them to the Hidden and Center layers.

Now, use one of your new tools to draw a line. Notice that when you select the tool the current layer switches to the one assigned to the tool. When you finish using the tool, the current layer switches back to the previous layer. Try each of the other new tools. Experiment with some of the other general properties, such as Color, Linetype, and Lineweight.

PROFESSIONAL TIP

Creating layer-specific drawing tools as outlined above can streamline your drafting. For example, in an architectural drawing you could have tools set to draw solid lines on a Walls layer, dashed lines on an Electrical layer, etc. This prevents you from needing to set a new layer current or changing properties after drawing lines.

Customizing Block Insertion Tools

When a block insertion tool is modified in the **Tool Properties** dialog box, properties specific to block insertion are displayed in a category labeled **Insert**, **Figure 24-20**. These properties are described as follows.

Figure 24-20.
The **Tool Properties** dialog box for a block insertion tool.

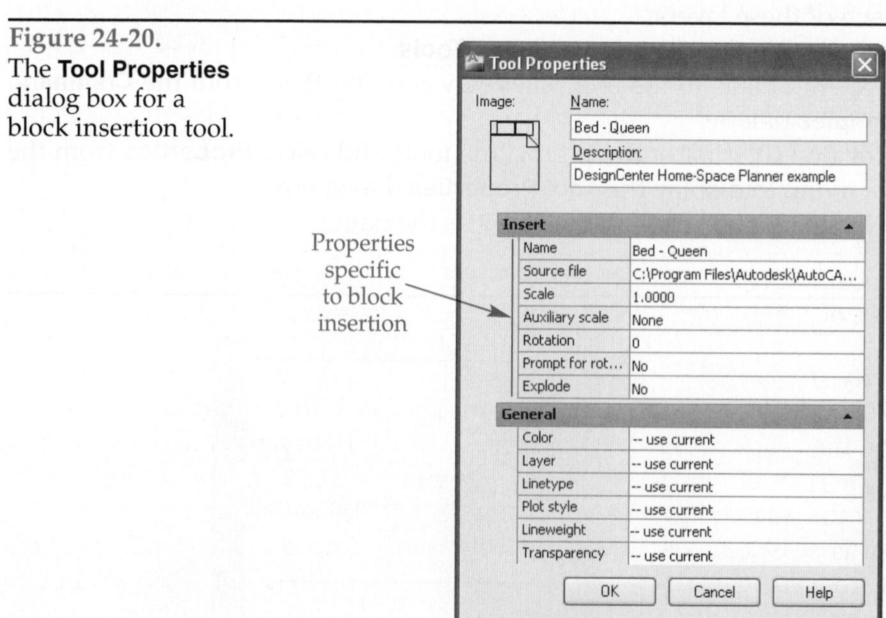

Name

The Name property contains the name of the block to be inserted. It is not usually modified.

Source file

The Source file property lists the name of and path to the source drawing containing the block. If the drawing file has been moved to a new location, specify the correct location in the text box for this property. When you pick in the text box, an ellipses button (...) appears at the right-hand side of the box. You can pick this button to browse for the drawing file.

Scale

The Scale property value is the scale that will be applied to the block when it is inserted into the drawing. The scale is applied equally in the X, Y, and Z directions. The default value is 1.000.

Auxiliary scale

The block is inserted at a scale calculated by multiplying the Auxiliary scale property value by the Scale property value. This drop-down list allows you to apply either the dimension scale or the plot scale to the insertion scale. The default value is None.

Rotation

The rotation angle for the block when it is inserted is set by the Rotation property. The default value is 0.

Prompt for rotation

The Prompt for rotation property determines whether or not the user is prompted for a rotation angle when the block is inserted. The default value in the drop-down list is No.

Explode

The Explode property determines whether or not the block is inserted as a block or as its component objects (exploded). The default value in the drop-down list is No, which means the block is inserted unexploded.

Customizing Hatch Pattern Tools

When a hatch pattern insertion tool is modified in the **Tool Properties** dialog box, properties specific to hatch patterns are displayed in the **Pattern** category, Figure 24-21. These properties are described as follows. Which properties are disabled or enabled is determined by the pattern type.

Tool type

The Tool type property determines if the hatch pattern is a standard hatch or a gradient fill. Choosing Gradient in the drop-down list changes the rest of the properties found in this area of the dialog box. Gradient fill properties are discussed later.

Type

The Type property determines the type of hatch pattern. To change the type, select the property and then pick the ellipses button (...) at the right-hand end of the entry. The **Hatch Pattern Type** dialog box is displayed, Figure 24-22. In the **Pattern type:** drop-down list of this dialog box, select User-defined, Predefined, or Custom.

If Predefined is selected, the **Pattern...** button and drop-down list are enabled. Select a pattern from the drop-down list or pick the button to select a pattern in the **Hatch Pattern Palette** dialog box. In this dialog box, you can select which hatch pattern to use, Figure 24-23.

Figure 24-21.
The **Tool Properties** dialog box for a hatch insertion tool.

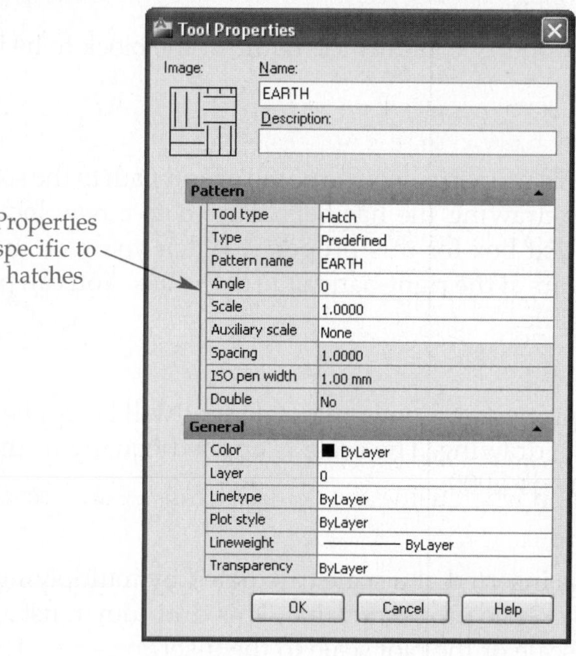

Figure 24-22.
The **Hatch Pattern Type** dialog box is used to determine the type of hatch. A predefined hatch pattern can also be selected in this dialog box.

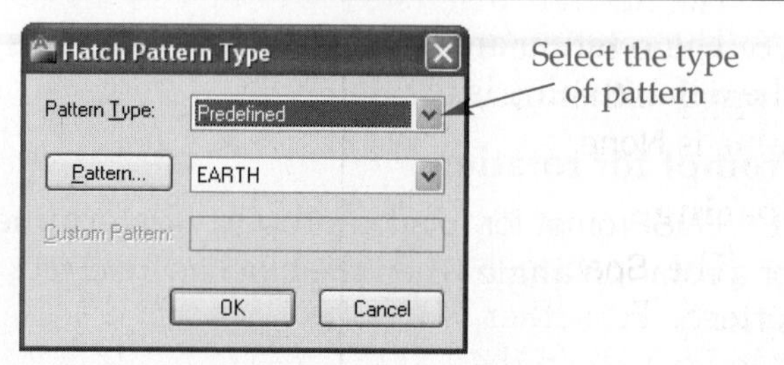

Figure 24-23.
Selecting a predefined hatch pattern.

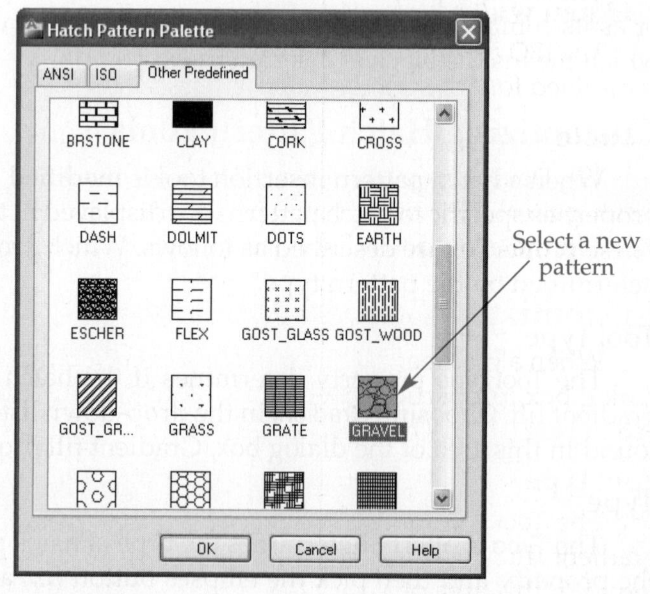

If User-defined is selected, the rest of the items in the dialog box are disabled. The properties for user-defined hatch patterns are set in the **Tool Properties** dialog box, as discussed next.

Selecting Custom disables the **Pattern...** button and drop-down list and enables the **Custom Pattern:** text box. In this text box, enter the name of the custom pattern to use.

Pattern name

If the pattern type is set to Predefined, you can change the pattern using the Pattern name property. Select the property and pick the ellipses button (...) to open the **Hatch Pattern Palette** dialog box, Figure 24-23. If you selected the pattern in the **Hatch Pattern Type** dialog box, you will not need to select it using this property. This property is read-only when the type is set to User-defined or Custom.

Angle

The Angle property allows you to rotate the hatch pattern. This is the same as entering a rotation when using the **BHATCH** command. The default value is 0.

Scale

The Scale property determines the scale to be applied to the hatch pattern. Pick in the text box and enter a scale factor for the pattern. The default value is 1.000.

Auxiliary scale

The hatch is inserted at a scale calculated by multiplying the Auxiliary scale property value by the Scale property value. The Auxiliary scale drop-down list allows you to apply either the dimension scale or the plot scale to the insertion scale. The default value is None.

Spacing

The Spacing property determines the spacing of lines in a user-defined hatch pattern. For other hatch pattern types, this property is read-only. To change the spacing, pick in the text box and enter the relative distance between lines. The default value is 1.000.

ISO pen width

The ISO pen width property allows you to set the pen width for ISO hatch patterns. It is disabled for other hatch patterns. The default value in the drop-down list is 1.00 mm.

Double

The Double property determines whether or not the user-defined hatch is a crosshatch pattern. For other hatch pattern types, this property is read-only. The default value in the drop-down list is No. To make a crosshatch pattern, select Yes.

Customizing Gradient Fill Tools

When a gradient fill hatch pattern insertion tool is modified in the **Tool Properties** dialog box, properties specific to gradient fills are displayed in the **Pattern** category. See Figure 24-17. These properties are described as follows.

Tool type

The Tool type property determines if the hatch pattern is a standard hatch or a gradient fill. Choosing Hatch in the drop-down list changes the rest of the properties found in this area of the dialog box, as described earlier.

Color 1 and Color 2

The Color 1 property is used to specify the first color of the gradient fill. The Color 2 property is used to specify the second color of the gradient fill. When you pick the property, a drop-down list is displayed. Pick a color from the drop-down list or choose Select Color... to pick a color in the **Color Selector** dialog box.

Gradient angle

The value of the Gradient angle property determines the rotation of the gradient fill. Pick in the text box and enter the number of degrees for the angle of the gradient. When the fill is created, the angle is relative to the current UCS. When you press [Enter] the **Image:** preview tile is updated to reflect the setting.

Centered

The Centered property determines whether or not the gradient fill is centered. The default value in the drop-down list is Yes. This creates a symmetrical fill. Selecting No shifts the gradient to the left. This is used to simulate a light source illuminating the fill from the left side.

Gradient name

The type of gradient fill is set by the Gradient name property. AutoCAD's preset gradient fills appear in the drop-down list. Select the type of gradient fill to use.

Customizing Command Tools

When a command tool is modified in the **Tool Properties** dialog box, properties specific to commands are displayed in the **Command** category, Figure 24-24. These properties are described as follows.

- **Use flyout.** This property determines whether or not the tool is a flyout. Flyouts are discussed later in this chapter.
- **Flyout options.** This property allows you to pick which commands are associated with the flyout, as discussed later.
- **Command string.** The command macro for the tool is entered in this text box, if the tool is not a flyout. Creating custom commands is discussed in Chapter 22. If the tool is a flyout, this text box is disabled.

For example, suppose you need a tool that will draw three concentric circles with diameters of .50, 1.00, and 1.50. This tool is shown in Figure 24-24. Use the following procedure.

1. In the current drawing, create a circle of any diameter at any location.
2. Drag and drop the circle onto a tool palette to create a new tool.
3. Right-click on the new **Circle** tool and select **Properties...** from the shortcut menu.

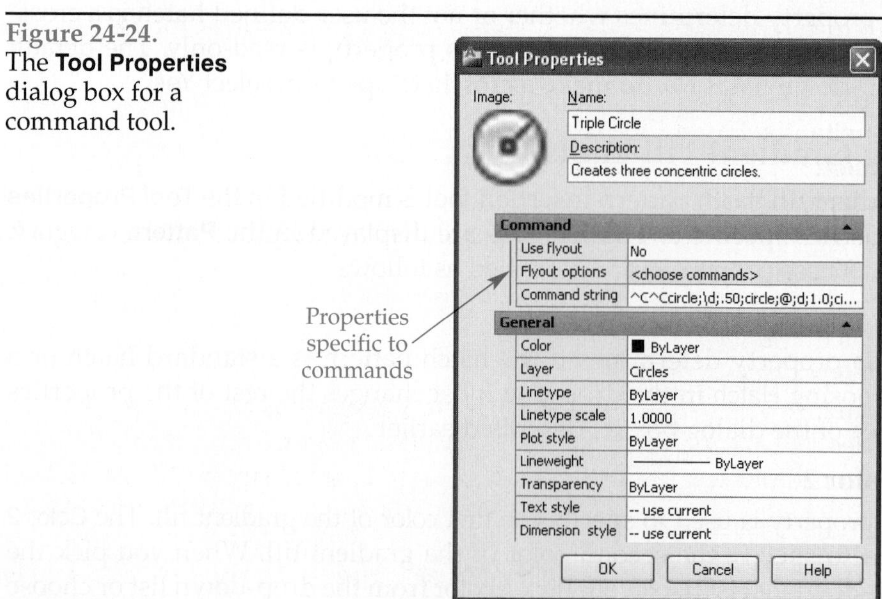

Figure 24-24.
The **Tool Properties** dialog box for a command tool.

4. In the **Tool Properties** dialog box, change the following properties. The Use flyout property must be set to No to enable the Command string property.
   Name: Triple Circle
   Description: Creates three concentric circles.
   Use Flyout: No
   Command string: ^C^Ccircle;\d;.50;circle;@;d;1.0;circle;@;d;1.5
5. Pick the **OK** button to close the **Tool Properties** dialog box.
6. Test the new tool.

Look closely at the command macro for the **Triple Circle** tool you just created. Can you identify each component of the macro? If not, use the tool and then display the **AutoCAD Text Window** by pressing [F2]. Using the text window, determine what function each component of the macro performs.

NOTE

The **VisualLisp Expression** tool included on the **Command Tool Samples** palette is not actually a "command" tool. It is merely there to let you know that you can use a command tool to execute Visual LISP expressions. Visual LISP functions can be added in the Command string property in the **Tool Properties** dialog box. For more information on AutoLISP and Visual LISP, refer to Chapters 27 and 28 of this text.

Exercise 24-3

Complete the exercise on the student website.
www.g-wlearning.com/CAD

Customizing Camera and Light Tools

Cameras and lights are stored in the drawing file. By placing camera and light tools on a tool palette, you can store your favorite settings for use in all drawings. Productivity is increased since you do not have to use **DesignCenter** to browse for your favorite cameras and lights within other drawings. In fact, AutoCAD provides several tool palettes with different light tools. These tool palettes are provided for lights:

- **Generic Lights**
- **Fluorescent**
- **High Intensity Discharge**
- **Incandescent**
- **Low Pressure Sodium**

AutoCAD also has a **Cameras** tool palette containing three camera tools. These tools are provided on the **Cameras** tool palette:

- **Normal Camera**
- **Wide-Angle Camera**
- **Extreme Wide-Angle Camera**

When customizing a camera or light tool, the **Tool Properties** dialog box provides all of the properties that are required to create a camera or light. See Figure 24-25. For detailed information about setting up lights and cameras, see Chapters 17 and 19.

Figure 24-25. A—The **Tool Properties** dialog box for a light tool. B—The **Tool Properties** dialog box for a camera tool.

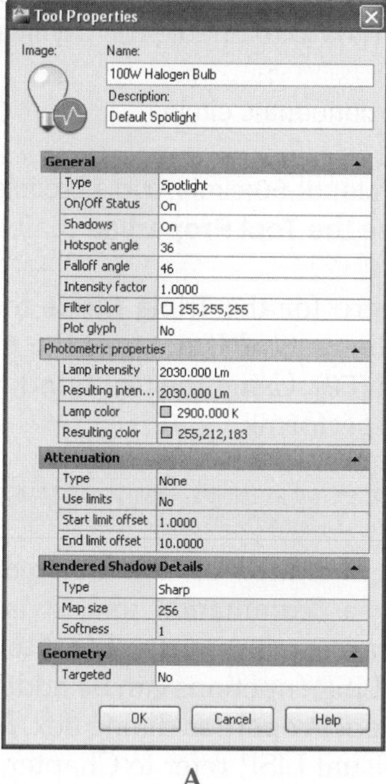

A

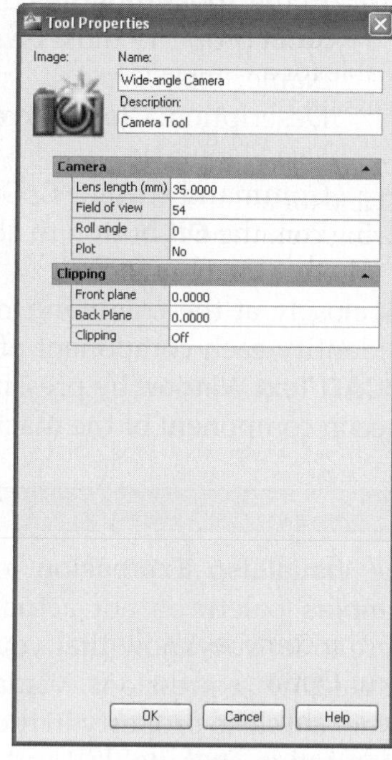

B

Changing a Tool Icon

You can change the icon associated with a tool. This can be done from within the **Tool Properties** dialog box or directly on the tool palette. Right-click on the icon, either in the dialog box or on the tool palette, and select **Specify Image...** from the shortcut menu. The **Select Image File** dialog box is displayed. This is a standard "open" dialog box. Navigate to the image file, select it, and pick the **Open** button. The icon displays the new image.

NOTE

Some tools have an icon image based on the specific settings for the tool. You should not specify an image for these tools. If the settings for the tool is redefined, right-click on the tool icon in the tool palette and select **Update tool image** from the shortcut menu.

Working with Flyouts

A command tool in a tool palette can be set to function as a ***flyout.*** A flyout tool is similar to a flyout button found on the ribbon or a toolbar. The tool icon for a flyout displays a small arrow to the right of the tool. When the arrow is picked, a graphic shortcut menu is displayed that contains the command tools in the flyout.

To set a command tool as a flyout, open the **Tool Properties** dialog box for the tool. Then, in the **Command** category, set the Use flyout property to Yes. The tool is now a flyout. Notice that the **Command** string text box in the dialog box is disabled. You must now select the commands that will be displayed in the flyout.

The Flyout options property is used to specify which commands will be displayed in the flyout for a tool. This property is disabled unless the Use flyout property is set to Yes. To choose the commands to appear in the flyout, select the property and pick the ellipses button (**...**) at the right-hand side. The **Flyout Options** dialog box appears, **Figure 24-26.**

Figure 24-26.
A—The commands available for a dimensioning command flyout tool.
B—The commands available for a drawing command flyout tool.

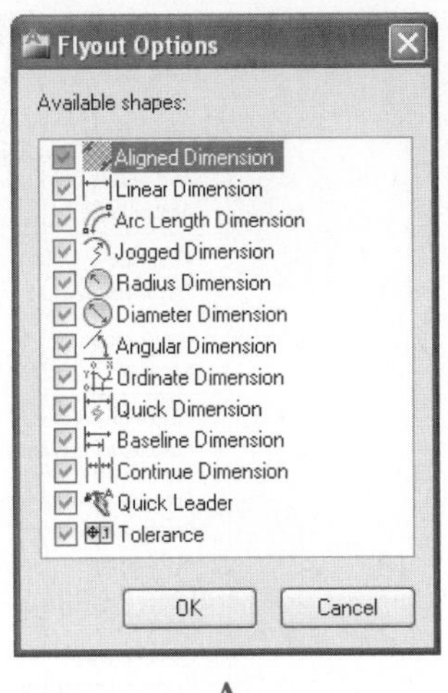

A

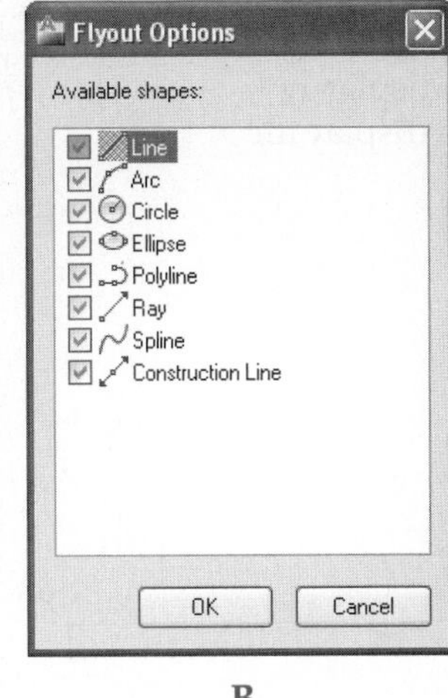

B

The commands that will appear in the flyout tool are indicated with a check in the **Flyout Options** dialog box. To prevent a command from being displayed in the flyout, remove the check mark next to its name. Then, close the **Flyout Options** dialog box to return to the **Tool Properties** dialog box. Finally, close the **Tool Properties** dialog box and test the tool.

The commands displayed in the **Flyout Options** dialog box depend on which command tool is being modified. **Figure 24-26A** shows the commands available when a dimensioning command tool is modified. When a drawing command tool is modified, the commands shown in **Figure 24-26B** are available. These also act as the default tools for other commands, such as modify commands or custom commands.

PROFESSIONAL TIP

When a drawing or dimensioning command is added to a tool palette, the tool is automatically a flyout. It can then be modified to customize or remove its flyout capabilities.

Organizing Tool Palettes into Groups

Tool palettes are very useful, productive tools. You may find it beneficial to create many tool palettes to meet your design needs. However, having too many palettes visible at once can be counterproductive. The tool palette names become abbreviated in the **Tool Palettes** window or the tabs may be stacked on top of each other. This is the case if all of the default AutoCAD tool palettes are displayed. Fortunately, tool palettes can be divided into named groups and then a single group can be displayed.

By default, all of the tool palettes are shown in the **Tool Palettes** window, but the existing tool palettes are divided into named groups. Right-click on the title bar of the **Tool Palettes** window and look at the shortcut menu. At the bottom of the shortcut menu, notice the entries below the last separator, **Figure 24-27.** These are the palette groups. The check mark next to **All Palettes** indicates that all of the defined tool palettes are being displayed. Select **Annotation and Design** from the shortcut menu and notice that only the block-related tool palettes are visible. This is indicated in the title bar. Right-click on the title bar again and select **3D Make** from the shortcut menu. Notice that three tool palettes that may be used in an initial 3D design are displayed.

Figure 24-27.
Select a palette group to display or choose to display all palettes.

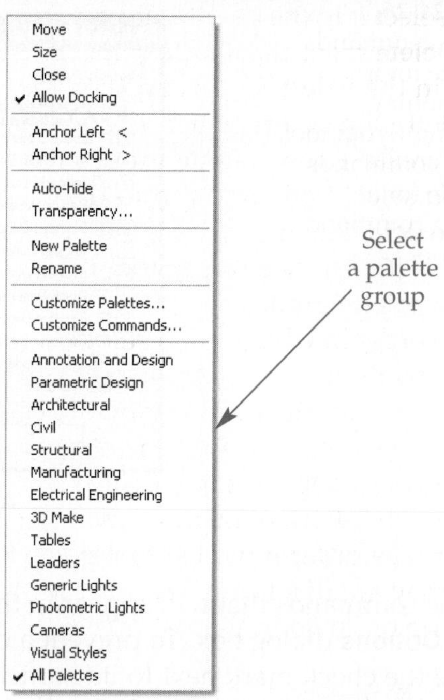

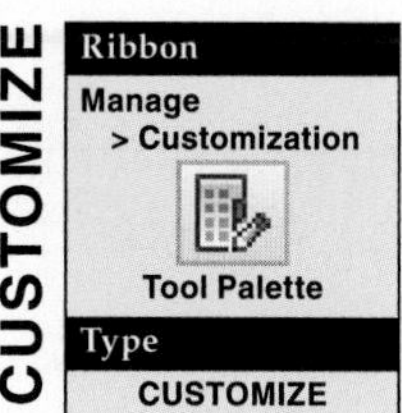

The **Customize** dialog box is used to make new palette groups or customize existing groups. To open this dialog box, right-click on the **Tool Palettes** window title bar and select **Customize Palettes...** from the shortcut menu. You can also use the **CUSTOMIZE** command. All of the currently defined tool palettes are listed in the **Palettes:** area on the left side of the dialog box, Figure 24-28. The currently defined palette groups are listed in the **Palette Groups:** area on the right side of the dialog box. Palette groups are shown as folders. The tool palettes contained within the group are shown in the tree below the folder. The current palette group is indicated at the bottom of the dialog box and by the bold folder name in the **Palette Groups:** area.

You can customize one of the existing palette groups by adding tool palettes to or removing tool palettes from the group. To remove a tool palette from a group, select the palette name under the group name in the **Palette Groups:** area. Then, press the [Delete] key or right-click and select **Remove** from the shortcut menu. To delete a palette

Figure 24-28.
The **Customize** dialog box is used to create and manage palette groups.

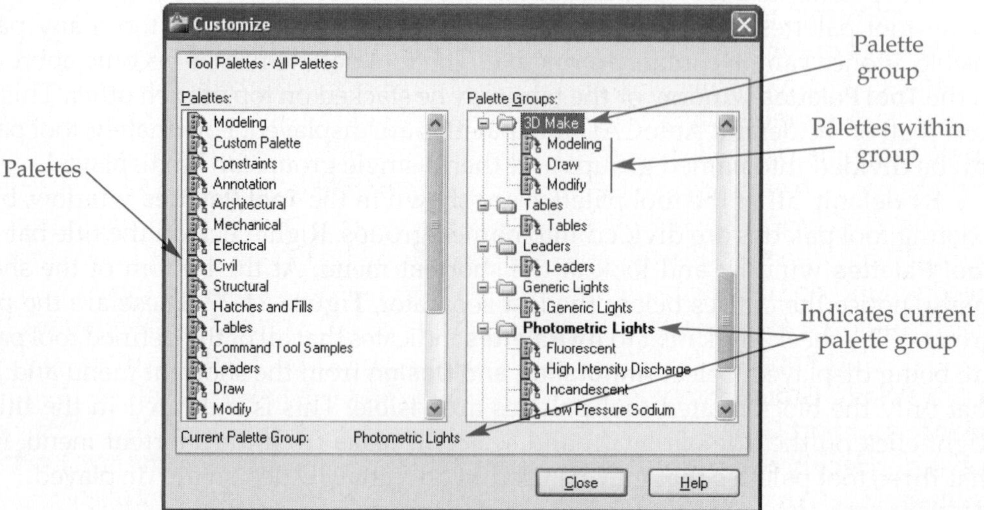

group, select it in the **Palette Groups:** area and press the [Delete] key or right-click and select **Delete** from the shortcut menu. To add a palette to a group, simply select the palette in the **Palettes:** area and drag it into the group in the **Palette Groups:** area. A tool palette can be a member of more than one palette group.

To create a new palette group, right-click in a blank area of the **Palette Groups:** area and select **New Group** from the shortcut menu. A new folder appears in the tree with the default name highlighted in a text box. Type a name for the new palette group and press [Enter]. Then, drag tool palettes from the **Palettes:** area and drop them into the new palette group.

The order in which tool palettes appear in the **Tool Palettes** window is determined by their positions in the tree. To reorder the tool palettes, simply drag them to different positions within the palette group. To reorder the palette groups, drag the folders to different positions within the tree in the **Palette Groups:** area. The order in which the palette groups appear in the **Palette Groups:** area determines the order in which they appear in the shortcut menu displayed by right-clicking on the **Tool Palettes** window title bar. The order in which tool palettes appear in the **Palettes:** area is the order in which they are displayed when the **All Palettes** option is selected. By dragging a tool palette up or down in the tree, you can reorder the list.

CAUTION

When adding a palette group, be careful where you right-click. If you right-click within a group, the new palette group is nested within that group. If you do not want the palette group nested, simply drag the group to the top of the tree in the **Customize** dialog box.

Saving and Sharing Tool Palettes and Palette Groups

Tool palettes and palette groups can be exported from and imported into AutoCAD. Both operations are performed in the **Customize** dialog box. Some precautions about tool palette files:

- Tool palette files can only be imported into the same version of AutoCAD as the version from which the file was exported.
- Tool palette files exported from AutoCAD and imported into AutoCAD LT may have tools that will not work or behave the same. For example, color property tools using a color other than an AutoCAD Color Index (ACI) color are converted to ByLayer in AutoCAD LT. Also, gradient fill tools convert to hatch tools in AutoCAD LT. Raster image tools do not work in AutoCAD LT.

To export a tool palette, right-click on the palette to be exported in the **Palettes:** area of the **Customize** dialog box and select **Export...** from the shortcut menu. To export a palette group, right-click on the name of the group in the **Palette Groups:** area and select **Export...** from the shortcut menu. The **Export Palette** or **Export Group** dialog box is displayed. These are standard Windows "save" dialog boxes. Name the file, navigate to the folder where you want to save it, and pick the **Save** button. Tool palettes are saved with a .xtp file extension. Palette groups are saved with a .xgp file extension.

To import a tool palette, right-click in the **Palettes:** area of the **Customize** dialog box and select **Import...** from the shortcut menu. To import a palette group, right-click in the **Palette Groups:** area and select **Import...** from the shortcut menu. The **Import Palette** or **Import Group** dialog box is displayed. Navigate to the folder where the file is saved, select it, and pick the **Open** button. The imported tool palette is added to the **Palettes:** area. An imported palette group is added to the **Palette Groups:** area.

Chapter Test

Answer the following questions. Write your answers on a separate sheet of paper or complete the electronic chapter test on the student website.
www.g-wlearning.com/CAD

1. Name three ways to open the **Tool Palettes** window.
2. How do you dock the **Tool Palettes** window?
3. When enabled, what does the autohide feature do to the **Tool Palettes** window? How is the **Tool Palettes** window accessed when autohide is enabled?
4. Name two ways to toggle the autohide feature on the **Tool Palettes** window.
5. Describe how anchoring the **Tool Palettes** window differs from using the autohide feature or docking it.
6. How do you activate the transparency feature for the **Tool Palettes** window?
7. Name the three view styles in which tools can be displayed in the **Tool Palettes** window.
8. How do you create a new, blank tool palette?
9. How do you add a tool palette that contains all of the blocks in a particular drawing?
10. Which dialog box must be open to create a tool on a tool palette from a toolbar button? How is it opened?
11. What are the names of the two files that store hatch pattern definitions?
12. How do you rearrange tools on a tool palette?
13. Name four of the general properties that can be customized on individual tools on a tool palette.
14. What is the purpose of the auxiliary scale property for block and hatch insertion tools?
15. What are the two types of patterns that can be inserted using a tool in a tool palette?
16. If you are creating a command tool that performs a custom function, where is the command macro entered?
17. Which two types of commands can be included in a flyout tool?
18. What is the purpose of creating tool palette groups?
19. What is the extension given to an exported tool palette file?
20. How do you import a tool palette file?

Drawing Problems

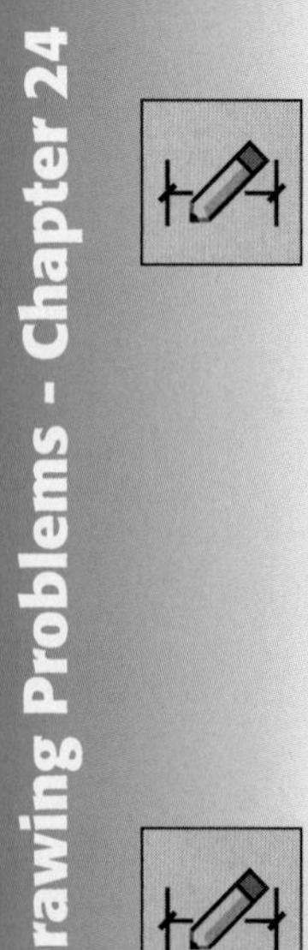

Before customizing or creating any tool palettes, check with your instructor or supervisor for specific instructions or guidelines.

1. Design a complete tool palette system for your chosen discipline. Incorporate:
   - Multiple tool palettes
   - Block tools
   - Hatch tools
   - Drawing command tools (with and without flyouts)
   - Modification command tools
   - Inquiry command tools
   - Custom macro tools
   - Dimensioning command tools (with and without flyouts)

   A. On the tool palettes that use multiple types of command tools, use separators and text labels to group the types of tools.
   B. Create groups for the multiple tool palettes.
2. Export the tool palette groups created in problem 1. Copy the files to removable media or an archive drive.

CHAPTER 25

User Profiles and Workspaces

Learning Objectives

After completing this chapter, you will be able to:

- ✓ Describe user profiles.
- ✓ Create user profiles.
- ✓ Restore a user profile.
- ✓ Describe workspaces.
- ✓ Explain the **Quick Access** toolbar.
- ✓ Create workspaces.
- ✓ Customize a workspace.
- ✓ Restore a workspace.
- ✓ Customize the **Quick Access** toolbar.

In a school or company, there is often more than one person who will use the same AutoCAD workstation. Each drafter has a unique style for creating a drawing. While there are often general rules to follow, many times the method used to arrive at the end result is not important. As you learned in previous chapters, there are many ways to customize AutoCAD. You can set screen colors and other features of the AutoCAD environment, create custom menus, and customize the ribbon. Many of these settings can be saved in a user profile or workspace. User profiles and workspaces allow you to quickly and easily restore a group of custom settings.

User Profiles

A ***user profile*** is a group of settings for devices and AutoCAD functions. Some of the settings and values a profile can contain are:

- Temporary drawing file location.
- Template drawing file location.
- Text display format.
- Startup dialog box display.
- Minutes between automatic saves.
- File extension for temporary files.
- AutoCAD screen or pull-down menu display (on/off).
- Color and font settings for AutoCAD's text and graphics screens.

- Type of pointer and length of crosshairs.
- Default locations of printers, plotters (PC3 files), and plot styles (CTB/STB files).

Multiple profiles can be saved by a single user for different applications or several users can create individual profiles for their own use. A user profile should not be confused with settings found in a drawing. Template files are used to save settings relating to a drawing session, such as units, limits, object snap settings, drafting settings, grip settings, arc and circle smoothness, dimension styles, and text styles. A user profile, on the other hand, saves settings related to the performance and appearance of the software and hardware.

Creating a User Profile

A user profile is basically a collection of all of the AutoCAD environmental settings you have customized in AutoCAD *except* ribbon tabs and panels, application and pull-down menus, tool palettes, and toolbars. These customizations are usually done to make AutoCAD easier for you to use. For example, as you gain experience in AutoCAD, you realize that you may:

- Prefer the crosshairs extending to the edges of the graphics window.
- Need to have the plotters and plot styles shared on a network.
- Prefer the graphics window background color to be gray.

Through the course of several drawing sessions, you have customized AutoCAD to reflect these preferences. Now, so you do not lose your preferred settings, you should create a user profile.

First, open the **Options** dialog box and pick the **Profiles** tab, Figure 25-1. Pick the **Add to List...** button on the right side of the tab. The **Add Profile** dialog box is opened, Figure 25-2. Enter a name and description. Then, pick the **Apply & Close** button to close the **Add Profile** dialog box. The current settings are saved to the user profile and the new user profile is now listed in the **Profiles** tab of the **Options** dialog box. The user profile is saved and will be available in the current and future AutoCAD drawing sessions.

Figure 25-1.
Settings can be saved in a profile.

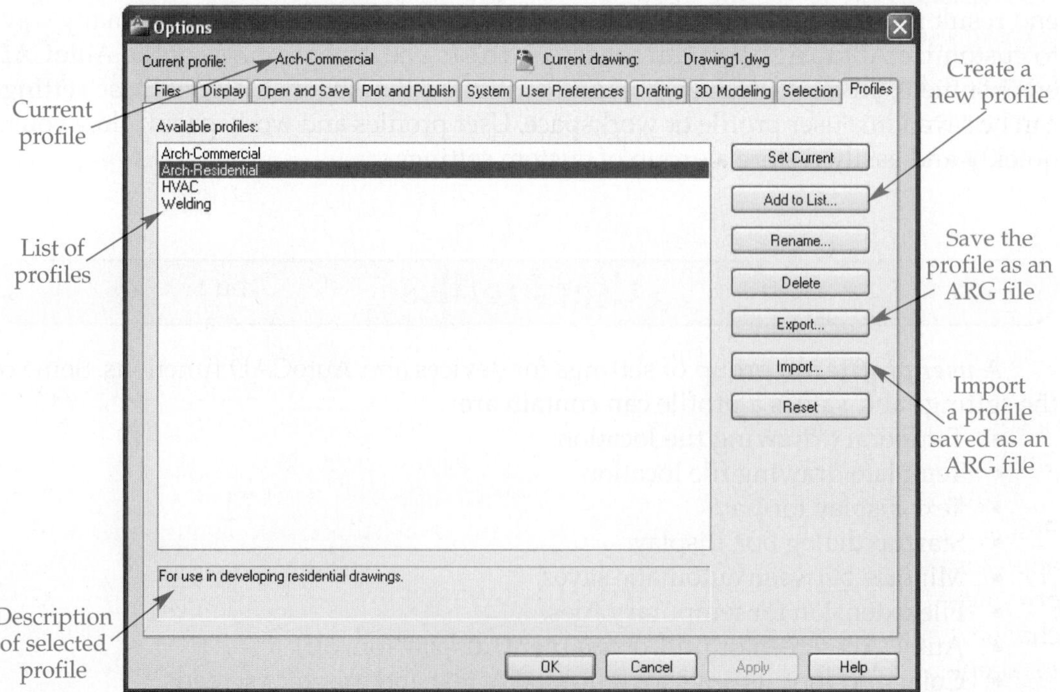

Figure 25-2.
The **Add Profile** dialog box is used to create a new profile.

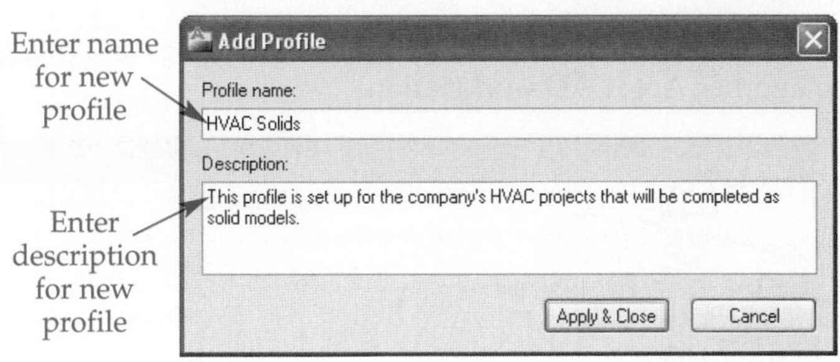

To change the name of a user profile, pick the **Rename...** button in the **Profiles** tab. In the **Change Profile** dialog box, enter a new name and description. To delete a user profile, highlight it in the **Profiles** tab and pick the **Delete** button. You cannot delete the current user profile. If you pick the **Reset** button, the highlighted user profile has all of its settings restored to AutoCAD defaults.

PROFESSIONAL TIP

Changes made to the environment (screen color, toolbar display, etc.) are automatically saved to the current profile and stored in the Windows registry.

Restoring a User Profile

Once a user profile is saved, it is available in the current and future AutoCAD drawing sessions. To set any saved user profile as the current user profile, first open the **Options** dialog box. The current profile is indicated at the top of the **Options** dialog box. Then, pick the **Profiles** tab. Highlight the name of the profile to restore in the **Available profiles:** list. Then, pick the **Set Current** button. You can also double-click on the name in the **Available profiles:** list to restore a profile. All of the settings in the user profile are applied while the **Options** dialog box is still open. Close the dialog box to return to the drawing editor.

Exercise 25-1

Complete the exercise on the student website.
www.g-wlearning.com/CAD

Importing and Exporting User Profiles

A user profile can be exported and imported. This allows you to take your user profile to a different AutoCAD workstation. A user profile is saved as an ARG file.

To export a user profile, open the **Options** dialog box and pick the **Profiles** tab. Highlight the profile to export and pick the **Export...** button. The **Export Profiles** dialog box is opened, Figure 25-3. Then, select a folder and name the file. When you pick the **Save** button, the user profile is saved with the .arg file extension.

To import a user profile, pick the **Import...** button in the **Profiles** tab. The **Import Profile** dialog box is displayed. This is a standard "open" dialog box. Then, navigate to the appropriate folder, select the proper ARG file, and pick the **Open** button. A second dialog box named **Import Profile** is displayed. See Figure 25-4. You can rename the user profile, change the description, and choose to include the file path. Pick the **Apply & Close** button to complete the process. The user profile is then available in the **Profiles** tab.

Figure 25-3.
The **Export Profile** dialog box is used to save a user profile as an ARG file, which can be transferred to another AutoCAD workstation.

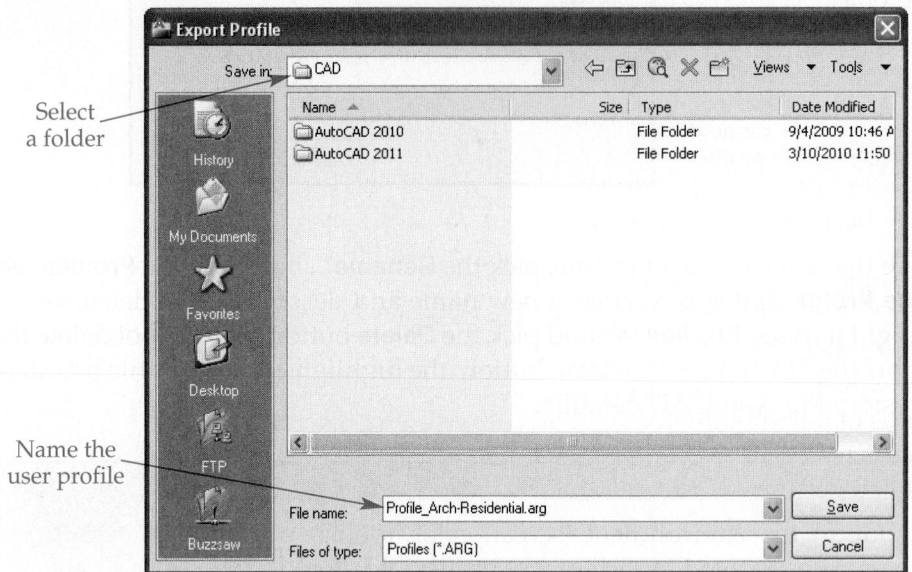

Figure 25-4.
Importing a user profile.

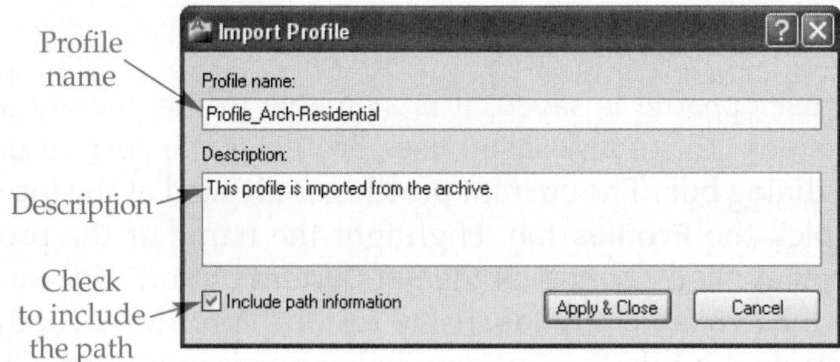

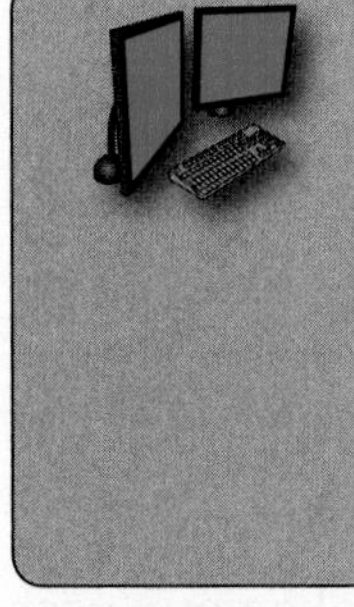

PROFESSIONAL TIP

A variety of settings can be changed in the **Options** dialog box. Settings that have the AutoCAD drawing icon located next to them can be stored in a template file (or drawing). Using the appropriate template file to begin a drawing automatically resets these settings. Settings within the **Options** dialog box that do not have the AutoCAD drawing icon next to them can typically be restored through the use of user profiles.

Exercise 25-2

Complete the exercise on the student website.
www.g-wlearning.com/CAD

Workspaces

As you have seen, a user profile allows you to set and restore settings for screen colors, drafting settings, and file locations. On the other hand, a workspace allows you to set and restore settings for the **Quick Access** toolbar, ribbon, menus, and palettes (**DesignCenter**, **Properties** palette, **Tool Palettes** window, etc.), but it does not contain environmental settings. A ***workspace*** is a collection of displayed toolbars, palettes, ribbon tabs and panels, and menus and their configurations. A workspace stores not only which of these graphic tools are visible, but also their on-screen locations.

PROFESSIONAL TIP

A user profile also stores which toolbars are displayed and their location. However, the configurations of the menus and ribbon and display status of palettes are not saved in a user profile. Therefore, if a workspace is restored, the settings for toolbar display and location override the current user profile settings.

Creating a Workspace

AutoCAD has four default workspaces—2D Drafting and Annotation, 3D Basics, 3D Modeling, and AutoCAD Classic. When AutoCAD is first installed, it also includes the Initial Setup workspace, but this workspace is available only until another workspace is selected. A workspace can be set current using the **Workspace Switching** button on the right-hand side of the status bar or the drop-down list on the **Quick Access** toolbar. See Figure 25-5. Picking this button displays a shortcut menu containing the names of all available workspaces. It also contains options for working with workspaces. When AutoCAD is launched, the workspace that was last active is restored. AutoCAD can be set up so that any changes made to the ribbon, toolbars, menus, and palettes are saved to this workspace. However, it is best to create your own workspaces.

The first step in setting up and storing your own workspace is to arrange the ribbon, toolbars, tool palettes, and menus to your liking. Refer to Chapter 22. Next, use the **WSSAVE** command to open the **Save Workspace** dialog box, Figure 25-6. You can easily access the **WSSAVE** command by selecting **Save Current As...** in the shortcut menu displayed by picking the **Workspace Switching** button or in the drop-down list on the **Quick Access** toolbar. In the **Save Workspace** dialog box, enter a name for the workspace, such as Normal Design or Standard Arrangement, and then pick the **Save** button.

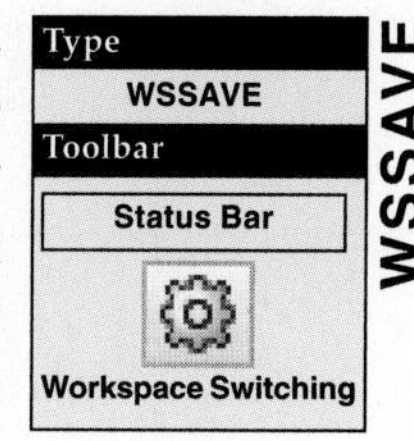

Figure 25-5.
Switching workspaces.

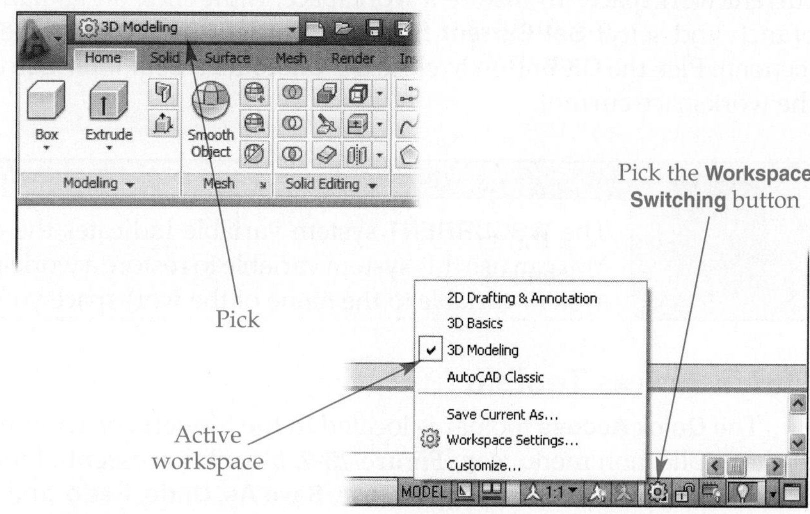

Figure 25-6.
Creating a new workspace.

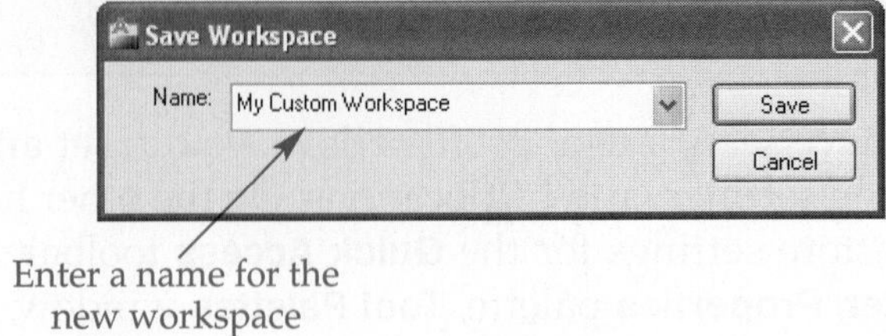

The current settings for toolbars, pull-down menus, and tool palettes are now stored in the new workspace, which is also made current. The workspace name also appears in the shortcut menu displayed by picking the **Workspace Switching** button and the drop-down list on the **Quick Access** toolbar. The current workspace is indicated in this menu by a check mark.

PROFESSIONAL TIP

Workspaces are saved in a CUIx file. By default, they are saved to the main CUIx file (acad.cuix). See Chapter 22 for details regarding CUIx files.

Restoring a Workspace

Once a workspace has been saved, it can easily be restored. A list of available workspaces appears in the shortcut menu displayed by picking the **Workspace Switching** button and the drop-down list on the **Quick Access** toolbar. To select a different workspace, simply pick the name of the workspace in the pop-up list. A workspace can also be restored using the command line:

```
Command: WORKSPACE↵
Enter workspace option [setCurrent/SAveas/Edit/Rename/Delete/SEttings/?]
   <setCurrent>: C↵
Enter name of workspace to make current [?] <current>: MY CUSTOM
   WORKSPACE↵
Command:
```

Notice that you can manage workspaces using this command.

You can also use the **Customize User Interface** dialog box to restore a workspace. First, open the dialog box. Then, expand the Workspaces branch in the **Customizations in All Files** pane. All of the workspaces defined in the default CUIx file and any open CUIx files are displayed in this branch. The label (current) follows the name of the current workspace. To restore a workspace, right-click on its name in the Workspaces branch and select **Set Current** from the shortcut menu. Its name is now followed by (current). Pick the **OK** button to close the **Customize User Interface** dialog box and make the workspace current.

PROFESSIONAL TIP

The **WSCURRENT** system variable indicates the current workspace. You can use this system variable to restore a workspace. Simply set the system variable to the name of the workspace you want to restore.

Quick Access Toolbar

The **Quick Access** toolbar is located in the top-left corner of the screen to the right of the application menu icon, Figure 25-7. It contains essential tools for AutoCAD. By default, it displays the **New**, **Open**, **Save**, **Save As**, **Undo**, **Redo**, and **Plot** tools along with

Figure 25-7.
The **Quick Access** toolbar is customized as part of the workspace.

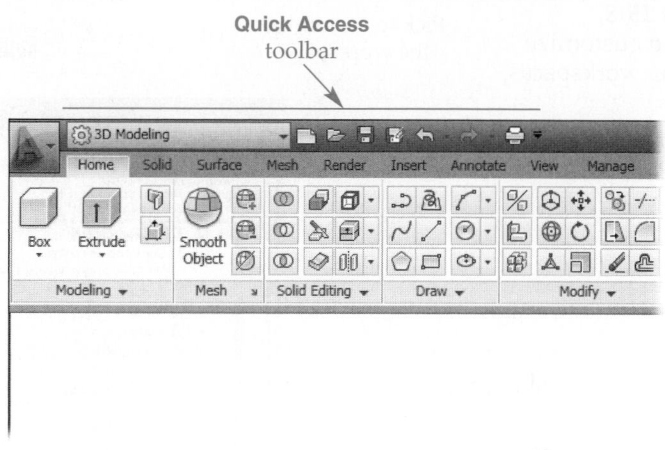

the **Workspace** drop-down list. In addition, you can display other tools on the toolbar by picking the down arrow button on right-hand end of the **Quick Access** toolbar to display a shortcut menu. In this menu, select the tools to show and hide. The additional available tools are **Match Properties**, **Batch Plot**, **Plot Preview**, **Properties**, **Sheet Set Manager**, and **Render**. Any other defined command can be added to the toolbar by selecting **More Commands...** in the drop-down list. This displayed the **Customize User Interface** dialog box.

Unlike the toolbars discussed in Chapter 23, the **Quick Access** toolbar is customized inside of the workspace. You may create more than one **Quick Access** toolbar, but only one may be active in a workspace. To create an additional **Quick Access** toolbar, open the **Customize User Interface** dialog box, right-click on the Quick Access Toolbars branch, and select **New Quick Access Toolbar** from the context menu. When creating a new **Quick Access** toolbar, it is automatically populated with the tools outlined above. Customizing the **Quick Access** toolbar is discussed in the next section.

PROFESSIONAL TIP

Any tool on a ribbon panel can be quickly added to the **Quick Access** toolbar. Right-click on the tool icon and select **Add to Quick Access Toolbar** from the shortcut menu.

Customizing a Workspace

An existing workspace can be customized using the **Customize User Interface** dialog box. First, select the workspace to be customized in the Workspaces branch of the **Customizations in All Files** pane. Remember, the name of this pane will change based on the selection in the drop-down list. The **Workspace Contents** pane at the upper-right corner of the dialog box displays the contents of the selected workspace. There are Quick Access Toolbar, Toolbars, Menus, Palettes, and Ribbon Tabs branches. Refer to Figure 25-8. Expand a branch to see which components the workspace contains.

To customize the workspace, pick the **Customize Workspace** button at the top of the **Workspace Contents** pane. The tree in the pane turns blue to indicate you are in customization mode and the button changes to the **Done** button. Also, notice that the tree in the **Customizations in All Files** pane has changed. Several branches have disappeared and the Menus branch has a green check mark next to it. If you expand the branches in the **Customizations in All Files** pane, you will see that a green check mark also appears next to the components that currently are in the workspace. See Figure 25-9.

To add a component to the workspace, pick the blank box in front of its name to place a check mark in the box. The component also appears in the **Workspace Contents** pane.

Figure 25-8.
You can customize existing workspaces.

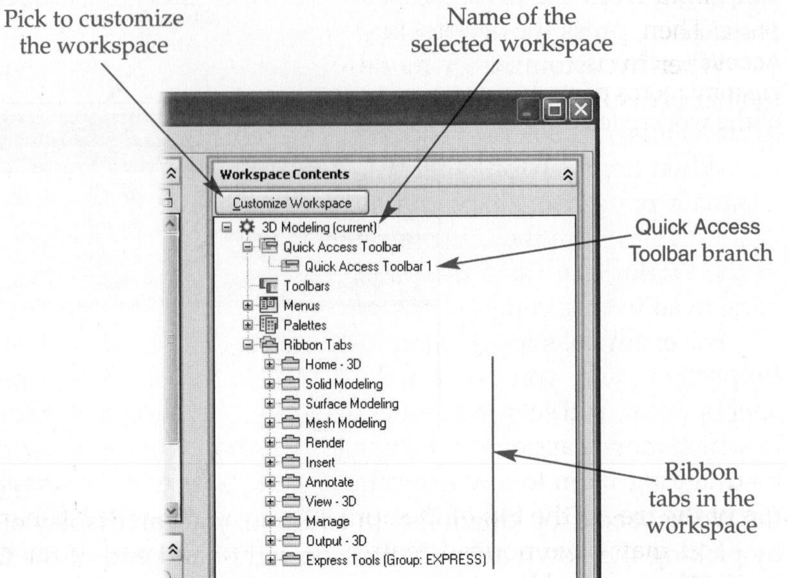

Pick to customize the workspace

Name of the selected workspace

Quick Access Toolbar branch

Ribbon tabs in the workspace

Figure 25-9.
Specifying which toolbars, menus, and ribbon tabs are included in the workspace.

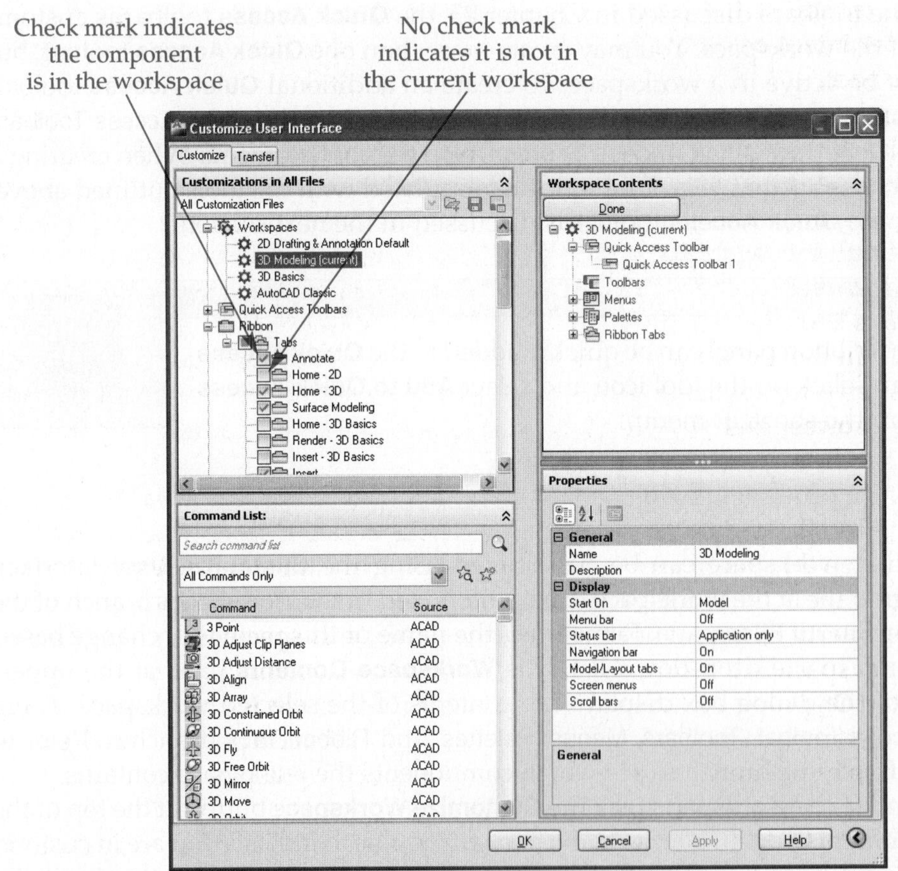

To remove a component from the workspace, pick the check mark in front of its name to clear the box. The component is also removed from the **Workspace Contents** pane.

The components in the Quick Access Toolbar branch are commands. To add a command to the Quick Access Toolbar branch, locate it in the **Command List:** pane. Then, drag and drop it into the branch in the **Workspace Contents** pane. To remove a command from the Quick Access Toolbar branch, select it in the **Workspace Contents** pane. Then, press the [Delete] key.

When in customization mode for workspaces, you may choose which **Quick Access** toolbar is displayed in the workspace. However, you cannot be in customization mode to customize a **Quick Access** toolbar.

When done customizing the workspace, pick the **Done** button in the **Workspace Contents** pane. The tree is no longer displayed in blue. Also, all branches are once again displayed in the **Customizations in All Files** pane. You can now expand a branch in the **Workspace Contents** pane, select a component name, and use the **Properties** pane to adjust the component's properties.

For example, suppose you have added the **Layers** toolbar to a workspace. In the **Properties** pane, you can set the orientation to floating, specify the location of its anchor point, and set the number of rows for the toolbar. You can also change the order in which menus appear in the application menu (menu browser) and on the menu bar by dragging them to a new location in the tree in the **Workspace Contents** pane. The top of the tree is the top of the application menu or the left-hand side of the menu bar.

You may have noticed that there is not a Palettes branch in the **Customizations in All Files** pane. All palettes are automatically available in all workspaces. You can, however, specify whether a palette is displayed or hidden in a workspace. You can also change other properties of a palette, such as floating/docked status, its size, and whether or not the autohide feature is enabled. To change the properties of a palette, first select it in the Palettes branch in the **Workspace Contents** pane. Then, in the **Properties** pane, adjust the properties as needed. When you pick **OK** to close the **Customize User Interface** dialog box, the new default properties of the palette are set for that workspace.

For example, the **Materials Browser** palette in the 3D Modeling workspace is, by default, displayed floating. You can hide the **Materials Browser** palette when in the drawing editor by simply picking the **Close** button (**X**). You can also drag the palette into the drawing area to a new position or dock it. However, these actions do not change the default setting for the workspace. If you restore the workspace in the future, the **Materials Browser** palette will again be floating. You must alter the default settings for the **Materials Browser** palette in the workspace. First, select Material Browser in the Palettes branch in the **Workspace Contents** pane. Then, in the **Properties** pane, change the Show property to No. See **Figure 25-10.** Now, when you restore the workspace, the **Materials Browser** palette will not be displayed. It can, of course, be manually displayed as needed.

You can also set up a workspace so that it displays model space or layout (paper) space when restored. By default, a workspace displays model space when it is set as current. To change this, highlight the workspace name in either the **Customizations in All Files** pane or the **Workspace Contents** pane. Then, in the **Properties** pane, change the Start On property to Layout or Do not change. If Model is specified for the Start On property, model space is displayed when the workspace is restored. If Layout is specified for the Start On property, the most recently active layout tab is displayed when the workspace is restored. If Do not change is specified for the Start On property, the current tab (model space or paper space) remains active when the workspace is restored.

PROFESSIONAL TIP

If the **Customize User Interface** dialog box is open, but you are not in workspace customization mode, you can also remove a component from the workspace by right-clicking on it in the tree in the **Workspace Contents** pane and selecting **Remove from Workspace** in the shortcut menu. This is true for all component types except palettes, as all palettes are always available to all workspaces.

Figure 25-10.
Changing the default properties of a palette for a given workspace. Click the property drop-down list to see the choices.

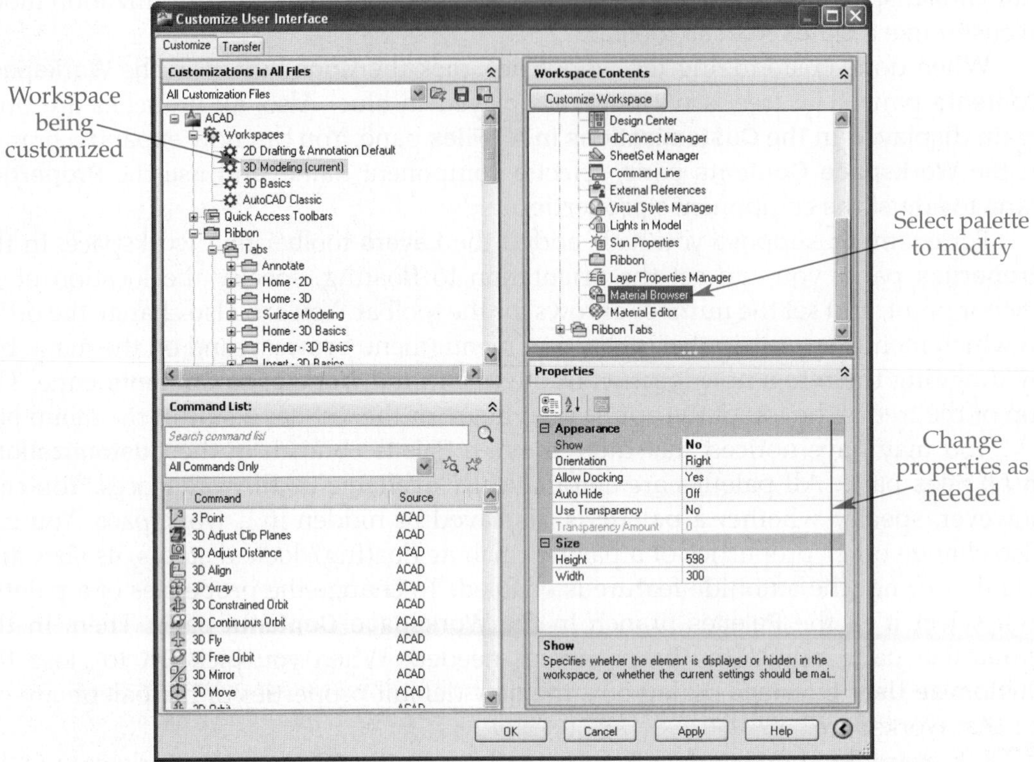

Workspace Settings

WSSETTINGS

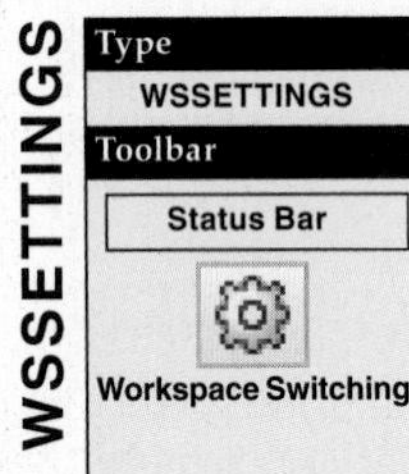

There are various settings related to workspaces. These are set in the **Workspace Settings** dialog box. See **Figure 25-11.** The **WSSETTINGS** command opens this dialog box. It can also be displayed by selecting **Workspace Settings...** from the shortcut menu displayed by picking the **Workspace Switching** button on the status bar or the drop-down list on the **Quick Access** toolbar.

At the top of the **Workspace Settings** dialog box is the **My Workspace =** drop-down list. All saved workspaces in the CUIx file appear in this list. The workspace that is selected in the list is defined as My Workspace. The workspace that is designated as My Workspace is restored when the **My Workspace** button on the **Workspaces** toolbar is picked. This can be useful if one person primarily uses a machine, but others may temporarily use the machine with their own workspace settings. You may also find this useful if you have more than one workspace, but use one more often than all of the others. This feature can only be accessed when the **Workspaces** toolbar is displayed, which, by default, it is not except in the AutoCAD Classic workspace.

The **Menu Display and Order** area of the **Workspace Settings** dialog box contains a list of all workspaces saved in the current CUIx file. The order of this list determines the order of the list that appears in the **Workspace Switching** shortcut menu, in the drop-down list on the **Quick Access** toolbar, and in the drop-down list on the **Workspaces** toolbar. The order of the list can be modified by highlighting one of the workspaces and using the **Move Up** and **Move Down** buttons. The **Add Separator** button is used to add a horizontal line, or menu separator, to the list. A separator is used to logically group workspace names within the list. The separator can be relocated within the list just like a workspace name.

Figure 25-11.
The **Workspace Settings** dialog box is used to set which workspace is My Workspace, specify which workspaces are shown in the menu and drop-down list and their order, and set whether or not changes are automatically saved when a different workspace is restored.

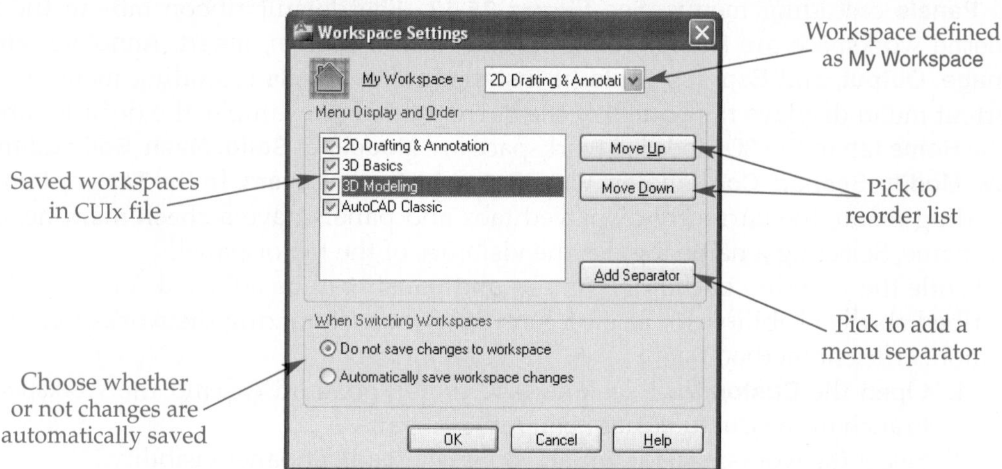

You can prevent a workspace name or separator from being displayed in the menu or drop-down list by removing the check box next to its name. The check box next to the current workspace and the workspace designated as My Workspace can be cleared. However, these workspaces will always be displayed in the menu and drop-down list.

At the bottom of the **Workspace Settings** dialog box is the **When Switching Workspaces** area. The radio buttons in this area determine whether or not changes you have made since you last saved the workspace, such as the visibility of toolbars, are saved when you switch to a different workspace. To retain the settings as you last saved them, pick the **Do not save changes to workspace** radio button. Changes made since the workspace was last saved are discarded when a different workspace is restored. If the **Automatically save workspace changes** radio button is on, any "as you work" changes are automatically saved to the workspace when a different workspace is restored.

At any time, you can manually save the settings to the current workspace. Pick the Save Current As… selection in the **Workspace Switching** shortcut menu. When the **Save Workspace** dialog box appears, select the current workspace from the drop-down list. Then, pick the **Save** button. An alert is displayed stating that a workspace with that name already exists and asking if you would like to replace it. Pick the **Yes** button to save the changes to the current workspace.

PROFESSIONAL TIP

Unlike a user profile, changes to the environment (toolbar display, menu configuration, etc.) are not necessarily automatically saved. To ensure the changes are only saved when you decide to save them, be sure the **Do not save workspace changes** radio button is on in the **Workspace Settings** dialog box.

Exercise 25-3

Complete the exercise on the student website.
www.g-wlearning.com/CAD

Controlling the Display of Ribbon Tabs and Panels

Right-click anywhere on the ribbon to display the shortcut menu. Notice the **Tabs** and **Panels** cascading menus. See **Figure 25-12.** The default ribbon tabs in the 3D Modeling workspace are **Home**, **Solid**, **Surface**, **Mesh**, **Render**, **Insert**, **Annotate**, **View**, **Manage**, **Output**, and **Express Tools** (if installed). The **Panels** cascading menu in the shortcut menu displays the panels of the current tab. For example, the default panels of the **Home** tab in the 3D Modeling workspace are **Modeling**, **Solid**, **Mesh**, **Solid Editing**, **Draw**, **Modify**, **Section**, **Coordinates**, **View**, **Subobject**, and **Layers**. In the **Tabs** or **Panels** cascading menu, the currently displayed tabs and panels have a check mark next to their name. Selecting a name toggles the visibility of the tab or panel.

While the visibility of the ribbon tabs and panels can be adjusted "on the fly" as described above, visibility can also be controlled by customizing the workspace. This is a more efficient method.

1. Open the **Customize User Interface** dialog box and expand the Workspaces branch in the **Customizations in All Files** pane.
2. Select the workspace for which you wish to adjust panel visibility.
3. In the **Workspace Contents** pane, pick the **Customize Workspace** button.
4. Expand the Ribbon Tabs branch in the **Workspace Contents** pane. The tabs visible in the workspace appear in the tree. The panels visible in the workspace appear as branches below the tabs.
5. In the **Customizations in All Files** pane, expand the Ribbon branch and then the Tabs branch. All available tabs appear in the branch. The currently displayed tabs have a check mark next to their name. See **Figure 25-13.**
6. Check the tabs to display and uncheck the tabs to hide.
7. To control which panels are displayed for a tab, expand the tab's branch in the **Workspace Contents** pane. Then, select the panel to hide and, in the **Properties** pane, change its Show property to No. To display a panel, change the property to Yes.
8. In the **Workspace Contents** pane, pick the **Done** button.
9. Exit the **Customize User Interface** dialog box.

Tabs and panels appear in the ribbon in the order in which they are displayed in the tree of the **Workspace Contents** pane. Top to bottom equals left to right when the ribbon is horizontal. As you add panels to the workspace, they are added to the bottom

Figure 25-12.
A shortcut menu is displayed when you right-click on the ribbon. A—The **Tabs** cascading menu. B—The **Panels** cascading menu. This is for the **Home** tab.

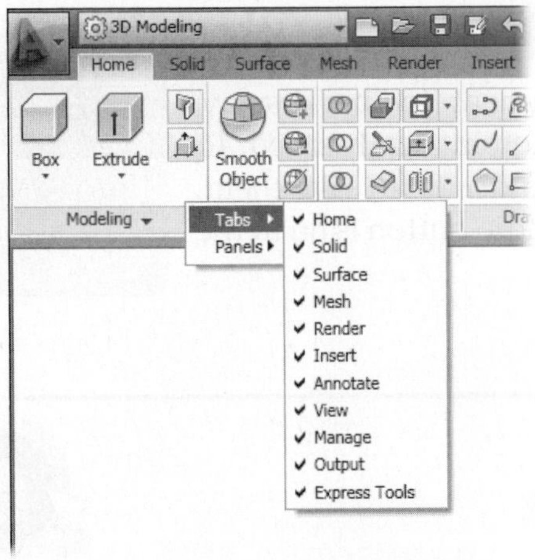

A

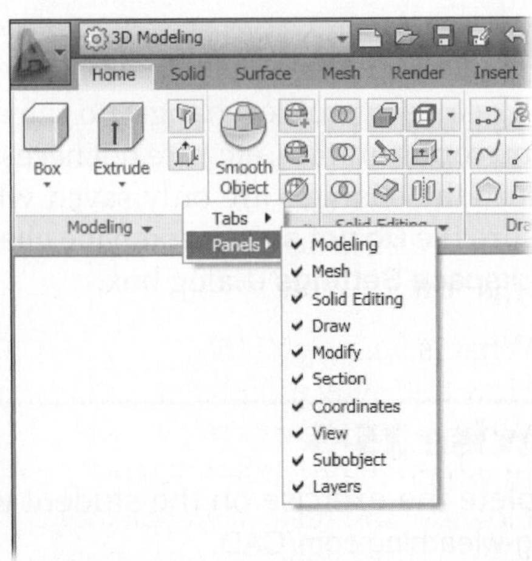

B

Figure 25-13.
Setting which ribbon tabs and panels are displayed for a workspace.

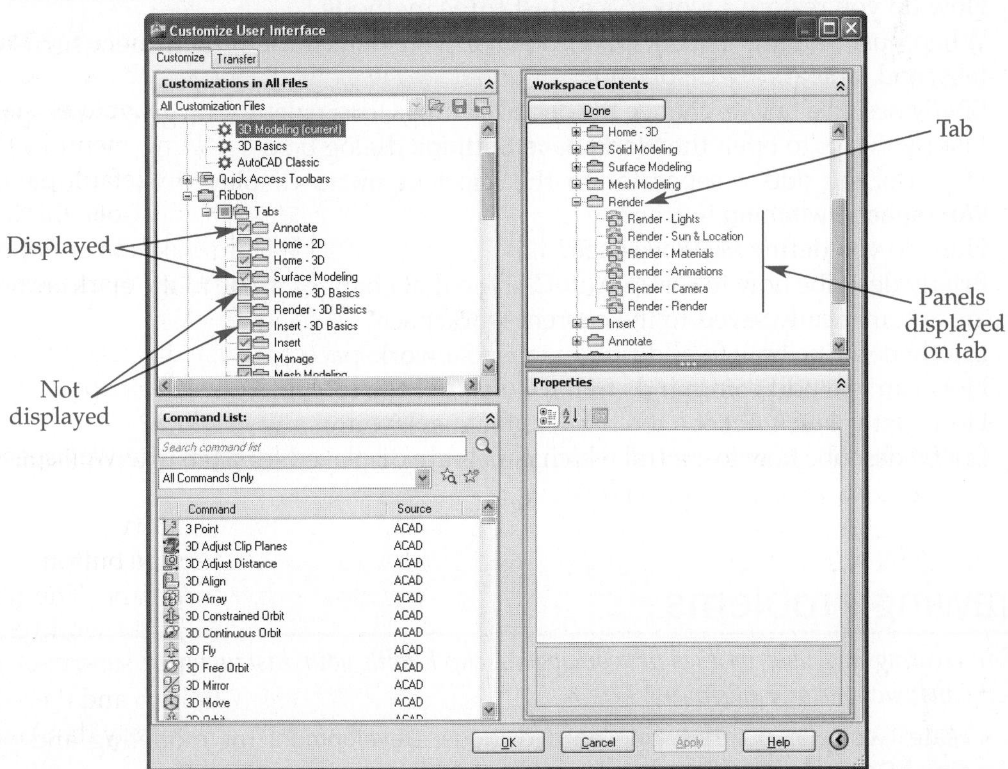

of the Ribbon Tabs branch. Refer to Chapter 22 for information on creating custom ribbon tabs and panels. To rearrange the tabs or panels, pick and drag them within the **Workspace Contents** pane. You do not need to be in customization mode. You can also remove a tab from the workspace by right-clicking on its name in the **Workspace Contents** pane and picking **Remove from Workspace** from the shortcut menu.

PROFESSIONAL TIP

You do not need to be in customization mode to change a panel's Show property. Also, when a panel's Show property is set to No, the panel is still available in the **Panels** cascading menu in the shortcut menu displayed by right-clicking on the ribbon.

Chapter Test

Answer the following questions. Write your answers on a separate sheet of paper or complete the electronic chapter test on the student website.
www.g-wlearning.com/CAD

1. What is a *user profile?*
2. How and why are profiles used?
3. What is the file extension used for a profile when it is exported?
4. In which dialog box is a user profile created?
5. How do you restore a user profile?
6. Why would you export a user profile?
7. Briefly describe how to import a user profile.

8. Define *workspace* as related to AutoCAD.
9. List three ways to open the **Save Workspace** dialog box.
10. How do you restore a workspace? List three methods.
11. When customizing a workspace, how do you determine which menus, ribbon tabs, and toolbars are displayed?
12. Briefly describe how to change the default settings for a palette for a given workspace.
13. List two ways to open the **Workspace Settings** dialog box.
14. How do you add a separator to the shortcut menu displayed by picking the **Workspace Switching** button?
15. How do you define My Workspace?
16. Briefly describe how to set up AutoCAD so that changes made to the environment are automatically saved to the current workspace.
17. Briefly describe how to add ribbon tabs to a workspace.
18. How can you add commands to the **Quick Access** toolbar?
19. How many **Quick Access** toolbars can be displayed in a workspace?
20. Briefly describe how to control which panels are displayed for a tab in a workspace.

Drawing Problems

Before creating any user profiles or workspaces, check with your instructor or supervisor for specific instructions or guidelines.

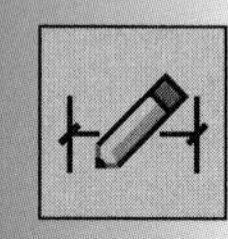

1. Create two user profiles, one named Model Development for modeling and one named Rendering for visual styles and rendering.
   A. Display the toolbars that contain the commands needed for each type of work.
   B. Change the color of the drawing area as needed. For example, some drafters prefer a white background when drawing. However, a black background is often desired when shading and rendering the model.

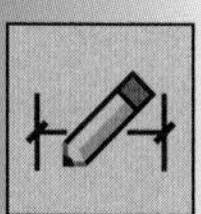

2. Export the user profiles created in problem 1 to ARG files. Then, delete each profile from AutoCAD. Restart AutoCAD and verify that the profiles are no longer available. Next, import each profile from file. Restore each profile to verify the settings.

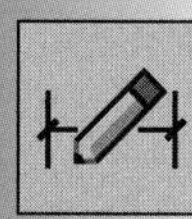

3. Create two workspaces, one named Design Development and one named Dimensioning.
   A. For the Design Development workspace, display the ribbon tabs and panels related to drawing and editing. Also, rearrange the menus to group the drawing and editing/modifying menus together. Remove any menus that are not needed.
   B. For the Dimensioning workspace, display the ribbon tabs and panels related to dimensioning the drawing. You may consider removing editing/modifying menus.
   C. Create custom ribbon tabs and panels as needed. Include flyouts when advantageous.

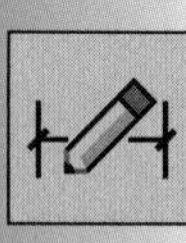

4. Customize the two workspaces created in problem 3.
   A. Change default on/off status of any palettes to suit the purpose of the workspace. For example, you may want the **DesignCenter**, **Tool Palette**, and **Properties** palettes displayed for the Design Development workspace.
   B. Enable the autohide status of all displayed palettes.
   C. Change the Dimensioning workspace so that it starts in a layout.

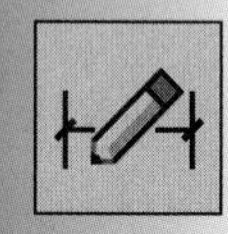

5. Set up the workspaces from problem 4 so that "on the fly" changes are automatically saved when a different workspace is made current.

CHAPTER 26
Recording and Using Actions

Learning Objectives

After completing this chapter, you will be able to:

- ✓ Describe the action recorder.
- ✓ Create and store an action.
- ✓ Run an action.
- ✓ Modify an action.
- ✓ Explain the limitations of the action recorder.

In schools and workplaces, certain design elements and customizing features can be enhanced by automation. The AutoCAD ***action recorder*** is a tool that allows the user to automate a task or function by saving the steps needed to complete it. These steps are called an ***action.*** Creating a library of recorded actions allows a user to increase productivity by more efficiently working in AutoCAD.

Action Recorder Overview

The **Action Recorder** panel is located in the **Manage** tab of the ribbon. It contains seven tools: **Record/Stop**, **Insert Message**, **Insert Base Point**, **Pause for User Input**, **Play**, **Preference**, and **Manage Action Macros**. In addition, a drop-down list displays all recorded actions. See Figure 26-1. The expanded **Action Recorder** panel displays the action tree. The ***action tree*** shows the steps used to create the current action that is selected in the drop-down list. The steps are listed in order from top to bottom. The first step is located at the top of the tree.

The action recorder saves all steps in an action to an ACTM file. By saving the action to a file, it can repeatedly be used. The ACTM file name cannot contain spaces or special characters. Also, the name cannot be the same as an existing AutoCAD command name.

The folder where actions are stored is defined in the **Files** tab of the **Options** dialog box. To change the location where actions are stored, expand the Action Recorder Settings branch in the **Files** tab, Figure 26-2. Then, pick the Actions Recording File Location branch, pick the **Browse...** button, and locate the new default folder. The Action Recorder Settings branch also contains the Additional Actions Reading File Locations branch. Use this to define network or shared folder connections for ACTM files.

Figure 26-1.
The **Action Recorder** panel on the **Manage** tab of the ribbon with the action tree displayed.

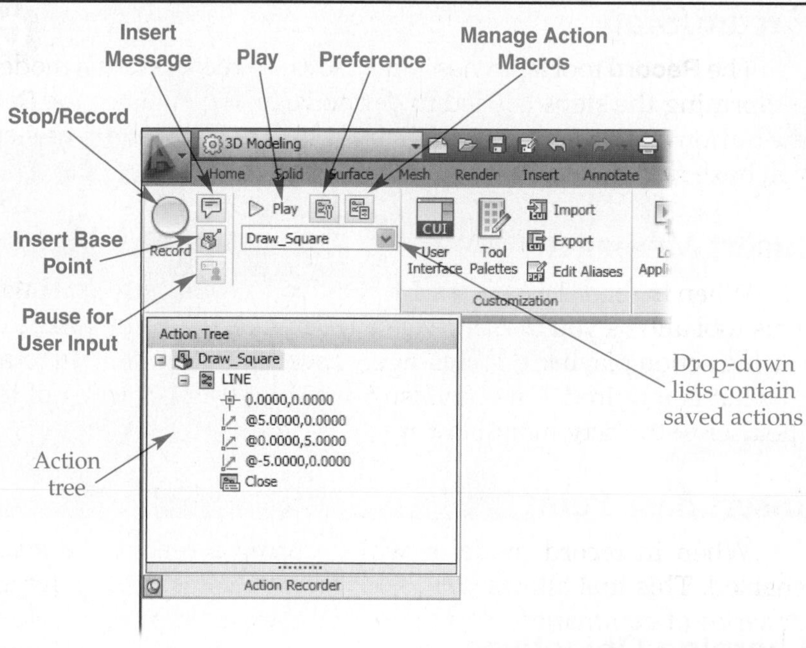

Figure 26-2.
The location where actions are stored is set in the **Files** tab of the **Options** dialog box.

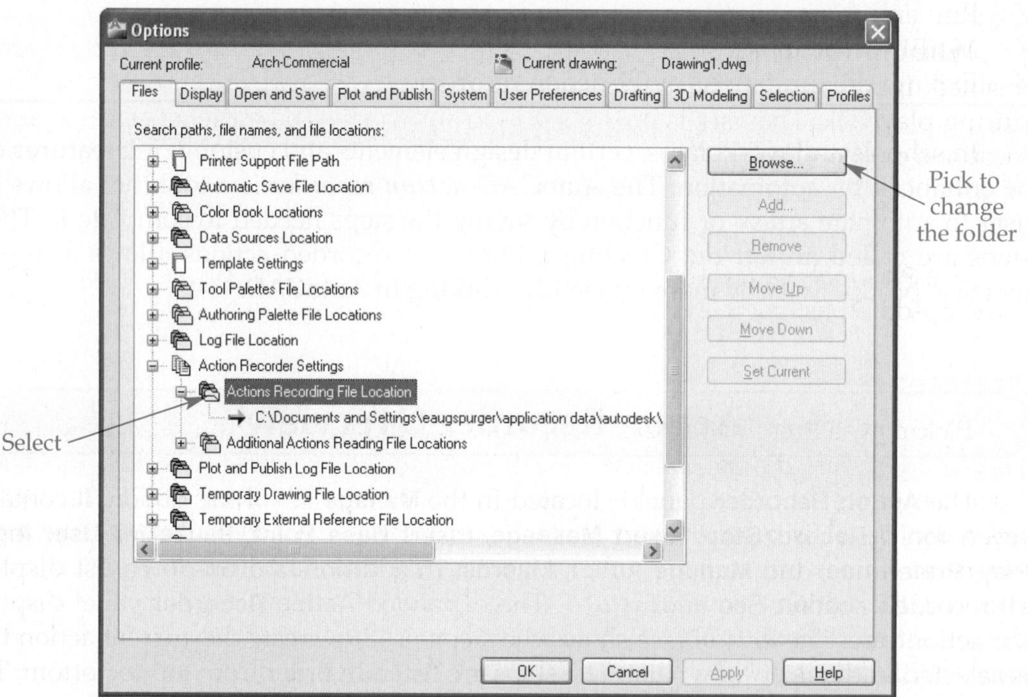

Record/Stop

Ribbon
Manage
> Action Recorder
Record
Type
ACTRECORD
ARR

ACTRECORD

The **Record** tool launches AutoCAD into record-action mode. Then, you can begin performing the steps needed to define your action. Once the **Record** tool is activated, the button changes to the **Stop** button. This is used when the steps are completed and you have finished recording your action.

Insert Message

Ribbon
Manage
> Action Recorder
Insert Message
Type
ACTUSERMESSAGE
ARM

ACTUSERMESSAGE

When in record mode or while editing a macro, the **Insert Message** tool is enabled. This tool allows you to insert a message dialog box that will be displayed for the user during action playback. A message is especially useful for directing the user's attention to what is required. One can also be used at the beginning of the action to introduce the user to the action and how it will perform.

Insert Base Point

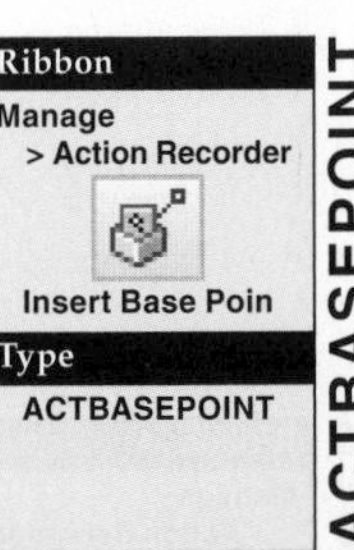

ACTBASEPOINT

When in record mode or while editing a macro, the **Insert Base Point** tool is enabled. This tool allows you to choose a starting location for a particular command or series of commands. This is used to override the default absolute coordinates that are automatically entered while creating an object. It can be used during recording, but only when applicable. The tool is disabled when not available. Additionally, when editing a macro, you may add this option in front of a particular command.

Pause for User Input

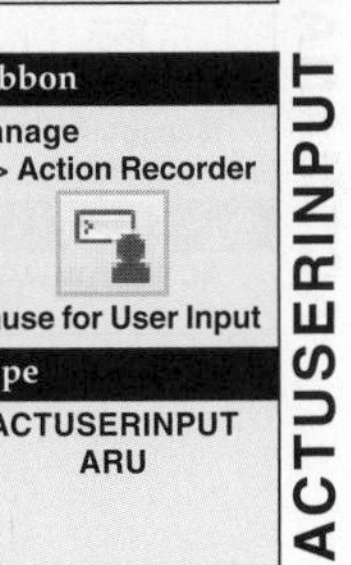

ACTUSERINPUT

When in record mode or while editing a macro, the **Pause for User Input** tool is enabled. Inserting a request for user input temporarily halts the action at that point during playback. The user is prompted for input, such as selecting objects or picking locations on the screen.

Play

The **Play** tool plays the current macro. The current macro is displayed at the top of the drop-down list in the **Action Recorder** panel. This drop-down list is discussed later.

Preferences

Preference

Picking the **Preference** tool displays the **Action Recorder Preferences** dialog box, Figure 26-3. This dialog box contains three check boxes for setting how the **Action Recorder** panel functions. When the **Expand on playback** check box is checked, **Action Recorder** panel is automatically expanded to show the action tree during action playback. This is unchecked by default. When the **Expand on recording** check box is checked, the **Action Recorder** panel automatically expands to display the action tree during action recording. This is checked by default. When the **Prompt for action macro name** check box is checked, the **Action Macro** dialog box is displayed after recording of the action is complete (the **Stop** button is picked). This dialog box prompts for the action name, which is the name of the ACTM file. See Figure 26-4.

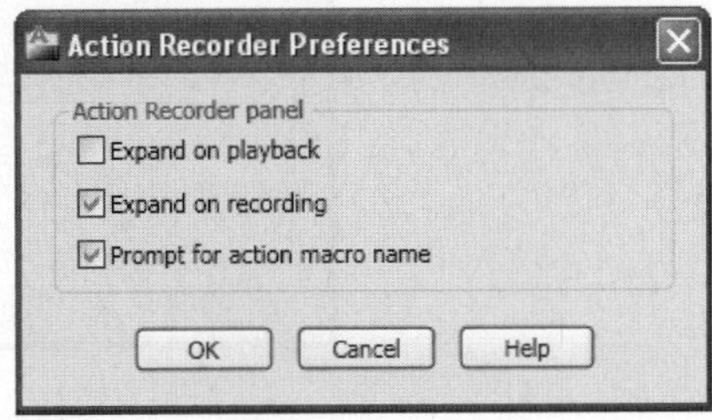

Figure 26-3.
The **Action Recorder Preferences** dialog box is used to set certain behavior of the action recorder.

Figure 26-4.
The **Action Macro** dialog box is used to save an action.

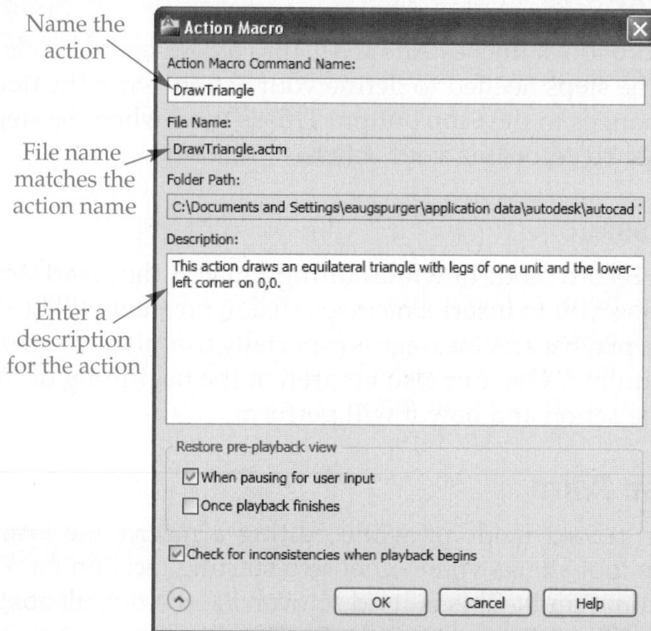

Manage Action Macros

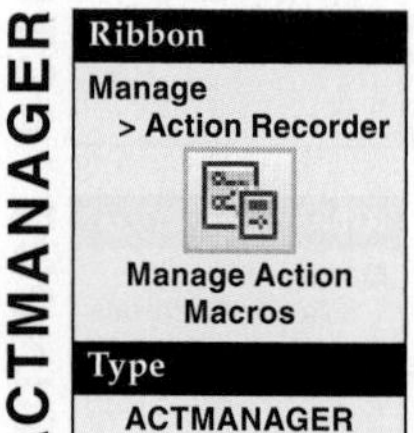

The **Action Macro Manager** dialog box allows you to rename saved actions or change their descriptions. Picking the **Manage Action Macros** button in the **Action Recorder** panel displays this dialog box. See Figure 26-5. The dialog box contains a list of actions saved in the action macro folder specified in the **Options** dialog box.

Picking the **Copy** button displays the **Action Macro** dialog box, which is also displayed by default when recording of an action is complete. Enter a name and description for the new action and set the other properties. When you pick the **OK** button, the steps in the action highlighted in the **Manage Action Macros** dialog box are saved under the new name. The original action is not altered.

Picking the **Rename** button replaces the name of the highlighted macro with an edit box. This allows you to enter a new name for the action. Renaming the action also changes the name of the ACTM file.

Picking the **Modify** button also displays the **Action Macro** dialog box. However, when you make changes to the name, description, or other properties, the action highlighted in the **Manage Action Macros** dialog box is altered.

Figure 26-5.
The **Action Macro Manager** dialog box is used to copy, rename, or change an existing macro.

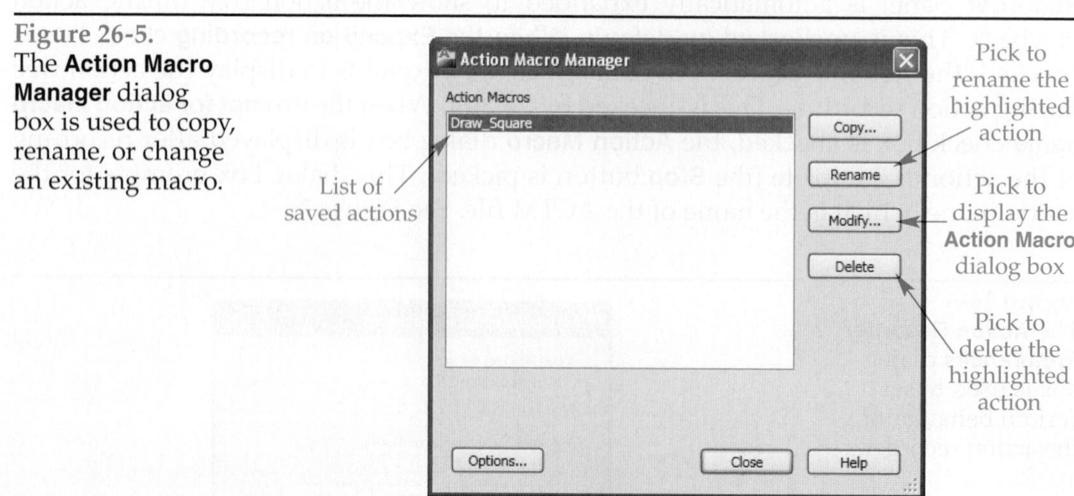

Picking the **Delete** button deletes the action and associated ACTM file. Picking the **Options...** button displays the **Options** dialog box with the Action Recorder Settings branch highlighted in the **Files** tab.

Action Tree and Available Actions

The **Available Action Macro** drop-down list displays all action files (ACTM) located in the action folder. As explained earlier, the action folder is set in the **Files** tab of the **Options** dialog box. The current action is displayed at the top of the list. To see all of the available actions, pick the drop-down list. To set a different action current, simply pick it in the drop-down list.

At the bottom of the expanded **Action Recorder** panel is the action tree. The action tree contains all of the steps in the action in graphic form. See **Figure 26-6.** The name of the action is the top of the tree. Each step is a branch in the tree. In addition, the command branches can be expanded to see the values entered into the action.

PROFESSIONAL TIP

Inserting messages is especially useful to direct the user about the steps of an action. Inserting requests for user input in an action pauses the action and allows the user to select objects or on-screen points. These features make actions more dynamic and help ensure proper usage.

Exercise 26-1

Complete the exercise on the student website.
www.g-wlearning.com/CAD

Figure 26-6.
The action tree contains all of the steps in the current action.

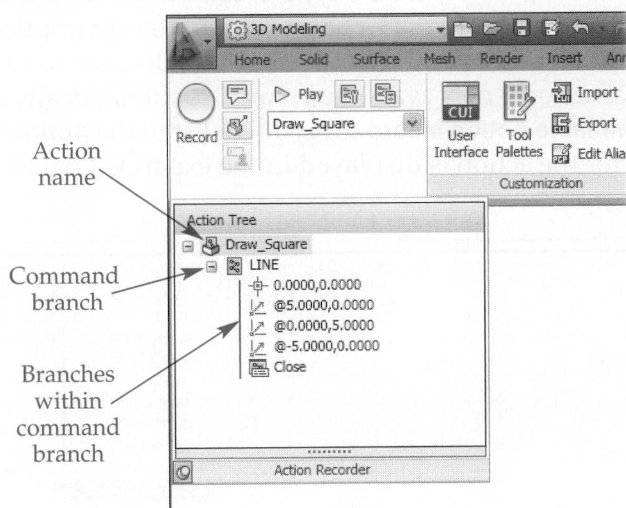

Creating and Storing an Action

Now that you have investigated the features of the action recorder, it is time to create an action. This example action will create a layer named A-Anno-Iden, set the layer color to orange (ACI = 30), and make the layer current. The action will also set the annotation scale to 1/4″ = 1′-0″. In this section, you will create the basic action. Later, you will modify the action to make it more interactive. Follow these steps to create the basic action named AnnoQtr.

1. Start a new drawing.
2. Pick the **Record** button in the **Action Recorder** panel on the **Manage** tab of the ribbon. Notice that the panel is shaded in red and a red dot appears next to the crosshairs. These are indications that you are in record mode.
3. Display the **Layer Properties Manager** palette.
4. Pick the **New Layer** button in the palette's toolbar.
5. In the edit box, name the new layer A-Anno-Iden.
6. Pick the color swatch for the layer and, in the **Select Color** dialog box, pick color 30 on the **Index Color** tab. Then, pick the **OK** button to close the **Select Color** dialog box.
7. Right-click on the newly created layer and pick **Set Current** from the shortcut menu.
8. Close the **Layer Properties Manager** palette.
9. Using the **Annotation Scale** button on the status bar, set the 1/4″ = 1′-0″ scale current. See **Figure 26-7**. If the drawing is based on the acad3D.dwt template, you must type the **CANNOSCALE** command since the status bar button is unavailable.
10. Pick the **Stop** button on the **Action Recorder** panel in the **Tools** tab of the ribbon. This ends recording of the action.
11. The **Action Macro** dialog box appears when you pick the **Stop** button, unless you have turned off the "prompt for name" option.
12. In the **Action Macro** dialog box, type AnnoQtr in the **Action Macro Command Name:** text box.
13. In the **Description:** text box, type Creates a new layer with color 30 and sets the 1/4″ = 1′-0″ annotation scale current or a similar description.
14. Pick the **OK** button in the **Action Macro** dialog box to save the action.

The AnnoQtr action is now available to run. It is automatically selected as the current action in the **Available Action Macro** drop-down list in the **Action Recorder** panel. Also, the action tree for the action is displayed in the expanded panel. See **Figure 26-8.**

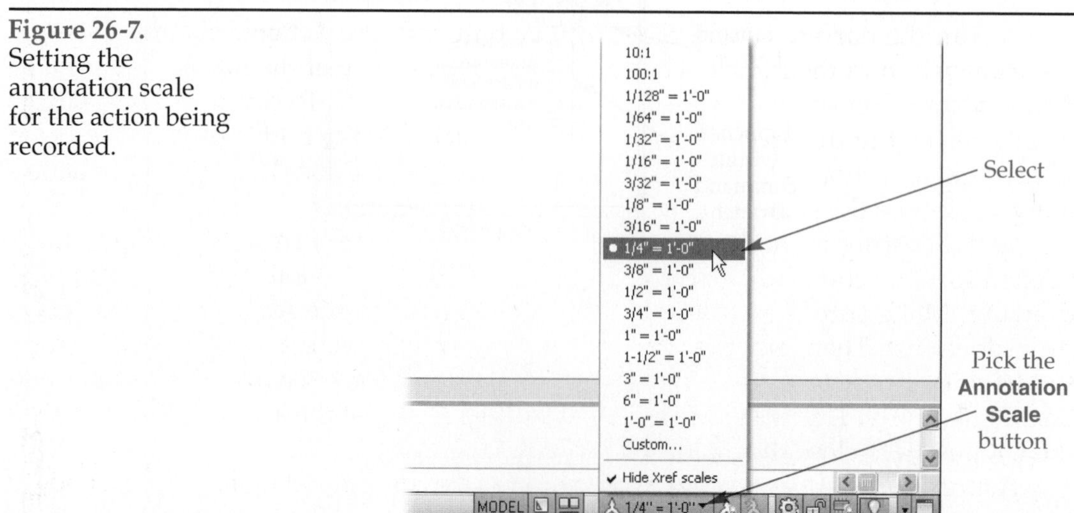

Figure 26-7.
Setting the annotation scale for the action being recorded.

Figure 26-8.
The new action is automatically set current. Its action tree is displayed in the expanded panel.

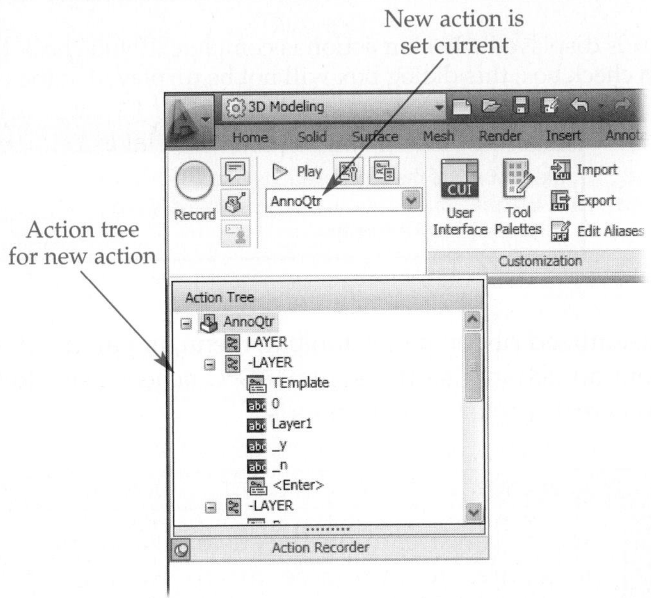

PROFESSIONAL TIP

In a design environment, saving actions in a shared folder on a network can streamline productivity. This creates an environment in which AutoCAD users can share their knowledge by creating actions drawn from their own design experiences. Use the Additional Actions Reading File Locations branch in the **Options** dialog box to establish the shared connection.

Exercise 26-2

Complete the exercise on the student website.
www.g-wlearning.com/CAD

Running an Action

To run the current action, pick the **Play** button in the **Action Recorder** panel on the **Manage** tab in the ribbon. The current action is shown at the top of the **Available Action Macro** drop-down list in the **Action Recorder** panel. To run a different action, simply select it in the drop-down list and then pick the **Play** button. Another method of running an action is to type its name on the command line. This is why the action name cannot be the same as an existing AutoCAD command.

Before testing an action, be sure to delete all graphics from the screen unless needed for the action. To test the action recorded in the previous section, start a new drawing. Make sure the AnnoQtr action is set current in the **Available Action Macro** drop-down list. Then, pick the **Play** button and watch the action run. By default, when the action is complete, AutoCAD displays the **Action Macro—Playback Complete** dialog box, Figure 26-9. This is simply a message to the user that the action is finished. Pick the **OK** button to close the dialog box.

Actions run just like regular AutoCAD commands. Since they can be executed on the command line, actions can be made into custom AutoCAD commands. They can be

Figure 26-9.
This dialog box is displayed when an action is complete. If you check the **Do not show this message again** check box, this dialog box will not be displayed at the end of action playback.

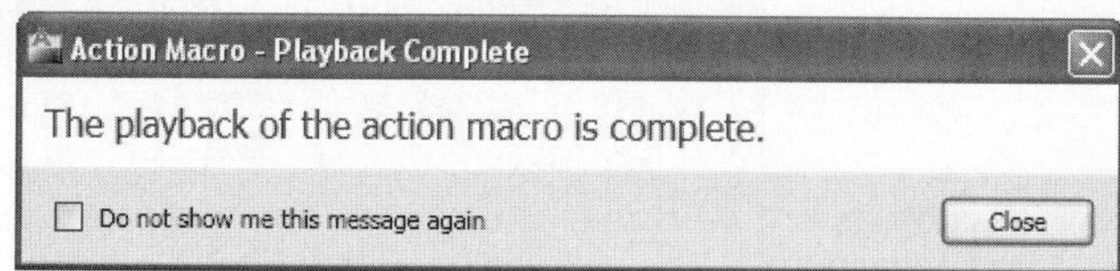

placed in a customized ribbon panel, toolbar, menu, or palette. When creating a custom command from an action, use the syntax ^C^C*_action_name*. Refer to Chapter 22 for information on creating custom commands.

PROFESSIONAL TIP

Actions can streamline the customization process by creating setup procedures to minimize macro writing when creating custom commands. For example, suppose you have an annotation block that you want inserted at a set scale and on a specific layer. You can create an action similar to the example above (AnnoQtr). Then, you can insert the action in the command string prior to the block insertion. This will streamline your customization process.

Modifying an Action

The action tree can be used to edit an action. Examine the action tree for the AnnoQtr example, as shown in **Figure 26-10.** At the top of the action tree is the action name. It is always the top branch of the tree. The icon to the left of the name indicates that it is an action. The branches below the action name contain the commands and steps in the action.

Notice the dash in front of the **LAYER** command. This dash indicates the command is a command-line command. Even though you used the **Layer Properties Manager** palette when recording the action, the action recorder translated your picks into the command-line version of the steps.

Branches below the command names contain the specific information used to create each item. Each branch is automatically converted to the command-line entry. The icon next to each branch indicates the type of action or input represented by the branch. These branches are also called *nodes.* For example, reviewing the **LAYER** command branches, you can see a text icon next to the name A-Anno-Iden. This icon indicates that the branch, or node, is a text entry. Additionally, the icon for the branch where the layer color is set to 30 is a color wheel. This indicates the branch is a change in color value.

An icon with a head-and-shoulders image in the lower-right corner indicates the action will pause at this point to allow the user to enter information. This icon is called a user icon. The AnnoQtr action does not currently contain any user icons.

Now, you will modify the action recorded earlier. You will change some of the steps to require user input. For example, you will have the user type the annotation scale instead of automatically setting it to 1/4″ = 1′-0″ scale. So the user knows what is required, you will add a message explaining the step. Since the scale is not automatically set, the action also needs to be renamed to remove the Qtr (quarter inch scale) designation. Complete the following steps.

1. Expand the **Action Recorder** panel to see the action tree. You may want to pin the panel in the expanded state.

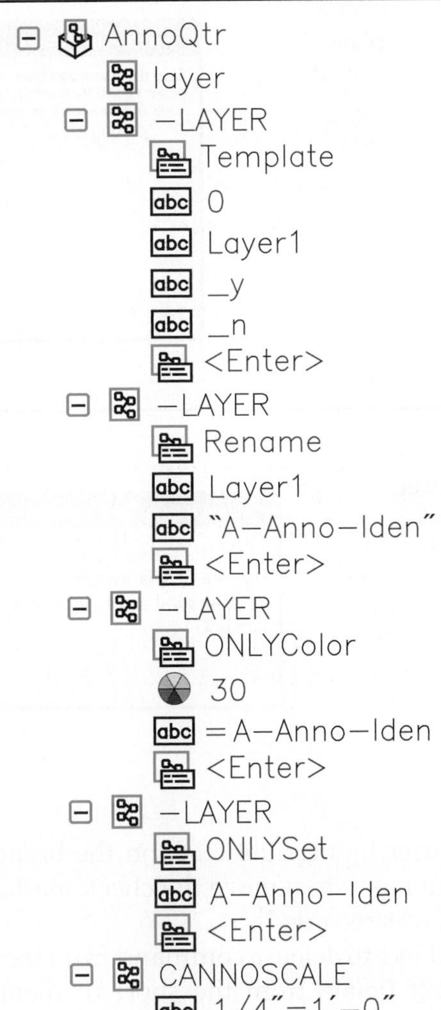

Figure 26-10.
The action tree for the recorded action prior to editing the action.

2. Right-click on the action name AnnoQtr and select **Rename...** from the shortcut menu. If you want to retain the original action instead of renaming it, select **Copy...** from the shortcut menu. In either case, the **Action Macro** dialog box is displayed.
3. In **Action Macro Command Name:** text box, type AnnoUserScale and then pick the **OK** button to rename the action. Note: the edits in steps 2 and 3 may also be accomplished using the **Action Macro Manager** dialog box.
4. In the action tree, locate the CANNOSCALE branch. Right-click on the 1/4″ = 1′-0″ branch below it and select **Pause for User Input** from the shortcut menu. The value in the branch (1/4″ = 1′-0″) is grayed out and the icon for the branch now contains the user icon.
5. To add a message describing how to type the correct scale, right-click on the 1/4″ = 1′-0″ branch and select **Insert User Message...** from the shortcut menu. The **Insert User Message** dialog box appears, Figure 26-11.
6. In the **Insert User Message** dialog box, type the message:

   Enter the desired annotation scale on the command line. Please use proper syntax. For example, type 1/8″ = 1′-0″ for the 1/8″ Architectural scale. Include a space on each side of the equals symbol.

Notice that a User Message branch is added to the action tree above the branch that was right clicked. The action is now altered. Start a new drawing and run the altered action. Notice the dialog box that appears to guide the user through the inputs. See Figure 26-12.

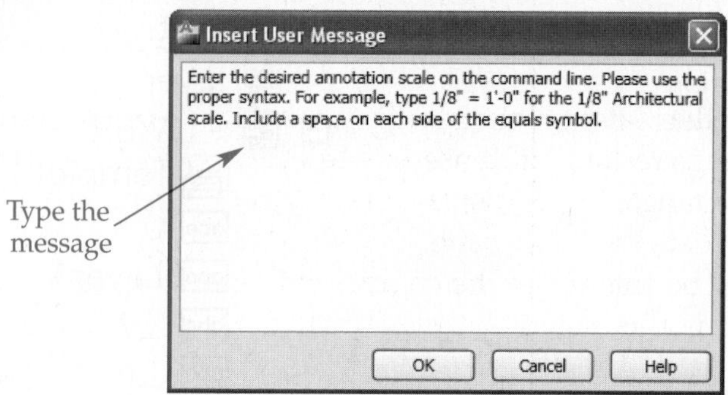

Figure 26-11.
Adding a user message to the action.

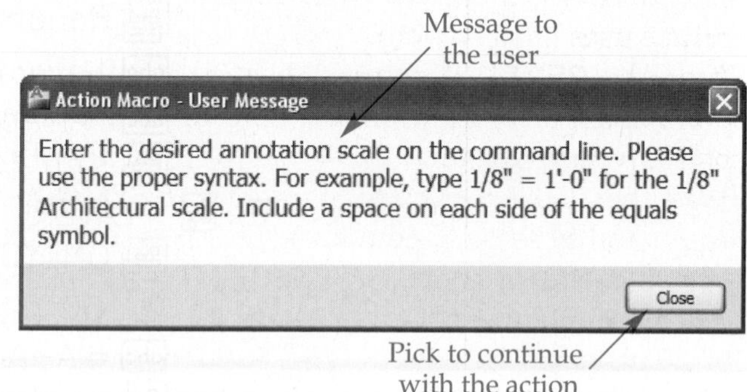

Figure 26-12.
The message displayed to the user during playback of the example action.

To remove user input, right-click on the branch and select **Pause for User Input** from the shortcut menu to remove the check mark. The action then uses the value set when the action was recorded.

If you would like to delete a command from the action, right-click on the command branch and select **Delete** from the shortcut menu. Once deleted, the command is permanently removed from the action. This cannot be undone. If zooming or panning is done while recording the action, the action tree will display a View Change node with its appropriate icon. You may right-click and select **Delete** or **Insert User Message** from the shortcut menu to edit this node.

To edit a value, right-click on the corresponding branch and select **Edit** from the shortcut menu. The value is replaced by a text box. Type the new value and press [Enter]. Careful; if you move the cursor off of the panel while the text box is displayed, whatever is displayed in the text box is entered as the value.

If you are creating objects such as a rectangle during recording of an action, the default coordinates are set relative to the previous point. If you wish to toggle to absolute coordinate entry, right-click on the coordinate in the action tree and uncheck **Relative to Base Point** in the shortcut menu. Furthermore, if you right-click in the action tree, you may uncheck **All Points are Relative** in the shortcut menu, which will edit the entire action.

NOTE

When recording the action, use the buttons on the **Action Recorder** panel to insert messages, base points, and user-input requests.

Exercise 26-3

Complete the exercise on the student website.
www.g-wlearning.com/CAD

Practical Example

This example creates a rectangle and fills it with a hatch pattern. First, you will record an action that draws a rectangle at a user-specified location and then uses the **ZOOM** command to center the rectangle in the display. The action will draw a brick hatch pattern inside of the rectangle. Then, the hatch pattern is edited to change/reset its origin. In this example, commands will be entered on the command line. Although it is more time-consuming to create an action in this way, it allows easy access to and editing of the action steps.

1. Start a new drawing.
2. Begin recording a new action.
3. Select the **Insert Base Point** button and choose any location on the screen. This allows the user to place the rectangle anywhere on the screen during future uses of the macro.
4. Type the **RECTANG** command. When prompted to specify the first corner point, pick anywhere on the screen. When prompted for the other corner point, type @120,60.
5. Type the **ZOOM** command and enter the **Extents** option. Enter the command again and type .7x to center the rectangle on the screen.
6. Type the **-HATCH** command. Use the **Select object** option to select the rectangle.
7. Use the **Properties** option and select the hatch AR-B816. This is the architectural brick pattern. Enter a scale of 1.000 and 0° rotation angle. Be sure to enter the values; do not press [Enter] to accept the defaults. Finish the **-HATCH** command.
8. Type the **-HATCHEDIT** command and select the brick pattern.
9. Select the **Origin** option, then select the **Set new origin** option. Pick the lower, right-hand corner of the rectangle. When prompted, store this location as the default origin.
10. Stop recording of the action and name it BrickHatchEdit.

Next, edit the action to allow user input and display messages. This will aid the user in selecting points.

11. For the <Select Objects> node of the **-HATCHEDIT** command, change it to pause for user input. This allows the user to pick the hatch to edit.
12. Insert a user message for the **-HATCHEDIT** command just above the <Select Objects> node. The message should read Select the brick hatch object. This gives instruction to the user.
13. For the point location under the **Set** option of the **-HATCHEDIT** command, change it to pause for user input. In this way, the user selects the origin point for where the hatch pattern begins.
14. Insert a user message for the point location under the **Set** option of the **-HATCHEDIT** command. The message should read Edit the hatch and select a new origin point when prompted. This gives instruction to the user.
15. Test the action. When prompted for the new origin, select the midpoint of the lower line of the rectangle. Notice the result is different from the pattern created when recording the action.

PROFESSIONAL TIP

The action recorder works best with command-line versions of commands. Many commands can be run on the command line only by placing a hyphen (-) in front of the command name. For example, the **LAYER** command displays the **Layer Properties Managers** palette. However, the **-LAYER** command displays all options and settings on the command line without displaying a palette or dialog box. Not all commands have a command-line version.

Limitations of the Action Recorder

As a customizing tool, the action recorder is very powerful. However, as in all customization, there is syntax to follow and limitations to address. For example, the AutoCAD commands shown in **Figure 26-13** cannot be recorded in an action.

The following chapters discuss AutoLISP and other programming features of AutoCAD. Actions are able to call and run AutoLISP routines during the recording process. However, the specific AutoLISP routines, Object ARX files, VBA macros, .NET assemblies, and associated files must all be loaded into AutoCAD in order for the programming feature to function during playback of the action. See Chapters 27 through 29 for further details on programming AutoCAD. These chapters are located on the student website.

When creating an action for use in a professional or school environment, you must always consider the source files. For example, if you create an action that reads from a template file or loads particular blocks, those external files must always be either loaded into AutoCAD or accessible to the user.

Finally, you cannot append an existing action. If you need to add steps to the action, you must re-record the existing steps along with the new steps. For this reason, it is very important to plan an action before recording it.

PROFESSIONAL TIP

You can create actions with other Autodesk applications, such as AutoCAD Architecture. If you create an action using AEC objects, for example, the action can only properly function if the action is played back in AutoCAD Architecture.

Figure 26-13.
These commands cannot be used in an action.

ACTBASEPOINT
ACTMANAGER
ACTSTOP
ACTUSERINPUT
ACTUSERMESSAGE
-ACTUSERMESSAGE
DXFIN
EXPORTLAYOUT
FILEOPEN
NEW
OPEN
PARTIALOPEN
PRESSPULL
QNEW
RECOVER
TABLEDIT
VBAIDE
VBALOAD
-VBALOAD
VBAMAN
VBANEW
VBAPREF
VBARUN
-VBARUN
VBASTMT
VBAUNLOAD
XOPEN

Chapter Test

Answer the following questions. Write your answers on a separate sheet of paper or complete the electronic chapter test on the student website.
www.g-wlearning.com/CAD

1. What is an *action?*
2. What is used to record an action?
3. In which file format is an action saved?
4. By default, where are action files stored?
5. What is the *action tree?*
6. How do you set an action current?
7. Name two ways to play an action.
8. How do you pause an action for user input?
9. How can you display a custom message in an action?
10. Briefly describe how to change the value associated with a branch in an action.

Drawing Problems

1. In this problem, you will create an action that sets up a drawing for architectural work. The action will create layers and set the units. The action should:
   A. Display a message to the user indicating what the action will do.
   B. Create a layer named A-Wall and set its color to ACI 113 (or the color of your choice).
   C. Create a layer named A-Door and set its color to ACI 31 (or the color of your choice).
   D. Create a layer named A-Demo and set its color to ACI 1 (or the color of your choice).
   E. Set the units to architectural with a precision of 1/4″.
   F. Save the action as ArchSetup.actm. If the file is saved in a folder other than the default, make sure that folder is added to the AutoCAD search path.

2. In this problem, you will create an action that allows the user to select objects and then places those objects at an elevation of 12 units. The action should:
   A. Display a message to the user indicating what the action will do.
   B. Prompt the user to select objects.
   C. Change the elevation property of all objects to 12 units.
   D. Save the action as 1FtElevChange.actm. If the file is saved in a folder other than the default, make sure that folder is added to the AutoCAD search path.

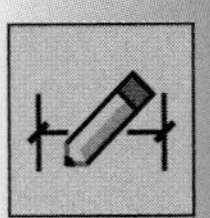

3. In this problem, you will modify the action created in problem 2. Edit the action to allow the user to enter the new elevation. Be sure to display a message to the user indicating what is being asked of them. This is a good tool for flattening linework or resetting solids. Save the action as ElevChange.actm. If the file is saved in a folder other than the default, make sure that folder is added to the AutoCAD search path.

The following chapters are available on the student website. These chapters discuss various aspects of programming AutoCAD, including working with AutoLISP and creating dialog boxes using DCL.
www.g-wlearning.com/CAD

Programming AutoCAD

Index

The index entries shown in black refer to chapters located in this printed book. The index entries shown in color or page number listings in color refer to chapters that are located on the student website.

 www.g-wlearning.com/CAD

The index entries shown in color refer to chapters that are located on the student website.

B

The index entries shown in color refer to chapters that are located on the student website.

C

The index entries shown in color refer to chapters that are located on the student website.

The index entries shown in color refer to chapters that are located on the student website.

The index entries shown in color refer to chapters that are located on the student website.

D

E

F

The index entries shown in color refer to chapters that are located on the student website.

G

H

The index entries shown in color refer to chapters that are located on the student website.

The index entries shown in color refer to chapters that are located on the student website.

The index entries shown in color refer to chapters that are located on the student website.

N

O

P

The index entries shown in color refer to chapters that are located on the student website.

Q

R

The index entries shown in color refer to chapters that are located on the student website.

S

The index entries shown in color refer to chapters that are located on the student website.

T

The index entries shown in color refer to chapters that are located on the student website.

U

The index entries shown in color refer to chapters that are located on the student website.

Index–Advanced

The index entries shown in color refer to chapters that are located on the student website.

Nursing considerations provide guidance throughout the nursing process

Alert icon highlights high-alert drugs and clinical considerations

2 abacavir

Drug/Lab Test
Increase: glucose, triglycerides, GGT, LFTs

NURSING CONSIDERATIONS

Assess:
- Symptoms of HIV and possible infections; increased temp

⚠ **Lactic acidosis** (elevated lactate levels, increased LFTs), severe hepatomegaly with steatosis, discontinue treatment and do not restart; may have large liver, elevated AST, ALT, lactate levels; women are at greater chance of lactic acidosis

Black Box Warning: Fatal hypersensitivity reactions: fever, rash, nausea, vomiting, fatigue, cough, dyspnea, diarrhea, abdominal discomfort; treatment should be discontinued and not restarted; those with HLA B5701 are at great risk for hypersensitivity; obtain genetic testing for HLA B5701 before starting treatment

⚠ **Blood dyscrasias** (anemia, granulocytopenia): bruising, fatigue, bleeding, poor healing

- Renal studies: BUN, serum uric acid, CCr before, during therapy; these may be elevated

Black Box Warning: Hepatic studies before and monthly during therapy: bilirubin, AST, ALT, amylase, alk phos, creatine phosphokinase, creatinine

- Blood counts; monitor viral load and CD4 counts during treatment; watch for decreasing granulocytes, Hgb; if low, therapy may have to be discontinued and restarted after hematologic recovery; blood transfusions may be required; perform hepatitis B virus (HBr) screening to confirm correct treatment

Evaluate:
- Therapeutic response: increased CD4 count, decrease viral load

Teach patient/family:
- That product is not a cure but will control symptoms; patient is still infective, may pass AIDS virus on to others, to carry emergency ID with condition, products taken
- That body fat distribution may occur; not to share product

⚠ To notify prescriber of sore throat, swollen lymph nodes, malaise, fever; other infections may occur; to stop product and to notify prescriber immediately if skin rash, fever, cough, shortness of breath, GI symptoms occur; advise all health care providers that allergic reaction has occurred with abacavir

- That follow-up visits must be continued because serious toxicity may occur; blood counts must be done
- To use contraception during treatment; if patient is pregnant, register with the Antiretroviral Pregnancy Registry at 1-800-258-4263
- Give patient Medication Guide and Warning Card, discuss points on guide
- That other products may be necessary to prevent other infections and that drug is taken with other antiretrovirals
- Not to drink alcohol while taking this product

⚠ Nurse Alert

Black Box Warnings identify serious and life-threatening adverse effects

Mosby's 2015
NURSING DRUG REFERENCE

Mosby's 2015 NURSING DRUG REFERENCE

Linda Skidmore-Roth, RN, MSN, NP
Consultant
Littleton, Colorado
Formerly, Nursing Faculty
New Mexico State University
Las Cruces, New Mexico
El Paso Community College
El Paso, Texas

ELSEVIER

ELSEVIER
MOSBY

3251 Riverport Lane
St. Louis, Missouri 63043

MOSBY'S 2015 NURSING DRUG REFERENCE,
TWENTY-EIGHTH EDITION

ISBN: 978-0-323-27801-0
ISSN: 1044-8470

Notices

Knowledge and best practice in this field are constantly changing. As new research and experience broaden our understanding, changes in research methods, professional practices, or medical treatment may become necessary.

Practitioners and researchers must always rely on their own experience and knowledge in evaluating and using any information, methods, compounds, or experiments described herein. In using such information or methods they should be mindful of their own safety and the safety of others, including parties for whom they have a professional responsibility.

With respect to any drug or pharmaceutical products identified, readers are advised to check the most current information provided (i) on procedures featured or (ii) by the manufacturer of each product to be administered, to verify the recommended dose or formula, the method and duration of administration, and contraindications. It is the responsibility of practitioners, relying on their own experience and knowledge of their patients, to make diagnoses, to determine dosages and the best treatment for each individual patient, and to take all appropriate safety precautions.

To the fullest extent of the law, neither the Publisher nor the authors, contributors, or editors, assume any liability for any injury and/or damage to persons or property as a matter of products liability, negligence or otherwise, or from any use or operation of any methods, products, instructions, or ideas contained in the material herein.

ISBN: 978-0-323-27801-0

Director, eContent Solutions: Robin Carter
Product Specialist: Shephali Graf
Content Coordinator: Samantha Taylor
Publishing Services Manager: Pat Joiner; Deborah L. Vogel
Senior Project Manager: Jodi Willard
Design Direction: Brian Salisbury

Printed in China

Last digit is the print number: 9 8 7 6 5 4 3 2 1

Consultants

Timothy Loren Brenner, PharmD, BCOP
Clinical Pharmacy Specialist
UPMC Cancer Centers
Pittsburgh, Pennsylvania

David S. Chun, PharmD, BCPS
Pharmacist
Richmond Heights, Missouri

Donna Ciulla, RPh
Faculty
Salem State University
Salem, Massachusetts
Staff Pharmacist
Lahey Health Systems
Beverly, Massachusetts

Amanda Gross, RPh
Clinical Pharmacist
Atrium Pharmacy
University of Colorado Hospital
Aurora, Colorado

Paul E. Milligan, PharmD
Clinical Lead
BJC Learning Institute
St. Louis, Missouri

Joshua J. Neumiller, PharmD, CDE, CGP, FASCP
Assistant Professor
Washington State University
Spokane, Washington

Kristin Oneail, MSN, RN
Assistant Professor
Lourdes University
Sylvania, Ohio

Sarah R. Pool, RN, MS
Nursing Education Specialist Cardiac Surgery
Mayo Clinic
Rochester, Minnesota

Sheila M. Seed, PharmD, MPH, RPh
Associate Professor and Vice Chair of Pharmacy Practice
MCPHS University
Worcester, Massachusetts

Melissa Sellers, MSN, BSN, RN
Nursing Instructor
Jackson State Community College
Jackson, Tennessee

Travis E. Sonnett, PharmD, FASCP
Clinical Pharmacy Specialist
Spokane VA Medical Center
Clinical Assistant Professor
Washington State University College of Pharmacy
Spokane, Washington

Patricia A. Talbert, RN
Preoperative/Pain Clinic Nurse
Aromatherapist
Herbalist
Horticulturist
Baxter Regional Medical Center
Mountain Home, Arkansas

Shamim Tejani, PharmD
Clinical Pharmacist
Adelante Healthcare
Phoenix, Arizona

Kristine Willett, PharmD
Assistant Professor of Pharmacy Practice
MCPHS University
Manchester, New Hampshire

Preface

Increasingly, patients are relying on nurses to know every detail of health care. More important, nurses are expected to have these answers, especially when it comes to medication. Let *Mosby's 2015 Nursing Drug Reference* be your answer. Our indispensable, yet compact, resource contains hundreds of monographs with several easy-to-use features.

NEW FEATURES

This edition features:

- A beautiful, full-color design to spotlight critical and safety-related information
- Twenty recent FDA-approved drugs located in **Appendix A** (see Contents for a complete list). Included are monographs for:
  - dapagliflozin (Farxiga)—for diabetes
  - levomilnacipran (Fetzima)—for depression
  - macitentan (Opsumit)—for hypertension
- Newly added complete monographs for select ophthalmic, otic, nasal, and topical products
- **A new ebook** with easy-to-use navigation for quick access to drug categories and monographs of your choice

NEW FACTS

This edition features more than 2000 new drug facts, including:

- New drugs and dosage information
- Newly researched side effects and adverse reactions
- New and revised Black Box Warnings
- The latest precautions, interactions, and contraindications
- IV therapy updates
- Revised nursing considerations
- Updated patient/family teaching guidelines

ORGANIZATION

This reference is organized into two main sections:

- Individual drug monographs (in alphabetical order by generic name)
- Appendixes (identified by the wide, blue thumb tabs on the edge)

The guiding principle behind this book is to provide fast, easy access to drug information and nursing considerations. Every detail—the paper, typeface, cover, binding, use of color, and appendixes—has been carefully chosen with the user in mind.

INDIVIDUAL DRUG MONOGRAPHS

This book contains monographs for more than 1300 generic and 4500 trade medications. Common trade names are given for all drugs regularly used in the United States and Canada, with drugs available only in Canada identified by a maple leaf 🍁.

The following information is provided, whenever possible, for safe, effective administration of each drug:

High-alert status: Identifies high-alert drugs with a label and icon. Visit the Institute for Safe Medication Practices (ISMP) at http://www.ismp.org/tools/highalertmedications.pdf for a list of medications and drug classes with the greatest potential for patient harm if they are used in error.

Tall man lettering: Uses the capitalization of distinguishing letters to avoid medication errors and is required by the FDA for drug manufacturers.

Pronunciation: Helps the nurse master complex generic names.

Rx/OTC: Identifies prescription or over-the-counter drugs.

Functional and chemical classifications: Allow the nurse to see similarities and dissimilarities among drugs in the same functional but different chemical classes.

Do not confuse: Present drug names that might easily be confused within each appropriate monograph.

Action: Describes pharmacologic properties concisely.

Uses: List the conditions the drug is used to treat.

Unlabeled uses: Describe drug uses that may be encountered in practice but are not yet FDA approved.

Dosages and routes: List all available and approved dosages and routes for adult, pediatric, and geriatric patients.

Available forms: Include tablets, capsules, extended-release, injectables (IV, IM, SUBCUT), solutions, creams, ointments, lotions, gels, shampoos, elixirs, suspensions, suppositories, sprays, aerosols, and lozenges.

Side effects: Groups these reactions by body system, with common side effects *italicized* and life-threatening reactions (those that are potentially fatal and/or permanently disabling) in **bold, red type** for emphasis.

Contraindications: List conditions under which the drug absolutely should not be given, including FDA pregnancy safety categories D or X.

Precautions: List conditions that require special consideration when the drug is prescribed, including FDA pregnancy safety categories A, B, or C.

Black Box Warnings: Identify FDA warnings that highlight serious and life-threatening adverse effects.

Pharmacokinetics: Outlines metabolism, distribution, and elimination.

Interactions: Include confirmed drug interactions, followed by the drug or nutrient causing that interaction, when applicable.

Drug/herb: Highlight potential interactions between herbal products and prescription or OTC drugs.

Drug/food: Identifies many common drug interactions with foods.

Drug/lab test: Identifies how the drug may affect lab test results.

Nursing considerations: Identify key nursing considerations for each step of the nursing process: Assess, Administer, Evaluate, and Teach Patient/Family. Instructions for giving drugs by various routes (e.g., PO, IM, IV) are included, with route subheadings in bold.

Compatibilities: List syringe, Y-site, and additive compatibilities and incompatibilities. If no compatibilities are listed for a drug, the necessary compatibility testing has not been done and that compatibility information is unknown. To ensure safety, assume that the drug may not be mixed with other drugs unless specifically stated.

"Nursing Alert" icon ⚠: Highlights a critical consideration.

Treatment of overdose: Provides drugs and treatment for overdoses where appropriate.

APPENDIXES

Selected new drugs: Includes comprehensive information on 20 key drugs approved by the FDA during the past 12 months.

Ophthalmic, otic, nasal, and topical products: Provides essential information for more than 80 ophthalmic, otic, nasal, and topical products commonly used today, grouped by chemical drug class.
Vaccines and toxoids: Features an easy-to-use table with generic and trade names, uses, dosages and routes, and contraindications for over 40 key vaccines and toxoids.

I am indebted to the nursing and pharmacology consultants who reviewed the manuscript and thank them for their criticism and encouragement. I would also like to thank Robin Carter and Shephali Graf, my editors, whose active encouragement and enthusiasm have made this book better than it might otherwise have been. I am likewise grateful to Jodi Willard and Graphic World Inc. for the coordination of the production process and assistance with the development of the new edition.

Linda Skidmore-Roth

FDA pregnancy categories

A	No risk demonstrated to the fetus in any trimester
B	No adverse effects in animals; no human studies available
C	Only given after risks to the fetus are considered; animal studies have shown adverse reactions; no human studies available
D	Definite fetal risks, may be given in spite of risks if needed in life-threatening conditions
X	Absolute fetal abnormalities; not to be used at any time during pregnancy

Note: **UK** = Unknown fetal risk (used in this text but not an official FDA pregnancy category)

Contents

abacavir (Rx)

(ah-bak′ah-veer)

Ziagen

Func. class.: Antiretroviral

Chem. class.: Nucleoside reverse transcriptase inhibitor (NRTI)

Do not confuse:
abacavir/amprenavir

ACTION: Inhibitory action against HIV-1; inhibits replication of the virus by incorporating into cellular DNA by viral reverse transcriptase, thereby terminating the cellular DNA chain

USES: In combination with other antiretroviral agents for HIV-1 infection (not to be used with lamivudine or tenofovir)
Unlabeled uses: HIV prophylaxis following occupational exposure

CONTRAINDICATIONS

Black Box Warning: Hypersensitivity, moderate severe hepatic disease

Precautions: Pregnancy (C), breastfeeding, children <3 mo, granulocyte count <1000/mm^3 or Hgb <9.5 g/dl, severe renal disease, impaired hepatic function, HLA B5701 (black, caucasian, Asian patients), abrupt discontinuation; Guillain Barré Syndrome, immune reconstitution syndrome, MI, obesity, polymyositis

Black Box Warning: Lactic acidosis

DOSAGE AND ROUTES

- **Adult and adolescent ≥16 yr: PO** 300 mg bid or 600 mg/day with other antiretrovirals
- **Adolescent <16 yr and child ≥3 mo: PO** 8 mg/kg bid, max 300 mg bid with other antiretrovirals

Hepatic dose
- **Adult: PO** (Child-Pugh 5-6) (oral sol) 200 mg bid; severe hepatic disease, do not use

HIV prophylaxis (unlabeled)
- **Adult: PO** 300 mg bid to be added to the basic 2-drug regimen ×4 wk

Available forms: Tabs 300 mg; oral sol 20 mg/ml

Administer:
- Give in combination with other antiretrovirals; do not use triple therapy as a beginning treatment, resistance may occur
- May give without regard to food
- Reduce dose in hepatic disease, use oral sol
- Storage in cool environment; protect from light; oral sol stored at room temperature; do not freeze

SIDE EFFECTS

CNS: *Fever, headache, malaise, insomnia,* paresthesia
GI: *Nausea, vomiting, diarrhea, anorexia,* cramps, abdominal pain, increased AST, ALT, hepatotoxicity, hepatomegaly with steatosis
HEMA: Granulocytopenia, anemia, lymphopenia
INTEG: *Rash,* urticaria, hypersensitivity reactions
META: Lactic acidosis
OTHER: Fatal hypersensitivity reactions, MI, fat redistribution, immune reconstitution
RESP: Dyspnea

PHARMACOKINETICS

Rapid/extensive absorption, distributed to extravascular space then erythrocytes; 50% protein binding; extensively metabolized to inactive metabolites; half-life 1½ hr; excreted in urine, feces (unchanged); onset, peak, duration unknown

INTERACTIONS

- Do not coadminister with abacavir-containing products

⚠ **Increase:** possible lactic acidosis—ribavirin
Increase: abacavir levels—alcohol
Decrease: abacavir levels—tipranavir
Decrease: levels of—methadone

Drug/Lab Test
Increase: glucose, triglycerides, GGT, LFTs

NURSING CONSIDERATIONS

Assess:

• Symptoms of HIV and possible infections; increased temp

⚠ **Lactic acidosis** (elevated lactate levels, increased LFTs), severe hepatomegaly with steatosis, discontinue treatment and do not restart; may have large liver, elevated AST, ALT, lactate levels; women are at greater chance of lactic acidosis

Black Box Warning: Fatal hypersensitivity reactions: fever, rash, nausea, vomiting, fatigue, cough, dyspnea, diarrhea, abdominal discomfort; treatment should be discontinued and not restarted; those with HLA B5701 are at great risk for hypersensitivity; obtain genetic testing for HLA B5701 before starting treatment

⚠ **Blood dyscrasias** (anemia, granulocytopenia): bruising, fatigue, bleeding, poor healing

• Renal studies: BUN, serum uric acid, CCr before, during therapy; these may be elevated

Black Box Warning: Hepatic studies before and monthly during therapy: bilirubin, AST, ALT, amylase, alk phos, creatine phosphokinase, creatinine

• Blood counts; monitor viral load and CD4 counts during treatment; watch for decreasing granulocytes, Hgb; if low, therapy may have to be discontinued and restarted after hematologic recovery; blood transfusions may be required; perform hepatitis B virus (HBr) screening to confirm correct treatment

Evaluate:

• Therapeutic response: increased CD4 count, decrease viral load

Teach patient/family:

• That product is not a cure but will control symptoms; patient is still infective, may pass AIDS virus on to others, to carry emergency ID with condition, products taken

• That body fat distribution may occur; not to share product

⚠ To notify prescriber of sore throat, swollen lymph nodes, malaise, fever; other infections may occur; to stop product and to notify prescriber immediately if skin rash, fever, cough, shortness of breath, GI symptoms occur; advise all health care providers that allergic reaction has occurred with abacavir

• That follow-up visits must be continued because serious toxicity may occur; blood counts must be done

• To use contraception during treatment; if patient is pregnant, register with the Antiretroviral Pregnancy Registry at 1-800-258-4263

• Give patient Medication Guide and Warning Card, discuss points on guide

• That other products may be necessary to prevent other infections and that drug is taken with other antiretrovirals

• Not to drink alcohol while taking this product

abatacept (Rx)

(ab-a-ta′sept)

Orencia

Func. class.: Antirheumatic agent (disease modifying); immunomodulator

ACTION: A selective costimulation modulator, inhibits T-lymphocytes, inhibits production of tumor necrosis factor (TNF-α), interferon-γ, interleukin-2, which are involved in immune and inflammatory reactions

USES: Polyarticular juvenile rheumatoid arthritis; moderate to severe rheumatoid arthritis; acute, chronic rheumatoid arthritis that has not responded to other disease-modifying agents; may use in combination with DMARDs; do not use with TNF antagonists (adalimumab, etanercept, infliximab), anakinra

CONTRAINDICATIONS:
Hypersensitivity

Precautions: Pregnancy (C), breastfeeding, children, geriatric patients, recurrent infections, COPD, TB, viral hepatitis, immunosuppression, neoplastic disease, respiratory infection

DOSAGE AND ROUTES

Rheumatoid arthritis

- **Adult: SUBCUT** 125 mg within 1 day after single IV loading dose, then 125 mg weekly; weekly subcut dose may be initiated without an IV loading dose for those unable to receive an infusion
- **Adult >100 kg (220 lb): IV INF** 1 g over 30 min, give at 2, 4 wk after first inf, then q4wk
- **Adult 60-100 kg (132-220 lb): IV INF** 750 mg over 30 min, give at 2, 4 wk after first inf, then q4wk
- **Adult <60 kg (132 lb): IV INF** 500 mg over 30 min, give at 2, 4 wk after first inf, then q4wk

Juvenile rheumatoid arthritis (JRA)/juvenile idiopathic arthritis (JIA)

- **Adolescent and child ≥6 yr and >100 kg: IV INF** 1 g over 30 min q2wk × 3 doses, then 1 g over 30 min q4wk starting at wk 8
- **Adolescent and child ≥6 yr and 75-100 kg: IV INF** 750 mg over 30 min q2wk × 2 doses, then 750 mg over 30 min q4wk starting at wk 8
- **Adolescent and child ≥6 yr and <75 kg: IV INF** 10 mg/kg over 30 min q2wk × 3 doses, then 10 mg/kg q4wk starting at wk 8

Available forms: Lyophilized powder, single-use vials 250 mg; sol for subcut inj 125 mg/ml

Administer:

- Storage in refrigerator; do not use expired vials, protect from light, do not freeze

Intermittent IV INF route

- To reconstitute, remove plastic flip top from vial and wipe the top with alcohol wipe; insert syringe needle into vial and direct stream of sterile water for inj on the wall of vial; rotate vial until mixed; vent with needle to rid foam after reconstitution (25 mg/ml); further dilute in 100 ml NS from a 100-ml inf bag/bottle; withdraw the needed volume (2 vials remove 20 ml; 3 vials remove 30 ml, 4 vials remove 40 ml); slowly add the reconstituted Orencia sol from each vial into the inf bag/bottle using the same disposable syringe supplied; mix gently; discard unused portions of vials; do not use if particulate is present or discolored; give over 30 min; use non–protein-binding filter (0.2-1.2 microns); protect from light
- Do not admix with other sol or medications

SUBCUT route

- Use prefilled syringe for subcut only (do not use for IV); only those trained should use this system; allow syringe to warm to room temp (30-60 min); do not speed up warming process in any way; the amount of liquid should be between the 2 lines on the barrel; do not use the syringe if there is more or less liquid; inject into fronts of thighs, outer area of upper arm, or abdomen except for 2-inch area around the navel; do not inject into tender, bruised area
- Gently pinch skin and hold firmly; insert needle at 45-degree angle; inject full amount in 125-mg syringe
- Rotate injection sites

SIDE EFFECTS

CNS: Headache, asthenia, dizziness

CV: *Hypo/hypertension*

GI: Abdominal pain, dyspepsia, nausea

INTEG: Rash, *inj site reaction,* flushing, urticaria, pruritus

RESP: *Pharyngitis, cough, URI,* non-URI, *rhinitis,* wheezing

SYST: **Anaphylaxis, malignancies, serious infections**

PHARMACOKINETICS

Terminal half-life IV 13 days, subcut 14.3 days, steady state 60 days; subcut half-life 85 days, clearance increases with increased body weight

INTERACTIONS

• Do not give concurrently with vaccines; immunizations should be brought up to date before treatment
• Do not use with TNF antagonists: adalimumab, etanercept, infliximab; anakinra
• Avoid use with corticosteroids, immunosuppressives, atropine, scopolamine, halothane, nitrous oxide

NURSING CONSIDERATIONS

Assess:

• **RA:** pain, stiffness, ROM, swelling of joints during treatment
• For latent/active TB, viral hepatitis before beginning treatment
• For inj site pain, swelling
• Patient's overall health at each visit; product should not be given with active infections; parenteral product contains maltose, glucose monitoring must be done with glucose-specific testing

⚠ **Infection:** sinusitis, urinary tract infection, influenza, bronchitis; serious infections have occurred

Evaluate:

• Therapeutic response: decreased inflammation, pain in joints

Teach patient/family:

• That product must be continued for prescribed time to be effective
• To use caution when driving; dizziness may occur
• Not to have vaccinations while taking this product
• About patient information included in packaging

RARELY USED
⚠ HIGH ALERT

abciximab (Rx)

(ab-six′i-mab)

ReoPro

Func. class.: Platelet aggregation inhibitor

USES: Used with heparin and aspirin to prevent acute cardiac ischemia after percutaneous transluminal coronary angioplasty in patients with NSTEMI or UA when percutaneous coronary intervention (PCI) is planned within 24 hr

Unlabeled uses: Acute MI, Kawasaki disease (child)

CONTRAINDICATIONS: Hypersensitivity to this product or murine protein; GI, GU bleeding; CVA within 2 yr, bleeding disorders, intracranial neoplasm, intracranial arteriovenous malformations, intracranial aneurysm, platelet count <100,000/mm^3, recent surgery, aneurysm, uncontrolled severe hypertension, vasculitis, coagulopathy

DOSAGE AND ROUTES

PCI

• **Adult: IV BOL** 250 mcg (0.25 mg)/kg 10-60 min before PCI followed by 0.125 mcg/kg/min **CONT INF** for 12 hr

MI (unlabeled)

• **Adult: IV BOL** 0.25 mg/kg over 5 min, then 0.125 mcg/kg/min (max 10 mcg/min); **IV INF** for 12 hr unless complications

abiraterone

(a′bir-a′ter-one)

Zytiga

Func. class.: Androgen inhibitor

ACTION: Converted to abiraterone, which inhibits CYP17, the enzyme required for androgen biosynthesis;

androgen-sensitive prostate cancer responds to treatment that decreases androgens

USES:
Metastatic castration-resistant prostate cancer in combination with predniSONE

CONTRAINDICATIONS:
Pregnancy (X), women, children

Precautions: Adrenal insufficiency, cardiac disease, MI, heart failure, hepatic disease, hypertension, hypokalemia, infection, surgery, ventricular dysrhythmia

DOSAGE AND ROUTES

- **Adult males: PO** 1000 mg/day with predniSONE 5 mg bid

Hepatic dose

- **Adult males (Child-Pugh B, 7-9): PO** 250 mg/day with predniSONE; (Child-Pugh C, >10) do not use

Available forms:

Tabs 250 mg

Administer:

PO route

- Give whole, on empty stomach 2 hr before or 1 hr after meals with full glass of water

⚠ Women who are pregnant or may become pregnant should not touch tabs without gloves

- Store tabs at room temperature

SIDE EFFECTS

CV: Angina, dysrhythmia exacerbation, atrial flutter/fibrillation/tachycardia, AV block, chest pain, edema, heart failure, MI, hypertension, QT prolongation, sinus tachycardia, supraventricular tachycardia, ventricular tachycardia

ENDO: Hot flashes

GI: Diarrhea, dyspepsia

GU: Increased urinary frequency, nocturia, urinary tract infection

META: Adrenocortical insufficiency, hyperbilirubinemia, hypertriglyceridemia, hypokalemia, hypophosphatemia

MS: Arthralgia, myalgia

RESP: Cough, upper respiratory infection

SYST: Infection

PHARMACOKINETICS

99% protein binding, converted to abiraterone (active metabolite), mean terminal half-life 7-17 hr; excreted 88% (feces), 5% (urine); high-fat food increases effect, give on empty stomach; increased effect in hepatic disease

INTERACTIONS

- Avoid use with: CYP3A4 inhibitors (clarithromycin, atazanavir, nefazodone, saquinavir, telithromycin, ritonavir, indinavir, nelfinavir, voriconazole, ketoconazole, itraconazole), loperamide, guinine, naprosen, pioglitazone, pitavastin, repaglinide, rosiglitazone
- Avoid use with: CYP3A4 inducers (carBAMazepine, phenytoin, rifampin, rifabutin, rifapentine, PHENobarbital)
- Increase action of CYP2D6 substrate—dextromethorphan, thioridazine; dose of these products should be reduced

Drug/Food

Increase: abiraterone action—must be taken on an empty stomach

NURSING CONSIDERATIONS

Assess:

- **Prostate cancer:** monitor prostate-specific antigen (PSA), serum potassium, serum bilirubin

⚠ **Hepatotoxicity:** monitor liver function tests (AST/ALT) at baseline, every 2 wk for 3 mo, monthly thereafter in patients with no known hepatic disease; interrupt treatment in patients without known hepatic disease at baseline who develop ALT/AST >5 × ULN or total bilirubin >3 × ULN; patients with moderate hepatic disease at baseline, measure ALT, AST, bilirubin before the start of treatment, every wk for 1 mo, every 2 wk for the following 2 mo, monthly thereafter; if elevations in ALT and/or AST >5 × ULN or total bilirubin >3 × ULN occur in patients with moderate hepatic impairment at baseline, discontinue and do NOT restart; measure serum total bilirubin, AST/ALT if hepatotoxicity is suspected; elevations of AST, ALT, bilirubin from baseline should prompt more frequent monitoring

- **Musculoskeletal pain, joint swelling, discomfort:** arthritis, arthralgia, joint swelling, and joint stiffness, some severe; muscle discomfort that includes muscle spasms, musculoskeletal pain, myalgia, musculoskeletal discomfort, and musculoskeletal stiffness may be relieved with analgesics
- Signs, symptoms of adrenocorticoid insufficiency; monthly for hypertension, hypokalemia, fluid retention

⚠ **QT prolongation:** Monitor ECG for QT prolongation, ejection fraction in patients with cardiac disease, small increases in the QTc interval such as <10 ms have occurred; monitor for arrhythmia exacerbation such as sinus tachycardia, atrial fibrillation, supraventricular tachycardia (SVT), atrial tachycardia, ventricular tachycardia, atrial flutter, bradycardia, AV block complete, conduction disorder, bradyarrhythmia

Teach patient/family:

⚠ That women must not come in contact with tabs; wear gloves if product needs to be handled, pregnancy (X)

- To report chest pain, swelling of joints, burning/pain when urinating

acamprosate
(a-kam-pro′sate)

Campral

Func. class.: Alcohol deterrent

Chem. class.: Synthetic amino acid neurotransmitter analog

ACTION: Not completely understood; in vitro data suggest it has affinity for type A and type B GABA receptors, lowers neuronal excitability, centrally mediated

USES: Alcohol abstinence management

CONTRAINDICATIONS:

Black Box Warning: Hypersensitivity to this product or sulfites, creatinine clearance ≤30 ml/min

Precautions: Pregnancy (C), breastfeeding, infants, children, ethanol intoxication, renal impairment, depression, suicidal ideation, driving or operating machinery, geriatric patients

DOSAGE AND ROUTES

- **Adult: PO** 666 mg tid

Renal dosage

- **Adult: PO** CCr 30-50 ml/min 333 mg tid; CCr <30 ml/min do not use

Available forms: Del-rel tabs 333 mg

Administer

- Without regard to food; do not crush, chew, break del-rel tab
- Use only after alcohol is stopped
- Store at room temperature

SIDE EFFECTS

CNS: Anxiety, depression, dizziness, headache, insomnia, paresthesias, suicidal ideation, tremors, abnormal thoughts, chills, drowsiness

CV: Palpitations, hypertension, peripheral edema

EENT: Rhinitis, pharyngitis, abnormal vision

GI: Anorexia, constipation, diarrhea, dry mouth, abdominal pain, flatulence, nausea, vomiting, taste change, weight gain

GU: Impotence

INTEG: Rash, pruritus, increased sweating

MISC: Infection, flulike symptoms

MS: Back pain, myalgias, arthralgia

RESP: Dyspnea, bronchitis

PHARMACOKINETICS

Peak 3-8 hr, half-life 20-33 hr

INTERACTIONS

Increase: LFTs, blood glucose, bilirubin, uric acid

Decrease: Hgb/Hct, platelets

NURSING CONSIDERATIONS

Assess:

- Mental status: depression, abnormal thinking, suicidal thoughts/behaviors, length of alcohol use, date of discontinuing alcohol use
- B/P baseline and periodically

Evaluate:

- Therapeutic response: continued alcohol abstinence

Teach patient/family:

- To notify prescriber of depression, abnormal thoughts, suicidal thoughts/behaviors
- To take without regard to food; not to break, crush, chew del-rel tabs
- To notify prescriber if pregnancy is planned or suspected; to use effective contraception
- Not to engage in hazardous activities until effect is known

acarbose (Rx)

(ay-car′bose)

Precose

Func. class.: Oral antidiabetic

Chem. class.: α-Glucosidase inhibitor

Do not confuse:
Precose/preCare

ACTION: Delays digestion/absorption of ingested carbohydrates by inhibiting α-glucosidase, results in smaller rise in postprandial blood glucose after meals; does not increase insulin production

USES: Type 2 diabetes mellitus, alone or in combination with a sulfonylurea, metformin, insulin

Unlabeled uses: Adjunct in type 1 diabetes mellitus

CONTRAINDICATIONS: Breastfeeding, hypersensitivity, diabetic ketoacidosis, cirrhosis, inflammatory bowel disease, ileus, colonic ulceration, partial intestinal obstruction, chronic intestinal disease, serum creatinine >2 mg/dl, CCr <25 ml/min

Precautions: Pregnancy (B), children, renal/hepatic disease

DOSAGE AND ROUTES

- **Adult >60 kg (132 lb): PO** 25 mg tid initially, with 1st bite of meal; maintenance dose may be increased to 50-100 mg tid; dosage adjustment at 4-8 wk intervals, individualized
- **Adult <60 kg (132 lb): PO** max 50 mg tid

Type 1 diabetes mellitus (unlabeled)

- **Adult: PO** 50 mg tid with meals × 2 wk, then 100 mg tid with meals

Available forms: Tabs 25, 50, 100 mg

Administer:

PO route

- With 1st bite of each meal 3 ×/day
- Store in a tight container, cool environment

SIDE EFFECTS

GI: *Abdominal pain, diarrhea, flatulence*

PHARMACOKINETICS

Poor systemic absorption, peak 1 hr, duration 2-4 hr, metabolized in GI tract, excreted as intact product in urine, half-life 2 hr

INTERACTIONS

- Do not use with gatifloxacin

Increase: acetaminophen toxicity—acetaminophen combined with alcohol

Increase or decrease: glycemic control—androgens, lithium, bortezomib, quinolones

Decrease: effect of digoxin

Increase: hypoglycemia—sulfonylureas, insulin, MAOIs, salicylates, fibric acid derivatives, bile acid sequestrants, ACE inhibitors, angiotensin II receptor antagonists, beta blockers

Decrease: effect, increase hyperglycemia—digestive enzymes, intestinal absorbents, thiazide diuretics, loop diuretics, corticosteroids, estrogen, progestins, oral contraceptives, sympathomimetics, isoniazid, phenothiazines; protease inhibitors, atypical antipsychotics, carbonic anhydrase inhibitors, cycloSPORINE, tacrolimus, baclofen

Drug/Herb

Increase: hypoglycemia—chromium, garlic, horse chestnut

Drug/Lab Test

Increase: ALT, AST

Decrease: calcium, vit B_6, Hgb, Hct

NURSING CONSIDERATIONS

Assess:

- **Hypoglycemia** (weakness, hunger, dizziness, tremors, anxiety, tachycardia, sweating), hyperglycemia; even though product does not cause hypoglycemia, if patient is on sulfonylureas or insulin, hypoglycemia may be additive; if hypoglycemia occurs, treat with dextrose, or, if severe, with IV glucose or glucagon
- For stress, surgery, or other trauma that may require change in dose
- Monitor AST, ALT q3mo × 1 yr and periodically thereafter; if elevated, dose may need to be reduced or discontinued, usually increased with doses ≥300 mg/day; A1c q3mo, monitor serum glucose, 1 hr PP throughout treatment
- GI side effects for tolerability/compliance

Evaluate:

- Therapeutic response: improved signs/symptoms of diabetes mellitus (decreased polyuria, polydipsia, polyphagia; clear sensorium, absence of dizziness, stable gait)

Teach patient/family:

- The symptoms of hypo/hyperglycemia; what to do about each
- That medication must be taken as prescribed; explain consequences of discontinuing medication abruptly; that insulin may need to be used for stress, including trauma, surgery, fever
- To avoid OTC medications and herbal supplements unless approved by health care provider
- That diabetes is a lifelong illness; that the diet and exercise regimen must be followed; that this product is not a cure
- To carry emergency ID and a glucose source; to avoid sugar, because sugar is blocked by acarbose
- That blood glucose monitoring is required to assess product effect
- To avoid breastfeeding if using acarbose with other antidiabetics
- That GI side effects may occur

acetaminophen (OTC)

(a-seat-a-mee′noe-fen)

Acephen, Apacet, APAP, Apra, Children's Feverall, Equaline Children's Pain Relief, Equaline Infant's Pain Relief, Genapap, GoodSense Acetaminophen, GoodSense Children's Pain Relief, Infantaire, Leader Children's Pain Reliever, Mapap, Neopap, Ofirmev, Q-Pap, Q-Pap Children's, Redutemp, Ridenol, Silapap, Tempra, T-Painol, Tylenol, Walgreen's Non-Aspirin, XS pain reliever

Func. class.: Nonopioid analgesic, antipyretic

Chem. class.: Nonsalicylate, paraaminophenol derivative

ACTION: May block pain impulses peripherally that occur in response to inhibition of prostaglandin synthesis; does not possess antiinflammatory properties; antipyretic action results from inhibition of prostaglandins in the CNS (hypothalamic heat-regulating center)

USES: Mild to moderate pain or fever, arthralgia, dental pain, dysmenorrhea, headache, myalgia, osteoarthritis

Unlabeled uses: Migraine

CONTRAINDICATIONS: Hypersensitivity

Precautions: Pregnancy (B), (C) IV; breastfeeding, geriatric patients, anemia, renal/hepatic disease, chronic alcoholism

DOSAGE AND ROUTES

• **Adult/child >12 yr: PO/RECT** 325-650 mg q4-6hr prn, max 4 g/day; weight ≥50 kg IV 1000 mg q6hr or 650 mg q4hr prn, max single dose 1000 mg, min dosing interval 4 hr; weight <50 kg IV 15 mg/kg/dose q6hr or 12.5 mg/kg/dose q4hr, max single dose 15 mg/kg, min dosing interval 4 hr, max 75 mg/kg/day from all sources, ext rel 650-1300 mg q8hr as needed, max 4 g/day

• **Child ≥2 yr and <50 kg: IV** 15 mg/kg/dose q6hr or 12.5 mg/kg/dose q4hr, max single dose 15 mg/kg, min dosing interval 4 hr, max 75 mg/kg/day from all sources

• **Child 1-12 yr: PO** 10-15 mg/kg q4-6hr, max 5 doses/24 hr

• **Child 1-12 yr: RECT** 10-20 mg/kg/dose q4-6hr

• **Neonate: RECT** 10-15 mg/kg/dose q6-8hr

Migraine (unlabeled)

• **Adult and adolescent: PO/RECT** 500-1000 mg, max 1 g/dose or max 4 g/day

Available forms: Rect supp 120, 325, 650 mg; soft chew tabs 80 mg; caps 500 mg; elix 120, 160, 325 mg/5 ml; oral disintegrating tab 80, 160 ml; oral drops 80 mg/0.8 ml, liquid 500 mg/5 ml, 160/5 ml, 1000/30 ml; ext rel 650 mg, 80 mg/ml; tabs 325, 500, 650 mg; sol for inj 1000 mg/100 ml

Administer:

PO route

• Crushed or whole, do not crush EXT REL product; chewable tabs may be chewed; give with full glass of water

• With food or milk to decrease gastric symptoms if needed

• Susp after shaken well; check elixir, liquid, suspension concentration carefully; susp and cups are bioequivalent

Rectal route

• Store suppositories <80° F (27° C)

Intermittent IV INF route

• No further dilution needed; do not add other medications to vial or inf device

• For doses equal to single vial, averted IV set may be used to deliver directly from vial; for doses less than a single vial, withdraw dose and place in an empty sterile syringe, plastic IV container, or glass bottle; infuse over 15 min

• Discard unused portion; if seal is broken, vial penetrated, or drug transferred to another container, give within 6 hr

SIDE EFFECTS

GI: Nausea, vomiting, abdominal pain; **hepatotoxicity, hepatic seizure (overdose), GI bleeding**

GU: Renal failure (high, prolonged doses)

HEMA: Leukopenia, neutropenia, hemolytic anemia (long-term use), thrombocytopenia, pancytopenia

INTEG: Rash, urticaria

SYST: Hypersensitivity

TOXICITY: Cyanosis, anemia, neutropenia, jaundice, pancytopenia, CNS stimulation, delirium followed by vascular collapse, seizures, coma, death

PHARMACOKINETICS

85%-90% metabolized by liver, excreted by kidneys; metabolites may be toxic if overdose occurs; widely distributed; crosses placenta in low concentrations; excreted in breast milk; half-life 1-4 hr

PO: Onset 10-30 min, peak ½-2 hr, duration 4-6 hr, well absorbed

RECT: Onset slow, peak 1-2 hr, duration 4-6 hr, absorption varies

INTERACTIONS

• Avoid use with salicylates

Increase: renal adverse reactions—NSAIDs, salicylates

Increase: hypoprothrombinemia—warfarin, long-term use, high doses of acetaminophen

Increase: hepatotoxicity—barbiturates, alcohol, carBAMazepine, hydantoins, ri-

fampin, rifabutin, isoniazid, diflunisal, zidovudine, lamoTRIgine, imatinib
Decrease: absorption—colestipol, cholestyramine
Drug/Herb
Increase: heptatotoxicity—St. John's wort, due to acetaminophen metabolism
Drug/Lab Test
Interference: 5-HIAA
Increase: LFTs, potassium, bilirubin, LDH, pro-time
Decrease: Hgb/Hct, WBC, RBC, platelets; albumin, magnesium, phosphate (pediatrics)

NURSING CONSIDERATIONS

Assess:
- **Hepatic studies:** AST, ALT, bilirubin, creatinine before therapy if long-term therapy is anticipated; may cause hepatic toxicity at doses >4 g/day with chronic use
- **Renal studies:** BUN, urine creatinine, occult blood, albumin, if patient is on long-term therapy; presence of blood or albumin indicates nephritis
- **Blood studies:** CBC, PT if patient is on long-term therapy
- I&O ratio; decreasing output may indicate renal failure (long-term therapy)
- **For fever and pain:** type of pain, location, intensity, duration
- **Chronic poisoning:** rapid, weak pulse; dyspnea; cold, clammy extremities; report immediately to prescriber
- **Hepatotoxicity:** dark urine; clay-colored stools; yellowing of skin, sclera; itching; abdominal pain; fever; diarrhea if patient is on long-term therapy
- **Allergic reactions:** rash, urticaria; if these occur, product may have to be discontinued

Evaluate:
- Therapeutic response: absence of pain using pain scoring; fever

Teach patient/family:
- ⚠ Not to exceed recommended dosage; the elixir, liquid, suspension come in several concentrations, read label carefully; acute poisoning with liver damage may result; tell parents of children to check products carefully; that acute toxicity includes symptoms of nausea, vomiting, abdominal pain and that prescriber should be notified immediately; that toxicity may occur when used with other combination products
- Not to use with alcohol, herbals, OTC products without approval of prescriber
- May be used when breastfeeding, short-term
- To recognize signs of chronic overdose: bleeding, bruising, malaise, fever, sore throat
- That those with diabetes may notice blood glucose monitoring changes
- To notify prescriber of pain or fever lasting more than 3 days, not to be used in patients <2 yr unless approved by prescriber

TREATMENT OF OVERDOSE:

Product level, gastric lavage, activated charcoal; administer oral acetylcysteine to prevent hepatic damage *(see acetylcysteine monograph)*; monitor for bleeding

acetaZOLAMIDE (Rx)

(a-set-a-zole′a-mide)

Diamox Sequels

Func. class.: Diuretic, carbonic anhydrase inhibitor, antiglaucoma agent, antiepileptic
Chem. class.: Sulfonamide derivative

Do not confuse:
acetaZOLAMIDE/acetoHEXAMIDE
Diamox/Trimox/Dobutrex

ACTION: Inhibits carbonic anhydrase activity in proximal renal tubules to decrease reabsorption of water, sodium, potassium, bicarbonate resulting in increased urine volume and alkalinization of urine; decreases carbonic anhydrase in CNS, increasing seizure threshold; able to decrease secretion of aqueous humor in eye, which lowers intraocular pressure

USES: Open-angle glaucoma, angle-closure glaucoma (preoperatively, if surgery delayed), seizures (petit mal, grand mal, mixed, absence), edema in CHF, product-induced edema, acute altitude sickness

Unlabeled uses: Urine alkalinization, metabolic alkalosis in mechanical ventilation, decrease CSF production in infants with hydrocephalus, familial periodic paralysis, nystagmus

CONTRAINDICATIONS: Hypersensitivity to sulfonamides, severe renal/hepatic disease, electrolyte imbalances (hyponatremia, hypokalemia), hyperchloremic acidosis, Addison's disease, long-term use for closed-angle glaucoma, adrenocortical insufficiency, metabolic acidosis, acidemia, anuria

Precautions: Pregnancy (C), breastfeeding, hypercalciuria, respiratory acidosis, pulmonary obstruction/emphysema, COPD

DOSAGE AND ROUTES

Angle-closure glaucoma

- **Adult: PO/IV** 250 mg q4hr or 250 mg bid for short-term therapy

Open-angle glaucoma

- **Adult: PO/IV** 250 mg/day in divided doses for amounts of more than 250 mg or 500 mg **EXT REL** bid, max 1 g/day

Edema in CHF

- **Adult: PO/IV** 250-375 mg/day or 5 mg/kg in AM, give for 2 days, then 1-2 days drug free
- **Child: PO/IV** 5 mg/kg/day or 150 mg/m^2 in AM

Seizures

- **Adult: PO/IV** 8-30 mg/kg/day, in 1-4 divided doses, usual range 375-1000 mg/day; **ER** not recommended with seizures
- **Child: PO/IV** 8-30 mg/kg/day in divided doses tid or qid, or 300-900 mg/m^2/day, not to exceed 1 g/day

Altitude sickness

- **Adult: PO** 250 mg q6-12hr; **EXT REL** 500 mg q12-24hr, start therapy 24-48 hr before ascent and give for ≥48 hr after arrival at high altitude
- **Geriatric: PO** 250 mg bid, use lowest effective dose

Renal dose

- **Adult: PO/IV** CCr 50-80 ml/min give dose ≥6 hr reg rel of IV; CCr 10-50 ml/min give dose q12hr; CCr <10 ml/min, avoid use

Infants with hydrocephalus (unlabeled)

- **Infant: PO/IV** 5 mg/kg q6hr, may increase by 25 mg/kg/day; max 100 mg/kg/day

Urine alkalinization (unlabeled)

- **Adult: IV** 5 mg/kg/dose, repeat 2-3× over 24 hr

Familial periodic paralysis (unlabeled)

- **Adult: PO** 250-375 mg/day in divided doses

Metabolic alkalosis in mechanical ventilation (unlabeled)

- **Adult: IV** 500 mg as a single dose or 250 mg q6hr × 4 doses

Vestibular nystagmus (unlabeled)

- **Adult: PO** 250 mg, increase by 250 mg q3days; max 3 g/day in divided doses

Available forms: Tabs 125, 250 mg; ext rel caps 500 mg; inj 500 mg

Administer:

- In AM to avoid interference with sleep if using product as diuretic
- Potassium replacement if potassium level is <3 mg/dl

PO route

- Do not break, crush, or chew ext rel caps; this product should be used for altitude sickness, glaucoma
- With food if nausea occurs; absorption may be decreased slightly

IV route

- After diluting 500 mg in ≥5 ml sterile water for inj; **direct IV:** give at 100-500 mg/min; **intermittent INF:** may be diluted further in LR, D_5W, $D_{10}W$, 0.45% NaCl, 0.9% NaCl, or Ringer's sol and infused over 15-30 min; use within 24 hr of dilution
- Store in cool, dark area; use reconstituted solution within 24 hr

Additive compatibilities: cimetidine, ranitidine

SIDE EFFECTS

CNS: Anxiety, *confusion,* seizures, *depression,* dizziness, *drowsiness, fatigue,* headache, *paresthesia,* stimulation

EENT: Myopia, tinnitus

ENDO: *Hyper/hypoglycemia*

GI: *Nausea, vomiting, anorexia, diarrhea,* melena, *weight loss,* hepatic insufficiency, cholestatic jaundice, fulminant hepatic necrosis, *taste alterations,* bleeding

GU: *Frequency, polyuria,* uremia, glucosuria, hematuria, dysuria, crystalluria, renal calculi

HEMA: Aplastic anemia, hemolytic anemia, leukopenia, thrombocytopenia, purpura, pancytopenia

INTEG: *Rash,* pruritus, urticaria, fever, Stevens-Johnson syndrome, photosensitivity, flushing, toxic epidermal necrolysis

META: *Hypokalemia, hyperchloremic acidosis,* hyponatremia, sulfonamide-like reactions, metabolic acidosis, growth inhibition in children, hyperuricemia, hypercalcemia

PHARMACOKINETICS

65% absorbed if fasting (oral), 75% absorbed if given with food; half-life $2^1/_2$-$5^1/_2$ hr; excreted unchanged by kidneys (80% within 24 hr), crosses placenta

PO: Onset 1-$1^1/_2$ hr, peak 2-4 hr, duration 8-12 hr

PO-EXT REL: Onset 2 hr, peak 8-12 hr, duration 18-24 hr

IV: Onset 2 min, peak 15 min, duration 4-5 hr

INTERACTIONS

Increase: action of—amphetamines, flecainide, memantine, phenytoin, procainamide, quiNIDine, anticholinergics, methenamine, mecamylamine, ePHEDrine, memantine, mexiletene, folic acid antagonists

Increase: excretion of lithium, primidone

Increase: osteomalacia—carBAMazepine, ethotoin

Increase: toxicity—salicylates, cycloSPORINE

Increase: hypokalemia—corticosteroids, amphotericin B, corticotropin, ACTH

Increase: cardiac toxicity if hypokalemia develops—arsenic trioxide, cardiac glycosides, levomethadyl

Increase: renal stone formation, heat stroke—topiramate (avoid concurrent use)

Decrease: primidone levels

Drug/Lab Test

Increase: glucose, bilirubin, calcium, uric acid

Decrease: thyroid iodine uptake

False positive: urinary protein, 17 hydroxysteroid

NURSING CONSIDERATIONS

Assess:

- Edema: weight daily, I&O daily to determine fluid loss; effect of product may be decreased if used daily; monitor geriatric patients for dehydration
- Ocular status: intraocular pressure, ophthalmologic examination
- For cross-sensitivity with other sulfonamides and this product
- B/P lying, standing; postural hypotension may occur
- Electrolytes: potassium, sodium, chloride; also BUN, blood glucose, CBC, serum creatinine, blood pH, ABGs, LFTs; I&O, platelet count, patient may need to be on a high-potassium diet; identify signs of hypokalemia (vomiting, fatigue, weakness)
- **Seizures:** neurologic status, provide seizure precaution

Evaluate:

- Therapeutic response: improvement in edema of feet, legs, sacral area daily if medication is being used for CHF; decrease in aqueous humor if medication is being used for glaucoma; decreased frequency of seizures, prevention of altitude sickness

A

Teach patient/family:

- To take exactly as prescribed; if dose is missed, take as soon as remembered; do not double dose
- Altitude sickness: to avoid rapid ascent
- Diabetic: that drug may alter blood glucose and to monitor blood glucose
- To use sunscreen to prevent photosensitivity
- To avoid prolonged sun exposure
- To avoid hazardous activities if drowsiness occurs
- To increase fluids to 2-3 L/day if not contraindicated

⚠ To report nausea, vertigo, rapid weight gain, change in stools, weakness, numbness, rash, sore throat, bleeding/bruising; Stevens-Johnson syndrome, toxic epidermal necrolysis (blistering, red rash that spreads)

TREATMENT OF OVERDOSE:

Lavage if taken orally; monitor electrolytes; administer dextrose in saline; monitor hydration, CV, renal status

acetylcholine ophthalmic

See Appendix B

acetylcysteine (Rx)

(a-se-teel-sis′tay-een)

Acetadote, Mucomyst ✱

Func. class.: Mucolytic; antidote—acetaminophen

Chem. class.: Amino acid L-cysteine

Do not confuse:

acetylcysteine/acetylcholine

Mucomyst/Mucinex

ACTION: Decreases viscosity of secretions by breaking disulfide links of mucoproteins; serves as a substrate in place of glutathione, which is necessary to inactivate toxic metabolites with acetaminophen overdose

USES: Acetaminophen toxicity; bronchitis; cystic fibrosis; COPD; atelectasis

Unlabeled uses: Prevention of contrast medium nephrotoxicity

CONTRAINDICATIONS: Hypersensitivity, increased intracranial pressure, status asthmaticus

Precautions: Pregnancy (B), breastfeeding, hypothyroidism, Addison's disease, CNS depression, brain tumor, asthma, renal/hepatic disease, COPD, psychosis, alcoholism, seizure disorders, bronchospasms, anaphylactoid reactions, fluid restriction, weight <40 kg

DOSAGE AND ROUTES

Acetaminophen toxicity

- **Adult and child: PO** 140 mg/kg, then 70 mg/kg q4hr × 17 doses to total of 1330 mg/kg; **IV** loading dose 150 mg/kg over 60 min (dilution 150 mg/kg in 200 ml of D_5); maintenance dose 1: 50 mg/kg over 4 hr (dilution 50 mg/kg in 500 ml D_5); maintenance dose 2: 100 mg/kg over 16 hr (dilution 100 mg/kg in 1000 ml D_5)

Mucolytic

- **Adult and child: INSTILL** 1-20 ml (10%-20% sol) q6-8hr prn or 3-5 ml (20% sol) or 6-10 ml (10% sol) tid or qid; nebulization (face mask, mouthpiece, tracheostomy) 1-10 ml of a 20% sol, or 2-20 ml of a 10% sol, q6-8hr; nebulization (tent, croupette) may require large dose, up to 300 ml/treatment

Nephrotoxicity prophylaxis (unlabeled)

- **Adult: PO** 600 mg bid, given day before and day of administration of contrast media or **IV** 150 mg/kg in 500 ml NS over 30 min before contrast, then 50 mg/kg in 500 ml NS over the next 4 hr; **IV BOL** 1200 mg before contrast medium and 1200 mg **PO** bid for 48 hr (MI undergoing angioplasty)

Available forms: Oral sol 10%, 20%; inj 20% (200 mg/ml)

Administer:

PO route

- **Antidotal use:** give within 24 hr; dilute 10% or 20% sol to a 5% sol with diet soda, may use water if giving via gastric tube; dilution of 10% sol 1:1, 20% sol 1:3, store open undiluted solution refrigerated ≤96 hr

Direct intratracheal instill route

- By syringe: 1-2 ml of 10%-20% sol up to q1hr
- Decreased dose to geriatric patients; metabolism may be slowed
- Only if suction machine is available
- Before meals ½-1 hr for better absorption, to decrease nausea
- 20% sol diluted with NS or water for inj; may give 10% sol undiluted
- Only after patient clears airway by deep breathing, coughing
- Assistance with inhaled dose: bronchodilator if bronchospasm occurs; mechanical suction if cough insufficient to remove excess bronchial secretions

IV route

- **21-hr regimen:** loading dose: dilute 150 mg/kg in 200 ml D_5W; maintenance dose 1: dilute 50 mg/kg in 500 ml D_5W; maintenance dose 2: dilute 100 mg/kg in 1000 ml D_5W; give loading dose over 15 min; give maintenance dose 1 over 4 hr; give maintenance dose 2 over 16 hr, administer sequentially without time between doses
- Store in refrigerator; use within 96 hr of opening

SIDE EFFECTS

CNS: *Dizziness, drowsiness,* headache, fever, chills

CV: Hypotension, flushing tachycardia

EENT: *Rhinorrhea,* tooth damage

GI: *Nausea,* stomatitis, constipation, vomiting, anorexia, hepatotoxicity, diarrhea

INTEG: Urticaria, rash, fever, clamminess, pruritus

RESP: Bronchospasm, burning, hemoptysis, chest tightness, cough, dyspnea

MISC: Anaphylaxis, angioedema

PHARMACOKINETICS

PO: (antidote), peak 1-2 hr, duration 4 hr

INH/INSTILL: Onset 5-10 min, peak 10 min, duration 1 hr

Excreted in urine, half-life 5.6 hr (adult), 11 hr (newborn)

IV: Protein binding 83%

INTERACTIONS

- Do not use with iron, copper, rubber, nickel, activated charcoal

Increase: effect—nitrates

NURSING CONSIDERATIONS

Assess:

- **Mucolytic use:** cough—type, frequency, character, including sputum
- Rate, rhythm of respirations, increased dyspnea; sputum; discontinue if bronchospasm occurs
- VS, cardiac status including checking for dysrhythmias, increased rate, palpitations
- ABGs for increased CO_2 retention in asthma patients
- **Antidotal use:** LFTs, PT, BUN, creatinine, glucose, electrolytes, acetaminophen levels; inform prescriber if dose is vomited or vomiting is persistent
- Nausea, vomiting, rash; notify prescriber if these occur

Evaluate:

- Therapeutic response: absence of purulent secretions when coughing, clear lung sounds (mucolytic use); absence of hepatic damage with acetaminophen toxicity

Teach patient/family:

- That foul odor and smell may be unpleasant
- To clear airway for inhalation
- That discoloration of sol after bottle is opened does not impair its effectiveness
- To report vomiting because dose may need to be repeated

aclidinium
(a'kli-din'ee-um)

Tudorza Pressair

Func. class.: Anticholinergic, bronchodilator

Chem. class.: Synthetic quaternary ammonium compound

ACTION: Inhibits interaction of acetylcholine at receptor sites on the bronchial smooth muscle, thereby resulting in decreased cGMP and bronchodilation

USES: Long-term maintenance treatment of bronchospasm in COPD, emphysema, chronic bronchitis, not indicated for initial treatment of acute episodes

CONTRAINDICATIONS: Hypersensitivity

Precautions: Pregnancy (C), breastfeeding, milk sensitivity, contact lenses, narrow-angle glaucoma, neonates, ocular exposure, prostatic hypertrophy, bladder obstruction

DOSAGE AND ROUTES

• **Adults, including geriatric patients:** Oral inhalation 400 μg (1 actuation) bid; doses should be 12 hours apart

Available forms: Powder for inhalation 400 μg/actuation

Administer

Oral inhalation route

• Before initial use, remove inhaler from pouch; remove the cap by squeezing the arrows marked on each side and pulling outward; instruct the patient to hold the inhaler with the mouthpiece facing patient but not inside the mouth; the green button should be facing straight up

• Before placing the inhaler into the mouth, the green button should be pushed all the way down and then released; patient should not continue to hold it down; once the green button is pressed, the control window changes from red to green, indicating that the medication is ready for inhalation; if the control window remains red, repeat the press-and-release actions until the control window is green

• Before inhaling the dose, have patient breathe out completely away from the inhaler; patient should never breathe out into the inhaler

• Instruct patient to put their lips tightly around the mouthpiece and to breathe in quickly and deeply through the mouth until a clicking sound is heard

• The patient should remove the inhaler from the mouth and hold breath for as long as is comfortable, then breathe out slowly through the nose

• The control window should turn red after the full dose has been inhaled. If it remains green after the dose is inhaled, the inhalation process should be repeated. If correct inhalation has not been achieved after several attempts, the doctor or health care professional should be contacted

• Once the window has turned red, the protective cap should be placed back onto the inhaler by pressing it back onto the mouthpiece

• The dose indicator displays how many doses are left. The first time the inhaler is used, the indicator displays the number 60. The indicator number counts down as the patient uses the inhaler

• Discard inhaler when the marking "0" with red background shows in the middle of the dose indicator, the device locks out after 45 days (whichever comes first); the inhaler does not need to be cleaned

SIDE EFFECTS

CNS: *Anxiety, dizziness, headache,* nervousness

CV: Palpitation

EENT: Dry mouth, blurred vision, nasopharyngitis congestion

GI: *Nausea, vomiting, diarrhea*

INTEG: Rash

RESP: *Cough, worsening of symptoms,* **bronchospasm**

PHARMACOKINETICS

Half-life 5-8 hr

INTERACTIONS

Increase: toxicity—other bronchodilators

NURSING CONSIDERATIONS

Assess:
- Tolerance over long-term therapy; dose might have to be increased or changed
- Respiratory status: rate, rhythm, auscultate breath sounds before and after administration; pulmonary function tests at baseline and periodically

Evaluate:
- Therapeutic response: ability to breathe adequately

Teach patient/family:
- That compliance is necessary with number of inhalations/24 hr or overdose can occur; about spacer device for geriatric patients; that max therapeutic effects can take 2-3 mo
- About the correct method of inhalation
- To report any visual effects or urinary retention

activated charcoal (OTC)

Actidose-Aqua, Actidose with Sorbitol, Charcoal Plus, Charcoal Plus DS, Charcocaps, EZ Char

Func. class.: Antiflatulent; antidote

Do not confuse:
Actidose/Actos

ACTION: Binds poisons, toxins, irritants; increases adsorption in GI tract; inactivates toxins and binds until excreted

USES: Poisoning, overdose
Unlabeled uses: Diarrhea, flatulence

CONTRAINDICATIONS: Hypersensitivity to this product, unconsciousness, semiconsciousness, cyanide poisoning, mineral acids, alkalis, gag reflex depression, ethanol intoxication, intestinal obstruction, absent bowel sounds
Precautions: Pregnancy (C), hypersensitivity to quiNIDine, quiNINE

DOSAGE AND ROUTES

Children should not get more than 1 dose of products that contain sorbitol

Poisoning
- Tabs/caps should not be used for poisonings
- **Adult/adolescents: PO** (activated charcoal aqueous susp) 5-10 × estimated weight of drug/chemical ingested or 50-100 g dose, may repeat q4-6hr as needed; (activated charcoal with sorbitol susp) 50 g as a single dose, do not use multiple dosing
- **Child: PO** (activated charcoal aqueous susp) 1-2 g/kg/dose or 25-50 g dose, may repeat as needed q4-6hr
- **Infant: PO** (activated charcoal aqueous susp) 1 g/kg/dose, may repeat as needed q4-6hr

Diarrhea/flatulance (unlabeled)
- **Adult: PO** (CharcoCaps) 520 mg (2 caps) after meals or prn, max 4.16 g (16 cap)/day

Available forms: Powder 15, 25 ✱, 30, 40, 120, 240 g/container; oral susp 12.5 g/60 ml, 15 g/72 ml, 15 g/120 ml, 25 g/120 ml, 30 g/120 ml, 50 g/240 ml; 15 g/120 ml ✱, 25 g/125 ml, 50 g/225 ml, 50 g/250 ml

Administer:
- Store in tightly closed container to prevent absorption of gases
- Use stool softener or laxative to lessen constipation unless sorbitol has been given

PO route
- Give orally to those with intact gag reflexes, protected airways
- After inducing vomiting unless vomiting contraindicated (i.e., cyanide or alkalis)
- After mixing with water or fruit juice to form thick syrup; do not use dairy products, chocolate syrup to mix charcoal

- Repeat dose if vomiting occurs soon after dose; give with a laxative to promote elimination
- After spacing at least 2 hr before or after other products or absorption will be decreased
- Do not use tabs, caps to treat overdose

NG route

- Through nasogastric tube if patient unable to swallow

SIDE EFFECTS

GI: *Nausea, black stools,* vomiting, constipation, diarrhea, abdominal pain
OTHER: **Pulmonary aspiration**

PHARMACOKINETICS

PO: Excreted in feces, not absorbed, excreted unchanged in feces

INTERACTIONS

- Inactivation of acetylcysteine

Decrease: effects of acarbose, carBAMazepine, digoxin, ipecac, phenytoin, acetaminophen, barbiturates, furosemide, methotrexate, phenothiazine, salicylates, tricyclics; use for overdose

NURSING CONSIDERATIONS

Assess:

- Respiration, pulse, B/P to determine charcoal effectiveness if taken for barbiturate/opiate poisoning; intact gag reflex, serum
- Not used for all types of overdose

Evaluate:

- Therapeutic response: LOC alert (poisoning)

Teach patient/family:

- That stools will be black
- How to prevent further poisonings
- To drink 8 glasses of water to prevent constipation

acyclovir (Rx)

(ay-sye′kloe-veer)

Zovirax

Func. class.: Antiviral
Chem. class.: Purine nucleoside analog

Do not confuse:
Zovirax/Zyvox/Valtrex/Zostrix

ACTION: Interferes with DNA synthesis by conversion to acyclovir triphosphate, thereby causing decreased viral replication

USES: Mucocutaneous herpes simplex virus, herpes genitalis (HSV-1, HSV-2), varicella infections, herpes zoster, herpes simplex encephalitis
Unlabeled uses: Bell's palsy, prevention of CMV, Epstein-Barr virus, esophagitis, hairy leukoplakia, prevention of herpes labialis, herpes simplex, herpes simplex ocular prophylaxis, keratoconjunctivitis, pharyngitis, pneumonitis, prevention of postherpetic neuralgia, proctitis, stomatitis, tracheobronchitis, varicella prophylaxis

CONTRAINDICATIONS: Hypersensitivity to this product, valacyclovir
Precautions: Pregnancy (B), breastfeeding, renal/hepatic/neurologic disease, electrolyte imbalance, dehydration, hypersensitivity to famciclovir, ganciclovir, penciclovir, valganciclovir

DOSAGE AND ROUTES

Herpes simplex (recurrent)

- **Adult: PO** 400 mg 3 ×/day for 5 days or 200 mg 5 ×/day × 5 days
- **Adult and child >12 yr: IV INF** 5 mg/kg over 1 hr q8hr × 7 days, use ideal body weight for patients with obesity
- **Infant >3 mo/child <12 yr: IV INF** 10 mg/kg q8hr × 7 day; if HIV infected 5-10 mg/kg q8hr (moderate to severe)
- **Neonate: IV INF** 10 mg/kg q8hr × 10 days, may use higher dose

Genital herpes
• **Adult: PO** 200 mg q4hr (5 ×/day while awake) for 5 days to 6 mo depending on whether initial, recurrent, or chronic; **IV** 5 mg/kg q8hr × 5 days

Genital herpes, initial limited, mucocutaneous HSV in immunocompromised patients, non-life-threatening
• **Adult/child ≥12 yr: TOP** cover lesions q3hr 6 times/day

Herpes simplex encephalitis
• **Adult: IV** 10 mg/kg over 1 hr q8hr × 10 days
• **Child 3 mo-12 yr: IV** 20 mg/kg q8hr × 10 days
• **Child birth-3 mo: IV** 10 mg/kg q8hr × 10 days

Herpes labialis, recurrent
• **Adult/child ≥12 yr:** top apply cream 5 ×/day for 4 days, start as soon as symptoms appear

Herpes zoster
• **Adult: PO** 800 mg q4hr while awake × 7-10 days; **IV** 10 mg/kg q8hr × 7 days

Herpes zoster (shingles) immunocompromised patients
• **Adult/adolescent: PO** 800 mg q4hr 5×/day for 7-10 days; **IV** 10 mg/kg q8hr × 7 days
• **Child ≥12 yr: IV** 10 mg/kg/dose q8hr × 7 days
• **Infant/child <12 yr: IV** 20 mg/kg/dose q8hr × 7-10 days

Herpes zoster (shingles) immunocompetent
• **Adult: PO** 800 mg q4hr 5×/day × 7-10 days; start within 48-72 hr of rash onset

Varicella (chickenpox) immunocompetent
• **Adult/adolescent/child >40 kg: PO** 800 mg 4×/day × 5 days
• **Child ≥2 yr and ≤40 kg: PO** 20 mg/kg/dose (max 800 mg) 4×/day × 5 days

Mucosal/cutaneous herpes simplex infections in immunosuppressed patients
• **Adult and child >12 yr: IV** 5 mg/kg q8hr × 7 days
• **Infant >3 mo/child <12: IV** 10 mg/kg q8hr × 7 days

Renal dose
• **Adult and child: PO/IV** CCr >50 ml/min 100% dose q8hr, CCr 25-50 ml/min 100% dose q12hr, CCr 10-25 ml/min 100% dose q24hr, CCr 0-10 ml/min 50% dose q24hr

Recurrent ocular herpes, prevention (unlabeled)
• **Adult/child ≥12 yr: PO** 600-800 mg every day × 8-12 mo

CMV prophylaxis (unlabeled)
• **Adult: IV** 500 mg/m^2 q8hr

Herpes simplex in pneumonitis/ esophagitis/tracheobronchitis/ proctitis/stomatitis/pharyngitis (unlabeled)
• **Adult and adolescent: IV** 5-10 mg/kg q8hr × 2-7 days or **PO** 200 mg q4hr 5×/day × 7-10 days or 400 mg 3-5 ×/day × ≥10 days
• **Child 6 mo-12 yr: IV** 1000 mg/day in 3-5 divided doses × 7-14 days

Herpes simplex prophylaxis for chronic suppression therapy (unlabeled)
• **Adult and adolescent: PO** 400 mg bid up to 12 mo
• **Child: PO** 800-1000 mg/day in 2-5 divided doses, max 80 mg/kg/day

Available forms: Caps 200 mg; tabs 400, 800 mg; powder for inj 500, 1000 mg; sol for inj 50 mg/ml; oral susp 200 mg/5 ml; ointment/cream 5%

Administer:

PO route
• Do not break, crush, or chew caps
• May give without regard to meals, with 8 oz of water
• Shake susp before use

Topical route
• Use finger cot or glove to cover all lesions completely; do not get in eye; wash hands after use

Intermittent IV INF route

- Increase fluids to 3 L/day to decrease crystalluria; most critical during first 2 hr after IV
- Reconstitute with 10 ml compatible sol/500 mg or 20 mg/1 g of product, conc of 50 mg/ml, shake, further dilute in 50-125 ml compatible sol; use within 12 hr; give over at least 1 hr (constant rate) by inf pump to prevent nephrotoxicity; do not reconstitute with sol containing benzyl alcohol in neonates
- Store at room temp for up to 12 hr after reconstitution; if refrigerated, sol may show a precipitate that clears at room temp; yellow discoloration does not affect potency

Solution compatibilities: D_5W, LR, or NaCl (D_5 0.9% NaCl, 0.9% NaCl) sol

Y-site compatibilities: Alemtuzumab, alfentanil, allopurinol, amikacin, aminophylline, amphotericin B cholesteryl, amphotericin B liposome, ampicillin, anidulafungin, argatroban, atracurium, bivalirudin, buprenorphine, busulfan, butorphanol, calcium chloride/gluconate, CARBOplatin, cefazolin, cefonicid, cefoperazone, cefotaxime, cefoxitin, ceftazidime, ceftizoxime, cefTRIAXone, cefuroxime, cephapirin, chloramphenicol, cholesteryl sulfate complex, cimetidine, clindamycin, dexamethasone sodium phosphate, dimenhyDRINATE, DOXOrubicin, doxycycline, erythromycin, famotidine, filgrastim, fluconazole, gallium, gentamicin, granisetron, heparin, hydrocortisone sodium succinate, hydromorphone, imipenem/cilastatin, LORazepam, magnesium sulfate, melphalan, methylPREDNISolone sodium succinate, metoclopramide, metroNIDAZOLE, multivitamin, nafcillin, oxacillin, paclitaxel, penicillin G potassium, PENTobarbital, perphenazine, piperacillin, potassium chloride, propofol, ranitidine, remifentanil, sodium bicarbonate, tacrolimus, teniposide, theophylline, thiotepa, ticarcillin, tobramycin, trimethoprim-sulfamethoxazole, vancomycin, vasopressin, voriconazole, zidovudine

SIDE EFFECTS

CNS: Tremors, confusion, lethargy, hallucinations, **seizures**, dizziness, headache, encephalopathic changes
EENT: Gingival hyperplasia
GI: Nausea, vomiting, diarrhea, increased ALT/AST, abdominal pain, glossitis, colitis
GU: **Oliguria, proteinuria, hematuria**, vaginitis, moniliasis, **glomerulonephritis, acute renal failure**, changes in menses, polydipsia
HEMA: **Thrombotic thrombocytopenia purpura, hemolytic uremic syndrome** (immunocompromised patients)
INTEG: Rash, urticaria, pruritus, pain or phlebitis at IV site, unusual sweating, alopecia, **Stevens-Johnson syndrome**
MS: Joint pain, leg pain, muscle cramps

PHARMACOKINETICS

Distributed widely; crosses placenta; CSF concentrations are 50% plasma; protein binding 9%-33%
PO: Absorbed minimally, onset unknown, peak 1.5-2 hr, terminal half-life 2.5-3.3 hr
IV: Onset immediate, peak immediate, duration unknown, half-life 20 min-3 hr (terminal); metabolized by liver, excreted by kidneys as unchanged product (95%)

INTERACTIONS

Increase: CNS side effects—zidovudine
Increase: levels, toxicity—probenecid
Increase: nephrotoxicity—aminoglycosides
Increase: concentrations of—entecavir, pemetrexed, tenofovir
Decrease: action of—hydantoins, valproic acid

Drug/Lab Test
Increase: BUN, creatinine

NURSING CONSIDERATIONS

Assess:

- Signs of infection, anemia
- **Toxicity: any patient with compromised renal system because product is**

excreted slowly with poor renal system function; toxicity may occur rapidly

• Hepatic, renal studies: AST, ALT; urinalysis, protein, BUN, creatinine, CCr, watch for increasing BUN and serum creatinine or decreased CCr; I&O ratio; report hematuria, oliguria, fatigue, weakness; may indicate **nephrotoxicity;** check for protein in urine during treatment

• Bowel pattern before, during treatment; if severe abdominal pain with bleeding occurs, product should be discontinued

• Skin eruptions: rash, urticaria, itching

• Allergies before treatment, reaction of each medication; place allergies on chart in bright red letters

• Neurologic status with herpes encephalitis

• Provide adequate intake of fluids (2 L) to prevent deposits in kidneys, more likely to occur with rapid administration or in dehydration

Evaluate:

• Therapeutic response: absence of itching, painful lesions; crusting and healed lesions; decreased symptoms of chickenpox; healing, decreased pain with herpes zoster

Teach patient/family:

• To take as prescribed; if dose is missed, take as soon as remembered up to 1 hr before next dose; do not double dose

• That product may be taken orally before infection occurs; product should be taken when itching or pain occurs, usually before eruptions

• That sexual partners need to be told that patient has herpes because they can become infected; condoms must be worn to prevent reinfections

• Not to touch lesions to avoid spreading infection to new sites

• That product does not cure infection, just controls symptoms and does not prevent infecting others

• That product must be taken in equal intervals around the clock to maintain blood levels for duration of therapy

• To seek dental care during treatment to prevent gingival hyperplasia

• That women with genital herpes are more likely to develop cervical cancer; to keep all gynecologic appointments

TREATMENT OF OVERDOSE:

Discontinue product, hemodialysis, resuscitate if needed

adalimumab (Rx)

(add-a-lim′yu-mab)

Humira

Func. class.: Antirheumatic agent (disease modifying), immunomodulator, anti-TNF

Chem. class.: Recombinant human IgG1 monoclonal antibody, DMARD

Do not confuse:
Humira/Humulin/Humalog

ACTION: A form of human IgG1 monoclonal antibody specific for human tumor necrosis factor (TNF); elevated levels of TNF are found in patients with rheumatoid arthritis

USES: Reduction of signs and symptoms and inhibition of progression of structural damage in patients with moderate to severe active rheumatoid arthritis who are ≥18 years of age and who have not responded to other disease-modifying agents, juvenile rheumatoid arthritis (JRA), psoriatic arthritis, Crohn's disease, moderate to severe plaque psoriasis, ankylosing spondylitis, ulcerative colitis

CONTRAINDICATIONS: Hypersensitivity

Precautions: Pregnancy (B), breastfeeding, children, geriatric patients, CNS demyelinating disease, lymphoma, CHF, hepatitis B carriers, manitol hypersensitivity, latex allergy, neoplastic disease

Black Box Warning: Active infections, risk of lymphomas/leukemias, TB

DOSAGE AND ROUTES

Rheumatoid arthritis/ankylosing spondylitis/psoriatic arthritis

• **Adult: SUBCUT** 40 mg every other wk or every wk if not combined with methotrexate; **IV** (unlabeled) 0.25-3 mg/kg q2-4wk

Juvenile rheumatoid arthritis

• **Child ≥4 yr/adolescent ≥30 kg: SUBCUT** 40 mg every other wk

• **Child ≥4 yr/adolescent ≥15 kg to <30 kg: SUBCUT** 20 mg every other wk

• **Child ≥4 yr/adolescent <15 kg: SUBCUT** 24 mg/m^2 BSA (up to 40 mg total) every other wk, then 20 mg every other wk

Crohn's disease/ulcerative colitis

• **Adult: SUBCUT** 160 mg given as 4 inj on day 1 or 2 inj each on days 1 and 2, then 80 mg at wk 2 and 40 mg every other wk starting at wk 4

Plaque psoriasis

• **Adult: SUBCUT** 80 mg baseline as 2 inj then 40 mg every other wk starting 1 wk after initial dose × 16 wk

Available form: Inj 40 mg/0.8 ml; 20 mg/0.4 ml (pediatric)

Administer:

SUBCUT route

• Do not admix with other sol or medications; do not use filter; protect from light; give at 45-degree angle using abdomen, thighs; rotate inj sites; discard unused portions

SIDE EFFECTS

CNS: *Headache,* Guillain-Barré syndrome

CV: *Hypertension,* CHF

EENT: *Sinusitis*

GI: Abdominal pain, nausea, hepatic damage, **GI bleeding**

HEMA: **Leukopenia pancytopenia, aplastic anemia, agranulocytopenia, thrombocytopenia**

INTEG: *Rash, inj site reaction*

MISC: Flulike symptoms, UTI, back pain, lupus-like syndrome, **increased cancer risk,** *antibody development to this drug;* **risk of infection (TB, invasive fungal infections, other opportunistic infections), may be fatal, Stevens-Johnson syndrome, anaphylaxis**

RESP: *URI,* **pulmonary fibrosis, bronchitis**

PHARMACOKINETICS

Absorption 65%, terminal half-life 2 wk, lower clearance with advancing age (40-75 yr)

INTERACTIONS

Increase: **serious infections—other TNF blockers, rilonacept**

• **Do not use with anakinra; serious infections may occur**

• Do not give concurrently with vaccines; immunizations should be brought up to date before treatment

Drug/Lab Test

Increase: ALT, cholesterol

NURSING CONSIDERATIONS

Assess:

• **RA:** pain, stiffness, ROM, swelling of joints before, during treatment

• For inj site pain, swelling, redness—usually occur after 2 inj (4-5 days)—use cold compress to relieve pain/swelling

Black Box Warning: For infections (fever, flulike symptoms, dyspnea, change in urination, redness/swelling around any wounds), stop treatment if present; some serious infections including sepsis may occur, may be fatal; patients with active infections should not be started on this product

• **May reactivate hepatitis B in chronic carriers, may be fatal**

Black Box Warning: Latent TB before therapy, treat before starting this product

- **Anaphylaxis, latex allergy:** stop therapy if lupus-like syndrome develops
- **Blood dyscrasias:** CBC, differential periodically

Black Box Warning: For neoplastic disease (lymphomas/leukemia) in children, adolescents, hepatosplenic T-cell lymphoma is more likely in adolescent males with Crohn's disease or ulcerative colitis

Evaluate:

- Therapeutic response: decreased inflammation, pain in joints, decreased joint destruction

Teach patient/family:

- About self-administration if appropriate: inj should be made in thigh, abdomen, upper arm; rotate sites at least 1 inch from old site; do not inject in areas that are bruised, red, hard
- That if medication is not taken when due, inject next dose as soon as remembered and inject next dose as scheduled
- Not to take any live virus vaccines during treatment
- To report signs of infection, allergic reaction, or lupus-like syndrome, immediately

adefovir (Rx)

(add-ee-foh′veer)

Hepsera

Func. class.: Antiviral

Chem. class.: Adenosine monophosphate analog

ACTION: Inhibits hepatitis B virus DNA polymerase by competing with natural substrates and by causing DNA termination after its incorporation into viral DNA; causes viral DNA death

USES: Chronic hepatitis B

CONTRAINDICATIONS: Hypersensitivity

Precautions: Pregnancy (C), labor, breastfeeding, children, geriatric patients, dialysis, females, obesity, organ transplant

Black Box Warning: Severe renal disease, impaired hepatic function, lactic acidosis, HIV

DOSAGE AND ROUTES

- **Adult/adolescent: PO** 10 mg/day, optimal duration unknown

Renal dose

- **Adult: PO** CCr ≥50 ml/min 10 mg q24hr; CCr 30-49 ml/min 10 mg q48hr; CCr 10-29 ml/min 10 mg q72hr; hemodialysis 10 mg q7days after dialysis

Available forms: Tabs 10 mg

Administer:

- By mouth without regard for food
- Take with full glass of water
- Store in cool environment; protect from light

SIDE EFFECTS

CNS: *Headache*

GI: *Dyspepsia,* abdominal pain, nausea, vomiting, diarrhea, hepatomegaly, flatulence, pancreatitis

GU: Hematuria, glycosuria, nephrotoxicity, Fanconi syndrome, renal failure

MISC: Fever, rash, weight loss, cough

PHARMACOKINETICS

PO: Rapidly absorbed from GI tract, peak 1¾ hr, excreted by kidneys 45%, terminal half-life 7.48 hr

INTERACTIONS

- Do not use in combination with emtricitabine/tenofovir, emtricitabine/rilpivirine, emtricitabine/efavirenz/tenofovir

Increase: serum conc and possible toxicity—aminoglycosides, memantine, emtricitabine, efavirenz, dofetilide, digoxin, cycloSPORINE, aMILoride, quiNINE, quiNIDine, procainamide, PEMEtrexed, midodrine, metFORMIN, NSAIDs, vancomycin, trospium, triamterene, tenofovir, tacrolimus, ranitidine, cimetidine, morphine

Black Box Warning: Increase: lactic acidosis, severe hepatomegaly, NNRTIs, NRTIs, antiretroviral protease inhibitors

A

Drug/Lab Test
Increase: ALT, AST, amylase, creatine kinase

NURSING CONSIDERATIONS

Assess:

Black Box Warning: Nephrotoxicity: increasing CCr, BUN

Black Box Warning: HIV antibody testing before beginning treatment, HIV resistance may occur in chronic hepatitis B patients

Black Box Warning: For lactic acidosis, severe hepatomegaly with stenosis; for use of NNRTIs, NRTIs, antiretroviral protease inhibitors (PIs), lactic acidosis with severe hepatomegaly is more common in females, obese patients, and with prolonged nucleoside use

- Geriatric patients more carefully; may develop renal, cardiac symptoms more rapidly

Black Box Warning: For exacerbations of hepatitis after discontinuing treatment, monitor LFTs, hepatitis B serology

- Pregnancy: If planned or suspected; if pregnant call the Pregnancy Registry: 800-258-4263

Evaluate:

- Therapeutic response: decreased symptoms of chronic hepatitis B, improving LFTs

Teach patient/family:

- That optimal duration of treatment is unknown; that product is not a cure; that transmission may still occur
- To avoid use with other medications unless approved by prescriber
- To notify prescriber of decreased urinary output
- Not to stop abruptly unless directed; worsening of hepatitis may occur
- To report immediately dyspnea, nausea, vomiting, abdominal pain, weakness, dizziness, cold extremities
- To notify prescriber if pregnancy is planned or suspected; avoid breastfeeding

⚠ HIGH ALERT

adenosine (Rx)

(a-den′oh-seen)

Func. class.: Antidysrhythmic

Chem. class.: Endogenous nucleoside

Do not confuse:
adenosine/adeosine phosphate

ACTION: Slows conduction through AV node, can interrupt reentry pathways through AV node, and can restore normal sinus rhythm in patients with paroxysmal supraventricular tachycardia; decreases cardiac oxygen demand, decreasing hypoxia

USES: PSVT, as a diagnostic aid to assess myocardial perfusion defects in CAD, Wolff-Parkinson-White syndrome
Unlabeled uses: Wide-complex tachycardia diagnosis

CONTRAINDICATIONS: Hypersensitivity, 2nd- or 3rd-degree AV block, sick sinus syndrome
Precautions: Pregnancy (C), breastfeeding, children, geriatric patients, asthma, atrial flutter, atrial fibrillation, ventricular tachycardia, bronchospastic lung disease, symptomatic bradycardia, bundle branch block, heart transplant, unstable angina, COPD, hypotension, hypovolemia, vascular heart disease, CV disease

DOSAGE AND ROUTES

Antidysrhythmic

- **Adult and child >50 kg: IV BOL** 6 mg; if conversion to normal sinus rhythm does not occur within 1-2 min, give 12 mg by rapid **IV BOL;** may repeat 12-mg dose again in 1-2 min
- **Infant and child <50 kg: IV BOL** 0.1 mg/kg; if not effective, increase dose by 0.05-0.1 mg/kg q2min to a max of 0.3 mg/kg/dose
- **Neonate: IV BOL** 0.05 mg/kg by rapid

IV BOL, may increase by 0.05 mg/kg q2min, max 0.3 mg/kg/dose

Available forms: 3 mg/ml

Administer:

IV, direct route

- Warm to room temp; crystals will dissolve
- Undiluted; give 6 mg or less by rapid inj over 1-2 sec; if using an IV line, use port near insertion site, flush with NS (20 ml), then elevate arm
- Store at room temperature; sol should be clear; discard unused product

Y-site compatibilities: abciximab

Solution compatibilities: D_5LR, D_5W, LR, 0.9% NaCl

SIDE EFFECTS

CNS: Lightheadedness, dizziness, arm tingling, numbness, headache

CV: Chest pain, pressure, atrial tachydysrhythmias, sweating, palpitations, hypotension, *facial flushing,* AV block, cardiac arrest, ventricular dysrhythmias, atrial fibrillation

GI: *Nausea,* metallic taste

RESP: *Dyspnea, chest pressure,* hyperventilation, bronchospasm (asthmatics)

PHARMACOKINETICS

Cleared from plasma in <30 sec, half-life 10 sec, converted to inosine/adenosine monophosphate

INTERACTIONS

Increase: risk for higher degree of heart block—carBAMazepine

Increase: risk for ventricular fibrillation—digoxin, verapamil

- Smoking: increase tachycardia

Increase: effects of adenosine—dipyridamole

Decrease: activity of adenosine—theophylline or other methylxanthines (caffeine)

Drug/Herb

Increase: adenosine effect—ginger

Decrease: adenosine effect—guarana, green tea

NURSING CONSIDERATIONS

Assess:

- **Cardiopulmonary status:** B/P, pulse, respiration, rhythm, ECG intervals (PR, QRS, QT); check for transient dysrhythmias (PVCs, PACs, sinus tachycardia, AV block)
- Respiratory status: rate, rhythm, lung fields for crackles; watch for respiratory depression; bilateral crackles may occur in **CHF** patient; increased respiration, increased pulse, product should be discontinued
- CNS effects: dizziness, confusion, psychosis, paresthesias, seizures; product should be discontinued

Evaluate:

- Therapeutic response: normal sinus rhythm or diagnosis of perfusion defect

Teach patient/family:

- To report facial flushing, dizziness, sweating, palpitations, chest pain; usually transient

TREATMENT OF OVERDOSE:

Defibrillation, vasopressor for hypotension, theophylline

aflibercept

(a-fli-ber′sept)

EYLEA

Func. class.: Biologic response modifier; signal transduction inhibitor (STIs) (Ophthalmic)

ACTION:

A recombinant fusion protein consisting of portions of human VEGF receptors 1 and 2 extracellular domains fused to human IgG1; acts as a soluble decoy receptor that binds vascular endothelial growth factor-A (VEGF-A) and placental growth factor (PIGF); can act as mitogenic, chemotactic, vascular permeability factors for endothelial cells; VEGF-A interacts with VEGFR-1 and VEGFR-2 on the surface of endothelial cells; results in neovascularization and vascular permeability; binding of aflibercept to VEGF-A and PIGF prevents activation of these receptors

USES:
For treatment of neovascular (wet) age-related macular degeneration (AMD), macular edema after central retinal vein

CONTRAINDICATIONS:
Hypersensitivity, ocular/periocular infection, active intraocular inflammation

Precautions: Neonates, infants, children, adolescents, pregnancy (C), breastfeeding; history of glaucoma, ocular surgery; driving or operating machinery

DOSAGE AND ROUTES

• **Adult: Intravitreal INJ** 2 mg (0.05 ml) into affected eye(s) q4wk × 12 wk, then 2 mg (0.05 ml) q8wk

Available forms: Solution for injection 2 mg/0.05 ml

Administer:

Intravitreal route

• Visually inspect for particulate matter, discoloration before use; do not use if particulates, cloudiness, discoloration are visible; only for use by physicians trained in administration

• Use controlled aseptic conditions (sterile gloves, sterile drape, sterile eyelid speculum); adequate anesthesia, topical broad-spectrum antiinfective should be given before use

• Use each vial for treatment of single eye only; if other eye is being treated, use new vial and change the sterile field, syringe, gloves, drapes, eyelid speculum, filter, injection needles before administering to the other eye

• Use aseptic technique: withdraw all of vial contents through a 5-micron 19-G filter needle attached to 1-ml syringe supplied by manufacturer; after vial contents are withdrawn, discard filter needle and replace with sterile 30-G × ½ inch needle for intravitreal injection; expel any air bubbles and contents of syringe until plunger tip is aligned with line that marks 0.05 ml

• Immediately after the intravitreal injection, monitor patient for elevation in intraocular pressure (IOP); appropriate monitoring may consist of a check for perfusion of optic nerve head or tonometry; sterile paracentesis needle should be available if required

• Storage: do not freeze, protect from light, refrigerate, store in original package until time of use

SIDE EFFECTS

CV: Thromboembolism, nonfatal stroke, nonfatal myocardial infarction, vascular death

EENT: *Ocular hemorrhage, ocular pain, cataracts, vitreous detachment, vitreous floaters,* conjunctival hyperemia, corneal erosion, detachment of retinal pigment epithelium, injection site pain, foreign body sensation, increased lacrimation, blurred vision, retinal pigment epithelium tear, injection site hemorrhage, blepharedema, corneal edema

SYST: Hypersensitivity

PHARMACOKINETICS

Absorbed into systemic circulation; present in its unbound form and stable inactive form bound with endogenous VEGF; elimination by binding to free endogenous VEGF; metabolism by proteolysis; terminal half-life in plasma 5-6 days

NURSING CONSIDERATIONS

Assess:

• **Infection:** monitor for infection during week after injection to permit early treatment of any ocular infection that may develop; proper aseptic injection technique should be used to minimize infection

• **Increased intraocular pressure:** monitor for acute increases in intraocular pressure within 60 min of injection; sustained increases in intraocular pressure have been reported after repeated intravitreal dosing; monitor intraocular pressure and optic nerve head perfusion, a sterile paracentesis needle should be available

Evaluate:
- Prevention of further vision loss

Teach patient/family:

⚠ To seek immediate care if symptoms of endophthalmitis or retinal detachment develop (ocular pain, hyperemia of the conjunctiva, photophobia, blurry vision)

albumin, human 5% (Rx)

(al-byoo′min)

Albumarc, Albuminar-5, Albutein 5%, Buminate 5%, Plasbumin-5

albumin, human 25%

Albuminar-25, Albutein 25%, Buminate 25%, Plasbumin-25

Func. class.: Plasma volume expander

ACTION:
Exerts oncotic pressure, which expands volume of circulating blood and maintains cardiac output

USES:
Restores plasma volume after burns, hyperbilirubinemia, shock, hypoproteinemia, prevention of cerebral edema, cardiopulmonary bypass procedures, ARDS, nephrotic syndrome

CONTRAINDICATIONS:
Hypersensitivity, CHF, severe anemia, renal insufficiency, pulmonary edema

Precautions: Pregnancy (C), decreased salt intake, decreased cardiac reserve, lack of albumin deficiency, renal/hepatic disease, chronic anemia

DOSAGE AND ROUTES

Burns
- **Adult:** **IV** dose to maintain plasma albumin at 3-4 mg/dl

Hypovolemic Shock
- **Adult:** **IV** rapidly give 5% sol, when close to normal inf at ≤2-4 ml/min (25% sol ≤1 ml/min)
- **Child:** **IV** 0.5-1 g/kg/dose 5% sol, may repeat as needed, max 6 g/kg/day

Nephrotic Syndrome
- **Adult:** **IV** 100-200 ml of 25% and loop diuretic × 7-10 days

Hypoproteinemia
- **Adult:** **IV** 25 g, may repeat in 15-30 min, or 50-75 g of 25% albumin infused at ≤2 ml/min
- **Child and infant:** **IV** 0.5-1 g/kg/dose over 2-4 hr, may repeat q1-2days

Hyperbilirubinemia/erythroblastosis fetalis
- **Infant:** **IV** 1 g/kg 1-2 hr before transfusion

Available forms: Inj (5%) 50 mg/ml, (25%) 250 mg/ml

Administer:

IV route
- Slowly, to prevent fluid overload; dilute with NS for injection or D_5W; 5% is given undiluted; 25% may be given diluted or undiluted; give over 30-60 min, use inf pump, use large-gauge needle; inf must be completed within 4 hr
- 5% solution may be used with hypovolemic/intravascular depletion
- 25% solution may be used with sodium/fluid restrictions

Solution compatibilities: LR, NaCl, Ringer's, D_5W, $D_{10}W$, $D_{2½}W$, NaCl 0.9%, dextrose/Ringer's, dextrose/LR

Y-site compatibilities: Diltiazem, LORazepam

SIDE EFFECTS

CNS: Fever, chills, flushing, headache

CV: Fluid overload, hypotension, erratic pulse, tachycardia

GI: Nausea, vomiting, increased salivation

INTEG: Rash, urticaria

RESP: Altered respirations, pulmonary edema

PHARMACOKINETICS

In hyponutrition states, metabolized as protein/energy source; terminal half-life 21 days

INTERACTIONS

Drug/Lab Test

Increase: serum albumin

NURSING CONSIDERATIONS

Assess:

- Blood studies Hct, Hgb; if serum protein declines, dyspnea, hypoxemia can result
- Decreased B/P, erratic pulse, respiration
- I&O ratio: urinary output may decrease
- Adequate hydration before, during administration
- Check type of albumin; some stored at room temp; some need to be refrigerated; use within 4 hr of opening

⚠ **Circulatory/pulmonary overload:** CVP, pulmonary wedge pressure distended neck veins indicate circulatory overload; shortness of breath, anxiety, insomnia, expiratory crackles, frothy blood-tinged cough, cyanosis indicate pulmonary overload

- Allergy: fever, rash, itching, chills, flushing, urticaria, nausea, vomiting, hypotension, requires discontinuation of inf, use of new lot if therapy reinstituted; premedicate with diphenhydrAMINE

Evaluate:

- Therapeutic response: increased B/P, decreased edema, increased serum albumin levels, increased plasma protein

Teach patient/family:

- Reason for product; to report hypersensitivity

albuterol (Rx)

(al-byoo′ter-ole)

Accuneb, Airomir ✤, Apo-Salvent, Gen-Salbutamol ✤, Proair HFA, Proventil, Proventil HFA, ReliOn, Ventolin HFA, VoSpire ER

Func. class.: Adrenergic β_2-agonist, sympathomimetic, bronchodilator

Do not confuse:

albuterol/atenolol/Albutein
Ventolin/Vantin
Proventil/Prinivil

ACTION: Causes bronchodilation by action on β_2 (pulmonary) receptors by increasing levels of cAMP, which relaxes smooth muscle; produces bronchodilation, CNS, cardiac stimulation as well as increased diuresis and gastric acid secretion; longer acting than isoproterenol

USES: Prevention of exercise-induced asthma, acute bronchospasm, bronchitis, emphysema, bronchiectasis, or other reversible airway obstruction

Unlabeled uses: Hyperkalemia in dialysis patients

CONTRAINDICATIONS: Hypersensitivity to sympathomimetics, tachydysrhythmias, severe cardiac disease, heart block

Precautions: Pregnancy (C), breastfeeding, cardiac/renal disease, hyperthyroidism, diabetes mellitus, hypertension, prostatic hypertrophy, angle-closure glaucoma, seizures, exercise-induced bronchospasm (aerosol) in children <12 yr, hypoglycemia

DOSAGE AND ROUTES

Bronchospasm prophylaxis

- **Adult and child ≥4 yr: INH** (metered-dose inhaler) 2 puffs q4-6hr as needed

Other respiratory conditions

• **Adult and child ≥12 yr: INH** (metered-dose inhaler) 1 puff q4-6hr; **PO** 2-4 mg tid-qid, max 32 mg; **NEB/IPPB** 2.5 mg tid-qid

• **Geriatric: PO** 2 mg tid-qid, may increase gradually to 8 mg tid-qid

• **Child 2-12 yr: INH** (metered-dose inhaler) 0.1 mg/kg tid (max 2.5 mg tid-qid); **NEB/IPPB** 0.1-0.15 mg/kg/dose tid-qid or 1.25 mg tid-qid for child 10-15 kg or 2.5 mg tid-qid for child >15 kg

Hyperkalemia (unlabeled)

• **Adult: ORAL INH** (albuterol nebulizer sol) 10-20 mg

Available forms: Aerosol 90 mcg/actuation; oral syr 2 mg/5 ml; tabs 2, 4 mg; ext rel 4, 8 mg; INH sol 0.5, 0.83, 1, 2, 5 mg/ml; powder for INH (Ventodisk) 200, 400 mcg; INH cap 200 mcg; 100 mcg/spray, 80 INH/canister, 200 INH/canister

Administer:

• Store in light-resistant container; do not expose to temperatures of more than 86° F (30° C)

PO route

• Do not break, crush, or chew ext rel tabs; give with meals to decrease gastric irritation

• **Oral sol** to children (no alcohol, sugar)

Inhalation route

• For geriatric patients and children, a spacing device is advised

• After shaking metered-dose inhaler, exhale, place mouthpiece in mouth, inhale slowly while depressing inhaler, hold breath, remove, exhale slowly; give INH at least 1 min apart

• **NEB/IPPB** diluting 5 mg/ml sol/2.5 ml 0.9% NaCl for INH; other sol do not require dilution; for neb O_2 flow or compressed air 6-10 L/min

SIDE EFFECTS

CNS: *Tremors, anxiety,* insomnia, headache, dizziness, stimulation, *restlessness,* hallucinations, flushing, irritability

CV: Palpitations, tachycardia, angina, hypo/hypertension, dysrhythmias

EENT: Dry nose, irritation of nose and throat

GI: Heartburn, nausea, vomiting

MISC: Flushing, sweating, anorexia, bad taste/smell changes, hypokalemia, metabolic acidosis

MS: Muscle cramps

RESP: Cough, wheezing, dyspnea, paradoxical bronchospasm, dry throat

PHARMACOKINETICS

Extensively metabolized in the liver and tissues, crosses placenta, breast milk, blood-brain barrier

PO: Onset ½ hr, peak 2-3 hr, duration 4-6 hr, half-life 2.7-6 hr, well absorbed

PO-ER: Onset ½ hr; peak 2-3 hr; duration 8-12 hr

INH: Onset 5-15 min, peak 1-1½ hr, duration 3-6 hr, half-life 4 hr

INTERACTIONS

Increase: QTc prolongation—other drugs that increase QT prolongation

Increase: digoxin level—digoxin

Increase: CNS stimulation—CNS stimulants

Increase: ECG changes/hypokalemia—potassium-losing diuretics

Increase: severe hypotension—oxytocics

Increase: toxicity—theophylline

Increase: action of aerosol bronchodilators

Increase: action of albuterol—tricyclics, MAOIs, other adrenergics; do not use together

Increase: CV effects—atomoxetine, selegiline

Decrease: albuterol—other β-blockers

Drug/Herb

Increase: stimulation—caffeine (cola nut, green/black tea, guarana, yerba maté, coffee, chocolate)

Drug/Lab Test

Decrease: potassium

NURSING CONSIDERATIONS

Assess:

- **Respiratory function:** vital capacity, forced expiratory volume, ABGs; lung sounds, heart rate and rhythm, B/P, sputum (baseline and peak); whether patient has not received theophylline therapy before giving dose
- Patient's ability to self-medicate
- For evidence of allergic reactions
- For paradoxical bronchospasm; hold medication, notify prescriber if bronchospasm occurs

Evaluate:

- Therapeutic response: absence of dyspnea, wheezing after 1 hr, improved airway exchange, improved ABGs

Teach patient/family:

- To use exactly as prescribed; to take missed dose when remembered, alter dosing schedule; not to use OTC medications; that excess stimulation may occur
- About use of inhaler: review package insert with patient; use demonstration; return demonstration; shake, prime before 1st use and when not used for >2 wk; release 4 test sprays into air, away from the face
- To avoid getting aerosol in eyes (blurring of vision may result) or using near flames or sources of heat
- To wash inhaler in warm water daily and dry; to track number of inhalations used and to discard product when labeled inhalations have been used
- To avoid smoking, smoke-filled rooms, persons with respiratory infections

⚠ That **paradoxical bronchospasm** may occur; to stop product immediately, call prescriber

- To limit caffeine products such as chocolate, coffee, tea, colas

TREATMENT OF OVERDOSE:

Administer β_1-adrenergic blocker, IV fluids

alendronate (Rx)

(al-en-drone′ate)

Binosto, Fosamax, Fosamax plus D

Func. class.: Bone-resorption inhibitor

Chem. class.: Bisphosphonate

Do not confuse:

Fosamax/Flomax

ACTION: Decreases rate of bone resorption and may directly block dissolution of hydroxyapatite crystals of bone; inhibits osteoclast activity

USES: Treatment and prevention of osteoporosis in postmenopausal women, treatment of osteoporosis in men, Paget's disease, treatment of corticosteroid-induced osteoporosis in postmenopausal women not receiving estrogen and in men who are on continuing corticosteroid treatment with low bone mass

CONTRAINDICATIONS: Hypersensitivity to bisphosphonates, delayed esophageal emptying, inability to sit or stand for 30 min, hypocalcemia

Precautions: Pregnancy (C), breastfeeding, children, CCr <35 ml/min, esophageal disease, ulcers, gastritis, poor dental health, increased esophageal cancer risk

DOSAGE AND ROUTES

Osteoporosis in postmenopausal women

- **Adult and geriatic: PO** 10 mg/day or 70 mg/wk

Osteoporosis in men

- **Adult: PO** 10 mg/day or 70 mg/wk

Paget's disease

- **Adult and geriatric: PO** 40 mg/day × 6 mo, consider retreatment for relapse

Prevention of osteoporosis

- **Adult/postmenopausal female: PO** 5 mg/day or 35 mg/wk

Corticosteroid-induced osteoporosis in postmenopausal women (not receiving estrogen)
- **Adult: PO** 10 mg/day

Corticosteroid-induced osteoporosis in men or premenopausal women
- **Adult: PO** 5 mg/day

Renal dose
- **Adult: PO** CCr ≤35 ml/min, not recommended

Available forms: Tabs 5, 10, 35, 40, 70 mg; tabs 70 mg with 2800 IU vit D_3, 70 mg with 5600 IU Vit D_3; oral sol 70 mg/75 ml, effervescent tab 70 mg

Administer:
- For 6 mo to be effective for Paget's disease
- Store in cool environment, out of direct sunlight
- **Tablet:** take with 8 oz of water 30 min before 1st food, beverage, or medication of the day
- Do not lie down for ≥30 min after dose; do not take at bedtime or before rising
- **Liquid:** use oral syringe or calibrated device; give in AM with ≥2 oz of water ≥30 min before food, beverage, or medication

SIDE EFFECTS

CNS: Headache
GI: Abdominal pain, constipation, nausea, vomiting, esophageal ulceration, acid reflux, dyspepsia, esophageal perforation, diarrhea, esophageal cancer
META: Hypophosphatemia, hypocalcemia
MS: Bone pain, osteonecrosis of the jaw, bone fractures
SYST: Angioedema, Stevens-Johnson syndrome, toxic epidermal necrolysis

PHARMACOKINETICS

Bioavailability 60%, protein binding 78%, rapidly cleared from circulation, taken up mainly by bones, eliminated primarily through kidneys, bound to bone, half-life >10 yr

INTERACTIONS

Increase: GI adverse reactions—NSAIDs, salicylates, H_2 blockers, proton pump inhibitors, gastric mucosal agents
Decrease: absorption—antacids, calcium supplements, aminoglycosides

Drug/Food
Decrease: absorption when used with caffeine, orange juice, food

Drug/Lab Test
Decrease: calcium, phosphate

NURSING CONSIDERATIONS

Assess:
⚠ Serious reactions: angioedema, Stevens-Johnson syndrome, toxic epidermal necrolysis, atrial fibrillation
- Hormonal status in women, before treatment
- **For osteoporosis:** bone density test before and during treatment
- **For Paget's disease:** increased skull size, bone pain, headache; decreased vision, hearing
- Electrolytes; BUN/creatinine; calcium, phosphorous, magnesium, potassium
- **For hypercalcemia:** paresthesia, twitching, laryngospasm; Chvostek's, Trousseau's signs
- Alk phos levels, baseline and periodically, 2 × upper limit of normal is indicative of Paget's disease
- Dental status: regular dental exams should be performed; dental extractions (cover with antiinfectives before procedure)

Evaluate:
- Therapeutic response: increased bone mass, absence of fractures

Teach patient/family:
- To remain upright for 30 min after dose to prevent esophageal irritation; if dose is missed, skip dose, do not double doses or take later in day; to take in AM before food, other meds; to take with 6-8 oz of water only (no mineral water)
- To take calcium, vit D if instructed by health care provider
- To perform weight-bearing exercise to increase bone density

• To let health care provider know if pregnant or if pregnancy is planned or if breastfeeding; to inform dentist of the use of this product
• To maintain good oral hygiene

alfuzosin (Rx)

(al-fyoo′zoe-sin)

Uroxatral

Func. class.: Urinary tract, antispasmodic, α_1-agonist

Chem. class.: Quinazolone

ACTION: Binds to α_{1A}-adrenoceptor subtype located mainly in the prostate, relaxing smooth muscles

USES: Symptoms of benign prostatic hyperplasia

Unlabeled uses: Lower urinary tract symptoms, erectile dysfunction with sildenafil

CONTRAINDICATIONS: Hypersensitivity, moderate to severe hepatic impairment; not indicated for use in women or children, breastfeeding

Precautions: Pregnancy (B) but not used in females, geriatric patients; CAD, coronary insufficiency, mild hepatic disease, mild/moderate/severe renal disease, history of QT prolongation or coadministration with meds known to prolong QT interval, torsades de pointes, syncope, surgery, prostate cancer, orthostatic hypotension, ocular surgery, dysrhythmias, angina

DOSAGE AND ROUTES

• **Adult: PO EXT REL** 10 mg/day, taken after same meal each day

Available forms: Ext rel tabs 10 mg

Administer:

PO route

• Do not break, crush, chew tabs; give with food; take at same time each day
• Store in tight container in cool environment

SIDE EFFECTS

CNS: *Dizziness, headache,* fatigue, flushing

CV: Postural hypotension (dizziness, lightheadedness, fainting) within a few hours of administration, chest pain, tachycardia, angina

GI: Nausea, abdominal pain, dyspepsia, constipation, diarrhea, liver injury, jaundice

GU: Impotence, priapism

HEMA: Thrombocytopenia

INTEG: Rash, urticaria, angioedema, pruritus

MISC: Body pain in general, xerostomia, rhinitis

RESP: Upper respiratory infection, pharyngitis, bronchitis, sinusitis

PHARMACOKINETICS

Peak 8 hr, elimination half-life 10 hr, extensively metabolized in liver by CYP3A4 enzyme, excreted via urine (11% unchanged), moderately protein binding (82%-90%)

INTERACTIONS

• Not to be taken with prazosin, terazosin, doxazosin

Increase: QT prolongation (slight)—class IA/III antidysrhythmics

Increase: effects of alfuzosin—alcohol

Increase: effects—CYP3A4 inhibitors (ketoconazole, itraconazole, and ritonavir); do not use together

Increase: hypotension—β-blockers, phosphodiesterase 5 inhibitors, nitrates

NURSING CONSIDERATIONS

Assess:

• **Prostatic hyperplasia:** change in urinary patterns (hesitancy, dribbling, dysuria, urgency), baseline and throughout treatment
• **Serious skin reactions:** angioedema

Evaluate:

• Therapeutic response: decreased symptoms of benign prostatic hyperplasia

Teach patient/family:
- To take at same time each day with food; to not double doses
- Not to drive or operate machinery for 4 hr after 1st dose or after dosage increase, dizziness may occur
- About orthostatic hypotension; to rise slowly from sitting or lying
- To avoid all OTC products unless approved by prescriber
- To notify prescriber of fainting, dizziness
- That erectile dysfunction is a side effect and is temporary
- Store in tight container at room temperature

SIDE EFFECTS

CV: Orthostatic hypotension, hypotension
CNS: Headache, dizziness, seizures, torsades de pointes
GI: *Diarrhea*
GU: Renal stones, increased uric acid
INTEG: Rash
META: Hyperkalemia
MISC: Angioedema, cough

PHARMACOKINETICS

Poorly absorbed, bioavailability 2.5%, peak 1-3 hr, steady state 7-8 days, 91% excreted unchanged in the feces, half-life 24 hr

INTERACTIONS

- Do not use with ACE inhibitors, angiotensin II receptor antagonists in diabetes mellitus

Increase: potassium levels—ACE inhibitors, angiotensin II receptor antagonists, potassium supplements, potassium-sparing diuretics
Increase: hypotension—other antihypertensives, diuretics
Increase: aliskiren levels—atorvastatin, itraconazole, ketoconazole, cycloSPORINE; concurrent use is not recommended
Decrease: levels of warfarin
Drug/Food
Decrease: absorption—high-fat meal
Drug/Lab Test
Increase: uric acid, CPK, BUN, serum creatinine, potassium
Decrease: Hct, Hgb

aliskiren (Rx)

(a-lis′kir-en)

Tekturna

Func. class.: Antihypertensive
Chem. class.: Direct renin inhibitor

ACTION: Renin inhibitor that acts on the renin-angiotensin system (RAS)

USES: Hypertension, alone or in combination with other antihypertensives

CONTRAINDICATIONS: Hypersensitivity

Black Box Warning: Pregnancy (D)

Precautions: Breastfeeding, children, geriatric patients, angioedema, aortic/renal artery stenosis, cirrhosis, CAD, dialysis, hyper/hypokalemia, hyponatremia, hypotension, hypovolemia, renal/hepatic disease, surgery, diabetes

DOSAGE AND ROUTES

- **Adult: PO** 150 mg/day, may increase to 300 mg/day if needed, max 300 mg/day

Available forms: Tabs 150, 300 mg
Administer:
- Do not use with a high-fat meal
- Daily with a full glass of water, titrate up to achieve correct dose
- Do not discontinue abruptly, correct electrolyte/volume depletion before treatment

NURSING CONSIDERATIONS

Assess:
- Renal studies: uric acid, serum creatinine, BUN may be increased; potassium, hyperkalemia may occur

⚠ **Allergic reactions: angioedema** may occur (swelling of face; trouble breathing, swallowing)

• Daily dependent edema in feet, legs; weight, B/P, orthostatic hypotension

• **Diabetes:** identify the use of ACE inhibitors, angiotensin II receptor antagonists if in use, do not use aliskiren

Evaluate:

• Therapeutic response: decrease in B/P

Teach patient/family:

• About the importance of complying with dosage schedule even if feeling better; that if dose is missed, take as soon as possible; that if it is almost time for the next dose, take only that dose; do not double dose

Black Box Warning: To notify if pregnancy is planned or suspected; if pregnant, product will need to be discontinued, pregnancy (D)

• How to take B/P and normal reading for age group

• Not to use OTC products including herbs, supplements unless approved by prescriber

• To report to prescriber immediately: dizziness, faintness, chest pain, palpitations, uneven or rapid heart beat, headache, severe diarrhea, swelling of tongue or lips, trouble breathing, difficulty swallowing, tightening of the throat

• Not to operate machinery or perform hazardous tasks if dizziness occurs

• To rise slowly; to avoid faintness

allopurinol (Rx)

(al-oh-pure′i-nole)

Aloprim, Zyloprim

Func. class.: Antigout drug, antihyperuricemic

Chem. class.: Xanthene oxidase inhibitor

Do not confuse:
Zyloprim/Zovirax

ACTION: Inhibits the enzyme xanthine oxidase, reducing uric acid synthesis

USES: Chronic gout, hyperuricemia associated with malignancies, recurrent calcium oxalate calculi, uric acid calculi

CONTRAINDICATIONS: Hypersensitivity

Precautions: Pregnancy (C), breastfeeding, children, renal/hepatic disease

DOSAGE AND ROUTES

Increased uric acid levels in malignancies

• **Adult: PO** 600-800 mg/day in divided doses for 2-3 days; start up to 1-2 days before chemotherapy; **IV INF** 200-400 mg/m^2/day, max 600 mg/day 24-48 hr before chemotherapy, may be divided at 6-, 8-, 12-hr intervals

• **Child 6-10 yr: PO** 300 mg/day, adjust dose after 48 hr

• **Child <6 yr: PO** 150 mg/day, adjust dose after 48 hr

• **Child: IV INF** 200 mg/m^2/day, initially as a single dose or divided q6-12hr

Recurrent calculi

• **Adult: PO** 200-300 mg/day in a single dose or divided bid-tid, max 300 mg/dose, 800 mg/day

Uric acid nephropathy prevention

• **Adult and child >10 yr: PO** 600-800 mg/day × 2-3 days

Gout (mild)

• **Adult: PO** 100 mg/day, increase weekly based on uric acid levels, max 800 mg/day; maintenance dose 100-200 mg bid-tid

Gout (moderate-severe)

• **Adult: PO** 400-600 mg/day in a single dose or divided bid-tid, max 800 mg/day, doses >300 mg should be given in divided doses

Renal dose

• **Adult: PO/IV** CCr 81-100 ml/min 300 mg/day; CCr 61-80 ml/min 250 mg/day; CCr 41-60 ml/min 200 mg/day; 21-40 ml/min 150 mg/day; CCr 10-20 ml/min 100-200 mg/day; CCr 3-9 ml/min 100 mg/day or 100 mg every other day; CCr <3 ml/min 100 mg q24hr or longer or 100 mg every 3rd day

Available forms: Tabs, scored 100, 300 mg; powder for inj 500 mg/vial

Administer:

PO route

- With meals to prevent GI symptoms; may crush, add to foods or fluids
- 1-2 days before antineoplastic therapy

Intermittent IV INF route

- Reconstitute 30-ml vial with 25 ml of sterile water for inj; dilute to desired conc (≤6 mg/ml) with 0.9% NaCl for inj or D_5 for inj; begin inf within 10 hr

Y-site compatibilities: Acyclovir, aminophylline, amphotericin B lipid complex, anidulafungin, argatroban, atenolol, aztreonam, bivalirudin, bleomycin, bumetanide, buprenorphine, butorphanol, calcium gluconate, CARBOplatin, caspofungin, ceFAZolin, cefoperazone, cefoTEtan, cefTAZidime, ceftizoxime, cefTRIAXone, cefuroxime, CISplatin, cyclophosphamide, DACTINomycin, DAUNOrubicin citrate liposome, dexamethasone, dexmedetomidine, DOCEtaxel, DOXOrubicin liposomal, enalaprilat, etoposide, famotidine, fenoldopam, filgrastim, fluconazole, fludarabine, fluorouracil, furosemide, gallium, ganciclovir, gatifloxacin, gemcitabine, gemtuzumab, granisetron hydrochloride, heparin, hydrocortisone phosphate, hydrocortisone succinate, HYDROmorphone, ifosfamide, linezolid injection, LORazepam, mannitol, mesna, methotrexate, metroNIDAZOLE, milrinone, mitoXANtrone, morphine, nesiritide, octreotide, oxytocin, PACLitaxel, pamidronate, pantoprazole, PEMEtrexed, piperacillin, piperacillin-tazobactam, plicamycin, potassium chloride, ranitidine, sodium acetate, sulfamethoxazole-trimethoprim, teniposide, thiotepa, ticarcillin, ticarcillin-clavulanate, tigecycline, tirofiban, vancomycin, vasopressin, vinBLAStine, vinCRIStine, voriconazole, zidovudine, zoledronic acid

SIDE EFFECTS

CNS: *Headache,* drowsiness, neuritis, paresthesia

GI: *Nausea, vomiting, anorexia, malaise,* metallic taste, cramps, peptic ulcer, diarrhea

HEMA: Agranulocytosis, thrombocytopenia, aplastic anemia, pancytopenia, leukopenia, bone marrow suppression, eosinophilia

INTEG: Dermatitis, pruritus, purpura, erythema, rash, Stevens-Johnson syndrome

MISC: Myopathy, arthralgia, hepatomegaly, cholestatic jaundice, renal failure, exfoliative dermatitis

PHARMACOKINETICS

Protein binding <1%, half-life 1-2 hr

PO: Peak 1.5 hr; excreted in feces, urine

IV: Peak up to 30 min

INTERACTIONS

Increase: kidney stone formation—ammonium chloride, vit C, potassium/sodium phosphate

Increase: rash—ampicillin, amoxicillin, avoid concurrent use

Increase: action of oral anticoagulants, oral antidiabetics, theophylline

Increase: hypersensitivity, toxicity—ACE inhibitors, thiazides

Increase: bone marrow depression—antineoplastics (mercaptopurine, azaTHIOprine)

Increase: xanthine nephropathy, calculi—rasburicase

NURSING CONSIDERATIONS

Assess:

- **For gout:** joint pain, swelling; may use with NSAIDs for acute gouty attacks; uric acid levels q2wk; uric acid levels should be ≤6 mg/dl, effect may take several wk
- CBC, AST, BUN, creatinine before starting treatment, periodically
- I&O ratio; increase fluids to 2 L/day to prevent stone formation and toxicity
- For rash, hypersensitivity reactions, discontinue allopurinol

Evaluate:

• Therapeutic response: decreased pain in joints, decreased stone formation in kidneys, decreased uric acid levels

Teach patient/family:

• To take as prescribed; if dose is missed, take as soon as remembered; do not double dose; tabs may be crushed

• To increase fluid intake to 2 L/day

• To report skin rash, stomatitis, malaise, fever, aching; product should be discontinued

• To avoid hazardous activities if drowsiness or dizziness occurs

• To avoid alcohol, caffeine; will increase uric acid levels

• To avoid large doses of vit C; kidney stone formation may occur

• To reduce dairy products, refined sugars, sodium, meat if taking for calcium oxalate stones

almotriptan (Rx)

(al-moh-trip′tan)

Axert

Func. class.: Antimigraine agent, abortive

Chem. class.: 5-HT_1-receptor agonist, triptan

Do not confuse:

Axert/Antivert

ACTION: Binds selectively to the vascular 5-$HT_{1B/1D/1F}$-receptors, exerts antimigraine effect

USES: Acute treatment of migraine with or without aura (adult/adolescent/child ≥12 yr)

CONTRAINDICATIONS: Hypersensitivity, acute MI, angina, CV disease, CAD, stroke, vasospastic angina, ischemic heart disease or risk for, peripheral vascular syndrome, uncontrolled hypertension, basilar or hemiplegic migraine

Precautions: Pregnancy (C), postmenopausal women, men >40 yr, breastfeeding, children <18 yr, geriatric patients, risk factors for CAD, MI; hypercholesterolemia, obesity, diabetes, impaired renal/hepatic function, sulfonamide hypersensitivity, cardiac dysrhythmias, Raynaud's disease, tobacco smoking, Wolff-Parkinson-White syndrome

DOSAGE AND ROUTES

• **Adult, adolescent, and child ≥12 yr: PO** 6.25-12.5 mg; may repeat dose after 2 hr; max 2 doses/24 hr, 25 mg/day or 4 treatment cycles within any 30-day period

Hepatic/renal dose CCr 10-30 ml/min

• **Adult: PO** 6.25 mg initially, max 12.5 mg

Available forms: Tabs 6.25, 12.5 mg

Administer:

• Avoid using more than 2 ×/wk; rebound headache may occur

• Swallow tabs whole; do not break, crush, chew, without regard to food

SIDE EFFECTS

CNS: *Tingling, hot sensation, burning, feeling of pressure, tightness, numbness, dizziness, sedation,* headache, anxiety, fatigue, cold sensation, **seizures**

CV: *Flushing,* palpitations, tachycardia, **coronary artery vasospasm, MI, ventricular fibrillation, ventricular tachycardia**

EENT: Throat, mouth, nasal discomfort; vision changes

GI: Nausea, xerostomia

INTEG: Sweating

MS: *Weakness, neck stiffness,* myalgia

RESP: Chest tightness, pressure

PHARMACOKINETICS

Onset of pain relief 2 hr; peak 1-3 hr; duration 3-4 hr; bioavailability 70%; protein binding 35%; metabolized in the liver (metabolite), metabolized by MAO-A, CYP2D6, CYP3A4; excreted in urine (40%), feces (13%); half-life 3-4 hr

INTERACTIONS

Increase: serotonin syndrome—SSRIs, SNRIs, serotonin-receptor agonists, sibutramine

Increase: vasospastic effects—ergot, ergot derivatives, other 5-HT_1 agonists; avoid concurrent use

⚠ **Increase:** almotriptan effect—MAOIs, do not use together

Increase: plasma concentration of almotriptan—(CYP3A4 inhibitors) itraconazole, ritonavir, erythromycin, ketoconazole, avoid concurrent use in renal/hepatic disease

Drug/Herb

- Avoid use with feverfew

Increase: serotonin syndrome—St. John's wort

NURSING CONSIDERATIONS

Assess:

- **Migraine:** pain location, aura, duration, intensity, nausea, vomiting
- **Serotonin syndrome:** occurs in those taking SSRIs, SNRIs; agitation, confusion, hallucinations, diaphoresis, hypertension, diarrhea, fever, tremors, usually occurs when dose is increased
- B/P; signs/symptoms of coronary vasospasms
- For stress level, activity, recreation, coping mechanisms
- Neurologic status: LOC, blurring vision, nausea, vomiting, tingling, hot sensation, burning, feeling of pressure, numbness, flushing preceding headache
- **Tyramine foods** (pickled products, beer, wine, aged cheese), food additives, preservatives, colorings, artificial sweeteners, chocolate, caffeine, which may precipitate these types of headaches
- Quiet, calm environment with decreased stimulation from noise, bright light, excessive talking

Evaluate:

- Therapeutic response: decrease in severity of migraine

Teach patient/family:

- To report chest pain, drowsiness, dizziness, tingling, flushing
- To use contraception while taking product, to notify prescriber if pregnancy is planned or suspected, to avoid breastfeeding
- That if one dose does not relieve migraine, to take another after 2 hr
- That product does not prevent or reduce number of migraine attacks; use to relieve attack only

alogliptin

(al'oh-glip'tin)

Nesina

Func. class.: Antidiabetic

Chem. class.: Dipeptidyl peptidase-4 (DPP-4) inhibitor

ACTION: A dipeptidyl-peptidase-IV (DDP-IV) inhibitor for the treatment of type 2 diabetes mellitus (monotherapy or in combination with other antidiabetic agents), potentiates the effects of the incretin hormones by inhibiting their breakdown by DDP-IV

USES: Type 2 diabetes mellitus (T2DM)

CONTRAINDICATIONS: Hypersensitivity

Precautions: Pregnancy (B), breastfeeding, hepatic disease, burns, ketoacidosis, diarrhea, fever, GI obstruction, hyper/hypoglycemia, hyper/hypothyroidism, type 1 diabetes, hypercortisolism, children, ileus, malnutrition, pancreatitis, surgery, trauma, vomiting, kidney disease, adrenal insufficiency

Black Box Warning: Angioedema

DOSAGE AND ROUTES

- **Adult: PO** 25 mg/day; when used in combination, a lower dose of the other antidiabetic may be needed

Renal dose

- **Adult: PO** CCr 30–59 ml/min: 12.5 mg q day; CCr <30 ml/min: 6.25 mg q day;

CONTRAINDICATIONS:

Pregnancy (D), breastfeeding, hypersensitivity to benzodiazepines, closed-angle glaucoma, psychosis, addiction

Precautions: Geriatric patients, debilitated patients, hepatic disease, obesity, severe pulmonary disease

DOSAGE AND ROUTES

Anxiety disorder

• **Adult: PO** 0.25-0.5 mg tid, may increase q3-4days if needed, max 4 mg/day in divided doses

• **Geriatric: PO** 0.125-0.25 mg bid; increase by 0.125 as needed

Panic disorder

• **Adult: PO** 0.5 mg tid, may increase up to 1 mg/day q3-4days, max 10 mg/day; **EXT REL** (Xanax XR) give daily in AM, 0.5-1 mg initially, maintenance 3-6 mg/day

Hepatic dose

• Reduce dose by 50%-60%

Premenstrual dysphoric disorders/PMS (unlabeled)

• **Adult: PO** 0.25 mg bid-qid starting on day 16-18 of menses, taper over 2-3 days when menses occurs, max 4 mg/day

Insomnia (unlabeled)

• **Adult: PO** 0.25-0.5 mg at bedtime

Available forms: Tabs 0.25, 0.5, 1, 2 mg; ext rel tabs (Xanax XR) 0.5, 1, 2, 3 mg; orally disintegrating tabs 0.25, 0.5, 1, 2 mg; oral sol 1 mg/ml

Administer:

• Tabs may be crushed, mixed with food, fluids if patient is unable to swallow medication whole; do not break, crush, chew ext rel (XR), give ext rel tab in AM

• With food or milk for GI symptoms; high-fat meal will decrease absorption

• To discontinue, decrease by 0.5 mg q3days

• May divide total daily doses into more times/day if anxiety occurs between doses

• **Orally disintegrating tabs** on tongue to dissolve and swallow, protect from moisture

SIDE EFFECTS

CNS: *Dizziness, drowsiness,* confusion, headache, anxiety, tremors, stimulation, fatigue, depression, insomnia, hallucinations, memory impairment, poor coordination, **suicide**

CV: *Orthostatic hypotension,* **ECG changes, tachycardia,** hypotension

EENT: *Blurred vision,* tinnitus, mydriasis

GI: Constipation, dry mouth, nausea, vomiting, anorexia, diarrhea, weight gain/loss, increased appetite

GU: Decreased libido

INTEG: Rash, dermatitis, itching, **angioedema**

PHARMACOKINETICS

PO: Well absorbed; widely distributed; onset 30 min; peak 1-2 hr; duration 4-6 hr; *oral disintegrating tab* peak 1.5-2 hr; therapeutic response 2-3 days; metabolized by liver (CYP3A4), excreted by kidneys; crosses placenta, breast milk; half-life 12-15 hr, protein binding 80%

INTERACTIONS

Increase: ALPRAZolam action—CYP3A4 inhibitors (cimetidine, disulfiram, erythromycin, FLUoxetine, isoniazid, itraconazole, ketoconazole, metoprolol, propoxyphene, propanolol, valproic acid)

Increase: CNS depression—anticonvulsants, alcohol, antihistamines, sedative/hypnotics, opioids

Decrease: sedation—xanthines

Decrease: ALPRAZolam action—CYP3A4 inducers (barbiturates, rifampin)

Decrease: action of levodopa

Decrease: product level—cigarette smoking

Drug/Herb

Increase: CNS depression—kava, melatonin, St. John's wort, valerian

Drug/Food

Increase: product level—grapefruit juice, avoid concurrent use

Drug/Lab Test

Increase: AST/ALT, alk phos

NURSING CONSIDERATIONS

Assess:

- Mental status: anxiety, mood, sensorium, orientation, affect, sleeping pattern, drowsiness, dizziness, especially in geriatric patients both before and during treatment, suicidal thoughts, behaviors
- B/P lying, standing; pulse; if systolic B/P drops 20 mm Hg, hold product, notify prescriber
- Hepatic, blood studies: AST, ALT, bilirubin, creatinine, LDH, alk phos, CBC; may cause neutropenia, decreased Hct, increased LFTs

⚠ **Physical dependency, withdrawal symptoms:** anxiety, panic attacks, agitation, seizures, headache, nausea, vomiting, muscle pain, weakness; withdrawal seizures may occur after rapid decrease in dose or abrupt discontinuation; because duration of action is short, considered to be the product of choice for geriatric patients

- Pregnancy: if planned or suspected, pregnancy (D), to avoid breastfeeding

Evaluate:

- Therapeutic response: decreased anxiety, restlessness, sleeplessness

Teach patient/family:

- Not to double doses; take exactly as prescribed; if dose is missed, take within 1 hr as scheduled; that product may be taken with food
- Not to use for everyday stress or for more than 4 mo unless directed by prescriber; not to take more than prescribed amount; that product may be habit forming; that memory impairment is a result of long-term use
- To avoid OTC preparations unless approved by prescriber
- Not to use during pregnancy (D), to avoid breastfeeding
- Not to discontinue medication abruptly after long-term use
- To avoid driving, activities that require alertness because drowsiness may occur
- To avoid alcohol, other psychotropic medications unless directed by prescriber
- To rise slowly or fainting may occur, especially among geriatric patients
- That drowsiness may worsen at beginning of treatment

TREATMENT OF OVERDOSE:

Lavage, VS, supportive care, flumazenil

RARELY USED

alprostadil (Rx)

(al-pros′ta-dil)

Caverject, Caverject Impulse, Edex, Muse, Prostin VR Pediatric

Func. class.: Hormone

USES:

To maintain patent ductus arteriosus (temporary treatment), erectile dysfunction

CONTRAINDICATIONS:

Hypersensitivity, respiratory distress syndrome, those at risk for priapism

DOSAGE AND ROUTES

Patent ductus arteriosus

- **Infant: IV INF** 0.05-0.1 mcg/kg/min until desired response, then reduce to lowest effective amount, max 0.4 mcg/kg/min

Erectile dysfunction of vasculogenic or mixed etiology, psychogenic

- **Men: INTRACAVERNOSAL** 2.5 mcg, may increase by 2.5 mcg, may then increase by 5-10 mcg until adequate response occurs (max 60 mcg/dose); **INTRAURETHRAL** 125-250 mcg, max 2 doses/24 hr, max dose 1000 mcg; administer as needed to achieve erection

⚠ HIGH ALERT

alteplase (tissue plasminogen activator, t-PA) (Rx)

(al-ti-plaze′)

Activase, Cathflo

Func. class.: Thrombolytic enzyme

Chem. class.: Tissue plasminogen activator (TPA)

Do not confuse:
alteplase/Altace

ACTION: Produces fibrin conversion of plasminogen to plasmin; able to bind to fibrin, convert plasminogen in thrombus to plasmin, which leads to local fibrinolysis, limited systemic proteolysis

USES: Lysis of obstructing thrombi associated with acute MI, ischemic conditions that require thrombolysis (i.e., PE, unclotting arteriovenous shunts, acute ischemic CVA), central venous catheter occlusion (Cathflo)

Unlabeled uses: Arterial thromboembolism, deep vein thrombosis (DVT), occlusion prophylaxis, percutaneous coronary intervention (PCI), subarachnoid hemorrhage

CONTRAINDICATIONS: Active internal bleeding, history of CVA, severe uncontrolled hypertension, intracranial/intraspinal surgery/trauma (within 3 mo), aneurysm, brain tumor, platelets <100,000 mm³, bleeding diathesis including INR >1.7 or PR >15 sec, arteriovenous malformation, subarachnoid hemorrhage, intracranial hemorrhage, uncontrolled hypertension, seizure at onset of stroke

Precautions: Pregnancy (C), breastfeeding, children, geriatric patients, neurologic deficits, mitral stenosis, recent GI/GU bleeding, diabetic retinopathy, subacute bacterial endocarditis, arrhythmias, diabetic hemorrhage retinopathy, CVA, recent major surgery, hypertension, acute pericarditis, hemostatic defects, significant hepatic disease, septic thrombophlebitis, occluded AV cannula at seriously infected site

DOSAGE AND ROUTES

MI (standard infusion)

• **Adult >65 kg:** 100 mg total given over 3 hr as follows: 6- to 10-mg IV BOL over 1-2 min; remaining 50-54 mg given over the remainder of the hr; during 2nd, 3rd hr 20 mg given by **CONT IV INF** (20 mg/hr)

• **Adult <65 kg:** 1.25 mg/kg over 3 hr: 60% during 1st hr (6%-10% as bolus); remaining 50% given over remainder of hr; during 2nd hr, 20% of dose given by CONT IV INF; 20% given during 3rd hr by CONT IV INF

MI (accelerated infusion)

• **Adult >67 kg:** 100 mg total dose: give 15-mg IV bolus, then 50 mg over 30 min, then 35 mg over 60 min

• **Adult <67 kg:** 15-mg IV bolus, then 0.75 mg/kg (max 50 mg) over 30 min, then 0.5 mg/kg (max 35 mg) over next 60 min

Pulmonary embolism

• **Adult: IV** 100 mg over 2 hr, then heparin

Acute ischemic stroke

• **Adult: IV** 0.9 mg/kg, max 90 mg; give as **INF** over 1 hr, give 10% of dose **IV BOL** over 1st min

Occluded venous access devices

• **Adult/child ≥30 kg: IV** 2 mg/2 ml instilled in occluded catheter, may repeat if needed after 2 hr

• **Child 10-29 kg: IV** 110% of lumen volume, max 2 mg/2 ml instilled in occluded catheter, may repeat if needed after 2 hr

Arterial thromboembolism (unlabeled)

• **Adult: IV** 2 mg/hr for up to 5 hr

Deep venous thrombosis (DVT) (unlabeled)

• **Adult: IV** 4 mcg/kg/min as 2-hr inf, then 1 mcg/kg/min × 33 hr

Percutaneous coronary intervention (PCI) (unlabeled)

• **Adult: INTRACARDIAC** 20 mg over 5 min then 50 mg over the next 60 min

Occlusion prophylaxis (unlabeled)

• **Adult >30 kg:** Do not exceed 2 mg in 2 ml; may use up to 2 doses 120 min apart

Available forms: Powder for inj 50 mg (29 million international units/vial), 100 mg (58 million international units/vial); Cathflo Actinase: lyophilized powder for inj 2 mg

Administer:

Intermittent IV INF route

• After reconstituting with provided diluent, add appropriate amount of sterile water for inj (no preservatives) 20-mg vial/20 ml or 50-mg vial/50 ml to make 1 mg/ml, mix by slow inversion or dilute with NaCl, D_5W to a conc of 0.5 mg/ml; 1.5 to <0.5 mg/ml may result in precipitation of product; use 18G needle; flush line with NaCl after administration, give over 3 hr for MI, 2 hr for PE

• Heparin therapy after thrombolytic therapy is discontinued, TT, ACT, or APTT less than 2× control (about 3-4 hr); treatment can be initiated before coagulation study results obtained, inf should be discontinued if pretreatment INR >1.7, PT >15 sec, or elevated aPTT is identified

• Use reconstituted IV solution within 8 hr or discard

• Avoid invasive procedures, inj, rectal temp

• Pressure for 30 sec to minor bleeding sites; 30 min to sites of atrial puncture followed by pressure dressing; inform prescriber if this does not attain hemostasis; apply pressure dressing

• Store powder at room temperature or refrigerate; protect from excessive light

Y-site compatibilities: Eptifibatide, lidocaine, metoprolol, propranolol

• **Cathflo Actinase:** reconstitute by using 2.2 ml of sterile water provided and injecting in vial, direct flow into powder (1 mg/ml), foam will disappear after standing; swirl, do not shake, sol will be pale yellow or clear, use within 8 hr, instill 2 ml of reconstituted sol into occluded catheter, try to aspirate after ½ hr; if unable to remove, allow 2 hr, a 2nd dose may be used; aspirate 5 ml of blood to remove clot and product, irrigate with normal saline

SIDE EFFECTS

CV: Sinus bradycardia, cholesterol microembolization, ventricular tachycardia, accelerated idioventricular rhythm, bradycardia, recurrent ischemic stroke, hypotension

EENT: Orolingual angioedema

INTEG: Urticaria, rash

SYST: GI, GU, intracranial, retroperitoneal bleeding, *surface bleeding*, anaphylaxis, fever

PHARMACOKINETICS

Cleared by liver, 80% cleared within 10 min of product termination, onset immediate, peak 45 min, duration 4 hr, half-life 35 min

INTERACTIONS

Increase: bleeding—anticoagulants, salicylates, dipyridamole, other NSAIDs, abciximab, eptifibatide, tirofiban, clopidogrel, ticlopidine, some cephalosporins, plicamycin, valproic acid

Increase: orolingual angioedema—ACE inhibitors

Decrease: effect—nitroglycerin

Drug/Herb

Increase: risk for bleeding—feverfew, garlic, ginger, ginkgo, ginseng, green tea

Drug/Lab Test

Increase: PT, APTT, TT

NURSING CONSIDERATIONS

Assess:

• Treatment is not recommended for patients with acute ischemic stroke >3 hr after symptom onset or with minor neurologic deficits or rapidly improving symptoms

• VS, B/P, pulse, respirations, neurologic signs, temp at least q4hr; temp >104° F (40° C) indicates internal bleeding; monitor rhythm closely; ventricular dysrhythmias may occur with hyperfusion; monitor heart, breath sounds, neurologic status, peripheral pulses; assess neurologic status, neurologic change may indicate intracranial bleeding; those with severe neurologic deficit (NIH SS >22) at presentation have increased risk of hemorrhage

⚠ **For bleeding** during 1st hour of treatment and 24 hr after procedure: hematuria, hematemesis, bleeding from mucous membranes, epistaxis, ecchymosis; guaiac all body fluids, stools; do not use 150 mg or more total dose because intracranial bleeding may occur

⚠ **Hypersensitivity:** fever, rash, itching, chills, facial swelling, dyspnea, notify prescriber immediately; stop product, keep resuscitative equipment nearby; mild reaction may be treated with antihistamines

• Previous allergic reactions or streptococcal infection; alteplase may be less effective

• Blood studies (Hct, platelets, PTT, PT, TT, APTT) before starting therapy; PT or APTT must be less than 2× control before starting therapy TT or PT q3-4hr during treatment

• **MI:** ECG continuously, cardiac enzymes, radionuclide myocardial scanning/coronary angiography; chest pain intensity, character; monitor those with major early infarct signs on CT scan with substantial edema, mass effect, midline shift

• PE: pulse, B/P, ABGs, rate/rhythm of respirations

• **Occlusion:** have patient exhale then hold breath when connecting/disconnecting syringe to prevent air embolism

• **Cholesterol embolism:** purple toe syndrome, acute renal failure, gangrenous digits, hypertension, livedo reticularis, pancreatitis, MI, cerebral infarction, spinal cord infarction, retinal artery occlusion, bowel infarction, rhabdomyolysis

Evaluate:

• Therapeutic response: lysis of thrombi

Teach patient/family:

• The purpose and expected results of the treatment; to report adverse reactions

aluminum hydroxide (OTC)

Func. class.: Antacid, hypophosphatemic

Chem. class.: Aluminum product, phosphate binder

ACTION: Neutralizes gastric acidity; binds phosphates in GI tract; these phosphates are then excreted

USES: Antacid, hyperphosphatemia in chronic renal failure; adjunct in gastric, peptic, duodenal ulcers; hyperacidity, reflux esophagitis, heartburn, stress ulcer prevention in critically ill, GERD

Unlabeled uses: GI bleeding

CONTRAINDICATIONS: Hypersensitivity to product or aluminum products

Precautions: Pregnancy (C), breastfeeding, geriatric patients, fluid restriction, decreased GI motility, GI obstruction, dehydration, renal disease, sodium-restricted diets

DOSAGE AND ROUTES

Antacid

• **Adult: PO** 600 mg 1 hr after meals, at bedtime, max 6 doses/day

Hyperphosphatemia

• **Adult: PO** 300-600 mg tid

• **Child: PO** 50-150 mg/kg/day in 4-6 divided doses

Available forms: Susp 320 mg/5 ml, 600 mg/5 ml

Administer:

• 2 tsp (10 ml) will neutralize 20 mEq of acid

PO route

• **Hyperphosphatemia:** give with 8 oz water, meals unless contraindicated

• Laxatives or stool softeners if constipation occurs, especially for geriatric patients
• After shaking susp
NG route: By nasogastric tube if patient unable to swallow

SIDE EFFECTS

GI: *Constipation,* anorexia, obstruction, fecal impaction
META: *Hypophosphatemia,* hypercalciuria

PHARMACOKINETICS

PO: Onset 20-40 min, duration 1-3 hr, excreted in feces

INTERACTIONS

Decrease: effectiveness of—allopurinol, amprenavir, cephalosporins, corticosteroids, delavirdine, digoxin, gabapentin, gatifloxacin, H_2-antagonists, iron salts, isoniazid, ketoconazole, penicillamine, phenothiazines, phenytoin, quiNIDine, quinolones, tetracyclines, thyroid hormones, ticlopidine, anticholinergics; separate by at least 4-6 hr
Drug/Food
Decrease: product effect—high-protein meal
Drug/Lab Test
Decrease: phosphate interference: Tc-99m

NURSING CONSIDERATIONS

Assess:
• **Pain:** location, intensity, duration, character, aggravating, alleviating factors
• Phosphate, calcium levels because product is bound in GI system
• **Hypophosphatemia:** anorexia, weakness, fatigue, bone pain, hyporeflexia
• Constipation; increase bulk in diet if needed, may use stool softeners or laxatives; record amount and consistency of stools
• **Aluminum toxicity:** severe renal disease, may also be used for hyperphosphatemia

Evaluate:
• Therapeutic response: absence of pain, decreased acidity, healed ulcers, decreased phosphate levels
Teach patient/family:
• Not to use for prolonged periods for patients with low serum phosphate or patients on low-sodium diets; to shake liquid well
• That stools may appear white or speckled
• To check with prescriber after 2 wk of self-prescribed antacid use
• To separate from other medications by 2 hr
• **Hyperphosphatemia:** to avoid phosphate foods (most dairy products, eggs, fruits, carbonated beverages) during product therapy
• To notify prescriber of black tarry stools, which may indicate bleeding

alvimopan

(al-vim′oh-pan)
Entereg
Func. class.: Functional GI disorder agent
Chem. class.: Peripheral mu-opioid receptor antagonist

ACTION: A peripherally selective mu-opioid receptor antagonist with activity restricted to the GI tract

USES: Prevention of postoperative ileus
Unlabeled uses: Opiate-agonist–induced constipation

CONTRAINDICATIONS: Those who have taken therapeutic doses of opioids for more than 7 consecutive days immediately before starting alvimopan, end-stage renal disease, Child–Pugh C
Precautions: Risk for MI, surgery for complete GI obstruction, hepatic disease, renal disease, pregnancy B, breastfeeding

DOSAGE AND ROUTES

- **Adult/geriatric patient: PO** 12 mg 30 min-5 hr before surgery, then 12 mg bid beginning the day after surgery; max 7 days or hospital discharge; max 15 doses

Opiate agonist–induced constipation (unlabeled)

- **Adult: PO** 0.5 mg or 1 mg every day

Renal/hepatic dosage

- **Adult: PO** Do not use in end-stage renal disease or Child–Pugh Class C

Available forms: Cap 12 mg

Administer

- Without regard to food

Black Box Warning: Only in a hospital setting approved for the ENTEREG Access Support and Education (E.A.S.E.) program

- Store at room temperature

SIDE EFFECTS

GI: *Constipation, dyspepsia*, flatulence, diarrhea, abdominal pain, cramping
HEMA: *Anemia*
META: Hypokalemia
MISC: Back pain, urinary retention, MI

PHARMACOKINETICS

Half-life 10-17 hr, extended in hepatic/renal disease, protein binding 80%-94%

INTERACTIONS

Increase: Alvimopan concentrations, toxicity—amiodarone, bepridil, cycloSPORINE, diltiazem, itraconazole, quiNIDine, quiNINE, spironolactone, verapamil

- Do not use if opiate agonists have been used for 7 consecutive days before alvimopan
- Do not use concurrently with other opiate antagonists

NURSING CONSIDERATIONS

Assess:

- Blood studies: Hgb/Hct, potassium; hyperkalemia occurs
- Recent opioid use, do not use within 7 consecutive days
- Opioid use for chronic pain; **MI** is more common in this population

Evaluate:

- Therapeutic response: absence of postoperative ileus

Teach patient/family:

- To notify prescriber if pregnancy is planned or suspected, avoid breastfeeding
- That product is only used for a limited time in a hospital setting
- To report constipation, abdominal pain, cramping

amantadine (Rx)

(a-man′ta-deen)

Func. class.: Antiviral, antiparkinsonian agent

Chem. class.: Tricyclic amine

Do not confuse:
amantadine/ranitidine/rimantidine

ACTION: Prevents uncoating of nucleic acid in viral cell, thereby preventing penetration of virus to host; causes release of DOPamine from neurons

USES: Prophylaxis or treatment of influenza type A, EPS, parkinsonism, Parkinson's disease

Unlabeled uses: Neuroleptic malignant syndrome, MS-associated fatigue

CONTRAINDICATIONS: Hypersensitivity, breastfeeding, children <1 eczematic rash

Precautions: Pregnancy (C), geriatric patients, epilepsy, CHF, orthostatic hypotension, psychiatric disorders, renal/hepatic disease, peripheral edema

DOSAGE AND ROUTES

Influenza type A

- **Adult and child >9 yr: PO** 200 mg/day in single dose or divided bid
- **Geriatric: PO** No more than 100 mg/day
- **Child 1-8 yr: PO** 4.4-8.8 mg/kg/day divided bid-tid, max 150 mg/day

Extrapyramidal reaction/ parkinsonism
• **Adult: PO** 100 mg bid up to 400 mg/day in EPS; give for 1 wk then 100 mg as needed up to 400 mg for parkinsonism
Renal dose
• **Adult: PO** CCr 30-50 ml/min 200 mg 1st day then 100 mg/day; CCr 15-29 ml/min 100 mg 1st day then 100 mg on alternate days; CCr 15 ml/min reduce dose and interval to 200 mg q7days
MS-associated fatigue (unlabeled)
• **Adult: PO** 200 mg/day or 100 mg bid
Neuroleptic malignant syndrome (unlabeled)
• **Adult: PO** 100 mg bid × 3 wk
Available forms: Caps 100 mg; oral sol 50 mg/5 ml; tab 100 mg
Administer:
• **Prophylaxis:** before exposure to influenza; continue for 10 days after contact; **treatment:** initiate within 24-48 hr of onset of symptoms, continue for 24-48 hr after symptoms disappear
• After meals for better absorption to decrease GI symptoms; at least 4 hr before bedtime to prevent insomnia
• In divided doses to prevent CNS disturbances: headache, dizziness, fatigue, drowsiness
• Store in tight, dry container

SIDE EFFECTS

CNS: *Headache, dizziness*, drowsiness, fatigue, *anxiety*, psychosis, *depression, hallucinations*, tremors, seizures, confusion, *insomnia*
CV: *Orthostatic hypotension*, CHF
EENT: Blurred vision
GI: *Nausea, vomiting*, constipation, dry mouth, anorexia
GU: *Frequency, retention*
HEMA: Leukopenia, agranulocytosis
INTEG: Photosensitivity, dermatitis, livedo reticularis

PHARMACOKINETICS

PO: Onset 48 hr, peak 1-4 hr, half-life 24 hr, not metabolized, excreted in urine (90%) unchanged, crosses placenta, excreted in breast milk

INTERACTIONS

Increase: anticholinergic response—atropine, other anticholinergics
Increase: CNS stimulation—CNS stimulants
Decrease: amantadine effect—metoclopramide, phenothiazines
Decrease: renal excretion of amantadine—triamterene, hydrochlorothiazide
Decrease: effect—S/B H1N1 influenza A virus vaccine; avoid use 2 wk before or 48 hr after amantadine
Drug/Lab Test
Increase: BUN, creatinine, alk phos, CK, LDH, bilirubin, AST, ALT, GGT

NURSING CONSIDERATIONS

Assess:
• Mental status: may cause increased psychiatric disorders especially in the elderly
• **CHF** (weight gain, jugular venous distention, dyspnea, crackles)
• Skin eruptions, photosensitivity after administration of product
• Reaction to each medication
• Signs of infection
• **Livedo reticularis:** mottling of the skin, usually red; edema; itching in lower extremities
• **Parkinson's disease:** gait, tremors, akinesia, rigidity, may be effective if anticholinergics have not been effective
⚠ **Toxicity:** confusion, behavioral changes, hypotension, seizures
Evaluate:
• Therapeutic response: absence of fever, malaise, cough, dyspnea with infection; tremors, shuffling gait with Parkinson's disease
Teach patient/family:
• To change body position slowly to prevent orthostatic hypotension
• About aspects of product therapy: to report dyspnea, weight gain, dizziness, poor concentration, dysuria, complex sleep behaviors
• To avoid hazardous activities if dizziness, blurred vision occurs

- To take product exactly as prescribed; parkinsonian crisis may occur if product is discontinued abruptly; not to double dose; if a dose is missed, not to take within 4 hr of next dose; caps may be opened and mixed with food
- To avoid alcohol

TREATMENT OF OVERDOSE:
Withdraw product, maintain airway, administer EPINEPHrine, aminophylline, O_2, IV corticosteroids, physostigmine

ambrisentan (Rx)
(am-bri-sen′tan)
Letairis
Func. class.: Antihypertensive
Chem class.: Vasodilator/endothelin receptor antagonist

ACTION: Endothelin-A receptor antagonist; endothelin-A is a vasoconstrictor

USES: Pulmonary arterial hypertension, alone or in combination with other antihypertensives

CONTRAINDICATIONS: Breastfeeding, hypersensitivity, idiopathic pulmonary fibrosis (IPF)

Black Box Warning: Pregnancy (X)

Precautions: Children, females, geriatric patients, hepatitis, anemia, heart failure, jaundice, peripheral edema, hepatic disease, pulmonary disease

DOSAGE AND ROUTES
- **Adult: PO** 5 mg/day; may increase to 10 mg/day if needed

Hepatic dose
- **Adult: PO Discontinue if AST/ALT >5× ULN, or if elevations are accompanied by bilirubin >2× ULN, or other signs of liver dysfunction**

Available forms: Tabs 5, 10 mg

Administer:
- Do not break, crush, chew tabs
- Daily with a full glass of water without regard to food
- Do not discontinue abruptly
- Only those facilities enrolled in the LEAP program (866-664-5327) may administer this product
- Store in tight container at room temp

SIDE EFFECTS
CNS: *Headache,* fever, flushing, fatigue
CV: Orthostatic hypotension, hypotension, *peripheral edema,* palpitations
EENT: Sinusitis, rhinitis
GI: Abdominal pain, constipation, anorexia, **hepatotoxicity**
GU: Decreased sperm counts
HEMA: *Anemia*
INTEG: Rash, **angioedema**
RESP: Pharyngitis, dyspnea, **pulmonary edema, veno-occlusive disease (VOD)**

PHARMACOKINETICS
Rapidly absorbed, peak 2 hr, protein binding 99%, metabolized by CYP3A4, CYP2C19, UGTa, terminal half-life 15 hr, effective half-life 9 hr

INTERACTIONS
- Possibly increase ambrisentan: cimetidine, clopidogrel, efavirenz, felbamate, FLUoxetine, modafinil, OXcarbazepine, ticlopidine

Increase: hypotension—other antihypertensives, diuretics, MAOIs
Increase: ambrisentan—CYP3A4 inhibitors (amprenavir, aprepitant, atazanavir, clarithromycin, conivaptan, cycloSPORINE, dalfopristin, danazol, darunavir, erythromycin, estradiol, imatinib, itraconazole, ketoconazole, nefazodone, nelfinavir, propoxyphene, quinupristin, ritonavir, RU-486, saquinavir, tamoxifen, telithromycin, troleandomycin, zafirlukast); CYP2C19/CYP3A4 (chloramphenicol, delavirdine, fluconazole, fluvoxaMINE, isoniazid, voriconazole)

Decrease: ambrisentan—CYP3A4 inducers (carBAMazepine, PHENobarbital, phenytoin, rifampin)
Decrease: ambrisentan absorption—mefloquine, niCARdipine, propafenone, quiNIDine, ranolazine, tacrolimus, testosterone
Drug/Herb
• Need for ambrisentan dosage change: St. John's wort, ephedra (ma huang)
Drug/Food
• Avoid use with grapefruit products
Drug/Lab Test
Increase: LFTs, bilirubin
Decrease: Hct, Hgb

NURSING CONSIDERATIONS

Assess:
• **Pulmonary status:** improvement in breathing, ability to exercise; pulmonary edema that may indicate veno-occlusive disease
• Blood studies: CBC with differential; Hct, Hgb may be decreased
• Liver function tests: AST, ALT, bilirubin

Black Box Warning: Assess pregnancy status before giving this product; pregnancy (X)

• Hepatotoxicity: nausea, vomiting, abdominal pain/cramping, jaundice, anorexia, itching
Evaluate:
• Therapeutic response: decrease in B/P; decreased shortness of breath; importance of follow up with labs
Teach patient/family:
• The importance of complying with dosage schedule even if feeling better

Black Box Warning: To notify if pregnancy is planned or suspected (if pregnant, product will need to be discontinued, pregnancy test done monthly); to use 2 contraception methods while taking this product

• Not to use OTC products including herbs, supplements unless approved by prescriber
• To report to prescriber immediately: dizziness, faintness, chest pain, palpitations, uneven or rapid heart rate, headache, edema, weight gain
• To report hepatic dysfunction: nausea/vomiting, anorexia, fatigue, jaundice, right upper quadrant abdominal pain, itching, fever, malaise

amifostine (Rx)
(a-mi-foss'teen)
Ethyol
Func. class.: Cytoprotective agent for cisplatin/radiation

ACTION: Binds and detoxifies damaging metabolites of cisplatin, alkylating agents, DNA-reactive agents, ionizing radiation by converting this product by alk phos in tissue to an active free thiol compound

USES: To reduce renal toxicity when cisplatin is given repeatedly for ovarian cancer; to reduce xerostomia (dry mouth) after radiation therapy for head, neck cancer
Unlabeled uses: To prevent or reduce cisplatin-induced neurotoxicity, cyclophosphamide-induced granulocytopenia; to prevent or reduce toxicity of radiation therapy; to reduce toxicity of PACLitaxel, myelodysplastic syndrome (MDS)

CONTRAINDICATIONS: Breastfeeding, hypersensitivity to mannitol, aminothiol; hypotension, dehydration
Precautions: Pregnancy (C), children, geriatric patients, CV disease, hypocalcemia, MI, radiation therapy, stroke, dehydration, hypotension, chemotherapy, exfoliative dermatitis

DOSAGE AND ROUTES

Reduction of renal damage with cisplatin
• **Adult: IV** 910 mg/m^2/day within $^1/_2$ hr before chemotherapy, give over 15 min; may reduce dose to 740 mg/m^2 if higher dose is poorly tolerated

A

Xerostomia prophylaxis
• **Adult: IV** 200 mg/m²/day over 3 min as inf 15-30 min before radiation therapy
Bone marrow suppression/ nephrotoxicity/neurotoxicity prophylaxis (unlabeled)
• **Adult: IV** 100-340 mg/m²/day over 15 min before each dose of chemotherapy/ radiation
Myelodyplastic syndrome (MDS) (unlabeled)
• **Adult: IV** 100 mg/m² 3×/wk, max 300 mg/m² 3×/wk
Available forms: Powder for inj, lyophilized 500 mg/vial
Administer:
Intermittent IV INF route
• Intermittent inf after reconstituting 500 mg/9.7 ml of sterile 0.9% NaCl, further dilute with 0.9% NaCl to concentration of 5-40 mg/ml, give over 15 min within ½ hr of chemotherapy, place patient in supine position during inf
• **Xerostomia:** Give over 3 min 15-30 min before radiation

Y-site compatibilities: Amikacin, aminophylline, amphotericin B liposome, ampicillin, ampicillin/sulbactam, aztreonam, bivalirudin, bleomycin, bumetanide, buprenorphine, butorphanol, calcium gluconate, CARBOplatin, carmustine, caspofungin, ceFAZolin, cefotaxime, cefoTEtan, cefOXitin, cefTAZidime, ceftizoxime, cefTRIAXone, cefuroxime, cimetidine, ciprofloxacin, clindamycin, cyclophosphamide, cytarabine, dacarbazine, DACTINomycin, DAPTOmycin, DAUNOrubicin, dexamethasone, dexmedetomidine, diltiazem, diphenhydrAMINE, DOBUTamine, DOCEtaxel, DOPamine, DOXOrubicin, doxycycline, droperidol, enalaprilat, epirubicin, ertapenem, etoposide, famotidine, fenoldopam, floxuridine, fluconazole, fludarabine, fluorouracil, furosemide, gallium, gentamicin, granisetron, haloperidol, heparin, hydrocortisone, HYDROmorphone, IDArubicin, ifosfamide, imipenem-cilastatin, leucovorin, levofloxacin, linezolid, LORazepam, magnesium sulfate, mannitol, mechlorethamine, meperidine, mesna, methotrexate, methylPREDNISolone, metoclopramide, metroNIDAZOLE, mezlocillin, milrinone, mitoMYcin, morphine, nalbuphine, nesiritide, netilmicin, octreotide, ondansetron, oxaliplatin, PACLitaxel, palonosetron, pantoprazole, PEMEtrexed, piperacillin, plicamycin, potassium chloride, promethazine, ranitidine, sodium bicarbonate, streptozocin, tacrolimus, teniposide, thiotepa, ticarcillin, ticarcillin/clavulanate, tigecycline, tirofiban, tobramycin, trastuzumab, trimethoprim-sulfamethoxazole, trimetrexate, vancomycin, vinBLAStine, vinCRIStine, voriconazole, zidovudine
Solution compatibility: 0.9% NaCl
Additive incompatibilities: Do not mix with other products

SIDE EFFECTS

CNS: Dizziness, somnolence, loss of consciousness, **seizures**
CV: *Hypotension*
EENT: *Sneezing*
GI: *Nausea, vomiting,* hiccups, *diarrhea*
INTEG: Flushing, feeling of warmth
MISC: *Hypocalcemia,* rash, chills, **anaphylaxis, toxic epidermal necrolysis, toxicoderma, Stevens-Johnson syndrome, exfoliative dermatitis, erythema multiforme**

PHARMACOKINETICS

Metabolized to free thiol compound, half-life 8 min, onset 5-8 min

INTERACTIONS

Increase: hypotension—antihypertensives
Drug/Lab
Decrease: calcium

NURSING CONSIDERATIONS

Assess:
• B/P before and q5min during inf; antihypertensives should be discontinued 24 hr before inf if severe hypotension occurs, give IV 0.9% NaCl to expand fluid

volume, place patient in modified Trendelenburg position

• Fluid status before administration; administer antiemetic before administration to prevent severe nausea and vomiting; may also give dexamethasone 20 mg IV and a serotonin antagonist such as ondansetron, dolasetron, granisetron

• Calcium levels before and during treatment; may cause hypocalcemia; calcium supplements may be needed

• **Xerostomia:** mouth lesions, dry mouth during therapy

• **Anaphylaxis:** rash, pruritus, wheezing, laryngeal edema, discontinue; give antihistamines, EPINEPHrine depending on severity; reactions may be delayed, if severe product should be permanently discontinued; other serious skin disorders (Stevens-Johnson syndrome, toxic epidermal necrolysis, toxicoderma, erythema multiforme, exfoliative dermatitis)

Evaluate:

• Therapeutic response: prevention of renal toxicity associated with cisplatin therapy; decreased xerostomia associated with radiation therapy of head, neck cancer

Teach patient/family:

• To stay supine during inf; B/P will be monitored often

• To drink fluids

• To report signs, symptoms, severe reactions

• The reason for the medication and expected results

• That side effects may include severe nausea, vomiting, decreased B/P, chills, dizziness, somnolence, hiccups, sneezing, skin reactions

• Not to breastfeed

amikacin (Rx)

(am-i-kay′sin)

Amikin

Func. class.: Antiinfective

Chem. class.: Aminoglycoside

Do not confuse:

Amikin/Amicar

Amikacin/anakinra

ACTION: Interferes with protein synthesis in bacterial cells by binding to ribosomal subunits, which causes misreading of genetic code; inaccurate peptide sequence forms in protein chain, thereby causing bacterial death

USES: Severe systemic infections of CNS, respiratory tract, GI tract, urinary tract, bone, skin, soft tissues caused by *Staphylococcus aureus (MSSA), Pseudomonas aeruginosa, Escherichia coli, Enterobacter, Acinetobacter, Providencia, Citrobacter, Serratia, Proteus, Klebsiella pneumoniae*

Unlabeled uses: *Mycobacterium avium* complex (MAC) (intrathecal or intraventricular) in combination; aerosolization, actinomycotic mycetoma, febrile neutropenia

CONTRAINDICATIONS: Pregnancy (D), mild to moderate infections, hypersensitivity to aminoglycosides, sulfites

Precautions: Breastfeeding, neonates, geriatric patients, myasthenia gravis, Parkinson's disease

Black Box Warning: Hearing impairment, renal/neuromuscular disease

DOSAGE AND ROUTES

Severe systemic infections

• **Adult and child: IV INF** 15 mg/kg/day in 2-3 divided doses q8-12hr in 100-200 ml D_5W over 30-60 min, max 1.5 g; **pulse dosing** (once-daily dosing) may be used with some infections; **IM** 10-15 mg/kg/day in divided doses q8-12hr;

daily or extended-interval dosing as an alternative dosing regimen

• **Infant: IV/IM** 10 mg/kg initially then 7.5 mg/kg q12hr

• **Neonate: IV/IM** 10 mg/kg initially then 7.5 mg/kg q8-12hr

Severe urinary tract infections

• **Adult: IM** 10-15 mg/kg/day divided q8-12hr

Hemodialysis

• **Adult: IM/IV** 7.5 mg/kg followed by 5 mg/kg 3×/wk after each dialysis session (for TIW dialysis)

Renal dose (extended interval dosing)

• **Adult: IV** CCr 40-59 ml/min 15 mg/kg q36hr; CCr 20-39 ml/min 15 mg/kg q48hr; <20 ml/min adjust based on serum concentrations and MIC (use traditional dosing)

Traditional dosing

• Decrease dose and maintain interval or maintain dose and decrease interval

Mycobacterium avium complex (MAC) (unlabeled)

• **Adult and adolescent: IM/IV** 15-20 mg/kg/day or 5× per wk

• **Child: IV** 15-30 mg/kg/day divided q12-24hr as part of multiple-drug regimen, max 1.5 g/day

Actinomycotic mycetoma (unlabeled)

• **Adult: IM/IV** 15 mg/kg/day in 2 divided doses × 3 wk with co-trimoxazole for 5 wk; repeat cycle once, may be repeated 2×

Available forms: Inj 50, 250 mg/ml

Administer:

• Obtain C&S before administration; begin treatment before results

IM route

• Inj in large muscle mass; rotate inj sites

• Obtain peak 1 hr after IM, trough before next dose

Intermittent IV INF route

• Dilute 500 mg of product/100-200 ml of D_5W, 0.9% NaCl and give over ½-1 hr; dilute in sufficient volume to allow for inf over 1-2 hr (infants); flush after administration with D_5W or 0.9% NaCl; sol clear or pale yellow; discard if precipitate or dark color develops

• In children, amount of fluid will depend on ordered dose; in infants, infuse over 1-2 hr

• In evenly spaced doses to maintain blood level, separate from penicillins by at least 1 hr

Y-site compatibilities: Acyclovir, alatrafloxacin, aldesleukin, alemtuzumab, alfentanil, amifostine, aminophylline, amiodarone, amsacrine, anidulafungin, argatroban, ascorbic acid, atracurium, atropine, aztreonam, benztropine, bivalirudin, bumetanide, buprenorphine, butorphanol, calcium chloride/gluconate, CARBOplatin, caspofungin, ceFAZolin, cefepime, cefonicid, cefotaxime, cefoTEtan, cefOXitin, cefTAZidime, ceftizoxime, cefTRIAXone, cefuroxime, chloramphenicol, chlorproMAZINE, cimetidine, cisatracurium, CISplatin, clindamycin, codeine, cyanocobalamin, cyclophosphamide, cycloSPORINE, cytarabine, DACTINomycin, DAPTOmycin, dexamethasone, dexmedetomidine, digoxin, diltiazem, diphenhydrAMINE, DOBUTamine, DOCEtaxel, DOPamine, doripenem, doxacurium, DOXOrubicin, doxycycline, eftifibatide, enalaprilat, ePHEDrine, EPINEPHrine, epirubicin, epoetin alfa, ertapenem, erythromycin, esmolol, etoposide, famotidine, fentaNYL, filgrastim, fluconazole, fludarabine, fluorouracil, foscarnet, furosemide, gemcitabine, gentamicin, glycopyrrolate, granisetron, hydrocortisone, HYDROmorphone, IDArubicin, ifosfamide, IL-2, imipenem/cilastin, isoproterenol, ketorolac, labetalol, levofloxacin, lidocaine, linezolid, LORazepam, magnesium sulfate, mannitol, mechlorethamine, melphalan, meperidine, metaraminol, methotrexate, methoxamine, methyldopate, methylPREDNISolone, metoclopramide, metoprolol, metroNIDAZOLE, midazolam, milrinone, mitoXANtrone, morphine, multivitamins, nafcillin, nalbuphine, naloxone, niCARdipine, nitroglycerin, nitroprusside, norepinephrine, octreotide, ondansetron, oxaliplatin, oxytocin, PACLitaxel, palonosetron, pantoprazole, papaverine, PEMEtrexed, penicillin G, pentazocine, perphenazine, PHENobarbital, phenylephrine, phytonadione, piperacillin/tazobactam, potassium chloride, procainamide, prochlorperazine, promethazine, propranolol, protamine, pyridoxime, quinupristin/dalfopristin, rantidine, remifentanil, riTUXimab, rocuromium, sargramostim, sodium acetate, sodium bicarbonate, succinylcholine, SUFentanil, tacrolimus, teniposide, theophylline, thiamine, thiotepa, ticarcillin/clavulanate, tigecycline, tirofiban, tobramycin, tolazoline, trimethaphan, urokinase, vancomycin, vasopressin, vecuronium, verapamil, vinCRIStine, vinorelbine, voriconazole, warfarin, zidovudine, zoledronic acid

SIDE EFFECTS

CNS: Confusion, depression, numbness, tremors, seizures, muscle twitching, neurotoxicity, dizziness, vertigo, tinnitus, neuromuscular blockade with respiratory paralysis
CV: Hypo/hypertension
EENT: *Ototoxicity,* deafness
GI: *Nausea, vomiting, anorexia;* bilirubin
GU: Oliguria, hematuria, renal damage, azotemia, renal failure, nephrotoxicity
HEMA: Eosinophilia, anemia
INTEG: *Rash,* burning, urticaria, dermatitis, alopecia

PHARMACOKINETICS

IM: Onset rapid, peak 1 hr, leads to unpredictable concentrations, IV preferred
IV: Onset immediate, peak 15-30 min, half-life 2 hr, prolonged up to 7 hr in infants; not metabolized; excreted unchanged in urine; crosses placenta; removed by hemodialysis

INTERACTIONS

Increase: masking ototoxicity: dimenhyDRINATE, ethacrynic acid
Increase: serum trough and peak—NSAIDs

Black Box Warning: Increase: nephrotoxicity—cephalosporins, acyclovir, vancomycin, amphotericin B, cycloSPORINE, loop diuretics, cidofovir

Black Box Warning: Increase: Ototoxicito-IV loop diuretics

Increase: neuromuscular blockade, respiratory depression—anesthetics, nondepolarizing neuromuscular blockers
Decrease: amikacin effect—parenteral penicillins, do not combine

Drug/Lab Test

Increase: BUN, creatinine, urea levels (urine)

NURSING CONSIDERATIONS

Assess:

- Weight before treatment; calculation of dosage is usually based on ideal body weight but may be calculated on actual body weight
- I&O ratio; urinalysis daily for proteinuria, cells, casts; report sudden change in urine output
- VS during infusion; watch for hypotension, change in pulse
- IV site for thrombophlebitis including pain, redness, swelling q30min; change site if needed; apply warm compresses to discontinued site

Black Box Warning: Renal impairment by securing urine for CCr, BUN, serum creatinine; lower dosage should be given with renal impairment; nephrotoxicity may be reversible if product stopped at 1st sign

Black Box Warning: Deafness: audiometric testing, ringing, roaring in ears, vertigo; assess hearing before, during, after treatment

- Dehydration: high specific gravity, decrease in skin turgor, dry mucous membranes, dark urine
- Overgrowth of infection: increased temp, malaise, redness, pain, swelling, perineal itching, diarrhea, stomatitis, change in cough, sputum
- Vestibular dysfunction: nausea, vomit-

ing, dizziness, headache; product should be discontinued if severe

Black Box Warning: Neuromuscular blockade; respiratory paralysis may occur

• Provide adequate fluids of 2-3 L/day, unless contraindicated, to prevent irritation of tubules

Evaluate:

• Therapeutic response: absence of fever, draining wounds, negative C&S after treatment

Teach patient/family:

• To report headache, dizziness, symptoms of overgrowth of infection, renal impairment, symptoms of neurotoxicity, hepatotoxicity

⚠ To report loss of hearing; ringing, roaring in ears; feeling of fullness in head

• To report hypersensitivity: rash, itching, trouble breathing, facial edema; notify health care provider

TREATMENT OF HYPERSENSITIVITY: Hemodialysis, exchange transfusion in the newborn, monitor serum levels of product, may give ticarcillin or carbenicillin

aMILoride (Rx)

(a-mill′oh-ride)

Midamor

Func. class.: Potassium-sparing diuretic

Chem. class.: Pyrazine

Do not confuse:

aMILoride/amLODIPine/amiodarone

ACTION: Inhibits sodium, potassium ion exchange in the distal tubule, cortical collecting duct resulting in inhibition of sodium reabsorption and decreasing potassium secretion

USES: Edema in CHF in combination with other diuretics, for hypertension, adjunct with other diuretics to maintain potassium

Unlabeled uses: Ascites

CONTRAINDICATIONS: Anuria, hypersensitivity, diabetic nephropathy

Black Box Warning: Hyperkalemia

Precautions: Pregnancy (B), breastfeeding, children, geriatric patients, dehydration, diabetes, respiratory acidosis, hyponatremia, impaired renal function

DOSAGE AND ROUTES

• **Adult: PO** 5-10 mg/day in 1-2 divided doses; may be increased to 10-20 mg/day if needed

• **Infant/child (6-20 kg): PO** 0.625 mg/kg/day

Renal dose

• **Adult: PO** CCr 10-50 ml/min reduce dose by 50%; however, avoid if possible; CCr <10 ml/min contraindicated

Ascites (unlabeled)

• **Adult: PO** 10 mg/day, max 40 mg

Available forms: Tabs 5 mg

Administer:

• In AM to avoid interference with sleep if using as diuretic; if 2nd daily dose is needed, give in late afternoon

• With food; if nausea occurs, absorption may be decreased slightly

SIDE EFFECTS

CNS: *Headache,* dizziness, fatigue, weakness, paresthesias, tremors, depression, anxiety, encephalopathy

CV: *Orthostatic hypotension,* dysrhythmias, chest pain

EENT: Blurred vision, increased intraocular pressure

ELECT: Hyperkalemia, dehydration, hyponatremia, hypochloremia

GI: *Nausea, diarrhea,* dry mouth, *vomiting, anorexia,* cramps, constipation, abdominal pain, jaundice

GU: *Polyuria,* dysuria, urinary frequency, impotence

HEMA: Aplastic anemia, neutropenia

INTEG: *Rash, pruritus,* alopecia, urticaria

MS: Cramps

RESP: *Cough, dyspnea,* shortness of breath

PHARMACOKINETICS

50% absorbed from GI tract; widely distributed; onset 2 hr; peak 3-4 hr; duration 24 hr; excreted in urine, feces; half-life 6-9 hr

INTERACTIONS

Black Box Warning: Hyperkalemia: other potassium-sparing diuretics, potassium products, ACE inhibitors, salt substitutes, cycloSPORINE, tacrolimus; avoid concurrent use, if using together monitor potassium level

Increase: lithium toxicity: lithium, monitor lithium levels
Increase: action of antihypertensives
Decrease: effect of aMILoride—NSAIDs, avoid concurrent use

Drug/Herb
Increase: effect—hawthorn, horse chestnut

Drug/Food
- Possible hyperkalemia: foods high in potassium, potassium-based salt substitutes

Drug/Lab Test
Increase: LFTs, BUN, potassium, sodium
Interference: GTT

NURSING CONSIDERATIONS

Assess:
- Heart rate, B/P lying, standing; postural hypotension may occur
- Electrolytes: potassium, sodium, chloride; glucose (serum), BUN, CBC, serum creatinine, blood pH, ABGs, periodic ECG
- Discontinue potassium-sparing diuretics 3 days before GTT, hyperkalemia may occur
- **Hypokalemia:** weakness, polyuria, polydipsia, fatigue, ECG U wave

Black Box Warning: Hyperkalemia: fatigue, weakness, paresthesia, confusion, dyspnea, dysrhythmias, ECG changes

Evaluate:
- Therapeutic response: improvement in edema of feet, legs, sacral area daily if medication is being used for CHF; decreased B/P; prevention of hypokalemia (diuretics)

Teach patient/family:
- To take as prescribed; if dose is missed, to take when remembered within 1 hr of next dose; to take with food or milk for GI symptoms; to take early in day to prevent nocturia; to avoid alcohol
- About adverse reactions: muscle cramps, weakness, nausea, dizziness, blurred vision
- To avoid potassium-rich foods: oranges, bananas, salt substitutes, dried fruits, potassium supplements
- To rise slowly from sitting to standing to avoid orthostatic hypotension
- To avoid hazardous activities if dizziness occurs

TREATMENT OF OVERDOSE:

Lavage if taken orally, monitor electrolytes, administer sodium bicarbonate for potassium >6.5 mEq/L, IV glucose, kayoxalate as needed; monitor hydration, CV, renal status

A

amino acids
(a-mee′noe)

amino acid infusions (crystalline)
Aminosyn, Amynosyn II, Aminosyn-PF, Clinisol, FreAmine III, Premasol, Travasol, TrophAmine

amino acid infusions/ dextrose
Aminosyn II with dextrose, Clinimix, Amino Acid Infusions/ electrolytes

aminosyn with electrolytes
Aminosyn II with electrolytes, FreAmine III with electrolytes, ProcalAmine with electrolytes, Travasol with electrolytes

amino acid infusions/ electrolytes/dextrose
Aminosyn II with electrolytes in dextrose

amino acid infusions (hepatic failure)
HepatAmine, Hepatasol

amino acid infusions (high metabolic stress)
Aminosyn-HBC, FreAmine HBC

amino acid infusions (renal failure)
Aminosyn-RF, NephrAmine

Func. class.: Nutritional supplement/ protein
Chem. class.: n/a

ACTION: Needed for anabolism to maintain structure, decrease catabolism, promote healing

Black Box Warning: Central infusions: Administration by central venous catheter should be used only by those familiar with this technique and its complications

USES: Hepatic encephalopathy, cirrhosis, hepatitis, nutritional support in cancer; burn or solid organ transplant patients; to prevent nitrogen loss when adequate nutrition by mouth, gastric, or duodenal tube cannot be obtained

CONTRAINDICATIONS: Hypersensitivity, severe electrolyte imbalances, anuria, severe liver damage, maple syrup urine disease, PKU, azotemia, genetic disease of amino acid metabolism

Precautions: Pregnancy (C), breastfeeding, children, renal disease, diabetes mellitus, CHF, sulfite sensitivity

DOSAGE AND ROUTES

Nutritional support (cirrhosis, hepatic encephalopathy, hepatitis)
- **Adult:** **IV** 80-120 g/day amino acids/12-18 g nitrogen of hepatic failure formula

TPN
- **Adult:** **IV** 1-1.5 g/kg/day

Metabolic stress (severe)
- **Adult:** **IV** 1.5 g/kg, use formula for high metabolic stress

Renal failure (nutritional support)
- **Adult:** **IV** Minosyn-RF 300-600 ml/70% dextrose/day; NephrAmine 250-500 ml/70% dextrose/day; the total daily dosage is calculated based on the daily protein requirements, as well as the patient's metabolic and clinical response. Check product instructions for specific directions
- **Child ≥10 kg (22 lb):** **IV** 20-25 g/day × first 10 kg, then 1-1.25 g/kg/day for each kg >10 kg
- **Child <10 kg (22 lb):** **IV** 2-4 kg/day
- **Child:** **IV** 1 g/kg/day max
- The dosage and constant infusion rate of IV dextrose must be selected with caution in children, particularly neonates

and low birth weight infants, because of the increased risk of hyper/hypoglycemia. Owing to their concentration, these solutions are not recommended for use in children younger than 1 year

• The total daily dosage is calculated based on the daily protein requirements, as well as the patient's metabolic and clinical response. Check each product's instructions for specific directions

Available forms:

Injection: 250, 500, 1000, 2000 ml, containing amino acids in various concentrations; amino acid infusions, crystalline: Aminosyn: 3.5%, 5%, 7%, 8.5%, 10%; Aminosyn II: 3.5%, 5%, 7%, 8.5%, 10%, 15%; Aminosyn-PF: 7%, 10%; Clinisol: 15%; FreAmine III: 8.5%, 10%; Premasol: 6%, 10%; Travasol: 10%; TrophAmine: 6%, 10%

Amino acid infusion/dextrose:

Aminosyn II: 3.5% in 5% dextrose, 4.25% in 20% dextrose, 4.25% in 10% dextrose 4.25% in 25% dextrose; Clinimix: 2.75% in 5% dextrose, 4.25% in 5% dextrose, 4.25% in 10% dextrose 4.25% in 20% dextrose, 4.25% in 25% dextrose, 5% in 10% dextrose, 5% in 15% dextrose, 5% in 20% dextrose, 5% in 25% dextrose, 5 Amino acid infusions/electrolytes: Aminosyn: 3.5%, 7%, 8.5%; Aminosyn II: 3.5%, 7%, 8.5%; FreAmine III: 3%, 8.5%; ProcalAmine: 3%; Travasol: 3.5%, 5.5%, 8.5%

Amino acid infusions/electrolytes/dextrose:

Aminosyn II: 3.5% with electrolytes in 5% dextrose, 3.5% with electrolytes in 25% dextrose, 4.25% with electrolytes in 10% dextrose, 4.25% with electrolytes in 20% dextrose, 4.25% with electrolytes in 25% dextrose; Amino acid infusions (hepatic failure): HepatAmine: 8%; Hepatasol: 8%

Amino acid infusions (high metabolic stress):

Aminosyn-HBC 7%; Freamine HBC: 6.9%; Amino acid (renal failure): Aminosyn-RF:5.2%; NephrAmine: 5.4%

Administer:

Continuous IV INF route

• Up to 40% protein and dextrose (up to 12.5%) via peripheral vein; stronger sol requires central IV administration

• TPN only mixed with dextrose to promote protein synthesis

• Immediately after mixing under strict aseptic technique, use inf pump, in-line filter (0.22 μm) unless mixed with fat emulsion and dextrose (3 in 1); using careful monitoring technique; do not speed up inf; pulmonary edema, glucose overload will result

SIDE EFFECTS

CNS: Dizziness, headache, confusion, loss of concentration, fever

CV: Hypertension, **CHF/pulmonary edema**, flushing, **thrombosis**

ENDO: Hyperglycemia, rebound hypoglycemia, electrolyte imbalances, hyperosmolar hyperglycemic nonketotic syndrome, alkalosis, hypophosphatemia, hyperammonemia, dehydration, hypocalcemia

GI: Nausea, abdominal pain

GU: Glycosuria osmotic dieresis

INTEG: **Extravasation necrosis, phlebitis at inj site**

INTERACTIONS

Individual Drugs

Decrease: Protein-sparing effects—tetracycline

Drug/Lab Test

Increase: LFTs, ammonia glucose

Decrease: Potassium, phosphate, glucose

NURSING CONSIDERATIONS

Assess:

• Electrolytes (potassium, sodium, phosphate, chloride, magnesium, bicarbonate, blood glucose, ammonia, ketones)

• Renal/hepatic studies: BUN, creatinine, ALT, AST, bilirubin

• Weight changes, triglycerides before and after inf; vit A level with renal disease

- Inj site for extravasation: redness along vein, edema at site, necrosis, pain, hard tender area; site should be changed immediately
- Discontinue inf, culture tubing and sol
- **Sepsis:** chills, fever, increased temperature, if sepsis is suspected
- For impending hepatic coma: asterixis, confusion, uremic fetor, lethargy
- Hyperammonemia: nausea, vomiting, malaise, tremors, anorexia, seizures
- Change of dressing and IV tubing to prevent infection every 24-48 hr if chills, fever, other signs of infection occur

Evaluate:

- Therapeutic response: weight gain, decrease in jaundice with liver disorders, increased LOC

Teach patient/family:

- Reason for use of TPN
- If chills, sweating are experienced to report at once
- About infusion pump and blood glucose

⚠ HIGH ALERT

amiodarone (Rx)

(a-mee-oh′da-rone)

Cordarone, Nexterone, Pacerone

Func. class.: Antidysrhythmic (class III)

Chem. class.: Iodinated benzofuran derivative

Do not confuse:
amiodarone/Inamrinone
Cordarone/Inocor

ACTION: Prolongs duration of action potential and effective refractory period, noncompetitive α- and β-adrenergic inhibition; increases PR and QT intervals, decreases sinus rate, decreases peripheral vascular resistance

USES: Hemodynamically unstable ventricular tachycardia, supraventricular tachycardia, ventricular fibrillation not controlled by 1st-line agents

Unlabeled uses: Atrial fibrillation treatment/prophylaxis, atrial flutter, cardiac arrest, cardiac surgery, CPR, heart failure, PSVT, Wolff-Parkinson-White (WPW) syndrome, supraventricular tachycardia

CONTRAINDICATIONS: Pregnancy (D), breastfeeding, neonates, infants, severe sinus node dysfunction, hypersensitivity to this product/iodine/benzyl alcohol, cardiogenic shock

Black Box Warning: 2nd-3rd degree AV block, bradycardia

Precautions: Children, goiter, Hashimoto's thyroiditis, electrolyte imbalances, CHF, respiratory disease, torsades de pointes

Black Box Warning: Severe hepatic disease, cardiac arrhythmias, pneumonitis, pulmonary fibrosis

DOSAGE AND ROUTES

Ventricular dysrhythmias

- **Adult: PO** Loading dose 800-1600 mg/day for 1-3 wk then 600-800 mg/day for 1 mo, maintenance 400 mg/day; **IV** loading dose (1st rapid) 150 mg over the first 10 min then slow 360 mg over the next 6 hr, maintenance 540 mg given over the remaining 18 hr, decrease rate of the slow inf to 0.5 mg/min
- **Child (unlabeled): PO** Loading dose 10-15 mg/kg/day in 1-2 divided doses for 4-14 days then 5 mg/kg/day
- **Child and infant: IV/INTRAOSSEOUS** (during CPR) 5 mg/kg as a bolus (PALS guidelines)

Supraventricular dysrhythmias (atrial fibrillation, atrial flutter, PSVT, WPW syndrome) (unlabeled)

- **Adult: PO** 1.2-1.8 g/day divided until a total of 10 g has been given, then 200-400 mg/day (class IIa recommendation); **IV** 5-7 mg/kg over 30-60 min, then 1.2-1.8 g as **CONT IV INF** or in divided PO doses until 10 g, then 200-400 mg/day (class IIa recommendation)
- **Child and infant: PO** 10-20 mg/kg/day in divided doses for 7-10 days, then 5-10 mg/kg/day once daily

Available forms: Tabs 100, 200, 400 mg; inj 50 mg/ml

Administer:

PO route

- May be used with/without food but be consistent

IV, direct route

- **Peripheral:** max 2 mg/ml for more than 1 hr; preferred through central venous line with in-line filter; concentrations of more than 2 mg/ml should be given by central line
- **Cardiac arrest:** give 300 bol; diluted to a total volume of 20 ml D5W; may repeat 150 mg after 3-5 min

Intermittent IV INF route

- **Rapid loading:** add 3 ml (150 mg), 100 ml D_5W (1.5 mg/ml), give over 10 min
- **Slow loading:** add 18 ml (900 mg), 500 ml D_5W (1.8 mg/ml), give over next 6 hr

Continuous IV INF route

- After 24 hr, dilute 50 ml to 1-6 mg/ml, give 1-6 mg/ml at 1 mg/min for the 1st 6 hr, then 0.5 mg/min

Y-site compatibilities: Amikacin, clindamycin, DOBUTamine, DOPamine, doxycycline, erythromycin, esmolol, gentamicin, insulin, isoproterenol, labetalol, lidocaine, metaraminol, metroNIDAZOLE, midazolam, morphine, nitroglycerin, norepinephrine, penicillin G potassium, phenylephrine, potassium chloride, procainamide, tobramycin, vancomycin

Solution compatibility: D_5W, 0.9% NaCl

SIDE EFFECTS

CNS: *Headache, dizziness,* involuntary movement, *tremors, peripheral neuropathy,* malaise, *fatigue,* ataxia, *paresthesias,* insomnia

CV: *Hypotension, bradycardia,* sinus arrest, CHF, dysrhythmias, SA node dysfunction, AV block, increased defibrillation energy requirement

EENT: Blurred vision, halos, photophobia, *corneal microdeposits,* dry eyes

ENDO: *Hypo*/hyperthyroidism

GI: *Nausea, vomiting,* diarrhea, abdominal pain, *anorexia, constipation,* hepatotoxicity

GU: Epididymitis, ED

INTEG: Rash, photosensitivity, blue-gray skin discoloration, alopecia, spontaneous ecchymosis, toxic epidermal necrolysis, urticaria, pancreatitis, phlebitis (IV)

MISC: Flushing, abnormal taste or smell, edema, abnormal salivation, coagulation abnormalities

MS: Weakness, pain in extremities

RESP: Pulmonary fibrosis/toxicity, pulmonary inflammation, ARDS; gasping syndrome if used with neonates

PHARMACOKINETICS

PO: Onset 1-3 wk, peak 2-7 hr, half-life 26-107 days, increased in geriatrics, metabolized by liver (CYP3A4, CYP2C8), excreted by kidneys, 99% protein binding

INTERACTIONS

Increase: QT prolongation—azoles, fluoroquinolones, macrolides

Increase: amiodarone concentrations, possible serious dysrhythmias-protease inhibitors, reduce dose

Increase: myopathy—HMG-CoA reductase inhibitors

Increase: bradycardia—β-blockers, calcium channel blockers

Increase: levels of cycloSPORINE, dextromethorphan, digoxin, disopyramide, flecainide, methotrexate, phenytoin, procainamide, quiNIDine, theophylline, class I antidysrhythmics

Increase: anticoagulant effects—warfarin

Drug/Food

- Toxicity: grapefruit juice

Drug/Lab Test

Increase: T_4, ALT, AST, GGT alk phos, cholesterol, lipids, PT, INR

Decrease: T_3

NURSING CONSIDERATIONS

Assess:

Black Box Warning: Pulmonary toxicity: dyspnea, fatigue, cough, fever, chest pain; product should be discontinued; for ARDS, pulmonary fibrosis, crackles, tachypnea, increased at higher doses, toxicity is common

Black Box Warning: ECG continuously to determine product effectiveness; measure PR, QRS, QT intervals; check for PVCs, other dysrhythmias, B/P continuously for hypo/hypertension; report dysrhythmias, slowing heart rate; monitor amiodarone level: therapeutic 1-2.5 mcg/ml; toxic >2.5 mcg/ml

- Electrolytes (sodium, potassium, chloride); hepatic studies: AST, ALT, bilirubin, alk phos; for dehydration, hypovolemia, monitor PT, INR if using warfarin
- Chest x-ray, thyroid function tests
- CNS symptoms: confusion, psychosis, numbness, depression, involuntary movements; product should be discontinued
- **Hypothyroidism:** lethargy; dizziness; constipation; enlarged thyroid gland; edema of extremities; cool, pale skin
- **Hyperthyroidism:** restlessness; tachycardia; eyelid puffiness; weight loss; frequent urination; menstrual irregularities; dyspnea; warm, moist skin
- Ophthalmic exams at baseline and periodically (PO); to prevent corneal deposits, use methylcellulose
- Cardiac rate, respirations: rate, rhythm, character, chest pain; start with patient hospitalized and monitored up to 1 wk; for rebound hypertension after 1-2 hr

Evaluate:

- Therapeutic response: decrease in ventricular tachycardia, supraventricular tachycardia, fibrillation

Teach patient/family:

- To take this product as directed; to avoid missed doses; not to use with grapefruit juice; not to discontinue abruptly
- To use sunscreen or stay out of sun to prevent burns; that dark glasses may be needed for photophobia
- To report side effects immediately; more common at high dose
- That skin discoloration is usually reversible

TREATMENT OF OVERDOSE:

O_2, artificial ventilation, ECG, administer DOPamine for circulatory depression, administer diazepam, thiopental for seizures, isoproterenol

amitriptyline (Rx)

(a-mee-trip′ti-leen)

Func. class.: Antidepressant—tricyclic

Chem. class.: Tertiary amine

Do not confuse:
amitriptyline/nortriptyline/aminophylline

ACTION: Blocks reuptake of norepinephrine, serotonin into nerve endings, thereby increasing action of norepinephrine, serotonin in nerve cells

USES: Major depression

Unlabeled uses: Neuropathic pain, prevention of cluster/migraine headaches, fibromyalgia, ADHD, bulimia nervosa, diabetic neuropathy, enuresis, insomnia, panic disorder, postherpetic neuralgia, hiccups, social phobia

CONTRAINDICATIONS: Hypersensitivity to tricyclics, carBAMazepine; recovery phase of myocardial infarction

Precautions: Pregnancy (C), breastfeeding, geriatric patients, seizure disorders, prostatic hypertrophy, schizophrenia, psychosis, severe depression, increased intraocular pressure, closed-angle glaucoma, urinary retention, renal/hepatic/cardiac disease, hyperthyroidism, electroshock therapy, elective surgery

Black Box Warning: Children <12 yr, suicidal patients

DOSAGE AND ROUTES

Depression

- **Adult/adolescent: PO** 25-75 mg/day as single dose at bedtime or in divided doses, may increase to 200 mg/day; max 300 mg/day (if hospitalized)
- **Geriatric: PO** 10-25 mg at bedtime, may be increased to 150 mg/day

Cluster/migraine headache (unlabeled)

- **Adult: PO** 10-300 mg/day

Pain (unlabeled)

- **Adult: PO** 75-300 mg/day

Fibromyalgia/insomnia (unlabeled)

- **Adult: PO** 10-50 mg nightly

Enuresis (unlabeled)

- **Child 11-14 yr: PO** 50 mg at bedtime
- **Child 6-10 yr: PO** 25 mg at bedtime

ADHD/bulimia nervosa (unlabeled)

- **Adult: PO** 25 mg tid, titrate to 200 mg/day by 25-50 mg at weekly intervals
- **Child 6-12 yr: PO** 10-30 mg/day or 1-5 mg/kg/day in divided doses

Available forms: Tabs 10, 25, 50, 75, 100, 150 mg

Administer:

- Increase fluids, bulk in diet if constipation, urinary retention occur, especially in geriatric patients
- With food, milk for GI symptoms
- Crushed if patient unable to swallow medication whole
- Dosage at bedtime if oversedation occurs during day; may take entire dose at bedtime; geriatric patients may not tolerate once daily dosing
- Store at room temperature; do not freeze

SIDE EFFECTS

CNS: *Dizziness, drowsiness,* confusion, headache, anxiety, tremors, stimulation, weakness, *insomnia,* nightmares, EPS (geriatric patients), increased psychiatric symptoms, seizures, suicidal thoughts, anxiety

CV: *Orthostatic hypotension,* ECG changes, tachycardia, hypertension, palpitations, dysrhythmias

EENT: *Blurred vision,* tinnitus, mydriasis, ophthalmoplegia, amblyopia

GI: *Constipation, dry mouth,* weight gain, nausea, vomiting, paralytic ileus, increased appetite, cramps, epigastric distress, jaundice, hepatitis, stomatitis

GU: *Urinary retention,* sexual dysfunction

HEMA: Agranulocytosis, thrombocytopenia, eosinophilia, leukopenia, aplastic anemia

INTEG: Rash, urticaria, sweating, pruritus, photosensitivity

RESP: Asthma, exacerbation, rhinitis

SYST: Neuroleptic malignant syndrome, serotonin syndrome

PHARMACOKINETICS

Onset 45 min; peak 2-12 hr; therapeutic response 4-10 days; metabolized by liver to nortriptyline; excreted in urine, feces; crosses placenta; excreted in breast milk; half-life 10-46 hr

INTERACTIONS

⚠ Hyperpyretic crisis, seizures, hypertensive episode: MAOIs, do not use within 14 days of MAOIs

Increase: risk for agranulocytosis—antithyroid agents

Increase: QT prolongation—procainamide, quiNIDine, amiodarone, tricyclics, class IA, III antidysrhythmics

Increase: amitriptyline levels, toxicity—cimetidine, FLUoxetine, phenothiazines, oral contraceptives, antidepressants, carBAMazepine, class IC antidysrhythmics

Increase: effects of direct-acting sympathomimetics (EPINEPHrine), alcohol, barbiturates, benzodiazepines, CNS depressants, opioids, sedative/hypnotics

Increase: serotonin syndrome, linezolid, methyline blue, use cautiously

Decrease: effects of guanethidine, cloNIDine, indirect-acting sympathomimetics (ePHEDrine)

Drug/Herb

Increase: serotonin syndrome—SAM-e, St. John's wort, yohimbe

Increase: CNS depression—kava, hops, chamomile, lavender, valerian

Drug/Lab Test

Increase: serum bilirubin, blood glucose, alk phos, LFTs

Decrease: WBCs, platelets, granulocytes

NURSING CONSIDERATIONS

Assess:

- B/P lying, standing; pulse q4hr; if systolic B/P drops 20 mm Hg, hold product, notify prescriber; take vital signs q4hr with CV disease; ECG for flattening of T wave, prolongation of QTc interval, bundle branch block, AV block, dysrhythmias in cardiac patients, avoid use immediately after MI
- **Blood studies:** CBC, leukocytes, differential, cardiac enzymes if patient is receiving long-term therapy, thyroid function tests
- **Hepatic studies:** AST, ALT, bilirubin
- Weight every wk; appetite may increase with product
- EPS primarily in geriatric patients: rigidity, dystonia, akathisia
- Paralytic ileus, glaucoma exacerbation

Black Box Warning: Mental status: mood, sensorium, affect, suicidal tendencies; increase in psychiatric symptoms: depression, panic; suicidal tendencies are higher in those ≤24 yr, restrict amount of product available

- **Serotonin syndrome:** may occur with other serotonergic products (hyperthermia, hypertension, rigidity, delirium)
- Urinary retention, constipation; constipation is most likely to occur in children and geriatric patients
- **Withdrawal symptoms:** headache, nausea, vomiting, muscle pain, weakness; do not usually occur unless product was discontinued abruptly
- **Alcohol consumption:** if alcohol is consumed, hold dose until morning
- **Pain syndromes (unlabeled):** intensity, location, severity; use pain scale; product may be taken for 1-2 mo before effective
- **Sexual dysfunction:** erectile dysfunction, decreased libido

Evaluate:

- Therapeutic response: decrease in depression, absence of suicidal thoughts

Teach patient/family:

- To take medication as directed; not to double dose; that therapeutic effects may take 2-3 wk; not to discontinue medication quickly after long-term use: may cause nausea, headache, malaise
- To use caution when driving, performing other activities that require alertness because of drowsiness, dizziness, blurred vision; to avoid rising quickly from sitting to standing (especially geriatric patients); how to manage anticholinergic effects
- The symptoms of serotonin syndrome
- To avoid alcohol, other CNS depressants
- To wear sunscreen or large hat when outdoors, photosensitivity occurs
- That contraception is recommended during treatment; to avoid breastfeeding
- To watch for suicide

TREATMENT OF OVERDOSE:

ECG monitoring, lavage; administer anticonvulsant, sodium bicarbonate

amLODIPine (Rx)

(am-loe′di-peen)

Norvasc

Func. class.: Antianginal, antihypertensive, calcium channel blocker
Chem. class.: Dihydropyridine

Do not confuse:
amLODIPine/aMILoride

ACTION: Inhibits calcium ion influx across cell membrane during cardiac depolarization; produces relaxation of coronary vascular smooth muscle, peripheral vascular smooth muscle; dilates coronary vascular arteries; increases myocardial O_2 delivery in patients with vasospastic angina

USES:
Chronic stable angina pectoris, hypertension, variant angina (Prinzmetal's angina); may coadminister with other antihypertensives, antianginals

Unlabeled uses: Hypertension (pediatric patients)

CONTRAINDICATIONS:
Hypersensitivity to this product, severe aortic stenosis, severe obstructive CAD

Black Box Warning: Hypersensitivity to dihydropyridine

Precautions: Pregnancy (C), breastfeeding, children, geriatric patients, CHF, hypotension, hepatic injury, GERD

DOSAGE AND ROUTES

Coronary artery disease
- **Adult: PO** 5-10 mg/day
- **Geriatric: PO** 5 mg/day, max 10 mg/day

Hypertension
- **Adult: PO** 2.5-5 mg/day initially, max 10 mg/day
- **Geriatric: PO** 2.5 mg/day, may increase to 5 mg/day, max 10 mg/day
- **Child 6-16 yr (unlabeled): PO** 2.5-5 mg/day
- **Child <6 yr (unlabeled): PO** 0.05-0.2 mg/kg/day in 1-2 divided doses

Hepatic dose
- **Adult: PO** 2.5 mg/day; may increase to 10 mg/day (antihypertensive); 5 mg/day, may increase to 10 mg/day (antianginal)

Available forms: Tabs 2.5, 5, 10 mg

Administer:
- Once a day without regard to meals

SIDE EFFECTS

CNS: *Headache,* fatigue, dizziness, asthenia, anxiety, depression, insomnia, paresthesia, somnolence

CV: *Peripheral edema,* bradycardia, hypotension, palpitations, syncope, chest pain

GI: Nausea, vomiting, diarrhea, gastric upset, constipation, flatulence, anorexia, gingival hyperplasia, dyspepsia

GU: Nocturia, polyuria, sexual difficulties

INTEG: Rash, pruritus, urticaria, alopecia

OTHER: Flushing, muscle cramps, cough, weight gain, tinnitus, epistaxis

PHARMACOKINETICS

Peak 6-12 hr; half-life 30-50 hr; increased in geriatric patients, hepatic disease; metabolized by liver (CYP3A4); excreted in urine (90% as metabolites); protein binding >93%

INTERACTIONS

Increase: neurotoxicity—lithium

Increase: hypotension—alcohol, antihypertensives, nitrates, fentaNYL, quiNIDine

Increase: amLODIPine level—diltiazem

Decrease: antihypertensive effect—NSAIDs

Drug/Herb

Decrease: effect—yohimbe

Drug/Food

Increase: hypotensive effect—grapefruit juice

NURSING CONSIDERATIONS

Assess:
- Cardiac status: B/P, pulse, respirations, ECG; some patients have developed severe angina, acute MI after calcium channel blockers if obstructive CAD is severe
- Peripheral edema, dyspnea, jugular vein distention, crackles
- **Angina:** intensity, location, duration of pain

Evaluate:
- Therapeutic response: decreased anginal pain, decreased B/P, increased exercise tolerance

Teach patient/family:
- To take product as prescribed, not to double or skip dose
- To avoid hazardous activities until stabilized on product, dizziness is no longer a problem

- To avoid OTC products unless directed by prescriber
- To comply in all areas of medical regimen: diet, exercise, stress reduction, product therapy, smoking cessation
- **To notify prescriber of irregular heartbeat; shortness of breath; swelling of feet, face, hands; severe dizziness; constipation; nausea; hypotension; if chest pain does not improve, use nitroglycerin when angina is severe**
- To use correct technique when monitoring pulse; to contact prescriber if pulse <50 bpm
- To avoid large amounts of grapefruit juice, alcohol
- To change positions slowly to prevent orthostatic hypotension
- To continue with good oral hygiene to prevent gingival disease
- To use sunscreen, protective clothing to prevent photosensitivity
- To notify all health care providers of use of this product

TREATMENT OF OVERDOSE:
Defibrillation, β-agonists, IV calcium inotropic agents, diuretics, atropine for AV block, vasopressor for hypotension

amoxicillin (Rx)
(a-mox-i-sill′in)

Moxatag

Func. class.: Antiinfective, antiulcer

Chem. class.: Aminopenicillin

Do not confuse:
amoxicillin/amoxapine/Amoxil
Trimox/Diamox/Tylox
Wymox/Tylox

ACTION: Interferes with cell wall replication of susceptible organisms; bactericidal: lysis mediated by bacterial cell wall autolysins

USES: Treatment of skin, respiratory, GI, GU infections, otitis media, gonorrhea; for gram-positive cocci *(Staphylococcus aureus, Streptococcus pyogenes, Streptococcus faecalis, Streptococcus pneumoniae)*, gram-negative cocci *(Neisseria gonorrhoeae, Neisseria meningitidis)*, gram-positive bacilli *(Corynebacterium diphtheriae, Listeria monocytogenes)*, gram-negative bacilli *(Haemophilus influenzae, Escherichia coli, Proteus mirabilis, Salmonella)*; gastric ulcer, β-lactase–negative organisms

Unlabeled uses: Lyme disease, anthrax treatment and prophylaxis, cervicitis, *Chlamydia trachomatis*, dental abscess/infection, dyspepsia, non-gonococcal urethritis, periodontitis, typhoid fever; prophylaxis of bacterial endocarditis; in combination with other products for treatment of *Helicobacter pylori*

CONTRAINDICATIONS: Hypersensitivity to penicillins

Precautions: Pregnancy (B), breastfeeding, neonates, hypersensitivity to cephalosporins, carbapenems; severe renal disease, mononucleosis, phenylketonuria, diabetes, geriatrics, asthma, child, colitis, dialysis, eczema, pseudomembranous colitis, syphilis

DOSAGE AND ROUTES

Upper respiratory infections

- **Adult/adolescent/child (≥40 kg): (mild-moderate infections) PO** 500 mg q12hr, or 250 mg q8hr; **(severe infections)** 875 mg q12hr or 500 mg q8hr
- **Child (<40 kg): (mild-moderate infections)** 45-90 mg/kg/day individual doses q12hr; **(severe infections)** 40 mg/kg/day divided q8hr or 45 mg/kg/day divided q12hr

***S. pyogenes* infection (pharyngitis/tonsillitis)**

- **Adult/child >12 yr: PO EXT REL** 775 mg every day with meal × 10 days

Gonorrhea (not CDC approved)

- **Adult: PO** 3 g given with 1 g probenecid as a single dose followed by tetracycline or erythromycin therapy

C. trachomatis

- **Adult: PO** 500 mg/tid × 1 wk

Duodenal/gastric ulcer/dyspepsia from *H. pylori* infection (unlabeled)
• **Adult: PO** 1000 mg bid with lansoprazole or clarithromycin/omeprazole
Bacterial endocarditis prophylaxis (unlabeled)
• **Adult: PO** 2 g 1 hr before procedure
• **Child: PO** 50 mg/kg 1 hr before procedure; max 2 g
Lyme disease (unlabeled)
• **Adult: PO** 500 mg tid × 14-21 days
• **Child: PO** 50 mg/kg/day in divided doses q8hr × 14-21 days
Anthrax treatment/prophylaxis (unlabeled)
• **Adult and child >20 kg: PO** 500 mg q8hr × 10-14 days (prophylaxis), 60 days (treatment)
• **Child <20 kg: PO** 80 mg/kg divided in 3 doses q8hr × 60 days (treatment)
Renal disease
• **Adult: PO** CCr 10-30 ml/min 250-500 mg q12hr; CCr <10 ml/min 250-500 mg q24hr; do not use 775, 875 mg strength if CCr <30 ml/min

Available forms: Caps 250, 500 mg; chew tabs 125, 200, 250, 400 mg; tabs 250, 500, 875 mg; ext rel tab (Moxatag) 775 mg; susp 125, 200, 250, 400 mg/5 ml

Administer:
PO route
• Identify allergies before use
• **Susp:** shake well before each dose; use calibrated spoon, oral syringe, or measuring cup; may be used alone, mixed in drinks; use immediately; discard unused portion after 14 days, store in refrigerator
• Give around the clock; caps may be emptied, mixed with liquids if needed without regard to food
• **Ext rel:** do not crush, chew, break; take with food

SIDE EFFECTS

CNS: Headache, seizures, agitation, confusion, dizziness, insomnia
GI: *Nausea, vomiting, diarrhea,* increased AST, ALT, abdominal pain, glossitis, colitis, pseudomembranous colitis, jaundice, cholestasis, eosinophils, thrombocytopenia, agranulocystosis
HEMA: Anemia, increased bleeding time, bone marrow depression, granulocytopenia, hemolytic anemia
INTEG: *Urticaria, rash*
SYST: Anaphylaxis, respiratory distress, serum sickness, Stevens-Johnson syndrome, toxic epidermal necrolysis, exfoliative dermatitis

PHARMACOKINETICS

PO: Peak 1-2 hr, duration 6-8 hr, half-life 1-1⅓ hr, metabolized in liver, excreted in urine, crosses placenta, enters breast milk

INTERACTIONS

Increase: rash—allopurinol
Increase: amoxicillin level—probenecid
Increase: anticoagulant action—warfarin
Increase: methotrexate levels—methotrexate
Decrease: effectiveness of—oral contraceptives
Drug/Lab Test
Increase: AST/ALT, alk phos, LDH, eosinophils
Decrease: Hgb, WBC, platelets
Interference: urine glucose test (Clinitest, Benedict's reagent, cupric SO_4)

NURSING CONSIDERATIONS

Assess:
• Report hematuria, oliguria, penicillin products, in high doses are nephrotoxic
• Hepatic studies: AST, ALT
• Blood studies: WBC, RBC, Hgb, Hct, bleeding time
• Renal studies: urinalysis, protein, blood, BUN, creatinine
• C&S before product therapy; product may be given as soon as culture is taken
⚠ **Pseudomembranous colitis:** bowel pattern before, during treatment; diarrhea, cramping, blood in stools; report to prescriber
• Skin eruptions after administration of

penicillin to 1 wk after discontinuing product; rash is more common if allopurinol is taken concurrently

⚠ **Anaphylaxis:** rash, itching, dyspnea, facial/laryngeal edema

Evaluate:

• Therapeutic response: absence of infection; prevention of endocarditis, resolution of ulcer symptoms

Teach patient/family:

• That caps may be opened, contents taken with fluids; that chewable form is available; to take as prescribed, not to double dose

• All aspects of product therapy: to complete entire course of medication to ensure organism death (10-14 days); that culture may be taken after completed course of medication

⚠ To report sore throat, fever, fatigue, diarrhea **(superinfection or agranulocytopenia),** blood in stool, abdominal pain **(pseudomembranous colitis),** decreased urinary output

• That product must be taken in equal intervals around the clock to maintain blood levels; to take without regard to food

• To wear or carry emergency ID if allergic to penicillins

TREATMENT OF ANAPHYLAXIS: Withdraw product, maintain airway; administer EPINEPHrine, aminophylline, O_2, IV corticosteroids

amoxicillin/clavulanate (Rx)

(a-mox-i-sill′in)

Amoclan, Apo-Amoxi Clav ♣, Augmentin, Augmentin XR, Clavulin ♣

Func. class.: Broad-spectrum antiinfective

Chem. class.: Aminopenicillin β-lactamase inhibitor

Do not confuse:

Augmentin/amoxicillin

ACTION: Bacteriocidal, interferes with cell wall replication of susceptible organisms; lysis mediated by bacterial cell wall autolytic enzymes, combination increases spectrum of activity against β-lactamase–resistant organisms

USES: Lower respiratory tract infections, sinus infections, pneumonia, otitis media, impetigo, skin infection, UTI; effective for *Actinomyces* sp., *Bacillus anthracis, Bacteroides* sp., *Bordetta pertassis, Borrelia burdofer; Brucella* sp., *Burkholderia pseudomallei, Clostridium perfringens/tetani, Corynebacterium diphtheriae, Eikenella corodens, Enterobacter* sp., *Enterococcus faecalis, Erysipelothrix rhusioputhae, Escherichiacoli, Eubacterium* sp., *Fusobacterium* sp., *Haemophilus ducrey/ parainfluenzal (positive/negative beta-lactamase), Heliobacter pylori, Klebsiella* sp., *Lactobacillus* sp., *Listeria monocytogenes, Moraxella catarrholis, Neisseria gonorrheal/meningitis, Nocardia brasilensis, Peptococcus* sp., *Peptostreptococcus* sp., *Prevotella melaninogemica, Propionibacterium* sp., *shigella* sp., *Staphylococcus aureus (MSSA)/epidermidis/ Saprophyticus, streptococcus agalactial (group B Streptococci)/dysglactial/ pneumonia/ pygenes (group A Streptococci) reponema pallidium, Vibrocholeral, viridans streptococci*

Unlabeled uses: Actinomycotic mycetoma, chancroid, dental infections, dentoalveolar infection, melioidosis, pericoronitis, SARS

CONTRAINDICATIONS: Hypersensitivity to penicillins, severe renal disease, dialysis, jaundice

Precautions: Pregnancy (B), breastfeeding, neonates, children, hypersensitivity to cephalosporins; renal/GI disease, asthma, colitis, diabetes, eczema, leukemia, mononucleosis, viral infections, phenylketonuria

DOSAGE AND ROUTES

- **Adult: PO** 250-500 mg q8hr or 500-875 mg q12hr, depending on severity of infection
- **Child ≤40 kg: PO** 20-90 mg/kg/day in divided doses q8-12hr

Renal disease

- **Adult: PO** CCr 10-30 ml/min dose q12hr; CCr <10 ml/min dose q24hr; do not use 875-mg strength or ext rel if CCr <30 ml/min; Augmentin XR is contraindicated with renal disease

Available forms: Tabs 250, 500, 875 mg/125 mg clavulanate; chew tabs 200/28.5, 400/57 mg; powder for oral susp 250/28.5, 200/28.5, 400/57, 600/42.9 mg/5 ml; ext rel tabs (XR) 1000 mg amoxicillin, 62.5 mg clavulanate; powder for oral susp (ES) 600 mg amoxicillin, 42.9 mg clavulanate

Administer:

PO route

- Do not break, crush, chew ext rel product

⚠ Only as directed; two 250-mg tabs not equivalent to one 500-mg tab due to strength of clavulanate

- Shake susp well before each dose; may be used alone, mixed in drinks; use immediately, discard unused portion of susp after 14 days, store in refrigerator
- Give around the clock
- Give with light meal for increased absorption, less GI effects, confusion, behavioral changes

SIDE EFFECTS

CNS: Headache, fever, seizures, agitation, insomnia

GI: *Nausea, diarrhea, vomiting,* increased AST/ALT, abdominal pain, glossitis, colitis, black tongue, pseudomembranous colitis

GU: Oliguria, proteinuria, hematuria, *vaginitis, moniliasis,* glomerulonephritis

HEMA: Anemia, bone marrow depression, granulocytopenia, leukopenia, eosinophilia, thrombocytopenic purpura

INTEG: *Rash,* urticaria, dermatitis, toxic epidermal necrolysis

META: Hypo/hyperkalemia, alkalosis, hypernatremia

SYST: Anaphylaxis, respiratory distress, serum sickness, superinfection, Stevens-Johnson syndrome, candidiasis

PHARMACOKINETICS

PO: Peak 1-2.5 hr, duration 6-8 hr, half-life 1-1⅓ hr, metabolized in liver, excreted in urine, crosses placenta, excreted in breast milk, removed by hemodialysis

INTERACTIONS

Increase: amoxicillin levels—probenecid

Increase: anticoagulant effect—warfarin; monitor closely, dose adjustment may be needed

Increase: skin rash—allopurinol

Decrease: action of oral contraceptives

Drug/Food

Decrease: absorption by a high-fat meal

Drug/Lab Test

Increase: AST/ALT, alk phos, LDH

Interference: urine glucose tests (Clinitest, Benedict's reagent, Cupric SO_4)

NURSING CONSIDERATIONS

Assess:

⚠ Nephrotoxicity at high doses: I&O ratio; report hematuria, oliguria

- Hepatic studies: AST, ALT
- Blood studies: WBC, RBC, Hgb, Hct, bleeding time
- Renal studies: urinalysis, protein, blood, BUN, creatinine; if urine output decreases, long-acting products should not be used
- C&S before product therapy; product may be given as soon as culture is taken

⚠ **Pseudomembranous colitis:** bowel pattern before, during treatment; diarrhea, cramping, blood in stools; report to prescriber

⚠ **Anaphylaxis:** rash, itching, dyspnea, facial/laryngeal edema; skin eruptions after administration of penicillin to 1 wk after discontinuing product

- Adrenaline, suction, tracheostomy set, endotracheal intubation equipment on unit
- Adequate intake of fluids (2 L) during diarrhea episodes

Evaluate:

- Therapeutic response: absence of infection

Teach patient/family:

- To take as prescribed, not to double dose
- All aspects of product therapy: to complete entire course of medication to ensure organism death (10-14 days); that culture may be taken after completed course of medication

⚠ To report sore throat, fever, fatigue **(superinfection or agranulocytosis);** diarrhea, cramping, blood in stools **(pseudomembranous colitis)**

- That product must be taken in equal intervals around the clock to maintain blood levels
- To wear or carry emergency ID if allergic to penicillins

TREATMENT OF HYPERSENSITIVITY: Withdraw product, maintain airway, administer EPINEPHrine, aminophylline, O_2, IV corticosteroids for anaphylaxis

amphotericin B lipid complex (ABLC)

(am-foe-ter'i-sin)

Abelcet

Func. class.: Antifungal

Chem. class.: Amphoteric polyene

ACTION: Increases cell membrane permeability in susceptible fungi by binding sterols; alters cell membrane, thereby causing leakage of cell components, cell death

USES: Indicated for the treatment of invasive fungal infections in patients who cannot tolerate or have failed conventional amphotericin B therapy; broad-spectrum activity against many fungal, yeast and mold pathogen infections, including *Aspergillus, Zygomycetes, Fusarium, Cryptococcus,* and many hard-to-treat *Candida* species; *Aspergillus fumigatus, Aspergillus* sp., *Blastomyces dermatitidis, Candida albicans, Candida guilliermondii, Candida* sp., *Candida stellatoidea, Candida tropicalis, Coccidioides immitis, Cryptococcus* sp., *Histoplasma* sp., *sporotiaichosis*

CONTRAINDICATIONS: Hypersensitivity

Precautions: Anemia, breastfeeding, cardiac disease, children, electrolyte imbalance, geriatric, hematological/hepatic/renal disease, hypotension pregnancy (B)

DOSAGE AND ROUTES

- **Adult: IV** 3-5 mg/kg/day as a single inf given at 2.5 mg/kg/hr

Available forms: Susp for inj 100 mg/20-ml vial

Administer:

⚠ Do not confuse four different types; these are not interchangeable: conventional amphotericin B, amphotericin B cholesteryl, amphotericin B lipid complex, amphotericin B liposome

- May premedicate with acetaminophin, diphenhydrAMINE

IV route

- Give product only after C&S confirms organism, product needed to treat condition; make sure product is used for life-threatening infections
- Handle with aseptic technique because amphotericin B lipid complex (ABLC) no preservatives; visually inspect parenteral products for particulate matter and discoloration before use

Filtration and dilution:

- Prior to dilution, store at 36°-46° F (2°-8° C), protected from moisture and light; do not freeze; the diluted, ready-for-use admixture is stable for up to 48 hours at 36°-46° F (2°-8° C) and an

additional 6 hr at room temperature; do not freeze
• Prepare the admixture for infusion by first shaking the vial until there is no evidence of yellow sediment on the bottom of the vial
• Transfer the appropriate amount of drug from the required number of vials into one or more sterile syringes using an 18-gauge needle
• Attach the provided 5-micron filter needle to the syringe; inject the syringe contents through the filter needle, into an IV bag containing the appropriate amount of D5W injection; each filter needle may be used on the contents of no more than four 100-mg vials
• The suspension must be diluted with D5W injection to a final concentration of 1 mg/ml; for pediatric patients and patients with cardiovascular disease, the final concentration may be 2 mg/ml; DO NOT USE SALINE SOLUTIONS OR MIX WITH OTHER DRUGS OR ELECTROLYTES
• The diluted ready-for-use admixture is stable for up to 48 hr at 2-8° C (36-46° F) and an additional 6 hr at room temperature; do not freeze

IV INF:
• Flush IV line with D_5W injection before use or use a separate IV line; DO NOT USE AN IN-LINE FILTER
• Before infusion, shake the bag until the contents are thoroughly mixed; max rate 2.5 mg/kg/hr; if the infusion time exceeds 2 hr, mix the contents by shaking the infusion bag every 2 hr

Y-site compatibilities: acyclovir, allopurinol, aminocaproic acid, aminophylline, amiodarone, anidulafungin, argatroban, arsenic trioxide, atracurium, azithromycin, aztreonam, bumetanide, buprenorphine, busulfan, butorphanol, CARBOplatin, carmustine, ceFAZolin, cefepime, cefotaxime, cefoTEtan, cefOXitin, cefTAZidime, ceftizoxime, cefTRIAXone, cefuroxime, chloramphenicol, chlorproMAZINE, cimetidine, cisatracurium, clindamycin, cyclophosphamide, cycloSPORINE, cytarabine, DACTINomycin, dexamethasone, digoxin, diphenhydrAMINE, DOCEtaxel, doxacurium, DOXOrubicin liposomal, enalaprilat, EPINEPHrine eptifibatide, ertapenem, etoposide, famotidine, fentaNYL, fludarabine, fluorouracil, fosphenytoin, furosemide, ganciclovir, granisetron, heparin, hydrocortisone, HYDROmorphone, ifosfamide, insulin, regular ketorolac, lepirudin, lidocaine, linezolid, LORazepam, mannitol, melphalan, meperidine, methotrexate, methylPREDNISolone, metoclopramide, mitoMYcin, mivacurium, nafcillin, nesiritide, mitroglycerin, nitroprusside, octreotide, oxaliplatin, PACLitaxel, pamidronate, pantoprazole, PEMEtrexed, pentazocine, PENTobarbital, PHENobarbital, phentolamine, piperacillin-tazobactam, procainamide, ranitidine, succinylcholine, SUFentanil, tacrolimus, telavancin, teniposide, theophylline, thiopental, thiotepa, ticarcillin, ticaracillin-clavulanate, trimethobenzamide, verapamil, vinBLAStine, vinCRIStine, zidovudine, zoledronic acid

SIDE EFFECTS

CNS: *Headache, fever, chills,* peripheral nerve pain, paresthesias, peripheral neuropathy, seizures, dizziness
CV: Bradycardia, hypotension, cardiac arrest, chest pain
EENT: Tinnitus, deafness, diplopia, blurred vision
GI: *Nausea, vomiting, anorexia,* diarrhea, cramps, hemorrhagic gastroenteritis, acute liver failure
GU: *Hypokalemia,* azotemia, hyposthenuria, renal tubular acidosis, nephrocalcinosis, permanent renal impairment, anuria, oliguria
HEMA: Normochromic, normocytic anemia, thrombocytopenia, agranulocytosis, leukopenia, eosinophilia
INTEG: *Burning, irritation,* pain, necrosis at inj site with extravasation, flushing, dermatitis
META: Hyponatremia, hypomagnesemia
MS: Arthralgia, myalgia, generalized pain, weakness, weight loss

RESP: Bronchospasm, dyspnea
SYST: Toxic epidermal neurolysis, exfoliative dermatitis, anaphylaxis

PHARMACOKINETICS

IV: 4-6 days; terminal half-life lipid complex mean 7 days

INTERACTIONS

- Do not use with cidofovir

Increase: nephrotoxicity—other nephrotoxic antibiotics (aminoglycosides, CISplatin, vancomycin, cycloSPORINE, polymyxin B), antineoplastics, pentamidine, salicylates, tacrolimus, tenofovir
Increase: hypokalemia—corticosteroids, digoxin, skeletal muscle relaxants, thiazides, loop diuretics
Decrease: amphotericin B lipid complex-azole antifungals may still be used concurrently in serious resistant infections
Drug/Lab Test
Increase: AST/ALT, alk phos, BUN, creatinine, LDH, bilirubin
Decrease: magnesium, potassium, Hgb, WBC, platelets

NURSING CONSIDERATIONS

Assess:

- VS every 15-30 min during first inf; note changes in pulse, B/P
- I&O ratio; watch for decreasing urinary output, change in specific gravity; discontinue product to prevent permanent damage to renal tubules
- Blood studies: CBC, potassium, sodium, calcium, magnesium every 2 wk; BUN, creatinine 2-3 ×/wk
- Weight weekly; if weight increases by more than 2 lb/wk, edema is present; renal damage should be considered
- **For renal toxicity:** increasing BUN, serum creatinine; if BUN is >40 mg/dl or if serum creatinine is >3 mg/dl, product may be discontinued, dosage reduced
- **For hepatotoxicity:** increasing AST, ALT, alk phos, bilirubin
- For allergic reaction: dermatitis, rash; product should be discontinued, antihistamines (mild reaction) or EPINEPHrine (severe reaction) should be administered
- For hypokalemia: anorexia, drowsiness, weakness, decreased reflexes, dizziness, increased urinary output, increased thirst, paresthesias
- Infusion reactions: fever, chills, pain, swelling at site
- For ototoxicity: tinnitus (ringing, roaring in ears), vertigo, loss of hearing (rare)

Evaluate:

- Therapeutic response: decreased fever, malaise, rash, negative C&S for infecting organism

Teach patient/family:

- That long-term therapy may be needed to clear infection (2 wk-3 mo, depending on type of infection)
- To notify prescriber of bleeding, bruising, or soft-tissue swelling

amphotericin B liposomal (LAmB)

(am-foe-ter'i-sin)

AmBisome

Func. class.: Antifungal
Chem. class.: Amphoteric polyene

ACTION: Increases cell membrane permeability in susceptible fungi by binding to membrane sterols; alters cell membrane, thereby causing leakage of cell components, cell death

THERAPEUTIC OUTCOME

Resolution of infection

USES: Empirical therapy for presumed fungal infection in febrile neutropenic patients; treatment of *Cryptococcal* Meningitis in HIV-infected patients; treatment of *Aspergillus* species, *Candida* species and/or *Cryptococcus* species infections refractory to amphotericin B deoxycholate, or in patients where renal impairment or unacceptable toxicity precludes the use of amphotericin B deoxycholate *(Aspergillus flavus, Aspergillus fumigatus, Blastomyces dermatitidis, Candida albicans, Candida krusei, Candida lusitaniae, Candida parapsilosis, Candida tropicalis, Cryptococcus neoformans)*; treatment of visceral leishmaniasis

Unlabeled uses: Coccidioidomycosis, histoplasmosis

CONTRAINDICATIONS:

Hypersensitivity, severe bone marrow depression

Precautions: Anemia, breastfeeding, cardiac disease, children, electrolyte imbalance, geriatric, hematological/hepatic/renal disease, hypotension pregnancy (B)

DOSAGE AND ROUTES

Visceral leishmaniasis

- **Adult and child ≥1 mo: IV** 3 mg/kg every 24 hr days 1-5, and days 14, 21 (immunocompetent), 4 mg/kg every 24 hr days 1-5, and days 10, 17, 24, 31, 38 (immunocompromised)

Cryptococcal meningitis in HIV

- **Adult and child ≥1 mo: IV** 6 mg/kg/day

Fungal infection, empirical

- **Adult and child ≥1 mo: IV** 3 mg/kg/day

Fungal infection, systemic

- **Adult and child ≥1 mo: IV** 3-5 mg/kg/day

Available forms: Powder for inj 50-mg vial

Administer

⚠ Do not confuse four different types; these are not interchangeable: conventional amphotericin B, amphotericin B cholesteryl, amphotericin B lipid complex, amphotericin B liposome

- May premedicate with acetaminophen, diphenhydrAMINE

IV route

- Give only after C&S confirms organism, product needed to treat condition; make sure product is used for life-threatening infections
- Administer by IV infusion only; handle with aseptic technique as LAmB does not contain any preservatives
- Visually inspect products for particulate matter and discoloration

Reconstitution

- LAmB *must* be reconstituted using sterile water for injection (without a bacteriostatic agent); DO NOT RECONSTITUTE WITH SALINE OR ADD SALINE TO THE RECONSTITUTED SUSPENSION, DO NOT MIX WITH OTHER DRUGS; doing so can cause a precipitate to form
- Reconstitute vials containing 50 mg of LAmB /12 ml of sterile water (4 mg/ml)
- Immediately after the addition of water, SHAKE THE VIAL VIGOROUSLY for 30 sec; the suspension should be yellow and translucent; visually inspect vial for particulate matter and continue shaking until product is completely dispersed
- Store suspension for up to 24 hours refrigerated if using sterile water for injection; do not freeze

Filtration and dilution

- Calculate the amount of reconstituted (4 mg/ml) suspension to be further diluted and withdraw this amount into a sterile syringe
- Attach the provided 5-micron filter to the syringe; inject the syringe contents through the filter, into the appropriate amount of D_5W injection; use only one filter per vial
- The suspension must be diluted with D_5W injection to a final concentration of 1-2 mg/ml before administration; for infants and small children, lower concentrations (0.2-0.5 mg/ml) may be appropriate to provide sufficient volume for infusion
- Use injection of LAmB within 6 hr of dilution with D5W

IV INF

- Flush intravenous line with D5W injection before infusion; if this cannot be done, then a separate IV line must be used
- An inline membrane filter may be used provided the mean pore diameter of the filter is not less than 1 micron

- Administer by IV infusion using a controlled infusion device over a period of approximately 120 min; infusion time may be reduced to approximately 60 min in patients who tolerate the infusion; if discomfort occurs during infusion, the duration of infusion may be increased

Acetaminophen and diphenhydrAMINE

- 30 min before inf to reduce fever, chills, headache
- Store protected from moisture and light; diluted solution is stable for 24 hr at room temp

Y-site compatibilities:
Acyclovir, amifostine, aminophylline, anidulafungin, atropine, azithromycin, bivalirudin, bumetanide, buprenorphine, busulfan, butorphanol, CARBOplatin, carmustine, ceFAZolin, ceFOXitin, ceftizoxime, cefTRIAXone, cefuroxime, cimetidine, clindamycin, cyclophosphamide, cytarabine, DACTINomycin, DAPTOmycin, dexamethasone, dexmedetomidine, diphenhydrAMINE, doxacurium, enalaprilat, ePHEDrine, EPINEPHrine, eptifibatide, ertapenem, esmolol, etoposide, famotidine, fenoldopam, fentaNYL, fludarabine, fluorouracil, fosphenytoin, furosemide, granisetron, haloperidol, heparin, hydrocortisone, HYDROmorphone, ifosfamide, isoproterenol, ketorolac, levorphanol, lidocaine, linezolid, mesna, methotrexate, methylPREDNISolone, metoprolol, milrinone, mitoMYcin, nesiritide, nitroglycerin, nitroprusside, octreotide, oxaliplatin, oxytocin, palonosetron, pancuronium, pantoprazole, PEMEtrexed, PENTobarbital, PHENObarbital, phenylephrine, piperacillin/tazobactam, potassium chloride, procainamide, ranitidine, SUFentanil, tacrolimus, theophylline, thiopental, thiotepa, ticarcillin/clavulanate, tigecycline, trimethoprim-sulfamethoxazole, vasopressin, vinCRIStine, voriconazole, zidovudine

SIDE EFFECTS

CNS: *Headache, fever, chills,* peripheral nerve pain, paresthesias, peripheral neuropathy, seizures, dizziness, insomnia

CV: Bradycardia, hypotension, **cardiac arrest**

EENT: Tinnitus, deafness, diplopia, blurred vision

ENDO: Hyperglycemia

GI: *Nausea, vomiting, anorexia,* diarrhea, cramps, **hemorrhagic gastroenteritis, acute liver failure**

GU: *Hypokalemia,* **azotemia, hyposthenuria,** renal tubular acidosis, **nephrocalcinosis,** permanent renal impairment, anuria, oliguria

HEMA: **Normochromic normocytic anemia, thrombocytopenia, agranulocytosis, leukopenia, eosinophilia,** hyponatremia, hypomagnesemia

INTEG: *Burning, irritation,* pain, **necrosis at inj site with extravasation,** flushing, dermatitis, skin rash (topical route)

MS: Arthralgia, myalgia, generalized pain, weakness, weight loss

RESP: Dyspnea

SYST: **Stevens–Johnson syndrome, toxic epidermal neurolysis, exfoliative dermatitis, anaphylaxis**

PHARMACOKINETICS

IV: Peak 1-2 hr; initial half-life (LAmB) mean 4-6 days; metabolized in liver; excreted in urine (metabolites), breast milk; protein binding 90%; poorly penetrates CSF, bronchial secretions, aqueous humor, muscle, bone; terminal half-life mean 4-6 days

INTERACTIONS

Increase: nephrotoxicity—other nephrotoxic antibiotics (aminoglycosides, CISplatin, vancomycin, cycloSPORINE, polymyxin B)

Increase: hypokalemia—corticosteroids, digoxin, skeletal muscle relaxants, thiazides

NURSING CONSIDERATIONS

Assess:

- VS every 15-30 min during first inf; note changes in pulse, B/P
- I&O ratio; watch for decreasing urinary output, change in specific gravity;

discontinue product to prevent permanent damage to renal tubules

• Blood studies: CBC, potassium, sodium, calcium, magnesium every 2 wk, BUN, creatinine 2-3 ×/wk

• Weight weekly; if weight increases by more than 2 lb/wk, edema is present; renal damage should be considered

• **For renal toxicity:** increasing BUN, serum creatinine; if BUN is >40 mg/dl or if serum creatinine is >3 mg/dl, product may be discontinued, dosage reduced

• **For hepatotoxicity:** increasing AST, ALT, alk phos, bilirubin, monitor LFTs

• **For allergic reaction:** dermatitis, rash; product should be discontinued, antihistamines (mild reaction) or EPINEPHrine (severe reaction) administered

• **For hypokalemia:** anorexia, drowsiness, weakness, decreased reflexes, dizziness, increased urinary output, increased thirst, paresthesias

• **For ototoxicity:** tinnitus (ringing, roaring in ears), vertigo, loss of hearing (rare)

• **Infusion reaction:** chills, fever, pain, swelling at site

Evaluate:

• Therapeutic response: decreased fever, malaise, rash, negative C&S for infecting organism

Teach patient/family:

• That long-term therapy may be needed to clear infection (2 wk-3 mo, depending on type of infection)

• To notify prescriber of bleeding, bruising, or soft-tissue swelling

ampicillin (Rx)

(am-pi-sill'in)

Func. class.: Antiinfective—broad-spectrum

Chem. class.: Aminopenicillin

Do not confuse:

Omnipen/imipenem

ACTION: Interferes with cell wall replication of susceptible organisms; the cell wall, rendered osmotically unstable, swells, bursts from osmotic pressure; lysis mediated by cell wall autolysins

USES: Effective for gram-positive cocci *(Staphylococcus aureus, Streptococcus pyogenes, Streptococcus faecalis, Streptococcus pneumoniae)*, gram-negative cocci *(Neisseria meningitidis)*, gram-negative bacilli *(Haemophilus influenzae, Proteus mirabilis, Salmonella, Shigella, Listeria monocytogenes)*, gram-positive bacilli; meningitis, GI/GU/respiratory infections, endocarditis, septicemia, otitis media, skin infection

Unlabeled uses: Biliary tract infection, shigellosis, typhoid fever, PID, OB/GYN infections, leptospirosis

CONTRAINDICATIONS: Hypersensitivity to penicillins, antimicrobial resistance

Precautions: Pregnancy (B), breastfeeding, neonates, hypersensitivity to cephalosporins, renal disease, mononucleosis

DOSAGE AND ROUTES

Systemic infections

• **Adult and child ≥40 kg: PO** 250-500 mg q6hr; **IV/IM** 2-8 g/day in divided doses q4-6hr

• **Child <40 kg: PO** 50-100 mg/kg/day in divided doses q6-8hr; **IV/IM** 50-500 mg/kg/day in divided doses q6-8hr

Bacterial meningitis

• **Adult and adolescent: IM/IV** 150-200 mg/kg/day in divided doses q3-4hr; IDSA dose IV 12 g in divided doses q4hr

• **Infant and child: IM/IV** 150-200 mg/kg/day in divided doses q3-4hr; IDSA dose IV 300 mg/kg/day in divided doses q6hr

• **Neonates >7 days and >2000 g: IM/IV** 200 mg/kg/day in divided doses q6hr; IDSA dose IV 200 mg/kg/day in divided doses q6-8hr

Prevention of bacterial endocarditis

• **Adult: IM/IV** 2 g 30 min before procedure

• **Child: IM/IV** 50 mg/kg 30 min before procedure, max 2 g

GI/GU infections other than *N. gonorrhoeae*

• **Adult and child >20 kg: PO** 500 mg q6hr, may use larger dose for more serious infections

• **Child <40 kg: PO** 50 mg/kg/day in divided doses q6-8hr

Renal disease

• **Adult and child:** CCr 10-50 ml/min extend to q6-12hr; CCr <10 ml/min extend to q12-16hr

Shigellosis in AIDS patients (unlabeled)

• **Adult: PO** 500 mg qid × 5 days (not recommended by CDC)

Typhoid fever (unlabeled)

• **Adult/adolescent/child: IV** 100 mg/kg/day in divided doses q6hr × 14 days or more

Available forms: Powder for inj 125, 250, 500 mg, 1, 2, 10 g; IV inj 500 mg, 1, 2 g; caps 250, 500 mg; powder for oral susp 125, 250/5 ml

Administer:

PO route

• On empty stomach with plenty of water for best absorption (1-2 hr before meals or 2-3 hr after meals)

• Shake susp well before each dose; store after reconstituting in refrigerator up to 2 wk, 1 wk room temperature

IM route (painful)

• **Reconstitute** by adding 0.9-1.2 ml/125-mg vial; 0.9-1.9 ml/250-mg vial; 1.2-1.8 ml/500-mg vial; 2.4-7.4 ml/1-g vial; 6.8 ml/2-g vial

IV route

• **IV direct:** after diluting with sterile water 0.9-1.2 ml/125 mg product, administer over 3-5 min (up to 500 mg), 10-15 min (>500 mg)

Intermittent IV INF route

• May be diluted in 50 ml or more of D_5W, D_5 0.45% NaCl to a conc of 30 mg/ml or less; IV sol is stable for 1 hr; give at prescribed rate, do not give in same tubing as aminoglycosides, separate by ≥1 hr

Y-site compatibilities: Acyclovir, alemtuzumab, alprostadil, amifostine, aminocaproic acid, anidulafungin, argatroban, atenolol, azithromycin, bivalirudin, bleomycin, CARBOplatin, carmustine, CISplatin, clarithromycin, cyclophosphamide, cytarabine, DACTINomycin, DAPTOmycin, DAUNOrubicin liposome, dexmedetomidine, dexrazoxane, DOCEtaxel, doxacurium, doxapram, DOXOrubicin liposome, eptifibatide, etoposide, etoposide phosphate, filgrastim, fludarabine, fluorouracil, foscarnet, gallium, gatafloxacin, gemcitabine, gemtuzumab, granisetron, hetastarch, ifosfamide, irinotecan, lepirudin, leucovorin, levofloxacin, linezolid, mannitol, mechlorethamine, melphalan, methotrexate, metroNIDAZOLE, milrinone, octreotide, ofloxacin, oxaliplatin, PACLitaxel, palonosetron, pamidronate, pancuronium, pantoprazole, PEMEtrexed, penicillin G potassium, perphenazine, potassium acetate, propofol, remifentanil, riTUXimab, rocuronium, sodium acetate, teniposide, thiotepa, tigecycline, tirofiban, TNA, trastuzumab, vecuronium, vinBLAStine, vinCRIStine, vit B/C, voriconazole, zoledronic acid

SIDE EFFECTS

CNS: Lethargy, hallucinations, anxiety, depression, twitching, **coma, seizures**

GI: *Nausea, vomiting, diarrhea,* **pseudomembranous colitis,** stomatitis

GU: Oliguria, proteinuria, hematuria, *vaginitis, moniliasis,* **glomerulonephritis**

HEMA: Anemia, increased bleeding time, **bone marrow depression, granulocytopenia,** leukopenia, eosinophilia, **hemolysis**

INTEG: *Rash, urticaria,* erythema multiforme

MISC: Anaphylaxis, serum sickness, Stevens-Johnson syndrome, toxic epidermal necrolysis

PHARMACOKINETICS

Half-life 50-110 min; metabolized in liver; excreted in urine, bile, breast milk; crosses placenta; removed by dialysis

PO: Peak 2 hr, duration 6-8 hr
IM: Peak 1 hr
IV: Peak 1 hr

INTERACTIONS

Increase: bleeding, oral anticoagulants, monitor INR/PIJ
Increase: ampicillin concentrations—probenecid
Increase: ampicillin-induced skin rash—allopurinol, monitor for rash
Decrease: ampicillin level—H2 antagonists, proton pump inhibitors
Decrease: oral contraceptive effect

Drug/Lab Test
Increase: eosinophil
Decrease: conjugated estrone during pregnancy, conjugated estriol, Hgb, WBC, platelets
False positive: urine glucose
Interference: urine glucose, (clinitest, Benedict's reagent, cupric SO4)

NURSING CONSIDERATIONS

Assess:
- **Infection:** characteristics of wound, sputum, WBC; baseline, periodically; C&S before product therapy, product may be taken as soon as culture is taken

⚠ **Nephrotoxicity:** I&O ratio; report hematuria, oliguria; renal studies: urinalysis, protein, blood, BUN, creatinine
- Hepatic studies: AST, ALT
- Blood studies: WBC, RBC, Hgb, Hct, bleeding time
- Bowel pattern before, during treatment
- Skin eruptions after administration of penicillin to 1 wk after discontinuing product; identify allergies before using
- Respiratory status: rate, character, wheezing, tightness in chest

⚠ **Anaphylaxis:** rash, itching, dyspnea, facial swelling; stop product, notify prescriber, have emergency equipment available
- Adequate intake of fluids (2 L) during diarrhea episodes
- Scratch test to assess allergy after securing order from prescriber; usually done when penicillin is only product of choice
- Store in tight container; after reconstituting, oral suspension refrigerated for 2 wk or stored at room temp for 1 wk

Evaluate:
- Therapeutic response: absence of fever, draining wounds, other symptoms of infection

Teach patient/family:
- That tabs may be crushed; caps may be opened, mixed with water
- To take oral ampicillin on empty stomach with full glass of water
- All aspects of product therapy: to complete entire course of medication to ensure organism death (10-14 days); that culture may be taken after completed course of medication

⚠ To report sore throat, fever, fatigue, diarrhea (may indicate **superinfection**); to report rash, other signs of allergy
- That product must be taken in equal intervals around the clock to maintain blood levels
- To wear or carry emergency ID if allergic to penicillins
- **Pseudomembranous colitis:** diarrhea with blood or pus; notify prescriber

TREATMENT OF ANAPHYLAXIS: Withdraw product, maintain airway; administer EPINEPHrine, aminophylline, O_2, IV corticosteroids

ampicillin, sulbactam (Rx)

Unasyn

Func. class.: Antiinfective—broad-spectrum
Chem. class.: Aminopenicillin with β-lactamase inhibitor

ACTION: Interferes with cell wall replication of susceptible organisms; the cell wall, rendered osmotically unstable, swells, bursts from osmotic pressure; lysis due to cell wall autolytic enzymes;

combination extends spectrum of activity by β-lactamase inhibition

USES:
Skin infections, intraabdominal infections, cellulitis, diabetic foot ulcer, nosocomial pneumonicus, gynecologic infections; *acinetobacter* sp., *actinomyces* sp., *Bacillus anthracias, Bacteroides* sp., *Bifidobacterium* sp., *Bordetella pertussis, Borrelia burgdorferi, Brucella* sp., *Clostridium* sp., *Corynebacterium diptherial/xerosis, Eikenella Corrodens, Entercoccus faecalis, Erysipelothrix rhusiopathial, Escherichia coli, Eubacterium* sp., *Fusobacterium* sp., *Gardnerella vaginalis, Haemophilas influenzal (beta-lactamase negative/ positive) Helicobacterpylori, Klebsiella* sp., *Lactobacillus* sp., *Leptospira* sp., *Listeria monocytogenes, Moraxella Catarrhalis, Morganella morganii, Neisseria gonorrhoeae, Pasteurella multocida Peptococcus* sp., *Pepto streptococcus* sp., *Porphyromonas* sp., *Prevotella* sp., *Propionibacterium Sp. Proteus mirabilis, Prteus Vulgaris, Provedencia rettgeri, Provedencia stuarti, Salmonella* sp., *Shigella* sp., *Staphylococcus aureus (MSSA)/ epidermidis/ saprophyticus, Streptococcus agalactinel, dysgalactial pneumonial/pyrogenes, Treponema pallidum, Viridans Streptococci*

Unlabeled uses: Aspiration pneumonia, bone/joint infections, asthma, diabetes, mellitus, dialysis, diarrhea, eczema, IBS, leukemia, pseudomembranous colitis, ulcerative colitis

CONTRAINDICATIONS:
Hypersensitivity to penicillins, sulbactam

Precautions: Pregnancy (B), breastfeeding, neonates, hypersensitivity to cephalosporins/carbapenems, renal disease, mononucleosis, viral infections, syphilis

DOSAGE AND ROUTES

- **Adult/adolescent/child ≥40 kg: IM/IV** 1.5-3 g q6hr, max 4 g/day sulbactam
- **Child ≤40 kg: IV** 150-300 mg/kg/day divided q6hr

Renal disease

- **Adult ≥40 kg: IM/IV** CCr 15-30 ml/min dose q12hr; CCr 5-15 ml/min dose q24hr

Available forms: Powder for inj 1.5 g (1 g ampicillin, 0.5 g sulbactam), 3 g (2 g ampicillin, 1 g sulbactam), 15 g (10 g ampicillin, 5 g sulbactam)

Administer:

- Scratch test to assess allergy after securing order from prescriber; usually done when penicillin is only product choice
- Store at room temperature

IM route

- Reconstitute by adding 3.2 ml sterile water/1.5-g vial; 6.4 ml/3-g vial, give deep in large muscle, aspirate
- Do not use IM in child

Direct IV route

- After diluting 1.5 g/3.2 ml sterile water for inj or 3 g/6.4 ml (250 mg ampicillin/ 125 mg sulbactam), allow to stand until foaming stops; may give over 15 min

Intermittent IV INF route

- Dilute further in 50 ml or more of D_5W, NaCl; administer within 1 hr after reconstitution; give over 15-30 min, separate doses from aminoglycosides by ≥1 hr

Y-site compatibilities: Alemtuzumab, amifostine, aminocaproic acid, anidulafungin, argatroban, atenolol, bivalirudin, bleomycin, CARBOplatin, carmustine, cefepime, CISplatin, codeine, cyclophosphamide, cytarabine, DAPTOmycin, DAUNOrubicin liposome, dexmedetomidine, DOCEtaxel, doxacurium, DOXOrubicin liposomal, eptifibatide, etoposide, fenoldopam, filgrastim, fludarabine, fluorouracil, foscarnet, gallium, gatifloxacin, gemcitabine, granisetron, hetastarch, irinotecan, levofloxacin, linezolid, methotrexate, metroNIDAZOLE, octreotide, oxaliplatin, PACLitaxel, palonosetron, pamidronate, pancuronium, pantoprazole, PEMEtrexed, remifentanil, riTUXimab, rocuronium, tacrolimus, teniposide, thiotepa, tigecycline, tirofiban, TNA, TPN, trastuzumab, vencuronium, vinCRIStine, voriconazole, zoledronic acid

SIDE EFFECTS

CNS: Lethargy, hallucinations, anxiety, depression, twitching, coma, seizures
GI: *Nausea, vomiting, diarrhea,* increased AST/ALT, abdominal pain, glossitis, colitis, pseudomembranous colitis, hepatic necrosis/failure, black hairy tongue
GU: Oliguria, proteinuria, hematuria, *vaginitis, moniliasis,* glomerulonephritis, dysuria
HEMA: Anemia, increased bleeding time, bone marrow depression, granulocytopenia, leukopenia, eosinophilia, hemolysis
INTEG: Injection site reactions, rash, edema, urticaria
MISC: Anaphylaxis, serum sickness, toxic epidermal necrolysis, Stevens-Johnson syndrome, hypoalbuminemia

PHARMACOKINETICS

IV: Peak 5 min, IM 1 hr; half-life 50-110 min, little metabolized in liver, 75%-85% of both products excreted in urine, diffuses to breast milk, crosses placenta

INTERACTIONS

Increase: bleeding risk—oral anticoagulants; check, INR, PT
Increase: ampicillin-induced skin rash—allopurinol, check for rash
Increase: ampicillin level—probenecid
Increase: methotrexate level—methotrexate

Drug/Lab Test
False positive: urine glucose, urine protein

NURSING CONSIDERATIONS

Assess:
- **Infection:** characteristics of wound, sputum; take temperature, WBC count; C&S before product therapy, product may be given as soon as culture is taken
- Bowel pattern before, during treatment
- I&O ratio; report hematuria, oliguria because penicillin in high doses is nephrotoxic

⚠ Any patient with compromised renal system, because product excreted slowly with poor renal system function; toxicity may occur rapidly

- Hepatic studies: AST, ALT if on long-term therapy
- Blood studies: WBC, RBC, Hct, Hgb, bleeding time
- Renal studies: urinalysis, protein, blood, BUN, creatinine

⚠ **Anaphylaxis:** skin eruptions after administration of ampicillin to 1 wk after discontinuing product

- Allergies before treatment; reaction to each medication; report allergies

Evaluate:
- Therapeutic response: absence of fever, draining wounds; negative C&S

Teach patient/family:
- To report superinfection: vaginal itching; loose, foul-smelling stools; black furry tongue

⚠ **Pseudomembranous colitis:** fever, diarrhea with pus, blood, mucus; may occur up to 4 wk after treatment; report immediately to health care provider

- To wear or carry emergency ID if allergic to penicillin products

TREATMENT OF ANAPHYLAXIS:
Withdraw product, maintain airway; administer EPINEPHrine, aminophylline, O_2, IV corticosteroids

anakinra (Rx)

(an-ah-kin′rah)

Kineret

Func. class.: Antirheumatic (DMARD), immunomodulator

Chem. class.: Recombinant form of human interleukin-1 receptor antagonist (IL-1Ra)

Do not confuse:
Anakinra/amikacin

ACTION:
A form of human interleukin-1 receptor antagonist (IL-1Ra) produced by DNA technology; blocks activity of IL-1, thereby resulting in decreased inflammation, cartilage degradation, bone resorption

USES:
Reduction in signs and symptoms of moderate to severe active rheumatoid arthritis in patients ≥18 yr who have not responded to other disease-modifying agents

Unlabeled uses: Cryopyrin-associated periodic syndromes

CONTRAINDICATIONS:
Hypersensitivity to *Escherichia-coli*–derived proteins, product, latex; sepsis

Precautions: Pregnancy (B), breastfeeding, children, geriatric patients, renal impairment, active infections, immunosuppression, neoplastic disease, asthma

DOSAGE AND ROUTES

- **Adult: SUBCUT** 100 mg/day

Renal dose

- **Adult:** CCr <30 ml/min **SUBCUT** 100 mg every other day

Available form: Inj 100 mg/0.67 ml prefilled glass syringe

Administer:

SUBCUT route

- Do not use if cloudy, discolored, if particulate is present; protect from light
- Do not admix with other sol or medications; do not use filter; give at same time each day
- Apply cold compress before, after inj, allow sol to warm to room temp before use
- Use middle thigh, abdomen (outside 2 inches from navel), upper outer buttocks, upper outer area of arm; rotate sites, give inj at least 1 inch from old site; do not give in skin that is bruised, red, tender, hard; remove needle cover immediately before use, pull gently back on plunger, if no blood appears, inject entire contents of prefilled syringe; discard any unused portion

SIDE EFFECTS

CNS: *Headache*
EENT: *Sinusitis*
GI: *Abdominal pain, nausea, diarrhea*
HEMA: Neutropenia
INTEG: Rash, *inj site reaction,* allergic reaction
MISC: Flulike symptoms, infection
MS: *Worsening of RA, arthralgia*
RESP: *URI*

PHARMACOKINETICS

Terminal half-life 4-6 hr; eliminated renally

INTERACTIONS

- Do not use rilonacept

Increase: risk for severe infection—TNF-blocking agents, do not use together

Decrease: antibody reactions—vaccines, avoid concurrent use

NURSING CONSIDERATIONS

Assess:

- **Rheumatoid arthritis:** pain, stiffness, ROM, swelling of joints, baseline, periodically during treatment
- For inj site pain, swelling; usually occurs after 2 inj (4-5 days)
- For infections (increased WBC, fever, flulike symptoms); stop treatment if present; do not start if patient has active infection

- CBC with differential, neutrophil counts before treatment, monthly × 3 mo, quarterly for up to 1 yr thereafter
- For allergic reactions (rash, dyspnea), discontinue if severe
- For urinary status: decreasing urinary output

Evaluate:
- Therapeutic response: decreased inflammation, pain in joints

Teach patient/family:
- Not to receive vaccines while taking this product; to update vaccines before treatment
- About self-administration, if appropriate: inj should be made in thigh, abdomen, upper arm; rotate sites at least 1 inch from old site; give at same time of day, store in refrigerator, do not freeze
- To notify prescriber if pregnancy is planned, suspected; to avoid breastfeeding; to notify prescriber of allergic reaction, decreasing urine output, signs/symptoms of infection

⚠ HIGH ALERT

anastrozole (Rx)

(an-a-stroh′zole)

Arimidex

Func. class.: Antineoplastic

Chem. class.: Aromatase inhibitor

ACTION: Highly selective nonsteroidal aromatase inhibitor that lowers serum estradiol concentrations; many breast cancers have strong estrogen receptors

USES: Advanced breast carcinoma not responsive to other therapy in estrogen-receptor–positive patients (postmenopausal); patients with advanced disease taking tamoxifen, adjunct therapy for early breast cancer

Unlabeled uses: Uterine leiomyomata, breast cancer (in those who have received tamoxifen for 2-3 yr)

CONTRAINDICATIONS: Pregnancy (X), breastfeeding, hypersensitivity

Precautions: Children, geriatric patients, premenopausal women, osteoporosis, hepatic/cardiac disease

DOSAGE AND ROUTES

- **Adult: PO** 1 mg/day

Available forms: Tabs 1 mg

Administer:
- Give without regard to meals at same time of day
- Store in light-resistant container at room temperature

SIDE EFFECTS

CNS: *Hot flashes, headache, lightheadedness,* depression, dizziness, confusion, insomnia, anxiety, fatigue, mood changes

CV: Chest pain, *hypertension,* **thrombophlebitis,** *edema,* angina, **MI, cerebral infarct, CVA,** *vasodilation*

GI: *Nausea, vomiting,* altered taste leading to anorexia, diarrhea, constipation, abdominal pain, dry mouth

GU: Vaginal bleeding, vaginal dryness, pelvic pain, pruritus vulvae, UTI

HEMA: **Leukopenia**

INTEG: *Rash,* **Stevens-Johnson syndrome,** anemia

MISC: Hypercholesterolemia

MS: Bone pain, myalgia, *asthenia,* bone loss/osteoporosis, arthralgia, fractures

RESP: Cough, sinusitis, dyspnea, **pulmonary embolism**

PHARMACOKINETICS

Peak 2 hr; half-life 50 hr; excreted in feces, urine, terminal half-life 50 hr

INTERACTIONS

- **Do not use with oral contraceptives, estrogen, tamoxifen, androstenedione, DHEA**

Drug/Lab Test

Increase: GGT, AST, ALT, alk phos, cholesterol, LDL

NURSING CONSIDERATIONS

Assess:

- Bone mineral density, cholesterol, lipid panel, periodically

⚠ **Serious skin reactions:** Stevens-Johnson syndrome

- Not effective in hormone-receptor–negative disease, use only in postmenopausal women

Evaluate:

- Therapeutic response: decreased tumor size, spread of malignancy

Teach patient/family:

- To report any complaints, side effects to prescriber
- That vaginal bleeding, pruritus, hot flashes reversible after discontinuing treatment
- To report continued vaginal bleeding immediately
- That **tumor flare**—increase in size of tumor, increased bone pain—may occur and will subside rapidly; may take analgesics for pain
- To take adequate calcium and vitamin D due to risk for bone loss/fractures

anidulafungin (Rx)

(a-nid-yoo-luh-fun′jin)

Eraxis

Func. class.: Antifungal, systemic

Chem. class.: Echinocandin

ACTION:
Inhibits fungal enzyme synthesis; causes direct damage to fungal cell wall

USES:
Esophageal candidiasis, *Candida albicans, C. glabrata, C. parapsilosis, C. tropicalis*

Unlabeled uses: Fungal prophylaxis, disseminated candidiasis, oropharyngeal candidiasis, *Aspergillosis* sp.

CONTRAINDICATIONS:
Hypersensitivity to product, other echinocandins

Precautions: Pregnancy (C), breastfeeding, children, severe hepatic disease

DOSAGE AND ROUTES

Candidemia and other *Candida* infections

- **Adult: IV** 200 mg loading dose on day 1 then 100 mg/day × 14 days or more until last positive culture

Esophageal candidiasis

- **Adult: IV** 100 mg loading dose on day 1 then 50 mg/day × 14 days, for at least 7 days after symptoms resolved

Fungal prophylaxis (unlabeled)

- **Adolescent and child 2-17 yr: IV** 1.5 mg/kg over 90 min on day 1 then 0.75 mg/kg/day over 45 min × 5-28 days

Available forms: Powder for inj, lyophilized 50, 100 mg

Administer:

IV route

- Visually inspect prepared infusions for particulate matter and discoloration, do not use if present; give by IV infusion only, after dilution
- **Reconstitution:** Reconstitute each 50 mg or 100 mg vial/15 ml or 30 ml of Sterile Water for Injection, respectively (3.33 mg/ml)
- **Storage:** Reconstituted solutions are stable for up to 24 hours at room temperature
- Dilution
- Do not use any other diluents besides Dextrose 5% in Water (D5W) or Sodium Chloride 0.9% (NS)
- *Preparation of the 200 mg loading dose infusion:* Withdraw the contents of either four 50 mg reconstituted vials OR two 100 mg reconstituted vials and add to an IV infusion bag or bottle containing 200 ml of D5W or NS to give a total infusion volume of 260 ml
- *Preparation of the 100 mg daily infusion:* Withdraw the contents of one 100 mg reconstituted vial OR two 50-mg reconstituted vials and add to an IV infusion bag or bottle containing 100 ml of D5W or NS to give a total infusion volume of 130 ml
- *Preparation of a 50 mg daily infusion:* Withdraw the contents of one 50 mg reconstituted vial and add to an IV

infusion bag or bottle containing 50 ml of D5W or NS to give a total infusion volume of 65 ml

- *Storage:* Diluted solutions are stable for up to 48 hr at temperatures up to 77° F (25° C) or for 72 hours if stored frozen

Intermittent IV Infusion

- Do not mix or co-infuse with other medications
- Administer as a slow IV infusion at a rate of 1.4 ml/min or 84 ml/hour; the minimum duration of infusion is 180 min for the 200 mg dose, 90 min for the 100 mg dose, and 45 min for the 50 mg dose
- Store reconstituted vials at 59° F-86° F for up to 24 hr, do not freeze (dehydrated alcohol); store reconstituted vials at 36° F-46° F (sterile water) for up to 24 hr, do not freeze

Y-site compatibilities: Acyclovir, alemtuzumab, alfentanil, allopurinol, amifostine, amikacin, aminocaproic acid, aminophylline, amiodarone, amphotericin B lipid complex, amphotericin B liposome, ampicillin, ampicillin sulbactam, argatroban, arsenic trioxide, atenolol, atracurium, azithromycin, aztreonam, bivalirudin, bleomycin, bumetanide, buprenorphine, busulfan, butorphanol, calcium chloride/gluconate, CARBOplatin, carmustine, caspofungin, ceFAZolin, cefepime, cefotaxime, cefoTEtan, cefOXitin, cefTAZidime, ceftizoxime, cefTRIAXone, cefuroxime, chloramphenicol, chlorproMAZINE, cimetidine, ciprofloxacin, cisatracurium, CISplatin, clindamycin, cyclophosphamide, cycloSPORINE, cytarabine, dacarbazine, DACTINomycin, DAUNOrubicin liposome, DAUNOrubicin hydrochloride, dexamethasone, dexmedetomidine, dexrazoxane, digoxin, diltiazem, diphenhydrAMINE, DOBUtamine, DOCEtaxel, dolasetron, DOPamine, doripenem, doxacurium, DOXOrubicin, DOXOrubicin liposomal, doxycycline, droperidol, enalaprilat, ePHEDrine, EPINEPHrine, epirubicin, eptifibatide, erythromycin, esmolol, etoposide, etoposide phosphate, famotidine, fenoldopam, fentaNYL, fluconazole, fludarabine, fluorouracil, foscarnet, fosphenytoin, furosemide, gallium nitrate, ganciclovir, gatifloxacin, gemcitabine, gentamicin, glycopyrrolate, granisetron, haloperidol, heparin, hydrALAZINE, hydrocortisone, HYDROmorphone, hydrOXYzine, IDArubicin, ifosfamide, imipenem-cilastatin, inamrinone, insulin (regular), irinotecan, isoproterenol, ketorolac, labetalol, leucovorin, levofloxacin, lidocaine, linezolid injection, LORazepam, mannitol, mechlorethamine, melphalan, meperidine, meropenem, mesna, metaraminol, methotrexate, methyldopate, methylPREDNISolone, metoclopramide, metoprolol, metroNIDAZOLE, midazolam, milrinone, mitoMYcin, mitoXANtrone, mivacurium, morphine, moxifloxacin, mycophenolate mofetil, nafcillin, naloxone, nesiritide, niCARdipine, nitroglycerin, nitroprusside, norepinephrine, octreotide, ondansetron, oxaliplatin, oxytocin, PACLitaxel, palonosetron, pamidronate, pancuronium, pantoprazole, PEMEtrexed, pentamidine, pentazocine, PENTobarbital, PHENobarbital, phentolamine, phenylephrine, piperacillin-tazobactam, polymyxin B, potassium acetate/chloride, procainamide, prochlorperazine, promethazine, propranolol, quiNIDine, quinupristin-dalfopristin, ranitidine, remifentanil, rocuronium, sodium acetate, streptozocin, succinylcholine, SUFentanil, sulfamethoxazole-trimethoprim, tacrolimus, teniposide, theophylline, thiopental, thiotepa, ticarcillin, ticarcillin-clavulanate, tirofiban, tobramycin, topotecan, trimethobenzamide, vancomycin, vasopressin, vecuronium, verapamil, vinBLAStine, vinCRIStine, vinorelbine, voriconazole, zidovudine, zoledronic acid

SIDE EFFECTS

Candidemia/other *Candida* infections

CNS: Seizures, dizziness, *headache*

CV: DVT, atrial fibrillation, right bundle branch block, hypotension, sinus arrhythmia, thrombophlebitis superficial, ven-

tricular extrasystoles, QT prolongation (rare)
GI: *Nausea, anorexia, vomiting, diarrhea, increased AST, ALT*
META: Hypokalemia
Esophageal candidiasis
CNS: *Headache*
GI: *Nausea, anorexia, vomiting, diarrhea,* **hepatic necrosis**
HEMA: **Neutropenia, thrombocytopenia, leukopenia, coagulopathy**
INTEG: *Rash,* urticaria, itching, flushing
META: Hypocalcemia, hyperglycemia, hyperkalemia, hypernatremia, hypomagnesium (rare)
MS: *Back pain, rigors*

PHARMACOKINETICS

Steady state after loading dose, distribution half-life 0.5-1 hr, terminal half-life 40-50 hr, protein binding 99%

INTERACTIONS

Increase: plasma concentrations—cycloSPORINE
Drug/Lab Test
Increase: amylase, bilirubin, CPK, creatinine, ECG, lipase, PT
Decrease: platelets, magnesium, potassium, transferase, urea

NURSING CONSIDERATIONS

Assess:
- **Infection:** clearing of cultures during treatment; obtain culture at baseline and throughout treatment; product may be started as soon as culture is taken, those with HIV pharyngeal candidiasis may need antifungals

⚠ **Blood dyscrasias (rare): CBC (RBC, Hct, Hgb), differential, platelet count periodically; notify prescriber of results**
- Hepatic studies before, during treatment: bilirubin, AST, ALT, alk phos, as needed; also uric acid

⚠ **Bleeding: hematuria, heme-positive stools, bruising/petechiae of mucosa or orifices; blood dyscrasias can occur**
- GI symptoms: frequency of stools, cramping; if severe diarrhea occurs, electrolytes may need to be given

Evaluate:
- Therapeutic response: decreased symptoms of *Candida* infection, negative culture

Teach patient/family:
- To notify prescriber if pregnancy is suspected, planned; use nonhormonal form of contraception while taking this product
- To avoid breastfeeding while taking this product
- To inform prescriber of renal/hepatic disease
- To report bleeding
- To report signs of infection: increased temp, sore throat, flulike symptoms
- To notify prescriber of nausea, vomiting, diarrhea, jaundice, anorexia, clay-colored stools, dark urine; hepatotoxicity may occur

antithymocyte

See lymphocyte immune globulin

apomorphine

(ah-poe-more′feen)

Apokyn

Func. class.: Antiparkinson agent
Chem. class.: Dopamine agonist, non-ergot

ACTION: Has similarities to dopamine that are thought to contribute to its central dopamine receptor agonist properties; exhibits a high affinity for dopamine D_4 receptors, a moderate affinity for dopamine D_2, D_3, and D_5 receptors, and a low affinity for D_1 receptors; treats erectile dysfunction by activating dopamine receptors in the hypothalamus and limbic and neural pathways to increase erections

USES:
For use as rescue of "off" episodes associated with advanced Parkinson's disease

CONTRAINDICATIONS:
Hypersensitivity to this product, sulfites, or benzyl alcohol, IV use, major psychotic disorder; concurrent treatment with drugs of the 5-HT_3 antagonist class (e.g., ondansetron, granisetron, alosetron)

Precautions: History of suicidal ideation or depression, CNS/respiratory depression, decreased alertness, gag reflex depression, seizures, seizure disorder, coma, asthma, emphysema, atrial fibrillation, hypertension, bradycardia, orthostatic hypotension, QT prolongation, vasovagal syncope, acute circulatory failure, orthostatic hypotension, cardiac disease, congenital heart disease, cardiac arrhythmias, cardiac depression, heart failure, shock, hepatic/renal disease, pregnancy C, breastfeeding, debilitated/geriatric patients, abrupt discontinuation

DOSAGE AND ROUTES

• **Adult: Test dose: SUBCUT** 0.2 ml (2 mg) (test dose) where B/P can be closely monitored (before dose and 20, 40, 60 min after); if tolerated and patient response, then begin with 0.2 ml (2 mg); may increase by 1 mg every few days, max 0.6 ml (6 mg); if the test dose of 0.2 ml (2 mg) is tolerated but the patient does not respond, give a test dose of 0.4 ml (4 mg) no sooner than 2 hr after the 0.2 ml (2 mg) test dose, where BP can be closely monitored (before dose and 20, 40, and 60 min after the dose); if the 0.4 ml (4 mg) test dose is tolerated, the starting dose should be 0.3 ml (3 mg); may be increased by 1 mg every few days as required, max 0.6 ml (6 mg). If the 0.4 ml (4 mg) test dose is not tolerated, administer a test dose of 0.3 ml (3 mg) no sooner than 2 hr after the 0.4 ml (4 mg) test dose where BP can be closely monitored (before dose and 20, 40, and 60 min after the dose); if the 0.3 ml (3 mg) test dose is tolerated, begin with 0.2 ml (2 mg); may be increased by 0.1 ml (1 mg) every few days as required, max 0.4 ml (4 mg; outpatient)

• **Usual dosage:** 0.3-0.6 ml (3 to 6 mg), average frequency three times a day

Available forms: Penfill sol for inj 10 mg/ml

Administer

Subcut route

• To reduce nausea/vomiting, give trimethobenzamide × 3 days before the first dose, and continued ≥2 months of treatment

• For subcut use only; do use IM or IV

• Wash hands before and after use, skin contact can cause an allergic skin reaction

• To be used by those qualified and who have received instruction on the use

• **Dosing pen (Apokyn Pen) is dosed in ml, not mg;** the pen allows the patient's usual dose to be dialed in, even if there is not enough medicine in the pen to do so; if there is not enough medicine for the usual dose, the device must be "rearmed" and the remaining amount needed should be dialed in; this process should be avoided if possible to avoid dosage errors; keep a record of doses used so the cartridges can be replaced as needed

• Do not inject into infected, erythematous, or irritated skin

• Double-check dosage in syringe before use

• Aspirate before injection to avoid injection into a blood vessel; inject subcut, taking care not to inject ID

• Rotate injection sites with each injection (abdomen, upper arm, thigh)

SIDE EFFECTS

CNS: Drowsiness, dizziness, flushing, hallucinations, confusion, worsening psychiatric/behavioral effects, dyskinesias, sweating

CV: Hypotension, CHF, edema, MI, cardiac arrest

GI: Nausea, vomiting, constipation, or diarrhea

INTEG: Contact dermatitis, ecchymosis, injection reaction
MISC: Yawning, rhinorrhea, respiratory depression
MS: Arthralgia, back pain

PHARMACOKINETICS

Subcut: Onset 20 min, peak 20-60 min, duration 2 hr, half-life 30-60 min

INTERACTIONS

Do not use with dronedarone
Increase: hypotension, unconsciousness—serotonin-receptor antagonists (alosetron, dolasetron, granisetron, ondansetron, palonosetron), do not use together
Increase: QT interval—antiemetics (dolasetron, ondansetron, palonosetron), other products that increase QT, can occur at high doses
Increase: hypotension—antihypertensives, vasodilators
Decrease: apomorphine effect—antipsychotics

NURSING CONSIDERATIONS

Assess:

- Assess for the acute, intermittent treatment of hypomobility, "off" episodes ("end-of-dose wearing off" and unpredictable "on/off" episodes) associated with advanced Parkinson's disease
- During the test dosing to determine dose titration, B/P should be closely monitored (before dose and 20, 40, and 60 min after the injection)
- Monitor for patient response to the injection (i.e., reduction in "off" episode symptoms)
- Contact the health care provider if patient's symptoms do not respond to the first dose for a particular "off" episode; do not use a second dose for that episode
- If the patient has not received an apomorphine injection in more than 1 wk, must reinstitute following the initial test dosage and titration instructions

Evaluate:

- Positive therapeutic outcome—reduction of "off" episode symptoms in Parkinson's disease

Teach patient/family:

- Not to rise quickly; fainting can occur
- To use with antiemetic; nausea and vomiting are common
- All persons involved in care, how to use subcut pen
- To report hallucinations, other psychiatric symptoms to prescriber immediately
- Not to consume alcohol or kava; CNS effects can worsen

apraclonidine ophthalmic

See Appendix B

aprepitant (Rx)

(ap-re′pi-tant)
Emend

fosaprepitant

Emend
Func. class.: Antiemetic
Chem. class.: Miscellaneous

ACTION:
Selective antagonist of human substance P/neurokinin 1 (NK_1) receptors that decreases emetic reflex

USES:
Prevention of nausea/vomiting associated with cancer chemotherapy (highly emetogenic/moderately emetogenic), including high-dose CISplatin; used in combination with other antiemetics; postoperative nausea/vomiting

CONTRAINDICATIONS:
Hypersensitivity
Precautions: Pregnancy (B), breastfeeding, children, geriatric patients, hepatic disease

DOSAGE AND ROUTES

Highly emetogenic (aprepitant)

• **Adult: PO** Day 1 (1 hr before chemotherapy) aprepitant 125 mg with 12 mg dexamethasone **PO**, with 32 mg ondansetron IV; day 2 aprepitant 80 mg with 8 mg dexamethasone **PO;** day 3 aprepitant 80 mg with 8 mg dexamethasone **PO;** day 4 only dexamethasone 8 mg **PO; IV INF** 115 mg over 15 min, 30 min before chemotherapy as alternative to 1st dose of aprepitant on day 1 of regimen (fosaprepitant)

Moderately emetogenic

• **Adult: PO** Day 1 125 mg aprepitant with dexamethasone 12 mg **PO**, with ondansetron 8 mg **PO** × 2; days 2 and 3 80 mg aprepitant only (aprepitant); **IV INF** 115 mg over 15 min, 30 min before chemotherapy as alternative to 1st dose of aprepitant on day 1 of regimen (fosaprepitant)

Prevention of postoperative nausea/vomiting

• **Adult: PO** 40 mg within 3 hr of induction of anesthesia

Available forms: Caps 40, 80, 125 mg; powder for inj 115, 150 mg; combo pack cap 80-125 mg

Administer:

PO route

• Do not break, crush, or chew
• PO on 3-day schedule, give with full glass of water 1 hr before chemotherapy, with or without food, given with other antiemetics
• Store at room temperature; keep in original bottles, blisters

Intermittent IV INF route

• Only approved as a substitute for the 1st dose of aprepitant in 3-day regimen
• **Reconstitution:** use aseptic technique; inject 5 ml 0.9% NaCl into the vial, directing stream to wall of vial to prevent foam; swirl (do not shake)
• Prepare inf bag with 110 ml NS/115 mg; 145 ml/150 mg; do not dilute, reconstitute with any divalent cations such as calcium, magnesium, including LR, Hartmann's sol
• Withdraw entire volume from vial, transfer to inf bag; total volume 115 ml (1 mg/1 ml)
• Gently invert bag 2-3 times; reconstituted sol stable for 24 hr at lower room temp or <25° C
• Visually inspect for particulates, discoloration
• Infuse over 15 min, stable for 24 hr at room temperature

SIDE EFFECTS

CNS: *Headache, dizziness,* insomnia, anxiety, depression, confusion, peripheral neuropathy
CV: Bradycardia, tachycardia, DVT, hypo/hypertension
GI: *Diarrhea, constipation,* abdominal pain, anorexia, gastritis, increased AST, ALT, *nausea,* vomiting, heartburn
GU: Increased BUN, serum creatine, proteinuria, dysuria
HEMA: Anemia, thrombocytopenia, neutropenia
INTEG: Pruritus, rash, urticaria
MISC: Asthenia, fatigue, dehydration, fever, hiccups, tinnitus, alopecia
SYST: Anaphylaxis

PHARMACOKINETICS

Absorption 60%-65%, peak 4 hr, metabolized in liver by CYP3A4 enzymes to active metabolite, half-life 9-12 hr, 95% protein bound, not excreted in kidneys, crosses blood-brain barrier

INTERACTIONS

Increase: aprepitant action—CYP3A4 inhibitors (ketoconazole, itraconazole, nefazodone, troleandomycin, clarithromycin, ritonavir, nelfinavir, diltiazem)
Increase: action of CYP3A4 substrates (pimozide, cisapride, dexamethasone, methylPREDNISolone, midazolam, ALPRAZolam, triazolam, DOCEtaxel, PACLitaxel, etoposide, irinotecan, imatinib, ifosfamide, vinorelbine, vinBLAStine, vinCRIStine)
Decrease: aprepitant action—CYP3A4 inducers (rifampin, carBAMazepine, phenytoin)

Decrease: action of CYP2C9 substrates (warfarin, TOLBUTamide, phenytoin), oral contraceptives
Decrease: action of both products—paroxetine
Drug/Food
Decrease: effect—grapefruit juice

NURSING CONSIDERATIONS

Assess:

⚠ **For hypersensitive reactions:** pruritus, rash, urticaria, anaphylaxis

- CV status: hypo/hypertension, bradycardia, tachycardia, DVT
- For absence of nausea, vomiting during chemotherapy
- CBC, LFTs, creatinine baseline and periodically

Evaluate:

- Therapeutic response: absence of nausea, vomiting during cancer chemotherapy

Teach patient/family:

- To report diarrhea, constipation
- To take only as prescribed; to take 1st dose 1 hr before chemotherapy
- To report all medications and herbals to prescriber before taking this medication
- To use nonhormonal form of contraception while taking this agent and for 1 mo thereafter; oral contraceptive effect may be decreased
- That those patients also taking warfarin should have clotting monitored closely during 2-wk period after administration of aprepitant
- To avoid breastfeeding

arformoterol (Rx)

(ar-for-moe′ter-ole)

Brovana

Func. class.: Long-acting adrenergic β_2-agonist, sympathomimetic, bronchodilator

Do not confuse:
Brovana/Boniva

ACTION: Causes bronchodilation by action on β_2 (pulmonary) receptors by increasing levels of cAMP, which relaxes smooth muscle; produces bronchodilation and CNS, cardiac stimulation as well as increased diuresis and gastric acid secretion; longer acting than isoproterenol

USES: Maintenance bronchospasm prevention in COPD, including chronic bronchitis, emphysema

CONTRAINDICATIONS: Hypersensitivity to sympathomimetics, product, racemic formoterol; tachydysrhyhmias, severe cardiac disease, heart block, children, monotherapy in asthma
Precautions: Pregnancy (C), breastfeeding, cardiac disorders, hyperthyroidism, diabetes mellitus, hypertension, prostatic hypertrophy, angle-closure glaucoma, seizures, hypoglycemia

Black Box Warning: Asthma-related death

DOSAGE AND ROUTES

COPD

- **Adult: NEB** 15 mcg, bid, AM, PM

Available forms: Inh sol 15 mcg/2 ml
Administer:

- By nebulization only; no dilution needed; give over 10-15 min; sol should be colorless
- Store in refrigerator; if stored at room temp, discard after 6 wk or if past expiration date, whichever is sooner

SIDE EFFECTS

CNS: *Tremors, anxiety,* insomnia, headache, dizziness, stimulation, *restlessness,* hallucinations, flushing, irritability
CV: Palpitations, tachycardia, hypertension, angina, hypotension, dysrhythmias, **AV block heart failure, prolonged QT supraventricular tachycardia**
EENT: Dry nose, irritation of nose, throat
GI: Heartburn, nausea, vomiting

MISC: Flushing, sweating, anorexia, bad taste/smell changes, hypokalemia, anaphylaxis, *hypoglycemia*
MS: Muscle cramps
RESP: Cough, wheezing, dyspnea, bronchospasm, dry throat

PHARMACOKINETICS

Onset 5 min; peak 1-1½ hr; duration 4-6 hr; terminal half-life (COPD) 26 hr; extensively metabolized by direct conjugation by CYP2D6, CYP2C19; crosses placenta; protein binding 52%-65%; excreted in urine 63%, feces 11%

INTERACTIONS

Increase: QT prolongation—Class IA/III antidysrthymics, MAOIs, tricyclics
Increase: severe hypotension—oxytocics
Increase: toxicity—theophylline
Increase: ECG changes/hypokalemia—potassium-losing diuretics
Increase: action of nebulized bronchodilators
Increase: action of arformoterol—tricyclics, MAOIs, other adrenergics; do not use together
Decrease: arformoterol action, asthma-related death—other β-blockers
Drug/Herb
Increase: stimulation—caffeine (cola nut, green/black tea, guarana, yerba maté, coffee, chocolate)

NURSING CONSIDERATIONS

Assess:

Black Box Warning: Respiratory function: vital capacity, forced expiratory volume, ABGs; lung sounds, heart rate, rhythm, B/P, sputum (baseline, peak); actively deteriorating COPD may occur; a rescue inhaler should be readily available

- Whether patient has received theophylline therapy, other bronchodilators before giving dose
- Patient's ability to self-medicate
- For evidence of allergic reactions; anaphylaxis may occur
- For **paradoxical bronchospasm;** hold medication, notify prescriber if bronchospasm occurs

Evaluate:

- Therapeutic response: absence of dyspnea, wheezing after 1 hr, improved airway exchange, improved ABGs

Teach patient/family:

- To use exactly as prescribed; that death has resulted from asthma with products similar to this one; to have a rescue inhaler always
- Not to use OTC medications because excess stimulation may occur
- That an opened unit-dose vial should be used right away

⚠ HIGH ALERT

argatroban (Rx)

(are-ga-troe′ban)

Func. class.: Anticoagulant
Chem. class.: Thrombin inhibitor

Do not confuse:
argatroban/Aggrastat

ACTION: Direct inhibitor of thrombin, it reversibly binds to thrombin active site

USES: Anticoagulation prevention/treatment of thrombosis in heparin-induced thrombocytopenia; adjunct to percutaneous coronary intervention (PCI) in those with history of HIT, deep vein thrombosis, pulmonary embolism
Unlabeled uses: Acute MI, DIC, use in infants/children/adolescents

CONTRAINDICATIONS: Hypersensitivity, overt major bleeding
Precautions: Pregnancy (C), breastfeeding, children, intracranial bleeding, renal function impairment, hepatic disease, severe hypertension, after lumbar puncture, spinal anesthesia, major surgery/trauma, congenital/acquired bleeding, GI ulcers

DOSAGE AND ROUTES

DVT, Pulmonary Embolism

• **Adult: CONT IV INF** 2 mcg/kg/min; adjust dose until steady-state aPTT is 1.5-3× initial baseline, max 100 sec, max dose 10 mcg/kg/min

• **Infant/child/adolescent (unlabeled): CONT IV INF** 0.75 mcg/kg/min, monitor aPTT q2hr until stable then at least daily

Hepatic dose

• **Adult: CONT INF** 0.5 mcg/kg/min, adjust rate based on aPTT

Percutaneous coronary intervention (PCI) in HIT

• **Adult: IV INF** 25 mcg/kg/min and bolus of 350 mcg/kg given over 3-5 min, check ACT 5-10 min after bolus completed, proceed if ACT >300 sec; if ACT <300 sec, give another 150 mcg/kg **BOL,** increase inf rate to 30 mcg/kg/min, recheck ACT in 5-10 min; if ACT >450 sec, decrease inf rate to 15 mcg/kg/min, recheck ACT in 5-10 min; when ACT is therapeutic, continue for duration of procedure

Acute MI (unlabeled)

• **Adult: IV** loading dose of 100 mcg/kg over 1 min, then 1-3 mcg/kg/min

DIC (unlabeled)

• **Adult: CONT IV** 0.7 mcg/kg/min

Available forms: Inj 100 mg/ml (2.5 ml; must dilute 100-fold), 50 mg/50 ml, 125 mg/125 ml

Administer:

• Avoid all IM inj that may cause bleeding

IV, direct route

• **For PCI:** 350 mg/kg bol and continuous inf of 25 mcg/kg/min; check ACT 5-10 min after bolus

Intermittent IV INF route

• **Dilute** in 0.9% NaCl, D_5, LR to a final conc of 1 mg/ml; **dilute** each 2.5-ml vial 100-fold by mixing with 250 ml of diluent, mix by repeated inversion of the diluent bag for 1 min; may briefly be slightly hazy

• Dosage adjustment may be made after review of aPTT, max 10 mcg/kg/min

SIDE EFFECTS

CNS: *Fever,* intracranial bleeding, headache

CV: Atrial fibrillation, coronary thrombosis, MI, myocardial ischemia, coronary occlusion, ventricular tachycardia, bradycardia, *chest pain, hypotension*

GI: *Nausea, vomiting, abdominal pain, diarrhea,* GI bleeding

GU: Hematuria, abnormal kidney function, UTI

HEMA: Hemorrhage

MISC: *Back pain,* infection

RESP: Dyspnea, coughing, hemoptysis, pulmonary edema

SYST: Sepsis

PHARMACOKINETICS

Metabolized in liver, distributed to extracellular fluid, 54% plasma protein binding, half-life 39-51 min, excreted in feces, steady state 1-3 hr

INTERACTIONS

Increase: bleeding risk—antiplatelets, NSAIDs, salicylates, dipyridamole, clopidogrel, ticlopidine, heparin, warfarin, glycoprotein IIb/IIIa antagonists (abciximab, tirofiban, eptifibatide), thrombolytics (alteplase, reteplase, urokinase, tenecteplase), other anticoagulants

Drug/Herb

Increase: bleeding risk—ginger, garlic, ginkgo, horse chestnut

NURSING CONSIDERATIONS

Assess:

• Baseline aPTT before treatment; do not start treatment if aPTT ratio is ≥2.5; then check aPTT 2 hr after initiation of treatment and at least daily thereafter

• aPTT, which should be 1.5-3 × control, draw blood for ACT every 20-30 min during long PCI

⚠ **Bleeding** gums: petechiae; ecchymosis; black, tarry stools; hematuria/epistaxis; B/P; vaginal bleeding, possible hemorrhage

⚠ **Anaphylaxis:** dyspnea, rash during treatment

• Fever, skin rash, urticaria

Evaluate:

• Therapeutic response: absence or decrease of thrombosis

Teach patient/family:

• To use a soft-bristle toothbrush to avoid bleeding gums; avoid contact sports; use electric razor; avoid IM inj

• To report any signs of bleeding: gums, under skin, urine, stools; trouble breathing wheezing, skin rash

• To notify prescriber if planning to become pregnant, breastfeeding

ARIPiprazole (Rx)

(a-rip-ip-pra′zol)

Abilify, Abilify Discmelt, Abilify Maintena

Func. class.: Antipsychotic

Chem. class.: Quinolinone

ACTION: Exact mechanism unknown; may be mediated through both DOPamine type 2 (D_2, D_3) and serotonin type 2 ($5\text{-}HT_{1A}$, $5\text{-}HT_{2A}$) antagonism

USES: Schizophrenia and bipolar disorder (adults and adolescents), agitation, mania, major depressive disorder, short-term mania or mixed episodes of bipolar disorder; irritability in patients with autism

Unlabeled uses: Psychosis in patients with dementia

CONTRAINDICATIONS: Breastfeeding, hypersensitivity, seizure disorders

Precautions: Pregnancy (C), geriatric patients, renal/hepatic/cardiac disease, neutropenia

Black Box Warning: Children, dementia, suicidal ideation

DOSAGE AND ROUTES

Major depressive disorder

• **Adult: PO** 2-5 mg/day as an adjunct to other antidepressant treatment; adjust by 5 mg at ≥1 wk (range, 2-15 mg/day)

Schizophrenia

• **Adult: PO** 10-15 mg/day; if needed, dosage may be increased to 30 mg/day after 2 wk; maintenance 15 mg/day; periodically reassess; IM/ext rel monthly inj susp) 400 mg monthly

• **Adolescent 13-17 yr: PO** 2 mg/day, may increase to 5 mg after 2 days, then 10 mg after 2 more days, max 30 mg/day

Bipolar disorder

• **Adult: PO** 15 mg/day, may increase to 30 mg if needed (monotherapy); adjunctive to lithium or valproate PO 10-15 mg daily, may increase to 30 mg if needed

• **Child >10 yr, adolescents: PO** 2 mg, titrate to 5 mg/day after 2 days to target of 10 mg/day after another 2 days

Agitation with bipolar disorder/schizophrenia

• **Adult: IM** 9.75 mg as a single dose, may start with a lower dose, max 30 mg/day

Irritability associated with autism

• **Child ≥6 yr, adolescents: PO** 2 mg/day, increase to 5 mg/day after 1 wk, may increase to 10-15 mg/day if needed; dose changes should not occur more frequently than q1wk

Potential CYP2D6 inhibitor, strong CYP3A4 inhibitors

• **Adult: PO** Reduce to 50% of usual dose; increase dose when CYP2D6, CYP3A4 inhibitor withdrawn

Combination of strong CYP3A4/CYP2D6 inhibitors

• **Adult: PO** reduce to 25% of usual dose

Acute psychosis (unlabeled)

• **Adults: PO** 15 mg/day, may increase to 20-30 mg/day after 3 wk

Available forms: Tabs 2, 5, 10, 15, 20, 30 mg; inj 9.75 mg/1.3 ml; orally disintegrating tab 10, 15 mg; oral sol 1 mg/ml; sol for inj 1.75 mg/1.3 ml; inj susp ext rel 300, 400 mg

Administer:

PO route

• Store in tight, light-resistant container

• May be given without regard to meals

• **Orally disintegrating tabs;** do not open blister until ready to use, do not

push tab through foil; place on tongue, allow to dissolve, swallow, do not divide
- **Oral liquid:** use calibrated measuring device
- **Oral solution:** can be substituted for tablet on a mg-per-mg, up to 25-mg dose. Patients receiving 30-mg tablets should receive 25 mg, immediate release of solution

IM route
- Give IM only; inject slowly, deeply into muscle mass; discard unused portion; do not give IV or subcut
- Available as ready to use
- Ext rel monthly (Abilify Maintena)

SIDE EFFECTS

CNS: *Drowsiness, insomnia, agitation, anxiety, headache,* **seizures, neuroleptic malignant syndrome,** *light-headedness, akathisia, asthenia, tremor,* **stroke, suicidal ideation,** dystonia
CV: Orthostatic hypotension, **tachycardia**
EENT: *Blurred vision, rhinitis*
GI: *Constipation, nausea, vomiting,* jaundice, *weight gain*
INTEG: *Rash*
META: Hyperglycemia, dyslipidemia
MS: Musculoskeletal pain/stiffness, myalgia
RESP: *Cough*
SYST: **Death among geriatric patients with dementia**

PHARMACOKINETICS

PO: Absorption 87%; extensively metabolized by liver to a major active metabolite; plasma protein binding >99%; terminal half-life 75-146 hr; excretion via urine 25%, feces 55%; clearance decreased in geriatric patients

INTERACTIONS

Increase: effects of ARIPiprazole—CYP3A4 inhibitors (ketoconazole, erythromycin), CYP2D6 inhibitors (quiNIDine, FLUoxetine, PARoxetine); reduce dose of ARIPiprazole
Increase: sedation—other CNS depressants, alcohol
Increase: EPS—other antipsychotics, lithium
Decrease: ARIPiprazole level—famotidine, valproate
Decrease: effects of ARIPiprazole—CYP3A4 inducers (carBAMazepine)

Drug/Herb
Decrease: ARIPiprazole effect—St. John's wort

NURSING CONSIDERATIONS

Assess:

Black Box Warning: **Mental status** before initial administration, children/young adults may exhibit suicidal thoughts/behaviors, therefore smallest amount of product should be given; elderly patients with dementia-related psychosis are at increased risk of death

- Swallowing of PO medication; check for hoarding, giving of medication to other patients
- I&O ratio; palpate bladder if urinary output is low
- Bilirubin, CBC, LFTs, weight, lipid profile, fasting blood glucose q mo
- Affect, orientation, LOC, reflexes, gait, coordination, sleep pattern disturbances
- B/P standing and lying; also pulse, respirations; take q4hr during initial treatment; establish baseline before starting treatment; report drops of 30 mm Hg; watch for ECG changes
- Dizziness, faintness, palpitations, tachycardia on rising
- **EPS,** including akathisia (inability to sit still, no pattern to movements), tardive dyskinesia (bizarre movements of the jaw, mouth, tongue, extremities), pseudoparkinsonism (rigidity, tremors, pill rolling, shuffling gait)

⚠ **Neuroleptic malignant syndrome:** hyperthermia, increased CPK, altered mental status, muscle rigidity; notify prescriber immediately

- Constipation, urinary retention daily; if these occur, increase bulk, water in diet; stool softeners, laxatives may be needed

• Supervised ambulation until patient is stabilized on medication; do not involve patient in strenuous exercise program because fainting is possible; patient should not stand still for a long time

Evaluate:

• Therapeutic response: decrease in emotional excitement, hallucinations, delusions, paranoia; reorganization of patterns of thought, speech

Teach patient/family:

• That orthostatic hypotension may occur; to rise from sitting or lying position gradually
• To avoid hot tubs, hot showers, tub baths; hypotension may occur
• To avoid abrupt withdrawal of this product; EPS may result; product should be withdrawn slowly
• To avoid OTC preparations (cough, hay fever, cold) unless approved by prescriber because serious product interactions may occur; to avoid use with alcohol, CNS depressants because increased drowsiness may occur
• To avoid hazardous activities if drowsy, dizzy
• About compliance with product regimen
• To report impaired vision, tremors, muscle twitching, urinary retention
• That heat stroke may occur in hot weather; to take extra precautions to stay cool
• To notify prescriber if pregnant or intending to become pregnant; not to breastfeed

Black Box Warning: To report suicidal thoughts/behaviors immediately

TREATMENT OF OVERDOSE: Lavage if orally ingested; provide airway; *do not induce vomiting*

RARELY USED

armodafinil (Rx)

(ar-moe-daf′in-il)

Nuvigil

Controlled Substance Schedule IV

USES: Narcolepsy, obstructive sleep apnea/hypoapnea syndrome, circadian rhythm disruption (shift-work sleep problems)

CONTRAINDICATIONS: Hypersensitivity to this product or modafinil

DOSAGE AND ROUTES

Narcolepsy, obstructive sleep apnea/hypoapnea syndrome

• **Adult and adolescent ≥17 yr: PO** 150-250 mg in AM

Circadian rhythm disruption (shift work sleep problems)

• **Adult and adolescent ≥17 yr: PO** 150 mg at start of shift

ascorbic acid (vit C) (OTC, Rx)

(a-skor′bic)

Acerola C, Apo-C ♣, Ascor L-500, Cenolate, Equaline Vitamin C, Walgreens Gold Seal, and many more

Func. class.: Vit C—water-soluble vitamin

ACTION: Wound healing, collagen synthesis, antioxidant, carbohydrate metabolism

USES: Vit C deficiency, scurvy; delayed wound, bone healing; chronic disease; urine acidification; before gastrectomy; dietary supplement

Unlabeled uses: Common cold prevention

CONTRAINDICATIONS:
Tartrazine, sulfite sensitivity; G6PD deficiency

Precautions: Pregnancy (C), gout, diabetes, renal calculi (large doses)

DOSAGE AND ROUTES

Dietary supplementation

- **Adult:** 50-500 mg/day
- **Child 14-18 yr: PO** 65 mg (female), 75 mg (male)
- **Child 9-13 yr: PO** 45 mg/day
- **Child 4-8 yr: PO** 25 mg/day
- **Child 1-3 yr: PO** 15 mg/day
- **Infant: PO** 40-50 mg/day

Scurvy

- **Adult: PO/SUBCUT/IM/IV** 100-250 mg/day × 2 wk then 50 mg or more daily
- **Child: PO/SUBCUT/IM/IV** 100-300 mg/day × 2 wk then 35 mg or more daily

Wound healing/chronic disease/ fracture (may be given with zinc)

- **Adult: SUBCUT/IM/IV/PO** 200-500 mg/day for 1-2 mo
- **Child: SUBCUT/IM/IV/PO** 100-200 mg added doses for 1-2 mo

Urine acidification

- **Adult:** 4-12 g/day in divided doses
- **Child:** 500 mg q6-8hr

Available forms: Tabs 25, 50, 100, 250, 500, 1000, 1500 mg; effervescent tabs 1000 mg; chewable tabs 100, 250, 500 mg; timed-release tabs 500, 750, 1000, 1500 mg; timed-release caps 500 mg; crys 4 g/tsp; powder 4 g/tsp; liq 35 mg/0.6 ml; sol 100 mg/ml; syr 20 mg/ml, 500 mg/5 ml; inj SUBCUT, IM, IV 100, 250, 500 mg/ml

Administer:

PO route

- Do not crush or chew ext rel tabs or caps
- Caps may be opened and contents mixed with jelly

IV, direct route

- 100 mg undiluted by direct IV over at least 1 min; rapid inf may cause fainting

Intermittent IV INF route

- Diluted with D_5W, D_5NaCl, NS, LR, Ringer's, sodium lactate and given over 15 min

Syringe compatibilities: Metoclopramide, aminophylline, theophylline

Y-site compatibilities: Warfarin

SIDE EFFECTS

CNS: Headache, insomnia, dizziness, fatigue, flushing

GI: Nausea, vomiting, diarrhea, anorexia, heartburn, cramps

GU: Polyuria, urine acidification, oxalate/urate renal stones, dysuria

HEMA: **Hemolytic anemia in patients with G6PD**

INTEG: Inflammation at inj site

PHARMACOKINETICS

PO/INJ: Readily absorbed PO, metabolized in liver; unused amounts excreted in urine (unchanged), metabolites; crosses placenta, breast milk

INTERACTIONS

Drug/Lab Test

False negative: occult blood, urine bilirubin, leukocyte determination

NURSING CONSIDERATIONS

Assess:

- I&O ratio; urine pH (acidification)
- Ascorbic acid levels throughout treatment if continued deficiency is suspected
- Nutritional status: citrus fruits, vegetables
- Inj sites for inflammation
- Thrombophlebitis if receiving large dose

Evaluate:

- Therapeutic response: absence of anorexia, irritability, pallor, joint pain, hyperkeratosis, petechiae, poor wound healing

Teach patient/family:
• Necessary foods to include in diet, such as citrus fruits
• That smoking decreases vit C levels; not to exceed prescribed dose; that excesses will be excreted in urine, except when taking timed-release forms

asenapine (Rx)
(a-sen′a-peen)
Saphris
Func. class.: Antipsychotic, atypical
Chem. class.: Dibenzapine

ACTION: Unknown; may be mediated through both DOPamine type 2 (D2) and serotonin type 2 (5-HT2A) antagonism

USES: Bipolar 1 disorder, schizophrenia
Unlabeled uses: Agitation

CONTRAINDICATIONS: Breastfeeding, hypersensitivity
Precautions: Pregnancy (C), children, geriatric patients, cardiac/renal/hepatic disease, breast cancer, Parkinson's disease, dementia, seizure disorder, CNS depression, agranulocytosis, QT prolongation, torsades de pointes, suicidal ideation, substance abuse, diabetes mellitus

Black Box Warning: Increased mortality in elderly patients with dementia-related psychosis

DOSAGE AND ROUTES
Schizophrenia
• **Adult: SL** 5 mg bid, max 20 mg/day
Bipolar 1 disorder
• **Adult: SL** 10 mg bid, may decrease to 5 mg bid as needed, max 20 mg/day
Available forms: SL tab 5, 10 mg
Administer:
• Anticholinergic agent to be used for EPS
• Store in tight, light-resistant container
• **SL tab:** remove tab; place tab under tongue; after it dissolves, swallow; advise patient not to chew, crush, swallow tabs, not to eat, drink for 10 min

SIDE EFFECTS
CNS: *EPS, pseudoparkinsonism, akathisia, dystonia, tardive dyskinesia; drowsiness, insomnia, agitation, anxiety, headache,* seizures, neuroleptic malignant syndrome, dizziness, suicidal ideation, drowsiness, depression
CV: Orthostatic hypotension, sinus tachycardia; heart failure, QT prolongation, stroke, bundle branch block
ENDO: Hyperglycemia, hyperlipidemia
GI: *Nausea,* vomiting, *constipation,* weight gain, increased appetite; oral hypoesthesia/parasthesia, mucosal ulcers, increased salivation (SL)
HEMA: Thrombocytopenia, agranulocytosis, anemia, leukopenia
INTEG: Serious allergic reactions (anaphylaxis, angioedema)

PHARMACOKINETICS
Extensively metabolized by liver, protein binding 95%, peak 0.5-1.5 hr, terminal half-life 24 hr

INTERACTIONS
Increase: sedation—other CNS depressants, alcohol
Increase: EPS—CYP2D6 inhibitors/substrates (SSRIs)
Increase: serotonin syndrome—SSRIs
Increase: Seizure risk—buPROPion
Increase: EPS—other antipsychotics
Increase: asenapine excretion—carBAMazepine
Increase: QT prolongation—class IA/III antidysrhythmics, some phenothiazines, β-agonists, local anesthetics, tricyclics, haloperidol, methadone, chloroquine, clarithromycin, droperidol, erythromycin, pentamidine
Decrease: asenapine action—CYP2D6 inducers (carBAMazepine, barbiturates, phenytoins, rifampin)
Drug/Herb
Increase: CNS depression—kava
Increase: EPS—betel palm, kava

Drug/Lab Test
Increase: prolactin levels, glucose LFTs, cholesterol, LFTs, lipids, triglycerides
Decrease: sodium

NURSING CONSIDERATIONS

Assess:

⚠ Mental status before initial administration; watch for suicidal thoughts and behaviors; dementia and death may occur among elderly patients

- Affect, orientation, LOC, reflexes, gait, coordination, sleep pattern disturbances
- B/P standing and lying; also pulse, respirations; take these q4hr during initial treatment; establish baseline before starting treatment; report drops of 30 mm Hg; watch for ECG changes; QT prolongation may occur
- Dizziness, faintness, palpitations, tachycardia on rising
- **EPS,** including akathisia, tardive dyskinesia (bizarre movements of the jaw, mouth, tongue, extremities), pseudoparkinsonism (rigidity, tremors, pill rolling, shuffling gait)
- **Neuroleptic malignant syndrome:** hyperthermia, increased CPK, altered mental status, muscle rigidity
- Constipation daily; increase bulk, water in diet if needed
- Weight gain, hyperglycemia, metabolic changes with diabetes
- Supervised ambulation until patient stabilized on medication; do not involve patient in strenuous exercise program because fainting is possible; patient should not stand still for a long time

Evaluate:

- Therapeutic response: decrease in emotional excitement, hallucinations, delusions, paranoia; reorganization of patterns of thought, speech

Teach patient/family:

- That orthostatic hypotension may occur; to rise from sitting or lying position gradually
- To avoid hot tubs, hot showers, tub baths; hypotension may occur
- To avoid abrupt withdrawal of this product; EPS may result; product should be withdrawn slowly
- To avoid OTC preparations (cough, hay fever, cold) unless approved by prescriber; serious product interactions may occur; to avoid use of alcohol; increased drowsiness may occur
- To avoid hazardous activities if drowsy, dizzy
- About compliance with product regimen
- That heat stroke may occur in hot weather; to take extra precautions to stay cool
- To use contraception; to inform prescriber if pregnancy is planned, suspected

Black Box Warning: To report suicidal thoughts/behaviors immediately

TREATMENT OF OVERDOSE:

Lavage if orally ingested; provide airway; *do not induce vomiting*

RARELY USED
⚠ HIGH ALERT

asparaginase (Rx)

(a-spare′a-gi-nase)

Elspar

Func. class.: Antineoplastic
Chem. class.: *Escherichia coli* enzyme

USES:

Acute lymphocytic leukemia in combination with other antineoplastics

CONTRAINDICATIONS:

Hypersensitivity to product or *E. coli* protein, thromboembolic disease, infants, breastfeeding, pancreatitis

DOSAGE AND ROUTES

In combination

- **Adult and child: IM/IV** 25,000 international units/m^2/wk × 2 wk or 6000 international units/m^2 every other day ×

3-4 wk or 1000-20,000 international units/m² for 10-12 days

asparaginase *Erwinia chrysanthemi*

Erwinaze

Func. class.: Antineoplastic, natural and semisynthetic

ACTION:

Contains an asparaginase-specific enzyme L-asparaginase derived from Erwinia chrysanthemi, which catalyzes of asparagine to aspartic acid and ammonia and causes reduced circulating concentrations of asparagine; efficacy of asparaginase *Erwinia chrysanthemi* may be leukemic cell cytotoxicity from asparagine deficiency

USES:

Treatment of acute lymphocytic leukemia (ALL) in combination with other chemotherapeutic agents in patients who have developed hypersensitivity to *E. coli*–derived asparaginase

CONTRAINDICATIONS:

Hypersensitivity, breastfeeding, history of serious pancreatitis, bleeding, or serious thrombosis with prior L-asparaginase therapy

Precautions: Pregnancy (C), children <2 yr, hepatic disease, diabetes mellitus

DOSAGE AND ROUTES

- **Adult, adolescent, child ≥2 yr (substitute for pegaspargase): IM** 25,000 IU/m² 3 ×/wk (Monday/Wednesday/Friday) × 6 doses for each planned dose of pegaspargase within a treatment
- **Adult (substitute for L-asparaginase *E. coli*): IM** 25,000 IU/m² for each scheduled dose of native *E. coli* asparaginase within a treatment

Available forms: Powder for inj 10,000 units

Administer:

- Slowly inject 1 or 2 ml of preservative-free sterile sodium chloride (0.9%) inj against inner vial wall; do not forcefully inject sol directly onto or into powder; if 1 ml of NS is used, conc is 10,000 IU/ml; if 2 ml of NS is used, conc is 5000 IU/ ml; dissolve contents by gentle mixing or swirling; do not shake or invert vial
- Reconstituted sol should be clear and colorless; discard if any visible particles or protein aggregates are present
- Calculate the volume needed to obtain dose; withdraw volume containing calculated dose from vial into polypropylene syringe within 15 min of reconstitution

IM route

Administer dose by IM inj within 4 hr of reconstitution; limit volume to 2 ml per inj site; multiple inj sites may be needed

- Do not freeze or refrigerate the reconstituted solution; discard any unused portions

SIDE EFFECTS

ENDO: Hyperglycemia
GI: Pancreatitis
HEMA: Bleeding, serious thrombosis
INTEG: Local inj site reaction
MISC: Hyperammonemia (seizure, headache), neurotoxicity, elevated hepatic enzymes, hyperbilirubinemia
SYST: Anaphylaxis

PHARMACOKINETICS

Achieved serum trough asparaginase activity concentrations ≥0.1 IU/ml by 72 hr after dose 3

INTERACTIONS

Synergistic or antagonistic action of: methotrexate
Increase: neurotoxicity—vinCRIStine
Increase: pancreatitis risk—cytarabine
Increase: hyperglycemia—corticosteroids
Increase: bleeding risk—NSAIDs, thrombolytics, salicylates, anticoagulants, platelet inhibitors

Drug/Lab Test

Increase: liver function tests, ammonia
Decrease: fibrinogen, protein C activity, protein S activity, anti-thrombin III

A

NURSING CONSIDERATIONS

Assess:

- **Anaphylaxis:** give only with resuscitation equipment and other agents necessary to treat anaphylaxis; if a serious hypersensitivity reaction occurs, discontinue product
- **Bleeding:** identify if NSAIDs, salicylates, thrombolytics, platelet inhibitors, anticoagulants have been used; grade 1 or 2 bleeding or coagulation abnormalities may occur (serious thrombosis, sagittal sinus thrombosis); discontinue for a thrombotic or hemorrhagic event until symptoms resolve; after resolution, treatment may be resumed; coagulation proteins were decreased in most patients after a 2-wk course: fibrinogen, protein C activity, protein S activity, antithrombin III

⚠ **Pancreatitis:** discontinue for severe or hemorrhagic pancreatitis (abdominal pain >72 hr and amylase elevation ≥2 × upper limit of normal); mild pancreatitis, hold until the signs, symptoms subside and amylase concentrations return to normal, then resume

- **Hyperglycemia:** glucose intolerance that is irreversible in some cases; monitor baseline and periodic glucose concentration; administer insulin therapy as needed

⚠ **Hyperammonemia/neurotoxicity:** seizures, headache; asparagine rescue may be used to treat acute CNS toxicity; doses of asparagine used were 1-2 mmol/kg/day continuous IV inf × 5 days Monitor liver function tests and for hyperbilirubinemia

Local injection site reaction: no more than 2 ml of the reconstituted product is to be administered at a single injection site

Evaluate:

- Decreased spread of malignancy

Teach patient/family:

⚠ To get immediate medical advice if they experience abdominal pain, nausea, vomiting, diarrhea; pancreatitis may occur

⚠ To get immediate medical advice if experiencing anaphylactic reactions

- To contact provider if patient experiences excessive thirst or any increase in volume or frequency of urination; hyperglycemia may occur

aspirin (OTC)

(as′pir-in)

APC-ASA Coated Aspirin ✤, A.S.A., Ascriptin Enteric, Aspergum, Aspirin ✤, Aspir-Low, Aspir-trin ✤, Bayer Aspirin, Bayer Children's Aspirin, Bufferin, Ecotrin, Equaline, Good Sense Aspirin, Halfprin, PMS-ASA ✤, St. Joseph Children's, St. Joseph's Adult, Walgreens Aspirin Adult

Func. class.: Nonopioid analgesic, nonsteroidal antiinflammatory, antipyretic, antiplatelet

Chem. class.: Salicylate

Do not confuse:

Ascendin/Afrin

ACTION: Blocks pain impulses in CNS, reduces inflammation by inhibition of prostaglandin synthesis; antipyretic action results from vasodilation of peripheral vessels; decreases platelet aggregation

USES: Mild to moderate pain or fever including RA, osteoarthritis, thromboembolic disorders; TIAs, rheumatic fever, post-MI, prophylaxis of MI, ischemic stroke, angina, acute MI

Unlabeled uses: Prevention of cataracts (long-term use), prevention of pregnancy loss in women with clotting disorders, bone pain, claudication, colorectal cancer prophylaxis, Kawasaki disease, PCI, preeclampsia/thrombosis prophylaxis, vernal keratoconjunctivitis, pericarditis

CONTRAINDICATIONS: Pregnancy (D) 3rd trimester, breastfeeding, children <12 yr, children with flulike symptoms, hypersensitivity to salicylates, tartrazine (FDC yellow dye #5), GI bleeding, bleeding disorders, vit K deficiency, peptic ulcer, acute bronchospasm, agranulocytosis, increased intracranial pressure, intracranial bleeding, nasal polyps, urticaria

Precautions: Abrupt discontinuation, acetaminophen/NSAIDs hypersensitivity, acid/base imbalance, alcoholism, ascites, asthma, bone marrow suppression in elderly patients, dehydration, G6PD deficiency, gout, heart failure, anemia, renal/hepatic disease, pre/postoperatively, gastritis, pregnancy (C) 1st trimester

DOSAGE AND ROUTES

Arthritis

- **Adult: PO** 3 g/day in divided doses q4-6hr
- **Child >25 kg (55 lb): PO/RECT** 90-130 mg/kg/day in divided doses

Pain/fever

- **Adult: PO/RECT** 325-650 mg q4hr prn, max 4 g/day
- **Child 2-11 yr: PO** 10-15 mg/kg/dose q4hr, max 4 g/day

Thromboembolic disorders

- **Adult: PO** 325-650 mg/day or bid

Transient ischemic attacks (risk)

- **Adult: PO** 50-325 mg/day (grade 1A)

Evolving MI with ST segment elevation (STEMI)

- **Adult: PO** 160-325 mg nonenteric, chewed and swallowed immediately, maintenance 75-162 mg daily

MI, stroke prophylaxis

- **Adult: PO** 50-325 mg/day

Prevention of recurrent MI

- **Adult: PO** 75-162 mg/day

CABG

- **Adult: PO** 75-325 mg/day starting 6 hr postprocedure, continue for 1 yr

PTCA

- **Adult: PO** 325 mg 2 hr before surgery

Thrombosis prophylaxis in ACS (unlabeled)

- **Adult: PO** 160-325 mg nonenteric, chewed and swallowed immediately

Idiopathic/viral pericarditis (unlabeled)

- **Adult: PO** 800 mg tid-qid × 7-10 days with gradual tapering to 800 mg/day q wk for an additional 2-3 wk

Colorectal cancer prophylaxis (unlabeled)

- **Adult: PO** 325 mg every other day

Kawasaki disease (unlabeled)

- **Child: PO** 80-100 mg/kg/day in 4 divided doses, maintenance 3-5 mg/kg/day

Available forms: Tabs 81, 325, 500, 650, 800 mg; chewable tabs 81 mg; supp 300, 600 mg; gum 227 mg; enteric-coated tabs 81, 325, 500, 975 mg; ext rel tabs 800 mg; del rel tabs 325, 500 mg; suppository 300, 600 mg

Administer:

PO route

- Do not break, crush, or chew enteric product
- Crushed or whole, chewable tablets may be chewed
- 1/2 hr before planned exercise
- With food or milk to decrease gastric symptoms; separate by 2 hr from enteric products
- With 8 oz of water; sit upright for 1/2 hr after dose to facilitate product passing into stomach

Rectal route

- Place suppository in refrigerator for at least 30 minutes before removing wrapper

SIDE EFFECTS

CNS: Stimulation, drowsiness, dizziness, confusion, seizures, headache, flushing, hallucinations, coma

CV: Rapid pulse, pulmonary edema

EENT: Tinnitus, hearing loss

ENDO: Hypoglycemia, hyponatremia, hypokalemia

GI: *Nausea, vomiting,* GI bleeding, diarrhea, heartburn, anorexia, hepatitis, GI ulcer

HEMA: Thrombocytopenia, agranulocytosis, leukopenia, neutropenia, hemolytic anemia, increased PT, aPTT, bleeding time

INTEG: *Rash*, urticaria, bruising
RESP: Wheezing, hyperpnea, **bronchospasm**
SYST: Reye's syndrome (children), anaphylaxis, laryngeal edema

PHARMACOKINETICS

Enteric metabolized by liver; inactive metabolites excreted by kidneys; crosses placenta; excreted in breast milk; half-life 15-20 min, up to 9 hr in large dose; rectal products may be erratic, protein binding 90%
PO: Onset 15-30 min, peak 1-2 hr, duration 4-6 hr, well absorbed
RECT: Onset slow, duration 4-6 hr

INTERACTIONS

Increase: gastric ulcer risk—corticosteroids, antiinflammatories, NSAIDs, alcohol
Increase: bleeding—alcohol, plicamycin, cefamandole, thrombolytics, ticlopidine, clopidogrel, tirofiban, eptifibatide, anticoagulants
Increase: effects of warfarin, insulin, methotrexate, thrombolytic agents, penicillins, phenytoin, valproic acid, oral hypoglycemics, sulfonamides
Increase: salicylate levels—urinary acidifiers, ammonium chloride, nizatidine
Increase: hypotension—nitroglycerin
Decrease: effects of aspirin—antacids (high doses), urinary alkalizers, corticosteroids
Decrease: antihypertensive effect—ACE inhibitors
Decrease: effects of probenecid, spironolactone, sulfinpyrazone, sulfonylamides, NSAIDs, β-blockers, loop diuretics

Drug/Herb
Increase: risk of bleeding—feverfew, garlic, ginger, ginkgo, ginseng *(Panax)*, horse chestnut

Drug/Food
Increase: risk of bleeding—fish oil (omega-3 fatty acids)
• Foods that acidify urine may increase aspirin level

Drug/Lab Test
Increase: coagulation studies, LFTs, serum uric acid, amylase, CO_2, urinary protein
Decrease: serum potassium, cholesterol
Interference: VMA, 5-HIAA, xylose tolerance test, TSH, pregnancy test

NURSING CONSIDERATIONS

Assess:
• **Pain:** character, location, intensity; ROM before and 1 hr after administration
• **Fever:** temp before and 1 hr after administration
• Hepatic studies: AST, ALT, bilirubin, creatinine if patient is receiving long-term therapy
• Renal studies: BUN, urine creatinine; I&O ratio; decreasing output may indicate renal failure (long-term therapy)
• Blood studies: CBC, Hct, Hgb, PT if patient is receiving long-term therapy
⚠ **Hepatotoxicity:** dark urine, clay-colored stools, yellowing of skin, sclera, itching, abdominal pain, fever, diarrhea if patient is receiving long-term therapy
• **Allergic reactions:** rash, urticaria; if these occur, product may have to be discontinued; patients with asthma, nasal polyps, allergies: severe allergic reaction may occur
• **Ototoxicity:** tinnitus, ringing, roaring in ears; audiometric testing needed before, after long-term therapy
• **Salicylate level:** therapeutic level 150-300 mcg/ml for chronic inflammation
• Edema in feet, ankles, legs
• Product history; many product interactions

Evaluate:
• Therapeutic response: decreased pain, inflammation, fever

Teach patient/family:
• To report any symptoms of hepatotoxicity, renal toxicity, visual changes, ototoxicity, allergic reactions, bleeding (long-term therapy)
• To avoid if allergic to tartrazine

- Not to exceed recommended dosage; acute poisoning may result
- To read labels on other OTC products because many contain aspirin, salicylates
- That the therapeutic response takes 2 wk (arthritis)
- To report tinnitus, confusion, diarrhea, sweating, hyperventilation
- To avoid alcohol ingestion; GI bleeding may occur
- That patients who have allergies, nasal polyps, asthma may develop allergic reactions
- To discard tabs if vinegar-like smell is detected
- That medication is not to be given to children or teens with flulike symptoms or chickenpox because Reye's syndrome may develop
- To take with a full glass of water
- Not to use during 3rd trimester of pregnancy (D)

TREATMENT OF OVERDOSE:

Lavage, activated charcoal, monitor electrolytes, VS

atazanavir (Rx)

(at-a-za-na'veer)

Reyataz

Func. class.: Antiretroviral

Chem. class.: Protease inhibitor

ACTION: Inhibits human immunodeficiency virus (HIV-1) protease, which prevents maturation of the infectious virus

USES: HIV-1 infection in combination with other antiretroviral agents

CONTRAINDICATIONS: Hypersensitivity

Precautions: Pregnancy (B), breastfeeding, children, geriatric patients, hepatic disease, alcoholism, drug resistance, AV block, diabetes, dialysis, elderly, females, hemophilia, hypercholesterolemia, immune reconstitution syndrome, lactic acidosis, pancreatitis, cholelithiasis, serious rash

DOSAGE AND ROUTES

Antiretroviral-naive patients

- **Adult: PO** 400 mg/day (unable to take ritonavir); 300 mg with ritonavir 100 mg/day
- **Child ≥6 yr/adolescent ≥40 kg: PO** 300 mg with ritonavir 100 mg daily
- **Child ≥6 yr/adolescent 20 to <40 kg: PO** 200 mg with ritonavir 100 mg daily
- **Child ≥6 yr/adolescent 15 to <20 kg: PO** 150 mg with ritonavir 80 mg daily

Antiretroviral-experienced patients

- **Adult: PO** 300 mg with ritonavir 100 mg daily
- **Pregnant adults/adolescents (2nd/3rd trimester) with H2 blocker or tenofovir: PO** 400 mg with ritonavir 100 mg daily
- **Child ≥6 yr/adolescent ≥40 kg: PO** 300 mg with ritonavir 100 mg daily
- **Child ≥6 yr/adolescent 20 to <40 kg: PO** 200 mg with ritonavir 100 mg daily

Hepatic dose

- **Adult: PO** Child-Pugh B: 300 mg/day; Child-Pugh C: do not use

Available forms: Caps 100, 150, 200, 300 mg

Administer:

- With food; 2 hr before or 1 hr after antacid or didanosine; swallow cap whole

SIDE EFFECTS

CNS: Headache, depression, dizziness, insomnia, peripheral neurologic symptoms

GI: Vomiting, *diarrhea, abdominal pain, nausea,* hepatotoxicity, cholelithiasis

INTEG: *Rash,* Stevens-Johnson syndrome, *photosensitivity,* DRESS

MISC: Fatigue, fever, arthralgia, back pain, cough, lipodystrophy, pain, gynecomastia, nephrolithiasis; lactic acidosis, hyperbilirubinemia (pregnancy, females, obesity)

A

PHARMACOKINETICS

Rapidly absorbed, absorption increased with food, peak 2½ hr, 86% protein bound, extensively metabolized in liver by CYP3A4, 27% excreted unchanged in urine/feces (minimal), half-life 7 hr

INTERACTIONS

⚠ **Increase:** levels, toxicity of immunosuppressants (cycloSPORINE, sirolimus, tacrolimus, sildenafil), tricylic antidepressants, warfarin, calcium channel blockers, clarithromycin, chlorazepate, diazepam, irinotecan, HMG-CoA reductase inhibitors, antidysrhythmics, midazolam, triazolam, ergots, pimozide, other protease inhibitors

Increase: effects of estrogens, oral contraceptives

Increase: atazanavir levels—CYP3A4 substrates, CYP3A4 inhibitors

Increase: hyperbilirubinemia—indinavir

Decrease: teleprevir level when used with atazanavir and ritonavir

Decrease: atazanavir levels—CYP3A4 inducers, rifampin, antacids, didanosine, efavirenz, proton-pump inhibitors, H_2-receptor antagonists

Drug/Herb

Decrease: atazanavir levels—St. John's wort, avoid concurrent use

Increase: myopathy, rhabdomyolysis—red yeast rice

Drug/Lab Test

Increase: AST, ALT, total bilirubin, amylase, lipase, CK

Decrease: Hgb, neurophils, platelets

NURSING CONSIDERATIONS

Assess:

⚠ For hepatic failure; hepatic studies: ALT, AST, bilirubin

• Immune reconstitution syndrome: when given with combination antiretroviral therapy

• For **lactic acidosis, hyperbilirubinemia** (females, pregnancy, obesity), if pregnant call Antiretroviral Pregnancy Registry 800-258-4263

• For signs of infection, anemia, nephrolithiasis

• Bowel pattern before, during treatment; if severe abdominal pain with bleeding occurs, product should be discontinued; monitor hydration

• Viral load, CD4 count throughout treatment

• **Serious rash (Stevens-Johnson syndrome, DRESS): most rashes last 1-4 wk; if serious, discontinue product**

• **Immune reconstitution syndrome: time of onset is variable**

Evaluate:

• Therapeutic response: increasing CD4 counts; decreased viral load, resolution of symptoms of HIV-1 infection

Teach patient/family:

• To take as prescribed with other antiretrovirals as prescribed; if dose is missed, to take as soon as remembered up to 1 hr before next dose; not to double dose, share with others

• That product must be taken daily to maintain blood levels for duration of therapy

• That product may cause photosensitivity; to use protective clothing, stay out of the sun

• To notify prescriber if diarrhea, nausea, vomiting, rash occurs; dizziness, lightheadedness may occur; ECG may be altered

• That product interacts with many products; St. John's wort; to advise prescriber of all products, herbal products used

• That redistribution of body fat may occur, the effect is not known

• That product does not cure HIV-1 infection, prevent transmission to others; only controls symptoms

• That, if taking phosphodiesterase type 5 inhibitor with atazanavir, there may be increased risk of phosphodiesterase type 5 inhibitor–associated adverse events (hypotension, prolonged penile erection); to notify physician promptly of these symptoms

atenolol (Rx)

(a-ten′oh-lole)

Tenormin

Func. class.: Antihypertensive, antianginal

Chem. class.: β-Blocker, β_1-, β_2-blocker (high doses)

Do not confuse:
atenolol/albuterol
Tenormin/thiamine/Imuran

ACTION:
Competitively blocks stimulation of β-adrenergic receptor within vascular smooth muscle; produces negative chronotropic activity (decreases rate of SA node discharge, increases recovery time), slows conduction of AV node, decreases heart rate, negative inotropic activity decreases O_2 consumption in myocardium; decreases action of renin-aldosterone-angiotensin system at high doses, inhibits β_2 receptors in bronchial system at higher doses

USES:
Mild to moderate hypertension, prophylaxis of angina pectoris; suspected or known MI (IV use); MI prophylaxis

Unlabeled uses: Migraine prophylaxis, supraventricular tachycardia prophylaxis (PSVT), unstable angina, alcohol withdrawal, lithium-induced tremor

CONTRAINDICATIONS:
Pregnancy (D), hypersensitivity to β-blockers, cardiogenic shock, 2nd- or 3rd-degree heart block, sinus bradycardia, cardiac failure

Precautions: Breastfeeding, major surgery, diabetes mellitus, thyroid/renal disease, CHF, COPD, asthma, well-compensated heart failure, dialysis, myasthenia gravis, Raynaud's disease, pulmonary edema

Black Box Warning: Abrupt discontinuation

DOSAGE AND ROUTES

- **Adult: PO** 25-50 mg/day, increasing q1-2wk to 100 mg/day; may increase to 200 mg/day for angina, up to 100 mg/day for hypertension
- **Child: PO** 0.8-1 mg/kg/dose initially; range, 0.8-1.5 mg/kg/day; max 2 mg/kg/day
- **Geriatric: PO** 25 mg/day initially

Chronic stable angina

- **Adult: PO** 50 mg/day, then 100 mg/day as needed after 7 days, max 200 mg/day

Post MI, MI prophylaxis

- **Adult: PO** 100 mg/day in 1-2 divided doses; may need for 1-3 yr after MI

Renal disease

- **Adult: PO** CCr 15-35 ml/min, max 50 mg/day; CCr <15 ml/min, max 25 mg/day; hemodialysis 25-50 mg after dialysis

PSVT prophylaxis (unlabeled)

- **Child: PO** 0.3-1.3 mg/kg/day

Ethanol withdrawal prevention (unlabeled)

- **Adult: PO** 50-100 mg/day

Migraine prophylaxis (unlabeled)

- **Adult: PO** 50-150 mg/day, titrate to response

Lithium-induced tremor (unlabeled)

- **Adult: PO** 50 mg/day

Available forms: Tabs 25, 50, 100 mg

Administer:

PO route

- Before meals, at bedtime; tab may be crushed, swallowed whole
- Reduced dosage with renal dysfunction
- Store protected from light, moisture; place in cool environment

SIDE EFFECTS

CNS: *Insomnia, fatigue, dizziness, mental changes,* memory loss, hallucinations, depression, lethargy, drowsiness, strange dreams, catatonia

CV: Profound hypotension, bradycardia, CHF, *cold extremities, postural hypotension, 2nd- or 3rd-degree heart block*

EENT: Sore throat; dry, burning eyes; blurred vision; stuffy nose

ENDO: Increased hypoglycemic response to insulin
GI: *Nausea, diarrhea,* vomiting, **mesenteric arterial thrombosis, ischemic colitis**
GU: Impotence, decreased libido
HEMA: **Agranulocytosis, thrombocytopenia purpura**
INTEG: Rash, fever, alopecia
RESP: **Bronchospasm,** dyspnea, wheezing, pulmonary edema

PHARMACOKINETICS

PO: Peak 2-4 hr; onset 1 hr; duration 24 hr; half-life 6-7 hr; excreted unchanged in urine, feces (50%); protein binding 5%-15%

INTERACTIONS

- Mutual inhibition: sympathomimetics (cough, cold preparations)

Increase: hypotension, bradycardia—reserpine, hydrALAZINE, methyldopa, prazosin, anticholinergics, digoxin, diltiazem, verapamil, cardiac glycosides, antihypertensives
Increase: hypoglycemia—insulins, oral antidiabetics
Increase: hypertension—amphetamines, ePHEDrine, pseudoephedrine
Decrease: effect—insulin, oral antidiabetic agents, theophylline, DOPamine, MAOIs

Drug/Herb
Increase: atenolol effect—hawthorn
Decrease: atenolol effect—ephedra (ma huang)

Drug/Lab Test
Increase: blood glucose, BUN, potassium, triglycerides, uric acid, ANA titer

NURSING CONSIDERATIONS

Assess:

- I&O, weight daily; watch for CHF (rales/crackles, jugular vein distention, weight gain, edema)
- **Hypertension:** B/P, pulse q4hr; note rate, rhythm, quality; apical/radial pulse before administration; notify prescriber of any significant changes (<50 bpm); ECG
- **Hypotension:** may be caused in hemodialysis
- **Hypoglycemia:** may be masked in diabetes mellitus
- Baselines in renal/hepatic studies before therapy begins

Black Box Warning: Taper gradually do not discontinue abruptly

Evaluate:

- Therapeutic response: decreased B/P after 1-2 wk, increased activity tolerance, decreased anginal pain

Teach patient/family:

Black Box Warning: Not to discontinue product abruptly, taper over 2 wk (angina); to take at same time each day as directed

- Not to use OTC products unless directed by prescriber
- To report bradycardia, dizziness, confusion, depression, fever
- To take pulse at home; advise when to notify prescriber
- To limit alcohol, smoking, sodium intake
- To comply with weight control, dietary adjustments, modified exercise program
- To carry emergency ID to identify product, allergies, conditions being treated
- To avoid hazardous activities if dizziness is present
- To change position slowly
- That product may mask symptoms of hypoglycemia in diabetic patients
- To use contraception while taking this product, pregnancy (D); to avoid breastfeeding

TREATMENT OF OVERDOSE:

Lavage, IV atropine for bradycardia, IV theophylline for bronchospasm, dextrose for hypoglycemia, digoxin, O_2, diuretic for cardiac failure, hemodialysis

atomoxetine (Rx)

(at-o-mox′eh-teen)

Strattera

Func. class.: Psychotherapeutic—miscellaneous

Chem. class.: Selective norepinephrine reuptake inhibitor

ACTION: Selective norepinephrine reuptake inhibitor; may inhibit the presynaptic norepinephrine transporter

USES: Attention deficit hyperactivity disorder

CONTRAINDICATIONS: Hypersensitivity, angle-closure glaucoma, arteriosclerosis, cardiac disease, cardiomyopathy, heart failure, jaundice, MAOI therapy, history of pleochromocytoma

Precautions: Pregnancy (C), breastfeeding, hepatic disease, angioedema, bipolar disorder, dysrhythmias, CAD, hypo/hypertension

Black Box Warning: Children <6 yr, suicidal ideation

DOSAGE AND ROUTES

- **Child ≤70 kg >6 yrs: PO** 0.5 mg/kg/day, increase after 3 days to target daily dose of 1.2 mg/kg in AM or evenly divided doses AM, late afternoon; max 1.4 mg/kg/day or 100 mg/day, whichever is less
- **Adult and child >70 kg: PO** 40 mg/day, increase after 3 days to target daily dose of 80 mg in AM or evenly divided doses AM, late afternoon; max 100 mg/day

Maintenance

- **Adolescent ≤15 yr and child ≥6 yr: PO** 1.2-1.8 mg/kg/day

Initial dose titration with strong CYP2D6 inhibitors

- **Adult and child >6 yr weighing >70 kg: PO** 40 mg/day each AM or 2 evenly divided doses, titrate to target of 80 mg/day if symptoms do not improve after 4 wk and dose is well tolerated

Hepatic dose

- Child-Pugh B: reduce dose by 50%; Child-Pugh C: reduce dose by 75%

Available forms: Caps 10, 18, 25, 40, 60, 80, 100 mg

Administer:

- Whole; do not break, crush, chew
- Gum, hard candy, frequent sips of water for dry mouth
- Without regard to food

SIDE EFFECTS

CNS: *Insomnia,* dizziness, headache, irritability, crying, mood swings, fatigue, hypoesthesia, lethargy, paresthesia

CV: *Palpitations,* hot flushes, tachycardia, increased B/P

ENDO: Growth retardation

GI: Dyspepsia, nausea, anorexia, dry mouth, weight loss, vomiting, diarrhea, constipation, hepatic injury

GU: Urinary hesitancy, retention, dysmenorrhea, erectile disturbance, ejaculation failure, impotence, prostatitis, abnormal orgasm, male pelvic pain

INTEG: Exfoliative dermatitis, sweating, rash

MISC: Cough, rhinorrhea, dermatitis, ear infection

PHARMACOKINETICS

Peak 1-2 hr, metabolized by liver, excreted by kidneys, 98% protein binding

INTERACTIONS

Increase: hypertensive crisis—MAOIs or within 14 days of MAOIs, vasopressors

Increase: cardiovascular effects of albuterol, pressor agents

Increase: effects of atomoxetine—CYP2D6 inhibitors (amiodarone, cimetidine [weak], clomipramine, delavirdine, gefitinib, imatinib, propafenone, quiNIDine [potent], ritonavir, citalopram, escitalopram, FLUoxetine, sertraline, PARoxetine, thioridazine, venlafaxine)

A

NURSING CONSIDERATIONS

Assess:

- VS, B/P; check patients with cardiac disease more often for increased B/P
- **Hepatic injury:** may cause liver failure: monitor LFT; assess for jaundice, pruritus, flulike symptoms, upper right quadrant pain

⚠ Mental status: mood, sensorium, affect, stimulation, insomnia, aggressiveness, suicidal ideation in children/young adults

- Appetite, sleep, speech patterns
- For increased attention span, decreased hyperactivity with ADHD, growth rate weight, therapy may need to be discontinued

Evaluate:

- Therapeutic response: decreased hyperactivity (ADHD)

Teach patient/family:

- To avoid OTC preparations unless approved by prescriber, no tapering needed when discontinuing product
- To avoid alcohol ingestion
- To avoid hazardous activities until stabilized on medication
- To get needed rest; patients will feel more tired at end of day; not to take dose late in day, insomnia may occur

Black Box Warning: To report suicidal ideation

- To notify prescriber immediately if erection >4 hr

atorvastatin (Rx)

(a-tore′va-stat-in)

Lipitor

Func. class.: Antilipidemic

Chem. class.: HMG-CoA reductase inhibitor (statin)

Do not confuse:

Lipitor/Levatol

ACTION: Inhibits HMG-CoA reductase enzyme, which reduces cholesterol synthesis; high doses lead to plaque regression

USES: As adjunct for primary hypercholesterolemia (types Ia, Ib), dysbetalipoproteinemia, elevated triglyceride levels, prevention of CV disease by reduction of heart risk in those with mildly elevated cholesterol

Unlabeled uses: Atherosclerosis

CONTRAINDICATIONS: Pregnancy (X), breastfeeding, hypersensitivity, active hepatic disease

Precautions: Previous hepatic disease, alcoholism, severe acute infections, trauma, severe metabolic disorders, electrolyte imbalance

DOSAGE AND ROUTES

- **Adult: PO** 10-20 mg/day, usual range 10-80 mg/day, dosage adjustments may be made in 2-4 wk intervals, max 80 mg/day; patients who require >45% reduction in LDL may be started at 40 mg/day

Atherosclerosis (unlabeled)

- **Adult: PO** 80 mg/day

Available forms: Tabs 10, 20, 40, 80 mg

Administer:

- Total daily dose at any time of day without regard to meals
- Store in cool environment in tight container protected from light

SIDE EFFECTS

CNS: Headache, asthenia

EENT: Lens opacities

GI: *Abdominal cramps, constipation, diarrhea, flatus, heartburn,* dyspepsia, **liver dysfunction, pancreatitis,** nausea, increased serum transaminase

GU: Impotence

INTEG: Rash, pruritus, alopecia; photosensitivity (rare)

MISC: Hypersensitivity

MS: Arthralgia, myalgia, **rhabdomyolysis**

RESP: Pharyngitis, sinusitis

PHARMACOKINETICS

Peak 1-2 hr, metabolized in liver, highly protein-bound, excreted primarily in urine, half-life 14 hr, protein binding 98%

INTERACTIONS

Increase: rhabdomyolysis—azole antifungals, cycloSPORINE, erythromycin, niacin, gemfibrozil, clofibrate
Increase: serum level of digoxin
Increase: levels of oral contraceptives
Increase: levels of atorvastatin—erythromycin
Increase: effects of warfarin
Decrease: atorvastatin levels—colestipol

Drug/Herb
Decrease: effect—St. John's wort

Drug/Food
• Possible toxicity when used with grapefruit juice; oat bran may reduce effectiveness

Drug/Lab Test
Increase: bilirubin, alk phos, ALT, AST, CK
Interference: thyroid function tests

NURSING CONSIDERATIONS

Assess:
• **Hypercholesterolemia:** diet, obtain diet history including fat, cholesterol in diet; cholesterol triglyceride levels periodically during treatment; check lipid panel 6-12 wk after changing dose
• Hepatic studies q1-2mo, at initiation, 6, 12 wk after initiation or change in dose, periodically thereafter; AST, ALT, LFTs may be increased
• Renal studies in patients with compromised renal system: BUN, I&O ratio, creatinine
• Bowel status: constipation, stool softeners may be needed; if severe, add fiber, water to diet
⚠ **Rhabdomyolysis:** for muscle pain, tenderness, obtain CPK baseline; if markedly increased, product may need to be discontinued

Evaluate:
• Therapeutic response: decrease in LDL, total cholesterol, triglycerides, CAD; increase in HDL

Teach patient/family:
• That blood work and eye exam will be necessary during treatment
• To report blurred vision, severe GI symptoms, headache, muscle pain, weakness
• That previously prescribed regimen will continue: low-cholesterol diet, exercise program, smoking cessation
• Not to take product if pregnant (X), breastfeeding; to avoid alcohol
• To stay out of the sun; to use sunscreen, protective clothing to prevent photosensitivity (rare)

atovaquone (Rx)

(a-toe′va-kwon)

Mepron

Func. class.: Antiprotozoal
Chem. class.: Aromatic diamide derivative, analog of ubiquinone

ACTION: Interferes with DNA/RNA synthesis in protozoa

USES: *Pneumocystis jiroveci* infections in patients intolerant of trimethoprim-sulfamethoxazole, prophylaxis, *Toxoplasma gondii,* toxoplasmosis
Unlabeled uses: Babesiosis, malaria treatment/prophylaxis, toxoplasmosis prophylaxis, *Plasmodium* sp.

CONTRAINDICATIONS: Hypersensitivity or history of developing life-threatening allergic reactions to any component of the formulation, benzyl alcohol sensitivity
Precautions: Pregnancy (C), breastfeeding, neonates, hepatic disease, GI disease, respiratory insufficiency

DOSAGE AND ROUTES

Acute, mild, moderate *Pneumocystis jiroveci* pneumonia
• **Adult and adolescent 13-16 yr: PO** 750 mg with food bid for 21 days

***Pneumocystis jiroveci* pneumonia, prophylaxis**

- **Adult and adolescent: PO** 1500 mg/day with meal

Babesiosis (unlabeled)

- **Adult: PO** 750 mg q12hr with azithromycin (500-1000 mg on day 1 then 250 mg/day × 7-14 days)

Toxoplasmosis prophylaxis in AIDS (unlabeled)

- **Adult: PO** 1500 mg alone or in combination

***Plasmodium falciparum* (unlabeled)**

- **Adult: PO** 250 mg with proguanil daily
- **Child: PO** 17 mg/kg with proguanil daily

Available forms: Susp 750 mg/5 ml

Administer:

- With high-fat food to increase absorption of product and higher plasma concentrations
- Oral susp; shake before using
- All contents of foil pouch

SIDE EFFECTS

CNS: *Dizziness, headache, anxiety, insomnia,* asthenia, fever

CV: Hypotension

GI: *Nausea, vomiting, diarrhea,* anorexia, increased AST/ALT, acute pancreatitis, constipation, abdominal pain

HEMA: Anemia, neutropenia

INTEG: Pruritus, urticaria, *rash*

META: Hyperkalemia, hypoglycemia, hyponatremia

OTHER: Cough, dyspnea

PHARMACOKINETICS

Excreted unchanged in feces (94%), highly protein bound (99%), half-life 2-3 days

INTERACTIONS

Increase: level of—zidovudine, monitor for toxicity

Decrease: effect of atovaquone—rifampin, rifabutin, tetracycline, avoid concurrent use

Drug/Lab Test

Increase: AST, ALT, alk phos

Decrease: glucose, neutrophils, Hgb, sodium

NURSING CONSIDERATIONS

Assess:

Infection: WBC, vital signs; sputum baseline, periodically; obtain specimens needed before giving 1st dose

- Bowel pattern before, during treatment
- Respiratory status: rate, character, wheezing, dyspnea; risk for respiratory infection
- Allergies before treatment, reaction to each medication

Evaluate:

- Therapeutic response: decreased temp, ability to breathe

Teach patient/family:

- To take with food to increase plasma concentrations

atovaquone/proguanil

(a-toe′va-kwon)

Malarone

Func. class.: Antiprotozoal

Chem. class.: Aromatic diamide derivative

ACTION:

The constituents of Malarone, atovaquone, and proguanil hydrochloride interfere with 2 different pathways involved in DNA/RNA synthesis in protozoa

USES:

Malaria, malaria prophylaxis

CONTRAINDICATIONS:

Hypersensitivity to this product, malaria prophylaxis in patients with severe renal impairment

Precautions: Pregnancy (C), breastfeeding, children, hepatic/GI/renal disease

DOSAGE AND ROUTES

Treatment of acute, uncomplicated *P. falciparum* malaria

Malarone adult strength tabs

- **Adult/adolescent/child >40 kg: PO** 4 adult strength tabs every day as a single dose × 3 consecutive days
- **Child 31-40 kg: PO** 3 adult strength tabs every day as a single dose × 3 consecutive days
- **Child 21-30 kg: PO** 2 adult strength tabs every day as a single dose × 3 consecutive days
- **Infant/child 11-20 kg: PO** 1 adult strength tab every day × 3 consecutive days

Malarone Pediatric Tabs

- **Infant/child 11-20 kg: PO** 4 pediatric tabs every day × 3 consecutive days
- **Infant/child 9-10 kg: PO** 3 pediatric tabs every day × 3 consecutive days
- **Infant/child 5-8 kg: PO** 2 pediatric tabs every day every 3 consecutive days

***P. falciparum* malaria prophylaxis, including chloroquine resistance areas**

- Each Malarone adult strength tab contains atovaquone 250 mg/proguanil 100 mg; Malarone Pediatric tablet contains atovaquone 62.5 mg/proguanil 25 mg

Malarone adult strength tabs

- **Adult/adolescent/child >40 kg: PO** 1 adult strength tab every day; begin prophylaxis 1-2 days before entering the endemic area; continue daily during the stay and for 7 days after leaving the area

Malarone Pediatric Tab

- **Child 31-40 kg: PO** 3 pediatric tabs every day; begin prophylaxis 1-2 days before entering the endemic area; continue daily during the stay and for 7 days after leaving the area
- **Child 21-30 kg: PO** 2 pediatric tabs every day; begin prophylaxis 1-2 days before entering the endemic area; continue daily during the stay and for 7 days after leaving the area
- **Infant/child 11-20 kg: PO** 1 pediatric tab every day; begin prophylaxis 1-2 days before entering the endemic area; continue daily during the stay and for 7 days after leaving the area

Available forms: Tabs (adult) 250 mg atovaquone/proguanil 100 mg; tabs (pediatric) 62.5 atovaquone/proguanil 25 mg

Administer

- Give with food or with milk or milk-based drink (nutritional supplement shake) to enhance oral absorption of atovaquone; food with high fat content is desired
- Give dose at the same time each day; administer a repeat dose if vomiting occurs within 1 hr after dosing
- Tabs may be crushed and mixed with condensed milk for children unable to swallow whole tablets

SIDE EFFECTS

CNS: *Dizziness, headache, anxiety, insomnia,* asthenia, fever

CV: Hypotension

GI: *Nausea, vomiting, diarrhea,* anorexia, increased AST/ALT, acute pancreatitis, constipation, abdominal pain

HEMA: Anemia, neutropenia

INTEG: Pruritus, urticaria, *rash*

META: Hyperkalemia, hypoglycemia, hyponatremia

OTHER: Cough, dyspnea

PHARMACOKINETICS

Atovaquone excreted unchanged in feces (94%), highly protein-bound (99%), Protaguanil 75% protein-bound, 40%-60% excreted in urine, hepatic metabolism; half-life 2-3 days

INTERACTIONS

Increase: level of indinavir

Decrease: effect of atovaquone-rifampin, rifabutin, tetracycline, metoclopramide

Increase: AST, ALT, alk phos

Decrease: glucose, neutrophils, Hgb, sodium

NURSING CONSIDERATIONS

Assess:

- **Malaria:** identify when the patient will be entering an area with malaria
- Bowel pattern before, during treatment
- Respiratory status: rate, character, wheezing, dyspnea; risk for respiratory infection
- Allergies before treatment, reaction to each medication
- CBC, LFTs, serum amylase, creatinine/BUN, sodium; increases in LFTs can persist for 4 wk after discontinuation of treatment

Evaluate:

Therapeutic response:

- Resolution/prevention of malaria

Teach patient/family:

- To take with food to increase plasma concentrations

RARELY USED

atracurium (Rx)

(a-tra-kyoor′ee-um)

Func. class.: Neuromuscular blocker (nondepolarizing)

USES: Facilitation of endotracheal intubation, skeletal muscle relaxation during mechanical ventilation, surgery, or general anesthesia

CONTRAINDICATIONS: Hypersensitivity

Black Box Warning: Respiratory insufficiency

DOSAGE AND ROUTES

- **Adult and child >2 yr: IV BOL** 0.4-0.5 mg/kg then 0.08-0.1 mg/kg 20-45 min after 1st dose if needed for prolonged procedures; give smaller doses with halothane
- **Child 1 mo-2 yr: IV BOL** 0.3-0.4 mg/kg

⚠ HIGH ALERT

atropine (Rx)

(a′troe-peen)

Atreza, AtroPen, Sal-Tropine

Func. class.: Antidysrhythmic, anticholinergic parasympatholytic, antimuscarinic

Chem. class.: Belladonna alkaloid

Do not confuse:

atropine/Akarpine

ACTION: Blocks acetylcholine at parasympathetic neuroeffector sites; increases cardiac output, heart rate by blocking vagal stimulation in heart; dries secretions by blocking vagus

USES: Bradycardia <40-50 bpm, bradydysrhythmia, reversal of anticholinesterase agents, insecticide poisoning, blocking cardiac vagal reflexes, decreasing secretions before surgery, antispasmodic with GU, biliary surgery, bronchodilator, AV heart block

Unlabeled uses: Cardiac arrest, CPR, diarrhea, pulseless electrical activity, ventricular asystole, asthma

CONTRAINDICATIONS: Hypersensitivity to belladonna alkaloids, closed-angle glaucoma, GI obstructions, myasthenia gravis, thyrotoxicosis, ulcerative colitis, prostatic hypertrophy, tachycardia/tachydysrhythmias, asthma, acute hemorrhage, severe hepatic disease, myocardial ischemia, paralytic ileus

Precautions: Pregnancy (C), breastfeeding, children <6 yr, geriatric patients, renal disease, CHF, hyperthyroidism, COPD, hypertension, intraabdominal infection, Down syndrome, spastic paralysis, gastric ulcer

DOSAGE AND ROUTES

Bradycardia/bradydysrhythmia

- **Adult: IV BOL** 0.5-1 mg given q3-5min, max 3 mg

• **Child: IV BOL** 0.01 mg/kg up to 0.4 mg or 0.3 mg/m^2; may repeat q4-6hr; min dose 0.1 mg to avoid paradoxical reaction, max single dose 0.5 mg

Organophosphate poisoning

• **Adult and child: IM/IV** 1-2 mg hourly until muscarinic symptoms disappear; may need 6 mg every hr

• **Adult and child >90 lb, usually >10 yr: AtroPen** 2 mg

• **Child 40-90 lb, usually 4-10 yr: AtroPen** 1 mg

• **Child 15-40 lb: AtroPen** 0.5 mg

• **Infant <15 lb: IM/IV** 0.05 mg/kg q5-20min as needed

Presurgery

• **Adult and child >20 kg: SUBCUT/IM/IV** 0.4-0.6 mg 30-60 min before anesthesia

• **Child <20 kg: IM/SUBCUT** 0.01 mg/kg up to 0.4 mg ½-1 hr preop, max 0.6 mg/dose

Available forms: Inj 0.05, 0.1, 0.4, 0.5, 0.8, 1 mg/ml; tabs 0.4 mg; AtroPen 0.5, 1, 2 mg inj prefilled autoinjectors

Administer:

PO route

• Increased bulk, water in diet if constipation occurs

• Without regard to meals

IM route

• Atropine flush may occur in children and is not harmful

AtroPen

• Use no more than 3 AtroPen inj unless under the supervision of trained medical provider

• Use as soon as symptoms appear (tearing, wheezing, muscle fasciculations, excessive oral secretions)

IV route

• Undiluted or diluted with 10 ml sterile water; give at 0.6 mg/min through Y-tube or 3-way stopcock; do not add to IV sol; may cause paradoxical bradycardia for 2 min

Y-site compatibilities: Amrinone, etomidate, famotidine, heparin, hydrocortisone, meropenem, nafcillin, potassium chloride, sufentanil, vit B/C

SIDE EFFECTS

CNS: Headache, dizziness, involuntary movement, confusion, psychosis, anxiety, coma, flushing, drowsiness, insomnia, weakness; delirium (geriatric patients)

CV: Hypo/hypertension, paradoxical bradycardia, angina, PVCs, tachycardia, ectopic ventricular beats, bradycardia

EENT: Blurred vision, photophobia, glaucoma, eye pain, pupil dilation, nasal congestion

GI: Dry mouth, nausea, vomiting, abdominal pain, anorexia, constipation, paralytic ileus, abdominal distention, altered taste

GU: Retention, hesitancy, impotence, dysuria

INTEG: Rash, urticaria, contact dermatitis, dry skin, flushing

MISC: Suppression of lactation, decreased sweating, anaphylaxis

PHARMACOKINETICS

Half-life 2-3 hr, terminal 12.5 hr, excreted by kidneys unchanged (70%-90% in 24 hr), metabolized in liver, 40%-50% crosses placenta, excreted in breast milk

PO: Onset ½-2 hr, peak ½-1 hr, duration 4-6 hr, well absorbed

IM/SUBCUT: Onset 15-50 min, peak 30 min, duration 4-6 hr, well absorbed

IV: Peak 2-4 min, duration 4-6 hr

INTERACTIONS

Increase: mucosal lesions—potassium chloride tab

Increase: anticholinergic effects—tricyclics, amantadine, antiparkinson agents

Decrease: absorption—ketoconazole, levodopa

Decrease: effect of atropine—antacids

NURSING CONSIDERATIONS

Assess:

• I&O ratio; check for urinary retention, daily output

• **ECG** for ectopic ventricular beats, PVC, tachycardia in cardiac patients

• For bowel sounds, constipation

- Respiratory status: rate, rhythm, cyanosis, wheezing, dyspnea, engorged neck veins
- **Increased intraocular pressure:** eye pain, nausea, vomiting, blurred vision, increased tearing
- Cardiac rate: rhythm, character, B/P continuously
- Allergic reaction: rash, urticaria

Evaluate:
- Therapeutic response: decreased dysrhythmias, increased heart rate, secretions; GI, GU spasms; bronchodilation

Teach patient/family:
- To report blurred vision, chest pain, allergic reactions, constipation, urinary retention, to use sunglasses to protect the eyes
- Not to perform strenuous activity in high temperatures; heat stroke may result
- To take as prescribed; not to skip or double doses
- Not to operate machinery if drowsiness occurs
- Not to take OTC products without approval of prescriber
- Not to freeze or expose to light (AtroPen)

TREATMENT OF OVERDOSE:
O_2, artificial ventilation, ECG; administer DOPamine for circulatory depression; administer diazepam or thiopental for seizures; assess need for antidysrhythmics

atropine ophthalmic
See Appendix B

RARELY USED

auranofin (Rx)
(au-rane′oh-fin)

Ridaura

Func. class.: Antiinflammatory, gold compound

Do not confuse:
Ridaura/Cardura

USES: RA; not for 1st-line therapy
Unlabeled uses: SLE, psoriatic arthritis, pemphigus

CONTRAINDICATIONS: Breastfeeding, children <6 yr, hypersensitivity to gold, necrotizing enterocolitis, pulmonary fibrosis, exfoliative dermatitis, recent radiation therapy, renal/hepatic disease, marked hypertension, uncontrolled CHF

Black Box Warning: Bone marrow suppression, blood dyscrasias, hematuria, anemia, diarrhea

DOSAGE AND ROUTES
- **Adult: PO** 6 mg/day or 3 mg bid; may increase to 9 mg/day after 3 mo

avanafil
(a-van′a-fil)

Stendra

Func. class.: Impotence agent
Chem. class.: Phosphodiesterase type 5 inhibitor

ACTION: Inhibits phosphodiesterase type 5 (PDE5); enhances erectile function by increasing the amount of cGMP causing smooth muscle relaxation and increasing blood flow to the corpus cavernosum

USES: Treatment of erectile dysfunction

CONTRAINDICATIONS:

Hypersensitivity, severe renal/hepatic disease, current nitrates/nitrites, patients <18 yr

Precautions: Pregnancy (C) although not indicated for women, anatomic penile deformities, sickle cell anemia, leukemia, multiple myeloma, renal/hepatic/CV disease, bleeding disorders, active peptic ulcer, prolonged erection, aortic stenosis, HIV, stroke, geriatric patients, tinnitus, MI, visual disturbances, retinitis pigmentosa

DOSAGE AND ROUTES

Erectile dysfunction

- **Adult: PO** 100 mg 30 min before sexual activity, dose may be reduced to 50 mg or increased to 200 mg; usual max dose frequency is 1 time/day

Potent CYP3A4 inhibitors/nitrates

- Do not use

Moderate CYP3A4 inhibitors/alpha-blockers

- **Adult: PO** Max 50 mg/day

Hepatic dosage/severe renal disease

- **Adult: Child–Pugh C: PO** not recommended

Available forms: Tabs 50, 100, 200 mg

Administer

PO route

- May be taken 30 min before sexual activity on an as-needed basis, but no more than once per day
- May be used without regard to meals
- Products should not be used with nitrates/nitrates or strong CYP3A4 inhibitors

SIDE EFFECTS

CNS: *Headache, flushing*

EENT: Nasal congestion, nasopharyngitis

MISC: Back pain

PHARMACOKINETICS

99% protein binding, metabolized by CYP3A4, excreted as metabolites; urine 62%; 21% feces, half-life 5 hr, peak 45 min

INTERACTIONS

Do not use with nitrates/nitrites because of unsafe drop in B/P, which could result in MI, stroke

Do not use with strong CYP3A4 inhibitors (ketoconazole, ritonavir, atazanavir, clarithromycin, indivinavir, itraconazole, nefazodone, nelfinavir, saquinavir, telithromycin, isoniazid, boceprevir, delavirdine, telaprevir, tipranavir)

Avoid use with other phosplie sterage inhibitors (vardenafil, sildenafil, tadalafil)

Increase: Avanafil level—moderate CYP3A4 inhibitors (erythromycin, amprenavir, aprepitant, diltiazem, fluconazole, fosamprenavir, verapamil, amiodarone, crizotinib, darunavir, dasantinib, dronedarone, imatinib, lapatinib, ticagrelor, voriconazole)

Decrease: B/P—alcohol, alpha-blockers, amLODIPine

Increase: Avanafil effect—grapefruit juice

NURSING CONSIDERATIONS

Assess:

Erectile dysfunction: Assess for underlying cause before treatment; use of organic nitrates that should not be used with this product; any loss of vision/hearing while taking this product, hypersensitivity reactions

Evaluate:

Therapeutic response: Ability to engage in sexual intercourse

Teach patient/family:

- Sexual dysfunction: may be taken 30 min before sexual activity on an as-needed basis, but no more than once per day
- May be used without regard to meals
- ⚠ That products should not be used with nitrates/nitrates, or strong CYP3A4 inhibitors
- That product does not protect against sexually transmitted disease including HIV
- That product has no effect in the absence of sexual stimulation; to seek help if erection lasts >4 hr

- To tell prescriber about all medication, vitamins, herbs being taken, especially ritonavir, indinavir, ketoconazole, itraconazole, erythromycin, nitrates, α-blockers
- Not to drink large amounts of alcohol

⚠ **To notify prescriber immediately and to stop taking product if vision or hearing loss occurs, if erection lasts >4 hr, or if chest pain occurs**

axitinib

Inlyta

Func. class.: Antineoplastics, biologic response modifiers, signal transduction inhibitors (STIs)

Chem. class.: Tyrosine kinase inhibitor

ACTION: Inhibits receptor tyrosine kinases including vascular endothelial growth factor receptors (VEGFR)-1, VEGFR-2, and VEGFR-3; inhibits tumor growth and phosphorylation of VEGFR-2 and VEGF-mediated endothelial cell proliferation

USES: Treatment of advanced renal cell cancer after failure of 1 prior systemic therapy

CONTRAINDICATIONS: Pregnancy (D), breastfeeding

Precautions Risk for or history of thromboembolic disease, recent bleeding, untreated brain metastasis, recent GI bleeding, GI perforation, fistula, surgery, moderate hepatic disease, uncontrolled hypertension, hyper/hypothyroidism, proteinuria, infertility, end-stage renal disease (CrCl <15 ml/min); not intended for use in adolescents, children, infants, neonates

DOSAGE AND ROUTES

- **Adult:** PO 5 mg bid (at 12 hr intervals), may increase to 7 mg bid and then to 10 mg bid in those not receiving antihypertensives who tolerate the lower dosage for at least 2 consecutive wk with no more than grade 2 adverse reactions; reduce to 3 mg bid if a dose reduction is needed; if further reduction is necessary, reduce to 2 mg bid
- **Adult receiving a strong CYP 3A4/5 inhibitor:** Reduce dose by 1/2, adjust as needed

Administer:

- Give with or without food; swallow tablet whole with a glass of water
- If patient vomits or misses a dose, an additional dose should not be taken; the next dose should be taken at the usual time
- Store at room temperature

SIDE EFFECTS

CNS: Dizziness, headache, **reversible posterior leukoencephalopathy syndrome (RPLS)**, fatigue

CV: Hypertension, **arterial thromboembolic events (ATE), venous thromboembolic events (VTE)**

ENDO: Hypothyroidism, hyperthyroidism

GI: **Lower GI bleeding/perforation/fistula,** abdominal pain, constipation, diarrhea, dysgeusia, dyspepsia, dysphonia, hemorrhoids, nausea, mucosal inflammation, stomatitis, vomiting, increased ALT/AST

GU: Proteinuria

HEMA: **Bleeding intracranial bleeding,** anemia, polycythemia, decreased/increased hemoglobin, **lymphopenia thrombocytopenia, neutropenia**

INTEG: Palmar-plantar erythrodysesthesia (hand and foot syndrome) rash, dry skin, pruritus, alopecia, erythema

MISC: Weight loss dehydration metabolic and electrolyte laboratory abnormalities included

MS: Asthenia, arthralgia, musculoskeletal pain, myalgia

RESP: Cough, dyspnea

PHARMACODYNAMICS

Absorption: bioavailability 58%; distribution: protein binding >99%; metabolized in liver by CYP3A4/5, CYP1A2,

CYP2C19, and UGT1A1; metabolites are carboxylic acid, sulfoxide, and N-glucuronide; excretion 41% in feces and 23% in urine, 12% unchanged; half-life: 2.5-6.1 hr; steady state 2-3 days; onset unknown, peak 2.5-4.1 hr, increased in moderate hepatic disease; duration unknown

INTERACTIONS

Increase: effect of axitinib—CYP3A4/5 inhibitor, strong, moderate (ketoconazole, boceprevir, chloramphenicol, conivaptan, delavirdine, fosamprenavir, imatinib, indinavir, isoniazid itraconazole, dalfopristin, quinupristin, posaconazole, ritonavir, telithromycin, tipranavir (boosted with ritonavir), darunavir (boosted with ritonavir), aldesleukin, IL-2, amiodarone, aprepitant, fosaprepitant atazanavir, bromocriptine, clarithromycin, crizotinib, danazol, diltiazem, dronedarone, erythromycin, fluvoxaMINE, lanreotide, lapatinib, miconazole, mifepristone, nefazodone, nelfinavir, niCARdipine, octreotide, pantoprazole, saquinavir, tamoxifen, verapamil, voriconazole, grapefruit juice)

Decrease: effect of axitinib—CYP3A4/5 inducers, strong/moderate (rifampin, carBAMazepine, dexamethasone, phenytoin, PHENobarbital, rifabutin, rifapentine, St. John's wort, ethanol, bexarotene, bosentan, efavirenz, etravirine, griseofulvin, metyrapone, modafinil, nafcillin, nevirapine, OXcarbazepine, vemurafenib, pioglitazone, topiramate)

Increase or decrease: effect of axitinib—CYP3A4/5 inhibitor and inducers (quiNINE)

Drug/Lab Test

Increase: creatinine, lipase, amylase, sodium, potassium, glucose

Decrease: bicarbonate, calcium, albumin, glucose, phosphate, sodium

Increase or decrease: sodium, glucose

Drug/Herb

Decrease: effect of axitinib—St. John's wort

NURSING CONSIDERATIONS

Assess:

- **Bleeding:** monitor for GI bleeding or perforation; temporarily discontinue therapy if a patient develops any bleeding that requires treatment
- **Surgery:** discontinue ≥24 hr before surgery; may be resumed after adequate wound healing
- **Hepatic/renal disease:** dosage should be reduced in patients with moderate (Child-Pugh Class B) hepatic disease; monitor liver function tests (ALT, AST, bilirubin) before and periodically during therapy; monitor CCr before and during treatment
- **Hypertension:** B/P should be well controlled before starting treatment; monitor patients for hypertension and administer antihypertensive therapy as needed before and during therapy; dose should be reduced for persistent hypertension; therapy should be discontinued if B/P remains elevated after a dosage reduction or if there is evidence of hypertensive crisis; after discontinuation monitor B/P for hypotension in those receiving antihypertensives
- **Hyper/hypothyroidism:** monitor thyroid function tests before and periodically during therapy; thyroid disease should be treated with thyroid medications
- Monitor for proteinuria before and periodically during therapy; product may need to be decreased or discontinued if moderate to severe proteinuria occurs

⚠ **Pregnancy/breastfeeding:** pregnancy (D); determine if the patient is pregnant or breastfeeding before using this product; may also cause infertility

Evaluate:

- Decreased spread of malignancy

Teach patient/family:

⚠ To use contraception during treatment (pregnancy [D]) or to avoid use of this product; to notify prescriber if pregnancy is planned to suspected, not to breastfeed

⚠ To notify prescriber of bleeding that is severe or that requires treatment

- That product will be discontinued ≥24 hr before surgery; may be resumed after adequate wound healing
- That laboratory testing will be required before and periodically during product use
- How to monitor B/P and that B/P products should be continued as directed by prescriber

⚠ HIGH ALERT

azaCITIDine (Rx)

(a-za-sie-ti′deen)

Vidaza

Func. class.: Antineoplastic-nucleoside analog

Chem. class.: DNA demethylation agent

Do not confuse:
azaCITIDine/azaTHIOprine

ACTION:
Cytotoxic by producing damage to double-strand DNA during DNA synthesis

USES:
Myelodysplastic syndrome (MDS)

Unlabeled uses: Acute myelogenous leukemia (AML), chronic myelogenous leukemia (CML)

CONTRAINDICATIONS:
Pregnancy (D), hypersensitivity to product or mannitol, advanced malignant hepatic tumors

Precautions: Breastfeeding, children, geriatric patients, renal/hepatic disease, baseline albumin <30 g/L; a man should not father a child while taking product

DOSAGE AND ROUTES

- **Adult: SUBCUT/IV** 75 mg/m²/day × 7 days q4wk, dose may be increased to 100 mg/m² if no response seen after 2 treatment cycles; minimum treatment, 4 cycles

Available forms: Powder for inj 100 mg

Administer:

- Use cytotoxic handling procedures

SUBCUT route

- **Reconstitute** with 4 ml sterile water for inj (25 mg/ml), inject diluents slowly into vial, invert vial 2-3 times, gently rotate; sol will be cloudy, use immediately; divide doses >4 ml into 2 syringes; invert contents 2-3 times, gently roll syringe between the palms for 30 sec immediately before administration, rotate inj site

Intermittent IV INF route

- **Reconstitute** each vial with 10 ml sterile water for inj, shake well until all solids are dissolved, withdraw sol (10 mg/ml), inject in 50-100 NS or LR inf run over 10-40 min

SIDE EFFECTS

CNS: *Anxiety, depression, dizziness, fatigue, headache, fever, insomnia*

CV: Cardiac murmur, hypotension, tachycardia, peripheral edema, *chest pain*

GI: *Diarrhea, nausea, vomiting,* anorexia, *constipation,* abdominal pain, distention, tenderness, hemorrhoids, mouth hemorrhage, tongue ulceration, stomatitis, dyspepsia, **hepatotoxicity, hepatic coma**

GU: **Renal failure, renal tubular acidosis,** dysuria, UTI

HEMA: **Leukopenia, anemia, thrombocytopenia, neutropenia, febrile neutropenia,** ecchymosis, petechiae

INTEG: *Irritation at site, rash,* sweating, pyrexia, pruritus

META: Hypokalemia

MS: Weakness, arthralgia, muscle cramps, myalgia, back pain

RESP: Cough, *dyspnea, pharyngitis,* **pleural effusion**

PHARMACOKINETICS

Rapidly absorbed, peak ½ hr, metabolized in the liver, half-life 4 hr, excreted in urine

INTERACTIONS

Increase: bone marrow depression—other antineoplastics

NURSING CONSIDERATIONS

Assess:

- For CNS symptoms: fever, headache, chills, dizziness
- **Bone marrow suppression/hematologic response:** CBC with differential, baseline WBC ≥3000/mm^3, absolute neutrophil count (ANC) ≥1500/mm^3, platelets >7500/mm^3, adjust dose based on nadir; ANC <500/mm^3, platelets <25,000/mm^3, give 50% dose next course; ANC 500-1500/mm^3, platelets 25,000-50,000/mm^3, give 67% next course; bruising, bleeding, blood in stools, urine, sputum, emesis; myelodysplastic syndrome (MDS), splenomegaly
- Buccal cavity q8hr for dryness, sores, or ulceration, white patches, oral pain, bleeding, dysphagia
- **Myelodysplastic syndrome:** severe anemia, cytopenias, splenomegaly
- Blood studies: BUN, bicarbonate, creatine, LFTs
- Increased fluid intake to 2-3 L/day to prevent dehydration unless contraindicated
- Rinsing of mouth tid-qid with water, club soda; brushing of teeth bid-tid with soft brush or cotton-tipped applicator for stomatitis; use unwaxed dental floss

Evaluate:

- Therapeutic response: improvement in blood counts with refractory anemia or refractory anemia with excess blasts

Teach patient/family:

- To avoid crowds, persons with known infections; not to receive immunizations
- To avoid foods with citric acid or hot or rough texture if stomatitis is present; to drink adequate fluids
- To report stomatitis; any bleeding, white spots, ulcerations in mouth; to examine mouth daily, report symptoms, infection site reactions, pruritus, fever
- To use contraception during and for several months after therapy (pregnancy [D]); not to breastfeed; not to father a child while receiving product

azaTHIOprine (Rx)

(ay-za-thye′oh-preen)

Azasan, Imuran

Func. class.: Immunosuppressant

Chem. class.: Purine antagonist

Do not confuse:

Imuran/Imferon/Elmiron/IMDUR/Enduron/Tenormin

azaTHIOprine/azaCITIDine

ACTION:

Produces immunosuppression by inhibiting purine synthesis in cells

USES:

Renal transplants to prevent graft rejection, refractory rheumatoid arthritis

Unlabeled uses: Myasthenia gravis, chronic ulcerative colitis, Crohn's disease, Behçet's disease, autoimmune hepatitis, dermatomyositis, thrombocytopenic purpura, lupus nephritis, polymyositis, pulmonary fibrosis, systemic lupus erythematosus (SLE), Wegener's granulomatosis, vasculitis, atopic dermatitis

CONTRAINDICATIONS:

Pregnancy (D), hypersensitivity, breastfeeding

Precautions: Severe renal/hepatic disease, geriatric patients, thiopurine methyltransferase deficiency, infection

Black Box Warning: Bone marrow suppression, neoplastic disease, must be used by experienced clinician

DOSAGE AND ROUTES

Prevention of rejection

- **Adult and child: IV** 3-5 mg/kg/day then maintenance **(PO)** of ≥ 1-3 mg/kg/day

Refractory rheumatoid arthritis

- **Adult: PO** 1 mg/kg/day, may increase dose after 2 mo by 0.5 mg/kg/day and then q4wk, max 2.5 mg/kg/day

Lupus nephritis/SLE/Wegener's granulomatosis/idiopathic pulmonary fibrosis (unlabeled)

• **Adult: PO** 2-3 mg/kg/day

Atopic dermatitis (unlabeled)

• **Adult/adolescent ≥16 yr: PO** 2.5 mg/kg/day

Available forms: Tabs 50, 75, 100 mg; inj 100 mg

Administer:

• For several days before transplant surgery

• All medications PO if possible; avoid IM inj because bleeding may occur

PO route

• With meals to reduce GI upset

IV route

• Prepare in biologic cabinet with gown, gloves, mask

Direct IV

• **Dilute** to 10 mg/ml with 0.9% NaCl, 0.45% NaCl, D_5W, **give** over 5 min

Intermittent IV INF route

• **Reconstitute** 100 mg/10 ml of sterile water for inj; rotate to dissolve; **further dilute** with 50 ml or more saline or glucose in saline, **give** over ½-1 hr

Y-site compatibilities: Alfentanil, atracurium, atropine, benztropine, calcium gluconate, cycloSPORINE, enalaprilat, epoetin alfa, erythromycin, fentaNYL, fluconazole, folic acid, furosemide, glycopyrrolate, heparin, insulin, mannitol, mechlorethamine, metoprolol, naloxone, nitroglycerin, oxytocin, penicillin G, potassium chloride, propranolol, protamine, SUFentanil, trimetaphan, vasopressin

Solution compatibilities: D_5W, NaCl 0.9%, NaCl 0.45%

SIDE EFFECTS

GI: *Nausea, vomiting,* stomatitis, esophagitis, **pancreatitis, hepatotoxicity, jaundice, hepatic veno-occlusive disease**

HEMA: **Leukopenia, thrombocytopenia, anemia, pancytopenia, bleeding**

INTEG: Rash, alopecia

MISC: **Serum sickness,** Raynaud's symptoms, **secondary malignancy, infection**

MS: Arthralgia, muscle wasting

PHARMACOKINETICS

Metabolized in liver, excreted in urine (active metabolite), crosses placenta, half-life 3 hr

INTERACTIONS

Increase: leukopenia—ACE inhibitors, sulfamethoxazole-trimethoprim

Increase: myelosuppression—cycloSPORINE, mercaptopurine

Increase: action of azaTHIOprine—allopurinol

Decrease: immune response—vaccines, toxoids

Decrease: action of warfarin—warfarin

• Do not admix with other products

Drug/Lab Test

Increase: LFTs

Decrease: uric acid

Interference: CBC, differential count

NURSING CONSIDERATIONS

Assess:

• **For infection: increased temp, WBC; sputum, urine**

• I&O, weight daily, report decreasing urine output; toxicity may occur

Black Box Warning: Bone marrow suppression: severe leukopenia, pancytopenia, thrombocytopenia; Hgb, WBC, platelets during treatment monthly; if leukocytes are <3000/mm³ or platelets <100,000/mm³, product should be discontinued, CBC

⚠ **Hepatotoxicity:** if dark urine, jaundice, itching, light-colored stools, increased LFTs, product should be discontinued; hepatic studies: alk phos, AST, ALT, bilirubin

• **Arthritis:** pain, ROM, swelling, mobility before, during treatment

Evaluate:

• Therapeutic response: absence of graft rejection, immunosuppression in autoimmune disorders

Teach patient/family:

- To take as prescribed; not to miss doses; if dose is missed on daily regimen, to skip dose; if taking multiple doses/day, to take as soon as remembered
- That therapeutic response may take 3-4 mo with RA; to continue with prescribed exercise, rest, other medications
- To report fever, rash, severe diarrhea, chills, sore throat, fatigue because **serious infections** may occur; report unusual bleeding, bruising; signs/symptoms of **renal/hepatic toxicity**
- To use contraceptive measures during treatment, for 16 wk after ending therapy (pregnancy [D]); to avoid vaccinations
- To avoid crowds to reduce risk for infection
- To take with food to decrease GI intolerance
- About multiple significant drug–drug interactions
- To use soft-bristled toothbrush to prevent bleeding
- That treatment is ongoing to prevent transplant rejection

azelaic acid

(aze-eh-lay′ik)

Azelex, Finacea

Func. class.: Antiacne agent

Chem. class.: Dicarboxylic acid

ACTION: Active for *Staphylococcus epidermidis, Propionibacterium acnes* by inhibiting microbial protein synthesis in the cell

USES: Mild to moderate inflammatory acne vulgaris, rosacea

CONTRAINDICATIONS: Hypersensitivity

Precautions: Pregnancy (B), breastfeeding; patients with dark complexions should be monitored for early signs of hypopigmentation

DOSAGE AND ROUTES

Adult/child ≥12 yr:

- Apply a thin film and massage into affected areas bid am and pm

Available forms:

Cream 20%, Gel 15%

Administer

- Before applying, the affected areas should be thoroughly washed and patted dry; then apply cream or gel and massage gently into the affected areas; occlusive dressings or wrappings should not be used; wash hands after applying
- Do not apply to the eye; avoid contact with the mouth, eyes, and other mucous membranes; if contact with the eye(s) occur, the eye(s) should be washed with large amounts of water; contact prescriber if ocular irritation persists
- Avoid concurrent use of alcoholic cleansers, astringents, abrasives, and peeling agents
- Store at room temperature, do not freeze

SIDE EFFECTS

MISC: Allergic reaction

SKIN: Itching, burning, stinging, dermatitis, peeling, erythema, edema, acne

PHARMACOKINETICS

Onset, peak, duration unknown; half-life 12 hr

NURSING CONSIDERATIONS

Assess

- Skin: assess for redness, irritation, stinging; if severe, product might need to be discontinued
- Hypopigmentation: in patients with dark complexion

Evaluate

- Therapeutic response: decreased acne lesions

Teach patient/family:

- To thoroughly wash and pat dry the affected areas before applying product; apply cream or gel and massage gently into the affected areas; occlusive dress-

ings or wrappings should not be used; wash hands after applying
- Not to apply to the eye; avoid contact with the mouth, eyes, and other mucous membranes; if contact with the eyes occurs, wash with large amounts of water; contact prescriber if ocular irritation persists
- To avoid concurrent use of alcoholic cleansers, astringents, abrasives, and peeling agents
- To store at room temperature; not to freeze
- To use for prescribed amount of time
- To report burning, stinging, irritation
- To report change in skin color

azelastine (ophthalmic)

(ah-zell′ah-steen)

Optivar

Func. class.: Antihistamine (ophthalmic)

Chem. class.: H_1 receptor antagonist

ACTION: Decreases the allergic response by inhibiting histamine release

USES: Pruritus from allergic conjunctivitis

CONTRAINDICATIONS: Hypersensitivity

Precautions: Pregnancy (C), breastfeeding, child <3 yr

DOSAGE AND ROUTES

- **Adult/child ≥3 yr: OPHTH** 1 drop into each affected eye bid

Available forms: Ophthalmic sol: 0.05%

Administer
- Tip of dropper should not touch the eye
- Store upright and tightly closed at room temperature

SIDE EFFECTS

CNS: Headache

EENT: Eye burning/stinging/irritation, blurred vision, rhinitis, bitter taste

INTEG: Pruritus

RESP: Asthma, dyspnea, wheezing

PHARMACOKINETICS

Onset 3 min, duration 8 hr, half-life 22 hr, protein binding 88%

NURSING CONSIDERATIONS

Assess:
- Eyes: for itching, redness, use of soft or hard contact lenses

Evaluate:
- Therapeutic response: absence of redness, itching in the eyes

Teach patient/family:
- To use in the eyes only; not to touch dropper to eye/eyelid
- Not to wear contact lenses if eyes are red and itching
- To wait at least 10 min before inserting contact lenses; soft contact lenses can absorb preservative

azelastine nasal agent

See Appendix B

azilsartan

Edarbi

Func. class.: Antihypertensive

Chem. class.: Angiotensin II receptor antagonist

ACTION: Antagonizes angiotensin II at the AT_1 receptor in tissues like vascular smooth muscle and the adrenal gland; two angiotensin II receptors, AT_1 and AT_2, have been identified; azilsartan exhibits more than 10,000-fold greater affinity for the AT_1 receptor than the AT_2 receptor

USES: Hypertension, alone or in combination with other antihypertensives

CONTRAINDICATIONS:

Black Box Warning: Pregnancy (D) 2nd/3rd trimesters

Precautions: Pregnancy (C) 1st trimester, breastfeeding, children, geriatric patients, angioedema, African descent, renal disease, renal artery stenosis, heart failure, hypovolemia

DOSAGE AND ROUTES:

- **Adult: PO** 80 mg/day, may give an initial dose of 40 mg/day in patients receiving high-dose diuretic therapy

Available forms: Tabs 40, 80 mg

Administer:

- May administer without regard to food
- Use original package to protect from light and moisture beat

SIDE EFFECTS

CNS: Dizziness, fatigue, asthenia, syncope
CV: Hypotension, orthostatic hypotension
GI: Nausea, diarrhea
HEMA: Anemia
INTEG: Angioedema
MS: Muscle cramps
RESP: Cough

PHARMACOKINETICS:

Protein binding >99% to serum albumin; metabolized by CYP2C9; elimination half-life 11 hr, steady state within 5 days and no accumulation in plasma occurs with once-daily dosing; elimination 55% in feces, 42% in urine; hydrolyzed to the active metabolite, azilsartan, in GI tract during absorption, rapidly absorbed, peak 1.5-3 hr; absolute bioavailability (60%) not affected by food

INTERACTIONS:

Increase: hypotensive effect—other antihypertensives, other angiotensin receptor antagonists, MAOIs
Increase: hypoglycemia—antidiabetics
⚠ **Increase:** renal failure risk—cyclosporine, diuretics, NSAIDs in those with poor renal function; monitor closely
⚠ **Increase:** lithium toxicity—lithium
⚠ **Increase:** phosphate nephropathy—sodium phosphate monobasic monohydrate; sodium phosphate dibasic anhydrous

Drug/Herb:
Increase: antihypertensive effect—Hawthorn
Decrease: antihypertensive effect—ephedra

NURSING CONSIDERATIONS

Assess:

⚠ **Angioedema:** Assess for facial swelling, difficulty breathing

Black Box Warning: For pregnancy (D) 2nd/3rd trimester, can cause fetal death

- Response and adverse reactions especially in renal disease
- B/P, pulse when beginning therapy and periodically thereafter; note rhythm, rate, quality; obtain electrolytes before beginning therapy

Evaluate:

- Therapeutic response: decreased B/P

Teach patient/family:

- To comply with dosage schedule even if feeling better

⚠ To notify prescriber of facial swelling; if pregnancy is planned or suspected (category D 2nd/3rd trimester)

- That diarrhea, dehydration, excessive perspiration, vomiting, may lead to fall in B/P; to consult prescriber if these occur
- To rise slowly from lying or sitting to minimize orthostatic hypotension; that product may cause dizziness
- To avoid OTC medications unless approved by prescriber; to inform all health care providers of product use
- To use proper technique for obtaining B/P

A

azithromycin (Rx)

(ay-zi-thro-my′sin)

Zithromax, Zmax, Zithromax Tri Pak, Zithromax Z Pak

Func. class.: Antiinfective

Chem. class.: Macrolide (azalide)

Do not confuse:
azithromycin/erythromycin
Zithromax/Zinacef

ACTION: Binds to 50S ribosomal subunits of susceptible bacteria and suppresses protein synthesis; much greater spectrum of activity than erythromycin; more effective against gram-negative organisms

USES: Mild to moderate infections of the upper respiratory tract, lower respiratory tract; uncomplicated skin and skin structure infections caused by *Bacillus anthracias, Bacteroides bivius, Bordetella paertassis, Borrelia burgdorferi, Campylobacter jun juni, CDC coryneform group G, Chlamydia trachomatis, Chlamydophila pneumonial, Clostridium perffringes, Gardnerella vaginalis, Haemophilisducreyi, influenzal* (betalactamase negative/positive) *Helicobacterphlori, Klebsiella granulomatis, Legionella pnemohila moraxella catarrhalis, Mycobacterium arium/ intracellulare, Mycoplasma genitalium/homonis/pneumonial, Neisseria gonorrhoel, Peptostreptococcus* sp., *Prevotella bivia, Rickettsia tsutsugamushi, Salmonella typhi, Staphylococcus aureus (MSSA)/epidermidis, Streptococcus* sp., *Toxoplasma gondi, Treponema pallidium, Ureaplasma urealytican, Vibriocholerae, Viridancs streptococci;* **PO;** acute pharyngitis/tonsillitis (group A streptococcal); acute skin/soft tissue infections; community-acquired pneumonia; pharyngitis/tonsillitis

Unlabeled uses: Babesiosis, cholera, cystic fibrosis, dental abscess/infection, endocarditis prophylaxis, granuloma inguinale, Legionnaire's disease, Lyme disease, lymphogranuloma venereum, MAC, periodontitis, pertussis, prostatitis, shigellosis, syphilis, toxoplasmosis, typhoid fever

CONTRAINDICATIONS: Hypersensitivity to azithromycin, erythromycin, any macrolide, hepatitis, jaundice

Precautions: Pregnancy (B), breastfeeding; geriatric patients; renal/hepatic/ cardiac disease; <6 mo for otitis media; <2 yr for pharyngitis, tonsillitis, QT prolongation, ulcerative colitis, torsades de pointes, sunlight exposure, sodium restriction, myasthenia gravis, *Pseudomembranous colitis,* contact lenses, hypokalemia, hypomagnesemia

DOSAGE AND ROUTES

Most infections

- **Adult: PO** 500 mg on day 1 then 250 mg/day on days 2-5 for a total dose of 1.5 g or 500 mg a day × 3 days
- **Child 2-15 yr: PO** 10 mg/kg on day 1 then 5 mg/kg × 4 days

Disseminated MAC infections

- **Adult: PO** 600 mg/day with ethambutol 15 mg/kg/day

Pelvic inflammatory disease

- **Adult: PO/IV** 500 mg **IV** q24hr × 2 doses then 250 mg **PO** q24hr × 7-10 days

Cervicitis, chlamydia, chancroid, nongonococcal urethritis, syphilis

- **Adult: PO** 1 g single dose

Gonorrhea

- **Adult: PO** 2 g single dose

Lower respiratory tract infections

- **Adult: PO** 500 mg day 1, then 250 mg × 4 days
- **Child: PO** 5-12 mg/kg/day × 5 days

Acute otitis media

- **Child >6 mo: PO** 30 mg/kg as a single dose or 10 mg/kg/day × 3 days or 10 mg/kg as a single dose on day 1 (max 500 mg/day) then 5 mg/kg on days 2-5 (max 250 mg/day)

Prevention of acute otitis media
• **Child: PO** 10 mg/kg q wk × 6 mo
MAC in HIV
• **Adult/adolescent: PO** 1.2 g q1wk, alone or with rifabutin
Legionnaire's disease/early Lyme disease (unlabeled)
• **Adult: PO** 500 mg/day
Pertussis (unlabeled)
• **Adult: PO** 500 mg on day 1 then 250 mg/day for 2-5 days
• **Infant ≥6 mo and child: PO** 10 mg/kg/day (max 500 mg) on day 1, then 5 mg/kg/day (max 250 mg) on days 2-5
• **Infant <6 mo: PO** 10 mg/kg/day × 5 days

Available forms: Tabs 250, 500, 600 mg; powder for inj 500 mg; susp 100, 200 mg/5 ml 1 g single-dose powder for susp; ext rel powder for susp 2 g; ophthalmic drops 1% solution

Administer:
PO route
• **Susp** 1 hr before meal or 2 hr after meal; reconstitute 1 g packet for susp with 60 ml water, mix, rinse glass with more water and have patient drink to consume all medication; packets not for pediatric use
• Store at room temperature

Intermittent IV INF route
• **Reconstitute** 500 mg of product with 4.8 ml sterile water for inj (100 mg/ml); shake, **dilute** with 250 or 500 ml 0.9% NaCl, 0.45% NaCl, or LR to 1-2 mg/ml; diluted sol stable for 24 hr or 7 days if refrigerated
• **Give** 1 mg/ml sol over 3 hr or 2 mg/ml sol over 1 hr; never give IM or as bolus
• Reconstituted product is stable for 24 hr at room temperature or 7 days refrigerated

Y-site compatibilities: Acyclovir, alatrofloxacin, alemtuzumab, alfentanol, aminocaproic acid, aminophyllum, amphotericin B liposome/complex, ampicillin, ampicillin-sulbactum, anidulafungin, atenolol, bivalirudin, bleomycin, bumetanide, buprenorphine, butorphanol, calcium chloride/gluconate, CARBOplatin, carmustine, ceFAZolin, cefepime, cefoperazone, cefoTEtan, cefoxitin, ceftaroline, cefTAZidime, ceftizoxime, ceftobiprole, cimetidine, cisatracurium, CISplatin, cyclophosphamide, cycloSPORINE, cytarabine, DAPTOmycin, DAUNOrubicin liposome, dexamethasome, dexmedetomidine, dexrazoxane, digoxin, diltiazem, diphenhydrAMINE, DOBUTamine, DOCEtaxel, dolasetron, doripenem, doxacurium, DOXOrubicin liposomal, doxycycline, droperidol, enalaprilat, EPINEPHrine, epirubicin, eptifibatide, ertapenem, esmolol, etoposide, etoposide phosphate, fenoldopam, fluconozole, fluorouracil, foscarnet, fosphenytoin, gallium, ganciclovir, gatifloxacin, gemcitabine, granisetron, haloperidol, heparin, hydrocortisone phosphate/succinate, HYDROmorphone, hydrOXYzine, IDArubicin, ifosfamide, inamrinone, irinotecan, isoproterenol, labetalol, lepirudin, magnesium sulfate, mannitol, meperidine, meropenem, mesna, methochlorethamine, methohexital, methotrexate, methylPREDNISolone, metoclopramide, metroNIDAZOLE, milrinone, minocycline, miralacurium, nalbuphine, naloxone, nesiritide, nitroglycerin, nitroprusside, octreotide, ofloxacin, ondansetron, oxaliplatin, oxytocin, PACLitaxel, palonosetron, pamidronate, pantoprazole, PEMEtrexed, PENTobarbital, phenylephrine, piperacillin, potassium acetate/phosphates, proclinimide, proclorperazine, promethazine, propranolol, ranitidine, remifentanil, rocuronium, sodium acetate, succinylcholine, SUFentanil, sulfamethoxazole-trimethoprim, tacrolimus, telavancin, teniposide, thiotepa, ticarcillin, tigecycline, tirofiban, TPN, trimethobenzamide, vancomycin, vasopressin, vecuronium, verapamil, vinCRIStine, voriconazole, zidovudine, zoledronic acid

SIDE EFFECTS

CNS: Dizziness, headache, vertigo, somnolence, fatigue

CV: Palpitations, chest pain, **QT prolongation, torsades de pointes (rare)**

EENT: Hearing loss, tinnitus, loss of smell (anosmia)

GI: *Nausea, diarrhea,* **hepatotoxicity,** abdominal pain, stomatitis, heartburn, dyspepsia, flatulence, melena, **cholestatic jaundice, pseudomembranous colitis,** tongue discoloration

GU: Vaginitis, moniliasis, nephritis

HEMA: Anemia

INTEG: Rash, urticaria, pruritus, photosensitivity, pain at inj site

SYST: **Angioedema, Stevens-Johnson syndrome, toxic epidermal necrolysis**

PHARMACOKINETICS

PO: Peak 2-4 hr, duration 24 hr

IV: Peak end of inf; duration 24 hr; half-life 11-57 hr; excreted in bile, feces, urine primarily as unchanged product; may be inhibitor of P-glycoprotein

INTERACTIONS

Increase: ergot toxicity—ergotamine

⚠ **Increase:** dysrhythmias—pimozide: fatal reaction, do not use concurrently

Increase: QT prolongation—amiodarone, quiNIDine, nilotinib, droperidol, methadone, propafenone, fluoroquinolones, lithium, paliperidone

Increase: effects of oral anticoagulants, digoxin, theophylline, methylPREDNISolone, cycloSPORINE, bromocriptine, disopyramide, triazolam, carBAMazepine, phenytoin, tacrolimus, nelfinavir

Decrease: clearance of triazolam

Decrease: absorption of azithromycin—aluminum, magnesium antacids, separate by ≥2 hr

Drug/Lab Test

Increase: CPK, ALT, AST, bilirubin, BUN, creatinine, alk phos, potassium, blood glucose

Drug/Food

Decrease: absorption—food (susp)

Decrease: blood glucose, potassium, sodium

NURSING CONSIDERATIONS

Assess:

- I&O ratio; report hematuria, oliguria with renal disease
- Hepatic studies: AST, ALT, CBC with differential
- Renal studies: urinalysis, protein, blood
- C&S before product therapy; product may be taken as soon as culture is taken; C&S may be repeated after treatment
- **QT prolongation, torsades de pointes:** assess for patients with serious bradycardia, ongoing proarrhythmic conditions, or elderly; more common in these patients
- **Serious skin reactions:** Stevens-Johnson syndrome, toxic epidermal necrolysis, angioedema; discontinue if rash develops, treat symptomatically
- **Superinfection:** sore throat, mouth, tongue; fever, fatigue, diarrhea, anogenital pruritus
- **Pseudomembranous colitis:** diarrhea, abdominal pain, fever, fatigue, anorexia; obtain CBC, serum albumin
- Bowel pattern before, during treatment
- Respiratory status: rate, character; wheezing, tightness in chest: discontinue product
- Cardiovascular death has occurred in those with serious bradycardia or ongoing hypokalemia, hypomagnesemia, avoid use

Evaluate:

- Therapeutic response: C&S negative for infection; decreased signs of infection

Teach patient/family:

⚠ To report sore throat, fever, fatigue, severe diarrhea, anal/genital itching (may indicate superinfection)

- Not to take aluminum-magnesium–containing antacids simultaneously with this product (PO)

⚠ To notify nurse of diarrhea, dark urine, pale stools; yellow discoloration of eyes, skin; severe abdominal pain

- To complete dosage regimen

• To take ZMAX 1 hr before or 2 hr after a meal; shake well before use
• To use protective clothing or stay out of the sun, photosensitivity may occur

TREATMENT OF HYPERSENSITIVITY: Withdraw product, maintain airway; administer EPINEPHrine, aminophylline, O_2, IV corticosteroids

azithromycin ophthalmic

See Appendix B

RARELY USED

aztreonam (Rx)

(az-tree'oh-nam)

Azactam, Cayston

Func. class.: Antibiotic—miscellaneous

USES: Urinary tract infection; septicemia; skin, muscle, bone infection; lower respiratory tract, intraabdominal infections; other infections caused by gram-negative organisms

CONTRAINDICATIONS: Hypersensitivity to product, severe renal disease

DOSAGE AND ROUTES

Urinary tract infections
• **Adult: IM/IV** 500 mg-1 g q8-12hr

Systemic infections
• **Adult: IM/IV** 1-2 g q8-12hr
• **Child: IM/IV** 90-120 mg/kg/day in divided doses q6-8hr; max 8 g/day **IV**

Severe systemic infections
• **Adult: IM/IV** 2 g q6-8hr; max 8 g/day; continue treatment for 48 hr after negative culture or until patient is asymptomatic

Cystic fibrosis with *Pseudomonas aeruginosa*
• **Adult, adolescent, child ≥7 yr: NEB** 75 mg tid × 28 days, then 28 days off; give q4hr or more; give bronchodilator before aztreonam

baclofen (Rx)

(bak'loe-fen)

Gablofen, Lioresal

Func. class.: Skeletal muscle relaxant, central acting

Chem. class.: GABA chlorophenyl derivative

Do not confuse:
Lioresal/Lotensin

ACTION: Inhibits synaptic responses in CNS by stimulating GABAb receptor subtype, which decreases neurotransmitter function; decreases frequency, severity of muscle spasms

USES: Spasticity with spinal cord injury, multiple sclerosis

Unlabeled uses: Neuropathic pain, hiccups, trigeminal neuralgia/nystagmus, recurrent priapism

CONTRAINDICATIONS: Hypersensitivity

Precautions: Pregnancy (C), breastfeeding, geriatric patients, peptic ulcer disease, renal/hepatic disease, stroke, seizure disorder, diabetes mellitus

Black Box Warning: Abrupt discontinuation

DOSAGE AND ROUTES

- **Adult: PO** 5 mg tid × 3 days, then 10 mg tid × 3 days, then 15 mg tid × 3 days, then 20 mg tid × 3 days, then titrated to response, max 80 mg/day; **INTRATHECAL** use implantable intrathecal inf pump, use screening trial of 3 separate bol doses if needed 24 hr apart (50 mcg, 75 mcg, 100 mcg); patients who do not respond to 100 mcg should not be considered for chronic IT therapy; initial: double screening dose that produced result, give over 24 hr, increase by 10%-30% q24hr only; maintenance: 1200-1500 mcg/day
- **Child >2-7 yr: PO** 10-15 mg/day divided q8hr; titrate q3days by 5-15 mg/day to max 40 mg/day
- **Child ≥8 yr:** As above; max 60 mg/day
- **Child: INTRATHECAL** initial test dose same as adult; for small children, initial dose of 25 mcg/dose may be used; 25-1200 mcg/day inf titrated to response in screening phase
- **Geriatric: PO** 5 mg bid-tid

Neuropathic pain including trigeminal neuralgia (unlabeled)

- **Adult: PO** 10 mg tid, may increase by 10 mg every other day; max 80 mg/day

Hiccups (unlabeled)

- **Adult: PO** 10 mg qid

Stuttering/recurrent priapism (unlabeled)

- **Adult: PO** 40 mg at bedtime

Available forms: Tabs 10, 20 mg; IT inj 10 mg/20 ml (500 mcg/ml), 10 mg/5 ml (2000 mcg/ml); pharmacy can prepare extemperaneous liquid preparations

Administer:

PO route

- With meals for GI symptoms
- Store in a tight container at room temperature

IT route

- **For screening,** dilute to a concentration of 50 mcg/ml with NaCl for inj (preservative free); give test dose over 1 min; watch for decreasing muscle tone, frequency of spasm; if inadequate, use 2 more test doses q24hr; **maintenance inf** via implantable pump of 500-2000 mcg/ml because individual titration is required
- Do not give IT dose by inj, IV, IM, SUBCUT, epidural

Additive compatibilities: CloNIDine, morphine, ziconotide

SIDE EFFECTS

CNS: *Dizziness, weakness, fatigue, drowsiness,* headache, *disorientation,* insomnia, paresthesias, tremors; **seizures, life-threatening CNS depression, coma; CNS infection** (IT)

CV: Hypotension, chest pain, palpitations, edema; cardiovascular collapse (IT)
EENT: Nasal congestion, blurred vision, mydriasis, tinnitus
GI: *Nausea,* constipation, *vomiting,* increased AST, alk phos, abdominal pain, dry mouth, anorexia
GU: Urinary frequency, hematuria
INTEG: Rash, pruritus
RESP: Dyspnea; respiratory failure (IT)

PHARMACOKINETICS

PO: Onset 3-4 days, peak 2-3 hr, duration >8 hr, half-life 2$^1/_2$-4 hr, partially metabolized in liver, excreted in urine (unchanged)
INTRATHECAL: CSF levels with plasma levels 100 × that of the oral route, peak 4 hr, duration 4-8 hr
BOLUS: Onset $^1/_2$-1 hr
CONT INF: Onset 6-8 hr, peak 24-48 hr

INTERACTIONS

Increase: CNS depression—alcohol, tricyclics, opiates, barbiturates, sedatives, hypnotics, MAOIs
Increase: hypotension—antihypertensives

Drug/Herb
Increase: CNS depression—kava, valerian

Drug/Lab Test
Increase: AST, alk phos, blood glucose

NURSING CONSIDERATIONS

Assess:

- **Multiple sclerosis:** spasms, spasticity, ataxia; improvement should occur with product
- B/P, weight, blood glucose, hepatic function periodically

⚠ **Seizures:** for increased seizure activity with seizure disorders; product decreases seizure threshold; EEG in epileptic patients

- I&O ratio; check for urinary frequency
- Allergic reactions: rash, fever, respiratory distress
- Severe weakness, numbness in extremities
- Tolerance: increased need for medication, more frequent requests for medication, increased pain
- **Withdrawal symptoms:** CNS depression, dizziness, drowsiness, psychiatric symptoms
- **Intrathecal:** have emergency equipment nearby; assess test dose and titration; if no response, check pump, catheter for proper functioning
- Assistance with ambulation if dizziness, drowsiness occurs

Evaluate:

- Therapeutic response: decreased pain, spasticity

Teach patient/family:

- Not to discontinue medication quickly; hallucinations, spasticity, tachycardia will occur; product should be tapered off over 1-2 wk
- Not to take with alcohol, other CNS depressants
- To avoid hazardous activities if drowsiness, dizziness occurs; to rise slowly to prevent orthostatic hypotension
- To avoid using OTC medications; not to take cough preparations, antihistamines unless directed by prescriber
- To notify prescriber if nausea, headache, tinnitus, insomnia, confusion, constipation, inadequate or painful urination continues
- **MS:** may require 1-2 mo for full response

TREATMENT OF OVERDOSE:

Induce emesis in conscious patient, activated charcoal, dialysis, physostigmine to reduce life-threatening CNS side effects

B

⚠ HIGH ALERT

basiliximab (Rx)

(bas-ih-liks′ih-mab)

Simulect

Func. class.: Immunosuppressant

Chem. class.: Murine/human monoclonal antibody (interleukin-2) receptor antagonist

ACTION: Binds to and blocks the IL-2 receptor, which is selectively expressed on the surface of activated T lymphocytes; impairs the immune system to antigenic challenges

USES: Acute allograft rejection in renal transplant patients when used with cycloSPORINE and corticosteroids

Unlabeled uses: Liver transplant rejection prophylaxis, graft-versus-host disease

CONTRAINDICATIONS: Breastfeeding, hypersensitivity to mannitol/murine, exposure to viral infections

Precautions: Pregnancy (B), children, geriatric patients, human anti-murine antibody

Black Box Warning: Infections

DOSAGE AND ROUTES

- **Adult/child ≥35 kg: IV** 20 mg × 2 doses; 1st dose within 2 hr before transplant surgery; 2nd dose 4 days after transplantation
- **Child/adolescent <35 kg: IV** 10 mg × 2 doses; 1st dose within 2 hr before transplant surgery; 2nd dose 4 days after transplantation

Available forms: Powder for inj 10, 20 mg

Administer:

Intermittent IV INF route

- **Reconstitute** 10-mg vial/2.5 ml or 20-mg vial in 5 ml sterile water for inj; shake gently to dissolve; **dilute** reconstituted sol in 25 ml (10 mg vial) or 50 ml (20-mg vial) with 0.9% NaCl or D_5; gently invert bag, do not shake; **give** over 1/2 hr, do not admix
- Store reconstituted sol refrigerated for up to 24 hr or at room temp for 4 hr

SIDE EFFECTS

CNS: *Pyrexia, chills, tremors, headache, insomnia, weakness, dizziness*

CV: *Chest pain,* angina, cardiac failure, hypotension, *hypertension, edema*

GI: *Vomiting, nausea, diarrhea, constipation, abdominal pain,* GI bleeding, *gingival hyperplasia, stomatitis*

INTEG: *Acne,* pruritus, impaired wound healing

META: *Acidosis, hypercholesterolemia, hyperuricemia, hypo/hyperkalemia, hypocalcemia, hypophosphatemia*

MISC: *Infection, moniliasis,* anaphylaxis, anemia, allergic reaction, dysuria, CMV infection, candidiasis

MS: *Arthralgia, myalgia*

RESP: *Dyspnea, wheezing,* pulmonary edema, *cough*

PHARMACOKINETICS

Peak 1/2 hr (adults); terminal half-life 7 days (adult), 9 1/2 days (children)

INTERACTIONS

Increase: immunosuppression—other immunosuppressants

Drug/Herb

Decrease: St. John's wort, turmeric

Drug/Lab Test

Increase: cholesterol, BUN, uric acid, creatinine, potassium, calcium blood glucose, Hgb, Hct

Decrease: Hgb, Hct, platelets, magnesium, phosphate

NURSING CONSIDERATIONS

Assess:

Black Box Warning: For infection: increased temp, WBC, sputum, urine; may be fatal (bacterial, protozoal, fungal)

- Blood studies: Hgb, WBC, platelets during treatment monthly; if leukocytes are <3000/mm³, product should be discontinued; electrolytes, B/P, edema assessment

- Hepatic studies: alk phos, AST, ALT, bilirubin

⚠ **Anaphylaxis, hypersensitivity:** dyspnea, wheezing, rash, pruritus, hypotension, tachycardia; if severe hypersensitivity reactions occur, product should not be used again

Evaluate:

- Therapeutic response: absence of graft rejection

Teach patient/family:

Black Box Warning: To report fever, chills, sore throat, fatigue; serious infection may occur

- To avoid crowds, persons with known upper respiratory tract infections
- To use contraception during treatment
- To report GI symptoms, bleeding, allergic reactions

beclomethasone (Rx)

(be-kloe-meth′a-sone)

QVAR

Func. class.: Corticosteroid, synthetic
Chem. class.: Glucocorticoid

Do not confuse:
beclomethasone/betamethasone

ACTION: Prevents inflammation by suppression of the migration of polymorphonuclear leukocytes, fibroblasts and the reversal of increased capillary permeability and lysosomal stabilization; does not suppress hypothalamus and pituitary function

USES: Chronic asthma, allergic/vasomotor rhinitis, nasal polyps

CONTRAINDICATIONS: Hypersensitivity, status asthmaticus (primary treatment)

Precautions: Pregnancy (C), breastfeeding, children <12 yr, nasal disease/surgery, nonasthmatic bronchial disease; bacterial, fungal, viral infections of mouth, throat, lungs; HPA suppression, osteoporosis, Cushing's syndrome, diabetes mellitus, measles, cataracts, corticosteroid hypersensitivity, glaucoma, herpes infection

DOSAGE AND ROUTES

- **Adult and child >12 yr: ORAL INH** 40-80 mcg bid (alone) or 40-160 mcg bid (previous inhaled corticosteroids); max 320 mcg bid
- **Child 5-11 yr: ORAL INH** 40 mcg bid; max 80 mcg bid

Available forms: Oral inh 40, 80, 250 ✦mcg/metered spray

Administer:

- Bronchodilator spray; if used, should be used 1st, then wait a few minutes, then use beclomethasone
- Prime before 1st use or if not used for 7-10 days; prime by spraying 2 actuations into the air, away from the face; do not share inhaler
- **Oral inhalation** (metered-dose non-CFC aerosol) (QVAR); shake well, use spacer; after using, rinse mouth, gargle if possible; clean weekly with dry cloth/tissue; do not wash inhaler
- Titrated dose, use lowest effective dose

SIDE EFFECTS

CNS: *Headache,* psychiatric/behavioral changes (child)

EENT: *Hoarseness, candidal infection of oral cavity, sore throat,* loss of taste/smell, dysgeusia

ENDO: HPA suppression

GI: Dry mouth, dyspepsia

MISC: Angioedema, adrenal insufficiency, facial edema, Churg-Strauss syndrome (rare)

RESP: Bronchospasm, wheezing, cough

PHARMACOKINETICS

INH: Onset 1-4 wk; excreted in feces, urine (metabolites); half-life 2.8 hr; crosses placenta; metabolized in lungs, liver (by CYP3A)

NURSING CONSIDERATIONS

Assess:

- For fungal infection in mucous membranes
- Adrenal function periodically for HPA axis suppression during prolonged therapy, monitor growth/development
- Gum, rinsing of mouth for dry mouth

Evaluate:

- Therapeutic response: decreased dyspnea, wheezing, dry crackles

Teach patient/family:

- To gargle/rinse mouth after each use to prevent oral fungal infections
- That during times of stress, systemic corticosteroids may be needed to prevent adrenal insufficiency; not to discontinue oral product abruptly, to taper slowly
- To notify prescriber if therapeutic response decreases; dosage adjustment may be needed
- Proper administration technique and cleaning technique
- About all aspects of product usage, including cushingoid symptoms
- About **adrenal insufficiency symptoms:** nausea, anorexia, fatigue, dizziness, dyspnea, weakness, joint pain, depression

beclomethasone (nasal)

(be-kloe-meth′a-sone)

Beconase AQ, Qnasal

Func. class.: Nasal corticosteroid

ACTION: Readily crosses cell membranes and binds with high affinity to specific cytoplasmic receptors; inhibition of leukocyte infiltration at the site of inflammation, interference in the function of mediators of inflammatory response, and suppression of humoral immune responses

USES: To relieve symptoms of seasonal/perennial rhinitis

CONTRAINDICATIONS: Hypersensitivity

Precautions: Child <6 yr, untreated fungal infections, glaucoma and/or cataracts, nasal septum ulcers/surgery/trauma

DOSAGE AND ROUTES

- **Adult/child ≥12 yr:** Nasal 1-2 sprays in each nostril bid (42 mcg/spray); 2 sprays in each nostril every day (80 mcg/actuation)
- **Child 6-12 yr:** Nasal 1 spray (42 mcg) in each nostril bid

Available forms: Nasal spray 42 mcg/metered spray; nasal aerosol 80 mcg/actuation

Administer

- Products are not always interchangeable owing to differences in route of administration and in the amount of active drug released per spray
- To avoid the spread of infection, do not use the container for more than one person
- Product's effectiveness depends on regular use

Nasal inhalation (metered-dose aerosol) (Qnasal):

- Instruct patient to shake the canister well before administering
- Before first use, instruct the patient to prime the pump by actuating 4 times; after the initial priming, the dose-counter should read 120
- If the canister is not used for 7 consecutive days, instruct the patient to prime by actuating 2 times
- Instruct patient on proper administration technique

Nasal inhalation (pump spray) (Beconase AQ):

- Instruct patient to shake the nasal sprayer well before use
- Before first use, instruct the patient to prime the pump by actuating 6 times
- If the pump is not used for 7 days, prime until a fine spray appears
- Instruct patient on proper administration technique
- After use, rinse the tip of the bottle with hot water, taking care not to suck water

into the bottle, and dry with a clean tissue; replace the cap

SIDE EFFECTS

CNS: Headache, dizziness
EENT: Nasal burning, epistaxis, nasal fungal infections, nasal congestion, sneezing
GI: Nausea

PHARMACOKINETICS

Onset 7 days, peak 21 days, half-life 15 hr

NURSING CONSIDERATIONS

Assess:
- Nasal symptoms: assess for sneezing, running of nose before and after use; avoid use longer than 3 wk; check for fungal infections, changes in vision

Evaluate:
- Decrease nasal running, sneezing, other symptoms of seasonal/perennial rhinitis

Teach patient/family:
- That products are not always interchangeable owing to differences in route of administration and in the amount of active drug released per spray
- Not to use the container for more than one person, to avoid the spread of infection
- That product effectiveness depends on regular use

Nasal inhalation (metered-dose aerosol) (Qnasal):
- Instruct patient to shake the canister well before administering
- Before first use, instruct the patient to prime the pump by actuating 4 times; after the initial priming, the dose-counter should read 120
- If the canister is not used for 7 consecutive days, instruct the patient to prime by actuating 2 times
- Instruct patient on proper administration technique

Nasal inhalation (pump spray) (Beconase AQ):
- Instruct patient to shake the nasal sprayer well before use
- Before first use, instruct the patient to prime the pump by actuating 6 times
- If the pump is not used for 7 days, prime until a fine spray appears
- Instruct patient on proper administration technique
- After use, rinse the tip of the bottle with hot water, taking care not to suck water into the bottle, and dry with a clean tissue; replace the cap

belatacept

(bel-a-ta′sept)

Nulojix

Func. class.: Biologic response modifier

Chem. class.: Fusion protein

ACTION:

Activated T lymphocytes are the mediators of immunologic rejection, this product is a selective T-cell costimulation blocker; blocks the CD28 mediated costimulation of T lymphocytes by binding to CD80 and CD86 on antigen-presenting cells; inhibits T lymphocyte proliferation and the production of the cytokines interleukin-2, interferon-gamma, interleukin-4, and TNF-alpha.

USES:

Kidney transplant rejection prophylaxis given with basiliximab induction, mycophenolate mofetil, corticosteroids

CONTRAINDICATIONS:

Hypersensitivity

Precautions: Breastfeeding, child/infant/neonate, pregnancy (C), diabetes mellitus, progressive multifocal leukoencephalopathy, immunosuppression, sunlight exposure, TB

Black Box Warning: Infection, organ transplant, requires an experienced clinician, secondary malignancy, posttransplant lymphoproliferation disorder

DOSAGE AND ROUTES

• **Adult:** IV 10 mg/kg rounded to the nearest 12.5 mg increment; give over 30 min the day of transplantation (day 1) but before transplantation, on day 5 approximately 96 hours after the day 1 dose 1, at the end of wk 2, at the end of wk 4, at the end of wk 8, and at the end of wk 12; maintenance dosage is 5 mg/kg rounded to the nearest 12.5 mg increment; give over 30 min at the end of wk 16 and every 4 wk ± 3 days thereafter; doses should be calculated on actual body weight on the transplantation day unless the patient's weight varies by >10%

Available forms: Powder for inj 250 mg

Administer:

Black Box Warning: Only providers skilled in the use of immunosuppressant and management of transplant should use these products

IV route

• Visually inspect product for particulate matter, discoloration whenever sol/container permit; discard if present

• Calculate the number of drug vials required to provide total inf dose

• Reconstitute each vial/10.5 ml of sterile water for injection, 0.9% sodium chloride, D_5W, using the silicone-free disposable syringe provided with each vial and an 18-21G needle; if silicone-free disposable syringe is dropped or becomes contaminated, use a new silicone-free disposable syringe from inventory; if you need additional silicone-free disposable syringes, call 1-888-685-6549; if the powder is accidentally reconstituted using a different syringe than the one provided, the sol may develop a few translucent particles; discard any sol prepared using siliconized syringes

• Using aseptic technique, inject the diluent into the vial and direct the stream of diluent to the glass wall of the vial; to minimize foaming, rotate the vial and invert with gentle swirling until the contents are dissolved; do not shake when reconstituted (25 mg/ml), should be clear to slightly opalescent and colorless to pale yellow; do not use if opaque particles, discoloration, or other foreign particles are present

• Calculate the total volume of the reconstituted 25 mg/ml sol required to provide the total inf dose; further dilute this volume with a volume of inf fluid equal to the volume of the reconstituted drug sol required to provide the prescribed dose; use either NS or D_5W if drug was reconstituted with SWFI; use NS if drug was reconstituted with NS; use D_5W if drug was reconstituted with D_5W; with the same silicone-free disposable syringe used for reconstitution, withdraw the required amount of belatacept sol from the vial, inject it into the inf container, gently rotate the inf container to ensure mixing; final conc in inf container should range (2-10 mg/ml); volume of 100 ml will be appropriate for most patients and doses, but total inf volumes ranging from 50-250 ml may be used; discard any unused drug sol remaining in the vials; after reconstitution, immediately transfer the reconstituted sol from the vial to the inf bag or bottle; complete within 24 hr

IV INF route

• Give over 30 min, use an inf set and a sterile, nonpyrogenic, low-protein-binding filter (0.2-1.2 mm), use a separate line

• Store refrigerated, protected from light ≤24 hr; max 4 hr of the total 24 hr can be at room temp and room light

SIDE EFFECTS

CNS: Guillain-Barré syndrome, anxiety, dizziness, headache, fever, insomnia, tremors

EENT: Pharyngitis, stomatitis

GI: Abdominal pain, constipation, diarrhea, nausea, vomiting

GU: **Renal tubular necrosis, renal failure,** proteinuria, urinary incontinence

HEMA: Anemia, neutropenia, leucopenia, leukoencephalopathy
INTEG: Acne, alopecia
META: Hypercholesterolemia, hyperglycemia, hyper/hypokalemia, hypocalcemia, hypophosphatemia, hypomagnesemia
MS: Arthralgia
SYST: Secondary malignancy, posttransplant lymphoproliferation disorder, wound dehiscence, BK-virus-associated neuropathy

PHARMACOKINETICS

Half-life, 6.1-15.1 days during receipt of 10 mg/kg IV doses; during receipt of 5 mg/kg IV doses, terminal half-life 3.1-11.9; steady-state by wk 8 after transplantation and by mo 6 during maintenance phase

INTERACTIONS

Increase: Basiliximab induction, mycophenolate mofetil

NURSING CONSIDERATIONS

Assess:

Black Box Warning: Transplant rejection: flulike symptoms, decreasing urinary output, malaise; some may experience pain in area (rare; monitor BUN/creatinine)

Black Box Warning: Infection: monitor for fever, chills, increased WBC, **wound dehiscences**

Black Box Warning: Posttransplant lymphoproliferation disorder: may lead to secondary malignancy (lymphoma) or infectious mononucleosis-like lesions; may be treated with antivirals or immunosuppressant; product may need to be discontinued

- Hyperlipidemia: monitor cholesterol, triglycerides; an antilipidemic may be needed

Evaluate:

- Therapeutic response: absence of renal transplant rejection

Teach patient/family:

- Reason for product and expected result
- To avoid exposure to sunlight, tanning beds, risk of secondary malignancy
- To avoid crowds, persons with known infections
- That repeated lab test will be needed
- To avoid with vaccines
- That immunosuppressants will be needed for life to prevent rejection; symptoms of rejection and to call provider immediately

belimumab

(be-lim′ue-mab)

Benlysta

Func. class.: Monoclonal antibody
Chem. class.: Disease-Modifying Antirheumatic Drugs (DMARDs)

ACTION: Inhibits B lymphocyte stimulator (BLyS), needed for B-cell survival; normally, soluble BLyS binds to its receptors on B cells and allows B-cell survival; binds BLyS and prevents binding to its receptors on B-cells

USES: Active, autoantibody-positive, systemic lupus erythematosus (SLE) in combination with standard therapy

CONTRAINDICATIONS: Hypersensitivity

Precautions: Pregnancy (C), breastfeeding, children/infants, geriatric patients, African descent patients, depression, immunosuppression, infection, suicidal ideation, vaccination, secondary malignancy, cardiac disease; requires experienced clinician

DOSAGE AND ROUTES

- **Adult: IV** 10 mg/kg over 1 hr q2wk for the first 3 doses then q4wk

Available forms: Powder for injection 120, 400 mg

Administer:

- Only health care providers prepared to manage anaphylaxis should administer this product; may give premedication for prophylaxis against infusion and hypersensitivity reactions

Intermittent IV INF route

- Visually inspect particulate matter and discoloration whenever sol and container permits
- Give as IV inf only; do not give IV bolus or push; give over 1 hr and slow or stop if inf reactions occur
- Do not give with any other agents in the same IV line
- Allow to stand at room temp for 10-15 min before using
- Reconstitute with the appropriate amount of sterile water for injection (80 mg/ml); add 1.5 ml of sterile water (120 mg/vial) or 4.8 ml of sterile water (400 mg/vial)
- Direct the stream of sterile water toward the side of the vial to minimize foaming; gently swirl for 60 sec, allow to sit during reconstitution, gently swirl for 60 sec q5min until the powder is dissolved; do not shake; reconstitution is complete in 10-30 min
- If a mechanical reconstitution device (swirler) is used, max 500 rpm swirled for ≤30 min
- Sol should be opalescent and colorless to pale yellow and without particles; small air bubbles are expected; protect from sunlight
- Dilution: only dilute in normal saline for injection; dilute reconstituted sol with enough normal saline to 250 ml; from a 250-ml inf bag or bottle of normal saline, withdraw and discard a volume equal to the volume of the reconstituted sol required for dose; add the required volume of the reconstituted sol the infusion bag/bottle; gently invert to mix
- Discard any unused sol
- Store in refrigerator or at room temp; total time from reconstitution to completion of inf max 8 hr

SIDE EFFECTS

CNS: Headache, dizziness, anxiety, depression, fever, insomnia, migraine, suicidal ideation

CV: Bradycardia, hypotension

GI: Nausea, diarrhea

MISC: Rash, dyspnea, cystitis, leukopenia, myalgia, rash, bronchitis, nasopharyngitis, pharyngitis

SYST: Anaphylaxis, angioedema, antibody formation, secondary malignancy, infection, influenza

PHARMACOKINETICS

Terminal half-life 19.4 days; distribution half-life 1.75 days

NURSING CONSIDERATIONS

Assess:

- **SLE:** monitor for decreasing fever, malaise, fatigue, joint pain, myalgias
- **Suicidal ideation:** more common in those with preexisting depression
- **Infection:** determine if a chronic or acute infection is present, may be fatal when used with this product; do not begin therapy if any products are being used for a chronic infection; leukopenia may occur with this product and susceptibility to infections increased
- **Anaphylaxis,** infusion site reactions: if these occur, stop infusion
- **African descent patients:** use cautiously in these patients, may not respond to this product
- Cardiac disease: monitor closely for cardiovascular side effects, bradycardia, hypotension
- Pregnancy: determine if pregnant or if pregnancy is planned or suspected; if pregnant, call 1-877-681-6269 to enroll in registry

Evaluate:

- Positive response: decreasing symptoms of SLE: decreasing fatigue, fever, malaise

Teach patient/family:

- To notify prescriber if pregnancy is planned or suspected; to use reliable contraception during and for 4 mo after final treatment; to avoid breastfeeding
- To seek treatment immediately for serious hypersensitive reactions
- Not to receive live vaccinations during treatment

benazepril (Rx)

(ben-aze′uh-pril)

Lotensin

Func. class.: Antihypertensive

Chem. class.: Angiotensin-converting enzyme (ACE) inhibitor

Do not confuse:
benazepril/Benadryl

ACTION:

Selectively suppresses renin-angiotensin-aldosterone system; inhibits ACE, thus preventing conversion of angiotensin I to angiotensin II

USES:

Hypertension, alone or in combination with thiazide diuretics

Unlabeled uses: CHF, diabetic nephropathy, proteinuria, renal impairment

CONTRAINDICATIONS:

Breastfeeding, children, hypersensitivity to ACE inhibitors, angioedema

Black Box Warning: Pregnancy (D)

Precautions: Geriatric patients, impaired renal/hepatic function, dialysis patients, hypovolemia, blood dyscrasias, CHF, asthma, bilateral renal artery stenosis

DOSAGE AND ROUTES

- **Adult: PO** 10 mg/day initially, then 20-40 mg/day divided bid or daily (without a diuretic); reduce initial dose to 5 mg **PO** daily (with a diuretic); max 80 mg/day
- **Geriatric: PO** used on the basis of the clinical response
- **Child ≥6 yrs: PO** 0.2 mg/kg/day max 5 mg/day

Renal dose

- **Adult: PO** CCr <30 ml/min 5 mg **PO** daily, max 40 mg/day

Renal impairment due to diabetic nephropathy (unlabeled)

- **Adult: PO** 5-10 mg/day

Heart failure (unlabeled)

- **Adult: PO** 2-20 mg/day

Available forms: Tabs 5, 10, 20, 40 mg

Administer:

- May give without regard to food
- Do not discontinue product abruptly
- Store in tight container at 86° F (30° C) or less

SIDE EFFECTS

CNS: Anxiety, hypertonia, insomnia, paresthesia, headache, dizziness, fatigue
CV: Hypotension, postural hypotension, syncope, palpitations, angina
GI: Nausea, constipation, vomiting, gastritis, melena, diarrhea, hepatotoxicity
GU: Increased BUN, creatinine, decreased libido, impotence, UTI
HEMA: Agranulocytosis, neutropenia
INTEG: Rash, flushing, sweating
META: Hyperkalemia, hyponatremia
MISC: Angioedema, Stevens-Johnson syndrome
MS: Arthralgia, arthritis, myalgia
RESP: Cough, asthma, bronchitis, dyspnea, sinusitis

PHARMACOKINETICS

Peak 1-2 hr fasting, 2-4 hr after food; protein binding 89%-95%; half-life 10-11 hr; metabolized by liver (metabolites); excreted in urine 33%

INTERACTIONS

Increase: hypotension—phenothiazines, nitrates, acute alcohol ingestion, diuretics, other antihypertensives
Increase: hyperkalemia—potassium-sparing diuretics, potassium supplements
Increase: myelosuppression—azaTHIOprine
Increase: serum levels of lithium, digoxin
Decrease: hypotensive effects—NSAIDs

Drug/Herb

Increase: antihypertensive effect—hawthorn
Decrease: antihypertensive effect—ephedra (Ma huang)

Drug/Lab Test

Increase: AST, ALT, alk phos, bilirubin, uric acid, blood glucose
Positive: ANA titer
False positive: ANA titer

NURSING CONSIDERATIONS

Assess:

- **Hypertension:** B/P, pulse at baseline, periodically; orthostatic hypotension, syncope when used with diuretic; notify prescriber of changes; monitor compliance
- **Blood dyscrasias:** neutrophils, decreased platelets; WBC with differential at baseline, q3mo; if neutrophils $<1000/mm^3$, discontinue treatment; recommended with **collagen-vascular disease**
- Renal studies: protein, BUN, creatinine; increased levels may indicate nephrotic syndrome; monitor urine for protein; LFTs, uric acid, glucose may be increased; diuretic should be discontinued 3 days before initiation of benazepril, if hypertension is not controlled, a diuretic can be added; measure B/P at peak 2-4 hr and trough (before next dose); this product is less effective in African descent patients
- Potassium levels, although hyperkalemia rarely occurs
- **Allergic reactions:** rash, fever, pruritus, urticaria; product should be discontinued if antihistamines fail to help; Stevens-Johnson syndrome; angioedema is more common in patients of African descent
- Renal symptoms: polyuria, oliguria, frequency, dysuria
- **CHF (unlabeled):** edema in feet, legs daily; weight daily

Evaluate:

- Therapeutic response: decrease in B/P

Teach patient/family:

- Not to use OTC products (cough, cold, allergy) unless directed by prescriber; not to use salt substitutes that contain potassium without consulting prescriber
- The importance of complying with dosage schedule, even if feeling better

Black Box Warning: To notify prescriber of pregnancy (D); product will need to be discontinued

- To rise slowly to sitting or standing position to minimize orthostatic hypotension
- To notify prescriber of mouth sores, sore throat, fever, swelling of hands or feet, irregular heartbeat, chest pain, bruising, bleeding, swelling of face, tongue, lips, difficulty breathing
- To report excessive perspiration, dehydration, vomiting, diarrhea; may lead to fall in B/P; to use caution in hot weather
- That product may cause dizziness, fainting, lightheadedness; that this may occur during first few days of therapy
- That product may cause skin rash or impaired perspiration
- How to take B/P, and normal readings for age-group
- To avoid potassium-containing products (salt substitutes)

TREATMENT OF OVERDOSE:
0.9% NaCl IV INF, hemodialysis

HIGH ALERT

bendamustine (Rx)

(ben-da-muss′teen)

Treanda

Func. class.: Antineoplastic alkylating agent

Chem. class.: Mechlorethamine derivative

ACTION: Cross-linking DNA that causes single strand and double strand breaks, inhibits several mitotic checkpoints, combines alkylating and antimetabolite properties

USES: Chronic lymphocytic leukemia, non-Hodgkin's lymphoma

Unlabeled uses: Mantle cell lymphoma (MCL)

CONTRAINDICATIONS: Pregnancy (D), fetal harm may occur; breastfeeding, children, hepatic disease, renal impairment, hypersensitivity to product or mannitol

Precautions: Hyperuricemia, infusion-related reactions, myelosuppression, infection, skin reactions

DOSAGE AND ROUTES

Chronic lymphocytic leukemia

• **Adult: IV INF** 100 mg/m² over 30 min on days 1, 2 q28days up to 6 cycles

Non-Hodgkin's lymphoma

• **Adult: IV INF** 120 mg/m² over 60 min on days 1, 2 q21days up to 8 cycles

Mantle cell lymphoma (unlabeled)

• **Adult: IV INF** 90 mg/m² on days 1, 2 with rituximab on day 1 q28days for 6 cycles

Renal/hepatic dose

• **Adult: IV INF** CCr <40 ml/min, do not use; AST or ALT 2.5-10 × upper limit normal or bilirubin 1.5-3 × ULN, do not use

Available forms: Powder for inj 25, 100 mg

Administer:

• Allopurinol for 1-2 wk to those at high risk for tumor lysis syndrome; usually develops in first treatment cycle

• Blood transfusions, RBC colony-stimulating factors to counter anemia

• Antiemetic 30-60 min before giving product to prevent vomiting

• All medications PO; if possible, avoid IM inj if platelets are <100,000/mm³

Intermittent IV INF route

• Prepare in biologic cabinet wearing gown, gloves, mask; avoid contact with skin, can cause burning, stain the skin brown; use cytotoxic handling procedures

• After **reconstituting** 100 mg product/20 ml or 25 mg/5 ml sterile water for inj (5 mg/ml), sol should be clear, colorless to pale yellow, completely dissolve in 5 min; if particulate is present, do not use

• Within 30 min of reconstitution, withdraw volume needed and **further dilute** in 500 ml NS or $D_{2.5/0.45}$%NS to a final conc of 0.2-0.6 mg/ml; doses of ≤100 mg/m², **give** over 30 min; doses of >100 mg/m², **give** over 60 min

• Monitor for inf reactions; may use antihistamines or corticosteroids for grade 1, 2 reactions; if grade 3 or 4 occurs, discontinue if needed

• Store reconstituted sol in refrigerator for 24 hr or at room temp for 3 hr; protect from light; store vials at room temp

SIDE EFFECTS

CNS: Asthenia, *fatigue,* fever, *headache,* chills, hypertension

CV: Hypertension, hypertensive crisis

GI: *Nausea, vomiting, diarrhea,* hyperbilirubinemia, *constipation,* stomatitis, *anorexia,* weight loss

GU: Renal failure

HEMA: Thrombocytopenia, leukopenia, anemia, lymphocytopenia, neutropenia, secondary malignancy, toxic epidermal necrolysis, tumor lysis syndrome

INTEG: *Bulbous rash, pruritus,* extravasation

META: Hyperuricemia

SYST: Anaphylaxis, infection, dehydration, severe skin toxicities, tumor lysis syndrome, Stevens-Johnson syndrome

PHARMACOKINETICS

95% protein binding, metabolized by hydrolysis via CYP450 1A2, 2 metabolites are produced, half-life 40 min, 90% excreted unchanged (feces)

INTERACTIONS

Increase: agranulocytosis risk—cloZAPine (do not use concurrently)

Increase: bleeding risk—aspirin, anticoagulants, NSAIDs, platelet inhibitors, thrombolytics

Increase: myelosuppression—myelosuppressive agents

Increase: toxicity—other antineoplastics, radiation

Increase: adverse reactions, decreased antibody reaction—live vaccines

Increase: bendamustine—CYP1A2 inhibitors (atazanavir, cimetidine, ciprofloxacin, enoxacin, ethyl estradiol, fluvoxaMINE, mexiletine, norfloxacin, tacrine, thiabendazole, zileuton)

Decrease: bendamustine—CYP1A2 inducers (barbiturates, carBAMazepine, rifampin)

Drug/Lab Test

Increase: LFTs

NURSING CONSIDERATIONS

Assess:

- **Blood dyscrasias:** CBC, differential, platelet count weekly; withhold product if WBC is <1000 or if platelet count is <75,000; notify prescriber of results
- Hepatic studies: AST, ALT, bilirubin
- Renal studies: BUN, serum uric acid, urine CCr before, during therapy; I&O ratio; report fall in urine output of 30 or 40 ml/hr; electrolytes
- Monitor for cold, cough, fever (may indicate beginning infection)
- For malignancy regression
- Bleeding: hematuria, guaiac, bruising, petechiae, mucosa, orifices q8hr
- **Serious skin toxicities:** toxic epidermal necrolysis, Stevens-Johnson syndrome; product should be discontinued
- **Tumor lysis syndrome:** monitor uric acid, potassium; may occur during 1st treatment cycle; use allopurinol for patients at high risk for this condition, usually during the 1st 2 wk; provide adequate hydration

Evaluate:

- Therapeutic response: improvement in blood counts, morphology

Teach patient/family:

- To avoid hazardous activity that requires mental alertness
- To avoid crowds, persons with upper respiratory infections
- To report immediately fever, sore throat, flulike symptoms; indicates infection
- To report immediately allergic reaction, facial swelling, difficulty breathing, itchy rash
- Until reaction is known, not to breastfeed; males should also use contraception during and for 3 mo after
- To use contraception during therapy and for 3 mo after pregnancy (D)
- To avoid use of aspirin, ibuprofen, razors, commercial mouthwash
- To report signs of anemia (fatigue, irritability, SOB, faintness)
- To report signs of infection, myelosuppression, skin toxicities, diarrhea, nausea, vomiting

benzocaine topical

See Appendix B

RARELY USED

benzonatate (Rx)

(ben-zoe′na-tate)

Tessalon Perles, Zonatuss

Func. class.: Antitussive, nonopioid

USES: Nonproductive cough

CONTRAINDICATIONS: Hypersensitivity

DOSAGE AND ROUTES

- **Adult and child: PO** 100 mg up to tid; max 600 mg/day

benztropine (Rx)

(benz′troe-peen)

Cogentin

Func. class.: Cholinergic blocker, antiparkinson's agent

Chem. class.: Tertiary amine

ACTION: Blockade of central acetylcholine receptors, balances cholinergic activity

USES: Parkinson's symptoms, EPS associated with neuroleptic products, acute dystonic reactions, hypersalivation

CONTRAINDICATIONS: Children <3 yr, hypersensitivity, closed-angle glaucoma, dementia, tardive dyskinesia

Precautions: Pregnancy (C), breastfeeding, geriatric patients, tachycardia, renal/hepatic disease, substance abuse history, dysrhythmias, hypo/hypertension, myasthenia gravis, GI/GU obstruction, peptic ulcer, megacolon, prostate hypertrophy, psychosis

DOSAGE AND ROUTES

Drug-induced EPS

- **Adult: IM/IV/PO** 1-4 mg daily/bid; give **PO** dose as soon as possible
- **Child >3 yr: IM/IV/PO** 0.02-0.05 mg/kg/dose 1-2×/day

Parkinson's symptoms

- **Adult: PO/IM** 0.5-1 mg at bedtime; increase by 0.5 mg q5-6days titrated to patient response, max 6 mg/day

Acute dystonic reactions

- **Adult: IM/IV** 1-2 mg, may increase to 1-2 mg bid **(PO)**

Available forms: Tabs 0.5, 1, 2 mg; inj 1 mg/ml

Administer:

PO route

- With or after meals to prevent GI upset; may give with fluids other than water
- At bedtime to avoid daytime drowsiness with parkinsonism
- Store at room temperature

IM route

- Inject deeply in muscle; use filtered needle to remove solution from ampule

IV, direct route

- **Undiluted** IV (1 mg 5 1 ml): **give** 1 mg/1 min (rarely used)

Syringe compatibilities: Metoclopramide, perphenazine

Y-site compatibilities: Fluconazole, tacrolimus

SIDE EFFECTS

CNS: Anxiety, restlessness, irritability, delusions, hallucinations, headache, sedation, depression, incoherence, dizziness, memory loss; *confusion;* delirium (geriatric patients)

CV: Palpitations, tachycardia, hypotension, bradycardia

EENT: Blurred vision, photophobia, dilated pupils, difficulty swallowing

GI: *Dryness of mouth, constipation,* nausea, vomiting, abdominal distress, paralytic ileus

GU: Urinary hesitancy/retention, dysuria

INTEG: Rash, urticaria, dermatoses

MISC: Increased temperature, flushing, decreased sweating, hyperthermia, heat stroke, numbness of fingers

PHARMACOKINETICS

PO: Onset 1 hr, duration 6-10 hr

IM/IV: Onset 15 min, duration 6-10 hr

INTERACTIONS

Increase: anticholinergic effect—amantadine; antihistamines, phenothiazines, tricyclics, disopyramide, quiNIDine; reduce dose

Decrease: absorption—antidiarrheals, antacids

Decrease: anticholinergic effect of—cholinergic

NURSING CONSIDERATIONS

Assess:

- **Parkinsonism:** EPS, shuffling gait, muscle rigidity, involuntary movements, loss of balance
- Paralytic ileus: abdominal pain, intermittent constipation/diarrhea
- I&O ratio; commonly causes decreased urinary output; urinary hesitancy, retention; palpate bladder if retention occurs
- Constipation: increase fluids, bulk, exercise if this occurs
- Mental status: affect, mood, CNS depression, worsening of mental symptoms during early therapy
- Use caution during hot weather; product may increase susceptibility to heat stroke by decreasing sweating
- With benztropine "buzz" or "high," patients may imitate EPS
- Hard candy, gum, frequent drinks to relieve dry mouth

Evaluate:
• Therapeutic response: absence of involuntary movements after 2 days of treatment
Teach patient/family:
• To report urinary hesitancy/retention, dysuria
• That tabs may be crushed, mixed with food; may take whole dose at bedtime if approved by prescriber
• Not to discontinue product abruptly; to taper off over 1 wk or withdrawal symptoms may occur (EPS, tremors, insomnia, tachycardia, restlessness); to take as directed; not to double dose
• To avoid driving, other hazardous activities; drowsiness/dizziness may occur
• To avoid OTC medications: cough, cold preparations with alcohol, antihistamines, antacids, antidiarrheals within 2 hr unless directed by prescriber
• To change positions slowly to prevent orthostatic hypotension
• To use good oral hygiene, frequent sips of water, sugarless gum for dry mouth

bepotastine
(beh-pot′uh-steen)
Bepreve
Func. class.: Antihistamine (ophthalmic)
Chem. class.: Histamine 1 receptor antagonist

ACTION: A topically active, direct H_1-receptor antagonist and mast cell stabilizer; by reducing these inflammatory mediators, relieves the ocular pruritus associated with allergic conjunctivitis

USES: Ocular pruritus associated with signs and symptoms of allergic conjunctivitis

CONTRAINDICATIONS: Hypersensitivity
Precautions: Pregnancy (C), breastfeeding, children, contact lenses

DOSAGE AND ROUTES
• **Adult/Child ≥2 yrs: OPHTH** Instill 1 drop in affected eye bid, max 2 drops/day in each eye
Available forms: Ophthalmic solution 1.5%
Administer
Ophthalmic route
• For topical ophthalmic use only
• Wash hands before and after use; tilt the head back slightly and pull the lower eyelid down with the index finger; squeeze the prescribed number of drops into the conjunctival sac and gently close eyes for 1-2 min; do not blink
• Do not touch the tip of the dropper to the eye, fingertips, or other surface
• Wait ≥10 min after instilling the ophthalmic solution before inserting contact lenses; contact lenses should not be worn if eye is red; the preservative in this product may be absorbed by soft contact lenses
• Do not share ophthalmic drops with others

SIDE EFFECTS
CNS: Headache
EENT: Taste change, ocular irritation, pharyngitis

NURSING CONSIDERATIONS
Assess:
• Eyes: for itching, redness, use of soft or hard contact lenses
Evaluate:
• Therapeutic response: absence of redness, itching in the eyes
Teach patient/family:
Ophthalmic route
• Product is for topical ophthalmic use only
• Wash hands before and after use; tilt the head back slightly and pull the lower eyelid down with the index finger; squeeze the prescribed number of drops into the conjunctival sac and gently close eyes for 1-2 min; do not blink
• Do not touch the tip of the dropper to the eye, fingertips, or other surface

• Wait ≥10 min after instilling the ophthalmic solution before inserting contact lenses; contact lenses should not be worn if eye is red
• Do not share ophthalmic drops with others
• Remove **contact lenses** before use because the preservative, benzalkonium chloride, may be absorbed by soft contact lenses; product should not be used to treat contact lens–related irritation

RARELY USED

beractant (Rx)

(ber-ak′tant)

Survanta

Func. class.: Natural lung surfactant

USES: Prevention and treatment (rescue) of respiratory distress syndrome in premature infants

DOSAGE AND ROUTES

• **Newborn: INTRATRACHEAL INSTILL** 4 doses can be administered during the 1st 48 hr of life; give doses no more frequently than q6hr; each dose is 100 mg of phospholipids/kg birth weight

betamethasone (topical)

(bay-ta-meth′a-sone)

betamethasone dipropionate

Diprolene, Diprolene AF

betamethasone augmented dipropionate

betamethasone valerate

Beta-Val, Luxiq

Func. class.: Corticosteroid, topical

ACTION: Crosses cell membrane to attach to receptors to decrease inflammation, itching; inhibits multiple inflammatory cytokines

USES: Inflammation/itching corticosteroid-responsive dermatoses on the skin/scalp

CONTRAINDICATIONS: Hypersensitivity, use of some preparations on face, axilla, groin

Precautions: Pregnancy (C), skin infections

DOSAGE AND ROUTES

• **Adult: TOP** 1-2 times/day (dipropionate) or 1-3 times/day (valerate)

Available forms: dipropionate: gel, lotion, ointment, cream 0.05%; valerate: cream, lotion, ointment 0.1%, foam 0.12%

SIDE EFFECTS

INTEG: Burning, folliculitis, pruritus, dermatitis, maceration, erythema

MISC: Hyperglycemia; glycosuria, Cushing syndrome, HPA axis suppression

PHARMACOKINETICS

Unknown

NURSING CONSIDERATIONS

Assess:

• Skin reactions: burning pruritus, folliculitis, dermatitis

Evaluate:

• Decreased itching, inflammation on the skin, scalp

Teach patient/family:

Topical route:

• That betamethasone valerate may be used with occlusive dressings for psoriasis or recalcitrant conditions; not to use dipropionate with occlusive dressings

Cream/ointment/lotion:

• To apply sparingly in a thin film, using gloves, and rub gently into the cleansed, slightly moist affected area

Gel:
• To apply sparingly in a thin film, using gloves, and rub gently into the cleansed, slightly moist affected area
Scalp foam:
• To invert can and dispense a small amount of foam onto a saucer or other cool surface; not to dispense directly onto hands; to pick up small amounts of foam with fingers and gently massage into affected area until foam disappears; repeat until entire affected scalp area is treated
• That treatment should be limited to 2 wk

betamethasone (augmented) topical

See Appendix B

betaxolol (ophthalmic)

(beh-tax′oh-lol)

Betoptic-S

Func. class.: Antiglaucoma
Chem. class.: Beta blocker

ACTION: Can decrease aqueous humor and increase outflows

USES: Treatment of chronic open-angle glaucoma and ocular hypertension

CONTRAINDICATIONS: Hypersensitivity, AV block, heart failure, bradycardia, sick sinus syndrome
Precautions: Abrupt discontinuation, children, pregnancy, breastfeeding, asthma, COPD, depression, diabetes mellitus, myasthenia gravis, hyperthyroidism, pulmonary disease, angle-closure glaucoma

DOSAGE AND ROUTES

• **Adult: IV INF** Instill 1-2 drops in the affected eye(s) bid
Available forms: Ophthalmic sol 0.5%; ophthalmic susp 0.25%

SIDE EFFECTS

CNS: Insomnia, headache, dizziness
CV: Palpitations
EENT: Eye stinging/burning, tearing, photophobia
MISC: Bronchospasm

PHARMACOKINETICS

Onset 30 min, peak 2 hr, duration ≥12 hr

NURSING CONSIDERATIONS

Assess:
⚠ **Systemic absorption:** When used in the eye, systemic absorption is common, with the same adverse reactions and interactions
• Glaucoma: monitor intraocular pressure
Evaluate:
• Decreasing intraocular pressure
Teach patient/family:
• That strength is expressed in betaxolol base
• That drug is for ophthalmic use only; shake the ophthalmic suspension well before use
• Not to touch the tip of the dropper to the eye, fingertips, or other surface to prevent contamination
• To wash hands before and after use; tilt the head back slightly and pull the lower eyelid down with the index finger to form a pouch; squeeze the prescribed number of drops into the pouch; close eyes to spread drops; to avoid excessive systemic absorption, apply finger pressure on the lacrimal sac for 1-2 min following use
• That if more than one topical ophthalmic drug product is being used, the drugs should be administered at least 5 min apart
• To avoid contamination or the spread of infection, do not use dropper for more than one person

bethanechol (Rx)

(be-than′e-kole)

Urecholine

Func. class.: Urinary tract stimulant, cholinergic

Chem. class.: Synthetic choline ester

ACTION: Stimulates muscarinic ACH receptors directly; mimics effects of parasympathetic nervous system stimulation; stimulates gastric motility, micturition; increases lower esophageal sphincter pressure

USES: Urinary retention (postoperative, postpartum), neurogenic atony of bladder with retention

Unlabeled uses: Ileus, GERD, anticholinergic syndrome

CONTRAINDICATIONS: Hypersensitivity, severe bradycardia, asthma, severe hypotension, hyperthyroidism, peptic ulcer, parkinsonism, seizure disorders, CAD, COPD, coronary occlusion, mechanical obstruction, peritonitis, recent urinary/GI surgery, GI/GU obstruction

Precautions: Pregnancy (C), breastfeeding, children <8 yr, hypertension

DOSAGE AND ROUTES

- **Adult: PO** 10-50 mg bid-qid
- **Child (unlabeled): PO** 0.6 mg/kg/day in 3-4 divided doses

Ileus (unlabeled)

- **Adult: PO** 10-20 mg tid-qid before meals

Available forms: Tabs 5, 10, 25, 50 mg

Administer:

- To avoid nausea, vomiting, take on an empty stomach
- Only after all other cholinergics have been discontinued
- Store at room temp

SIDE EFFECTS

CNS: Dizziness, headache, malaise

CV: Hypotension, bradycardia, reflex tachycardia, cardiac arrest, circulatory collapse

EENT: Miosis, increased salivation, lacrimation, blurred vision

GI: *Nausea, bloody diarrhea, belching, vomiting, cramps, fecal incontinence*

GU: Urgency

INTEG: Rash, urticaria, flushing, increased sweating

RESP: Acute asthma, dyspnea, bronchoconstriction

PHARMACOKINETICS

PO: Onset 30-90 min, duration 6 hr

INTERACTIONS

Increase: severe hypotension—ganglionic blockers

Increase: action or toxicity—cholinergic agonists, anticholinesterase agents

Decrease: action of anticholinergics, procainamide, quiNIDine

Drug/Lab Test

Increase: AST, lipase/amylase, bilirubin

NURSING CONSIDERATIONS

Assess:

- **Urinary patterns:** retention, urgency
- B/P, pulse: observe after parenteral dose for 1 hr; may need to use atropine subcut 0.6 mg or IV push slowly for bronchoconstriction
- I&O ratio: check for urinary retention, urge incontinence
- **Toxicity:** bradycardia, hypotension, bronchospasm, headache, dizziness, seizures, respiratory depression; product should be discontinued if toxicity occurs

Evaluate:

- Therapeutic response: absence of urinary retention, abdominal distention

Teach patient/family:

- To take product exactly as prescribed; 1 hr before meals or 2 hr after meals
- To make position changes slowly; orthostatic hypotension may occur

• To avoid driving, hazardous activities until effects are known

TREATMENT OF OVERDOSE:
Administer atropine 0.6-1.2 mg IV or IM (adult)

⚠ HIGH ALERT

bevacizumab (Rx)
(beh-va-kiz′you-mab)

Avastin

Func. class.: Antineoplastic—miscellaneous

Chem. class.: Monoclonal antibody

Do not confuse:
Avastin/Astelin

ACTION:
DNA-derived monoclonal antibody selectively binds to and inhibits activity of human vascular endothelial growth factor (VEGF) to reduce microvascular growth and metastatic disease progression

USES:
Non–small-cell lung cancer (NSCLC), metastatic carcinoma of the colon or rectum, renal cell carcinoma, glioblastoma

Unlabeled uses: Adjunctive for pancreatic/neovascular/ovarian cancer; (wet) macular degeneration

CONTRAINDICATIONS:
Hypersensitivity

Precautions: Pregnancy (C), breastfeeding, children, geriatric patients, CHF, blood dyscrasias, CV disease, hypertension, surgery, thromboembolic disease, hamster protein/murine hypersensitivity

Black Box Warning: GI perforation, wound dehiscence

DOSAGE AND ROUTES

Non–small-cell lung cancer

• **Adult: IV** 15 mg/kg over 60-90 min with CARBOplatin and paclitaxel q3wk

Colorectal cancer

• **Adult: IV INF** in combination with 5-fluorouracil 5-10 mg/kg q14days given over 90 min; if well tolerated, next inf may be given over 60 min; if 60-min infs well tolerated, subsequent inf may be given over 30 min; (second line) 5 mg/kg q2wk or 7.5 mg/kg q3wk with fluoropyrimidine and irinotecan or fluoropyramide and oxaliplatin-based agent

Metastatic renal cell carcinoma

• **Adult: IV** 10 mg/kg q2wk with interferon alfa 9 million units SUBCUT 3×/wk up to 52 wk

Glioblastoma single agent

• **Adult: IV** 10 mg/kg q2wk given over 60-90 min; 28-day cycle

Metastatic Breast Cancer (unlabeled)

• **Adult: IV** 15 mg/kg on day 1 with docetaxel 100 mg/m^2 q3wk, up to 9 cycles (those who have not received previous chemotherapy)

Metastatic renal cell cancer (unlabeled)

• **Adult (single agent): IV** 10 mg/kg over 60-90 min q2wk; may be given in combination with other products

Ovarian cancer (unlabeled)

• **Adult: IV** 15 mg/kg q21days until unacceptable toxicity, disease progression

Neovascular (wet) macular degeneration (unlabeled)

• **Adult: INTRAVITREOUS INJ** 1.25 mg monthly

Available forms: Inj 25 mg/ml

Administer:

Intermittent IV INF route

• Do not give by IV bolus, IV push; do not shake vial

• Withdraw amount of product to be given, dilute in 100 ml 0.9% NaCl, discard any unused portion

Black Box Warning: Wound dehiscence: do not give for ≥28 days after surgery; make sure wound is healed before giving product

• Give as IV inf over 90 min for 1st dose and 60 min thereafter if well tolerated; subsequent inf may be given over 30 min; do not admix with dextrose

• **Rapid infusion rate (unlabeled):** give at rate of 0.5 mg/kg/min for all doses including initial inf (5 mg/kg over 10 min; 10 mg/kg over 20 min; 15 mg/kg over 30 min)

SIDE EFFECTS

CNS: *Asthenia, dizziness,* intracranial hemorrhage (malignant glioma), headache, fatigue, confusion
CV: Deep vein thrombosis, hypo/hypertension, hypertensive crisis, heart failure
GI: Nausea, vomiting, *anorexia, diarrhea,* constipation, *abdominal pain,* colitis, stomatitis, GI hemorrhage/perforation
GU: Proteinuria, urinary frequency/urgency, nephrotic syndrome, ovarian failure
HEMA: Leukopenia, neutropenia, thrombocytopenia, microangiopathic hemolytic anemia, thromboembolism, bleeding
META: Bilirubinemia, hypokalemia
MISC: Exfoliative dermatitis, hemorrhage, non-GI fistula formation, *alopecia, impaired wound healing,* osteonecrosis of the jaw, antibody formation
RESP: Dyspnea, upper respiratory tract infection

PHARMACOKINETICS

Half-life 20 days, steady-state 100 days

INTERACTIONS

• Avoid concurrent use with SUNItinib; microangiopathic hemolytic anemia may occur

NURSING CONSIDERATIONS

Assess:

• B/P; take frequently if hypertension develops
• For symptoms of infection; may be masked by product
• **CNS reaction:** dizziness, confusion
• **CHF:** crackles, jugular venous distention, dyspnea during treatment
⚠ **GU status** (proteinuria): nephrotic syndrome may occur; monitor urinalysis for increasing protein level; product should be held if protein ≥2 g/24 hr; resume when <2 g/24 hr

Black Box Warning: Wound dehiscence: Hold for ≥28 days until incision is healed

Black Box Warning: **GI perforation, serious bleeding, nephrotic syndrome, hypertensive crisis;** product should be discontinued permanently, surgery should be postponed

• Reversible posterior leukoencephalopathy syndrome (RPLS): discontinue if this disorder develops

Evaluate:

• Therapeutic response: decrease in size of tumors

Teach patient/family:

• To avoid hazardous tasks because confusion, dizziness may occur
• To report signs of infection: sore throat, fever, diarrhea, vomiting
• Not to become pregnant while taking this product or for several months after discontinuing treatment
• To report bleeding, changes in urinary patterns, edema, abdominal pain
• To avoid immunizations

⚠ *HIGH ALERT*

bicalutamide (Rx)

(bye-kal-u'ta-mide)

Casodex

Func. class.: Antineoplastic hormone
Chem. class.: Nonsteroidal antiandrogen

Do not confuse:
Casodex/Kapidex

ACTION: Binds to cytosolic androgen in target tissue, which competitively inhibits the action to androgens

USES: Stage D-2 metastatic prostate cancer in combination with luteinizing-hormone releasing hormone (LHRH) analog
Unlabeled uses: Recurrent priapism

CONTRAINDICATIONS: Pregnancy (X), women, hypersensitivity

Precautions: Breastfeeding, geriatric patients, renal/hepatic disease, diabetes mellitus

DOSAGE AND ROUTES

- **Adult: PO** 50 mg/day with LHRH analog

Recurrent priapism (unlabeled)

- **Adult: PO** 50 mg every other day

Available forms: Tabs 50 mg

Administer:

- At same time each day, either AM or PM, with/without food
- With LHRH treatment; start both products at same time

SIDE EFFECTS

CNS: Dizziness, paresthesia, insomnia, anxiety, neuropathy, headache

CV: **CHF**, *edema, hot flashes,* hypertension, chest pain, **MI**

GI: *Diarrhea, constipation, nausea,* vomiting, increased hepatic enzymes, anorexia, dry mouth, melena, *abdominal pain,* **hepatitis, hepatotoxicity**

GU: *Nocturia, hematuria,* UTI, impotence, gynecomastia; urinary incontinence, frequency, dysuria, retention, urgency; breast tenderness, decreased libido

INTEG: Rash, sweating, dry skin, pruritus, alopecia

MISC: *Infection,* anemia, dyspnea, bone pain, headache, *asthenia, back pain,* flulike symptoms

PHARMACOKINETICS

Well absorbed; peak 31½ hr; metabolized by liver; excreted in urine, feces; half-life 5.8 days; 96% protein binding

INTERACTIONS

Increase: anticoagulation—anticoagulants

Increase: bicalutamide effects—CYP3A4 inhibitors (amiodarone, antiretrovirals, protease inhibitors, clarithromycin, dalfopristin, quinupristin, delavirdine, efavirenz, erythromycin, FLUoxetine, fluvoxaMINE, imatinib, mifepristone, RU-486, nefazodone, some azole antifungals)

Decrease: bicalutamide effects—CYP3A4 inducers (barbiturates, bosentan, carBAMazepine, dexamethasone, nevirapine, OXcarbazepine, phenytoins, rifabutin, rifampin, rifapentine)

Drug/Herb

- May require dosage change when used with St. John's wort

Drug/Food

- Do not use with grapefruit juice

Drug/Lab Test

Increase: AST, ALT, bilirubin, BUN, creatinine

Decrease: Hgb, WBC

NURSING CONSIDERATIONS

Assess:

- For diarrhea, constipation, nausea, vomiting
- For hot flashes, gynecomastia; assure patient that these are common side effects
- Prostate-specific antigen, LFTs

Evaluate:

- Therapeutic response: decreased tumor size, decreased spread of malignancy

Teach patient/family:

- To recognize, report signs of anemia, hepatoxicity, renal toxicity
- That hair may be lost but this is reversible after therapy is completed
- Not to use other products unless approved by prescriber
- To report severe diarrhea
- To use contraception while taking this product

bimatoprost (ophthalmic/topical)

(by-mat′oh-prost)

Latisse, Lumigan

Func. class.: Antiglaucoma agent

Chem. class.: Prostaglandin agonist

Do not confuse:

latanoprost/travoprost

ACTION:

Latisse

Promotion of eyelash growth, thickness, and darkness: unknown; possible increase in the percent of hairs and an increase in the duration of the hair growth (anagen) phase

Lumigan

Reduction of intraocular pressure (IOP) in patients with ocular hypertension or open-angle glaucoma; selectively mimics endogenous prostamides to produce ocular hypotension

USES:
Increased intraocular pressure in those with open-angle glaucoma/ocular hypertension (Lumigan); eyelash hypotrichosis (Latisse)

CONTRAINDICATIONS:
Hypersensitivity to this product, benzalkonium chloride

Precautions: Children, intraocular inflammation, closed-angle glaucoma, macular edema, contact lenses

DOSAGE AND ROUTES

Increased intraocular pressure/ ocular hypertension (Lumigan)

- **Adult: OPHTH Instill** 1 drop in each affected eye (conjunctival sac) every night

Eyelash Hypotrichosis (Latisse)

- **Adult: Apply** 1 drop to skin of upper eyelid margin at base of eyelashes every night using a new supplied disposable sterile applicator

Available forms: Ophthalmic solution 0.01%, 0.03%; topical solution 0.03%

SIDE EFFECTS

EENT: *Conjunctival hyperemia, growth of eyelashes (hypertrichosis), ocular pigment changes ocular pruritus*, xerophthalmia, visual disturbance, ocular irritation/burning, foreign body sensation, ocular pain, blepharitis, cataracts, superficial punctate keratitis

INTEG: Hyperpigmentation of the periocular skin, eyelash darkening, lacrimation, photophobia, conjunctivitis, asthenopia, iritis, macular edema

MISC: Influenza, upper respiratory tract infections, asthenia, headache, hirsutism

PHARMACOKINETICS

Ophthalmic: Onset 4 hr, peak 8-12 hr; half-life 45 min

INTERACTIONS

Decrease: Intraocular pressure lowering effect—latanoprost, travoprost (no longer available in the U.S.)

Drug/Lab Test

Increase: LFTs

NURSING CONSIDERATIONS

Assess:

- **Intraocular pressure:** in those with ongoing increased IOP or those using latanoprost, travoprost (no longer available in the U.S.)

Evaluate:

- Decreasing IOP or increased growth of eyelashes

Teach patient/family:

Ophthalmic route (Lumigan):

- To wash hands before and after use; remove contact lenses before use and reinsert 15 min after use; Lumigan contains benzalkonium chloride, which can be absorbed by soft contact lenses
- To tilt the head back slightly and pull the lower eyelid down with the index finger to form a pouch; squeeze the prescribed number of drops into the pouch and gently close the eyes for 1-2 min; do not blink; to avoid contamination, do not touch the tip of the dropper to the eye, fingertips, or other surface
- That the solution may be used concomitantly with other topical ophthalmic drug products to lower IOP; if more than one topical ophthalmic drug is being used, the drugs should be administered at least 5 min apart

Topical route (Latisse):

• To ensure the patient's face is clean and makeup is removed before using Latisse; the disposable sterile applicator is the only applicator that should be used; each applicator should be used for 1 eye only; dispose of the applicator after each use; after applying 1 drop of solution to the applicator, apply evenly along the skin of the upper eyelid margin at the base of the eyelashes; blot excess solution runoff outside the upper eyelid margin with a tissue or other absorbent cloth; do not apply to the lower eyelash line

bisacodyl (Rx, OTC)

(bis-a-koe′dill)

Alophen, Correctol, Dacodyl, Dulcolax, Ex-Lax Ultra Tab, Femilax, Feminine, Good Sense Women's, Leader Laxative, Top Care Laxative, Veracolate, Walgreens Gentle, Walgreens Women's

Func. class.: Laxative, stimulant

Chem. class.: Diphenylmethane

ACTION: Acts directly on intestine by increasing motor activity; thought to irritate colonic intramural plexus

USES: Short-term treatment of constipation; bowel or rectal preparation for surgery, examination

CONTRAINDICATIONS: Hypersensitivity, rectal fissures, abdominal pain, nausea, vomiting, appendicitis, acute surgical abdomen, ulcerated hemorrhoids, acute hepatitis, fecal impaction, intestinal/biliary tract obstruction

Precautions: Pregnancy (C), breastfeeding

DOSAGE AND ROUTES

• **Adult and child ≥12 yr: PO** 10-15 mg in PM or AM; may use up to 30 mg for bowel or rectal preparation; **RECT** 10 mg as a single dose; 30-ml enema

• **Child 3-11 yr: PO** 5-10 mg as a single dose; **RECT** 5-10 mg as a single dose

Available forms: Tabs del rel 5, 10 mg; enteric-coated tabs 5 mg; supp 5, 10 mg; enema 10 mg/30 ml

Administer:

PO route

• Swallow tabs whole with full glass of water; do not break, crush, chew tabs

• Alone only with water for better absorption; do not take within 1 hr of other products or within 1 hr of antacids, milk, H_2 antagonists; do not take enteric product with proton-pump inhibitors

• In AM or PM

Rectal route

• Insert high in rectum

SIDE EFFECTS

CNS: Muscle weakness

GI: *Nausea, vomiting, anorexia, cramps,* diarrhea, rectal burning (suppositories)

META: Protein-losing enteropathy, alkalosis, hypokalemia, tetany; electrolyte, fluid imbalances

PHARMACOKINETICS

Small amounts metabolized by liver; excreted in urine, bile, feces, breast milk

PO: Onset 6-10 hr

RECT: Onset 15-60 min

INTERACTIONS

Increase: gastric irritation—antacids, milk, H_2-blockers, gastric acid pump inhibitors

Drug/Food

• Increase irritation—dairy products separate by 2 hr

Drug/Lab

Increase: Sodium phosphate

Decrease: Calcium, magnesium

NURSING CONSIDERATIONS

Assess:

- Blood, urine electrolytes if product is used often by patient
- I&O ratio to identify fluid loss
- Cause of constipation; identify whether fluids, bulk, exercise missing from lifestyle; determine use of constipating products
- **GI symptoms:** cramping, rectal bleeding, nausea, vomiting; if these symptoms occur, product should be discontinued
- Multiple products/routes may be used for bowel prep

Evaluate:

- Therapeutic response: decrease in constipation

Teach patient/family:

- Not to use laxatives for long-term therapy because bowel tone will be lost; 1-wk use is usually sufficient
- That normal bowel movements do not always occur daily
- Not to use in presence of abdominal pain, nausea, vomiting
- To notify prescriber if constipation is unrelieved or if symptoms of electrolyte imbalance occur: muscle cramps, pain, weakness, dizziness
- To take with a full glass of water; not to take with dairy products
- Identify bulk, water, constipating products, exercise in patient's life

bismuth subsalicylate (OTC)

(bis′muth sub-sal-iss′uh-late)

Bismatrol, Equaline Stomach Relief, Good Sense Stomach Relief, GNP Pink Bismuth, Leader Pink Bismuth, Peptic Relief, Pepto-Bismol, Pink Bismuth, Top Care Stomach Relief, Walgreens Soothe

Func. class.: Antidiarrheal, weak antacid

Chem. class.: Salicylate

Do not confuse:

Kaopectate/Kayoxalate

ACTION: Inhibits the prostaglandin synthesis responsible for GI hypermotility, intestinal inflammation; stimulates absorption of fluid and electrolytes; binds toxins produced by *Escherichia coli*

USES: Diarrhea (cause undetermined), prevention of diarrhea when traveling; may be included to treat *Helicobacter pylori,* heartburn, indigestion, nausea

CONTRAINDICATIONS: Children <3 yr, children with chickenpox, history of GI bleeding, renal disease, flu-like symptoms, hypersensitivity to product or salicylates

Precautions: Pregnancy (C), breastfeeding, geriatric patients, anticoagulant therapy, immobility, gout, diabetes mellitus, bleeding disorders, previous hypersensitivity to NSAIDs, *Clostridium-difficile*–associated diarrhea when used with antiinfectives for *H. pylori*

DOSAGE AND ROUTES

Antidiarrheal/Gastric Distress

- **Adult: PO** 2 tabs or 30 ml (15 ml extra/max strength) q30min or 2 tabs q60min, max 4.2 g/24 hr

Antiulcer (unlabeled)

• **Adult/adolescent: PO** 525 mg qid, max 4.2 g/24 hr; given with metroNIDAZOLE or tetracycline and acid-suppressive therapy × 14 days

Available forms: Tabs 262 mg; chewable tabs 262 mg; susp 87 mg/5 ml, 130 mg/15 ml, 262 mg/15 ml, 525 mg/15 ml

Administer:

PO route

• Increased fluids to rehydrate patient
• **Susp:** shake liquid before using
• Tabs can be chewed, dissolved in mouth; caplets to be swallowed whole with water

SIDE EFFECTS

CNS: Confusion, twitching, **neurotoxicity (high doses)**

EENT: Hearing loss, tinnitus, metallic taste, blue gums, black tongue

GI: Increased fecal impaction (high doses), dark stools, constipation, diarrhea, nausea

HEMA: Increased bleeding time

PHARMACOKINETICS

PO: Onset 1 hr, peak 2 hr, duration 4 hr

INTERACTIONS

Increase: toxicity—salicylates, methotrexate

Increase: effects of oral anticoagulants, oral antidiabetics

Decrease: absorption of tetracycline, quinolones, separate for ≥2 hr

Drug/Lab Test

Interference: radiographic studies of GI system

NURSING CONSIDERATIONS

Assess:

• **Diarrhea:** bowel pattern before product therapy, after treatment
• Electrolytes potassium, sodium, chloride if diarrhea is severe or continues long term; assess skin turgor, other signs of dehydration

Evaluate:

• Therapeutic response: decreased diarrhea, absence of diarrhea when traveling; resolution of ulcers

Teach patient/family:

• To chew, dissolve medication in mouth; not to swallow whole; to shake liquid before using
• To avoid other salicylates unless directed by prescriber; not to give to children, possibility of Reye's syndrome
• That stools may turn black; that tongue may darken; that impaction may occur in debilitated patients
• To stop use if symptoms do not improve within 2 days or become worse or if diarrhea is accompanied by high fever

bisoprolol (Rx)

(bis-oh′pro-lole)

Zebeta

Func. class.: Antihypertensive
Chem. class.: β_1-Blocker

Do not confuse:

Zebeta/DiaBeta/Zetia

ACTION:

Preferentially and competitively blocks stimulation of β_1-adrenergic receptors within cardiac muscle (decreases rate of SA node discharge, increases recovery time), slows conduction of AV node, decreases heart rate, which decreases O_2 consumption in myocardium; decreases renin-angiotensin-aldosterone system; inhibits β_2-receptors in bronchial and vascular smooth muscle at high doses

USES:

Mild to moderate hypertension

Unlabeled uses: Stable angina, stable CHF

CONTRAINDICATIONS:

Hypersensitivity to β-blockers, cardiogenic shock, heart block (2nd, 3rd degree), sinus bradycardia, CHF, cardiac failure

Precautions: Pregnancy (C), breastfeeding, children, major surgery, diabetes mellitus, thyroid/renal/hepatic disease, COPD, asthma, well-compensated heart failure, aortic or mitral valve disease, peripheral vascular disease, myasthenia gravis

Black Box Warning: Abrupt discontinuation

DOSAGE AND ROUTES

Hypertension

• **Adult: PO** 5 mg/day; reduce to 2.5 mg in bronchospastic disease; may increase to 20 mg/day if necessary; max 20 mg/day

Renal/hepatic dose

• **Adult: PO** CCr <40 ml/min 2.5 mg, titrate upward

Angina (unlabeled)

• **Adult: PO** 5-20 mg/day

Heart failure (unlabeled)

• **Adult: PO** 1.25 mg/day × 48 hr, then 2.5 mg/day for 1st mo, then 5 mg/day; max 10 mg/day

Available forms: Tabs 5, 10 mg

Administer:

• Tab may be crushed, swallowed whole; may give without regard to meals

• Reduced dosage with renal/hepatic dysfunction

• Store protected from light, moisture; place in cool environment

SIDE EFFECTS

CNS: Vertigo, headache, insomnia, fatigue, dizziness, mental changes, memory loss, hallucinations, depression, lethargy, drowsiness, strange dreams, catatonia, peripheral neuropathy

CV: Ventricular dysrhythmias, profound hypotension, bradycardia, CHF, cold extremities, postural hypotension, 2nd- or 3rd-degree heart block

EENT: Sore throat; dry, burning eyes

ENDO: Increased hypoglycemic response to insulin

GI: Nausea, diarrhea, vomiting, mesenteric arterial thrombosis, ischemic colitis, flatulence, gastritis, gastric pain

GU: Impotence, decreased libido

HEMA: Agranulocytosis, thrombocytopenia, purpura, eosinophilia

INTEG: Rash, flushing, alopecia, pruritus, sweating

MISC: Facial swelling, weight gain, decreased exercise tolerance

MS: Joint pain, arthralgia

RESP: Bronchospasm, dyspnea, wheezing, cough, nasal stuffiness

PHARMACOKINETICS

Peak 2-4 hr, half-life 9-12 hr, 50% excreted unchanged in urine, protein binding 30%-36%, metabolized in liver to inactive metabolites

INTERACTIONS

Increase: hypotension—reserpine, guanethidine

Increase: myocardial depression—calcium channel blockers

Increase: antihypertensive effect—ACE inhibitors, α-blockers, calcium channel blockers, diuretics

Increase: bradycardia—digoxin, amiodarone

Increase: peripheral ischemia—ergots

Increase: antidiabetic effect—antidiabetics; may make hypoglycemic symptoms

Decrease: antihypertensive effect—NSAIDs, salicylates

Drug/Herb

Increase: β-blocking effect—hawthorn

Decrease: β-blocking effect—ephedra

Drug/Lab Test

Increase: AST, ALT, ANA titer, blood glucose, BUN, uric acid, potassium, lipoprotein

Interference: glucose/insulin tolerance tests

NURSING CONSIDERATIONS

Assess:

• **Hypertension:** B/P during beginning treatment, periodically thereafter; pulse q4hr: note rate, rhythm, quality; apical/radial pulse before administration; notify prescriber of any significant changes (pulse <50 bpm)

• Baselines of renal, hepatic studies before therapy begins

• **CHF:** I&O, weight daily; increased weight, jugular venous distention, dyspnea, crackles, edema in feet, legs daily
• Skin turgor, dryness of mucous membranes for hydration status, especially for geriatric patients

Evaluate:

• Therapeutic response: decreased B/P after 1-2 wk

Teach patient/family:

Black Box Warning: Not to discontinue product abruptly; may cause precipitate angina, rebound hypertension; evaluate noncompliance

• Not to use OTC products that contain α-adrenergic stimulants (e.g., nasal decongestants, OTC cold preparations) unless directed by prescriber
• To report bradycardia, dizziness, confusion, depression, fever, cold extremities
• To take pulse at home; advise when to notify prescriber
• To avoid alcohol, smoking, sodium intake
• To comply with weight control, dietary adjustments, modified exercise program
• To carry emergency ID to identify product, allergies
• To avoid hazardous activities if dizziness is present

⚠ To report symptoms of CHF: difficulty breathing, especially on exertion or when lying down, night cough, swelling of extremities

• That if diabetic, product may mask signs of hypoglycemia or alter blood glucose levels

TREATMENT OF OVERDOSE:

Lavage, IV atropine for bradycardia; IV theophylline for bronchospasm; digoxin, O_2, diuretic for cardiac failure; hemodialysis, IV glucose for hypoglycemia; IV diazepam or phenytoin for seizures

⚠ HIGH ALERT

bivalirudin (Rx)

(bye-val-i-rue′din)

Angiomax

Func. class.: Anticoagulant

Chem. class.: Thrombin inhibitor

ACTION:

Direct inhibitor of thrombin that is highly specific; able to inhibit free and clot-bound thrombin

USES:

Unstable angina in patients undergoing percutaneous transluminal coronary angioplasty (PTCA), used with aspirin; heparin-induced thrombocytopenia, with/without thrombosis syndrome

Unlabeled uses: Acute MI, DVT prophylaxis

CONTRAINDICATIONS:

Hypersensitivity, active bleeding, cerebral aneurysm, intracranial hemorrhage, recent surgery, CVA

Precautions: Pregnancy (B), breastfeeding, children, geriatric patients, renal function impairment, hepatic disease, asthma, blood dyscrasias, thrombocytopenia, GI ulcers, hypertension, inflammatory bowel disease, vitamin K deficiency

DOSAGE AND ROUTES

• **Adult: IV BOL** 0.75 mg/kg then **IV INF** 1.75 mg/kg/hr for 4 hr; another **IV INF** may be used at 0.2 mg/kg/hr for ≤20 hr; this product is intended to be used with aspirin (325 mg/day) adjusted to body weight

Renal dose

• **Adult: IV** CCr ≥30 ml/min no adjustment; CrL 10-29 ml/min consider reducing to 1 mL/kg/hr

Acute MI (unlabeled)

• **Adult: IV BOL** 0.25 mg/kg then **CONT IV INF** 0.5 mg/kg/hr × 12 hr

DVT prophylaxis (unlabeled)

• **Adult: SUBCUT** 1 mg/kg q8hr for those undergoing orthopedic surgery

Available forms: Inj, lyophilized 250 mg/vial

Administer:

• Before PTCA; give with aspirin (325 mg)

• Store reconstituted vials in refrigerator for up to 24 hr; store diluted conc at room temp for 24 hr

IV, direct route

• Dilute by adding 5 ml of sterile water for inj/250 mg bivalirudin, swirl until dissolved, further dilute in 50 ml of D_5W or 0.9% NaCl (5 mg/ml), give by bolus inj 0.75 mg/kg, then intermittent inf

Continuous IV INF route

• To each 250-mg vial add 5 ml of sterile water for inj, swirl until dissolved, further dilute in 500 ml D_5W or 0.9% NaCl (0.5 mg/ml); give inf after bolus dose at a rate of 1.75 mg/kg/hr; may give an additional inf at 0.2 mg/kg/hr

• Do not mix other IV medications with bivalirudin or provide via the same IV line as bivalirudin

Y-site compatibilities: Abciximab, acyclovir, alfentanil, allopurinol, amifostine, amikacin, aminocaproic acid, aminophylline, amphotericin B liposome, ampicillin, ampicillin-sulbactam, anidulafungin, argatroban, arsenic trioxide, atenolol, atracurium, atropine, azithromycin, aztreonam, bleomycin, bumetanide, buprenorphine, busulfan, butorphanol, calcium chloride/gluconate, capreomycin, CARBOplatin, carmustine, ceFAZolin, cefepime, cefoperazone, cefotaxime, cefoTEtan, cefOXitin, cefTAZidime, ceftizoxime, cefTRIAXone, cefuroxime, chloramphenicol, cimetidine, ciprofloxacin, cisatracurium, CISplatin, clindamycin, cyclophosphamide, cycloSPORINE, cytarabine, dacarbazine, DACTINomycin, DAPTOmycin, DAUNOrubicin, DAUNOrubicin liposome, dexamethasone, dexmedetomidine, dexrazoxane, digoxin, diltiazem, diphenhydrAMINE, DOCEtaxel, dolasetron, DOPamine, DOXOrubicin, DOXOrubicin liposomal, doxycycline, droperidol, enalaprilat, ePHEDrine, EPINEPHrine, epirubicin, epoprostenol, eptifibatide, ertapenem, erythromycin, esmolol, etoposide, etoposide phosphate, famotidine, fenoldopam, fentaNYL, fluconazole, fludarabine, fluorouracil, foscarnet, fosphenytoin, furosemide, gallium, ganciclovir, gatifloxacin, gemcitabine, gentamicin, glycopyrrolate, granisetron, haloperidol, heparin, hydrALAZINE, hydrocortisone, HYDROmorphone, hydrOXYzine, IDArubicin, ifosfamide, imipenem-cilastatin, inamrinone, insulin (regular), irinotecan, isoproterenol, ketorolac, labetalol, leucovorin, levofloxacin, lidocaine, linezolid, LORazepam, magnesium, mannitol, mechlorethamine, melphalan, meperidine, meropenem, mesna, methohexital, methotrexate, methyldopate, methylPREDNISolone, metoclopramide, metoprolol, metroNIDAZOLE, midazolam, milrinone, mitoMYcin, mitoXANtrone, mivacurium, morphine, moxifloxacin, mycophenolate mofetil, nafcillin, nalbuphine, naloxone, nesiritide, niCARdipine, nitroglycerin, nitroprusside, norepinephrine, octreotide, ofloxacin, ondansetron, oxaliplatin, oxytocin, PACLitaxel, palonosetron, pamidronate, pancuronium, PEMEtrexed, PENTobarbital, PHENobarbital, phenylephrine, piperacillin, piperacillin-tazobactam, polymyxin B, potassium acetate/chloride/phosphates, procainamide, promethazine, propranolol, ranitidine, remifentanil, rocuronium, sodium acetate/bicarbonate/phosphates, streptozocin, succinylcholine, SUFentanil, sulfamethoxazole-trimethoprim, tacrolimus, teniposide, theophylline, thiopental, thiotepa, ticarcillin, ticarcillin-clavulanate, tigecycline, tirofiban, tobramycin, topotecan, vasopressin, vecuronium, verapamil, vinBLAStine, vinCRIStine, vinorelbine, voriconazole, warfarin, zidovudine, zoledronic acid

SUBCUT injection (unlabeled)

• May be used for DVT prophylaxis

SIDE EFFECTS

CNS: *Headache, insomnia, anxiety, nervousness*
CV: *Hypo/hypertension, bradycardia,* ventricular fibrillation
GI: *Nausea, vomiting, abdominal pain, dyspepsia*
HEMA: **Hemorrhage, thrombocytopenia**
MISC: Pain at inj site, pelvic pain, urinary retention, fever, **anaphylaxis,** infection
MS: *Back pain*

PHARMACOKINETICS

Excreted in urine, half-life 25 min, duration 1 hr, no protein binding

INTERACTIONS

Increase: bleeding risk—anticoagulants, aspirin, treprostinil, thrombolytics

Drug/Herb

Increase: bleeding risk—agrimony, alfalfa, angelica, anise, bilberry, black haw, bogbean, buchu, cat's claw, chamomile, chondroitin, devil's claw, dong quai, evening primrose, fenugreek, feverfew, fish oil, garlic, ginger, ginkgo, ginseng, horse chestnut, Irish moss, kava, kelp, kelpware, khella, licorice, lovage, lungwort, meadowsweet, motherwort, mugwort, nettle, papaya, parsley (large amts), pau d'arco, pineapple, poplar, prickly ash, red clover, safflower, saw palmetto, senega, skullcap, tonka bean, turmeric, wintergreen, yarrow
Decrease: anticoagulant effect—coenzyme Q10, flax, glucomannan, goldenseal, guar gum

NURSING CONSIDERATIONS

Assess:

- Baseline and periodic ACT, APTT, PT, INR, TT, platelets, Hgb, Hct

⚠ **Bleeding: check arterial and venous sites, IM inj sites, catheters; all punctures should be minimized; fall in B/P or Hct may indicate hemorrhage, hematoma, hemorrhage at puncture site are more common in the elderly**

- Fever, skin rash, urticaria
- CV status: B/P, watch for hypo/hypertension, bradycardia
- Neurologic status: any focal or generalized deficits should be reported immediately
- PCI use: possible thrombosis, stenosis, unplanned stent, prolonged ischemia, decreased reflow

Evaluate:

- Therapeutic response: anticoagulation with PTCA; resolution of heparin-induced thrombocytopenia, thrombosis syndrome

Teach patient/family:

- About the reason for the product and expected results
- To report black, tarry stools; blood in urine; difficulty breathing
- Not to use any OTC, herbal products unless approved by prescriber
- Not to use hard-bristle toothbrush or regular razor to avoid any injury; hemorrhage may result

⚠ ***HIGH ALERT***

bleomycin (Rx)

(blee-oh-mye′sin)

Blenoxane

Func. class.: Antineoplastic, antibiotic
Chem. class.: Glycopeptide

ACTION: Inhibits synthesis of DNA, RNA, protein; derived from *Streptomyces verticillus;* phase specific to the G_2 and M phases; a nonvesicant, sclerosing agent

USES: Cancer of head, neck, penis, cervix, vulva of squamous cell origin; Hodgkin's/non-Hodgkin's disease; testicular carcinoma; as a sclerosing agent for malignant pleural effusion

Unlabeled uses: Cutaneous T-cell lymphoma (CTCL), hemangioma, Kaposi's sarcoma, malignant ascites, verruca plantaris/vulgaris, osteogenic sarcoma

CONTRAINDICATIONS:

Pregnancy (D), breastfeeding, hypersensitivity, prior idiosyncratic reaction

Precautions: Patients >70 yr old, renal/hepatic disease, respiratory disease

Black Box Warning: Idiosyncratic reaction, pulmonary fibrosis

DOSAGE AND ROUTES

Test dose

- **Adult and child (unlabeled): IM/IV/SUBCUT** ≤2 units for first 2 doses followed by 24 hr of observation
- **Adult and child: SUBCUT/IV/IM** 0.25-0.5 units/kg 1-2 ×/wk or 10-20 units/m^2, then 1 unit/day or 5 units/wk; may also be given by **CONT INF**; do not exceed total dose of 400 units during lifetime

Renal dose

- **Adult/child:** CCr 40-50 ml/min reduce dose by 30%; CCr 30-39 ml/min reduce dose by 40%; CCr 20-29 ml/min reduce dose by 45%; CCr 10-19 ml/min reduce dose by 55%; CCr 5-10 ml/min reduce dose by 60%

Cutaneous T-cell lymphoma (CTCL) (unlabeled)

- **Adult: IV** 15 units twice weekly with vinBLAStine and predniSONE

Kaposi's sarcoma (unlabeled)

- **Adult: IV** 15 units q2wk with DOXOrubicin and vinCRIStine

Available forms: Powder for inj, 15, 30 units/vial

Administer:

- Antiemetic 30-60 min before giving product to prevent vomiting; continue antiemetics 6-10 hr after treatment
- Topical or systemic analgesics for pain of stomatitis as ordered; antihistamines and antipyretics for fever, chills
- May be given IM, subcut, IV, intrapleurally, intralesionally, intraarterially

IM/SUBCUT route

- After reconstituting 15 units/1-5 ml or 30 mg/2-10 ml of 0.9% NaCl or bacteriostatic water for inj, rotate inj sites; do not use products that contain benzyl alcohol when giving to neonates or that contain dextrose because of loss of potency

Intrapleural route

- 60 units/100 ml of 0.9% NaCl administered by MD through thoracostomy tube

IV route

- Use cytotoxic handling procedures
- After reconstituting 15- or 30-unit vial with 5 or 10 ml of NS, respectively, inj slowly over 10 min or after further dilution with 50-100 ml 0.9% NaCl; give 15 units or less over 10 min through Y-tube or 3-way stopcock
- For patients with lymphoma, give 2 test doses of 2-5 units before initial dose; monitor for anaphylaxis
- Store for 2 wk after reconstituting if refrigerated or for 24 hr at room temp; discard unused portions

Y-site compatibilities: Acyclovir, alfentanil, allopurinol, amifostine, amikacin, aminocaproic acid, aminophylline, amiodarone, ampicillin, ampicillin-sulbactam, anidulafungin, atenolol, atracurium, azithromycin, aztreonam, bivalirudin, bumetanide, buprenorphine, busulfan, butorphanol, calcium chloride/gluconate, CARBOplatin, carmustine, caspofungin, ceFAZolin, cefepime, cefotaxime, cefoTEtan, cefOXitin, cefTAZidime, ceftizoxime, cefTRIAXone, cefuroxime, chloramphenicol, chlorproMAZINE, cimetidine, ciprofloxacin, cisatracurium, CISplatin, clindamycin, codeine, cyclophosphamide, cycloSPORINE, cytarabine, dacarbazine, DACTINomycin, DAPTOmycin, DAUNOrubicin, dexamethasone, dexmedetomidine, dexrazoxane, digoxin, diltiazem, diphenhydrAMINE, DOBUTamine, DOCEtaxel, DOPamine, doxacurium, DOXOrubicin, DOXOrubicin liposomal, doxycycline, droperidol, enalaprilat, ePHEDrine, EPINEPHrine, epirubicin, ertapenem, erythromycin, esmolol, etoposide, famotidine, fenoldopam, fentaNYL, filgrastim, fluconazole, fludarabine, fluorouracil, foscarnet, fosphenytoin, furosemide, ganciclovir, gatifloxacin, gemcitabine, gentamicin, glycopyrrolate, granisetron, haloperidol, heparin, hydrALAZINE, hydrocortisone sodium succinate, HYDROmorphone, hydrOXYzine, IDArubicin, ifosfamide, imipenem-cilastatin, inamrinone, insulin (regular), irinotecan, isoproterenol, ketorolac, labetalol, leucovorin, levofloxacin, levorphanol, lidocaine, linezolid, LORazepam, magnesium sulfate, mannitol, mechlorethamine, melphalan, meperidine, meropenem, mesna, metaraminol, methohexital, methotrexate, methyldopate, methylPREDNISolone, metoclopramide, metoprolol, metroNIDAZOLE, midazolam, milrinone, minocycline, mitoMYcin, mitoXANtrone, mivacurium, morphine, nafcillin, nalbuphine, naloxone, nesiritide, niCARdipine, nitroglycerin, nitroprusside, norepinephrine, octreotide, ondansetron, oxaliplatin, palonosetron, pamidronate, pancuronium, pantoprazole, PEMEtrexed, pentamidine, pentazocine, PENTobarbital, PHENobarbital, phenylephrine, piperacillin, piperacillin-tazobactam, polymyxin B, potassium chloride, potassium phosphates, procainamide, prochlorperazine, promethazine, propranolol, quiNIDine, ranitidine, remifentanil, riTUXimab, rocuronium, sargramostim, sodium acetate, sodium bicarbonate, sodium phosphates, succinylcholine, SUFentanil, sulfamethoxazole-trimethoprim, tacrolimus, teniposide, theophylline, thiopental, thiotepa, ticarcillin, ticarcillin-clavulanate, tirofiban, tobramycin, tolazoline, trastuzumab, trimethobenzamide, vancomycin, vasopressin, vecuronium, verapamil, vinBLAStine, vinCRIStine, vinorelbine, voriconazole, zidovudine

SIDE EFFECTS

CNS: Pain at tumor site, headache, confusion
CV: **MI, stroke**
GI: *Nausea, vomiting, anorexia, stomatitis, weight loss,* ulceration of mouth, lips
GU: **Hemolytic-uremic syndrome**
IDIOSYNCRATIC REACTION: Hypotension, *confusion, fever,* chills, wheezing
INTEG: *Rash, hyperkeratosis, nail changes, alopecia,* pruritus, acne, striae, peeling, hyperpigmentation, phlebitis
RESP: **Fibrosis, pneumonitis,** wheezing, **pulmonary toxicity**
SYST: **Anaphylaxis,** radiation recall, Raynaud's phenomenon

PHARMACOKINETICS

Half-life 2 hr; when CCr is >35 ml/min, half-life is increased with lower clearance; metabolized in liver; 50% excreted in urine (unchanged)

INTERACTIONS

- Avoid live virus vaccines concurrently

Increase: toxicity—other antineoplastics, radiation therapy, general anesthesia, filgrastim, sargramostim
Decrease: serum phenytoin levels—phenytoin, fosphenytoin
Drug/Lab Test
Increase: uric acid

NURSING CONSIDERATIONS

Assess:

- IM test dose in patients with lymphoma of 1-2 units before 1st 2 doses

Black Box Warning: Pulmonary toxicity/fibrosis: pulmonary function tests; chest x-ray before, during therapy, should be obtained q2wk during treatment; pulmonary diffusion capacity for carbon monoxide (DLCO) monthly, if <40% of pretreatment value, stop treatment; treat pulmonary infection before treatment; dyspnea, crackles, unproductive cough, chest pain, tachypnea, fatigue, increased pulse, pallor, lethargy, more common in the elderly, radiation therapy, pulmonary disease

- Temp; fever may indicate beginning infection
- Renal status: serum creatinine/BUN; CBC
- Effects of alopecia, skin color alterations on body image; discuss feelings about body changes

• Buccal cavity q8hr for dryness, sores, ulceration, white patches, oral pain, bleeding, dysphagia
• Local irritation, pain, burning, discoloration at inj site
⚠ **Anaphylaxis:** rash, pruritus, urticaria, purpuric skin lesions, itching, flushing, wheezing, hypotension; have emergency equipment available

Black Box Warning: Idiosyncratic reaction: hypotension, mental confusion, fever, chills, wheezing

• Rinsing of mouth tid-qid with water, club soda; brushing of teeth with baking soda bid-tid with soft brush or cotton-tipped applicators for stomatitis; use unwaxed dental floss

Evaluate:
• Therapeutic response: decrease in size of tumor

Teach patient/family:
• To report any changes in breathing, coughing, fever
• That hair may be lost during treatment and that wig or hairpiece may make patient feel better; that new hair may be different in color, texture
• To avoid foods with citric acid, hot or rough texture
• To report any bleeding, white spots, ulcerations in mouth; to examine mouth daily and report symptoms
• To use contraception during treatment (pregnancy [D]), to avoid breastfeeding
• Not to receive vaccines during treatment

boceprevir

(boe-se′pre-vir)

Victrelis

Func. class.: Antiviral, anti-hepatitis agents

ACTION: Prevents hepatitis C viral (HCV) replication by blocking the activity of HCV NS3/4A serine protease. Hepatitis C virus NS3/4A serine protease is an enzyme responsible for the conversion of HCV encoded polyproteins to mature/functioning viral proteins

USES: Hepatitis C infection in combination with peginterferon alfa and ribavirin

CONTRAINDICATIONS: Pregnancy (X), male partners of women who are pregnant

Precautions: Breastfeeding, neonates, infants, children, adolescents <18 years of age, anemia, neutropenia, thrombocytopenia, HIV, hepatitis B, decompensated hepatic disease, in liver or other organ transplants, hypersensitivity

DOSAGE AND ROUTES

Chronic hepatitis C infection (genotype 1) compensated liver disease (without cirrhosis, previously untreated with interferon and ribavirin therapy)/(Partial responders/relapsers/null responders)

• **Adults: PO** Before starting therapy peginterferon alfa and ribavirin must be given 4 wk, then add boceprevir 800 mg (four 200 mg caps) PO TID (7-9 hours); treatment length is determined by HCV RNA conc at treatment wk 4, 8, 12, and 24; if patient has undetectable HCV RNA conc at wk 8 and 24, discontinue all 3 medications at wk 28 (previously untreated); 36 wk (partial responders/relapsers); if HCV RNA is detectable at wk 8 but undetectable at wk 24, the 3-drug regimen through wk 36, then give only peginterferon alfa and ribavirin through treatment wk 48; if the patient has a poor response to peginterferon alfa and ribavirin during the initial 4 wk, continue treatment with all 3 medications for a total of 48 wk; discontinue the 3-drug regimen if the HCV RNA conc >100 international units/ml at treatment wk 12 or a detectable HCA RNA conc at treatment wk 24

Chronic hepatitis C infection (genotype 1) compensated liver disease with cirrhosis

• **Adults: PO** Before starting therapy with boceprevir; peginterferon alfa and ribavirin must be given 4 wk; then add boceprevir 800 mg (four 200 mg caps) PO TID (7-9 hr) to peginterferon alfa and ribavirin for an additional 44 wk (48 wk total)

Available forms: Caps 200 mg

Administer:

• Only use in combination with peginterferon alfa and ribavirin; never give as monotherapy

• Discontinue in hepatitis C virus (HCV) if RNA concentrations ≥100 international units/ml at wk 12 or a confirmed detectable HCV RNA concentrations at wk 24

• Any contraindication to peginterferon alfa or ribavirin also applies to boceprevir

• Give with food

SIDE EFFECTS

When used in combination with peginterferon/ribavarin

CNS: *Fatigue, chills, asthenia*, insomnia, irritability, dizziness

GI: Nausea, vomiting, diarrhea, dysgeusia, decreased appetite, xerostomia

HEMA: **Anemia (Hgb <10 g/dL), neutropenia, thrombocytopenia**

INTEG: *Alopecia, rash*, xerosis

MISC: *Arthralgia, exertional dyspnea,* **drug rash with eosinophilia and systemic symptoms (DRESS) syndrome, exfoliative dermatitis, Stevens-Johnson syndrome, toxic epidermal necrolysis**

INTERACTIONS

⚠ **Increase: life-threatening reactions of each product: alfuzosin, ergots (dihydroergotamine, ergotamine, ergonovine, methylergonovine), cisapride, pimozide, lovastatin, simvastatin, ezetimibe, niacin with simvastatin and boceprevir; triazolam, oral midazolam; sildenafil, tadalafil (pulmonary arterial hypertension); do not use concurrently**

Increase: adverse reactions of each product—phosphodiesterase type 5 (PDE5) inhibitors (for erectile dysfunction), acetaminophen, alfentanil, aliskiren, almotriptan, alosetron, ALPRAZolam, aminophylline, amiodarone, amitriptyline, amLODIPine, ARIPiprazole, astemizole, atorvastatin, atorvastin, bepridil, boceprevir, bosentan, budesonide, bupivacaine, buprenorphine, busPIRone, carvedilol, cevimeline, chloroquine, cilostazol, cinacalcet, citalopram, clarithromycin, clomiPRAMINE, clonazePAM, clopidogrel, cloZAPine, colchicine, cyclobenzaprine, cycloSPORINE, dapsone, DAUNOrubicin, desipramine, desloratadine, dexamethasone, dexlansoprazole, dextromethorphan, diazepam, diclofenac, digoxin, diltiazem, disopyramide, disulfiram, DOCEtaxel, dolasetron, donepezil, DOXOrubicin, droperidol, dutasteride, ebastine, eletriptan, eplerenone, erlotinib, erythromycin, estazolam, eszopiclone, ethosuximide, etoposide, exemestane, felodipine, fentaNYL, fexofenadine, finasteride, flecainide, flunitrazepam, flurazepam, galantamine, gefitinib, glyburide, granisetron, halofantrine, haloperidol, HYDROcodone, ifosfamide, imipramine, indiplon, irinotecan, isradipine, itraconazole, ivermectin, ixabepilone, ketoconazole, lansoprazole, lidocaine, loperamide, loratadine, losartan, maraviroc, mefloquine, meloxicam, mirtazapine, mitoMYcin, montelukast, morphine, nateglinide, niCARdipine, NIFEdipine, nisoldipine, nortriptyline, omeprazole, ondansetron, oxybutynin, oxyCODONE, PACLitaxel, palonosetron, paricalcitol, plicamycin, posaconazole, prasugrel, praziquantel, propafenone, quazepam, QUEtiapine, quinacrine, quiNIDine, ramelteon, repaglinide, rifabutin, risperiDONE, ropivacaine, salmeterol, selegiline, sertraline, sibutramine, silodosin, sirolimus, sitaxsentan, solifenacin, SUFentanil, SUNItinib, systemic corticosteroids, tacrolimus, telithromycin, teniposide, terfenadine, testosterone, theophylline, tiaGABine, ti-

nidazole, tolterodine, tolvaptan, traMADol, traZODone, vardenafil, venlafaxine, verapamil, vinBLAStine, vinCRIStine, voriconazole, warfarin, and others; use cautiously, may need to reduce dose
Increase: hyperkalemia—drospirenone
Decrease: estrogen levels—ethinyl estradiol
Decrease: boceprevir effect—CYP3A4 inhibitors (phenytoin, carBAMazepine, PHENobarbital, rifampin)
Decrease: effect of—methadone
Possible treatment failure: efavirenz, ritonavir, atazanavir, lopinavir with ritonavir

Drug/Herb

• Do not use with St. John's wort

Drug/Lab Test

Decrease: Hgb

NURSING CONSIDERATIONS

Assess:

⚠ **Pregnancy:** obtain a pregnancy test before, monthly during, and for 6 mo after treatment is completed; those who are not willing to practice strict contraception should not receive treatment; report any cases of prenatal ribavirin exposure to the Ribavirin Pregnancy Registry at (800) 593-2214

• **Anemia:** monitor Hgb CBC with differential before, at treatment wk 2, 8, 12, and as needed. If Hgb is <10 g/dL, decrease ribavirin dosage; if Hgb is <8.5 g/dL, discontinuation of therapy is recommended; dosage should not be altered based on adverse reactions; anemia may be managed through ribavirin dose modifications; never alter the dose of boceprevir; if anemia persists despite a reduction in ribavirin dose, consider discontinuing boceprevir; if management of anemia requires permanent discontinuation of ribavirin, treatment with boceprevir MUST also be permanently discontinued; once boceprevir has been discontinued, it must not be restarted; monitor CBC with differential at treatment wk 4, 8, 12, and at other treatment points as needed

• **Serious skin disorders (DRESS, Stevens-Johnson syndrome, toxic epidermal necrolysis, exfoliative dermatitis):** These reactions may be due to combination use with peginterferon alfa, ribavirin; if serious skin reactions occur, discontinue all 3 products

Teach patient/family:

• To take with food to increase absorption

• To use precautions to prevent transmission of hepatitis C

• To inform prescriber of all medications, herbs, supplements used

• To use 2 forms of effective contraception (intrauterine devices and barrier methods) during treatment and for 6 mo after treatment (pregnancy [X]); to avoid breastfeeding

boric acid otic

See Appendix B

⚠ *HIGH ALERT*

bortezomib (Rx)

(bor-tez′oh-mib)

Velcade

Func. class.: Antineoplastic—miscellaneous
Chem. class.: Proteasome inhibitor

ACTION: Reversible inhibitor of chymotrypsin-like activity in mammalian cells; causes delay in tumor growth by disrupting normal homeostatic mechanisms

USES: Multiple myeloma previously untreated or when at least 2 other treatments have failed; mantle cell lymphoma who have received ≥1 prior therapy
Unlabeled uses: Non-Hodgkin's lymphoma (NHL)

CONTRAINDICATIONS:

Pregnancy (D), breastfeeding; hypersensitivity to product, boron, mannitol

Precautions: Children, geriatric patients, peripheral neuropathy, renal/hepatic disease, hypotension, tumor lysis syndrome, thrombocytopenia, infection, diabetes mellitus, bone marrow suppression

DOSAGE AND ROUTES

Multiple myeloma (previously untreated)

- **Adult: IV BOL/SUBCUT** Give for 9 6-wk cycles; cycles 1-4, 1.3 mg/m^2/dose given on days 1, 4, 8, 11, then a 10-day rest period (days 12-21), then give again on days 22, 25, 29, 32, then a 10-day rest period (days 33-42); given with melphalan (9 mg/m^2/day on days 1-4) and predniSONE (60 mg/m^2/day on days 1-4); during cycles 5-9, give bortezomib 1.3 mg/m^2/dose on days 1, 8, 22, 29 with melphalan (9 mg/m^2/day on days 1-4) and predniSONE (60 mg/m^2/day on days 1-4); this 6-wk cycle is considered 1 course; at least 72 hr should elapse between consecutive doses

Mantle cell lymphoma

- **Adult: IV BOL/SUBCUT** 1.3 mg/m^2/dose on days 1, 4, 8, 11 followed by a 10-day rest period (days 12 to 21); max 8 cycles

Hepatic dose

- **Adult: IV** bilirubin >1.5 × ULN, reduce to 0.7 mg/m^2 during cycle 1; consider dose escalation to 1 mg/m^2 or further reduction to 0.5 mg/m^2 during next cycles based on tolerability

Non-Hodgkin's Lymphoma (unlabeled)

- **Adult: IV BOL** 1.3 mg/m^2 on days 1, 4, 8, 11 repeated every 21 days

Available forms: Lyophilized powder for inj 3.5 mg

Administer:

IV bolus route

- **Reconstitute** each vial with 3.5 ml of 0.9% NaCl (1 mg/ml); sol should be clear/colorless; **inj** as bolus over 3-5 sec
- Store unopened product at room temperature, protect from light
- Wear protective clothing during handling, preparation; avoid contact with skin
- Check for extravasation at inj site

SIDE EFFECTS

CNS: Anxiety, insomnia, dizziness, headache, *peripheral neuropathy,* rigors, paresthesia, fever

CV: *Hypotension,* edema, **CHF**

GI: Abdominal pain, *constipation, diarrhea,* dyspepsia, *nausea, vomiting,* anorexia

HEMA: *Anemia,* **neutropenia, thrombocytopenia**

MISC: Dehydration, weight loss, herpes zoster, *rash,* pruritus, blurred vision

MS: *Fatigue, malaise, weakness,* arthralgia, bone pain, muscle cramps, myalgia, back pain, **tumor lysis syndrome**

RESP: Cough, pneumonia, dyspnea, URI, **ARDs, pneumonitis, interstitial pneumonia, lung infiltration**

PHARMACOKINETICS

Half-life 9-15 hr, protein binding 83%, metabolized by CYP450 enzymes (3A4, 2D6, 2C19, 2C9, 1A2)

INTERACTIONS

- **Do not use hematopoietic progenitor cells (sargramostim, filgrastim) within 24 hr of chemotherapy**
- Oral hypoglycemics: may result in hypo/hyperglycemia

Increase: risk for bleeding—anticoagulants, NSAIDs, platelet inhibitors, salicylates, thrombolytics

Increase: hypotension—antihypertensives

Increase: peripheral neuropathy—amiodarone, antivirals (amprenavir; atazanavir; didanosine, lamiVUDine, 3TC; ritonavir; stavudine, zidovudine), chloramphenicol, CISplatin, colchicine, cycloSPORINE, dapsone, disulfiram, DOCEtaxel, gold salts, HMG-CoA reductase inhibitors, iodoquinol, INH, metro-

NIDAZOLE, nitrofurantoin, oxaliplatin, PACLitaxel, penicillamine, phenytoin, sulfaSALAzine, thalidomide, vinBLAStine, vinCRIStine, zalcitabine ddc, isoniazid, statins, others
Increase: toxicity or decrease efficacy when administered with products that induce or inhibit CYP3A4
Decrease: effect of norethindrone, estradiol, combination oral contraceptives, another non-hormonal contraceptive should be used
Drug/Herb
Increase: toxicity or decrease efficacy—St. John's wort

NURSING CONSIDERATIONS

Assess:
Fatal pulmonary toxicity: asses for risk factors or new worsening pulmonary symptoms
Tumor lysis syndrome: usually with those with a high tumor burden
• Hematologic status: platelets, CBC throughout treatment; platelets ≥70 × 10^9/L and ANC ≥1.0 × 10^9/L before any cycle; nonhematologic toxicities should be grade 1 or baseline before any cycle
• B/P, fluid status, peripheral neuropathy symptoms
Evaluate:
• Therapeutic response: improvement of multiple myeloma symptoms
Teach patient/family:
• To use contraception while taking this product (pregnancy [D]); to avoid breastfeeding
• To monitor blood glucose levels if diabetic
• To contact prescriber about new or worsening peripheral neuropathy, severe vomiting, diarrhea, easy bruising, bleeding, infection
• To avoid driving, operating machinery until effect is known
• To avoid using other medications unless approved by prescriber

bosentan (Rx)

(boh′sen-tan)
Tracleer
Func. class.: Vasodilator
Chem. class.: Endothelin receptor antagonist

ACTION: Peripheral vasodilation occurs via the antagonism of the effect of endothelin on endothelium and vascular smooth muscle

USES: Pulmonary arterial hypertension with WHO class III, IV symptoms
Unlabeled uses: Septic shock to improve microcirculatory blood flow, functional class II pulmonary arterial hypertension

CONTRAINDICATIONS: Hypersensitivity, CVA, CAD
Black Box Warning: Pregnancy X
Precautions: Breastfeeding, children, geriatric patients, mitral stenosis
Black Box Warning: Hepatic disease

DOSAGE AND ROUTES

• **Adult and adolescent >40 kg: PO** 62.5 mg bid × 4 wk then 125 mg bid
• **Adult and adolescent <40 kg: PO** 62.5 mg bid
Hepatic dose
• **Adult: PO** baseline AST/ALT<3×ULN no dosage change, monitor LFTs q mo reduce or interrupt if elevated; AST/ALT>3 and ≤5 × ULN repeat test, if confirmed reduce to 62.5 mg BID or interrupt; monitor LFTs q2wks, if interrupted, restart when LFTs <3× ULN, check LFTs within 3 days; increase in AST/ALT>5 and ≤5 × ULN; during treatment repeat test to confirm, discontinue, monitor LFTs q2wk until LFTs <3× ULN, restart at starting dose; AST/ALT>8× ULN **discontinue permanently**
Available forms: Tabs 62.5, 125 mg

Administer:
- Give without regard to meals
- Only available through the TAP program; 866-228-3546
- Do not stop product abruptly; taper
- Store at room temp

SIDE EFFECTS

CNS: Headache, flushing, fatigue, fever
CV: *Hypo/hypertension,* palpitations, edema of lower limbs, fluid retention
GI: Abnormal hepatic function, diarrhea, dyspepsia, **hepatotoxicity**
HEMA: **Anemia, leukopenia, neutropenia, lymphopenia, thrombocytopenia**
INTEG: Pruritus, **anaphylaxis,** rash, **Stevens-Johnson syndrome, toxic epidermal necrolysis**
MISC: Oligospermia, **tumor lysis syndrome**
SYST: **Secondary malignancy**

PHARMACOKINETICS

Metabolized by inducer of CYP2C9, CYP3A4, possibly CYP2C19; metabolized by the liver; terminal half-life 5 hr; steady state 3-5 days

INTERACTIONS

- **Do not coadminister cycloSPORINE with bosentan; bosentan is increased, cycloSPORINE is decreased**
- **Do not coadminister glyBURIDE with bosentan; glyBURIDE is decreased significantly, bosentan is also decreased, hepatic enzymes may be increased**

Increase: bosentan effects—CYP2C9, CYP3A4 inhibitors
Increase: bosentan level—ketoconazole
Decrease: effects of warfarin, hormonal contraceptives, statins

Drug/Lab Test
Increase: ALT, AST
Decrease: Hgb, Hct

NURSING CONSIDERATIONS

Assess:
- **Serious skin toxicities: Angioedema occurring 8-21 days after initiating therapy**
- B/P, pulse during treatment until stable
- Blood studies: Hct, Hgb after 1 mo, 3 mo, then every 3 mo may be decreased
- Pulmonary hypertension/CHF: Fluid retention, weight gain, increased leg edema; may occur within weeks

Black Box Warning: Hepatic toxicity: vomiting, jaundice; product should be discontinued; hepatic studies: AST, ALT, bilirubin; hepatic enzymes may increase; if ALT/AST >3× and ≤5× ULN, decrease dose or interrupt treatment and monitor AST/ALT q2wk; if bilirubin >2× ULN or signs of hepatitis or hepatic disease are present, stop treatment

Evaluate:
- Therapeutic response: decrease in pulmonary hypertension

Teach patient/family:
- To report jaundice, dark urine, joint pain, fatigue, malaise, bruising, easy bleeding

Black Box Warning: Pregnancy (X), monitor pregnancy test monthly

- To avoid pregnancy; to use nonhormonal form of contraception, hormonal contraceptives may not be effective
- That lab work will be required periodically
- To take without regard to food

bosutinib

(boe-sue'ti-nib)

Bosulif

Func. class.: Antineoplastic biologic response modifiers
Chem. class.: Signal transduction inhibitors (STIs), tyrosine kinase inhibitor

ACTION: Inhibits bcr-abl tyrosine kinase created in patients with chronic myeloid leukemia (CML)

USES: Treatment of CML; Philadelphia-chromosome–positive patients in blast-cell crisis

CONTRAINDICATIONS:
Pregnancy (D), hypersensitivity

Precautions: Breastfeeding, children, diarrhea, geriatric patients, hepatic disease, bone marrow suppression, infection, thrombocytopenia, neutropenia, immunosuppression

DOSAGE AND ROUTES

• **Adult: PO** 500 mg daily with food, may increase to 600 mg/day in those who have not developed grade 3 toxicity or in patients who do not reach complete hematological response by wk 8 or complete cytogenic response (CCyR) by wk 12

Hepatic dosage

• **Adult: PO** Any baseline hepatic impairment: Start at 200 mg/day; liver transaminase >5 × ULN, hold dose until levels are ≤2.5 × ULN, then resume at 400 mg/day; liver transaminase level ≥3 × ULN and bilirubin >2 × ULN and alk phos <2 × ULN, discontinue

Dosage adjustments for treatment-related toxicity

Hematologic toxicity:

⚠ ***ANC <1000 × 10⁶/L or platelet count <50,000 × 10⁶/L:*** hold dose until ANC is ≥1000 × 10⁶/L and platelets are ≥50,000 × 10⁶/L; if recovery within 2 wk, resume therapy at the same dosage; if blood counts remain low after 2 wk, upon recovery, resume at 100 mg/day less than the previous dosage

Diarrhea:

⚠ ***Grade 3 or 4 diarrhea (≥7 stools/day compared with baseline):*** hold therapy until recovery to grade 1 toxicity or lower; resume therapy at 400 mg/day

Other nonhematologic toxicity:

⚠ ***Significant or moderate or severe toxicity:*** hold therapy until toxicity resolves; resume therapy at 400 mg/day

Available forms: Tabs 100, 500 mg

Administer:

PO route

• Give with food; swallow whole

• If dose is missed, take within 12 hr of missed dose; if >12 hr have passed, skip dose

• Follow cytotoxic handling procedures

SIDE EFFECTS

CNS: Headache, dizziness

GI: Nausea, vomiting, anorexia, abdominal pain, diarrhea

HEMA: Neutropenia, thrombocytopenia, bleeding

INTEG: Rash, pruritus

MS: Arthralgia, myalgia

RESP: Cough, dyspnea, pleural effusion, edema

OTHER: Elevated LFTs

PHARMACOKINETICS

Protein binding 96%; metabolized by CYP3A4; half-life 22.5 hr

INTERACTIONS

Increase: bosutinib concentrations—CYP3A4 inhibitors (ketoconazole, itraconazole, erythromycin, clarithromycin)

Increase: plasma concentrations of simvastatin, calcium channel blockers, ergots

Decrease: bosutinib concentrations—CYP3A4 inducers (dexamethasone, phenytoin, carBAMazepine, rifampin, PHENobarbital), antacids, proton-pump inhibitors

Drug/Food

Increase: increase bosutinib effect—grapefruit juice; avoid use while taking product

Drug/Herb

Decrease: bosutinib concentration—St. John's wort

NURSING CONSIDERATIONS

Assess:

⚠ **Myelosuppression: anemia, thrombocytopenia, neutropenia; obtain a CBC weekly × 1 mo, then monthly as needed**

• LFTs every mo × 3 mo, then as clinically indicated

Evaluate:

• Therapeutic response: decrease in leukemic cells or size of tumor

Teach patient/family:
- To report adverse reactions immediately, bleeding
- About reason for treatment, expected results
- To use effective contraception during treatment and up to 30 days after discontinuing treatment

⚠ HIGH ALERT

brentuximab
(bren-tuk′see-mab)

Adcetris

Func. class.: Monoclonal antibody; antineoplastic

ACTION: The anticancer activity is due to the binding of the ADC to CD30-expressing cells, followed by the internalization and transportation of the ADC-CD30 complex to lysosomes and the release of MMAE via selective proteolytic cleavage; MMAE binds to tubulin and disrupts the microtubule network within the cell, inducing cell cycle arrest and apoptotic death of the cells

USES: For the treatment of Hodgkin's disease after failure of autologous stem cell transplant (ASCT) or after failure of at least 2 prior multiagent chemotherapy regimens in patients who are not ASCT candidates; for the treatment of non-Hodgkin's lymphoma (NHL); for the treatment of systemic anaplastic large cell lymphoma (sALCL) after failure of at least 1 prior multiagent chemotherapy regimen

CONTRAINDICATIONS Hypersensitivity, pregnancy (category D)

Precautions: Breastfeeding, children, infants, neonates, neutropenia, peripheral neuropathy, tumor lysis syndrome (TLS)

Black Box Warning: Progressive multifocal leukoencephalopathy

DOSAGE AND ROUTES
- **Adult: IV** 1.8 mg/kg over 30 min every 3 wk until a maximum of 16 cycles, disease progression, or unacceptable toxicity; for patients >100 kg, max weight used for dosage calculation should be 100 kg, which translates to no more than 180 mg/dose

Dose adjustments for toxicity due to peripheral neuropathy:
- For grade <3: no dosage adjustments are recommended; for new or worsening grade 2-3: interrupt treatment until toxicity resolves to grade ≤1; when resuming treatment, reduce dosage to 1.2 mg/kg IV q3wk; for grade 4: discontinue treatment

Dose adjustments for toxicity due to neutropenia:
- For neutropenia grade <3: no dosage adjustments; for grade 3-4 neutropenia: interrupt treatment until toxicity resolves to baseline or grade ≤2; consider the use of growth factors (CSFs) for subsequent cycles of therapy; for grade 4 neutropenia despite the use of growth factors: discontinue treatment or reduce the dose to 1.2 mg/kg IV q3wk

Available forms: Powder for inj 50 mg

Administer:

Intermittent IV INF route
- Visually inspect for particulate matter and discoloration whenever sol and container permit
- Only as an IV inf, do not give as an IV push or bolus
- Use cytotoxic handling procedures
- Do not mix, or administer as an infusion, with other IV products
- Calculate the dose (mg) and the number of vials required. For patients weighing >100 kg, use 100 kg to calculate the dose; reconstitute each 50-mg vial per 10.5 ml of sterile water for inj (5 mg/ml)
- Direct the stream of sterile water toward the wall of the vial and not directly at the cake or powder; gently swirl the vial to aid in dissolution, do not shake
- Discard any unused portion left in the vial

- After reconstitution, dilute immediately with ≥100 ml of 0.9% sodium chloride, 5% dextrose, or lactated Ringer's solution to a final concentration (0.4 mg/ml-1.8 mg/ml)
- Infuse over 30 min
- Use the diluted sol immediately or store in refrigerator for ≤24 hr after reconstitution; do not freeze

SIDE EFFECTS

CNS: Headache, dizziness, *fever,* peripheral neuropathy, anxiety, chills, confusion, *fatigue,* paresthesias, insomnia, night sweats, progressive multifocal leukoencephalopathy
CV: Peripheral edema, supraventricular arrhythmia
GI: *Abdominal pain, nausea, vomiting,* constipation, *diarrhea,* weight loss
INTEG: *Rash,* pruritus, alopecia, xerosis
RESP: Pneumothorax, pneumonitis, pulmonary embolism, dyspnea, *cough*
SYST: Anaphylaxis, tumor lysis syndrome, antibody formation, Stevens-Johnson syndrome
HEMA: Anemia, neutropenia, thrombocytopenia

PHARMACOKINETICS

Protein binding is 68%-82%, only a small amount is metabolized; potent inhibitors or inducers of CYP3A4 may alter action; ADC peak at end of infusion; MME peak 1-3 days; terminal half-life is 4-6 days; 3 components are released: MMAE (monomethyl auristatin E), ADC, and the total antibody; the half-life of MMAE a component is 3.43-3.6 days

INTERACTIONS

Increase: brentuximab component action: ketoconazole, boceprevir, delavirdine, isoniazid, indinavir, itraconazole, dalfopristin; quinupristin, telithromycin, tipranavir, rifampin, ritonavir
Increase noninfectious pulmonary toxicity: bleomycin, do not use together

Drug/Herb
- Increased brentuximab component action: St. John's wort

NURSING CONSIDERATIONS

Assess:
- **Tumor lysis syndrome (TLS):** assess for hyperkalemia, hypophosphatemia, hypocalcemia; may develop renal failure may use allopurinol or rasburicase to prevent TLS; monitor serum BUN/creatinine
- **Pregnancy: determine if pregnancy is planned or suspected, pregnancy category D**
- **Peripheral neuropathy: progressive multifocal leukoencephalopathy:** assess for weakness or paralysis, vision loss, impaired speech, and cognitive deterioration; often fatal
- Monitor CBC, and differential

Evaluate:
- Decreasing symptoms of Hodgkin's disease (increased lymph nodes, night sweats, weight loss, splenomegaly, hepatomegaly)

Teach patient/family:
- To report immediately weakness, change in vision, impaired speech
- To use reliable contraception (pregnancy [D]); to avoid breastfeeding

brimonidine (ophthalmic)

(bri-moe′ni-deen)

Alphagan P

Func. class.: Antiglaucoma
Chem. class.: Selective alpha-2 agonist

Do not confuse:
brimonidine/bimatoprost

ACTION: Decrease aqueous humor and increase outflows

USES: Treatment of chronic open-angle glaucoma and ocular hypertension

CONTRAINDICATIONS: Hypersensitivity, AV block, heart failure, bradycardia, sick sinus syndrome, within 14 days of MAOIs therapy

Precautions: Breastfeeding, depression, cerebrovascular disease, hepatic/renal impairment, Raynaud's phenomenon, orthostatic hypotension, thromboangiitis obliterans

DOSAGE AND ROUTES

- **Adult/child >2 yr:** Instill 1 drop in the affected eye(s) tid

Available forms: Ophthalmic solution 0.1%, 0.15%, 0.2%

SIDE EFFECTS

CNS: Headache, dizziness, somnolence
CV: Hyper/hypotension
EENT: Eye stinging/burning, tearing, photophobia, change in vision, sinus infection

PHARMACOKINETICS

Peak ½-2 hr, half-life 2 hr

INTERACTIONS

Increase: intraocular pressure reduction—apraclonidine, dorzolamide, pilocarpine, timolol
Increase: effects of—CNS depressants
Decrease: B/P—beta blockers, antihypertensives
Decrease: brimonidine effect—tricyclic antidepressants, may cause HTN crisis MAOIs, linezolid

NURSING CONSIDERATIONS

Assess:

- Glaucoma: monitor intraocular pressure

Evaluate:

- Decreasing intraocular pressure

Teach patient/family:

- That drug is for ophthalmic use only
- Not to touch the tip of the dropper to the eye, fingertips, or other surface to prevent contamination
- To wash hands before and after use; to tilt the head back slightly and pull the lower eyelid down with the index finger to form a pouch; squeeze the prescribed number of drops into the pouch; close eyes to spread drops; to avoid excessive systemic absorption, apply finger pressure on the lacrimal sac for 1-2 min following use
- That if more than one topical ophthalmic drug product is being used, the drugs should be administered at least 5 min apart
- To avoid contamination or the spread of infection, do not use dropper for more than one person

brinzolamide ophthalmic

See Appendix B

bromfenac (ophthalmic)

(brom′fen-ak)

Bromday, Prolensa

Func. class.: Antiinflammatory (ophthalmic)

Chem. class.: Nonsteroidal antiflammatory drug

ACTION: The mechanism of action is thought to be ability to block prostaglandin synthesis by inhibiting cyclooxygenase 1 and 2. In studies performed in animal eyes, prostaglandins have been shown to produce disruption of the blood–aqueous humor barrier, vasodilation, increased vascular permeability, leukocytosis, and increased intraocular pressure

USES: To reduce pain and inflammation after cataract surgery

CONTRAINDICATIONS: Hypersensitivity to this product, sulfites, NSAIDs, salicylates

Precautions: Bleeding disorders, complicated ocular surgery, corneal denervation, diabetes mellitus, rheumatoid arthritis, dry eye syndrome, pregnancy (C), breastfeeding
Do not administer while wearing contact lenses

DOSAGE AND ROUTES

• **Adult:** Instill 1 drop (0.09 mg) into affected eye twice daily beginning 24 hr before cataract surgery, continued on the day of surgery and through the first 14 days of the postoperative period
Available forms: Ophthalmic solution 0.07%, 0.09%
Administer
• Apply topically to the eye
• Remove contact lenses before instilling solution; contact lenses should not be worn during use of this product
• Instruct patient on proper instillation of eye solution
• Do not touch the tip of the dropper to the eye, fingertips, or other surface
• Do not share bottle with other patients
• If more than one ophthalmic medication is being used, the medications should be administered at least 5 min apart

SIDE EFFECTS

CNS: Headache
EENT: Abnormal sensation in eye, conjunctival hyperemia, ocular irritation, ocular pain, ocular pruritus, conjunctival hyperemia, iritis, keratitis

PHARMACOKINETICS

Unknown

INTERACTIONS

Increase: corneal erosion, poor healing—topical corticosteroids
Increase: bleeding—anticoagulants

NURSING CONSIDERATIONS

Assess:
Eyes:
• For pain, inflammation, burning, redness after cataract surgery
• Identify if patient is using topical corticosteroids, anticoagulants; use cautiously in those using these products
Evaluate:
• Decreased pain and inflammation after cataract surgery
Teach patient/family:
• To apply topically to the eye
• To remove contact lenses before instilling solution; contact lenses should not be worn during use of this product
• Proper instillation of eye solution
• Not to touch the tip of the dropper to the eye, fingertips, or other surface
• Not to share bottle with other patients
• That if more than one ophthalmic medication is being used, the medications should be administered at least 5 min apart

bromocriptine (Rx)

(broe-moe-krip′teen)

Cycloset, Parlodel

Func. class.: Dopamine receptor agonist, antiparkinson agent
Chem. class.: Ergot alkaloid derivative

Do not confuse:
Parlodel/pindolol/Provera
bromocriptine/benztropine/brimonidine

ACTION: Inhibits prolactin release by activating postsynaptic dopamine receptors; activation of striatal dopamine receptors may be reason for improvement in Parkinson's disease

USES: Parkinson's disease, amenorrhea/galactorrhea caused by hyperprolactinemia, infertility, acromegaly, pituitary adenomas, adjunct for type 2 diabetes

Unlabeled uses: Neuroleptic malignant syndrome, alcoholism, premenstrual syndrome, mastalgia, cocaine withdrawal, premenstrual breast symptoms

CONTRAINDICATIONS:
Severe ischemic disease, uncontrolled hypertension, severe peripheral vascular disease; hypersensitivity to ergot, bromocriptine; migraine, preeclampsia

Precautions: Pregnancy (B), breastfeeding, children, renal/hepatic disease, pituitary tumors, peptic ulcer disease, sulfite hypersensitivity, pulmonary fibrosis, dementia, GI bleeding, bipolar disorder

DOSAGE AND ROUTES

Hyperprolactinemia

• **Adult: PO** 1.25-2.5 mg with meals; may increase by 2.5 mg q3-7days, usual range 2.5-15 mg/day, max 30 mg/day

Acromegaly

• **Adult: PO** 1.25-2.5 mg × 3 days at bedtime; may increase by 1.25-2.5 mg q3-7days; usual range 20-30 mg/day, max 100 mg/day

Parkinson's disease

• **Adult: PO** 1.25 mg bid with meals; may increase q2-4wk by 2.5 mg/day, max 100 mg/day; levodopa should be continued while bromocriptine is being instituted

Pituitary adenoma

• **Adult: PO** 1.25 mg bid-tid; may increase over several wk to 10-20 mg/day

Type 2 diabetes (Cycloset only)

• **Adult: PO** (initially) 0.8 mg daily in AM within 2 hr of waking; titrate by 0.8 mg/day no more than weekly to max 1.6-4.8 mg/day

Neuroleptic malignant syndrome (unlabeled)

• **Adult: PO** 2.5-10 mg tid

Cocaine withdrawal (unlabeled)

• **Adult: PO** 0.625 mg qid × 42 days

Alcoholism (unlabeled)

• **Adult: PO** 7.5 mg/day

Mastalgia (unlabeled)

• **Adult: PO** 2.5-7.5 bid starting 10-14 days before menses; discontinue when menses begins

Available forms: Caps 5 mg; tabs 2.5 mg; Cycloset tabs 0.8 mg

Administer:

• With meal to prevent GI symptoms

• At bedtime so that dizziness, orthostatic hypotension do not occur

• Store at room temp in tight, light-resistant container

SIDE EFFECTS

CNS: *Headache,* depression, restlessness, anxiety, nervousness, confusion, **seizures,** *hallucinations,* dizziness, fatigue, drowsiness, abnormal involuntary movements, psychosis

CV: Orthostatic hypotension, decreased B/P, palpitations, extrasystole, **shock,** dysrhythmias, bradycardia, **MI**

EENT: Blurred vision, diplopia, burning eyes, nasal congestion

GI: *Nausea, vomiting, anorexia,* cramps, constipation, diarrhea, dry mouth, GI hemorrhage

GU: Frequency, retention, incontinence, diuresis

INTEG: *Rash on face, arms;* alopecia; coolness, pallor of fingers, toes; peripheral edema

PHARMACOKINETICS

Peak 1-3 hr, duration 4-8 hr, 90%-96% protein bound, half-life 3 hr, metabolized by liver (inactive metabolites), 85%-98% of dose excreted in feces, >90% of absorbed dose undergoes 1st-pass metabolism

INTERACTIONS

• Disulfiram-like reaction: alcohol

Increase: action of antihypertensives, levodopa, chloramphenicol, probenecid, salicylates, sulfonamides

Decrease: action of bromocriptine—phenothiazines, oral contraceptives, progestins, estrogens, haloperidol, loxapine, methyldopa, metoclopramide, MAOIs, reserpine

Decrease/increase: effect of Cyclosert-CYP3A4 inhibitors/inducers

Decrease: effect of Cyclosert-butyrophenones, metoclopramide, phenothiazine, thioxanthenes

Drug/Lab Test

Increase: growth hormone, AST, ALT, CK, BUN, uric acid, alk phos

NURSING CONSIDERATIONS

Assess:

- B/P; establish baseline, compare with other readings; this product decreases B/P and causes orthostatic hypotension
- **Parkinson's symptoms:** pill rolling, shuffling gait, restlessness, tremors, postural instability before and during treatment
- **Neuroleptic malignant syndrome:** decreased temp, seizures, sweating, pulse indicates resolution of symptoms
- Change in size of soft-tissue volume with acromegaly
- **Pregnancy:** may cause postpartum conception, use pregnancy testing q4wk or if menstruation does not occur

Evaluate:

- Therapeutic response (Parkinson's disease): decreased dyskinesia, slow movements, drooling

Teach patient/family:

- That tabs may be crushed, mixed with food; Cyclosert to be taken within 2 hr of rising
- To change position slowly to prevent orthostatic hypotension
- To use contraceptives during treatment with this product; that pregnancy may occur; to use methods other than oral contraceptives/subdermal implants
- That therapeutic effect for Parkinson's disease may take 2 mo
- To avoid hazardous activity if dizziness occurs
- To report symptoms of MI immediately
- To take with food, avoid alcohol

budesonide (Rx)

(byoo-des′oh-nide)

Entocort EC, Pulmicort, Pulmicort Flexhaler, Rhinocort Aqua, Uceris

Func. class.: Glucocorticoid

Chem. class.: Nonhalogenated

ACTION: Prevents inflammation by depressing migration of polymorphonuclear leukocytes and fibroblasts, reversal of increased capillary permeability, and lysosomal stabilization; does not suppress hypothalamus or pituitary function

USES: Rhinitis; prophylaxis for asthma; Crohn's disease, ulcerative colitis

Unlabeled uses: Microscopic colitis, laryngotrache obronchitis (croup)

CONTRAINDICATIONS: Hypersensitivity, status asthmaticus

Precautions: Pregnancy (C), inhaled form (B); breastfeeding; children; TB; fungal, bacterial, systemic viral infections; ocular herpes simplex; nasal septal ulcers; hepatic disease (caps)

DOSAGE AND ROUTES

Rhinitis (Rhinocort Aqua)

- **Adult and child >12 yr: SPRAY/INH** 256 mcg/day (2 sprays in each nostril AM, PM or 4 sprays in each nostril AM)

Asthma

- **Adult: INH** 360 mcg bid, max 720 mcg bid

Crohn's disease/ulcerative colitis (Uceris)

- **Adult: PO** 9 mg/day AM × 8 wk

Laryngotracheobronchitis (croup) (unlabeled)

- **Infant ≥3 mo-child ≤5 yr: NEB** (Pulmicort Respules INH susp) 2 mg inhaled as a single dose

Available forms: Dry powder for INH 90, 180, 200 mcg/actuation (Pulmicort Flexhaler); 32 mcg/actuation (Rhinocort

Aqua) nasal spray; susp for INH 0.5 mg/2 ml, 0.25 mg/2 ml; cap 3 mg; ext rel tab (Uceris) 9 mg

Administer:

PO route (Crohn's disease/ulcerative colitis)

- Swallow caps whole; do not break, crush, chew
- May repeat 8-wk course if needed; may taper to 6 mg/day for 2 wk before cessation
- Store at 59° F-86° F (15° C-30° C); keep away from heat, open flame

SIDE EFFECTS

CNS: *Headache,* insomnia, hypertonia, syncope, dizziness, drowsiness

CV: Chest pain, hypertension, sinus tachycardia, palpitation

EENT: *Sinusitis, pharyngitis,* rhinitis, oral candidiasis

ENDO: Adrenal insufficiency, growth suppression in children

GI: Dry mouth, dyspepsia, nausea, vomiting, abdominal pain

MISC: Ecchymosis, fever, *hypersensitivity,* flulike symptoms, epistaxis, dysuria

MS: Back pain, myalgias, fractures

RESP: Nasal irritation, cough, nasal bleeding, *respiratory infections,* **bronchospasm**

PHARMACOKINETICS

Peak: Respules 4-6 wk, Rhinocort Aqua 2 wk, half-life 2-3.6 hr

Onset: Respules 2-8 days, Rhinocort Aqua 10 hr

Enters breast milk

INTERACTIONS

Increase: budesonide effect, CYP3A inhibitors, dose adjustment may be needed

- Avoid concurrent use of varicella live vaccine in pediatric patients

NURSING CONSIDERATIONS

Assess:

- Respiratory status: rate, rhythm, increase in bronchial secretions, wheezing, chest tightness; provide fluids to 2 L/day to decrease thickness of secretions; check for oral candidiasis
- **Bronchospasm:** stop treatment, give bronchodilator
- Viral infections: corticosteroid use can mask infections
- Increased intraocular pressure: discontinue use if this occurs

Evaluate:

- Therapeutic response: absence of asthma, rhinitis

Teach patient/family:

- To notify prescriber of pharyngitis, nasal bleeding, oral candidiasis
- Not to exceed recommended dose because adrenal suppression may occur
- To carry emergency ID that identifies steroid use
- To read and follow package directions
- To prevent exposure to infections (especially viral)
- To use good oral hygiene if using nebulizer or inhaler
- To avoid breastfeeding
- That burning or stinging may occur with first few doses of inhalation use
- That product is not a bronchodilator and not to be used for asthma; to use regularly
- Teach how to use as described in "administer"
- To notify prescriber if symptoms persist after wks, that results usually take 2 wk
- To notify prescriber if exposure to measles, chickenpox occurs

budesonide nasal agent

See Appendix B

bumetanide (Rx)

(byoo-met'a-nide)

Func. class.: Loop diuretic, antihypertensive

Chem. class.: Sulfonamide derivative

ACTION: Acts on ascending loop of Henle by inhibiting reabsorption of chloride, sodium

USES: Edema in CHF, renal/hepatic disease, heart failure

Unlabeled uses: Hypercalcemia, hypertension, ascites

CONTRAINDICATIONS: Hypersensitivity to sulfonamides, anuria, hepatic coma

Black Box Warning: Electrolyte imbalance

Precautions: Pregnancy (C), breastfeeding, neonates, ascites, severe renal disease, hepatic cirrhosis, blood dyscrasias, ototoxicity, hyperuricemia, hypokalemia, hyperglycemia, oliguria, hypomagnesemia, hypovolemia

Black Box Warning: Dehydration

DOSAGE AND ROUTES

- **Adult and adolescent: PO** 0.5-2.0 mg/day; may give 2nd or 3rd dose at 4-5 hr intervals, max 10 mg/day; may be given on alternate days or intermittently; **IV/IM** 0.5-1.0 mg; may give 2nd or 3rd dose at 2-3 hr intervals, not to exceed 10 mg/day
- **Child and infant (unlabeled): PO/IM/IV** 0.015-0.1 mg/kg daily or every other day, max 10 mg/day

Hypercalcemia (unlabeled)

- **Adult: IV** 1-2 mg q1-4hr to maintain urine output of 200-250 ml/hr; give saline before 1st dose of this product

Hypertension (unlabeled)

- **Adult and adolescent: PO** 0.5-2 mg/day, max 10 mg/day in 2 divided doses

Available forms: Tabs 0.5, 1, 2 mg; inj 0.25 mg/ml

Administer:

- In AM to avoid interference with sleep if using product as a diuretic; without regard to meals
- Potassium replacement if potassium is <3.0

PO route

- With food if nausea occurs; absorption may be decreased slightly

IV, direct route

- Direct IV undiluted slowly over 1-2 min through Y-tube, 3-way stopcock, or heplock

Intermittent IV INF route

- Dilute in LR, D_5W, 0.9% NaCl (rarely given by this method), give over 12 hr with renal disease

Syringe compatibilities: Doxapram

Y-site compatibilities: Acyclovir, alfentanil, allopurinol, amifostine, amikacin, aminocaproic acid, aminophylline, amiodarone, amoxicillin, amphotericin B lipid complex (Abelcet), amphotericin B liposome (AmBisome), anidulafungin, ascorbic acid injection, atenolol, atracurium, atropine, aztreonam, benztropine, bivalirudin, bleomycin, buprenorphine, butorphanol, calcium chloride/gluconate, CARBOplatin, caspofungin, cefamandole, ceFAZolin, cefepime, cefmetazole, cefonicid, cefoperazone, cefotaxime, cefoTEtan, cefOXitin, cefTAZidime, ceftizoxime, ceftobiprole, cefTRIAXone, cefuroxime, cephalothin, cephapirin, chloramphenicol, cimetidine, cisatracurium, CISplatin, cladribine, clarithromycin, clindamycin, codeine, cyanocobalamin, cyclophosphamide, cycloSPORINE, cytarabine, DACTINomycin, DAPTOmycin, dexamethasone, dexmedetomidine, digoxin, diltiazem, diphenhydrAMINE, DOBUTamine, DOCEtaxel, DOPamine, doripenem, doxacurium, DOXOrubicin, doxycycline, enalaprilat, ePHEDrine, EPINEPHrine, epirubicin, epoetin alfa, eptifibatide, ertapenem, erythromycin, esmolol, etoposide, famotidine, fentaNYL, filgrastim, fluconazole, fludarabine, fluorouracil, folic acid, furosemide, gatifloxacin, gemcitabine, gentamicin, glycopyrrolate, granisetron, heparin, hydrocortisone sodium succinate, HYDROmorphone, hydrOXYzine, IDArubicin, ifosfamide, imipenem-cilastatin, indomethacin, insulin (regular), irinotecan, isoproterenol, ketorolac, labetalol, levofloxacin, lidocaine, linezolid, LORazepam, magnesium sulfate, mannitol, mechlorethamine, melphalan, meperidine, metaraminol, methotrexate, methoxamine, methyldopate, methylPREDNISolone, metoclopramide, metoprolol, metroNIDAZOLE, mezlocillin, micafungin, miconazole, milrinone, mitoXANtrone, morphine, moxalactam, multiple vitamins injection, mycophenolate, nafcillin, nalbuphine, naloxone, netilmicin, nitroglycerin, nitroprusside, norepinephrine, octreotide, ondansetron, oxacillin, oxaliplatin, oxytocin, palonosetron, pamidronate, pancuronium, pantoprazole, PEMEtrexed, penicillin G potassium/sodium, pentazocine, PENTobarbital, PHENobarbital, phenylephrine, phytonadione, piperacillin, piperacillin-tazobactam, polymyxin B, potassium chloride, procainamide, promethazine, propofol, propranolol, protamine, pyridoxine, quiNIDine, ranitidine, remifentanil, rifampin, ritodrine, riTUXimab, rocuronium, sodium acetate, sodium bicarbonate, succinylcholine, SUFentanil, tacrolimus, teniposide, theophylline, thiamine, thiotepa, ticarcillin, ticarcillin-clavulanate, tigecycline, tirofiban, TNA, tobramycin, tolazoline, TPN, traMADol, trastuzumab, trimetaphan, urokinase, vancomycin, vasopressin, vecuronium, verapamil, vinCRIStine, vinorelbine, voriconazole

SIDE EFFECTS

CNS: *Headache,* fatigue, weakness, *dizziness,* encephalopathy
CV: **Chest pain,** *hypotension,* **circulatory collapse,** ECG changes, dehydration
EENT: *Loss of hearing*
ELECT: *Hypokalemia, hypochloremic alkalosis, hypomagnesemia, hyperuricemia, hypocalcemia, hyponatremia*
ENDO: *Hyperglycemia*
GI: *Nausea,* diarrhea, dry mouth, vomiting, anorexia, cramps, upset stomach, abdominal pain, **acute pancreatitis, jaundice**
GU: *Polyuria,* **renal failure,** glycosuria, premature ejaculation, hypercholesterolemia
HEMA: **Thrombocytopenia, leukopenia, granulocytopenia, hemoconcentration**
INTEG: *Rash, pruritus,* purpura, **Stevens-Johnson syndrome,** sweating, photosensitivity
MS: Muscular cramps, arthritis, stiffness

PHARMACOKINETICS

Excreted by kidneys (50% unchanged), feces (20%); crosses placenta; excreted in breast milk; protein binding >96%; half-life 1-1½ hr, 6-15 hr in neonates
PO: Onset ½-1 hr, peak 1-2 hr, duration 3-6 hr
IM: Onset 40 min, peak 1-2 hr, duration 4-6 hr
IV: Onset 5 min, peak 15-30 min, duration 3-6 hr

INTERACTIONS

- Ototoxicity: aminoglycosides
- Hypokalemia: potassium-wasting products

Increase: toxicity—lithium, digoxin
Increase: diuresis, electrolyte loss—metolazone
Decrease: diuretic effect—indomethacin, NSAIDs, probenecid
Decrease: antidiabetic effects—antidiabetics

Drug/Herb
Increase: effect—hawthorn, horse chestnut
Decrease: effect of bumetanide—ginseng, ephedra

NURSING CONSIDERATIONS

Assess:
- For tinnitus; obtain audiometric testing for long-term IV treatment
- Weight, I&O daily to determine fluid loss; if urinary output decreases or azote-

mia occurs, product should be discontinued; safest dosage schedule is alternate days

• B/P lying, standing; postural hypotension may occur

Black Box Warning: Electrolyte imbalances: Potassium, sodium, calcium; include BUN, blood glucose, CBC, serum creatinine, blood pH, ABGs, uric acid, calcium, magnesium; severe electrolyte imbalances should be corrected before starting treatment

• Blood glucose if patient is diabetic; blood uric acid levels in those with gout
• Improvement in edema of feet, legs, sacral area daily if medication is being used for CHF
• Signs of metabolic alkalosis: drowsiness, restlessness
• **Hypokalemia:** postural hypotension, malaise, fatigue, tachycardia, leg cramps, weakness
• Rashes, temp elevation daily
• Confusion, especially in geriatric patients; take safety precautions if needed
• **Digoxin toxicity** in patients taking digoxin products: anorexia, nausea, vomiting, confusion, paresthesia, muscle cramps; **lithium toxicity** in those taking lithium

Evaluate:

• Therapeutic response: decreased edema, B/P

Teach patient/family:

• To increase fluid intake to 2-3 L/day unless contraindicated; to take potassium supplement; to rise slowly from lying or sitting position
• To recognize adverse reactions: muscle cramps, weakness, nausea, dizziness
• To take with food, milk for GI symptoms; to avoid alcohol
• To take early in day to prevent nocturia
• To use sunscreen to prevent photosensitivity

TREATMENT OF OVERDOSE:

Lavage if taken orally; monitor electrolytes; administer dextrose in saline; monitor hydration, CV, renal status

buprenorphine (Rx)

(byoo-pre-nor′feen)

Buprenex, Butrans

Func. class.: Opioid analgesic, partial agonist

Chem. class.: Thebaine derivative

Controlled Substance Schedule V (Parenteral); Schedule III (Tablet, TD)

Do not confuse:
Buprenex/Bumex

ACTION:

Depresses pain impulse transmission at the spinal cord level by interacting with opioid receptors

USES:

Moderate to severe pain, opiate agonist withdrawal

Unlabeled uses: Cocaine withdrawal

CONTRAINDICATIONS:

Hypersensitivity, ileus, status asthmaticus

Black Box Warning: Respiratory depression

Precautions: Pregnancy (C), breastfeeding, substance abuse/alcoholism, increased intracranial pressure, MI (acute), severe heart disease, respiratory depression, renal/hepatic/pulmonary disease, hypothyroidism, Addison's disease

Black Box Warning: QT prolongation, accidental exposure, potential for overdose/poisoning, substance abuse

DOSAGE AND ROUTES

• **Adult: IM/IV** 0.3 mg q6hr prn, reduce dosage in geriatric patients, may repeat after 30-60 min; **EPIDURAL** (unlabeled) 4 mcg/kg or 2 mcg/kg (epidural inj); **TD** each patch is worn for 7 days (moderate-severe pain); **opioid-naive patients** (those taking <30 mg of oral morphine or equivalent before beginning treatment with TD buprenorphine), 5 mcg/hr q7days, overestimating dose can be fatal;

conversion from other opiate agonist therapy, titrate from other opioids for up to 7 days to no more than 30 mg oral morphine or equivalent before beginning TD therapy, begin with 5 mcg/hr q7days; for those with daily dose of 30-80 mg oral morphine or equivalent, start with 10 mcg/hr q7days; for those taking >80 mg oral morphine or equivalent, start with 20 mcg/hr q7days

• **Child 2-12 yr: IM/IV** 2-6 mcg/kg q4-6hr

Available forms: Inj 0.3 mg/ml (1-ml vials); SL tab 2, 8 mg as base; TD system 5, 10, 20 mcg/hr (weekly)

Administer:

SL route

• Do not chew, dissolve under tongue, use 2 or more at same time

Transdermal route

• Apply to clean, dry, intact skin; each patch should be worn for 7 days; do not apply direct heat source to patch, will increase absorption of product, may use first aid tape if edge of patch is not adhering

• Apply to upper outer arm, upper chest/back, or side of chest

IM route

• In deep muscle mass

IV, direct route

• **Give** undiluted over ≥2 min, titrate to patient response; rapid injection will increase side effects

• With antiemetic if nausea, vomiting occur

• When pain is beginning to return; determine dosage interval by patient response

Additive compatibilities: Bupivacaine, glycopyrrolate, haloperidol

Syringe compatibilities: Glycopyrrolate, haloperidol, heparin, midazolam

Y-site compatibilities: Acyclovir, alfentanil, allopurinol, amifostine, amikacin, aminocaproic acid, amphotericin B liposome (AmBisone), anidulafungin, ascorbic acid injection, atenolol, atracurium, atropine, aztreonam, benztropine, bivalirudin, bleomycin, bumetanide, butorphanol, calcium chloride/gluconate, CARBOplatin, cefamandole, ceFAZolin, cefepime, cefoperazone, cefotaxime, cefoTEtan, cefOXitin, cefTAZidime, ceftizoxime, cefTRIAXone, cefuroxime, chloramphenicol, chlorproMAZINE, cimetidine, cisatracurium, CISplatin, cladribine, clindamycin, cyanocobalamin, cyclophosphamide, cycloSPORINE, cytarabine, D_5W-dextrose 5%, DACTINomycin, DAPTOmycin, dexamethasone, dexmedetomidine, digoxin, diltiazem, diphenhydrAMINE, DOBUTamine, DOCEtaxel, DOPamine, doxacurium, DOXOrubicin HCl, doxycycline, enalaprilat, ePHEDrine, EPINEPHrine, epirubicin, epoetin alfa, eptifibatide, ertapenem, erythromycin, esmolol, etoposide, famotidine, fenoldopam, fentaNYL, filgrastim, fluconazole, fludarabine, gatifloxacin, gemcitabine, gentamicin, glycopyrrolate, granisetron, heparin, hydrocortisone, hydrOXYzine, IDArubicin, ifosfamide, imipenem-cilastatin, inamrinone, insulin (regular), irinotecan, isoproterenol, ketorolac, labetalol, lactated Ringer's injection, levofloxacin, lidocaine, linezolid, LORazepam, magnesium sulfate, mannitol, mechlorethamine, melphalan, meperidine, metaraminol, methicillin, methotrexate, methoxamine, methyldopate, methylPREDNISolone, metoclopramide, metoprolol, metroNIDAZOLE, mezlocillin, miconazole, midazolam, milrinone, minocycline, mitoXANtrone, morphine, moxalactam, multiple vitamins injection, mycophenolate mofetil, nafcillin, nalbuphine, naloxone, nesiritide, netilmicin, nitroglycerin, nitroprusside, norepinephrine, octreotide, ondansetron, oxacillin, oxaliplatin, oxytocin, palonosetron, pamidronate, pancuronium, papaverine, PEMEtrexed, penicillin G potassium/sodium, pentamidine, pentazocine, phenylephrine, phytonadione, piperacillin, piperacillin-tazobactam, polymyxin B, potassium chloride, procainamide, prochlorperazine, promethazine, propofol, propranolol, protamine, pyridoxine, quiNIDine, ranitidine, remifentanil, Ringer's injection, riTUXimab, rocuronium, sodium acetate, succinylcholine, SUFentanil, tacrolimus, teniposide, theophylline, thiamine, thiotepa, ticarcillin, ticarcillin-clavulanate, tigecycline, tirofiban, TNA (3-in-1), tobramycin, tolazoline, TPN, trastuzumab, trimetaphan, urokinase, vancomycin, vasopressin, vecuronium, verapamil, vinCRIStine, vinorelbine, voriconazole

SIDE EFFECTS

CNS: *Drowsiness, dizziness, confusion, headache, sedation, euphoria,* increased intracranial pressure, amnesia
CV: Palpitations, bradycardia, change in B/P, tachycardia
EENT: Tinnitus, blurred vision, *miosis,* diplopia
GI: *Nausea,* vomiting, anorexia, constipation, cramps, dry mouth
GU: Dysuria, urinary retention
INTEG: *Rash,* urticaria, bruising, flushing, diaphoresis, pruritus
RESP: Respiratory depression, dyspnea, hypo/hyperventilation

PHARMACOKINETICS

Metabolized in liver by CYP3A4, excreted by kidneys and in feces, crosses placenta, excreted in breast milk, half-life 2½-3½ hr, 96% bound to plasma proteins
IM: Onset 15 min, peak 1 hr, duration 6 hr
SL: Onset, peak, duration unknown, half-life 37 hr
IV: Onset 1 min, peak 5 min, duration 6 hr, half-life 2.2 hr
TD: Half-life 26 hr
Epidural: Duration dose dependent

INTERACTIONS

Increase: effect with other CNS depressants—alcohol, opioids, sedative/hypnotics, antipsychotics, skeletal muscle relaxants, MAOIs
Increase: buprenorphine effect—CYP3A4 inhibitors (erythromycin, indinavir, ketoconazole, ritonavir, saquinavir)
Increase: QT prolongation—class IA, III antidysrhythmics
Decrease: buprenorphine effect—CYP3A4 inducers (carBAMazepine, PHENobarbital, phenytoin, rifampin)
Drug/Herb
Increase: CNS depression—St. John's wort

NURSING CONSIDERATIONS

Assess:
- **Pain:** intensity, location, type before treatment, after 5 min (IV); need for pain medication, tolerance
- I&O ratio; check for decreasing output; may indicate urinary retention
- Bowel pattern; severe constipation can occur
- CNS changes, dizziness, drowsiness, hallucinations, euphoria, LOC, pupil reaction; withdrawal in opioid-dependent persons; if dependence occurs, within 2 wk of discontinuing product **withdrawal symptoms** will occur
- Allergic reactions: rash, urticaria
- Respiratory dysfunction: respiratory depression, character, rate, rhythm; notify prescriber if respirations are <12/min

Black Box Warning: Potential for overdose may occur from chewing, swallowing, snorting, or injecting extracted product from TD formulation

Black Box Warning: QT prolongation: in those taking class Ia, III antidysrhythmics; patients with hypokalemia, cardiac instability (TD), max TD 20 mcg/hr q7day

Evaluate:
- Therapeutic response: decrease in pain, absence of grimacing

Teach patient/family:
- To report any symptoms of CNS changes, allergic reactions

Black Box Warning: That psychologic dependence leading to substance abuse may result when used for extended periods; that long-term use not recommended

- To avoid hazardous activities such as driving unless reaction known

B

TREATMENT OF OVERDOSE:
Naloxone 0.4 mg ampule diluted in 10 ml 0.9% NaCl given by direct IV push 0.02 mg q2min (adult)

buPROPion (Rx)
(byoo-proe′pee-on)

Aplenzin, Budeprion SR, Budeprion XL, Buproban, Wellbutrin, Wellbutrin SR, Wellbutrin XL, Zyban

Func. class.: Antidepressant—miscellaneous smoking deterrent

Chem. class.: Aminoketone

Do not confuse:
buPROPion/busPIRone
Zyban/Diovan/Zagam

ACTION: Inhibits reuptake of DOPamine, norepinephrine, serotonin

USES: Depression (Wellbutrin), smoking cessation (Zyban); seasonal affective disorder

Unlabeled uses: Neuropathic pain, enhancement of weight loss, ADHD (attention-deficit/hyperactivity disorder)

CONTRAINDICATIONS: Hypersensitivity, head trauma, stroke, intracranial mass, eating disorders, seizure disorders

Precautions: Pregnancy (C), breastfeeding, geriatric patients, renal/hepatic disease, recent MI, cranial trauma, seizure disorder

Black Box Warning: Children <18 yr, suicidal thinking/behavior (young adults)

DOSAGE AND ROUTES

Depression

• **Adult: PO** 100 mg bid initially then increase after 3 days to 100 mg tid if needed, max 150 mg single dose; **ER/SR** initially 150 mg AM, increase to 300 mg/day if initial dose is tolerated, after no less than 4 days; after several wk, titrate to 200 mg bid; Aplenzin 174 mg q AM, may increase to 348 mg q AM on day 4, may increase to 522 mg after several weeks if needed

• **Geriatric: PO** 50-100 mg/day, may increase by 50-100 mg q3-4days

Smoking cessation (Zyban)

• **Adult: SR** 150 mg q day × 3 days then 150 mg bid for remainder of treatment, initiate 1-2 wk before targeted "quit day," continue for 7-12 wk; in combination with nicotine TD, 150 mg q day × 3 days then 150 mg bid for remainder of treatment, give ≥8 hr apart, max 300 mg/day, initiate 1-2 wk before targeted "quit day," continue for 7-12 wk, may be continued for 8-20 wk

Seasonal affective disorder

• **Adult: PO** (Wellbutrin XL) 150 mg as a single dose in the AM, after 1 wk may be increased to 300 mg/day; (Aplenzin) 174 mg q day in AM, after 7 days may increase to 348 mg q day

ADHD (unlabeled) (Wellbutrin)

• **Adult: PO** 100 mg bid, after ≥3 days titrate to 100 mg tid; **SR** 300 mg/day, 200 mg 8 AM, 100 mg 4 PM

Diabetic neuropathy/postherpetic neuralgia (unlabeled)(Wellbutrin SR)

• **Adult: PO** SR 150-300 mg/day

Available forms: Tabs 75, 100 mg; sus rel tabs (SR) 150; ext rel tab (XL) 100, 150, 300 mg; (SR-12 hr, XL-24 hr); tab ext rel (Aplenzin) 174, 348, 522 mg

Administer:

PO route

⚠ When switching to Aplenzin from Wellbutrin, Wellbutrin SR or XL, use these equivalents: 174 mg buPROPion HBr = 150 mg buPROPion HCl; 348 mg buPROPion HBr = 300 mg buPROPion HCl; 522 mg buPROPion HBr = 450 mg buPROPion HCl

• **Wellbutrin immediate rel,** separate by ≥6 hr, give in 3 divided doses; **Wellbutrin SR,** if multiple doses are used, separate by ≥8 hr; **Wellbutrin XL,** give daily in AM; **Zyban SR,** give in 2 divided doses, ≥8 hr apart; **Aplenzin ER,** give daily in AM, a larger dose of Aplenzin is needed because these products are not equivalent

- Do not break, crush, chew sus rel, ext rel tab
- At evenly spaced times to prevent seizures; seizure risk increases with high doses
- Increase fluids, bulk in diet if constipation occurs
- With food, milk for GI symptoms
- Sugarless gum, hard candy, frequent sips of water for dry mouth
- Avoid giving at night to prevent insomnia

SIDE EFFECTS

CNS: *Headache, agitation, dizziness, akinesia, bradykinesia, confusion,* seizures, delusions, *insomnia, sedation, tremors,* suicidal ideation, mania, hot flashes

CV: *Dysrhythmias, hypertension,* palpitations, *tachycardia,* hypotension, complete AV block; QRS prolongation (overdose)

EENT: *Blurred vision, auditory disturbance,* tinnitus

GI: *Nausea, vomiting,* anorexia, diarrhea, *dry mouth,* increased appetite, *constipation,* altered taste

GU: Impotence, urinary frequency, retention, *menstrual irregularities,* nocturia, altered libido

INTEG: *Rash,* pruritus, *sweating,* Stevens-Johnson syndrome

MISC: *Weight loss or gain*

PHARMACOKINETICS

Onset 1-4 wk, half-life 14 hr, extensively metabolized by liver, some conversion to active metabolites, steady state 5-8 days, protein binding 84%, excreted in urine and feces

INTERACTIONS

- Do not use within 14 days of MAOIs

⚠ **Increase:** adverse reactions, seizures—levodopa, MAOIs, phenothiazines, antidepressants, benzodiazepines, alcohol, theophylline, systemic steroids

Increase: buPROPion toxicity—ritonavir

Increase: buPROPion level—cimetidine

Increase: buPROPion effect—CYP2D6/CYP2B6 inhibitors

Decrease: effect of tamoxifen

Decrease: buPROPion effect—carBAMazepine, cimetidine, PHENobarbital, phenytoin or other products (CYP2D6); CYP2B6 inducers

Drug/Herb

Increase: CNS depression—kava, valerian

Drug/Lab Test

Positive: urine drug screen for amphetamine possible

NURSING CONSIDERATIONS

Assess:

- Hepatic/renal function in patients with hepatic, kidney impairment
- For increased risk of seizures; if patient has excessively used CNS depressants and OTC stimulants, dosage of buPROPion should not be exceeded
- Monitor weight regularly
- For smoking cessation after 7-12 wk; if progress has not been made, product should be discontinued

Black Box Warning: Mental status: mood, sensorium, affect, suicidal tendencies, increase in psychiatric symptoms

- Assistance with ambulation during beginning therapy because sedation occurs
- Safety measures, primarily for geriatric patients

Evaluate:

- Therapeutic response: decreased depression, ability to perform daily activities, ability to sleep throughout the night, smoking cessation

Teach patient/family:

- That therapeutic effects may take 2-4 wk; not to increase dose without prescriber's approval; that treatment for smoking cessation lasts 7-12 wk
- To use caution when driving, performing other activities that require alertness; sedation, blurred vision may occur

B

• To avoid alcohol, other CNS depressants; alcohol may increase risk of seizures
• Not to use with nicotine patches unless directed by prescriber; may increase B/P
• To notify prescriber immediately if urinary retention occurs
• **That risk of seizures increased when dose exceeded, if patient has seizure disorder**

Black Box Warning: That suicidal ideas, behaviors, hostility, depression may occur in children or young adults

• To notify prescriber if pregnancy is suspected, planned

TREATMENT OF OVERDOSE:

ECG monitoring; lavage, activated charcoal; administer anticonvulsant

busPIRone (Rx)

(byoo-spye′rone)

BuSpar, BuSpar Dividose

Func. class.: Antianxiety, sedative

Chem. class.: Azaspirodecanedione

Do not confuse:
busPIRone/buPROPion

ACTION: Acts by inhibiting the action of serotonin (5-HT); has shown little potential for abuse; a good choice with substance abuse

USES: Management and short-term relief of generalized anxiety disorders

Unlabeled uses: Autism

CONTRAINDICATIONS: Children <18 yr, hypersensitivity

Precautions: Pregnancy (B), breastfeeding, geriatric patients, impaired hepatic/renal function

DOSAGE AND ROUTES

• **Adult: PO** 7.5 mg bid; may increase by 5 mg/day q2-3days, max 60 mg/day

Autism with anxiety (unlabeled)

• **Adult: PO** 5-15 mg tid after titration, max 60 mg/day
• **Child ≥5 yr: PO** 0.2-0.6 mg/kg/day, max 60 mg/day; titrate to higher dose

Hepatic/renal dose

• **Adult: PO** reduce by 25%-50% for mild-moderate hepatic disease; do not use for severe hepatic disease; CCr 11-70 ml/min reduce by 25%-50%, CCr <10 ml/min do not use

Available forms: Tabs 5, 7.5, 10, 15, 30 mg

Administer:

• With food, milk for GI symptoms; avoid grapefruit juice; give drug at same time of day, with/without food consistently
• Crushed if patient unable to swallow medication whole
• Sugarless gum, hard candy, frequent sips of water for dry mouth

SIDE EFFECTS

CNS: *Dizziness, headache, depression, stimulation, insomnia, nervousness, lightheadedness, numbness, paresthesia, incoordination,* nightmares, *tremors,* excitement, involuntary movements, confusion, akathisia, hostility

CV: *Tachycardia, palpitations,* hypo/hypertension, **CVA, CHF, MI**

EENT: *Sore throat, tinnitus, blurred vision, nasal congestion;* red, itching eyes; change in taste, smell

GI: *Nausea, dry mouth, diarrhea, constipation,* flatulence, increased appetite, rectal bleeding

GU: Frequency, hesitancy, menstrual irregularity, change in libido

INTEG: *Rash,* edema, pruritus, alopecia, dry skin

MISC: *Sweating,* fatigue, weight gain, fever, **serotonin syndrome**

MS: *Pain, weakness,* muscle cramps, spasms

RESP: Hyperventilation, chest congestion, shortness of breath

PHARMACOKINETICS

Peak 40-90 min, half-life 2-3 hr, rapidly absorbed, metabolized by liver (CYP3A4), excreted in feces, protein binding 86%

INTERACTIONS

Increase: busPIRone—product metabolized by CYP3A4 (erythromycin, itraconazole, nefazodone, ketoconazole, ritonavir, several other protease inhibitors)
Increase: B/P—procarbazine, MAOIs; do not use together
Increase: CNS depression—psychotropic products, alcohol (avoid use)
Increase: serotonin syndrome—SSRIs, SNRIs, serotonin receptor agonists
Decrease: busPIRone effects—rifampin
Decrease: busPIRone action—products induced by CYP3A4 (rifampin, phenytoin, PHENobarbital, carBAMazepine, dexamethasone)
Drug/Food
Increase: peak concentration of busPIRone—grapefruit juice

NURSING CONSIDERATIONS

Assess:
- B/P lying, standing; pulse; if systolic B/P drops 20 mm Hg, hold product, notify prescriber
- CNS reactions because some may be unpredictable
- Mental status: mood, sensorium, affect, sleeping pattern, drowsiness, dizziness; withdrawal symptoms when dose reduced, product discontinued
- Safety measures if drowsiness, dizziness occurs

Evaluate:
- Therapeutic response: decreased anxiety, restlessness, sleeplessness

Teach patient/family:
- That product may be taken consistently with/without food
- To avoid OTC preparations, alcohol ingestion, other psychotropic medications unless approved by prescriber; to avoid large amounts of grapefruit juice
- To avoid activities that require alertness because drowsiness may occur
- Not to discontinue medication abruptly after long-term use; if dose missed, do not double
- To rise slowly because fainting may occur, especially among geriatric patients
- That drowsiness may worsen at beginning of treatment; that 1-2 wk of therapy may be required before therapeutic effects occur
- Serotonin syndrome: to report immediately (fever, tremor, sweating, diarrhea, delirium)

HIGH ALERT

busulfan (Rx)

(byoo-sul′fan)

Busulfex, Myleran

Func. class.: Antineoplastic alkylating agent

Chem. class.: Bifunctional alkylating agent

Do not confuse:
Myleran/Leukeran

ACTION: Changes essential cellular ions to covalent bonding with resultant alkylation; this interferes with the normal biological function of DNA; activity is not phase-specific; action is due to myelosuppression

USES: Chronic myelocytic leukemia, bone marrow ablation, stem cell transplant preparation with CML

CONTRAINDICATIONS: Pregnancy (D) 3rd trimester, breastfeeding, radiation, chemotherapy, blastic phase of chronic myelocytic leukemia, hypersensitivity

Precautions: Women of childbearing age, leukopenia, anemia, hepatotoxicity, renal toxicity, seizures, tumor lysis syndrome, hyperkalemia, hyperphosphatemia, hypocalcemia, hyperuricemia

Black Box Warning: Thrombocytopenia, neutropenia, secondary malignancy

DOSAGE AND ROUTES

Chronic myelocytic leukemia

- **Adult: PO** 4-8 mg/day or 1.8-4 mg/m^2/day initially, reduce dose if WBC reaches 30,000-40,000/mm^3, discontinue if WBC ≤20,000/mm^3, maintenance 1-3 mg/day
- **Child: PO** 0.06-0.12 mg/kg/day or 1.8-4.6 mg/m^2/day; reduce if WBC reaches 30,000-40,000/mm^3, discontinue if WBC ≤20,000/mm^3

Allogenic hemopoietic stem cell transplantation with chronic myelogenous leukemia

- **Adult: IV** 0.8 mg/kg over 2 hr, q6hr × 4 days (total 16 doses); give cyclophosphamide **IV** 60 mg/kg over 1 hr daily for 2 days starting after 16th dose of busulfan; **PO** (unlabeled) 1 mg/kg q6hr × 16 doses
- **Adolescent and child (unlabeled) >12 kg: IV** 0.8 mg/kg over 2 hr q6hr × 16 doses (4 days) then high-dose cyclophosphamide 50 mg/kg/day × 4 days
- **Infant/child ≤12 kg (unlabeled): IV** 1.1 mg/kg over 2 hr q6hr × 16 doses (4 days) then high-dose cyclophosphamide 50 mg/kg/day × 4 days

Available forms: Tabs 2 mg; inj 6 mg/ml

Administer:

- Store in tight container

PO route

- Give at same time daily on empty stomach

Intermittent IV INF route

- Prepare in biologic cabinet while wearing gloves, gown, mask; **dilute** with 10 times volume of product with D_5W, 0.9% NaCl, (0.5 mg/ml); when withdrawing product, use needle with 5-micron filter provided, remove amount needed, remove filter, and **inject** product into diluent; always add product to diluent (not vice versa); stable for 8 hr at room temp (using D_5W) or 12 hr refrigerated; **give** by central venous catheter over 2 hr q6hr × 4 days, use inf pump, do not admix
- Give antiemetics before IV route on schedule
- In those with history of seizures, give phenytoin before IV drug to prevent seizures (using 0.9% NaCl)

Y-site compatibilities: Acyclovir, amphotericin B lipid complex, amphotericin B liposome, anidulafungin, atenolol, bivalirudin, bleomycin, caspofungin, codeine, DAPTOmycin, dexmedetomidine, diltiazem, DOCEtaxel, ertapenem, fenoldopam, gatifloxacin, granisetron, HYDROmorphone, levofloxacin, linezolid, LORazepam, meperidine, metroNIDAZOLE, milrinone, nesiritide, octreotide acetate, ondansetron, palonosetron, pancuronium, piperacillin-tazobactam, riTUXimab, sodium acetate, tacrolimus, tigecycline, tirofiban, trastuzumab, vasopressin

SIDE EFFECTS

PO route

CV: *Hypotension,* thrombosis, *chest pain,* tachycardia, atrial fibrillation, heart block, pericardial effusion, cardiac tamponade (high dose with cyclophosphamide)

GI: *Anorexia, constipation, diarrhea, dry mouth, nausea, vomiting*

RESP: Alveolar hemorrhage, atelectasis, cough, hemoptysis, hypoxia, pleural effusion, pneumonia, sinusitis, pulmonary fibrosis

IV route

CNS: Cerebral hemorrhage, coma, seizures, *anxiety, depression, dizziness, headache,* encephalopathy, *weakness,* mental changes

EENT: *Pharyngitis, epistaxis,* cataracts

GI: Nausea, vomiting, *diarrhea, weight loss*

GU: Impotence, sterility, amenorrhea, gynecomastia, renal toxicity, hyperuremia, adrenal-insufficiency–like syndrome

HEMA: Thrombocytopenia, leukopenia, pancytopenia, severe bone marrow depression

INTEG: Dermatitis, hyperpigmentation, alopecia
OTHER: Chromosomal aberrations
RESP: Irreversible pulmonary fibrosis, pneumonitis

PHARMACOKINETICS

Well absorbed orally, excreted in urine, crosses placenta, excreted in breast milk, half-life 2.5 hr

INTERACTIONS

Increase: hepatotoxicity—thioguanine
Increase: cardiac tamponade—cyclophosphamide
Increase: toxicity—other antineoplastics, radiation
Increase: risk for bleeding—anticoagulants, salicylates
Increase: antibody response—live virus vaccines
Decrease: busulfan level—phenytoin
Decrease: busulfan clearance—acetaminophen, itraconazole
Drug/Lab Test
False positive: breast, bladder, cervix, lung cytology tests

NURSING CONSIDERATIONS

Assess:

Black Box Warning: CBC, differential, platelet count weekly; withhold product if WBC is <15,000/mm^3 or platelet count is <150,000/mm^3; notify prescriber of results; institute thrombocytopenia precautions; levels for withholding product will be different for children

Black Box Warning: Bone marrow status before chemotherapy; seizure history; bone marrow suppression may be prolonged (up to 2 mo)

- **Pulmonary fibrosis:** pulmonary function tests, chest x-ray films before, during therapy; chest film should be obtained q2wk during treatment; pulmonary fibrosis may occur up to 10 yr after treatment with busulfan
- Renal studies: BUN, serum uric acid, urine CCr before, during therapy; monitor ALT, alk phos, bilirubin, uric acid before and during treatment; I&O ratio; report fall in urine output of <30 ml/hr; hyperuricemia

Black Box Warning: For secondary malignancy within 5-8 yr of chronic oral therapy, long-term follow-up may be required

- Monitor for cold, fever, sore throat (may indicate beginning infection)
- Bleeding: hematuria, guaiac, bruising, petechiae; mucosa, orifices q8hr; no rectal temps
- Dyspnea, crackles, nonproductive cough, chest pain, tachypnea
- Inflammation of mucosa, breaks in skin; use viscous xylocaine for oral pain
- Comprehensive oral hygiene
- Strict medical asepsis, protective isolation if WBC levels low
- Increased fluid intake to 2-3 L/day to prevent urate deposits, calculi formation

Evaluate:

- Therapeutic response: decreased exacerbations of chronic myelocytic leukemia

Teach patient/family:

- To avoid use of products that contain aspirin, ibuprofen; razors; commercial mouthwash
- To use effective contraception during and for at least 3 mo after treatment; to avoid breastfeeding; may cause infertility, discuss family planning before initiating therapy, pregnancy (D)
- To report signs of **anemia** (fatigue, headache, irritability, faintness, shortness of breath); symptoms of **infection;** jaundice; persistent cough, congestion, skin pigmentation, darkening of skin; sudden weakness, weight loss (may resemble adrenal insufficiency)
- To report symptoms of bleeding (hematuria, tarry stools)
- To avoid vaccinations, crowds, persons with known infections
- That impotence, amenorrhea can occur; that these are reversible after discontinuing treatment

butoconazole vaginal antifungal

See Appendix B

butorphanol (Rx)

(byoo-tor′fa-nole)

Func. class.: Opioid analgesic

Chem. class.: Mixed opioid antagonist, partial agonist

Controlled Substance Schedule IV

ACTION:

Depresses pain impulse transmission at the spinal cord level by interacting with opioid receptors

USES:

Moderate to severe pain, general anesthesia induction/maintenance, headache, migraine, preanesthesia

Unlabeled uses: Pruritus

CONTRAINDICATIONS:

Hypersensitivity to product, preservative; addiction (opioid)

Precautions: Pregnancy (C), breastfeeding, children <18 yr, addictive personality, increased intracranial pressure, respiratory depression, renal/hepatic disease, ileus, COPD

DOSAGE AND ROUTES

Moderate-severe pain

- **Adult: IM** 1-4 mg q3-4hr prn; **IV** 0.5-2 mg q3-4hr prn; **INTRANASAL** 1 spray in 1 nostril, may give another dose 1-1½ hr later; repeat if needed 3-4hr after last dose
- **Geriatric: IV** ½ adult dose at 2× the interval; **INTRANASAL** if no relief after 90-120 min, may repeat with 1 spray

Renal/hepatic dose

- **Adult: INTRANASAL** max 1 mg followed by 1 mg after 90-120 min; **IM/IV** give 50% of dose (0.5 mg **IV**, 1 mg **IM**), do not repeat within 6 hr

Opioid-induced pruritus (unlabeled)

- **Adult: INTRANASAL** 1 mg (1 spray) in each nostril q4-6hr

Intractable pruritus with inflammatory skin or systemic disease (unlabeled)

- **Adult: INTRANASAL** 1-4 mg/day

Available forms: Inj 1, 2 mg/ml; nasal spray 10 mg/ml

Administer:

- With antiemetic if nausea, vomiting occur
- When pain beginning to return; determine dosage interval according to patient response
- Store in light-resistant container at room temp

Nasal route

- Prime before first use, point sprayer away from the face, pump activator 7 × until a fine, wide spray occurs if not used for 48 hr, reprime by pumping 1-2 ×
- If more than 1 spray is needed, use other nostril
- Do not share with others
- Nasal congestion/irritation may occur

IM route

- Deeply in large muscle mass

IV direct route

- Undiluted at a rate of <2 mg/>3-5 min, titrate to patient response; inject directly in vein or tubing of free-flowing compatible IV inf

Syringe compatibilities: Atropine, chlorproMAZINE, cimetidine, diphenhydrAMINE, droperidol, fentaNYL, hydrOXYzine, meperidine, methotrimeprazine, metoclopramide, midazolam, morphine, pentazocine, perphenazine, prochlorperazine, promethazine, scopolamine, thiethylperazine

Y-site compatibilities: Acyclovir, alfentanil, allopurinol, amifostine, amikacin, aminocaproic acid, aminophylline, amphotericin B liposome (AmBisome), anidulafungin, ascorbic acid injection, atenolol, atracurium, atropine, aztreonam, benztropine, bivalirudin, bleomycin, bumetanide, buprenorphine, calcium chloride/gluconate, CARBOplatin, caspofungin, cefamandole, ceFAZolin, cefepime, cefoperazone, cefotaxime, cefoTEtan, cefOXitin, cefTAZidime, ceftizoxime, cefTRIAXone, cefuroxime, cephalothin, chlorproMAZINE, cimetidine, cisatracurium, CISplatin, cladribine, clindamycin, cyanocobalamin, cyclophosphamide, cycloSPORINE, cytarabine, DACTINomycin, DAPTOmycin, dexamethasone phosphate, dexmedetomidine, digoxin, diltiazem, diphenhydrAMINE, DOBUTamine, DOCEtaxel, DOPamine, doxacurium, DOXOrubicin, DOXOrubicin liposomal, doxycycline, enalaprilat, ePHEDrine, EPINEPHrine, epirubicin, epoetin alfa, eptifibatide, ertapenem, erythromycin, esmolol, etoposide, famotidine, fenoldopam, fentaNYL, filgrastim, fluconazole, fludarabine, fluorouracil, gatifloxacin, gemcitabine, gentamicin, glycopyrrolate, granisetron, heparin, hydrocortisone, hydrOXYzine, IDArubicin, ifosfamide, imipenem-cilastatin, irinotecan, isoproterenol, ketorolac, labetalol, lactated Ringer's injection, levofloxacin, lidocaine, linezolid injection, LORazepam, magnesium, mannitol, mechlorethamine, melphalan, meperidine, metaraminol, methicillin, methotrexate, methoxamine, methyldopate, methylPREDNISolone, metoclopramide, metoprolol, metroNIDAZOLE, mezlocillin, milrinone, minocycline, mitoXANtrone, morphine, multiple vitamins injection, mycophenolate mofetil, nafcillin, nalbuphine, naloxone, nesiritide, netilmicin, niCARdipine, nitroglycerin, nitroprusside, norepinephrine, octreotide, ondansetron, oxacillin, oxaliplatin, oxytocin, palonosetron, pamidronate, pancuronium, papaverine, PEMEtrexed, penicillin G potassium/sodium, pentazocine, PHENobarbital, phenylephrine, phytonadione, piperacillin, piperacillin-tazobactam, polymyxin B, potassium chloride, procainamide, prochlorperazine, promethazine, propofol, propranolol, protamine, pyridoxine, quiNIDine, ranitidine, remifentanil, Ringer's injection, riTUXimab, rocuronium, sargramostim, sodium acetate, succinylcholine, SUFentanil, tacrolimus, teniposide, theophylline, thiamine, thiotepa, ticarcillin, ticarcillin-clavulanate, tigecycline, tirofiban, TNA, tobramycin, tolazoline, TPN, trastuzumab, urokinase, vancomycin, vasopressin, vecuronium, verapamil, vinCRIStine, vinorelbine, voriconazole

SIDE EFFECTS

CNS: *Drowsiness, dizziness, confusion, headache, sedation, euphoria, weakness, hallucinations,* insomnia (nasal)
CV: Palpitations, bradycardia, hypotension
EENT: Tinnitus, blurred vision, miosis, diplopia, nasal congestion, unpleasant taste
GI: *Nausea, vomiting, anorexia, constipation, cramps*
GU: Urinary retention
INTEG: Rash, urticaria, bruising, flushing, diaphoresis, pruritus
RESP: Respiratory depression

PHARMACOKINETICS

Metabolized by liver, excreted by kidneys, crosses placenta, excreted in breast milk, half-life 2-9 hr, protein binding 80%
IM: Onset 5-15 min, peak 30-60 min, duration 3-4 hr
INTRANASAL: Onset within 15 min, peak 1-2 hr, duration 4-5 hr
IV: Onset 1 min, peak 4-5 min, duration 2-4 hr

INTERACTIONS

Increase: CNS effects—alcohol, opioids, sedative/hypnotics, antipsychotics, skeletal muscle relaxants, other CNS depressants, MAOIs

NURSING CONSIDERATIONS

Assess:

- For decreasing output; may indicate urinary retention
- ⚠ For withdrawal symptoms in opioid-dependent patients; PE, vascular occlusion, abscesses, ulcerations

• CNS changes: dizziness, drowsiness, hallucinations, euphoria, LOC, pupil reaction
• Allergic reactions: rash, urticaria
• Respiratory dysfunction: respiratory depression, character, rate, rhythm; notify prescriber if respirations are <10/min
• Need for pain medication, physical dependence
• Safety measures: night-light, call bell within easy reach, assistance with ambulation, especially for geriatric patients

Evaluate:
• Therapeutic response: decrease in pain

Teach patient/family:
• To report any symptoms of CNS changes, allergic reactions
• That physical dependency may result when used for extended periods
• That withdrawal symptoms may occur: nausea, vomiting, cramps, fever, faintness, anorexia
• How to use nasal product
• To avoid hazardous activities

TREATMENT OF OVERDOSE:

Naloxone 0.4-2 mg IV, O_2, IV fluids, vasopressors

cabozantinib

(ka'boe-zan'ti-nib)

Cometriq

Func. class.: Antineoplastic biologic response modifiers

Chem. class.: Signal transduction inhibitor (STI)

ACTION: Inhibits abnormal tyrosine kinase associated with growth and development of metastatic medullary thyroid cancer

USES: Treatment of progressive metastatic medullary thyroid cancer

CONTRAINDICATIONS: Pregnancy (D), hypersensitivity

Precautions: Breastfeeding, children, geriatric patients, cardiac/renal/hepatic/dental disease, bone marrow suppression, wound dehiscence, skin disease MI, hypertension, proteinuria, infertility, surgery, thrombocytopenia, neutropenia, immunosuppression

Black Box Warning: Bleeding, fistula, GI bleeding/perforations

DOSAGE AND ROUTES

- **Adult: PO** 140 mg/day until disease progression or unacceptable toxicity
- Avoid the concomitant use of strong 3A4 inhibitors or inducers if possible; temporary interruption of therapy and a dosage reduction may be necessary in patients who develop toxicity or intolerable side effects

Available forms: Tabs 60, 100, 140 mg

Administer:

- Take on an empty stomach; do not eat for ≥2 hr before and ≥1 hr after use; swallow whole; do not open or crush caps
- Do not take with grapefruit juice
- Do not take a missed dose within 12 hr of the next dose; if the next dose is in 12 hr or more, take the missed dose; if the next dose is in less than 12 hr, skip the missed dose and take the next dose at the scheduled time
- Store at room temperature

Dosage adjustments for treatment-related toxicity:

⚠ ***Grade 4 hematologic toxicity, grade 3 or higher nonhematologic toxicity, or intolerable grade 2 toxicity:*** Hold until resolution/improvement, then reduce the daily dosage by 40 mg (eg, 140 mg/day to 100 mg/day or 100 mg/day to 60 mg/day)

Dosage guidance in patients on strong CYP3A4 inducers/inhibitors:

⚠ ***Strong CYP3A4 inhibitors:*** Avoid use if possible; if a strong CYP3A4 inhibitor is used, reduce the daily cabozantinib dosage by 40 mg; resume the prior dosage after 2 or 3 days if the strong CYP3A4 inhibitor is discontinued

⚠ ***Strong CYP3A4 inducers:*** Avoid chronic concomitant use if possible; if a strong CYP3A4 inducer is required, increase the daily cabozantinib dosage by 40 mg, max 180 mg/day; resume the prior dosage after 2 or 3 days if the strong CYP3A4 inducer is discontinued

SIDE EFFECTS

CNS: Headache, dizziness

GI: Nausea, vomiting, dyspepsia, anorexia, abdominal pain, diarrhea, GI bleeding/fistula/perforation

HEMA: Neutropenia, thrombocytopenia, bleeding

INTEG: Rash

META: Hypokalemia

MISC: Fatigue, hypothyroidism, elevated liver enzymes

MS: Arthralgia, myalgia

PHARMACOKINETICS

Protein binding 99%, metabolized by CYP3A4

INTERACTIONS

Increase: cabozantinib concs—CYP3A4 inhibitors (ketoconazole, itraconazole, erythromycin, clarithromycin)

Increase: plasma concentrations of simvastatin, calcium channel blockers, ergots

Decrease: cabozantinib conc—CYP3A4 inducers (dexamethasone, phenytoin, carBAMazepine, rifampin, PHENobarbital)

Drug/Food

Increase: cabozantinib effect—grapefruit juice; avoid use while taking product

Drug/Herb

Cabozantinib conc—St. John's wort

NURSING CONSIDERATIONS

Assess:

⚠ **Proteinuria:** monitor urine protein regularly; discontinue if nephritic syndrome occurs

⚠ **GI bleeding/fistula/perforation:** can be fatal

⚠ **Wound dehiscence:** discontinue product ≥20 days before surgery

Evaluate:

- Therapeutic response: decrease in size, spread of tumor

Teach patient/family:

- To report adverse reactions immediately: bleeding
- About reason for treatment, expected results
- That effect on male infertility is unknown

calcitonin (rDNA) (Rx)

(kal-sih-toh′nin)

Fortical

calcitonin (salmon) (Rx)

Miacalcin

Func. class.: Parathyroid agents (calcium regulator)

Chem. class.: Polypeptide hormone

ACTION: Decreases bone resorption, blood calcium levels; increases deposits of calcium in bones; opposes parathyroid hormone

USES: Paget's disease, postmenopausal osteoporosis, hypercalcemia

Unlabeled uses: Bone/neuropathic pain, diabetic neuropathy, osteolytic metastases, osteoporosis prophylaxis, phantom limb pain

CONTRAINDICATIONS: Hypersensitivity to this product, fish

Precautions: Pregnancy (C), breastfeeding, children, renal disease, osteogenic sarcoma, pernicious anemia

DOSAGE AND ROUTES

rDNA

Paget's disease

- **Adult: SUBCUT** 0.5 mg/day initially; may require 0.5 mg bid × 6 mo, then decrease until symptoms reappear

Salmon

Postmenopausal osteoporosis

- **Adult: SUBCUT/IM** 100 international units every other day; **INTRANASAL** 200 international units (1 spray) daily, alternating nostrils daily, activate pump before 1st dose

Paget's disease

- **Adult: SUBCUT/IM** 100 international units/day, maintenance 50 international units daily, every other day, or 3 × per wk

Hypercalcemia

- **Adult: SUBCUT/IM** 4 international units/kg q12hr, increase to 8 international units/kg q12hr if response is unsatisfactory

Neuropathic pain/phantom limb pain/diabetic neuropathy (unlabeled)

- **Adult: IV/SUBCUT** 100-200 international units/day for phantom limb pain **IV** 200 international units over 20 min and 2nd inf given

Bone pain due to osteoporosis, osteolytic metastases (unlabeled)

- **Adult: SUBCUT** 50-100 international units/day or **INTRANASAL** 200 international units in 1 nostril/day

Available forms: Inj 200 international units/ml; nasal spray 200 international units/actuation

Administer:
- Store at <77° F (25° C); protect from light

SUBCUT route (rDNA)
- By SUBCUT route only; rotate inj sites; use within 6 hr of reconstitution; **give** at bedtime to minimize nausea, vomiting, rotate sites

IM route (Salmon)
- After test dose of 10 international units/ml, 0.1 ml intradermally; watch for 15 min; **give** only with EPINEPHrine, emergency meds available
- IM inj slowly into deep muscle mass; rotate sites; preferred route if volume is >2 ml

SIDE EFFECTS

CNS: Headache, tetany, chills, weakness, dizziness, fever, tremors
CV: Chest pressure
EENT: Nasal congestion, eye pain
GI: Nausea, diarrhea, vomiting, anorexia, abdominal pain, salty taste, epigastric pain
GU: Diuresis, nocturia, urine sediment, frequency
INTEG: Rash, flushing, pruritus of earlobes, edema of feet, reaction at inj site
MS: Swelling, tingling of hands, backache
RESP: Dyspnea
SYST: Anaphylaxis

PHARMACOKINETICS

IM/SUBCUT: Onset 15 min, peak 4 hr, duration 8-24 hr, metabolized by kidneys, excreted as inactive metabolites via kidneys

INTERACTIONS

Decrease: lithium effect
Decrease: effect of nasal spray—bisphosphonates (Paget's disease)

NURSING CONSIDERATIONS

Assess:
- **Anaphylaxis, hypersensitivity reaction** (rash, fever, inability to breathe); emergency equipment should be nearby
- GI symptoms, polyuria, flushing, head swelling, tingling, headache; may indicate hypercalcemia
- Nutritional status; diet for sources of vit D (milk, some seafood), calcium (dairy products, dark green vegetables), phosphates
- Urinalysis (calcium should be kept at 9-10 mg/dl, vit D 50-135 international units/dl), alk phos baseline, q3-6mo; monitor urine hydroproline with Paget's disease, biochemical markers of bone formation/absorption, radiologic evidence of fracture; bone density (osteoporosis)
- **Toxicity (can occur rapidly),** increased drug level; have parenteral calcium on hand if calcium level drops too low; check for tetany (irritability, paresthesia, nervousness, muscle twitching, seizures, tetanic spasms)

Evaluate:
- Therapeutic response: calcium levels 9-10 mg/dl, decreasing symptoms of Paget's disease

Teach patient/family:
- About the method of inj if patient will be responsible for self-medication
- To report difficulty swallowing, any changes in side effects to prescriber immediately

Nasal
- To use alternating nostrils for nasal spray; use after allowing to warm to room temperature; prime to get full spray

calcitriol vitamin D_3 (Rx)

(kal-sih-try′ole)

Calcijex, Rocaltrol, Vectical

Func. class.: Parathyroid agent (calcium regulator)
Chem. class.: Vit D hormone

Do not confuse:
calcitriol/Calciferol

ACTION: Increases intestinal absorption of calcium; provides calcium for bones; increases renal tubular resorption of phosphate

USES:
Hypocalcemia with chronic renal disease, hyperparathyroidism pseudohypoparathyroidism

Unlabeled uses: Osteopetrosis, osteoporosis, osteoporosis prophylaxis, rickets, familial hypophosphatemia

CONTRAINDICATIONS:
Hypersensitivity, hyperphosphatemia, hypercalcemia, vit D toxicity

Precautions: Pregnancy (C), breastfeeding, renal calculi, CV disease

DOSAGE AND ROUTES

Hypocalcemia (stage 5 chronic kidney disease, on dialysis)

- **Adult and child ≥6 yr: PO** 0.25 mcg/day **IV** 0.5 mcg tid initially; may increase by 0.25 every other day q4-8wk
- **Child 1-5 yr: PO** 0.25-2 mcg/day

Renal osteodystrophy

- **Adult and child ≥3 yr: PO** 0.25 mcg/day, may increase to 0.5 mcg/day
- **Child <3 yr: PO** 0.01-0.015 mcg/kg/day

Hypoparathyroidism

- **Adult and child ≥6 yr: PO** 0.25 mcg/day, may increase q2-4wk, maintenance 0.5-2 mcg/day
- **Child 1-5 yr: PO** 0.25-0.75 mcg daily
- **Child <1 yr: PO** 0.04-0.08 mcg/kg/day

Rickets (unlabeled)

- **Adult and child: PO** 1 mcg/day

Familial hypophosphatemia (unlabeled)

- **Adult: PO** 2 mcg/day
- **Child: PO** 0.015-0.02 mcg/kg/day, maintenance 0.03-0.06 mcg/kg/day; max 2 mcg/day

Postmenopausal osteoporosis (unlabeled)

- **Adult: PO** 0.25 mcg bid, adjust to serum calcium levels

Osteopetrosis (unlabeled)

- **Child: PO** high-dose calcitriol 1-2 mcg/kg/day given in 4-6 divided doses

Osteoporosis prophylaxis in corticosteroid therapy (unlabeled)

- **Adult: PO** 0.5-1 mcg/day

Available forms: Caps 0.25, 0.5 mcg; inj 1 mcg/ml, 2 mcg/ml; oral sol 1 mcg/ml; top 3 mcg/g

Administer:

PO route

- Do not break, crush, chew caps
- Give without regard to meals
- Store protected from light, heat, moisture

IV route

- Give by direct IV over 1 min

SIDE EFFECTS

CNS: Drowsiness, headache, vertigo, fever, lethargy, hallucinations

CV: Palpitations, hypertension

EENT: Blurred vision, photophobia

GI: Nausea, diarrhea, vomiting, jaundice, anorexia, dry mouth, constipation, cramps, metallic taste

GU: Polyuria, hypercalciuria, hyperphosphatemia, hematuria, thirst

MS: Myalgia, arthralgia, decreased bone development, weakness

SYST: Anaphylaxis

PHARMACOKINETICS

PO: Absorbed readily from GI tract, peak 10-12 hr, duration 3-5 days, half-life 3-6 hr, undergoes hepatic recycling, excreted in bile

INTERACTIONS

- Hypercalcemia: thiazide diuretics, calcium supplements
- Cardiac dysrhythmias: cardiac glycosides, verapamil
- Hypermagnesemia: magnesium antacids
- Toxicity: other vit D products

Increase: metabolism of vit D—phenytoin

Decrease: absorption of calcitriol—cholestyramine, mineral oil, fat-soluble vitamins

Drug/Food
- Large amounts of high-calcium foods may cause hypercalcemia

Drug/Lab Test
False increase: Cholesterol
Interference: Alk phos, electrolytes

NURSING CONSIDERATIONS

Assess:
- BUN, urinary calcium, PTH, creatinine, chloride, magnesium, electrolytes, phosphate; may increase calcium; should be kept at 9-10 mg/dl, vit D 50-135 international units/dl, phosphate 70 mg/dl; toxic reactions may occur rapidly
- **Hypercalcemia:** dry mouth, metallic taste, polyuria, bone pain, muscle weakness, headache, fatigue, change in level of consciousness, dysrhythmias, increased respirations, anorexia, nausea, vomiting, cramps, diarrhea, constipation; paresthesia, twitching, dysrhythmias, Chvostek's sign, Trousseau's sign **(hypocalcemia)**
- Renal status: decreased urinary output (oliguria, anuria), edema in extremities, weight gain 5-7 lb, periorbital edema
- Nutritional status, diet for sources of vit D (milk, some seafood); calcium (dairy products, dark green vegetables), phosphates (dairy products) must be avoided

Perform/provide:
- Restriction of sodium, potassium if required; restriction of fluids if required for chronic renal failure

Evaluate:
- Therapeutic response: calcium 9-10 mg/dl; decreasing symptoms of hypocalcemia, hypoparathyroidism

Teach patient/family:
- About symptoms of hypercalcemia: renal stones, nausea, vomiting, anorexia, lethargy, thirst, bone, or flank pain
- To avoid products with sodium: cured meats, dairy products, cold cuts, olives, beets, pickles, soups, meat tenderizers with chronic renal failure; products with potassium: oranges, bananas, dried fruit, peas, dark green leafy vegetables, milk, melons, beans
- To avoid OTC products that contain calcium, potassium, sodium with chronic renal failure
- To monitor weight weekly; to maintain fluid intake

calcium carbonate (PO-OTC)

AcidFree, Alka-Mints, Amitone, Apo-Cal ♣, Calcarb, Calci-Chew, Calci-Mix, Calcite ♣, Cal-Gest, Caltrate, Equaline Calcium, Leader Calcium, Maalox Antacid, Os-Cal 500, Rolaids Extra Strength Softchew, Tums, Tums E-X, Walgreens Calcium

calcium acetate (OTC)

(kal′see-um ass′e-tate)

Calphron, Eliphos, PhosLo

Func. class.: Antacid, calcium supplement

Chem. class.: Calcium product

Do not confuse:
Os-Cal/Asacol

ACTION: Neutralizes gastric acidity

USES: Antacid, calcium supplement; not suitable for chronic therapy, hyperphosphatemia, hypertension during pregnancy, osteoporosis, prevention, treatment of hypocalcemia, hypoparathyroidism

Unlabeled uses: Duodenal ulcer, PMS, stress gastritis

CONTRAINDICATIONS: Hypersensitivity, hypercalcemia, hyperparathyroidism, bone tumors

Precautions: Pregnancy (C), breastfeeding, geriatric patients, fluid restriction, decreased GI motility, GI obstruction, dehydration, renal disease

DOSAGE AND ROUTES

Nutritional supplement including osteoporosis prophylaxis

- **Adult ≥51 yr: PO** 1000-1500 mg/day elemental calcium (2500-3750 mg/day calcium carbonate)
- **Adult 19-50 yr: PO** 1000 mg/day elemental calcium (2500 mg/day calcium carbonate)

Chronic hypocalcemia

- **Adult: PO** 2-4 g/day elemental calcium (5-10 g/day calcium carbonate) in 3-4 divided doses
- **Child: PO** 45-65 mg/kg/day elemental calcium (112.5-162.5 mg/kg/day calcium carbonate) in 4 divided doses
- **Neonate: PO** 50-150 mg/kg/day elemental calcium (125-375 mg/kg/day in 4-6 divided doses, max 1 g/day)

Supplementation

- **Adolescent and child 9-18 yr: PO** 1300 mg elemental calcium (3250 mg calcium carbonate)
- **Child 4-8 yr: PO** 800 mg/day elemental calcium (2000 mg/day calcium carbonate)
- **Child 1-3 yr: PO** 500 mg/day elemental calcium (1250 mg/day calcium carbonate)
- **Infant 6-12 mo: PO** 270 mg/day elemental calcium based on total intake
- **Neonates and infants <6 mo: PO** 210 mg/day elemental calcium based on total intake

Hyperphosphatemia

- **Adult: PO** Individualized on response

Heartburn, dyspepsia, hyperacidity (OTC)

- **Adult: PO** 1-2 tabs q2hr, max 9 tabs/24 hr (Alka-mints); chew 2-4 tab q1hr prn, max 16 tabs (Tums regular strength); chew 2-4 tab q1hr prn, max 10 tabs (Tums E-X); chew 2-3 tabs q1hr prn, max 10 tabs/24 hr (Tums Ultra); chew 2 tabs q2-3hr, max 19 tabs/24 hr (Titralac Extra Strength)

Duodenal ulcer/stress gastritis (unlabeled)

- **Adult: PO** 80-140 mEq q1-3hr

PMS (unlabeled)

- **Adult: PO** (Tums E-X, Tums Calcium for Life PMS) Chew 2 tabs bid

Available forms: Calcium carbonate: chewable tabs 350, 420, 450, 500, 750, 1000, 1250 mg; tabs 500, 600, 650, 667, 1000, 1250, 1500 mg; gum 300, 450, 500 mg; susp 1250 mg/5 ml; caps 1250 mg; powder 6.5 g/packet; **calcium acetate:** tabs 250 mg (65 mg Ca), 667 mg (169 mg Ca), 668 mg (169 mg Ca), 1 g (250 mg Ca); caps 500 mg (125 mg Ca)

Administer:

PO route

- Do not give enteric-coated within 1 hr of calcium carbonate
- For ulcer treatment (adjunct): give 1 and 3 hr after meals and at bedtime
- For a phosphate binder: give 1 hr after each meal or snack and at bedtime
- For supplement: give 1-1½ hr after meals; avoid oxalic acid foods (spinach, rhubarb), phytic acid (brans, cereals) or phosphorus (milk, dairy), may decrease calcium absorption
- Suspension: shake well; use calibrated measuring device
- Laxatives or stool softeners if constipation occurs

SIDE EFFECTS

GI: *Constipation,* anorexia, nausea, vomiting, flatulence, diarrhea, rebound hyperacidity, eructation

GU: Calculi, hypercalciuria

PHARMACOKINETICS

⅓ of dose absorbed by small intestine, onset 20 min, duration 20-180 min, excreted in feces and urine, crosses placenta, must have adequate vit D for absorption

INTERACTIONS

Increase: digoxin toxicity—hypercalcemia

Increase: plasma levels of quiNIDine, amphetamines

Increase: hypercalcemia—thiazide diuretics
Decrease: levels of salicylates, calcium channel blockers, ketoconazole, iron salts, tetracyclines, fluoroquinolones, phenytoin, etidronate, risedronate, atenolol
Drug/Lab Test
False increase: chloride
False positive: benzodiazepines
False decrease: magnesium, oxylate, lipase

NURSING CONSIDERATIONS

Assess:

- Calcium (serum, urine), calcium should be 8.5-10.5 mg/dl, urine calcium should be 150 mg/day, monitor weekly; serum phosphate

⚠ **Milk-alkali syndrome:** nausea, vomiting, disorientation, headache

- Constipation; increase bulk in the diet if needed
- **Hypercalcemia:** headache, nausea, vomiting, confusion; hypocalcemia: paresthesia, twitching colic, dysrhythmias, Chvostek's sign, Trousseau's sign
- Those taking digoxin for toxicity
- Antacid—for abdominal pain, heartburn, indigestion before, after administration

Evaluate:

- Therapeutic response: absence of pain, decreased acidity; decreased hyperphosphatemia with renal failure

Teach patient/family:

- To increase fluids to 2 L unless contraindicated; to add bulk to diet for constipation; to notify prescriber of constipation
- Not to switch antacids unless directed by prescriber; not to use as antacid for >2 wk without approval by prescriber
- That therapeutic dose recommendations are figured as elemental calcium
- To avoid excessive use of alcohol, caffeine, tobacco
- To avoid spinach, cereals, dairy products in large amounts

CALCIUM SALTS

calcium chloride (Rx)
calcium gluceptate (Rx)
calcium gluconate (Rx)
calcium lactate (Rx)

Func. class.: Electrolyte replacement—calcium product

ACTION: Caution needed for maintenance of nervous, muscular, skeletal function; enzyme reactions; normal cardiac contractility; coagulation of blood; affects secretory activity of endocrine, exocrine glands

USES: Prevention and treatment of hypocalcemia, hypermagnesemia, hypoparathyroidism, neonatal tetany, cardiac toxicity caused by hyperkalemia, lead colic, hyperphosphatemia, vit D deficiency, osteoporosis prophylaxis, calcium antagonist toxicity (calcium channel blocker toxicity)
Unlabeled uses: Electrolyte abnormalities in cardiac arrest, CPR

CONTRAINDICATIONS: Hypercalcemia, digoxin toxicity, ventricular fibrillation, renal calculi
Precautions: Pregnancy (C), breastfeeding, children, respiratory/renal disease, cor pulmonale, digitalized patient, respiratory failure, diarrhea, dehydration

DOSAGE AND ROUTES

Calcium chloride

- **Adult: IV** 500 mg-1 g q1-3days as indicated by serum calcium levels, give at <1 ml/min; **IV** 200-800 mg injected in ventricle of heart

Calcium gluceptate

- **Adult: IV** 5-20 ml; **IM** 2-5 ml

Calcium gluconate

- **Adult: PO** 0.5-2 g bid-qid; **IV** 0.5-2 g at

0.5 ml/min (10% solution); max **IV** dose 3 g
- **Child: PO/IV** 500 mg/kg/day in divided doses

Calcium lactate
- **Adult: PO** 325 mg-1.3 g tid with meals
- **Child: PO** 500 mg/kg/day in divided doses

Available forms: Many; check product listings

Administer:

PO route (only acetate, carbonate, citrate, glubionate, lactate, phosphate)
- Give in 3-4 divided doses with or 1 hr after meals, follow with full glass of water; if using as phosphate binder in renal dialysis, do not follow with water, do not give oral medications within 1 hr of oral calcium; *chew tab:* chew thoroughly; *effervescent tab:* dissolve in full glass of water; *oral powder:* mix and give with food; *oral solution:* give before meals; *oral suspension:* shake well
- Store at room temperature

IM route
- Glycerophosphate, lactate may be given IM
- Do not give chloride, gluconate IM
- Use only if IV is not feasible
- Inject into gluteal region (adult), lateral thigh (child)
- Aspirate before inj
- Do not give chloride subcut

IV route
- Undiluted or diluted with equal amounts of NS to a 5% sol for inj, give 0.5-1 ml/min
- Through small-bore needle into large vein; if extravasation occurs, necrosis will result (IV)
- Remain recumbent ½ hr after IV dose

Calcium chloride

Additive compatibilities: Amikacin, amphotericin B, ampicillin, ascorbic acid, cefTRIAXone, cephapirin, chloramphenicol, DOPamine, hydrocortisone, isoproterenol, lidocaine, methicillin, norepinephrine, penicillin G potassium, penicillin G sodium, PENTobarbital, PHENobarbital, verapamil, vit B/C

Syringe compatibilities: Milrinone

Y-site compatibilities: Inamrinone, DOBUTamine, EPINEPHrine, esmolol, morphine, PACLitaxel

Calcium gluceptate

Additive compatibilities: Ascorbic acid inj, isoproterenol, lidocaine, norepinephrine, phytonadione, sodium bicarbonate

Calcium gluconate

Additive compatibilities: Amikacin, aminophylline, ascorbic acid inj, cephapirin, chloramphenicol, cisatracurium, corticotropin, dimenhyDRINATE, DOXOrubicin liposome, erythromycin, furosemide, heparin, hydrocortisone, lidocaine, magnesium sulfate, methicillin, norepinephrine, penicillin G potassium, penicillin G sodium, PHENobarbital, potassium chloride, remifentanil, tobramycin, vancomycin, verapamil, vit B/C

Syringe compatibilities: Aldesleukin, allopurinol, amifostine, aztreonam, ceFAZolin, cefepime, ciprofloxacin, cladribine, DOBUTamine, enalaprilat, EPINEPHrine, famotidine, filgrastim, granisetron, heparin/hydrocortisone, labetalol, melphalan, midazolam, netilmicin, piperacillin/tazobactam, potassium chloride, prochlorperazine, propofol, sargramostim, tacrolimus, teniposide, thiotepa, tolazoline, vinorelbine, vit B/C

SIDE EFFECTS

CV: Shortened QT, heart block, hypotension, bradycardia, dysrhythmias; cardiac arrest (IV)

GI: Vomiting, nausea, constipation

HYPERCALCEMIA: Drowsiness, lethargy, muscle weakness, headache, constipation, coma, anorexia, nausea, vomiting, polyuria, thirst

INTEG: Pain, burning at IV site, severe venous thrombosis, necrosis, extravasation

PHARMACOKINETICS

Crosses placenta, enters breast milk, excreted via urine and feces, half-life unknown, protein binding 40%-50%

PO: Onset, peak, duration unknown, absorption from GI tract
IV: Onset immediate, duration ½-2 hr

INTERACTIONS

Increase: milk-alkali syndrome—antacids
Increase: dysrhythmias—digoxin glycosides
Increase: toxicity—verapamil
Increase: hypercalcemia—thiazide diuretics
Decrease: absorption of fluoroquinolones, tetracyclines, iron salts, phenytoin, thyroid hormones when calcium is taken PO
Decrease: effects of atenolol, verapamil

Drug/Herb
Increase: action/side effects—lily of the valley, pheasant's eye, shark cartilage, squill

Drug/Lab Test
Increase: calcium

NURSING CONSIDERATIONS

Assess:
- **ECG for decreased QT and T wave inversion:** hypercalcemia, product should be reduced or discontinued, consider cardiac monitoring
- Calcium levels during treatment (8.5-11.5 g/dl is normal level); urine calcium if hypercaluria occurs
- Cardiac status: rate, rhythm, CVP (PWP, PAWP if being monitored directly)
- **Hypocalcemia:** muscle twitching, paresthesia, dysrhythmias, laryngospasm
- Digitalized patients frequently; an increase in calcium increases digoxin toxicity risk
- Seizure precautions: padded side rails, decreased stimuli (noise, light); place airway suction equipment, padded mouth gag if calcium levels are low
- Store at room temperature

Evaluate:
- Therapeutic response: decreased twitching, paresthesias, muscle spasms; absence of tremors, seizures, dysrhythmias, dyspnea, laryngospasm; negative Chvostek's sign, negative Trousseau's sign

Teach patient/family:
- To add foods high in vit D
- To add calcium-rich foods to diet: dairy products, shellfish, dark green leafy vegetables; to decrease oxalate- and zinc-rich foods: nuts, legumes, chocolate, spinach, soy
- To prevent injuries; to avoid immobilization

RARELY USED

calfactant (Rx)

(cal-fak′tant)

Infasurf

Func. class.: Natural lung surfactant extract

USES: Prevention and treatment (rescue) of respiratory distress syndrome in premature infants

DOSAGE AND ROUTES

- **Newborn: INTRATRACHEAL INSTILL:** 3 ml/kg of birth weight given as 2 doses of 1.5 ml/kg; repeat doses of 3 ml/kg of birth wt until up to 3 doses 12 hr apart have been given

candesartan (Rx)

(can-deh-sar′tan)

Atacand

Func. class.: Antihypertensive
Chem. class.: Angiotensin II receptor (type AT_1) antagonist

ACTION: Blocks the vasoconstrictor and aldosterone-secreting effects of angiotensin II; selectively blocks the binding of angiotensin II to the AT_1 receptor found in tissues

C

USES:
Hypertension, alone or in combination; CHF NYHA Class II-IV and ejection fraction ≤40%

CONTRAINDICATIONS:
Hypersensitivity

Black Box Warning: Pregnancy (D) 2nd/3rd trimesters

Precautions: Pregnancy (C) 1st trimester, breastfeeding, children, geriatric patients, hypersensitivity to ACE inhibitors, volume depletion, renal/hepatic impairment, renal artery stenosis, hypotension

DOSAGE AND ROUTES

Hypertension

- **Adult: PO** single agent 16 mg/day initially in patients who are not volume depleted, range 8-32 mg/day, with diuretic or volume depletion 2-32 mg/day as single dose or divided bid
- **Adolescent and child ≥6 yr and weight >50 kg: PO** 8-16 mg/day or divided bid, adjust to B/P; usual range 4-32 mg/day, max 32 mg/day
- **Child ≥6 yr, weight <50 kg: PO** 4-8 mg/day or divided bid, adjust to B/P
- **Child ≥1 yr and <6 yr: PO** 0.2 mg/kg/day in 1 dose or in 2 divided doses, adjust to B/P, max 0.4 mg/kg/day

Heart failure

- **Adult: PO** 4 mg/day, may be doubled ≥2 wk, target dose 32 mg/day

Renal/hepatic disease

- **Adult: PO** ≤8 mg/day for severe renal disease/moderate hepatic disease, adjust dose as needed

Available forms: Tabs 4, 8, 16, 32 mg

Administer:

- Without regard to meals
- Oral liquid (compounded): shake well, do not freeze

SIDE EFFECTS

CNS: *Dizziness,* fatigue, headache, syncope

CV: Chest pain, peripheral edema, hypotension, palpitations

EENT: Sinusitis, rhinitis, pharyngitis

GI: *Diarrhea,* nausea, abdominal pain, vomiting

GU: **Renal failure**

MS: Arthralgia, pain

RESP: *Cough, upper respiratory infection*

SYST: **Angioedema**

PHARMACOKINETICS

Peak 3-4 hr, protein binding 99%, half-life 9-12 hr, extensively metabolized, excreted in urine (33%) and feces (67%)

INTERACTIONS

Increase: lithium level—lithium

Increase: hyperkalemia—potassium, potassium-sparing diuretics

Increase: hypotension—ACE inhibitors, β-blockers, calcium channel blockers, α-blockers

Decrease: effect—salicylates, NSAIDs

Drug/Herb

Increase: antihypertensive effect—hawthorn

Decrease: antihypertensive effect—ephedra

NURSING CONSIDERATIONS

Assess:

⚠ **Serious hypersensitivity reaction:** angioedema, anaphylaxis: facial swelling, difficulty breathing (rare)

Black Box Warning: For pregnancy; this product can cause fetal death when given during pregnancy (D), 2nd/3rd trimester

- Response and adverse reactions, especially with renal disease
- B/P, pulse q4hr; note rate, rhythm, quality; electrolytes: potassium, sodium, calcium; baselines of renal/hepatic studies before therapy begins
- **Heart failure:** jugular venous distention, weight, edema, dyspnea, crackles

Evaluate:

- Therapeutic response: decreased B/P

Teach patient/family:

- To comply with dosage schedule, even if feeling better

• To notify prescriber of mouth sores, fever, swelling of hands or feet, irregular heartbeat, chest pain
• That excessive perspiration, dehydration, vomiting, diarrhea may lead to fall in B/P; to consult prescriber if these occur

Black Box Warning: To notify prescriber immediately if pregnant (D) 2nd/3rd trimester, (C) 1st trimester; to use if breastfeeding

• To avoid all OTC medications unless approved by prescriber; to inform all health care providers of medication use
• To use proper technique for obtaining B/P; to understand acceptable parameters; to rise slowly to sitting or standing position to minimize orthostatic hypotension; that product may cause dizziness, fainting, lightheadedness

capecitabine (Rx)

(cap-eh-sit′ah-bean)

Xeloda

Func. class.: Antineoplastic, antimetabolite

Chem. class.: Fluoropyrimidine carbamate

Do not confuse:
Xeloda/Xenical

ACTION: Competes with physiologic substrate of DNA synthesis, thereby interfering with cell replication in the S phase of cell cycle (before mitosis); also interferes with RNA and protein synthesis; product is converted to 5-FU

USES: Monotherapy for PACLitaxel; anthracycline-resistant, metastatic breast, colorectal cancer when 5-FU monotherapy is preferred; treatment of colorectal cancer patients who have undergone complete resection of their primary tumors

Unlabeled uses: Biliary tract, ovarian cancer

CONTRAINDICATIONS: Pregnancy (D), hypersensitivity to 5-FU, infants, severe renal impairment (CCr <30 ml/min), DPD deficiency

Precautions: Breastfeeding, children, geriatric patients, renal/hepatic disease

DOSAGE AND ROUTES

Metastatic breast cancer resistant to both PACLitaxel and anthracycline or resistant to PACLitaxel and when further anthracycline therapy is not indicated

• **Adult:** **PO** 2500 mg/m²/day divided q12hr after a meal × 2 wk, repeat q3wk

Breast cancer (locally advanced/metastatic) with DOCEtaxel, previously treated with anthracycline

• **Adult:** **PO** 2500 mg/m²/day divided q12hr after a meal on days 1-14, with DOCEtaxel 75 mg/m² IV on day 1

Advanced/metastatic breast cancer (HER2 positive) previously treated with anthracycline, taxane, and trastuzumab

• **Adult:** **PO** 2000 mg/m²/day divided q12hr after a meal on days 1-14 with lapatinib 1250 mg/day on days 1-21, repeat q21days

Metastatic/locally advanced breast cancer resistant to anthracycline, previously treated with a taxane or taxane resistant, and when further anthracycline is contraindicated

• **Adult:** **PO** 2000 mg/m²/day divided q12hr on days 1-14 with ixabepilone 40 mg/m² IV over 3 hr, repeat q3wk

As an adjuvant for Dukes C colorectal cancer with a complete resection when fluoropyrimidine alone is preferred

• **Adult:** **PO** 2500 mg/m²/day divided q12hr within 30 min of a meal × 2 wk, repeat q3wk for 8 cycles

First-line treatment of metastatic colorectal cancer when fluoropyrimidine alone is preferred

• **Adult:** **PO** 2500 mg/m²/day divided q12hr after a meal × 2 wk, repeat q3wk

First-line treatment of metastatic colorectal cancer with oxaliplatin with or without bevacizumab (unlabeled)

• **Adult: PO** 2000 mg/m^2/day divided q12hr after a meal on days 1-14 with oxaliplatin on day 1, repeat q3wk

First-/second-line treatment of advanced colorectal cancer with oxaliplatin (unlabeled)

• **Adult: PO** 2000 mg/m^2/day divided q12hr on days 1-14 and oxaliplatin 130 mg/m^2 IV on day 1, repeat q3wk

Unresectable advanced/metastatic biliary tract cancer (unlabeled)

• **Adult: PO** 2500 mg/m^2 divided q12hr on days 1-14, then 7-day rest period; given with CISplatin 60 mg/m^2 IV over 1 hr on day 1, repeat q21days

Renal dose

• **Adult: PO** CCr 30-50 ml/min, decrease initial dose to 75% of usual dose; CCr <30 ml/min, contraindicated

Available forms: Tabs 150, 500 mg

Administer:

• **Dosage adjustments of capecitabine monotherapy based on most severe toxicity OR when used in combination with ixabepilone based on nonhematologic toxicity: Grade 1 toxicity:** maintain current dosage; **Grade 2 toxicity (1st appearance):** interrupt therapy until toxicity is resolved to Grade 0-1; do not replace missed doses, begin the next cycle with 100% of the starting dose; **Grade 2 toxicity (2nd appearance):** interrupt therapy until toxicity is resolved to Grade 0-1; do not replace missed doses, begin the next cycle with 75% of the starting dose; **Grade 2 toxicity (3rd appearance):** interrupt therapy until toxicity is resolved to Grade 0-1; do not replace missed doses, begin the next cycle with 50% of the starting dose; **Grade 2 toxicity (4th appearance):** discontinue treatment permanently; **Grade 3 toxicity (1st appearance):** interrupt therapy until toxicity is resolved to Grade 0-1

SIDE EFFECTS

CNS: Dizziness, *headache, paresthesia, fatigue,* insomnia

CV: **Venous thrombosis**

GI: *Nausea, vomiting, anorexia, diarrhea, stomatitis, abdominal pain, constipation, dyspepsia,* **intestinal obstruction, necrotizing enterocolitis,** hyperbilirubinemia, **hepatic failure**

HEMA: **Neutropenia, lymphopenia, thrombocytopenia,** anemia

INTEG: *Hand and foot syndrome, dermatitis,* nail disorders

OTHER: *Eye irritation, edema, myalgia,* limb pain, *pyrexia,* dehydration

RESP: *Cough, dyspnea,* **pulmonary embolism**

PHARMACOKINETICS

Readily absorbed, peak 1½ hr, food decreases absorption, extensively metabolized in the liver, elimination half-life 45 min

INTERACTIONS

Increase: toxicity—leucovorin

Increase: capecitabine levels—antacids (aluminum, magnesium)

Increase: phenytoin level—phenytoin

Black Box Warning: Increase: bleeding risk—anticoagulants

Drug/Food

Increase: absorption; give within 30 min of a meal

Drug/Lab Test

Increase: bilirubin

Decrease: Hgb/HcT/RBC, neutrophils, platelets, WBC

NURSING CONSIDERATIONS

Assess:

• **Bone marrow suppression,** CBC (RBC, Hct, Hgb), differential, platelet count weekly; withhold product if WBC is <4000/mm^3, platelet count is <75,000/mm^3, or RBC, Hct, Hgb low; notify prescriber of these results; frequently monitor INR in those receiving warfarin concurrently

- Renal studies: BUN, serum uric acid, urine CCr, electrolytes before, during therapy
- Monitor temperature q4hr; fever may indicate beginning infection; no rectal temperatures
- Hepatic studies before, during therapy: bilirubin, ALT, AST, alk phos as needed or monthly

Black Box Warning: Bleeding: hematuria, heme-positive stools, bruising or petechiae of mucosa or orifices q8hr, monitor INR and PT in those taking anticoagulants

- Dyspnea, crackles, unproductive cough, chest pain, tachypnea, fatigue, increased pulse, pallor, lethargy; personality changes with high doses
- **Hand and foot syndrome:** paresthesia, tingling, painful/painless swelling, blistering, erythema with severe pain of hands or feet
- **Toxicity:** severe diarrhea (multiple times/day or at night), nausea, vomiting, stomatitis, fever
- Buccal cavity q8hr for dryness, sores, ulceration, white patches, oral pain, bleeding, dysphagia
- **GI symptoms:** frequency of stools, cramping; if severe diarrhea occurs, fluid, electrolytes may need to be given
- Rinsing of mouth tid-qid with water, club soda; brushing of teeth bid-tid with soft brush or cotton-tipped applicators for stomatitis; use unwaxed dental floss

Evaluate:

- Therapeutic response: decreased tumor size, spread of malignancy

Teach patient/family:

- To avoid foods with citric acid, hot or rough texture if stomatitis is present; take with water within 30 min of end of meal
- To notify prescriber if pregnancy is planned or suspected, pregnancy (D); to avoid pregnancy while taking this product; not to breastfeed
- Not to double dose if dose is missed
- ⚠ To immediately report severe diarrhea, vomiting, stomatitis, fever of more than 100° F (37.8° C), hand and foot syndrome, anorexia
- To report signs of **infection:** increased temperature, sore throat, flulike symptoms; signs of **anemia:** fatigue, headache, faintness, shortness of breath, irritability; **bleeding;** to avoid use of razors, commercial mouthwash
- OTC antidiarrheals for mild diarrhea (4-6 stools/day or diarrhea at night)

captopril (Rx)

(kap'toe-pril)

Apo-Capto 🍁

Func. class.: Antihypertensive

Chem. class.: Angiotensin-converting enzyme (ACE) inhibitor

Do not confuse:
captopril/Capitrol/carvedilol

ACTION: Selectively suppresses renin-angiotensin-aldosterone system; inhibits ACE; prevents conversion of angiotensin I to angiotensin II

USES: Hypertension, CHF, left ventricular dysfunction after MI, diabetic nephropathy, proteinuria

Unlabeled uses: Acute MI, hypertensive emergency/urgency, scleroderma renal crisis (SRC)

CONTRAINDICATIONS: Breastfeeding, children, hypersensitivity, heart block, potassium-sparing diuretics, bilateral renal artery stenosis, angioedema

Black Box Warning: Pregnancy (D) 2nd/3rd trimester

Precautions: Dialysis patients, hypovolemia, leukemia, scleroderma, SLE, blood dyscrasias, CHF, diabetes mellitus, thyroid/renal/hepatic disease, African descent, pregnancy (C) 1st trimester, collagen vascular disease, hyperkalemia, hyponatremia

DOSAGE AND ROUTES

Hypertension

• **Adult: PO** initial dose: 12.5-25 mg bid-tid; may increase to 50 mg bid-tid at 1-2 wk intervals; usual range: 25-150 mg bid-tid; max 450 mg/day

• **Child: PO** 0.3-0.5 mg/kg/dose, titrate up to 6 mg/kg/day in 1-4 divided doses

• **Neonate: PO** 0.01-0.1 mg/kg/dose, may increase as needed

CHF

• **Adult: PO** 25 mg bid; may increase to 50 mg tid; after 14 days, may increase to 150 mg tid if needed

Diabetic nephropathy

• **Adult: PO** 25 mg tid

Renal dose

• **Adult: PO** CCr >50 ml/min, no change; CCr 10-50 ml/min, decrease dose by 25%; CCr <10 ml/min, decrease dose by 50%

Acute MI (unlabeled) or post-MI

• **Adult: PO** 6.25-12.5 mg tid, increase to 25 mg tid gradually

Hypertensive emergency/urgency (unlabeled)

• **Adult: PO** 25 mg, may repeat q30min

Available forms: Tabs 12.5, 25, 50, 100 mg

Administer:

• Store in tight container at 86° F (30° C) or less

• 1 hr before or 2 hr after meals

• **Oral sol:** may crush tab, dissolve in water; give within ½ hr; make sure tab completely dissolved

SIDE EFFECTS

CNS: Fever, chills

CV: *Hypotension,* postural hypotension, *tachycardia,* angina

GI: Loss of taste, increased LFTs

GU: Impotence, dysuria, nocturia, proteinuria, nephrotic syndrome, acute reversible renal failure, polyuria, oliguria, urinary frequency

HEMA: Neutropenia, agranulocytosis, pancytopenia, thrombocytopenia, anemia

INTEG: Rash, pruritus

MISC: Angioedema, hyperkalemia

RESP: Bronchospasm, *dyspnea, cough*

PHARMACOKINETICS

Peak 1 hr; duration 2-6 hr; half-life <2 hr, increased in renal disease; metabolized by liver (metabolites); excreted in urine; crosses placenta; excreted in breast milk, small amounts; protein binding 25%-30%

INTERACTIONS

• Do not use with potassium-sparing diuretics, sympathomimetics, potassium supplements

Increase: possible toxicity—lithium, digoxin

Increase: hypoglycemia—insulin, oral antidiabetics

Increase: hypotension—diuretics, other antihypertensives, phenothiazines, nitrates, acute alcohol ingestion

Decrease: captopril effect—antacids, NSAIDs, salicylates

Drug/Herb

Increase: antihypertensive effect—hawthorn

Decrease: antihypertensive effect—ephedra

Drug/Food

Decrease: absorption of captopril

Drug/Lab Test

Increase: AST, ALT, alk phos, bilirubin, uric acid, potassium

Decrease: platelets, WBC, RBC, Hgb/HcT

False positive: urine acetone, ANA titer

NURSING CONSIDERATIONS

Assess:

• **Blood dyscrasias:** blood studies: decreased platelets; WBC with differential at baseline and periodically q3mo; if neutrophils <1000/mm^3, discontinue treatment (recommended with collagen-vascular or renal disease)

• **Hypertension:** B/P, pulse rates at baseline, frequently

• Renal studies: protein, BUN, creatinine; watch for raised levels, may indicate nephrotic syndrome; increased LFTs, uric acid; glucose, potassium

- **Allergic reaction:** rash, fever, pruritus, urticaria; discontinue product if antihistamines fail to help
- **CHF:** edema, dyspnea, wet crackles, increased B/P, weight gain

Evaluate:
- Therapeutic response: decrease in B/P with hypertension; edema, moist crackles (CHF)

Teach patient/family:
- To take 1 hr before or 2 hr after meals; not to discontinue product abruptly; if dose is missed, take as soon as remembered but not if almost time for next dose; not to double doses
- Not to use OTC products (cough, cold, or allergy) unless directed by prescriber; avoid salt substitutes, high-potassium or high-sodium foods
- To adhere to dosage schedule, even if feeling better
- To rise slowly to sitting or standing position to minimize orthostatic hypotension
- To notify prescriber of mouth sores, sore throat, fever, swelling of hands or feet, irregular heartbeat, chest pain, signs of angioedema
- That excessive perspiration, dehydration, vomiting, diarrhea may lead to fall in B/P; to consult prescriber if these occur
- That dizziness, fainting, lightheadedness may occur during first few days of therapy; to avoid activities that require concentration
- That skin rash or impaired perspiration may occur
- How to take B/P and when to notify prescriber

Black Box Warning: To report if pregnancy is suspected or planned, pregnancy (D)

TREATMENT OF OVERDOSE:
0.9% NaCl IV/INF; hemodialysis

carbachol ophthalmic
See Appendix B

carBAMazepine (Rx)
(kar-ba-maz′e-peen)

Carbatrol, Equetro, Mazepine ♣, Novo-Carbamaz ♣, TEGretol, TEGretol-XR

Func. class.: Anticonvulsant
Chem. class.: Iminostilbene derivative

Do not confuse:
TEGretol/Toradol

ACTION: Exact mechanism unknown; appears to decrease polysynaptic responses and block posttetanic potentiation

USES: Tonic-clonic, complex-partial, mixed seizures; trigeminal neuralgia; bipolar disorder
Unlabeled uses: Neurogenic pain, psychotic behavior with dementia, diabetic neuropathy, agitation, hiccups

CONTRAINDICATIONS: Pregnancy (D), hypersensitivity to carBAMazepine or tricyclics

Black Box Warning: Bone marrow suppression

Precautions: Breastfeeding, children <6 yr, glaucoma, AV or bundle branch block, cardiac/renal/hepatic disease, psychosis, alcoholism, hepatic porphyria

Black Box Warning: Hematologic disease, Asian patients, agranulocytosis, leukopenia, neutropenia, thrombocytopenia

DOSAGE AND ROUTES
Seizures
- **Adult and child >12 yr: PO** 200 mg bid, may be increased by 200 mg/day in weekly intervals, give in divided doses q6-8hr; maintenance 800-1200 mg/day, max 1600 mg/day (adult); max child 12-15 yr 1000 mg/day; max child >15 yr 1200 mg/day; adjust to minimum dose to control seizures; **EXT REL** give bid; rec-

tal administration of **ORAL SUSP** 200 mg/10 ml or 6 mg/kg as a single dose

- **Child 6-12 yr: PO** tabs 100 mg bid or susp 50 mg qid; may increase by <100 mg weekly; max 1000 mg/day, usual dose 15-30 mg/kg/day; **EXT REL** tabs daily-bid
- **Child <6 yr: PO** 10-20 mg/kg/day in 2-3 divided doses or 4 divided dose (susp), may increase every wk

Trigeminal neuralgia

- **Adult: PO** 100 mg bid with meals; may increase 100 mg q12hr until pain subsides, max 1200 mg/day; maintenance 200-400 mg bid

Diabetic neuropathy (unlabeled)

- **Adult: PO** 100 mg bid or 50 mg qid, titrate to 600-800 mg/day

Bipolar disorder

- **Adult: PO** (Equetro only [regular release]) 200 mg bid, may adjust dose q3-4days to achieve carBAMazepine level to 8-12 mcg/ml micro response, max 1600 mg/day

Agitation due to dementia (unlabeled)

- **Adult: PO** 100 mg bid, may increase to 250-300 mg/day

Hiccups (unlabeled)

- **Adult: PO** 200 mg tid

Available forms: Chewable tabs 100 mg; tabs 200 mg; ext rel tabs (XR) 100, 200, 400 mg; oral susp 100 mg/5 ml; ext rel caps 100, 200, 300 mg

Administer:

PO route

- Do not crush, chew ext rel tab; ext rel cap may be opened and beads sprinkled over food; patient should chew chewable tab, not swallow it whole
- With food, milk to decrease GI symptoms; shake oral susp before use
- **(Susp):** Turn off NG/enteral feeding 15 min before and hold for 15 min after
- Mix an equal amount of water, D_5W, 0.9% NaCl when giving by NG tube; flush tube with 15-30 ml of above sol
- Store at room temperature

SIDE EFFECTS

CNS: *Drowsiness,* dizziness, unsteadiness, confusion, fatigue, **paralysis,** headache, hallucinations, **worsening of seizures,** speech disturbance, **suicidal thoughts/behaviors, neuroleptic malignant syndrome (when used with psychotropics)**

CV: Hypertension, CHF, dysrhythmias, AV block, hypotension, aggravation of cardiac artery disease

EENT: Tinnitus, dry mouth, blurred vision, diplopia, nystagmus, conjunctivitis

ENDO: SIADH (geriatric patients)

GI: *Nausea, constipation, diarrhea,* anorexia, vomiting, abdominal pain, stomatitis, glossitis, increased hepatic enzymes, **hepatitis, hepatic porphyria, hypercholesterolemia, pancreatitis**

GU: Frequency, retention, albuminuria, glycosuria, impotence, increased BUN, **renal failure**

HEMA: Thrombocytopenia, leukopenia, agranulocytosis, leukocytosis, aplastic anemia, eosinophilia, increased PT, lymphadenopathy

INTEG: *Rash,* **Stevens-Johnson syndrome,** urticaria, photosensitivity, **toxic epidermal necrolysis, DRESS, alopecia**

MS: Osteoporosis

RESP: Pulmonary hypersensitivity (fever, dyspnea, pneumonitis)

PHARMACOKINETICS

Onset slow; peak 4-5 hr (PO), 1.5 hr (susp) metabolized by liver; excreted in urine, feces; crosses placenta, blood-brain barrier; excreted in breast milk; half-life 18-65 hr then 8-29 hr after 1st month; protein binding 76%; metabolized by CYP3A4

INTERACTIONS

- CNS toxicity: lithium

⚠ **Fatal reaction: MAOIs**

- **Do not use with: NNRTIs (non-nucleoside reverse transcriptase inhibitors nefazodone)**

Increase: carBAMazepine levels—CYP3A inhibitors (cimetidine, clarithromycin,

danazol, diltiazem, erythromycin, FLUoxetine, fluvoxamine, isoniazid, propoxyphene, valproic acid, verapamil, voriconazole)
Increase: effects of desmopressin, lithium, lypressin, vasopressin
Decrease: carBAMazepine effect—CYP1A2, CYP2C9 substrates
Decrease: effect of CYP3A inducers
Decrease: effects of benzodiazepines, doxycycline, felbamate, haloperidol, oral contraceptives, PHENobarbital, phenytoin, primidone, theophylline, thyroid hormones, warfarin
Decrease: carBAMazepine levels—CYP3A4 inducers (CISplatin, darunavir, delavirdine, DOXOrubicin, felbamate, nefazodone, OXcarbazepine, PHENobarbital, phenytoin, primidone, rifampin, theophylline)

Drug/Herb

Decrease: carBAMazepine metabolism, increased levels—echinacea
Decrease: anticonvulsant effect—St. John's wort

Drug/Food

Increase: peak conc of carBAMazepine—grapefruit juice

Drug/Lab Test

Decrease: serum calcium, sodium
Increase: cholesterol

NURSING CONSIDERATIONS

Assess:

Black Box Warning: Asian patients for serious skin reaction; genetic test before administration

- **Seizures:** character, location, duration, intensity, frequency, presence of aura
- **Trigeminal neuralgia:** facial pain including location, duration, intensity, character, activity that stimulates pain
- Renal studies: urinalysis, BUN, urine creatinine q3mo
- **Serious skin, multi organ hypersensitivity (Stevens-Johnson syndrome, toxic epidermal necrolysis, DRESS):** may be increased in HLA-A 3101 gene and may be fatal

Black Box Warning: **Bone marrow depression:** blood studies: CBC reticulocyte counts every wk for 4 wk then q3-6mo if on long-term therapy; if myelosuppression occurs, product should be discontinued; blood dyscrasias: fever, sore throat, bruising, rash, jaundice

- Blood studies: ALT, AST, bilirubin; serum calcium, may be decreased and lead to osteoporosis; cholesterol periodically
- Product levels during initial treatment or when changing dose; should remain at 4-12 mcg/ml; anorexia may indicate increased blood levels

⚠ Mental status: mood, sensorium, affect, behavioral changes, **suicidal thoughts/behaviors;** if mental status changes, notify prescriber

- Eye problems: need for ophthalmic examinations before, during, after treatment (slit lamp, funduscopy, tonometry)
- Allergic reaction: purpura, red, raised rash; if these occur, product should be discontinued; increased risk if past hypersensitivity to hydantoins

⚠ Toxicity: bone marrow depression, nausea, vomiting, ataxia, diplopia, CV collapse

- Hard candy, gum, frequent rinsing for dry mouth

Evaluate:

- Therapeutic response: decreased seizure activity; document on patient's chart

Teach patient/family:

- To carry emergency ID stating patient's name, products taken, condition, prescriber's name, and phone number
- To avoid driving, other activities that require alertness usually for the first 3 days of treatment
- Not to discontinue medication quickly after long-term use
- To immediately report chills, rash, light-colored stools, dark urine, yellowing of skin and eyes, abdominal pain, sore throat, mouth ulcers, bruising, blurred vision, dizziness, skin rash, fever
- That urine may turn pink to brown
- To notify if pregnancy is planned or suspected, pregnancy (D), avoid breastfeeding

TREATMENT OF OVERDOSE:
Lavage, VS

⚠ HIGH ALERT

CARBOplatin (Rx)
(kar-boe-pla′-tin)

Func. class.: Antineoplastic alkylating agent

Chem. class.: Platinum coordination compound

Do not confuse:
CARBOplatin/CISplatin

ACTION: Produces interstrand DNA cross-links and, to a lesser extent, DNA-protein cross-links; activity is not cell-cycle–phase specific

USES: Initial treatment of advanced ovarian cancer in combination with other agents; palliative treatment of ovarian carcinoma recurrent after treatment with other antineoplastic agents

Unlabeled uses: Acute lymphocytic leukemia (ALL), acute myelogenous leukemia (AML), bladder/breast/head/neck/lung/testicular cancer, bone marrow ablation, malignant glioma, neuroblastoma, non-Hodgkin's lymphoma, osteogenic sarcoma, soft-tissue sarcoma, stem-cell transplant preparation, Wilms' tumor, stage I seminoma

CONTRAINDICATIONS: Pregnancy (D), breastfeeding, hypersensitivity, significant bleeding, aluminum products used to prepare or administer CARBOplatin

Black Box Warning: Severe bone marrow depression, platinum compound hypersensitivity

Precautions: Geriatric patients, radiation therapy within 1 mo, other cancer chemotherapy within 1 mo, renal/hepatic disease

Black Box Warning: Anemia, infection

DOSAGE AND ROUTES

Dosing with the Calvert equation

- **Adult:** The total CARBOplatin dose in mg for adults may be calculated using the Calvert equation:

Total dose (mg) = target AUC × (GFR + 25)

- **Children:** Calculate CARBOplatin dosage (mg/m^2) in children:

Total dose (mg/m^2) = target AUC × [(0.93 × GFR) + 15]

Advanced ovarian cancer

- **Adult (single agent): IV INF** initially 300 mg/m^2 on day 1 with cyclophosphamide, 600 mg/m^2 **IV** on day 1, repeat q4wk × 6 cycles; refractory tumors 360 mg/m^2 single dose, may repeat q4wk as needed, do not repeat until neutrophils >2000/mm^3 and platelets >100,000/mm^3

Renal dose

- **Adult (single agent): IV INF** CCr 41-59 ml/min, 250 mg/m^2, CCr 16-40 ml/min, 200 mg/m^2, do not use if CCr <15 ml/min

ALL (unlabeled)

- **Adult <21 yr/adolescent/child: IV** 635 mg/m^2 on day 3 with ifosfamide, mesna, etoposide

Relapsed Wilms' tumor (unlabeled)

- **Child: IV** 400 mg/m^2 × 2 days with etoposide, ifosfamide

Osteogenic sarcoma (unlabeled)

- **Adult/adolescent/child: IV** 560 mg/m^2 on day 1 with ifosfamide 2.65 g/m^2/day **IV** on days 1-3

Neuroblastoma/soft-tissue sarcoma (unlabeled)

- **Child: IV** 300-600 mg/m^2 q4wk or 400 mg/m^2/day for 2 days q4wk or 160 mg/m^2/day × 5 days q4wk

Available forms: Lyophilized powder for inj 50-, 150-, 450-mg vials; aqueous sol for inj 50 mg/5-ml vial, 150 mg/15-ml vial, 450 mg/45-ml vial, 600 mg/60-ml vial

Administer:

- Antiemetic 30-60 min before product and prn for vomiting

IV route

- Do not use needles or IV administration sets that contain aluminum; may cause precipitate or loss of potency
- Use cytotoxic handling procedures
- Store protected from light at room temperature; reconstituted vials stable for 24 hr at room temperature, Paraplatin multidose (10 mg/ml) vials stable for up to 14 days after entry into vial
- **Reconstitute** CARBOplatin 50, 150, or 450 mg with 5, 15, or 45 ml, respectively, of sterile water for inj, D_5W, or NaCl (10 mg/ml); then further **dilute** with the same sol to 0.5-4 mg/ml; **give** over 15 min or more (intermittent INF)
- **Continuous IV INF** over 24 hr; max dose based on (GFR=125 mg/ml)

Solution compatibilities: D_5/0.2% NaCl, D_5/0.45% NaCl, D_5/0.9% NaCl, 0.9% NaCl, D_5W, sterile water for inj

Y-site compatibilities: Acyclovir, alfentanil, allopurinol, amifostine, amikacin, aminocaproic acid, aminophylline, amiodarone, amphotericin B lipid complex, amphotericin B liposome, ampicillin, ampicillin-sulbactam, anidulafungin, atenolol, atracurium, azithromycin, aztreonam, bivalirudin, bleomycin, bumetanide, buprenorphine, butorphanol, calcium chloride/gluconate, caspofungin, ceFAZolin, cefepime, cefoperazone, cefotaxime, cefoTEtan, cefOXitin, cefTAZidime, ceftizoxime, cefTRIAXone, cefuroxime, cimetidine, ciprofloxacin, cisatracurium, CISplatin, cladribine, clindamycin, codeine, cyclophosphamide, cycloSPORINE, cytarabine, DAPTOmycin, DAUNOrubicin, dexamethasone, dexmedetomidine, dexrazoxane, digoxin, diltiazem, diphenhydrAMINE, DOBUTamine, DOCEtaxel, DOPamine, doripenem, doxacurium, DOXOrubicin, DOXOrubicin liposomal, doxycycline, droperidol, enalaprilat, ePHEDrine, EPINEPHrine, epirubicin, ertapenem, erythromycin, esmolol, etoposide, famotidine, fenoldopam, fentaNYL, filgrastim, fluconazole, fludarabine, fluorouracil, foscarnet, fosphenytoin, furosemide, ganciclovir, gatifloxacin, gemcitabine, gentamicin, granisetron, haloperidol, heparin, hydrocortisone, HYDROmorphone, hydrOXYzine, IDArubicin, ifosfamide, imipenem-cilastatin, inamrinone, insulin (regular), irinotecan, isoproterenol, ketorolac, labetalol, levofloxacin, levorphanol, lidocaine, linezolid injection, LORazepam, magnesium sulfate, mannitol, melphalan, meperidine, meropenem, mesna, methohexital, methotrexate, methylPREDNISolone, metoclopramide, metoprolol, metroNIDAZOLE, micafungin, midazolam, milrinone, minocycline, mitoXANtrone, mivacurium, morphine, nafcillin, nalbuphine, naloxone, nesiritide, niCARdipine, nitroglycerin, nitroprusside, norepinephrine, octreotide, ofloxacin, ondansetron, oxaliplatin, PACLitaxel, palonosetron, pamidronate, pancuronium, pantoprazole, PEMEtrexed, pentamidine, PENTobarbital, PHENobarbital, phenylephrine, piperacillin, piperacillin-tazobactam, potassium chloride, potassium phosphates, prochlorperazine, promethazine, propofol, propranolol, ranitidine, remifentanil, riTUXimab, rocuronium, sargramostim, sodium acetate, sodium bicarbonate, sodium phosphates, succinylcholine, SUFentanil, sulfamethoxazole-trimethoprim, tacrolimus, teniposide, theophylline, thiotepa, ticarcillin, ticarcillin-clavulanate, tigecycline, tirofiban, TNA, tobramycin, topotecan, TPN, trastuzumab, trimethobenzamide, vancomycin, vasopressin, vecuronium, verapamil, vinBLAStine, vinCRIStine, vinorelbine, voriconazole, zidovudine

SIDE EFFECTS

CNS: Seizures, central neurotoxicity, *peripheral neuropathy,* dizziness, confusion

CV: Cardiac abnormalities (fatal CV events), stroke

EENT: Tinnitus, hearing loss, *vestibular toxicity,* visual changes

GI: *Severe nausea, vomiting,* diarrhea, weight loss, mucositis, anorexia, constipation, taste change

HEMA: Thrombocytopenia, leukopenia, pancytopenia, neutropenia, anemia, bleeding
INTEG: *Alopecia,* dermatitis, rash, erythema, pruritus, urticaria
META: Hypomagnesemia, hypocalcemia, hypokalemia, hyponatremia, hyperuremia
SYST: Anaphylaxis

PHARMACOKINETICS

Initial half-life 1-2 hr, postdistribution half-life $2^{1}/_{2}$-6 hr, not bound to plasma proteins, excreted by the kidneys

INTERACTIONS

Increase: nephrotoxicity or ototoxicity—aminoglycosides, amphotericin B
Increase: bleeding risk—aspirin, NSAIDs, thrombolytics, anticoagulants, platelet inhibitors
Increase: toxicity—radiation, bone marrow suppressants
Increase: myelosuppression—myelosuppressives
Decrease: phenytoin levels, monitor levels

Drug/Lab Test
Increase: AST, BUN, alk phos, bilirubin, creatinine
Decrease: platelets, neutrophils, WBC, RBC, Hgb/HCT, calcium, potassium, magnesium, phosphate

NURSING CONSIDERATIONS

Assess:

Black Box Warning: To only be used by person experienced in the use of chemotherapeutic products, in a specialized care setting

Black Box Warning: **Bone marrow depression:** CBC, differential, platelet count weekly; withhold product if neutrophil count is <2000/mm^3 or platelet count is <100,000/mm^3; notify prescriber of results; calcium, magnesium, phosphate, potassium, sodium, uric acid, CCR, bilirubin; CCr <60 ml/min may be responsible for increased bone marrow suppression; assess frequently for infection

- Renal studies: BUN, creatinine, serum uric acid; urine CCr before, during therapy; I&O ratio; report fall in urine output to <30 ml/hr
- Monitor temperature q4hr (may indicate beginning infection)
- Hepatic studies before, during therapy (bilirubin, AST, ALT, LDH) as needed or monthly; jaundice of skin, sclera; dark urine, clay-colored stools, itchy skin, abdominal pain, fever, diarrhea

⚠ **Anaphylaxis:** hypotension, rash, pruritus, wheezing, tachycardia may occur within a few mins of use; notify prescriber after discontinuing product; resuscitation equipment should be available

- Delay dental work until blood counts have returned to normal; regular toothbrushes, dental floss, and toothpicks should not be used; use soft bristle toothbrush
- **Peripheral neuropathy:** may be increased in geriatric patients
- **Bleeding;** hematuria, stool guaiac, bruising, petechiae, mucosa or orifices; avoid all IM injections if platelets <50,000/mm^3
- Effects of alopecia on body image; discuss feelings about body changes

Evaluate:

- Therapeutic response: decreasing size of tumor, spread of malignancy

Teach patient/family:

- To report ringing/roaring in the ears; numbness, tingling in face, extremities; weight gain
- That impotence or amenorrhea can occur; that this is reversible after treatment is discontinued; to notify prescriber if pregnancy is suspected or planned; pregnancy (D), that contraception should be used if patient is fertile
- Not to breastfeed during treatment
- To avoid OTC products with aspirin, NSAIDs, alcohol; not to receive vaccinations during treatment

⚠ To notify prescriber immediately of fever, fatigue, sore throat, bleeding, bruising, chills, back pain, blood in stools, dyspnea

- That hair may be lost during treatment; that a wig or hairpiece may make the patient feel better; that new hair may be different in color, texture

• To avoid crowds, persons with known infections; to avoid the use of razors, stiff-bristle toothbrushes

carboprost (Rx)

(kar'boe-prost)

Hemabate

Func. class.: Oxytocic, abortifacient

Chem. class.: Prostaglandin

ACTION: Stimulates uterine contractions, causes complete abortion in approximately 16 hr

USES: Abortion at 13-20 wk gestation, postpartum hemorrhage caused by uterine atony not controlled by other methods

Unlabeled uses: Hemorrhagic cystitis

CONTRAINDICATIONS: Hypersensitivity to this product or benzyl alcohol, severe CV/respiratory/renal/hepatic disease, PID

Precautions: Pregnancy (C), asthma, anemia, jaundice, diabetes mellitus, hypo/hypertension, seizure disorders, past uterine surgery

DOSAGE AND ROUTES

Pregnancy termination between 13-20 wk gestation

• **Adult: IM** 100 mcg (0.4 ml) test dose, then 250 mcg, then 250 mcg q1½-3½hr; may increase to 500 mcg if no response, max 12 mg total dose

Postpartum hemorrhage

• **Adult: IM** 250 mcg, repeat at 15- to 90-min intervals; max total dosage 2 mg

Hemorrhagic cystitis (unlabeled)

• **Adult: INTRAVESICULAR** 0.8 mg/dl in 50 ml of saline instilled into the bladder for 60 min, q6hr × 4 doses

Available forms: Inj 250 mcg/ml

Administer:

• Only by trained personnel in a hospital that can provide emergency services

• Incomplete abortion may occur in 20% of patients

• Give antiemetics to prevent nausea/vomiting

• Store in refrigerator

• In deep muscle mass; aspirate before inj, rotate inj sites if additional doses given

SIDE EFFECTS

CNS: *Fever, chills,* headache

GI: *Nausea, vomiting, diarrhea*

PHARMACOKINETICS

Peak 15-60 min, excreted in urine (major metabolites)

INTERACTIONS

Increase: action—other oxytocics

NURSING CONSIDERATIONS

Assess:

• B/P, pulse; watch for change that may indicate hemorrhage

• For length, duration of contractions; notify prescriber of contractions that last more than 1 min or absence of contractions; watch for signs of uterine rupture

• For incomplete abortion, pregnancy must be terminated by another method; product is teratogenic

Evaluate:

• Therapeutic response: expulsion of fetus, control of bleeding

Teach patient/family:

• To report increased blood loss, abdominal cramps, increased temperature, foul-smelling lochia

carfilzomib

(car-fil'zoe-mib)

Kyprolis

Func. class.: Antineoplastic biologic response modifiers

Chem. class.: Signal transduction inhibitors (STIs)

ACTION: Antiproliferative and pro-apoptotic activity

USES: Multiple myeloma in those who have received ≥2 therapies (including bortezomib and immunomodulatory agents)

CONTRAINDICATIONS: Pregnancy (D), hypersensitivity

Precautions: Breastfeeding, children, cardiac disease, cardiac arrest, dysrhythmias, MI, infusion-related reactions, pulmonary/hepatic disease, edema, thrombocytopenia, neutropenia, tumor lysis syndrome

DOSAGE AND ROUTES

- **Adult: IV** 20 mg/m^2 over 2-10 min on days 1, 2, 8, 9, 15, 16, then 12 day rest (days 17-28), then may increase to 27 mg/m^2 on days 1, 2, 8, 9, 15, 16 repeated every 28 days

⚠ Refer to package insert for dosage adjustments for treatment-related toxicity

Available forms: Powder for injection 60 mg

Administer:

- Premedicate with dexamethasone 4 mg PO/IV before all carfilzomib 20-mg/m^2 doses during cycle 1 and before all carfilzomib 27-mg/m^2 doses in cycle 2; dexamethasone may be given in subsequent cycles if infusion-related reactions occur
- Hydration with 250–500 ml of NS or other IV fluids before each dose in cycle 1; additional hydration with 250–500 ml may be given after the carfilzomib infusion in cycle 1, continue hydration as needed
- Do not mix with other products
- Flush IV line with NS or D$_5$ for injection, before and after use

Reconstitution:

- Add 29 ml of sterile water for injection to the inside wall of the vial to minimize foaming (2 mg/mL); to mix, gently swirl and/or invert the vial slowly for about 1 min or until the cake or powder completely dissolves; do not shake; if foaming occurs, allow the solution to rest for 2 to 5 min or until foaming subsides; visually inspect for particulate and discoloration before use

IV injection route

- Give over 2-10 min; do not give as an IV bolus; the reconstituted sol may be stored in the vial/syringe at room temperature × 4 hr or ≤24 hr refrigerated

IV infusion route

- May further dilute in D$_5$W; measure and inject the correct dose from the reconstituted vial into 50 ml D$_5$W
- Administer IV over 2-10 min
- The diluted solution may be stored at room temperature × 4 hr or ≤24 hr refrigerated

SIDE EFFECTS

CNS: Headache, dizziness, insomnia

CV: **Heart failure**

GI: Nausea, vomiting, dyspepsia, anorexia, diarrhea

HEMA: **Neutropenia, thrombocytopenia**

META: Hyperglycemia, hypercalcemia, hypomagnesemia, hyponatremia, hypophosphatemia

MISC: Fatigue

MS: Arthralgia, myalgia

PHARMACOKINETICS

Protein binding 97%

NURSING CONSIDERATIONS

Assess:

⚠ **Tumor lysis syndrome (TLS):** hydrate well; assess for hyperuricemia, hyperkalemia, hyperphosphatemia, hypocalcemia, uremia

⚠ **Hematologic toxicity** grade 3 and 4 neutropenia and thrombocytopenia; platelet nadirs occur day 8 of each 28-day cycle; counts return to baseline before the start of the next cycle; monitor blood and platelet counts frequently; hold dose for grade 3 or 4 neutropenia or grade 4 thrombocytopenia, may require dosage reduction

⚠ **Serious liver toxicity:** AST/ALT and bilirubin elevations and rare cases of fatal hepatic failure have occurred; monitor hepatic enzymes frequently; withhold doses until resolution or return to baseline in grade 3 or 4AST/ALT or bilirubin elevations

⚠ **Serious cardiac toxicity:** fatal cardiac arrest, CHF with decreased left ventricular function/ejection fraction, myocardial ischemia, and pulmonary edema; those with NYHA class III/ IV CHF, MI within 6 mo, cardiac arrhythmias (conduction abnormalities) may be at increased risk; monitor for cardiac complications; withhold doses until resolution or return to baseline for grade 3 or 4 cardiac toxicity

⚠ **Infusion-related reactions:** may occur ≤24 hr after dose; premedication with dexamethasone is recommended; assess for fever, chills, arthralgia, myalgia, facial flushing, facial edema, vomiting, weakness, shortness of breath, hypotension, syncope, chest tightness, angina

Evaluate:

- Decreased spread of multiple myeloma

Teach patient/family:

- To promptly report infusion-related symptoms (fever, chills, arthralgia, myalgia, facial flushing, facial edema, vomiting, weakness, shortness of breath, hypotension, syncope, chest tightness, angina)

carglumic acid

(kar-gloo′mik)

Carbaglu

Func. class.: Antihyperammonemic agent

ACTION: The enzyme *N*-acetylglutamate synthase (NAGS) produces *N*-acetylglutamate (NAG), which is an essential activator of carbamoyl phosphate synthetase 1 (CPS 1); CPS 1 is the first enzyme of the urea cycle, and the enzyme converts ammonia to urea; patients with NAGS deficiency do not produce enough NAG, resulting in hyperammonemia; carglumic acid is a synthetic structural analog of NAG that works to activate CPS 1 and thus convert ammonia to urea

USES: Acute or chronic hyperammonemia in persons with *N*-acetylglutamate synthetase deficiency

Precautions: Breastfeeding, pregnancy (C), geriatric patients

DOSAGE AND ROUTES

- **Adult/child:** PO 100-250 mg/kg/day (rounded to the nearest 100 mg) initially, divided; bid-qid, given immediately before meals, titrate to ammonia level; usual maintenance dose <100 mg/kg/day

Available forms: Tabs 200 mg

Administer

- Do not give whole or crushed tablets, disperse each tab in ≥2.5 ml of water immediately before use; the tabs do not dissolve completely in water, and undissolved tab particles can remain; rinse the mixing container with water and give to the patient, do not mix with foods or other fluids

Nasogastric tube

- Mix each tab in ≥2.5 ml of water (80 mg/ml), shake gently, give the correct volume immediately through the NG tube, flush with additional water, discard any unused portion

Oral syringe

• Mix each tab in ≥2.5 ml of water (80 mg/ml), shake gently, draw up the appropriate volume in an oral syringe and give immediately, refill syringe with ≥1-2 ml and give immediately, discard any unused portion

SIDE EFFECTS

CNS: Headache, fever, asthenia
EENT: Tonsillitis, ear infections
GI: Abdominal pain, vomiting, taste change, anorexia, diarrhea
OTHER: Infection, anemia, weight loss, rash

PHARMACOKINETICS

Peak 3 hr, half-life 5.6 hr

INTERACTIONS

Drug/Lab Test
Decrease: Hgb level

NURSING CONSIDERATIONS

Assess:

Neurologic symptoms and ammonia levels: Headache, fever, change in level of consciousness
Infection: Assess for upper respiratory infections, influenza, pneumonia

Evaluate:

• Decreasing ammonia levels

Teach patient/family:

• Not to swallow whole or crush, but disperse in water
• That blood for lab tests will be drawn regularly
• To advise prescriber if pregnancy is planned or suspected, not to breastfeed
• That a high-calorie, low-protein diet is necessary when ammonia levels are elevated
• To report adverse reactions of vomiting, infection, ear pain, headache
• To store unopened containers in refrigerator; opened bottle may be stored at room temperature for up to 1 mo; product stored at controlled room temperature should not be returned to a refrigerator; protect from moisture; discard opened bottle after 30 days

carisoprodol (Rx)

(kar-eye-soe-proe′dole)

Soma

Func. class.: Skeletal muscle relaxant, central acting
Chem. class.: Meprobamate congener

Controlled Substance Schedule IV

Do not confuse:
Soma/Soma Compound

ACTION: Depresses CNS by blocking interneuronal activity in descending reticular formation, spinal cord, thereby producing sedation

USES: Relieving pain, stiffness with musculoskeletal disorders

CONTRAINDICATIONS: Hypersensitivity to these products or carbamates, intermittent porphyria
Precautions: Pregnancy (C), breastfeeding, geriatric patients, Asian patients, renal/hepatic disease, substance abuse, seizure disorder, CNS depression, abrupt discontinuation

DOSAGE AND ROUTES

• **Adult/adolescent ≥16 yr: PO** 250-350 mg tid and at bedtime, max 3 wk of use

Available forms: Tabs 350 mg

Administer:

• With meals for GI symptoms
• For short term (2-3 wk), potential for habituation
• Store in tight container at room temperature

SIDE EFFECTS

CNS: *Dizziness, weakness, drowsiness,* headache, tremor, depression, insomnia, ataxia, irritability, **seizures**
CV: Postural hypotension, tachycardia

EENT: Diplopia, temporary loss of vision
GI: *Nausea*, vomiting, hiccups, epigastric discomfort
HEMA: Eosinophilia
INTEG: Rash, pruritus, fever, facial flushing, erythema multiforme
RESP: Asthmatic attacks
SYST: Angioedema, anaphylaxis

PHARMACOKINETICS

PO: Onset ½ hr; peak 4 hr; duration 4-6 hr; extensively metabolized by liver, substrate of CYP2C19; excreted in urine; crosses placenta; excreted in breast milk (large amounts); half-life 8 hr

INTERACTIONS

- Do not use together with meprobamate

Increase: CNS depression—alcohol, tricyclics, opioids, barbiturates, sedatives, hypnotics
Increase: carisoprodol effect—CYP2C19 inhibitors (FLUoxetine, fluvoxaMINE, isoniazid, modafinil)
Decrease: carisoprodol effect—CYP219 inducers (rifampin)
Drug/Herb
Increase: CNS depression—kava, valerian
Increase: metabolism of carisoprodol—St. John's wort
Drug/Lab Test
Increase: eosinophils
Increase: RBC, WBC, platelets

NURSING CONSIDERATIONS

Assess:

- **Pain**, stiffness, mobility, activities of daily living at baseline and throughout treatment
- **ECG in seizure patients:** poor seizure control has occurred among patients taking this product
- BUN, creatinine at baseline and periodically
- **Idiosyncratic reaction:** (weakness, dizziness, blurred vision, confusion, euphoria), anaphylaxis within a few minutes or hours of 1st to 4th dose, withhold and notify prescriber
- **Allergic reactions:** rash, fever, respiratory distress, anaphylaxis, angioedema
- **CNS depression:** dizziness, drowsiness, psychiatric symptoms, abuse potential
- **Abrupt discontinuation:** withdrawal reactions do occur but may be mild, dependence may occur
- Assistance with ambulation if dizziness, drowsiness occurs, especially for geriatric patients

Evaluate:

- Therapeutic response: decreased pain, spasticity; increased ROM

Teach patient/family:

- To avoid hazardous activities if drowsiness, dizziness occur; not to drive while taking product; to avoid rapid position changes, postural hypotension occurs
- To avoid using OTC medications (cough preparations, antihistamines) unless directed by prescriber; not to take with alcohol, other CNS depressants
- To report allergic reaction immediately: rash, swelling of tongue/lips, hives, dyspnea
- To take with food for GI symptoms

TREATMENT OF OVERDOSE:

Activated charcoal, dialysis, lavage

⚠ HIGH ALERT

carmustine (Rx)

(kar-mus'teen)

BiCNU, Gliadel

Func. class.: Antineoplastic alkylating agent
Chem. class.: Nitrosourea

ACTION: Alkylates DNA, RNA; able to inhibit enzymes that allow for the synthesis of amino acids in proteins; activity is not cell-cycle–phase specific

USES: Brain tumors such as glioblastoma, medulloblastoma, brain stem glioma, astrocytoma, ependymoma, metastatic brain tumors; multiple myeloma (with predniSONE), non-Hodgkin's disease, Hodgkin's disease, other lymphomas; GI, breast, bronchogenic, renal carcinomas; wafer, as adjunct to surgery/radiation for patients newly diagnosed with high-grade malignant glioma

Unlabeled uses: Ablation, mycosis fungoides, stem cell transplant preparation

CONTRAINDICATIONS:

Pregnancy (D), breastfeeding, hypersensitivity, leukopenia, thrombocytopenia

Precautions: Dental disease, extravasation, females, infection, leukopenia, neutropenia, secondary malignancy, thrombocytopenia

Black Box Warning: Bone marrow suppression, pulmonary fibrosis

DOSAGE AND ROUTES

Brain tumors, Hodgkin's disease, malignant lymphoma, multiple myeloma

- **Adult: IV** 75-100 mg/m^2 over 1-2 hr × 2 days or 150-200 mg/m^2 × 1 dose q6-8wk or 40 mg/m^2/day × 5 days q6wk or 150-200 mg/m^2 q6wk; **INTRACAVITARY** up to 8 wafers inserted into resection cavity
- **Child (unlabeled): IV** 200-250 mg/m^2 as a single dose q4-6wks

Stem cell transplant/bone marrow ablation (unlabeled)

- **Adult: IV** 450-600 mg/m^2 as a single dose or 2 divided doses q12hr at a rate of no more than 3 mg/m^2/min

Available forms: Powder for inj 100 mg; wafer 7.7 mg (intracavitary)

Administer:

- Store reconstituted sol in refrigerator for 24 hr or at room temperature for 8 hr; protect from light
- Blood transfusions, RBC colony-stimulating factors to counter anemia
- Antiemetic, serotonin antagonists, dexamethasone, nausea, vomiting can begin 2 hr after a dose and last ≤6 hr and may be severe
- All medications PO, if possible; avoid IM inj if platelets are <100,000/mm^3

Black Box Warning: Carmustine should not be given until platelets >100,000/mm^3 and WBC >4000/mm^2

Wafer route

- If wafers are broken into several pieces, they should not be used
- Foil pouches may be kept at room temperature for 6 hr if unopened

Intermittent IV INF route

- Use cytotoxic handling procedures; do not use if an oil film appears on vial (decomposition)
- Do not use with PVC IV tubing, do not admix
- After **diluting** 100 mg product/3 ml ethyl alcohol (provided), **further dilute** with 27 ml sterile water for inj; **then dilute** with 100-500 ml 0.9% NaCl or D_5W, **give** over 1 hr or more, reduce rate if discomfort is felt; use only glass containers, protect from light
- **Flush** IV line after carmustine with 10 ml 0.9% NaCl to prevent irritation at site

Y-site compatibilities: Amifostine, amphotericin B lipid complex, amphotericin B liposome, anidulafungin, aztreonam, bivalirudin, bleomycin, caspofungin, cefepime, codeine, DAPTOmycin, dexmedetomidine, DOCEtaxel, ertapenem, etoposide, fenoldopam, filgrastim, fludarabine, gemcitabine, granisetron, levofloxacin, melphalan, meperidine, mitoXANtrone, nesiritide, octreotide, ondansetron, PACLitaxel, palonosetron, pamidronate, pantoprazole, PEMEtrexed, piperacillin-tazobactam, riTUXimab, sargramostim, sodium acetate, tacrolimus, teniposide, thiotepa, tigecycline, tirofiban, trastuzumab, vinCRIStine, vinorelbine, voriconazole

SIDE EFFECTS

GI: *Nausea, vomiting, anorexia, stomatitis,* hepatotoxicity

GU: *Azotemia,* renal failure

HEMA: Thrombocytopenia, leukopenia, myelosuppression, anemia

INTEG: Pain, burning, hyperpigmentation at inj site, alopecia

RESP: Fibrosis, pulmonary infiltrate

SYST: Secondary malignant neoplastic disease

PHARMACOKINETICS

Degraded within 15 min; crosses blood-brain barrier; 70% excreted in urine within 96 hr; 10% excreted as CO_2, fate of 20% is unknown

INTERACTIONS

Increase: bleeding risk—aspirin, anticoagulants
Increase: myelosuppression—myelosuppressive agents
Increase: toxicity: other antineoplastics, radiation, cimetidine
Increase: adverse reactions, decreased antibody reaction—live vaccines
Decrease: effects of digoxin, phenytoins
Drug/Lab Test
Increase: bilirubin, prolactin, uric acid, LFTs
Decrease: platelets, WBC, neutrophils, HCT

NURSING CONSIDERATIONS

Assess:

Black Box Warning: **Bone marrow suppression:** CBC, differential, platelet count weekly; withhold product if WBC is <4000 or platelet count is <100,000; notify prescriber of results

- Hepatic studies: AST, ALT, bilirubin, monitor regularly, hepatotoxicity occurs rarely

Black Box Warning: **Pulmonary fibrosis/infiltrate:** pulmonary function tests, chest x-ray films before, during therapy; chest film should be obtained q2wk during treatment; monitor for dyspnea, cough, pulmonary fibrosis; infiltrate occurs after high doses or several low-dose courses (>1400 mg/m^2 cumulative dose), may occur months or years after treatment

Black Box Warning: Only to be used by an experienced clinician in cases of cancer, immune suppression

- Renal studies: BUN, serum uric acid, urine CCr before, during therapy; I&O ratio; report fall in urine output of 30 ml/hr, may use allopurinol for hyperuricemia with increased fluids
- Monitor for cold, cough, fever (may indicate beginning infection)
- **Bleeding:** hematuria, guaiac, bruising, petechiae, mucosa, orifices q8hr
- Rinsing of mouth tid-qid with water or club soda; use of sponge brush for stomatitis
- Warm compresses at inj site for inflammation; reduce flow rate if patient complains of burning at inf site

Evaluate:

- Therapeutic response: decreasing size of tumor, spread of malignancy

Teach patient/family:

Black Box Warning: To report any changes in breathing or coughing; to avoid smoking

- To avoid foods with citric acid, hot or rough texture if stomatitis is present; to report any bleeding, white spots, ulceration in mouth to prescriber; to examine mouth daily
- To avoid aspirin, ibuprofen, razors, commercial mouthwash
- To report signs of anemia (fatigue, irritability, shortness of breath, faintness); to report signs of infection (sore throat, fever); pulmonary toxicity can occur up to 15 yr after treatment
- To use contraception during treatment; to avoid breastfeeding, pregnancy (D)
- Not to receive live vaccines during treatment

carteolol (ophthalmic)

(kar′tee-oh-lol)
Func. class.: Antiglaucoma
Chem. class.: Beta-blocker

ACTION: May decrease aqueous humor and increase outflows

USES: Treatment of chronic open-angle glaucoma and ocular hypertension

CONTRAINDICATIONS:
Hypersensitivity, AV block, heart failure, bradycardia, sick sinus syndrome

Precautions: Abrupt discontinuation, children, pregnancy, breastfeeding, COPD, depression, diabetes mellitus, myasthenia gravis, hyperthyroidism, pulmonary disease, angle-closure glaucoma

DOSAGE AND ROUTES

- **Adult:** instill 1 drop in the affected eye(s) bid

Available forms: Ophthalmic solution 1%

Administer

- For ophthalmic use only; shake the ophthalmic suspension well before use
- Do not touch the tip of the dropper to the eye, fingertips, or other surface to prevent contamination
- Wash hands before and after use; tilt head back slightly and pull the lower eyelid down with the index finger to form a pouch; squeeze the prescribed number of drops into the pouch; close eyes to spread drops; to avoid excessive systemic absorption, apply finger pressure on the lacrimal sac for 1-2 min after use
- If more than one topical ophthalmic drug product is being used, the drugs should be administered at least 5 min apart
- To avoid contamination or the spread of infection, do not use dropper for more than one person, do not touch the eyelids or surrounding areas with the dropper tip of the bottle
- Decreased intraocular pressure can take several weeks; monitor IOP after 1 mo

SIDE EFFECTS

CNS: Insomnia, headache, dizziness
CV: Palpitations
EENT: Eye stinging/burning, tearing, photophobia

PHARMACOKINETICS

Onset 30 min, peak 2 hr, duration ≥12 hr

INTERACTIONS

Increase: beta blocking effect—oral beta blockers

Increased: intraocular pressure reduction—topical miotics, dipivefrin, EPINEPHrine, carbonic anhydrase inhibitors, this may be beneficial

Increase: B/P, severe—when abruptly stopping cloNIDine

Increase: depression of AV nodal conduction, bradycardia, or hypotension—adenosine, cardiac glycosides, disopyramide, other antiarrhythmics, class 1C antiarrhythmic drugs (flecainide, propafenone, moricizine, encainide, quiNIDine or drugs that significantly depress AV nodal conduction)

Increase: AV block nodal conduction, induce AV block—high doses of procainamide

Increase: antihypertensive effect—other antihypertensives

NURSING CONSIDERATIONS

Assess:

- **Systemic absorption:** when used in the eye, systemic absorption is common with the same adverse reactions and interactions
- Glaucoma: monitor intraocular pressure

Evaluate:

- Decreasing intraocular pressure

Teach patient/family:

- For ophthalmic use only; shake the ophthalmic suspension well before use
- Not to touch the tip of the dropper to the eye, fingertips, or other surface to prevent contamination
- To wash hands before and after use; tilt the head back slightly and pull the lower eyelid down with the index finger to form a pouch; squeeze the prescribed number of drops into the pouch; close eyes to spread drops; to avoid excessive systemic absorption, apply finger pressure on the lacrimal sac for 1-2 min following use
- If more than one topical ophthalmic drug product is being used, the drugs

should be administered at least 5 min apart
• To avoid contamination or the spread of infection, do not use dropper for more than one person, do not touch the eyelids or surrounding areas with the dropper tip of the bottle

carvedilol (Rx)

(kar-ved'i-lole)

Coreg, Coreg CR

Func. class.: Antihypertensive, α-/β-adrenergic blocker

Do not confuse:
carvedilol/captopril/carteolol

ACTION: A mixture of nonselective α-/β-adrenergic blocking activity; decreases cardiac output, exercise-induced tachycardia, reflex orthostatic tachycardia; causes vasodilation, reduction in peripheral vascular resistance

USES: Essential hypertension alone or in combination with other antihypertensives, CHF, LV dysfunction after MI, cardiomyopathy

Unlabeled uses: Angina, pediatric patients

CONTRAINDICATIONS: Hypersensitivity, asthma, class IV decompensated cardiac failure, 2nd- or 3rd-degree heart block, cardiogenic shock, severe bradycardia, pulmonary edema, severe hepatic disease

Precautions: Pregnancy (C), breastfeeding, children, geriatric patients, cardiac failure, hepatic injury, peripheral vascular disease, anesthesia, major surgery, diabetes mellitus, thyrotoxicosis, emphysema, chronic bronchitis, renal disease

Black Box Warning: Abrupt discontinuation

DOSAGE AND ROUTES

Essential hypertension
• **Adult: PO** 6.25 mg bid × 7-14 days; if tolerated well, then increase to 12.5 mg bid × 7-14 days; if tolerated well, may be increased (if needed) to 25 mg bid; not to exceed 50 mg/day; **EXT REL** cap 20 mg/day, may increase after 7-14 days to 40 mg/day, max 80 mg/day

Congestive heart failure
• **Adult: PO** 3.125 mg bid × 2 wk; if tolerated well, give 6.25 mg bid × 2 wk, then double q2wk to max dose of 25 mg bid <85 kg or 50 mg bid >85 kg; **EXT REL** caps (Coreg CR) 10 mg/day × 2 wk, increase to 20, 40, 80 mg/day over successive intervals of 2 wk

Postmyocardial infarction
• **Adult: PO** 6.25 mg bid with food × 3-10 days, lower starting dose may be used if indicated; titrate upward as tolerated; may increase to 12.5 mg bid then titrate to 25 mg bid; **PO EXT REL** 20 mg daily with food, lower starting dose of 10 mg/day may be used, titrate upward after 3-10 days, increase to 40 mg daily as required

Angina (unlabeled)
• **Adult: PO** 25-50 mg bid

Available forms: Tabs 3.125, 6.25, 12.5, 25 mg; ext rel cap 10, 20, 40, 80 mg

Administer:
• With food; tabs may be crushed or swallowed whole; give ext rel every AM with food; do not break, crush, chew ext rel cap; separate alcohol (including OTC products that contain ethanol) by ≥2 hr; caps may be opened and sprinkled over applesauce

Black Box Warning: Do not discontinue before surgery

SIDE EFFECTS

CNS: *Dizziness, fatigue,* weakness, somnolence, insomnia, ataxia, hyperesthesia, paresthesia, vertigo, depression, headache

CV: Bradycardia, *postural hypotension,* dependent edema, *peripheral edema,* AV block, extrasystoles, hypo/hypertension, palpitations, peripheral ischemia, **CHF, pulmonary edema**

GI: *Diarrhea,* abdominal pain, increased alk phos, ALT, AST

GU: Decreased libido, *impotence,* UTI

INTEG: Rash, Stevens-Johnson syndrome

MISC: Injury, back pain, viral infection, hypertriglyceridemia, **thrombocytopenia,** *hyperglycemia,* abnormal weight gain, aplastic anemia

RESP: Rhinitis, pharyngitis, dyspnea, **bronchospasm,** cough, **lung edema**

PHARMACOKINETICS

Peak 1-2 hr; readily and extensively absorbed PO; >98% protein binding; extensively metabolized by liver; excreted through bile into feces; terminal half-life 7-10 hr with increases in geriatric patients, hepatic disease

INTERACTIONS

Increase: conduction disturbances—calcium channel blockers

Increase: bradycardia, hypotension—levodopa, MAOIs, reserpine

Increase: hypoglycemia—antidiabetic agents

Increase: concs of digoxin, cycloSPORINE, CYP2D6 inhibitors (FLUoxetine, quiNIDine)

Increase: toxicity of carvedilol—cimetidine, other antihypertensives, nitrates, acute alcohol ingestion

Decrease: heart rate, B/P—cloNIDine

Decrease: carvedilol levels—rifampin, NSAIDs, thyroid medications

Drug/Herb

Increase: antihypertensive effect—hawthorn

Decrease: antihypertensive effect—ephedra (ma huang)

Drug/Lab Test

Increase: blood glucose, BUN, potassium, triglycerides, uric acid, bilirubin, cholesterol, creatinine

Decrease: sodium, HDL

NURSING CONSIDERATIONS

Assess:

- **Hypertension:** B/P when beginning treatment, periodically thereafter; pulse: note rate, rhythm, quality; apical/radial pulse before administration; notify prescriber of significant changes
- **CHF:** edema in feet, legs daily; fluid overload: dyspnea, weight gain, jugular venous distention, fatigue, crackles

Evaluate:

- Therapeutic response: decreased B/P with hypertension; decrease anginal pain

Teach patient/family:

- To comply with dosage schedule even if feeling better; that improvement may take several weeks
- To rise slowly to sitting or standing position to minimize orthostatic hypotension
- **To report bradycardia, dizziness, confusion, depression, fever, weight gain, SOB, cold extremities, rash, sore throat, bleeding, bruising**
- To weigh, take pulse, B/P at home; to advise if weight gain of >2 lb/day or 5 lb/wk and when to notify prescriber

Black Box Warning: Not to discontinue product abruptly; to taper over 1-2 wk; life-threatening dysrhythmias may occur

- To avoid hazardous activities until stabilized on medication; dizziness may occur
- To avoid all OTC medications unless approved by prescriber
- To carry emergency ID with product name, prescriber information at all times
- To carry emergency ID with product name, prescriber information at all times
- To inform all health care providers of products, supplements taken
- To report if pregnancy is planned or suspected, pregnancy (C), avoid breastfeeding
- That product may mask hypoglycemia

caspofungin (Rx)

(cas-po-fun′gin)

Cancidas

Func. class.: Antifungal, systemic

Chem. class.: Echinocandin

ACTION: Inhibits an essential component in fungal cell walls; causes direct damage to fungal cell wall

USES: Treatment of invasive aspergillosis and candidemia that has not responded to other treatment, including peritonitis, intraabdominal abscesses; susceptible species: *Aspergillus flavus, A. fumigatus, A. terreus, Candida albicans, C. glabrata, C. krusei, C. lusitaniae, C. parapsilosis, C. tropicalis,* esophageal candidiasis; empirical therapy for presumed fungal infection in febrile, neutropenic patients

Unlabeled uses: *Aspergillus niger,* fungal infections in premature neonates, neonates, infants, children <2 yr

CONTRAINDICATIONS: Hypersensitivity

Precautions: Pregnancy (C), breastfeeding, children, geriatric patients, severe hepatic disease

DOSAGE AND ROUTES

- **Adult: IV** loading dose 50-70 mg on day 1 then 50 mg/day maintenance dose, depending on condition; max 70 mg/day
- **Adolescent/child/infant ≥3 mo: IV INF** 70 mg/m^2 loading dose then 50 mg/m^2/day; max 70 mg/day
- **Neonate and infant <3 mo (unlabeled): IV** 25 mg/m^2/day

Esophageal candidiasis

- **Adult: IV** 50 mg × 7-14 days over 1 hr
- **Child:** 3 mo-17 yr: **IV** 70 mg/m^2 loading dose, then 50 mg/m^2 daily; max 70 mg/m^2

Available forms: Powder for inj 50, 70 mg

Administer:

- Do not mix or coinfuse with other medications, do not use dextrose-containing products to dilute, do not give as bolus
- Store at room temperature for up to 24 hr or refrigerated for 48 hr; store reconstituted sol at room temperature for 1 hr before preparation of sol for administration

Intermittent IV INF route

- Allow to warm to room temperature
- May administer loading dose on day 1
- **Reconstitute** 50-mg vial or 70-mg vial with 10.8 ml 0.9% NaCl, sterile water for inj or bacteriostatic water for inj (5 mg/ml or 7 mg/ml), respectively; **swirl** to dissolve, withdraw 10 ml reconstituted sol, and **further dilute** with 250 ml 0.9% NaCl, 0.45% NaCl, 0.225% NaCl, RL; **run** over 1 hr or more

SIDE EFFECTS

CNS: Dizziness, *headache,* fever

CV: Sinus tachycardia, hypertension

GI: Abdominal pain, *nausea, anorexia, vomiting, diarrhea, increased AST/ALT, alk phos*

GU: Renal failure

HEMA: Thrombophlebitis, vasculitis, anemia

INTEG: *Rash, pruritus, inj site pain*

META: Hypokalemia

MS: Myalgia

RESP: Acute respiratory distress syndrome (ARDS), pleural effusions

SYST: Anaphylaxis, histamine-related reactions, Stevens-Johnson syndrome

PHARMACOKINETICS

Metabolized in liver to inactive metabolites; excretion in feces, urine; phase II terminal half-life 9-11 hr; phase III terminal half-life 40-50 hr; protein binding 97%

INTERACTIONS

Increase: caspofungin levels—cycloSPORINE; may need dosage reduction

Decrease: levels of tacrolimus, sirolimus

Decrease: caspofungin levels—carBAMazepine, dexamethasone, efavirenz, nelfinavir, nevirapine, phenytoin, rifampin

Drug/Lab Test

Increase: AST, ALT, RBC, eosinophils

Decrease: HCT/Hgb, WBC, potassium

NURSING CONSIDERATIONS

Assess:

- **Infection;** clearing of cultures during treatment; obtain culture at baseline, throughout treatment; product may be started as soon as culture is taken (esophageal candidiasis); monitor cultures during HSCT for prevention of *Candida* infections
- Blood studies before, during treatment: bilirubin, AST, ALT, alk phos, as needed; obtain baseline renal studies; CBC with differential, serum potassium
- **Hypersensitivity:** rash, pruritus, facial swelling; also for phlebitis; anaphylaxis (rare)
- GI symptoms: frequency of stools, cramping; if severe diarrhea occurs, electrolytes may need to be given

Evaluate:

- Therapeutic response: decreased symptoms of *Candida, Aspergillus* infections

Teach patient/family:

- To notify prescriber if pregnancy is suspected or planned
- To inform prescriber of renal/hepatic disease
- To report bleeding, facial swelling, wheezing, difficulty breathing, itching, rash, hives, increasing warmth, flushing; anaphylaxis can occur

cefaclor

See cephalosporins—2nd generation

cefadroxil

ceFAZolin

See cephalosporins—1st generation

cefdinir

cefditoren pivoxil

cefepime

cefixime

cefotaxime

See cephalosporins—3rd generation

cefoTEtan

cefOXitin

See cephalosporins—2nd generation

cefpodoxime

See cephalosporins—3rd generation

cefprozil

See cephalosporins—2nd generation

cefTAZidime

ceftibuten

ceftizoxime

cefTRIAXone

See cephalosporins—3rd generation

cefuroxime

See cephalosporins—2nd generation

ceftaroline (Rx)

(sef-tar′oh-leen)

Teflaro

Func. class.: Cephalosporin action: Inhibits cell wall synthesis through binding to essential penicillin-binding protein (PBPs)

USES: Acute bacterial skin/skin structure infections (ABSSI), bacterial community-acquired pneumonia

CONTRAINDICATIONS: Cephalosporin hypersensitivity

Precautions: Child/infant/neonate, breastfeeding, elderly patients, antimicrobial resistance, carbapenem/penicillin hypersensitivity, coagulopathy, colitis, dialysis, diarrhea, GI disease, hypoprothrombinemia, IBS, pregnancy (B), pseudomembranous colitis, renal disease, ulcerative colitis, viral infection, vit K deficiency

DOSAGE AND ROUTES

• **Adult: IV** 600 mg q12hr × 5-14 days (skin/skin-structure infections) or × 5-7 days (bacterial community-acquired pneumonia)

Renal dose

• **Adult: IV** CCr >30-≤50 ml/min, 400 mg q12hr; CCr ≥15-≤30 ml/min, 300 mg q12hr, CCr <15 ml/min, 200 mg q12hr

Available forms: Powder for inj 400 mg, 600 mg

Administer:

• Obtain C&S before use

Intermittent IV INF route

• Visually inspect for particulate matter, discoloration if sol or container permits

• **Reconstitute:** add 20 mg of sterile water to 400- or 600-mg vial (20 mg/ml for 400 mg; 30 mg/ml for 600 mg), mix gently until dissolved; **dilute** in 250 ml of 0.9% NaCl, 0.45% NaCl, LR, D_5, D2.5, **give** over 1 hr, do not admix, use within 6 hr at room temperature or 24 hr refrigerated

SIDE EFFECTS

CNS: Dizziness, seizures

CV: Phlebitis, palpitations, bradycardia

ENDO: Hypo/hyperkalemia

GI: Diarrhea, nausea, vomiting, constipation, abdominal pain, pseudomembranous colitis (rare)

HEMA: Thrombocytopenia, neutropenia, anemia, eosinophilia

INTEG: Rash, anaphylaxis

PHARMACOKINETICS

Protein binding 20%; excreted in urine 88%, feces 6%; half-life 2.66 hr; not hepatically metabolized

INTERACTIONS

Increase: prothrombin time risk—anticoagulants

Drug/Lab Test

Increase: LFTs

Decrease: potassium, eosinophils, platelets

NURSING CONSIDERATIONS

Assess:

• **Infection:** vital signs, sputum, WBC before, during therapy

• **Hypersensitivity:** before use, obtain a history of hypersensitivity reactions to cephalosporins, carbapenems, penicillins; cross-sensitivity may occur

• **Anaphylaxis (rare):** rash, pruritus, laryngeal edema, dyspnea, wheezing; discontinue product, notify health care provider immediately, keep emergency equipment nearby

• **Pseudomembranous colitis:** diarrhea, abdominal pain, fever, bloody stools; report immediately if these occur; may occur several weeks after terminating therapy

• Monitor BUN, creatinine baseline, periodically; elderly and those with renal disease are at greater risk of renal dysfunction

Evaluate:

• Therapeutic response: negative C&S, resolution of symptoms of infection

Teach patient/family:
- About the reason for treatment and expected result
- To immediately report rash, itching, difficulty breathing, bloody diarrhea, fever, abdominal pain

HIGH ALERT

celecoxib (Rx)

(sel-eh-cox′ib)

CeleBREX

Func. class.: Nonsteroidal antiinflammatory, antirheumatic

Chem. class.: COX-2 inhibitor

Do not confuse:
CeleBREX/CeleXA/Cerebra/Cerebyx

ACTION: Inhibits prostaglandin synthesis by selectively inhibiting cyclooxygenase-2 (COX-2), an enzyme needed for biosynthesis

USES: Acute, chronic rheumatoid arthritis, osteoarthritis, acute pain, primary dysmenorrhea, ankylosing spondylitis, juvenile rheumatoid arthritis (JRA)

Unlabeled uses: Colorectal adenoma prophylaxis

CONTRAINDICATIONS: Pregnancy (D) 3rd trimester; hypersensitivity to salicylates, iodides, other NSAIDs, sulfonamides

Black Box Warning: CABG

Precautions: Pregnancy (C) 1st/2nd trimesters, breastfeeding, children <18 yr, geriatric patients, bleeding, GI/renal/hepatic/cardiac disorders, PVD, hypertension, severe dehydration, asthma

Black Box Warning: GI bleeding/perforation, peptic ulcer disease, MI, stroke

DOSAGE AND ROUTES

Do not exceed recommended dose; deaths have occurred

Acute pain/primary dysmenorrhea
- **Adult: PO** 400 mg initially, then 200 mg if needed on 1st day, then 200 mg bid prn on subsequent days; start with ½ dose for poor CYP2C9 metabolizers

Osteoarthritis
- **Adult: PO** 200 mg/day as a single dose or 100 mg bid; start with ½ dose for poor CYP2C9 metabolizers
- **Geriatric: PO** use lowest possible dose

Rheumatoid arthritis
- **Adult: PO** 100-200 mg bid; start with ½ dose for poor CYP2C9 metabolizers

Ankylosing spondylitis
- **Adult: PO** 200 mg/day or in divided doses (bid); start with ½ dose for poor CYP2C9 metabolizers

Juvenile rheumatoid arthritis (JRA)
- **Adolescent and child ≥2 yr (>25 kg): PO** 100 mg bid; start with ½ dose for poor CYP2C9 metabolizers
- **Child ≥2 yr (10-25 kg): PO** 50 mg bid; start with ½ dose for poor CYP2C9 metabolizers

Hepatic disease
- **Adult: PO** (Child-Pugh B) reduce dose by 50%

Colorectal adenoma prophylaxis (unlabeled)
- **Adult: PO** 200-400 mg BID for up to 3 yr

Available forms: Caps 50, 100, 200, 400 mg

Administer:
- Do not break, crush, chew, or dissolve caps; give with a full glass of water to enhance absorption; caps may be opened into applesauce or soft food, ingest immediately with water
- With food, milk to decrease gastric symptoms (with higher doses [400 mg bid]); do not increase dose

SIDE EFFECTS

CNS: *Fatigue, anxiety, depression, nervousness, paresthesia,* dizziness, insomnia, headache

CV: Stroke, MI, tachycardia, CHF, angina, palpitations, dysrhythmias, hypertension, fluid retention

EENT: Tinnitus, hearing loss, blurred vision, glaucoma, cataract, conjunctivitis, eye pain
GI: Nausea, anorexia, vomiting, constipation, dry mouth, diverticulitis, gastritis, gastroenteritis, hemorrhoids, hiatal hernia, stomatitis, GI bleeding/ulceration
GU: Nephrotoxicity: *dysuria,* hematuria, azotemia, cystitis, UTI, renal papillary necrosis
HEMA: Blood dyscrasias, epistaxis, anemia
INTEG: Serious (sometimes fatal) Stevens-Johnson syndrome, toxic epidermal necrolysis, purpura, rash, pruritus, sweating, erythema, petechiae, photosensitivity, alopecia, bruising, hot flashes
RESP: Pharyngitis, SOB, pneumonia, coughing

PHARMACOKINETICS

Well absorbed, crosses placenta, bound to plasma proteins, metabolized by CYP2C9 in liver, very little excreted by kidneys/in feces, peak 3 hr, half-life 11 hr, protein binding ~97%

INTERACTIONS

Increase: bleeding risk—anticoagulants, SSRIs, antiplatelets, thrombolytics, salicylates, alcohol
Increase: adverse reactions—glucocorticoids, NSAIDs, aspirin
Increase: toxicity—lithium, antineoplastics, bisphosphonates, cidofovir
Increase: celecoxib blood level—fluconazole
Decrease: effect of aspirin, ACE inhibitors, thiazide diuretics, furosemide

Drug/Herb
Decrease: effect of feverfew
Increase: bleeding risk—garlic, ginger, ginkgo

Drug/Lab Test
Increase: ALT, AST, BUN, cholesterol, glucose, potassium, sodium
Decrease: glucose, sodium, WBC, platelets

NURSING CONSIDERATIONS

Assess:
- **Pain** of rheumatoid arthritis, osteoarthritis; check ROM, inflammation of joints, characteristics of pain

Black Box Warning: For cardiac disease that may be worse after taking product; MI, stroke; do not use with coronary artery bypass graft (CABG)

- CBC during therapy; watch for decreasing platelets; if low, therapy may need to be discontinued, restarted after hematologic recovery; LFTs, serum creatinine/BUN, stool guaiac

Black Box Warning: For blood dyscrasias (thrombocytopenia): bruising, fatigue, bleeding, poor healing

- **GI toxicity:** black, tarry stools; abdominal pain
- ⚠ **Serious skin disorders:** Stevens-Johnson syndrome, toxic epidermal necrolysis; may be fatal

Evaluate:
- Therapeutic response: decreased pain, inflammation in arthritic conditions; decreased number of polyps

Teach patient/family:

Black Box Warning: Not to exceed recommended dose; to notify prescriber immediately of chest pain, skin eruptions; to stop product if these occur

- To check with prescriber to determine when product should be discontinued before surgery
- That product must be continued for prescribed time to be effective; to avoid other NSAIDs, aspirin, sulfonamides

Black Box Warning: To notify prescriber of GI symptoms: black, tarry stools; cramping or rash; edema of extremities; weight gain

Black Box Warning: To report bleeding, bruising, fatigue, malaise because blood abnormalities do occur

- To report possible respiratory infection: fever, SOB, coughing, painful swallowing

• To report if pregnancy is planned or suspected, pregnancy (C) before 30 wk, (D) after 30 wk

cephalexin

See cephalosporins—1st generation

CEPHALOSPORINS—1ST GENERATION

cefadroxil (Rx)
(sef-a-drox′ill)
Apo-Cefadroxil ♣
ceFAZolin (Rx)
(sef-a′zoe-lin)
cephalexin (Rx)
(sef-a-lex′in)
Keflex, Panixine
Func. class.: Antiinfective
Chem. class.: Cephalosporin (1st generation)

Do not confuse:
cephalexin/cefaclor

ACTION: Inhibits bacterial cell wall synthesis; renders cell wall osmotically unstable, leads to cell death; lysis mediated by cell wall autolytic enzymes

USES:

cefadroxil: Gram-negative bacilli: *Escherichia coli, Proteus mirabilis, Klebsiella* (UTI only); gram-positive organisms: *Streptococcus pneumoniae, Streptococcus pyogenes, Staphylococcus aureus;* upper, lower respiratory tract, urinary tract, skin infections; otitis media; tonsillitis; UTIs
ceFAZolin: Gram-negative bacilli: *Haemophilus influenzae, Escherichia coli, Proteus mirabilis, Klebsiella, pneumonial;* gram-positive organisms: *Staphylococcus aureus/epidermidis;* upper, lower respiratory tract, urinary tract, skin infections; bone, joint, biliary, genital infections; endocarditis, surgical prophylaxis, septicemia; *Streptococcus Sp.*
cephalexin: Gram-negative bacilli: *Haemophilus influenzae, Escherichia coli, Proteus mirabilis, Klebsiella pneumoniae;* gram-positive organisms: *Streptococcus pneumoniae, Streptococcus pyogenes, Streptococcus agalactial, Staphylococcus aureus;* upper, lower respiratory tract, urinary tract, skin, bone infections; otitis media

CONTRAINDICATIONS: Hypersensitivity to cephalosporins, infants <1 mo
Precautions: Pregnancy (B), breastfeeding, hypersensitivity to penicillins, renal disease

DOSAGE AND ROUTES

cefadroxil
• **Adult: PO** 1-2 g/day or divided q12hr; loading dose of 1 g initially
• **Child: PO** 30 mg/kg/day in divided doses bid, max 2 g/day
Renal dose
• **Adult: PO** CCr 25-50 ml/min, 1g, then 500 mg q12hr; CCr 10-24 ml/min, 1g, then 500 mg q24hr; CCr <10 ml/min, 1g, then 500 mg q36hr
Available forms: Caps 500 mg; tabs 1 g; oral susp 250, 500 mg/5 ml

ceFAZolin
Life-threatening infections
• **Adult: IM/IV** 1-2 g q6-8hr; max 12 g/day
• **Child >1 mo: IM/IV** 75-100 mg/kg/day in 3-4 divided doses; max 6 g/day
Mild/moderate infections
• **Adult: IM/IV** 250 mg-1 g q8hr, max 12 g/day
• **Child >1 mo: IM/IV** 50 mg/kg in 3-4 equal doses, max 6 g/day, or 2 g as a single dose
Renal dose
• **Adult: IM/IV** after loading dose, CCr 35-54 ml/min, dose q8hr; CCr 10-34 ml/min, 50% of dose q12hr; CCr <10 ml/min, 50% of dose q18-24hr

• **Child: IM/IV** CCr >70 ml/min, no dosage adjustment; CCr 40-70 ml/min after loading dose, reduce dose to 7.5-30 mg/kg q12hr; CCr 20-39 ml/min, give 3.125-12.5 mg/kg after loading dose q12hr; CCr 5-19 ml/min, 2.5-10 mg/kg after loading dose q24hr

Available forms: Inj 500 mg, 1, 5, 10, 20 g; inf 500 mg, 1 g/50-ml, 50 mg/500 ml vial

cephalexin

Moderate infections

• **Adult: PO** 250-500 mg q6hr, max 4 g/day

• **Child: PO** 25-100 mg/kg/day in 4 equal doses, max 4 g/day

Moderate skin infections

• **Adult: PO** 500 mg q12hr

Endocarditis prophylaxis

• 2 g 1 hr before procedure

Severe infections

• **Adult: PO** 500 mg-1 g q6hr, max 4 g

• **Child: PO** 50-100 mg/kg/day in 4 equal doses, max 4 g/day

Renal dose

• **Adult: PO** CCr 10-40 ml/min, 250-500 mg then 250-500 mg q8-12hr; CCr <10 ml/min, 250-500 mg then 250-500 mg q12-24hr

Available forms: Caps 250, 500 mg; tabs 250, 500 mg; oral susp 125 mg, 250 mg/5 ml

Administer:

cefadroxil

• For 10-14 days to ensure organism death, prevent superinfection

• With food if needed for GI symptoms

• Shake susp, refrigerate, discard after 2 wk

• After C&S completed

ceFAZolin

IV route

• Check for irritation, extravasation often; dilute in 2 ml/50 mg or 2.5 ml/1 g sterile water for inj, inject over 3-5 min; may be further diluted with 50-100 ml of NS, D_5W sol, run over 10 min-1 hr by Y-tube or 3-way stopcock

• After C&S completed

Syringe compatibilities: DimenhyDRINATE, heparin, salbutamol, vit B complex

Y-site compatibilities: Acyclovir, alfentanil, allopurinol, alprostadil, amifostine, amikacin, aminocaproic acid, aminophylline, amphotericin B liposome, anidulafungin, ascorbic acid injection, atenolol, atracurium, atropine, aztreonam, benztropine, bivalirudin, bleomycin, bumetanide, buprenorphine, butorphanol, calcium gluconate, CARBOplatin, cefamandole, cefmetazole, cefonicid, cefoperazone, cefoTEtan, cefOXitin, cefpirome, cefTAZidime, ceftizoxime, cefTRIAXone, cefuroxime, cephalothin, cephapirin, chloramphenicol, cimetidine, CISplatin, clindamycin, codeine, cyanocobalamin, cyclophosphamide, cycloSPORINE, cytarabine, DACTINomycin, DAPTOmycin, dexamethasone, dexmedetomidine, digoxin, diltiazem, DOCEtaxel, doxacurium, doxapram, DOXOrubicin liposomal, enalaprilat, ePHEDrine, EPINEPHrine, epirubicin, epoetin alfa, eptifibatide, esmolol, etoposide, fenoldopam, fentaNYL, filgrastim, fluconazole, fludarabine, fluorouracil, folic acid (as sodium salt), foscarnet, furosemide, gallium, gatifloxacin, gemcitabine, gentamicin, glycopyrrolate, granisetron, heparin, hydrocortisone, hydrOXYzine, IDArubicin, ifosfamide, imipenem-cilastatin, indomethacin, insulin (regular), irinotecan, isoproterenol, ketorolac, lidocaine, linezolid, LORazepam, LR's injection, mannitol, mechlorethamine, melphalan, meperidine, metaraminol, methicillin, methotrexate, methoxamine, methyldopate, methylPREDNISolone, metoclopramide, metoprolol, metroNIDAZOLE, mezlocillin, miconazole, midazolam, milrinone, morphine, moxalactam, multiple vitamins injection, nafcillin, nalbuphine, naloxone, nesiritide, niCARdipine, nitroglycerin, nitroprusside, norepinephrine, octreotide, ondansetron, oxacillin, oxaliplatin, oxytocin, PACLitaxel, palonosetron, pamidronate, pancuronium, pantoprazole, penicillin G potassium/sodium, peritoneal dialysis solution, perphenazine, PHENobarbital, phenylephrine, phytonadione, piperacillin, Plasma-Lyte M in dextrose 5%, polymyxin B, potassium chloride, procainamide, propofol, propranolol, ranitidine, remifentanil, Ringer's injection, ritodrine, riTUXimab, sargramostim, sodium acetate, sodium bicarbonate, succinylcholine, SUFentanil, tacrolimus, teniposide, tenoxicam, theophylline, thiamine, thiotepa, ticarcillin, ticarcillin-clavulanate, tigecycline, tirofiban, TNA, tolazoline, trastuzumab, trimetaphan, urokinase, vasopressin, vecuronium, verapamil, vinCRIStine, vitamin B complex with C, voriconazole, warfarin, zoledronic acid

cephalexin

- Shake susp, refrigerate, discard after 2 wk; use calibrated oral syringe, spoon, or measuring cup
- With food if needed for GI symptoms
- After C&S

SIDE EFFECTS

CNS: Headache, dizziness, weakness, paresthesia, fever, chills, **seizures** (with high doses)

GI: Nausea, vomiting, *diarrhea, anorexia,* pain, glossitis, bleeding; increased AST, ALT, bilirubin, LDH, alk phos; abdominal pain, **pseudomembranous colitis**

GU: Proteinuria, vaginitis, pruritus, candidiasis, increased BUN, **nephrotoxicity, renal failure**

HEMA: **Leukopenia, thrombocytopenia, agranulocytosis,** anemia, **neutropenia, lymphocytosis, eosinophilia, pancytopenia, hemolytic anemia**

INTEG: Rash, urticaria, dermatitis

MS: Arthralgia, arthritis

RESP: Dyspnea

SYST: **Anaphylaxis, serum sickness,** superinfection, **Stevens-Johnson syndrome**

PHARMACOKINETICS

cefadroxil: Peak 1-1½ hr, duration 12-24 hr, half-life 1-2 hr, 20% bound by plasma proteins, crosses placenta, excreted in breast milk

ceFAZolin

IV: Peak 10 min, duration 6-12 hr, eliminated unchanged in urine, 75%-85% protein bound

IM: Peak 1-2 hr, duration 6-12 hr, half-life 1½-2 hr

cephalexin: Peak 1 hr, duration 6-12 hr, half-life 30-72 min, 5%-15% bound by plasma proteins, 80%-100% eliminated unchanged in urine, crosses placenta, excreted in breast milk

INTERACTIONS

Increase: prothrombin time—anticoagulants; use cautiously

Increase: toxicity—aminoglycosides, loop diuretics, probenecid

Decrease: oral contraceptives possible, use another form of contraception

Drug/Lab Test

Increase: AST, ALT, alk phos, LDH, BUN, creatinine, bilirubin

False positive: urinary protein, direct Coombs' test, urine glucose

Interference: cross-matching

NURSING CONSIDERATIONS

Assess:

- Sensitivity to penicillin and other cephalosporins
- ⚠ **Nephrotoxicity: increased BUN, creatinine; urine output: if decreasing, notify prescriber**
- I&O daily
- Blood studies: AST, ALT, CBC, Hct, bilirubin, LDH, alk phos, Coombs' test monthly if patient is on long-term therapy
- Electrolytes: potassium, sodium, chlorine monthly if patient is on long-term therapy
- **Pseudomembranous colitis: bowel pattern daily; if severe diarrhea occurs, product should be discontinued**
- ⚠ **Anaphylaxis: rash, urticaria, pruritus, chills, fever, joint pain; angioedema; may occur a few days after therapy begins; discontinue product, notify prescriber im-**

mediately, keep emergency equipment nearby

• Bleeding: ecchymosis, bleeding gums, hematuria, stool guaiac daily

⚠ **Overgrowth of infection:** perineal itching, fever, malaise, redness, pain, swelling, drainage, rash, diarrhea, change in cough, sputum

Evaluate:

• Therapeutic response: decreased symptoms of infection, negative C&S

Teach patient/family:

• To use yogurt or buttermilk to maintain intestinal flora, decrease diarrhea

• To take all medication prescribed for length of time ordered

⚠ To report sore throat, bruising, bleeding, joint pain (may indicate **blood dyscrasias** [rare]); diarrhea with mucus, blood (may indicate **pseudomembranous colitis**)

TREATMENT OF ANAPHYLAXIS: EPINEPHrine, antihistamines; resuscitate if needed

CEPHALOSPORINS—2ND GENERATION

cefaclor (Rx)
(sef′a-klor)
Ceclor, Raniclor

cefoTEtan (Rx)
(sef′oh-tee-tan)
Cefotan

cefOXitin (Rx)
(se-fox′i-tin)
Mefoxin

cefprozil (Rx)
(sef-proe′zill)
Cefzil

cefuroxime (Rx)
(sef-yoor-ox′eem)
Ceftin, Zinacef

Func. class.: Antiinfective
Chem. class.: Cephalosporin (2nd generation)

Do not confuse:
cefaclor/cephalexin
Cefotan/Ceftin
cefprozil/ceFAZolin/cefuroxime
Cefzil/Ceftin

ACTION: Inhibits bacterial cell wall synthesis, renders cell wall osmotically unstable, leads to cell death by binding to cell wall membrane

USES:

cefaclor: Gram-negative bacilli: *Haemophilus influenzae, Escherichia coli, Proteus mirabilis, Klebsiella;* gram-positive organisms: *Streptococcus pneumoniae, Streptococcus pyogenes, Staphylococcus aureus;* respiratory tract, urinary tract, skin, bone, joint infections; otitis media

cefoTEtan: Gram-negative organisms: *Haemophilus influenzae, Escherichia coli, Enterobacter aerogenes, Proteus mirabilis, Klebsiella, Citrobacter, Sal-*

monella, Shigella, Acinetobacter, Bacteroides fragilis, Neisseria, Serratia; gram-positive organisms: *Streptococcus pneumoniae, Streptococcus pyogenes, Staphylococcus aureus;* upper and lower, serious respiratory tract, urinary tract, skin, bone, joint, gynecologic, gonococcal, intraabdominal infections

cefOXitin: Gram-negative bacilli: *Haemophilus influenzae, Escherichia coli, Proteus, Klebsiella, Bacteroides fragilis, Neisseria gonorrhoeae;* gram-positive organisms: *Streptococcus pneumoniae, Streptococcus pyogenes, Staphylococcus aureus;* anaerobes including *Clostridium;* lower respiratory tract, urinary tract, skin, bone, gynecologic, gonococcal infections; septicemia, peritonitis

cefprozil: Pharyngitis/tonsillitis; otitis media; secondary bacterial infection of acute bronchitis; acute bacterial exacerbation of chronic bronchitis; uncomplicated skin and skin-structure infections; acute sinusitis

cefuroxime: Gram-negative bacilli: *Haemophilus influenzae, Escherichia coli, Neisseria, Proteus mirabilis, Klebsiella;* gram-positive organisms: *Streptococcus pneumoniae, Streptococcus pyogenes, Staphylococcus aureus;* serious lower respiratory tract, urinary tract, skin, bone, joint, gonococcal infections; septicemia, meningitis

CONTRAINDICATIONS:

Hypersensitivity to cephalosporins or related antibiotics; seizures

Precautions: Pregnancy (B), breastfeeding, children, GI/renal disease, diabetes mellitus, coagulopathy

DOSAGE AND ROUTES

cefaclor

- **Adult: PO** 250-500 mg q8hr, max 4 g/day; ext rel 375 mg q12hr; max 1.5 mg/day (cup, oral susp); 1 g/day (ext rel)
- **Child >1 mo: PO** 20-40 mg/kg/day in divided doses q8hr or total daily dose may be divided and given q12hr, max 1 g/day

Available forms: Caps 250, 500 mg; oral susp 125, 187, 250, 375 mg/5; ext rel tab 500 mg

cefoTEtan

- **Adult: IM/IV** 1-2 g q12hr × 5-10 days
- **Adult: CCr** 30-50 ml/min, 1-2 g, then 1-2 g q8-12hr; **IM/IV** CCr 10-29 ml/min 1-2 g, then 1-2 g q12-24hr; **CCr** 5-9 ml/min 1-2 g, then 0.5-1 g q12-24hr; **CCr** <5 ml/min 1-2 g, then 0.5-1 g q24-48hr

Perioperative prophylaxis

- **Adult: IV** 1-2 g ½-1 hr before surgery

Available forms: Inj 1, 2, 10 g

cefOXitin

- **Adult: IM/IV** 1-2 g q6-8hr

Renal dose

- **Adult: IM/IV** after loading dose, CCr 30-50 ml/min, 1-2 g q8-12hr; CCr 10-29 ml/min, 1-2 g q12-24hr; CCr <10 ml/min, 0.5-1 g q12-24hr

Uncomplicated gonorrhea (outpatient)

Renal dose

- **Adult: IM/IV** CCr 10-30 ml/min, give dose q24hr or ½ dose q12hr; CCr <10 ml/min, give dose q48hr or ½ dose q24hr
- **Adult/adolescent/child ≥45 kg: IM** 2 g as single dose with 1 g **PO** probenecid at same time

Severe infections

- **Adult: IM/IV** 2 g q4hr
- **Child ≥3 mo: IM/IV** 80-160 mg/kg/day divided q4-6hr; max 12 g/day

Available forms: Powder for inj 1, 2, 10 g

cefprozil

Upper respiratory infections

- **Adult: PO** 500 mg q24hr × 10 days

Otitis media

- **Child 6 mo-12 yr: PO** 15 mg/kg q12hr × 10 days

Lower respiratory infections

- **Adult: PO** 500 mg q12hr × 10 days

Skin/skin-structure infections

- **Adult: PO** 250-500 mg q12hr × 10 days

Renal dose

- CCr <30 ml/min, 50% of dose

Available forms: Tabs 250, 500 mg; susp 125, 250 mg/5 ml

cefuroxime

- **Adult and child: PO** 250 mg q12hr; may increase to 500 mg q12hr for serious infections
- **Adult: IM/IV** 750 mg-1.5 g q8hr for 5-10 days

Urinary tract infections

- **Adult: PO** 125 mg q12hr; may increase to 250 mg q12hr if needed

Otitis media

- **Child <2 yr: PO** 125 mg bid
- **Child >2 yr: PO** 250 mg bid

Surgical prophylaxis

- **Adult: IV** 1.5 g ½-1 hr before surgery

Severe infections

- **Adult: IM/IV** 1.5 g q6hr; may give up to 3 g q8hr for bacterial meningitis
- **Child >3 mo: IM/IV** 50-100 mg/kg/day or IM in divided doses q6-8hr

Uncomplicated gonorrhea

- **Adult: IM** 1.5 g as single dose in 2 separate sites with oral probenecid

Renal dose

- Dosage reduction indicated with severe renal impairment (CCr <20 ml/min)

Available forms: Tabs 250, 500 mg; inj 1, 750 mg, 1.5, 7.5 g; inj 750 mg; 1.5 g powder; susp 125, 250 mg/5 ml

Administer:

- Do not break, crush, or chew ext rel tabs or caps
- On an empty stomach 1 hr before or 2 hr after a meal

cefaclor

- Shake susp, refrigerate, discard after 2 wk
- For 10-14 days to ensure organism death, prevent superinfection
- With food if needed for GI symptoms
- After C&S completed
- Swallow ext rel whole

cefoTEtan

- IV direct after diluting 1 g/10 ml sterile water for inj, give over 3-5 min; may be diluted further with 50-100 ml NS or D_5W; shake; run over ½-1 hr by Y-tube or 3-way stopcock; discontinue primary inf during administration
- May be stored 96 hr refrigerated or 24 hr at room temperature

Y-site compatibilities: Allopurinol, amifostine, aztreonam, diltiazem, famotidine, filgrastim, fluconazole, fludarabine, heparin, insulin (regular), melphalan, meperidine, morphine, PACLitaxel, remifentanil, sargramostim, tacrolimus, teniposide, theophylline, thiotepa

cefOXitin

IV route

- After diluting 1 g or less/10 ml or more D_5W, NS and give over 3-5 min; may be diluted further with 50-100 ml NS or D_5W; run over ½-1 hr by Y-tube or 3-way stopcock; discontinue primary inf during administration; give by cont inf at prescribed rate; may store 96 hr refrigerated or 24 hr at room temperature
- For 10-14 days to ensure organism death, prevent superinfection
- After C&S completed

Syringe compatibilities: Heparin, insulin

Y-site compatibilities: Acyclovir, amifostine, amphotericin B cholesteryl sulfate complex, aztreonam, cyclophosphamide, diltiazem, DOXOrubicin liposome, famotidine, fluconazole, foscarnet, HYDROmorphone, magnesium sulfate, meperidine, morphine, ondansetron, perphenazine, remifentanil, teniposide, thiotepa

cefprozil

- For 10-14 days to ensure organism death, prevent superinfection
- After C&S
- Refrigerate/shake susp before use

cefuroxime

- For 10-14 days to ensure organism death, prevent superinfection
- With food if needed for GI symptoms
- After C&S obtained

Y-site compatibilities: Acyclovir, allopurinol, amifostine, atracurium, aztreonam, cyclophosphamide, diltiazem, famotidine, fludarabine, foscarnet, HYDROmorphone, melphalan, meperidine, morphine, ondansetron, pancuronium, perphenazine, remifentanil, sargramostim, tacrolimus, teniposide, thiotepa, vecuronium

SIDE EFFECTS

CNS: Dizziness, headache, fatigue, paresthesia, fever, chills, confusion
GI: *Diarrhea,* nausea, vomiting, anorexia, dysgeusia, glossitis, bleeding; increased AST, ALT, bilirubin, LDH, alk phos; abdominal pain, loose stools, flatulence, heartburn, stomach cramps, colitis, jaundice, **pseudomembranous colitis**
GU: Vaginitis, pruritus, candidiasis, increased BUN, **nephrotoxicity, renal failure,** pyuria, dysuria, reversible interstitial nephritis
HEMA: **Leukopenia, thrombocytopenia, agranulocytosis,** anemia, **neutropenia, lymphocytosis, eosinophilia, pancytopenia, hemolytic anemia, leukocytosis, granulocytopenia**
INTEG: Rash, urticaria, dermatitis, **Stevens-Johnson syndrome**
RESP: Dyspnea
SYST: Anaphylaxis, **serum sickness,** superinfection

PHARMACOKINETICS

cefaclor
PO: Peak ½-1 hr, half-life 36-54 min, 25% bound by plasma proteins, 60%-85% eliminated unchanged in urine in 8 hr, crosses placenta, excreted in breast milk (low concs)
cefoTEtan
IM/IV: Peak 1½-3 hr, half-life 3-5 hr, 75%-90% bound by plasma proteins, 50%-80% eliminated unchanged in urine, crosses placenta, excreted in breast milk
cefOXitin
Half-life 0.75-1 hr; 65%-80% bound by plasma proteins; 90%-100% eliminated unchanged in urine; crosses placenta, blood-brain barrier; eliminated in breast milk; not metabolized
IM: Peak 15-30 min
IV: Peak 5 min
cefprozil
PO: Peak 1.5 hr, protein binding 36%, elimination half-life 1.3 hr (normal renal function), 2 hr (hepatic disease), 5½-6 hr (end-stage renal disease), extensively metabolized to an active metabolite, eliminated in urine 60%
cefuroxime
Peak PO 2 hr, IM 45 min, IV 2-3 min, 66% excreted unchanged in urine, half-life 1-2 hr in normal renal function

INTERACTIONS

Increase: effect/toxicity—aminoglycosides, furosemide, probenecid
Increase: bleeding risk (cefoTEtan)—anticoagulants, thrombolytics, NSAIDs, antiplatelets, plicamycin, valproic acid
Decrease: oral contraceptive, possible—use additional form of contraception
Decrease: absorption of cephalosporin—antacids
Decrease: effect of cephalosporin—H_2-blockers
Drug/Lab Test
False increase: creatinine (serum urine), urinary 17-KS
False positive: urinary protein, direct Coombs' test, urine glucose testing (Clinitest)
Interference: cross-matching

NURSING CONSIDERATIONS

Assess:
⚠ **Nephrotoxicity:** increased BUN, creatinine
- I&O ratio
- Blood studies: AST, ALT, CBC, Hct, bilirubin, LDH, alk phos, Coombs' test monthly if patient is on long-term therapy
- Electrolytes: potassium, sodium, chlorine monthly if patient is on long-term therapy
- Bowel pattern daily; if severe diarrhea occurs, product should be discontinued; may indicate pseudomembranous colitis
- Urine output; if decreasing, notify prescriber (may indicate nephrotoxicity)

⚠ **Anaphylaxis:** rash, flushing, urticaria, pruritus, dyspnea; discontinue product, notify prescriber, have emergency equipment available
- **Bleeding:** ecchymosis, bleeding gums, hematuria, stool guaiac daily

⚠ **Overgrowth of infection:** perineal itching, fever, malaise, redness, pain, swelling, drainage, rash, diarrhea, change in cough, sputum

Evaluate:

- Therapeutic response: negative C&S

Teach patient/family:

- If diabetic, to use blood glucose testing
- To complete full course of product therapy; to report persistent diarrhea
- To use yogurt, buttermilk to maintain intestinal flora, decrease diarrhea
- To notify prescriber if breastfeeding or of any side effects

⚠ To report sore throat, bruising, bleeding, joint pain (may indicate blood dyscrasias [rare]); diarrhea with mucus, blood (pseudomembranous colitis); symptoms of hypersensitivity

TREATMENT OF ANAPHYLAXIS: EPINEPHrine, antihistamines; resuscitate if needed

CEPHALOSPORINS—3RD/4TH GENERATION

cefdinir (Rx)
(sef′dih-ner)
Omnicef

cefditoren pivoxil (Rx)
(sef-dit′oh-ren pih-vox′il)
Spectracef

cefepime (Rx)
(sef′e-peem)
Maxipime (4th generation)

cefixime (Rx)
(sef-icks′ime)
Cefixime, Suprax

cefotaxime (Rx)
(sef-oh-taks′eem)
Claforan

cefpodoxime (Rx)
(sef-poe-docks′eem)
Vantin

cefTAZidime (Rx)
(sef′tay-zi-deem)
Ceptaz, Fortaz, Tazicef, Tazidime

ceftibuten (Rx)
(sef-ti-byoo′tin)
Cedax

ceftizoxime (Rx)
(sef-ti-zox′eem)
Cefizox

cefTRIAXone (Rx)
(sef-try-ax′one)
Rocephin

Func. class.: Broad-spectrum antibiotic

Chem. class.: Cephalosporin (3rd generation)

Do not confuse:
cefTAZidime/ceftizoxime
Vantin/Ventolin

ACTION: Inhibits bacterial cell wall synthesis, renders cell wall osmotically unstable, leads to cell death

USES:

cefdinir: Community-acquired pneumonia, otitis media, sinusitis, pharyngitis, skin and skin-structure infections, acute exacerbations of chronic bronchitis, pneumonia, tonsillitis, citrobacter diversus, escherichia coli, Klebsiella pneumonial, proteus mirabilis, Staphylococcus epidermidis, streptococcus agalactial (group B), viridans Streptococci alpha: *Haemophilus influenzae, Haemophilus parainfluenzae, Moraxella catarrhalis;* gram-positive organisms: *Streptococcus pneumoniae, Streptococcus pyogenes, Staphylococcus aureus* (MSSA)

cefditoren pivoxil: Acute bacterial exacerbations of chronic bronchitis caused by *Haemophilus influenzae, Haemophilus parainfluenzae, Streptococcus pneumoniae, Moraxella catarrhalis*; pharyngitis/tonsillitis caused by *Streptococcus pyogenes*; uncomplicated skin and skin-structure infections caused by *Staphylococcus aureus, Streptococcus pyogenes;* community-acquired pneumonia, viridans streptococci

cefepime: *Escherichia coli, Proteus, Klebsiella;* Acinetobacter calcoaceticus, acinetobacter lwoffii, aeromonas hydrophilia, citrobacter diversus, citrobacter freundii, enterobacter sp., Escherichia coli, gardnerella vaginalis, hafnia alvei, moraxella catarrhalis, morganella morganii, neisseria gonorrhoeal, meningitidis, providencia rettgeri, stuartii, pseudomonas aeruginosa, Salmonella sp., Serratia liquefaciens, Serratia marcescens, shigella sp., Staphylococcus epidermidis, staphylococcus saprophyticus, streptococcus agalactial, bovis, viridans streptococci, yersinia entercolitica: *Streptococcus pneumoniae, Streptococcus pyogenes, Staphylococcus aureus;* lower respiratory tract, urinary tract, skin, bone infections; febrile neutropenia intraabdominal infection

cefixime: Uncomplicated UTI *(Escherichia coli, Proteus mirabilis)*, pharyngitis and tonsillitis *(Streptococcus pyogenes)*, otitis media *(Haemophilus influenzae), Moraxella catarrhalis,* acute bronchitis and acute exacerbations of chronic bronchitis *(Streptococcus pneumoniae, H. influenzae)*, uncomplicated gonorrhea

cefotaxime: *Haemophilus influenzae, Haemophilus parainfluenzae, Escherichia coli, Enterococcus faecalis, Neisseria gonorrhoeae, Neisseria meningitidis, Proteus mirabilis, Klebsiella, Citrobacter, Serratia, Salmonella, Shigella Pseudomonas; Streptococcus pneumoniae, Streptococcus pyogenes, Staphylococcus aureus;* serious lower respiratory tract, urinary tract, skin, bone, gonococcal infections; bacteremia, septicemia, meningitis, skin, skin-structure infections; CNS infections; perioperative prophylaxis, intraabdominal infections, PID, UTI, ventriculitis

cefpodoxime: Bacteroides, *Neisseria gonorrhoeae, Haemophilus influenzae, Escherichia coli, Proteus mirabilis, Klebsiella;* gram-positive organisms: *Streptococcus pneumoniae, Streptococcus pyogenes, Staphylococcus aureus;* upper and lower respiratory tract, urinary tract, skin infections; otitis media; sexually transmitted diseases

cefTAZidime: *Haemophilus influenzae, Escherichia coli, Enterobacter aerogenes, Pseudomonas aeruginosa, Proteus mirabilis, Klebsiella, Citrobacter, Enterobacter, Salmonella, Shigella, Acinetobacter, Bacteroides fragilis, Neisseria, Serratia; Streptococcus pneumoniae, Streptococcus pyogenes, Staphylococcus aureus;* serious lower respiratory tract, urinary tract, skin, gynecologic, bone, joint, intraabdominal infections; septicemia, meningitis

ceftibuten: Pharyngitis/tonsillitis, otitis media, secondary bacterial infection of acute bronchitis

ceftizoxime: *Bacteroides, Haemophilus influenzae, Escherichia coli, Enterobacter aerogenes, Proteus mirabilis,*

Klebsiella, Enterobacter; Streptococcus pneumoniae, Streptococcus pyogenes, Staphylococcus aureus; serious lower respiratory tract, urinary tract, skin, intraabdominal infections; septicemia, meningitis, bone, joint infections; PID caused by *Neisseria gonorrhoeae*

cefTRIAXone: Gram-negative bacilli: *Haemophilus influenzae, Escherichia coli, Enterobacter aerogenes, Proteus mirabilis, Klebsiella, Citrobacter, Enterobacter, Salmonella, Shigella, Acinetobacter, Bacteroides fragilis, Neisseria, Serratia;* gram-positive organisms: *Streptococcus pneumoniae, Streptococcus pyogenes, Staphylococcus aureus;* serious lower respiratory tract, urinary tract, skin, gonococcal, intraabdominal infections; septicemia, meningitis, bone, joint infections; otitis media; PID

CONTRAINDICATIONS:

Hypersensitivity to cephalosporins, infants <1 mo

Precautions: Pregnancy (B), breastfeeding, children, hypersensitivity to penicillins, GI/renal disease, geriatrics, pseudomembranous colitis, viral infection, vit K deficiencies

DOSAGE AND ROUTES

cefdinir

Uncomplicated skin and skin-structure infections/community-acquired pneumonia

- **Adult and child ≥13 yr: PO** 300 mg q12hr × 10 days
- **Child 6 mo-12 yr: PO** 7 mg/kg q12hr or 14 mg/kg q24hr × 10 days

Acute exacerbations of chronic bronchitis/acute maxillary sinusitis

- **Adult and child ≥13 yr: PO** 300 mg q12hr or 600 mg q24hr × 10 days

Pharyngitis/tonsillitis

- **Adult and child ≥13 yr: PO** 300 mg q12hr or 600 mg q24hr × 5-10 days
- **Child 6 mo-12 yr: PO** 7 mg/kg q12hr × 5-10 days or 14 mg/kg q24hr × 5-10 days

Renal dose

- **Adult: PO** CCr <30 ml/min, 300 mg/day (adult); 7 mg/kg/day (child)

Available forms: Caps 300 mg; susp 125 mg, 250 mg/5 ml

cefditoren pivoxil

- **Adult: PO** 200-400 mg bid × 10 days

Renal dose

- **Adult: PO** CCr 30-49 ml/min, max 200 mg bid; CCr <30 ml/min, max 200 mg daily

Available forms: Tabs 200 mg

cefepime

Febrile neutropenia

- **Adult/adolescent >16 yrs/child ≥ 40 kg: IV** 2 g q8hr × 7 days or until neutropenia resolves
- **Infant ≥2 mo/child/adolescent ≤16 yr and ≤40 kg: IV** 50 mg/kg/dose q8hr × 7-10 days or until neutropenia resolves

Urinary tract infections (mild to moderate)

- **Adult: IV/IM** 0.5-1 g q12hr × 7-10 days

Urinary tract infections (severe)

- **Adult/adolescent >16 yr/child ≥40 kg: IV** 2 g q12hr × 10 days

Pneumonia (moderate to severe)

- **Adult: IV** 1-2 g q12hr × 10 days

Available forms: Powder for inj 500 mg, 1, 2 g; 1 g/50 ml, 2 g/100 ml

cefixime

Mild to moderate pharyngitis, tonsillitis, bronchitis

- **Adult/adolescent/child >45 kg: PO** 400 mg/day divided q12-24hr
- **Children ≤45 kg infants ≥6 months: PO** 8 mg/kg/day divided q12-24hr

Uncomplicated urinary tract infection (UTI)

- **Adult/adolescent/child >45 kg: PO** 400 mg/day divided q12-24hr
- **Children ≤45 kg and infants ≥6 mo: PO** 8 mg/kg/day divided q12-24hr max: 400 mg/day × 7-14 days is recommended by the American Academy of Pediatrics (AAP) for the treatment of initial UTI in febrile infants and young children 2-24 mo

- **Infants 2-5 mo (unlabeled): PO** 8 mg/kg/day × 7-14 days is recommended by the American Academy of Pediatrics (AAP) for the treatment of initial UTI in febrile infants and young children

Mild to moderate otitis media

- **Adult/adolescent/child >45 kg: PO** 400 mg/day divided q12-24hr
- **Children ≤45 kg: PO** 8 mg/kg/day divided q12-24hr, max 400 mg/day
- **Infants ≥6 months: PO** 8 mg/kg/day divided q12-24hr

Gonorrhea of uncomplicated cervicitis, or urethritis due to *N. gonorrhoeae*

- **Adult/adolescent: PO** as alternative therapy, 400 mg as a single dose with a regimen effective against uncomplicated genital *C. trachomatis* infection (e.g., azithromycin as a single dose or doxycycline for 7 days) if chlamydial infection is not ruled out; the CDC states that cefixime is only acceptable if IM cefTRIAXone is not an option because of rising cefixime MICs for gonorrhea; if cefixime is used, test-of-cure should be done at the infected site 1 wk after treatment; cefixime is not recommended for infections of the pharynx
- **Children ≥45 kg: PO** 400 mg as a single dose with a regimen effective against uncomplicated genital *C. trachomatis* infection (e.g., azithromycin as a single dose or doxycycline for 7 days) if chlamydial infection is not ruled out; cefixime is not recommended for infections of the pharynx; cefixime is an alternative to cefTRIAXone per AAP

Gonorrhea prophylaxis (victims of sexual assault) (unlabeled)

- **Adult/adolescent:** The CDC recommends 400 mg as a single dose with metroNIDAZOLE (for trichomoniasis and bacterial vaginosis prophylaxis) plus either azithromycin or doxycycline (for chlamydia prophylaxis); the CDC states that cefixime is only acceptable if IM cefTRIAXone is not an option because of rising cefixime MICs for gonorrhea

Typhoid fever caused by multidrug-resistant *Salmonella typhi* (unlabeled)

- **Adult/adolescent/child:** 100-200 mg **PO BID** or 15-20 mg/kg/day **PO** in 2 divided doses for 7-14 days; for quinolone-resistant organisms, 20 mg/kg/day **PO** in 2 divided doses for 7-14 days should be used
- **Children <50 kg:** should receive weight-based dosing, max adult dosages

Acute bacterial sinusitis (unlabeled)

- **Infants ≥6 mo/child/adolescent: PO** 8 mg/kg/day **PO** divided q12hr, max 400 mg/day) with clindamycin for 10-14 days

Renal dose

- **Adult: PO** CCr 21-59 ml/min, give 65% of dose; CCr <20 ml/min, give 50% of dose

Available forms: Tabs 400 mg; powder for oral susp 100 mg/5 ml, chew tabs 100, 200 mg, 200 mg/5 ml, cap 400 mg

cefotaxime

- **Adult/adolescent/child ≥50 kg: IV/IM** (uncomplicated infections) 1 g q12hr, (moderate-severe infection) 1-2 g q8hr, (severe infections) 2 g q6-8hr, (life-threatening infections) 2 g q4hr, max 12 g/day
- **Adolescent/child <50 kg and infants: IV/IM** 50-180 mg/kg/day divided q6-8hr, max 2 g/dose, (severe infections) 200-225 mg/kg/day divided q4-6hr max 12 hr
- **Neonates >7 days: IV/IM** 50 mg/kg/dose q8-12hr

Uncomplicated gonorrhea

- **Adult: IM** 500 mg as a single dose
- Dosage reduction indicated for severe renal impairment (CCr <30 ml/min)

Available forms: Powder for inj 500 mg, 1, 2, 10 g; inj 1, 2 g premixed frozen

cefpodoxime

Pneumonia

- **Adult >12 yr: PO** 200 mg q12hr × 14 days

Uncomplicated gonorrhea

- **Adult >13 yr: PO** 200 mg as a single dose

Skin and skin structure
- **Adult >13 yr: PO** 400 mg q12hr × 7-14 days

Pharyngitis and tonsillitis
- **Adult >13 yr: PO** 100 mg q12hr × 5-10 days
- **Child 5 mo-12 yr: PO** 5 mg/kg q12hr (max 100 mg/dose or 200 mg/day) × 5-10 days

Uncomplicated UTI
- **Adult >13 yr: PO** 100 mg q12hr × 7 days; dosing interval increased with severe renal impairment

Acute otitis media
- **Child 5 mo-12 yr: PO** 5 mg/kg q12hr × 10 days

Available forms: Tabs 100, 200 mg; granules for susp 50 mg, 100 mg/5 ml

cefTAZidime
- **Adult: IV/IM** 1-2 g q8
- **Child: IV** 30-50 mg/kg q8hr, max 6 g/day
- **Neonate: IV** 30-50 mg/kg q12hr

Renal dose
- **Adult: IM/IV** CCr 31-50 ml/min 1 g q12hr; CCr 16-30 ml/min 1 g q24hr; CCR 6-15 ml/min 1 g loading dose, then 0.5 mg q24hr; CCr <5 ml/min 1 g loading dose, then 0.5 g q48hr

Available forms: Inj 250, 500 mg, 1, 2, 6 g

ceftibuten
- **Adult: PO** 400 mg/day × 10 days
- **Child 6 mo-12 yr: PO** 9 mg/kg/day × 10 days

Renal dose
- **Adult: PO** CCr 30-49 ml/min, give 200 mg q24hr; CCr 5-29 ml/min, give 100 mg q24hr

Available forms: Caps 400 mg; susp 90 mg, 180 mg/5 ml

ceftizoxime
- **Adult: IM/IV** 1-2 g q8-12hr, may give up to 4 g q8hr in life-threatening infections
- **Child >6 mo: IM/IV** 50 mg/kg q6-8hr

Renal dose
- **Adult: IM/IV** CCr 50-79 ml/min, give 500-1500 mg q8hr; CCr 5-49 ml/min, give 250-1000 mg q12hr

PID
- **Adult: IV** 2 g q8hr, may increase to 4 g q8hr in severe infections

Available forms: Premixed 1 g, 2 g/50 ml

cefTRIAXone
- **Adult: IM/IV** 1-2 g/day, max 4 g q12-24hr
- **Child: IM/IV** 50-75 mg/kg/day in equal doses q12-24hr

Uncomplicated gonorrhea
- **Adult:** 250 mg **IM** as single dose
- Reduce dosage in severe renal impairment (CCr <10 ml/min)

Available forms: Inj 250, 500 mg, 1, 2, 10 g

Administer:
- Change IV site q72hr

cefdinir
- Oral susp after adding 39 ml water to the 60-ml bottle or 65 ml water to the 120-ml bottle; discard unused portion after 10 days; give without regard to food, do not give within 2 hr of antacids, iron supplements
- After C&S completed

cefditoren pivoxil
- For 10 days to ensure organism death, prevent superinfection
- With food; do not give with antacids
- After C&S completed

cefepime
Intermittent IV INF route
- IV after diluting in 50-100 ml or more D_5, NS; give over 30 min
- For 7-10 days to ensure organism death, prevent superinfection

Solution compatibilities: 0.9% NaCl, D_5, D_5W, 0.5%, 10% lidocaine, bacteriostatic water for inj with parabens/benzyl alcohol

Y-site compatibilities: DOXOrubicin liposome

cefixime
- For 10-14 days to ensure organism death, prevent superinfection
- Without regard to food

cefotaxime

IV route

- IV after **diluting** 1 g/10 ml D_5W, NS, sterile water for inj, **give** over 3-5 min by Y-tube or 3-way stopcock; may be **diluted further** with 50-100 ml NS or D_5W; **run** over ½-1 hr; discontinue primary inf during administration; may be **diluted** in larger vol of sol, given as a cont inf
- For 10-14 days to ensure organism death, prevent superinfection
- Thaw frozen container at room temperature or refrigeration; do not force thaw by immersion or microwave; visually inspect container for leaks

Syringe compatibilities: Caffeine, diphenhyDRINATE, heparin, ofloxacin

Y-site compatibilities: Acyclovir, alfentanil, alprostadil, amifostine, amikacin, aminocaproic acid, aminophylline, anidulafungin, ascorbic acid injection, atenolol, atracurium, atropine, aztreonam, benztropine, bivalirudin, bleomycin, bumetanide, buprenorphine, butorphanol, caffeine, calcium chloride/gluconate, CARBOplatin, cefamandole, cefmetazole, cefonicid, cefoperazone, cefoTEtan, cefOXitin, cefTAZidime (L-arginine), cefTRIAXone sodium, cefuroxime, cimetidine, CISplatin, clindamycin, codeine, cyanocobalamin, cyclophosphamide, cycloSPORINE, cytarabine, DACTINomycin, DAPTOmycin, dexamethasone, dexmedetomidine, digoxin, diltiazem, DOCEtaxel, DOPamine, doxacurium, doxycycline, enalaprilat, ePHEDrine, EPINEPHrine, epirubicin, epoetin alfa, eptifibatide, erythromycin, esmolol, etoposide, famotidine, fenoldopam, fentaNYL, fludarabine, fluorouracil, folic acid, furosemide, gatifloxacin, gentamicin, glycopyrrolate, granisetron, heparin, hydrocortisone, HYDROmorphone, ifosfamide, imipenem-cilastatin, insulin (regular), isoproterenol, ketorolac, lidocaine, linezolid, LORazepam, LR, magnesium sulfate, mannitol, mechlorethamine, melphalan, meperidine, metaraminol, methicillin, methotrexate, methoxamine, methyldopate, metoclopramide, metoprolol, metroNIDAZOLE, mezlocillin, miconazole, midazolam, milrinone, minocycline, mitoXANtrone, morphine, moxalactam, multiple vitamins, mycophenolate, nafcillin, nalbuphine, naloxone, nesiritide, netilmicin, nitroglycerin, nitroprusside, norepinephrine, normal saline, octreotide, ofloxacin, ondansetron, ornidazole, oxacillin, oxaliplatin, oxytocin, PACLitaxel, palonosetron, pamidronate, pancuronium, pantoprazole, papaverine, pefloxacin, PEMEtrexed, penicillin G potassium/sodium, pentamidine, pentazocine, PENTobarbital, peritoneal dialysis solution, perphenazine, PHENobarbital, phenylephrine, phenytoin, phytonadione, piperacillin, polymyxin B, potassium chloride, procainamide, prochlorperazine, promethazine, propofol, propranolol, protamine, pyridoxine, quiNIDine, quinupristin, ranitidine, remifentanil, Ringer's injection, ritodrine, riTUXimab, rocuronium, sargramostim, sodium acetate/bicarbonate, sodium fusidate, sodium lactate, succinylcholine, SUFentanil, sulfamethoxazole-trimethoprim, tacrolimus, teniposide, theophylline, thiamine, thiotepa, ticarcillin, ticarcillin-clavulanate, tigecycline, tirofiban, TNA, tobramycin, tolazoline, TPN, trastuzumab, trimetaphan, urokinase, vancomycin, vasopressin, vecuronium, verapamil, vinorelbine, voriconazole

cefpodoxime

- Do not break, crush, or chew tabs due to taste
- For 10-14 days to ensure organism death, prevent superinfection
- With food for better absorption; do not give within 2 hr of antacids, H_2-receptor antagonists
- Shake susp well, refrigerate, discard after 2 wk

cefTAZidime

IM route

- **Fortaz, Tazidime vials:** reconstitute 500 mg or 1 g with 1.5 or 3 ml, respectively, of sterile or bacteriostatic water for inj or 0.5%-1% lidocaine (approx 280 mg/ml)

- **Tazicef vials:** reconstitute 1 g/3 ml sterile water for inj (approx 280 mg/ml)
- **Ceptaz vials:** reconstitute 1 g/3 ml sterile or bacteriostatic water for inj or 0.5%-1% lidocaine (approx 250 mg/ml)
- **Withdraw** dose while making sure needle remains in vial; **ensure** no CO_2 bubbles present; **inject** deeply in large muscle mass, **aspirate** before injection

IV route

- Visually inspect for particulate matter, discoloration, if possible
- **Fortaz, Tazicef, Tazidime packs: reconstitute** 1 or 2 g/100 ml sterile water for inj or other compatible IV sol (10 or 20 mg/ml, respectively); reconstitution is done in two stages: first, **inject** 10 ml of the diluent into the pack and **shake** well to dissolve and become clear; CO_2 pressure inside container will occur, **insert** vent needle to release pressure; **add** remaining diluents, **remove** vent needle
- **Fortaz, Tazicef, Tazidime vials: reconstitute** 500 mg, 1 g, 2 g with 5, 10, 10 ml, respectively, of sterile water for inj or other compatible IV solution (100, 95-100, or 170-180 mg/ml, respectively); **shake** well to dissolve
- **Fortaz, Tazidime ADD-Vantage vials (for IV only): reconstitute** 1 or 2 g with NS, ½ NS, D_5W in either 50- or 100-ml flexible diluent container; to release CO_2 pressure, **insert** vent needle after dissolving, **remove** vent before using
- **Ceptaz packs: reconstitute** 1 or 2 g/100 ml sterile water for inj or compatible IV sol (10 or 20 mg/ml, respectively); reconstitution is done in two stages: first, **inject** 10 ml of the diluent into the pack and **shake** well to dissolve, **add** the remaining diluent, **insert** vent needle before giving
- **Ceptaz vials: reconstitute** 1 or 2 g/10 ml of sterile water for inj or compatible IV sol (90-95, or 170-180 mg/ml, respectively)
- **Ceptaz ADD-Vantage vials (for IV only): reconstitute** 1 or 2 g with NS, ½ NS, or D_5W in either 50- or 100-ml diluent container

Direct Intermittent IV INF route

- **Vials: withdraw** dose while making sure needle remains in sol; make sure there are no CO_2 bubbles in syringe before inj; **inject** directly over 3-5 min or slowly into tubing of a free-flowing compatible IV solution

Intermittent IV INF route

- **Vials: withdraw** dose while making sure needle opening remains in sol; make sure there are no CO_2 bubbles in syringe before inj; inf packs and ADD-Vantage systems ready for inf after reconstitution, **infuse** over 15-30 min

Syringe compatibilities: Cimetidine, dimenhyDRINATE, HYDROmorphone

Y-site compatibilities: Acyclovir, alfentanil, allopurinol, amifostine, amikacin, aminocaproic acid, aminophylline, amphotericin B lipid complex, anakinra, anidulafungin, atenolol, atropine sulfate, aztreonam, benzotropine, bivalirudin, bleomycin, bumetanide, buprenorphine, butorphanol, calcium gluconate, CARBOplatin, cefamandole, ceFAZolin, cefonicid, cefoperazone, cefoTEtan, cefOXitin, cefTAZidime, ceftizoxime, cefTRIAXone, cefuroxime, cephalothin, cephapirin, cimetidine, ciprofloxacin, CISplatin, clindamycin, codeine, cyanocobalamin, cyclophosphamide, cycloSPORINE, cytarabine, DACTINomycin, DAPTOmycin, dexamethasone, dexmedetomidine, digoxin, diltiazem, DOCEtaxel, DOPamine, doxacurium, doxapram, enalaprilat, ePHEDrine, EPINEPHrine, epoetin alfa, eptifibatide, esmolol, etoposide, famotidine, fenoldopam, fentaNYL, filgrastim, fludarabine, fluorouracil, folic acid, foscarnet, furosemide, gallium, gatifloxacin, gemcitabine, gentamicin, glycopyrrolate, granisetron, heparin, HYDROmorphone, ifosfamide, imipenem-cilastatin, indomethacin, insulin (regular), irinotecan, isepamicin, isoproterenol, isosorbide, ketamine, ketorolac, labetalol, levofloxacin, lidocaine, linezolid, LORazepam, LR, magnesium sulfate, mannitol, mechlorethamine, melphalan, meperidine, metaraminol, methicillin, methotrexate, methoxamine, methyldopate, methylPREDNISolone, metoclopramide, metoprolol, metroNIDAZOLE, miconazole, milrinone, morphine, moxalactam, multiple vitamin inj, nafcillin, nalbuphine, PACLitaxel, ranitidine, remifentanil, tacrolimus, teniposide, theophylline, thiotepa, vinorelbine, zidovudine

ceftibuten

- For 10 days to ensure organism death, prevent superinfection
- On empty stomach

ceftizoxime

IV route

- IV after **diluting** 1 g/10 ml sterile water, shake and **give** over 3-5 min; may be **diluted further** with 50-100 ml NS or D_5W; give through Y-tube or 3-way stopcock; **run** over ½-1 hr
- For 10-14 days to ensure organism death, prevent superinfection

Y-site compatibilities: Acyclovir, alfentanil, allopurinol, amifostine, amikacin, aminocaproic acid, aminophylline, amiodarone, amphotericin B cholesteryl, amphotericin B liposome, anidulafungin, ascorbic acid injection, atenolol, atracurium, atropine, aztreonam, benztropine, bivalirudin, bleomycin, bumetanide, buprenorphine, butorphanol, calcium chloride/gluconate, CARBOplatin, caspofungin, cefamandole, ceFAZolin, cefmetazole, cefoperazone, cefoTEtan, cefTAZidime, cefTRIAXone, cefuroxime, cephalothin, cephapirin, cimetidine, CISplatin, clindamycin, codeine, cyanocobalamin, cyclophosphamide, cycloSPORINE, cytarabine, DACTINomycin, DAPTOmycin, dexamethasone, dexmedetomidine, digoxin, diltiazem, diphenhydrAMINE, DOCEtaxel, DOPamine, doxacurium, DOXOrubicin, DOXOrubicin liposomal, enalaprilat, ePHEDrine, EPINEPHrine, epirubicin, epoetin alfa, eptifibatide, esmolol, etoposide, fenoldopam, fentaNYL, fluconazole, fludarabine, fluorouracil, folic acid, foscarnet, furosemide, gatifloxacin, gemcitabine, gentamicin, glycopyrrolate, granisetron, heparin, hydrocortisone, HYDROmorphone, IDArubicin, ifosfamide, imipenem-cilastatin, insulin (regular), irinotecan, isoproterenol, ketorolac, labetalol, levofloxacin, lidocaine, linezolid, LORazepam, LR, magnesium sulfate, mannitol, mechlorethamine, melphalan, meperidine, methotrexate, methoxamine, methyldopate, metoclopramide, metoprolol, metroNIDAZOLE, mezlocillin, midazolam, milrinone, mitoXANtrone, morphine, multiple vitamins injection, mycophenolate mofetil, nafcillin, naloxone, nesiritide, netilmicin, niCARdipine, nitroglycerin, nitroprusside, norepinephrine, octreotide, ondansetron, oxaliplatin, oxytocin, PACLitaxel, palonosetron, pamidronate, pancuronium, pantoprazole, PEMEtrexed, penicillin G potassium/sodium, PHENobarbital, phenylephrine, phytonadione, piperacillin, polymyxin B, potassium chloride, propofol, propranolol, ranitidine, remifentanil, Ringer's, ritodrine, riTUXimab, rocuronium, sargramostim, sodium acetate/bicarbonate, succinylcholine, SUFentanil, tacrolimus, teniposide, theophylline, thiotepa, ticarcillin-clavulanate, tigecycline, tirofiban, TNA, tobramycin, TPN, trastuzumab, trimetaphan, vasopressin, verapamil, vinCRIStine, vinorelbine, voriconazole

cefTRIAXone

- For 10-14 days to ensure organism death, prevent superinfection
- **IM** inj deeply in large muscle mass

IV route

- **IV** after **diluting** 250 mg/2.4 ml, 500 mg/4.8 ml, 1 g/9.6 ml, 2 g/19.2 ml D_5W, water for inj, 0.9% NaCl; may be **further diluted** with 50-100 ml NS, D_5W, $D_{10}W$; shake; **run** over ½ hr
- Do not mix with calcium salts

Y-site compatibilities: Acyclovir, alfentanil, allopurinol, amifostine, amikacin, aminocaproic acid, aminophylline, amiodarone, amphotericin B liposome, anidulafungin, atenolol, atracurium, atropine, aztreonam, benztropine, bivalirudin, bleomycin, bumetanide, buprenorphine, butorphanol, CARBOplatin, cefamandole, ceFAZolin, cefmetazole, cefonicid, cefoperazone, cefoTAXime, cefoTEtan, cefOXitin, cefTAZidime, ceftizoxime, cefuroxime, cephalothin, cephapirin, cimetidine, cisatracurium, CISplatin, codeine, cyanocobalamin, cyclophosphamide, cycloSPORINE, cytarabine, DACTINomycin, DAPTOmycin, dexamethasone, dexmedetomidine, digoxin, diltiazem, DOCEtaxel, DOPamine, doxacurium, DOXOrubicin liposomal, doxycycline, drotrecogin alfa, enalaprilat, ePHEDrine, EPINEPHrine, epoetin alfa, eptifibatide, erythromycin, esmolol, etoposide, fenoldopam, fludarabine, fluorouracil, folic acid, foscarnet, furosemide, gallium, gatifloxacin, gemcitabine, gentamicin, glycopyrrolate, granisetron, heparin, hydrocortisone, HYDROmorphone, ifosfamide, indomethacin, insulin (regular), isoproterenol, ketorolac, lansoprazole, levofloxacin, lidocaine, linezolid, LORazepam, mannitol, mechlorethamine, melphalan, meperidine, metaraminol, methicillin, methotrexate, methoxamine, methyldopate, methylPREDNISolone, metoclopramide, metoprolol, metroNIDAZOLE, mezlocillin, miconazole, midazolam, milrinone, morphine, moxalactam, multiple vitamins injection, nafcillin, nalbuphine, naloxone, nesiritide, netilimicin, nitroglycerin, nitroprusside, norepinephrine, octreotide, oxacillin, oxaliplatin, oxytocin, PACLitaxel, palonosetron, pamidronate, pancuronium, pantoprazole, PEMEtrexed, penicillin G potassium/sodium, PHENobarbital, phenylephrine, phytonadione, piperacillin, polymyxin B, potassium chloride, procainamide, propofol, propranolol, pyridoxine, ranitidine, remifentanil, ritodrine, riTUXimab, rocuronium, sargramostim, sodium acetate/bicarbonate, succinylcholine, SUFentanil, tacrolimus, teniposide, theophylline, thiamine, thiotepa, ticarcillin, ticarcillin-clavulanate, tigecycline, tirofiban, tolazoline, trastuzumab, trimetaphan, urokinase, vasopressin, vecuronium, verapamil, vinCRIStine, voriconazole, warfarin, zidovudine

SIDE EFFECTS

CNS: Headache, dizziness, weakness, paresthesia, fever, chills, seizures, dyskinesia (cefdinir); neurotoxicity (renal disease) cefepime
CV: Heart failure, syncope (cefdinir)
EENT: *Oral candidiasis*
GI: *Nausea, vomiting, diarrhea, anorexia,* pain, glossitis, bleeding; increased AST, ALT, bilirubin, LDH, alk phos; abdominal pain, pseudomembranous colitis; cholestasis (cefotaxime)
GU: Proteinuria, vaginitis, pruritus, *candidiasis,* increased BUN, nephrotoxicity, renal failure
HEMA: Leukopenia, thrombocytopenia, agranulocytosis, anemia, neutropenia, lymphocytosis, eosinophilia, pancytopenia, hemolytic anemia
INTEG: Rash, urticaria, dermatitis
MS: Arthralgia (cefditoren)
RESP: Dyspnea
SYST: Anaphylaxis, serum sickness, Stevens-Johnson syndrome, toxic epidermal necrolysis

PHARMACOKINETICS

cefdinir
Unchanged in urine; crosses placenta, blood-brain barrier; eliminated in breast milk, not metabolized; 60%-70% protein binding, half-life 1.7 hr
cefditoren pivoxil
Well absorbed when broken down (prodrug), wide distribution, half-life 100 min, onset rapid, peak 0.5-3 hr, duration 12 hr, 88% protein binding
cefepime
Peak 79 min; half-life 2 hr; 20% bound by plasma proteins; 90% excreted unchanged in urine; crosses placenta, blood-brain barrier; excreted in breast milk, not metabolized
cefixime
PO: Peak 2-8 hr, half-life 3-4 hr, 65%-70% protein binding, 50% eliminated unchanged in urine, crosses placenta, excreted in breast milk

cefotaxime
Half-life 1 hr, 35%-65% is bound by plasma proteins, 40%-65% is eliminated unchanged in urine in 24 hr, 25% metabolized in the liver to active metabolites, excreted in breast milk (small amounts)
IM: Onset 30 min
IV: Onset 5 min
cefpodoxime
Half-life 1-1.5 hr, 13%-38% bound by plasma proteins, 30% eliminated unchanged in urine in 8 hr, crosses placenta, excreted in breast milk
cefTAZidime
IM/IV:
Half-life $1^1/_2$-2 hr, 10% bound by plasma proteins, 80% eliminated unchanged in urine, crosses placenta, excreted in breast milk
ceftibuten
PO: Peak 2-3 hr; plasma protein binding 65%, elimination half-life 2 hr, extensively metabolized to an active metabolite
ceftizoxime
Half-life 1.6 hr, 30% bound by plasma proteins, 36%-60% eliminated unchanged in urine, crosses placenta, excreted in breast milk
IM: Peak 1 hr
IV: Onset 5 min
cefTRIAXone
Half-life 6-9 hr, 58%-96%, eliminated unchanged in urine, crosses placenta, excreted in breast milk
IM: Peak $1^1/_2$-4 hr
IV: Peak 30 min

INTERACTIONS

Many products should not be used with calcium salts (mixed or administered) or H2 blockers antacids (PO)
Increase: cycloSPORINE levels—cycloSPORINE
Increase: bleeding—anticoagulants, thrombolytics, plicamycin, valproic acid, NSAIDs
Increase: toxicity—aminoglycosides, furosemide, probenecid
Decrease: absorption of cefdinir—iron

Drug/Food
Decrease: absorption—iron-rich cereal, infant formula
Drug/Lab Test
Increase: ALT, AST, alk phos, LDH, bilirubin, BUN, creatinine
False increase: creatinine (serum urine), urinary 17-KS
False positive: urinary protein, direct Coombs' test, urine glucose
Interference: cross-matching

NURSING CONSIDERATIONS

Assess:
- Sensitivity to penicillin, other cephalosporins

⚠ **Nephrotoxicity:** increased BUN, creatinine; urine output: if decreasing, notify prescriber
- Blood studies: AST, ALT, CBC, Hct, bilirubin, LDH, alk phos, Coombs' test monthly if patient is on long-term therapy
- Electrolytes: potassium, sodium, chloride monthly if patient is on long-term therapy
- **Pseudomembranous colitis:** bowel pattern daily; if severe diarrhea occurs, product should be discontinued
- IV site for extravasation, phlebitis

⚠ **Anaphylaxis:** rash, urticaria, pruritus, chills, fever, joint pain, angioedema; may occur a few days after therapy begins
- Bleeding: ecchymosis, bleeding gums, hematuria, stool guaiac

⚠ **Overgrowth of infection:** perineal itching, fever, malaise, redness, pain, swelling, drainage, rash, diarrhea, change in cough, sputum
- Monitor heart rate during direct IV inf (cefotaxime)

Evaluate:
- Therapeutic response: decreased symptoms of infection; negative C&S

Teach patient/family:
- If diabetic, to check blood glucose

⚠ To report sore throat, bruising, bleeding, joint pain, may indicate **blood dyscrasias (rare);** diarrhea with mucus, blood, may indicate **pseudomembranous colitis**

- That cefditoren can be taken with oral contraceptives

TREATMENT OF ANAPHYLAXIS: EPINEPHrine, antihistamines; resuscitate if needed

cephradine

See cephalosporins—1st generation

certolizumab pegol (Rx)

(ser'tue-liz'oo-mab pegh'ol)

Cimzia

Func. class.: Biologic response modifier

Chem. class: Anti-tissue necrosis factor (anti-TNF) agent

ACTION: Monoclonal antibody that neutralizes the activity of tumor necrosis factor α (TNF-α) found in Crohn's disease; decreases infiltration of inflammatory cells

USES: Crohn's disease (moderate to severe) that has not responded to conventional therapy, rheumatoid arthritis (moderate to severe)

Unlabeled uses: Moderate to severe chronic plaque psoriasis, fistulizing Crohn's disease

CONTRAINDICATIONS: Influenza, IV administration, sepsis, hypersensitivity

Black Box Warning: Infection, neoplastic disease

Precautions: Pregnancy (B), breastfeeding, children, geriatric patients, AIDS, coagulopathy, diabetes, fungal infection, heart failure, hepatitis, human antichimeric antibody, immunosuppression, leukopenia, MS, cancer, neurologic/renal disease, surgery, thrombocytopenia, TB, vaccinations

DOSAGE AND ROUTES

Crohn's disease (moderate to severe)

- **Adult: SUBCUT** 400 mg given as 2 inj at wk 0, 2, 4; if clinical response occurs, give 400 mg q4wk

Rheumatoid arthritis (moderate to severe)

- **Adult: SUBCUT** 400 mg q2wk × 3 doses then 200 mg q2wk; given with methotrexate

Crohn's disease (fistulizing)/intolerant to infliximab (unlabeled)

- **Adult: SUBCUT** 400 mg wk 0, 2, 4 then 400 mg q4wk

Available forms: Powder for inj 400-mg kit

Administer:

SUBCUT route

- Give by subcut inj only
- Reconstitution: allow to warm to room temperature; add 1 ml sterile water for inj to each vial; 2 vials will be needed for patients with Crohn's disease
- Gently swirl; do not shake; full reconstitution may take up to 30 min; reconstituted product may remain at room temperature for up to 2 hr or refrigerated up to 24 hr
- If reconstituted product has been refrigerated, allow to warm to room temperature
- Use 2 syringes and two 20G needles
- Withdraw reconstituted sol from each vial into separate syringes; each will contain 200 mg; switch 20G to 23G needle; inject into 2 separate sites in abdomen or thigh
- Store in refrigerator; do not freeze

SIDE EFFECTS

CNS: *Dizziness,* syncope, peripheral neuropathy, fever, **seizures, demyelinating disease of CNS**

CV: Hypotension, **heart failure, MI, cardiac dysrhythmia**

EENT: Optic neuritis, retinal hemorrhage, uveitis

GI: Increased LFTs, **hepatitis, bowel obstruction**

GU: UTI, renal disease
HEMA: **Anemia, aplastic anemia, pancytopenia, thrombocytopenia**
INTEG: *Rash, urticaria,* **angioedema**
MISC: **Anaphylaxis,** antibody formation, arthralgia, bleeding, infection, lupuslike symptoms, lymphadenopathy, **malignancies, serum sickness, suicidal ideation**
RESP: Dyspnea, upper respiratory tract infection

PHARMACOKINETICS

Peak 54-171 hr, terminal half-life 14 days

INTERACTIONS

• **Do not administer live vaccines, toxoids concurrently**

Increase: possible infections—abatacept, adalimumab, anakinra, etanercept, immunosuppressive agents, infliximab, rilonacept; do not use concurrently
Increase: possible malignancies—adalimumab, etanercept, infliximab

NURSING CONSIDERATIONS

Assess:

• Antinuclear antibody test (ANA), hepatitis B serology, CBC
• For rheumatoid arthritis, ROM, pain
• GI symptoms: nausea, vomiting, abdominal pain, hepatitis, increased LFTs
• Periodic blood counts (CBC)
• CV status: B/P, pulse, chest pain

⚠ **Allergic reaction, anaphylaxis: rash, dermatitis, urticaria, dyspnea, hypotension, fever, chills; discontinue if severe; administer EPINEPHrine, corticosteroids, antihistamines; assess for allergies to murine proteins before starting therapy**

Black Box Warning: Infection: discontinue if infection occurs; do not administer to patients with active infection

Black Box Warning: Identify TB, risk for HBV before beginning treatment; TB test should be obtained; if present, TB should be treated before certolizumab treatment

Evaluate:

• Therapeutic response: absence of fever, mucus in stools

Teach patient/family:

• Not to breastfeed while taking this product

Black Box Warning: To notify prescriber of GI symptoms, hypersensitivity reactions, infections, fluid retention; redness, pain, swelling at inj site

Black Box Warning: Not to operate machinery, drive if dizziness, vertigo occur

cetirizine (Rx, OTC)

(se-teer'i-zeen)

All Day Allergy, All Day Allergy Children's, GNP All Day Allergy, GNP Children's All Day Allergy, Good Sense All Day Allergy, Good Sense Children's All Day Allergy, Publix Allergy Children's, Reactine ✦, Top Care Children's All Day Allergy, ZyrTEC, ZyrTEC Children's

Func. class.: Antihistamine (2nd generation, peripherally selective)
Chem. class.: Piperazine, H_1-histamine antagonist

Do not confuse:
ZyrTEC/Xanax/Zantac

ACTION: Acts on blood vessels, GI, respiratory system by competing with histamine for H_1-receptor site; decreases allergic response by blocking pharmacologic effects of histamine; minimal anticholinergic, sedative action

USES: Rhinitis, allergy symptoms, chronic idiopathic urticaria
Unlabeled uses: Asthma, atopic dermatitis

CONTRAINDICATIONS: Breastfeeding, newborn or premature infants, hypersensitivity to this product or hydrOXYzine, severe hepatic disease
Precautions: Pregnancy (B), children, geriatric patients, respiratory disease,

angle-closure glaucoma, prostatic hypertrophy, bladder neck obstruction, asthma

DOSAGE AND ROUTES

Perennial/seasonal allergic rhinitis or idiopathic urticaria

- **Adult and child ≥6 yr: PO** 5-10 mg/day
- **Child 2-5 yr: PO** 2.5 mg/day, may increase to 5 mg/day or 2.5 mg bid
- **Child 1-2 yr: PO** 2.5 mg/day, may increase to 2.5 mg q12hr
- **Geriatric: PO** 5 mg/day, may increase to 10 mg/day

Self-treatment of hay fever/other respiratory allergies

- **Adult/adolescent/child ≥6 yr: PO** 10 mg/day; **ORAL SOL** 5-10 mg/day

Renal dose/hemodialysis/hepatic dose

- **Adult: PO** CCr 11-31 ml/min, 5 mg/day

Atopic dermatitis (unlabeled)

- **Child 6-12 yr: PO** 5-10 mg/day
- **Child 1-2 yr: PO** 0.25 mg/kg bid

Available forms: Tabs 5, 10 mg; syr 5 mg/5 ml, prefilled spoons 1 mg/ml; oral sol 5 mg/ml; liquid-filled caps 10 mg; chew tabs 5, 10 mg

Administer:

- Without regard to meals
- Store in tight, light-resistant container
- **Caps:** swallow whole; do not break, cut, chew, crush
- **Chew tabs:** chew before swallowing; may use with or without water
- **Oral liquid:** use calibrated measuring device

SIDE EFFECTS

CNS: *Headache,* stimulation, *drowsiness,* sedation, *fatigue,* confusion, blurred vision, tinnitus, restlessness, tremors; paradoxical excitation in children, geriatric patients

GI: *Dry mouth,* increased LFTs, constipation

INTEG: Rash, eczema, photosensitivity, urticaria

RESP: *Thickening of bronchial secretions,* dry nose, throat

PHARMACOKINETICS

Absorption rapid; onset ½ hr; peak 1-2 hr; duration 24 hr; protein binding 93%; half-life decreased in children, increased in renal/hepatic disease

INTERACTIONS

Increase: CNS depression—alcohol, opiates, sedative/hypnotics, other CNS depressants

Increase: anticholinergic/sedative effect—MAOIs

Increase: cetirizine effect—ritonavir

Drug/Food

- Food prolongs absorption by 1.7 hr

Drug/Lab Test

False negative: skin allergy tests

NURSING CONSIDERATIONS

Assess:

- **Allergy symptoms:** pruritus, urticaria, watering eyes at baseline and during treatment
- Respiratory status: rate, rhythm, increase in bronchial secretions, wheezing, chest tightness
- Blood studies: LFTs, BUN, creatinine at baseline, periodically
- Hard candy, gum, frequent rinsing of mouth for dryness

Evaluate:

- Therapeutic response: absence of running or congested nose, rashes

Teach patient/family:

- About all aspects of product use; to notify prescriber if confusion, sedation, hypotension occur
- To avoid driving, other hazardous activity if drowsiness occurs
- To avoid alcohol, other CNS depressants, OTC antihistamines
- To avoid exposure to sunlight; burns may occur
- To use sugarless gum, candy, frequent sips of water to minimize dry mouth
- Not to breastfeed

TREATMENT OF OVERDOSE:

Administer diazepam, vasopressors, phenytoin IV

cetrorelix (Rx)

(set-roe-ree′lix)

Cetrotide

Func. class.: Gonadotropin-releasing hormone antagonist

Chem. class.: Synthetic decapeptide

ACTION: Inhibitor of pituitary gonadotropin secretion; initially increases LH and FSH; induces a rapid suppression of gonadotropin secretion

USES: For inhibition of premature LH surges in women undergoing controlled ovarian hyperstimulation

Unlabeled uses: Benign prostatic hyperplasia (BPH), endometriosis

CONTRAINDICATIONS: Pregnancy (X), breastfeeding, hypersensitivity, latex allergy, renal disease

Precautions: Geriatric patients

DOSAGE AND ROUTES

Single-dose regimen

- **Adult: SUBCUT** 3 mg when serum estradiol level at appropriate stimulation response, usually on stimulation day 7; if hCG not given within 4 days after inj of 3 mg cetrorelix, give 0.25 mg daily until day of hCG administration

Multiple-dose regimen

- **Adult: SUBCUT** 0.25 mg given on stimulation day 5 (either morning or evening) or 6 (morning) and continued daily until day hCG is given

BPH (unlabeled)

- **Adult (male): SUBCUT** 5 mg bid × 2 days then 1 mg/day

Endometriosis (unlabeled)

- **Adult (female): SUBCUT** 3 mg weekly

Available forms: Inj 0.25, 3 mg

Administer:

SUBCUT route

- Attach yellow-marked needle to diluent syringe; dilute powder by injecting liquid from syringe into vial, leaving syringe on vial; gently swirl until clear, avoid bubbles; withdraw contents of vial back into syringe, replace yellow-marked needle with gray-marked needle
- SUBCUT using abdomen, 1 inch away from navel or upper thigh; swab inj area with disinfectant; clean a 2-inch circle and allow to dry; pinch up area between thumb and finger; insert needle 45-90 degrees to surface; if positioned correctly, no blood will be drawn back into syringe; reposition needle without removing it; rotate inj sites
- Protect from light

SIDE EFFECTS

CNS: Headache

CV: Edema

ENDO: Ovarian hyperstimulation syndrome, abdominal pain (gyn)

GI: Nausea, vomiting, diarrhea

INTEG: Pain on inj; local site reactions, bruising, pruritus

Other: Rapid weight gain

RESP: Shortness of breath

SYST: Fetal death, anaphylaxis

PHARMACOKINETICS

Excreted in feces/urine, half-life depends on dosage, metabolized to metabolites, protein binding 86%

NURSING CONSIDERATIONS

Assess:

- Serum progesterone, LH; ovarian ultrasound day 7-14; pelvic exam, serum estradiol/gonadotropin; weight
- **For suspected pregnancy, product should not be used, pregnancy (X)**
- **For latex allergy, product should not be used**

⚠ **For anaphylaxis during first inf**

Evaluate:

- Therapeutic response: pregnancy

Teach patient/family:

- To report abdominal pain, vaginal bleeding, nausea, vomiting, diarrhea, SOB, peripheral edema
- How to perform self-administration technique if needed

cetuximab (Rx)

(se-tux′i-mab)

Erbitux

Func. class.: Antineoplastic—miscellaneous, monoclonal antibody

Chem. class.: Epidermal growth factor receptor inhibitor

ACTION: Not fully understood; binds to epidermal growth factor receptors (EGFRs); inhibits phosphorylation and activation of receptor-associated kinase, thereby resulting in inhibition of cell growth

USES: Alone or in combination with irinotecan for EGFRs expressing metastatic colorectal carcinoma, head/neck cancer

Unlabeled uses: Front-line use for non–small-cell lung cancer in combination with CISplatin and vinorelbine

CONTRAINDICATIONS: Hypersensitivity to this product, murine proteins

Precautions: Pregnancy (C), breastfeeding, children, geriatric patients; CV/renal/hepatic disease; ocular, pulmonary disorders

Black Box Warning: Arrhythmias, CAD, inf-related reactions, radiation/platinum-based therapy, cardiac, respiratory arrest

DOSAGE AND ROUTES

• **Adult: IV INF** 400 mg/m^2 loading dose given over 120 min, max inf rate 5 ml/min; weekly maintenance dose (all other inf) is 250 mg/m^2 given over 60 min, max inf rate 5 ml/min (10 mg/min); premedicate with an H_1-antagonist (diphenhydrAMINE 50 mg IV); dosage adjustments made for inf reactions or dermatologic toxicity; other protocols used

Non–small-cell lung cancer (NSCLC) (unlabeled)

• **Adult: IV** 400 mg/m^2 over 120 min (max 5 ml/min) week 1 with weekly inf of 250 mg/m^2 over 60 min (max 5 ml/min) with CISplatin 80 mg/m^2 on day 1 and vinorelbine 25 mg/m^2 on days 1, 8

Available forms: Sol for inj 100 mg/50 ml, 200 mg/100 ml

Administer:

Intermittent IV INF route

• Use cytoxic handling procedures
• By IV inf only; do not give by IV push or bolus; do not shake or dilute
• Do not dilute with other products
• Store refrigerated at 36° F-46° F, discard unused portions
• **Inf pump:** draw up volume of vial using appropriate syringe/needle (vented spike or other appropriate transfer device); fill Erbitux into sterile evacuated container/bag, repeat until calculated volume put into the container; use new needle for each vial; give through in-line filter (low protein binding 0.22 micrometer); affix inf line and prime before starting inf, max rate 5 ml/min; flush line at end of inf with 0.9% NaCl, use a low protein binding 0.22-micrometer in-line filter
• **Syringe pump:** draw up volume of vial using appropriate syringe/needle (vented spike); place syringe into syringe driver of syringe pump and set rate; use in-line filter (low protein binding 0.22-micrometer); connect inf line, start inf after priming; repeat until calculated volume given
• Use new needle and filter for each vial, max 5 ml/min rate; use 0.9% NaCl to flush line after inf
• Do not piggyback to patient inf line
• Observe patient for adverse reactions for 1 hr after inf

Black Box Warning: Inf reactions: if mild (grade 1 or 2), reduce all doses by 50%; if severe (grade 3 or 4), permanently discontinue

SIDE EFFECTS

CNS: *Headache, insomnia, depression,* aseptic meningitis

CV: Cardiac arrest

GI: *Nausea, diarrhea, vomiting, anorexia, mouth ulceration, dehydration, constipation, abdominal pain*
HEMA: Leukopenia, anemia
INTEG: Rash, pruritus, acne, dry skin, toxic epidermal necrolysis, angioedema, *blepharitis, cheilitis, cellulitis, cysts, alopecia, skin/nail disorder,* acute inf reactions, other skin toxicities
MISC: *Conjunctivitis, asthma, malaise, fever,* renal failure, hypomagnesemia
MS: *Back pain*
RESP: Interstitial lung disease, *cough, dyspnea,* pulmonary embolus, *peripheral edema,* respiratory arrest
SYST: Anaphylaxis, sepsis, infection, mucosal inflammation

PHARMACOKINETICS

Half-life 114 hr, steady state by 3rd wkly inf, peak 168-235 g/ml, trough 41-85 g/ml

INTERACTIONS

Drug/Lab:
Increase: LFTs

NURSING CONSIDERATIONS

Assess:

Black Box Warning: **Pulmonary changes:** lung sounds, cough, dyspnea; interstitial lung disease may occur, may be fatal; discontinue therapy if confirmed

• Cardiac arrest: monitor electrolytes, in those undergoing radiation therapy, electrolytes may be decreased; monitor cardiac patients that receive this product and radiation therapy or platinum based therapy with 5-FU (head, neck, cancer)

⚠ **Serious hypersensitivity reactions:** toxic epidermal necrosis, angioedema, anaphylaxis

• GI symptoms: frequency of stools, dehydration, abdominal pain, stomatitis

• **K-RAS mutations** with metastatic colorectal carcinoma; if K-RAS mutation on codon 12 or 13 detected, patient should not receive anti-EGFR antibody therapy

Evaluate:

• Therapeutic response: decreased growth, spread of EGFR-expressing metastatic colorectal, head/neck carcinoma

Teach patient/family:

Black Box Warning: To report adverse reactions immediately: shortness of breath, severe abdominal pain, skin eruptions

• About the reason for treatment, expected results

Black Box Warning: To use contraception during treatment (pregnancy [C]), not to breastfeed

• To wear sunscreen and hats to limit sun exposure; sun exposure can exacerbate any skin reactions

• To avoid crowds, persons with known infections

chlordiazePOXIDE (Rx)

(klor-dye-az-e-pox′ide)

Librium

Func. class.: Antianxiety

Chem. class.: Benzodiazepine, long-acting

Controlled Substance Schedule IV

Do not confuse:
Librium/Librax

ACTION: Potentiates the actions of GABA, especially in the limbic system, reticular formation

USES: Short-term management of anxiety, acute alcohol withdrawal, preoperatively for relaxation

CONTRAINDICATIONS: Pregnancy (D), breastfeeding, children <6 yr, hypersensitivity to benzodiazepines, closed-angle glaucoma, psychosis

Precautions: Geriatric patients, debilitated, renal/hepatic disease, suicidal ideation, abrupt discontinuation

DOSAGE AND ROUTES

Mild anxiety

- **Adult: PO** 5-10 mg tid-qid
- **Geriatric: PO** 5 mg bid initially, increase as needed
- **Child >6 yr: PO** 5 mg bid-qid, max 10 mg bid-tid

Severe anxiety

- **Adult: PO** 25-50 mg tid-qid

Preoperatively

- **Adult: PO** 5-10 mg tid-qid on day before surgery

Alcohol withdrawal

- **Adult: PO** 50-100 mg q4-6hr prn, max 300 mg/day

Renal disease

- **Adult: PO** CCr <10 ml/min, give 50% dose

Available forms: Caps 5, 10, 25 mg

Administer:

PO route

- With food or milk for GI symptoms

SIDE EFFECTS

CNS: *Dizziness, drowsiness,* confusion, headache, anxiety, tremors, stimulation, fatigue, depression, insomnia, hallucinations

CV: *Orthostatic hypotension,* edema, ECG changes, tachycardia, hypotension

EENT: *Blurred vision,* tinnitus, mydriasis

GI: Constipation, dry mouth, nausea, vomiting, anorexia, diarrhea

GU: Irregular periods, decreased libido

HEMA: Agranulocytosis

INTEG: Rash, dermatitis, itching

PHARMACOKINETICS

PO: Onset 30 min, peak within 2 hr, duration 4-6 hr, metabolized by liver, excreted by kidneys, crosses placenta, excreted in breast milk, half-life 5-30 hr (increased in geriatric patients)

INTERACTIONS

Increase: CNS depression—CNS depressants, alcohol

Increase: chlordiazePOXIDE—cimetidine, disulfiram, FLUoxetine, isoniazid, ketoconazole, metoprolol, oral contraceptives, propranolol, valproic acid

Decrease: action of levodopa

Decrease: action of chlordiazePOXIDE—CYP3A4 inhibitors (protease inhibitors, barbiturates, rifamycins)

Drug/Lab Test

Increase: LFTs

False increase: 17-OHCS

False positive: pregnancy test (some methods)

NURSING CONSIDERATIONS

Assess:

- B/P (lying, standing), pulse; if systolic B/P drops 20 mm Hg, hold product, notify prescriber
- Blood studies: CBC during long-term therapy; **blood dyscrasias** have occurred rarely
- Hepatic studies: AST, ALT, bilirubin, creatinine, LDH, alk phos during long-term therapy
- For ataxia, oversedation of geriatric patients, debilitated patients
- Physical dependency, withdrawal symptoms: headache, nausea, vomiting, muscle pain, weakness after long-term use
- Mental status: mood, sensorium, affect, sleeping pattern, drowsiness, dizziness; suicidal tendencies; paradoxic reactions such as excitement, stimulation, acute rage
- For pregnancy; product should not be used during pregnancy (D)
- Assistance with ambulation during beginning therapy because drowsiness, dizziness occur
- Check to confirm that PO medication has been swallowed if patient is depressed, suicidal
- Sugarless gum, hard candy, frequent sips of water for dry mouth

Evaluate:

- Therapeutic response: decreased anxiety, restlessness, sleeplessness

Teach patient/family:

- That product may be taken with food
- Not to use product for everyday stress or use for more than 4 mo unless directed by prescriber
- Not to take more than prescribed amount; may be habit forming
- To avoid OTC preparations unless approved by prescriber
- To avoid driving, activities that require alertness because drowsiness may occur
- To avoid alcohol ingestion, other psychotropic medications unless directed by prescriber
- Not to discontinue medication abruptly after long-term use because this may precipitate seizures
- To rise slowly because fainting may occur, especially among geriatric patients
- That drowsiness may be worse at beginning of treatment
- To notify prescriber if pregnancy is suspected or planned
- To immediately report suicidal thoughts/behaviors

TREATMENT OF OVERDOSE: Lavage, VS, supportive care, give flumazenil

chloroquine (Rx)

(klor′oh-kwin)

Aralen

Func. class.: Antimalarial

Chem. class.: Synthetic 4-aminoquinoline derivative

ACTION: Inhibits parasite replication, transcription of DNA to RNA by forming complexes with DNA of parasite

USES: Malaria of *Plasmodium vivax, P. malariae, P. ovale, P. falciparum* (some strains); amebiasis

Unlabeled uses: Discoid lupus erythematosus, polymorphous light eruption, rheumatoid arthritis, ulcerative colitis

CONTRAINDICATIONS: Hypersensitivity, retinal field changes

Precautions: Pregnancy (C), breastfeeding, children, blood dyscrasias, severe GI disease, neurologic disease, alcoholism, hepatic disease, G6PD deficiency, psoriasis, eczema, seizures, preexisting auditory damage, torsades de pointes

Black Box Warning: Infection

DOSAGE AND ROUTES

Acute malaria attacks

- **Adult: PO** 1000 mg (600-mg base), then 500 mg (300-mg base) in 6-8 hr, then 500 mg (300-mg base) daily × 2 days for a total of 2.5 g (1.5-g base) in 3 days
- **Adult/adolescent of low body weight, child/infant: PO** 16.5 mg (10-mg base)/kg, max 600-mg base; then 8.3 mg (5-mg base)/kg, max 300-mg base 6 hr after 1st dose; then 8.3 mg (5-mg base)/kg, max 300-mg base 24 hr after 1st dose; then 8.3 mg (5-mg base)/kg, max 300-mg base 36 hr after 1st dose

Malaria prophylaxis (in areas with chloroquine–sensitive *P. falciparum*)

- Adult: **PO** 500 mg (300-mg base) weekly on same day of each wk starting 2 wk before travel and for 8 wk after leaving

Extraintestinal amebiasis

- **Adult: PO** 1 g (600-mg base) daily × 2 days, then 500 mg (300-mg base) for ≥2-3 wk
- **Child (unlabeled): PO** 16.6 mg (10-mg base)/kg (max 300-mg base) daily × 2-3 wk

Rheumatoid arthritis/discoid lupus erythematosus (unlabeled)

- **Adult: PO** 250 mg (150-mg base) daily

Available forms: Tabs 250 mg (150-mg base), 500 mg (300-mg base) phosphate

Administer:

- Product in mg or base; they are different

PO route

- Before or after meals at same time each day to maintain product level

• Store in tight, light-resistant container at room temperature; keep inj in cool environment

SIDE EFFECTS

CNS: Headache, stimulation, fatigue, seizures, psychosis, hallucinations, insomnia
CV: Hypotension, heart block, asystole with syncope, ECG changes, cardiomyopathy
EENT: *Blurred vision, corneal changes, retinal changes, difficulty focusing,* tinnitus, vertigo, deafness, photophobia, corneal edema
GI: *Nausea, vomiting, anorexia,* diarrhea, cramps
HEMA: Thrombocytopenia, agranulocytosis, hemolytic anemia, leukopenia
INTEG: Pruritus, pigmentary changes, skin eruptions, lichen-planus–like eruptions, eczema, exfoliative dermatitis

PHARMACOKINETICS

Metabolized in liver; excreted in urine, feces, breast milk; crosses placenta
PO: Peak 1-3 hr, half-life 3-5 days

INTERACTIONS

• Reduced oral clearance and metabolism of chloroquine: cimetidine
Increase: QT prolongation, torsades de pointes—class IA, III antidysrhythmics
Increase: effects—2D6 inhibitors (amiodarone, chlorpheniramine, FLUoxetine, haloperidol, ritonavir, PARoxetine, terbinafine, ticlopidine); CYP3A4 inhibitors (diltiazem, verapamil, itraconazole, ketoconazole, erythromycin, doxycycline, clarithromycin)
Decrease: action of chloroquine—magnesium, aluminum compounds, kaolin; do not use concurrently
Decrease: effects of ampicillin, rabies vaccine (ID)
Drug/Lab Test
Decrease: Hgb, platelets, WBC

NURSING CONSIDERATIONS

Assess:

Black Box Warning: Infection: resistance is common, not to be used for *P. falciparum* acquired in areas of resistance or where prophylaxis has failed

• Ophthalmic test if long-term treatment or dosage of >150 mg/day, baseline and periodically
• Blood studies: CBC, as blood dyscrasias occur
• **ECG** during therapy; watch for depression of T waves, widening of QRS complex
• **Allergic reactions:** pruritus, rash, urticaria
• **Blood dyscrasias:** malaise, fever, bruising, bleeding (rare)
• **For ototoxicity** (tinnitus, vertigo, change in hearing); audiometric testing should be done before, after treatment
For toxicity: blurring vision; difficulty focusing; headache; dizziness; decreased knee, ankle reflexes; seizures, CV collapse; product should be discontinued immediately and IV fluids given

Evaluate:

• Therapeutic response: decreased symptoms of infection

Teach patient/family:

• To take with meals or immediately after meals
• To use sunglasses in bright sunlight to decrease photophobia
• That urine may turn rust or brown color
• To report hearing, visual problems; fever, fatigue, bruising, bleeding; may indicate blood dyscrasias

TREATMENT OF OVERDOSE:

Administer barbiturate (ultrashort-acting), vasopressor; tracheostomy may be necessary
• Keep away from pets, children; overdose is fatal

chlorpheniramine (OTC, Rx)

(klor-fen-ir'a-meen)

AHIST, Aller-Chlor, Allergy, Chlor-Pheniton, Chlor-Trimeton, Diabetic Tussin Allergy Relief, ED-Chlor-Tann, Equaline Allergy, Equate Chlortabs Allergy, Good Sense Allergy, Leader Allergy, P-Tann, Tana Hist-PD, Teldrin, Top Care Allergy, Wal-finate Allergy

Func. class.: Antihistamine (1st generation, nonselective)

Chem. class.: Alkylamine, H_1-receptor antagonist

Do not confuse:
Teldrin/Tedral

ACTION: Acts on blood vessels, GI system, respiratory system by competing with histamine for H_1-receptor site; decreases allergic response by blocking histamine

USES: Allergy symptoms, rhinitis, conjunctivitis (allergic)

Unlabeled uses: Nausea, vomiting due to motion sickness, pruritus, urticaria

CONTRAINDICATIONS: Newborns/neonates

Precautions: Pregnancy (B), breastfeeding, children, geriatric patients, increased intraocular pressure, cardiac/renal disease, hypertension, asthma, seizure disorder, hyperthyroidism, prostatic hypertrophy, GI obstruction, peptic ulcer disease, emphysema, hypersensitivity to H_1-receptor antagonists, lower respiratory tract disease, stenosed peptic ulcers, bladder neck obstruction, closed-angle glaucoma

DOSAGE AND ROUTES

- **Adult and child ≥12 yr: PO** 4 mg tid-qid, max 24 mg/day; **EXT REL** 8-12 mg bid-tid, max 24 mg/day
- **Child 6-12 yr: PO** 2 mg q4-6hr, max 12 mg/day; **EXT REL** 8 mg bedtime or daily; **EXT REL** not recommended for child <6 yr

Available forms: Chewable tabs 2 mg; tabs 4, 8, 12 mg; ext rel tabs 8, 12 mg; ext rel caps 8, 12 mg; syr 1 mg/5 ml, 2 mg/5 ml, 2.5 mg/5 ml

Administer:

- Avoid concurrent use with other CNS depressants
- Do not break, crush, or chew ext rel forms
- Without regard to meal
- Avoid use in children <6 yr
- Store in tight container at room temperature
- **Syrup:** Use dosing utensil to measure correct dose

SIDE EFFECTS

CNS: *Dizziness, drowsiness,* poor coordination, fatigue, anxiety, euphoria, confusion, paresthesia, neuritis

EENT: Blurred vision; dilated pupils; tinnitus; nasal stuffiness; dry nose, throat, mouth

GI: Nausea, anorexia, diarrhea

GU: *Retention,* dysuria, urinary frequency

HEMA: Thrombocytopenia, agranulocytosis, hemolytic anemia

INTEG: Photosensitivity

RESP: Increased thick secretions, wheezing, chest tightness

PHARMACOKINETICS

Detoxified in liver, excreted by kidneys (metabolites/free drug), half-life 12-15 hr

PO: Onset ½ hr, duration 4-12 hr

PO-ER: Duration 8-24 hr

INTERACTIONS

Increase: CNS depression—barbiturates, opiates, hypnotics, tricyclics, alcohol

Increase: effect of chlorpheniramine—MAOIs
Increase: anticholinergic action—atropine, phenothiazines, quiNIDine, haloperidol
Drug/Lab Test
False negative: skin allergy tests

NURSING CONSIDERATIONS

Assess:

- Be alert for urinary retention, frequency, dysuria; product should be discontinued
- Respiratory status: rate, rhythm, increase in bronchial secretions, wheezing, chest tightness
- Hard candy, gum, frequent rinsing of mouth for dryness

Evaluate:

- Therapeutic response: absence of running, congested nose, rashes, conjunctivitis

Teach patient/family:

- About all aspects of product use; to notify prescriber of confusion, sedation, hypotension, difficulty voiding
- To avoid driving, other hazardous activity if drowsiness occurs, especially geriatric patients
- To avoid concurrent use of alcohol

TREATMENT OF OVERDOSE:
Administer diazepam, vasopressors, phenytoin IV

chlorproMAZINE (Rx)
(klor-proe′ma-zeen)
Func. class.: Antipsychotic/antiemetic
Chem. class.: Phenothiazine-aliphatic

Do not confuse:
chlorproMAZINE/chlorproPAMIDE/prochlorperazine

ACTION: Depresses cerebral cortex, hypothalamus, limbic system, which control activity aggression; blocks neurotransmission produced by DOPamine at synapse; exhibits a strong α-adrenergic, anticholinergic blocking action; mechanism for antipsychotic effects is unclear

USES: Psychotic disorders, mania, schizophrenia, anxiety, intractable hiccups in adults, nausea, vomiting; preoperatively for relaxation; acute intermittent porphyria, behavioral problems in children, nonpsychotic, demented patients, Tourette's syndrome
Unlabeled uses: Vascular headache, agitation, dementia, neonatal abstinence syndrome

CONTRAINDICATIONS: Children <6 mo, hypersensitivity, circulatory collapse, liver damage, cerebral arteriosclerosis, coronary disease, severe hypo/hypertension, blood dyscrasias, coma, brain damage, bone marrow depression, alcohol/barbiturate withdrawal, closed-angle glaucoma
Precautions: Pregnancy (C), breastfeeding, geriatric patients, seizure disorders, hypertension, hepatic/cardiac disease, prostatic enlargement, Parkinson's disease, pulmonary disease

Black Box Warning: Dementia; increased mortality in geriatric patients with dementia-related psychosis

DOSAGE AND ROUTES

Psychosis

- **Adult: PO** 10-50 mg q1-4hr initially then increase up to 2 g/day if necessary; **IM** 10-50 mg q1-4hr, usual dose 300-800 mg/day
- **Geriatric:** 10-25 mg daily-bid, increase by 10-25 mg/day q4-7days, max 800 mg/day
- **Child >6 mo: PO** 0.55 mg/kg q4-6hr; **IM** 0.5 mg/kg q6-8hr

Nausea and vomiting

- **Adult: PO** 10-25 mg q4-6hr prn; **IM** 12.5-25 mg then 25-50 mg if no hypotension q3hr prn; q6-8hr prn, max 400 mg/day; **IV** 25-50 mg daily-qid

• **Child ≥6 mo: PO** 0.55 mg/kg q4-6hr; **IM** q6-8hr; **IM** ≤5 yr or ≤22.7 kg, 40 mg; max **IM** 5-10 yr or 22.7-45.5 kg, 75 mg

Intractable hiccups/acute intermittent porphyria

• **Adult: PO** 25-50 mg tid-qid; **IM** 25-50 mg (only if PO dose does not work); **IV** 25-50 mg in 500-1000 ml **NS** (only for severe hiccups)

Available forms: Tabs 10, 25, 50, 100, 200 mg; inj 25 mg/ml

Administer:

• Anticholinergic agent for EPS if ordered

PO route

• With full glass of water, milk or with food to decrease GI upset

• Periodically attempt dosage reduction in those with behavioral problems

• Store in tight, light-resistant container

IM route

• Use gloves to prepare product; if product touches skin, wash with soap and water to prevent contact dermatitis

• IM, inject in deep muscle mass, do not give SUBCUT, no dilution needed; if irritation occurs, may dilute in NS or procaine 2%

• Avoid skin contact with injection solution—may cause contact dermatitis

IV route

Direct IV: After **diluting** 1 mg/1 ml with NS, **give** 1 mg or less/2 min or more, never give undiluted

Intermittent IV INF: Dilute 25-50 mg/500-1000 NS or other compatible large IV sol, **give** over ½ hour, protect from light

Syringe compatibilities: Atropine, benztropine, butorphanol, diphenhydrAMINE, doxapram, droperidol, fentaNYL, glycopyrrolate, HYDROmorphone, hydrOXYzine, meperidine, metoclopramide, midazolam, morphine, pentazocine, perphenazine, prochlorperazine, promazine, promethazine, scopolamine

Y-site compatibilities: Alfentanil, amikacin, amphotericin B lipid complex, amsacrine, anidulafungin, ascorbic acid injection, atenolol, atracurium, atropine, benztropine, bleomycin sulfate, buprenorphine, butorphanol, calcium chloride/gluconate, caspofungin, cimetidine, cisatracurium, CISplatin, cladribine, codeine, cyanocobalamin, cyclophosphamide, cycloSPORINE, cytarabine, DACTINomycin, DAPTOmycin, dexmedetomidine, digoxin, diltiazem, diphenhydrAMINE, DOBUTamine, DOCEtaxel, DOPamine, doxacurium, DOXOrubicin, DOXOrubicin liposomal, doxycycline, enalaprilat, ePHEDrine, EPINEPHrine, epirubicin, erythromycin, esmolol, etoposide, famotidine, fenoldopam, fentaNYL, filgrastim, fluconazole, gatifloxacin, gemcitabine, gentamicin, glycopyrrolate, granisetron, hydrocortisone, HYDROmorphone, hydrOXYzine, IDArubicin, ifosfamide, isoproterenol, labetalol, levofloxacin, lidocaine, LORazepam, LR, magnesium sulfate, mannitol, mechlorethamine, meperidine, methicillin, methoxamine, methyldopate, methylPREDNISolone, metoclopramide, metoprolol, metroNIDAZOLE, miconazole, midazolam, milrinone, minocycline, mitoXANtrone, morphine, multiple vitamins injection, mycophenolate mofetil, nafcillin, nalbuphine, naloxone, netilmicin, nitroglycerin, norepinephrine, octreotide, ondansetron, oxacillin, oxaliplatin, palonosetron, pamidronate, pancuronium, papaverine, penicillin G potassium, pentamidine, pentazocine, phytonadione, polymyxin B, potassium chloride, procainamide, prochlorperazine, promethazine, propofol, propranolol, protamine sulfate, pyridoxine, quiNIDine, quinupristin-dalfopristin, ranitidine, Ringer's injection, ritodrine, riTUXimab, rocuronium, sodium acetate, succinylcholine, SUFentanil, tacrolimus, teniposide, theophylline, thiamine, thiotepa, tirofiban, TNA, tolazoline, TPN, trimetaphan, vancomycin, vasopressin, vecuronium, verapamil, vinCRIStine, vinorelbine, vitamin B complex with C, voriconazole, zoledronic acid

SIDE EFFECTS

CNS: *EPS: pseudoparkinsonism, akathisia, dystonia, tardive dyskinesia,* seizures, *headache,* neuroleptic malignant syndrome, dizziness
CV: *Orthostatic hypotension,* hypertension, cardiac arrest, ECG changes, tachycardia
EENT: Blurred vision, glaucoma, dry eyes
ENDO: SIADH
GI: *Dry mouth, nausea, vomiting, anorexia, constipation,* diarrhea, cholestatic jaundice, weight gain
GU: Urinary retention, enuresis, impotence, amenorrhea, gynecomastia, breast engorgement
HEMA: Anemia, leukopenia, leukocytosis, agranulocytosis
INTEG: *Rash,* photosensitivity, dermatitis
RESP: Laryngospasm, dyspnea, respiratory depression
SYST: Death in geriatric patients with dementia

PHARMACOKINETICS

Metabolized by liver, excreted in urine (metabolites), crosses placenta, enters breast milk, 95% bound to plasma proteins, elimination half-life 10-30 hr
PO: Absorption variable, widely distributed, onset erratic 30-60 min, duration 4-6 hr
PO-ER: Onset 30-60 min, peak unknown, duration 10-12 hr
IM: Well absorbed, peak 15-20 min, duration 4-8 hr
IV: Onset 5 min, peak 10 min, duration unknown

INTERACTIONS

Increase: CNS depression—other CNS depressants, alcohol, barbiturate anesthetics, antihistamines, sedatives/hypnotics, antidepressants
Increase: toxicity—EPINEPHrine
Increase: agranulocystosis—antithyroid agents
Increase: effects of both products—β-adrenergic blockers, alcohol
Increase: anticholinergic effects—anticholinergics, antidepressants, antiparkinsonian agents
Increase: valproic acid level
Decrease: seizure threshold—anticonvulsants
Decrease: absorption—aluminum hydroxide, magnesium hydroxide antacids
Decrease: antiparkinson activity—levodopa, bromocriptine
Decrease: serum chlorproMAZINE—lithium, barbiturates
Decrease: anticoagulant effect—warfarin

Drug/Lab Test
Increase: hepatic studies
Decrease: WBC, platelets, Hgb/HcT
False positive: pregnancy tests, PKU
False negative: urinary steroids, 17-OHCS

NURSING CONSIDERATIONS

Assess:

- Mental status: orientation, mood, behavior, presence and type of hallucinations before initial administration and monthly
- Any potentially reversible causes of behavior problems in geriatric patients before and during therapy
- Swallowing of PO medication; check for hoarding or giving of medication to other patients
- I&O ratio; palpate bladder if low urinary output occurs, especially in geriatric patients
- Bilirubin, CBC, LFTs, ocular exam; agranulocytosis, glaucoma, cholestatic jaundice may occur
- Respirations q4hr during initial treatment; establish baseline before starting treatment; report drops of 30 mm Hg; obtain baseline ECG; Q-wave and T-wave changes
- Dizziness, faintness, palpitations, tachycardia on rising

⚠ **Neuroleptic malignant syndrome:** hyperpyrexia, muscle rigidity, increased CPK, altered mental status, for acute dystonia (check chewing, swallowing, eyes, pill rolling)

• **EPS:** akathisia (inability to sit still, no pattern to movements), tardive dyskinesia (bizarre movements of the jaw, mouth, tongue, extremities), pseudoparkinsonism (rigidity, tremors, pill rolling, shuffling gait)
• Constipation, urinary retention daily; increase bulk, water in diet
• Supervised ambulation until stabilized on medication; do not involve in strenuous exercise program because fainting is possible; patient should not stand still for long periods
• Increased fluids, roughage to prevent constipation
• Candy, gum, sips of water for dry mouth

Evaluate:
• Therapeutic response: decrease in emotional excitement, hallucinations, delusions, paranoia; reorganization of patterns of thought, speech; increase in target behaviors

Teach patient/family:
• To use good oral hygiene; to use frequent rinsing of mouth, sugarless gum, candy, ice chips for dry mouth
• To avoid hazardous activities until product response is determined
• That orthostatic hypotension occurs often; to rise gradually from sitting or lying position
• To remain lying down for at least 30 min after IM inj
• To avoid hot tubs, hot showers, tub baths because hypotension may occur; that, during hot weather, heat stroke may occur; to take extra precautions to stay cool
• To avoid abrupt withdrawal of product or EPS may result; product should be withdrawn slowly
• To avoid OTC preparations (cough, hay fever, cold) unless approved by prescriber since serious product interactions may occur; avoid use with alcohol, increased drowsiness may occur
• To use a sunscreen and sunglasses to prevent burns
• To take antacids 2 hr before or after this product
• To report sore throat, malaise, fever, bleeding, mouth sores; CBC should be drawn and product discontinued
• To employ contraceptive measures
• That urine may turn pink or reddish brown

TREATMENT OF OVERDOSE:
Lavage if orally ingested; provide airway; *do not induce vomiting or use EPINEPHrine*

cholestyramine (Rx)
(koe-less-tir′a-meen)

Prevalite, Questran, Questran Light

Func. class.: Antilipemic
Chem. class.: Bile acid sequestrant

Do not confuse:
Questran/Quarzan

ACTION: Adsorbs, combines with bile acids to form insoluble complex that is excreted through feces; loss of bile acids lowers LDL, cholesterol levels

USES: Primary hypercholesterolemia (esp. type IIa/IIb hyperlipoproteinemia), pruritus associated with biliary obstruction

Unlabeled uses: Diarrhea caused by excess bile acid

CONTRAINDICATIONS: Hypersensitivity; complete biliary obstruction; hyperlipidemia III, IV, V

Precautions: Pregnancy (C), breastfeeding, children, PKU

DOSAGE AND ROUTES
• **Adult: PO** 4 g/day or bid, max 24 g/day
• **Child: PO** 240 mg/kg/day in 3 divided doses with food or drink, max 8 g/day titrated up over several weeks to decrease GI effects

Available forms: Powder for susp 4 g cholestyramine/packet or scoop; tab 1 g

Administer:

- Product daily or bid; give all other medications 1 hr before or 4-6 hr after cholestyramine to avoid poor absorption
- Product mixed with applesauce or stirred into beverage (2-6 oz), let stand for 2 min; do not take dry, avoid inhaling powder, avoid GI tube administration, take with food
- Supplemental doses of vit A, D, K if levels are low
- Doses are expressed in anhydrous cholestyramine resin; amount of resin varies with each product

SIDE EFFECTS

CNS: Headache, dizziness, drowsiness, vertigo, tinnitus, anxiety
GI: *Constipation, abdominal pain, nausea,* fecal impaction, hemorrhoids, flatulence, vomiting, steatorrhea, peptic ulcer
HEMA: Bleeding, increased PT
INTEG: Rash, irritation of perianal area, tongue, skin
META: Decreased vit A, D, K, red cell folate content; hyperchloremic acidosis
MS: Muscle, joint pain

PHARMACOKINETICS

PO: Excreted in feces, LDL lowered within 4-7 days, serum cholesterol lowered within 1 mo, duration 2-4 wk

INTERACTIONS

Decrease: absorption of warfarin; thiazides; cardiac glycosides; propranolol; corticosteroids; iron; thyroid hormones; acetaminophen; amiodarone; penicillin G; tetracyclines; clofibrate; gemfibrozil; oral vancomycin, glipiZIDE; vit A, D, E, K

Drug/Lab Test
Increase: AST, ALT, alk phos
Decrease: sodium, potassium

NURSING CONSIDERATIONS

Assess:

- Cardiac glycoside level if both products administered, may need to adjust dose of cardiac glycoside, if this product is increased or decreased
- For signs of vit A, D, K deficiency
- **Hypercholesterolemia:** fasting LDL, HDL, total cholesterol, triglyceride levels, electrolytes if receiving extended therapy; diet history
- **Pruritus:** for signs of itching
- Bowel pattern daily; increase bulk, water in diet for constipation; diarrhea may also occur

Evaluate:

- Therapeutic response: decreased LDL, cholesterol level (hyperlipidemia); diarrhea, pruritus (excess bile acids)

Teach patient/family:

⚠ About the symptoms of hypoprothrombinemia: bleeding mucous membranes, dark tarry stools, hematuria, petechiae; report immediately

- To take with food, never use dry
- That PKU patients should avoid Questran Light (contains aspartame and phenylalanine)
- About the importance of compliance
- That risk factors should be decreased: high-fat diet, smoking, alcohol consumption, absence of exercise
- That GI side effects will resolve with continued use

cidofovir (Rx)

(si-doh-foh′veer)

Vistide

Func. class.: Antiviral
Chem. class.: Nucleotide analog

ACTION: Suppresses cytomegalovirus (CMV) replication by selective inhibition of viral DNA synthesis

USES: CMV retinitis in patients with HIV; used with probenecid
Unlabeled uses: Adenovirus, condylomata acuminata, eczema vaccination, Epstein-Barr virus, generalized vaccinia, herpes genitalis/simplex, HPV, molluscum contagiosum, vaccinia necrosum, vaccinia, varicella-zoster, variola

CONTRAINDICATIONS:

Hypersensitivity to this product, probenecid, sulfa products; direct intraocular injection

Black Box Warning: Proteinuria, renal disease/failure

Precautions: Pregnancy (C), breastfeeding, children <6 mo, geriatric patients, preexisting cytopenias, renal function impairment, platelet count <25,000/mm^3

Black Box Warning: Neutropenia, infertility, secondary malignancy

DOSAGE AND ROUTES

- **Adult: IV** 5 mg/kg weekly × 2 wk then 3 mg/kg q2wk, give with probenecid

Renal dose

- **Adult: IV** CCr ≤55 ml/min, do not use; SCr increase of 0.3-0.4 mg/dl above baseline, decrease dose to 3 mg/kg; SCr increase of ≥0.5 mg/dl above baseline or ≥21 proteinuria, discontinue

Available forms: Inj 75 mg/ml

Administer:

Black Box Warning: Use cytotoxic handling procedures

- Allow to warm to room temperature
- If product comes in contact with skin, wash with soap and water immediately
- If zidovudine is used, reduce dose to 50% on cidofovir treatment days

Intermittent IV INF route

- **Dilute** in 100 ml 0.9% saline sol before administration; probenecid must be given PO 2 g 3 hr before the cidofovir inf and 1 g at 2 and 8 hr after ending the cidofovir inf; **give** 1 L of 0.9% saline sol IV with each INF of cidofovir, give saline INF over 1-2 hr period immediately before cidofovir; patient should be given a 2nd L if the patient can tolerate the fluid load (2nd L given at time of cidofovir or immediately afterward, should be given over 1-3 hr)
- **Give** slowly; do not give by bolus IV, SUBCUT inj
- Use diluted sol within 24 hr, do not freeze; do not use sol with particulate matter or discoloration
- Do not admix

SIDE EFFECTS

CNS: *Fever, chills,* **coma,** confusion, abnormal thoughts, *dizziness*, bizarre dreams, *headache*, psychosis, tremors, somnolence, paresthesia, *amnesia, anxiety, insomnia*, **seizures**

CV: Dysrhythmias, hypo/hypertension

EENT: Retinal detachment with CMV retinitis

GI: Abnormal LFTs, *nausea, vomiting, anorexia, diarrhea*, abdominal pain, **hemorrhage**

GU: **Hematuria,** increased creatinine, BUN, **nephrotoxicity**

HEMA: **Granulocytopenia, thrombocytopenia, irreversible neutropenia, anemia, eosinophilia**

INTEG: *Rash, alopecia, pruritus, acne*, urticaria, pain at inj site, phlebitis

RESP: Dyspnea

PHARMACOKINETICS

Terminal half-life 2.6 hr

INTERACTIONS

Black Box Warning: **Nephrotoxicity:** amphotericin B, foscarnet, aminoglycosides, pentamidine IV, NSAIDs, salicylates; wait 7 days after use to begin cidofovir

NURSING CONSIDERATIONS

Assess:

- Culture before treatment is initiated; cultures of blood, urine, and throat may all be taken; CMV not confirmed by this method; diagnosis made by ophthalmic exam

Black Box Warning: Renal, hepatic, increased hemopoietic studies, BUN; serum creatinine, AST, ALT, creatinine, CCr, A-G ratio, baseline and drip treatment, blood counts should be done q2wk; watch for decreasing granulocytes, Hgb; if low, therapy may have to be discontinued and restarted after hematologic recovery; blood transfusions may be required, renal failure can occur also, Fancom syndrome

- For GI symptoms: severe nausea, vomiting, diarrhea; severe symptoms may necessitate discontinuing product
- Electrolytes and minerals: calcium, phosphorus, magnesium, sodium, potassium; watch closely for tetany during 1st administration

Black Box Warning: **Blood dyscrasias** (anemia, granulocytopenia); bruising, fatigue, bleeding, poor healing; leukopenia, neutropenia, thrombocytopenia: WBCs, platelets q2days during 2 ×/day dosing and every wk thereafter; check for leukopenias with daily WBC count in patients with prior leukopenia, with other nucleoside analogs, or for whom leukopenia counts are <1000 cells/mm³ at start of treatment

- Allergic reactions: flushing, rash, urticaria, pruritus
- Monitor serum creatinine or CCr at least q2wk; give only to those with creatinine levels ≤1.5 mg/dl, CCr >55 ml/min, urine protein <100 mg/dl

Evaluate:

- Therapeutic response: decreased symptoms of CMV

Teach patient/family:

- To notify prescriber if sore throat, swollen lymph nodes, malaise, fever occur; may indicate other infections
- To report perioral tingling, numbness in extremities, paresthesias; report rash immediately
- That serious product interactions may occur if OTC products are ingested; check with prescriber
- That product is not a cure but will control symptoms
- That regular ophthalmic exams, renal studies must be continued
- That major toxicities may necessitate discontinuing product
- To use contraception during treatment, that infertility may occur, and that men should use barrier contraception for 90 days after treatment

TREATMENT OF OVERDOSE:

Discontinue product; use hemodialysis; increase hydration

cilostazol (Rx)

(sih-los′tah-zol)

Pletal

Func. class.: Platelet aggregation inhibitor

Chem. class.: Quinolinone derivative

Do not confuse:
Pletal/Plendil

ACTION: Multifactorial effects (antithrombotic, antiplatelet vasodilation)

USES: Intermittent claudication associated with PVD

Unlabeled uses: Buerger's disease, percutaneous coronary intervention (PCI)

CONTRAINDICATIONS: Hypersensitivity, acute MI, active bleeding conditions, hemostatic conditions

Black Box Warning: CHF

Precautions: Pregnancy (C), breastfeeding, children, geriatric patients, previous hepatic disease, cardiac/renal disease, increased bleeding risk, low platelet count, platelet dysfunction, smoking

DOSAGE AND ROUTES

- **Adult: PO** 100 mg bid taken ≥30 min before or 2 hr after breakfast and dinner or 50 mg bid if using products that inhibit CYP3A4 and CYP2C19; 12 wk of treatment may be needed for beneficial effect

PCI to prevent acute coronary thrombosis/Buerger's disease (unlabeled)

• **Adult: PO** 100 mg bid

Available forms: Tabs 50, 100 mg

Administer:

• Give bid 30 min before or 2 hr after meals with a full glass of water; do not give with grapefruit juice

SIDE EFFECTS

CNS: *Dizziness, headache*

CV: *Palpitations, tachycardia,* nodal dysrhythmia, postural hypotension

EENT: Blindness, diplopia, ear pain, tinnitus, retinal hemorrhage

GI: *Nausea,* vomiting, *diarrhea,* GI discomfort, colitis, cholelithiasis, ulcer, esophagitis, gastritis, anorexia, *flatulence, dyspepsia*

GU: Cystitis, frequency, vaginitis, **vaginal hemorrhage**, hematuria

HEMA: **Bleeding (epistaxis, hematuria, retinal hemorrhage, GI bleeding), thrombocytopenia**, anemia, **polycythemia, aplastic anemia**

INTEG: *Rash,* urticaria, dry skin, **Stevens-Johnson syndrome**

MISC: *Back pain, headache, infection, myalgia, peripheral edema,* chills, fever, malaise, diabetes mellitus

RESP: *Cough, pharyngitis, rhinitis,* asthma, pneumonia

PHARMACOKINETICS

95%-98% protein binding; metabolism: hepatic extensively by CYP3A4, 2C19 enzymes (active metabolite); excreted in urine (74%), feces (20%); half-life 11-13 hr

INTERACTIONS

Increase: bleeding tendencies—anticoagulants, NSAIDs, thrombolytics, abciximab, eptifibatide, tirofiban, ticlopidine

Increase: cilostazol levels—CYP3A4, CYP2C19 inhibitors; diltiazem, erythromycin, clarithromycin, verapamil, protease inhibitors, omeprazole; exercise caution when coadministering with fluvoxaMINE, FLUoxetine, ketoconazole, isoniazid, gemfibrozil, omeprazole, itraconazole, voriconazole, fluconazole; reduce dose to 50 mg bid

Decrease: cilostazol levels—CYP3A4 inducers

Drug/Herb

Decrease: action—chamomile, coenzyme Q10, feverfew, garlic, ginger, ginko biloba, flax, goldenseal, St. John's wort

Drug/Food

• Do not use with grapefruit juice, toxicity may occur

Increase: cilostazol action—fatty meal, avoid giving with food

NURSING CONSIDERATIONS

Assess:

Black Box Warning: For underlying CV disease because CV risk is great; for CV lesions with repeated oral administration; do not administer to patients with CHF of any severity; for severe headache, signs of toxicity

• Blood studies: CBC q2wk, Hct, Hgb, PT

Evaluate:

• Therapeutic response: improved walking distance, duration; decreased pain

Teach patient/family:

• To avoid hazardous activities until effect is known

• To report any unusual bleeding

• To report side effects such as diarrhea, skin rashes, subcutaneous bleeding

• That effects may take 2-4 wk; treatment of up to 12 wk may be required for necessary effect

• That reading the patient package insert is necessary

• That it is best to discontinue tobacco use, not to use grapefruit juice

• That there are many drug and herb interactions; obtain approval from prescriber before use

cimetidine (OTC, Rx)

(sye-met'i-deen)

Acid Reducer, Equaline Acid Reducer, Good Sense Heartburn Relief, Nu-Cimet ✦, Tagamet, Tagamet HB

Func. class.: H_2-histamine receptor antagonist

Chem. class.: Imidazole derivative

ACTION:
Inhibits histamine at H_2-receptor site in the gastric parietal cells, which inhibits gastric acid secretion

USES:
Short-term treatment of duodenal and gastric ulcers and maintenance; management of GERD (PO) and Zollinger-Ellison syndrome; prevention of upper GI bleeding; prevent, relieve heartburn, acid indigestion, upper GI bleeding

Unlabeled uses: Prevention of aspiration pneumonitis, stress ulcers, angioedema, molluscum contagiosum, NSAID-induced ulcer prophylaxis, verruca vulgaris

CONTRAINDICATIONS:
Hypersensitivity

Precautions: Pregnancy (B), breastfeeding, children <16 yr, geriatric patients, organic brain syndrome, renal/hepatic disease

DOSAGE AND ROUTES

Short-term treatment of active ulcers

- **Adult/adolescents ≥16 yr: PO** 300 mg qid with meals, at bedtime × 8 wk or 400 mg bid, 800 mg at bedtime; after 8 wk, give bedtime dose only; **IV BOL** 300 mg/20 ml 0.9% NaCl over 1-2 min q6hr; **IV INF** 300 mg/50 ml D_5W over 15-20 min; **IM** 300 mg q6hr, max 2400 mg/day
- **Child: PO** 20-40 mg/kg/day; **IM/IV** 5-10 mg/kg q6-8hr

Prophylaxis of duodenal ulcer

- **Adult and child >16 yr:** 400 mg at bedtime or 300 mg bid

GERD

- **Adult: PO** 800-1600 mg/day in divided doses × up to 12 wk

Hypersecretory conditions (Zollinger-Ellison syndrome)

- **Adult: PO/IM/IV** 300-600 mg q6hr; may increase to 12 g/day if needed; OTC use ≤200 mg daily or bid, max 2×/wk

Upper GI bleeding prophylaxis

- **Adult: IV** 50 mg/hr; lowered in renal disease

Heartburn

- **Adult/child ≥12 yrs: PO** 200 mg Tagamet Hb up to BID, may use before eating, max 400 mg/day, max daily use up to 2 wk

Renal disease

- **Adult: PO/IV** CCr <30 ml/min, 300 mg q12hr

Verruca vulgaris (unlabeled)

- **Child: PO** 30-40 mg/kg/day × 2 mo

Available forms: Tabs 100, 200, 300, 400, 800 mg; liq 200, 300 mg/5 ml; inj 300 mg/2 ml, 300 mg/50 ml 0.9% NaCl

Administer:

PO route

- With meals for prolonged product effect; antacids 1 hr before or 1 hr after cimetidine

IM route

- May give undiluted
- Give at end of dialysis
- Inject deeply in large muscle mass, aspirate

IV route

- After **diluting** 300 mg/20 ml of 0.9% NaCl for inj; give by **direct IV over** ≥5 min; **Intermittent IV INF** may be **diluted** 300 mg/50 ml of D_5W; **run** ≥30 min; or total daily dose (900 mg) diluted in 100-1000 ml D_5W given over 24 hr

Continuous IV INF

- Storage of diluted sol at room temperature up to 48 hr

Y-site compatibilities: Acyclovir, alfentanil, amifostine, amikacin, aminocaproic acid, aminophylline, amphotericin

B lipid complex/liposome, anakinra, anidulafungin, ascorbic acid injection, atenolol, atracurium, atropine, aztreonam, benztropine, bivalirudin, bleomycin, bumetanide, buprenorphine, butorphanol, calcium chloride/gluconate, CARBOplatin, caspofungin, cefamandole, ceFAZolin, cefmetazole, cefonicid, cefotaxime, cefoTEtan, cefOXitin, cefTAZidime, ceftizoxime, cefTRIAXone, cefuroxime, cephalothin, cephapirin, chlorproMAZINE, cisatracurium, CISplatin, cladribine, clarithromycin, clindamycin, codeine, cyanocobalamin, cyclophosphamide, cycloSPORINE, cytarabine, DACTINomycin, DAPTOmycin, dexamethasone, dexmedetomidine, digoxin, diltiazem, diphenhydrAMINE, DOBUTamine, DOCEtaxel, DOPamine, doripenem, doxacurium, doxapram, DOXOrubicin, DOXOrubicin liposome, enalaprilat, ePHEDrine, EPINEPHrine, epirubicin, epoetin alfa, eptifibatide, ertapenem, erythromycin, esmolol, etoposide, famotidine, fenoldopam, fentaNYL, filgrastim, fluconazole, fludarabine, fluorouracil, folic acid, foscarnet, gallium, gatifloxacin, gemcitabine, gentamicin, gycopyrrolate, granisetron, heparin, hydrocortisone, HYDROmorphone, hydrOXYzine, IDArubicin, ifosfamide, imipenem-cilastatin, irinotecan, isoproterenol, ketorolac, labetalol, levofloxacin, lidocaine, linezolid, LORazepam, LR, magnesium sulfate, mannitol, mechlorethamine, melphalan, meperidine, metaraminol, meropenem, methicillin, methotrexate, methoxamine, methyldopate, methylPREDNISolone, metoclopramide, metoprolol, metroNIDAZOLE, mezlocillin, miconazole, midazolam, milrinone, minocycline, mitoXANtrone, morphine, moxalactum, multiple vitamin injection, mycophenolate, nafcillin, nalbuphine, naloxone, nesiritide, netilmicin, niCARdipine, nitroglycerin, nitroprusside, norepinephrine, octreotide, ondansetron, oxacillin, oxaliplatin, oxytocin, PACLitaxel, palonosetron, pamidronate, pancuronium, pantoprazole, papaverine, PEMEtrexed, penicillin G sodium/potassium, pentamidine, pentazocine, phenylephrine, phytonadione, pipercillin, piperacillin/tazobactam, polymyxin B, potassium chloride, procainamide, prochlorperazine, promethazine, propofol, propranolol, protamine, pyridoxine, quiNIDine, quinupristin-dalfopristin, ranitidine, remifentanil, Ringers' ritodrine, riTUXimab, rocuronium, sargramostim, sodium acetate/bicarbonate, succinylcholine, SUFentanil, tacrolimus, teniposide, theophylline, thiamine, thiotepa, ticarcillin, ticarcillin-clavulanate, tigecycline, tirofiban, TNA, tobramycin, tolazoline, topotecan, TPN, trastuzumab, trimetaphan, urokinase, vancomycin, vasopressin, vecuronium, verapamil, vinCRIStine, vinorelbine, voriconazole, zidovudine, zoledronic acid

SIDE EFFECTS

CNS: *Confusion, headache,* depression, dizziness, anxiety, weakness, psychosis, tremors, **seizures**

CV: Bradycardia, tachycardia, **dysrhythmias**

GI: *Diarrhea,* abdominal cramps, **paralytic ileus, jaundice**

GU: Gynecomastia, galactorrhea, impotence, increase in BUN, creatinine

HEMA: **Agranulocytosis, thrombocytopenia, neutropenia, aplastic anemia, increase in PT**

INTEG: Urticaria, rash, alopecia, sweating, flushing, **exfoliative dermatitis**

RESP: **Pneumonia**

PHARMACOKINETICS

Half-life $1\frac{1}{2}$-2 hr; 30%-40% metabolized by liver, excreted in urine (unchanged), crosses placenta, enters breast milk

PO: Onset 30 min, peak 45-90 min; duration 4-5 hr, well absorbed

IM/IV: Onset 10 min, peak ½ hr, duration 4-5 hr, well absorbed (IM)

INTERACTIONS

Increase: toxicity due to CYP450 pathway—benzodiazepines, β-blockers, calcium channel blockers, carBAMazepine, chloroquine, lidocaine, metroNIDAZOLE, moricizine, phenytoin, quiNIDine, quiNINE, sulfonylureas, theophylline, tricyclics, valproic acid, warfarin

Increase: bone marrow suppression-carmustine

Decrease: absorption of cimetidine—antacids, sucralfate

Decrease: absorption—ketoconazole, itraconazole

Drug/Lab Test

Increase: alk phos, AST, creatinine, prolactin

False positive: gastroccult, hemoccult tests

False negative: TB skin tests

NURSING CONSIDERATIONS

Assess:

- **Ulcer symptoms:** epigastric pain, duration, intensity; aggravating, ameliorating factors

Evaluate:

- Therapeutic response: decreased pain in abdomen; healing of ulcers; absence of gastroesophageal reflux, gastric pH of 5

Teach patient/family:

- That gynecomastia, impotence may occur, are reversible
- To avoid driving, other hazardous activities until stabilized on this medication; drowsiness or dizziness may occur
- To avoid OTC preparations: aspirin; cough, cold preparations; condition may worsen
- That smoking decreases effectiveness of product
- That product must be taken exactly as prescribed and continued for prescribed time to be effective; not to double dose
- To report bruising, fatigue, malaise; blood dyscrasias may occur
- To report diarrhea, black tarry stools, sore throat, rash to prescriber

cinacalcet (Rx)

(sin-a-kal'set)

Sensipar

Func. class.: Calcium receptor agonist

Chem. class.: Polypeptide hormone

ACTION: Directly lowers PTH levels by increasing sensitivity of calcium-sensing receptors to extracellular calcium

USES: Hypercalcemia with parathyroid carcinoma, secondary hyperparathyroidism with chronic kidney disease for patient on dialysis, primary hyperparathyroidism

CONTRAINDICATIONS: Hypersensitivity, hypocalcemia

Precautions: Pregnancy (C), breastfeeding, children, seizure disorders, hepatic disease

DOSAGE AND ROUTES

Parathyroid carcinoma

- **Adult: PO** 30 mg bid, titrate q2-4wk, with sequential doses of 30 mg bid, 60 mg bid, 90 mg bid, 90 mg tid-qid to normalize calcium levels

Secondary hyperparathyroidism

- **Adult: PO** 30 mg/day, titrate no more frequently than q2-4wk with sequential doses of 30, 60, 90, 120, 180 mg/day

Available forms: Tabs 30, 60, 90 mg

Administer:

- Swallow tabs whole; do not break, crush, or chew; use with food or right after a meal
- Can be used alone or in combination with vit D sterols, phosphate binders
- Storage at <77° F (25° C)

Secondary hyperthyroidism

- Titrate q2-4wk to target iPTH consistent with National Kidney Foundation–Kidney Disease Outcomes Quality Initiative (NKF-K/DOQI) for chronic kidney disease patient on dialysis of 150-300 pg/

ml; if iPTH <150-300 pg/ml, reduce dose of cinacalcet and/or vit D sterols or discontinue treatment

SIDE EFFECTS

CNS: Dizziness, asthenia, seizures, tetany, hallucinations, depression
CV: Hypertension, dysrhythmia exacerbation
GI: Nausea, diarrhea, vomiting, anorexia
MISC: Access infection, noncardiac chest pain, hypocalcemia
MS: Myalgia

PHARMACOKINETICS

93%-97% bound to plasma; proteins metabolized by CYP3A4, 2D6, 1A2; half-life 30-40 hr; renal excretion of metabolites (80% renal, 15% in feces)

INTERACTIONS

Increase: cinacalcet levels: CYP3A4 inhibitors (ketoconazole, erythromycin, itraconazole)
Increase: levels of CYP2D6 inhibitors (flecainide, vinBLAStine, thioridazine, tricyclics)
Drug/Food
Increase: action by high-fat meal

NURSING CONSIDERATIONS

Assess:
- **Hypocalcemia:** cramping, seizures, tetany, myalgia, paresthesia; calcium, phosphorous within 1 wk and iPTH 1-4 wk after initiation or dosage adjustment when maintenance established; measure calcium, phosphorus monthly; iPTH q1-3mo, target range 150-300 pg/ml for iPTH level; biochemical markers of bone formation/resorption; radiologic evidence of fracture; serum testosterone
- **Renal disease (without dialysis): these patients should not receive treatment with this product, high risk of hypocalcemia**
- If calcium <8.4 mg/dl, do not start therapy

Evaluate:
- Therapeutic response: calcium levels 9-10 mg/dl, decreasing symptoms of hypercalcemia

Teach patient/family:
- To take with food or shortly after a meal; to take tabs whole
- To report cramping, seizures, muscle pain, tingling, tetany immediately

ciprofloxacin (Rx)

(sip-ro-floks′a-sin)

Cipro, Cipro XR

Func. class.: Antiinfective—broad spectrum
Chem. class.: Fluoroquinolone

Do not confuse:
ciprofloxacin/cephalexin

ACTION: Interferes with conversion of intermediate DNA fragments into high-molecular-weight DNA in bacteria; DNA gyrase inhibitor

USES: Infection caused by susceptible *Escherichia coli, Enterobacter cloacae, Proteus mirabilis, Klebsiella pneumoniae, Proteus vulgaris, Citrobacter freundii, Serratia marcescens, Pseudomonas aeruginosa, Staphylococcus aureus, Staphylococcus epidermidis, Enterobacter, Campylobacter jejuni, Salmonella;* chronic bacterial prostatitis, acute sinusitis, postexposure inhalation anthrax, infectious diarrhea, typhoid fever, complicated intraabdominal infections, nosocomial pneumonia, urinary tract infections
Unlabeled uses: *Acinetobacter/woffii, Aeromonas hydrophila,* brucellosis, *Burkholderia, pseudomallei,* chancroid, cholera, dental infection, *Edwardsiella tarda,* endocarditis, *Enterobacter aerogenes,* granuloma inguinale, *Klebsiella oxytoca,* Legionnaire's disease, melioidosis, meningococcal infection prophylaxis, *Pasteurella multocida,* PID, periodontitis, pharyngitis, *Salmonella sp., Stenotrophomonas malto-*

philia, tularemia, *Vibrio cholerae/ parahaemolyticus/vulnificus, Yersinia enterocolitica*

CONTRAINDICATIONS:

Hypersensitivity to quinolones

Precautions: Pregnancy (C), breastfeeding, children, geriatric patients, renal disease, seizure disorder, stroke, CV disease, hepatic disease, QT prolongation, hypokalemia

Black Box Warning: Tendon pain/ rupture, tendinitis, myasthenia gravis

DOSAGE AND ROUTES

Uncomplicated urinary tract infections

- **Adult: PO** 250 mg q12hr × 3 days or XL 500 mg q24hr × 3 days

Complicated/severe urinary tract infections

- **Adult: PO** 500 mg q12hr or XL 1000 mg q24hr × 7-14 days; **IV** 400 mg q12hr

Pyelonephritis, acute uncomplicated/UTI

- **Adult: PO** 250 mg q12hr; 500 mg q12hr (severe) × 7-14 days q4-6wk

Respiratory, bone, skin, joint infections (mild-moderate)

- **Adult: PO** 500-750 mg q12hr × 7-14 days; **IV** 400 mg q12hr

Nosocomial pneumonia

- **Adult: IV** 400 mg q8hr × 10-14 days

Intraabdominal infections, complicated

- **Adult: PO** 500 mg q12hr × 7-14 days; **IV** 400 mg q12hr × 7-14 days, usually given with metroNIDAZOLE

Acute sinusitis, mild/moderate

- **Adult: PO** 500 mg q12hr × 10 days; **IV** 400 mg q12hr × 10 days

Inhalational anthrax (postexposure)

- **Adult: PO** 500 mg q12hr × 60 days; **IV** 400 mg q12hr × 60 days
- **Child: PO** 15 mg/kg/dose q12hr × 60 days, max 500 mg/dose; **IV** 10 mg/kg q12hr, max 400 mg/dose

Infectious diarrhea

- **Adult: PO** 500-750 mg q12hr × 5-7 days

Chronic bacterial prostatis

- **Adult: PO** 500 mg q12hr × 28 days; **IV** 400 mg q12hr × 28 days

Renal disease

- **Adult:** CCr 30-50 ml/min, **PO** 250-500 mg q12hr; CCr 5-29 ml/min, **PO** 250-500 mg q18hr; **IV** 200-400 mg q18-24hr

Available forms: Tabs 250, 500, 750 mg; ext rel tabs (XR) 500, 1000 mg; inj 10 mg/ml, 200 mg/20 ml, 400 mg/40 ml, 200 mg/100 ml D_5, 400 mg/200 ml D_5; oral susp 250 mg, 500 mg/5 ml; powder for oral susp 5%, 10%

Administer:

- Obtain C&S before use
- Use caution when giving with antidysrhythmics IA, III

PO route

- Do not break, crush, chew XR (ext rel) product, use adequate fluids to prevent crystalluria
- 2 hr before or 6 hr after antacids, zinc, iron, calcium
- Do not give oral susp by GI tube

IV route

- Over 1 hr as an inf, comes in premixed plastic inf container or diluted 20- or 40-ml vial to a final conc of 0.5-2 mg/ml of NS or D_5W; give through Y-tube or 3-way stopcock, diluted vials can be stored for 14 days at room temperature or refrigerator; do not freeze

Y-site compatibilities: Amifostine, anakinra, anidulafungin, atenolol, aztreonam, bivalirudin, bleomycin, calcium gluconate, CARBOplatin, caspofungin, cefTAZidime, cisatracurium, CISplatin, clarithromycin, codeine, cytarabine, DACTINomycin, DAPTOmycin, dexmedetomidine, digoxin, diltiazem, diphenhydrAMINE, DOBUTamine, DOCEtaxel, doripenem, DOPamine, doxacurium, DOXOrubicin, epirubicin, eptifibatide, ertapenem, etoposide, fenoldopam, fludarabine, gallium, gemcitabine, gentamicin, granisetron, HYDROmorphone, hydrOXYzine, IDArubicin, ifosfamide, irinotecan, lidocaine, linezolid, LORazepam, LR, mechlorethamine, meperidine, methotrexate, metoclopramide, metroNIDAZOLE, midazolam, midodrine, milrinone, mitoXANtrone, mycophenolate, nesiritide, octreotide, ondansetron, oxaliplatin, oxytocin, PACLitaxel, palonosetron, pamidronate, pancuronium, piperacillin, potassium acetate/chloride, promethazine, ranitidine, remifentanil, rocuronium, sodium chloride, tacrolimus, teniposide, thiotepa, tigecycline, tirofiban, TNA, tobramycin, trastuzumab, vasopressin, vecuronium, verapamil, vinCRIStine, vinorelbine, voriconazole

SIDE EFFECTS

CNS: *Headache*, dizziness, fatigue, insomnia, depression, *restlessness*, **seizures**, confusion, hallucinations

GI: *Nausea, diarrhea*, increased ALT/AST, dry mouth, flatulence, heartburn, *vomiting*, oral candidiasis, dysphagia, **pseudomembranous colitis**, abdominal pain, **pancreatitis**

GU: Crystalluria, interstitial neuritis

HEMA: **Bone marrow depression, agranulocytosis**, eosinophils

INTEG: *Rash*, pruritus, urticaria, photosensitivity, flushing, fever, chills, **toxic epidermal necrolysis**, injection site reactions

MISC: **Anaphylaxis, Stevens-Johnson syndrome**, visual impairment, **QT prolongation, pseudotumor cerebri**

MS: Tremor, arthralgia, tendinitis, **tendon rupture**

PHARMACOKINETICS

PO: Peak 1-2 hr; half-life 1-2 hr; excreted in urine as active product, metabolites 35%-40%, 20%-40% protein binding

INTERACTIONS

> **Black Box Warning: Increase:** tendonitis, tendon rupture—Corticosteroids

Increase: nephrotoxicity—cycloSPORINE

Increase: ciprofloxacin levels—probenecid; monitor for toxicity

Increase: levels of theophylline, warfarin, monitor blood levels, reduce dose

Increase: levels of CYP1A2 inhibitors

Increase: QT prolongation—astemizole, droperidol, class IA/III antidysrhythmics, tricyclics, tetracyclines, local anesthetics, phenothiazines, haloperidol, risperiDONE, sertindole, ziprasidone, alfuzosin, arsenic trioxide, β-agonists, chloroquine, cloZAPine, cyclobenzapine, dasatinib, dolasetron, droperidol, flecainide, halogenated anesthetics, lapatinib, levomethadyl, macrolides, methadone, octreotide, ondansetron, paliperidone, palonosetron, pentamidine, propafenone, ranolazine, SUNItinib, tacrolimus, terfenadine, vardenafil, vorinostat; less likely than other quinolones

Decrease: ciprofloxacin absorption—antacids that contain magnesium, aluminum; zinc, iron, sucralfate, enteral feedings, calcium

Drug/Food

Increase: effect of caffeine

Decrease: absorption—dairy products, food

Drug/Lab Test

Increase: AST, ALT, BUN, creatinine, LDH, bilirubin, alk phos, glucose, proteinuria, albuminuria

Decrease: WBC, glucose

NURSING CONSIDERATIONS

Assess:

- **Infection:** WBC, temperature before treatment, periodically
- **QT Prolongation:** Monitor for changes in QTc if taking other products that increase QT

> **Black Box Warning: Myasthenia gravis:** Avoid use in these patients, increases muscle weakness

- **CNS symptoms:** headache, dizziness, fatigue, insomnia, depression, seizures
- Renal, hepatic studies: BUN, creatinine, AST, ALT
- I&O ratio, urine pH <5.5 is ideal

Black Box Warning: **Tendonitis, tendon rupture:** discontinue at first sign of tendon pain, inflammation; increased in those >60 yr, those taking corticosteroids, organ transplants

Anaphylaxis: fever, flushing, rash, urticaria, pruritus, dyspnea

- **Pseudomotor Cerebri:** may occur at excessive doses
- Limited intake of alkaline foods, products: milk, dairy products, alkaline antacids, sodium bicarbonate; caffeine intake if excessive cardiac or CNS stimulation
- Increase fluids to 3 L/day to avoid crystallization in kidneys

Evaluate:

- Therapeutic response: decreased pain, frequency, urgency, C&S; absence of infection

Teach patient/family:

- Not to take any products that contain magnesium, calcium (such as antacids), iron, aluminum with this product or 2 hr before, 6 hr after product; to drink fluids to prevent crystals in urine

Black Box Warning: To report tendon pain, chest pain, palpitations

- To ambulate, perform activities with assistance if dizziness occurs
- To complete full course of product therapy; not to double or miss doses
- To notify prescriber if rash occurs, discontinue product
- To notify prescriber if pregnancy is planned or suspected pregnancy (C); not to breastfeed
- To frequently rinse mouth; use sugarless candy, gum for dry mouth
- To contact prescriber if taking theophylline
- Extended release and regular release are not interchangeable
- Not to add or stop products without prescriber's approval
- To use calibrated measuring device for suspension

ciprofloxacin (ophthalmic)

(sip-roe-flox′a-sin)

Ciloxan

Func. class.: ophthalmic anti-infective

Chem. class.: fluoroquinolone

Do not confuse:
ciprofloxacin/gatifloxacin/levofloxacin/moxifloxacin/ofloxacin

ACTION: Inhibits DNA gyrase, thereby decreasing bacterial replication

USES: Corneal ulcers, bacterial conjunctivitis

CONTRAINDICATIONS: Hypersensitivity to this product or fluoroquinolones

Precautions: Pregnancy (C), breastfeeding

DOSAGE AND ROUTES

Bacterial conjunctivitis

- **Adult/adolescent/child ≥1 yr:** Ophthalmic (sol): 1-2 drops in affected eye(s) every 2 hr while awake × 2 days, then every 4 hr while awake for the next 5 days
- **Adult/adolescent/child ≥2 yr:** Ophthalmic (ointment): 1/2-inch ribbon to conjunctival sac tid × 2 days, then 1/2 inch bid for next 5 days

Ophthalmic infection associated with corneal ulcer

- **Adult/adolescent/child ≥1 yr:** Ophthalmic (solution): 2 drops in affected eye(s) every 15 min × 6 hr, then every 30 min for the remainder of the first day; for the second day, 2 drops every hr; for days 3-14, 2 drops every 4 hr

Available forms: Ophthalmic ointment, solution 0.3%

Administer

- Commercially available ophthalmic solutions are not for injection subconjunctivally or into the anterior chamber of the eye

Ophthalmic route

- Apply topically to the eye, taking care to avoid contamination
- Do not touch the tip of the dropper to the eye, fingertips, or other surface
- Apply pressure to lacrimal sac for 1 min after instillation
- Avoid wearing contact lens(es) while treating eye infection

SIDE EFFECTS

EENT: Burning, hypersensitivity, pruritus, precipitate in those with corneal ulcers

NURSING CONSIDERATIONS

Assess:

Allergic reaction:

- Assess for hypersensitivity, discontinue product

Evaluate:

- Decreased ophthalmic infection

Teach patient/family:

Ophthalmic route:

- To apply topically to the eye, taking care to avoid contamination; for ophthalmic use only
- Not to touch the tip of the dropper to the eye, fingertips, or other surface
- To apply pressure to lacrimal sac for 1 min after instillation
- To avoid wearing contact lens(es) while treating eye infection

cisatracurium

(sis-ah-trah-kyoo′ree-um)

Nimbex

Func. class.: Skeletal muscle relaxant

Chem. class.: Nondepolarizing neuromuscular blocker

ACTION: Antagonizes acetylcholine by binding to cholinergic receptors on the motor end plate, resulting in neuromuscular blockade

USES: To maintain neuromuscular blockade during mechanical ventilation and as an adjunct to general anesthesia

CONTRAINDICATIONS: Hypersensitivity

Precautions: Pregnancy (B), breastfeeding, children, benzyl alcohol hypersensitivity, electrolyte imbalances, long-term use in ICU, trauma, or burns

DOSAGE AND ROUTES

For neuromuscular blockade, as an adjunct to general anesthesia, to facilitate endotracheal intubation, and to provide skeletal muscle relaxation during surgery or mechanical ventilation in the ICU

Endotracheal intubation

- **Adult/adolescent (healthy): IV** 0.15-0.2 mg/kg, one time
- **Adult with myasthenia gravis: IV** use peripheral nerve stimulator monitoring and an initial dose ≤0.02 mg/kg
- **Child 2-12 yr: IV** 0.1-0.15 mg/kg over 5-10 sec during either halothane or opioid anesthesia
- **Infant/child ≤23 mo: IV** 0.15 mg/kg over 5-10 sec during either halothane or opioid anesthesia

To maintain neuromuscular blockade during prolonged surgical procedures:

- **Adult/adolescent/child ≥2 yr (healthy): IV** Maintenance dose 0.03 mg/kg; maintenance dosing is generally required 40-50 min after an initial dose of 0.15 mg/kg IV or 50-60 min after an initial dose of 0.2 mg/kg; the need for maintenance doses should be determined by clinical criteria

Available forms: Injection solution 2, 10 mg/ml

Administer

IV route

- Visually inspect for particulate matter and discoloration before use
- Only experienced clinicians, familiar with the use of neuromuscular blocking drugs, should administer or supervise the use this product

- Use by rapid IV injection or by continuous IV inf

IV INJ route

- Inject IV over 5-10 sec

Continuous IV INF route

- Dilute with NS, D_5W, or D_5NS (0.1-0.4 mg/ml); adjust the rate of inf according to peripheral nerve stimulation
- The amount of inf sol required per minute depends on the conc of cisatracurium in the inf sol, the desired dose of cisatracurium, and the patient's weight
- Store Nimbex injection diluted to 0.1 mg/ml either under refrigeration or at room temperature for 24 hr; dilutions to 0.1 mg/ml or 0.2 mg/ml in D_5W/LR injection may be stored under refrigeration for 24 hr

SIDE EFFECTS

CV: Bradycardia, flushing, hypotension
RESP: Apnea, bronchospasm, prolonged neuromuscular block

PHARMACOKINETICS

Onset 2 min, peak 3-5 min, duration 25-44 min, half-life 22-30 min

INTERACTIONS

Increase: neuromuscular blockade—aminoglycosides, clindamycin, lithium, local anesthetics, magnesium salts, colistin, colistimethate, procainamide, quiNIDine, tetracyclines, bacitracin, capreomycin, polymyxin B, vancomycin; amphotericin B, CISplatin, corticosteroids, loop/thiazide diuretics (if hypokalemia is present)
Decrease: neuromuscular blockade—carBAMazepine, phenytoin

NURSING CONSIDERATIONS

Assess:

Neuromuscular function:

- Use nerve stimulator to monitor neuromuscular function; if no response, stop until response; not to be used for rapid-sequence endotracheal intubation

Electrolyte:

- Electrolytes and acid-base balance may be altered

Malignant hyperthermia: Assess for malignant hyperthermia; those with a family history of this condition should not receive this product or it should be used cautiously

Evaluate:

- Maintenance of neuromuscular blockade

Teach patient/family:

- Reason for product and expected results

⚠ HIGH ALERT

CISplatin (Rx)

(sis′pla-tin)

Func. class.: Antineoplastic alkylating agent

Chem. class.: Platinum complex

Do not confuse:
CISplatin/CARBOplatin

ACTION:

Alkylates DNA, RNA; inhibits enzymes that allow for the synthesis of amino acids in proteins; activity not cell-cycle–phase specific

USES:

Advanced bladder cancer; adjunct in metastatic testicular cancer; osteosarcoma; soft-tissue sarcomas; adjunct in metastatic ovarian cancer; head, neck cancer; esophagus, prostate, lung, cervical cancer; lymphoma

Unlabeled uses: Astrocytoma; breast, gastric, head, neck, hepatocellular, lung, penile cancer; carcinoid, desmoid tumor; Hodgkin's disease, malignant glioma, malignant melanoma, neuroblastoma, non-Hodgkin's lymphoma (NHL), osteogenic sarcoma

CONTRAINDICATIONS:

Pregnancy (D), breastfeeding

Black Box Warning: Preexisting hearing impairment, bone marrow suppression, platinum compound hypersensitivity, renal disease/failure

Precautions: Geriatric patients, vaccination

DOSAGE AND ROUTES

Dosage protocols may vary

Metastatic testicular cancer

- **Adult: IV** 20 mg/m²/day × 5 days, repeat q3wk for 2 cycles or more, depending on response

Advanced bladder cancer

- **Adult: IV** 50-70 mg/m² q3-4wk

Metastatic ovarian cancer

- **Adult: IV** 100 mg/m² q4wk or 75-100 mg/m² q3wk with cyclophosphamide

Hodgkin's/non-Hodgkin's lymphoma (unlabeled)

- **Adult and child: IV INF** 100 mg/m² 24 hr continuous inf day 1 of 4-day regimen with cytarabine/dexamethasone q3-4wk

Gastric cancer (unlabeled)

- **Adult: IV** 75 mg/m² on day 1 with DOCEtaxel 75 mg/m² and fluorouracil 750 mg/m² on days 1-5, q21days

Available forms: Inj 0.5 ✱, 1 mg/ml

Administer:

IV route

- Do not use aluminum equipment during any preparation or administration, will form precipitate; do not refrigerate unopened powder or solution; protect from sunlight
- Prepare in biologic cabinet using gown, gloves, mask; do not allow product to come in contact with skin; use soap and water if contact occurs; use cytotoxic handling procedures
- Hydrate patient with 0.9% NaCl over 8-12 hr before treatment
- EPINEPHrine, antihistamines, corticosteroids for hypersensitivity reaction
- Antiemetic 30-60 min before product and prn; allopurinol to maintain uric acid levels, alkalinization of urine; diuretic (furosemide 40 mg IV) or mannitol after INF

Intermittent IV INF route

- **Dilute** 10 mg/10 ml or 50 mg/50 ml sterile water for inj, withdraw prescribed dose; **dilute** ½ dose with 1000 ml D_5 0.2 NaCl, D_5 0.45 NaCl with 37.5 g mannitol; IV INF is **given** over 3-4 hr; use a 0.45-μm filter; total dose 2 L over 6-8 hr; check site for irritation, phlebitis

Continuous IV INF route

- **Give** over 24 hr × 5 days

Additive compatibilities: CARBOplatin, cyclophosphamide, floxuridine, hydrOXYzine, ifosfamide, leucovorin, magnesium sulfate, mannitol, ondansetron, potassium chloride

Solution compatibilities: D_5/0.225% NaCl, D_5/0.45% NaCl, D_5/0.9% NaCl, D_5/0.45% NaCl with mannitol 1.875%, D_5/0.33% NaCl with KCl 20 mEq and mannitol 1.875%, 0.9% NaCl, 0.45% NaCl, 0.3% NaCl, 0.225% NaCl

Syringe compatibilities: Bleomycin, cyclophosphamide, doxapram, droperidol, fluorouracil, furosemide, heparin, leucovorin, methotrexate, metoclopramide, vinBLAStine, vinCRIStine

Y-site compatibilities: Acyclovir, alfentanil, allopurinol, amikacin, aminophylline, amiodarone, ampicillin, ampicillin-sulbactam, anidulafungin, atenolol, atracurium, azithromycin, aztreonam, bivalirudin, bleomycin, bumetanide, buprenorphine, butorphanol, calcium chloride/gluconate, carmustine, caspofungin, ceFAZolin, cefoperazone, cefotaxime, cefoTEtan, cefOXitin, cefTAZidime, ceftizoxime, cefTRIAXone, cefuroxime, chlorproMAZINE, cimetidine, ciprofloxacin, cisatracurium, cladribine, clindamycin, codeine, cyclophosphamide, cycloSPORINE, cytarabine, DACTINomycin, DAPTOmycin, DAUNOrubicin, dexamethasone, dexmedetomidine, dexrazoxane, digoxin, diltiazem, diphenhydrAMINE, DOBUTamine, DOCEtaxel, DOPamine, doripenem, doxacurium, DOXOrubicin, DOXOrubicin liposomal, doxycycline, droperidol, enalaprilat, ePHEDrine, EPINEPHrine, epirubicin, ertapenem, erythromycin, esmolol, etoposide, famotidine, fenoldopam, fentaNYL, filgrastim, fluconazole, fludarabine, fluorouracil, foscarnet, fosphenytoin, furosemide, ganciclovir, gatifloxacin, gemcitabine, gentamicin, glycopyrrolate, granisetron, haloperidol, heparin, hydrocortisone, HYDROmorphone, IDArubicin, ifosfamide, imipenem-cilastatin, inamrinone, indomethacin, irinotecan, isoproterenol, ketorolac, labetalol, leucovorin, levofloxacin, levorphanol, lidocaine, linezolid, LORazepam, magnesium sulfate, mannitol, melphalan, meperidine, meropenem, methohexital, methotrexate, methylPREDNISolone, metoclopramide, metoprolol, metroNIDAZOLE, midazolam, milrinone, minocycline, mitoMYcin, mitoXANtrone, mivacurium, nafcillin, naloxone, nesiritide, niCARdipine, nitroglycerin, nitroprusside, norepinephrine, octreotide, ofloxacin, ondansetron, oxaliplatin, PACLitaxel, palonosetron, pamidronate, pancuronium, PEMEtrexed, pentamidine, pentazocine, PENTobarbital, PHENobarbital, phenylephrine, phenytoin, piperacillin, polymyxin B, potassium chloride/phosphates, procainamide, prochlorperazine, promethazine, propofol, propranolol, quiNIDine, quinupristin-dalfopristin, ranitidine, remifentanil, riTUXimab, sargramostim, sodium acetate/bicarbonate/phosphates, succinylcholine, SUFentanil, sulfamethoxazole-trimethoprim, tacrolimus, teniposide, theophylline, thiopental, ticarcillin, ticarcillin-clavulanate, tigecycline, tirofiban, TNA, tobramycin, topotecan, trastuzumab, vancomycin, vasopressin, vecuronium, verapamil, vinBLAStine, vinCRIStine, vinorelbine, voriconazole, zidovudine, zoledronic acid

SIDE EFFECTS

CNS: Seizures, *peripheral neuropathy*
CV: Cardiac abnormalities
EENT: *Tinnitus, hearing loss, vestibular toxicity,* blurred vision, altered color perception
GI: *Severe nausea, vomiting, diarrhea, weight loss*
GU: Renal tubular damage, *renal insufficiency,* impotence, sterility, amenorrhea, gynecomastia, hyperuremia
HEMA: Thrombocytopenia, leukopenia, pancytopenia
INTEG: *Alopecia,* dermatitis
META: *Hypomagnesemia,* hypocalcemia, hypokalemia, hypophosphatemia
RESP: Fibrosis
SYST: Anaphylaxis

PHARMACOKINETICS

Absorption complete, metabolized in liver, excreted in urine, half-life 20 min first phase, 48-70 min second phase, 24 hr terminal phase, accumulates in body tissues for several months, enters breast milk

INTERACTIONS

Increase: bleeding risk—aspirin, NSAIDs, alcohol
Increase: ototoxicity—bumetanide, ethacrynic acid, furosemide
Increase: myelosuppression—myelosuppressive agents, radiation
Increase: nephrotoxicity—aminoglycosides, loop diuretics, salicylates
Decrease: effects of phenytoin
Decrease: antibody response—live virus vaccines

Drug/Lab Test
Increase: uric acid, BUN, creatinine
Decrease: CCr, calcium, phosphate, potassium, magnesium
Positive: Coombs' test

NURSING CONSIDERATIONS

Assess:

Black Box Warning: **Bone marrow depression:** CBC, differential, platelet count weekly; withhold product if WBC is <4000 or platelet count is <100,000; notify prescriber of results

Black Box Warning: **Renal toxicity:** BUN, creatinine, serum uric acid, urine CCr before, electrolytes during therapy; dose should not be given if BUN <25 mg/dl; creatinine <1.5 mg/dl; I&O ratio; report fall in urine output of <30 ml/hr

⚠ **Anaphylaxis:** wheezing, tachycardia, facial swelling, fainting; discontinue product, report to prescriber; resuscitation equipment should be nearby, may occur within minutes; often EPINEPHrine, corticosteroids, antihistamines may alleviate symptoms

- Monitor temperature q4hr; may indicate beginning infection
- Hepatic studies before, during therapy (bilirubin, AST, ALT, LDH) as needed

⚠ **Bleeding:** hematuria, guaiac, bruising, petechiae, mucosa, or orifices q8hr; obtain prescription for viscous lidocaine (Xylocaine)

Black Box Warning: **Ototoxicity:** more common in genetic variants TPMT 3B and 3C in children; use audiometric testing baseline and before each dose

- Effects of alopecia on body image; discuss feelings about body changes
- Comprehensive oral hygiene
- All medications PO, if possible; avoid IM inj when platelets <100,000/mm^3
- **Hyperuricemia:** Increase fluid intake to 2-3 L/day to prevent urate deposits, calculi formation; promote elimination of product, usually occurs between 3-5 days after a dose, may use allopurinol

Evaluate:

- Therapeutic response: decreased tumor size, spread of malignancy

Teach patient/family:

- To report signs of infection: increased temperature, sore throat, flulike symptoms
- To report signs of anemia: fatigue, headache, faintness, SOB, irritability
- To report bleeding, bruising, petechiae; to avoid use of razors, commercial mouthwash; to avoid aspirin, ibuprofen, NSAIDs, alcohol; may cause GI bleeding
- That impotence or amenorrhea can occur but is reversible after discontinuing treatment

Black Box Warning: **Ototoxicity:** to report loss of hearing, ringing or roaring in the ears

- To maintain adequate fluids; to report decreased urine output, flank pain
- That hair may be lost during treatment; a wig or hairpiece may make patient feel better; new hair may be different in color, texture
- To report numbness, tingling in face or extremities; poor hearing; joint pain, swelling
- Not to receive vaccinations during treatment
- To use contraception during treatment and for 4 mo after (pregnancy [D]); product may cause infertility; to avoid breastfeeding

citalopram (Rx)

(sigh-tal′oh-pram)

CeleXA

Func. class.: Antidepressant

Chem. class.: Selective serotonin reuptake inhibitor (SSRI)

Do not confuse:
CeleXA/CeleBREX/Cerebyx/Cerebra/Zyprexa

ACTION: Inhibits CNS neuron uptake of serotonin but not of norepinephrine; weak inhibitor of CYP450 enzyme system, thus making it more appealing than other products

USES: Major depressive disorder
Unlabeled uses: Premenstrual disorders, panic disorder, social phobia, obsessive-compulsive disorder in adolescents; anxiety, hot flashes, menopause; adjunct in schizophrenia, PTSD

CONTRAINDICATIONS: Hypersensitivity
Precautions: Pregnancy (C), breastfeeding, geriatric patients, renal/hepatic disease, seizure disorder, hypersensitivity to escitalopram, bradycardia, recent MI

Black Box Warning: Children, suicidal ideation

DOSAGE AND ROUTES

Depression

- **Adult: PO** 20 mg/day AM or PM, may increase if needed to 40 mg/day after 1 wk; maintenance: after 6-8 wk of initial treatment, continue for 24 wk (32 wk total), reevaluate long-term usefulness (max 40 mg/day)

Hepatic dose/geriatric

- **Adult: PO** 20 mg/day

Panic disorder (unlabeled)

- **Adult: PO** 20-40 mg/day

Premenstrual dysphoria/social phobia (unlabeled)

- **Adult: PO** 10-30 mg/day, used intermittently in premenstrual dysphoria

Available forms: Tabs 10, 20, 40 mg; oral sol 10 mg/5 ml

Administer:

- With food or milk for GI symptoms
- Crushed if patient is unable to swallow medication whole
- Dosages at bedtime if oversedation occurs during the day; may take entire dose at bedtime
- Do not give within 14 days of MAOIs
- Store at room temperature; do not freeze

SIDE EFFECTS

CNS: *Headache, nervousness, insomnia, drowsiness, anxiety, tremor, dizziness, fatigue, sedation, poor conc, abnormal dreams, agitation,* seizures, apathy, euphoria, hallucinations, delusions, psychosis, suicidal attempts, neuroleptic malignant-like syndrome reactions

CV: *Hot flashes, palpitations,* angina pectoris, hemorrhage, hypertension, tachycardia, 1st-degree AV block, bradycardia, MI, thrombophlebitis, QT prolongation, orthostatic hypotension, torsades de pointes

EENT: Visual changes, ear/eye pain, photophobia, tinnitus

GI: *Nausea, diarrhea, dry mouth, anorexia, dyspepsia, constipation, cramps, vomiting, taste changes, flatulence, decreased appetite*

GU: *Dysmenorrhea, decreased libido, urinary frequency, UTI,* amenorrhea, cystitis, impotence, urine retention

INTEG: *Sweating, rash, pruritus,* acne, alopecia, urticaria, photosensitivity

MS: *Pain,* arthritis, twitching

RESP: *Infection, pharyngitis, nasal congestion, sinus headache, sinusitis, cough, dyspnea, bronchitis,* asthma, hyperventilation, pneumonia

SYST: *Asthenia, viral infection, fever, allergy, chills;* hyponatremia (geriatric patients), serotonin syndrome, neonatal abstinence syndrome

PHARMACOKINETICS

Metabolized in liver by CYP3A4, CYP2C19; excreted in urine; steady state 1 wk; peak 4 hr; half-life 35 hr

INTERACTIONS

⚠ Fatal reactions: do not use with MAOIs

⚠ **Increase:** QTc interval—dofetilide, halofantrine, probucol, pimoside, quinolones, ziprasidone; do not use together

Increase: effect of tricyclics; use cautiously

Increase: serotonin syndrome—serotonin receptor agonists, SSRIs, traMADol, lithium, MAOIs, traZODone, SNRIs (venlafaxine, DULoxetine)

Increase: bleeding risk—NSAIDs, salicylates, thrombolytics, anticoagulants, antiplatelets

Increase: CNS effects—barbiturates, sedative/hypnotics, other CNS depressants

Increase: citalopram levels—macrolides, azole antifungals

Increase: plasma levels of β-blockers

Decrease: citalopram levels—carBAMazepine, cloNIDine

Drug/Herb

⚠ **Increase:** serotonin syndrome—St. John's wort, SAM-e; fatal reaction may occur; do not use concurrently

Increase: CNS stimulation—yohimbe

Drug/Lab Test

Increase: serum bilirubin, blood glucose, alk phos

Decrease: VMA, 5-HIAA

False increase: urinary catecholamines

NURSING CONSIDERATIONS

Assess:

Black Box Warning: Mental status: mood, sensorium, affect, suicidal tendencies, increase in psychiatric symptoms, depression, panic

⚠ Serotonin syndrome: increased heart rate, sweating, dilated pupils, tremors, twitching, hyperthermia, agitation

- B/P lying, standing, pulse q4hr; if systolic B/P drops 20 mm Hg, hold product, notify prescriber; take vital signs q4hr in patients with CV disease
- Weight weekly; appetite may decrease or increase with product
- **Torsades de pointes, QT prolongation:** is dose-dependent, ECG for flattening of T wave, bundle branch, AV block, dysrhythmias in cardiac patients
- Alcohol consumption; if alcohol is consumed, hold dose until AM
- Sexual dysfunction: erectile dysfunction, decreased libido
- Assistance with ambulation during therapy, as drowsiness, dizziness occur
- Safety measures, primarily for geriatric patients
- Sugarless gum, hard candy, frequent sips of water for dry mouth

Evaluate:

- Therapeutic response: decreased depression

Teach patient/family:

- That therapeutic effect may take 4-6 wk; that patient may have increased anxiety 1st 5-7 days of therapy; not to discontinue abruptly
- To use caution when driving, performing other activities that require alertness because of drowsiness, dizziness, blurred vision; to report signs, symptoms of bleeding
- To avoid alcohol, other CNS depressants

Black Box Warning: That suicidal ideas, behavior may occur in children or young adults

- To notify prescriber if pregnant, planning to become pregnant, or breastfeeding

⚠ About the effects of serotonin syndrome: nausea/vomiting, tremors; if symptoms occur, to discontinue immediately, notify prescriber

clarithromycin (Rx)

(klare-ith′row-my-sin)

Biaxin, Biaxin Filmtab, Biaxin XL

Func. class.: Antiinfective

Chem. class.: Macrolide

ACTION: Binds to 50S ribosomal subunits of susceptible bacteria and suppresses protein synthesis

USES: Mild to moderate infections of the upper and lower respiratory tract, uncomplicated skin and skin-structure infections caused by *Streptococcus pneumoniae, Mycoplasma pneumoniae, Legionella pneumophila, Moraxella catarrhalis, Neisseria gonorrhoeae, Corynebacterium diphtheriae, Listeria monocytogenes, Haemophilus influenzae, Streptococcus pyogenes, Staphylococcus aureus, Mycobacterium avium* complex (MAC); complex infection in AIDS patients; *Mycobacterium avium intracellulare, Helicobacter pylori* in combination with omeprazole, *H. parainfluenzae*

Unlabeled uses: Endocarditis prophylaxis, dyspepsia, gastric ulcer, Legionnaire's disease, pertussis, SARS

CONTRAINDICATIONS: Hypersensitivity to this product or macrolide antibiotics, torsades de pointes, QT prolongation

Precautions: Pregnancy (C), breastfeeding, geriatric patients, renal/hepatic disease

DOSAGE AND ROUTES

Acute exacerbation of chronic bronchitis

• **Adult: PO** 250-500 mg q12hr × 7-14 days or 1000 mg/day × 7 days (XL)

Pharyngitis/tonsillitis

• **Adult: PO** 250 mg q12hr × 10 days

Community-acquired pneumonia

• **Adult: PO** 250 mg q12hr × 7-14 days or 1000 mg/day × 7 days (XL)

MAC prophylaxis/treatment

• **Adult: PO** 500 mg bid; will require an additional antiinfective for active infection

***H. pylori* infection**

• **Adult: PO** 500 mg bid plus omeprazole 2 × 20 mg every AM (days 1-14), then omeprazole 20 mg every AM (days 15-28)

Acute maxillary sinusitis

• **Adult: PO** 500 mg q12hr × 14 days

Most infections

• **Child: PO** 7.5 mg/kg q12hr × 10 days, max 500 mg/dose for MAC

Renal dose

• **Adult and child: PO** CCr 30-60 ml/min decrease dose by 50% if using with ritonavir; <30 ml/min, reduce dose by 50%, if used with ritonavir reduce by 75%

Legionnaire's disease/SARS/whooping cough/gastric ulcer/dyspepsia *(H. pylori)* (unlabeled)

• **Adult: PO** 500 mg q12hr; may be used in combination for some of these conditions

Endocarditis prophylaxis (unlabeled)

• **Adult: PO** 500 mg 1 hr before procedure

Available forms: Tabs 250, 500 mg; oral susp 125 mg/5 ml, 250 mg/5 ml; ext rel tab (XL) 500 mg

Administer:

• Do not break, crush, or chew ext rel
• Adequate intake of fluids (2 L) during diarrhea episodes
• q12hr to maintain serum levels
• Store at room temperature

Susp: Shake well, store at room temperature, discard after 2 wk

SIDE EFFECTS

CV: Ventricular dysrhythmias, QT prolongation

GI: *Nausea, vomiting, diarrhea,* hepatotoxicity, *abdominal pain,* stomatitis, heartburn, anorexia, *abnormal taste,* pseudomembranous colitis, tooth/tongue discoloration, pancreatitis

GU: Vaginitis, moniliasis, interstitial nephritis, azotemia

HEMA: Leukopenia, thrombocytopenia, increased INR

INTEG: Rash, urticaria, pruritus, Stevens-Johnson syndrome, toxic epidermal necrolysis

MISC: *Headache,* hearing loss

PHARMACOKINETICS

Peak 2-2.5 hr; duration 12 hr; half-life 3-4 hr; metabolized by liver; excreted in bile, feces; possible inhibition of P-glycoprotein

INTERACTIONS

Increase: dysrhythmias—cisapride, pimozide

Increase: levels, increase toxicity—ALPRAZolam, busPIRone, carBAMazepine, cycloSPORINE, digoxin, disopyramide, ergots, felodipine, fluconazole, omeprazole, tacrolimus, theophylline, antidiabetics

Increase: oral anticoagulants effect—digoxin, theophylline, carBAMazepine

Increase: levels of HMG-CoA reductase inhibitors

Increase: action, risk for toxicity—all products metabolized by CYP3A enzyme system

Increase: effect of calcium channel blockers, midazolam, benzodiazepines, tacrolimus

Increase: QT prolongation—class IA, III antidysrhythmics or other products that prolong QT

Increase or decrease action: zidovudine

Decrease: clarithromycin action—CYP3A4 substrates

Decrease: levels—rifampin, rifabutin, nevirapine, etravirine, benzodiazepine

Drug/Food

• Do not use with grapefruit juice

Drug/Lab Test

Increase: 17-OHCS/17-KS, AST, ALT, BUN, creatinine, LDH, total bilirubin

Decrease: folate assay, WBC

NURSING CONSIDERATIONS

Assess:

• **Infection:** wound characteristics, urine, stool, sputum, WBC, temperature; C&S before product therapy; product may be given as soon as culture is taken; C&S may be repeated after treatment

• For ulcers: abdominal pain, bleeding in stools, emesis

• Renal, hepatic studies; report hematuria, oliguria

• Bowel pattern before, during treatment

• Respiratory status: rate, character, wheezing, tightness in chest; discontinue product

• Bleeding: check INR if anticoagulants are taken

• Allergies before treatment, reaction to each medication

• **QT prolongation, ventricular dysrhythmias:** monitor ECG, cardiac status in those with underlying cardiac abnormalities

• **Serious skin reaction:** Stevens-Johnson syndrome, toxic epidermal necrolysis; product should be discontinued immediately

Evaluate:

• Therapeutic response: C&S negative for infection

Teach patient/family:

• To take with full glass of water; may give with food to decrease GI symptoms

⚠ To report sore throat, fever, fatigue; may indicate superinfection

⚠ To notify prescriber of diarrhea, dark urine, pale stools, yellow discoloration of eyes or skin, severe abdominal pain

• To take at evenly spaced intervals; to complete dosage regimen; to notify prescribers of all products used

• To notify prescriber if pregnancy is suspected or planned

TREATMENT OF HYPERSENSITIVITY: Withdraw product, maintain airway, administer EPINEPHrine, aminophylline, O_2, IV corticosteroids

C

clevidipine (Rx)

(klev-id'i-peen)

Cleviprex

Func. class.: Calcium channel blocker (L-type)

Chem. class.: Dihydropyridine

ACTION: L-type calcium channels mediate the influx of calcium during depolarization in arterial smooth muscle; reduces mean arterial B/P by decreasing systemic vascular resistance

USES: Reduction of B/P when oral therapy not feasible

CONTRAINDICATIONS: Hypersensitivity to this product, eggs, soya lecithin; defective lipid metabolism; severe aortic stenosis, pancreatitis

Precautions: Pregnancy (C), labor, breastfeeding, children <18 yr, heart failure, hyperlipidemia, chronic hypertension, pheochromocytoma

DOSAGE AND ROUTES

• **Adult: CONT IV** 1-2 mg/hr; dose may be doubled q90sec initially; as B/P reaches goal, adjust dose less frequently (q5-10min) with smaller increases in dose; most patients require 4-6 mg/hr, max 32 mg/hr; no more than 1000 ml should be infused per 24-hr period due to lipid load restrictions

Available forms: Single-dose vial 50, 100 ml (0.5 mg/ml); IV emulsion

Administer:

Intermittent IV INF route

- Do not give through same line as other medications; do not dilute
- Gently invert several times before use; do not use if discolored or if particulate matter is present
- Give through central or peripheral line at 1-2 mg/hr; use inf device
- Store vials in refrigerator; do not freeze; leave vials in carton until use; product is photosensitive, but protection from light during administration is not required

SIDE EFFECTS

CNS: Headache

CV: MI, sinus tachycardia, syncope, reflex tachycardia, atrial fibrillation

GI: Nausea, vomiting

PHARMACOKINETICS

Onset 2-4 min; half-life initially 1 min, terminal 15 min; metabolized via esterases in blood, extravascular tissues; excreted in urine 63%-74%, feces 7%-22%; protein binding >99%

NURSING CONSIDERATIONS

Assess:

- Cardiac status: B/P, pulse, respiration, ECG; some patients have developed severe angina, acute MI after calcium channel blockers if obstructive CAD is severe; if not transitioned to other antihypertensive therapies after clevidipine inf, patients should be monitored ≥8 hr for rebound hypertension; monitor for rebound hypertension after product stoppage

Evaluate:

- Therapeutic response: decreased B/P

Teach patient/family:

- To notify prescriber immediately if neurological symptoms, visual changes, or symptoms of CHF occur
- To continue follow-up for hypertension
- To notify prescriber if pregnancy is planned, suspected, or if breastfeeding

clindamycin HCl (Rx)

(klin-da-my′sin)

Cleocin HCl, Dalacin C ✦

clindamycin palmitate (Rx)

Cleocin Pediatric, Dalacin C Flavored Granules ✦

clindamycin phosphate (Rx)

Cleocin Phosphate, Dalacin C Phosphate Sterile Solution ✦

Func. class.: Antiinfective—miscellaneous

Chem. class.: Lincomycin derivative

ACTION: Binds to 50S subunit of bacterial ribosomes, suppresses protein synthesis

USES: Infections caused by staphylococci, streptococci, *Rickettsia, Fusobacterium, Actinomyces, Peptococcus, Bacteroides, Pneumocystis jiroveci*

Unlabeled uses: Acne rosacea, *Bacillus anthracis,* dental infections, folliculitis, malaria, pemphigus, periodontitis, *Pneumocystis jiroveci* pneumonia (PCP), toxoplasmosis

CONTRAINDICATIONS: Hypersensitivity to this product or lincomycin, tartrazine dye; ulcerative colitis/enteritis

Black Box Warning: Pseudomembranous colitis

Precautions: Pregnancy (B), breastfeeding, GI/hepatic disease, asthma, allergy

Black Box Warning: Diarrhea

DOSAGE AND ROUTES

- **Adult: PO** 150-450 mg q6hr, max 1.8 g/day; **IM/IV** 1.2-2.7 g/day in 2-4 divided doses, max 2700 mg/day
- **Child >1 mo: PO** 8-25 mg/kg/day in divided doses q6-8hr; **IM/IV** 20-40 mg/kg/day in 3-4 equal divided doses q6-8hr

• **Neonate: IM/IV** 15-20 mg/kg/day divided doses q6-8hr

PID

• **Adult: IV** 900 mg q8hr plus gentamicin

Bacterial endocarditis prophylaxis (unlabeled)

• **Adult:** 600 mg 1 hr before procedure

***P. jiroveci pneumonia* (unlabeled)**

• **Adult: PO** 1200-1800 mg/day in divided doses with 15-30 mg primaquine/day × 21 days

Acne Vulgaris

• **Adult/adolescent:** top gel, lotion, sol, apply a thin film to affected areas bid for up to 12 wk

• **Adult/adolescents/child ≥2 yr:** top (foam), apply to affected areas daily; if no improvement in 8 wk, discontinue

Bacteria vaginosis nonpregnant

• **Adult/adolescent females:** intravaginal 1 applicator full (100 mg clindamycin/5 g cream) at bedtime × 7 days (CDC)

Available forms: HCl: caps 75, 150, 300 mg; **palmitate:** oral sol 75 mg/5 ml; **phosphate:** inj 150, 300, 600 mg base/4 ml; 900 mg base/ml; inj inf in D_5 300 mg, 600 mg, 900 mg; top pledget/foam/gel/lotion/solution/1%; Vag cream 2%; Vag suppository 100 mg

Administer:

• In equal intervals around the clock to maintain blood levels

• Obtain C&S before use

PO route

• Do not break, crush, chew caps

• Orally with at least 8 oz of water

Oral solution

• Do not refrigerate reconstituted product; store at room temperature ≤2 wk

• Reconstitute granules with most of 75 ml of water, shake well, add remaining water, shake well (75 mg/5 ml)

IM route

• IM deep inj; rotate sites; do not give >600 mg in single IM inj

IV route

• Visually inspect parenteral products for particulate matter and discoloration before use

• **Vials:** dilute 300 and 600 mg doses with 50 ml of a compatible diluent; dilute 900 mg doses with 50-100 ml of a compatible diluent; dilute 1200 mg doses with 100 ml of a compatible diluent, final conc max 18 mg/ml

• **ADD-vantage vials:** dilute 600 and 900 mg ADD-vantage containers with 50 or 100 mg, respectively, of NS or D5W

• **Storage:** when diluted in D5W, NS, or LR, solutions with concs of 6, 9, or 12 mg/ml are stable for 16 days at room temperature or 32 days under refrigeration when stored in glass bottles or minibags; when diluted in D5W, solutions with a conc of 18 mg/ml are stable for 16 days at room temperature

Intermittent IV inf route

• Infuse over at least 10-60 min, infusion rates max 30 mg/min and ≤1.2 g should be infused in a 1 hr period

• Infuse 300 mg doses over 10 min; 600 mg doses over 20 min, 900 mg doses over 30 min, and 1200 mg doses over 40 min

Continuous IV inf route

• Give first dose rapidly, then follow with continuous inf

• Rate is based on desired serum clindamycin levels:

• To maintain serum concs above 4 mcg/ml, use a rapid infusion rate of 10 mg/min for 30 min and a maintenance rate of 0.75 mg/min; to maintain serum concs above 5 mcg/ml, use a rapid infusion rate of 15 mg/min for 30 min and a maintenance rate of 1 mg/min; to maintain serum concs above 5 mcg/ml, use a rapid infusion rate of 20 mg/min for 30 min and a maintenance rate of 1.25 mg/min

Syringe compatibilities: Amikacin, aztreonam, gentamicin, heparin

Y-site compatibilities: Acyclovir, alfentanil, amifostine, amikacin, aminocaproic acid, aminophylline, amiodarone, amphotericin B cholesteryl, amphotericin B lipid complex, amsacrine, anakinra, anidulafungin, ascorbic acid injection, atenolol, atracurium, atropine, aztreonam, benztropine, bivalirudin, bleomycin, bumetanide, buprenorphine, butorphanol, calcium chloride/gluconate, CARBOplatin, cefamandole, ceFAZolin, cefmetazole, cefonicid, cefoperazone, cefotaxime, cefoTEtan, cefOXitin, cefpirome, cefTAZidime, ceftizoxime, ceftobiprole, cefuroxime, cephalothin, cephapirin, chloramphenicol, cimetidine, cisatracurium, CISplatin, codeine, cyanocobalamin, cyclophosphamide, cycloSPORINE, cytarabine, DACTINomycin, DAPTOmycin, dexamethasone, dexmedetomidine, digoxin, diltiazem, diphenhydrAMINE, DOCEtaxel, DOPamine, doxacurium, DOXOrubicin, DOXOrubicin liposomal, doxycycline, enalaprilat, ePHEDrine, EPINEPHrine, epirubicin, epoetin alfa, eptifibatide, esmolol, etoposide, famotidine, fenoldopam, fentaNYL, fludarabine, fluorouracil, folic acid, foscarnet, furosemide, gatifloxacin, gemcitabine, gemtuzumab, gentamicin, glycopyrrolate, granisetron, heparin, hydrocortisone, HYDROmorphone, ifosfamide, imipenem-cilastatin, indomethacin, insulin (regular), irinotecan, isoproterenol, ketorolac, levofloxacin, lidocaine, linezolid, LORazepam, LR, magnesium sulfate, mannitol, mechlorethamine, melphalan, meperidine, metaraminol, methicillin, methotrexate, methoxamine, methyldopate, methylPREDNISolone, metoclopramide, metoprolol, metroNIDAZOLE, mezlocillin, miconazole, milrinone, morphine, moxalactam, multiple vitamins injection, nafcillin, nalbuphine, naloxone, nesiritide, netilmicin, niCARdipine, nitroglycerin, nitroprusside, norepinephrine, octreotide, ondansetron, oxacillin, oxaliplatin, oxytocin, PACLitaxel, palonosetron, pamidronate, pancuronium, pantoprazole, PEMEtrexed, penicillin G potassium/sodium, pentazocine, perphenazine, PHENobarbital, phenylephrine, phytonadione, piperacillin, piperacillin-tazobactam, potassium chloride, procainamide, propofol, propranolol, protamine, pyridoxine, ranitidine, remifentanil, Ringer's, ritodrine, riTUXimab, rocuronium, sargramostim, sodium acetate/bicarbonate, succinylcholine, SUFentanil, tacrolimus, teniposide, theophylline, thiamine, thiotepa, ticarcillin, ticarcillin-clavulanate, tigecycline, tirofiban, TNA, tobramycin, tolazoline, TPN, trimetaphan, urokinase, vancomycin, vasopressin, vecuronium, verapamil, vinCRIStine, vinorelbine, vitamin B complex/C, voriconazole, zidovudine, zoledronic acid

SIDE EFFECTS

GI: *Nausea, vomiting, abdominal pain, diarrhea,* pseudomembranous colitis, anorexia, weight loss, increased AST/ALT, bilirubin, alk phos; jaundice
GU: *Vaginitis,* urinary frequency
INTEG: Rash, urticaria, pruritus, erythema, pain, abscess at inj site
SYST: Stevens-Johnson syndrome, exfoliative dermatitis

PHARMACOKINETICS

PO: Peak 45 min, duration 6 hr
IM: Peak 3 hr (adult), 1 hr (child); duration 8-12 hr; half-life 2½ hr; metabolized in liver; excreted in urine, bile, feces as inactive metabolites; crosses placenta; excreted in breast milk

INTERACTIONS

• May block clindamycin effect: erythromycin, chloramphenicol
Increase: neuromuscular blockade—neuromuscular blockers
Decrease: absorption—kaolin
Drug/Lab Test
Increase: alk phos, bilirubin, CPK, AST, ALT

NURSING CONSIDERATIONS

Assess:

• **Infection:** C&S before product therapy; product may be given as soon as culture is taken
• VS, urine, stools, sputum
• Hepatic studies: AST, ALT if on long-term therapy
• B/P, pulse in patient receiving product parenterally

Black Box Warning: Pseudomembranous colitis: bowel pattern before, during treatment; if severe diarrhea occurs, product should be discontinued, may occur several wk after therapy is terminated

• Respiratory status: rate, character, wheezing, tightness in chest
• **Serious skin reactions:** Stevens-Johnson syndrome, exfoliative dermatitis

Evaluate:

• Therapeutic response: decreased temperature, negative C&S

Teach patient/family:

• To take oral product with full glass of water; that antiperistaltic products may worsen diarrhea
• About all aspects of product therapy; to complete entire course of medication to ensure organism death (10-14 days); culture may be taken after medication course completed
• ⚠ To report sore throat, fever, fatigue; may indicate **superinfection**
• To take with food to reduce GI symptoms
• To notify nurse or prescriber of diarrhea with pus, mucous

TREATMENT OF HYPERSENSITIVITY:

• Withdraw product; maintain airway; administer EPINEPHrine, aminophylline, O_2, IV corticosteroids

clindamycin (topical, vaginal)

(klin-da-mye′sin)

Cleocin, Cleocin-T, Clindacin-P, Clindagel, ClindaMax, Clindasol ✦, Clinda-T ✦, Clindesse, Clindets, Dalacin ✦, Evoclin

Func. class.: Topical anti-infective
Chem. class.: Lincosamide derivative

ACTION: Antibacterial activity results from inhibition of protein synthesis; bacteriostatic

USES: For the treatment of acne vulgaris; treatment of bacterial vaginosis and anaerobic bacteria

CONTRAINDICATIONS: Hypersensitivity to this product or lincomycin, history of antibiotic-associated colitis or ulcerative colitis

Precautions: Breastfeeding, children <12 yr

DOSAGE AND ROUTES

Acne vulgaris

• **Adult/adolescent: TOP** (gel, lotion, solution) Apply a thin film of 1% to affected areas bid
• **Adult/adolescent/child ≥12 yr: TOP (foam)** Apply 1% topical foam to affected areas once daily; if there is no improvement after 6-8 weeks or if the condition worsens, discontinue treatment
• **TOP** (medicated pledgets) Use a pledget to apply a thin film to the affected area bid; more than one pledget may be used, but each pledget should be used only once and then discarded

Bacterial vaginosis and anaerobic bacteria

Nonpregnant adult/adolescent/postmenarchal females:

• **Intravaginal cream:** one applicatorful (100 mg clindamycin/5 g cream) intravaginally, preferably at bedtime, for 3 or 7 consecutive days in nonpregnant women and for 7 consecutive days in pregnant women; clindesse is administered as a single dose at any time of the day
• **Intravaginal ovules/suppositories:** as an alternative to first-line therapies, the CDC recommends one ovule (100 mg clindamycin) inserted intravaginally at bedtime for 3 days

Available forms: Topical gel, foam, lotion, pledget, solution 1%; vaginal cream 2%, vaginal suppositories 100 mg

Administer:

Topical route:

• Improvement occurs after 6 wk but can require 8-12 wk

• Topical skin products are not for intravaginal therapy and are for external use only; do not use skin products near the eyes, nose, or mouth
• Wash hands before and after use; wash affected area and gently pat dry

Cream/ointment/lotion:

• Shake well before use (lotion); apply a thin film to the cleansed affected area; massage gently into affected areas

Foam formulations:

• Do not dispense foam directly onto hands or face; the warmth of the skin will cause the foam to melt; instead, dispense desired amount directly into the cap or onto a cool surface; make sure enough foam is dispensed to cover the affected area(s); if the can feels warm or the foam seems runny, run the can under cold water; to apply, pick up small amounts of the foam with the fingertips and gently massage into the affected areas until the foam disappears

Solution formulations:

• Shake well before use; apply a thin film to the cleansed affected area; massage gently into affected areas; if using a solution-soaked pledget, patient may use more than 1 pledget per application as needed to treat affected areas, but each pledget should be used only once and then discarded

Intravaginal route:

• Only use dosage formulations specified for intravaginal use; intravaginal dosage forms are not for topical therapy; do not ingest

Suppository:

• Unwrap vaginal ovule (suppository) before insertion; use applicator(s) supplied by the manufacturer

Cream:

• Use applicator(s) supplied by the manufacturer

SIDE EFFECTS

GU: Colitis, diarrhea, overgrowth, vaginitis, vaginal moniliasis, UTI

INTEG: Redness, burning, dermatitis, rash, pruritus

NURSING CONSIDERATIONS

Assess:

Allergic reaction: Assess for hypersensitivity; product may need to be discontinued

• Contact prescriber immediately if severe diarrhea, stomach cramps/pain, or bloody stools occur

Infection: Assess for number of lesions, severity in acne, itching in vaginosis

Evaluate:

• Decreased lesions in acne, infection in vaginosis

Teach patient/family:

Topical route:

• That improvement occurs after 6 wk but can require 8-12 wk
• That topical skin products are not for intravaginal therapy and are for external use only; not to use skin products near the eyes, nose, or mouth
• To wash hands before and after use; wash affected area and gently pat dry

Cream/ointment/lotion:

• Shake well before use; **lotion:** apply a thin film to the cleansed affected area; massage gently into affected areas

Foam formulations:

• Do not dispense foam directly onto hands or face; the warmth of the skin will cause the foam to melt; instead, dispense desired amount directly into the cap or onto a cool surface; make sure enough foam is dispensed to cover the affected area(s); if the can feels warm or the foam seems runny, run the can under cold water; to apply, pick up small amounts of the foam with the fingertips and gently massage into the affected areas until the foam disappears

Solution formulations:

• Shake well before use; apply a thin film to the cleansed affected area; massage gently into affected areas; if using a solution-soaked pledget, patient may use more than one pledget per application as needed to treat affected areas, but each pledget should be used only once and then discarded

Intravaginal route:
• Only use dosage formulations specified for intravaginal use; intravaginal dosage forms are not for topical therapy; do not ingest
Suppository:
• Unwrap vaginal ovule (suppository) before insertion; use applicator(s) supplied by the manufacturer
Cream:
• Use applicator(s) supplied by the manufacturer

clobetasol

(kloe-bay'ta-sol)
Clobex, Cormax, Olux, Olux-E, Temovate, Temovate-E
Func. class.: Corticosteroid, topical

ACTION: Crosses cell membrane to attach to receptors to decrease inflammation, itching

USES: Inflammation/itching in corticosteroid-responsive dermatoses on the skin/scalp

CONTRAINDICATIONS: Hypersensitivity, use of some preparations on face, axilla, groin; monotherapy for primary bacterial infections
Precautions: Pregnancy (C), breastfeeding, children <12 yr

DOSAGE AND ROUTES

• **Adult: TOP** Apply to infected areas bid (shampoo: daily up to 4×/wk
Available forms: Gel, lotion, ointment, cream, shampoo, solution, spray, foam 0.05%
Administer
Topical route
⚠ Do not use with occlusive dressings
⚠ Treatment should be limited to 2 wk
• **Cream/Ointment/Lotion:** using gloves, apply sparingly in a thin film and rub gently into the cleansed, slightly moist affected area
• **Gel:** using gloves, apply sparingly in a thin film and rub gently into the cleansed, slightly moist affected area
• **Scalp foam:** invert can and dispense a small amount of foam onto a saucer or other cool surface; do not dispense directly onto hands; pick up small amounts of foam with fingers and gently massage into affected area until foam disappears; repeat until entire affected scalp area is treated
• **Shampoo:** apply onto dry scalp in thin film, leave lather on scalp for 15 min, rinse off

SIDE EFFECTS

INTEG: Burning, folliculitis, pruritus, dermatitis, irritation, erythema, hypertrichosis, acne
MISC: Hyperglycemia

NURSING CONSIDERATIONS

Assess:
• Skin reactions: burning, pruritus, folliculitis, dermatitis
Evaluate:
• Decreasing itching, inflammation on the skin, scalp
Teach patient/family:
Topical route
⚠ Do not use with occlusive dressings
⚠ Treatment should be limited to 2 wk
• **Cream/Ointment/Lotion:** to apply sparingly in a thin film and rub gently into the cleansed, affected area
• **Gel:** to apply sparingly in a thin film and rub gently into the cleansed, slightly moist affected area
• **Scalp foam:** to invert can and dispense a small amount of foam onto a saucer or other cool surface; not to dispense directly onto hands; to pick up small amounts of foam with fingers and gently massage into affected area until foam disappears; repeat until entire affected scalp area is treated

clomiPHENE (Rx)

(kloe′mi-feen)

Clomid, Serophene

Func. class.: Ovulation stimulant

Chem. class.: Nonsteroidal antiestrogenic

Do not confuse:
clomiPHENE/clomiPRAMINE
Serophen/Sarafem

ACTION: Increases LH, FSH release from the pituitary, which increases the maturation of the ovarian follicle, ovulation, and the development of the corpus luteum

USES: Female infertility (ovulatory failure)
Unlabeled uses: Oligospermia

CONTRAINDICATIONS: Pregnancy (X), hypersensitivity, hepatic disease, undiagnosed uterine bleeding, uncontrolled thyroid or adrenal dysfunction, intracranial lesion, ovarian cysts, endometrial carcinoma
Precautions: Hypertension, depression, seizures, diabetes mellitus, abnormal ovarian enlargement, ovarian hyperstimulation

DOSAGE AND ROUTES

- **Adult: PO** 50 mg/day × 5 days or 50 mg/day beginning on day 5 of menstrual cycle, may increase to 100 mg daily × 5 days with next cycle; may be repeated until conception occurs or max 6 cycles of therapy

Oligospermia (unlabeled)
- **Adult (men): PO** 25 mg/day × 25 days then 5 days off cycle each mo

Available forms: Tabs 50 mg
Administer:
- After discontinuing estrogen therapy
- At same time daily to maintain product level, without regard to food
- Avoid heat, moisture, light; store at room temperature

SIDE EFFECTS

CNS: *Headache, depression,* restlessness, anxiety, nervousness, fatigue, insomnia, dizziness, flushing
CV: Vasomotor flushing, phlebitis, deep venous thrombosis
EENT: Blurred vision, diplopia, photophobia
GI: *Nausea, vomiting, constipation,* abdominal pain, bloating, hepatitis
GU: Polyuria, urinary frequency, birth defects, spontaneous abortions, multiple ovulation, breast pain, oliguria, abnormal uterine bleeding, ovarian cyst, hypertrophy of ovary
INTEG: *Rash, dermatitis,* urticaria, alopecia

PHARMACOKINETICS

Metabolized in liver, excreted in feces

INTERACTIONS

Drug/Herb
Decrease: clomiPHENE effect—DHEA, black cohosh, chaste tree fruit
Drug/Food
Decrease: clomiPHENE effect—soy
Drug/Lab Test
Increase: LFTs

NURSING CONSIDERATIONS

Assess:
- LFTs before therapy: AST, ALT, alk phos
- Serum progesterone, urinary excretion of pregnanediol to identify occurrence of ovulation
- Ovarian size, cervical condition by pelvic examination
- Rule out endometrial carcinoma in women >35 yr by endometrial biopsy

Evaluate:
- Therapeutic response: fertility

Teach patient/family:
- That multiple births are common
- To notify prescriber immediately if low abdominal pain occurs; may indicate ovarian cyst, cyst rupture
- To notify prescriber of photophobia, blurred vision, diplopia, abnormal bleeding

- That, if dose is missed, to double it next time; if more than one dose is missed, to call prescriber
- That response usually occurs 4-10 days after last day of treatment
- About the method for taking, recording basal body temperature to determine whether ovulation has occurred
- If ovulation can be determined (there is a slight decrease in temperature then a sharp increase with ovulation), to attempt coitus 3 days before and every other day until after ovulation
- **If pregnancy is suspected, to notify prescriber immediately (X)**

clomiPRAMINE (Rx)

(kloe-mip′ra-meen)

Anafranil

Func. class.: Antidepressant, tricyclic
Chem. class.: Tertiary amine

Do not confuse:
clomiPRAMINE/clomiPHENE/chlorproMAZINE/desipramine/Norpramin

ACTION:
Potentiates serotonin and norepinephrine; moderate anticholinergic effect

USES:
Obsessive-compulsive disorder
Unlabeled uses: Autism, depression, premature ejaculation

CONTRAINDICATIONS:
Hypersensitivity to this product, carBAMazepine, tricyclics, immediate post-MI
Precautions: Pregnancy (C), breastfeeding, geriatric patients, seizures, cardiac disease, glaucoma, prostatic hypertrophy, urinary retention

Black Box Warning: Children, suicidal ideation

DOSAGE AND ROUTES

Obsessive-compulsive disorder
- **Adult: PO** 25 mg at bedtime, increase gradually over 4 wk to 75-250 mg/day in divided doses
- **Child 10-18 yr: PO** 25 mg/day gradually increase over 2 wk; max 3 mg/kg/day or 200 mg/day, whichever is smaller

Autism (unlabeled)
- **Adult: PO** 25 mg/day, may increase to 75-100 mg/day, max 250 mg/day
- **Child: PO** 25 mg/day, may increase if needed

Premature ejaculation (unlabeled)
- **Adult: PO** 25-50 mg/day

Depression (unlabeled)
- **Adult: PO** 25 mg at bedtime and increase gradually over 4 wk to 75-250 mg/day in divided doses
- **Child 10-18 yr: PO** 25-50 mg/day gradually increased; max 3 mg/kg/day or 200 mg/day, whichever is smaller

Available forms: Caps 25, 50, 75 mg
Administer:
- Do not break, crush, or chew caps
- Increased fluids, bulk in diet for constipation, especially for geriatric patients
- Without regard to food, during initial dosing and titration give with meals
- After titration, may be given as a single dose at bedtime to reduce daytime sedation
- Store in tight container, at room temperature; do not freeze

SIDE EFFECTS

CNS: *Dizziness, tremors, mania,* **seizures**, aggressiveness, EPS, drowsiness, headache, **neuroleptic malignant syndrome**, insomnia, agitation, anxiety, impaired memory
CV: Hypotension, tachycardia, **cardiac arrest**, hypertension, palpitations
EENT: Blurred vision, altered taste, tinnitus, increased intraocular pressure
ENDO: Galactorrhea, hyperprolactinemia
GI: *Constipation, dry mouth, nausea, dyspepsia,* weight gain, **hepatic toxicity**
GU: *Delayed ejaculation, anorgasmia,* urinary retention, decreased libido
HEMA: **Agranulocytosis, neutropenia, pancytopenia**
INTEG: Diaphoresis, photosensitivity, abnormal skin order, flushing, rash, pruritus

META: Hyponatremia
RESP: Pharyngitis, rhinitis, bronchospasm
SYST: Suicide in children, adolescents

PHARMACOKINETICS

Onset ≥2 wk (depression), 4-10 wk (OCD); peak 2-6 hr; extensively bound to tissue and plasma proteins; demethylated in liver; active metabolites excreted in urine (50%-60%), feces (24%-32%); half-life 20-30 hr; steady state 1-2 wk

INTERACTIONS

Increase: hypertensive crisis, seizures, hypertensive episode—MAOIs
Increase: serotonin syndrome-SSRIs, SNRIs serotonin syndrome, linezolid, methylene blue IV
Increase: clomiPRAMINE levels—cimetidine, FLUoxetine, fluvoxaMINE, sertraline; do not use together
Increase: hypertensive effect—cloNIDine, EPINEPHrine, norepinephrine
Increase: clomiPRAMINE level—CYP1A2, CYP2D6
Increase: CNS depression—alcohol, CNS depressants, general anesthetics
Increase: QT prolongation—other tricyclics, phenothiazines, quinolones
Decrease: effect of cloNIDine, levodopa, skeletal muscle relaxants, haloperidol, opiates
Decrease: clomiPRAMINE levels—barbiturates, carBAMazepine, phenytoin

Drug/Herb

Increase: serotonin syndrome—St. John's wort; do not use concurrently
Increase: CNS depression—hops, kava, valerian

Drug/Lab Test

Increase: prolactin, TBG, AST, ALT, blood glucose
Decrease: serum thyroid hormone (T_3, T_4)

NURSING CONSIDERATIONS

Assess:

- B/P lying, standing; pulse q4hr; if systolic B/P drops 20 mm Hg, withhold product, notify prescriber; take VS q4hr in patients with CV disease
- **Serotonin syndrome:** hyperpyrexia, rigidity, irregular pulse, diaphoresis
- **ECG** for flattening of T wave, QTc prolongation, bundle branch block, AV block, dysrhythmias in cardiac patients, may lead to cardiac collapse
- Blood studies: CBC, leukocytes, differential, cardiac enzymes if patient is receiving long-term therapy and signs of blood dyscrasias
- Hepatic studies: AST, ALT, bilirubin

Black Box Warning: Mental status: mood, sensorium, affect, suicidal tendencies; increase in psychiatric symptoms: depression, panic, frequency of obsessive-compulsive behaviors; watch closely for evidence of suicidal thoughts in children, adolescents; seizure disorders

- Urinary retention, constipation; constipation more likely in children
- **Withdrawal symptoms:** headache, nausea, vomiting, muscle pain, weakness; not usual unless product discontinued abruptly
- Alcohol consumption; if alcohol consumed, withhold dose until AM
- Assistance with ambulation during beginning therapy, since drowsiness, dizziness occurs
- Gum, hard candy, or frequent sips of water for dry mouth

Evaluate:

- Therapeutic response: decreased anxiety, depression

Teach patient/family:

- That the effects may take 4-6 wk to appear
- About risk for seizures
- To use caution when driving, performing other activities that require alertness because drowsiness, dizziness, blurred vision may occur
- To avoid alcohol, other CNS depressants

- Not to discontinue medication quickly after long-term use because this may cause nausea, headache, malaise

Black Box Warning: That suicidal thoughts/behaviors may occur in children, young adults; report immediately

- To wear sunscreen, protective clothing to prevent photosensitivity
- To notify prescriber if pregnancy is planned, suspected
- That men may experience a high incidence of sexual dysfunction
- **Serotonin syndrome:** report immediately sweating, diarrhea, twitching
- **Abrupt discontinuation:** do not stop abruptly

TREATMENT OF OVERDOSE:

ECG monitoring; induce emesis; lavage, activated charcoal; anticonvulsant; diazepam IV

clonazePAM (Rx)

(kloe-na′zi-pam)

KlonoPIN

Func. class.: Anticonvulsant

Chem. class.: Benzodiazepine derivative

Controlled Substance Schedule IV

Do not confuse:
clonazePAM/LORazepam/clorazepate/cloNIDine
KlonoPIN/cloNIDine

ACTION: Inhibits spike, wave formation during absence seizures (petit mal); decreases amplitude, frequency, duration, spread of discharge during minor motor seizures

USES: Absence, atypical absence, akinetic, myoclonic seizures; Lennox-Gastaut syndrome, panic disorder

Unlabeled uses: Anxiety, insomnia, nystagmus, restless leg syndrome

CONTRAINDICATIONS: Pregnancy (D), hypersensitivity to benzodiazepines, acute closed-angle glaucoma, psychosis, severe hepatic disease

Precautions: Breastfeeding, geriatric patients, open-angle glaucoma, chronic respiratory disease, renal/hepatic disease

DOSAGE AND ROUTES

Lennox-Gastaut syndrome/atypicals absence seizures/akinetic and myclonic seizures

- **Adult: PO** up to 1.5 mg/day in 3 divided doses; may be increased 0.5-1 mg q3days until desired response, max 20 mg/day
- **Geriatric: PO** 0.25 daily-bid initially, increase by 0.25/day q7-14days as needed
- **Child <10 yr or <30 kg: PO** 0.01-0.03 mg/kg/day in divided doses q8hr, max 0.05 mg/kg/day; may be increased 0.25-0.5 mg q3days until desired response, max 0.1-0.2 mg/kg/day

Panic disorder

- **Adult: PO** 0.25 mg bid increase to 1 m daily after 3 days, max 4 mg/day

Restless leg syndrome (RLS) (unlabeled)

- **Adult: PO** 0.5 mg tid or 0.5 mg in the evening and 30 min before bedtime

Insomnia/anxiety (unlabeled)

- **Adult: PO** 0.125-0.25 mg at bedtime, titrate up q3-4days as needed

Available forms: Tabs 0.5, 1, 2 mg; orally disintegrating tabs 0.125, 0.25, 0.5, 1, 2 mg

Administer:

PO route

- With food, milk for GI symptoms
- **Orally disintegrating tablets:** open pouch by peeling back foil on blister pack (do not push tab through foil), place on tongue, allow to dissolve; may be swallowed with/without water
- Store at room temperature

SIDE EFFECTS

CNS: *Drowsiness,* dizziness, confusion, behavioral changes, tremors, insomnia, headache, suicidal tendencies, slurred speech, anterograde amnesia

CV: Palpitations, bradycardia, tachycardia
EENT: *Increased salivation, nystagmus, diplopia,* abnormal eye movements
GI: *Nausea, constipation,* polyphagia, anorexia, xerostomia, diarrhea, gastritis, sore gums
GU: Dysuria, enuresis, nocturia, retention, libido changes
HEMA: Thrombocytopenia, leukocytosis, eosinophilia
INTEG: Rash, alopecia, hirsutism
RESP: Respiratory depression, dyspnea, congestion

PHARMACOKINETICS

PO: Peak 1-2 hr, metabolized by liver, excreted in urine, half-life 18-50 hr, duration 6-12 hr, protein binding 85%

INTERACTIONS

Increase: clonazePAM effects—CYP3A4 inhibitors (azoles, cimetidine, clarithromycin, diltiazem, erythromycin, FLUoxetine), oral contraceptives
Increase: CNS depression—alcohol, barbiturates, opiates, antidepressants, other anticonvulsants, general anesthetics, hypnotics, sedatives
Decrease: clonazePAM effect—CYP3A4 inducers (carBAMazepine, PHENobarbital, phenytoin)

Drug/Herb
Increase: CNS depression—kava, chamomile, valerian
Increase: clonazePAM effect—ginkgo, melatonin
Decrease: clonazePAM effect—ginseng, St. John's wort

Drug/Lab Test
Increase: AST, alk phos, bilirubin
Decrease: platelets, WBC

NURSING CONSIDERATIONS

Assess:

- **Seizures:** duration, type, intensity, with/without aura
- Blood studies: RBC, Hct, Hgb, reticulocyte counts every wk for 4 wk then monthly
- Hepatic studies: ALT, AST, bilirubin, creatinine
- **Abrupt discontinuation:** do not discontinue abruptly; seizures may increase
- Signs of physical withdrawal if medication suddenly discontinued
- ⚠ **Mental status:** mood, sensorium, affect, oversedation, behavioral changes, **suicidal thoughts/behaviors;** if mental status changes, notify prescriber
- Eye problems: need for ophthalmic exam before, during, after treatment (slit lamp, funduscopy, tonometry)
- Allergic reaction: red, raised rash; product should be discontinued
- ⚠ **Blood dyscrasias:** fever, sore throat, bruising, rash, jaundice
- **Toxicity:** bone marrow depression, nausea, vomiting, ataxia, diplopia, CV collapse; drug levels during initial treatment (therapeutic 20-80 ng/ml)
- Assistance with ambulation during early part of treatment; dizziness occurs, especially among geriatric patients

Evaluate:

- Therapeutic response: decreased seizure activity

Teach patient/family:

- To carry emergency ID bracelet stating name, products taken, condition; prescriber's name, phone number
- To avoid driving, other activities that require alertness
- To avoid alcohol, other CNS depressants; increased sedation may occur
- Not to discontinue medication quickly after long-term use; to taper off over several wk
- To notify prescriber of yellowing of skin/eyes, clay-colored stools, bleeding, fever, extreme fatigue, sore throat, suicidal thoughts/behaviors

TREATMENT OF OVERDOSE:

Lavage, activated charcoal, flumazenil, monitor electrolytes, VS, administer vasopressors

cloNIDine (Rx)

(klon'i-deen)

Catapres, Catapres-TTS, Duraclon, Kapvay, Nexiclon

Func. class.: Antihypertensive

Chem. class.: Central α-adrenergic agonist

Do not confuse:
cloNIDine/KlonoPIN/clonazePAM
Catapres/Cataflam/Catarase

ACTION: Inhibits sympathetic vasomotor center in CNS, which reduces impulses in sympathetic nervous system; blood pressure, pulse rate, cardiac output decrease, prevents pain signal transmission in CNS by α-adrenergic receptor stimulation of the spinal cord

USES: Mild to moderate hypertension, used alone or in combination; severe pain in cancer patients (epidural), attention-deficit/hyperactivity disorder (ADHD)

Unlabeled uses: Opioid withdrawal, prevention of vascular headaches, treatment of menopausal symptoms, dysmenorrhea, autism, cycloSPORINE nephrotoxicity prophylaxis, diabetic neuropathy, ethanol/nicotine/opiate agonist withdrawal, Tourette's syndrome, hypertensive emergency, neonatal abstinence syndrome, scleroderma renal crisis

CONTRAINDICATIONS: Hypersensitivity; (epidural) bleeding disorders, anticoagulants

Precautions: Pregnancy (C), breastfeeding, children <12 yr (transdermal), geriatric patients, noncompliant patients, MI (recent), diabetes mellitus, chronic renal failure, Raynaud's disease, thyroid disease, depression, COPD, asthma, pheochromocytoma

Black Box Warning: Labor

DOSAGE AND ROUTES

Hypertension

- **Adult: PO/TRANSDERMAL** 0.1 mg bid then increase by 0.1-0.2 mg/day at weekly intervals until desired response; range 0.2-0.6 mg/day in divided doses
- **Geriatric: PO** 0.1 mg at bedtime; may increase gradually
- **Child: PO** 5-10 mcg/kg/day in divided doses q8-12hr, max 0.9 mg/day

Severe pain

- **Adult: CONT EPIDURAL INF** 30 mcg/hr
- **Child: CONT EPIDURAL INF** 0.5 mcg/kg/hr then titrate to response

Opioid withdrawal (unlabeled)

- **Adult: PO** 0.3-1.2 mg/day; may decrease by 50% × 3 days then decrease by 0.1-0.2 mg/day or discontinue

ADHD

- **Adolescent/child ≥6 yr: PO** 0.05 mg/kg/day in 3-4 divided doses may increase in 0.1 mg/day weekly up to 0.4 mg/day; ext rel 0.1 mg at bedtime, increase dose by 0.1 mg/day up to 0.4 mg/day

Menopausal symptoms (unlabeled)

- **Adult: TRANSDERMAL** 0.1-mg patch q1wk; **PO** 0.05-0.4 mg/day

Tourette's syndrome (unlabeled)

- **Adult: PO** 0.15-0.2 mg/day

Hypertensive emergency (unlabeled)

- **Adult: PO** 0.1-0.2 mg q1hr to a total of 0.6 mg

Available forms: Tabs 0.025 ✦, 0.1, 0.2, 0.3 mg; transdermal 2.5, 5, 7.5 mg delivering 0.1, 0.2, 0.3 mg/24 hr, respectively; inj 100, 500 mcg/ml; ext rel tab 0.1 mg (Kapvay)

Administer:

- Store patches in cool environment, tablets in tight container

PO route

- Give last dose at bedtime
- Do not crush, cut, chew, or break ext rel tabs; Kapvay is not interchangeable with other products

Transdermal route

- Once weekly; apply to site without hair; best absorption over chest or upper arm;

rotate sites with each application; clean site before application; apply firmly, especially around edges; may secure with adhesive tape if loosen, fold sticky sides together and discard

- Should be removed before MRI

Epidural route

- Used for severe cancer pain
- May be used with opiates
- Use only if familiar with epidural inf devices

SIDE EFFECTS

CNS: *Drowsiness, sedation, headache, fatigue,* nightmares, insomnia, mental changes, anxiety, depression, hallucinations, delirium, syncope, dizziness

CV: *Orthostatic hypotension, palpitations,* CHF, ECG abnormalities, sinus tachycardia

EENT: Taste change, parotid pain

ENDO: Hyperglycemia

GI: *Nausea, vomiting, malaise,* constipation, *dry mouth*

GU: Impotence, dysuria, nocturia, gynecomastia

INTEG: *Rash,* alopecia, facial pallor, pruritus, hives, edema, burning papules, excoriation (transdermal patches)

MISC: Withdrawal symptoms

MS: Muscle, joint pain; leg cramps

PHARMACOKINETICS

Absorbed well

PO: Onset ½ to 1 hr, peak 2-4 hr, duration 8-12 hr, half-life 6-12 hr

TRANSDERMAL: Onset 3 days; duration 1 wk; metabolized by liver (metabolites); excreted in urine (45% unchanged, inactive metabolites), feces; crosses blood-brain barrier; excreted in breast milk

INTERACTIONS

- AV block: verapamil, diltiazem

⚠ Life-threatening elevations of B/P: tricyclics, β-blockers

Increase: CNS depression—opiates, sedatives, hypnotics, anesthetics, alcohol

Increase: hypotensive effects—diuretics, other antihypertensive nitrates

Decrease: hypotensive effects—tricyclics, MAOIs, appetite suppressants, amphetamines, prazosin, antipsychotics

Decrease: effect of levodopa

Drug/Herb

Increase: antihypertensive effect—hawthorn

Decrease: antihypertensive effect—ephedra, ginseng

Drug/Lab Test

Increase: blood glucose

Decrease: VMA, urinary catecholamines, aldosterone

NURSING CONSIDERATIONS

Assess:

- **Hypertension:** B/P, pulse; report significant changes
- **Opiate withdrawal (unlabeled):** fever, diarrhea, nausea, vomiting, cramps, insomnia, shivering, dilated pupils
- **Cancer pain:** location, intensity, character; alleviating, aggravating factors at baseline and frequently
- Edema in feet, legs daily; monitor I&O; check for falling output
- **Allergic reaction:** rash, fever, pruritus, urticaria; product should be discontinued if antihistamines fail to help
- **CHF:** edema, dyspnea, wet crackles, B/P, more common in geriatric patients
- **ADHD:** B/P, pulse, palpitations, syncope
- Renal symptoms: polyuria, oliguria, frequency

Evaluate:

- Therapeutic response: decrease in B/P with hypertension, decrease in withdrawal symptoms (opioid), decrease in pain

Teach patient/family:

- To avoid hazardous activities, since product may cause drowsiness
- To notify all health care providers of medication use
- Not to discontinue product abruptly or **withdrawal symptoms** may occur: anxiety, increased B/P, headache, insomnia, increased pulse, tremors, nausea, sweating; to comply with dosage schedule even if feeling better

⚠ Nurse Alert

- Not to use OTC (cough, cold, or allergy), alcohol or CNS depressant products unless directed by prescriber
- To rise slowly to sitting or standing position to minimize orthostatic hypotension, especially among geriatric patients
- To notify prescriber of mouth sores, sore throat, fever, swelling of hands or feet, irregular heartbeat, chest pain, signs of **angioedema**
- About excessive perspiration, dehydration, vomiting; diarrhea may lead to fall in B/P; consult prescriber if these occur; that product may cause dizziness, fainting; that lightheadedness may occur during first few days of therapy
- That product may cause dry mouth; to use hard candy, saliva product, sugarless gum, or frequent rinsing of mouth
- That compliance is necessary; not to skip or stop product unless directed by prescriber; tolerance may develop with long-term use
- **Transdermal:** how to use patch; that patch comes in two parts: product patch and overlay to keep patch in place; not to trim or cut patch; that response may take 2-3 days if product is given transdermally; on administration of patch, if switching from tabs to patch, to taper tabs to avoid withdrawal

TREATMENT OF OVERDOSE:
Supportive treatment; administer tolazoline, atropine, DOPamine prn

clopidogrel (Rx)
(klo-pid'oh-grel)

Plavix

Func. class.: Platelet aggregation inhibitor

Chem. class.: Thienopyridine derivative

Do not confuse:
Plavix/Paxil/Elavil

ACTION:
Inhibits ADP-induced platelet aggregation

USES:
Reducing the risk of stroke, MI, vascular death, peripheral arterial disease in high-risk patients, acute coronary syndrome, transient ischemic attack (TIA), unstable angina

Unlabeled uses: Cardiac surgery (infant and child), Kawasaki disease

CONTRAINDICATIONS:
Hypersensitivity, active bleeding

Precautions: Pregnancy (B), breastfeeding, children, previous hepatic disease, increased bleeding risk, neutropenia, agranulocytosis, renal disease, Asian/Black/Caucasian patients

Black Box Warning: CYP2C19 allele (poor metabolizers)

DOSAGE AND ROUTES

Recent MI, stroke, peripheral arterial disease, TIA
- **Adult: PO** 75 mg/day with/without aspirin

Acute coronary syndrome
- **Adult: PO** loading dose 300 mg then 75 mg/day with aspirin

Cardiac surgery/other cardiac conditions (unlabeled)
- **Child ≤2 yr, infant, neonate: PO** 0.2 mg/kg/day for platelet inhibition

Available forms: Tabs 75, 300 mg

Administer:
- Without regard to food
- Should be discontinued 5 days before elective surgery if an antiplatelet action is not desired

SIDE EFFECTS

CNS: Headache, dizziness, depression, syncope, hypesthesia, neuralgia, confusion, hallucinations

CV: Edema, hypertension, chest pain

GI: Nausea, vomiting, diarrhea, constipation, GI discomfort, **GI bleeding, pancreatitis, hepatic failure**

GU: **Glomerulonephritis**

HEMA: Epistaxis, purpura, **bleeding (major/minor from any site), neutropenia, aplastic anemia, agranulocytosis, thrombotic thrombocytopenic purpura**

INTEG: Rash, pruritus, anaphylaxis
MISC: UTI, hypercholesterolemia, chest pain, fatigue, intracranial hemorrhage, toxic epidermal necrolysis, Stevens-Johnson syndrome, flulike syndrome
MS: Arthralgia, back pain
RESP: Upper respiratory tract infection, dyspnea, rhinitis, bronchitis, cough, bronchospasm

PHARMACOKINETICS

Rapidly absorbed; peak 1-3 hr; metabolized by liver (CYP3A4); excreted in urine, feces; half-life 6 hr; plasma protein binding 95%; effect on platelets after 3-7 days

INTERACTIONS

Black Box Warning: Avoid use with CYP2C19 inhibitors (omeprazole, esomeprazole)

Increase: bleeding risk—anticoagulants, aspirin, NSAIDs, abciximab, eptifibatide, tirofiban, thrombolytics, ticlopidine, SSRIs, treprostinil, rifampin
Increase: action of some NSAIDs, phenytoin, TOLBUTamide, tamoxifen, torsemide, fluvastatin, warfarin
Decrease: clopidogrel effect—proton pump inhibitor (PPIs)
Decrease: CYP3A4 inhibitors/substrates—atorvastatin, simvastatin, cerivastatin

Drug/Herb

Increase: clopidogrel effect—feverfew, fish oil, omega-3 fatty acid, garlic, ginger, ginkgo biloba, green tea, horse chestnut
Decrease: clopidogrel effect—bilberry, saw palmetto

Drug/Lab Test

Increase: AST, ALT, bilirubin, uric acid, total cholesterol, nonprotein nitrogen (NPN)

NURSING CONSIDERATIONS

Assess:

⚠ **Thrombotic/thrombocytic purpura,** fever, thrombocytopenia, neurolytic anemia

Black Box Warning: CYP2C19 Allele (poor metabolizers): Consider using another antiplatelet product, higher CV reaction occurs after acute coronary syndrome or PCI, tests are available to determine CYP2C19 allele

- Symptoms of stroke, MI during treatment
- Hepatic studies: AST, ALT, bilirubin, creatinine (long-term therapy)
- Blood studies: CBC, differential, Hct, Hgb, PT, cholesterol (long-term therapy)

Evaluate:

- Therapeutic response: absence of stroke, MI

Teach patient/family:

- That blood work will be necessary during treatment
- To report any unusual bruising, bleeding to prescriber; that it may take longer to stop bleeding
- To take without regard to food
- To report diarrhea, skin rashes, subcutaneous bleeding, chills, fever, sore throat
- To tell all health care providers that clopidogrel is being used; may be held for 5 days before surgery

clotrimazole (topical, vaginal, oral)

(kloe-trim′a-zole)

Cruex, Gyne-Lotrimin, Lotrimin, Lotrimin AF, MyCelex, MyCelex-7, Trivagizole 3, Desenex

Func. class.: Topical antifungal
Chem. class.: Imidazole derivative

Do not confuse: clotrimazole/miconazole/clobetasol

ACTION: Antifungal activity results from altering cell wall permeability

USES: Vulvovaginal, oropharyngeal candidiasis; topical fungal infections

CONTRAINDICATIONS: Hypersensitivity, ophthalmic use
Precautions: Hepatic impairment (oral)

DOSAGE AND ROUTES

Tinea corporis, cruris, pedis, versicolor; candidiasis

• **Adult/child ≥2 yr:** Apply to affected area and rub into area AM/PM × 2-4 wk

Vulvovaginal candidiasis

• **Adult/child ≥12 yr: Vag cream** 1 applicator at bedtime × 3 days (2%) or 7 days (1%)

Oropharyngeal candidiasis

• **Adult/child ≥3 yr: Lozenge** 1 PO dissolved 5 × /day × 2 wk or adults 1 lozenge dissolved tid (prevention)

Available forms: Topical cream, solution, 1%; vaginal cream 1%, 2%; Lozenges, troches 10 mg

Administer:

Topical route

• Topical skin products are not for intravaginal therapy and are for external use only; do not use skin products near the eyes, nose, or mouth

• Wash hands before and after use; wash affected area and gently pat dry

• **Cream/solution:** apply to the cleansed affected area; massage gently into affected areas

• **Troches:** allow to dissolve; do not chew or swallow whole

Intravaginal route

• Only use dosage formulations specified for intravaginal use; intravaginal dosage forms are not for topical therapy; do not ingest

• **Cream:** use applicator(s) supplied by the manufacturer

SIDE EFFECTS

GI: Nausea, vomiting
GU: Vaginal burning, irritation
INTEG: Burning, peeling, rash, pruritus

PHARMACOKINETICS

PO duration 3 hr

INTERACTIONS

Drug/Lab Test

Increase: LFTs

NURSING CONSIDERATIONS

Assess:

• **Allergic reaction:** assess for hypersensitivity; product might need to be discontinued

• **Infection:** assess for severity of infection, itching

• Hepatic function studies periodically if using oral troches

Evaluate:

• Decreased infection, itching

Teach patient/family:

Topical route:

• That topical skin products are not for intravaginal therapy and are for external use only; do not use skin products near the eyes, nose, or mouth; do not use occlusive dressings

• To wash hands before and after use; wash affected area and gently pat dry

• **Cream:** to shake well before use; apply a thin film to the cleansed affected area; massage gently into affected areas

PO route:

• **Troches**: allow to dissolve; do not chew or swallow whole

Intravaginal route: To use only dosage formulations specified for intravaginal use; intravaginal dosage forms are not for topical therapy; do not ingest; abstain from sexual contact; may damage condoms, diaphragms

• **Cream:** to use applicator(s) supplied by the manufacturer

cloZAPine (Rx)

(kloz′a-peen)

Clozaril, FazaClo, Versacloz

Func. class.: Antipsychotic

Chem. class.: Tricyclic dibenzodiazepine derivative

Do not confuse:

Clozaril/Clinoril/Colazal

ACTION: Interferes with DOPamine receptor binding with lack of EPS; also acts as an adrenergic, cholinergic, histaminergic, serotonergic antagonist

USES:
Management of psychotic symptoms for schizophrenic patients for whom other antipsychotics have failed; recurrent suicidal behavior

Unlabeled uses: Agitation, bipolar disorder, psychosis in dementia, psychosis in Parkinson's disease

CONTRAINDICATIONS:
Hypersensitivity, severe granulocytopenia (WBC <3500 before therapy), coma

Black Box Warning: Myeloproliferative disorders, severe CNS depression, agranulocytosis, leukopenia, neutropenia, seizure disorder

Precautions: Pregnancy (B), breastfeeding, children <16 yr, geriatric patients; CV, pulmonary, cardiac, renal, hepatic disease; seizures, prostatic enlargement, closed-angle glaucoma, stroke

Black Box Warning: Bone marrow suppression, hypotension, myocarditis, orthostatic hypotension, seizures, geriatric patients with dementia-related psychosis

DOSAGE AND ROUTES

Schizophrenia

• **Adult: PO** 12.5 mg daily or bid; may increase by 25-50 mg/day; normal range 300-450 mg/day after 2 wk; dose >500 mg require 3 divided doses; do not increase dose more than 2×/wk; max 900 mg/day; use lowest dose to control symptoms; if dose is to be discontinued, taper over 1-2 wk

• **Child ≥9 yr, adolescents (unlabeled): PO** 6.25-12.5mg initially slowly titrate

Dementia with multiple behavioral disturbances (unlabeled)

• **Geriatric: PO** 12.5 mg daily at bedtime, may increase by 12.5 mg every other day, max 50 mg/day

Available forms: Tabs 25, 50, 100, 200 mg; orally disintegrating tabs 25, 100 mg (Fazaclo: 12.5, 25, 100, 150, 200 mg); oral suspension 50 mg/ml

Administer:

• May be taken with or without food

• Patient-specific registration required before administration; if WBC <3500 cells/mm^3 or ANC <2000 cells/mm^3, therapy should not be started; may only dispense the 7, 14, 28 day supply upon receipt of lab report that is appropriate

• Check to confirm PO medication swallowed; monitor for hoarding or giving of medication to other patients, if hospitalized; avoid giving patient >7 days' worth of medication if outpatient

• Store in tight, light-resistant container

• **Orally disintegrating tab:** do not push through foil; leave in foil blister until ready to take, peel back foil, place tab in mouth; allow to dissolve, swallow; water is not needed

• **Oral suspension**

SIDE EFFECTS

CNS: Neuroleptic malignant syndrome, *sedation, salivation, dizziness, headache, tremors, sleep problems, akinesia, fever,* seizures, *sweating, akathisia, confusion, fatigue, insomnia,* depression, slurred speech, anxiety, *agitation,* dystonia, obsessive-compulsive symptoms

CV: *Tachycardia, hypo/hypertension,* chest pain, ECG changes, orthostatic hypotension

EENT: *Blurred vision*

GI: *Drooling or excessive salivation, constipation, nausea, abdominal discomfort, vomiting, diarrhea,* anorexia, *weight gain, dry mouth,* heartburn, *dyspepsia, gastroesophageal reflux*

GU: *Urinary abnormalities,* incontinence, ejaculation dysfunction; frequency, urgency, retention, dysuria

HEMA: Leukopenia, agranulocytosis, eosinophilia

MS: Weakness; pain in back, neck, legs; spasm, *rigidity*

OTHER: *Diaphoresis*

RESP: Dyspnea, nasal congestion, lower respiratory tract infection

SYST: Death among geriatric patients with dementia, aggravation of diabetes mellitus

PHARMACOKINETICS

Bioavailability 27%-47%; 97% protein bound; completely metabolized by liver enzymes involved in metabolism CYP1A2, 2D6, 3A4; excreted in urine (50%), feces (30%) (metabolites); half-life 8-12 hr

INTERACTIONS

Increase: CNS depression—CNS depressants, psychoactives, alcohol

Increase: cloZAPine level—caffeine, citalopram, FLUoxetine, sertraline, ritonavir, risperiDONE, CYP1A2 inhibitors (fluvoxaMINE), CYP3A4 inhibitors (ketoconazole, erythromycin)

Increase: plasma conc—warfarin, digoxin, other highly protein-bound products

Increase: QT prolongation—β blockers, class IA/III antidysrhythmias, and other drugs that increase QT

Increase: hypotension, respiratory, cardiac arrest, collapse—benzodiazepines

Decrease: cloZAPine level—CYP1A2 inducers (carBAMazepine, omeprazole, rifampin); PHENobarbital

Drug/Lab Test

Increase: LFTs, cardiac enzymes, cholesterol, blood glucose, bilirubin, PBI, cholinesterase, ^{131}I, Hct/Hgb, erythrocyte sedimentation rate

Decrease: WBC

False positive: pregnancy tests, PKU

False negative: urinary steroids, 17-OHCS

NURSING CONSIDERATIONS

Assess:

Black Box Warning: Myocarditis; if suspected, discontinue use; myocarditis usually occurs during 1st month of treatment

Black Box Warning: Seizures; usually occur with higher doses >600 mg/day or dosage change >100 mg/day; do not use in uncontrolled seizure disorder, use cautiously in those with a predisposition to seizures

- I&O ratio; obtain baseline before treatment begins; palpate bladder if low urinary output occurs

Black Box Warning: **Bone marrow depression:** bilirubin, CBC, LFTs monthly; discontinue treatment if WBC <3000/mm³ or ANC <1500/mm³; test weekly; may resume when normal; if WBC <2000/mm³ or ANC <1000/mm³, discontinue; if agranulocytosis develops, never restart product

- Urinalysis recommended before, during prolonged therapy
- Affect, orientation, LOC, reflexes, gait, coordination, sleep pattern disturbances

Black Box Warning: **Hypotension:** B/P standing and lying; take pulse, respirations q4hr during initial treatment; establish baseline before starting treatment; report drops of 30 mm Hg

- Dizziness, faintness, palpitations, tachycardia on rising
- **EPS** including akathisia (inability to sit still, no pattern to movements), tardive dyskinesia (bizarre movements of the jaw, mouth, tongue, extremities), pseudoparkinsonism (rigidity, tremors, pill rolling, shuffling gait)

⚠ **Neuroleptic malignant syndrome:** tachycardia, seizures, fever, dyspnea, diaphoresis, increased/decreased B/P; notify prescriber immediately

- Constipation, urinary retention daily; if these occur, increase bulk, water in diet, especially for geriatric patients; stool softeners, laxatives may be needed
- If diabetic, check blood glucose levels
- Supervised ambulation until stabilized on medication; do not involve patient in strenuous exercise program because fainting is possible; patient should not stand still for long periods

Evaluate:

- Therapeutic response: decrease in emotional excitement, hallucinations, delusions, paranoia, reorganization of patterns of thought, speech

Teach patient/family:

- About symptoms of agranulocytosis and need for blood tests weekly for 6 mo, then q2wk; to report flulike symptoms
- That orthostatic hypotension often occurs; to rise gradually from sitting or lying position; to avoid hot tubs, hot showers, tub baths; hypotension may occur
- To avoid abrupt withdrawal of this product because EPS may result; that product should be withdrawn over 1-2 wk
- To avoid OTC preparations (cough, hay fever, cold) unless approved by prescriber, since serious product interactions may occur; to avoid use with alcohol or CNS depressants, increased drowsiness may occur
- About compliance with product regimen
- About EPS and necessity for meticulous oral hygiene, since oral candidiasis may occur

Black Box Warning: To report sore throat, malaise, fever, bleeding, mouth sores; if these occur, CBC should be drawn and product discontinued

- That heat stroke may occur in hot weather; to take extra precautions to stay cool
- To avoid driving, other hazardous activities; seizures may occur
- To notify prescriber if pregnant or if pregnancy is intended; not to breastfeed

TREATMENT OF OVERDOSE:

Lavage, activated charcoal; provide an airway; do not induce vomiting

codeine (Rx)

(koe′deen)

Func. class.: Opiate analgesic, antitussive

Chem. class.: Opiate, phenathrene derivative

Controlled Substance Schedule II, III, IV, V (depends on content)

Do not confuse:

codeine/Lodine/iodine/Cardene

ACTION: Depresses pain impulse transmission at the spinal cord level by interacting with opioid receptors; decreases cough reflex, GI motility

USES: Moderate to severe pain

Unlabeled uses: Diarrhea, arthralgia, bone/dental pain, headache, migraine, myalgia, nonproductive cough

CONTRAINDICATIONS: Breastfeeding, hypersensitivity to opiates, respiratory depression, increased intracranial pressure, seizure disorders, severe respiratory disorders

Black Box Warning: Children (tonsillectomy/adenoidectomy)

Precautions: Pregnancy (C), geriatric patients, cardiac dysrhythmias, prostatic hypertrophy, bowel impaction

DOSAGE AND ROUTES

Pain

- **Adult: PO** 15-60 mg q4hr prn
- **Child 6-17 yr: PO** 3 mg/kg/day in divided doses q4hr prn

Cough

- **Adult: PO** 10-20 mg q4-6hr, max 120 mg/day

Renal disease

- **Adult: PO** CCr 10-50 ml/min, 75% of dose; CCr <10 ml/min, 50% of dose

Diarrhea (unlabeled)

- **Adult: PO** 30 mg; may repeat qid prn

Arthralgia/bone pain/back pain/dental pain/headache/migraine/myalgia (unlabeled)

• **Adult: PO** 15-60 mg q4-6hr

• **Child ≥3 yr: PO** 0.5-1 mg/kg or 15 mg/m^2 (max 60 mg/dose) q4-6hr

Available forms: Tabs 15, 30, 60 mg; inj 15, 30 mg/ml; oral sol 30 mg/5 ml, 30 mg/5 ml sol

Administer:

Discontinue gradually after long-term use

• Store in light-resistant container at room temperature

SIDE EFFECTS

CNS: *Drowsiness, sedation,* dizziness, agitation, dependency, lethargy, restlessness, euphoria, **seizures,** hallucinations, headache, confusion

CV: Bradycardia, palpitations, orthostatic hypotension, tachycardia, **circulatory collapse**

GI: *Nausea, vomiting, anorexia, constipation,* dry mouth

GU: Urinary retention

INTEG: Flushing, rash, urticaria, pruritus

RESP: **Respiratory depression, respiratory paralysis,** dyspnea

SYST: **Anaphylaxis**

PHARMACOKINETICS

Bioavailability 60%-90%; peak ½-1 hr; duration 4-6 hr; metabolized by liver (CYP3A4 to morphine); excreted by kidneys, in breast milk; crosses placenta; half-life 3 hr; protein binding 7%; altered codeine metabolism occurs in different ethnic groups

PO: Onset 30-60 min

INTERACTIONS

Increase: CNS depression—CYP2D6, alcohol, opiates, sedative/hypnotics, antipsychotics, skeletal muscle relaxants

⚠ **Increase: toxicity—MAOIs; use cautiously**

Drug/Lab Test

Increase: lipase, amylase

NURSING CONSIDERATIONS

Assess:

• **Pain:** intensity, type, location, aggravating, alleviating factors; need for pain medication, tolerance; use pain scoring

• I&O ratio; check for decreasing output; may indicate urinary retention, especially among geriatric patients

• GI function: nausea, vomiting, constipation

• **Cough:** type, duration, ability to raise secretion for productive cough; do not use to suppress productive cough

• CNS changes, dizziness, drowsiness, hallucinations, euphoria, LOC, pupil reaction

• Allergic reactions: rash, urticaria

Black Box Warning: Children (tonsillectomy/adenoidectomy): Deaths have occurred; use is contraindicated

⚠ **Respiratory dysfunction: respiratory depression, character, rate, rhythm; notify prescriber if respirations are <10/min, shallow; obstructive sleep apnea (children) (tonsillectomy/adenoidectomy)**

Evaluate:

• Therapeutic response: decrease in pain, absence of grimacing, decreased cough, decreased diarrhea

Teach patient/family:

• Not to breastfeed

• To report any symptoms of CNS changes, allergic reactions

• That physical dependency may result after extended periods

• To decrease dry mouth use sugarless gum, rinse mouth often

• To change position slowly; orthostatic hypotension may occur

• To avoid hazardous activities if drowsiness, dizziness occurs

• To avoid alcohol, other CNS depressants unless directed by prescriber

TREATMENT OF OVERDOSE:

Naloxone 0.4-mg ampule diluted in 10 ml 0.9% NaCl and given by direct IV push, 0.02 mg q2min (adult)

colchicine (Rx)

(kol′chih-seen)

Colcrys

Func. class.: Antigout agent

Chem. class.: *Colchicum autumnale* alkaloid

ACTION: Inhibits microtubule formation of lactic acid in leukocytes, which decreases phagocytosis and inflammation in joints

USES: Gout, gouty arthritis (prevention, treatment); to arrest the progression of neurologic disability in those with MS, Mediterranean fever

Unlabeled uses: Hepatic cirrhosis, pericarditis, amyloidosis, Behçet's syndrome, biliary cirrhosis, dermatitis herpetiformis, idiopathic thrombocytopenic purpura, Paget's disease, pseudogout, pulmonary fibrosis

CONTRAINDICATIONS: Pregnancy (D) (injectable), serious GI, severe cardiac/renal/hepatic disorders, hypersensitivity

Precautions: Pregnancy (C) (PO), breastfeeding, children, geriatric patients, blood dyscrasias, hepatic disease

DOSAGE AND ROUTES

Gout prevention

- **Adult: PO** 0.6-1.2 mg/day, depending on severity

Gout treatment

- **Adult: PO** 1.2 mg initially, then 0.6 mg 1 hr later (1.8 mg); for those on strong CYP3A4 inhibitor (during past 14 days), 0.6 mg initially, then 0.3 mg 1 hr later

Renal dose

- **Adult: PO** CCr <30 ml/min, for acute gout, do not repeat course for 2 wk; for familial Mediterranean fever, 0.3 mg daily, increase cautiously

Mediterranean fever (unlabeled)

- **Adult on no interacting products: PO** 1.2-2.4 mg/day in 1-2 divided doses; strong CYP3A4 inhibitor; P-glycoprotein inhibitors within 14 day: max 0.6 mg/day in 1-2 divided doses; moderate CYP3A4 inhibitors with 14 day: max 1.2 mg/day in 1-2 divided doses
- **Adolescent: PO** 1.2-2.4 mg/day in 1-2 divided doses, titrate by 0.3 mg/day
- **Child > 6-12 yr: PO** 0.9-1.8 mg/day in 1-2 divided doses
- **Child 4-6 yr: PO** 0.3-1.8 mg/day in 1-2 divided doses

Amyloidosis/biliary cirrhosis/dermatitis herpetiformis/Paget's disease/Behçet's syndrome/chronic idiopathic thrombocytopenic purpura/pulmonary fibrosis (unlabeled)

- **Adult: PO** 0.5-0.6 mg bid-tid

Available forms: Tabs 0.5, 0.6, 1 ✦ mg

Administer:

PO route

- Without regard to food
- Cumulative doses ≤4 mg, renal patients ≤2 mg; when reached, administer only for 3 wk

SIDE EFFECTS

GI: *Nausea, vomiting, anorexia, malaise,* metallic taste, cramps, peptic ulcer, diarrhea

GU: Hematuria, oliguria, renal damage

HEMA: Agranulocytosis, thrombocytopenia, aplastic anemia, pancytopenia

INTEG: Chills, dermatitis, pruritus, purpura, erythema

MISC: Myopathy, alopecia, reversible azoospermia, peripheral neuritis

PHARMACOKINETICS

PO: Peak ½-2 hr, half-life 4.4 hr, deacetylates in liver, excreted in feces (metabolites/active product)

INTERACTIONS

Increase: colchicine level/toxicity—moderate/strong CYP3A4 inhibitors, reduce dose

Increase: GI effects—NSAIDs, ethanol

Increase: bone marrow depression—radiation, bone marrow depressants, cycloSPORINE
Decrease: action of vit B_{12}; may cause reversible malabsorption
Drug/Food
Increase: colchicine level—grapefruit juice
Drug/Lab Test
Increase: alk phos, AST
Decrease: platelets, WBC, granulocytes
False positive: urine Hgb
Interference: urinary 17-hydroxycorticosteroids

NURSING CONSIDERATIONS

Assess:
• Relief of pain, uric acid levels returning to normal
• I&O ratio; observe for decrease in urinary output
⚠ CBC, platelets, reticulocytes before, during therapy (q3mo); may cause aplastic anemia, agranulocytosis, decreased platelets
• **Toxicity:** weakness, abdominal pain, nausea, vomiting, diarrhea; product should be discontinued, report symptoms immediately
Evaluate:
• Therapeutic response: decreased stone formation, decreased pain in kidney region, absence of hematuria, decreased pain in joints
Teach patient/family:
• To avoid alcohol, OTC preparations that contain alcohol
• To report any pain, redness, hard areas, usually in legs; rash, sore throat, fever, bleeding, bruising, weakness, numbness, tingling, nausea, vomiting, abdominal pain
• About the importance of complying with medical regimen (diet, weight loss, product therapy); about the possibility of bone marrow depression occurring
• Advise all providers of product use; surgery may increase possibility of acute gout symptoms

TREATMENT OF OVERDOSE:

D/C medication; may need opioids to treat diarrhea

colesevelam (Rx)

(koe-leh-seve′eh-lam)
WelChol
Func. class.: Antilipemic
Chem. class.: Bile acid sequestrant

ACTION: Adsorbs, combines with bile acids to form insoluble complex excreted through feces; loss of bile acids lowers cholesterol levels

USES: Elevated LDL cholesterol, alone or in combination with HMG-COA reductase inhibitor; type 2 diabetes (adjunct)

CONTRAINDICATIONS: Hypersensitivity, bowel disease, primary biliary cirrhosis, triglycerides >500 mg/dl, bowel obstruction, pancreatitis, biliary obstruction, dysphagia, fat-soluble vitamin deficiency
Precautions: Pregnancy (B), breastfeeding, children

DOSAGE AND ROUTES

Monotherapy
• **Adult: PO** 3 625-mg tabs bid with meals or 6 tabs daily with meal; may increase to 7 tabs if needed
Combination therapy
• **Adult: PO** 3 tabs bid with meals or 6 tabs daily with meal given with an HMG-CoA reductase inhibitor
Type 2 diabetes, adjunct (to improve glycemic control)
• **Adult and geriatric: PO** Approx 3.8 g (6 tabs)/day or approx 1.9 g (3 tabs) bid
Heterozygous familial hypercholesterolemia
• **Females (postmenarchal and >10 yr) and males ≥10 yr: PO** 1.875-g packet bid or 3.75-g packet daily dissolved in 4-8 oz of water with meal
Available forms: Tabs 625 mg; powder for oral susp 3.75 g/packet

Administer:

- Swallow tabs whole; do not break, crush, or chew
- Give product daily or bid with meals; give all other medications 4 hr before colesevelam; with liquid to avoid poor absorption
- **Powder for oral susp:** empty contents of packet into a cup/glass, add 1/2-1 cup (4-8 oz) of water, fruit juice, diet soda; stir well before drinking

SIDE EFFECTS

GI: *Constipation, abdominal pain, nausea,* fecal impaction, hemorrhoids, flatulence, vomiting, GI obstruction
MISC: Hypertriglycerides, hypoglycemia
MS: Muscle, joint pain

PHARMACOKINETICS

Excreted in feces, peak response 2 wk

INTERACTIONS

Decrease: absorption of diltiazem, gemfibrozil, mycophenolate, phenytoin, propranolol, warfarin, thiazides, digoxin, penicillin G, tetracyclines, corticosteroids, iron, thyroid, fat-soluble vitamins, glyBURIDE, fluoroquinolones
Decrease: action of—oral contraceptives
Drug/Lab Test
Increase: LFTs

NURSING CONSIDERATIONS

Assess:

- Cardiac glycoside level, if both products administered
- Fasting LDL, HDL, total cholesterol, triglyceride levels, electrolytes if on extended therapy
- Bowel pattern daily; increase bulk, water in diet for constipation

Evaluate:

- Therapeutic response: decreased total cholesterol level, LDL cholesterol, apolipoproteins

Teach patient/family:

- About the importance of compliance; toxicity may result if doses missed; timing of dose 4 hr after other meds
- That risk factors should be decreased: high-fat diet, smoking, alcohol consumption, absence of exercise

conivaptan (Rx)

(kon-ih-vap′tan)

Vaprisol

Func. class.: Vasopressin receptor antagonist

ACTION:

Dual arginine vasopressin (AVP) antagonist with affinity for V_{1A}, V_2 receptors; level of AVP in circulating blood is critical for regulation of water, electrolyte balance and is usually elevated in euvolemic/hypervolemic hyponatremia

USES:

Euvolemia hyponatremia in hospitalized patients; not indicated for CHF, hypervolemic hyponatremia

CONTRAINDICATIONS:

Hypersensitivity, hypovolemia
Precautions: Pregnancy (C), breastfeeding, orthostatic disease, renal disease, heart failure, rapid correction of serum sodium

DOSAGE AND ROUTES

- **Adult: IV INF** loading dose 20 mg given over 30 min then **CONT IV** over 24 hr; after 1 day, give for an additional 1-3 days as a **CONT INF** of 20 mg/day total; can be titrated up to 40 mg/day if serum sodium is not rising at desired rate; max time 4 days

Hepatic/renal dose

- **Adult: IV Child-Pugh A-C or CCr 30-60 ml/min:** give IV loading dose over 10 min then **CONT IV INF** 10 mg over 24 hr × 2-4 days

Available forms: 5 mg/ml (20 mg/4 ml) in single-use ampule; 20 mg/100 ml in D_5 for inj

Administer:

IV route

- Withdraw 4 ml (20 mg), add to 100 ml D_5W, gently invert several times to mix, give over 30 min; in large vein, change site every 24 hr to minimize vascular irritation

Continuous IV INF route

- Withdraw 4 ml (20 mg), add to 250 ml D_5W, gently invert several times to mix, give over 24 hr; or give 40 mg in 250 ml D_5W, gently invert several times to mix, give over 24 hr

SIDE EFFECTS

CNS: Headache, confusion, insomnia
CV: Atrial fibrillation, hypo/hypertension, *orthostatic hypotension,* phlebitis
GI: Nausea, vomiting, constipation, dry mouth
GU: Hematuria, polyuria, UTI, pollakiuria
HEMA: Anemia
INTEG: Erythemia, inj site reaction
META: Dehydration, hypo/hyperglycemia, hypokalemia, hypomagnesia, hyponatremia
MISC: Oral candidiasis, pain, peripheral edema, pneumonia

PHARMACOKINETICS

Protein binding 99%, metabolized by CYP3A4, terminal half-life 5 hr

INTERACTIONS

Increase: effect of—CYP3A4 substrates (alfuzosin, ARIPiprazole, bexarolene, bortezomib, bosentan, bupivacaine, buprenorphine, carBAMazepine, cevimeline, cilostazol, cinacalcet, clopidogrel, colchicine, cyclobenzaprine, dapsone, darifenacin, disopyramide, DOCEtaxel, donepezil, DOXOrubicin, dutasteride, eletriptan, eplerenone, ergots, erlotinib, eszopiclone, ethinyl estradiol, ethosuximide, etoposide, fentaNYL, galantamine, gefitinib, halofantrine, ifosfamide, irinotecan, levobupivacaine, levomethadyl, lidocaine, loperimide, loratadine, mefloquine, methadone, modafinil, PACLitaxel, paricalcitrol, pimozide, praziquantel, quiNIDine, quiNINE, ramelteon, reboxetine, repaglinide, rifabutin, sibutramine, sildenafil, sirolimus, SUFentanil, SUNItinib, tacrolimus, tamoxifen, teniposide, testosterone, tiaGABine, tinidazole, trimetrexate, vardenafil, vinca alkaloids, ziprasidone, zolpidem, zonisamide); do not use concurrently

NURSING CONSIDERATIONS

Assess:

- Renal, hepatic function
- Frequent sodium volume status; overly rapid correction of sodium conc (>12 mEq/L per 24 hr) may result in osmotic demyelination syndrome
- Neurologic status: confusion, headache
- CV status: atrial fibrillation, hypo/hypertension, orthostatic hypotension; monitor B/P, pulse
- Monitor other electrolytes (magnesium and potassium)

Evaluate:

- Therapeutic response: correction of serum sodium levels

Teach patient/family:

- To report neurologic changes: headache, insomnia, confusion
- About administration procedure and expected results
- To report inj site pain, redness, swelling

CONTRACEPTIVES, HORMONAL

Monophasic, Oral

ethinyl estradiol/ desogestrel (Rx)

Apri, Cesia, Desogen, Kariva, Mircette, Ortho-Cept, Reclipsen, Solia, Velivet

ethinyl estradiol/ drospirenone (Rx)

Yasmin, Yaz 28, Ocella

ethinyl estradiol/ ethynodiol (Rx)

Kelnor 1/35, Zovia 1/35, Zovia 1/50

ethinyl estradiol/ levonorgestrel (Rx)

Alesse, Aviane-28, Enpresse, Jolessa, Lessina, Levlen, Levlite, Levora, Lutera, Nordette, Portia, Quasense, Seasonique, Sronyx

ethinyl estradiol/ norethindrone (Rx)

Brevicon, Genora 1/35, Junel 21 1/20, Junel 21 1.5/20, Loestrin 21 1.5/30, Loestrin 21 1/20, Microgestin, Modicon, N.E.E 1/35, Necon 0.5/35, Norcept-E 1/35, Norinyl 1+35, Nortrel 1/35, Nortrel 7/7/7

ethinyl estradiol/ norgestimate (Rx)

MonoNessa, Ortho-Cyclen, Previfem, Sprintec, Tri-Sprintec

ethinyl estradiol/ norgestrel (Rx)

Cryselle, Lo/Ovral, Low-Ogestrel, Ogestrel, Ovral

mestranol/ norethindrone (Rx)

Norinyl 1+50

Biphasic, Oral

ethinyl estradiol/ norethindrone (Rx)

Ortho-Novum 10/11

Triphasic, Oral

ethinyl estradiol/ desogestrel (Rx)

Cyclessa

ethinyl estradiol/ norethindrone (Rx)

Nortrel 7/7/7, Ortho-Novum 7/7/7, Tri-Norinyl

ethinyl estradiol/ norgestimate (Rx)

Ortho Tri-Cyclen, Ortho Tri-Cyclen Lo

ethinyl estradiol/ levonorgestrel (Rx)

Enpresse, Tri-Levlen, Triphasil

Extended Cycle, Oral

ethinyl estradiol/ levonorgestrel (Rx)

Seasonale

Progestin, Oral

norethindrone (Rx)

Errin, Ortho Micronor, Camila, Jolivette, Nor-Q D

Progressive Estrogen, Oral

ethinyl estradiol/ norethindrone acetate (Rx)

Estrostep, Estrostep Fe

Emergency

levonorgestrel/ethinyl estradiol (Rx)

Preven

levonorgestrel (Rx)

Plan B

medroxyPROGESTERone (Rx)

Depo-Provera

Intrauterine

levonorgestrel (Rx)

Mirena

Implant

etonogestrel (Rx)

Implanon

Vaginal Ring

ethinyl estradiol/ etonogestrel (Rx)

Nuva Ring

Transdermal

ethinyl estradiol/ norelgestromin (Rx)

Ortho Evra

ACTION: Prevents ovulation by suppressing FSH and LH; *monophasic:* estrogen/progestin (fixed dose) used during a 21-day cycle; ovulation is inhibited by suppression of FSH and LH; thickness of cervical mucus and endometrial lining prevents pregnancy; *biphasic:* ovulation is inhibited by suppression of FSH and LH; alteration of cervical mucus, endometrial lining prevents pregnancy; *triphasic:* ovulation is inhibited by suppression of FSH and LH; change of cervical mucus, endometrial lining prevents pregnancy; variable doses of estrogen/progestin combinations may be similar to natural hormonal fluctuations; *extended cycle:* estrogen/progestin continuous for 84 days, off for 7 days, results in 4 menstrual periods/yr; *progressive estrogen:* constant progestin with 3 progressive doses of estrogen; *progestin-only pill, implant, intrauterine:* change of cervical mucus and endometrial lining prevents pregnancy; ovulation may be suppressed

USES: To prevent pregnancy, regulation of menstrual cycle, treatment of acne in women >14 yr for whom other treatment has failed, emergency contraception; *injection:* inhibits gonadotropin secretion, ovulation, follicular maturation; *emergency:* inhibits ovulation and fertilization, decreases transport of sperm and egg from fallopian tube to uterus; *vaginal ring, transdermal:* inhibits ovulation, prevents sperm entry into uterus; *antiacne:* may decrease sex hormone binding globulin, results in decreased testosterone

CONTRAINDICATIONS: Pregnancy (X), breastfeeding, women ≥40 yr, reproductive cancer, thrombophlebitis, MI, hepatic tumors, hepatic disease, CAD, CVA, breast cancer, jaundice, stroke, vaginal bleeding

Precautions: Depression, hypertension, renal disease, seizure disorders, lupus erythematosus, rheumatic disease, migraine headache, amenorrhea, irregular menses, gallbladder disease, diabetes mellitus, heavy smoking, acute mononucleosis, sickle cell disease

Black Box Warning: Tobacco smoking

DOSAGE AND ROUTES

Monophasic

- **Adult: PO** Take first tab on Sunday after start of menses × 21 days; skip 7 days, then repeat cycle; start on 1st day of menses × 21 days; skip 7 days, then repeat cycle; may contain 7 placebo tabs when 1 tab is taken daily

Biphasic

- **Adult: PO** Take 10 days of small progestin, then large progestin; estrogen is the same during cycle; skip 7 days, then repeat cycle; may contain 7 placebo tabs when 1 tab is taken daily

Triphasic

• **Adult: PO** Estrogen dose remains constant; progestin changes throughout 21-day cycle; some products contain 28 tabs per month

Extended cycle

• **Adult: PO** Start taking on 1st day of menses; continue for 84 days of active tab, then 7 days of placebo; repeat cycle

Progestin

• **Adult: PO** Start on 1st day of menses, then daily and continuously

Progressive estrogen

• **Adult: PO** Progestin dose remains constant; estrogen increases q7days throughout 21-day cycle; may include 7 placebo tabs for 28-day cycle

Emergency

• **Adult/adolescent:** Give within 72 hr of intercourse, repeat 12 hr later; **Plan B** 1 tab, then 1 tab 12 hr later; **Preven** 2 tab, then 2 tab 12 hr later; **Ovral (unlabeled)** 2 white tabs; **Lo/Ovral (unlabeled)** 4 white tabs; **Levlen (unlabeled), Nordette (unlabeled)** 4 orange tabs; **Triphasil (unlabeled), Tri-Levlen (unlabeled)** 4 yellow tabs

Injectable

• **Adult: IM (Depo-Provera)** 150 mg within 5 days of start of menses or within 5 days postpartum (must not be breastfeeding); if breastfeeding, give 6 wk postpartum, repeat q3mo

Intrauterine

• **Adult:** To be inserted using the levonorgestrel-releasing intrauterine system (LRIS) by those trained in procedure; inserted into uterine cavity within 7 days of the onset of menstruation; use should not exceed 5 yr per implant

Vaginal ring

• **Adult: VAG** Insert 1 ring on or before day 5 of cycle; leave in place 3 wk; remove for 1 wk, then repeat

Transdermal

• **Adult: TD** Apply patch within 7 days of menses; change weekly × 3 wk; no patch wk 4; repeat cycle

Implant

• **Adult: SUBDERMAL** In inner side of upper arm on days 1-5 of menses, replace q3yr

Acne

• **Adult: PO (Ortho Tri-Cyclen)** Take daily × 21 days, off 7 days

Administer:

• PO with food for GI symptoms; give at same time each day

• Subdermal implant of 6 caps effective for 5 yr, then should be removed

• IM inj deep in large muscle mass after shaking suspension; ensure patient not pregnant if inj are 2 wk or more apart

SIDE EFFECTS

CNS: Depression, fatigue, dizziness, nervousness, anxiety, headache

CV: Increased B/P, cerebral hemorrhage, thrombosis, pulmonary embolism, fluid retention, edema, MI

EENT: Optic neuritis, retinal thrombosis, cataracts

ENDO: Decreased glucose tolerance, increased TBG, PBI, T_4, T_3, temporary infertility

GI: *Nausea,* vomiting, cramps, diarrhea, bloating, constipation, change in appetite, cholestatic jaundice, weight change

GU: Breakthrough bleeding, amenorrhea, spotting, dysmenorrhea, galactorrhea, endocervical hyperplasia, vaginitis, cystitis-like syndrome, breast changes

HEMA: Increased fibrinogen, clotting factor

INTEG: *Chloasma, melasma,* acne, rash, urticaria, erythema, pruritus, hirsutism, alopecia, photosensitivity

PHARMACOKINETICS

Excreted in breast milk

INTERACTIONS

Decrease: oral contraceptives effectiveness—anticonvulsants, rifampin, analgesics, antibiotics, antihistamines, griseofulvin

Decrease: oral anticoagulants action

Drug/Herb
- Altered action: black cohosh

Decrease: oral contraceptives effect—saw palmetto, St. John's wort

Drug/Food

Increase: peak level—grapefruit juice

Drug/Lab Test

Increase: PT; clotting factors VII, VIII, IX, X; TBG, PBI, T_4, platelet aggregability, BSP, triglycerides, bilirubin, AST, ALT

Decrease: T_3, antithrombin III, folate, metyrapone test, GTT, 17-OHCS

NURSING CONSIDERATIONS

Assess:
- Glucose, thyroid function, LFTs, BP
- Reproductive changes: changes in breasts, tumors; positive Pap smear; product should be discontinued

Evaluate:
- Therapeutic response: absence of pregnancy, endometriosis, hypermenorrhea

Teach patient/family:
- About detection of clots using Homan's sign
- To use sunscreen or avoid sunlight; photosensitivity can occur
- To take at same time each day to ensure equal product level
- To report GI symptoms that occur after 4 mo
- To use another birth control method during 1st week of oral contraceptive use
- To take another tablet as soon as possible if one is missed
- That, after product is discontinued, pregnancy may not occur for several months
- To report abdominal pain, change in vision, shortness of breath, change in menstrual flow, spotting, breakthrough bleeding, breast lumps, swelling, headache, severe leg pain
- That continuing medical care is needed: Pap smear and gynecologic examinations q6mo
- To notify health care providers and dentists of oral contraceptive use

Black Box Warning: Do not smoke; increased risk of CV side effects

⚠ HIGH ALERT

crizotinib
(kriz-oh'ti-nib)

XALKORI

Func. class.: Antineoplastic; biologic response modifiers

Chem. class.: Signal transduction inhibitors (STIs)

ACTION: Inhibits receptor tyrosine kinases (anaplastic lymphoma kinase (ALK), Hepatocyte Growth Factor Receptor (HGFR, c-Met), Recepteur d'Origine Nantais (RON)

USES: Locally advanced or metastatic non–small-cell lung cancer (NSCLC) that is anaplastic lymphoma kinase (ALK)-positive as detected by an FDA-approved test

CONTRAINDICATIONS: Pregnancy (D), breastfeeding, hypersensitivity

Precautions: Neonates, infants, children, adolescents, pneumonitis, severe hepatic disease, congenital long QT syndrome, severe renal impairment, end-stage renal disease, vision disorders

DOSAGE AND ROUTES
- **Adult: PO** 250 mg bid, continue as long beneficial

Dose adjustments for hematologic toxicities
- **For Grade 1-2:** no dosage adjustment needed; **Grade 3:** interrupt treatment until toxicity resolves to grade ≤2, then continue with the same dosage schedule; in case of recurrence after a grade 4 event with dose reduction, interrupt treatment until toxicity resolves to grade ≤2; when resuming treatment, reduce dosage to 250 mg PO daily; **Grade 4:** interrupt

treatment until toxicity resolves to grade ≤2; when resuming treatment, reduce dosage to 200 mg PO bid; in case of grade 4 recurrence, permanently discontinue treatment

Dose adjustments for hepatic laboratory abnormalities

• **For Grade 1:** No dosage adjustment necessary; **Grade 2 ALT/AST elevations with grade ≤1 total bilirubin elevations:** no dosage adjustment necessary; **Grade 3-4 ALT/AST elevations with grade ≤1 total bilirubin elevations:** interrupt treatment until toxicity resolves to grade ≤1 or baseline; when resuming treatment, reduce dosage to 200 mg PO bid; in case of recurrence, interrupt treatment until toxicity resolves to grade ≤1, and when resuming treatment, reduce dosage to 250 mg PO daily; permanently discontinue treatment in case of further recurrence; **Grade 2-4 ALT/AST elevations with concurrent Grade 2-4 total bilirubin elevations (in the absence of cholestasis or hemolysis):** permanently discontinue treatment

Dose adjustments for pneumonitis not attributable to NSCLC progression, other pulmonary disease, infection, or radiation effect

• **For any grade pneumonitis:** permanently discontinue

Dose adjustment for QTc prolongation:

• **For Grade 1-2 QTc prolongation:** no dosage adjustment necessary; **For Grade 3 QTc prolongation:** interrupt treatment until toxicity resolves to grade ≤1; when resuming treatment, reduce dosage to 200 mg PO bid; in case of recurrence, interrupt treatment until toxicity resolves to grade ≤1 and when resuming treatment, reduce dosage to 250 mg PO daily; permanently discontinue in case of further recurrence; **For Grade 4 QTc prolongation:** permanently discontinue

Available forms: Cap 200, 250 mg

Administer:

• May be taken orally with or without food

• Have the patient swallow capsule whole; do not crush or chew

• If a dose is missed, it can be taken up to 6 hr before the next dose is due to maintain the twice daily regimen; do not take both doses at the same time

• Store at room temperature

SIDE EFFECTS

CNS: Dizziness, balance disorder, presyncope, neuropathy (motor and sensory), burning sensation, dysesthesia, hyperesthesia, hypoesthesia, neuralgia, paresthesias, peripheral neuropathy (motor and sensory), headache, insomnia

CV: QT prolongation, disseminated intravascular coagulation (DIC), septic shock, bradycardia

EENT: *Diplopia, photopsia,* photophobia, *blurred vision,* visual field defect, *vitreous floaters,* visual brightness, reduced *visual acuity;* esophageal disorders

GI: *Nausea, diarrhea, vomiting, constipation,* decreased appetite, dysgeusia, abdominal pain, abdominal discomfort/pain, stomatitis, oral ulceration, glossodynia, glossitis, cheilitis, mucosal inflammation, oropharyngeal pain/discomfort, oral pain, esophageal disorder, elevated hepatic enzymes and hyperbilirubinemia, hepatotoxicity, dyspepsia, dysphagia, epigastric discomfort/pain, burning, esophagitis, esophageal obstruction/pain/spasm, esophageal ulceration, gastroesophageal reflux, odynophagia, and reflux esophagitis

HEMA: Grade 3/4 neutropenia, thrombocytopenia, lymphopenia

MISC: Fatigue, fever, *edema,* localized/peripheral edema, chest pain (unspecified), chest discomfort, musculoskeletal chest pain, arthralgia, back pain, *rash*

RESP: Severe, life-threatening pneumonitis, pneumonia, hypoxia, acute respiratory distress syndrome (ARDS), dyspnea, empyema, pulmonary hemorrhage, pulmonary embolism, upper respiratory tract infection (nasopharyngitis, pharyngitis, rhinitis), cough

PHARMACOKINETICS

Protein binding 91%; distribution into the tissues and plasma; metabolized by the CYP3A4/5; primary metabolic pathways are oxidation to metabolites; terminal half-life 42 hr; excreted 63% feces, 22% urine; unchanged drug 53% feces, 2.3% urine; absolute bioavailability is 43%; peak is 4-6 hr; steady state is reached within 15 days; dosage adjustments may need to be made in hepatic/renal disease and Asian patients

INTERACTIONS

Increase: CYP2B6 substrates (prasugrel, selegiline, cyclophosphamide)
Increase: CYP3A4 inhibitors (ketoconazole, atazanavir, indinavir, itraconazole, nefazodone, nelfinavir, ritonavir, voriconazole, boceprevir, delavirdine, isoniazid, dalfopristin; quinupristin, tipranavir)
Decrease: CYP3A4 inducers (rifampin, carBAMazepine, PHENobarbital, phenytoin, rifabutin); antacids, H2-blockers, proton pump inhibitors (PPIs)
Increase: action of—midazolam
Avoid use with CYP3A4 substrates(alfentanil, cycloSPORINE, ergotamine, dihydroergotamine fentaNYL, sirolimus, colchicine)
Increase: QT prolongation, torsades de pointes—arsenic trioxide, certain phenothiazines (chlorproMAZINE, mesoridazine, thioridazine), grepafloxacin, levomethadyl, pentamidine, probucol, sparfloxacin, troleandomycin, class IA antiarrhythmics (disopyramide, procainamide, quiNIDine), class III antiarrhythmics (amiodarone, dofetilide, ibutilide, sotalol), clarithromycin, ziprasidone, pimozide, haloperidol, halofantrine, quiNIDine, chloroquine, dronedarone, droperidol, erythromycin, methadone, posaconazole, propafenone, saquinavir, abarelix, amoxapine, apomorphine, asenapine, β-agonists, ofloxacin, eribulin, ezogabine, flecainide, gatifloxacin, gemifloxacin, halogenated anesthetics, iloperidone, levofloxacin, local anesthetics, magnesium sulfate, potassium sulfate, sodium, maprotiline, moxifloxacin, nilotinib, norfloxacin, ciprofloxacin, OLANZapine, paliperidone, some phenothiazines (fluPHENAZine, perphenazine, prochlorperazine, trifluoperazine), telavancin, tetrabenazine, tricyclic antidepressants, venlafaxine, vorinostat, citalopram, alfuzosin, cloZAPine, cyclobenzaprine, dolasetron, palonosetron, QUEtiapine, rilpivirine, SUNItinib, tacrolimus, tacrolimus, vardenafil, indacaterol, dasatinib, fluconazole, lapatinib, lopinavir/ritonavir, mefloquine, octreotide, ondansetron, ranolazine, risperiDONE, telithromycin, vemurafenib

Drug/Herb
Do not use with St. John's wort

Drug/Food
Do not use with grapefruit juice

NURSING CONSIDERATIONS

Assess:

- **Severe, life-threatening, or fatal treatment-related** pneumonitis: all cases occurred within 2 mo of treatment initiation; monitor for pulmonary symptoms that may indicate pneumonitis, other causes of pneumonitis should be excluded; permanently discontinue in patients with treatment-related pneumonitis I
- **Hepatic disease:** liver function test (LFT) abnormalities, altered bilirubin levels may occur during treatment; monitor LFTs and bilirubin levels before treatment, then monthly; more frequent testing is needed in those presenting with grade 2 or greater toxicities; laboratory alterations should be managed with dose reduction, treatment interruption, or discontinuation
- **QT prolongation** has been reported with use of product; therefore, avoid crizotinib use in those patients; monitor ECG and electrolytes in patients with congestive heart failure, bradycardia, electrolyte imbalance (hypokalemia, hypomagnesemia), or in patients taking concomitant medications known to prolong the QT interval; treatment interruption, dosage adjustment, treatment dis-

continuation may be needed in patients who develop QT prolongation

- **Vision disorders,** generally start within 2 wk of the start of therapy; ophthalmological evaluation should be considered, particularly if patients experience photopsia or new or increased vitreous floaters; caution should be used when driving or operating machinery by patients who experience vision disorders
- **Pregnancy/breastfeeding:** identify if pregnancy is planned or suspected (pregnancy category D), do not breastfeed
- CBC with differential; BUN/creatinine

Evaluate:

- Decreasing spread of malignancy

Teach patient/family:

- That missed doses can be taken up to 6 hr before the next dose is due to maintain the twice daily regimen
- To use reliable contraception; both women and men of childbearing age should use adequate contraceptive methods during therapy and for at least 90 days after completing treatment, pregnancy category D
- To report immediately shortness of breath, cough, fatigue, visual changes
- Not to take with grapefruit juice
- To avoid activities requiring mental alertness until effects are known
- To report signs of QT prolongation (abnormal heartbeats, dizziness, syncope)
- To swallow caps whole and avoid contact with broken cap

cromolyn (OTC, Rx)

(kroe'moe-lin)

Gastrocrom, Nasalcrom

Func. class.: Antiasthmatic

Chem. class.: Mast cell stabilizer

Do not confuse:
Nasalcrom/Nasalide

ACTION:
Stabilizes the membrane of the sensitized mast cell, thus preventing the release of chemical mediators after an antigen-IgE interaction

USES:
Severe perennial bronchial asthma; prevention of exercise-induced bronchospasm; acute bronchospasm induced by environmental pollutants, mastocytosis, allergic rhinitis

Unlabeled uses: Food allergy, ulcerative colitis

CONTRAINDICATIONS:
Hypersensitivity to this product or lactose; status asthmaticus, acute asthma

Precautions: Pregnancy (B), breastfeeding, children <5 yr (aerosol); <2 yr (nebulizer); <2 yr (nasal sol); oral <2 yr; renal/hepatic disease, safety not established; cardiac dysrhythmias, CAD

DOSAGE AND ROUTES

Allergic rhinitis

- **Adult and child >2 yr: NASAL SOL** 1 spray in each nostril tid-qid, max 6 doses/day

To prevent exercise-induced bronchospasm

- **Adult and child >5 yr: INH** 2 metered sprays inhaled ≤1 hr before exercise

Bronchial asthma

- **Adult and child >5 yr: INH** 2 metered sprays using inhaler qid; **NEB** 20 mg qid by nebulization

Systemic mastocytosis

- **Adult and child >12 yr: PO** 200 mg qid ½ hr before meals and at bedtime
- **Child 2-12 yr: PO** 100 mg qid ½ before meals and at bedtime

Ulcerative colitis (unlabeled)

- **Adult: PO** (Gastrocrom) 200 mg qid 20 min before meals and at bedtime, may double dose after 2 wk
- **Child 2-14 yr: PO** (Gastrocrom) 100 mg qid 20 min before meals and at bedtime, may double dose after 2-3 wk

Available forms: Nasal sol 5.2 mg/metered spray (40 mg/ml); neb sol 20 mg/2 ml; aerosol 800 mcg/actuation; oral conc 100 mg/5 ml

Administer:

- For oral conc: break open ampule, squeeze contents into glass of water, stir, drink

SIDE EFFECTS

CNS: *Headache, dizziness,* neuritis, confusion, drowsiness
EENT: Throat irritation, cough, nasal congestion, burning eyes, nasal stinging/irritation, sneezing
GI: Nausea, vomiting, anorexia, dry mouth, bitter taste
GU: Urinary frequency, dysuria
INTEG: Rash, urticaria, angioedema
MS: Joint pain/swelling
Oral conc:
CNS: Dizziness, headache, paresthesia, migraine, **seizures,** psychosis, anxiety, depression, hallucinations, insomnia
CV: Tachycardia, PVCs, palpitations
GI: Diarrhea, nausea, abdominal pain, constipation, dyspepsia, stomatitis, vomiting
HEMA: Polycythemia, neutropenia, pancytopenia
INTEG: Pruritus, rash, flushing, photosensitivity

PHARMACOKINETICS

Excreted unchanged in feces, half-life 80 min, 63%-76% protein binding

NURSING CONSIDERATIONS

Assess:
- Eosinophil count during treatment
- Respiratory status: rate, rhythm, characteristics, cough, wheezing, dyspnea
- Gargle, sip of water to decrease irritation in throat (INH/Neb)

Evaluate:
- Therapeutic response: decrease in asthmatic symptoms; congested, runny nose

Teach patient/family:
Nasal sol: Blow nose, hold pump between fingers; if 1st use, spray in air until fine mist occurs, insert nozzle in nostril, spray and breathe in through nose, repeat in other nostril
Aerosol (not for acute asthma): Take cover off mouthpiece, shake gently, breathe out slowly, place mouthpiece in mouth, close mouth around it, tilt head back, breathe in as the inhaler is depressed, remove, hold breath, then breathe out slowly
Inhalation: Do not swallow sol, empty ampule into power-driven nebulizer as directed; do not combine different meds
Oral: Take ½ hr before meals and at bedtime

C

crotamiton

(kroe-tam′-ih-tuhn)
Eurax
Func. class.: Scabicide/pediculicide

ACTION: Scabicidal and antipruritic properties; the mechanisms of these actions are not known, but the drug is toxic to the scabies mite; antipruritic effect may be due to counterirritation mechanisms

USES: Infestations of scabies, lice

CONTRAINDICATIONS: Hypersensitivity; raw, inflamed skin
Precautions: Pregnancy (C), children

DOSAGE AND ROUTES

- **Adult: TOP** After routine bath, apply over the entire body from the chin to the soles; do not apply to the face or head; repeat in 24 hr; patient may take a cleansing bath 48 hr after the second dose

Treatment of pruritus
- **Adult: TOP** Apply topically by massaging gently into affected area until medication is completely absorbed; repeat if needed

Available forms: Lotion, cream 10%
Administer:
- The cream and lotion are for topical use only to the skin; shake lotion well before use; do not apply in the eyes or mouth; do not apply to raw, weeping, or inflamed skin until acute inflammation has subsided

Scabies: After taking a bath or shower, apply by massaging into the skin from the chin to the toes, including folds and creases; ensure use under fingernails (they should be trimmed short), because scabies are likely to remain under nails; a toothbrush may be used to apply under fingernails; if a toothbrush is used to apply, wrap it in paper and discard immediately after use; accidental use of the same toothbrush in the mouth could cause poisoning; a second application 24 hr later is recommended; a 60-g tube of cream or bottle of lotion is sufficient for 2 applications for most patients; clothing and bed linen should be changed the next morning; contaminated clothing and bed linen may be washed in the hot cycle of the washing machine or dry cleaned; a cleansing bath should be taken 48 hours after the last application

Pruritus: Massage cream or lotion into affected areas until the medication is absorbed; repeat as needed

SIDE EFFECTS

INTEG: Irritation, rash

NURSING CONSIDERATIONS

Assess:

Scabies/lice: Area of infestation, irritation, scales; itching, raw, peeling, inflamed skin; contamination of other family members

Evaluate:

- Decreasing infestation, itching

Teach patient/family:

- To stop using if irritation, rash occurs; wash off completely; not to use on raw, inflamed skin
- How to use (see dosage and routes)

cyanocobalamin (vit B_{12}) (OTC, Rx)

(sye-an-oh-koe-bal′a-min)

Anacobin ♣, Bedoz ♣, Cyanabin ♣, Cyanoject, Cyomin, Ener-B

hydroxocobalamin (OTC, Rx)

LA-12, Nascobal, Neuroforte-R, Vitamin B_{12}

Func. class.: Vit B_{12}, water-soluble vitamin

ACTION: Needed for adequate nerve functioning, protein and carbohydrate metabolism, normal growth, RBC development, cell reproduction

USES: Vit B_{12} deficiency, pernicious anemia, vit B_{12} malabsorption syndrome, Schilling test, increased requirements with pregnancy, thyrotoxicosis, hemolytic anemia, hemorrhage, renal/hepatic disease, nutritional supplementation

CONTRAINDICATIONS: Hypersensitivity, optic nerve atrophy

Precautions: Pregnancy (A), breastfeeding, children

DOSAGE AND ROUTES

Cyanocobalamin

- **Adult: PO** Up to 1000 mcg/day **SUBCUT/IM** 30-100 mcg/day × 1 wk, then 100-200 mcg/mo

Schilling test

- **Adult and child: IM** 1000 mcg in 1 dose
- **Child: PO** Up to 1000 mcg/day **SUBCUT/IM** 30-50 mcg/day × 2 wk, then 100 mcg/mo; **NASAL** 500 mcg weekly

Hydroxocobalamin

- **Adult: SUBCUT/IM** 30-50 mcg/day × 5-10 days then 100-200 mcg/mo
- **Child: SUBCUT/IM** 30-50 mcg/day × 5-10 days then 30-50 mcg/mo

Available forms: *Cyanocobalamin:* tabs 25, 50, 100, 250, 500, 1000, 5000 mcg; ext rel tabs 100, 200, 500, 1000 mcg; lozenges 100, 250, 500 mcg; nasal 500 mcg/spray; inj 100, 1000 mcg/ml; *hydroxocobalamin:* inj 1000 mcg/ml

Administer:

PO route

- With fruit juice to disguise taste; immediately after mixing
- With meals if possible for better absorption; large doses should not be used because most is excreted
- Protect from light, heat

IM route

- By IM inj for pernicious anemia for life unless contraindicated

Intranasal route

- Avoid use within 1 hr of hot fluids, food

IV route

- IV route not recommended but may be admixed in TPN solution

Additive compatibilities: Ascorbic acid, chloramphenicol, hydrocortisone, metaraminol, vit B/C

Solution compatibilities: Dextrose/Ringer's or LR combinations, dextrose/saline combinations, D_5W, $D_{10}W$, 0.45% NaCl, Ringer's or LR sol

Y-site compatibilities: Alfentanil, amikacin, aminophylline, ascorbic acid, atracurium, atropine, azaTHIOprine, aztreonam, benztropine, bretylium, bumetanide, buprenorphine, butorphanol, calcium chloride/gluconate, cefamandole, ceFAZolin, cefmetazole, cefonicid, cefoperazone, cefotaxime, cefoTEtan, cefOXitin, cefTAZidime, ceftizoxime, cefTRIAXone, cefuroxime, cephalothin, cephapirin, chloramphenicol, chlorproMAZINE, cimetidine, clindamycin, dexamethasone, digoxin, diphenhydrAMINE, DOBUTamine, DOPamine, doxycycline, enalaprilat, ePHEDrine, EPINEPHrine, epoetin alfa, erythromycin, esmolol, famotidine, fentaNYL, fluconazole, folic acid, furosemide, ganciclovir, gentamicin, glycopyrrolate, heparin, hydrocortisone, hydrOXYzine, imipenem-cilastatin, indomethacin, insulin (regular), isoproterenol hydrochloride, ketorolac, labetalol, lidocaine, magnesium, mannitol, meperidine, metaraminol, methicillin, methoxamine, methyldopate, methylPREDNISolone, metoclopramide, metoprolol, mezlocillin, miconazole, midazolam, minocycline, morphine, moxalactam, multiple vitamins injection, nafcillin, nalbuphine, naloxone, netilmicin, nitroglycerin, nitroprusside, norepinephrine, ondansetron, oxacillin, oxytocin, papaverine, penicillin G potassium/sodium, pentamidine, pentazocine, PENTobarbital, PHENobarbital, phentolamine, phenylephrine, phytonadione, piperacillin, polymyxin B, potassium chloride, procainamide, prochlorperazine, promethazine, propranolol, protamine, pyridoxine, quiNIDine, ranitidine, ritodrine, sodium bicarbonate, succinylcholine, SUFentanil, theophylline, thiamine, ticarcillin, ticarcillin-clavulanate, tobramycin, tolazoline, trimetaphan, urokinase, vancomycin, vasopressin, verapamil, vitamin B complex with C

SIDE EFFECTS

CNS: Flushing, optic nerve atrophy
CV: **CHF**, peripheral vascular thrombosis, **pulmonary edema**
GI: *Diarrhea*
INTEG: Itching, rash, pain at inj site
META: Hypokalemia
SYST: **Anaphylactic shock**

PHARMACOKINETICS

Gastric intrinsic factor must be present for absorption to occur; stored in liver, kidneys, stomach; 50%-90% excreted in urine; crosses placenta; excreted in breast milk

INTERACTIONS

Increase: absorption—predniSONE
Decrease: absorption—aminoglycosides, anticonvulsants, colchicine, chloramphenicol, aminosalicylic acid, potassium preparations, cimetidine

Drug/Herb
Decrease: vit B_{12} absorption—goldenseal
Drug/Lab Test
False positive: intrinsic factor

NURSING CONSIDERATIONS

Assess:
- For vit B_{12} deficiency: red, beefy tongue; psychosis; pallor; neuropathy
- GI function: diarrhea, constipation
- Potassium levels during beginning treatment in megaloblastic anemia; q6mo in pernicious anemia; folic acid, plasma vit B_{12} (after 1 wk), reticulocyte counts
- Nutritional status: egg yolks, fish, organ meats, dairy products, clams, oysters: good sources of vit B_{12}
- For pulmonary edema, worsening of CHF in cardiac patients

Evaluate:
- Therapeutic response: decreased anorexia, dyspnea on exertion, palpitations, paresthesias, psychosis, visual disturbances

Teach patient/family:
- That treatment must continue for life for pernicious anemia
- To eat a well-balanced diet
- To avoid contact with persons with infection; that infections are common

TREATMENT OF OVERDOSE:
Discontinue product

cyclobenzaprine (Rx)
(sye-kloe-ben′za-preen)
Amrix, Fexmid, Flexeril
Func. class.: Skeletal muscle relaxant, central acting
Chem. class.: Tricyclic amine salt

Do not confuse:
cyclobenzaprine/cyproheptadine

ACTION: Reduces tonic muscle activity at the brain stem; may be related to antidepressant effects

USES: Adjunct for relief of muscle spasm and pain in musculoskeletal conditions
Unlabeled uses: Fibromyalgia

CONTRAINDICATIONS: Children <12 yr, acute recovery phase of MI, dysrhythmias, heart block, CHF, hypersensitivity, intermittent porphyria, thyroid disease, QT prolongation
Precautions: Pregnancy (B), breastfeeding, geriatric patients, renal/hepatic disease, addictive personality

DOSAGE AND ROUTES

Muscloskeletal disorders
- **Adult/adolescent ≥15 yr: PO** 5 mg tid × 1 wk, max 30 mg/day × 3 wk
- **Adult: EXT REL** 15 mg/day, max 30 mg/day × 3 wk
- **Geriatric: PO** 5 mg tid

Hepatic dose
- **Adult (mild hepatic disease): PO** 5 mg, titrate slowly

Fibromyalgia (unlabeled)
- **Adult: PO** 10 mg at bedtime, titrated up

Available forms: Tabs 5, 10 mg; ext rel tab 15, 30 mg
Administer:
- Without regard to meals, give with food for GI symptoms
- Do not crush, break, chew ext rel cap
- Store in tight container at room temperature

SIDE EFFECTS

CNS: *Dizziness, weakness, drowsiness,* headache, tremor, depression, insomnia, confusion, paresthesia, nervousness
CV: Postural hypotension, tachycardia, dysrhythmias
EENT: Diplopia, temporary loss of vision
GI: *Nausea,* vomiting, hiccups, dry mouth, constipation, hepatitis
GU: Urinary retention, frequency, change in libido
INTEG: Rash, pruritus, fever, facial flushing, sweating

PHARMACOKINETICS

PO: Onset 1 hr, peak 3-8 hr, duration 12-24 hr, half-life 1-3 days, metabolized by liver, excreted in urine, crosses placenta, excreted in breast milk

INTERACTIONS

- Do not use within 14 days of MAOIs, traMADol

Increase: QT interval—Class IA/III antidysrhythmics, and other products that increase QT interval

Increase: serotonin syndrome—SSRIs, SNRIs

Increase: CNS depression—alcohol, tricyclics, opiates, barbiturates, sedatives, hypnotics

Drug/Herb

Increase: CNS depression—kava

NURSING CONSIDERATIONS

Assess:

- **Serotonin syndrome:** If using with SSRIs, SNRIs, monitor closely, if syndrome occurs, discontinue both products immediately
- **Pain:** location, duration, mobility, stiffness at baseline, periodically
- **Allergic reactions:** rash, fever, respiratory distress
- Severe weakness, numbness in extremities
- Assistance with ambulation if dizziness, drowsiness occur, especially for geriatric patients

Evaluate:

- Therapeutic response: decreased pain, spasticity; muscle spasms of acute, painful musculoskeletal conditions generally short term; long-term therapy seldom warranted

Teach patient/family:

- Not to discontinue medication abruptly; that insomnia, nausea, headache, spasticity, tachycardia will occur; that product should be tapered off over 1-2 wk
- Not to take with alcohol, other CNS depressants
- To avoid hazardous activities if drowsiness, dizziness occur
- To avoid using OTC medication (cough preparations, antihistamines) unless directed by prescriber
- To use gum, frequent sips of water for dry mouth

TREATMENT OF OVERDOSE:

Administer activated charcoal; use anticonvulsants if indicated; monitor cardiac function

cyclopentolate ophthalmic

See Appendix B

⚠ HIGH ALERT

cyclophosphamide (Rx)

(sye-kloe-foss′fa-mide)

Cytoxan, Procytox ❖

Func. class.: Antineoplastic alkylating agent

Chem. class.: Nitrogen mustard

Do not confuse:

cyclophosphamide/cycloSPORINE

Cytoxan/Cytosar/Cytotec/cytarabine

ACTION: Alkylates DNA is responsible for cross-linking DNA strands; activity is not cell-cycle–phase specific

USES: Hodgkin's disease, lymphomas, leukemia; cancer of female reproductive tract, breast, lung, prostate; multiple myeloma; neuroblastoma; retinoblastoma; Ewing's sarcoma; disseminated neuroblastoma, nephrotic syndrome

Unlabeled uses: Aplastic anemia, chronic idiopathic thrombocytopenic purpura, dermatomyositis, pneumonitis, polymyositis, SLE, scleroderma, RA, Behçet's syndrome, Churg-Strauss syndrome, polyarteritis nodosa, Wegener's granulomatosis, idiopathic pulmonary fibrosis, localized neuroblastoma, CLL

CONTRAINDICATIONS:

Pregnancy (D), hypersensitivity, prostatic hypertrophy, bladder neck obstruction

Precautions: Radiation therapy, cardiac disease, anemia, dysrhythmias, child, dental disease/work, dialysis, geriatrics, heart failure, hematuria, infections, leukopenia QT prolongation, secondary malignancy surgery, tumor lysis syndrome, vaccinations, breastfeeding, severely depressed bone marrow function

DOSAGE AND ROUTES

Acute lymphocytic leukemia (ALL) (induction therapy)

• **Adult: PO** Initially 1-5 mg/kg over 2-5 days, maintenance is 1-5 mg/kg; **IV** 300-1500 mg/m^2 (total dose); 60-250 mg/m^2 in divided doses for 6 or more days; maintenance 10-15 mg/kg q7-10 days or 30 mg/kg q3-4wk; dose should be reduced by half when bone marrow depression occurs

Neuroblastoma

• **Child and infant: PO** 150 mg/m^2/day, days 1-7 with DOXOrubicin (**IV** 35 mg/m^2 on day 8) q21days × 5 cycles

• **Child: IV** 70 mg/kg/day with hydration on days 1, 2 with DOXOrubicin and vinCRIStine q21days for courses 1, 2, 4, 6 alternating with CISplatin and etoposide q21days for courses 3, 5, 7

Breast cancer

• **Adult: PO** 100-200 mg/m^2/day or 2 mg/kg/day × 4-14 days; **IV** 500-1000 mg/m^2 on day 1 in combination with fluorouracil and methotrexate or DOXOrubicin or DOXOrubicin alone, also cyclophosphamide 600 mg/m^2; may be given dose-dense on day 1 of q14days with DOXOrubicin (60 mg/m^2) with growth-factor support

Operable node-positive breast cancer IV (TAC regimen)

• **Adult: IV** 500 mg/m^2 with DOXOrubicin (50 mg/m^2 **IV**) then DOCEtaxel (75 mg/m^2) **IV** given 1 hr later q3wk × 6 cycles

Nephrotic syndrome

• **Adult: PO** 2-3 mg/kg/day for up to 12 wk when corticosteroids are unsuccessful

Aplastic anemia (unlabeled)

• **Adult: IV** 45-50 mg/kg divided over 4 days

Behçet's syndrome/Churg-Strauss syndrome/polyarteritis nodosa/uveitis/Wegener's granulomatosis (unlabeled)

• **Adult: PO** 1-2 mg/kg/day, **IV** 0.5-1 g/m^2

Rheumatoid arthritis (unlabeled)

• **Adult and child: PO** 1.5-2.5 mg/kg/day

CLL (unlabeled)

• **Adult: IV** 250 mg/m^2/day on days 1-3 with fludarabine 30 mg/m^2/day on days 1-3

Available forms: Inj 100, 200, 500 mg, 1, 2 g; tabs 25, 50 mg

Administer:

• Use cytotoxic handling procedures

• In AM so product can be eliminated before bedtime

• Fluids IV or PO before chemotherapy to hydrate patient

• Antacid before oral agent; give after evening meal, before bedtime

• Antiemetic 30-60 min before product and prn

• Allopurinol or sodium bicarbonate to maintain uric acid levels, alkalinization of urine

PO route

• Take on empty stomach; do not crush, break, chew tabs; wash hands immediately if in contact with tab

• May be taken as a single dose or divided doses

• Take in AM or afternoon, avoid evening

• Store in tight container at room temperature

Direct IV route

• Reconstitute with NS only

Intermittent IV INF route

• Use cytotoxic handling procedures

• IV after diluting 100 mg/5 ml of sterile water or bacteriostatic water; shake; let stand until clear; may be further diluted in ≤250 ml D_5 NS D_5/NS; 0.45% NaCl

give 100 mg or less/min through 3-way stopcock of glucose or saline inf
- Use 21, 23, 25G needle; check site for irritation, phlebitis

Solution compatibilities: Amino acids 4.25%/D_{25}, D_5/0.9% NaCl, D_5W, 0.9% NaCl
Syringe compatibilities: Bleomycin, CISplatin, doxapram, DOXOrubicin, droperidol, fluorouracil, furosemide, heparin, leucovorin, methotrexate, metoclopramide, mitoMYcin, vinBLAStine, vinCRIStine
Y-site compatibilities: Acyclovir, alfentanil, allopurinol, amifostine, amikacin, aminocaproic acid, aminophylline, amiodarone, amphotericin B lipid complex, amphotericin B liposome, ampicillin, ampicillin-sulbactam, anidulafungin, atenolol, atracurium, azlocillin, aztreonam, bivalirudin, bleomycin, bumetanide, buprenorphine, butorphanol, calcium chloride/gluconate, CARBOplatin, caspofungin, cefamandole, ceFAZolin, cefepime, cefoperazone, cefotaxime, cefoTEtan, cefOXitin, cefTAZidime, ceftizoxime, cefTRIAXone, cefuroxime, chloramphenicol, chlorproMAZINE, cimetidine, ciprofloxacin, cisatracurium, CISplatin, cladribine, clindamycin, codeine, cycloSPORINE, cytarabine, DACTINomycin, DAPTOmycin, DAUNOrubicin, dexamethasone, dexmedetomidine, dexrazoxane, digoxin, diltiazem, diphenhydrAMINE, DOBUTamine, DOCEtaxel, dolasetron, DOPamine, doripenem, doxacurium, DOXOrubicin, DOXOrubicin liposomal, doxycycline, droperidol, enalaprilat, ePHEDrine, EPINEPHrine, epirubicin, ertapenem, erythromycin, esmolol, etoposide, famotidine, fenoldopam, fentaNYL, filgrastim, fluconazole, fludarabine, fluorouracil, foscarnet, fosphenytoin, furosemide, gallium, ganciclovir, gatifloxacin, gemcitabine, gentamicin, granisetron, haloperidol, heparin, hydrocortisone, HYDROmorphone, hydrOXYzine, IDArubicin, imipenem-cilastatin, inamrinone, insulin (regular), irinotecan, isoproterenol, kanamycin, ketorolac, labetalol, leucovorin, levofloxacin, levorphanol, lidocaine, linezolid, LORazepam, magnesium sulfate, mannitol, melphalan, meperidine, meropenem, mesna, methohexital, methotrexate, methylPREDNISolone, metoclopramide, metoprolol, metroNIDAZOLE, midazolam, milrinone, minocycline, mitoMYcin, mitoXANtrone, mivacurium, morphine, nafcillin, nalbuphine, naloxone, nesiritide, nitroglycerin, nitroprusside, norepinephrine, octreotide, ondansetron, oxacillin, oxaliplatin, PACLitaxel, palonosetron, pamidronate, pancuronium, pantoprazole, PEMEtrexed, penicillin G potassium, pentamidine, PENTobarbital, PHENobarbital, phenylephrine, piperacillin, piperacillin-tazobactam, potassium chloride/phosphates, procainamide, prochlorperazine, promethazine, propofol, propranolol, quinupristin-dalfopristin, ranitidine, rapacuronium, remifentanil, riTUXimab, rocuronium, sargramostim, sodium acetate/bicarbonate/phosphates, succinylcholine, SUFentanil, sulfamethoxazole-trimethoprim, tacrolimus, teniposide, theophylline, thiopental, thiotepa, ticarcillin, ticarcillin-clavulanate, tigecycline, tirofiban, TNA, tobramycin, topotecan, TPN, trastuzumab, vancomycin, vasopressin, vecuronium, verapamil, vinBLAStine, vinCRIStine, vinorelbine, voriconazole, zidovudine, zoledronic acid

SIDE EFFECTS

CNS: Headache, dizziness
CV: **Cardiotoxicity (high doses), myocardial fibrosis, congestive heart failure, pericarditis**
ENDO: SIADH, gonadal suppression
GI: *Nausea, vomiting, diarrhea, weight loss,* colitis, **hepatotoxicity**
GU: **Hemorrhagic cystitis,** *hematuria, neoplasms, amenorrhea, azoospermia, sterility, ovarian fibrosis,* **renal tubular fibrosis**
HEMA: **Thrombocytopenia, leukopenia, pancytopenia; myelosuppression**
INTEG: *Alopecia,* dermatitis

META: Hyperuricemia
MISC: Secondary neoplasms, anaphylaxis
RESP: Pulmonary fibrosis, interstitial pneumonia

PHARMACOKINETICS

Metabolized by liver, excreted in urine, half-life 4-6½ hr, 50% bound to plasma proteins

INTERACTIONS

Increase: neuromuscular blockade—succinylcholine
Increase: cyclophosphamide toxicity—barbiturates
Increase: action of warfarin
Increase: bone marrow depression—allopurinol, thiazides
Increase: hypoglycemia—insulin
Decrease: digoxin levels—digoxin
Decrease: cyclophosphamide effect—chloramphenicol, corticosteroids
Decrease: antibody response—live virus vaccines

Drug/Herb
- Toxicity: St. John's wort

Drug/Lab Test
Increase: uric acid
False positive: Pap smear
False negative: PPD, mumps, trichophytin, *Candida, Trichophyton,* Pap smear

NURSING CONSIDERATIONS

Assess:
- **Hemorrhagic cystitis;** renal studies: BUN, serum uric acid, urine CCr before, during therapy; I&O ratio; report fall in urine output <30 ml/hr
- **Bone marrow depression:** CBC, differential, platelet count baseline, weekly; withhold product if WBC is <2500 or platelet count is <75,000; notify prescriber of results
- Pulmonary function tests, chest x-ray films before, during therapy; chest film should be obtained q2wk during treatment
- Monitor temperature q4hr; elevated temperature may indicate beginning infection
- **Hepatotoxicity:** hepatic studies before, during therapy (bilirubin, AST, ALT, LDH), as needed; jaundice of skin, sclera; dark urine, clay-colored stools; itchy skin; abdominal pain; fever; diarrhea
- **Bleeding:** hematuria, guaiac, bruising or petechiae, mucosa or orifices q8hr
- Dyspnea, crackles, unproductive cough, chest pain, tachypnea
- Effects of alopecia on body image, discuss feelings about body changes
- Buccal cavity q8hr for dryness, sores or ulceration, white patches, oral pain, bleeding, dysphagia; obtain prescription for viscous lidocaine (Xylocaine)

⚠ Symptoms that indicate severe allergic reaction: rash, pruritus, urticaria, purpuric skin lesions, itching, flushing

- Increase fluid intake to 2-3 L/day to prevent urate deposits, calculi formation, reduce incidence of hemorrhagic cystitis
- Rinsing of mouth tid-qid with water, club soda; brushing of teeth bid-tid with soft brush or cotton-tipped applicators for stomatitis; use unwaxed dental floss
- Warm compresses at inj site for inflammation

Evaluate:
- To take adequate fluids to eliminate product
- Therapeutic response: decreased tumor size, spread of malignancy

Teach patient/family:
- That amenorrhea can occur and may last up to 1 yr after therapy but is reversible after stopping treatment
- To report any changes in breathing or coughing
- That hair may be lost during treatment; a wig or hairpiece may make patient feel better; new hair may be different in color, texture
- To avoid foods with citric acid, hot, or rough texture
- To report signs of infection: increased temperature, sore throat, flulike symptoms
- To report signs of anemia: fatigue, headache, faintness, SOB, irritability

- To report bleeding (bruising, hematuria, petechiae); to avoid use of razors, commercial mouthwash
- To use reliable contraception during and for 4 mo after treatment; not to breastfeed
- To avoid use of aspirin products, ibuprofen
- To avoid vaccinations during therapy
- Advise patient on proper handling and disposal of chemotherapy drugs

cycloSPORINE (Rx)

(sye′kloe-spor-een)

CycloSPORINE, modified Gengraf, Neoral, Pulminiq, SandIMMUNE

Func. class.: Immunosuppressant

Chem. class.: Fungus-derived peptide

Do not confuse:
cycloSPORINE/cycloSERINE/cyclophosphamide
SandIMMUNE/SandoSTATIN

ACTION: Produces immunosuppression by inhibiting lymphocytes (T)

USES: Organ transplants (liver, kidney, heart) to prevent rejection, rheumatoid arthritis, psoriasis

Unlabeled uses: Recalcitrant ulcerative colitis, aplastic anemia, Crohn's disease, GVHD, thrombocytopenia purpura, lupus, nephritis, myasthenia gravis, psoriatic arthritis, atopic dermatitis

CONTRAINDICATIONS: Breastfeeding, hypersensitivity to polyxyethylated castor oil (inj only); psoriasis or RA in renal disease (Neoral/Gengraf); Gengraf/Neoral used with PUVA/UVB, methotrexate, coal tar; ocular infections

Black Box Warning: Uncontrolled, malignant hypertension; radiation in psoriasis, neoplastic disease, sunlight (UV) exposure, renal disease/failure

Precautions: Pregnancy (C), geriatric patients, severe hepatic disease

DOSAGE AND ROUTES

Prevention of transplant rejection (nonmodified)

- **Adult and child: PO** 15 mg/kg several hr before surgery, daily for 2 wk, reduce dosage by 2.5 mg/kg/wk to 5-10 mg/kg/day; **IV** 5-6 mg/kg several hr before surgery, daily, switch to PO form as soon as possible

Prevention of transplant rejection (modified)

- **Adult and child: PO** 4-12 mg/kg/day divided q12hr, depends on organ transplanted

Rheumatoid arthritis (Neoral/Gengraf)

- **Adult: PO** 2.5 mg/kg/day divided bid, may increase 0.5-0.75 mg/kg/day after 8-12 wk, max 4 mg/kg/day

Psoriasis (Neoral/Gengraf)

- **Adult: PO** 2.5 mg/kg/day divided bid, × 4 wk, then increase by 0.5 mg/kg/day q2wk, max 4 mg/kg/day

Idiopathic thrombocytopenia purpura (unlabeled)

- **Adult: PO** 1.25-2.5 mg/kg bid

Severe aplastic anemia (unlabeled)

- **Adult and child: PO** 12 mg/kg/day or 15 mg/kg/day (child) with antithymocyte globulin (ATG)

Atopic dermatitis (unlabeled)

- **Adult/adolescent/child ≥2 yr: PO** 5 mg/kg/day

Crohn's disease that is resistant to/intolerant of corticosteroids (unlabeled)

- **Adult: PO** 2.5-15 mg/kg/day (nonmodified)

Available forms: Oral sol 100 mg/ml; soft gel cap 25, 50, 100 mg; inj 50 mg/ml; sol for inh 300 mg/4.8 ml

Administer:

PO route

- Do not break, crush, or chew caps
- Use pipette provided to draw up oral sol; may mix with milk or juice; wipe pipette, do not wash (Neoral)
- For several days before transplant surgery; give at same time of day
- With corticosteroids

• With meals for GI upset or in chocolate milk, milk, or orange juice (SandIMMUNE)
• With oral antifungal for candida infections

Rheumatoid arthritis

• Give Neoral or Gengraf 2.5 mg/kg/day divided bid; may use with salicylates, NSAIDs, PO corticosteroids
• Always give the daily dose of Neoral/Gengraf in 2 divided doses on consistent schedule
• Give initial SandIMMUNE PO dose 4-12 hr before transplantation as a single dose of 15 mg/kg, continue the single daily dose for 1-2 wk, then taper 5%/wk to a maintenance dose of 5-10 mg/kg/day

Intermittent IV INF route

• After diluting each 50 mg/20-100 ml of 0.9% NaCl or D_5W; run over 2-6 hr, use an inf pump, glass inf bottles only

Continuous IV INF route

• May run over 24 hr
• **For SandIMMUNE parenteral,** give 1/3 of PO dose, initial dose 4-12 hr before transplantation as a single IV dose 5-6 mg/kg/day, continue the single daily dose until PO can be used

Solution compatibilities: D_5W, NaCl 0.9%

Y-site compatibilities: Abciximab, alatrofloxacin, alfentanil, amikacin, aminocaproic acid, aminophylline, amphotericin B lipid complex, anidulafungin, argatroban, ascorbic acid injection, atenolol, atracurium, atropine, azaTHIOprine, aztreonam, benztropine, bivalirudin, bleomycin, bretylium, bumetanide, buprenorphine, butorphanol, calcium chloride/gluconate, CARBOplatin, carmustine, caspofungin, cefamandole, ceFAZolin, cefmetazole, cefonicid, cefoperazone, cefotaxime, cefoTEtan, cefOXitin, ceftaroline, cefTAZidime, ceftizoxime, ceftobiprole, cefTRIAXone, cefuroxime, cephalothin, cephapirin, chloramphenicol, chlorproMAZINE, cimetidine, ciprofloxacin, CISplatin, clindamycin, codeine, cyanocobalamin, cyclophosphamide, cytarabine, DACTINomycin, DAPTOmycin, DAUNOrubicin, dexamethasone, dexmedetomidine, digoxin, diltiazem, diphenhydrAMINE, DOBUTamine, DOCEtaxel, DOPamine, doripenem, doxacurium, DOXOrubicin, doxycycline, enalaprilat, ePHEDrine, EPINEPHrine, epirubicin, epoetin alfa, eptifibatide, ertapenem, erythromycin, esmolol, etoposide, famotidine, fenoldopam, fentaNYL, fluconazole, fludarabine, fluorouracil, folic acid, furosemide, gallium, ganciclovir, gatifloxacin, gemcitabine, gentamicin, glycopyrrolate, granisetron, heparin, hydrocortisone, HYDROmorphone, hydrOXYzine, ifosfamide, imipenem-cilastatin, indomethacin, irinotecan, isoproterenol, ketorolac, labetalol, lansoprazole, levofloxacin, lidocaine, linezolid, LORazepam, mannitol, mechlorethamine, meperidine, meropenem, methicillin, methotrexate, methoxamine, methyldopate, methylPREDNISolone, metoclopramide, metoprolol, metroNIDAZOLE, mezlocillin, micafungin, miconazole, midazolam, milrinone, minocycline, mitoXANtrone, morphine, moxalactam, multiple vitamins injection, nafcillin, naloxone, nesiritide, netilmicin, nitroglycerin, nitroprusside, norepinephrine, octreotide, ondansetron, oxacillin, oxaliplatin, oxytocin, PACLitaxel, palonosetron, pamidronate, pancuronium, pantoprazole, papaverine, PEMEtrexed, penicillin G potassium/sodium, pentamidine, pentazocine, phentolamine, phenylephrine, phytonadione, piperacillin, piperacillin-tazobactam, polymyxin B, potassium acetate/chloride, procainamide, prochlorperazine, promethazine, propofol, propranolol, protamine, pyridoxine, quiNIDine, quinupristin-dalfopristin, ranitidine, ritodrine, sargramostim, sodium acetate/bicarbonate, succinylcholine, SUFentanil, tacrolimus, teniposide, theophylline, thiamine, thiotepa, ticarcillin, ticarcillin-clavulanate, tigecycline, tirofiban, tobramycin, trimetaphan, urokinase, vancomycin, vasopressin, vecuronium, verapamil, vinCRIStine, vinorelbine, zoledronic acid

SIDE EFFECTS

CNS: *Tremors, headache,* seizures, confusion, encephalopathy, migraine
GI: Nausea, vomiting, diarrhea, *oral candida, gum hyperplasia,* hepatotoxicity, pancreatitis
GU: Albuminuria, hematuria, proteinuria, renal failure, hemolytic uremic syndrome, nephrotoxicity
INTEG: Rash, acne, *hirsutism,* pruritus
META: Hyperkalemia, hypomagnesemia, hyperlipidemia, hyperuricemia
MISC: *Infection, hypertension*

PHARMACOKINETICS

Peak 4 hr; highly protein bound; half-life (biphasic) 1.2 hr, 25 hr; metabolized in liver; excreted in feces, 6% in urine; crosses placenta; excreted in breast milk

INTERACTIONS

Increase: action, toxicity of cycloSPORINE—allopurinol, amiodarone, amphotericin B, androgens, azole antifungals, β-blockers, bromocriptine, calcium channel blockers, carvedilol, cimetidine, colchicine, corticosteroids, fluoroquinolones, foscarnet, imipenem-cilastatin, macrolides, metoclopramide, oral contraceptives, NSAIDs, melphalan, SSRIs
Increase: effects of aliskiren, digoxin, etoposide, HMG-CoA reductase inhibitors, methotrexate, potassium-sparing diuretics, sirolimus, tacrolimus
Increase: action toxicity of—digoxin, colchicine
Decrease: cycloSPORINE action—anticonvulsants, nafcillin, orlistat, PHENObarbital, phenytoin, rifamycins, sulfamethoxazole-trimethoprim, terbinafine, ticlopidine
Decrease: antibody reaction—live virus vaccines
Drug/Food
• Slowed metabolism of product: grapefruit juice, food

NURSING CONSIDERATIONS

Assess:
• Renal studies: BUN, creatinine at least monthly during treatment, 3 mo after treatment
• Product blood level during treatment 12 hr after dose, toxic >400 ng/ml
• Hepatic studies: alk phos, AST, ALT, bilirubin; hepatotoxicity: dark urine, jaundice, itching, light-colored stools; product should be discontinued
• Serum lipids, magnesium, potassium, cycloSPORINE blood concs
⚠ **Encephalopathy:** impaired cognition, seizures, visual changes including blindness, loss of motor function, movement disorders and psychiatric changes; dosage reduction or discontinuation may be needed in severe cases
⚠ **Nephrotoxicity:** 6 wk after surgery, acute tubular necrosis, CyA trough level >200 ng/ml, gradual rise in creatinine (0.15 mg/dl/day), creatinine plateau <25% above baseline, intracapsular pressure <40 mm Hg
⚠ Signs/symptoms of encephalopathy, lymphoma
Evaluate:
• Therapeutic response: absence of rejection
Teach patient/family:
• To report fever, chills, sore throat, fatigue, since serious infections may occur; tremors, bleeding gums, increased B/P
• To use contraceptive measures during treatment, for 12 wk after ending therapy; to notify prescriber if pregnancy is planned or suspected
• To take at same time of day, every day; not to skip doses or double dose; not to use with grapefruit juice or receive vaccines; that there are many drug interactions; not to add new or discontinued products without approval of prescriber

Black Box Warning: To limit UV exposure

• That treatment is lifelong to prevent rejection; to identify signs of rejection
• To report severe diarrhea because drug loss may result

• About the signs of nephrotoxicity: increased B/P, tremors of the hands, changes in gums, increased hair on body, face
• To continue with all lab work and follow-up appointments
• That types of products are not interchangeable
• Not to wash syringe/container with water; variation in dose may result

⚠ HIGH ALERT

cytarabine (Rx)

(sye-tare′a-been)

Ara-C, Cytosar ✱

cytarabine liposomal (Rx)

DepoCyt

Func. class.: Antineoplastic, antimetabolite

Chem. class.: Pyrimidine nucleoside analog

Do not confuse:
Cytosar/Cytoxan/Cytovene

ACTION: Competes with physiologic substrate of DNA synthesis, thus interfering with cell replication in the S phase of the cell cycle (before mitosis)

USES: Acute myelocytic leukemia, acute nonlymphocytic leukemia, chronic myelocytic leukemia; lymphomatous meningitis (intrathecal/intraventricular)
Unlabeled uses: Hodgkin's/non-Hodgkin's lymphoma, malignant meningitis, mantle cell lymphoma

CONTRAINDICATIONS: Pregnancy (D), hypersensitivity
Precautions: Breastfeeding, children, renal/hepatic disease, tumor lysis syndrome, infection, hyperkalemia, hyperphosphatemia, hyperuricemia, hypocalcemia

Black Box Warning: Bone marrow suppression

DOSAGE AND ROUTES

Acute myelogenous leukemia (AML)
• **Adult: CONT IV INF** 100 mg/m²/day × 7 days q2wk as single agent or 2-6 mg/kg/day (100-200 mg/m²/day) as a single dose or 2-3 divided doses for 5-10 days until remission, used in combination; maintenance 70-200 mg/m²/day for 2-5 days monthly; **SUBCUT/IM** maintenance 100 mg/m²/day × 5 days q28days

Meningeal leukemia
• **Adult/child: INTRATHECAL** 5-70 mg/m² variable daily × 4 days to q2-7days

Refractory acute Hodgkin's/ refractory non-Hodgkin's lymphoma (unlabeled)
• **Adult/child: IV** 2 g/m²/day; on day 5 q21days, with etoposide, methylPREDNISolone, and CISplatin

Carcinomatous meningitis (liposoma)
• **Adult: IT** 50 mg over 1-5 min q14days, during induction and consolidation wk 1, 3, 5, 7, 9, give another 50 mg **IT** wk 13; maintenance 50 mg q28days on wk 17, 21, 25, 29 use with dexamethasone 4 mg **PO/IV** × 5 day on each day of cytarabine

Renal dose
• **Adult CCr ≤60 ml/min, serum creatinine 1.5-1.9 mg/dl or increase of 0.5-1.2 mg/dl from baseline during treatment:** reduce to 1 g/m²/dose; **serum creatinine ≥2 mg/dl or change from baseline serum creatinine was 1.2 mg/dl:** reduce to 100 mg/m²/day

Available forms: Powder for inj 100, 500 mg, 1, 2 g; sus rel, (DepoCyt) liposomal for intrathecal use 10 mg/ml

Administer:
• Antiemetic 30-60 min before product and prn
• Allopurinol to maintain uric acid levels and alkalinization of the urine
• Topical or systemic analgesics for pain

IT route
• Use preservative-free NS, add 5 ml/100-mg vial or 10 ml/500-mg vial; use immediately, discard unused product
• Use dexamethasone with IT administration

IV route

- Use cytotoxic handling precautions

Direct IV route

- After diluting 100 mg/5 ml of sterile water for inj; given by direct IV over 1-3 min through free-flowing tubing (IV)

Intermittent IV INF route

- May be further diluted in 50-100 ml NS or D_5W, given over 30 min to 24 hr, depending on dose

Continuous IV INF route

- May also be given by continuous inf

Additive compatibilities: Corticotropin, DAUNOrubicin with etoposide, etoposide, hydrOXYzine, lincomycin, mitoXANtrone, ondansetron, potassium chloride, prednisoLONE, sodium bicarbonate, vinCRIStine

Solution compatibilities: Amino acids, D_5/LR, D_5/0.2% NaCl, D_5/0.9% NaCl, D_{10}/0.9% NaCl, D_5W, invert sugar 10% in electrolyte #1, Ringer's, LR, 0.9% NaCl, sodium lactate 1/6 mol/L, TPN #57

Syringe compatibilities: Metoclopramide

Y-site compatibilities: Acyclovir, alfentanil, amifostine, amikacin, aminocaproic acid, aminophylline, amphotericin B lipid complex, amphotericin B liposome, ampicillin, ampicillin-sulbactam, amsacrine, anidulafungin, atenolol, atracurium, azithromycin, aztreonam, bivalirudin, bleomycin, bumetanide, buprenorphine, butorphanol, calcium chloride/gluconate, CARBOplatin, ceFAZolin, cefepime, cefoperazone, cefotaxime, cefoTEtan, cefOXitin, cefTAZidime, ceftizoxime, cefTRIAXone, cefuroxime, chlorproMAZINE, cimetidine, ciprofloxacin, cisatracurium, CISplatin, cladribine, clindamycin, codeine, cyclophosphamide, cycloSPORINE, DAUNOrubicin, dexamethasone, dexmedetomidine, dexrazoxane, digoxin, diltiazem, diphenhydrAMINE, DOBUTamine, DOCEtaxel, dolasetron, DOPamine, doxacurium, DOXOrubicin, DOXOrubicin liposomal, doxycycline, droperidol, enalaprilat, ePHEDrine, EPINEPHrine, ertapenem, erythromycin, esmolol, etoposide, famotidine, fenoldopam, fentaNYL, filgrastim, fluconazole, fludarabine, foscarnet, fosphenytoin, furosemide, gatifloxacin, gemcitabine, gemtuzumab, gentamicin, granisetron, haloperidol, heparin, hydrocortisone, HYDROmorphone, hydrOXYzine, IDArubicin, ifosfamide, imipenem-cilastatin, inamrinone, insulin (regular), irinotecan, isoproterenol, ketorolac, labetalol, leucovorin, levofloxacin, levorphanol, lidocaine, linezolid, LORazepam, magnesium sulfate, mannitol, melphalan, meperidine, meropenem, mesna, methohexital, methotrexate, methylPREDNISolone, metoclopramide, metoprolol, metroNIDAZOLE, midazolam, milrinone, minocycline, mitoXANtrone, mivacurium, morphine, nalbuphine, naloxone, nesiritide, niCARdipine, nitroglycerin, nitroprusside, norepinephrine, octreotide, ofloxacin, ondansetron, oxaliplatin, PACLitaxel, palonosetron, pamidronate, pancuronium, pantoprazole, PEMEtrexed, pentamidine, PENTobarbital, PHENobarbital, phenylephrine, piperacillin, piperacillin-tazobactam, potassium chloride/phosphates, procainamide, prochlorperazine, promethazine, propofol, propranolol, quinupristin-dalfopristin, ranitidine, rapacuronium, remifentanil, riTUXimab, rocuronium, sargramostim, sodium acetate/bicarbonate/phosphates, succinylcholine, SUFentanil, sulfamethoxazole-trimethoprim, tacrolimus, teniposide, theophylline, thiopental, thiotepa, ticarcillin, ticarcillin-clavulanate, tigecycline, tirofiban, TNA, tobramycin, trastuzumab, trimethobenzamide, vancomycin, vasopressin, vecuronium, verapamil, vinCRIStine, vinorelbine, voriconazole, zidovudine, zoledronic acid

SIDE EFFECTS

CNS: Neuritis, dizziness, headache, cerebellar syndrome, personality changes, ataxia, mechanical dysphasia, coma; chemical arachnoiditis (IT)

CV: Chest pain, cardiopathy

CYTARABINE SYNDROME: *Fever,* myalgia, bone pain, chest pain, *rash,* conjunc-

tivitis, malaise (6-12 hr after administration)
EENT: Sore throat, conjunctivitis
GI: *Nausea, vomiting, anorexia, diarrhea, stomatitis,* hepatotoxicity, abdominal pain, hematemesis, GI hemorrhage
GU: Urinary retention, renal failure, hyperuricemia
HEMA: Thrombophlebitis, bleeding, thrombocytopenia, leukopenia, myelosuppression, anemia
INTEG: *Rash, fever,* freckling, cellulitis
META: Hyperuricemia
RESP: Pneumonia, dyspnea, pulmonary edema (high doses)
SYST: Anaphylaxis, tumor lysis syndrome

PHARMACOKINETICS

INTRATHECAL: Half-life 100-236 hr; metabolized in liver; excreted in urine (primarily inactive metabolite); crosses blood-brain barrier, placenta
IV/SUBCUT: Distribution half-life 10 min, elimination half-life 1-3 hr

INTERACTIONS

- Do not use with live virus vaccines
- Do not use within 24 hr of chemotherapy—sargramostim, GM-CSF, filgrastim, G-CSF

Increase: toxicity—immunosuppressants, methotrexate, flucytosine, radiation, or other antineoplastics
Increase: bleeding risk—anticoagulants, platelet inhibitors, salicylates, thrombolytics, NSAIDs
Decrease: effects of oral digoxin, gentamicin

NURSING CONSIDERATIONS

Assess:

Black Box Warning: **Bone marrow suppression:** CBC (RBC, Hct, Hgb), differential, platelet count weekly; withhold product if WBC is <1000/mm^3, platelet count is <50,000/mm^3, or RBC, Hct, Hgb low; notify prescriber of these results

- Renal studies: BUN, serum uric acid, urine CCr, electrolytes before and during therapy
- I&O ratio; report fall in urine output to <30 ml/hr
- Monitor temperature; fever may indicate beginning infection; no rectal temperatures
- **Hepatotoxicity:** hepatic studies before and during therapy: bilirubin, ALT, AST, alk phos, as needed or monthly; check for jaundice of skin, sclera; dark urine; clay-colored stools; pruritus; abdominal pain; fever; diarrhea
- Blood uric acid during therapy

⚠ **For anaphylaxis:** rash, pruritus, facial swelling, dyspnea; resuscitation equipment should be nearby

⚠ **Chemical arachnoiditis (IT):** headache, nausea, vomiting, fever; neck rigidity pain, meningism, CSF pleocytosis; may be decreased by dexamethasone

- Cytarabine syndrome 6-12 hr after inf: fever, myalgia, bone pain, chest pain, rash, conjunctivitis, malaise; corticosteroids may be ordered
- Bleeding: hematuria, heme-positive stools, bruising or petechiae, mucosa or orifices q8hr

⚠ Dyspnea, crackles, unproductive cough, chest pain, tachypnea, fatigue, increased pulse, pallor, lethargy; personality changes, with high doses; pulmonary edema may be fatal (rare)

- Buccal cavity q8hr for dryness, sores or ulceration, white patches, oral pain, bleeding, dysphagia
- Local irritation, pain, burning, discoloration at inj site
- GI symptoms: frequency of stools, cramping; antispasmodic may be used
- Acidosis, signs of dehydration: rapid respirations, poor skin turgor, decreased urine output, dry skin, restlessness, weakness
- Increased fluid intake to 2-3 L/day to prevent urate deposits and calculi formation unless contraindicated
- Rinsing of mouth tid-qid with water, club soda; brushing of teeth bid-tid with

soft brush or cotton-tipped applicators for stomatitis; use unwaxed dental floss

Evaluate:

- Therapeutic response: improvement of hematologic parameters

Teach patient/family:

- To report any coughing, chest pain, changes in breathing; may indicate beginning **pneumonia, pulmonary edema**
- To avoid foods with citric acid, spicy, or rough texture if stomatitis is present; use sponge brush and rinse with water after each meal; to report stomatitis: any bleeding, white spots, ulcerations in mouth; to examine mouth daily, report any symptoms
- To report signs of **infection:** increased temperature, sore throat, flulike symptoms; to avoid crowds, persons with infections
- To report signs of **anemia:** fatigue, headache, faintness, SOB, irritability
- To report bleeding; to avoid use of razors, commercial mouthwash, salicylates, NSAIDs, anticoagulants
- To use thrombocytopenia precautions
- To take fluids to 3 L/day to prevent renal damage
- To use reliable contraception during treatment and for 4 mo thereafter; not to breastfeed
- To avoid receiving vaccines during treatment
- That fever, headache, nausea, vomiting likely to occur

dabigatran

(da-bye-gat′ran)

Pradaxa

Func. class.: Anticoagulant-thrombin inhibitor

ACTION: Direct thrombin inhibitor that inhibits both free and clot-bound thrombin, prevents thrombin-induced platelet aggregation and thrombus formation by preventing conversion of fibrinogen to fibrin

USES: Stroke/systemic embolism prophylaxis with nonvalvular atrial fibrillation

Unlabeled uses: Deep venous thrombus (DVT), pulmonary embolism prophylaxis

CONTRAINDICATIONS: Hypersensitivity, bleeding, prosthetic heart valves

Precautions: Pregnancy (C), labor, obstetric delivery, breastfeeding, children, geriatric patients, abrupt discontinuation, anticoagulant therapy, renal disease, surgery

Black Box Warning: Abrupt discontinuation

DOSAGE AND ROUTES

Stroke prophylaxis

- **Adult: PO** 150 mg bid

For conversion from an alternative anticoagulant to dabigatran

- When converting from warfarin to dabigatran, discontinue warfarin and initiate dabigatran therapy when the INR is <2.0; when converting from a parenteral anticoagulant to dabigatran, initiate dabigatran 0-2 hr before the time of the next scheduled anticoagulant dose or at the time of discontinuation of a continuously administered anticoagulant (e.g., intravenous unfractionated heparin)

For conversion from dabigatran to warfarin

- **Adult:** CCr >50 ml/min, start warfarin 3 days before discontinuing dabigatran; CCr 31-50 ml/min, start warfarin 2 days before discontinuing dabigatran; CCr 15-30 ml/min, start warfarin 1 day before discontinuing dabigatran

For conversion from dabigatran to parenteral anticoagulants

- **Adult: PO** discontinue dabigatran; start parenteral anticoagulant 12 hr (CCr ≥30 ml/min) or 24 hr (CCr <30 ml/min) after the last dabigatran dose

Renal dose

- **Adult: PO** CCr 15-30 ml/min, 75 mg bid

Deep venous thrombus (DVT)/pulmonary embolism prophylaxis (unlabeled)

- **Adult: PO** 220 mg or 150 mg/day × 28-35 days, starting with 1/2 dose 1-4 hr after surgery

Available forms: Caps 75, 150 mg

Administer:

- Do not crush, break, chew, or empty contents of capsule
- Without regard to food
- Store in original package at room temperature until time of use; discard after 30 days; protect from moisture

SIDE EFFECTS

CNS: Intracranial bleeding

CV: Myocardial infarction

GI: Abdominal pain, dyspepsia, peptic ulcer, esophagitis, GERD, gastritis, GI bleeding

HEMA: Bleeding (any site), hemorrhagic erosive gastritis

INTEG: Rash, pruritus

SYST: Anaphylaxis (rare)

PHARMACOKINETICS

Protein binding 35%, half-life 12-17 hr (extended in renal disease), peak 1 hr, high-fat meal delays peak

INTERACTIONS

Increase: bleeding risk—amiodarone, other anticoagulants, clopidogrel, ketoconazole, quiNIDine, thrombolytics, verapamil

Decrease: dabigatran effect—rifampin

Decrease: dabigatran effect—P-glycoprotein inducers (carBAMazepine, rifampin, tipranovir

Drug/Herb

Decrease: dabigatran—St. John's wort

Drug/Lab Test

Increase: thrombin time, aPTT

NURSING CONSIDERATIONS

Assess:

- **Bleeding:** blood in urine or emesis, dark tarry stools, lower back pain; caution with arterial/venous punctures, catheters, NG tubes; monitor vital signs frequently; elderly patients more prone to serious bleeding, monitor aPTT, ecarin clotting time baseline and during treatment
- **Thrombosis/MI/emboli:** swelling, pain, redness, difficulty breathing, chest pain, tachypnea, cough, coughing up blood, cyanosis
- **Postthrombotic syndrome:** pain, heaviness, itching/tingling, swelling, varicose veins, brownish/reddish skin discoloration, ulcers; use of ambulation, compression stockings, adequate anticoagulation can prevent this syndrome
- **Surgery:** discontinue 24-48 hr before surgery in those with CCr ≥50 ml/min, 72-96 hr in those with CCr <50 ml/min; longer times may be needed in major surgery

Evaluate:

- Therapeutic response: decreased thrombus formation/extension, absence of emboli, postthrombotic effects

Teach patient/family:

- About the purpose and expected results of this product; to take at same time of day; not to skip or double doses
- To take without regard to food, swallow cap whole
- To notify all providers that this product is being used
- To report any bleeding including blood in stool, emesis, urine; nosebleeds
- If dose is missed, take as soon as remembered if on the same day; do not administer if <6 hr before next dose

⚠ HIGH ALERT

dacarbazine (Rx)

(da-kar′ba-zeen)

DTIC ✦, DTIC-Dome

Func. class.: Antineoplastic alkylating agent

Chem. class.: Cytotoxic triazine

ACTION: Alkylates DNA, RNA; inhibits DNA, RNA synthesis; also responsible for breakage, cross-linking of DNA strands; activity is not cell-cycle–phase specific

USES: Hodgkin's disease, malignant melanoma

Unlabeled uses: Malignant pheochromocytoma in combination with cyclophosphamide and vinCRIStine, metastatic soft-tissue sarcoma in combination with other agents, carcinoma meningitis, neuroblastoma

CONTRAINDICATIONS: Breastfeeding, hypersensitivity

Precautions: Renal disease

Black Box Warning: Pregnancy (C) 1st trimester, radiation therapy, hepatic disease, bone marrow suppression, secondary malignancy

DOSAGE AND ROUTES

Metastatic malignant melanoma

- **Adult:** IV 2-4.5 mg/kg/day × 10 days or 100-250 mg/m²/day × 5 days; repeat q3-4wk depending on response

Hodgkin's disease

- **Adult:** IV 150 mg/m²/day × 5 days with other agents, repeat q4wk; or 375 mg/m² on days 1 and 15 when given in combination, repeat q28days

Osteogenic sarcoma (unlabeled)

- **Adult and child:** IV 250 mg/m²/day as continuous inf × 4 days q28days

Soft-tissue sarcoma (unlabeled)
• **Adult and child:** IV 250-300 mg/m²/day as continuous inf × 3 days q21-28days

Carcinoma meningitis (unlabeled)
• **Adult:** INTRATHECAL 5-30 mg in a fixed dose 2-3 ×/wk until disease controlled

Available forms: Powder for inj 10, 100, 200 mg

Administer:
• Antiemetic 30-60 min before giving product to prevent vomiting, nausea; vomiting may subside after several doses
• Antibiotics for prophylaxis of infection

IV route
• Use cytotoxic handling precautions

Direct IV route
• After diluting 100 mg/9.9 or 200 mg/19.7 ml of sterile water for inj (10 mg/ml), give by direct IV over 1 min through Y-tube or 3-way stopcock

Intermittent IV INF route
• May be further diluted in 50-250 ml D_5W or NS for inj, given as an inf over ½ hr
• Watch for extravasation; stop inf, apply ice to area
• Store in light-resistant container in a dry area

Y-site compatibilities: Amifostine, anidulafungin, atenolol, aztreonam, bivalirudin, bleomycin, caspofungin, DAPTOmycin, dexmedetomidine, DOCEtaxel, DOXOrubicin, ertapenem, etoposide, fenoldopam, filgrastim, fludarabine, gemtuzumab, granisetron, levofloxacin, mechlorethamine, melphalan, nesiritide, octreotide, ondansetron, oxaliplatin, PACLitaxel, palonosetron, pamidronate, quinupristin-dalfopristin, sargramostim, teniposide, thiotepa, tigecycline, tirofiban, vinorelbine, voriconazole, zoledronic acid

SIDE EFFECTS

CNS: Facial paresthesia, flushing, fever, malaise; confusion, headache, seizures, cerebral hemorrhage, blurred vision (high doses)

GI: *Nausea, anorexia, vomiting,* hepatotoxicity (rare)

HEMA: Thrombocytopenia, leukopenia, anemia

INTEG: *Alopecia,* dermatitis, pain at inj site, photosensitivity; severe sun reactions (high doses)

MISC: Flulike symptoms, malaise, fever, myalgia, hypotension

SYST: Anaphylaxis

PHARMACOKINETICS

Metabolized by liver; excreted in urine; half-life 35 min, terminal 5 hr, 5% protein bound

INTERACTIONS

Black Box Warning: **Toxicity, bone marrow suppression:** bone marrow suppressants, radiation, other antineoplastics

• Bleeding: salicylates, anticoagulants

Increase: adverse reaction; decrease antibody reaction—live virus vaccines

Increase: nephrotoxicity—aminoglycosides

Increase: ototoxicity—loop diuretics

Decrease: dacarbazine effect—phenytoin, PHENobarbital

NURSING CONSIDERATIONS

Assess:

Black Box Warning: **Bone marrow suppression:** CBC, differential, platelet count weekly; notify prescriber of results

• Monitor temperature; may indicate beginning infection

Black Box Warning: Secondary malignancy: assess for secondary malignancy that may occur with this product

• Bleeding: hematuria, guaiac, bruising, petechiae of mucosa or orifices q8hr
• Effects of alopecia on body image, discuss feelings about body changes

Black Box Warning: Hepatic disease: jaundice of skin, sclera; dark urine; clay-colored stools; itchy skin; abdominal pain; fever; diarrhea; hepatic studies before, during therapy (bilirubin, AST, ALT, LDH) as needed or monthly

- Inflammation of mucosa, breaks in skin

⚠ **Hypersensitivity reactions, anaphylaxis,** discontinue product, administer meds for anaphylaxis

- Increased fluid intake to 2-3 L/day to prevent urate deposits, calculi formation

Evaluate:

- Therapeutic response: decreased tumor size, spread of malignancy

Teach patient/family:

- That patient should avoid prolonged exposure to sun, wear sunscreen
- That hair may be lost during treatment; that a wig or hairpiece may make the patient feel better; that new hair may be different in color, texture
- To report signs of **infection:** fever, sore throat, flulike symptoms
- To report signs of **anemia:** fatigue, headache, faintness, SOB, irritability
- To report bleeding; to avoid use of razors, commercial mouthwash
- **Pregnancy:** to notify prescriber if pregnancy is planned or suspected pregnancy (D)
- To avoid aspirin products or ibuprofen

Black Box Warning: To use reliable contraceptives during and for several mo after therapy; not to breastfeed

dalfampridine (Rx)

(dal-fam′pri-deen)

Ampyra

Func. class.: Neurological agent—MS

Chem. class.: Broad spectrum potassium channel blocker

ACTION: Mechanism of action is not fully understood; a broad-spectrum potassium channel blocker that inhibits potassium channels and increased action potential conduction in demyelinated axions

USES: For improved walking in patients with multiple sclerosis

CONTRAINDICATIONS: Renal failure (CCr <50 ml/min), seizures

Precautions: Pregnancy (C), breastfeeding, geriatric patients, renal disease

DOSAGE AND ROUTES

- **Adult: PO** 10 mg q12hr

Renal dose

- **Adult: PO** CCr 51-80 ml/min, no dosage adjustment needed but seizure risk unknown; CCr ≤50 ml/min, do not use

Available forms: Ext rel tab 10 mg

Administer:

- Do not break, crush, or chew; give without regard to meals
- Do not give closer together than q12hr; seizures may occur
- Do not double doses; if a dose is missed, skip it

SIDE EFFECTS

CNS: Seizures, paresthesias, headache, dizziness, asthenia, insomnia

GI: Nausea, constipation, dyspepsia

GU: Urinary tract infection

MS: Back pain

PHARMACOKINETICS

Bioavailability 96%; peak 3-4 hr (fasting), longer if taken with food; largely unbound to plasma proteins; 96% recovered in urine

INTERACTIONS

- Do not use with fampridine, other 4-aminopyridine (4-AP)–containing products

NURSING CONSIDERATIONS

Assess:

- **Multiple sclerosis:** improved walking, including speed
- **Seizures:** more common in those with previous seizure disorder

Evaluate:
- Therapeutic response: ability to walk at improved speed in MS

Teach patient/family:
- To notify prescriber if pregnancy is planned or suspected; not to breastfeed
- Expected results; side effects, including seizures

⚠ HIGH ALERT

dalteparin (Rx)

(dahl'ta-pear-in)

Fragmin

Func. class.: Anticoagulant

Chem. class.: Low–molecular-weight heparin

ACTION: Inhibits factor Xa/IIa (thrombin), resulting in anticoagulation

USES: Unstable angina/non–Q-wave MI; prevention/treatment of deep venous thrombosis in abdominal surgery, hip replacement, or in those with restricted mobility during acute illness, pulmonary embolism

Unlabeled uses: Antiphospholipid antibody, arterial thromboembolism (after heart valve surgery), cerebral thromboembolism, acute MI

CONTRAINDICATIONS: Hypersensitivity to this product, heparin, or pork products; active major bleeding, hemophilia, leukemia with bleeding, thrombocytopenic purpura, cerebrovascular hemorrhage, cerebral aneurysm; those undergoing regional anesthesia for unstable angina, non–Q-wave MI, dalteparin-induced thrombocytopenia

Precautions: Hypersensitivity to benzyl alcohol, pregnancy (B), breastfeeding, children, recent childbirth, geriatric patients; hepatic disease; severe renal disease; blood dyscrasias; bacterial endocarditis; acute nephritis; uncontrolled hypertension; recent brain, spine, eye surgery; congenital or acquired disorders; severe cardiac disease; peptic ulcer disease; hemorrhagic stroke; history of HIT; pericarditis; pericardial effusion; recent lumbar puncture; vasculitis; other diseases in which bleeding is possible

Black Box Warning: Epidural anesthesia, lumbar puncture

DOSAGE AND ROUTES

DVT/pulmonary embolism
- **Adult:** SUBCUT 200 international units/kg daily during 1st mo (max single dose 18,000 international units), then 150 international units/kg daily for mo 2-6 (max single dose 18,000 international units), use prefilled syringe that is closest to calculated dose; if platelets are 50,000-100,000/mm^3, reduce dose by 2500 international units until platelets ≥100,000 mm^3; if platelets <50,000/mm^3, discontinue until >50,000/mm^3

Hip replacement surgery/DVT prophylaxis
- **Adult:** SUBCUT 2500 international units 2 hr before surgery and 2nd dose in the evening on the day of surgery (4-8 hr postop), then 5000 international units SUBCUT 1st postop day and daily × 5-10 days

Unstable angina/non–Q-wave MI
- **Adult:** SUBCUT 120 international units/kg q12hr × 5-8 days, max 10,000 international units q12hr × 5-8 days with concurrent aspirin; continue until stable

DVT, prophylaxis for abdominal surgery
- **Adult:** SUBCUT 2500 international units 1-2 hr before surgery; repeat daily × 5-10 days; for high-risk patients, 5000 international units should be used

Renal dose
- **Adult:** SUBCUT cancer patient with CCr <30 ml/min, monitor anti-factor Xa during extended treatment

APLA (unlabeled)
- **Adult (female):** SUBCUT Antepartum 5000 international units/day with aspirin;

maintain anti-factor Xa of 0.2-0.6 international units/ml

Arterial thromboembolism prophylaxis (unlabeled)

• **Adult: SUBCUT** LMWH in combination with oral anticoagulants until INR is in therapeutic range × 2 consecutive days

Available forms: Prefilled syringes, 2500, 5000 international units/0.2 ml; 7500 international units/0.3 ml; 10,000, 15,000, 18,000, 25,000 international units/ml

Administer:

• Cannot be used interchangeably (unit for unit) with unfractionated heparin or other LMWHs

• Do not give IM or IV product route; approved is SUBCUT only; do not mix with other inj or sol

• Have patient sit or lie down; SUBCUT inj may be 2 inches from umbilicus in a U-shape, upper outer side of thigh, around navel, or upper outer quadrangle of the buttocks; rotate inj sites

• Change inj site daily; use at same time of day

SIDE EFFECTS

CNS: Intracranial bleeding

HEMA: Thrombocytopenia

INTEG: Pruritus, superficial wound infection, skin necrosis, inj site reaction, alopecia

SYST: Hypersensitivity, hemorrhage, anaphylaxis possible, hematoma

PHARMACOKINETICS

87% absorbed, excreted by kidneys, elimination half-life 3-5 hr, peak 2-4 hr, onset 1-2 hr, duration >12 hr

INTERACTIONS

Increase: bleeding risk—aspirin, oral anticoagulants, platelet inhibitors, NSAIDs, salicylates, thrombolytics

Drug/Herb

Increase: bleeding risk—feverfew, garlic, ginger, ginkgo, horse chestnut

Drug/Lab Test

Increase: AST, ALT

NURSING CONSIDERATIONS

Assess:

• Blood studies (Hct/Hgb, CBC, platelets, anti-Xa, stool guaiac) during treatment because bleeding can occur

⚠ **Bleeding:** Bleeding gums, petechiae, ecchymosis, black tarry stools, hematuria, epistaxis, decrease in Hct, B/P; may indicate bleeding, possible hemorrhage; notify prescriber immediately, product should be discontinued

Black Box Warning: **Epidural anesthesia:** Neurologic impairment frequently in those when neuraxial anesthesia has been used, spinal/epidural hematomas can occur, with paralysis

• **Hypersensitivity:** fever, skin rash, urticaria; notify prescriber immediately

• Needed dosage change q1-2wk; dose may need to be decreased if bleeding occurs

Evaluate:

• Therapeutic response: absence of DVT

Teach patient/family:

• To avoid OTC preparations that contain aspirin; other anticoagulants, serious product interaction may occur unless approved by prescriber

• To use soft-bristle toothbrush to avoid bleeding gums; to avoid contact sports; to use electric razor; to avoid IM inj

• To report any signs of bleeding (gums, under skin, urine, stools), unusual bruising

TREATMENT OF OVERDOSE:

Protamine sulfate 1% given IV; 1 mg protamine/100 anti-Xa international units of dalteparin given

dantrolene (Rx)

(dan′troe-leen)

Dantrium, Revonto

Func. class.: Skeletal muscle relaxant, direct acting

Chem. class.: Hydantoin

Do not confuse:

Dantrium/danazol

ACTION: Interferes with intracellular release of calcium from the sarcoplasmic reticulum necessary to initiate contraction; slows catabolism in malignant hyperthermia

USES: Spasticity in multiple sclerosis, stroke, spinal cord injury, cerebral palsy, malignant hyperthermia

Unlabeled uses: Neuroleptic malignant syndrome

CONTRAINDICATIONS: Hypersensitivity, compromised pulmonary function, impaired myocardial function

Black Box Warning: Active hepatic disease

Precautions: Pregnancy (C), breastfeeding, geriatric patients, peptic ulcer disease, cardiac/renal/hepatic disease, stroke, seizure disorder, diabetes mellitus, ALS, COPD, MS, mannitol/gelatin hypersensitivity, labor, lactase deficiency, extravasation

Black Box Warning: Females >35 yr, with MS, or taking estrogens

DOSAGE AND ROUTES

Spasticity

- **Adult: PO** 25 mg/day; may increase to 25-100 mg bid-qid, max 400 mg/day
- **Child: PO** 0.5 mg/kg/day given in divided doses bid; dosage may increase gradually, max 400 mg/day

Prevention of malignant hyperthermia

- **Adult and child: PO** 4-8 mg/kg/day in 3-4 divided doses × 1-2 days prior to procedure, give last dose 4 hr preop; **IV** 2.5 mg/kg before anesthesia

Malignant hyperthermia

- **Adult and child: IV** 1-2.5 mg/kg, may repeat to total dose of 10 mg/kg; **PO** 4-8 mg/kg/day in 4 divided doses × 1-3 days

Neuroleptic malignant syndrome (unlabeled)

- **Adult: PO** 100-300 mg/day in divided doses; **IV** 1.25-1.5 mg/kg

Available forms: Caps 25, 50, 100 mg; powder for inj 20 mg/vial

Administer:

- Avoid use with other CNS depressants

PO route

- Do not crush or chew caps
- Caps may be opened, mixed with juice and swallowed
- With meals for GI symptoms

IV route

- IV after diluting 20 mg/60 ml sterile water for inj without bacteriostatic agent (333 mcg/ml); shake until clear; give by rapid IV push through Y-tube or 3-way stopcock; follow with prescribed doses immediately; may also give by intermittent inf over 1 hr before anesthesia
- Considered incompatible in sol or syringe; compatibility unknown
- Store in tight container at room temperature; protect diluted sol from light, use reconstituted sol within 6 hr

SIDE EFFECTS

CNS: *Dizziness, weakness, fatigue, drowsiness,* headache, disorientation, insomnia, paresthesias, tremors, seizures

CV: Hypotension, chest pain, palpitations

EENT: Nasal congestion, blurred vision, mydriasis

GI: Hepatic injury, *nausea,* constipation, vomiting, increased AST, alk phos, abdominal pain, dry mouth, anorexia, hepatitis, dyspepsia

GU: Urinary frequency, nocturia, impotence, crystalluria, hepatitis

HEMA: Eosinophilia, aplastic anemia, leukopenia, thrombocytopenia/lymphoma

INTEG: Rash, pruritus, photosensitivity, extravasation (tissue necrosis)

RESP: Pleural effusion, pulmonary edema

PHARMACOKINETICS

PO: Peak 5 hr, highly protein bound, half-life 8 hr, metabolized in liver, excreted in urine (metabolites), absorption poor (35%)

INTERACTIONS

Increase: dysrhythmias—verapamil
Increase: hepatotoxicity—estrogens, other hepatotoxics
Increase: CNS depression—alcohol, tricyclics, opiates, barbiturates, sedatives, hypnotics, antihistamines

NURSING CONSIDERATIONS

Assess:

- **Seizures:** increased seizure activity, ECG in epilepsy patient; poor seizure control has occurred
- I&O ratio; check for urinary retention, frequency, hesitancy, especially geriatric patients

Black Box Warning: Active hepatic disease: hepatic function by frequent determination of AST, ALT, bilirubin, alk phos, GGTP; renal function studies, BUN, creatinine, CBC

- **Allergic reactions:** rash, fever, respiratory distress
- Severe weakness, numbness in extremities; prescriber should be notified and product discontinued
- Tolerance: increased need/more frequent requests for medication, increased pain
- CNS depression: dizziness, drowsiness, insomnia, psychiatric symptoms

⚠ **Signs of hepatotoxicity:** jaundice, yellow sclera, pain in abdomen, nausea, fever; prescriber should be notified, product discontinued

Evaluate:

- Therapeutic response: decreased pain, spasticity

Teach patient/family:

- Not to discontinue medication quickly because hallucinations, spasticity, tachycardia will occur; product should be tapered off over 1-2 wk; to notify prescriber of abdominal pain, jaundiced sclera, clay-colored stools, change in color of urine
- That if improvement does not occur within 6 wk, prescriber may discontinue product
- To avoid hazardous activities if drowsiness, dizziness occurs
- To avoid using OTC medications: cough preparations, antihistamines, other CNS depressants, alcohol unless directed by prescriber
- To use sunscreen or stay out of the sun to prevent burns

TREATMENT OF OVERDOSE:

Activated charcoal, supportive care

dapiprazole ophthalmic

See Appendix B

DAPTOmycin (Rx)

(dap′toe-mye-sin)
Cubicin
Func. class.: Antiinfective—miscellaneous
Chem. class.: Lipopeptides

ACTION: A new class of antiinfective; it binds to the bacterial membrane and results in a rapid depolarization of the membrane potential, thereby leading to inhibition of DNA, RNA, and protein synthesis

USES: Bacteremia, endocarditis, UTI, complicated skin, skin-structure infections caused by *Staphylococcus aureus* including methicillin-resistant strains, *Streptococcus pyogenes, Streptococcus agalactiae, Streptococcus dysgalactiae, Enterococcus faecalis* (vancomycin-susceptible strains), *Streptococcus pyogenes* (group A beta hemolytic), *Staphylococcus aureus, Staphylococcus epidermidis* (MRSA, MSSA), *Corynebacterium jeikeium, Staphylococcus haemolyticus*

Unlabeled uses: Vancomycin-resistant enterococci (VRE), bone, joint infection, infectious arthritis, orthopedic device-related infection, osteomyelitis

CONTRAINDICATIONS: Hypersensitivity

Precautions: Pregnancy (B), breast-feeding, children, geriatric patients, GI/renal disease, myopathy, ulcerative/pseudomembranous colitis, rhabdomyolysis, eosinophilic pneumonia

DOSAGE AND ROUTES

- **Adult: IV INF** 4 mg/kg over $^{1}/_{2}$ hr diluted in 0.9% NaCl, give q24hr × 7-14 days; some indications may use up to 6 mg/kg
- **Adolescent/child/infant ≥5 mo (unlabeled): IV** 4-6 mg/kg/day

***Staphylococcus aureus* bacteremia, including right-sided infective endocarditis**

- **Adult: IV INF** 6 mg/kg daily × 2-6 wk, up to 8-10 mg/kg daily; treatment failures should use another agent

Renal dose

- **Adult: IV INF** CCr <30 ml/min, hemodialysis, CAPD 4 mg/kg q48hr

Bacteremia, endocarditis, UTI (unlabeled)

- **Adult: IV** 6 mg/kg/day

VRE (unlabeled)

- **Adult: IV** 4 mg/kg/day

Available forms: Lyophilized powder for inj 500 mg

Administer:

IV route

- After reconstitution with 10 ml 0.9% NaCl (500 mg/10 ml); further dilution is needed with 0.9 NaCl; infuse over $^{1}/_{2}$ hr or give reconstituted sol (50 mg/ml) by IV inj over 2 min; do not use dextrose-containing solutions

Y-site compatibilities: Alfentanil, amifostine, amikacin, aminocaproic acid, aminophylline, amiodarone, amphotericin B liposome, ampicillin, ampicillin-sulbactam, argatroban, arsenic trioxide, atenolol, atracurium, azithromycin, aztreonam, bivalirudin, bleomycin, bumetanide, buprenorphine, busulfan, butorphanol, calcium chloride/gluconate, CARBOplatin, carmustine, caspofungin, ceFAZolin, cefepime, cefotaxime, cefoTEtan, cefOXitin, cefTAZidime, ceftizoxime, cefTRIAXone, cefuroxime, chloramphenicol, chlorproMAZINE, cimetidine, ciprofloxacin, cisatracurium, CISplatin, clindamycin, cyclophosphamide, cycloSPORINE, dacarbazine, DACTINomycin, DAUNOrubicin, dexamethasone, dexmedetomidine, dexrazoxane, diazepam, digoxin, diltiazem, diphenhydrAMINE, DOBUTamine, DOCEtaxel, DOPamine, doripenem, doxacurium, DOXOrubicin, DOXOrubicin liposomal, doxycycline, droperidol, enalaprilat, ePHEDrine, EPINEPHrine, epirubicin, eptifibatide, ertapenem, erythromycin, esmolol, etoposide, famotidine, fenoldopam, fentaNYL, fluconazole, fludarabine, fluorouracil, foscarnet, fosphenytoin, furosemide, ganciclovir, gentamicin, glycopyrrolate, granisetron, haloperidol, heparin, hydrALAZINE, hydrocortisone, HYDROmorphone, hydrOXYzine, IDArubicin, ifosfamide, inamrinone, insulin (regular), irinotecan, isoproterenol, ketorolac, labetalol, lepirudin, leucovorin, levofloxacin, lidocaine, linezolid, LORazepam, magnesium sulfate, mannitol, mechlorethamine, melphalan, meperidine, meropenem, mesna, metaraminol, methyldopate, methylPREDNISolone, metoclopramide, metoprolol, midazolam, milrinone, mitoXANtrone, mivacurium, morphine, moxifloxacin, mycophenolate mofetil, nafcillin, nalbuphine, naloxone, niCARdipine, nitroprusside, norepinephrine, octreotide, ondansetron, oxaliplatin, oxytocin, PACLitaxel, palonosetron, pamidronate, pancuronium, PEMEtrexed, pentamidine, PHENobarbital, phenylephrine, piperacillin-tazobactam, polymyxin B, potassium acetate/chloride/phosphates, procainamide, prochlorperazine, promethazine, propranolol, quinupristin-dalfopristin, ranitidine, rocuronium, sodium acetate/bicarbonate/citrate/phosphates, succinylcholine, sulfamethoxazole-trimethoprim, tacrolimus, teniposide, theophylline, thiotepa, ticarcillin, ticarcillin-clavulanate, tigecycline, tirofiban, tobramycin, topotecan, trimethobenzamide, vasopressin, vecuronium, verapamil, vinBLAStine, vin-

CRIStine, vinorelbine, voriconazole, zidovudine, zoledronic acid

Solution compatibilities: 0.9% NaCl, LR

SIDE EFFECTS

CNS: Headache, insomnia, dizziness, confusion, anxiety, fatigue, fever

CV: Hypo/hypertension, heart failure, chest pain

GI: Nausea, constipation, diarrhea, vomiting, dyspepsia, pseudomembranous colitis, abdominal pain, stomatitis, xerostomia, anorexia

GU: Nephrotoxicity

HEMA: Leukocytosis, anemia, thrombocytopenia

INTEG: Rash, pruritus

META: Electrolyte imbalances

MISC: Fungal infections, UTI, anemia, hypoglycemia

MS: Muscle pain or weakness, arthralgia, pain, myopathy, rhabdomyolysis

RESP: Cough, eosinophilic pneumonia, dyspnea

SYST: Anaphylaxis, DRESS, Stevens-Johnson syndrome

PHARMACOKINETICS

Site of metabolism unknown, protein binding 92%, terminal half-life 8-9 hr, 78% excreted unchanged (urine), breast milk

INTERACTIONS

Increase: myopathy—HMG-CoA reductase inhibitors

Drug/Lab Test

Increase: CPK, AST, ALT, BUN, creatinine, albumin, LDH

Increase/Decrease: glucose

Decrease: alkaline phosphatase, magnesium, phosphate, bicarbonate

NURSING CONSIDERATIONS

Assess:

- **Eosinophilic pneumonia:** dyspnea, fever, cough, shortness of breath; if left untreated, can lead to respiratory failure and death

⚠ **Nephrotoxicity:** any patient with compromised renal system, toxicity may occur; BUN, creatinine

- **Rhabdomyolysis:** Check for myopathy CPK > 1000 U/L (5 × ULN), discontinue product
- I&O ratio: report hematuria, oliguria, nephrotoxicity may occur
- Blood studies: CBC, CPK
- C&S; product may be given as soon as culture taken
- B/P during administration; hypo/hypertension may occur
- Signs of infection
- Allergies before treatment, reaction of each medication

Evaluate:

- **Myopathy:** assess for muscle pain, weakness; CPK >10× ULN; symptoms resolve after 3 days and CPK return to normal within 7-10 days after stopping product
- **Pseudomembranous colitis:** assess for diarrhea with mucous, pus, product should be discontinued; if moderate to severe, provide fluids, electrolytes, protein supplements, and an antibacterial
- Therapeutic response: negative culture

Teach patient/family:

- About allergies before treatment, reaction to each medication
- To report sore throat, fever, fatigue; could indicate superinfection; diarrhea; muscle weakness, pain, shortness of breath
- Avoid breastfeeding

darbepoetin (Rx)

(dar′bee-poh′eh-tin)

Aranesp

Func. class.: Hematopoietic agent

Chem. class.: Recombinant human erythropoietin

ACTION: Stimulates erythropoiesis by the same mechanism as endogenous erythropoietin; in response to hypoxia, erythropoietin is produced in the kidney and released into the bloodstream, where it interacts with progenitor stem cells to increase red-cell production

USES:
Anemia associated with chronic renal failure, in patients on and not on dialysis, and anemia in nonmyeloid malignancies for patients receiving coadministered chemotherapy

CONTRAINDICATIONS:
Hypersensitivity to mammalian-cell–derived products, human albumin, polysorbate 80; uncontrolled hypertension; red-cell aplasia

Precautions: Pregnancy (C), breastfeeding, children, seizure disorder, porphyria, hypertension, sickle cell disease; vit B_{12}, folate deficiency; chronic renal failure, dialysis; latex hypersensitivity, CABG, angina, anemia

Black Box Warning: Hgb >12 g/dl, neoplastic disease

DOSAGE AND ROUTES

Correction of anemia in chronic renal failure

• **Adult: SUBCUT/IV** 0.45 mcg/kg as a single inj; every wk, titrate max target Hgb of 12 g/dl

Chemotherapy treatment

• **Adult: SUBCUT** 2.25 mcg/kg/wk or 500 mcg q3wk

Epoetin alfa to darbepoetin conversion

• **Adult: SUBCUT/IV** (epoetin alfa <2500 units/wk) 6.25 mcg/wk; (epoetin alfa 2500-4999 units/wk) 12.5 mcg/wk; (epoetin alfa 5000-10,999 units/wk) 25 mcg/wk; (epoetin alfa 11,000-17,999 units/wk) 40 mcg/wk; (epoetin alfa 18,000-33,999 units/wk) 60 mcg/wk; (epoetin alfa 34,000-89,999 units/wk) 100 mcg/wk; (epoetin alfa >90,000 units/wk) 200 mcg/wk

Available forms: Sol for inj 25, 40, 60, 100, 150, 200, 300, 500 mcg/ml

Administer:

SUBCUT/IV route

• Without shaking; check for discoloration, particulate matter, do not use if present; do not dilute, do not mix with other products or sol, discard unused portion, do not pool unused portions
• Subcut typically used for those not requiring dialysis
• IV given direct undiluted or bolus into IV tubing or venous line after completion of dialysis; watch for clotting of line
• Adjust dosage every mo or more
• Store refrigerated, do not freeze; protect from light

SIDE EFFECTS

CNS: Seizures, sweating, headache, dizziness, stroke

CV: *Hypo/hypertension,* cardiac arrest, *angina pectoris,* thrombosis, CHF, acute MI, dysrhythmias, chest pain, transient ischemic attacks, edema

GI: *Diarrhea, vomiting, nausea, abdominal pain, constipation*

HEMA: Red-cell aplasia

MISC: *Infection, fatigue, fever,* death, *fluid overload,* vascular access hemorrhage, dehydration, sepsis

MS: *Bone pain, myalgia, limb pain, back pain*

RESP: *URI, dyspnea, cough, bronchitis,* PE

SYST: Allergic reactions, anaphylaxis

PHARMACOKINETICS

IV: Onset of increased reticulocyte count 2-6 wk; distributed to vascular space; absorption slow and rate limiting; terminal half-life 49 hr (SUBCUT), 21 hr (IV); peak conc at 34 hr; increased Hgb levels not generally observed until 2-6 wk after treatment initiated

INTERACTIONS

⚠ Do not use epoetin alfa with product

Increase: darbepoetin-alfa effect—androgens

Drug/Lab Test

Increase: WBC, platelets

Decrease: bleeding time

NURSING CONSIDERATIONS

Assess:

• Symptoms of anemia: fatigue, dyspnea, pallor

⚠ **Serious allergic reactions:** rash, urticaria; if anaphylaxis occurs, stop product, administer emergency treatment (rare)

• Renal studies: urinalysis, protein, blood, BUN, creatinine; monitor dialysis shunts; during dialysis, heparin may need to be increased

Black Box Warning: Blood studies: ferritin, transferrin monthly; transferrin saturation ≥20%, ferritin ≥100 ng/ml; Hgb 2×/wk until stabilized in target range (30%-33%) then at regular intervals; those with endogenous erythropoietin levels of <500 units/L respond to this agent; iron stores should be corrected before beginning therapy; if there is lack of response, obtain folic acid, iron, B_{12} levels

Black Box Warning: Neoplastic disease: breast, non–small-cell lung, head and neck, lymphoid, or cervical cancers, increased tumor progression; use lowest dose to avoid RBC transfusion

• B/P: check for rising B/P as Hgb rises; antihypertensives may be needed

⚠ CV status: hypertension may occur rapidly, leading to **hypertensive encephalopathy;** Hgb >12 g/dl may lead to death, do not administer

• I&O; report drop in output to <50 ml/hr

• **Seizures:** if Hgb is increased by 4 pts within 2 wk, institute seizure precautions

• CNS symptoms: sweating, pain in long bones

• **Dialysis patients:** thrill, bruit of shunts, monitor for circulation impairment

Evaluate:

• Therapeutic response: increase in reticulocyte count, Hgb/Hct; increased appetite, enhanced sense of well-being

Teach patient/family:

• To avoid driving or hazardous activity during beginning of treatment

• To monitor B/P, Hgb

• To take iron supplements, vit B_{12}, folic acid as directed

• To report side effects to prescriber; to comply with treatment regimen

• That menses and fertility may return; to use contraception

• About home administration procedures, if appropriate

darifenacin

(da-ree-fen′ah-sin)

Enablex

Func. class.: Antispasmodic/Guanticholinergic

ACTION: Bladder smooth muscle relaxation by decreasing the action of muscarinic receptors, thereby relieving overactive bladder

USES: Urge incontinence, frequency, urgency in overactive bladder

CONTRAINDICATIONS: Hypersensitivity, urinary retention, narrow-angle glaucoma (uncontrolled)

Precautions: Severe hepatic disease (Child-Pugh C), GI/GU obstruction, controlled narrow-angle glaucoma, ulcerative colitis, myasthenia gravis, moderate hepatic disease (Child-Pugh B), elderly

DOSAGE AND ROUTES

• **Adult: PO** 7.5 mg/day, initially; may increase to 15 mg/day after 14 days if needed

Those taking a CYP3A4 inhibitor

• **Adult: PO** max 7.5 mg/day

Hepatic dosage

• **Adult: PO** (Child-Pugh B) max 7.5 mg/day

Available forms: Tabs, ext rel 7.5, 15 mg

Administer:
- Without regard to meals, do not crush, break, chew extended-release tabs
- Store at room temperature

SIDE EFFECTS

CNS: Dizziness, headache
EENT: Blurred vision, drying eyes, sinusitis, rhinitis
GI: Constipation, dry mouth, abdominal pain, nausea, vomiting, dyspepsia
GU: UTI, urine retention, vaginosis
INTEG: Rash, pruritus, skin drying
MISC: Bronchitis, flulike symptoms

PHARMACOKINETICS

Peak 7 hr, half-life 12-19 hr

INTERACTIONS

Increase: level of—digoxin
Increase: anticholinergic effect—anticholinergics
Increase: darifenacin level—CYP3A4 inhibitors
Increase: levels of—drugs metabolized by CYP2D6

NURSING CONSIDERATIONS

Assess:
- Urinary function: urgency, frequency, retention in bladder-outflow obstruction
- Constipation, may add bulk in the diet

Evaluate:
- Decreasing urgency, frequency of urination

Teach patient/family:
- To take without regard to meals; do not crush, break, chew extended-release tabs
- To store at room temperature
- To avoid breastfeeding; to notify prescriber if pregnancy (C) is planned or suspected
- About anticholinergic symptoms (dry mouth, constipation, dry eyes, heat prostration); not to become overheated
- To avoid hazardous activities until reaction is known; dizziness, blurred vision can occur

darunavir (Rx)

(dar-ue′na-vir)

Prezista

Func. class.: Antiretroviral
Chem. class.: Protease inhibitor

ACTION: Inhibits human immunodeficiency virus (HIV-1) protease; this prevents maturation of the virus

USES: HIV-1 in combination with ritonavir and other antiretrovirals

CONTRAINDICATIONS: Hypersensitivity

Precautions: Pregnancy (B), breastfeeding, children, geriatric patients, renal/hepatic disease, history of renal stones, diabetes, hypercholesterolemia, sulfonamide hypersensitivity, antimicrobial resistance, bleeding, immune reconstitution syndrome, pancreatitis

DOSAGE AND ROUTES

Treatment-naive patients
- **Adult: PO** 800 mg with ritonavir 100 mg daily

Treatment-experienced patients
- **Adult: PO** 600 mg bid; with ritonavir 100 mg bid with food; 800 mg daily with ritonavir 100 mg with food (without darunavir resistance)
- **Child ≥3 yr and adolescents ≥40 kg: PO** 800 mg/dose daily with ritonavir 100 mg/dose daily (no darunavir) resistance-associated substitutions
- **Adolescent ≥30 kg, <40 kg: PO** 450 mg bid with ritonavir 60 mg bid
- **Adolescent ≥20 kg, <30 kg: PO** 375 mg bid with ritonavir 50 mg bid
- **Child 3 to <6 yr (14 kg to <15 kg: PO** 280 mg (with ritonavir 48 mg) bid with food
- **Child 3 to <6 yr (13 kg to <14 kg: PO** 260 mg (with ritonavir 40 mg) bid with food
- **Child 3 to <6 yr (12 kg to <13 kg: PO** 240 mg (with ritonavir 40 mg) bid with food

• **Child 3 to <6 yr (11 kg to <12 kg: PO** 220 mg (with ritonavir 32 mg) bid with food
• **Child 3 to <6 yr (10 kg to <11 kg: PO** 200 mg (with ritonavir 32 mg) bid with food

Available forms: Tabs 75, 150, 400, 600 mg

Administer:
• With food and ritonavir
• Tab should be swallowed whole

SIDE EFFECTS

CNS: *Headache, insomnia,* dizziness, somnolence
GI: *Diarrhea, abdominal pain, nausea, vomiting,* anorexia, dry mouth, **hepatitis, hepatotoxicity**
GU: Nephrolithiasis
INTEG: Rash, **angioedema, Stevens-Johnson syndrome, toxic epidermal necrolysis, exanthematous pustulosis**
MS: Pain
OTHER: Asthenia, **insulin-resistant hyperglycemia,** hyperlipidemia, **ketoacidosis,** lipodystrophy

PHARMACOKINETICS

95% protein binding; metabolized by CYP3A; peak 2.5-4 hr; terminal half-life 15 hr; excreted in feces 79.5%, urine 13.9%

INTERACTIONS

⚠ Life-threatening dysrhythmias: ergots, midazolam, rifampin, pimozide, triazolam; do not use concurrently
Increase: myopathy, rhabdomyolysis—HMG-CoA reductase inhibitors (atorvastatin, lovastatin, simvastatin)
Increase: darunavir levels—CYP3A4 inhibitors: (ketoconazole, itraconazole)
Increase: levels of both products—clarithromycin, zidovudine
Increase: levels of telapravir, rilpivarine, monitor for adverse reactions
Increase: side effects—(YP3A4 substrate) artemether/lumefantrine
Decrease: darunavir levels—CYP3A4 inducers (carBAMazepine, phenytoin, fosphenytoin, PHENobarbital), rifamycins, fluconazole, nevirapine, efavirenz
Decrease: levels of oral contraceptives
Decrease: levels of both products—tenofovir, avoid concurrent use

Drug/Herb
Decrease: darunavir levels—St. John's wort; avoid concurrent use
Increase: myopathy, rhabdomyolysis risk—red yeast rice

Drug/Food
Increase: darunavir absorption

Drug/Lab Test
Increase: LFTs, bilirubin, uric acid
Decrease: WBC, neutrophils, platelets

NURSING CONSIDERATIONS

Assess:
• Complaints of lower back, flank pain; indicates kidney stones
• Signs of infection, anemia, the presence of other sexually transmitted diseases
• **Serious skin reactions: angioedema, Stevens-Johnson syndrome, toxic epidermal necrolysis; discontinue immediately, notify prescriber**
• **Hepatoxicity: hepatic studies (ALT, AST, bilirubin, amylase); all may be elevated in those with underlying liver disease; product should be discontinued in those with increased LFTs**
• Viral load, CD4, HIV RNA during treatment
• Bowel pattern before, during treatment; if severe abdominal pain with bleeding occurs, product should be discontinued; monitor hydration
• **Hyperlipidemia:** cholesterol, triglycerides, LDL may be elevated; monitor serum cholesterol, lipid panel throughout treatment

Evaluate:
• Therapeutic response: decreased viral load, increased CD4 count

Teach patient/family:
• To use nonhormonal birth control; not to breastfeed
• To take as prescribed; if dose is missed, to take as soon as remembered up to 1 hr before next dose; not to double dose

• That product must be taken at same time of day to maintain blood levels for duration of therapy

⚠ That **hyperglycemia** may occur; to watch for increased thirst, weight loss, hunger, dry, itchy skin; to notify prescriber if these occur

• To increase fluids to prevent kidney stones; that if stone formation occurs, treatment may need to be interrupted

• That product does not cure AIDS, only controls symptoms; not to donate blood; not to share medication; to notify all health care providers of use; not to use with any other products without prescriber's approval

dasatinib (Rx)

(da-si′ti-nib)

Sprycel

Func. class.: Antineoplastic—miscellaneous

Chem. class.: Protein-tyrosine kinase inhibitor

ACTION: Inhibits a tyrosine kinase enzyme, thereby reducing cell growth in leukemia

USES: Treatment of accelerated, chronic blast phase CML or acute lymphoblastic leukemia (ALL); chronic phase CML with resistance or intolerance to prior therapy; Philadelphia chromosome–positive CML in chronic phase

CONTRAINDICATIONS: Pregnancy (D), hypersensitivity

Precautions: Breastfeeding, children, geriatric patients, QT prolongation, infection, thrombocytopenia, accidental exposure, edema, infertility, lactase deficiency, neutropenia, anemia, autoimmune disease with immune reconstitution syndrome

DOSAGE AND ROUTES

Accelerated or myeloid/lymphoid blast phase CML with resistance/intolerance to prior therapy

• **Adult: PO** 140 mg daily titrated up to 180 mg daily in those resistant to therapy

Chronic phase CML with resistance/intolerance to prior therapy

• **Adult: PO** 100 mg daily either AM or PM

Dosage reduction for those taking a strong CYP3A4 inhibitor

• **Adult: PO** 20-40 mg daily

Available forms: Tabs 20, 50, 70, 80, 100 mg

Administer:

• Do not break, crush, or chew tab
• After meal and with large glass of water

SIDE EFFECTS

CNS: CNS hemorrhage, headache, dizziness, insomnia, neuropathy, asthenia

CV: Dysrhythmias, chest pain, CHF, pericardial effusion, congestive cardiomyopathy, decreased injection fraction, QT prolongation

GI: *Nausea,* vomiting, *anorexia, abdominal pain,* constipation, diarrhea, GI bleeding, muscositis, stomatitis

HEMA: Neutropenia, thrombocytopenia, bleeding

INTEG: *Rash, pruritus,* alopecia

META: Fluid retention, edema, hypocalcemia, hypophosphatemia

MISC: Increased/decreased weight

MS: Pain, arthralgia, myalgia

RESP: Cough, dyspnea, pulmonary edema/hypertension, pneumonia, upper respiratory tract infection, pleural effusion

PHARMACOKINETICS

Metabolized by CYP3A4; 96% protein bound; peak 0.5-6 hr; excreted in feces (85%), small amount in urine (4%); terminal half-life 1.3-5 hr

INTERACTIONS

• Altered action of CYP3A4 substrates: alfentanil, cycloSPORINE, ergots, fentaNYL, pimozide, quiNIDine, sirolimus, tacrolimus

Increase: dasatinib concs—CYP3A4 inhibitors: ketoconazole, itraconazole, erythromycin, clarithromycin, nefazodone, protease inhibitors, telithromycin

Increase: myopathy, rhabdomyolysis-HMG-CoA reductase inhibitors (rare)

Increase: plasma concs of simvastatin

Increase: QT prolongation—class IA/III antidysrhythmics and other products that increase QT prolongation

Decrease: dasatinib concs—CYP3A4 inducers (dexamethasone, phenytoin, carBAMazepine, rifampin, PHENobarbital), H_2 blockers (famotidine), proton pump inhibitors (omeprazole)

Drug/Food

• Grapefruit: do not use

Drug/Herb

Decrease: dasatinib conc—St. John's wort

NURSING CONSIDERATIONS

Assess:

• **Myelosuppression:** ANC, platelets; in chronic phase, if ANC <1 × 10^9/L and/or platelets <50 × 10^9/L, stop until ANC >1.5 × 10^9/L and platelets >75 × 10^9/L; in accelerated phase/blast crisis, if ANC <0.5 × 10^9/L and/or platelets <10 × 10^9/L, determine whether cytopenia is related to biopsy/aspirate, if not, reduce dose by 200 mg, if cytopenia continues, reduce dose by another 100 mg; if cytopenia continues for 4 wk, stop product until ANC ≥1 × 10^9/L, monitor CBC weekly × 8 wk, then monthly

• **Hepatotoxicity:** monitor LFTs before treatment and monthly; if liver transaminases >5 × IULN, withhold until transaminase levels return to <2.5 × IULN

• **Signs of fluid retention, edema:** weigh, monitor lung sounds, assess for edema; some fluid retention is dose dependent, may result in CHF, congestive cardiomyopathy, decreased injection failure

• **QT Prolongation:** more common in those with hypokalemia, hypomagnesemia, congenital long QT syndrome, those taking products that prolong QT; correct electrolyte imbalances before use

Evaluate:

• Therapeutic response: decrease in leukemic cells or size of tumor

Teach patient/family:

• To report adverse reactions immediately: SOB, swelling of extremities, bleeding, bruising

• About reason for treatment, expected results

• To use contraception (pregnancy [D]); to avoid breastfeeding; men should use condoms

• To take at same time of day; not to crush or chew; not to use grapefruit juice

D

⚠ HIGH ALERT

DAUNOrubicin (Rx)

(daw-noe-roo′bi-sin)

Cerubidine

DAUNOrubicin citrate liposomal (Rx)

DaunoXome

Func. class.: Antineoplastic, antibiotic
Chem. class.: Anthracycline glycoside

Do not confuse:
DAUNOrubicin/DOXOrubicin

ACTION: Inhibits DNA synthesis, primarily; derived from *Streptomyces coerulorubidus;* replication is decreased by binding to DNA, which causes strand splitting; cell-cycle specific (S phase); a vesicant

USES: Acute lymphocytic leukemia (ALL), acute myelogenous leukemia (AML); *liposomal:* Kaposi's sarcoma

Unlabeled uses: *Liposomal:* multiple myeloma, AML, breast cancer, non-Hodgkin's lymphoma

CONTRAINDICATIONS: Pregnancy (D), breastfeeding, hypersensitivity, systemic infections, cardiac disease, bone marrow depression

Black Box Warning: IM/SUBCUT use

Precautions: Tumor lysis syndrome, MI, infection, thrombocytopenia, renal/hepatic disease; gout

Black Box Warning: Bone marrow suppression, cardiac disease, extravasation, renal failure, hepatic disease; requires a specialized care setting and an experienced clinician

DOSAGE AND ROUTES

Use decreased dose for those >60 yr

DAUNOrubicin

• **Adult: IV** 45-60 mg/m^2/day × 3 days then 2 days of subsequent courses in combination, max 400-600 mg/m^2 total cumulative dose

• **Child: IV** 30-40 mg/m^2/day depending on cycle (AML); ≤2 yr or <0.5 m^2: 1 mg/kg on day 1 weekly in combination with vinCRIStine and predniSONE, base dose on body weight not surface area (ALL); >2 yr or 0.5 m^2: 25 mg/m^2 day 1 weekly in combination with vinCRIStine and predniSONE (ALL)

DAUNOrubicin citrate liposomal

• **Adult: IV** 40 mg/m^2 q2wk (Kaposi's sarcoma); 100 mg/m^2 q3wk (multiple myeloma, unlabeled); **IV** 100-140 mg/m^2 q3wk (non-Hodgkin's lymphoma, unlabeled; metastatic breast cancer, unlabeled); IV escalating doses of 75, 100, 125, 135, 150 mg/m^2/day × 3 days (AML, unlabeled)

Renal dose

• **Adult: IV** serum CCr >3 mg/dl, reduce dose by 50%

Hepatic dose

• **Adult: IV** serum bilirubin 1.2-3 mg/dl, reduce dose by 50%; bilirubin >3 mg/dl, reduce dose by 75%; bilirubin >5 mg/dl, omit dose

Available forms: Inj 20 mg powder/vial, *liposomal:* solution for inj 2 mg/ml

Administer:

• Antiemetic 30-60 min before giving product and 6-10 hr after treatment to prevent vomiting

Black Box Warning: To be used in a care setting with emergency equipment available

Black Box Warning: To be used by a clinician knowledgeable in cytotoxic therapy

Black Box Warning: Do not give by IM/subcut injection

IV route (Cerubidine)

Do not confuse with liposome

• Use cytotoxic handling precautions

• After diluting 20 mg/4 ml sterile water for inj (5 mg/ml), rotate, further dilute in 10-15 ml 0.9% NaCl; give over 3-5 min by direct IV through Y-tube or 3-way stopcock of inf of D_5 or 0.9% NaCl; or dilute in 50 ml 0.9% NaCl and give over 10-15 min; or dilute in 100 ml and give over 30 min

• Hydrocortisone for extravasation; apply ice compress after stopping inf

Solution compatibilities: $D_{3.3}$%/0.3% NaCl, D_5W, Normosol R, LR, 0.9% NaCl

Y-site compatibilities: Amifostine, anidulafungin, atenolol, bivalirudin, bleomycin, CARBOplatin, caspofungin, CISplatin, codeine, cyclophosphamide, cytarabine, DACTINomycin, DAPTOmycin, dexmedetomidine, etoposide, fenoldopam, filgrastim, gemcitabine, gemtuzumab, granisetron, melphalan, meperidine, methotrexate, nesiritide, octreotide, ondansetron, oxaliplatin, PACLitaxel, palonosetron, quinupristin-dalfopristin, riTUXimab, sodium acetate/bicarbonate, teniposide, thiotepa, tigecycline, trastuzumab, vinCRIStine, vinorelbine, voriconazole, zoledronic acid

IV route (DaunoXome)

• Dilute with D_5W to (1 mg/ml), give over 1-2 hr, do not use in-line filter, reconstituted sol may be stored ≤6 hr refrigerated; do not admix

IV compatibilities: Anidulafungin, bivalirudin, meperidine, octreotide, sodium acetate, tirofiban, trastuzumab

SIDE EFFECTS

DAUNOrubicin

CNS: Fever, chills

CV: CHF, pericarditis, myocarditis, peripheral edema, fatal myocarditis, left ventricular failure, QT prolongation, ST-T wave changes, QRS voltage changes, tachycardia, SVT, PVCs

GI: *Nausea, vomiting, anorexia, mucositis,* hepatotoxicity

GU: Impotence, sterility, amenorrhea, gynecomastia

HEMA: Thrombocytopenia, leukopenia, anemia

INTEG: *Rash,* extravasation, dermatitis, reversible alopecia, cellulitis, thrombophlebitis at inj site

SYST: Anaphylaxis, tumor lysis syndrome

DAUNOrubicin citrate liposomal

CNS: *Fatigue, headache,* depression, insomnia, dizziness, *malaise, neuropathy*

CV: Chest pain, edema

GI: Abdominal pain, stomatitis, *nausea, vomiting, diarrhea,* constipation

INTEG: *Alopecia, pruritus,* sweating

MISC: *Allergic reactions, chest pain, fever,* edema, flulike symptoms

MS: *Rigors,* arthralgia, back pain

RESP: *Cough, dyspnea, rhinitis, sinusitis*

PHARMACOKINETICS

Half-life $18^1/_2$ hr, liposome $55^1/_2$ hr; metabolized by liver; crosses placenta; excreted in breast milk, urine, bile

INTERACTIONS

Increase: QT prolongation, torsades de pointes—arsenic trioxide, chloroquine, clarithromycin, class IA, class III antidysrhythmics, dasatinib, dolasetron, droperidol, erythromycin, flecainide, halofantrine, haloperidol, levomethadyl, methadone, ondansetron, palonosetron, pentamidine, some phenothiazines, propafenone, risperiDONE, sparfloxacin; tricyclic antidepressants (high doses); vorinostat, ziprasidone

Increase: bleeding risk—NSAIDs, salicylates, anticoagulants, platelet inhibitors, thrombolytics

Increase: toxicity—other antineoplastics, radiation, cyclophosphamide

Decrease: DAUNOrubicin effects—hematopoietic progenitor cells given within 24 hr

Decrease: antibody reaction—live virus vaccines

Drug/Lab Test

Increase: uric acid

NURSING CONSIDERATIONS

Assess:

Black Box Warning: **Bone marrow suppression:** CBC, differential, platelet count weekly, leukocyte nadir within 2 wk after administration, recovery within 3 wk; do not administer if absolute granulocyte count is $<750/\text{mm}^3$ (liposome)

- **Acute renal failure, uric acid nephropathy:** renal studies: BUN, urine CCr, electrolytes, uric acid baseline before each dose; I&O ratio; report fall in urine output to <30 ml/hr; provide aggressive alkalinization of urine and use of allopurinol; can prevent urate nephropathy
- Monitor temperature q4hr; fever may indicate beginning infection
- **Hepatotoxicity:** monitor hepatic studies baseline before each dose: bilirubin, AST, ALT, alk phos; check for jaundice of skin, sclera; dark urine, clay-colored stools; itchy skin, abdominal pain, fever; diarrhea

Black Box Warning: **Cardiac toxicity:** chest x-ray, echocardiography, radionuclide angiography, MUGA, ECG; watch for ST-T wave changes, low QRS and QT prolongation, possible dysrhythmias (sinus tachycardia, heart block, PVCs); watch for CHF (jugular vein distention, weight gain, edema, crackles), may occur after 2-6 mo of treatment, cumulative dose (400-550 mg/m^2), 450 mg/m^2 if used in combination with radiation, cyclophosphamide

- Bleeding: hematuria, guaiac stools, bruising, petechiae, mucosa, or orifices q8hr

• Effects of alopecia on body image; discuss feelings about body changes
• Buccal cavity q8hr for dryness, sores, ulceration, white patches, oral pain, bleeding, dysphagia, rinse mouth tid-qid with water, club soda; brush teeth bid-qid with soft brush or cotton-tipped applicators for stomatitis; use unwaxed dental floss
• **Tumor lysis syndrome:** hyperkalemia, hyperphosphatemia, hyperuricemia, hypocalcemia

Black Box Warning: Extravasation: swelling, pain, decreased blood return; if extravasation occurs, stop inf, remove tubing, attempt to aspirate the drug prior to removing the needle, elevate area, treat with ice pack

• GI symptoms: frequency of stools, cramping
• Increase fluid intake to 2-3 L/day to prevent urate and calculi formation

Evaluate:
• Therapeutic response: decreased tumor size, spread of malignancy

Teach patient/family:
• To report signs of infection, bleeding, bruising, SOB, swelling, change in heart rate
• That hair may be lost during treatment; that wig or hairpiece may make patient feel better; that new hair may be different in color, texture
• To avoid pregnancy (D) while taking product and for 4 mo thereafter; not to breastfeed
• To avoid foods with citric acid, hot, or rough texture if stomatitis is present
• To report any bleeding, white spots, ulcerations in mouth; to examine mouth daily
• That urine and other body fluids may be red-orange for 48 hr
• To avoid vaccines, alcohol, aspirin, NSAIDs while taking this product
• To avoid crowds, those with known infections

RARELY USED

deferasirox (Rx)
(def-a′sir-ox)
Exjade
Func. class.: Heavy metal chelating agent

USES: Chronic iron overload, transfusion hemosiderosis

CONTRAINDICATIONS: Breastfeeding, children, hypersensitivity, severe renal/hepatic disease, GI hemorrhage

DOSAGE AND ROUTES
• **Adult and child >2 yr: PO** 20-30 mg/kg/day; oral dispersion tablet is dissolved in water <1 g in 3.5 oz; >1 g in 7 oz or more; give on empty stomach at least 30 min before meals

degarelix
(day-gah-rel′iks)
Firmagon
Func. class.: Antineoplastic
Chem. class.: GnRH-receptor antagonist

ACTION: Reduces gonadotropins release and testicular steroidogenesis by reversibly binding to GnRH receptors

USES: Advanced prostate cancer

CONTRAINDICATIONS: Hypersensitivity

DOSAGE AND ROUTES
• **Adult (male): SUBCUT** 240 mg given as 2 120 mg injections; (40 mg/ml conc) maintenance 80 mg (20 mg/ml conc) every 28 days, starting 28 days after first dose

Available forms: Injection 80, 120 mg vial

Administer:
• Do not give IV, subcut only

General reconstitution information:

• Use gloves, aseptic technique during preparation and administration

• Keep vials vertical at all times; do *not* shake the vials; give reconstituted drug within 1 hr after addition of sterile water for injection

Reconstitution of 120-mg vial (240 mg dose *only*):

• To achieve a 240-mg dose, 2 120-mg vials of degarelix must be used; repeat for each 120 mg vial: draw up exactly 3 ml of sterile water for injection with a 2-inch, 21-G needle; do not use bacteriostatic water for injection; inject the sterile water slowly into vial containing 120 mg; to maintain sterility, do not remove the syringe or the needle from the vial; keep the vial in an upright position and swirl gently until clear; avoid shaking; reconstitution can take up to 15 min; tilt the vial slightly and withdraw 3 ml (40 mg/ml); avoid turning the vial upside down; repeat with a new vial, needle, and syringe for the second 120-mg dose (total dose = 240 mg)

Reconstitution of 80-mg vial:

• Draw up 4.2 ml of sterile water for injection with a 2-inch, 21-G needle; do not use bacteriostatic water for injection; inject the sterile water slowly into vial containing 80 mg; do not remove the syringe or the needle from the vial; keep the vial upright, swirl gently until clear; avoid shaking; reconstitution can take up to 15 min; tilt vial slightly and withdraw 4 ml (20 mg/ml); avoid turning upside down during withdrawal

Subcut injection:

• Exchange the reconstitution needle with a 1.25-inch, 27-G needle; remove air bubbles; give in the abdominal region; rotate injection site periodically; use area not exposed to pressure; grasp the skin of abdomen, elevate the subcutaneous tissue, and insert the needle deeply at an angle ≥45 degrees; aspirate before injection; inject the dose subcut; when giving the loading dose of 2 120 mg doses, the second dose should be injected at a different site

SIDE EFFECTS

CNS: Chills, dizziness, fatigue, fever, headache, insomnia

CV: **Increased QT prolongation**, hypotension, hot flashes

GI: Diarrhea, constipation, nausea

GU: ED, UTI, gynecomastia

INTEG: **Injection site reactions**, pain at site, redness, swelling

MS: Back pain, decreased bone density

SYST: **Hypersensitivity, anaphylaxis, angioedema**

PHARMACOKINETICS

Peak 2 days, duration 50 days, half-life 53 days

INTERACTIONS

Increase: QT prolongation—Class IA/III antidysrhythmics

Drug/Lab Test

Increase: PSA, LFTs

Decrease: bone density test

NURSING CONSIDERATIONS

Assess:

QT prolongation: More common in those taking Class IA/III antidysrhythmics, heart failure, congenital long QT syndrome; monitor cardiac status at baseline and often thereafter

Anaphylaxis angioedema: Discontinue treatment and do not restart in serious reactions

• Liver function studies, PSA, GGT that may be elevated; bone density that may be decreased; electrolytes

Evaluate:

• Decreasing spread, size of tumor

Teach patient/family:

• To notify all prescribers of cardiac disease or use of cardiac products

• Injection technique if patient/family will be giving product (provide patient information)

delavirdine (Rx)

(de-la-veer′deen)

Rescriptor

Func. class.: Antiretroviral

Chem. class.: Nonnucleoside reverse transcriptase inhibitor (NNRTI)

ACTION: Binds directly to reverse transcriptase; blocks RNA-, DNA-dependent polymerase activities, causing a disruption of the enzyme's site

USES: HIV-1 in combination with at least 2 other antiretrovirals

CONTRAINDICATIONS: Hypersensitivity

Precautions: Pregnancy (C), breastfeeding, children, hepatic disease, achlorhydria, antimicrobial resistance, exfoliative dermatitis, hepatitis, immune reconstitution syndrome

DOSAGE AND ROUTES

• **Adult and adolescent ≥16 yr: PO** 400 mg tid, max 1200 mg/day

Available forms: Tabs 100, 200 mg

Administer:

• 100-mg tab: dispersion by adding 4 tab/3-4 oz water, let stand, stir, swallow, rinse glass, swallow; use only 100-mg tabs for dispersion; 200-mg tab take as intact tab

• Do not give within 1 hr of antacids or didanosine

• Always use as combination therapy; this product is not recommended for initial treatment; due to inferior, virologic effect, it is no longer listed as part of any preferred regimens

SIDE EFFECTS

CNS: Headache, fatigue, anxiety, insomnia, fever

GI: Diarrhea, abdominal pain, nausea, anorexia, vomiting, dyspepsia, **hepatotoxicity**

GU: **Nephrotoxicity**

HEMA: **Neutropenia, leukopenia, thrombocytopenia, anemia, granulocytopenia**

INTEG: Rash, pruritus

MISC: Cough

MS: Pain, myalgia

SYST: **Stevens-Johnson syndrome; immune reconstitution syndrome (combination therapy)**

PHARMOCOKINETICS

98% protein bound, half-life 5.8 hr, peak 1 hr, duration 8 hr, extensively metabolized by CYP3A4, excreted in urine, feces

INTERACTIONS

Do not coadminister with nevirapine, efavirenz, rilpivirine, combined use not beneficial

⚠ **Serious life-threatening adverse reaction: amphetamines, ergots, benzodiazepines, calcium channel blockers, sedative/hypnotics, antidysrhythmics, sildenafil, pimozide, cisapride, ALPRAZolam, astemizole, midazolam, terfenadine, opiates, triazolam**

Increase: levels of ALPRAZolam, clarithromycin, dapsone, ergots, felodipine, midazolam, NIFEdipine, indinavir, saquinavir, lovastatin, simvastatin, atorvastatin, other CYP3A4, 2D6 inhibitors

Increase: delavirdine levels—FLUoxetine, ketoconazole

Increase: levels of both products—quiNIDine, warfarin, clarithromycin

Decrease: delavirdine levels—antacids, anticonvulsants, rifamycins, protease inhibitors, didanosine

Decrease: action of oral contraceptives, didanosine

Drug/Herb

Decrease: delavirdine level—St. John's wort

NURSING CONSIDERATIONS

Assess:

• **HIV:** obtain hepatitis B virus (HBV) screening to ensure proper treatment, if coinfected, a fully suppressive antiretroviral regimen with products against both; CBC, blood chemistry, plasma HIV RNA, absolute CD41/CD81/cell counts/%, serum β-2 microglobulin, serum ICD124 antigen levels

- **Immune reconstitution syndrome:** when treated with combination therapy; development of opportunistic infections (*Mycobacterium avium complex* [MAC], cytomegalovirus [CMV], *Pneumocystis carinii* pneumonia [PCP], TB)
- Signs of infection, anemia
- Hepatic studies: ALT, AST; renal studies
- Bowel pattern before, during treatment; if severe abdominal pain with bleeding occurs, product should be discontinued; monitor hydration
- Allergies before treatment, reaction to each medication; place allergies on chart
- Plasma delavirdine concs (trough 10 micromolar)
- **Toxicity:** severe nausea/vomiting, maculopapular rash
- **Serious skin reactions:** Stevens-Johnson syndrome; rash may occur within 1-3 wk of beginning treatment; if rash is not severe, manage with diphenhydrAMINE, hydrOXYzine, topical corticosteroids

Evaluate:

- Therapeutic response: increased CD4 cell count, decreased viral load, improvement in symptoms of HIV

Teach patient/family:

- To take as prescribed; if dose is missed, to take as soon as remembered up to 1 hr before next dose; not to double dose
- That tabs may be dissolved in ½ cup of water; to stir; when dissolved, drink right away; to rinse cup with water and drink to get all medication
- To make sure health care provider knows about all medications being taken
- That if severe rash, mouth sores, swelling, aching muscles/joints, or eye redness occur, to notify health care provider
- Not to breastfeed if taking this product
- That this product is not a cure, only controls symptoms

demecarium ophthalmic

See Appendix B

denosumab (Rx)

(den-oh′sue-mab)

Prolia, Xgeva

Func. class.: Bone resorption inhibitor
Chem. class.: Monoclonal antibody, bone resorption

ACTION: Neutralizes activity of receptor activator nuclear factor kappa-B ligand (RANKL) by binding to it and blocking its interaction with cell-surface receptors; use of a RANKL inhibitor may reduce bone turnover and decrease tumor burden

USES: Osteoporosis in postmenopausal women or men at high risk for fractures; increase bone mass in men who are receiving androgen deprivation therapy for prostate cancer and women receiving aromastase inhibitor therapy for breast cancer at high risk for fractures; prevention of skeletal-related events in bone metastases from solid tumors (Xgeva), giant cell tumor of bone

CONTRAINDICATIONS: Hypersensitivity, hypocalcemia

Precautions: Pregnancy (C), breastfeeding, child/infant/neonate, anemia, coagulopathy, diabetes mellitus, dialysis, eczema, hypoparathyroidism, immunosuppression, latex hypersensitivity, malabsorption syndrome, neoplastic disease, pancreatitis, parathyroid disease, dental/renal/thyroid disease, TB, vit D deficiency

DOSAGE AND ROUTES

Postmenopausal osteoporosis

- **Adult female: SUBCUT** 60 mg q6mo with 1000 mg calcium and 400 international units vit D, max 60 mg q6mo

Bone metastases from solid tumors (Xgeva)

- **Adult: SUBCUT** 120 mg q4wk, max 120 mg q4wk; administer with calcium and vit D as necessary to prevent hypocalcemia

Giant cell tumor (bone) (Xgeva)
• **Adult: SUBCUT** 120 mg q4wk, then 2 more doses on day 8,15 of 1st mo only, use calcium, vit D as needed
Prevention of skeletal-related events with bone metastases from solid tumors
• **Adult: SUBCUT** (Xgeva): 120 mg q4wk
Available forms: Sol for inj 60 mg/ml (Prolia); 120 mg/1.7 ml (Xgeva)
Administer:
SUBCUT route
• Give acetaminophen before and for 72 hr after to decrease pain
• Do not use if particulate matter or discoloration is present; sol is clear and colorless to slightly yellow with small white/opalescent particles; remove from refrigerator and allow to warm to room temperature (15-30 min)
• **Use of prefilled syringe with needle safety guard:** leave green guard in original position until after administration; remove and discard needle cap immediately before inj; give by subcut inj in upper arm/thigh or abdomen; after inj, point needle away from people and slide green guard over needle
• **Use of single-use vials:** use 27-G needle; give in upper arm/thigh or abdomen; do not reinsert needle in vial; discard supplies as appropriate
• Store and use out of direct sunlight/heat; do not freeze; use within 14 days after removal from refrigerator; store unopened containers in refrigerator

SIDE EFFECTS

CNS: Chills, fever, flushing, headache, vertigo, neuropathic pain
CV: Angina, atrial fibrillation
GI: Abdominal pain, constipation, *diarrhea,* flatulence, GERD, *vomiting, nausea*
GU: Cystitis, lactation suppression
HEMA: Anemia, neutropenia
INTEG: Atopic dermatitis, pruritus
META: Hypercholesterolemia, hypocalcemia, hypophosphatemia
MS: Back, bone pain; MS pain, myalgia, osteonecrosis of the jaw, atypical low trauma femoral fractures
RESP: Cough, *dyspnea*
SYST: Infection, secondary malignancy

PHARMACOKINETICS

Terminal half-life 25.4 days, bioavailability 62%, max serum conc 3-21 days, steady state 6 mo

INTERACTIONS

Increase: infection, possible—immunosupressives, corticosteroids
Increase: osteonecrosis of the jaw, possible—antineoplastics, corticosteroids

NURSING CONSIDERATIONS

Assess:
• **Acute acute-phase reaction:** fever, myalgia, headache, flulike symptoms for 72 hr after inj; usually resolves after 72 hr
• **Blood tests:** serum calcium, creatitine, BUN, magnesium, phosphate
• **Hypocalcemia (may be fatal):** paresthesia, twitching, laryngospasm, Chvostek's and Trouseau's signs; preexisting hypocalcemia should be corrected before treatment; patient with vit D deficiency may require higher doses of vit D
• **Hypercalcemia:** nausea, vomiting, anorexia, weakness, thirst, constipation, dysrhythmias
• **Dental status:** correct dental complications before product use; good oral hygiene should be maintained; if dental work is to be performed, antiinfectives should be given to prevent osteonecrosis of the jaw
• **Infection:** do not start treatment in patients with active infections; infections should be resolved first
Evaluate:
• Therapeutic response: increased/maintained bone density, decreased calcium levels
Teach patient/family:
• To report hypercalcemic relapse: nausea, vomiting, bone pain, thirst

- To continue with dietary recommendations, including additional calcium and vit D
- To avoid use during pregnancy and breastfeeding; to notify prescriber if pregnancy is planned, suspected
- To use acetaminophen before and for 72 hr after inj to lessen bone pain
- About the purpose of this product and its expected results
- To avoid OTC, Rx medications and herbs and supplements unless approved by prescriber
- To exercise regularly, stop smoking, and avoid alcohol to maintain bone health
- To inform all health care providers of product use; to avoid dental procedures/ surgery if possible; to practice good oral hygiene
- That lab tests and follow-up exams will be required

desipramine (Rx)

(dess-ip′ra-meen)

Norpramin

Func. class.: Antidepressant, tricyclic

Chem. class.: Dibenzazepine, secondary amine

ACTION: Blocks reuptake of norepinephrine, serotonin into nerve endings, thereby increasing action of norepinephrine, serotonin in nerve cells

USES: Depression

Unlabeled uses: Chronic pain, postherpetic neuralgia, ADHD, bulimia, diabetic neuropathy, panic disorder, social phobia

CONTRAINDICATIONS: Hypersensitivity to tricyclics, carBAMazepine, closed-angle glaucoma, acute MI, MAOIs

Precautions: Pregnancy (C), breastfeeding, geriatric patients, severe depression, increased intraocular pressure, seizure disorder, CV disease, urinary retention, cardiac dysrhythmias, cardiac conduction disturbances, family history of sudden death, prostatic hypertrophy, thyroid disease

Black Box Warning: Children <18 yr, suicidal patients

DOSAGE AND ROUTES

Major depression

- **Adult: PO** 50-75 mg/day in 1-4 divided doses; titrate by 25-50 mg weekly up to 300 mg/day in single or divided doses (inpatient), 200 mg/day (outpatient)
- **Geriatric: PO** 25 mg/day at bedtime, titrate weekly; may increase to 150 mg/day
- **Adolescent: PO** 25-50 mg/day in divided doses; max 150 mg/day
- **Child 6-12 yr: PO** 1-3 mg/kg/day in divided doses; give >3 mg/kg/day with close medical monitoring; max 5 mg/kg/ day

Bulimia nervosa (unlabeled)

- **Adult: PO** 25 mg tid; may titrate to 200 mg/day by 25-50 mg/day at weekly intervals

ADHD (unlabeled)

- **Child ≥5 yr: PO** up to 3.5 mg/kg/day divided bid

Neuropathic pain/postherpetic neuralgia (unlabeled)

- **Adult: PO** 10-25 mg at bedtime, titrate to relief

Diabetic neuropathy (unlabeled)

- **Adult: PO** 75-150 mg

Available forms: Tabs 10, 25, 50, 75, 100, 150 mg

Administer:

- Increased fluids, bulk in diet for constipation, especially in geriatric patients; with food or milk for GI symptoms; crushed if patient is unable to swallow medication whole
- Dosage at bedtime if oversedation occurs during day; may take entire dose at bedtime; geriatric patients may not tolerate once-daily dosing

SIDE EFFECTS

CNS: *Dizziness, drowsiness,* confusion, headache, anxiety, tremors, stimulation, weakness, insomnia, nightmares, EPS (geriatric patients), increased psychiatric symptoms, paresthenia, suicidal ideation, impaired memory, seizures, serotonin syndrome

CV: *Orthostatic hypotension,* ECG changes, *tachycardia, hypertension,* palpitations

EENT: *Blurred vision,* tinnitus, mydriasis, ophthalmoplegia

ENDO: SIADH

GI: *Diarrhea, dry mouth,* nausea, vomiting, paralytic ileus, increased appetite, cramps, epigastric distress, jaundice, hepatitis, stomatitis, constipation, weight gain

GU: *Retention,* acute renal failure

HEMA: Agranulocytosis, thrombocytopenia, eosinophilia, leukopenia

INTEG: Rash, urticaria, sweating, pruritus, photosensitivity

PHARMACOKINETICS

Well absorbed, widely distributed, protein binding 92%, extensively metabolized in the liver to active metabolite of imipramine, half-life 7-60 hr

INTERACTIONS

Increase: serotonin syndrome, neuroleptic malignant syndrome—SSRIs, SNRIs, serotonin-receptor agonists, other tricyclic antidepressants

Increase: CNS depression—alcohol, barbiturates, opioids, CNS depressants

Increase: desipramine level—cimetidine, diltiazem, fluvoxaMINE, FLUoxetine, PARoxetine, sertraline, verapamil

Increase: life-threatening B/P elevations; do not use concurrently—cloNIDine

Increase: hypertension—EPINEPHrine, norepinephrine

Increase: hyperpyrexia, seizures, excitation; do not use within 14 days of MAOIs

Increase: QT interval—tricyclics, SUNItinib, vorinostat, ziprasidone, gatifloxacin, levofloxacin, moxifloxacin, sparfloxacin, class IA/III antidysrhythmics

Drug/Herb

Increase: serotonin syndrome, avoid concurrent use—St. John's wort

Increase: CNS depression—kava, valerian

Drug/Lab Test

Increase: serum bilirubin, blood glucose, alk phos

Decrease: sodium

NURSING CONSIDERATIONS

Assess:

- B/P (lying, standing), pulse q4hr; if systolic B/P drops 20 mm Hg, hold product, notify prescriber; take VS q4hr in patients with cardiovascular disease
- Hepatic studies: AST, ALT, bilirubin; thyroid function studies
- Weight weekly; appetite may increase with product
- **ECG** for flattening T wave, bundle branch block, AV block, dysrhythmias in cardiac patients
- **EPS** primarily in geriatric patients: rigidity, dystonia, akathisia
- **Seizure activity** in those with a history of seizures

Black Box Warning: Mental status: mood, sensorium, affect, **suicidal tendencies,** increase in psychiatric symptoms (depression, panic); this product is not indicated for children

- Urinary retention, constipation; constipation most likely in children
- **Withdrawal symptoms:** headache, nausea, vomiting, muscle pain, weakness; not usual unless product discontinued abruptly
- Alcohol consumption; if consumed, hold dose until morning
- Check to confirm that PO medication is swallowed

Evaluate:

- Therapeutic response: decreased depression

Teach patient/family:

- That therapeutic effects may take 2-3 wk

Black Box Warning: That suicidal thoughts/behaviors may occur; to notify prescriber immediately

- To use caution when driving, performing other activities requiring alertness because of drowsiness, dizziness, blurred vision

• To avoid alcohol, other CNS depressants
• Not to discontinue medication abruptly after long-term use because this may cause nausea, headache, malaise
• To wear sunscreen or large hat because photosensitivity occurs

TREATMENT OF OVERDOSE:
ECG monitoring; lavage, activated charcoal; administer anticonvulsant

desirudin
(deh-sihr′uh-din)
Iprivask
Func. class.: Anticoagulant
Chem. class.: Thrombin inhibitor

ACTION: Selectively inhibits free and clot-bound thrombin, prevents activation of clotting factors

USES: Prevents DVT in hip-replacement surgery
Unlabeled uses: Acute MI, PTCA, unstable angina

CONTRAINDICATIONS: Hypersensitivity to this product, mannitol (diluent), hirudin, active bleeding, coagulation disorders
Precautions: Renal disease (CCr <60 ml/min), hepatic disease, GI/respiratory bleeding ≤3 mo, severe uncontrolled hypertension, spinal/epidural anesthesia, bacterial endocarditis, children, geriatric patients, inflammatory bowel disease, labor, organ biopsy, trauma, pregnancy (C)
Black Box Warning: Epidural anesthesia, lumbar puncture

DOSAGE AND ROUTES
• **Adult: SUBCUT** 15 mg every 12 hr × 9-12 days; give first dose 5-15 min before surgery and after induction if a regional block is used

Renal dose
• **Adult: SUBCUT** CCr 31-60 ml/min 5 mg q12hr; CCr < 31 ml/min 1.7 mg q12hr
Available forms: INJ 15 mg and mannitol 3% in water
Administer:
• Visually inspect particulate matter and discoloration prior to use; do not use sols that are cloudy or contain particles
• Do not use IM
Subcut route
• Do not mix with other injections, solvents, or parenteral fluids
• **Reconstitute** each vial with 0.5 ml of provided diluent; shake gently until the drug is fully reconstituted; the injection should be clear, colorless; once reconstituted, each 0.5 ml contains 15.75 mg desirudin; use immediately; however, it remains stable ≤24 hr at room temperature and protected from light; discard any unused solution
• **Subcut injection:** have patient sit or lie down; using a syringe with a 26- or 27-G needle, which is approximately 0.5-inch long, withdraw the entire reconstituted sol into the syringe; inject total volume subcut (unless renal dose); alternate between the left and right anterolateral and left and right posterolateral thigh or abdominal wall; insert whole length of the needle in a skin fold held between the thumb and forefinger; the skin fold should be held throughout the injection; to minimize bruising, do not rub the site

SIDE EFFECTS
CNS: Dizziness, fever
CV: **Thrombosis, thrombophlebitis**, hypotension
EENT: Nosebleeds
GI: **Hematemesis**, nausea, vomiting
GU: Hematuria
HEMA: **Hemorrhage**, anemia
MISC: **Anaphylaxis**, impaired healing, edema

PHARMACOKINETICS
Onset ½ hr, 1-1½ hr, half-life 2 hr (12 hr in severe renal disease)

INTERACTIONS

Increase: bleeding risk—other anticoagulants, salicylates, NSAIDs, thrombolytics, glycoprotein IIb/IIIa antagonists, corticosteroids, dextran 40

Drug/Herb

Increase: bleeding risk—feverfew, garlic, ginger, ginkgo, horse chestnut

Drug/Lab Test

Decrease: Hct/Hgb

NURSING CONSIDERATIONS

Assess:

Bleeding:

- Gums, black tarry stools, hematuria, epistaxis, decreased Hct/Hgb, guiac-positive stools, bleeding from hip-replacement site, notify prescriber if these occur; observe for thrombosis, ecchymosis
- Monitor aPTT daily in those with bleeding risk, CCr <60 ml/min, aPTT should not be >2× control

Black Box Warning: Epidural/spinal anesthesia sites for hematomas; can result in irreversible paralysis

- CCr baseline and qd if CCr >60 ml/min

Evaluate:

- Decreased occurrence of DVT in hip-replacement surgery

Teach patient/family:

- To report any signs of bleeding
- To use a soft-bristle toothbrush to avoid bleeding gums, to use an electric razor

desloratadine (Rx)

(des′lor-at′ah-deen)

Clarinex, Clarinex RediTabs

Func. class.: Antihistamine, 2nd generation

Chem. class.: Selective histamine (H_1)-receptor antagonist

ACTION: Binds to peripheral histamine receptors, thus providing antihistamine action without sedation

USES: Seasonal/perennial allergic rhinitis, chronic idiopathic urticaria, pruritus

CONTRAINDICATIONS: Hypersensitivity, infants/neonates

Precautions: Pregnancy (C), breastfeeding, child, asthma, renal/hepatic impairment

DOSAGE AND ROUTES

- **Adult and child ≥12 yr: PO** 5 mg/day
- **Child 6-11 yr: PO** 2.5 mg/day
- **Child 2-5 yr: PO** 1.25 mg/day
- **Child 6-11 mo: PO** 1 mg/day (urticaria, only)

Hepatic/renal dose

- **Adult: PO** 5 mg every other day

Available form: Tabs 5 mg; orally disintegrating tabs 2.5, 5 mg (RediTabs); syr 0.5 mg/ml

Administer:

- Without regard to meals
- Do not remove RediTabs from blister until ready to use
- RediTabs directly on tongue; may take with or without water

SIDE EFFECTS

CNS: Sedation (more common with increased doses), headache, psychomotor hyperactivity, seizures, fatigue

GI: Hepatitis, nausea, dry mouth

MISC: Flulike symptoms

PHARMACOKINETICS

Onset antihistamine effect 1 hr, relief as early as 1 day, duration up to 24 hr, peak 1½ hr, elimination half-life 8½-28 hr, metabolized in liver to active metabolites, excreted in urine

INTERACTIONS

Increase: CNS depression (rare)—alcohol, opiates, sedative/hypnotics, H_1 blockers, antipsychotics, tricyclic antidepressants, anxiolytics

Increase: desloratadine—nilotinib, etravirine

NURSING CONSIDERATIONS

Assess:

- **Allergy:** hives, rash, rhinitis; monitor respiratory status; stop product 4 days before antigen skin test

Evaluate:

- Therapeutic response: absence of running or congested nose, other allergy symptoms

Teach patient/family:

- To avoid driving, other hazardous activities if drowsiness occurs; to use caution until product's effects are known
- That product may cause photosensitivity; to use sunscreen or stay out of the sun to prevent burns
- Not to exceed max dose

desmopressin (Rx)

(des-moe-press′in)

DDAVP, Minirin, Octostim ✦, Stimate

Func. class.: Pituitary hormone

Chem. class.: Synthetic antidiuretic hormone

ACTION: Promotes reabsorption of water by action on renal tubular epithelium; causes smooth muscle constriction, increase in plasma factor VIII levels, which increases platelet aggregation, thereby resulting in vasopressor effect; similar to vasopressin

USES: Hemophilia A, von Willebrand's disease type 1, nonnephrogenic diabetes insipidus, symptoms of polyuria/polydipsia caused by pituitary dysfunction, nocturnal enuresis

Unlabeled uses: Cardiopulmonary bypass, sickle cell disease, uremic bleeding

CONTRAINDICATIONS: Hypersensitivity, nephrogenic diabetes insipidus, severe renal disease

Precautions: Pregnancy (B), breastfeeding, coronary artery disease, hypertension, cystic fibrosis, thrombus

DOSAGE AND ROUTES

Primary nocturnal enuresis

- **Adult and child ≥6 yr: PO** 0.2 mg at bedtime, max 0.6 mg at bedtime; intranasal 0.2 ml at bedtime, half in each nostril

Diabetes insipidus

- **Adult: INTRANASAL** 0.1-0.4 ml/day in divided doses (1-4 sprays with pump); **IV/SUBCUT** 2-4 micromcg/day in 2 divided doses
- **Child 3 mo to 12 yr: INTRANASAL** 0.05-0.3 ml/day in divided doses

Hemophilia/von Willebrand's disease

- **Adult and child >3 mo: IV** 0.3 mcg/kg in 0.9% NaCl over 15-30 min; may repeat if needed

Antihemorrhagic

- **Adult and child >3 mo: IV** 0.3 mcg/kg
- **Adult and child <50 kg: INTRANASAL** 1 spray in 1 nostril
- **Adult and child >50 kg:** 1 spray in each nostril

Cardiopulmonary bypass (unlabeled)

- **Adult: IV** 0.3 mcg/kg with aminocaproic acid given as a single postop dose

Sickle cell disease (unlabeled)

- **Adult: SUBCUT/IV** 0.3 mcg/kg with a high fluid intake

Uremic bleeding (unlabeled)

- **Adult: SUBCUT/IV** 0.3-0.4 mg/kg as a single inj

Available forms: Inj 4, 15 mg/ml; rhinal tube delivery 2.5 mg/vial (0.1 mg/ml); tabs 0.1, 0.2 mg; nasal spray pump 10 mcg/spray (0.1 mg/ml); nasal sol 1.5 mg/ml (150 mcg/dose)

Administer:

PO route

- Store at room temperature

Direct IV route

- Undiluted over 1 min for diabetes insipidus

Intermittent IV INF route

- Diluted single dose/50 ml of 0.9% NaCl (adult and child >10 kg), single dose/10 ml as IV inf over 15-30 min for von Willebrand's disease or hemophilia A
- Store in refrigerator

SIDE EFFECTS

CNS: Drowsiness, headache, lethargy, flushing, seizures
CV: Increased B/P, palpitations, tachycardia
EENT: Nasal irritation, congestion, rhinitis
GI: Nausea, heartburn, cramps
GU: Vulval pain
META: Hyponatremia, hyponatremia-induced seizures
SYST: Anaphylaxis (IV)

PHARMACOKINETICS

PO: Onset 1 hr, peak 4-7 hr
INTRANASAL: Onset 1 hr; peak 1-4 hr; duration 8-20 hr; half-life 8 min, 76 min (terminal)
IV: Onset 1 min, peak ½ hr, duration >3 hr

INTERACTIONS

Increase: antidiuretic action—carBAMazepine, chlorproPAMIDE, clofibrate
Increase: pressor effect—pressor products
Decrease: antidiuretic action—lithium, alcohol, demeclocycline, heparin, large doses of EPINEPHrine

NURSING CONSIDERATIONS

Assess:
- Pulse, B/P when giving IV or SUBCUT
- I&O ratio, weight daily; check for edema in extremities; if water retention severe, diuretic may be prescribed
- **Water intoxication:** lethargy, behavioral changes, disorientation, neuromuscular excitability
- Intranasal use: nausea, congestion, cramps, headache; usually decreased with decreased dose; for nasal mucosa changes: congestion, edema, discharge, scarring (nasal route)

⚠ For severe allergic reaction, including **anaphylaxis (IV route)**
- Urine volume/osmolality and plasma osmolality (diabetes insipidus)
- Factor VIII coagulant activity, bleeding time before using for hemostasis
- **Nocturnal enuresis:** frequency of enuresis before and during treatment

Evaluate:
- Therapeutic response: absence of severe thirst, decreased urine output, decreased osmolality

Teach patient/family:
- About the proper technique for nasal instillation: to insert tube into nostril to instill product
- To avoid OTC products (cough, hay fever) because these preparations may contain EPINEPHrine, decrease product response; not to use with alcohol because adverse reactions may occur
- To wear emergency ID specifying therapy
- That, if dose is missed, to take when remembered up to 1 hr before next dose; not to double dose; to avoid fluids from 1 hr to up to 8 hr after PO dose
- To report upper respiratory infection, nasal congestion to prescriber

desonide topical

See Appendix B

desoximetasone

(dess-ox′ee-met′ah-sone)

Topicort, Topicort Spray

Func. class.: Corticosteroid, topical

ACTION: Crosses cell membrane to attach to receptors to decrease inflammation, itching; inhibits multiple inflammatory cytokines

USES: Inflammation/itching of corticosteroid-responsive dermatoses on the skin; spray-plaque psoriasis

CONTRAINDICATIONS: Hypersensitivity, use of some preparations on face, axilla, groin, intertriginous areas; monotherapy in primary bacterial infection, TB

Precautions: Pregnancy (C), breastfeeding, children, skin infections, Cushing syndrome

DOSAGE AND ROUTES

Adult/child >10 yr:

- Apply to affected areas 2 times/day

Available forms: Cream 0.05%, 0.25%; ointment 0.05%, 0.25%; gel 0.05%; spray 0.25%

Administer:

Topical route

⚠ Do not use with occlusive dressings

- **Cream/ointment/lotion:** apply sparingly in a thin film and rub gently into the cleansed, affected area
- **Gel:** apply sparingly in a thin film and rub gently into the cleansed, affected area
- **Spray:** discard after 30 days; keep away from heat/flame; store at room temperature

SIDE EFFECTS

INTEG: Burning, folliculitis, pruritus, dermatitis, maceration

MISC: Hyperglycemia, glycosuria, systemic absorption, HPA axis suppression, Cushing syndrome

NURSING CONSIDERATIONS

Assess:

- Skin reactions: burning pruritus, folliculitis, dermatitis
- **Systemic absorption:** hypothalamic-pituitary-adrenal (HPA) suppression and possible adrenocortical insufficiency after stopping treatment

Evaluate:

- Decreasing itching, inflammation on the skin

Teach patient/family:

Topical route

⚠ Not to use with occlusive dressings

- **Cream/ointment/lotion:** apply sparingly in a thin film and rub gently into the cleansed, affected area
- **Gel:** apply sparingly in a thin film and rub gently into the cleansed, affected area
- **Spray:** discard after 30 days; keep away from heat/flame; store at room temperature

desoxyephedrine nasal agent

See Appendix B

desvenlafaxine

Pristiq

Func. class.: Antidepressant, serotonin-receptor norepinephrine reuptake inhibitor (SNRI)

ACTION: May work by blocking the central presynaptic reuptake of 5-HT and NE, resulting in an increased sustained level of these neurotransmitters

USES: Major depressive disorder

Unlabeled uses: Vasomotor symptoms (hot flashes) associated with menopause

CONTRAINDICATIONS: Hypersensitivity to this product or venlafaxine, MAOI therapy

Precautions: CNS depression, abrupt discontinuation, hypertension, hepatic/renal disease, hyponatremia, geriatric patients, pregnancy (C), labor and delivery, breastfeeding, angina, bleeding, cardiac dysrhythmias, MI, stroke, mania, hypovolemia, dehydration, increased intraocular pressure

Black Box Warning: Children, suicidal ideation

DOSAGE AND ROUTES

- **Adult: PO** Initially, 50 mg/day; max 400 mg/day with adjustments as needed; **SPRAY** apply sparingly to area bid, rub gently (plaque psoriasis only)

Renal dose

- **Adult: PO** CCr 30-50 ml/min 50 mg daily; CCr < 30 ml/min or end stage renal disease 50 mg q other day

Vasopastic effects of menopause (unlabeled)

- **Adult: PO** 100-150 mg/day

Available forms:
Extended release tabs 50, 100 mg

Administer:
- Without regard to food; food may minimize GI symptoms
- Extended release tab: do not crush, break, or chew
- Store at room temperature

SIDE EFFECTS

CNS: *Dizziness,* drowsiness, *headache*, tremor, paresthesias, asthenia, suicidal thoughts/behaviors, seizures, fatigue, chills, yawning, hot flashes, flushing, *irritability*, *insomnia*, *anxiety, abnormal dreams, fatigue*

CV: Palpitations, sinus tachycardia, increased blood pressure, orthostatic hypotension

EENT: Blurred vision, mydriasis, tinnitus, bruxism

GI: *Nausea,* xerostomia, *diarrhea*, constipation, vomiting, anorexia, weight loss, dysgeusia, hypercholesterolemia, hypertriglyceridemia

GU: Urinary retention/hesitancy, orgasm dysfunction, decreased libido, impotence, proteinuria

HEMA: Impaired platelet aggregation

INTEG: Photosensitivity, hyperhidrosis, diaphoresis, rash

SYST: Serotonin syndrome, neuroleptic malignant syndrome–like symptoms, toxic epidermal necrolysis, Stevens-Johnson syndrome, erythema multiforme, angioedema; neonatal abstinence syndrome (fetal exposure)

PHARMACOKINETICS

Protein binding 30%, elimination half-life 11 hr; elimination half-life is increased (hepatic/renal disease)

INTERACTIONS

Increase: serotonin syndrome, neuroleptic malignant syndrome–like reactions—SSRIs, other SNRIs, serotonin receptor agonists (almotriptan, eletriptan, frovatriptan, naratriptan, rizatriptan, SUMAtriptan, ZOLMitriptan), tricyclics, traZODone, sibutramine, SUMAtriptan, ergots, dexfenfluramine, fenfluramine, lithium, nefazodone, meperidine, phentermine, MAOIs, dextromethorphan, linezolid, promethazine, methylphenidate, dexmethylphenidate, mirtazapine, pentazocine, tryptophan, methylene blue IV; do not administer concurrently

Increase: bleeding risk—salicylates, thrombolytics, NSAIDs, platelet inhibitors, anticoagulants

Increase: CNS depression—alcohol, opioids, antihistamines, sedatives/hypnotics

Increase: hallucinations, delusions, disorientation—zolpidem

Drug/Herb

Increase: desvenlafaxine action—kava, valerian

Drug/Lab Test

Increase: sodium, cholesterol, triglycerides

False positive: amphetamine, phencyclidine

NURSING CONSIDERATIONS

Assess:

Black Box Warning: Suicidal thoughts/behaviors: mental status and mood; identify suicidal ideation

- **Serotonin syndrome, neuroleptic malignant syndrome–like symptoms:** assess for nausea/vomiting, sedation, dizziness, diaphoresis (sweating), facial flush, hallucinations, mental status changes, myoclonia, restlessness, shivering, elevated blood pressure, hyperthermia, muscle rigidity, autonomic instability, mental status changes; if serotonin syndrome occurs, discontinue desvenlafaxine and any other serotonergic agents
- Monitor B/P baseline and periodically during treatment, lipid levels, signs of glaucoma
- Appetite and nutritional intake, weight loss is common, change diet as need to support weight

Evaluate:
- Decreased depression, increased sense of well-being, renewed interest in activities

Teach patient/family:

• To take as directed, not to double or skip doses; if a dose is missed, take as soon as remembered unless close to next dose; do not discontinue abruptly, decrease gradually

Black Box Warning: To report immediately suicidal thoughts/behaviors; have family members look for symptoms of suicidal ideation

• Not to operate machinery or engage in hazardous activities until reaction is known, may cause dizziness, drowsiness

• To avoid all other products unless approval by prescriber

• To report if pregnancy is planned or suspected (pregnancy [C]) or if breastfeeding

• To report immediately allergic reactions, including rash, hives, difficulty breathing, or swelling of face, lips

• That continuing follow-up exams will be needed

dexamethasone (Rx)

(dex-ah-meth′a-sone)

Baycadron, Decadron, Dexasone ✦, DexPak, Maxidex ✦

dexamethasone sodium phosphate (Rx)

Func. class.: Corticosteroid, synthetic
Chem. class.: Glucocorticoid, long acting

Do not confuse:
Decadron/Percodan

ACTION: Decreases inflammation by suppression of migration of polymorphonuclear leukocytes, fibroblasts, reversal of increased capillary permeability and lysosomal stabilization, suppresses normal immune response, no mineralocorticoid effects

USES: Inflammation, allergies, neoplasms, cerebral edema, septic shock, collagen disorders, dexamethasone suppression test for Cushing syndrome, adrenocortical insufficiency, TB, meningitis, acute exacerbations of MS

CONTRAINDICATIONS: Hypersensitivity to corticosteroids, sulfites, or benzyl alcohol; fungal infections, abrupt discontinuation, coagulopathy, ulcerative colitis, seizure disorders

Precautions: Pregnancy (C), breastfeeding, diabetes mellitus, osteoporosis, seizure disorders, ulcerative colitis, CHF, myasthenia gravis, renal disease, peptic ulcer, esophagitis, recent MI, hypertension, TB, active hepatitis, psychosis, sulfite hypersensitivity, thromboembolic disorders

DOSAGE AND ROUTES

Inflammation

• **Adult: PO** 0.75-9 mg/day in divided doses q6-12hr or phosphate **IM** 0.5-9 mg/day divided q6-12hr

• **Child: PO** 0.024-0.34 mg/kg/day in divided doses q6-12hr

Anaphylactic shock

• **Adult: IV** (phosphate) single dose 1-6 mg/kg or **IV** 40 mg q2-6hr as needed up to 72 hr

Cerebral edema

• **Adult: IV** (phosphate) 10 mg, then 4-6 mg **IM** q6hr × 2-4 days, then taper over 1 wk

• **Child:** loading dose 1-2 mg/kg **(PO/IM/IV)** then 1-1.5 mg/kg/day, max 16 mg/day divided q4-6hr for 2-4 days, then taper down weekly

Adrenocortical insufficiency

• **Adult: PO** 0.75-9 mg/day in divided doses

• **Child: PO** 0.03-0.3 mg/kg/day in 2-4 divided doses

Suppression test

• **Adult: PO** 1 mg at 11 PM or 0.5 mg q6hr × 48 hr

ARDS (unlabeled)

• **Adult: IM/IV** (phosphate) 0.5-9 mg/day in 2-4 divided doses

• **Child: IM/IV** 0.06-0.3 mg/kg/day or 1.2-10 mg/m^2 in divided doses q6-12hr

Bone pain (unlabeled)
- **Adult: PO/IV** 12-20 mg/day in divided doses

Pediatric bacterial meningitis (unlabeled)
- **Child and infant >2 mo: IV** 0.15 mg/kg qid × first 2 days of antibiotics

Available forms: Dexamethasone: tabs 0.5, 0.75, 1, 1.5, 2, 4, 6 mg; elix 0.5 mg/5 ml; oral sol 0.5 mg/5 ml, 1 mg/1 ml; **sodium phosphate:** 4, 10 mg/ml; ophth implant 0.7 mg; ophth susp drops/solution 0.1%

Administer:

PO route
- Titrated dose; use lowest effective dose
- With food or milk to decrease GI symptoms

IM route
- IM inj deeply in large muscle mass; rotate sites; avoid deltoid; use 21-G needle
- In 1 dose in AM to prevent adrenal suppression; avoid SUBCUT administration, may damage tissue

Direct IV route (sodium phosphate)
- Undiluted direct over ≤1 min

Intermittent IV INF route
- Diluted with 0.9% NaCl or D_5W, give as IV inf at prescribed rate

Dexamethasone sodium phosphate

Syringe compatibilities: Acetaminophen, caffeine, dimenhydrAMINE, furosemide, granisetron, hyaluronidase, ketamine, metoclopramide, octreotide, oxyCODONE, palonosetron, ranitidine, salbutamol, SUFentanil, traMADol

Y-site compatibilities: Acetaminophen, acyclovir, alfentanil, allopurinol, amifostine, amikacin, aminocaproic acid, aminophylline, amphotericin B cholesteryl, amphotericin B lipid complex, amphotericin B liposome, amsacrine, anidulafungin, argatroban, ascorbic acid injection, atenolol, atracurium, atropine, aztreonam, benztropine, bivalirudin, bleomycin, bumetanide, buprenorphine, butorphanol, caffeine, CARBOplatin, carmustine, cefamandole, ceFAZolin, cefepime, cefmetazole, cefonicid, cefoperazone, cefotaxime, cefoTEtan, cefOXitin, cefpirome, ceftaroline, cefTAZidime, ceftizoxime, ceftobiprole, cefTRIAXone, chloramphenicol, cimetidine, cisatracurium, CISplatin, cladribine, clindamycin, codeine, cyanocobalamin, cyclophosphamide, cycloSPORINE, cytarabine, DACTINomycin, DAPTOmycin, DAUNOrubicin, dexmedetomidine, digoxin, diltiazem, DOCEtaxel, DOPamine, doripenem, doxacurium, DOXOrubicin, DOXOrubicin liposomal, enalaprilat, ePHEDrine, EPINEPHrine, epoetin alfa, eptifibatide, ertapenem, etoposide, etoposide phosphate, famotidine, fentaNYL, filgrastim, fluconazole, fludarabine, fluorouracil, folic acid, fosaprepitant, foscarnet, furosemide, ganciclovir, gatifloxacin, gemcitabine, glycopyrrolate, granisetron, heparin, hydrocortisone, HYDROmorphone, ifosfamide, imipenem-cilastatin, indomethacin, insulin (regular), irinotecan, isoproterenol, ketorolac, lansoprazole, leucovorin, levofloxacin, lidocaine, linezolid, liposome, LORazepam, LR, mannitol, mechlorethamine, melphalan, meropenem, metaraminol, methadone, methicillin, methoxamine, methyldopate, methylPREDNISolone, metoclopramide, metoprolol, metroNIDAZOLE, mezlocillin, miconazole, milrinone, morphine, moxalactam, multiple vitamins injection, nafcillin, nalbuphine, naloxone, nitroglycerin, nitroprusside, norepinephrine, octreotide, ondansetron, oxacillin, oxaliplatin, oxyCODONE, oxytocin, PACLitaxel, palonosetron, pamidronate, pancuronium, PEMEtrexed, penicillin G potassium/sodium, PENTobarbital, PHENObarbital, phenylephrine, phytonadione, piperacillin, piperacillin-tazobactam, potassium chloride, procainamide, propofol, propranolol, pyridoxine, ranitidine, remifentanil, Ringer's, ritodrine, riTUXimab, sargramostim, sodium acetate/bicarbonate, succinylcholine, SUFentanil, tacrolimus, telavancin, teniposide, theophylline, thiamine, thiotepa, ticarcillin, ticarcillin-clavulanate, tigecycline, tirofiban, TNA, tolazoline, topotecan, trastuzumab, trimetaphan, urokinase, vancomycin, vasopressin, vecuronium, verapamil, vinCRIStine, vinorelbine, vitamin B complex/C, voriconazole, zidovudine, zoledronic acid

SIDE EFFECTS

CNS: *Depression, flushing, sweating,* headache, mood changes, euphoria, psychosis, **seizures**, insomnia, **pseudotumor cerebri**

CV: *Hypertension,* **circulatory collapse**, tachycardia, edema, cardiomyopathy, **thromboembolism, heart failure, dysrhythmias**

EENT: Fungal infections, increased intraocular pressure, blurred vision, cataracts, glaucoma

ENDO: HPA suppression, hyperglycemia, sodium, fluid retention

GI: *Diarrhea, nausea, abdominal distention,* **GI hemorrhage**, *increased appetite,* **pancreatitis**

HEMA: **Thrombocytopenia**, transient leukocytosis, **thromboembolism**

INTEG: Acne, poor wound healing, ecchymosis, petechiae, hirsutism, **angioedema**

META: Hypokalemia

MS: Fractures, osteoporosis, weakness, arthralgia, myopathy

PHARMACOKINETICS

Half-life 1-2 days

PO: Onset 1 hr, peak 1-2 hr, duration 2½ days

IM: Duration 2 days-3 wk

INTERACTIONS

Increase: toxicity—cycloSPORINE

Increase: side effects—alcohol, salicylates, amphotericin B, digoxin, cycloSPORINE, diuretics, NSAIDs

Increase: dexamethasone action—salicylates, estrogens, indomethacin, hormonal contraceptives, ketoconazole, macrolide antiinfectives, NSAIDs

Increase: tendinitis, tendon rupture risk—quinolones

Increase: effect of—antidiabetics

Decrease: dexamethasone action—cholestyramine, colestipol, barbiturates, rifampin, ePHEDrine, phenytoin, theophylline, antacids, bosentan, carBAMazepine, ethotoin

Decrease: anticoagulant effect—anticonvulsants, antidiabetics, ambenonium, neostigmine, isoniazid, toxoids, vaccines, anticholinesterases, salicylates, somatrem

Decrease: potassium levels—thiazide diuretics

Drug/Lab Test

Increase: cholesterol, sodium, blood glucose, uric acid, calcium

Decrease: calcium, potassium, T_4, T_3, thyroid ^{131}I uptake test, urine 17-OHCS, 17-KS, PBI

False negative: skin allergy tests

NURSING CONSIDERATIONS

Assess:

- Potassium, blood, urine glucose while receiving long-term therapy; hypo/hyperglycemia
- Weight daily; notify prescriber of weekly gain >5 lb
- B/P, pulse; notify prescriber of chest pain
- I&O ratio; be alert for decreasing urinary output, increasing edema
- **Cushingoid symptoms: assess for buffalo hump, moon face, increased B/P; monitor; plasma cortisol levels during long-term therapy (normal: 138-635 nmol/L SI units when drawn at 8 AM); prolonged use can cause cushingoid symptoms**
- **Infection:** fever, WBC even after withdrawal of medication; product masks infection
- **Potassium depletion:** paresthesias, fatigue, nausea, vomiting, depression, polyuria, dysrhythmias, weakness
- Edema, hypertension, cardiac symptoms
- **Mental status:** affect, mood, behavioral changes, aggression
- **Abrupt withdrawal: acute adrenal insufficiency and death may occur follow-**

ing abrupt discontinuation of systemic therapy; withdraw gradually

Evaluate:

- Therapeutic response: decreased inflammation

Teach patient/family:

- That ID as corticosteroid user should be carried
- To contact prescriber if surgery, trauma, stress occurs because dose may need to be adjusted
- To notify prescriber if therapeutic response decreases because dosage adjustment may be needed
- To take with food or milk
- That bruising may occur easily
- That if on long-term therapy, a high protein diet may be needed

⚠ Not to discontinue abruptly because **adrenal crisis** can result

- About symptoms of adrenal insufficiency: nausea, anorexia, fatigue, dizziness, dyspnea, weakness, joint pain
- To avoid OTC products: salicylates, alcohol in cough products, cold preparations unless directed by prescriber
- About all aspects of product usage, including cushingoid symptoms; to notify health care provider of infection
- To avoid exposure to chickenpox or measles, persons with infection

dexamethasone (ophthalmic)

(dex-a-meth′a-sone)

Maxidex

Func. class.: Ophthalmic antiinflammatory

Chem. class.: Corticosteroid

Do not confuse:

dexamethasone/desoximetasone

ACTION: Exact mechanism of antiinflammatory action unknown; inhibits multiple inflammatory cytokines; decreases inflammation, collagen deposits, capillary dilation, edema

USES: Treatment of corticosteroid-responsive ophthalmic disorders

CONTRAINDICATIONS: Hypersensitivity to this product or sulfites, ocular TB, acute herpes simplex (superficial), fungal/viral infections of the eye, posterior lens capsule rupture

Precautions: Corneal infected abrasions, glaucoma, pregnancy (C), breastfeeding, migration of intravitreal implant risk, children

DOSAGE AND ROUTES

Corticosteroid-responsive ophthalmic disorders including allergic conjunctivitis (not controlled topically), allergic marginal corneal ulcer, anterior segment inflammation, chorioretinitis, cyclitis, Graves' ophthalmopathy, giant papillary conjunctivitis (GPC), ophthalmic herpes zoster (herpes zoster ophthalmicus, herpes zoster keratitis), iritis, keratitis, superficial punctate keratitis, postoperative ocular inflammation, optic neuritis, diffuse choroiditis, sympathetic ophthalmia, vernal keratoconjunctivitis, corneal injury (corneal abrasion):

- **Adult:** instill 1 or 2 drops of 0.1% ophthalmic sol or susp every hr during the day and every 2 hr at night; reduce application to every 4 hr after response occurs

Available forms: Ophthalmic solution, suspension 0.1%

Administer:

- For ophthalmic use only
- Instruct patient on proper instillation of eye ointment or solution; do not touch the tip of the dropper to the eye, fingertips, or other surface; wait ≥15 min before inserting soft contact lens

SIDE EFFECTS

EENT: Burning, stinging, poor vision, corneal ulcerations, increased IOP, optic nerve damage

NURSING CONSIDERATIONS

Assess:
- **Corneal effects:** ulcerations, infections can worsen with this product

Evaluate:
- Decreased corneal inflammation

Teach patient/family:
- How to use products
- Not to share with others or use for other conditions
- **To notify prescriber immediately if vision changes or if condition worsens**
- To take as prescribed

dexamethasone topical

See Appendix B

dexlansoprazole (Rx)

(dex-lan-so-prey′zole)

Dexilant

Func. class.: Antiulcer, proton-pump inhibitor

Chem. class.: Benzimidazole

ACTION: Suppresses gastric secretion by inhibiting hydrogen/potassium ATPase enzyme system in gastric parietal cell; characterized as gastric acid pump inhibitor because it blocks final step of acid production

USES: Gastroesophageal reflux disease (GERD), severe erosive esophagitis, heartburn

CONTRAINDICATIONS: Hypersensitivity

Precautions: Pregnancy (B), breastfeeding, children, proton-pump hypersensitivity, gastric cancer, hepatic disease, vit B_{12} deficiency

DOSAGE AND ROUTES

Erosive esophagitis
- **Adult: PO** 60 mg daily for up to 8 wk; maintenance: **PO** 30 mg daily for up to 6 mo

GERD
- **Adult: PO** 30 mg daily × 4 wk

Hepatic disease
- **Adult: PO (Child-Pugh B): max 30 mg/day**

Available forms: Del rel caps 30, 60 mg

Administer:
- Swallow caps whole; do not crush, chew caps; caps may be opened, contents sprinkled on food, use immediately; do not chew contents of capsule; give without regard to food

SIDE EFFECTS

CNS: Headache, dizziness, confusion, agitation, amnesia, depression, **anxiety, seizures,** insomnia, migraine

CV: Chest pain, angina, bradycardia, palpitations, **CVA,** hypertension, **MI**

EENT: Tinnitus

GI: Diarrhea, abdominal pain, vomiting, nausea, constipation, flatulence, colitis, dysgeusia

HEMA: Anemia, **neutropenia, thrombocytopenia, pernicious anemia, thrombosis**

INTEG: Rash, urticaria, pruritus

META: Gout

MS: Arthralgia, mylagia

RESP: Upper respiratory infections, cough, epistaxis, dyspnea, **pneumonia**

SYST: **Anaphylaxis, Stevens-Johnson syndrome, toxic epidermal necrolysis, exfoliative dermatitis**

PHARMACOKINETICS

Absorption 57%-64%; plasma half-life 1-2 hr; protein binding 96.1%-98.8%; extensively metabolized in liver; excreted in urine, feces; clearance decreased in geriatric patients, renal/hepatic impairment; peak dual 1-2 hr, 4-5 hr

INTERACTIONS

Increase: dexlansoprazole effect—CYP2C19, 3A4 inhibitors (fluvoxaMINE, voriconazole)
- Dexlansoprazole absorption: sucralfate

Decrease: absorption of ketoconazole, itraconazole, iron, delavirdine, ampicillin, calcium carbonate

Drug/Lab Test

Increase: LFTs, bilirubin, creatinine, glucose, lipids

Decrease: platelets, magnesium

NURSING CONSIDERATIONS

Assess:

- GI system: bowel sounds q8hr, abdomen for pain, swelling, anorexia, monitor serum magnesium
- **Hepatotoxicity (rare):** hepatitis, jaundice, monitor hepatic studies (AST, ALT, alk phos) if hepatic adverse reactions occur
- **Hypomagnesemia:** usually 3 mo to 1 yr after beginning therapy; monitor magnesium level, assess for irregular heart beats, muscle spasms; in children, fatigue, upset stomach, dizziness; magnesium supplement may be used
- **Anaphylaxis (rare), serious skin disorders:** require emergency intervention

Evaluate:

- Therapeutic response: absence of epigastric pain, swelling, fullness

Teach patient/family:

- To report severe diarrhea; product may have to be discontinued
- That diabetic patient should know that hypoglycemia may occur
- To avoid hazardous activities; dizziness may occur
- To avoid alcohol, salicylates, ibuprofen; may cause GI irritation
- To report allergic reactions, symptoms of low magnesium levels
- To notify prescriber if pregnancy is planned or suspected; not to breastfeed
- To swallow cap whole, not to chew, crush

dexmethylphenidate (Rx)

(dex′meth-ul-fen′ih-dayt)

Focalin, Focalin XR

Func. class.: Central nervous system (CNS) stimulant, psychostimulant

Controlled Substance Schedule II

Do not confuse:
dexmethylphenidate/methylphenidate

ACTION: Increases release of norepinephrine and DOPamine into the extraneuronal space; also blocks the reuptake of norepinephrine and DOPamine into the presynaptic neuron; mode of action for treating attention-deficit/hyperactivity disorder (ADHD) is unknown

USES: ADHD

CONTRAINDICATIONS: Breastfeeding, children <6 yr, hypersensitivity to methylphenidate, anxiety, history of Gilles de la Tourette's syndrome, tics, psychosis, glaucoma, concurrent treatment with MAOIs or within 14 days of discontinuing treatment with MAOIs

Precautions: Pregnancy (C), hypertension, depression, seizures, CV disorders, alcoholism

Black Box Warning: Substance abuse

DOSAGE AND ROUTES

- **Adult/adolescent/child >6 yr: PO** 2.5 mg bid with doses at least 4 hr apart, gradually increase to a maximum of 20 mg/day (10 mg bid); for those taking methylphenidate, use 1/2 of methylphenidate dose initially then increase as needed to a max of 20 mg/day
- **Adolescent and child ≥6 yr: EXT REL** 5 mg/day, may adjust to 20 mg/day in 5-mg increments, max 30 mg/day

• **Adult: PO EXT REL** 10 mg/day, may adjust to 20 mg/day in 10-mg increments, max 40 mg/day

Available forms: Tabs 2.5, 5, 10 mg; ext rel caps 5, 10, 15, 20, 25, 30, 35, 40 mg (Focalin XR)

Administer:

• Twice daily at least 4 hr apart; ext rel once a day; in the morning, ext rel cap may be opened and contents sprinkled onto applesauce and consumed without chewing

• Without regard to meals

• Do not break, crush, or chew ext rel product

• Med guide should be provided by dispenser

SIDE EFFECTS

CNS: Dizziness, headache, drowsiness, nervousness, insomnia, **toxic psychosis, neuroleptic malignant syndrome (rare)**, Tourette's syndrome

CV: Palpitations, B/P changes, angina, **dysrhythmias, tachycardia, MI, stroke**

GI: *Nausea, anorexia*, abnormal hepatic function, **hepatic coma,** *abdominal pain*

HEMA: **Leukopenia, anemia, thrombocytopenic purpura**

INTEG: **Exfoliative dermatitis,** urticaria, rash, erythema multiforme

MISC: *Fever*, arthralgia, scalp hair loss

PHARMACOKINETICS

Readily absorbed, elimination half-life 2.2 hr, metabolized by liver, excreted by kidneys

PO: Peak 1½ hr, onset ½-1 hr, duration 4 hr

PO-ER: Onset unknown, peak 4 hr, duration 8 hr

INTERACTIONS

⚠ **Increase: hypertensive crisis—MAOIs or within 14 days of MAOIs, vasopressors**

Increase: sympathomimetic effect—decongestants, vasoconstrictors

Increase: effects of anticonvulsants, tricyclics, SSRIs, coumarin

Decrease: effects of antihypertensives

Drug/Herb

• Synergistic effect: melatonin

NURSING CONSIDERATIONS

Assess:

Black Box Warning: Substance abuse, past or current; psychotic episodes may occur, especially with parental abuse

• VS, B/P; may reverse antihypertensives; check patients with cardiac disease more often for increased B/P

• CBC, differential platelet counts during long-term therapy, urinalysis; with diabetes: blood glucose, urine glucose; insulin changes may have to be made because eating will decrease; platelets

• Height, growth rate q3mo in children; growth rate may be decreased

• Mental status: mood, sensorium, affect, stimulation, insomnia, aggressiveness, hostility

⚠ **Withdrawal symptoms: headache, nausea, vomiting, muscle pain, weakness**

• Appetite, sleep, speech patterns

• For attention span, decreased hyperactivity in persons with ADHD

Evaluate:

• Therapeutic response: decreased hyperactivity or ability to stay awake

Teach patient/family:

• To decrease caffeine consumption (coffee, tea, cola, chocolate); may increase irritability, stimulation

• To take early in day to prevent insomnia

• To avoid OTC preparations unless approved by prescriber; to avoid alcohol ingestion

• To taper off product over several wk to avoid depression, increased sleeping, lethargy

• To avoid hazardous activities until stabilized on medication

• To get needed rest; patients will feel more tired at end of day

• To notify all health care workers, including school nurse, of medication and schedule

• About information, instructions provided in patient information section

• To notify prescriber if pregnancy is planned or suspected; to avoid breast-feeding

TREATMENT OF OVERDOSE: Administer fluids; hemodialysis or peritoneal dialysis; antihypertensive for increased B/P; administer short-acting barbiturate before lavage

dextroamphetamine (Rx)

(dex-troe-am-fet′a-meen)

Dexedrine, ProCentra

Func. class.: Cerebral stimulant

Chem. class.: Amphetamine

Controlled Substance Schedule II

ACTION: Increases release of norepinephrine, DOPamine in cerebral cortex to reticular activating system

USES: Narcolepsy, attention-deficit/hyperactivity disorder (ADHD)

Unlabeled uses: Obesity

CONTRAINDICATIONS: Hypersensitivity to sympathomimetic amines, hyperthyroidism, hypertension, glaucoma, severe arteriosclerosis, anxiety, anorexia nervosa, tartrazine dye hypersensitivity

Black Box Warning: Symptomatic CV disease, substance abuse

Precautions: Pregnancy (C), breast-feeding, children <3 yr, depression, Gilles de la Tourette's disorder, cardiomyopathy, bipolar disorder, abrupt discontinuation, acute MI; benzyl alcohol, salicylate hypersensitivity; hypercortisolism, obesity, psychosis, seizure disorder

DOSAGE AND ROUTES

Narcolepsy

• **Adult: PO** 5 mg bid, titrate daily dose by no more than 10 mg/wk, max 60 mg/day

• **Child 6-12 yr: PO** 5 mg/day, titrate daily dose by no more than 5 mg/day at weekly intervals

ADHD

• **Adult: PO** 5-60 mg/day in divided doses

• **Child 3-5 yr: PO** 2.5 mg/day increasing by 2.5 mg/day at weekly intervals, max 40 mg/day

• **Child >6-12 yr: PO** 5 mg daily-bid increasing by 5 mg/day at weekly intervals

Obesity, exogenous (unlabeled)

• **Adult and adolescent: PO** 5-30 mg/dose given 30-60 min before meals, use for 3-6 wk only

Available forms: Tabs 5, 10 mg; oral sol 5 mg/5 ml

Administer:

• At least 6 hr before bedtime to avoid sleeplessness

• Use calibrated measuring device for oral sol

• Store all forms at room temperature

SIDE EFFECTS

CNS: *Hyperactivity, insomnia, restlessness, talkativeness,* dizziness, headache, chills, stimulation, dysphoria, irritability, aggressiveness, tremor, dependence, addiction

CV: *Palpitations, tachycardia,* hypertension, decrease in heart rate, **dysrhythmias**

GI: *Anorexia,* dry mouth, diarrhea, constipation, weight loss, metallic taste

GU: Impotence, change in libido

INTEG: Urticaria

PHARMACOKINETICS

Onset 1 hr; peak 2 hr; duration 4-20 hr; ext rel onset 1 hr, peak 2 hr, duration 8 hr; metabolized by liver; urine excretion pH dependent; crosses placenta, breast milk; half-life 6-8 hr (child), 10-12 hr (adult)

INTERACTIONS

⚠ Hypertensive crisis: MAOIs or within 14 days of MAOIs

Increase: serotonin syndrome, neuroleptic malignant syndrome: SSRIs, SNRIs, serotonin-receptor agonists; do not use concurrently

Increase: dextroamphetamine effect—acetaZOLAMIDE, antacids, sodium bicarbonate

Increase: CNS effect—haloperidol, tricyclics, phenothiazines

Decrease: absorption of barbiturates, phenytoin

Decrease: dextroamphetamine effect—ascorbic acid, ammonium chloride, guanethidine

Decrease: effect of adrenergic blockers, antidiabetics, antihypertensives, antihistamines

Drug/Herb

- Serotonin syndrome: St. John's wort

Decrease: stimulant effect—eucalyptus

Drug/Food

Increase: amine effect—caffeine (cola, coffee, tea [green/black])

Drug/Lab Test

Increase: plasma corticosteroids, urinary steroids

NURSING CONSIDERATIONS

Assess:

Black Box Warning: **Cardiac disease:** VS, B/P; product may reverse antihypertensives; check patients with cardiac disease often

Black Box Warning: **Substance abuse:** use for prolonged periods may lead to dependence; sudden death or serious CV events can occur from misuse; chronic intoxication (insomnia, irritability, personality changes)

- CBC, urinalysis; with diabetes: blood glucose, urine glucose; insulin changes may be required because eating will decrease
- Height, growth rate in children; growth rate may be decreased, weight
- Mental status: mood, sensorium, affect, stimulation, insomnia, irritability
- Tolerance or dependency: increased amount may be used to get same effect; will develop after long-term use
- **Overdose:** pain, fever, dehydration, insomnia, hyperactivity

Evaluate:

- Therapeutic response: increased CNS stimulation, decreased drowsiness

Teach patient/family:

- To take before meals (obesity)
- To decrease caffeine consumption (coffee, tea, cola, chocolate); may increase irritability, stimulation
- To avoid OTC preparations unless approved by prescriber
- To taper product over several wk; depression, increased sleeping, lethargy may occur
- To avoid alcohol ingestion
- To avoid hazardous activities until stabilized on medication
- To get needed rest; patient will feel more tired at end of day

TREATMENT OF OVERDOSE:

Administer fluids, hemodialysis, or peritoneal dialysis; antihypertensive for increased B/P, ammonium Cl for increased excretion

dextromethorphan (OTC)

(dex-troe-meth-or′fan)

Balminil ♣, Benylin ♣, Buckley's DM, Buckley's Mixture, Delsym 12-Hour, ElixSure Cough, Koffex ♣, Robitussin, Robitussin Cough with honey, Robitussin Long Acting Strength, Scot-Tussin Diabetes CF, Silphen-DM, Top Care Day Time Cough, Top Care Tussin Cough Suppressant Long-Acting, Triaminic Long Acting Cough, Tylenol Children's Simply Cough, Vicks DayQuil Cough, Vicks Formula 44, Wal-Tussin

Func. class.: Antitussive, nonopioid
Chem. class.: Levorphanol derivative

ACTION: Depresses cough center in medulla by direct effect

USES: Nonproductive cough caused by colds or inhaled irritants

CONTRAINDICATIONS: Hypersensitivity

Precautions: Pregnancy (C), fever, hepatic disease, asthma/emphysema, chronic cough

DOSAGE AND ROUTES

- **Adult and child ≥12 yr: PO** 10-20 mg q4hr or 30 mg q6-8hr, max 120 mg/day; **SUS-REL LIQ** 60 mg q12hr, max 120 mg/day
- **Child 6-12 yr: PO** 5-10 mg q4hr; **SUS REL LIQ** 30 mg bid, **LOZ** 5-10 mg q1-4hr; max 60 mg/day
- **Child 4-6 yr: PO** 2.5-7.5 mg q4-8hr, max 30 mg/day; **SUS REL LIQ** 15 mg bid
- **Child <2-3 yr: PO** 2.5-5 mg q4hr or 7.5 mg q6-8hr, max 30 mg/day; **EXT REL** not recommended

Available forms: Liq 7.5, 15 mg/5 ml; syr 10 mg/5 ml, 15 mg/5 ml, 30 mg/15 ml; gel caps 15 mg; caps 15 mg; ext rel susp: 30 mg/5 ml

Administer:

- **Chew tabs:** chew well; **syrup:** use calibrated measuring device; **ext rel susp:** shake well, use calibrated measuring device
- Decreased dose for geriatric patients; metabolism may be slowed

SIDE EFFECTS

CNS: *Dizziness,* sedation, confusion, ataxia, fatigue
GI: *Nausea*

PHARMACOKINETICS

PO: Onset 15-30 min, duration 3-6 hr
SUS: Duration 12 hr, terminal half-life 11 hr, metabolized by the liver, excreted via kidneys

INTERACTIONS

- Do not give with MAOIs or within 2 wk of MAOIs; avoid furazolidone, linezolid, procarbazine (MAOI activity)

Increase: CNS depression—alcohol, antidepressants, antihistamines, opioids, sedative/hypnotics
Increase: adverse reactions—amiodarone, quiNIDine, serotonin receptor agonist, sibutramine, SSRI

NURSING CONSIDERATIONS

Assess:

- **Cough:** type, frequency, character, including sputum
- Increase fluids to liquify secretions

Evaluate:

- Therapeutic response: absence of cough

Teach patient/family:

- To avoid driving, other hazardous activities until stabilized on medication
- To avoid smoking, smoke-filled rooms, perfumes, dust, environmental pollutants, cleaners that increase cough
- To avoid alcohol, CNS depressants
- To notify prescriber if cough persists over a few days

dextrose (D-glucose) (Rx)

Func. class.: Caloric, parenteral solution

ACTION: Needed for adequate utilization of amino acids; decreases protein, nitrogen loss; prevents ketosis

USES: Increases intake of calories; increases fluids in patients unable to take adequate fluids, calories orally; acute hypoglycemia

CONTRAINDICATIONS: Hyperglycemia, delirium tremens, hemorrhage (cranial/spinal), CHF, anuria, allergy to corn products (concentrated products)
Precautions: Cardiac/renal/hepatic disease, diabetes mellitus, carbohydrate intolerance

DOSAGE AND ROUTES

Hypoglycemia
- **Adult: PO/IV** 10-25 mg g/dose (20-50 ml of a 50% sol), may need subsequent continuous IV inf of 10% dextrose

Acute symptomatic hypoglycemia (infants/neonates)
- **Neonate/infant: IV** 250-500 mg/kg/dose (25% sol)

Available forms: Inj 2.5%, 5%, 10%, 20%, 25%, 30%, 38.5%, 40%, 50%, 60%, 70%; oral gel 40%; chew tab 5 g

Administer:
- Do not use concentrated solutions IM/IV; 25%, 50% may be used IV
- Only (4%) protein and dextrose (up to 12.5%) via peripheral vein; stronger sol: central IV administration
- May be given undiluted via prepared sol; give 10% sol, 5 ml/15 sec; 10% sol, 1000 ml/3 hr or more; 20% sol, 500 ml/½-1 hr; 50% sol, 10 ml/min; control rate, rapid inf may cause fluid shifts; do not use same inf set as used for blood; never discontinue hypertonic products abruptly
- To reduce contamination, all IV sets should be replaced ≤24 hr
- Oral glucose preparations (gel, chew tabs) to be used in conscious patients only; check serum blood glucose 10 min after 1st dose

SIDE EFFECTS

CNS: Confusion, loss of consciousness, dizziness
CV: Hypertension, CHF, pulmonary edema, intracranial hemorrhage
ENDO: Hyperglycemia, rebound hypoglycemia, hyperosmolar syndrome, hyperglycemic nonketotic syndrome, aluminum toxicity, hypokalemia, hypomagnesium
GI: Nausea
GU: Glycosuria, osmotic diuresis
INTEG: Chills, flushing, warm feeling, rash, urticaria, extravasation necrosis
RESP: Pulmonary edema

INTERACTIONS

Increase: fluid retention/electrolyte excretion—corticosteroids

Drug/Lab Test
Increase: glucose

NURSING CONSIDERATIONS

Assess:
- Electrolytes (potassium, sodium, calcium, chlorine, magnesium), blood glucose, ammonia, phosphate
- Inj site for extravasation: redness along vein, edema at site, necrosis, pain; hard, tender area; site should be changed immediately
- Monitor temperature for increased fever, indicating infection; if infection suspected, discontinue inf, culture tubing, bottle, catheter tip cultured
- Monitor glucose level, I&O, weight, fluid overload
- Nutritional status: calorie count by dietitian

Evaluate:
- Therapeutic response: increased weight

Teach patient/family:
- About the reason for dextrose inf

• To review hypoglycemia/hyperglycemia symptoms
• To review blood glucose monitoring procedures

diazepam (Rx)

(dye-az′-e-pam)

Diazemuls ♣, Diastat, Valium

Func. class.: Antianxiety, anticonvulsant, skeletal muscle relaxant, central acting

Chem. class.: Benzodiazepine, long-acting

Controlled Substance Schedule IV

Do not confuse:
diazepam/Ditropan/LORazepam

ACTION: Potentiates the actions of GABA, especially in the limbic system, reticular formation; enhances presympathetic inhibition, inhibits spinal polysynaptic afferent paths

USES: Anxiety, acute alcohol withdrawal, adjunct for seizure disorders; preoperatively as a relaxant for skeletal muscle relaxation; rectally for acute repetitive seizures

Unlabeled uses: Agitation, benzodiazepine withdrawal, chloroquine overdose, insomnia, seizure prophylaxis

CONTRAINDICATIONS: Pregnancy (D), hypersensitivity to benzodiazepines, closed-angle glaucoma, coma, myasthenia gravis, ethanol intoxication, hepatic disease, sleep apnea

Precautions: Breastfeeding, children <6 mo, geriatric patients, debilitation, renal disease, asthma, bipolar disorder, COPD, CNS depression, labor, Parkinson's disease, neutropenia, psychosis, seizures, substance abuse, smoking

DOSAGE AND ROUTES

Anxiety/seizure disorders
• **Adult: PO** 2-10 mg bid-qid; **IM/IV** 2-5 mg q3-4hr
• **Geriatric: PO** 1-2 mg daily-bid, increase slowly as needed
• **Child >6 mo: PO** 1-2.5 mg tid/qid; **IM/IV** 0.04-0.3 mg/kg/dose q2-4hr, max 0.6 mg/kg in an 8-hr period

Precardioversion
• **Adult: IV** 5-15 mg 5-10 min precardioversion

Preendoscopy
• **Adult: IV** 2.5-20 mg; **IM** 5-10 mg ½ hr preendoscopy

Muscle relaxation
• **Adult: PO** 2-10 mg tid-qid or **EXT REL** 15-30 mg/day; **IV/IM** 5-10 mg, repeat in 2-4 hr
• **Geriatric: PO** 2-5 mg bid-qid; **IV/IM** 2-5 mg, may repeat in 2-4 hr

Tetanic muscle spasms
• **Child >5 yr: IM/IV** 5-10 mg q3-4hr prn
• **Infant >30 days: IM/IV** 1-2 mg q3-4hr prn

Status epilepticus
• **Adult: IV/IM** 5-10 mg, 2 mg/min, may repeat q10-15min, max 30 mg; may repeat in 2-4 hr if seizures reappear
• **Child >5 yr: IM** 1 mg q2-5min; **IV** 1 mg slowly
• **Child 1 mo-5 yr: IV** 0.2-0.5 mg slowly; **IM** 0.2-0.5 mg slowly q2-5min up to 5 mg, may repeat in 2-4 hr prn

Seizures other than status epilepticus
• **Adult: RECT** 0.2 mg/kg, may repeat in 4-12 hr
• **Child 6-11 yr: RECT** 0.3 mg/kg, may repeat in 4-12 hr
• **Child 2-5 yr: RECT** 0.5 mg/kg, may repeat in 4-12 hr

Alcohol withdrawal
• **Adult: IV** 10 mg initially then 5-10 mg q3-4hr prn

Benzodiazepine withdrawal (unlabeled)
• **Adult: PO** Taper 0.5-2 mg over 4-16 wk

Febrile seizure prophylaxis (unlabeled)
- **Child 6 mo-5 yr: PO** 0.33 mg/kg q8hr until afebrile for ≥24 hr

Available forms: Tabs 2, 5, 10 mg; inj 5 mg/ml; oral sol 5 mg/5 ml, rectal 2.5 (pediatric), 10, 20 mg, twin packs; ext rel cap 15 mg

Administer:
- With food or milk for GI symptoms; crushed if patient is unable to swallow medication whole
- Reduced opioid dose by 1/3 if given concomitantly with diazepam
- **Concentrate:** use calibrated dropper only; mix with water, juice, pudding, applesauce; to be consumed immediately

Rectal route
- Do not use more than 5×/mo or for an episode q5days (Diastat)

Direct IV route
- Into large vein; give IV 5 mg or less/1 min or total dose over 3 min or more (children, infants); continuous inf is not recommended; inject as close to vein insertion as possible; **do not dilute or mix with other products**

SIDE EFFECTS

CNS: *Dizziness, drowsiness,* confusion, headache, anxiety, tremors, stimulation, fatigue, depression, insomnia, hallucinations, ataxia, fatigue
CV: *Orthostatic hypotension,* **ECG changes, tachycardia,** hypotension
EENT: *Blurred vision,* tinnitus, mydriasis, nystagmus
GI: Constipation, dry mouth, nausea, vomiting, anorexia, diarrhea
HEMA: Neutropenia
INTEG: Rash, dermatitis, itching
RESP: Respiratory depression

PHARMACOKINETICS

Metabolized by liver via CYP2C19, CYP3A4; excreted by kidneys; crosses placenta; excreted in breast milk; crosses the blood-brain barrier; half-life 20-50 hr; more reliable by mouth; 99% protein binding

PO: Rapidly absorbed, onset 1/2 hr, duration 2-3 hr
IM: Onset 15-30 min, duration 1-1 1/2 hr, absorption slow and erratic
RECT: Peak 1.5 hr
IV: Onset immediate, duration 15 min-1 hr

INTERACTIONS

Increase: diazepam effect—amiodarone, protease inhibitors, diltiazem, cimetidine, clarithromycin, dalfopristin, quinupristin, delavirdine, disulfiram, efavirenz, erythromycin, fluconazole, fluvoxaMINE, imatinib, itraconazole, ketoconazole, IV miconazole, nefazodone, niCARdipine, ranolazine, troleandomycin, verapamil, voriconazole, zafirlukast, zileauton, valproic acid
Increase: toxicity—barbiturates, SSRIs, cimetidine, CNS depressants, valproic acid, CYP3A4 inhibitors
Increase: CNS depression—CNS depressants, alcohol
Decrease: diazepam metabolism—oral contraceptives, valproic acid, disulfiram, isoniazid, propranolol
Decrease: diazepam effect—CYP3A4 inducers (rifampin, barbiturates, carBAMazepine, ethotoin, phenytoin, fosphenytoin), smoking

Drug/Lab Test
Increase: AST/ALT, serum bilirubin

NURSING CONSIDERATIONS

Assess:
- B/P (lying, standing), pulse; respiratory rate; if systolic B/P drops 20 mm Hg, hold product, notify prescriber; respirations q5-15min if given IV
- Blood studies: CBC during long-term therapy; blood dyscrasias (rare); hepatic studies: AST, ALT, bilirubin, creatinine, LDH, alk phos
- **Degree of anxiety;** what precipitates anxiety and whether product controls symptoms
- **Alcohol withdrawal symptoms,** including hallucinations (visual, auditory),

delirium, irritability, agitation, fine to coarse tremors

- Seizure control and type, duration, intensity of seizures
- For muscle spasms; pain relief
- IV site for thrombosis or phlebitis, which may occur rapidly
- Mental status: mood, sensorium, affect, sleeping pattern, drowsiness, dizziness, suicidal tendencies
- **Physical dependency, withdrawal symptoms:** headache, nausea, vomiting, muscle pain, weakness after long-term use

Evaluate:

- Therapeutic response: decreased anxiety, restlessness, insomnia

Teach patient/family:

- That product may be taken with food
- That product not to be used for everyday stress or for >4 mo unless directed by prescriber; to take no more than prescribed amount; that product may be habit forming
- To avoid OTC preparations unless approved by prescriber
- To avoid driving, activities that require alertness; drowsiness may occur
- To avoid alcohol, other psychotropic medications unless directed by prescriber; that smoking may decrease diazepam effect by increasing diazepam metabolism
- Not to discontinue medication abruptly after long-term use; to gradually taper
- To rise slowly or fainting may occur, especially in geriatric patients
- That drowsiness may worsen at beginning of treatment
- To notify prescriber if pregnancy is planned or suspected (D), avoid breastfeeding

TREATMENT OF OVERDOSE:

Lavage, VS, supportive care, flumazenil

dibucaine topical

See Appendix B

diclofenac epolamine (Rx)

(dye-kloe′fen-ak)

Flector

diclofenac potassium (Rx)

Cambia, Cataflam, Rapide ✱, Zipsor

diclofenac sodium (Rx)

Apo-Diclo ✱, Novo-Difenac ✱, Nu-Diclo ✱, PENNSAID, Sandoz Diclofenac ✱, Solaraze Topical Gel, Voltaren, Voltaren Topical Gel, Voltaren XR

Func. class.: Nonsteroidal antiinflammatory products (NSAIDs), nonopioid analgesic

Chem. class.: Phenylacetic acid

Do not confuse:
Cataflam/Catapres

ACTION: Inhibits COX-1, COX-2 by blocking arachidonate resulting in analgesic, antiinflammatory, antipyretic effects

USES: Acute, chronic RA; osteoarthritis; ankylosing spondylitis; analgesia; primary dysmenorrhea; patch: mild to moderate pain

Unlabeled uses: Arthralgia, headache, migraine, bone pain, myalgia

CONTRAINDICATIONS: Hypersensitivity to aspirin, iodides, other NSAIDs, bovine protein, asthma, serious CV disease; eczema, exfoliative dermatitis, skin abrasions (gel, patch)

Black Box Warning: Treatment of perioperative pain (CABG), surgery

Precautions: Pregnancy (C) (tabs, del rel tab, ext rel tab, top gel), (volteren, ophthalmic sol); pregnancy (B) (top gel)(solaraze); top patch, top sol, cap, powder for oral sol (C <30 wk, D >30

wk); breastfeeding, children, bleeding disorders, GI disorders, cardiac disorders, hypersensitivity to other antiinflammatory agents, CCr <30 ml/min, accidental exposure, acute bronchospasm, hypersensitivity to benzyl alcohol

Black Box Warning: GI bleeding, MI, stroke

DOSAGE AND ROUTES

Osteoarthritis

- **Adult: PO** (Cataflam) 50 mg bid-tid, max 150 mg/day; **DEL REL** (Voltaren) 50 mg bid-tid or 75 mg bid, max 150 mg/day; **EXT REL** (Voltaren-XR) 100 mg daily, max 150 mg/day; **TOP GEL** 1% (Voltaren gel) 4 g for each of lower extremities qid, max 16 g/day; 2 g for each of upper extremities qid, max 8 g/day; **TOP SOL** (Pennsaid) apply 40 drops to each affected knee qid; apply 10 drops at a time, spread over entire knee

Rheumatoid arthritis

- **Adult: PO** (Cataflam) 50 mg tid-qid, max 200 mg/day; **DEL REL** (Voltaren) 50 mg tid-qid or 75 mg bid, max 200 mg/day; **EXT REL** (Voltaren-XR) 100 mg daily, may increase to 200 mg/day, max 200 mg/day

Ankylosing spondylitis

- **Adult: PO DEL REL** (Voltaren) 25 mg qid and 25 mg at bedtime, max 125 mg/day

Acute migraine with/without aura

- **Adult: PO** (powder for oral sol) (Cambia) 50 mg as a single dose, mix contents of packet in 1-2 oz water

Mild to moderate pain

- **Adult: PO** (Zipsor) 25 mg qid

Dysmenorrhea or nonrheumatic inflammatory conditions

- **Adult: PO** (Cataflam) 50 mg tid or 100 mg initially then 50 mg tid, max 200 mg 1st day then 150 mg/day, immediate release only

Pain of strains/sprains

- **Adult: TOP PATCH** (Flector) apply patch to area bid

Actinic keratosis

- **Adult: TOP GEL** (Solaraze) apply to area bid

Prevention of heterotropic ossification (unlabeled)

- **Adult: PO** 50 mg tid × 3 wk

Renal dose

- **Avoid:** Use of topical gel, patch, sol, potassium oral tab for advanced renal disease

Available forms: Epolamine: topical patch 1.3%; **potassium:** tabs 50 mg; tabs liquid filled 25 mg; oral powder for sol 50 mg; **sodium:** delayed rel tabs (enteric-coated) 25, 50, 75 mg; Pennsaid: top sol 1.5%, ext rel 100 mg; topical gel 1%, 3%

Administer:

PO route

- Do not break, crush, or chew enteric products
- Take with a full glass of water to enhance absorption, remain upright for ½ hr; if dose missed, take as soon as remembered within 2 hr if taking 1-2×/day; do not double doses
- Store at room temperature

Topical patch route (Flector)

- Wash hands before handling patch
- Remove and release liner before administration
- Use only on normal, intact skin
- Remove before bath, shower, swimming, do not use heat or occlusive dressings
- Discard removed patch in trash away from children, pets
- Store at room temperature

Topical gel route

- Apply to intact skin, do not use heat or occlusive dressings
- Use only for osteoarthritis, mild to moderate pain
- Store at room temperature, avoid heat, do not freeze

Topical solution route

- Apply to clean, dry skin
- Wait until dry before applying clothing, other creams/lotions
- Wait ≥30 min after use before bathing, swimming
- Store at room temperature

SIDE EFFECTS

CNS: *Dizziness, headache,* drowsiness, fatigue, tremors, confusion, insomnia, anxiety, depression, nervousness, paresthesia, muscle weakness
CV: CHF, tachycardia, peripheral edema, palpitations, dysrhythmias, hypo/hypertension, fluid retention, MI, stroke
EENT: Tinnitus, hearing loss, blurred vision, laryngeal edema
GI: Nausea, anorexia, vomiting, diarrhea, jaundice, cholestatic hepatitis, constipation, flatulence, cramps, dry mouth, peptic ulcer, GI bleeding, hepatotoxicity, hematemesis
GU: Nephrotoxicity: dysuria, hematuria, oliguria, azotemia, cystitis
HEMA: Blood dyscrasias, epistaxis, anemia
INTEG: Purpura, rash, pruritus, sweating, erythema, petechiae, photosensitivity, alopecia
META: Hyperglycemia, hypoglycemia
RESP: Dyspnea, bronchospasm
SYST: Anaphylaxis, Stevens-Johnson syndrome

PHARMACOKINETICS

PO: Peak 2-3 hr; **TOP Patch:** peak 12 hr; elimination half-life 1-2 hr, patch 12 hr, 99% bound to plasma proteins, metabolized in liver to metabolite, excreted in urine

INTERACTIONS

• Need for dosage adjustment: antidiabetics
Increase: hyperkalemia—potassium-sparing diuretics
Increase: anticoagulant effect—anticoagulants, NSAIDs, platelet inhibitors, salicylates, thrombolytics, SSRIs
Increase: toxicity—phenytoin, lithium, cycloSPORINE, methotrexate, digoxin, lithium, cidofovir
Increase: GI side effects—aspirin, other NSAIDs, bisphosphonates, corticosteroids
Decrease: antihypertensive effect—β-blockers, diuretics, ACE inhibitors
Decrease: effect of diuretics
Drug/Herb
Increase: bleeding risk—garlic, ginger, ginkgo; monitor for bleeding

NURSING CONSIDERATIONS

Assess:

Black Box Warning: CABG: do not use oral, top, gel, patch, in perioperative pain in coronary artery bypass graft surgery for 10-14 days

Black Box Warning: Stroke/MI: may increase CHF and hypertension, increased CV thrombotic events that may be fatal; those with CV disease may be at greater risk

• **Pain:** location, character, aggravating/alleviating factors, ROM before and 1 hr after dose
• Patients with asthma, aspirin hypersensitivity, nasal polyps; may develop hypersensitivity
• LFTs (may be elevated), uric acid (may be decreased—serum; increased—urine) periodically; also BUN, creatinine, electrolytes (may be elevated)
⚠ **Blood dyscrasias (thrombocytopenia):** bruising, fatigue, bleeding, poor healing; blood counts during therapy; watch for decreasing platelets; if low, therapy may need to be discontinued, restarted after hematologic recovery; stool guaiac

Evaluate:
• Therapeutic response: decreased inflammation in joints, after cataract surgery

Teach patient/family:
• That product must be continued for prescribed time to be effective; to contact prescriber before surgery regarding when to discontinue this product
• To report bleeding, bruising, fatigue, malaise; **blood dyscrasias** do occur
• To avoid aspirin, alcoholic beverages, NSAIDs, or other OTC medications unless approved by prescriber
• To take with food, milk, or antacids to avoid GI upset; to swallow whole

- To use caution when driving; drowsiness, dizziness may occur
- To report **hepatotoxicity:** flulike symptoms, nausea, vomiting, jaundice, pruritus, lethargy
- To use sunscreen to prevent photosensitivity
- To notify all providers of product use
- To notify prescriber if pregnancy is planned or suspected (C, tabs) (C, <30 wk, D, >30 wk caps, topical patch/solution, powder for oral solution)

diclofenac ophthalmic

See Appendix B

RARELY USED

dicyclomine (Rx)

(dye-sye′kloe-meen)

Bentyl, Bentylol ✱, Formulex ✱, Lomine ✱

Func. class.: Gastrointestinal anticholinergic

USES: Treatment of peptic ulcer disease in combination with other products; infant colic, urinary incontinence, IBS

CONTRAINDICATIONS: Hypersensitivity to anticholinergics, closed-angle glaucoma, GI obstruction, myasthenia gravis, paralytic ileus, GI atony, toxic megacolon, dementia

DOSAGE AND ROUTES

- **Adult: PO** 10-20 mg tid-qid; **IM** 20 mg q4-6hr; max 160 mg/day
- **Child >2 yr: PO** 10 mg tid-qid
- **Child 6 mo-2 yr: PO** 5 mg tid-qid

didanosine (Rx)

(dye-dan′oh-seen)

ddI, Videx Pediatric Powder, Videx EC

Func. class.: Antiretroviral

Chem. class.: Nucleoside reverse transcriptase inhibitor (NRTI)

ACTION: Nucleoside analog incorporating into cellular DNA by viral reverse transcriptase, thereby terminating the cellular DNA chain

USES: HIV-1 infection in combination with at least 2 other antiretrovirals

Unlabeled uses: HIV prophylaxis

CONTRAINDICATIONS: Hypersensitivity, lactic acidosis, pancreatitis, phenylketonuria

Precautions: Pregnancy (B), breastfeeding, children, renal disease, sodium-restricted diets, elevated amylase, preexisting peripheral neuropathy, hyperuricemia, gout, CHF, noncirrhotic portal hypertension

Black Box Warning: Hepatic disease, lactic acidosis, pancreatitis

DOSAGE AND ROUTES

- **Adult/adolescent/child ≥6 yr and ≥60 kg: PO EXT REL CAP** 400 mg/day; if used with tenofovir, reduce to 250 mg/day
- **Adult/adolescent/child ≥6 yr and 25 kg to <60 kg: PO EXT REL CAP** 250 mg/day; if used with tenofovir, reduce to 200 mg/day
- **Adolescent 20 kg to <25 kg: PO EXT REL CAP** 200 mg/day
- **Adult ≥60 kg: PO ORAL SOL** 200 mg bid or 400 mg/day; if used with tenofovir, reduce to 250 mg/day
- **Adult <60 kg: PO ORAL SOL** 125 mg bid or 250 mg/day; if used with tenofovir, reduce to 200 mg/day

• **Adolescent/child/infant >8 mo: PO ORAL SOL** 120 mg/m^2 every 12 hr, max adult dosing
• **Infant <8 mo/neonate ≥2 wk: PO ORAL SOL** 100 mg/m^2 every 12 hr

Renal dose

• **Adult: PO** CrCl ≥60 ml/min: No change
• **Adult/adolescent ≥60 kg: PO** CCr 30-59 ml/min: reduce oral sol to 100 mg every 12 hr or 200 mg every 24 hr; reduce EXT-REL caps to 200 mg/day; CCr 10-29 ml/min: reduce oral sol to 150 mg every 24 hr; reduce EXT-REL caps to 125 mg/day; CCr <10 ml/min: reduce oral sol to 100 mg every 24 hr; reduce EXT-REL caps to 125 mg/day
• **Adult/adolescent <60 kg: PO** CCr 30-59 ml/min: reduce oral sol to 75 mg every 12 hr or to 150 mg every 24 hr; reduce EXT-REL caps to 125 mg/day; CCr 10-29 ml/min: reduce oral sol to 100 mg every 24 hr; reduce EXT-REL caps to 125 mg/day; CCr <10 ml/min: reduce oral sol to 75 mg every 24 hr; EXT-REL caps are not recommended

Intermittent hemodialysis/ continuous ambulatory peritoneal dialysis

• **Adult/adolescent >60 kg:** Give 100 mg oral sol or 125 mg EXT-REL CAPS every 24 hr
• **Adult/adolescent <60 kg:** Give 75 mg oral sol every 24 hr, EXT-REL CAPS are not recommended

Available forms: Powder for oral sol 10 mg/ml; del rel caps 125, 200, 250, 400 mg

Administer:

• Pediatric powder for oral sol after preparation by pharmacist; dilution required using purified USP water then antacid (10 mg/ml), refrigerate, shake before use
• On an empty stomach ≥30 min before or 2 hr after meals
• Adjust dose with renal impairment
• Store tabs, caps in tightly closed bottle at room temperature; store oral sol after dissolving at room temperature ≤4 hr

SIDE EFFECTS

CNS: Peripheral neuropathy, seizures, confusion, *anxiety,* hypertonia, abnormal thinking, asthenia, *insomnia,* CNS depression, pain, dizziness, chills, fever
CV: Hypertension, vasodilation, dysrhythmia, syncope, CHF, palpitation
EENT: Ear pain, otitis, photophobia, visual impairment, retinal depigmentation, optic neuritis
GI: Pancreatitis, *diarrhea, nausea,* vomiting, *abdominal pain,* constipation, stomatitis, dyspepsia, liver abnormalities, flatulence, taste perversion, dry mouth, oral thrush, melena, increased ALT/AST, alk phos, amylase, hepatic failure, noncirrhotic portal hypertension
GU: Increased bilirubin, uric acid
HEMA: Leukopenia, granulocytopenia, thrombocytopenia, anemia
INTEG: *Rash, pruritus,* alopecia, ecchymosis, hemorrhage, petechiae, sweating
MS: Myalgia, arthritis, myopathy, muscular atrophy
RESP: Cough, pneumonia, dyspnea, asthma, epistaxis, hypoventilation, sinusitis
SYST: Lactic acidosis, anaphylaxis

PHARMACOKINETICS

PO: Peak 0.67 hr, del rel 2 hr; elimination half-life 1.62 hr; extensive metabolism thought to occur; administration within 5 min of food will decrease absorption (50%); excreted urine, feces

INTERACTIONS

Increase: didanosine level—allopurinol, tenofovir
Increase: side effects from magnesium, aluminum antacids
Increase: pancreatitis risk—stavudine
Decrease: absorption—ketoconazole, dapsone
Decrease: concs of fluoroquinolones, other antiretrovirals, itraconazole, tetracyclines
Decrease: didanosine level—methadone
• Do not use with these products PO: gatifloxacin, gemifloxacin, grepafloxacin,

levofloxacin, lomefloxacin, moxifloxacin, norfloxacin, sparfloxacin, trovafloxacin

Drug/Food

• Any food decreases rate of absorption 50%, do not use with food
• Do not use with acidic juices

NURSING CONSIDERATIONS

Assess:

Black Box Warning: **Pancreatitis:** do not use in those with symptoms of pancreatitis; may be dose related in advanced HIV, alcoholism, history of pancreatitis

• **Peripheral neuropathy:** tingling or pain in hands and feet, distal numbness; onset usually occurs 2-6 mo after beginning treatment, may persist if product not discontinued

Black Box Warning: **Lactic acidosis, severe hepatomegaly, pancreatitis:** abdominal pain, nausea, vomiting, elevated hepatic enzymes; product should be discontinued because condition can be fatal

• Children by dilated retinal exam q6mo to rule out retinal depigmentation
• CBC, differential, platelet count monthly; notify prescriber of results; alk phos, monitor amylase; viral load, CD4 count
• Renal studies: BUN, serum uric acid, urine CCr before, during therapy
• Temperature may indicate beginning infection
• Hepatic studies before, during therapy (bilirubin, AST, ALT) as needed, monthly
• Clean up of powdered products; use wet mop or damp sponge

Evaluate:

• Therapeutic response: absence of infection; symptoms of HIV

Teach patient/family:

• To avoid use with alcohol; not to take with food
• To report numbness/tingling in extremities
• To take on an empty stomach; not to take dapsone at same time as ddI; not to mix powder with fruit juice; chew tab or crush and dissolve in water; to drink powder immediately after mixing
• To report signs of **infection:** increased temperature, sore throat, flulike symptoms
• To report signs of **anemia:** fatigue, headache, faintness, SOB, irritability
• To report **bleeding;** to avoid use of razors, commercial mouthwash
• That hair may be lost during therapy (rare); that a wig or hairpiece may make patient feel better
• That product does not cure, only controls symptoms

RARELY USED

diflunisal

(dye-floo′ni-sal)

Dolobid

Func. class.: Nonsteroidal antiinflammatory drug

USES: Mild to moderate pain including osteoarthritis, rheumatoid arthritis

CONTRAINDICATIONS: Hypersensitivity to this product, NSAIDs, salicylates, severe renal disease

Black Box Warning: Perioperative pain after CABG

DOSAGE AND ROUTES

Osteoarthritis, rheumatoid arthritis

• **Adult: PO** 500-1500 mg/day in 2 divided doses every 12 hr
• **Child ≥12 yr: PO** 250-1000 mg/day in 2 divided doses every 12 hr, max 1500 mg

Pain

• **Adult/child ≥12 yr: PO** 1000 mg, then 250-500 mg every 8-12 hr; a lower dose may be used

Available forms: Tab 500 mg

difluprednate (ophthalmic)

(die-flu′pred-nate)

Durezol

Func. class.: Ophthalmic antiinflammatory

Chem. class.: Corticosteroid

ACTION: Exact mechanism of antiinflammatory action unknown; inhibits multiple inflammatory cytokines; decreases release of arachidonic acid, which increases in inflammation

USES: For the treatment of postoperative ocular pain and postoperative ocular inflammation; for the treatment of endogenous anterior uveitis

CONTRAINDICATIONS: Hypersensitivity to this product, glycerin, polysorbate, ocular TB, acute herpes simplex (superficial), fungal/viral infections of the eye

Precautions: Pregnancy (C), breastfeeding, children, corneal infected abrasions, glaucoma

DOSAGE AND ROUTES

Postoperative ocular pain, postoperative ocular inflammation

- **Adult/geriatric/adolescents/children/infants: OPHTH** Instill 1 drop into the conjunctival sac of the affected eye(s) qid beginning 24 hr after surgery; continue giving 4 ×/day for the first 2 wk of the postoperative period, then administer bid × 1 wk; at the end of the third wk, taper dosage based on response

Endogenous anterior uveitis:

- **Adult: OPHTH** Instill 1 drop into the conjunctival sac of the affected eye(s) qid × 14 days, followed by tapering based on response

Available forms: Ophthalmic emulsion 0.05%

Administer:

- Apply topically to the eye, shake well before use
- Do not touch the tip of the dropper to the eye, fingertips, or other surface
- Instruct patient on proper instillation of eye sol
- When using this product, the patient should not wear contact lenses

SIDE EFFECTS

EENT: Burning, stinging, poor vision, corneal ulcerations, increased IOP, optic nerve damage

NURSING CONSIDERATIONS

Assess:

- **Corneal effects:** ulcerations, infections can worsen with this product, monitor IOP used IO over ten days

Evaluate:

- **Therapeutic response:** Decreased corneal inflammation

Teach patient/family:

- How to use product
- Not to share with others or use for other conditions
- To notify prescriber immediately if vision changes or if condition worsens
- To take as prescribed

HIGH ALERT

digoxin (Rx)

(di-jox′in)

APO-Digoxin ♣, Lanoxin

Func. class.: Cardiac glycoside, inotropic, antidysrhythmic

Chem. class.: Digoxin preparation

Do not confuse:

Lanoxin/Lasix/Lonox/Lomotil/Xanax/Levoxine

ACTION: Inhibits the sodium-potassium ATPase pump, which makes more calcium available for contractile proteins, thereby resulting in increased cardiac output (positive inotropic effect); increases force of contractions; decreases heart rate (negative chronotropic effect); decreases AV conduction speed

USES:
Heart failure, atrial fibrillation, atrial flutter, atrial tachycardia, cardiogenic shock, paroxysmal atrial tachycardia, rapid digitalization in these disorders

Unlabeled uses: Atrial flutter, paroxysmal supraventricular tachycardia (PSVT) treatment/prophylaxis

CONTRAINDICATIONS:
Hypersensitivity to digoxin, ventricular fibrillation, ventricular tachycardia

Precautions: Pregnancy (C), breastfeeding, geriatric patients, renal disease, acute MI, AV block, severe respiratory disease, hypothyroidism, sinus nodal disease, hypokalemia, carotid sinus syndrome, 2nd- or 3rd-degree heart block, electrolyte disturbances, hypertension, cor pulmonale, Wolff-Parkinson-White syndrome

DOSAGE AND ROUTES

Loading dose, IV route

- **Adult/adolescent/child >10 yr: IV** 8-12 mcg/kg , divided into ≥3 doses, with the first dose equaling one-half the total, give subsequent doses every 6-8 hr
- **Child 5-10 yr: IV** 15-30 mcg/kg divided into ≥3 doses, with the first dose equaling one-half the total, give subsequent doses every 6-8 hr
- **Child 2-4 yr: IV** 25-35 mcg/kg, divided into ≥3 doses, with the first dose equaling one-half the total, give subsequent doses every 6-8 hr
- **Infant/child <2 yr: IV** 30-50 mcg/kg, divided into ≥3 doses, with the first dose equaling one-half the total, give subsequent doses every 6-8 hr
- **Full-term neonate: IV** 20-30 mcg/kg, divided into ≥3 doses, with the first dose equaling one-half the total, give subsequent doses every 6-8 hr
- **Premature neonate: IV** 15-25 mcg/kg, divided into ≥3 doses, with the first dose equaling one-half the total, give subsequent doses every 6-8 hr

Loading dose, PO (tablets): Tablets are 60%-80% bioavailable; oral elixir should be used to obtain the appropriate dose in infants, young pediatric patients, or patients with very low body weight

- **Adult/adolescent/child >10 yr: PO** Total dosage of 10-15 mcg/kg in 3 divided doses, give one-half the total loading dose initially, then one-fourth the loading dose every 4-8 hr × 2 doses
- **Child 5-10 yr: PO** Total dosage of 20-45 mcg/kg in 3 divided doses, give one-half the total loading dose initially, then one-fourth the loading dose every 4-8 hr × 2 doses

Loading dose, PO (elixir): Elixir is approximately 70%-85% bioavailable

- **Adult/adolescent/child >10 yr: PO** Total dosage of 10-15 mcg/kg, give one-half the total loading dose initially, then additional fractions of the planned total dose at 4-8 hr
- **Child 5-10 yr: PO** Total dosage of 20-35 mcg/kg, give one-half the total loading dose initially, then additional fractions of the planned total dose at 4-8 hr
- **Child 2-4 yr: PO** Total dosage of 30-45 mcg/kg, give one-half the total loading dose initially, then additional fractions of the planned total dose at 4-8 hr
- **Infant/child <2 yr: PO** Total dosage of 35-60 mcg/kg, give one-half the total loading dose initially then additional fractions of the planned total dose at 4-8 hr
- **Full-term neonate: PO** Total dosage of 25-35 mcg/kg, give one-half the total loading dose initially, then additional fractions of the planned total dose at 4-8 hr
- **Premature neonate: PO** Total dosage of 20-30 mcg/kg, give one-half the total loading dose initially, then additional fractions of the planned total dose at 4-8 hr

Maintenance dose, IV

- **Adult: IV** 125-350 mcg/day, depending on CrCl, usual daily maintenance dosage for CHF based on corrected CrCl (ml/min/70 kg) and lean body weight (LBW) are listed below

• LBW 50-59 kg: CrCl ≥100 ml/min: 175 mcg//day; CrCl 70-99 ml/min: 150 mcg/day; CrCl 60-69 ml/min: 125 mcg/day
• LBW 60-69 kg: CrCl ≥90 ml/min: 200 mcg/day; CrCl 70-89 ml/min: 175 mcg/day; CrCl 60-69 ml/min: 150 mcg/day
• LBW 70-79 kg: CrCl ≥100 ml/min: 250 mcg/day; CrCl 90-99 ml/min: 225 mcg/day; CrCl 70-89 ml/min: 200 mcg/day; CrCl 60-69 ml/min: 175 mcg/day
• LBW 80-89 kg: CrCl ≥100 ml/min: 275 mcg/day; CrCl 80-99 ml/min: 250 mcg/day; CrCl 70-79 ml/min: 225 mcg/day; CrCl 60-69 ml/min: 200 mcg/day
• LBW 90-99 kg: CrCl ≥90 ml/min: 300 mcg/day; CrCl 80-89 ml/min: 275 mcg/day; CrCl 70-79 ml/min: 250 mcg/day; CrCl 60-69 ml/min: 225 mcg/day
• LBW ≥100 kg: CrCl ≥100 ml/min: 350 mcg/day; CrCl 90-99 ml/min: 325 mcg IV/day; CrCl 80-89 ml/min: 300 mcg/day; CrCl 70-79 ml/min: 275 mcg IV/day; CrCl 60-69 ml/min: 250 mcg/day
• **Child >10 yr:** 25%-35% of the IV digitalizing dose/day
• **Child 5-10 yr:** 25%-35% of the IV digitalizing dose bid
• **Children 2-4 yr:** 25%-35% of the IV digitalizing dose bid
• **Infant, child <2 yr:** 25%-35% of the IV digitalizing dose bid
• **Full term neonate:** 25%-35% of the IV digitalizing dose bid
• **Preterm neonate:** 20%-30% of the IV digitalizing dose bid

Maintenance dose, PO (tablets)
• **Adult/adolescent/child >10 yr: PO** 3.4-5.1 mcg/kg/day
• LBW 40-49 kg: CrCl ≥70 ml/min: 187.5 mcg/day; CrCl ≥60-69 ml/min: 125 mcg/day
• LBW 50-59 kg: CrCl ≥90 ml/min: 250 mcg/day; CrCl 60-89 ml/min: 187.5 mcg/day
• LBW 60-69 kg: CrCl ≥100 ml/min: 312.5 mcg/day; CrCl 60-99 ml/min: 250 mcg/day
• LBW 70-79 kg: CrCl ≥80 ml/min: 312.5 mcg/day; CrCl 60-79 ml/min: 250 mcg/day
• LBW 80-89 kg: CrCl ≥90 ml/min: 375 mcg/day; CrCl 60-89 ml/min: 312.5 mcg/day
• LBW 90-99 kg: CrCl ≥90 ml/min: 437.5 mcg/day; CrCl 70-89 ml/min: 375 mcg/day; CrCl 60-69 ml/min: 312.5 mcg/day
• LBW ≥100 kg: CrCl ≥100 ml/min: 500 mcg/day; CrCl 80-99 ml/min: 437.5 mcg/day; CrCl 60-79 ml/min: 375 mcg/day
• **Child 5-10 yr: PO** 6.4-12.9 mcg/kg/day in 2 divided doses is recommended starting maintenance dose

Maintenance dose, PO (elixir)
• **Adult/adolescent/child >10 yr: PO** 3-4.5 mcg/kg/day
• LBW 40-49 kg: CrCl ≥100 ml/min: 170 mcg/day; CrCl 90-99 ml/min: 160 mcg/day; CrCl 80-89 ml/min: 150 mcg/day; CrCl 70-79 ml/min: 140 mcg/day; CrCl 60-69 ml/min: 130 mcg/day
• LBW 50-59 kg: CrCl ≥100 ml/min: 213 mcg/day; CrCl 90-99 ml/min: 200 mcg/day; CrCl 80-89 ml/min: 188 mcg/day; CrCl 70-79 ml/min: 175 mcg/day; CrCl 60-69 ml/min: 163 mcg/day
• LBW 60-69 kg: CrCl ≥100 ml/min: 255 mcg/day; CrCl 90-99 ml/min: 240 mcg/day; CrCl 80-89 ml/min: 225 mcg/day; CrCl 70-79 ml/min: 210 mcg/day; CrCl 60-69 ml/min: 195 mcg/day
• LBW 70-79 kg: CrCl ≥100 ml/min: 298 mcg/day; CrCl 90-99 ml/min: 280 mcg/day; CrCl 80-89 ml/min: 263 mcg/day; CrCl 70-79 ml/min: 245 mcg/day; CrCl 60-69 ml/min: 228 mcg/day
• LBW 80-89 kg: CrCl ≥100 ml/min: 340 mcg/day; CrCl 90-99 ml/min: 320 mcg/day; CrCl 80-89 ml/min: 300 mcg/day; CrCl 70-79 ml/min: 280 mcg/day; CrCl 60-69 ml/min: 260 mcg/day
• LBW 90-99 kg: CrCl ≥100 ml/min: 383 mcg/day; CrCl 90-99 ml/min: 360 mcg/day; CrCl 80-89 ml/min: 338 mcg/day; CrCl 70-79 ml/min: 315 mcg/day; CrCl 60-69 ml/min: 293 mcg/day
• LBW ≥100 kg: CrCl ≥100 ml/min: 425 mcg/day; CrCl 90-99 ml/min: 400 mcg/day; CrCl 80-89 ml/min: 375 mcg/

day; CrCl 70-79 ml/min: 350 mcg/day; CrCl 60-69 ml/min: 325 mcg/day

• **Child 5-10 yr: PO** 5.6-11.3 mcg/kg/day in 2 divided doses

• **Child 2-4 yr: PO** 9.4-13.1 mcg/kg/day in 2 divided doses

• **Infant/child <2 yr: PO** 11.3-18.8 mcg/kg/day in 2 divided doses

• **Full-term neonate: PO** 7.5-11.3 mcg/kg/day in 2 divided doses

• **Preterm neonate: PO** 4.7-7.8 mcg/kg/day in 2 divided doses

Renal dose, IV route

• **CrCl 50-59 ml/min:** LBW 50-59 kg: 125 mcg 1× daily; LBW 60-69 kg: 150 mcg 1× daily; LBW 70-79 kg: 175 mcg 1× daily; LBW 80-89 kg: 200 mcg 1× daily; LBW 90-99 kg: 225 mcg 1× daily; LBW ≥100 kg: 250 mcg 1× daily

• **CrCl 40-49 ml/min:** LBW 50-59 kg: 100 mcg 1× daily; LBW 60-69 kg: 125 mcg 1× daily; LBW 70-79 kg: 150 mcg 1× daily; LBW 80-89 kg: 175 mcg 1× daily; LBW 90-99 kg: 200 mcg 1× daily; LBW ≥100 kg: 225 mcg 1× daily

• **CrCl 30-39 ml/min:** LBW 50-59 kg: 100 mcg 1× daily; LBW 60-69 kg: 125 mcg 1× daily; LBW 70-89 kg: 150 mcg 1× daily; LBW 90-99 kg: 175 mcg 1× daily; LBW ≥100 kg: 200 mcg 1× daily

• **CrCl 20-29 ml/min:** LBW 50-69 kg: 100 mcg 1× daily; LBW 70-79 kg: 125 mcg 1× daily; LBW 80-99 kg: 150 mcg 1× daily; LBW ≥100 kg: 175 mcg 1× daily

• **CrCl 10-19 ml/min:** LBW 50-59 kg: 75 mcg 1× daily; LBW 60-79 kg: 100 mcg 1× daily; LBW 80-89 kg: 125 mcg 1× daily; LBW ≥90 kg: 150 mcg 1× daily

• **CrCl <10 ml/min** LBW 50-69 kg: 75 mcg 1× daily; LBW 70-89 kg: 100 mcg 1× daily; LBW 90-99 kg: 125 mcg 1× daily; LBW ≥100 kg: 150 mcg 1× daily

Renal dose, PO (tablets)

• **CrCl 40-59 ml/min:** LBW 50-69 kg: 187.5 mcg 1× daily

• **CrCl 50-59 ml/min:** LBW 70-89 kg: 250 mcg 1× daily; LBW ≥90 kg: 312.5 mcg 1× daily

• **CrCl 30-39 ml/min:** LBW 50-59 kg: 125 mcg 1× daily; LBW 60-79 kg: 187.5 mcg 1× daily; LBW 80-99 kg: 250 mcg 1× daily; LBW ≥100 kg: 312.5 mcg 1× daily

• **CrCl 20-29 ml/min:** LBW 50-69 kg: 125 mcg 1× daily; LBW 70-89 kg: 187.5 mcg 1× daily; LBW ≥90 kg: 250 mcg 1× daily

• **CrCl <20 ml/min:** LBW 50-69 kg: 125 mcg 1× daily; LBW 70-99 kg: 187.5 mcg 1× daily; LBW ≥100 kg: 250 mcg 1× daily

Renal dose, PO (elixir)

• **CrCl 50-59 ml/min:** LBW 50-59 kg: 150 mcg 1× daily; LBW 60-69 kg: 180 mcg 1× daily; LBW 70-79 kg: 210 mcg 1× daily; LBW 80-89 kg: 240 mcg 1× daily; LBW 90-99 kg: 270 mcg 1× daily; LBW ≥100 kg: 300 mcg 1× daily

• **CrCl 40-49 ml/min:** LBW 50-59 kg: 138 mcg 1× daily; LBW 60-69 kg: 165 mcg 1× daily; LBW 70-79 kg: 193 mcg 1× daily; LBW 80-89 kg: 220 mcg 1× daily; LBW 90-99 kg: 248 mcg 1× daily; LBW ≥100 kg: 275 mcg 1× daily

• **CrCl 30-39 ml/min:** LBW 50-59 kg: 125 mcg 1× daily; LBW 60-69 kg: 150 mcg 1× daily; LBW 70-79 kg: 175 mcg PO 1× daily; LBW 80-89 kg: 200 mcg PO 1× daily; LBW 90-99 kg: 225 mcg 1× daily; LBW ≥100 kg: 250 mcg PO 1× daily

• **CrCl 20-29 ml/min:** LBW 50-59 kg: 113 mcg 1× daily; LBW 60-69 kg: 135 mcg 1× daily; LBW 70-79 kg: 158 mcg 1× daily; LBW 80-89 kg: 180 mcg PO 1× daily; LBW 90-99 kg: 203 mcg 1× daily; LBW ≥100 kg: 225 mcg PO 1× daily

• **CrCl <20 ml/min:** LBW 50-59 kg: 100 mcg 1× daily; LBW 60-69 kg: 120 mcg 1× daily; LBW 70-79 kg: 140 mcg 1× daily; LBW 80-89 kg: 160 mcg 1× daily; LBW 90-99 kg: 180 mcg 1× daily; LBW ≥100 kg: 200 mcg 1× daily

The daily maintenance dose can also be estimated using patient CrCl and loading dose (LD) according to the method of Jelliffe and Brooker: Daily % loss = 14 + CrCl/5

Available forms: Caps 0.05, 0.1, 0.2 mg; elix 0.05 mg/ml; tabs 0.125, 0.25, 0.5 mg; inj 0.5 ✦, 0.25 mg/ml; pediatric inj 0.1 mg/ml

Administer:

PO route

⚠ Bioavailability varies among different oral dosage forms of digoxin and among different brands of the same dosage form; changing from one preparation to another might require dosage adjustments

- **All dosage forms:** may be administered without regard to meals
- **Tab:** may be crushed and administered with food or fluids
- **Pediatric elixir:** administer using a calibrated measuring device

Injectable

- IV is preferred over IM because it is less painful and more rapid action
- PO should replace parenteral therapy as soon as possible
- Visually inspect parenteral products for particulate matter and discoloration before use

IM route

- Do not administer >2 ml at any one IM injection site
- Inject deeply into gluteal muscle, then massage area

IV route

- May be given undiluted or each 1 ml may be diluted in 4 ml of sterile water for injection, NS, D_5W, or LR; diluent volumes <4 ml will cause precipitation; use diluted solutions immediately
- Inject over ≥5 min via Y-site or 3-way stopcock; in patients with pulmonary edema, administer over 10-15 min; to avoid inadvertent overdosage, do not flush the syringe following administration

Additive compatibilities: Cimetidine, floxacillin, furosemide, lidocaine, potassium chloride, ranitidine, verapamil

Syringe compatibilities: Heparin, milrinone

Y-site compatibilities: Acyclovir, alfentanil, amikacin, aminocaproic acid, aminophylline, amphotericin B lipid complex, anidulafungin, ascorbic acid injection, atenolol, atracurium, atropine, aztreonam, benztropine, bivalirudin, bleomycin, bumetanide, buprenorphine, butorphanol, calcium chloride/gluconate, CARBOplatin, cefamandole, ceFAZolin, cefmetazole, cefonicid, cefoperazone, cefotaxime, cefoTEtan, cefOXitin, cefTAZidime, ceftizoxime, ceftobiprole, cefTRIAXone, cefuroxime, cephalothin, cephapirin, chloramphenicol, chlorproMAZINE, cimetidine, ciprofloxacin, cisatracurium, CISplatin, clindamycin, codeine, cyanocobalamin, cyclophosphamide, cycloSPORINE, cytarabine, DACTINomycin, DAPTOmycin, dexamethasone, dexmedetomidine, diltiazem, diphenhydrAMINE, DOBUTamine, DOCEtaxel, DOPamine, doripenem, doxacurium, doxycycline, enalaprilat, ePHEDrine, EPINEPHrine, epirubicin, epoetin alfa, eptifibatide, ertapenem, erythromycin, esmolol, etoposide, famotidine, fenoldopam, fentaNYL, fludarabine, fluorouracil, folic acid, furosemide, ganciclovir, gatifloxacin, gemcitabine, gentamicin, glycopyrrolate, granisetron, heparin, hydrocortisone, HYDROmorphone, hydrOXYzine, ifosfamide, imipenem-cilastatin, indomethacin, irinotecan, isoproterenol, ketorolac, labetalol, levofloxacin, lidocaine, linezolid, LORazepam, LR, magnesium sulfate, mannitol, mechlorethamine, meperidine, meropenem, metaraminol, methicillin, methotrexate, methoxamine, methyldopate, methylPREDNISolone, metoclopramide, metoprolol, metroNIDAZOLE, mezlocillin, miconazole, midazolam, milrinone, morphine, moxalactam, multiple vitamins injection, mycophenolate mofetil, nafcillin, nalbuphine, naloxone, nesiritide, metilmicin, nitroglycerin, nitroprusside, norepinephrine, octreotide, ondansetron, oxacillin, oxaliplatin, oxytocin, palonosetron, pamidronate, pancuronium, pantoprazole, papaverine, PEMEtrexed, penicillin G potassium/sodium, pentazocine, PENTobarbital, PHENobarbital, phenylephrine, phytonadione, piperacillin, piperacillin-tazobactam, polymyxin B, potassium chloride, procainamide, prochlorperazine, promethazine, propranolol, protamine, pyridoxine, quiNIDine, ranitidine, remifentanil, Ringer's, ritodrine, riTUXimab, rocuronium, sodium acetate/bicarbonate, succinylcholine, SUFentanil, tacrolimus, teniposide, theophylline, thiamine, thiotepa, ticarcillin, ticarcillin-clavulanate, tigecycline, tirofiban, TNA, tobramycin, tolazoline, TPN, trastuzumab, trimetaphan, urokinase, vancomycin, vasopressin, vecuronium, verapamil, vinCRIStine, vinorelbine, vitamin B complex, voriconazole, zoledronic acid

SIDE EFFECTS

CNS: *Headache,* drowsiness, apathy, confusion, disorientation, fatigue, depression, hallucinations
CV: **Dysrhythmias,** *hypotension,* bradycardia, **AV block**
EENT: Blurred vision, yellow-green halos, photophobia, diplopia
GI: Nausea, vomiting, anorexia, abdominal pain, diarrhea

PHARMACOKINETICS

Half-life 1.5 days, excreted in urine, protein binding 20%-30%
PO: Onset ½-2 hr, peak 6-8 hr, duration 3-4 days
IV: Onset 5-30 min, peak 1-5 hr, duration variable

INTERACTIONS

Increase: toxicity—azole antifungals, macrolides, tetracyclines, ritonavir
Increase: hypercalcemia, hypomagnesemia, digoxin toxicity—thiazides, parenteral calcium
Increase: hypokalemia, digoxin toxicity—diuretics, amphotericin B, carbenicillin, ticarcillin, corticosteroids
Increase: digoxin levels—propantheline, quiNIDine, verapamil, amiodarone, anticholinergics, diltiazem, NIFEdipine, indomethacin
Increase: bradycardia—β-adrenergic blockers, antidysrythmics
Increase: cardiac dysrhythmia risk—sympathomimetics
Decrease: digoxin absorption—antacids, kaolin/pectin, cholestyramine, metoclopramide
Decrease: digoxin level—thyroid agents, cholestyramine, colestipol, metoclopramide, aMILoride

Drug/Food
Decrease: digoxin
Decrease: GI absorption—flaxseed, psyllium

Drug/Herb
Decrease: product effect—St. John's wort

Drug/Lab Test
Increase: CPK

NURSING CONSIDERATIONS

Assess:
- Apical pulse for 1 min before giving product; if pulse <60 in adult or <90 in infant, take again in 1 hr; if <60 in adult, call prescriber; note rate, rhythm, character; monitor ECG continuously during parenteral loading dose
- Electrolytes: potassium, sodium, chloride, magnesium, calcium; renal function studies: BUN, creatinine; blood studies: ALT, AST, bilirubin, Hct, Hgb before initiating treatment and periodically thereafter
- Monitor product levels; therapeutic level 0.5-2 ng/ml

Evaluate:
- Therapeutic response: decreased weight, edema, pulse, respiration, crackles; increased urine output; serum digoxin level (0.5-2 ng/ml)

Teach patient/family:
- Not to stop product abruptly; about all aspects of product; to take exactly as ordered; how to monitor heart rate
- To avoid OTC medications, herbal remedies because many adverse product interactions may occur; not to take antacid at same time
- **To notify prescriber of loss of appetite, lower stomach pain, diarrhea, weakness, drowsiness, headache, blurred or yellow vision, rash, depression, toxicity**
- About the toxic symptoms of this product; when to notify prescriber
- To maintain a sodium-restricted diet as ordered
- To use one brand consistently

TREATMENT OF OVERDOSE:
Discontinue product; give potassium; monitor ECG; give adrenergic-blocking agent, digoxin immune FAB

digoxin immune FAB (ovine) (Rx)

(di-jox′in im-myoon′ FAB)

DigiFab

Func. class.: Antidote—digoxin specific

ACTION:
Antibody fragments bind to free digoxin or digitoxin to reverse toxicity by not allowing digoxin or digitoxin to bind to sites of action

USES:
Life-threatening digoxin toxicity

CONTRAINDICATIONS:
Mild digoxin toxicity, hypersensitivity to this product, papain or ovine protein

Precautions: Pregnancy (C), breastfeeding, children, geriatric patients, renal/cardiac disease, allergy to ovine proteins, hypocalcemia, heart failure

DOSAGE AND ROUTES

1 (38-mg) vial binds 0.5 mg digoxin; 1 (40 mg) DigiFab binds 0.5 mg digoxin

Digoxin toxicity (known amount) (tabs, oral sol, IM)

- **Adult and child: IV** dose (mg) = dose ingested (mg) × 0.8/1000 × 38- or 40-mg vial

Toxicity (known amount) (cap, IV)

- **Adult and child: IV** dose = dose ingested (mg)/0.5 × 38- or 40-mg vial

Toxicity (known amount) by serum digoxin concs (SDCs)

- **Adult and child: IV** SDC (ng/ml) × kg of weight/100 × 38- or 40-mg vial

Digoxin toxicity (unknown amount)

- **Adult and child >20 kg: IV** 228 mg (6 vials)
- **Infant and child <20 kg: IV** 38 mg (1 vial)

Acute ingestion

- **Adult: IV** 380 mg (10 vials)

Life-threatening ingestion

- **Adult: IV** 760 mg (20 vials)

Skin test

- **Adult: ID** 9.5 mcg

Available forms: Inj 38 mg/vial (binds 0.5 mg digoxin), 40 mg/vial (binds 0.5 mg digoxin)

Administer:

- Test doses proven to be ineffective in the general population; only use test dose in those with known allergies or those previously treated with digoxin immune FAB
- Store reconstituted sol for up to 4 hr in refrigerator
- Do not freeze DigiFab
- **For test dose** dilute 0.1 ml of reconstituted product (9.5 mg/ml) in 9.9 ml sterile isotonic saline, inj 0.1 ml (1:100 dilution) ID and observe for wheal with erythema; read in 20 min
- **For scratch test,** place 1 gtt of sol on skin, make a scratch through the drop with a sterile needle; read in 20 min
- After diluting 38 mg/4 ml of sterile water for inj 10 mg/ml mix; may be further diluted with normal saline; sol should be clear, colorless
- By bolus if cardiac arrest is imminent or IV over 30 min using a 0.22-μm filter

SIDE EFFECTS

CV: CHF, ventricular rate increase, atrial fibrillation, low cardiac output, hypotension

INTEG: *Hypersensitivity,* allergic reactions, facial swelling, redness, phlebitis

META: Hypokalemia

MISC: Anaphylaxis (rare)

RESP: Impaired respiratory function, rapid respiratory rate

PHARMACOKINETICS

IV: Peaks after completion of inf; onset 30 min (variable); not known if crosses placenta, breast milk; half-life biphasic: 14-20 hr, prolonged with renal disease; excreted by kidneys

INTERACTIONS

• Considered incompatible with all products in syringe or sol

Drug/Lab Test

Interference: immunoassay digoxin

NURSING CONSIDERATIONS

Assess:

• **Hypokalemia:** ST depression, flat T waves, presence of U wave, ventricular dysrhythmia; potassium levels may decrease rapidly

• **CHF:** dyspnea, crackles, peripheral edema, B/P, volume overload

Evaluate:

• Therapeutic response: correction of digoxin toxicity; check digoxin levels 0.5-2 ng/ml; digitoxin level 9-25 ng/ml

Teach patient/family:

• About the purpose of medication; to report delayed hypersensitivity: fever, chills, itching, swelling, dyspnea

⚠ HIGH ALERT

diltiazem (Rx)

(dil-tye′a-zem)

Cardizem, Cardizem CD, Cardizem LA, Cartia XT, Dilacor-XR, Dilt-CD, Diltia XR, Diltia XT, Diltzac, Taztia XT, Tiazac

Func. class.: Calcium channel blocker, antiarrhythmic class IV, antihypertensive

Chem. class.: Benzothiazepine

Do not confuse:
Cardizem/Cardene

ACTION:
Inhibits calcium ion influx across cell membrane during cardiac depolarization; produces relaxation of coronary vascular smooth muscle, dilates coronary arteries, slows SA/AV node conduction times, dilates peripheral arteries

USES:
PO angina pectoris due to coronary artery spasm, hypertension, **IV** atrial fibrillation, flutter, paroxysmal supraventricular tachycardia

Unlabeled uses: Unstable angina, proteinuria, cardiomyopathy, diabetic neuropathy

CONTRAINDICATIONS:
Sick sinus syndrome, AV heart block, hypotension <90 mm Hg systolic, acute MI, pulmonary congestion, cardiogenic shock

Precautions: Pregnancy (C), breastfeeding, children, geriatric patients, CHF, aortic stenosis, bradycardia, GERD, hepatic disease, hiatal hernia, ventricular dysfunction

DOSAGE AND ROUTES

Prinzmetal's or variant angina, chronic stable angina

• **Adult: PO** 30 mg qid, increasing dose gradually to 180-360 mg/day in divided doses or (SR) 60-120 mg bid; may increase to 240-360 mg/day or 120 or 180 mg **EXT REL** (LA, CD, XT, XR products) **PO** daily

Atrial fibrillation/flutter, paroxysmal supraventricular tachycardia

• **Adult: IV BOL** 0.25 mg/kg over 2 min initially then 0.35 mg/kg may be given after 15 min; if no response, may give **CONT INF** 5-15 mg/hr for up to 24 hr

Hypertension

• **Adult: PO** or 120-240 mg **(EXT REL once-daily dosing)** daily or 60-120 mg bid **(SUS REL twice-daily dosing)**

Rapid ventricular rate secondary to dysrhythmias (unlabeled)

• **Adolescent/child/infant >7 mo: IV BOL** 0.25 mg/kg over 5 min then **CONT IV INF** 0.11 mg/kg/hr

Available forms: Tabs 30, 60, 90, 120 mg; ext rel tabs 120, 180, 240, 300, 360, 420 mg; ext rel caps 60, 90, 120, 180, 240, 300, 360, 420 mg; inj 5 mg/ml (5, 10 ml)

Administer:

PO route

- Store at room temperature
- **Cardiazem LA** ext rel tab 24 hr: give daily, either AM or PM, without regard to meals
- **Dilacor XR/Diltia XT** ext rel cap 24 hr: give daily; take on empty stomach; swallow whole; do not cut, crush, chew, open
- **Tiazac, Tiztia XT:** give daily without regard to meals
- **Conventional regular-rel tab:** give before meals, at bedtime
- **Cardizem CD or equivalent (Cartia XT):** generic ext rel cap 24 hr: give daily, without regard to meals
- May crush, sprinkle regular tab on applesauce for administration

Oral suspension (unlabeled)

- Grind 16 90-mg diltiazem regular rel tab into fine powder
- In separate container, mix 60 ml Ora-Sweet and 60 ml Ora-Plus
- Add small amount of sol to powder to form paste, add geometric amounts of base to achieve desired vol, place in amber container

Direct IV route

- IV undiluted over 2 min

Continuous IV INF route

- Diluted 125 mg/100 ml, 250 mg/250 ml of D_5W, 0.9% NaCl, $D_5/0.45\%$ NaCl, give 10 mg/hr, may increase by 5 mg/hr to 15 mg/hr, continue inf up to 24 hr

Y-site compatibilities: Albumin, amikacin, amphotericin B, aztreonam, bumetanide, ceFAZolin, cefotaxime, cefoTEtan, cefOXitin, cefTAZidime, cefTRIAXone, cefuroxime, cimetidine, ciprofloxacin, clindamycin, digoxin, DOBUTamine, DOPamine, doxycycline, EPINEPHrine, erythromycin, esmolol, fentaNYL, fluconazole, gentamicin, hetastarch, HYDROmorphone, imipenem-cilastatin, labetalol, lidocaine, LORazepam, meperidine, metoclopramide, metroNIDAZOLE, midazolam, milrinone, morphine, multivitamins, niCARdipine, nitroglycerin, norepinephrine, oxacillin, penicillin G potassium, pentamidine, piperacillin, potassium chloride, potassium phosphates, ranitidine, sodium nitroprusside, theophylline, ticarcillin, ticarcillin/clavulanate, tobramycin, trimethoprim-sulfamethoxazole, vancomycin, vecuronium

SIDE EFFECTS

CNS: *Headache, fatigue, drowsiness,* dizziness, depression, weakness, insomnia, tremor, paresthesia

CV: Dysrhythmia, *edema,* CHF, bradycardia, hypotension, palpitations, heart block

GI: *Nausea,* vomiting, diarrhea, gastric upset, *constipation*, increased LFTs

GU: Nocturia, polyuria, acute renal failure

INTEG: *Rash,* flushing, photosensitivity, burning, pruritus at inj site

RESP: Rhinitis, dyspnea, pharyngitis

PHARMACOKINETICS

Onset 30-60 min; peak 2-3 hr immediate rel, 10-14 hr ext rel, 6-11 hr sus rel; half-life $3^1/_2$-9 hr; metabolized by liver; excreted in urine (96% as metabolites)

INTERACTIONS

Increase: effect, toxicity—theophylline

Increase: effects of β-blockers, digoxin, lithium, carBAMazepine, cycloSPORINE, anesthetics, HMG-CoA reductase inhibitors, benzodiazepines, lovastatin, methylPREDNISolone

Increase: effects of diltiazem—cimetidine

NURSING CONSIDERATIONS

Assess:

- **CHF:** dyspnea, weight gain, edema, jugular venous distention, rales; monitor I&O ratios daily, weight
- **Angina:** location, duration, alleviating factors, activity when pain starts
- **Dysrhythmias:** cardiac status: B/P, pulse, respiration, ECG and intervals PR, QRS, QT; if systolic B/P $<$90 mm Hg or HR $<$50 bpm, hold dose, notify prescriber

Evaluate:

- Therapeutic response: decreased anginal pain, decreased B/P

Teach patient/family:

- How to take pulse, B/P before taking product; that a record or graph should be kept
- To avoid hazardous activities until stabilized on product, dizziness is no longer a problem
- To limit caffeine consumption; to avoid grapefruit juice
- To avoid OTC products unless directed by prescriber
- About the importance of complying with all areas of medical regimen: diet, exercise, stress reduction, product therapy
- To change position slowly
- ⚠ To report dizziness, SOB, palpitations
- Not to discontinue abruptly

TREATMENT OF OVERDOSE:
Atropine for AV block, vasopressor for hypotension

dimenhyDRINATE (OTC, Rx)

(dye-men-hye′dri-nate)

Apo-DimenhyDRINATE ✱, Dramamine, Driminate, Gravol ✱

Motion Sickness

TripTone, Wal-Dram

Func. class.: Antiemetic, antihistamine, anticholinergic

Chem. class.: H_1-receptor antagonist, ethanolamine derivative

Do not confuse:
dimenhyDRINATE/diphenhydrAMINE

ACTION: Competes with histamine for H_1 receptors in GI tract, blood vessels, respiratory tract; central anticholinergic activity, which results in decreased vestibular stimulation and blockade of chemoreceptor trigger zone

USES: Motion sickness, nausea, vomiting, vertigo

Unlabeled uses: Hyperemesis gravidarum, Ménière's syndrome

CONTRAINDICATIONS: Hypersensitivity, infants, neonates, tartrazine dye hypersensitivity

Precautions: Pregnancy (B), breastfeeding, children, geriatric patients, cardiac dysrhythmias, asthma, prostatic hypertrophy, bladder-neck obstruction, closed-angle glaucoma, stenosing peptic ulcer, pyloroduodenal obstruction

DOSAGE AND ROUTES

- **Adult: PO** 50-100 mg q4hr; **IM/IV** 50 mg q4hr as needed (Canada only)
- **Child 6-12 yr: PO** 25-50 mg q6-8hr prn, max 150 mg/day
- **Child 2-5 yr: PO** 12.5-25 mg q6-8hr, max 75 mg/day

Available forms: Tabs 50 mg; inj 50 mg/ml ✱; elixir 15 mg/5 ml ✱; chew tabs 50 mg

Administer:

- IM inj in large muscle mass; aspirate to avoid IV administration (Canada only)
- Tablets may be swallowed whole, chewed, or allowed to dissolve

IV route (Canada only)

- After diluting 50 mg/10 ml of NaCl inj, give ≤50 mg over 2 min

SIDE EFFECTS

CNS: *Drowsiness,* restlessness, headache, dizziness, insomnia, confusion, nervousness, tingling, vertigo

CV: Hypertension, *hypotension*, palpitation

EENT: *Dry mouth,* blurred vision, diplopia, nasal congestion, photosensitivity, xerostomia

GI: Nausea, anorexia, vomiting, *constipation*

INTEG: Rash, urticaria, fever, chills, flushing

MISC: **Anaphylaxis**

PHARMACOKINETICS

PO/IM: Onset 15-30 min, duration 4-6 hr

INTERACTIONS

Increase: effect—alcohol, anticholinergic, tricyclics, MAOIs, opiates, sedative/hypnotics, other CNS depressants

Drug/Lab Test

False negative: Allergy skin testing

NURSING CONSIDERATIONS

Assess:

- VS, B/P; check patients with cardiac disease more often
- **Signs of toxicity** of other products or masking of symptoms of disease: brain tumor, intestinal obstruction
- Observe for drowsiness, dizziness

Evaluate:

- Therapeutic response: absence of nausea, vomiting, or vertigo

Teach patient/family:

- To avoid hazardous activities, activities requiring alertness because dizziness may occur; to request assistance with ambulation
- To avoid alcohol, other CNS depressants

dinoprostone (Rx)

(dye-noe-prost′one)

Cervidil, Prepidil, Prostin E-2

Func. class.: Oxytocic, abortifacient

Chem. class.: Prostaglandin E_2

Do not confuse:

Prepidil/bepridil

ACTION:

Stimulates uterine contractions, causing abortion; acts within 30 hr for complete abortion

USES:

Abortion during 2nd trimester, benign hydatidiform mole, expulsion of uterine contents in fetal deaths to 28 wk, missed abortion, to efface and dilate the cervix in pregnancy at term

CONTRAINDICATIONS:

Hypersensitivity, C-section, surgery, fetal distress, multiparity, vaginal bleeding, cephalopelvic disproportion

Precautions: Pregnancy (C), cardiac/renal/hepatic disease, asthma, anemia, jaundice, diabetes mellitus, seizure disorders, hypertension, glaucoma, uterine fibrosis, cervical stenosis, pelvic surgery, pelvic inflammatory disease, respiratory disease

Black Box Warning: Requires a specialized setting and an experienced clinician

DOSAGE AND ROUTES

Abortifacient/2nd trimester/missed abortion/benign hydatidiform mole/intrauterine fetal death

- **Adult: VAG SUPP** 20 mg, repeat q3-5hr until abortion occurs, max dose is 240 mg

Cervical ripening

- **Adult: GEL** 0.5 mg vag gel placed in cervical canal, may repeat after 6 hr, max 1.5 mg/24 hr; vag insert 10 mg high in vagina, remove at onset of active labor or within 12 hr

Available forms: VAG SUPP 20 mg; gel 0.5 mg/3 g (prefilled syringe); vag insert 10 mg

Administer:

- **By gel:** after warming to room temperature, remove seal from end of syringe, remove protective end cap and insert into plunger stopper assembly; make sure patient is in dorsal position; **insert:** must be kept frozen until use
- Antiemetic/antidiarrheal before administration of this product

SIDE EFFECTS

CNS: *Headache*, dizziness, chills, fever, flushing

CV: Hypotension, dysrhythmias, DIC

EENT: Blurred vision

SYST: Anaphylactoid syndrome of pregnancy

FETAL: Bradycardia (i.e., deceleration)

GI: *Nausea, vomiting, diarrhea*

GU: Vaginitis, vaginal pain, vulvitis, vaginismus

INTEG: Rash, skin color changes
MS: *Leg cramps, joint swelling,* weakness
GEL: Uterine contractile abnormality, GI side effects, back pain, fever
INSERT: Uterine hyperstimulation, fever, nausea, vomiting, diarrhea, abdominal pain
SUPPOSITORY: Uterine rupture, anaphylaxis

INTERACTIONS

Increase: effect—other oxytocics
Decrease: oxytocic effect—alcohol

PHARMACOKINETICS

Metabolized in spleen, kidney, lungs; excreted in urine
GEL: Onset 10 min, peak 30-45 min
SUPP: Onset 10 min, duration 2-3 hr

NURSING CONSIDERATIONS

Assess:

Black Box Warning: Specialized setting, specialized clinician: use only with emergency equipment nearby, by a clinician experienced when used in pregnancy termination; complete abortion should result within 17 hr

- **Cervical ripening:** dilation, effacement of cervix and uterine contraction, fetal heart tones, check for contractions over 1 min
- For fever that occurs 1/2 hr after suppository insertion (abortion)
- Respiratory rate, rhythm, depth; notify prescriber of abnormalities, pulse, B/P, temperature
- **Vaginal discharge:** check for itching, irritation; indicates vaginal infection

Evaluate:

- Therapeutic response: expulsion of fetus

Teach patient/family:

- To remain supine for 10-15 min after insertion of supp, 2 hr after insert, 15-30 min after gel
- To report excessive cramping, bleeding, chills, fever
- About some methods of pain, comfort control
- To avoid intercourse, tub baths, douches, tampon use for at least 2 wk

diphenhydrAMINE (OTC, Rx)

(dye-fen-hye′dra-meen)

Allerdryl ♣, AllerMax ♣, Altaryl, Banophen, Benadryl, Benadryl Allergy, Benadryl Allergy Dye Free, Benadryl Children's Allergy, Buckley's Bedtime, Diphedryl, Diphenhist, Dytan, ElixSure Allergy, Equaline Allergy, Equaline Children's Allergy, Equate Allergy, Equate Children's Allergy, Genahist, Good Sense Children's Allergy Relief, Good Sense Diphedryl, Leader Complete Allergy, Nytol, PediaCare Children's Allergy, PediaCare Nighttime Cough, Q-Dryl Allergy, Select Brand Allergy, Siladryl, Silphen, Simply Sleep, Sleepinal, Sleep Tabs, Sominex, Top Care Allergy, Top Care Children's Allergy, Unisom ♣, Valu-Dryl, Wal-dryl Allergy, Wal-dryl Allergy Dye Free, Wal-dryl Children's Allergy, Walgreen's Sleep Aid, Walgreen's Sleep II, Wal-Som

Func. class.: Antihistamine (1st generation, nonselective)
Chem. class.: Ethanolamine derivative, H_1-receptor antagonist

Do not confuse:
diphenhydrAMINE/dicyclomine
diphenhydrAMINE/dimenhyDRINATE

ACTION: Acts on blood vessels, GI, respiratory system by competing with histamine for H_1-receptor site; decreases allergic response by blocking histamine

USES: Allergy symptoms, rhinitis, motion sickness, antiparkinsonism, nighttime sedation, infant colic, nonproductive cough, insomnia in children

Unlabeled uses: Nystagmus

CONTRAINDICATIONS: Hypersensitivity to H_1-receptor antagonist, acute asthma attack, lower respiratory tract disease, neonates

Precautions: Pregnancy (B), breastfeeding, children <2 yr, increased intraocular pressure, cardiac/renal disease, hypertension, bronchial asthma, seizure disorder, stenosed peptic ulcers, hyperthyroidism, prostatic hypertrophy, bladder neck obstruction

DOSAGE AND ROUTES

- **Adult and child >12 yr: PO** 25-50 mg q4-6hr, max 300 mg/day; **IM/IV** 10-50 mg, max 300 mg/day
- **Child 6-12 yr: PO/IM/IV** 5 mg/kg/day in 4 divided doses, max 300 mg/day

Nighttime sleep aid

- **Adult and child ≥12 yr: PO** 25-50 mg at bedtime

Antitussive (syrup only)

- **Adult and child ≥12 yr: PO** 25 mg q4hr, max 150 mg/24 hr
- **Child 6-12 yr: PO** 12.5 mg q4hr, max 75 mg/24 hr

Renal disease

- CCr >50 ml/min, dose q6hr; CCr 10-50 ml/min, dose q6-12hr; CCr <10 ml/min, dose q12-18hr

Peripheral vestibular nystagmus (unlabeled)

- **Adult: PO** 25-50 mg q4-6hr up to 48 hr

Available forms: Caps 25, 50 mg; tabs 25, 50 mg; chew tabs 12.5 mg; elix 12.5 mg/5 ml; syr 12.5 mg/5 ml; inj 10, 50 mg/ml; orally disintegrating tabs 12.5, 25 mg

Administer:

⚠ Avoid use in children <2 yr; death has occurred; overdose has occurred with topical gel taken orally (adult/child)

- With meals for GI symptoms; absorption rate may slightly decrease
- At bedtime only if using for sleep aid

IM route

- Deep IM in large muscle; rotate site

Direct IV route

- Undiluted; give 25 mg/min or less

Intermittent IV INF route

- Dilute with 0.9% NaCl, 0.45% NaCl, D_5W, 0.9% NaCl, $D_{10}W$, LR, Ringer's

Syringe compatibilities: Atropine, butorphanol, chlorproMAZINE, cimetidine, dimenhyDRINATE, droperidol, fentaNYL, fluPHENAZine, glycopyrrolate, HYDROmorphone, hydrOXYzine, meperidine, metoclopramide, midazolam, morphine, nalbuphine, pentazocine, perphenazine, prochlorperazine, promazine, promethazine, ranitidine, scopolamine, SUFentanil

Y-site compatibilities: Abciximab, aldesleukin, alfentanil hydrochloride, amifostine, amikacin sulfate, aminocaproic acid, amphotericin B lipid complex (Abelcet), amphotericin B liposome (AmBisome), amsacrine, anidulafungin, argatroban, ascorbic acid injection, atenolol, atracurium besylate, atropine sulfate, azithromycin, benztropine mesylate, bivalirudin, bleomycin, bumetanide, buprenorphine, butorphanol, calcium chloride/gluconate, CARBOplatin, caspofungin, cefTAZidime, ceftizoxime, chlorproMAZINE, cimetidine, ciprofloxacin, cisatracurium, CISplatin, cladribine, clindamycin, codeine, cyanocobalamin, cyclophosphamide, cycloSPORINE, cytarabine, DACTINomycin, DAPTOmycin, digoxin, diltiazem, DOBUTamine, DOCEtaxel, DOPamine, doripenem, doxacurium, DOXOrubicin, DOXOrubicin liposomal, doxycycline, enalaprilat, ePHEDrine, EPINEPHrine, epirubicin, epoetin alfa, eptifibatide, ertapenem, erythromycin, esmolol, etoposide, famotidine, fenoldopam, fentaNYL, filgrastim, fluconazole, fludarabine, folic acid, gallium, gatifloxacin, gemcitabine, gemtuzumab, gentamicin, glycopyrrolate, granisetron, HYDROmorphone, hydrOXYzine, IDArubicin, ifosfamide, imipenem-cilastatin, irinotecan, isoproterenol, labetalol, levofloxacin, lidocaine, linezolid, LORazepam, LR, magnesium sulfate, mannitol, mechlorethamine, melphalan, meperidine, meropenem, metaraminol, methadone, methicillin, methotrexate, methoxamine, methyldopate, metoclopramide, metoprolol, metroNIDAZOLE, miconazole, midazolam, minocycline, mitoXANtrone, morphine, multiple vitamins injection, mycophenolate, nalbuphine, naloxone, nesiritide, netilmicin, nitroglycerin, norepinephrine, octreotide, ondansetron, oxaliplatin, oxytocin, PACLitaxel, palonosetron, pamidronate, pancuronium, papaverine, PEMEtrexed, penicillin G potassium/sodium, pentamidine, pentazocine, phenylephrine, phytonadione, piperacillin, piperacillin-tazobactam, polymyxin B, potassium chloride, procainamide, prochlorperazine, promethazine, propofol, propranolol, protamine, pyridoxine, quiNIDine, quinupristin-dalfopristin, ranitidine, remifentanil, Ringer's, ritodrine, riTUXimab, rocuronium, sargramostim, sodium acetate, succinylcholine, SUFentanil, tacrolimus, teniposide, theophylline, thiamine, thiotepa, ticarcillin, ticarcillin-clavulanate, tigecycline, tirofiban, TNA, tobramycin, tolazoline, TPN, trastuzumab, trimetaphan, urokinase, vancomycin, vasopressin, vecuronium, verapamil, vinCRIStine, vinorelbine, vitamin B complex/C, voriconazole, zoledronic acid

SIDE EFFECTS

CNS: *Dizziness, drowsiness,* poor coordination, fatigue, anxiety, euphoria, confusion, paresthesia, neuritis, **seizures**
CV: Hypotension, palpitations
EENT: Blurred vision, dilated pupils, tinnitus, nasal stuffiness, dry nose, throat, mouth
GI: Nausea, anorexia, diarrhea
GU: *Retention,* dysuria, frequency
HEMA: **Thrombocytopenia, agranulocytosis, hemolytic anemia**
INTEG: Photosensitivity
MISC: **Anaphylaxis**
RESP: Increased thick secretions, wheezing, chest tightness

PHARMACOKINETICS

Metabolized in liver, excreted by kidneys, crosses placenta, excreted in breast milk, half-life 2-7 hr
PO: Peak 1-3 hr, duration 4-7 hr
IM: Onset ½ hr, peak 1-4 hr, duration 4-7 hr
IV: Onset immediate, duration 4-7 hr

INTERACTIONS

Increase: CNS depression—barbiturates, opiates, hypnotics, tricyclics, alcohol
Increase: diphenhydrAMINE effect—MAOIs
Drug/Lab Test
False negative: skin allergy tests

NURSING CONSIDERATIONS

Assess:
- Urinary retention, frequency, dysuria; product should be discontinued
- CBC during long-term therapy; blood dyscrasias may occur
- Respiratory status: rate, rhythm, increase in bronchial secretions, wheezing, chest tightness
- Product should be discontinued 4 days before skin allergy tests
- Store in tight container at room temperature

Evaluate:
- Therapeutic response: absence of running or congested nose or rashes, improved sleep

Teach patient/family:
- About all aspects of product use; to notify prescriber of confusion, sedation, hypotension
- To avoid driving, other hazardous activity if drowsiness occurs
- That photosensitivity may occur
- To avoid concurrent use of alcohol, other CNS depressants
- To avoid breastfeeding; **not to breastfeed (injectable)**

• To use hard candy, gum, frequent rinsing of mouth for dryness

TREATMENT OF OVERDOSE:
Administer diazepam, vasopressors, phenytoin IV

diphenoxylate/atropine (Rx)
(dye-fen-ox′ee-late/a′troe-peen)

Lomotil, Lonox

difenoxin/atropine (Rx)
(dye-fen-ox′in/a′troe-peen)

Motofen

Func. class.: Antidiarrheal

Chem. class.: Phenylpiperidine derivative opiate agonist

Controlled Substance Schedule V

diphenoxylate/atropine

Controlled Substance Schedule IV

difenoxin/atropine (US)

Do not confuse:
Lomotil/LaMICtal/LamISIL/Lanoxin/Lasix/Ludomil

ACTION: Inhibits gastric motility by acting on mucosal receptors responsible for peristalsis

USES: Acute nonspecific and acute exacerbations of chronic functional diarrhea

CONTRAINDICATIONS: Children <2 yr, hypersensitivity, pseudomembranous colitis, severe electrolyte imbalances, diarrhea associated with organisms that penetrate intestinal mucosa

Precautions: Pregnancy (C), breastfeeding, hepatic disease, ulcerative colitis, severe hepatic disease, substance abuse, dehydration

DOSAGE AND ROUTES
Diphenoxylate/atropine

• **Adult: PO** 5 mg qid titrated to patient response needed, max 8 tabs/day

• **Child 2-12 yr: PO** (liquid only) 0.3-0.4 mg/kg/day in 4 divided doses

Difenoxin/atropine

• **Adult: PO** 2 tabs then 1 tab after each loose stool or q3-4hr prn, max 8 tabs/day

Available forms: Diphenoxylate/atropine: tabs 2.5 mg with atropine 0.025 mg; liquid 2.5 mg with atropine 0.025 mg/5 ml; **difenoxin/atropine:** tabs 1 mg difenoxin/0.025 atropine

Administer:

• For 48 hr only; if no response, product should be discontinued

SIDE EFFECTS
CNS: *Dizziness, drowsiness, lightheadedness, headache,* fatigue, nervousness, insomnia, confusion

EENT: Burning eyes, blurred vision

GI: *Nausea, vomiting, dry mouth, epigastric distress,* constipation, paralytic ileus, toxic megacolon

MISC: Anaphylaxis, angioedema

RESP: Respiratory depression

PHARMACOKINETICS
PO: Onset 40-60 min, peak 2 hr, duration 3-4 hr, terminal half-life 12-14 hr, metabolized in liver to active metabolite; excreted in urine and feces

INTERACTIONS
• Do not use with MAOIs; hypertensive crisis may occur

Increase: action of alcohol, opioids, barbiturates, other CNS depressants, anticholinergics

Decrease: GI motility, possible toxic megacolon—amantadine, antimuscarinics, amoxapine, diphenhydrAMINE, cloZAPine, clemastine, cyclobenzaprine, loperamide, maprotiline, phenothiazines, tricyclics, disopyramide, OLANZapine

NURSING CONSIDERATIONS

Assess:

- Electrolytes (potassium, sodium, chlorine) if receiving long-term therapy
- Bowel pattern before; for rebound constipation after termination of medication; bowel sounds
- Response after 48 hr; if none, product should be discontinued
- **Abdominal distention, toxic megacolon;** may occur in ulcerative colitis
- Hepatic studies if receiving long-term therapy

Evaluate:

- Therapeutic response: decreased diarrhea

Teach patient/family:

- To avoid OTC products unless directed by prescriber (may contain alcohol); not to use alcohol or CNS depressants
- Not to exceed recommended dose
- That product may be habit forming
- Not to engage in hazardous activities; that drowsiness may occur; not to use for longer than 48 hr for acute diarrhea

dipivefrin ophthalmic

See Appendix B

dipyridamole (Rx)

(dye-peer-id′a-mole)

Persantine

Func. class.: Coronary vasodilator, antiplatelet agent

Chem. class.: Nonnitrate

ACTION: Inhibits adenosine uptake, which produces coronary vasodilation; increases oxygen saturation in coronary tissues, coronary blood flow; acts on small resistance vessels with little effect on vascular resistance; may increase development of collateral circulation; decreases platelet aggregation by the inhibition of phosphodiesterase (an enzyme)

USES: Prevention of transient ischemic attacks, inhibition of platelet adhesion to prevent myocardial reinfarction, thromboembolism, with warfarin in prosthetic heart valves, prevention of coronary bypass graft occlusion with aspirin; IV form used to evaluate CAD; used as alternative to exercise with thallium myocardial perfusion imaging to evaluate CAD

Unlabeled uses: Cardiomyopathy, MI prophylaxis, proteinuria, TIA, valvular heart disease

CONTRAINDICATIONS: Hypersensitivity

Precautions: Pregnancy (B), breastfeeding, hypotension, unstable angina, asthma, hepatic disease, labor

DOSAGE AND ROUTES

Inhibition of platelet adhesion

- **Adult: PO** 75-100 mg qid in combination with aspirin or warfarin

Thallium myocardial perfusion imaging

- **Adult: IV** 570 mcg/kg

TIA with aspirin (unlabeled)

- **Adult: PO** 225-400 mg/day max 400 mg/day

Available forms: Tabs 25, 50, 75 mg; inj 10 mg/2 ml

Administer:

PO route

- On empty stomach: 1 hr before meals or 2 hr after; give with 8 oz water for better absorption
- Store at room temperature

IV route

- IV after diluting to at least 1:2 ratio using D_5W, 0.45% NaCl, or 0.9% NaCl to a total vol of 20-50 ml; give over 4 min; do not give undiluted
- Inject thallium 201 within 5 min after product inf
- Do not admix

SIDE EFFECTS

CNS: *Headache, dizziness, weakness, fainting, syncope;* IV: transient cerebral ischemia, weakness
CV: *Postural hypotension;* IV: MI
GI: *Nausea, vomiting,* anorexia, diarrhea
INTEG: *Rash,* flushing

PHARMACOKINETICS

PO: Peak 1.25 hr, duration 6 hr, therapeutic response may take several mo, metabolized in liver, excreted in bile, undergoes enterohepatic recirculation, protein binding 91%-99%, terminal half-life 12 hr

INTERACTIONS

• Prevention of coronary vasodilation: theophylline
Increase: digoxin effect—digoxin
Increase: bleeding risk—NSAIDs, cefamandole, cefoTEtan, cefoperazone, plicamycin, valproic acid, salicylates, sulfinpyrazole, anticoagulants, thrombolytics

NURSING CONSIDERATIONS

Assess:
• B/P, pulse during treatment until stable; take B/P lying, standing; orthostatic hypotension is common
• Cardiac status: chest pain; what aggravates, ameliorates condition
Evaluate:
• Therapeutic response: decreased platelet adhesion
Teach patient/family:
• That medication is not a cure; may have to be taken continuously in evenly spaced doses only as directed
• To avoid hazardous activities until stabilized on medication; dizziness may occur
• To rise slowly from sitting or lying to prevent orthostatic hypotension
• Not to use alcohol or OTC medications unless approved by prescriber

DOBUTamine (Rx)

(doe-byoo′ta-meen)
Func. class.: Adrenergic direct-acting β_1-agonist, cardiac stimulant
Chem. class.: Catecholamine

Do not confuse:
DOBUTamine/DOPamine

ACTION: Causes increased contractility, increased cardiac output without marked increase in heart rate by acting on β_1-receptors in heart; minor α and β_2 effects

USES: Cardiac decompensation due to organic heart disease or cardiac surgery
Unlabeled uses: Cardiogenic shock in children; congenital heart disease in children undergoing cardiac catheterization

CONTRAINDICATIONS: Hypersensitivity, idiopathic hypertrophic subaortic stenosis
Precautions: Pregnancy (B), breastfeeding, children, hypertension, CAD, MI, hypovolemia, dysrhythmias, sulfite hypersensitivity, renal failure, geriatric patients

DOSAGE AND ROUTES

• **Adult and child: IV INF** 0.5-1 mcg/kg/min; titrate to 2-20 mcg/kg/min may increase to 40 mcg/kg/min if needed
Available forms: Inj 12.5 mg/ml, 250 mg/20 ml
Administer:
Injectable
• Visually inspect parenteral products for particulate matter and discoloration before administration whenever solution and container permit
• Store reconstituted sol for 24 hr if refrigerated
IV route
Note: Inf ≤72 hr have been given without development of tolerance; however, beta-receptor desensitization can occur with prolonged inf of any beta-adrenergic agonist, including DOBUTamine, or as a

consequence of sympathetic compensatory mechanisms associated with advanced congestive heart failure, resulting in alterations in DOBUTamine pharmacodynamics; experience with intravenous DOBUTamine in controlled trials does not extend beyond 48 hr of repeated boluses and/or continuous inf

- Must be diluted before administration
- Infuse into a large vein

Dilution

- Concentrate for injection must be diluted with ≥50 ml of a compatible IV solution (strongly alkaline [e.g., sodium bicarbonate] solutions are incompatible); a common dilution is 500 mg (40 ml) in 210 ml D_5W or NS (withdraw 40 ml from a 250 ml bag) to produce a final conc of 2000 mcg/ml; or 1000 mg (80 ml) in 170 ml D_5W or NS (withdraw 80 ml from a 250 ml bag) to produce a final conc of 4000 mcg/ml; maximum conc should not exceed 5000 mcg/ml and should be adjusted according to the patient's fluid requirements

Infusion

- Administer diluted solution by IV inf using a controlled-infusion device
- Premixed bags of DOBUTamine in D_5W solutions can exhibit a pink color that increase with time; this color change is due to slight oxidation of the drug, but there is no significant loss of potency
- Do not administer DOBUTamine simultaneously with solutions containing sodium bicarbonate or strong alkaline solutions (incompatible)
- Inf of DOBUTamine should be started at a low rate and titrated frequently to reach the optimal dosage (see Dosage and Routes); dosage titration is guided by the patient's response, including systemic blood pressure, urine flow, frequency of ectopic activity, heart rate, and (whenever possible) measurements of cardiac output, central venous pressure, and/or pulmonary capillary wedge pressure

Y-site compatibilities: Alfentanil, alprostadil, amifostine, amikacin, aminocaproic acid, amiodarone, anidulafungin, argatroban, ascorbic acid injection, atenolol, atracurium, atropine, aztreonam, benztropine, bleomycin, bumetanide, buprenorphine, butorphanol, calcium chloride/gluconate, CARBOplatin, caspofungin, chlorproMAZINE, cimetidine, ciprofloxacin, cisatracurium, CISplatin, cladribine, clarithromycin, cloNIDine, codeine, cyanocobalamin, cyclophosphamide, cycloSPORINE, cytarabine, DACTINomycin, DAPTOmycin, dexmedetomidine, digoxin, diltiazem, diphenhydrAMINE, DOCEtaxel, DOPamine, doripenem, doxacurium, DOXOrubicin, DOXOrubicin liposomal, doxycycline, enalaprilat, ePHEDrine, EPINEPHrine, epirubicin, epoetin alfa, eptifibatide, erythromycin, esmolol, etoposide, famotidine, fenoldopam, fentaNYL, fluconazole, fludarabine, gatifloxacin, gemcitabine, gentamicin, glycopyrrolate, granisetron, HYDROmorphone, hydrOXYzine, IDArubicin, ifosfamide, irinotecan, isoproterenol, labetalol, levofloxacin, lidocaine, linezolid, LORazepam, LR, magnesium sulfate, mannitol, mechlorethamine, meperidine, meropenem, metaraminol, methoxamine, methyldopate, methylPREDNISolone, metoclopramide, metoprolol, metroNIDAZOLE, miconazole, milrinone, minocycline, mitoXANtrone, morphine, multiple vitamins injection, mycophenolate mofetil, nafcillin, nalbuphine, naloxone, netilmicin, niCARdipine, nitroglycerin, norepinephrine, octreotide, ondansetron, oxaliplatin, oxytocin, PACLitaxel, palonosetron, pamidronate, pancuronium, papaverine, pentamidine, pentazocine, phenylephrine, polymyxin B, potassium chloride, procainamide, prochlorperazine, promethazine, propofol, propranolol, protamine, pyridoxine, quiNIDine, ranitidine, remifentanil, Ringer's, ritodrine, riTUXimab, rocuronium, sodium acetate, succinylcholine, SUFentanil, tacrolimus, temocillin, teniposide, theophylline, thiamine, thiotepa, tigecycline, tirofiban, TNA, tobramycin, tolazoline, TPN, trastuzumab, trimetaphan, urokinase, vancomycin, vasopressin, vecuronium, verapamil, vinCRIStine, vinorelbine, voriconazole, zidovudine, zoledronic acid

SIDE EFFECTS

CNS: *Anxiety,* headache, dizziness, fatigue
CV: Palpitations, tachycardia, hypo/hypertension, PVCs, angina
ENDO: Hypokalemia
GI: Heartburn, nausea, vomiting
MS: Muscle cramps (leg)
RESP: Dyspnea

PHARMACOKINETICS

IV: Onset 1-2 min, peak 10 min, half-life 2 min, metabolized in liver (inactive metabolites), excreted in urine

INTERACTIONS

Increase: severe hypertension—guanethidine
Increase: dysrhythmias—general anesthetics
Increase: pressor effect, dysrhythmias—atomoxetine, COMT inhibitors, tricyclics, MAOIs, oxytocics
Decrease: DOBUTamine action—other β-blockers

NURSING CONSIDERATIONS

Assess:

- **Hypovolemia;** if present, correct first; administer cardiac glycoside before DOBUTamine
- **Oxygenation/perfusion deficit:** check B/P, chest pain, dizziness, loss of consciousness
- **Heart failure:** S_3 gallop, dyspnea, neck venous distention, bibasilar crackles in patients with CHF, cardiomyopathy, palpate peripheral pulses; report if extremities become cold or mottled or if peripheral pulses decrease
- **ECG** during administration continuously; if B/P increases, product is decreased; CVP or PCWP, cardiac output during inf; report changes
- Serum electrolytes, urine output

⚠ **Sulfite sensitivity,** which may be life threatening

Evaluate:

- Therapeutic response: increased B/P with stabilization, increased urine output

Teach patient/family:

- About the reason for product administration; to report dyspnea, chest pain, numbness of extremities, headache, IV site discomfort

TREATMENT OF OVERDOSE:

Administer a β_1-adrenergic blocker; reduce IV or discontinue, ensure oxygenation/ventilation; for severe tachydysrhythmias (ventricular), give lidocaine or propranolol

DOCEtaxel (Rx)

(doe-se-tax'el)

Taxotere

Func. class.: Antineoplastic—miscellaneous
Chem. class.: Taxane

Do not confuse:
Taxotere/Taxol

ACTION:

Inhibits reorganization of microtubule network needed for interphase and mitotic cellular functions; also causes abnormal bundles of microtubules during cell cycle and multiple esters of microtubules during mitosis

USES:

Locally advanced or metastatic breast cancer, non–small-cell lung cancer, androgen-independent metastatic prostate cancer, postsurgery operable node-positive breast cancer, induction treatment of locally advanced squamous cell of the head and neck

Unlabeled uses: Malignant melanoma, ovarian cancer, front-line use with bevacizumab for metastatic breast cancer, adjuvant treatment of breast cancer with CARBOplatin and trastuzumab, gastric adenocarcinoma

CONTRAINDICATIONS:

Pregnancy (D), breastfeeding, hypersensitivity to this product, bilirubin exceeding upper normal limit

Black Box Warning: Other products with polysorbate 80, neutropenia of <1500/mm^3

Precautions: Children, cardiovascular disease, pulmonary disorders, bone marrow depression, herpes zoster, pleural effusion

Black Box Warning: Edema, hepatic disease, lung cancer, taxane hypersensitivity

DOSAGE AND ROUTES:

- Other regimens are used

Locally advanced or metastatic breast cancer after failure of other chemotherapy

- **Adult: IV** 60-100 mg/m^2 given over 1 hr q3wk; if neutrophil count is <500 cells/mm^3 for >1 wk, reduce dose by 25%

Operable node-positive breast cancer, adjuvant postsurgery treatment of operable node-positive breast cancer

- **Adult: IV** (TAC regimen) 75 mg/m^2 1 hr after DOXOrubicin 50 mg/m^2 and cyclophosphamide 500 mg/m^2 q3wk × 6 cycles

Adjuvant treatment of operable stage I-III invasive breast cancer in combination with cyclophosphamide

- **Adult: IV** (TAC regimen) DOCEtaxel 75 mg/m^2 with cyclophosphamide 600 mg/m^2 q21days × 4 cycles

Locally advanced or metastatic non–small-cell lung cancer after failure of CISplatin chemotherapy

- **Adult: IV** 75 mg/m^2 over 1 hr q3wk; if neutrophil count is <500 cells/mm^3 for >1 wk, reduce dose to 55 mg/m^2; if patient develops grade 3 peripheral neuropathy, stop product

Unreactable, locally advanced, or metastatic non–small-cell lung cancer previously treated with chemotherapy

- **Adult: IV** 75 mg/m^2 over 1 hr then CISplatin 75 mg/m^2 **IV** given over 30-60 min q3wk; reduce dose to 65 mg/m^2 in those with hematologic or non-hematologic toxicities

Androgen-independent metastatic prostate cancer

- **Adult: IV** 75 mg/m^2 given over 1 hr q3wk with 5 mg predniSONE **PO** bid continuously; give dexamethasone 8 mg **PO** at 12 hr, 3 hr, and 1 hr prior to DOCEtaxel; if neutrophil count is <500 cells/mm^3 for more than 1 wk or other toxicities occur, reduce dose to 60 mg/m^2

Squamous cell of head and neck

- **Adult: IV** 75 mg/m^2 over 1 hr, then CISplatin 100 mg/m^2 over 1 hr on day 1, then 5FU 1000 mg/m^2/day **CONT INF** × 5 days, repeat cycle q3wk

Gastric adenocarcinoma

- **Adult: IV** 75 mg/m^2 q3wk, given with CISplatin, fluorouracil

Advanced ovarian cancer/metastatic melanoma (unlabeled)

- **Adult: IV** 100 mg/m^2 over 1 hr q3wk

Available forms: Inj 20 mg/0.5 ml, 20 mg/ml, 80 mg/2 ml, 80 mg/4 ml

Administer:

- Premedicate with dexamethasone 8 mg **PO** bid × 3 days starting 1 day prior to treatment
- Antiemetic 30-60 min before product and prn
- Confirmation that dexamethasone was given 12 hr and 6 hr before inf begins
- Store prepared sol up to 27 hr in refrigerator

Intermittent IV INF route

- Use cytotoxic handling procedures
- Use non-PVC bag and use non-DEHP tubing
- Allow vials to warm to room temperature; withdraw all diluent, inject in vial of DOCEtaxel; rotate gently to mix; allow to stand to decrease foaming, then with-

draw the required amount (10 mg/ml), inject in 250 ml of 0.9% NaCl, D_5W; mix gently; give over 1 hr

Y-site compatibilities: Acyclovir, alfentanil, allopurinol, amifostine, amikacin, aminocaproic acid, aminophylline, amiodarone, amphotericin B lipid complex, ampicillin, ampicillin-sulbactam, anidulafungin, atenolol, atracurium, azithromycin, aztreonam, bivalirudin, bleomycin, bumetanide, buprenorphine, busulfan, butorphanol, calcium chloride/gluconate, CARBOplatin, carmustine, caspofungin, ceFAZolin, cefepime, cefonicid, cefoperazone, cefotaxime, cefoTEtan, cefOXitin, cefTAZidime, ceftizoxime, cefTRIAXone, cefuroxime, cephapirin, chloramphenicol, chlorproMAZINE, cimetidine, ciprofloxacin, cisatracurium, CISplatin, clindamycin, codeine, cyclophosphamide, cycloSPORINE, cytarabine, dacarbazine, DACTINomycin, DAPTOmycin, dexamethasone, dexmedetomidine, dexrazoxane, diazepam, digoxin, diltiazem, diphenhydrAMINE, DOBUTamine, DOPamine, doripenem, doxacurium, DOXOrubicin HCL, doxycycline, droperidol, enalaprilat, ePHEDrine, EPINEPHrine, epirubicin, ertapenem, erythromycin, esmolol, etoposide, famotidine, fenoldopam, fentaNYL, fluconazole, fludarabine, fluorouracil, foscarnet, fosphenytoin, furosemide, ganciclovir, gatifloxacin, gemcitabine, gentamicin, glycopyrrolate, granisetron, haloperidol, heparin, hydrALAZINE, hydrocortisone, HYDROmorphone, hydrOXYzine, ifosfamide, imipenem-cilastatin, inamrinone, insulin (regular), irinotecan, isoproterenol, ketorolac, labetalol, leucovorin, levofloxacin, levorphanol, lidocaine, linezolid, LORazepam, LR, magnesium sulfate, mannitol, meperidine, meropenem, mesna, methotrexate, methyldopate, metoclopramide, metoprolol, metroNIDAZOLE, midazolam, milrinone, minocycline, mitoXANtrone, mivacurium, morphine, nafcillin, naloxone, nesiritide, netilmicin, niCARdipine, nitroglycerin, nitroprusside, norepinephrine, octreotide, ofloxacin, ondansetron, oxaliplatin, palonosetron, pamidronate, pancuronium, pantoprazole, PEMEtrexed, pentamidine, pentazocine, PENTobarbital, PHENobarbital, phenylephrine, piperacillin, piperacillin-tazobactam, polymyxin B, potassium chloride/phosphates, procainamide, prochlorperazine, promethazine, propranolol, quiNIDine, quinupristin-dalfopristin, ranitidine, remifentanil, riTUXimab, rocuronium, sodium acetate/bicarbonate/phosphates, succinylcholine, SUFentanil, sulfamethoxazole-trimethoprim, tacrolimus, teniposide, theophylline, thiopental, thiotepa, ticarcillin, ticarcillin-clavulanate, tigecycline, tirofiban, tobramycin, tolazoline, trastuzumab, trimethobenzamide, vancomycin, vasopressin, vecuronium, verapamil, vinCRIStine, vinorelbine, voriconazole, zidovudine, zoledronic acid

SIDE EFFECTS

CNS: Seizures
CV: *Hypotension, fluid retention, peripheral edema,* flushing, MI, sinus tachycardia
GI: *Nausea, vomiting, diarrhea,* hepatotoxicity, stomatitis, colitis
HEMA: Neutropenia, leukopenia, thrombocytopenia, anemia, bleeding, infections, myelosuppression
INTEG: *Alopecia,* nail pain, rash, skin eruptions
MISC: Amenorrhea, fever of unknown origin, secondary malignancy, Stevens-Johnson syndrome, epiphora
MS: *Arthralgia, myalgia,* back pain, weakness
NEURO: *Peripheral neuropathy*
RESP: Dyspnea, pulmonary edema, fibrosis, embolism
SYST: *Hypersensitivity reactions*

PHARMACOKINETICS

Metabolized in liver, excreted in feces, terminal half-life 11.1 hr

INTERACTIONS

Increase: CYP3A inhibition: anastrozole (high doses), aprepitant, fosaprepitant, clarithromycin, conivaptan, delavirdine, efavirenz (induces or inhibits), erythromycin, fluconazole, FLUoxetine, fluvoxaMINE, imatinib, itraconazole, ketoconazole, mibefradil, nefazodone, voriconazole, and others

Increase: CYP3A induction: barbiturates, bosentan, carBAMazepine, nevirapine, phenytoin, fosphenytoin, rifabutin, rifampin, rifapentine, troglitazone

Increase: myelosuppression—other antineoplastics, radiation

Decrease: immune response—live virus vaccines

NURSING CONSIDERATIONS

Assess:

Black Box Warning: CBC, differential, platelet count before treatment and weekly; withhold product if WBC is <1500/mm^3 or platelet count is <100,000/mm^3; notify prescriber

Black Box Warning: DOCEtaxel, polysorbate 80 hypersensitivity: contraindicated

Black Box Warning: Edema: oral corticosteroids should be given as premedication, assess for fluid retention

Black Box Warning: Lung cancer: increased mortality in those with increased LFTs and a history of platinum-based products

- Monitor temperature; fever may indicate beginning of infection

Black Box Warning: **Hepatic disease:** hepatic studies before, during therapy (bilirubin, AST, ALT, LDH) prn or monthly; check for jaundiced skin and sclera, dark urine, clay-colored stools, itchy skin, abdominal pain, fever, diarrhea

- **CNS changes:** confusion, paresthesias, peripheral neuropathy, dysethenia, pain, weakness; if severe, product should be discontinued
- VS during 1st hr of inf, check IV site for signs of infiltration

Black Box Warning: **Hypersensitive reactions, anaphylaxis,** including hypotension, dyspnea, angioedema, generalized urticaria; discontinue inf immediately

- **Bone marrow depression/bleeding:** hematuria, guaiac, bruising or petechiae, mucosa or orifices q8hr; obtain prescription for viscous lidocaine (Xylocaine); avoid invasive procedures
- Effects of alopecia on body image; discuss feelings about body changes

Evaluate:

- Therapeutic response: decreased tumor size, spread of malignancy

Teach patient/family:

- To report signs of **infection:** fever, sore throat, flulike symptoms
- To report signs of **anemia:** fatigue, headache, faintness, SOB, irritability
- To report **bleeding;** to avoid use of razors, commercial mouthwash
- To avoid use of aspirin, ibuprofen
- That hair may be lost during treatment; that a wig or hairpiece may make patient feel better; that new hair may be different in color and texture
- That pain in muscles and joints 2-5 days after inf is common
- To use barrier contraception during and for several mo after treatment, pregnancy (D); to avoid breastfeeding
- To avoid receiving vaccinations while taking product

docosanol topical

See Appendix B

docusate calcium (OTC)

(dok′yoo-sate cal′see-um)

Kao-Tin, Stool Softener DC, Sur-Q-Lax, Walgreen's Stool Softener

docusate sodium (OTC)

Colace, Correctol, Diocto, Doc-Q-Lace, Docu DOK, Doculace, Enemeez, Equaline Stool Softener, Good Sense Stool Softener, Leader Stool Softener, Phillip's Stool Softener, Regulex ✱, Select Brand Docusate Sodium, Selex ✱, Silace, Soflax ✱, Top Care Stool Softener, Walgreen's Stool Softener

Func. class.: Laxative, emollient; stool softener

Chem. class.: Anionic surfactant

ACTION: Increases water, fat penetration in intestine; allows for easier passage of stool

USES: Prevention of dry, hard stools

CONTRAINDICATIONS: Hypersensitivity, obstruction, fecal impaction, nausea/vomiting

Precautions: Pregnancy (C), breastfeeding

DOSAGE AND ROUTES

- **Adult: PO** 50-300 mg/day (sodium) or 240 mg (calcium) prn; **ENEMA** 4 ml
- **Child >12 yr: ENEMA** 2 ml
- **Child 6-12 yr: PO** 40-150 mg/day (sodium) in divided doses
- **Child 3-6 yr: PO** 20-60 mg/day (sodium) in divided doses
- **Child <3 yr: PO** 10-40 mg/day (sodium) in divided doses

Available forms: *Calcium:* 240 mg; *sodium:* caps 50, 100, 250 mg; tabs 100 mg; syr 20 mg/5 ml, 50 mg/15 ml, 100 mg/30 ml, 150 mg/15 ml; oral sol 10, 50 mg/ml; enema 283 mg/3.9 cap

Administer:
- Swallow tabs whole; do not break, crush, or chew
- Oral sol: diluted in milk, fruit juice to decrease bitter taste
- In morning or evening (oral dose)
- Store in cool environment; do not freeze

SIDE EFFECTS

EENT: Bitter taste, throat irritation

GI: Nausea, anorexia, cramps, diarrhea

INTEG: Rash

PHARMACOKINETICS

Onset 12-72 hr

INTERACTIONS

- **Toxicity:** mineral oil

Drug/Herb

Increase: laxative action—flax, senna

NURSING CONSIDERATIONS

Assess:
- **Cause of constipation;** identify whether fluids, bulk, or exercise missing from lifestyle; constipating products
- Cramping, rectal bleeding, nausea, vomiting; if these occur, product should be discontinued

Evaluate:
- Therapeutic response: decrease in constipation

Teach patient/family:
- That normal bowel movements do not always occur daily
- Not to use in presence of abdominal pain, nausea, vomiting
- To notify prescriber if constipation unrelieved or if symptoms of electrolyte imbalance occur: muscle cramps, pain, weakness, dizziness, excessive thirst
- That product may take up to 3 days to soften stools
- To take oral preparations with a full glass of water (unless on fluid restrictions) and to increase fluid intake

dofetilide (Rx)

Tikosyn

Func. class.: Antidysrhythmic (Class III)

ACTION: Blocks cardiac ion channel carrying the rapid component of delayed potassium current; no effect on sodium channels

USES: Atrial fibrillation, flutter, maintenance of normal sinus rhythm

CONTRAINDICATIONS: Children, hypersensitivity, digoxin toxicity, aortic stenosis, pulmonary hypertension, severe renal disease

Black Box Warning: QT prolongation, torsades de pointes, renal failure

Precautions: Pregnancy (C), breastfeeding, AV block, bradycardia, electrolyte imbalance

Black Box Warning: Renal disease, arrhythmias, ventricular arrhythmias/tachycardia

DOSAGE AND ROUTES

Conversion of atrial fibrillation/atrial flutter to normal sinus rhythm; maintenance therapy with highly symptomatic atrial fibrillation/atrial flutter of ≥1 wk duration

• **Adult: PO** Individualize dosage based on renal function and QTc in a monitored facility; refer to the step-by-step procedure for determining the initial dosage of dofetilide

Maintenance therapy of atrial fibrillation/atrial flutter after hospital discharge

• **Adult: PO** Continue dosage at discharge as from initial dosage titration; individualize dosage based on renal function and QTc, which should be re-evaluated every 3 mo or as medically warranted; if the QTc >500 msec (550 msec in patients with ventricular conduction abnormalities) at any time, discontinue; carefully monitor until QTc returns to baseline; if renal function deteriorates, adjust the dosage as described in the dosage guidelines for patients with renal impairment

Discontinuation of dofetilide before use of interacting drugs

• **Adult: PO** Discontinue dofetilide for ≥2 days before starting a potentially interacting drug

Renal dose

• **Adult: PO** CCr >60 ml/min, 500 mcg bid; CCr 40-60 ml/min, 250 mcg bid; CCr 20-39 ml/min, 125 mcg bid; CCr <20 ml/min, do not use

Available forms: Caps 125, 250, 500 mcg

Administer:

Black Box Warning: Step 1: Assess cardiac conduction: Before first dose, the QTc interval must be determined using an average of 5-10 beats; if the QTc interval is >440 msec (or >500 msec in ventricular conduction abnormalities), do not use; if baseline heart rate is <60 bpm, then the QT interval should be used

Black Box Warning: Step 2: Assess renal function: Before first dose, determine renal function using the Cockroft-Gault equation, use actual body weight to calculate creatinine clearance

Black Box Warning: Step 3: Adjust starting dose according to renal function: Refer to the renal dose section above to determine the appropriate initial dosage

Black Box Warning: Step 4: ECG monitoring: Begin continuous ECG monitoring starting with the first dose

Black Box Warning: Step 5: Dosage adjustments: Approximately 2-3 hr after the first dose, determine the QTc interval; if the QTc interval has increased by >15% (compared to baseline), or if the QTc interval is >500 msec (>550 msec in patients with ventricular conduction abnormalities), the initial dosage should be reduced by half as follows:

- Decrease an initial dose of 500 mcg bid to 250 mcg bid
- Decrease an initial dose of 250 mcg bid to 125 mcg bid
- Decrease an initial dose of 125 mcg bid to 125 mcg/day

Black Box Warning: **Step 6: Reassess QTc interval:** Reassess the QTc interval 2-3 hr after each subsequent dose; if the QTc interval lengthens to >500 msec (or >550 msec in patients with ventricular conduction abnormalities), *discontinue*

Black Box Warning: **Step 7: ECG monitoring:** Monitor continuous ECG for a minimum of 3 days or for 12 hr after conversion to normal sinus rhythm, whichever is greater

SIDE EFFECTS

CNS: *Syncope, dizziness,* headache, stroke

CV: *Hypotension, postural hypotension, bradycardia,* angina, PVCs, substernal pressure, transient hypertension, precipitation of angina, QT prolongation, torsades de pointes, ventricular dysrhythmias, chest pain

GI: *Nausea, vomiting,* severe diarrhea, anorexia

MISC: Angioedema

RESP: Dyspnea, respiratory infections

PHARMACOKINETICS

Well absorbed, max plasma conc 2-3 hr, steady state 2-3 days, half-life 10 hr, metabolized by liver, excreted by kidneys

INTERACTIONS

- Do not use with cimetidine, ketoconazole, verapamil, prochlorperazine, trimethoprim-sulfamethoxazole, megestrol, hydrochlorothiazide

Increase: QT prolongation, torsades de pointes—class IA/III antidysrhythmics, arsenic trioxide, chloroquine, clarithromycin, droperidol, erythromycin, halofantrine, haloperidol, levomethadyl, methadone, pentamidine, some phenothiazines, ziprasidone, ciprofloxacin

Increase: hypokalemia—potassium-depleting diuretics

Increase: toxicity—aMILoride metFORMIN, entecavir, lamiVUDine, memantine, triamterene, procainamide, trospium

Increase: dofetilide levels—antiretroviral protease inhibitors

Drug/Food

- Do not use with grapefruit juice

NURSING CONSIDERATIONS

Assess:

- AF patients should receive anticoagulation prior to cardioversion
- Cardiac status: rate, rhythm, character, continuously; B/P

Black Box Warning: **Severe renal impairment CCr <20 ml/min:** do not use for mild to moderate renal disease; monitor BUN/creatinine; adjust dose based on creatinine clearance

Evaluate:

- Therapeutic response: control of atrial fibrillation

Teach patient/family:

- To make position changes slowly; orthostatic hypotension may occur
- To notify prescriber if fast heartbeats with fainting or dizziness occur
- To notify all prescribers of all medications, supplements taken
- That if dose is missed, not to double; to take next dose at usual time
- To avoid breastfeeding

dolasetron (Rx)

(do-la'se-tron)

Anzemet

Func. class.: Antiemetic

Chem. class.: 5-HT3 receptor antagonist

ACTION: Prevents nausea, vomiting by blocking serotonin peripherally, centrally, and in the small intestine

USES:
Prevention of nausea, vomiting associated with cancer chemotherapy, radiotherapy; prevention of postoperative nausea, vomiting

Unlabeled uses: Radiotherapy-induced nausea/vomiting

CONTRAINDICATIONS:
Hypersensitivity

Precautions: Pregnancy (B), breastfeeding, children, geriatric patients, hypokalemia, electrolyte imbalances; granisetron/ondansetron/palonosetron hypersensitivity, QT prolongation

DOSAGE AND ROUTES

Prevention of nausea and vomiting during cancer chemotherapy

- **Adult: PO** 100 mg 1 hr before chemotherapy
- **Child 2-16 yr: PO** 1.8 mg/kg before chemotherapy; max 100 mg

Prevention of postoperative nausea and vomiting

- **Adult: IV** 12.5 mg as single dose 15 min before cessation of anesthesia; **PO** 100 mg 2 hr before surgery (prevention only)
- **Child 2-16 yr: IV** 0.35 mg/kg as single dose 15 min before cessation of anesthesia; **PO** 1.2 mg/kg 2 hr before surgery (prevention only)

Available forms: Tabs 50, 100 mg; inj 20 mg/ml (12.5 mg/0.625 ml)

Administer:

PO route

- Do not mix product for oral administration in apple or apple-grape juice until immediately before administration; diluted product can be kept for 2 hr at room temperature
- Store at room temperature 48 hr after dilution

Intermittent IV INF route

- By inj 100 mg/30 sec or more or diluted in 50 ml compatible sol; give over 15 min
- Do not admix

SIDE EFFECTS

CNS: *Headache,* dizziness, fatigue, drowsiness

CV: Dysrhythmias, ECG changes, hypo/hypertension, tachycardia, bradycardia; **ventricular tachycardia/fibrillation, QT prolongation, torsades de pointes, cardiac arrest (IV)**

GI: *Diarrhea,* constipation, increased AST/ALT, abdominal pain, anorexia

GU: Urinary retention, oliguria

MISC: Rash, **bronchospasm**

PHARMACOKINETICS

Well absorbed, metabolized to active metabolite, half-life of active metabolite 8 hr, max concs after 1 hr

INTERACTIONS

- **QT Prolongation:** QRS, PR prolongation; do not use in those with congenital long QT syndrome, hypokalemia, hypomagnesemia, complete heart block (unless a pacemaker is in place), correct electrolytes before use, monitor ECG in elderly patients, renal cardiac disease

Increase: dysrhythmias—antidysrhythmics

Increase: dolasetron levels—cimetidine

Increase: QT prolongation—thiazide/loop diuretics, antidysrhythmics—class IA, III, arsenic trioxide, chloroquine, clarithromycin, droperidol, erythromycin, halofantrine, haloperidol, levomethadyl, methadone, pentamidine, some phenothiazines, ziprasidone; occurs at higher dose of dolasetron

Decrease: dolasetron levels—rifampin

NURSING CONSIDERATIONS

Assess:

- For absence of nausea, vomiting during chemotherapy
- **Hypersensitivity reaction:** rash, bronchospasm
- Cardiac conduction conditions, electrolyte imbalances, dysrhythmias, heart rate

Evaluate:

- Therapeutic response: absence of nausea, vomiting during cancer chemotherapy

Teach patient/family:

- To report diarrhea, constipation, nausea, vomiting, rash, or changes in respirations
- May cause headache; use analgesic

donepezil (Rx)

(don-ep-ee′zill)

Aricept, Aricept ODT

Func. class.: Anti-Alzheimer's agent

Chem. class.: Reversible cholinesterase inhibitor

ACTION: Elevates acetylcholine concs (cerebral cortex) by slowing degradation of acetylcholine released in cholinergic neurons; does not alter underlying dementia

USES: Mild to severe dementia with Alzheimer's disease

Unlabeled uses: Subcortical, vascular dementia; dementia with Lewy bodies

CONTRAINDICATIONS: Hypersensitivity to this product or piperidine derivatives

Precautions: Pregnancy (C), breastfeeding, children, sick sinus syndrome, history of ulcers, GI bleeding, hepatic disease, bladder obstruction, asthma, seizures, COPD, abrupt discontinuation, AV block, GI obstruction, Parkinson's disease, surgery

DOSAGE AND ROUTES

- **Adult: PO** 5 mg/day at bedtime; may increase to 10 mg/day after 4-6 wk, may increase to 23 mg/day after 3 mo of 10 mg/day

Available forms: Tabs 5, 10, 23 mg; orally disintegrating tabs (Aricept ODT) 5, 10 mg

Administer:

- Daily in the evening before bedtime; swallow whole; do not cut, break, chew, or crush tab
- Dosage adjusted to response no more than q4-6wk; oral dosage forms are interchangeable
- **Orally disintegrating tabs:** allow to dissolve on tongue before swallowing; may be given with/without water

SIDE EFFECTS

CNS: Dizziness, *insomnia,* somnolence, *headache,* fatigue, abnormal dreams, syncope, seizures, drowsiness, agitation, depression, confusion, fever, hallucinations

CV: Atrial fibrillation, hypo/hypertension, sinus bradycardia, AV block

GI: *Nausea, vomiting,* anorexia, *diarrhea,* abdominal pain, GI bleeding, weight loss

GU: Urinary frequency, UTI, incontinence

INTEG: Rash, flushing, diaphoresis, bruising

META: Hyperlipidemia

MS: Cramps, arthritis, arthralgia, back pain

RESP: Rhinitis, URI, cough, pharyngitis, dyspnea

PHARMACOKINETICS

Well absorbed PO; metabolized by CYP2D6, CYP3A4; elimination half-life 10 hr single dose, 70 hr multiple doses; protein binding 96%

INTERACTIONS

Increase: donepezil effects—CYP2D6, CYP3A4 inhibitors

Increase: synergistic effect—succinylcholine, cholinesterase inhibitors, cholinergic agonists

Increase: GI intolerance—NSAIDs

Decrease: donepezil effects—CYP2D6, CYP3A4 inducers

Decrease: action of anticholinergics

Decrease: donepezil effect—carBAMazepine, dexamethasone, phenytoin, PHENObarbital, rifampin

Drug/Herb
Decrease: donepezil—St. John's wort

NURSING CONSIDERATIONS

Assess:
- B/P: hypo/hypertension, heart rate
- Mental status: affect, mood, behavioral changes, depression, complete suicide assessment; neurologic status
- GI status: nausea, vomiting, anorexia, diarrhea; monitor weight
- GU status: urinary frequency, incontinence, I&O
- Assistance with ambulation during beginning therapy; dizziness, ataxia may occur

Evaluate:
- Therapeutic response: decrease in confusion, improved mood

Teach patient/family:
- To report side effects: twitching, nausea, vomiting, sweating, dizziness; indicates cholinergic crisis or overdose
- To use product exactly as prescribed
- To notify prescriber of nausea, vomiting, diarrhea (dose increase or beginning treatment), or rash
- Not to increase or abruptly decrease dose; serious consequences may result
- That product is not a cure, relieves symptoms

⚠ HIGH ALERT

DOPamine (Rx)

(doe′pa-meen)

Func. class.: Adrenergic
Chem. class.: Catecholamine

Do not confuse:
DOPamine/DOBUTamine

ACTION: Causes increased cardiac output; acts on β_1- and α-receptors, causing vasoconstriction in blood vessels; low dose causes renal and mesenteric vasodilation; β_1 stimulation produces inotropic effects with increased cardiac output

USES: Shock, increased perfusion, hypotension, cardiogenic/septic shock
Unlabeled uses: Bradycardia, cardiac arrest, CPR, acute renal failure, cirrhosis, barbiturate intoxication

CONTRAINDICATIONS: Hypersensitivity, ventricular fibrillation, tachydysrhythmias, pheochromocytoma, hypovolemia
Precautions: Pregnancy (C), breastfeeding, geriatric patients, arterial embolism, peripheral vascular disease, sulfite hypersensitivity, acute MI

Black Box Warning: Extravasation

DOSAGE AND ROUTES

Shock
- **Adult: IV INF** 2-5 mcg/kg/min, titrate upward in 5-10 mcg/kg/min increments, max 50 mcg/kg/min; titrate to patient's response
- **Child: IV** 1-5 mcg/kg/min initially; usual dosage range, 2-20 mcg/kg/min

COPD
- **Adult: IV** 4 mcg/kg/min

CHF
- **Adult: IV** 3-10 mcg/kg/min

Bradycardia (unlabeled)
- **Adult: IV** 2-10 mcg/kg/min, titrate as needed

Available forms: Inj 40 mg, 80 mg, 160 mg/ml; conc for IV inf 0.8, 1.6, 3.2 mg/ml in 250, 500 ml D_5W

Administer:
- Store reconstituted sol for up to 24 hr if refrigerated
- Do not use discolored sol; protect from light

IV route
- IV after diluting 200-400 mg/250-500 ml of D_5W, D_5 0.45% NaCl, D_5 0.9% NaCl, D_5LR, LR; use large vein
- After reconstituting, use inf pump; give at rate of 0.5-5 mcg/kg/min, increase by 1-4 mcg/kg/min at 10-30 min intervals until desired response

Black Box Warning: Extravasation: if extravasation occurs, stop inf, may inject area with phentolamine 10 mg/15 ml of NS

Y-site compatibilities: Alfentanil, alprostadil, amifostine, amikacin, aminocaproic acid, aminophylline, amiodarone, anidulafungin, argatroban, ascorbic acid injection, atenolol, atracurium, atropine, aztreonam, benztropine, bivalirudin, bleomycin, bumetanide, buprenorphine, butorphanol, calcium chloride/gluconate, CARBOplatin, caspofungin, cefamandole, cefmetazole, cefonicid, cefotaxime, cefoTEtan, cefOXitin, cefpirome, cefTAZidime, ceftizoxime, cefTRIAXone, cefuroxime, chlorproMAZINE, cimetidine, ciprofloxacin, cisatracurium, CISplatin, cladribine, clarithromycin, clindamycin, cloNIDine, codeine, cyanocobalamin, cyclophosphamide, cycloSPORINE, cytarabine, DACTINomycin, DAPTOmycin, dexamethasone, dexmedetomidine, digoxin, diltiazem, diphenhydrAMINE, DOBUTamine, DOCEtaxel, doripenem, doxacurium, DOXOrubicin, DOXOrubicin liposomal, doxycycline, droperidol, enalaprilat, ePHEDrine, EPINEPHrine, epirubicin, epoetin alfa, eptifibatide, ertapenem, erythromycin, esmolol, etoposide, famotidine, fenoldopam, fentaNYL, fluconazole, fludarabine, fluorouracil, folic acid, foscarnet, gatifloxacin, gemcitabine, gemtuzumab, gentamicin, glycopyrrolate, granisetron, heparin, hydrocortisone, HYDROmorphone, hydrOXYzine, IDArubicin, ifosfamide, imipenem-cilastatin, irinotecan, isoproterenol, ketorolac, labetalol, levofloxacin, lidocaine, linezolid, LORazepam, LR, magnesium sulfate, mannitol, mechlorethamine, meperidine, metaraminol, methicillin, methoxamine, methyldopate, methylPREDNISolone, metoclopramide, metoprolol, metroNIDAZOLE, mezlocillin, micanfungin, miconazole, midazolam, milrinone, minocycline, mitoXANtrone, morphine, moxalactam, multiple vitamins injection, mycophenolate, nafcillin, nalbuphine, naloxone, netilmicin, niCARdipine, nitroglycerin, nitroprusside, norepinephrine, octreotide, ondansetron, oxacillin, oxaliplatin, oxytocin, PACLitaxel, palonosetron, pamidronate, pancuronium, pantoprazole, papaverine, PEMEtrexed, penicillin G potassium/sodium, pentamidine, pentazocine, PENTobarbital, PHENobarbital, phenylephrine, phytonadione, piperacillin, piperacillin-tazobactam, polymyxin B, potassium chloride, procainamide, prochlorperazine, promethazine, propofol, propranolol, protamine, pyridoxine, quiNIDine, ranitidine, remifentanil, Ringer's, ritodrine, riTUXimab, rocuronium, sargramostim, sodium acetate, succinylcholine, SUFentanil, tacrolimus, temocillin, teniposide, theophylline, thiamine, thiotepa, ticarcillin, ticarcillin-clavulanate, tigecycline, tirofiban, TNA, tobramycin, tolazoline, TPN, trastuzumab, trimetaphan, urokinase, vancomycin, vasopressin, vecuronium, verapamil, vinCRIStine, vinorelbine, vitamin B complex/C, voriconazole, warfarin, zidovudine, zoledronic acid

SIDE EFFECTS

CNS: *Headache*, anxiety
CV: *Palpitations*, tachycardia, *hypertension, ectopic beats, angina,* wide QRS complex, peripheral vasoconstriction, hypotension
GI: *Nausea, vomiting, diarrhea*
INTEG: Necrosis, tissue sloughing with extravasation, gangrene
RESP: Dyspnea

PHARMACOKINETICS

IV: Onset 5 min; duration <10 min; metabolized in liver, kidney, plasma; excreted in urine (metabolites); half-life 2 min

INTERACTIONS

- Do not use within 2 wk of MAOIs; hypertensive crisis may result

Increase: bradycardia, hypotension—phenytoin

Increase: dysrhythmias—general anesthetics
Increase: severe hypertension—ergots
Increase: B/P—oxytocics
Increase: pressor effect—tricyclics, MAOIs
Decrease: DOPamine action—β-/α-blockers

Drug/Lab Test
Increase: urinary catecholamine, serum glucose

NURSING CONSIDERATIONS

Assess:

- Hypovolemia; if present, correct first
- **Oxygenation/perfusion deficit:** check B/P, chest pain, dizziness, loss of consciousness
- **Heart failure:** S_3 gallop, dyspnea, neck venous distention, bibasilar crackles in patients with CHF, cardiomyopathy, palpate peripheral pulses
- I&O ratio: if urine output decreases without decrease in B/P, product may need to be reduced
- **ECG** during administration continuously; if B/P increases, product should be decreased; PCWP, CVP during inf
- B/P, pulse q5min
- Paresthesias and coldness of extremities; peripheral blood flow may decrease
- Inj site: tissue sloughing; if this occurs, administer phentolamine mixed with NS

Evaluate:

- Therapeutic response: increased B/P with stabilization; increased urine output

Teach patient/family:

- About the reason for product administration

TREATMENT OF OVERDOSE:

Discontinue IV, may give a short-acting α-adrenergic blocker

doripenem (Rx)

(dore-i-pen′em)

Doribax

Func. class.: Antiinfective—miscellaneous
Chem. class.: Carbapenem

D

ACTION: Bactericidal; interferes with cell-wall replication of susceptible organisms; osmotically unstable cell wall swells, bursts from osmotic pressure

USES: Serious infections caused by *Acinetobacter baumannii, Bacteroides caccae, Bacteroides fragilis, Bacteroides thetaiotaomicron, Bacteroides uniformis, Bacteroides vulgatus, Citrobacter freundii, Escherichia coli, Klebsiella pneumoniae, Peptostreptococcus micros, Proteus mirabilis, Pseudomonas aeruginosa, Serratia marcescens, Staphylococcus aureus, Streptococcus constellatus, Streptococcus intermedius*; complicated urinary tract infections, pyelonephritis, complicated intraabdominal infections

CONTRAINDICATIONS: Hypersensitivity to carbapenems (meropenem, doripenem, imipenem), penicillin, β-lactam; viral infection

Precautions: Pregnancy (B), breastfeeding, geriatric patients, renal disease, seizure disorder, pseudomembranous colitis, nebulizer or inhalation use, hypersensitivity to cephalosporins, children/adolescents

DOSAGE AND ROUTES

- **Adult: IV** 500 mg q8hr × 5-14 days; if improvement occurs after 3 days, switch to appropriate oral product

Renal dose

- **Adult: IV** CCr 30-50 ml/min, 250 mg over 1 hr, q8hr; CCr >10 to <30 ml/min, 250 mg over 1 hr q12hr; CCr ≤10 ml/min, no data

Available forms: Powder for inj 250, 500 mg

Administer:

IV route

- Visually inspect parenteral products for particulate matter and discoloration before use, diluted range in color from clear, colorless solutions to solutions that are clear and slightly yellow
- **Reconstitution:** no bacteriostatic preservative is present; observe aseptic technique while preparing the inf
- **500-mg dose using the 500-mg vial**: constitute the vial with 10 ml of sterile water for injection or sodium chloride 0.9% (normal saline); gently shake (50 mg/ml); *the constituted suspension is not for direct injection; further dilution is required;* using a syringe with a 21-G needle, withdraw the suspension and add it to an inf bag containing 100 ml of NS or D_5W; gently shake until clear: final conc: 4.5 mg/ml
- **250-mg dose using the 500 mg vial**: constitute the vial with 10 ml of sterile water for injection or sodium chloride 0.9% (normal saline); gently shake (50 mg/ml); *the constituted suspension is not for direct injection; further dilution is required;* using a syringe with a 21-G needle, withdraw the suspension and add it to an inf bag containing 100 ml of normal saline or D_5W; gently shake until clear; remove 55 ml of this solution and discard; the remaining inf sol contains 250 mg (4.5 mg/ml)
- **250-mg dose using the 250 mg vial**: constitute the vial with 10 ml of sterile water for injection or sodium chloride 0.9% (normal saline); gently shake (25 mg/ml); *the constituted suspension is not for direct injection; further dilution is required;* using a syringe with a 21-G needle, withdraw the suspension and add it to an inf bag containing 50 or 100 ml of normal saline or D_5W; gently shake until clear; final conc 4.2 mg/ml (50 ml inf bag) or 2.3 mg/ml (100 ml inf bag)
- **Storage:** constituted suspensions may be held in vial for up to 1 hr before transfer and dilution in the inf bag; including storage and inf time, diluted inf sols are stable for up to 12 hr (NS) or 4 hr (D_5W) at controlled room temperature; diluted inf sols are stable for up to 72 hr (NS) or 24 hr (D_5W) refrigerated; do not freeze constituted solutions
- If Baxter Minibag Plus inf bags are to be used, consult the instructions provided by the inf bag manufacturer

Intermittent IV INF route

- After C&S is taken
- Do not mix with or physically add to solutions containing other drugs; infuse over 1 hr

Y-site compatibilities: Acyclovir, amikacin, aminophylline, amiodarone, anidulafungin, atropine, azithromycin, bumetanide, calcium gluconate, CARBOplatin, caspofungin, ceftaroline, ceftobiprole, cimetidine, ciprofloxacin, CISplatin, cyclophosphamide, cycloSPORINE, DAPTOmycin, dexamethasone, digoxin, diltiazem, diphenhydrAMINE, DOBUTamine, DOCEtaxel, DOPamine, DOXOrubicin, enalaprilat, esmolol, esomeprazole, etoposide, famotidine, fentaNYL, fluconazole, fluorouracil, foscarnet, furosemide, gemcitabine, gentamicin, granisetron, heparin, hydrocortisone, HYDROmorphone, ifosfamide, insulin (regular), labetalol, levofloxacin, linezolid, LORazepam, magnesium sulfate, mannitol, meperidine, methotrexate, methylPREDNISolone, metoclopramide, metroNIDAZOLE, micafungin, midazolam, milrinone, morphine, moxifloxacin, norepinephrine, ondansetron, PACLitaxel, pantoprazole, PHENObarbital, phenylephrine, potassium chloride, ranitidine, sodium bicarbonate/phosphates, tacrolimus, telavancin, tigecycline, tobramycin, vancomycin, voriconazole, zidovudine

Solution compatibilities: D_5W, 0.9% NaCl, sterile water for inj

SIDE EFFECTS

CNS: Seizures, headache
GI: *Diarrhea, nausea,* vomiting, pseudomembranous colitis, hepatitis
GU: Renal impairments/failure
HEMA: Neutropenia, leukopenia, anemia
INTEG: *Rash,* urticaria, phlebitis, erythema at inj site, Stevens-Johnson syndrome, toxic epidermal necrolysis, pruritus
RESP: Pneumonitis (inhalation)
SYST: Anaphylaxis, Stevens-Johnson syndrome, toxic epidermal necrolysis

PHARMACOKINETICS

IV: Distributed to most body fluids/tissue, excreted mainly unchanged in urine, 70% recovered in 48 hr, half-life 1 hr, half-life extended in renal disease

INTERACTIONS

Increase: doripenem plasma levels—probenecid
Decrease: effect of valproic acid, divalproex sodium

Drug/Lab Test

Increase: AST, ALT, LDH, BUN, alk phos, bilirubin, creatinine
False positive: direct Coombs' test

NURSING CONSIDERATIONS

Assess:

- Sensitivity to carbapenem antibiotics, penicillins, cephalosporins, other beta lactams
- Renal disease: lower dose may be required
- Bowel pattern daily; if severe diarrhea occurs, product should be discontinued; may indicate pseudomembranous colitis
- For infection: temperature; sputum; characteristics of wound before, during, and after treatment

⚠ **Allergic reactions, anaphylaxis:** rash, urticaria, pruritus; may occur few days after therapy begins

- **Overgrowth of infection:** perineal itching, fever, malaise, redness, pain, swelling, drainage, rash, diarrhea, change in cough, sputum

Evaluate:

- Therapeutic response: negative C&S; absence of symptoms and signs of infection

Teach patient/family:

- To report severe diarrhea; may indicate pseudomembranous colitis
- To report sore throat, bruising, bleeding, joint pain; may indicate blood dyscrasias (rare)
- To report overgrowth of infection: black, furry tongue; vaginal itching; foul-smelling stools
- To avoid breastfeeding; product is excreted in breast milk

TREATMENT OF HYPERSENSITIVITY: EPINEPHrine, antihistamines; resuscitate if needed (anaphylaxis)

dorzolamide (ophthalmic)

(dor-zole′ah-mide)

Trusopt

Func. class.: Antiglaucoma
Chem. class.: Carbonic anhydrase inhibitor

ACTION: Decreases aqueous humor secretion by decreasing bicarbonate, thus decreasing IOP

USES: For the treatment of elevated intraocular pressure in patients with ocular hypertension or open-angle glaucoma

CONTRAINDICATIONS: Hypersensitivity
Precautions: Hypersensitivity to sulfonamides, hepatic/renal disease, angle-closure glaucoma, electrolyte disturbances

DOSAGE AND ROUTES

Elevated intraocular pressure in patients with ocular hypertension or open-angle glaucoma

- **Adult/adolescent/child/infant/neonate ≥1 wk: Ophthalmic** Instill 1 drop of a 2% solution into the affected eye(s) tid

Available forms: Ophthalmic solution 2%

Administer:

- Wash hands before and after use, tilt the head back slightly and pull the lower eyelid down with the index finger to form a pouch, squeeze the prescribed number of drops into the pouch and gently close eyes for 1-2 min; do not blink
- Care should be taken to avoid contamination; do not touch the tip of the dropper to the eye, fingertips, or other surface
- The sol may be used concomitantly with other topical ophthalmic drug products to lower IOP; if more than one topical ophthalmic drug is being used, administer ≥10 min apart

SIDE EFFECTS

CNS: Headache

EENT: Blurred vision, tearing, allergy, burning/stinging, photophobia

GI: Bitter taste

PHARMACOKINETICS

Onset 1-2 hr, peak 3 hr, duration 8 hr, half-life 4 mo

INTERACTIONS

Increase: effects—carbonic anhydrase inhibitors (PO), salicylates

NURSING CONSIDERATIONS

Assess:

- Hypersensitivity
- Monitor IOP during treatment

Evaluate:

- Decreasing IOP

Teach patient/family:

- How to use product
- Not to share with others or use for other conditions
- To notify prescriber immediately if vision changes or if condition worsens
- To take as prescribed

doxapram (Rx)

(dox′a-pram)

Dopram

Func. class.: Analeptic

ACTION:

Respiratory stimulation through activation of peripheral carotid chemoreceptor; with higher doses, medullary respiratory centers are stimulated; with progressive CNS stimulation

USES:

Chronic obstructive pulmonary disease (COPD), postanesthesia respiratory depression, prevention of acute hypercapnia, product-induced CNS depression

Unlabeled uses: Neonatal apnea

CONTRAINDICATIONS:

Hypersensitivity, seizure disorders, severe hypertension, severe bronchial asthma, severe dyspnea, severe cardiac disorders, flail chest, pneumothorax, PE, severe respiratory disease

Precautions: Pregnancy (B), breastfeeding, children, bronchial asthma, pheochromocytoma, severe tachycardia, dysrhythmias, hypertension, hyperthyroidism

DOSAGE AND ROUTES

Postanesthesia

- **Adult:** **IV** inj 0.5-1 mg/kg, not to exceed 1.5 mg/kg total as a single inj; **IV INF** 250 mg in 250 ml sol, not to exceed 4 mg/kg; run at 1-3 mg/min

Drug-induced CNS depression

- **Adult:** **IV** Priming dose of 2 mg/kg, repeated in 5 min; repeat q1-2hr until patient awakes; **IV INF** priming dose 2 mg/kg at 1-3 mg/min, not to exceed 3 g/day

COPD (hypercapnia)

- **Adult:** **IV INF** 1-2 mg/min, not to exceed 3 mg/min for no longer than 2 hr

Available forms: Inj 20 mg/ml

SIDE EFFECTS

CNS: **Seizures (clonus/generalized),** *headache,* restlessness, dizziness, confusion, paresthesias, flushing, sweating, bilateral Babinski's sign, rigidity, depression

CV: *Chest pain, hypertension, change in heart rate,* lowered T waves, tachycardia, **dysrhythmias**

EENT: Pupil dilation, sneezing

GI: Nausea, vomiting, diarrhea, desire to defecate

GU: Retention, incontinence, elevation of BUN, albuminuria

INTEG: Pruritus, irritation at inj site

RESP: **Laryngospasm, bronchospasm,** rebound hypoventilation, dyspnea, cough, tachypnea, hiccups

PHARMACOKINETICS

IV: Onset 20-40 sec, peak 1-2 min, duration 5-10 min, metabolized by liver, excreted by kidneys (metabolites), half-life 2.5-4 hr

INTERACTIONS

- **Synergistic pressor effect: MAOIs, sympathomimetics**
- **Cardiac dysrhythmias: halothane, cyclopropane, enflurane; delay use of doxapram for at least 10 min after inhalation anesthetics**

NURSING CONSIDERATIONS

Assess:

- BP, heart rate, deep tendon reflexes, ABGs, LOC before administration, q30min
- Po_2, Pco_2, O_2 saturation during treatment
- **Hypertension, dysrhythmias, tachycardia, dyspnea, skeletal muscle hyperactivity; may indicate overdosage; discontinue product**
- Respiratory stimulation: increased rate, abnormal rhythm
- Extravasation; change IV site q48hr
- Patient closely for 1/2-1 hr

Administer:

IV route

- Undiluted or diluted with equal parts of sterile water for inj; may be diluted 250 mg/250 ml of D_5W, $D_{10}W$ and run as inf; rapid inf may cause hemolysis
- IV undiluted over 5 min; IV inf at 1-3 mg/min; adjust for desired respiratory response, using inf pump IV; if an inf is used after initial dose, start at 1-3 mg/min depending on patient response; D/C after 2 hr; wait 1-2 hr and repeat
- Only after adequate airway is established
- After O_2, IV barbiturates, resuscitative equipment available
- Discontinue inf if side effects occur; narrow margin of safety

Syringe compatibilities: Amikacin, bumetadine, chlorproMAZINE, cimetidine, CISplatin, cyclophosphamide, DOPamine, doxycycline, EPINEPHrine, hydrOXYzine, imipramine, isoniazid, lincomycin, methotrexate, netilmicin, phytonadione, pyridoxine, terbutaline, thiamine, tobramycin, vinCRIStine

Evaluate:

- Therapeutic response: increased breathing capacity

Teach patient/family:

- Purpose of medication

doxazosin (Rx)

(dox-ay′zoe-sin)

Cardura, Cardura XL

Func. class.: Peripheral α_1-adrenergic receptor blocker

Chem. class.: Quinazoline

Do not confuse:

Cardura/Coumadin/Cardene/Ridaura

ACTION: Dilates peripheral blood vessels, lowers peripheral resistance; reduction in B/P results from peripheral α_1-adrenergic receptors being blocked

USES:
Hypertension, urinary outflow obstruction, symptoms of benign prostatic hyperplasia

CONTRAINDICATIONS:
Hypersensitivity to quinazolines

Precautions: Pregnancy (C), breastfeeding, children, hepatic disease, geriatric patients

DOSAGE AND ROUTES

BPH

- **Adult: PO** 1 mg/day at bedtime; increase in stepwise manner to 2, 4, 8 mg/day as needed at 1-2 wk intervals, max 8 mg; ext rel tab (Cardura XL) 4 mg daily with breakfast, adjust dose q3-4wk, up to 8 mg daily

Hypertension

- **Adult: PO** 1 mg/day at bedtime; increasing up to 16 mg/day if required; usual range 4-16 mg/day
- **Geriatric: PO** 0.5 mg nightly, gradually increase

Available forms: Tabs 1, 2, 4, 8 mg; ext rel tabs 4, 8 mg

Administer:

- Store in tight container at room temperature
- **Tabs** broken, crushed, or chewed; if chewed, will be bitter; do not break, crush, chew XL tabs
- **Immediate release tab:** without regard to meals; **ext rel tabs:** give with breakfast; when switching from immediate release to ext rel, the final evening dose of immediate release should not be taken

SIDE EFFECTS

CNS: *Dizziness, headache,* drowsiness, anxiety, depression, *vertigo,* weakness, fatigue, asthenia, syncope

CV: Palpitations, *orthostatic hypotension*, tachycardia, *edema,* dysrhythmias, chest pain

EENT: Epistaxis, tinnitus, dry mouth, red sclera, pharyngitis, rhinitis

GI: *Nausea,* vomiting, diarrhea, constipation, abdominal pain, hepatitis

GU: Incontinence, polyuria, priapism, impotence

PHARMACOKINETICS

PO: Onset 2 hr, peak 5-6 hr, duration up to 24 hr, half-life 22 hr, metabolized in liver, excreted via bile/feces (<63%) and in urine (9%), extensively protein bound (98%)

INTERACTIONS

Increase: hypotensive effects—alcohol, other antihypertensives, nitrates, PDE-5 inhibitors

Decrease: antihypertensive effects of cloNIDine

NURSING CONSIDERATIONS

Assess:

- **Hypertension:** B/P (lying, standing), pulse 2-6 hr after each dose, with each increase; postural effects may occur, crackles, dyspnea, orthopnea with B/P; pulse; jugular venous distention during beginning treatment
- **BPH:** urinary pattern changes (hesitancy, dribbling, incomplete bladder emptying, dysuria, urgency, nocturia, urgency incontinence, intermittency) before and during treatment
- I&O, weight daily; edema in feet, legs daily

Evaluate:

- Therapeutic response: decreased B/P; decreased symptoms of BPH

Teach patient/family:

- That fainting occasionally occurs after 1st dose; not to drive, operate machinery for 4 hr after 1st dose, after dosage increase; to take 1st dose at bedtime; may take 1-2 wk to respond with BPH
- Rise slowly from sitting position

TREATMENT OF OVERDOSE:
Administer volume expanders or vasopressors; discontinue product; place patient in supine position

doxepin (Rx)

(dox′e-pin)

Prudoxin Cream, Silenor, Zonalon Topical Cream

Func. class.: Antidepressant, tricyclic, antihistamine (topical)

Chem. class.: Dibenzoxepin, tertiary amine

ACTION: Blocks reuptake of norepinephrine, serotonin into nerve endings, increasing action of norepinephrine, serotonin in nerve cells

USES: Major depression, anxiety; *topical:* lichen simplex, atopic dermatitis, eczema

Unlabeled uses: Topical pruritus, insomnia, migraine prophylaxis

CONTRAINDICATIONS: Hypersensitivity to tricyclics, urinary retention, closed-angle glaucoma, prostatic hypertrophy, acute recovery from MI

Precautions: Pregnancy (C) (PO) (B) (topical), breastfeeding, geriatric patients, seizures

Black Box Warning: Children, suicidal patients

DOSAGE AND ROUTES

Depression/anxiety

- **Adult: PO** 50-75 mg/day, may increase to 300 mg/day for severely ill; give in divided doses if >150 mg/day
- **Geriatric: PO** 25-50 mg at bedtime, increase weekly by 25-50 mg to desired dose, max 150 mg/day

Pruritus

- **Adult: PO** 10 mg at bedtime, may increase to 25 mg at bedtime; **TOP** apply thin film qid at least 3 hr apart

Insomnia (Silenor)

- **Adult: PO** 6 mg 30 min before bedtime, 3 mg may be sufficient, max 6 mg/night

Available forms: Caps 10, 25, 50, 75, 100, 150 mg; oral conc 10 mg/ml; cream 5%; tabs (Silenor) 3, 6 mg

Administer:

- **Oral conc:** should be diluted with 120 ml water, milk or orange, grapefruit, tomato, prune, or pineapple juice; do not mix with grape juice
- Increased fluids, bulk in diet for constipation
- With food, milk for GI symptoms; do not give with carbonated beverages
- Dosage at bedtime to avoid oversedation during day; may take entire dose at bedtime; geriatric patients may not tolerate daily dosing
- Gum, hard candy, or frequent sips of water for dry mouth
- Store in tight container protected from direct sunlight
- **Topical:** by applying to affected area, rub slightly; do not use occlusive dressings

SIDE EFFECTS

CNS: *Dizziness, drowsiness,* confusion, headache, anxiety, tremors, stimulation, weakness, insomnia, nightmares, EPS (geriatric patients), increased psychiatric symptoms, paresthesia, **suicidal ideation**

CV: *Orthostatic hypotension, ECG changes, tachycardia,* **hypertension,** palpitations, **dysrhythmias**

EENT: *Blurred vision,* tinnitus, mydriasis, ophthalmoplegia, glossitis

GI: *Diarrhea, dry mouth,* nausea, vomiting, **paralytic ileus,** increased appetite, cramps, epigastric distress, jaundice, **hepatitis,** stomatitis, constipation

GU: *Urinary retention,* **acute renal failure**

HEMA: **Agranulocytosis, thrombocytopenia, eosinophilia, leukopenia,** pancytopenia, purpuric disorder

INTEG: Rash, urticaria, sweating, pruritus, photosensitivity

PHARMACOKINETICS

PO: Steady state 2-8 days, metabolized by liver, excreted by kidneys, crosses placenta, excreted in breast milk, half-life 8-24 hr

INTERACTIONS

Increase: hyperpyretic crisis, seizures, hypertensive episode—MAOIs

Increase: hypertensive action—EPINEPHrine, norepinephrine

Increase: hypertensive crisis—cloNIDine; do not use together

Increase: doxepin effect—cimetidine, FLUoxetine, fluvoxaMINE, PARoxetine, sertraline

Increase: CNS depression—barbiturates, benzodiazepines, sedative/hypnotics, alcohol, other CNS depressants

Increase: QT interval: class IC/III antiarrhythmics (propafenone, flecainide), quinolones

Serotonin syndrome

Increase: toxicity—SSRIs, SNRIs, serotonin-receptor agonists

Increase: anticholinergic effects—anticholinergics

Drug/Herb

- Serotonin syndrome: St. John's wort

Drug/Lab Test

Increase: serum bilirubin, blood glucose, alk phos, LFTs

NURSING CONSIDERATIONS

Assess:

- B/P (lying, standing), pulse q4hr; if systolic B/P drops 20 mm Hg, hold product, notify prescriber; VS q4hr in patients with CV disease
- Blood studies: CBC, leukocytes, differential, cardiac enzymes if patient is receiving long-term therapy
- Hepatic studies: AST, ALT, bilirubin
- Weight weekly; appetite may increase with product
- **ECG** for flattening of T wave, bundle branch block, AV block, dysrhythmias in cardiac patients; product should be discontinued gradually several days before surgery
- **EPS** primarily in geriatric patients: rigidity, dystonia, akathisia
- **Depression:** mood, sensorium, affect, suicidal tendencies, increase in psychiatric symptoms
- **Chronic pain:** location, severity, type before and during treatment, alleviating/aggravating factors
- Urinary retention, constipation; constipation most likely in children, geriatric patients
- **Withdrawal symptoms:** headache, nausea, vomiting, muscle pain, weakness; not usual unless product is discontinued abruptly
- Alcohol consumption; if alcohol is consumed, hold dose until morning
- Assistance with ambulation during beginning therapy because drowsiness/dizziness occurs; safety measures primarily for geriatric patients

Evaluate:

- Therapeutic response: decreased anxiety, depression

Teach patient/family:

- That therapeutic effect (depression) may take 2-3 wk, antianxiety effects sooner
- To use caution when driving, during other activities requiring alertness because of drowsiness, dizziness, blurred vision
- To avoid alcohol, other CNS depressants; may potentiate effects
- Not to discontinue medication abruptly after long-term use; may cause nausea, headache, malaise
- To wear sunscreen or large hat; photosensitivity occurs
- That clinical worsening and suicide may occur
- To immediately report urinary retention

TREATMENT OF OVERDOSE:

ECG monitoring; lavage, activated charcoal; administer anticonvulsant, sodium bicarbonate

DOXOrubicin

(dox-oh-roo'bi-sin)

Adriamycin

Func. class.: Antineoplastic, antibiotic

Chem. class.: Anthracycline glycoside

Do not confuse:
DOXOrubicin /DOXOrubicin liposomal/ DAUNOrubicin

ACTION: Inhibits DNA synthesis primarily; replication is decreased by binding to DNA, which causes strand splitting; active throughout entire cell cycle; a vesicant

USES: Wilms' tumor; bladder, breast, lung, ovarian, stomach, thyroid cancer; Hodgkin's/non-Hodgkin's disease; acute lymphoblastic leukemia; myeloblastic leukemia; neuroblastomas; lymphomas; soft tissue/bone sarcomas

CONTRAINDICATIONS: Pregnancy (D) 1st trimester, breastfeeding, hypersensitivity, systemic infections, cardiac disorders, severe myelosuppression, lifetime dose of 550 mg/m^2

Black Box Warning: Hepatic disease

Precautions: Accidental exposure, cardiac disease, dental work, electrolyte imbalance, infection, hyperuricemia

Black Box Warning: Bone marrow suppression, extravasation, heart failure, secondary malignancy; requires an experienced clinician

DOSAGE AND ROUTES

- **Adult: IV** 60-75 mg/m^2 every 3 wk, or may be used in combination with other antineoplastics with 40-60 mg/m^2 every 21-28 days, max cumulative dose 550 mg/m^2 or 450 mg/m^2 if prior DAUNOrubicin, cyclophosphamide, mediastinal XRT

Hepatic dose

⚠ **Adult: IV** Bilirubin 1.2-3 mg/dl, give 50% of dose; bilirubin 3.1-5 mg/dl, give 25% of dose

Renal dose

⚠ **Adult: IV** CCr $<$ 10 ml/min give 75% of dose

Available forms: Powder for inj 10, 20, 50 mg; inj 2 mg/ml

Administer:

IV route

- Give antiemetic 30-60 min before product to prevent vomiting
- Give allopurinol or sodium bicarbonate to maintain uric acid levels, alkalization of urine
- Use cytotoxic handling procedures: inspect for particulate and discoloration before use

Black Box Warning: Do not give IM, subcut

Black Box Warning: If extravasation occurs, stop inf and complete via another vein, preferably in another limb

- Aluminum needles may be used during administration; avoid aluminum during storage
- Rapid injection can cause facial flushing or erythema along the vein

Reconstitution:

- To avoid risks with reconstitution, the commercially available injection may be used; there are still risks involved in handling the injection
- Do not use diluents containing preservatives to reconstitute powder for injection
- Reconstitute 10, 20, 50, 100 mg of DOXOrubicin with 5, 10, 25, 50 ml, respectively, of nonbacteriostatic NS injection (2 mg/ ml), shake until completely dissolved; use reconstituted solution within 24 hr; do not expose to sunlight
- **IV injection** Inject reconstituted solution over $>$3-5 min via Y-site or 3-way stopcock into a free-flowing IV inf of NS or D_5W; a butterfly needle inserted into a large vein is preferred

Black Box Warning: Care should be taken to avoid extravasation because the drug is extremely irritating to extravascular tissue

- Increased fluid intake to 2-3 L/day to prevent urate, calculi formation
- Store at room temperature for 24 hr after reconstituting

Y-site compatibilities: Alemtuzumab, alfentanil, amifostine, amikacin, anidulafungin, argatroban, aztreonam, bivalirudin, bleomycin, bumetanide, buprenorphine, butorphanol, calcium chloride/gluconate, CARBOplatin, carmustine, caspofungin, ceftizoxime, chlorproMAZINE, cimetidine, ciprofloxacin, CISplatin, cladribine, clindamycin, cyclophosphamide, cycloSPORINE, cytarabine, DACTINomycin, DAPTOmycin, dexamethasone, diltiazem, diphenhydrAMINE, DOBUTamine, DOCEtaxel, dolasetron, DOPamine, doripenem, doxycycline, droperidol, enalaprilat, ePHEDrine, EPINEPHrine, erythromycin, esmolol, etoposide, etoposide phosphate, famotidine, fenoldopam, fentaNYL, filgrastim, fluconazole, fludarabine, gemcitabine, gentamicin, granisetron, haloperidol, hydrocortisone, HYDROmorphone, ifosfamide, imipenem cilastatin, inamrinone, isoproterenol, ketorolac, labetalol, leucovorin, levorphanol, lidocaine, linezolid, LORazepam, mannitol, mechlorethamine, melphalan, meperidine, mesna, methotrexate, metoclopramide, metoprolol, metroNIDAZOLE, midazolam, milrinone, mitoMYcin, morphine, nalbuphine, naloxone, nesiritide, niCARdipine, nitroglycerin, nitroprusside, octreotide, ofloxacin, ondansetron, oxaliplatin, PACLitaxel, palonosetron, pancuronium, phenylephrine, potassium chloride, procainamide, prochlorperazine, promethazine, propranolol, quinupristin-dalfopristin, ranitidine, sargramostim, sodium acetate, tacrolimus, teniposide, theophylline, thiotepa, ticarcillin/clavulanate, tigecycline, tirofiban, tobramycin, topotecan, trastuzumab, trimethobenzamide, vancomycin, vasopressin, vecuronium, verapamil, vinBLAStine, vinCRIStine, vinorelbine, zidovudine, zoledronic acid

SIDE EFFECTS

CV: Increased B/P, sinus tachycardia, PVCs, chest pain, bradycardia, extrasystoles, irreversible cardiomyopathy, acute left ventricular failure
GI: *Nausea, vomiting,* anorexia, *mucositis,* hepatotoxicity
GU: Impotence, sterility, amenorrhea, gynecomastia, hyperuricemia, urine discoloration
HEMA: Thrombocytopenia, leukopenia, anemia
INTEG: *Rash,* necrosis at inj site, dermatitis, reversible *alopecia,* cellulitis, thrombophlebitis at inj site
SYST: Anaphylaxis

PHARMACOKINETICS

Half-life 30 min, terminal 16.5 hr; metabolized by liver; crosses placenta; excreted in urine, bile, breast milk

INTERACTIONS

Increase: life-threatening dysrhythmias—posaconazole, fluconazole, do not use together
Increase: QT prolongation—other drugs that increase QT prolongation
Increase: neutropenia, thrombocytopenia—progesterone
Increase: cardiomyopathy—calcium-channel blockers
Increase: toxicity—other antineoplastics, cycloSPORINE, radiation, mercaptopurine
Increase: hemorrhagic cystitis risk, cardiac toxicity—cyclophosphamide
Increase: effect of phenytoin, fosphenytoin
Increase: DOXOrubicin effect—streptozocin
Decrease: DOXOrubicin effect—PHENObarbital
Decrease: antibody response—live virus vaccine
Decrease: antineoplastic effect—hematopoietic progenitor cell; do not use 24 hr before or after treatment

Decrease: clearance of DOXOrubicin—PACLitaxel

Drug/Lab Test

Increase: uric acid

NURSING CONSIDERATIONS

Assess:

Black Box Warning: **Bone marrow depression:** CBC, differential, platelet count weekly; withhold or reduce dose of product if WBC is <1500/mm^3 or platelet count is <50,000/mm^3; notify prescriber of these results

Renal studies: BUN, serum uric acid, urine CCr, electrolytes before, during therapy

I&O ratio: Report fall in urine output to <30 ml/hr

Monitor temperature: Fever might indicate beginning infection

Black Box Warning: **Hepatotoxicity:** hepatic studies before, during therapy: bilirubin, AST, ALT, alk phos as needed or monthly; check for jaundice of skin and sclera, dark urine, clay-colored stools, itchy skin, abdominal pain, fever, diarrhea

Black Box Warning: **Dysrhythmias:** ECG; watch for ST-T wave changes, low QRS and T, possible dysrhythmias (sinus tachycardia, heart block, PVCs), ejection fraction before treatment, signs of irreversible cardiomyopathy, can occur up to 6 mo after treatment begins

- Bleeding: hematuria, guaiac, bruising, petechiae of mucosa or orifices every 8 hr
- Effects of alopecia on body image; discuss feelings about body changes; almost total alopecia is expected
- Buccal cavity every 8 hr for dryness, sores, ulceration, white patches, oral pain, bleeding, dysphagia
- Alkalosis if severe vomiting is present

Black Box Warning: **Extravasation:** local irritation, pain, burning at inj site; a vesicant; if extravasation occurs, stop drug, restart at another site, apply ice, elevate extremity to reduce swelling; if resolution does not occur, surgical debridement may be required

- GI symptoms: frequency of stools, cramping
- Rinsing of mouth tid-qid with water, club soda; brushing of teeth bid-tid with soft brush or cotton-tipped applicators for stomatitis; use unwaxed dental floss

Evaluate:

- Therapeutic response: decreased tumor size, spread of malignancy

Teach patient/family:

- To add 2-3 L of fluids unless contraindicated before and for 24-48 hr after to decrease possible hemorrhagic cystitis
- To report any complaints, side effects to nurse or prescriber
- That hair may be lost during treatment; that wig or hairpiece might make patient feel better; that new hair might be different in color, texture
- To avoid foods with citric acid, hot or rough texture
- To report any bleeding, white spots, ulcerations in mouth to prescriber; to examine mouth daily
- That urine, other body fluids may be red-orange for 48 hr
- To avoid crowds and persons with infections when granulocyte count is low
- That barrier contraceptive measures are recommended during therapy and for 4 mo after (pregnancy [D]); to avoid breastfeeding
- To avoid vaccinations

DOXOrubicin liposomal

(dox-oh-roo′bi-sin)

Doxil, Lipodex

Func. class.: Antineoplastic, antibiotic

Chem. class.: Anthracycline glycoside

Do not confuse: DOXOrubicin/DOXOrubicin liposomal/DAUNOrubicin

ACTION: Inhibits DNA synthesis primarily; replication is decreased by binding to DNA, which causes strand splitting; active throughout entire cell cycle; a vesicant

USES:
AIDS-related Kaposi's sarcoma, multiple myeloma, metastatic ovarian carcinoma

CONTRAINDICATIONS:
Pregnancy (D), breastfeeding, hypersensitivity, systemic infections, cardiac disorders

Black Box Warning: Cardiotoxicity, inf reactions, myelosuppression, hepatic disease

Precautions: Children, infection, leukopenia, stomatitis, thrombocytopenia

DOSAGE AND ROUTES

Max lifetime cumulative dose 550 mg/m^2; 400 mg/m^2 for those that have received other cardiotoxics or mediastinal radiation

Kaposi's sarcoma

- **Adult:** IV 20 mg/m^2 every 3 wk

Multiple myeloma

- **Adult:** IV 30 mg/m^2 IV inf on day 4 every 3 wk plus bortezomib 1.3 mg/m^2/dose IV bolus on days 1, 4, 8, 11 of each cycle; give DOXOrubicin liposomal after bortezomib receipt on day 4; administer up to 8 treatment cycles or until disease progression or unacceptable toxicity occurs

Ovarian cancer that has progressed or recurred after platinum-based chemotherapy

- **Adult:** IV 50 mg/m^2 every 4 wk; continue treatment for as long as the patient shows benefit and does not progress; a minimum of 4 courses is recommended

Available forms: Liposomal dispersion for inj: 2 mg/ml

Administer:

- Prepared liposomal DOXOrubicin is a translucent, red liposomal dispersion; visually inspect for particulate matter and discoloration before use
- Pegylated liposomal DOXOrubicin (Doxil) is for IV INF use only and should not be given IM/subcut, give under the supervision of a physician who is experienced in cancer chemotherapy

Black Box Warning: Care should be taken to avoid extravasation because the drug is irritating to extravascular tissue

- Premedication with antiemetics is recommended

IV route

- **Reconstitution (Doxil):** dilute the appropriate dose, not to exceed 90 mg/250 ml D_5W; do not mix with any other diluent, drugs, or bacteriostatic agent, use aseptic technique; product contains no preservative or bacteriostatic agent; diluted solution must be refrigerated and used within 24 hr
- **IV INF (Doxil):** do not administer as a bolus injection or an undiluted solution; rapid injection can increase the risk of an inf-related reaction

Black Box Warning: Care should be taken to avoid extravasation because the drug is irritating to extravascular tissue

- An acute inf reaction can occur during the first inf and is usually resolved by slowing the rate of inf; most patients can tolerate subsequent inf
- **Rate:** infuse at an initial rate of 1 mg/min; if no inf-related action, the rate can be increased to complete the inf over 1 hr; do not filter
- **For hematologic toxicity in patients with ovarian cancer or HIV-related Kaposi's sarcoma: Grade 1 (ANC of 1500-1900/mm^3, platelets ≥75,000/mm^3):** No dose reduction; **Grade 2 (ANC of 1000-1499/mm^3, platelets ≥50,000/mm^3 and < 75,000/mm^3:** Wait until ANC ≥1500 cells/mm^3 and platelets ≥75,000 cells/mm^3; redose with no dose reduction; **Grade 3 (ANC of 500-999/mm^3, platelets ≥25,000/mm^3 and <50,000/mm^3):** Wait until ANC ≥1500 cells/mm^3 and platelets ≥75,000 cells/mm^3; redose with no dose reduction; **Grade 4 (ANC <500/mm^3, platelets <25,000/mm^3):** Wait until ANC ≥1500 cells/mm^3 and platelets ≥75,000 cells/mm^3; reduce dose by 25% or continue with full dose with colony-stimulating factor

- Give antiemetic 30-60 min before product to prevent vomiting
- Use allopurinol or sodium bicarbonate to maintain uric acid levels, alkalinization of urine
- Avoid mixing with other products
- Increased fluid intake to 2-3 L/day to prevent urate, calculi formation
- Store refrigerated for 24 hr after reconstituting

SIDE EFFECTS

CNS: Paresthesias, headache, depression, insomnia, fatigue, fever
CV: Chest pain, decreased B/P, cardiomyopathy, heart failure, dysrhythmias, tachycardia
EENT: Optic neuritis, rhinitis, pharyngitis, stomatitis
GI: *Nausea, vomiting,* anorexia, *mucositis,* hepatotoxicity, constipation, oral candidiasis, abdominal pain
HEMA: Thrombocytopenia, leukopenia, anemia, neutropenia
INTEG: *Rash,* necrosis at inj site, dermatitis, reversible *alopecia,* exfoliative dermatitis, palmar-plantar erythrodysesthesia, thrombophlebitis at inj site
RESP: Dyspnea, cough, respiratory infections

PHARMACOKINETICS

Half-life 55 hr; metabolized by liver; crosses placenta; excreted in urine, bile, breast milk

INTERACTIONS

Increase: life-threatening dysrhythmias—posaconazole, fluconazole, do not use together
Increase: QT prolongation—other drugs that increase QT prolongation
Increase: neutropenia, thrombocytopenia—progesterone
Increase: cardiomyopathy—calcium-channel blockers
Increase: toxicity—other antineoplastics, cycloSPORINE, radiation, mercaptopurine
Increase: hemorrhagic cystitis risk, cardiac toxicity—cyclophosphamide
Increase: effect of—phenytoin, fosphenytoin
Increase: DOXOrubicin effect—streptozocin
Decrease: DOXOrubicin effect—PHENObarbital
Decrease: antibody response—live virus vaccine
Decrease: antineoplastic effect—hematopoietic progenitor cell; do not use 24 hr before or after treatment
Decrease: clearance of DOXOrubicin—PACLitaxel

Drug/Lab Test
Increase: uric acid

NURSING CONSIDERATIONS

Assess:

Black Box Warning: **Bone marrow depression:** CBC, differential, platelet count weekly; withhold product if WBC is <4000/mm³ or platelet count is <75,000/mm³; notify prescriber of these results

- Renal studies: BUN, serum uric acid, urine CCr, electrolytes before, during therapy
- I&O ratio: report fall in urine output to <30 ml/hr
- Monitor temperature; fever can indicate beginning infection

Black Box Warning: **Hepatotoxicity:** Hepatic studies before, during therapy: bilirubin, AST, ALT, alk phos as needed or monthly; check for jaundice of skin and sclera, dark urine, clay-colored stools, itchy skin, abdominal pain, fever, diarrhea

Black Box Warning: **Dysrhythmias:** ECG: watch for ST-T wave changes, low QRS and T, possible dysrhythmias (sinus tachycardia, heart block, PVCs), ejection fraction before treatment, signs of irreversible cardiomyopathy, can occur up to 6 mo after treatment begins

- Bleeding: hematuria, guaiac, bruising, petechiae of mucosa or orifices every 8 hr
- Effects of alopecia on body image;

discuss feelings about body changes; almost total alopecia is expected
• Inflammation of mucosa, breaks in skin
• Buccal cavity every 8 hr for dryness, sores, ulceration, white patches, oral pain, bleeding, dysphagia
• Alkalosis if severe vomiting is present

Black Box Warning: Extravasation: Local irritation, pain, burning at inj site; a vesicant; if extravasation occurs, stop drug, restart at another site, apply ice, elevate extremity to reduce swelling; if resolution does not occur, surgical debridement may be required

• GI symptoms: frequency of stools, cramping
• Rinsing of mouth tid-qid with water, club soda; brushing of teeth bid-tid with soft brush or cotton-tipped applicators for stomatitis; use unwaxed dental floss

Evaluate:
• Therapeutic response: decreased tumor size, spread of malignancy

Teach patient/family:
• To add 2-3 L of fluids unless contraindicated before and for 24-48 hr after to decrease possible **hemorrhagic cystitis**
• To report any complaints, side effects to nurse or prescriber
• That hair may be lost during treatment; that wig or hairpiece might make patient feel better; that new hair may be different in color, texture
• To avoid foods with citric acid, hot or rough texture
• To report any bleeding, white spots, ulcerations in mouth to prescriber; to examine mouth daily
• That urine, other body fluids may be red-orange for 48 hr
• To avoid crowds and persons with infections when granulocyte count is low
• That barrier contraceptive measures are recommended during therapy and for 4 mo after (pregnancy [D]); to avoid breastfeeding
• To avoid vaccinations because reactions can occur; to avoid alcohol

doxycycline (Rx)
(dox-i-sye′kleen)

Oracea

doxycycline calcium
Vibramycin

doxycycline hyclate
Adoxa, Apo-Doxy ✱, Doryx, Doxy, Doxycaps, Doxycin ✱, Periostat, Vibramycin, Vibra-Tabs

doxycycline monohydrate
Adoxa, Monodox, Vibramycin

Func. class.: Antiinfective
Chem. class.: Tetracycline

Do not confuse:
doxycycline/doxepin/dicyclomine

ACTION: Inhibits protein synthesis, phosphorylation in microorganisms by binding to 30S ribosomal subunits, reversibly binding to 30S ribosomal subunits; bacteriostatic

USES: *Acinetobacter* sp., *Actinomyces israelii, Bacillus anthracis, Bacteroides* sp., *Balantidium coli, Bartonella bacilliformis, Borrelia recurrentis, Brucella* sp., *Campylobacter fetus, Chlamydia psittaci, Chlamydia trachomatis, Clostridium* sp., *Entamoeba histolytica, Enterobacter aerogenes, Enterococcus* sp., *Escherichia coli, Francisella tularensis, Fusobacterium fusiforme, Haemophilus ducreyi, Haemophilus influenzae* (beta-lactamase negative), *Haemophilus influenzae* (beta-lactamase positive), *Klebsiella granulomatis, Klebsiella* sp., *Leptospira* sp., *Listeria monocytogenes, Mycoplasma pneumoniae, Neisseria gonorrhoeae, Neisseria meningitidis, Orientia tsutsugamushi, Plasmodium falciparum, Propionibacterium acnes, Rickettsia akari, Rickettsia prowazekii, Rickettsia rickettsii, Shigella* sp., *Staphylococcus aureus* (MSSA), *Strep-*

tococcus pneumoniae, Streptococcus pyogenes (group A beta-hemolytic streptococci), *Streptococcus* sp., *Treponema pallidum, Treponema pertenue, Ureaplasma urealyticum, Vibrio cholerae, Viridans streptococci, Yersinia pestis;* syphilis, gonorrhea, lymphogranuloma venereum, uncommon gram-negative/gram-positive organisms, malaria prophylaxis

Unlabeled uses: Enterocolitis, biliary tract, intraabdominal infections; epididymitis *(Chlamydia trachomatis);* chronic prostatitis *(Ureaplasma urealyticum);* traveler's diarrhea (enterotoxigenic *Escherichia coli*); Legionnaire's disease *(Legionella pneumophila);* Lyme disease *(Borrelia burgdorferi),* Lyme disease (erythema migrans); Lyme arthritis; Lyme carditis; pleural effusion; malaria (chloroquine-resistant *Plasmodium falciparum*); pelvic inflammatory disease (PID), tubo-ovarian abscess in combination; acute dental infection, dentoalveolar infection, endodontic infection; aggressive juvenile periodontitis, plaque prophylaxis *(Yersinia pestis);* tularemia prophylaxis *(Francisella tularensis);* Bancroft's filariasis (elephantiasis) *(Wuchereria bancrofti):* melioidosis due to *Burkholderia pseudomallei; Burkholderia pseudomallei* bacteremia melioidosis; leptospirosis (*Leptospira* sp); infection prophylaxis for gynecologic procedures/surgical infection prophylaxis hysterosalpingogram or chromotubation/induced abortion/dilation and evacuation; methicillin-resistant *Staphylococcus aureus* (MRSA)-associated bone and joint infections

CONTRAINDICATIONS:

Pregnancy (D), children <8 yr, hypersensitivity to tetracyclines, esophageal ulceration

Precautions: Breastfeeding, hepatic disease, pseudomembranous colitis, ulcerative colitis, sulfite hypersensitivity, excessive sunlight

DOSAGE AND ROUTES

Most infections

- **Adult: PO/IV** 100 mg q12hr on day 1 then 100 mg/day; **IV** 200 mg in 1-2 inf on day 1 then 100-200 mg/day
- **Child >8 yr, ≥45 kg: PO** 100 mg q12hr on day 1 then 100 mg daily; severe infections 100 mg q12hr; **IV** 200 mg on day 1 then 100-200 mg daily, give 200 mg dose as 1 or 2 inf
- **Child ≥8 yr, <45 kg: PO** 2.2 mg/kg q12hr on day 1 then 2.2 mg/kg daily, severe infections 2.2 mg/kg q12hr; **IV** 4.4 mg/kg on day 1 then 2.2-4.4 mg/kg daily in 1 to 2 divided doses

Gonorrhea (uncomplicated) in patients allergic to penicillin

- **Adult: PO** 100 mg q12hr × 7 days or 300 mg followed 1 hr later by another 300 mg

Malaria prophylaxis

- **Adult: PO** 100 mg/day 1-2 days before travel, daily during travel, and for 4 wk after return
- **Adolescent/child ≥8 yr, <45 kg: PO** 2 mg/kg/day (up to 100 mg/day) begin 1-2 days before travel, continue for 4 wk after return

C. trachomatis

- **Adult: PO** 100 mg bid × 7 days

Syphilis

- **Adult: PO** 100 mg bid × 14 days

Anthrax

- **Adult and child >8 yr and ≥45 kg: IV** 100 mg q12hr; change to **PO** when able × 60 days
- **Adolescent/child ≥8 yr and <45 kg: PO** 2.2 mg/kg q12hr × 60 days; **IV** 100 mg q12hr, change to **PO** when able × 60 days

Lyme disease

- **Adult/adolescent/child ≥8 yr: PO** 100 mg bid × 10-21 days

Periodontitis

- **Adult:** 20 mg bid after scaling and root planing for ≤9 mo; give close to meal time AM or PM

Pleural effusion (unlabeled)

• **Adult: INTRACAVITARY** 500 mg diluted with 250 ml 0.9% NaCl given by chest tube lavage and drainage

Available forms: Doxycycline: cap 40 mg; doxycycline calcium: susp 50 mg/5 ml; doxycycline hyclate: cap 50, 100 mg; del rel tabs 75, 100, 150 mg; del rel cap 75, 100 mg; inj 42.5, 100, 200 mg; tabs 20, 100 mg; doxycycline monohydrate: caps 50, 100, 150 mg; tabs 50, 75, 100 mg; oral susp 25 mg/5 ml

Administer:

PO route

• Do not break, crush, or chew caps; may crush tabs and mix with food

• On empty stomach or with full glass of water 2 hr before or after meals; avoid dairy products, antacids, laxatives, iron-containing products; if these must be taken, give 2 hr before or after product; avoid giving oral products within 1 hr of bedtime, esophageal uleration may occur

• **Delayed release cap:** Swallow whole or open and sprinkle on applesauce

• **Susp:** Shake well, use calibrated device, may give with food/milk for GI irritation

Intermittent IV INF route

• After diluting 100 mg or less/10 ml or 200 mg/20 ml of sterile water or NS for inj, each 100 mg must be further diluted with 100-1000 ml of NaCl, D_5, Ringer's, LR D_5LR, Normosol-M, Normosol-R in D_5W; run 100 mg or less over 1-4 hr; inf must be completed in 6 hr when diluted in LR sol or 12 hr with other sol; protect from light, heat

• Avoid rapid use, extravasation

• Store in tight, light-resistant container at room temperature; IV stable for 12 hr at room temperature, 72 hr refrigerated; discard if precipitate forms

Y-site compatibilities: Acyclovir, alemtuzumab, alfentanil, amifostine, amikacin, aminophylline, amiodarone, anidulafungin, ascorbic acid, atracurium, atropine, aztreonam, bivalirudin, bumetanide, buprenorphine, butorphanol, calcium chloride/gluconate, CARBOplatin, caspofungin, cefonicid, cefotaxime, cefTRIAXone, chlorproMAZINE, cimetidine, cisatracurium, CISplatin, clindamycin, codeine, cyanocobalamin, cyclophosphamide, cycloSPORINE, cytarabine, DACTINomycin, DAPTOmycin, dexmedetomidine, digoxin, diltiazem, diphenhydrAMINE, DOBUTamine, DOCEtaxel, DOPamine, doxacurium, DOXOrubicin, enalaprilat, ePHEDrine, EPINEPHrine, epirubicin, epoetin alfa, eptifibatide, ertapenem, esmolol, etoposide, etoposide phosphate, famotidine, fenoldopam, fentaNYL, filgrastim, fluconazole, fludarabine, gemcitabine, gemtuzumab, gentamicin, glycopyrrolate, granisetron, HYDROmorphone, IDArubicin, ifosfamide, imipenem/cilastatin, insulin, isoproterenol, labetalol, levofloxacin, lidocaine, linezolid, LORazepam, magnesium sulfate, mannitol, mechlorethamine, melphalan, meperidine, metaraminol, methoxamine, methyldopate, metoclopramide, metoprolol, metroNIDAZOLE, miconazole, midazolam, milrinone, mitoXANtrone, morphine, multivitamins, nalbuphine, naloxone, nesiritide, netilmicin, nitroglycerin, nitroprusside, norepinephrine, octreotide, ondansetron, oxaliplatin, oxytocin, PACLitaxel, pancuronium, pantoprazole, papaverine, pentamidine, pentazocine, perphenazine, phentolamine, phenylephrine, phytonadione, potassium chloride, procainamide, prochlorperazine, promethazine, propofol, propranolol, protamine, pyridoxime, quinupristin/dalfopristin, ranitidine, remifentanil, ritodrine, riTUXimab, rocuronium, sargramostim, sodium acetate, succinylcholine, SUFentanil, tacrolimus, telavancin, teniposide, theophylline, thiamine, thiotepa, tirofiban, tobramycin, tolazoline, TPN (2 in 1), trastuzumab, trimetaphan, urokinase, vancomycin, vasopressin, vecuronium, verapamil, vinCRIStine, vinorelbine, voriconazole, zoledronic acid

SIDE EFFECTS

CNS: Fever, headache

CV: Pericarditis

EENT: Dysphagia, glossitis, decreased calcification of deciduous teeth, oral candidiasis, tooth discoloration
GI: *Nausea, abdominal pain, vomiting, diarrhea,* anorexia, enterocolitis, **hepatotoxicity**, flatulence, abdominal cramps, gastric burning, stomatitis
GU: *Increased BUN*
HEMA: **Eosinophilia, neutropenia, thrombocytopenia, hemolytic anemia**
INTEG: *Rash, urticaria, photosensitivity, increased pigmentation,* **exfoliative dermatitis,** pruritus, phlebitis, injection, site reaction
MS: Bone growth retardation (<8 yr old), muscle, joint pain
RESP: Cough
SYST: **Stevens-Johnson syndrome, angioedema, anaphylaxis,** toxic epidermal necrolysis

PHARMACOKINETICS

PO: Well absorbed; widely distributed; peak $1^1/_2$-4 hr; half-life 14-17 hr; excreted in urine, feces, bile; 90% protein bound; crosses placenta; enters breast milk

INTERACTIONS

Increase: effect of—anticoagulants, digoxin, methotrexate
Decrease: doxycycline effect—antacids, $NaHCO_3$, dairy products, alkali products, iron, kaolin/pectin, barbiturates, carBAMazepine, phenytoin, cimetidine sucralfate, cholestyramine, colestipol, rifampin, bismuth; iron, magnesium, zinc, calcium, aluminum salts, Sevelamer
Decrease: effects—penicillins, oral contraceptives, digoxin

Drug/Lab Test
Increase: BUN, alk phos, bilirubin, amylase, ALT, AST, eosinophils, WBC
Decrease: Hgb
False increase: urinary catecholamines

NURSING CONSIDERATIONS

Assess:
- I&O ratio
- Blood studies: PT, CBC, AST, ALT, BUN, creatinine
- Signs of infection
- **Allergic reactions: rash, itching, pruritus, angioedema**
- Nausea, vomiting, diarrhea; administer antiemetic, antacids as ordered
- **Overgrowth of infection: fever, malaise, redness, pain, swelling, drainage, perineal itching, diarrhea, changes in cough or sputum**
- IV site for phlebitis/thrombosis; product is highly irritating
- After C&S is obtained, do not wait for results

Evaluate:
- Therapeutic response: decreased temperature, absence of lesions, negative C&S

Teach patient/family:
- To avoid sun because burns may occur; that sunscreen does not seem to decrease photosensitivity
- **That all prescribed medication must be taken to prevent superinfection; not to use outdated products because Fanconi syndrome may occur (reversible nephrotoxicity)**
- **That if children ≤8 yr old are undergoing tooth development, teeth will be permanently discolored**
- Not to use with antacids, iron products, H2 blockers, Sevelamer
- To take with full glass of water; if nausea occurs take with food

⚠ HIGH ALERT

dronedarone (Rx)

(drone′da′rone)

Multaq

Func. class.: Antidysrhythmic (class III)
Chem. class.: Iodinated benzofuran derivative

ACTION: Prolongs duration of action potential and effective refractory period, noncompetitive α- and β-adrenergic inhibition; increases RR and QT intervals, decreases sinus rate, decreases peripheral vascular resistance

USES: Atrial fibrillation, atrial flutter

CONTRAINDICATIONS: Pregnancy (X), breastfeeding; 2nd-, 3rd-degree AV block; bradycardia, severe sinus node dysfunction, hypersensitivity, heart failure, hepatic disease, QT prolongation, aminodarone-induced lung/liver toxicity

Black Box Warning: NYHA Class IV heart failure or Class II-III with recent decompensation requiring hospitalization, permanent atrial fibrillation (cannot restore sinus rhythm)

Precautions: Children, geriatric patients, Asian patients, females, electrolyte imbalances, atrial fibrillation/flutter

DOSAGE AND ROUTES

- **Adult: PO** 400 mg bid; discontinue class I, III antidysrhythmics or strong CYP3A4 inhibitors before beginning treatment; max 800 mg/day

Available forms: Tabs 400 mg

Administer:

PO route

- Give bid with morning, evening meals
- Give MedGuide; should be dispensed with each prescription, refill

SIDE EFFECTS

CNS: Weakness

CV: *Bradycardia,* heart failure, QT prolongation, torsades de pointes, atrial flutter

ENDO: Hypo/hyperthyroidism

GI: Nausea, vomiting, diarrhea, abdominal pain, severe hepatic injury, hepatic failure

INTEG: Rash, photosensitivity, anaphylaxis, angioedema

RESP: Interstitial pneumonitis, pulmonary fibrosis

PHARMACOKINETICS

Peak 3-6 hr, half-life 13-19 hr, metabolized by liver, excreted in feces (84%), via kidneys (6%), protein binding >98%

INTERACTIONS

Increase: dronedarone levels: CYP3A inhibitors/2D6 inhibitors

Decrease: dronedarone levels: 3A/2D6 inducers

Increase: 3A/2D6 substrate levels

Increase: bradycardia—β-blockers, calcium channel blockers

Increase: levels of cycloSPORINE, dextromethorphan, digoxin, disopyramide, flecainide, methotrexate, phenytoin, procainamide, quiNIDine, theophylline

Increase: anticoagulant effects—dabigatran, warfarin

Drug/Herb

Increase: anticoagulant effect—yohimbine

Decrease: dronedarone effect—St. John's wort

Drug/Food:

Increase: dronedarone effect, grapefruit; avoid use

Drug/Lab Test

Increase: T_4, creatinine, LFTs, bilirubin

Decrease: potassium, magnesium

NURSING CONSIDERATIONS

Assess:

Black Box Warning: NYHA Class IV heart failure or symptomatic heart failure with recent decomposition requiring hospitalization doubles risk of death

- **ECG** to determine product effectiveness; measure PR, QRS, QT intervals; check for PVCs, other dysrhythmias, B/P continuously for hypo/hypertension; report dysrhythmias, slowing heart rate
- Serum creatinine, potassium, magnesium
- I&O ratio; electrolytes (potassium, creatinine, magnesium)
- Dehydration or hypovolemia
- Rebound hypertension after 1-2 hr
- **Hypothyroidism:** lethargy, dizziness, constipation, enlarged thyroid gland, edema of extremities; cool, pale skin
- **Hyperthyroidism:** restlessness, tachycardia, eyelid puffiness, weight loss, frequent urination, menstrual irregularities, dyspnea; warm, moist skin
- Cardiac rate, respiration: rate, rhythm, character, chest pain; start with patient hospitalized and monitored up to 1 wk

Evaluate:

- Therapeutic response: atrial fibrillation, flutter

Teach patient/family:

- To take this product as directed; to avoid missed doses; not to use with grapefruit juice, to avoid all other products without approval of provider
- To immediately report weight gain, edema, difficulty breathing
- To use effective contraception during treatment (pregnancy [X]) not to breastfeed

TREATMENT OF OVERDOSE:

O_2, artificial ventilation, ECG, administer DOPamine for circulatory depression; administer diazepam or thiopental for seizures, isoproterenol

HIGH ALERT

droperidol (Rx)

(droe-per'i-dole)

Func. class.: Sedative/hypnotic

Chem. class.: Butyrophenone

ACTION: Acts on CNS at subcortical levels, produces tranquilization, sleep; antiemetic; mild α-blockade

USES: Premedication for surgery; induction, maintenance in general anesthesia; postoperatively for nausea, vomiting

Unlabeled uses: Anxiety, general anesthesia induction/maintenance, preanesthesia, sedation induction

CONTRAINDICATIONS: Breastfeeding, children <2 yr, hypersensitivity

Black Box Warning: QT prolongation, torsades de pointes

Precautions: Pregnancy (C), geriatric patients, CV disease (hypotension, bradydysrhythmias), renal/hepatic disease, Parkinson's disease, pheochromocytoma, CHF, hypokalemia, hypomagnesemia, cardiac hypertrophy

DOSAGE AND ROUTES

Induction, adjunct

- **Adult: IV/IM** 1.25-2.5 mg, may give additional 1.25 mg with caution
- **Child 2-12 yr: IV** 0.05-0.1 mg/kg, titrate to response

Premedication

- **Adult: IM** 2.5 mg ½-1 hr before surgery, may give 1.25-2.5 mg additionally
- **Child 2-12 yr: IM** 0.05-0.1 mg/kg

Available forms: Inj 2.5 mg/ml

Administer:

- Protect sol from light
- Anticholinergics (benztropine, diphenhydrAMINE) for EPS
- Only with crash cart, resuscitative equipment nearby
- IM deep in large muscle mass

Direct IV route

- Undiluted; give through Y-tube at 10 mg or less/min; titrate to patient response

Intermittent IV INF route

- May be given as inf by adding dose to 250 ml LR, D_5W, 0.9% NaCl; give slowly, titrate to patient response

Syringe compatibilities: Atropine, bleomycin, butorphanol, chlorproMAZINE, cimetidine, CISplatin, cyclophosphamide, dimenhyDRINATE, diphenhydrAMINE, DOXOrubicin, fentaNYL, glycopyrrolate, hydrOXYzine, meperidine, metoclopramide, midazolam, mitoMYcin, morphine, nalbuphine, pentazocine, perphenazine, prochlorperazine, promazine, promethazine, scopolamine, vinBLAStine, vinCRIStine

Y-site compatibilities: Amifostine, aztreonam, bleomycin, cisatracurium, CISplatin, cladribine, cyclophosphamide, cytarabine, DOXOrubicin, DOXOrubicin liposome, famotidine, filgrastim, fluconazole, fludarabine, granisetron, hydrocortisone, IDArubicin, melphalan, meperidine, metoclopramide, mitoMYcin, ondansetron, PACLitaxel, potassium chloride, propofol, remifentanil, sargramostim, teniposide, thiotepa, vinBLAStine, vinCRIStine, vinorelbine, vit B/C

SIDE EFFECTS

CNS: EPS (dystonia, akathisia, flexion of arms, fine tremors); dizziness, anxiety, drowsiness, restlessness, hallucination, depression, seizures, neuroleptic malignant syndrome
CV: *Tachycardia, hypotension,* prolonged QT, torsades de pointes
EENT: Upward rotation of eyes, oculogyric crisis
INTEG: *Chills, facial sweating, shivering*
RESP: Laryngospasm, bronchospasm

PHARMACOKINETICS

IM/IV: Onset 3-10 min, peak ½ hr, duration 3-6 hr, metabolized in liver, excreted in urine as metabolites, crosses placenta, half-life 2-3 hr

INTERACTIONS

Increase: CNS depression—alcohol, opiates, barbiturates, antihistamines, antipsychotics, or other CNS depressants
Increase: hypotension—nitrates, antihypertensives
Increase: side effects of lithium

Black Box Warning: Increase: QT prolongation—class IA/III antiarrhythmics, some phenothiazines, tricyclics, some quinolones, others

Drug/Herb
Increase: action—kava, valerian, hops, chamomile

NURSING CONSIDERATIONS

Assess:
- VS q10min during IV administration, q30min after IM dose
- EPS: dystonia, akathisia

⚠ For increasing heart rate or decreasing B/P, notify prescriber at once; do not place patient in Trendelenburg position because sympathetic blockade may occur, thereby causing respiratory arrest

Black Box Warning: QT prolongation, torsades de pointes: ECG before and 2-3 hr after administration for serious arrhythmias

Evaluate:
- Therapeutic response: decreased anxiety, absence of vomiting during and after surgery

Teach patient/family:
- To rise slowly from sitting or standing to minimize orthostatic hypotension
- To avoid ambulation without assistance; drowsiness, dizziness may occur

DULoxetine (Rx)

(du-lox′uh-teen)

Cymbalta

Func. class.: Antidepressant
Chem. class.: Serotonin-norepinephrine reuptake inhibitor (SNRI)

ACTION: May potentiate serotonergic, noradrenergic activity in the CNS; in studies, DULoxetine is a potent inhibitor of neuronal serotonin and norepinephrine reuptake

USES: Major depressive disorder (MDD), neuropathic pain associated with diabetic neuropathy, generalized anxiety disorder, fibromyalgia, chronic low back pain, osteoarthritis pain
Unlabeled uses: Stress, urinary incontinence

CONTRAINDICATIONS: Alcohol intoxication, alcoholism, closed-angle glaucoma, hepatic disease, hepatitis, jaundice, hypersensitivity
Precautions: Pregnancy (C), breastfeeding, geriatric patients, mania, hypertension, renal/cardiac disease, seizures, increased intraocular pressure, anorexia nervosa, bleeding, dehydration, diabetes, hyponatremia, hypotension, hypovolemia, orthostatic hypotension, abrupt product withdrawal

Black Box Warning: Children, suicidal ideation

DOSAGE AND ROUTES

Depression

- **Adult: PO** 40-60 mg/day as single dose or 2 divided doses

Diabetic neuropathy

- **Adult: PO** 60 mg/day

Generalized anxiety disorder

- **Adult: PO** 60 mg/day, may start with 30 mg/day × 1 wk then increase to 60 mg/day; maintenance 60-120 mg/day

Fibromyalgia

- **Adult: PO** 30 mg/day × 1 wk then 60 mg/day

Musculoskeletal pain

- **Adult: PO** 60 mg/day or 30 mg/day × 1 wk then 60 mg/day

Renal dose

- **Adult: PO** Start with 20 mg, gradually increase; avoid use in severe renal disease

Available forms: Caps 20, 30, 60 mg

Administer:

- Swallow cap whole; do not break, crush, or chew; do not sprinkle on food or mix with liquid
- Without regard to food
- Store in tight container at room temperature; do not freeze

SIDE EFFECTS

CNS: Insomnia, anxiety, dizziness, tremor, somnolence, fatigue, decreased appetite, decreased weight, agitation, diaphoresis, hallucinations, neuroleptic malignant–like syndrome reaction, aggression, seizures, *headache,* abnormal dreams, flushing, hot flashes, chills

CV: Thrombophlebitis, peripheral edema, hypertension, palpitations, supraventricular dysrhythmia, orthostatic hypotension

EENT: *Abnormal vision*

ENDO: Hypo/hyperglycemia, SIADH

GI: Constipation, diarrhea, dysphagia, *nausea,* vomiting, anorexia, dry mouth, colitis, gastritis, abdominal pain, hepatic failure

GU: Abnormal ejaculation, urinary hesitation/retention/frequency, ejaculation delayed, erectile dysfunction, gynecologic bleeding

INTEG: Photosensitivity, bruising, sweating, Stevens-Johnson syndrome

MS: Gait disturbance, muscle spasm, restless leg syndrome, myalgia

SYST: Anaphylaxis, angioedema, serotonin syndrome, Stevens-Johnson syndrome

PHARMACOKINETICS

Well absorbed; extensively metabolized (CYP2D6, CYP1A2) in the liver to an active metabolite; 70% of product recovered in urine, 20% in feces; 90% protein binding; elimination half-life 9.2-19.1 hr

INTERACTIONS

- Do not use with linezolid or methylene blue IV
- Narrow therapeutic index: CYP2D6 extensively metabolized products (flecainide, phenothiazines, propafenone, tricyclics, thioridazine)

⚠ Hyperthermia, rigidity, rapid fluctuations of vital signs, mental status changes, neuroleptic malignant syndrome—MAOIs, coadministration contraindicated within 14 days of MAOI use

Increase: CNS depression—opioids, antihistamines, sedative/hypnotics

Increase: serotonin syndrome, neuroleptic malignant syndrome—SSRIs, serotonin-receptor agonists

Increase: bleeding risk—anticoagulants, antiplatelets, salicylates, NSAIDs

Increase: action of DULoxetine—CYP1A2 inhibitors (fluvoxamine, quinolone anti-infectives); CYP2D6 inhibitors (FLUoxetine, quiNIDine, PARoxetine)

Increase: ALT, bilirubin—alcohol

Drug/Herb

- Serotonin syndrome: St. John's wort

Increase: CNS depression—kava, valerian

Drug/Lab Test

Increase: blood glucose

NURSING CONSIDERATIONS

Assess:

Black Box Warning: Depression: mood, sensorium, affect, **suicidal tendencies,** increase in psychiatric symptoms; depression, panic, monitor children weekly face to face during first 4 wk, or dosage change, then every other wk for next 4 wk, then at 12 wk

- B/P lying, standing; pulse q4hr; if systolic B/P drops 20 mm Hg, hold product, notify prescriber; take VS q4hr in patients with CV disease
- Hepatic studies: AST, ALT, bilirubin
- Weight weekly; weight loss or gain; appetite may increase; peripheral edema may occur
- Sugarless gum, hard candy, frequent sips of water for dry mouth
- **Withdrawal symptoms:** headache, nausea, vomiting, muscle pain, weakness; not common unless product is discontinued abruptly

⚠ **Malignant neuroleptic-like syndrome reaction**

⚠ **Serotonin syndrome:** nausea/vomiting, dizziness, facial flush, shivering, sweating

- **Sexual dysfunction:** ejaculation dysfunction, erectile dysfunction, decreased libido, orgasm dysfunction
- Assistance with ambulation during beginning therapy; drowsiness, dizziness occur

Evaluate:

- Therapeutic response: decreased depression

Teach patient/family:

- To report urinary retention; about signs and symptoms of bleeding (GI bleeding, nosebleed, ecchymoses, bruising)
- To use with caution when driving, performing other activities requiring alertness because of drowsiness, dizziness, blurred vision
- To avoid alcohol ingestion, MAOIs, other CNS depressants
- Not to discontinue medication quickly after long-term use; may cause nausea, headache, malaise; taper

Black Box Warning: That clinical worsening and suicide risk may occur

- To wear sunscreen or large hat; photosensitivity may occur
- To notify prescriber if pregnancy planned or suspected, or if breastfeeding
- Improvement may occur in 4-8 wk or in up to 12 wk (geriatric patients)

dutasteride (Rx)

(doo-tass′ter-ide)

Avodart

Func. class.: Androgen inhibitor

Chem. class.: Synthetic 5α-reductase inhibitor, 4-azasteroid compound

ACTION: Inhibits both type 1 and type 2 forms of a steroid enzyme that converts testosterone to 5α-dihydrotestosterone (DHT), which is responsible for the initial growth of prostatic tissue

USES: Treatment of benign prostatic hyperplasia (BPH) in men with an enlarged prostate gland; may be used in combination with tamsulosin

Unlabeled uses: Alopecia

CONTRAINDICATIONS: Pregnancy (X), breastfeeding, women, children, hypersensitivity

Precautions: Hepatic disease

DOSAGE AND ROUTES

Benign prostatic hyperplasia (BPH)

- **Adult: PO** 0.5 mg/day

Alopecia (unlabeled)

- **Adult: PO** 0.5-2.5 mg/day

Available form: Caps 0.5 mg

Administer:

- Swallow caps whole; do not break, crush, chew
- Without regard to meals

SIDE EFFECTS

GU: Decreased libido, impotence, gynecomastia, ejaculation disorders (rare), mastalgia, teratogenesis

INTEG: Serious skin infections

PHARMACOKINETICS

Peak 2-3 hr, protein binding 99%, metabolized in liver by CYP3A4, excreted in feces, half-life 5 wk at steady state

INTERACTIONS

Increase: dutasteride concs—ritonavir, ketoconazole, verapamil, diltiazem, cimetidine, ciprofloxacin, antiretroviral protease inhibitors, or other products metabolized by CYP3A4

Drug/Lab Test

Decrease: PSA

NURSING CONSIDERATIONS

Assess:

- **For decreasing symptoms of BPH:** decreasing urinary retention, frequency, urgency, nocturia
- PSA levels; digital rectal, urinary obstruction; determine the absence of urinary cancer before starting treatment
- Blood studies: ALT, AST, bilirubin, CBC with differential, serum creatinine, serum electrolytes

Evaluate:

- Therapeutic response: decreasing symptoms of BPH; decreased urinary retention, frequency, urgency, nocturia

Teach patient/family:

- To read patient information leaflet before starting therapy; to reread it upon prescription renewal
- To notify prescriber if therapeutic response decreases, if edema occurs
- Not to discontinue product abruptly
- About changes in sex characteristics
- That men taking dutasteride should not donate blood for at least 6 mo after last dose to prevent blood administration to pregnant female
- That caps should not be handled by a woman who is pregnant or who may become pregnant because product can be absorbed through skin
- That ejaculate volume may decrease during treatment; that product rarely interferes with sexual function
- Should not be used or handled by breastfeeding women
- Swallow whole; do not crush, chew or open
- May increase risk for developing high-grade prostate cancer

ecallantide

(ee-kal'an-tide)

Kalbitor

Func. class.: Hematological agents

Chem. class.: Kallikrein inhibitor

ACTION: Selectively, reversibly inhibits kallikrein within the inflammatory pathways, preventing excess production of bradykinin

USES: Acute attacks of hereditary angioedema (≥16 yr)

CONTRAINDICATIONS: Hypersensitivity to the drug or its components

Black Box Warning: Anaphylaxis has occurred after administrations (usually within first hour after dosing); drug should be administered only by health care provider with medical support available to treat anaphylaxis and hereditary angioedema; monitor patient closely

Precautions: Pregnancy (C), breastfeeding, geriatric patients

DOSAGE AND ROUTES

- **Adult/adolescent ≥16 yr: SUBCUT** 30 mg given as three 10-mg injections; give additional 30-mg dose within 24 hours if attack persists

Available forms: Solution 10 mg/ml

Administer:

- Using aseptic technique, withdraw 1 ml (10 mg) using a large bore needle; change the needle on the syringe to a needle suitable for subcut inj (27 gauge recommended)
- Inject into the skin of the abdomen, thigh, or upper arm
- Repeat the procedure for each of the three vials (30 mg total dose); the injection site for each of the three injections may be in the same or different anatomic location (abdomen, thigh, or upper arm); there is no need for site rotation; individual injections should be separated by at least 2 inches and away from the anatomical site of attack
- The same directions for administration apply if an additional dose is required within 24 hr; different injection sites or the same anatomical location as used for the first dose may be used
- Visually inspect parenteral products for particulate matter and discoloration prior to administration

SIDE EFFECTS

CNS: Fatigue, fever, headache

EENT: Nasopharyngitis

GI: Abdominal pain, diarrhea, nausea, vomiting

INTEG: Injection-site reactions, pruritus, rash, urticaria

RESP: Upper respiratory tract infection

Black Box Warning: Assess: anaphylaxis; usually within first hr after dosing

NURSING CONSIDERATIONS

Teach patient/family:

- To immediately report wheezing, cough, chest tightness, trouble breathing, dizziness, fainting, throat tightness, itchiness, hives, and swelling of tongue or throat.
- Inform patient that drug must be given by health care provider in a health care setting in case serious allergic reaction occurs

econazole

(ee-koe'na-zole)

Func. class.: Topical antifungal

Chem. class.: Imidazole derivative

ACTION: Antifungal activity results from inhibiting cell wall permeability

USES: Tinea corporis, cruris, pedis, versicolor; cutaneous candidiasis

CONTRAINDICATIONS: Hypersensitivity

Precautions: Pregnancy (C)

DOSAGE AND ROUTES

Tinea corporis, cruris, pedis, versicolor

• **Adult/child: TOP** rub into affected areas daily × 2 wk, or × 4 wk (pedis)

Cutaneous candidiasis

• **Adult/child:** rub into affected areas bid × 2 wk

Available forms: Topical cream 1%

Administer:

Topical route

• Do not use products near the eyes, nose, or mouth

• Wash hands before and after use; wash affected area and gently pat dry

• **Cream:** apply to the cleansed affected area; massage gently into affected areas

SIDE EFFECTS

INTEG: Burning, rash, pruritus, erythema

PHARMACOKINETICS

Unknown

NURSING CONSIDERATIONS

Assess:

• **Allergic reaction**: assess for hypersensitivity, product might need to be discontinued

• **Infection**: assess for itching, peeling

Evaluate:

• Decreased itching, peeling

Teach patient/family:

Topical route

• That product is for external use only; do not use skin products near the eyes, nose, or mouth

• To wash hands before and after use; to wash affected area and gently pat dry

• **Cream:** to apply a thin film to the cleansed affected area; massage gently into affected areas

econazole topical

See Appendix B

ecothiophate ophthalmic

See Appendix B

RARELY USED

edetate calcium disodium (Rx)

(ee′de-tate)

Calcium Disodium Versenate

Func. class.: Heavy metal antagonist (antidote)

Do not confuse:

edetate calcium disodium/edetate disodium

USES: Lead poisoning, acute lead encephalopathy

CONTRAINDICATIONS: Hypersensitivity, anuria, poisoning of other metals, severe renal disease, hepatitis

Black Box Warning: Child <3 yr

DOSAGE AND ROUTES

Lead mobilization test (lead toxicity 25-45 mcg/dl)

• **Adult/adolescent: IV INF** 500 mg/m^2 over 1 hr or **IM**

• **Child: IV INF** 500 mg/m^2 over 1 hr or **IM** as single dose or 2 divided doses

Acute lead encephalopathy (blood levels >70 mcg/dl)

• **Adult/adolescent/child/infant: IM/IV** 1500 mg/m^2 as **IV INF** over 12-24 hr in combination with dimercaprol **IM**, give 1st dose ≥4 hr after initial dimercaprol, when urine flow established

efavirenz (Rx)

(ef-ah-veer′enz)

Sustiva

Func. class.: Antiretroviral

Chem. class.: Nonnucleoside reverse transcriptase inhibitor (NNRTI)

ACTION: Binds directly to reverse transcriptase and blocks RNA, DNA polymerase, thus causing a disruption of the enzyme's site

USES: HIV-1 in combination with at least 2 other antivirals

Unlabeled uses: HIV prophylaxis

CONTRAINDICATIONS: Pregnancy (D), hypersensitivity

Precautions: Breastfeeding, children <3 yr, renal/hepatic disease, myelosuppression, depression, seizures

DOSAGE AND ROUTES

Given in combination with protease inhibitor or nucleoside analog reverse transcriptase inhibitors (NARTIs)

- **Adult and child >40 kg: PO** 600 mg/day at bedtime
- **Child ≥3 mo, 32.5-39.9 kg: PO** 400 mg/day at bedtime
- **Child ≥3 mo, 25-32.4 kg: PO** 350 mg/day at bedtime
- **Child ≥3 mo, 20-24.9 kg: PO** 300 mg/day at bedtime
- **Child ≥3 mo, 15-19.9 kg: PO** 250 mg/day at bedtime
- **Child ≥3 mo, 10-14.9 kg: PO** 200 mg/day at bedtime
- **Child ≥3 mo, 5-<7.5 kg: PO** 150 mg/day at bedtime
- **Child 3.5-≤5 kg:** 100 mg/day at bedtime

Available forms: Caps 50, 100, 200 mg; 600-mg tabs

Administer:

- Give on empty stomach; give at bedtime to decrease CNS side effects
- Caps may be opened, added to grape jelly to disguise peppery taste

SIDE EFFECTS

CNS: Fatigue, impaired cognition, insomnia, abnormal dreams, depression, headache, dizziness, anxiety, drowsiness

GI: *Diarrhea,* abdominal pain, *nausea,* hyperlipidemia, constipation, increased LFTs

GU: Hematuria, kidney stones

INTEG: Rash, erythema multiforme, Stevens-Johnson syndrome, toxic epidermal necrolysis, exfoliative dermatitis

PHARMACOKINETICS

Peak 3-5 hr, well absorbed, metabolized by liver; terminal half-life 52-76 hr; >99% protein binding, excreted in urine, feces; concs higher in females and those of African, Asian, and Hispanic descent

INTERACTIONS

- Avoid use with boceprevir, delavirdine, rilpivirine; dosage change may be needed if given with telaprevir
- Do not give together with benzodiazepines, ergots, midazolam, triazolam, CISapride

Increase: CNS depression—alcohol, antidepressants, antihistamines, opioids

Increase: levels of both products—ritonavir, estrogens, anticonvulsants

Increase: levels of warfarin, statins (except pravastatin, fluvastatin)

Decrease: levels of indinavir, amprenavir, lopinavir, oral contraceptives, nonoral contraceptives, ketoconazole, itraconazole, posaconazole, voriconazole, saquinavir, clarithromycin, methadone

Decrease: efavirenz metabolism—CYP3A4 inhibitors (conivaptan, ambrisentan, SORAfenib)

Decrease: efavirenz effect—CYP3A4 inducers (carBAMazepine, rifamycins)

Drug/Herb

Decrease: efavirenz level—St. John's wort; do not use together

Drug/Food

Increase: absorption—high-fat foods

Drug/Lab Test
Increase: ALT
False positive: cannibinoids

NURSING CONSIDERATIONS

Assess:

- **Pregnancy:** Rule out pregnancy (D) before starting treatment; a type of contraception is needed, oral/nonoral contraceptives are decreased
- Bowel pattern before, during treatment; if severe abdominal pain with bleeding occurs, product should be discontinued; monitor hydration
- **Serious skin reactions:** Stevens-Johnson syndrome, toxic epidermal necrolysis
- **HIV:** Monitor CBC, blood chemistry, plasma HIV RNA, absolute CD4+/CD8+ cell counts/%, serum β_2 microglobulin, serum ICD+24 antigen levels, cholesterol, hepatic enzymes
- **Signs of toxicity:** severe nausea/vomiting, maculopapular rash

Evaluate:

- Therapeutic response: increased CD4 cell counts; decreased viral load; slowing progression of HIV

Teach patient/family:

- To take as prescribed; if dose is missed, to take as soon as remembered; not to double dose; to take with water, juice; to take on empty stomach at bedtime
- To make sure health care provider knows all medications, supplements, OTC products taken
- That, if severe rash occurs, to notify health care provider; that adverse reactions (rash, dizziness, abnormal dreams, insomnia) lessen after 1 mo
- Not to breastfeed or become pregnant if taking this product; to use nonhormonal contraception because serious birth defects have occurred (pregnancy [D])
- To avoid hazardous activities if dizziness, drowsiness occur
- That product does not cure disease but controls symptoms; that HIV can be transmitted to others even while taking this product; to continue with safe-sex practices

eletriptan (Rx)

(el-ee-trip′tan)
Relpax
Func. class.: Antimigraine agent, abortive
Chem. class.: 5-HT_1-1B/1D receptor agonist, triptan

ACTION: Binds selectively to the vascular 5-HT_1-receptor subtype; causes vasoconstriction in cranial arteries

USES: Acute treatment of migraine with/without aura

CONTRAINDICATIONS: Hypersensitivity, coronary artery vasospasm, peripheral vascular disease, hemiplegic/basilar migraine, uncontrolled hypertension; ischemic bowel, heart disease; severe renal/hepatic disease, acute MI, stroke, angina, CV disease

Precautions: Pregnancy (C), breastfeeding, children, geriatric patients, postmenopausal women, men >40 yr; risk factors of CAD, MI, or other cardiac disease; hypercholesterolemia, obesity, diabetes, impaired renal/hepatic function

DOSAGE AND ROUTES

- **Adult: PO** 20 or 40 mg, may increase if needed, max 40 mg (single dose); may repeat in 2 hr if headache improves but returns, max 80 mg/24 hr

Available forms: Tabs 20, 40 mg

Administer:

- Swallow tabs whole; do not break, crush, or chew; use with 8 oz of water
- At beginning of headache; if headache returns, repeat dose after 2 hr of 1st dose if 1st dose is ineffective

SIDE EFFECTS

CNS: *Dizziness,* headache, anxiety, paresthesia, asthenia, somnolence, flushing, fatigue, hot/cold sensation, chills, vertigo, hypertonia, seizures, serotonin syndrome

CV: Chest pain, palpitations, hypertension, MI, sinus tachycardia, stroke, ventricular fibrillation/tachycardia, atrial fibrillation, AV block, bradycardia, chest pressure syndrome, coronary vasospasm

GI: Nausea, dry mouth, vomiting

MS: *Weakness,* back pain

RESP: Chest tightness, pressure

PHARMACOKINETICS

Onset of pain relief 2 hr, metabolized in the liver, 70% excreted in urine and feces

INTERACTIONS

Increase: plasma conc of eletriptan—CYP3A4 inhibitors (clarithromycin, erythromycin, itraconazole, ketoconazole, nelfinavir, ritonavir), propranolol, ergots

Increase: serotonin syndrome—SSRIs, SNRIs, serotonin-receptor agonists

NURSING CONSIDERATIONS

Assess:

- **Migraine:** pain location, character, intensity, nausea, vomiting, aura; quiet, calm environment with decreased stimulation from noise, bright light, excessive talking
- B/P; signs, symptoms of coronary vasospasms, geriatric patients may be at higher risk
- Tingling, hot sensation, burning, feeling of pressure, numbness, flushing
- Stress level, activity, recreation, coping mechanisms
- Neurologic status: LOC, blurring vision, nausea, vomiting, tingling in extremities preceding headache
- Ingestion of tyramine foods (pickled products, beer, wine, aged cheese), food additives, preservatives, colorings, artificial sweeteners, chocolate, caffeine, which may precipitate these types of headaches
- Patients with CAD risk factors; 1st dose should be administered in prescriber's office or medical facility

Evaluate:

- Therapeutic response: decrease in frequency, severity of migraine

Teach patient/family:

- To report any side effects to prescriber
- To use contraception while taking product; to inform prescriber if pregnant or intending to become pregnant
- To provide dark, quiet environment
- That product does not prevent or reduce number of migraine attacks

RARELY USED

eltrombopag

(ell-trom-bow′pag)

Promacta

Func. class.: Hematopoietic

USES:

Thrombocytopenia in chronic immune thrombocytopenic purpura when unresponsive to other treatment, chronic hepatitis C-associated thrombocytopenia

CONTRAINDICATIONS:

Hypersensitivity

Black Box Warning: Hepatotoxicity

DOSAGE AND ROUTES

- **Adult: PO** 50 mg/day, adjust dosage to maintain platelets at $\geq 50 \times 10^9$, max 75 mg/day; 100 mg/day chronic hepatitis C

emedastine ophthalmic

See Appendix B

emtricitabine (Rx)

(em-tri-sit'uh-bean)

Emtriva

Func. class.: Antiretroviral

Chem. class.: Nucleoside reverse transcriptase inhibitor (NRTI)

ACTION: A synthetic nucleoside analog of cytosine; inhibits replication of HIV virus by competing with the natural substrate and then becoming incorporated into cellular DNA by viral reverse transcriptase, thereby terminating cellular DNA chain

USES: HIV-1 infection with other antiretroviral

Unlabeled uses: HBV (hepatitis B virus) infection with HIV, HIV prophylaxis

CONTRAINDICATIONS: Hypersensitivity

Black Box Warning: Lactic acidosis

Precautions: Pregnancy (B), breastfeeding, children, geriatric patients, renal disease

Black Box Warning: Hepatic insufficiency, chronic hepatitis B virus (HPV)

DOSAGE AND ROUTES

Oral cap and solution are not interchangeable

- **Adult: PO** Caps 200 mg/day; oral sol 240 mg (24 ml)/day
- **Adolescent/child >33 kg: PO** Caps 200 mg/day; **child 3 mo-17 yr:** oral sol 6 mg/kg/day, max 240 mg (24 ml)
- **Infants <3 mo: PO oral sol** 3 mg/kg daily, do not use caps

Renal dose

- **Adult: PO** Caps CCr 30-49 ml/min, 200 mg q48hr; oral sol 120 mg q24hr; caps CCr 15-29 ml/min, 200 mg q72hr; oral sol 80 mg q24hr; caps CCr <15 ml/min, 200 mg q96hr; oral sol 60 mg q24hr

Available forms: Cap 200 mg; oral sol 10 mg/ml

Administer:

- Give without regard to meals
- Oral cap and solution not interchangeable
- Store (caps) at 25° C (77° F); (oral sol) refrigerated, use within 3 mo

SIDE EFFECTS

CNS: *Headache,* abnormal dreams, *depression,* dizziness, *insomnia,* neuropathy, paresthesia, *asthenia*

GI: *Nausea, vomiting, diarrhea, anorexia, abdominal pain, dyspepsia,* **hepatomegaly with steatosis (may be fatal)**

INTEG: *Rash,* skin discolorization

MS: Arthralgia, myalgia

RESP: *Cough*

SYST: Change in body fat distribution, **lactic acidosis**

PHARMACOKINETICS

Rapidly, extensively absorbed; peak 1-2 hr; protein binding <4%; excreted unchanged in urine (86%), feces (14%); half-life 10 hr

INTERACTIONS

- Do not use with efavirenz, tenofovir, lamiVUDine, treatment duplication

Decrease: emtricitabine level—interferons

- Complex interactions—ribavirin, cautious use

NURSING CONSIDERATIONS

Assess:

- Renal/hepatic function tests: AST, ALT, bilirubin, amylase, lipase, triglycerides periodically during treatment

Black Box Warning: Lactic acidosis, severe hepatomegaly with steatosis; if lab reports confirm these conditions, discontinue treatment; may be fatal

Black Box Warning: Hepatotoxicity: do not use in those with risk factors such as alcoholism, discontinue if hepatotoxicity occurs

Black Box Warning: Hepatitis B and HIV coinfection (unlabeled); perform HBV screening in any patient who has HIV to ensure appropriate treatment; avoid single-drug treatments in HBV

Evaluate:

- Therapeutic response: decreased signs, symptoms of HIV; decreased viral load, increased CP4 counts

Teach patient/family:

- That GI complaints resolve after 3-4 wk of treatment
- To report planned or suspected pregnancy; not to breastfeed while taking product
- That product must be taken at same time of day to maintain blood level
- That product will control symptoms but is not a cure for HIV; patient still infectious, may pass HIV virus on to others; that other products may be necessary to prevent other infections
- That changes in body fat distribution may occur

Black Box Warning: Lactic acidosis: to notify prescriber immediately if fatigue, muscle aches/pains, abdominal pain, difficulty breathing, nausea, vomiting, change in heart rhythm occur

Black Box Warning: Hepatotoxicity: to notify prescriber of dark urine, yellowing of skin/eyes, clay-colored stools, anorexia, nausea, vomiting

enalapril/enalaprilat (Rx)

(e-nal′a-pril)/(e-nal′a-pril-at)

Vasotec

Func. class.: Antihypertensive

Chem. class.: Angiotensin-converting enzyme (ACE) inhibitor

Do not confuse:

enalapril/ramipril/Anafranil/Eldepryl

ACTION: Selectively suppresses renin-angiotensin-aldosterone system; inhibits ACE; prevents conversion of angiotensin I to angiotensin II, dilation of arterial, venous vessels

USES: Hypertension, CHF, left ventricular dysfunction

Unlabeled uses: Diabetic nephropathy, hypertensive emergency/urgency, post-MI, proteinuria, renal crisis in scleroderma

CONTRAINDICATIONS: Hypersensitivity, history of angioedema

Black Box Warning: Pregnancy (D)

Precautions: Breastfeeding, renal disease, hyperkalemia, hepatic failure, dehydration, bilateral renal artery/aortic stenosis

DOSAGE AND ROUTES

Hypertension

- **Adult: PO** 2.5-5 mg/day, may increase or decrease to desired response, range 10-40 mg/day in 1-2 divided doses; **IV** 0.625-1.25 mg q6hr over 5 min
- **Child: PO** 0.08 mg/kg/day in 1-2 divided doses, max 0.58 mg/kg/day
- **Child: IV** 5-10 mcg/kg/dose q8-24hr

CHF

- **Adult: PO** 2.5-20 mg/day in 2 divided doses, max 40 mg/day in divided doses

Renal disease

- **Adult: PO** 2.5 mg/day (CCr <30 ml/min), increase gradually; **IV** CCr >30 ml/min, 1.25 mg q6hr; CCr <30 ml/min, 0.625 mg as one-time dose, increase as per B/P

Hypertensive emergency/urgency (unlabeled)

- **Adult: IV** 1.25-5 mg q6hr

Available forms: *Enalapril:* tabs 2.5, 5, 10, 20 mg; *enalaprilat:* inj 1.25 mg/ml

Administer:

PO route

- Tab may be crushed, given without regard to meals

IV route

- Prepare in sterile environment using aseptic technique
- Dilute each dose with ≤50 ml compatible sol
- For 25 mcg/ml dilution often used for neonatal or pediatric patients, combine 1

ml enalaprilat 1.25 mg/ml and 49 ml compatible sol for IV

IV, Direct/Intermittent IV INF route

• Undiluted over ≥5 min, use diluent provided or 50 ml D_5W, 0.9% NaCl, 0.9% NaCl in D_5W or LR, Isolyte E; give through Y-tube of free-flowing inf of 0.9% NaCl, D_5W, LR, Isolyte E

Y-site compatibilities: Acyclovir, alemtuzumab, alfentanil, allopurinol, amifostine, amikacin, aminophylline, amphotericin B liposome, anidulafungin, ascorbic acid, atracurium, atropine, azaTHIOprine, aztreonam, benztropine, bivalirudin, bretylium, bumetanide, buprenorphine, butorphanol, calcium chloride/gluconate, CARBOplatin, ceFAZolin, cefonicid, cefoperazone, cefotaxime, cefoTEtan, cefOXitin, cefTAZidime, ceftizoxime, cefTRIAXone, cefuroxime, cephalothin, cephapirin, chloramphenicol, cimetidine, cisatracurium, cladribine, clindamycin, cyanocobalamin, cyclophosphamide, cycloSPORINE, cytarabine, DACTINomycin, DAPTOmycin, dexamethasone, dexmedetomidine, dextran 40, digoxin, diltiazem, diphenhydrAMINE, DOBUTamine, DOCEtaxel, DOPamine, doripenem, doxacurium, DOXOrubicin, DOXOrubicin liposome, doxycycline, ePHEDrine, EPINEPHrine, epirubicin, epoetin, ertapenem, erythromycin, esmolol, etoposide, etoposide phosphate, famotidine, fenoldopam, fentaNYL, filgrastim, fluconazole, fludarabine, fluorouracil, folic acid, furosemide, ganciclovir, gemcitabine, gentamicin, granisetron, heparin, hydrocortisone, HYDROmorphone, ifosfamide, imipenem-cilastatin, indomethacin, insulin, isoproterenol, ketorolac, labetalol, levofloxacin, lidocaine, linezolid, LORazepam, magnesium sulfate, mannitol, mechlorethamine, melphalan, meperidine, meropenem, metaraminol, methicillin, methotrexate, methoxamine, methyldopate, methylPREDNISolone, metoclopramide, metoprolol, metroNIDAZOLE, mezlocillin, miconazole, midazolam, milrinone, minocycline, mitoXANtrone, morphine, moxalactam, multiple vitamin injection, nafcillin, nalbuphine, naloxone, netilmicin, niCARdipine, nitroglycerin, nitroprusside, norepinephrine, octreotide, ondansetron, oxacillin, oxaliplatin, oxytocin, PACLitaxel, palonosetron, papaverine, PEMEtrexed, penicillin G potassium, pentamidine, pentazocine, PENTobarbital, PHENobarbital, phentolamine, phenylephrine, phytonadione, piperacillin-tazobactam, potassium chloride/phosphate, procainamide, prochlorperazine, promethazine, propofol, propranolol, protamine, pyridoxime, quinupristin-dalfopristin, ranitidine, remifentanil, ritodrine, riTUXimab, rocuronium, sodium acetate, sodium bicarbonate, succinylcholine, SUFentanil, tacrolimus, teniposide, tetracycline, theophylline, thiamine, thiotepa, ticarcillin/clavulanate, tigecycline, tirofiban, tobramycin, tolazoline, trastuzumab, trimetaphan, urokinase, vancomycin, vasopressin, vecuronium, verapamil, vinCRIStine, vinorelbine, voriconazole, zoledronic acid

Additive compatibilities: DOBUTamine, DOPamine, heparin, meropenem, nitroglycerin, nitroprusside, potassium chloride

SIDE EFFECTS

CNS: *Insomnia, dizziness,* paresthesias, headache, fatigue, anxiety

CV: *Hypotension,* chest pain, tachycardia, **dysrhythmias**, syncope, angina, **MI**, orthostatic hypotension

EENT: *Tinnitus;* visual changes; sore throat; double vision; dry, burning eyes

GI: Nausea, vomiting, colitis, cramps, diarrhea, constipation, flatulence, dry mouth, loss of taste, **hepatotoxicity**

GU: **Proteinuria, renal failure,** increased frequency of polyuria or oliguria

HEMA: **Agranulocytosis, neutropenia**

INTEG: Rash, purpura, alopecia, hyperhidrosis, photosensitivity

META: Hyperkalemia

RESP: Dyspnea, dry cough, crackles

SYST: **Toxic epidermal necrolysis, Stevens-Johnson syndrome, angioedema**

PHARMACOKINETICS

Enalapril: PO: Onset 1 hr, peak 4-6 hr, duration ≥24 hr
Enalaprilat: IV: Onset 5-15 min, peak up to 4 hr, duration 4-6 hr, half-life 11 hr
Metabolized by liver to active metabolite, excreted in urine

INTERACTIONS

Increase: hypersensitivity—allopurinol
Increase: hypotension—diuretics, other antihypertensives, phenothiazines, nitrates, acute alcohol ingestion, general anesthesia
Increase: potassium levels—salt substitutes, potassium-sparing diuretics, potassium supplements, cycloSPORINE, NSAIDs
Increase: levels of lithium, digoxin
Decrease: effects of enalapril—antacids, rifampin

Drug/Lab Test
Increase: ALT, AST, bilirubin, alk phos, glucose, uric acid, BUN, creatine
False positive: ANA titer

NURSING CONSIDERATIONS

Assess:
- **Bone marrow depression (rare):** neutrophils, decreased platelets; WBC with differential baseline, q3mo; if neutrophils <1000/mm^3, discontinue treatment (recommended with collagen-vascular disease)
- **Hypertension:** B/P, peak/trough level, orthostatic hypotension, syncope when used with diuretic, pulse q4hr; note rate, rhythm, quality
- Baselines of renal, hepatic studies before therapy begins and 1 wk into therapy; electrolytes: potassium, sodium, chloride during 1st 2 wk of therapy
- Skin turgor, dryness of mucous membranes for hydration status; edema in feet, legs daily
- **Symptoms of CHF:** edema, dyspnea, wet crackles, weight gain, jugular venous distension, difficulty breathing

Evaluate:
- Therapeutic response: decreased B/P

Teach patient/family:
- Not to use OTC (cough, cold, or allergy) products unless directed by prescriber; to avoid potassium, salt substitutes
- To avoid sunlight or wear sunscreen for photosensitivity
- To comply with dosage schedule even if feeling better
- To notify prescriber of mouth sores, sore throat, fever, swelling of hands or feet, irregular heartbeat, chest pain, signs of angioedema
- That excessive perspiration, dehydration, vomiting, diarrhea may lead to fall in blood pressure; to consult prescriber if these occur, maintain adequate hydration
- That product may cause dizziness, fainting; that lightheadedness may occur during 1st few days of therapy, avoid activities requiring coordination
- That product may cause skin rash, impaired perspiration or angioedema; to discontinue if angioedema occurs
- Not to discontinue product abruptly
- That CV adverse reactions may reoccur
- To rise slowly to sitting or standing position to minimize orthostatic hypotension

Black Box Warning: To notify prescriber if pregnancy is planned or suspected; to use contraception during treatment; pregnancy (D)

TREATMENT OF OVERDOSE:

Lavage, IV atropine for bradycardia, IV theophylline for bronchospasm, digoxin, O_2, diuretic for cardiac failure

enfuvirtide (Rx)

(en-fyoo′vir-tide)

Fuzeon

Func. class.: Antiretroviral
Chem. class.: Fusion Inhibitor

ACTION: Inhibitor of the fusion of HIV-1 with CD4+ cells

USES: Treatment of HIV-1 infection in combination with other antiretrovirals
Unlabeled uses: HIV prophylaxis after occupational exposure

CONTRAINDICATIONS: Breastfeeding, hypersensitivity
Precautions: Pregnancy (B), children <6 yr, liver disease, myelosuppression, infections

DOSAGE AND ROUTES

- **Adult: SUBCUT** 90 mg (1 ml) bid
- **Child 6-16 yr and <42.6 kg: SUBCUT** 2 mg/kg bid, max 90 mg bid; 11-15.5 kg 27 mg/0.3 ml bid; 15.6-20 kg 36 mg/0.4 ml bid; 20.1-24.5 kg 45 mg/0.5 ml bid; 24.6-29 kg 54 mg/0.6 ml bid; 29.1-33.5 kg 63 mg/0.7 ml bid; 33.6-38 kg 72 mg/0.8 ml bid; 38.1-42.5 kg 81 mg/0.9 ml bid

HIV prophylaxis (unlabeled)

- **Adult: SUBCUT** 90 mg bid added to PEP regimen

Available forms: Powder for inj, lyophilized 108 mg (90 mg/ml when reconstituted)

Administer:

SUBCUT route

- **Reconstitute** vial with 1.1 ml sterile water for inj; tap and roll to mix; allow to stand until completely dissolved, may take up to 45 min; after dissolved, immediately **inject** or refrigerate up to 24 hr
- Do not mix with other medications
- SUBCUT: give bid, rotate sites; preferred sites: upper arm, anterior thigh, abdomen

SIDE EFFECTS

CNS: Anxiety, peripheral neuropathy, taste disturbance, **Guillain-Barré syndrome**, insomnia, depression
GI: Abdominal pain, anorexia, constipation, **pancreatitis**
GU: **Glomerulonephritis, renal failure**
HEMA: **Thrombocytopenia, neutropenia**
INTEG: *Inj site reactions*
MISC: Influenza, cough, conjunctivitis, lymphadenopathy, myalgia, hyperglycemia, **pneumonia**, rhinitis, fatigue, hypersensitivity

PHARMACOKINETICS

Peak 8 hr, terminal half-life 3.8 hr, well absorbed, undergoes catabolism, 92% protein binding

NURSING CONSIDERATIONS

Assess:

- **Signs of infection, inj site reactions**
- **Glomerulonephritis/renal failure:** BUN, creatinine, renal failure may occur
- Bowel pattern before, during treatment; if severe abdominal pain or constipation occurs, notify prescriber; monitor hydration
- Skin eruptions, rash, urticaria, itching
- Allergies before treatment, reaction to each medication
- **Immune reconstitution syndrome:** with combination theory
- **HIV:** CBC, blood chemistry, plasma HIV RNA, absolute CD4+/CD8+ cell counts/%, serum β_2 microglobulin, serum ICD+24 antigen levels, cholesterol

Evaluate:

- Therapeutic response: increased CD4 cell counts; decreased viral load; slowing progression of HIV-1 infection

Teach patient/family:

- To notify prescriber if pregnancy is suspected or if breastfeeding
- That pneumonia may occur; to contact prescriber if cough, fever occur
- That hypersensitive reactions may occur; rash, pruritus; to stop product, contact prescriber
- That product is not a cure for HIV-1 infection but controls symptoms; HIV-1 can still be transmitted to others; that product is to be used in combination only with other antiretrovirals

⚠ HIGH ALERT

enoxaparin (Rx)

(ee-nox′a-par-in)

Lovenox

Func. class.: Anticoagulant, antithrombotic

Chem. class.: Low-molecular-weight heparin (LMWH)

Do not confuse:
enoxaparin/enoxacin
Lovenox/Lotronex

ACTION: Binds to antithrombin III inactivating factors Xa/IIa, thereby resulting in a higher ratio of anti-factor Xa to IIa

USES: Prevention of DVT (inpatient or outpatient), PE (inpatient) in hip and knee replacement, abdominal surgery at risk for thrombosis; unstable angina, acute MI, coronary artery thrombosis

Unlabeled uses: Antiphospholipid antibody syndrome, arterial thromboembolism prophylaxis, cerebral thromboembolism, percutaneous coronary intervention

CONTRAINDICATIONS: Hypersensitivity to this product, heparin, pork; active major bleeding, hemophilia, leukemia with bleeding, thrombocytopenic purpura, heparin-induced thrombocytopenia

Precautions: Pregnancy (B), breastfeeding, children, geriatric patients, low weight men (<57 kg), women (<45 kg), severe renal/hepatic disease, severe hypertension, subacute bacterial endocarditis, acute nephritis, recent burn, spinal surgery, indwelling catheters, hypersensitivity to benzyl alcohol

Black Box Warning: Lumbar puncture, aneurysm, coagulopathy, epidural anesthesia, spinal anesthesia

DOSAGE AND ROUTES

DVT/PE prophylaxis

- **Adult:** (moderate risk—general surgery, non surgery 40-60 yr major surgery <40 yr with no risk factors)
- **Adult: SUBCUT** 20 mg daily; higher risk adults—abdominal surgery, geriatric—general surgery, major surgery <40 yr no risk factors; **SUBCUT** 30 mg q12hr or 40 mg daily

DVT prevention before hip or knee surgery

- **Adult: SUBCUT** 30 mg bid given 12-24 hr postop for 7-10 days until DVT risk is diminished

DVT prevention before hip replacement

- **Adult: SUBCUT** 40 mg/day started 9-15 hr preop or 30 mg q12hr started 12-24 hr postop, continued until DVT risk diminished or patient adequately on anticoagulant

DVT prophylaxis before abdominal surgery

- **Adult: SUBCUT** 40 mg/day starting 24 hr prior to surgery × 7-10 days to prevent thromboembolic complications

Treatment of DVT or PE

- **Adult: SUBCUT** 1 mg/kg q12hr (without PE, outpatient); 1 mg/kg q12hr or 1.5 mg/kg/day (with or without PE, inpatient); warfarin should be started within 72 hr, continued ≥5 days until INR is 2-3 (usually 7 days)

Prevention of ischemic complications in unstable angina or non–Q-wave MI/non-ST

- **Adult: SUBCUT/IV** 1 mg/kg q12hr until stable with aspirin 100-325 mg/day × ≥2 days

Renal dose

- **Adult: SUBCUT** <30 ml/min 30 mg/day (thrombosis prophylaxis)

Available forms: Prefilled syringes/inj 30 mg/0.3 ml, 40 mg/0.4 ml, 60 mg/0.6 ml, 80 mg/0.8 ml, 100 mg/1 ml, 120 mg/0.8 ml, 150 mg/ml; multidose vials 100 mg/ml (3 ml)

Administer:

- **Only after screening patient for bleeding disorders**
- **Do not mix with other products or inf fluids**

⚠ **Only this product when ordered; not interchangeable with heparin or other LMWHs**

- At same time each day to maintain steady blood levels
- Avoid all IM inj that may cause bleeding
- Prepare in a sterile environment using aseptic technique
- Dilution may be stored for up to 4 wk in glass vial at room temperature, up to 2 wk in TB syringes with rubber stoppers at room temperature or refrigerated

SUBCUT route

- Do not give IM; begin 1 hr prior to surgery; do not aspirate; rotate sites; do not expel bubble from syringe before administration
- To recumbent patient, give SUBCUT; rotate inj sites (left/right anterolateral, left/right posterolateral abdominal wall)
- Insert whole length of needle into skin fold held with thumb and forefinger
- If withdrawing from multidose vial, use TB syringe for proper measurement
- Prefilled syringes (30, 40 mg) not graduated; do not use for partial doses
- Do not administer if particulate is present

Direct IV route

- Use multidose vial for IV administration; use TB syringe, other graduated syringe to measure dose; give IV BOL through IV line, flush after

SIDE EFFECTS

CNS: Fever, confusion
GI: Nausea
HEMA: **Hemorrhage from any site, hypochromic anemia, thrombocytopenia, bleeding**
INTEG: Ecchymosis, inj site hematoma
META: Hyperkalemia in renal failure
MS: Osteoporosis
SYST: Edema, peripheral edema

PHARMACOKINETICS

SUBCUT: 90% absorbed, maximum antithrombin activity (3-5 hr), elimination half-life 4½ hr, excreted in urine

INTERACTIONS

Increase: enoxaparin action—anticoagulants, salicylates, NSAIDs, antiplatelets, thrombolytics, RU-486

Drug/Lab Test

Increase: bleeding risk—feverfew, garlic, ginger, ginkgo, horse chestnut
Increase: AST, ALT
Decrease: platelet count

NURSING CONSIDERATIONS

Assess:

- Blood studies (Hct/Hgb, CBC, coagulation studies, platelets, occult blood in stools), anti-factor Xa (should be checked 4 hr after inj); thrombocytopenia may occur
- Renal studies: BUN/creatinine baseline and periodically
- **Bleeding:** gums, petechiae, ecchymosis, black tarry stools, hematuria; notify prescriber

Black Box Warning: Neurologic symptoms in patients who have received spinal anesthesia

Evaluate:

- Therapeutic response: prevention of DVT

Teach patient/family:

- To use soft-bristle toothbrush to avoid bleeding gums; to use electric razor
- To report any signs of bleeding: gums, under skin, urine, stools
- To avoid OTC products containing aspirin unless approved by prescriber

TREATMENT OF OVERDOSE:

Protamine SO_4 1% sol; dose should equal dose of enoxaparin

entacapone (Rx)

(en'ta-kah-pone)

Comtan

Func. class.: Antiparkinson agent

Chem. class.: COMT inhibitor

ACTION: Inhibits COMT (catechol *O*-methyltransferase) and alters the plasma pharmacokinetics of levodopa; given with levodopa/carbidopa

USES: Parkinson's disease for those experiencing end of dose; decreased effect as adjunct to levodopa/carbidopa

CONTRAINDICATIONS: Hypersensitivity

Precautions: Pregnancy (C), breastfeeding, children, renal/hepatic disease, affective disorders, psychosis

DOSAGE AND ROUTES

- **Adult: PO** 200 mg given with carbidopa/levodopa, max 1600 mg/day; may allow for 25% dosage reduction in levodopa therapy

Available forms: Tabs, film coated 200 mg

Administer:

- Only after MAOIs have been discontinued for 2 wk
- Give with dose of levodopa/carbidopa; product has no effect on its own

SIDE EFFECTS

CNS: *Involuntary choreiform movements, hand tremors, fatigue, headache, anxiety, twitching, numbness, dyskinesia, hypokinesia, hyperkinesia, weakness, confusion, agitation, nightmares,* psychosis, hallucination, hypomania, severe depression, dizziness, neuroleptic malignant syndrome

CV: *Orthostatic hypotension*

GI: *Nausea, vomiting, anorexia, abdominal distress, dry mouth, flatulence, bitter taste, diarrhea, constipation, dyspepsia,* gastritis, GI disorder

INTEG: Rash, sweating, alopecia

MISC: Dark urine and other body fluids, back pain, dyspnea, purpura, fatigue, asthenia, bacterial infection, rhabdomyolysis

PHARMACOKINETICS

Duration up to 8 hr; excreted in urine, feces; well absorbed; protein binding 98%; metabolized in liver extensively; enters breast milk; half-life of levodopa extended, half-life 0.5 hr initial, 2.5 hr second

INTERACTIONS

- Prevents catecholamine metabolism—nonselective MAOIs; do not use together

Increase: B/P, tachycardia, dysrhythmias, avoid use—bitolterol, DOPamine, DOBUTamine, EPINEPHrine, methyldopa, isoetharine, norepinephrine

Decrease: excretion of entacapone—ampicillin, chloramphenicol, probenecid, erythromycin, rifampin

Drug/Herb

Increase: B/P—ma huang

Decrease: effect—kava

NURSING CONSIDERATIONS

Assess:

- ⚠ **Neuroleptic malignant syndrome:** high temperature, increased CPK, rigidity, change in LOC usually during rapid withdrawal
- **Involuntary movements of Parkinson's disease:** akinesia, tremors, staggering gait, muscle rigidity, drooling when given with levodopa/carbidopa
- B/P, respirations during initial treatment
- Mental status: affect, mood, behavioral changes, depression; complete suicide assessment
- **Rhabdomyolysis:** muscle pain, tenderness, weakness; swelling of affected muscles; may lead to decreased B/P, shock
- Assistance with ambulation during beginning therapy

Evaluate:

- Therapeutic response: decrease in akathisia, increased mood when given with levodopa/carbidopa

Teach patient/family:

- That hallucinations, mental changes, nausea, dyskinesia can occur; may mean patient is overmedicated
- To change positions slowly to prevent orthostatic hypotension; not to drive, operate machinery until stabilized on medication and mental performance not affected
- To use product exactly as prescribed; if dose is missed, to take as soon as remembered up to 2 hr before next dose; not to discontinue abruptly; to withdraw gradually
- That urine, sweat may darken
- To notify prescriber if pregnancy is suspected; if lactating, that product excreted in breast milk

entecavir (Rx)

(en-te′ka-veer)

Baraclude

Func. class.: Antiretroviral nucleoside reverse transcriptase inhibitor (NRTIs)

Chem. class.: Guanosine nucleoside analog

ACTION: Inhibits hepatitis B virus DNA polymerase by competing with natural substrates and by causing DNA termination after its incorporation into viral DNA; causes viral DNA death

USES: Chronic hepatitis B (HBV)

CONTRAINDICATIONS: Hypersensitivity

Precautions: Pregnancy (C), breastfeeding, children, geriatric patients, severe renal disease

Black Box Warning: Hepatic disease, hepatitis, HIV, lactic acidosis

DOSAGE AND ROUTES

Chronic hepatitis B (nucleoside treatment naive)

- **Adult and adolescent ≥16 yr: PO** 0.5 mg/day

Chronic hepatitis B with compensated liver disease and history of hepatitis B viremia while receiving lamiVUDine or known lamiVUDine/telbivudine-resistant mutations

- **Adult and adolescent ≥16 yr: PO** 1 mg/day

Renal dose

- **Adult: PO** CCr ≥50 ml/min, 0.5 mg/day; CCr 30-49 ml/min, 0.25 mg/day, 0.5 mg/day or 1 mg q48hr for lamiVUDine-refractory patient; CCr 10-29 ml/min, 0.15/day, 0.3 mg or 1 mg q72hr for lamiVUDine-refractory patient; CCr <10 ml/min, 0.05 mg/day, 0.1 mg/day or 1 mg q7days for lamiVUDine-refractory patient

Available forms: Tabs, film coated 0.5, 1 mg; oral sol 0.05 mg/ml

Administer:

- After hemodialysis
- By mouth on empty stomach 2 hr before or after food
- Store at room temperature
- **Oral liquid:** use calibrated oral dosing spoon provided; may be used interchangeably with tabs

SIDE EFFECTS

CNS: *Headache,* fatigue, dizziness, insomnia

ENDO: Hyperglycemia

GI: *Dyspepsia,* nausea, vomiting, diarrhea, elevated liver function enzymes

INTEG: Alopecia, rash

SYST: Lactic acidosis, severe hepatomegaly with steatosis

PHARMACOKINETICS

Peak 0.5-1.5 hr, steady state 6-10 days, 100% bioavailability, extensively distributed to tissues, protein binding 13%, terminal half-life 128-149 hr, excreted unchanged (62%-73%) via kidneys

INTERACTIONS

Drug/Food

Decrease: absorption—high-fat meal

Drug/Lab Test
Increase: ALT, AST, total bilirubin, amylase, lipase, creatinine, blood glucose, urine glucose
Decrease: platelets, albumin

NURSING CONSIDERATIONS

Assess:

- For nephrotoxicity: increasing CCr, BUN

Black Box Warning: For HIV before beginning treatment because HIV resistance may occur in chronic hepatitis B patients; monitor HIV RNA

Black Box Warning: For lactic acidosis, severe hepatomegaly with stenosis; increased serum lactate, increased hepatic enzymes, palpate line; discontinue if present

- Geriatric patients more carefully; may develop renal, cardiac symptoms more rapidly

Black Box Warning: For exacerbations of hepatitis (jaundice, pruritus, fatigue) after discontinuing treatment, monitor LFTs

Evaluate:

- Therapeutic response: decreased symptoms of chronic hepatitis B, improving LFTs

Teach patient/family:

- Not to take with food
- To take exactly as prescribed
- Not to stop medication without approval of prescriber
- That optimal duration of treatment is unknown
- To avoid use with other medications unless approved by prescriber
- To notify prescriber of decreased urinary output, blood in urine

Black Box Warning: Symptoms of lactic acidosis: muscle pain, severe tiredness, weakness, trouble breathing, stomach pain with nausea/vomiting, coldness in arms/legs, fast/irregular heartbeat, dizziness

Black Box Warning: Symptoms of hepatotoxicity: eyes/skin turning yellow, dark urine, light bowel movements, no appetite for days, nausea, stomach pain

- That product does not cure but lowers amount of HBV in body
- That product does not stop spread of HBV to others by sex, sharing needles, or being exposed to blood
- Not to breastfeed

enzalutamide

(en-zal-u'ta-mide)

Xtandi

Func. class.: Antineoplastic hormone
Chem. class.: Nonsteroidal antiandrogen

ACTION: Binds to cytosolic androgen receptors in target tissue, which competitively inhibits the action to androgens

USES: Metastatic castration-resistant prostate cancer in those who have received DOCEtaxel

CONTRAINDICATIONS: Pregnancy (X), women, hypersensitivity
Precautions: Breastfeeding, brain tumor, head trauma, infertility, seizures, stroke, male-mediated teratogenicity

DOSAGE AND ROUTES

- **Adult: PO** 160 mg (4 × 40-mg caps) daily
- If a patient experiences a grade 3 or higher toxicity or an intolerable adverse effect, withhold dosing for 1 wk or until symptoms improve to grade 2 or less, then resume at the same or a reduced dosage (120 or 80 mg), if warranted
- The concomitant use of strong cytochrome P450 (CYP-450) 2C8 inhibitors should be avoided if possible; if a strong CYP2C8 inhibitor must be coadministered, reduce the enzalutamide dosage to 80 mg once daily

Available forms: Tabs 40 mg

Administer:
- At same time each day, without regard to food
- Swallow whole

SIDE EFFECTS

CNS: Dizziness, paresthesias, insomnia, anxiety, headache
CV: Hot flashes, hypertension
GU: Urinary frequency
INTEG: Dry skin, pruritus
MISC: Infection, asthenia

PHARMACOKINETICS

Terminal half-life 2.8 to 10.2 days, 97% to 98% protein binding

INTERACTIONS

Increase: anticoagulation—anticoagulants
Increase: enzalutamide effects—CYP3A4 inhibitors (amiodarone, antiretrovirals, protease inhibitors, clarithromycin, dalfopristin, quinupristin, delavirdine, efavirenz, erythromycin, FLUoxetine, fluvoxaMINE, imatinib, mifepristone, RU-486, nefazodone, some azole antifungals)
Decrease: enzalutamide effects—CYP3A4 inducers (barbiturates, bosentan, carBAMazepine, dexamethasone, nevirapine, OXcarbazepine, phenytoin, rifabutin, rifampin, rifapentine)
Drug/Lab Test
Might require dosage change when used with St. John's wort

NURSING CONSIDERATIONS

Assess:
⚠ **Male-mediated teratogenicity:** a condom and another effective birth control method should be used if patient is having sex with a woman of childbearing potential during and 3 mo after end of therapy
⚠ **Seizures:** those with a history of seizures, underlying head trauma with loss of consciousness, TIA within 12 mo, stroke, brain tumor, or drugs lowering seizure threshold, should be told about danger of loss of consciousness when engaging in activities
⚠ For hot flashes, assure patient that these are common side effects
Evaluate:
- Therapeutic response: decreased tumor size, decreased spread of malignancy

Teach patient/family:
- That infertility can occur

⚠ To use a condom and another form of birth control during and for 3 mo after end of therapy if female sex partner is of child bearing potential
⚠ That those with seizure disorders might experience seizures because seizure threshold is decreased
- Not to use other products unless approved by prescriber

ePHEDrine nasal agent

See Appendix B

epinastine (ophthalmic)

(ep-ih-nas′teen)

Elestat

Func. class.: Antihistamine (ophthalmic)
Chem. class.: Histamine 1 receptor antagonist/mast cell stabilizer

ACTION: A topically active, direct H_1-receptor antagonist and mast cell stabilizer; by reducing these inflammatory mediators, it relieves the ocular pruritus associated with allergic conjunctivitis

USES: Prevention of ocular pruritus associated with signs and symptoms of allergic conjunctivitis

CONTRAINDICATIONS: Hypersensitivity
Precautions: Pregnancy (C), breastfeeding, children, contact lenses

DOSAGE AND ROUTES

• **Adult/child ≥3 yr: OPHTH** Instill 1 drop in each eye bid

Available forms: Ophthalmic sol 0.5%

Ophthalmic route

• For topical ophthalmic use only
• Wash hands before and after use; tilt the head back slightly and pull the lower eyelid down with the index finger; squeeze the prescribed number of drops into the conjunctival sac and gently close eyes for 1-2 min; do not blink
• Do not touch the tip of the dropper to the eye, fingertips, or other surface
• The preservative benzalkonium chloride may be absorbed by soft contact lenses; wait ≥10 min after instilling the ophthalmic solution before inserting contact lenses; contact lenses should not be worn if eye is red
• Do not share ophthalmic drops with others
• Keep bottle tightly closed when not in use
• Treatment should be continued throughout the period of exposure (i.e., until the pollen season is over or until exposure to the offending allergen is terminated), even when symptoms are absent

SIDE EFFECTS

EENT: Ocular irritation, folliculosis, hyperemia, ocular pruritus

MISC: Infection (including cold symptoms and upper respiratory infections), headache, rhinitis, sinusitis, increased cough, pharyngitis

PHARMACOKINETICS

Onset 3-5 min, peak 2 hr, duration 8 hr

NURSING CONSIDERATIONS

Assess:

• Eyes: for itching, redness, use of soft or hard contact lenses

Evaluate:

• Therapeutic response: absence of redness, itching in the eyes

Teach patient/family:

Ophthalmic route

• Product is for topical ophthalmic use only
• Wash hands before and after use; tilt the head back slightly and pull the lower eyelid down with the index finger; squeeze the prescribed number of drops into the conjunctival sac and gently close eyes for 1-2 min; do not blink
• Do not touch the tip of the dropper to the eye, fingertips, or other surface
• Wait ≥10 min after instilling the ophthalmic solution before inserting contact lenses; contact lenses should not be worn if eye is red
• Keep bottle tightly closed when not in use
• Do not share ophthalmic drops with others
• Remove contact lenses before use because the preservative benzalkonium chloride may be absorbed by soft contact lenses; product should not be used to treat contact lens–related irritation

HIGH ALERT

EPINEPHrine (Rx)

(ep-i-nef′rin)

Adrenaclick, Primatene Mist, Twinject, Walgreens Bronchial Mist

EPINEPHrine HCL

Adrenalin, EpiPen, EpiPen Jr.

Func. class.: Bronchodilator nonselective adrenergic agonist, vasopressor

Chem. class.: Catecholamine

Do not confuse:

EPINEPHrine/ePHEDrine

ACTION: β_1- and β_2-agonist causing increased levels of cAMP, thereby producing bronchodilation, cardiac, and CNS stimulation; high doses cause vasoconstriction via α-receptors; low doses can cause vasodilation via β_2-vascular receptors

USES:
Acute asthmatic attacks, hemostasis, bronchospasm, anaphylaxis, allergic reactions, cardiac arrest, adjunct in anesthesia, shock

Unlabeled uses: Bradycardia, chloroquine overdose

CONTRAINDICATIONS:
Hypersensitivity to sympathomimetics, sulfites, closed-angle glaucoma, nonanaphylactic shock during general anesthesia

Precautions: Pregnancy (C), breastfeeding, cardiac disorders, hyperthyroidism, diabetes mellitus, prostatic hypertrophy, hypertension, organic brain syndrome, local anesthesia of certain areas, labor, cardiac dilation, coronary insufficiency, cerebral arteriosclerosis, organic heart disease

DOSAGE AND ROUTES

Anaphylaxis/severe asthma exacerbation

• **Adult: IM/SUBCUT** 0.3-0.5 mg, may repeat q10-15min (anaphylaxis) or q20min-4 hr (asthma)

Severe allergic reactions type I

• **Adult/child ≥30 kg: IM** 0.3 mg (EpiPen/EpiPen 2-Pak, 1:1000)

• **Child <30 kg: IM** 0.15 mg (EpiPen Jr/EpiPen Jr 2-Pak 1:2000)

• **Adult/child ≥66 lb: IM/SUBCUT** 0.3 mg (0.3 ml) initially (Twinject 1.1 ml 1:1000, 1 mg/ml, containing 2 doses of 0.3 mg)

• **Adult/child 33-66 lb: IM/SUBCUT** 0.15 mg (0.15 ml) initially, may give another 0.15 mg after 10 min (Twinject 1.1 ml 1:1000 [1 mg/ml] containing 2 doses of 0.15 mg)

Status asthmaticus

• **Adult/adolescent:** SUBCUT 0.3-0.5 mg (0.3-0.5 ml of the 1:1000 injection) q20min × 3 doses

• **Infant/child:** SUBCUT 0.01 mg/kg-0.5 mg q20min × 3 doses

Available forms: Nasal spray (sol) 1 mg/ml; sol for inj 1 mg/ml, 1:10,000, 1:1000; inh vapor (sol) 0.22 mg/actuation; pressurized inh (sol) 0.22 mg/actuation; sol for inj 0.15 mg/0.15 ml autoinjector, 0.3 mg/0.3 ml autoinjector, 0.15 mg/0.3 ml

Administer:

• Increased dose of insulin for diabetic patients if glucose is elevated

• Check for correct conc, route, dosage before administering

• Give subcut, IM, intraosseously, IV; suspensions are for subcut use only; do not give IV

• Visually inspect parenteral products for particulate matter and discoloration before use; do not use sols that are pinkish to brownish or that contain a precipitate

• Avoid extravasation during parenteral administration; if extravasation occurs, infiltrate the affected area with phentolamine diluted in NS

• Store reconstituted sol refrigerated ≤24 hr

Inhalation route

• Place in nebulizer (10 gtt of a 1% base sol)

• Dilute racepinephrine 2.25% sol

IM route

• EPINEPHrine injection should preferably be into the deltoid or anterior thigh (vastus lateralis); do not administer into the gluteal muscle

• Twinject is light sensitive and should be stored in the carrying case provided; do not refrigerate; protect from freezing; replace if solution is discolored or contains a precipitate

SUBCUT route

• Inject, taking care not to inject intradermally; massage injection site well after use to enhance absorption and to decrease local vasoconstriction; injection can cause tissue irritation

Intraosseous INF route (unlabeled)

• During CPR, the same EPINEPHrine dosage may be given via the intraosseous route when IV access is not available

Intracardiac route

• Intracardiac route should be reserved for extreme emergencies; intracardiac injection should only be performed by properly trained medical personnel

Endotracheal route

• Per the ACLS or PALS guidelines, the EPINEPHrine parenteral product is administered via endotracheal (ET) route; ET administration should only be used if access to IV or intraosseous routes is not possible

• **Adult:** Dilute dose in 5-10 ml of NS or sterile distilled water; administer via ET tube; endotracheal absorption of EPINEPHrine may be improved by diluting with water instead of NS

• **Child:** after dose administration, flush the ET tube with a minimum of 5 ml NS

Direct IV INJ route

• Inject EPINEPHrine directly into a vein over 5-10 min for adults or 1-3 min for children; may be given IV push in cardiac arrest

• In neonates, may administer via the umbilical vein

• **During adult cardiopulmonary resuscitation (CPR):** Resuscitation drugs may be given IV by bolus injection into a peripheral vein, followed by an injection of 20 ml IV fluid; elevate the extremity for 10-20 sec to facilitate drug delivery to the central circulation

Continuous IV INF route

• Dilute 1 mg EPINEPHrine in 250 or 500 ml of a compatible IV inf sol to provide a conc of 4 or 2 mcg/ml, respectively; give into a large vein, if possible; more-concentrated sols (16-32 mcg/ml) may be used in fluid-restricted patients when administered through a central line

Y-site compatibilities: Alfentanil, amikacin, amiodarone, amphotericin B liposome, anidulafungin, ascorbic acid, atracurium, atropine, aztreonam, benztropine, bivalirudin, bleomycin, bumetanide, buprenorphine, butorphanol, calcium chloride/gluconate, CARBOplatin, caspofungin, ceFAZolin, cefoperazone, cefotaxime, cefoTEtan, cefOXitin, cefTAZidime, ceftizoxime, cefTRIAXone, cefuroxime, chloramphenicol, chlorproMAZINE, cimetidine, cisatracurium, CISplatin, clindamycin, cyanocobalamin, cyclophosphamide, cycloSPORINE, cytarabine, DACTINomycin, DAPTOmycin, dexamethasone, dexmedetomidine, digoxin, diltiazem, diphenhydrAMINE, DOBUTamine, DOCEtaxel, DOPamine, DOXOrubicin, doxycycline, enalaprilat, epirubicin, epoetin alfa, ertapenem, erythromycin, esmolol, etoposide, etoposide phosphate, famotidine, fenoldopam, fentaNYL, fluconazole, fludarabine, folic acid, furosemide, gemcitabine, gentamicin, glycopyrrolate, granisetron, heparin, hydrocortisone, HYDROmorphone, ifosfamide, imipenem-cilastatin, isoproterenol, ketorolac, labetalol, levofloxacin, lidocaine, linezolid, LORazepam, magnesium sulfate, mannitol, mechlorethamine, meperidine, metaraminol, methicillin, methotrexate, methoxamine, methyldopa, methylPREDNISolone, metoclopramide, metoprolol, metroNIDAZOLE, midazolam, milrinone, minocycline, mitoXANtrone, morphine, multiple vitamins, nafcillin, nalbuphine, naloxone, niCARdipine, nitroglycerin, nitroprusside, norepinephrine, octreotide, ondansetron, oxacillin, oxaliplatin, oxytocin, PACLitaxel, palonosetron, pancuronium, pantoprazole, PEMEtrexed, penicillin G potassium, pentamidine, pentazocine, phentolamine, phenylephrine, phytonadione, piperacillin/tazobactam, potassium chloride, procainamide, prochlorperazine, promethazine, propofol, propranolol, protamine, pyridoxime, quinupristin/dalfopristin, ranitidine, remifentanil, ritodrine, rocuronium, sodium acetate, streptomycin, succinylcholine, SUFentanil, tacrolimus, teniposide, theophylline, thiamine, thiotepa, ticarcillin/clavulanate, tigecycline, tirofiban, tobramycin, tolazoline, trimethaphan, urokinase, vancomycin, vasopressin, vecuronium, verapamil, vinCRIStine, vinorelbine, vitamin B complex with C, voriconazole, warfarin, zoledronic acid

SIDE EFFECTS

CNS: *Tremors, anxiety,* insomnia, headache, *dizziness,* confusion, hallucinations, **cerebral hemorrhage**, weakness, drowsiness
CV: *Palpitations, tachycardia,* hypertension, **dysrhythmias, increased T wave**
GI: *Anorexia, nausea, vomiting*
MISC: Sweating, dry eyes
RESP: *Dyspnea*

PHARMACOKINETICS

Crosses placenta, metabolized in liver
IM: Onset variable, duration 1-4 hr
SUBCUT: Onset 5-15 min, duration 20 min-4 hr
INH: Onset 1-5 min, duration 1-3 hr

INTERACTIONS

- **Do not use with MAOIs or tricyclics; hypertensive crisis may occur**
- **Toxicity: other sympathomimetics**

Decrease: hypertensive effects—β-adrenergic blockers
Drug/Herb
- Increased stimulation: coffee, tea, guarana, yerba maté

NURSING CONSIDERATIONS

Assess:
- **Asthma:** auscultate lungs, pulse, B/P, respirations, sputum (color, character); monitor pulmonary function studies before and during treatment
- **ECG during administration continuously; if B/P increases, decrease dose; B/P, pulse q5min after parenteral route; CVP, ISVR, PCWP during inf if possible; inadvertent high arterial B/P can result in angina, aortic rupture, cerebral hemorrhage**
- Inj site: tissue sloughing; administer phentolamine with NS
- **Sulfite sensitivity; may be life-threatening**
- Cardiac status, I&O; blood glucose in diabetes
- **Allergic reactions, bronchospasms: withhold dose, notify prescriber**

Evaluate:
- Therapeutic response: increased B/P with stabilization or ease of breathing

Teach patient/family:
- About the reason for product administration; how to administer
- To rinse mouth after use to prevent dryness after inhalation
- Not to take OTC preparations

TREATMENT OF OVERDOSE:

Administer α-blocker and β-blocker

EPINEPHrine/ epinephryl borate ophthalmic

See Appendix B

EPINEPHrine nasal agent

See Appendix B

HIGH ALERT

epirubicin (Rx)

(ep-ih-roo′bi-sin)

Ellence, Pharmorubicin ✦

Func. class.: Antineoplastic, antibiotic
Chem. class.: Anthracycline

Do not confuse:
DOXOrubicin DAUNOrubicin, eribulin, IDArubicin
epirubicin/eribulin

ACTION:

Inhibits DNA synthesis primarily; replication is decreased by binding to DNA, which causes strand splitting; maximum cytotoxic effects at S and for G_2 phases; a vesicant

USES:

Adjuvant therapy for breast cancer with axillary node involvement after resection
Unlabeled uses: Used in combination for treatment of advanced forms of can-

cer: bladder, gastric, head and neck, hepatocellular, lung, ovarian, multiple myeloma, soft-tissue sarcoma

CONTRAINDICATIONS:

Pregnancy (D), breastfeeding; hypersensitivity to product, anthracyclines, anthracenediones; baseline neutrophil count <1500 cell/mm^3, severe myocardial insufficiency, recent MI, heart failure, cardiomyopathy

Black Box Warning: Severe hepatic disease, IM/SUBCUT use

Precautions: Children, geriatric patients, cardiac/renal/hepatic disease, accidental exposure, angina, dental disease, herpes, hyperkalemia, hyperphosphatemia, hypertension, hyperuricemia, hypocalcemia, infection, infertility, tumor lysis syndrome, ventricular dysfunction, previous anthracycline use

Black Box Warning: Bone marrow depression (severe), heart failure, extravasation, secondary malignancy, requires an experienced clinician

DOSAGE AND ROUTES

Breast cancer with axillary node involvement following resection of the primary tumor in combination with cyclophosphamide and fluorouracil

• **Adult: IV** 100 mg/m^2 on day 1 with fluorouracil and cyclophosphamide (FEC regimen) every 21 days × 6 cycles or 60 mg/m^2 on days 1 and 8 with oral cyclophosphamide and fluorouracil every 28 days × 6 cycles

Breast cancer in combination with cyclophosphamide

• **Adult: IV** 60 mg/m^2 day 1 with cyclophosphamide (500 mg/m^2 IV day 1), repeated every 21 days × 8 cycles or a higher-dose regimen of epirubicin 100 mg/m^2 IV day 1 with cyclophosphamide (830 mg/m^2 IV day 1), every 21 days × 8 cycles

Dosage adjustments based upon hematologic and non-hematologic toxicities

• **Nadir platelet counts <50,000/mm^3, absolute neutrophil counts (ANC) <250/mm^3, neutropenic fever, or grades 3/4 non-hematologic toxicities:** day 1 dose in subsequent cycles should be reduced by 25% of the previous dose

• **For patients receiving divided-dose epirubicin (i.e., day 1 and 8):** Day 8 dose should be reduced by 25% of the day 1 dose if the platelet counts are 75,000-100,000/mm^3 and the ANC is 1000-1499/mm^3; if day 8 platelet counts are <75,000/mm^3, ANC <1000/mm^3, or Grade 3/4 non-hematologic toxicity has occurred, omit the day 8 dose

Hepatic dose

• **Adult: IV** Bilirubin 1.2-3 mg/dl or AST 2-4× normal upper limit, 50% of starting dose; bilirubin >3-5 mg/dl or AST >4 × normal upper limit, 25% of starting dose

Available forms: Inj (2 mg/ml) 10 mg/5 ml, 50 mg/25 ml, 150 mg/75 ml, 200 mg/100 ml

Administer:

• Antiemetic 30-60 min before product to prevent vomiting

• Allopurinol or sodium bicarbonate to maintain uric acid levels, alkalinization of urine, monitor electrolytes

Black Box Warning: To be used by a clinician experienced in giving cytotoxic products

Black Box Warning: Do not use IM/SUBCUT because of severe tissue necrosis

• Give IV, do not give IM, subcut; a vesicant; if extravasation occurs stop and complete via another vein, preferably in another limb; avoid inf into veins over joints or in extremities with compromised venous or lymphatic drainage

• Rapid injection can cause facial flushing or erythema along the vein; avoid administration time of <3 min

• Product should be given to those with neutrophils ≥1500/mm^3, platelet count

≥100,000/mm³, and non-hematologic toxicities recovered to ≤grade 1
- When refrigerated, the preservative-free, ready-to-use solution can form a gelled product and will return to solution after 2-4 hr at room temperature
- Visually inspect for particulate matter and discoloration before use

IV route
- Use cytotoxic handling procedures; pregnant women must not handle product
- Reconstitute 50 mg and 200 mg powder for injection vials with 25 ml and 100 ml, respectively, of sterile water for injection (2 mg/ml), shake vigorously for up to 4 min; reconstituted sols are stable for 24 hr when stored refrigerated and protected from light or at room temperature in normal light
- Solution can be further diluted with sterile water for injection

IV INJ route
- Give doses of 100-120 mg/m² into tubing of a freely flowing 0.9% sodium chloride (NS) or D_5W IV inf over 15-20 min; the inf time may be decreased, proportionally, in those who require lower doses; inf times <3 min are not recommended
- Direct injection into the vein is not recommended because of the risk of extravasation; avoid use with any solution of alkaline pH because hydrolysis will occur

IV INF route
- Dilute dose in 0.9% sodium chloride (NS) or D_5W, infuse over 30-60 min, avoid use with any solution of alkaline pH because hydrolysis will occur

Y-site compatibilities: Alemtuzumab, alfentanil, amifostine, amikacin, aminocaproic acid, anidulafungin, atracurium, aztreonam, bivalirudin, bleomycin, bumetanide, buprenorphine, butorphanol, calcium chloride/gluconate, CARBOplatin, caspofungin, ceFAZolin, cefotaxime, ceftizoxime, chlorproMAZINE, cimetidine, ciprofloxacin, cisatracurium, CISplatin, clindamycin, cyclophosphamide, cycloSPORINE, DAPTOmycin, dexrazoxane, digoxin, diltiazem, diphenhydrAMINE, DOBUTamine, DOCEtaxel, dolasetron, DOPamine, doxacurium, doxycycline, droperidol, enalaprilat, ePHEDrine, EPINEPHrine, ertapenem, erythromycin, etoposide, famotidine, fenoldopam, fentaNYL, fluconazole, gatifloxacin, gemcitabine, gentamicin, granisetron, haloperidol, hydrocortisone, HYDROmorphone, hydrOXYzine, ifosfamide, imipenem-cilastatin, inamrinone, insulin (regular), isoproterenol, labetalol, levofloxacin, levorphanol, lidocaine, linezolid, LORazepam, mannitol, meperidine, mesna, methotrexate, metoclopramide, metoprolol, metroNIDAZOLE, midazolam, milrinone, minocycline, mitoMYcin, mivacurium, morphine, nalbuphine, naloxone, nesiritide, niCARdipine, nitroglycerin, nitroprusside, norepinephrine, octreotide, ofloxacin, ondansetron, oxaliplatin, PACLitaxel, palonosetron, pamidronate, pancuronium, pentamidine, pentazocine, phenylephrine, potassium chloride, procainamide, prochlorperazine, promethazine, propranolol, quinupristin-dalfopristin, ranitidine, remifentanil, rocuronium, sodium acetate, succinylcholine, SUFentanil, tacrolimus, teniposide, theophylline, thiotepa, tigecycline, tirofiban, tobramycin, trimethobenzamide, vancomycin, vasopressin, vecuronium, verapamil, vinBLAStine, vinCRIStine, vinorelbine, voriconazole, zidovudine, zoledronic acid

SIDE EFFECTS

CV: Increased B/P, **sinus tachycardia, PVCs,** chest pain, **bradycardia, extrasystoles,** cardiomyopathy

GI: *Nausea, vomiting, anorexia, mucositis, diarrhea*

GU: *Amenorrhea, hot flashes, hyperuricemia*

HEMA: **Thrombocytopenia, leukopenia, anemia, neutropenia, secondary AML**

INTEG: *Rash,* **necrosis,** *pain at inj site, reversible alopecia*

MISC: *Infection, febrile neutropenia, lethargy, fever, conjunctivitis,* tumor lysis syndrome

PHARMACOKINETICS

Triphasic pattern of elimination; half-life 3 min, 2.5 hr, 33 hr; metabolized by liver; crosses placenta; excreted in urine, bile, breast milk

INTERACTIONS

• Give epirubicin before PACLitaxel if both are given

Increase: toxicity—other antineoplastics or radiation, cimetidine

Increase: ventricular dysfunction, CHF—trastuzumab

Increased: heart failure—calcium channel blockers

Decrease: antibody response—live virus vaccine

NURSING CONSIDERATIONS

Assess:

Black Box Warning: **Bone marrow depression (severe):** CBC, differential, platelet count weekly; withhold product if baseline neutrophil ≤1500/mm³; leukocyte nadir occurs 10-14 days after administration, recovery by day 21; notify prescriber of results

• **Infection:** treat before receiving this product if regimens >120 mg/m², prophylactic antibiotics should be given (trimethaprin-sulfamethoxazole or a quinolone)

• Blood, urine uric acid levels; swelling, joint pain primarily in extremities; patient should be well hydrated to prevent urate deposits

• Renal disease: BUN, serum uric acid, urine CCr, electrolytes before, during therapy; I&O ratio; report fall in urine output to <30 ml/hr; dosage adjustment needed if serum creatinine >5 mg/dl

• Increased fluid intake to 2-3 L/day to prevent urate, calculi formation

Black Box Warning: Hepatic studies before, during therapy: bilirubin, AST, ALT, alk phos as needed or monthly

Black Box Warning: **Heart failure:** B/P, pulse, character, rhythm, rate, ABGs, ECG, LVEF, MUGA scan, or ECHO; watch for ST-T wave changes, low QRS and T, possible dysrhythmias (sinus tachycardia, heart block, PVCs); identify cumulative amount of anthracycline received (lifetime)

• Bleeding: hematuria, guaiac, bruising, or petechiae, mucosa or orifices q8hr

• Effects of alopecia on body image; discuss feelings about body changes

Black Box Warning: **Extravasation (vesicant):** local irritation, pain, burning, necrosis at inj site, discontinue and start at another site

• GI symptoms: frequency of stools, cramping

Evaluate:

• Therapeutic response: decreased tumor size, spread of malignancy

Teach patient/family:

• That hair may be lost during treatment; that wig or hairpiece may make patient feel better; that new hair may be different in color, texture; new hair growth occurs in ≤3 mo after treatment

• To avoid crowds, persons with infections when granulocyte count is low

• That contraceptive measures recommended during therapy and for 4 mo thereafter for men and women; pregnancy (D)

• To avoid vaccinations because reactions may occur; to avoid cimetidine during therapy

• That urine may appear red for 2 days

• To avoid OTC medications, supplements unless approved by prescriber

Black Box Warning: That irreversible myocardial damage, leukopenia, menopause may occur

• To report rapid heart beat, trouble breathing, fever, nausea, vomiting, oral sores

eplerenone (Rx)

(ep-ler-ee′known)

Inspra

Func. class.: Antihypertensive

Chem. class.: Selective aldosterone receptor antagonist

Do not confuse:
Inspra, Spiriva

ACTION: Binds to mineralocorticoid receptor and blocks the binding of aldosterone, a component of the renin-angiotensin-aldosterone system (RAAS)

USES: Hypertension, alone or in combination with thiazide diuretics, CHF, post-MI

CONTRAINDICATIONS: Hypersensitivity; increased serum creatinine >2 mg/dl (male), >1.8 mg/dl (female); potassium >5.5 mEq/L, type 2 diabetes with microalbuminuria, hepatic disease, CCr <30 ml/min; CCr <50 ml/min in hypertension

Precautions: Pregnancy (B), breastfeeding, child, geriatric patients, impaired renal/hepatic function, hyperkalemia

DOSAGE AND ROUTES

- **Adult: PO** 50 mg/day initially, may increase to 50 mg bid after 4 wk; start dose at 25 mg/day if patient is taking CYP3A4 inhibitors

CHF or post-MI

- **Adult: PO** 25 mg/day initially, may increase to 50 mg/day max after 4 wk

Available forms: Tabs 25, 50 mg

Administer:

- Without regard to food
- Do not use salt substitutes containing potassium
- Store in tight container at ≤86° F (30° C)

SIDE EFFECTS

CNS: Headache, *dizziness, fatigue*

CV: Angina, MI

GI: Increased GGT *diarrhea,* abdominal pain, increased ALT

GU: Gynecomastia, mastodynia (males), abnormal vaginal bleeding

META: *Hyperkalemia,* hyponatremia, hypercholesteremia, hypertriglyceridemia, increased uric acid

RESP: *Cough*

PHARMACOKINETICS

Peak 1½ hr; serum protein binding 50%; half-life 4-6 hr; metabolized in liver by CYP3A4 inhibitor; excreted in urine, feces

INTERACTIONS

Increase: hyperkalemia—ACE inhibitors, angiotensin II antagonists, NSAIDs, potassium supplements, potassium-sparing diuretics

Increase: serum levels of lithium

Increase: eplerenone levels—erythromycin, fluconazole, verapamil; reduce dose of eplerenone

Increase: levels of eplerenone—CYP3A4 inhibitors (ketoconazole, itraconazole, saquinavir, clarithromycin, imatinib, nelfinavir, nefazodone, ritonavir, troleandomycin; do not use concurrently); reduce dose of eplerenone

Decrease: antihypertensive effect—NSAIDs

Drug/Herb

Decrease: antihypertensive effect—ephedra

Drug/Food

- Grapefruit, grapefruit juice increase product level by 25%
- Do not use salt substitutes containing potassium

Drug/Lab Test

Increase: BUN, creatinine, potassium, cholesterol, lipids, uric acid

Decrease: sodium

NURSING CONSIDERATIONS

Assess:

- **Hypertension:** B/P at peak/trough level of product, orthostatic hypotension, syncope when used with diuretic; monitor lithium level in those also taking lithium
- Renal studies: protein, BUN, creatinine; increased LFTs, uric acid may be increased
- Potassium levels, hyperkalemia may occur

Evaluate:

- Therapeutic response: decreased B/P

Teach patient/family:

- Not to discontinue product abruptly
- Not to use OTC products (cough, cold, allergy) unless directed by prescriber; not to use salt substitutes containing potassium without consulting prescriber
- To comply with dosage schedule, even if feeling better
- That product may cause dizziness, fainting, lightheadedness; may occur during first few days of therapy
- How to take B/P; and about normal readings for age group
- Avoid activities that require coordination

epoetin alfa (Rx)

(ee-poe′e-tin)

Epogen, Eprex ✱, Procrit

Func. class.: Antianemic, biologic modifier, hormone

Chem. class.: Amino acid polypeptide

ACTION: Erythropoietin is 1 factor controlling the rate of red cell production; product is developed by recombinant DNA technology

USES: Anemia caused by reduced endogenous erythropoietin production, primarily end-stage renal disease; to correct hemostatic defect in uremia; anemia due to AZT treatment in patients with HIV or those receiving chemotherapy; reduction of allogenic blood transfusion in surgery patients

Unlabeled uses: Anemia in premature preterm infants, anemia due to ribavirin and interferon-alfa therapy in hepatitic C

CONTRAINDICATIONS: Hypersensitivity to mammalian-cell–derived products, human albumin; uncontrolled hypertension

Precautions: Pregnancy (C), breastfeeding, children <1 mo, seizure disorder; multidose preserved formulation contains benzyl alcohol and should not be used in premature infants; porphyria, CV disease, hemodialysis, latex allergy, hypertension, history of CABG

Black Box Warning: Hgb >12 g/dl, surgery, neoplastic disease

DOSAGE AND ROUTES

Anemia (chronic kidney disease including dialysis-dependent and dialysis-independent patients to decrease the need for red blood cell transfusion)

- **Adult/adolescent ≥17 yr: SUBCUT/IV** Initially, 50–100 units/kg 3×/wk; for patients on dialysis, administer IV; for patients on dialysis, initiate treatment when hemoglobin (Hgb) is <10 g/dl; if Hgb approaches or exceeds 11 g/dl, reduce or interrupt the dose; for patients not on dialysis, consider initiating treatment only when Hgb is <10 g/dl and the rate of Hgb decline indicates the likelihood of requiring RBC transfusion and reducing the risk of alloimmunization and/or other RBC transfusion–related risks is a goal; if Hgb is >10 g/dl, reduce or interrupt the dose, and use the lowest dose sufficient to reduce the need for RBC transfusions; if the Hgb rises >1 g/dl in any 2-wk period, reduce dose by 25% or more as needed to reduce rapid responses; in contrast, if Hgb has not increased >1 g/dl after 4 wk of therapy, increase the dose by 25%; for patients who do not respond adequately over a 12-wk escalation period, increasing the dose further is unlikely to improve response and can increase risks; use the lowest dose that will maintain a Hgb conc sufficient to reduce the need for RBC transfusions; evaluate other causes of anemia, and discontinue if responsiveness does not improve

• **Infant/child/adolescent ≤16 yr: SUBCUT/IV** 50 units/kg 3×/wk initially; for dosage adjustments, see adult dosage

Zidovudine-induced anemia in HIV-infected patients with circulating endogenous erythropoietin concs ≤500 mUnits/ml who are receiving a dose of zidovudine ≤4200 mg/wk

• **Adult: SUBCUT/IV** Initially, 100 units/kg 3×/wk; if Hgb does not increase after 8 wk, increase by 50-100 units/kg at 4-8-wk intervals until Hgb is at a conc to avoid RBC transfusions or a dose of 300 units/kg is reached; if the Hgb is >12 g/dl, withhold, once Hgb is <11 g/dl resume at a dose 25% below the previous dose

Anemia (non-myeloid malignancies where the anemia is due to the effect of concomitantly administered chemotherapy and at least 2 additional mo of chemotherapy is planned)

• **Adult: SUBCUT** 150 units/kg 3×/wk or 40,000 units 1×/wk only when the hemoglobin is <10 g/dl and only until the chemotherapy course is completed; adjust the dosage to maintain the lowest Hgb conc sufficient to avoid RBC transfusions; if no rise in Hgb ≥1 g/dl after 4 wk of therapy and Hgb is <10 g/dl, the dosage may be increased to 300 units/kg subcut 3×/wk or 60,000 units 1×/wk; discontinue if after 8 wk of therapy there is no response as measured by Hgb concs or if transfusions are still required; reduce the dosage by approximately 25% if Hgb increases by >1 g/dl in any 2-wk period or if Hgb reaches a conc needed to avoid RBC inf; if Hgb is increasing and exceeds a conc necessary to avoid blood transfusions, hold therapy and reinstitute at a dose that is 25% lower when the Hgb reaches a conc where transfusions may be needed

• **Adolescent/child ≥5 yr: IV** 600 units/kg/wk only when the hemoglobin is <10 g/dl and only until the chemotherapy course is completed; adjust the dosage to maintain the lowest Hgb conc sufficient to avoid RBC transfusions; if no rise in Hgb ≥1 g/dl after 4 wk of therapy and Hgb is <10 g/dl, the dosage may be increased to 900 units/kg (up to 60,000 units)/wk IV; discontinue if after 8 wk there is no response as measured by Hgb concs or if transfusions are still required; reduce the dosage by approximately 25% if Hgb increases by more than 1 g/dl in any 2-wk period or if Hgb reaches a conc needed to avoid RBC inf; if the Hgb is increasing and exceeds a conc necessary to avoid blood transfusions, hold therapy and reinstitute at a dose that is 25% lower when the Hgb reaches a conc where transfusions may be needed

To reduce the need for allogenic blood transfusions in anemic patients (hemoglobin >10 and ≤13 g/dl) scheduled to undergo elective, noncardiac, nonvascular surgery

• **Adult: SUBCUT** 300 units/kg/day × 10 days before surgery, on the day of surgery, and for 4 days after surgery (14 days total) or 600 units/kg 1×/wk, 21, 14, and 7 days before surgery plus 1 dose on the day of surgery

Available forms: Inj 2000, 3000, 4000, 10,000, 20,000, 40,000 units/ml

Administer:

• Do not shake vial

• Use 1 single-use vial/dose, once syringe has entered single dose vial, sterility cannot be guaranteed, do not administer with other product, multidose vials can be stored in refrigerator up to 21 days once opened

SUBCUT route

• Before injecting preservative free-, single-dose formulation may be admixed using 0.9% NaCl with benzyl alcohol 0.9% at a 1:1 ratio to reduce inj site discomfort

Direct IV route

- Additional heparin to lower chance of clots
- By direct inj or bolus into IV tubing or venous line at end of dialysis
- Decrease dose by 25 units/kg if Hct increases by 4% in 2 wk; increase dose if Hct does not increase by 5-6 pts after 8 wk of therapy; suggested target Hct range 30%-36%

Solution compatibilities: Do not dilute or administer with other sol

SIDE EFFECTS

CNS: Seizures, coldness, sweating, headache
CV: *Hypertension,* hypertensive encephalopathy, CHF, edema, DVT
INTEG: Pruritus, rash, inj site reaction
MISC: Iron deficiency
MS: Bone pain
RESP: Cough

PHARMACOKINETICS

IV: Metabolized in body, extent of metabolism unknown, onset of increased reticulocyte count 2-6 wk, peak immediate

INTERACTIONS

- Need for increased heparin during hemodialysis

NURSING CONSIDERATIONS

Assess:

- Renal studies: urinalysis, protein, blood, BUN, creatinine; I&O, report drop in output <50 ml/hr

Black Box Warning: Blood studies: ferritin, transferrin, serum iron monthly; transferrin sat ≥20%, ferritin ≥100 ng/ml; Hct 2×/wk until stabilized in target range (30%-36%) then at regular intervals; those with endogenous erythropoietin levels of <500 units/L respond to product; monitor Hct 2×/wk with chronic renal failure; patients treated with zidovudine or patients with cancer should be monitored weekly then periodically after stabilization; death may occur with Hgb >12 g/dl

- B/P; check for rising B/P as Hct rises, antihypertensives may be needed; hypertension may occur rapidly, leading to hypertensive encephalopathy
- CNS symptoms: coldness, sweating, pain in long bones; for seizures if Hct is increased within 2 wk by 4 pts
- Hypersensitivity reactions: skin rashes, urticaria (rare), antibody development does not occur

⚠ **Pure cell aplasia (PRCA)** in absence of other causes; evaluate by testing sera for recombinant erythropoetin antibodies; any loss of response to epoetin should be evaluated

- Dialysis patients: thrill, bruit of shunts; monitor for circulation impairment
- **Seizures:** place patient on seizure precautions if increase of ≥4 points HCT in 2 wk, increased B/P; more common in chronic renal failure during the first 90 days of treatment

Evaluate:

- Therapeutic response: increase in reticulocyte count in 2-6 wk, Hgb/Hct; increased appetite, enhanced sense of well-being

Teach patient/family:

- To avoid driving or hazardous activities during beginning of treatment
- To monitor B/P
- To take iron supplements, vit B_{12}, folic acid as directed

eprosartan (Rx)

(ep-roh-sar′tan)

Teveten

Func. class.: Antihypertensive
Chem. class.: Angiotensin II–receptor antagonist (Subtype AT_1)

ACTION: Blocks the vasoconstrictive and aldosterone-secreting effects of angiotensin II; selectively blocks the binding of angiotensin II to the AT_1 receptor found in tissues

USES: Hypertension, alone or with other antihypertensives

⚠ Nurse Alert

CONTRAINDICATIONS:

Hypersensitivity

Black Box Warning: Pregnancy (D)

Precautions: Breastfeeding, children, geriatric patients, hypersensitivity to ACE inhibitors; renal/hepatic disease, angioedema, hyperkalemia

DOSAGE AND ROUTES

- **Adult: PO** 600 mg/day; dose may be divided, given bid, with total daily doses from 400-800 mg, max 900 mg/day

Renal dose

- **Adult: PO** CCr ≤30 ml/min, max 600 mg/day

Available forms: Tabs 400, 600 mg

Administer:

- Without regard to meals

SIDE EFFECTS

CNS: *Dizziness,* depression, *fatigue,* headache

CV: Chest pain, hypotension, palpitations

EENT: Sinusitis

GI: *Diarrhea, dyspepsia, abdominal pain*

GU: UTI

HEMA: **Neutropenia**

INTEG: Pruritus, **angioedema**

META: Hypertriglyceridemia

MS: *Myalgia,* arthralgia, **rhabdomyolysis**

RESP: *Cough, upper respiratory infection,* rhinitis, pharyngitis, viral infection

SYST: **Anaphylaxis**

PHARMACOKINETICS

Peak 1-2 hr, food delays absorption; protein binding 98%; moderate renal impairment increases product levels by 30%, hepatic impairment increases levels by 40%; excreted in urine and feces; half-life 5-9 hr

INTERACTIONS

Increase: antihypertensive effect—other antihypertensives

Increase: hyperkalemia—ACE inhibitors, angiotensin II receptor antagonists, potassium sparing diuretics, potassium supplements

Increase: lithium toxicity—lithium

Decrease: antihypertensive effect—NSAIDs, salicylates

Drug/Herb

Decrease: antihypertensive effect—ephedra

Increase: antihypertensive effect—hawthorn

Drug/Lab Test

Increase: ALT, AST, alk phos, potassium

Decrease: Hgb

NURSING CONSIDERATIONS

Assess:

- B/P with position changes, pulse q4hr; note rate, rhythm, quality

⚠ Hypersensitivity reactions, including anaphylaxis

⚠ Myalgia, arthralgia; may cause rhabdomyolysis

- Baselines of renal, hepatic studies before therapy begins
- Edema in feet, legs daily
- Skin turgor, dryness of mucous membranes for hydration status

Evaluate:

- Therapeutic response: decreased B/P

Teach patient/family:

- To comply with dosage schedule, even if feeling better
- To notify prescriber of fever; chest pain; swelling of hands, feet, face, lip, or tongue
- That excessive perspiration, dehydration, diarrhea may lead to fall in B/P; consult prescriber if these occur, maintain adequate hydration
- That product may cause dizziness; to avoid hazardous activities until effect is known; to rise slowly from sitting

Black Box Warning: Not to take this product if pregnant or breastfeeding, or if have had an allergic reaction to product

- To take missed dose as soon as possible unless within 1 hr before next dose
- That therapeutic effect may take 2-3 wk

⚠ HIGH ALERT

eptifibatide (Rx)

(ep-tih-fib′ah-tide)

Integrilin

Func. class.: Antiplatelet agent

Chem. class.: Glycoprotein IIb/IIIa inhibitor

ACTION: Platelet glycoprotein antagonist; this agent reversibly prevents fibrinogen, von Willebrand's factor from binding to the glycoprotein IIb/IIIa receptor, thus inhibiting platelet aggregation

USES: Acute coronary syndrome including those undergoing percutaneous coronary intervention (PCI)

CONTRAINDICATIONS: Hypersensitivity, active internal bleeding; recent history of bleeding, stroke within 30 days or any hemorrhagic stroke; major surgery with severe trauma, severe hypertension, current or planned use of another parenteral GP IIb/IIIa inhibitor, dependence on renal dialysis, coagulopathy, AV malformation, aneurysm

Precautions: Pregnancy (B), breastfeeding, children, geriatric patients, bleeding, impaired renal function

DOSAGE AND ROUTES

Acute coronary syndrome

- **Adult: IV BOL** 180 mcg/kg as soon as diagnosed, max 22.6 mg, then **IV CONT** 2 mcg/kg/min until discharge or CABG up to 72 hr, max 15 mg/hr

PCI in patients without acute coronary syndrome

- **Adult: IV BOL** 180 mcg/kg given immediately before PCI then 2 mcg/kg/min × 18 hr CONT IV INF and a second 180-mcg/kg bolus by 10 min after 1st bolus; continue inf for up to 18-24 hr at rate of 1 mcg/kg/min

Renal dose

- **Adult: IV** maintenance CCr <50 ml/min, 1 mcg/kg/min, max rate 7.5 mg/hr; CCr <10 ml/min, contraindicated

Available forms: Sol for inj 2 mg/ml (10 ml), 0.75 mg/ml (100 ml)

Administer:

- Aspirin and heparin may be given with this product; check for bleeding
- D/C heparin before removing femoral artery sheath, after PCI
- Do not give discolored solutions, those with particulates; discard unused amount
- Discontinue product prior to CABG
- All medications PO if possible; avoid IM inj, all catheters

Direct IV route

- After withdrawing bolus dose from 10-ml vial, give IV push over 1-2 min

Continuous IV INF route

- Follow bolus dose with continuous inf using pump; give product undiluted directly from 100-ml vial, spike 100-ml vial with vented inf set, use caution when centering spike on circle of stopper top

Y-site compatibilities: Alfentanil, alteplase, amikacin, aminophylline, amphotericin B lipid complex, amphotericin B liposome, ampicillin, ampicillin-sulbactam, anidulafungin, argatroban, atenolol, atracurium, atropine, azithromycin, aztreonam, bivalirudin, bumetanide, buprenorphine, butorphanol, calcium chloride/gluconate, ceFAZolin, cefepime, cefoperazone, cefotaxime, cefoTEtan, cefOXitin, cefTAZidime, ceftizoxime, cefTRIAXone, cefuroxime, cimetidine, ciprofloxacin, cisatracurium, clindamycin, cycloSPORINE, DAPTOmycin, dexamethasone, D_5/NaCl 0.9%, diazepam, diltiazem, diphenhydrAMINE, DOBUTamine, dolasetron, DOPamine, doxycycline, droperidol, enalaprilat, ePHEDrine, EPINEPHrine, ertapenem, erythromycin, esmolol, famotidine, fentaNYL, fluconazole, fosphenytoin, ganciclovir, gatifloxacin, gentamicin, granisetron, haloperidol, heparin, hydrocortisone, HYDROmorphone, hydrOXYzine, imipenem-cilastatin, inamrinone, isoproterenol, ketorolac, labetalol, leucovorin, levofloxacin, levorphanol, lidocaine, linezolid, LORazepam, magnesium sulfate, mannitol, meperidine, meropenem, methylPREDNISolone, metoclopramide, metoprolol, metroNIDAZOLE, micafungin, midazolam, milrinone, minocycline, mivacurium, morphine, nalbuphine, naloxone, niCARdipine, nitroglycerin, nitroprusside, NS, octreotide, ofloxacin, ondansetron, oxytocin, palonosetron, pancuronium, PEMEtrexed, PENTobarbital, PHENobarbital, phenylephrine, piperacillin, piperacillin-tazobactam, potassium chloride/phosphates, procainamide, prochlorperazine, promethazine, propranolol, ranitidine, remifentanil, rocuronium, sodium bicarbonate/phosphates, succinylcholine, SUFentanil, sulfamethoxazole-trimethoprim, teniposide, theophylline, ticarcillin, ticarcillin-clavulanate, tigecycline, tirofiban, tobramycin, trimethobenzamide, vancomycin, vecuronium, verapamil, zidovudine, zoledronic acid

Solution compatibilities: 0.9% NaCl, D_5/0.9% NaCl

SIDE EFFECTS

CV: Stroke, hypotension
GU: Hematuria
HEMA: Thrombocytopenia
SYST: Major/minor bleeding from any site, anaphylaxis

PHARMACOKINETICS

Onset within 1 hr, protein binding 25%, half-life 1.5-2 hr, steady state 4-6 hr, metabolism limited, excretion via kidneys

INTERACTIONS

- Do not give with glycoprotein inhibitors IIb, IIIa

Increase: bleeding—aspirin, heparin, NSAIDs, anticoagulants, ticlopidine, clopidogrel, dipyridamole, thrombolytics, valproate, abciximab, SSRIs, SNRIs

Drug/Herb
Increase: Bleeding risk—feverfew, garlic, ginger, ginkgo, ginseng

NURSING CONSIDERATIONS

Assess:

⚠ **Thrombocytopenia:** platelets, Hgb, Hct, creatinine, APTT baseline within 6 hr of loading dose, daily therafter, patients undergoing PCI should have ACT monitored; maintain APTT 50-70 sec unless PCI to be performed; during PCI, ACT should be 200-300 sec; if platelets drop $<100,000/mm^3$, obtain additional platelet counts; if thrombocytopenia is confirmed, discontinue product; draw Hct, Hgb, serum creatinine

⚠ **Bleeding:** gums, bruising, ecchymosis, petechiae; from GI, GU tract, cardiac cath sites, IM inj sites

Teach patient/family:

- About reason for medication and expected results
- To report bruising, bleeding, chest pain immediately

eribulin
(er′i-bu′lin)
Halaven
Func. class.: Antineoplastics—non-taxane

Do not confuse:
eribulin/epirubicin/erlotinib

ACTION: Potent antimitotic agent, different from taxanes, vinca alkaloids, epothilones; blocks cell progression during G2-M phase; inhibits the growth phase of microtubules and sequesters tubules, leading to the disruption of mitotic spindles and apoptotic cell death

USES: Metastatic breast cancer in patients who have received at least 2 chemotherapy regimens

CONTRAINDICATIONS: Hypersensitivity, pregnancy (D)

Precautions: Breastfeeding, neonates, infants, children, bradycardia, electrolyte imbalances, heart failure, hypokalemia, hypomagnesemia, infertility, neutropenia, peripheral neuropathy, QT prolongation, hepatic/renal disease

DOSAGE AND ROUTES

• **Adult: IV** 1.4 mg/m^2 over 2-5 min on days 1 and 8, repeat q21days

• **Recommendations for dose delay:** for ANC <1000/mm^3, platelets <75,000/mm^3, or grade 3 or 4 nonhematologic toxicities: do not administer; the day 8 dose may be delayed a maximum of 1 wk; for the day 8 dose, if toxicities do not resolve to ≤ Grade 2 by day 15: omit the dose; for the day 8 dose, if toxicities resolve or improve to ≤ Grade 2 by day 15: administer eribulin at reduced dose (see below), initiate the next cycle no sooner than 2 wk later

• **Dose adjustments for hematologic toxicity:** ANC <500/mm^3 for >7 days or ANC <1000/mm^3 with fever or infection: permanently reduce dose to 1.1 mg/m^2; platelets <25,000/mm^3 or <50,000/mm^3 requiring transfusion: permanently reduce dose to 1.1 mg/m^2; if day 8 of previous cycle omitted or delayed: permanently reduce dose to 1.1 mg/m^2; while receiving 1.1 mg/m^2, if recurrence of hematologic event occurs, or if day 8 of previous cycle omitted or delayed: permanently reduce dose to 0.7 mg/m^2; while receiving 0.7 mg/m^2, if recurrence of hematologic event occurs, or if day 8 of previous cycle omitted or delayed: discontinue

• **Dose adjustments of eribulin for nonhematologic toxicity during treatment:** any Grade 3 or 4 nonhematologic toxicity: permanently reduce dose to 1.1 mg/m^2; if day 8 of previous cycle omitted or delayed: permanently reduce dose to 1.1 mg/m^2; while receiving 1.1 mg/m^2, if recurrence of Grade 3 or 4 nonhematologic toxicity occurs, or if day 8 of previous cycle omitted or delayed: permanently reduce dose to 0.7 mg/m^2; while receiving 0.7 mg/m^2, if recurrence of Grade 3 or 4 nonhematologic toxicity occurs, or if day 8 of previous cycle omitted or delayed: discontinue

Available forms: Sol for inj 1 mg/2 ml

Administer:

IV direct, intermittent route

• Visually inspect for particulate matter, discoloration as solution and container permit; withdraw required amount (0.5 mg/ml) from single-use vial, give undiluted over 2-5 min or diluted in 100 ml 0.9% NaCl and give as intermittent inf; do not give through line with dextrose or any other product

• Store at room temperature for 4 hr or 24 hr refrigerated

SIDE EFFECTS

CNS: Depression, dizziness, *fatigue*, fever, headache, insomnia, *peripheral neuropathy*

CV: QT prolongation, peripheral edema

GI: Abdominal pain, anorexia, constipation, diarrhea, dyspepsia, nausea, vomiting, weight loss

HEMA: Anemia, neutropenia, thrombocytopenia

INTEG: *Alopecia*, rash, stomatitis

META: Hypokalemia

MS: Arthralgia, myalgia, bone/back pain

RESP: Cough, dyspnea

SYST: Infection

PHARMACOKINETICS

Protein binding 49%-65%; inhibits CYP3A4; excreted in feces 82%; urine 9%; elimination half-life 40 hr; increased levels in hepatic/renal disease

INTERACTIONS

Increase: QT prolongation—arsenic trioxide, astemizole, bepridil, chloroquine, certain phenothiazines (chlorproMAZINE, mesoridazine, thioridazine), cisapride, clarithromycin, class IA antiarrhythmics (disopyramide, procainamide, quiNIDine), class III antiarrhythmics (amiodarone, bretylium, dofetilide, ibutilide, sotalol), dextromethorphan; quiNIDine, dronedarone, droperidol, erythromycin, halofantrine, haloperidol, levomethadyl, methadone, pentamidine, pimozide, posaconazole, probucol, propafenone, saquinavir, sparfloxacin terfenadine, troleandomycin, and ziprasidone; also to a lesser degree abarelix, alfuzosin, amoxapine, apomorphine, artemether; lumefantrine, asenapine, β-agonists, ofloxacin, cloZAPine, cyclobenzaprine, dasatinib, dolasetron, flecainide, gatifloxacin, gemifloxacin, halogenated anesthetics, iloperidone, lapatinib, levofloxacin, local anesthetics, lopinavir; ritonavir, magnesium sulfate; potassium sulfate; sodium sulfate, maprotiline, mefloquine, moxifloxacin, nilotinib, norfloxacin, octreotide, ciprofloxacin, OLANZapine, ondansetron, paliperidone, palonosetron, some phenothiazines (fluPHENAZine, perphenazine, prochlorperazine, trifluoperazine), QUEtiapine, ranolazine, risperiDONE, sertindole, SUNItinib, tacrolimus, telavancin, telithromycin, tetrabenazine, tricyclic antidepressants, venlafaxine, vardenafil, vorinostat

NURSING CONSIDERATIONS

Assess:

- Peripheral neuropathy: pain, numbness in extremities
- Infection: increased temperature, sore throat, flulike symptoms
- **QT prolongatation:** assess for drug interactions that may occur; monitor ECG, heart rate
- **Bone marrow depression:** CBC, differential, serum creatinine, BUN, electrolytes, LFTs at baseline, periodically; increased AST/ALT >3 × ULN or total bilirubin >1.5 × ULN involve greater chance of Grade 4 or febrile neutropenia

Teach patient/family:

- **Infection:** to notify prescriber of increased temperature, sore throat, fatigue, flulike symptoms
- **QT prolongation:** to report extra heart beats
- **Peripheral neuropathy:** to report tingling, pain in extremities
- About reason for product and expected results
- To avoid other medications, supplements unless approved by provider; serious drug interactions may occur
- About hair loss, use of wig or hairpiece
- To notify prescriber if pregnancy is planned or suspected (pregnancy [D]), to avoid breastfeeding

E

erlotinib (Rx)

(er-loe′tye-nib)

Tarceva

Func. class.: Antineoplastic—miscellaneous

Chem. class.: Epidermal growth factor receptor inhibitor

ACTION: Not fully understood; inhibits intracellular phosphorylation of cell-surface receptors associated with epidermal growth factor receptors

USES: Non–small-cell lung cancer (NSCLC) including EGFR ex on 19 deletions or ex on 21 substitution mutations, pancreatic cancer

Unlabeled uses: Squamous cell head and neck cancer

CONTRAINDICATIONS: Pregnancy (D), breastfeeding

Precautions: Children, geriatric patients, ocular/pulmonary/renal/hepatic disorders, bradycardia, heart failure, hypokalemia, infertility, QT prolongation

DOSAGE AND ROUTES

Non–small-cell lung cancer (NSCLC)

- **Adult: PO** 150 mg/day

Pancreatic cancer

- **Adult: PO** 100 mg/day in combination with gemcitabine 1000 mg/m^2 cycle 1, days 1, 8, 15, 22, 29, 36, 43 of 8-wk cycle; cycle 2 and subsequent cycle, days 1, 8, 15 of 4-wk cycle

CYP3A4 inducers concurrently (e.g., rifampin, phenytoin)
- Dosage increase is advised

CYP3A4 inhibitors (atazanavir, clarithromycin, indinavir, itraconazole, ketoconazole, telithromycin, ritonavir, saquinavir, troleandomycin, nelfinavir)
- Dosage reduction may be needed

Hepatic dose
- **Adult: PO** interrupt if total bilirubin >3 times ULN and/or transaminases >5 times ULN

Head or neck cancer (unlabeled)
- **Adult: PO** 150 mg daily

Available forms: Tabs 25, 100, 150 mg

Administer:
- 1 hr before or 2 hr after food; at same time of day

SIDE EFFECTS

CNS: CVA, anxiety, depression, headache, rigors

CV: MI/ischemia

EENT: Ocular changes, *conjunctivitis, eye pain*

GI: *Nausea, diarrhea, vomiting, anorexia, mouth ulceration,* hepatic failure, GI perforation

GU: Renal impairment/failure

HEMA: Deep vein thrombosis

INTEG: *Rash,* Stevens-Johnson–like skin reaction, toxic epidermal necrolysis

MISC: *Fatigue, infection*

RESP: Interstitial lung disease, *cough, dyspnea,* ARDS, pulmonary fibrosis

SYST: Hepatorenal syndrome

PHARMACOKINETICS

Slowly absorbed (60%); peak 3-7 hr; excreted in feces (86%), urine (<4%); metabolized by CYP3A4; terminal half-life 36 hr; protein binding 93%

INTERACTIONS

Increase: GI bleeding, may be fatal—warfarin

Increase: erlotinib concs—CYP3A4 inhibitors (ketoconazole, itraconazole, erythromycin, clarithromycin, telithromycin)

Increase: plasma conc of warfarin, metoprolol

Increase: myopathy—HMG-CoA reductase inhibitors

Decrease: erlotinib levels—CYP3A4 inducers (phenytoin, rifampin, carBAMazepine, PHENobarbital), proton-pump inhibitors

Drug/Herb

Decrease: erlotinib levels—St. John's wort

Drug/Smoking

Decrease: erlotinib level; dose may need to be increased

Drug/Food

Increase: effect of erlotinib—grapefruit juice

NURSING CONSIDERATIONS

Assess:
- Serious skin toxicities: toxic epidermal necrolysis, Stevens-Johnson syndrome

⚠ **MI/ischemia, CVA** in patients with pancreatic cancer

⚠ **Pulmonary changes:** lung sounds, cough, dyspnea; interstitial lung disease may occur, may be fatal; discontinue therapy if confirmed

- **Ocular changes:** eye irritation, corneal erosion/ulcer, aberrant eyelash growth
- GI symptoms: frequency of stools; if diarrhea is poorly tolerated, therapy may be discontinued for ≤14 days, monitor for dehydration, fluid status during period of vomiting and diarrhea
- Blood studies: INR, LFTs, PT
- **Hepatic failure:** interrupt dosing if severe changes to liver function occur (total bilirubin >3× ULN and/or transaminases >5× ULN when normal pretreatment LFTs)
- **GI perforation/bleeding:** some cases have been fatal

Evaluate:
- Therapeutic response: decrease in NSCLC cells, pancreatic cancer cells

Teach patient/family:

⚠ To report adverse reactions immediately: SOB, severe abdominal pain, persistent diarrhea or vomiting, ocular changes, skin eruptions

• About reason for treatment, expected results

• To use reliable contraception during treatment (pregnancy D); to avoid breastfeeding

• To avoid use with other products, herbs, supplements unless approved by provider

• To avoid smoking, decreases effect of this product

ertapenem (Rx)

(er-tah-pen'em)

INVanz

Func. class.: Antiinfective—miscellaneous

Chem. class.: Carbapenem

Do not confuse:
INVanz/AVINza

ACTION: Interferes with cell wall replication of susceptible organisms; bactericidal

USES: *Bacteroides distasonis, Bacteroides fragilis, Bacteroides ovatus, Bacteroides thetaiotaomicron, Bacteroides uniformis, Bacteroides vulgatus, Citrobacter freundii, Citrobacter koseri, Clostridium clostridioforme, Clostridium perfringens, Enterobacter aerogenes, Enterobacter cloacae, Escherichia coli, Eubacterium lentum, Fusobacterium* sp., *Haemophilus influenzae* (beta-lactamase negative), *Haemophilus influenzae* (beta-lactamase positive), *Haemophilus parainfluenzae, Klebsiella oxytoca, Klebsiella pneumoniae, Moraxella catarrhalis, Morganella morganii, Peptostreptococcus* sp., *Porphyromonas asaccharolytica, Prevotella bivia, Proteus mirabilis, Proteus vulgaris, Providencia rettgeri, Providencia stuartii, Serratia marcescens, Staphylococcus aureus* (MSSA), *Staphylococcus epidermidis, Streptococcus agalactiae* (group B streptococci), *Streptococcus pneumoniae, Streptococcus pyogenes* (group A beta-hemolytic streptococci); bacteremia, community-acquired pneumonia, diabetic foot ulcer, endometritis, gyn/intraabdominal skin/skin structure/urinary tract infections, surgical infection prophylaxis

CONTRAINDICATIONS: Hypersensitivity to this product, its components, amide-type local anesthetics (IM only); anaphylactic reactions to β-lactams

Precautions: Pregnancy (B), breastfeeding, children, geriatric patients, GI/renal/hepatic disease, seizures

DOSAGE AND ROUTES

Complicated intraabdominal infections

• **Adult/adolescent: IM/IV** 1 g/day × 5-14 days

• **Infant ≥3 mo/child: IM/IV** 15 mg/kg bid (max 1 g/day) × 5-14 days

Complicated skin/skin-structure infections

• **Adult/adolescent: IM/IV** 1 g/day × 7-14 days

• **Child 3 mo-12 yr: IM/IV** 15 mg/kg bid × 7-14 days

Community-acquired pneumonia

• **Adult/adolescent: IM/IV** 1 g/day × 10-14 days

• **Child 3 mo-12 yr: IM/IV** 15 mg/kg bid × 10-14 days, max 1 g/day

Complicated UTI

• **Adult/adolescent: IM/IV** 1 g/day × 10-14 days

• **Child 3 mo-12 yr: IM/IV** 15 mg/kg bid × 10-14 days

Acute pelvic infections

• **Adult/adolescent: IM/IV** 1 g/day × 3-10 days

• **Child 3 mo-12 yr: IM/IV** 15 mg/kg bid 3-10 days

Surgical infection prophylaxis (unlabeled)

• **Adult:** **IV** 1 g as a single dose 1 hr prior to surgical incision

Renal dose

• **Adult: IM/IV** CCr ≤30 ml/min 500 mg daily

Available form: Powder, lyophilized, 1 g

Administer:

IM route

• Reconstitute the 1 g vial of ertapenem with 3.2 ml of 1% lidocaine HCl injection (without EPINEPHrine) (280 mg/ml), agitate well to form a solution; the IM reconstituted formulation is not for IV use

• IM administration may be used as an alternative to IV administration in the treatment of infections where IM therapy is appropriate; only give via IM injection × 7 days

• **For a 1 g dose:** immediately withdraw the contents of the vial and inject deeply into a large muscle, aspirate before injection to avoid injection into a blood vessel

• **For a dose** <1 g (i.e., for pediatric patients 3 mo-12 yr): immediately withdraw a volume equal to 15 mg/kg (max 1 g/day) and inject deeply into a large muscle, aspirate before injection to avoid injecting into a blood vessel; use the reconstituted IM sol within 1 hr after preparation

IV route

• Visually inspect for particulate matter and discoloration before use, may be colorless to pale yellow; do not mix with other products; dextrose sols are not compatible

• **1 g vial:** For each gram reconstitute with 10 ml of either NS injection, sterile water for injection, or bacteriostatic water for injection to 100 mg/ml, shake

• **1 g dose**: immediately transfer contents of the reconstituted vial to 50 ml of NS injection; for a dose <1 g (pediatric patients 3 mo-12 yr): from the reconstituted vial, immediately withdraw a volume equal to 15 mg/kg of body weight (max 1 g/day) and dilute in NS injection to a conc of 20 mg/ml or less

IV INF route

• Complete the inf within 6 hr of reconstitution, infuse over 30 min; do not coinfuse with other medications

• The reconstituted IV sol may be stored at room temperature if used within 6 hr, or store under refrigeration for 24 hr and use within 4 hours after removal from refrigeration; do not freeze

Y-site compatibilities: Acyclovir, alfentanil, amifostine, amikacin, aminocaproic acid, aminophylline, amphotericin B lipid complex, amphotericin B liposome, argatroban, arsenic trioxide, atenolol, atracurium, azithromycin, aztreonam, bivalirudin, bleomycin, bumetanide, buprenorphine, busulfan, butorphanol, calcium chloride/gluconate, CARBOplatin, carmustine, chloramphenicol, cimetidine, ciprofloxacin, cisatracurium, CISplatin, cyclophosphamide, cycloSPORINE, cytarabine, dacarbazine, DACTINomycin, DAPTOmycin, dexamethasone, dexmedetomidine, dexrazoxane, digoxin, diltiazem, diphenhydrAMINE, DOCEtaxel, dolasetron, DOPamine, doxacurium, doxycycline, enalaprilat, ePHEDrine, EPINEPHrine, eptifibatide, erythromycin, esmolol, etoposide, etoposide phosphate, famotidine, fenoldopam, fluconazole, fludarabine, fluorouracil, foscarnet, fosphenytoin, furosemide, ganciclovir, gatifloxacin, gemcitabine, gemtuzumab, gentamicin, glycopyrrolate, granisetron, haloperidol, heparin, hydrocortisone, HYDROmorphone, ifosfamide, inamrinone, insulin (regular), irinotecan, isoproterenol, ketorolac, labetalol, lepirudin, leucovorin, levofloxacin, lidocaine, linezolid, LORazepam, magnesium sulfate, mannitol, mechlorethamine, melphalan, meperidine, mesna, metaraminol, methotrexate, methyldopate, methylPREDNISolone, metoclopramide, metroNIDAZOLE, milrinone, mitoMYcin, mivacurium, morphine, moxifloxacin, nalbuphine, naloxone, nesiritide, nitroglycerin, nitroprusside, norepinephrine, octreotide, oxaliplatin, oxytocin, PACLitaxel, pamidronate, pancuronium, pantoprazole, PEMEtrexed, PENTobarbital, PHENobarbital, phentolamine, phenylephrine, polymyxin B, potassium acetate/chloride/phosphates, procainamide, propranolol, ranitidine, remifentanil, rocuronium, sodium acetate/bicarbonate/phosphates, streptozocin, succinylcholine, SUFentanil, sulfamethoxazole-trimethoprim, tacrolimus, telavancin, teniposide, theophylline, thiotepa, tigecycline, tirofiban, tobramycin, trimethobenzamide, vancomycin, vasopressin, vecuronium, vinBLAStine, vinCRIStine, vinorelbine, voriconazole, zidovudine, zoledronic acid

SIDE EFFECTS

CNS: Insomnia, seizures, dizziness, *headache,* agitation, confusion, somnolence, disorientation, edema, hypotension

CV: Tachycardia, seizures

GI: *Diarrhea, nausea, vomiting,* pseudomembranous colitis, cholelithiasis, jaundice, abdominal pain

GU: *Vaginitis,* dysuria

INTEG: *Rash,* urticaria, *pruritus,* pain at inj site, *infused vein complication, phlebitis/thrombophlebitis,* erythema at inj site, dermatitis

RESP: Dyspnea, cough, pharyngitis, crackles, respiratory distress

SYST: Anaphylaxis, angioedema

PHARMACOKINETICS

IV: Onset immediate; peak dose dependent; half-life 4 hr; metabolized by liver; excreted in urine, feces, breast milk

INTERACTIONS

Increase: ertapenem levels—probenecid; do not coadminister

Decrease: effect of valproic acid

Drug/Lab Test

Increase: hepatic enzymes

NURSING CONSIDERATIONS

Assess:

- Renal disease: lower dose may be required
- **Pseudomembranous colitis:** bowel pattern daily: if severe diarrhea occurs, product should be discontinued
- For infection: temperature, sputum, characteristics of wound before, during, after treatment
- ⚠ **Allergic reactions, anaphylaxis;** rash, urticaria, pruritus; may occur a few days after therapy begins; sensitivity to carbapenem antibiotics, other β-lactam antibiotics, penicillins
- **Overgrowth of infection:** perineal itching, fever, malaise, redness, pain, swelling, drainage, rash, diarrhea, change in cough or sputum

Evaluate:

- Therapeutic response: negative C&S; absence of signs, symptoms of infection

Teach patient/family:

- To report severe diarrhea (may indicate **pseudomembranous colitis**), CNS side effects
- To report overgrowth of infection: black, furry tongue; vaginal itching; foul-smelling stools
- To avoid breastfeeding; product is excreted in breast milk

TREATMENT OF OVERDOSE:

EPINEPHrine, antihistamines; resuscitate if needed (anaphylaxis)

erythromycin

(e-rith′roe-mye′sin)

Ilotycin, Romycin

Func. class.: Ophthalmic anti-infective

Chem. class.: Macrolide

ACTION:

Inhibits protein synthesis, thereby decreasing bacterial replication

USES:

Conjunctivitis, eye infections, prevention of ophthalmic neonatorum

CONTRAINDICATIONS:
Hypersensitivity to this product or macrolides
Precautions: Pregnancy (B), breastfeeding

DOSAGE AND ROUTES

Bacterial conjunctivitis

- **Adult/adolescent/child:** apply 1 cm of ointment directly to the eye up to 6 times a day ×7-10 days depending on severity of infection

Prevention of ophthalmic neonatorum

Neonate: Ointment apply 1 cm ribbon to lower conjunctival sac of each eye once after birth

Administer:

Ophthalmic route

- Apply ribbon of ointment directly to the eye; for ophthalmic use only

SIDE EFFECTS

EENT: Hypersensitivity, irritation, redness

PHARMACOKINETICS

Unknown

NURSING CONSIDERATIONS

Assess:

- **Allergic reaction:** assess for hypersensitivity, discontinue product

Evaluate:

- Decreased ophthalmic infection

Teach patient/family:

Ophthalmic route:

- Apply ribbon of ointment directly to the eye; for ophthalmic use only

erythromycin base (Rx)

(eh-rith-roh-my'sin)

Apo-Erythro ♣, Ery-Tab, Novo-Rythro Encap ♣, PCE

erythromycin ethylsuccinate (Rx)

Apo-Erythro-Es ♣, E.E.S., Ery Ped, Novo-Rythro ♣

erythromycin lactobionate (Rx)

Erythrocin

erythromycin stearate (Rx)

Apo-Erythro-S ♣, Erythrocin, My-E, Novo-Rythro ♣

Func. class.: Antiinfective
Chem. class.: Macrolide

Do not confuse:
erythromycin/azithromycin

ACTION:
Binds to 50S ribosomal subunits of susceptible bacteria and suppresses protein synthesis

USES:
Infections caused by *Neisseria gonorrhoeae;* mild to moderate respiratory tract, skin, soft-tissue infections caused by *Bordetella pertussis, Borrelia burgdorferi, Chlamydia trachomatis; Corynebacterium diphtheriae, Haemophilus influenzae* (when used with sulfonamides); *Legionella pneumophila,* Legionnaire's disease, *Listeria monocytogenes; Mycoplasma pneumoniae, Streptococcus pneumoniae,* syphilis: *Treponema pallidum; Staphylococcus* sp.

Unlabeled uses: Bartonellosis, burn wound infection, chancroid, cholera, diabetic gastroparesis, endocarditis prophylaxis, gastroenteritis, granuloma inguinale, Lyme disease, tetanus

CONTRAINDICATIONS:
Hypersensitivity, preexisting hepatic disease (estolate)

Precautions: Pregnancy (B), breast-feeding, geriatric patients, hepatic disease, GI disease, QT prolongation, seizure disorder, myasthenia gravis

DOSAGE AND ROUTES

Acne vulgaris

- **Adult: PO** 250 mg qid

Mild to moderately severe upper respiratory tract infections (otitis media, sinusitis) or lower respiratory tract infections (pneumonia, bronchitis) caused by susceptible organisms

- **Adult: PO** 250-500 mg (of base, estolate, or stearate) every 6 hr or 400-800 mg (ethylsuccinate) every 6 hr; **IV** 15-20 mg/kg/day in divided doses every 4-6 hr, max 4 g/day
- **Adolescent/child/infant: PO** 20-50 mg/kg/day divided every 6 hr, max adult doses; **IV** 15-20 mg/kg/day in divided doses every 4-6 hr, or as a continuous inf, max dose 4 g/day
- **Neonate >7 days, ≥1200 g: PO** 30 mg/kg/day in divided doses every 8 hr
- **Neonates >7 days, <1200 g: PO** 20 mg/kg/day in divided doses every 12 hr
- **Neonates ≤7 days: PO** 20 mg/kg/day in divided doses every 12 hr

Pneumonia caused by *Chlamydia trachomatis*

- **Infant/neonate: PO** CDC recommends 50 mg/kg/day in 4 divided doses × 14 days (erythromycin base or ethylsuccinate)

Mycoplasma infection such as *Mycoplasma pneumoniae* pneumonia

- **Adult: PO** 250-500 mg tid
- **Adult/adolescent/child/infant: IV** 15-20 mg/kg/day, given in divided doses every 4-6 hr, or as a continuous inf, max dose 4 g/day; replace by oral dosage as soon as possible

Legionnaire's disease (caused by *Legionella pneumophila*)

- **Adult: PO/IV** 0.5-1 g every 6 hr × 21 days

Treatment of group A β-hemolytic streptococcal (GAS) pharyngitis (primary rheumatic fever prophylaxis)

- **Adult: PO** 250-500 mg (base, estolate, or stearate) every 6 hr or 400-800 mg (ethylsuccinate) every 6 hr × 10 days
- **Adolescent/child/infant: PO** 20-50 mg/kg/day, divided every 6 hr × 10 days, max adult dose

Secondary prevention of rheumatic fever (prevention of recurrent attacks of rheumatic fever)

- **Adult/adolescent/child: PO** 250 mg bid in patients allergic to penicillin and sulfADIAZINE for 10 yr or age 40 whichever is longer, secondary prophylaxis (American Heart Association)

Listeriosis

- **Adult: PO** 250-500 mg (base, estolate or stearate) every 6 hr or 400-800 mg (ethylsuccinate) every 6 hr
- **Adolescent/child/infant: PO** 20-50 mg/kg/day, divided every 6 hr, max adult doses

Cervicitis caused by *Chlamydia trachomatis*

- **Adult/adolescent: PO** CDC recommends erythromycin base 500 mg qid or erythromycin ethylsuccinate 800 mg qid × 7 days as alternatives to first-line agents doxycycline or azithromycin
- **Pregnant females: PO** As alternatives to first-line agents azithromycin or amoxicillin, CDC recommends base 500 mg q12hr, 333 mg q8hr, or 250 mg qid × 14 days
- **Child ≤45 kg: PO** CDC recommends base or ethylsuccinate 50 mg/kg/day in 4 doses × 14 days

Chlamydial conjunctivitis caused by *Chlamydia trachomatis* including trachoma and inclusion conjunctivitis

- **Pregnant/lactating woman/child <8 yr: PO** 250-500 mg qid × 10-14 days

Infant pneumonia caused by *Chlamydia trachomatis*

- **Infant/neonate: PO** (base or ethylsuccinate) CDC recommends 50 mg/kg/day 4 divided doses × 14 days

- **Pregnant female: PO** CDC recommends base 400 mg qid × 14 days

Non-gonococcal urethritis (NGU) caused by *Chlamydia trachomatis* or *Ureaplasma urealyticum*

- **Adult/adolescent: PO** CDC recommends 500 mg (base) qid or 800 mg (ethylsuccinate) qid × 7 days as alternatives to first-line agents doxycycline or azithromycin
- **Child <45 kg: PO** CDC recommends base 50 mg/kg/day in 4 divided doses × 14 days, second course of therapy may be required

Ophthalmia neonatorum caused by *Chlamydia trachomatis*

- **Neonate: PO** (erythromycin base or ethylsuccinate) CDC recommends 50 mg/kg/day in qid × 14 days, may repeat if condition returns

Lymphogranuloma venereum caused by *Chlamydia trachomatis*

- **Adult: PO** (base) CDC recommends 500 mg qid × 21 days as an alternative to doxycycline

Adjunctive treatment of diphtheria to prevent establishment of carrier state and to eradicate *Corynebacterium diphtheriae* in carriers

- **Adult: PO** 500 mg every 6 hr × 10 days

Intestinal amebiasis (unable to take metroNIDAZOLE)

- **Adult: PO** 250 mg every 6 hr × 10-14 days
- **Adolescent/child: PO** 30-50 mg/kg/day, divided every 6 hr × 10-14 days, max adult dose

Acute pelvic inflammatory disease (PID) caused by *Neisseria gonorrhoeae*

- **Adult: IV** 500 mg (lactobionate) every 6 hr × 3 days, then PO 250 mg (base, estolate, stearate) or **PO** 400 mg (ethylsuccinate) every 6 hr × 7 days

Pertussis (whooping cough) caused by *Bordetella pertussis* or for postexposure pertussis prophylaxis:

- **Adult: PO** 500 mg qid (2 g total) × 14 days
- **Adolescent/child/infant: PO** 40-50 mg/kg/day (max 2 g/day) in 4 divided doses × 14 days

Primary or secondary syphilis (caused by *Treponema pallidum*) in penicillin-allergic nonpregnant patients

- **Adult: PO** 48-64 g (ethylsuccinate) or 30-40 g (base or stearate) in divided doses × 10-15 days

Surgical infection prophylaxis as a bowel preparation in combination with neomycin

- **Adult:** It is generally recommended that if surgery is scheduled for 8 am, 1 g of erythromycin PO with neomycin sulfate PO should be given at 1 pm, 2 pm, and 11 pm on the day before surgery

Impetigo, burn wound infection (unlabeled)

- **Adult: PO** 250-500 mg q6hr (base, estolate, stearate) or 400-800 mg (ethylsuccinate) q6hr
- **Infant/child/adolescent:** 20-50 mg/kg/day divided q6hr

Available forms: *Base:* enteric-coated tabs 250, 333, 500 mg; film-coated tabs 250, 500 mg; enteric-coated caps 250, 333 mg; *stearate:* film-coated tabs 250 mg; *ethylsuccinate:* granules for oral susp 200, 400 mg/5 ml; powder for inj 500 mg, 1 g (lactobionate), 1 g (as gluceptate)

Administer:

- Do not break, crush, or chew time rel cap or tab; chew only chewable tabs; enteric-coated tablets may be given with food
- Do not give by IM or IV push
- Oral product with full glass of water; do not give with fruit juice
- Give 1 hr before or 2 hr after meals
- Store at room temperature; store susp in refrigerator

- Adequate intake of fluids (2 L) during diarrhea episodes

IV route

- After **reconstituting** 500 mg or less/10 ml sterile water without preservatives; dilute further in 100-250 ml of 0.9% NaCl, LR, Normosol-R; may be **further diluted** to 1 mg/ml and **given** as cont inf; run 1 g or less/100 ml over ½-1 hr; cont inf over 6 hr, may require buffers to neutralize pH if dilution is <250 ml, use inf pump

Lactobionate

Y-site compatibilities: Acyclovir, alfentanil, amikacin, aminocaproic acid, aminophylline, amiodarone, anidulafungin, argatroban, atenolol, atosiban, atracurium, atropine, azaTHIOprine, benztropine, bivalirudin, bleomycin, bumetanide, buprenorphine, butorphanol, calcium chloride/gluconate, CARBOplatin, caspofungin, cefotaxime, cefTRIAXone, cefuroxime, chlorproMAZINE, cimetidine, CISplatin, cyanocobalamin, cyclophosphamide, cycloSPORINE, cytarabine, DACTINomycin, DAPTOmycin, dexmedetomidine, digoxin, diltiazem, diphenhydrAMINE, DOBUTamine, DOCEtaxel, DOPamine, doxacurium, doxapram, DOXOrubicin, enalaprilat, ePHEDrine, EPINEPHrine, epirubicin, epoetin alfa, eptifibatide, ertapenem, esmolol, etoposide, famotidine, fenoldopam, fentaNYL, fluconazole, fludarabine, fluorouracil, folic acid, foscarnet, gatifloxacin, gemcitabine, gentamicin, glycopyrrolate, granisetron, hydrocortisone, HYDROmorphone, hydrOXYzine, IDArubicin, ifosfamide, imipenem-cilastatin, insulin (regular), irinotecan, isoproterenol, labetalol, levofloxacin, lidocaine, LORazepam, LR, mannitol, mechlorethamine, meperidine, methicillin, methotrexate, methoxamine, methyldopate, methylPREDNISolone, metoclopramide, metroNIDAZOLE, miconazole, midazolam, milrinone, mitoXANtrone, morphine, multiple vitamins injection, mycophenolate, nafcillin, nalbuphine, naloxone, nesiritide, netilmicin, niCARdipine, nitroglycerin, norepinephrine, octreotide, ondansetron, oxacillin, oxaliplatin, oxytocin, PACLitaxel, palonosetron, pamidronate, pancuronium, papaverine, pentamidine, pentazocine, perphenazine, phenylephrine, phytonadione, piperacillin, piperacillin-tazobactam, polymyxin B, procainamide, prochlorperazine, promethazine, propranolol, protamine, pyridoxine, quiNIDine, ranitidine, Ringer's, ritodrine, sodium acetate/bicarbonate, succinylcholine, SUFentanil, tacrolimus, temocillin, teniposide, theophylline, thiamine, thiotepa, tigecycline, tirofiban, TNA, tobramycin, tolazoline, TPN, trimetaphan, urokinase, vancomycin, vasopressin, vecuronium, verapamil, vinCRIStine, vinorelbine, vitamin B complex/C, voriconazole, zidovudine, zoledronic acid

E

SIDE EFFECTS

CNS: **Seizures**
CV: **Dysrhythmias, QT prolongation**
EENT: Hearing loss, tinnitus
GI: *Nausea, vomiting, diarrhea,* **hepatotoxicity,** abdominal pain, stomatitis, heartburn, anorexia, **pseudomembranous colitis, esophagitis, hepatotoxicity**
GU: *Vaginitis, moniliasis*
INTEG: Rash, urticaria, pruritus, thrombophlebitis (IV site)
SYST: **Anaphylaxis**

PHARMACOKINETICS

Peak 1-4 hr (base); ½-2½ hr (ethylsuccinate); half-life 1-2 hr; metabolized in liver; excreted in bile, feces; protein binding 75%-90%; inhibitor of CYP3A4 and P-glycoprotein

INTERACTIONS

⚠ **Serious dysrhythmias—diltiazem, itraconazole, ketoconazole, nefazodone, pimozide, protease inhibitors, verapamil**
Increase: **QT prolongation—products that increase QT prolongation**
Increase: action, toxicity of alfentanil, ALPRAZolam, bromocriptine, busPIRone, carBAMazepine, cilostazol, clinda-

mycin, cloZAPine, cycloSPORINE, diazepam, digoxin, disopyramide, ergots, felodipine, HMG-CoA reductase inhibitors, methylPREDNISolone, midazolam, quiNIDine, rifabutin, sildenafil, tacrolimus, tadalafil, theophylline, triazolam, vardenafil, vinBLAStine, warfarin

Drug/Lab Test

Increase: AST/ALT

Decrease: folate assay

False increase: 17-OHCS/17-KS

NURSING CONSIDERATIONS

Assess:

- **Infection:** temperature, characteristics of wounds, urine, stools, sputum, WBCs at baseline and periodically
- I&O ratio; report hematuria, oliguria in renal disease
- Hepatic studies: AST, ALT if patient is receiving long-term therapy
- Hearing at baseline and after treatment
- Renal studies: urinalysis, protein, blood
- C&S before product therapy; product may be given as soon as culture is taken; C&S may be repeated after treatment
- **Pseudomembranous colitis:** diarrhea with blood, mucus; abdominal pain, fever; product should be discontinued immediately, notify prescriber
- **Anaphylaxis:** generalized hives, itching, flushing, swelling of lips, tongue, throat, wheezing; have emergency equipment nearby
- **QT prolongation:** may occur (IV >15 mg/min); those with electrolyte imbalances, congenital QT prolongation, elderly at greater risk; correct electrolyte imbalances before treatment, ECG

Evaluate:

- Therapeutic response: decreased symptoms of infection

Teach patient/family:

- To report sore throat, fever, fatigue (could indicate superinfection), rhythm changes in the heart, hearing loss
- To notify nurse of diarrhea stools, dark urine, pale stools, jaundice of eyes or skin, severe abdominal pain
- To take at evenly spaced intervals; to complete dosage regimen; to take without food
- To avoid use with other products unless approved by prescriber

TREATMENT OF HYPERSENSITIVITY: Withdraw product; maintain airway; administer EPINEPHrine, aminophylline, O_2, IV corticosteroids

erythromycin (topical)

(e-rith-roe-mye′sin)

Akne-mycin, Ery-sol ✦

Func. class.: Topical anti-infective, anti-acne

Chem. class.: Macrolide

ACTION: Antibacterial activity results from inhibition of protein synthesis; bacteriostatic

USES: Treatment of acne vulgaris

CONTRAINDICATIONS: Hypersensitivity, children

DOSAGE AND ROUTES

Acne vulgaris

- **Adult/adolescent: TOP** Apply to affected areas bid, AM, PM

Available forms: Topical gel, ointment, pledget, solution 2%

Administer:

Topical route

- For external use only; do not use skin products near the eyes, nose, or mouth
- Wash hands before and after use. Wash affected area and gently pat dry before using
- **Gel/ointment/pledget/solution:** Apply to the cleansed affected area. Massage gently into affected areas
- Each pledget should be used once and discarded

SIDE EFFECTS

INTEG: Burning, rash, pruritus, peeling, irritation

NURSING CONSIDERATIONS

Assess:

• **Allergic reaction:** assess for hypersensitivity, product might need to be discontinued

• **Infection**: assess for number of lesions, severity in acne

Evaluate:

• Decreased lesions in acne

Teach patient/family:

Topical Route:

• That product is for external use only; do not use skin products near the eyes, nose, or mouth

• To wash hands before and after use, wash affected area and gently pat dry before using

• **Gel/pledget/solution/ointment/lotion:** To apply to the cleansed affected area; massage gently into affected areas

• That each pledget should be used once and discarded

escitalopram (Rx)

(es-sit-tal′oh-pram)

Lexapro

Func. class.: Antidepressant, SSRI (selective serotonin reuptake inhibitor)

ACTION: Inhibits CNS neuron uptake of serotonin but not of norepinephrine

USES: General anxiety disorder; major depressive disorder in adults/adolescents

Unlabeled uses: Panic disorder, social phobia, autism

CONTRAINDICATIONS: Hypersensitivity to this product, citalopram

Precautions: Pregnancy (C), breastfeeding, geriatric patients, renal/hepatic disease, history of seizures

Black Box Warning: Children/adolescents ≤12 yr, suicidal ideation

DOSAGE AND ROUTES

• **Adult: PO** 10 mg/day in AM or PM; after 1 wk, if no clinical improvement is noted, dose may be increased to 20 mg/day PM; maintenance 10-20 mg/day; reassess to determine need for treatment

Hepatic dose/geriatric

• **Adult: PO** 10 mg/day

Available forms: Tabs 5, 10, 20 mg; oral sol 5 mg (as base)/5 ml (contains sorbitol)

Administer:

• With food or milk for GI symptoms, give with full glass of water

• Crushed if patient is unable to swallow medication whole

• Dosage at bedtime if oversedation occurs during the day

• Gum, hard candy, frequent sips of water for dry mouth

• **Oral sol:** measure with calibrated device

• Store at room temperature; do not freeze

SIDE EFFECTS

CNS: *Headache, nervousness, insomnia,* **suicidal ideation**, *drowsiness, anxiety, tremor, dizziness, fatigue, sedation, poor conc, abnormal dreams, agitation,* **seizures**, apathy, euphoria, hallucinations, delusions, psychosis, **neuroleptic malignant-like syndrome,** ataxia

CV: *Hot flashes, palpitations,* angina pectoris, **hemorrhage,** hypertension, **tachycardia,** 1st-degree AV block, **bradycardia, MI, thrombophlebitis,** postural hypotension

EENT: Visual changes, ear/eye pain, photophobia, tinnitus, pupil dilation

GI: *Nausea, diarrhea, dry mouth, anorexia, dyspepsia, constipation, cramps, vomiting, taste changes, flatulence, decreased appetite,* **hepatitis**

GU: *Dysmenorrhea, decreased libido, urinary frequency, UTI,* amenorrhea, cystitis, impotence, urine retention, ejaculation disorder

INTEG: *Sweating, rash, pruritus,* acne, alopecia, urticaria, photosensitivity, bruising
MS: *Pain,* arthritis, twitching, osteopenia
RESP: *Infection, pharyngitis, nasal congestion, sinus headache, sinusitis, cough, dyspnea, bronchitis,* asthma, hyperventilation, pneumonia
SYST: *Asthenia, viral infection, fever, allergy, chills,* serotonin syndrome, neonatal abstinence syndrome, Stevens-Johnson syndrome

PHARMACOKINETICS

PO: Metabolized in liver; excreted in urine; 56% protein binding; metabolized by CYP2C19, 3A4, half-life 27-32 hr; half-life increased by 50% in geriatric patients

INTERACTIONS

• Paradoxical worsening of OCD: busPIRone
Increase: serotonin syndrome—tryptophan, amphetamines, busPIRone, lithium, amantadine, bromocriptine, SSRI, SNRIs, serotonin-receptor agonists, traMADol
⚠ Do not use pimozide, MAOIs, with or 14 days before escitalopram
Increase: CNS depression—alcohol, antidepressants, opioids, sedatives
Increase: side effects of escitalopram—highly protein-bound products
Increase: levels or toxicity of carBAMazepine, lithium, warfarin, phenytoin, antipsychotics, antidysrhythmics
Increase: levels of tricyclics, phenothiazines, haloperidol, diazepam
Increase: bleeding risk—NSAIDs, salicylates, anticoagulants, SSRIs, platelet inhibitors
Decrease: escitalopram effect—cyproheptadine

Drug/Herb
• St. John's wort: do not use together, serotonin syndrome may occur
Increase: CNS effect—kava, valerian

Drug/Food
• Grapefruit juice—increased escitalopram effect

Drug/Lab Test
Increase: serum bilirubin, blood glucose, alk phos
Decrease: VMA, 5-HIAA
False increase: urinary catecholamines

NURSING CONSIDERATIONS

Assess:

Black Box Warning: Mental status: mood, sensorium, affect, **suicidal tendencies,** increase in psychiatric symptoms, depression, panic, not approved for use in children

• Appetite with bulimia nervosa, weight daily; increase nutritious foods in diet, watch for binging and vomiting
• **Allergic reactions:** itching, rash, urticaria; product should be discontinued, may need to give antihistamine
• B/P (lying/standing), pulse q4hr; if systolic B/P drops 20 mm Hg, hold product, notify prescriber
• Blood studies: CBC, leukocytes, differential, cardiac enzymes if patient receiving long-term therapy; check platelets; bleeding can occur
• **Serotonin syndrome:** nausea, vomiting, sedation, dizziness, sweating, facial flushing, mental changes, shivering, increased B/P; discontinue product, notify prescriber
• Hepatic studies: AST, ALT, bilirubin, creatinine; thyroid function studies
• Weight weekly; appetite may decrease with product
• **ECG** for flattening of T wave, bundle branch, AV block, dysrhythmias in cardiac patients
• Alcohol consumption; if alcohol is consumed, hold dose until AM
• **Sexual dysfunction:** ejaculation dysfunction, erectile dysfunction, decreased libido, orgasm dysfunction, priapism
• Assistance with ambulation during therapy, since drowsiness, dizziness occur; safety measures primarily for geriatric patients

Evaluate:
• Therapeutic response: decreased depression

Teach patient/family:

- That therapeutic effect may take 1-4 wk, may have increased anxiety for first 5-7 days
- To use caution when driving, performing other activities requiring alertness because drowsiness, dizziness, blurred vision may occur
- To avoid alcohol, other CNS depressants; to avoid all OTC products unless approved by prescriber
- To notify prescriber if pregnant or planning to become pregnant or breastfeeding
- To change positions slowly, orthostatic hypotension may occur
- To report signs of urinary retention immediately

Black Box Warning: That clinical worsening and suicide risk may occur

- To use MedGuide provided

TREATMENT OF OVERDOSE:

Activated charcoal, supportive care, serotonin antagonist

esmolol (Rx)

(ez′moe-lole)

Brevibloc

Func. class.: β-Adrenergic blocker (antidysrhythmic II)

Do not confuse:
esmolol/Osmitrol
Brevibloc/Brevital

ACTION: Competitively blocks stimulation of β_1-adrenergic receptors in the myocardium; produces negative chronotropic, inotropic activity (decreases rate of SA node discharge, increases recovery time), slows conduction of AV node, decreases heart rate, decreases O_2 consumption in myocardium; also decreases renin-aldosterone-angiotensin system at high doses; inhibits β_2-receptors in bronchial system at higher doses

USES: Supraventricular tachycardia, noncompensatory sinus tachycardia, hypertensive crisis, intraoperative and postoperative tachycardia and hypertension, atrial fibrillation/flutter

Unlabeled uses: Acute MI, ECT, thyroid storm, pheochromocytoma, hypertensive crisis/urgency, unstable angina

CONTRAINDICATIONS: 2nd- or 3rd-degree heart block; cardiogenic shock, CHF, cardiac failure, hypersensitivity, severe bradycardia

Precautions: Pregnancy (C), breastfeeding, geriatric patients, hypotension, peripheral vascular disease, diabetes, hypoglycemia, thyrotoxicosis, renal disease, atrial fibrillation, bronchospasms, hyperthyroidism, myasthenia gravis

Black Box Warning: Abrupt discontinuation

DOSAGE AND ROUTES

Atrial fibrillation/flutter

- **Adult: IV** loading dose 500 mcg/kg/min over 1 min; maintenance 50 mcg/kg/min for 4 min; if no response after 5 min, give 2nd loading dose then increase inf to 100 mcg/kg/min for 4 min; if no response, repeat loading dose then increase maintenance inf by 50 mcg/kg/min (max of 200 mcg/kg/min); titrate to patient response
- **Child: IV** total loading dose of 600 mcg/kg over 2 min, maintenance **IV INF** 200 mcg/kg/min, titrate upward by 50-100 mcg/kg/min q5-10min until B/P, heart rate reduced by >10%

Perioperative Hypertension/ Tachycardia

- **Adult: IV** Immediate control 80 mcg (Bolus) over 30 seconds, then 150 mcg/kg/min, adjust to response, max 300 mcg/kg/min

Hypertensive Emergency (Unlabeled)

- **Adult: IV** 250-500 mcg/kg over 1 min, then **IV INF** 50-100 mcg/kg/min × 4 min

Available forms: Inj 10 mg, 20 mg/ml

Administer:
- Do not discontinue product suddenly
- Store protected from light, moisture; in cool environment

IV route
- Check that correct conc being given
- 10 mg/ml inj sol needs no dilution, may be used as an IV loading dose using a handheld syringe

Continuous IV INF route
- Ready-to-use bags of premixed isotonic sol of 10 mg/ml and 20 mg/ml available in 100-, 250-ml bags; use controlled inf device, central line preferred; rate is based on patient's weight

Additive compatibilities: Aminophylline, atracurium, heparin, potassium chloride, sodium bicarbonate

Y-site compatibilities: Amikacin, aminophylline, amiodarone, atracurium, butorphanol, calcium chloride, ceFAZolin, cefTAZidime, ceftizoxime, chloramphenicol, cimetidine, cisatracurium, clindamycin, diltiazem, DOPamine, enalaprilat, erythromycin, famotidine, fentaNYL, gentamicin, insulin (regular), labetalol, magnesium sulfate, methyldopate, metroNIDAZOLE, midazolam, morphine, nitroglycerin, nitroprusside, norepinephrine, pancuronium, penicillin G potassium, piperacillin, polymyxin B, potassium chloride, potassium phosphate, propofol, ranitidine, remifentanil, streptomycin, tacrolimus, tobramycin, trimethoprim-sulfamethoxazole, vancomycin, vecuronium, voriconazole, zoledronic acid

SIDE EFFECTS

CNS: Confusion, lightheadedness, paresthesia, somnolence, fever, dizziness, fatigue, headache, depression, anxiety, seizures

CV: Hypotension, bradycardia, chest pain, peripheral ischemia, SOB, CHF, conduction disturbances; 1st-, 2nd-, 3rd-degree heart block

GI: *Nausea,* vomiting, anorexia, gastric pain, flatulence, constipation, heartburn, bloating

GU: Urinary retention, impotence, dysuria

INTEG: *Induration, inflammation at site,* discoloration, edema, erythema, burning pallor, flushing, rash, pruritus, dry skin, alopecia

RESP: Bronchospasm, dyspnea, cough, wheeziness, nasal stuffiness, pulmonary edema

PHARMACOKINETICS

Onset very rapid, duration short, half-life 9 min, metabolized by hydrolysis of ester linkage, excreted via kidneys

INTERACTIONS

- Avoid use with MAOIs, Sotalol

Increase: antihypertensive effect—general anesthetics

Increase: digoxin levels—digoxin

Increase: α-adrenergic stimulation—ePHEDrine, EPINEPHrine, amphetamine, norepinephrine, phenylephrine, pseudoePHEDrine

Decrease: action of thyroid hormones

Decrease: action of esmolol—thyroid hormone

Drug/Herb

Increase: β-blocking effect—hawthorn

Decrease: antihypertensive effect—ephedra

Drug/Lab Test

Interference: glucose/insulin tolerance test

NURSING CONSIDERATIONS

Assess:
- **CHF:** I&O ratio, weight daily, jugular venous distention, weight gain, crackles, edema
- **Dysrhythmias:** B/P, pulse q4hr; note rate, rhythm, quality; rapid changes can cause shock; if systolic <100 or diastolic <60, notify prescriber before giving product; ECG continuously during inf, hypotension common, if severe, slow or stop inf
- Baselines in renal/hepatic studies, blood glucose before therapy begins
- **Bronchospasm:** breath sounds, respiratory pattern

Evaluate:
- Therapeutic response: lower B/P immediately, lower heart rate

Teach patient/family:
- About reason for use, expected results
- To notify prescriber if chest pain, SOB, wheezing, hypotension, bradycardia, pain, swelling at IV site occurs

TREATMENT OF OVERDOSE:
Discontinue product

esomeprazole (Rx)
(es′oh-mep′rah-zohl)
NexIUM
Func. class.: Antiulcer
Chem. class.: Proton-pump inhibitor, benzimidazole

Do not confuse:
NexIUM/NexAVAR

ACTION: Suppresses gastric secretions by inhibiting hydrogen/potassium ATPase enzyme system in gastric parietal cell; characterized as gastric acid pump inhibitor because it blocks the final step of acid production

USES: Gastroesophageal reflux disease (GERD), adult/child/infant; severe erosive esophagitis, adult/child; treatment of active duodenal ulcers in combination with antiinfectives for *Helicobacter pylori* infection; long-term use for hypersecretory conditions

CONTRAINDICATIONS: Hypersensitivity to proton-pump inhibitors (PPIs)

Precautions: Pregnancy (B), breastfeeding, children, geriatric patients

DOSAGE AND ROUTES

Active duodenal ulcers associated with *H. pylori*
- **Adult: PO** 40 mg/day × 10 days in combination with clarithromycin 500 mg bid × 10 days and amoxicillin 1000 mg bid × 10 days

Hepatic dose
- **Adult: PO/IV** max 20 mg/day (severe hepatic disease)

GERD/erosive esophagitis
- **Adult: PO** 20 or 40 mg/day × 4-8 wk; no adjustment needed in renal/liver failure, geriatric patients; **IV** 20 or 40 mg/day up to 10 days
- **Adolescent and child 12-17 yr: PO** 20 or 40 mg/day 1 hr before meals for ≤8 wk
- **Child 1-11 yr and ≥20 kg: PO** 10 mg/day 1 hr before meals for ≤8 wk
- **Infant ≥1 mo: IV** 0.5 mg/day over 10-30 min
- **Infant 1-11 mo (>7.5-12 kg): PO** 10 mg daily × up to 6 wk
- **Infant 1-11 mo (>5-7.5 kg): PO** 5 mg daily × up to 6 wk
- **Infant 1-11 mo (3-5 kg): PO** 2.5 mg daily × up to 6 wk

Available forms: Del rel caps 20, 40 mg; powder for IV inj 20, 40 mg/vial; del rel powder for oral susp 10, 20, 40 mg

Administer:

PO route
- Swallow caps whole; do not crush or chew; cap may be opened and sprinkled over Tbsp of applesauce
- Same time daily, 1 hr before meal
- **Oral susp (del rel):** empty contents of packet into container with 1 Tbsp of water, let stand 2-3 min to thicken, restir, give within 30 min of mixing; any residual product should be flushed with more water, taken immediately
- **NG tube (del rel oral susp):** add 15 ml water to contents of packet in syringe, shake, leave 2-3 min to thicken, shake, inject through NG tube within 30 min

IV, direct route
- Reconstitute each vial with 5 ml 0.9% NaCl, D_5W, LR; give over 3 min

Intermittent IV INF route
- Dilute reconstituted sol to 50 ml, give over 30 min, do not admix, flush line with D_5W, 0.9% NaCl, LR after inf

Solution compatibilities: D_5W, LR, 0.9% NaCl

SIDE EFFECTS

CNS: *Headache, dizziness*
GI: *Diarrhea, flatulence,* abdominal pain, constipation, dry mouth, **hepatic failure, hepatitis,** microscopic colitis
INTEG: *Rash,* dry skin
MISC: Heart failure
RESP: *Cough,* pneumonia
SYST: **Stevens-Johnson syndrome, toxic epidermal necrolysis, exfoliative dermatitis**

PHARMACOKINETICS

Well absorbed 90%; protein binding 97%; extensively metabolized in liver (CYP2C19); terminal half-life 1-1.5 hr; eliminated in urine as metabolites and in feces; in geriatric patients, elimination rate decreased, bioavailability increased

INTERACTIONS

Increase: effect, toxicity of diazepam, digoxin, penicillins, saquinavir
Decrease: effect—atazanavir, nelfinavir, dapsone, iron, itraconazole, ketoconazole, indinavir, calcium carbonate, vit B_{12}, clopidogrel
Drug/Lab Test
Interference: sodium, Hgb, WBC, platelets, magnesium

NURSING CONSIDERATIONS

Assess:
- GI system: bowel sounds q8hr, abdomen for pain, swelling, anorexia
- **Hepatic failure, hepatitis:** AST, ALT, alk phos at baseline and periodically during treatment
- **Serious skin disorders:** Stevens-Johnson syndrome, toxic epidermal necrolysis, exfoliative dermatitis

Evaluate:
- Therapeutic response: absence of epigastric pain, swelling, fullness

Teach patient/family:
- **To report severe diarrhea; abdominal pain; black, tarry stools, rash; product may have to be discontinued**
- That hypoglycemia may occur if diabetic
- To avoid hazardous activities; dizziness may occur
- To avoid alcohol, salicylates, NSAIDs; may cause GI irritation
- To take ≥1 hr before meal; not to crush, chew del rel product
- If cap is unable to be swallowed, whole contents may be mixed with a Tbsp of applesauce

estradiol (Rx)
(es-tra-dye′ole)
Estrace
estradiol cypionate (Rx)
Depo-Estradiol
estradiol gel (Rx)
Divigel, Elestrin, Estrogel
estradiol spray (Rx)
Evamist
estradiol topical emulsion (Rx)
Estrasorb
estradiol valerate (Rx)
Delestrogen
estradiol transdermal system (Rx)
Alora, Climara, Menostar, Vivelle-Dot
estradiol vaginal tablet (Rx)
Vagifem
estradiol vaginal ring (Rx)
Estring, Femring
Func. class.: Estrogen, progestins

ACTION: Needed for adequate functioning of female reproductive system; affects release of pituitary gonadotropins; inhibits ovulation, adequate calcium use in bone

USES: Vasomotor symptoms (menopause), inoperable breast cancer (selected cases), prostatic cancer, atrophic vaginitis, kraurosis vulvae, hypogonadism, primary ovarian failure, prevention of osteoporosis, castration

CONTRAINDICATIONS: Pregnancy (X), breastfeeding, reproductive cancer, genital bleeding (abnormal, undiagnosed), protein C, protein S, antithrombin deficiency

Black Box Warning: Breast/endometrial cancer, thromboembolic disorders, MI, stroke

Precautions: Hypertension, asthma, blood dyscrasias, gallbladder/bone/renal/hepatic disease, CHF, diabetes mellitus, depression, migraine headache, seizure disorders, family history of cancer of breast or reproductive tract, smoking, uterine fibroids, vaginal irritation/infection, accidental exposure of children, pets to topical product, history of angioedema

Black Box Warning: Cardiac disease, dementia

DOSAGE AND ROUTES

Hormone replacement/menopause symptoms

• **Adult: TRANSDERMAL** 1 patch delivering 0.025, 0.0375, 0.05, 0.075, or 0.1 mg/day 2×/wk (Alora, Estraderm, Vivelle-Dot); 1 patch delivering 0.025, 0.0375, 0.05, 0.06, 0.075, or 0.1 mg/day replace q7days (Climara); 1 patch delivering 0.025 mg/day, replace q7days, may increase to 2 patches after 4-6 wk; **GEL** apply entire unit-dose packet to 5 × 7-inch area of upper thigh/day, alternate thighs; **SPRAY** (Evamist) 1 spray to inner surface of forearm/day in AM

Menopause/hypogonadism/castration/ovarian failure

• **Adult: PO** 1-2 mg/day, 3 wk on, 1 wk off or 5 days on, 2 days off; **IM** (cypionate) 1-5 mg q3-4wk; (valerate) 10-20 mg q4wk

• **Adult: TOP** (Estraderm) 0.05 mg/24 hr applied 2×/wk; (Climara) 0.05 mg/hr applied 1×/wk in cyclic regimen; women with hysterectomy may use continuously

Prostatic cancer (inoperable)

• **Adult: IM** (valerate) 30 mg q1-2wk;-**PO** (oral estradiol) 1-2 mg bid-tid

Breast cancer (palliative treatment)

• **Adult: PO** 10 mg tid × 3 mo or longer

Atropic vaginitis/kraurosis vulvae

• **Adult: VAG CREAM** 2-4 g/day × 1-2 wk, then 1 g 1-3×/wk cycled; vag tab 1/day × 2 wk, maintenance 1 tab 2×/wk; **VAG RING** inserted, left in place continuously for 3 mo

Vasomotor symptoms

• **Adult: TOP** after cleaning and drying skin on left thigh, calf, rub in contents of pouch using both hands until completely absorbed; wash hands

Available forms: ***Estradiol:*** tabs 0.5, 1, 2 mg; ***valerate:*** inj 10, 20, 40 mg/ml; ***transdermal:*** 0.025, 0.0375, 0.05, 0.06, 0.075, 0.1 mg/24 hr release rate; ***vag cream:*** 100 mcg/g; ***vag tab:*** 10 mcg; ***vag ring:*** 2 mg/90 days; ***topical emulsion:*** 2.5 mg; ***gel*** (Divigel) 0.1%; ***spray*** (Evamist) 1.53 mg/acuation

Administer:

• Titrated dose; use lowest effective dose
• IM inj deeply in large muscle mass

PO route

• With food or milk to decrease GI symptoms

Transdermal route

• May contain aluminum or other metals in backing of patch, can overheat in MRI scan and burn patients
• Apply to trunk of body 2×/wk; press firmly, hold in place for 10 sec to ensure good contact; do not apply to breasts
• On intermittent cycle schedule: 3 wk on then 1 wk off; if patch falls off, reapply

Topical route

• Use Evamist daily; spray to inner upper arm; may increase to 2-3×/day based on response; allow to dry for 2 min, avoid secondary exposure to children, pets, caregivers

Vaginal route

- Use a new applicator daily, provided

SIDE EFFECTS

CNS: Dizziness, headache, migraines, depression, seizures
CV: Hypertension, thrombophlebitis, edema, thromboembolism, stroke, pulmonary embolism, MI, chest pain
EENT: Contact lens intolerance, increased myopia, astigmatism, throat swelling, eyelid edema
GI: *Nausea,* vomiting, diarrhea, anorexia, pancreatitis, cramps, constipation, increased appetite, increased weight, cholestatic jaundice, hepatic adenoma
GU: Amenorrhea, cervical erosion, breakthrough bleeding, dysmenorrhea, vaginal candidiasis, breast changes, *gynecomastia, testicular atrophy, impotence,* increased risk of breast cancer, endometrial cancer, changes in libido; toxic shock, vaginal wall ulceration/erosion (vag ring)
INTEG: Rash, urticaria, acne, hirsutism, alopecia, oily skin, seborrhea, purpura, erythema, pruritus, melasma; site irritation (transdermal)
META: Folic acid deficiency, hypercalcemia, hyperglycemia

PHARMACOKINETICS

PO/INJ/TRANSDERMAL: Degraded in liver, excreted in urine, crosses placenta, excreted in breast milk

INTERACTIONS

Increase: action of corticosteroids
Increase: toxicity—cycloSPORINE, dantrolene
Decrease: action of anticoagulants, oral hypoglycemics, tamoxifen
Decrease: estradiol action—anticonvulsants, barbiturates, phenylbutazone, rifampin, calcium

Drug/Herb

- Altered estrogen effect: black cohosh, DHEA

Decrease: estrogen effect—saw palmetto, St. John's wort

Drug/Food

Increase: estrogen level—grapefruit juice

Drug/Lab Test

Increase: BSP retention test, PBI, T_4, serum sodium, platelet aggregation, thyroxine-binding globulin (TBG), prothrombin; factors VII, VIII, IX, X; triglycerides
Decrease: serum folate, serum triglyceride, T_3 resin uptake test, glucose tolerance test, antithrombin III, pregnanediol, metyrapone test
False positive: LE prep, ANA

NURSING CONSIDERATIONS

Assess:

Black Box Warning: For previous breast/endometrial cancer, thrombo-embolic disorders, MI, stroke, dementia

- Blood glucose of diabetic patient; hyperglycemia may occur
- Weight daily; notify prescriber of weekly weight gain >5 lb; if increase, diuretic may be ordered
- B/P q4hr; watch for increase caused by water and sodium retention
- I&O ratio; decreasing urinary output, increasing edema, report changes
- Hepatic studies, including AST, ALT, bilirubin, alk phos at baseline, periodically; periodic folic acid level
- Hypertension, cardiac symptoms, jaundice, hypercalcemia
- Mental status: affect, mood, behavioral changes, aggression
- Female patient for intact uterus; if so, progesterone should be added to estrogen therapy to decrease risk of endometrial cancer

Evaluate:

- Therapeutic response: reversal of menopause symptoms; decrease in tumor size in prostatic, breast cancer

Teach patient/family:

- To weigh weekly; to report gain >5 lb
- ⚠ To report breast lumps, vaginal bleeding, edema, jaundice, dark urine, clay-

colored stools, dyspnea, headache, blurred vision, abdominal pain, numbness or stiffness in legs, chest pain; tenderness, redness, and swelling in extremities; males to report impotence, gynecomastia; to report dermal rash with transdermal patch

- To avoid grapefruit or grapefruit juice (PO)

RARELY USED

estramustine (Rx)

(ess-tra-muss'teen)

Emcyt

Func. class.: Antineoplastic alkylating agent

USES: Metastatic prostate cancer

CONTRAINDICATIONS: Pregnancy (D), hypersensitivity to estradiol, thromboembolic disorders, stroke, thrombophlebitis

DOSAGE AND ROUTES

- **Adult: PO** 14-16 mg/kg/day or 600 mg/m²/day in 3-4 divided doses; treatment may continue for ≥3 mo

estrogens, conjugated (Rx)

Cenestin, Premarin

estrogens, conjugated synthetic B (Rx)

Enjuvia

Func. class.: Estrogen, hormone

Do not confuse:
Premarin/Provera

ACTION: Needed for adequate functioning of female reproductive system; affects release of pituitary gonadotropins, inhibits ovulation, adequate calcium use in bone

USES: Vasomotor symptoms (menopause), inoperable breast cancer, prostatic cancer, abnormal uterine bleeding, hypogonadism, primary ovarian failure, prevention of osteoporosis, castration, atrophic vaginitis

Unlabeled uses: hyperparathyroidism, infertility

CONTRAINDICATIONS: Pregnancy (X), breastfeeding, thromboembolic disorders, reproductive cancer, genital bleeding (abnormal, undiagnosed), hypersensitivity, MI, stroke, thrombophlebitis

Black Box Warning: Endometrial, breast cancer, thromboembolic diseases

Precautions: Hypertension, asthma, blood dyscrasias, CHF, diabetes mellitus, depression, migraine headache, seizure disorders, gallbladder/bone/hepatic/renal disease, family history of cancer of breast or reproductive tract, smoking, dementia, hypothyroidism, obesity, SLE

DOSAGE AND ROUTES

Estrogens conjugated

Vasomotor symptoms (menopause)

- **Adult: PO** 0.3-1.25 mg/day 3 wk on, 1 wk off

Prevention of osteoporosis

- **Adult: PO** 0.3 mg/day or in cycle

Atrophic vaginitis

- **Adult: VAG CREAM** 0.5 g/day × 21 days, off 7 days, repeat

Prostatic cancer

- **Adult: PO** 1.25-2.5 mg tid

Advanced inoperable breast cancer

- **Adult: PO** 10 mg tid × ≥3 mo

Abnormal uterine bleeding

- **Adult: IV/IM** 25 mg q6-12hr

Castration/primary ovarian failure

- **Adult: PO** 1.25 mg/day 3 wk on, 1 wk off

Hypogonadism

- **Adult: PO** 2.5-7.5 mg/day × 20 days/mo

Estrogens conjugated synthetic B
Vasomotor symptoms (menopause)
• **Adult: PO** 0.625 mg/day initially; may increase based on response
Available forms: Tabs 0.3, 0.45, 0.625, 0.9, 1.25, 2.5 mg; inj 25 mg/vial; vag cream 0.625 mg/g; *synthetic B:* tabs 0.625, 1.25 mg
Administer:
• Titrated dose; use lowest effective dose
PO route
• Give with or immediately after food to reduce nausea
IM route
• IM reconstitute after withdrawing 5 ml of air from container, inject sterile diluent on vial side, rotate to dissolve; give inj deep in large muscle mass, aspirate before inj
Vaginal route
• Use applicator provided, wash after use
Direct IV route
• IV, after reconstituting as for IM, inject into distal port of running IV line of D_5W, 0.9% NaCl at ≤5 mg/min

Y-site compatibilities: Heparin, hydrocortisone, potassium chloride, vit B/C

SIDE EFFECTS

CNS: Dizziness, headache, migraine, depression, seizures, mood disturbances
CV: Hypertension, thrombophlebitis, edema, thromboembolism, stroke, pulmonary embolism, MI, chest pain
EENT: Contact lens intolerance, increased myopia, astigmatism
GI: *Nausea,* vomiting, diarrhea, anorexia, pancreatitis, cramps, constipation, increased appetite, cholestatic jaundice, hepatic adenoma, weight gain/loss
GU: Amenorrhea, cervical erosion, breakthrough bleeding, dysmenorrhea, vaginal candidiasis, breast changes, *gynecomastia, testicular atrophy, impotence,* increased risk of breast cancer, endometrial cancer, libido changes
INTEG: Rash, urticaria, acne, hirsutism, alopecia, oily skin, seborrhea, purpura, melasma
META: Folic acid deficiency, hypercalcemia, hyperglycemia

PHARMACOKINETICS

PO/IM/IV: Degraded in liver, excreted in urine, crosses placenta, excreted in breast milk

INTERACTIONS

Increase: toxicity—cycloSPORINE, dantrolene
Increase: action of corticosteroids
Decrease: action of estrogens—anticonvulsants, barbiturates, phenylbutazone, rifampin
Decrease: action of anticoagulants, oral hypoglycemics, tamoxifen, thyroid
Drug/Food
Increase: estrogen level—grapefruit juice
Drug/Lab Test
Increase: BSP retention test, PBI, T_4, serum sodium, platelet aggregation, thyroxine-binding globulin (TBG), prothrombin; factors VII, VIII, IX, X; triglycerides
Decrease: serum folate, serum triglyceride, T_3 resin uptake test, glucose tolerance test, antithrombin III, pregnanediol, metyrapone test
False positive: LE prep, antinuclear antibodies

NURSING CONSIDERATIONS

Assess:

Black Box Warning: Breast, endometrial cancer: estrogens should not be used in known, suspected, or history of these disorders

Black Box Warning: Stroke, thromboembolic disease of MI: Should not be used in these conditions or known protein C deficiency, protein S deficiency, or antithrombin deficiency

• Blood glucose if diabetic patient; hyperglycemia may occur
• Weight daily; notify prescriber of weekly weight gain >5 lb; if increase,

diuretic may be ordered; check for edema; B/P baseline and periodically
• Hepatic studies: AST, ALT, bilirubin, alk phos
• Hypertension, cardiac symptoms, jaundice, hypercalcemia
• Mental status: affect, mood, behavioral changes, aggression
• Female patient for intact uterus; if so, progesterone should be added to estrogen therapy to decrease risk of endometrial cancer; abnormal uterine bleeding, breast exam; Pap smear

Evaluate:
• Therapeutic response: absence of breast engorgement, reversal of menopause symptoms, decrease in tumor size with prostatic cancer

Teach patient/family:
• To avoid breastfeeding, since product excreted in breast milk
• To weigh weekly; to report gain >5 lb

Black Box Warning: To report breast lumps, vaginal bleeding, edema, jaundice, dark urine, clay-colored stools, dyspnea, headache, blurred vision, abdominal pain; leg pain and redness, numbness or stiffness; chest pain; males to report impotence or gynecomastia

• To avoid sunlight or wear sunscreen; burns may occur
• To notify prescriber if pregnancy is suspected
• That vasomotor symptoms improve in 2 wk, max relief in 8 wk

RARELY USED

estropipate

(ess-troe-pip′ate)
Func. class.: Estrogen

USES: Primary ovarian failure, menopausal symptoms, prevention of osteoporosis, vaginal atrophy

CONTRAINDICATIONS: Hypersensitivity, coagulation disorders, breast and reproductive cancers, undiagnosed genital bleeding

Black Box Warning: Pregnancy (X)

DOSAGE AND ROUTES

Vaginal atrophy/vulval atrophy/menopausal symptoms
• **Adult (women): PO** 0.75-6 mg/day × 3 wk on 1 wk off, may be used continuously in menopausal symptoms

Primary ovarian failure
• **Adult (women): PO** 1.5-9 mg/day × 3 wk, then off for 8-10 days; if bleeding does not occur, repeat cycle

Osteoporosis prevention
• **Adult (women): PO** 0.75 mg/day × 25 days, 6 days off, repeat as needed

eszopiclone (Rx)

(es-zop′i-klone)
Lunesta
Func. class.: Sedative/hypnotic, nonbenzodiazepine
Chem. class.: Cyclopyrrolone

Controlled Substance Schedule IV

ACTION: Interacts with GABA receptors

USES: Insomnia

CONTRAINDICATIONS: Hypersensitivity

Precautions: Pregnancy (C), breastfeeding, children, geriatric patients, severe hepatic disease, abrupt discontinuation, COPD, depression, labor, sleep apnea, substance abuse, suicidal ideation, ethanol intoxication

DOSAGE AND ROUTES

• **Adult: PO** 2 mg immediately before bed, may increase to 3 mg if needed

Hepatic dose/CYP3A4 inhibitors
• **Adult: PO** 1 mg immediately before bed with severe hepatic disease

Available forms: Tabs 1, 2, 3 mg

Administer:

- Do not break, crush, or chew tab
- Immediately before bedtime; avoid use with food; for short-term use only

SIDE EFFECTS

CNS: Worsening depression, hallucinations, headache, daytime drowsiness, suicidal thoughts/actions, migraine, restlessness, anxiety, sleep driving, sleepwalking

CV: Peripheral edema, chest pain

GI: Dry mouth, bitter taste (dysgeusia)

GU: Gynecomastia, dysmenorrhea

INTEG: Rash, angioedema

PHARMACOKINETICS

Onset rapid; peak 1 hr; duration 6 hr; extensively metabolized in the liver by CYP3A4, CYP2E1; excreted via kidneys; half-life 6 hr, geriatric patients 9 hr, protein binding 52%-59%

INTERACTIONS

Increase: CNS depression—CNS depressants

Increase: toxicity due to decreased eszopiclone elimination—CYP3A4 inhibitors (clarithromycin, itraconazole, ketoconazole, nefazodone, nelfinavir, ritonavir, troleandomycin, SSRIs)

Drug/Food

Decrease: product action—food

NURSING CONSIDERATIONS

Assess:

- **Sleep pattern:** ability to go to sleep, stay asleep, early morning awakenings, conservative methods used
- For abuse of this product, other products
- **Anaphylaxis, angioedema:** monitor during first dose
- Alternative methods to improve sleep: reading, quiet environment, warm bath, milk
- Assistance with ambulation, night light, call bell within reach

Evaluate:

- Therapeutic response: ability to fall asleep and stay asleep throughout the night

Teach patient/family:

- That daytime drowsiness may occur; not to engage in hazardous activities until effect is known; that memory problems may occur
- That all other medications and supplements should be avoided unless approved by prescriber; to avoid alcohol
- To notify prescriber if pregnancy is suspected or planned
- To avoid use after a high fat meal
- To swallow tab whole
- Not to stop drug abruptly, tolerance may occur

etanercept (Rx)

(eh-tan′er-sept)

Enbrel

Func. class.: Antirheumatic agent (disease modifying) (DMARDs)

Chem. class.: Anti-TNF agent

ACTION: Binds tumor necrosis factor (TNF), which is involved in immune and inflammatory reactions

USES: Acute, chronic rheumatoid arthritis that has not responded to other disease-modifying agents, polyarticular course of juvenile rheumatoid arthritis (JRA), ankylosing spondylitis, plaque psoriasis, psoriatic arthritis

Unlabeled uses: Crohn's disease; plaque psoriasis (child ≥4 yr)

CONTRAINDICATIONS: Sepsis

Precautions: Pregnancy (B), breastfeeding, children <4 yr, geriatric patients, malignancies, CHF, seizures, multiple sclerosis, latex hypersensitivity

Black Box Warning: Infection, lymphoma, neoplastic disease

DOSAGE AND ROUTES

Rheumatoid/psoriatic arthritis, ankylosing spondylitis

- **Adult: SUBCUT** 50 mg/wk or 25 mg 2×/wk, 3-4 days apart
- **Child 2-17 yr: SUBCUT** 0.8 mg/kg/wk, max 50 mg/wk

Plaque psoriasis

- **Adult: SUBCUT** 50 mg 2×/wk × 3 mo
- **Adolescent/child 4-17 yr (unlabeled): SUBCUT** 0.8 mg/kg/wk, max 50 mg/wk

Juvenile rheumatoid arthritis (JRA)

- **Adolescent/child 2-17 yr: SUBCUT** 0.8 mg/kg/wk, max 50 mg/wk

Available forms: Powder for inj 25 mg; inj 50 mg/ml; autoinjector, single use

Administer:

- If appropriate, may be administered by the patient or a caregiver after thorough instruction of proper injection preparation and administration. Assess the patient's or caregiver's ability to inject subcut and observe the first injection
- Administration of one 50 mg/ml prefilled syringe or autoinjector provides a dose equivalent to two 25 mg prefilled syringes or two 25 mg vials of lyophilized when vials are reconstituted and administered as recommended
- The needle cap on the prefilled syringe and on the SureClick autoinjector contain dry natural rubber (latex) and should not be handled by persons sensitive to this product

Route-Specific Use

Injectable Use

- Visually inspect parenteral products for particulate matter and discoloration prior to use, solution should be clear and colorless, although small white particles in solution may be noted in the autoinjector or prefilled syringe

Subcut Use

- Injection sites include front of the thigh, abdomen except the 2 in around the navel, or outer area of the upper arm; rotate injection sites; do not administer where skin is tender, bruised, red, or hard; do not inject directly into any raised, thick, red, or scaly skin patches or lesions related to psoriasis

Reconstitution and administration of the vial:

- Do not mix or transfer the contents of one vial into another vial; do not filter reconstituted product during preparation or administration; do not add other medications to solutions containing etanercept; ONLY use the supplied diluent
- A vial adaptor is supplied for use when reconstituting the powder; however, the adaptor should not be used if multiple doses are going to be withdrawn from the vial; to reconstitute using the vial adaptor, slide the plunger into the flange end of the syringe; attach the plunger to the gray rubber stopper in the syringe by turning the plunger clockwise until a slight resistance is felt; remove the twist-off cap from the prefilled diluent syringe by turning counterclockwise; once the twist-off cap is removed, twist the vial adapter onto the syringe clockwise until a slight resistance is felt; place the vial adapter over the top of the vial, being careful not to bump or touch the plunger, the plastic spike inside the vial adapter should puncture the gray stopper; push the plunger down until all the liquid from the syringe is in the vial and gently swirl to dissolve the powder; after the diluent is added, some foaming may occur; do not shake; generally, dissolution takes less than 10 min; the solution should be clear and colorless. Each reconstituted vial contains 25 mg/ml of etanercept; turn the vial upside down and slowly pull the plunger down to the unit markings on the side of the syringe that correspond with the needed dose; gently tap the syringe to make any air bubbles rise to the top of the syringe, and slowly push the plunger up to remove them; remove the syringe from the vial adapter by turning the syringe counterclockwise and attach the 27 gauge needle
- If the vial will be used for multiple doses, use a 25-gauge needle for reconstituting and withdrawing the solution;

insert the 25 gauge needle or the vial adapter straight into the center of the gray stopper; a "pop" will be felt; inject the diluent very slowly; after the diluent is added, some foaming may occur; do not shake; swirl contents gently during dissolution; generally, dissolution takes less than 10 minutes; the solution should be clear and colorless; write the mixing date on the supplied sticker and attach to the vial; each reconstituted vial contains 25 mg/ml of etanercept; withdraw the correct dose of the solution into the syringe; remove any air bubbles; remove the 25-gauge needle from the syringe; attach a 27-gauge needle

- Hold the barrel of the syringe with one hand and pull the needle cover straight off; hold the syringe in one hand like a pencil and use the other hand to gently pinch a fold of skin at the cleaned injection site; insert the needle at a 45 degree angle to the skin; let go of the skin and hold the syringe near its base to stabilize it; push the plunger to inject all of the solution at a slow, steady rate; withdraw the needle at the same angle as insertion; do NOT rub the site
- Use as soon as possible after reconstitution; place reconstituted vials for multiple doses in the refrigerator at 36°-46° F (2°-8° C) within 4 hr of reconstitution and may be stored up to 14 days; DO NOT FREEZE
- **Use of the SureClick autoinjector:** allow to reach room temperature, do not shake; immediately before use, remove the needle shield by pulling it straight off
- Stretch the skin under and around the prefilled autoinjector, place the open end against the injection site at a 90 degree angle; without pushing the purple button on top, push the autoinjector firmly against the skin to unlock; press the purple button on top once and release the button; listen for the first click; wait for the second click or wait 15 seconds, and remove the autoinjector from injection site; do NOT rub the site
- Look at the inspection window; if it is not purple, call 1-888-436-2735; do not try to reuse the autoinjector

Use of the prefilled syringe:

- **Single-use:** allow to room temperature, do not shake; immediately before use, remove the needle shield by pulling it straight off; do not twist off or recap; check to see if the amount of liquid in the prefilled syringe falls between the two purple fill level indicator lines on the syringe; hold the prefilled syringe with the covered needle pointing down; if bubbles are seen in the syringe, very gently tap the prefilled syringe to allow any bubbles to rise to the top of the syringe; turn the syringe so that the purple horizontal lines on the barrel are directly facing you; do not use if the syringe does not have the right amount of liquid
- Hold the barrel of the prefilled syringe with one hand and pull the needle cover straight off; holding the syringe with the needle pointing up, check the syringe for air bubbles; if there are bubbles, gently tap until the air bubbles rise to the top of the syringe; slowly push the plunger up to force the air bubbles out of the syringe
- Hold the syringe in one hand like a pencil and use the other hand to gently pinch a fold of skin at the cleaned injection site; insert the needle at a 45 degree angle to the skin; let go of the skin and hold the syringe near its base to stabilize it; push the plunger to inject all of the solution at a slow, steady rate; withdraw the needle at the same angle as insertion; do NOT rub the site

SIDE EFFECTS

CNS: *Headache,* asthenia, dizziness, seizures

CV: Heart failure

GI: Abdominal pain, dyspepsia, vomiting, hepatitis

HEMA: Pancytopenia, anemia, thrombocytopenia, leukopenia, neutropenia

INTEG: Rash, *inj site reaction,* keratoderma blenorrhagicum

RESP: *Pharyngitis, cough, URI, non-URI,* sinusitis, *rhinitis*

SYST: Serious infections, sepsis, death, malignancies, Stevens-Johnson syndrome

PHARMACOKINETICS

Elimination half-life 102 hr, 60% absorbed (SUBCUT)

INTERACTIONS

- **Increase:** neutropenia—sulfaSALAzine
- Do not give concurrently with live vaccines; immunizations should be brought up to date before treatment
- Avoid use with anakinra, cyclophosphamide, rilonacept

Drug/Lab Test
Increase: LFTs

NURSING CONSIDERATIONS

Assess:

- **RA:** pain, stiffness, ROM, swelling of joints before, during, after treatment
- For inj site pain, swelling; usually occurs after 2 inj (4-5 days)

Black Box Warning: **Infection:** patients using immunosuppressives, corticosteroids, methotrexate at greater risk; assess for fever

- **Hypersensitivity:** to this product, latex needle cap, benzyl alcohol; usual reactions to product last 3-5 days

Evaluate:

- Therapeutic response: decreased inflammation, pain in joints

Teach patient/family:

- That product must be continued for prescribed time to be effective
- To use caution when driving; dizziness may occur
- Not to receive live vaccinations during treatment
- About self-administration if appropriate: inj should be made in thigh, abdomen, upper arm; rotate sites at least 1 in from previous site
- To notify prescriber of possible infection (upper respiratory, other)

ethambutol (Rx)

(e-tham′byoo-tole)

Etibi ✦, Myambutol

Func. class.: Antitubercular

Chem. class.: Diisopropylethylene diamide derivative

Do not confuse:
ethambutol/Ethmozine

ACTION:
Inhibits RNA synthesis, decreases tubercle bacilli replication

USES:
Pulmonary TB as an adjunct, other mycobacterial infections

CONTRAINDICATIONS:
Children <13 yr, hypersensitivity, optic neuritis

Precautions: Pregnancy (B), breastfeeding, renal disease, diabetic retinopathy, cataracts, ocular defects, hepatic and hematopoietic disorders

DOSAGE AND ROUTES

- **Adult/child >13 yr: PO** 15-25 mg/kg/day as single dose, 50 mg/kg 2×/wk, or 25-30 mg/kg 3×/wk

Renal disease

- CCr 10-50 ml/min, dose q24-36hr; CCr <10 ml/min, dose q48hr

Retreatment

- **Adult: PO** 25 mg/kg/day as single dose × 2 mo with at least 1 other product then decrease to 15 mg/kg/day as single dose, max 2.5 g/day
- **Child: PO** 15 mg/kg/day

Available forms: Tabs 100, 400 mg

Administer:

- With meals to decrease GI symptoms
- Antiemetic if vomiting occurs
- After C&S completed; monthly to detect resistance
- 2 hr before antacids

SIDE EFFECTS

CNS: *Headache, confusion,* fever, malaise, dizziness, ***disorientation,*** hallucinations, peripheral neuropathy

EENT: Blurred vision, optic neuritis, photophobia, decreased visual acuity
GI: *Abdominal distress, anorexia, nausea, vomiting*
INTEG: Dermatitis, pruritus, toxic epidermal necrolysis, erythema multiforme
META: *Elevated uric acid, acute gout,* impaired hepatic function
MISC: Thrombocytopenia, joint pain, anaphylaxis

PHARMACOKINETICS

Peak 2-4 hr, half-life 3 hr, metabolized in liver, excreted in urine (unchanged product/inactive metabolites, unchanged product in feces)

INTERACTIONS

- Delayed absorption of ethambutol: aluminum salts, separate by 4 hr
- Neurotoxicity: other neurotoxics

NURSING CONSIDERATIONS

Assess:

- Hepatic studies weekly × 2 wk then q2mo: ALT, AST, bilirubin; decreased appetite, jaundice, dark urine, fatigue
- Signs of anemia: Hct, Hgb, fatigue
- Mental status often: affect, mood, behavioral changes; psychosis may occur
- C&S, including sputum, before treatment
- Visual status: decreased activity, altered color perception
- **Serious skin reactions:** toxic epidermal necrolysis

Evaluate:

- Therapeutic response: decreased symptoms of TB, decrease in acid-fast bacteria

Teach patient/family:

- To avoid alcohol products
- That compliance with dosage schedule, duration is necessary
- That scheduled appointments must be kept or relapse may occur
- To report to prescriber any visual changes; rash; hot, swollen, painful joints; numbness, tingling of extremities

etodolac

(ee-toe′doe-lak)

Lodine, Lodine XL

Func. class.: Nonsteroidal antiinflammatory/nonopioid analgesic

Do not confuse: Lodine/codeine/iodine

ACTION:
Inhibits COX1,2; analgesic, antiinflammatory

USES:
Mild to moderate pain, osteoarthritis

CONTRAINDICATIONS:
Hypersensitivity; patients in whom aspirin, iodides, or NSAIDs have produced asthma; urticaria

Black Box Warning: Coronary artery bypass graft surgery (CABG)

Precautions: Edema, renal/hepatic disease, children, GI ulcers, geriatric patients, bronchospasm, nasal polyps

Black Box Warning: GI bleeding, perforation, MI, stroke

DOSAGE AND ROUTES

Osteoarthritis

- **Adult: PO** 300 mg BID-TID, or 400-500 mg BID initially, then adjust dosage to 600-1200 mg/day in divided doses; max 1200 mg/day; patients <60 kg max 20 mg/kg; ext rel 400-1200 mg daily; ext rel 400-1000 mg daily

Analgesia

- **Adult: PO** 200-400 mg every 6-8 hr up to 1000 mg daily; max 1200 mg/day; patients <60 kg, max 20 mg/kg; ext rel 400-1000 mg daily

Available forms: Caps 200, 300 mg; tabs 400, 500 mg; ext rel 400, 500, 600 mg

Administer:

- Without regard to food
- Store at room temperature

SIDE EFFECTS

CNS: Dizziness, headache, drowsiness, fatigue, tremors, confusion, insomnia, anxiety, depression, lightheadedness, vertigo

CV: Tachycardia, peripheral edema, fluid retention, palpitations, dysrhythmias, CHF

EENT: Tinnitus, hearing loss, blurred vision, photophobia

GI: Nausea, anorexia, vomiting, diarrhea, jaundice, cholestatic hepatitis, constipation, flatulence, cramps, dry mouth, peptic ulcer, dyspepsia, GI bleeding

GU: Nephrotoxicity: dysuria, hematuria, oliguria, azotemia, cystitis, urinary tract infection

HEMA: Blood dyscrasias

INTEG: Erythema, urticaria, purpura, rash, pruritus, sweating, Stevens-Johnson syndrome

SYST: Angioedema, anaphylaxis

PHARMACOKINETICS

Peak $1^1/_2$-2 hr, serum protein binding >99%, half-life 7 hr; metabolized by liver (metabolites excreted in urine)

INTERACTIONS

Increase: toxicity—cycloSPORINE, digoxin, lithium, methotrexate, phenytoin, cidofovir

Increase: GI toxicity—aspirin

Decrease: effect of—etodolac: antacids

Decrease: effect of—beta blockers, diuretics

NURSING CONSIDERATIONS

Assess:

- **Pain:** location, frequency, characteristics; relief after medication
- Blood, renal, liver tests: BUN, creatinine, AST, ALT, Hgb, platelets before treatment, periodically thereafter
- For GI bleeding: black stools, hematemesis
- Audiometric, ophthalmic examination before, during, after treatment
- For eye, ear problems: blurred vision, tinnitus; can indicate toxicity
- For asthma, aspirin hypersensitivity, nasal polyps that may be hypersensitive to etodolac

Evaluate:

- Therapeutic response: decreased pain, stiffness, swelling in joints, ability to move more easily

Teach patient/family:

- To report blurred vision or ringing, roaring in the ears; might indicate toxicity
- Not to break, crush, or chew ext rel tabs
- To report change in urine pattern, weight increase, edema, pain increase in joints, fever, blood in urine; indicates nephrotoxicity
- That therapeutic effects can take up to 1 mo
- To avoid aspirin, NSAIDs, acetaminophen, alcoholic beverages while taking this medication

RARELY USED

etomidate (Rx)

(e-tom'i-date)

Amidate

Func. class.: General anesthetic

USES: Induction of general anesthesia

CONTRAINDICATIONS: Hypersensitivity, labor/delivery

DOSAGE AND ROUTES

- **Adult/child >10 yr: IV** 0.2-0.6 mg/kg over $^1/_2$-1 min

⚠ HIGH ALERT

etoposide (Rx)

(e-toe-poe′side)

Toposar

etoposide phosphate (Rx)

Etopophos

Func. class.: Antineoplastic—miscellaneous

Chem. class.: Semisynthetic podophyllotoxin

ACTION:
Inhibits cells from entering mitosis, depresses DNA/RNA synthesis, cell-cycle–specific S and G_2; binds to a complex of DNA and topoisomerase II leading to DNA strand breaks

USES:
Testicular cancer

Unlabeled uses: Leukemias (ALL, AML), desmoid tumor, gastric/ovarian cancer, bone marrow ablation, Hodgkin's/non-Hodgkin's lymphoma, malignant glioma, neuroblastoma, stem cell transplant preparation, trophoblastic disease

CONTRAINDICATIONS:
Pregnancy (D), breastfeeding, hypersensitivity

Precautions: Children, renal/hepatic disease, gout

Black Box Warning: Bone marrow depression, infection, bleeding, requires an experienced clinician

DOSAGE AND ROUTES

Testicular cancer

- **Adult: IV** 100 mg/m²/day on days 1, 2 in combination with methotrexate, leucovorin, actinomycin D, cyclophosphamide and vinCRIStine (EMA-CO regimen) q2-3wk or 1-5 with bleomycin, CISplatin (BEP regimen) or 100 mg/m²/day on days 1, 3, 5, repeat q3-4wk

Renal dose

- **Adult: IV** CCr 45-60 ml/min, reduce dose by 15%; CCr 30-44 ml/min, reduce dose by 20%; CCr <30 ml/min, reduce dose by 25%

Hepatic dose

- **Adult: IV/PO** total bilirubin 1.5-3 mg/dl reduce dose by 50% total bilirubin 3-5 mg/dl: reduce dose by 75%; total bilirubin >5 mg/dl: hold

Available forms: Inj 20 mg/ml; caps 50 mg

Administer:

- Antiemetic 30-60 min before product and prn to prevent vomiting
- **Urate nephropathy:** Allopurinol, aggressive alkalinization to maintain uric acid levels
- Antispasmodic, EPINEPHrine, corticosteroids, antihistamines for reactions

PO route

- Give without regard to food
- Increase fluid intake to 2-3 L/day to prevent urate deposits, calculi formation
- Refrigerate oral product; do not freeze

IV route

- Do not use acrylic or ABS plastic devices; may crack, leak

Intermittent IV INF route (etoposide)

- Use cytotoxic handling procedures, use Luer-Lok fittings to prevent leakage
- After **diluting** 5-ml vial with 100 mg/250 ml or more D_5W or NaCl to 0.2-0.4 mg/ml, **infuse** over 30-60 min

Additive compatibilities: CARBOplatin, cytarabine, floxuridine, fluorouracil, hydrOXYzine, ifosfamide, ondansetron

Y-site compatibilities: Acyclovir, alfentanil, allopurinol, amifostine, amikacin, aminocaproic acid, aminophylline, amiodarone, amphotericin B colloidal, amphotericin B lipid complex, amphotericin B liposome, ampicillin, ampicillin-sulbactam, anidulafungin, atenolol, atracurium, aztreonam, bivalirudin, bleomycin, bumetanide, buprenorphine, butorphanol, calcium chloride/gluconate, CARBOplatin, caspofungin, ceFAZolin, cefoperazone, cefotaxime, cefoTEtan, cefOXitin, cefTAZidime, ceftizoxime, cefTRIAXone, cefuroxime, chloramphenicol, chlorproMAZINE, cimetidine, ciprofloxacin, cisatracurium, CISplatin, cladribine, clindamycin, codeine, cyclophosphamide, cycloSPORINE, cytarabine, DACTINomycin, DAPTOmycin, DAUNOrubicin, dexamethasone, dexmedetomidine, dexrazoxane, digoxin, diltiazem, diphenhydrAMINE, DOBUTamine, DOCEtaxel, DOPamine, doxacurium, DOXOrubicin, DOXOrubicin liposomal, doxycycline, droperidol, enalaprilat, ePHEDrine, EPINEPHrine, epirubicin, ertapenem, erythromycin, esmolol, famotidine, fenoldopam, fentaNYL, floxuridine, fluconazole, fludarabine, fluorouracil, foscarnet, fosphenytoin, furosemide, ganciclovir, gatifloxacin, gemcitabine, gentamicin, glycopyrrolate, granisetron, haloperidol, heparin, hydrALAZINE, hydrocortisone, HYDROmorphone, hydrOXYzine, ifosfamide, imipenem-cilastatin, inamrinone, insulin (regular), irinotecan, isoproterenol, ketorolac, labetalol, lansoprazole, leucovorin, levofloxacin, levorphanol, lidocaine, linezolid, LORazepam, magnesium sulfate, mannitol, mechlorethamine, melphalan, meperidine, meropenem, mesna, methohexital, methotrexate, methyldopate, methylPREDNISolone, metoclopramide, metoprolol, metroNIDAZOLE, micafungin, midazolam, milrinone, minocycline, mitoXANtrone, mivacurium, morphine, nafcillin, nalbuphine, naloxone, nesiritide, nitroglycerin, nitroprusside, norepinephrine, NS, octreotide, ofloxacin, ondansetron, oxaliplatin, PACLitaxel, palonosetron, pamidronate, pancuronium, PEMEtrexed, pentamidine, pentazocine, PENTobarbital, PHENobarbital, phenylephrine, piperacillin, piperacillin-tazobactam, polymyxin B, potassium chloride/phosphates, procainamide, prochlorperazine, promethazine, propranolol, quinupristin-dalfopristin, ranitidine, remifentanil, rocuronium, sargramostim, sodium acetate/bicarbonate/phosphates, succinylcholine, SUFentanil, sulfamethoxazole-trimethoprim, tacrolimus, teniposide, theophylline, thiotepa, ticarcillin, ticarcillin-clavulanate, tigecycline, tirofiban, tobramycin, topotecan, trimethobenzamide, vancomycin, vasopressin, vecuronium, verapamil, vinBLAStine, vinCRIStine, vinorelbine, voriconazole, zidovudine, zoledronic acid

Intermittent IV INF route (etoposide phosphate)

- **Reconstitute** each vial with 5 or 10 ml of D_5W, 0.9% NaCl for a conc of 20 mg/ml or 10 mg/ml, respectively; may give diluted or undiluted to conc of as little as 0.1 mg/ml, **give** over 5-210 min

Y-site compatibilities: Acyclovir, alfentanil, amifostine, amikacin, aminocaproic acid, aminophylline, amiodarone, ampicillin, ampicillin-sulbactam, anidulafungin, atenolol, atracurium, aztreonam, bivalirudin, bleomycin, bumetanide, buprenorphine, butorphanol, calcium acetate/chloride/gluconate, CARBOplatin, carmustine, caspofungin, ceFAZolin, cefonicid, cefoperazone, cefotaxime, cefoTEtan, cefOXitin, cefTAZidime, ceftizoxime, cefTRIAXone, cefuroxime, chloramphenicol, cimetidine, ciprofloxacin, cisatracurium, CISplatin, clindamycin, codeine, cyclophosphamide, cycloSPORINE, cytarabine, dacarbazine, DACTINomycin, DAPTOmycin, DAUNOrubicin, dexamethasone, digoxin, diltiazem, diphenhydrAMINE, DOBUTamine, DOCEtaxel, DOPamine, doripenem, doxacurium, DOXOrubicin, HCL/liposome, doxycycline, enalaprilat, ePHEDrine, EPINEPHrine, epirubicin, ertapenem, erythromycin, esmolol, famotidine, fenoldopam, fentaNYL, floxuridine, fluconazole, fludarabine, fluorouracil, foscarnet, fosphenytoin, furosemide, ganciclovir, gatifloxacin, gemcitabine, gentamicin, glycopyrrolate, granisetron, haloperidol, heparin, hydrALAZINE, hydrocortisone, HYDROmorphone, hydrOXYzine, IDArubicin, ifosfamide, inamrinone, insulin (regular), irinotecan, isoproterenol, ketorolac, labetalol, leucovorin, levofloxacin, levorphanol, lidocaine, linezolid, LORazepam, magnesium sulfate, mannitol, mechlor-

E

ethamine, meperidine, meropenem, mesna, metaraminol, methotrexate, methyldopate, metoclopramide, metoprolol, metroNIDAZOLE, midazolam, milrinone, minocycline, mitoXANtrone, mivacurium, morphine, nafcillin, nalbuphine, naloxone, nesiritide, netilmicin, nitroglycerin, nitroprusside, norepinephrine, octreotide, ofloxacin, ondansetron, oxaliplatin, PACLitaxel, palonosetron, pamidronate, pancuronium, PEMEtrexed, pentamidine, pentazocine, PENTobarbital, PHENobarbital, phenylephrine, piperacillin, piperacillin-tazobactam, plicamycin, polymyxin B, potassium chloride/phosphates, procainamide, promethazine, propranolol, quiNIDine, quinupristin-dalfopristin, ranitidine, remifentanil, riTUXimab, rocuronium, sodium acetate/bicarbonate/phosphates, streptozocin, succinylcholine, SUFentanil, sulfamethoxazole-trimethoprim, tacrolimus, teniposide, theophylline, thiopental, thiotepa, ticarcillin, ticarcillin-clavulanate, tigecycline, tirofiban, tobramycin, tolazoline, trastuzumab, trimethobenzamide, vancomycin, vasopressin, vecuronium, verapamil, vinBLAStine, vinCRIStine, vinorelbine, voriconazole, zidovudine, zoledronic acid

SIDE EFFECTS

CNS: Headache, *fever,* peripheral neuropathy, paresthesias, confusion, chills, fever
CV: *Hypotension,* MI, dysrhythmias
GI: *Nausea, vomiting, anorexia,* hepatotoxicity, dyspepsia, diarrhea, constipation
GU: Nephrotoxicity
HEMA: Thrombocytopenia, leukopenia, myelosuppression, anemia
INTEG: *Rash, alopecia,* phlebitis at IV site, Stevens-Johnson syndrome
RESP: Bronchospasm
SYST: Anaphylaxis, secondary malignancy

PHARMACOKINETICS

Half-life 7 hr, metabolized in liver, excreted in urine, feces, crosses placental barrier, protein binding 95%

INTERACTIONS

Increase: bone marrow depression—other antineoplastics, radiation, immunosuppressives
Increase: adverse reactions—live virus vaccines, toxoids
Increase: effect of etoposide, toxicity—voriconazole, conivaptan, cycloSPORINE, imatinib, nilotinib, etravirine, telithromycin
Increase: risk of bleeding—anticoagulants, NSAIDs, platelet inhibitors, thrombolytics, salicylate
Decrease: etoposide effect—sargramostim, filgrastim, separate by ≥24 hr

Drug/Food
- Decreased oral etoposide—grapefruit juice

Drug/Lab Test
Decrease: platelets, RBC, WBC, neutrophils, Hgb, calcium, phosphate
Increase: uric acid, potassium

NURSING CONSIDERATIONS

Assess:

Black Box Warning: **Bone marrow depression:** CBC, differential, platelet count weekly; withhold product if WBC is <500 or platelet count is <50,000; notify prescriber, treatment should be delayed

- **Nephrotoxicity:** BUN; serum uric acid; urine CCr; electrolytes before, during therapy; I&O ratio; report fall in urine output to <30 ml/hr; check B/P bid, report any significant decrease

Black Box Warning: **Infection:** Monitor temperature; fever may indicate beginning infection, treat active infection before treatment

Black Box Warning: **Infection:** Hepatic studies before, during therapy (bilirubin, AST, ALT, LDH) as needed or monthly

Black Box Warning: **Bleeding:** hematuria, guaiac stools, bruising or petechiae, mucosa or orifices q8hr

- Effects of alopecia on body image; discuss feelings about body changes

• **Hepatotoxicity:** Jaundice of skin and sclera, dark urine, clay-colored stools, itchy skin, abdominal pain, fever, diarrhea
• B/P q15min during inf; if systolic reading <90 mm Hg, discontinue inf, notify prescriber
• Buccal cavity q8hr for dryness, sores or ulceration, white patches, oral pain, bleeding, dysphagia
• **Injection site reaction:** monitor site closely for infiltration
• Local irritation, pain, burning, discoloration at inj site
⚠ **Symptoms indicating severe allergic reaction:** rash, pruritus, urticaria, purpuric skin lesions, itching, flushing, restlessness, coughing, difficulty breathing
• **Frequency of stools, characteristics:** cramping, acidosis; **signs of dehydration:** rapid respirations, poor skin turgor, decreased urine output, dry skin, restlessness, weakness
• **Geriatric patients:** increased alopecia, GI effects, infection, nephrotoxicity, myelosuppression

Black Box Warning: An experienced clinician knowledgable in the use of cytotoxic products

Evaluate:
• Therapeutic response: decreased tumor size, spread of malignancy

Teach patient/family:
• To report any changes in breathing or coughing
• That hair may be lost during treatment; that a wig or hairpiece may make patient feel better; that new hair may be different in color, texture
• That metallic taste may occur

Black Box Warning: **Infection:** To report symptoms of infection (fever, sore throat); to avoid crowds, persons with known infections

• To avoid immunizations
• **Pregnancy:** To notify prescriber if pregnancy is planned or suspected pregnancy (D); to use reliable contraception during and for several mo after therapy; to avoid breastfeeding
• To take as prescribed (PO), not to double dose
• To report signs of infection (flu-like symptoms, fever, fatigue, sore throat)
• To take B/P often, hypotension occurs

etravirine (Rx)
(e-tra′veer-een)
INTELENCE
Func. class.: Antiretroviral
Chem. class.: Nonnucleoside reverse transcriptase inhibitor (NNRTI)

ACTION: Binds directly to reverse transcriptase, thus blocking the RNA- and DNA-dependent DNA polymerase action and causing a disruption of the enzyme's catalytic site

USES: In combination with other antiretroviral agents for HIV infection in treatment-experienced patients with evidence of HIV replication despite ongoing antiretroviral therapy

CONTRAINDICATIONS: Breastfeeding, hypersensitivity
Precautions: Pregnancy (B), children, geriatric patients, impaired hepatic function, antimicrobial resistance, hepatitis, hypercholesterolemia, hypertriglycerides, immune reconstitution syndrome

DOSAGE AND ROUTES
• **Adult: PO** 200 mg bid after a meal, max 400 mg/day; not established for treatment-naive patients

Available forms: Tabs 100, 200 mg

Administer:
• In combination with other antiretrovirals with food or after a meal
• Tabs may be dispersed in water; once dispersed, stir well, give immediately, rinse glass, have patient drink to ensure all medication taken
• Store in cool environment; protect from light

SIDE EFFECTS

CNS: *Headache, insomnia,* amnesia, anxiety, confusion, fatigue, nightmares, peripheral neuropathy, seizures, stroke, tremor
CV: Atrial fibrillation, hypertension, MI
EENT: Blurred vision
GI: *Nausea, vomiting, diarrhea, anorexia,* abdominal pain, increased AST/ALT, constipation, flatulence, gastritis, GERD, hematemesis, hepatitis, hepatomegaly, pancreatitis
GU: Renal failure
HEMA: Hemolytic anemia, neutropenia, thrombocytopenia, anemia
INTEG: *Rash,* erythema multiforme, angioedema, Stevens-Johnson syndrome
MS: Rhabdomyolysis
OTHER: Diabetes mellitus, gynecomastia, hyperamylasemia, hypercholesterolemia, hyperglycemia, hyperlipidemia
RESP: Dyspnea, bronchospasm
SYST: DRESS

PHARMACOKINETICS

99.9% plasma protein binding; metabolized by CYP3A4, 2C9, 2C19; half-life 21-61 hr; excreted in feces

INTERACTIONS

- Do not use concurrently with atazanavir, carBAMazepine, delavirdine, fosamprenavir, fosphenytoin, phenytoin, PHENobarbital, rifapentine, rifampin, tipranavir, rilpivirine
- Altered effect of cycloSPORINE, tacrolimus, sirolimus

Increase: myopathy, rhabdomyolysis—HMG-CoA reductase inhibitors
Increase: etravirine levels—CYP3A4 inhibitors (fluconazole, itraconazole, ketoconazole, lopinavir, posaconazole, ritonavir, voriconazole)
Increase: withdrawal symptoms—methadone
Increase: levels of diazepam, rifampin, voriconazole, warfarin
Decrease: levels of CYP3A4 inducers (amiodarone, atazanavir, clarithromycin, flecainide, fosamprenavir, lidocaine, mexiletine, propafenone, quiNIDine, sildenafil, tadalafil, vardenafil)
Decrease: etravirine levels—darunavir, dexamethasone, disopyramide, efavirenz, nevirapine, ritonavir, saquinavir, tipranavir

Drug/Herb
Decrease: etravirine—St. John's wort

NURSING CONSIDERATIONS

Assess:

- **Symptoms of HIV, possible infections;** increased temperature
- ⚠ **Fatal hypersensitivity reactions:** fever, rash, nausea, vomiting, fatigue, cough, dyspnea, diarrhea, abdominal discomfort; treatment should be discontinued and not restarted; incidence of rash may be worse in women
- **Blood dyscrasias** (anemia, granulocytopenia): bruising, fatigue, bleeding, poor healing
- **Renal failure:** BUN, serum uric acid, CCr before, during therapy; may be elevated throughout treatment
- **Hepatitis/pancreatitis:** hepatic studies before and during therapy: bilirubin, AST, ALT, amylase, alk phos, creatine phosphokinase, creatinine, monthly
- HIV: monitor viral load, CD4 counts, plasma HIV RNA during treatment; watch for decreasing granulocytes, Hgb; if low, therapy may have to be discontinued and restarted after hematologic recovery; blood transfusions may be required; cholesterol/lipid profile

Evaluate:

- Therapeutic response: increased CD4 count, decreased viral load

Teach patient/family:

- That product is not a cure but will control symptoms; that patient is still infective, may pass AIDS virus to others
- To notify prescriber of sore throat, swollen lymph nodes, malaise, fever; other infections may occur; to stop product and notify prescriber immediately if skin rash, fever, cough, SOB, GI symptoms occur; to advise all health care providers

that allergic reaction has occurred with etravirine

- That follow-up visits must be continued, since serious toxicity may occur; blood counts must be performed
- To use contraception during treatment; that patient still able to transmit disease
- About information on medication guide and warning card; discuss points on guide
- That other products may be necessary to prevent other infections
- To take medication after a meal

everolimus (Rx)

(e-ve-ro′li-mus)

Afinitor, Afinitor Disperz, Zortress

Func. class.: Antineoplastic—miscellaneous

Chem. class.: Immunosuppressant, macrolide

Do not confuse:
everolimus/sirolimus/tacrolimus/temsirolimus

ACTION: Proliferation signal inhibitor that inhibits mammalian target of rapamycin (mTOR); this pathway is dysregulated in cancer

USES: Renal cell cancer in those with failed treatment with SORAfenib or SUNItinib kidney transplant rejection prophylaxis with cycloSPORINE, subependymal giant cell astrocytoma, progressive pancreatic neuroendocrine tumor (PNET) with unresectable locally advanced/metastatic disease, breast cancer hormone receptor positive/HER-2 negative, renal angiomyolipoma, tuberous sclerosis complex

CONTRAINDICATIONS: Breastfeeding; hypersensitivity to this product, Rapamune, torisel, pregnancy (D)

Precautions: Children, renal/hepatic disease; diabetes mellitus, hyperlipidemia, plural effusion

Black Box Warning: Immunosuppression, infection, renal artery thrombosis, renal impairment, renal vein thrombosis

DOSAGE AND ROUTES

Kidney transplant rejection prophylaxis (Zortress)

- **Adult: PO** 0.75 mg q12hr with cycloSPORINE in combination with basiliximab, corticosteroids, reduced doses of cycloSPORINE

Advanced renal cancer (Afinitor)

- **Adult: PO** 10 mg daily as long as clinically beneficial; with strong 3A4 inducers 10 mg daily then may increase by 5 mg increments to 20 mg daily

Progressive neuroendocrine tumor (PNET) (Afinitor only)

- **Adult: PO** 10 mg daily, reduce dose to 5 mg daily if intolerable adverse reactions occur

Subependymal giant-cell astrocytoma (SEGA) (Afinitor only)

- **Adult/adolescent/child:** 4.5 mg/m^2 daily, then titrate to a target trough of 5-15 ng/ml

Hepatic dose

- **Adult: PO** (Child-Pugh A): Afinitor 7.5 mg/day; (Child-Pugh B) Afinitor: 5 mg/day; Zortress: 0.75 mg/day divided q12hr; (Child-Pugh C) Afinitor 2.5 mg/day

Available forms: Tabs 0.25, 0.5, 0.75 (Zortess); 2.5, 5, 7.5, 10 mg (Afinitor)

Administer:

- Follow procedure for proper handling of antineoplastics
- Swallow tabs whole with a full glass of water; do not chew, crush, or break
- **Afinitor:** take at same time of day; if unable to swallow, consistently with or without food, disperse in 30 ml of water
- **Zortress:** must take consistently with/without food, give at same time of day q12hr with cycloSPORINE
- Store protected from light at room temperature

Afinitor adjustments for toxicity

Noninfectious pneumonitis

- Grade 1, asymptomatic with radiographic findings only: no dosage change

- Grade 2, symptomatic but no interference with activities of daily living (ADL): consider withholding therapy, resume Afinitor at a lower dosage when symptoms improve to ≤ grade 1; discontinue Afinitor if symptoms do not improve within 4 wk
- Grade 3, symptomatic and interfering with ADL and oxygen therapy indicated: hold therapy; consider resuming Afinitor at a lower dosage when symptoms improve to ≤ grade 1; consider discontinuing Afinitor if grade 3 toxicity recurs
- Grade 4, life-threatening and ventilator support indicated: discontinue therapy

Stomatitis

- Grade 1, minimum symptoms and normal diet: no dosage adjustment required
- Grade 2, symptomatic but can eat and swallow modified diet: hold therapy until symptoms improve to ≤ grade 1 and resume Afinitor at the same dosage; if grade 2 toxicity recurs, hold therapy and resume Afinitor at a lower dosage when symptoms improve to ≤ grade 1
- Grade 3, symptomatic and unable to adequately eat or hydrate orally: hold therapy; resume Afinitor at a lower dosage when symptoms improve to ≤ grade 1
- Grade 4, symptomatic and life-threatening: discontinue therapy

Other nonhematologic toxicity (excluding metabolic events)

- **Grade 1:** No dosage adjustment required if toxicity is tolerable
- **Grade 2:** No dosage adjustment required if toxicity is tolerable; if toxicity is intolerable, hold therapy until symptoms improve to ≤ grade 1 and resume Afinitor at the same dosage; if grade 2 toxicity recurs, hold therapy and resume Afinitor at a lower dosage when symptoms improve to ≤ grade 1
- **Grade 3:** Hold therapy; consider resuming Afinitor at a lower dosage when symptoms improve to ≤ grade 1; if grade 3 toxicity recurs, consider discontinuing therapy
- **Grade 4:** Discontinue Afinitor therapy

Metabolic events (hyperglycemia, dyslipidemia)

- **Grade 1 or 2:** No dose adjustment required
- **Grade 3:** Temporarily withhold therapy; resume Afinitor at a lower dosage
- **Grade 4:** Discontinue Afinitor therapy

SIDE EFFECTS

CNS: *Headache, insomnia, paresthesia,* chills, fever, seizure/personality changes/dizziness

CV: *Hypertension, CHF, peripheral edema*

EENT: Blurred vision, photophobia

GI: Nausea, vomiting, diarrhea, constipation, stomatitis

GU: Renal failure

HEMA: Anemia, leukopenia, thrombocytopenia

INTEG: *Rash, acne*

META: Hyperglycemia, increased creatinine, *hyperlipemia,* hypophosphatemia, weight loss

RESP: Pleural effusion, *dyspnea,* noninfectious pneumonitis, pulmonary embolism

PHARMACOKINETICS

Rapidly absorbed; peak 1-2 hr; protein binding 74%; extensively metabolized by CYP3A4 enzyme system; half-life 30 hr; reduced by high-fat meal; excreted in feces (80%), urine (5%)

INTERACTIONS

Increase: Everolimus effect—CYP3A4 inhibitors (strong, moderate), antifungals, calcium channel blockers, cimetidine, danazol, erythromycin, cycloSPORINE, HIV-protease inhibitors

Decrease: blood levels of everolimus—carBAMazepine, PHENobarbital, phenytoin, rifamycin, rifapentine

Increase: nephrotoxicity—immunosuppressants

Decrease: effect of live vaccines

Drug/Herb

- St. John's wort: may decrease effect of everolimus

Drug/Food
• Alters bioavailability; use consistently with/without food; do not use with grapefruit juice

Drug/Lab Test
Increase: bilirubin, calcium, cholesterol, glucose, potassium, lipids, phosphate, triglycerides, uric acid
Decrease: calcium, glucose, potassium, magnesium, phosphate

NURSING CONSIDERATIONS

Assess:
• Lipid profile: cholesterol, triglycerides, lipid-lowering agent may be needed; blood glucose

Black Box Warning: **Immunosuppression:** CBC with differential during treatment monthly; if leukocytes <3000/mm³ or platelets <100,000/mm³, product should be discontinued or reduced; decreased hemoglobulin level may indicate bone marrow suppression

• Hepatic/renal studies: AST, ALT, amylase, bilirubin, creatinine, phosphate, and for hepatotoxicity: dark urine, jaundice, itching, light-colored stools; product should be discontinued

Black Box Warning: **Infection:** bacterial fungal infections can occur and are more common with combination immunosuppression therapy

• **Renal artery/vein thrombosis (Zortress):** May result in graft loss within 30 days after transplantation
• Obtain everolimus blood levels in kidney transplant, hepatic disease, CYP3A4 inducers, inhibitors

Evaluate:
• Therapeutic response

Teach patient/family:

Black Box Warning: To report fever, rash, severe diarrhea, chills, sore throat, fatigue; serious infections may occur; to report clay-colored stools, cramping (hepatotoxicity)

Black Box Warning: To avoid crowds, persons with known infections to reduce risk for infection

• **Pregnancy:** to notify prescriber if pregnancy is planned or suspected, pregnancy (D)
• To use contraception before, during, and 12 wk after product discontinued; to avoid breastfeeding
• Not to use with grapefruit juice
• To avoid live vaccines
• That product may decrease male, female fertility
• That drinking alcohol is not recommended
• To take consistently with or without food
• To report vision changes, weight gain, edema, shortness of breath

exemestane (Rx)

(ex-em′eh-stane)

Aromasin

Func. class.: Antineoplastic
Chem. class.: Aromatase inhibitor

Do not confuse:
exemestane/ezetimibe/estramustine

ACTION: Lowers serum estradiol concs; many breast cancers have strong estrogen receptors

USES: Advanced breast carcinoma not responsive to other therapy (postmenopausal)

CONTRAINDICATIONS: Pregnancy (X), breastfeeding, premenopausal women, hypersensitivity
Precautions: Children, geriatric patients, renal/hepatic disease

DOSAGE AND ROUTES

• **Adult: PO** 25 mg/day after meals; may need 50 mg/day if taken with a potent CYP3A4 inducer

Available forms: Tabs 25 mg

Administer:
• After meals at same time of day
• Store at room temperature

SIDE EFFECTS

CNS: *Headache, depression, insomnia, anxiety, fatigue, hot flashes, diaphoresis,* dizziness
CV: Hypertension, edema
GI: *Nausea,* vomiting, diarrhea, constipation, abdominal pain, increased appetite
HEMA: *Lymphopenia*
MS: Fracture, bone loss
RESP: Cough, *dyspnea*

PHARMACOKINETICS

Half-life 24 hr; excreted in feces, urine

INTERACTIONS

Decrease: exemestane action—CYP3A4 inducers, estrogens

NURSING CONSIDERATIONS

Assess:

- B/P; hypertension may occur
- Bone mineral density, x-ray of thoracic or lumbar spine if bone changes suspected

Evaluate:

- Therapeutic response: decreased tumor size, spread of malignancy

Teach patient/family:

- To report any complaints, side effects to prescriber
- That hot flashes are reversible after discontinuing treatment
- To use reliable contraception (pregnancy [X]); not to breastfeed
- That vit D, calcium may be used for bone loss

exenatide (Rx)

(ex-en′a-tide)

Bydureon, Byetta

Func. class.: Antidiabetic
Chem. class.: Incretin mimetic

ACTION: Binds and activates known human GLP-1 receptor, mimics natural physiology for self-regulating glycemic control

USES: Type 2 diabetes mellitus given in combination with metFORMIN, a sulfonylurea, thiazolidinedione, insulin glargine; once-weekly dosing

CONTRAINDICATIONS: Hypersensitivity

Black Box Warning: Medullary thyroid carcinoma, multiple endocrine neoplasia syndrome type 2 (men-2), thyroid cancer

Precautions: Pregnancy (C), geriatric patients, severe renal/hepatic/GI disease, pancreatitis, vit D deficiency

DOSAGE AND ROUTES

- **Adult: SUBCUT** 5 mcg bid 1 hr before morning and evening meal; may increase to 10 mcg bid after 1 mo of therapy; ext rel SUBCUT (Bydureon) 2 mg q7days

Available forms: Inj 5, 10 mcg pen; ext rel powder for susp for inj 2 mg

Administer:

- Store in refrigerator for unopened pen; may store at room temperature after opening for up to 30 days

SUBCUT route (regular release—Byetta)

- May be used as monotherapy or combined with other products
- SUBCUT only, do not give IV/IM
- Pen needles must be purchased separately, compatible; prime before use; inject into thigh, abdomen, upper arm; rotate sites
- Product 1 hr before meals, approximately 6 hr apart; if patient is NPO, may need to hold dose to prevent hypoglycemia
- If added to insulin glargine, insulin, or detemir, a dosage reduction in these products may be required

SUBCUT route (ext rel—Bydureon)

- Give every 7 days (weekly); the dose can be given at any time of day without regard to meals
- Available as a single-dose tray containing a vial of 2 mg, a pre-filled syringe delivering 0.65 ml diluent, a vial connector, and two custom needles (23GX 5/16′) specific to this delivery system

(one is a spare needle); do not substitute needles or any other components

- Inject immediately after the white to off-white powder is suspended in the diluent and transferred to the syringe
- Inject subcut into the thigh, abdomen, or upper arm, rotate sites to prevent lipodystrophy

SIDE EFFECTS

CNS: *Headache, dizziness,* feeling jittery, restlessness, weakness
ENDO: **Hypoglycemia,** thyroid hyperplasia
GI: Nausea, vomiting, diarrhea, dyspepsia, anorexia, gastroesophageal reflux, weight loss, **pancreatitis**
SYST: **Angioedema, anaphylaxis,** inj site reactions

PHARMACOKINETICS

Immediate release: Peak 2.1 hr, elimination by glomerular filtration
Ext Rel: Peak 2 wk

INTERACTIONS

- May decrease effect of acetaminophen
- **Do not use with erythromycin, metoclopramide**

Increase: hypoglycemia—ACE inhibitors, disopyramide, sulfonylureas, androgens, fibric acid derivatives, alcohol
Increase: hyperglycemia—phenothiazines, corticosteroids, anabolic steroids
Decrease: action of digoxin, lovastatin, acetaminophen (elixir)
Decrease: efficacy—niacin, dextrothyroxine, thiazide diuretics, triamterene, estrogens, progestins, oral contraceptives, MAOIs

NURSING CONSIDERATIONS

Assess:

- Fasting blood glucose, A1c levels, postprandial glucose during treatment to determine diabetes control
- **Pancreatitis:** severe abdominal pain, with or without vomiting, product should be discontinued
- **Anaphylaxis, Angioedema:** product should be discontinued immediately
- Renal studies: urinalysis, creatinine
- Hypo/hyperglycemic reaction that can occur soon after meals; for severe hypoglycemia, give IV $D_{50}W$ then IV dextrose solution
- Nausea, vomiting, diarrhea, ability to tolerate product, may cause dehydration

Evaluate:

- Therapeutic response: decrease in polyuria, polydipsia, polyphagia, clear sensorium, improving A1c, weight; absence of dizziness, stable gait

Teach patient/family:

- About the symptoms of hypo/hyperglycemia, what to do about each; to have glucagon emergency kit available; to carry a glucose source (candy, sugar cube) to treat hypoglycemia
- That product must be continued on a daily or weekly basis (ext rel); about consequences of discontinuing product abruptly
- That diabetes is a lifelong illness; product will not cure disease; to carry emergency ID with prescriber and medication information
- To continue weight control, dietary restrictions, exercise, hygiene
- That regular blood glucose monitoring and A1c testing is needed
- To notify prescriber if pregnant or intending to become pregnant (C)
- About the importance of reading "Information for the Patient" and "Pen User Manual;" about self-injection
- **Pancreatitis:** If severe abdominal pain with or without vomiting occurs seek medical attention immediately
- To review injection procedure; to store product in refrigerator, room temperature after first use; discard 30 days after first use; do not freeze; protect from light (Byetta)

ezetimibe (Rx)

(ehz-eh-tim′bee)

Ezetrol ✦, Zetia

Func. class.: Antilipemic; cholesterol absorption inhibitor

ACTION: Inhibits absorption of cholesterol by the small intestine, causes reduced hepatic cholesterol stores

USES: Hypercholesterolemia, homozygous familial hypercholesterolemia (HoFH), homozygous sitosterolemia

CONTRAINDICATIONS: Hypersensitivity, severe hepatic disease

Precautions: Pregnancy (C), breastfeeding, children, hepatic disease

DOSAGE AND ROUTES

- **Adult/adolescent/child >10 yr: PO** 10 mg/day; may be given with HMG-CoA reductase inhibitor at same time; may be given with bile acid sequestrant; give ezetimibe 2 hr before or 4 hr after bile acid sequestrant

Available forms: Tabs 10 mg

Administer:

- Without regard to meals

SIDE EFFECTS

CNS: Fatigue, dizziness, headache

GI: Diarrhea, abdominal pain

MISC: Chest pain

MS: *Myalgias, arthralgias,* back pain, myopathy, rhabdomyolysis

RESP: Pharyngitis, sinusitis, cough, *URI*

SYST: Angioedema

PHARMACOKINETICS

Metabolized in small intestine, liver; excreted in feces 78%, urine 11%; peak 4-12 hr; half-life 22 hr

INTERACTIONS

Increase: action of ezetimibe—fibric acid derivatives, cycloSPORINE

Decrease: action of ezetimibe—antacids, cholestyramine

Drug/Lab Test

Increase: LFTs

NURSING CONSIDERATIONS

Assess:

- **Hypercholestrolemia:** diet history: fat content, lipid levels (triglycerides, LDL, HDL, cholesterol); LFTs at baseline, periodically during treatment
- **Myopathy/rhabdomyolysis:** increased CPK, myalgia, muscle cramps, musculoskeletal pain, lethargy, fatigue, fever; more common when combined with statins

Evaluate:

- Therapeutic response: decreased cholesterol, LDL; increased HDL

Teach patient/family:

- That compliance is needed
- That risk factors should be decreased: high-fat diet, smoking, alcohol consumption, absence of exercise
- To notify prescriber if pregnancy suspected, planned, or if breastfeeding
- To notify prescriber if unexplained weakness, muscle pain present
- To notify prescriber of dietary/herbal supplements

ezogabine

(e-zog′a-been)

Potiga

Func. class.: Anticonvulsant

ACTION: The exact mechanism of anticonvulsant effects is not fully known; however, studies indicate that the drug enhances transmembrane potassium currents, which may stabilize the resting membrane potential and reduce brain excitability; may also augment GABA-mediated currents

USES: Partial seizures, migraines

CONTRAINDICATIONS:
Hypersensitivity

Precautions: Suicidal ideation/behavior, prostatic hypertrophy, dementia, psychotic disorders, QT prolongation, congestive heart failure, ventricular hypertrophy, hypokalemia, hypomagnesemia, abrupt discontinuation, renal impairment, hepatic disease, geriatric patients, pregnancy category C, breastfeeding, neonates, infants, children, adolescents

DOSAGE AND ROUTES

- **Adult/geriatric patient ≤65 yr: PO** initially, 100 mg tid, increase by ≤50 mg tid per day at weekly intervals depending on response, up to a maintenance dose of 200-400 mg tid depending on response; max is 400 mg tid (1200 mg/day).
- **Geriatric patient >65 yr: PO** initially, 50 mg tid, increase by ≤50 mg tid per day at weekly intervals depending on response; max 250 mg tid (750 mg/day).

Available forms: Film-coated tabs 50, 200, 300, 400 mg

Administer:

- Tab should be swallowed whole without regard to meals
- Give in 3 equally divided doses

SIDE EFFECTS

CNS: Dizziness, drowsiness, memory impairment, tremor, vertigo, abnormal coordination, disturbance in attention, gait disturbance, aphasia, dysarthria, balance disorder, paresthesias, amnesia, dysphagia, myoclonia, hypokinesia, confusion, anxiety, hallucinations, **suicidal thoughts/behaviors,** fatigue, asthenia, malaise, euphoria

EENT: Diplopia, blurred vision, retinal pigment change

GI: Nausea, constipation, dyspepsia, xerostomia, constipation, weight gain, appetite stimulation

GU: Urinary retention, hydronephrosis, dysuria, urinary hesitation, hematuria, chromaturia

HEMA: **Thrombocytopenia, leukopenia, neutropenia**

INTEG: Rash, alopecia, blue skin discoloration

MISC: Influenza, dyspnea, **QT prolongation**

MS: Muscle spasms, weakness

PHARMACOKINETICS

80% protein bound; extensively distributed in the body; extensively metabolized by glucuronidation and acetylation; inactive N-glucuronides are the primary metabolites; elimination half-lives of ezogabine and its N-acetyl metabolite are 7 hr and 11 hr, respectively; 36% excreted renally as exogabine, 18% as NAMR, 24% as the N-glururonides of ezogabine and NAMR, 14% fecal excretion; rapidly absorbed, peak 0.5-2 hr, bioavailability 60%; high-fat food increases peak conc; increased in hepatic/renal disease, geriatric patients, young adult females

INTERACTIONS:

Increase: QT prolongation—arsenic trioxide, chloroquine, chlorproMAZINE, mesoridazine thioridazine, clarithromycin, Class IA antiarrhythmics (disopyramide, procainamide, quiNIDine), Class III antiarrhythmics (amiodarone, dofetilide, ibutilide, sotalol), dextromethorphan, dronedarone, droperidol, erythromycin, grepafloxacin, halofantrine, levomethadyl, methadone, pentamidine, pimozide, posaconazole, probucol, propafenone, quiNIDine, saquinavir, sparfloxacin, terfenadine, troleandomycin, ziprasidone

Decrease: effect of phenytoin

Decrease: effect of ezogabine—carBAMazepine

Increase: effect of digoxin

Increase: urinary retention—antimuscarinics, amantadine, H1-blockers

Increase: CNS depression—anxiolytics, sedatives, hypnotics, buprenorphine, butorphanol, dronabinol, mirtazapine, nabilone, nalbuphine, opiate agonists, pentazocine, pregabalin, skeletal muscle relaxants, traMADol, traZODone, ethanol

Drug/Lab Test
Increase: LFTs

NURSING CONSIDERATIONS

Assess:

⚠ **Seizures:** Assess for type, duration, location, activity, presence of aura

⚠ **QT prolongation:** Monitor in those with known QT prolongation, congestive heart failure, ventricular hypertrophy, hypokalemia, hypomagnesemia, and in patients receiving medications known to cause QT prolongation; QT prolongation can occur within 3 hr of dose

⚠ **Abrupt withdrawal:** Withdraw gradually to minimize increased seizure frequency

⚠ **Suicidal thoughts/behaviors:** Assess for any unusual changes in moods or behaviors, including emotional lability or emerging or worsening depression and suicidal ideation

- **Vision changes:** Obtain a baseline eye exam, then periodic eye exams; discontinue if ophthalmic changes occur, unless no other treatment is available

Teach patient/family:

- To avoid driving or operating machinery or performing other tasks that require mental alertness until reaction is known
- To avoid concurrent use of alcohol
- To avoid abruptly discontinuing this product

⚠ **Suicidal thought/behaviors:** Advise patient to notify prescriber immediately of suicidal thought/behaviors

- That skin may turn blue; it is unknown if it is reversible

⚠ HIGH ALERT

factor IX complex (human) (Rx)

AlphaNine SD, Bebulin, BeneFIX, Mononine, Profilnine SD

Func. class.: Hemostatic

Chem. class.: Factors II, VII, IX, X

ACTION: Causes an increase in blood levels of clotting factors II, VII, IX, X; factor IX (human) has IX activity

USES: Hemophilia B (Christmas disease), factor IX deficiency, anticoagulant reversal, control of bleeding in patients with factor VIII inhibitors, reversal of overdose of anticoagulants in emergencies

CONTRAINDICATIONS: Hypersensitivity to mouse or hamster protein, DIC, mild factor IX deficiency

Precautions: Pregnancy (C), neonates/infants, hepatic disease, elective surgery

DOSAGE AND ROUTES

Factor IX complex (human) bleeding in hemophilia B

- **Adult and child: IV** establish 25% of normal factor IX or 60-75 units/kg then 10-20 units/kg/day 1-2×/wk

Prophylaxis for bleeding in hemophilia B (long term)

- **Adult and child: IV** 25-40 units/kg 2×/wk

Bleeding in hemophilia A/inhibitors of factor VIII (Proplex T, Konyne 80)

- **Adult and child: IV** 75 units/kg, repeat after 12 hr

Oral anticoagulant reversal (Factor IX complexer only) (unlabeled)

- **Adult and child: IV** 20-50 units/kg

Factor VII deficiency (use Proplex T only)

- **Adult and child: IV** 0.5 units/kg × weight (kg) × desired factor IX increase (% of normal); repeat q4-6hr if needed

Factor IX (human) minor to moderate hemorrhage

Use only Alpha-Nine, Alpha-Nine SD

- **Adult and child: IV** dose to increase factor IX level to 20%-30% in one dose

Serious hemorrhage

- **Adult and child: IV** dose to increase factor IX to 30%-50% as daily inf

Minor hemorrhage (mononine only)

- **Adult and child: IV** dose to increase factor IX to 15%-25% (20-30 units/kg), repeat after 24 hr if needed

Major hemorrhage

- **Adult and child: IV** dose to increase factor IX to 25%-50% (75 units/kg) q18-30hr × ≤10 days

Available forms: Inj (number of units noted on label)

Administer:

- Hepatitis B vaccine before administration
- IV after warming to room temperature ≤3 ml/min, with plastic syringe only; do not admix
- After dilution with provided diluent, 50 units/ml or 25 units/ml; do not exceed 10 ml/min; decrease rate if fever, headache, flushing, tingling occur
- After cross-matching if patient has blood type A, B, AB to determine incompatibility with factor
- Store reconstituted sol for 3 hr at room temperature or for ≤2 yr with refrigeration (powder); check expiration date

BeneFIX

- Allow vials of concentrate/diluent to warm to room temp
- After removing flip-top cap from vial, wipe top of vial with alcohol swab; allow to dry
- Peel back cover from vial adapter package; do not remove
- Place vial adapter over vial; press firmly until it snaps; attach plunger rod to diluent syringe and break plastic tip cap from diluent syringe
- Lift package away from adapter and connect diluent syringe; depress plunger; swirl contents

SIDE EFFECTS

CNS: Headache, dizziness, malaise, paresthesia, *lethargy, chills, fever, flushing*
CV: *Hypotension,* tachycardia, *MI,* venous thrombosis, pulmonary embolism
GI: Nausea, vomiting, abdominal cramps, jaundice, viral hepatitis
HEMA: Thrombosis, hemolysis, AIDS, DIC
INTEG: Rash, flushing, *urticaria,* inj site reactions
RESP: Bronchospasm

PHARMACOKINETICS

IV: Half-life factor IX: 22 hr; rapidly cleared from plasma

INTERACTIONS

- Incompatible with protein products

⚠ **Increase:** thrombosis risk—aminocaproic acid; do not administer
Decrease: effect of warfarin

NURSING CONSIDERATIONS

Assess:

- Blood studies: coagulation factor assays by % normal: 5% prevents spontaneous hemorrhage, 30%-50% for surgery, 80%-100% for severe hemorrhage; APTT, clotting inhibitor titers, D-dimer, factor IX, VII conc, fibrin degradation products, fibrinogen, platelets, PT, thrombin time
- Increased B/P, pulse
- For bleeding q15-30min; immobilize and apply ice to affected joints
- I&O; if urine becomes orange or red, notify prescriber
- **Allergic or pyrogenic reaction:** fever, chills, rash, itching, slow inf rate if not severe

⚠ **DIC:** bleeding, ecchymosis, hypersensitivity, changes in coagulation tests

- For tingling sensation; if present, reduce rate

Evaluate:

- Therapeutic response: prevention of hemorrhage

Teach patient/family:

- To report any signs of bleeding: gums, under skin, urine, stools, emesis; calf pain, joint pain, yellowing of eyes/skin
- About the risk for viral hepatitis, AIDS; to be tested q2-3mo for HIV, although low risk
- That immunization for hepatitis B may be given first
- To carry emergency ID identifying disease; to avoid salicylates, NSAIDs; to inform other health professionals of condition

famciclovir (Rx)

(fam-cy′clo-veer)

Famvir

Func. class.: Antiviral
Chem. class.: Guanosine nucleoside

ACTION: Inhibits DNA polymerase and viral DNA synthesis by conversion of this guanosine nucleoside to penciclovir

USES: Treatment of acute herpes zoster (shingles), genital herpes; recurrent mucocutaneous herpes simplex virus (HSV) in patients with HIV; initial episodes of herpes genitalis; herpes labialis in the immunocompromised
Unlabeled uses: Bell's palsy, herpes labialis prophylaxis, postherpetic neuralgia prophylaxis

CONTRAINDICATIONS: Hypersensitivity to this product, penciclovir, acyclovir, ganciclovir, valacyclovir, valganciclovir
Precautions: Pregnancy (B), breastfeeding, renal disease

DOSAGE AND ROUTES

Herpes zoster

- **Adult: PO** 500 mg q8hr for 7 days

Renal dose

- **Adult: PO** CCr ≥60 ml/min, 500 mg q8hr; CCr 40-59 ml/min, 500 mg q12hr; CCr 20-39 ml/min, 500 mg q24hr; CCr <20 ml/min, 250 mg q24hr

Recurrent herpes simplex virus
• **Adult: PO** 125 mg q12hr × 5 days
Renal dose
• **Adult: PO** 40-59 ml/min 500 mg bid; CCr 20-39 ml/min 500 daily; CCr <20 ml/min 250 mg daily
Suppression of recurrent herpes simplex virus
• **Adult: PO** 250 mg q12hr up to 1 yr
Renal dose
• **Adult: PO** CCr 20-39 ml/min, 125 mg q12hr × 5 days; CCr <20 ml/min, 125 mg q24hr × 5 days
Genital herpes/herpes labialis (recurrent)
• **Adult: PO** 125 mg bid × 5 days or 1000 mg bid for 1 day; begin treatment at 1st sign of recurrence; immunocompetent herpes labialis 1500 mg as a single dose
Suppression of recurrent genital herpes
• **Adult: PO** 250 mg bid for up to 1 yr
Herpes genitalis initial episodes
• **Adult: PO** 250 mg tid × 7-10 days
Bell's palsy (unlabeled)
• **Adult: PO** 750 mg tid × 7 days with predniSONE
Varicella-zoster virus (shingles); chickenpox (unlabeled)
• **Adult: PO** 500 mg q8hr × 7 days, preferably within 48 hr of onset
Herpes zoster in HIV (unlabeled)
• **Adult/adolescent: PO** 500 mg tid × 7-10 days
Available forms: Tabs 125, 250, 500 mg
Administer:
• Without regard to meals
• As soon as diagnosed; for herpes zoster within 72 hr

SIDE EFFECTS

CNS: *Headache, fatigue, dizziness,* paresthesia, somnolence, fever
GI: Nausea, vomiting, diarrhea, constipation, abdominal pain, anorexia
GU: Decreased sperm count
INTEG: *Pruritus*
MS: Back pain, arthralgia
RESP: Pharyngitis, sinusitis

PHARMACOKINETICS

Bioavailability 77%, 20% protein binding, 73% excreted via kidneys, terminal plasma half-life 2-3 hr

INTERACTIONS

Decrease: renal excretion—theophylline, probenecid, digoxin
Decrease: metabolism—cimetidine

NURSING CONSIDERATIONS

Assess:
• **Herpes zoster:** number, distribution of lesions; burning, itching, pain, which are early symptoms of herpes infection; assess daily during therapy
• Renal studies: urine CCr; BUN before and during treatment if decreased renal function; dose may have to be lowered
• Bowel pattern before, during treatment; diarrhea may occur
• Posttherapeutic neuralgia during and after treatment
Evaluate:
• Therapeutic response: decreased size, spread of lesions
Teach patient/family:
• How to recognize beginning infection
• How to prevent spread of infection; that this medication does not prevent spread to others; that condoms should be used
• About the reason for medication, expected results
• That women with genital herpes should have yearly Pap smears; that cervical cancer is more likely

famotidine (OTC, Rx)

(fa-moe′ti-deen)

Pepcid, Pepcid AC

Func. class.: H_2-histamine receptor antagonist

ACTION: Competitively inhibits histamine at histamine H_2-receptor site, thus decreasing gastric secretion while pepsin remains at a stable level

USES:
Short-term treatment of active duodenal ulcer, maintenance therapy for duodenal ulcer, Zollinger-Ellison syndrome, multiple endocrine adenomas, gastric ulcers; gastroesophageal reflux disease, heartburn

Unlabeled uses: GI disorders in those taking NSAIDs; urticaria; prevention of stress ulcers, aspiration pneumonitis, inactivation of oral pancreatic enzymes in pancreatic disorders

CONTRAINDICATIONS:
Hypersensitivity

Precautions: Pregnancy (B), breastfeeding, children <12 yr, geriatric patients, severe renal/hepatic disease

DOSAGE AND ROUTES

Active ulcer

- **Adult: PO** 40 mg/day at bedtime × 4-8 wk then 20 mg/day at bedtime if needed (maintenance); **IV** 20 mg q12hr if unable to take **PO**
- **Child 1-16 yr: PO** 0.5 mg/kg/day at bedtime or divided bid, max 40 mg/day

Hypersecretory conditions

- **Adult: PO** 20 mg q6hr; may give 160 mg q6hr if needed; **IV** 20 mg q6hr if unable to take **PO**

Heartburn relief/prevention

- **Adult: PO** 10 mg with water or 15 min-1 hr before eating

Renal disease

- **Adult: PO** CCr <50 ml/min, decrease dose by 50% or extend interval to 36-48 hr

Available forms: Tabs 10, 20, 40 mg; gel cap 10 mg; powder for oral susp 40 mg/5 ml; inj 10 mg/ml, 20 mg/50 ml; chew tabs 10 mg

Administer:

- Store in cool environment (oral); IV sol stable for 48 hr at room temperature; do not use discolored sol; discard unused oral sol after 1 mo

PO route

- Antacids 1 hr before or 2 hr after famotidine; may be given with foods or liquids
- After shaking oral suspension

Direct IV route

- After diluting 2 ml of product (10 mg/ml) in 0.9% NaCl to total volume of 5-10 ml; inject over 2 min to prevent hypotension

Intermittent IV INF route

- After diluting 20 mg (2 ml) of product in 100 ml of LR, 0.9% NaCl, D_5W, $D_{10}W$; run over 15-30 min

Continuous IV INF route

- *Adults:* Dilute 40 mg of product in 250 ml D_5W, NS; infuse over 24 hr, run at 11 ml/hr, use inf device

Y-site compatibilities: Acyclovir, alfentanil, allopurinol, amifostine, amikacin, aminocaproic acid, aminophylline, amiodarone, amphotericin B lipid complex, amphotericin B liposome, amsacrine, anakinra, anidulafungin, ascorbic acid injection, atenolol, atracurium, atropine, aztreonam, benztropine, bivalirudin, bleomycin, bumetanide, buprenorphine, butorphanol, calcium chloride/gluconate, CARBOplatin, caspofungin, cefonicid, cefotaxime, cefTAZidime, cefuroxime, chlorproMAZINE, cimetidine, cisatracurium, CISplatin, cladribine, clindamycin, codeine, cyanocobalamin, cyclophosphamide, cycloSPORINE, cytarabine, DACTINomycin, DAPTOmycin, dexamethasone, dexmedetomidine, digoxin, diltiazem, diphenhydrAMINE, DOBUTamine, DOCEtaxel, DOPamine, doripenem, doxacurium, DOXOrubicin, DOXOrubicin liposomal, doxycycline, droperidol, enalaprilat, ePHEDrine, EPINEPHrine, epirubicin, epoetin alfa, eptifibatide, ertapenem, erythromycin, esmolol, etoposide, fenoldopam, fentaNYL, filgrastim, fluconazole, fludarabine, fluorouracil, folic acid, gatifloxacin, gemcitabine, gentamicin, glycopyrrolate, granisetron, heparin, hydrocortisone, HYDROmorphone, hydrOXYzine, IDArubicin, ifosfamide, imipenem-cilastatin, irinotecan, isoproterenol, ketorolac, labetalol, levofloxacin, lidocaine, linezolid, LORazepam, LR, magnesium sulfate, mannitol, mechlorethamine, melphalan, meperidine, metaraminol, methicillin, methotrexate, methoxamine, methyldopate, methylPREDNISolone, metoclopramide, metoprolol, metroNIDAZOLE, miconazole, midazolam, milrinone, mitoXANtrone, morphine, moxalactam, multiple vitamins injection, mycophenolate, nafcillin, nalbuphine, naloxone, nesiritide, netilmicin, niCARdipine, nitroglycerin, nitroprusside, norepinephrine, 0.9% NaCl, octreotide, ondansetron, oxacillin, oxaliplatin, oxytocin, PACLitaxel, palonosetron, pamidronate, pancuronium, papaverine, PEMEtrexed, penicillin G potassium/sodium, pentamidine, pentazocine, PENTobarbital, perphenazine, PHENobarbital, phenylephrine, phytonadione, polymyxin B, potassium chloride/phosphates, procainamide, prochlorperazine, promethazine, propofol, propranolol, protamine, pyridoxine, quiNIDine, ranitidine, remifentanil, Ringer's, ritodrine, riTUXimab, sargramostim, sodium acetate/bicarbonate, succinylcholine, SUFentanil, tacrolimus, teniposide, theophylline, thiamine, thiotepa, ticarcillin, ticarcillin-clavulanate, tigecycline, tirofiban, TNA, tobramycin, tolazoline, TPN, trastuzumab, trimetaphan, urokinase, vancomycin, vasopressin, vecuronium, verapamil, vinCRIStine, vinorelbine, voriconazole, zoledronic acid

SIDE EFFECTS

CNS: *Headache, dizziness,* paresthesia, depression, anxiety, somnolence, insomnia, fever, **seizures in renal disease**

CV: **Dysrhythmias, QT prolongation (impaired renal functioning)**

EENT: Taste change, tinnitus, orbital edema

GI: *Constipation,* nausea, vomiting, anorexia, cramps, abnormal hepatic enzymes, diarrhea

INTEG: Rash, **toxic epidermal necrolysis, Stevens-Johnson syndrome**

MS: Myalgia, arthralgia

RESP: **Pneumonia**

PHARMACOKINETICS

Plasma protein binding 15%-20%, metabolized in liver 30% (active metabolites), 70% excreted by kidneys, half-life 2½-3½ hr

PO: Onset 30-60 min, duration 6-12 hr, peak 1-3 hr, absorption 50%

IV: Onset immediate, peak 30-60 min, duration 8-15 hr

INTERACTIONS

Decrease: absorption—ketoconazole, itraconazole, cefpodoxime, cefditoren

Decrease: famotidine absorption—antacids

Decrease: effect of—atazanavir, delavirdine

NURSING CONSIDERATIONS

Assess:

- **Ulcers:** epigastric pain, adominal pain, frank or occult blood in emesis, stools
- Intragastric pH, serum creatinine/BUN baseline and periodically
- **For bleeding, hematuria, hematuresis, occult blood in stools; abdominal pain**
- Increase in bulk and fluids in diet to prevent constipation

Evaluate:

- Therapeutic response: decreased abdominal pain

Teach patient/family:

- That product must be continued for prescribed time in prescribed method to be effective; not to double dose
- About possibility of decreased libido; that this is reversible after discontinuing therapy
- To avoid irritating foods, alcohol, aspirin, extreme-temperature foods that may irritate GI system
- That smoking should be avoided because it diminishes effectiveness of product
- To avoid tasks requiring alertness because dizziness, drowsiness may occur

fat emulsions (Rx)

Intralipid 10%, Intralipid 20%, Liposyn II 10%, Liposyn II 20%, Liposyn III 10%, Liposyn III 20%

Func. class.: Caloric

Chem. class.: Fatty acid, long chain; nutritional supplement

ACTION:
Needed for energy, heat production; consist of neutral triglycerides, primarily unsaturated fatty acids

USES:
Increase calorie intake, fatty acid deficiency, prevention

CONTRAINDICATIONS:
Hypersensitivity to this product or eggs, soybeans, legumes; hyperlipemia, lipid necrosis, acute pancreatitis accompanied by hyperlipemia, hyperbilirubinemia of the newborn; renal insufficiency, hepatic damage

Precautions: Pregnancy (C), premature/term newborns, severe hepatic disease, diabetes mellitus, thrombocytopenia, gastric ulcers, sepsis

Black Box Warning: Preterm infants

DOSAGE AND ROUTES

Deficiency

- **Adult and child:** IV 8%-10% of required calorie intake (intralipid)

Adjunct to TPN

- **Adult:** IV 1 ml/min over 15-30 min (10%) or 0.5 ml/min over 15-30 min (20%); may increase to 500 ml over 4-8 hr if no adverse reactions occur; max 2.5 g/kg
- **Child:** IV 0.1 ml/min over 10-15 min (10%) or 0.05 ml/min over 10-15 min (20%); may increase to 1 g/kg over 4 hr if no adverse reactions occur; max 4 g/kg

Prevention of deficiency

- **Adult:** IV 500 ml 2×/wk (10%), given 1 ml/min for 30 min, max 500 ml over 6 hr
- **Child:** IV 5-10 ml/kg/day (10%), given 0.1 ml/min for 30 min, max 100 ml/hr

Available forms: Inj 10% (50, 100, 200, 250, 500 ml), 20% (50, 100, 200, 250, 500 ml)

Administer:

Intermittent IV INF route

- At 10% (1 ml/min) or 20% (0.5 ml/min) initially × 15-30 min, may increase 10% (120 ml/hr) or 20% (62.5 ml/hr) if no adverse reaction; do not give more than 500 ml on 1st day
- After changing IV tubing at each inf; infection may occur with old tubing
- With inf pump at prescribed rate; do not use in-line filter sized for lipid emulsion; clogging will occur
- Do not use mixed sol if separated or oily looking

SIDE EFFECTS

CNS: Dizziness, headache, drowsiness, focal seizures

CV: Shock

GI: Nausea, vomiting, hepatomegaly

HEMA: Hyperlipemia, hypercoagulation, thrombocytopenia, leukopenia, leukocytosis

RESP: Dyspnea, fat in lung tissue

PHARMACOKINETICS

Completely absorbed, distributed to intravascular space, converted to triglycerides then to free fatty acids

NURSING CONSIDERATIONS

Assess:

- Triglycerides, free fatty acid levels, platelet counts daily to prevent fat overload, thrombocytopenia
- Hepatic studies: AST, ALT, Hct, Hgb; notify prescriber if abnormal
- Nutritional status: calorie count by dietitian; monitor weight daily

Evaluate:

- Therapeutic response: increased weight

Teach patient/family:

- About the reason for use of lipids

febuxostat (Rx)

(feb-ux'oh-stat)

Uloric

Func. class.: Antigout drug, antihyperuricemic

Chem. class.: Xanthene oxidase inhibitor

ACTION: Inhibits the enzyme xanthine oxidase, thereby reducing uric acid synthesis; more selective for xanthine oxidase than allopurinol

USES: Chronic gout, hyperuricemia

CONTRAINDICATIONS: Hypersensitivity

Precautions: Pregnancy (C), breastfeeding, children, renal/hepatic/cardiac/neoplastic disease, stroke, MI, organ transplant, Lesch-Nyhan syndrome

DOSAGE AND ROUTES

- **Adult: PO** 40 mg daily, may increase to 80 mg daily if uric acid levels are >6 mg/dl after 2 wk of therapy

Available forms: Tabs 40, 80 mg

Administer:

PO route

- Without regard to meals or antacids; may crush and add to foods or fluids
- A few days before antineoplastic therapy

SIDE EFFECTS

CNS: Weakness, flushing

CV: **MI, atrial fibrillation, atrial flutter, AV block**, bradycardia, hypo/hypertension, palpitations, **sinus tachycardia, stroke**, angina

EENT: Retinopathy, cataracts, epistaxis

GI: *Nausea, vomiting, anorexia,* constipation, diarrhea, dyspepsia, hematemesis, **hepatitis**, hepatomegaly, weight gain/loss, cholecystitis, cholelithiasis, melena

GU: **Renal failure**, urinary urgency/frequency/incontinence, nephrolithiasis, hematuria

HEMA: **Thrombocytopenia, anemia, pancytopenia, leukopenia, bone marrow suppression**

INTEG: Rash

MISC: Arthralgia, gout flare

PHARMACOKINETICS

Peak 1-1.5 hr; excreted in feces, urine; half-life 5-8 hr; protein binding 99.2%

INTERACTIONS

Increase: toxicity—azaTHIOprine

Increase: xanthine nephropathy, calculi—rasburicase, antineoplastics

Increase: myelosuppression—mercaptopurine, theophylline

NURSING CONSIDERATIONS

Assess:

- **Hyperuricemia:** uric acid levels q2wk; uric acid levels should be ≤6 mg/dl, flares may occur during first 6 wk of treatment
- Hepatic studies before use, then 2, 4 mo and then periodically; assess for fatigue, anorexia, right upper abdominal discomfort, dark urine, jaundice
- CBC, AST, BUN, creatinine before starting treatment, periodically
- **Renal disease:** I&O ratio; increase fluids to 2 L/day to prevent stone formation and toxicity
- For rash, hypersensitivity reactions; discontinue
- **Gout:** joint pain, swelling; may use with NSAIDs for acute gouty attacks and gout flare (first 6 wk)

Evaluate:

- Therapeutic response: decreased pain in joints, decreased stone formation in kidneys, decreased uric acid levels

Teach patient/family:

- That tabs may be crushed
- To take as prescribed; if dose is missed, to take as soon as remembered; not to double dose
- To increase fluid intake to 2 L/day unless contraindicated
- To avoid alcohol, caffeine because they will increase uric acid levels

F

- To report cardiovascular events to prescriber immediately
- **Gout:** that flares may occur during first 6 wk of treatment

felodipine (Rx)

(fe-loe′-di-peen)

Plendil ✦, Renedil ✦

Func. class.: Antihypertensive, calcium channel blocker, antianginal

Chem. class.: Dihydropyridine

Do not confuse:
Plendil/pindolol/Pletal/PriLOSEC/Prinivil/Isordil

ACTION: Inhibits calcium ion influx across cell membrane, resulting in the inhibition of the excitation and contraction of vascular smooth muscle

USES: Essential hypertension alone or with other antihypertensives
Unlabeled uses: Hypertension in adolescents and children, angina pectoris; Prinzmetal's angina (vasospastic)

CONTRAINDICATIONS: Hypersensitivity to this product or dihydropyridines, sick sinus syndrome, 2nd- or 3rd-degree heart block, hypotension <90 mm Hg systolic
Precautions: Pregnancy (C), breastfeeding, children, geriatric patients, CHF, hepatic injury, renal disease, coronary artery disease

DOSAGE AND ROUTES

- **Adult: PO** 5 mg/day initially; usual range 2.5-10 mg/day; max 10 mg/day; do not adjust dosage at intervals of <2 wk
- **Geriatric: PO** 2.5 mg/day

Hepatic disease
- **Adult: PO** 2.5-5 mg, max 10 mg/day

Hypertension in adolescent/child (unlabeled)
- **Adolescent and child: PO** 2.5 mg initially, titrate upward, max 10 mg/day

Available forms: Ext rel tabs 2.5, 5, 10 mg

Administer:
PO route
- Swallow whole; do not break, crush, or chew ext rel products
- Once daily with light meal; avoid grapefruit juice

SIDE EFFECTS

CNS: *Headache,* fatigue, drowsiness, dizziness, anxiety, depression, nervousness, insomnia, light-headedness, paresthesia, tinnitus, psychosis, somnolence, flushing
CV: Dysrhythmia, *edema,* CHF, hypotension, palpitations, MI, pulmonary edema, tachycardia, syncope, AV block, angina
GI: Nausea, vomiting, diarrhea, gastric upset, constipation, increased LFTs, dry mouth
GU: Nocturia, polyuria
HEMA: Anemia
INTEG: Rash, pruritus, peripheral edema
MISC: Flushing, sexual difficulties, cough, nasal congestion, SOB, wheezing, epistaxis, respiratory infection, chest pain, angioedema, gingival hyperplasia

PHARMACOKINETICS

Peak plasma levels 2.5-5 hr, highly protein bound >99%, metabolized in liver, 0.5% excreted unchanged in urine, elimination half-life 11-16 hr

INTERACTIONS

Increase: bradycardia, CHF—β-blockers, digoxin, phenytoin, disopyramide
Increase: toxicity, hypotension—nitrates, alcohol, quiNIDine, zileuton, miconazole, diltiazem, delavirdine, quinupristin, dalfopristin, conivaptan, cycloSPORINE, cimetidine, clarithromycin, antiretroviral protease inhibitors, other antihypertensives, MAOIs, ketoconazole, erythromycin, itraconazole, propranolol
Decrease: antihypertensive effects—NSAIDs, carBAMazepine, barbiturates, phenytoin
Drug/Herb
Increase: antihypertensive effect—ginseng, ginkgo, hawthorn

Decrease: antihypertensive effect—ephedra, St. John's wort

Drug/Food

Increase: felodipine level—grapefruit juice

NURSING CONSIDERATIONS

Assess:

- **CHF:** I&O, weight daily; weight gain, crackles, dyspnea, edema, jugular venous distention
- Cardiac status: B/P, pulse, respiration; ECG periodically
- **Angina pain:** location, duration, intensity; ameliorating, aggravating factors

Evaluate:

- Therapeutic response: decreased B/P, decreased anginal attacks, increased activity tolerance

Teach patient/family:

- To avoid hazardous activities until stabilized on product, dizziness no longer a problem
- To avoid OTC products, alcohol unless directed by prescriber; to limit caffeine consumption
- About the importance of complying with all areas of medical regimen: diet, exercise, stress reduction, product therapy
- That tablets may appear in stools but are insignificant
- To report dyspnea, palpitations, irregular heart beat, swelling of extremities, nausea, vomiting, severe dizziness, severe headache
- To change positions slowly to prevent orthostatic hypotension
- To obtain correct pulse; to contact prescriber if pulse <50 bpm
- Use good oral hygiene to prevent gingival hyperplasia
- Do not stop abruptly
- To avoid grapefruit juice

TREATMENT OF OVERDOSE:

Atropine for AV block, vasopressor for hypotension

fenofibrate (Rx)

(fen-oh-fee′brate)

Antara, Apo-Feno-Micro ✤, Fenoglide, Lipidil Micro ✤, Lipidil Supra Cap ✤, Lipofen, Lofibra, Tricor, Triglide

Func. class.: Antilipemic

Chem. class.: Fibric acid derivative

F

ACTION:

Increases lipolysis and elimination of triglyceride-rich particles from plasma by activating lipoprotein lipase, thereby resulting in changes in triglyceride size and composition of LDL, leading to rapid breakdown of LDL; mobilizes triglycerides from tissue; increases excretion of neutral sterols

USES:

Hypercholesterolemia; types IV, V hyperlipidemia that do not respond to other treatment and that increase risk for pancreatitis; Fredrickson type IV, V hypertriglyceridemia

CONTRAINDICATIONS:

Hypersensitivity, severe renal/hepatic disease, primary biliary cirrhosis, preexisting gallbladder disease, breastfeeding

Precautions: Pregnancy (C), geriatric patients, peptic ulcer, pancreatitis, renal/hepatic disease, diabetes mellitus

DOSAGE AND ROUTES

Hypertriglyceridemia

- **Adult: PO** (Antara) 43-130 mg/day; (Lofibra) 67-200 mg/day; (Tricor) 48-145 mg/day; (Triglide) 50-160 mg/day

Primary hypercholesterolemia/mixed hyperlipidemia

- **Adult: PO** (Antara) 130 mg/day; (Lofibra) 200 mg/day; (Tricor) 145 mg/day; (Triglide) 160 mg/day

Renal dose (geriatric)

- **Adult: PO** (Tricor) CCr 30-80 ml/min, 48 mg/day; CCr <30 ml/min, contraindicated
- **Adult: PO** CCr 30-80 ml/min, 50 mg/day (Triglide, Liprofen), 43 mg/day (Antara), 40 mg daily (fenoglide), 67 mg/day

(Lofibra caps), 54 mg/day (Lofibra tabs); CCr <30 ml/min, contraindicated (Antara, Lipofen, Lofibra, Triglide)

Available forms: Tabs (Triglide) 50, 160 mg; (Tricor) 48, 145 mg; micronized cap (Antara) 43, 130 mg; (Lofibra) 54, 160; (Lofibra micronized) 67, 134, 200 mg; cap; (Lipofen) 50, 150 mg; tabs; (Fenoglide) 40, 120 mg

Administer:

• Product with meals (Lipofen, Lofibra); Triglide, Antara without regard to food; may increase q4-8wk; brands are not interchangeable; therapy should be discontinued if there is not adequate response after 2 mo

SIDE EFFECTS

CNS: *Fatigue, weakness,* drowsiness, dizziness, insomnia, depression, vertigo

CV: Angina, hypo/hypertension

GI: *Nausea,* vomiting, dyspepsia, increased liver enzymes, flatulence, hepatomegaly, gastritis, pancreatitis, cholelithiasis

GU: Dysuria, urinary frequency

HEMA: Anemia, leukopenia, thrombosis/pulmonary embolism

INTEG: *Rash,* urticaria, pruritus, photosensitivity

MISC: Polyphagia, weight gain, infection, flulike syndrome

MS: *Myalgias, arthralgias,* myopathy, rhabdomyolysis

RESP: Pharyngitis, bronchitis, cough

PHARMACOKINETICS

Peak 6-8 hr, protein binding 99%, converted to fenofibric acid, metabolized in liver, excreted in urine (60%), feces (25%), half-life 20 hr

INTERACTIONS

Increase: myopathy—colchicine

Increase: nephrotoxicity—cycloSPORINE

• Avoid use with HMG-CoA reductase inhibitors; rhabdomyolysis may occur

Increase: anticoagulant effects—oral anticoagulants

Increase: effects of—antidiabetics

Decrease: absorption of fenofibrate—bile acid sequestrants

Drug/Herb

Increase: effect—red yeast rice

Drug/Food

Increase: absorption

Drug/Lab Test

Increase: ALT, AST, BUN, CK, creatinine

Decrease: WBC, uric acid, Hgb, paradoxical effect in HDL

NURSING CONSIDERATIONS

Assess:

• **Diet history:** fat content; lipid levels (triglycerides, LDL, HDL, cholesterol), may cause a paradoxical decrease in HDL, LFTs at baseline, periodically during treatment, CPK if muscle pain occurs, CBC, Hct, Hgb; PT with anticoagulant therapy

• **Pancreatitis, cholelithiasis renal failure, rhabdomyolysis** (when combined with HMG Co-A reductase inhibitors), myositis; product should be discontinued

Evaluate:

• Therapeutic response: decreased triglycerides, cholesterol levels

Teach patient/family:

• That compliance is needed; not to consume chipped or broken tabs (Triglide)

• That risk factors should be decreased: high-fat diet, smoking, alcohol consumption, absence of exercise

• To notify prescriber if pregnancy is suspected or planned

• To report GU symptoms: decreased libido, impotence, dysuria, proteinuria, oliguria, hematuria

• To notify prescriber of muscle pain, weakness, fever, fatigue, epigastric pain

⚠ HIGH ALERT

fentaNYL (Rx)

(fen′ta-nill)

ABSTRAL, Actiq, Fentora, Onsolis, RAN-Fentanyl ♣

fentaNYL transdermal (Rx)

Duragesic

fentaNYL nasal spray (Rx)

Lazanda

fentaNYL SL spray

Subsys

Func. class.: Opioid analgesic

Chem. class.: Synthetic phenylpiperidine

Controlled Substance Schedule II

Do not confuse:
fentaNYL/Sufenta

ACTION: Inhibits ascending pain pathways in CNS, increases pain threshold, alters pain perception by binding to opiate receptors

USES: Controls moderate to severe pain; preoperatively, postoperatively; adjunct to general anesthetic, adjunct to regional anesthesia; **FentaNYL:** anesthesia as premedication, conscious sedation; **Actiq:** breakthrough cancer pain

CONTRAINDICATIONS: Hypersensitivity to opiates, myasthenia gravis

Black Box Warning: Headache, migraine (Actiq, ABSTRAL, Fentora, Lazanda, Onsolis); emergency room use (ABSTRAL, Lazanda); outpatient surgeries (Duragesic TD); opioid-naive patients, respiratory depression

Precautions: Pregnancy (C), breastfeeding, geriatric patients, increased intracranial pressure, seizure disorders, severe respiratory disorders, cardiac dysrhythmias

Black Box Warning: Accidental exposure, ambient temperature increase, fever, skin abrasion (TD patch), substance abuse, surgery, requires an experienced clinician

DOSAGE AND ROUTES

FentaNYL

Anesthetic

- **Adult: IV** 50-100 mcg/kg over 1-2 min

Anesthesia supplement

- **Adult and child >12 yr: IM/IV** 2-20 mcg/kg **IV INF** 0.025-0.25 mcg/kg/min

Induction and maintenance

- **Child 2-12 yr: IV** 2-3 mcg/kg

Preoperatively

- **Adult and child >12 yr: IM/IV** 50-100 mcg q30-60min before surgery

Postoperatively

- **Adult and child >12 yr: IM/IV** 0.05-0.1 mg q1-2hr prn

Moderate/severe pain

- **Adult: IV/IM** 50-100 mcg q1-2hr

Actiq

- **Adult: TRANSMUCOSAL** 200 mcg; redose if needed 15 min after completion of 1st dose; do not give more than 2 doses during titration period, max 4 doses/day

Fentora

- **Adult: BUCCAL/SL** 100 mcg placed above rear molar between upper cheek and gum, a second 100 mcg dose, if needed, may be started 30 min after 1st dose

Onsolis

- **Adult: TRANSMUCOSAL** 200 mcg, titrate, max 1200 mcg/dose or 4 doses/day

ABSTRAL

- **Adult: SL** 100 mcg; another dose may be taken 30 min after 1st, max 2 doses per episode of breakthrough pain; ≥2 hr must elapse before treating again; titrate stepwise over consecutive episodes

FentaNYL transdermal

• **Adult:** 25 mcg/hr; may increase until pain relief occurs; apply patch to flat surface on upper torso and wear for 72 hr; apply new patch on different site; may use 12.5 mcg/hr if <60 mg/day morphine equivalent

FentaNYL nasal spray

• **Adult:** 100 mcg (1 spray in 1 nostril), may retreat after ≥2 hr, titrate upward until adequate analgesia; treat a max of 4 episodes daily

FentaNYL SL spray

• **Adult:** 100 mcg sprayed under tongue, titrate stepwise carefully

Available forms: Inj 0.05 mg/ml; lozenges 100, 200, 300, 400, 600, 800, 1200, 1600 mcg; lozenges on a stick 200, 400, 600, 800, 1200, 1600 mcg; buccal tab 100, 200, 400, 600, 800 mcg; oral dissolving film (Onsolis) 200, 400, 600, 800, 1200 mcg; SL tab (ABSTRAL) 100, 200, 300, 400, 600, 800 mcg; *transdermal:* patch 12, 25, 50, 75, 100 mcg/hr; *SL spray:* 100, 200, 400, 600, 800, 1200, 1600 mcg/spray

Administer:

• By inj (IM, IV); give slowly to prevent rigidity

• Must have emergency equipment available, opioid antagonists, O_2; to be used only by those appropriately trained; IV products to be used in OR, ER, ICU

Transmucosal route

• Remove foil just before administration; instruct patient to place product between cheek and lower gum, moving it back and forth and sucking, not chewing (Actiq); place above rear molar (Fentora); place film on the inside of the cheek (Onsolis); all products not used or only partially used should be flushed down the toilet; this product may be used SL

Transdermal route

• q72hr for continuous pain relief; dosage adjusted after at least 2 applications; apply to clean, dry skin and press firmly

• Give short-acting analgesics until patch takes effect (8-24 hr); when reducing dosage or switching to alternative IV treatment, withdraw gradually; serum levels drop gradually, give ½ the equianalgesic dose of new analgesic 12-18 hr after removal as ordered

SL spray

• Open blister package with scissors immediately before use; spray contents of unit under tongue; dispose of each used unit after use by placing it into one of the disposable bags provided; seal bag, discard into a trash container out of reach of children

IV route

• IV undiluted by anesthesiologist or diluted with 5 ml or more sterile water or 0.9% NaCl given through Y-tube or 3-way stopcock at 0.1 mg or less/1-2 min. Muscular rigidity may occur with rapid IV administration

Additive compatibilities: Bupivacaine, caffeine citrate, cloNIDine, droperidol, EPINEPHrine, ketamine, lidocaine, ziconotide

Solution compatibilities: D_5W, 0.9% NaCl

Syringe compatibilities: Alprostadil, atracurium, atropine, bupivacaine/ketamine, butorphanol, chlorproMAZINE, cimetidine, cloNIDine/lidocaine, diphenhyDRINATE, diphenhydrAMINE, droperidol, heparin, HYDROmorphone, hydrOXYzine, lidocaine, meperidine, metoclopramide, midazolam, morphine, pentazocine, perphenazine, prochlorperazine, promazine, promethazine, ranitidine, scopolamine

Y-site compatibilities: Abciximab, acyclovir, alfentanil, alprostadil, amikacin, aminocaproic acid, aminophylline, amiodarone, amphotericin B cholesteryl, amphotericin B lipid complex, amphotericin B liposome, anidulafungin, argatroban, ascorbic acid injection, atenolol, atracurium, atropine, azaTHIOprine, aztreonam, benztropine, bivalirudin, bleomycin, bumetanide, buprenorphine, butorphanol, calcium chloride/gluconate, CARBOplatin, caspofungin, cefamandole, ceFAZolin, cefmetazole, cefonicid, cefoperazone, cefotaxime, cefoTEtan, cefOXitin, cefTAZidime, ceftizoxime, ceftobiprole, cefTRIAXone, cefuroxime, cephalothin, chloramphenicol, chlorproMAZINE, cimetidine, cisatracurium, CISplatin, clindamycin, cloNIDine, cyanocobalamin, cyclophosphamide, cycloSPORINE, cytarabine, DACTINomycin, DAPTOmycin, dexamethasone, dexmedetomidine, digoxin, diltiazem, diphenhydrAMINE, DOBUTamine, DOCEtaxel, DOPamine, doripenem, doxacurium, doxapram, DOXOrubicin, doxycycline, enalaprilat, ePHEDrine, EPINEPHrine, epirubicin, epoetin alfa, eptifibatide, erythromycin, esmolol, etomidate, etoposide, famotidine, fenoldopam, fluconazole, fludarabine, fluorouracil, folic acid, furosemide, ganciclovir, gatifloxacin, gemcitabine, gentamicin, glycopyrrolate, granisetron, heparin, hydrocortisone, HYDROmorphone, hydrOXYzine, IDArubicin, ifosfamide, imipenem-cilastatin, inamrinone, insulin (regular), irinotecan, isoproterenol, ketorolac, labetalol, lansoprazole, levofloxacin, lidocaine, linezolid, LORazepam, LR, magnesium sulfate, mannitol, mechlorethamine, meperidine, metaraminol, methicillin, methotrexate, methotrimeprazine, methoxamine, methyldopate, methylPREDNISolone, metoclopramide, metoprolol, metroNIDAZOLE, mezlocillin, miconazole, midazolam, milrinone, minocycline, mitoXANtrone, mivacurium, morphine, moxalactam, multiple vitamins injection, mycophenolate, nafcillin, nalbuphine, naloxone, nesiritide, netilmicin, niCARdipine, nitroglycerin, nitroprusside, norepinephrine, octreotide, ondansetron, oxacillin, oxaliplatin, oxytocin, PACLitaxel, palonosetron, pamidronate, pancuronium, papaverine, PEMEtrexed, penicillin G potassium/sodium, pentamidine, pentazocine, PENTobarbital, PHENObarbital, phenylephrine, phytonadione, piperacillin, piperacillin-tazobactam, polymyxin B, potassium chloride, procainamide, prochlorperazine, promethazine, propofol, propranolol, protamine, pyridoxine, quiNIDine, quinupristin-dalfopristin, ranitidine, remifentanil, Ringer's, ritodrine, riTUXimab, rocuronium, sargramostim, scopolamine, sodium acetate/bicarbonate, succinylcholine, SUFentanil, tacrolimus, teniposide, theophylline, thiamine, thiopental, thiotepa, ticarcillin, ticarcillin-clavulanate, tigecycline, tirofiban, TNA, tobramycin, tolazoline, TPN, trastuzumab, trimetaphan, urokinase, vancomycin, vasopressin, vecuronium, verapamil, vinCRIStine, vinorelbine, vitamin B complex/C, voriconazole, zoledronic acid

F

SIDE EFFECTS

CNS: Dizziness, delirium, euphoria, sedation
CV: **Bradycardia, arrest**, hypo/hypertension
EENT: Blurred vision, miosis
GI: Nausea, vomiting, constipation
GU: Urinary retention
INTEG: Rash, diaphoresis
MS: Muscle rigidity
RESP: **Respiratory depression, arrest, laryngospasm**

PHARMACOKINETICS

Metabolized by liver, excreted by kidneys, crosses placenta, excreted in breast milk, Half-life: IV: 2-4 hr, Transdermal: 13-22 hr, Transmucosal: 7 hr, Buccal: 4-12 hr, 80% bound to plasma proteins
IM: Onset 7-8 min, peak 30 min, duration 1-2 hr
IV: Onset 1 min, peak 3-5 min, duration ½-1 hr

INTERACTIONS

Black Box Warning: Increase: fentaNYL effect, fetal respiratory depression: (cycloSPORINE, ketoconazole, itraconazole, cimetidine, conivaptan, fluconazole, nefazodine, ranolazine, zafirlukast, zileuton

Increase: hypotension—droperidol
Increase: CV depression—diazepam
Increase: fentaNYL effect with other CNS depressants—alcohol, opioids, sedative/hypnotics, antipsychotics, skeletal muscle relaxants, protease inhibitors
Decrease: fentaNYL effect—CYP3A4 inducers (carBAMazepine, phenytoin, PHENObarbital, rifampin)

Drug/Herb
Increase: action of fentaNYL—St. John's wort
Decrease: effect of fentaNYL—echinacea
Drug/Lab Test
Increase: amylase, lipase

NURSING CONSIDERATIONS

Assess:
- VS after parenteral route; note muscle rigidity, drug history, hepatic/renal function tests
- CNS changes: dizziness, drowsiness, hallucinations, euphoria, LOC, pupil reaction
- Allergic reactions: rash, urticaria

Black Box Warning: **Respiratory dysfunction: respiratory depression,** character, rate, rhythm; notify prescriber if respirations are <10/min

Black Box Warning: **Headache/migraine:** ABSTRAL, Actiq, Fentora, Lazanda, Onsolis not to be used for this condition; ABSTRAL, Lazanda not to be used in ER; Duragesic TD not to be used for outpatient surgery patients

Black Box Warning: **Apnea, respiratory arrest in opioid-naive patients:** do not use ABSTRAL, Actiq, Duragesic, Fentora, Lazanda, Onsolis; opioid-tolerant patients are those using ≥60 mg/day oral morphine, ≥30 mg/day oxyCODONE PO, 8 mg/day HYDROmorphone, 25 mcg TD fentaNYL/hr

Evaluate:
- Therapeutic response: induction of anesthesia, relief of breakthrough cancer pain, general pain relief
- Cancer pain, general pain relief

Teach patient/family:
- About CNS changes: physical dependence; not to use with alcohol, other CNS depressants

Black Box Warning: **Accidental exposure:** Discuss the dangers of children or pets ingesting or coming in contact with product

Transdermal route

Black Box Warning: **Ambient temperature:** that excessive heat may increase absorption; do not use with heating pads, electric blankets, heat/tanning lamps, saunas, hot tubs, heated waterbeds, when sunbathing

- That excessive perspiration may alter adhesiveness
- To dispose of patch by placing sticky sides together and flushing down toilet
- That patient may need to clip hair before applying to ensure adhesion
- May add first aid tape around the edges if there is a problem with adhesion

ferrous fumarate (Rx)

Ferretts, Ferrimin, Ferro-Sequels, Hemocyte, Palafer ✦, Walgreens Finest Iron

ferrous gluconate (Rx)

Apo-Ferrous Gluconate ✦, Ferate, Walgreens Gold Seal Ferrous Gluconate

ferrous sulfate (Rx)

Apo-Ferrous Sulfate ✦, Equaline Ferrous Sulfate, Leader Ferrous Sulfate, Slow Release Iron, Walgreens Gold Seal Ferrous Sulfate

ferrous sulfate, dried (Rx)

Slow Fe

carbonyl iron (OTC)

(kar′bo-nil)

ICAR Pediatric, Iron Chews

iron polysaccharide (OTC)

iFerex, Niferex, Nu-Iron

Func. class.: Hematinic
Chem. class.: Iron preparation

ACTION: Replaces iron stores needed for red blood cell development as well as energy and O_2 transport and use; fumarate contains 33% elemental iron; gluconate, 12%; sulfate, 20%; iron, 30%; ferrous sulfate exsiccated

USES:
Iron deficiency anemia, prophylaxis for iron deficiency in pregnancy, nutritional supplementation

CONTRAINDICATIONS:
Sideroblastic anemia, thalassemia, hemosiderosis/hemochromatosis

Precautions: Pregnancy (B) (ferric gluconate complex), (C) (iron dextran, oral products), anemia (long term), ulcerative colitis/regional enteritis, peptic ulcer disease, hemolytic anemia, cirrhosis, sulfite sensitivity

Black Box Warning: Accidental exposure

DOSAGE AND ROUTES

Fumarate

- **Adult: PO** 200-325 mg tid
- **Child: PO** 3 mg/kg/day (elemental iron) tid-qid
- **Infant: PO** 10-25 mg/day (elemental iron) in 3-4 divided doses, max 15 mg/day

Gluconate

- **Adult: PO** 50-100 mg elemental iron tid
- **Child: PO** 3 mg/kg/day in divided doses

Sulfate

- **Adult: PO** 0.75-1.5 g/day in divided doses tid
- **Child 6-12 yr: PO** 3 mg/kg/day in divided doses

Pregnancy

- **Adult: PO** 300-600 mg/day in divided doses

Iron polysaccharide

- **Adult: PO** 100-200 mg tid
- **Child: PO** 4-6 mg/kg/day in 3 divided doses (severe iron deficiency)

Available forms: Fumarate: tabs 63, 195, 200, 324, 325 mg; chewable tabs 100 mg; cont rel tabs 300 mg; oral susp 100 mg/5 ml, 45 mg/0.6 ml; **gluconate:** tabs 300, 320, 325 mg; caps 86, 325, 435 mg; film-coated tabs 300 mg; elix 300 mg/5 ml; **sulfate:** tabs 195, 300, 325 mg; enteric-coated tabs 325 mg; ext rel tabs, time-rel caps, 525 mg; **dried:** tabs 200 mg; ext rel tabs 160 mg; ext rel caps 160 mg; **iron polysaccharide:** tabs 50 mg; caps 150 mg; sol 100 mg/5 ml

Administer:

PO route

- Swallow tabs whole; do not break, crush, or chew unless labeled as chewable
- Between meals for best absorption; may give with juice; do not give with antacids or milk, delay at least 1 hr; if GI symptoms occur, give after meals even if absorption is decreased; eggs, milk products, chocolate, caffeine interfere with absorption
- Store in tight, light-resistant container
- **Liquid** through plastic straw to avoid discoloration of tooth enamel; dilute thoroughly
- At least 1 hr before bedtime; corrosion may occur in stomach; ferrous gluconate is less irritating of GI tract than ferrous sulfate
- For <6 mo for anemia

SIDE EFFECTS

GI: *Nausea, constipation, epigastric pain, black and red tarry stools,* vomiting, diarrhea

INTEG: Temporarily discolored tooth enamel and eyes

SYST: **Hypersensitivity reactions (Ferrlecit)**

PHARMACOKINETICS

PO: Excreted in feces, urine, skin, breast milk; enters bloodstream; bound to transferrin; crosses placenta

INTERACTIONS

Increase: action of iron preparation—ascorbic acid, chloramphenicol

Decrease: absorption of penicillamine, levodopa, methyldopa, fluoroquinolones, L-thyroxine, tetracycline

Decrease: absorption of iron preparations—antacids, H_2-antagonists, proton pump inhibitors, cholestyramine, vit E

Drug/Food

Decrease: absorption—dairy products, caffeine, eggs

Drug/Lab Test

False positive: occult blood

NURSING CONSIDERATIONS

Assess:

- Blood studies: Hct, Hgb, reticulocytes, bilirubin before treatment, at least monthly; iron studies (Iron, TIBC, ferritin)

⚠ **Toxicity:** nausea, vomiting, diarrhea (green then tarry stools), hematemesis, pallor, cyanosis, shock, coma

- Elimination: if constipation occurs, increase water, bulk, activity
- **Nutrition:** amount of iron in diet (meat, dark green leafy vegetables, dried beans, dried fruits, eggs)
- Cause of iron loss or anemia, including salicylates, sulfonamides, antimalarials, quiNIDine

Evaluate:

- Therapeutic response: improvement in Hct, Hgb, reticulocytes; decreased fatigue, weakness

Teach patient/family:

- That iron will turn stools black or dark green
- **Accidental exposure:** to keep out of reach of children, pets; iron poisoning may occur if increased beyond recommended level
- Not to substitute 1 iron salt for another; that elemental iron content differs (e.g., 300 mg ferrous fumarate contains about 100 mg elemental iron; 300 mg ferrous gluconate contains only about 30 mg elemental iron)
- To avoid reclining position for 15-30 min after taking product to avoid esophageal corrosion
- To follow a diet high in iron; to avoid taking iron, dairy products, calcium supplements, and vit C together because they compete for absorption

TREATMENT OF OVERDOSE:

Induce vomiting; give eggs, milk until lavage can be done

fesoterodine (Rx)

(fess′oh-ter-oh-deen)

Toviaz

Func. class.: Overactive bladder product

Chem. class.: Muscarinic receptor antagonist

ACTION: Relaxes smooth muscles in urinary tract by inhibiting acetylcholine at postganglionic sites

USES: Overactive bladder (urinary frequency, urgency), urinary incontinence

CONTRAINDICATIONS: GI obstruction, ileus, pyloric stenosis, urinary retention, gastric retention, hypersensitivity, closed-angle glaucoma

Precautions: Pregnancy (C), breastfeeding, children, renal/hepatic disease, urinary tract obstruction, ambient temperature increase, autonomic neuropathy, constipation, contact lenses, hazardous activity, GERD, gastroparesis, myasthenia gravis, prostatic hypertrophy, toxic megacolon, ulcerative colitis, possible cross-sensitivity with tolterodine

DOSAGE AND ROUTES

- **Adult and geriatric: PO EXT REL** 4 mg/day, may increase to 8 mg/day based on response, max 4 mg/day in those taking potent CYP3A4 inhibitors

Renal dose

- **Adult: PO EXT REL** CCr <30 ml/min, max 4 mg/day in severe renal impairment

Available forms: EXT REL TABS 4, 8 mg

Administer:

- Do not break, crush, or chew ext rel product
- Give without regard to meals

• Store at room temperature; protect from moisture

SIDE EFFECTS

CV: Chest pain, angina, **QT prolongation**
EENT: Xerophthalmia
GI: *Nausea, vomiting,* abdominal pain, constipation, dry mouth
GU: Dysuria, urinary retention
INTEG: Rash, **angioedema**
MISC: Peripheral edema, insomnia
MS: Back pain
RESP: Cough
SYST: Infection

PHARMACOKINETICS

Peak 5 hr, rapidly absorbed, protein binding 50%, excreted in urine/feces, half-life 7 hr

INTERACTIONS

Increase: action of fesoterodine—CYP3A4 inhibitors (antiretroviral protease inhibitors, macrolide antiinfectives, azole antifungals)
Increase: anticholinergic effect—antimuscarinics, anticholinergics
Increase: urinary frequency—diuretics
Drug/Herb
Decrease: fesoterodine—caffeine, green tea, guarana
Drug/Food
Increase: fesoterodine level—grapefruit juice
Decrease: fesoterodine level—cola, coffee, tea

NURSING CONSIDERATIONS

Assess:
• **Urinary patterns:** distention, nocturia, frequency, urgency, incontinence
• **Allergic reactions:** rash; if this occurs, product should be discontinued
Evaluate:
• Therapeutic response: absence of urinary frequency, urgency, incontinence
Teach patient/family:
• Not to drink liquids before bedtime
• About the importance of bladder maintenance

fexofenadine (Rx, OTC)

(fex-oh-fi′na-deen)

Allegra, Allegra ODT, GNP Allergy Relief, Premier Value Allergy Relief

Func. class.: Antihistamine—2nd generation
Chem. class.: Piperidine, peripherally selective

Do not confuse:
Allegra/Viagra

ACTION: Acts on blood vessels, GI, respiratory system by competing with histamine for H_1-receptor site; decreases allergic response by blocking pharmacologic effects of histamine, less sedating

USES: Rhinitis, allergy symptoms, chronic idiopathic urticaria

CONTRAINDICATIONS: Breastfeeding, newborn or premature infants, hypersensitivity, severe hepatic disease
Precautions: Pregnancy (C), children, geriatric patients, respiratory disease, closed-angle glaucoma, prostatic hypertrophy, bladder neck obstruction, asthma

DOSAGE AND ROUTES

• **Adult and child >12 yr: PO** Rx only, 60 mg bid or 180 mg/day; PO OTC only, 60 mg bid or 180 mg/day (self-treatment of allergic rhinitis)
• **Child 2-11 yr: PO** 30 mg bid; **ORALLY DISINTEGRATING TAB Child 6-11 yr:** 30 mg bid dissolved on tongue

Renal dose
• **Adult and child ≥12 yr: PO** CCr <80 ml/min, 60 mg/day
• **Child 2-11 yr: PO** CCr <80 ml/min, 30 mg daily
• **Child <2 yr: PO** CCr <80 ml/min, 15 mg daily

Available forms: Tabs 30, 60, 180 mg; oral susp 6 mg/ml, orally disintegrating tab 30 mg

Administer:
- Without regard to meals; caps/tabs should not be given with or right before grapefruit, orange, or apple juice
- Store in tight, light-resistant container
- **Orally disintegrating tab:** allow to dissolve, swallow; do not remove from blister pack until time of administration
- **Oral susp:** shake well; use calibrated measuring device

SIDE EFFECTS

CNS: Headache, stimulation, drowsiness, sedation, fatigue, confusion, blurred vision, tinnitus, restlessness, tremors, paradoxical excitation in children or geriatric patients
CV: Hypotension, palpitations, bradycardia, tachycardia, dysrhythmias (rare)
GI: Nausea, diarrhea, abdominal pain, vomiting, constipation
GU: Frequency, dysuria, urinary retention, impotence
HEMA: Hemolytic anemia, thrombocytopenia, leukopenia, agranulocytosis, pancytopenia
INTEG: Rash, eczema, photosensitivity, urticaria
RESP: Thickening of bronchial secretions, dry nose, throat

PHARMACOKINETICS

Well absorbed; onset 1 hr; peak 2-3 hr; duration 12-24 hr; 80% excreted in urine; half-life 14.5 hr, increased in renal disease

INTERACTIONS

Decrease: effect—magnesium-aluminum-containing antacids
Drug/Food
Decrease: absorption of product—apple, orange, grapefruit juice
Drug/Lab Test
False negative: skin allergy tests

NURSING CONSIDERATIONS

Assess:
- **Allergy:** itchy, runny, watery eyes; congested nose; before and during treatment
- I&O ratio: be alert for urinary retention, frequency, dysuria, especially among geriatric patients; product should be discontinued if these occur; serum creatinine/BUN at baseline and periodically during treatment
- Respiratory status: rate, rhythm, increase in bronchial secretions, wheezing, chest tightness

Evaluate:
- Therapeutic response: absence of running or congested nose or rashes

Teach patient/family:
- About all aspects of product use; to notify prescriber if confusion, sedation, hypotension occur
- To avoid driving, other hazardous activity if drowsiness occurs
- To avoid alcohol, other CNS depressants
- Not to exceed recommended dose; that dysrhythmias may occur

TREATMENT OF OVERDOSE:

Lavage, diazepam, vasopressors, IV phenytoin

fidaxomicin

(fye-dax′oh-mye′sin)

Dificid

Func. class.: Antiinfective-macrolide

ACTION: Bactericidal against *Clostridium difficile;* is a fermentation product obtained from *Dactylosporangium aurantiacum;* inhibits RNA synthesis by inhibiting transcription of bacterial RNA polymerases; may act at the early stages of transcription

USES: Pseudomembraneous colitis *Clostridium difficile*-associated diarrhea

CONTRAINDICATIONS: Hypersensitivity

Precautions: Pregnancy (B), breastfeeding, children

DOSAGE AND ROUTES

- **Adult: PO** 200 mg bid × 10 days

Available forms: Tab 200 mg

Administer:

- Without regard to food
- Store at room temperature

SIDE EFFECTS

GI: Nausea, vomiting, abdominal pain, GI bleeding

HEMA: Anemia, neutropenia

INTEG: Rash, pruritus

META: **Metabolic acidosis,** hyperglycemia

PHARMACOKINETICS

Half-life 12 hr, onset 1 hr, peak 1-5 hr, excreted in feces 92%, minimal absorption, substrate of PGP efflux transporter

INTERACTIONS

Increase: fidaxomicin action—cycloSPORINE

Drug/Lab Test

Increase: Glucose, LFTs, alk phos

Decrease: sodium bicarbonate

NURSING CONSIDERATIONS

Assess:

- **Pseudomembraneous colitis:** for diarrhea, abdominal pain, fever, fatigue, anorexia, anemia, elevated WBC and low serum albumin; product may be used in place of vancomycin; monitor CBC with differential and stool culture (*Clostridium difficile*), not to be used for systemic infection, obtain C&S before use, monitor glucose (diabetic patients), monitor fluid, electrolyte depletion

Evaluate:

- Positive therapeutic response: resolution of *Clostridium difficile*

Teach patient/family:

- To report GI bleeding or severe abdominal pain
- To report if pregnancy is planned or suspected or if breastfeeding
- May take without regard to food

filgrastim (Rx)

(fill-grass′stim)

G-CSF, granulocyte colony stimulator, Neupogen

Func. class.: Biologic modifier

Chem. class.: Granulocyte colony-stimulating factor

F

ACTION:

Stimulates proliferation and differentiation of neutrophils

USES:

To decrease infection in patients receiving antineoplastics that are myelosuppressive; to increase WBC in patients with product-induced neutropenia; bone marrow transplantation

Unlabeled uses: Neutropenia with HIV infection, aplastic anemia, ganciclovir-induced neutropenia, zidovudine-induced neutropenia

CONTRAINDICATIONS:

Hypersensitivity to proteins of *Escherichia coli*

Precautions: Pregnancy (C), breastfeeding, cardiac conditions, children, myeloid malignancies, radiation therapy, sepsis, sickle cell disease, chemotherapy, respiratory disease

DOSAGE AND ROUTES

After myelosuppressive chemotherapy

- **Adult and child: IV/SUBCUT** 5 mcg/kg/day in a single dose × 14 days; may increase by 5 mcg/kg with each cycle

After bone marrow transplantation

- **Adult: IV/SUBCUT** 10 mcg/kg/day as **INF (IV)** over 4 hr or 24 hr, begin 24 hr after chemotherapy and 24 hr after bone marrow transplantation

Peripheral blood progenitor cell collection/therapy

- **Adult:** 10 mcg/kg/day as bolus or **CONT INF** × ≥4 days before leukapheresis, continue to last leukapheresis; may alter dose if WBC >100,000 cells/mm^3

Severe neutropenia (chronic), idiopathic/cyclical

• **Adult: SUBCUT** 5 mcg/kg daily

Available forms: Inj 300 mcg/ml, 480 mcg/1.6 ml, 480 mcg/0.8 ml, 3000 mcg/0.5 ml

Administer:

• Given by subcut inj, short IV inf, or continuous SC or IV inf

• Avoid use within 24 hr before or after chemotherapy

• Do not shake commercial single-dose vials before withdrawing the dose; if the vial is shaken and froth or bubbles form, allow the vial to stand undisturbed for a few minutes until the froth or bubbles dissipate

• Before injection, filgrastim may be allowed to reach room temperature for a maximum of 24 hr; any vial or syringe exposed to room temperature for more than 24 hr should be discarded

• Visually inspect for particulate matter and discoloration before use

• Store in refrigerator; do not freeze; may store at room temperature up to 24 hr

SUBCUT route

• **Subcut inj:** no dilution is necessary; inject by rapid subcut inj, taking care not to inject intradermally

• **Subcut continuous inf:** infuse subcut at a rate not to exceed 2 ml/hr

IV route

• May be diluted with 5% dextrose; do not dilute with NS; product can precipitate

• May be diluted to concs 5-15 mcg/ml; should be protected from adsorption to plastic by the addition of albumin to a final albumin conc of 2 mg/ml; do not dilute filgrastim to a conc <5 mcg/ml

• **IV inf:** infuse IV over 15-30 min or as a continuous inf over 24 hr

Y-site compatibilities: Acyclovir, allopurinol, amikacin, aminophylline, ampicillin, ampicillin/sulbactam, aztreonam, bleomycin, bumetanide, buprenorphine, butorphanol, calcium gluconate, CARBOplatin, carmustine, ceFAZolin, cefoTEtan, cefTAZidime, chlorproMAZINE, cimetidine, CISplatin, cyclophosphamide, cytarabine, dacarbazine, DAUNOrubicin, dexamethasone, diphenhydrAMINE, DOXOrubicin, doxycycline, droperidol, enalaprilat, famotidine, floxuridine, fluconazole, fludarabine, gallium, ganciclovir, granisetron, haloperidol, hydrocortisone, HYDROmorphone, hydrOXYzine, IDArubicin, ifosfamide, leucovorin, LORazepam, mechlorethamine, melphalan, meperidine, mesna, methotrexate, metoclopramide, miconazole, minocycline, mitoXANtrone, morphine, nalbuphine, netilmicin, ondansetron, plicamycin, potassium chloride, promethazine, ranitidine, sodium bicarbonate, streptozocin, ticarcillin, ticarcillin/clavulanate, tobramycin, trimethoprim-sulfamethoxazole, vancomycin, vinBLAStine, vinCRIStine, vinorelbine, zidovudine

SIDE EFFECTS

CNS: Fever, headache

GI: *Nausea,* vomiting, diarrhea, mucositis, anorexia

HEMA: Thrombocytopenia, excessive leukocytosis

INTEG: Alopecia, exacerbation of skin conditions, urticaria, cutaneous vasculitis

MS: Osteoporosis, skeletal pain

OTHER: Chest pain, hypotension

RESP: Acute respiratory distress syndrome, wheezing, alveolar hemorrhage

PHARMACOKINETICS

SUBCUT: Onset 5-60 min, peak 2-8 hr, duration up to 1 wk

IV: Onset 5-60 min, peak 24 hr, duration up to 1 wk

INTERACTIONS

Increase: adverse reactions—do not use this product concomitantly with antineoplastics, lithium

Drug/Lab Test

Increase: uric acid, LDH, alk phos, WBC

NURSING CONSIDERATIONS

Assess:

- Blood studies: CBC, platelet count before treatment and twice weekly; neutrophil counts may be increased for 2 days after therapy
- B/P, respirations, pulse before and during therapy
- Bone pain; give mild analgesics
- CBC with differential, platelets

Evaluate:

- Therapeutic response: absence of infection

Teach patient/family:

- About the technique for self-administration: dose, side effects, disposal of containers and needles; provide instruction sheet

finasteride (Rx)

(fin-ass′te-ride)

Propecia, Proscar

Func. class.: Hormone, androgen inhibitor, hair stimulant

Chem. class.: 5-α-Reductase inhibitor

Do not confuse:

finasteride/furosemide

Proscar/ProSom/PROzac

ACTION:

Inhibits 5-α-reductase and reduction in DHT; DHT induces androgenic effects by binding to androgen receptors in the cell nuclei of the prostate gland, liver, skin; prevents development of BHP

USES:

Symptomatic benign prostatic hyperplasia (Proscar); male-pattern baldness (Propecia)

Unlabeled uses: Hirsutism, prostate cancer prophylaxis

CONTRAINDICATIONS:

Pregnancy (X), breastfeeding, children, women who are pregnant or who may become pregnant should not handle tabs, hypersensitivity

Precautions: Large residual urinary volume, severely diminished urinary flow, hepatic function abnormalities

DOSAGE AND ROUTES

BPH

- **Adult: PO** 5 mg/day × 6-12 mo (Proscar)

Male-pattern baldness

- **Adult: PO** 1 mg/day for 3 mo or more for results (Propecia)

Hirsutism (unlabeled)

- **Adult (nonpregnant): PO** 5 mg/day alone or in combination with oral contraceptives

Prostate cancer prophylaxis (unlabeled)

- **Adult (male): PO** 5 mg/day

Available forms: Tabs (Propecia) 1 mg, (Proscar) 5 mg

Administer:

- Without regard to meals
- For a minimum of 6 mo; not all patients will respond
- Store <86° F (30° C); protect from light; keep container tightly closed

SIDE EFFECTS

GU: Impotence, decreased libido, decreased volume of ejaculate, sexual dysfunction

INTEG: Rash

MISC: Breast tenderness, **secondary malignancy**

PHARMACOKINETICS

Bioavailability 63%; readily absorbed from GI tract; plasma protein binding 90%; metabolized in the liver; excreted in urine (metabolites) 39%, feces (57%); crosses blood-brain barrier; peak 1-2 hr; duration 24 hr

INTERACTIONS

Decrease: finasteride effect—theophylline, adrenergic bronchodilators, anticholinergics

Drug/Lab Test

Decrease: PSA levels

NURSING CONSIDERATIONS

Assess:

- **BPH:** urinary patterns, residual urinary volume, severely diminished urinary flow
- PSA levels and digital rectal exam before initiating therapy and periodically thereafter
- Hepatic studies before treatment; extensively metabolized in liver

Evaluate:

- Therapeutic response: increased urinary flow; decreased postvoiding dribbling, frequency, nocturia or hair growth within 3-6 mo; regression of prostate size

Teach patient/family:

⚠ That pregnant women or women who may become pregnant should not touch crushed tabs or come into contact with the semen of a patient taking this product; that product may adversely affect developing male fetus

- That volume of ejaculate may be decreased during treatment; that impotence and decreased libido may also occur and may continue after discontinuing treatment
- That Propecia results may not occur for 3 mo
- That Proscar results may not occur for 6-12 mo

fingolimod (Rx)

(fin-gol'i-mod)

Gilenya

Func. class.: Biologic response modifier

Chem. class.: Sphingosine 1-phosphate receptor modulator

ACTION: Binds with high affinity to sphingosine 1-phosphate receptors; blocks lymphocyte egress to lymph nodes, thereby reducing the number of peripheral blood lymphocytes; may reduce lymphocyte migration into the CNS

USES: To reduce frequency of exacerbation, to delay physical disability of relapsing forms of MS

CONTRAINDICATIONS: Hypersensitivity

Precautions: Pregnancy (C), breastfeeding, neonates/infants/children, AIDS, asthma, AV block, bradycardia, cardiac disease, COPD, diabetes mellitus, dysrhythmias, heart failure, hepatic disease, HIV, hypertension, immunosuppression, leukemia, lymphoma, QT prolongation, respiratory insufficiency, sick sinus syndrome, syncope, uveitis

DOSAGE AND ROUTES

- **Adult: PO** 0.5 mg/day

Hepatic dose

- **Adult: PO** Child-Pugh C, total score >10: closely monitor, fingolimod exposure is doubled

Available forms: Caps 0.5 mg

Administer:

PO route

- Watch patient for 6 hr after initial dose or if product not given for >2 wk for development of bradycardia
- Give without regard to food
- Store at room temperature, protect from moisture

SIDE EFFECTS

CNS: Asthenia, depression, fatigue, headache, dizziness, encephalopathy, migraine, paresthesias, stroke

CV: AV block, bradycardia, chest pain, hypertension, palpitations

EENT: Blurred vision, vision impairment, ocular pain, macular edema

GI: Abdominal pain, anorexia, diarrhea, jaundice, vomiting, weight loss

HEMA: Leukopenia, lymphopenia, neutropenia

INTEG: Alopecia, pruritus

MS: Back pain

RESP: Dyspnea, cough

SYST: Infection, influenza, secondary malignancy

PHARMACOKINETICS

Protein binding (99.7%), distributed to RBCs (86%), steady state 1-2 mo, metabolized by CYP4F2 and CYP2D6 to a lesser extent, terminal half-life 6-9 days, excreted in urine (81% inactive metabolites), peak 12-16 hr

INTERACTIONS

Increase: immunosuppression—antineoplastics, immunosuppressants, immune modulating therapies

Increase: fingolimod effect—ketoconazole

Increase: infection risk—live vaccines

Decrease: effect of—inactive vaccines, toxoids

Increase: risk of torsades de pointes—Class Ia/III antidysrhythmics

NURSING CONSIDERATIONS

Assess:

- **Multiple sclerosis:** improving paresthesia, muscle weakness, clonus, muscle spasms, difficulty with moving, difficulty with coordination of balance, speech, swallowing, vision problems, fatigue; prevention of increasing disability
- **Laboratory monitoring:** obtain before initial dose; CBC, LFTs, serum bilirubin, ophthalmologic exam, antibodies to VZV; if there is no history of chickenpox or no vaccination, may give VZV vaccination to antibody-negative patient before giving product, postpone for 1 mo after vaccination; obtain ECG for evidence of bradycardia, AV block

Evaluate:

- Therapeutic response: improved symptoms of multiple sclerosis and prevention of increasing disability

Teach patient/family:

- About use of product and expected results; provide med guide to patient
- That continuing follow-up exams and laboratory tests will be required on a regular basis
- To report any side effects of therapy
- To protect product from moisture
- To report chest pain, palpitations, jaundice

flecainide (Rx)

(flek-ay′nide)

Tambocor

Func. class.: Antidysrhythmic (Class IC)

F

ACTION: Decreases conduction in all parts of the heart, with greatest effect on the His-Purkinje system, which stabilizes cardiac membrane

USES: Life-threatening ventricular dysrhythmias, sustained ventricular tachycardia, supraventricular tachydysrhythmias, paroxysmal atrial fibrillation/flutter associated with disabling symptoms

Unlabeled uses: Atrial fibrillation, single dose

CONTRAINDICATIONS: Hypersensitivity, AV bundle branch block, cardiogenic shock

Precautions: Pregnancy (C), breastfeeding, children, geriatric patients, renal/hepatic disease, CHF, respiratory depression, myasthenia gravis, electrolyte abnormalities, atrial fibrillation, sick sinus syndrome, torsades de pointes, MI, bundle branch block

Black Box Warning: MI, cardiac arrhythmias, atrial fibrillation

DOSAGE AND ROUTES

PSVT/PAT

- **Adult: PO** 50 mg q12hr; may increase q4days by 50 mg q12hr to desired response, max 300 mg/day

Life-threatening ventricular dysrhythmias

- **Adult: PO** 100 mg q12hr; may increase by 50 mg q12hr q4days, max 400 mg/day

Renal dose

- **Adult: PO** CCr <35 ml/min, 100 mg daily or 50 mg bid initially

Available forms: Tabs 50, 100, 150 mg

Administer:

PO route

- Reduced dosage as soon as dysrhythmia is controlled
- May give with meals for GI upset
- May adjust dose q4days
- Therapeutic trough serum concs for adults range from 200 to 1000 ng/ml (average, 500 ng/ml); toxicity is more common with trough serum concs >1000 ng/mL; usual therapeutic range in children is 200-500 ng/mL; in some cases, up to 800 ng/mL is required
- Adjust dosage at intervals of ≥4 days (approximate plateau effects) following dosage adjustments; however, longer intervals are needed in patients with renal or hepatic impairment
- Frequent serum drug conc monitoring is required for patients with severe renal (CrCl <35 ml/min) or hepatic disease and may also be helpful in patients with CHF or in patients with moderate renal disease
- Monitoring of flecainide serum concs is strongly recommended in patients receiving amiodarone therapy

SIDE EFFECTS

CNS: *Headache, dizziness,* involuntary movement, confusion, psychosis, restlessness, irritability, paresthesias, ataxia, flushing, somnolence, depression, anxiety, malaise, fatigue, asthenia, tremors

CV: *Hypotension, bradycardia,* angina, PVCs, heart block, cardiovascular collapse, arrest, dysrhythmias, CHF, fatal ventricular tachycardia, palpitations, QT prolongation, torsades de pointes

EENT: Tinnitus, *blurred vision,* hearing loss, corneal deposits, dry eyes

GI: Nausea, vomiting, anorexia, constipation, abdominal pain, flatulence, change in taste, diarrhea

GU: Impotence, decreased libido, polyuria, urinary retention

HEMA: Leukopenia, thrombocytopenia

INTEG: Rash, urticaria, edema, swelling

RESP: Dyspnea, respiratory depression

PHARMACOKINETICS

Peak 3 hr, half-life 12-27 hr, metabolized by liver, excreted unchanged by kidneys (10%), excreted in breast milk

INTERACTIONS

Increase: of both products—propranolol

Increase: CV depressant action—β-blockers, disopyramide, verapamil

Increase: flecainide level—amiodarone, cimetidine, ritonavir

Increase: digoxin level—digoxin

Increase or decrease: effect—urinary, alkalinizing agents, acidifying agents

Drug/Herb

- Do not use with hawthorn

Drug/Lab Test

Increase: CPK

NURSING CONSIDERATIONS

Assess:

Black Box Warning: **MI, CHF, cardiogenic shock:** should not be used in these conditions

Black Box Warning: **Atrial fibrillation:** avoid use, risk of ventricular dysrhythmias

Black Box Warning: **Cardiac dysrhythmias:** discontinue in those with prolonged QRS >180 ms or prolonged PR >300 ms

- Monitor ECG, B/P, pulse
- **Electrolyte imbalances:** hypo/hyperkalemia before administration; correct electrolytes before use
- CNS effects: dizziness, confusion, psychosis, paresthesias, seizures; product should be discontinued
- **Flecainide level:** Monitor level in those with CHF or renal failure; peak, trough

Evaluate:

- Therapeutic response: decreased dysrhythmias

Teach patient/family:

- To change position slowly from lying or sitting to standing to minimize orthostatic hypotension

- To take as prescribed; not to skip or double dose
- To avoid hazardous activities that require alertness until response is known
- To carry emergency ID with disorder, medications taken
- To notify all health care providers of treatment
- To report new or worsening cardiac symptoms

TREATMENT OF OVERDOSE:
O_2, artificial ventilation, ECG, DOPamine for circulatory depression, diazepam or thiopental for seizures, treat ventricular dysrhythmias

fluconazole (Rx)
(floo-kon′a-zole)

Diflucan

Func. class.: Antifungal, systemic; azole

Chem class: Triazole

Do not confuse:
Diflucan/Diprivan

ACTION: Inhibits ergosterol biosynthesis, causes direct damage to fungal membrane phospholipids

USES: Oropharyngeal candidiasis, chronic mucocutaneous candidiasis; systemic, vaginal, urinary candidiasis; cryptococcal meningitis; prevention of candidiasis in bone marrow transplant in those who receive chemotherapy and/or radiation therapy; cystitis, fungal prophylaxis, peritonitis, pneumonia, pyelonephritis

Unlabeled uses: Prophylaxis, systemic candidiasis in very-low-birthweight premature infants, blastomycosis, chemotherapy-induced neutropenia, coccidioidomycosis cryptococcosis prophylaxis, endocarditis, endophthalmitis, histoplasmosis infectious arthritis, myocarditis, osteomyelitis, pericarditis

CONTRAINDICATIONS: Hypersensitivity to this product or azoles, pregnancy (D)

Precautions: Breastfeeding, renal/hepatic disease, torsades de pointes

DOSAGE AND ROUTES
Vulvovaginal candidiasis
- **Adult: PO** 150 mg as a single dose

Serious fungal infections
- **Adult: PO/IV** 50-400 mg initially then 200 mg/day for 4 wk
- **Child:** 6-12 mg/kg/day

Oropharyngeal candidiasis
- **Adult: PO/IV** 200 mg initially then 100 mg/day for ≥2 wk
- **Child: PO/IV** 6 mg/kg initially then 3 mg/kg/day for ≥2 wk

Esophageal candidiasis
- **Adult: PO/IV** 200 mg on 1st day, then 100 mg daily × ≥3 wk and for ≥2 wk after resolution of symptoms
- **Child: PO/IV** 6 mg/kg on 1st day, then 3 mg/kg × ≥3 wk and for ≥2 wk after resolution of symptoms

Cryptococcal meningitis
- **Adult: PO/IV** 400 mg on 1st day, then 200 mg daily × 10-12 wk after CSF culture negative
- **Child/infants/neonates ≥14 days: PO/IV** 12 mg/kg on 1st day, then 6-12 mg/kg daily × 10-12 wk after CSF culture negative, max 600 mg/day
- **Neonates 0-14 days: PO/IV** 12 mg/kg on 1st day, then 6-12 mg/kg q72hr × 10-12 wk after CSF culture negative

Prevention of candidiasis in bone marrow transplant
- **Adult: PO/IV** 400 mg/day, those anticipated to have neutrophils <500/mm^3, start several days before anticipated onset of neutropenia and continue for 7 days after rise of neutrophils >1000/mm^3

Renal disease
- **Adult: PO/IV** CCr ≤50 ml/min, after loading dose, give 50% of usual dose

Available forms: Tabs 50, 100, 150, 200 mg; inj 2 mg/ml; powder for oral susp 10 mg, 40 mg

Administer:

PO route

- Tap bottle to loosen powder, add water in 2 portions, review manufacturer reconstitution instructions
- Shake oral susp before each use, use within 2 wk

Intermittent IV INF route

- After diluting according to package directions; run at ≤200 mg/hr; do not use plastic containers in connections; check for bag leaks
- Use inf pump; check for extravasation and necrosis q2hr
- Do not use if cloudy or precipitated
- Do not admix; do not refrigerate
- Store protected from moisture and light; diluted sol stable 24 hr; do not freeze

Y-site compatibilities: Acyclovir, aldesleukin, alfentanil, allopurinol, amifostine, amikacin, aminocaproic acid, aminophylline, amiodarone, anidulafungin, ascorbic acid injection, atenolol, atracurium, atropine, azaTHIOprine, aztreonam, benztropine, bivalirudin, bleomycin, bumetanide, buprenorphine, butorphanol, calcium chloride, CARBOplatin, caspofungin, cefamandole, ceFAZolin, cefepime, cefmetazole, cefonicid, cefoperazone, cefoTEtan, cefOXitin, cefpirome, cefTAZidime, ceftizoxime, ceftobiprole, cephalothin, cephapirin, chlorproMAZINE, cimetidine, cisatracurium, CISplatin, codeine, cyanocobalamin, cyclophosphamide, cycloSPORINE, cytarabine, DACTINomycin, DAPTOmycin, dexamethasone, diltiazem, dimenhyDRINATE, diphenhydrAMINE, DOBUTamine, DOCEtaxel, DOPamine, doripenem, doxacurium, DOXOrubicin, DOXOrubicin liposomal, doxycycline, droperidol, drotrecogin alfa, enalaprilat, ePHEDrine, EPINEPHrine, epirubicin, epoetin alfa, eptifibatide, ertapenem, erythromycin, esmolol, etoposide, famotidine, fenoldopam, fentaNYL, filgrastim, fludarabine, fluorouracil, folic acid, foscarnet, gallium, ganciclovir, gatifloxacin, gemcitabine, gentamicin, glycopyrrolate, granisetron, heparin, hydrocortisone, HYDROmorphone, IDArubicin, ifosfamide, IV immune globulin, inamrinone, indomethacin, insulin (regular), irinotecan, isoproterenol, ketorolac, labetalol, lansoprazole, leucovorin, levofloxacin, lidocaine, linezolid, LORazepam, LR, magnesium sulfate, mannitol, mechlorethamine, melphalan, meperidine, meropenem, metaraminol, methicillin, methotrexate, methoxamine, methyldopate, methylPREDNISolone, metoclopramide, metoprolol, metroNIDAZOLE, mezlocillin, miconazole, midazolam, milrinone, minocycline, mitoXANtrone, morphine, moxalactam, multiple vitamins injection, mycophenolate, nafcillin, nalbuphine, naloxone, nesiritide, nitroglycerin, nitroprusside, norepinephrine, octreotide, ondansetron, oxacillin, oxaliplatin, oxytocin, PACLitaxel, palonosetron, pamidronate, pancuronium, papaverine, PEMEtrexed, penicillin G potassium/sodium, pentazocine, PENTobarbital, PHENobarbital, phenylephrine, phenytoin, phytonadione, piperacillin-tazobactam, polymyxin B, potassium chloride, procainamide, prochlorperazine, promethazine, propofol, propranolol, protamine, pyridoxine, quiNIDine, quinupristin-dalfopristin, ranitidine, remifentanil, Ringer's, ritodrine, riTUXimab, rocuronium, sargramostim, sodium acetate/bicarbonate, succinylcholine, SUFentanil, tacrolimus, temocillin, teniposide, theophylline, thiotepa, ticarcillin-clavulanate, tigecycline, tirofiban, TNA, tobramycin, tolazoline, TPN, trastuzumab, trimetaphan, urokinase, vancomycin, vasopressin, vecuronium, verapamil, vinCRIStine, vinorelbine, voriconazole, zidovudine, zoledronic acid

SIDE EFFECTS

CNS: *Headache,* seizures

CV: QT prolongation, torsades de pointes

GI: *Nausea, vomiting,* diarrhea, cramping, flatus, increased AST, ALT, hepatotoxicity

HEMA: Agranulocytosis, eosinophilia, leukopenia, neutropenia, thrombocytopenia
INTEG: Stevens-Johnson syndrome, angioedema, anaphylaxis, exfoliative dermatitis, toxic epidermal necrolysis

PHARMACOKINETICS

Peak 2-4 hr, bioavailability (PO) >90%, excreted unchanged in urine 80%, metabolized by CYP3A enzyme system at dose >200 mg/day, elimination half-life 30 hr

INTERACTIONS

Increase: hypoglycemia—oral sulfonyl/ureas (glipiZIDE)
Increase: anticoagulation—warfarin
Increase: plasma concs—cycloSPORINE, phenytoin, theophylline, rifabutin, tacrolimus, sirolimus, zidovudine, zolpidem
Increase: myopathy, rhabdomyolysis risk—HMG-CoA reductase inhibitors: lovastatin, simvastatin
Increase: effect of zidovudine, methadone, SUFentanil, alfentanil, buprenorphine, saquinavir, fentaNYL, ergots
Decrease: effect of oral contraceptives, calcium channel blockers
Decrease: fluconazole effect—proton pump inhibitors
Drug/Lab Test
Increase: alk phos, LFTs
Decrease: WBC, platelets

NURSING CONSIDERATIONS

Assess:
- **Infection:** clearing of CSF and other culture during treatment, obtain C&S baseline and throughout treatment, product may be started as soon as culture is taken
- ⚠ **Hepatotoxicity:** increasing AST, ALT, periodically alk phos, bilirubin; for renal status: BUN, creatinine
- **Skin symptoms:** color, lesions, injection site reactions; if lesions progress, stop product

Evaluate:
- Therapeutic response: decreasing oral candidiasis, fever, malaise, rash; negative C&S for infection organism

Teach patient/family:
- That long-term therapy may be needed to clear infection
- That medication may be taken with food to reduce GI effects
- To notify prescriber of nausea, vomiting, diarrhea, jaundice, anorexia, clay-colored stools, dark urine, skin rash
- To use alternative method of contraception while taking this product, pregnancy (D)

RARELY USED

fludarabine (Rx)

(floo-dar′a-been)

Fludara, Oforta

Func. class.: Antineoplastic, antimetabolite

USES: Chronic lymphocytic leukemia, non-Hodgkin's lymphoma

CONTRAINDICATIONS: Pregnancy (D), breastfeeding, hypersensitivity

Black Box Warning: Hemolytic anemia, bone marrow suppression, coma, seizures, visual disturbances

DOSAGE AND ROUTES

- **Adult: IV** 25 mg/m² over 30 min × 5 days, may repeat q28days; reconstitute with 2 ml of sterile water for inj; dissolution should occur in <15 sec, adjust dose based on toxicity; **PO** 40 mg/m² × 5 days q28days

fludrocortisone (Rx)

(floo-droe-kor′ti-sone)

Func. class.: Corticosteroid, synthetic

Chem. class.: Mineralocorticoid

ACTION: Promotes increased reabsorption of sodium and loss of potassium, water, hydrogen from distal renal tubules

USES: Adrenal insufficiency, salt-losing adrenogenital syndrome, Addison's disease

Unlabeled uses: Renal tubular acidosis (type IV), idiopathic orthostatic hypotension

CONTRAINDICATIONS: Children <2 yr, hypersensitivity, acute glomerulonephritis, amebiasis, psychoses, Cushing syndrome, fungal infections

Precautions: Pregnancy (C), breastfeeding, children >2 yr, osteoporosis, CHF, hypertension, diabetes

DOSAGE AND ROUTES

- **Adult: PO** 100-200 mcg/day
- **Child: PO** 50-100 mcg/day

Idiopathic hypotension (unlabeled)

- **Adult: PO** 50-200 mcg/day

Available forms: Tabs 100 mcg (0.1 mg)

SIDE EFFECTS

CNS: *Flushing, sweating,* headache, paralysis, dizziness, seizures

CV: *Hypertension,* circulatory collapse, thrombophlebitis, embolism, tachycardia, CHF, *edema*

ENDO: Weight gain, adrenal suppression, hyperglycemia

META: *Hypokalemia*

MISC: Hypersensitivity, cataracts, GI ulcers, anaphylaxis

MS: Fractures, osteoporosis, weakness

PHARMACOKINETICS

Peak 1.5 hr, half-life 18-36 hr, metabolized by liver, excreted in urine

INTERACTIONS

Increase: B/P—sodium-containing food or medication

Decrease: fludrocortisone action—barbiturates, rifampin, phenytoin

Decrease: potassium levels—thiazides, potassium-wasting products, loop diuretics, amphotericin B, piperacillin, mezlocillin

Drug/Herb

Increase: hypokalemia—aloe, buckthorn, cascara sagrada, Chinese rhubarb, senna

Increase: corticosteroid effect—aloe, licorice, perilla

Drug/Lab Test

Increase: potassium, sodium

Decrease: Hct

NURSING CONSIDERATIONS

Assess:

- Weight daily; notify prescriber of weekly gain >5 lb
- I&O ratio; be alert for decreasing urinary output, increasing edema
- B/P q4hr, pulse; notify prescriber if chest pain occurs
- Potassium depletion: paresthesias, fatigue, nausea, vomiting, depression, polyuria, dysrhythmias, weakness
- Electrolytes: sodium, potassium, chloride, hypokalemia is common

Administer:

- Titrated dose; use lowest effective dose
- With food or milk to decrease GI symptoms
- Assistance with ambulation in patient with bone tissue disease to prevent fractures

Evaluate:

- Therapeutic response: correction of adrenal insufficiency

Teach patient/family:

- That emergency ID as steroid user should be carried

• Not to discontinue this medication abruptly
• To notify health care provider of muscle cramps, weight gain, edema, nausea, infection, trauma, stress
• Not to breastfeed while taking this medication
• Avoid exposure to disease, trauma

flumazenil (Rx)

(flu-maz′e-nill)

Anexate ✦, Romazicon

Func. class.: Antidote: benzodiazepine receptor antagonist

Chem. class.: Imidazobenzodiazepine derivative

ACTION: Antagonizes actions of benzodiazepines on CNS, competitively inhibits activity at benzodiazepine recognition site on GABA/benzodiazepine receptor complex

USES: Reversal of sedative effects of benzodiazepines

CONTRAINDICATIONS: Hypersensitivity to this product or benzodiazepines, serious cyclic antidepressant overdose, patients given benzodiazepine for control of life-threatening conditions

Precautions: Pregnancy (C), breastfeeding, children, geriatric patients, status epilepticus, head injury, labor/delivery, renal/hepatic disease, hypoventilation, panic disorder, drug and alcohol dependency, ambulatory patients

Black Box Warning: Benzodiazepine dependence, seizures

DOSAGE AND ROUTES

Reversal of conscious sedation or general anesthesia

• **Adult:** IV 0.2 mg given over 15 sec; wait 45 sec then give 0.2 mg if consciousness does not occur; may be repeated at 60-sec intervals prn (max 3 mg/hr) or 1 mg/5 min
• **Child:** IV 10 mcg (0.01 mg)/kg; cumulative dose of 1 mg or less

Management of suspected benzodiazepine overdose

• **Adult:** IV 0.2 mg given over 30 sec; wait 30 sec then give 0.3 mg over 30 sec if consciousness does not occur; further doses of 0.5 mg can be given over 30 sec at intervals of 1 min up to cumulative dose of 3 mg
• **Child:** IV 10 mcg (0.01 mg/kg), cumulative dose of <1 mg

Available forms: Inj 0.1 mg/ml

Administer:

• Check airway and IV access before administration
• Use large vein

Direct IV route

• Give undiluted or diluted with 0.9% NaCl, D_5W, LR; give over 15-30 sec into running IV
• Stable for 24 hr if drawn into a syringe or mixed with other solutions

SIDE EFFECTS

CNS: Dizziness, agitation, emotional lability, confusion, **seizures**, somnolence, panic attacks

CV: Hypertension, palpitations, cutaneous vasodilation, **dysrhythmias**, bradycardia, tachycardia, chest pain

EENT: Abnormal vision, blurred vision, tinnitus

GI: Nausea, vomiting, hiccups

SYST: Headache, inj site pain, increased sweating, fatigue, rigors

PHARMACOKINETICS

Terminal half-life 41-79 min, metabolized in liver, onset 1-2 min

INTERACTIONS

• Toxicity: mixed product overdosage
• Antagonize action of benzodiazepines, zaleplon, zolpidem

NURSING CONSIDERATIONS

Assess:

• Cardiac status using continuous monitoring

Black Box Warning: For seizures; protect patient from injury; most likely among those who are withdrawing from benzodiazepines; flumazenil can precipitate benzodiazepine withdrawal and onset of seizures, seizures are increased in head trauma

• GI symptoms: nausea, vomiting; place patient in side-lying position to prevent aspiration
• **Allergic reactions:** flushing, rash, urticaria, pruritus

Black Box Warning: Seizures/benzodiazepine dependence: do not use in those who have used these products for IIP or status epilepticus; use in intensive care setting cautiously, there may be unrecognized benzodiazepine dependence

Evaluate:
• Therapeutic response: decreased sedation, respiratory depression, toxicity

Teach patient/family:
• That amnesia may continue
• Not to engage in hazardous activities for 18-24 hr after discharge
• Not to take any alcohol or nonprescription products for 18-24 hr

flunisolide nasal agent
See Appendix B

fluocinolone ophthalmic
See Appendix B

fluocinolone topical
See Appendix B

fluorometholone ophthalmic
See Appendix B

⚠ HIGH ALERT

fluorouracil (Rx)
(flure-oh-yoor'a-sil)

Adrucil, Carac, Efudex, 5-FU

Func. class.: Antineoplastic, antimetabolite

Chem. class.: Pyrimidine analog

Do not confuse:
fluorouracil/flucytosine

ACTION: Inhibits DNA, RNA synthesis; interferes with cell replication by competitively inhibiting thymidylate production, specific for S phase of cell cycle

USES: **Systemic:** cancer of breast, colon, rectum, stomach, pancreas; **topical:** multiple actinic keratoses, superficial basal cell carcinomas

Unlabeled uses: Anal, biliary tract, cervical, head and neck, hepatocellular, ovarian cancer

CONTRAINDICATIONS: Pregnancy (X), breastfeeding, hypersensitivity, poor nutritional status, serious infections

Black Box Warning: Bone marrow suppression

Precautions: Children, renal/hepatic disease, angina

Black Box Warning: GI bleeding

DOSAGE AND ROUTES

Doses vary widely, based on actual body weight unless obese, then based on lean body weight

Advanced colorectal cancer
• **Adult: IV bolus** 300-500 mg/m^2/day × 4-5 days q28days or 600-1500 mg/m^2 weekly or every other wk; continuous IV inf: 300-1000 mg/m^2/day × 4-5 days q4wk or 300 mg/m^2/day indefinitely, high dose 3000-3400 mg/m^2 over 24-72 hr

Breast cancer

• **Adult: IV bolus** 400-600 mg/m² on days 1 and 8 of every cycle with cyclophosphamide and methotrexate or 600 mg/m² day 1 with cyclophosphamide and methotrexate q21-28days

Pancreatic cancer

• **Adult: IV bolus** 600 mg/m² on days 1, 8, 29, 36 with DOXOrubicin and mitoMYcin q8wk

Actinic/solar keratoses

• **Adult: TOP** 1% cream/sol 1-2×/day or 2-5% sol for hands

Superficial basal cell carcinoma

• **Adult: TOP** 5% sol or cream 2×/day × 3-12 wk

Available forms: Inj 50 mg/ml; cream 0.5%, 1%, 5%; sol 1%, 2%, 5%

Administer:

• Antiemetic 30-60 min before product to prevent vomiting, for several days thereafter

Topical route

• The 1% strength is used on face; higher strengths are used on other parts of the body

• Wear gloves when applying; may use with a loose dressing; use plastic or wooden applicator

IV route

• Prepared in biologic cabinet using gloves, gown, mask; use cytotoxic handling procedures

• Undiluted; may inject through Y-tube or 3-way stopcock; give over 1-3 min; may be diluted in NS, D_5W; given as a continuous inf in plastic containers; do not refrigerate/freeze; protect from light; discard unused portion

Y-site compatibilities: Acyclovir, alatrofloxacin, alfentanil, allopurinol, amifostine, amikacin, amphotericin B lipid complex, amphotericin B liposome, ampicillin, ampicillin-sulbactam, anidulafungin, argatroban, atenolol, atracurium, azithromycin, aztreonam, bivalirudin, bleomycin, bumetanide, butorphanol, calcium gluconate, CARBOplatin, ceFAZolin, cefepime, cefoperazone, cefotaxime, cefoTEtan, cefOXitin, cefTAZidime, ceftizoxime, cefTRIAXone, cefuroxime, cimetidine, cisatracurium, CISplatin, clindamycin, codeine, cyclophosphamide, cycloSPORINE, DAPTOmycin, dexamethasone, digoxin, DOCEtaxel, DOPamine, doripenem, DOXOrubicin liposomal, enalaprilat, ePHEDrine, ertapenem, erythromycin, esmolol, etoposide phosphate, famotidine, fenoldopam, fentaNYL, fluconazole, fludarabine, foscarnet, fosphenytoin, furosemide, ganciclovir, gatifloxacin, gemcitabine, gentamicin, granisetron, heparin, hydrocortisone, HYDROmorphone, ifosfamide, imipenem-cilastatin, inamrinone, isoproterenol, ketorolac, labetalol, leucovorin, levorphanol, lidocaine, linezolid, magnesium sulfate, mannitol, melphalan, meperidine, meropenem, mesna, methohexital, methotrexate, methylPREDNISolone, metoprolol, metroNIDAZOLE, milrinone, mitoMYcin, mitoXANtrone, morphine sulfate, nalbuphine, naloxone, nesiritide, nitroglycerin, nitroprusside, octreotide, ofloxacin, PACLitaxel, palonosetron, pamidronate, pancuronium, pantoprazole, PEMEtrexed, PENTobarbital, PHENobarbital, phenylephrine, piperacillin, piperacillin-tazobactam, potassium chloride/phosphates, procainamide, propofol, propranolol, ranitidine, remifentanil, riTUXimab, sargramostim, sodium acetate/bicarbonate/phosphates, succinylcholine, SUFentanil, sulfamethoxazole-trimethoprim, teniposide, theophylline, thiopental, thiotepa, ticarcillin, ticarcillin-clavulanate, tigecycline, tirofiban, tobramycin, trastuzumab, vasopressin, vecuronium, vinBLAStine, vinCRIStine, vitamin B complex/C, voriconazole, zidovudine, zoledronic acid

SIDE EFFECTS

Systemic use

CNS: Lethargy, malaise, weakness, acute cerebellar dysfunction

CV: Myocardial ischemia, angina

EENT: Light intolerance, lacrimation

GI: *Anorexia, stomatitis,* diarrhea, nausea, vomiting, hemorrhage, enteritis, glossitis
HEMA: Thrombocytopenia, leukopenia, myelosuppression, anemia, agranulocytosis
INTEG: *Rash,* fever, photosensitivity, anaphylaxis

PHARMACOKINETICS

Half-life 20 hr terminal; metabolized in liver; excreted in urine; crosses blood-brain barrier

INTERACTIONS

Increase: bleeding—anticoagulants, NSAIDs, platelet inhibitors, thrombolytics
Increase: toxicity—metroNIDAZOLE
Increase: toxicity, bone marrow depression—radiation or other antineoplastics, leucovorin
Decrease: antibody response—live virus vaccines
Decrease: effect of phenytoin
Drug/Lab Test
Increase: AST, ALT, LDH, serum bilirubin, Hct, Hgb, WBC, platelets, 5-HIAA
Decrease: albumin

NURSING CONSIDERATIONS

Assess:

Black Box Warning: **Bone marrow suppression:** CBC, differential, platelet count daily (IV); withhold product if WBC is <3500/mm^3 or platelet count is <100,000/mm^3; notify prescriber of results; product should be discontinued; nadir of leukopenia within 2 wk, recovery 1 mo, if pretreatment of WBC <2,000/mm^3 or platelets <100,000/mm^3, delay until recovery of counts above this level

- Renal studies: BUN, serum uric acid, urine CCr, electrolytes before, during therapy
- Hepatic studies before, during therapy: bilirubin, alk phos, AST, ALT, LDH before, during therapy
- **Bleeding:** hematuria, guaiac, bruising, petechiae, mucosa or orifices q8hr
- Inflammation of mucosa, breaks in skin; buccal cavity q8hr for dryness, sores or ulceration, white patches, oral pain, bleeding, dysphagia
- GI symptoms: frequency of stools, cramping, intractable vomiting, stomatitis
- **Infection:** those with current infections should be treated before receiving 5-FU, the dose reduced or discontinued if infection occurs

Evaluate:

- Therapeutic response: decreased tumor size, spread of malignancy

Teach patient/family:

- To avoid crowds, persons with known infection
- To avoid foods with citric acid, hot or rough texture if stomatitis is present; to drink adequate fluids
- To report stomatitis: any bleeding, white spots, ulcerations in mouth; that patient should examine mouth daily, report symptoms; viscous lidocaine may be used; rinsing of mouth tid-qid with water, club soda; brushing of teeth bid-tid with soft brush or cotton-tipped applicator for stomatitis; use unwaxed dental floss, give ice chips for mucositis
- To report signs of **infection:** fever, sore throat, flulike symptoms
- To report signs of **anemia:** fatigue, headache, faintness, shortness of breath, irritability
- To report **bleeding:** to avoid razors, commercial mouthwash, IM inj if counts are low
- Not to use aspirin products or NSAIDs
- To use contraception during therapy (men and women), pregnancy (X); to avoid breastfeeding (topical use)
- Not to receive vaccinations during therapy
- To use sunscreen or stay out of the sun to prevent photosensitivity
- About hair loss; to explore use of wigs or other products until hair regrowth occurs
- To apply topically only to affected areas, being careful around mouth, nose, eyes

FLUoxetine (Rx)

(floo-ox′eh-teen)

PROzac, PROzac Weekly, Sarafem

Func. class.: Antidepressant, SSRI (selective serotonin reuptake inhibitor)

Do not confuse:
PROzac/Proscar/ProSom/PriLOSEC
Sarafem/Serophene

ACTION: Inhibits CNS neuron uptake of serotonin but not of norepinephrine

USES: Major depressive disorder, obsessive-compulsive disorder (OCD), bulimia nervosa, generalized anxiety disorder social phobia; *Sarafem:* premenstrual dysphoric disorder (PMDD), panic disorder

Unlabeled uses: Alcoholism, anorexia nervosa, borderline personality disorder, obesity, posttraumatic stress disorder, autism, fibromyalgia, orthostatic hypotension, premature ejaculation

CONTRAINDICATIONS: Hypersensitivity

Precautions: Pregnancy (C), breastfeeding, geriatric patients, diabetes mellitus, narrow-angle glaucoma, cardiac malformations in infants (exposed to FLUoxetine in utero)

Black Box Warning: Children, suicidal ideation

DOSAGE AND ROUTES

Depression/obsessive-compulsive disorder

- **Adult: PO** 20 mg/day in AM; after 4 wk, if no clinical improvement is noted, dose may be increased to 20 mg bid in AM, PM, max 80 mg/day; **PO** 90 mg/wk
- **Geriatric: PO** 5-10 mg/day, increase as needed
- **Child 8-18 yr: PO** 10-20 mg/day, max 20 mg/day

Premenstrual dysphoric disorder (Sarafem)

- **Adult: PO** 20 mg/day, may be taken daily 14 days before menses

Alcoholism (unlabeled)

- **Adult: PO** 20-80 mg/day

Anorexia nervosa (unlabeled)

- **Adult: PO** 10 mg daily, max 60 mg/day

Borderline personality disorder/fibromyalgia/autism/orthostatic hypotension/premature ejaculation/hot flashes (unlabeled)

- **Adult: PO** 20 mg/day, max 80 mg/day

Obesity (unlabeled)

- **Adult: PO** 60 mg/day after titration

Posttraumatic stress disorder (unlabeled)

- **Adult: PO** 10-80 mg/day

Available forms: Caps 10, 20, 40 mg; tabs 10, 20 mg; oral sol 20 mg/5 ml; del rel caps (PROzac Weekly) 90 mg; tab 10, 15, 20 mg (Sarafem)

Administer:

- Without regard to meals
- Crushed if patient is unable to swallow medication whole (tab only)
- Gum, hard candy, frequent sips of water for dry mouth
- Store at room temperature; do not freeze
- **PROzac Weekly** on the same day each wk, swallow whole; do not crush, cut, chew
- **Oral sol:** use oral syringe or calibrated measuring device

SIDE EFFECTS

CNS: *Headache, nervousness, insomnia, drowsiness, anxiety, tremor, dizziness, fatigue, sedation, poor conc, abnormal dreams, agitation,* **seizures,** apathy, euphoria, hallucinations, delusions, psychosis, **suicidal ideation, neuroleptic malignant syndrome–like reactions**

CV: *Hot flashes, palpitations,* angina pectoris, hypertension, **tachycardia, 1st-degree AV block, bradycardia, MI, thrombophlebitis,** generalized edema

EENT: Visual changes, ear/eye pain, photophobia, tinnitus, increased intraocular pressure

GI: *Nausea, diarrhea, dry mouth, anorexia, dyspepsia, constipation,* taste changes, flatulence, decreased appetite
GU: *Dysmenorrhea, decreased libido, urinary frequency, UTI,* amenorrhea, cystitis, impotence, urine retention
HEMA: Hemorrhage
INTEG: *Sweating, rash, pruritus,* acne, alopecia, urticaria, angioedema, exfoliative dermatitis, Stevens-Johnson syndrome, toxic epidermal necrolysis
META: Hyponatremia
MS: *Pain,* arthritis, twitching
RESP: *Pharyngitis, cough, dyspnea, bronchitis,* asthma, hyperventilation, pneumonia
SYST: *Asthenia,* serotonin syndrome, flu-like symptoms, neonatal abstinence syndrome

PHARMACOKINETICS

PO: Peak 6-8 hr, metabolized in liver, excreted in urine, terminal half-life 2-3 days, norfluoxetine active metabolite half-life 4-16 days, steady state 28-35 days, protein binding 94%

INTERACTIONS

Increase: serotonin syndrome—SSRIs, SNRIs, serotonin-receptor agonists, selegiline, busPIRone, tryptophan, phenothiazines, haloperidol, loxapine, thiothixene, tricyclics; do not use concurrently
Increase: bleeding risk—platelet inhibitors, thrombolytics, NSAIDs, salicylates, anticoagulants
⚠ Do not use MAOIs with or 14 days before FLUoxetine
Increase: levels or toxicity of carBAMazepine, lithium, digoxin, warfarin, phenytoin, diazepam, vinBLAStine, donepezil, antidiabetics, dorifenacin, paricalcitrol, budesonide, bosentan, thioridazine
Increase: CNS depression—alcohol, antidepressants, opioids, sedatives
Decrease: FLUoxetine effect—cyproheptadine

Drug/Herb
⚠ Do not use together; increased risk of serotonin syndrome: St. John's wort, SAM-e
Increase: CNS effect—hops, kava, lavender, valerian

Drug/Lab Test
Increase: serum bilirubin, blood glucose, alk phos, BUN, CK

NURSING CONSIDERATIONS

Assess:

Black Box Warning: Mental status: mood, sensorium, affect, suicidal tendencies (child/young adult), increase in psychiatric symptoms, depression, panic; monitor for seizures, seizure potential increased, sarafem is not approved for children

- **Bulimia nervosa:** appetite, weight daily, increase nutritious foods in diet, watch for bingeing and vomiting

⚠ **Allergic reactions/serious skin reactions:** angioedema, exfoliative dermatitis, Stevens-Johnson syndrome, toxic epidermal necrolysis, itching, rash, urticaria; product should be discontinued, may need to give antihistamine

- B/P (lying/standing), pulse q4hr; if systolic B/P drops 20 mm Hg, hold product, notify prescriber; ECG for flattening of T wave, bundle branch, AV block, dysrhythmias in cardiac patients
- Blood studies: CBC, leukocytes, differential, cardiac enzymes if patient is receiving long-term therapy; check platelets; bleeding can occur, thyroid function, growth rate (children), weight
- Hepatic studies: AST, ALT, bilirubin, creatinine, weight weekly; appetite may decrease with product
- Alcohol consumption; if alcohol is consumed, hold dose until AM
- Safety measures, primarily for geriatric patients

Evaluate:

- Therapeutic response: decreased depression, symptoms of OCD

Teach patient/family:

- That therapeutic effect may take 1-4 wk, not to discontinue abruptly
- To use caution when driving, performing other activities requiring alertness because of drowsiness, dizziness, blurred vision
- To avoid alcohol, other CNS depressants
- To notify prescriber if pregnant, planning to become pregnant, or breastfeeding
- To change positions slowly because orthostatic hypotension may occur
- To avoid all OTC products unless approved by prescriber

Black Box Warning: That suicidal thoughts/behaviors may occur in young adults, children

fluPHENAZine decanoate (Rx)

(floo-fen′a-zeen)

Modecate ♣

fluPHENAZine hydrochloride (Rx)

Func. class.: Antipsychotic

Chem. class.: Phenothiazine, piperazine

Do not confuse:
Prolixin/Proloid

ACTION: Depresses cerebral cortex, hypothalamus, limbic system, which control activity and aggression; blocks neurotransmission produced by DOPamine at synapse; exhibits strong α-adrenergic and anticholinergic blocking action; mechanism for antipsychotic effects is unclear

USES: Schizophrenia

Unlabeled use: Agitation

CONTRAINDICATIONS: Hypersensitivity, blood dyscrasias, coma, bone marrow depression

Precautions: Pregnancy (C), breastfeeding, children <12 yr, geriatric patients, seizure disorders, hypertension, cardiac/hepatic disease, abrupt discontinuation; accidental exposure, agranulocytosis, ambient temperature increase, angina, hypersensitivity to benzyl alcohol/parabens/sesame oil/tartrazine dye, QT prolongation, suicidal ideation, renal failure, Parkinson's disease, hypocalcemia, head trauma, prostatic hypertrophy, pulmonary disease, infection, ileus, chemotherapy, breast cancer

Black Box Warning: Increased mortality in elderly patients with dementia-related psychosis

DOSAGE AND ROUTES

Decanoate

- **Adult and child >12 yr: IM/SUBCUT** 12.5-25 mg q1-3wk, may increase slowly, max 100 mg/dose

HCl

- **Adult: PO** 2.5-10 mg in divided doses q6-8hr, max 40 mg/day; **IM** initially 1.25 mg then 2.5-10 mg in divided doses q6-8hr

Available forms: *Decanoate:* inj 25 mg/ml; *HCl:* tabs 1, 2.5, 5, 10 mg; inj 2.5 mg/ml; elixir 2.5 mg/5 ml; oral solution 5 mg/ml

Administer:

PO route

- Give with food, milk, or a full glass of water to minimize gastric irritation
- **Oral concentrate:** Give using a calibrated measuring device; dilute just before use with 120-240 ml of water, saline, milk, 7-Up, carbonated orange beverage, or apricot, orange, pineapple, prune, tomato, or V-8 juice; do not mix with beverages containing caffeine (coffee, cola), tannics (tea), or pectinates (apple juice) or with other liquid medications; avoid spilling the solution on the skin and clothing
- **Oral elixir:** Give using a calibrated measuring device; avoid spilling the solution on the skin and clothing

Injectable routes

- Visually inspect for particulate matter and discoloration before use, slight yellow to amber color does not alter potency, markedly discolored solutions should be discarded

IM route (fluPHENAZine HCl only)

- No dilution necessary; if irritation occurs, subsequent IM doses may be diluted with NS for injection or 2% procaine
- Inject slowly and deeply into the upper outer quadrant of the gluteal muscle using a dry syringe and needle, aspirate before injection
- Keep patient in a recumbent position for at least 30 min following injection to minimize hypotensive effects
- Rotate the site of injection to avoid irritation or sterile abscess formation with repeat use

IM depot injection (fluPHENAZine decanoate or fluPHENAZine enanthate)

- FluPHENAZine decanoate or enanthate injections are administered IM or subcut, do not give IV
- FluPHENAZine depot injections must be drawn up using a dry syringe and needle of at least 21-G, do not dilute
- Inject slowly and deeply into the upper outer quadrant of the gluteal muscle, aspirate before injection
- Keep patient in a recumbent position for at least 30 min following the initial injection to minimize hypotensive effects; once patient is dose-stabilized on depot formulation, this precaution may be modified; rotate the site of injection to avoid irritation or sterile abscess formation with repeat administration

Subcut injection route (fluPHENAZine decanoate or fluPHENAZine enanthate)

- FluPHENAZine decanoate or enanthate injections are administered via IM or SC injection only, do not administer IV
- FluPHENAZine depot injection solutions must be drawn up using a dry syringe and a needle of at least 21-G, do not dilute
- Inject subcut, taking care not to inject intradermally
- Keep patient in a recumbent position for at least 30 min following the initial injection to minimize hypotensive effects. Once patient is dose-stabilized on depot formulation, this precaution may be modified for ambulatory patients; rotate the injection sites

SIDE EFFECTS

CNS: *EPS: pseudoparkinsonism, akathisia, dystonia, tardive dyskinesia, drowsiness, headache,* seizures, neuroleptic malignant syndrome

CV: *Orthostatic hypotension,* hypertension, cardiac arrest, ECG changes, tachycardia

EENT: Blurred vision, glaucoma, dry eyes, nasal congestion

GI: *Dry mouth, nausea, vomiting, anorexia, constipation,* diarrhea, jaundice, weight gain, paralytic ileus, hepatitis, cholecystic jaundice

GU: Urinary retention, urinary frequency, enuresis, impotence, amenorrhea, gynecomastia

HEMA: Anemia, leukopenia, leukocytosis, agranulocytosis, aplastic anemia, thrombocytopenia

INTEG: *Rash,* photosensitivity, dermatitis, hyperpigmentation (long-term use)

RESP: Laryngospasm, dyspnea, respiratory depression

PHARMACOKINETICS

Metabolized by liver, excreted in urine (metabolites), crosses placenta, enters breast milk, protein binding >90%, not dialyzable

PO/IM (HCl): Onset 1 hr, peak 2-4 hr, duration 6-8 hr, half-life 3.5-4 days

IM/SUBCUT (decanoate): Onset 1-3 days; peak 1-2 days, duration over 4 wk, single-dose half-life 6.8-9.6 days, multiple dose 14.3 days

INTERACTIONS

Increase: serotonin syndrome, neuroleptic malignant syndrome—SSRIs, SNRIs, serotonin-receptor agonists

Increase: QT prolongation, torsades de pointes (at higher doses)—amiodarone, arsenic trioxide, astemizole, dasatinib, disopyramide, dofetilide, droperidol, erythromycin, flecainide, gatifloxacin, ibutilide, levomethadyl, ondansetron, paliperidone, palonosetron, some antidepressants, vorinostat, ziprasidone, haloperidol, phenothiazines, ARIPiprazole, lurasidone

Increase: sedation—other CNS depressants, alcohol, barbiturate anesthetics, haloperidol, metyrosine, risperiDONE

Increase: toxicity—EPINEPHrine

Increase: anticholinergic effects—anticholinergics

Decrease: effects of levodopa, lithium

Decrease: fluPHENAZine effects—smoking, barbiturates

Drug/Lab Test

Increase: LFTs, cardiac enzymes, cholesterol, blood glucose, prolactin, bilirubin, cholinesterase

Decrease: hormones (blood and urine)

False positive: pregnancy tests, PKU urinary steroids, 17-OHCS

NURSING CONSIDERATIONS

Assess:

Black Box Warning: Increased mortality in elderly patients with dementia-related psychosis

- **Serotonin syndrome, neuroleptic malignant syndrome:** severe EPS, increased CPK, altered mental state, sinus tachycardia, change in B/P, sweating, often occurs in young men, heat stress, physical exhaustion, dehydration, organic brain disease

⚠ **QT prolongation, torsades de pointes:** ECG for changes

- Bilirubin, CBC, LFTs monthly; ophthalmic exams periodically
- Urinalysis recommended before and during prolonged therapy
- Affect, orientation, LOC, reflexes, gait, coordination, sleep pattern disturbances
- B/P standing and lying; pulse and respirations q4hr during initial treatment; establish baseline before starting treatment; report drops of 30 mm Hg
- Dizziness, faintness, palpitations, tachycardia on rising
- **EPS** including akathisia (inability to sit still, no pattern to movements), tardive dyskinesia (bizarre movements of jaw, mouth, tongue, extremities), pseudoparkinsonism (rigidity, tremors, pill rolling, shuffling gait)
- Constipation, urinary retention daily; if these occur, increase bulk, water in diet
- Supervised ambulation until stabilized on medication; do not involve patient in strenuous exercise; fainting possible; patient should not stand still for long periods

Evaluate:

- Therapeutic response: decrease in emotional excitement, hallucinations, delusions, paranoia, reorganization of patterns of thought, speech

Teach patient/family:

- That orthostatic hypotension occurs often; to rise from sitting or lying position gradually; to avoid hazardous activities until stabilized on medication
- To avoid hot tubs, hot showers, tub baths because hypotension may occur; that in hot weather, heat stroke may occur; to take extra precautions to stay cool

⚠ To avoid abrupt withdrawal of this product or EPS may result; that product should be withdrawn slowly

- To avoid OTC preparations (cough, hay fever, cold) unless approved by prescriber; that serious product interactions may occur; to avoid use with alcohol, CNS depressants; that increased drowsiness may occur
- To use a sunscreen to prevent burns
- About the importance of compliance with product regimen, follow up, lab, ophthalmic exams
- About EPS and the need for meticulous oral hygiene because oral candidiasis may occur
- To report sore throat, malaise, fever, bleeding, mouth sores; if these occur, CBC should be drawn, product discontinued

- That urine may turn pink to reddish brown

TREATMENT OF OVERDOSE:
Lavage; if orally ingested, provide an airway; *do not induce vomiting*

flurandrenolide topical
See Appendix B

flurbiprofen ophthalmic
See Appendix B

flutamide (Rx)
(floo′ta-mide)

Euflex ♣

Func. class.: Antineoplastic, hormone

Chem. class.: Antiandrogen

ACTION: Interferes with androgen uptake in the nucleus or androgen activity in target tissues; arrests tumor growth in androgen-sensitive tissue (i.e., prostate gland)

USES: Metastatic prostatic carcinoma, stage D_2 in combination with LHRH agonistic analogs (leuprolide), B_2-C in combination with goserelin and radiation

CONTRAINDICATIONS: Pregnancy (D), hypersensitivity

Black Box Warning: Severe hepatic disease

Precautions: G6PD deficiency, hemoglobinopathy, lactase deficiency, polycystic ovary syndrome, tobacco smoking

DOSAGE AND ROUTES
- **Adult: PO** 250 mg q8hr for a daily dosage of 750 mg

Available forms: Caps 125, 250 ♣ mg

Administer:
- Do not break, crush, chew caps
- Give without regard to food with a full glass of water
- Use cytotoxic handling procedures

SIDE EFFECTS
CNS: *Hot flashes,* drowsiness, confusion, depression, anxiety, paresthesia

GI: *Diarrhea, nausea, vomiting,* increased levels in hepatic studies, hepatitis, anorexia, hepatotoxicity, abdominal pain, cholestasis, **hepatic necrosis/failure**

GU: *Decreased libido, impotence, gynecomastia*

HEMA: Hemolytic anemia, leukopenia, thrombocytopenia

INTEG: Irritation at site, rash, photosensitivity

MISC: Edema, neuromuscular and pulmonary symptoms, hypertension, secondary malignancy

PHARMACOKINETICS
Rapidly and completely absorbed; excreted in urine and feces as metabolites; half-life 6 hr, geriatric half-life 8 hr; 94% protein binding, peak 2 hr

INTERACTIONS
Increase: PT—warfarin

Decrease: flutamide action—LHRH analog (leuprolide)

Drug/Lab Test

Increase: LFTs, BUN, creatinine

Decrease: WBC, platelets

NURSING CONSIDERATIONS
Assess:

Black Box Warning: Severe hepatic disease: AST, ALT, alk phos, which may be elevated; if LFTs elevated, product may need to be discontinued; monitor CBC, bilirubin, creatinine

- CNS symptoms, including drowsiness, confusion, depression, anxiety

Evaluate:
- Therapeutic response: decrease in prostatic tumor size, decrease in spread of cancer

Teach patient/family:

- To report side effects: decreased libido, impotence, breast enlargement, hot flashes, diarrhea

Black Box Warning: **Hepatotoxicity:** to report nausea, vomiting, yellow eyes or skin, dark urine, clay-colored stools; hepatotoxicity may be the cause

- Notify of yellow, green urine discoloration
- Avoid sun exposure, tanning beds
- To use contraception during treatment; pregnancy category (D)

fluticasone (Rx)

(floo-tic′a-sone)

Flonase, Flovent HFA, Flovent Diskus, Veramyst (nasal spray)

Func. class: Corticosteroids, inhalation; antiasthmatic

ACTION: Decreases inflammation by inhibiting mast cells, macrophages, and leukotrienes; antiinflammatory and vasoconstrictor properties

USES: Prevention of chronic asthma during maintenance treatment in those requiring oral corticosteroids; nasal symptoms of seasonal/perennial, allergic/nonallergic rhinitis

Unlabeled uses: COPD

CONTRAINDICATIONS: Hypersensitivity to this product or milk protein, primary treatment in status asthmaticus, acute bronchospasm

Precautions: Pregnancy (C), breastfeeding, active infections, glaucoma, diabetes, immunocompromised patients, Cushing syndrome

DOSAGE AND ROUTES

Prevention of chronic asthma during maintenance treatment in those requiring oral corticosteroids

Flovent HFA

- **Adult and child ≥12 yr: INH** 88-440 mcg bid (in those previously taking bronchodilators alone); **INH** 88-220 mcg bid, max 440 mcg bid (in those previously taking inhaled corticosteroids); **INH** 440 mcg bid, max 880 mcg bid (in those previously taking oral corticosteroids)
- **Child 4-11 yr: INH** 88 mcg bid

Flovent Diskus

- **Adult and child ≥12 yr: INH** 100 mcg bid, max 500 mcg bid (in those previously taking bronchodilators alone); **INH** 100-250 mcg bid, max 500 mcg bid (in those previously taking inhaled corticosteroids); **INH** 500-1000 mcg bid, max 1000 mcg bid (in those previously taking oral corticosteroids)
- **Child 4-11 yr: INH** Initially 50 mcg bid, max 100 mcg bid (in those previously taking bronchodilators alone or inhaled corticosteroids)

Nasal symptoms of seasonal, perennial allergic, nonallergic rhinitis

Flonase

- **Adult:** nasal 2 sprays initially, in each nostril daily or 1 spray bid, when controlled, lower to 1 spray in each nostril daily
- **Adolescent/child >4 yr:** nasal 1 spray in each nostril daily, may increase to 2 sprays in each nostril daily, when controlled lower to 1 spray in each nostril daily

Veramyst

- **Adult/child ≥12 yr:** nasal 2 sprays in each nostril daily
- **Child 2-11 yr:** nasal 1 spray in each nostril daily

Available forms: Oral inhalation aerosol 44, 110, 220 mcg; oral inhalation powder 50, 100, 250 mcg; nasal spray (Veramyst) 27.5 mcg/actuation, (Flonase) 50 mcg/actuation

Administer:

- Give at 1-min intervals; if a bronchodilator aerosol spray is used, use bronchodilator first, wait 5-15 min, then use fluticasone
- Decrease dose to lowest effective dose after desired effect; decrease dose at 2-4 wk intervals

Inhalation route (aerosol)

- Shake well, prime before 1st use, release 4 sprays into air away from face, prime using 1 spray if not used for ≥7 days; when the counter reads 000, discard; clean mouthpiece daily in warm water, dry; do not share inhaler with others
- In general, child <4 yr requires a face mask with spacer/VHC device for delivery; allow 3-5 INH per actuation; do not use spacer with Flovent Diskus

Inhalation route: Powder for oral inhalation (Flovent Diskus)

- Before initial use, remove the diskus foil pouch and throw away the foil pouch; the diskus will be in the closed position; fill in the "Pouch opened" and "Use by" dates in the blank lines on the label; the "Use by" date for Flovent Diskus 50 mcg is 6 wk from the date the pouch is opened; the "Use by" date for Diskus 100 mcg and 250 mcg is 2 mo from the date the pouch is opened
- Open the diskus by holding in one hand and using the thumb of the other to push the thumb grip away as far as it will go until the mouthpiece shows and snaps into place
- Hold the diskus in a level, flat position with the mouthpiece toward the patient and slide the lever away from the patient as far as it will go until it clicks; the number on the dose counter will count down by 1; the diskus is now ready to use
- Before inhaling the dose, have patient breathe out as much as possible while holding the diskus level and away from the mouth; the patient should never breathe out into the mouthpiece
- Instruct the patient to put the mouthpiece to the lips and breathe in through the mouth quickly and deeply through the diskus; remove the diskus from the mouth, hold breath for about 10 sec, and breathe out slowly
- After taking a dose, close the diskus by sliding the thumb grip back as far as it will go; the diskus will click shut; the lever automatically returns to its original position
- The counter displays how many doses are left; the counter number counts down each time the patient uses the diskus; after 55 doses (23 doses from the sample pack), numbers 5 to 0 are red to warn that there are only a few doses left
- After use, patient should rinse mouth with water and spit out the water, not swallow it
- To avoid the spread of infection, do not use the inhaler for more than one person

Intranasal

- Prime before first use
- Shake bottle gently before each use
- Rinse tip after use, dry with tissue
- Blow nose before use

SIDE EFFECTS

CNS: Fatigue, fever, headache, nervousness, dizziness, migraines

EENT: *Pharyngitis*, sinusitis, rhinitis, laryngitis, hoarseness, dry eyes, cataracts, nasal discharge, epistaxis, blurred vision

GI: Diarrhea, abdominal pain, nausea, vomiting, *oral candidiasis*

INTEG: Urticaria, dermatitis

META: Hyperglycemia, growth retardation in children, cushingoid features

MISC: Influenza, eosinophilic conditions, angioedema, Churg-Strauss syndrome, anaphylaxis, adrenal insufficiency (high doses), bone mineral density reduction

MS: Osteoporosis, muscle soreness, joint pain, arthralgia

RESP: *Upper respiratory infection*, dyspnea, cough, bronchitis, bronchospasm

PHARMACOKINETICS

Absorption 30% aerosol, 13.5% powder; protein binding 91%; metabolized in liver after absorption in lung; half-life 7.8 hr; <5% excreted in urine and feces

Oral INH: Onset 24 hr, peak several days, duration 1-2 wk
Intranasal: Onset 12 hr, peak several days

INTERACTIONS

• Avoid use with live virus vaccines
Increase: fluticasone levels—CYP3A4 inhibitors (ketoconazole, itraconazole), darunavir, nelfinavir, ritonavir, amprenavir, fosamprenavir, atazanavir, delavirdine, saquinavir
Increase: tendonitis, tendon rupture—quinolones
Increase: cardiac toxicity—isoproterenol (asthma patients)
Decrease: effects of growth hormones—mecasermin
Drug/Lab Test
Increase: urine/serum glucose

NURSING CONSIDERATIONS

Assess:
• **Respiratory status:** lung sounds, pulmonary function tests during, for several mo after change from systemic to inhalation corticosteroids
• Withdrawal symptoms from oral corticosteroids: depression, pain in joints, fatigue
⚠ **Adrenal insufficiency:** nausea, weakness, fatigue, hypotension, hypoglycemia, anorexia; may occur when changing from systemic to inhalation corticosteroids; may be life-threatening; adrenal function tests periodically: hypothalamic–pituitary–adrenal axis suppression in long-term treatment
• Growth rate in children; blood glucose, serum potassium for all patients
Evaluate:
• Therapeutic response: decreased severity of asthma, COPD, allergies
Teach patient/family:
• To use bronchodilator 1st, before using inhalation, if taking both
• Not to use for acute asthmatic attack; acute asthma may require oral corticosteroids
• To avoid smoking, smoke-filled rooms, those with URIs, those not immunized against chickenpox or measles
• To rinse mouth after inhaled product to decrease risk of oral candidiasis
• To report immediately cushingoid symptoms: no appetite, nausea, weakness, fatigue, decreased B/P

fluticasone (topical)

(floo-tic′a-sone)
Cutivate
Func. class.: Corticosteroid, topical

Do not confuse:
fluticasone/mometasone/fludrocortisone

ACTION: Crosses cell membrane to attach to receptors to decrease inflammation, itching; inhibits multiple inflammatory cytokines

USES: Inflammation/itching of corticosteroid-responsive dermatoses on the skin

CONTRAINDICATIONS: Hypersensitivity to this product or milk protein, status asthmaticus, monotherapy in primary infections, Cushing syndrome, rosacea, perioral dermatits, diabetes mellitus
Precautions: Pregnancy (C), children, breastfeeding, skin infections, skin atrophy

DOSAGE AND ROUTES

• **Adult:** Apply to affected areas bid (cream/ointment) or daily (lotion) × 4 wk
Available forms: Lotion, cream 0.05%, ointment 0.005%
Administer:
Topical route
⚠ Do not use with occlusive dressings
• **Cream/Ointment/Lotion:** Apply sparingly in a thin film and rub gently into the cleansed, affected area, wash

hands, use gloves, avoid use on face, groin, underarms

• Reassess treatment after 2 uses

SIDE EFFECTS

INTEG: Burning, pruritus, dermatitis, hypertrichosis, hives, rash, xerosis, irritation, hyperpigmentation, miliaria

META: Hyperglycemia, glycosuria

MISC: HPA axis suppression, Cushing syndrome

PHARMACOKINETICS

Absorption 5% but variable; half life 7 hr

INTERACTIONS

Increase: Blood glucose

NURSING CONSIDERATIONS

Assess:

• Skin reactions: burning, pruritus, dermatitis

Evaluate:

• Decreasing itching, inflammation on the skin

Teach patient/family:

Topical route:

⚠ Not to use with occlusive dressings

• **Cream/Ointment/Lotion:** To apply sparingly in a thin film and rub gently into the cleansed, affected area; avoid use on face, groin, underarms; wash hands; use gloves

• To reassess treatment after 2 uses

fluticasone/salmeterol

(floo-tic'a-sone) (sal-mee'ter-ol)

Advair Diskus, Advair HFA

Func. class.: Corticosteroid, long-acting/β_2-adrenergic agonist

ACTION:
Decreased inflammation in inhibiting mast cells, macrophages and leukotrienes; anti-inflammatory and vasoconstrictor properties relaxes bronchial smooth muscles

USES:
Maintenance of asthma (long term), COPD

CONTRAINDICATIONS:
Hypersensitivity, acute asthma/COPD episodes, severe hypersensitivity to milk proteins

Black Box Warning: Asthma-related deaths

Precautions: Pregnancy (C), breastfeeding, active infections, diabetes mellitus, glaucoma, immunosuppression, hyperthyroidism, Cushing syndrome, hypertension, QT prolongation, pheochromocytoma, seizures, MAOIs, other long-acting B_2 agonists, inhaled corticosteroid

DOSAGE AND ROUTES

Asthma maintenance

• **Adult/adolescent ≥12 yr: INH** 1 inhalation of Advair diskus q12hr, or 2 inhalations of Advair HFA q12hr

• **Child 4-11 yr: INH** 1 inhalation of fluticasone 100 mcg/salmeterol 50 mcg (Advair Diskus) q12hr, must be 12 hr apart

COPD

• **Adult: INH** 1 inhalation of Advair 250/50 diskus q12hr, must be 12 hr apart

Available forms: Inhalation 100/50, 250/50, 500/50 mcg fluticasone/salmeterol; aerosol spray 45/21, 115/21, 230/21 mcg fluticasone/salmeterol

Administer:

Oral inhalation route

Powder for oral inhalation (Diskus):

• Most children <4 years of age do not generate sufficient inspiratory flow to activate dry powder inhalers

• Give with the Diskus device: Ask the patient to open and prepare mouthpiece, slide device lever to activate the first dose, do not advance the lever >1 time; holding the Diskus mouthpiece level to, but away from the mouth, exhale; then put the mouthpiece to the lips and breathe in the dose deeply and slowly; remove the Diskus from the mouth, hold breath for at least 10 sec, and then exhale slowly; close the Diskus, which also resets the dose lever for the next scheduled dose

- Mouth should be rinsed, spit out water
- Discard device after 1 mo or when counter reads 0 (whichever comes first)

HFA aerosol:

- Shake canister; prime the inhaler before first use with 4 test sprays away from face or with 2 test sprays (away from the face) if it has not been used for more than 4 wk, or after dropping; use spacer if unable to coordinate inhalation/activation
- Rinse mouth with water after use, spit out water; clean inhaler mouthpiece at least every day; discard inhaler after 120 sprays or when the counter reads 000

SIDE EFFECTS

CNS: Fever, headache, nervousness, dizziness, migraines, insomnia, tremors, agitation, anxiety, depression, hyperactivity, irritability

EENT: Pharyngitis, sinusitis, rhinitis, laryngitis, hoarseness, dry eyes, cataracts, nasal discharge, epistaxis, hypersalivation, eye edema dysphonia, conjunctivitis

GI: Diarrhea, abdominal pain, nausea, vomiting, oral candidiasis

GU: UTI

INTEG: Urticaria, dermatitis

META: Hyperglycemia, growth retardation in children, cushingoid features

MISC: Influenza, eosinophilic conditions, angioedema, Churg–Strauss syndrome, anaphylaxis, adrenal insufficiency (high doses), reduced bone mineral density, HPA axis suppression

MS: Osteoporosis, muscle soreness, joint pain, decreased growth velocity

RESP: Upper respiratory infection, dyspnea, cough, bronchitis, bronchospasm

PHARMACOKINETICS

Fluticasone: half-life 8 hr, peak 1-2 hr; Salmeterol: half-life 5.5 hr, peak 5 min

INTERACTIONS

Increase: CNS stimulation—theophylline

Increase: Fluticasone levels—CYP3A4 inhibitors (ketoconazole, itraconazole), darunavir, nelfinavir, ritonavir, amprenavir, fosamprenavir, atazanavir, delavirdine, saquinavir, MAOIs, linezolid

Increase: Tendinitis, tendon rupture—quinolones

Increase: Hypokalemia—loop diuretics, thiazides, theophylline

Drug/Lab Test

Increase: LFTs

Decrease: Potassium

NURSING CONSIDERATIONS

Assess:

Respiratory status: Lung status, pulmonary function tests during, for several mo after change from systemic to inhalation corticosteroid

- Withdrawal symptoms from oral corticosteroids: depression, pain in joints, fatigue

Black Box Warning: **Adrenal Insufficiency:** nausea, weakness, fatigue, hypotension, hypoglycemia, anorexia; can occur when changing from systemic to inhalation corticosteroids; may be life threatening; adrenal function tests periodically: hypothalamic–pituitary–adrenal axis suppression in long-term treatment

- Growth rate in children; blood glucose, serum potassium for all patients

Evaluate:

- Therapeutic response: decreased severity of asthma

Teach patient/family:

- To use bronchodilator first, before using inhalation, if taking both; may use spacer or valved chamber
- Not to use for acute asthmatic attack; acute asthma might require oral corticosteroids; may use short-acting B_2 agonists for rescue
- To avoid smoking, smoke-filled rooms, those with URIs, those not immunized against chickenpox or measles

F

• To rinse mouth after inhaled product to reduce the risk of oral candidiasis; not to swallow

fluvastatin (Rx)

(flu′vah-stay-tin)

Lescol, Lescol XL

Func. class.: Antilipemic

Chem. class.: HMG-CoA reductase inhibitor

Do not confuse:
fluvastatin/FLUoxetine

ACTION:
Inhibits HMG-CoA reductase enzyme, which reduces cholesterol synthesis

USES:
As an adjunct for primary hypercholesterolemia (types Ia, Ib), coronary atherosclerosis in CAD; to reduce the risk for secondary prevention of coronary events in patients with CAD; as an adjunct to diet to reduce LDL, total cholesterol, apo B levels in heterozygous familial hyper-cholesterolemia (LDL-C ≥190 mg/dl) or LDL-C ≥160 mg/dl with history of premature CV disease

CONTRAINDICATIONS:
Pregnancy (X), breastfeeding, hypersensitivity, active hepatic disease

Precautions: Previous hepatic disease, alcoholism, severe acute infections, trauma, hypotension, uncontrolled seizure disorders, severe metabolic disorders, electrolyte imbalance, myopathy, rhabdomyolysis

DOSAGE AND ROUTES

• **Adult: PO** 20-40 mg/day in PM initially, usual range 20-80 mg, max 80 mg; may be given in 2 doses (40 mg AM, 40 mg PM); dosage adjustments may be made at ≥4-wk intervals

Heterozygous familial hypercholesterolemia

• **Adolescent ≥1 yr postmenarche (10-16 yr): PO** 20 mg daily at bedtime, may increase q6wk, max 40 mg bid (cap) or 80 mg (ext rel)

Available forms: Caps 20, 40 mg; ext rel tab 80 mg

Administer:

• Do not break, crush, or chew ext rel tabs, use at any time of day (tab), in the evening (cap)
• Bile acid sequestrant should be given at least 2 hr before fluvastatin
• Give without regard to food
• Store at room temperature, protected from light

SIDE EFFECTS

CNS: Headache, dizziness, insomnia, confusion

EENT: Lens opacities

GI: *Abdominal pain, cramps, nausea, constipation, diarrhea, dyspepsia, flatus,* hepatic dysfunction, pancreatitis

HEMA: Thrombocytopenia, hemolytic anemia, leukopenia

INTEG: Rash, pruritus

MISC: Fatigue, influenza, photosensitivity

MS: Myalgia, myositis, rhabdomyolysis, *arthritis, arthralgia*

PHARMACOKINETICS

Peak response 3-4 wk, metabolized in liver, >98% protein bound, excreted primarily in feces, enters breast milk, half-life 1.9 hr, steady state 4-5 wk

INTERACTIONS

Increase: effects of warfarin, digoxin, phenytoin

Increase: myopathy—cycloSPORINE, niacin, colchicine, protease inhibitors, fibric acid derivatives

Increase: effects of fluvastatin—alcohol, cimetidine, ranitidine, omeprazole, phenytoin

Decrease: fluvastatin effect—cholestyramine, colestipol, separate by ≥4 hr

Drug/Herb

Increase: adverse reactions—red yeast rice

Drug/Food

• Grapefruit juice: possible increased toxicity

Drug/Lab Test

Increase: LFTs, CK

Decrease: platelets, WBC

NURSING CONSIDERATIONS

Assess:

• **Hypercholesterolemia:** diet history: fats, fasting lipid profile (cholesterol, LDL, HDL, TG) before and q4-6wk, then q3-6mo when stable

• **Hepatotoxicity/pancreatitis:** monitor hepatic studies before, q12wk after dosage change, then q6mo; AST, ALT, LFTs may be increased

• Renal studies in patients with compromised renal system: BUN, I&O ratio, creatinine

⚠ **Myopathy, rhabdomyolysis:** muscle pain, tenderness; obtain baseline CPK if elevated; if these occur, product should be discontinued

Evaluate:

• Therapeutic response: decrease in sLDL, VLDL, total cholesterol; increased HDL, decreased triglycerides, slowing of CAD

Teach patient/family:

• That blood work will be necessary during treatment; to take product as prescribed; that effect may take ≥4 wk

• To report severe GI symptoms, headache, muscle pain, weakness, tenderness

• That previously prescribed regimen will continue: low-cholesterol diet, exercise program, smoking cessation

• To report suspected pregnancy; not to use during pregnancy (X), breastfeeding

• To take without regard to meals; take immediate release product in the evening; separate by ≥4 hr from bile-acid product

fluvoxaMINE (Rx)

(flu-vox′a-meen)

Luvox CR

Func. class.: Antidepressant SSRI (selective serotonin reuptake inhibitor)

Do not confuse:

Luvox/Levoxyl/Lasix/Lovenox

FluvoxaMINE/FLUoxetine/FluPHENAZine

ACTION: Inhibits CNS neuron uptake of serotonin but not of norepinephrine

USES: Obsessive-compulsive disorder, social phobia

Unlabeled uses: Depression, bulimia nervosa, panic disorder, autism, anxiety, posttraumatic stress disorder (PTSD), premenstrual dysphoric disorder (PMDD)

CONTRAINDICATIONS: Hypersensitivity

Precautions: Pregnancy (C), breastfeeding, geriatric patients, hepatic/cardiac disease, abrupt discontinuation, dehydration, ECT, hyponatremia, hypovolemia, bipolar disorder, seizure disorder

Black Box Warning: Children <8 yr, suicidal ideation

DOSAGE AND ROUTES

Obsessive-compulsive disorder (OCD)

• **Adult: PO** 50 mg at bedtime, increase by 50 mg at 4-7 day intervals, max 300 mg; doses over 100 mg should be divided; **EXT REL** 100 mg at bedtime, may titrate upward by 50 mg/wk, max 300 mg/day

• **Child 12-17 yr: PO** 25 mg at bedtime, increase by 25 mg/day q4-7days, max 300 mg/day; doses over 50 mg should be divided

• **Child 8-11 yr: PO** 25 mg/day at bedtime, may increase q4-7days, max 200 mg/day

Social anxiety disorder
• **Adult: PO EXT REL CAP** (Luvox CR) 100 mg at bedtime initially, titrate as needed by 50 mg/wk to 100-300 mg/day; **PO** 50 mg at bedtime, titrate as needed by 50 mg q4-7days to 50-300 mg/day
• **Child/adolescent 12-17 yr: PO** 25 mg at bedtime, titrate by 25-50 mg q4-7days, max 300 mg/day; if total daily dose >50 mg, divide equally
Hepatic dose/geriatric
• **Adult: PO** 25 mg at bedtime, may titrate upward slowly
Bulimia nervosa, depression (unlabeled)
• **Adult: PO** 50 mg at bedtime × 4-7 days, titrate by 25-50 mg/dose q4-7days as needed
Premenstrual dysphoric disorder (unlabeled)
• **Adult: PO** 50 mg/day, may titrate to 100 mg/day
Posttraumatic stress disorder (PTSD) (unlabeled)
• **Adult: PO** 25-50 mg at bedtime × 4-7 days then titrate by 25-50 mg/dose q4-7days, range 25-300 mg/day single or divided dose × 3-12 wk
Available forms: Tabs 25, 50, 100 mg; ext rel cap 100, 150 mg
Administer:
• With food, milk for GI symptoms
• Store at room temperature; do not freeze
• **Immediate release:** give at bedtime; doses >100 mg/day (or >50 mg/day in those aged 8-17 yr) in 2 divided doses; if doses are not equal, give larger dose at bedtime
• **Ext rel:** give at bedtime; do not break, crush, chew ext rel product

SIDE EFFECTS

CNS: *Headache, drowsiness, dizziness, seizures,* sleep disorders, insomnia, suicidal ideation (children/adolescents), neuroleptic malignant syndrome–like reactions, *weakness,* neuroleptic malignant syndrome, tremors
CV: Palpitation, chest pain, syncope, nervousness, agitation
GI: *Nausea, anorexia, constipation,* hepatotoxicity, *vomiting, diarrhea,* dry mouth, altered taste
GU: *Decreased libido,* anorgasmia, urinary frequency priapism
INTEG: *Rash, sweating*
MS: Myalgia
SYST: Neonatal abstinence syndrome

PHARMACOKINETICS

Crosses blood-brain barrier, 77% protein binding, metabolism by the liver, terminal half-life 15.6 hr, peak 2-8 hr

INTERACTIONS

⚠ Fatal reaction—MAOIs
Increase: CNS depression—alcohol, barbiturates, benzodiazepines
Increase: effect of—ramelteon, thioridazine; do not use together
⚠ **Increase:** QT prolongation, death—pimozide, do not use together
Increase: fluvoxaMINE, toxicity levels—tricyclics, cloZAPine, alosetron, tiZANidine; do not use together
Increase: metabolism, decrease effects—smoking
Increase: serotonin syndrome, neuroleptic malignant syndrome: SSRIs, SNRIs, serotonin-receptor agonists, atypical antipsychotics
Increase: bleeding risk—anticoagulants, NSAIDs, salicylates, thrombolytics
• Avoid use with clopidogrel
Decrease: metabolism, increase action of propranolol, diazepam, lithium, theophylline, carBAMazepine, warfarin
Drug/Herb
Increase: CNS effect—kava, valerian
Increase: serotonin syndrome—tryptophan, St. John's wort; do not use together

NURSING CONSIDERATIONS

Assess:
• Hepatic studies: AST, ALT, bilirubin
• Mental status: mood, sensorium, affect, **suicidal tendencies;** increase in psychiatric symptoms: depression, panic, obsessive-compulsive symptoms

- Constipation; most likely in geriatric patients
- Growth rate (children), bone density (post-menopausal females), glucose (diabetes)

⚠ For toxicity: nausea, vomiting, diarrhea, syncope, increased pulse, seizures

Evaluate:

- Therapeutic response: decrease in depression

Teach patient/family:

- That therapeutic effects may take 2-3 wk; not to discontinue abruptly
- To use caution when driving, performing other activities requiring alertness because drowsiness, dizziness may occur
- Not to use other CNS depressants, alcohol, barbiturates, benzodiazepines, St. John's wort, kava
- To notify prescriber if pregnancy is suspected, planned
- To notify prescriber of allergic reaction
- To increase bulk in diet if constipation occurs, especially in geriatric patients

Black Box Warning: That suicidal thoughts/behaviors may occur

- To stop taking MAOIs at least 14 days before starting product

TREATMENT OF OVERDOSE:

Activated charcoal, gastric lavage

folic acid (vit B_9) (OTC)

(foe′lik a′sid)

Equaline Folic Acid, Folacin, Vitamin B_9, Walgreens Gold Seal Folic Acid

Func. class.: Vit B complex group, water-soluble vitamin

ACTION: Needed for erythropoiesis; increases RBC, WBC, platelet formation with megaloblastic anemias

USES: Megaloblastic or macrocytic anemia caused by folic acid deficiency; hepatic disease, alcoholism, hemolysis, intestinal obstruction, pregnancy to reduce risk for neural tube defects

Unlabeled uses: Reduce risk for heart disease, stroke, methotrexate toxicity prophylaxis

CONTRAINDICATIONS: Hypersensitivity

Precautions: Pregnancy (A), anemias other than megaloblastic/macrocytic anemia, vit B_{12} deficiency anemia, uncorrected pernicious anemia

DOSAGE AND ROUTES

RDA

- **Adult and child ≥14 yr: PO** 400 mcg
- **Adult (pregnant/lactating): PO** 600 mcg/day
- **Child 9-13 yr: PO** 300 mcg
- **Child 4-8 yr: PO** 200 mcg
- **Child 1-3 yr: PO** 150 mcg
- **Infant 6 mo-1 yr: PO** 80 mcg
- **Neonate/infant <6 mo: PO** 65 mcg

Megaloblastic/macrocytic anemia due to folic acid or nutritional deficiency

- **Pregnant/lactating: PO** 800-1000 mcg

Therapeutic dose

- **Adult and child: PO/IM/SUBCUT/IV** up to 1 mg/day

Maintenance dose

- **Adult and child >4 yr: PO/IM/SUBCUT/IV** 0.4 mg/day
- **Pregnant and lactating: PO/IM/SUBCUT/IV** 0.8-1 mg/day
- **Child <4 yr: PO/IM/SUBCUT/IV** up to 0.3 mg/day
- **Infant: PO/IM/SUBCUT/IV** up to 0.1 mg/day

Prevention of neural tube defects during pregnancy

- **Adult: PO** 0.6 mg/day

Prevention of megaloblastic anemia during pregnancy

- **Adult: PO/IM/SUBCUT** up to 1 mg/day during pregnancy

Tropical sprue

- **Adult: PO** 3-15 mg/day

Available forms: Tabs 0.1, 0.4, 0.8, 1, 5 mg; inj 5, 10 mg/ml

Administer:

SUBCUT route

- Do not inject intradermally

IM route

- Inject deeply in large muscle mass, aspirate

Direct IV route

- Direct undiluted ≤5 mg/1 min or more

Continuous IV INF route

- May be added to most IV sol or TPN

Y-site compatibilities: Alfentanil, aminophylline, ascorbic acid injection, atracurium, atropine, azaTHIOprine, aztreonam, benztropine, bumetanide, calcium gluconate, cefamandole, ceFAZolin, cefmetazole, cefonicid, cefoperazone, cefotaxime, cefoTEtan, cefOXitin, cefTAZidime, ceftizoxime, cefTRIAXone, cefuroxime, cephalothin, cephapirin, chloramphenicol, cimetidine, clindamycin, cyanocobalamin, cycloSPORINE, dexamethasone, digoxin, diphenhydrAMINE, DOPamine, enalaprilat, ePHEDrine, EPINEPHrine, epoetin alfa, erythromycin, esmolol, famotidine, fentaNYL, fluconazole, furosemide, ganciclovir, glycopyrrolate, heparin, hydrocortisone, hydrOXYzine, imipenem-cilastatin, indomethacin, insulin (regular), ketorolac, labetalol, lidocaine, LR, magnesium sulfate, mannitol, meperidine, methicillin, methylPREDNISolone, metoclopramide, metoprolol, mezlocillin, midazolam, moxalactam, multiple vitamins injection, naloxone, nitroglycerin, nitroprusside, ondansetron, oxacillin, oxytocin, penicillin G potassium/sodium, PENTobarbital, PHENobarbital, phenylephrine, phytonadione, piperacillin, potassium chloride, procainamide, propranolol, ranitidine, Ringer's, ritodrine, sodium bicarbonate, succinylcholine, SUFentanil, theophylline, ticarcillin, ticarcillin-clavulanate, TPN, trimetaphan, urokinase, vancomycin, vasopressin

SIDE EFFECTS

CNS: Confusion, depression, excitability, irritability

GI: Anorexia, nausea

INTEG: Pruritus, rash, erythema

RESP: Bronchospasm

SYST: Anaphylaxis (rare)

PHARMACOKINETICS

PO: Peak ½-1 hr, bound to plasma proteins, excreted in breast milk, metabolized by liver, excreted in urine (small amounts)

INTERACTIONS

Increase: need for folic acid—estrogen, hydantoins, carBAMazepine, glucocorticoids

Decrease: folate levels—methotrexate, sulfonamides, sulfaSALAzine, trimethoprim

Decrease: phenytoin levels, fosphenytoin, may increase seizures

NURSING CONSIDERATIONS

Assess:

- **Megaloblastic anemia:** fatigue, dyspnea, weakness
- Hgb, Hct, reticulocyte count
- Nutritional status: bran, yeast, dried beans, nuts, fruits, fresh vegetables, asparagus
- Products currently taken: estrogen, carBAMazepine, glucocorticoids, hydantoins; these products may cause increased folic acid use by body and contribute to a deficiency if taking other neurotoxic products

Evaluate:

- Therapeutic response: increased weight, oriented, well-being; absence of fatigue; increase in reticulocyte count within 5 days of beginning treatment, absence of neural tube defect

Teach patient/family:

- To take product exactly as prescribed; that periodic lab work is required
- To alter nutrition to include high–folic-acid foods: organ meats, vegetables, fruit
- That urine will turn bright yellow
- To notify prescriber of allergic reaction
- To avoid breastfeeding

⚠ HIGH ALERT

fondaparinux (Rx)

(fon-dah-pair′ih-nux)

Arixtra

Func. class.: Anticoagulant, antithrombotic

Chem. class.: Synthetic, selective factor Xa inhibitor

Do not confuse:
Arixtra/Anti-Xa

ACTION: Inhibits factor Xa; neutralization of factor Xa interrupts blood coagulation and thrombin formation

USES: Prevention/treatment of deep venous thrombosis, PE in hip and knee replacement, hip fracture or abdominal surgery

Unlabeled uses: Acute coronary syndrome, acute MI, unstable angina

CONTRAINDICATIONS: Hypersensitivity to this product; hemophilia, leukemia with bleeding, peptic ulcer disease, hemorrhagic stroke, surgery, thrombocytopenic purpura, weight <50 kg, severe renal disease (CCr <30 ml/min), active major bleeding, bacterial endocarditis

Precautions: Pregnancy (B), breastfeeding, children, geriatric patients, alcoholism, hepatic disease (severe), blood dyscrasias, heparin-induced thrombocytopenia, uncontrolled severe hypertension, acute nephritis, mild to moderate renal disease

Black Box Warning: Spinal/epidural anesthesia, lumbar puncture

DOSAGE AND ROUTES

Deep venous thrombosis/PE

- **Adult <50 kg: SUBCUT** 5 mg/day × ≥5 days until INR 2-3; give warfarin within 72 hr of fondaparinux
- **Adult 50-100 kg: SUBCUT** 7.5 mg/day × ≥5 days until INR 2-3; give warfarin within 72 hr of fondaparinux
- **Adult >100 kg: SUBCUT** 10 mg/day × ≥5 days until INR 2-3; give warfarin within 72 hr of fondaparinux

Prevention of deep venous thrombosis

- **Adult: SUBCUT** 2.5 mg/day given 6 hr after surgery; (hemostasis established) continue for 5-9 days; for hip surgery, up to 32 days; for abdominal surgery, up to 24 days

Coronary artery thrombosis prophylaxis/acute coronary syndrome (unlabeled)

- **Adult: SUBCUT** 2.5 mg until hospital discharge or ≤8 days with standard treatment

Renal disease

- **Adult: SUBCUT** CCr 30-50 ml/min, use cautiously; CCr <30 ml/min, do not use

Available forms: Inj 2.5 mg/0.5 ml, 5 mg/0.4 ml, 7.5 mg/0.6 ml, 10 mg/0.8 ml prefilled syringes

Administer:

- Alone; do not mix with other products or solutions; cannot be used interchangeably (unit to unit) with other anticoagulants
- Only after screening patient for bleeding disorders

SUBCUT route

- SUBCUT only; do not give IM; do not give <6 hr after surgery
- Check for discolored sol or sol with particulate; if present, do not give
- Administer 6-8 hr after surgery; administer to recumbent patient, rotate inj sites (left/right anterolateral, left/right posterolateral abdominal wall)
- Wipe surface of inj site with alcohol swab, twist plunger cap and remove, remove rigid needle guard by pulling straight off needle; do not aspirate, do not expel air bubble from surface
- Insert whole length of needle into skinfold held with thumb and forefinger
- When product is injected, a soft click may be felt or heard

- Give at same time each day to maintain steady blood levels; observe inj site
- Avoid all IM inj that may cause bleeding

⚠ Administer only this product when ordered; not interchangeable with heparin

- Store at 77° F (25° C); do not freeze

SIDE EFFECTS

CNS: *Fever*, confusion, headache, dizziness, *insomnia*
GI: *Nausea, vomiting*, diarrhea, dyspepsia, *constipation*, increased AST, ALT
GU: UTI, urinary retention
HEMA: *Anemia*, minor bleeding, purpura, hematoma, thrombocytopenia, major bleeding (intracranial, cerebral, retroperitoneal hemorrhage), postoperative hemorrhage, heparin-induced thrombocytopenia
INTEG: Increased wound drainage, bullous eruption, local reaction—*rash*, pruritus, inj site bleeding
META: Hypokalemia
OTHER: Hypotension, pain, *edema*

PHARMACOKINETICS

Rapidly, completely absorbed; peak steady state 3 hr; distributed primarily in blood; does not bind to plasma proteins except 94% to ATIII; eliminated unchanged in urine within 72 hr with normal renal function; terminal half-life 17-21 hr

INTERACTIONS

Increase: bleeding risk—salicylates, NSAIDs, abciximab, eptifibatide, tirofiban, clopidogrel, dipyridamole, quiNIDine, valproic acid
Drug/Herb
Increase: bleeding risk—feverfew, garlic, ginger, ginkgo, ginseng, green tea, horse chestnut, kava

NURSING CONSIDERATIONS

Assess:

Black Box Warning: Monitor patients who have received epidural/spinal anesthesia or lumbar puncture for neurological impairment

- Blood studies (CBC, anti-Xa, Hgb/Hct, prothrombin time, platelets, occult blood in stools), thrombocytopenia may occur; if platelets $<100,000/mm^3$, treatment should be discontinued; renal studies: BUN, creatinine
- For bleeding: gums, petechiae, ecchymosis, black tarry stools, hematuria; decreased Hct, notify prescriber
- For risk of hemorrhage if coadministering with other products that may cause bleeding
- For hypersensitivity: rash, fever, chills; notify prescriber

Evaluate:

- Therapeutic response: prevention of DVT

Teach patient/family:

- To use soft-bristle toothbrush to avoid bleeding gums; to use electric razor
- To report any signs of bleeding: gums, under skin, urine, stools
- To avoid OTC products containing aspirin, NSAIDs

formoterol (Rx)

(for-moh′ter-ahl)

Foradil Aerolizer, Oxeze ✱, Perforomist

Func. class.: Bronchodilator
Chem. class.: β-Adrenergic agonist

Do not confuse:
Foradil/Toradol

ACTION: Has β_1 and β_2 action; relaxes bronchial smooth muscle and dilates the trachea and main bronchi by increasing levels of cAMP, which relaxes smooth muscles; causes increased contractility and heart rate by acting on β-receptors in heart

USES: Maintenance, treatment of asthma, COPD; prevention of exercise-induced bronchospasm

CONTRAINDICATIONS: Hypersensitivity to sympathomimetics, monotherapy for asthma, COPD, status asthmaticus

Precautions: Pregnancy (C), geriatric patients, cardiac disorders, hyperthyroidism, diabetes mellitus, prostatic hypertrophy, hypertension, African descendants, aneurysm

Black Box Warning: Asthma-related death

DOSAGE AND ROUTES

Maintenance, treatment of asthma

- **Adult and child ≥5 yr: INH** AM and PM long-term 1 cap (12 mcg) q12hr using aerolizer inhaler

Maintenance of COPD

- **Adult: INH** 12 mcg q12hr

Prevention of exercise-induced bronchospasm

- **Adult and child ≥12 yr: INH** prn occasionally 1 cap (12 mcg) ≥15 min before exercise, do not use additional doses for ≥12 hr

Available form: INH powder in cap 12 mcg; nebulizer sol for INH 20 mcg/2 ml

Administer:

Inhalation route

- Place cap in aerolizer inhaler; cap is punctured; do not wash aerolizer inhaler
- Pull off cover, twist mouthpiece to open, push buttons in; make sure the 4 pins are visible; remove cap from blister pack, place cap in chamber; twist to close, press (a click will be heard), release; patient should exhale, place inhaler in mouth, inhale rapidly
- Store at room temperature; protection from heat, moisture

SIDE EFFECTS

CNS: *Tremors*, *anxiety*, insomnia, headache, dizziness, stimulation

CV: Palpitations, tachycardia, hypertension, chest pain

GI: Nausea, vomiting, xerostomia

RESP: Bronchial irritation, dryness of oropharynx, **bronchospasms** (overuse), infection, inflammatory reaction (child)

PHARMACOKINETICS

Onset 15 min; peak 1-3 hr; duration 12 hr; metabolized in liver, lungs, GI tract; half-life 10 hr

INTERACTIONS

⚠ **Increase: serious dysrhythmias—MAOIs, tricyclics**

Increase: hypokalemia—loop/thiazide diuretics

Increase: effects of both products—other sympathomimetics, thyroid hormones

Increase: QT prolongation—class IA/III antiarrhythmics, phenothiazines, pimozide, haloperidol, risperiDONE, sertindole, ziprasidone, amoxapine, arsenic trioxide, chloroquine, clarithromycin, dasatinib, dolasetron, droperidol, erythromycin, halofantrine, halogenated anesthetics, levomethadyl, maprotiline, methadone, some quinolones, ondansetron, paliperidone, palonosetron, pentamidine, probucol, ranolazine, SUNItinib, tricyclics, vorinostat

Decrease: action when used with β-blockers

NURSING CONSIDERATIONS

Assess:

- Respiratory function: B/P, pulse, lung sounds; note sputum color, character; respiratory function tests before, during treatment
- Cardiac status: hypertension, palpitations, tachycardia
- For paresthesias, coldness of extremities; peripheral blood flow may decrease

Evaluate:

- Therapeutic response: ease of breathing

Teach patient/family:

Black Box Warning: Asthma-related death, severe asthma exacerbations; if wheezing worsens and cannot be relieved during an acute asthma attack, immediate medical attention should be sought

- To rinse mouth after use
- About correct use of inhaler/nebulizer (review package insert with patient); to avoid getting aerosol in eyes
- About all aspects of product; to avoid smoking, smoke-filled rooms, persons with respiratory infections; not to swallow caps

TREATMENT OF OVERDOSE:

Administration of β-blocker

fosamprenavir (Rx)

(fos-am-pren′a-veer)

Lexiva, Telzir ♣

Func. class.: Antiretroviral

Chem. class.: Protease inhibitor

ACTION: A prodrug of amprenavir; inhibits human immunodeficiency virus (HIV) protease, which prevents maturation of the infectious virus

USES: HIV-1 infection in combination with antiretrovirals

CONTRAINDICATIONS: Hypersensitivity to protease inhibitors

Precautions: Pregnancy (C), breastfeeding, geriatric patients, hepatic disease, hemolytic anemia, diabetes, sulfa sensitivity, autoimmune disease with immune reconstitution

DOSAGE AND ROUTES

Therapy-naive patients

- **Adult: PO** 1400 mg bid without ritonavir or fosamprenavir 1400 mg/day and with ritonavir 200 mg/day or fosamprenavir 700 mg bid and ritonavir 100 mg bid
- **Infant/child ≥4 wk, adolescents: PO** 30 mg/kg bid max 1400 mg/dose, without ritonavir

Protease-experienced patients (PI)

- **Adult: PO** 700 mg bid and ritonavir 100 mg bid
- **Child/adolescent ≥20 kg: PO** 18 mg/kg bid with ritonavir 3 mg/kg bid
- **Child/adolescent 15 kg to <20 kg: PO** 23 mg/kg bid with ritonavir 3 mg/kg bid
- **Child/adolescent 11 kg to <15 kg: PO** 30 mg/kg bid with ritonavir 3 mg/kg bid
- **Child/adolescent <11 kg: PO** 45 mg bid with ritonavir 7 mg bid
- **Infant ≥6 mo: 15 kg to <20 kg: PO** susp 23 mg/kg bid with ritonavir 3 mg/kg bid
- **Infant ≥6 mo: 11 kg to <15 kg: PO** susp 30 mg/kg bid with ritonavir 3 mg/kg bid
- **Infant ≥6 mo: <11 kg: PO** susp 45 mg/kg bid with ritonavir 7 mg/kg bid

Combination with efavirenz

- **Adult: PO** add another 100 mg/day of ritonavir for a total of 300 mg/day when all 3 products given

Hepatic dose

- **Adult: PO** (Child-Pugh 5-6) 700 mg bid without ritonavir (treatment-naive patients) or 700 mg bid with ritonavir 100 mg daily (treatment-naive or experienced patients); (Child-Pugh 7-9) 700 mg bid without ritonavir (treatment-naive patients) or 450 mg bid with ritonavir 100 mg daily (treatment-naive or experienced patients); (Child-Pugh 10-15) 350 mg bid without ritonavir (treatment-naive patients) or 300 mg bid with ritonavir 100 mg daily (treatment-naive or experienced patients)

Available forms: Tabs 700 mg (equivalent to 600 mg amprenavir); oral susp 50 mg/ml

Administer:

- **Tab:** without regard to food
- **Oral susp:** give without food (adult), with food (pediatric); if vomiting occurs within 30 min of dose, readminister; shake vigorously before dose, use calibrated device

SIDE EFFECTS

CNS: Headache, fatigue, depression, oral paresthesia

GI: *Nausea, diarrhea, vomiting, abdominal pain*

INTEG: Rash, pruritus

MISC: Redistribution or accumulation of body fat, hyperglycemia, **Stevens-Johnson syndrome**

PHARMACOKINETICS

Prodrug of amprenavir, peak 1½-4 hr; 90% protein binding; metabolized in liver by cytochrome P4503AY (CYP3A4); excretion of unchanged product minimal; half-life 7.7 hr

INTERACTIONS

Increase: treatment failure—ritonavir with boceprevir or ritonavir with telaprevir, not recommended

Increase: effect of—warfarin

⚠ **Serious life-threatening reactions: amiodarone, calcium channel blockers, lidocaine, pimozide, ergots, midazolam, triazolam, flecainide, lovastatin, simvastatin, propafenone**

Increase: effect of—ARIPiprazole, rifbutin, ketoconazole, itraconazole, sildenafil, vardenafil

Increase: toxicity—HMG-CoA reductase inhibitors

Decrease: dose of maraviroc with product if used with ritonavir; do not use maraviroc with unboosted fosamprenavir

Decrease: effect of oral contraceptives, methadone

Decrease: fosamprenavir levels—nevirapine, antacids, efavirenz, saquinavir, ranitidine, carBAMazepine, phenytoin, lopinavir/ritonavir, barbiturates, proton-pump inhibitors, H_2-receptor antagonists, dexamethasone; avoid concurrent use

Drug/Herb

- Avoid use with St. John's wort

Drug/Lab Test

Increase: serum glucose, AST, ALT, triglycerides

NURSING CONSIDERATIONS

Assess:

- **HIV:** monitor CD4 + T cell count, plasma HIV RNA, serum cholesterol, serum lipid profile
- Bowel pattern before, during treatment; monitor hydration
- Skin eruptions, rash, urticaria, itching; allergy to sulfonamides; cross-sensitivity may occur

⚠ **Stevens-Johnson syndrome; skin reactions, report immediately**

- **Immune reconstitution syndrome: may occur with combination antiretroviral therapy, autoimmune disease may also develop, up to months after treatment starts**

Teach patient/family:

- To avoid taking with other medications unless directed by provider
- That product does not cure but does manage symptoms; that product does not prevent transmission of HIV to others
- To use a nonhormonal form of birth control while taking this product
- That if dose is missed, to take as soon as remembered up to 1 hr before next dose; not to double dose
- Not to alter dose or stop therapy without talking to physician
- To advise physician if patient has sulfa allergy
- To report all medications, including herbal supplements, to physician
- That patients receiving phosphodiesterase type 5 inhibitors may be at increased risk for PDE5 inhibitor adverse effects

foscarnet (Rx)

(foss-kar′net)

Foscavir

Func. class.: Antiviral

Chem. class.: Inorganic pyrophosphate organic analog

ACTION: Antiviral activity is produced by selective inhibition at the pyrophosphate binding site on virus-specific DNA polymerases and reverse transcriptases at concs that do not affect cellular DNA polymerases

USES:
Treatment of CMV retinitis, HSV infections; used with ganciclovir for relapsing patients

CONTRAINDICATIONS:
Hypersensitivity, CCr <0.4 ml/min/kg

Precautions: Pregnancy (C), breastfeeding, children, geriatric patients, seizure disorders, severe anemia

Black Box Warning: Renal disease, electrolyte/mineral imbalances

DOSAGE AND ROUTES

Acyclovir-resistant mucocutaneous herpes simplex virus infection (herpes labialis, herpes febrilis, herpes genitalis) in HIV-infected patients

- **Adult/adolescent (unlabeled):** IV 40 mg/kg every 8-12 hr × 2-3 wk or until lesions are healed

Cytomegalovirus (CMV) encephalitis (unlabeled); CMV neurological disease (unlabeled) (including encephalitis) in HIV-infected patients

- **Adult/adolescent:** IV 90 mg/kg every 12 hr or 60 mg/kg every 8 hr × 3 wk (or until symptomatic improvement) with ganciclovir 5 mg/kg IV every 12 hr × 2-3 wk

Encephalitis (unlabeled) caused by human herpesvirus 6 (HHV-6) in immunocompromised patients

- **Adult:** IV 60 mg/kg every 8 hr or 90 mg/kg every 12 hr alone or in combination with ganciclovir 5 mg/kg IV every 12 hr

CMV retinitis or disseminated disease (unlabeled) in HIV-infected patients, recurrent or relapsed CMV retinitis in HIV-infected patients

- **Adult:** IV Induction with 90 mg/kg every 12 hr or 60 mg/kg IV every 8 hr × 14-21 days, depending upon the clinical response

Renal dose

- HSV induction dosage equivalent to 80 mg/kg/day (40 mg/kg IV every 12 hr)
- **Adult:** IV CCr >1.4 ml/min/kg: no change; CCr >1-1.4 ml/min/kg: decrease to 30 mg/kg every 12 hr; CCr >0.8-1 ml/min/kg: decrease to 20 mg/kg every 12 hr; CCr >0.6-0.8 ml/min/kg: decrease to 35 mg/kg every 24 hr; CCr >0.5-0.6 ml/min/kg: decrease to 25 mg/kg every 24 hr; CCr ≥0.4-0.5 ml/min/kg: decrease to 20 mg/kg every 24 hr; CCr <0.4 ml/min/kg: not recommended

HSV induction dosage equivalent to 120 mg/kg/day (40 mg/kg IV every 8 hr)

- **Adult:** IV CCr >1.4 ml/min/kg: no change; CCr >1-1.4 ml/min/kg: decrease to 30 mg/kg every 8 hr; CCr >0.8-1 ml/min/kg: decrease to 35 mg/kg every 12 hr; CCr >0.6-0.8 ml/min/kg: decrease to 25 mg/kg every 12 hr; CCr 0.5-0.6 ml/min/kg: decrease to 40 mg/kg every 24 hr; CCr ≥0.4-0.5 ml/min/kg: decrease to 35 mg/kg every 24 hr; CCr <0.4 ml/min/kg: not recommended

CMV induction dosage equivalent to 180 mg/kg/day (60 mg/kg every 8 hr)

- **Adult:** IV CCr >1.4 ml/min/kg: no change; CCr >1-1.4 ml/min/kg: decrease to 45 mg/kg every 8 hr; CCr >0.8-1 ml/min/kg: decrease to 50 mg/kg every 12 hr; CCr >0.6-0.8 ml/min/kg: decrease to 40 mg/kg every 12 hr; CCr >0.5-0.6 ml/min/kg: decrease to 60 mg/kg every 24 hr; CCr ≥0.4-0.5 ml/min/kg: decrease to 50 mg/kg every 24 hr; CCr <0.4 ml/min/kg: not recommended

CMV induction dosage equivalent to 180 mg/kg/day (90 mg/kg IV every 12 hr)

- **Adult:** IV CCr >1.4 ml/min/kg: no change; CCr >1-1.4 ml/min/kg: decrease to 70 mg/kg every 12 hr; CCr >0.8-1 ml/min/kg: decrease to 50 mg/

kg every 12 hr; CCr >0.6-0.8 ml/min/kg: decrease to 80 mg/kg every 24 hr; CCr >0.5-0.6 ml/min/kg: decrease to 60 mg/kg every 24 hr; CCr ≥0.4-0.5 ml/min/kg: decrease to 50 mg/kg every 24 hr; CCr <0.4 ml/min/kg: not recommended

Available forms: Inj 6000 mg/250 ml, 12,000 mg/500 ml (24 mg/ml)

Administer:

- Increased fluids before and during product administration to induce diuresis, minimize renal toxicity

Intermittent IV INF route

- Using inf device at no more than 1 mg/kg/min; do not give by rapid or bolus IV; give by CVP or peripheral vein; standard 24 mg/ml sol may be used without dilution if using by CVP; dilute the 24 mg/ml sol to 12 mg/ml with D_5W or NS if using peripheral vein
- Manufacturer recommends product not be given with other medications in syringe or admixed

SIDE EFFECTS

CNS: *Fever*, dizziness, *headache*, seizures, *fatigue*, neuropathy, asthenia, encephalopathy, malaise, meningitis, *paresthesia*, depression, *confusion*, *anxiety*

CV: ECG abnormalities, 1st-degree AV block, nonspecific ST-T segment changes, cerebrovascular disorder, cardiomyopathy, cardiac arrest, atrial fibrillation, CHF, sinus tachycardia

GI: *Nausea*, *vomiting*, *diarrhea*, *anorexia*, abdominal pain, pancreatitis

GU: Acute renal failure, decreased CCr, increased serum creatinine, azotemia, diabetes insipidus, renal tubular disorders

HEMA: *Anemia*, granulocytopenia, leukopenia, thrombocytopenia, thrombosis, neutropenia, lymphadenopathy

INTEG: *Rash*, sweating, pruritus, skin discoloration

RESP: *Coughing*, *dyspnea*, pneumonia, pulmonary infiltration, pneumothorax, hemoptysis

SYST: *Hypokalemia*, *hypocalcemia*, *hypomagnesemia*; hypophosphatemia

PHARMACOKINETICS

14%-17% protein bound, half-life 18-88 hr in normal renal function, 79%-92% excreted via kidneys

INTERACTIONS

Black Box Warning: **Increase:** nephrotoxicity—acyclovir, cidofovir, CISplatin, gold compounds, penicillamine, tacrolimus, tenofovir, vancomycin, aminoglycosides, amphotericin B, NSAIDs, lithium, cycloSPORINE

Increase: hypocalcemia—pentamidine

NURSING CONSIDERATIONS

Assess:

Black Box Warning: **Renal tubular disorders:** I&O ratio, urine pH, serum creatinine at baseline, 3×/wk during initial therapy then 2×/wk thereafter; CCr at baseline, throughout treatment; if CCr <0.4 ml/min/kg, discontinue

- Blood counts q2wk; watch for decreasing granulocytes, Hgb; if low, therapy may have to be discontinued and restarted after hematologic recovery; blood transfusions may be required
- Lesions in HSV
- Electrolytes and minerals (calcium, phosphate, magnesium, potassium); watch closely for tetany during 1st administration
- GI symptoms: nausea, vomiting, diarrhea; severe symptoms may necessitate discontinuing product

⚠ **Blood dyscrasias** (anemia, granulocytopenia): bruising, fatigue, bleeding, poor healing

- **Allergic reactions:** flushing, rash, urticaria, pruritus

CMV retinitis

- Culture should be performed before treatment (blood, urine, throat); a negative culture does not rule out CMV
- Ophthalmic exam should confirm diagnosis

• Close monitoring during therapy for tingling, numbness, paresthesias; if these occur, stop inf, obtain lab sample for electrolytes

Evaluate:

• Therapeutic response: improvement in CMV retinitis

Teach patient/family:

• To call prescriber if sore throat, swollen lymph nodes, malaise, fever occur, since other infections may occur
• To report perioral tingling, numbness in extremities, and paresthesias
• That serious product interactions may occur if OTC products are ingested; check first with prescriber
• That product is not a cure but will control symptoms

fosinopril (Rx)

(foss′in-oh-pril)

Monopril

Func. class.: Antihypertensive

Chem. class.: Angiotensin-converting enzyme (ACE) inhibitor

Do not confuse:
Monopril/minoxidil/Accupril/Monoket

ACTION: Selectively suppresses renin-angiotensin-aldosterone system; inhibits ACE; prevents conversion of angiotensin I to angiotensin II; results in dilation of arterial, venous vessels

USES: Hypertension, alone or in combination with thiazide diuretics, systolic CHF

Unlabeled uses: Proteinuria in nondiabetic nephropathy

CONTRAINDICATIONS: Breastfeeding, children, hypersensitivity to ACE inhibitors, history of ACE-inhibitor–induced angioedema

Black Box Warning: Pregnancy (D)

Precautions: Geriatric patients, impaired hepatic function, hypovolemia, blood dyscrasias, CHF, COPD, asthma, angioedema, hyperkalemia, renal artery stenosis, renal disease, aortic stenosis, autoimmune disorders, collagen vascular disease, febrile illness

DOSAGE AND ROUTES

CHF

• **Adult: PO** 5-10 mg/day then up to 40 mg/day increased over several wk; use lower dose for those diuresed before fosinopril

Hypertension

• **Adult: PO** 10 mg/day initially then 20-40 mg/day divided bid or daily, max 80 mg/day

Available forms: Tabs 10, 20, 40 mg

Administer:

• May be taken without regard to meals
• Store in tight container at ≤86° F (30° C)

SIDE EFFECTS

CNS: *Headache, dizziness,* fatigue, syncope

CV: *Hypotension,* orthostatic hypotension, tachycardia

GI: *Nausea,* constipation, *vomiting,* diarrhea, hepatotoxicity, cholestatic jaundice, fulminant hepatic necrosis, hepatic failure, death

GU: Increased BUN, creatinine, azotemia, renal artery stenosis

HEMA: Decreased Hct, Hgb; eosinophilia, leukopenia, neutropenia, agranulocytosis

META: *Hyperkalemia*

RESP: *Cough*

SYST: Anaphylaxis, angioedema

PHARMACOKINETICS

Peak 3 hr, serum protein binding 99%, half-life 11.5-14 hr, metabolized by liver (metabolites excreted in urine, feces)

INTERACTIONS

Increase: hyperkalemia risk—potassium-sparing diuretics, potassium supplements

Increase: hypotension—diuretics, other antihypertensives, ganglionic blockers, adrenergic blockers, nitrates, acute alcohol ingestion

Increase: toxicity—vasodilators, hydrALAZINE, prazosin, potassium-sparing diuretics, sympathomimetics, digoxin, lithium, NSAIDs

Decrease: absorption—antacids

Decrease: antihypertensive effect—salicylates

Drug/Herb

Increase: antihypertensive effect—hawthorn

Decrease: antihypertensive effect—ephedra

Drug/Lab Test

Increase: AST, ALT, alk phos, glucose, bilirubin, uric acid

False positive: urine acetone

Positive: ANA titer

NURSING CONSIDERATIONS

Assess:

- **Hypertension:** B/P, orthostatic hypotension, syncope
- **Collagen vascular disease:** neutrophils, decreased platelets; obtain WBC with differential baseline and monthly × 6 mo then q2-3mo × 1 yr; if neutrophils <1000/mm³, discontinue
- Renal studies: protein, BUN, creatinine; increased levels may indicate nephrotic syndrome
- Baselines of renal, hepatic studies before therapy begins
- Potassium levels, although hyperkalemia rarely occurs
- **CHF:** edema in feet, legs daily; weigh daily
- **Allergic reactions:** rash, fever, pruritus, urticaria; product should be discontinued if antihistamines fail to help
- Supine position for severe hypotension

Evaluate:

- Therapeutic response: decrease in B/P

Teach patient/family:

- Not to discontinue product abruptly; to take at same time of day
- Not to use OTC products (cough, cold, allergy) unless directed by prescriber; not to use salt substitutes containing potassium without consulting prescriber
- About the importance of complying with dosage schedule, even if feeling better
- To rise slowly to sitting or standing position to minimize orthostatic hypotension
- To notify prescriber of mouth sores, sore throat, fever, swelling of hands or feet, irregular heartbeat, chest pain, nonproductive cough
- To report excessive perspiration, dehydration, vomiting, diarrhea; may lead to fall in B/P
- That product may cause dizziness, fainting, lightheadedness during first few days of therapy
- That product may cause skin rash or impaired perspiration
- How to take B/P; normal readings for age group

Black Box Warning: To notify prescriber if pregnancy is planned or suspected, pregnancy (D)

TREATMENT OF OVERDOSE:

0.9% NaCl IV inf, hemodialysis

fosphenytoin (Rx)

(foss-fen′i-toy-in)

Func. class.: Anticonvulsant

Chem. class.: Hydantoin, phosphate phenytoin ester

ACTION: Inhibits spread of seizure activity in motor cortex by altering ion transport; increases AV conduction, prodrug of phenytoin

USES: Generalized tonic-clonic seizures, status epilepticus, partial seizures

CONTRAINDICATIONS: Pregnancy (D), hypersensitivity, bradycardia, SA and AV block, Stokes-Adams syndrome

Precautions: Breastfeeding, allergies, renal/hepatic disease, myocardial insufficiency, hypoalbuminemia, hypothyroidism, Asian patients positive for HLA-B 1502, abrupt discontinuation, agranulocytosis, alcoholism, carBAMazepine/barbiturate hypersensitivity, bone marrow suppression, CAD, geriatric patients, hemolytic anemia, hyponatremia, methemoglobinemia, myasthenia gravis, psychosis, suicidal ideation

Black Box Warning: Rapid IV inf

DOSAGE AND ROUTES

All doses in PE (phenytoin sodium equivalent)

Status epilepticus

- **Adult and adolescent: IV** 15-20 mg PE/kg
- **Child <12 yr (unlabeled): IV** 15-20 mg PE/kg

Nonemergency/maintenance dosing

- **Adult and adolescent >16 yr: IM/IV** 10-20 mg PE/kg; 4-6 mg PE/kg/day (maintenance); start maintenance 12 hr after loading dose; give in 2-3 divided doses

Available forms: Inj 150 mg (100 mg phenytoin equiv), 750 mg (500 mg phenytoin equiv), 50-mg/ml vials

Administer:

Injectable routes

- Give IM/IV; the dosage, conc, and inf rate of fosphenytoin should always be expressed, prescribed, and dispensed in phenytoin sodium equivalents (PE); exercise extreme caution when preparing and administering fosphenytoin; the conc and dosage should be carefully confirmed; fatal overdoses have occurred in children when the per ml conc of the product (50 mg PE/mL) was misinterpreted as the total amount of drug in the vial
- Visually inspect for particulate matter and discoloration before use

IV INF route

- Before inf, dilute in 5% dextrose or 0.9% saline solution to a conc ranging from 1.5 to 25 mg PE/ml
- Because of the risk of hypotension, do not exceed recommended inf rates; continuous monitoring of ECG, B/P, and respiratory function is recommended, especially throughout the period where phenytoin concs peak (about 10-20 min after the end of the inf)
- Loading doses should always be followed by maintenance doses of oral or parenteral phenytoin or parenteral fosphenytoin
- **Adult: IV** Infuse at a max rate of 150 mg PE/min
- **Elderly or debilitated adults: IV** Infuse at a max 3 mg PE/kg/min or 150 mg PE/min, whichever is less
- **Child: IV** Infuse at a rate of 0.5-3 mg PE/kg/min or max 150 mg PE/min, whichever is less
- **Infant/neonate: IV** Infuse at a rate max 0.5-3 mg PE/kg/min

Y-site compatibilities: Aminocaproic acid, amphotericin B lipid complex, amphotericin B liposome, anidulafungin, atenolol, bivalirudin, bleomycin, CARBOplatin, CISplatin, cyclophosphamide, cytarabine, DACTINomycin, DAPTOmycin, dexmedetomidine, diltiazem, DOCEtaxel, doxacurium, eptifibatide, ertapenem, etoposide, fludarabine, fluorouracil, gatifloxacin, gemcitabine, gemtuzumab, granisetron, ifosfamide, levofloxacin, linezolid, LORazepam, mechlorethamine, meperidine, methotrexate, metroNIDAZOLE, nesiritide, octreotide, oxaliplatin, oxytocin, PACLitaxel, palonosetron, pamidronate, pantoprazole, PEMEtrexed, PHENobarbital, piperacillin-tazobactam, rocuronium, sodium acetate, tacrolimus, teniposide, thiotepa, tigecycline, tirofiban, vinCRIStine, vinorelbine, voriconazole, zoledronic acid

SIDE EFFECTS

CNS: *Drowsiness,* dizziness, insomnia, paresthesias, depression, suicidal tendencies, aggression, headache, confusion, paresthesia, emotional liability, syncope, cerebral edema

CV: Hypo/hypertension, CHF, shock, dysrhythmias

EENT: Nystagmus, diplopia, blurred vision

GI: Nausea, vomiting, diarrhea, constipation, anorexia, weight loss, hepatitis, jaundice, gingival hyperplasia
HEMA: Agranulocytosis, leukopenia, aplastic anemia, thrombocytopenia, megaloblastic anemia
INTEG: Rash, lupus erythematosus, Stevens-Johnson syndrome, hirsutism, hypersensitivity, pruritus
RESP: Bronchospasm, cough
SYST: Hyperglycemia, hypokalemia, SJS/TEN Asian patients positive for HLA-B 1502; drug reaction with eosinophilia and systemic symptoms (DRESS), purple glove syndrome, anaphylaxis

PHARMACOKINETICS

Metabolized by liver, excreted by kidneys, protein binding 95%-99%, rapidly converted to phenytoin

INTERACTIONS

Increase: fosphenytoin level—cimetidine, amiodarone, chloramphenicol, estrogens, H_2 antagonists, phenothiazines, salicylates, sulfonamides, tricyclics, CYP1A2 inhibitors
Decrease: fosphenytoin effects—alcohol (chronic use), antihistamines, antacids, traMADol, antineoplastics, rifampin, folic acid, carBAMazepine, theophylline, CYP1A2 inducers
Decrease: virologic response, resistance—delavirdine, do not use concurrently

Drug/Herb

Increase: anticonvulsant effect—ginkgo
Decrease: anticonvulsant effect—ginseng, valerian

Drug/Lab Test

Increase: glucose, alk phos
Decrease: dexamethasone, metyrapone test serum, PBI, urinary steroids, potassium

NURSING CONSIDERATIONS

Assess:

• Drug level: target level 10-20 mcg/ml, toxic level 30-50 mcg/ml, wait >2 hr after dose before testing, 4 hr after IM dose
• Blood studies: CBC, platelets q2wk until stabilized, then monthly × 12 mo, then q3mo; discontinue product if neutrophils <1600/mm³; serum calcium, albumin, phosphorus, potassium
⚠ Mental status: mood, sensorium, affect, memory (long, short), **suicidal thoughts/behaviors**
• **Serious skin reactions:** usually occurring within 28 days of treatment; if a rash develops, patient should be evaluated for DRESS
• Seizure activity including type, location, duration, character; provide seizure precaution
• Renal studies: urinalysis, BUN, urine creatinine
• Hepatic studies: ALT, AST, bilirubin, creatinine
• Allergic reaction: red, raised rash; product should be discontinued
• **Toxicity/bone marrow depression:** nausea, vomiting, ataxia, diplopia, cardiovascular collapse, slurred speech, confusion
• Respiratory depression: rate, depth, character of respirations
⚠ **Blood dyscrasias:** fever, sore throat, bruising, rash, jaundice
• Continuous monitoring of ECG, B/P, respiratory function
• Rash; discontinue as soon as rash develops, serious adverse reactions such as **Stevens-Johnson syndrome** can occur

Evaluate:

• Therapeutic response: decrease in severity of seizures

Teach patient/family:

• About the reason for, expected outcomes of treatment
• Not to use machinery or engage in hazardous activity, since drowsiness, dizziness may occur
• To carry emergency ID denoting product use, name of prescriber
• To notify prescriber of rash, bleeding, bruising, slurred speech, jaundice of skin or eyes, joint pain, nausea, vomiting,

severe headaches, depression, suicidal thought
• To keep all medical appointments, including those for lab work, physical assessment
• To notify prescriber if pregnancy is planned, suspected
• To use contraception while using this product

⚠ HIGH ALERT

fospropofol (Rx)

(fos-proe′poe-fol)

Lusedra

Func. class.: General anesthetic

ACTION: Produces dose-dependent CNS depression by activation of GABA receptor

USES: Induction or maintenance of anesthesia as part of balanced anesthetic technique; sedation in mechanically ventilated patients

CONTRAINDICATIONS: Hypersensitivity to product

Precautions: Pregnancy (B), breastfeeding, children, geriatric patients, respiratory depression, severe respiratory disorders, cardiac dysrhythmias, labor/delivery, renal disease, hyperlipidemia

DOSAGE AND ROUTES

Induction/Maintenance

• **Adult <65 (healthy), >90 kg:** 577.5 **IV BOL,** may give supplemental doses up to max of 140 mg, give no more frequently than q4min; 61-89 kg 6.5 mg/kg (max 577.5 mg) **IV BOL** may give supplemental doses up to 1.6 mg/kg/dose (max 140 mg/dose; give no more frequently than q4min); <60 kg 385 mg **IV BOL,** may give supplemental doses up to max 105 mg/dose, give no more frequently than q4min

• **Adult: <65 yr (severe systemic disease), ≥90 kg** 437.5 mg **IV BOL,** give supplemental doses up to max 105 mg/dose, give no more frequently than q4min; 61-89 kg 4.875 mg/kg (max 437.5 mg) give supplemental doses of 75% of standard dose/up to 1.2 mg/kg/dose (max 105 mg/dose) give no more frequently than q4min; <60 kg 297.5 **IV BOL,** give supplemental doses up to max 70 mg/dose, give no more frequently than q4min

• **Geriatric ≥90 kg: IV BOL** 437.5 then supplemental doses up to max 105 mg; give no more frequently than q4min; 61-89 kg, **IV BOL** give 75% standard dosing regimen (4.875 mg/kg), give no more frequently than q4min (max 105 mg/dose)

Available forms: Inj 1050 mg/30 ml

Administer:

IV Bolus route

• Visually inspect for particulate matter and discoloration
• Each vial is single patient/single use
• Draw from vial, discard unused portion
• Do not mix with other drugs before use
• Give by IV BOL in free-flowing peripheral IV line of D_5W, 5% dextrose/0.2% NaCl, 5% dextrose/0.45% NaCl D_5LR, LR, 0.45% NaCl, NS, 5% dextrose/0.45% NaCl/20 mEq KCl; do not mix with other fluids, flush line with NS before and after administration
• No filtration needed
• Give only with resuscitative equipment available
• Only qualified persons trained in anesthesia should administer
• Store unopened vials at room temperature

SIDE EFFECTS

CNS: Involuntary movement, headache, jerking, fever, dizziness, shivering, tremor, confusion, somnolence, paresthesia, agitation, abnormal dreams, euphoria, fatigue, increased intracranial pressure, impaired cerebral flow, seizures

CV: *Bradycardia, hypotension,* hypertension, PVC, PAC, tachycardia, abnormal ECG, ST segment depression, **asystole, bradydysrhythmias**

EENT: Blurred vision, tinnitus, eye pain, strange taste, diplopia

GI: *Nausea, vomiting, abdominal cramping,* dry mouth, swallowing, hypersalivation, **pancreatitis**

GU: Urine retention, green urine, cloudy urine, oliguria

INTEG: *Flushing, phlebitis, hives, burning/stinging at inj site,* rash, pain of extremities

MS: Myalgia

RESP: **Apnea,** *cough, hiccups,* dyspnea, hypoventilation, sneezing, wheezing, tachypnea, hypoxia, respiratory acidosis

PHARMACOKINETICS

Onset 15-30 sec, rapid distribution, half-life 1-8 min, terminal half-life .81-.88 hr; 70% excreted in urine; metabolized in liver by conjugation to inactive metabolites, 95%-99% protein binding

INTERACTIONS

⚠ **Do not use within 10 days of MAOIs**

Increase: CNS depression—alcohol, opioids, sedative/hypnotics, antipsychotics, skeletal muscle relaxants, inhalational anesthetics

Drug/Herb

Increase: fospropofol effect—St. John's wort

Drug/Lab Test

Increase: LFTs, bilirubin

NURSING CONSIDERATIONS

Assess:

- Inj site: phlebitis, burning, stinging
- **ECG** for changes: PVC, PAC, ST segment changes; monitor VS
- CNS changes: movement, jerking, tremors, dizziness, LOC, pupil reaction
- Allergic reactions: hives
- **Respiratory depression,** character, rate, rhythm; notify prescriber if respirations are <10/min; hypoxemia detectable by pulse oximetry

Evaluate:

- Therapeutic response: induction of anesthesia

Teach patient/family:

- That this medication will cause dizziness, drowsiness, sedation

TREATMENT OF OVERDOSE:

Discontinue product; administer vasopressor agents or anticholinergics, artificial ventilation

F

frovatriptan (Rx)

(froh-vah-trip′tan)

Frova

Func. class.: Antimigraine agent

Chem. class.: 5-HT_1-Receptor agonist

ACTION: Binds selectively to the vascular 5-HT_{1B}, 5-HT_{1D} receptor subtypes; exerts antimigraine effect; binds to benzodiazepine receptor sites, causes vasoconstriction in cranium

USES: Acute treatment of migraine with/without aura

CONTRAINDICATIONS: Hypersensitivity, angina pectoris, history of MI, documented silent ischemia, Prinzmetal's angina, ischemic heart disease; concurrent ergotamine-containing preparations; uncontrolled hypertension; basilar or hemiplegic migraine; ischemic bowel disease; peripheral vascular disease, severe hepatic disease, prophylactic migraine treatment

Precautions: Pregnancy (C), breastfeeding, children, geriatric patients, postmenopausal women, men >40 yr, risk factors for CAD, hypercholesterolemia, obesity, diabetes, impaired hepatic function, seizure disorder

DOSAGE AND ROUTES

- **Adult: PO** 2.5 mg; a 2nd dose may be taken after ≥2 hr; max 3 tabs (7.5 mg/day)

Available form: Tabs 2.5 mg
Administer:
• Swallow tabs whole; do not break, crush, or chew
• With fluids
• 2 days/wk or less; rebound headache may occur

SIDE EFFECTS

CNS: *Hot/cold sensation*, paresthesia, *dizziness*, headache, fatigue, insomnia, anxiety, somnolence, seizures
CV: *Flushing*, chest pain, palpitation
GI: Dry mouth, dyspepsia, abdominal pain, diarrhea, vomiting, nausea
MS: Skeletal pain

PHARMACOKINETICS

Onset of pain relief 2-3 hr, terminal half-life 25-29 hr, protein binding 15%, metabolized liver by CYP1A2

INTERACTIONS

Increase: frovatriptan levels—CYP1A2 inhibitors (cimetidine, ciprofloxacin, erythromycin), estrogen, propranolol, oral contraceptives
Increase: toxicity—SSRIs, other serotonin agonists (dextromethorphan, traMADol, antidepressants)

NURSING CONSIDERATIONS

Assess:
• **Migraine symptoms:** aura, unable to view light; ingestion of tyramine-containing foods (pickled products, beer, wine, aged cheese), food additives, preservatives, colorings, artificial sweeteners, chocolate, caffeine, which may precipitate these types of headaches
• B/P; signs, symptoms of coronary vasospasms
• For stress level, activity, recreation, coping mechanisms
• Quiet, calm environment with decreased stimulation from noise, bright light, excessive talking
Evaluate:
• Therapeutic response: decrease in frequency, severity of migraine

Teach patient/family:
• To report any side effects to prescriber
• To use contraception while taking product; to inform prescriber if pregnant or planning to become pregnant
• To consult prescriber if breastfeeding

fulvestrant (Rx)

(full-vess'trant)
Faslodex
Func. class.: Antineoplastic
Chem. class.: Estrogen-receptor antagonist

ACTION:
Inhibits cell division by competitive binding to cytoplasmic estrogen receptors, downregulates estrogen receptors

USES:
Advanced breast carcinoma in estrogen-receptor–positive patients (usually postmenopausal)
Unlabeled uses: Loading dose for metastatic breast cancer

CONTRAINDICATIONS:
Pregnancy (D), breastfeeding, children, hypersensitivity
Precautions: Hepatic disease, jaundice, thrombocytopenia, biliary tract disease, coagulopathy

DOSAGE AND ROUTES

• **Adult: IM** 500 mg as 2, 5-ml injections on days 1, 15, 29 and monthly thereafter
Available forms: Inj 50 mg/ml
Administer:
IM route
• IM 5 ml give one inj in each buttock slowly
• Antiemetic 30-60 min before product to prevent vomiting prn
• Store in refrigerator; protect from light

SIDE EFFECTS

CNS: *Headache,* depression, dizziness, insomnia, paresthesia, anxiety

GI: *Nausea, vomiting,* anorexia, constipation, diarrhea, abdominal pain, **hepatitis, hepatic failure, hyperbilirubinemia**
HEMA: **Anemia**
INTEG: *Rash,* sweating, *hot flashes,* inj site pain
MS: Bone pain, arthritis, back pain
RESP: Pharyngitis, dyspnea, cough
SYST: **Angioedema**

PHARMACOKINETICS

Half-life 40 days, metabolized by CYP3A4, excretion in feces 90%

INTERACTIONS

Increase: Bleeding—anticoagulants, do not use concurrently
Drug/Lab Test
Increase: LFTs

NURSING CONSIDERATIONS

Assess:
- For anticoagulant use
- For side effects; report to prescriber

Evaluate:
- Therapeutic response: decreased tumor size, spread of malignancy

Teach patient/family:
- To report any complaints, side effects to prescriber
- To report vaginal bleeding immediately
- That tumor flare (increase in size of tumor, increased bone pain) may occur, will subside rapidly; that analgesics may be taken for pain
- That premenopausal women must use mechanical birth control because ovulation may be induced; not to breastfeed
- **To use contraception to prevent pregnancy; pregnancy category (D)**

furosemide (Rx)

(fur-oh′se-mide)
Lasix
Func. class.: Loop diuretic
Chem. class.: Sulfonamide derivative

Do not confuse:
furosemide/torsemide
Lasix/Luvox/Lomotil/Lanoxin/Losec

ACTION:
Inhibits reabsorption of sodium and chloride at proximal and distal tubule and in the loop of Henle

USES:
Pulmonary edema; edema with CHF, hepatic disease, nephrotic syndrome, ascites, hypertension
Unlabeled uses: Hypercalcemia with malignancy, hypertensive emergency/urgency, pulmonary edema or prevention of hemodynamic effects associated with blood product transfusion, ascites

CONTRAINDICATIONS:
Anuria
Precautions: Pregnancy (C), breastfeeding, diabetes mellitus, dehydration, severe renal disease, cirrhosis, ascites, hypersensitivity to sulfonamides/thiazides, infants, hypovolemia, electrolyte depletion, hypersensitivity

DOSAGE AND ROUTES

Edema
- **Adult: PO** 20-80 mg/day in AM; may give another dose after 6 hr up to 600 mg/day; **IM/IV** 20-40 mg; increase by 20 mg q2hr until desired response
- **Child: PO/IM/IV** 2 mg/kg; may increase by 1-2 mg/kg q6-8hr up to 6 mg/kg

Antihypercalcemia
- **Adult: IM/IV** 80-100 mg q1-4hr or **PO** 120 mg/day or divided bid
- **Child: IM/IV** 25-50 mg, repeat q4hr if needed

Acute/chronic renal failure
- **Adult: PO** 80 mg/day, increase by 80-120 mg/day to desired response; **IV** 100-200 mg, max 600-800 mg

Hypertensive emergency/urgency (unlabeled)
- **Adult: IV** 40-80 mg

Pulmonary edema/prevention of adverse hemodynamic effects associated with blood product transfusion (unlabeled)
- **Adult: IV** 40 mg injected slowly then 80 mg injected slowly after 2 hr if needed
- **Child: IM/IV** 1-2 mg/kg q6-12hr
- **Premature neonate >32 wk postconceptional age: IM/IV** 1-2 mg/kg q12-24hr
- **Premature neonate ≤32 wk postconceptional age: IM/IV** max 1 mg/kg, give no more frequently than q24hr

Available forms: Tabs 20, 40, 80 mg; oral sol 8 mg/ml, 10 mg/ml; inj 10 mg/ml

Administer:
- In AM to avoid interference with sleep if using product as diuretic
- Potassium replacement if potassium <3 mg/dl

PO route
- PO with food if nausea occurs; absorption may be decreased slightly; tabs may be crushed

IV route
- Undiluted; may be given through Y-tube or 3-way stopcock; give ≤20 mg/min

Intermittent IV Inf route
- May be added to NS or D_5W; if large doses required and given as IV inf, max 4 mg/min; use inf pump

Y-site compatibilities: Acyclovir, alfentanil, allopurinol, alprostadil, amifostine, amikacin, aminocaproic acid, aminophylline, amphotericin B cholesteryl/lipid complex/liposome, anidulafungin, argatroban, ascorbic acid, atenolol, atropine, azaTHIOprine, aztreonam, bivalirudin, bleomycin, bumetanide, calcium chloride/gluconate, CARBOplatin, cefamandole, ceFAZolin, cefepime, cefmetazole, cefonicid, cefoperazone, cefotaxime, cefoTEtan, cefOXitin, cefTAZidime, ceftizoxime, ceftobiprole, cefTRIAXone, cefuroxime, cephalothin, cephapirin, chloramphenicol, CISplatin, cladribine, clindamycin, cyanocobalamin, cyclophosphamide, cycloSPORINE, cytarabine, DACTINomycin, DAPTOmycin, dexamethasone, dexmedetomidine, digoxin, DOCEtaxel, doripenem, doxacurium, DOXOrubicin liposome, enalaprilat, ePHEDrine, EPINEPHrine, etoposide, fentaNYL, fludarabine, fluorouracil, folic acid, foscarnet, gallium nitrate, ganciclovir, granisetron, heparin, hydrocortisone, HYDROmorphone, ifosfamide, imipenem-cilastatin, indomethacin, insulin (regular), isosorbide, kanamycin, leucovorin, lidocaine, linezolid, LORazepam, LR, mannitol, mechlorethamine, melphalan, meropenem, methicillin, methotrexate, methylPREDNISolone, metoprolol, metroNIDAZOLE, mezlocillin, micafungin, miconazole, mitoMYcin, moxalactam, multiple vitamin injection, nafcillin, naloxone, nitroprusside, octreotide, oxacillin, oxaliplatin, oxytocin, PACLitaxel, palonosetron, pamidronate, pantoprazole, PEMEtrexed, penicillin G, PENTobarbital, PHENobarbital, phytonadione, piperacillin, piperacillin-tazobactam, potassium chloride, procainamide, propofol, propranolol, ranitidine, remifentanil, Ringer's, ritodrine, sargramostim, sodium acetate/bicarbonate, succinylcholine, SUFentanil, temocillin, teniposide, theophylline, thiopental, thiotepa, ticarcillin, ticarcillin-clavulanate, tigecycline, tirofiban, TNA, tobramycin, urokinase, vit B/C, voriconazole, zoledronic acid

SIDE EFFECTS

CNS: Headache, fatigue, weakness, vertigo, paresthesias

CV: Orthostatic hypotension, chest pain, ECG changes, circulatory collapse

EENT: Loss of hearing, ear pain, tinnitus, blurred vision

ELECT: *Hypokalemia, hypochloremic alkalosis, hypomagnesemia, hyperuricemia, hypocalcemia, hyponatremia,* metabolic alkalosis

ENDO: *Hyperglycemia*

GI: *Nausea,* diarrhea, dry mouth, vomiting, anorexia, cramps, oral, gastric irritations, pancreatitis
GU: *Polyuria,* renal failure, glycosuria, bladder spasms
HEMA: Thrombocytopenia, agranulocytosis, leukopenia, neutropenia, anemia
INTEG: *Rash, pruritus,* purpura, Stevens-Johnson syndrome, sweating, photosensitivity, urticaria
MS: Cramps, stiffness
SYST: Toxic epidermal necrolysis

PHARMACOKINETICS

PO: Onset 1 hr, peak 1-2 hr, duration 6-8 hr, absorbed 70%
IV: Onset 5 min; peak ½ hr; duration 2 hr (metabolized by the liver 30%); excreted in urine, some as unchanged product, and feces; crosses placenta; excreted in breast milk; half-life ½-1 hr

INTERACTIONS

Increase: toxicity—lithium, nondepolarizing skeletal muscle relaxants, digoxin
Increase: hypotensive action of antihypertensives, nitrates
Increase: ototoxicity—aminoglycosides, CISplatin, vancomycin
Increase: effects of anticoagulants, salicylates
Decrease: furosemide effect—probenecid
Drug/Lab Test
Interference: GTT
Increase: LDL

NURSING CONSIDERATIONS

Assess:
- **CHF:** weight, I&O daily to determine fluid loss; effect of product may be decreased if used daily
- **Hypertension:** B/P lying, standing; postural hypotension may occur
- Metabolic alkalosis: drowsiness, restlessness
- **Hypokalemia:** postural hypotension, malaise, fatigue, tachycardia, leg cramps, weakness
- Rashes, temperature elevation daily
- Confusion, especially in geriatric patients; take safety precautions if needed
- **Hearing,** including tinnitus and hearing loss, when giving high doses for extended periods or rapid inf
- Rate, depth, rhythm of respiration, effect of exertion, lung sounds
- Electrolytes (potassium, sodium, chloride); include BUN, blood glucose, CBC, serum creatinine, blood pH, ABGs, uric acid
- Glucose in urine if patient diabetic
- Allergies to sulfonamides, thiazides

Evaluate:
- Therapeutic response: improvement in edema of feet, legs, sacral area (CHF); increase urine output, decreased B/P; decreased calcium levels (hypercalcemia)

Teach patient/family:
- To discuss the need for a high-potassium diet or potassium replacement with prescriber
- To rise slowly from lying or sitting position because orthostatic hypotension may occur
- To recognize adverse reactions that may occur: muscle cramps, weakness, nausea, dizziness
- About the entire treatment regimen, including exercise, diet, stress relief for hypertension
- To take with food or milk for GI symptoms
- To use sunscreen or protective clothing to prevent photosensitivity
- To take early in the day to prevent sleeplessness
- To avoid OTC medications unless directed by prescriber

TREATMENT OF OVERDOSE:

Lavage if taken orally; monitor electrolytes; administer dextrose in saline; monitor hydration, CV, renal status

gabapentin (Rx)

(gab′a-pen-tin)

Gralise, Horizant, Neurontin

Func. class.: Anticonvulsant

Do not confuse:
Neurontin/Noroxin/Neoral

ACTION: Mechanism unknown; may increase seizure threshold; structurally similar to GABA; gabapentin binding sites in neocortex, hippocampus

USES: Adjunct treatment of partial seizures, with/without generalization in patients >12 yr; adjunct for partial seizures in children 3-12 yr, postherpetic neuralgia, primary restless leg syndrome in adults

Unlabeled uses: Tremors with multiple sclerosis (MS), neuropathic pain, bipolar disorder, migraine prophylaxis, diabetic neuropathy, nystagmus, pruritus, spasticity, menopause, hot flashes, ALS, MS

CONTRAINDICATIONS: Hypersensitivity to this product

Precautions: Pregnancy (C), breastfeeding, children <3 yr, geriatric patients, renal disease, hemodialysis, suicidal thoughts, depression

DOSAGE AND ROUTES

- **Adult and child >12 yr: PO** 900-1800 mg/day in 3 divided doses; may titrate by giving 300 mg on 1st day, 300 mg bid on 2nd day, 300 mg tid on 3rd day; may increase to 1800-2400 mg/day by adding 300 mg on subsequent days
- **Child 3-12 yr: PO** 10-15 mg/kg/day in 3 divided doses, initially titrate dose upward over approximately 3 days; if > syringe 25-35 mg/kg/day; if 3-4 yr old, 40 mg/kg/day divided in 3 doses

Postherpetic neuralgia

- **Adult: PO** 300 mg on day 1, 600 mg/day divided bid on day 2, 900 mg/day divided tid, may titrate to 1800-3600 mg divided tid if needed; **EXT REL** (Gralise only) 300 mg on day 1, 600 mg on day 2, 900 mg on days 3-6, 1200 mg on days 7-10, 1500 mg on days 11-14, 1800 mg on day 15 and thereafter

Moderate to severe restless leg syndrome (RLS) (Horizant only)

- **Adult: PO EXT REL** 600 mg daily with food at about 5 PM; if dose missed, take next day at 5 PM

Renal dose

- **Adult and child >12 yr:** CCr 30-59 ml/min, 200-700 mg bid, CCr 15-29 ml/min, 200-700 mg/day, CCr <15 ml/min, 100-300 mg/day

Uremic pruritus in hemodialysis (unlabeled)

- **Adult: PO** 300 mg 3×/wk

Brachioradical pruritus (unlabeled)

- **Adult: PO** 300-1800 mg/day

ALS (unlabeled)

- **Adult: PO** 1000 mg/day in divided doses × 6 mo

Pendular/congenital nystagmus (unlabeled)

- **Adult: PO** 900 mg/day in divided doses initially, up to 2400 mg/day in divided doses

Spasticity in MS (unlabeled)

- **Adult: PO** 600-1200 mg/day in divided doses

Available forms: Caps 100, 300, 400 mg; tabs 600, 800 mg; Horizant: ext rel tab 600 mg; oral sol 250 mg/5 ml; ext rel tab (Gralise): 300, 600 mg

Administer:

- Do not crush or chew caps, ext rel tabs; caps may be opened and contents put in applesauce or dissolved in juice; scored tabs may be cut in half
- 2 hr apart when giving antacids
- Give without regard to meals
- Gradually withdraw over 7 days; abrupt withdrawal may precipitate seizures
- Beginning dose at bedtime to minimize daytime drowsiness
- **Oral sol:** measure with calibrated device
- **Ext rel:** give with food at about 5 PM; bioavailability increased with food; swal-

low whole; do not interchange Gralise with Horizant

• Store at room temperature away from heat and light

SIDE EFFECTS

CNS: *Drowsiness, confusion,* dizziness, fatigue, anxiety, somnolence, ataxia, amnesia, abnormal thinking, unsteady gait, *depression;* children 3-12 yr old, emotional lability, aggression, thought disorder, hyperkinesia, hostility, seizures, suicidal ideation, impaired cognition, euphoria

CV: Vasodilation, peripheral edema, hypotension, hypertension

EENT: Dry mouth, blurred vision, *diplopia,* nystagmus, conjunctivitis; otitis media (child 3-12 yr)

GI: Constipation/diarrhea, weight gain, increased appetite, dental abnormalities, nausea, vomiting

GU: Impotence, bleeding, *UTI*

HEMA: Leukopenia, thrombocytopenia

INTEG: Pruritus, abrasion, Stevens-Johnson syndrome, acne vulgaris

MS: Myalgia, back pain, gout

RESP: *Rhinitis,* pharyngitis, cough, upper respiratory infection (child 3-12 yr)

SYST: Drug reaction with eosinophilia and systemic symptoms (DRESS); dehydration (child 3-12 yr)

PHARMACOKINETICS

Protein binding <3% not metabolized; excreted in urine (unchanged); elimination half-life 5-7 hr, prolonged to 130 hr in ESRD

INTERACTIONS

Increase: CNS depression—alcohol, sedatives, antihistamines, all other CNS depressants

Increase: gabapentin levels—morphine

Decrease: gabapentin levels—antacids, sevelamar, cimetidine

Decrease: effect of—HYDROcodone

Drug/Lab Test

Increase: LFTs

False positive: urinary protein using Ames N-multistix SG

NURSING CONSIDERATIONS

Assess:

• **Seizures:** aura, location, duration, frequency, activity at onset

• **Pain:** location, duration, characteristics if using for chronic pain, migraine

⚠ Mental status: mood, sensorium, affect, behavioral changes, **suicidal thoughts/behaviors;** if mental status changes, notify prescriber

• Eye problems, need for ophthalmic exam before, during, after treatment (slit lamp, funduscopy, tonometry)

• WBC, gabapentin level (therapeutic 5.9-21 mcg/ml, toxic >85 mcg/ml), serum creatinine/BUN, weight

• Drug reaction with eosinophilia and systemic symptoms

• **Seizure precautions:** padded side rails; move objects that may harm patient

• Increased fluids, bulk in diet for constipation

Evaluate:

• Therapeutic response: decreased seizure activity; decrease in chronic pain

Teach patient/family:

• To carry emergency ID stating patient's name, products taken, condition, prescriber's name and phone number

• To avoid driving, other activities that require alertness because dizziness, drowsiness may occur

• Not to discontinue medication quickly after long-term use; to taper over ≥1 wk because withdrawal-precipitated seizures may occur; not to double doses if dose is missed; to take if 2 hr or more before next dose

• To notify prescriber if pregnancy planned or suspected; to avoid breastfeeding

• Not to use within 2 hr of antacid, may take regular release without regard to meals; ext rel should be taken with food; to take as directed, doses interval should not be ≥12 hr

galantamine (Rx)

(gah-lan′tah-meen)

Razadyne, Razadyne ER, Reminyl ✱

Func. class.: Anti-Alzheimer agent

Chem. class.: Centrally acting cholinesterase inhibitor

ACTION: Enhances cholinergic functioning by increasing acetylcholine in cerebral cortex

USES: Mild to moderate dementia of Alzheimer's disease, dementia with Lewy bodies

Unlabeled uses: Vascular dementia, Pick's disease

CONTRAINDICATIONS: Hypersensitivity to this product, GI bleeding, jaundice, renal failure, children

Precautions: Pregnancy (B), geriatric patients, respiratory/renal/hepatic/cardiac disease, seizure disorder, peptic ulcer, asthma, bradycardia, heart block, surgery, urinary tract obstruction, breastfeeding

DOSAGE AND ROUTES

- **Adult: PO** 4 mg bid with morning and evening meals; after 4 wk or more, may increase to 8 mg bid; may increase to 12 mg bid after another 4 wk, usual dose 16-24 mg/day in 2 divided doses; **EXT REL** 8 mg/day in AM; may increase to 16 mg/day after 4 wk and 24 mg/day after another 4 wk

Hepatic dose

- **Adult: PO** Child-Pugh 7-9, max 16 mg/day; Child-Pugh 10-15, avoid use

Renal dose

- **Adult: PO** CCr 10-70 ml/min, max 16 mg/day; CCr <9 ml/min, avoid use

Available forms: Tabs 4, 8, 12 mg; ext rel tabs 8, 16, 24 mg; oral sol 4 mg/ml

Administer:

PO route

- With meals; take with morning and evening meal (immediate rel); morning (ext rel)
- Dose increase after minimum of 4 wk at prior dose; if dose is interrupted for ≥3 days, restart at lower dose, titrate to current dose
- Ext rel in AM with food; do not crush, open, or chew
- **Oral sol:** use pipette provided, put in liquid and have patient consume

SIDE EFFECTS

CNS: *Tremors, insomnia,* depression, dizziness, headache, somnolence, fatigue

CV: Bradycardia, chest pain, AV block

GI: *Nausea, vomiting, anorexia, abdominal distress, flatulence,* diarrhea

GU: Urinary incontinence, bladder outflow obstruction, hematuria

HEMA: Anemia

META: Weight decrease

MS: Asthenia

RESP: Upper respiratory tract infection, rhinitis

PHARMACOKINETICS

Rapidly and completely absorbed; metabolized by CYP2D6, 3A4; excreted via kidneys; clearance is lower in geriatric patients, hepatic disease; clearance is 20% lower in females; elimination half-life 7 hr; 18% protein binding

INTERACTIONS

Increase: synergistic effect—cholinomimetics, other cholinesterase inhibitors

Increase: galantamine effect—CYP3A4, CYP2D6 inhibitors (antiretroviral protease inhibitors, ketoconazole, erythromycin, conivaptan, delaviridine, diltiazem, efavirenz, fluconazole, fluvoxaMINE, imatinib, itraconazole, clarithromycin, troleandomycin, nefazodone, niCARdipine, verapamil, voriconazole, zafirlukast)

Increase: GI effects—NSAIDs

Decrease: galantamine effect—CYP3A4, CYP2D6 inducers (bosentan, carBAMazepine, nevirapine, OXcarbazepine, phenytoin, fosphenytoin/rifabutin, rifampin, rifapentine, troglitazone), anticholinergics

Drug/Herb

Decrease: galantamine effect—St. John's wort

NURSING CONSIDERATIONS

Assess:

Alzheimer's disease: mental status: affect, mood, behavioral changes, depression, attention, confusion; neurologic status: long- and short-term memory, cognitive functioning

- Hepatic/renal studies: AST, ALT, alk phos, LDH, bilirubin, CBC; BUN, creatinine
- For severe GI effects: nausea, vomiting, anorexia, weight loss
- B/P, heart rate, respiration during initial treatment: bradycardia/AV block may occur
- Fluid status: ensure adequate hydration
- Assistance with ambulation during beginning therapy

Evaluate:

- Therapeutic response: decreased confusion

Teach patient/family:

- To notify all prescribers of use
- About correct procedure for giving oral sol using instruction sheet provided
- To notify prescriber of severe GI effects; hypo/hypertension, slow heart rate
- That product is not a cure but relieves symptoms
- That results may take several wk or mo to occur
- To take with food to prevent intolerance

ganciclovir (Rx)

(gan-sye′kloe-vir)

Cytovene, Vitrasert, Zirgan

Func. class.: Antiviral

Chem. class.: Synthetic nucleoside analog

Do not confuse:

Cytovene/Cytosar

ACTION: Inhibits replication of herpesviruses; competitively inhibits human CMV DNA polymerase and is incorporated, resulting in termination of DNA elongation

USES: Cytomegalovirus (CMV) retinitis in immunocompromised persons, including those with AIDS, after indirect ophthalmoscopy confirms diagnosis; prophylaxis for CMV in transplantation; ophthalmic: acute herpes keratitis

Unlabeled uses: CMV pneumonia in organ transplant patients; CMV gastroenteritis, esophagitis, colitis; CMV pneumonitis, congenital CMV disease; Epstein-Barr virus; herpes simplex types 1, 2; varicella-zoster, hepatitis B

CONTRAINDICATIONS: Hypersensitivity to acyclovir, ganciclovir

Black Box Warning: Absolute neutrophil count <500, platelet count <25,000 (intravitreal)

Precautions: Pregnancy (C), breastfeeding, children <6 mo, geriatric patients, preexisting cytopenias, renal function impairment, radiation therapy, hypersensitivity to famciclovir, penciclovir, valacyclovir, valganciclovir

Black Box Warning: Secondary malignancy, bone marrow suppression, anemia, infertility, neutropenia

DOSAGE AND ROUTES

Prevention of CMV

- **Adult/adolescent (unlabeled): IV** 5 mg/kg/dose over 1 hr q12hr × 1-2 wk, then 5 mg/kg/day 7 day/wk, then 6 mg/kg/day × 5 days/wk; **PO** 1000 mg tid starting 10 days posttransplant × 14 wk

Induction treatment

- **Adult: IV** 5 mg/kg/dose given over 1 hr, q12hr × 2-3 wk

Maintenance treatment

- **Adult: IV INF** 5 mg/kg given over 1 hr, daily × 7 days/wk or 6 mg/kg/day × 5 days/wk; **PO** 1000 mg tid with food or 500 mg q3hr while awake for 6 doses; **INTRAVITREAL** 4.5 mg implant

Acute herpes keratitis

- **Adult/adolescent/child ≥2 yr: OPHTH** 1 drop in affected eye 5 times daily

Renal dose

- **Adult: IV** CCr 50-69 ml/min, reduce to 2.5 mg/kg q12hr (induction), 2.5 mg/kg q24hr (maintenance); **PO** 1500 mg/day or 500 mg tid; **IV** CCr 25-49 ml/min, reduce to 2.5 mg q24hr (induction); 1.25 mg/kg q24hr (maintenance); **PO** 1000 mg/day or 500 mg bid; **IV** CCr 10-24 ml/min, reduce to 1.25 mg/kg q24hr (induction); 0.625 mg/kg q24hr (maintenance); **PO** 500 mg/day; **IV** CCr <10 ml/min, reduce to 1.25 mg/kg 3×/wk after hemodialysis (induction); 0.625 mg/kg 3×/wk after hemodialysis (maintenance); **PO** 500 mg 3×/wk after hemodialysis

Available forms: Powder for inj 500 mg/vial; caps 250, 500 mg; implant, intraviteral 4.5 mg; ophth gel 0.15% (Zirgan)

Administer:

PO route

- With food and a full glass of water
- Do not open, crush capsules

Intravitreal implant route

- Implanted by surgeon only
- Handle carefully to prevent damage to coating

Intravitreal inj route (unlabeled)

- Reconstitute and dilute IV powder to 2 mg/0.1 ml or 5 mg/0.1 ml, depending on dose; injected using TB syringe

IV route

- Mixed in biologic cabinet using gown, gloves, mask; use cytotoxic handling procedures

Intermittent IV INF route

- IV after reconstituting 500 mg/10 ml sterile water for inj (50 mg/ml); shake; further dilute in 50-250 ml D_5W, 0.9% NaCl, LR, Ringer's and run over 1 hr; use inf pump, in-line filter
- Do not give by bolus IV, IM, SUBCUT inj
- Use reconstituted sol within 24 hr; do not refrigerate or freeze

Y-site compatibilities: Allopurinol, amphotericin B cholesteryl, CISplatin, cyclophosphamide, DOXOrubicin liposome, enalaprilat, etoposide, filgrastim, fluconazole, gatifloxacin, granisetron, linezolid, melphalan, methotrexate, PACLitaxel, propofol, remifentanil, tacrolimus, teniposide, thiotepa

SIDE EFFECTS

CNS: *Fever*, chills, coma, *confusion*, abnormal thoughts, dizziness, bizarre dreams, *headache*, psychosis, tremors, somnolence, *paresthesia*, *weakness*, seizures, peripheral neuropathy
CV: Dysrhythmia, hypo/hypertension
EENT: Retinal detachment in CMV retinitis, ocular hypertension, ocular pain, conjunctival scarring, cataracts
GI: *Abnormal LFTs, nausea, vomiting, anorexia, diarrhea, abdominal pain,* hemorrhage, perforation, pancreatitis
GU: Hematuria, *increased creatinine*, BUN, infertility, decreased sperm count
HEMA: Granulocytopenia, thrombocytopenia, irreversible neutropenia, anemia, eosinophilia, pancytopenia
INTEG: *Rash*, alopecia, *pruritus*, urticaria, pain at site, phlebitis
RESP: Dyspnea

PHARMACOKINETICS

Half-life 3-4½ hr; excreted by kidneys (unchanged); crosses blood-brain barrier, CSF, increased bioavailability with fatty foods

INTERACTIONS

Increase: severe granulocytopenia—zidovudine, antineoplastics, radiation; do not give together
Increase: ganciclovir toxicity—adriamycin, amphotericin B, cycloSPORINE, dapsone, DOXOrubicin, flucytosine, pentamidine, probenecid, trimethoprim-sulfamethoxazole combinations, vinBLAStine, vinCRIStine, other nucleoside analogs, mycophenolate, tenofovir, tacrolimus, aminoglycosides, NSAIDs
Increase: seizures—imipenem/cilastatin
Increase: didanosine effect—didanosine
Drug/Lab Test
Increase: LFTs, creatinine
Decrease: Hgb, WBC, platelets, neutrophils, granulocytes

NURSING CONSIDERATIONS

Assess:
- **CMV retinitis:** culture should be completed before starting treatment (urine, blood, throat), ophthalmic exam
- **Infection:** increased temperature, sore throat, chills, fever; report to prescriber

Black Box Warning: **Leukopenia/neutropenia/thrombocytopenia:** CBC, WBCs, platelets q2days during 2×/day dosing and then q1wk for leukopenia with daily WBC count in patients with prior leukopenia with other nucleoside analogs or for whom leukopenia counts are <1000 cells/mm^3 at start of treatment

- Serum creatinine or CCr ≥q2wk, BUN; LFTs; ophthalmic exam
- For seizures, dysrhythmias

Evaluate:

Black Box Warning: Secondary malignancy: avoid direct contact with powder in caps/solution, if skin contact occurs, wash thoroughly with soap and water; do not get in the eyes

- Therapeutic response: decreased symptoms or prevention of CMV

Teach patient/family:
- Do not wear contact lenses while using gel
- That product does not cure condition; that regular blood tests, ophthalmologic exams are necessary
- That major toxicities may necessitate discontinuing product

Black Box Warning: To use contraception during treatment and that infertility may occur; to use barrier contraception for 90 days after treatment; may cause reversible infertility at lower doses, irreversible infertility at higher doses

- To take PO with food

⚠ **To report infection:** fever, chills, sore throat; blood dyscrasias: bruising, bleeding, petechiae; to avoid crowds, persons with respiratory infections

ganciclovir ophthalmic

See Appendix B

gatifloxacin

(ga-ti-floks′a-sin)

Zymaxid

Func. class.: Ophthalmic anti-infective

Chem. class.: Fluoroquinolone

Do not confuse: gatifloxacin/levofloxacin/moxifloxacin

ACTION: Inhibits DNA gyrase, thereby decreasing bacterial replication

USES: Bacterial conjunctivitis

CONTRAINDICATIONS: Hypersensitivity to this product or fluoroquinolones, infants <1 yr
Precautions: Pregnancy (C), breastfeeding

DOSAGE AND ROUTES

Bacterial conjunctivitis

• **Adult/adolescent/child ≥1 yr: OPHTH SOL** instill 1 drop (0.5% sol) q2hr while awake; on days 2-7 instill 1 drop in eye 2-4×/day while awake

Available forms: Ophthalmic solution 0.5%

Administer:

Ophthalmic route

• Commercially available ophthalmic solutions are not for injection subconjunctivally or into the anterior chamber of the eye

• Apply topically to the eye, taking care to avoid contamination

• Do not touch the tip of the dropper to the eye, fingertips, or other surface

• Apply pressure to lacrimal sac for 1 min after instillation

• Avoid wearing contact lenses during treatment

SIDE EFFECTS

EENT: Hypersensitivity, irritation, redness, tearing, keratitis, blepharitis, taste changes, ocular discharge/hemorrhage/irritation/pain

PHARMACOKINETICS

Unknown

NURSING CONSIDERATIONS

Assess:

• **Allergic reaction**: hypersensitivity, discontinue product

Evaluate:

• Decreased ophthalmic infection

Teach patient/family:

Ophthalmic route

• To apply topically to the eye, taking care to avoid contamination; for ophthalmic use only

• Not to touch the tip of the dropper to the eye, fingertips, or other surface

• To apply pressure to lacrimal sac for 1 min after instillation

• To avoid wearing contact lenses during treatment

gefitinib (Rx)

(ge-fi′tye-nib)

Iressa

Func. class.: Antineoplastic—miscellaneous

Chem. class.: Epidermal growth factor receptor inhibitor

ACTION:

Not fully understood; inhibits intracellular phosphorylation of cell surface receptors associated with epidermal growth factor receptors

USES:

Advanced/metastatic non-small-cell lung cancer (NSCLC) in those who have not responded to platinum or DOCEtaxel products

CONTRAINDICATIONS:

Pregnancy (D), breastfeeding, children, hypersensitivity

Precautions: Geriatric patients, ocular/pulmonary/renal/hepatic disorders

DOSAGE AND ROUTES

• **Adult: PO** 250 mg/day, max 500 mg/day

Available forms: Tabs 250 mg

SIDE EFFECTS

EENT: Amblyopia, conjunctivitis, eye pain, corneal erosion/ulcer

GI: *Nausea, diarrhea, vomiting,* anorexia, pancreatitis, mouth ulceration, hepatotoxicity

INTEG: *Rash,* pruritus, *acne, dry skin,* toxic epidermal necrolysis, angioedema

MISC: Peripheral edema, hemorrhage

RESP: Interstitial lung disease, cough, dyspnea, pneumonia

PHARMACOKINETICS

Slowly absorbed; excreted in feces (86%), urine (<4%); elimination half-life 48 hr; metabolism by CYP3A4

INTERACTIONS

Increase: gefitinib concs—ketoconazole, itraconazole, erythromycin, clarithromycin

Increase: bone marrow suppression—cloZAPine

Increase: plasma conc of warfarin, metoprolol

Decrease: gefitinib levels—phenytoin, rifampin, cimetidine, ranitidine, sodium bicarbonate

Drug/Herb

Decrease: gefitinib levels—St. John's wort

NURSING CONSIDERATIONS

Assess:

⚠ Pulmonary changes: lung sounds, cough, dyspnea; interstitial lung disease may occur, may be fatal; discontinue therapy if confirmed

- Ocular changes: eye irritation, corneal erosion/ulcer, aberrant eyelash growth

⚠ Pancreatitis: abdominal pain, levels of amylase, lipase

⚠ Toxic epidermal necrolysis, angioedema

- GI symptoms: frequency of stools, if diarrhea is poorly tolerated, therapy may be discontinued for up to 14 days; LFTs baseline and periodically

Administer:

- Without regard to food; avoid grapefruit juice
- May be dispersed in half-glass of water (noncarbonated); rinse glass and drink liquid; may be given through NG tube

Evaluate:

- Therapeutic response: decreased NSCLC

Teach patient/family:

⚠ To report adverse reactions immediately: shortness of breath, severe abdominal pain, ocular changes, skin eruptions

- Reason for treatment, expected results
- Use contraception during treatment; avoid breastfeeding
- Avoid persons with infections

gemcitabine (Rx)

(jem-sit′a-been)

Gemzar

Func. class.: Antineoplastic—miscellaneous

Chem. class.: Nucleoside analog

Do not confuse:

Gemzar/Zinecard

ACTION: Exhibits antitumor activity by killing cells undergoing DNA synthesis (S phase) and blocking G1/S-phase boundary

USES: Adenocarcinoma of the pancreas (nonresectable stage II, III, or metastatic stage IV); in combination with CISplatin for inoperable, advanced, or metastatic non–small-cell lung cancer; advanced breast cancer in combination with PACLitaxel; with CARBOplatin for ovarian cancer

Unlabeled uses: Bladder cancer, mesothelioma, adjuvant treatment of pancreatic cancer, ovarian cancer single agent, biliary tract cancer, advanced T-cell lymphoma

CONTRAINDICATIONS: Pregnancy (D), breastfeeding, hypersensitivity

Precautions: Children, geriatric patients, myelosuppression, radiation therapy, renal/hepatic disease, accidental exposure, alcoholism, dental disease, infection

DOSAGE AND ROUTES

Pancreatic carcinoma (nonresectable stage II, III, IV)

- **Adult: IV** 1000 mg/m^2 given over 30 min/wk up to 7 wk then 1 wk rest period; subsequent cycles should be infused 1×/wk × 3 wk out of every 4 wk depending on hematologic toxicity

Non–small-cell lung cancer

- **Adult: IV** (4-wk schedule) 1000 mg/m^2 given over 30 min on days 1, 8, 15, of each 28-day cycle; give CISplatin **IV** 100 mg/m² on day 1 after gemcitabine; 3-wk schedule: 1250 mg/m² given over 30 min on days 1, 8 of each 21-day cycle; give CISplatin **IV** 100 mg/m² after the inf of gemcitabine on day 1

Advanced breast cancer

• **Adult: IV** 1250 mg/m² over 30 min on days 1 and 8 of 21-day cycle; give with PACLitaxel 175 mg/m² over 3 hr before gemcitabine on day 1

Recurrent ovarian cancer (single agent) (unlabeled)

• **Adult: IV** 1 g/m², days 1, 8, 15 of 28-day cycle

Available forms: Lyophilized powder for inj 20 mg/ml (10-, 50-ml vials)

Administer:

IV route

• Prepare in biologic cabinet using gown, mask, gloves; use cytotoxic handling procedures

• After reconstituting with 0.9% NaCl 5 ml/200-mg vial of product or 25 ml/1-g vial of product (38 mg/ml) shake; may be further diluted with 0.9% NaCl to conc as low as 0.1 mg/ml; discard unused portions, give over 30 min, do not admix

• Diluted solution stable at room temperature for 24 hr, do not refrigerate

• Inf longer than 60 min increase toxicity

• **Bone marrow depression:** CBC, differential, platelet count before each dose; single agent: absolute granulocyte count >1000 and platelets >100,000, give complete dose; absolute granulocyte count 500-999, platelets 50,000-99,999, give 75%; absolute granulocyte count <500 or platelets <50,000, do not give; combination with PACLitaxel for breast cancer: absolute granulocyte count >1200 and platelets >75,000, give complete dose; absolute granulocyte count 1000-1199 or platelets 50,000-75,000, give 75%; absolute granulocyte 700-999 or platelets ≥50,000, give 50%; granulocyte count <700 or platelet <50,000, do not give; combination with CARBOplatin for ovarian cancer: absolute granulocyte count >1500 and platelet count >100,000, give complete dose; absolute granulocyte count 1000-1499 or platelets 75,000-99,000, give 75%; absolute granulocyte count <1000 or platelets <75,000, do not give

Y-site compatibilities: Amifostine, amikacin, aminophylline, ampicillin, aztreonam, bleomycin, bumetanide, butorphanol, calcium gluconate, cefOXitin, cefTAZidime, ceftizoxime, cefTRIAXone, chlorproMAZINE, cimetidine, ciprofloxacin, CISplatin, clindamycin, cyclophosphamide, cytarabine, DACTINomycin, DAUNOrubicin, diphenhydrAMINE, DOBUTamine, DOCEtaxel, DOPamine, DOXOrubicin, droperidol, enalaprilat, etoposide, famotidine, floxuridine, fluconazole, fludarabine, fluorouracil, gentamicin, granisetron, haloperidol, heparin, hydrocortisone, HYDROmorphone, IDArubicin, ifosfamide, leucovorin, linezolid, LORazepam, mannitol, meperidine, mesna, metoclopramide, metroNIDAZOLE, minocycline, mitoXANtrone, morphine, nalbuphine, ondansetron, PACLitaxel, promethazine, ranitidine, streptozocin, teniposide, thiotepa, ticarcillin, tobramycin, topotecan, trimethoprim/sulfamethoxazole, vancomycin, vinBLAStine, vinCRIStine, vinorelbine, zidovudine, zoledronic acid

SIDE EFFECTS

ENDO: Hyperglycemia

GI: Diarrhea, *nausea, vomiting,* anorexia, constipation, stomatitis

GU: *Proteinuria,* hematuria

HEMA: Leukopenia, anemia, neutropenia, thrombocytopenia

INTEG: Irritation at site, *rash, alopecia*

META: Hypocalcemia, hypokalemia, hypomagnesemia

OTHER: Dyspnea, *fever,* hemorrhage, infection, flulike symptoms, paresthesia, *peripheral edema,* myalgia, capillary leak syndrome

PHARMACOKINETICS

Half-life 42-379 min, crosses placenta, excretion: renal, 92%-98%

INTERACTIONS

Increase: bleeding—NSAIDs, alcohol, salicylates, anticoagulants

Increase: myelosuppression, diarrhea—other antineoplastics, radiation

Decrease: antibody response—live virus vaccines

Drug/Lab Test

Increase: BUN, AST, ALT, alk phos, bilirubin, creatinine

Decrease: Hgb, WBC, neutrophils, platelets

NURSING CONSIDERATIONS

Assess:

- **Blood dyscrasias:** bruising, bleeding, petechiae
- I&O, nutritional intake; food preferences: list likes, dislikes
- Renal, hepatic studies before and during treatment; may increase AST, ALT, alk phos, bilirubin, BUN, creatinine, calcium, potassium, glucose, magnesium, urine protein
- Buccal cavity for dryness, sores, ulceration, white patches, oral pain, bleeding, dysphagia
- GI symptoms: frequency of stools, cramping
- Signs of dehydration: rapid respirations, poor skin turgor, decreased urine output, dry skin, restlessness, weakness
- Increased fluid intake to 2-3 L/day to prevent dehydration unless contraindicated
- Rinsing of mouth tid-qid with water, club soda; brushing of teeth bid-tid with soft brush or cotton-tipped applicator for stomatitis; use unwaxed dental floss
- Antiemetic agents

Evaluate:

- Therapeutic response: decrease in tumor size; decrease in spread of cancer; symptom relief

Teach patient/family:

- To avoid foods with citric acid or hot or rough texture if stomatitis is present; to drink adequate fluids; to avoid use with NSAIDs, alcohol, salicylates
- To report stomatitis and any bleeding, white spots, ulcerations in mouth; to examine mouth daily, report symptoms
- To report signs of anemia: fatigue, headache, faintness, SOB, irritability; hematuria, dysuria
- To use contraception during therapy and for 4 mo after; pregnancy (D), do not breastfeed
- Not to receive vaccinations during treatment
- About possible hair loss and what can be done
- To report flulike symptoms, swelling of feet/legs
- To report bruising, bleeding: gums, blood in urine, stool, emesis
- To avoid crowds, persons with known upper-respiratory infections
- To avoid use of hard-bristle toothbrush, electric razor
- Infection: to report sore throat, fever, flulike symptoms immediately

gemfibrozil (Rx)

(jem-fi′broe-zil)

Lopid

Func. class.: Antilipemic

Chem. class.: Fibric acid derivative

Do not confuse:
Lopid/Levbid/Slo-bid

ACTION: Inhibits biosynthesis of VLDL, decreases triglycerides, increases HDL

USES: Type IIb, IV, V hyperlipidemia as adjunct with diet therapy, hypertriglyceridemia

CONTRAINDICATIONS: Severe renal/hepatic disease, preexisting gallbladder disease, primary biliary cirrhosis, hypersensitivity

Precautions: Pregnancy (C), breastfeeding, renal disease, cholelithiasis

DOSAGE AND ROUTES

- **Adult: PO** 600 mg bid 30 min before AM, PM meal

Hepatic/renal dose

- Avoid use

Available forms: Tabs 600 mg; caps 300 mg ✱

Administer:

PO route

- 30 min before AM, PM meals
- Discontinue product if response does not occur within 3 mo

SIDE EFFECTS

CNS: Fatigue, vertigo, headache, paresthesia, dizziness, somnolence

GI: *Dyspepsia, diarrhea, abdominal pain,* nausea, vomiting

HEMA: Leukopenia, anemia, eosinophilia, thrombocytopenia

INTEG: Rash, urticaria, pruritus

MISC: Taste perversion

MS: Myopathy, rhabdomyolysis

SYST: Angioedema, exfoliative dermatitis

PHARMACOKINETICS

Peak 1-2 hr; plasma protein binding >95%; half-life $1^1/_2$ hr; 70% excreted in urine mostly unchanged; metabolized in liver (minimal)

INTERACTIONS

- Do not use with repaglinide, simvastatin

Increase: hypoglycemic effect—sulfonylureas, repaglinide

Increase: anticoagulant properties—warfarin

Increase: risk of myositis, myalgia—HMG-CoA reductase inhibitors

Increase: nephrotoxicity of cycloSPORINE

Decrease: effect of gemfibrozil—bile acid sequestrants, separate by >2 hr

Drug/Lab Test

Increase: LFTs, CK

Decrease: Hgb, Hct, WBC, potassium, eosinophils, platelets

NURSING CONSIDERATIONS

Assess:

- Hyperlipidemia: diet history: fats, triglycerides, cholesterol; if lipids increase, product should be discontinued; LDL, VLDL baseline and periodically
- **Myopathy, rhabdomyolysis:** For muscle pain, tenderness; obtain baseline CPK; if elevated or if these occur, product should be discontinued; at greater risk if combined with HMG-Co-A reductase inhibitors
- Renal, hepatic studies, CBC, blood glucose if patient is receiving long-term therapy; if LFTs increase, therapy should be discontinued; monitor hematologic and hepatic functions
- Bowel pattern daily; watch for increasing diarrhea (common)

Evaluate:

- Therapeutic response: decreased cholesterol, triglyceride levels; HDL, cholesterol ratios improved

Teach patient/family:

- That compliance is needed for positive results; not to double or skip dose
- That risk factors should be decreased: high-fat diet, smoking, alcohol consumption, absence of exercise
- To notify prescriber of diarrhea, nausea, vomiting, chills, fever, sore throat, muscle cramps, abdominal cramps, severe flatulence, tendon pain
- To avoid driving, hazardous activities if dizziness, blurred vision occur

gemifloxacin (Rx)

(gem-ah-flox′a-sin)

Factive

Func. class.: Antiinfective

Chem. class.: Fluoroquinolone

ACTION: Inhibits DNA gyrase, which is an enzyme involved in replication, transcription, and repair of bacterial DNA

USES:
Acute bacterial exacerbation of chronic bronchitis caused by *Streptococcus pneumoniae, Haemophilus influenzae, Haemophilus parainfluenzae, Moraxella catarrhalis;* community-acquired pneumonia caused by *Streptococcus pneumoniae* including multiproduct-resistant strains, *H. influenzae, M. catarrhalis, Mycoplasma pneumoniae, Chlamydia pneumoniae, Klebsiella pneumoniae*

CONTRAINDICATIONS:
Hypersensitivity to quinolones

Precautions: Pregnancy (C), breastfeeding, children, geriatric patients, hypokalemia, hypomagnesium, renal disease, seizure disorders, excessive exposure to sunlight, psychosis, increased intracranial pressure, history of QT interval prolongation, dysrhythmias, myasthenia gravis, torsades de pointes

Black Box Warning: Tendon pain/rupture, tendinitis, myasthenia gravis

DOSAGE AND ROUTES
• **Adult: PO** 320 mg/day × 5-7 days depending on type of infection

Renal dose

• **Adult: PO** CCr ≤40 ml/min, 160 mg q24hr

Available forms: Tabs 320 mg

Administer:

• 2 hr before or 3 hr after aluminum/magnesium antacids, iron, zinc products, multivitamins, buffered products 2 hr before sucralfate

SIDE EFFECTS
CNS: *Dizziness, headache,* somnolence, depression, insomnia, nervousness, confusion, agitation, **seizures, pseudotumor cerebri**

CV: QT prolongation, vasodilation

EENT: Visual disturbances, retinal detachment

GI: Diarrhea, *nausea,* vomiting, anorexia, flatulence, heartburn, dry mouth; increased AST, ALT; constipation, abdominal pain, oral thrush, glossitis, stomatitis, **pseudomembranous colitis**

GU: Crystalluria (rare), vaginitis, interstitial nephritis

HEMA: **Thrombocytopenia, neutropenia,** anemia, agranulocytosis, **thrombotic thrombocytopenic purpura,** muscle weakness

INTEG: Rash, pruritus, urticaria, *photosensitivity*

MS: Tendinitis, **tendon rupture**

SYST: **Anaphylaxis, Stevens-Johnson syndrome, toxic epidermal necrolysis, exfoliative dermatitis**

G

PHARMACOKINETICS
Rapidly absorbed; bioavailability 71%; peak 1-2 hr; half-life 4-12 hr; excreted in urine as active product, metabolites

INTERACTIONS
Increase: CNS stimulation—NSAIDs

Increase: toxicity of gemifloxacin—probenecid

⚠ **Increase: QT prolongation—class IA, III antidysrhythmics, tricyclics, amoxapine, maprotiline, phenothiazines, haloperidol, pimozide, risperiDONE, sertindole, ziprasidone, β-blockers, chloroquine, cloZAPine, dasatinib, dolasetron, droperidol, dronedarone, flecainide, halogenated/local anesthetics, local anesthetics, lapatinib, methadone, erythromycin, telithromycin, troleandomycin, octreotide, ondansetron, palonosetron, pentamidine, propafenone, ranolazine, SUNItinib, tacrolimus, vardenafil, vorinostat**

Decrease: absorption antacids containing aluminum, magnesium, sucralfate, zinc, iron; give 2 hr before or 3 hr after meals

NURSING CONSIDERATIONS
Assess:

• Renal, hepatic studies: BUN, creatinine, AST, ALT; I&O ratio, electrolytes

• CNS symptoms: insomnia, vertigo, headache, agitation, confusion

⚠ **Allergic reactions and anaphylaxis: rash, flushing, urticaria, pruritus, chills,** fever, joint pain; may occur a few days after therapy begins; EPINEPHrine and resuscitation equipment should be available for anaphylactic reaction

• **Pseudomembranous colitis:** bowel pattern daily; if severe diarrhea, fever, abdominal pain occur, product should be discontinued

• **QT prolongation:** avoid use of quinolones in patients with known QT prolongation, females and those with ongoing proarrhythmic conditions (TdP) are at a greater risk; monitor ECG and/or holter monitoring if product is used

• **Overgrowth of infection:** perineal itching, fever, malaise, redness, pain, swelling, drainage, rash, diarrhea, change in cough, sputum

Black Box Warning: **Tendon rupture:** tendon pain, inflammation; if present, discontinue use; more common when used with corticosteroids; discontinue immediately if tendon pain, inflammation occurs

• **Toxic psychosis/pseudotumor cerebri:** headache, blurred vision, neck/shoulder pain, nausea, vomiting, dizziness, tinnitus; discontinue immediately

Evaluate

• Therapeutic response: negative C&S, absence of signs, symptoms of infection

Teach patient/family:

• To take with/without food
• That fluids must be increased to 2 L/day to avoid crystallization in kidneys
• That if dizziness or lightheadedness occurs, to perform activities with assistance
• To complete full course of product therapy
• To contact prescriber if adverse reactions occur
• To avoid iron- or mineral-containing supplements or aluminum/magnesium antacids, buffered products within 2 hr before and 3 hr after dosing, 2 hr before sucralfate
• That photosensitivity may occur and sunscreen should be used
• To use frequent rinsing of mouth, sugarless candy or gum for dry mouth
• To avoid other medication unless approved by prescriber

gentamicin (Rx)

(jen-ta-mye′sin)

Func. class.: Antiinfective
Chem. class.: Aminoglycoside

Do not confuse:

gentamicin/kanamycin

ACTION: Interferes with protein synthesis in bacterial cell by binding to 30S ribosomal subunit, thus causing misreading of genetic code; inaccurate peptide sequence forms in protein chain, thereby causing bacterial death

USES: Severe systemic infections of CNS, respiratory, GI, urinary tract, bone, skin, soft tissues caused by susceptible strains of *Pseudomonas aeruginosa, Proteus, Klebsiella, Serratia, Escherichia coli, Enterobacter, Citrobacter, Staphylococcus, Shigella, Salmonella, Acinetobacter, Bacillus anthracis*

Unlabeled uses: Bartonellosis, bronchiectasis, cystic fibrosis, endocarditis prophylaxis, febrile neutropenia, gonorrhea, granuloma inguinale, PID, plaque, surgical infection prophylaxis, tubo-ovarian abscess, tularemia

CONTRAINDICATIONS: Hypersensitivity to this product, other aminoglycosides

Black Box Warning: Pregnancy (D)

Precautions: Breastfeeding, neonates, geriatric patients, pseudomembranous colitis

Black Box Warning: Renal disease, hearing deficits, myasthenia gravis, Parkinson's disease, infant botulism, neuromuscular disease, tinnitus

DOSAGE AND ROUTES

Severe systemic infections

- **Adult: IV INF** 3-5 mg/kg/day in divided doses q8hr; **IV** (pulse dosing, once-daily dosing) (unlabeled) 5-7 mg/kg; **IM** 3-5 mg/kg/day in divided doses q8hr
- **Child: IM/IV** 2-2.5 mg/kg q8hr; **IV** (pulse dosing, once daily dosing) (unlabeled) 5 mg/kg
- **Neonate and infant: IM/IV** 2.5 mg/kg q8-12hr
- **Neonate <1 wk: IV** 2.5 mg/kg q12

Renal dose-Regular dosing

- **Adult: IM/IV** CCr 70-100 ml/min, reduce dose by multiplying maintenance dose by 0.85, give q8-12hr; CCr 50-69 ml/min, reduce as above, give q12hr; CCr 25-49 ml/min, reduce as above, give q24hr; CCr <25 ml/min, reduce as above, give based on serum concs, give doses after dialysis

Renal dose: extend interval (unlabeled)

- **Adult: IV** CCr 40-59 ml/min 5-7 mg/kg q36hr; CCr 20-39 ml/min 5-7 mg/kg q48hr; CCr <20 ml/min 5-7 mg/kg once, then base on serial levels

Available forms: Inj 10, 40, 60, 80, 100 mg/ml; premixed inj 40, 60, 70, 80, 90, 100, 120/100 ml NS

Administer:

IM route

- IM inj in large muscle mass; rotate inj sites
- Product in evenly spaced doses to maintain blood level

Intermittent IV INF route

- After diluting in 50-200 ml NS, D_5W; decrease vol of diluent in child; maintain 0.1% sol run over ½-1 hr (adults) or up to 2 hr (children); flush IV line with NS, D_5W after administration

Syringe compatibilities: Clindamycin

Y-site compatibilities: Amifostine, amiodarone, amsacrine, atracurium, aztreonam, cefpirome, ciprofloxacin, cyclophosphamide, cytarabine, diltiazem, enalaprilat, esmolol, famotidine, filgrastim, fluconazole, fludarabine, foscarnet, granisetron, HYDROmorphone, IL-2, insulin, labetalol, LORazepam, magnesium sulfate, melphalan, meperidine, meropenem, midazolam, morphine, multivitamins, ondansetron, PACLitaxel, pancuronium, perphenazine, sargramostim, tacrolimus, teniposide, theophylline, thiotepa, tolazine, vecuronium, vinorelbine, vit B/C, zidovudine, and many more

SIDE EFFECTS

CNS: Confusion, depression, numbness, tremors, **seizures**, muscle twitching, **neurotoxicity**, dizziness, vertigo, **encephalopathy**, fever, headache, lethargy

CV: Hypo/hypertension, palpitations, edema

EENT: **Ototoxicity**, *deafness*, visual disturbances, tinnitus

GI: *Nausea, vomiting, anorexia;* increased ALT, AST, bilirubin; hepatomegaly, **hepatic necrosis**, splenomegaly

GU: **Oliguria, hematuria, renal damage, azotemia, renal failure, nephrotoxicity,** proteinuria

HEMA: **Agranulocytosis, thrombocytopenia, leukopenia**, eosinophilia, anemia

INTEG: *Rash*, burning, urticaria, dermatitis, alopecia, photosensitivity, **anaphylaxis**

MS: Twitching, myasthenia gravis-like symptoms

RESP: **Apnea**

PHARMACOKINETICS

Not metabolized, excreted unchanged in urine, crosses placental barrier

IM: Onset rapid, peak 30-60 min

IV: Onset immediate; peak 30-90 min; plasma half-life 1-2 hr, infants 6-7 hr; duration 6-8 hr

INTERACTIONS

- Do not use at the same time or physically mix with penicillins

Black Box Warning: Increase: ototoxicity, neurotoxicity, nephrotoxicity—other aminoglycosides, amphotericin B, polymyxin, vancomycin, ethacrynic acid, furosemide, mannitol, methoxyflurane, CISplatin, cephalosporins, penicillins, cidofovir, acyclovir, foscarnet, cycloSPORINE, tacrolimus, ganciclovir, zoledronic acid, pamidronate

Increase: effects—nondepolarizing neuromuscular blockers, digoxin, entecavir

Drug/Lab Test

Increase: LDH, AST, ALT, bilirubin, BUN, creatinine, eosinophils

Decrease: Hgb, WBC, platelet, granulocytes

NURSING CONSIDERATIONS

Assess:

Black Box Warning: Neuromuscular disease (myasthenia gravis, Parkinson's disease, infant botulism): paresthesias, tetany, positive Chvostek's/Trousseau's signs, confusion (adults), tetany, muscle weakness (infants); correct electrolyte imbalance

- Weight before treatment; calculation of dosage is usually based on ideal body weight but may be calculated on actual body weight

Black Box Warning: Renal disease: I&O ratio, urinalysis daily for proteinuria, cells, casts; report sudden change in urine output; urine pH if product is used for UTI; urine should be kept alkaline; urine for CCr testing, BUN, serum creatinine; lower dosage should be given with renal impairment (CCr <80 ml/min); toxicity is increased in patients with decreased renal function if high doses are given

- VS during inf; watch for hypotension, change in pulse
- IV site for thrombophlebitis, including pain, redness, swelling q30min, change site if needed; discontinue, apply warm compresses to site
- Serum peak drawn at 30-60 min after IV inf or 60 min after IM inj and trough level drawn just before next dose; blood level should be 2-4 times bacteriostatic level; peak (5-10 mcg/ml), trough (0.5-2 mcg/ml), depending on type of infection (based on traditional dosing)

Black Box Warning: Hearing deficits: eighth cranial nerve dysfunction by audiometric testing; also ringing, roaring in ears, vertigo; assess hearing before, during, after treatment

- Dehydration: high specific gravity, decrease in skin turgor, dry mucous membranes, dark urine
- **Overgrowth of infection:** fever, malaise, redness, pain, swelling, perineal itching, diarrhea, stomatitis, change in cough or sputum
- C&S before starting treatment to identify infecting organism
- **Vestibular dysfunction:** nausea, vomiting, dizziness, headache; product should be discontinued if severe
- Inj sites for redness, swelling, abscesses; use warm compresses at site
- Adequate fluids of 2-3 L/day unless contraindicated to prevent irritation of tubules
- Supervised ambulation, other safety measures with vestibular dysfunction

Evaluate:

- Therapeutic response: absence of fever, draining wounds, negative C&S after treatment

Teach patient/family:

- To report headache, dizziness, symptoms of overgrowth of infection, renal impairment

Black Box Warning: To report loss of hearing; ringing, roaring in ears; feeling of fullness in head

- To drink adequate fluids
- To avoid hazardous activities until reaction is known

gentamicin (ophthalmic)

(jen-ta-mye′sin)

Gentak, Garamycin

Func. class.: Ophthalmic anti-infective

Chem. class.: Aminoglycoside

Do not confuse: clindamycin/tobramycin/erythromycin/vancomycin/GenTeal

ACTION: Inhibits protein synthesis, thereby decreasing bacterial replication

USES: External ocular infections

CONTRAINDICATIONS: Hypersensitivity to this product or aminoglycosides

Precautions: Pregnancy (C), breastfeeding

DOSAGE AND ROUTES

Ophthalmic (solution)

- **Adult/adolescent/child ≥1 mo: SOL** 1-2 drops in affected eye(s) every 4 hr while awake × 2 days, then every 4 hr; severe injections ≤2 drops every 1 hr
- **Ointment:** apply a small amount (1/2″) to lower conjunctival sac bid or tid

Available forms: Ophthalmic ointment, ophthalmic solution 0.3%

Administer:

Ophthalmic route

- Commercially available ophthalmic solutions are not for injection subconjunctivally or into the anterior chamber of the eye
- Apply topically to the eye, taking care to avoid contamination
- Do not to touch the tip of the dropper to the eye, fingertips, or other surface
- Apply pressure to lacrimal sac for 1 min after instillation
- To apply the ointment, pull down gently on lower eyelid and apply a thin film of the ointment

SIDE EFFECTS

EENT: Burning, hypersensitivity, stinging, blurred vision, hyperemia, corneal ulcers

PHARMACOKINETICS

Unknown

NURSING CONSIDERATIONS

Assess:

- **Allergic reaction**: hypersensitivity, discontinue product

Evaluate:

- Decreased ophthalmic infection

Teach patient/family:

Ophthalmic route

- To apply topically to the eye, taking care to avoid contamination; for ophthalmic use only
- Not to touch the tip of the dropper to the eye, fingertips, or other surface
- To apply pressure to lacrimal sac for 1 min after instillation
- To apply the ointment by pulling down gently on lower eyelid and applying a thin film of the ointment

gentamicin (topical)

(jen-ta-mye′sin)

Func. class.: Topical anti-infective

Chem. class.: Aminoglycoside

Do not confuse: gentamicin/clindamycin

ACTION: Antibacterial activity results from inhibition of protein synthesis; bactericidal

USES: Superficial infections

CONTRAINDICATIONS: Hypersensitivity to this product or other aminoglycosides

Precautions: Infections, local sensitivity

DOSAGE AND ROUTES

- **Adult/child >1 yr:** apply to affected areas tid-qid

Available forms: Topical cream, ointment 0.1%
Administer:
• For external use only; do not use skin products near the eyes, nose, or mouth
• Wash hands before and after use; wash affected area and gently pat dry
• **Cream/Ointment:** Apply to the cleansed affected area, massage gently into affected areas

SIDE EFFECTS

INTEG: Rash, irritation, erythema, pruritus

PHARMACOKINETICS

Unknown

NURSING CONSIDERATIONS

Assess:
• **Allergic reaction**: hypersensitivity, product may need to be discontinued
• **Infection**: skin infection
Evaluate:
• Decreased skin infection
Teach patient/family:
• To use for external use only; do not use skin products near the eyes, nose, or mouth
• To wash hands before and after use; wash affected area and gently pat dry
• **Cream/Ointment:** to apply to the cleansed affected area and massage gently into affected areas

glatiramer (Rx)

(glah-tear'a-meer)
Copaxone
Func. class.: Multiple sclerosis agent
Chem.. class.: Biological response modifier

ACTION: Unknown; may modify the immune responses responsible for multiple sclerosis (MS)

USES: Reduction of the frequency of relapses in patients with relapsing or remitting MS after first clinical episode with MRI results consistent with MS

CONTRAINDICATIONS: Hypersensitivity to this product or mannitol, IV use
Precautions: Pregnancy (B), breastfeeding, children <18 yr, immune disorders, renal disease, infection, vaccinations, geriatric patients

DOSAGE AND ROUTES

• **Adult: SUBCUT** 20 mg/day
Available forms: Inj, premixed 20 mg/ml and 40 mg mannitol in single use syringe
Administer:
SUBCUT route
• If refrigerated, allow to warm for 20 min; visually inspect for particulate or cloudiness, if present discard; prefilled syringe contents is for single use; administer SUBCUT into hip, thigh, arm; discard unused portion
• Use SUBCUT route only; do not give IM or IV

SIDE EFFECTS

CNS: *Anxiety, hypertonia, tremor, vertigo,* speech disorder, *agitation,* confusion, flushing
CV: *Migraine, palpitations, syncope, tachycardia, vasodilation,* chest pain, hypertension
EENT: *Ear pain, blurred vision*
GI: *Nausea, vomiting, diarrhea, anorexia, gastroenteritis*
GU: *Urinary urgency, dysmenorrhea, vaginal moniliasis*
HEMA: *Ecchymosis, lymphadenopathy*
INTEG: *Pruritus, rash, sweating, urticaria, erythema,* inj site reaction
META: Edema, weight gain
MS: *Arthralgia, back pain, neck pain,* increased muscle tone
RESP: *Bronchitis, dyspnea, laryngismus, rhinitis*

PHARMACOKINETICS

May be hydrolyzed locally, may reach regional lymph nodes

NURSING CONSIDERATIONS

Assess:

• CNS symptoms: anxiety, confusion, vertigo

• GI status: diarrhea, vomiting, abdominal pain, gastroenteritis

• Cardiac status: tachycardia, palpitations, vasodilation, chest pain

Evaluate:

• Therapeutic response: decreased symptoms of MS

Teach patient/family:

• With written, detailed instructions about product; provide initial and return demonstrations on inj procedure; give information about use and disposal of product, inj site reaction (hives, rash, irritation, severe pain, flushing, chest pain)

• To notify prescribers of allergic reactions including itching, trouble breathing; chest pain, dizziness, sweating

• That irregular menses, dysmenorrhea, metrorrhagia, breast pain may occur; to use contraception during treatment

• That if pregnancy is suspected or if nursing to notify prescriber

• Not to change dosing or stop taking product without advice of prescriber

• About **immediate postinjection reaction:** flushing, chest pain, palpitations, anxiety, dyspnea, laryngeal constriction, urticaria; does not usually require treatment, may occur months after beginning treatment

• To take as directed; not to stop product or change schedule; teach on self-injection technique

glimepiride (Rx)

(glye-me′pi-ride)

Amaryl

glipiZIDE (Rx)

(glip-i′zide)

Glucotrol, Glucotrol XL

Func. class.: Antidiabetic

Chem. class.: Sulfonylurea (2nd generation)

Do not confuse:

glipiZIDE/Glucotrol/glyBURIDE

ACTION: Causes functioning β cells in pancreas to release insulin, leading to drop in blood glucose levels; may improve insulin binding to insulin receptors or increase the number of insulin receptors with prolonged administration; may also reduce basal hepatic glucose secretion; not effective if patient lacks functioning β cells

USES: Type 2 diabetes mellitus

CONTRAINDICATIONS: Hypersensitivity to sulfonylureas, type 1 diabetes, diabetic ketoacidosis

Precautions: Pregnancy (C), geriatric patients, cardiac disease, severe renal/hepatic disease, G6PD deficiency

DOSAGE AND ROUTES

Glimepiride

• **Adult: PO** 1-2 mg/day with breakfast, then increase by ≤2 mg/day q1-2wk, max 8 mg/day

• **Geriatric: PO** 1 mg/day; may increase if needed

Renal dose

• **Adult: PO** 1 mg/day with breakfast; may titrate upward as needed

GlipiZIDE

• **Adult: PO** 5 mg initially before breakfast then increase by 2.5-5 mg after several days to desired response; max 40 mg/day in divided doses or 15 mg/dose;

PO (XL) 5 mg/day with breakfast, may increase to 10 mg/day, max 20 mg/day
- **Geriatric: PO** 2.5 mg/day; may increase if needed

Hepatic disease
- **Adult: PO** 2.5 mg initially then increase to desired response; max 40 mg/day in divided doses or 15 mg/dose

Available forms: *Glimepiride:* tabs 1, 2, 4 mg; *glipiZIDE:* tabs, scored 5, 10 mg; ext rel tab (XL) 2.5, 5, 10 mg

Administer:
- Do not break, crush, or chew ext rel tabs; may crush tabs and mix with fluids if unable to swallow whole
- **GlipiZIDE:** product 30 min before meals (regular release); with breakfast (ext rel); **Glimepiride:** with breakfast; if patient is NPO, may need to hold dose to prevent hypoglycemia
- Gradual conversion from other oral hypoglycemics to these products is not needed; insulin ≥20 units/day, convert using 25% reduction in insulin dose every day or every other day
- Store in tight, light-resistant container at room temperature

SIDE EFFECTS

CNS: *Headache, weakness, dizziness, drowsiness,* tinnitus, fatigue, vertigo
ENDO: Hypoglycemia
GI: Hepatotoxicity, cholestatic jaundice, nausea, vomiting, diarrhea, heartburn
HEMA: Leukopenia, thrombocytopenia, agranulocytosis, aplastic anemia; increased AST, ALT, alk phos; pancytopenia, hemolytic anemia
INTEG: Rash, allergic reactions, pruritus, urticaria, eczema, photosensitivity, erythema, allergic vasculitis

PHARMACOKINETICS

PO: Completely absorbed by GI route; **glipiZIDE:** onset 1-1½ hr, peak 2-3 hr, duration 12-24 hr, half-life 2-4 hr; **glimepiride:** peak 2-3 hr, half-life 5 hr; metabolized in liver, excreted in urine, 90%-95% plasma-protein bound

INTERACTIONS

- May mask symptoms of hypoglycemia: β-blockers

Increase: action of digoxin, glycosides
Increase: hypoglycemic effects—insulin, MAOIs, cimetidine, chloramphenicol, guanethidine, methyldopa, NSAIDs, salicylates, probenecid, androgens, anticoagulants, clofibrate, fenfluramine, fluconazole, gemfibrozil, histamine H_2 antagonists, magnesium salts, phenylbutazone, sulfinpyrazone, sulfonamides, tricyclics, urinary acidifiers, clarithromycin, fibric acid derivatives, voriconazole
Decrease: hypoglycemic effect—thiazide diuretics, rifampin, isoniazid, cholestyramine, diazoxide, hydantoins, urinary alkalinizers, charcoal, corticosteroids

Drug/Herb
Increase: antidiabetic effect—garlic, horse chestnut
Decrease: hypoglycemic effect—green tea

Drug/Lab Test
Increase: AST, ALT, LDH, BUN, creatinine

NURSING CONSIDERATIONS

Assess:
- **Hypo/hyperglycemic reaction** that can occur soon after meals; for severe hypoglycemia give IV $D_{50}W$, then IV dextrose solution
- Blood glucose, A1c levels during treatment to determine diabetes control

⚠ Blood dyscrasias: CBC at baseline and throughout treatment; report decreased blood count

Evaluate:
- Therapeutic response: decrease in polyuria, polydipsia, polyphagia; clear sensorium; absence of dizziness; stable gait; improved serum glucose, A1c

Teach patient/family:
- Not to drink alcohol; about disulfiram reaction (nausea, headache, cramps, flushing, hypoglycemia)
- To report bleeding, bruising, weight gain, edema, SOB, weakness, sore throat

- To check for symptoms of cholestatic jaundice: dark urine, pruritus, yellow sclera; prescriber should be notified
- About symptoms of hypo/hyperglycemia, what to do about each; to have glucagon emergency kit available; to carry sugar packets
- That product must be continued on daily basis; about consequences of discontinuing product abruptly; to take product in morning to prevent hypoglycemic reactions at night
- To use sunscreen or stay out of the sun, wear protective clothing (photosensitivity)
- To avoid OTC medications unless ordered by prescriber
- That diabetes is a lifelong illness; product will not cure disease
- That all food in diet plan must be eaten to prevent hypoglycemia; to continue weight control, dietary restrictions, exercise, hygiene
- To carry emergency ID with prescriber and medication information
- To test using blood glucose meter while taking this product
- That ext rel tab may appear in stool

TREATMENT OF OVERDOSE:

Glucose 25 g IV via dextrose 50% sol, 50 ml, 1 mg glucagon, oral carbohydrate depending on severity

glucagon

(gloo′ka-gon)

GlucaGen

Func. class.: Antihypoglycemic

ACTION: Increases in blood glucose, relaxation of smooth muscle of the GI tract, and a positive inotropic and chronotropic effect on the heart; increases in blood glucose are secondary to stimulation of glycogenolysis

USES: Hypoglycemia, used to temporarily inhibit movement of GI tract as a diagnostic test

CONTRAINDICATIONS: Hypersensitivity, pheochromocytoma, insulinoma

Precautions: Pregnancy (B), breastfeeding, cardiac disease, adrenal insufficiency

DOSAGE AND ROUTES

Hypoglycemia in those with diabetes mellitus

- **Adult/adolescent/child ≥55 lb (25 kg): IM/IV/SUBCUT (GlucaGen)** 1 mg (1 IU)
- **Child <55 lb (25 kg) or <6-8 yr: IM/IV/SUBCUT** 0.5 mg (0.5 IU) (Glucagon)
- **Adult/adolescent/child ≥44 lb (20 kg): IM/IV/SUBCUT** (Glucagon) 1 mg (1IU)
- **Child <44 lb (20 kg):** (Glucagon) 0.5 mg (0.5 IU) or 0.02-0.03 mg/kg (IU/kg) max/mg

Available forms: Powder for injection 1-mg vial

Administer:

- Visually inspect for particulate matter and discoloration before use

Reconstitution: Reconstitute with 1 ml of sterile water for injection or with diluent supplied by the manufacturer; the reconstituted injection should be clear and of water-like consistency (1 mg/ml); discard any unused portion

IM route

- Inject into a large muscle mass; aspirate before injection to avoid injection into a blood vessel

SUBCUT route

- Inject, taking care not to inject intradermally

IV route

- Inject directly into a vein at a rate ≤1 IU/min (mg/min); may be given through line running D_5W or given at the same time as a bolus of dextrose

SIDE EFFECTS

CNS: Dizziness, headache
CV: Hypotension
GI: Nausea, vomiting
SYST: Hypersensitivity

PHARMACOKINETICS

IV: Onset immediate, peak 30 min, duration 1-1½ hr
IM/subcut: Onset 5-10 min, peak 13-20 min, duration 12-30 min

INTERACTIONS

Increase: bleeding risk—anticoagulants

NURSING CONSIDERATIONS

Assess:

- **Hypoglycemia:** monitor glucose levels before and after product use; use other products to control hypoglycemia if patient is conscious

Evaluate:

- Decreased hypoglycemia

Teach patient/family:

- How to use this product, symptoms of hypoglycemia

glucarpidase

(glu-car-pie′dase)

Voraxaze

Func. class.: Recombinant bacterial enzyme

ACTION: A recombinant bacterial enzyme that hydrolyzes glutamate residue from methotrexate and provides an alternate nonrenal pathway for methotrexate elimination in patients with renal dysfunction during high-dose methotrexate treatment; as a result of its enzymatic action, converts methotrexate to its inactive metabolites

USES: The treatment of methotrexate toxicity (methotrexate plasma conc >1 micromole/liter) in those with delayed methotrexate clearance due to impaired renal function

Precautions: Pregnancy (C), breastfeeding

DOSAGE AND ROUTES

- **Adult/adolescent/child: IV** 50 units/kg over 5 min; continue to give leucovorin, IV hydration, urinary alkalinization as needed

Available forms: Sol for IV

Administer:

IV direct route

- Visually inspect parenteral products for particulate matter and discoloration before administration
- Reconstitution: add 1 ml of sterile saline for injection to vial, roll and tilt the vial gently to mix; do not shake; flush the IV line then inject as a bolus inj over 5 min; flush the IV line again
- Do not administer to patients who exhibit the expected clearance of methotrexate (plasma methotrexate concs within 2 standard deviations of the mean methotrexate excretion curve specific for the dose of methotrexate administered)
- Do not administer to patients with normal or mildly impaired renal function because of the potential risk of subtherapeutic exposure to methotrexate
- Continue to administer leucovorin after glucarpidase, but do not administer leucovorin within 2 hr before or after a glucarpidase; for the first 48 hr after glucarpidase administration, administer the same leucovorin dose as given before glucarpidase; after 48 hr after glucarpidase, administer leucovorin based on the measured methotrexate conc; do not discontinue therapy with leucovorin based on the determination of a single methotrexate conc below the leucovorin treatment threshold; continue leucovorin until the methotrexate conc has been maintained below the leucovorin treatment threshold for a minimum of 3 days; use of a chromatographic method to determine methotrexate concs is needed for the first 48 hr after glucarpidase receipt
- Continue hydration and alkalinization of the urine as needed
- Store reconstituted solution refrigerated ≤4 hr; discard any unused product, no preservative is present

SIDE EFFECTS

CNS: Paresthesias, headache
CV: Hypo/hypertension
EENT: Throat irritation or tightness
GI: Nausea, vomiting
INTEG: Flushing, hot, burning sensation; rash, hypersensitivity, serious allergic reactions
SYST: Antibody formation

PHARMACOKINETICS

Elimination half-life of 5.6 hr, methotrexate conc reduced 97% within 15 min

INTERACTIONS

Decrease: elucarpidase level—PEMEtrexed, PRALAtrexate; avoid concurrent use

NURSING CONSIDERATIONS

Assess:

- Continue monitoring of methotrexate blood levels, renal status (creatinine/BUN) after glucarpidase is given; methotrexate concs within 48 hr of glucarpidase are only reliable measure by a chromatographic method; leucovorin receipt is needed until methotrexate level has been maintained below leucovorin treatment threshold for a minimum of 3 days

Evaluate:

- For resolution of methotrexate toxicity

Teach patient/family:

- To immediately report any inf-related reactions: fever, chills, flushing, feeling hot, rash, hives, itching, throat tightness or breathing problems, tingling, numbness, or headache

glyBURIDE (Rx)

(glye′byoor-ide)

DiaBeta, Glynase PresTab

Func. class.: Antidiabetic

Chem. class.: Sulfonylurea (2nd generation)

Do not confuse:
glyBURIDE/Glucotrol/glipiZIDE
DiaBeta/Zebeta

ACTION: Causes functioning β cells in pancreas to release insulin, thereby leading to a drop in blood glucose levels; may improve insulin binding to insulin receptors and increase number of insulin receptors with prolonged administration; may also reduce basal hepatic glucose secretion; not effective if patient lacks functioning β cells

USES: Type 2 diabetes mellitus
Unlabeled use: Gestational diabetes not controlled by diet

CONTRAINDICATIONS: Hypersensitivity to sulfonylureas, type 1 diabetes, diabetic ketoacidosis, renal failure
Precautions: Pregnancy (B) Glynase; (C) DiaBeta, geriatric patients, cardiac/thyroid disease, severe renal/hepatic disease, severe hypoglycemic reactions, sulfonamide/sulfonylurea hypersensitivity, G6PD deficiency

DOSAGE AND ROUTES

DiaBeta (nonmicronized)

- **Adult: PO** 1.25-5 mg initially then increase to desired response at weekly intervals up to 20 mg/day; may be given as a single or divided dose
- **Geriatric: PO** 1.25 mg initially then increase to desired response; max 20 mg/day, maintenance 1.25-20 mg/day

Glynase PresTab (micronized)

- **Adult: PO** 1.5-3 mg/day initially, may increase by 1.5 mg/wk, max 12 mg/day
- **Geriatric: PO** 0.75-3 mg/day, may increase by 1.5 mg/wk

Gestational diabetes (unlabeled)

- **Adult (pregnant female): PO** 2.5 mg/day titrated up to 20 mg/day (conventional glyBURIDE)

Available forms: Tabs (DiaBeta) 1.25, 2.5, 5 mg (nonmicronized); (Glynase PresTab) 1.5, 3, 6 mg (micronized)

Administer:

- With breakfast as single or divided dose; hold dose if patient NPO to avoid hypoglycemia; take at same time each day
- Gradual conversion from other oral hypoglycemics to product is not needed; insulin ≥20 units/day; convert using 25% reduction in insulin dose every day
- Micronized glyBURIDE/nonmicronized glyBURIDE are not equivalent
- Store in tight container in cool environment

SIDE EFFECTS

CNS: *Headache, weakness,* paresthesia, tinnitus, fatigue, vertigo

EENT: Blurred vision

ENDO: Hypoglycemia

GI: Nausea, fullness, heartburn, hepatotoxicity, cholestatic jaundice, vomiting, diarrhea, hepatic failure

HEMA: Leukopenia, thrombocytopenia, agranulocytosis, aplastic anemia (rare), increased AST, ALT, alk phos

INTEG: Rash, allergic reactions, pruritus, urticaria, eczema, photosensitivity, erythema

MISC: Angioedema

MS: Joint pain, vasculitis

PHARMACOKINETICS

PO: Completely absorbed by GI route; onset 2 hr; peak 2-4 hr; duration 24 hr; half-life 10 hr; metabolized in liver; excreted in urine, feces (metabolites); crosses placenta; 99% plasma-protein bound

INTERACTIONS

Increase: masking symptoms of hypoglycemia—β-blockers

Increase: level—digoxin

Increase: hypoglycemic effects—insulin, MAOIs, oral anticoagulants, chloramphenicol, guanethidine, methyldopa, NSAIDs, salicylates, probenecid, androgens, fenfluramine, fluconazole, gemfibrozil, histamine H_2 antagonists, magnesium salts, phenylbutazone, sulfinpyrazone, sulfonamides, tricyclics, urinary acidifiers, β-blockers, clarithromycin

Increase: LFTs—bosentan; avoid concurrent use

Increase: triglyceride levels—colesevelam

Increase: action of—cycloSPORINE

Decrease: both products' effects—diazoxide

Decrease: glyBURIDE action—thiazide diuretics, rifampin, isoniazid, cholestyramine, hydantoins, urinary alkalinizers, charcoal, corticosteroids, phenothiazines; oral contraceptives, estrogens, thyroid, voriconazole

Drug/Herb

Increase: antidiabetic effect—garlic, horse chestnut

Decrease: hypoglycemic effect—green tea

Drug/Lab Test

Increase: AST, ALT, LDH, BUN, creatinine

Decrease: Hgb, sodium, glucose

NURSING CONSIDERATIONS

Assess:

- Hypo/hyperglycemic reaction that can occur soon after meals; for severe hypoglycemia, give IV $D_{50}W$ then IV dextrose sol
- Blood glucose; A1c levels during treatment

⚠ **Blood dyscrasias:** CBC at baseline, throughout treatment; report decreased blood counts

Evaluate:

- Therapeutic response: decrease in polyuria, polydipsia, polyphagia; clear sensorium; absence of dizziness; stable gait; improved serum glucose, A1c

Teach patient/family:

- To check for symptoms of cholestatic jaundice: dark urine, pruritus, jaundiced sclera; if these occur, notify prescriber
- To use a blood glucose meter for testing while taking this product
- About the symptoms of hypo/hyperglycemia, what to do about each
- That product must be continued on a daily basis; about consequences of discontinuing product abruptly; that in times of stress, infection, surgery, trauma, a higher dose may be needed
- To take product in morning to prevent hypoglycemic reactions at night if taking once a day
- To avoid OTC medications unless ordered by prescriber
- To report bleeding, bruising, weight gain, edema, shortness of breath, weakness, sore throat
- That diabetes is a lifelong illness; that product will not cure disease
- That all food included in diet plan must be eaten to prevent hypoglycemia; to have glucagon emergency kit, sugar packets always available
- To use sunscreen or stay out of the sun, wear protective clothing (photosensitivity)
- To carry an emergency ID with prescriber and medication information

TREATMENT OF OVERDOSE:

Glucose 25 g IV via dextrose 50% sol, 50 ml, 1 mg glucagon, oral carbohydrate depending on severity

RARELY USED

glycerin (OTC)

(gli′ser-in)

Sani-Supp

Func. class.: Laxative, hyperosmotic

USES: Constipation

CONTRAINDICATIONS: Hypersensitivity

DOSAGE AND ROUTES

Laxative

- **Adult and child >6 yr: RECT SUPP** 3 g; **ENEMA** 5-15 ml
- **Child <6 yr: RECT SUPP** 1-1.8 g; **ENEMA** 2-5 ml

golimumab (Rx)

(goal-lim′yu-mab)

Simponi, Simponi Aria

Func. class.: Antirheumatic agent (disease modifying), immunomodulator

Chem. class.: Monoclonal antibody, DMARD, tumor necrosis factor (TNF-α) modifier

ACTION: Monoclonal antibody specific for human tumor necrosis factor (TNF); elevated levels of TNF are found in patients with rheumatoid arthritis

USES: Rheumatoid arthritis (RA), ankylosing spondylitis, psoriatic arthritis, ulcerative colitis

CONTRAINDICATIONS: Hypersensitivity, active infections

Precautions: Pregnancy (B), breastfeeding, children, geriatric patients, CNS demyelinating disease, Guillain-Barré syndrome, CHF, hepatitis B carriers, blood dyscrasias, surgery, MS, neurologic disease, diabetes, immunosuppression

Black Box Warning: Neoplastic disease, TB; fungal, bacterial, viral infections

DOSAGE AND ROUTES

Rheumatoid arthritis:

- **Adult: SUBCUT** 50 mg monthly; for RA, give with methotrexate; IV (Simponi Aria only) 2 mg/kg over 30 min, repeat 4 wk later, then q8wk give with methotrexate

Ulcerative colitis:
• **Adult: SUBCUT** 200 mg for 1 dose, then 100 mg in 2 wk, maintenance 100 mg q4wk starting at wk 6
Available form: Inj 50 mg/0.5 ml prefilled syringe, SmartJect Auto Injector
Administer:
SUBCUT route
• Refrigerate, do not freeze; allow to warm to room temperature before using
• Visually inspect sol for particulate or discoloration; sol should be clear to slightly opalescent and colorless to slightly yellow; there may be tiny white particles; do not shake; rotate injection sites
• **SmartJect Auto Injector:** allow to reach room temperature for 30 min before use; remove cap, and inject within 5 min of removing cap; do not put cap back on; place open end against the inj site at a 90-degree angle without pushing button; push the injector firmly against the skin; press the button once and release; listen for the first click; wait for the second click or 15 sec, then remove injector; do not rub site
• **Prefilled syringe:** allow to warm to room temperature for 30 min; remove needle cover by pulling straight off; do not twist or recap; inject within 5 min of needle cover removal; hold syringe in one hand like a pencil and use other hand to pinch the skin; inject needle at a 45-degree angle; push plunger down as far as it will go; keep pressure on plunger head and remove needle from skin; remove pressure from plunger head; needle guard will cover needle; do not rub site
IV route
(Simponi Aria)
• Calculate number of vials needed; do not shake; dilute total volume of product in NS to yield 100 ml for inf; slowly add product, mix gently
• Infuse over 30 min use inf set with in-line, sterile, nonpyrogenic, low-protein binding filter (≤0.22 mm pore size)

SIDE EFFECTS

CNS: Dizziness, paresthesia, CNS demyelinating disorder, weakness
CV: Hypertension, CHF
GI: Hepatitis
HEMA: Agranulocytosis, aplastic anemia, leukopenia, polycythemia, thrombocytopenia, pancytopenia
INTEG: Psoriasis
MISC: Increased cancer risk, antibody development to this drug; risk for infection (TB, invasive fungal infections, other opportunistic infections), may be fatal, inj site reactions

PHARMACOKINETICS

Terminal half-life 2 wk

INTERACTIONS

• Do not give concurrently with live vaccines; immunizations should be brought up to date before treatment
• Dosage change may be needed: warfarin, cycloSPORINE, theophylline
Increase: infection—abatacept, etanercept, rilonacept, riTUXimab, adalimumab, anakinra, immunosuppressants, inFLIXimab

NURSING CONSIDERATIONS

Assess:
• **Pain,** stiffness, ROM, swelling of joints during treatment
• Inj site pain, swelling; usually occur after 2 inj (4-5 days)

Black Box Warning: TB: obtain TB skin test before starting treatment, treat latent TB before starting therapy

⚠ **Blood dyscrasias:** CBC, differential before and periodically during treatment

Black Box Warning: Infection: fever, flulike symptoms, dyspnea, change in urination, redness/ swelling around any wounds; stop treatment if present; some serious infections including sepsis may occur, may be fatal; patients with active infections should not be started on this product; obtain a chest x-ray, fungal serology, TB testing before starting treatment

- LFTs; hepatitis B serology, may reactivate HBV
- **CHF:** B/P, pulse, edema, SOB
- **Psoriasis:** may occur or worsen

Black Box Warning: Neoplastic disease: may occur in those <18 yr, avoid use in those with known malignancies

Evaluate:

- Therapeutic response: decreased inflammation, pain in joints, decreased joint destruction

Teach patient/family:

- About self-administration if appropriate: inj should be made in thigh, abdomen, upper arm; rotate sites at least 1 inch from old site; do not inject in areas that are bruised, red, hard
- That, if medication not taken when due, to inject next dose as soon as remembered and then the following dose as scheduled
- Not to receive any live virus vaccines during treatment

Black Box Warning: To report signs, symptoms of infection, allergic reaction, or lupuslike syndrome

- To notify prescriber if pregnancy is planned or suspected; not to breastfeed

goserelin (Rx)

(goe′se-rel-lin)

Zoladex

Func. class.: Gonadotropin-releasing hormone, antineoplastic (hormone)

Chem. class.: Synthetic decapeptide analog of LHRH

ACTION: Inhibitor of pituitary gonadotropin secretion; initially increases LH and FSH, with increases in testosterone, reduction in sex steroid levels (substitute serum testosterone levels)

USES: Advanced and locally confined prostate cancer stage B2-C (10.8 mg), endometriosis, advanced breast cancer, endometrial thinning (3.6 mg)

CONTRAINDICATIONS: Hypersensitivity; pregnancy D (breast cancer), X (endometriosis); breastfeeding, children, nondiagnosed vaginal bleeding; hypersensitivity to LHRH, LHRH-agonist analogs; 10.8 mg dose contraindicated in women

Precautions: Spinal cord decompression, renal disease, bone mineral density loss, hyperglycemia, diabetes mellitus, CV disease

DOSAGE AND ROUTES

Breast cancer

- **Adult: SUBCUT** 3.6 mg q28days or 10.8 mg q12wk

Endometrial thinning

- **Adult: SUBCUT** 3.6 mg 1-2 depot inj, usually 1 depot, surgery performed at 4 wk; if 2 depots, surgery performed 2-4 wk after 2nd depot

Available forms: Depot inj 3.6, 10.8 mg

Administer:

Depot

- SUBCUT using implant, inserted by qualified person into upper subcutaneous tissue in abdominal wall q28days or q12wk (10.8 mg); do not attempt to remove air bubbles from syringe

SIDE EFFECTS

CNS: *Headaches,* **spinal cord compression,** *anxiety, depression, dizziness, insomnia, lethargy,* hot flashes, emotional lability

CV: **Dysrhythmia, cerebrovascular accident,** hypertension, chest pain, **CHF;** MI, sudden cardiac death, stroke (men), *peripheral edema*

ENDO: Gynecomastia, breast tenderness, breast enlargement, *hot flashes;* hyperglycemia, diabetes (men)

GI: *Nausea,* vomiting, constipation, diarrhea, ulcer

GU: *Spotting, breakthrough bleeding, decreased libido,* renal insufficiency, urinary obstruction, urinary tract infection, *impotence*

INTEG: Rash, pain on inj, diaphoresis

MS: Osteoneuralgia
RESP: COPD, URI

PHARMACOKINETICS

Peak 12-15 days, half-life 4½ hr, 30% protein bound

INTERACTIONS

Drug/Lab Test
Increase: alk phos, estradiol, FSH, LH, testosterone levels, triglycerides
Decrease: testosterone levels, progesterone

NURSING CONSIDERATIONS

Assess:

- **Reproductive studies:** pelvic ultrasound, pelvic exam, PSA, serum estradiol/testosterone, pregnancy test before therapy
- I&O ratios; palpate bladder for distention in urinary obstruction
- **Cancer metastases:** for relief of bone pain (back pain), change in motor function
- Blood studies: lipid profile, acid phosphatase; calcium; hypercalcemia may occur

Evaluate:

- Therapeutic response: more normal levels of PSA, acid phosphatase, alk phos; testosterone level of <25 ng/dl; thinning of endometrial lining

Teach patient/family:

- To continue with appointments monthly
- That hyperglycemia may occur in diabetic patients
- That gynecomastia and postmenopausal symptoms may occur but will decrease when treatment is discontinued
- That bone pain may increase then decrease
- To notify prescriber of difficulty urinating, hot flashes
- Not to breastfeed; to use effective nonhormonal contraception; to notify prescriber if menstrual period continues

granisetron (Rx)

(grane-iss′e-tron)

Granisol, Kytril, Sancuso

Func. class.: Antiemetic

Chem. class.: 5-HT_3 receptor antagonist

ACTION: Prevents nausea, vomiting by blocking serotonin peripherally, centrally, and in the small intestine

USES: Prevention of nausea, vomiting associated with cancer chemotherapy, including high-dose CISplatin, radiation
Unlabeled uses: Acute nausea, vomiting after surgery

CONTRAINDICATIONS: Hypersensitivity to this product, benzyl alcohol
Precautions: Pregnancy (B), breastfeeding, children, geriatric patients, ondansetron/palonosetron/dolasetron hypersensitivity, cardiac dysrhythmias, cardiac/hepatic disease, electrolyte imbalances

DOSAGE AND ROUTES

Nausea, vomiting in chemotherapy

- **Adult and child ≥2 yr: IV** 10 mcg/kg over 5 min, 30 min before the start of cancer chemotherapy; **TD** apply 1 patch (3.1 mg/24 hr) to upper outer arm 24-48 hr before chemotherapy, patch may be worn up to 7 days
- **Adult: PO** 1 mg bid, give 1st dose 1 hr before chemotherapy and next dose 12 hr after 1st or 2 mg as a single dose anytime within 1 hr before chemotherapy

Nausea, vomiting in radiation therapy

- **Adult: PO** 2 mg/day 1 hr before radiation

Available forms: Inj 1 mg/ml; tab 1 mg; oral sol 2 mg/10 ml; patch TD 3.1 mg/24 hr

Administer:

- Chemotherapy/radiation: given on day of chemotherapy or radiation

PO route

• Give dose 1 hr before chemotherapy/radiation and another 12 hr after 1st dose

Direct IV route

• May give undiluted over 30 sec via Y-site

Intermittent IV INF route

• Dilute in 0.9% NaCl for inj or D_5W (20-50 ml); give over 5-15 min 30 min before chemotherapy

• Store at room temperature for 24 hr after dilution

Additive compatibilities: Dexamethasone, methylPREDNISolone

Solution compatibilities: D_5W, 0.9% NaCl

Y-site compatibilities: Acyclovir, allopurinol, amifostine, amikacin, aminophylline, amphotericin B cholesteryl, ampicillin, ampicillin/sulbactam, amsacrine, aztreonam, bleomycin, bumetanide, buprenorphine, butorphanol, calcium gluconate, CARBOplatin, carmustine, ceFAZolin, cefepime, cefonicid, cefoperazone, cefotaxime, cefoTEtan, cefOXitin, cefTAZidime, ceftizoxime, cefTRIAXone, cefuroxime, chlorproMAZINE, cimetidine, ciprofloxacin, CISplatin, cladribine, clindamycin, cyclophosphamide, cytarabine, dacarbazine, DACTINomycin, DAUNOrubicin, dexamethasone, diphenhydrAMINE, DOBUTamine, DOPamine, DOXOrubicin, DOXOrubicin liposome, doxycycline, droperidol, enalaprilat, etoposide, famotidine, filgrastim, fluconazole, fluorouracil, floxuridine, fludarabine, furosemide, gallium, ganciclovir, gentamicin, haloperidol, heparin hydrocortisone, HYDROmorphone, hydrOXYzine, IDArubicin, ifosfamide, imipenem-cilastatin, leucovorin, LORazepam, magnesium sulfate, melphalan, meperidine, mesna, methotrexate, methylPREDNISolone, metoclopramide, metroNIDAZOLE, mezlocillin, miconazole, minocycline, mitoMYcin, mitoXANtrone, morphine, nalbuphine, netilmicin, ofloxacin, PACLitaxel, piperacillin, piperacillin/tazobactam, plicamycin, potassium chloride, prochlorperazine, promethazine, propofol, ranitidine, sargramostim, sodium bicarbonate, streptozocin, teniposide, thiotepa, ticarcillin, ticarcillin/clavulanate, tobramycin, trimethoprim-sulfamethoxazole, vancomycin, vinBLAStine, vinCRIStine, vinorelbine, zidovudine

Transdermal route

• Apply to dry, clean, intact skin of upper outer arm 24-48 hr before chemotherapy, firmly press on skin, keep on during chemotherapy; can bathe, avoid swimming, whirlpool; remove ≥24 hr after chemotherapy completion; do not cut patch

SIDE EFFECTS

CNS: *Headache, asthenia,* anxiety, dizziness

CV: Hypertension, **QT prolongation**

GI: Diarrhea, *constipation,* increased AST, ALT, *nausea*

HEMA: **Leukopenia**, anemia, **thrombocytopenia**

MISC: Rash, **bronchospasm**

PHARMACOKINETICS

Metabolized in liver to an active metabolite, half-life 10-12 hr, protein binding 65%

INTERACTIONS

Increase: EPS—antipsychotics

Increase: QT prolongation—amoxapine, β-blockers, chloroquine, class IA, III antidysrhythmics, cloZAPine, dasatinib, dolasetron, dronedarone, droperidol, erythromycin, flecainide, halogenated/local anesthetics, haloperidol, lapatinib, maprotiline, methadone, octreotide, ondansetron, palonosetron, pentamidine, phenothiazines, pimozide, propafenone, ranolazine, risperiDONE, sertindole, SUNItinib, tacrolimus, telithromycin, tricyclics, troleandomycin, vardenafil, vorinostat, ziprasidone

NURSING CONSIDERATIONS

Assess:

- For absence of nausea, vomiting during chemotherapy
- **Hypersensitivity reaction:** rash, bronchospasm

Evaluate:

- Therapeutic response: absence of nausea, vomiting during cancer chemotherapy

Teach patient/family:

- To report diarrhea, constipation, rash, changes in respirations
- That headache requiring an analgesic is common

guaiFENesin (OTC, Rx)

(gwye-fen′e-sin)

Alfen, Altarussin, Balminil ✤, Bidex, Diabetic Tussin, Elix Sure EX, Equaline Non-Drowsy Tussin, Equate Tussin Cough, Good Sense Mucus Relief, Good Sense Tussin Chest Congestion, GuaiFENesin NR, Guiatuss, Humibid, Iophen NR, Liquibid, Miltuss EX, Mucinex, Mucinex Children's Mucus Relief, Mucinex Junior Strength, Naldecon Senior EX, Organ-1NR, Organidin NR, Q-Tussin, Robafen, Scot-Tussin Expectorant, Siltussin DAS, Siltussin SA, Top Care Children's Mucus Relief, Top Care Tussin Chest Congestion Solution, Wal-Tussin Expectorant, XPect

Func. class.: Expectorant

ACTION: Increases the volume and reduces the viscosity of secretions in the trachea and bronchi to facilitate secretion removal

USES: Productive and nonproductive cough

CONTRAINDICATIONS: Hypersensitivity; chronic, persistent cough

Precautions: Pregnancy (C), breastfeeding, CHF, asthma, emphysema, fever

DOSAGE AND ROUTES

- **Adult and adolescent: PO** 200-400 mg q4hr; **EXT REL** 600-1200 mg q12hr, max 2.4 g/day
- **Child 6-11 yr: PO** 100-200 mg q4hr; **EXT REL** 600 mg q12hr, max 1.2 g/day
- **Child 2-5 yr: PO** 50-100 mg q4hr; max 600 mg/day; ext rel 300 mg q12hr, max 600 mg/day

Available forms: Tabs 200, 400 mg; oral sol 100 mg/5 ml; ext rel tabs 600, 1200 mg; syrup 100 mg/5 ml; oral granules 50, 100 mg/packet

Administer:

- Do not break, crush, chew ext rel tabs
- Store at room temperature

SIDE EFFECTS

CNS: Drowsiness, headache, dizziness

GI: Nausea, anorexia, vomiting, diarrhea

PHARMACOKINETICS

Half-life 1 hr, excreted in urine (metabolites)

NURSING CONSIDERATIONS

Assess:

- **Cough:** type, frequency, character, including sputum; fluids should be increased to 2 L/day
- Increased fluids, room humidification to liquefy secretions

Evaluate:

- Therapeutic response: productive cough, thinner secretions

Teach patient/family:

- To avoid driving, other hazardous activities if drowsiness occurs (rare)
- To avoid smoking, smoke-filled room, perfumes, dust, environmental pollutants, cleansers
- To consult health provider if cough lasts >7 days

halcinonide topical

See Appendix B

haloperidol (Rx)

(hal-oh-pehr′ih-dol)

haloperidol decanoate (Rx)

Haldol Decanoate

haloperidol lactate (Rx)

Haldol

Func. class.: Antipsychotic, neuroleptic

Chem. class.: Butyrophenone

Do not confuse:
haloperidol/Halotestin

ACTION: Depresses cerebral cortex, hypothalamus, limbic system, which control activity and aggression; blocks neurotransmission produced by DOPamine at synapse; exhibits strong α-adrenergic, anticholinergic blocking action; mechanism for antipsychotic effects unclear

USES: Psychotic disorders, control of tics, vocal utterances in Gilles de la Tourette's syndrome, short-term treatment of hyperactive children showing excessive motor activity, prolonged parenteral therapy in chronic schizophrenia, organic mental syndrome with psychotic features, hiccups (short-term), emergency sedation of severely agitated or delirious patients, ADHD

Unlabeled uses: Nausea, vomiting during surgery; autism; migraine

CONTRAINDICATIONS: Hypersensitivity, coma, Parkinson's disease

Precautions: Pregnancy (C), breastfeeding, geriatric patients, seizure disorders, hypertension, pulmonary/cardiac/hepatic disease, QT prolongation, torsades de pointes, prostatic hypertrophy, hyperthyroidism, thyrotoxicosis, children, blood dyscrasias, brain damage, bone marrow depression, alcohol and barbiturate withdrawal states, angina, epilepsy, urinary retention, closed-angle glaucoma, CNS depression

Black Box Warning: Increased mortality in elderly patients with dementia-related psychosis

DOSAGE AND ROUTES

Acute psychosis

- **Adult: IM/IV** (lactate) 2-10 mg, may repeat q1hr, convert to **PO** as soon as possible, **PO** should be 150% of total parenteral dose required
- **Child 6-12 yr: IM/IV** (lactate) (unlabeled) 1-3 mg q4-8hr, max 0.15 mg/kg/day, switch to **PO** as soon as possible

Chronic schizophrenia

- **Adult: IM** (decanoate) 50-100 mg q4wk, max 100 mg for 1st inj
- **Child 3-12 yr: PO/IM** 0.05-0.15 mg/kg/day

Tourette's syndrome

- **Adult and adolescent: PO** 0.5-2 mg bid-tid, increase until desired response occurs
- **Child 3-12 yr or weighing 15-40 kg: PO** 0.25-0.5 mg/day in 2-3 divided doses, increase by 0.25-0.5 mg q5-7days, max 0.15 mg/kg/day

ADHD

- **Child 3-12 yr: PO** 0.25-0.5 mg/day in 2-3 divided doses, may increase by 0.025-0.5 mg q5-7days; maintenance 0.01-0.03 mg/kg/day as a single dose

Autism (unlabeled)

- **Child: PO** 0.04 mg/kg/day or 1-3 mg/day, max 4 mg/day

Migraine (unlabeled)

- **Adult: PO/IM** (lactate) 5 mg at onset, may repeat once

Available forms: Tabs 0.5, 1, 2, 5, 10, 20 mg; **lactate:** oral sol 2 mg/ml; inj 5 mg/ml, **decanoate:** 50 mg/ml, 100 mg/ml

Administer:

- Reduced dose to geriatric patients
- Antiparkinsonian agent if EPS occurs
- Avoid use with CNS depressants

PO route

- Store in tight, light-resistant container

- **Oral liquid:** use calibrated device; do not mix in coffee or tea
- PO with food or milk
- Avoid skin contact with oral suspension or solution—may cause contact dermatitis

IM route

- IM inj into large muscle mass, use 21-G, 2-in needle; give no more than 3 ml/inj site; patient should remain recumbent for 30 min

IV route (lactate)

- Give undiluted for psychotic episode at 5 mg/min
- Only use lactate for IV

Intermittent IV INF route

- Dilute in 30-50 ml of D_5W, give over 30 min

Y-site compatibilities: Amifostine, amsacrine, cisatracurium, cladribine, DOXOrubicin liposome, filgrastim, fludarabine, granisetron, LORazepam, melphalan, milrinone, PACLitaxel, palonosetron, pamidronate, pancuronium, propofol, remifentanil, riTUXimab, rocuronium, tacrolimus, teniposide, thiotepa, tigecycline, tirofiban, TPN, trastuzumab, vecuronium, vinCRIStine, vinorelbine, voriconazole, zoledronic acid

SIDE EFFECTS

CNS: *EPS: pseudoparkinsonism, akathisia, dystonia, tardive dyskinesia, drowsiness, headache,* seizures, neuroleptic malignant syndrome, confusion

CV: *Orthostatic hypotension,* hypertension, cardiac arrest, ECG changes, tachycardia, QT prolongation, sudden death, torsades de pointes

EENT: Blurred vision, glaucoma, dry eyes

GI: *Dry mouth, nausea, vomiting, anorexia, constipation,* diarrhea, jaundice, weight gain, ileus, hepatitis

GU: Urinary retention, dysuria, urinary frequency, enuresis, impotence, amenorrhea, gynecomastia

INTEG: *Rash,* photosensitivity, dermatitis

RESP: Laryngospasm, dyspnea, respiratory depression

SYST: Risk for death (dementia)

PHARMACOKINETICS

Metabolized by liver; excreted in urine, bile; crosses placenta; enters breast milk; protein binding 92%; terminal half-life 12-36 hr (metabolites)

PO: Onset erratic, peak 2-6 hr, half-life 24 hr

IM: Onset 15-30 min, peak 15-20 min, half-life 21 hr

IM (Decanoate): Peak 4-11 days, half-life 3 wk

INTERACTIONS

Increase: serotonin syndrome, neuroleptic malignant syndrome—SSRIs, SNRIs

Increase: QT prolongation—class IA, III antidysrhythmics, tricyclics, amoxapine, maprotiline, phenothiazines, pimozide, risperiDONE, sertindole, ziprasidone, β-blockers, chloroquine, cloZAPine, dasatinib, dolasetron, droperidol, dronedarone, flecainide, halogenated/local anesthetics, lapatinib, methadone, erythromycin, telithromycin, troleandomycin, octreotide, ondansetron, palonosetron, pentamidine, propafenone, ranolazine, SUNItinib, tacrolimus, vardenafil, vorinostat, usually with IV use

Increase: oversedation—other CNS depressants, alcohol, barbiturate anesthetics

Increase: toxicity—EPINEPHrine, lithium

Increase: both drugs effects—β-adrenergic blockers, alcohol

Increase: anticholinergic effects—anticholinergics

Decrease: effects—lithium, levodopa

Decrease: haloperidol effects—PHENObarbital, carBAMazepine

Drug/Lab Test

Increase: LFTs

NURSING CONSIDERATIONS

Assess:

- Prolactin, CBC, urinalysis, ophthalmic exam before and during prolonged therapy

Black Box Warning: Dementia, affect, orientation, LOC, reflexes, gait, coordination, sleep pattern disturbances, risk for death in dementia-related psychosis

- B/P standing, lying; take pulse, respirations q4hr during initial treatment; establish baseline before starting treatment; report drops of 30 mm Hg
- Dizziness, faintness, palpitations, tachycardia on rising
- **EPS** including akathisia (inability to sit still, no pattern to movements), tardive dyskinesia (bizarre movements of jaw, mouth, tongue, extremities), pseudoparkinsonism (rigidity, tremors, pill rolling, shuffling gait)

⚠ **Neuroleptic malignant syndrome/serotonin syndrome:** hyperthermia, muscle rigidity, altered mental status, increased CPK, seizures, hypo/hypertension, tachycardia; notify prescriber immediately

- Constipation, urinary retention daily; if these occur, increase bulk, water in diet
- **Abrupt discontinuation:** do not withdraw abruptly, taper
- **QT prolongation:** more common with IV use at high doses; monitor ECG in those with CV disease
- Supervised ambulation until patient stabilized on medication; do not involve patient in strenuous exercise program, fainting is possible; patient should not stand still for long periods
- Sips of water, sugarless candy, gum for dry mouth

Evaluate:

- Therapeutic response: decrease in emotional excitement, hallucinations, delusions, paranoia, reorganization of patterns of thought, speech; improvement in specific behaviors

Teach patient/family:

- That orthostatic hypotension occurs often; to rise from sitting or lying position gradually; to remain lying down after IM inj for at least 30 min
- To avoid hazardous activities until stabilized on medication and effects are known
- To avoid abrupt withdrawal of this product because EPS may result; to withdraw slowly
- To avoid OTC preparations (cough, hay fever, cold) unless approved by prescriber, serious product interactions may occur; to avoid use with alcohol, increased drowsiness may occur
- About EPS and necessity of meticulous oral hygiene because oral candidiasis may occur
- To report impaired vision, jaundice, tremors, muscle twitching

TREATMENT OF OVERDOSE:

Activated charcoal, lavage if orally ingested; provide an airway; do not induce vomiting

⚠ *HIGH ALERT*

heparin (Rx)

(hep′a-rin)

Hepalean ✦, Heparin Leo ✦, Hep-Lock, Hep-Lock U/P, Monoject Prefill

Func. class.: Anticoagulant, antithrombotic

Do not confuse:
heparin/Hespan

ACTION: Prevents conversion of fibrinogen to fibrin and prothrombin to thrombin by enhancing inhibitory effects of antithrombin III

USES: Prevention treatment of deep venous thrombosis, PE, MI, open heart surgery, disseminated intravascular clotting syndrome, atrial fibrillation with embolization, as an anticoagulant in transfusion and dialysis procedures, to maintain patency of indwelling venipuncture devices; diagnosis, treatment of DIC

CONTRAINDICATIONS: Bleeding, hypersensitivity

Precautions: Pregnancy (C), children, geriatric patients, alcoholism, hyperlipidemia, diabetes, renal disease, heparin-induced thrombocytopenia (HIT), hemophilia, leukemia with bleeding, peptic ulcer disease, severe thrombocytopenic purpura, severe renal/hepatic disease, blood dyscrasias, severe hypertension, subacute bacterial endocarditis, acute nephritis; benzyl alcohol products in neonates/infants/pregnancy/lactation

DOSAGE AND ROUTES

Deep venous thrombosis/ pulmonary embolism

- **Adult: IV BOL** 80 international units/kg then maintenance **IV INF** 18 international units/kg/hr; if aPTT <35 (1.2× normal), increase **IV INF** rate by 4 international units/kg/hr and rebolus with 80 international units/kg; if aPTT 35-45 (1.2-1.5× normal), increase **IV INF** by 2 international units/kg/hr and rebolus with 40 international units/kg; if aPTT 46-70 (1.5-2.3× normal), maintain **IV INF**; if aPTT 71-90 (2.3-3× normal), decrease **IV INF** by 2 international units/kg/hr; if aPTT >90 (>3× normal), hold **IV INF** for 1 hr then decrease rate 3 international units/kg/hr
- **Child/infant/neonate: IV** loading dose 75 international units/kg
- **Child >1 yr:** 20 international units/kg/hr
- **Infant/neonate <1 yr:** 28 international units/kg/hr as initial maintenance dose

Thrombosis prophylaxis (open heart/CV surgery)

- **Adult: IV** ≥150 international units/kg; procedures <60 min, up to 300 international units/kg; procedures >60 min, up to 400 international units/kg based on ACT

Thrombosis prophylaxis (PCI, not receiving abciximab)

- **Adult: IV BOL** weight adjusted with 60-100 international units/kg, maintain ACT within 250-300 sec (HemoTec) or 300-350 sec (Hemochron)
- **Child/infant/neonate: IV BOL** 100-150 international units/kg

Prophylaxis for DVT/PE

- **Adult: SUBCUT** 5000 units q8-12hr

IV catheter occlusion prophylaxis

- **Adult/child: IV** 10-100 units/ml
- **Infant <10 kg: IV** 10 units/ml

Available forms: Sol for inj 10, 100, 1000, 2000, 5000, 7500, 10,000, 20,000 units/ml; premixed 1000 units/500 ml, 2000 units/1000 ml, 12,500 units/250 ml, 25,000 units/250 ml, 25,000 units/500 ml; lock flush preparations 10 units/ml

Administer:

- Cannot be used interchangeably (unit for unit) with LMWHs or heparinoids
- At same time each day to maintain steady blood levels
- Store at room temperature

Heparin Lock route

⚠ Do not mistake heparin sodium inj 10,000 units/ml and Hep-Lock U/P 10 units/ml; they have similar blue labeling; deaths in pediatric patients have occurred when heparin sodium inj vials were confused with heparin flush vials

SUBCUT route

- Give deeply with 25-G 3/8-1/2-in needle; do not massage area or aspirate when giving SUBCUT inj; give in abdomen between pelvic bones, rotate sites; do not pull back on plunger; leave in for 10 sec; apply gentle pressure for 1 min
- Changing needles is not recommended
- Avoid all IM inj that may cause bleeding, hematoma

Direct IV route

- Give loading dose undiluted, over ≥1 min, use before continuous inf

Continuous IV INF route

- Dilute 25,000 units/250-500 ml 0.9 NaCl or D_5W, (50-100 units) solutions are premixed and ready for use
- When product is added to inf sol for cont IV, invert container at least 6 times to ensure adequate mixing

Additive compatibilities: Avoid adding to sol, even if compatible; rate of heparin inf times may need to be changed

Y-site compatibilities: Acyclovir, allopurinol, amifostine, aminophylline, atro-

⚠ Nurse Alert

pine, aztreonam, betamethasone, bleomycin, calcium gluconate, ceFAZolin, cefoTEtan, cefTAZidime, cefTRIAXone, chlordiazePOXIDE, cimetidine, CISplatin, cladribine, clindamycin, conjugated estrogens, cyanocobalamin, cyclophosphamide, cytarabine, dexamethasone, digoxin, DOPamine, DOXOrubicin liposome, edrophonium, enalaprilat, EPINEPHrine, esmolol, ethacrynate, etoposide, famotidine, fentaNYL, fluconazole, fludarabine, fluorouracil, foscarnet, furosemide, gallium, gemcitabine, granisetron, hydrocortisone, HYDROmorphone, insulin (regular), isoproterenol, kanamycin, leucovorin, linezolid, lidocaine, LORazepam, magnesium sulfate, melphalan, menadiol, meropenem, methotrexate, methoxamine, methyldopate, methylergonovine, metoclopramide, metroNIDAZOLE, midazolam, milrinone, minocycline, mitoMYcin, morphine, nafcillin, neostigmine, nitroglycerin, nitroprusside, norepinephrine, ondansetron, oxacillin, oxytocin, PACLitaxel, pancuronium, penicillin G potassium, phytonadione, piperacillin, piperacillin/tazobactam, potassium chloride, prednisoLONE, procainamide, propofol, propranolol, pyridostigmine, ranitidine, remifentanil, sargramostim, scopolamine, sodium bicarbonate, succinylcholine, tacrolimus, theophylline, thiopental, thiotepa, ticarcillin, ticarcillin/clavulanate, tirofiban, trimethobenzamide, trimethaphan, vecuronium, vinBLAStine, warfarin, zidovudine, zoledronic acid

SIDE EFFECTS

CNS: *Fever,* chills, headache

GU: **Hematuria**

HEMA: **Hemorrhage, thrombocytopenia, anemia, HIT**

INTEG: *Rash,* dermatitis, urticaria, pruritus, alopecia, hematoma, cutaneous necrosis (SUBCUT), inj site reactions

META: Hyperkalemia, hypoaldosteronism, hyperlipidemia

SYST: **Anaphylaxis**

PHARMACOKINETICS

Half-life $1^1/_2$ hr; excreted in urine; 95% bound to plasma proteins; does not cross placenta or alter breast milk; removed from the system via the lymph and spleen; partially metabolized in kidney, liver; excreted in urine (<50% unchanged)

SUBCUT: Onset 20-60 min, duration 8-12 hr, well absorbed >35,000 international units/24 hr

IV: Peak 5 min, duration 2-6 hr

INTERACTIONS

Increase: heparin action—oral anticoagulants, salicylates, dextran, NSAIDs, platelet inhibitors, cephalosporins, penicillins, ticlopidine, dipyridamole, antineoplastics, clopidogrel, SSRIs, SNRIs

Decrease: heparin action—digoxin, tetracyclines, antihistamines, cardiac glycosides, nicotine, nitroglycerin

Drug/Herb

Increase: bleeding risk—garlic, ginger, ginkgo, feverfew, green tea, horse chestnut

Drug/Lab Test

Increase: ALT, AST, INR, PT, PTT, potassium

Decrease: platelets

NURSING CONSIDERATIONS

Assess:

⚠ **Bleeding, hemorrhage: gums, petechiae, ecchymosis, black tarry stools, hematuria, epistaxis, decrease in Hct, B/P; HIT may occur after product discontinuation**

- Blood studies (Hct, occult blood in stools) q3mo
- Partial prothrombin time, which should be 1.5-2.5× control; for continuous IV inf, check aPTT baseline 6 hr after initiation and 6 hr after any dose change; use aPTT for dosing adjustments; after therapeutic aPTT has been measured 2×, check aPTT daily
- Platelet count q2-3days; thrombocytopenia may occur on 4th day of treatment
- **Hypersensitivity:** rash, chills, fever, itching; report to prescriber

Evaluate:

- Therapeutic response: decrease of DVT, adequate anticoagulation based on aPTT

Teach patient/family:

- To avoid OTC preparations that may cause serious product interactions unless directed by prescriber
- That product may be held during active bleeding (menstruation), depending on condition
- To use soft-bristle toothbrush to avoid bleeding gums; to avoid contact sports; to use an electric razor, to avoid IM inj
- To carry emergency ID identifying product taken
- To report to prescriber any signs of bleeding: gums, under skin, urine, stools
- To report to prescriber any signs of hypersensitivity: rash, chills, fever, itching

TREATMENT OF OVERDOSE:

Withdraw product, protamine 1 mg protamine/100 units heparin

hepatitis B immune globulin (HBIG) (Rx)

HepaGam B, Hyper HEP B S/D, Nabi-HB

Func. class.: Immune globulin

ACTION: Provides passive immunity to hepatitis B

USES: Prevention of hepatitis B virus in exposed patients, including passive immunity in neonates born to HBsAg-positive mother, prevention of hepatitis B recurrence after liver transplant in HBsAg-positive patients

CONTRAINDICATIONS: Hypersensitivity to immune globulins, coagulation disorders

Precautions: Pregnancy (C), breastfeeding, children, geriatric patients, hemophilia, active infection, IgA deficiency

DOSAGE AND ROUTES

Hepatitis B exposure in those at high risk

- **Adult and child: IM** 0.06 ml/kg (usual 3-5 ml) within 7 days of exposure; repeat 28 days after exposure if patient wishes to not receive hepatitis B vaccine

Neonates born to hepatitis B surface-antigen–positive persons

- **Neonate: IM** 0.5 ml within 12 hr of birth

Prevention of hepatitis B infection recurrence after liver transplant

- **Adult: IV** (HepaGam B only) 20,000 international units concurrent with grafting transplanted liver, then 20,000 international units/day on days 1-7, then 20,000 international units q2wk starting on day 14, then 20,000 international units/mo starting with mo 4

Available forms: Inj 1-, 4-, 5-ml vials; neonatal syringe 0.5 ml; HepaGam B sol for inj 312 units/ml; Hyper HEP B S/D 217 units/ml

Administer:

IM route

- After rotating vial; do not shake
- Only with EPINEPHrine 1:1000 on unit to treat laryngospasm
- In deltoid for better absorption (adult)

IV route (HepaGam B only)

- Calculate volume needed for each 20,000 international-unit dose using measured potency of each lot; HBIG potency stamped on label
- Promptly use after vial entered; discard unused product
- Give at 2 ml/min through separate IV line, use inf pump, decrease to 1 ml/min if inf-related event occurs, patient becomes uncomfortable
- Do not use HyperHEP B BS/D or Nabi-HB IV

SIDE EFFECTS

CNS: Headache, dizziness, fever

GI: Nausea, vomiting

INTEG: Soreness at inj site, urticaria, erythema, swelling

SYST: Induration, anaphylaxis, angioedema

INTERACTIONS

• Do not use within 3 mo of hepatitis B immune globulin, MMR, varicella, or rotavirus vaccines even after discontinuing product

NURSING CONSIDERATIONS

Assess:

• History of allergies, skin conditions (eczema, psoriasis, dermatitis), reactions to vaccinations

• Skin reactions: rash, induration, urticaria

⚠ **Anaphylaxis:** inability to breathe, bronchospasm, hypotension, wheezing, diaphoresis, fever, flushing

• Can be used with hepatitis B vaccine in cases of direct contact

• Written record of immunization

Evaluate:

• Prevention of hepatitis B

Teach patient/family:

• That discomfort may occur at site

• To report any rash, wheezing, inability to breathe immediately

homatropine ophthalmic

See Appendix B

hydrALAZINE (Rx)

(hye-dral′a-zeen)

Apresoline ✱

Func. class.: Antihypertensive, direct-acting peripheral vasodilator

Chem. class.: Phthalazine

Do not confuse:

hydrALAZINE/hydrOXYzine

Apresoline/allopurinol

ACTION: Vasodilates arteriolar smooth muscle by direct relaxation; reduction in blood pressure with reflex increases in heart rate, stroke volume, cardiac output

USES: Essential hypertension; hypertensive emergency/urgency

Unlabeled uses: CHF, preeclampsia

CONTRAINDICATIONS: Hypersensitivity to hydrALAZINEs, mitral valvular rheumatic heart disease, CAD

Precautions: Pregnancy (C), breastfeeding, geriatric patients, CVA, advanced renal disease, hepatic disease, SLE, dissecting aortic aneurysm

DOSAGE AND ROUTES

Hypertension

• **Adult: PO** 10 mg qid 2-4 days, then 25 mg for rest of 1st wk, then 50 mg qid individualized to desired response, max 300 qid mg/day

• **Child: PO** 0.75-1 mg/kg/day in 2-4 divided doses, max 25 mg/dose, increase over 3-4 wk to max 7.5 mg/kg/day or 200 mg, whichever is less

Hypertensive crisis

• **Adult: IV BOL** 10-20 mg q4-6hr, administer **PO** as soon as possible; **IM** 10-50 mg q4-6hr

• **Child: IV BOL** 0.1-0.6 mg/kg q4hr; **IM** 0.1-0.6 mg/kg q4-6hr, max 1.7-3.5 mg/kg/day

CHF

• **Adult: PO** 10-25 mg tid, max 100 mg tid

Available forms: Inj 20 mg/ml; tabs 10, 25, 50, 100 mg

Administer:

PO route

• Give with meals (PO) to enhance absorption

IM route

• Do not admix, switch to PO as soon as possible

• No dilution needed, inject deeply in large muscle, aspirate

Direct IV route

• IV undiluted; give through Y-tube or 3-way stopcock, give each 10 mg over ≥1 min

• To recumbent patient, keep recumbent for 1 hr after administration

Y-site compatibilities: Alemtuzumab, anidulafungin, argatroban, atenolol, bivalirudin, bleomycin, DACTINomycin, DAPTOmycin, dexrazoxone, diltiazem, DOCEtaxel, etoposide, fludarabine, gatifloxacin, gemcitabine, granisetron, HYDROmorphone, IDArubicin, irinotecan, leucovorin, linezolid, mechlorethamine, metroNIDAZOLE, milrinone, mitoXANtrone, octreotide, oxaliplatin, PACLitaxel, palonosetron, pancuronium, potassium chloride, tacrolimus, teniposide, thiotepa, tirofiban, vecuronium, vinorelbine, vitamin B/C, voriconazole

Solution compatibilities: D_5LR, D_5W, $D_{10}W$, $D_{10}LR$, 0.45% NaCl, 0.9% NaCl, Ringer's, LR

SIDE EFFECTS

CNS: *Headache, tremors, dizziness, anxiety,* peripheral neuritis, depression, fever, chills

CV: *Palpitations, reflex tachycardia, angina,* shock, rebound hypertension, orthostatic hypotension

GI: *Nausea, vomiting, anorexia, diarrhea,* constipation, paralytic ileus, hepatotoxicity

GU: Urinary retention, glomerulonephritis, hematuria

HEMA: Leukopenia, agranulocytosis, anemia, thrombocytopenia

INTEG: Rash, pruritus, urticaria

MISC: Nasal congestion, muscle cramps, *lupuslike symptoms,* flushing, edema, dyspnea

PHARMACOKINETICS

Half-life 3-7 hr, metabolized by liver, 12%-14% excreted in urine, protein binding 89%

PO: Onset 20-30 min, peak 1-2 hr, duration 2-4 hr

IM: Onset 10-30 min, peak 1 hr, duration up to 12 hr

IV: Onset 5-30 min, peak 10-80 min, duration up to 12 hr

INTERACTIONS

Increase: severe hypotension—MAOIs

Increase: tachycardia, angina—sympathomimetics (EPINEPHrine, norepinephrine)

Increase: hypotension—other antihypertensives, alcohol, levodopa, thiazide diuretics

Increase: effects of β-blockers

Decrease: hydrALAZINE effects—NSAIDs, estrogens

Drug/Lab Test

Decrease: Hgb, WBC, RBC, platelets, neutrophils

Positive: ANA titer

NURSING CONSIDERATIONS

Assess:

- Cardiac status: B/P q5min × 2 hr, then q1hr × 2 hr, then q4hr; pulse, jugular venous distention q4hr
- Electrolytes, blood studies: potassium, sodium, chloride, carbon dioxide, CBC, serum glucose, LE prep, ANA titer before, during treatment; assess for fever, joint pain, rash, sore throat (lupuslike symptoms); notify prescriber
- Weight daily, I&O
- Edema in feet, legs daily, skin turgor, dryness of mucous membranes for hydration status
- Crackles, dyspnea, orthopnea
- IV site for extravasation, rate
- Mental status: affect, mood, behavior, anxiety; check for personality changes

Evaluate:

- Therapeutic response: decreased B/P

Teach patient/family:

- To take with food to increase bioavailability (PO)
- To avoid OTC preparations unless directed by prescriber
- To notify prescriber if chest pain, severe fatigue, fever, muscle or joint pain occurs
- To rise slowly to prevent orthostatic hypotension
- To notify prescriber if pregnancy is suspected

TREATMENT OF OVERDOSE:

Administer vasopressors, volume expanders for shock; if PO, lavage or give activated charcoal, digitalization

hydrochlorothiazide (Rx)

(hye-droe-klor-oh-thye′a-zide)

Apo-Hydro ✱, Ezide, Neo-Codema ✱

Func. class.: Thiazide diuretic, antihypertensive

Chem. class.: Sulfonamide derivative

ACTION:
Acts on distal tubule and ascending limb of loop of Henle by increasing excretion of water, sodium, chloride, potassium

USES:
Edema, hypertension, diuresis, CHF; edema in corticosteroid, estrogen, NSAIDs; idiopathic lower extremity edema therapy

Unlabeled uses: Diabetes insipidus, hypercalciuria, nephrolithiasis, premenstrual syndrome, renal calculus

CONTRAINDICATIONS:
Hypersensitivity to thiazides or sulfonamides, pregnancy (D) preeclampsia, anuria, renal decompensation

Precautions: Pregnancy (B), breastfeeding, hypokalemia, renal/hepatic disease, gout, COPD, LE, diabetes mellitus, hyperlipidemia, CCr <30 ml/min, hypomagnesemia

DOSAGE AND ROUTES

Hypertension:

- **Adult/adolescent: PO** 12.5-25 mg/day, may increase to 50 mg/day in 1-2 divided doses
- **Child >6 mo: PO** 1-2 mg/kg/day in divided doses, max 37.5 mg for 6 mo-2 yr; max 100 mg/day for 2-12 yr
- **Child <6 mo: PO** up to 2-3.3 mg/kg/day in divided doses

Renal dose:

- **Adult: PO** CCr <30 ml/min, do not use; not effective

Available forms: Tabs 12.5, 25, 50 mg; caps 12.5 mg

Administer:

PO route

- In AM to avoid interference with sleep if using product as a diuretic; tab may be crushed, mixed with food
- Potassium replacement if potassium <3 mg/dl
- With food; if nausea occurs, absorption may be decreased slightly

SIDE EFFECTS

CNS: Drowsiness, paresthesia, depression, headache, *dizziness, fatigue, weakness,* fever

CV: Irregular pulse, *orthostatic hypotension,* palpitations, volume depletion, allergic myocarditis

EENT: Blurred vision

ELECT: *Hypokalemia,* hypercalcemia, hyponatremia, hypochloremia, hypomagnesemia

GI: *Nausea, vomiting, anorexia,* constipation, diarrhea, cramps, **pancreatitis,** GI irritation, **hepatitis,** jaundice

GU: *Urinary frequency,* polyuria, **uremia, glucosuria,** hyperuricemia, **renal failure**

HEMA: **Aplastic anemia, hemolytic anemia, leukopenia, agranulocytosis, thrombocytopenia, neutropenia**

INTEG: *Rash,* urticaria, purpura, photosensitivity, alopecia, erythema multiforme

META: *Hyperglycemia, hyperuricemia,* increased creatinine, BUN

SYST: **Stevens-Johnson syndrome**

PHARMACOKINETICS

PO: Onset 2 hr, peak 4 hr, duration 6-12 hr, half-life 6-15 hr, excreted unchanged by kidneys, crosses placenta, enters breast milk

INTERACTIONS

Increase: hyperglycemia, hyperuricemia, hypotension—diazoxide

Increase: hypokalemia—glucocorticoids, amphotericin B

Increase: toxicity—lithium, non-depolarizing skeletal muscle relaxants, cardiac glycosides

Increase: renal failure risk—NSAIDs
Increase: effects—loop diuretics
Decrease: antidiabetics effects
Decrease: thiazides absorption—cholestyramine, colestipol
Drug/Food
Increase: severe hypokalemia—licorice
Drug/Lab Test
Increase: parathyroid test, uric acid, calcium, glucose, cholesterol, triglycerides
Decrease: potassium, sodium, Hgb, WBC, platelets

NURSING CONSIDERATIONS

Assess:
- Weight, I&O daily to determine fluid loss; effect of product may be decreased if used daily
- Rate, depth, rhythm of respiration, effect of exertion
- B/P lying, standing; postural hypotension may occur
- Blood studies: BUN, blood glucose, CBC, serum creatinine, blood pH, ABGs, uric acid, electrolytes
- **Signs of metabolic alkalosis:** drowsiness, restlessness
- **Signs of hypokalemia:** postural hypotension, malaise, fatigue, tachycardia, leg cramps, weakness, dehydration
- Confusion, especially in geriatric patients; take safety precautions if needed

Evaluate:
- Therapeutic response: improvement in edema of feet, legs, sacral area daily, decreased B/P

Teach patient/family:
- To rise slowly from lying or sitting position
- To notify prescriber of muscle weakness, cramps, nausea, dizziness; hypokalemia is common
- That product may be taken with food or milk
- To use sunscreen for photosensitivity
- That blood glucose may be increased in diabetics
- To take early in day to avoid nocturia
- To avoid alcohol, OTC meds unless approved by prescriber
- To rise slowly to prevent dizziness

TREATMENT OF OVERDOSE: Lavage if taken orally; monitor electrolytes; administer dextrose in saline; monitor hydration, CV, renal status

HYDROcodone (Rx)

(hye-droe-koe′done)

Hycodan ✱, Tussigon

HYDROcodone/acetaminophen (Rx)

Co-Gesic, Dolorex Forte, Duocet, Hycet, Lorcet, Lortab, Margesic H, Liquicet, Maxidone, Norco, Polygesic, Stagesic, Vanacet, Vicodin, Vicodin ES, Vicodin HP, Xodol, Zamicet, Zydone

HYDROcodone/ibuprofen (Rx)

Ibudone, Reprexain, Vicoprofen

Func. class.: Antitussive opioid analgesic/nonopioid analgesic

Controlled Substance Schedule III

Do not confuse:
HYDROcodone/hydrocortisone
Hycodan/Vicodin

ACTION: Acts directly on cough center in medulla to suppress cough; binds to opiate receptors in CNS to reduce pain

USES: Hyperactive and nonproductive cough, mild to moderate pain

CONTRAINDICATIONS: Acne rosacea/vulgaris, Cushing syndrome, measles, perioral dermatitis, varicella, abrupt discontinuation; hypersensitivity to this product, benzyl

Precautions: Pregnancy (C), breastfeeding, neonates, addictive personality, increased intracranial pressure, MI (acute), severe heart disease, respiratory depression, renal/hepatic disease, bowel impaction, urinary retention, viral infection, ulcerative colitis, seizures, sulfite hypersensitivity, psychosis, hypertension, hyperthyroidism

DOSAGE AND ROUTES

Analgesic

• **Adult: PO** 2.5-10 mg q3-6hr prn, max 60 mg/day

Antitussive

• **Adult: PO** 5 mg q4-6hr prn, max 30 mg/24 hr

Available forms: ***HYDROcodone:*** bulk powder; ***HYDROcodone/acetaminophen:*** 5 mg HYDROcodone/500 mg acetaminophen (Co-Gesic, Lorcet, Lortab 5/500, Stagesic, Vicodin); 7.5 mg HYDROcodone/400 mg acetaminophen (Zydone), 7.5 mg HYDROcodone/500 mg acetaminophen (Lortab 7.5/500), 7.5 mg HYDROcodone/750 mg acetaminophen (Vicodin ES), 5 mg HYDROcodone/325 acetaminophen, 10 mg HYDROcodone/325 acetaminophen (Norco), 10 mg HYDROcodone/500 mg acetaminophen (Lortab 10/500), 10 mg HYDROcodone/650 mg acetaminophen (Lorcet 10/650, Vicodin HP), 10 mg HYDROcodone/660 acetaminophen (Vicodin HP); caps 5 mg HYDROcodone/500 mg acetaminophen (Stagesic, Zydone); ***HYDROcodone/ibuprofen:*** tabs 7.5 mg HYDROcodone/200 mg ibuprofen (Vicoprofen)

Administer:

• Do not break, crush, or chew tabs; only scored tabs can be broken

• With antiemetic after meals if nausea or vomiting occurs

• Do not exceed 4 g acetaminophen with combination product

• Give with food or milk to prevent gastric upset

• Store in light-resistant area at room temperature

SIDE EFFECTS

CNS: *Drowsiness*, dizziness, light-headedness, confusion, headache, sedation, euphoria, dysphoria, weakness, hallucinations, disorientation, mood changes, dependence, **seizures**

CV: Palpitations, tachycardia, bradycardia, change in B/P, **circulatory depression**, syncope, **cardiac arrest (children)**

EENT: Tinnitus, blurred vision, miosis, diplopia

GI: *Nausea, vomiting, anorexia, constipation*, cramps, dry mouth, ulcers

GU: Increased urinary output, dysuria, urinary retention

INTEG: Rash, urticaria, flushing, pruritus

RESP: **Respiratory depression; pulmonary edema, bronchopneumonia, respiratory arrest (children)**

PHARMACOKINETICS

Onset 10-20 min, duration 4-6 hr, half-life $3^1/_2$-$4^1/_2$ hr, metabolized in liver, excreted in urine, crosses placenta

INTERACTIONS

Increase: CNS depression—alcohol, opioids, sedative/hypnotics, phenothiazines, skeletal muscle relaxants, general anesthetics, tricyclics

Increase: severe reactions—MAOIs

Drug/Herb

Increase: CNS depression—lavender, valerian

Drug/Lab Test

Increase: amylase, lipase

NURSING CONSIDERATIONS

Assess:

• **Pain:** intensity, type, location, other characteristics before, 1 hr after giving product; titrate upward by 25% until pain reduced by half; need for pain medication, physical dependence; opioid is more effective before pain is severe

• **CNS changes:** dizziness, drowsiness, hallucinations, euphoria, LOC, pupil reaction

• B/P, pulse, respirations before, periodically; if respirations <10/min, dose may need to be reduced, oversedation may occur

• Bowel status: constipation; provide fluids, fiber in diet, may need stimulate laxatives

• **Allergic reactions:** rash, urticaria

• Cough and respiratory dysfunction: respiratory depression, character, rate, rhythm
• History of ulcers if using ibuprofen combination product
• Safety measures: night-light, call bell within easy reach; assistance with ambulation

Evaluate:
• Therapeutic response: decrease in pain or cough

Teach patient/family:
• To report any symptoms of CNS changes, allergic reactions
• That physical dependency may result when used for extended periods
• That withdrawal symptoms may occur: nausea, vomiting, cramps, fever, faintness, anorexia
• To avoid driving, other hazardous activities because drowsiness occurs
• To avoid other CNS depressants; they will enhance sedating properties of this product
• To change positions slowly to reduce orthostatic hypotension

TREATMENT OF OVERDOSE:
Naloxone HCl (Narcan) 0.2-0.8 mg IV, O_2, IV fluids, vasopressors

hydrocortisone (Rx)
(hy-dro-kor′tih-sone)
Cortef, Colocort, Cortenema

hydrocortisone acetate (Rx)
Anucort, Anusol, Cortifoam, Hemril, Proctocort, Rectasol

hydrocortisone sodium succinate (Rx)
A-hydroCort, Solu-CORTEF
Func. class.: Corticosteroid
Chem. class.: Short-acting glucocorticoid

Do not confuse:
hydrocortisone/HYDROcodone

ACTION: Decreases inflammation by suppression of migration of polymorphonuclear leukocytes, fibroblasts, reversal of increased capillary permeability, and lysosomal stabilization

USES: Severe inflammation, adrenal insufficiency, ulcerative colitis, collagen disorders, asthma, COPD, Hodgkin's disease, SLE, Stevens-Johnson syndrome, ulcerative, TB

Unlabeled uses: Carpal tunnel syndrome, Churg-Strauss syndrome, endophthalmitis, mixed connective-tissue disease, multiple myeloma, polyarteritis nodosa, polychondritis, pulmonary edema, temporal arteritis, Wegener's granulomatosis, septic shock

CONTRAINDICATIONS: Fungal infection, hypersensitivity

Precautions: Pregnancy (C), breastfeeding, children <2 yr, diabetes mellitus, glaucoma, osteoporosis, seizure disorders, ulcerative colitis, CHF, myasthenia gravis, renal disease, esophagitis, peptic ulcer, metastatic carcinoma, psychosis, idiopathic thrombocytopenia (IM), acute glomerulonephritis, amebiasis, nonasthmatic bronchial disease, AIDS, TB, recent MI (associated with left ventricular rupture), Cushing syndrome, hepatic disease, hypothyroidism, coagulopathy, thromboembolism

DOSAGE AND ROUTES

Adrenal insufficiency/inflammation
• **Adult: PO** 20-240 mg daily; in divided doses **IM/IV** 100-500 mg (succinate), may repeat q2-6hr

Shock prevention
• **Adult: IM/IV** (succinate) 50 mg/kg repeated after 4 hr, repeat q24hr as needed

Colitis
• **Adult: PO** 20-240 mg (base)/day in 2-4 divided doses; **ENEMA** 100 mg nightly for 21 days
• **Child: PO** 2-8 mg (base)/kg/day or 60-240 mg (base)/m^2/day in 3-4 divided doses

Available forms: Hydrocortisone: enema 100 mg/60 ml; tabs 5, 10, 20 mg; **Acetate: rectal aerosol foam:** 10%; **Cypionate:** tabs 5, 10, 20 mg; **Succinate:** injection 100, 250, 500, 1000 mg vial

Administer:

- Daily dose in AM for better results
- In one dose in AM to prevent adrenal suppression; avoid SUBCUT administration, may damage tissue
- Do not use acetate or susp for IV

PO route

- With food or milk for GI symptoms

Rectal route

- Tell patient to retain for 1 hr if possible

IM route

- IM inj deep in large muscle mass; rotate sites; avoid deltoid; use 21-G needle

IV route

- **Succinate:** IV in mix-o-vial or reconstitute ≤250 mg/2 ml bacteriostatic water for inj; mix gently; give direct IV over ≥1 min; may be further diluted in 100, 250, 500, or 1000 ml of D_5W, D_5 0.9%, NaCl 0.9% given over ordered rate

Sodium succinate preparations

Y-site compatibilities: Acyclovir, allopurinol, amifostine, amphotericin B cholesteryl, ampicillin, amrinone, amsacrine, atracurium, atropine, aztreonam, betamethasone, calcium gluconate, cefepime, cefmetazole, cephalothin, cephapirin, chlordiazePOXIDE, chlorproMAZINE, cisatracurium, cladribine, cyanocobalamin, cytarabine, dexamethasone, digoxin, diphenhydrAMINE, DOPamine, DOXOrubicin liposome, droperidol, edrophonium, enalaprilat, EPINEPHrine, esmolol, estrogens conjugated, ethacrynate, famotidine, fentaNYL, fentaNYL/droperidol, filgrastim, fludarabine, fluorouracil, foscarnet, furosemide, gallium, granisetron, heparin, hydrALAZINE, insulin (regular), isoproterenol, kanamycin, lidocaine, LORazepam, magnesium sulfate, melphalan, menadiol, meperidine, methicillin, methoxamine, methylergonovine, minocycline, morphine, neostigmine, norepinephrine, ondansetron, oxacillin, oxytocin, PACLitaxel, pancuronium, penicillin G potassium, pentazocine, phytonadione, piperacillin/tazobactam, prednisoLONE, procainamide, prochlorperazine, propofol, propranolol, pyridostigmine, remifentanil, scopolamine, sodium bicarbonate, succinylcholine, tacrolimus, teniposide, theophylline, thiotepa, trimethaphan, trimethobenzamide, vecuronium, vinorelbine

SIDE EFFECTS

CNS: *Depression, flushing, sweating,* headache, mood changes, **pseudotumor cerebri, euphoria, insomnia, seizures**

CV: *Hypertension,* **circulatory collapse, thrombophlebitis, embolism,** tachycardia, edema, heart failure

EENT: Fungal infections, increased intraocular pressure, blurred vision, cataracts, glaucoma

GI: *Diarrhea, nausea,* abdominal distention, **GI hemorrhage,** increased appetite, **pancreatitis,** vomiting

HEMA: **Thrombocytopenia**

INTEG: Acne, poor wound healing, ecchymosis, petechiae

MISC: Adrenal insufficiency (after stress/withdrawal)

MS: Fractures, osteoporosis, weakness

PHARMACOKINETICS

Metabolized by liver, excreted in urine (17-OHCS, 17-KS), crosses placenta

PO: Peak 1-2 hr, duration 1-1½ days

IM/IV: Onset 20 min, peak 4-8 hr, duration 1-1½ days

IV: Peak 1-2 hr

RECT: Onset 3-5 days

INTERACTIONS

Increase: GI bleeding risk—salicylates, NSAIDs, acetaminophen

Increase: side effects—alcohol, amphotericin B, digoxin, cycloSPORINE, diuretics

Increase: neurologic reactions—live virus vaccines/toxoids

Decrease/Increase: anticoagulation—oral anticoagulants

Decrease: hydrocortisone action—bosentan, cholestyramine, colestipol, barbiturates, rifampin, phenytoin, theophylline, carBAMazepine
Decrease: anticoagulant effects, anticonvulsants, antidiabetics, calcium supplements, toxoids, vaccines

Drug/Herb
Decrease: hydrocortisone levels—ephedra

Drug/Lab Test
Increase: cholesterol, sodium, blood glucose, uric acid, calcium, glucose
Decrease: calcium, potassium, T_4, T_3, thyroid ^{131}I uptake test, urine 17-OHCS, 17-KS
False negative: skin allergy tests

NURSING CONSIDERATIONS

Assess:

- Potassium, blood glucose, urine glucose while patient receiving long-term therapy; hypokalemia and hyperglycemia; potassium depletion: paresthesias, fatigue, nausea, vomiting, depression, polyuria, dysrhythmias, weakness
- B/P, pulse; notify prescriber of chest pain
- I&O ratio; be alert for decreasing urinary output, increasing edema; weight daily, notify prescriber of weekly gain >5 lb
- **Adrenal insufficiency (cushingoid symptoms):** nausea, anorexia, SOB, moon face, fatigue, dizziness, weakness, joint pain before and during treatment; plasma cortisol levels during long-term therapy (normal level: 138-635 nmol/L SI units when drawn at 8 AM)
- **Infection:** increased temperature, WBC even after withdrawal of medication; product masks infection
- Mental status: affect, mood, behavioral changes, aggression
- **GI effects:** nausea, vomitting, anorexia or appetite stimulation, diarrhea, constipation, abdominal pain, hiccups, gastritis, pancreatitis, GI bleeding/perforation with long-term treatment
- Assistance with ambulation for patient with bone-tissue disease to prevent fractures

Evaluate:

- Therapeutic response: decreased inflammation, GI symptoms

Teach patient/family:

- That emergency ID as corticosteroid user should be carried
- To immediately report abdominal pain, black tarry stools because GI bleeding/perforation can occur
- To notify prescriber if therapeutic response decreases; that dosage adjustment may be needed; about signs of infection
- Not to discontinue abruptly because adrenal crisis can result; that product should be tapered
- That supplemental calcium/vit D may be needed if patient receiving long-term therapy
- That product can mask infection and cause hypo/hyperglycemia (diabetic)
- To avoid OTC products: salicylates, alcohol in cough products, cold preparations unless directed by prescriber
- About cushingoid symptoms of adrenal insufficiency: nausea, anorexia, fatigue, dizziness, dyspnea, weakness, joint pain, moon face
- To avoid live-virus vaccines if using steroids long term

hydrocortisone nasal

See Appendix B

hydrocortisone (topical)

(hye-droe-kor′ti-sone)

Ala-Cort, Ala-Scalp, Anusol HC, Caldecort, Cetacort, Cortizone-5, Cortizone-10, Cortizone-10 Dermolate, Hytone, LactiCare-HC, Maximum Strength Cortaid Faststick, Procort, Texacort

hydrocortisone acetate

Anusol HC, Cortaid, Cortef, Corticaine, Cortifoam, ProctoCream-HC, ProctoFoam-HC, Tucks, U-cort

hydrocortisone butyrate

Locoid, Locoid Lipocream

hydrocortisone probutate

Pandel

hydrocortisone valerate

Func. class.: Corticosteroid, topical

ACTION: Crosses cell membrane to attach to receptors to decrease inflammation, itching, inhibits multiple inflammatory cytokines

USES: Inflammation/itching in corticosteroid-responsive dermatoses on the skin, rectal area

CONTRAINDICATIONS: Hypersensitivity

Precautions: Pregnancy (C), breastfeeding, children

DOSAGE AND ROUTES

Corticosteroid-responsive dermatoses, inflammation, pruritus

- **Adult/child: TOP** apply to affected area 1 to 4 times per day

Inflammation from proctitis

- **Adult: Rectal** 1 applicator full of foam once or twice a day × 2-3 wk, then every other day as needed; suppository: 1 bid × 2 wk

Available forms:

Hydrocortisone: cream 0.5%, 1%, 2.5%; gel 1%, 2%; lotion 0.25%, 1%, 2%, 2.5%; ointment 0.5%, 1%, 2.5%; rectal cream 1%; rectal ointment 1%; spray 1%; solution 1%, 2.5%; **Hydrocortisone acetate:** cream 0.5%, 1%, 2%, 2.5%; lotion 0.5%; ointment 0.5%, 1%; rectal foam 90 mg/application; suppositories 25 mg, 30 mg; **Hydrocortisone butyrate:** cream 0.1%; ointment 0.1%; lotion; **Hydrocortisone probutate:** cream 0.1%; **Hydrocortisone valerate:** cream 0.2%, ointment 0.2%

Administer:

Topical route

- May be used with dressings

Cream/Ointment/Lotion

- Apply sparingly in a thin film and rub gently

SIDE EFFECTS

CNS: Seizures, **increased intracranial pressure**, headache

CV: Hypertension

EENT: Cataracts, glaucoma

INTEG: Burning, folliculitis, pruritus, dermatitis, maceration

MISC: Hyperglycemia; glycosuria, HPA suppression

PHARMACOKINETICS

Unknown, minimally absorbed

NURSING CONSIDERATIONS

Assess:

- Skin reactions: burning, pruritus, folliculitis, dermatitis

Evaluate:

- Decreasing itching, inflammation on the skin, rectal area

Teach patient/family:

Topical route

- That product may be used with dressings

Cream/Ointment/Lotion

- To apply sparingly in a thin film and rub gently into the affected area

Gel
• To apply sparingly in a thin film and rub gently
Rectal
• To remove wrapper and insert suppository

⚠ HIGH ALERT

HYDROmorphone (Rx)

(hye-droe-mor′fone)
Dilaudid, Dilaudid HP, Exalgo, Hydromorph Contin ♣
Func. class.: Opiate analgesic
Chem. class.: Semisynthetic phenanthrene

Controlled Substance Schedule II

Do not confuse:
HYDROmorphone/meperidine/morphine
Dilaudid/Demerol

ACTION: Inhibits ascending pain pathways in CNS, increases pain threshold, alters pain perception

USES: Moderate to severe pain
Unlabeled uses: Arthralgia, bone, dental pain, headache, migraine, myalgia

CONTRAINDICATIONS: Hypersensitivity to this product/sulfite, COPD, cor pulmonale, emphysema, GI obstruction, ileus, increased intracranial pressure, obstetric delivery, status asthmaticus

Black Box Warning: Respiratory depression, opioid-naive patients

Precautions: Pregnancy (C), breastfeeding, children <18 yr, addictive personality, renal/hepatic disease, abrupt discontinuation, adrenal insufficiency, angina, asthma, biliary tract disease, bladder obstruction, hypothyroidism, hypovolemia, hypoxemia, IBD, IV use, lactase/paraben deficiency, labor, latex hypersensitivity, myxedema, seizure disorders, sleep apnea

Black Box Warning: Substance abuse, accidental exposure, potential for overdose/poisoning

DOSAGE AND ROUTES

Analgesic
• **Adult: PO** (oral solution) 2.5-10 mg q3-6hr or (tabs) 2-4 mg q4-6hr; **EXT REL** (Exalgo): convert to **EXT REL** by giving total daily dose of immediate release/day, in 1 daily dose, if needed titrate **EXT REL** q3-4days until adequate pain relief; use 25%-50% increase for each titration step, if more than 2 doses of rescue medication needed in 24 hr consider titration; **IV** 0.2-1 mg q2-3hr given over 2-3 min; **IM/SUBCUT** 1-2 mg q4-6hr prn, may be increased (opioid-naive patients may require lower dose); **RECT** 3 mg q6-8hr prn
• **Geriatric: PO** 1-2 mg q4-6hr
• **Child >50 kg (unlabeled): PO** 2-4 mg q3-4hr in opioid-naive patients, titrate; **IV** 0.2-1 mg q2-4hr or 0.3 mg/kg inf (opioid-naive will require lower dose)
• **Infant >6 mo/child <50 kg (unlabeled): PO** 0.04-0.08 mg/kg q3-4hr in opioid-naive patients, titrate; **IV** 0.015-0.02 mg/kg q2-4hr or 0.006 mg/kg/hr inf (opioid-naive)
Hepatic disease
• **Adult: Child-Pugh B, or C** (oral liquid, immediate rel tab, supp) give reduced dose based on response, impairment (parental) give 25%-50% of dose (moderate impairment)
Available forms: Powder for inj 250 mg; inj 1, 2, 4, 10 mg/ml; tabs 2, 4, 8 mg; supp 3 mg; oral sol 5 mg/5 ml; ext rel tab 8, 12, 16, 32 mg
Administer:
PO route
• Give with food or milk for GI irritation
• **Ext rel (Exalgo):** discontinue all other ext rel opioids, give q24hr; do not crush, break, chew
• When pain is beginning to return; determine interval by response
• Store in light-resistant area at room temperature

Black Box Warning: Do not use ext rel products in opioid-naive patients

Extended release

• **Converting from oral opioids:** conversion ratios are approximate; initiate ext rel tabs at 50% of calculated total daily equivalent dose of ext rel, give q24hr; max increase q3-4days, consider titration increases of 25%-50% with each step

• **Converting from transdermal patch (fentaNYL):** initiate ext rel tabs 18 hr after removal of patch; for each 25 mcg/hr dose of transdermal fentaNYL dose is 12 mg q24hr, start dose at 50% of calculated HYDROmorphone ext rel dose q24hr; titrate no more often than q3-4days, consider dose increases of 25%-50% with each step; if more than 2 rescue doses are required in 24 hr, consider titration

SUBCUT route

• Use short 30-G needle; make sure not to inject ID

• Rotate inj sites

IV route

• **Direct,** diluted with 5 ml sterile water or NS; give through Y-connector or 3-way stopcock; give ≤2 mg over 3-5 min

• **IV INF:** Dilute each 0.1-1 mg/ml NS (0.1-1 mg/ml), deliver by opioid syringe infusor; may be diluted in D_5W, D_5/NaCl, 0.45% NaCl, NS for larger amounts, delivery through inf pump

Additive compatibilities: Bupivacaine, cloNIDine, fluorouracil, heparin, midazolam, ondansetron, potassium chloride, promethazine, verapamil, ziconotide

Solution compatibilities: D_5W, D_5/0.45% NaCl, D_5/0.9% NaCl, D_5/LR, D_5/Ringer's sol, 0.45% NaCl, 0.9% NaCl, Ringer's and lactated Ringer's sol

Syringe compatibilities: Atropine, bupivacaine, cefTAZidime, chlorproMAZINE, cimetidine, diphenhydrAMINE, fentaNYL, glycopyrrolate, hydrOXYzine, LORazepam, midazolam, pentazocine, promethazine, ranitidine, scopolamine, thiethylperazine, trimethobenzamide

Y-site compatibilities: Acyclovir, allopurinol, amifostine, amikacin, amsacrine, aztreonam, cefamandole, cefepime, cefoperazone, cefotaxime, cefOXitin, cefTAZidime, ceftizoxime, cefuroxime, chloramphenicol, cisatracurium, CISplatin, cladribine, clindamycin, cyclophosphamide, cytarabine, diltiazem, DOBUTamine, DOPamine, DOXOrubicin, DOXOrubicin liposome, doxycycline, EPINEPHrine, erythromycin lactobionate, famotidine, fentaNYL, filgrastim, fludarabine, foscarnet, furosemide, gentamicin, granisetron, heparin, kanamycin, labetalol, LORazepam, magnesium sulfate, melphalan, methotrexate, metroNIDAZOLE, midazolam, milrinone, morphine, nafcillin, niCARdipine, nitroglycerin, norepinephrine, ondansetron, oxacillin, PACLitaxel, penicillin G potassium, piperacillin, piperacillin/tazobactam, propofol, ranitidine, remifentanil, teniposide, thiotepa, ticarcillin, tobramycin, trimethoprim-sulfamethoxazole, vancomycin, vecuronium, vinorelbine

SIDE EFFECTS

CNS: *Drowsiness, dizziness, confusion, headache, sedation, euphoria,* mood changes, **seizures**

CV: Palpitations, bradycardia, change in B/P, hypotension, tachycardia, peripheral vasodilation

EENT: Tinnitus, blurred vision, miosis, diplopia

GI: *Nausea, vomiting, anorexia, constipation, cramps,* dry mouth, paralytic ileus

GU: Increased urinary output, dysuria, urinary retention

INTEG: *Rash,* urticaria, bruising, flushing, diaphoresis, pruritus

RESP: **Respiratory depression,** dyspnea

PHARMACOKINETICS

IM: Onset 15-30 min, peak ½-1 hr, duration 4-5 hr, metabolized by liver, excreted by kidneys, crosses placenta, excreted in breast milk, half-life 2-3 hr

INTERACTIONS

Increase: effects—other CNS depressants (alcohol, opiates, sedative/hypnotics, antipsychotics, skeletal muscle relaxants)
Increase: CNS, respiratory depression—MAOIs
Decrease: HYDROmorphone effects—opiate antagonists
Drug/Herb
Increase: action—chamomile, hops, kava, lavender, St. John's wort, valerian
Drug/Lab Test
Increase: amylase

NURSING CONSIDERATIONS

Assess:

Black Box Warning: **Respiratory dysfunction:** respiratory depression, character, rate, rhythm; notify prescriber if respirations <10/min

- I&O ratio; check for decreasing output; may indicate urinary retention
- CNS changes: dizziness, drowsiness, hallucinations, euphoria, LOC, pupil reaction
- Bowel function, constipation
- Allergic reactions: rash, urticaria
- Need for pain medication, physical dependence
- **Pain:** control, sedation by scoring on 0-10 scale, ATC dosing is best for pain control
- Assistance with ambulation
- Safety measures: side rails, nightlight, call bell within easy reach

Evaluate:

- Therapeutic response: decrease in pain

Teach patient/family:

- To report any symptoms of CNS changes, allergic reactions
- That physical dependency may result when used for extended periods; that withdrawal symptoms may occur: nausea, vomiting, cramps, fever, faintness, anorexia
- To avoid driving, other hazardous activities because drowsiness occurs

TREATMENT OF OVERDOSE:

Naloxone (Narcan) 0.2-0.8 mg IV (nontolerant patients), O_2, IV fluids, vasopressors

hydroxychloroquine (Rx)

(hye-drox-ee-klor'oh-kwin)

Apo-Hydroxyquine ♣, Plaquenil

Func. class.: Antimalarial, antirheumatic (DMARDs)

Chem. class.: 4-Aminoquinoline derivative

ACTION:

Impairs complement-dependent antigen–antibody reactions

USES:

Malaria caused by susceptible strains of *Plasmodium vivax, P. malariae, P. ovale, P. falciparum* (some strains); SLE, rheumatoid arthritis
Unlabeled uses: SLE in children

CONTRAINDICATIONS:

Hypersensitivity to this product or chloroquine; retinal field changes

Black Box Warning: Children (long term), ocular disease

Precautions: Pregnancy (C), breastfeeding, blood dyscrasias, severe GI disease, neurologic disease, alcoholism, hepatic disease, G6PD deficiency, psoriasis, eczema

DOSAGE AND ROUTES

Malaria

- **Adult: PO Suppression or prevention:** 400 mg/wk, begin 1-2 wk before travel, continue 4 wk after returning; **treatment:** 800 mg, then 400 mg after 6-8 hr, then 400 mg/day on 2nd and 3rd day, total dose 2 g
- **Child: PO Suppression or prevention:** 6.4 mg/kg (5 mg/kg base) weekly, begin 1-2 wk before travel, continue 4 wk after returning; **treatment:** 10 mg/kg, 6.4 mg/kg (5 mg/kg base) at 6, 18, 24 hr after 1st dose

Lupus erythematosus

- **Adult: PO** 400 mg (310 mg base) daily-bid; length depends on patient response; **maintenance** 200-400 mg/day
- **Child (unlabeled): PO** 5 mg/kg/day, max 400 mg/day; long-term therapy is contraindicated

Rheumatoid arthritis

- **Adult: PO** 400-600 mg/day for 4-12 wk, then 200-300 mg/day after good response

Available forms: Tabs 200 mg

Administer:

PO route

- Tabs may be crushed and mixed with food, fluids
- With food or milk; at same time each day to maintain product level
- For malaria, prophylaxis should be started 2 wk before exposure, continued for 4-6 wk after leaving exposure area
- Store in tight, light-resistant container at room temperature; keep inj in cool environment

SIDE EFFECTS

CNS: Headache, stimulation, fatigue, irritability, **seizures**, bad dreams, dizziness, confusion, psychosis, decreased reflexes

CV: Hypotension, heart block, **asystole with syncope**

EENT: *Blurred vision, corneal changes, retinal changes, difficulty focusing*, tinnitus, vertigo, deafness, photophobia, corneal edema

GI: *Nausea, vomiting, anorexia*, diarrhea, cramps

HEMA: **Thrombocytopenia, agranulocytosis, leukopenia, aplastic anemia**

INTEG: Pruritus, pigmentation changes, skin eruptions, lichen-planus–like eruptions, eczema, **exfoliative dermatitis**, alopecia, **Stevens-Johnson syndrome**, photosensitivity

PHARMACOKINETICS

Peak 3 hr; terminal half-life 32-50 days; metabolized in liver; excreted in urine, feces, breast milk; crosses placenta, protein binding 45%

INTERACTIONS

Increase: digoxin, methotrexate levels

Increase: antibody titer—rabies vaccine

Decrease: hydroxychloroquine action—magnesium or aluminum compounds

Decrease: effect of—live virus vaccines, botulinum toxoids

NURSING CONSIDERATIONS

Assess:

- **SLE, malaria symptoms:** before treatment and daily
- **Rheumatoid arthritis:** pain, swelling, ROM, temperature of joints; for decreased reflexes: knee, ankle
- Ophthalmic exam at baseline and q6mo if long-term treatment or product dosage >150 mg/day
- Hepatic studies weekly: AST, ALT, bilirubin if patient receiving long-term treatment
- **Blood dyscrasias:** blood studies: CBC, platelets; WBC, RBC, platelets may be decreased; if severe, product should be discontinued; assess for malaise, fever, bruising, bleeding (rare)
- **ECG** during therapy: watch for depression of T waves, widening of QRS complex
- **Allergic reactions:** pruritus, rash, urticaria
- **Ototoxicity** (tinnitus, vertigo, change in hearing); audiometric testing should be done before, after treatment

⚠ **For toxicity:** blurring vision, difficulty focusing, headache, dizziness, knee, ankle reflexes; product should be discontinued immediately

Evaluate:

- Therapeutic response: decreased symptoms of malaria, SLE, rheumatoid arthritis

Teach patient/family:

- To use sunglasses in bright sunlight to decrease photophobia; to wear protective clothing (photosensitivity)
- That urine may turn rust or brown; that skin may become blue-black
- To report hearing, visual problems, fever, fatigue, bruising, bleeding, which may indicate blood dyscrasias

H

TREATMENT OF OVERDOSE:

Induce vomiting; gastric lavage; administer barbiturate (ultrashort acting), vasopressor, ammonium chloride; tracheostomy may be necessary

hydroxyurea (Rx)

(hye-drox'ee-yoo-ree-ah)

Droxia, Hydrea

Func. class.: Antineoplastic, antimetabolite

Chem. class.: Synthetic urea analog

Do not confuse:
hydroxyurea/hydrOXYzine

ACTION: Acts by inhibiting DNA synthesis without interfering with RNA or protein synthesis; incorporates thymidine into DNA, thereby causing direct damage to DNA strands; specific for S phase of cell cycle

USES: Melanoma, chronic myelogenous leukemia, recurrent or metastatic ovarian cancer, squamous cell carcinoma of the head and neck, sickle cell anemia

Unlabeled uses: Psoriasis, acute myelogenous leukemia (AML), astrocytoma, HIV, lung cancer, malignant glioma, polycythemia vera, thrombocytosis

CONTRAINDICATIONS: Pregnancy (D), breastfeeding, hypersensitivity, leukopenia (<2500/mm³), thrombocytopenia (<100,000/mm³), anemia (severe)

Precautions: Renal disease (severe), anemia, bone marrow suppression, dental disease, geriatric patients, HIV, hyperkalemia, hyperphosphatemia, hyperuricemia, hypocalcemia, infection, infertility, IM injection, tumor lysis syndrome, vaccinations

Black Box Warning: Requires an experienced clinician, secondary malignancy

DOSAGE AND ROUTES

Ovarian cancer, malignant melanoma

• **Adult: PO** 80 mg/kg as a single dose q3days or 20-30 mg/kg as a single dose daily

Ovarian cancer in combination with radiation

• **Adult: PO** 80 mg/kg as a single dose q3days

Chronic myelogenous leukemia (CML)/acute myelogenous leukemia (unlabeled)

• **Adult: PO** WBC >100,000/mm³, 50-75 mg/kg/day; WBC <100,000/mm³, 10-30 mg/kg/day; adjust for WBCs

• **Child: PO** 10-20 mg/kg/day, adjust to hematologic response

Sickle cell anemia

• **Adult: PO** 15 mg/kg/day, may increase by 5 mg/kg/day q12wk, max 35 mg/kg/day

Renal disease

• **Adult:** CCr <59 ml/min use 50% of dose

Available forms: Caps 200, 300, 400, 500 mg

Administer:

• Gloves should be worn when handling bottles or caps, including by caregivers; wash hands immediately and thoroughly

• Do not crush or chew caps; caps can be opened and contents mixed with water

• Antiemetic 30-60 min before product and prn

SIDE EFFECTS

CNS: Headache, confusion, hallucinations, dizziness, seizures

CV: Angina, ischemia

GI: Nausea, vomiting, anorexia, diarrhea, stomatitis, constipation, hepatotoxicity, pancreatitis

GU: Increased BUN, uric acid, creatinine, temporary renal function impairment

HEMA: Leukopenia, anemia, thrombocytopenia, megaloblastic erythropoiesis

INTEG: *Rash,* urticaria, pruritus, dry skin, facial erythema

META: Hyperphosphatemia, hyperuricemia, hypocalcemia
MISC: Fever, chills, malaise, secondary cancers, tumor lysis syndrome
RESP: Pulmonary fibrosis, diffuse pulmonary infiltrates

PHARMACOKINETICS

Readily absorbed when taken orally; peak level in 1-4 hr; degraded in liver; excreted in urine; almost totally eliminated within 24 hr; readily crosses blood-brain barrier; eliminated as CO_2; terminal half-life 3.5-4.5 hr

INTERACTIONS

Increase: pancreatitis/hepatotoxicity—didanosine, stavudine
Increase: toxicity—radiation or other antineoplastics
Increase: bleeding risk—NSAIDs, anticoagulants, thrombolytics, salicylates, platelet inhibitors
Increase: uric acid levels—probenecid, sulfinpyrazone
• Do not use with live virus vaccines
• Do not use hematopoietic progenitor cells (sargramostim, filgrastim) 24 hr before or after antineoplastic

Drug/Lab Test
Increase: BUN, creatinine, LFTs, uric acid
False increase: urea, uric acid, lactic acid
Decrease: Hgb, WBC, platelets

NURSING CONSIDERATIONS

Assess:
• **Bone marrow suppression:** determine the hemoglobin conc, total leukocyte count, and platelet count at least once a week during entire course; if the WBC is ≤2500/mm³ or platelets are ≤100,000/mm³, interrupt Hydrea until the values rise significantly toward normal concs; if severe anemia occurs, manage it without interrupting Hydrea receipt; for Droxia, monitor blood counts q2wk and interrupt drug receipt if neutrophils are <2000/mm³, platelets are <80,000/mm³, hemoglobin is <4.5 g/dl, or reticulocytes are <80,000/mm³ when the hemoglobin conc is <9 g/dl; after recovery, Droxia may be resumed at lower dosage; Droxia therapy requires an experienced clinician knowledgeable in the use of this medication for the treatment of sickle cell anemia
• Renal studies: BUN, serum uric acid, urine CCr, electrolytes before, during therapy
• **Tumor lysis syndrome;** hyperkalemia, hyperphosphatemia, hyperuricemia, hypocalcemia; uric acid nephropathy, acute renal failure, metabolic acidosis can also occur; aggressive alkalinization of urine, allopurinol can prevent this
• I&O ratio; report fall in urine output to <30 ml/hr
• Monitor temperature; fever may indicate beginning infection
• Hepatic studies before, during therapy: bilirubin, alk phos, AST, ALT, LDH; prn or monthly
• **Cutaneous vasculitic toxicity and gangrene:** more common in those who are receiving interferon
• **Bleeding:** hematuria, guaiac, bruising or petechiae, mucosa or orifices q8hr
• Buccal cavity for dryness, sores or ulceration, white patches, oral pain, bleeding, dysphagia
• **Pulmonary reactions:** assess for pulmonary fibrosis, fever, dyspnea, diffuse pulmonary infiltrates
• **Symptoms indicating severe allergic reaction:** rash, urticaria, itching, flushing
• **Neurotoxicity:** headaches, hallucinations, seizures, dizziness
• Rinsing of mouth tid-qid with water, club soda; brushing of teeth bid-tid with soft brush or cotton-tipped applicators for stomatitis; use unwaxed dental floss

Evaluate:
• Therapeutic response: decreased tumor size, spread of malignancy

Teach patient/family:
• To report signs of infection: elevated temperature, sore throat, flulike symptoms

- To report signs of anemia: fatigue, headache, faintness, SOB, irritability
- To report bleeding; to avoid use of razors, commercial mouthwash
- To avoid use of aspirin products, ibuprofen (thrombocytopenia)
- To avoid foods with citric acid, hot or rough texture if stomatitis is present
- To report stomatitis: any bleeding, white spots, ulcerations in the mouth; to examine mouth daily, report symptoms
- To notify prescriber if pregnancy is planned or suspected, pregnancy (D)
- To notify prescriber of fever, chills, sore throat, nausea, vomiting, anorexia, diarrhea, bleeding, bruising; may indicate blood dyscrasias; mental status changes

hydrOXYzine (Rx)

(hye-drox′i-zeen)

Vistaril

Func. class.: Antianxiety/antihistamine/sedative/hypnotic, antiemetic

Chem. class.: Piperazine derivative

Do not confuse:
hydrOXYzine/hydrALAZINE
Vistaril/Versed

ACTION: Depresses subcortical levels of CNS, including limbic system, reticular formation; competes with H_1-receptor sites

USES: Anxiety preoperatively, postoperatively to prevent nausea, vomiting; to potentiate opioid analgesics; sedation; pruritus, ethanol withdrawal
Unlabeled uses: Insomnia, allergic rhinitis, generalized anxiety disorder

CONTRAINDICATIONS: Pregnancy 1st trimester, breastfeeding; hypersensitivity to this product or cetirizine; acute asthma
Precautions: Pregnancy (C) (2nd/3rd trimester), geriatric patients, debilitated, renal/hepatic disease, closed-angle glaucoma, COPD, prostatic hypertrophy, asthma

DOSAGE AND ROUTES

Anxiety
- **Adult: PO** 50-100 mg qid, max 600 mg/day; **IM** 50-100 mg q4-6hr
- **Geriatric: PO** max 50 mg/day
- **Child >6 yr: PO** 50-100 mg/day in divided doses
- **Child <6 yr: PO** 50 mg/day in divided doses

Alcohol withdrawal
- **Adult: IM** 50-100 mg q4-6hr

Preoperatively/postoperatively (nausea/vomiting)
- **Adult: IM** 25-100 mg q4-6hr
- **Child: IM** 1.1 mg/kg as a single dose

Pruritus
- **Adult: PO** 25 mg tid-qid; **IM** 50-100 mg then q4-6hr prn, switch to **PO** as soon as feasible
- **Child ≥6 yr: PO** 50-100 mg/day in divided doses; **IM** 0.5-1 mg/kg/dose q4-6hr prn, use **PO** when possible
- **Child <6 yr: PO** 50 mg/day in divided doses

Insomnia (unlabeled)
- **Adult: PO** 50-100 mg 30-60 min before bedtime; **IM** 50 mg 30-60 min before bedtime

Renal dose
- **Adult: PO** CCr <50 ml/min, give 50% of dose

Available forms: Tabs 10, 25, 50 mg; caps 25, 50, 100 mg; oral sol 10 mg/ 5 ml; inj 25, 50 mg/ml; oral susp 25 mg/ 5 ml

Administer:

PO route
- Without regard to meals
- Crushed if patient is unable to swallow medication whole
- Shake oral susp before giving

IM route
- Does not need to be diluted; give by Z-track inj in large muscle to decrease pain, chance of necrosis; never give IV/ SUBCUT (HCl)

Additive compatibilities: CISplatin, cyclophosphamide, cytarabine, dimenhyDRINATE, etoposide, lidocaine, mesna, methotrexate, nafcillin

Syringe compatibilities: Atropine, butorphanol, chlorproMAZINE, cimetidine, codeine, diphenhydrAMINE, doxapram, droperidol, fentaNYL, fluPHENAZine, glycopyrrolate, HYDROmorphone, lidocaine, meperidine, metoclopramide, midazolam, morphine, nalbuphine, oxymorphone, pentazocine, perphenazine, procaine, prochlorperazine, scopolamine, SUFentanil

SIDE EFFECTS

CNS: *Dizziness, drowsiness,* confusion, headache, tremors, fatigue, depression, **seizures**
CV: Hypotension
GI: Dry mouth, increased appetite, nausea, diarrhea, weight gain

PHARMACOKINETICS

PO: Onset 15-60 min, duration 4-6 hr, half-life 3 hr, metabolized by liver, excreted by kidneys

INTERACTIONS

Increase: CNS depressant effect—barbiturates, opioids, analgesics, alcohol, sedative/hypnotics, other CNS depressants
Increase: anticholinergic effects—phenothiazines, quiNIDine, disopyramide, antihistamines, antidepressants, atropine, haloperidol, MAOIs
Drug/Lab Test
False negative: skin allergy testing
False increase: 17-hydroxycorticosteroids

NURSING CONSIDERATIONS

Assess:
- Anticholinergic effects: dry mouth, dizziness, confusion, hypotension, increased sedation; monitor B/P
- Assistance with ambulation during beginning therapy, since drowsiness, dizziness occurs

Evaluate:
- Therapeutic response: decreased anxiety

Teach patient/family:
- To avoid OTC preparations (cold, cough, hay fever) unless approved by prescriber
- To avoid driving, activities that require alertness
- To avoid alcohol, psychotropic medications
- Not to discontinue medication quickly after long-term use
- To rise slowly because fainting may occur

TREATMENT OF OVERDOSE:
Lavage if orally ingested; VS, supportive care; IV norepinephrine for hypotension

H

hyoscyamine (Rx)
(hye-oh-sye′a-meen)

Anaspaz, Colidrops, Colytrol Pediatric, Cystospaz-M, ED-SPAZ, HyoMax, HyoMax SL, Hyosyne, Levisin SL, NuLev, Oscimin, Spasdel, Symax
Func. class.: Anticholinergic/antispasmodics
Chem. class.: Belladonna alkaloid

ACTION: Inhibits muscarinic actions of acetylcholine at postganglionic parasympathetic neuroeffector sites; reduces rigidity, tremors, hyperhidrosis of parkinsonism

USES: Treatment of peptic ulcer disease in combination with other products; other GI disorders, other spastic disorders, IBS, urinary incontinence

CONTRAINDICATIONS: Hypersensitivity to anticholinergics, closed-angle glaucoma, GI obstruction, myasthenia gravis, paralytic ileus, GI atony, toxic megacolon, prostatic hypertrophy, urinary tract obstruction
Precautions: Pregnancy (C), geriatric patients, hyperthyroidism, dysrhythmias, CHF, ulcerative colitis, hypertension, hia-

tal hernia, renal/hepatic disease, urinary retention, CAD

DOSAGE AND ROUTES

- **Adult/adolescent/child ≥12 yr: PO/SL** 0.125-0.25 mg q4hr; **EXT REL** 0.375-0.75 mg q12hr
- **Adult: IM/SUBCUT/IV** 0.25-0.5 mg in a single dose or 2-4×/day q6hr
- **Geriatric:** Max 1.5 mg/day in divided doses or max 4 biphasic tabs
- **Child 2-12 yr: PO** SL 0.0625-0.125 q4hr

Available forms: Tabs 0.125, 0.15 mg; ext rel caps 0.375 mg; sol 0.125 mg/ml; elix 0.125 mg/5 ml; sol for inj 0.5 mg/ml; SL tab 0.125 mg; tab, biphasic 0.125, 0.375 mg; orally disintegrating tab 0.125 mg

Administer:

PO route

- Do not break, crush, or chew ext rel caps
- ½ hr before meals for better absorption
- Decreased dose to geriatric patients; metabolism may be slowed
- Store in tight container protected from light

IV route

- Use undiluted, inject slowly

SIDE EFFECTS

CNS: *Confusion, stimulation in geriatric patients,* headache, insomnia, dizziness, drowsiness, anxiety, weakness, hallucination

CV: *Palpitations,* tachycardia

EENT: *Blurred vision,* photophobia, mydriasis, cycloplegia, increased ocular tension

GI: *Dry mouth, constipation, paralytic ileus,* heartburn, nausea, vomiting, dysphagia, absence of taste

GU: *Urinary hesitancy, retention,* impotence

INTEG: Urticaria, rash, pruritus, anhidrosis, fever, allergic reactions

PHARMACOKINETICS

PO: Duration 4-6 hr, metabolized by liver, excreted in urine, half-life 3.5 hr

INTERACTIONS

Increase: anticholinergic effect—amantadine, tricyclics, MAOIs, H_1-antihistamines

Decrease: hyoscyamine effect—antacids

Decrease: effect of phenothiazines, levodopa, ketoconazole

NURSING CONSIDERATIONS

Assess:

- VS, cardiac status: checking for dysrhythmias, increased rate, palpitations
- I&O ratio; check for urinary retention or hesitancy
- GI complaints: pain, nausea, vomiting, anorexia
- Increased fluids, bulk, exercise to decrease constipation

Evaluate:

- Therapeutic response: absence of epigastric pain, bleeding, nausea, vomiting

Teach patient/family:

- To avoid driving, other hazardous activities until stabilized on medication
- To avoid alcohol or other CNS depressants; they will enhance sedating properties of this product
- To avoid hot environments because heat stroke may occur; that product suppresses perspiration
- To use sunglasses when outside to prevent photophobia; that product may cause blurred vision
- To notify prescriber if pregnancy is planned or suspected (C)

ibandronate (Rx)

(eye-ban′dro-nate)

Boniva

Func. class.: Bone-resorption inhibitor, electrolyte modifier

Chem. class.: Bisphosphonate

ACTION: Inhibits bone resorption, apparently without inhibiting bone formation and mineralization; absorbs calcium phosphate crystals in bone and may directly block dissolution of hydroxyapatite crystals of bone; more potent than other products

USES: Osteoporosis and prophylaxis

Unlabeled uses: Hypercalcemia of malignancy, osteolytic metastases, Paget's disease, osteoporosis (treatment/prevention) in those taking anastrozole

CONTRAINDICATIONS: Achalasia, esophageal stricture, hypocalcemia, intraarterial administration, renal failure, hypersensitivity to bisphosphonates

Precautions: Pregnancy (C), breastfeeding, children, geriatric patients, anemia, chemotherapy, coagulopathy, dental disease, diabetes mellitus, dysphagia, GI/renal disease, GERD, hypertension, infection, multiple myeloma, phosphate hypersensitivity, vit D deficiency

DOSAGE AND ROUTES

Postmenopausal osteoporosis

- **Adult: PO** 2.5 mg/day or 150 mg/mo; **IV BOL** 3 mg q3mo

Prophylaxis

- **Adult: PO** 2.5 mg/day or 150 mg/mo

Paget's disease (unlabeled)

- **Adult: IV** 2 mg as a single dose

Osteoporosis in those taking anastrozole (unlabeled)

- **Postmenopausal women: PO** 150 mg/mo

Osteolytic metastases (unlabeled)

- **Adult: IV** 6 mg over 1 hr × 3 days, repeat q4wk

Hypercalcemia (unlabeled)

- **Adult: IV INF** 2-4 mg over 2 hr

Renal dose

- **Adult: PO** CCr <30 ml/min, avoid use

Available forms: Tabs 2.5, 150 mg; sol for inj 1 mg/ml

Administer:

PO route

- Give early AM with a glass of water; if monthly, give on same day of each month
- Store at room temperature

Direct IV route

- Use single-dose prefilled syringe; discard unused portion; give over 15-30 sec; give q3mo; do not use if discolored or contains particulates

SIDE EFFECTS

CNS: Fever, insomnia, dizziness, headache

CV: Hypertension, atrial fibrillation

EENT: Ocular pain/inflammation, uveitis

GI: Constipation, nausea, vomiting, diarrhea, dyspepsia

INTEG: Rash, inj site reaction

META: *Hypomagnesemia, hypophosphatemia, hypocalcemia,* hypercholesterolemia

MS: Bone pain, myalgia, osteonecrosis of the jaw

PHARMACOKINETICS

Half-life 5-60 hr, 86%-99% protein binding; taken up mainly by bones, primarily in areas of high bone turnover; eliminated primarily by kidneys

INTERACTIONS

Increase: neurotoxicity—aminoglycosides, cycloSPORINE, tacrolimus, NSAIDs, radiopaque contrast agents, vancomycin

Increase: hypocalcemia—loop diuretics

Decrease: ibandronate effect—calcium/vit D/iron/aluminum/magnesium salts; separate by 1 hr

Drug/Food

- Do not take with food, calcium

Increase: cholesterol

Drug/Lab Test
Decrease: Alk phos, magnesium, calcium, phosphate

NURSING CONSIDERATIONS

Assess:
- **Osteoporosis:** before and during treatment; DEXA scan for bone mineral density, correct electrolyte imbalances (calcium, magnesium, phosphate) before starting therapy
- Atrial fibrillation
- **Dental health:** before dental extraction, give antiinfectives, osteonecrosis of the jaw may occur
- Blood studies: electrolytes, creatinine/BUN, vit D: correct deficiencies before treatment
- For bone pain; use analgesics; may begin within 24 hr or even years after treatment; pain usually subsides after treatment is discontinued

Evaluate:
- Therapeutic response: increased bone mineral density

Teach patient/family:
- **To report hypercalcemic relapse:** nausea, vomiting, bone pain, thirst, unusual muscle twitching, muscle spasms, severe diarrhea, constipation
- To continue with dietary recommendations, including calcium, vit D
- To obtain an analgesic from provider for bone pain
- That, if nausea, vomiting occur, small, frequent meals may help
- To report vision symptoms: blurred vision, edema, inflammation
- To report if pregnancy is planned or suspected or if planning to breastfeed, pregnancy (C)
- To exercise regularly, stop smoking, decrease alcohol intake
- To take PO first thing in AM at least 60 min before other medications, food, beverages, to take monthly dose on same day
- To sit upright for ≥60 min after PO

ibritumomab (Rx)

(ee-brit-u-moe′mab)

Zevalin

Func. class.: Radiopharmaceutical
Chem. class.: Monoclonal antibody

ACTION: High affinity for indium-111, yttrium-90; induces CD20+ B-cell lines

USES: Non-Hodgkin's lymphoma, B-cell NHL

CONTRAINDICATIONS: Pregnancy (D), hypersensitivity to murine proteins, prior murine antibody exposure

Black Box Warning: Hypersensitivity to this product, neutropenia, thrombocytopenia

Precautions: Breastfeeding, children, geriatric patients, cardiac conditions, immunizations after therapy

Black Box Warning: Altered biodistribution, inf-related reaction

DOSAGE AND ROUTES

Step 1
- **Adult: IV** riTUXimab 250 mg/m^2 given first; premedicate with acetaminophen 650 mg and diphenhydrAMINE 50 mg

Step 2
- **Adult (baseline platelets ≥150,000/mm^3):** 7-9 days after step 1, give a second dose of riTUXimab 250 mg/m^2; within 4 hr of completing riTUXimab inf, give Y-90 ibritumomab tiuxetan 0.4 mCi/kg (14.8 MBq/kg) IV over 10 min, the dose given should be within 10% of actual prescribed dose of Y-90 ibritumomab tiuxetan, total dose given max 32 mCi (1184 MBq) regardless of patient's weight
- **Adult (baseline platelets 100,000-149,000/mm^3):** 7-9 days after step 1, give a second dose of riTUXimab 250 mg/m^2 IV; within 4 hr of completing riTUXimab inf, give Y-90 ibritumomab tiuxetan 0.3 mCi/kg (11.1 MBq/kg) IV over 10 min;

the dose given should be within 10% of actual prescribed dose of Y-90 ibritumomab tiuxetan, total dose given max 32 mCi (1184 MBq) regardless of patient's weight

- **Adult (baseline platelets <100,000/mm^3):** do not use

Available forms: Inj 3.2 mg/2 ml

Administer:

- Do not use as bolus or direct IV
- See manufacturer's product labeling for preparation
- Increased fluid intake to 2-3 L/day to prevent dehydration unless contraindicated
- Emergency equipment nearby with EPINEPHrine, antihistamines, corticosteroids

SIDE EFFECTS

CV: Cardiac dysrhythmias

GI: *Nausea, vomiting, anorexia,* abdominal pain, diarrhea

GU: Renal failure

HEMA: Leukopenia, neutropenia, thrombocytopenia, anemia

INTEG: *Irritation at site, rash,* fatal mucocutaneous infections (rare)

OTHER: Fever, chills, asthenia, headache, angioedema, hypotension, myalgia, bronchospasm, hemorrhage, infections, cough, dyspnea, dizziness, anxiety

SYST: Stevens-Johnson syndrome, secondary malignancies (AML, MDS), fatal infections

PHARMACOKINETICS

Half-life 30 hr

NURSING CONSIDERATIONS

Assess:

Black Box Warning: **Fatal inf reaction:** hypoxia, pulmonary infiltrates, ARDS, MI, ventricular fibrillation, cardiogenic shock; most fatal inf reactions occur with 1st inf; potentially fatal

- Biodistribution: 1st image 2-24 hr, 2nd image 48-72 hr, 3rd image 90-120 hr (optimal)
- For infection; murine antibody titers

Black Box Warning: **Severe mucocutaneous reactions:** Stevens-Johnson syndrome, lichenoid dermatitis, toxic epidermal lysis; occur 1-13 wk after product given

Black Box Warning: **Tumor lysis syndrome:** acute renal failure requiring hemodialysis, hyperkalemia, hypocalcemia, hyperuricemia, hyperphosphatemia

Black Box Warning: **Bone marrow depression:** CBC, differential, platelet count weekly; withhold product if WBC <3500/mm^3, or platelet count <150,000/mm^3; notify prescriber

- GI symptoms: frequency of stools
- Signs of dehydration: rapid respirations, poor skin turgor, decreased urine output, dry skin, restlessness, weakness

Evaluate:

- Therapeutic response: improvement in blood counts, decreased evidence of disease

Teach patient/family:

- To report adverse reactions
- To use contraception during and for 12 mo after therapy; pregnancy (D)
- About radiation safety precautions, disposal of bodily fluids
- About symptoms of infection
- About neutropenia and bleeding precautions

ibuprofen (OTC, Rx)

(eye-byoo-proe′fen)

Advil, Advil Children's, Advil Infants' Concentrated, Advil Junior Strength, Caldolor, ElixSure IB, Genpril, Ibren, IBU, Infants Leader Ibu-drops, Midol Cramps and Body Aches Formula, Motrin ✱, Motrin IB, Motrin Infant's, Motrin Junior Strength, Pamprin IB ✱, Samson-8 Select Brand, Top Care Infant, Wal-Profen

ibuprofen lysine (Rx)

NeoProfen

Func.class.: Nonsteroidal antiinflammatory, antipyretic, nonopioid analgesics

Chem. class.: Propionic acid derivative

Do not confuse:
Nuprin/Lupron

ACTION: Inhibits COX-1, COX-2 by blocking arachidonate; analgesic, antiinflammatory, antipyretic

USES: Rheumatoid arthritis, osteoarthritis, primary dysmenorrhea, dental pain, musculoskeletal disorders, fever, migraine

Unlabeled uses: Ankylosing spondylitis, bone pain, cystic fibrosis, gouty arthritis, psoriatic arthritis, patent ductus arteriosus

CONTRAINDICATIONS: Pregnancy (D) IV after 30 wk; hypersensitivity to this product, NSAIDs, salicylates; asthma; severe renal/hepatic disease

Black Box Warning: Perioperative pain in CABG

Precautions: Pregnancy (C) 1st and 2nd trimesters, breastfeeding, children, geriatric patients, bleeding disorders, GI disorders, cardiac disorders, hypersensitivity to other antiinflammatory agents, CHF, CCr <25 ml/min

Black Box Warning: GI bleeding, MI, stroke

DOSAGE AND ROUTES

Self-treatment of minor aches/pains

- **Adult/adolescent: PO** (OTC product) 200 mg q4-6hr, may increase to 400 mg q4-6hr if needed, max 1200 mg/day

Analgesic

- **Adult: PO** 200-400 mg q4-6hr, max 3.2 g/day; OTC use max 1200 mg/day
- **Child: PO** 4-10 mg/kg/dose q6-8hr

Moderate to severe pain (hospitalized patients) (Caldolor)

- **Adult: IV** 400-800 mg q6hr as an adjunct to opiate-agonist therapy

Dysmenorrhea

- **Adult: PO** 400 mg q4hr, max 1200 mg/day

Antipyretic

- **Child 6 mo-12 yr: PO** 5 mg/kg (temperature <102.5° F or 39.2° C), 10 mg/kg (temperature >102.5° F), may repeat q6-8hr, max 40 mg/kg/day

Antiinflammatory

- **Adult: PO** 400-800 mg tid-qid, max 3.2 g/day
- **Child: PO** 30-40 mg/kg/day in 3-4 divided doses, max 50 mg/kg/day

Patent ductus arteriosus (PDA) (NeoProfen)

- **Premature neonate ≤32 wk gestation who weighs 500-1500 g: IV** 10 mg/kg initially, then, if needed, 2 doses of 5 mg/kg at 24-hr intervals; if oliguria occurs, hold dose

Available forms: Tabs 100, 200, 400, 600, 800 mg; cap, liq gels 200 mg; oral susp 100 mg/5 ml; liq 100 mg/5 ml; chew tabs 50, 100 mg; drops 50 mg/1.25 ml; inj 10 mg/ml (NeoProfen); inj (Caldolor) 100 mg/ml

Administer:

PO route

- With food, milk, or antacid to decrease GI symptoms; if nausea and vomiting occur or persist, notify prescriber
- Shake susp well before use
- Do not use in pregnancy after 30 wk gestation
- Store at room temperature

IV route

- Patient must be well hydrated before administration
- Dilute to ≤4 mg/ml with 0.9% NaCl, LR, D_5; infuse over ≥30 min; do not give IM
- Discard unused portion
- Visually inspect for particulate
- **Ibuprofen lysine:** dilute with dextrose or saline to appropriate volume (10 mg/ml of ibuprofen is recommended); give within 30 min of preparation; give via IV port nearest insertion site; give over 15 min
- Check for extravasation; do not give in same line with TPN; interrupt TPN for 15 min before and after product administration

SIDE EFFECTS

CNS: *Headache,* dizziness, drowsiness, fatigue, tremors, confusion, insomnia, anxiety, depression

CV: Tachycardia, peripheral edema, palpitations, dysrhythmias, **CV thrombotic events, MI, stroke**

EENT: Tinnitus, hearing loss, blurred vision

GI: Nausea, *anorexia,* vomiting, diarrhea, jaundice, **hepatitis,** constipation, flatulence, cramps, dry mouth, peptic ulcer, **GI bleeding, ulceration, necrotizing enterocolitis, GI perforation**

GU: **Nephrotoxicity:** dysuria, hematuria, oliguria, azotemia

HEMA: **Blood dyscrasias,** increased bleeding time

INTEG: Purpura, rash, pruritus, sweating, urticaria, **nectrotizing fasciitis,** photosensitivity, photophobia

META: Hyperkalemia, hyperuricemia, hypoglycemia, hyponatremia

SYST: **Anaphylaxis, Stevens-Johnson syndrome**

PHARMACOKINETICS

PO: Onset ½ hr, peak 1-2 hr, half-life 1.8-2 hr, metabolized in liver (inactive metabolites), excreted in urine (inactive metabolites), 90%-99% plasma protein binding, does not enter breast milk, well absorbed

INTERACTIONS

Increase: bleeding risk—valproic acid, thrombolytics, antiplatelets, anticoagulants, salicylates

Increase: blood dyscrasia risk—antineoplastics, radiation

Increase: toxicity—lithium, oral anticoagulants, cycloSPORINE, methotrexate

Increase: GI reactions—aspirin, corticosteroids, NSAIDs, alcohol, tobacco

Increase: hypoglycemia—oral antidiabetics

Decrease: effect of antihypertensives, thiazides, furosemide

Decrease: ibuprofen action—aspirin

Drug/Herb

Increase: bleeding risk—feverfew, garlic, ginger, ginkgo, ginseng *(Panax)*

Drug/Lab Test

Increase: BUN, creatinine, LFTs

Decrease: Hgb/Hct, blood glucose, WBC, platelets

NURSING CONSIDERATIONS

Assess:

- Renal, hepatic, blood studies: BUN, creatinine, AST, ALT, Hgb, stool guaiac, before treatment, periodically thereafter, monitor electrolytes as needed

Black Box Warning: Perioperative pain in CABG: MI, and stroke can result for 10-14 days; can be fatal

- **Pain:** note type, duration, location, intensity with ROM 1 hr after administration

• Audiometric, ophthalmic exam before, during, after long-term treatment; for eye, ear problems: blurred vision, tinnitus; may indicate toxicity
• **Infection,** may mask symptoms; fever: temperature before and 1 hr after administration
• Cardiac status: edema (peripheral), tachycardia, palpitations; monitor B/P, pulse for character, quality, rhythm, especially in patients with cardiac disease, geriatric patients
• For history of peptic ulcer disorder; asthma, aspirin, hypersensitivity; check closely for hypersensitivity reactions

Black Box Warning: GI bleeding/perforation: chronic use can cause gastritis with or without bleeding, those with a prior history of peptic ulcer disease or GI bleeding, initiate treatment at lower dose; geriatric patients are at greater risk and those who consume >3 alcohol drinks/day

Evaluate:
• Therapeutic response: decreased pain, stiffness in joints; decreased swelling in joints; ability to move more easily; reduction in fever or menstrual cramping

Teach patient/family:
• To use sunscreen, sunglasses, and protective clothing to prevent photosensitivity, photophobia
• To report blurred vision, ringing, roaring in ears (may indicate toxicity); that eye and hearing tests should be done during long-term therapy
• To avoid driving, other hazardous activities if dizziness or drowsiness occurs
⚠ **Nephrotoxicity:** to report change in urinary pattern, increased weight, edema, increased pain in joints, fever, blood in urine; monitor fluid status, BUN, creatinine
• That therapeutic antiinflammatory effects may take up to 1 mo
⚠ To avoid alcohol, NSAIDs, salicylates; bleeding may occur
• To report use of this product to all health care providers
• **Pregnancy:** notify prescriber if pregnancy (C) is planned or suspected; avoid during 3rd trimester pregnancy (D) IV after 30 wk

TREATMENT OF OVERDOSE:
Lavage, activated charcoal, induce diuresis

⚠ HIGH ALERT

ibutilide (Rx)
(eye-byoo′tih-lide)
Corvert
Func. class.: Antidysrhythmic (class III)
Chem. class.: Methane sulfonamide

ACTION:
Prolongs duration of action potential and effective refractory period

USES:
For rapid conversion of atrial fibrillation/flutter, including within 1 wk of coronary artery bypass or valve surgery

CONTRAINDICATIONS:
Hypersensitivity
Precautions: Pregnancy (C), breastfeeding, children <18 yr, geriatric patients, sinus node dysfunction, 2nd- or 3rd-degree AV block, electrolyte imbalances, bradycardia, renal/hepatic disease, CHF

Black Box Warning: QT prolongation, torsades de pointes, ventricular arrhythmias, ventricular tachycardia, cardiac dysrhythmias

DOSAGE AND ROUTES
Atrial fibrillation/flutter
• **Adult ≥60 kg: IV INF** 1 vial (1 mg) given over 10 min, may repeat same dose after 10 min
• **Adult <60 kg: IV INF** 0.01 mg/kg given over 10 min, may repeat same dose after 10 min
Available forms: Inj 1 mg/10 ml

Administer:

- Ice compress after stopping inf for extravasation; tubing should be removed and attempt to aspirate product; elevate affected areas

IV route

- Undiluted or diluted in 50 ml 0.9% NaCl, or D_5W (0.017 mg/ml); give over 10 min
- Solution is stable for 48 hr refrigerated or 24 hr at room temperature
- Do not admix with other sol, products
- Reduce dosage slowly with ECG monitoring
- Stop inf as soon as arrhythmia is controlled
- Do not use if discolored or particulate is present

SIDE EFFECTS

CNS: *Headache*

CV: *Hypotension, bradycardia,* **sinus arrest, CHF, dysrhythmias, torsades de pointes,** hypertension, extrasystoles, ventricular tachycardia, bundle branch block, AV block, palpitations, supraventricular extrasystoles, syncope, **prolonged QT interval**

GI: Nausea

PHARMACOKINETICS

Elimination half-life 6 hr, metabolized by liver, excreted by kidneys

INTERACTIONS

Increase: prodysrhythmia—phenothiazines, tricyclics, tetracyclics, antidepressants, H_1-receptor antagonists, antihistamines

Increase: masking of cardiotoxicity—digoxin

- Do not use within 5 hr of ibutilide: class Ia antidysrhythmics (disopyramide, quiNIDine, procainamide), class III agents (amiodarone, sotalol)

Drug/Herb

- Do not use with hawthorn

Increase: antiarrhythmic action—ginger

NURSING CONSIDERATIONS

Assess:

Black Box Warning: ECG continuously for ≥4 hr to determine product effectiveness; measure PR, QRS, QT intervals, check for PVCs, other dysrhythmias; discontinue if atrial fibrillation/flutter ceases; continue until QT interval corrected for heart rate (QTc) returned to baseline; if used ≥2 days, anticoagulation must be adequate

- I&O ratio; electrolytes: potassium, sodium, chlorine
- Hepatic studies: AST, ALT, bilirubin, alk phos
- Dehydration or hypovolemia
- Rebound hypertension after 1-2 hr
- Cardiac rate, respiration: rate, rhythm, character, chest pain

Evaluate:

- Therapeutic response: decrease in atrial fibrillation/flutter

Teach patient/family:

- To report side effects immediately
- About reason for medication

icosapent

(eye-koe'sa-pent)

Vascepa

Func. class.: Antilipidemic

Chem. class.: Omega-3 fatty acid ethylester

ACTION: Inhibits hepatic very low density lipoprotein (VLDL) and triglyceride synthesis; enhances clearance of triglycerides from circulating VLDL particles

USES: As adjunct to diet in adults with severe hypertriglyceridemia (≥500 mg/dl)

CONTRAINDICATIONS: Hypersensitivity to icosapent ethyl

Precautions: Hepatic disease, bleeding, breastfeeding, children, fish/shellfish hypersensitivity, thrombolytic/anticoagulation therapy, pregnancy (C)

DOSAGE AND ROUTES

- **Adult: PO** 2 g bid

Available forms: Soft gel cap 1 g
Administer:

- Give with food, swallow whole
- Store at room temperature

SIDE EFFECTS

HEMA: Ecchymosis, epistaxis
MS: Arthralgia

PHARMACOKINETICS

Peak 5 hr, protein binding 99%, half-life 89 hr

INTERACTIONS

Increase: effects of anticoagulants, platelet inhibitors, thrombolytics
Drug/Lab Test
Increase: bleeding time

NURSING CONSIDERATIONS

Assess:
Hypertriglyceridemia: Obtain diet history including fat, cholesterol in diet; cholesterol, triglyceride levels periodically during treatment
Liver disease: Monitor LFTs baseline, periodically in those with liver disease
Bleeding: Monitor ecchymosis, epistaxis, bleeding time; use cautiously in anticoagulant/thrombolytic therapy
Fish/shellfish hypersensitivity: Identify fish hypersensitivity, use cautiously
Evaluate:

- Therapeutic response: decrease in triglycerides

Teach patient/family:

- That blood work may be necessary during treatment
- To report bleeding if anticoagulants or thrombolytics are used
- Not to break, crush, open, or dissolve caps
- That previously prescribed regimen will continue: low-cholesterol diet, exercise program, smoking cessation

⚠ HIGH ALERT

IDArubicin (Rx)

(eye-dah-roob′ih-sin)

Idamycin PFS

Func. class.: Antineoplastic, antibiotic
Chem. class.: Anthracycline glycoside

Do not confuse:
IDArubicin/DOXOrubicin/DAUNOrubicin/epirubicin
Idamycin/Adriamycin

ACTION: Non–cell-cycle specific; topoisomerase II inhibitor; vesicant; intercalcalating between DNA base pairs, causing shape change, low free radicals

USES: Used in combination with other antineoplastics for acute myelocytic leukemia in adults
Unlabeled uses: Breast cancer, liquid tumors, non-Hodgkin's lymphoma, ALL, CLL, AML

CONTRAINDICATIONS: Pregnancy (D), breastfeeding, hypersensitivity

Black Box Warning: Myelosuppression, bilirubin >5 mg/dl

Precautions: Children, gout, bone marrow depression, preexisting CV disease

Black Box Warning: Renal/hepatic disease, heart failure

DOSAGE AND ROUTES

- **Adult: IV** 8-12 mg/m^2/day × 3 days in combination with cytarabine (induction)
- **Adolescent/child (unlabeled): IV** 10-12 mg/m^2/day × 3 days

Renal/hepatic dose

- **Adult: IV** CCr >2.5 mg/dl, reduce dose; bilirubin 2.5-5 mg/dl, reduce dose by 50%; bilirubin >5 mg/dl, do not use

Available forms: Inj 1 mg/ml
Administer:

- Ice compress after stopping inf for extravasation

• Store at room temperature for 3 days after reconstituting or 7 days refrigerated

Intermittent IV INF route

• Do not give IM/SUBCUT
• Use cytotoxic handling procedures after preparing in biologic cabinet wearing gown, gloves, mask
• Antiemetic 30-60 min before product and 6-10 hr after treatment to prevent vomiting
• After reconstituting 5-mg vial with 5 ml 0.9% NaCl (1 mg/1 ml); give over 10-15 min through Y-tube or 3-way stopcock of inf of D_5 or NS; discard unused portion; use caution when needle inserted into vial (negative pressure)

Y-site compatibilities: Amifostine, amikacin, aztreonam, cimetidine, cladribine, cyclophosphamide, cytarabine, diphenhydrAMINE, droperidol, erythromycin, filgrastim, granisetron, imipenem/CISplatin, magnesium sulfate, mannitol, melphalan, metoclopramide, potassium chloride, ranitidine, sargramostim, thiotepa, vinorelbine

SIDE EFFECTS

CNS: Fever, chills, *headache,* **seizures**
CV: **Dysrhythmias, CHF, pericarditis, myocarditis,** peripheral edema, angina, **MI, myocardial toxicity**
GI: *Nausea, vomiting, abdominal pain, mucositis, diarrhea,* **hepatotoxicity**
GU: **Nephrotoxicity,** red urine
HEMA: **Thrombocytopenia, leukopenia, anemia**
INTEG: Rash, **extravasation,** dermatitis, *reversible alopecia,* urticaria; thrombophlebitis and tissue necrosis at inj site; radiation recall
SYST: **Infection,** tumor lysis syndrome

PHARMACOKINETICS

Half-life 22 hr; metabolized by liver; crosses placenta; excreted in bile, urine (primarily as metabolites); 97% protein binding

INTERACTIONS

Increase: bleeding risk—anticoagulants, salicylates, NSAIDs, thrombolytics; avoid concurrent use
Decrease: IDArubicin effect—corticosteroids
Increase: CHF, ventricular dysfunction—trastuzumab
Increase: ECG changes (QT prolongation, changes in QRS voltage)—class IA/III antidysrhythmias, some phenothiazines, and other products that increase QT prolongation
Increase: cardiotoxicity—cyclophosphamide
Increase: toxicity—other antineoplastics or radiation
Decrease: antibody response—live virus vaccines

Drug/Lab Test
Decrease: calcium, platelets, neutrofils
Increase: uric acid, phosphate, potassium

NURSING CONSIDERATIONS

Assess:

Black Box Warning: CBC, differential, platelet count weekly; notify prescriber of results, severe myelosuppression can occur

• Renal studies: BUN, serum uric acid, urine CCr, electrolytes before, during therapy
⚠ **Tumor lysis syndrome: hyperkalemia, hyperphosphatemia, hyperuricemia, hypocalcemia**
• I&O ratio; report fall in urine output to <30 ml/hr
• Monitor temperature; fever may indicate beginning infection

Black Box Warning: Hepatic studies before, during therapy: bilirubin, AST, ALT, alk phos prn or monthly; check for jaundice of skin and sclera, dark urine, clay-colored stools, itchy skin, abdominal pain, fever, diarrhea

Black Box Warning: Cardiac toxicity: CHF, dysrythmias, cardiomyopathy; cardiac studies before and periodically during treatment: ECG, chest x-ray, MUGA; ECG: watch for ST-T wave changes, low QRS and T, possible dysrhythmias (sinus tachycardia, heart block, PVCs)

- **Bleeding:** hematuria, guaiac stools, bruising or petechiae, mucosa or orifices
- Effects of alopecia on body image; discuss feelings about body changes
- Inflammation of mucosa, breaks in skin
- Buccal cavity for dryness, sores, ulceration, white patches, oral pain, bleeding, dysphagia

Black Box Warning: Local irritation, pain, burning at inj site; extravasation (vesicant)

- GI symptoms: frequency of stools, cramping
- Increase fluid intake to 2-3 L/day to prevent urate and calculi formation

Evaluate:

- Therapeutic response: decreased liquid tumor, spread of malignancy

Teach patient/family:

- To report signs of CHF, cardiac toxicity, beginning infection
- That hair may be lost during treatment; that wig or hairpiece may make patient feel better; that new hair may be different in color, texture
- To avoid foods with citric acid, hot or rough texture
- To avoid crowds, persons with upper respiratory illness
- To report any bleeding, white spots, ulcerations in mouth; to examine mouth daily
- That urine may be red-orange for 48 hr
- To report if pregnancy is planned or suspected, pregnancy (D)
- To use contraception during treatment, for ≥4 mo after treatment
- That all body fluids will change color

⚠ HIGH ALERT

ifosfamide (Rx)

(i-foss′fa-mide)

Ifex

Func. class.: Antineoplastic alkylating agent

Chem. class.: Nitrogen mustard

Do not confuse:
ifosfamide/cyclophosphamide

ACTION: Alkylates DNA, RNA, inhibits enzymes that allow synthesis of amino acids in proteins; also responsible for cross-linking DNA strands; activity is not cell-cycle–stage specific

USES: Testicular cancer

Unlabeled uses: Soft-tissue sarcoma, Ewing's sarcoma, non-Hodgkin's lymphoma, lung/pancreatic sarcoma, bladder, breast, cervical, thymic cancer, desmoid tumor, Ewing's sarcoma, rhabdomyosarcoma

CONTRAINDICATIONS: Pregnancy (D), hypersensitivity

Black Box Warning: Bone marrow suppression

Precautions: Breastfeeding, children, renal/hepatic disease, accidental exposure, dehydration, dental disease, infection, IM injection, ocular exposure varicella

Black Box Warning: Coma, hemorrhagic cystitis

DOSAGE AND ROUTES

- **Adult: IV** 1.2-2 g/m²/day × 5 days, repeat course q3wk, given with mesna, epirubicin, ondansetron alpha

Renal dose

- **Adult: IV** CCr 31-60 ml/min, give 75% of dose; CCr 10-30 ml/min, give 50% of dose; CCr <10 ml/min, do not give

Available forms: Inj 1-, 3-g vials

Administer:

- Antiemetic 30-60 min before product to prevent vomiting
- Visually inspect parenteral products for particulate matter and discoloration before use
- Store powder at room temperature

IV route

- Give as an intermittent inf or continuous inf
- Well hydrate with ≥2 L/day of oral or IV fluids to prevent bladder toxicity
- Must be given in combination with mesna to prevent hemorrhagic cystitis
- Close hematologic monitoring is recommended; WBC count, platelet count, and hemoglobin should be obtained before each use and periodically thereafter
- A urinalysis should be performed before each dose to monitor for hematuria

Reconstitution and further dilution:

- Reconstitute 1 or 3 g with 20 or 60 ml, respectively, of sterile water for injection or bacteriostatic water for injection containing parabens or benzyl alcohol to give IV solutions containing 50 mg/ml
- Solutions may be diluted further to achieve concs of 0.6-20 mg/ml in the following solutions: 5% dextrose for injection, 0.9% sodium chloride for injection, lactated Ringer's for injection, or sterile water for injection
- Infuse slowly over at least 30 min
- Diluted and reconstituted solutions must be refrigerated and used within 24 hr

Additive compatibilities: CARBOplatin, CISplatin, etoposide, fluorouracil, mesna

Y-site compatibilities: Allopurinol, amifostine, amphotericin B cholesteryl, aztreonam, DOXOrubicin liposome, filgrastim, fludarabine, gallium, granisetron, melphalan, ondansetron, PACLitaxel, piperacillin/tazobactam, propofol, sargramostim, sodium bicarbonate, teniposide, thiotepa, vinorelbine

SIDE EFFECTS

CNS: Facial paresthesia, fever, malaise, somnolence, confusion, depression, hallucinations, dizziness, disorientation, **seizures, coma,** cranial nerve dysfunction, **encephalopathy**

GI: Nausea, vomiting, anorexia, **hepatotoxicity,** stomatitis, dyslipidemia, hyperglycemia, constipation, diarrhea

GU: **Hematuria, nephrotoxicity, hemorrhagic cystitis,** dysuria, urinary frequency

HEMA: **Thrombocytopenia, leukopenia, anemia,** retrograde ejaculation

INTEG: Dermatitis, alopecia, pain at inj site, hyperpigmentation

META: Metabolic acidosis

PHARMACOKINETICS

Metabolized by liver, saturation occurs at high doses, excreted in urine, half-life 7-15 hr, depends on dose

INTERACTIONS

Increase: myelosuppression—other antineoplastics, radiation

Increase: toxicity—CYP3A4, a weak P-gp inhibitor, inducers, barbiturates, allopurinol

Increase: bleeding risk—NSAIDs, anticoagulants, salicylates, thrombolytics

Decrease: antibody response—live virus vaccines

Decrease: effect of ifosfamide—CYP3A4 inhibitors

- Do not use within 24 hr of hematopoietic progenitor cells

NURSING CONSIDERATIONS

Assess:

- Hepatic studies before, during therapy (bilirubin, AST, ALT, LDH) monthly or as needed; jaundice of skin and sclera, dark urine, clay-colored stools, itchy skin, abdominal pain, fever, diarrhea

Black Box Warning: CBC, differential, platelet count weekly; withhold product if WBC <2000 or platelet count <50,000; notify prescriber; severe myelosuppression may occur

- Monitor temperature (may indicate beginning infection)

• Blood dyscrasias (anemia, granulocytopenia); bruising, fatigue, bleeding, poor healing
• Allergic reactions: dermatitis, exfoliative dermatitis, pruritus, urticaria

Black Box Warning: I&O ratio; monitor for hematuria; hemorrhagic cystitis can occur; increase fluids to 3 L/day; urinalysis before each dose, not to give at night

Black Box Warning: Neurologic symptoms: hallucinations, confusion, disorientation, coma; product should be discontinued

• Bleeding: hematuria, guaiac, bruising or petechiae, mucosa or orifices

Evaluate:
• Therapeutic response: decrease in size and spread of tumor

Teach patient/family:
• To notify prescriber of sore throat, swollen lymph nodes, malaise, fever; other infections may occur
• Not to have vaccinations during or after treatment
• That hair may be lost during treatment; that wig or hairpiece may make the patient feel better; that new hair may be different in color, texture
• To report signs of anemia: fatigue, headache, faintness, SOB, irritability
• To report bleeding; to avoid use of razors, commercial mouthwash
• To avoid use of aspirin products, NSAIDs, ibuprofen because hemorrhage can occur
• To notify prescriber if pregnancy is planned or suspected, pregnancy (D)
• To use contraceptive measures during therapy; not to breastfeed
• To avoid crowds, persons with infections
• To report confusion, hallucinations, extreme drowsiness, numbness, tingling; to avoid alcohol use for ≥4 mo after treatment
• To avoid driving, hazardous activities until reaction is known
• To use excessive fluids and urinate often to prevent hemorrhagic cystitis

iloperidone (Rx)

(ill-o-pehr′ih-dohn)

Fanapt

Func. class.: Antipsychotic

Chem. class.: Benzisoxazole derivative

ACTION: Unknown; may be mediated through both DOPamine type 2 (D2) and serotonin type 2 (5-HT2) antagonism, high receptor binding affinity for norepinephrine (alpha 1)

USES: Schizophrenia

Unlabeled uses: Agitation

CONTRAINDICATIONS: Breastfeeding, hypersensitivity

Precautions: Pregnancy (C), children, geriatric patients, renal/hepatic disease, breast cancer, Parkinson's disease, dementia with Lewy bodies, seizure disorder, QT prolongation, bundle branch block, acute MI, ambient temperature increase, AV block, stroke, substance abuse, suicidal ideation, tardive dyskinesia, torsade de pointes, blood dyscrasias, dysphagia

Black Box Warning: Increased mortality in elderly patients with dementia-related psychosis

DOSAGE AND ROUTES

• **Adult: PO** 1 mg bid day 1, 2 mg bid day 2, 4 mg bid day 3, 6 mg bid day 4, 8 mg bid day 5, 10 mg bid day 6, 12 mg bid day 7; max 24 mg/day in 2 divided doses; reduce dose by 50% in patient who is a poor metabolizer of CYP2D6 or when used with strong CYP2D6/CYP3A4 inhibitors

Available forms: Tabs 1, 2, 4, 6, 8, 10, 12 mg; titration pack

Administer:
• Use without regard to meals
• Reduced dose in geriatric patients
• Anticholinergic agent for EPS

- Avoid use with CNS depressants
- Store in tight, light-resistant container

SIDE EFFECTS

CNS: *EPS, pseudoparkinsonism, akathisia, dystonia, tardive dyskinesia; drowsiness,* **seizures, neuroleptic malignant syndrome,** dizziness, delirium, depression, paranoia, fatigue, hostility, lethargy, restlessness, vertigo, tremor

CV: Orthostatic hypotension, **heart failure, AV block, QT prolongation,** tachycardia

EENT: Blurred vision, cataracts, nystagmus, tinnitus

GI: *Nausea,* vomiting, *anorexia, constipation,* jaundice, weight gain/loss, abdominal pain, stomatitis, xerostomia

GU: Hyperprolactinemia, urinary retention/incontinence, testicular pain, **renal failure**

HEMA: Agranulocytosis, leukopenia, neutropenia

MISC: Renal artery occlusion, hyperglycemia, dyslipidemia

PHARMACOKINETICS

PO: Extensively metabolized by liver to major active metabolite by CYP2D6, CYP3A4; protein binding 95%; peak 2-4 hr; excreted in urine and feces; terminal half-life 18 hr in extensive metabolizers; 33 hr in poor metabolizers

INTERACTIONS

Increase: serotonin syndrome, neuroleptic malignant syndrome—SSRIs, SNRI

Increase: sedation—other CNS depressants, alcohol

Increase: iloperidone effect, decreased clearance—CYP2D6, CYP3A4 inhibitors (delavirdine, indinavir, isoniazid, itraconazole, dalfopristin, ritonavir, tipranavir), reduce dose

Increase: QT prolongation—class IA/ III antidysrhythmics, some phenothiazines, β-agonists, local anesthetics, tricyclics, haloperidol, methadone, chloroquine, clarithromycin, droperidol, erythromycin, pentamidine

Decrease: iloperidone action—CYP2D6, CYP3A4 inducers (carBAMazepine, barbiturates, phenytoins, rifampin)

Drug/Lab Test

Increase: prolactin levels, cholesterol, glucose, lipids, triglycerides

Decrease: potassium

NURSING CONSIDERATIONS

Assess:

- Mental status before initial administration
- I&O ratio; palpate bladder if urinary output is low
- Bilirubin, CBC, hepatic studies monthly, monitor glucose, hyperglycemia may occur
- Urinalysis before, during prolonged therapy, monitor potassium, magnesium and lipid panel
- Affect, orientation, LOC, reflexes, gait, coordination, sleep-pattern disturbances
- B/P standing and lying; pulse, respirations; q4hr during initial treatment; establish baseline before starting treatment; report drops of 30 mm Hg; watch for ECG changes; QT prolongation may occur; dizziness, faintness, palpitations, tachycardia on rising
- EPS, including akathisia, tardive dyskinesia (bizarre movements of the jaw, mouth, tongue, extremities), pseudoparkinsonism (rigidity, tremors, pill rolling, shuffling gait)

Black Box Warning: **Serious reactions in the geriatric patient:** fatal pneumonia, heart failure, sudden death, not to be used in the elderly with dementia

Black Box Warning: **Neuroleptic malignant syndrome:** hyperthermia, increased CPK, altered mental status, muscle rigidity

- Constipation, urinary retention daily; if these occur, increase bulk and water in diet
- Weight gain, hyperglycemia, metabolic changes in diabetes
- Supervised ambulation until patient is stabilized on medication; do not involve

patient in strenuous exercise program because fainting is possible; patient should not stand still for long periods
- Sips of water, candy, gum for dry mouth

Evaluate:
- Therapeutic response: decrease in emotional excitement, hallucinations, delusions, paranoia; reorganization of patterns of thought, speech

Teach patient/family:
- That orthostatic hypotension may occur; to rise from sitting or lying position gradually
- To avoid hot tubs, hot showers, tub baths because hypotension may occur; that heat stroke may occur in hot weather; to take extra precautions to stay cool
- To avoid abrupt withdrawal of product because EPS may result; that product should be withdrawn slowly, to review symptoms of neuroleptic malignant syndrome
- To avoid OTC preparations (cough, hay fever, cold) unless approved by prescriber because serious product interactions may occur; to avoid use of alcohol because increased drowsiness may occur
- To avoid hazardous activities if drowsy or dizzy
- To comply with product regimen
- To report impaired vision, tremors, muscle twitching
- To use contraception; to inform prescriber if pregnancy is planned or suspected

TREATMENT OF OVERDOSE:

Lavage if orally ingested; provide airway; *do not induce vomiting*

imatinib (Rx)

(im-ah-tin′ib)

Gleevec

Func. class.: Antineoplastic—miscellaneous

Chem. class.: Protein-tyrosine kinase inhibitor

ACTION:
Inhibits Bcr-Abl tyrosine kinase created in patients with chronic myeloid leukemia (CML), also inhibits tyrosine kinases including EGF, FGF, PDGF, SCF, VEGF, NFG

USES:
Treatment of chronic myeloid leukemia (CML); Philadelphia-chromosome–positive patients in blast-cell crisis or patients in chronic failure after treatment failure with interferon alfa; gastrointestinal stromal tumors (GIST) positive for c-Kit; chronic eosinophilic leukemia, acute lymphocytic leukemia, dermatofibrosarcoma protuberans, myelodysplastic syndrome, systemic mastocytosis

Unlabeled uses: Desmoid tumor

CONTRAINDICATIONS:
Pregnancy (D), hypersensitivity

Precautions: Breastfeeding, children, geriatric patients, cardiac/renal/hepatic/dental disease, GI bleeding, bone marrow suppression, infection, thrombocytopenia, neutropenia, immunosuppression

DOSAGE AND ROUTES

For the treatment of Philadelphia chromosome positive (Ph+) chronic myelogenous leukemia (CML) chronic phase as initial therapy

- **Adult: PO** 400 mg/day, continue as long as beneficial; may increase to 600 mg/day in the absence of severe adverse reactions and severe non–leukemia-related neutropenia or thrombocytopenia
- **Adolescent/child >2 yr: PO** 340 mg/m^2/day, max 600 mg/day; the daily dose

may be given as a single dose or split into 2 doses given once in the morning and once in the evening

Adult with Ph+ CML in chronic phase after the failure of interferon-alfa therapy

• **Adult: PO** 400 mg every day; continue as long as beneficial, may increase to 600 mg/day

Pediatric patients with Ph+ chronic phase CML whose disease has recurred after hematopoietic stem cell transplant or who are resistant to interferon-alfa therapy

• **Adolescent/child >3 yr: PO** 260 mg/m^2/day as a single daily dose, or the dose may be divided given once in the morning and once in the evening; may increase to 340 mg/m^2/day

Patients with Ph+ CML in accelerated phase or blast crisis

• **Adult: PO** 600 mg every day, continue as long as beneficial, may increase to 800 mg/day (400 mg bid)

Resistant or relapsed PH+ acute lymphocytic leukemia (ALL)

• **Adult: PO** 600 mg every day, continue as long beneficial

***Kit* (CD117)-positive unresectable and/or metastatic GIST**

• **Adult: PO** 400-600 mg every day, may increase to 400 mg bid

Adjuvant treatment of *Kit* (CD117)-positive GIST after complete gross resection

• **Adult: PO** 400 mg/day

Patients with hypereosinophilic syndrome (HES) and/or chronic eosinophilic leukemia (CEL) who have the FIPL1L1-PDGFR α-fusion kinase (mutational analysis or FISH demonstration of CHIC2 allele deletion) and for patients with HES and/or CEL who are FIPL1L-PDGFR α-fusion kinase negative or unknown

• **Adult: PO** 400 mg/day in those who are FIPL1L-PDGFR α-fusion kinase negative or unknown; for HES/CEL patients with demonstrated FIP1L1-PDGFR α-fusion kinase, 100 mg/day, may increase to 400 mg

Myelodysplastic syndrome (MDS)/myeloproliferative disease (MPD) associated with the PDGFR (platelet-derived growth factor receptor) gene rearrangements

• **Adult: PO** 400 mg/day

Aggressive systemic mastocytosis (ASM) without D816V c-Kit mutation or with c-Kit mutation status unknown

• **Adult: PO** 400 mg/day in those without the FIP1L1-PDGFR-α c-Kit mutation; if c-Kit status is unknown or not available, give 400 mg/day

Unresectable, recurrent, and/or metastatic dermatofibrosarcoma protuberans (DFSP)

• **Adult: PO** 400 mg bid (800 mg/day)

Renal dose

• **Adult: PO** CCr 40-59 ml/min, max 600 mg/day; CCr 20-39 ml/min, decrease initial dose by 50%, max 400 mg/day; CCr <20 ml/min, use with caution, 100 mg/day

Hepatic dose

• **Adult: PO** Total bilirubin 1.5-3 × ULN and any AST, decrease initial dose to 400 mg/day; total bilirubin >3 × ULN and any AST, decrease initial dose to 300 mg/day

Available forms: Tabs 100, 400 mg

Administer:

• With meal and large glass of water to decrease GI symptoms; doses of 800 mg should be given as 400 mg bid
• Tab may be dispersed in a glass of water or apple juice, use 50 ml of liquid for 100-mg tab, 200 ml liquid for 400-mg tab
• Continue as long as beneficial
• Store at 77° F (25° C)

SIDE EFFECTS

CNS: **CNS hemorrhage,** headache, dizziness, insomnia, **subdermal hematoma**

CV: **Hemorrhage, heart failure, cardiac tamponade, hypereosinophilia, cardiac toxicity**

EENT: Blurred vision, conjunctivitis

GI: *Nausea,* hepatotoxicity, vomiting, dyspepsia, GI hemorrhage, *anorexia, abdominal pain,* GI perforation, diarrhea
HEMA: Neutropenia, thrombocytopenia, bleeding
INTEG: *Rash, pruritus,* alopecia, photosensitivity
META: Fluid retention, hypokalemia, edema
MISC: Fatigue, epistaxis, pyrexia, night sweats, increased weight, flulike symptoms, hypothyroidism
MS: Cramps, pain, arthralgia, myalgia
RESP: Cough, dyspnea, nasopharyngitis, pneumonia, upper respiratory tract infection, pleural effusion, edema

PHARMACOKINETICS

Well absorbed (98%); protein binding 95%; metabolized by CYP3A4; excreted in feces, small amount in urine; peak 2-4 hr; duration 24 hr (imatinib), 40 hr (metabolite); half-life 18-40 hr

INTERACTIONS

Increase: hepatotoxicity—acetaminophen
Increase: imatinib concs—CYP3A4 inhibitors (ketoconazole, itraconazole, erythromycin, clarithromycin)
Increase: plasma concs of simvastatin, calcium channel blockers, ergots
Increase: plasma conc of warfarin; avoid use with warfarin; use low-molecular-weight anticoagulants instead
Decrease: imatinib concs—CYP3A4 inducers (dexamethasone, phenytoin, carBAMazepine, rifampin, PHENobarbital)

Drug/Herb

Decrease: imatinib conc—St. John's wort

Drug/Lab Test

Increase: bilirubin, amylase, LFTs
Decrease: albumin, calcium, potassium, sodium, phosphate, platelets, neutrophils, leukocytes, lymphocytes

NURSING CONSIDERATIONS

Assess:

- **Bone marrow suppression:** ANC, platelets; during chronic phase, if ANC $<1 \times 10^9$/L and/or platelets $<50 \times 10^9$/L, stop until ANC $>1.5 \times 10^9$/L and platelets $>75 \times 10^9$/L; during accelerated phase/blast crisis, if ANC $<0.5 \times 10^9$/L and/or platelets $<10 \times 10^9$/L, determine whether cytopenia related to biopsy/aspirate; if not, reduce dose by 200 mg; if cytopenia continues, reduce dose by another 100 mg; if cytopenia continues for 4 wk, stop product until ANC $\geq 1 \times 10^9$/L
- **Renal toxicity:** if bilirubin $>3 \times$ IULN, withhold imatinib until bilirubin levels return to $<1.5 \times$ IULN
- **Hepatotoxicity:** monitor LFTs, before treatment monthly; if liver transaminases $>5 \times$ IULN, withhold imatinib until transaminase levels return to $<2.5 \times$ IULN
- Monitor CBC for first mo, biweekly next mo and periodically thereafter; neutropenia (2-3 wk) and thrombocytopenia (3-4 wk) and anemia may occur; may need dosage decrease or discontinuation
- Signs of fluid retention, edema: weigh, monitor lung sounds, assess for edema; some fluid retention is dose dependent

Evaluate:

- Therapeutic response: decrease in leukemic cells or size of tumor

Teach patient/family:

- To report adverse reactions immediately: shortness of breath, swelling of extremities, bleeding
- About reason for treatment, expected results
- That effect on male infertility is unknown
- Not to stop or change dose—to avoid hazardous activities until response is known, dizziness may occur
- To take with food and water; for those unable to swallow tabs, to mix in liquid (30 m for 100 mg) or 200 ml for 400 mg) after dissolved, stir and consume
- To avoid OTC products unless approved by prescriber

• To notify prescriber if pregnancy is planned or suspected, pregnancy (D), do not breastfeed

imipenem/cilastatin (Rx)

(i-me-pen′em sye-la-stat′in)

Primaxin IM, Primaxin IV

Func. class.: Antiinfective—miscellaneous

Chem. class.: Carbapenem

Do not confuse:
imipenem/Omnipen
Primaxin/Premarin

ACTION: Interferes with cell-wall replication of susceptible organisms; osmotically unstable cell wall swells, bursts from osmotic pressure; addition of cilastatin prevents renal inactivation that occurs with high urinary concs of imipenem

USES: Serious infections caused by gram-positive *Streptococcus pneumoniae,* group A β-hemolytic streptococci, *Staphylococcus aureus,* enterococcus; gram-negative *Klebsiella, Proteus, Escherichia coli, Acinetobacter, Serratia, Pseudomonas aeruginosa, Salmonella, Shigella, Haemophilus influenzae, Listeria* sp.

CONTRAINDICATIONS: Hypersensitivity to this product, amide local anesthetics, or carbapenems; AV block, shock (IM)

Precautions: Pregnancy (C), breastfeeding, children, geriatric patients, seizure disorders, renal disease, head trauma; hypersensitivity to cephalosporins, penicillins; pseudomembranous colitis, ulcerative colitis, diabetes mellitus

DOSAGE AND ROUTES

Intraabdominal, gynecologic, lower respiratory tract, skin and skin structure, bone and joint infections; septicemia, endocarditis, febrile neutropenia (unlabeled), and polymicrobial infections for fully susceptible organisms including gram-positive or gram-negative aerobes and anaerobes

• **Adult ≥70 kg: IV** 250 mg every 6 hr (mild infections); 500 mg every 6-8 hr (moderate infections); 500 mg every 6 hr (severe life-threatening infections)
• **Adult 60 kg: IV** 250 mg IV every 8 hr (mild infections); 250 mg every 6 hr (moderate or severe life-threatening infections)
• **Adult 50 kg: IV** 125 mg every 6 hr (mild infections) 250 mg every 6 hr (moderate or severe life-threatening infections)
• **Adult 40 kg: IV** 125 mg every 6 hr (mild infections); 250 mg every 6-8 hr (moderate infections); 250 mg every 6 hr (severe life-threatening infections)
• **Adult 30 kg: IV** 125 mg every 8 hr (mild infections); 125 mg every 6 hr or 250 mg every 8 hr (moderate infections); 250 mg every 8 hr (severe life-threatening infections)
• **Adolescent/child/infant ≥3 mo: IV** 15-25 mg/kg every 6 hr
• **Infant 1-3 mo and ≥1500 g: IV** 25 mg/kg every 6 hr
• **Neonate 1-4 wk and ≥1500 g: IV** 25 mg/kg every 8 hr
• **Neonate <7 days and ≥1500 g: IV** 25 mg/kg every 12 hr

Moderately susceptible organisms, primarily some strains of *P. aeruginosa*

• **Adult ≥70 kg: IV** 500 mg every 6 hr (mild infections); 500 mg every 6 hr or 1 g every 8 hr (moderate infections); 1 g every 6-8 hr (life-threatening infections)
• **Adult 60 kg: IV** 500 mg every 8 hr (mild infections); 500 mg every 8 hr or 750 mg every 8 hr (moderate infec-

I

tions); 0.75-1 g every 8 hr (life-threatening infections)

- **Adult 50 kg: IV** 250 mg every 6 hr (mild infections); 250-500 mg every 6 hr (moderate infections); 500 mg every 6 hr or 750 mg every 8 hr (life-threatening infections)
- **Adult 40 kg: IV** 250 mg every 6 hr (mild infections); 250 mg every 6 hr or 500 mg every 8 hr (moderate infections); 500 mg every 6-8 hr (life-threatening infections)
- **Adult 30 kg: IV** 250 mg every 8 hr (mild infections); 250 mg every 6-8 hr; 250 mg every 6 hr or 500 mg every 8 hr (life-threatening infections)
- **Adolescent/child/infant ≥3 mo: IV** 15-25 mg/kg every 6 hr
- **Infant 1-3 mo weighing ≥1500 g: IV** 25 mg/kg every 6 hr
- **Neonate 1-4 wk weighing ≥1500 g: IV** 25 mg/kg every 8 hr
- **Neonate <7 days weighing ≥1500 g: IV** 25 mg/kg every 12 hr

Mild to moderate lower respiratory tract, skin and skin structure, or gynecologic infections

- **Adult/adolescent/child ≥12 yr: IM** 500 or 750 mg every 12 hr, max 1.5 g/day

Mild to moderate intraabdominal infections, including acute gangrenous or perforated appendicitis and appendicitis with peritonitis

- **Adult/adolescent/child ≥12 yr: IM** 750 mg every 12 hr, max 1.5 g/day

Community-acquired pneumonia (CAP) in ICU patients with risk factors for *Pseudomonas* infection

- Imipenem; cilastatin in combination with ciprofloxacin or an aminoglycoside plus a respiratory fluoroquinolone or an advanced macrolide
- **Adult ≥70 kg: IV** 500 mg every 6-8 hr
- **Adult 60 kg: IV** 250 mg every 6 hr
- **Adult 50 kg: IV** 250 mg every 6 hr
- **Adult 40 kg: IV** 250 mg every 6-8 hr
- **Adult 30 kg: IV** 125 mg every 6 hr or 250 mg every 8 hr

Empiric treatment of aspiration pneumonia

- **Adult: IV** 500-1000 mg every 6 hr × 10 days

Renal dose

- **Adult ≥70 kg (reduce normal dose of 1 g/day to): IV** CCr 41-70 ml/min, 250 mg q8hr; CCr 6-40 ml/min, 250 mg q12hr; *(reduce normal dose of 1.5 g/day to):* CCr 41-70 ml/min, 250 mg q6hr; CCr 21-40 ml/min, 250 mg q8hr; CCr 6-20 ml/min, 250 mg q12hr; *(reduce normal dose of 2 g/day to):* CCr 41-70 ml/min, 500 mg q8hr; CCr 21-40 ml/min, 250 mg q6hr; CCr 6-20 ml/min, 250 mg q12hr

Available forms: Powder for sol inj 250, 500 mg; powder for susp 500 mg

Administer:

- After C&S is taken

IM route

- Reconstitute 500 mg/2 ml lidocaine without EPINEPHrine; shake
- Inject deeply in large muscle, aspirate, **product for IM is not for IV use**

IV route

- After reconstitution of 250 or 500 mg with 10 ml of diluent and shake; add to ≥100 ml of same inf sol
- 250-500 mg over 20-30 min; ≥750 mg over 40-60 min; give through Y-tube or 3-way stopcock; do not give by IV bolus or if cloudy

Y-site compatibilities: Acyclovir, alfentanil, amifostine, amikacin, aminocaproic acid, anidulafungin, argatroban, ascorbic acid, atenolol, atracurium, atropine, benztropine, bivalirudin, bleomycin, bumetanide, buprenorphine, butorphanol, CARBOplatin, carmustine, caspofungin, cefamandole, ceFAZolin, cefoperazone, cefotaxime, cefoTEtan, cefOXitin, ceftazidime, cefuroxime, chloramphenicol, cimetidine, cisatracurium, CISplatin, clindamycin, codeine, cyanocobalamin, cyclophosphamide, cycloSPORINE, cytarabine, DACTINomycin, dexamethasone, dexrazoxane, digoxin, diltiazem, diphenhydrAMINE, DOCEtaxel, dolasetron, DOPamine, doxacurium, DOXOrubicin, DOXOrubicin liposomal, doxycycline, enalaprilat, famotidine, fludarabine, foscarnet, granisetron, IDArubicin, insulin (regular), melphalan, methotrexate, ondansetron, propofol, remifentanil, tacrolimus, teniposide, thiotepa, vinorelbine, zidovudine

SIDE EFFECTS

CNS: Fever, somnolence, **seizures**, confusion, dizziness, weakness, myoclonus, drowsiness
CV: Hypotension, palpitations, tachycardia
GI: *Diarrhea, nausea, vomiting,* **pseudomembranous colitis, hepatitis,** glossitis, gastroenteritis, abdominal pain, jaundice
GU: **Renal toxicity/failure**
HEMA: **Agranulocytosis, eosinophilia, neutropenia,** decreased Hgb, Hct
INTEG: Rash, urticaria, pruritus, pain at inj site, phlebitis, erythema at inj site, erythema multiform
MISC: Hearing loss, tinnitus, electrolyte abnormalities
RESP: Chest discomfort, dyspnea, hyperventilation
SYST: **Anaphylaxis, Stevens-Johnson syndrome, toxic epidermal necrolysis, angioedema**

PHARMACOKINETICS

IV: Onset immediate, peak 20 min-1 hr, half-life 1 hr, 70%-80% excreted unchanged in urine

INTERACTIONS

Increase: imipenem plasma levels—probenecid
Increase: antagonistic effect—β-lactam antibiotics
Increase: seizure risk—ganciclovir, theophylline, aminophylline, cycloSPORINE
Decrease: effect of valproic acid
Drug/Lab Test
Increase: AST, ALT, LDH, BUN, alk phos, bilirubin, creatinine, potassium, chloride
Decrease: sodium
False positive: direct Coombs' test

NURSING CONSIDERATIONS

Assess:

- Renal studies: creatinine/BUN, electrolytes
- **Infection:** increased temperature, WBC, characteristics of wounds, sputum, urine or stool culture
- Sensitivity to penicillin, other Beta lactams—may have sensitivity to this product
- Renal disease: lower dose may be required
- Bowel pattern daily; if severe diarrhea occurs, product should be discontinued; may indicate pseudomembranous colitis

⚠ **Allergic reactions, anaphylaxis:** **rash, urticaria, pruritus, wheezing, laryngeal edema; may occur a few days after therapy begins; have EPINEPHrine, antihistamine, emergency equipment available**

- **Overgrowth of infection:** perineal itching, fever, malaise, redness, pain, swelling, drainage, rash, diarrhea, change in cough, sputum

Evaluate:

- Therapeutic response: negative C&S; absence of signs and symptoms of infection

Teach patient/family:

⚠ **To report severe diarrhea; may indicate pseudomembranous colitis**

⚠ **To report sore throat, bruising, bleeding, joint pain; may indicate blood dyscrasias (rare)**

- **To report seizures, immediately**

TREATMENT OF ANAPHYLAXIS: EPINEPHrine, antihistamines; resuscitate if needed

imipramine (Rx)

(im-ip′ra-meen)

Tofranil, Tofranil PM

Func. class.: Antidepressant, tricyclic
Chem. class.: Dibenzazepine, tertiary amine

Do not confuse:
imipramine/desipramine

ACTION: Blocks reuptake of norepinephrine, serotonin into nerve endings, thereby increasing action of norepinephrine, serotonin in nerve cells

USES: Depression, enuresis in children
Unlabeled uses: Chronic pain, migraine headaches, cluster headaches as adjunct, incontinence, ADHD, neuralgia, bulimia, neuropathic pain, social phobia

CONTRAINDICATIONS: Pregnancy (D), hypersensitivity to this product or carBAMazepine; acute MI
Precautions: Breastfeeding, geriatric patients, suicidal patients, severe depression, increased intraocular pressure, closed-angle glaucoma, urinary retention, cardiac/hepatic disease, hyperthyroidism, electroshock therapy, elective surgery, seizure disorders, prostatic hypertrophy, MI, AV block, bundle branch block, ileus, QT prolongation, hypersensitivity to tricyclics

Black Box Warning: Children other than for enuresis; suicidal ideation

DOSAGE AND ROUTES

Depression

- **Adult: PO** 75-100 mg/day in divided doses, may increase by 25-50 mg to 200 mg/day (outpatients), 300 mg/day (inpatients); may give daily dose at bedtime
- **Geriatric: PO** 25-50 mg at bedtime, may increase to 100 mg/day in divided doses
- **Child ≥6 yr (unlabeled): PO** 1.5 mg/kg/day in divided doses, max 2.5 mg/kg/day

Enuresis

- **Child 6-12 yr: PO** 10-25 mg at bedtime, max 50 mg

Neuropathic pain (unlabeled)

- **Adult: PO** 10-150 mg/day

Social phobia/panic disorder (unlabeled)

- **Adult: PO** 10 mg at bedtime, titrate q2-4days to 100-200 mg/day

Overactive bladder (OAB) (unlabeled)

- **Adult: PO** 10-50 mg daily, may titrate to 150 mg/day

Available forms: Tabs 10, 25, 50 mg; caps 75, 100, 125, 150 mg

Administer:

PO route

- Not to break, crush, or chew caps
- With food or milk for GI symptoms
- Dosage at bedtime if oversedation occurs during day; may take entire dose at bedtime; geriatric patients may not tolerate once daily dosing
- Sugarless gum, hard candy, or frequent sips of water for dry mouth
- Store in tight container at room temperature; do not freeze

SIDE EFFECTS

CNS: *Dizziness, drowsiness,* confusion, seizures, headache, anxiety, tremors, stimulation, weakness, insomnia, nightmares, EPS (geriatric patients), increased psychiatric symptoms, paresthesia, ataxia
CV: *Orthostatic hypotension, ECG changes, tachycardia,* hypertension, palpitations, dysrhythmias
EENT: Blurred vision, tinnitus, mydriasis
ENDO: Hyperglycemia, hypo/hyperthyroidism, goiter, SIADH
GI: *Diarrhea, dry mouth,* nausea, vomiting, paralytic ileus; increased appetite; cramps, epigastric distress, jaundice, hepatitis, stomatitis, constipation, taste change, weight gain
GU: *Retention,* acute renal failure, impotence, decreased libido

HEMA: **Agranulocytosis, thrombocytopenia, eosinophilia, leukopenia**
INTEG: Rash, urticaria, sweating, pruritus, photosensitivity; hyperpigmentation (rare)

PHARMACOKINETICS

Steady state 2-5 days; metabolized to desipramine by liver; excreted in urine, breast milk, feces; crosses placenta; half-life 6-20 hr

INTERACTIONS

⚠ **Hyperpyretic crisis, seizures, hypertensive episode: MAOIs, cloNIDine**
⚠ **Increase: serotonin syndrome, neuroleptic malignant syndrome—SSRIs, SNRIs, serotonin-receptor agonists; avoid concurrent use, linezolid, methylene blue IV**
Increase: QT interval—class IA/III antidysrhythmics, tricyclics, gatifloxacin, levofloxacin, moxifloxacin, ziprasidone
Increase: effects of direct-acting sympathomimetics (EPINEPHrine), alcohol, barbiturates, benzodiazepines, CNS depressants
Decrease: effects of guanethidine, cloNIDine, indirect-acting sympathomimetics (ePHEDrine)

Drug/Herb
Increase: serotonin syndrome—SAM-e, St. John's wort

Drug/Lab Test
Increase: serum bilirubin, alk phos, blood glucose, LFTs
Decrease: 5-HIAA, VMA, urinary catecholamines

NURSING CONSIDERATIONS

Assess:

- B/P (lying, standing), pulse q4hr; if systolic B/P drops 20 mm Hg, hold product, notify prescriber; take vital signs q4hr in patients with CV disease
- Blood studies: CBC, leukocytes, differential, cardiac enzymes, serum imipramine levels (125-250 ng/ml) if patient is receiving long-term therapy
- Hepatic studies: AST, ALT, bilirubin
- Weight weekly; appetite may increase with product

⚠ **QT prolongation:** **ECG for flattening of T wave, bundle branch block, AV block, dysrhythmias in cardiac patients**

- EPS primarily in geriatric patients: rigidity, dystonia, akathisia
- Mental status: mood, sensorium, affect, **suicidal tendencies**; increase in psychiatric symptoms: depression, panic
- Urinary retention, constipation; constipation is more likely to occur in children, geriatric patients; increase fluids, bulk in diet

⚠ **Withdrawal symptoms: headache, nausea, vomiting, muscle pain, weakness, diarrhea, insomnia, restlessness; not usual unless product is discontinued abruptly**

- **Serotonin syndrome, hypertensive episodes, identify drug interactions before use of product**
- Alcohol consumption; if alcohol is consumed, hold dose until morning
- Assistance with ambulation during beginning therapy because drowsiness, dizziness, orthostatic hypotension occurs
- Safety measures, primarily for geriatric patients

Evaluate:

- Therapeutic response: decreased depression, enuresis, pain

Teach patient/family:

- That therapeutic effects may take 2-3 wk
- That product is dispensed in small amounts because of suicide potential, especially at beginning of therapy
- To use caution when driving, performing other activities requiring alertness because of drowsiness, dizziness, blurred vision
- To report urinary retention immediately
- To avoid alcohol, other CNS depressants during treatment
- Not to discontinue medication abruptly after long-term use; may cause nausea, headache, malaise
- To wear sunscreen or large hat because photosensitivity occurs

• To rise slowly, orthostatic hypotension may occur

TREATMENT OF OVERDOSE: ECG monitoring; lavage, activated charcoal; administer anticonvulsant

immune globulin IM (IMIG/IGIM) (Rx)

Bay Gam 15%, Flebogamma 5%, Flebogamma DIF 5%, Gammagard, Gamunex 10%, Privigen 10%, Vivaglobin 10%

immune globulin IV (IGIV, IVIG) (Rx)

Bay Gam 15%, Carimune NF, Flebogamma 5%, Flebogamma 10% DIF, Gammagard S/D, Gammagard Liquid 10%, Gammaked, Gammaplex, Gammar-P IV, Gamunex, Iveegam EN, Polygam S/D, Privigen, Vivaglobin

immune globulin SC (SCIG/IGSC)

Bay Gam 15%, Flebogamma 5%, Flebogamma DIF 5%, Gammagard 10%, Gammaked, Gammaplex, Gamunex 10%, Privigen 10%, Vivaglobin, Hizentra

Func. class.: Immune serum
Chem. class.: IgG

ACTION: Provides passive immunity to hepatitis A, measles, varicella, rubella, immune globulin deficiency; contains gamma globulin antibodies (IgG)

USES: Immunodeficiency syndrome; B-cell chronic lymphocytic leukemia; Kawasaki syndrome; bone marrow transplantation; pediatric HIV infection; agammaglobulinemia; hepatitis A, B exposure; measles exposure; measles vaccine complications; purpura; rubella exposure; chickenpox exposure; chronic inflammatory demyelinating polyneuropathy, multifocal motor neuropathy

Unlabeled uses: IV posttransfusion purpura, Guillain-Barré syndrome, refractory pemphigus vulgaris, West Nile virus, meningitis, myasthenia gravis, thrombocytopenia encephalitis, HIV, cytomegalovirus, neonatal jaundice, RSV infection, primary humoral immunodeficiency, refractory pemphigus vulgaris

CONTRAINDICATIONS: Coagulopathy, hemophilia, IgA deficiency, thrombocytopenia

Precautions: Pregnancy (C), breastfeeding, children, agammaglobulinemia, bleeding, hypogammaglobulinemia, infection, IV, viral infection

DOSAGE AND ROUTES

Immune globulin IM (IMIG, IGIM)

Hepatitis A prophylaxis

• **Adult/geriatric/adolescent/child/infant (unlabeled): IM** 0.02 ml/kg for those who have not received hepatitis A vaccine and been exposed during the prior 2 wk

Measles prophylaxis (exposed during prior 6 days)

• **Adult: IM** 0.25 ml/kg (immunocompetent)

• **Child (unlabeled): IM** 0.5 ml/kg as a single dose, max 15 ml (immunocompromised)

Varicella prophylaxis

• **Adult: IM** 0.6-1.2 ml/kg as soon as possible and if varicella-zoster immune globulin is not available

Rubella prophylaxis in exposed/susceptible individual who will not consider a therapeutic abortion

• **Adult pregnant women: IM** 0.55 ml/kg

Immunoglobulin deficiency

• **Adult: IM** 1.32 ml/kg then 0.66 ml/kg (≥100 mg/kg) q3-4wk

Immune globulin IV (IVIG, IGIV)

Primary immunodeficiency

Gammagard S/D

- **Adult/adolescent/child:** IV 300-600 mg/kg q3-4wk

Polygam S/D

- **Adult/adolescent/child:** IV 100 mg/kg/mo; initially 200-400 mg/kg may be used

Gammar-P IV

- **Adult:** IV 200-400 mg/kg q3-4wk
- **Adolescent/child:** IV 200 mg/kg q3-4wk

Gamunex

- **Adult/adolescent/child:** IV INF 300-600 mg/kg (3-6 ml/kg) q3-4wk, initial inf rate 1 mg/kg/min (max 8 mg/kg/min)

Iveegam EN

- **Adult/adolescent/child:** IV 200 mg/kg monthly, max 800 mg/kg/mo

Panglobulin NF/Carimune NF

- **Adult/adolescent/child:** IV 200 mg/kg/mo

Gamagard liquid/Flebogamma 5%

- **Adult/adolescent/child:** IV 300-600 mg/kg q3-4wk

Privigen

- **Adult/adolescent/child ≥3 yr:** IV 200-800 mg q3wk

Idiopathic thrombocytopenic purpura (ITP)

Panglobulin NF/Carimune NF

- **Adult/child:** IV 400 mg/kg daily × 2-5 days; with acute ITP of childhood, only 2 of 5 days are needed if initial platelets are 30,000-50,000 mcl after 2 doses

Gammagard S/D/Polygam S/D

- **Adult/adolescent/child:** IV 1000 mg/kg as a single dose; may give on alternate days for up to 3 doses

Gamunex

- **Adult/adolescent/child:** IV INF total dose of 2000 mg/kg on 2 consecutive days; initial rate is 1 mg/kg/min (max 8 mg/kg/min); if after 1st dose adequate platelets are observed after 24 hr, may withhold 2nd dose

Privigen

- **Adult/adolescent ≥15 yr:** IV 1 g/kg/day × 2 days

Kawasaki disease

Iveegam EN

- **Child:** IV 400 mg/kg daily × 4 consecutive days or a single dose of 2000 mg/kg over 10 hr, given with aspirin 100 mg/kg/day through 14th day of illness, then 3-5 mg/kg each day thereafter for 5 wk

Gammagard S/D/Polygam S/D

- **Infant/child:** IV 1000 mg/kg as a single dose or 400 mg/kg/day × 4 days beginning within 7 days of fever onset, with aspirin 80-100 mg/kg/day × 4 divided doses

Immune globulin SC (SCIG/IGSC)

- **Adult/child >2 yr:** SUBCUT INF 100-200 mg/kg weekly, **Vivaglobin** brand of SCIG 160 mg IgG/ml, **SUBCUT** inj 15 ml/inj site, given at max of 20 ml/hr

Hizentra

- **Adult/child:** 1.53 times previous IGIV dose in grams divided by number of wk between IGIV dose, multiply dose in q × 5 for dose in ml, adjust by 1.3 trough before last IGIV treatment, give by subcut inf weekly

Available forms: IM: Inj 2-, 10-ml vial (Bay Gam); IV: 5%, 10% sol (Gamimune N); powder for inj 1-, 3-, 6-, 12-g vials (Carimune NF); 50 mg protein/ml in 2.5-, 5-, 10-g vials (Gammagard S/D); 1-, 2.5-, 5-, 10-g vials (Gammar-P IV); 500 mg, 1-, 2.5-, 5-g vials (Iveegam); 6-, 12-g vials (Panglobulin); 2.5-, 5-, 10-g vials (Polygam S/D); sol for inj 1-, 2.5-, 5-, 10-, 20-g vials (Gamunex); Human sol for inj: Flebogamma 5%, 10% DIF; subcut inj 10, 16, 20% single use

Administer:

- IM ≤3 ml at 1 site; use large muscle mass
- Only with EPINEPHrine 1:1000, resuscitative equipment available
- Only within 2 wk of exposure to hepatitis A
- Store at 36° F-46° F (2° C-8° C)

IV route

- **Gamimune N:** IV undiluted or dilute with D_5; give 0.01 ml/kg/min; may increase to 0.02-0.04 ml/kg/min

- **Venoglobulin-I:** (50 mg/ml sol) give 0.01-0.02 ml/kg/min; if no adverse reaction within ½ hr, increase to 0.04 ml/kg/min; store at room temperature
- **Gammagard:** reconstitute with sterile water for inj (50 mg protein/ml); give 0.5 ml/kg/hr, may increase to 4 ml/kg/hr; use inf set provided
- **Gammar-IV:** give 0.01 ml/kg/min (50 mg/ml sol) × 15-30 min, may increase to 0.02 ml/kg/min, may increase to 0.03-0.06 ml/kg/min

Y-site compatibilities: Fluconazole, sargramostim

SIDE EFFECTS

CNS: Headache, fatigue, malaise
GI: Abdominal pain
HEMA: Thromboembolism (Vivaglobin)
INTEG: Pain at inj site, rash, pruritus, chills
MS: Arthralgia, chest pain
SYST: Lymphadenopathy, anaphylaxis

INTERACTIONS

- Do not administer live virus vaccines within 3 mo of this product

Drug/Lab Test
Interference: glucose testing system

NURSING CONSIDERATIONS

Assess:

- **Exposure date:** this product should be given within 6 days of measles, 7 days of hepatitis B, 14 days of hepatitis A
- **Anaphylaxis:** diaphoresis, wheezing, chest tightness, hypotension

Evaluate:

- Prevention of infection, increased platelets

Teach patient/family:

- That passive immunity is temporary
- About the treatment of anaphylaxis: EPINEPHrine, diphenhydrAMINE, oxygen, vasopressors, corticosteroids

indacaterol

(in-da-kat'er-ol)

Arcapta Neohaler

Func. class.: Beta-2 agonist, long acting respiratory

ACTION: An agonist at β-2 receptors. These receptors are present in large numbers in the lungs and are located on bronchiolar smooth muscle. Stimulation of β-2 receptors in the lung causes relaxation of bronchial smooth muscle, which produces bronchodilation and an increase in bronchial airflow. These effects may be mediated, in part, by increased activity of adenyl cyclase, an intracellular enzyme responsible for the formation of cyclic-3′,5′-adenosine monophosphate (cAMP); has >24-fold agonist activity at β-2 receptors (primarily in the lung) compared to β-1 receptors (primarily in the heart)

USES: Bronchitis, chronic obstructive pulmonary disease (COPD), emphysema

CONTRAINDICATIONS: Acute bronchospasm, acute asthma attack, status asthmaticus, acute respiratory insufficiency, monotherapy of asthma, hypersensitivity

Precautions: Ischemic cardiac disease (coronary artery disease), hypertension, cardiac arrhythmias, tachycardia, QT prolongation, congenital long QT syndrome, torsades de pointes history, hyperthyroidism (thyrotoxicosis, thyroid disease), pheochromocytoma, unusual responsiveness to other sympathomimetic amines, seizure disorder, diabetes mellitus, hypokalemia, milk protein hypersensitivity, severe hepatic disease; not indicated for neonates, infants, children, or adolescents under the age of 18 yr, pregnancy (C), breastfeeding

Black Box Warning: Asthma-related deaths

DOSAGE AND ROUTES

• **Adult: NH** 75 mcg (contents of 1 capsule) inhaled once daily; administer at the same time every day, max 1 dose in 24 hours

Available forms: Powder for inhalation 75 mcg

Administer:

Inhalation route

• For oral inhalation use only; DO NOT swallow the capsules; always use the Neohaler Inhaler; this inhaler should not be used with any other products; DO NOT use with a spacer

• Use dry hands to remove a capsule from the blister pack immediately before use and place into the capsule chamber of the Neohaler Inhaler; click closed; do not place capsule into the mouthpiece; then, holding the inhaler upright, depress buttons fully once to pierce capsule, a click will be heard; have patient breathe out fully away from inhaler, place inhaler in the mouth with buttons positioned to the left and right (not up and down), close lips around mouthpiece, then breathe deeply, rapidly, and steadily in through the inhaler; a whirring sound should be heard; if no sound is heard, check the chamber because the capsule may be stuck; gently tap the base of the device to loosen capsule, if necessary; after inhalation, patient should hold the breath as long as comfortable while removing inhaler from the mouth; check the chamber to see if any powder remains in the capsule; repeat inhalation steps until no powder remains; most patients empty the capsule in 1 or 2 inhalations; after administration, open chamber and discard empty capsule

• Advise coughing after administration is not problematic; as long as the capsule is empty the full dose has been administered

• The gelatin capsule might break into small pieces that pass through the inhaler screen and reach the mouth; accidental inhalation or ingestion of these pieces is harmless; piercing capsule more than once increases risk of shattering capsule: do not wash inhaler; keep it dry; a clean, dry, lint-free cloth may be used to wipe out the inhaler

• Use the new inhaler provided with each new prescription

• To avoid the spread of infection, do not share inhaler

SIDE EFFECTS

CNS: Headache, tremor

CV: **Sinus tachycardia, hypertension, QT prolongation and ST-T wave changes, prolonged QTc (an increase of >60 ms from baseline), nonsustained ventricular tachycardia, supraventricular tachycardia (SVT) episodes, intermittent ectopic atrial rhythm**

ENDO: Hyperglycemia

GI: Nausea, dry mouth (xerostomia)

META: Hypokalemia

MS: Muscle cramps/spasm, musculoskeletal pain

RESP: **Paradoxical bronchospasm,** cough, dyspnea, sputum purulence or volume, wheezing, nasopharyngitis, pneumonia, sinusitis and upper respiratory tract infection

PHARMACOKINETICS:

Protein binding 94%-96%, approximately 1/3 of total drug-related AUC over 24 hours, a hydroxylated derivative, glucuronate and dealkylate metabolites are present; metabolized by CYP3A4, CYP1A1, CYP2D6, UGT1A1; elimination multiphasic with terminal half-life of 45.5-126 hours; excreted renally (2%-6%), fecally (>90%); onset 5 min, peak 15 min, steady state 12-15 days

INTERACTIONS:

Increase QT prolongation: **Class IA/III antiarrhythmics, flecainide, propafenone some antipsychotics (phenothiazines, pimozide, haloperidol risperiDONE, sertindole ziprasidone), amoxapine, arsenic trioxide, astemizole, bepridil, cisapride, citalopram, chloroquine, clarithromycin, dasatinib, dolasetron, dronedarone, dro-**

peridol, erythromycin, halofantrine, halogenated anesthetics, levomethadyl, maprotiline, methadone, some quinolones (ofloxacin, gatifloxacin, gemifloxacin, grepafloxacin, levofloxacin, moxifloxacin, sparfloxacin), ondansetron, paliperidone, palonosetron, pentamidine, probucol, propafenone, ranolazine, SUNItinib, terfenadine, thioridazine, tricyclic antidepressants, troleandomycin, vorinostat, tetrabenazine
Increase hypokalemia: Theophylline, aminophylline, corticosteroids
Increase cardiovascular reactions: MAOIs, furazolidone, procarbazine, rasagiline

NURSING CONSIDERATIONS

Assess

Black Box Warning: Asthma-related death; not to be used in asthma

- COPD, emphysema, bronchospasm: monitor pulmonary function tests
- **QT prolongation:** monitor ECG, ejection fraction for QT prolongation
- **Paradoxical bronchospasm:** if paradoxical bronchospasm occurs, discontinue product immediately, use a short-acting β-agonist for rescue therapy, as appropriate

Teach patient/family

- To report dyspnea, wheezing, bronchospasm
- Not to use with other products unless approved by prescriber; there are many interactions

Black Box Warning: Do not use for asthma

indapamide (Rx)

(in-dap′a-mide)

Lozide ♣

Func. class.: Diuretic—thiazide-like, antihypertensive
Chem. class.: Indoline

ACTION:
Acts on proximal section of distal renal tubule by inhibiting reabsorption of sodium; may act by direct vasodilation caused by blocking of calcium channels

USES:
Edema of CHF, hypertension

CONTRAINDICATIONS:
Hypersensitivity to this product or sulfonamides; anuria, hepatic coma
Precautions: Breastfeeding, hypokalemia, dehydration, ascites, hepatic disease, severe renal disease, CCr <30 ml/min (not effective), diabetes mellitus, gout, pregnancy (C), cardiac dysrhythmias

DOSAGE AND ROUTES

Edema

- **Adult: PO** 2.5 mg/day in AM; may be increased to 5 mg/day if needed after 1 wk

Antihypertensive

- **Adult: PO** 1.25-5 mg/day; may increase to 5 mg/day over 8 wk

Available forms: Tabs 1.25, 2.5 mg
Administer:

- In AM to avoid interference with sleep
- With food if nausea occurs; absorption may be decreased slightly

SIDE EFFECTS

CNS: *Headache, dizziness, fatigue,* weakness, nervousness, agitation, extremity numbness, depression
CV: Orthostatic hypotension, volume depletion, palpitations, dysrhythmias, PVCs, vasculitis
EENT: Blurred vision, nasal congestion, increased intraocular pressure

ELECT: *Hypochloremic alkalosis, hypomagnesemia, hyperuricemia, hypercalcemia, hyponatremia, hypokalemia,* hyperglycemia
GI: *Nausea,* diarrhea, dry mouth, vomiting, anorexia, cramps, constipation, abdominal pain, hypercholesterolemia
GU: *Polyuria,* nocturia, urinary frequency, impotence
HEMA: **Agranulocytosis, anemia**
INTEG: *Rash, pruritus,* **Stevens-Johnson syndrome**
MS: *Cramps*

PHARMACOKINETICS

Well absorbed (PO); widely distributed; metabolized by liver; excreted by kidney (small amounts); onset 1-2 hr; peak 2 hr; duration up to 36 hr; excreted in urine, feces; half-life 14-18 hr

INTERACTIONS

Increase: hyperglycemia—diazoxide
Increase: toxicity of muscle relaxants, steroids, lithium, digoxin
Increase: hypokalemia—corticosteroids, amphotericin B, loop diuretics, thiazide diuretics
Decrease: effects—antidiabetics, antigout agents, anticoagulants
Decrease: absorption—cholestyramine, colestipol
Decrease: hypotensive effect—indomethacin, NSAIDs

Drug/Food
Increase: severe hypokalemia—licorice

Drug/Herb
Increase: antihypertensive effect—hawthorn

Drug/Lab Test
Increase: calcium, parathyroid test, glucose, uric acid

NURSING CONSIDERATIONS

Assess:
- Weight, I&O daily to determine fluid loss; effect of product may be decreased if used daily
- Rate, depth, rhythm of respirations, effect of exertion
- B/P lying, standing; postural hypotension may occur
- Electrolytes: potassium, magnesium, sodium, chloride: include BUN, CBC, serum creatinine, blood pH, ABGs, uric acid, Ca, glucose
- Signs of metabolic alkalosis, hypokalemia
- Rashes, fever daily; allergy to sulfa products
- Confusion, especially in geriatric patients; take safety precautions if needed
- Hydration: skin turgor, thirst, dry mucous membranes

Evaluate:
- Therapeutic response: improvement in edema of feet, legs, sacral area daily; decreased B/P

Teach patient/family:
- To consume diet high in potassium; to rise slowly from lying or sitting position
- To recognize adverse reactions: muscle cramps, weakness, nausea, dizziness
- To take with food, milk for GI symptoms; to take early in day to prevent nocturia; to avoid alcohol
- **To notify prescriber if urinary output decreases; to monitor daily weight**
- Not to stop product abruptly

TREATMENT OF OVERDOSE:

Lavage if taken orally; monitor electrolytes, administer IV fluids; monitor hydration, CV, renal status

indinavir (Rx)

(en-den′a-veer)

Crixivan

Func. class.: Antiretroviral
Chem. class.: Protease inhibitor

Do not confuse: indinavir/Denavir

ACTION: Inhibits human immunodeficiency virus (HIV-1) protease; this prevents the maturation of the virus

USES: HIV-1 in combination with at least 2 other antiretrovirals

Unlabeled uses: Prevention of HIV-1 after exposure

CONTRAINDICATIONS:

Hypersensitivity, breastfeeding

Precautions: Pregnancy (C), children, renal/hepatic disease, history of renal stones, diabetes, hypercholesterolemia, hemophilia, autoimmune disease, immune reconstitution syndrome

DOSAGE AND ROUTES

• **Adult: PO** 800 mg q8hr; 400 mg bid with ritonavir 400 mg bid or 800 mg bid with ritonavir 100-200 mg bid; decrease dose to 600 mg bid when given with lopinavir, ritonavir

Mild to moderate hepatic impairment

• **Adult: PO** 600 mg q8h

Available forms: Caps 200, 400 mg

Administer:

• Do not break, crush, or chew caps

• With water, 1 hr before or 2 hr after meals; may be given with other liquids or small meal; do not give with high-fat, high-protein meals

• Dosage adjustment will need to be considered when given with efavirenz

• Increase water to 1.5 L/day minimum to prevent nephrolithiasis

SIDE EFFECTS

CNS: *Headache, insomnia,* dizziness, somnolence

GI: *Diarrhea, abdominal pain, nausea, vomiting,* anorexia, dry mouth

GU: Nephrolithiasis

INTEG: Rash

MS: Pain

OTHER: Asthenia, insulin-resistant hyperglycemia, hyperlipidemia, ketoacidosis, lipodystrophy

PHARMACOKINETICS

Terminal half-life 2 hr; 60% protein binding; metabolized liver; excreted <20% unchanged in urine, 83% in feces

INTERACTIONS

⚠ Life-threatening dysrhythmias: ergots, midazolam, rifampin, triazolam, amiodarone, pimozide, alfazosin

Increase: myopathy—statins (atorvastatin, lovastatin, simvastatin)

Increase: indinavir levels—CYP3A4 inhibitors (arepitant, protease inhibitors, azole antifungals, nefazodone, verapamil); phosphodiesterase-5 inhibitors (sildenafil, tadalafil, vardenafil)

Increase: levels of both products—clarithromycin, zidovudine

Increase: levels of isoniazid, oral contraceptives

Decrease: indinavir levels—CYP3A4 inducers (barbiturates, carBAMazepine, nonnucleoside reverse transcriptase inhibitors, phenytoins, rifamycins, modafinil)

Decrease: effect of both products—anticonvulsants

Decrease: effect—CYP3A4 substrates (calcium channel blockers, immunosuppressants, benzodiazepines, azole antifungals, macrolides, SSRIs, statins)

Drug/Herb

Decrease: indinavir levels—St. John's wort; avoid concurrent use

Drug/Food

Decrease: indinavir absorption—grapefruit juice; high-fat, high-protein foods

Drug/Lab Test

Increase: AST, ALT, amylase, total bilirubin

NURSING CONSIDERATIONS

Assess:

• Complaints of lower back, flank pain; indicates kidney stones

• Signs of infection, anemia, presence of other sexually transmitted diseases

• Blood/hepatic studies: ALT, AST; total bilirubin, amylase, blood glucose, serum cholesterol/lipid profile, may be elevated

• Plasma HIV RNA, viral load, CD4 during treatment

• Bowel pattern before, during treatment; if severe abdominal pain with bleeding occurs, product should be discontinued; monitor hydration

- Skin eruptions; rash, urticaria, itching
- Allergies before treatment, reaction of each medication; place allergies on chart

Teach patient/family:

- To take as prescribed; if dose is missed, to take as soon as remembered up to 1 hr before next dose; not to double dose
- That product must be taken in equal intervals around the clock to maintain blood levels for duration of therapy
- ⚠ That hyperglycemia may occur; to watch for increased thirst, weight loss, hunger, and dry, itchy skin; to notify prescriber
- To increase fluids to prevent kidney stones; if stone formation occurs, that treatment may need to be interrupted
- That product does not cure AIDS, only controls symptoms; not to donate blood

indomethacin (Rx)

(in-doe-meth′a-sin)

Indocin, Nu-Indo ♣

Func. class.: Nonsteroidal antiinflammatory product (NSAID), antirheumatic

Chem. class.: Acetic acid derivative

Do not confuse:
Indocin/Endocet/minocin/Vicodin

ACTION: Inhibits prostaglandin synthesis by decreasing enzyme needed for biosynthesis; analgesic, antiinflammatory, antipyretic

USES: RA, ankylosing spondylitis, osteoarthritis, bursitis, tendinitis, acute gouty arthritis; closure of patent ductus arteriosus in premature infants (IV)

Unlabeled uses: Bone pain, headache, heterotopic ossification, juvenile rheumatoid arthritis, pericarditis

CONTRAINDICATIONS: Pregnancy (D) 3rd trimester, aortic coarctation, bleeding salicylate/NSAID hypersensitivity, GI bleeding

Black Box Warning: Perioperative pain in CABG

Precautions: Pregnancy (B) 1st trimester, breastfeeding, children, bleeding disorders, GI disorders, cardiac disorders, depression, renal/hepatic disease, asthma, diabetes, acute bronchospasm, ulcerative colitis, seizures, Parkinson's disease, neonates

Black Box Warning: Stroke, GI bleeding, MI

DOSAGE AND ROUTES

Arthritis/antiinflammatory

- **Adult: PO** 25-50 mg bid-tid; max 200 mg/day; **EXT REL** 75 mg/day, may increase to 75 mg bid

Acute gouty arthritis

- **Adult: PO** 50 mg tid; use only for acute attack then reduce dose

Patent ductus arteriosus

Longer or repeated treatment courses may be necessary for very premature infants

- **Infant <2 days: IV** 0.2 mg/kg then 0.1 mg/kg × 2 doses at 12, 24 hr
- **Infant 2-7 days: IV** 0.2 mg/kg then 0.2 mg/kg × 2 doses at 12, 24 hr
- **Infant >7 days: IV** 0.2 mg/kg then 0.25 mg/kg × 2 doses at 12, 24 hr

Available forms: Caps 25, 50 mg; ext rel caps 75 mg; inj 1-mg vial; supp 50 mg; oral susp 5 mg/ml

Administer:

PO route

- Do not break, crush, or chew sus rel cap or reg caps
- With food to decrease GI symptoms and prevent ulcerations
- Shake susp; do not mix with other liquids
- Store at room temperature

IV route

- After reconstituting 1 mg with 1 or 2 ml NS or sterile water for inj without preservative; to give 1 or 0.5 mg/ml, respectively; do not dilute further

• Infuse over 20-35 min; avoid extravasation
• Do not inj/inf via umbilical catheter to avoid dramatic shift in cerebral blood flow

Y-site compatibilities: Furosemide, insulin (regular), potassium chloride, sodium bicarbonate, sodium nitroprusside

SIDE EFFECTS

CNS: Dizziness, drowsiness, fatigue, confusion, insomnia, anxiety, depression, *headache*

CV: Tachycardia, peripheral edema, palpitations, dysrhythmias, hypertension, CV thrombotic events, MI, stroke

EENT: Tinnitus, hearing loss, blurred vision

GI: *Nausea,* anorexia, *vomiting,* diarrhea, jaundice, cholestatic hepatitis, *constipation,* flatulence, cramps, peptic ulcer, ulceration, perforation, GI bleeding

GU: Nephrotoxicity: dysuria, hematuria, oliguria, azotemia

HEMA: Blood dyscrasias, prolonged bleeding

INTEG: Purpura, rash, pruritus, sweating

PHARMACOKINETICS

PO: Onset 30 min; peak 2 hr; duration 4-6 hr; metabolized in liver, kidneys; excreted in urine 60%, feces 33%; crosses placenta; excreted in breast milk; 99% protein binding; half-life 1 hr 1st pass, 2.6-11.2 hr 2nd pass

INTERACTIONS

Increase: hyperkalemia—potassium-sparing diuretics

Increase: toxicity—lithium, methotrexate, cycloSPORINE, probenecid, cidofovir

Increase: effect of—digoxin, phenytoin, aminoglycosides

Increase: bleeding risk—anticoagulants, abciximab, clopidogrel, eptifibatide, plicamycin, ticlopidine, tirofiban, thrombolytics, aspirin, SSRIs, SNRIs

Decrease: effect of—antihypertensives

NURSING CONSIDERATIONS

Assess:

• **Arthritis symptoms:** ROM, pain, swelling before and 2 hr after treatment

Black Box Warning: Cardiac disease, CV, thrombotic events (MI, stroke) before administration, not to be used in perioperative pain in CABG surgery

• Patent ductus arteriosus: respiratory rate, character, heart sounds
• Renal, hepatic, blood studies: BUN, creatinine, AST, ALT, Hgb before treatment, periodically thereafter; if renal function has decreased, do not give subsequent doses
• Eye/ear problems: blurred vision, tinnitus; may indicate toxicity; audiometric, ophthalmic exam before, during, after treatment if patient receiving long-term therapy
• Confusion, mood changes, hallucinations, especially among geriatric patients
• Asthma, nasal polyps, aspirin sensitivity, may develop hypersensitivity to indomethacin

Black Box Warning: GI bleeding/perforation: chronic use can lead to GI bleeding, use cautiously in those with a history of active GI disease

Black Box Warning: MI, stroke: may be greater with longer-term use and those with CV risk factors

Evaluate:

• Therapeutic response: decreased pain, stiffness, swelling in joints; ability to move more easily

Teach patient/family:

• To report blurred vision, ringing, roaring in ears; may indicate toxicity
• To avoid driving, other hazardous activities if dizziness, drowsiness occurs
• To report change in urine pattern, increased weight, edema, increased pain in joints, fever, blood in urine; may indicate nephrotoxicity
• To report mood changes: anxiety, depression
• That therapeutic antiinflammatory effects may take up to 1 mo

• To avoid alcohol, NSAIDs, salicylates because bleeding may occur
• To report use to all health care providers

inFLIXimab (Rx)

(in-fliks′ih-mab)

Remicade

Func. class.: Biologic response modifiers

Chem. class.: Tumor necrosis factor modifiers

Do not confuse: Remicade/Renacidin/inFLIXimab/riTUXimab

ACTION: Monoclonal antibody that neutralizes the activity of tumor necrosis factor alpha (TNFα) found in Crohn's disease; decreased infiltration of inflammatory cells

USES: Crohn's disease, fistulizing (moderate to severe); RA, given with methotrexate; plaque psoriasis, ankylosing spondylitis, ulcerative colitis, psoriasis

Unlabeled uses: Psoriatic arthritis, Behçet's syndrome, uveitis, juvenile arthritis

CONTRAINDICATIONS: Hypersensitivity to murines, moderate to severe CHF (NYHA class III/IV)

Precautions: Pregnancy (B), breastfeeding, children, geriatric patients, COPD, hepatotoxicity, hemalologic abnormalities, hepatitis B, Guillain-Barré syndrome, seizures, multiple sclerosis

Black Box Warning: Infection, neoplastic disease, TB

DOSAGE AND ROUTES

Crohn's disease (moderate to severe)/(fistulizing)

• **Adult/adolescent/child ≥6 yr: IV INF** 5 mg/kg initially then at 2 wk, 6 wk, q8wk thereafter; may increase to 10 mg/kg/dose if needed (adults)

Rheumatoid arthritis

• **Adult: IV** 3 mg/kg initially then at 2 wk, 6 wk, q8wk thereafter; max 10 mg/kg/dose

Available forms: Powder for inj 100 mg

Administer:

Intermittent IV INF route

• Pretreat with diphenhydrAMINE, acetaminophen, predniSONE if a reaction is inf related
• Give immediately after reconstitution; reconstitute each vial with 10 ml sterile water for inj; further dilute total dose/250 ml of 0.9% NaCl inj to a total conc of 0.4-4 mg/ml; use 21-G or smaller needle for reconstitution; direct sterile water at glass wall of vial; gently swirl; do not shake; may foam; allow to stand for 5 min, give within 3 hr
• Give over ≥2 hr, use polyethylene-lined inf with in-line, sterile, low-protein-bind filter
• Do not admix
• Refrigerated storage; do not freeze

SIDE EFFECTS

CNS: *Headache, dizziness, depression, vertigo, fatigue, anxiety, fever,* **seizures,** *chills, flulike symptoms,* demyelinating disease

CV: Chest pain, hypo/hypertension, **tachycardia,** CHF, **acute coronary syndrome**

GI: *Nausea, vomiting, abdominal pain, stomatitis, constipation, dyspepsia, flatulence*

GU: Dysuria, urinary frequency

HEMA: **Anemia, leukopenia, thrombocytopenia, pancytopenia**

INTEG: *Rash, dermatitis, urticaria,* dry skin, sweating, flushing, hematoma, pruritus, keratoderma blenorrhagicum

MS: Myalgia, back pain, arthralgia

RESP: URI, pharyngitis, bronchitis, cough, dyspnea, sinusitis

SYST: **Anaphylaxis, fatal infections, sepsis, malignancies, immunogenicity, Stevens-Johnson syndrome, toxic epidermal necrolysis**

PHARMACOKINETICS

Distributed to vascular compartment, half-life 9.5 days

INTERACTIONS

Increase: infections neutropenia—TNF blockers (abatacept, anakinra, golimumab, rilonacept), avoid concurrent use
- Do not administer live vaccines concurrently

NURSING CONSIDERATIONS

Assess:
- For RA, ROM, pain
- GI symptoms: nausea, vomiting, abdominal pain
- Periodic blood counts (CBC), ANA titer, LFTs
- CV status: B/P, pulse, chest pain

⚠ **Allergic reaction, anaphylaxis:** rash, dermatitis, urticaria, dyspnea, hypotension, fever, chills; discontinue if severe, administer EPINEPHrine, corticosteroids, antihistamines; assess for allergies to murine proteins before starting therapy

Black Box Warning: Fatal infections: discontinue if infection occurs, do not administer to patients with active infection; identify TB before beginning treatment; a TB test should be obtained; if present, TB should be treated before patient receives product; exercise caution when switching from 1 DMARD to another

- Report suspected adverse reactions to the FDA (1-800-FDA-1088)

Black Box Warning: For neoplastic disease in those <18 yr, including hepatosplenic T-cell lymphoma, usually occurs in those with inflammatory bowel disease

Evaluate:
- Therapeutic response: absence of fever, mucus in stools

Teach patient/family:
- Inf reaction should be reported immediately
- If infection occurs, notify prescriber immediately
- Not to breastfeed while taking this product
- To notify prescriber of GI symptoms, hypersensitivity reactions, heart symptoms
- Not to operate machinery, drive if dizziness, vertigo occur
- To avoid live virus vaccinations; bring up to date before use

⚠ **HIGH ALERT**

INSULINS

Rapid Acting

insulin glulisine (Rx)

Apidra, Apidra SoloStar

insulin aspart (Rx)

NovoLOG, NovoLOG Flexpen, NovoLOG Pen Fill, NovoMix 30 ✦, Novo Rapid ✦

insulin lispro (Rx)

HumaLOG

Short Acting

insulin, regular (OTC)

HumuLIN R, NovoLIN R, ReliOn R

insulin, regular concentrated (Rx)

HumuLIN R U-500

Intermediate Acting

insulin, isophane suspension (NPH) (OTC)

HumuLIN N, NovoLIN ge NPH ✦, NovoLIN N, NovoLIN N Prefilled, ReliOn N

Long Acting

insulin detemir (Rx)

Levemir

insulin glargine (Rx)

Lantus

Mixtures

insulin, isophane suspension and regular insulin (Rx)

HumuLIN 70/30, HumuLIN 30/70 ✦, NovoLIN 70/30 Prefilled, ReliOn 70/30

isophane insulin suspension (NPH) and insulin mixtures (Rx)

HumuLIN 50/50

insulin lispro mixture (Rx)

HumaLOG KwikPen Mix 50/50, HumaLOG Mix 25 ✦, HumaLOG Mix 50 ✦, HumaLOG Mix 75/25, HumaLOG Mix 50/50

insulin aspart mixture (Rx)

NovoLOG 70/30, NovoLOG Mix Flexpen Prefilled Syringe 70/30

Func.class.: Antidiabetic, pancreatic hormone

Chem. class.: Modified structures of endogenous human insulin

Do not confuse:
Lantus/lente
NovoLIN 70/30 PenFill/NovoLIN 70/30 Prefilled

ACTION: Decreases blood glucose; by transport of glucose into cells and the conversion of glucose to glycogen, indirectly increases blood pyruvate and lactate, decreases phosphate and potassium; insulin may be human (processed by recombinant DNA technologies)

USES: Type 1 diabetes mellitus, type 2 diabetes mellitus, gestational diabetes; insulin lispro may be used in combination with sulfonylureas in children >3 yr

CONTRAINDICATIONS: Hypersensitivity to protamine; creosol (aspart)

Precautions: Pregnancy (B) lispro, detemir, regular aspart; (C) all others

DOSAGE AND ROUTES

Insulin glulisine

- **Adult/adolescent/child ≥4 yr: SUBCUT** dosage individualized, give within 15 min before or 20 min after starting a meal; **Adult: IV** dilute to 1 unit/ml in inf systems with 0.9% NaCl, use PVC Viaflex inf bags and PVC tubing, use dedicated line

Insulin aspart

- **Adult/adolescent/child ≥6 yr: INTERMITTENT SUBCUT** Total daily dose is given as 2-4 inj/day just before beginning of meal; in general, 50%-70% of total daily insulin may be given as insulin aspart, remainder should be intermediate- or long-acting insulin; **CONTINUOUS SUBCUT** used with external insulin pump via cont SUBCUT insulin inf (CSII), insulin dose should be based on insulin dose from previous regimen

Insulin lispro

- **Adult/adolescent/child ≥3 yr: SUBCUT** 15 min before meals; **CONT SUBCUT INF (external insulin pump):** total daily dose should be based on insulin dose from previous regimen, 50% of total dose can be given as meal-related boluses, remainder as basal inf

Human regular

- **Adult: SUBCUT** ½-1 hr before meals

Insulin, isophane suspension

- **Adult: SUBCUT** dosage individualized by blood, urine glucose; usual dose 7-26 units; may increase by 2-10 units/day if needed

Insulin detemir

- **Adult/adolescent/child ≥2 yr: SUBCUT** 1-2×/day; if 1×, give with evening meal

Insulin glargine

- **Adult and child ≥6 yr: SUBCUT** 10 international units/day, range 2-100 international units/day

Regular insulin (ketoacidosis)

- **Adult: IV** 5-10 units, then 5-10 units/hr until desired response, then switch to

SUBCUT dose; **IV/INF** 2-12 units (50 units/500 ml of normal saline)

- **Child: IV** 0.1 units/kg

Replacement

- **Adult and child: SUBCUT** 0.5-1 units/kg/day qid given 30 min before meals
- **Adolescent: SUBCUT** 0.8-1.2 mg/kg/day; this dosage is used during rapid growth

Available forms: ***NPH*** Inj 100 units/ml; ***regular*** inj 100 units/ml, cartridges 100 units/ml; ***insulin analog*** inj 100 units/ml; ***isophane insulin*** inj 100 units/ml, cartridges 100 units/ml; ***insulin lispro*** 100 units/ml, 1.5-ml cartridges, ***insulin lispro*** HumaLOG Pen sol for inj 100 units/ml; ***insulin glulisine*** inj 100 units/ml; ***insulin glargine*** inj 100 units/ml; ***insulin detemir*** inj 100 units/ml in 10 vials, 3-ml cartridges; ***insulin aspart*** inj 100 mg/ml (Flexpen, Pen Fill)

Administer:

- Store at room temperature for <1 mo (some insulins); keep away from heat and sunlight; refrigerate all other supply; NPH, premixed insulins are cloudy; regular, rapid-acting analogs, long-acting analogs are clear; do not freeze—IV route, regular only

SUBCUT route

- After warming to room temperature by rotating in palms to prevent injecting cold insulin; use only insulin syringes with markings or syringe matching units/ml; rotate inj sites within one area: abdomen, upper back, thighs, upper arm, buttocks; keep record of sites
- Increased dosages if tolerance occurs
- Premixed insulins, NPH are cloudy suspensions
- Regular human insulin, rapid-acting analogs, long-acting analogs are clear; do not use if cloudy, thick, or discolored

CONT SUBCUT route (insulin inf CSII)

- Do not mix with other insulins when using a pump
- Insulin lispro 3-ml cartridges to be used in Disetronic H-TRON plus V100 pump using Disetronic rapid inf sets; inf set and cartridge adapter should be changed q3days; replace 3-ml cartridge q6days

IV route (insulin glulisine only)

- Dilute to 1 international unit/ml in inf systems with 0.9% NaCl using PVC viaflex inf bags and PVC tubing; use dedicated line; do not admix

IV route (regular only)

⚠ When regular insulin is administered IV, monitor glucose, potassium often to prevent fatal hypoglycemia, hypokalemia

- IV direct, undiluted via vein, Y-site, 3-way stopcock; give at ≤50 units/min
- By cont inf after diluting with IV sol and run at prescribed rate; use IV inf pump for correct dosing; give reduced dose at serum glucose level of 250 mg/100 ml

Additive compatibilities: Cimetidine, lidocaine, meropenem, ranitidine, verapamil

Y-site compatibilities: Amiodarone, ampicillin, ampicillin/sulbactam, aztreonam, ceFAZolin, cefoTEtan, DOBUTamine, esmolol, famotidine, gentamicin, heparin, heparin/hydrocortisone, imipenem/ cilastatin, indomethacin, magnesium sulfate, meperidine, meropenem, midazolam, morphine, nitroglycerin, oxytocin, PENTobarbital, potassium chloride, propofol, ritodrine, sodium bicarbonate, sodium nitroprusside, tacrolimus, terbutaline, ticarcillin, ticarcillin/clavulanate, tobramycin, vancomycin, vit B/C

SIDE EFFECTS

EENT: Blurred vision, dry mouth
INTEG: Flushing, rash, urticaria, warmth, lipodystrophy, lipohypertrophy, swelling, redness
META: *Hypoglycemia,* rebound hyperglycemia (Somogyi effect 12-72 hr or longer)
MISC: Peripheral edema
SYST: Anaphylaxis

PHARMACOKINETICS

Rapid acting

Insulin glulisine: Onset 15-30 min, peak ½-1½ hr, duration 3-4 hr

Insulin aspart: Onset 10-20 min, peak 1-3 hr, duration 3-5 hr

Insulin lispro: Onset 15-30 min, peak ½-1½ hr, duration 3-5 hr

Short acting

Insulin regular: Onset 30 min, peak 2.5-5 hr, duration up to 7 hr

Intermediate acting

Insulin, isophane suspension (NPH): Onset 1.5-4 hr, peak 4-12 hr, duration ≤24 hr

Long acting

Insulin detemir: Onset 0.8-2 hr, peak unknown, duration ≤24 hr (conc dependent)

Insulin glargine: Onset 1.5 hr, no peak identified, duration ≥24 hr

Mixtures

Insulin, isophane suspension and regular insulin (70/30): Onset 10-20 min, peak 2.4 hr, duration ≤24 hr

Insophane insulin suspension (NPH) and insulin mixtures (50/50): Onset ½-1 hr, peak dual, duration 10-16 hr

INTERACTIONS

Increase: hypoglycemia—salicylate, alcohol, β-blockers, anabolic steroids, fenfluramine, phenylbutazone, sulfinpyrazone, guanethidine, oral hypoglycemics, MAOIs, tetracycline

Decrease: hypoglycemia—thiazides, thyroid hormones, oral contraceptives, corticosteroids, estrogens, DOBUTamine, EPINEPHrine

Drug/Lab Test

Increase: VMA

Decrease: potassium, calcium

Interference: LFTs, thyroid function studies

NURSING CONSIDERATIONS

Assess:

- Fasting blood glucose; A1c may be drawn to identify treatment effectiveness q3mo
- Urine ketones during illness; insulin requirements may increase during stress, illness, surgery
- Hypoglycemic reaction that can occur during peak time (sweating, weakness, dizziness, chills, confusion, headache, nausea, rapid weak pulse, fatigue, tachycardia, memory lapses, slurred speech, staggering gait, anxiety, tremors, hunger)
- **Hyperglycemia:** acetone breath; polyuria; fatigue; polydipsia; flushed, dry skin; lethargy

Evaluate:

- Therapeutic response: decrease in polyuria, polydipsia, polyphagia; clear sensorium; absence of dizziness; stable gait

Teach patient/family:

- That blurred vision occurs; not to change corrective lens until vision is stabilized after 1-2 mo
- To keep insulin, equipment available at all times; to carry a glucagon kit, candy, or lump of sugar to treat hypoglycemia
- That product does not cure diabetes but controls symptoms
- To carry emergency ID as diabetic
- To recognize hypoglycemia reaction: headache, tremors, fatigue, weakness
- To recognize hyperglycemia reaction: frequent urination, thirst, fatigue, hunger
- About the dosage, route, mixing instructions, diet restrictions (if any), disease process
- About the symptoms of **ketoacidosis:** nausea; thirst; polyuria; dry mouth; decreased B/P; dry, flushed skin; acetone breath; drowsiness; Kussmaul respirations
- That a plan is necessary for diet, exercise; that all food on diet should be eaten; that exercise routine should not vary
- About blood glucose testing; how to determine glucose level
- To avoid OTC products unless directed by prescriber

TREATMENT OF OVERDOSE:

Glucose 25 g IV, via dextrose 50% sol, 50 ml or glucagon 1 mg

interferon alfacon-1 (Rx)

(in-ter-feer′on al′fa-kon)

Infergen

Func. class.: Recombinant type 1 interferon

ACTION: Induces biologic responses and has antiviral, antiproliferative, and immunomodulatory effects

USES: Chronic hepatitis C infections in those ≥18 yr with compensated liver disease who have anti-HCV antibodies or HCV RNA, may use in combination with ribavirin

Unlabeled uses: Hairy-cell leukemia when used with G-CSF

CONTRAINDICATIONS: Hypersensitivity to α-interferons or products from *Escherichia coli;* decompensated hepatic disease, autoimmune hepatitis

Precautions: Pregnancy (C), breastfeeding, children <18 yr, geriatric patients, thyroid disorders, myelosuppression, hepatic disease, seizure disorder, alcoholism, hepatitis

Black Box Warning: Cardiac disease, autoimmune disease, infection, depression

DOSAGE AND ROUTES

- **Adult: SUBCUT monotherapy** 9 mcg as single inj 3×/wk × 24 wk; leave ≥48 hr between injections; for patients who did not respond or relapsed after discontinuation, give 15 mcg 3×/wk × 48 wk; **combination: SUBCUT** 15 mcg daily with ribavirin 1000 mg/day PO (<75 kg), 1200 mg/day PO (≥75 kg); give in 2 divided doses for up to 48 wks, use stepwise dose reduction of the interferon dose from 15 mcg to 9 mcg to 6 mcg for serious adverse reactions

Available forms: Inj 9 mcg/0.3 ml, 15 mcg/0.5 ml

Administer:

- Premedicate with acetaminophen or ibuprofen
- Do not shake vial; use 1 dose per vial; discard unused portion; use proper inj sites; rotate sites, discard unused product
- Do not miss doses

SIDE EFFECTS

CNS: Depression, headache, fatigue, fever, rigors, insomnia, dizziness, agitation, nervousness, anxiety, lability, abnormal thinking

CV: Hypertension, palpitation, tachycardia

EENT: Tinnitus, earache, conjunctivitis, eye pain

GI: Abdominal pain, nausea, diarrhea, anorexia, dyspepsia, vomiting, constipation, flatulence, hemorrhoids, decreased salivation

GU: Dysmenorrhea, vaginitis, menstrual disorders

HEMA: Granulocytopenia, thrombocytopenia, leukopenia, ecchymosis, aplastic anemia

INTEG: Alopecia, pruritus, rash, erythema, dry skin

MISC: Anaphylaxis, angioedema, flulike illness

MS: Back, limb, neck skeletal pain; rigors

RESP: Pharyngitis, upper respiratory infection, cough, sinusitis, rhinitis, respiratory tract congestion, epistaxis, dyspnea, bronchitis

PHARMACOKINETICS

Peak 1-4 hr, peak biologic response 24-36 hr

INTERACTIONS

Increase: hearing loss—eflornithine (systemic), consider serial audiograms

Increase: NRTI, use together cautiously

Increase: myelosuppression—myelosuppressives

Increase: toxicity—aldesleukin, IL-2 eflornithine, theophylline

Decrease: effect of—antiretrovirals (NNRTs, NRTIs, protease inhibitors)

NURSING CONSIDERATIONS

Assess:

Black Box Warning: Past or present history of depression, seizures; use with caution with these disorders

Black Box Warning: Hemolytic anemia: ribavirin may be used with this product and cause hemolytic anemia, cardiac symptoms

- Ophthalmologic status; report periodically
- **Pregnancy:** determine if patient is pregnant before use with ribavirin, pregnancy (X), when used with ribavirin
- CBC, LFTs, ECG, platelet counts, heme conc, ANC, serum creatinine, albumin, bilirubin, TSH, T_4, triglycerides at baseline and periodically
- Myelosuppression: hold dose if neutrophil count is $<500 \times 10^6$/L or if platelets are $<50 \times 10^9$/L, monitor for infection
- For hypersensitivity: discontinue immediately if hypersensitivity occurs

Evaluate:

- Therapeutic response: decreased chronic hepatitis C signs/symptoms

Teach patient/family:

- With detailed written information about product
- To report signs, symptoms of infection, thyroid/liver dysfunction, changes in behavior
- To report if pregnancy is planned or suspected (C), when combined with ribavirin pregnancy (X)
- To use OTC analgesics to decrease flu-like symptoms
- If patient will be self-administering injection, teach all aspects of product use, administration, disposal, provide medication guide

interferon beta-1a (Rx)

(in-ter-feer′on)

Avonex, Rebif

interferon beta-1b (Rx)

Betaseron, Extavia

Func. class.: Multiple sclerosis agent, immune modifier

Chem. class.: Interferon, *Escherichia coli* derivative

ACTION: Antiviral, immunoregulatory; action not clearly understood; biologic response-modifying properties mediated through specific receptors on cells, inducing expression of interferon-induced gene products

USES: Ambulatory patients with relapsing or remitting MS

Unlabeled uses: May be useful for treatment of AIDS, AIDS-related Kaposi's sarcoma, malignant melanoma, metastatic renal cell carcinoma, cutaneous T-cell lymphoma, acute non-A/non-B hepatitis, chronic hepatitis C

CONTRAINDICATIONS: Hypersensitivity to natural or recombinant interferon-β or human albumin, hamster protein, rotovirus vaccine

Precautions: Pregnancy (C), breastfeeding, children <18 yr, chronic progressive MS, depression, mental disorders, seizure disorder, latex allergy, autoimmune disorders, bone marrow suppression, hepatotoxicity, cardiac disease, alcoholism, chickenpox, herpes zoster

DOSAGE AND ROUTES

Interferon beta-1a

Remitting or relapsing multiple sclerosis

- **Adult: IM** (Avonex) 30 mcg/wk
- **Adult: SUBCUT** (Rebif) 22 or 44 mcg 3×/wk with each dose 48 hr apart, titrate to full dose over 4-wk period

Chronic hepatitis C (unlabeled)
- **Adult: SUBCUT** (Rebif) 44 mcg 3×/wk × 24 wk

Interferon beta-1b

Relapsing or remitting multiple sclerosis
- **Adult: SUBCUT** 0.0625 mg every other day for wk 1 and 2, then 0.125 mg every other day for wk 3 and 4, then 0.1875 mg every other day for wk 5 and 6, then 0.25 mg every other day thereafter; higher doses should not be used

Available forms: *beta-1a:* (Avonex) 30 mcg (6.6 million international units/vial) (autoinjector pen); (Rebif) 22 mcg, 44 mcg/0.5 ml; *beta-1b:* powder for inj 0.3 mg (9.6 m international units) Kit

Administer:

Interferon beta-1a

- Visually inspect parenteral products for particulate matter and discoloration before use
- Store in refrigerator; do not freeze

IM route
- Premedicate with acetaminophen or ibuprofen and give at bedtime to lessen flulike symptoms
- Interferon β1a (Avonex) 30 mcg is equivalent to 6 million IU
- If a dose is missed, give it as soon as possible; continue the regular schedule but do not give 2 injections within 2 days of each other; all products are for single-use only; do not reuse needles, syringes, prefilled syringes, or autoinjectors
- Injection sites should be rotated to minimize the likelihood of injection site reactions
- Do not inject into an area of the body where skin is irritated, reddened, bruised, infected, or scarred
- The injection site should be checked after 2 hr for redness, edema, or tenderness
- The manufacturer of Avonex offers free training on IM use for patients and health care partners; contact MS ActiveSource for more information (800-456-2255)

Reconstitution and administration of Avonex lyophilized powder for IM route
- Use appropriate aseptic technique for preparation of solution
- Sites for injection include the thigh or upper arm
- Slowly add 1.1 ml sterile water for injection, USP, preservative-free (supplied by manufacturer) to the vial; rapid addition of the diluent can cause foaming, making it difficult to withdraw the solution
- Gently swirl the vial to aid in dissolution; do not shake; final conc should be 30 mcg/ml (6 million IU/ml)
- The reconstituted solution should be clear to slightly yellow without particles; discard if the reconstituted product contains particulate matter or is discolored
- Withdraw 1 ml of reconstituted solution into a syringe; attach the sterile needle and inject IM
- A 25-G 1-inch needle for intramuscular injection may be substituted for the 23-G 1¼-inch needle provided by the manufacturer, if needed
- **Storage:** Use within 6 hr of reconstitution; store reconstituted solution in refrigerator; *do not freeze;* discard any unused solution; both drug and diluent vials are single-use only

Administration of Avonex prefilled syringe
- Patients may self-inject only if provider determines that it is appropriate, and with medical follow-up, and after proper training in IM injection technique
- The first injection should be performed under the supervision of an appropriately qualified person
- If self-injecting, rotate injection site between thighs; with help from another person, may rotate injection site between thighs and upper arms
- Wash hands before handling the Dose Pack
- Remove prefilled syringe from the refrigerator to warm to room temperature (usually 30 min before use); do not use external heat sources such as hot water to warm the syringe

• Hold the syringe so the cap is facing down and the 0.5 ml mark is at eye level; be sure the amount of liquid in the syringe is the same or very close to the 0.5 ml mark; if the syringe does not contain the correct amount of liquid, do not use it and call the pharmacist
• Hold syringe upright so that the rubber cap faces up; remove the cap by bending it at a 90 degree angle until it snaps free
• Attach the needle by pressing it onto the syringe and turning it clockwise until it locks in place; be careful not to push the plunger while attaching the needle
• Use the alcohol wipe to clean the skin at the injection site you choose; then, pull the protective cover straight off the needle; do not twist the cover off
• Inject intramuscularly at a 90-degree angle into the thigh or upper arm as directed by the provider
• Use gauze pad to apply pressure for a few seconds after the injection
• Dispose of used needles and syringes in a puncture-resistant container and discard appropriately
• Instruct patients to contact the health care provider if a skin reaction occurs that does not resolve in a few days
• A 25-G 1-inch needle for intramuscular injection may be substituted for the 23-G 1¼-inch needle provided by the manufacturer, if deemed appropriate by the physician
• Refer to the Patient Medication Guide for detailed instructions for preparing and giving a dose
• **Storage:** Store refrigerated; if refrigeration is unavailable, may store at 77° F or less for up to 7 days; after removing from refrigerator, do not store product above 25° C; if the product has been exposed to conditions other than recommended, discard the product; do not expose to high temperatures; do not freeze; protect from light

Administration of Avonex prefilled auto-injector

• Patients may self-inject only if their provider determines that it is appropriate, and with medical follow-up, and after proper training in IM technique
• The first injection should be performed under the supervision of provider
• Remove one Administration Dose Pack from the refrigerator to warm to room temperature (about 30 min before use); do not use external heat sources such as hot water to warm the syringe Dose Pack
• Wash hands before handling Dose Pack contents
• Ensure tamper-evident cap has not been removed or is loose; then grasp the cap and bend it at a 90-degree angle until it snaps off; pull off the sterile foil from the needle cover
• Hold the Avonex Pen with the glass syringe tip pointing up; press the needle onto the glass syringe tip; gently turn the needle clockwise until firmly attached; do not remove plastic cover from the needle
• Hold Pen with one hand and, using other hand, hold onto the injector shield (grooved area) tightly and quickly pull up on the injector shield until the injector shield covers the needle all the way; the plastic needle cover will pop off after the injector shield has been fully extended
• When the injector shield is extended the right way, there will be a small blue rectangular area next to the oval medication display window; check the display window and make sure the Avonex is clear and colorless
• Do not use the injection if the liquid is colored, cloudy, or has lumps or particles; air bubbles will not affect the dose
• Do not push down on the injector shield and the blue activation button at the same time until you are ready to give injection
• Avonex Pen should be injected into the upper outer thigh
• Use the alcohol wipe to clean the skin at the injection site and allow it to dry before injection
• Hold Pen at 90-degree angle to the injection site; firmly push the body of the pen down against the thigh to release the

safety lock; safety lock is released when blue rectangle area above the oval medication display window is gone, push down on blue activation button with thumb and count to 10, you will hear a click if the injection is given the right way
• After counting to 10, pull the Pen straight out of the skin, use gauze pad to apply pressure for a few seconds
• The circular display window on the Pen is yellow if the full dose is received
• Cover exposed needle with Pen cover; do not hold the Pen cover with your hands while inserting the needle
• Dispose of used needles and syringes in a puncture-resistant container and discard appropriately
• Instruct patients to contact the health care provider if a skin reaction occurs that does not resolve in a few days
• Refer to the Patient Medication Guide for detailed instructions for preparing and giving a dose
• **Storage:** Store at 36°-46° F (2°-8° C); if refrigeration is unavailable, may store at 77° F or less for up to 7 days; after removing from refrigerator, do not store product above 25° C; if the product has been exposed to conditions other than recommended, discard the product and do not use; do not expose to high temperatures; do not freeze; protect from light

Subcutaneous administration

• Give at the same time (preferably late in the afternoon or evening) on the same days of the week at least 48 hours apart
• Do not give on two consecutive days; if a dose is missed, administer the dose as soon as possible, then skip the following day; return to the regular schedule the following week
• Premedication with acetaminophen or ibuprofen can lessen the severity of flu-like symptoms
• Interferon β1a (Rebif) 44 mcg is equivalent to 12 million IU
• Rotate injection sites; appropriate injection sites include thigh, outer surface of upper arm, stomach, or buttocks; do not inject into an area where the skin is irritated, reddened, bruised, or infected
• A Starter Pack containing a lower dose of Rebif syringes is available for the initial titration period; patients and/or their caregivers should be trained and understand appropriate preparation and administration
• The manufacturer offers complimentary services including injection training and reimbursement support; contact MS LifeLines at 877-44-REBIF

Injection (Rebif)

• Interferon β1a (Rebif) is available in a prefilled syringe with a 29-G needle
• Inject subcutaneously into the outer surface of the upper arm, abdomen, thigh, or buttock do not inject the area near the navel or waistline; take care not to inject intradermally
• Discard any unused solution; prefilled syringes do not contain preservatives and are single-use only

Interferon beta-1b

SUBCUT route

• The manufacturers of Betaseron and of Extavia offer materials to assist with training on subcut use, call 1-800-788-1467 (Betaseron), 1-888-669-6682 (Extavia)
• Premedication with acetaminophen or ibuprofen and use of product at bedtime can lessen the severity of flulike symptoms
• Visually inspect parenteral products for particulate matter and discoloration before use; do not use if particulate matter is present

Reconstitution

• Add 1.2 ml of 0.54% sodium chloride injection (supplied by the manufacturer) to the vial by using the vial adapter to attach the prefilled syringe that contains the diluent; keep the plunger depressed, and gently swirl, do not shake; if you take your thumb off the plunger, the solution can come back into the syringe before the product is fully reconstituted; if foaming occurs, allow the vial to sit until the foam settles; final conc (250 mcg interferon β1b/ml, which corresponds to 8 million IU/ml)

- If not used immediately, store in the refrigerator for up to 3 hr; do not freeze; discard any unused portion after 3 hr

Injection

- Withdraw the desired amount of the reconstituted solution into the syringe by turning the vial and syringe to get the vial on top; pull the plunger back to get the desired amount of product; turn the syringe to point the needle upward, and tap the syringe and release any air bubbles; twist the vial adapter to remove it and the vial
- Choose an injection site on the upper back arm; abdomen; buttock; or front thigh; do not inject within 2 in of the navel or in a site where the skin is red, bruised, infected, broken, painful, uneven, or scabbed; rotate injection sites to minimize injection site reactions such as necrosis or localized infection
- Inject subcutaneously; take care not to inject intradermally

SIDE EFFECTS

CNS: *Headache, fever, pain, chills, mental changes, depression,* hypertonia, **suicide attempts, seizures**
CV: *Migraine, palpitations, hypertension,* tachycardia, peripheral vascular disorders
EENT: *Conjunctivitis,* blurred vision
GI: *Diarrhea, constipation, vomiting, abdominal pain*
GU: *Dysmenorrhea, irregular menses, metrorrhagia,* cystitis, breast pain
HEMA: **Decreased lymphocytes, ANC, WBC;** *lymphadenopathy,* anemia
INTEG: *Sweating, inj site reaction*
MS: *Myalgia,* **myasthenia**
RESP: *Sinusitis,* dyspnea

PHARMACOKINETICS

β-1a: Onset ≤12 hr, peak 48 hr, duration 4 days, half-life 8.6 hr
β-1b: Onset rapid, peak 2-8 hr, duration unknown, half-life 8 min-4.3 hr

INTERACTIONS

Increase: hepatic damage—antiretrovirals (NNRTIs, NRTIs, protease inhibitors)
Increase: myelosuppression—antineoplastics
Decrease: clearance of zidovudine

Drug/Herb

- Change in immunomodulation: astragalus, echinacea, melatonin

Drug/Lab Test

Interference: vaccines, toxoids; avoid concurrent use
Increase: LFTs

NURSING CONSIDERATIONS

Assess:

- Blood, hepatic studies: CBC, differential, platelet counts, BUN, creatinine ALT, urinalysis; if absolute neutrophil count $<750/mm^3$ or if AST/ALT is 10× normal, discontinue product
- CNS symptoms: headache, fatigue, depression
- GI status: diarrhea or constipation, vomiting, abdominal pain
- Cardiac status: increased B/P, tachycardia
- Mental status: depression, depersonalization, **suicidal thoughts,** insomnia
- Multiple sclerosis symptoms

Evaluate:

- Therapeutic response: decreased symptoms of multiple sclerosis

Teach patient/family:

- With written, detailed information about product
- That blurred vision, sweating may occur
- That female patients may experience irregular menses, dysmenorrhea or metrorrhagia, breast pain
- To use sunscreen to prevent photosensitivity
- To notify prescriber if pregnancy is suspected
- About inj technique, care of equipment
- **To notify prescriber of increased temperature, chills, muscle soreness, fatigue, depression, symptoms of hepatotoxicity**

interferon gamma-1b (Rx)

(in-ter-feer'on)

Actimmune

Func. class.: Biologic response modifier

Chem. class.: Lymphokine, interleukin type

ACTION: Species-specific protein synthesized in response to viruses, effects; can mediate killing of *Staphylococcus aureus, Toxoplasma gondii, Leishmania donovani, Listeria monocytogenes, Mycobacterium avium intracellulare;* enhances oxidative metabolism of macrophages, enhances antibody-dependent cellular cytotoxicity

USES: Serious infections associated with chronic granulomatous disease, osteopetrosis

Unlabeled uses: *Mycobacterium avium* complex (MAC), pulmonary fibrosis

CONTRAINDICATIONS: Hypersensitivity to interferon-γ, *Escherichia coli*–derived products

Precautions: Pregnancy (C), breastfeeding, children <1 yr, cardiac disease, seizure disorders, CNS disorders, myelosuppression

DOSAGE AND ROUTES

- **Adult: SUBCUT** 50 mcg/m² (1.5 million units/m²) for patients with surface area >0.5 m²; 1.5 mcg/kg/dose for patients with surface area <0.5 m²; give Monday, Wednesday, Friday for 3×/wk dosing

Available forms: Inj 100 mcg (2 million units)/single-dose vial

Administer:

- At bedtime to minimize adverse reactions; give acetaminophen for fever, headache
- 50% of dose if severe reactions occur or discontinue treatment until reactions subside
- In right and left deltoid and anterior thigh
- Warm to room temperature before use; do not leave at room temperature >12 hr (unopened vial)
- Store in refrigerator upon receipt; do not freeze; do not shake

SIDE EFFECTS

CNS: *Headache, fatigue,* depression, fever, chills

GI: *Nausea, anorexia,* abdominal pain, weight loss, diarrhea, vomiting, colitis

HEMA: Leukopenia, thrombocytopenia, neutropenia

INTEG: Rash, pain at inj site, Stevens-Johnson syndrome

MS: Myalgia, arthralgia

PHARMACOKINETICS

SUBCUT: Dose absorbed 89%, elimination half-life 5.9 hr, peak 7 hr

INTERACTIONS

Increase: myelosuppression—other myelosuppressive agents

Increase: level of theophylline, aminophylline

Increase: liver toxicity—protease inhibitors, nucleoside reverse transcriptase inhibitors (NRTIs), nonnucleoside reverse transcriptase inhibitors (NNRTIs)

NURSING CONSIDERATIONS

Assess:

- Blood, renal, hepatic studies: CBC, differential, platelet counts, BUN, creatinine, ALT, urinalysis
- CNS symptoms: headache, fatigue, depression

Evaluate:

- Therapeutic response: decreased serious infections; improvement in existing infections and inflammatory conditions

Teach patient/family:

- About the method of administration if family members will be giving medication

• With written, detailed information about product

ipilimumab

(ip-i-lim′ue-mab)

Yervoy

Func. class.: Antineoplastic; biologic response modifier

ACTION:
A recombinant, human monoclonal antibody that binds to the cytotoxic T-lymphocyte-associated antigen 4 (CTLA-4); action is indirect, possibly through T-cell–mediated anti-tumor immune responses

USES:
Treatment of unresectable or metastatic malignant melanoma

CONTRAINDICATIONS
Hypersensitivity

Precautions: Pregnancy, breastfeeding, Crohn's disease, hepatitis, immunosuppression, inflammatory bowel disease, iritis, ocular disease, organ transplant, pancreatitis, renal disease, rheumatoid arthritis, sarcoidosis, systemic lupus erythematosus, thyroid disease, ulcerative colitis, uveitis

Black Box Warning: Adrenal insufficiency, diarrhea, Guillain-Barré syndrome, hepatic disease, myasthenia gravis, hypo/hyperthyroidism, hypopituitarism, peripheral neuropathy, serious rash

DOSAGE AND ROUTES

• **Adult/geriatric: IV** 3 mg/kg over 90 min q3wk × 4 doses; permanently discontinue if the full treatment course is not completed within 16 wk from 1st dose or for severe or life-threatening adverse reactions; withhold a dose for any moderate endocrine or immune-mediated adverse reactions; if the moderate adverse reaction completely or partially resolves (Grade 0-1) and if the patient is receiving <7.5 mg predniSONE or equivalent/day, resume at a dose of 3 mg/kg IV q3wk until all 4 planned doses or 16 wk from 1st dose, whichever occurs earlier; if moderate adverse reactions are persistent or if the corticosteroid dose cannot be reduced to 7.5 mg predniSONE or equivalent/day, permanently discontinue

Available forms: Sol for inj 50 mg/10 ml, 200 mg/40 ml

Administer:

Intermittent IV INF route

• Visually inspect parenteral products for particulate matter and discoloration before using whenever sol and container permit; sol may have a pale yellow color and have translucent to white, amorphous particles; discard the vial if sol is cloudy, if there is pronounced discoloration, or if particulate matter is present

• Allow vials to stand at room temperature for 5 min before inf preparation; withdraw the required volume and transfer into an IV bag; discard partially used vials or empty vials; dilute with 0.9% sodium chloride injection or 5% dextrose injection to a final conc (1-2 mg/ml); mix diluted sol by gentle inversion; do not admix

• Give inf over 90 min through an IV line with a low-protein binding in-line filter, do not give with other products; after each inf, flush the line with 0.9% sodium chloride injection or 0.5% dextrose injection

• Store once diluted for no more than 24 hr refrigerated or at room temperature

SIDE EFFECTS

CNS: Severe and fatal immune-mediated neuropathies, fatigue, headache, fever

EENT: Uveitis, iritis, episcleritis

ENDO: Severe and fatal immune-mediated endocrinopathies

GI: Severe and fatal immune-mediated enterocolitis, hepatitis, pancreatitis, abdominal pain, nausea, diarrhea, appetite decreased, vomiting, constipation, colitis

INTEG: Severe and fatal immune-mediated dermatitis pruritus, rash, urticaria

MISC: Cough, dyspnea, anemia, eosinophilia, nephritis
SYST: Antibody formation, Stevens-Johnson syndrome, toxic epidermal necrolysis

PHARMACOKINETICS:

Steady state by 3rd dose; terminal half-life 14.7 days

NURSING CONSIDERATIONS

- **Serious skin disorders: Stevens-Johnson syndrome, toxic epidermal necrolysis:** permanently discontinue in these or rash complicated by full thickness dermal ulceration or necrotic, bullous, or hemorrhagic manifestations like bullous rash; give systemic corticosteroids at a dose of 1-2 mg/kg/day of predniSONE or equivalent; when dermatitis is controlled, taper corticosteroids over a period of at least 1 mo, withhold in patients with moderate to severe reactions; for mild to moderate dermatitis (localized rash and pruritus), give topical or systemic corticosteroids
- **Hepatotoxicity:** monitor liver functions tests baseline and before each dose to rule out infectious or malignant causes, increase the frequency of liver function test monitoring until resolution, permanently discontinue in patients with Grade 3-5 toxicity, give systemic corticosteroids at a dose of 1-2 mg/kg/day of predniSONE or equivalent
- **Neuropathy:** monitor for motor or sensory neuropathy (unilateral or bilateral weakness, sensory alterations, or paresthesias) before each dose; permanently discontinue if severe neuropathy (interfering with daily activities), such as Guillain-Barré–like syndromes, occur
- **Endocrinopathy:** monitor thyroid function tests at baseline and before each dose; monitor hypophysitis, adrenal insufficiency, adrenal crisis, hypo/hyperthyroidism (fatigue, headache, mental status changes, abdominal pain, unusual bowel habits, hypotension, or nonspecific symptoms that may resemble other causes)

Evaluate:
Decreasing spread of malignant melanoma

Teach patient/family:

- To immediately report allergic reactions, skin rash, severe abdominal pain, yellowing of skin or eyes, tingling of extremities, change in bowel habits
- About the reason for treatment and expected results

ipratropium (Rx)

(i-pra-troe′pee-um)

Atrovent HFA

Func. class.: Anticholinergic, bronchodilator
Chem. class.: Synthetic quaternary ammonium compound

Do not confuse:
Atrovent/Alupent

ACTION: Inhibits interaction of acetylcholine at receptor sites on the bronchial smooth muscle, thereby resulting in decreased cGMP and bronchodilation

USES: COPD; rhinorrhea (nasal spray)

CONTRAINDICATIONS: Hypersensitivity to this product, atropine, bromide, soybean or peanut products
Precautions: Breastfeeding, children <12 yr, angioedema, heart failure, surgery, acute bronchospasm, bladder obstruction, closed-angle glaucoma, prostatic hypertrophy, urinary retention, pregnancy (B)

DOSAGE AND ROUTES

Bronchospasm in chronic bronchitis/emphysema

- **Adult: INH** 2 sprays (17 mcg/spray) 3-4×/day, max 12 **INH**/24 hr; **SOL** 500 mcg (1 unit dose) given 3-4×/day by nebulizer; nasal spray: 2 sprays (42 mcg/spray) 3-4×/day

• **Child 5-11 yr: INH** 4-8 inhalations q20min as needed for ≤3 hr (asthma, unlabeled); **NEB** 250-500 mcg q20min as needed for ≤3 hr (asthma, unlabeled)

Rhinorrhea perennial rhinitis

• **Adult/child ≥6 yr: INTRANASAL** 2 sprays (43 mcg)/nostril bid or tid

• **Child 5-12 yr: INTRANASAL** 2 sprays (0.03%) in each nostril 3×/day

Available forms: Aerosol 17 mcg/actuation; nasal spray 0.03%, 0.06%; sol for inh 0.0125% ♣, 0.02%

Administer:

• Store at room temperature

Nebulizer route

• Use sol in nebulizer with a mouthpiece rather than a face mask

Intranasal route

• Priming pump initially requires 7 actuations of pump; priming again is not necessary if used regularly, tilt head backward after dose

SIDE EFFECTS

CNS: *Anxiety, dizziness, headache,* nervousness

CV: Palpitation

EENT: Dry mouth, blurred vision, nasal congestion

GI: *Nausea, vomiting, cramps*

INTEG: Rash

RESP: *Cough, worsening of symptoms,* **bronchospasms**

PHARMACOKINETICS

Half-life 2 hr, does not cross blood-brain barrier

INTERACTIONS

Increase: toxicity—other bronchodilators (INH)

Increase: anticholinergic action—phenothiazines, antihistamines, disopyramide

Drug/Herb

Increase: anticholinergic effect—belladonna

Increase: bronchodilator effect—green tea (large amts), guarana

NURSING CONSIDERATIONS

Assess:

• Palpitations; if severe, product may have to be changed

• Tolerance over long-term therapy; dose may have to be increased or changed

• Atropine sensitivity; patient may also be sensitive to this product

• Respiratory status: rate, rhythm, auscultate breath sounds before and after administration

• Hard candy, frequent drinks, sugarless gum to relieve dry mouth

Evaluate:

• Therapeutic response: ability to breathe adequately

Teach patient/family:

• That compliance is necessary with number of inhalations/24 hr or overdose may occur; about spacer device for geriatric patients; that max therapeutic effects may take 2-3 mo

• To shake before using

• About the correct method of inhalation; how to clean equipment daily

• Patients should prime the inhaler before using for the first time by releasing 2 test sprays into the air, away from the face

• Each inhaler has 200 actuations or sprays

irbesartan (Rx)

(er-be-sar′tan)

Avapro

Func. class.: Antihypertensive

Chem. class.: Angiotensin II receptor blocker (Type AT_1)

Do not confuse:

Avapro/Anaprox

ACTION: Blocks the vasoconstrictor and aldosterone-secreting effects of angiotensin II; selectively blocks the binding of angiotensin II to the AT_1 receptor found in tissues

USES:
Hypertension, alone or in combination; nephropathy in type 2 diabetic patients; proteinuria
Unlabeled uses: Heart failure

CONTRAINDICATIONS:
Hypersensitivity

Black Box Warning: Pregnancy (D) 2nd/3rd trimester

Precautions: Pregnancy (C) 1st trimester, breastfeeding, children <6 yr, geriatric patients, hypersensitivity to ACE inhibitors; hepatic/renal disease; renal artery stenosis, African descent, angioedema

DOSAGES AND ROUTES

Hypertension
- **Adult: PO** 150 mg/day; may be increased to 300 mg/day, volume-depleted patients: start with 75 mg/day

Nephropathy in type 2 diabetic patients
- **Adult: PO** maintenance dose 300 mg/day, start 75 mg/day

Available forms: Tabs 75, 150, 300 mg
Administer:
- Without regard to meals
- May be used with other antihypertensives, diuretic

SIDE EFFECTS
CNS: *Dizziness,* anxiety, *headache, fatigue,* syncope
CV: Hypotension
GI: *Diarrhea, dyspepsia,* hepatitis, cholestasis
HEMA: Thrombocytopenia
MISC: Edema, chest pain, rash, tachycardia, UTI, angioedema, hyperkalemia
RESP: *Cough, upper respiratory tract infection,* sinus disorder, pharyngitis, rhinitis

PHARMACOKINETICS
Peak 1.5-2 hr, extensively metabolized by CYP2C9, half-life 11-15 hr, highly bound to plasma proteins, excreted in urine and feces, protein binding 90%

INTERACTIONS
Increase: hyperkalemia: potassium-sparing diuretics, potassium salt substitutes, ACE inhibitors
Increase: irbesartan level—CYP2C9 inhibitors (amiodarone, delavirdine, fluconazole, FLUoxetine, fluvastatin, fluvoxaMINE, imatinib, sulfonamides, sulfinpyrazone, voriconazole, zafirlukast)
Decrease: antihypertensive effect—NSAIDs
Drug/Herb
Increase: antihypertensive effect—black cohosh, goldenseal, hawthorn, kelp
Increase or decrease: antihypertensive effect—astragalus, cola tree
Decrease: antihypertensive effect—guarana, khat, licorice, yohimbe

NURSING CONSIDERATIONS
Assess:
- **Hypotension:** for severe hypotension, place in supine position and give IV inf of NS, drug may be continued after B/P is restored
- B/P, pulse q4hr; note rate, rhythm, quality
- Baselines of renal/hepatic studies before therapy begins; periodically monitor LFTs, total/direct bilirubin
- Skin turgor, dryness of mucous membranes for hydration status; edema in feet, legs daily

Evaluate:
- Therapeutic response: decreased B/P

Teach patient/family:
- To comply with dosage schedule, even if feeling better; that max therapeutic effects may take 2-3 mo, to take without regard to food
- That product may cause dizziness, fainting, lightheadedness
- To rise slowly to sitting or standing position to minimize orthostatic hypotension
- Do not stop product abruptly

Black Box Warning: To notify prescriber if pregnancy is suspected; discontinue if pregnant; pregnancy (D) 2nd/3rd trimester, (C) 1st trimester

⚠ HIGH ALERT

irinotecan (Rx)

(ear-een-oh-tee′kan)

Camptosar

Func. class.: Antineoplastic

Chem. class.: Camptothecin analog

ACTION: Cytotoxic by producing damage to single-strand DNA during DNA synthesis; binds to topoisomerase I

USES: Metastatic carcinoma of the colon or rectum or 1st-line treatment in combination with 5-FU and leucovorin for metastatic colon or rectal carcinomas

Unlabeled uses: Cervical, gastric, lung, ovarian, pancreatic cancer, malignant glioma, rhabdomyosarcoma

CONTRAINDICATIONS: Pregnancy (D), hypersensitivity

Precautions: Breastfeeding, children, geriatric patients, irradiation, hepatic disease

Black Box Warning: Myelosuppression, diarrhea

DOSAGE AND ROUTES

First-line treatment of colorectal cancer in combination with 5-fluorouracil (5-FU):

Intravenous dosage (with bolus 5-FU/leucovorin)

- **Adult:** IV 125 mg/m^2 over 90 min followed by leucovorin 20 mg/m^2 IV bolus and then 5-FU 500 mg/m^2 IV bolus on days 1, 8, 15, and 22; the next course begins on day 43 or when toxicity has recovered to NCI grade 1 or less

Intravenous dosage (with infusional 5-FU/leucovorin)

- **Adult:** IV 180 mg/m^2 over 90 min followed by leucovorin 200 mg/m^2 IV over 2 hr then 5-FU bolus and continuous inf 400 mg/m^2 IV bolus, then 600 mg/m^2 IV inf over 22 hr on days 1, 15, and 29; leucovorin and 5-FU are given on days 1, 2, 15, 16, 29, and 30; the next course begins on day 43 or when toxicity has recovered to NCI grade 1 or less

Single-agent treatment of metastatic colorectal cancer that has recurred or progressed after 5-FU–based therapy

Intravenous dosage (weekly dosage schedule)

- **Adult:** IV 125 mg/m^2 over 90 min weekly × 4 wk, every 6 wk

Intravenous dosage (once-every-3-weeks dosage schedule)

- **Adult:** IV 350 mg/m^2 over 90 min every 3 wk

Available forms: Inj 20 mg/ml

Administer:

- Use cytotoxic handling precautions

IV route

- Premedicate with antiemetic dexamethasone 10 mg plus another antiemetic agent, such as a 5-HT$_3$ blocker, given at least 30 min before use
- **Before beginning a course of therapy, the granulocyte count should be ≥1500, the platelet count should be ≥100,000, and treatment-related diarrhea should be fully resolved**

Dilution

- Dilute appropriate dose in D$_5$W (preferred) or NS injection to a final conc of 0.12-2.8 mg/ml
- Store up to 24 hr at room temperature and room lighting; however, because of possible microbial contamination during preparation, an admixture prepared with D$_5$W or NS should be used within 6 hr, solutions prepared with D$_5$W, refrigerated, protected from light must be used within 48 hr; avoid refrigeration if prepared with NS

Intravenous inf

- Infuse intravenously over 90 min

SIDE EFFECTS

CNS: Fever, headache, chills, dizziness

CV: Vasodilation, edema, **thromboembolism**

GI: **Severe diarrhea,** *nausea, vomiting,* anorexia, constipation, cramps, flatus, stomatitis, dyspepsia, **hepatotoxicity**

HEMA: Leukopenia, anemia, neutropenia
INTEG: Irritation at site, rash, sweating, alopecia
MISC: Edema, asthenia, weight loss, back pain
RESP: Dyspnea, increased cough, rhinitis

PHARMACOKINETICS

Rapidly and completely absorbed, excreted in urine and bile as metabolites, half-life 6-12 hr, bound to plasma proteins 30%-68%, increased risk for toxicity in patients homozygous for UGT1A1 28

INTERACTIONS

Increase: toxicity—fluorouracil
Increase: bleeding risk—NSAIDs, anticoagulants
Increase: irinotecan levels—some CYP3A4 inhibitors (ketoconazole)
Increase: myelosuppression, diarrhea—other antineoplastics, radiation
Increase: lymphocytopenia, hyperglycemia—dexamethasone
Increase: akathisia—prochlorperazine
Increase: dehydration—diuretics
Decrease: irinotecan levels—CYP3A4 inducers (phenytoin, carBAMazepine, PHENobarbital)

Drug/Herb
Decrease: product level—St. John's wort; avoid concurrent use

Drug/Lab Test
Increase: alk phos, LFTs, bilirubin
Decrease: platelets, WBC, neutrophils, Hgb/HcT

NURSING CONSIDERATIONS

Assess:
- CNS symptoms: fever, headache, chills, dizziness

Black Box Warning: CBC, differential, platelet count weekly; use colony-stimulating factor if WBC $<2000/mm^3$ or platelet count $<100,000/mm^3$, Hgb ≤ 9 g/dl, neutrophil $\leq 1000/mm^3$; notify prescriber of results; product should be discontinued and colony-stimulating factor given

- Buccal cavity for dryness, sores or ulceration, white patches, oral pain, bleeding, dysphagia

Black Box Warning: GI symptoms: frequency of stools; cramping; severe, life-threatening diarrhea may occur with fluid and electrolyte imbalances, treat diarrhea within 24 hr of use with 0.25-1 mg atropine IV; treat diarrhea >24 hr of use with loperamide, diarrhea >24 hr (late diarrhea) can be fatal

- Signs of dehydration: rapid respirations, poor skin turgor, decreased urine output, dry skin, restlessness, weakness
- Bone marrow depression: bruising, bleeding, blood in stools, urine, sputum, emesis
- Increased fluid intake to 2-3 L/day to prevent dehydration unless contraindicated

Evaluate:
- Therapeutic response: decrease in tumor size, spread of cancer

Teach patient/family:
- To avoid foods with citric acid or hot or rough texture if stomatitis is present; to drink adequate fluids
- To report stomatitis; any bleeding, white spots, ulcerations in mouth; to examine mouth daily, report symptoms
- To report signs of anemia: fatigue, headache, faintness, SOB, irritability, infection, rash
- To use contraception during therapy
- To report if pregnancy is planned or suspected pregnancy (D)
- To avoid salicylates, NSAIDs, alcohol because bleeding may occur; to avoid all products unless approved by prescriber
- About alopecia; that, when hair grows back, it will be different texture, thickness
- To avoid vaccinations while taking this product
- ⚠ To report diarrhea that occurs 24 hr after administration; severe dehydration can occur rapidly

⚠ Nurse Alert

iron dextran (Rx)

DexFerrum, INFeD

Func. class.: Hematinic

Chem. class.: Ferric hydroxide complex with dextran

ACTION: Iron is carried by transferrin to the bone marrow, where it is incorporated into hemoglobin

USES: Iron-deficiency anemia

CONTRAINDICATIONS:

Black Box Warning: Hypersensitivity

Precautions: Pregnancy (C), breastfeeding, neonates, infants <4 mo, children, acute renal disease, asthma, rheumatoid arthritis (IV), ankylosing spondylitis, lupus, hypotension, all anemias excluding iron-deficiency anemia, hepatic/cardiac/renal disease

DOSAGE AND ROUTES

- Total dose in ml = [0.0442 × (desired Hb − observed Hb) × LBW] + (0.26 × LBW)
- **Adult and child: IM** 0.5 ml as a test dose by Z-track then no more than the total dose including test dose following/day:
- **Adult/adolescent/child >15 kg: IM** 100 mg/day
- **Child (10-15 kg): IM** 100 mg/kg
- **Child/infant >4 mo (5-9.9 kg): IM** 50 mg/day
- **Infant >4 mo (<5 kg): IM** 25 mg/day

Available forms: Inj 50 mg/ml (2-ml, 10-ml vials)

Administer:

- D/C oral iron before parenteral; give only after test dose of 25 mg by preferred route; wait at least 1 hr before giving remaining portion
- Store at room temperature in cool environment
- Recumbent position 30 min after IV inj to prevent orthostatic hypotension

IM route

- IM deeply in large muscle mass; use Z-track method, 19- to 20-G 2- to 3-in needle; ensure needle long enough to place product deep in muscle; change needles after withdrawing product and before injecting to prevent skin, tissue staining

⚠ Only with EPINEPHrine available in case of anaphylactic reaction during dose

IV route

- IV after flushing with 10 ml 0.9% NaCl; give undiluted; may be diluted in 50-250 ml NS for inf; give ≤1 ml (50 mg) over ≥1 min; flush line after use with 10 ml 0.9% NaCl; patient should remain recumbent for ½-1 hr
- IV inj requires single-dose vial without preservative; verify on label that IV use approved

SIDE EFFECTS

CNS: Headache, paresthesia, dizziness, shivering, weakness, **seizures**

CV: Chest pain, **shock**, hypotension, tachycardia

GI: *Nausea,* vomiting, metallic taste, abdominal pain

HEMA: **Leukocytosis**

INTEG: Rash, pruritus, urticaria, fever, sweating, chills, brown skin discoloration, pain at inj site, necrosis, sterile abscesses, phlebitis

OTHER: **Anaphylaxis**

RESP: Dyspnea

PHARMACOKINETICS

IM: Excreted in feces, urine, bile, breast milk; crosses placenta; most absorbed through lymphatics; can be gradually absorbed over weeks/months from fixed locations

INTERACTIONS

Increase: toxicity—oral iron; do not use

Decrease: reticulocyte response—chloramphenicol

Drug/Lab Test

False increase: serum bilirubin

False decrease: serum calcium

False positive: ^{99m}Tc diphosphate bone scan, iron test (large doses >2 ml)

NURSING CONSIDERATIONS

Assess:

- Observe for 1 hr after test dose
- Blood studies: Hct, Hgb, reticulocytes, transferrin, plasma iron concs, ferritin, total iron binding, bilirubin before treatment, at least monthly

Black Box Warning: Allergy: anaphylaxis, rash, pruritus, fever, chills, wheezing; notify prescriber immediately, keep emergency equipment available

- Cardiac status: anginal pain, hypotension, tachycardia
- Nutrition: amount of iron in diet (meat, dark green leafy vegetables, dried beans, dried fruits, eggs)
- Cause of iron loss or anemia, including use of salicylates, sulfonamides
- **Toxicity:** nausea, vomiting, diarrhea, fever, abdominal pain (early symptoms), cyanotic-looking lips, nailbeds, seizures, CV collapse (late symptoms)
- Therapeutic response: increased serum iron levels, Hct, Hgb

Teach patient/family:

- That iron poisoning may occur if increased beyond recommended level; not to take oral iron preparation or vitamins containing iron
- That delayed reaction may occur 1-2 days after administration and last 3-4 days (IV), 3-7 days (IM); to report fever, chills, malaise, muscle, joint aches, nausea, vomiting, backache
- To avoid breastfeeding
- That stools may become dark

TREATMENT OF OVERDOSE:

Discontinue product, treat allergic reaction, give diphenhydrAMINE or EPINEPHrine as needed, give iron-chelating product for acute poisoning

iron sucrose (Rx)

Venofer

Func. class.: Hematinic

Chem. class.: Ferric hydroxide complex with dextran

ACTION: Iron is carried by transferrin to the bone marrow, where it is incorporated into hemoglobin

USES: Iron-deficiency anemia

Unlabeled uses: Dystrophic epidermolysis bullosa (DEB)

CONTRAINDICATIONS: Hypersensitivity, all anemias excluding iron-deficiency anemia, iron overload

Precautions: Pregnancy (B), breastfeeding, children, geriatric patients, abdominal pain, anaphylactic shock, arthralgia, chest pain, cough, diarrhea, dizziness, dyspnea, edema, increased LFTs, fever, headache, heart failure, hypo/hypertension, infection, MS pain nausea/vomiting, seizures, weakness

DOSAGE AND ROUTES

- **Adult:** IV 5 ml (100 mg of elemental iron) given during dialysis; most will need 1000 mg of elemental iron over 10 sequential dialysis sessions
- **Child ≥2 yr/adolescents:** IV 0.5 mg/kg by slow IV inj over 5 min (undiluted) or diluted in 25 ml of 0.9% NaCl, give over 5-60 min, max 100 mg every 2 wk × 12 wk

Available forms: Inj 20 mg/ml

Administer:

⚠ Only with EPINEPHrine, Solu-MEDROL available in case of anaphylactic reaction during dose

IV route

- Do not use if particulate is present or if discolored
- Give directly in dialysis line by slow inj or inf; give by slow inj at 1 ml/min (5 min/vial); inf dilute each vial exclusively in ≤100 ml 0.9% NaCl, give at 100 mg of iron/15 min; discard unused portions

- Do not use with IV products
- Store at room temperature in cool environment; do not freeze

SIDE EFFECTS

CNS: Headache, dizziness
CV: Chest pain, hypo/hypertension, hypervolemia, **heart failure**
GI: *Nausea, vomiting, abdominal pain*
INTEG: Rash, pruritus, urticaria, fever, sweating, chills
OTHER: **Anaphylaxis**, hyperglycemia
RESP: Dyspnea, pneumonia, cough

PHARMACOKINETICS

Excreted in urine, half-life 6 hr

INTERACTIONS

Increase: toxicity—oral iron, dimercaprol, do not use
Decrease: iron sucrose effect—chloramphenicol
Drug/Lab Test
Increase: glucose

NURSING CONSIDERATIONS

Assess:

- Blood studies: Hct, Hgb, reticulocytes, transferrin, plasma iron concs, ferritin, total iron binding; bilirubin before treatment, at least monthly
- Allergy, anaphylaxis: rash, pruritus, fever, chills, wheezing; notify prescriber immediately, keep emergency equipment available
- Cardiac status: hypo/hypertension, hypervolemia
- **Toxicity:** nausea, vomiting, diarrhea, fever, abdominal pain (early symptoms), cyanotic-looking lips, nailbeds, seizures, CV collapse (late symptoms)

Evaluate:

- Therapeutic response: increased serum iron levels, Hct, Hgb

Teach patient/family:

- To report itching, rash, chest pain, headache, vertigo, nausea, vomiting, abdominal pain, joint/muscle pain, numbness, tingling
- That iron poisoning may occur if dosage increased beyond recommended level; not to take oral iron preparation

TREATMENT OF OVERDOSE:

Discontinue product, treat allergic reaction, give diphenhydrAMINE or EPINEPHrine as needed, give iron-chelating product for acute poisoning

isoflurophate ophthalmic

See Appendix B

isoniazid (Rx)

(eye-soe-nye′a-zid)
Isotamine ♣
Func. class.: Antitubercular
Chem. class.: Isonicotinic acid hydrazide

ACTION: Bactericidal interference with lipid, nucleic acid biosynthesis

USES: Treatment, prevention of TB

CONTRAINDICATIONS: Hypersensitivity

Black Box Warning: Acute hepatic disease

Precautions: Pregnancy (C), renal disease, diabetic retinopathy, cataracts, ocular defects, IV drug users, >35 yr, postpartum, HIV, neuropathy

Black Box Warning: Alcoholism, females (African descent/Hispanic patients)

DOSAGE AND ROUTES

- **Adult/adolescent with/without HIV: PO/IM** 5 mg/kg/day ≤300 mg/day or 15 mg/kg 2-3×/wk, max 900 mg 2-3×/wk
- **Child/infant with HIV: PO/IM** 10-15 mg/kg/day, max 300 mg/day

Available forms: Tabs 100, 300 mg; inj 100 mg/ml

Administer:

PO route

• PO with meals to decrease GI symptoms; better to take on empty stomach 1 hr before or 2 hr after meals

IM route

• IM deep in large muscle mass; massage; rotate inj site; warm inj to room temperature to dissolve crystals

SIDE EFFECTS

CNS: *Peripheral neuropathy, dizziness,* memory impairment, toxic encephalopathy, seizures, psychosis, slurred speech

EENT: Blurred vision, optic neuritis

GI: *Nausea, vomiting,* epigastric distress, jaundice, fatal hepatitis

HEMA: Agranulocytosis, hemolytic, aplastic anemia, thrombocytopenia, eosinophilia, methemoglobinemia

Hypersensitivity: Fever, skin eruptions, lymphadenopathy, vasculitis

MISC: Dyspnea, B_6 deficiency, pellagra, hyperglycemia, metabolic acidosis, gynecomastia, rheumatic syndrome, SLE-like syndrome

PHARMACOKINETICS

Metabolized in liver, excreted in urine (metabolites), crosses placenta, excreted in breast milk, half life 1-4 hr

PO: Peak 1-2 hr

IM: Peak 45-60 min

INTERACTIONS

Increase: toxicity—tyramine foods, alcohol, cycloSERINE, ethionamide, rifampin, carBAMazepine, phenytoin, benzodiazepines, meperidine

Increase: serotonin syndrome—SSRIs, SNRIs

Decrease: absorption—aluminum antacids

Decrease: effectiveness of BCG vaccine, ketoconazole

Drug/Food

• Do not give with high-tyramine foods, alcohol

Drug/Lab Test

Increase: LFTs, bilirubin, glucose

Decrease: platelets granulocytes

NURSING CONSIDERATIONS

Assess:

Black Box Warning: Hepatic studies weekly: baseline in all patients, those >35 yr and all women should be monitored periodically; ALT, AST, bilirubin; increased test results may indicate hepatitis; hepatic status: decreased appetite, jaundice, dark urine, fatigue, those with fast acetylation (genetic) may metabolize product more than 5 times faster (black, Asian patients are at greater risk) some Caucasian patients; fatal hepatitis is at greater risk in black/Hispanic patients after birth

• Mental status often: affect, mood, behavioral changes; psychosis may occur

• Paresthesia in hands, feet

Evaluate:

• Therapeutic response: decreased symptoms of TB

Teach patient/family:

• That compliance with dosage schedule, duration is necessary; not to skip or double dose

• That scheduled appointments must be kept or relapse may occur

⚠ To avoid alcohol while taking product; may increase risk for hepatic injury

• That, if diabetic, to use blood glucose monitor to obtain correct result

⚠ To report weakness, fatigue, loss of appetite, nausea, vomiting, jaundice of skin or eyes, tingling/numbness of hands/feet

Black Box Warning: Fatal hepatitis: to notify prescriber immediately of yellow skin/eyes, dark urine, loss of appetite

TREATMENT OF OVERDOSE:

Pyridoxine

isosorbide dinitrate (Rx)

(eye-soe-sor'bide)

Apo-ISDN ✱, Dilatrate-SR, Isochron, IsoDitrate, Isordil

isosorbide mononitrate (Rx)

Apo-ISMN ✱, Imdur

Func. class.: Antianginal, vasodilator

Chem. class.: Nitrate

Do not confuse:
Imdur/Imuran/Inderal/K-Dur

ACTION: Relaxation of vascular smooth muscle, which leads to decreased preload, afterload, which is responsible for decreasing left ventricular end-diastolic pressure, systemic vascular resistance, and reducing cardiac oxygen demand

USES: Treatment, prevention of chronic stable angina pectoris

Unlabeled uses: Diffuse esophageal spasm, heart failure (dinitrate)

CONTRAINDICATIONS: Hypersensitivity to this product or nitrates; severe anemia, closed-angle glaucoma

Precautions: Pregnancy (C), breastfeeding, children, orthostatic hypotension, MI, CHF, severe renal/hepatic disease, increased intracranial pressure, cerebral hemorrhage, acute MI, geriatric patients, GI disease, syncope

DOSAGE AND ROUTES

Dinitrate

- **Adult: PO** 5-20 mg bid-tid initially, maintenance 10-40 mg bid-tid; **SL,** buccal 2.5-5 mg, may repeat q5-10min × 3 doses; **EXT REL** 40-80 mg q8-12hr, max 160 mg/day

Mononitrate

- **Adult: PO** (Monoket) 10-20 mg bid, 7 hr apart; (Imdur) initiate at 30-60 mg/day as a single dose, increase q3days as needed, may increase to 120 mg/day, max 240 mg/day

Available forms: *Dinitrate:* sus rel caps (SR) 40 mg, SR tabs 40 mg; tabs 5, 10, 20, 30, 40 mg; SL tabs 2.5, 5 mg; *mononitrate:* tabs (Monoket) 10, 20 mg; ext rel (Imdur) 30, 60, 120 mg

Administer:

- Do not break, crush, or chew sus rel caps, SL tabs
- After checking expiration date
- PO with 8 oz water on empty stomach
- SL tabs should be placed under the tongue until dissolved; avoid smoking, eating, drinking until dissolved

SIDE EFFECTS

CNS: *Vascular headache, flushing, dizziness,* weakness

CV: *Orthostatic hypotension,* tachycardia, **collapse**, syncope

GI: Nausea, vomiting

INTEG: Pallor, sweating, rash

MISC: Twitching, **hemolytic anemia, methemoglobinemia**, tolerance, xerostomia

PHARMACOKINETICS

Dinitrate

Metabolized by liver, excreted in urine as metabolites (80%-100%)

PO: Onset 15-30 min, duration 4-6 hr, half-life 5-6 hr

SUS REL: Onset ≤4 hr, duration 6-8 hr

SL: Onset 2-5 min, duration 1-4 hr, half-life 2 hr

Mononitrate

SUS REL: Onset 30-60 min, peak 1-4 hr, duration 6-8 hr, half-life 5 hr

INTERACTIONS

⚠ **Fatal hypotension: sildenafil, tadalafil, vardenafil, do not use together**

Increase: hypotension—β-blockers, diuretics, antihypertensives, alcohol, calcium channel blockers, phenothiazines

Increase: heart rate, B/P—sympathomimetics

Increase: myocardial ischemia—rosiglitazone; avoid concurrent use

NURSING CONSIDERATIONS

Assess:

- **Anginal pain:** duration, time started, activity being performed, character
- **Methemoglobinemia (rare):** Cyanosis of lips, nausea/vomiting, coma, shock, usually caused by high dose of product but may occur with normal dosing
- B/P, pulse, respirations during beginning therapy and periodically thereafter
- Tolerance if taken over long period, to prevent, allow intervals of 12-14 hr/day without product
- Headache, lightheadedness, decreased B/P; may indicate a need for decreased dosage, treat headache with OTC analgesics

Evaluate:

- Therapeutic response: decrease or prevention of anginal pain

Teach patient/family:

- To leave tabs in original container
- To avoid alcohol, OTC products unless approved by prescriber
- That product may cause headache; that taking with meals may reduce or eliminate headache; to take no later than 7 PM (last dose)
- To avoid hazardous activities if dizziness occurs
- About the importance of complying with complete medical regimen
- To make position changes slowly to prevent orthostatic hypotension
- Not to use with sildenafil, tadalafil, vardenafil with nitrates, may cause serious drop in B/P
- Not to discontinue abruptly, may cause heart attack
- To use at beginning of angina symptoms, may repeat every 15 mins; if no relief, seek medical attention immediately

RARELY USED

ISOtretinoin (Rx)

(eye-soe-tret′i-noyn)

Amnesteem, Claravis, Sotret

Func. class.: Antiacne agent, retinoid

USES: Severe recalcitrant nodulocystic acne

CONTRAINDICATIONS: Hypersensitivity to this product, parabens, retinoids, inflamed skin, blood donation

Black Box Warning: Pregnancy (X)

DOSAGE AND ROUTES

- **Adult: PO** 0.5-2 mg/kg/day in 2 divided doses × 15-20 wk; if relapse occurs, repeat after 2 mo off product

itraconazole (Rx)

(it-ra-con′a-zol)

Onmel, Sporanox

Func. class.: Antifungal, systemic

Chem. class.: Triazole derivative

ACTION: Alters cell membranes; inhibits several fungal enzymes

USES: Histoplasmosis, blastomycosis (pulmonary and extrapulmonary), aspergillosis, onychomycosis of toenail/fingernail

Unlabeled uses: Dermatomycosis, histoplasmosis, chromoblastomycosis, coccidioidomycosis, pityriasis versicolor, sebopsoriasis, vaginal candidiasis, cryptococcus, subcutaneous mycoses, dimorphic infections, fungal keratitis, zygomycosis, superficial mycoses (dermatophytoses), chronic mucocutaneous candidiasis

CONTRAINDICATIONS: Hypersensitivity, fungal meningitis; onychomycosis or dermatomycosis with cardiac dysfunction, in pregnant women

Black Box Warning: Heart failure, ventricular dysfunction, coadministration with other products

Precautions: Pregnancy (C), breastfeeding, children, cardiac/renal/hepatic disease, achlorhydria or hypochlorhydria (product-induced), dialysis, hearing loss, cystic fibrosis neuropathy

DOSAGE AND ROUTES

Dose varies with type of infection

- **Adult: PO** 200 mg/day with food; may increase to 200 mg bid if needed; life-threatening infections may require a loading dose of 200 mg tid × 3 days
- **Child: PO** 2-10 mg/kg/day

Available forms: Caps 100 mg; oral sol 10 mg/ml; tab 200 mg

Administer:

- In the presence of acid products only; do not use alkaline products, antacids within 2 hr of product; may give coffee, tea, acidic fruit juices

PO route

- Swallow caps whole; do not break, crush, or chew caps
- Give caps after full meal to ensure absorption
- Oral sol: patient should swish in mouth vigorously, use on empty stomach
- Oral sol and caps are not interchangeable on mg/mg basis
- Store in tight container at room temperature, do not freeze

SIDE EFFECTS

CNS: *Headache, dizziness,* insomnia, somnolence, depression

CV: Hypertension, CHF

GI: *Nausea, vomiting, anorexia, diarrhea,* cramps, *abdominal pain,* flatulence, **GI bleeding, hepatotoxicity**

GU: Gynecomastia, impotence, decreased libido

INTEG: *Pruritus,* fever, *rash,* **toxic epidermal necrolysis, Stevens-Johnson syndrome**

MISC: *Edema, fatigue,* malaise, hypokalemia, tinnitus, **rhabdomyolysis**

RESP: Rhinitis, sinusitis, upper respiratory infection, pulmonary edema

PHARMACOKINETICS

PO: Peak 3-5 hr; half-life 21 hr, IV 35.4 hr; metabolized in liver; excreted in bile, feces, urine 40%; requires acid pH for absorption; distributed poorly to CSF; 99.8% protein bound; inhibits CYP4503A4

INTERACTIONS

⚠ Life-threatening CV reactions: pimozide, quiNIDine, dofetilide, levomethadyl, dronedarone

Increase: tinnitus, hearing loss—quiNIDine

Increase: hepatotoxicity—other hepatotoxic products

Increase: edema—calcium channel blockers

Increase: severe hypoglycemia—oral hypoglycemics

Increase: sedation—ALPRAZolam, clorazepate, diazepam, estazolam, flurazepam, triazolam, oral midazolam

Increase: levels, toxicity—busPIRone, busulfan, clarithromycin, cycloSPORINE, diazepam, digoxin, felodipine, fentaNYL, atorvastatin, carBAMazepine, disopyramide, indinavir, isradipine, niCARdipine, NiFEdipine, niMODipine, phenytoin, quiNIDine, QUEtiapine, ritonavir, saquinavir, tacrolimus, warfarin

Decrease: itraconazole action—antacids, H_2-receptor antagonists, rifamycins, didanosine, carBAMazepine, isoniazid, proton pump inhibitors

Drug/Food

- Food increases absorption
- Grapefruit juice decreases itraconazole level

Drug/Lab

Increase: LFTs

NURSING CONSIDERATIONS

Assess:

- **CHF:** if present, discontinue product
- Type of infection; may begin treatment before obtaining results
- **Infection:** temperature, WBC, sputum at baseline and periodically
- I&O ratio, potassium levels
- Hepatic studies (ALT, AST, bilirubin) if patient receiving long-term therapy
- Allergic reaction: rash, photosensitivity, urticaria, dermatitis

⚠ **Hepatotoxicity:** nausea, vomiting, jaundice, clay-colored stools, fatigue

Evaluate:

- Therapeutic response: decreased fever, malaise, rash, negative C&S for infecting organism

Teach patient/family:

- That long-term therapy may be needed to clear infection (1 wk-6 mo, depending on infection)
- To avoid hazardous activities if dizziness occurs
- To take 2 hr before administration of other products that increase gastric pH (antacids, H_2-blockers, omeprazole, sucralfate, anticholinergics); to avoid grapefruit juice; to notify health care provider of all medications taken; to take after a full meal (caps) or on empty stomach (oral sol)
- About the importance of compliance with product regimen; to use alternative methods of contraception
- To notify prescriber of GI symptoms; signs of hepatic dysfunction (fatigue, jaundice, nausea, anorexia, vomiting, dark urine, pale stools); heart failure (trouble breathing, unusual weight gain, fatigue, swelling); hearing changes

RARELY USED

ivacaftor

(eye′va-kaf′tor)

Kalydeco

Func. class.: Respiratory agent

USES: Cystic fibrosis in those with G551D mutation in the *CFTR* gene

DOSAGE AND ROUTES

- **Adult/adolescent/child ≥6 yr: PO** 150 mg every 12 hr with fat-containing food

ixabepilone (Rx)

(ix-ab-ep′i-lone)

Ixempra

Func. class.: Antineoplastic—miscellaneous

Chem. class.: Epothilone

ACTION: Microtubule stabilizing agent; microtubules are needed for cell division

USES: Breast cancer

CONTRAINDICATIONS: Pregnancy (D), breastfeeding, hypersensitivity to products with polyoxyethylated castor oil, neutropenia of <1500/mm^3, thrombocytopenia

Black Box Warning: Hepatic disease

Precautions: Children, geriatric patients, alcoholism, bone marrow suppression, cardiac dysrhythmias, cardiac/renal disease, diabetes mellitus, peripheral neuropathy, ventricular dysfunction

DOSAGE AND ROUTES

Breast cancer, metastatic or locally advanced given with capecitabine and resistant to anthracycline, taxane

- **Adult: IV INF** 40 mg/m^2 over 3 hr q3wk plus capecitabine **PO** 2000 mg/m^2/

day in 2 divided doses on days 1-14 q21days; in those with BSA >2.2 m^2, dose should be calculated for a BSA of 2.2 m^2

Breast cancer, metastatic or locally advanced resistant/refractory to anthracycline, taxane, capecitabine

• **Adult:** **IV INF** 40 mg/m^2 over 3 hr q3wk; in those with BSA >2.2 m^2, dose should be calculated for a BSA of 2.2 m^2

Dosage reduction in those taking a strong CYP3A4 inhibitor

• **Adult:** **IV INF** 20 mg/m^2 over 3 hr q3wk

Available forms: Powder for inj 15, 45 mg

Administer:

• Premedicate with histamine antagonists 1 hr before use, prevents hypersensitivity

• Antiemetic 30-60 min before product and prn

IV route

• Let kit stand at room temperature for 30 min; to reconstitute, withdraw supplied diluent (8 ml for 15-mg vials, 23.5 ml for 45-mg vials); slowly inject sol into vial; gently swirl and invert to mix, final conc 2 mg/ml; further dilute in LR in DEHP-free bags, final conc should be between 0.2 and 0.6 mg/ml; after added, mix by manual rotation

• Diluted sol stable for 6 hr at room temperature; inf must be completed within 6 hr

• Use in-line filter, 0.2-1.2 micron

• Give over 3 hr

SIDE EFFECTS

CNS: *Peripheral neuropathy,* impaired cognition, chills, fatigue, fever, flushing, headache, insomnia, *asthenia*

CV: Bradycardia, *hypotension,* **abnormal ECG,** angina, **atrial flutter, cardiomyopathy,** chest pain, edema, **MI,** vasculitis

GI: *Nausea, vomiting, diarrhea,* abdominal pain, anorexia, colitis, constipation, gastritis, jaundice, GERD, hepatic failure, trismus

GU: **Renal failure**

HEMA: **Neutropenia, thrombocytopenia, anemia,** infections, coagulopathy

INTEG: *Alopecia,* rash, hot flashes

META: Hypokalemia, metabolic acidosis

MS: *Arthralgia, myalgia*

RESP: **Bronchospasm,** cough, dyspnea

SYST: *Hypersensitivity reactions,* **anaphylaxis,** dehydration, **radiation recall reaction**

PHARMACOKINETICS

Metabolized in liver by P45CYP3A4; excreted in feces (65%) and urine (21%); terminal half-life 52 hr

INTERACTIONS

Increase: ixabepilone level—CYP3A4 inhibitors (amiodarone, amprenavir, aprepitant, atazanavir, chloramphenicol, clarithromycin, conivaptan, cycloSPORINE, danazol, darunavir, dalforpistin, delavirdine, diltiazem, erythromycin, estradiol, fluconazole, fluvoxaMINE, fosamprenavir, imatinib, indinavir, isoniazid, itraconazole, ketoconazole, lopinavir, miconazole, nefazodone, nelfinavir, propoxyphene, ritonavir, RU-486, saquinavir, tamoxifen, telithromycin, troleandomycin, verapamil, voriconazole, zafirlukast)

Decrease: ixabepilone levels—CYP3A4 inducers (aminoglutethimide, barbiturates, bexarotene, bosentan, carBAMazepine, dexamethasone, efavirenz, griseofulvin, modafinil, nafcillin, nevirapine, OXcarbazepine, phenytoin, rifamycin, topiramate)

Drug/Herb

• Avoid use with St. John's wort

Drug/Food

• Avoid use with grapefruit products

NURSING CONSIDERATIONS

Assess:

• CBC, differential, platelet count before treatment and weekly; withhold product if WBC is <1500/mm^3 or platelet count is <100,000/mm^3, notify prescriber

• Monitor temperature q4hr (may indicate beginning infection)

Black Box Warning: Liver function tests before, during therapy (bilirubin, AST, ALT, LDH) prn or monthly; check for jaundiced skin and sclera, dark urine, clay-colored stools, itchy skin, abdominal pain, fever, diarrhea

• VS during 1st hr of inf; check IV site for signs of infiltration

⚠ **Hypersensitivity reactions, anaphylaxis** including hypotension, dyspnea, angioedema, generalized urticaria; discontinue inf immediately; keep emergency equipment available

• Effects of alopecia on body image; discuss feelings about body changes

Evaluate:

• Therapeutic response: decreased tumor size, spread of malignancy

Teach patient/family:

• To report signs of infection: fever, sore throat, flulike symptoms

• To report signs of anemia: fatigue, headache, faintness, SOB

• To report any complaints or side effects to nurse or prescriber

• That hair may be lost during treatment; that a wig or hairpiece may make patient feel better; that new hair may be different in color, texture

• That pain in muscles and joints 2-5 days after inf is common

• To use nonhormonal type of contraception

• To avoid receiving vaccinations while receiving product

ketoconazole (Rx)

(kee-toe-koe′na-zole)

Func. class.: Antifungal

Chem. class.: Imidazole derivative

ACTION: Alters cell membrane permeability and inhibits several fungal enzymes, thereby leading to cell death

USES: Systemic candidiasis, chronic mucocandidiasis, oral thrush, candiduria, coccidioidomycosis, histoplasmosis, chromomycosis, paracoccidioidomycosis, blastomycosis; tinea cruris, tinea corporis, tinea versicolor, *Pityrosporum ovale*

Unlabeled uses: Cushing syndrome, advanced prostatic cancer, candidiasis/fungal prophylaxis, fungal keratitis, leishmaniasis

CONTRAINDICATIONS: Breastfeeding, hypersensitivity, fungal meningitis

Black Box Warning: Coadministration with other products (ergot derivatives, cisapride, or triazolam) may cause fatal cardiac arrhythmias due to inhibition of CYP3A4 enzyme system

Precautions: Pregnancy (C), children <2 yr, renal disease, achlorhydria (product-induced)

Black Box Warning: Hepatic disease

DOSAGE AND ROUTES

- **Adult: PO** 200-400 mg/day for 1-2 wk (candidiasis), 6 wk (other infections)
- **Child ≥2 yr: PO:** 3.3-6.6 mg/kg/day as a single daily dose

Prostate cancer (unlabeled)

- **Adult: PO** 400 mg tid

Available forms: Tabs 200 mg; oral susp 100 mg/5 ml ✤

Administer:

- In the presence of acid products only; do not use alkaline products, proton pump inhibitors, H_2-antagonists, antacids within 2 hr of product; may give coffee, tea, acidic fruit juices, cola
- With food to decrease GI symptoms
- With HCl if achlorhydria is present; dissolve tab/4 ml of aqueous sol 0.2 N hydrochloric acid; use straw to avoid contact; rinse with water afterward and swallow
- Store in tight container at room temperature

SIDE EFFECTS

CNS: Headache, dizziness, somnolence

GI: Nausea, vomiting, anorexia, diarrhea, abdominal pain, **hepatotoxicity**

GU: Gynecomastia, impotence

HEMA: **Thrombocytopenia, leukopenia, hemolytic anemia**

INTEG: Pruritus, fever, chills, photophobia, rash, dermatitis, purpura, urticaria

SYST: **Anaphylaxis, angioedema**

PHARMACOKINETICS

PO: Peak 1-2 hr; half-life 2 hr, terminal 8 hr; metabolized in liver; excreted in bile, feces; requires acid pH for absorption; distributed poorly to CSF; highly protein bound

INTERACTIONS

Increase: effect of CNS depression—chlordiazePOXIDE, clonazePAM, clorazepate, diazepam, estazolam, flurazepam, prazepam, quazepam, avoid concurrent use

Decrease: effect of—theophylline

Decrease: metabolism of PACLitaxel

Increase: hepatotoxicity—other hepatotoxic products, alcohol

Increase: anticoagulant effect—warfarin, anticoagulants

Increase: effect of toxicity: alfentanil, ALPRAZolam, amprenavir, ARIPiprazole, atorvastatin, calcium channel blockers, carBAMazepine, clarithromycin, corticosteroids, cyclophosphamide, cycloSPORINE, donepezil, eletriptan, erythromycin, fentaNYL, ifosfamide, indinavir, lovastatin, midazolam, nelfinavir, nisoldipine, quiNIDine, ritonavir, saquinavir, sildenafil, sim-

K

vastatin, SUFentanil, tamoxifen, triazolam, vinBLAStine, vinca alkaloids, vinCRIStine, zolpidem
Decrease: action of ketoconazole—antacids, H_2-receptor antagonists, anticholinergics, phenytoin, isoniazid, rifampin, ddI, gastric acid pump inhibitors
Drug/Lab Test
Increase: alkaline phosphatase, LFTs, lipids
Decrease: Hgb, WBC, platelets

NURSING CONSIDERATIONS

Assess:

- Infection symptoms before and after treatment
- C&S before starting treatment
- Allergic reaction: rash, photosensitivity, urticaria, dermatitis

Black Box Warning: Hepatotoxicity: nausea, vomiting, jaundice, clay-colored stools, fatigue; hepatic studies (ALT, AST, bilirubin) if patient receiving long-term therapy, use cautiously in those with hepatic disease

- Using with CYP3A4 inhibitors can lead to increased product level, toxicity

Evaluate:

- Therapeutic response: decreased fever, malaise, rash, negative C&S for infectious organism, absence of scaling

Teach patient/family:

- That long-term therapy may be needed to clear infection (1 wk-6 mo, depending on infection)
- To avoid hazardous activities if dizziness occurs
- To take 2 hr before administration of other products that increase gastric pH (antacids, H_2-blockers, omeprazole, sucralfate, anticholinergics)
- About the importance of compliance with product regimen; that testing will be done periodically to indicate if infection has been eradicated
- To decrease nausea, divide dose with meals
- Not to use any products (OTC, herbal) unless approved by prescriber

Black Box Warning: To notify prescriber of GI symptoms, signs of hepatic dysfunction (fatigue, nausea, anorexia, vomiting, dark urine, pale stools)

- To use sunglasses to prevent photophobia (rare)
- To notify prescriber if pregnancy is planned or suspected; not to breastfeed

ketoconazole (topical)

(kee-toe-koe′na-zole)

Extina, Ketoderm ♣, Ketodan, Nizoral, Nizoral A-D, Xolegel

Func. class.: Topical antifungal
Chem. class.: Imidazole derivative

ACTION: Antifungal activity results from altering cell membrane permeability

USES: Seborrheic dermatitis (immunocompromised), tinea corporis, cruris, pedis, versicolor, dandruff

CONTRAINDICATIONS: Hypersensitivity, sulfite allergy
Precautions: Pregnancy (C), breastfeeding, children

DOSAGE AND ROUTES

Seborrheic dermatitis:

- **Adult/child ≥12 yr: TOP FOAM** apply to affected areas bid × 4 wk
- **GEL** apply to affected areas daily × 2 wk

Tinea corporis, cruris, pedis, versicolor

- **Adult: TOP** cover areas daily × 2 wk

Dandruff:

- **Adult: SHAMPOO** wet hair, lather, massage for 1 min, rinse, repeat 2×/wk spaced by 3 days, for up to 8 wk, then as needed

Available forms: Topical gel, foam, cream 2%; shampoo 1%, 2%

Administer:

Topical route

- For external use only; do not use skin products near the eyes, nose, or mouth, wash hands before and after use
- **Cream/lotion:** Apply to the cleansed affected area, massage gently into affected areas, do not use on skin that is broken or irritated

SIDE EFFECTS

INTEG: Irritation, stinging, pustules, pruritus

NURSING CONSIDERATIONS

Assess allergic reaction:

- Assess for hypersensitivity, product might need to be discontinued

⚠ Assess for sulfite allergy; may be life threatening

Evaluate:

- Decreased itching scaling

Teach patient/family:

Topical route

- These products are not for intravaginal therapy and are for external use only; do not use skin products near the eyes, nose, or mouth; wash hands before and after use; do not wash affected area for ≥3 hr after application
- **Cream/Ointment/Lotion:** apply a thin film to the cleansed affected area, massage gently
- **Foam Formulations:** do not dispense foam directly onto hands or face; the warmth of the skin will cause the foam to melt; dispense desired amount directly into the cap or onto a cool surface; make sure enough foam is dispensed to cover the affected area(s); if the can feels warm or the foam seems runny, run the can under cold water; to apply, pick up small amounts of the foam with the fingertips and gently massage into the affected areas until the foam disappears
- To continue for prescribed time, tinea corporis/cruris ≥2 wk

ketoprofen (OTC, Rx)

(ke-toe-proe′fen)

Apo-Keto ✦

Func. class.: Nonsteroidal antiinflammatory product (NSAID), antirheumatic

Chem. class.: Propionic acid derivative

ACTION:
Inhibits COX-1, COX-2; analgesic, antiinflammatory, antipyretic

USES:
Mild to moderate pain, osteoarthritis, rheumatoid arthritis, dysmenorrhea; OTC relief of minor aches, pains

Unlabeled uses: Ankylosing spondylitis, bone pain, gouty arthritis

CONTRAINDICATIONS:
Pregnancy (D) 2nd/3rd trimester, hypersensitivity to this product, NSAIDs, salicylates

Black Box Warning: Perioperative pain with CABG

Precautions: Pregnancy (C) 1st trimester, breastfeeding, children, geriatric patients, bleeding, GI/cardiac disorders, hypersensitivity to other antiinflammatory agents; asthma, severe renal/hepatic disease, ulcer disease

Black Box Warning: GI bleeding, MI, stroke

DOSAGE AND ROUTES

Antiinflammatory

- **Adult: PO** 50 mg qid or 75 mg tid, max 300 mg/day or **EXT REL** 200 mg/day

Analgesic

- **Adult: PO** 25-50 mg q6-8hr, max 300 mg/day

Available forms: Caps 50, 75 mg; ext rel cap 200 mg

Administer:

- Do not break, crush, or chew ext rel caps
- Store at room temperature
- Give with antacids, milk, or food for GI upset

SIDE EFFECTS

CNS: Dizziness, drowsiness, fatigue, confusion, insomnia, depression, headache
CV: Tachycardia, peripheral edema, palpitations, hypertension, CV thrombotic events, MI, stroke
EENT: Tinnitus, hearing loss, blurred vision
GI: *Nausea, anorexia, vomiting, diarrhea,* jaundice, hepatitis, constipation, flatulence, cramps, dry mouth, peptic ulcer, GI bleeding
GU: Nephrotoxicity: dysuria, hematuria, azotemia
HEMA: Blood dyscrasias
INTEG: Purpura, rash, pruritus
SYST: Anaphylaxis

PHARMACOKINETICS

PO: Peak 1.2 hr; ext rel 6.8 hr, half-life 2-4 hr; 5.4 hr ext rel, metabolized in liver, urine (metabolites), breast milk; 99% plasma protein binding

INTERACTIONS

Increase: toxicity—cycloSPORINE, lithium, methotrexate, phenytoin, alcohol, cidofovir
Increase: bleeding risk—anticoagulants, clopidogrel, eptifibatide, thrombolytics, ticlopidine, tirofiban, SSRIs
Increase: ketoprofen levels—aspirin, probenecid
Increase: adverse GI reactions—aspirin, corticosteroids, NSAIDs, alcohol
Increase: hematologic toxicity, antineoplastics
Decrease: effect of diuretics, antihypertensives

Drug/Herb
Increase: bleeding risk—feverfew, garlic, ginger, ginkgo

Drug/Lab Test
Increase: BUN, alk phos, AST, ALT, LDH, creatinine, bleeding time

NURSING CONSIDERATIONS

Assess:

- **Pain:** type, location, intensity, ROM before and 1-2 hr after treatment
- Renal, hepatic, blood studies: BUN, creatinine, AST, ALT, Hgb before treatment, periodically thereafter

⚠ Aspirin sensitivity, asthma; these patients may be more likely to develop hypersensitivity to NSAIDs

- Audiometric, ophthalmic exam before, during, after treatment
- For eye/ear problems: blurred vision, tinnitus; may indicate toxicity

Black Box Warning: GI bleeding: blood in sputum, emesis, stools

Black Box Warning: CV thrombotic events: MI, stroke

Evaluate:

- Therapeutic response: decreased pain, stiffness, swelling in joints; ability to move more easily; decreased fever

Teach patient/family:

- To report blurred vision, ringing, roaring in ears; may indicate toxicity
- To avoid driving, other hazardous activities if dizziness, drowsiness occurs, especially in geriatric patients
- To report immediately change in urine pattern, increased weight, edema, increased pain in joints, fever, blood in urine (indicate nephrotoxicity); rash, itching, blurred vision, ringing in ears, flulike symptoms; blood in urine, vomit, or stools (bleeding)
- That therapeutic effects may take up to 1 mo; to take with 8 oz water; to sit upright for 1/2 hr after administration to prevent GI irritation, not to crush, chew ext rel products
- To avoid aspirin, alcohol, corticosteroids, acetaminophen, other medications, supplements unless approved by prescriber
- To wear sunscreen, protective clothing to prevent photosensitivity
- To report product use to all health care providers
- To report planned or suspected pregnancy (D) (2nd/3rd trimester), (C) (1st trimester); avoid breastfeeding

ketorolac (ophthalmic)

(kee′toe-role-ak)

Acular, Acular LS, Acuvail

Func. class.: Antiinflammatory (ophthalmic)

Chem. class.: Nonsteroidal antiinflammatory drug

ACTION: Inhibits miosis by inhibiting the biosynthesis of ocular prostaglandins; prostaglandins play a role in the miotic response produced during ocular surgery by constricting the iris sphincter independently of cholinergic mechanisms

USES: Pain and inflammation after cataract surgery, refractive surgery, seasonal allergic conjunctivitis

CONTRAINDICATIONS: Hypersensitivity to this product, NSAIDs, salicylates

Precautions: Bleeding disorders, complicated ocular surgery, corneal denervation, diabetes mellitus, rheumatoid arthritis, dry eye syndrome, pregnancy (C), breastfeeding, children, contact lens

DOSAGE AND ROUTES

Seasonal Allergic Conjunctivitis (Acular)

- **Adult/child ≥2 yr:** Instill 1 drop into affected eye qid

Inflammation after cataract extraction (Acular)

- **Adult: OPHTH** 1 drop in affected eye qid beginning 24 hr after surgery × 2 wk

Corneal refracture surgery, pain, burning (Acular LS)

- **Adult: OPHTH** 1 drop in affected eye qid × ≤4 days

Incision refraction surgery, pain, photophobia (Acular PF)

- **Adult: OPHTH** 1 drop qid in affected eye × 3 days

Cataract surgery, pain, inflammation (Acuvail)

- **Adult: OPHTH** 1 drop bid in affected eye, starting 1 day before surgery, on the day of surgery, × 2 wks after surgery

Available forms: Ophthalmic solution Acular (0.5%), Acular LS (0.4%), Acuvail (0.45%)

Administer:

- Apply topically to the eye, separate by ≥5 min when using with other ophthalmics
- Remove contact lenses before instillation of solution
- Instruct patient on proper instillation of eye solution
- Do not touch the tip of the dropper to the eye, fingertips, or other surface
- Do not share bottle with other patients

SIDE EFFECTS

CNS: Headache

EENT: Abnormal sensation in eye, conjunctival hyperemia, ocular irritation, ocular pain, ocular pruritus, conjunctival hyperemia, iritis, keratitis, blurred vision, transient burning/stinging

NURSING CONSIDERATIONS

Assess:

- Eyes: for pain, inflammation, burning, redness after cataract surgery

Evaluate:

- Decreased pain and inflammation after cataract surgery, refractive surgery, seasonal allergic conjunctivitis

Teach patient/family:

- To apply topically to the eye
- To remove contact lenses before instillation of solution
- Proper instillation of eye solution
- Not to touch the tip of the dropper to the eye, fingertips, or other surface
- Not to share bottle with other patients

K

ketorolac (systemic, nasal) (Rx)

(kee-toe′role-ak)

Toradol ✦, Sprix

Func. class.: Nonsteroidal antiinflammatory/nonopioid analgesic

Chem. class.: Acetic acid

Do not confuse:
Ketorolac/Ketalar

ACTION: Inhibits prostaglandin synthesis by decreasing an enzyme needed for biosynthesis; analgesic, antiinflammatory, antipyretic effects

USES: Mild to moderate pain (short term)

CONTRAINDICATIONS: Pregnancy (D) 3rd trimester, hypersensitivity to this product, salicylates, asthma, hepatic disease, peptic ulcer disease, CV bleeding, C-section intracranial bleeding

Black Box Warning: Severe renal disease, L&D, perioperative pain in CABG, before major surgery, epidural/intrathecal administration, GI bleeding, hypovolemia, NSAID hypersensitivity

Precautions: Pregnancy (C), breastfeeding, GI/cardiac disorders, hypersensitivity to other antiinflammatory agents, CCr <25 ml/min

Black Box Warning: Children, geriatric patients, bleeding, MI, stroke

DOSAGE AND ROUTES

For the short-term treatment of moderate pain

Intranasal dosage:

- **Adults ≥50 kg with normal renal function:** 1 spray (15.75 mg/spray) in each nostril (total dose of 31.5 mg) q6-8hr; max 4 doses (8 sprays/126 mg) daily; total systemic therapy max 5 days
- **Adults <50 kg or who have renal impairment, and geriatric patients:** 1 spray (15.75 mg/spray) in one nostril q6-8hr max 4 doses (4 sprays/63 mg) daily; total systemic therapy max 5 days

For the short-term treatment of moderately severe pain

IM/IV (single-dose treatment):

- **Adults/adolescents 17 yr and ≥50 kg and have normal renal function:** 60 mg IM, or 30 mg IV
- **Adults/adolescents 17 yr and <50 kg or have renal impairment, and geriatric patients:** 30 mg IM or 15 mg IV
- **Child ≥2 yr/adolescents ≤16 yr:** 1 mg/kg IM; max 30 mg or 0.5 mg/kg IV to a max of 15 mg

IM/IV (multiple-dose treatment):

- **Adults/adolescents 17 yr and weigh ≥50 kg and have normal renal function:** 30 mg IM/IV q6hr; total systemic therapy max 5 days
- **Adults/adolescents 17 yr and weigh <50 kg or have renal impairment, and geriatric patients:** 15 mg IM/IV q6hr; total systemic therapy max 5 days

PO (continuation therapy from IM/IV only):

- **Adults/adolescents 17 yr and weigh ≥50 kg and have normal renal function:** 20 mg initially following IV/IM therapy, then 10 mg q4-6hr; max 40 mg daily; total systemic therapy max 5 days
- **Adults/adolescents 17 yr and weigh <50 kg or have renal impairment, and geriatric patients:** 10 mg initially following IV/IM therapy, then 10 mg q4-6hr; max 40 mg daily; total systemic therapy should not exceed 5 days

Renal dose

- Do not use in advanced renal disease

Available forms: Inj 15, 30 mg/ml (prefilled syringes), 60 mg/2 ml; tab 10 mg; nasal spray 15.75 mg/spray; multi dose vial 300 mg/10 ml

Administer:

- Not to exceed 5 days
- Store at room temperature, protect from light

IM route

- IM inj deeply and slowly in large muscle mass

Nasal route

- Prime pump before using for the first time, point away from person/pets, pump activator 5 times, no need to reprime
- For single use only, discard 24 hr after opening if not used
- Do not share with others
- Have patient blow nose, sit upright to spray

IV route

- Give undiluted over ≥15 sec

Solution compatibility: D_5W, 0.9% NaCl, LR, D_5, Plasma-Lyte A

Syringe compatibilities: SUFentanil

Y-site compatibilities: Cisatracurium, remifentanil, SUFentanil

SIDE EFFECTS

CNS: Dizziness, *drowsiness,* tremors, seizures, headache

CV: Hypertension, pallor, edema, CV thrombotic events, MI, stroke

EENT: Tinnitus, hearing loss, blurred vision

GI: Nausea, anorexia, vomiting, diarrhea, constipation, flatulence, cramps, dry mouth, peptic ulcer, GI bleeding, perforation, taste change, hepatic failure

GU: Nephrotoxicity: dysuria, hematuria, oliguria

HEMA: Blood dyscrasias, prolonged bleeding

INTEG: Purpura, rash, pruritus, sweating, angioedema, Stevens-Johnson syndrome, toxic epidermal necrolysis

PHARMACOKINETICS

Half-life 6 hr, enters breast milk, metabolized by liver, excreted by kidneys

PO: Onset 30-60 min, duration 4-6 hr

IM: Onset 30 min, duration 4-6 hr

INTERACTIONS

Increase: toxicity—methotrexate, lithium, cycloSPORINE, pentoxifylline, probenecid, cidofovir

Increase: bleeding risk—anticoagulants, clopidogrel, eptifibatide, salicylates, ticlopidine, tirofiban, thrombolytics, SSRIs, SNRIs

Increase: renal impairment—ACE inhibitors

⚠ **Increase:** ketorolac levels—aspirin, other NSAIDs; contraindicated

Increase: GI effects—corticosteroids, alcohol, aspirin, NSAIDs

Decrease: effects—antihypertensives, diuretics

Drug/Lab Test

Increase: AST, ALT, LDH, bleeding time

NURSING CONSIDERATIONS

Assess:

- **Aspirin sensitivity, asthma:** patients may be more likely to develop hypersensitivity to NSAIDs; monitor for hypersensitivity
- **Pain:** type, location, intensity, ROM before and 1 hr after treatment

Black Box Warning: Renal, hepatic, blood studies: BUN, creatinine, AST, ALT, Hgb before treatment, periodically thereafter; check for dehydration

Black Box Warning: Bleeding times; check for bruising, bleeding, occult blood in urine

Black Box Warning: Do not use epidurally, intrathecally, alcohol is present in the solution

- Eye/ear problems: blurred vision, tinnitus (may indicate toxicity)

⚠ Hepatic dysfunction: jaundice, yellow sclera and skin, clay-colored stools

Black Box Warning: CV thrombotic events: MI, stroke, do not use in perioperative pain in CABG

- Audiometric, ophthalmic exam before, during, after treatment

Evaluate:

- Therapeutic response: decreased pain, stiffness, swelling in joints, ability to move more easily

Teach patient/family:

- To report blurred vision, ringing/roaring in ears (may indicate toxicity)
- To avoid driving, other hazardous activities if dizziness or drowsiness occurs

Black Box Warning: To report change in urine pattern, weight increase, edema; pain increase in joints, fever, blood in urine (indicates nephrotoxicity); bruising, black tarry stools (indicates bleeding); pruritus, jaundice, nausea, right upper quadrant pain, abdominal pain (hepatotoxicity); to notify prescriber immediately

- To avoid alcohol, salicylates, other NSAIDs
- To report product use to all health care providers, not to use with other products unless approved by prescriber; use for ≤5 days
- **Nasal:** to discard within 24 hr of opening; may cause irritation, may drink water after dose
- To report if pregnancy is planned or suspected pregnancy (C) systemic; not to breastfeed

ketotifen (ophthalmic)

(kee-toe-tye′fen)

Alaway, Zaditor, ZyrTEC Itchy Eye, Claritin Eye

Func. class.: Antihistamine (ophthalmic)

Chem. class.: Histamine 1 receptor antagonist/mast cell stabilizer

ACTION: A topically active, direct H_1-receptor antagonist and mast cell stabilizer; by reducing these inflammatory mediators, relieves the ocular pruritus associated with allergic conjunctivitis

USES: For the temporary relief of ocular pruritus due to ragweed, pollen, grass, animal hair, animal dander

CONTRAINDICATIONS: Hypersensitivity

Precautions: Pregnancy (C), breastfeeding, children, contact lenses

DOSAGE AND ROUTES

- **Adult/child ≥3 yr:** OPHTH instill 1 drop in affected eye(s) every 8-12 hr

Available forms: Ophthalmic solution 0.025%

Administer:

Ophthalmic route

- For topical ophthalmic use only
- Wash hands before and after use; tilt the head back slightly and pull the lower eyelid down with the index finger; squeeze the prescribed number of drops into the conjunctival sac and gently close eyes for 1-2 min, do not blink
- Do not touch the tip of the dropper to the eye, fingertips, or other surface
- Wait ≥10 min after instilling the ophthalmic solution before inserting contact lenses; contact lenses should not be worn if eye is red
- Do not share ophthalmic drops with others

SIDE EFFECTS

CNS: Headache

EENT: Conjunctival hyperemia, rhinitis, allergic reactions, ocular irritation consisting of burning or stinging, conjunctivitis, eyelid disorder, flu syndrome, keratitis, lacrimation disorder, mydriasis, ocular discharge, ocular pain, pharyngitis, photophobia, pruritus, rash, xerophthalmia (dry eyes)

NURSING CONSIDERATIONS

Assess:

- Eyes: for itching, redness, use of soft or hard contact lens

Evaluate:

- Absence of redness, itching in the eyes

Teach patient/family:

Ophthalmic route

- That product is for topical ophthalmic use only
- To wash hands before and after use; tilt the head back slightly and pull the lower eyelid down with the index finger; squeeze the prescribed number of drops into the conjunctival sac and gently close eyes for 1-2 min; not to blink

- Not to touch the tip of the dropper to the eye, fingertips, or other surface
- To wait ≥10 min after instilling the ophthalmic solution before inserting contact lenses; contact lenses should not be worn if eye is red
- Not to share ophthalmic drops with others
- To remove contact lenses before use because the preservative benzalkonium chloride may be absorbed by soft contact lenses; that product should not be used to treat contact lens-related irritation

labetalol (Rx)

(la-bet′a-lole)

Trandate

Func. class.: Antihypertensive, antianginal

Chem. class.: α-1/β-Blocker

Do not confuse:
Trandate/Tridrate

ACTION: Produces decreases in B/P without reflex tachycardia or significant reduction in heart rate through mixture of α-blocking, β-blocking effects; elevated plasma renins are reduced

USES: Mild to moderate hypertension; treatment of severe hypertension (IV)
Unlabeled uses: Hypertension in patients with pheochromocytoma, hypertension during cloNIDine withdrawal, pediatric hypertension

CONTRAINDICATIONS: Hypersensitivity to β-blockers, cardiogenic shock, heart block (2nd or 3rd degree), sinus bradycardia, CHF, bronchial asthma
Precautions: Pregnancy (C), breastfeeding, geriatric patients, major surgery, diabetes mellitus, thyroid/renal/hepatic disease, COPD, well-compensated heart failure, nonallergic bronchospasm, peripheral vascular disease

Black Box Warning: Abrupt discontinuation

DOSAGE AND ROUTES

Hypertension

- **Adult: PO** 100 mg bid; may be given with diuretic; may increase to 200 mg bid after 2 days; may continue to increase q1-3days; max 2400 mg/day in divided doses
- **Child/adolescent (unlabeled): PO** 1-3 mg/kg/day, titrate to max 10-20 mg/kg/day based on B/P; **IV** 0.2-1 mg/kg over 2 min, max 40 mg/dose; **IV INF** 0.25-3 mg/kg/hr

Hypertensive crisis

- **Adult: IV Intermittent** 20 mg over 2 min; may repeat 20-80 mg over 2 min q10min, 200 mg; **IV Cont INF** after loading dose give 1-2 mg/min until desired response or max 300 mg

Available forms: Tabs 100, 200, 300 mg; inj 5 mg/ml

Administer:

- PO before meals, at bedtime; tab may be crushed or swallowed whole; give with meals to increase absorption
- Do not discontinue before surgery
- Store in dry area at room temperature; do not freeze

Direct IV route

- Give undiluted (5 mg/ml) over 2 min

Continuous IV INF route

- Diluted in LR, D_5W, D_5 in 0.2%, 0.9%, 0.33% NaCl, Ringer's inj; inf is titrated to patient response; 200 mg of product/160 ml sol = 1 mg/ml; 300 mg of product/240 ml sol = 1 mg/ml; 200 mg of product/250 ml sol = 2 mg/3 ml; use inf pump
- Keep patient recumbent during and for 3 hr after administration; monitor VS q5-15min

Y-site compatibilities: Alemtuzumab, alfentanil, amikacin, aminocaproic acid, aminophylline, amiodarone, anidulafungin, argatroban, arsenic trioxide, ascorbic acid injection, atracurium, atropine, azithromycin, aztreonam, benztropine, bivalirudin, bleomycin, bretylium, bumetanide, buprenorphine, butorphanol, calcium chloride/gluconate, CARBOplatin, carmustine, caspofungin, cefamandole, ceFAZolin, cefoperazone, cefotaxime, cefoTEtan, cefOXitin, ceftaroline, cefTAZidime, ceftizoxime, chlorproMAZINE, cimetidine, CISplatin, cloNIDine, cyanocobalamin, cyclophosphamide, cycloSPORINE, cytarabine, DACTINomycin, DAPTOmycin, DAUNOrubicin liposome, dexmedetomidine, dexrazoxane, digoxin, diltiazem, diphenhydrAMINE, DOBUTamine, DOCEtaxel, dolasetron, DOPamine, doripenem, doxacurium, DOXOrubicin, DOXOrubicin liposomal, doxycycline, enalaprilat, ePHEDrine, EPINEPHrine, epirubicin, epoetin alfa, eptifibatide, ertapenem, erythromycin lactobionate, esmolol, etoposide, etoposide phosphate, famotidine, fenoldopam, fentaNYL, fluconazole, fludarabine, fluorouracil, folic acid, gallium, ganciclovir, gatifloxacin, gemcitabine, gentamicin, glycopyrrolate, granisetron, HYDROmorphone, hydroxyzine, IDArubicin, ifosfamide, imipenem-cilastatin, inamrinone, irinotecan, isoproterenol, lactated Ringer's injection, lepirudin, leucovorin, levofloxacin, lidocaine, linezolid injection, LORazepam, magnesium sulfate, mannitol, mechlorethamine, meperidine, metaraminol, methyldopate, methylPREDNISolone, metoclopramide, metoprolol, metroNIDAZOLE, midazolam, milrinone, minocycline, mitoXANtrone, morphine, moxifloxacin, multiple vitamins injection, mycophenolate, nalbuphine, naloxone, netilmicin, niCARdipine, nitroglycerin, nitroprusside, norepinephrine, octreotide, ondansetron, oxacillin, oxaliplatin, oxytocin, palonosetron, pamidronate, pancuronium, papaverine, PEMEtrexed, pentamidine, pentazocine, PENTobarbital, PHENobarbital, phentolamine, phenylephrine, phytonadione, polymyxin B, potassium acetate/chloride/phosphates, procainamide, prochlorperazine, promethazine, propofol, propranolol, protamine, pyridoxine, quiNIDine, quinupristin-dalfopristin, ranitidine, Ringer's injection, rocuronium, sodium acetate/bicarbonate, succinylcholine, SUFentanil, tacrolimus, telavancin, teniposide, theophylline, thiamine, thiotepa, ticarcillin-clavulanate, tigecycline, tirofiban, tobramycin, tolazoline, urokinase, vancomycin, vasopressin, vecuronium, verapamil, vinBLAStine, VinCRIStine, vinorelbine, voriconazole, zoledronic acid

SIDE EFFECTS

CNS: *Dizziness,* mental changes, drowsiness, *fatigue,* headache, catatonia, depression, anxiety, nightmares, paresthesias, lethargy

CV: *Orthostatic hypotension, bradycardia,* **CHF**, chest pain, **ventricular dysrhythmias**, AV block, scalp tingling

EENT: *Tinnitus,* visual changes; sore throat; double vision; dry, burning eyes, floppy iris syndrome; nasal congestion

ENDO: Hyperkalemia

GI: *Nausea, vomiting, diarrhea,* dyspepsia, taste distortion, **hepatotoxicity**

GU: Impotence, dysuria, ejaculatory failure

HEMA: **Agranulocytosis, thrombocytopenia, purpura** (rare)

INTEG: Rash, alopecia, urticaria, pruritus, fever, **exfoliative dermatitis**

RESP: **Bronchospasm**, dyspnea, wheezing

PHARMACOKINETICS

Half-life 2.5-8 hr, metabolized by liver (metabolites inactive), excreted in urine, crosses placenta, excreted in breast milk, protein binding 50%

PO: Onset 30 min, peak 1-4 hr, duration 8-24 hr

IV: Onset 2-5 min, peak 5-15 min, duration 2-4 hr

INTERACTIONS

- **Do not use within 2 wk of MAOIs**

Increase: **myocardial depression—hydantoins, general anesthetics, verapamil, class I antidysrhythmics**

Increase: tremor—tricyclic antidepressants

Increase: hypotension—diuretics, other antihypertensives, cimetidine, nitroglycerin, alcohol, nitrates

Decrease: effects of—sympathomimetics, lidocaine, theophylline, β-blockers, bronchodilators, xanthines

Decrease: antihypertensive effect—NSAIDs, salicylates

Increase or decrease: effects of—antidiabetics

Drug/Herb

Increase: antihypertensive effect—hawthorn

Decrease: antihypertensive effect—ephedra

Drug/Lab Test
Increase: ANA titer, blood glucose, alk phos, LDH, AST, ALT, BUN, potassium, triglyceride, uric acid, serum lipoprotein
False increase: urinary catecholamines

NURSING CONSIDERATIONS

Assess:

- **Hypertension:** B/P during beginning treatment, periodically thereafter; note pulse, rate, rhythm, quality; apical/radial pulse before administration; notify prescriber of any significant changes

⚠ CHF: I&O, weight daily; fluid overload: weight gain, jugular venous distention, edema, crackles in lungs

Black Box Warning: Abrupt discontinuation: product should be tapered to prevent adverse reactions

- Baselines of renal/hepatic studies before therapy begins

Evaluate:

- Therapeutic response: decreased B/P after 1-2 wk

Teach patient/family:

Black Box Warning: Not to discontinue product abruptly; to taper over 2 wk; may cause precipitate angina

- Not to use OTC products containing α-adrenergic stimulants (nasal decongestants, OTC cold preparations) unless directed by prescriber
- To report bradycardia, dizziness, confusion, depression, fever
- To take pulse at home; advise when to notify prescriber
- May mask symptoms of hypoglycemia; monitor blood glucose closely
- To avoid alcohol, smoking, increased sodium intake
- To comply with weight control, dietary adjustments, modified exercise program
- To carry emergency ID to identify product, allergies
- To avoid hazardous activities if dizziness is present
- **To report symptoms of CHF:** difficulty breathing, especially on exertion or when lying down; night cough; swelling of extremities
- To take medication at bedtime to prevent effect of orthostatic hypotension; to rise slowly
- To avoid driving or other hazardous activities until response is known; dizziness, drowsiness, may occur
- To wear support hose to minimize effects of orthostatic hypotension

TREATMENT OF OVERDOSE:

Lavage, IV glucagonor atropine for bradycardia, IV theophylline for bronchospasm; digoxin, O_2, diuretic for cardiac failure; hemodialysis useful for removal/hypotension; administer vasopressor

lacosamide (Rx)

(la-koe′sa-mide)

Vimpat

Func. class.: Anticonvulsant
Chem. class.: Functionalized amino acid

ACTION: May act through action at sodium channels; exact action is unknown

USES: Adjunctive therapy for partial seizures

CONTRAINDICATIONS: Hypersensitivity

Precautions: Pregnancy (C), breastfeeding, children <17 yr, geriatric patients, allergies, cardiac/renal/hepatic disease, acute MI, atrial fibrillation/flutter, AV block, bradycardia, CHD, dehydration, depression, dialysis, hazardous activity, electrolyte imbalance, heart failure, labor, PR prolongation, sick sinus syndrome, substance abuse, suicidal ideation, syncope, torsades de pointes

DOSAGE AND ROUTES

- **Adult and adolescent ≥17 yr: PO** 50 mg bid, may increase weekly by 100 mg bid to 200-400 mg/day; **IV** 50 mg bid, infuse over 30-60 min, may be increased by 100 mg/day weekly up to 200-400 mg/day maintenance

Renal/hepatic dose

• **Adult: PO/IV** max 300 mg/day for mild to moderate hepatic disease or CCr ≤30 ml/min

Available forms: Film-coated tabs 50, 100, 150, 200 mg; IV 20 ml single-use vials (200 mg/20 ml); oral sol 10 mg/ml

Administer:

• Store PO products/IV vials at room temperature; sol is stable for 24 hr when mixed with compatible diluents in glass or PVC bags at room temperature

PO route

• **Tablet:** give without regard to meals
• **Oral sol:** measure with calibrated measuring device

IV route

• May give undiluted or mixed in 0.9%NaCl, D_5W, or LR
• Infuse over 30-60 min
• Do not use if discolored or if particulates are present; discard unused portions

SIDE EFFECTS

CNS: Dizziness, syncope, tremor, vertigo, ataxia, drowsiness, fever, hypoesthesia, paresthesias, depression, fatigue, headache, confusion, irritability, psychologic dependence, suicidal ideation, euphoria

CV: Atrial fibrillation/flutter, AV block, bradycardia, myocarditis, orthostatic hypotension, palpitations, PR prolongation

EENT: Diplopia, blurred vision, nystagmus, tinnitus

GI: Nausea, constipation, vomiting, hepatitis, diarrhea, dyspepsia

HEMA: Anemia, neutropenia, agranulocytosis

INTEG: Rash, erythema, inj site reaction, pruritus, xerostomia

MS: Asthenia, dysarthria, muscle cramps

SYST: Drug reaction with eosinophilia, systemic symptoms (DRESS)

PHARMACOKINETICS

Metabolized by liver; excreted by kidneys, 95%; protein binding <15%

PO: Peak 1-4 hr

IV: Peak 30-60 min; half-life 13 hr; elimination half-life 15-23 hr

INTERACTIONS

⚠ **Increase:** PR prolongation—beta-blockers, calcium channel blockers, atazanavir, dronedarone, digoxin, lopinavir, ritonavir

⚠ **Increase:** lincosamide effect—CYP2C19 inhibitors (fluconazole, isoniazid, miconazole)

Drug/Lab Test

Increase: LFTs

NURSING CONSIDERATIONS

Assess:

• **Seizures:** duration, type, intensity precipitating factors
• Renal function: albumin conc
• CV status: orthostatic hypotension, PR prolongation; monitor cardiac status throughout treatment
• Mental status: mood, sensorium, affect, memory (long, short term), depression, suicidal ideation, psychologic dependence
• Rash, hypersensitivity reactions
• **Pregnancy:** Enroll in UCB Antiepileptic Drugs Registry 1-888-537-7734

Evaluate:

• Therapeutic response: decrease in severity of seizures

Teach patient/family:

• Not to discontinue product abruptly; to taper over 1 week because seizures may occur
• To avoid hazardous activities until stabilized on product
• To carry emergency ID stating product use
• To notify prescriber of suicidal thoughts/behaviors, syncope, cardiac changes
• To notify prescriber if pregnancy is planned or suspected
• That interactions with other medications may occur
• To consult MedGuide for proper use, risks

lactulose (Rx)

(lak′tyoo-lose)

Constulose, Enulose, Generlac, Kristalose

Func. class.: Laxative; ammonia detoxicant (hyperosmotic)

Chem. class.: Lactose synthetic derivative

ACTION: Prevents absorption of ammonia in colon by acidifying stool; increases water, softens stool

USES: Chronic constipation, portal-systemic encephalopathy in patients with hepatic disease

CONTRAINDICATIONS: Hypersensitivity, low-galactose diet

Precautions: Pregnancy (B), breastfeeding, geriatric patients, debilitated patients, diabetes mellitus

DOSAGE AND ROUTES

Constipation

- **Adult: PO** 15-30 ml/day (10-20 g), may increase to 60 ml/day prn
- **Child: PO** 7.5 ml/day

Hepatic encephalopathy

- **Adult: PO** 30-45 ml (20-30 g) tid or qid until stools soft; **RETENTION ENEMA** 300 ml (200 g) diluted
- **Child: PO** 40-90 ml/day in 3-4 divided doses
- **Infant: PO** 2.5-10 ml/day in divided doses

Available forms: Oral sol (encephalopathy) 10 g/15 ml; oral sol (constipation) 10 g/15 ml

Administer:

PO route

- With 8 oz fruit juice, water, milk to increase palatability of oral form; for rapid effect, give on empty stomach
- Increased fluids to 2 L/day; do not give with other laxatives; if diarrhea occurs, reduce dosage
- **Kristalose:** dissolve contents of packet/4 oz water

Rectal route

- **Retention enema** by diluting 300 ml lactose/700 ml of water; administer by rectal balloon catheter

SIDE EFFECTS

GI: *Nausea, vomiting, anorexia, abdominal cramps,* diarrhea, flatulence, distention, belching

META: Hypernatremia

PHARMACOKINETICS

Metabolized in colon, excreted by kidneys, onset 1-2 days, peak unknown, duration unknown

INTERACTIONS

- Do not use with other laxatives (hepatic encephalopathy)

Increase: GI obstruction—NIFEdipine ext-rel tabs

Decrease: lactulose effects—neomycin, other oral antiinfectives, antacids

Drug/Herb

Increase: laxative action—flax, senna

Drug/Lab Test

Increase: blood glucose (diabetic patients)

Decrease: blood ammonia

NURSING CONSIDERATIONS

Assess:

- **Stool:** amount, color, consistency
- **Cause of constipation;** determine whether fluids, bulk, or exercise is missing from lifestyle; use of constipating products
- **Hepatic encephalopathy:** blood ammonia level (30-70 mg/100 ml); may decrease ammonia level by 25%-50%; clearing of confusion, lethargy, restlessness, irritability if portal-systemic encephalopathy
- Blood, urine electrolytes if product used often; may cause diarrhea, hypokalemia, hyponatremia
- I&O ratio to identify fluid loss
- Cramping, rectal bleeding, nausea,

vomiting; if these symptoms occur, product should be discontinued

Evaluate:

• Therapeutic response: decreased constipation, decreased blood ammonia level, clearing of mental state

Teach patient/family:

• Not to use laxatives long term
• To dilute with water or fruit juice to counteract sweet taste
• To store in cool environment; not to freeze
• To take on an empty stomach for rapid action
• To report diarrhea; may indicate overdose

lamiVUDine 3TC (Rx)

(lam-i-voo'deen)

Epivir, Epivir HBV

Func. class.: Antiretroviral

Chem. class.: Nucleoside reverse transcriptase inhibitor (NRTI)

Do not confuse:
lamiVUDine/lamoTRIgine

ACTION: Inhibits replication of HIV virus by incorporating into cellular DNA by viral reverse transcriptase, thereby terminating cellular DNA chain

USES: HIV-1 infection in combination with at least 2 other antiretrovirals; chronic hepatitis B (Epivir-HBV)

Unlabeled uses: Prophylaxis of HIV: postexposure with indinavir and zidovudine

CONTRAINDICATIONS: Hypersensitivity

Black Box Warning: Lactic acidosis

Precautions: Pregnancy (C), breastfeeding, children, geriatric patients, granulocyte count $<1000/mm^3$ or Hgb <9.5 g/dl, renal disease, pancreatitis, peripheral neuropathy

Black Box Warning: Severe hepatic dysfunction

DOSAGE AND ROUTES

HIV

• **Adult and adolescent >16 yr: PO** 150 mg bid or 300 mg/day
• **Child 3 mo-16 yr: PO** 4 mg/kg bid, max 150 mg bid

Chronic hepatitis B

• **Adult: PO** 100 mg/day
• **Child and adolescent 2-17 yr: PO** 3 mg/kg/day, max 100 mg

Renal dose

• **Adult: PO** CCr 30-49 ml/min: Epivir 150 mg/day; Epivir HBV 100 mg 1st dose then 50 mg/day; CCr 15-29 ml/min: Epivir 150 mg 1st dose then 100 mg/day; Epivir HBV 100 mg 1st dose then 25 mg/day; CCr 5-14 ml/min: Epivir 150 mg 1st dose then 50 mg/day; Epivir HBV 35 mg 1st dose then 15 mg/day; CCr <5 ml/min: Epivir 50 mg 1st dose then 25 mg/day; Epivir HBV 35 mg 1st dose then 10 mg/day

Available forms: (Epivir) oral sol 10 mg/ml; tabs 150, 300 mg; **(Epivir HBV)** oral sol 5 mg/ml; tabs 100 mg

Administer:

• PO daily or bid, without regard to meals
• Epivir and Epivir HBV are not interchangeable
• With other antiretrovirals only
• Store in cool environment; protect from light

SIDE EFFECTS

CNS: *Fever, headache, malaise, dizziness, insomnia, depression, fatigue, chills,* **seizures**, peripheral neuropathy, paresthesias

EENT: Taste change, hearing loss, photophobia

GI: *Nausea, vomiting, diarrhea,* anorexia, cramps, dyspepsia, **hepatomegaly with steatosis, pancreatitis**

HEMA: **Neutropenia, anemia, thrombocytopenia**

INTEG: *Rash*

MS: *Myalgia, arthralgia, pain*

RESP: *Cough*

SYST: **Lactic acidosis, anaphylaxis, Stevens-Johnson syndrome**

L

PHARMACOKINETICS

Rapidly absorbed, distributed to extravascular space, excreted unchanged in urine, protein binding <36%, terminal half-life 5-7 hr

INTERACTIONS

Decrease: both products—zalcitabine; avoid concurrent use

Increase: lamiVUDine level—trimethoprim-sulfamethoxazole, aMILoride, dofetilide, entecavir, metFORMIN, memantine, procainamide, trospium

• Do not use with emtricitabine, duplication

Decrease: lamiVUDine effect—interferons

Drug/Lab Test

Increase: ALT, bilirubin

Decrease: Hgb, neutrophil, platelet count

NURSING CONSIDERATIONS

Assess:

• **HIV:** Test for HIV before starting treatment, blood counts q2wk; watch for neutropenia, thrombocytopenia, Hgb, CD4, viral load; if low, therapy may have to be discontinued and restarted after hematologic recovery; blood transfusions may be required; assess for lessening of symptoms; if HBV is present, a higher dose of Epivir-HBV is needed

• **Hepatitis B:** fatigue, anorexia, pruritus, jaundice during and for several months after discontinuation; AST, ALT, bilirubin; amylase, lipase, triglycerides, periodically during treatment

• **Children for pancreatitis:** abdominal pain, nausea, vomiting, neuropathy

Black Box Warning: Lactic acidosis, severe hepatomegaly with steatosis: obtain baseline LFTs, if elevated, discontinue treatment; discontinue even if LFTs are normal if lactic acidosis, severe hepatomegaly develops, may be fatal

Evaluate:

• Decreasing symptoms of HIV, CD4, viral load

Teach patient/family:

• That GI complaints, insomnia resolve after 3-4 wk of treatment

• That product is not a cure for HIV but will control symptoms; that compliance is necessary; to take as directed; to complete full course of treatment even if feeling better

• To notify prescriber of sore throat, swollen lymph nodes, malaise, fever, peripheral neuropathy; other infections may occur

• That patient is still infective, may pass HIV virus on to others

• That follow-up visits must be continued since serious toxicity may occur; that blood counts must be done

• That other products may be necessary to prevent other infections

• That product may cause fainting or dizziness

• Not to breastfeed, excreted in breast milk

lamoTRIgine (Rx)

(la-moe′tri-geen)

LaMICtal, LaMICtal CD, LaMICtal ODT, LaMICtal XR

Func. class.: Anticonvulsant—miscellaneous

Chem. class.: Phenyltriazine

Do not confuse:

lamoTRIgine/lamiVUDine

LaMICtal/Lomotil/LamISIL

ACTION: Inhibits voltage-sensitive sodium channels, thus decreasing seizures

USES: Adjunct for the treatment of partial, tonic-clonic seizures; children with Lennox-Gastaut syndrome, bipolar disorder

Unlabeled uses: Absence, seizures

CONTRAINDICATIONS: Hypersensitivity

Precautions: Pregnancy (C) (cleft lip/palate during 1st trimester), breastfeeding, geriatric patients, cardiac/renal/hepatic disease, severe depression, suicidal, blood dyscrasias, children <16 yr, serious rash

DOSAGE AND ROUTES

Seizures: monotherapy

- **Adult and adolescent ≥16 yr: PO** 50 mg/day while receiving 1 enzyme-inducing AED (carBAMazepine, PHENobarbital, phenytoin, primidone but not valproic acid) wk 1-2, then increase to 100 mg divided bid wk 3-4; maintenance 300-500 mg/day; EXT REL 50 mg/day × 1-2 wk, then 100 mg/day wk 3-4, then 200 mg/day wk 5, then 300 mg/day wk 6, then 400 mg/day wk 7; after wk 7, range is 400-600 mg/day
- **Adolescent <16 yr and child: PO** 0.3 mg/kg/day wk 1-2, then 0.6 mg/kg/day wk 3-4; depends on use of AED; usual dose 4.5-7.5 mg/kg/day, max 300 mg/day

Monotherapy for patients taking valproate

- **Adult and adolescent ≥16 yr receiving lamoTRIgine and valproate without enzyme-inducing drug: PO** (immediate release) stabilize on valproate, target dose of 200 mg/day lamoTRIgine; if patient is not taking lamoTRIgine 200 mg/day, increase dose by 25-50 mg/day q1-2wk to reach 200 mg/day; while maintaining lamoTRIgine 200 mg/day, decrease valproate to 500 mg/day by ≤500 mg/day/wk, maintain valproate at 500 mg/day × 1 wk, then increase lamoTRIgine to 300 mg/day while decreasing valproate 250 mg/day × 1 wk, then discontinue valproate and increase lamoTRIgine by 100 mg/day/wk to maintenance of 500 mg/day

Seizures: multiple therapy with valproate

- **Adult and adolescent ≥16 yr: PO** 25 mg every other day then 25 mg/day wk 3-4, increase by 25-50 mg q1-2wk, maintenance 100-400 mg/day
- **Adolescent <16 yr and child: PO** 0.1-0.2 mg/kg/day initially then increase q2wk as needed to 1-5 mg/kg/day or 200 mg/day

Bipolar disorder (escalation regimen for those not taking carBAMazepine, other enzyme-inducing drugs, or valproate)

- **Adult and adolescent ≥16 yr: PO** wk 1-2, 25 mg/day; wk 3-4, 50 mg/day; wk 5, 100 mg/day; wk 6-7, 200 mg/day; for patients taking valproic acid: wk 1-2, 25 mg every other day; wk 3-4, 25 mg/day; wk 5, 50 mg/day; wk 6, 100 mg/day; wk 7, 100 mg/day

Hepatic dose

- **Adult: PO** moderate hepatic impairment or severe without ascites: reduce by 25%; severe hepatic impairment with ascites: reduce by 50%

Absence seizures (unlabeled)

- **Adolescent and child 3-13 yr: PO** 0.5 mg/kg/day in 2 divided doses × 2 wk then 1 mg/kg/day in 2 divided doses × 2 wk, adjusted q5days

Available forms: Tabs 25, 100, 150, 200 mg; PO ext rel 25-50-100, 50-100-200 mg titration kit; PO 25-100 mg starter kit; ext rel 25, 50, 100, 250 mg; chew dispersible tabs 5, 25 mg; oral disintegrating tab 25, 50, 100, 200 mg; oral disintegrating tab 25-50, 50-100 mg, 25-50-100 mg titration kit

Administer:

- Correct starter kit; severe side effects here occurred from incorrect starter kit
- **Orange starter kit:** for those **NOT** taking carBAMazepine, phenytoin, PHENobarbital, primidone, rifampin, valproate
- **Green starter kit:** for those taking carBAMazepine, phenytoin, PHENobarbital, primidone, rifampin but **NOT** valproate
- **Blue starter kit:** for those taking valproate
- Discontinue all products gradually over ≥2 wk; abrupt discontinuation can increase seizures
- All forms may be given without regard to meals

• **Chewable dispersible tab:** may be swallowed whole, chewed, mixed in water or fruit juice; to mix, add to small amount of liquid in glass or spoon; tabs will dissolve in 1 min, then mix in more liquid and swirl and swallow immediately; do not cut tabs in half
• **Orally disintegrating tabs:** place on tongue, move around in mouth, when disintegrated, swallow; examine blister pack before use, do not use if blisters are torn or missing
• **Extended-release tabs:** swallow whole, do not cut, break, chew; without regard to food

SIDE EFFECTS

CNS: *Dizziness,* ataxia, *headache,* fever, insomnia, tremor, depression, anxiety, suicidal ideation
EENT: Nystagmus, *diplopia, blurred vision*
GI: *Nausea, vomiting, anorexia, abdominal pain,* hepatotoxicity
GU: *Dysmenorrhea*
HEMA: Anemia, DIC, leukopenia, thrombocytopenia
INTEG: Rash (potentially life-threatening), alopecia
SYST: Stevens-Johnson syndrome, angioedema, toxic epidermal necrolysis, DRESS

PHARMACOKINETICS

Half-life varies depending on dose; terminal half-life 24 hr, 15 hr with enzyme inducers; rapidly, completely absorbed; metabolized by glucuronic acid conjunction; protein binding 55%; peak 1.4-2.3 hr; crosses placenta; excreted in breast milk

INTERACTIONS

Decrease: metabolic clearance of lamoTRIgine—valproic acid, CYP3A4 inhibitors
Decrease: lamoTRIgine conc—carBAMazepine, rifamycins, oral contraceptives, acetaminophen, phenytoin, primodone, PHENobarbital, OXcarbazepine, succinimides, estrogen
Drug/Herb
Increase: anticonvulsant effect—ginkgo
Decrease: anticonvulsant effect—ginseng

NURSING CONSIDERATIONS

Assess:
• **Seizure:** duration, type, intensity, halo before seizure
⚠ Rash (Stevens-Johnson syndrome, toxic epidermal necrolysis) in pediatric patients; product should be discontinued at first sign of rash
⚠ **Bipolar disorder:** suicidal thoughts/behaviors

Evaluate:
• Therapeutic response: decrease in severity of seizures or of bipolar symptoms

Teach patient/family:
• To take PO doses divided, with or after meals to decrease adverse effects; not to discontinue product abruptly because seizures may occur
• To avoid hazardous activities until stabilized on product
• To carry emergency ID; to notify prescriber of skin rash, increased seizure activity; to use sunscreen, protective clothing if photosensitivity occurs
• To notify prescriber if pregnant, intending to become pregnant
• To notify prescriber immediately of suicidal thoughts/behaviors
• To notify prescriber if pregnancy is planned or suspected, pregnancy (C), product decreases folate, avoid breastfeeding

lansoprazole (Rx, OTC)

(lan-so-prey′zole)

Prevacid, Prevacid SoluTab

Func. class.: Antiulcer, proton pump inhibitor
Chem. class.: Benzimidazole

Do not confuse:
Prevacid/Pravachol/Prinivil

ACTION: Suppresses gastric secretion by inhibiting hydrogen/potassium ATPase enzyme system in gastric parietal cell; characterized as gastric acid pump inhibitor because it blocks the final step of acid production

USES: Gastroesophageal reflux disease (GERD), severe erosive esophagitis, poorly responsive systemic GERD, pathologic hypersecretory conditions (Zollinger-Ellison syndrome, systemic mastocytosis, multiple endocrine adenomas); possibly effective for treatment of duodenal, gastric ulcers, maintenance of healed duodenal ulcers
Unlabeled uses: GERD (infants/neonates)

CONTRAINDICATIONS: Hypersensitivity
Precautions: Pregnancy (B), breastfeeding, children

DOSAGE AND ROUTES

Frequent heartburn
- **Adult: PO (OTC)** 15 mg daily up to 14 days

Duodenal ulcer
- **Adult: PO** 15 mg/day before eating for 4 wk then 15 mg/day to maintain healing of ulcers; associated with *Helicobacter pylori*: 30 mg lansoprazole, 500 mg clarithromycin, 1 g amoxicillin bid × 14 days or 30 mg lansoprazole, 1 g amoxicillin tid × 14 days

Pathologic hypersecretory conditions
- **Adult: PO** 60 mg/day, may give up to 90 mg bid, administer doses of >120 mg/day in divided doses

GERD/esophagitis
- **Adult and adolescent: PO** 15-30 mg/day × 8 wk
- **Child 1-11 yr (>30 kg): PO** 30 mg/day ≤12 wk
- **Child 1-11 yr (≤30 kg): PO** 15 mg/day ≤12 wk
- **Infant (unlabeled): PO** 1-1.74 mg/kg/day; limited data available
- **Neonate (unlabeled): PO** 0.5-1 mg/kg/day

Stress gastric prophylaxis
- **Adult: NG** Use 30 mg oral cap or 30 mg disintegrating tab

Available forms: Del rel caps 15, 30 mg; orally disintegrating tabs 15, 30 mg
Administer:
PO route
- Swallow caps whole 30 min before eating; do not crush or chew caps; caps may be opened and contents sprinkled on food

NG route
- **Oral cap:** open cap and pour ¼ of granules into NG feeding syringe with plunger removed, slowly add water and depress plunger, repeat until all granules used; flush tube with 15 ml water
- Place on tongue, allow to dissolve, use without regard to water
- **Oral syringe:** dissolve 15 mg/4 ml or 30 mg/10 ml water, use extra water in syringe to remove all of the product

NG tube
- **Oral disintegrating tab:** mix 30 mg tab in 10 ml water, give via NG tube, flush tube with 10 ml sterile water, clamp for 60 min

SIDE EFFECTS

CNS: *Headache*
GI: Diarrhea, abdominal pain, vomiting, nausea, *constipation,* flatulence, acid regurgitation, anorexia, irritable colon, microscopic colitis
GU: **Hematuria,** glycosuria, impotence, kidney calculus, breast enlargement

PHARMACOKINETICS

Absorption after granules leave stomach 80%; plasma half-life 1½-2 hr; protein binding 97%; extensively metabolized in liver; excreted in urine, feces; clearance decreased in geriatric patients, renal/hepatic impairment

INTERACTIONS

Decrease: antiplatelets, effect of—clopidogrel

Increase: lansoprazole, toxicity—fluvoxaMINE, voriconazole
Decrease: lansoprazole absorption—sucralfate
Decrease: absorption of ketoconazole, itraconazole, iron salts, calcium carbonate, atazanavir, ampicillin
Increase: hypomagnesemia—loop/thiazide diuretics
Decrease: lansoprazole effect—antimuscarinics, octreotide, H_2-blockers, misoprostol
Decrease: release of ext rel amphetamine/dextroamphetamine
- Avoid use with dasatinib, delavirdine

Drug/Herb
- Avoid use with red yeast rice, St. John's wort

NURSING CONSIDERATIONS

Assess:
- **GI system:** bowel sounds q8hr, abdomen for pain, swelling, anorexia, blood in stool, emesis
- **Hepatic studies:** AST, ALT, alk phos during treatment
- INR and prothrombin time when taking warfarin
- Magnesium: low magnesium may occur

Evaluate:
- Therapeutic response: absence of epigastric pain, swelling, fullness

Teach patient/family:
- To report severe diarrhea; product may have to be discontinued
- That hypoglycemia may occur if diabetic
- To avoid hazardous activities; that dizziness may occur
- To avoid alcohol, salicylates, ibuprofen; may cause GI irritation

RARELY USED

lanthanum (Rx)
(lan′-tha-num)
Fosrenol
Func. class.: Phosphate binder

USES: End-stage renal disease

CONTRAINDICATIONS: Hypophosphatemia, hypersensitivity

DOSAGE AND ROUTES
- **Adult: PO** 750-1500 mg/day in divided doses with meals; titrate dose q2-3wk until an acceptable phosphate level is reached; tabs should be chewed completely before swallowing; intact tabs should not be swallowed; maintenance dose 1500-3000 mg/day divided with meals; max 3750 mg/day

⚠ HIGH ALERT

lapatinib (Rx)
(la-pa′tin-ib)
Tykerb
Func. class.: Antineoplastic—miscellaneous
Chem. class.: Biologic response modifier, signal transduction inhibitor (STIs)

ACTION: Reversibly binds tyrosine kinase of both the epidermal growth factor receptor (ERbB1) and the human epidermal receptor type 2 (HER2) (ERbB2), blocking downstream signaling for cell proliferation

USES: Advanced metastatic breast cancer patients with tumor that overexpresses HER2 protein and who have received previous chemotherapy

CONTRAINDICATIONS:

Pregnancy (D), breastfeeding, hypersensitivity

Precautions: Geriatric patients, cardiac disease, bradycardia, hypertension, hypokalemia, hypomagnesemia, QT prolongation, pneumonitis, interstitial lung disease

Black Box Warning: Hepatic disease

DOSAGE AND ROUTES

Advanced/metastatic breast cancer with HER2 overexpression who have received previous therapy

- **Adult: PO** 1250 mg (5 tabs)/day 1 hr before or after food on days 1-21 plus capecitabine 2000 mg/m^2/day in 2 divided doses on days 1-14 in a repeating 21-day cycle; continue until therapeutic response or toxicity occurs

Metastatic breast cancer with HER2 overexpression for whom hormonal therapy is indicated

- **Adult: PO** 1500 mg (6 tabs) 1 hr before food with letrozole 2.5 mg/day

Hepatic dose

- **Adult: PO** (Child-Pugh C) 750 mg/day (with capecitabine); 1000 mg/day (with letrozole)

Available forms: Tabs 250 mg

Administer:

- Once a day with water on an empty stomach, 1 hr before or after food
- Do not use with grapefruit products
- Store at room temperature, away from heat

SIDE EFFECTS

CNS: Fatigue, insomnia, palmar-plantar erythrodysesthesia (hand/foot syndrome)

CV: Heart failure, palpitations, QT prolongation

GI: Anorexia, diarrhea, dyspepsia, mouth ulcerations, nausea, vomiting, xerosis

HEMA: Anemia, neutropenia, thrombocytopenia

INTEG: Rash

RESP: Dyspnea, pneumonitis

PHARMACOKINETICS

Bioavailability incomplete; peak 4 hr; >99% protein bound; extensively metabolized in the liver by P450 enzymes CYP3A4, CYP3A5; elimination half-life 24 hr; steady state 6-7 days; increased half-life in hepatic disease

INTERACTIONS

⚠ **Increase:** effect of lapatinib—CYP3A4 inhibitors (amiodarone, amprenavir, aprepitant, atazanavir, chloramphenicol, clarithromycin, conivaptan, dalfopristin, danazol, darunavir, delavirdine, diltiazem, efavirenz, erythromycin, estradiol, fluconazole, fluvoxaMINE, imatinib, indinavir, isoniazid, itraconazole, ketoconazole, miconazole, mifepristone, nefazodone, nelfinavir, propoxyphene, quinupristin, ritonavir, RU-486, saquinavir, telithromycin, troleandomycin, verapamil, voriconazole, zafirlukast); avoid concurrent use

⚠ **Increase:** QT prolongation—CYP3A4 inhibitors (amiodarone, clarithromycin, erythromycin, telithromycin, troleandomycin); class IA/III antidysrhythmics, arsenic trioxide, chlorproMAZINE, chloroquine, haloperidol, levomethadyl, mesoridazine, pentamine, thioridazine, CYP3A4 substrates (methadone, pimozide, QUEtiapine, quiNIDine, risperiDONE, terfenadine, ziprasidone)

NURSING CONSIDERATIONS

Assess:

- **Cardiac status:** ECG for QT prolongation, ejection fraction; chest pain, palpitations, dyspnea
- **Hepatic status:** liver function tests; jaundice of sclera, skin; dose should be reduced in hepatic disease

⚠ **Skin toxicities** NCI CTC grade ≥2 discontinue use in those with decreased left ventricular ejection fraction (LVEF) or for a LVEF that drops below institution's lower limit of normal; product may be restarted after 2 wk if LVEF recovers to normal at 1000 mg/day; restart at 1250 mg/day when toxicity improves to grade 1 or better

Evaluate:
• Therapeutic response: decrease in breast cancer progression

Teach patient/family:
• To take with a full glass of water, once a day, 1 hr before or after food; not to take with food or grapefruit products
• To take as directed only; if a dose is missed, to take as soon as remembered; if it is close to the next dose, to take only that dose; not to double
⚠ To report chest pain, difficulty breathing, fever, chills, sore throat, bleeding, bruising, yellow skin or eyes, severe fatigue, dizziness, palpitations to prescriber
• That other side effects that may occur, but do not need to be reported: nausea, diarrhea, heartburn, mouth sores, rash, numbness/pain in hands/feet
⚠ To use adequate contraception because the fetus could be damaged by this product (D)

latanoprost (ophthalmic)

(lah-tan′oh-prost)

Xalatan

Func. class.: Antiglaucoma agent
Chem. class.: Prostaglandin agonist

Do not confuse:
latanoprost/bimatoprost

ACTION: Increases aqueous humor outflow

USES: Increased intraocular pressure in those who have open-angle glaucoma/ocular hypertension and who do not respond to other IOP-lowering products

CONTRAINDICATIONS: Hypersensitivity to this product, benzalkonium chloride

Precautions: Eye infections, angle-closure glaucoma, renal/hepatic function impairment, children

DOSAGE AND ROUTES

• **Adult: OPHTH** Instill 1 drop in each affected eye (conjunctival sac) every night

Available forms: Ophthalmic solution 0.005%

Administer:

Ophthalmic route
• Wash hands before and after use; contact lenses should be removed before using the product, reinsert 15 min after use; contains benzalkonium chloride, which may be absorbed by soft contact lenses
• Tilt the head back slightly and pull the lower eyelid down with the index finger to form a pouch; squeeze the prescribed number of drops into the pouch and gently close the eyes for 1-2 min; do not blink; to avoid contamination, do not touch the tip of the dropper to the eye, fingertips, or other surface
• The solution may be used concomitantly with other topical ophthalmic drug products to lower IOP; if more than one topical ophthalmic drug is being used, the drugs should be administered at least 5 min apart
• Store unopened bottle refrigerated; once opened, it may be stored at room temperature, protected from light, for up to 6 wk

SIDE EFFECTS

EENT: *Conjunctival hyperemia, iris color change, ocular pruritus*, xerophthalmia, visual disturbance, ocular irritation burning, foreign body sensation, ocular pain, blepharitis, cataracts, and superficial punctate keratitis

INTEG: Rash

MISC: Flulike symptoms

PHARMACOKINETICS

Ophthalmic: Onset 3-4 hr, peak 8-12 hr; half-life 3 hr

NURSING CONSIDERATIONS

Assess:

- **Intraocular pressure:** in those with on-going increased IOP

Evaluate:

- Decreasing IOP

Teach patient/family:

Ophthalmic route

- To wash hands before and after use; that contact lenses should be removed before using product, reinsert 15 min after use; contains benzalkonium chloride, which may be absorbed by soft contact lenses
- To tilt the head back slightly and pull the lower eyelid down with the index finger to form a pouch; squeeze the prescribed number of drops into the pouch and gently close the eyes for 1-2 min; not to blink; to avoid contamination, do not touch the tip of the dropper to the eye, fingertips, or other surface
- That the solution may be used concomitantly with other topical ophthalmic drug products to lower IOP; if more than one topical ophthalmic drug is being used, the drugs should be administered at least 5 min apart
- To store unopened bottle refrigerated; that once opened, it may be stored at room temperature, protected from light, for up to 6 wk

leflunomide (Rx)

(leh-floo′noh-mide)

Arava

Func. class.: Antirheumatic (DMARDs)

Chem. class.: Immune modulator, pyrimidine synthesis inhibitor

ACTION:
Inhibits an enzyme involved in pyrimidine synthesis; has antiproliferative, antiinflammatory effect

USES:
RA: to reduce disease process and symptoms

Unlabeled uses: Juvenile RA

CONTRAINDICATIONS:
Breastfeeding, hypersensitivity, jaundice, lactase deficiency, hepatic disease

Black Box Warning: Pregnancy (X)

Precautions: Children, renal disorders, vaccinations, infection, alcoholism, immunosuppression

Black Box Warning: Hepatic disease

DOSAGE AND ROUTES

Rheumatoid arthritis

- **Adult: PO** Loading dose 100 mg/day × 3 days, maintenance 20 mg/day; may be decreased to 10 mg/day if not well tolerated

Juvenile rheumatoid arthritis (unlabeled)

- **Adolescent and child >40 kg: PO** 20 mg
- **Adolescent and child 20-40 kg: PO** 15 mg
- **Adolescent and child 10-19.9 kg: PO** 10 mg

Available forms: Tabs 10, 20 mg

Administer:

- Give loading dose of 100 mg/day × 3, then 20 mg/day; decrease to 10 mg/day if poorly tolerated
- With food for GI upset
- **Drug elimination:** give cholestyramine 8 g tid × 11 days, check levels

SIDE EFFECTS

CNS: *Headache,* dizziness, insomnia, depression, paresthesia, anxiety, migraine, neuralgia

CV: Palpitations, hypertension, chest pain, angina pectoris, peripheral edema

EENT: Pharyngitis, oral candidiasis, stomatitis, dry mouth, blurred vision

GI: *Nausea, anorexia, vomiting, constipation, flatulence, diarrhea, elevated LFTs,* **hepatotoxicity**

HEMA: Anemia, ecchymosis, hyperlipidemia

INTEG: Rash, pruritus, alopecia, acne, hematoma, herpes infections

RESP: Pharyngitis, rhinitis, bronchitis, cough, respiratory infection, pneumonia, sinusitis, interstitial lung disease
SYST: Opportunistic/fatal infections, Stevens-Johnson syndrome

PHARMACOKINETICS

Metabolized in liver to active metabolite, half-life of metabolite 2 wk, excreted in urine

INTERACTIONS

Increase: NSAID effect—NSAIDs
Increase: hepatotoxicity—hepatotoxic agents, methotrexate
Increase: leflunomide levels—rifampin
Decrease: antibody response—live virus vaccines
Decrease: leflunomide effect—activated charcoal, cholestyramine, use for overdose

NURSING CONSIDERATIONS

Assess:
• Screen for latent TB before starting treatment; if TB is present, pretreat before using product
• **Interstitial lung disease:** increased or worsening cough, SOB, fever; product may need to be discontinued
• **Arthritic symptoms:** ROM, mobility, swelling of joints at baseline and during treatment

Black Box Warning: Hepatic studies: if ALT elevations are > 2× ULN, reduce dose to 10 mg/day, monitor monthly or more frequently

• CBC with differential monthly × 6 mo, then q 6-8 wk thereafter; pregnancy test; serum electrolytes
⚠ **Infections:** fatal infections can occur
• B/P, weight; edema can occur

Black Box Warning: Pregnancy (X): determine that patient is not pregnant before treatment; not to be given to women of child-bearing potential that are not using reliable contraception

Evaluate:
• Therapeutic response: decreased inflammation, pain in joints
Teach patient/family:
• That product must be continued for prescribed time to be effective, that up to a month may be required for improvement
• To take with food, milk, or antacids to avoid GI upset; to take at same time of day
• To use caution when driving because drowsiness, dizziness may occur
• To take with a full glass of water to enhance absorption, may continue with correct prescribed treatment with other anti-inflammatories

Black Box Warning: Not to become pregnant (X) while taking this product; not to breastfeed while taking this product; men should also discontinue product and begin leflunomide removal protocol if pregnancy is planned

• That hair may be lost; review alternatives
• To avoid live virus vaccinations during treatment
• To notify prescriber of weight loss
• Overdose treatment: give cholestyramine 8 g tid × 11 days

letrozole (Rx)

(let′tro-zohl)

Femara

Func. class.: Antineoplastic, nonsteroidal aromatase inhibitor

ACTION: Binds to the heme group of aromatase; inhibits conversion of androgens to estrogens to reduce plasma estrogen levels

USES: Early, advanced, or metastatic breast cancer in postmenopausal women who are hormone receptor positive
Unlabeled uses: Infertility, idiopathic short stature, constitutional delayed puberty

CONTRAINDICATIONS:
Pregnancy (D), premenopausal females, hypersensitivity

Precautions: Respiratory/hepatic disease, osteoporosis

DOSAGE AND ROUTES

- **Adult: PO** 2.5 mg/day

Infertility (unlabeled)

- **Adult: PO** 2.5, 5, 7.5 mg/day × 5 days, usually days 3-7 of menstrual cycle

Idiopathic short stature, constitutional delayed puberty (unlabeled)

- **Adolescent and child ≥9 (male): PO** 2.5 mg/day; use with testosterone for delayed puberty

Available forms: Tabs 2.5 mg

Administer:

- Without regard to meals; with small glass of water
- May administer biphosphates to increase bone density

SIDE EFFECTS

CNS: *Headache, lethargy,* somnolence, dizziness, depression, anxiety

CV: Angina, MI, CVA, thromboembolic events, hypertension, peripheral edema

GI: *Nausea, vomiting, anorexia,* constipation, heartburn, diarrhea

GU: Endometrial cancer, vaginal bleeding, endometrial proliferation disorders

INTEG: *Rash, pruritus,* alopecia, sweating

MISC: Hot flashes, night sweats, **second malignancies, anaphylaxis, angioedema**

MS: Arthralgia, arthritis, bone fracture, myalgia, osteoporosis

RESP: Dyspnea, cough

PHARMACOKINETICS

Metabolized in liver, excreted in urine, peak 2 days, terminal half-life 48 hr, steady state 2-6 wk

INTERACTIONS

Decrease: letrozole effect—estrogens, oral contraceptives

NURSING CONSIDERATIONS

Assess:

- Hepatic studies before, during therapy (bilirubin, AST, ALT, LDH) as needed or monthly

Evaluate:

- Therapeutic response: decrease in size of tumor

Teach patient/family:

- **To report allergic reactions (rash; hives; difficulty breathing; tightness in chest; swelling of mouth, face, lips, tongue)**
- **To report vaginal bleeding, diarrhea, chest/bone pain**
- **To use adequate contraception in perimenopausal, recently postmenopausal women; pregnancy (D)**

leucovorin (Rx)

(loo-koe-vor′in)

Func. class.: Vitamin, folic acid/methotrexate antagonist antidote

Chem. class.: Tetrahydrofolic acid derivative

Do not confuse:

leucovorin/Leukeran/leukine

folinic acid/folic acid

ACTION:
Needed for normal growth patterns; prevents toxicity during antineoplastic therapy by protecting normal cells

USES:
Megaloblastic or macrocytic anemia caused by folic acid deficiency, overdose of folic acid antagonist, methotrexate/pyrimethamine/ trimetrexate/trimethoprim toxicity, pneumocystosis, toxoplasmosis

CONTRAINDICATIONS:
Hypersensitivity to this product or folic acid, benzyl alcohol; anemias other than megaloblastic not associated with vit B_{12} deficiency

Precautions: Pregnancy (C), neonates, breastfeeding, geriatric patients, seizures, stomatitis, vomiting

DOSAGE AND ROUTES

Megaloblastic anemia caused by enzyme deficiency

• **Adult and child: PO/IV/IM** up to 6 mg/day

Megaloblastic anemia caused by deficiency of folate

• **Adult and child: IM** ≤1 mg/day until adequate response

Methotrexate toxicity/leucovorin rescue

• **Adult and child: PO/IM/IV Normal elimination** given 6 hr after dose of methotrexate (10 mg/m^2) until methotrexate $<5 \times 10^{-8}$ m, CCr >50% above prior level, or methotrexate level 5×10^{-8} m at 24 hr or $>9 \times 10^{-8}$ m at 48 hr; give leucovorin 100 mg/m^2 q3hr until level drops to $<10^{-8}$ m

Pyrimethamine/trimethoprim toxicity

• **Adult and child: PO/IM** 5-15 mg/day

Advanced colorectal cancer

• **Adult: IV** 200 mg/m^2, then 5-FU 370 mg/m^2 or leucovorin 20 mg/m^2, then 5-FU 425 mg/m^2; give daily × 5 days q4-5wk

Available forms: Tabs 5, 10, 15, 25 mg; inj 3, 5 mg/ml; powder for inj 10 mg/ml

Administer:

• Within 1 hr of folic acid antagonist
• Do not give concurrently with systemic methotrexate
• Increased fluid intake if used to treat folic acid inhibitor overdose
• Protection from light and heat

IM route

• No reconstitution needed
• For treatment of megaloblastic anemia

IV route

• Reconstitute 50 mg/5 ml bacteriostatic or sterile water for inj (10 mg/ml) or (100 mg/10 ml); use immediately if sterile water used

Direct IV route

• Give over 160 mg/min or less (16 ml of 10 mg/ml sol/min)

Intermittent IV INF route

• Give after diluting in 100-500 ml of 0.9% NaCl, D_5W, $D_{10}W$, LR, Ringer's sol

Y-site compatibilities: Acyclovir, alemtuzumab, alfentanil, allopurinol, amifostine, amikacin, aminocaproic acid, aminophylline, ampicillin, ampicillin-sulbactam, anidulafungin, argatroban, atenolol, atracurium, azithromycin, aztreonam, bivalirudin, bleomycin, bumetanide, buprenorphine, busulfan, butorphanol, calcium acetate, chloride/gluconate, carmustine, caspofungin, ceFAZolin, cefepime, cefotaxime, cefoTEtan, cefOXitin, cefTAZidime, ceftizoxime, cefuroxime, chloramphenicol, cimetidine, ciprofloxacin, cisatracurium, CISplatin, cladribine, clindamycin, codeine, cyclophosphamide, cycloSPORINE, cytarabine, DACTINomycin, DAPTOmycin, DAUNOrubicin liposome, dexamethasone, dexmedetomidine, dexrazoxane, digoxin, diltiazem, diphenhydrAMINE, DOBUTamine, DOCEtaxel, dolasetron, DOPamine, doxacurium, DOXOrubicin, DOXOrubicin liposomal, doxycycline, enalaprilat, ePHEDrine, EPINEPHrine, eptifibatide, ertapenem, erythromycin, esmolol, etoposide, etoposide phosphate, famotidine, fenoldopam, fentaNYL, filgrastim, fluconazole, fludarabine, fluorouracil, fosphenytoin, furosemide, gallium nitrate, ganciclovir, gatifloxacin, gemcitabine, gentamicin, glycopyrrolate, granisetron, haloperidol, heparin, hydrALAZINE, hydrocortisone, HYDROmorphone, hydrOXYzine, IDArubicin, ifosfamide, imipenem-cilastatin, inamrinone, insulin, regular, irinotecan, isoproterenol, ketorolac, labetalol, lactated Ringer's injection, lepirudin, levofloxacin, lidocaine, linezolid injection, LORazepam, magnesium sulfate, mannitol, mechlorethamine, melphalan, meperidine, meropenem, mesna, metaraminol, methotrexate, methyldopate, metoclopramide, metoprolol, metroNIDAZOLE, midazolam, milrinone, minocycline, mitoMYcin, mitoXANtrone, mivacurium, morphine, mycophenolate mofetil, nafcillin, nalbuphine, nesiritide, niCARdipine, nitroglycerin, nitroprusside, norepinephrine, octreotide, ondansetron, oxaliplatin, oxytocin, PACLitaxel (solvent/

surfactant), palonosetron, pancuronium, PEMEtrexed, pentazocine, PENTobarbital, PHENobarbital, phentolamine, phenylephrine, piperacillin-tazobactam, polymyxin B, potassium acetate/chloride, procainamide, prochlorperazine, promethazine, propranolol, quiNIDine gluconate, ranitidine, remifentanil, riTUXimab, rocuronium, sodium acetate, sodium phosphates, streptozocin, succinylcholine, SUFentanil, sulfamethoxazole-trimethoprim, tacrolimus, teniposide, theophylline, thiotepa, ticarcillin, ticarcillin-clavulanate, tigecycline, tirofiban, TNA (3-in-1) total nutrient admixture, tobramycin, TPN (2-in-1) total parenteral nutrition admixture, trastuzumab, trimethobenzamide, vasopressin, vecuronium, verapamil, vinBLAStine, vinCRIStine, vinorelbine, voriconazole, zidovudine

SIDE EFFECTS

HEMA: Thrombocytosis (intraarterial)
INTEG: Rash, pruritus, erythema, urticaria
RESP: Wheezing

INTERACTIONS

Increase: metabolism of barbiturates, hydantoins, primidone
Decrease: effect of—sulfamethoxazole, trimethoprim

NURSING CONSIDERATIONS

Assess:

- CCr, creatinine before leucovorin rescue, daily to detect nephrotoxicity; methotrexate level
- I&O; urine pH q6hr, maintain >7 to prevent neurotoxicity; watch for nausea and vomiting
- **Other products taken:** alcohol, hydantoins, trimethoprim may cause increased folic acid use by body
- Neurologic status (rescue): weakness, fatigue
- Monitor calcium levels
- Megaloblastic anemia, plasma lactic acid, leticulocyte count, Hct, Hgb
- **Hypersensitivity:** rash, urticaria, wheezing; notify prescriber

Evaluate:

- Therapeutic response: increased weight; improved orientation, well-being; absence of fatigue; reversal of toxicity (methotrexate, folic acid antagonist overdose)

Teach patient/family:

- **For leucovorin rescue** to drink 3 L fluid/day of rescue
- **For folic acid deficiency,** to eat folic-acid–rich foods: bran; yeast; dried beans; nuts; fresh, green leafy vegetables
- To take product exactly as prescribed
- To notify prescriber of side effects
- To report signs of hyposensitivity reaction immediately
- To avoid breastfeeding

⚠ HIGH ALERT

leuprolide (Rx)

(loo-proe′lide)

Eligard, Lupron Depot, Lupron Depot-Ped, Viadur

Func. class.: Antineoplastic hormone
Chem. class.: Gonadotropin-releasing hormone

Do not confuse:
Lupron/Nuprin/Lopurin

ACTION: Causes initial increase in circulating levels of LH, FSH; continuous administration results in decreased LH, FSH; in men, testosterone is reduced to castrate levels; in premenopausal women, estrogen is reduced to menopausal levels

USES: Metastatic prostate cancer (inj implant), management of endometriosis, central precocious puberty, uterine leiomyomata (fibroids)
Unlabeled uses: Breast cancer, recurrent priapism, benign prostatic hyperplasia

CONTRAINDICATIONS:

Pregnancy (X), breastfeeding, hypersensitivity to GnRH or analogs, thromboembolic disorders, undiagnosed vaginal bleeding; Viadur implant or Eligard should not be used in women, children

Precautions: Edema, hepatic disease, CVA, MI, seizures, hypertension, diabetes mellitus, CHF, depression, osteoporosis, spinal cord compression, urinary tract obstruction

DOSAGE AND ROUTES

Prostate cancer

- **Adult: SUBCUT** 1 mg/day; **IM** 7.5 mg/dose monthly; Viadur implant (72 mg) yearly; or **IM** 22.5 mg q3mo; or **IM** 30 mg q4mo; or IM 45 mg q6mo

Endometriosis/fibroids

- **Adult: IM** 3.75 mg monthly for 6 mo or 11.25 q3mo for 6 mo or 30 mg q4mo

Central precocious puberty

- **Child: SUBCUT** 50 mcg/kg/day; may increase by 10 mcg/kg/day as needed
- **Child >37.5 kg: IM** 15 mg q4wk
- **Child 25-37.5 kg: IM** 11.25 mg q4wk
- **Child ≤25 kg:** 7.5 mg q4wk

Benign prostatic hyperplasia (BPH) (unlabeled)

- **Adult: SUBCUT** (sol for inj) 1 mg/day, must be cont; IM (inj susp) 3.75 mg q28day × 24 wk

Available forms: Powder for inj depot 1 mo 3.75 mg, depot 3 mo/1.25 mg, 22.5 mg; depot 4 mo 30 mg; depot 6 mo 45 mg; depot-ped 7.5, 11.25, 15 mg; sol for inj 1 mg/0.2 ml; susp for inj 7.5, 22.5, 30, 45 mg; implant (Viadur) 65 mg

Administer:

- Store in tight container at room temperature
- **SUBCUT:** No dilution needed if patient self-administering; make sure patient using syringes provided by manufacturer
- **Viadur DUROS Implant:** Insert in inner aspect of arm, remove after 12 mo
- **SUBCUT: Eligard:** bring to room temperature, once mixed, give within 30 min, prepare the 2 syringes for mixing, join the 2 syringes together by pushing in and twisting until secure; mix the product by pushing the contents of both syringes back and forth between syringes until uniform; should be light tan to tan, hold syringes vertically with syringe B on the bottom, draw entire mixed product into syringe B (short, wide syringe) by depressing the syringe A plunger and slightly withdrawing syringe B plunger, uncouple syringe A, while pushing down on syringe A plunger, small air bubbles will remain, hold syringe B upright, remove pink cap, attach needle cartridge to the end of syringe B, remove needle cover, give by subcut

IM route

- **Monthly:** reconstitute single-use vial with 1 ml of diluent; if multiple vials used, withdraw 0.5 ml, inject into each vial (1 ml); withdraw all, inject at 90-degree angle (3.75 mg)
- **3-mo:** reconstitute microspheres using 1.5 ml of diluent, inject into vial; shake, withdraw, inject
- **12-mo:** insert into upper arm; at the end of 12 mo, implant must be removed

SIDE EFFECTS

CNS: Memory impairment, depression, seizures

CV: MI, PE, dysrhythmias, peripheral edema

GI: Nausea, vomiting, anorexia, diarrhea, GI bleeding

GU: Edema, hot flashes, impotence, decreased libido, amenorrhea, vaginal dryness, gynecomastia, profuse vaginal bleeding

INTEG: Alopecia

MS: Bone pain

RESP: Dyspnea, pulmonary fibrosis, interstitial lung disease

PHARMACOKINETICS

SUBCUT: Onset 1-2 wk; peak 2-4 wk; absorbed rapidly (SUBCUT), slowly (IM depot); half-life 3 hr

INTERACTIONS

Increase: antineoplastic action—flutamide, megestrol

Drug/Herb

- Do not use with black cohosh or chaste tree fruit, may interfere with treatment

NURSING CONSIDERATIONS

Assess:

- **Prostate cancer:** increased bone pain for first 4 wk of treatment; those with metastases in spinal column may exhibit severe back pain
- **Symptoms of endometriosis** (lower abdominal pain)/**fibroids** (pelvic pain, excessive vaginal bleeding, bloating) before, during, after treatment
- **Central precocious puberty (CPP)** diagnosis should have been confirmed by secondary S_4 characteristics in children <9 yr, estradiol/testosterone levels, GnRH test, tomography of head, adrenal steroids, chorionic gonadotropin, wrist x-ray, height, weight
- Hepatic studies before, during therapy (bilirubin, AST, ALT, LDH) monthly, as needed; PSA, calcium, testosterone with prostate cancer; bone mineral density; blood glucose, HbA1c
- Pituitary gonadotropic and gonadal function during therapy and 4-8 wk after therapy decreased
- **Tumor flare:** worsening of signs and symptoms; normal during beginning therapy
- Fatigue, increased pulse, pallor, lethargy; edema in feet, joints; stomach pain

⚠ **Severe allergic reaction:** rash, pruritus, urticaria, purpuric skin lesions, itching, flushing

Evaluate:

- Therapeutic response: decreased tumor size and spread of malignancy; decrease in lesions, pain with endometriosis, fibroids, correction of CPP; increased follicle maturation

Teach patient/family:

- To notify prescriber if menstruation continues; menstruation should stop
- To notify prescriber if pregnancy is planned or suspected (X), avoid breastfeeding
- That bone pain will disappear after 1 wk
- To report any complaints, side effects to nurse, prescriber; hot flashes may occur; record weight, report gain of >2 lb/day
- How to prepare, give; to rotate sites for SUBCUT inj
- To keep accurate records of dose
- **That tumor flare may occur:** increase in size of tumor, increased bone pain, will subside rapidly; may take analgesics for pain
- That voiding problems may increase during beginning of therapy but will decrease in several weeks

levalbuterol (Rx)

(lev-al-byoo′ter-ole)

Xopenex, Xopenex HFA

Func. class.: Bronchodilator, adrenergic β_2-agonist

ACTION: Causes bronchodilation by action on β_2 (pulmonary) receptors by increasing levels of cAMP, which relaxes smooth muscle; produces bronchodilation, CNS, cardiac stimulation as well as increased diuresis and gastric acid secretion

USES: Treatment or prevention of bronchospasm (reversible obstructive airway disease)

CONTRAINDICATIONS: Hypersensitivity to sympathomimetics, this product, albuterol

Precautions: Pregnancy (C), breastfeeding, hyperthyroidism, diabetes mellitus, hypertension, prostatic hypertrophy, angle-closure glaucoma, seizures, renal disease, QT prolongation, tachydysrhythmias, severe cardiac disease, hypokalemia

DOSAGE AND ROUTES

- **Adult/child ≥12 yr: INH** 0.63 mg tid q6-8hr by nebulization, may increase 1.25 mg q8hr
- **Adult/adolescent/child >4 yr:** (HFA, metered dose) 90 mcg (2 **INH**) q4-6hr
- **Child 6-11 yr: INH** 0.31 mg tid by nebulization, max 0.63 mg tid

Available forms: Sol, inh pediatric 0.31/3 ml; 0.63 mg, 1.25 mg/3 ml; 45 mcg per actuation (HFA)

Administer:

- Every 6-8hr; wait ≥1 min between inhalation of aerosols

Inhalation route

- Shake well before use, use a spacer device, prime with 4 test sprays in new canister or when not used for >3 days

SIDE EFFECTS

CNS: *Tremors, anxiety,* insomnia, *headache,* dizziness, stimulation, *restlessness,* irritability, weakness

CV: Palpitations, tachycardia, hypertension, angina, hypotension, dysrhythmias, QT prolongation

EENT: Dry nose, irritation of nose and throat, rhinitis

GI: Heartburn, nausea, vomiting, diarrhea

INTEG: Rash

META: *Hypokalemia, hyperglycemia*

MS: Muscle cramps

RESP: Cough

SYST: Anaphylaxis, angioedema

PHARMACOKINETICS

Metabolized in the liver and tissues; crosses placenta, breast milk, blood-brain barrier; half-life 3.3-4 hr

INH sol: Onset 10-17 min, peak 1½ hr, duration 5-6 hr; **INH aerosol:** onset 4.5-10.2 min, peak 76-78 min, duration ≤6 hr

INTERACTIONS

Increase: QT prolongation—Class IA, III antidysrhythmics, usually at high doses or hypokalemia

Increase: action of aerosol bronchodilators

Increase: levalbuterol action—tricyclics, MAOIs, other adrenergics

Decrease: levalbuterol action—other β-blockers, severe bronchospasm may occur

Drug/Herb

Increase: stimulation—black/green tea, coffee, cola nut, guarana, yerba maté

NURSING CONSIDERATIONS

Assess:

- **Respiratory function:** vital capacity, forced expiratory volume, ABGs, lung sounds, heart rate and rhythm (baseline); character of sputum: color, consistency, amount
- Cardiac status: palpitations, increase/decrease in B/P, dysrhythmias
- ⚠ For evidence of allergic reactions, paradoxic bronchospasm, anaphylaxis, angioedema

Evaluate:

- Therapeutic response: absence of dyspnea, wheezing after 1 hr; improved airway exchange, ABGs

Teach patient/family:

- Not to use OTC medications because excess stimulation may occur
- To avoid getting aerosol in eyes because blurring may result
- To avoid smoking, smoke-filled rooms, persons with respiratory infections
- ⚠ That paradoxic bronchospasm may occur; to stop product immediately, contact prescriber
- To limit caffeine products such as chocolate, coffee, tea, colas, and herbs such as cola nut, guarana, yerba maté
- To use this product first if using other inhalers; to wait 5 min or more between products; to rinse mouth with water after each dose to prevent dry mouth

TREATMENT OF OVERDOSE:

Administer a β_1-adrenergic blocker

levETIRAcetam (Rx)

(lev-eh-teer-ass′eh-tam)

Keppra, Keppra XR

Func. class.: Anticonvulsant

Do not confuse:
Keppra/Kaletra

ACTION: Unknown; may inhibit nerve impulses by limiting influx of sodium ions across cell membrane in motor cortex

USES: Adjunctive therapy for partial-onset seizures, primary generalized tonic-clonic seizures, myoclonic seizures in juvenile patients
Unlabeled use: Pediatrics

CONTRAINDICATIONS: Hypersensitivity, breastfeeding
Precautions: Pregnancy (C), children, geriatric patients, renal/cardiac disease, psychosis

DOSAGE AND ROUTES

Adjunctive treatment of partial seizures

• **Adult/adolescent ≥16 yr: IV** 500 mg bid, may be titrated by 1000 mg/day q2wk, max 3000 mg/day in divided doses; **EXT REL** 1000 mg/day, may increase q2wk, max 3000 mg/day

• **Adolescent <16 yr/child/infant: PO** 10 mg/kg bid, increase daily dose q2wk by 20 mg/kg to dose of 30 mg/kg bid; if patient unable to tolerate, may reduce dose

Myoclonic seizures/tonic-clonic seizures/partial seizures

• **Adult/adolescent ≥16 yr: PO/IV** 500 mg bid, may increase by 1000 mg/day q2wk, max 3000 mg/day

Renal dose

• **Adult: PO** CCr 50-80 ml/min, 500-1000 mg q12hr or **EXT REL** 1000-2000 q24hr, max 2000 mg/day; CCr 30-49 ml/min, 250-750 mg q12hr or **EXT REL** 500-1500 mg q24hr, max 1500 mg/day; CCr <30 ml/min, 250-500 mg q12hr or **EXT REL** 500-1000 q24hr, max 1000 mg/day

Available forms: Tabs 250, 500, 750, 1000 mg; oral sol 100 mg/ml; sol for inj 100 mg/ml; ext rel tab 500, 750 mg; 1000 mg/100 ml 0.75% NaCl, 1500 mg/100 ml 0.54% NaCl, 500 mg/100 ml 0.82% NaCl

Administer:

PO route

• Extended release product should not be used in dialysis patients

• Swallow tab whole; do not break, crush, or chew

• With food, milk to decrease GI symptoms (rare)

• Store at room temperature (PO)

• **Child:** <20 kg should be given oral solution; use calibrated device

Intermittent IV INF route

• Single-use vials: dilute in 100 ml of 0.9% NaCl, D_5, LR; give over 15 min, discard unused vial contents, do not use product with particulates or discoloration

• Diluted preparation stable for 24 hr at room temperature in polyvinyl bags

Additive compatibilities: diazepam, LORazepam, valproate

SIDE EFFECTS

CNS: Dizziness, somnolence, asthenia, psychosis, **suicidal ideation**, nonpsychotic behavioral symptoms, headache, ataxia

EENT: Diplopia, conjunctivitis

GI: Nausea, vomiting, anorexia, diarrhea, constipation, **hepatitis**

HEMA: Lowered Hct, Hgb, RBC, infection, leukopenia

INTEG: Pruritus, rash

MISC: Infection, abdominal pain, pharyngitis

SYST: **Stevens-Johnson syndrome, toxic epidermal necrolysis**; dehydration (child <4 yr)

PHARMACOKINETICS

Rapidly absorbed; not protein bound; excreted via kidneys 66% unchanged;

half-life 6-8 hr, longer in geriatric patients or with renal disease

INTERACTIONS

- Avoid use with alcohol
- Possible increased carBAMazepine toxicity: carBAMazepine

Decrease: levETIRAcetam absorption—sevelamer; separate by 1 hr before, 3 hr after sevelamer

Drug/Lab Test

Increase: eosinophils

NURSING CONSIDERATIONS

Assess:

- **Seizures:** type, location, duration, character; provide seizure precautions
- Renal studies: urinalysis, BUN, urine creatinine q3mo
- Blood studies: CBC, LFTs

⚠ Mental status: mood, sensorium, affect, behavioral changes, **suicidal thoughts/behaviors;** if mental status changes, notify prescriber

- Assistance with ambulation during early part of treatment; dizziness occurs

Evaluate:

- Therapeutic response: decreased seizure activity; document on patient's chart

Teach patient/family:

- To carry emergency ID stating patient's name, products taken, condition, prescriber's name, phone number
- How to use oral sol; if trouble swallowing, measure oral sol in medicine cup or dropper, do not use teaspoon
- To notify prescriber if pregnant, intending to become pregnant
- To avoid driving, other activities that require alertness
- Not to discontinue medication quickly after long-term use because withdrawal seizure may occur
- Not to breastfeed, excreted in breast milk
- To report suicidal thoughts

levobetaxolol ophthalmic

See Appendix B

levobunolol (ophthalmic)

(lee′voe-byoo′no-lahl)

Betagan

Func. class.: Antiglaucoma

Chem. class.: β-Blocker

ACTION: Can decrease aqueous humor and increase outflows

USES: Treatment of chronic open-angle glaucoma and ocular hypertension

CONTRAINDICATIONS: Hypersensitivity, AV block, heart failure, bradycardia, sick sinus syndrome, asthma

Precautions: Abrupt discontinuation, children, pregnancy, breastfeeding, COPD, depression, diabetes mellitus, myasthenia gravis, hyperthyroidism, pulmonary disease, sulfite sensitivity, angle-closure glaucoma

DOSAGE AND ROUTES

- **Adult:** Instill 1-2 drops in the affected eyes once or twice a day (0.5% solution), bid (0.25% solution)

Available forms: Ophthalmic solution 0.25%, 0.5%

Administer:

- For ophthalmic use only
- Do not touch the tip of the dropper to the eye, fingertips, or other surface to prevent contamination
- Wash hands before and after use; tilt head back slightly and pull the lower eyelid down with the index finger to form a pouch; squeeze the prescribed number of drops into the pouch; close eyes to spread drops; to avoid excessive systemic absorption, apply finger pressure on the lacrimal sac for 1-2 min after use

• If more than one topical ophthalmic drug product is being used, the drugs should be administered at least 5 min apart
• To avoid contamination or the spread of infection, do not use dropper for more than one person
• Decreased intraocular pressure can take several weeks, monitor IOP after a month

SIDE EFFECTS

CNS: Insomnia, headache, dizziness
CV: Palpitations
EENT: Eye stinging/burning, tearing, photophobia
PULM: Bronchospasm

PHARMACOKINETICS

Onset 60 min, peak 2-6 hr, duration 24 hr

INTERACTIONS

Increase: β-blocking effect—oral β-blockers
Increase: Intraocular pressure reduction—topical miotics, dipivefrin, EPINEPHrine, carbonic anhydrase inhibitors; this may be beneficial
Increase: B/P, severe—when abruptly stopping cloNIDine
Increase: Depression of AV nodal conduction, bradycardia, or hypotension—adenosine, cardiac glycosides, disopyramide, other antiarrhythmics, class 1C antiarrhythmic drugs (flecainide, propafenone, moricizine, encainide, quiNIDine, or drugs that significantly depress AV nodal conduction
Increase: AV block nodal conduction, induce AV block—high doses of procainamide
Increase: Antihypertensive effect—other antihypertensives

NURSING CONSIDERATIONS

Assess:
⚠ **Systemic absorption:** When used in the eye, systemic absorption is common with the same adverse reactions and interactions
• **Glaucoma:** Monitor intraocular pressure

Evaluate:
• Decreasing intraocular pressure

Teach patient/family:
• That product is for ophthalmic use only
• Not to touch the tip of the dropper to the eye, fingertips, or other surface to prevent contamination
• To wash hands before and after use; tilt the head back slightly and pull the lower eyelid down with the index finger to form a pouch; squeeze the prescribed number of drops into the pouch; close eyes to spread drops; to avoid excessive systemic absorption by applying finger pressure on the lacrimal sac for 1-2 min following use
• That if more than one topical ophthalmic drug product is being used, the drugs should be administered at least 5 min apart
• To avoid contamination or the spread of infection by not using dropper for more than one person

levocabastine ophthalmic

See Appendix B

levocetirizine (Rx)

(lee-voh-she-teer′ah-zeen)

Xyzal

Func. class.: Antihistamine, low sedating
Chem. class.: H_1 histamine blocker, low sedating

ACTION:

Acts on blood vessels, GI, respiratory system by competing with histamine for H_1-receptor site; decreases allergic response by blocking pharmacologic effects of histamine; minimal anticholinergic action

USES:
Perennial or seasonal rhinitis, allergy symptoms, chronic idiopathic urticaria

CONTRAINDICATIONS:
Breastfeeding; children 6-11 yr with renal disease; end-stage renal disease; dialysis; hypersensitivity to this product, cetirizine, hydrOXYzine

Precautions: Pregnancy (B), driving, renal disease

DOSAGE AND ROUTES

- **Adult and child ≥12 yr: PO** 2.5-5 mg/day in the evening
- **Child 6-11 yr: PO** (oral solution) 2.5 mg/day in the evening
- **Child 2-5 yr: PO** (oral solution) 1.25 mg/day in the evening
- **Geriatric: PO** 2.5-5 mg/day in the evening

Renal dose

- **Adult: PO** CCr 50-80 ml/min, 2.5 mg/day; CCr 30-50 ml/min, 2.5 mg every other day; CCr 10-30 ml/min, 2.5 mg 2×/wk; CCr <10 ml/min, do not use

Available forms: Tabs 5 mg; oral sol 2.5 mg/15 ml

Administer:

- Without regard to meals in the evening; tabs scored, may be broken in half
- Store in tight, light-resistant container

SIDE EFFECTS

CNS: *Drowsiness, fatigue,* asthenia, dizziness

GI: Dry mouth, increase LFTs, hepatitis

MISC: Urinary retention

PHARMACOKINETICS

Rapid absorption; peak 0.9 hr; protein binding 91%-92%; half-life 8 hr; excreted in urine 85.4%, feces 12.9%

INTERACTIONS

Increase: CNS depression—alcohol, other CNS depressants

Increase: anticholinergic/sedative effect—MAOIs, phenothiazines, tricyclics

Decrease: clearance of levocetirizine—ritonavir

Drug/Lab Test

False negative: Skin allergy tests

NURSING CONSIDERATIONS

Assess:

- **Allergy symptoms:** pruritus, urticaria, watering eyes at baseline, during treatment
- **Respiratory status:** rate, rhythm, increase in bronchial secretions, wheezing, chest tightness
- Liver function tests, serum creatinine, BUN

Evaluate:

- Therapeutic response: absence of running or congested nose or rashes

Teach patient/family:

- About all aspects of product use; to notify prescriber if confusion, sedation, hypotension occur, not to exceed recommended dose
- To avoid driving, other hazardous activities if drowsiness occurs
- To avoid alcohol, other CNS depressants
- That product not recommended while breastfeeding

TREATMENT OF OVERDOSE:
Administer diazepam, vasopressors, IV phenytoin

levodopa-carbidopa (Rx)

(lee-voe-doe′pa)-(kar-bi-doe′pa)

Apo-Levocarb ✱, Atamet, Parcopa, Sinemet, Sinemet CR

Func. class.: Antiparkinson agent

Chem. class.: Catecholamine

ACTION:
Decarboxylation of levodopa in periphery is inhibited by carbidopa; more levodopa is made available for transport to the brain and for conversion to DOPamine in the brain

USES:
Parkinson's disease, parkinsonism resulting from carbon monoxide, chronic manganese intoxication, cerebral arteriosclerosis

Unlabeled uses: Restless leg syndrome

CONTRAINDICATIONS:
Hypersensitivity, malignant melanoma, history of malignant melanoma or undiagnosed skin lesions resembling melanoma

Precautions: Pregnancy (C), breastfeeding, diabetes, closed-angle glaucoma, respiratory/cardiac/renal/hepatic disease, MI with dysrhythmias, seizures, peptic ulcer, depression

DOSAGE AND ROUTES

Beginning therapy for those not taking levodopa

• **Adult: PO** 25 mg carbidopa/100 mg levodopa tid, may increase daily or every other day by 1 tab to desired response (8 tabs/day); **EXT REL** tabs 50 mg carbidopa/200 mg levodopa bid

For those not taking levodopa ER

• 50 mg carbidopa/200 mg levodopa bid

For those taking levodopa ER

• Begin treatment with 10% more levodopa/day given q4-8hr, may increase or decrease dose q3days

For those taking levodopa <1.5 g/day

• **Adult: PO** 25 mg carbidopa/100 mg levodopa tid-qid, may increase daily to desired response

For those taking levodopa >1.5 g/day

• **Adult: PO** 25 mg carbidopa/250 mg levodopa tid-qid, may increase daily to desired response

Restless leg syndrome (RLS) (unlabeled)

• **Adult: PO** 25 mg carbidopa/100 mg levodopa, 1 tab at bedtime, may repeat if awakening within 2 hr or 50 mg carbidopa/200 mg levodopa sus rel tab 1-2 tabs 1 hr before bedtime

Available forms: Tabs 10 mg carbidopa/100 mg levodopa, 25 mg carbidopa/100 mg levodopa, 25 mg carbidopa/250 mg levodopa; ext rel tab 25 mg/100 mg, 50 mg carbidopa/200 mg levodopa (Sinemet CR); oral disintegrating tab (Parcopa) 10 mg carbidopa/100 mg levodopa, 25 mg carbidopa/100 mg levodopa; 25 mg carbidopa/250 mg levodopa

Administer:

• Pyridoxine (B_6) not effective for reversing Sinemet or Sinemet CR

PO route

• Do not crush or chew **ext rel tabs;** they may be broken in half; adjust dosage to response

• **Oral disintegrating tab** by gently removing from bottle, placing on tongue and swallowing with saliva; after tab dissolves, liquid is not necessary

• With meals if GI symptoms occur; limit protein taken with product

• Only after nonselective MAOIs have been discontinued for 2 wk; if patient has been previously treated with levodopa, discontinue for at least 12 hr before change to carbidopa-levodopa

SIDE EFFECTS

CNS: *Involuntary choreiform movements, hand tremors, fatigue, headache, anxiety, twitching, numbness, weakness, confusion, agitation, insomnia, nightmares,* psychosis, hallucination, hypomania, severe depression, dizziness, impulsive behaviors, **neuroleptic malignant syndrome**

CV: *Orthostatic hypotension,* tachycardia, hypertension, palpitation

EENT: Blurred vision, diplopia, dilated pupils

GI: *Nausea, vomiting, anorexia, abdominal distress, dry mouth, flatulence, dysphagia,* bitter taste, diarrhea, constipation

HEMA: **Hemolytic anemia, leukopenia, agranulocytosis**

INTEG: Rash, sweating, alopecia

MISC: Urinary retention, incontinence, weight change, dark urine

L

PHARMACOKINETICS

PO: Onset 30 min, peak 1-3 hr, excreted in urine (metabolites)
EXT REL: Onset 4-6 hr

INTERACTIONS

- Hypertensive crisis: nonselective MAOIs

Increase: effects of levodopa—antacids, metoclopramide
Decrease: effects of levodopa—anticholinergics, hydantoins, papaverine, pyridoxine, benzodiazepines, antipsychotics

Drug/Lab Test
Increase: BUN, AST, ALT, bilirubin, alk phos, LDH, serum glucose
Decrease: BUN, creatinine, uric acid
False positive: urine ketones (dipstick), Coombs' test
False negative: urine glucose
False increase: urine protein

Drug/Food
Decrease: absorption of levodopa—protein

NURSING CONSIDERATIONS

Assess:

- **Parkinson's symptoms:** tremors, pill rolling, drooling, akinesia, rigidity, shuffling gait before, during treatment
- B/P, respiration; orthostatic B/P
- Mental status: affect, mood, behavioral changes, depression, complete suicide assessment
- **Toxicity:** muscle twitching, blepharospasm
- Renal, hepatic, hematopoietic tests; also for diabetes, acromegaly if on long-term therapy

Evaluate:

- Therapeutic response: decrease in akathisia/bradykinesis, tremor, rigidity, improved mood

Teach patient/family:

- To change positions slowly to prevent orthostatic hypotension
- To report side effects: twitching, eye spasms because these indicate overdose
- To use product as prescribed; if discontinued abruptly, parkinsonian crisis, neuroleptic malignant syndrome (NMS) may occur; to gradually taper
- That urine, sweat may darken
- To use physical activities to maintain mobility, lessen spasms
- That improvement may not occur for 2-4 mo; about "on-off phenomenon"

levofloxacin (Rx)

(lee-voh-floks'a-sin)

Levaquin

Func. class.: Antiinfective
Chem. class.: Fluoroquinolone

ACTION:
Interferes with conversion of intermediate DNA fragments into high-molecular-weight DNA in bacteria; DNA gyrase inhibitor; inhibits topoisomerase IV

USES:
Acute sinusitis, acute chronic bronchitis, community-acquired pneumonia, uncomplicated skin infections, UTI, cellulitis, prostatitis, inhalational anthrax (postexposure); acute pyelonephritis caused by *Streptococcus pneumoniae, Haemophilus influenzae, Haemophilis parainfluenzae, Moraxella catarrhalis, Escherichia coli, Serratia marcescens, Klebsiella pneumoniae, Chlamydia pneumoniae, Legionella pneumophilia, Mycoplasma pneumoniae, Enterococcus faecalis, Staphylococcus epidermidis, Staphylococcus pyogenes;* inhalation anthrax in children

Unlabeled uses: Adnexitis, Bartholin abscess, bartholinitis, cervicitis, epididymis, gastroenteritis, *H. pylori* eradication, mastitis, MAC, nongonococcal urethritis, obstetric infections, PID, plague, SARS, shigellosis, TB, typhoid fever, disseminated; otitis media, otitis externa, tonsillitis, pharyngitis, sialadenitis

CONTRAINDICATIONS:
Hypersensitivity to quinolones

Precautions: Pregnancy (C), breastfeeding, children, photosensitivity, acute MI, atrial fibrillation, colitis, dehydration, diabetes, QT prolongation, myasthenia gravis, renal disease, seizure disorder, syphilis

Black Box Warning: Tendon pain/rupture, tendinitis

DOSAGE AND ROUTES

Acute bacterial exacerbation of chronic bronchitis

• **Adult: PO/IV** 500 mg q24hr × 7 days

Acute bacterial sinusitis

• **Adult: PO** 500 mg q24hr × 10-14 days or 750 mg q24hr × 5 days

Mild-moderate UTI/acute pyelonephritis

• **Adult: PO/IV** 750 mg q24hr × 5 days or 250 mg q24hr × 10 days

Chronic bacterial prostitis

• **Adult: PO** 500 mg q24hr × 28 days

Postexposure inhalational anthrax

• **Adult/adolescent/child >50 kg: PO/IV** 500 mg q24hr × 60 days

• **Infant >6 mo/child <50 kg: IV** 8 mg/kg q12hr × 60 days, max 250 mg/dose

Pneumonia, community acquired

• **Adult: PO/IV** 500 mg q24hr × 7-14 days or 750 mg q24hr × 5 days

Pneumonia, nosocomial

• **Adult: PO/IV** 750 mg q24hr × 7-14 days

Skin/skin-structure infections, complicated

• **Adult: PO/IV** 750 mg q24hr × 7-14 days

Skin/skin-structure infections, uncomplicated

• **Adult: PO** 500 mg q24hr × 7-10 days

UTI, complicated

• **Adult: PO/IV** 750 mg q24hr × 5 days or 250 mg q24hr × 10 days

UTI, uncomplicated

• **Adult: PO** 250 mg q24hr × 3 days

Plague *(Y. pestis)*

• **Adult: PO/IV** 500 mg q24hr × 10-14 days, with pneumonia 750 mg q24hr

• **Child/adolescent <50 kg: PO/IV** 8 mg/kg (max 250 mg/dose) q12hr × 10-14 days

Otitis media (unlabeled)

• **Adult: PO** 100-200 mg bid-tid × 3-14 days

• **Child 6 mo-14 yr: PO** 10 mg/kg bid × ≥10 days

Renal disease

• **Adult: PO/IV** CCr 20-49 ml/min for 750 mg doses, give 750 mg q48hr; for 500 mg doses, give 500 mg once, then 250 mg q24hr; for 250 mg doses, no adjustment; CCr 10-19 ml/min for 750 mg dose, give 750 mg once, then 500 mg q48hr; for 500 mg dose, give 500 mg once, then 250 mg q48hr; for 250 mg dose, give 250 mg q48hr, except when treating complicated UTI, then no dose adjustment

Available forms: Single-use vials 500, 750 mg; premixed flexible containers 250 mg/50 ml D_5W, 500 mg/100 ml D_5W, 750 mg/150 ml D_5W; tabs 250, 500, 750 mg; oral sol 25 mg/ml

Administer:

• Obtain C&S before treatment and periodically

• PO 2 hr before or after antacids, iron, calcium, zinc, give fluids

• **Oral solution:** Give 1 hr before or 2 hr after food

Intermittent IV INF route

• Discard any unused sol in single-dose vial

• Visually inspect for particulate matter/discoloration prior to use

IV (Single use vial)

• **500 mg/20 ml vials:** To prepare a dose of 500 mg withdraw 10 ml from a 20 ml vial and dilute with a compatible IV solution (D_5W, NS) to a total volume of 50 ml; to prepare a 500 mg dosage, withdraw all 20 ml from the vial and dilute with a compatible intravenous solution to a total volume of 100 ml

• **750 mg/30 ml vials:** To prepare a dose of 750 mg, withdraw 30 ml from a 30 ml vial and dilute with a compatible intravenous solution (D_5W, NS) to a total volume of 150 ml

- The concentration of the diluted solution should be 5 mg/ml prior to administration; solutions contain no preservatives; any unused portions must be discarded
- **Storage:** The diluted solution may be stored for up to 72 hr when stored at or below 25°C (77°F) or for 14 days when stored under refrigeration at 5°C (41°F) in plastic containers; solutions may be frozen for up to 6 mo (−20°C or −4°F) in glass bottles of plastic containers; thaw frozen solutions at room temperature (25°C or 77°F) or in a refrigerator (8°C or 46°F); do not force thaw by microwave or water bath immersion; do not refreeze after initial thawing

Premixed IV solution

- No dilution is necessary

Intermittent IV injection

- Infuse doses of ≤500 mg IV over 60 min and doses of 750 mg IV over 90 min; shorter infusions or bolus inj should be avoided because of the risk of hypotension

Y-site compatibilities: Alemtuzumab, alfentanil, amifostine, amikacin, aminocaproic acid, aminophylline, ampicillin, ampicillin-sulbactam, anidulafungin, argatroban, atenolol, atracurium, aztreonam, bivalirudin, bleomycin, bumetanide, buprenorphine, busulfan, butorphanol, caffeine citrate, calcium gluconate, CARBOplatin, carmustine, caspofungin, cefepime, cefoTEtan, ceftaroline, cefTAZidime, ceftizoxime, cefTRIAXone, cefuroxime, chlorproMAZINE, cimetidine, cisatracurium, CISplatin, clindamycin, codeine, cyclophosphamide, cycloSPORINE, cytarabine, dacarbazine, DACTINomycin, DAPTOmycin, DAUNOrubicin liposomal, dexamethasone, dexrazoxane, digoxin, diltiazem, diphenhydrAMINE, DOBUTamine, DOCEtaxel, dolasetron, DOPamine, doripenem, doxacurium, doxycycline, droperidol, enalaprilat, ePHEDrine, EPINEPHrine, epirubicin, ertapenem, erythromycin, esmolol, etoposide, etoposide phosphate, famotidine, fenoldopam, fentaNYL, filgrastim, floxuridine, fluconazole, fludarabine, foscarnet, fosphenytoin, gallium, gemcitabine, gemtuzumab, gentamicin, granisetron, haloperidol, hydrocortisone, HYDROmorphone, IDArubicin, ifosfamide, imipenem-cilastatin, irinotecan, isoproterenol, labetalol, lepirudin, leucovorin, levorphanol, lidocaine, linezolid, mannitol, mechlorethamine, meperidine, mesna, methylPREDNISolone, metoclopramide, metroNIDAZOLE, midazolam, milrinone, minocycline, mitoMYcin, mitoXANtrone, mivacurium, morphine, mycophenolate mofetil, nalbuphine, naloxone, nesiritide, netilmicin, niCARdipine, octreotide, ondansetron, oxacillin, oxaliplatin, oxytocin, PACLitaxel, palonosetron, pamidronate, pancuronium, PEMEtrexed, penicillin G sodium, pentamidine, phenylephrine, plicamycin, potassium acetate/chloride, promethazine, propranolol, quinupristin-dalfopristin, ranitidine, remifentanil, rocuronium, sargramostim, sodium bicarbonate, succinylcholine, SUFentanil, sulfamethoxazole-trimethoprim, tacrolimus, teniposide, theophylline, thiotepa, ticarcillin, ticarcillin-clavulanate, tigecycline, tirofiban, tobramycin, topotecan, trimethobenzamide, vancomycin, vasopressin, vecuronium, verapamil, vinBLAStine, vinCRIStine, vinorelbine, voriconazole, zidovudine, zoledronic acid

Solution compatibilities: 0.9% NaCl, D_5W, D_5/0.9% NaCl, D_5LR, D_5/0.45% NaCl, sodium lactate, plasma-lyte 56/D_5W

SIDE EFFECTS

CNS: *Headache*, dizziness, *insomnia*, anxiety, seizures, *encephalopathy*, paresthesia, pseudotumor cerebri

CV: Chest pain, palpitations, vasodilation, QT prolongation, hypotension (rapid inf)

EENT: Dry mouth, visual impairment, tinnitus

GI: *Nausea*, flatulence, *vomiting*, diarrhea, abdominal pain, pseudomembranous colitis, hepatotoxicity, esophagitis, pancreatitis

GU: Vaginitis, crystalluria

HEMA: Eosinophilia, hemolytic anemia, lymphopenia
INTEG: Rash, pruritus, *photosensitivity*, epidermal necrolysis, injection site reaction, edema
MISC: Hypoglycemia, hypersensitivity, tendinitis, tendon rupture, rhabdomyolysis
RESP: Pneumonitis
SYST: Anaphylaxis, multisystem organ failure, Stevens-Johnson syndrome, angioedema, toxic epidermal necrolysis

PHARMACOKINETICS

Excreted in urine unchanged, half-life 6-8 hr, peak 1-2 hr

INTERACTIONS

Black Box Warning: **Increase:** tendon rupture—corticosteroids

• Do not use with magnesium in same IV line
Increase: QT prolongation products causing a QT prolongation
Increase: levofloxacin levels—probenecid
Increase: CNS stimulation, seizures—NSAIDs, foscarnet, cycloSPORINE
Increase: bleeding risk—warfarin
Decrease: levofloxacin absorption—antacids containing aluminum, magnesium; sucralfate, zinc, iron, calcium
Decrease: clearance of theophylline; toxicity may result

Drug/Lab Test
Increase: PT, INR
Decrease: glucose, lymphocytes

NURSING CONSIDERATIONS

Assess:

• Previous sensitivity reaction to quinolones
• **Signs, symptoms of infection:** characteristics of sputum, WBC >10,000/mm^3, fever; obtain baseline information before, during treatment
• C&S before beginning product therapy to identify if correct treatment initiated
⚠ **Allergic reactions, anaphylaxis:** rash, urticaria, pruritus, chills, fever, joint pain; may occur a few days after therapy begins; EPINEPHrine and resuscitation equipment should be available for anaphylactic reaction
• **Pseudomembranous colitis:** bowel pattern daily; if severe diarrhea, fever occurs, product should be discontinued
⚠ **Overgrowth of infection:** perineal itching, fever, malaise, redness, pain, swelling, drainage, rash, diarrhea, change in cough, sputum
• Renal function (BUN/creatinine)

Black Box Warning: **Tendon rupture:** discontinue product at first sign of tendon pain or inflammation, usually the Achilles tendon is affected; can occur up to few months after treatment and may require surgical repair; steroids may increase risk

• Increased fluid intake to 2 L/day to prevent crystalluria

Evaluate:

• Therapeutic response: absence of signs, symptoms of infection (WBC <10,000/mm^3, temperature WNL)

Teach patient/family:

• To contact prescriber if vaginal itching; loose, foul-smelling stools; furry tongue occur (may indicate superinfection); to report itching, rash, pruritus, urticaria
• To notify prescriber of diarrhea with blood or pus
• To take product 2 hr before or after antacids, iron, calcium, zinc products
• To complete full course of therapy
• To avoid hazardous activities until response is known
• To use frequent rinsing of mouth, sugarless candy or gum for dry mouth
• To avoid other medication unless approved by prescriber
• To prevent sun exposure or to use sunscreen to prevent phototoxicity
• To monitor glucose (diabetes)

Black Box Warning: To notify prescriber of tendon pain, inflammation; avoid corticosteroids with this product

levofloxacin ophthalmic

See Appendix B

levothyroxine (T_4) (Rx)

(lee-voe-thye-rox′een)

Eltroxin ✦, Levothroid, Levoxyl, Synthroid, Tirosint, Unithroid

Func. class.: Thyroid hormone

Chem. class.: Levoisomer of thyroxine

Do not confuse:
Synthroid/Symmetrel

ACTION: Increases metabolic rate; controls protein synthesis; increases cardiac output, renal blood flow, O_2 consumption, body temperature, blood volume, growth, development at cellular level via action on thyroid hormone receptors

USES: Hypothyroidism, myxedema coma, thyroid hormone replacement, thyrotoxicosis, congenital hypothyroidism, some types of thyroid cancer, pituitary TSH suppression

CONTRAINDICATIONS: Adrenal insufficiency, recent MI, thyrotoxicosis, hypersensitivity to beef, alcohol intolerance (inj only)

Black Box Warning: Obesity treatment

Precautions: Pregnancy (A), breastfeeding, geriatric patients, angina pectoris, hypertension, ischemia, cardiac disease, diabetes

DOSAGE AND ROUTES

Hypothyroidism

- **Adult ≤50 yr: PO** 1.7 mcg/kg/day, 6-8 wk, average dose 100-200 mcg/day; max 200 mcg/day **IM/IV** 50-100 mcg/day as single dose or 50% of usual oral dosage
- **Adult >50 yr without heart disease or <50 yr with heart disease: PO** 25-50 mcg/day, titrate q6-8wk
- **Adult >50 yr with heart disease: PO** 12.5-25 mcg/day, titrate by 12.5-25 mcg q6-8wk
- **Child >12 yr: PO** 2-3 mcg/kg/day as single dose in AM
- **Child 6-12 yr: PO** 4-5 mcg/kg/day as single dose in AM
- **Child 1-5 yr: PO** 5-6 mcg/kg/day as single dose in AM
- **Child 6-12 mo: PO** 6-8 mcg/kg/day as single dose in AM
- **Child to 6 mo: PO** 8-10 mcg/kg/day as single dose in AM

Myxedema coma

- **Adult: IV** 200-500 mcg, may increase by 100-300 mcg after 24 hr; give oral medication as soon as possible

Subclinical hypothyroidism

- **Adult: PO** 1 mcg/kg/day may be sufficient

Available forms: Powder for inj 200, 500 mcg/vial; tabs 25, 50, 88, 100, 112, 125, 137, 150, 175, 200, 300 mcg; cap (liquid filled) 13, 25, 50, 75, 88, 100, 112, 125, 137, 150 mcg

Administer:

- Store in tight, light-resistant container; sol should be discarded if not used immediately
- Withdrawal of medication 4 wk before RAIU test

PO route

- In AM if possible as single dose to decrease sleeplessness; at same time each day to maintain product level; take on empty stomach
- Only for hormone imbalances; not to be used for obesity, male infertility, menstrual conditions, lethargy
- Lowest dose that relieves symptoms; lower dose to geriatric patients and for those with cardiac diseases
- Crush and mix with water; nonsoy formula or breast milk for infants, children
- Separate antacids, iron, calcium products by 4 hr

Direct IV route

- IV after diluting with provided diluent 500 mcg/5 ml, 200 mcg/2 ml; shake; give through Y-tube or 3-way stopcock; give ≤100 mcg/1 min; do not add to IV inf

• Considered to be incompatible in syringe with all other products

SIDE EFFECTS

CNS: *Anxiety, insomnia, tremors,* headache, **thyroid storm**, excitability
CV: *Tachycardia, palpitations, angina, dysrhythmias,* hypertension, **cardiac arrest**
GI: Nausea, diarrhea, increased or decreased appetite, cramps
MISC: Menstrual irregularities, weight loss, sweating, heat intolerance, fever, alopecia, decreased bone mineral density

PHARMACOKINETICS

Half-life euthyroid 6-7 days, hypothyroid 9-10 days, hyperthyroid 3-4 days, distributed throughout body tissues
PO: Onset 3-5 days, peak 6-8 wk, duration 1-3 wk
IV: Onset 6-8 hr, peak 24 hr, duration unknown

INTERACTIONS

Increase: cardiac insufficiency risk—EPINEPHrine products
Increase: effects of anticoagulants, sympathomimetics, tricyclics
Decrease: levothyroxine absorption—bile acid sequestrants, orlistat, ferrous sulfate
Decrease: levothyroxine effect—estrogens, antacids, sucralfate, aluminum, magnesium, calcium, iron, rifampin, rifabutin
Drug/Herb
Decrease: thyroid hormone effect—soy
Drug/Lab Test
Increase: CPK, LDH, AST, blood glucose
Decrease: thyroid function tests

NURSING CONSIDERATIONS

Assess:
• B/P, pulse periodically during treatment
• Weight daily in same clothing, using same scale, at same time of day
• Height, growth rate of child
• T_3, T_4, FTIs, which are decreased; radioimmunoassay of TSH, which is increased; radio uptake, which is increased if patient receiving too low a dose of medication
• Patient may require decreased anticoagulant; check for bleeding, bruising
• Increased nervousness, excitability, irritability, which may indicate too high a dose of medication, usually after 1-3 wk of treatment
• Cardiac status: angina, palpitation, chest pain, change in VS

Evaluate:
• Therapeutic response: absence of depression; increased weight loss, diuresis, pulse, appetite; absence of constipation, peripheral edema, cold intolerance; pale, cool, dry skin; brittle nails, alopecia, coarse hair, menorrhagia, night blindness, paresthesias, syncope, stupor, coma, rosy cheeks

Teach patient/family:
• That hair loss will occur in child, is temporary; that hypothyroid child will show almost immediate behavior/personality change
• To report excitability, irritability, anxiety, which indicate overdose
• Not to switch brands unless approved by prescriber
• That product may be discontinued after giving birth; that thyroid panel should be evaluated after 1-2 mo
• That product is not to be taken to reduce weight
• To avoid OTC preparations with iodine; to read labels; to separate antacids, iron, calcium products by 4 hr
• To avoid iodine-rich food, iodized salt, soybeans, tofu, turnips, high-iodine seafood, some bread
• That product is not a cure but controls symptoms; that treatment is lifelong, full effect may take up to 6 wk

L

⚠ HIGH ALERT

lidocaine (parenteral) (Rx)

(lye′doe-kane)

LidoPen Auto-Injector, Xylocaine, Xylocard ♣

Func. class.: Antidysrhythmic (Class Ib)

Chem. class.: Aminoacyl amide

ACTION: Increases electrical stimulation threshold of ventricle, His-Purkinje system, which stabilizes cardiac membrane, decreases automaticity

USES: Ventricular tachycardia, ventricular dysrhythmias during cardiac surgery, digoxin toxicity, cardiac catheterization

Unlabeled uses: Attenuation of intracranial pressure increased during intubation/endotracheal tube suctioning

CONTRAINDICATIONS: Hypersensitivity to amides, severe heart block, supraventricular dysrhythmias, Adams-Stokes syndrome, Wolff-Parkinson-White syndrome

Precautions: Pregnancy (B), breastfeeding, children, geriatric patients, renal/hepatic disease, CHF, respiratory depression, malignant hyperthermia, myasthenia gravis, weight <50 kg

DOSAGE AND ROUTES

- **Adult: IV BOL** 50-100 mg (1-1.5 mg/kg) 25-50 mg/min, repeat q3-5min, max 300 mg in 1 hr; begin **IV INF; IV INF** 1-4 mg/min (20-50 mcg/kg/min); **IM** 200-300 mg (4.3 mg/kg) in deltoid muscle, may repeat after 1-1½ hr if needed

Available forms: ***IV INF*** 0.2% (2 mg/ml), 0.4% (4 mg/ml), 0.8% (8 mg/ml); IV ad 4% (40 mg/ml), 10% (100 mg/ml), 20% (200 mg/ml); ***IV dir*** 1% (10 mg/ml), 2% (20 mg/ml); ***IM*** 10% 300 mg/ml

Administer:

- IM inj in deltoid; aspirate to avoid intravascular administration; check IV site daily for infiltration or extravasation

IV route

- Bolus undiluted (1%, 2% only), give ≤50 mg/1 min or dilute 1 g/250-500 ml D_5W; titrate to patient response; use inf pump; pediatric inf 120 mg lidocaine/100 ml D_5W; 1-2.5 ml/kg/hr = 20-50 mcg/kg/min; use only 1%, 2% sol for IV bol

Y-site compatibilities: Acetaminophen, alemtuzumab, alfentanil, alteplase, amikacin, aminocaproic acid, aminophylline, amiodarone, amphotericin B lipid/liposome, anidulafungin, argatroban, ascorbic acid injection, atenolol, atropine, atracurium, azithromycin, aztreonam, benztropine, bivalirudin, bleomycin, bumetanide, buprenorphine, butorphanol, calcium chloride/gluconate, CARBOplatin, carmustine, cefamandole, ceFAZolin, cefoperazone, cefotaxime, cefoTEtan, cefOXitin, ceftaroline, cefTAZidime, ceftizoxime, cefTRIAXone, cefuroxime, chloramphenicol, chlorproMAZINE, cimetidine, ciprofloxacin, cisatracurium, CISplatin, clarithromycin, clindamycin, cyanocobalamin, cyclophosphamide, cycloSPORINE, cytarabine, DACTINomycin, DAPTOmycin, DAUNOrubicin, dexamethasone, dexmedetomidine, dexrazoxane, digoxin, diltiazem, diphenhydrAMINE, DOBUTamine, DOCEtaxel, dolasetron, DOPamine, doxacurium, DOXOrubicin, DOXOrubicin liposomal, doxycycline, enalaprilat, EPINEPHrine, epirubicin, epoetin alfa, eptifibatide, ertapenem, erythromycin, esmolol, etomidate, etoposide, etoposide phosphate, famotidine, fenoldopam, fentaNYL, fluconazole, fludarabine, fluorouracil, folic acid, furosemide, gentamicin, granisetron, haloperidol, heparin, hydrocortisone, imipenem/cilastatin, inamrinone, insulin, isoproterenol, ketorolac, labetalol, levofloxacin, linezolid, LORazepam, magnesium sulfate, meperidine, methylPREDNISolone sodium succinate, metoclopramide, metoprolol, metroNIDAZOLE, micafungin, midazolam, morphine, nafcillin, niCARdipine, nitroglycerin, nitroprusside, norepinephrine, ondansetron, palonosetron, penicillin G potassium, phenylephrine, phytonadione, piperacillin/tazobactam, potassium chloride, procainamide, prochlorperazine, promethazine, propofol, propranolol, protamine, quinupristin/dalfopristin, ranitidine, remifentanil, sodium bicarbonate, streptokinase, tacrolimus, theophylline, ticarcillin/clavulanate, tigecycline, tirofiban, tobramycin, vancomycin, vasopressin, verapamil, vitamin B complex with C, voriconazole, warfarin

SIDE EFFECTS

CNS: *Headache, dizziness,* involuntary movement, confusion, tremor, drowsiness, euphoria, seizures, shivering
CV: *Hypotension, bradycardia,* heart block, CV collapse, arrest
EENT: Tinnitus, blurred vision
GI: Nausea, vomiting, anorexia
HEMA: Methemoglobinemia
INTEG: Rash, urticaria, edema, swelling, petechiae, pruritus
MISC: Febrile response, phlebitis at inj site
RESP: Dyspnea, respiratory depression

PHARMACOKINETICS

Half-life 8 min, 1-2 hr (terminal); metabolized in liver; excreted in urine; crosses placenta
IM: Onset 5-15 min, duration $1^1/_2$ hr
IV: Onset 2 min, duration 20 min

INTERACTIONS

Increase: cardiac depression, toxicity—amiodarone, phenytoin, procainamide, propranolol
Increase: hypotensive effects—MAOIs, antihypertensives
Increase: neuromuscular blockade—neuromuscular blockers, tubocurarine
Increase: lidocaine effects—cimetidine, beta blockers, protease inhibitors, ritonavir
Decrease: lidocaine effects—barbiturates, ciprofloxacin, voriconazole
Decrease: effect of—cycloSPORINE
Drug/Lab Test
Increase: CPK

NURSING CONSIDERATIONS

Assess:
⚠ ECG continuously to determine increased PR or QRS segments; if these develop, discontinue or reduce rate; watch for increased ventricular ectopic beats, may have to rebolus; B/P
• **Blood levels:** therapeutic level, 1.5-5 mcg/ml
• I&O ratio, electrolytes (potassium, sodium, chlorine)
⚠ **Malignant hyperthermia:** tachypnea, tachycardia, changes in B/P, increased temperature
• **Respiratory status:** rate, rhythm, lung fields for crackles, watch for respiratory depression; lung fields, bilateral crackles may occur with CHF; increased respiration, pulse; product should be discontinued
• **CNS effects:** dizziness, confusion, psychosis, paresthesias, convulsions; product should be discontinued
Evaluate:
• Therapeutic response: decreased dysrhythmias
Teach patient/family:
• About the use of automatic lidocaine injection device if ordered for personal use
• To report signs of toxicity

TREATMENT OF OVERDOSE:

O_2, artificial ventilation, ECG; administer DOPamine for circulatory depression, diazepam or thiopental for seizures; decrease product if needed

lidocaine topical

See Appendix B

L

linaclotide

(lin′a-kloe′tide)

Linzess

Func. class.: Functional GI disorder agent

Chem. class.: Selective guanylate cyclase C agonist

ACTION: Acts locally on the luminal surface of the intestinal epithelium; stimulates secretion of chloride and bicarbonate into the intestinal lumen; this action results in increased intestinal fluid and accelerated GI transit

USES: Irritable bowel syndrome (IBS) with constipation, chronic constipation not associated with IBS

CONTRAINDICATIONS: Hypersensitivity, urticaria, infection, sinusitis, GI obstruction

Black Box Warning: Children <6 yr, infants, neonates

Precautions: Pregnancy C, breastfeeding, diarrhea, abdominal pain, cholelithiasis, Crohn's disease, diverticulitis, fecal impaction, gastric cancer, GI disease, IBS

DOSAGE AND ROUTES

Constipation

- **Adult: PO** 145 mcg/day on empty stomach, AM

Irritable bowel syndrome

- **Adult: PO** 290 mcg/day on empty stomach, AM

Available forms: Caps 145, 290 mcg

Administer:

- Give on empty stomach, ≥30 min before first meal of day, swallow whole
- Store at room temperature

SIDE EFFECTS

CNS: Headache, fatigue

GI: Nausea, abdominal pain, diarrhea, flatulence, fecal incontinence, vomiting, GERD, **GI bleeding**

PHARMACOKINETICS

Minimally absorbed, metabolized within the GI tract

INTERACTIONS

Do not use with antidiarrheals

Decrease: linaclotide effect—antimuscarinics

NURSING CONSIDERATIONS

Assess:

Black Box Warning: Neonates/infants/children <6 yr: Can cause serious effects, including death

- GI symptoms: nausea, abdominal pain, evidence of mechanical GI obstruction; check for other sources of constipation, determine stool quality and frequency

Evaluate:

- Decreased constipation in IBS

Teach patient/family:

- To notify prescriber of GI symptoms, avoid antidiarrheals, if severe diarrhea occurs, stop product
- Do not crush, break, chew
- Pregnancy C: Avoid use in pregnancy, breastfeeding

linagliptin

(lin′a-glip′tin)

Tradjenta

Func. class.: Antidiabetic

Chem. class.: Didipeptidyl peptidase-4 inhibitor

ACTION: Slows the inactivation of incretin hormones; concs of the active, intact hormones are increased, thereby increasing and prolonging the action of these hormones; incretin hormones are released by the intestine throughout the day, and levels are increased in response to a meal

USES: Type 2 diabetes mellitus

CONTRAINDICATIONS:
Hypersensitivity to linagliptin, type 1 diabetes mellitus, diabetic ketoacidosis (DKA)

Precautions: Pregnancy (category B), breastfeeding, adolescents or children <18 yr, debilitated physical condition, malnutrition, uncontrolled adrenal insufficiency, pituitary insufficiency, hypo/hyperthyroidism, diarrhea, gastroparesis, GI obstruction, ileus, female hormonal changes, high fever, severe psychological stress, uncontrolled hypercortisolism

DOSAGE AND ROUTES

• **Adult: PO** 5 mg daily; when used with a sulfonylurea or insulin; a lower dose of the sulfonylurea may be necessary to minimize the risk of hypoglycemia

Available forms: Tab 5 mg

Administer:

• Once daily; may give without regard to food

• Store at room temperature

SIDE EFFECTS

CNS: Headache

EENT: Nasopharyngitis

ENDO: Hypoglycemia, hyperuricemia

GI: Body weight loss, **pancreatitis**

INTEG: Hypersensitivity reactions, urticaria, **angioedema, exfoliative dermatitis**

MISC: Arthralgia, back pain

RESP: Bronchial hyperreactivity (with **bronchospasm**), nasopharyngitis, cough

PHARMACOKINETICS

Extensively distributed in the tissues, protein binding is conc-dependent, a weak to moderate inhibitor of CYP3A4, plasma terminal half life of >100 hr; effective half-life 12 hr, 90% excreted unchanged, 85% excreted via the enterohepatic system (80%) or in urine (5%) within 4 days of dosing, rapidly absorbed, peak in 1.5 hr; bioavailability 30%

INTERACTIONS

Increase: hypoglycemia: sulfonylureas, beta blockers, ACE inhibitors, angiotensin II receptor antagonists, disopyramide, guanethidine, cloNIDine, octreotide, fenfluramine, dexfenfluramine, fibric acid derivatives, monoamine oxidase inhibitors (MAOIs), FLUoxetine, salicylates

Increase: masking of the signs and symptoms of hypoglycemia: reserpine, beta blockers

Increase: need for dosing change: cisapride, metoclopramide, tegaserod, androgens, alcohol, lithium, quinolones

Decrease: hypoglycemic effect—dextrothyroxine, bumetanide, furosemide, ethacrynic acid, torsemide, estrogens, progestins, oral contraceptives, thyroid hormones, glucocorticoids, glucagon, carbonic anhydrase inhibitors, phenytoin, fosphenytoin, or ethotoin; atypical antipsychotics (ARIPiprazole, cloZAPine, OLANZapine, QUEtiapine, risperiDONE, and ziprasidone), phenothiazine, niacin (nicotinic acid), triamterene, thiazide diuretics

Decrease: effect of linagliptin—CYP3A4 inducers (topiramate, rifabutin, pioglitazone, OXcarbazepine, carBAMazepine, nevirapine, modafinil, metyrapone, etravirine, efavirenz, bosentan, barbiturates, aprepitant, fosaprepitant)

Drug/Herb:

Decrease: linagliptin effect—St. John's wort

Drug/Lab Test:

Increase: uric acid

NURSING CONSIDERATIONS

Assess

• Monitor blood glucose, A1c, during treatment to determine diabetes control

• CBC baseline and periodically during treatment, report decreased blood counts

Evaluate:

• Improving blood glucose level, A1c; decreasing polydipsia, polyphagia, polyuria, clear sensorium, absence of dizziness

Teach patient/family:

• About the symptoms of hypo/hyperglycemia and what to do about each; to have

L

glucagon emergency kit available, to carry sugar packets

- That product must be continued on a daily basis, about the consequences of discontinuing product abruptly; to take only as directed
- To avoid OTC products unless approved by prescriber
- That diabetes is a life-long illness, that product will not cure diabetes
- To carry emergency ID with prescriber, condition and medications taken
- To immediately report skin disorders, swelling, difficulty breathing, or severe abdominal pain

lindane (Rx)

(lin′dane)

Hexit ✦

Func. class.: Scabicide, pediculicide

Chem. class.: Chlorinated hydrocarbon (synthetic)

ACTION: Stimulates nervous system of arthropods, resulting in seizures, death

USES: Scabies, lice (head/pubic/body), nits in those intolerant to or who do not respond to other agents

CONTRAINDICATIONS: Hypersensitivity, patients with known seizure disorders, Norwegian (crusted) scabies

Black Box Warning: Premature neonate; inflammation of skin, abrasions, skin breaks; seizure disorder

Precautions: Pregnancy (C), breastfeeding, infants, children <10 yr, avoid contact with eyes

DOSAGE AND ROUTES

Lice

- **Adult and child: CREAM/LOTION** shampoo using 30 ml: work into lather, rub for 5 min, rinse, dry with towel; comb with fine-toothed comb to remove nits; most require 1 oz, max 2 oz

Scabies

- **Adult and child: TOP** apply 1% cream/lotion to skin, from neck to bottom of feet, toes; wash area with soap, water; remove visible crusts; apply to skin surfaces; remove with soap, water in 8-12 hr; repeat after 1 wk prn; most require 1 oz, max 2 oz

Available forms: Lotion, shampoo, cream (1%)

Administer:

- Caregivers should wear gloves less permeable to lindane, thoroughly clean hands after application; avoid natural latex gloves
- **Cream/ointment/lotion:** use for scabies only; skin should be clean without other products on it, wait 1 hr after bathing or showering before application, shake well, apply under fingernails after trimming; toothbrush can be used to apply; after application wrap toothbrush in paper and throw away; use only a single application, apply as thin layer over all skin from neck down, close bottle containing leftover lotion, throw away
- Do not cover, wash off after 8-12 hr
- Use warm, not hot, water; do not leave on >12hr
- **Shampoo:** for lice only; do not use other hair products before use; shake well; hair should be completely dry; use only enough shampoo to lightly coat hair and scalp, work into hair, do not use water; allow to remain only 4 min, rinse, lather away, towel briskly
- To scalp only; do not apply to face, lips, mouth, eyes, any mucous membrane, anus, or meatus
- Topical corticosteroids as ordered to decrease contact dermatitis; antihistamines
- Lotions of menthol or phenol to control itching
- Topical antibiotics for infection

SIDE EFFECTS

CNS: Seizures, CNS toxicity, stimulation, dizziness

INTEG: *Pruritus, rash, irritation, contact dermatitis*

PHARMACOKINETICS

Peak 6 hr, half-life 18-22 hr

INTERACTIONS

- Oils may increase absorption; if oil-based hair dressing used, shampoo, rinse, dry hair before applying lindane shampoo

NURSING CONSIDERATIONS

Assess:

Black Box Warning: Skin for abrasions, breaks, inflammation; do not use on these areas

- **Infestation:** head, hair for lice, nits before and after treatment; if scabies present, check all skin surfaces; identify source of infection: school, family, sexual contacts
- Isolation until areas on skin, scalp have cleared, treatment completed
- Removal of nits with fine-toothed comb rinsed in vinegar after treatment; use gloves

Evaluate:

- Therapeutic response: decreased crusts, nits, brownish trails on skin, itching papules in skin folds, decreased itching after several weeks

Teach patient/family:

- To wash all inhabitants' clothing using insecticide; that preventive treatment may be required of all persons living in same house using lotion or shampoo to decrease spread of infection; to use rubber gloves when applying product
- That itching may continue for 4-6 wk
- That product must be reapplied if accidently washed off or treatment will be ineffective
- Not to apply to face; if accidental contact with eyes occurs, flush with water
- To remove product after specified time to prevent toxicity
- To treat sexual contacts simultaneously
- To check for CNS toxicity: dizziness, cramps, anxiety, nausea, vomiting, seizures

linezolid (Rx)

(line-zoe′lide)

Zyvox

Func. class.: Broad-spectrum antiinfective

Chem. class.: Oxazolidinone

Do not confuse:

Zyvox/Ziox/Zosyn

ACTION:

Inhibits protein synthesis by interfering with translation; binds to bacterial 23S ribosomal RNA of the 50S subunit, thus preventing formation of the bacterial translation process in primarily gram-positive organisms

USES:

Vancomycin-resistant *Enterococcus faecium* infections, nosocomial pneumonia caused by *Staphylococcus aureus* or *Streptococcus* pneumonia, uncomplicated or complicated skin and skin-structure infections, community-acquired pneumonia, *Pasteurella multocida,* viridans streptococci

CONTRAINDICATIONS:

Hypersensitivity

Precautions: Pregnancy (C), breastfeeding, children, thrombocytopenia, bone marrow suppression, hypertension, hyperthyroidism, pheochromocytoma, seizure disorder, ulcerative colitis, MI, PKU, renal/GI disease

DOSAGE AND ROUTES

Vancomycin-resistant *Enterococcus faecium* infections

- **Adult/adolescent/child ≥12 yr: IV/PO** 600 mg q12hr × 14-28 days; max 1200 mg/day
- **Child <12 yr/infant/term neonate: IV/PO** 10 mg/kg q8hr × 14-28 days

Nosocomial pneumonia/ complicated skin infections/ community-acquired pneumonia/ concurrent bacterial infection

- **Adult: IV/PO** 600 mg q12hr × 10-14 days; max 1200 mg/day

• **Child: birth-11 yr: PO** 10 mg/kg q8hr × 10-14 days

Uncomplicated skin infections

• **Adult: PO** 400 mg q12hr × 10-14 days; max 1200 mg/day

• **Adolescent: PO** 600 mg q12hr × 10-14 days; max 1200 mg/day

• **Infant preterm <7 days old: PO** 10 mg/kg q12hr × 10-14 days

Available forms: Tabs 600 mg; oral sus 100 mg/5 ml; inj 2 mg/ml

Administer:

PO route

• With/without food

• Store reconstituted oral susp at room temperature, use within 3 wk

Intermittent IV INF route

• Do not use if particulate is present, yellow color is normal

• Premixed sol ready to use (2 mg/ml), give over 30-120 min; do not use IV inf bag in series connections; do not use with additives in sol; do not use with another product, administer separately, flush line before and after use

Y-site compatibilities: Acyclovir, alfentanil, amikacin, aminophylline, ampicillin, aztreonam, buprenorphine, butorphanol, calcium gluconate, CARBOplatin, ceFAZolin, cefoperazone, cefoTEtan, cefOXitin, cefTAZidime, ceftizoxime, cefuroxime, cimetidine, ciprofloxacin, cisatracurium, CISplatin, clindamycin, cyclophosphamide, cycloSPORINE, cytarabine, digoxin, furosemide, ganciclovir, gemcitabine, gentamicin, heparin, HYDROmorphone, ifosfamide, labetalol, leucovorin, levofloxacin, lidocaine, LORazepam, magnesium sulfate, mannitol, meperidine, meropenem, mesna, methotrexate, methylPREDNISolone, metoclopramide, metroNIDAZOLE, midazolam, minocycline, mitoXANtrone, morphine, nalbuphine, naloxone, nitroglycerin, ofloxacin, ondansetron, PACLitaxel, PENTobarbital, PHENobarbital, piperacillin, potassium chloride, prochlorperazine, promethazine, propranolol, ranitidine, remifentanil, SUFentanil, theophylline, ticarcillin, tobramycin, vancomycin, vecuronium, verapamil, vinCRIStine, zidovudine

Solution compatibilities: D_5, 0.9% NaCl, LR

SIDE EFFECTS

CNS: *Headache,* dizziness, insomnia

GI: *Nausea, diarrhea,* **pseudomembranous colitis**, increased ALT/AST, *vomiting,* taste change, tongue-color change

HEMA: **Myelosuppression**

MISC: Vaginal moniliasis, fungal infection, oral moniliasis, **lactic acidosis, anaphylaxis, angioedema, Stevens-Johnson syndrome**

PHARMACOKINETICS

Peak 1-2 hr, terminal half-life 4-5 hr, rapidly and extensively absorbed, protein binding 31%, metabolized by oxidation of the morpholine ring

INTERACTIONS

⚠ **Do not use with MAOIs (or within 2 wk) or with products that possess MAOI-like action (furazolidone, isoniazid, INH, procarbazine); hypertensive crisis may occur**

⚠ **Increase: hypertensive crisis, seizures, coma—amoxapine, maprotiline, mirtazapine, traZODone, cyclobenzaprine, tricyclics, methyldopa**

Increase: serotonin syndrome—SSRIs, SNRIs, serotonin receptor agonists

Increase: effects of adrenergic agents (DOPamine, EPINEPHrine, pseudoephedrine)

Drug/Herb

• Avoid use with green tea, valerian, ginseng, yohimbine, kava, guarana, St. John's wort

Drug/Food

• Tyramine foods: avoid; increased pressor response

NURSING CONSIDERATIONS

Assess:

• CBC with differential weekly, assess for myelosuppression (anemias, leukopenia, pancytopenia, thrombocytopenia)

• **Serotonin syndrome:** at least 2 wk should elapse between discontinuing linezolid and starting serotonergic agents; assess for increased heart rate, shivering, sweating, dilated pupils, tremor, high B/P, hyperthermia, headache, confusion; if these occur, stop linezolid, administer a serotonin antagonist if needed
• **Lactic acidosis:** repeated nausea/vomiting, unexplained acidosis, low bicarbonate level; notify prescriber immediately
• **Anaphylaxis/angioedema/Stevens-Johnson syndrome:** rash, pruritus, difficulty breathing, fever; have emergency equipment nearby
• CNS symptoms: headache, dizziness
• Hepatic studies: AST, ALT
• **Diabetes mellitus:** monitor those receiving insulin or oral antidiabetics for increased hypoglycemia
⚠ **Pseudomembranous colitis:** diarrhea, abdominal pain, fever, fatigue, anorexia, possible anemia, elevated WBC, low serum albumin; stop product, usually either vancomycin or IV metroNIDAZOLE given

Evaluate:
• Therapeutic response: decreased symptoms of infection, blood cultures negative

Teach patient/family:
• If dizziness occurs, to ambulate, perform activities with assistance
• To complete full course of product therapy
• To contact prescriber if adverse reaction occurs
• To inform prescriber if SSRIs or cold products, decongestants being used
• To inform prescriber of history of hypertension
• To avoid large amounts of high-tyramine foods, drinks; provide list

liothyronine (T_3) (Rx)

(lye-oh-thye′roe-neen)

Cytomel, Triostat

Func. class.: Thyroid hormone
Chem. class.: Synthetic T_3

Do not confuse:
Cytomel/Cytotec

ACTION:
Increases metabolic rates, cardiac output, O_2 consumption, body temperature, blood volume, growth, development at cellular level; exact mechanism unknown

USES:
Hypothyroidism, myxedema coma, thyroid hormone replacement, congenital hypothyroidism, nontoxic goiter, T_3 suppression test

CONTRAINDICATIONS:
Adrenal insufficiency, MI, thyrotoxicosis, untreated hypertension

Black Box Warning: Obesity treatment

Precautions: Pregnancy (A), breastfeeding, geriatric patients, angina pectoris, hypertension, ischemia, cardiac disease, diabetes

DOSAGE AND ROUTES
• **Adult: PO** 25 mcg/day, increase by 12.5-25 mcg q1-2wk until desired response, maintenance dose 25-75 mcg/day, max 100 mcg/day
• **Geriatric: PO** 5 mcg/day, increase by 5 mcg/day q1-2wk, maintenance 25-75 mcg/day

Congenital hypothyroidism
• **Child >3 yr: PO** 50-100 mcg/day
• **Child <3 yr: PO** 5 mcg/day, increase by 5 mcg q3-4days titrated to response, infant maintenance 20 mcg/day; 1-3 yr 50 mcg/day

Myxedema, severe hypothyroidism
• **Adult: PO** 25-50 mcg then may increase by 5-10 mcg q1-2wk; maintenance dose 50-100 mcg/day

Myxedema coma/precoma
- **Adult: IV** 25-50 mcg initially, 5 mcg in geriatric patients, 10-20 mcg with cardiac disease; give doses q4-12hr

Nontoxic goiter
- **Adult: PO** 5 mcg/day, increase by 12.5-25 mcg q1-2wk; maintenance dose 75 mcg/day

Suppression test
- **Adult: PO** 75-100 mcg/day × 1 wk; radioactive ^{131}I given before and after 1-wk dose

Available forms: Tabs 5, 25, 50 mcg; inj 10 mcg/ml

Administer:
- In AM if possible as a single dose to decrease sleeplessness
- At same time each day to maintain product level
- Only for hormone imbalances; do not use for obesity, male infertility, menstrual conditions, lethargy
- Lowest dose that relieves symptoms
- Liothyronine after discontinuing other thyroid preparations
- Do not give with calcium, iron, aluminum, magnesium, soy products

SIDE EFFECTS

CNS: *Insomnia, tremors,* headache, thyroid storm

CV: *Tachycardia, palpitations, angina, dysrhythmias,* hypertension, cardiac arrest

GI: Nausea, diarrhea, increased or decreased appetite, cramps

MISC: Menstrual irregularities, weight loss, sweating, heat intolerance, fever, alopecia

PHARMACOKINETICS

PO/IV: Peak 2-3 days, duration 72 hr, half-life 2.5 days

INTERACTIONS

Increase: effects of—anticoagulants, sympathomimetics, tricyclics, amphetamines, decongestants, vasopressors

Decrease: absorption of liothyronine—cholestyramine; colestipol; calcium, iron, aluminum, magnesium products

Decrease: effects of liothyronine—estrogens

Drug/Herb

Decrease: thyroid hormone effect—soy

Drug/Lab Test

Increase: CPK, LDH, AST, PBI, blood glucose

Decrease: thyroid function tests

NURSING CONSIDERATIONS

Assess:
- B/P, pulse periodically during treatment
- Weight daily in same clothing, using same scale, at same time of day
- Height, growth rate of child
- T_3, T_4, which are decreased; radioimmunoassay of TSH, which is increased; radio uptake, which is increased if patient receiving too low a dose of medication
- Patient may require decreased anticoagulant; check for bleeding, bruising
- Increased nervousness, excitability, irritability, which may indicate too high a dose of medication, usually after 1-3 wk of treatment
- Cardiac status: angina, palpitation, chest pain, change in VS
- Remove medication 4 wk before RAIU test

Evaluate:
- Therapeutic response: absence of depression; increased weight loss, diuresis, pulse, appetite; absence of constipation, peripheral edema, cold intolerance; pale, cool, dry skin; brittle nails; alopecia, coarse hair; menorrhagia; night blindness; paresthesia; syncope, stupor, coma

Teach patient/family:
- That hair loss will occur in child but is temporary
- To report excitability, irritability, anxiety, which indicate overdose
- Not to switch brands unless approved by prescriber

- That hypothyroid child will show almost immediate behavior/personality change
- That product is not to be taken to reduce weight
- To avoid OTC preparations with iodine; to read labels; not to take with calcium, iron, aluminum, magnesium products
- To avoid iodine-rich food, iodized salt, soybeans, tofu, turnips, high iodine seafood, some bread
- That product controls symptoms but does not cure; that treatment is lifelong

liotrix (Rx)

(lye'oh-trix)

Thyrolar, T_3/T_4

Func. class.: Thyroid hormone

Chem. class.: Levothyroxine/liothyronine (synthetic T_4, T_3)

Do not confuse:
Thyrolar/Thyrar

ACTION:
Increases metabolic rates, cardiac output, O_2 consumption, body temperature, blood volume, growth, development at cellular level; exact mechanism unknown

USES:
Hypothyroidism, thyroid hormone replacement

CONTRAINDICATIONS:
Adrenal insufficiency, MI, thyrotoxicosis

Black Box Warning: Obesity treatment

Precautions: Pregnancy (A), breastfeeding, geriatric patients, angina pectoris, hypertension, ischemia, cardiac disease, diabetes

DOSAGE AND ROUTES

- **Adult: PO** single dose of Thyrolar, 1/4 or 1/2 of adult dose, adjust as needed at 2-wk intervals
- **Geriatric: PO** 1/4 tab initially, adjust q6-8wk

Available forms: Tabs: Levothyroxine 12.5 mcg/liothyronine 3.1 mcg; levothyroxine 25 mcg/liothyronine 6.25 mcg (Thyrolar-1/2); levothyroxine 50 mcg/liothyronine 12.5 mcg (Thyrolar-1); levothyroxine 100 mcg/liothyronine 25 mcg (Thyrolar-2); levothyroxine 150 mcg/liothyronine 37.5 mcg (Thyrolar-3)

Administer:

- Separate products containing calcium, iron by ≥4 hr
- In AM if possible as a single dose to decrease sleeplessness
- At same time each day to maintain product level
- Only for hormone imbalances; do not use for obesity, male infertility, menstrual conditions, lethargy
- Lowest dose that relieves symptoms
- Store in airtight, light-resistant container

SIDE EFFECTS

CNS: *Insomnia, tremors,* headache, **thyroid storm**, nervousness

CV: *Tachycardia, palpitations, angina, dysrhythmias,* hypertension, **cardiac arrest**

GI: Nausea, vomiting, diarrhea, increased or decreased appetite, cramps

MISC: Menstrual irregularities, weight loss, sweating, heat intolerance, fever

PHARMACOKINETICS

PO (T_4): Onset unknown, peak 1-3 wk, duration 1-3 wk, terminal half-life 6-7 days

PO (T_3): Onset unknown, peak 24-72 hr, duration 72 hr, terminal half-life 1-2 days

INTERACTIONS

Increase: effects of amphetamines, decongestants, vasopressors, anticoagulants, sympathomimetics, tricyclics, catecholamines

Decrease: absorption of liotrix—cholestyramine, colestipol

Decrease: effects of liotrix—estrogens, phenytoin, carBAMazepine, rifampin

Drug/Herb

Decrease: thyroid hormone effect—soy

Drug/Lab Test

Increase: CPK, LDH, AST, PBI, blood glucose

Decrease: thyroid function tests

NURSING CONSIDERATIONS

Assess:

- B/P, pulse periodically during treatment
- Weight daily in same clothing, using same scale, at same time of day
- Height, growth rate of child
- T_3, T_4, FTIs, which are decreased; radioimmunoassay of TSH, which is increased; radio uptake, which is increased if patient is receiving too low a dose of medication
- May require decreased anticoagulant; check for bleeding, bruising
- Increased nervousness, excitability, irritability, which may indicate too high a dose of medication, usually after 1-3 wk of treatment
- Cardiac status: angina, palpitation, chest pain, change in VS
- Withdraw medication 4 wk before RAIU test

Evaluate:

- Therapeutic response: absence of depression; increased weight loss, diuresis, pulse, appetite; absence of constipation, peripheral edema, cold intolerance; pale, cool, dry skin; brittle nails; coarse hair; menorrhagia; night blindness; paresthesias; syncope, stupor, coma; rosy cheeks

Teach patient/family:

- That hair loss will occur in child, is temporary
- To report excitability, irritability, chest pain, increased pulse rate, palpitations, excessive sweating, heat intolerance, nervousness, anxiety, which indicate overdose
- Not to switch brands unless approved by prescriber
- That hypothyroid child will show almost immediate behavior/personality change
- That product is not to be taken to reduce weight
- To avoid OTC preparations with iodine; read labels; separate products containing calcium, iron by ≥4 hr
- To avoid iodine food, iodized salt, soybeans, tofu, turnips, high iodine seafood, some bread
- That product does not cure but controls symptoms; treatment is lifelong

liraglutide (Rx)

(lir′a-gloo′tide)

Victoza

Func. class.: Antidiabetic agent

Chem. class.: Incretin mimetics

ACTION: Improved glycemic control and potential weight loss via activation of the glucagon-like peptide-1 (GLP-1) receptor

USES: Type 2 diabetes mellitus in combination with diet/exercise

CONTRAINDICATIONS: Hypersensitivity, pancreatitis

Black Box Warning: Medullary thyroid carcinoma (MTC), multiple endocrine neoplasia syndrome type 2 (MEN 2), thyroid cancer

Precautions: Breastfeeding, children, geriatric patients, alcoholism, cholelithiasis, ketoacidosis, diarrhea, fever, gastroparesis, hepatic/renal disease, hypoglycemia, infection, surgery, thyroid disease, trauma, vomiting

DOSAGES AND ROUTES

- **Adult: SUBCUT** 0.6 mg/day × 1 wk, then increase to 1.2 mg/day, max 1.8 mg/day

Available forms: Solution for injection 18 mg/3 ml pre-filled pen

Administer:

SUBCUT route

- Give subcut only, inspect for particulate matter, discoloration; do not use if unusually viscous, cloudy, discolored, or if particles present; give daily anytime, without regard to meals; pen needles must be purchased separately, use Novo Nordisk needle; before first use, prime, see manual for directions; give in thigh, abdomen, or upper arm; lightly pinch fold of skin, insert needle at 90-degree angle (45-degree angle if thin), release skin; aspiration is not needed, give over 6 sec, rotate injection sites

- If dose is missed, resume once-daily dosing at next scheduled dose; if >3 days have elapsed since last dose, reinitiate at 0.6 mg, titrate
- Storage: do not store pen with needle attached; avoid direct heat and sunlight; discard 30 days after first use; after first use may be stored at room temperature or refrigerated; do not freeze

SIDE EFFECTS

CNS: Dizziness, headache
CV: Hypertension
ENDO: Hypoglycemia
EENT: Sinusitis
GI: Abdominal pain, anorexia, constipation, diarrhea, dyspepsia, nausea, vomiting, pancreatitis
INTEG: Angioedema, erythema, injection site reaction, urticaria
MS: Back pain
SYST: Antibody formation, infection, influenza, secondary thyroid malignancy, anaphylaxis, angioedema

INTERACTIONS

Increase: hypoglycemic reactions—angiotensin II receptor antagonists, ACE inhibitors, other antidiabetics, β-blockers, dexfenfluramine, fenfluramine, disopyramide, FLUoxetine, fibric acid derivatives, mecasermin, MAOIs, octreotide, pegvisomant, salicylates
Decrease: liraglutide effect—protease inhibitors, phenothiazines, baclofen, atypical antipsychotics, corticosteroids, cycloSPORINE, tacrolimus, carbonic anhydrase inhibitors, dextrothyroxine, diazoxide, phenytoin, fosphenytoin, ethotoin, isoniazid, INH, niacin, nicotine, estrogens, progestins, oral contraceptives, growth hormones, sympathomimetics
Increase or decrease: hypoglycemic reactions—androgens, bortezomib, quinolones, cloNIDine, alcohol, lithium, pentamidine
Increase or decrease: effects of—atorvastatin, acetaminophen, griseofulvin
Decrease: levels of digoxin

PHARMACOKINETICS

Protein binding (98%); half-life 12-13 hr; binds to albumin, then released into circulation; peak 8-12 hr; body weight significantly affects pharmacokinetics

NURSING CONSIDERATIONS

Assess:

Black Box Warning: Thyroid C-cell tumors; monitor during treatment

- **Hypoglycemic reactions** that may occur soon after meals: hunger, sweating, weakness, dizziness, tremors, restlessness, tachycardia
- Hypersensitivity to this product
- Serum glucose, A1c, CBC during treatment
- **Stress:** those diabetic patients exposed to stress, surgery, fever, infections may require insulin administration temporarily

⚠ **Serious skin reactions:** angioedema; also pancreatitis

Evaluate:

- Therapeutic response: stable and improved serum glucose, A1C, weight loss

Teach patient/family:

- About symptoms of hypo/hyperglycemia, what to do for each; to have glucagon emergency kit available; to carry a carbohydrate source at all times
- About side effects associated with therapy, such as nausea and vomiting; that upward dose titration can be delayed or ignored, depending on tolerance
- That diabetes is a lifelong illness; that product does not cure disease and must be continued on a daily basis

- To carry emergency ID with prescriber's phone number and medications taken
- To continue with other recommendations: diet, exercise, hygiene
- To test blood glucose using a blood glucose meter
- To avoid other medications, herbs, supplements unless approved by prescriber

⚠ To report serious skin effects, abdominal pain with nausea/vomiting

- Provide patient with written instructions if self-administration is ordered

lisdexamfetamine (Rx)

(lis-dex′am-fet′a-meen)

Vyvanse

Func. class.: CNS stimulant

Chem. class.: Amphetamine

Controlled Substance Schedule II

ACTION: Increases release of norepinephrine, DOPamine in cerebral cortex to reticular activating system

USES: Attention-deficit/hyperactivity disorder (ADHD)

CONTRAINDICATIONS: Breastfeeding, hyperthyroidism, hypertension, glaucoma, severe arteriosclerosis, CV disease, hypersensitivity to sympathomimetic amines

Black Box Warning: Substance abuse

Precautions: Pregnancy (C), children <6 yr, Gilles de la Tourette's disorder, depression, anorexia nervosa, psychosis, seizure disorder, suicidal ideation, MI, heart failure, alcoholism, aortic stenosis, bipolar disorder

DOSAGE AND ROUTES

- **Child 6-12 yr: PO** 30 mg/day, may increase by 10-20 mg/day at weekly intervals, max 70 mg/day

Available forms: Caps 30, 50, 70 mg

Administer:

- Give daily in AM
- Without regard to meals
- Caps: may take whole or opened and contents dissolved in water

SIDE EFFECTS

CNS: *Hyperactivity, insomnia, restlessness, talkativeness,* dizziness, headache, dysphoria, irritability, aggressiveness, CNS tumor, dependence, addiction, mild euphoria, somnolence, lability, psychosis, mania, hallucinations, aggression

CV: *Palpitations, tachycardia,* hypertension, decrease in heart rate, dysrhythmias, MI, cardiomyopathy

EENT: Blurred vision, mydriasis, dyplopia

ENDO: Growth inhibition

GI: *Anorexia,* dry mouth, diarrhea, weight loss

GU: Impotence, change in libido

INTEG: Urticaria, angioedema, Stevens-Johnson syndrome, toxic epidermal necrolysis

PHARMACOKINETICS

Metabolized by liver; urine excretion pH dependent; crosses placenta, breast milk; half-life <1 hr

INTERACTIONS

⚠ Hypertensive crisis: MAOIs or within 14 days of MAOIs

Increase: serotonin syndrome, neuroleptic malignant syndrome—SSRIs, SNRIs, serotonin-receptor agonists

Increase: lisdexamfetamine effect—acetaZOLAMIDE, antacids, sodium bicarbonate, urinary alkalinizers

Increase: CNS effect—haloperidol, tricyclics, phenothiazines, modafinil, meperidine, PHENobarbital, phenytoin

Increase: CNS stimulation—melatonin

Decrease: absorption of phenytoin

Decrease: lisdexamfetamine effect—ascorbic acid, ammonium chloride, urinary acidifiers

Decrease: effect of—adrenergic blockers, antidiabetics

Drug/Herb

- Serotonin syndrome: St. John's wort

Increase: stimulant effect—khat, melatonin, green tea, guarana

Decrease: stimulant effect—eucalyptus

Drug/Food

Increase: amine effect—caffeine

NURSING CONSIDERATIONS

Assess:

- VS, B/P; product may reverse antihypertensives; check patients with cardiac disease often
- CBC, urinalysis; in diabetes: blood glucose; insulin changes may be required because eating may decrease
- Height, growth rate in children; growth rate may be decreased
- Mental status: mood, sensorium, affect, stimulation, insomnia, irritability

⚠ **Serotonin syndrome, neuroleptic malignant syndrome:** increased heart rate, shivering, sweating, dilated pupils, tremors, high B/P, hyperthermia, headache, confusion; if these occur, stop product, administer serotonin antagonist if needed; at least 2 wk should elapse between discontinuation of serotonergic agents and start of product

- **Tolerance or dependency:** increased amount of product may be used to get same effect; will develop after long-term use
- Overdose: pain, fever, dehydration, insomnia, hyperactivity

Black Box Warning: Before giving this product, identify presence of substance abuse; high potential for abuse

Perform/provide:

- Gum, hard candy, frequent sips of water for dry mouth

Evaluate:

- Therapeutic response: increased CNS stimulation, decreased drowsiness

Teach patient/family:

- **Seizures:** product may decrease seizure threshold; those with a seizure disorder should notify prescriber if seizure occurs
- To decrease caffeine consumption (coffee, tea, cola, chocolate); may increase irritability, stimulation
- To avoid OTC preparations unless approved by prescriber
- To taper product over several weeks; depression, increased sleeping, lethargy may occur
- To avoid alcohol ingestion
- To avoid breastfeeding
- To avoid hazardous activities until stabilized on medication
- To get needed rest; patient will feel more tired at end of day
- To use as part of a comprehensive treatment program

Black Box Warning: Serious CV effects may occur from increasing dose

TREATMENT OF OVERDOSE:

Administer fluids, antihypertensive for increased B/P, ammonium chloride for increased excretion, chlorproMAZINE for antagonizing CNS effects

lisinopril (Rx)

(lyse-in′oh-pril)

Prinivil, Zestril

Func. class.: Antihypertensive, angiotensin-converting enzyme 1 (ACE) inhibitor

Chem. class.: Enalaprilat lysine analog

Do not confuse:

lisinopril/RisperDAL/Lipitor
Prinivil/Plendil/Proventil/PriLOSEC
Zestril/Zetia/Zyprexa

ACTION: Selectively suppresses renin-angiotensin-aldosterone system; inhibits ACE, thereby preventing conversion of angiotensin I to angiotensin II

USES: Mild to moderate hypertension, adjunctive therapy of systolic CHF, acute MI

Unlabeled uses: Diabetic nephropathy/retinopathy, proteinuria, post MI

CONTRAINDICATIONS:
Hypersensitivity, angioedema

Black Box Warning: Pregnancy (D), 2nd/3rd trimesters

Precautions: Breastfeeding, renal disease, hyperkalemia, renal artery stenosis, CHF, pregnancy (C) 1st trimester, aortic stenosis

DOSAGE AND ROUTES

Hypertension

- **Adult: PO** 10-40 mg/day; max 80 mg/day
- **Child ≥6 yr: PO** 0.07 mg/kg/day up to 5 mg/day; titrate q1-2wk up to 0.6 mg/kg/day or 40 mg/day
- **Geriatric: PO** 2.5-5 mg/day, increase q7days

CHF

- **Adult: PO** 5 mg initially with diuretics, range 5-40 mg
- **Acute myocardial infarction in adults who are hemodynamically stable: PO** give 5 mg within 24 hr of onset of symptoms, then 5 mg after 24 hr, 10 mg after 48 hr, then 10 mg daily

Renal dose

- **Adult: PO** CCr <30 ml/min, reduce dose by 50%, initially 5 mg/day, max 40 mg/day; CCr <10 ml/min, 2.5 mg/day, max 40 mg/day

Available forms: Tabs 2.5, 5, 10, 20, 30, 40 mg

Administer:

- Severe hypotension may occur after 1st dose of product; may be prevented by reducing or discontinuing diuretic therapy 3 days before beginning lisinopril therapy

SIDE EFFECTS

CNS: *Vertigo,* depression, stroke, insomnia, paresthesias, *headache, fatigue,* asthenia, *dizziness*

CV: Chest pain, *hypotension,* sinus tachycardia

EENT: Blurred vision, nasal congestion

GI: Nausea, vomiting, anorexia, constipation, flatulence, GI irritation, diarrhea, hepatic failure, hepatic necrosis

GU: Proteinuria, renal insufficiency, sexual dysfunction, impotence

HEMA: Neutropenia, agranulocytosis

INTEG: Rash, pruritus

MISC: Muscle cramps, *hyperkalemia*

RESP: Dry cough, dyspnea

SYST: Angioedema, anaphylaxis, toxic epidermal necrolysis

PHARMACOKINETICS

Onset 1 hr, peak 6-8 hr, duration 24 hr, excreted unchanged in urine, half-life 12 hr

INTERACTIONS

Increase: hyperkalemia—potassium salt substitutes, potassium-sparing diuretics, potassium supplements, cycloSPORINE

Increase: possible toxicity—lithium

Increase: hypotensive effect—diuretics, other antihypertensives, probenecid, phenothiazines, nitrates, acute alcohol ingestion

Increase: hypersensitivity—allopurinol

Decrease: lisinopril effects—aspirin, indomethacin, NSAIDs

Drug/Food

- High-potassium diet (bananas, orange juice, avocados, nuts, spinach) should be avoided; hyperkalemia may occur

Drug/Lab Test

Interference: glucose/insulin tolerance tests, ANA titer

NURSING CONSIDERATIONS

Assess:

⚠ Blood studies, platelets; WBC with differential at baseline, periodically q3mo; if neutrophils <1000/mm³, discontinue treatment (recommended with collagen-vascular disease)

- Baselines of renal, hepatic studies before therapy begins, periodically; LFTs, uric acid, glucose may be increased
- **Angioedema: anaphylaxis, toxic epidermal necrolysis,** facial swelling, dyspnea, tongue swelling (rare)

Black Box Warning: Pregnancy before starting treatment; pregnancy (D)

• **Hypertension:** B/P, pulse q4hr during beginning treatment and periodically thereafter; note rate, rhythm, quality; apical/pedal pulse before administration; notify prescriber of any significant changes
• Electrolytes: potassium, sodium, chlorine
• **CHF:** edema in feet, legs daily; weight daily; dyspnea, wet crackles
• Skin turgor, dryness of mucous membranes for hydration status

Evaluate:
• Therapeutic response: decreased B/P, CHF symptoms

Teach patient/family:
• Not to discontinue product abruptly; to taper
• To rise slowly to sitting or standing position to minimize orthostatic hypotension
• To avoid increasing potassium in the diet
• To report dry cough

Black Box Warning: To report if pregnancy is planned or suspected; pregnancy (D) 2nd/3rd trimesters

TREATMENT OF OVERDOSE:

Lavage, IV atropine for bradycardia, IV theophylline for bronchospasm, digoxin, O_2, diuretic for cardiac failure

lithium (Rx)

(li'thee-um)

Carbolith ✱, Duralith ✱, Lithobid

Func. class.: Psychotropic agent—antimanic

Chem. class.: Alkali metal ion salt

ACTION:

May alter sodium, potassium ion transport across cell membrane in nerve, muscle cells; may balance biogenic amines of norepinephrine, serotonin in CNS areas involved in emotional responses

USES:

Bipolar disorders (manic phase), prevention of bipolar manic-depressive psychosis

CONTRAINDICATIONS:

Pregnancy (D), breastfeeding, children <12 yr, hepatic disease, brain trauma, organic brain syndrome, schizophrenia, severe cardiac/renal disease, severe dehydration

Precautions: Geriatric patients, thyroid disease, seizure disorders, diabetes mellitus, systemic infection, urinary retention

Black Box Warning: Lithium level >1.5 mmol/L

DOSAGE AND ROUTES

• **Adult: PO** 300-600 mg tid, maintenance 300 mg tid or qid; **SLOW REL TABS** q300 mg bid; dose should be individualized to maintain blood levels at 1-1.5 mEq/L
• **Geriatric: PO** 300 mg bid, increase q7days by 300 mg to desired dose
• **Child: PO** 15-20 mg/kg/day in 3-4 divided doses; increase as needed; do not exceed adult doses; maintain blood levels at 0.4-0.5 mEq/L

Available forms: Caps 150, 300, 600 mg; tabs 300 mg; ext rel tabs 300, 450 mg; syr 300 mg/5 ml (8 mEq/5 ml); slow rel caps 150, 300 mg ✱

Administer:
• Do not break, crush, chew caps, ext rel tabs
• Reduced dose to geriatric patients
• With meals to avoid GI upset
• Adequate fluids (2-3 L/day) to prevent dehydration during initial treatment, 1-2 L/day during maintenance

SIDE EFFECTS

CNS: *Headache, drowsiness, dizziness,* tremors, twitching, ataxia, **seizure,** slurred speech, restlessness, confusion, stupor, memory loss, clonic movements, fatigue

CV: *Hypotension,* ECG changes, dysrhythmias, circulatory collapse, edema, Brugada syndrome
EENT: Tinnitus, blurred vision
ENDO: Hyponatremia, goiter, hyperglycemia, hypo/hyperthyroidism
GI: *Dry mouth, anorexia, nausea, vomiting, diarrhea,* incontinence, abdominal pain, metallic taste
GU: Polyuria, glycosuria, proteinuria, albuminuria, urinary incontinence, polydipsia
HEMA: Leukocytosis
INTEG: Drying of hair, alopecia, rash, pruritus, hyperkeratosis, acneiform lesions, folliculitis
MS: Muscle weakness

PHARMACOKINETICS

PO: Onset rapid, peak ½-12 hr, half-life 18-36 hr depending on age, crosses blood-brain barrier, 80% of filtered lithium reabsorbed by renal tubules, excreted in urine, crosses placenta, enters breast milk, well absorbed by oral method

INTERACTIONS

- Neurotoxicity: haloperidol, thioridazine

Increase: hypothyroid effects—antithyroid agents, calcium iodide, potassium iodide, iodinated glycerol
Increase: effects of neuromuscular blocking agents, phenothiazines
Increase: renal clearance—sodium bicarbonate, acetaZOLAMIDE, mannitol, aminophylline
Increase: masking of lithium toxicity—beta-blockers used for lithium tremor
Increase: toxicity—indomethacin, diuretics, NSAIDs, losartan
Increase: lithium effect/toxicity—carBAMazepine, FLUoxetine, methyldopa, thiazide diuretics, probenecid
Decrease: lithium effects—theophyllines, urea, urinary alkalinizers

Drug/Herb

- Avoid use with kava, St. John's wort, valerian

Decrease: lithium levels—black/green tea, guarana

Drug/Food

- Significant changes in sodium intake alter lithium excretion

Drug/Lab Test

Increase: potassium excretion, urine glucose, blood glucose, protein, BUN
Decrease: VMA, T_3, T_4, ^{131}I

NURSING CONSIDERATIONS

Assess:

- **Mental status:** manic symptoms, mood, behavior before, during treatment

Black Box Warning: **Lithium toxicity:** diarrhea, vomiting, tremor, twitching; serum lithium levels 2×/wk initially then q2mo (therapeutic level: 0.5-1.5 mEq/L); toxic level >1.5 mcg/L

- Weight daily; check for, report edema in legs, ankles, wrists
- Sodium intake; decreased sodium intake with decreased fluid intake may lead to lithium retention; increased sodium, fluids may decrease lithium retention
- Skin turgor at least daily
- Urine for albuminuria, glycosuria, uric acid during beginning treatment, q2mo thereafter
- Neurologic status: LOC, gait, motor reflexes, hand tremors
- ECG in those >50 yr with CV disease, cardiology consult is recommended in those with risk factor

Evaluate:

- Therapeutic response: decrease in excitement, manic phase

Teach patient/family:

- About **the symptoms of minor toxicity:** vomiting, diarrhea, poor coordination, fine motor tremors, weakness, lassitude; **major toxicity:** coarse tremors, severe thirst, tinnitus, diluted urine
- To monitor urine specific gravity, emphasize need for follow-up care to determine lithium levels; to monitor lithium levels to ensure effective levels and treatment
- That contraception is necessary because lithium may harm fetus (pregnancy [D]); not to breastfeed

• Not to operate machinery until lithium levels stable
• That beneficial effects may take 1-3 wk
• About products that interact with lithium (provide list); about need for adequate, stable intake of salt and fluids; not to use OTC products unless approved by prescriber

TREATMENT OF OVERDOSE:
Induce emesis or lavage, maintain airway, respiratory function; dialysis for severe intoxication

lodoxamide ophthalmic
See Appendix B

loperamide (OTC, Rx)
(loe-per′a-mide)

Anti-Diarrheal, Equaline Anti-Diarrheal, Good Sense Anti-Diarrheal, Imodium ✦, Imodium A-D, Select Brand Anti-Diarrheal, Top Care Anti-Diarrheal, Walgreens Anti-Diarrheal

Func. class.: Antidiarrheal
Chem. class.: Piperidine derivative

Do not confuse:
Imodium/Indocin
Loperamide/furosemide

ACTION: Direct action on intestinal muscles to decrease GI peristalsis; reduces volume, increases bulk; electrolytes not lost

USES: Diarrhea (cause undetermined), travelers' diarrhea, chronic diarrhea, to decrease amount of ileostomy discharge
Unlabeled uses: Irinotecan-induced diarrhea, irritable bowel syndrome

CONTRAINDICATIONS: Hypersensitivity, pseudomembranous colitis, constipation, dysentery, GI bleeding/obstruction/perforation, ileus, vomiting

Precautions: Pregnancy (C), breastfeeding, children <2 yr, hepatic disease, dehydration, gastroenteritis, toxic megacolon, geriatric patients, severe ulcerative colitis

DOSAGE AND ROUTES
• **Adult: PO** 4 mg then 2 mg after each loose stool, max 16 mg/day
• **Child 9-11 yr: PO** 2 mg then 1 mg after each loose stool, max 6 mg/24 hr
• **Child 6-8 yr: PO** 2 mg then 0.1 mg/kg after each loose stool, max 4 mg/day
• **Child 2-5 yr: PO** 1 mg then 0.1 mg/kg after each loose stool, max 4 mg/24 hr

Irinotecan-induced diarrhea (unlabeled)
• **Adult: PO** 4 mg at first sign of late diarrhea (≥24 hr after irinotecan) then 2 mg q2hr × ≥12 hr; at night, 4 mg q4hr

Available forms: Caps 2 mg; liq 1 mg/5 ml; tabs 2 mg, chew tabs 2 mg

Administer:
• Do not break, crush, or chew caps
• For 48 hr only
• Do not mix oral sol with other sol

SIDE EFFECTS
CNS: Dizziness, drowsiness, fatigue
GI: *Nausea, dry mouth, vomiting, constipation,* abdominal pain, anorexia, **toxic megacolon**, bacterial enterocolitis, flatulence
INTEG: Rash
MISC: Hyperglycemia
SYST: **Anaphylaxis, angioedema, toxic epidermal necrolysis**

PHARMACOKINETICS
PO: Onset 1-3 hr, duration 4-5 hr, half-life 9-14 hr, metabolized in liver, excreted in feces as unchanged product, small amount in urine

INTERACTIONS
Increase: CNS depression—alcohol, antihistamines, analgesics, opioids, sedative/hypnotics

Drug/Herb
Increase: CNS depression—chamomile, hops, kava, valerian

NURSING CONSIDERATIONS

Assess:
- **Stools:** volume, color, characteristics, frequency; bowel pattern before product; rebound constipation
- Electrolytes (potassium, sodium, chlorine) if receiving long-term therapy
- Skin turgor q8hr if dehydration is suspected; fluid replacement
- Response after 48 hr; if no response, product should be discontinued
- **Dehydration, CNS problems in children, those with hepatic disease**
- **Abdominal distention, toxic megacolon; may occur with ulcerative colitis**

Evaluate:
- Therapeutic response: decreased diarrhea (48 hr); decreased chronic diarrhea (10 days)

Teach patient/family:
- To avoid OTC products unless directed by prescriber
- That ileostomy patient may take product for extended time
- If drowsiness occurs, not to operate machinery
- To use hard candy, sips of water for dry mouth

lopinavir/ritonavir
(low-pin′ah-ver/ri-toe′na-veer)
Kaletra
Func. class.: Antiretroviral
Chem. class.: Protease inhibitor

ACTION: Inhibits human immunodeficiency virus (HIV-1) protease and prevents maturation of the infectious virus

USES: HIV-1 in combination with or without other antiretrovirals

CONTRAINDICATIONS: Hypersensitivity to this product or polyoxyethylated castor oil (oral solution), CYP3A4 metabolized products

Precautions: Pregnancy (C), breastfeeding, hepatic disease, pancreatitis, diabetes, hemophilia, AV block, hypercholesterolemia, immune reconstitution syndrome, neonates, cardiomyopathy, congenital long-QT prolongation, hypokalemia, elderly patients, Grave's disease, polymyositis, Guillain-Barré syndrome, children, HBV/HCVCO-infection

DOSAGE AND ROUTES

HIV infection in combination with other antiretroviral agents
- **Adult: PO** 400 mg lopinavir/100 mg ritonavir bid or 800 mg lopinavir/200 mg ritonavir per day may be administered to patients with <3 lopinavir resistance–associated substitutions (L10F/I/R/V, K20M/N/R, L24I, L33F, M36I, I47V, G48V, I54L/T/V, V82A/C/F/S/T, and I84V); do not give once-daily dosing with concomitant carBAMazepine, PHENobarbital, or phenytoin; do not give as once daily dosing with efavirenz, nevirapine, (fos)amprenavir, nelfinavir
- **Pregnant adults: PO** 400 mg lopinavir/100 mg ritonavir bid, may need 600 mg lopinavir/150 mg ritonavir bid in the 2nd/3rd trimesters; once-daily dosing is not recommended; a preferred PI-based regimen for pregnant patients combines twice-daily lopinavir; ritonavir with zidovudine and either lamiVUDine or emtricitabine
- **Adult receiving concomitant efavirenz, nelfinavir, or nevirapine: PO TABS,** 500 mg lopinavir/125 mg ritonavir bid; **CAPS/SOL,** 533 mg lopinavir/133 mg ritonavir bid
- **Adolescent/child/infant >6 mo:** 300 mg lopinavir/75 mg ritonavir/m^2/dose bid. The once-daily regimen is not recommended in pediatric patients; capsules are not recommended for use in patients ≤40 kg; a preferred PI-based regimen for all children combines lopinavir/ritonavir with a dual NRTI; a lopinavir/ritonavir plus dual NRTI regimen is an alternative option in adolescents

Available forms: Oral solution 400 mg lopinavir/100 mg ritonavir/5 ml; tablets 100 mg lopinavir/25 mg ritonavir, 200 mg lopinavir/50 mg ritonavir, cup 133.3/33.3 mg

Administer:

PO route

- **TAB:** take without regard to food; swallow whole, do not crush, break, chew
- **CAP:** must take with food
- **ORAL SOL:** shake well, use calibrated measuring device
- Drug resistance testing should be done before beginning therapy in antiretroviral-naive patients and before changing therapy for treatment failure
- For adults and adolescents, therapy is recommended in those with a CD4 ≤500/mm³, who is pregnant, who has HIV-associated nephropathy, or who is being treated for hepatitis B (HBV) infection; therapy should be offered to patients at risk of transmitting HIV to sexual partners; treatment in those with a CD4 count >500/mm³ may be considered
- For children, use is recommended in any symptomatic patient; for asymptomatic or mildly symptomatic children ≥5 years, therapy is recommended for those with HIV RNA ≥100,000 copies/ml or CD4 <500/mm³; for asymptomatic or mildly symptomatic children 1-4 yr, therapy is recommended for patients with HIV RNA ≥100,000 copies/ml or CD4 <25%
- For infants <12 mo, therapy is recommended regardless of clinical status, CD4 percentage, or viral load

SIDE EFFECTS

CNS: Paresthesia, headache, **seizures**, fever, dizziness, insomnia, asthenia, **intracranial bleeding, encephalopathy**

CV: **QT, PR interval prolongation, deep vein thrombosis**

EENT: Blurred vision, otitis media, tinnitus

GI: Diarrhea, buccal mucosa ulceration, abdominal pain, nausea, taste perversion, dry mouth, vomiting, anorexia

INTEG: Rash

MISC: Asthenia, **angioedema, anaphylaxis, Stevens-Johnson syndrome**, increase lipids, lipodystrophy

MS: Pain, **rhabdomyolysis**, myalgias

PHARMACOKINETICS

Well absorbed, 98% protein binding, hepatic metabolism, peak 2-4 hr, terminal half-life 6 hr

INTERACTIONS

Increase: Toxicity—amiodarone, astemizole, avanafil, azole antifungals, benzodiazepines, buPROPion, cisapride, cloZAPine, desipramine, dihydroergotamine, encainide, ergotamine, flecainide, HMG-CoA reductase inhibitors, interleukins, meperidine, midazolam, pimozide, piroxicam, propafenone, propoxyphene, quiNIDine, ranolazine, rivaroxaban, saquinavir, terfenadine, triazolam, zolpidem

Increase: QT prolongation—class 1A/III antidysrhythmics, some phenothiazines, β-agonists, local anesthetics, tricyclics, haloperidol, chloroquine, droperidol, pentamidine, CYP3A4 inhibitors (amiodarone, clarithromycin, erythromycin, telithromycin, troleandomycin), arsenic trioxide, levomethadyl, CYP3A4 substrates (methadone, pimozide, QUEtiapine, quiNIDine, risperiDONE, ziprasidone)

Increase: Ritonavir levels—fluconazole

Increase: Level of both products—clarithromycin, ddI

Increase: Levels of bosentan

Decrease: Ritonavir levels—rifamycins, nevirapine, barbiturates, phenytoin, budesonide, predniSONE

Decrease: Levels of anticoagulants, atovaquone, divalproex, ethinyl estradiol, lamoTRIgine, phenytoin, sulfamethoxazole, theophylline, voriconazole, zidovudine

Drug/Lab Test

Increase: AST, ALT, CPK, cholesterol, GGT, triglycerides, uric acid

Decrease: Hct, Hgb, RBC, neutrophils, WBC

L

Drug/Herb

Decrease: Ritonavir levels—St. John's wort; avoid concurrent use

Avoid use with red yeast rice, evening primrose oil

NURSING CONSIDERATIONS

Assess:

- HIV: viral load, CD4 at baseline, throughout therapy; blood glucose, plasma HIV RNA, serum cholesterol/lipid profile; resistance testing before starting therapy and after treatment failure
- Signs of infection, anemia
- Hepatic studies: ALT, AST
- Bowel pattern before, during treatment; if severe abdominal pain with bleeding occurs, discontinue product; monitor hydration
- Skin eruptions; rash

⚠ **Rhabdomyolysis:** Muscle pain, increased CPK, weakness, swelling of affected muscles; if these occur and if confirmed by CPK, product should be discontinued

⚠ **QT prolongation:** ECG for QT prolongation, ejection fraction; assess for chest pain, palpitations, dyspnea

⚠ **Serious skin disorders:** Stevens-Johnson syndrome, angioedema, anaphylaxis

Evaluate:

- Therapeutic response: improvement in HIV symptoms; improving viral load, CD4+ T cells

Teach patient/family:

- To take as prescribed; if dose is missed, to take as soon as remembered up to 1 hr before next dose; not to double dose
- That product is not a cure for HIV; that opportunistic infections can continue to be acquired
- That redistribution of body fat or accumulation of body fat may occur
- That others can continue to contract HIV from patient
- To avoid OTC, prescription medications, herbs, supplements unless approved by prescriber; not to use St. John's wort because it decreases product's effect
- That regular follow-up exams and blood work will be required

loratadine (OTC, Rx)

(lor-a′ti-deen)

Alavert, Claritin, Claritin Children's, Claritin RediTabs, Clear-Atadine, Dimetapp, Equaline Non-Drowsy, Equate Allergy Relief, Good Sense Non-Drowsy, Leader Allergy Relief, Tavist ND, Wal-itin Aller-Melts, Wal-vert

Func. class.: Antihistamine, 2nd generation

Chem. class.: Selective histamine (H_1)-receptor antagonist

Do not confuse:

loratadine/lovastatin/LORazepam/losartan

ACTION: Binds to peripheral histamine receptors, thereby providing antihistamine action without sedation

USES: Seasonal rhinitis, chronic idiopathic urticaria for those ≥2 yr

CONTRAINDICATIONS: Hypersensitivity, acute asthma attacks, lower respiratory tract disease

Precautions: Pregnancy (B), breastfeeding, increased intraocular pressure, bronchial asthma, hepatic/renal disease

DOSAGE AND ROUTES

- **Adult and child ≥6 yr: PO** 10 mg/day
- **Child 2-5 yr: PO** 5 mg/day

Renal/hepatic dose

- **Adult: PO** CCr <30 ml/min or hepatic disease, 10 mg every other day

Available forms: Tabs 10 mg; rapid-disintegrating tabs 10 mg; orally disintegrating tabs 10 mg; syr 1 mg/ml; susp 5 mg/ml, ext rel tab 10 mg

Administer:

- **Rapid-disintegrating tabs** by placing on tongue, to be swallowed after disintegrated with/without water
- Use within 6 mo of opening pouch and immediately after opening blister pack
- On empty stomach daily

Ext Rel Tab

- Do not break, crush, or chew

SIDE EFFECTS

CNS: Sedation (more common with increased doses), headache, fatigue, restlessness

EENT: Dry mouth

PHARMACOKINETICS

Onset 1-3 hr, peak 8-10 hr, duration 24 hr, metabolized in liver to active metabolites, excreted in urine, active metabolite desloratadine half-life 20 hr

INTERACTIONS

Increase: antihistamine effects—MAOIs

Increase: CNS depressant effects—alcohol, antidepressants, other antihistamines, sedative/hypnotics

Increase: loratadine level—cimetidine, ketoconazole, macrolides (clarithromycin, erythromycin)

Drug/Lab Test

False negative: skin allergy tests (discontinue antihistamine 3 days before testing)

NURSING CONSIDERATIONS

Assess:

- **Allergy:** hives, rash, rhinitis; monitor respiratory status

Evaluate:

- Therapeutic response: absence of running or congested nose, other allergy symptoms

Teach patient/family:

- To avoid driving, other hazardous activities if drowsiness occurs
- To avoid use of other CNS depressants

LORazepam (Rx)

(lor-a′ze-pam)

Ativan

Func. class.: Sedative, hypnotic; antianxiety

Chem. class.: Benzodiazepine, short acting

Controlled Substance Schedule IV

Do not confuse:
LORazepam/ALPRAZolam/clonazePAM

ACTION:
Potentiates the actions of GABA, especially in the limbic system and the reticular formation

USES:
Anxiety, irritability with psychiatric or organic disorders, preoperatively; insomnia; adjunct for endoscopic procedures, status epilepticus

Unlabeled uses: Antiemetic before chemotherapy, rectal use, alcohol withdrawal, seizure prophylaxis, agitation, insomnia, sedation maintenance

CONTRAINDICATIONS:
Pregnancy (D), breastfeeding, hypersensitivity to benzodiazepines, benzyl alcohol; closed-angle glaucoma, psychosis, history of drug abuse, COPD, sleep apnea

Precautions: Children <12 yr, geriatric patients, debilitated, renal/hepatic disease, addiction, suicidal ideation, abrupt discontinuation

DOSAGE AND ROUTES

Anxiety

- **Adult/adolescent: PO** 2-3 mg/day in divided doses, max 10 mg/day
- **Geriatric: PO** 1-2 mg/day in divided doses or 0.5-1 mg at bedtime

Preoperatively

- **Adult: IM** 50 mcg/kg 2 hr before surgery; **IV** 44 mcg/kg 15-20 min before surgery, max 2 mg 15-20 min before surgery
- **Child ≥12 yr: IV** 0.05 mg/kg, max 4 mg

L

Status epilepticus

• **Neonate: IV** 0.05 mg/kg

• **Child: IV** 0.1 mg/kg up to 4 mg/dose; **RECT** (unlabeled) 0.05-0.1 mg × 2; wait 7 min before giving 2nd dose

Alcohol withdrawal (unlabeled)

• **Adult: PO** 2 mg q6hr × 4 doses then 1 mg q6hr for 8 doses

Insomnia (unlabeled)

• **Adult: PO** 2-4 mg at bedtime; only minimally effective after 2 wk continuous therapy

• **Geriatric: PO** 0.5-1 mg initially

Available forms: Tabs 0.5, 1, 2 mg; inj 2, 4 mg/ml; conc oral sol 2 mg/ml

Administer:

PO route

• With food or milk for GI symptoms; crushed if patient is unable to swallow medication whole

• Sugarless gum, hard candy, frequent sips of water for dry mouth

• Give largest dose before bedtime if giving in divided doses

• **Concentrate:** use calibrated dropper; add to food/drink; consume immediately

IM route

• Deep into large muscle mass

• Use this route when IV is not feasible

Direct IV route

• Prepare immediately before use; short stability time

• IV after diluting in equal vol sterile water, 5% dextrose, or 0.9% NaCl for inj; give through Y-tube or 3-way stopcock; give at ≤2 mg/1 min; do not give rapidly

• Do not use in neonates (benzyl alcohol)

Y-site compatibilities: Acyclovir, albumin, allopurinol, amifostine, amikacin, amoxicillin, amoxicillin/clavulanate, amphotericin B cholesteryl, amsacrine, atenolol, atracurium, bivalirudin, bleomycin, bumetanide, butorphanol, calcium chloride/gluconate, CARBOplatin, ceFAZolin, cefepime, cefotaxime, cefoTEtan, cefOXitin, cefTAZidime, ceftizoxime, ceftobiprole, cefTRIAXone, cefuroxime, chloramphenicol, chlorproMAZINE, cimetidine, ciprofloxacin, cisatracurium, CISplatin, cladribine, clindamycin, cloNIDine, cyclophosphamide, cycloSPORINE, cytarabine, DACTINomycin, DAPTOmycin, dexamethasone, dexmedetomidine, diltiazem, DOBUTamine, DOCEtaxel, DOPamine, doripenem, DOXOrubicin, DOXOrubicin liposomal, droperidol, enalaprilat, ePHEDrine, EPINEPHrine, epirubicin, eptifibatide, erythromycin, esmolol, etomidate, famotidine, fenoldopam, fentaNYL, filgrastim, fluconazole, fludarabine, fosphenytoin, furosemide, ganciclovir, gatifloxacin, gemcitabine, gentamicin, glycopyrrolate, granisetron, haloperidol, heparin, hydrocortisone, HYDROmorphone, hydrOXYzine, ifosfamide, inamrinone, insulin (regular), irinotecan, isoproterenol, ketorolac, labetalol, lidocaine, linezolid, magnesium sulfate, mannitol, mechlorethamine, melphalan, meropenem, metaraminol, methadone, methotrexate, methyldopate, methylPREDNISolone, metoclopramide, metoprolol, metroNIDAZOLE, micafungin, midazolam, milrinone, minocycline, mitoXANtrone, morphine, mycophenolate, nafcillin, nalbuphine, naloxone, nesiritide, niCARdipine, nitroglycerin, nitroprusside, norepinephrine, octreotide, oxaliplatin, oxytocin, PACLitaxel, palonosetron, pamidronate, pancuronium, PEMEtrexed, pentamidine, PENTobarbital, PHENobarbital, piperacillin, piperacillin-tazobactam, polymyxin B, potassium chloride, propofol, ranitidine, remifentanil, tacrolimus, teniposide, theophylline, thiotepa, ticarcillin, ticarcillin-clavulanate, tigecycline, tirofiban, tobramycin, TPN, trastuzumab, trimethobenzamide, trimethoprim-sulfamethoxazole, vancomycin, vasopressin, vecuronium, verapamil, vinCRIStine, vinorelbine, voriconazole, zidovudine

SIDE EFFECTS

CNS: *Dizziness, drowsiness,* confusion, headache, anxiety, tremors, stimulation, fatigue, depression, insomnia, hallucinations, weakness, unsteadiness

CV: *Orthostatic hypotension,* ECG changes, tachycardia, hypotension; apnea, cardiac arrest (IV, rapid)

EENT: *Blurred vision,* tinnitus, mydriasis
GI: Constipation, dry mouth, nausea, vomiting, anorexia, diarrhea
INTEG: Rash, dermatitis, itching
MISC: Acidosis

PHARMACOKINETICS

Metabolized by liver; excreted by kidneys; crosses placenta, excreted in breast milk; half-life 42 hr (neonates), 10.5 hr (older child), 12 hr (adult), 91% protein bound
PO: Onset 1/2 hr, peak 1-6 hr, duration 12-24 hr
IM: Onset 15-30 min, peak 1-1 1/2 hr, duration 6-8 hr
IV: Onset 5-15 min, peak unknown, duration 6-8 hr

INTERACTIONS

Increase: LORazepam effects—CNS depressants, alcohol, disulfiram
Decrease: LORazepam effects—valproic acid, oral contraceptives
Drug/Herb
Increase: CNS depression—chamomile, kava, valerian
Drug/Lab Test
Increase: AST, ALT

NURSING CONSIDERATIONS

Assess:
⚠ **Anxiety:** decrease in anxiety; mental status: mood, sensorium, affect, sleeping pattern, drowsiness, dizziness, suicidal tendencies
- Renal/hepatic/blood status if receiving high-dose therapy
- **Physical dependency, withdrawal symptoms:** headache, nausea, vomiting, muscle pain, weakness, tremors, seizures, after long-term, excessive use

Perform/provide:
- Assistance with ambulation during beginning therapy, since drowsiness, dizziness occurs
- Check to confirm that PO medication has been swallowed
- Refrigerate parenteral form

Evaluate:
- Therapeutic response: decreased anxiety, restlessness, insomnia

Teach patient/family:
- That product may be taken with food
- Not to use product for everyday stress or for >4 mo unless directed by prescriber
- Not to take more than prescribed amount; may be habit forming
- To avoid OTC preparations (cough, cold, hay fever) unless approved by prescriber
- To avoid driving, activities that require alertness, since drowsiness may occur
- To avoid alcohol, other psychotropic medications unless directed by prescriber
- Not to discontinue medication abruptly after long-term use
- To rise slowly because fainting may occur, especially among geriatric patients
- That drowsiness may worsen at beginning of treatment
- To notify prescriber if pregnancy is planned or suspected, pregnancy (D), do not breastfeed

TREATMENT OF OVERDOSE:

Lavage, VS, supportive care, flumazenil

lorcaserin

(lor-ca-ser'in)

Belviq

Func. class.: Weight-control agent (anorexiant)

Chem. class.: Serotonin 2C (5-HT_{2C}) receptor agonist

ACTION: Decreases food consumption and decreases hunger by selectively activating 5-HT_{2C} receptors

USES: Obesity management

CONTRAINDICATIONS: Pregnancy (X), breastfeeding, hypersensitivity, severe renal impairment

Precautions: Children, other organic causes of obesity, anemia, AV block, bradycardia, bundle branch block, depression, dialysis, liver/kidney disease, multiple myeloma, neutropenia, suicidal ideation, Peyronie's disease, pulmonary hypertension, sick sinus syndrome

DOSAGE AND ROUTES

Adult:

- **PO** 10 mg bid; do not exceed recommended dosage

Available forms: Tabs, film-coated 10 mg

Administer:

- For obesity if patient is on weight reduction program that includes dietary changes, exercise
- May give without regard to food

SIDE EFFECTS

CNS: Insomnia, depression, serotonin syndrome, anxiety, suicidal ideation, dizziness, headache, fatigue

CV: Bradycardia, hypertension

GI: Diarrhea, constipation, nausea

HEMA: Neutropenia, leukopenia, lymphopenia

INTEG: Rash

MS: Back pain

PHARMACOKINETICS

70% protein binding, half-life 11 hr

INTERACTIONS

Increase: life-threatening serotonin syndrome or NMS—SSRIs, SNRIs, serotonin-receptor agonists, sibutramine, MAOIs, linezolid, tricyclic antidepressants, buPROPion, lithium, DOPamine antagonist, traMADol

Increase: risk of hypoglycemia with sulfonylureas and insulin

Drug/Herb

Increase: Serotonin syndrome—St. John's wort

NURSING CONSIDERATIONS

Assess:

- Weight weekly; oral hypoglycemic dosage might need to be reduced in diabetic patients
- Monitor blood glucose, CBC with differential, Hct/Hgb, serum prolactin

⚠ Pregnancy (X): do not use in pregnancy

Suicidal Ideation:

⚠ Use caution in psychiatric disorders with emotional lability; assess for depression, suicidal thoughts/behaviors

Evaluate:

- Therapeutic response: decrease in weight

Teach patient/family:

- To avoid hazardous activities until stabilized on medication
- To discuss unpleasant side effects

⚠ To notify prescriber if pregnancy is planned or suspected, pregnancy X

losartan

(lo-zar'tan)

Cozaar

Func. class.: Antihypertensive

Chem. class.: Angiotensin II receptor (type AT_1) antagonist

Do not confuse:

losartan/valsartan

Cozaar/Zocor

ACTION: Blocks the vasoconstrictor and aldosterone-secreting effects of angiotensin II; selectively blocks the binding of angiotensin II to the AT_1 receptor found in tissues

USES: Hypertension, alone or in combination; nephropathy in type 2 diabetes; proteinuria; stroke prophylaxis for hypertensive patients with left ventricular hypertrophy

Unlabeled uses: Heart failure

CONTRAINDICATIONS: Hypersensitivity

Black Box Warning: Pregnancy (D) 2nd/3rd trimesters

Precautions: Pregnancy (C) 1st trimester, breastfeeding, children, geriatric patients; hypersensitivity to ACE inhibitors; hepatic disease, angioedema, renal artery stenosis, African descent, hyperkalemia, hypotension

DOSAGE AND ROUTES

Hypertension

- **Adult: PO** 50 mg/day alone or 25 mg/day in combination with diuretic; maintenance 25-100 mg/day
- **Child ≥6 yr: PO** 0.7 mg/kg/day, max 50 mg/day

Hypertension with left ventricular hypertrophy (benefit does not apply to those of African American descent)

- **Adult: PO** 50 mg/day; add hydrochlorothiazide 12.5 mg/day and/or increase losartan to 100 mg/day then increase hydrochlorothiazide to 25 mg/day

Nephropathy in type 2 diabetic patients/proteinuria

- **Adult: PO** 50 mg/day, may increase to 100 mg/day

Hepatic dose/volume/depletion

- **Adult: PO** 25 mg/day as starting dose

Available forms: Tabs 25, 50, 100 mg

Administer:

- Without regard to meals

SIDE EFFECTS

CNS: *Dizziness, insomnia,* anxiety, confusion, abnormal dreams, migraine, tremor, vertigo, headache, malaise, depression, fatigue

CV: Angina pectoris, 2nd-degree AV block, **cerebrovascular accident**, *hypotension,* **MI, dysrhythmias**

EENT: Blurred vision, burning eyes, conjunctivitis

GI: *Diarrhea, dyspepsia,* anorexia, constipation, dry mouth, flatulence, gastritis, vomiting

GU: Impotence, nocturia, urinary frequency, UTI, **renal failure**

HEMA: Anemia, **thrombocytopenia**

INTEG: Alopecia, dermatitis, dry skin, flushing, photosensitivity, rash, pruritus, sweating, **angioedema**

META: Gout, hyperkalemia, hypoglycemia

MS: Cramps, myalgia, pain, stiffness

RESP: *Cough, upper respiratory infection,* congestion, dyspnea, bronchitis

PHARMACOKINETICS

Peak 1-4 hr, extensively metabolized, half-life 2 hr, metabolite 6-9 hr, excreted in urine/feces, protein binding 98.7%

INTERACTIONS

Increase: lithium toxicity—lithium

Increase: antihypertensive effect—fluconazole

Increase: hyperkalemia—potassium-sparing diuretics, potassium supplements, ACE inhibitors

Decrease: antihypertensive effect—NSAIDs, PHENobarbital, rifamycin, salicylates

NURSING CONSIDERATIONS

Assess:

- B/P with position changes, pulse before and periodically during treatment; note rate, rhythm, quality
- Baselines of renal, hepatic, electrolyte studies before therapy begins and periodically thereafter
- Skin turgor, dryness of mucous membranes for hydration status

⚠ Angioedema: facial swelling, dyspnea, wheezing; may occur rapidly; tongue swelling (rare)

- **CHF:** jugular venous distention; edema in feet/legs, weight daily
- **Blood dyscrasias:** thrombocytopenia, anemia (rare)

Black Box Warning: Pregnancy before starting treatment; pregnancy (D) 2nd/3rd trimester

Evaluate:

- Therapeutic response: decreased B/P, slowing diabetic neuropathy

Teach patient/family:

- To avoid sunlight or to wear sunscreen if in sunlight; that photosensitivity may occur
- To comply with dosage schedule, even if feeling better; not to discontinue abruptly
- To notify prescriber of mouth sores, fever, swelling of hands or feet, irregular heartbeat, chest pain
- That excessive perspiration, dehydration, vomiting, diarrhea may lead to fall in B/P; to consult prescriber if these occur
- That product may cause dizziness, fainting, lightheadedness; to avoid hazardous activities until reaction is known
- To rise slowly to sitting or standing position to minimize orthostatic hypotension

Black Box Warning: To use contraception while taking this product; pregnancy (D) 2nd/3rd trimesters

- To avoid salt substitutes, alcohol, grapefruit juice, OTC products unless approved by prescriber

loteprednol ophthalmic

See Appendix B

lovastatin (Rx)

(loh-vah-stat'in)

Altoprev, Mevacor

Func. class.: Antilipemic

Chem. class.: HMG-CoA reductase inhibitor

Do not confuse:

lovastatin/Lotensin/Leustatin

Mevacor/mivacron

ACTION: Inhibits HMG-CoA reductase enzyme, which reduces cholesterol synthesis

USES: As an adjunct for primary hypercholesterolemia (types IIa, IIb), atherosclerosis; heterozygous familial hypercholesterolemia (adolescents)

CONTRAINDICATIONS: Pregnancy (X), breastfeeding, hypersensitivity, active hepatic disease

Precautions: Children, past hepatic disease, alcoholism, severe acute infections, trauma, hypotension, uncontrolled seizure disorders, severe metabolic disorders, electrolyte imbalances, visual disorder

DOSAGE AND ROUTES

- **Adult: PO** 20 mg/day with evening meal; may increase to 20-80 mg/day in single or divided doses, max 80 mg/day; **EXT REL** 20-60 mg/day at bedtime, max 60 mg/day

Heterozygous familial hypercholesterolemia

- **Adolescent 10-17 yr: PO** 10-40 mg with evening meal

Renal dose

- **Adult: PO** CCr <30 mg/min, max 20 mg/day unless titrated

Available forms: Tabs 10, 20, 40 mg; ext rel tab (Altocor) 10, 20, 40, 60 mg

Administer:

- In evening with meal; if dose is increased, take with breakfast and evening meal
- Altroprev not equivalent to Mevacor
- Do not crush, chew ext rel tab
- Store in cool environment in airtight, light-resistant container

SIDE EFFECTS

CNS: *Dizziness, headache, tremor,* insomnia, paresthesia

EENT: *Blurred vision,* lens opacities

GI: *Flatus, nausea, constipation, diarrhea, dyspepsia, abdominal pain, heartburn,* hepatic dysfunction, vomiting, acid regurgitation, dry mouth, dysgeusia

HEMA: Thrombocytopenia, hemolytic anemia, leukopenia

INTEG: *Rash, pruritus,* photosensitivity

MS: *Muscle cramps, myalgia,* myositis, rhabdomyolysis; leg, shoulder, or localized pain

PHARMACOKINETICS

PO: Peak 2 hr; peak response 4-6 wk, ext rel peak 14 hr; metabolized in liver (metabolites); highly protein bound; excreted in urine 10%, feces 83%; crosses blood-brain barrier, placenta; excreted in breast milk; half-life 1 hr

INTERACTIONS

Increase: myalgia, myositis, rhabdomyolysis—azole antifungals, clarithromycin, clofibrate, cycloSPORINE, dalfopristin, danazol, diltiazem, erythromycin, gemfibrozil, niacin, protease inhibitors, quinupristin, telithromycin, verapamil, avoid concurrent use
Increase: bleeding—warfarin
Increase: lovastatin effects—diltiazem
Decrease: effects of lovastatin—bile acid sequestrants, exonatide, bosentan

Drug/Herb

Decrease: effect—pectin, St. John's wort
Increase: adverse reactions—red yeast rice

Drug/Food

- Possible toxicity: grapefruit juice

Increase: levels of lovastatin with food; must be taken with food
Decrease: absorption—oat bran

Drug/Lab Test

Increase: CK, LFTs

NURSING CONSIDERATIONS

Assess:

- **Diet;** obtain diet history including fat, cholesterol in diet
- Fasting cholesterol, LDL, HDL, triglycerides periodically during treatment
- Hepatic studies at initiation, 6 wk, 12 wk after initiation or change in dose, periodically thereafter; AST, ALT, LFTs may increase
- Renal function in patients with compromised renal system: BUN, creatinine, I&O ratio

Evaluate:

- Therapeutic response: decreased triglycerides, sLDL, total cholesterol; increased HDL; slowing CAD

Teach patient/family:

- To report suspected pregnancy (pregnancy [X]); not to breastfeed
- That blood work, ophthalmic exam will be necessary during treatment
- To report blurred vision, severe GI symptoms, dizziness, headache, muscle pain, weakness
- To use sunscreen or to stay out of the sun to prevent photosensitivity
- That previously prescribed regimen will continue: low-cholesterol diet, exercise program, smoking cessation
- That product should be taken with food, not to crush, chew ext rel product

loxapine (Rx)

(lox′a-peen)

Adasuve, Loxapac ✱, Loxitane

Func. class.: Antipsychotic, neuroleptic

Chem. class.: Dibenzoxazepine

Do not confuse:
Loxitane/Soriatane

ACTION: Depresses cerebral cortex, hypothalamus, limbic system, which control activity and aggression; blocks neurotransmission produced by DOPamine at synapse; exhibits strong α-adrenergic, anticholinergic blocking action; mechanism for antipsychotic effects is unclear

USES: Schizophrenia, bipolar disorder

Unlabeled uses: Anxiety

CONTRAINDICATIONS: *Hypersensitivity, coma*

Black Box Warning: Acute bronchospasm, asthma, COPD, emphysema

Precautions: Pregnancy (C), breastfeeding, children <16 yr, geriatric patients, seizure disorders, cardiac/renal/hepatic disease, prostatic hypertrophy, cardiac conditions, blood dyscrasias, brain damage, bone marrow depression, alcohol and barbiturate withdrawal states, severe CNS depression, closed-angle glaucoma

Black Box Warning: Increased mortality in elderly patients with dementia-related psychosis

DOSAGE AND ROUTES

- **Adult: PO** 10 mg bid-qid initially, may be rapidly increased depending on severity of condition, maintenance 60-100 mg/day; inhalation powder 10 mg as a single dose in 24 hr
- **Geriatric: PO** 5-10 mg daily-bid, increase q4-7days by 5-10 mg, max 250 mg/day

Available forms: Caps 5, 10, 25, 50 mg; tabs 5, 10, 25, 50 mg; inhalation powder

Administer:

- Store in airtight, light-resistant container
- Reduced dose to geriatric patients
- Anticholinergic agent if EPS symptoms occur

PO route

- Give with food
- Concentrate mixed in orange or grapefruit juice

Inhalation route

- Due to the risk of bronchospasm, loxapine for oral inhalation (Adasuve) has restricted availability only through the Adasuve Risk Evaluation and Mitigation Strategy (ADASUVE REMS) program
- Only use by a healthcare professional in a healthcare facility enrolled in the ADASUVE REMS program and with immediate on-site access to equipment and personnel trained to manage acute bronchospasm, including advanced airway management (intubation and mechanical ventilation)
- Prior to administration, all patients must be screened for history of asthma, COPD, or other pulmonary disease; use of medications to treat airway disease; and examined (including chest auscultation) for respiratory abnormalities such as wheezing
- Each inhaler is for single use only, use only once in a 24 hr period
- *Step 1:* When ready to use, tear open pouch and remove inhaler; the indicator light on the inhaler should be off
- *Step 2:* Firmly pull the plastic tab from rear of inhaler and ensure the green light turns on, which indicates the inhaler is ready to use
- *Step 3:* Inform the patient that the inhaler may produce a flash of light and clicking sound and may become warm during use
- *Step 4:* Instruct the patient to hold the inhaler away from the mouth and breathe out fully to empty the lungs
- *Step 5:* The patient should place the mouthpiece of the inhaler between the lips, close the lips, and inhale through the mouthpiece with a steady deep breath
- *Step 6:* Instruct the patient to remove the inhaler from the mouth and hold their breath for as long as possible, up to 10 sec
- *Step 7:* Check that green light turns off, indicating that the dose was delivered; if the light remains on, then steps 4-6 should be repeated up to 2 additional times; if the green light does not turn off, discard the inhaler and use a new one
- Use inhaler within 15 min after removing the tab to prevent automatic deactivation
- The green light will turn off after use, at which time the inhaler is no longer usable
- After use, the patient must be monitored for signs and symptoms of bronchospasm or other respiratory distress, including a physical exam with chest auscultation, at least every 15 minutes for a minimum of 1 hr
- Refer to the Important Administration Instructions section

SIDE EFFECTS

CNS: *EPS: pseudoparkinsonism, akathisia, dystonia, tardive dyskinesia, drowsiness, headache,* **seizures**, confusion, **neuroleptic malignant syndrome**

CV: *Orthostatic hypotension,* **cardiac arrest,** ECG changes, tachycardia
EENT: Blurred vision, glaucoma
ENDO: Hyperprolactinemia, galactorrhea, gynecomastia, menstrual irregularities
GI: *Dry mouth, nausea, vomiting, anorexia, constipation,* diarrhea, jaundice, weight gain
GU: Urinary retention, urinary frequency, enuresis, impotence, amenorrhea, gynecomastia
HEMA: **Anemia, leukopenia, leukocytosis, agranulocytosis**
INTEG: *Rash,* photosensitivity, dermatitis
RESP: **Laryngospasm,** dyspnea, **respiratory depression**

PHARMACOKINETICS

96.6% protein binding, metabolized by liver, excreted in urine, crosses placenta, enters breast milk, initial half-life 5 hr
PO: Onset 20-30 min, peak 2-4 hr, duration 12 hr

INTERACTIONS

Increase: toxicity—EPINEPHrine, lithium
Increase: EPS—other antipsychotics, metoclopramide
Increase: CNS depression—MAOIs, antidepressants, alcohol
Increase: anticholinergic effects—anticholinergics
Decrease: effects—guanadrel, guanethidine, levodopa, carBAMazepine
Drug/Herb
Increase: CNS depression—chamomile, cola tree, hops, kava, nettle, nutmeg, skullcap, valerian
Increase: EPS—betel palm, kava

NURSING CONSIDERATIONS

Assess:
- Mental status before initial administration
- Swallowing of PO product; check for hoarding, giving of product to other patients
- I&O ratio; palpate bladder if low urinary output occurs, urinary retention may be present
- Bilirubin, CBC, LFTs monthly
- Urinalysis recommended before and during prolonged therapy
- Affect, orientation, LOC, reflexes, gait, coordination, sleep-pattern disturbances
- B/P standing and lying; pulse, respirations q4hr during initial treatment; establish baseline before starting treatment; report drops of 30 mm Hg
- Dizziness, faintness, palpitations, tachycardia on rising
- **EPS** including akathisia (inability to sit still, no pattern to movements), tardive dyskinesia (bizarre movements of jaw, mouth, tongue, extremities), pseudoparkinsonism (rigidity, tremors, pill rolling, shuffling gait)

⚠ **Neuroleptic malignant syndrome:** muscle rigidity, increased CPK, altered mental status, hyperthermia
- **Bronchospasm:** monitor closely after use of inhalation powder
- Constipation, urinary retention daily; if these occur, increase bulk, water in diet
- Supervised ambulation until stabilized on medication; do not involve patient in strenuous exercise program because fainting is possible; patient should not stand still for long periods

Evaluate:
- Therapeutic response: decrease in emotional excitement, hallucinations, delusions, paranoia; reorganization of patterns of thought, speech

Teach patient/family:
- That orthostatic hypotension may occur; to rise from sitting or lying position gradually
- To avoid hot tubs, hot showers, tub baths because hypotension may occur; that, in hot weather, heat stroke may occur; to take extra precautions to stay cool
- To avoid abrupt withdrawal of product because EPS may result; that product should be withdrawn slowly
- To avoid OTC preparations (cough, hay fever, cold) unless approved by prescriber; that serious product interactions

may occur; to avoid use with alcohol, CNS depressants; that increased drowsiness may occur
• To avoid hazardous activities until stabilized on medication
• To use sunscreen during sun exposure to prevent burns
• About necessity for meticulous oral hygiene because oral candidiasis may occur
• To report impaired vision, jaundice, tremors, muscle twitching

TREATMENT OF OVERDOSE:

Lavage; provide an airway, do not use EPINEPHrine

lubiprostone (Rx)

(loo-bee-pros'tone)

Amitiza

Func. class.: Gastrointestinal agent—miscellaneous

ACTION: Locally acting chloride channel activator; enhances a chloride-rich intestinal fluid secretion without altering other electrolytes; increases motility in the intestine, thereby increasing softening and passage of stool

USES: Chronic idiopathic constipation, constipation-predominant irritable bowel syndrome in women >18 yr, opiate agonist–induced constipation with chronic non-cancer pain

CONTRAINDICATIONS: Hypersensitivity, GI obstruction

Precautions: Pregnancy (C), breastfeeding, children, diarrhea, inflammatory bowel disease, abdominal pain, cholelithiasis, fecal impaction, GI/hepatic disease

DOSAGE AND ROUTES

Chronic idiopathic constipation/opiate agonist–induced constipation
• **Adult: PO** 24 mcg bid with food, water

IBS with constipation (females)
• **Adult and adolescent ≥18 yr: PO** 8 mcg bid with food, water

Hepatic dose
• **Adult: PO For chronic constipation:** 16 mcg bid (Child-Pugh B); 8 mcg bid (Child-Pugh C); **for irritable bowel:** 8 mcg daily (Child-Pugh C), may be increased if tolerated

Available forms: Caps 8, 24 mcg

Administer:
• With food bid
• Store at room temperature

SIDE EFFECTS

CNS: *Headache,* dizziness, depression, fatigue, insomnia
CV: Hypertension, chest pain
GI: *Nausea, abdominal pain, eructation,* abdominal distention, constipation, diarrhea, dry mouth, dyspepsia, flatulence, gastroenteritis viral, gastroesophageal reflux disease, vomiting, fecal incontinence, fecal urgency
GU: UTI
MISC: Chest pain, peripheral edema, influenza, pyrexia, viral infection
MS: Back pain, arthralgia, muscle cramps, pain in extremities
RESP: Bronchitis, cough, dyspnea, nasopharyngitis, sinusitis, upper respiratory tract infection

PHARMACOKINETICS

Peak 1.14 hr; 94% protein binding; half-life 0.9-1.4 hr; metabolism rapid, extensively in stomach, jejunum

INTERACTIONS

Decrease: effect by antidiarrheals and anticholinergics
• Do not use with sodium phosphate monobasic monohydrate, sodium phosphate dibasic anhydrous or with other laxative, purgatives when evacuating bowel before radiologic exam or surgery
• Possible GI obstruction: NIFEdipine ext rel tab

NURSING CONSIDERATIONS

Assess:

- **GI symptoms:** nausea, abdominal pain, diarrhea
- Need for continued treatment periodically

Evaluate:

- Therapeutic response: decreased constipation

Teach patient/family:

- To notify prescriber of GI symptoms, diarrhea, hypersensitivity reactions

lucinactant

(loo′sin-ak′tant)

Surfaxin

Func. class.: Synthetic lung surfactant

USES: Prevention of respiratory distress syndrome in premature neonates (RDS)

DOSAGE AND ROUTES

Premature neonate

Intratracheal 5.8 ml/kg birth weight divided in 4 doses; give each dose with neonate in a different position; provide positive pressure ventilation when stable; dosage may be repeated 4 times in first 48 hr

lurasidone (Rx)

(loo-ras′i-done)

Latuda

Func. class.: Atypical antipsychotic

Chem. class.: Benzoisothiazol derivative

ACTION: May modulate central DOPaminergic and serotonergic activity; high affinity for DOPamine-D2 receptors, serotonin 5-HT2A receptors; partial agonist at serotonin 5-HT1A receptor

USES: Schizophrenia

Unlabeled uses: Agitation

CONTRAINDICATIONS: Hypersensitivity

Precautions: Pregnancy (B), breastfeeding, children, geriatric patients, abrupt discontinuation, ambient temperature increase, breast cancer, cardiac disease, dehydration, diabetes, ketoacidosis, driving, operating machinery, dysphagia, heart failure, hematologic/hepatic/renal disease, hypotension, hypovolemia, MI, infertility, obesity, Parkinson's disease, seizures, strenuous exercise, stroke, substance abuse, suicidal ideation, syncope, tardive dyskinesia

Black Box Warning: Dementia: antipsychotics (e.g., as lurasidone) not approved for treatment of dementia-related psychosis in geriatric patients; may increase risk of death in this population

DOSAGE AND ROUTES

- **Adult: PO** 40 mg/day, range 40-160 mg/day; for those receiving CYP3A4 inhibitors (max 80 mg/day), do not use with strong CYP3A4 inducers/inhibitors

Hepatic/renal dose

- **Adult: PO** Child-Pugh class B/C; CCr ≥10 ml/min, ≤50 ml/min, max 40 mg/day

Available forms: Tabs 20, 40, 80, 120 mg

Administer:

- Give with meal of ≥350 calories

SIDE EFFECTS

CNS: Agitation, akathisia, anxiety, dizziness, drowsiness, fatigue, hyperthermia, insomnia, dystonic reactions; **neuroleptic malignant syndrome (rare)**, pseudoparkinsonism, restlessness, **seizures, suicidal ideation**, syncope, tardive dyskinesia, vertigo

CV: Angina, **AV block, bradycardia,** hypertension, orthostatic hypotension, **sinus tachycardia, stroke**

EENT: Blurred vision

ENDO: Diabetes mellitus, ketoacidosis, hyperglycemia, hyperprolactinemia
GI: Abdominal pain, diarrhea, dyspepsia, nausea, vomiting, gastritis, weight gain/loss
GU: Amenorrhea, breast enlargement, dysmenorrhea, impotence, dysuria, renal failure
HEMA: Agranulocytosis, anemia, leucopenia, neutropenia
INTEG: Pruritus, rash
MS: Back pain, dysarthria; rhabdomyolysis (rare)
SYST: Angioedema

PHARMACOKINETICS

99% protein binding; excreted 80% in feces, 9% in urine; elimination half-life 18 hr; 9%-19% absorbed; peak 1-3 hr, steady state 7 days

INTERACTIONS

- Do not use with metoclopramide

Increase: lurasidone effect—strong CYP3A4 inhibitors; do not use concurrently
Increase: serotonin syndrome, neuroleptic malignant syndrome—SSRIs, SNRIs
Increase: sedation—other CNS depressants, alcohol
Decrease: lurasidone effect—CYP3A4 inducers; do not use concurrently

NURSING CONSIDERATIONS

Assess:

- **Schizophrenia:** hallucinations, delusions, agitation, social withdrawal; monitor orientation, behavior, mood before and periodically during therapy
- **Neuroleptic malignant syndrome (rare):** fever, dyspnea, tachycardia, seizures, sweating, hypo/hypertension, muscle stiffness, pallor; report immediately
- **Blood dyscrasias:** CBC periodically; blood dyscrasias may occur
- **Serious cardiac symptoms:** AV block, stroke, bradycardia may occur
- **EPS:** restlessness, difficulty speaking, loss of balance, pill rolling, masklike face, shuffling gait, rigidity, tremors, muscle spasms; monitor before and periodically during therapy; report tardive dyskinesia immediately

Black Box Warning: Dementia: this product is not approved for geriatric patients with dementia-related psychosis

- Weight gain, hyperglycemia, metabolic changes in diabetes

Evaluate:

- Therapeutic response: decreasing hallucinations, delusions, agitation, social withdrawal

Teach patient/family:

- About reason for treatment, expected results
- That lab work will be needed regularly
- To avoid hazardous activities until response is known
- To avoid OTC products unless approved by prescriber
- To report fast heartbeat, extra beats
- To report EPS symptoms, blood dyscrasias: sore throat, fever, unusual bleeding/bruising

lymphocyte immune globulin (antithymocyte, equine) (Rx)

Atgam

Func. class.: Immune globulins—immunosuppressant

ACTION:

Produces immunosuppression by inhibiting the function of lymphocytes (T)

USES:

Renal organ transplants to prevent rejection, aplastic anemia
Unlabeled uses: Immunosuppressant in liver, bone marrow, heart, and other organ transplants; stem-cell transplant preparations

CONTRAINDICATIONS:

Hypersensitivity to this product or equine/leporine protein; acute viral illness

Precautions: Pregnancy (C), breastfeeding, children, severe renal/hepatic disease, leukopenia, thrombocytopenia

Black Box Warning: Infection, neoplastic disease

DOSAGE AND ROUTES

Renal allograft rejection

- **Adult/child:** **IV** 10-15 mg/kg/day × 14 days then every other day for 14 days if needed up to 21 total

Prevent renal allograft rejection

- **Adult:** **IV** 15 mg/kg/day × 7-14 days then every other day × 14 days for total of 21 doses in 28 days

Aplastic anemia

- **Adult:** **IV** 10-20 mg/kg/day × 8-14 days then every other day for ≤21 total doses

Available forms: Inj 50 mg equine gamma globulin/ml

Administer:

- Do not infuse <4 hr; usually given over 4-8 hr
- Use 0.2-1 micron in-line filter
- Skin testing must be completed before treatment; use intradermal inj of 0.1 ml of a 1:1000 dilution (5 mcg horse IgG) in 0.9% NaCl; if wheal or rash >10 mm or both, use caution during inf
- Dilute in saline sol before inf; invert IV bag so undiluted product does not contact air inside; conc should not be >1 mg/ml; do not shake
- Keep emergency equipment nearby for severe allergic reaction

SIDE EFFECTS

Renal transplant

CNS: Fever, chills, headache, dizziness, weakness, faintness, seizures

CV: Chest pain, hypo/hypertension, tachycardia

GI: Diarrhea, nausea, vomiting, epigastric pain, GI bleeding

INTEG: Rash, pruritus, urticaria, wheal

SYST: Anaphylaxis

Aplastic anemia

CNS: Fever, chills, headache, seizures, lightheadedness, encephalitis, postviral encephalopathy

CV: Bradycardia, myocarditis, irregularity

GI: Nausea, LFTs abnormality

HEMA: Thrombocytopenia, leukopenia

MS: Backache

RESP: Dyspnea, pulmonary edema

PHARMACOKINETICS

Onset rapid, half-life 5-7 days

NURSING CONSIDERATIONS

Assess:

- **Infection:** if infection occurs, evaluation will be needed to continue therapy
- Renal studies: BUN, creatinine at least monthly during treatment, for 3 mo after treatment
- Hepatic studies: alk phos, AST, ALT, bilirubin
- CBC with differential

Evaluate:

- Therapeutic response: absence of rejection; hematologic recovery (aplastic anemia)

Teach patient/family:

- To report fever, chills, sore throat, fatigue, since serious infections may occur
- To use contraceptive measures during treatment, for 12 wk after therapy

L

mafenide topical

See Appendix B

MAGNESIUM SALTS

magnesium chloride (Rx)

Mag-64

magnesium citrate (OTC)

magnesium gluconate (OTC)

Mag-G, Magtrate

magnesium oxide (OTC)

Mag-Ox 400, Uro-Mag

magnesium hydroxide (OTC)

Freelax, MOM, Phillips' Milk of Magnesia

magnesium sulfate (OTC, Rx)

epsom salts; magnesium sulfate (IV)—HIGH ALERT

Func. class.: Electrolyte; anticonvulsant; saline laxative, antacid

ACTION: Increases osmotic pressure, draws fluid into colon, neutralizes HCl

USES: Constipation; bowel preparation before surgery or exam; anticonvulsant for preeclampsia, eclampsia (magnesium sulfate); electrolyte

Unlabeled uses: *Magnesium sulfate:* persistent pulmonary hypertension of the newborn (PPHN), cardiac arrest, CPR, digitoxin/digoxin toxicity, premature labor, seizure prophylaxis, status asthmaticus, torsades de pointes, ventricular fibrillation/tachycardia

CONTRAINDICATIONS: Hypersensitivity, abdominal pain, nausea/vomiting, obstruction, acute surgical abdomen, rectal bleeding, heart block, myocardial damage

Precautions: Pregnancy (A); (B) (magnesium sulfate), renal/cardiac disease

DOSAGE AND ROUTES

Laxative

- **Adult: PO** (milk of magnesia) 15-60 ml at bedtime
- **Adult and child >12 yr: PO** (magnesium sulfate) 15 g in 8 oz water; **PO** (Concentrated milk of magnesia) 5-30 ml; **PO** (magnesium citrate) 5-10 oz at bedtime
- **Child 2-6 yr: PO** (milk of magnesia) 5-15 ml/day

Prevention of magnesium deficiency

- **Adult and child ≥10 yr: PO** (male) 350-400 mg/day; (female) 280-300 mg/day; (breastfeeding) 335-350 mg/day; (pregnancy) 320 mg/day
- **Child 8-10 yr: PO** 170 mg/day
- **Child 4-7 yr: PO** 120 mg/day

Magnesium sulfate deficiency

- **Adult: PO** 200-400 mg in divided doses tid-qid; **IM** 1 g q6hr × 4 doses; **IV** 5 g (severe)
- **Child 6-12 yr: PO** 3-6 mg/kg/day in divided doses tid-qid

Pre-eclampsia/eclampsia (magnesium sulfate)

- **Adult: IM/IV** 4-5 g IV inf; with 5 g **IM** in each gluteus, then 5 g q4hr or 4 g **IV INF**, then 1-2 g/hr **CONT INF**, max 40 g/day or 20 g/48 hr in severe renal disease

Persistent pulmonary hypertension of the newborn (PPHN) in mechanically ventilated neonates (unlabeled)

- **Premature infants >33 wk and term neonates: IV** (magnesium sulfate) 200 mg/kg over 20-30 min then **CONT IV INF** 20-150 mg/kg/hr to maintain blood magnesium levels at 3.5-5.5 mmol/L

Status asthmaticus (unlabeled)

- **Adult: IV** (magnesium sulfate) 2 g
- **Child: IV INF** (PALS) (magnesium sulfate) 25-50 mg/kg diluted in D_5W, given over 10-20 min, max 2 g/dose

Premature labor (unlabeled)
• **Adult: IV INF** (magnesium sulfate) 4-6 g given as a loading dose over 20-30 min then 2-4 g/hr **CONT INF;** use inf pump until contractions cease; continue inf at lowest dose over 12-24 hr; **PO** (magnesium chloride/gluconate/oxide) 648-1200 mg/day elemental magnesium in divided doses

Torsades de pointes/cardiac dysrhythmias with hypomagnesemia (unlabeled)
• **Adult: IV** (magnesium sulfate) use ACLS guidelines or 1-2 g in 50-100 ml D_5W given over 5-20 min in emergent cases or over 5-60 min

Available forms: Chloride: sus rel tabs 535 mg (64 mg Mg); enteric tabs 833 mg (100 mg Mg); **citrate:** oral sol 240-, 296-, 300-ml bottles (77 mEq/100 ml); **oxide:** tabs 400 mg; caps 140 mg; **hydroxide:** liq 400 mg/5 ml; conc liq 800 mg/5 ml; chew tabs 300, 600 mg; **sulfate:** powder for oral, bulk packages; epsom salts, bulk packages; inj 10%, 12.5%, 25%, 50%

Administer:

PO route
• With 8 oz water
• Refrigerate magnesium citrate before giving
• Shake susp before using as antacid at least 2 hr after meals
• Tablets should be chewed thoroughly before patient swallows; give 4 oz of water afterwards
• **Laxative:** give on empty stomach

IM route (magnesium sulfate)
• Give deeply in gluteal site

IV route (magnesium sulfate)
• Only when calcium gluconate available for magnesium toxicity

Direct IV route
• Dilute 50% sol to ≤20%, give at ≤150 mg/min

Continuous IV INF route
• May dilute to 20% sol, infuse over 3 hr
• IV at less than 125 mg/kg/hr; circulatory collapse may occur; use inf pump

Y-site compatibilities: Acyclovir, aldesleukin, amifostine, amikacin, ampicillin, aztreonam, ceFAZolin, cefoperazone, cefotaxime, cefOXitin, cephalothin, cephapirin, chloramphenicol, cisatracurium, DOBUTamine, doxycycline, DOXOrubicin liposome, enalaprilat, erythromycin, esmolol, famotidine, fludarabine, gallium, gentamicin, granisetron, heparin, HYDROmorphone, IDArubicin, insulin, kanamycin, labetalol, meperidine, metroNIDAZOLE, minocycline, morphine, moxalactam, nafcillin, ondansetron, oxacillin, PACLitaxel, penicillin G potassium, piperacillin, piperacillin/tazobactam, potassium chloride, propofol, remifentanil, sargramostim, thiotepa, ticarcillin, tobramycin, trimethoprim-sulfamethoxazole, vancomycin, vit B complex/C

SIDE EFFECTS

CNS: Muscle weakness, flushing, sweating, confusion, sedation, depressed reflexes, **flaccid paralysis**, hypothermia
CV: Hypotension, heart block, **circulatory collapse**, vasodilation
GI: *Nausea, vomiting, anorexia, cramps,* diarrhea
HEMA: Prolonged bleeding time
META: Electrolyte, fluid imbalances
RESP: Respiratory depression/paralysis

PHARMACOKINETICS

PO: Onset 1-2 hr
IM: Onset 1 hr, duration 4 hr
IV: Duration ½ hr
Excreted by kidney, effective anticonvulsant serum levels 2.5-7.5 mEq/L

INTERACTIONS

Increase: effect of neuromuscular blockers
Increase: hypotension—antihypertensives
Decrease: absorption of tetracyclines, fluoroquinolones, nitrofurantoin
Decrease: effect of digoxin

M

NURSING CONSIDERATIONS

Assess:

- **Laxative:** cause of constipation; lack of fluids, bulk, exercise; cramping, rectal bleeding, nausea, vomiting; product should be discontinued
- ⚠ **Eclampsia:** seizure precautions, B/P, ECG (magnesium sulfate); **magnesium toxicity:** thirst, confusion, decrease in reflexes; I&O ratio; check for decrease in urinary output

Evaluate:

- Therapeutic response: decreased constipation; absence of seizures (eclampsia), normal serum calcium levels

Teach patient/family:

- Not to use laxatives for long-term therapy because bowel tone will be lost
- That chilling improves taste of magnesium citrate
- To shake suspension well
- Not to give at bedtime as a laxative, may interfere with sleep; that MOM is usually given at bedtime
- To give citrus fruit after administering to counteract unpleasant taste
- About reason for product, expected results

mannitol (Rx)

(man'i-tole)

Osmitrol, Resectisol

Func. class.: Diuretic, osmotic

Chem. class.: Hexahydric alcohol

ACTION: Acts by increasing osmolarity of glomerular filtrate, which inhibits reabsorption of water and electrolytes and increases urinary output

USES: Edema; promotion of systemic diuresis in cerebral edema; decrease in intraocular/intracranial pressure; improved renal function in acute renal failure, chemical poisoning, urinary bladder irrigation

CONTRAINDICATIONS: Active intracranial bleeding, hypersensitivity, anuria, severe pulmonary congestion, edema, severe dehydration, progressive heart, renal failure, acute MI, aneurysm, stroke

Precautions: Pregnancy (C), breastfeeding, geriatric patients, dehydration, severe renal disease, CHF, electrolyte imbalances

Black Box Warning: Acute bronchospasm asthma

DOSAGE AND ROUTES

Oliguria, prevention

- **Adult:** **IV** after initial test dose; if urine output is 30-50 ml/hr × 2 hr, give 20-100 g over a 24-hr period of 15% or 20% sol

Oliguria, treatment

- **Adult:** **IV** after initial test dose; give balance of 50 g of a 20% sol over 1 hr then 5% via **CONT IV INF** to maintain output at 50 ml/hr
- **Child (unlabeled):** **IV** 0.5-2 g/kg as 15%-20% sol, run over 30-60 min; maintenance 0.25-0.5 g/kg q4-6hr

Edema

- **Adult:** **IV** after test dose, use product 10%-20% at a rate of 25-75 ml/hr; give loop diuretics before mannitol
- **Child (unlabeled):** **IV** 0.5-2 g/kg of 15%-20% mannitol over 2-6 hr

Intraocular pressure

- **Adult:** **IV** 1.5-2 g/kg of 15%-25% sol over 30-60 min

ICP

- **Adult:** **IV** 1-2 g/kg then 0.25-1 g/kg q4hr

Diuresis with drug intoxication

- **Adult and child >12 yr:** 5%-10% sol continuously up to 200 g **IV** while maintaining 100-500 ml urine output/hr

Available forms: Inj 5%, 10%, 15%, 20%, 25%; GU irrigation: 5%

Administer:

Intermittent/continuous IV route

- Precipitate may occur with PVC
- Change IV q24hr
- In 15%-25% sol with filter; rapid inf may worsen CHF; warm in hot water, shake to dissolve if crystals are present, use in-line filter, do not give as direct injection

• **Test dose** with severe oliguria, 0.2 g/kg over 3-5 min; if continued oliguria, give 2nd test dose; if no response, reassess patient

Y-site compatibilities: Acyclovir, aldesleukin, alemtuzumab, amifostine, amikacin, ampicillin, atropine, aztreonam, bivalirudin, bumetanide, calcium gluconate, caspofungin, ceFAZolin, cefotaxime, cefOXitin, cefTAZidime, ceftizoxime, chloramphenicol, cimetidine, cisatracurium, clindamycin, DAPTOmycin, dexmedetomidine, digoxin, diltiazem, diphenhydrAMINE, DOBUTamine, DOCEtaxel, DOPamine, DOXOrubicin liposome, doxycycline, enalaprilat, EPINEPHrine, ertapenem, esmolol, famotidine, fenoldopam, fentaNYL, fluconazole, fludarabine, gentamicin, granisetron, heparin, HYDROmorphone, hydrOXYzine, IDArubicin, imipenem/cilastatin, insulin, isoproterenol, ketorolac, labetalol, levofloxacin, lidocaine, linezolid, LORazepam, meperidine, metoclopramide, metoprolol, metroNIDAZOLE, micafungin, midazolam, milrinone, morphine, nafcillin, niCARdipine, nitroglycerin, nitroprusside, norepinephrine, ondansetron, oxaliplatin, PACLitaxel, palonosetron, pantoprazole, penicillin G potassium, phenylephrine, piperacillin/tazobactam, potassium chloride, procainamide, prochlorperazine, promethazine, propofol, propranolol, protamine, quinupristin/dalfopristin, ranitidine, remifentanil, sargramostim, sodium bicarbonate, tacrolimus, thiotepa, ticarcillin/clavulanate, tirofiban, tobramycin, trimethoprim/sulfamethoxazole, vancomycin, vasopressin, verapamil, vit B complex with C, voriconazole

SIDE EFFECTS

CNS: *Dizziness, headache,* **seizures, rebound increased ICP,** confusion

CV: Edema, thrombophlebitis, hypo/hypertension, **tachycardia,** angina-like chest pains, fever, chills, **CHF, circulatory overload,** PVCs

EENT: Loss of hearing, blurred vision, nasal congestion, decreased intraocular pressure

ELECT: Fluid, electrolyte imbalances, *acidosis,* electrolyte loss, dehydration, hypo/hyperkalemia

GI: *Nausea, vomiting,* dry mouth, diarrhea

GU: Marked diuresis, urinary retention, thirst

RESP: Pulmonary congestion, *cough,* dyspnea

PHARMACOKINETICS

IV: Onset 1-3 hr for diuresis, ½-1 hr for intraocular pressure, 15 min for cerebrospinal fluid; duration 4-6 hr for intraocular pressure, 3-8 hr for cerebrospinal fluid; excreted in urine; half-life 100 min

INTERACTIONS

Increase: elimination of mannitol—lithium

Increase: excretion of salicylates, barbiturates, imipramine, bromides

Increase: hypokalemia—arsenic trioxide, cardiac glycosides, levomethadyl

Drug/Lab Test

Interference: inorganic phosphorus, ethylene glycol

NURSING CONSIDERATIONS

Assess:

• Weight, I&O daily to determine fluid loss; effect of product may be decreased if used daily; output every hr prn

• B/P lying, standing; postural hypotension may occur

• Electrolytes: potassium, sodium, chloride; include BUN, CBC, serum creatinine, blood pH, ABGs, CVP, PAP

• **Metabolic acidosis:** drowsiness, restlessness

• **Hypokalemia:** postural hypotension, malaise, fatigue, tachycardia, leg cramps, weakness, or hyperkalemia

• Rashes, temperature daily

• Confusion, especially in geriatric patients; take safety precautions if needed

M

• Hydration including skin turgor, thirst, dry mucous membranes, provide adequate fluids
• Blurred vision, pain in eyes before, during treatment **(increased intraocular pressure)**; neurologic checks, intracranial pressure during treatment **(increased intracranial pressure)**

Evaluate:
• Therapeutic response: improvement in edema of feet, legs, sacral area daily if medication being used with CHF; decreased intraocular pressure, prevention of hypokalemia, increased excretion of toxic substances; decreased ICP

Teach patient/family:
• To rise slowly from lying or sitting position
• About the reason for, method of treatment
• To report signs of electrolyte imbalance, confusion

TREATMENT OF OVERDOSE:
Discontinue inf; correct fluid, electrolyte imbalances; hemodialysis; monitor hydration, CV status, renal function

maraviroc (Rx)
(mah-rav′er-rock)
Selzentry
Func. class.: Antiretroviral
Chem. class.: Fusion inhibitor, CCR5-receptor antagonist

ACTION: Interferes with entry into HIV-1 by inhibiting the fusion of the virus and the cell membrane

USES: CCR5-tropic HIV in combination with other antiretroviral agents for treating experienced patients

CONTRAINDICATIONS: Hypersensitivity, dialysis, renal impairment
Precautions: Pregnancy (B), Asian patients, breastfeeding, renal/hepatic/cardiac disease, electrolyte imbalance, dehydration, immune reconstitution syndrome, infection, MI, orthostatic hypotension, children, geriatric patients, Graves' disease, Guillain-Barré syndrome, polymyositis

Black Box Warning: Hepatitis, fever, serious rash

DOSAGE AND ROUTES
Those not taking any CYP3A inducers/inhibitors
• **Adult/adolescent ≥16 yr: PO** 300 mg bid

Those taking CYP3A4 inhibitors with/without a CYP3A inducer
• **Adult/adolescent ≥16 yr: PO** 150 mg bid

Those taking CYP3A4 inducers without a strong CYP3A inhibitor
• **Adult/adolescent ≥16 yr: PO** 600 mg bid

Renal dose
• **Adult: PO** ≤30 ml/min, reduce dose to 150 mg bid

Available forms: Tabs 150, 300 mg

Administer:
• May give without regard to meals, with 8 oz water; swallow whole; do not crush, chew, break
• Store at room temperature

SIDE EFFECTS
CV: MI, cardiac ischemia, orthostatic hypotension
CNS: Dizziness, depression, viral meningitis, disturbances in consciousness, peripheral neuropathy, paresthesia, dysesthesia, fever
EENT: Gingival hyperplasia
GI: Diarrhea, constipation, dyspepsia, pseudomembranous colitis, hepatotoxicity
INTEG: Rash, urticaria, pruritus, folliculitis
MS: Joint pain, leg pain, muscle cramps
RESP: Cough, upper respiratory tract infection, sinusitis, bronchitis, pneumonia, bronchospasm, obstruction
SYST: Herpes virus

PHARMACOKINETICS

Metabolized by P450 system; CYP3A metabolism; excreted 20% urine, 76% feces; protein binding 76%; terminal half-life 14-18 hr

INTERACTIONS

Increase: maraviroc levels—CYP3A inhibitors (amiodarone, aprepitant, chloramphenicol, clarithromycin, conivaptan, cycloSPORINE, dalfopristin, danazol, diltiazem, erythromycin, estradiol, fluconazole, fluvoxaMINE, imatinib, isoniazid, itraconazole, ketoconazole, miconazole, nefazodone, niCARdipine, propoxyphene, RU-486, tamoxifen, telithromycin, troleandomycin, verapamil, voriconazole, zafirlukast); reduce dose

Decrease: maraviroc levels—CYP3A4 inducers (efavirenz, aminoglutethimide, barbiturates, bexaroten, bosentan, carBAMazepine, dexamethasone, griseofulvin, modafinil, nafcillin, OXcarbazepine, phenytoin, fosphenytoin, rifabutin, rifampin, rifapentine, topiramate, tipranavir); increase dose

Drug/Herb

• Decreased maraviroc effect: St. John's wort

Drug/Food

• High-fat meal decreases absorption 33%

NURSING CONSIDERATIONS

Assess:

• **HIV:** CD_4, T-cell count, plasma HIV RNA, CCR5-tropic HIV-1; assess for changes in symptoms, other infections during treatment

• Renal studies: serum creatinine

• Bowel pattern before, during treatment

• **Allergies:** skin eruptions: rash, urticaria, itching; discontinue product

Black Box Warning: **Hepatitis/hepatotoxicity:** dark urine; abdominal pain, vomiting; yellowing of skin, eyes; hepatomegaly; discontinue product; monitor liver function tests; serious rash, fever, eosinophilia, or elevated IgE before hepatotoxicity may occur

Evaluate:

• Therapeutic response: improvement in CD4, viral load, T-cell count

Teach patient/family:

• To take as prescribed; if dose is missed, to take as soon as remembered up to 1 hr before next dose; not to double dose; that product does not cure condition, should not be shared with others

• That product does not cure infection, just controls symptoms and does not prevent infecting others

⚠ To report sore throat, fever, fatigue **(may indicate superinfection);** yellow skin/eyes, abdominal pain, vomiting **(hepatitis);** itching, SOB **(allergic reaction)**

• That product must be taken in equal intervals around the clock to maintain blood levels for duration of therapy

• To avoid all OTC products unless approved by prescriber

• To avoid driving, other hazardous activities until reaction is known; that dizziness may occur

• To make position changes slowly to prevent postural hypotension

• To notify prescriber if pregnancy is planned or suspected; not to breastfeed

mebendazole (RX)

(me-ben′da-zole)

Func. class.: Anthelmintic

Chem. Class.: Carbamate

ACTION: Inhibits glucose uptake, degeneration of cytoplasmic microtubules in the cell; interferes with absorption, secretory function

M

USES: Pinworms, roundworms, hookworms, whipworms, thread-worms, pork tapeworms, dwarf tapeworms, beef tapeworms, hydatid cyst

CONTRAINDICATIONS: Hypersensitivity

Precautions: Pregnancy (C) (1st trimester), breastfeeding, children <2 yr, Crohn's disease, hepatic disease, inflammatory bowel disease, ulcerative colitis

DOSAGE AND ROUTES

- **Adult and child >2 yr: PO** 100 mg as a single dose (pinworms) or bid ×3 days (whipworms, roundworms, or hookworms); course may be repeated in 3 wk if needed, max 200 mg/day

Available forms: Chew tabs 100 mg

Administer:

- May be crushed, chewed, swallowed whole, mixed with food
- PO after meals to avoid GI symptoms
- Second course after 3 wk if needed; usually recommended
- Store in tight container

SIDE EFFECTS

CNS: Dizziness, fever, headache, seizures (rare)

GI: Transient diarrhea, abdominal pain, nausea, vomiting, constipation, hepatitis

INTEG: Rash

PHARMACOKINETICS

PO Peak ½-7 hr; excreted in feces primarily (metabolites), small amount in urine (unchanged); highly bound to plasma proteins 95%

INTERACTIONS

Decrease: mebendazole effect—carBAMazepine, hydantoins

Drug/Food

Increase: absorption—high-fat meal

NURSING CONSIDERATIONS

Assess:

- Stools during entire treatment; specimens must be sent to lab while still warm, also 1-3 wk after treatment is completed
- For allergic reaction: rash (rare)
- For diarrhea during expulsion of worms; avoid self-contamination with patient's feces
- For infection in other family members, since infection from person to person is common
- Blood studies: AST, ALT, alk phos, BUN, CBC during treatment

Evaluate:

- Therapeutic response: expulsion of worms and 3 negative stool cultures after completion of treatment

Teach patient/family:

- Proper hygiene after BM, including hand-washing technique; tell patient to avoid putting fingers in mouth; clean fingernails
- That infected person should sleep alone; not to shake bed linen, change bed linen daily, wash in hot water, change and wash undergarments daily
- To clean toilet daily with disinfectant (green soap solution)
- The need for compliance with dosage schedule, duration of treatment
- To wear shoes, wash all fruits and vegetables well before eating; use commercial fruit/vegetable cleaner
- That all members of the family should be treated (pinworms)
- To report jaundice, liver pain

mecasermin (Rx)

(mec-a'sir-men)

Increlex

Func. class.: Biologic response modifier; insulin-like growth factor

ACTION: Stimulates growth; IGF-1 is the principal hormonal mediator of statural growth; GH binds to its receptor in the liver and other tissues

USES:
Growth failure in children with severe primary insulin-like growth factor-1 (IGF-1) deficiency (primary IGFD) or with growth hormone (GH) gene deletion who have developed neutralizing antibodies to GH

Unlabeled uses: ALS

CONTRAINDICATIONS:
Hypersensitivity, benzyl alcohol, closed epiphyses, active/suspected neoplasia, IV use

Precautions: Pregnancy (C), breastfeeding, children <2 yr, diabetes mellitus, hypothyroidism, lymphoid tissue hypertrophy, increased intracranial pressure, malnutrition, scoliosis, sleep apnea

DOSAGE AND ROUTES

- **Child ≥2 yr: SUBCUT** 0.04-0.08 mg/kg (40-80 mcg/kg) bid; if well tolerated for 1 wk, may increase by 0.04 mg/kg/dose, max 0.12 mg/kg bid

Available forms: Inj 10 mg/ml

Administer:

SUBCUT route

- Give within 20 min of meal or snack; do not give if unable to eat or if vomiting
- Rotate inj site; use sterile, disposable syringe/needles; use small-volume syringe for accurate measurement; if using syringe (units), convert to mg/kg (weight [kg] × dose [mg/kg] × 1 ml/mg × 100 units/1 ml = units/injection)
- Do not double dose if a dose is missed
- Store in refrigerator before opening, avoid freezing; after opening, stable for 30 days after initial vial entry, store in refrigerator; do not use if particulate matter is present, avoid direct light, do not use after expiration date

SIDE EFFECTS

CNS: Headache, **seizures,** dizziness, cardiac valvulopathy, **increased intracranial pressure**

CV: Cardiac murmur

EENT: Ear pain, otitis media, abnormal tympanometry, papilledema, visual impairment, tonsillar hypertrophy

ENDO: **Hypoglycemia, ketosis, hypothyroidism, hypercholesterolemia, hypertriglyceridemia**

INTEG: Pruritus, urticaria, **anaphylaxis, angioedema**

GI: Vomiting, nausea

HEMA: Thymus hypertrophy

MISC: Bruising, lipohypertrophy, inj site reaction

MS: Arthralgia, joint pain, slipped upper femoral epiphysis

RESP: Snoring, **apnea**

SYST: **Antibodies to growth hormone, secondary malignancy**

PHARMACOKINETICS

Bioavailability almost 100%, metabolized in liver/kidney, half-life 5.8 hr

INTERACTIONS

Increase: hypoglycemia—antidiabetics, corticosteroids

Decrease: growth suppression possible—psychostimulants

M

NURSING CONSIDERATIONS

Assess:

- Monitor preprandial glucose at beginning of treatment and until well tolerated
- By funduscopic exam at beginning and periodically during treatment
- **Allergic reactions;** if present, interrupt treatment and notify prescriber
- Growth rate of child at intervals during treatment
- **Serious skin disorders: angioedema, anaphylaxis**

Evaluate:

- Therapeutic response: growth in children

Teach patient/family:

- That treatment may continue for years; that regular assessments are required
- To avoid hazardous activities, driving within 2-3 hr of dosing
- About correct administration, needle disposal

meclizine (OTC, Rx)

(mek'li-zeen)

Antivert, Bonamine ✱, Bonine, Dramamine Less Drowsy Formula, Medivert, Travel Sickness, Wal-Dram II

Func. class.: Antiemetic, antihistamine, anticholinergic

Chem. class.: H_1-receptor antagonist, piperazine derivative

ACTION:
Acts centrally by blocking chemoreceptor trigger zone, which in turn acts on vomiting center

USES:
Vertigo, motion sickness

CONTRAINDICATIONS:
Hypersensitivity to cyclizines

Precautions: Pregnancy (B), breastfeeding, children, geriatric patients, closed-angle glaucoma, prostatic hypertrophy, hepatic/renal disease

DOSAGE AND ROUTES

Vertigo

- **Adult/adolescent: PO** 25-100 mg/day in divided doses

Motion sickness

- **Adult/adolescent: PO** 25-50 mg 1 hr before traveling, repeat dose q24hr prn

Available forms: Tabs 12.5, 25, 50 mg; chew tabs 25 mg; caps 25, 30 mg

Administer:

PO route

- May give without regard to food
- **Chew tab:** give without regard to water or may be swallowed whole with water
- Lowest possible dose for geriatric patients; anticholinergic effects

SIDE EFFECTS

CNS: *Drowsiness,* fatigue, restlessness, headache, insomnia

CV: Hypotension

EENT: Dry mouth, blurred vision

GI: Nausea, anorexia, constipation, increased appetite

GU: Urinary retention

SYST: Anaphylaxis

PHARMACOKINETICS

PO: Onset 1 hr, duration 8-24 hr, half-life 6 hr

INTERACTIONS

Increase: sedation—CYP2D6, inhibitors

Increase: anticholinergic effects—other antihistamines, atropine, antidepressants, phenothiazines

Increase: effect of alcohol, opioids, other CNS depressants

Drug/Lab Test

False negative: allergy skin testing (allergen extracts)

NURSING CONSIDERATIONS

Assess:

- **Vertigo/motion sickness:** nausea, vomiting after 1 hr; assess vertigo periodically

⚠ Signs of toxicity of other products, masking of symptoms of disease: brain tumor, intestinal obstruction

- Observe for drowsiness, dizziness, level of consciousness

Evaluate:

- Therapeutic response: absence of dizziness, vomiting

Teach patient/family:

- That a false-negative result may occur with skin testing for allergies; that these procedures should not be scheduled for ≤4 days after discontinuing use
- To avoid hazardous activities, activities requiring alertness because dizziness may occur; to request assistance with ambulation
- To avoid alcohol, other depressants; not to breastfeed

medroxyPROGESTERone (Rx)

(me-drox′ee-proe-jess′te-rone)

Depo-Provera, Depo-sub Q Provera, Gen-Medroxy ✦, Provera

Func. class.: Antineoplastic, hormone, contraceptive

Chem. class.: Progesterone derivative

Do not confuse:
medroxyPROGESTERone/methylPREDNISolone
Provera/Premarin/Covera

ACTION: Inhibits secretion of pituitary gonadotropins, which prevents follicular maturation and ovulation; antineoplastic action against endometrial cancer

USES: Uterine bleeding (abnormal); secondary amenorrhea; prevention of endometrial changes associated with estrogen replacement therapy (ERT); contraceptive; inoperable, recurrent, metastatic endometrial/ renal cancer

Unlabeled uses: Hot flashes; symptoms of menopause; paraphilia (men); hot flashes (men) with prostate cancer

CONTRAINDICATIONS: Pregnancy (X), hypersensitivity, reproductive cancer, genital bleeding (abnormal, undiagnosed), missed abortion, stroke, cerebrovascular disease, cervical cancer, hepatic disease, uterine/vaginal cancer

Black Box Warning: Breast cancer, MI, stroke, thromboembolic disease, thrombophlebitis

Precautions: Breastfeeding, hypertension, asthma, blood dyscrasias, gallbladder disease, CHF, diabetes mellitus, bone disease, depression, migraine headache, seizure disorders, renal/hepatic disease, family history of cancer of breast or reproductive tract, bone mineral density loss, ocular disorders, AIDS/HIV, alcoholism, children, hyperlipidemia

Black Box Warning: Cardiac disease, dementia, osteoporosis

DOSAGE AND ROUTES

Secondary amenorrhea
- **Adult: PO** 5-10 mg/day × 5-10 days

Uterine bleeding
- **Adult: PO** 5-10 mg/day × 5-10 days starting on 16th or 21st day of menstrual cycle

With ERT
- **Adult: PO** 5-10 mg daily × 10-14 or more days/mo (sequential estrogen); 2.5-5 mg daily (continuous estrogen)

Contraceptive
- **Adult: IM** (contraceptive inj) 150 mg q12wk; **SUBCUT** (depot SUBCUT Provera 104 inj) 104 mg q3mo

Endometrial/renal cancer
- **Adult: IM** 400 mg-1 g (using 400 mg/ml depot inj susp) weekly

Hot flashes/symptoms of menopause (unlabeled)
- **Adult (female): PO** 20 mg/day; **IM** 150 mg monthly

Hot flashes (men) in prostate cancer (unlabeled)
- **Adult (male): IM** Depot 150 or 400 mg

Available forms: Tabs 2.5, 5, 10 mg; inj susp, 150, 400 mg/ml; depot-SUBCUT inj: 104 mg/0.65 ml

Administer:
- Store in dark area

PO route
- Give without regard to food

IM route
- Visually inspect particulate matter and discoloration before use

Depo-Provera contraceptive injection suspension
- IM only, *never* IV
- Instruct patient on risks and warnings associated with hormonal contraceptives (see Patient Information)
- The possibility of pregnancy should be excluded before giving the first dose of medroxyPROGESTERone or whenever >14 wk has passed since the last dose

M

- Do not dilute
- Shake vigorously immediately before administration
- Inject deeply into the gluteal or deltoid muscle; aspirate before injection to avoid injection into a blood vessel

Depo-Provera sterile aqueous suspension, preserved

- IM only, *never* IV
- Instruct patient on risks and warnings associated with progestin use (see Patient Information)
- Shake vigorously immediately before use
- When multidose vials are used, take special care to prevent contamination
- Inject deeply into the gluteal or deltoid muscle; aspirate before injection

SUBCUT route

Depo-subQ provera 104 contraceptive injection suspension *only*

- For subcut only, *never* give IM or IV
- Instruct patient on risks and warnings associated with hormonal contraceptives (see Patient Information)
- Shake vigorously for at least 1 min before use
- Inject the entire contents of the prefilled syringe subcut into the anterior thigh or abdomen, avoiding bony areas and the umbilicus; gently grasp and squeeze a large area of skin in the chosen injection area, ensuring that the skin is pulled away from the body; insert the needle at a 45-degree angle; inject until the syringe is empty; this usually requires 5-7 sec; following use, press lightly on the injection site with a clean cotton pad for a few seconds; do not rub the area

SIDE EFFECTS

CNS: Dizziness, headache, migraines, depression, fatigue, nervousness

CV: Hypotension, thrombophlebitis, edema, thromboembolism, stroke, PE, MI

EENT: Diplopia

GI: *Nausea,* vomiting, anorexia, cramps, increased weight, cholestatic jaundice, abdominal pain

GU: Amenorrhea, cervical erosion, breakthrough bleeding, dysmenorrhea, vaginal candidiasis, breast changes, *gynecomastia, testicular atrophy, impotence,* endometriosis, spontaneous abortion, vaginitis, libido increased/decreased

INTEG: Rash, urticaria, acne, hirsutism, alopecia, oily skin, seborrhea, purpura, melasma, photosensitivity, injection site reaction

META: Hyperglycemia

MS: Decreased bone density

SYST: Angioedema, anaphylaxis

PHARMACOKINETICS

PO: Duration 24 hr; excreted in urine and feces; metabolized in liver

INTERACTIONS

Decrease: medroxyPROGESTERone action—aminoglutethimide, carBAMazepine, phenytoins, PHENobarbital, rifampin

Decrease: bone mineral density—anticoagulants, corticosteroids

Drug/Lab Test

Increase: LFTs, HDL, triglycerides

Decrease: GTT, HDL

NURSING CONSIDERATIONS

Assess:

- Pelvic exam, Pap smear before treatment, periodically
- ⚠ **Severe allergic reaction, angioedema;** have EPINEPHrine and rescusitative equipment available
- Weight daily; notify prescriber of weekly weight gain >5 lb; bone mineral density
- B/P at beginning of treatment and periodically
- I&O ratio; be alert for decreasing urinary output, increasing edema
- Hepatic studies: ALT, AST, bilirubin periodically during long-term therapy
- Edema, hypertension, cardiac symptoms, jaundice
- Mental status: affect, mood, behavioral changes, depression

Black Box Warning: This product should not be given to those with breast cancer, MI, stroke, thromboembolic disorders

• **Bone mineral density loss:** in those taking anticoagulants, corticosteroids with Depo-Provera or Depo-sub Q Provera

Black Box Warning: Use of product shown to increase dementia in women ≥65 yr old; use may increase osteoporosis in long-term treatment; those who smoke also at greater risk; adequate calcium and vit D should be taken

Evaluate:

• Therapeutic response: decreased abnormal uterine bleeding, absence of amenorrhea

Teach patient/family:

• To avoid sunlight or to use sunscreen; photosensitivity can occur

⚠ To report breast lumps, vaginal bleeding, edema, jaundice, dark urine, clay-colored stools, dyspnea, headache, blurred vision, abdominal pain, sudden change in speech/coordination, numbness or stiffness in legs, chest pain; males to report impotence, gynecomastia

• To report suspected pregnancy (X); fertility returns 6-12 mo after discontinuing

Black Box Warning: Long-term use decreases bone density; exercise, calcium supplements can help lessen osteoporosis

medrysone ophthalmic

See Appendix B

megestrol (Rx)

(me-jess′trole)

Megace, Megace ES

Func. class.: Antineoplastic hormone

Chem. class.: Progestin

Do not confuse:

Megace/Reglan

ACTION: Affects endometrium with antiluteinizing effect; thought to bring about cell death

USES: Breast, endometrial cancer; cachexia, anorexia, weight loss with AIDS

Unlabeled uses: Hot flashes in women (menopause) or men (prostate cancer), unexplained weight loss in geriatric patients, endometriosis, renal cell cancer (palliative), endometrial cancer, breast cancer (metastatic)

CONTRAINDICATIONS: Pregnancy (D) tabs, (X) susp; hypersensitivity

Precautions: Diabetes, thrombosis, adrenal insufficiency

DOSAGE AND ROUTES

Endometrial/ovarian carcinoma

• **Adult: PO** 40-320 mg/day in divided doses

Breast carcinoma

• **Adult: PO** 40 mg qid or 160 mg/day

Cachexia (AIDS)

• **Adult: PO** 800 mg/day (oral susp) or 625 mg/day (ES)

Hot flashes (unlabeled)

• **Adult: PO** 20 mg bid

Metastatic endometrial cancer

• **Adult: PO** 40-320 mg/day in divided doses × ≥2 mo

Available forms: Tabs 20, 40 mg; oral susp 40, 125 mg/ml

Administer:

• Oral susp for AIDS patients; shake well
• Tablets for carcinoma
• Without regard to food
• Store in tight container at room temperature

SIDE EFFECTS

CNS: Mood swings, insomnia, fever, lethargy, depression

CV: **Thrombophlebitis, thromboembolism,** hypertension

ENDO: Adrenal insufficiency

GI: Nausea, vomiting, diarrhea, abdominal cramps, weight gain, flatus, indigestion

GU: Gynecomastia, fluid retention, hypercalcemia, vaginal bleeding, discharge, impotence, decreased libido, menstruation disorders

M

INTEG: Alopecia, rash, pruritus, purpura, itching, sweating
META: Hyperglycemia
MISC: Tumor flare
RESP: Dyspnea

PHARMACOKINETICS

PO: Half-life 13-105 hr; metabolized in liver; excreted in feces, urine, breast milk; food increases bioavailability of oral sol

INTERACTIONS

• Do not use with dofetilide
Decrease: megestrol effect—antidiabetics
Drug/Lab Test
Increase: glucose

NURSING CONSIDERATIONS

Assess:
• PSA levels in men (prostate cancer); blood glucose, LFTs, serum calcium, weight
• Effects of alopecia on body image; feelings about body changes
• Frequency of stools, characteristics: cramping, acidosis, signs of dehydration (rapid respirations, poor skin turgor, decreased urine output, dry skin, restlessness, weakness)
• Anorexia, nausea, vomiting, constipation, weakness, loss of muscle tone
⚠ **Thrombophlebitis:** Homans' sign, edema; pain in calf, thigh; notify prescriber immediately
Evaluate:
• Therapeutic response: decreased tumor size, spread of malignancy; weight gain in AIDS patients; resolved dysfunctional uterine bleeding
Teach patient/family:
• To report vaginal bleeding
• That nonhormonal contraception should be used during and for 4 mo after treatment; pregnancy (D) tabs, (X) susp
• That gynecomastia, alopecia can occur; reversible after discontinuing treatment
⚠ To recognize signs of fluid retention, thromboemboli; to report these immediately
• To monitor blood glucose if diabetic

⚠ HIGH ALERT

melphalan (Rx)

(mel′fa-lan)
Alkeran
Func. class.: Antineoplastic, alkylating agent
Chem. class.: Nitrogen mustard

Do not confuse:
melphalan/Myleran

ACTION: Responsible for cross-linking DNA strands, thereby leading to cell death; activity is not cell-cycle–phase specific

USES: Multiple myeloma, malignant melanoma, advanced ovarian cancer
Unlabeled uses: Breast, testicular, prostate carcinoma; osteogenic sarcoma, amyloidosis, chronic myelogenous leukemia, non-Hodgkin's lymphoma, pediatric rhabdomyosarcoma, stem cell transplant, bone marrow ablation, AML, myelodysplastic syndrome

CONTRAINDICATIONS: Pregnancy (D), breastfeeding, other nitrogen mustards

Black Box Warning: Hypersensitivity to this product

Precautions: Children, radiation therapy, infections, renal disease

Black Box Warning: Bone marrow depression, secondary malignancy, radiation therapy; requires an experienced clinician

DOSAGE AND ROUTES

Multiple myeloma
• **Adult: PO** 6 mg daily × 2-3 wk, adjust dose based on blood counts or 10 mg daily × 7-10 day

• **Adult: IV INF** 16 mg/m^2, reduce with renal insufficiency, give over 15-20 min, give at 2-wk intervals × 4 doses then at 4-wk intervals

Ovarian carcinoma

• **Adult: PO** 200 mcg/kg/day × 5 days q4-5wk

Testicular cancer/breast cancer/ non-Hodgkin's lymphoma/ osteogenic sarcoma (unlabeled)

• **Adult: PO** 150 mcg/kg/day × 7 days q4wk; when leukocytes normal, give 50 mcg/kg/day maintenance

Pediatric rhabdomyosarcoma (unlabeled)

• **Child: IV** 10-35 mg/m^2 q21-28days

Available forms: Tabs 2 mg, powder for inj 50 mg

Administer:

• Antiemetic 30-60 min before product to prevent vomiting

PO route

• Give on empty stomach

• Protect from light; store refrigerated

Intermittent IV INF route

• Use gloves during administration; if skin exposure occurs, wash immediately with soap and water; use cytotoxic handling procedures

• Give after reconstituting with 10 ml diluent provided (5 mg/ml); shake, dilute dose with 0.9% NaCl (≤0.45 mg/ml); give within 1 hr, run over ≥15 min; degrading of product occurs rapidly; make sure inf is completed in time specified

Y-site compatibilities: Acyclovir, amikacin, aminophylline, ampicillin, aztreonam, bleomycin, bumetanide, buprenorphine, butorphanol, calcium gluconate, CARBOplatin, carmustine, ceFAZolin, cefepime, cefoperazone, cefotaxime, cefoTEtan, cefTAZidime, ceftizoxime, cefTRIAXone, cefuroxime, cimetidine, CISplatin, clindamycin, cyclophosphamide, cytarabine, dacarbazine, DACTINomycin, DAUNOrubicin, dexamethasone, diphenhydrAMINE, DOXOrubicin, doxycycline, droperidol, enalaprilat, etoposide, famotidine, floxuridine, fluconazole, fludarabine, fluorouracil, furosemide, gallium, ganciclovir, gentamicin, granisetron, haloperidol, heparin, hydrocortisone, hydrocortisone sodium phosphate, HYDROmorphone, hydrOXYzine, IDArubicin, ifosfamide, imipenem-cilastatin, LORazepam, mannitol, mechlorethamine, meperidine, mesna, methotrexate, methylPREDNISolone, metoclopramide, metroNIDAZOLE, miconazole, minocycline, mitoMYcin, mitoXANtrone, morphine, nalbuphine, netilmicin, ondansetron, pentostatin, piperacillin, plicamycin, potassium chloride, prochlorperazine, promethazine, ranitidine, sodium bicarbonate, streptozocin, teniposide, thiotepa, ticarcillin, ticarcillin/ clavulanate, tobramycin, trimethoprim-sulfamethoxazole, vancomycin, vinBLAStine, vinCRIStine, vinorelbine, zidovudine

SIDE EFFECTS

GI: *Nausea, vomiting,* stomatitis, diarrhea, **hepatitis**

GU: Amenorrhea, hyperuricemia, gonadal suppression, hyperuricemia

HEMA: **Thrombocytopenia, neutropenia, leukopenia,** anemia

INTEG: Rash, urticaria, alopecia, pruritus

RESP: **Fibrosis, dysplasia,** dyspnea, pneumonitis

SYST: **Anaphylaxis,** allergic reactions, **secondary malignancies,** edema

PHARMACOKINETICS

Metabolized in liver, excreted in urine, half-life 1½ hr, protein binding 80%-90%

INTERACTIONS

• Avoid administration of sargramostim, GM-CSF, filgrastim, G-CSF 14 hr before or 24 hr after product

Increase: toxicity—antineoplastics, radiation

Increase: pulmonary toxicity—carmustine

Increase: renal failure risk—cycloSPORINE

Increase: enterocolitis risk—nalidixic acid
Increase: bleeding risk—NSAIDs, anticoagulants, salicylates, thrombolytics, platelet inhibitors
Decrease: antibody response—live virus vaccines

NURSING CONSIDERATIONS

Assess:

Black Box Warning: Bone marrow depression: CBC, differential, platelet count weekly; withhold product if WBC is <3000/mm³ or platelet count is <100,000/mm³; notify prescriber; recovery usually occurs in 6 wk

- Renal studies: BUN, serum uric acid before, during therapy
- I&O ratio; report fall in urine output to 30 ml/hr
- **Infection:** fever, cough, temperature, chills, sore throat; notify prescriber
- Hepatic studies before, during therapy (bilirubin, AST, ALT, LDH) as needed; jaundiced skin and sclera, dark urine, clay-colored stools, itchy skin, abdominal pain, fever, diarrhea
- **Bleeding:** hematuria, guaiac, bruising or petechiae, mucosa or orifices q8hr
- Buccal cavity q8hr for dryness, sores, ulceration, white patches, oral pain, bleeding, dysphagia
- Local irritation, pain, burning, discoloration at inj site
- **Hyperuricemia:** joint pain, edema, increased uric acid, increase fluids to >2 L unless contraindicated

⚠ **Severe allergic reaction:** rash, pruritus, urticaria, purpuric skin lesions, itching, flushing; assess allergy to chlorambucil; cross-sensitivity may occur

- Increase fluid intake to 2-3 L/day to prevent urate deposits, calculi formation
- Rinsing of mouth tid-qid with water, club soda; brushing of teeth bid-tid with soft brush or cotton-tipped applicators for stomatitis; use unwaxed dental floss

Evaluate:

- Therapeutic response: decreased tumor size, spread of malignancy

Teach patient/family:

- That usually sterility, amenorrhea occur; reversible after discontinuing treatment
- To avoid foods with citric acid, hot or rough texture
- To report any bleeding, white spots, or ulcerations in mouth to prescriber; to examine mouth daily
- To report signs of infection: fever, sore throat, flulike symptoms
- To report suspected pregnancy; to use contraception during treatment; pregnancy (D)
- To report signs of anemia: fatigue, headache, faintness, SOB, irritability
- To avoid use of razors, commercial mouthwash
- To avoid use of aspirin products, NSAIDs, alcohol

memantine (Rx)

(me-man′teen)

Ebixa ✱, Namenda, Namenda XR

Func. class.: Anti-Alzheimer agent

Chem. class.: NMDA receptor antagonist

ACTION: Antagonist action of CNS NMDA receptors that may contribute to the symptoms of Alzheimer's disease

USES: Moderate to severe dementia in Alzheimer's disease

Unlabeled uses: Vascular dementia, acquired pendular nystagmus

CONTRAINDICATIONS: Children, hypersensitivity

Precautions: Pregnancy (B), breastfeeding, renal disease, GU conditions that raise urine pH, seizures, severe hepatic disease, renal failure

DOSAGE AND ROUTES

- **Adult: PO** 5 mg/day, may increase dose in 5-mg increments ≥1 wk intervals over a 3-wk period; recommended target dose as 10 mg bid; at wk 4; ext rel 7 mg daily, increased by 7 mg ≤1 wk up to target dose of 28 mg daily

Available forms: Tabs 5, 10 mg; tab titration pak 5, 10 mg; oral sol 2 mg/ml (10 mg/5 ml); cap ext rel 7, 14, 21, 28 mg

Administer:

- Can be taken without regard to meals
- Twice a day if dose >5 mg
- Dosage adjusted to response no more than q1wk
- **Ext rel caps:** do not crush, chew, divide; swallow whole or open and sprinkle on applesauce
- **Oral sol** using device provided; remove dosing syringe, green cap, plastic tube from plastic; attach tube to green cap; open cap by pushing down on cap, turning counterclockwise; remove unscrewed cap; carefully remove seal from bottle, discard; insert plastic tube fully into bottle, screw green cap tightly onto bottle by turning cap clockwise; keeping bottle upright on table, remove lid; with plunger fully depressed, insert tip of syringe into cap; while holding syringe, gently pull up on plunger; remove syringe; invert syringe, slowly press plunger to level that removes large air bubbles; keep plunger in inverted position; few small air bubbles may be present

SIDE EFFECTS

CNS: *Dizziness, confusion,* somnolence, headache, hallucinations, **stroke,** insomnia, depression, anxiety

CV: Hypertension, **heart failure, CHF**

GI: Vomiting, constipation

HEMA: Anemia

INTEG: Rash

MISC: Back pain, fatigue, pain, influenza-like symptoms

RESP: Coughing, dyspnea

PHARMACOKINETICS

Rapidly absorbed PO, 44% protein binding, very little metabolism, 57%-82% excreted unchanged in urine, terminal elimination half-life 60-80 hr

INTERACTIONS

Increase/decrease: both products: hydrochlorothiazide, triamterene, cimetidine, quiNIDine, ranitidine, nicotine

Increase: effect—levodopa, some ergots

Decrease: clearance of memantine—products that make urine alkaline (sodium bicarbonate, carbonic anhydrase inhibitors)

- Use cautiously with amantadine, dextromethorphan, ketamine; reaction unknown

NURSING CONSIDERATIONS

Assess:

- **Alzheimer's dementia:** affect, mood, behavioral changes; hallucinations, confusion, attention, orientation, memory; monitor serum creatinine
- Provide assistance with ambulation during beginning therapy; dizziness may occur

Evaluate:

- Therapeutic response: decrease in confusion, improved mood, maintenance of function, even with no improvement in symptoms

Teach patient/family:

- To report side effects: restlessness, psychosis, visual hallucinations, stupor, LOC; may indicate overdose
- To use product exactly as prescribed; that product not a cure; to avoid alcohol, nicotine
- To use oral sol dispenser
- To avoid OTC, herbal products unless approved by prescriber

M

⚠ HIGH ALERT

meperidine (Rx)

(me-per'i-deen)

Demerol, Meperitab

Func. class.: Opioid analgesic

Chem. class.: Phenylpiperidine derivative

Controlled Substance Schedule II

Do not confuse:
meperidine/HYDROmorphone/meprobamate/morphine
Demerol/Dilaudid/Desyrel/Demulen

ACTION: Depresses pain impulse transmission at the spinal cord level by interacting with opioid receptors

USES: Moderate to severe pain preoperatively, postoperatively

Unlabeled uses: Obstetric/regional analgesic, acute severe headache/migraine, shaking chills induced by IV amphotericin B or postoperative shivering

CONTRAINDICATIONS: Hypersensitivity, severe respiratory insufficiency, MAOI therapy

Precautions: Pregnancy (C), breastfeeding, children, geriatric patients, addictive personality, increased intracranial pressure, respiratory depression, renal/hepatic disease, seizure disorder, abrupt discontinuation, chronic pain, cardiac disease, adrenal insufficiency, alcoholism, angina, anticoagulant therapy, asthma, atrial flutter, biliary tract disease, bladder obstruction, cardiac dysrhythmias, COPD, CNS depression, coagulopathy, constipation, cor pulmonale, dehydration, diarrhea, driving epidural use, geriatric patients, GI obstruction, head trauma, heart failure, hypotension, hypothyroidism, ileus, IBS, IM/intrathecal/IV use, labor, myxedema, thrombocytopenia

DOSAGE AND ROUTES

Moderate to severe pain

- **Adult: PO/SUBCUT/IM** 50-150 mg q3-4hr prn
- **Child: PO/SUBCUT/IM** 1-1.8 mg/kg q3-4hr prn, max single dose 150 mg

Labor analgesia

- **Adult: SUBCUT/IM** 50-100 mg given when contractions regularly spaced, repeat q1-3hr prn

Preoperatively

- **Adult: IM/SUBCUT** 50-100 mg 30-90 min before surgery
- **Child: IM/SUBCUT** 1-2.2 mg/kg 30-90 min before surgery, max 100 mg

Shaking chills (unlabeled)

- **Adult: IV** 25-50 mg as a single dose
- **Child/adolescent: IV** 0.35-1 mg/kg, max adult dose

Renal dose

- Avoid or modify dose

Available forms: Inj 10, 25, 50, 75, 100 mg/ml; tabs 50, 100 mg; oral sol 50 mg/5 ml

Administer:

PO route

- May give with food or milk to decrease GI irritation
- **Oral liquid:** dilute in 4 oz water
- Store in light-resistant container at room temperature

IM/SUBCUT route

- Patient should remain recumbent for 1 hr after IM/SUBCUT route
- With antiemetic for nausea, vomiting
- When pain beginning to return; determine dosage interval by patient response
- In gradually decreasing dose after long-term use; withdrawal symptoms may occur
- Inject IM into large muscle mass; IM preferred route for multiple inj

Direct IV route

- Dilute to conc of 10 mg/ml with sterile water for inj or NS
- Inject slowly ≤25 mg/min
- Have emergency equipment and opiate antagonist on hand

Continuous IV INF route
- Dilute to conc of 1 mg/ml
- Infuse using inf pump

Syringe compatibilities: Butorphanol, chlorproMAZINE, cimetidine, dimenhyDRINATE, diphenhydrAMINE, droperidol, fentaNYL, glycopyrrolate, hydrOXYzine, ketamine, metoclopramide, midazolam, pentazocine, perphenazine, prochlorperazine, promazine, ranitidine, scopolamine

Y-site compatibilities: Amifostine, amikacin, atenolol, aztreonam, bumetanide, ceFAZolin, cefotaxime, cefOXitin, cefTAZidime, ceftizoxime, cefTRIAXone, cefuroxime, cisatracurium, cladribine, clindamycin, diltiazem, diphenhydrAMINE, DOBUTamine, DOPamine, DOXOrubicin HCL, doxycycline, droperidol, erythromycin, famotidine, filgrastim, fluconazole, fludarabine, gallium, gentamicin, granisetron, hydrocortisone, insulin (regular), kanamycin, labetalol, lidocaine, methyldopate, melphalan, metoclopramide, metoprolol, metroNIDAZOLE, ondansetron, oxytocin, PACLitaxel, penicillin G potassium, piperacillin, potassium chloride, propofol, propranolol, ranitidine, remifentanil, sargramostim, teniposide, thiotepa, ticarcillin, ticarcillin/clavulanate, tobramycin, vancomycin, verapamil, vinorelbine

Continuous intrathecal INF route
- Use controlled inf device; implantable controlled micro inf device used for highly concentrated inf; monitor for several days after implantation
- Filling of inf reservoir should only be done by those fully qualified
- To prevent pain, depletion of reservoir should be avoided

SIDE EFFECTS

CNS: *Drowsiness, dizziness, confusion, headache, sedation, euphoria,* **increased intracranial pressure, seizures,** serotonin syndrome

CV: Palpitations, bradycardia, hypotension, change in B/P, tachycardia (IV)

EENT: Tinnitus, blurred vision, miosis, diplopia, depressed corneal reflex

GI: Nausea, vomiting, anorexia, constipation, cramps, biliary spasm, paralytic ileus

GU: Urinary retention, dysuria

INTEG: Rash, urticaria, bruising, flushing, diaphoresis, pruritus

RESP: **Respiratory depression**

SYST: **Anaphylaxis**

PHARMACOKINETICS

Metabolized by liver (to active/inactive metabolites), excreted by kidneys; crosses placenta, excreted in breast milk; half-life 3-4 hr; toxic by-product accumulation can result from regular use or renal disease; protein binding 65%-75%

PO: Onset 15 min, peak 1-1.5 hr, duration 2-4 hr, absorption 50%

SUBCUT/IM: Onset 10 min, peak 1 hr, duration 2-4 hr, well absorbed

IV: Onset 5 min, duration 2 hr

INTERACTIONS

⚠ **May cause fatal reaction: MAOIs, procarbazine**

Increase: serotonin syndrome, neuroleptic malignant syndrome, SSRIs, SNRIs, serotonin-receptor agonists

Increase: effects with other CNS depressants, alcohol, opioids, sedative/hypnotics, antipsychotics, skeletal muscle relaxants

Increase: adverse reactions—protease inhibitor antiretrovirals

Decrease: meperidine effect—phenytoin

Drug/Herb

Increase: CNS depression—St. John's wort

Drug/Lab Test

Increase: amylase, lipase

NURSING CONSIDERATIONS

Assess:

- **Pain:** location, type, character; give product before pain becomes extreme; reassess after 60 min (IM, SUBCUT, PO) and 5-10 min (IV)

- Renal function before initiating therapy; poor renal function can lead to accumulation of toxic metabolite and seizures
- I&O ratio; check for decreasing output; may indicate urinary retention
- For constipation; increase fluids, bulk in diet; give stimulant laxatives if needed
- CNS changes: dizziness, drowsiness, hallucinations, euphoria, LOC, pupil reactions with chronic or high-dose use
- Allergic reactions: rash, urticaria

⚠ **Respiratory dysfunction:** depression, character, rate, rhythm; notify prescriber if respirations are <12/min

- CNS stimulation: with chronic or high doses

Evaluate:

- Therapeutic response: decrease in pain

Teach patient/family:

- To report any symptoms of CNS changes, allergic reactions
- That physical dependency may result from extended use; use should be short term only
- That drowsiness, dizziness may occur
- That withdrawal symptoms may occur: nausea, vomiting, cramps, fever, faintness, anorexia
- To make position changes slowly; orthostatic hypotension can occur
- To avoid OTC medications, alcohol unless directed by prescriber

TREATMENT OF OVERDOSE:

Naloxone (Narcan) 0.2-0.8 mg IV, caution in physically dependent patients, O_2, IV fluids, vasopressors

mercaptopurine (6-MP) (Rx)

(mer-kap-toe-pyoor'een)

Purinethol

Func. class.: Antineoplastic-antimetabolite

Chem. class.: Purine analog

ACTION: Inhibits purine metabolism at multiple sites, which inhibits DNA and RNA synthesis; specific for S phase of cell cycle

USES: Acute lymphocytic leukemia

Unlabeled uses: Ulcerative colitis, Crohn's disease

CONTRAINDICATIONS: Pregnancy (D), breastfeeding, patients with prior product resistance, hypersensitivity, requires an experienced clinician, secondary malignancy, infection, hypocalcemia, hyperuricemia, hyperphosphatemia, hyperkalemia

Precautions: Renal/hepatic disease, tumor lysis syndrome, dental disease, herpes, radiation therapy, leukopenia, thrombocytopenia, anemia

DOSAGE AND ROUTES

Acute lymphocytic leukemia

- **Adult: PO** 2.5-5 mg/kg/day or 80-100 mg/m²/day, maintenance 1.5-2.5 mg/kg/day
- **Child: PO** 2.5-5 mg/kg/day, maintenance 1.5-2.5 mg/kg/day or 70-100 mg/m²/day

Crohn's disease/ulcerative colitis (unlabeled)

- **Adult: PO** 1.5-2 mg/kg/day

Available forms: Tabs 50 mg

Administer:

- Store in tightly closed container in cool environment
- Give product after evening meal, before bedtime, on an empty stomach
- Sodium bicarbonate to maintain uric acid levels, alkalinization of urine

SIDE EFFECTS

CNS: Weakness
GI: *Nausea, vomiting, anorexia, diarrhea, stomatitis,* hepatotoxicity (high doses), jaundice, gastritis, pancreatitis
GU: Renal failure, hyperuricemia, oliguria, crystalluria, hematuria
HEMA: Thrombocytopenia, leukopenia, myelosuppression, anemia
INTEG: *Rash,* dry skin, urticaria, alopecia

PHARMACOKINETICS

Incompletely absorbed when taken orally, metabolized in liver, excreted in urine, peak 1-2 hr, terminal half-life 1-1.5 hr

INTERACTIONS

Increase: effects of mercaptopurine—allopurinol, avoid use of decrease dose
Increase: effects—radiation or other antineoplastics, immunosuppressants
Increase: bone marrow depression—azaTHIOprine, sulfamethoxazole-trimethoprim, avoid concurrent use
Increase: anticoagulant action—anticoagulants, NSAIDs, thrombolytics, platelet inhibitors, salicylates
Decrease: antibodies—live virus vaccines
Decrease: TPMT, rapid bone marrow suppression—balzalazide, olsalazine, mesalamine, sulfaSALAzine, use cautiously

NURSING CONSIDERATIONS

Assess:

⚠ **Bone marrow suppression:** CBC, differential, platelet count weekly; withhold product at first sign of abnormally large decrease in blood counts, unless bone marrow aplasia is the goal

- **Thiopurine methyltransferase (TPMT) deficiency:** individuals are prone to rapid bone marrow suppression; dosage reduction may be required in homozygous-TPMT-deficient persons
- **Tumor lysis syndrome:** monitor for increased potassium, uric acid, phosphate, decreased urine output, calcium
- Renal studies: BUN, serum uric acid, urine CCr, electrolytes before, during therapy
- I&O ratio; report fall in urine output to <30 ml/hr
- Monitor temperature; fever may indicate beginning infection; no rectal temperature
- **Hepatotoxicity:** Hepatic studies before, during therapy: bilirubin, alk phos, AST, ALT, weekly during beginning therapy, hepatic encephalopathy, toxic hepatitis, ascites can be fatal
- **Bleeding:** hematuria, guaiac, bruising, petechiae, mucosa or orifices, avoid IM inj if platelets are low; blood transfusions may be needed
- **Stomatitis:** buccal cavity for dryness, sores, ulceration, white patches, oral pain, bleeding, dysphagia
- Increase fluid intake to 2-3 L/day to prevent urate deposits, calculi formation, unless contraindicated
- Rinsing of mouth tid-qid with water, club soda; brushing of teeth bid-tid with soft brush or cotton-tipped applicators for stomatitis; use unwaxed dental floss

Evaluate:

- Therapeutic response: decreased size of tumor, spread of malignancy

Teach patient/family:

- To avoid foods with citric acid, hot or rough texture for stomatitis; to report stomatitis: any bleeding, white spots, ulcerations in mouth; to examine mouth daily, report symptoms
- **Pregnancy:** that contraceptive measures recommended during therapy (D); to avoid breastfeeding
- To drink 10-12 8-oz glasses of fluid/day
- To notify prescriber of fever, chills, sore throat, nausea, vomiting, anorexia, diarrhea, bleeding, bruising, all of which may indicate blood dyscrasias/infection
- To report signs of infection: fever, sore throat, flulike symptoms
- To report signs of anemia: fatigue, headache, faintness, SOB, irritability
- To report bleeding; to avoid use of razors, commercial mouthwash

- To avoid use of aspirin products, NSAIDs
- To take entire dose at one time

meropenem (Rx)

(mer-oh-pen′em)

Merrem

Func. class.: Antiinfective—miscellaneous

Chem. class.: Carbapenem

ACTION: Bactericidal; interferes with cell-wall replication of susceptible organisms

USES: *Acinetobacter* sp., *Aeromonas hydrophila, Bacteroides distasonis, Bacteroides fragilis, Bacteroides ovatus, Bacteroides thetaiotaomicron, Bacteroides uniformis, Bacteroides ureolyticus, Bacteroides vulgatus, Campylobacter jejuni, Citrobacter diversus, Citrobacter freundii, Clostridium difficile, Clostridium perfringens, Enterobacter cloacae, Enterococcus faecalis, Escherichia coli, Eubacterium lentum, Fusobacterium* sp., *Haemophilus influenzae* (beta-lactamase negative), *Haemophilus influenzae* (beta-lactamase positive), *Hafnia alvei, Klebsiella oxytoca, Klebsiella pneumoniae, Moraxella catarrhalis, Morganella morganii, Neisseria meningitidis, Pasteurella multocida, Peptostreptococcus* sp., *Porphyromonas asaccharolytica, Prevotella bivia, Prevotella intermedia, Prevotella melaninogenica, Propionibacterium acnes, Proteus mirabilis, Proteus vulgaris, Pseudomonas aeruginosa, Salmonella* sp., *Serratia marcescens, Shigella* sp., *Staphylococcus aureus* (MSSA), *Staphylococcus epidermidis, Streptococcus agalactiae* (group B streptococci), *Streptococcus pneumoniae, Streptococcus pyogenes* (group A beta-hemolytic streptococci), *Viridans streptococci, Yersinia enterocolitica;* appendicitis, bacteremia, intraabdominal infections, meningitis, peritonitis, skin/skin structure infections, pseudomembranous colitis

Unlabeled uses: Febrile, neutropenic, community-acquired pneumonia

CONTRAINDICATIONS: Hypersensitivity to this product, carbapenems, cephalosporins, penicillins

Precautions: Pregnancy (B), breastfeeding, geriatric patients, renal disease, seizure disorder, gram-negative infection, nosocomial pneumonia, pneumonia

DOSAGE AND ROUTES

Intraabdominal infections (complicated appendicitis, peritonitis)

- **Adult/child/adolescent >50 kg: IV** 1 g q8hr or adult 500 mg q6hr
- **Infant ≥3 mo/child/adolescent ≤50 kg: IV** 20 mg/kg q8hr
- **Term neonates/infants <3 mo (unlabeled): IV** 30 mg/kg/dose q8hr

Extended dose (3-hr inf) (unlabeled)

- **Adult: IV** 1 g over 3 hr, q8hr

Complicated skin and skin structure infections

- **Adult/adolescents/child >50 kg: IV** 500 mg q8hr over 15-30 min
- **Infants ≥3 mo/children/adolescents ≤50 kg: IV** 10 mg/kg q8hr over 15-30 min

Bacterial meningitis

- **Adult/child/adolescent >50 g: IV** 2 g q8hr
- **Infant/child/adolescent ≤50 kg: IV** 40 mg/kg q8hr

Renal disease

- **Adult: IV** CCr 26-50 ml/min, give dose q12hr; CCr 10-25 ml/min, give 1/2 dose q12hr; CCr <10 ml/min, give 1/2 dose q24hr

Febrile neutropenia (unlabeled)

- **Adult: IV** 1 g q8hr

Community-acquired pneumonia (CAP) (unlabeled)

- **Adult: IV** 1 g q8hr with ciprofloxacin or with aminoglycoside plus fluoroquinolone

Available forms: Powder for inj 500 mg, 1 g

Administer:

- After C&S is taken

Direct IV route

- Reconstitute 500-mg or 1-g vials with 10, 20 ml of sterile water for inj, respectively; shake to dissolve; let stand until clear (average conc 50 mg/ml); reconstituted sol may be stored for 2 hr at room temperature or for 12 hr refrigerated; inject up to 1 g in 5-20 ml over 3-5 min

Intermittent IV INF route

- Vials may be directly reconstituted with compatible inf fluid (NS, D_5W) to 2.5-50 mg/ml; vials with NS can be stored 2 hr at room temperature or for ≤18 hr refrigerated, (D_5W solutions) may be stored for up to 1 hr at room temperature or ≤8 hr refrigerated; infuse over 15-30 min

Continuous IV Infusion (unlabeled)

- **3 g/day continuous IV infusion:** Constitute a 1 g vial according to manufacturer recommendations; further dilute in 50 ml or 250 ml of NS and run over 8 hr; for continuous infusion, administer a new infusion bag q8hr
- **4 g/day continuous IV infusion:** Constitute a 1 g vial according to manufacturer recommendations; further dilute in 100 ml of NS and administer over 6 hr; for continuous infusion, administer a new infusion bag q6hr
- **3 g/day IV continuous infusion in ambulatory infusion pump with freezer packs:** Reconstitute 1 g vial according to manufacturer recommendations by adding 20 ml of NS into each vial; add 3 g (60 ml) to a 100 ml medication cassette reservoir and bring the final volume to 100 ml (final concentration, 30 mg/ml) run over 24 hr

Y-site compatibilities: Alemtuzumab, aminocaproic acid, aminophylline, anidulafungin, argatroban, atenolol, atropine, azithromycin, bivalirudin, bleomycin, CARBOplatin, carmustine, caspofungin, cimetidine, CISplatin, cyclophosphamide, cycloSPORINE, cytarabine, DACTINomycin, DAPTOmycin, dexamethasone, dexmedetomidine, dexrazoxane, digoxin, diltiazem, diphenhydrAMINE, DOCEtaxel, doxacurium, DOXOrubicin liposomal, enalaprilat, eptifibatide, etoposide, etoposide phosphate, fluconazole, fludarabine, fluorouracil, foscarnet, furosemide, gallium, gatifloxacin, gemcitabine, gemtuzumab, gentamicin, granisetron, heparin sodium, HYDROmorphone, ifosfamide, insulin, regular, irinotecan, lepirudin, leucovorin, linezolid injection, LORazepam, mechlorethamine, methotrexate, metoclopramide, metroNIDAZOLE, milrinone, mitoXANtrone, morphine, nesiritide, norepinephrine, octreotide, oxaliplatin, oxytocin, PACLitaxel, palonosetron, pamidronate, pancuronium, PEMEtrexed, PHENobarbital, potassium acetate/chloride, rocuronium, teniposide, thiotepa, tigecycline, tirofiban, TNA (3-in-1) total nutrient admixture, vancomycin, vasopressin, vecuronium, vinBLAStine, vinCRIStine, vinorelbine, voriconazole, zoledronic acid

M

SIDE EFFECTS

CNS: **Seizures,** dizziness, weakness, *headache,* insomnia, agitation, confusion, drowsiness

CV: Hypotension, tachycardia

ENDO: Hypoglycemia

GI: Diarrhea, nausea, vomiting, **pseudomembranous colitis, hepatitis,** glossitis, jaundice

INTEG: *Rash,* urticaria, *pruritus,* pain at inj site, phlebitis, erythema at inj site

RESP: Dyspnea, hyperventilation

SYST: **Anaphylaxis, Stevens-Johnson syndrome, angioedema**

PHARMACOKINETICS

IV: Onset immediate, peak dose dependent, half-life 1 hr, excreted unchanged in urine (70%)

INTERACTIONS

Increase: meropenem plasma levels—probenecid

Decrease: effect of valproic acid
Drug/Lab Test
Increase: AST, ALT, LDH, BUN, alk phos, bilirubin, creatinine
Decrease: prothrombin time
False positive: direct Coombs' test

NURSING CONSIDERATIONS

Assess:

- Sensitivity to carbapenem antibiotics, penicillins, cephalosporins
- Renal disease: lower dose may be required; monitor serum creatinine/BUN before, during therapy
- **Pseudomembranous colitis:** bowel pattern daily; if severe diarrhea, fever, abdominal pain, fatigue occurs, product should be discontinued
- **Infection:** temperature, sputum, characteristics of wound before, during, and after treatment

⚠ **Allergic reactions, anaphylaxis:** rash, laryngeal edema, wheezing, urticaria, pruritus; may occur immediately or several days after therapy begins; identify if there has been hypersensitivity to penicillins, cephalosporins, beta-lactams, cross-sensitivity may occur

- **Seizures:** may occur in those with brain lesions, seizure disorder, bacterial meningitis, or renal disease; stop product, notify prescriber if seizures occur
- Overgrowth of infection: perineal itching, fever, malaise, redness, pain, swelling, drainage, rash, diarrhea, change in cough, sputum

Evaluate:

- Therapeutic response: negative C&S; absence of symptoms and signs of infection

Teach patient/family:

- **Pseudomembranous colitis:** to report severe diarrhea
- To report sore throat, bruising, bleeding, joint pain; may indicate blood dyscrasias (rare)
- To report overgrowth of infection: black, furry tongue; vaginal itching; foul-smelling stools; seizures
- To avoid breastfeeding; product is excreted in breast milk

TREATMENT OF ANAPHYLAXIS: EPINEPHrine, antihistamines; resuscitate if necessary

mesalamine, 5-ASA (Rx)

(mez-al'a-meen)

Apriso, Asacol, Asacol HD, Canasa, Delzicol, Lialda, Pentasa, Rowasa Salofalk ♣, sf Rowasa

Func. class.: GI antiinflammatory
Chem. class.: 5-Aminosalicylic acid

Do not confuse:
Asacol/Ansaid/Os-Cal

ACTION: May diminish inflammation by blocking cyclooxygenase, inhibiting prostaglandin production in colon; local action only

USES: Mild to moderate active distal ulcerative colitis, proctosigmoiditis, proctitis
Unlabeled uses: Crohn's disease

CONTRAINDICATIONS: Hypersensitivity to this product or salicylates, 5-aminosalicylates
Precautions: Pregnancy (B), breastfeeding, children, geriatric patients, renal disease, sulfite sensitivity, pyloric stenosis

DOSAGE AND ROUTES

Treatment of ulcerative colitis

- **Adult: RECT** 60 ml (4 g) at bedtime, retained for 8 hr × 3-6 wk; **DEL REL TAB (Lialda)** 2.4-4.8 g/day × 8 wk; **DEL REL TAB (Asacol)** 800 mg tid × 6 wk; **CONTROLLED REL CAP (Pentasa)** 1 g qid up to 8 wk; **EXT REL** cap **(Apriso)** 1500 mg (4 caps) qAM daily up to 6 mo, **RECT SUPP** 500 mg bid retained for 1-3 hr × 3-6 wk until remission, may increase tid if needed; del rel cap (Delzicol) 800 mg tid × 6 wk

Maintenance of remission

- **Adult: PO** (del rel tab: Asacol) 800 mg bid or 400 mg qid; **PO** (del rel cap: Apriso) 1500 mg (4 caps) each AM; **PO** (del rel tab: Lialda) 2.4 g (2 tabs) daily with meal; del rel cap (Delzicol) 800 mg bid

Available forms: Enema 4 g/60 ml (Rowasa, sf Rowasa); ext rel tab 500 mg; ext rel cap 250, 500 mg (Pentasa); 0.375 g (Apriso); del rel tab 400 mg (Asacol), 800 mg (Asacol HD); del rel tab (Lialda) 1.2 g; rectal supp 1000 mg (Canasa); del rel cap (Delzicol) 400 mg

Administer:

PO route

- Swallow tabs whole; do not break, crush, or chew tabs
- **Lialda:** take with meal
- **Apriso caps:** take without regard to meals in AM
- **Delzicol caps:** give ≥1 hr before a meal or 2 hr after a meal

Rectal suspension

- Product should be given at bedtime, retained until morning (8 hr); empty bowel before insertion, shake well

Rectal suppository

- Moisten before insertion; suppository should be retained for 1-3 hr

SIDE EFFECTS

CNS: *Headache, fever, dizziness,* insomnia, asthenia, weakness, fatigue
CV: Pericarditis, myocarditis, chest pain, palpitations
EENT: Sore throat, cough, pharyngitis, rhinitis
GI: *Cramps, gas, nausea, diarrhea,* rectal pain, constipation
GU: Nephrotoxicity, interstitial nephritis
INTEG: *Rash, itching,* acne
SYST: *Flulike symptoms, malaise,* back pain, peripheral edema, leg and joint pain, arthralgia, dysmenorrhea, anaphylaxis, acute intolerance syndrome

PHARMACOKINETICS

RECT: Primarily excreted in feces but some in urine as metabolite; half-life 1 hr, metabolite half-life 5-10 hr

INTERACTIONS

- Do not give H_2 blockers with Apriso

Increase: nephrotoxicity—NSAIDs
Increase: action, adverse reactions of azaTHIOprine, mercaptopurine
Decrease: mesalamine absorption—lactulose, antacids
Decrease: effect of—warfarin

Drug/Lab Test

Increase: AST, ALT, alk phos, LDH, GGTP, amylase, lipase

NURSING CONSIDERATIONS

Assess:

- **Allergy to salicylates, sulfonamides;** if allergic reactions occur, discontinue product
- Renal studies: BUN, creatinine before, periodically during treatment; renal toxicity may occur
- **Bowel disorders:** cramps, gas, nausea, diarrhea, rectal pain; if severe, product should be discontinued

Evaluate:

- Therapeutic response: absence of pain, bleeding from GI tract, decrease in number of diarrhea stools

Teach patient/family:

- That usual course of therapy is 3-6 wk
- To shake bottle well (rectal susp)
- About method of rectal administration
- To inform prescriber of GI symptoms
- To report abdominal cramping, pain, diarrhea with blood, headache, fever, rash, chest pain; product should be discontinued

M

metFORMIN (Rx)

(met-for'min)

Fortamet, Glucophage, Glucophage XR, Glumetza, Riomet

Func. class.: Antidiabetic, oral
Chem. class.: Biguanide

ACTION:

Inhibits hepatic glucose production and increases sensitivity of peripheral tissue to insulin

USES:
Type 2 diabetes mellitus

Unlabeled uses: Precocious puberty or early-normal onset of puberty to delay menarche; polycystic ovary syndrome, infertility

CONTRAINDICATIONS:
Creatinine ≥1.5 mg/ml (males); diabetic ketoacidosis

Black Box Warning: History of lactic acidosis

Precautions: Pregnancy (B), breastfeeding, geriatric patients, previous hypersensitivity, thyroid disease, CHF, type 1 diabetes mellitus, hypersensitivity; hepatic disease; creatinine ≥1.4 (females); alcoholism; cardiopulmonary disease; acidemia; acute MI; cardiogenic shock; metabolic acidosis, renal disease

DOSAGE AND ROUTES

Type 2 diabetes mellitus

- **Adult: PO** 500 mg bid or 850 mg/day initially then 500 mg weekly or 850 mg q2wk up to 2000 mg/day in divided doses with morning meal, with dosage increased every other wk, max 2550 mg/day, **EXT REL** (Glucophage XR) 500 mg daily with evening meal, may increase by 500 mg per wk, max 2000 mg/day; (Glumetza) 1000 mg daily with food, preferably with PM meal, may increase by 500 mg per wk, max 2000 mg daily; (Fortamet) 500-1000 mg daily with PM meal, may increase by 500 mg per wk, max 2550 mg daily; regular rel or oral sol 2000-2500 mg/day for ext rel tab, depending on formulation
- **Geriatric: PO** Use lowest effective dose

To delay early menarche and to prolong pubertal growth with early onset of puberty (unlabeled)

- **Child 8-9 yr: PO** 825 mg/day with PM meal

To delay clinical puberty and early menarche in precocious puberty (unlabeled)

- **Child >6 yr: PO** 425 mg/day with PM meal

Polycystic ovary syndrome/ infertility related to hyperinsulinemia secondary to polycystic ovary syndrome (unlabeled)

- **Adult (female): PO** 500 mg tid

Available forms: Tabs 500, 850, 1000 mg; ext rel tab 500, 850, 1000 mg; oral sol 500 mg/5 ml

Administer:

PO route

- **Immediate rel product:** twice a day given with meals to decrease GI upset, and provide the best absorption; immediate rel tabs crushed, mixed with meal, fluids for patients with difficulty swallowing
- **Ext rel product** may also be taken as single dose; titrate slowly to therapeutic response, side effect tolerance
- Ext rel tabs: do not chew, break, crush

SIDE EFFECTS

CNS: *Headache, weakness, dizziness, drowsiness,* tinnitus, fatigue, vertigo, *agitation*

ENDO: Lactic acidosis, hypoglycemia

GI: *Nausea, vomiting, diarrhea,* heartburn, anorexia, metallic taste

HEMA: Thrombocytopenia, decreased vit B_{12} levels

INTEG: Rash

PHARMACOKINETICS

Excreted by kidneys unchanged 35%-50%, half-life 6 hr, peak 2-3 hr (immediate release); 7 hr (ext release); 2½ hr (solution)

INTERACTIONS

- Do not give with radiologic contrast media; may cause renal failure
- Do not use with dofetilide; may cause lactic acidosis

Increase: digoxin levels—digoxin

Increase: metFORMIN level—cimetidine, digoxin, morphine, procainamide, quiNIDine, ranitidine, triamterene, vancomycin

Increase: hyperglycemia—calcium channel blockers, corticosteroids, estrogens, oral contraceptives, phenothiazines, sympathomimetics, diuretics, phenytoin, β-blockers

Drug/Herb

Increase: hyperglycemia—glucosamine

Increase: hypoglycemia—garlic, green tea, horse chestnut

Drug/Lab Test

Decrease: vit B12

NURSING CONSIDERATIONS

Assess:

- **Hypoglycemic reactions** (sweating, weakness, dizziness, anxiety, tremors, hunger); hyperglycemic reactions soon after meals; these occur rarely with product, may occur when product combined with sulfonylureas
- CBC (baseline, q3mo) during treatment; check LFTs periodically, AST, LDH, renal studies: BUN, creatinine during treatment; glucose, A1c; folic acid, vit B_{12} q1-2yr
- **Surgery:** product should be discontinued temporarily for surgical procedures when patient is NPO or if contrast media is used, resume when patient is eating

Black Box Warning: **Lactic acidosis:** malaise, myalgia, abdominal distress; risk increases with age, poor renal function; monitor electrolytes, lactate, pyruvate, blood pH, ketones, glucose; suspect in any diabetic patient with metabolic acidosis, with ketoacidosis; immediately stop product if hypoxemia or significant renal dysfunction occurs

Perform/provide:

- Conversion from other oral hypoglycemic agents; change may be made without gradual dosage change; monitor serum glucose, urine ketones tid during conversion
- Store in tight container in cool environment

Evaluate:

- Therapeutic response: decrease in polyuria, polydipsia, polyphagia; clear sensorium; absence of dizziness; stable gait; blood glucose, A1c at normal level

Teach patient/family:

Black Box Warning: **Lactic acidosis:** hyperventilation, fatigue, malaise, chills, myalgia, somnolence; to notify prescriber immediately; stop product

- To regularly self-monitor blood glucose with blood-glucose meter
- About signs, symptoms of hypo/hyperglycemia; what to do about each (rare)
- That product must be continued on daily basis; about consequences of discontinuing product abruptly
- To avoid OTC medications, alcohol unless approved by prescriber
- That diabetes is a lifelong illness; that product is not a cure, only controls symptoms
- To carry emergency ID and glucagon emergency kit
- That Glucophage XR tab may appear in stool
- To take with meals; not to break, crush, chew ext rel product

M

⚠ HIGH ALERT

methadone (Rx)

(meth′a-done)

Dolophine, Metadol ✦, Methadose

Func. class.: Opioid analgesic

Chem. class.: Synthetic diphenylheptane derivative

Controlled Substance Schedule II

Do not confuse:

methadone/methylphenidate

ACTION: Depresses pain impulse transmission at the spinal cord level by interacting with opioid receptors; produces CNS depression

USES: Severe pain, opioid withdrawal

Unlabeled uses: Bone pain

CONTRAINDICATIONS: Hypersensitivity to this product or chlorobutanol (inj); asthma, ileus

Black Box Warning: Respiratory depression

Precautions: Pregnancy (C), breastfeeding, children <18 yr, geriatric patients, addictive personality, increased intracranial pressure, MI (acute), severe heart disease, respiratory depression, pulmonary/renal/hepatic disease, respiratory insufficiency, torsades de pointes, COPD, seizures

Black Box Warning: QT prolongation, pain, substance abuse, potential for overdose, poisoning, accidental exposure

DOSAGE AND ROUTES

Severe pain

- **Adult: PO/SUBCUT/IM/IV** 2.5-10 mg q8-12hr prn; IV dose approximately 1/2 PO dose

Opioid withdrawal

- **Adult including pregnant woman:** 20-30 mg initially unless low opioid tolerance expected; additional 5-10 mg q2-4hr as needed after initial dose; if symptoms continue, may give for ≤5 days

Renal/hepatic disease

- **Adult:** may need to be modified

Available forms: Inj 10 mg/ml; tabs 5, 10 mg; oral sol 5, 10 mg/5 ml, 10 mg/ml

Administer:

PO route

- When using during a methadone maintenance program, use only PO according to NATA guidelines

IM route

- Rotating inj sites, give deep in large muscle mass (IM)

SUBCUT route

- Pain and induration may occur at site

SIDE EFFECTS

CNS: *Drowsiness, dizziness, confusion, headache, sedation,* euphoria, seizures

CV: Palpitations, bradycardia, change in B/P, cardiac arrest, shock, hypotension, torsades de pointes, QT prolongation

EENT: Tinnitus, blurred vision, miosis, diplopia

GI: *Nausea, vomiting, anorexia, constipation, cramps,* biliary tract spasm

GU: Increased urinary output, dysuria, urinary retention, impotence

INTEG: *Rash,* urticaria, bruising, flushing, diaphoresis, pruritus

RESP: Respiratory depression, respiratory arrest

PHARMACOKINETICS

Metabolized by liver; excreted by kidneys; crosses placenta; excreted in breast milk; half-life 8-59 hr, extended interval with continued dosing; 90% bound to plasma proteins

PO: Onset 30-60 min, peak 1-1.5 hr, duration 6-8 hr, cumulative 22-48 hr; PO half as active as INJ

SUBCUT/IM: Onset 10-20 min, peak 1 1/2-2 hr, duration 4-6 hr, cumulative 22-48 hr

INTERACTIONS

⚠ Unpredictable reactions: MAOIs; do not use together

- Do not use within 2 wk of selegiline

Increase: effects with other CNS depressants—alcohol, opiates, sedative/hypnotics, antipsychotics, skeletal muscle relaxants

Increase: toxicity—CYP3A4 inhibitors (aprepitant, antiretroviral protease inhibitors, clarithromycin, danazol, delavirdine, diltiazem, erythromycin, fluconazole, FLUoxetine, fluroxamine, imatinib, ketoconazole, mibefradil, nefazodone, telithromycin, voriconazole)

Increase: QT prolongation—class IA antiarrhythmics (disopyramide, procainamide, quiNIDine), class III antiarrhythmics (amiodarone, dofetilide, ibutilide, sotalol), astemizole, arsenic trioxide, cisapride, chloroquine, clarithromycin, levomethadye, pentamidine, some phenothiazines, pimozide, terfenadine

Decrease: analgesia—rifampin, phenytoin, nalbuphine, pentazine

Decrease: methadone effect—CYP3A4 inducers (barbiturates, bosentan, carBAMazepine, efavirenz, phenytoins, nevirapine, rifabutin, rifampin), withdrawal symptoms may occur

Drug/Food

- Avoid use with grapefruit juice

Drug/Herb

- Avoid use with St. John's wort; withdrawal may result

Increase: CNS depression—chamomile, hops, kava, valerian

Drug/Lab Test

Increase: amylase, lipase

NURSING CONSIDERATIONS

Assess:

- **Pain:** type, location, intensity, grimacing before, 1½-2 hr after administration; use pain scoring
- I&O ratio; check for decreasing output; may indicate urinary retention
- CNS changes: dizziness, drowsiness, hallucinations, euphoria, LOC, pupil reaction
- Allergic reactions: rash, urticaria

Black Box Warning: **Respiratory dysfunction:** respiratory depression, character, rate, rhythm; notify prescriber if respirations are <10/min

Black Box Warning: **QT prolongation:** may be dose related or use with other products that increase QT

Black Box Warning: Accidental exposure: make sure product is not accessible to children, pets

Black Box Warning: Overdose, poisoning: advise persons involved in correct use

Black Box Warning: Substance abuse: may occur but has less psychological dependence than other opiate agonists

- Opioid detoxification: no analgesia occurs, only prevention of withdrawal symptoms

Black Box Warning: B/P, pulse, ECG; hypotension, palpitations may occur

- Bowel changes, bulk, fluids, laxatives should be used for constipation

Evaluate:

- Therapeutic response: decrease in pain, successful opioid withdrawal

Teach patient/family:

- To report any symptoms of CNS changes, allergic reactions
- That physical dependency may result from extended use

⚠ That withdrawal symptoms may occur: nausea, vomiting, cramps, fever, faintness, anorexia

- To maintain proper hydration; to avoid alcohol use
- To avoid use with other products without approval of prescriber; many drug interactions

TREATMENT OF OVERDOSE:

Naloxone (Narcan) 0.2-0.8 mg IV, O_2, IV fluids, vasopressors

methimazole (Rx)

(meth-im′a-zole)

Tapazole

Func. class.: Thyroid hormone antagonist (antithyroid)

Chem. class.: Thioamide

Do not confuse:

methimazole/metoprolol/minoxidil

ACTION: Inhibits synthesis of thyroid hormones by decreasing iodine use in manufacture of thyroglobin and iodothyronine; does not affect circulatory T_4, T_3

USES: Hyperthyroidism, preparation for thyroidectomy, thyrotoxic crisis; thyroid storm when PTU is contraindicated

CONTRAINDICATIONS:

Pregnancy (D), breastfeeding, hypersensitivity

Precautions: Infection, bone marrow suppression, hepatic disease, bleeding disorders

DOSAGE AND ROUTES

Hyperthyroidism

- **Adult: PO** 15 mg/day (mild hyperthyroidism); 30-40 mg/day (moderate to severe); 60 mg/day (severe); maintenance 5-15 mg/day; may be divided
- **Child: PO** 0.4 mg/kg/day in divided doses q8hr; continue until euthyroid; maintenance dose 0.2 mg/kg/day in divided doses q8hr, max 30 mg/24 hr; may be divided

Preparation for thyroidectomy

- **Adult and child: PO** same as above; iodine may be added × 10 days before surgery

Thyrotoxic crisis

- **Adult and child: PO** same as hyperthyroidism with iodine and propranolol

Available forms: Tabs 5, 10, 15, 20 mg

Administer:

- With meals to decrease GI upset
- At same time each day to maintain product level
- Lowest dose that relieves symptoms; discontinue before RAIU

SIDE EFFECTS

CNS: *Drowsiness, headache, vertigo, fever,* paresthesias, neuritis

ENDO: *Enlarged thyroid*

GI: *Nausea, diarrhea, vomiting,* jaundice, hepatitis, loss of taste

GU: Nephritis

HEMA: Agranulocytosis, leukopenia, thrombocytopenia, hypothrombinemia, lymphadenopathy, bleeding, vasculitis

INTEG: *Rash, urticaria, pruritus, alopecia, hyperpigmentation,* lupuslike syndrome

MS: Myalgia, arthralgia, nocturnal muscle cramps

PHARMACOKINETICS

Onset 12-18 hr; duration 36-72 hr; half-life 4-12 hr; excreted in urine, breast milk; crosses placenta

INTERACTIONS

Increase: bone marrow depression—radiation, antineoplastic agents

Increase: response to digoxin

Decrease: effectiveness—amiodarone, potassium iodide

Decrease: Anticoagulant effect—warfarin

Drug/Lab Test

Increase: PT, AST, ALT, alk phos

NURSING CONSIDERATIONS

Assess:

- **Hyperthyroidism:** palpitation, nervousness, loss of hair, insomnia, heat intolerance, weight loss, diarrhea
- **Hypothyroidism:** constipation, dry skin, weakness, fatigue, headache, intolerance to cold, weight gain; adjustment may be needed
- Pulse, B/P, temperature
- I&O ratio; check for edema: puffy hands, feet, periorbits; these indicate hypothyroidism
- Weight daily; same clothing, scale, time of day
- T_3, T_4, which are increased; serum TSH, which is decreased; free thyroxine index, which is increased if dosage too low; discontinue product 3-4 wk before RAIU

⚠ **Blood dyscrasias:** CBC, leukopenia, thrombocytopenia, agranulocytosis; if these occur, product should be discontinued and other treatment initiated; may occur at higher doses

- **Hypersensitivity:** rash, enlarged cervical lymph nodes; product may have to be discontinued
- **Hypoprothrombinemia:** bleeding, petechiae, ecchymosis
- **Clinical response:** after 3 wk should include increased weight; decreased T_4, pulse

⚠ Bone marrow suppression: sore throat, fever, fatigue
• Increased fluids to 3-4 L/day unless contraindicated

Evaluate:
• Therapeutic response: weight gain, decreased pulse, decreased T_4, B/P

Teach patient/family:
• Not to breastfeed
• To take pulse daily
• To report redness, swelling, sore throat, mouth lesions, fever, which indicate blood dyscrasias
• To keep graph of weight, pulse, mood
• To avoid OTC products, seafood that contains iodine, other iodine products
• Not to discontinue product abruptly because thyroid crisis may occur; stress patient response
• That response may take several months if thyroid is large
• **Symptoms and signs of overdose:** periorbital edema, cold intolerance, mental depression
• **Symptoms of inadequate dose:** tachycardia, diarrhea, fever, irritability
• To take medication as prescribed; not to skip or double dose

⚠ HIGH ALERT

methotrexate (Rx)

(meth-oh-trex′ate)

Rheumatrex, Trexall

Func. class.: Antineoplastic-antimetabolite (vesicant)

Chem. class.: Folic acid antagonist

Do not confuse:
methotrexate/metolazone/mitoXANtrone

ACTION: Inhibits an enzyme that reduces folic acid, which is needed for nucleic acid synthesis in all cells; specific to S phase of cell cycle; immunosuppressive

USES: Acute lymphocytic leukemia; in combination for breast, lung, head, neck carcinoma; lymphosarcoma, gestational choriocarcinoma, hydatidiform mole, psoriasis, RA, mycosis fungoides, osteosarcoma

Unlabeled uses: Burkitt's lymphoma, bladder or ovarian cancer, carcinomatous meningitis, desmoid tumor, fibromatosis, asthma, active Crohn's disease, ulcerative colitis, GVHD prophylaxis, ectopic pregnancy, pregnancy termination, psoriatic arthritis, pruritus due to cholestasis or primary biliary cirrhosis, SLE, sarcoidosis

CONTRAINDICATIONS: Hypersensitivity, leukopenia ($<3500/mm^3$), thrombocytopenia ($<100,000/mm^3$), anemia; psoriatic patients with severe renal disease, alcoholism, HIV

Black Box Warning: Pregnancy (X), hepatic disease

Precautions: Breastfeeding, children

Black Box Warning: Renal disease, ascites, diarrhea, exfoliative dermatitis, infection, intrathecal administration, lymphoma, pleural effusion, pulmonary disease, radiation therapy, stomatitis, tumor lysis syndrome, renal impairment

M

DOSAGE AND ROUTES

Acute lymphocytic leukemia
• **Adult and child: PO/IM/IV** 3.3 mg/m^2/day × 4-6 wk until remission then 20-30 mg/m^2 **PO/IM** weekly in 2 divided doses or 2.5 mg/kg **IV** × 2 wk

Choriocarcinoma
• **Adult and child: PO/IM** 15-30 mg/day × 5 days then off 1 wk; may repeat

Meningeal leukemia
• **Adult and child:** 12 mg/m^2 **INTRATHECALLY** q2-5days until CSF is normal then 1 additional dose, max 15 mg

Lymphosarcoma (stage III)
• **Adult: PO/IM/IV** 0.625-2.5 mg/kg/day

Osteosarcoma
• **Adult and child: IV** 12 g/m^2 given over 4 hr then leucovorin rescue

Mycosis fungoides
• **Adult: PO** 5-50 mg weekly or 15-37.5 mg twice weekly; **IV/IM** 50 mg weekly or 15-37.5 mg twice weekly

Psoriasis
• **Adult: PO/IM/IV** 10-25 mg/wk or 2.5 mg **PO** q12hr × 3 doses/wk, may increase to 25 mg/wk

Breast cancer
• **Adult: IV** 40-60 mg/m^2 on day 1 of every 21-28 days with other antineoplastics

Epidermal head/neck cancer
• **Adult/child: IV** 40 mg/m^2 on days 1 and 15, q21days alone or in combination with bleomycin, CISplatin
• **Adult: PO** 25-50 mg/m^2 q7days
• **Child: PO** 7.5-30 mg/m^2 q7-14days

Rheumatoid arthritis
• **Adult: PO** 7.5 mg/wk or in divided doses of 2.5 mg q12hr × 3 given per wk; max 20 mg/wk

Polyarticular-course juvenile RA
• **Child: PO/IM** 10 mg/m^2/wk

Burkitt's lymphoma (stages I, II, III)
• **Adult/adolescent/child: IV** 200 mg/m^2 days 8 and 15 q21days with bleomycin, cyclophosphamide, vinCRIStine, dexamethasone

Bladder cancer (unlabeled)
• **Adult: IV** 30 mg/m^2 on days 1, 15, 22 q28days in combination with vinBLAStine, DOXOrubicin, CISplatin (MVAC) regimen

Active Crohn's disease/ulcerative colitis (unlabeled)
• **Adult: IM** 25 mg/wk; **SUBCUT** 15 mg/kg/wk × 16 wk

GVHD prophylaxis (unlabeled)
• **Adult and child: IV** 15 mg/m^2 on day 1 after transplant then 10 mg/m^2 on days 3, 6, 11

Ectopic pregnancy (unlabeled)
• **Adult: IM** 50 mg/m^2 may be used in combination with mifepristone

Pregnancy termination before 63rd day of pregnancy (unlabeled)
• **Adult: IM** 50 mg/m^2 then intravaginal misoprostol 5-7 days later

Psoriatic arthritis (unlabeled)
• **Adult: PO** 5-7.5 mg weekly

Available forms: Tabs 2.5, 5, 7.5, 10, 15 mg; inj 25 mg/ml; powder for inj 1 g

Administer:
• Using chemotherapeutic handling
• Antiemetic 30-60 min before product
• Allopurinol or sodium bicarbonate to maintain uric acid levels, alkalinization of urine (pH >7.5), adequate fluids
• Store in tightly closed container in cool environment; store injection, powder for inj in dark, dry area

Direct IV route
• After diluting 5 mg/2 ml of sterile water for inj; give through Y-tube or 3-way stopcock

Intermittent/Continuous IV INF route
• Further dilute in D_5W, D_5NS, NS, before inf check patency of vein; flush with 5-10 ml of D_5W, NS; infuse at 4-20 mg/hr or prescribed rate

⚠ **Leucovorin rescue:** leucovorin calcium within 24 hr of product to prevent tissue damage; check agency policy; continue until methotrexate level $<10^{-8}$ m

IV INF intermediate or high dose (500 mg/m^2 over <4 hr or >1 g/m^2 over >4 hr): confirm WBC >1500/mm^3, neutrophils >200/mm^3, platelets >75,000/mm^3, serum bilirubin <1.2 mg/dl, serum creatinine WNL, SGPT <450 U, creatinine clearance >60 ml/min

⚠ Give sodium bicarbonate tabs or IV fluids to prevent precipitation of product at high doses; urine pH should be >7; may need to reduce dosage if BUN 20-30 mg/dl or creatinine is 1.2-2 mg/dl; stop product if BUN >30 mg/dl or creatinine >2 mg/dl

Additive compatibilities: Cephalothin, cyclophosphamide, cytarabine, fluorouracil, hydrOXYzine, mercaptopurine, ondansetron, sodium bicarbonate, vinCRIStine

Solution compatibilities: Amino acids, 4.25%/D_{25}, D_5W, sodium bicarbonate 0.05 mol/L, sodium chloride 0.9%

Syringe compatibilities: Bleomycin, CISplatin, cyclophosphamide, doxapram, DOXOrubicin, fluorouracil, furosemide, heparin, leucovorin, mitoMYcin, vinBLAStine, vinCRIStine

Y-site compatibilities: Allopurinol, amifostine, amphotericin B cholesteryl, asparaginase, aztreonam, bleomycin, cefepime, cefTRIAXone, cimetidine, CISplatin, cyclophosphamide, cytarabine, DAUNOrubicin, dexchlorpheniramine, diphenhydrAMINE, DOXOrubicin, DOXOrubicin liposome, etoposide, famotidine, filgrastim, fludarabine, fluorouracil, furosemide, gallium, ganciclovir, granisetron, heparin, HYDROmorphone, imipenem-cilastatin, leucovorin, LORazepam, melphalan, mesna, methylPREDNISolone, metoclopramide, mitoMYcin, morphine, ondansetron, oxacillin, PACLitaxel, piperacillin/tazobactam, prochlorperazine, ranitidine, sargramostim, teniposide, thiotepa, vinBLAStine, vinCRIStine, vinorelbine

Intrathecal route

Black Box Warning: Use preservative-free sol, reconstitute with NS; dose should be drawn into 5- to 10-ml syringe after LP, vol of CSF should be withdrawn equal to vol of methotrexate; allow CSF to flow into syringe and mix, inject over 15-30 sec with bevel of needle upward

SIDE EFFECTS

CNS: Dizziness, **seizures, leukoencephalopathy,** headache, confusion, **encephalopathy,** hemiparesis, malaise, fatigue, chills, fever; **arachnoiditis** (intrathecal)

EENT: Blurred vision, optic neuropathy

GI: *Nausea, vomiting, anorexia, diarrhea, ulcerative stomatitis,* **hepatotoxicity,** cramps, ulcer, gastritis, **GI hemorrhage,** abdominal pain, hematemesis, **hepatic fibrosis, acute toxicity**

GU: Urinary retention, **renal failure,** menstrual irregularities, defective spermatogenesis, **hematuria, azotemia, uric acid nephropathy**

HEMA: **Leukopenia, thrombocytopenia, myelosuppression, anemia**

INTEG: *Rash, alopecia,* dry skin, urticaria, photosensitivity, folliculitis, vasculitis, petechiae, ecchymosis, acne, alopecia, **severe fatal skin reaction**

RESP: **Methotrexate-induced lung disease**

SYST: **Sudden death,** ***Pneumocystis jiroveci,*** **tumor lysis syndrome**

PHARMACOKINETICS

Not metabolized; excreted in urine (unchanged); crosses placenta, blood-brain barrier; 50% plasma protein bound; terminal half-life 10-12 hr

PO: Readily absorbed

PO/IM/IV: Onset, duration unknown

IT: Onset, peak, duration unknown

INTERACTIONS

Do not use with proton pump inhibitors

Increase: **toxicity—salicylates, sulfa products, other antineoplastics, radiation, alcohol, probenecid, NSAIDs, phenylbutazone, theophylline, penicillins**

Increase: hypoprothrombinemia—oral anticoagulants

Increase: **hepatitis—acitretin; avoid concurrent use**

Decrease: effect of oral digoxin, vaccines, phenytoin, fosphenytoin

Decrease: effect of methotrexate—folic acid supplements

NURSING CONSIDERATIONS

Assess:

• Make sure product is taken weekly in RA, JRA

Black Box Warning: **Infection:** those with active infections should be treated for infection before product use; monitor temperature, fever may indicate beginning of infection

• Make sure drug–drug interacting products are discontinued before therapy, and do not resume until methotrexate level is safe

⚠ CBC, differential, platelet count weekly; avoid use until WBC is >1500/mm^3 or platelet count is >75,000/mm^3, neutrophils >200/mm^3; notify prescriber; WBC, platelet nadirs occur on day 7; monitor

M

Black Box Warning: Renal disease: avoid use in renal failure, BUN, serum uric acid, urine CCr, electrolytes before, during therapy; I&O ratio; report fall in urine output to <30 ml/hr

• Bleeding time, coagulation time during treatment; bleeding: hematuria, guaiac, bruising, or petechiae in mucosa or orifices
• Effects of alopecia on body image; discuss feelings about body changes

Black Box Warning: Pulmonary toxicity: those with ascites or pleural effusion at greater risk for toxicity; fluid should be removed before treatment; monitor plasma methotrexate levels

Black Box Warning: Tumor lysis syndrome: hyperkalemia, hyperphosphatemia, hyperuricemia, hypocalcemia, decreased urine output; use aggressive hydration, allopurinol to correct severe electrolyte imbalances, renal toxicity

Black Box Warning: Hepatotoxicity: jaundiced skin and sclera, dark urine, clay-colored stools, pruritus, abdominal pain, fever, diarrhea, hepatic studies before and during therapy: bilirubin, alk phos, AST, ALT; liver biopsy should be done before start of therapy (psoriasis)

• Monitor methotrexate levels, adjust leucovorin dose based on level
• Buccal cavity for dryness, sores, ulceration, white patches, oral pain, bleeding, dysphagia

Black Box Warning: Serious skin reaction: Stevens-Johnson syndrome, exfoliate dermatitis, skin necrosis, erythema multiform may occur within days of receiving product by any route; product should be discontinued

• **Strokelike encephalopathy:** common in high-dose therapy; assess for confusion, hemiparesis, seizures, coma; usually transient
• **Rheumatoid arthritis:** ROM, pain, joint swelling before, during treatment
• **Psoriasis:** skin lesions before, during treatment
• Increased fluid intake to 2-3 L/day to prevent urate deposits, calculi formation unless contraindicated
• Rinsing of mouth tid-qid with water, club soda; brushing of teeth bid-tid with soft brush or cotton-tipped applicators for stomatitis; use unwaxed dental floss

Evaluate:

• Therapeutic response: decreased tumor size, spread of malignancy; decreased joint inflammation, pain in RA

Teach patient/family:

Black Box Warning: To report any complaints, side effects to nurse or prescriber: black tarry stools, chills, fever, sore throat, bleeding, bruising, cough, SOB, dark or bloody urine, seizures

• That hair may be lost during treatment; that wig or hairpiece may make patient feel better; that new hair may be different in color, texture (alopecia rare)
• To avoid foods with citric acid, hot or rough texture if stomatitis is present
• To report stomatitis and any bleeding, white spots, ulcerations in mouth to prescriber; to examine mouth daily; to report symptoms to nurse; to use good oral hygiene

Black Box Warning: That contraceptive measures are recommended during therapy and for at least 8 wk after cessation of therapy for women and men; to discontinue breastfeeding; that toxicity to infant may occur; pregnancy (X)

• To drink 10-12 glasses of fluid/day
• To avoid alcohol, salicylates, live vaccines
• To avoid use of razors, commercial mouthwash
• To use sunblock to prevent burns
• To use good dental care to prevent overgrowth of infection in the mouth
• How to use this product with leucovorin rescue
• To continue leucovorin until told it is safe to stop
• To report CNS symptoms, vision changes
• To report fever, other symptoms of infection
• To report decreased urine output

methylcellulose (OTC)

(meth-ill-sell'yoo-lose)

Citrucel, Equaline Fiber Therapy, Fiber Therapy, Leader Fiber Therapy, Walgreens Fiber Therapy

Func. class.: Laxative, bulk forming

Chem. class.: Hydrophilic semisynthetic cellulose derivative

Do not confuse:
Citrucel/Citracal

ACTION: Attracts water, expands in intestine to increase peristalsis; absorbs excess water in stool; decreases diarrhea

USES: Chronic constipation

CONTRAINDICATIONS: Hypersensitivity, GI obstruction, hepatitis

DOSAGE AND ROUTES

- **Adult: PO** ≤6 g/day in divided doses
- **Child 6-12 yr: PO** 3 g/day in divided doses

Available forms: Powder 105 mg/g, 196 mg/g; tab 500 mg

Administer:

PO route

- Alone for better absorption; do not take within 1 hr of other products
- In AM or PM (oral dose)

SIDE EFFECTS

GI: Obstruction, abdominal distention

PHARMACOKINETICS

PO: Onset 12-24 hr, peak 1-3 days

INTERACTIONS

Decrease: absorption—digoxin, nitrofurantoin, salicylates, tetracyclines, oral anticoagulants

Drug/Herb

Increase: laxative action—flax senna

NURSING CONSIDERATIONS

Assess:

- Blood, urine electrolytes if used often
- I&O ratio to identify fluid loss
- Cause of constipation; lack of fluids, bulk, exercise, constipating products
- Cramping, rectal bleeding, nausea, vomiting; product should be discontinued

Evaluate:

- Therapeutic response: decrease in constipation

Teach patient/family:

- To mix powder in water, take with full glass of water
- To increase fluid intake
- That normal bowel movements do not always occur daily
- Not to use in presence of abdominal pain, nausea, vomiting
- To notify prescriber if constipation unrelieved or if symptoms of electrolyte imbalance occur: muscle cramps, pain, weakness, dizziness, excessive thirst

M

methyldopa/methyldopate (Rx)

(meth-ill-doe'pa)

Func. class.: Antihypertensive

Chem. class.: Centrally acting α-adrenergic inhibitor

Do not confuse:
methyldopa/L-dopa/levodopa

ACTION: Stimulates central inhibitory α-adrenergic receptors or acts as false transmitter, resulting in reduction of arterial pressure

USES: Hypertension, hypertensive crisis

CONTRAINDICATIONS: Active hepatic disease, hypersensitivity, MAOI therapy

Precautions: Pregnancy (B), geriatric patients, cardiac disease, autoimmune disease, depression, dialysis, hemolytic anemia, Parkinson's disease, pheochromocytoma, sulfite hypersensitivity

DOSAGE AND ROUTES

• **Adult: PO** 250-500 mg bid or tid then adjusted q2days as needed, 0.5-2 g/day in 2-4 divided doses (maintenance), max 3 g/day; **IV** 250-500 mg in 100 ml D_5W q6hr, run over 30-60 min, max 1 g q6hr; switch to oral as soon as possible
• **Geriatric: PO** 125 mg bid-tid, increase q2days as needed, max 3 g/day
• **Child: PO** 10 mg/kg/day in 2-4 divided doses, max 65 mg/kg or 3 g/day, whichever is less; **IV** 20-40 mg/kg/day in 4 divided doses, max 65 mg/kg or 3 g, whichever is less

Renal dose

• **Adult: PO** CCr 10-50 ml/min dose q8-12hr; CCr <10 ml/min dose q12-24hr

Available forms: *Methyldopa:* tabs 250, 500 mg; *methyldopate:* inj 50 mg/ml

Administer:

PO route

• Increase in dose should be done in the evening to minimize drowsiness

Intermittent IV INF route

• After diluting with 100 ml D_5W; run over 1/2-1 hr

Y-site compatibilities: Alfentanil, amikacin, aminophylline, anidulafungin, ascorbic acid, atenolol, atracurium, atropine, aztreonam, benztropine, bivalirudin, bleomycin, bumetanide, buprenorphine, butorphanol, calcium chloride/gluconate, caspofungin, cefamandole, ceFAZolin, cefmetazole, cefonicid, cefotaxime, cefoTEtan, cefOXitin, cefTAZidime, ceftizoxime, cefTRIAXone, cefuroxime, cephalothin, chlorproMAZINE, cimetidine, clindamycin, cyanocobalamin, cycloSPORINE, DACTINomycin, DAPTOmycin, dexamethasone, digoxin, diltiazem, diphenhydrAMINE, DOCEtaxel, DOPamine, doxycycline, enalaprilat, ePHEDrine, EPINEPHrine, epoetin alfa, ertapenem, erythromycin, esmolol, etoposide, etoposide phosphate, famotidine, fenoldopam, fentaNYL, fluconazole, fludarabine, gatifloxacin, gemcitabine, gentamicin, glycopyrrolate, granisetron, heparin, hydrocortisone, HYDROmorphone, hydrOXYzine, IDArubicin, insulin (regular), irinotecan, isoproterenol, labetalol, lidocaine, linezolid, LORazepam, magnesium sulfate, mannitol, mechlorethamine, meperidine, metaraminol, methicillin, methoxamine, methylPREDNISolone, metoclopramide, metoprolol, metroNIDAZOLE, mezlocillin, miconazole, midazolam, milrinone, minocycline, mitoXANtrone, morphine, moxalactam, multiple vitamins, mycophenolate mofetil, nafcillin, nalbuphine, naloxone, netilmicin, nitroglycerin, nitroprusside, norepinephrine, octreotide, ondansetron, oxacillin, oxaliplatin, oxytocin, PACLitaxel, palonosetron, pamidronate, pancuronium, pantoprazole, papaverine, PEMEtrexed, penicillin G potassium/sodium, pentazocine, phentolamine, phenylephrine, phytonadione, piperacillin, polymyxin B, potassium chloride, procainamide, prochlorperazine, promethazine, propranolol, protamine, pyridoxine, quiNIDine, ranitidine, ritodrine, sodium bicarbonate, succinylcholine, SUFentanil, tacrolimus, teniposide, theophylline, thiamine, thiotepa, ticarcillin, ticarcillin-clavulanate, tigecycline, tirofiban, tobramycin, tolazoline, trimetaphan, urokinase, vancomycin, vasopressin, vecuronium, verapamil, vinorelbine, voriconazole, zoledronic acid

SIDE EFFECTS

CNS: *Drowsiness, weakness, dizziness, sedation, headache,* depression, psychosis paresthesias, parkinsonism, Bell's palsy, nightmares
CV: Bradycardia, myocarditis, orthostatic hypotension, angina, edema, weight gain, CHF, paradoxic pressor response (IV)
EENT: Nasal congestion
ENDO: Breast enlargement, gynecomastia, amenorrhea
GI: Nausea, vomiting, diarrhea, constipation, hepatic dysfunction, sore or "black" tongue, pancreatitis, colitis, flatulence
GU: Impotence, failure to ejaculate

HEMA: Leukopenia, thrombocytopenia, hemolytic anemia, granulocytopenia, positive Coombs' test
INTEG: Rash, toxic epidermal necrolysis, lupuslike syndrome

PHARMACOKINETICS

PO: Peak 4-6 hr, duration 12-24 hr
IV: Peak 2 hr, duration 10-16 hr
Metabolized by liver, excreted in urine, half-life 2 hr

INTERACTIONS

• Lithium toxicity: lithium
⚠ **Increase:** pressor effect—sympathomimetic amines, MAOIs; do not use concurrently with MAOIs
Increase: hypotension, CNS toxicity—levodopa
Increase: hypotension—diuretics, other antihypertensives
Increase: psychosis—haloperidol
Increase: CNS depression—alcohol, antihistamines, antidepressants, analgesics, sedative/hypnotics
Increase: B/P—phenothiazines, β-blockers, amphetamines, NSAIDs, tricyclics, barbiturates
Increase: hypoglycemia—TOLBUTamide
Decrease: methyldopa absorption—iron

Drug/Lab Test
Increase: creatinine, LFTs
Decrease: platelets, WBC, Hgb/HcT
Interference: urinary uric acid, serum creatinine, AST
False increase: urinary catecholamines

NURSING CONSIDERATIONS

Assess:
• Blood studies: neutrophils, decreased platelets, CBC
• **Hemolytic anemia:** Direct Coombs' test before, after 6, 12 mo of therapy; a positive test may indicate hemolytic anemia; usually reverses within weeks to months after discontinuing treatment; monitor Hgb/HcT and RBC; do not start therapy in those with hemolytic anemia
• Baselines of renal, hepatic studies before therapy begins
• **Hypertension:** B/P when beginning treatment, periodically thereafter; report significant changes
• **Allergic reaction:** rash, fever, pruritus, urticaria; product should be discontinued if antihistamines fail to help
• CNS symptoms, especially in geriatric patients; depression, change in mental status
• **CHF:** edema, dyspnea, wet crackles, B/P
• Renal symptoms: polyuria, oliguria, urinary frequency; I&O ratio, weight; report weight gain >5 lb
• **Product tolerance:** may occur within 3 mo of starting treatment; a dosage change and other products may be needed

Evaluate:
• Therapeutic response: decrease in B/P with hypertension

Teach patient/family:
• To avoid hazardous activities
• Not to discontinue product abruptly because withdrawal symptoms may occur: anxiety, increased B/P, headache, insomnia, increased pulse, tremors, nausea, sweating
• To rise slowly to sitting or standing position to minimize orthostatic hypotension
• To notify prescriber of mouth sores, sore throat, fever, swelling of hands or feet, irregular heartbeat, chest pain, signs of angioedema
• That excessive perspiration, dehydration, vomiting, diarrhea may lead to fall in B/P; to consult prescriber
• That dizziness, fainting, lightheadedness may occur during first few days of therapy
• Not to use OTC (cough, cold, allergy) products unless directed by prescriber; that compliance is necessary; not to skip or stop product unless directed by prescriber
• That product may cause skin rash or impaired perspiration

M

TREATMENT OF OVERDOSE:

Gastric evacuation, sympathomimetics may be indicated; if severe, hemodialysis

methylergonovine (Rx)

(meth-ill-er-goe-noe′veen)

Func. class.: Oxytocic

Chem. class.: Ergot alkaloid

ACTION: Stimulates uterine, vascular, and smooth muscle, thereby causing contractions; decreases bleeding; arterial vasoconstriction

USES: Treatment of hemorrhage postpartum or postabortion, uterine contractions

CONTRAINDICATIONS: Pregnancy (other than obstetric delivery/abortion), hypertension, preeclampsia, eclampsia, elective induction of labor, hypersensitivity to ergot preparations

Precautions: Severe renal/hepatic disease, jaundice, diabetes mellitus, seizure disorders, sepsis, CAD, last stage of labor

DOSAGE AND ROUTES

• **Adult: PO** 200 mcg tid-qid × ≤7 days; **IM/IV** 200 mcg q2-4hr × 1-5 doses

Available forms: Inj 200 mcg/ml; tabs 200 mcg

Administer:

PO route

• Do not exceed dosage limits

• Store tabs at room temperature

• Give with water

• Only during 4th stage of labor; not to be used to augment labor

IM route

• Protect from light

• IM in deep muscle mass; rotate inj sites for additional doses, aspirate

Direct IV route

• Undiluted through Y-tube or 3-way stopcock; give ≤0.2 mg/min or diluted in 5 ml 0.9% NaCl given through Y-site

• With crash cart available on unit; IV route used only in emergencies

• Refrigerated storage of ampules; protect from light; give only if solution is clear; colorless

Y-site compatibilities: Heparin, hydrocortisone sodium succinate, potassium chloride, vit B/C

SIDE EFFECTS

CNS: *Headache, dizziness,* seizures, hallucinations; stroke (IV)

CV: Hypotension, chest pain, palpitation, hypertension, dysrhythmias, CVA (IV)

EENT: Tinnitus

GI: *Nausea, vomiting*

GU: Cramping

INTEG: Sweating, rash, allergic reactions

MS: Leg cramps

RESP: Dyspnea

PHARMACOKINETICS

Metabolized in liver, excreted in urine

PO: Onset 5-25 min, duration 3 hr

IM: Onset 2-5 min, duration 3 hr

IV: Onset immediate, duration 45 min

INTERACTIONS

Increase: vasoconstriction—DOPamine, ergots, anesthetics (regional), vasopressors, nicotine

Increase: ergot toxicity—CYP3A4 inhibitors, do not use together

NURSING CONSIDERATIONS

Assess:

• B/P, pulse, character and amount of vaginal bleeding; watch for indications of hemorrhage

• Uterine relaxation; observe for severe cramping

⚠ **Ergot toxicity:** tinnitus, hypertension, palpitations, chest pain, nausea, vomiting, weakness; cold, numb extremities

Evaluate:

• Therapeutic response: absence of hemorrhage

Teach patient/family:
- To report increased blood loss, severe abdominal cramps, fever, or foul-smelling lochia

methylnaltrexone (Rx)
(meth-il-nal-trex'one)
Relistor
Func. class.: Opioid antagonist

ACTION: Peripheral μ-opioid receptor antagonist that reduces constipation associated with opiate agonists

USES: Treatment of opioid-induced constipation in patients with advanced illness who are receiving palliative care when response to laxative therapy has been insufficient
Unlabeled uses: Pruritus; nausea, vomiting related to morphine; urinary retention from opioids

CONTRAINDICATIONS: Hypersensitivity, GI obstruction, IV route
Precautions: Pregnancy (B), breastfeeding, children, geriatric patients, renal disease, diarrhea, driving, operating machinery, neoplastic disease, Crohn's disease, peptic ulcer, ulcerative colitis

DOSAGE AND ROUTES
Opiate-agonist–induced constipation
- **Adult >114 kg: SUBCUT** 0.15 mg/kg every other day prn
- **Adult 62-114 kg: SUBCUT** 12 mg every other day prn, max 12 mg/34 hr
- **Adult 38-<62 kg: SUBCUT** 8 mg every other day prn, max 8 mg/24 hr
- **Adult <38 kg: SUBCUT** 0.15 mg/kg every other day prn, max 0.15 mg/kg/24 hr

Renal dose
- **Adult: SUBCUT** CCr <30 ml/min, reduce normal adult dose by 50%

Pruritus, nausea, and vomiting related to morphine (unlabeled) (oral dose investigational)
- **Adult: PO** 19.2 mg/kg 20 min before morphine

Available forms: Sol for inj 12 mg/0.6 ml, 8 mg/0.4 ml
Administer:
- SUBCUT only; oral dose investigational, not currently available
- Do not give IV; IV dosing for urinary retention investigational
- Store at 59° F-86° F (15° C-30° C); do not freeze
- Store away from light

SUBCUT route
- Inspect sol before use; should be clear, colorless to pale yellow aqueous sol; do not use if particulate matter or discoloration are present
- Withdraw needed amount of sol into sterile syringe; if immediate administration is impossible, syringe may be kept at room temperature for ≤24 hr; syringe does not need to be kept away from light during the 24-hr period; immediately discard any unused portion in vial; no preservatives are present
- Administer into upper arm, abdomen, or thigh ≤1×/24 hr; rotate inj sites; do not inject same spot each time; do not inject into areas where skin is tender, bruised, red, or hard; avoid areas with scars or stretch marks
- If using with retractable needle, slowly push down on plunger past resistance point until the syringe is empty and click is heard

M

SIDE EFFECTS
CNS: Dizziness
GI: Nausea, vomiting, diarrhea, flatulence, abdominal pain, **GI perforation**
INTEG: Hyperhidrosis

PHARMACOKINETICS
Terminal half-life 8 hr, protein binding 11%-15.3%; renal impairment has marked effect on renal excretion of methylnaltrexone; dose adjustment is re-

quired for patients with CCr <30 ml/min; renal clearance decreased and total systemic exposure increased in patients with severe renal impairment who receive single SUBCUT dose of 0.3 mg/kg
SUBCUT: Peak 30 min

NURSING CONSIDERATIONS

Assess:
- Serum creatinine
- **Opioid-induced constipation:** stool characteristics, bowel sounds during treatment

Evaluate:
- Therapeutic response: decreasing constipation

Teach patient/family:
- That, after 30 min, to remain near toilet facilities because bowel relaxation occurs; not to use more than 1 dose in 24 hr
- To notify prescriber of abdominal pain, continuous or severe diarrhea, nausea, vomiting
- To avoid use in pregnancy unless absolutely necessary; avoid in breastfeeding

methylphenidate (Rx)

(meth-ill-fen'i-date)

Biphentin ♣, Concerta, Daytrana, Metadate CD, Metadate ER, Methylin, Quillivant XR, Ritalin, Ritalin LA, Ritalin SR

Func. class.: Cerebral stimulant
Chem. class.: Piperidine derivative

Controlled Substance Schedule II

Do not confuse:
methylphenidate/methadone

ACTION: Increases release of norepinephrine, DOPamine in cerebral cortex to reticular activating system; exact action not known

USES: Attention deficit disorder (ADD), attention-deficit/hyperactivity disorder (ADHD); narcolepsy (except Concerta, Metadate CD, Ritalin LA)

CONTRAINDICATIONS: Children <6 yr, hypersensitivity, anxiety, history of Gilles de la Tourette's syndrome; glaucoma, anorexia nervosa, tartrazine dye hypersensitivity

Precautions: Pregnancy (C), breastfeeding, hypertension, depression, seizures

Black Box Warning: Substance abuse

DOSAGE AND ROUTES

Attention-deficit/hyperactivity disorder (ADHD) initial treatment, not currently on methylphenidate

Regular release: Ritalin, Methylin, Methylin oral sol, Methylin chew tabs
- **Adult: PO** 20-30 mg/day, range 10-60 mg/day in 2-3 divided doses, 30-45 min before meals
- **Child ≥6 yr: PO** 5 mg bid initially, increase 5-10 mg/day weekly, usual dose 0.3-2 mg/kg/day, max 60 mg/day

Extended release: Ritalin SR, Metadate ER, Methylin ER
- **Adult/adolescent/child ≥6 yr: PO** max 20-30 mg tid

Extended-release once-daily tabs: Concerta
- **Adult: PO** 18-36 mg/day initially then adjust by 18 mg/wk, max 72 mg/day
- **Adolescent: PO** 18 mg/day initially then adjust by 18 mg/wk, max 72 mg/day
- **Child ≥6 yr: PO** 18 mg/day initially then adjust by 18 mg/wk, max 54 mg/day

Extended-release once-daily capsules: Ritalin LA
- **Adult/adolescent/child ≥6 yr: PO** 20 mg daily in AM initially, adjust by 10 mg/wk, max 60 mg/day

Extended-release once-daily capsules: Metadate CD
- **Adult/adolescent/child ≥6 yr: PO** 20 mg/day in AM, adjust by 20 mg/wk, max 60 mg/day

Transdermal: Daytrana
- **Adolescent/child ≥6 yr: TD** wk 1: 10 mg/day (9-mg patch); wk 2: 15 mg/day

(9-mg patch); wk 3: 20 mg/day (9-mg patch); wk 4: 30 mg/day (9-mg patch)

Conversion to once-daily treatment from other forms for ADHD

Extended-release once-daily capsules: Metadate CD

• **Adult/adolescent/child ≥6 yr: PO** give no more than total daily dose of other forms, may adjust by 20 mg/wk, max 60 mg/day

Extended-release once-daily capsules: Ritalin LA

• **Adult/adolescent/child ≥6 yr: PO** give no more than total daily dose of other forms, may adjust by 10 mg/wk, max 60 mg/day

Extended-release once-daily tablets: Concerta

• **Adult/adolescent/child ≥6 yr (currently on 10-15 mg/day): PO** 18 mg every AM initially, adjust by 18 mg/wk, max 72 mg/day (adult); max 72 mg/day, 2 mg/kg/day (adolescent); 54 mg/day (child) mg/day

• **Adult/adolescent/child ≥6 yr (currently receiving 20-30 mg/day): PO** 36 mg every AM, adjust by 18 mg/wk, max 72 mg/day (adult); 72 mg/day, 2 mg/kg/day (adolescent); 54 mg/day (child)

• **Adult/adolescent/child ≥6 yr (currently receiving 30-45 mg/day): PO** 54 mg every AM, adjust by 18 mg/wk, max 72 mg/day (adult); 72 mg/day, 2 mg/kg/day (adolescent); 54 mg/day (child)

• **Adult/adolescent/child ≥6 yr (currently receiving 40-60 mg/day): PO** 72 mg every AM, 72 mg/day

Extended-release once-daily suspension: Quillivant XR

• **Adolescents/child ≥6 yr: PO** give in AM, increase in 10-20 mg increments weekly

Narcolepsy

Immediate release: Ritalin, Methylin oral sol, Methylin chew tabs

• **Adult: PO** 20-30 mg/day, range 10-60 mg/day in 2-3 divided doses

• **Child ≥6 yr: PO** 5 mg bid, may increase by 5-10 mg/wk, max 60 mg/day

Extended-release tabs: Ritalin SR, Metadate ER

• **Adult/adolescent/child ≥6 yr: PO** max 20 mg tid

Poststroke depression; major depression (unlabeled)

• **Adult and geriatric: PO** (immediate rel tabs) 2.5 mg bid, may increase by 2.5-5 mg q2-3days

Available forms: Tabs 5, 10, 20 mg; ext rel tabs 10, 20, mg; ext rel tabs (Concerta) 18, 27, 36, 54 mg; ext rel caps 10, 20, 30, 40 mg; oral sol 5 mg, 10 mg/ml; chew tabs (Methylin) 2.5, 5, 10 mg; transdermal patch 12.5 cm^2 (10 mg), 18.75 cm^2 (15 mg), 25 cm^2 (20 mg), 37.5 cm^2 (30 mg); ext rel oral susp 300 mg/60 ml, 600 mg/120 ml, 750 mg/150 ml, 900 mg/180 ml

Administer:

PO route

• Do not crush, chew ext rel product; caps may be opened, beads sprinkled over spoonful of applesauce; give without regard to meals

• Gum, hard candy, frequent sips of water for dry mouth

• Give immediate rel dose 30-45 min before meals

• **Chew tab** with adequate water to prevent choking; contains phenylalanine

• Avoid metadate CD on day of surgery

Extended Release Oral Suspension (Quillivant XR)

• Shake bottle well; use calibrated dosing dispenser provided; give in AM without regard to meals; reconstitute as per manufacturer

Transdermal route

• Place on clean, dry area of hip; avoid waist; remove 9 hr after application; fold after removal; flush down toilet

• If patch falls off, apply new patch to different site; total wear time should be 9 hr

SIDE EFFECTS

CNS: *Hyperactivity, insomnia, restlessness, talkativeness,* dizziness, drowsiness, toxic psychosis, headache, akathi-

M

sia, dyskinesia, masking or worsening of Tourette's syndrome, seizures, hallucinations, malignant neuroleptic syndrome, aggression; cerebral vasculitis, hemorrhage, stroke (rare)
CV: *Palpitations, tachycardia,* B/P changes, angina, dysrhythmias
ENDO: Growth retardation
GI: Nausea, anorexia, dry mouth, weight loss, abdominal pain
HEMA: Leukopenia, anemia, thrombocytopenic purpura
INTEG: Exfoliative dermatitis, urticaria, rash, erythema multiforme, hypersensitivity reactions
MISC: Fever, arthralgia, scalp hair loss

PHARMACOKINETICS

PO: Onset $^{1}/_{2}$-1 hr, duration 4-6 hr, metabolized by liver, excreted by kidneys, half-life 3-4 hr

INTERACTIONS

- Hypertensive crisis: MAOIs or within 14 days of MAOIs, vasopressors

Increase: effects of tricyclics, SSRIs, anticonvulsants, SNRIs, CNS stimulants
Decrease: effect of antihypertensives
Drug/Herb
Increase: CNS stimulation—cola nut, guarana, horsetail, yerba maté, yohimbe
Drug/Food
Increase: stimulation—caffeine

NURSING CONSIDERATIONS

Assess:
- **ADHD:** attention span, decreased hyperactivity

Black Box Warning: Substance abuse: there is a high potential for abuse; use caution in those with history of substance abuse

- VS, B/P; may reverse antihypertensives; check patients with cardiac disease more often for increased B/P
- CBC with differential, platelets, LFTs, urinalysis; in diabetes: blood glucose, urine glucose; insulin changes may have to be made because eating will decrease
- Height, growth rate q3mo in children; growth rate may be decreased, but normal growth will resume when product is discontinued
- Mental status: mood, sensorium, affect, stimulation, insomnia, aggressiveness

⚠ **Withdrawal symptoms:** headache, nausea, vomiting, muscle pain, weakness, usually not associated with drug holidays
- Appetite, sleep, speech patterns
- **Narcolepsy:** identify frequency, length of narcoleptic episodes

Evaluate:
- Therapeutic response: decreased hyperactivity (ADHD); increased ability to stay awake (narcolepsy)

Teach patient/family:
- To decrease caffeine consumption (coffee, tea, cola, chocolate), may increase irritability, stimulation; not to use guarana, yerba maté, cola nut
- To avoid OTC preparations unless approved by prescriber
- To taper off product over several weeks because depression, increased sleeping, lethargy will occur
- To avoid driving, hazardous activities if dizziness, blurred vision occur
- To avoid alcohol
- To avoid hazardous activities until stabilized on medication
- To get needed rest; patients will feel more tired at end of day
- That shell of Concerta tab may appear in stools
- To take regular tab at least 6 hr before sleep, 10 hr for ext rel, use dosing syringe to measure liquid; not to use household teaspoon
- **Transdermal:** after tray is opened, use within 2 mo; do not store patches without protective patch; notify prescriber if skin irritation or rash occurs

TREATMENT OF OVERDOSE:

Administer fluids; hemodialysis or peritoneal dialysis; antihypertensive for increased B/P; administer short-acting barbiturate before lavage

methylPREDNISolone (Rx)

(meth-il-pred-niss'oh-lone)

A-Methapred, Depo-Medrol, Medrol, Solu-MEDROL

Func. class.: Corticosteroid, synthetic

Chem. class.: Glucocorticoid, immediate acting

Do not confuse:
methylPREDNISolone/predniSONE/medroxyPROGESTERone/methylTESTOSTERone

ACTION: Decreases inflammation by suppression of migration of polymorphonuclear leukocytes, fibroblasts; reversal of increased capillary permeability and lysosomal stabilization

USES: Severe inflammation, shock, adrenal insufficiency, collagen disorders, management of acute spinal cord injury, multiple sclerosis

Unlabeled uses: Multiple myeloma, bronchospasm prophylaxis, airway-obstructing hemangioma, noncardiogenic pulmonary edema, idiopathic pulmonary fibrosis, carpal tunnel syndrome, temporal arteritis, Churg-Strauss syndrome, mixed connective-tissue disease, polyarteritis nodosa, relapsing polychondritis, polymyalgia rheumatica, vasculitis, Wegener's granulomatosis, *Pneumocystis jiroveci* pneumonia in AIDS patients, acute spinal cord injury, severe acute respiratory syndrome (SARS), acute interstitial nephritis

CONTRAINDICATIONS: Hypersensitivity, intrathecal use, neonates

Precautions: Pregnancy (C), breastfeeding, diabetes mellitus, glaucoma, osteoporosis, seizure disorders, ulcerative colitis, CHF, myasthenia gravis, renal disease, esophagitis, peptic ulcer, tartrazine, benzyl alcohol, corticosteroid hypersensitivity, viral infection, TB, traumatic brain injury, Cushing syndrome, measles, varicella, fungal infections

DOSAGE AND ROUTES

Adrenal insufficiency/inflammation
- **Adult: PO** 4-48 mg in 4 divided doses; **IM** 10-80 mg (acetate); **IM/IV** 10-250 mg (succinate); **INTRAARTICULAR** 4-80 mg (acetate)
- **Child: IV** 0.5-1.7 mg/kg in 3-4 divided doses (succinate)

Multiple sclerosis
- **Adult: PO** 160 mg/day × 1 wk then 64 mg every other day × 30 days

Multiple myeloma/temporal arteritis/Churg-Strauss syndrome/mixed connective-tissue disease/polyarteritis nodosa/relapsing polychondritis/polymyalgia rheumatica/vasculitis/Wegener's granulomatosis (unlabeled)
- **Adult: PO** 4-48 mg/day in 4 divided doses; **IM** 10-120 mg (acetate); **IV** 10-40 mg over several min (sodium succinate)
- **Child: PO/IM** 0.5-1.7 mg/kg or 5-25 mg/m^2/day in divided doses q6-12hr

Bronchospasm prophylaxis (unlabeled)
- **Adult and adolescent: PO/IV** 40-80 mg/day in 1-2 divided doses
- **Child: PO/IV** 1 mg/kg in 2 divided doses (max 60 mg)

Airway-obstructing hemangioma (unlabeled)
- **Child: PO** 0.5-1.7 mg/kg or 5-25 mg/m^2/day in divided doses q6-12hr

Idiopathic pulmonary fibrosis (unlabeled)
- **Adult: IV** 1-2 g/wk or every other week

Carpal tunnel syndrome (unlabeled)
- **Adult: INJ** (local) 40-80 mg as a single inj

Available forms: Tabs 2, 4, 8, 16, 24, 32 mg; inj 20, 40, 80 mg/ml acetate; inj 40, 125, 500, 1000, 2000 mg/vial succinate

Administer:
- Titrated dose; use lowest effective dose

PO route
- With food or milk to decrease GI symptoms (PO)

M

• Once-a-day dose should be given in AM to coincide with body's normal cortisol secretion

IM route

• IM inj deep in large muscle mass; rotate sites; avoid deltoid; use 21-G needle; after shaking suspension (parenteral); inj site reaction may occur (induration, pain at site, atrophy)

• In one dose in AM to prevent adrenal suppression; avoid SUBCUT administration; may damage tissue

⚠ Do not give Solu-MEDROL intrathecally

IV route

• Use only Solu-MEDROL; never use acetate product

Direct IV route

• Use only methylPREDNISolone sodium succinate (Solu-MEDROL) IV; never use methylPREDNISolone acetate suspension IV

• After diluting with diluent provided; agitate slowly; give ≤500 mg/≥1 min directly over 3-15 min; doses ≥2 mg/kg or 250 mg should be given by intermittent IV inf unless potential benefits outweigh risks

Intermittent/continuous INF route

• Dilute further in D5W, NS, D5NS; haze may form; give over 15-60 min; large dose (≥500 mg) should be given over 30-60 min

Syringe compatibilities: Granisetron, metoclopramide

Y-site compatibilities: Acyclovir, amifostine, amphotericin B cholesteryl, amrinone, aztreonam, cefepime, CISplatin, cladribine, cyclophosphamide, cytarabine, DOPamine, DOXOrubicin, enalaprilat, famotidine, fludarabine, granisetron, heparin, melphalan, meperidine, methotrexate, metroNIDAZOLE, midazolam, morphine, piperacillin/tazobactam, remifentanil, sodium bicarbonate, tacrolimus, teniposide, theophylline, thiotepa

SIDE EFFECTS

CNS: Depression, flushing, sweating, headache, mood changes

CV: Hypertension, circulatory collapse, thrombophlebitis, embolism, tachycardia

EENT: Fungal infections, increased intraocular pressure, blurred vision, cataracts

GI: Diarrhea, nausea, abdominal distention, GI hemorrhage, increased appetite, pancreatitis

HEMA: Thrombocytopenia

INTEG: Acne, poor wound healing, ecchymosis, petechiae

MS: Fractures, osteoporosis, weakness

PHARMACOKINETICS

Half-life >3½ hr (plasma), 18-36 hr (tissue); crosses placenta, enters breast milk in small amounts; metabolized in liver; excreted by kidneys (unchanged)

PO: Peak 1-2 hr, duration 1½ days, well absorbed

IM: Peak 4-8 days, duration 1-4 wk, well absorbed

Intraarticular: Peak 1 wk

INTERACTIONS

Increase: side effects—amphotericin B, diuretics

Increase: methylPREDNISolone action—oral contraceptives, estrogens

Increase: adrenal suppression—CYP3A4 inhibitors (aprepitant, antiretroviral protease inhibitors, clarithromycin, danazol, delavirdine, diltiazem, erythromycin, fluconazole, FLUoxetine, fluvoxaMINE, imatinib, ketoconazole, mibefradil, nefazodone, telithromycin, voriconazole)

Decrease: methylPREDNISolone effect—CYP3A4 inducers (barbiturates, bosentan, carBAMazepine, efavirenz, phenytoins, nevirapine, rifabutin, rifampin)

Decrease: effects of antidiabetics, vaccines, somatrem

Drug/Herb

• Avoid use with St. John's wort

Drug/Food

• Do not use with grapefruit juice; level of methylPREDNISolone will be increased

Drug/Lab Test

Increase: cholesterol, blood glucose

Decrease: Calcium, potassium, T_4, T_3, thyroid ^{131}I uptake test, urine 17-OHCS, 17-KS
False negative: skin allergy tests

NURSING CONSIDERATIONS

Assess:

- **Potassium depletion:** parethesias, fatigue, nausea, vomiting, depression, polyuria, dysrhythmias, weakness
- Edema, hypertension, cardiac symptoms
- Mental status: affect, mood, behavioral changes, aggression
- Potassium, blood glucose, urine glucose while receiving long-term therapy; hypokalemia and hyperglycemia
- Joint mobility, pain, edema if product given intraarticularly
- B/P q4hr, pulse; notify prescriber of chest pain, crackles
- I&O ratio; be alert for decreasing urinary output, increasing edema; weight daily; notify prescriber of weekly gain >5 lb

⚠ **Adrenal insufficiency:** weight loss, nausea, vomiting, confusion, anxiety, hypotension, weakness; plasma cortisol levels during long-term therapy (normal level: 138-635 nmol/L SI units when drawn at 8 AM)

- Growth in children receiving long-term treatment
- **Infection:** increased temperature, WBC, even after withdrawal of product; product masks infection

Evaluate:

- Therapeutic response: ease of respirations, decreased inflammation; decreased symptoms of adrenal insufficiency

Teach patient/family:

- To increase intake of potassium, calcium, protein
- To carry emergency ID as corticosteroid user
- To notify prescriber if therapeutic response decreases; that dosage adjustment may be needed

⚠ Not to discontinue abruptly because adrenal crisis can result

- To take PO with food, milk to decrease GI symptoms
- To avoid OTC products: salicylates, alcohol in cough products, cold preparations unless directed by prescriber; to avoid vaccinations because immunosuppression occurs
- **Adrenal insufficiency:** nausea, anorexia, fatigue, dizziness, dyspnea, weakness, joint pain
- **Cushingoid symptoms:** buffalo hump, moon face, rapid weight gain, excess sweating
- **Infection:** to avoid persons with known infections; corticosteroids can mask symptoms of infection

metipranolol ophthalmic

See Appendix B

metoclopramide (Rx)

(met-oh-kloe-pra′mide)

Apo-Metoclop ♣, Metozolv ODT, Reglan

Func. class.: Cholinergic, antiemetic
Chem. class.: Central dopamine receptor antagonist

Do not confuse:
metoclopramide/metolazone
Reglan/Megace/Renagel

ACTION: Enhances response to acetylcholine of tissue in upper GI tract, which causes the contraction of gastric muscle; relaxes pyloric, duodenal segments; increases peristalsis without stimulating secretions; blocks dopamine in chemoreceptor trigger zone of CNS

USES: Prevention of nausea, vomiting induced by chemotherapy, radiation, delayed gastric emptying, gastroesophageal reflux
Unlabeled uses: Hiccups, migraines, breastfeeding induction, lung cancer

CONTRAINDICATIONS:
Hypersensitivity to this product, procaine, or procainamide; seizure disorder, pheochromocytoma, breast cancer (prolactin dependent), GI obstruction

Precautions: Pregnancy (B), breastfeeding, GI hemorrhage, CHF, Parkinson's disease

Black Box Warning: Tardive dyskinesia

DOSAGE AND ROUTES

Nausea/vomiting (chemotherapy)

- **Adult:** IV 1-2 mg/kg 30 min before administration of chemotherapy, then q2hr × 2 doses, then q3hr × 3 doses
- **Child (unlabeled):** IV 1-2 mg/kg/dose

Facilitate small-bowel intubation for radiologic exams

- **Adult and child >14 yr:** IV 10 mg over 1-2 min
- **Child <6 yr:** IV 0.1 mg/kg
- **Child 6-14 yr:** IV 2.5-5 mg

Diabetic gastroparesis

- **Adult:** PO 10 mg 30 min before meals, at bedtime × 2-8 wk
- **Geriatric:** PO 5 mg 30 min before meals, at bedtime, increase to 10 mg if needed

Gastroesophageal reflux

- **Adult:** PO 10-15 mg qid 30 min before meals and at bedtime
- **Child:** PO 0.4-0.8 mg/kg/day in 4 divided doses

Renal dose

- **Adult:** CCr <40 ml/min 50% of dose

Lactation induction (unlabeled)

- **Adult:** PO 10 mg bid-tid, may increase to 20-45 mg/day in divided doses

Non–small-cell lung cancer (NSCLC) radiation sensitizer (unlabeled)

- **Adult:** IV (Sensamide IV) 2 mg/kg given 1 hr before radiation therapy 3×/wk

Hiccups (unlabeled)

- **Adult:** PO/IM/IV 10 mg q6hr

Available forms: Tabs 5, 10 mg; syr 5 mg/5 ml; inj 5 mg/ml; conc sol 10 mg/ml; orally disintegrating tab 5, 10 mg

Administer:

PO route

- ½-1 hr before meals for better absorption
- Gum, hard candy, frequent rinsing of mouth for dry oral cavity
- **Oral disintegrating:** place on tongue, allow to dissolve, swallow, remove from bottle immediately before use

IM route

- Give for post-operative nausea, vomiting before end of surgery

Direct IV route

- DiphenhydrAMINE IV or benztropine IM for EPS
- Undiluted if dose ≤10 mg; give over 2 min

Intermittent IV INF route

- >10 mg may be diluted in ≥50 ml D_5W, NaCl, Ringer's, LR, given over ≥15 min

Y-site compatibilities: Alfentanil, amifostine, amikacin, aminophylline, ascorbic acid, atracurium, atropine, azaTHIOprine, aztreonam, bivalirudin, bleomycin, bumetanide, buprenorphine, butorphanol, calcium chloride/gluconate, CARBOplatin, caspofungin, ceFAZolin, cefonicid, cefoperazone, cefotaxime, cefoTEtan, cefOXitin, cefTAZidime, ceftizoxime, cefTRIAXone, cefuroxime, chloramphenicol, chlorproMAZINE, cimetidine, ciprofloxacin, cisatracurium, CISplatin, cladribine, clindamycin, cyanocobalamin, cyclophosphamide, cycloSPORINE, cytarabine, DACTINomycin, DAPTOmycin, dexamethasone, dexmedetomidine, digoxin, diltiazem, diphenhydrAMINE, DOBUTamine, DOCEtaxel, DOPamine, doripenem, doxapram, DOXOrubicin hydrochloride, doxycycline, droperidol, enalaprilat, ePHEDrine, EPINEPHrine, epirubicin, epoetin alfa, ertapenem, erythromycin, esmolol, etoposide, etoposide phosphate, famotidine, fenoldopam, fentaNYL, filgrastim, fluconazole, fludarabine, folic acid, foscarnet, gallium nitrate, gemcitabine, gentamicin, glycopyrrolate, granisetron, heparin, hydrocortisone, HYDROmorphone, IDArubicin, ifosfamide, imipenem/cilastatin, indomethacin, insulin, isoproterenol, ketorolac, labetalol, leucovorin, levofloxacin, lidocaine, linezolid, LORazepam, magnesium sulfate, mannitol, mechlorethamine, melphalan, meperidine, meropenem, metaraminol, methadone, methotrexate, methoxamine, methyldopate, methylPREDNISolone, metoprolol, metroNIDAZOLE, miconazole, midazolam, milrinone, minocycline, mitoMYcin, morphine, moxalactam, multiple vitamins, nafcillin, nalbuphine, naloxone, nesiritide, nitroglycerin, nitroprusside, norepinephrine, octreotide, ondansetron, oxaliplatin, oxytocin, PACLitaxel, palonosetron, pantoprazole, papaverine, PEMEtrexed, penicillin G, pentamidine, pentazocine, PENTobarbital, PHENobarbital, phentolamine, phenylephrine, phytonadione, piperacillin/tazobactam, potassium chloride, procainamide, prochlorperazine, promethazine, propranolol, protamine, pyridoxine, quinupristin/dalfopristin, ranitidine, remifentanil, riTUXimab, rocuronium, sargramostim, sodium acetate/bicarbonate, succinylcholine, SUFentanil, tacrolimus, teniposide, theophylline, thiamine, thiotepa, ticarcillin/clavulanate, tigecycline, tirofiban, tobramycin, tolazoline, topotecan, trastuzumab, trimethaphan, urokinase, vancomycin, vasopressin, vecuronium, verapamil, vinBLAStine, vinCRIStine, vinorelbine, voriconazole, zidovudine

SIDE EFFECTS

CNS: *Sedation, fatigue, restlessness, headache, sleeplessness, dystonia,* dizziness, drowsiness, **suicidal ideation, seizures,** EPS, **neuroleptic malignant syndrome; tardive dyskinesia (>3 mo, high doses)**

CV: Hypotension, supraventricular tachycardia

GI: Dry mouth, constipation, nausea, anorexia, vomiting, diarrhea

GU: Decreased libido, prolactin secretion, amenorrhea, galactorrhea

HEMA: **Neutropenia, leukopenia, agranulocytosis**

INTEG: Urticaria, rash

PHARMACOKINETICS

Metabolized by liver, excreted in urine, half-life 4 hr

PO: Onset 1/2-1 hr, duration 1-2 hr

IM: Onset 10-15 min, duration 1-2 hr

IV: Onset 1-3 min, duration 1-2 hr

INTERACTIONS

- Avoid use with MAOIs

Increase: sedation—alcohol, other CNS depressants

Increase: risk for EPS—haloperidol, phenothiazines

Decrease: action of metoclopramide—anticholinergics, opiates

Drug/Lab Test

Increase: prolactin, aldosterone, thyrotropin

NURSING CONSIDERATIONS

Assess:

Black Box Warning: EPS, tardive dyskinesia; more likely to occur in treatment >3 mo, geriatric patients and may be irreversible; assess for involuntary movements often

- **Neuroleptic malignant syndrome:** **hyperthermia, change in B/P, pulse, tachycardia, sweating, rigidity, altered consciousness (rare)**
- Mental status: depression, anxiety, irritability
- GI complaints: nausea, vomiting, anorexia, constipation

Perform/provide:

- Protect from light with aluminum foil during inf
- Discard open ampules

Evaluate:

- Therapeutic response: absence of nausea, vomiting, anorexia, fullness

Teach patient/family:

- To avoid driving, other hazardous activities until stabilized on product

• To avoid alcohol, other CNS depressants that will enhance sedating properties of this product

metolazone (Rx)

(me-tole'a-zone)

Zaroxolyn

Func. class.: Diuretic, antihypertensive

Chem. class.: Thiazide-like quinazoline derivative

Do not confuse:
metolazone/methotrexate/metoclopramide

ACTION: Acts on distal tubule by increasing excretion of water, sodium, chloride, potassium, magnesium, bicarbonate; decreases GFR

USES: Edema, hypertension

Unlabeled uses: Heart failure, nephrotic syndrome

CONTRAINDICATIONS: Pregnancy (D) (preeclampsia, intrauterine growth retardation), hypersensitivity to thiazides, sulfonamides; anuria, coma, hepatic encephalopathy

Precautions: Pregnancy (B), breastfeeding, geriatric patients, hypokalemia, renal/hepatic disease, gout, COPD, lupus erythematosus, diabetes mellitus, hypotension, history of pancreatitis; hypersensitivity to sulfonamides, thiazides; electrolyte imbalance

DOSAGE AND ROUTES

Edema

• **Adult: PO** 5-10 mg/day; max 20 mg/day

Hypertension

• **Adult: PO** 2.5-5 mg/day

• **Child: PO** 0.2-0.4 mg/kg/day in divided doses q12-24hr

Available forms: Tabs 2.5, 5, 10 mg

Administer:

• In AM to avoid interference with sleep if using product as diuretic

• Potassium replacement if potassium <3 mg/dl

• With food if nausea occurs; absorption may be decreased slightly

SIDE EFFECTS

CNS: Anxiety, depression, *headache, dizziness, fatigue, weakness*

CV: *Orthostatic hypotension,* palpitations, volume depletion, hypotension, chest pain

EENT: Blurred vision

ELECT: *Hypokalemia,* hypercalcemia, hyponatremia

GI: *Nausea, vomiting, anorexia,* constipation, diarrhea, cramps, pancreatitis, GI irritation, dry mouth, jaundice, **hepatitis**

GU: *Urinary frequency,* polyuria, **uremia, glucosuria,** nocturia, impotence

HEMA: **Aplastic anemia, hemolytic anemia, leukopenia, agranulocytosis, neutropenia**

INTEG: *Rash,* urticaria, purpura, photosensitivity, fever, dry skin, **toxic epidermal necrolysis, Stevens-Johnson syndrome**

META: *Hyperglycemia,* increased creatinine, BUN

MS: Muscle cramps, joint pain, swelling

PHARMACOKINETICS

Protein binding 33%, peak 8 hr, duration 12-24 hr, excreted unchanged by kidneys, crosses placenta, enters breast milk, half-life 14 hr

INTERACTIONS

Increase: hyperglycemia—antidiabetics

Increase: hypokalemia—mezlocillin, piperacillin, amphotericin B, glucocorticoids, digoxin, stimulant laxatives

Increase: hypotension—alcohol (large amounts), nitrates, antihypertensives, barbiturates, opioids

Increase: toxicity—lithium

Increase: metolazone effects—loop diuretics

Decrease: action of metolazone—NSAIDs, salicylates

Drug/Food

Increase: severe hypokalemia—licorice

Drug/Herb

Decrease: antihypertensive effect—ephedra (Ma huang)

Increase: antihypertensive effect—hawthorn

NURSING CONSIDERATIONS

Assess:

- Weight, I&O daily to determine fluid loss; effect of product may be decreased if used daily
- **CHF:** improvement in edema of feet, legs, sacral area daily if product being used
- **Hypertension:** B/P lying, standing; postural hypotension may occur
- Electrolytes: potassium, magnesium, sodium, chloride; include BUN, blood glucose, CBC, serum creatinine, blood pH, ABGs, uric acid, calcium
- **Hypokalemia:** postural hypotension, malaise, fatigue, tachycardia, leg cramps, weakness
- Rashes, fever daily
- Confusion, especially among geriatric patients; take safety precautions if needed

Black Box Warning: **Hepatic encephalopathy:** do not use in hepatic coma or precoma; fluctuations in electrolytes can occur rapidly and precipitate hepatic coma; use caution in patients with impaired hepatic function

Evaluate:

- Therapeutic response: decreased edema, B/P

Teach patient/family:

- To increase fluid intake to 2-3 L/day unless contraindicated; to rise slowly from lying or sitting position
- To notify prescriber of muscle weakness, cramps, nausea, dizziness
- That product may be taken with food or milk
- To use sunscreen for photosensitivity
- That blood glucose may be increased in diabetics
- To take early in day to avoid nocturia
- To avoid alcohol
- To avoid sodium foods; to increase potassium foods in diet
- Not to stop product abruptly

TREATMENT OF OVERDOSE:

Lavage if taken orally; monitor electrolytes; administer dextrose in saline; monitor hydration, CV, renal status

metoprolol (Rx)

(meh-toe′proe-lole)

Lopressor, Nu-Metop ♣, Toprol-XL

Func. class.: Antihypertensive, antianginal

Chem. class.: β_1-Blocker

Do not confuse:
metoprolol/misoprostol

ACTION: Lowers B/P by β-blocking effects; reduces elevated renin plasma levels; blocks β_2-adrenergic receptors in bronchial, vascular smooth muscle only at high doses; negative chronotropic effect

USES: Mild to moderate hypertension, acute MI to reduce cardiovascular mortality, angina pectoris, NYHA class II, III heart failure, cardiomyopathy

Unlabeled uses: Migraine prevention, heart rate control for atrial fibrillation/flutter without accessory pathway, essential tremor, unstable angina

CONTRAINDICATIONS: Hypersensitivity to β-blockers, cardiogenic shock, heart block (2nd, 3rd degree), sinus bradycardia, pheochromocytoma, sick sinus syndrome

Precautions: Pregnancy (C), breastfeeding, geriatric patients, major surgery, diabetes mellitus, thyroid/renal/hepatic disease, COPD, CAD, nonallergic bronchospasm, bronchial asthma, CVA, children, depression, vasospastic angina

Black Box Warning: Abrupt discontinuation

DOSAGE AND ROUTES

Hypertension

- **Adult: PO** 50 mg bid or 100 mg/day; may give up to 200-450 mg in divided doses; **EXT REL** 25-100 mg daily, titrate at weekly intervals; max 400 mg/day
- **Geriatric: PO** 25 mg/day initially, increase weekly as needed
- **Child and adolescent 6-16 yr: PO ER** 1 mg/kg up to 50 mg daily

Myocardial infarction

- **Adult: IV BOL** (early treatment) 5 mg q2min × 3 then 50 mg **PO** 15 min after last dose and q6hr × 48 hr; (late treatment) **PO** maintenance 50-100 mg bid for 1-3 yr

Heart failure (NYHA class II/III)

- **Adult: PO EXT REL** 25 mg daily × 2 wk (class II); 12.5 mg daily (class III)

Angina

- **Adult: PO** 100 mg/day as a single dose or in 2 divided doses, increase weekly prn or 100 mg **EXT REL** daily, max 400 mg/day ext rel

Migraine prevention (unlabeled)

- **Adult: PO** 25-100 mg bid-qid; 50-200 mg daily (XL)

Heart rate control for atrial fibrillation/flutter without accessory pathway (unlabeled)

- **Adult: IV BOL** (acute setting) 2.5-5 mg over 2 min, may repeat dose × 3; **PO** (nonacute setting) 25-100 mg bid

Essential tremor (unlabeled)

- **Adult: PO** 50 mg/day, may increase, max 300 mg/day in divided doses; **EXT REL** 100 mg/day, max 400 mg/day

Available forms: Tabs 25, 50, 100 mg; inj 1 mg/ml; ext rel tab (succinate) (XL) 25, 50, 100, 200 mg; ext rel tabs, tartrate: 100 mg

Administer:

PO route

- Do not break, crush, or chew ext rel tabs
- Regular release tab after meals, at bedtime; tab may be crushed or swallowed whole; take at same time each day
- Store in dry area at room temperature; do not freeze

Direct IV route

- IV, undiluted, give over 1 min × 3 doses at 2 to 5-min intervals; start **PO** 15 min after last IV dose

Y-site compatibilities: Abciximab, acyclovir, alemtuzumab, alfentanil, alteplase, amikacin, aminophylline, amiodarone, amphotericin B liposome, anidulafungin, argatroban, ascorbic acid, atracurium, atropine, azaTHIOprine, aztreonam, benztropine, bivalirudin, bleomycin, bumetanide, buprenorphine, butorphanol, calcium chloride/gluconate, CARBOplatin, caspofungin, ceFAZolin, cefonicid, cefoperazone, cefotaxime, cefoTEtan, cefOXitin, cefTAZidime, ceftizoxime, cefTRIAXone, cefuroxime, chloramphenicol, chlorproMAZINE, cimetidine, CISplatin, clindamycin, cyanocobalamin, cyclophosphamide, cycloSPORINE, cytarabine, DACTINomycin, DAPTOmycin, dexamethasone, dexmedetomidine, digoxin, diltiazem, diphenhydrAMINE, DOBUTamine, DOCEtaxel, DOPamine, doxacurium, DOXOrubicin, doxycycline, enalaprilat, ePHEDrine, EPINEPHrine, epirubicin, epoetin alfa, eptifibatide, esmolol, etoposide, etoposide phosphate, famotidine, fenoldopam, fentaNYL, fluconazole, fludarabine, fluorouracil, folic acid, furosemide, ganciclovir, gemcitabine, gentamicin, glycopyrrolate, granisetron, heparin, hydrocortisone, HYDROmorphone, IDArubicin, ifosfamide, imipenem/cilastatin, indomethacin, insulin, isoproterenol, ketorolac, labetalol, linezolid, LORazepam, magnesium sulfate, mannitol, mechlorethamine, meperidine, metaraminol, methotrexate, methoxamine, methyldopate, methylPREDNISolone, metoclopramide, metroNIDAZOLE, midazolam, milrinone, mitoXANtrone, morphine, multivitamins, nafcillin, nalbuphine, naloxone, nitroprusside, norepinephrine, octreotide, ondansetron, oxacillin, oxaliplatin, oxytocin, PACLitaxel, palonosetron, pancuronium, papaverine, PEMEtrexed, penicillin G, pentamidine, pentazocine, PENTobarbital, PHENobarbital, phentolamine, phenylephrine, phytonadione, piperacillin/tazobactam, potassium chloride, procainamide, prochlorperazine, promethazine, propranolol, protamine, pyridoxime, quinupristin/dalfopristin, ranitidine, rocuronium, sodium bicarbonate, succinylcholine, SUFentanil, tacrolimus, teniposide, theophylline, thiamine, thiotepa, ticarcillin/clavulanate, tigecycline, tirofiban, tobramycin, tolazoline, trimetaphan, urokinase, vancomycin, vasopressin, vecuronium, verapamil, vinCRIStine, vinorelbine, voriconazole

SIDE EFFECTS

CNS: *Insomnia, dizziness,* mental changes, hallucinations, depression, anxiety, headaches, nightmares, confusion, fatigue
CV: *Hypotension,* **bradycardia**, CHF, *palpitations,* dysrhythmias, **cardiac arrest, AV block, pulmonary/peripheral edema, chest pain**
EENT: Sore throat; dry, burning eyes
GI: *Nausea, vomiting,* colitis, cramps, *diarrhea,* constipation, flatulence, dry mouth, *hiccups*
GU: Impotence
HEMA: **Agranulocytosis, eosinophilia, thrombocytopenia, purpura**
INTEG: Rash, purpura, alopecia, dry skin, urticaria, pruritus
RESP: **Bronchospasm**, dyspnea, wheezing

PHARMACOKINETICS

Half-life 3-4 hr, metabolized in liver (metabolites), excreted in urine, crosses placenta, enters breast milk
PO: Peak 2-4 hr, duration 13-19 hr
PO-ER: Peak 6-12 hr, duration 24 hr
IV: Onset immediate, peak 20 min, duration 6-8 hr

INTERACTIONS

- **Do not use with MAOIs**

Increase: hypotension, bradycardia—reserpine, hydrALAZINE, methyldopa, prazosin, amphetamines, EPINEPHrine, H_2-antagonists, calcium channel blockers
Increase: hypoglycemic effects—insulin, oral antidiabetics
Increase: metoprolol level—cimetidine
Increase: effects of benzodiazepines
Decrease: antihypertensive effect—salicylates, NSAIDs
Decrease: metoprolol level—barbiturates
Decrease: effects of—xanthines

Drug/Food

Increase: absorption with food

Drug/Lab Test

Increase: BUN, potassium, ANA titer, serum lipoprotein, triglycerides, uric acid, alk phos, LDH, AST, ALT

NURSING CONSIDERATIONS

Assess:

Black Box Warning: Abrupt withdrawal: may cause MI, ventricular dysrhythmias, myocardial ischemia; taper dose over 7-14 days

- **Hypertension/angina:** ECG directly when giving IV during initial treatment
- I&O, weight daily; check for CHF (weight gain, jugular venous distention, crackles, edema, dyspnea)
- Monitor B/P during initial treatment, periodically thereafter; pulse q4hr; note rate, rhythm, quality; apical/radial pulse before administration; notify prescriber of any significant changes or pulse <60 bpm
- Baselines of renal, hepatic studies before therapy begins

Evaluate:

- Therapeutic response: decreased B/P after 1-2 wk, decreased anginal pain

Teach patient/family:

- To take immediately after meals; to take medication at bedtime to prevent effect of orthostatic hypotension

Black Box Warning: Not to discontinue product abruptly; to taper over 2 wk; may cause angina

- Not to use OTC products containing α-adrenergic stimulants (nasal decongestants, OTC cold preparations) unless directed by prescriber; to avoid alcohol, smoking, sodium intake

- To report bradycardia, dizziness, confusion, depression, fever, sore throat, SOB, decreased vision to prescriber
- To take pulse, B/P at home; when to notify prescriber
- To comply with weight control, dietary adjustments, modified exercise program
- To carry emergency ID to identify product, allergies
- To monitor blood glucose closely if diabetic
- To avoid hazardous activities if dizziness is present
- To report symptoms of CHF: difficult breathing, especially on exertion or when lying down; night cough; swelling of extremities
- To wear support hose to minimize effects of orthostatic hypotension
- To report Raynaud's symptoms

TREATMENT OF OVERDOSE:

Lavage, IV atropine for bradycardia, IV theophylline for bronchospasm, digoxin, O_2, diuretic for cardiac failure, hemodialysis, administer vasopressor

metroNIDAZOLE (Rx)

(me-troe-ni'da-zole)

Flagyl, Flagyl ER, Flagyl IV, Flagyl IV RTU, Florazone ER ♣, Novo-Nidazol ♣

Func. class.: Antiinfective—miscellaneous

Chem. class.: Nitroimidazole derivative

ACTION:

Direct-acting amebicide/trichomonacide binds and disrupts DNA structure, thereby inhibiting bacterial nucleic acid synthesis

USES:

Intestinal amebiasis, amebic abscess, trichomoniasis, refractory trichomoniasis, bacterial anaerobic infections, giardiasis, septicemia, endocarditis; bone, joint, lower respiratory tract infections; rosacea

Unlabeled uses: Crohn's disease, urethritis, amebiasis due to *Dientamoeba fragilis, Entamoeba polecki,* giardiasis, pruritus; gastric ulcer, dyspepsia *(H. pylori),* pseudomembranous colitis, guinea worm disease, periodontitis

CONTRAINDICATIONS:

Pregnancy 1st trimester, breastfeeding, hypersensitivity to this product

Precautions: Pregnancy (B) 2nd/3rd trimesters, geriatric patients, *Candida* infections, heart failure, fungal infection, dental disease, bone marrow suppression, hematologic disease, GI/renal/hepatic disease, contracted visual or color fields, blood dyscrasias, CNS disorders

Black Box Warning: Secondary malignancy

DOSAGE AND ROUTES

Trichomoniasis

- **Adult: PO** 500 mg bid × 7 days or 2 g as single dose; do not repeat treatment for 4-6 wk
- **Child ≥45 kg (unlabeled): PO** 2 g once
- **Child <45 kg (unlabeled): PO** 15 mg/kg/day in 3 divided doses × 7-10 days
- **Infant (unlabeled): PO** 15 mg/kg/day divided in 3 doses × 7 days

Amebic hepatic abscess

- **Adult: PO** 750 mg tid × 7-10 days
- **Child: PO** 35-50 mg/kg/day in 3 divided doses × 7-10 days

Intestinal amebiasis

- **Adult: PO** 750 mg tid × 5-10 days
- **Child: PO** 35-50 mg/kg/day in 3 divided doses × 7-10 days then oral iodoquinol

Anaerobic bacterial infections

- **Adult: IV INF** 15 mg/kg over 1 hr then 7.5 mg/kg **IV** or **PO** q6hr, not to exceed 4 g/day; 1st maintenance dose should be administered 6 hr after loading dose

Bacterial vaginosis
- **Adult: PO** regular rel 500 mg bid or 250 mg tid × 7 days; ext rel 750 mg/day × 7 days

Persistent urethritis (unlabeled)
- **Adult and adolescent: PO** 2 g as single dose with azithromycin

Dientamoeba fragilis **(unlabeled)**
- **Child: PO** 250 mg tid × 7 days

Entamoeba polecki **(unlabeled)**
- **Adult: PO** 750 mg tid × 10 days then diloxanide furoate (500 mg tid × 10 days
- **Child: PO** 30-50 mg/kg/day in 3 divided doses × 5-10 days

Crohn's disease (unlabeled)
- **Adult: PO** 250 mg tid-qid

Giardiasis (unlabeled)
- **Adult: PO** 250 mg tid × 5-7 days
- **Child: PO** 15 mg/kg/day in 3 divided doses × 5 days

Antibiotic-associated pseudomembranous colitis (unlabeled)
- **Adult: PO** 500 mg 3-4×/day × 10-14 days; **IV** 500 mg q8hr
- **Child: PO/IV** 30 mg/kg/day (max 2 g) divided q6hr × 10 days

Available forms: Tabs 250, 500 mg; ext rel tab (ER) 750 mg; caps 375 mg; inj 500 mg/100 ml; injection solution 5 mg/ml

Administer:
- Store in light-resistant container; do not refrigerate

PO route
- Do not break, crush, or chew ext rel product, give on empty stomach
- PO with or after meals to avoid GI symptoms, metallic taste; crush tabs if needed

IV route

Intermittent INF ([Flagyl] IV RTU)
- Prediluted, ready to use; inf over 30-60 min
- Lyophilized vials: dilute with 4.4 ml sterile water, 0.9% NaCl; must be diluted further with ≤8 mg/ml with 0.9% NaCl, D_5W, or LR; must neutralize with 5 mEq $NaCO_3$/500 mg; CO_2 gas will be generated and may require venting; run over ≥1 hr; primary IV must be discontinued; may be given as cont inf; do not use aluminum products; IV may require venting
- Do not use aluminum needles or other products to prepare product

Y-site compatibilities: Acyclovir, alemtuzumab, alfentanil, allopurinol, amifostine, amikacin, aminophylline, amiodarone, ampicillin, ampicillin/sulbactam, anidulafungin, atracurium, bivalirudin, bumetanide, buprenorphine, busulfan, butorphanol, calcium acetate/chloride/gluconate, CARBOplatin, ceFAZolin, cefepime, cefoperazone, cefotaxime, cefoTEtan, cefTRIAXone, cefuroxime, chloramphenicol, chlorproMAZINE, cimetidine, ciprofloxacin, cisatracurium, CISplatin, clindamycin, codeine, cyclophosphamide, cycloSPORINE, cytarabine, DACTINomycin, dexamethasone, dexmedetomidine, dexrazoxane, digoxin, diltiazem, dimenhyDRINATE, diphenhydrAMINE, DOBUTamine, DOCEtaxel, DOPamine, doripenem, doxacurium, doxapram, DOXOrubicin, DOXOrubicin liposome, doxycycline, droperidol, enalaprilat, ePHEDrine, EPINEPHrine, epirubicin, eptifibatide, ertapenem, erythromycin, esmolol, etoposide, etoposide phosphate, famotidine, fenoldopam, fentaNYL, fluconazole, fludarabine, fluorouracil, foscarnet, fosphenytoin, furosemide, gemcitabine, gentamicin, glycopyrrolate, granisetron, haloperidol, heparin, hydrALAZINE, hydrocortisone, HYDROmorphone, IDArubicin, ifosfamide, imipenem/cilastatin, inamrinone, insulin, isoproterenol, ketorolac, labetalol, leucovorin, levofloxacin, lidocaine, linezolid, LORazepam, magnesium sulfate, mannitol, mechlorethamine, melphalan, meperidine, meropenem, mesna, metaraminol, methotrexate, methyldopate, methylPREDNISolone, metoclopramide, metoprolol, midazolam, milrinone, mitoXANtrone, morphine, nafcillin, nalbuphine, naloxone, nesiritide, niCARdipine, nitroglycerin, nitroprusside, norepinephrine, octreotide, ondansetron, oxaliplatin, oxytocin, PACLitaxel, palonosetron, pancuronium, pentamidine, pentazocine, PENTobarbital, perphenazine, PHENobarbital, phentolamine, phenylephrine, piperacillin/tazobactam, potassium chloride/phosphates, prochlorperazine, promethazine, propranolol, ranitidine, remifentanil, riTUXimab, rocuronium, sargramostim, sodium acetate/bicarbonate/phosphates, streptozocin, succinylcholine, SUFentanil, tacrolimus, teniposide, theophylline, thiopental, thiotepa, ticarcillin/clavulanate, tigecycline, tirofiban, tobramycin, trastuzumab, trimethobenzamide, trimethoprim/sulfamethoxazole, vancomycin, vasopressin, vecuronium, verapamil, vinCRIStine, vinorelbine, voriconazole, zidovudine, zoledronic acid

SIDE EFFECTS

CNS: *Headache, dizziness,* confusion, irritability, restlessness, ataxia, depression, fatigue, drowsiness, insomnia, paresthesia, peripheral neuropathy, seizures, incoordination, depression, encephalopathy, aseptic meningitis

CV: Flattening of T waves

EENT: Blurred vision, sore throat, retinal edema, dry mouth, metallic taste, furry tongue, glossitis, stomatitis, photophobia, optic neuritis

GI: *Nausea, vomiting, diarrhea,* epigastric distress, *anorexia,* constipation, *abdominal cramps,* pseudomembranous colitis, xerostomia, metallic taste, abdominal pain, pancreatitis

GU: Darkened urine, vaginal dryness, polyuria, albuminuria, dysuria, cystitis, decreased libido, nephrotoxicity, incontinence, dyspareunia, candidiasis, increased urinary frequency

HEMA: Leukopenia, bone marrow, depression, aplasia, thrombocytopenia

INTEG: Rash, pruritus, urticaria, flushing, Stevens-Johnson syndrome, phlebitis at injection site, toxic epidermal necrolysis

PHARMACOKINETICS

Crosses placenta, enters breast milk, metabolized by liver 30%-60%, excreted in urine (60%-80%), half-life 6-8 hr

PO: Peak 1-2 hr, absorbed 80%-85%

IV: Onset immediate, peak end of inf

INTERACTIONS

- Avoid use with zalcitabine, bortezomib, norfloxacin, disulfiram
- Do not use with amprenavir

Decrease: metroNIDAZOLE half-life—barbiturates

Decrease: metroNIDAZOLE—cholestyramine

Increase: disulfiram reaction—alcohol, oral ritonavir, any product with alcohol

Increase: busulfan toxicity—busulfan; avoid concurrent use

Increase: metroNIDAZOLE level, toxicity—cimetidine

Increase: lithium, CYP3A4 substrates

Increase: action of warfarin, phenytoin, lithium, fosphenytoin

Increase: leukopenia—azaTHIOprine, fluorouracil

Drug/Lab Test

Altered: AST, ALT, LDH

Decrease: WBC, neutrophils

False decrease: triglycerides

NURSING CONSIDERATIONS

Assess:

- **Infection:** WBC, wound symptoms, fever, skin or vaginal secretions; start treatment after C&S; for opportunistic fungal infections; superinfection: fever, monilial growth, fatigue, malaise
- Stools during entire treatment; should be clear at end of therapy; stools should be free of parasites for 1 yr before patient considered cured (amebiasis)
- Vision by ophthalmic exam during, after therapy; vision problems often occur

⚠ **Neurotoxicity:** peripheral neuropathy, seizures, dizziness, uncoordination, pruritus, joint pain; product may be discontinued

- **Allergic reaction:** fever, rash, itching, chills; product should be discontinued if these occur
- Renal, reproductive dysfunction: dysuria, polyuria, impotence, dyspareunia, decreased libido, I&O; weight daily

Black Box Warning: Secondary malignancy: use only when indicated; avoid unnecessary use

Evaluate:
- Therapeutic response: decreased symptoms of infection

Teach patient/family:
- That urine may turn dark reddish brown; that product may cause metallic taste
- About proper hygiene after bowel movement; handwashing technique
- To notify physician about numbness or tingling of extremities
- To avoid hazardous activities because dizziness can occur
- About need for compliance with dosage schedule, duration of treatment
- To use condoms if treatment for trichomoniasis or cross-contamination may occur; to notify prescriber if pregnant or planning to become pregnant; that treatment of both partners is necessary for trichomoniasis
- To use frequent sips of water, sugarless gum, candy for dry mouth
- Not to drink alcohol or use preparations containing alcohol during use or for 48 hr after use of product; disulfiram-like reaction can occur
- To notify if pregnancy is planned or suspected, pregnancy (B) 2nd/3rd trimester in trichomoniasis

metroNIDAZOLE (topical, vaginal)

(met-roe-ni′da-zole)

MetroCream, MetroGel, MetroGel Vaginal, MetroLotion, Noritate, Rosasol ♣, Rosadan, Vandazole

Func. class.: Antiprotozoal, antibacterial

Chem. class.: Nitroimidazole

ACTION: Antibacterial and antiprotozoal activity may result from interacting with DNA

USES: Acne rosacea, bacterial vaginosis

CONTRAINDICATIONS: Hypersensitivity to this product or nitroimidazoles, parabens

Precautions: Hepatic disease, blood dyscrasias; CNS conditions (vaginal), children

DOSAGE AND ROUTES

Acne rosacea
- **Adult: TOP** apply to affected areas bid (0.75%) daily (1%); adjust therapy based on response

Bacterial vaginosis
- **Adult: VAG** 1 applicatorful bid × 5 days

Available forms: Topical cream 0.75%, 1%; gel 0.75%, 1%; lotion 0.75%; vaginal gel 0.75%

Administer:

Topical route
- Topical skin products are not for intravaginal therapy and are for external use only; do not use skin products near the eyes, nose, or mouth
- Wash hands before and after use; wash affected area and gently pat dry
- **Cream/Gel/Lotion:** Apply a thin film to the cleansed affected area; massage gently into affected areas

Intravaginal route
- Only use dosage formulations specified for intravaginal use; intravaginal dosage forms are not for topical therapy; do not ingest
- Avoid vaginal intercourse during treatment
- **Cream:** Use applicator(s) supplied by the manufacturer

SIDE EFFECTS

GU: Vaginitis, cervicitis; nausea, vomiting, cramping

INTEG: Redness, burning, dermatitis, rash, pruritus

M

INTERACTIONS

- MetroNIDAZOLE may increase warfarin anticoagulant effect
- Caution with drinking alcohol or using disulfiram while using metroNIDAZOLE products(vaginal gel)
- Possible lithium toxicity (vaginal gel)

NURSING CONSIDERATIONS

Assess:

- **Allergic reaction:** assess for hypersensitivity; product may need to be discontinued
- **Infection:** assess for number of lesions and severity of acne rosacea, itching in vaginosis

Evaluate:

- Decreased severity of acne rosacea, infection in vaginosis

Teach patient/family:

- That topical skin products are not for intravaginal therapy and are for external use only; not to use skin products near the eyes, nose, or mouth
- To wash hands before and after use; to wash affected area and gently pat dry
- **Cream/Gel/Lotion:** to apply a thin film to the cleansed affected area and massage gently into affected areas
- **Intravaginal Route:** to use only dosage formulations specified for intravaginal use; not to ingest intravaginal dosage forms because these are not for topical therapy; to avoid vaginal intercourse during treatment
- **Cream:** to use applicator(s) supplied by the manufacturer

micafungin (Rx)

(my-ca-fun′gin)

Mycamine

Func. class.: Antifungal, systemic

Chem. class.: Echinocandin

ACTION: Inhibits an essential component of fungal cell walls; causes direct damage to fungal cell wall

USES: Treatment of esophageal candidiasis; prophylaxis for *Candida* infections in patients undergoing hematopoietic stem-cell transplantation (HSCT); susceptible *Candida* sp.: *C. albicans, C. glabrata, C. krusei, C. parapsilosis, C. tropicalis*

Unlabeled uses: *Aspergillus* sp., pediatrics to prevent candidiasis, endocarditis, endophthalmitis, infectious arthritis, myocarditis, osteomyelitis, pericarditis, pneumonia, sinusitis, tracheobronchitis, prophylaxis of HIV-related esophageal candidiasis

CONTRAINDICATIONS: Hypersensitivity to this product or other echinocandins

Precautions: Pregnancy (C), breastfeeding, children, geriatric patients, severe hepatic disease, renal impairment, hemolytic anemia

DOSAGE AND ROUTES

Esophageal candidiasis

- **Adult: IV INF** 150 mg/day given over 1 hr

Candidemia/acute disseminated candidiasis, abscess/peritonitis

- **Adult: IV** 100 mg/day over 1 hr

Prophylaxis for *Candida* infections

- **Adult: IV INF** 50 mg/day given over 1 hr
- **Adolescent/child/infant ≥6 mo (unlabeled): IV INF** 1 mg/kg/day, max 50 mg/day

***Aspergillus* sp. (unlabeled)**

- **Adult: IV INF** 25-150 mg/day × ≥30 days

Available forms: Powder for inj 50 mg, in single-dose vials 50-, 100-mg vial

Administer:

- Do not use if cloudy or precipitated; do not admix; by IV inf only; product diluted, solution protected from light
- Store at room temperature, away from light, do not freeze; discard unused sol

IV route

- Flush line before and after use with 0.9% NaCl
- **For *Candida* prevention,** reconstitute with provided diluent 0.9% NaCl without bacteriostatic product; 50-mg vial/5 ml (10 mg/ml), swirl to dissolve, do not shake; further dilute with 100 ml 0.9% NaCl only; run over 1 hr
- **For *Candida* infection:** reconstitute with provided diluent 50 mg/5 ml (10 mg/ml); further dilute 3 reconstituted vials in 100 ml 0.9% NaCl, run over 1 hr

Y-site compatibilities: Aminophylline, bumetanide, calcium chloride/gluconate, cycloSPORINE, DOPamine, eptifibatide, esmolol, fenoldopam, furosemide, heparin, HYDROmorphone, lidocaine, LORazepam, magnesium sulfate, milrinone, nitroglycerin, nitroprusside, norepinephrine, phenylephrine, potassium chloride, potassium phosphate, tacrolimus, vasopressin

SIDE EFFECTS

CNS: **Seizures**, dizziness, *headache, somnolence,* fever, anxiety
CV: Flushing, hypertension, phlebitis, tachycardia, **atrial fibrillation**
GI: Abdominal pain, *nausea, anorexia, vomiting, diarrhea, hyperbilirubinemia,* **hepatitis**
GU: **Renal failure**
HEMA: **Neutropenia, thrombocytopenia, leukopenia, coagulopathy, anemia, hemolytic anemia**
INTEG: *Rash, pruritus, inj site pain*
META: Hypokalemia, hypocalcemia, hypomagnesemia
MS: *Rigors*

PHARMACOKINETICS

Metabolized in liver; excreted in feces, urine; terminal half-life 14-17.2 hr; protein binding 99%

INTERACTIONS

Increase: plasma concs—itraconazole, sirolimus, NIFEdipine; may need dosage reduction

Drug/Lab Test
Increase: ALT/AST, alk phos, bilirubin, potassium, sodium
Decrease: blood glucose

NURSING CONSIDERATIONS

Assess:

- **Infection,** clearing of cultures during treatment; obtain culture at baseline and during treatment; product may be started as soon as culture is taken (esophageal candidiasis); monitor cultures during HSCT for prevention of *Candida* infections
- CBC (RBC, Hct, Hgb), differential, platelet count periodically; notify prescriber of results
- Renal studies: BUN, urine CCr, electrolytes before and during therapy
- Hepatic studies before and during treatment: bilirubin, AST, ALT, alk phos as needed
- **Bleeding:** hematuria, heme-positive stools, bruising, or petechiae, mucosa or orifices; blood dyscrasias can occur
- **For hypersensitivity:** rash, pruritus, facial swelling, phlebitis
- For hemolytic anemia
- **GI symptoms:** frequency of stools, cramping; if severe diarrhea occurs, electrolytes may need to be given

Evaluate:

- Therapeutic response: prevention of *Candida* infection with HSCT or decreased symptoms of *Candida* infection, negative culture

Teach patient/family:

- To notify prescriber if pregnancy is suspected or planned
- To avoid breastfeeding while taking this product
- To inform prescriber of kidney or liver disease
- To report bleeding, facial swelling, wheezing, difficulty breathing, itching, rash, hives, increasing warmth, flushing
- To report signs of infection: increased temperature, sore throat, flulike symptoms
- To notify prescriber of nausea, vomiting, diarrhea, jaundice, anorexia, clay-

M

colored stools, dark urine; heptatotoxicity may occur

miconazole

(mi-kon′a-zole)

Oravig

miconazole nitrate

Desenex, Femizol-M, Fungoid, Tincture, Lotrimin AF, Micatin, Micozole, Monistat-1, Monistat-3, Monistat-7, M-Zole 3, M-Zole 7, Tetterine, Ting, Vagistat-1, Zeasorb-AF

Func. class.: Antifungal

Chem. class.: Imidazole

Do not confuse: miconazole/clotrimazole/metroNIDAZOLE

ACTION:

Antifungal activity results from disruption of cell membrane permeability

USES:

Treatment of topical fungal infection, vulvovaginal candidiasis; athlete's foot (tinea pedis), jock itch (tinea cruris), and ringworm (tinea corporis)

CONTRAINDICATIONS:

Hypersensitivity to this product or imidazoles; pregnancy first trimester (vaginal)

Precautions: Breastfeeding, children

DOSAGE AND ROUTES

Oropharyngeal candidiasis (thrush)

- **Adult/adolescent ≥16 yr: BUCCAL** apply 1 tab (50 mg) to upper gum region, just above incisor daily × 14 days

Tinea corporis, cruris, pedis; cutaneous candidiasis

- **Adult/child >2 yr: TOP** apply bid × 2-4 wk

Tinea versicolor

- **Adult/child >2 yr: TOP** use bid × 2 wk, apply sparingly every day

Vulvovaginal candidiasis

- **Adult/child ≥12 yr: VAG** 1 applicatorful of Monistat-7 (100 mg) or 1 supp (100 mg) at bedtime × 7 days, repeat if needed, or Monistat-3 (200 mg) × 3 days or a 1200 mg supp × 1 day

Available forms: Topical cream, ointment, solution, lotion, powder, aerosol, powder 2%; aerosol spray 2%; vag cream 2, 4%; vag supp 100, 200, 1200 mg; buccal tab 50 mg

Administer:

Transmucosal Use (adhesive buccal tablet)

- Apply tab in the morning after brushing the teeth, use dry hands
- Place the rounded surface of the tab against the upper gum just above the incisor tooth, hold in place with slight pressure over the upper lip for 30 sec to assure adhesion
- Although the tab is rounded on one side for comfort, the flat side may also be applied to the gum
- The tab will gradually dissolve
- Administration of subsequent tab should be made to alternate sides of the mouth
- Before applying the next tab, clear away any remaining tab material
- Do not crush, chew, or swallow; food and drink can be taken normally; avoid chewing gum
- If tab does not adhere or falls off within the first 6 hr, the same tab should be repositioned immediately; if the tab still does not adhere, a new tab should be used
- If the tab falls off or is swallowed after it was in place for 6 hr or more, a new tab should not be applied until the next regularly scheduled dose

Topical route

- Topical skin products are not for intravaginal therapy and are for external use only; do not use skin products near the eyes, nose, or mouth
- Wash hands before and after use; wash affected area and gently pat dry
- **Cream/ointment/lotion/solution:** apply a thin film to the cleansed affected area; massage gently into affected areas
- **Solution formulations:** apply a thin film to the cleansed affected area; massage gently into affected areas; if using a

solution-soaked pledget, patient may use more than 1 pledget per application as needed to treat affected areas, but each pledget should be used only once and then discarded

• **Intravaginal route:** only use dosage formulations specified for intravaginal use; intravaginal dosage forms are not for topical therapy; do not ingest

• **Suppository:** unwrap vaginal ovule (suppository) before insertion; use applicator(s) supplied by the manufacturer

• **Cream:** use applicator(s) supplied by the manufacturer

SIDE EFFECTS

CNS: Headache
GI: Diarrhea, nausea
GU: Pruritus, irritation, vaginal burning
INTEG: Burning, dermatitis, rash

NURSING CONSIDERATIONS

Assess:

Allergic reaction:

• Assess for hypersensitivity; product may need to be discontinued

Infection:

• Assess for severity of infection

Evaluate:

• Decreasing severity of infection

Teach patient/family:

Topical route

• That topical skin products are not for intravaginal therapy and are for external use only; not to use skin products near the eyes, nose, or mouth

• To wash hands before and after use; wash affected area and gently pat dry

• **Cream/ointment/lotion/solution:** to apply a thin film to the cleansed affected area and massage gently into affected areas

• **Solution formulations:** to shake well before use, apply a thin film to the cleansed affected area, and massage gently into affected areas

Intravaginal route

• To only use dosage formulations specified for intravaginal use; not to ingest intravaginal dosage forms; not to use tampons, douches, spermicides; not to engage in sexual activity; product may damage condoms, diaphragms, cervical caps

• **Suppository:** to unwrap vaginal ovule (suppository) before inserting; to use applicator(s) supplied by the manufacturer

• **Cream:** to use applicator(s) supplied by the manufacturer

midazolam (Rx)

(mid′ay-zoe-lam)

Func. class.: Sedative, hypnotic, antianxiety

Chem. class.: Benzodiazepine, short-acting

Controlled Substance Schedule IV

ACTION: Depresses subcortical levels in CNS; may act on limbic system, reticular formation; may potentiate γ-aminobutyric acid (GABA) by binding to specific benzodiazepine receptors

USES: Preoperative sedation, general anesthesia induction, sedation for diagnostic endoscopic procedures, intubation, anxiety

Unlabeled uses: Refractory status epilepticus, alcohol withdrawal, agitation

CONTRAINDICATIONS: Pregnancy (D), hypersensitivity to benzodiazepines, acute closed-angle glaucoma, epidural/intrathecal use

Precautions: Breastfeeding, children, geriatric patients, COPD, CHF, chronic renal failure, chills, debilitated, hepatic disease, shock, coma, alcohol intoxication, status asthmaticus

Black Box Warning: Neonates (contains benzyl alcohol), IV administration, respiratory depression/insufficiency

M

DOSAGE AND ROUTES

Preoperative sedation/amnesia induction

- **Adult and child ≥12 yr: IM** 0.07-0.08 mg/kg 1/2-1 hr before general anesthesia
- **Child 6 yr-12 yr: IV** 0.025-0.05 mg/kg; total dose of 0.4 mg/kg may be necessary
- **Child 6 mo-5 yr: IV** 0.05-0.1 mg/kg; total dose of 0.6 mg/kg may be necessary

Induction of general anesthesia

- **Adult >55 yr:** (ASA I/II) **IV** 150-300 mcg/kg over 30 sec; (ASA III/IV) limit dose to 250 mcg/kg (nonpremedicated) or 150 mcg/kg (premedicated)
- **Adult <55 yr: IV** 200-350 mcg/kg over 20-30 sec; if patient has not received premedication, may repeat by giving 20% of original dose; if patient has received premedication, reduce dose by 50 mcg/kg
- **Child:** no safe and effective dose established; however, doses of 50-200 mcg/kg **IV** have been used

Continuous inf for mechanical ventilation (critical care)

- **Adult: IV** 0.01-0.05 mg/kg over several min; repeat at 10- to 15-min intervals until adequate sedation then 0.02-0.10 mg/kg/hr maintenance; adjust as needed
- **Child: IV** 0.05-0.2 mg/kg over 2-3 min then 0.06-0.12 mg/kg/hr by cont inf; adjust as needed
- **Neonate: IV** 0.03-0.06 mg/kg/hr, titrate using lowest dose

Status epilepticus (unlabeled)

- **Child and infant >2 mo: IV** 0.15 mg/kg then **CONT IV** 1 mcg/kg/min, titrate upward q5min until seizures controlled

Alcohol withdrawal (unlabeled)

- **Adult: IV** 1-5 mg q1-2hr (mild-moderate symptoms); **CONT IV INF** 1-20 mg q1-2hr (delirium tremens)

Available forms: Inj 1, 5 mg/ml, 25 mg/5 ml, 50 mg/10 ml; syr 2 mg/ml

Administer:

- Store at room temperature; protect from light

PO route

- Remove cap of press-in bottle adaptor, push adaptor into neck of bottle; close with cap; remove cap, insert tip of dispenser, insert into adaptor; turn upside-down, withdraw correct dose; place in mouth

IM route

- IM deep into large muscle mass

IV route

- May be given diluted or undiluted
- After diluting with D_5W or 0.9% NaCl to 0.25 mg/ml; give over 2 min (conscious sedation) or over 30 sec (anesthesia induction)

Y-site compatibilities: Abciximab, alfentamil, amikacin, amiodarone, argatroban, atracurium, atropine, aztreonam, benzotropine, calcium gluconate, ceFAZolin, cefotaxime, cefOXitine, cefTRIAXone, cimetidine, ciprofloxacin, CISplatin, clindamycin, cloNIDine, cyanocobalamin, cycloSPORINE, DACTINomycin, digoxin, diltiazem, diphenhydrAMINE, DOCEtaxal, DOPamine, doxycyclin, enalaprilat, EPINEPHrine, erythromycin, esmolol, etomidate, etoposide, famotidine, fentaNYL, fluconazole, folic acid, gatifloxacin, gemcitabine, gentamicin, glycopyrrolate, granisetron, heparin, hetastarch, HYDROmorphone, hydrOXYzine, inamrinone, isoproterenol, labetalol, lactated Ringer's, levofloxacin, lidocaine, linezolid, LORazepam, magnesium, mannitol, meperdine, methadone, methyldopa, methylPREDNISolone, metoclopromide, metomolol, metroNIDAZOLE, milrinone, morphine, nalbuphine, naloxone, niCARdipine, nitroglycerin, nitroprusside, norepinephrine, ondansetron, oxacillin, oxytocin, PACLitaxel, palonasetron, pancuronium, papaverin, phytonadione, piperacillin, potassium chloride, propanolol, protamine, pyridoxine, ranitidine, remifentanil, sodium nitroprusside, succinylcholine, SUFentanil, teniposide, theophylline, thiotepa, ticarcillin, tobramycin, vancomycin, vasopressin, vecuronium, verapamil, voriconazole

SIDE EFFECTS

CNS: Retrograde amnesia, euphoria, confusion, headache, anxiety, insomnia, slurred speech, paresthesia, tremors, weakness, chills, agitation, paradoxic reactions

CV: Hypotension, PVCs, tachycardia, bigeminy, nodal rhythm, **cardiac arrest**

EENT: Blurred vision, nystagmus, diplopia, loss of balance

GI: *Nausea, vomiting,* increased salivation, hiccups

INTEG: Urticaria; pain, swelling, pruritus at inj site; rash

RESP: Coughing, **apnea, bronchospasm, laryngospasm,** dyspnea, **respiratory depression**

PHARMACOKINETICS

Protein binding 97%; half-life 1-5 hr, metabolized in liver; by CYP3A4 to metabolites excreted in urine; crosses placenta, blood-brain barrier

PO: Onset 10-30 min

IM: Onset 15 min, peak ½-1 hr, duration 2-3 hr

IV: Onset 1.5-5 min, onset of anesthesia 1½-2½ min, duration 2 hr

INTERACTIONS

Increase: hypotension—antihypertensives, opiates, alcohol, nitrates

Increase: extended half-life—CYP3A4 inhibitors (cimetidine, erythromycin, ranitidine)

Increase: respiratory depression—other CNS depressants, alcohol, barbiturates, opiate analgesics, verapamil, ritonavir, indinavir, fluvoxaMINE, protease inhibitors

Decrease: midazolam metabolism—CYP3A4 inducers (azole antifungals, theophylline)

Drug/Herb

Increase: sedation—kava, valerian

Decrease: midazolam effect—St. John's wort

Drug/Food

Increase: (PO) midazolam effect—grapefruit juice

NURSING CONSIDERATIONS

Assess:

- B/P, pulse, respirations during IV; emergency equipment should be nearby
- Inj site for redness, pain, swelling
- Degree of amnesia in geriatric patients; may be increased
- Anterograde amnesia
- Vital signs for recovery period in obese patients; half-life may be extended

Black Box Warning: **Respiratory depression insufficiency:** apnea, respiratory depression that may be increased in geriatric patients

- Assistance with ambulation until drowsy period ends
- **Immediate availability of resuscitation equipment, O_2 to support airway; do not give by rapid bolus**

Evaluate:

- Therapeutic response: induction of sedation, general anesthesia

Teach patient/family:

- That amnesia occurs; that events may not be remembered

TREATMENT OF OVERDOSE:

Flumazenil, O_2

miglitol (Rx)

(mig'lih-tol)

Glyset

Func. class.: Oral hypoglycemic

Chem. class.: α-Glucosidase inhibitor

ACTION: Delays digestion and absorption of ingested carbohydrates, which results in a smaller rise in blood glucose after meals; does not increase insulin production

USES: Type 2 diabetes mellitus

Unlabeled uses: Type 1 diabetes mellitus

CONTRAINDICATIONS: Hypersensitivity, diabetic ketoacidosis, cirrhosis, inflammatory bowel disease, colonic

M

ulceration, partial intestinal obstruction, chronic intestinal disease, ileus
Precautions: Pregnancy (B), breastfeeding, children, diarrhea, hiatal hernia, hypoglycemia, renal disease, Type 1 diabetes, vomiting

DOSAGE AND ROUTES

- **Adult: PO** 25 mg tid initially, with 1st bite of meal; maintenance dose may be increased to 50 mg tid; may be increased to 100 mg tid if needed with dosage adjustment at 4- to 8-wk intervals

Available forms: Tabs 25, 50, 100 mg
Administer:
- Tid with first bite of each meal
- Store in tight container at room temperature

SIDE EFFECTS

GI: *Abdominal pain, diarrhea, flatulence,* hepatotoxicity
HEMA: Low iron
INTEG: Rash
RESP: Pneumatosis cystoides, intestinalis

PHARMACOKINETICS

Peak 2-3 hr, not metabolized, excreted in urine as unchanged product, half-life 2 hr

INTERACTIONS

Decrease: levels of digoxin, propranolol, ranitidine
Decrease: miglitol levels—digestive enzymes, intestinal adsorbents; do not use together
Drug/Food
Increase: diarrhea—carbohydrates

NURSING CONSIDERATIONS

Assess:
- **Hypo/hyperglycemia;** even though product does not cause hypoglycemia, if patient receiving sulfonylureas or insulin, hypoglycemia may be additive (rare)
- Blood glucose levels, hemoglobin, A1c LFTs; if hypoglycemia occurs with monotherapy, treat with glucose

Evaluate:
- Therapeutic response: decreased signs, symptoms of diabetes mellitus (polyuria, polydipsia, polyphagia; clear sensorium, absence of dizziness; stable gait); improved blood glucose, A1c

Teach patient/family:
- About the symptoms of hypo/hyperglycemia; what to do about each; that, during periods of stress, infection, or surgery, insulin may be required
- That medication must be taken as prescribed; about consequences of discontinuing medication abruptly
- To avoid OTC medications unless approved by health care provider
- That diabetes is lifelong; that product is not a cure
- To carry ID for emergency purposes
- That diet and exercise regimen must be followed
- About GI side effects

⚠ HIGH ALERT

milrinone (Rx)

(mill'rih-nohn)
Func. class.: Inotropic/vasodilator agent with phosphodiesterase activity
Chem. class.: Bipyridine derivative

ACTION: Positive inotropic agent; increases contractility of cardiac muscle with vasodilator properties; reduces preload and afterload by direct relaxation on vascular smooth muscle

USES: Short-term management of advanced heart failure that has not responded to other medication
Unlabeled uses: Adolescents, children, infants

CONTRAINDICATIONS: Hypersensitivity to this product, severe aortic disease, severe pulmonic valvular disease, acute MI
Precautions: Pregnancy (C), breastfeeding, children, geriatric patients,

renal/hepatic disease, atrial flutter/fibrillation

DOSAGE AND ROUTES

- **Adult: IV BOL** 50 mcg/kg given over 10 min; start inf of 0.375-0.75 mcg/kg/min
- **Adolescent/child/infant (unlabeled): IV** 50-75 mcg/kg over 10-60 min then 0.5-0.75 mcg/kg/min

Renal dose

- **Adult: IV** CCr 41-50 ml/min, 0.43 mcg/kg/min, titrate up; CCr 31-40 ml/min, 0.38 mcg/kg/min, titrate up; CCr 21-30 ml/min, 0.33 mcg/kg/min, titrate up; CCr 11-20 ml/min, 0.08 mcg/kg/min; CCr 6-10 ml/min, 0.23 mcg/kg/min; CCr ≤5 ml/min, 0.20 mcg/kg/min; max for all doses 0.75 mcg/kg/min

Available forms: Inj 1 mg/ml; premixed inj 200 mcg/ml in D_5W

Administer:

- Potassium supplements if ordered for potassium levels <3 mg/dl

Direct IV route

- Give IV loading dose undiluted over 10 min, use inf device

Continuous IV route

- Dilute 20-mg vial with 80, 113, 180 ml of 0.45% NaCl, 0.9% NaCl, or D_5W to a conc of 200, 150, 100 mcg/ml respectively
- Titrate rate based on hemodynamic and clinical response, use inf device
- Precipitation will form when furosemide is injected into line with milrinone

Y-site compatibilities: Acyclovir, alfentanil, allopurinol, amifostine, amikacin, aminocaproic acid, aminophylline, amiodarone, amphotericin B liposome, ampicillin, ampicillin-sulbactam, anidulafungin, argatroban, atenolol, atracurium, aztreonam, bivalirudin, bleomycin, bumetanide, buprenorphine, busulfan, butorphanol, calcium chloride/gluconate, CARBOplatin, caspofungin, ceFAZolin, cefepime, cefotaxime, cefoTEtan, cefOXitin, cefTAZidime, ceftizoxime, cefTRIAXone, cefuroxime, chloramphenicol, chlorproMAZINE, cimetidine, ciprofloxacin, cisatracurium, CISplatin, clindamycin, cyclophosphamide, cycloSPORINE, cytarabine, DACTINomycin, DAPTOmycin, dexamethasone, digoxin, diltiazem, DOBUTamine, DOCEtaxel, DOPamine, doripenem, doxacurium, DOXOrubicin, doxycycline, droperidol, enalaprilat, ePHEDrine, EPINEPHrine, epirubicin, eptifibatide, ertapenem, erythromycin, etoposide, famotidine, fenoldopam, fentaNYL, fluconazole, fludarabine, fluorouracil, gallium, ganciclovir, gatifloxacin, gemcitabine, gentamicin, glycopyrrolate, granisetron, haloperidol, heparin, hydrALAZINE, hydrocortisone, HYDROmorphone, IDArubicin, ifosfamide, insulin (regular), irinotecan, isoproterenol, ketorolac, labetalol, levofloxacin, linezolid, LORazepam, magnesium sulfate, mannitol, mechlorethamine, melphalan, meperidine, meropenem, methohexital, methotrexate, methyldopate, methylPREDNISolone, metoclopramide, metoprolol, metroNIDAZOLE, micafungin, midazolam, mitoXANtrone, morphine, mycophenolate, nafcillin, nalbuphine, naloxone, nesiritide, niCARdipine, nitroglycerin, nitroprusside, norepinephrine, octreotide, oxacillin, oxaliplatin, oxytocin, PACLitaxel, palonosetron, pamidronate, pancuronium, PEMEtrexed, pentamidine, pentazocine, PENTobarbital, PHENobarbital, phenylephrine, piperacillin, piperacillin-tazobactam, polymyxin B, potassium chloride/phosphates, prochlorperazine, promethazine, propofol, propranolol, quiNIDine, quinupristin-dalfopristin, ranitidine, remifentanil, rocuronium, sodium acetate/bicarbonate/phosphates, streptozocin, succinylcholine, SUFentanil, sulfamethoxazole-trimethoprim, tacrolimus, teniposide, theophylline, thiopental, thiotepa, ticarcillin, ticarcillin-clavulanate, tigecycline, tirofiban, tobramycin, torsemide, vancomycin, vasopressin, vecuronium, verapamil, vinCRIStine, vinorelbine, voriconazole, zidovudine, zoledronic acid

SIDE EFFECTS

CV: Dysrhythmias, hypotension, chest pain, *PVCs*
GI: Nausea, vomiting, anorexia, abdominal pain, hepatotoxicity, jaundice
HEMA: Thrombocytopenia
MISC: Headache, hypokalemia, tremor, inj site reactions

PHARMACOKINETICS

IV: Onset 2-5 min, peak 10 min, duration variable; terminal half-life 2.3 hr; metabolized in liver; excreted in urine as product (83%), metabolites (12%)

INTERACTIONS

Increase: effects of antihypertensives, diuretics

NURSING CONSIDERATIONS

Assess:

⚠ ECG continuously during IV; ventricular dysrhythmia can occur

- B/P, pulse q5min during inf; if B/P drops 30 mm Hg, stop inf, call prescriber
- Electrolytes: potassium, sodium, chloride, calcium; renal studies: BUN, creatinine; blood studies: platelet count
- ALT, AST, bilirubin daily
- I&O ratio, weight daily; diuresis should increase with continuing therapy
- If platelets are <150,000/mm³, product is usually discontinued and another product started

Evaluate:

- Therapeutic response: increased cardiac output, decreased PCWP, adequate CVP; decreased dyspnea, fatigue, edema, ECG

Teach patient/family:

- To report angina immediately during inf
- To report headache, which can be treated with analgesics

TREATMENT OF OVERDOSE:

Discontinue product, support circulation

minocycline (Rx)

(min-oh-sye′kleen)

Arestin, Dynacin, Minocin, Solodyn

Func. class.: Broad-spectrum antiinfective

Chem. class.: Tetracycline

ACTION:

Inhibits protein synthesis, phosphorylation in microorganisms by binding to ribosomal subunits, reversibly binding to ribosomal subunits; bacteriostatic

USES:

Syphilis, *Chlamydia trachomatis,* gonorrhea, lymphogranuloma venereum, rickettsial infections, inflammatory acne, *Neisseria meningitidis, Neisseria gonorrhoeae, Treponema pallidum, Chlamydia trachomatis, Ureaplasma urealyticum, Mycoplasma pneumoniae, Nocardia,* periodontitis, methicillin-resistant *S. aureus* (MRSA) infection, nonnodular moderate to severe acne vulgaris, *Rickettsia* sp.

Unlabeled uses: Rheumatoid arthritis, bullous pemphigoid, dental infection, prostatis, pleural effusion

CONTRAINDICATIONS:

Pregnancy (D), children <8 yr, hypersensitivity to tetracyclines

Precautions: Hepatic disease, breastfeeding

DOSAGE AND ROUTES

- **Adult: PO/IV** 200 mg then 100 mg q12hr, max 400 mg/24 hr **IV; SUBGINGIVAL** inserted into periodontal pocket
- **Child >8 yr: PO/IV** 4 mg/kg then 4 mg/kg/day **PO** in divided doses q12hr

Rickettsial infections

- **Adult: PO/IV** 200 mg, then 100 mg q12hr
- **Child ≥8 yr/adolescent: PO/IV** 4 mg/kg, then 2 mg/kg q12hr, max adult dose

Gonorrhea
- **Adult: PO** 200 mg then 100 mg q12hr × ≥4 days

Chlamydia trachomatis
- **Adult: PO** 100 mg bid × 7 days

Syphilis
- **Adult: PO** 200 mg then 100 mg q12hr × 10-15 days

Uncomplicated gonococcal urethritis in men
- **Adult: PO** 100 mg q12hr × 5 days

Acne vulgaris (Solodyn only)
- **Adult/adolescent/child ≥12 yr:** ext rel 1 mg/kg/day × 12 wk or those weighing 126-136 kg—135 mg/day; 111-125 kg—115 mg/day; 97-110 kg—105 mg/day; 85-96 kg—90 mg/day; 72-84 kg—80 mg/day; 60-71 kg—65 mg/day; 50-59 kg—55 mg/day; 45-49 kg—45 mg/day

Acne vulgaris (all except Solodyn)
- **Adult/adolescent/child ≥12 yr:** ext rel 1 mg/kg × 12 wk or 91-136 kg, 135 mg/day; 60-90 kg, 90 mg/day; 45-59 kg, 45 mg/day

Rheumatoid arthritis (unlabeled)
- **Adult: PO** 100 mg bid for ≤48 wk

Bullous pemphigus (unlabeled)
- **Adult: PO** 50 mg/day; may increase to 100 mg/day after 1-2 wk

Available forms: Caps 50, 75, 100 mg; powder for inj 100 mg; caps, pellet filled 50, 100 mg; tabs 50, 75, 100 mg; ext rel tabs 45, 55, 65, 80, 90, 105, 115, 135 mg

Administer:
- After C&S obtained
- Store in airtight, light-resistant container at room temperature

PO route
- With full glass of water; with food for GI symptoms
- 2 hr before or after laxative or ferrous products; 3 hr after antacid

IV route
- After diluting 100 mg/5 ml sterile water for inj; further dilute in 500-1000 ml of NaCl, dextrose sol, LR, Ringer's sol; run 100 mg/6 hr

Y-site compatibilities: Alfentanil, amikacin, atracurium, benztropine, buprenorphine, butorphanol, calcium chloride, CARBOplatin, caspofungin, cefonicid, chlorproMAZINE, cimetidine, cisatracurium, codeine, cyclophosphamide, cycloSPORINE, cytarabine, DACTINomycin, dexmedetomidine, diltiazem, diphenhydrAMINE, DOBUTamine, DOCEtaxel, doxacurium, doxycycline, enalaprilat, ePHEDrine, EPINEPHrine, eptifibatide, etoposide, fenoldopam, fentaNYL, filgrastim, fludarabine, gatafloxacin, gemcitabine, gentamicin, glycopyrrolate, granisetron, heparin, hetastarch, IDArubicin, ifosfamide, inamrinone, isoproterenol, labetalol, levofloxacin, lidocaine, linezolid, LORazepam, magnesium sulfate, mannitol, melphalan, metaraminol, methotrexate, methyldopa, metoclopramide, metoprolol, midazolam, mitoXANtrone, nalbuphine, naloxone, perphenazine, potassium chloride, remifentanil, sargramostim, teniposide, vinorelbine, vit B/C

SIDE EFFECTS

CNS: *Dizziness*, fever, lightheadedness, vertigo, **seizures, increased intracranial pressure**

CV: Pericarditis

EENT: Dysphagia, glossitis, decreased calcification of deciduous teeth, permanent discoloration of teeth, oral candidiasis

GI: *Nausea,* abdominal pain, *vomiting, diarrhea,* anorexia, enterocolitis, **hepatotoxicity,** flatulence, abdominal cramps, epigastric burning, stomatitis, **pseudomembranous colitis**

GU: *Increased BUN,* polyuria, polydipsia, **renal failure, nephrotoxicity**

HEMA: **Eosinophilia, neutropenia, thrombocytopenia, hemolytic anemia, pancytopenia**

INTEG: *Rash, urticaria, photosensitivity, increased pigmentation,* **exfoliative dermatitis,** pruritus, blue-gray color of skin, mucous membranes

MS: Myalgia, arthritis, bone discoloration, joint stiffness

SYST: **Angioedema, Stevens-Johnson syndrome**

M

PHARMACOKINETICS

PO: Peak 1-4 hr, half-life 11-22 hr; excreted in urine, feces, breast milk; crosses placenta; 70%-75% protein bound

INTERACTIONS

Increase: effect of warfarin, digoxin, insulin, oral anticoagulants, theophylline, neuromuscular blockers

Increase: chance of pseudomotor cerebri—retinoids; do not use concurrently

Decrease: effect of minocycline—antacids, sodium bicarbonate, alkali products, iron, kaolin/pectin, cimetidine, quinapril, sucralfate

Decrease: effect of barbiturates, carBAMazepine, phenytoin, penicillins, oral contraceptives, calcium

Drug/Lab Test

False negative: urine glucose with Clinistix or Tes-Tape

NURSING CONSIDERATIONS

Assess:

⚠ **Pseudomembranous colitis:** diarrhea, abdominal cramps, fever; may start up to 2 mo after treatment ends

- I&O ratio
- Age and tooth development
- Blood tests: PT, CBC, AST, ALT, BUN, creatinine
- Signs of anemia: Hct, Hgb, fatigue

⚠ **Allergic reactions:** rash, itching, pruritus, angioedema

- Nausea, vomiting, diarrhea; administer antiemetic, antacids as ordered

⚠ **Overgrowth of infection:** fever, malaise, redness, pain, swelling, drainage, perineal itching, diarrhea; changes in cough or sputum; black, furry tongue

Evaluate:

- Therapeutic response: decreased temperature, absence of lesions, negative C&S

Teach patient/family:

- To avoid sunlight, wear protective clothing; sunscreen does not seem to decrease photosensitivity
- That all prescribed medication must be taken to prevent superinfection; not to use outdated product because Fanconi's syndrome may occur
- To avoid taking antacids, iron, cimetidine; use 2 hr before, 6 hr after this product; absorption may be decreased
- That teeth discoloration, joint or muscle pain may occur

minoxidil (Rx, OTC)

(mi-nox'i-dill)

Rogaine (topical)

Func. class.: Antihypertensive

Chem. class.: Vasodilator, peripheral

Do not confuse:

minoxidil/Monopril

ACTION: Directly relaxes arteriolar smooth muscle, causing vasodilation; reduces peripheral vascular resistance, decreases B/P

USES: Severe hypertension unresponsive to other therapy (use with diuretic and β-blocker); topically to treat alopecia

Unlabeled uses: Scleroderma renal crisis (SRC) to control hypertension, anal fissures

CONTRAINDICATIONS: Dissecting aortic aneurysm, hypersensitivity, pheochromocytoma

Black Box Warning: Acute MI

Precautions: Pregnancy (C), breastfeeding, children, geriatric patients, renal disease, CVD

Black Box Warning: CAD, CHF, cardiac disease, cardiac tamponade, edema, hypotension, orthostatic hypotension, pericardial effusion

DOSAGE AND ROUTES

Severe hypertension

- **Adult: PO** 2.5-5 mg/day in 1-2 divided doses; max 100 mg/day; usual range 10-40 mg/day in single dose
- **Geriatric: PO** 2.5 mg/day, may be increased gradually

• **Child <12 yr: PO** (initial) 0.1-0.2 mg/kg/day; (effective range) 0.25-1 mg/kg/day; (max) 50 mg/day

Alopecia

• **Adult: TOP** 1 ml bid, rub into scalp daily, max 2 ml/day

Renal dose

• **Adult: PO** CCr 10-15 ml/min extend interval to q24hr; CCr <10 ml/min not recommended

Anal fissures (unlabeled)

• **Adult/adolescent: TOP** (0.5% minoxidil in white paraffin base) 0.5 g each compounded minoxidil and lignocaine ointment q8hr

Scleroderma renal crisis (unlabeled)

• **Adult: PO** 5 mg/day in 1-2 divided doses, increase after 3 days by 10-20 mg/day to reach desired B/P, max 100 mg/day

Available forms: Tabs 2.5, 10 mg; topical 2%, 5% sol; topical foam 5%

Administer:

• Store protected from light and heat

PO route

• Without regard to meals

• With β-blocker and/or diuretic for hypertension

Topical route

• 1 ml no matter how much balding has occurred; increasing dosage does not speed growth

SIDE EFFECTS

Systemic

CNS: Headache, fatigue

CV: *Severe rebound hypertension on withdrawal in children,* tachycardia, angina, increased T wave, CHF, pulmonary edema, pericardial effusion, edema, sodium, water retention, *hypotension*

GI: Nausea, vomiting

GU: Breast tenderness

HEMA: Hct, Hgb; erythrocyte count may decrease initially, leukopenia

INTEG: Pruritus, Stevens-Johnson syndrome, rash, hirsutism, contact dermatitis

PHARMACOKINETICS

PO: Onset 30 min, peak 2-3 hr, duration 48-120 hr; half-life 4.2 hr; metabolized in liver; metabolites excreted in urine, feces; protein binding minimal

INTERACTIONS

Increase: hypotension—antihypertensives, MAOIs

Decrease: antihypertensive effect—NSAIDs, salicylates, estrogens

Drug/Herb

Increase: antihypertensive effect—hawthorn

Drug/Lab Test

Increase: renal studies

Decrease: Hgb/Hct/RBC

NURSING CONSIDERATIONS

Assess:

⚠ Monitor closely; usually given with β-blocker to prevent tachycardia and increased myocardial workload; usually given with diuretic to prevent serious fluid accumulation; patient should be hospitalized during beginning treatment

• Nausea, edema in feet, legs daily

• Skin turgor, dryness of mucous membranes for hydration status

• Crackles, dyspnea, orthopnea

• Electrolytes: potassium, sodium, chloride, CO_2

• Renal studies: catecholamines, BUN, creatinine

• Hepatic studies: AST, ALT, alk phos

• B/P, pulse

• Weight daily, I&O

Evaluate:

• Therapeutic response: decreased B/P, increased hair growth

Teach patient/family:

• That body hair will increase but is reversible after discontinuing treatment

• Not to discontinue product abruptly

• To report pitting edema, dizziness, weight gain >5 lb, SOB, bruising or bleeding, heart rate >20 beats/min over normal, severe indigestion, dizziness, lightheadedness, panting, new or aggravated symptoms of angina

• To take product exactly as prescribed because serious side effects may occur

Topical

• That for topical use, treatment must continue for the long term, or new hair will be lost

• Not to use except on scalp; use on clean, dry scalp before styling aids; wash hands after each use

TREATMENT OF OVERDOSE:

Administer normal saline IV, vasopressors

mirabegron

(mir′a-beg′ron)

Myrbetriq

Func. class.: Bladder antispasmodic

Chem. class.: β_3 Adrenergic receptor agonist

ACTION: Relaxes smooth muscles in urinary tract

USES: Overactive bladder (urinary frequency, urgency), urinary incontinence

CONTRAINDICATIONS: Hypersensitivity

Precautions: Pregnancy (C), breastfeeding, children, kidney/liver disease, bladder obstruction, dialysis, hypertension

DOSAGE AND ROUTES

• **Adult: PO** 25 mg/day, may increase to 50 mg/day if needed

Hepatic/renal dose

⚠ **PO** Child–Pugh B or (CCr 15-29 ml/min, max 25 mg/day; Child–Pugh C or CCr <15 ml/min, not recommended

Available forms: Tabs ext rel 25, 50 mg

Administer:

• Give whole; take with liquids; do not crush, chew, or break ext rel product; use without regard to meals

SIDE EFFECTS

CNS: Fatigue, dizziness, headache

CV: Hypertension

EENT: Xerophthalmia, blurred vision

GI: Nausea, vomiting, anorexia, abdominal pain, constipation, diarrhea, dyspepsia

GU: Dysuria, urinary retention, frequency, UTI, bladder discomfort

INTEG: Rash, pruritus

MISC: Arthralgia, back pain

RESP: Pharyngitis

SYST: Stevens–Johnson syndrome

PHARMACOKINETICS

71% protein binding, excretion 25% unchanged in urine, terminal half-life 50 hr, peak 3.5 hr

INTERACTIONS

Increase: Effect of digoxin, warfarin, desipramine, thioridazine, flecainide, propafenone

Increase: risk of urinary retention with antimuscarinic agents (eg, atropine, scopolamine)

NURSING CONSIDERATIONS

Assess:

• Urinary patterns: distention, nocturia, frequency, urgency, incontinence

• LFTs at baseline, periodically

• Monitor B/P

Evaluate:

• Decreasing dysuria, frequency, nocturia, incontinence

Teach patient/family:

• To avoid hazardous activities; dizziness can occur

• Not to drink liquids before bedtime

• About the importance of bladder maintenance

mirtazapine (Rx)

(mer-ṭa'za-peen)

Remeron, Remeron Soltab

Func. class.: Antidepressant

Chem. class.: Tetracyclic

ACTION: Blocks reuptake of norepinephrine and serotonin into nerve endings, thereby increasing action of norepinephrine and serotonin in nerve cells; antagonist of central α_2-receptors; blocks histamine receptors

USES: Depression; dysthymic disorder; bipolar disorder: depressed, agitated depression

Unlabeled uses: Resting tremor, benign familial tremor, levodopa-induced dyskinesias, pruritus

CONTRAINDICATIONS: Hypersensitivity to tricyclics, recovery phase of MI, agranulocytosis, jaundice

Precautions: Pregnancy (C), geriatric patients, suicidal patients, severe depression, increased intraocular pressure, closed-angle glaucoma, urinary retention, cardiac/renal/hepatic disease, hypo/hyperthyroidism, electroshock therapy, elective surgery, seizure disorder, bone marrow suppression, thrombocytopenia

Black Box Warning: Suicidal ideation, children

DOSAGE AND ROUTES

- **Adult: PO** 15 mg/day at bedtime, maintenance to continue for 6 mo, titrate up to 45 mg/day; **ORALLY DISINTEGRATING** tabs: open blister pack, place tab on tongue, allow to disintegrate, swallow
- **Geriatric: PO** 7.5 mg at bedtime, increase by 7.5 mg q1-2wk to desired dose, max 45 mg/day

Resting tremor/benign familial tremor/levodopa-induced dyskinesias (unlabeled)

- **Adult: PO** titrate up to 30 mg at bedtime

Pruritus (unlabeled)

- **Adult: PO** 15-30 mg/day

Available forms: Tabs 7.5, 15, 30, 45 mg; orally disintegrating tab (Soltab) 15, 30, 45 mg

Administer:

- Increased fluids, bulk in diet for constipation, especially for geriatric patients
- Without regard to meals
- Dosage at bedtime if oversedation occurs during day; may take entire dose at bedtime; geriatric patients may not tolerate once daily dosing
- Gum, hard candy, or frequent sips of water for dry mouth
- Store in tight container at room temperature; do not freeze
- **Orally disintegrating tab:** no water needed; allow to dissolve on tongue, do not split; contains phenylalanine

SIDE EFFECTS

CNS: *Dizziness, drowsiness,* confusion, headache, anxiety, tremors, stimulation, weakness, nightmares, EPS (geriatric patients), increased psychiatric symptoms, **seizures**

CV: *Orthostatic hypotension, ECG changes,* tachycardia, **hypertension**, palpitations

EENT: *Blurred vision,* tinnitus, mydriasis

GI: *Diarrhea, dry mouth,* nausea, vomiting, **paralytic ileus**, increased appetite, cramps, epigastric distress, constipation, **jaundice, hepatitis**, stomatitis, weight gain

GU: *Urinary retention,* **acute renal failure**, urinary frequency

HEMA: **Agranulocytosis, thrombocytopenia, eosinophilia, leukopenia**

INTEG: Rash, urticaria, sweating, pruritus, photosensitivity

MS: Back pain, myalgia

RESP: Cough

SYST: Flulike symptoms, increased lipid levels

PHARMACOKINETICS

PO: Peak 2 hr, metabolized by CYP1A2, 2D6, 3A4 in liver; excreted in urine, feces; crosses placenta; half-life 20-40 hr, protein binding 85%

M

INTERACTIONS

⚠ Increase: hyperpyretic crisis, seizures, hypertensive episode—MAOIs

Increase: CNS depression—alcohol, barbiturates, benzodiazepines, other CNS depressants

Increase: serotonin syndrome—SSRIs, SNRIs, serotonin-receptor agonists, fenfluramine, dexfenfluramine, sibutramine, nefazodone

Decrease: effects of cloNIDine, indirect-acting sympathomimetics (ePHEDrine)

Drug/Herb

- Serotonin syndrome: St. John's wort

Increase: CNS depression—kava

Drug/Lab Test

Increase: serum bilirubin, blood glucose, alk phos

Decrease: VMA, 5-HIAA

False increase: urinary catecholamines

NURSING CONSIDERATIONS

Assess:

- B/P (lying, standing), pulse q4hr; if systolic B/P drops 20 mm Hg, hold product, notify prescriber; vital signs q4hr in patients with CV disease
- Blood studies: CBC, leukocytes, differential, cardiac enzymes, lipid profile, blood glucose, LFTs, serum creatinine/BUN if patient is receiving long-term therapy
- Weight weekly; appetite may increase with product
- ECG for flattening of T wave, bundle branch block, AV block, dysrhythmias in cardiac patients

Black Box Warning: Mental status: mood, sensorium, affect, suicidal tendencies (especially among adolescents, young adults), increase in psychiatric symptoms: depression, panic; EPS primarily in geriatric patients: rigidity, dystonia, akathisia

⚠ **Serotonin syndrome:** hyperthermia, hypertension, myoclonus, rigidity, delirium, coma; if using other serotonergic products

- Alcohol consumption; if alcohol consumed, hold dose until morning
- Assistance with ambulation during beginning therapy, since drowsiness, dizziness occurs

Evaluate:

- Therapeutic response: decreased depression

Teach patient/family:

- That therapeutic effects may take 2-3 wk; to take at bedtime; that there is decreased sedation with increased doses; not to discontinue abruptly
- To use caution when driving, performing other activities requiring alertness because of drowsiness, dizziness, blurred vision
- To avoid alcohol, other CNS depressants
- About how to take orally disintegrating tabs; dissolve on tongue, swallow
- Not to use within 14 days of MAOIs

Black Box Warning: To notify prescriber of suicidal thoughts, behavior

TREATMENT OF OVERDOSE:

ECG monitoring, lavage, activated charcoal; administer anticonvulsant, IV fluids

misoprostol (Rx)

(mye-soe-prost'ole)

Cytotec

Func. class.: Gastric mucosa protectant, antiulcer

Chem. class.: Prostaglandin E_1 analog

Do not confuse:

misoprostol/metoprolol

Cytotec/Cytoxan

ACTION: Inhibits gastric acid secretion; may protect gastric mucosa; can increase bicarbonate, mucus production

USES: Prevention of NSAID-induced gastric ulcers

Unlabeled uses: Pregnancy termination, postpartum hemorrhage, cervical ripening/labor induction (vaginal), active duodenal/gastric ulcer, kidney transplant rejection prophylaxis

CONTRAINDICATIONS: Hypersensitivity to this product or prostaglandins

Black Box Warning: Pregnancy (X), females

Precautions: Breastfeeding, children, geriatric patients, renal/CV disease, abnormal fetal position, cardiac/renal/inflammatory bowel disease, C-section, dehydration, diarrhea, fever, ectopic pregnancy, fetal distress, sepsis, vaginal bleeding

DOSAGE AND ROUTES

- **Adult: PO** 200 mcg qid with food for duration of NSAID therapy, with last dose given at bedtime; if 200 mcg is not tolerated, 100 mcg may be given

Active duodenal/gastric ulcer (unlabeled)

- **Adult: PO** 100-200 mcg qid with meals at bedtime × 4-8 wk

Pregnancy termination before 63rd day (unlabeled)

- **Adult: INTRAVAGINALLY** 800 mcg 5-7 days after methotrexate IM

Cervical ripening induction for term pregnancy (unlabeled)

- **Adult: INTRAVAGINALLY** 25 mcg q3-6hr

Available forms: Tabs 100, 200 mcg

Administer:

- PO with meals for prolonged product effect; avoid use of magnesium antacids
- Store at room temperature

SIDE EFFECTS

GI: *Diarrhea,* nausea, vomiting, flatulence, constipation, dyspepsia, abdominal pain

GU: Spotting, cramps, hypermenorrhea, menstrual disorders

PHARMACOKINETICS

PO: Peak 12 min, plasma steady state achieved within 2 days, excreted in urine

INTERACTIONS

Drug/Food

Decrease: maximum concs when taken with food

NURSING CONSIDERATIONS

Assess:

- **NSAID-induced ulcer prophylaxis:** monitor GI symptoms (hematemesis, occult or frank blood in stools, gastric aspirate, cramping, severe diarrhea)

Black Box Warning: Pregnancy (X): Obtain a negative pregnancy test; miscarriages are common

Evaluate:

- Therapeutic response: absence of pain or GI complaints; prevention of ulcers

Teach patient/family:

- To take only as directed; to read patient information leaflet

Black Box Warning: Not to take if pregnant (can cause miscarriage) (X); not to become pregnant while taking product; if pregnancy occurs during therapy, discontinue product, notify prescriber; not to breastfeed

- Not to give product to anyone else or to take for more than 4 wk unless directed by prescriber
- To avoid OTC preparations: aspirin, cough, cold products; condition may worsen

⚠ HIGH ALERT

mitoMYcin (Rx)

(mye-toe-mye′sin)

Mitosol

Func. class.: Antineoplastic, antibiotic

Do not confuse:
mitoMYcin/mithramycin/mitotane/mitoXANtrone

ACTION: Inhibits DNA synthesis, primarily; derived from *Streptomyces caespitosus;* appears to cause cross-linking of DNA; vesicant

M

USES:
Pancreatic, stomach, colorectal, bladder cancer

Unlabeled uses: Palliative treatment of anal, bladder, head, neck, colon, breast, biliary, cervical, lung malignancies; bone marrow ablation, desmoid tumor, mesothelioma, stem cell transplant preparation

CONTRAINDICATIONS:
Pregnancy (D) 1st trimester, breastfeeding, hypersensitivity, as single agent, coagulation disorders

Black Box Warning: Thrombocytopenia

Precautions: Accidental exposure, acute bronchospasm, anemia, children, dental disease/work, extravasation, females, infection, radiation therapy, surgery, vaccines, renal/respiratory disease

Black Box Warning: Bone marrow suppression, hemolytic-uremic syndrome

DOSAGE AND ROUTES
- **Adult: IV** 20 mg/m^2 q6-8wk

Available forms: Inj 5, 20, 40 mg/vial

Administer:

Direct IV route
- Use port or central line if possible
- Antiemetic 30-60 min before product to prevent vomiting
- IV after diluting 5 mg/10 ml, 20 mg/ 40 ml, 40 mg/80 ml (0.5 mg/ml) sterile water for inj; shake, allow to stand, give through Y-tube or 3-way stopcock; give slow IV push or infuse over 15-30 min; color of reconstituted sol is gray
- Avoid excessive heat; store reconstituted product in refrigerator; discard after 2 wk; store unreconstituted product at room temperature

Y-site compatibilities: Amifostine, amphotericin B lipid complex, amphotericin B liposome, anidulafungin, argatroban, atenolol, bivalirudin, bleomycin, caspofungin, CISplatin, cyclophosphamide, DACTINomycin, dolasteron, DOXOrubicin, droperidol, epirubicin, ertapenem, fluorouracil, furosemide, granisetron, heparin, leucovorin, melphalan, methotrexate, metoclopramide, nesiritide, octreotide, ondansetron, oxaliplatin, PACLitaxel, palonosetron, PEMEtrexed, riTUXimab, teniposide, thiotepa, tigecycline, tirofiban, trastuzumab, vinBLAStine, vinCRIStine, voriconazole, zoledronic acid

SIDE EFFECTS
CNS: Fever, headache, confusion, drowsiness, syncope, fatigue

EENT: Blurred vision

GI: *Nausea, vomiting, anorexia, stomatitis,* hepatotoxicity, diarrhea

GU: Urinary retention, renal failure, edema

HEMA: Thrombocytopenia, leukopenia, anemia

INTEG: *Rash,* alopecia, extravasation, nail discoloration

MISC: Hemolytic uremic syndrome, CHF

RESP: Fibrosis, pulmonary infiltrate, dyspnea

PHARMACOKINETICS
Half-life 1 hr, metabolized in liver, 10% excreted in urine (unchanged)

INTERACTIONS
Increase: toxicity—other antineoplastics, radiation

Increase: bleeding risk—NSAIDs, anticoagulants
- Avoid use with vaccines

Drug/Herb
- Avoid use with black cohosh

NURSING CONSIDERATIONS
Assess:

Black Box Warning: Bone marrow suppression: CBC, differential, platelet count weekly; withhold product if WBC is <4000/mm^3, serum creatinine >1.7 mg/dl, or platelet count is <100,000/mm^3; notify prescriber; bleeding: hematuria, guaiac, bruising, petechiae, mucosa, or orifices

Black Box Warning: Fatal hemolytic-uremic syndrome: hypertension, thrombocytopenia, microangiopathic hemolytic anemia; occurs in those receiving long-term therapy

- Renal studies: BUN, serum uric acid, urine CCr, electrolytes before, during therapy, adjust dose based on renal function
- I&O ratio; report fall in urine output to <30 ml/hr
- Monitor temperature q4hr; fever may indicate beginning infection
- Hepatic studies before, during therapy: bilirubin, AST, ALT, alk phos as needed or monthly; check for jaundiced skin and sclera, dark urine, clay-colored stools, itchy skin, abdominal pain, fever, diarrhea

⚠ **Pulmonary fibrosis:** bronchospasm, dyspnea, crackles, unproductive cough; chest pain, tachypnea, fatigue, increased pulse, pallor, lethargy; pulmonary function tests; chest x-ray before, during therapy; chest x-ray should be obtained q2wk during treatment

- Effects of alopecia on body image; discuss feelings about body changes
- Buccal cavity q8hr for dryness, sores, ulceration, white patches, oral pain, bleeding, dysphagia
- Local irritation, pain, burning at inj site
- GI symptoms: frequency of stools, cramping
- Adequate fluids 2-3 L/day unless contraindicated
- Rinsing of mouth tid-qid with water; brushing of teeth with baking soda bid-tid with soft brush or cotton-tipped applicators for stomatitis; use unwaxed dental floss

Evaluate:

- Therapeutic response: decreased tumor size, spread of malignancy

Teach patient/family:

- That hair may be lost during treatment; that wig or hairpiece may make patient feel better; that new hair may be different in color, texture
- To avoid foods with citric acid, hot or rough texture
- To report any bleeding, white spots, ulcerations in mouth; to examine mouth daily
- To report sign of IV site reaction, redness, inflammation, burning, pain
- To avoid crowds, persons with infections if granulocyte count is low
- **Infection:** report fever, flulike symptoms, sore throat
- To immediately report urine retention, absence of urine, dyspnea, bleeding, jaundice, signs of pulmonary toxicity
- To report if pregnancy is planned or suspected; pregnancy (D) 1st trimester, do not breastfeed

⚠ ***HIGH ALERT***

mitoXANtrone (Rx)

(mye-toe-zan′trone)

Func. class.: Antineoplastic, antiinfective, immunomodulator

Chem. class.: Synthetic anthraquinone

Do not confuse:
mitoXANtrone/mitoMYcin/mithramycin/mitotane

ACTION: DNA reactive agent; cytocidal effect on both proliferating and nonproliferating cells; topoisomerase II inhibitor (vesicant)

USES: Acute myelogenous leukemia (adult), relapsed leukemia; used with steroids to treat bone pain (advanced prostate cancer), multiple sclerosis (MS)

Unlabeled uses: Liver malignancies, non-Hodgkin's lymphoma, breast cancer, ALL, bone marrow ablation, CLL, hepatocellular cancer, ovarian cancer, pleural effusion, stem cell transplant preparation

CONTRAINDICATIONS: Pregnancy (D), hypersensitivity

M

Precautions: Breastfeeding, children; myelosuppression, renal/cardiac/hepatic disease; gout

Black Box Warning: Secondary malignancy, neutropenia, intrathecal administration, extravasation, heart failure

DOSAGE AND ROUTES

Acute myelogenous leukemia/induction

- **Adult: IV INF** 12 mg/m^2/day on days 1-3 and 100 mg/m^2 cytarabine × 7 days as continuous 24-hr inf

Consolidation

- **Adult: IV** 12 mg/m^2 given as short 5- to 15-min inf for 2 days with cytarabine × 5 days

Advanced prostate cancer

- **Adult: IV** 12-14 mg/m^2 as single dose or short inf q21days

Multiple sclerosis, relapsing

- **Adult: IV INF** 12 mg/m^2 as 5- to 15-min inf q3mo, cumulative lifetime dose 140 mg/m^2

Available forms: Inj 2, 10, 12.5, 15 mg/ml

Administer:

- Other medications by oral route if possible; avoid IM, SUBCUT, IV routes
- Antiemetic 30-60 min before product to prevent vomiting

Direct IV route

- IV after diluting with ≥50 ml NS or D_5W; give over 3-5 min into running IV of D_5W or NS; check for extravasation; do not give IM, SUBCUT, or intraarterially

Intermittent IV INF route

- May be diluted further in D_5W, NS, run over 15-30 min

Continuous IV INF route

- Give over 24 hr

Y-site compatibilities: Acyclovir, alemtuzumab, alfentanil, allopurinol, amikacin, aminocaproic acid, aminophylline, amiodarone, anidulafungin, argatroban, atracurium, bivalirudin, bleomycin, bretylium, bumetanide, buprenorphine, butorphanol, calcium chloride, calcium gluconate, CARBOplatin, carmustine, caspofungin, cefoTEtan, ceftizoxime, chloramphenicol, chlorproMAZINE, cimetidine, ciprofloxacin, cisatracurium, CISplatin, cladribine, codeine, cyclophosphamide, cycloSPORINE, cytarabine, DACTINomycin, DAPTOmycin, DAUNOrubicin citrate liposome, dexmedetomidine, dexrazoxane, diltiazem, diphenhydrAMINE, DOBUTamine, DOCEtaxel, dolasetron, DOPamine, doxacurium, doxycycline, droperidol, enalaprilat, ePHEDrine, EPINEPHrine, erythromycin, esmolol, etoposide, etoposide phosphate, famotidine, fenoldopam, fentaNYL, filgrastim, fluconazole, fludarabine, fluorouracil, ganciclovir, gatifloxacin, gemcitabine, gentamicin, glycopyrrolate, granisetron, haloperidol, hydrALAZINE, hydrocortisone sodium succinate, HYDROmorphone, hydrOXYzine, ifosfamide, imipenem-cilastatin, inamrinone, insulin regular, irinotecan, isoproterenol, ketorolac, labetalol, leucovorin, levofloxacin, levorphanol, lidocaine, linezolid, LORazepam, magnesium sulfate, mannitol, melphalan, meperidine, meropenem, mesna, metaraminol, methohexital, methotrexate, methyldopate, metoclopramide, metoprolol, metroNIDAZOLE, midazolam, milrinone, minocycline, mivacurium, morphine sulfate, nalbuphine, naloxone, nesiritide, niCARdipine, nitroglycerin, norepinephrine, octreotide, ondansetron, oxaliplatin, palonosetron, pamidronate, pancuronium, pentamidine, pentazocine, PENTObarbital, PHENObarbital, phentolamine, phenylephrine, polymyxin b, potassium acetate, potassium chloride, procainamide, prochlorperazine, promethazine hydrochloride, propranolol, quiNIDine gluconate, quinupristin-dalfopristin, ranitidine, remifentanil, riTUXimab, rocuronium, sargramostim, sodium acetate, sodium bicarbonate, succinylcholine, SUFentanil, sulfamethoxazole-trimethoprim, tacrolimus, teniposide, theophylline, thiopental, thiotepa, tigecycline, tirofiban, tobramycin, tolazoline, trastuzumab, trimethobenzamide, vancomycin, vasopressin, vecuronium, verapamil, vinCRIStine, vinorelbine, zidovudine, zoledronic acid

SIDE EFFECTS

CNS: Headache, **seizures**, fatigue
CV: **CHF, cardiopathy, dysrhythmias**
EENT: Conjunctivitis, blue/green sclera, blurred vision
GI: *Nausea, vomiting, diarrhea, anorexia, mucositis,* **hepatotoxicity**, abdominal pain, constipation, jaundice
GU: Amenorrhea, menstrual disorders
HEMA: **Thrombocytopenia, leukopenia, myelosuppression, anemia, secondary leukemia**
INTEG: *Rash, necrosis at inj site,* dermatitis, thrombophlebitis at inj site, alopecia
MISC: Fever, hyperuricemia, infections
RESP: Cough, dyspnea
SYST: **Tumor lysis syndrome, sepsis**

PHARMACOKINETICS

Protein binding 78%; metabolized in liver; excreted via renal, hepatobiliary systems; half-life 23-215 hr

INTERACTIONS

Increase: **bone marrow depression toxicity—radiation, other antineoplastics**
Increase: adverse reactions—live virus vaccines, trastuzumab
Increase: bleeding risk—NSAIDs, anticoagulants
Increase: mitoXANtrone effects of—digoxin, phenytoin

Drug/Lab Test

Increase: LFTs, uric acid
Decrease: HcT/Hgb, platelets, WBC

NURSING CONSIDERATIONS

Assess:

- CBC, differential, platelet count weekly; withhold product if WBC is <1500/mm^3; leukopenia, neutropenia, thrombocytopenia are expected, leukocyte nadir 10-14 days, recovers in 2-3 wk
- **Hepatotoxicity: hepatic studies before, during therapy: bilirubin, AST, ALT, alk phos prn or monthly; dose reduction needed with hepatic disease; jaundiced skin and sclera, dark urine, clay-colored stools, itchy skin, abdominal pain, fever, diarrhea**
- Renal studies: BUN, serum uric acid, urine CCr, electrolytes before, during therapy
- Bleeding, hematuria, guaiac, bruising or petechiae, mucosa or orifices q8hr

Black Box Warning: ECG, ECHO, chest x-ray, MUGA, RAI angiography; assess ejection fraction before and during treatment; cardiotoxic may develop during treatment or months to years after treatment; use vigilant cardiac monitoring in MS

Black Box Warning: Secondary acute myelogenous leukemia (AML) that can develop after taking this product

Black Box Warning: Multiple sclerosis: obtain MUGA, LVEF baselines; repeat LVEF if symptoms of CHF occur or if cumulative dose is >100 mg/m^2; do not administer to patients who have received lifetime dose of ≥140 mg/m^2 or if LVEF <50% or significant LVEF

- Obtain pregnancy test for all women of childbearing age, even if birth control is used
- Rinsing of mouth tid-qid with water, club soda; brushing of teeth bid-qid with soft brush or cotton-tipped applicators for stomatitis; use unwaxed dental floss
- Provide increased fluids to 2-3 L/day unless contraindicated

Evaluate:

- Therapeutic response: decreased tumor size, spread of malignancy

Teach patient/family:

- To immediately report bleeding, dyspnea, possible infections, seizure, jaundice, fever, cough, or dyspnea
- To avoid hot foods or those with citric acid, rough texture
- To report any bleeding, white spots, ulcerations in mouth; to examine mouth daily
- To avoid crowds, persons with infections
- That sclera, urine may turn blue or green; that hair loss may occur

Black Box Warning: To notify prescriber if pregnancy is suspected or planned; to use effective contraception

modafinil (Rx)

(mo-daf'i-nil)

Alertec ✱, Provigil

Func. class.: CNS stimulant

Chem. class.: Racemic compound

Controlled Substance IV

ACTION:

Similar action as that of sympathomimetics; does not alter release of DOPamine, norepinephrine

USES:

Narcolepsy, shift-work sleep disturbance, obstructive sleep apnea

Unlabeled uses: Fatigue in MS, ADHD

CONTRAINDICATIONS:

Hypersensitivity, ischemic heart disease, left ventricular hypertrophy, chest pain, dysrhythmias

Precautions: Pregnancy (C), breastfeeding, child <16 yr, geriatric patients, unstable angina, history of MI, severe hepatic disease

DOSAGE AND ROUTES

To improve wakefulness with daytime sleepiness

- **Adult and adolescent ≥16 yr: PO** 200 mg daily

Hepatic dose (severe hepatic disease)

- **Adult: PO** 100 mg daily

Multiple sclerosis (unlabeled)

- **Adult/geriatric/adolescents ≥6 yr: PO** 200-400 mg daily in the AM

Available forms: Tabs 100, 200 mg

Administer:

- Give 1 hr before start of shift work or in AM for those with narcolepsy or sleep apnea
- Store at room temperature

SIDE EFFECTS

CNS: *Headache,* anxiety, cataplexy, depression, dizziness, insomnia, amnesia, confusion, ataxia, tremors, paresthesia, dyskinesia, suicidal ideation

CV: Dysrhythmias, hypo/hypertension, chest pain, vasodilation

EENT: Change in vision, *rhinitis,* pharyngitis, epistaxis

GI: Nausea, vomiting, changes in LFTs, anorexia, diarrhea, thirst, mouth ulcers

GU: Ejaculation disorder, urinary retention, albuminuria

HEMA: Eosinophilia

INTEG: Rash, dry skin, herpes simplex, Stevens-Johnson syndrome

MISC: Infection, hyperglycemia, neck pain

RESP: *Dyspnea*, lung changes

PHARMACOKINETICS

Absorbed rapidly, 60% protein binding, metabolized by the liver (90%), half-life 15 hr, peak 2-4 hr

INTERACTIONS

Increase: Altered levels of CYP3A4 inhibitors (azole antibiotics some SSRIs), reaction difficult to predict

Increase: levels of CYP2C19 substrates (diazepam, phenytoin some tricyclics)

Decrease: effects of—cycloSPORINE, hormonal contraceptives, theophylline, estrogens

Delayed effect modafinil by 1 hr: methylphenidate

Altered: Levels of CYP3A4 inducers (carBAMazepine, phenytoin, rifampin, cycloSPORINE, theophylline)

Drug/Herb

Increase: stimulation—cola nut, guarana, yerba maté, coffee, tea

Drug/Lab Test

Increase: LFTs, glucose, eosinophils

NURSING CONSIDERATIONS

Assess:

- Narcolepsy, shift work, history of sleep apnea
- Depression, suicidal ideation

• Monitor B/P in those with hypertension

Evaluate:

• Ability to stay awake

Teach patient/family:

• To take only as directed; that product may be taken with/without food

• **To use other form of contraception during and for ≥30 days after discontinuing medication if using hormonal birth control; to notify prescriber if pregnancy is planned or suspected or if breastfeeding**

• To notify prescriber of allergic reaction, tremors, confusion

• To avoid all OTC medications unless approved by prescriber

• To avoid hazardous activities until drug effect is known

montelukast (Rx)

(mon-teh-loo'kast)

Singulair

Func. class.: Bronchodilator

Chem. class.: Leukotriene receptor antagonist, cysteinyl

ACTION:
Inhibits leukotriene (LTD_4) formation; leukotrienes exert their effects by increasing neutrophil, eosinophil migration; aggregation of neutrophils, monocytes; smooth muscle contraction, capillary permeability; these actions further lead to bronchoconstriction, inflammation, edema

USES:
Chronic asthma in adults and children, seasonal allergic rhinitis, bronchospasm prophylaxis

CONTRAINDICATIONS:
Hypersensitivity

Precautions: Pregnancy (B), breastfeeding, children <6 yr, acute attacks of asthma, alcohol consumption, severe hepatic disease, corticosteroid withdrawal, phenylketonuria, suicidal ideation, depression

DOSAGE AND ROUTES

• **Adult and child ≥15 yr: PO** 10 mg/day in PM

• **Child 6-14 yr: PO** 5-mg chew tab/day in PM

• **Child 2-5 yr: PO** (chew tab/granules) 4 mg/day

Asthma

• **Child 12-23 mo: PO** 1 packet (4 mg) granules taken in PM

Exercise-induced bronchoconstriction

• **Adult/child ≥6 yr: PO** 10 mg 2 hr before exercise; do not take another dose within 24 hr

Available forms: Tabs 10 mg; chew tabs 4, 5 mg; oral granules 4 mg/packet

Administer:

PO route

• In PM daily for all uses except exercise-induced bronchoconstriction; then take 2 hr before exercise

• Granules directly in mouth or mixed with spoonful of soft food (carrots, applesauce, ice cream, rice)

• Do not open granules packet until ready to use; mix whole dose; give within 15 min

SIDE EFFECTS

CNS: *Dizziness, fatigue, headache,* behavior changes, hallucinations, **seizures,** agitation, anxiety, depression, fever, drowsiness, **suicidal ideation, memory impairment, hostility, somnambulism**

GI: *Abdominal pain,* dyspepsia, nausea, vomiting, diarrhea, **pancreatitis**

HEMA: Thrombocytopenia

INTEG: Rash, pruritus, erythema

MS: Asthenia, myalgia, muscle cramps

RESP: *Influenza, cough,* nasal congestion

SYST: Anaphylaxis, angioedema, Churg-Strauss syndrome, Stevens-Johnson syndrome, toxic epidermal necrolysis

PHARMACOKINETICS

Rapidly absorbed; peak 3-4 hr, chew tab (5 mg) 2-2.5 hr; half-life 2.7-5.5 hr, extended in hepatic disease; protein bind-

M

ing 99%; metabolized by liver; excreted via bile

INTERACTIONS

Increase: adverse reactions of CYP2C8 substrates
Decrease: montelukast levels—barbiturates, rifabutin, rifapentine, carBAMazepine, fosphenytoin, phenytoin, rifampin
Drug/Herb
Increase: stimulation—black, green tea, guarana
Drug/Lab Test
Increase: ALT, AST

NURSING CONSIDERATIONS

Assess:
⚠ **Churg-Strauss syndrome:** rare adult patients carefully for symptoms: eosinophilia, vasculitic rash, worsening pulmonary symptoms, cardiac complications, neuropathy
• Allergic reactions: rash, urticaria; product should be discontinued
⚠ For behavior changes and suicidal ideation, other neuropsychiatric reactions
• **Severe hepatic disease:** use cautiously
Evaluate:
• Therapeutic response: ability to breathe more easily
Teach patient/family:
• To check OTC medications, current prescription medications for ePHEDrine, which will increase stimulation; to avoid alcohol
• To avoid hazardous activities; dizziness may occur
• That product is not to be used for acute asthma attacks
• If aspirin sensitivity is known, not to take NSAIDs while taking this product
• To continue to use inhaled β-agonists if exercise-induced asthma occurs
• **Granules:** to give directly in mouth or mixed in a spoonful of room temperature soft food (use only applesauce, carrots, rice, or ice cream); use within 15 min of opening packets; discard unused portions

⚠ HIGH ALERT

morphine (Rx)

(mor'feen)

Astramorph PF, AVINza, Depo Dur, Infumorph PF, Kadian, M.O.S. ✦, MS Contin, MSIR ✦, Oramorph SR

Func. class.: Opioid analgesic
Chem. class.: Alkaloid

Controlled Substance Schedule II

Do not confuse:
morphine/HYDROmorphone
MS Contin/OxyCONTIN

ACTION: Depresses pain impulse transmission at the spinal cord level by interacting with opioid receptors

USES: Moderate to severe pain
Unlabeled uses: Agitation, bone/dental pain, dyspnea in end stage cancer or pulmonary disease, sedation induction, rapid-sequence intubation

CONTRAINDICATIONS: Hypersensitivity, addiction (opioid/alcohol), hemorrhage, bronchial asthma, increased intracranial pressure, paralytic ileus, hypovolemia, shock, MAOI therapy

Black Box Warning: Respiratory depression

Precautions: Pregnancy (C), breastfeeding, children <18 yr, geriatric patients, addictive personality, acute MI, severe heart disease, renal/hepatic disease, bowel impaction, abrupt discontinuation, seizures

Black Box Warning: Accidental exposure, epidural/intrathecal/IM/subcut administration, opioid-naive patients, substance abuse

DOSAGE AND ROUTES

Acute moderate to severe pain

PO regular-release route

- **Adult ≥50 kg:** initially, 10-30 mg every 3-4 hr as needed
- **Adult <50 kg/geriatric patients:** might require lower doses and/or extended dosing intervals; doses should be titrated carefully
- **Child/infant ≥6 mo:** 0.2-0.5 mg/kg every 4-6 hr as needed
- **Infants <6 mo/neonate: PO** 0.1 mg/kg every 3-4 hr

IV/IM/SUBCUT route

- **Adult ≥50 kg:** 2.5-15 mg every 2-6 hr as needed, titrate; or a loading dose of 0.05-0.1 mg/kg IV, followed by 0.8-10 mg/hr IV, titrate
- **Adult <50 kg/geriatric patients:** might require lower doses and/or extended dosing intervals 0.1 mg/kg every 3-4 hr, titrate
- **Child/infant ≥6 mo:** 0.05-0.2 mg/kg every 2-4 hr, titrate to relief; max initial dose 15 mg/dose, avoid IM when possible
- **Infants <6 mo/neonate:** 0.03-0.05 mg/kg every 3-8 hr, titrate to relief, avoid IM when possible

Epidural (morphine sulfate injection, but *not* DepoDur)

- **Adult:** initially, 5 mg in the lumbar region; if pain relief does not occur in 1 hr, give 1-2 mg epidurally; max 10 mg/24 hr; **continuous epidural inf** 2-4 mg/24 hr; may give another 1-2 mg

Intrathecal route (morphine sulfate injection, but *not* DepoDur)

- Do not inject >2 ml of the 0.5 mg/ml or 1 ml of the 1 mg/ml ampule
- **Adult:** 0.2-1 mg in the lumbar area as a single dose or to establish dosage for continuous intrathecal inf; repeated injections are not recommended

Rectal route

- **Adult: PR** 10-20 mg every 4 hr, as needed
- **Child:** Individualize

Chronic moderate and severe pain:

⚠ Do not use extended-release cap or tab as prn analgesics, for acute pain, or if the pain is mild or not expected to persist for an extended period of time; use for postoperative pain only if the patient is receiving chronic opioid therapy before surgery or if the postoperative pain is expected to be moderate to severe and expected to persist for an extended period of time; do not use controlled-release tablets (MS Contin) immediately after surgery (for the first 24 hr) in patients not previously taking the drug

⚠ Do not use in opioid-naïve patients: 90 mg, 120 mg morphine biphasic-release capsules (AVINza); 100 mg, 130 mg, 150 mg, 200 mg morphine extended-release capsules (Kadian), 100 mg, 200 mg morphine control-release tablets (MS Contin); patients considered opioid tolerant are those who are taking at least 60 mg/day oral morphine, 30 mg/day oral oxyCODONE, 8 mg/day oral HYDROmorphone, or an equal dose of another opioid, for 1 wk or longer

PO (extended-release tab [MS Contin, Oramorph SR] or caps [Kadian, AVINza]) in opiate agonist naive patients

- **Adult:** 15-30 mg every 12 hr (tabs); 10 mg bid or 20 mg daily (Kadian); or 30 mg daily (AVINza), titrate; AVINza should be adjusted in increments ≤30 mg every 4 days; Kadian should be increased ≤20 mg every 1-2 days, taper gradually; to discontinue, gradually decrease AVINza and Kadian every 2-4 days
- **Child (unlabeled):** 0.3-0.6 mg/kg every 12 hr (tabs)

IV/SUBCUT route (opiate)

- **Adult: IV** 2-10 mg loading dose, then 0.8-10 mg/hour IV, titrate; maintenance 0.8-80 mg/hour IV
- **Children/infants ≥6 mo: IV** initially, 0.04-0.07 mg/kg/hr (range: 0.025-2.6 mg/kg/hr)
- **Subcut inf** 0.025-1.79 mg/kg/hour
- Infants <6 mo/neonate: **IV** 0.01 mg/kg/hour initially; inf rates max 0.015-0.02 mg/kg/hour IV

M

Breakthrough pain in patients receiving long-acting or continuous-inf morphine

PO (regular-release) route

- **Adult/child:** the dose is usually 1/4 to 1/3 the 8- to 12-hour extended-release dose every 4-6 hr as needed
- **Adult/child:** For PCA, intermittent dosage is usually 25%-30% of the hourly rate IV/SC every 6-15 min as needed; intermittent IV injection dosage is 25%-30% of the hourly rate given IV/SC every 1-2 hr as needed

Available forms: Inj 0.5, 1, 2, 4, 5, 8, 10, 15, 25, 50 mg/ml; oral sol 10, 20 mg/5 ml, 10 mg/0.5 ml, 100 mg/5 ml; oral tabs 15, 30 mg; rect supp 5, 10, 20, 30 mg; ext rel tabs 15, 30, 60, 100, 200 mg; cont rel cap pellets (Kadian) 10, 20, 30, 50, 60, 80, 100, 200 mg; ext rel caps (AVINza) 30, 45, 60, 75, 90, 120 mg

Administer:

PO route

- Give with food or milk to minimize GI effects
- Begin with immediate-release products and titrate to correct dose and convert to a sustained-release product
- **Immediate-release cap:** may swallow whole, or cap may be opened and contents sprinkled on cool food (pudding or applesauce) or added to juice; give immediately or delivered via gastric or NG tube by either adding to or following with liquid
- **Extended-release and controlled-release tabs:** swallow whole; do not crush, break, dissolve, or chew
- The use of MS Contin 100 mg or 200 mg tabs should be limited to opioid-tolerant patients requiring oral doses equivalent to ≥200 mg/day; use of the 100-mg or 200-mg tablet is only recommended for patients who have already been titrated to a stable analgesic regimen using lower strengths of MS Contin or other opioids
- **Sustained-release caps:** swallow; do not chew, crush, or dissolve; caps may be opened and contents sprinkled on applesauce (at room temperature or cooler) immediately before ingestion; do not chew, crush, or dissolve the pellets/beads inside the cap; the applesauce should be swallowed without chewing; if the pellets/beads are chewed, an immediate release of a potentially fatal morphine dose may be delivered; rinse mouth to ensure all the pellets/beads have been swallowed; do not separate applesauce into separate doses; the entire portion should be taken; discard unused portion
- **Kadian caps:** may be given through a 16-F gastrostomy tube; flush with water, and sprinkle the cap contents into 10 ml of water; using a funnel and a swirling motion, pour the pellets and water into the tube; rinse the beaker with 10 ml of water, and pour the water into the funnel; repeat until no pellets remain in the beaker; do not administer AVINza tabs through a gastrostomy tube; *do not* administer Kadian or AVINza through a nasogastric tube
- Avoid concurrent administration of AVINza with prescription or nonprescription medications that contain alcohol; consumption of alcohol while taking the extended-release capsules can result in the rapid release and absorption of a potentially fatal dose of morphine
- AVINza ≥90 mg or Kadian 100 mg, 130 mg, 150 mg, or 200 mg caps are given only to opioid-tolerant patients

Oral liquid

- Check dose before use because many concs of oral solution are available; may be diluted in fruit juice; protect from light

Injectable administration

- Visually inspect for particulate matter, discoloration before use; do not use if a precipitate is present after shaking; do not use the Duramorph solution if a precipitate is present or if the color is darker than pale yellow

SUBCUT route

- Inject, taking care not to inject intradermally
- **Continuous SC inf:** morphine is not approved by the FDA for subcut use; di-

lute to an appropriate conc in D_5W; give using a portable, controlled, subcut device; adjust rate based on patient response and tolerance; max subcut rate is 2 ml/hour/site

Intrathecal/epidural route

⚠ Morphine sulfate injection is not interchangeable with morphine sulfate extended-release liposome injection (DepoDur); DepoDur is only for epidural administration

⚠ Do not use Infumorph (10 mg/ml or 25 mg/ml) for single-dose neuraxial injection because lower doses can be more reliably administered with Duramorph (0.5 mg/ml or 1 mg/ml

Rectal route

• Moisten the suppository with water before insertion; if suppository is too soft, chill in the refrigerator for 30 min or run cold water over it before removing the wrapper

IV route

• Before use, an opiate antagonist and emergency facilities should be available

⚠ Do not use the highly concentrated morphine injections (i.e., 10-25 mg/ml) for IV, IM, or SC administration of single doses; these injection solutions are intended for use via continuous, controlled microinfusion devices

• **Direct IV route:** dilute dose with ≥5 ml of sterile water for injection or NS injection; inject 2.5-15 mg directly into a vein or into the tubing of a freely flowing IV solution over 4-5 min; do not give rapidly

• **Continuous IV inf:** dilute in 5% dextrose; use a controlled-inf device; adjust dosage and rate based on patient response

• **Patient-controlled analgesia (PCA):** a compatible patient-controlled inf device must be used; dilute solutions to obtain a conc of 1 or 10 mg/ml for ease in calculations and programming of PCA pumps; adjust dosage and rate based on patient response; consult the patient-controlled inf device manual for directions on rate of inf

Syringe compatibilities: Atropine, bupivacaine, butorphanol, cimetidine, dimenhyDRINATE, diphenhydrAMINE, droperidol, fentaNYL, glycopyrrolate, hydrOXYzine, ketamine, metoclopramide, milrinone, pentazocine, perphenazine, promazine, ranitidine, scopolamine

Y-site compatibilities: Allopurinol, amifostine, amikacin, aminophylline, amiodarone, amsacrine, atenolol, atracurium, aztreonam, bumetanide, calcium chloride, cefamandole, ceFAZolin, cefotaxime, cefoTEtan, cefOXitin, cefTAZidime, ceftizoxime, cefTRIAXone, cefuroxime, cephalothin, chloramphenicol, cisatracurium, cladribine, clindamycin, cyclophosphamide, cytarabine, dexamethasone, digoxin, diltiazem, DOBUTamine, DOPamine, doxycycline, enalaprilat, EPINEPHrine, erythromycin, esmolol, etomidate, famotidine, fentaNYL, filgrastim, fluconazole, fludarabine, foscarnet, gentamicin, granisetron, heparin, hydrocortisone, HYDROmorphone, kanamycin, labetalol, lidocaine, LORazepam, magnesium sulfate, melphalan, meropenem, methotrexate, methyldopate, methylPREDNISolone, metoclopramide, metoprolol, metroNIDAZOLE, midazolam, milrinone, nafcillin, niCARdipine, nitroglycerin, norepinephrine, ondansetron, oxacillin, oxytocin, PACLitaxel, pancuronium, penicillin G potassium, piperacillin, piperacillin/tazobactam, potassium chloride, propranolol, ranitidine, remifentanil, sodium bicarbonate, teniposide, thiotepa, ticarcillin, ticarcillin/clavulanate, tobramycin, vancomycin, vecuronium, vinorelbine, vit B/C, warfarin, zidovudine

SIDE EFFECTS

CNS: Drowsiness, dizziness, confusion, headache, sedation, euphoria, insomnia, **seizures**

CV: Palpitations, **bradycardia**, change in B/P, **shock**, **cardiac arrest**, chest pain, hypo/hypertension, edema, **tachycardia**

EENT: Blurred vision, miosis, diplopia

ENDO: Gynecomastia

M

GI: Nausea, vomiting, anorexia, constipation, cramps, biliary tract pressure
GU: Urinary retention, impotence, gonadal suppression
HEMA: Thrombocytopenia
INTEG: Rash, urticaria, bruising, flushing, diaphoresis, pruritus
RESP: Respiratory depression, respiratory arrest, apnea

PHARMACOKINETICS

PO: Onset variable, peak variable, duration variable
IM: Onset 1/2 hr, peak 50-90 min, duration 3-7 hr
SUBCUT: Onset 15-20 min, peak 50-90 min, duration 3-5 hr
IV: Peak 20 min
RECT: Peak 1/2-1 hr, duration 4-5 hr
Intrathecal: Onset rapid, duration ≤24 hr
Metabolized by liver, crosses placenta; excreted in urine, breast milk; half-life IM 3-4 hr; AVINza 24 hr; Kadian 11-13 hr

INTERACTIONS

- Unpredictable reaction, avoid use: MAOIs

Increase: effects with other CNS depressants—alcohol, opiates, sedative/hypnotics, antipsychotics, skeletal muscle relaxants
Decrease: morphine action—rifampin
Drug/Herb
Increase: CNS depression—chamomile, hops, kava, St. John's wort, valerian
Drug/Lab Test
Increase: amylase

NURSING CONSIDERATIONS

Assess:

- **Pain:** location, type, character; give dose before pain becomes severe
- Bowel status; constipation common, use stimulant laxative if needed
- I&O ratio; check for decreasing output; may indicate urinary retention; monitor serum sodium
- B/P, pulse, respirations (character, depth, rate)
- CNS changes: dizziness, drowsiness, hallucinations, euphoria, LOC, pupil reaction
- **Abrupt discontinuation:** gradually taper to prevent withdrawal symptoms; decrease by 50% q1-2days; avoid use of narcotic antagonists
- Allergic reactions: rash, urticaria

Black Box Warning: Accidental exposure: if Duramorph or Infumorph gets on skin, remove contaminated clothing, rinse affected area with water

- **Respiratory dysfunction:** depression, character, rate, rhythm; notify prescriber if respirations are <12/min; accidental overdose has occurred with high-potency oral sols
- Gradual withdrawal after long-term use

Evaluate:

- Therapeutic response; decrease in pain intensity

Teach patient/family:

- To change position slowly; orthostatic hypotension may occur
- To report any symptoms of CNS changes, allergic reactions
- That physical dependency may result from long-term use
- To avoid use of alcohol, CNS depressants
- That withdrawal symptoms may occur: nausea, vomiting, cramps, fever, faintness, anorexia

TREATMENT OF OVERDOSE:

Naloxone (Narcan) 0.2-0.8 mg IV, (caution with opioid-tolerant individuals) O_2, IV fluids, vasopressors

moxifloxacin (Rx)

Avelox, Avelox IV
Func. class.: Antiinfective
Chem. class.: Fluoroquinolone

ACTION: Interferes with conversion of intermediate DNA fragments into high-molecular-weight DNA in bacteria; DNA gyrase inhibitor

USES:
Acute bacterial sinusitis: *Streptococcus pneumoniae, Haemophilus influenzae, Moraxella catarrhalis;* acute bacterial exacerbation of chronic bronchitis: *S. pneumoniae, H. influenzae, Haemophilus parainfluenzae, Klebsiella pneumoniae, Staphylococcus aureus, M. catarrhalis;* community-acquired pneumonia: *S. pneumoniae, H. influenzae, Mycoplasma pneumoniae, Chlamydia pneumoniae, M. catarrhalis;* uncomplicated skin/skin-structure infections: *S. aureus, Streptococcus pyogenes;* complicated intraabdominal infections including polymicrobial infections: *E. coli, Bacteroides fragilis, S. anginosus, S. constellatus, Enterococcus faecalis, Proteus mirabilis, Clostridium perfringens, Bacteroides thetaiotaomicron, Peptostreptococcus* sp; complicated skin, skin-structure infections caused by methicillin-susceptible: *S. aureus, E. coli, K. pneumoniae, Enterobacter cloacae*

Unlabeled uses: Anthrax treatment/prophylaxis, gastroenteritis, MAC, nongonococcal urethritis, shigellosis, surgical infection prophylaxis, TB

CONTRAINDICATIONS:
Hypersensitivity to quinolones

Precautions: Pregnancy (C), breastfeeding, children, hepatic/cardiac/renal/GI disease, epilepsy, uncorrected hypokalemia, prolonged QT interval; patients receiving class IA, III antidysrhythmics; seizure disorder, pseudomembranous colitis, diabetes mellitus

Black Box Warning: Tendon pain, rupture; tendinitis, myasthenia gravis

DOSAGE AND ROUTES

Acute bacterial sinusitis
- **Adult: PO/IV** 400 mg q24hr × 10 days

Acute bacterial exacerbation of chronic bronchitis
- **Adult: PO/IV** 400 mg q24hr × 5 days

Community-acquired pneumonia
- **Adult: PO/IV** 400 mg q24hr × 7-14 days

Uncomplicated skin/skin-structure infections
- **Adult: PO/IV** 400 mg q24hr × 7 days

Complicated intraabdominal infections
- **Adult: IV** 400 mg/day × 5-14 days

Complicated skin, skin-structure infections
- **Adult: PO/IV** 400 mg/day × 7-21 days

Available forms: Tabs 400 mg; inj premix 400 mg/250 ml

Administer:

PO route
- 4 hr before or 8 hr after antacids, zinc, iron, calcium
- Without regard to food
- Store at room temperature

IV route
- Discontinue primary IV while administering moxifloxacin, give over 60 min
- Do not give SUBCUT, IM
- Available as premixed sol; may be diluted at ratios from 1:10 to 10:1; do not refrigerate; give by direct inf or through Y-type inf set; do not add other medications to sol or inf through same IV line at same time
- Flush line with compatible sol before and after use
- Do not admix

Solution compatibilities: 0.9% NaCl, D_5, D_{10}, LR, sterile water for inj

SIDE EFFECTS

CNS: *Headache,* dizziness, fatigue, insomnia, depression, *restlessness,* **seizures,** confusion, **increased intracranial pressure,** peripheral neuropathy, **pseudotumor cerebri,** fever

CV: **Prolonged QT interval, dysrhythmias, torsades de pointes,** tachycardia

EENT: Blurred vision, tinnitus, taste changes

GI: *Nausea, diarrhea,* increased ALT, AST, flatulence, heartburn, *vomiting,* oral candidiasis, dysphagia, **pseudomembranous colitis,** abdominal pain, dyspepsia, constipation, gastroenteritis, xerostomia

GU: Renal failure

INTEG: *Rash,* pruritus, urticaria, photosensitivity, flushing, fever, chills, injection site reactions

MISC: Candidiasis vaginitis

MS: Tremor, arthralgia, tendinitis, tendon rupture, myalgia

SYST: Anaphylaxis, Stevens-Johnson syndrome, angioedema, toxic epidermal necrolysis

PHARMACOKINETICS

Excreted in urine as active product, metabolites; parent product excreted in urine (20%), feces (25%); terminal half-life PO 12-16 hr, IV 8-15 hr

INTERACTIONS

Increase: QT prolongation—drugs that interval QT

Increase: moxifloxacin serum levels—probenecid

Increase: warfarin, cycloSPORINE effect

Increase: seizure risk—NSAIDs

Black Box Warning: Increase: tendon rupture—corticosteroids

Decrease: moxifloxacin absorption—magnesium antacids, aluminum hydroxide, zinc, iron, sucralfate, calcium, enteral feeding, didanosine

Drug/Lab Test

Increase: glucose, amylase, lipids, triglycerides, uric acid, LDH

Decrease: potassium

NURSING CONSIDERATIONS

Assess:

- CNS symptoms: headache, dizziness, fatigue, insomnia, depression, seizures
- Renal, hepatic studies: BUN, creatinine, AST, ALT, electrolytes
- I&O ratio, urine pH <5.5 is ideal

⚠ **Allergic reactions, Stevens-Johnson syndrome, toxic epidermal necrolysis, anaphylaxis:** fever, flushing, rash, urticaria, pruritus, sore throat, fatigue, ulcers, other lesions; keep EPINEPHrine, emergency equipment nearby for anaphylaxis

Black Box Warning: Tendon pain, rupture, tendinitis; if tendon becomes inflamed, product should be discontinued; more common in Achilles tendon

⚠ **Cardiac status:** prolonged QT or use of products that increase QT prolongation

⚠ **Pseudomembranous colitis:** assess for diarrhea, abdominal pain, fever, fatigue, anorexia; possible anemia, elevated WBC, low serum albumin; stop product; usually either vancomycin or IV metroNIDAZOLE given

- Increased fluids to 3 L/day to avoid crystallization in kidneys

Evaluate:

- Therapeutic response: decreased pain, C&S; absence of infection

Teach patient/family:

Black Box Warning: To notify prescriber of tendon pain, inflammation; stop drug

- Not to take any products containing magnesium or calcium (such as antacids), iron, or aluminum with this product or 4 hr before or 8 hr after
- That photosensitivity may occur; to avoid sunlight or use sunscreen to prevent burns
- To use frequent rinsing of mouth, sugarless candy or gum for dry mouth
- To take as prescribed; not to double or miss doses
- If dizziness occurs, to ambulate, perform activities with assistance
- To complete full course of product therapy
- To contact prescriber if abnormal heart rhythm or seizures occur

moxifloxacin (ophthalmic)

(mocks-ih-floks′a-sin)

Vigamox, Moxeza

Func. class.: Ophthalmic anti-infective

Chem. class.: Fluoroquinolone

Do not confuse: ciprofloxacin/gatifloxacin/levofloxacin

ACTION: Inhibits DNA gyrase, thereby decreasing bacterial replication

USES: Bacterial conjunctivitis (aerobic gram-positive/negative organisms), *Chlamydia trachomatis*

CONTRAINDICATIONS: Hypersensitivity to this product or fluoroquinolones

Precautions: Pregnancy (C), breastfeeding

DOSAGE AND ROUTES

Bacterial conjunctivitis

- **Adult/adolescent/child ≥1 yr: ophthalmic SOL** 1 drop in affected eye(s) bid (Moxeza) or tid (Vigamox) × 7 days

Available forms: Ophthalmic solution 0.5%

Administer:

Ophthalmic route

- Commercially available ophthalmic solutions are not for injection subconjunctivally or into the anterior chamber of the eye
- Apply topically to the eye, taking care to avoid contamination
- Do not touch the tip of the dropper to the eye, fingertips, or other surface
- Apply pressure to lacrimal sac for 1 min after instillation
- Avoid wearing contact lenses during treatment

SIDE EFFECTS

EENT: Hypersensitivity, pruritus, blurred vision, tearing

PHARMACOKINETICS

Half-life 13 hr

NURSING CONSIDERATIONS

Assess:

- **Allergic reaction:** assess for hypersensitivity, discontinue product

Evaluate:

- Decreased ophthalmic infection

Teach patient/family:

Ophthalmic route

- To apply topically to the eye, taking care to avoid contamination
- That product is for ophthalmic use only
- Not to touch the tip of the dropper to the eye, fingertips, or other surface
- To apply pressure to lacrimal sac for 1 min after installation
- To avoid wearing contact lenses during treatment

mupirocin (topical, nasal)

(myoo-pihr′oh-sin)

Bactroban, Centany

Func. class.: Topical anti-infective

ACTION: Antibacterial activity results from inhibition of protein synthesis; bacteriostatic at low conc, bactericidal at high conc

USES: Impetigo, skin lesions *(Staphylococcus aureus/Streptococcus pyogenes);* nasal: methicillin-resistant *S. aureus*

CONTRAINDICATIONS: Hypersensitivity to this product

Precautions: Open wounds, burns, severe kidney disease, children, pregnancy (B), breastfeeding

DOSAGE AND ROUTES

Impetigo

- **Adult/child: TOP** apply to affected area tid × 1-2 wk

Skin lesions
• **Adult/child: TOP** apply to affected area tid × 10 days
Methicillin-resistant *S. Aureus* in the nose
• **Adult/child ≥12 yr: NASAL** divide ointment in single use tube in half; use in each nostril bid × 5 days
Available forms: Topical cream, ointment 2%; intranasal ointment 2%
Administer:
Topical route
• Do not use skin products near the eyes, nose, or mouth
• Wash hands before and after use; wash affected area and gently pat dry
• May cover treated areas with gauze dressing
Cream/ointment
• Apply a thin film to the cleansed affected area; massage gently into affected areas, do not use near eyes, mouth
• **Nasal:** Close nostrils by squeezing and releasing and gently massaging over 1 min

SIDE EFFECTS

CNS: Headache
EENT: Burning, pharyngitis, rhinitis (nasal)
GI: Taste change, nausea
INTEG: Burning, rash, pruritus

INTERACTIONS

Decrease: Effect of other nasal products

NURSING CONSIDERATIONS

Assess:
• **Allergic reaction:** assess for hypersensitivity; product may need to be discontinued
• **Infection:** assess for number of lesions, severity in impetigo, other skin disorders
Evaluate:
• Decreased lesions in impetigo, other skin disorders

Teach patient/family:
Topical route
• Not to use skin products near the eyes, nose, or mouth
• To wash hands before and after use and to wash affected area and gently pat dry
• **Cream/Ointment:** to apply a thin film to the cleansed affected area; to cover treated areas with gauze dressing if desired
• **Nasal:** to close nostrils by squeezing and releasing and gently massaging over 1 min

mycophenolate mofetil (Rx)

(mye-koe-phen′oh-late)
CellCept, Myfortic
Func. class.: Immunosuppressant

ACTION: Inhibits inflammatory responses that are mediated by the immune system

USES: Prophylaxis for organ rejection in allogenic cardiac, hepatic, renal transplants
Unlabeled uses: Refractory uveitis, second-line therapy for Churg-Strauss syndrome, diffuse proliferative lupus nephritis (in combination), rheumatoid arthritis, psoriasis, GVHD, kidney disease, myasthenia gravis, atopic dermatitis

CONTRAINDICATIONS: Hypersensitivity to this product or mycophenolic acid

Black Box Warning: Pregnancy (D)

Precautions: Breastfeeding, lymphomas, neutropenia, renal disease, accidental exposure, anemia

Black Box Warning: Infection, neoplastic disease

DOSAGE AND ROUTES

Renal transplant

To prevent organ rejection

- **Adult: PO** mycophenolate mofetil 1 g or 720 mg mycophenolate sodium; 1 g or 720 mg bid given to renal transplant patients in combination with corticosteroids, cycloSPORINE
- **Child: PO-ER** 400 mg/m² bid, max 720 mg bid

Renal dose

- **Adult: PO/IV** GFR <25 ml/min, max 2 g/day

Cardiac transplant

To prevent organ rejection

- **Adult: PO/IV** 1.5 g bid, **IV** can be started ≤24 hr after transplant, switch to **PO** when able

Hepatic transplant

To prevent organ rejection

- **Adult: PO** 1.5 g bid; **IV** 1 g over ≥2 hr

Refractory acute kidney transplant rejection (unlabeled)

- **Adult: PO** 1.5 g (Mofetil) bid

Rheumatoid arthritis (unlabeled)

- **Adult: PO** 250 mg-2 g/day (Mofetil)

GVHD (unlabeled)

- **Adult: PO** 2 g/day (Mofetil) with cycloSPORINE and prednisoLONE

Diffuse proliferative lupus nephritis (unlabeled)

- **Adult: PO** 1 g/day (Mofetil)

Uveitis (unlabeled)

- **Adult: PO** 1 g (Mofetil) bid × 6-41 mo

Atopic dermatitis (unlabeled)

- **Adult: PO** 1 g (Mofetil) bid × 4 wk, then 500 mg bid × 4 wk
- **Adolescent/child ≥2 yr: PO** 30-50 mg/kg/day (Mofetil) in 2 divided doses

Available forms: Caps 250 mg; tabs 500 mg; inj (powder) 500 mg/20-ml vial; powder for oral susp 200 mg/ml; ext rel tab (Myfortic) 180, 360 mg

Administer:

- May be given in combination with corticosteroids, cycloSPORINE
- Del rel tab, cap, oral susp, tab are not interchangeable

PO route

- Do not break, crush, or chew tabs; do not open caps
- Give at same time each day
- Avoid inhalation or direct contact with skin, mucous membranes; teratogenic in animals
- **Oral susp:** tap closed bottle several times to loosen powder; use 94 ml of water in graduated cylinder; add ½ total amount of water for constitution and shake the closed bottle; add remaining water and shake again; remove child-resistant cap; push adapter into neck of bottle; close tightly
- Give alone for better absorption

Intermittent IV INF route

- Reconstitute each vial with 14 ml D_5W; shake gently; further dilute to 6 mg/ml; dilute 1 g/140 ml D_5W, 1.5 g/210 ml D_5W; give by slow IV inf ≥2 hr; never give by bolus or rapid IV inj
- Do not give with other medications or sol

Y-site compatibilities: Alemtuzumab, alfentanil, amikacin, anidulafungin, argatroban, bivalirudin, caspofungin, cefepime, DAPTOmycin, DOPamine, norepinephrine, octreotide, oxytocin, tacrolimus, tigecycline, tirofiban, vancomycin

SIDE EFFECTS

CNS: *Tremor, dizziness, insomnia, headache, fever,* anxiety, pain, **progressive multifocal leukoencephalopathy,** asthenia, paresthesia

CV: *Hypertension, chest pain,* hypotension, edema

GI: *Diarrhea, constipation, nausea, vomiting,* stomatitis, **GI bleeding,** abdominal pain, anorexia, dyspepsia

GU: *UTI, hematuria,* **renal tubular necrosis, polyomavirus-associated nephropathy**

HEMA: **Leukopenia, thrombocytopenia, anemia, pancytopenia, pure red cell aplasia,** neutropenia

INTEG: *Rash*

META: *Peripheral edema, hypercholesterolemia, hypophosphatemia, edema, hyperkalemia, hypokalemia, hyperglycemia,* hypocalcemia, hypomagnesemia
MS: Arthralgia, muscle wasting, back pain, weakness
RESP: *Dyspnea, respiratory infection, increased cough, pharyngitis, bronchitis, pneumonia,* plural effusion, pulmonary fibrosis
SYST: Lymphoma, *nonmelanoma skin carcinoma,* sepsis

PHARMACOKINETICS

Rapidly and completely absorbed; metabolized to active metabolite (MPA); excreted in urine, feces; protein binding (MPA) 97%; half-life (MPA) 17.9 hr

INTERACTIONS

Increase: bone marrow suppression—azaTHIOprine
• Increased bleeding risk: anticoagulants, NSAIDs, thrombolytics, salicylates
Increase: toxicity—acyclovir, ganciclovir, valacyclovir
Increase: effects of phenytoin, theophylline
Increase: mycophenolate levels—probenecid, immunosuppressives, salicylates
Decrease: mycophenolate levels—antacids (magnesium, aluminum), cholestyramine, cycloSPORINE, rifamycin
Decrease: protein binding of phenytoin, theophylline
Decrease: effect of live attenuated vaccines, oral contraceptives

Drug/Herb
Interference with immunosuppression: astragalus, echinacea, melatonin

Drug/Food
Decrease: absorption if taken with food

Drug/Lab Test
Increase: serum creatitine, BUN
• Abnormal LFTs

NURSING CONSIDERATIONS

Assess:
⚠ **Progressive multifocal leukoencephalopathy,** may be fatal; ataxia, confusion, apathy, hemiparesis, visual problems, weakness; side effects should be reported to FDA
• Blood studies: CBC during treatment monthly
• Hepatic studies: alk phos, AST, ALT, bilirubin
• Renal studies: BUN, CCr, electrolytes

Black Box Warning: Pregnancy test within 1 wk before initiation of treatment; confirm negative pregnancy test

Evaluate:
• Therapeutic response: absence of graft rejection

Teach patient/family:
• To report fever, rash, severe diarrhea, chills, sore throat, fatigue because serious infections may occur
• To reduce risk of infection by avoiding crowds
• About the need for repeated lab tests

Black Box Warning: To use 2 forms of contraception before, during, and for 6 wk after therapy

• To take at same time each day

nabumetone (Rx)

(na-byoo′me-tone)

Relafen

Func. class.: Nonsteroidal antiinflammatory

Chem. class.: Acetic acid derivative

ACTION: Metabolite inhibits COX-1, COX-2 by blocking arachidonate; analgesic, antiinflammatory, antipyretic

USES: Osteoarthritis, rheumatoid arthritis, acute or chronic treatment

CONTRAINDICATIONS: Hypersensitivity to this product or aspirin, NSAIDs

Black Box Warning: Perioperative pain with CABG surgery

Precautions: Pregnancy (C), breastfeeding, children, geriatric patients, bleeding/GI/cardiac/renal disorders, hepatic dysfunction, asthma, bone marrow suppression, lupus (SLE), ulcerative colitis, blood dyscrasias

Black Box Warning: MI, stroke, GI bleeding

DOSAGE AND ROUTES

- **Adult: PO** 1 g as single dose or divided bid; max 2 g/day if needed

Renal dose

- **Adult: PO** CCr 31-49 ml/min, 750 mg daily, max 1500 mg/day; CCr <30 ml/min 500 mg daily, max 1000 mg/day

Available forms: Tabs 500, 750 mg

Administer:

- With food or antacids
- Store at room temperature

SIDE EFFECTS

CNS: Dizziness, headache, drowsiness, fatigue, tremors, confusion, insomnia, anxiety, depression, nervousness

EENT: Tinnitus

CV: Tachycardia, peripheral edema, palpitations, dysrhythmias, **CHF, MI, stroke**

GI: Nausea, anorexia, vomiting, diarrhea, jaundice, **cholestatic hepatitis,** constipation, flatulence, cramps, dry mouth, peptic ulcer, gastritis, **ulceration, perforation, bleeding**

GU: **Nephrotoxicity, dysuria, hematuria, azotemia**

HEMA: **Thrombocytopenia**

INTEG: Purpura, rash, pruritus, sweating, photosensitivity

RESP: Dyspnea, pharyngitis, **bronchospasm**

SYST: **Anaphylaxis, angioedema, Stevens-Johnson syndrome**

PHARMACOKINETICS

PO: Peak $2^1/_2$-4 hr; protein binding >99%; half-life 24 hr; metabolized in liver to active metabolite; excreted in urine (metabolites), breast milk

INTERACTIONS

Increase: bleeding risk—anticoagulants, thrombolytics, clopidogrel, eptifibatide, ticlopidine, SSRIs, SNRIs

Increase: hematologic toxicity—antineoplastics

Increase: GI reactions—salicylates, NSAIDs, alcohol, corticosteroids

Increase: toxicity—lithium, methotrexate, cidofovir

Decrease: effect of diuretics, antihypertensives

Drug/Lab Test

Increase: bleeding time, potassium, BUN, AST, ALT, LDH, alk phos, creatinine

Decrease: CCr, blood glucose, Hct, Hgb

NURSING CONSIDERATIONS

Assess:

Black Box Warning: Cardiac status: CV thrombotic events, MI, stroke; may be fatal; not to be used in perioperative pain after CABG

Black Box Warning: GI status: ulceration, bleeding, perforation; may be fatal

- **Pain:** frequency, intensity, characteristics; relief of pain after taking product

N

- Asthma, aspirin sensitivity, or nasal polyps; increased hypersensitivity reactions
- Renal, hepatic studies: BUN, creatinine, AST, ALT, Hgb, LDH, blood glucose, WBC, platelets; CCr before treatment, periodically thereafter
- Audiometric, ophthalmic exam before, during, after treatment
- For eye, ear problems: blurred vision, tinnitus; may indicate toxicity

Evaluate:

- Therapeutic response: decreased pain and stiffness in joints

Teach patient/family:

- To avoid alcoholic beverages and aspirin
- ⚠ To report blurred vision, ringing, roaring in ears; may indicate toxicity
- To avoid driving, other hazardous activities if dizziness, drowsiness occur
- ⚠ To report change in urine pattern, increased weight, edema, increased pain in joints, fever, blood in urine; indicates nephrotoxicity
- That therapeutic effects may take up to 1 mo
- To take with a full glass of water and food or antacids to reduce GI upset
- To report dark stools; may indicate GI bleeding
- To use sunscreen, wear protective clothing if in the sun
- To report use to all health care providers, to avoid all other products unless approved by prescriber

nadolol (Rx)

(nay-doe′lole)

Corgard

Func. class.: Antihypertensive, antianginal

Chem. class.: β-Adrenergic receptor blocker

Do not confuse:

Corgard/Cognex/Coreg

Nadolol/Mandol

ACTION:

Long-acting, nonselective β-adrenergic receptor blocking agent, blocks β_1 in the heart and β_2 in the lungs, uterus, and circulatory system; mechanism is similar to that of propranolol

USES:

Chronic stable angina pectoris, mild to moderate hypertension

Unlabeled uses: Tachydysrhythmias, anxiety, tremors, esophageal varices (rebleeding only), prophylaxis of migraine headaches, portal hypertension, atrial fibrillation

CONTRAINDICATIONS:

Hypersensitivity to this product, cardiac failure, cardiogenic shock, 2nd/3rd degree heart block, bronchospastic disease, sinus bradycardia, CHF, COPD

Precautions: Pregnancy (C), breastfeeding, diabetes mellitus, renal disease, hyperthyroidism, peripheral vascular disease, myasthenia gravis, major surgery, nonallergic bronchospasm

Black Box Warning: Abrupt discontinuation

DOSAGE AND ROUTES

- **Adult: PO** 40 mg/day, increase by 40-80 mg q2-14days; maintenance 40-240 mg/day for angina, 40-320 mg/day for hypertension
- **Geriatric: PO** 20 mg/day, may increase by 20 mg until desired dose

Renal dose

- **Adult: PO** CCr 31-50 ml/min, give q24-36hr; CCr 10-30 ml/min, give q24-48hr; CCr <10 ml/min, give q40-60hr

Migraine prevention (unlabeled)

- **Adult: PO** 40-240 mg/day × 2-18 mo

Available forms: Tabs 20, 40, 80 mg

Administer:

- With 8 oz water, check apical pulse before use, if <60 bpm, withhold; call prescriber
- Give without regard to food

SIDE EFFECTS

CNS: Depression, *dizziness, fatigue,* lethargy, paresthesias, headache, *weakness,* insomnia, memory loss, nightmares

CV: *Bradycardia, hypotension,* **CHF,** palpitations, **AV block,** chest pain, peripheral ischemia, flushing, edema, vasodilation, conduction disturbances

EENT: Blurred vision, dry eyes, nasal congestion

ENDO: Hyperglycemia, hypoglycemia

GI: Nausea, vomiting, diarrhea, colitis, constipation, cramps, dry mouth, flatulence, hepatomegaly, **pancreatitis,** taste distortion

GU: *Impotence,* decreased libido

HEMA: **Agranulocytosis, thrombocytopenia**

INTEG: Rash, pruritus, fever, alopecia

RESP: Dyspnea, respiratory dysfunction, **bronchospasm,** cough, wheezing, **pulmonary edema,** pharyngitis, **laryngospasm**

PHARMACOKINETICS

PO: Onset variable, peak 3-4 hr, duration 10-24 hr; half-life 20-24 hr; not metabolized; excreted in urine (unchanged), bile, breast milk; protein binding 30%

INTERACTIONS

- **Do not use with MAOIs; bradycardia may occur**
- Peripheral ischemia: ergots

Increase: bradycardia—digoxin

Increase: hypotension, bradycardia—cloNIDine, EPINEPHrine

Increase: hypotensive effects—other hypotensive agents, phenothiazines

Decrease: β-blocking effect—thyroid hormones

Decrease: antihypertensive effect—NSAIDs

Drug/Lab Test

Increase: serum potassium, serum uric acid, ALT, AST, alk phos, LDH, blood glucose, cholesterol, ANA, triglycerides

NURSING CONSIDERATIONS

Assess:

- B/P, pulse, respirations during beginning therapy; orthostatic hypotension may occur
- Weight daily; report gain of 5 lb
- I&O ratio, CCr if kidney damage diagnosed; crackles, jugular venous distention, fatigue, dyspnea
- **Pain:** duration, time started, activity being performed, character
- Headache, lightheadedness, decreased B/P; may indicate a need for decreased dosage

Black Box Warning: Abrupt discontinuation: can result in MI, myocardial ischemia, ventricular dysrhythmias, severe hypertension; withdraw slowly by tapering

Evaluate:

- Therapeutic response: decreased B/P, heart rate, symptoms of angina

Teach patient/family:

- That product may mask signs of hypoglycemia or alter blood glucose in diabetics
- To avoid OTC products unless prescriber approves
- To avoid hazardous activities if dizziness occurs
- **Hypertension:** to comply with complete medical regimen; to report weight gain of >5 lb, swelling, unusual bruising, bleeding
- To rise slowly to prevent orthostatic hypotension
- About how and when to check B/P, pulse; to hold dose, contact prescriber if pulse ≤60 bpm, systolic B/P <90 mm Hg; to take missed dose as soon as possible if less than 8 hr

Black Box Warning: Not to discontinue abruptly; may cause life-threatening cardiac changes

N

nafcillin (Rx)

(naf-sill'in)

Func. class.: Antiinfective, broad-spectrum

Chem. class.: Penicillinase-resistant penicillin

ACTION: Bacteriocidal, interferes with cell wall replication of susceptible organisms; cell lysis mediated by cell wall autolytic enzymes

USES: Effective for gram-positive cocci *(Staphylococcus aureus, Streptococcus viridans, Streptococcus pneumoniae)*, infections caused by penicillinase-producing *Staphylococcus*

CONTRAINDICATIONS: Hypersensitivity to penicillins or corn

Precautions: Pregnancy (B), breastfeeding, neonates; hypersensitivity to cephalosporins or carbapenems; GI disease, asthma, electrolyte imbalances, hepatic/renal disease, pseudomembranous colitis

DOSAGE AND ROUTES

- **Adult: IV** 500-2000 mg q4hr; IM 500 mg q6-8hr, max 12 g/day
- **Child and infant >1 mo: IV** 150-200 mg/kg/day in divided doses
- **Neonates >7 days (weight >2000 g): IV** 25 mg/kg q6hr
- **Neonates ≤7 days (weight <2000 g): IV** 25 mg/kg q8hr

Available forms: Powder for inj 1, 2, 10 g, premixed or Add-Vantage vials (1-2 g)

Administer:

- Product after C&S has been drawn, begin therapy while waiting for results

IM route

- Reconstitute vials: add 1.7 (1.8 nafcil), 3, 4, or 6.4 ml (6.6 ml NaCl) (sterile water for inj, 0.9% NaCl, bacteriostatic water for inj with benzyl alcohol or parabens) to vials with 1 g, 2 g of nafcillin, respectively (250 mg/ml)
- No further dilution needed; after reconstitution, inject in deep muscle mass

IV route

- Reconstitute vials: add 1.7 (1.8 nafcil), 3, 4, or 6.4 ml (6.6 ml NaCl) sterile water for inj, 0.9% NaCl, bacteriostatic water for inj with benzyl alcohol or parabens) to vials with 1 g, 2 g of nafcillin, respectively (250 mg/ml); pharmacy bulk pack reconstitute 10 g/93 ml sterile water inj or 0.9% NaCl (100 mg/ml)
- **Nallpen piggyback units:** reconstitute 1 or 2 g with 50-100 ml or 99 ml, respectively, of sterile water for inj, 0.45% NaCl, 0.9% NaCl
- **Unipen piggyback units:** reconstitute according to manufacturer

Direct Intermittent IV INJ route

- Further dilute reconstituted sol in 15-30 ml sterile water inj, 0.45% NaCl, 0.9% NaCl; inj slowly over 5-10 min into tubing of free-flowing compatible IV solution

Intermittent IV INF route

- Vials, further dilute reconstituted sol to 2-40 mg/ml for peripheral vein inf ≤20 mg/ml (preferred); piggyback unit, no further dilution needed; infuse ≥30-60 min, make sure entire dose is given before ≥10% of sol is inactivated
- Extravasation management with cold packs, hyaluronidase

Y-site compatibilities: Acyclovir, alfentanil, amikacin, aminophylline, amphotericin B lipid complex (Abelcet), anidulafungin, argatroban, ascorbic acid injection, atenolol, atracurium, atropine, aztreonam, benztropine, bivalirudin, bleomycin, bretylium, bumetanide, buprenorphine, butorphanol, calcium chloride/gluconate, CARBOplatin, carmustine, cefamandole, ceFAZolin, cefoperazone, cefotaxime, cefoTEtan, cefOXitin, cefTAZidime, ceftizoxime, cefTRIAXone, cefuroxime, chlorproMAZINE, cimetidine, CISplatin, clindamycin, cyanocobalamin, cyclophosphamide, cycloSPORINE, DACTINomycin, DAPTOmycin, DAUNOrubicin liposome, dexamethasone, digoxin, DOBUTamine, DOCEtaxel, DOPamine, DOXOrubicin liposomal, enalaprilat, EPHEDrine, EPINEPHrine, epoetin alfa, erythromycin, etoposide, etoposide phosphate, famotidine, fenoldopam, fentaNYL, fluconazole, fludarabine, foscarnet, furosemide, gallium, ganciclovir, gatifloxacin, gemtuzumab, gentamicin, glycopyrrolate, granisetron, heparin, hydrocortisone, HYDROmorphone, imipenem-cilastatin, indomethacin, isoproterenol, ketorolac, lactated Ringer's, lepirudin, leucovorin, lidocaine, linezolid injection, LORazepam, magnesium sulfate, mannitol, methyldopate, methylPREDNISolone, metoclopramide, metoprolol, metroNIDAZOLE, milrinone, morphine, multiple vitamins injection, naloxone, niCARdipine, nitroglycerin, nitroprusside, norepinephrine, octreotide, ondansetron, oxacillin, oxaliplatin, oxytocin, PACLitaxel (solvent/surfactant), pamidronate, pancuronium, pantoprazole, PEMEtrexed, penicillin G potassium/sodium, PENTobarbital, perphenazine, PHENobarbital, phentolamine, phenylephrine, phytonadione, piperacillin, polymyxin B, potassium acetate/chloride, procainamide, prochlorperazine, propofol, propranolol, ranitidine, Ringer's injection, sodium bicarbonate, SUFentanil, tacrolimus, teniposide, theophylline, thiamine, thiotepa, ticarcillin, ticarcillin-clavulanate, tigecycline, tirofiban, TNA (3-in-1), tobramycin, tolazoline, TPN (2-in-1), urokinase, vasopressin, vinBLAStine, voriconazole, zidovudine, zoledronic acid

SIDE EFFECTS

CNS: Lethargy, hallucinations, anxiety, depression, twitching, **coma, seizures**

GI: *Nausea, vomiting, diarrhea,* increased AST, ALT, abdominal pain, glossitis, **pseudomembranous colitis**

GU: Oliguria, **proteinuria, hematuria,** vaginitis, moniliasis, **glomerulonephritis, interstitial nephritis**

HEMA: Anemia, increased bleeding time, **bone marrow depression, neutropenia, agranulocytosis**

INTEG: Tissue necrosis, extravasation injury at inj site, **rash, pruritus, exfoliative dermatitis**

SYST: **Anaphylaxis, serum sickness, Stevens-Johnson syndrome**

PHARMACOKINETICS

Half-life 30-90 min; metabolized by liver; excreted in bile, urine; 70%-90% protein bound; peak 30-120 min (PO); peak 30-60 min (IM)

INTERACTIONS

- Avoid use with tetracyclines

Increase: nafcillin concs—probenecid

Decrease: effect of cycloSPORINE—warfarin

Decrease: nafcillin effect—chloramphenicol, macrolides, sulfonamides, tetracyclines, aminoglycosides

Drug/Food

Decrease: absorption—food, carbonated drinks, citrus fruit juices

Decrease: Hgb/HcT, neutrophils

Drug/Lab Test

False positive: urine glucose, urine protein

Decrease: potassium

N

NURSING CONSIDERATIONS

Assess:

- I&O ratio; report hematuria, oliguria, high doses are nephrotoxic
- **Pseudomembranous colitis: assess for diarrhea, abdominal pain, fever, fatigue, anorexia; possible anemia, elevated WBC, low serum albumin; stop product; usually either vancomycin or IV metroNIDAZOLE given**
- Hepatic studies: AST, ALT
- Blood studies: CBC with differential bleeding time, electrolytes
- ⚠ **Renal studies: urinalysis, BUN, creatinine; abnormal urinalysis may indicate nephrotoxicity**
- C&S before product therapy; product may be given as soon as culture is taken
- Respiratory status: rate, character, wheezing, tightness in chest

⚠ Allergies before initiation of treatment; monitor for anaphylaxis, dyspnea, rash, laryngeal edema; stop product; keep emergency equipment nearby; skin eruptions after administration of penicillin to 1 wk after discontinuing product; cross-sensitivity with cephalosporins may occur
- Differential WBC 2 × per wk in patients receiving long-term therapy
- IV site: for redness, swelling, pain at site

Evaluate:
- Therapeutic response: absence of fever, draining wounds

Teach patient/family:

⚠ To report sore throat, fever, fatigue (may indicate superinfection); CNS reactions; pseudomembraneous colitis (diarrhea, fever, abdominal pain, fatigue)
- To wear or carry emergency ID if allergic to penicillins
- To avoid use with other products unless approved by prescriber

TREATMENT OF ANAPHYLAXIS: Withdraw product; maintain airway; administer EPINEPHrine, aminophylline, O_2, IV corticosteroids

⚠ HIGH ALERT

nalbuphine (Rx)

(nal′byoo-feen)

Func. class.: Opioid analgesic

Chem. class.: Synthetic opioid agonist, antagonist

ACTION: Depresses pain impulse transmission at the spinal cord level by interacting with opioid receptors

USES: Moderate to severe pain, supplement to anesthesia

CONTRAINDICATIONS: Hypersensitivity to this product or parabens, addiction (opiate)

Precautions: Pregnancy (B), breastfeeding, addictive personality, increased intracranial pressure, MI (acute), severe heart disease, respiratory depression, renal/hepatic disease, bowel impaction, abrupt discontinuation

DOSAGE AND ROUTES

Analgesic
- **Adult: SUBCUT/IM/IV** 10 mg q3-6hr prn (based on 70-kg body weight), max 160 mg/day (IV/IM/SUBCUT); max 20 mg/dose if opiate naïve (IV/IM/SUBCUT)

Balanced anesthesia adjunct
- **Adult: IV** 0.3-3 mg/kg given over 10-15 min; may give 0.25-0.5 mg/kg as needed for maintenance

Available forms: Inj 10, 20 mg/ml

Administer:
- With antiemetic if nausea, vomiting occur
- When pain beginning to return; determine dosage interval by response
- Store in light-resistant area at room temperature

IM route
- IM deep in large muscle mass, rotate inj sites, protect from light

Direct IV route
- Undiluted ≤10 mg over 3-5 min into free-flowing IV line of D_5W, NS, or LR

Syringe compatibilities: Atropine, cimetidine, diphenhydrAMINE, droperidol, glycopyrrolate, hydrOXYzine, lidocaine, midazolam, prochlorperazine, ranitidine, scopolamine, trimethobenzamide

Y-site compatibilities: Amifostine, aztreonam, cisatracurium, cladribine, filgrastim, fludarabine, granisetron, melphalan, PACLitaxel, propofol, remifentanil, teniposide, thiotepa, vinorelbine

SIDE EFFECTS

CNS: *Drowsiness, dizziness, confusion, headache, sedation, euphoria,* dysphoria (high doses), hallucinations, dreaming, tolerance, physical, psychologic dependency

CV: Bradycardia, change in B/P, cardiac arrest

EENT: Blurred vision, miosis, diplopia

GI: *Nausea, vomiting, anorexia, constipation, cramps,* abdominal pain, dyspepsia, xerostomia, bitter taste
GU: Urinary urgency
INTEG: *Rash,* urticaria, flushing, diaphoresis, pruritus
RESP: Respiratory depression/arrest, pulmonary edema

PHARMACOKINETICS

SUBCUT/IM/IV: Peak 30 min, onset 2-15 min, IV 2-3 min, duration 3-6 hr, metabolized by liver, excreted by kidneys, half-life 3-6 hr

INTERACTIONS

Increase: effects with other CNS depressants—alcohol, opiates, sedative/hypnotics, antipsychotics, skeletal muscle relaxants

NURSING CONSIDERATIONS

Assess:

- **Pain:** type, location, intensity before and 30-60 min after administration; titrate upward with 25%-50% until 50% of pain reduced; need for pain medication by pain sedation scoring, physical dependency
- Bowel status; constipation is common; may need laxative or stool softener

⚠ **Withdrawal reactions** in opiate-dependent individuals: PE, vascular occlusion; abscesses, ulcerations, nausea, vomiting, seizures; low potential for dependence

- **CNS changes:** dizziness, drowsiness, hallucinations, euphoria, LOC, pupil reaction
- Allergic reactions: rash, urticaria
- **Respiratory dysfunction: respiratory depression,** character, rate, rhythm; notify prescriber if respirations are <10/min

Evaluate:

- Therapeutic response: decrease in pain without respiratory depression

Teach patient/family:

- To report any symptoms of CNS changes, allergic reactions
- That physical dependency may result from long-term use; low potential for dependency
- That withdrawal symptoms may occur: nausea, vomiting, cramps, fever, faintness, anorexia, profuse sweating, twitching; without treatment symptoms resolve in 5-14 days, chronic abstinence syndrome may last 2-6 mo
- To avoid CNS depressants, alcohol
- To avoid driving, operating machinery if drowsiness occurs

TREATMENT OF OVERDOSE:

Naloxone (Narcan) 0.2-0.8 mg IV, O_2, IV fluids, vasopressors

naloxone (Rx)

(nal-oks′one)

Func. class.: Opioid antagonist, antidote

Chem. class.: Thebaine derivative

Do not confuse:
naloxone/naltrexone

ACTION:

Competes with opioids at opiate receptor sites

USES:

Respiratory depression induced by opioids, opiate agonist overdose

Unlabeled uses: IBS, opiate agonist dependence, opiate agonist-induced constipation, pruritus, urinary retention, coma, nausea, vomiting

CONTRAINDICATIONS:

Hypersensitivity

Precautions: Pregnancy (B), breastfeeding, children, neonates, CV disease, opioid dependency, seizure disorder, drug dependency, hepatic disease

DOSAGE AND ROUTES

Opioid-induced respiratory depression (known for suspected opiate agonist overdose)

- **Adult: IV/SUBCUT/IM** 0.4-2 mg, repeat q2-3min if needed, max 10 mg; **IV**

N

inf loading dose 0.005 mg/kg, then 0.0025 mg/kg/hr

• **Child <5 yr or ≤20 kg: IV/INTRAOSSEOUS** 0.01 mg/kg slowly followed by 0.1 mg/kg if needed; IV inf (PALS) 0.04-0.16 mg/kg/hr, titrate

Postoperative opioid-induced respiratory depression

• **Adult: IV** 0.1-0.2 mg q2-3min prn

• **Child: IV** 0.005-0.01 mg/kg q2-3min prn

Diagnosis of opiate-agonist dependence (unlabeled)

• **Adult: IM** 0.16 mg; if no withdrawal symptoms after 20-30 min, give 0.24 mg **IV**

Nausea/vomiting from continuous morphine inf/urinary retention (unlabeled)

• **Adult: IV** 0.2 mg

Available forms: Inj 0.02, 0.4, 1 mg/ml

Administer:

• Only with resuscitative equipment, O_2 nearby

• Only sol prepared within 24 hr

• Dark storage at room temperature

Direct IV route

• Undiluted with sterile water for inj; give ≤0.4 mg over 15 sec

Continuous IV INF route

• Dilute 2 mg/500 ml 0.9% NaCl or D_5W (4 mcg/ml), titrate to response

Y-site compatibilities: Acyclovir, alfentanil, amikacin, aminocaproic acid, aminophylline, anidulafungin, ascorbic acid, atenolol, atracurium, atropine, azaTHIOprine, aztreonam, benztropine, bivalirudin, bleomycin, bumetanide, buprenorphine, butorphanol, calcium chloride/gluconate, CARBOplatin, caspofungin, cefamandole, ceFAZolin, cefmetazole, cefonicid, cefoperazone, cefotaxime, cefoTEtan, cefOXitin, cefTAZidime, ceftizoxime, cefTRIAXone, cefuroxime, cephalothin, cephapirin, chloramphenicol, chlorproMAZINE, cimetidine, CISplatin, clindamycin, cyanocobalamin, cyclophosphamide, cycloSPORINE, cytarabine, DACTINomycin, DAPTOmycin, dexamethasone, digoxin, diltiazem, diphenhydrAMINE, DOBUTamine, DOCEtaxel, DOPamine, doxacurium, DOXOrubicin, doxycycline, enalaprilat, ePHEDrine, EPINEPHrine, epirubicin, epoetin alfa, eptifibatide, ertapenem, erythromycin, esmolol, etoposide, etoposide phosphate, famotidine, fenoldopam, fentaNYL, fluconazole, fludarabine, fluorouracil, folic acid, furosemide, ganciclovir, gatifloxacin, gemcitabine, gentamicin, glycopyrrolate, granisetron, heparin, hydrocortisone, hydrOXYzine, IDArubicin, ifosfamide, imipenem-cilastatin, inamrinone, indomethacin, insulin (regular), irinotecan, isoproterenol, ketorolac, labetalol, levofloxacin, lidocaine, linezolid, LORazepam, mannitol, mechlorethamine, meperidine, metaraminol, methicillin, methotrexate, methoxamine, methyldopate, methylPREDNISolone, metoclopramide, metoprolol, metroNIDAZOLE, mezlocillin, miconazole, midazolam, milrinone, minocycline, mitoXANtrone, morphine, moxalactam, multiple vitamins, mycophenolate, nafcillin, nalbuphine, nesiritide, netilmicin, nitroglycerin, nitroprusside, norepinephrine, octreotide, ondansetron, oxacillin, oxaliplatin, oxytocin, PACLitaxel, palonosetron, pamidronate, pancuronium, papaverine, PEMEtrexed, penicillin G potassium/sodium, pentamidine, pentazocine, PENTobarbital, PHENobarbital, phentolamine, phenylephrine, phytonadione, piperacillin, piperacillin-tazobactam, polymyxin B, potassium chloride, procainamide, prochlorperazine, promethazine, propofol, propranolol, protamine, pyridoxine, quiNIDine, quinupristin-dalfopristin, ranitidine, ritodrine, rocuronium, sodium acetate/bicarbonate, succinylcholine, SUFentanil, tacrolimus, teniposide, theophylline, thiamine, ticarcillin, ticarcillin-clavulanate, tigecycline, tirofiban, tobramycin, tolazoline, trimetaphan, urokinase, vancomycin, vasopressin, vecuronium, verapamil, vinCRIStine, vinorelbine, voriconazole, zoledronic acid

SIDE EFFECTS

CNS: Nervousness, seizures, tremor, opioid withdrawal symptoms
CV: Rapid pulse, increased systolic B/P (high doses), ventricular tachycardia, fibrillation, hypo/hypertension, cardiac arrest, sinus tachycardia
GI: Nausea, vomiting
RESP: Pulmonary edema, dyspnea

PHARMACOKINETICS

Well absorbed IM, SUBCUT; metabolized by liver, crosses placenta; excreted in urine, breast milk; half-life 30-81 min
IM/SUBCUT: Onset 2-5 min, duration 45-60 min
IV: Onset 1 min, duration 45 min

INTERACTIONS

Increase: seizures—traMADol overdose
Decrease: effect of opioid analgesics

NURSING CONSIDERATIONS

Assess:

- **Withdrawal:** cramping, hypertension, anxiety, vomiting, signs of withdrawal in drug-dependent individuals may occur ≤2 hr after administration
- VS q3-5min
- ABGs including Po_2, Pco_2
- Cardiac status: tachycardia, hypertension; monitor ECG
- **Respiratory dysfunction:** respiratory depression, character, rate, rhythm; if respirations are <10/min, administer naloxone; probably due to opioid overdose; monitor LOC
- **Pain:** duration, intensity, location before and after administration; may be used for respiratory depression

Evaluate:

- Therapeutic response: reversal of respiratory depression; LOC—alert

Teach patient/family:

- When patient is lucid, about reasons for, expected results of product

naltrexone (Rx)

(nal-trex′one)

ReVia, Vivitrol

Func. class.: Opioid antagonist
Chem. class.: Thebaine derivative

Do not confuse:
naloxone/naltrexone

ACTION:
Competes with opioids at opioid-receptor sites

USES:
Blockage of opioid analgesics; used for treatment of opiate addiction, alcoholism, opiate agonist overdose
Unlabeled uses: Nicotine withdrawal, opiate-agonist withdrawal, pruritus

CONTRAINDICATIONS:
Hypersensitivity, opioid dependence

Black Box Warning: Hepatic failure, hepatitis

Precautions: Pregnancy (B), breastfeeding, children, renal disease, depression, suicidal ideation, coagulopathy, respiratory depression, IV use

Black Box Warning: Hepatic disease

DOSAGE AND ROUTES

Adjunct in opiate-agonist dependence

- **Adult: PO** 25 mg; if no withdrawal symptoms in 1 hr, then 25 mg additionally; if no withdrawal symptoms, then 50-150 mg/day or in divided doses

Adjunct in alcoholism treatment

- **Adult: PO** 50 mg/day with food × 12 wk; **IM** (Vivitrol) 380 mg q4wk

Pruritus (unlabeled)

- **Adult: PO** 50 mg/day × 7 days to 4 wk

Nicotine withdrawal (unlabeled)

- **Adult: PO** 50 mg/day

Ultrarapid opiate detoxification (unlabeled)

- **Adult: PO** 50 mg before sedation with midazolam

Available forms: Tabs 50 mg; powder for inj 380 mg

Administer:

PO route

- Give with food or after meals, antacid to prevent nausea, vomiting
- Store in tight container

IM route

- Do not give until opioid-free for 7-10 days to prevent opioid withdrawal (relapse only)
- IM deep in gluteal, alternate inj sites; use supplied needle to prevent inj site reaction; aspirate before inj
- Only if resuscitative equipment is nearby
- Not to use IV or SUBCUT

SIDE EFFECTS

CNS: *Stimulation, drowsiness,* dizziness, confusion, seizures, headache, flushing, hallucinations, nervousness, irritability, suicidal ideation, syncope, anxiety

CV: Rapid pulse, pulmonary edema, hypertension, DVT

EENT: Tinnitus, hearing loss, blurred vision

GI: *Nausea, vomiting, diarrhea, heartburn,* anorexia, hepatitis, constipation, abdominal pain

GU: Delayed ejaculation, impotence

INTEG: *Rash,* urticaria, bruising, oily skin, acne, pruritus, inj site reactions

MISC: Increased thirst, chills, fever

MS: Joint and muscle pain

RESP: Wheezing, hyperpnea, nasal congestion, rhinorrhea, sneezing, sore throat, pneumonia

PHARMACOKINETICS

Metabolized by liver, excreted by kidneys; crosses placenta, excreted in breast milk; half-life 4 hr; IM half-life 5-10 days; extensive first-pass metabolism; protein binding 21%-28%

PO: Onset 15-30 min, peak 1-2 hr, duration is dose dependent

IM: Peak 2 hr

INTERACTIONS

Increase: lethargy—phenothiazines

Increase: hepatotoxicity—disulfiram

Increase: bleeding risk—anticoagulants

NURSING CONSIDERATIONS

Assess:

Black Box Warning: Hepatic status: LFTs, jaundice, hepatitis, hepatic failure

- ABGs including Po_2, Pco_2, LFTs, VS q3-5min
- Signs of withdrawal in drug-dependent individuals, use naltrexone challenge to test opioid dependence; must be free of opioids for 7-10 days before using this product, or withdrawal symptoms can occur
- Cardiac status: tachycardia, hypertension
- **Respiratory dysfunction: respiratory depression,** character, rate, rhythm; if respirations $<$10/min, respiratory stimulant should be administered
- Mental status: depression, suicidal ideation

Evaluate:

- Therapeutic response: blocking opiate ingestion; successful nicotine, alcohol withdrawal

Teach patient/family:

- That patient must be drug-free to start treatment
- ⚠ That using opioid while taking this product could prove fatal because high dose is needed to overcome this antagonist; not to self-dose with OTC products unless approved by prescriber
- To carry emergency ID stating product used
- That, if surgery is needed, all involved should be aware of this product
- To use caution while driving or performing other hazardous tasks until effect is known
- ⚠ That suicidal thoughts/behaviors may occur; to report these immediately

naphazoline nasal agent

See Appendix B

naphazoline (ophthalmic)

(na-faz′oh-leen)

Advanced Eye Relief, Redness Maximum Relief, Ak-Con, All Clear, Clear Eyes, Naphcon Forte ✦, Vasoclear

Func. class.: Ophthalmic vasoconstrictor

Chem. class.: Sympathomimetic

ACTION: Acts on the blood vessels in the eye to produce vasoconstriction

USES: Ocular congestion, irritation, itching of the eye

CONTRAINDICATIONS: Hypersensitivity, acute angle-closure glaucoma, 0.1% solution in children/infants

Precautions: Hyperthyroidism, diabetes mellitus, hypertension, cardiac conditions

DOSAGE AND ROUTES

- **Adult: OPHTH** instill 1-2 drops in affected eye in the conjunctival sac every 3-4 hr, up to qid

Available forms: ophthalmic solution 0.012%, 0.1%, 0.02%

Administer:

- Store at room temperature; keep tightly closed

SIDE EFFECTS

CNS: Headache

EENT: Blurred vision, irritation, photophobia, dilation, stinging, elevated IOP, keratitis

PHARMACOKINETICS

Onset 10 min, duration up to 6 hr

INTERACTIONS

Increase: Systemic effects—β-blockers

- Do not use within 14 days of MAOIs

NURSING CONSIDERATIONS

Assess:

- Ocular itching, congestion, irritation: should show improvement quickly; avoid using more than 3 days; long-term use or exceeding dosage can lead to rebound congestion; report eye pain, blurred vision

Evaluate:

- Decreasing ocular itching, congestion, irritation

Teach patient/family:

- Method for instilling drops
- To notify prescriber of eye pain, blurred vision
- That ocular itching, congestion, irritation should show improvement quickly
- To avoid using longer than 3 days; that long-term use or exceeding dosage can lead to rebound congestion
- To wait for at least 15 min before wearing contact lenses

naproxen (Rx, OTC)

Aleve, Anaprox, Anaprox DS, Apo-Napro-Na ✦, EC-Naprosyn, Equaline All Day Relief, Good Sense All Day Pain Relief, Midol Extended Relief, Naprelan, Novo-Naprox ✦, Novo-Naprox Sodium ✦, Nu-Naprox ✦, Top Care All Day Pain Relief, Wal-Proxen

Func. class.: Nonsteroidal antiinflammatory, nonopioid analgesic

Chem. class.: Propionic acid derivative

Do not confuse:

Naprosyn/Natacyn/Naprelan

ACTION: Inhibits COX-1, COX-2 by blocking arachidonate; analgesic, antiinflammatory, antipyretic

USES: Osteoarthritis; rheumatoid, gouty arthritis; primary dysmenorrhea; ankylosing spondylitis, bursitis, tendini-

N

tis, myalgia, dental pain, juvenile rheumatoid arthritis

Unlabeled uses: Bone pain, migraine/migraine prophylaxis, heterotropic ossification

CONTRAINDICATIONS:

Pregnancy (D) 3rd trimester, hypersensitivity to NSAIDs, salicylates

Black Box Warning: Perioperative pain in CABG surgery

Precautions: Pregnancy (C) breastfeeding, children <2 yr, geriatric patients, bleeding disorders, GI disorders, cardiac disorders, hypersensitivity to other antiinflammatory agents, CCr <30 ml/min, asthma, renal failure, hepatic disease

Black Box Warning: MI, GI bleeding, stroke

DOSAGE AND ROUTES

200 mg base = 220 mg naproxen sodium

Antiinflammatory/analgesic/antidysmenorrheal

- **Adult: PO** 250-500 mg bid, max 1500 mg/day; **DEL REL** 375-500 mg bid
- **Child ≥2 yr: PO** 7 mg/kg q12hr

Antigout

- **Adult: PO** 750 mg, then 250 mg q8hr

OTC use

- **Adult: PO** 220 mg q8-12hr or 440 mg then 220 mg q12hr; max 660 mg/24hr taken ≤10 days
- **Geriatric >65 yr: PO** max 220 mg q12hr

Available forms: Naproxen: tabs 250, 375, 500 mg; del rel tabs (EC-Naprosyn, Naprosyn-E) 250 ♣, 375, 500 mg; oral susp 125 mg/5 ml; ext rel tabs (CR) 375, 500, 750 mg; **naproxen sodium:** tabs 220, 275, 550 mg tab, ext rel 220 mg

Administer:

- Store at room temperature
- With food to decrease GI symptoms; take on empty stomach to facilitate absorption; give with full glass of liquid
- Do not crush, break, or chew ext rel tabs
- OTC for ≤10 days unless approved by prescriber
- Adequately hydrate in those taking angiotensin receptor blockers/angiotensin converting enzyme inhibitors
- **Oral susp:** shake well; use measuring cup provided or other calibrated device

SIDE EFFECTS

CNS: Dizziness, drowsiness, fatigue, tremors, confusion, insomnia, anxiety, depression

CV: Tachycardia, peripheral edema, palpitations, dysrhythmias, MI, stroke

EENT: Tinnitus, hearing loss, blurred vision

GI: Nausea, anorexia, vomiting, diarrhea, jaundice, hepatitis, constipation, flatulence, cramps, peptic ulcer, GI ulceration, bleeding, perforation

GU: Nephrotoxicity: dysuria, hematuria, oliguria, azotemia

HEMA: Blood dyscrasias

INTEG: Purpura, rash, pruritus, sweating

SYST: Anaphylaxis, Stevens-Johnson syndrome

PHARMACOKINETICS

PO: Peak 2-4 hr, half-life 12-17 hr; metabolized in liver; excreted in urine (metabolites), breast milk; 99% protein binding

INTERACTIONS

- Do not use with adefovir, cidofovir; nephrotoxicity is increased

Increase: renal impairment—ACE inhibitors

⚠ **Increase:** toxicity risk—methotrexate, lithium, antineoplastics, probenecid, radiation treatment

Increase: bleeding risk—oral anticoagulants, thrombolytic agents, eptifibatide, tirofiban, clopidogrel, ticlopidine, plicamycin, SSRIs, SNRIs, tricyclics

Increase: GI side effects risk—aspirin, corticosteroids, alcohol, NSAIDs

Decrease: effect of antihypertensives, diuretics

Decreased: absorption of naproxen—antacids, sucralfate, cholestyramine

Drug/Herb

• Bleeding risk: feverfew, garlic, ginger, ginkgo, ginseng *(Panax)*

Drug/Lab Test

Increase: BUN, alk phos, LFTs, potassium, glucose, cholesterol

Decrease: potassium, sodium

False increase: 5-HIAA, 17KS

NURSING CONSIDERATIONS

Assess:

Black Box Warning: Cardiac status: CV thrombotic events, MI, stroke; may be fatal; not to be used with CABG

Black Box Warning: GI status: ulceration, bleeding, perforation; may be fatal; obtain stool guaiac

• **Pain:** frequency, characteristics, intensity; relief before and 1-2 hr after product

• **Arthritis:** range of motion, pain, swelling before and 1-2 hr after use

• **Fever:** before, 1 hr after use

⚠ Asthma, aspirin hypersensitivity or nasal polyps, increased risk of hypersensitivity

• **Renal, hepatic, blood studies:** BUN, creatinine, AST, ALT, Hgb, LDH, blood glucose, Hct, WBC, platelets, CCr before treatment, periodically thereafter during long-term therapy

• Audiometric, ophthalmic exam before, during, after treatment; if taking long term, eye, ear problems: blurred vision, tinnitus (may indicate toxicity)

Evaluate:

• Therapeutic response: decreased pain, stiffness, swelling in joints; ability to move more easily

Teach patient/family:

• To report blurred vision, ringing, roaring in ears (may indicate toxicity)

• To avoid driving, other hazardous activities if dizziness or drowsiness occurs

⚠ To report change in urine pattern, weight increase, edema (face, lower extremities), pain increase in joints, fever, blood in urine (indicates nephrotoxicity); black stools, flulike symptoms

• That therapeutic effects may take up to 1 mo in arthritis

• To avoid ASA, alcohol, steroids, or other OTC medications without prescriber approval

• To report use to all health care providers

• To notify prescriber if pregnancy is planned or suspected, pregnancy (C); to avoid breastfeeding

naratriptan (Rx)

(nair'ah-trip-tan)

Amerge

Func. class.: Antimigraine agent

Chem. class.: 5-HT_1 receptor agonist

Do not confuse:

Amerge/Altace/Amaryl

ACTION: Binds selectively to the vascular 5-HT_1 B/D receptor subtype; exerts antimigraine effect; causes vasoconstriction in cranial arteries

USES: Acute treatment of migraine with/without aura

CONTRAINDICATIONS: Hypersensitivity, angina pectoris, history of MI, documented silent ischemia, ischemic heart disease, concurrent ergotamine-containing preparations, uncontrolled hypertension, CV syndromes, hemiplegic or basilar migraines, severe renal disease (CCr <15 ml/min); severe hepatic disease (Child-Pugh grade C)

Precautions: Pregnancy (C), breastfeeding, children, geriatric patients, postmenopausal women, men >40 yr, CAD risk, hypercholesterolemia, obesity, diabetes, impaired renal/hepatic function, peripheral vascular disease overuse

DOSAGE AND ROUTES

• **Adult: PO** 1 or 2.5 mg with fluids; if headache returns, repeat 1× after 4 hr; max 5 mg/24 hr

N

Hepatic/renal dose

- **Adult: PO** CCr 15-39 ml/min or mild to moderate hepatic disease max 2.5 mg/24 hr

Available forms: Tabs 1, 2.5 mg

Administer:

- With fluids as soon as symptoms appear; may take another dose after 4 hr; do not take >5 mg during any 24-hr period

SIDE EFFECTS

CNS: Dizziness, sedation, fatigue
CV: Increased B/P, palpitations, tachydysrhythmias, PR, QTc prolongation, ST/T wave changes, PVCs, atrial flutter/fibrillation, coronary vasospasm
EENT: EENT infections, photophobia
GI: *Nausea, vomiting*
MISC: Temperature change sensations; tightness, pressure sensations
MS: *Weakness, neck stiffness,* myalgia

PHARMACOKINETICS

Onset 2-3 hr; peak 2-3 hr; 28%-31% protein binding; half-life 6 hr; metabolized in liver (metabolite); excreted in urine, feces; may be excreted in breast milk

INTERACTIONS

Increase: serotonin syndrome, neuroleptic malignant syndrome—SSRIs (FLUoxetine, fluvoxaMINE, PARoxetine, sertraline), SNRIs, serotonin receptor agonists, sibutramine
Increase: vasospastic effects—ergot, ergot derivatives, other 5-HT_1 agonists
Increase: adverse reactions risk—MAOIs; do not use together

Drug/Herb

- Serotonin syndrome: SAM-e, St. John's wort

NURSING CONSIDERATIONS

Assess:

- **Migraine symptoms:** aura, duration, effect on lifestyle, aggravating/alleviating factors

⚠ **Serotonin syndrome, neuroleptic malignant syndrome:** increased heart rate, shivering sweating, dilated pupils, tremors, high B/P, hyperthermia, headache, confusion; if these occur, stop product, administer serotonin antagonist if needed; at least 2 wk should elapse between discontinuation of serotonergic agents and start of product

⚠ Cardiac status: ECG, increased B/P, dysrhythmias, monitor for PR, QT prolongation, ST-T wave changes, PVCs in those with cardiac disease

- Stress level, activity, recreation, coping mechanisms
- Neurologic status: LOC blurred vision, nausea, vomiting, tingling in extremities preceding headache
- Quiet, calm environment with decreased stimulation (noise, bright light, excessive talking)

Evaluate:

- Therapeutic response: decrease in frequency, severity of headache

Teach patient/family:

⚠ To report pain, tightness in chest, neck, throat, or jaw; to notify prescriber immediately if sudden, severe abdominal pain occurs

- Not to use if another 5-HT_1 agonist or ergot preparation has been used during past 24 hr; to avoid using >2 days/wk because rebound headache may occur
- To notify prescriber if pregnancy is planned or suspected; to avoid breastfeeding

natalizumab (Rx)

(na-ta-liz'u-mab)

Tysabri

Func. class.: Biologic response modifier, immunoglobulins, monoclonal antibody

ACTION: Biologic-response-modifying properties mediated through specific receptors on cells; may be secondary to blockade of the interaction of inflammatory cells with vascular endothelial cells

USES:
Ambulatory patients with relapsing/remitting MS who have not responded to other treatment; those with moderate to severe Crohn's disease

CONTRAINDICATIONS:
Hypersensitivity, immunocompromised individuals (HIV, AIDS, leukemia, lymphoma, transplants), PML, murine (mouse) protein allergy

Black Box Warning: Progressive multifocal leukoencephalopathy

Precautions: Pregnancy (C), breastfeeding, geriatric patients, chronic progressive MS, depression, mental disorders, diabetes, TB, active infections, hepatotoxicity

DOSAGE AND ROUTES

- **Adult: IV INF** 300 mg q4wk; give over 1 hr; observe during and for 1 hr after inf
- **Adolescent and child ≥11 yr (unlabeled): IV INF** pediatric Crohn's disease activity index (PCDAI) >30, 3 mg/kg q4wk

Available forms: Single-use vial, 300 mg/100 ml 0.9% NaCl

Administer:

- Acetaminophen for fever, headache
- Only after being enrolled in the TOUCH Prescribing Program

Intermittent IV INF route

- Use only clear, colorless solution, without particulates
- Withdraw 15 ml from the vial using aseptic technique: inj conc into 100 ml 0.9% NaCl; do not use other diluents; mix completely; do not shake; inf immediately or refrigerate for ≤8 hr; warm to room temperature before using; flush with 0.9% NaCl before, after inf; do not admix or use in same line with other agents
- Withhold product at first sign of PML
- Prescribers must be registered in the TOUCH Prescribing Program (1-800-456-2255)
- Store sol in refrigerator; do not freeze or shake; protect from light

SIDE EFFECTS

CNS: *Headache, fatigue,* rigors, syncope, tremors, *depression,* **progressive multifocal leukoencephalopathy (PML), suicidal ideation**, anxiety

CV: Chest discomfort, hypo/hypertension, tachycardia

GI: *Abdominal discomfort,* abnormal LFT, gastroentritis, **severe hepatic injury**

GU: Amenorrhea, *UTI, irregular menses,* vaginitis, urinary frequency

INTEG: *Rash,* dermatitis, pruritus, **skin melanoma**, inf-related reactions

MS: *Arthralgia,* myalgia

RESP: *Lower respiratory tract infection,* dyspnea

SYST: **Anaphylaxis, angioedema**

PHARMACOKINETICS

Half-life approximately 11 days

INTERACTIONS

- Do not use with vaccines

Increase: infection—immunosuppressants, antineoplastics, immunomodulators, tumor necrosis factors

NURSING CONSIDERATIONS

Assess:

Black Box Warning: Progressive multifocal leukoencephalopathy (weakness, paralysis, vision loss, impaired speech, cognitive deterioration; obtain gadolinium-enhanced MRI scan of the brain, possibly cerebrospinal fluid for JC viral DNA; signs, symptoms of PML (decreased cognition, vision; ataxia, dysphagia), incidences increase with number of doses, over 2 yr immunosuppressants and anti-JC virus antibody; consider testing for the anti-JC virus and periodically retest

- **Infection:** report serious opportunistic infections to the manufacturer; those with Crohn's disease and chronic oral corticosteroids may be at greater risk of infection

• Blood, renal, hepatic studies: CBC, differential, platelet counts, BUN, creatinine, ALT, urinalysis, antibody testing
• CNS symptoms: headache, fatigue, depression, rigors, tremors
• GI status: abdominal discomfort, gastroenteritis, severe hepatic injury, abnormal LFTs
• Mental status: depression, depersonalization, suicidal thoughts, insomnia
• **MS symptoms;** product should only be used by patients who have not responded to other treatments
Anaphylaxis: SOB, hives; swelling, tightness in throat, chest pain; usually within 2 hr of inf

Evaluate:
• Therapeutic response: decreased symptoms of MS, Crohn's disease

Teach patient/family:
• Provide patient or family member with written, detailed information about product (med guide)
• That female patients may experience irregular menses, amenorrhea; may worsen over several days; to notify prescriber if pregnancy is suspected; to avoid breastfeeding while taking this product; if pregnant, call the Tysabri Pregnancy Exposure Registry (1-800-456-2255)
To notify prescriber of possible infection: sore throat, cough, increased temperature, inf site reactions
• That continuing follow-up will be needed at 3, 6 mo after first dose, then every 6 mo
• To inform all prescribers of product use

natamycin ophthalmic

See Appendix B

nebivolol (Rx)

(ne-biv'oh-lol)

Bystolic

Func. class.: Antihypertensive
Chem. class.: β_1-blocker

ACTION: Competitively blocks stimulation of β-adrenergic receptors within vascular smooth muscle; decreases rate of SA node discharge; increases recovery time; slows conduction of AV node, thereby resulting in decreased heart rate (negative chronotropic effect), which decreases O_2 consumption in myocardium due to β_1-receptor antagonism; decreases renin-aldosterone-angiotensin system at high doses; inhibits β_2-receptors in bronchial system at high doses

USES: Hypertension alone or in combination
Unlabeled uses: Heart failure

CONTRAINDICATIONS: Cardiogenic shock, acute heart failure, severe hepatic disease, severe bradycardia, sick sinus syndrome, AV heart block; hypersensitivity to product, β-blockers
Precautions: Pregnancy (C), breastfeeding, children, major surgery, peripheral vascular disease, diabetes mellitus, thyrotoxicosis disease, COPD, asthma, well-compensated heart failure, renal/hepatic disease, abrupt discontinuation, acute bronchospasm

DOSAGE AND ROUTES

Hypertension
• **Adult: PO** 5 mg/day, may be increased to desired response q2wk; max 40 mg/day
• **Geriatric: PO** max 40 mg/day

Renal/hepatic dose
• **Adult: PO** CCr $<$30 ml/min, 2.5 mg/day; may increase cautiously; (Child-Pugh class B) 2.5 mg daily, use dose escalation cautiously

Heart failure (unlabeled)

• **Adult: PO** 1.25 mg titrated to max 10 mg/day

Available forms: Tabs 2.5, 5, 10, 20 mg

Administer:

PO route

• Without regard to meals; tab may be crushed or swallowed whole; give with food to prevent GI upset

• Store protected from light, moisture; place in cool environment

SIDE EFFECTS

CNS: *Insomnia, fatigue, dizziness, mental changes,* drowsiness, *headache*

CV: Bradycardia, MI, AV heart block, edema

GI: *Nausea, diarrhea,* vomiting, abdominal pain

GU: *Impotence*

HEMA: Thrombocytopenia

INTEG: Rash, pruritus, vasculitis, urticaria, psoriasis, angioedema

MISC: Renal failure, pulmonary edema, hyperuricemia, hypercholesterolemia, withdrawal symptoms

RESP: Bronchospasm, dyspnea

PHARMACOKINETICS

Peak 1.5-4 hr; half-life 12 hr; metabolized in liver by CYP2D6; 38% excreted in urine, 44% in feces

INTERACTIONS

⚠ Do not give with other β-blockers, mefloquine

Increase: nebivolol action—CYP2D6 inhibitors (amiodarone, buPROPion, chloroquine, chlorpheniramine, chlorproMAZINE, cinacalcet, diphenhydrAMINE, DULoxetine, FLUoxetine, haloperidol, imatinib, PARoxetine, promethazine, propoxyphene, quiNIDine, quiNINE, ritonavir, terbinafine, thioridazine), cimetidine, calcium channel blockers (nondihydropyridine)

Decrease: nebivolol action—CYP2D6 inducers (rifampin), sildenafil

Drug/Herb

• May increase nebivolol effect—hawthorn

• May decrease nebivolol effect—ephedra

Drug/Lab Test

Increase: serum lipoprotein levels, BUN, potassium, triglycerides, uric acid, LDH, AST, ALT, alk phos

Decrease: platelets

NURSING CONSIDERATIONS

Assess:

• **Hypertension:** B/P during beginning treatment, periodically thereafter; apical/radial pulse before administration; notify prescriber of any significant changes (pulse <50 bpm); **signs of CHF** (dyspnea, crackles, weight gain, jugular venous distention)

• Blood glucose in diabetics

• Baselines of renal, hepatic function tests before therapy begins and periodically, do not use in Child-Pugh class >B

• Edema in feet, legs daily: monitor I&O

Evaluate:

• Therapeutic response: decreased B/P after 1-2 wk; decreased dysrhythmias

Teach patient/family:

⚠ Not to discontinue product abruptly because severe cardiac reactions may occur; to taper over 2 wk; not to double dose; if dose is missed, to take as soon as remembered up to 4 hr before next dose

• That product may mask signs of hypoglycemia or alter blood glucose levels

• Not to use OTC products containing α-adrenergic stimulants (nasal decongestants, OTC cold preparations) unless directed by prescriber

• To report low pulse, dizziness, confusion, depression, fever

• To take pulse, B/P at home; advise patient when to notify prescriber

• To comply with weight control, dietary adjustments, modified exercise program

• To carry emergency ID to identify product, allergies

• To avoid hazardous activities if dizziness, drowsiness present

N

- **To report symptoms of CHF:** difficulty breathing, especially on exertion or when lying down, night cough, swelling of extremities
- To continue with required lifestyle changes (exercise, diet, weight loss, stress reduction)

TREATMENT OF OVERDOSE:

Lavage, IV atropine for bradycardia, IV theophylline for bronchospasm, digoxin, O_2, diuretic for cardiac failure, IV glucose for hypoglycemia, IV diazepam (or phenytoin) for seizures, IV fluids, IV pressors

nelfinavir (Rx)

(nell-fin′a-ver)

Viracept

Func. class.: Antiretroviral

Chem. class.: Protease inhibitor

ACTION: Inhibits human immunodeficiency virus (HIV-1) protease, which prevents maturation of the infectious virus

Uses: HIV-1 in combination with other antiretrovirals

CONTRAINDICATIONS: Hypersensitivity to protease inhibitors

Precautions: Pregnancy (B), breastfeeding, renal/hepatic disease, hemophilia, PKU, pancreatitis, diabetes, infection, immune reconstitution syndrome

DOSAGE AND ROUTES

HIV infection

- **Adult and child >13 yr: PO** 750 mg tid or 1250 mg bid
- **Child 2-13 yr: PO** 25-35 mg/kg tid, max 2500 mg/day

Prevention of HIV infection after exposure (unlabeled)

- **Adult: PO** 1250 mg bid with 2 other antiretroviral agents × 4 wk

Available forms: Tabs 250, 625 mg; powder, oral 50 mg/g/scoop

Administer:

PO route

- Do not mix with juice or acidic fluids
- **Oral powder:** mixed with fluids if desired; stable mixed for 6 hr; may use in child unable to take tabs; do not mix with water in original bottle
- **Tabs** may be crushed and dispersed in water or mixed with food; consume immediately

SIDE EFFECTS

CNS: Headache, asthenia, poor conc, seizures, suicidal ideation

CV: Bleeding

ENDO: Hyperglycemia, hyperlipidemia

GI: *Diarrhea,* anorexia, dyspepsia, *nausea, flatulence,* hepatitis, pancreatitis

HEMA: Anemia, leukopenia, thrombocytopenia, Hgb abnormalities

INTEG: *Rash,* dermatitis, anaphylaxis

MS: Pain, arthralgia, myalgia, myopathy

Other: Hypoglycemia, redistribution/accumulation of body fat, immune reconstitution syndrome

PHARMACOKINETICS

Half-life 3½-5 hr, excreted in feces (87%), peak 2-4 hr, 98% protein binding; metabolized by CYP3A4 enzyme system; potent inhibitor of CYP3A4

INTERACTIONS

Drug/Herb

⚠ **Increase:** serious dysrhythmias: amiodarone, ergots, lovastatin, midazolam, pimozide, quiNIDine, simvastatin, triazolam, salmeterol

Increase: effect of—atorvastatin, azithromycin, rifabutin, indinavir, saquinavir, cycloSPORINE, tacrolimus, sirolimus, sildenafil, alfentanil, alosetron, buprenorphine, busPIRone, bortezomib, calcium channel blockers, cilostazol, disopyramide, dofetilide, DOCEtaxel, donepezil, ethosuximide, fentaNYL, galantamine, gefitinib, levomethadyl, systemic lidocaine, PACLitaxel, sibutramine, SUFentanil, vinca alkaloids, ziprasidone,

zonisamide, traZODone, tricyclic antidepressants, sildenafil
Increase: nelfinavir levels—ketoconazole, indinavir, ritonavir; delavirdine, other HIV protease inhibitors
Decrease: nelfinavir levels—rifamycins, nevirapine, PHENobarbital, phenytoin, carBAMazepine
Decrease: effect of—didanosine, methadone, oral contraceptives, phenytoin
Drug/Herb
⚠ **Decrease:** antiretroviral effect—St. John's wort; do not use concurrently
Drug/Food
Increase: absorption with food
Drug/Lab Test
Increase: AST, ALT, alk phos, total bilirubin, CPK, LDH, lipids, uric acid
Decrease: WBC, platelets

NURSING CONSIDERATIONS

Assess:

- Resistance testing at initiation, with failure of treatment
- Signs of infection, anemia
- Hepatic studies: ALT, AST
- Bowel pattern before, during treatment; if severe abdominal pain with bleeding occurs, product should be discontinued; monitor hydration

⚠ **Immune reconstitution syndrome:** occurs with combination therapy, including MAC, CMV, PCP TB requiring treatment
⚠ **Anaphylaxis, hypersensitivity reaction:** wheezing, flushing; swelling of lips, tongue, throat, skin eruptions, rash, urticaria, itching

- **HIV:** serum lipid profile, plasma HIV RNA, blood glucose, viral load, CD4 cell counts at baseline and throughout treatment

Teach patient/family:

- To avoid taking with other medications unless directed by prescriber
- **Diarrhea** is most common side effect; may use loperamide to control
- That product does not cure but does manage symptoms; that product does not prevent transmission of HIV to others
- To use a nonhormonal form of birth control while taking this product if using contraceptives
- If dose is missed, to take as soon as remembered up to 1 hr before next dose; not to double dose
- To take with food
- To report symptoms of hyperglycemia, bleeding, abdominal pain; yellowing of skin, eyes
- **Phenylketonuria:** powder contains phenylalanine

RARELY USED

neomycin (Rx)

(nee-oh-mye′sin)

Func. class.: Antiinfective—aminoglycoside

USES: Severe systemic infections of CNS, respiratory, GI, urinary tract, eye, bone, skin, soft tissues; hepatic coma, preoperatively to sterilize bowel, infectious diarrhea caused by enteropathogenic *E. coli, Enterobacter* sp., *Escherichia coli, Klebsiella* sp. May be effective for *Acinetobacter* sp., *Bacillus anthracis, Citrobacter* sp., *Haemophilus influenzae* (beta-lactamase negative), *Haemophilus influenzae* (beta-lactamase positive), *Neisseria* sp., *Proteus mirabilis, Proteus vulgaris, Providencia* sp., *Salmonella* sp., *Serratia* sp., *Shigella* sp., *Staphylococcus aureus* (MSSA), *Staphylococcus epidermidis*

CONTRAINDICATIONS: Infants, children, bowel obstruction (oral use), severe renal disease, hypersensitivity, GI disease

Precautions: Dehydration, geriatric patients, respiratory insufficiency

Black Box Warning: Hearing impairment, neuromuscular disease, renal disease

N

DOSAGE AND ROUTES

Hepatic encephalopathy

- **Adult: PO** 4-12 g/day in divided doses q6hr × 5-6 days
- **Child: PO** 50-100 mg/kg/day in divided doses q6hr × 5-6 days

Preoperative intestinal antisepsis

- **Adult: PO** 1 g/hr × 4 hr then 1 g q4hr for remaining 24 hr

neomycin topical

See Appendix B

nepafenac ophthalmic

See Appendix B

HIGH ALERT

nesiritide (Rx)

(neh-seer′ih-tide)

Natrecor

Func. class.: Vasodilator

Chem. class.: Human B-type natriuretic peptide

ACTION: Uses DNA technology; human B-type natriuretic peptide binds to the receptor in vascular smooth muscle and endothelial cells, thereby leading to smooth muscle relaxation

USES: Acutely decompensated CHF

CONTRAINDICATIONS: Hypersensitivity to this product or *Escherichia coli* protein; cardiogenic shock or B/P <90 mm Hg as primary therapy

Precautions: Pregnancy (C); breastfeeding; children; mitral stenosis; significant valvular stenosis, restriction, or obstructive cardiomyopathy, or any condition dependent on venous return; renal disease; constrictive pericarditis

DOSAGE AND ROUTES

- **Adult: IV BOL** 2 mcg/kg then **CONT IV INF** 0.01 mcg/kg/min

Available forms: Powder for inj, 1.5-mg single-use vial

Administer:

IV route

- Do not give through a central catheter containing other products; administer other products through separate catheter or central line heparin-coated catheter because nesiritide binds to heparin
- Reconstitute one 1.5-mg vial/5 ml of diluent from prefilled 250-ml plastic IV bag with diluent of choice (preservative free D_5, 0.9% NaCl, $D_5/^1/_2$ NaCl, D_5/0.2% NaCl); do not shake vial; roll gently; use only clear sol
- Withdraw all contents of reconstituted vial and add to 250-ml plastic IV bag (6 mcg/ml), invert bag several times
- Use within 24 hr of reconstituting
- Prime IV fluid with inf of 5 ml before connecting to patient's vascular access port and before bolus dose or IV inf

Direct IV route

- Prime tubing with 5 ml inf sol; calculate dose based on patient's weight, 0.33 × patient weight (kg) = bolus vol (ml) (6 mcg/ml); withdraw prescribed bolus dose (volume) from prepared inf bag; give over 1 min through IV port

Intermittent IV INF route

- After bolus dose, use inf, give at 0.1 ml/kg/hr (0.01 mcg/kg/min)

Y-site compatibilities: Acyclovir, alfentanil, allopurinol, amifostine, aminocaproic acid, aminophylline, amiodarone, amphotericin B colloidal, amphotericin B lipid complex, amphotericin B liposome, anidulafungin, argatroban, atenolol, atracurium, azithromycin, aztreonam, bivalirudin, bleomycin, buprenorphine, busulfan, butorphanol, calcium acetate/chloride/gluconate, CARBOplatin, carmustine, ceFAZolin, cefotaxime, cefoTEtan, cefOXitin, cefTAZidime, ceftizoxime, cefTRIAXone, cefuroxime, chloramphenicol, cimetidine, ciprofloxacin, cisatracurium, CISplatin, clindamycin, cyclophosphamide, cycloSPORINE, cytarabine, dacarbazine, DACTINomycin, DAUNOrubicin, digoxin, diltiazem, diphenhydrAMINE, DOCEtaxel, dolasetron, doxacurium, DOXOrubicin, doxycycline, droperidol, ePHEDrine, epirubicin, ertapenem, erythromycin, esmolol, etoposide, etoposide phosphate, famotidine, fenoldopam, fentaNYL, filgrastim, fluconazole, fludarabine, fluorouracil, foscarnet, fosphenytoin, ganciclovir, gatifloxacin, gemcitabine, gemtuzumab, glycopyrrolate, granisetron, haloperidol, hydrocortisone, HYDROmorphone, hydrOXYzine, IDArubicin, ifosfamide, imipenem-cilastatin, irinotecan, ketorolac, leucovorin, levofloxacin, lidocaine, linezolid, LORazepam, magnesium sulfate, mannitol, mechlorethamine, melphalan, meropenem, mesna, metaraminol, methohexital, methotrexate, methylPREDNISolone, metoclopramide, metroNIDAZOLE, midazolam, milrinone, minocycline, mitoMYcin, mitoXANtrone, mivacurium, moxifloxacin, mycophenolate, nalbuphine, naloxone, niCARdipine, nitroglycerin, nitroprusside, octreotide, ondansetron, oxaliplatin, oxytocin, PACLitaxel, palonosetron, pamidronate, pancuronium, PEMEtrexed, pentamidine, PENTobarbital, PHENobarbital, phentolamine, phenylephrine, polymyxin B sulfate, potassium chloride/phosphates, prochlorperazine, propranolol, quiNIDine, quinupristin-dalfopristin, ranitidine, remifentanil, rocuronium, sodium acetate/bicarbonate/phosphates, streptozocin, succinylcholine, SUFentanil, tacrolimus, teniposide, theophylline, thiotepa, ticarcillin, tigecycline, tirofiban, tolazoline, topotecan, torsemide, trimethobenzamide, vancomycin, vasopressin, vecuronium, verapamil, vinBLAStine, vinCRIStine, vinorelbine, zidovudine, zoledronic acid

SIDE EFFECTS

CNS: Headache, insomnia, dizziness, anxiety, confusion, paresthesia, tremor

CV: *Hypotension,* tachycardia, dysrhythmias, bradycardia, **ventricular tachycardia, ventricular extrasystoles, atrial fibrillation**

GI: Vomiting, nausea

INTEG: Rash, sweating, pruritus, inj site reaction

MISC: Abdominal pain, back pain

RESP: Increased cough, hemoptysis, **apnea**

PHARMACOKINETICS

Half-life 18 min

INTERACTIONS

Increase: hypotension—ACE inhibitors, antihypertensives, inotropes, IV nitrates

NURSING CONSIDERATIONS

Assess:

• PCWP, RAP, cardiac index, MPAP, B/P, pulse during treatment until stable

• Daily serum creatinine, BUN

• **CHF:** weight gain, dyspnea, crackles, I&O ratios, peripheral edema

Evaluate:

• Therapeutic response: improvement in CHF with improved PCWP, RAP, MPAP

Teach patient/family:

• About purpose of medication, expected results; to report pain at IV site

• To report dizziness, blurred vision, lightheadedness, sweating

nevirapine (Rx)

(ne-veer′a-peen)

Viramune, Viramune XR

Func. class.: Antiretroviral

Chem. class.: Nonnucleoside reverse transcriptase inhibitor (NNRTI)

Do not confuse:

nevirapine/nelfinavir

Viramune/Viracept

ACTION: Binds directly to reverse transcriptase and blocks RNA, DNA, thus causing a disruption of the enzyme's site

USES:
HIV-1 in combination with other highly active antiretroviral therapy (HAART)

CONTRAINDICATIONS:
Black Box Warning: Hypersensitivity, hepatic disease

Precautions: Pregnancy (B), breast-feeding, children, renal disease, Hispanic patients

Black Box Warning: Females, hepatitis

DOSAGE AND ROUTES

Treatment of HIV infection in combination with other antiretrovirals

- **Adult and adolescent: PO** 200 mg/day × 2 wk then 200 mg bid in combination; **EXT REL** tab (adults not currently taking immediate rel nevirapine) 200 mg/day (immediate rel tab) × 14 days with other antiretrovirals; if rash develops during lead-in periods and persists beyond 14 days, do not use ext rel tab; if no consistent rash present, then give 400 mg/day ext rel tab with other antiretrovirals; if interrupted >7 days, restart 14 day lead-in dosing; for adults switched from immediate rel tab, give 400 mg/day ext rel tab
- **Child/adolescent ≥6 yr: PO EXT REL** not currently taking immediate release 150 mg/m^2 (immediate release) daily (max 200 mg/day) × 14 days then BSA 0.58-0.83 m^2 200 mg/day; BSA 0.84-1.16 m^2 300 mg/day
- **Child/infant/neonate ≥15 days old: PO** 150 mg/m^2/day × 14 days then 150 mg/m^2 bid, max 400 mg/day

Perinatal transmission prophylaxis (unlabeled)

- **Females with no previous antiretroviral therapy: PO** 200 mg as a single dose at onset of labor with zidovudine 2 mg/kg over 1 hr followed by zidovudine 1 mg/kg/hr until delivery
- **Neonate ≥34 wk gestation: PO** Nevirapine 12 mg (>2 kg) or 8 mg (1.5-2 kg) × 3 doses; 1st dose 48 hr after birth, 2nd dose 48 hr after 1st dose, 3rd dose 96 hr after 2nd dose and **PO** zidovudine

Hepatic Dose

- **Adult: PO** do not use with Child-Pugh grade B or C

Available forms: Tabs 200 mg; oral susp 50 mg/5 ml; ext rel 400 mg

Administer:

- Do not initiate treatment in females when CD4 counts >250 cells/mm^3 or in males when >400 cells/mm^3 unless benefits outweigh risks
- Without regard to meals
- Use in combination with at least 1 other antiretroviral
- **Oral susp** should be shaken before giving

SIDE EFFECTS

CNS: *Paresthesia, headache, fever, peripheral neuropathy*

GI: *Diarrhea,* abdominal pain, *nausea, stomatitis,* hepatotoxicity, hepatic failure

HEMA: Neutropenia, anemia, thrombocytopenia

INTEG: *Rash,* toxic epidermal necrolysis

MISC: Stevens-Johnson syndrome, anaphylaxis

MS: Pain, myalgia, rhabdomyolysis

PHARMACOKINETICS

Rapidly absorbed, peak 4 hr, 60% bound to plasma proteins, metabolized by liver; metabolized by hepatic P450 enzyme system, excreted 91% in urine, terminal half-life 25-30 hr, 50% removed by peritoneal dialysis; with hepatic disease and in Hispanic patients, African American patients, slower rate of clearance

INTERACTIONS

Increase: nevirapine levels—cimetidine, macrolide antiinfectives

Decrease: effects of protease inhibitors, oral contraceptives, ketoconazole, methadone, itraconazole

Decrease: nevirapine levels—rifamycins, anticonvulsants, clonazePAM, diazepam, warfarin

Drug/Herb
Decrease: action of antiretroviral—St. John's wort; do not use concurrently
Drug/Lab Test
Increase: ALT, AST, GGT, bilirubin, Hgb
Decrease: neutrophil count

NURSING CONSIDERATIONS

Assess:

- Resistance testing before therapy and when therapy fails

⚠ Signs of infection, anemia, hepatotoxicity, immune reconstitution syndrome; hepatitis B or C, liver toxicity may occur

- **HIV:** blood studies during treatment: ALT, AST, viral load, CD4, plasma HIV RNA, renal studies; if LFTs elevated significantly, product should be withheld; glucose levels in diabetic patients, if treatment is interrupted by >1 wk, restart at initial dose

⚠ **Rhabdomyolysis:** pain, tenderness, weakness, edema; product should be discontinued

- Bowel pattern before, during treatment; if severe abdominal pain with bleeding occurs, product should be discontinued; monitor hydration

⚠ **Stevens-Johnson syndrome, toxic epidermal necrolysis,** allergies before treatment, reaction to each medication; skin eruptions; rash, urticaria, itching; if rash is severe or systemic symptoms occur, discontinue immediately

Evaluate:

- Therapeutic response: absence of AIDS-defining symptoms, improvement in quality of life; decreased viral load, increase in CD4 count

Teach patient/family:

⚠ To report any right quadrant pain, yellowing of eyes or skin, dark urine, nausea, anorexia, muscle pain or tenderness, rash immediately

- That product may be taken with food, antacids
- To take as prescribed; if dose is missed, to take as soon as remembered up to 1 hr before next dose; not to double dose
- That product is not a cure, does not prevent transmission; controls symptoms of HIV
- To avoid OTC agents unless approved by prescriber
- To use a nonhormonal form of contraception during treatment in those using contraceptives

niacin (OTC, Rx)

(nye′a-sin)

Equaline Niacin, Niaspan, Ni-Odan ✦, Slo-Niacin

niacinamide (OTC, Rx)

Func. class.: Vit B_3, antihyperlipidemic
Chem. class.: Water-soluble vitamin

ACTION:
Needed for conversion of fats, protein, carbohydrates by oxidation reduction; acts directly on vascular smooth muscle, thus causing vasodilation; reduces total cholesterol, LDL, VLDL, triglycerides; increases HDL

USES:
Pellagra, hyperlipidemias (types 4, 5), peripheral vascular disease that presents a risk for pancreatitis

CONTRAINDICATIONS:
Breastfeeding, hypersensitivity, peptic ulcer, hepatic disease, hemorrhage, severe hypotension

Precautions: Pregnancy (C), breastfeeding, glaucoma, cardiovascular disease, CAD, diabetes mellitus, gout, schizophrenia

DOSAGE AND ROUTES

Niacin deficiency

- **Adult: PO** 100-500 mg/day in divided doses; **IM/SUBCUT** 5-100 mg ≥5×/day; **IV** 25-100 mg bid or tid
- **Child: PO** ≤300 mg/day in divided doses

Adjunct in hyperlipidemia

- **Adult: PO** 250 mg after evening meal; may increase dose at 1-4 wk intervals to 1-2 g tid, max 6 g/day; **EXT REL** 500 mg at bedtime × 4 wk then 1000 mg at bedtime for wk 5-8; do not increase by >500 mg q4wk, max 2000 mg/day

Pellagra

• **Adult: PO** 300-500 mg/day in divided doses; **IM** 50-100 mg 5×/day or **IV** 25-100 mg bid by slow **IV INF**

• **Child: PO** 100-300 mg/day in divided doses; **IV** ≤300 mg/day by slow **IV/INF**

Peripheral vascular disease

• **Adult: PO** 250-800 mg/day in 3-5 divided doses

Available forms: *Niacin:* tabs 50, 100, 250, 500, mg; ext rel caps 250, 500 mg; ext rel tab 250, 500, 750, 1000 mg; *niacinamide:* tab 100, 500 mg

Administer:

• Do not break, crush, or chew ext rel tabs, caps

• With meals for GI symptoms; with 81-325 mg aspirin or NSAIDs 30 min before dose to decrease flushing

SIDE EFFECTS

CNS: Paresthesias, headache, dizziness, anxiety

CV: Postural hypotension, vasovagal attacks, dysrhythmias, vasodilation

EENT: Blurred vision, ptosis

GI: Nausea, vomiting, anorexia, jaundice, hepatotoxicity, diarrhea, peptic ulcer, dyspepsia, hepatitis

GU: Hyperuricemia, glycosuria, hypoalbuminemia

INTEG: Flushing, dry skin, rash, pruritus, itching, tingling

PHARMACOKINETICS

PO: Peak 30-70 min (depends on formulation), half-life 45 min; metabolized in liver; 30% excreted unchanged in urine

INTERACTIONS

Increase: postural hypotension—ganglionic blockers

⚠ **Increase:** myopathy, rhabdomyolysis—HMG-CoA reductase inhibitors

Increase: flushing, pruritus—alcohol, avoid use

Drug/Herb

⚠ **Increase:** myopathy, rhabdomyolysis—red yeast rice

Drug/Lab Test

Increase: bilirubin, alk phos, hepatic enzymes, LDH, uric acid, glucose

Decrease: cholesterol

False increase: urinary catecholamines

False positive: urine glucose

NURSING CONSIDERATIONS

Assess:

• Cardiac status: rate, rhythm, quality; postural hypotension, dysrhythmias

• Nutritional status: liver, yeast, legumes, organ meat, lean poultry; fat in diet

⚠ **Hepatotoxicity:** clay-colored stools, itching, dark urine, jaundice; hepatic studies: AST, ALT, bilirubin, uric acid, alk phos; blood glucose before and during treatment

• CNS symptoms: headache, paresthesias, blurred vision

• **Niacin deficiency:** nausea, vomiting, anemia, poor memory, confusion, dermatitis

• **Hyperlipidemia:** for lipid, triglyceride, cholesterol level; obtain diet history

Evaluate:

• Therapeutic response: decreased lipids, warm extremities, absence of numbness in extremities

Teach patient/family:

• That flushing and increase in feelings of warmth will occur several hr after taking product (PO); after 2 wk of therapy, these side effects diminish

• To remain recumbent if postural hypotension occurs; to rise slowly to prevent orthostatic hypotension

• To abstain from alcohol if product is prescribed for hyperlipidemia

• To avoid sunlight if skin lesions are present

⚠ **Hepatotoxicity:** to report clay-colored stools, anorexia, yellow eyes or skin, dark urine

niCARdipine (Rx)

(nye-card'i-peen)

Cardene IV, Cardene SR

Func. class.: Calcium channel blocker, antianginal, antihypertensive

Chem. class.: Dihydropyridine

Do not confuse:
niCARdipine/NIFEdipine
Cardene/Cardizem
Cardene SR/Cardizem SR

ACTION: Inhibits calcium ion influx across cell membrane during cardiac depolarization; produces relaxation of coronary vascular smooth muscle, peripheral vascular smooth muscle; dilates coronary vascular arteries; increases myocardial oxygen delivery in patients with vasospastic angina

USES: Chronic stable angina pectoris, hypertension

CONTRAINDICATIONS: Sick sinus syndrome, 2nd-/3rd-degree heart block; hypersensitivity to this product or dihydropyridine; advanced aortic stenosis

Precautions: Pregnancy (C), breastfeeding, children, geriatric patients, CHF, hypotension, hepatic injury, renal disease

DOSAGE AND ROUTES

Hypertension

- **Adult: PO** 20 mg tid initially; may increase after 3 days (range 20-40 mg tid) or 30 mg bid **SUS REL;** may increase to 60 mg bid or **IV** 5 mg/hr; may increase by 2.5 mg/hr q15min; max 15 mg/hr

Angina

- **Adult: PO** 20 mg tid; may be adjusted q3days; may use 20-40 mg tid

Renal dose

- **Adult: PO** adjust based on response

Hepatic dose

- **Adult: PO** 20 mg bid

Available forms: Caps 20, 30 mg; sus rel caps 30, 45, 60 mg; inj 2.5 mg/ml, premixed 20 mg/200 ml, 40 mg/200 ml

Administer:

PO route

- Do not break, crush, chew, or open sus rel cap
- Without regard to meals

IV route

- Dilute each 25 mg/240 ml of compatible sol (0.1 mg/ml), give slowly, titrate to patient response, change IV site q12hr
- Stable at room temperature for 24 hr

Solution compatibilities: D_5W, $D_5/0.45\%$ NaCl, $D_5/0.9\%$ NaCl

Y-site compatibilities: Alemtuzumab, amikacin, aminophylline, aztreonam, bivalirudin, butorphanol, calcium gluconate, CARBOplatin, caspofungin, ceFAZolin, ceftizoxime, chloramphenicol, cimetidine, CISplatin, clindamycin, cytarabine, DAPTOmycin, dexmedetomidine, diltiazem, DOBUTamine, DOCEtaxel, DOPamine, DOXOrubicin hydrochloride, enalaprilat, EPINEPHrine, epirubicin, erythromycin, esmolol, famotidine, fenoldopam, fentaNYL, gentamicin, hydrocortisone, HYDROmorphone, labetalol, lidocaine, linezolid, LORazepam, magnesium sulfate, mechlorethamine, methylPREDNISolone, metroNIDAZOLE, midazolam, milrinone, morphine, nafcillin, nesiritide, nitroglycerin, nitroprusside, norepinephrine, octreotide, oxaliplatin, oxytocin, palonosetron, penicillin G potassium, potassium chloride/phosphate, quinupristin/dalfopristin, ranitidine, rocuronium, tacrolimus, tirofiban, tobramycin, trimethoprim/sulfamethoxazole, vancomycin, vasopressin, vecuronium, vinCRIStine, voriconazole, zoledronic acid

SIDE EFFECTS

CNS: *Headache, dizziness,* anxiety, depression, confusion, paresthesia, somnolence, *flushing*

CV: Edema, bradycardia, hypotension, palpitations, pulmonary edema, chest

N

pain, tachycardia, increased angina, arrhythmias, CHF

GI: Nausea, vomiting, gastric upset, constipation, hepatitis, abdominal cramps, dry mouth, sore throat

GU: Nocturia, polyuria

INTEG: Rash, inf site discomfort, Stevens-Johnson syndrome

OTHER: Blurred vision, flushing, sweating, SOB, impotence

PHARMACOKINETICS

Metabolized by liver, excreted in urine 60%, feces 35%

PO: Onset 30 min, peak 1-2 hr, duration 8 hr

PO-SR: Onset unknown, peak 2-6 hr, duration 10-12 hr, half-life 2-5 hr

INTERACTIONS

Increase: effects of digoxin, neuromuscular blocking agents, theophylline, other antihypertensives, nitrates, alcohol, quiNIDine

Increase: niCARdipine effects—cimetidine

Increase: toxicity risk—cycloSPORINE, prazosin, carBAMazepine, quiNIDine, propranolol

Decrease: antihypertensive effect—NSAIDs, rifampin

Drug/Herb

Increase: effect—ginkgo, ginseng, hawthorn

Decrease: effect—ephedra, melatonin, St. John's wort, yohimbe

Drug/Food

Increase: hypotensive effect—grapefruit juice

Increase: LFTs

Drug/Lab Test

Decrease: potassium (IV), phosphate, platelets

NURSING CONSIDERATIONS

Assess:

⚠ Cardiac status: B/P often, pulse, respiration, ECG during long-term treatment

- **Anginal pain:** intensity, location, duration; alleviating, precipitating factors
- Potassium, renal, hepatic studies periodically

⚠ CHF: weight gain, crackles, jugular venous distention, dyspnea, I&O

- **Hypertension:** decreasing B/P; assess salt in diet, smoking, exercise, weight, monitor B/P often

Evaluate:

- Therapeutic response: decreased anginal pain, decreased B/P

Teach patient/family:

- To avoid hazardous activities until stabilized on product, dizziness is no longer a problem
- To limit caffeine consumption; to avoid alcohol products; to take without regard to food, avoid high-fat foods; to swallow sus rel product whole
- To avoid OTC products, grapefruit juice unless directed by prescriber
- **Hypertension:** comply in all areas of medical regimen: diet, exercise, stress reduction, product therapy

⚠ To notify prescriber of irregular heartbeat, SOB, swelling of feet and hands, pronounced dizziness, constipation, nausea, hypotension, change in severity/pattern/incidence of angina

TREATMENT OF OVERDOSE:

Defibrillation, β-agonists, IV calcium, diuretics, atropine for AV block, vasopressor for hypotension

nicotine

(nik'o-teen)

nicotine chewing gum

CVS Nicotine Polacrilex, Equate Nicotine, GNP Nicotine, Good Sense Nicotine, Leader Nicotine Gum, NICOrelief, Nicorette, Publix Stop Smoking Aid, TopCare Nicotine, Walgreens Nicotine

nicotine inhaler (OTC, Rx)

Nicotrol

nicotine lozenge (OTC)

CVS Nicotine Polacrilex, GNP Nicotine Polacrilex, Good Sense Nicotine, Polacrilex, TopCare Nicotine Polacrilex, Walgreens Nicotine Polacrilex

nicotine nasal spray (Rx)

Nicotrol NS

nicotine transdermal (OTC, Rx)

CVS Nicotine Transdermal System, Equate Nicotine Transdermal System, Habitrol ✦, Leader Nicotine Transdermal Patch, Nicoderm CQ, Sunmark Nicotine Transdermal System, Walgreens Nicotine Transdermal Patch

Func. class.: Smoking deterrent
Chem. class.: Ganglionic cholinergic agonist

ACTION: Agonist at nicotinic receptors in peripheral, central nervous systems; acts at sympathetic ganglia, on chemoreceptors of aorta, carotid bodies; also affects adrenalin-releasing catecholamines

USES: Deter cigarette smoking
Unlabeled uses: Gilles de la Tourette's syndrome, ulcerative colitis

CONTRAINDICATIONS: Pregnancy (D) (transdermal, inhaler); hypersensitivity, immediate post-MI recovery period, severe angina pectoris
Precautions: Pregnancy (C) (gum); breastfeeding, vasospastic disease, dysrhythmias, diabetes mellitus, hyperthyroidism, pheochromocytoma, esophagitis, peptic ulcer, coronary/renal/hepatic disease; MRI (patch)

DOSAGE AND ROUTES

Nicotine chewing gum
- **Adult:** chew 1 piece of gum (2 mg nicotine) whenever urge to smoke occurs; dose varies; usually 20 mg/day during first mo, max 24 pieces/day, max 3 mo

Nicotine inhaler
- **Adult: INH** 6 cartridges/day for first 3-6 wk, max 16 cartridges/day × 12 wk

Nicotine lozenge
- **Adult:** if cigarette is desired >30 min after awakening, start with 2-mg lozenge; if <30 min after awakening, start with 4-mg lozenge then again q1-2hr, max 20 lozenges/day or 5 lozenges/6 hr × 6 wk, then 1 lozenge q2-4hr × 2 wk, then 1 lozenge q4-8hr × 2 wk, then discontinue

Nicotine nasal spray
- **Adult:** 1 spray in each nostril 1-2×/hr, max 5×/hr or 40×/day, max 3 mo

Nicotine transdermal/inhaler system
- **Habitrol ✦, Nicoderm:** 21 mg/day × 4-8 wk; 14 mg/day × 2-4 wk; 7 mg/day × 2-4 wk
- **Nicotrol:** 15 mg/day × 12 wk; 10 mg/day × 2 wk; 5 mg/day × 2 wk
- **Nicotrol inhaler:** delivers 30% of nicotine that smoker receives from an actual cigarette

N

Gilles de la Tourette's syndrome (unlabeled)
• **Adult and child: chewing gum** 2 mg chewed × 30 min bid for 1-6 mo; **TRANSDERMAL** 7- or 10-mg patch daily × 2 days

Available forms: Transdermal patch (Habitrol ✱, Nicoderm, nicotine transdermal system) delivering 7, 14, 21 mg/day; (Nicoderm) 5, 10, 15 mg/day; **nicotine inhaler** 4 mg delivered; **nasal spray** 0.5 mg nicotine/actuation; **gum** 2, 4 mg/piece; **lozenge** 2 mg, 4 mg

Administer:
• **Gum:** chew gum slowly for 30 min to promote buccal absorption of product; do not chew >45 min
• Begin product withdrawal after 3 mo of use; do not exceed 6 mo
• **Transdermal patch:** 1 × day to nonhairy, clean, dry area of skin on upper body or upper outer arm; rotate sites to prevent skin irritation
• **Inhaler:** puffing on mouthpiece delivers nicotine through mouth

SIDE EFFECTS

CNS: Dizziness, vertigo, insomnia, headache, confusion, seizures, depression, euphoria, numbness, tinnitus, strange dreams
CV: Dysrhythmias, tachycardia, palpitations, edema, flushing, hypertension
EENT: Jaw ache, irritation in buccal cavity
GI: *Nausea, vomiting, anorexia, indigestion,* diarrhea, abdominal pain, constipation, eructation, irritation
RESP: Breathing difficulty, cough, hoarseness, sneezing, wheezing, bronchial spasm

PHARMACOKINETICS

Onset 15-30 min, metabolized in liver, excreted in urine, half-life 2-3 hr, 30-120 hr (terminal)

INTERACTIONS

Increase: vasoconstriction—ergots, bromocriptine, cabergoline
Increase: effect of—adenosine
Increase: B/P—buPROPion
Decrease: effect of—α-blockers, insulin
Decrease: nicotine clearance—cimetidine

Drug/Food
• Avoid use of gum with acidic foods (colas, coffee) and for 15 min after

NURSING CONSIDERATIONS

Assess:
• **Smoking:** number of cigarettes smoked, years used; **withdrawal:** headache, cravings, restlessness, irritation, drowsiness, insomnia, sore throat, increased appetite
• **Adverse reaction:** irritation of buccal cavity, dislike of taste, jaw ache

Evaluate:
• Therapeutic response: decrease in urge to smoke, decreased need for gum after 3-6 mo

Teach patient/family:
• **Gum:** about all aspects of product use; give package insert to patient and explain
• That gum will not stick to dentures, dental appliances
• That gum is as toxic as cigarettes; that it is to be used only to deter smoking, call prescriber immediately, stop use if difficulty breathing or rash occur
• To avoid use during pregnancy
• **Transdermal patch:** that patch is as toxic as cigarettes; to be used only to deter smoking
• Not to use during pregnancy because birth defects may occur; not to breastfeed
• To keep used and unused system out of reach of children and pets
• To stop smoking immediately when beginning patch treatment
• To apply promptly after removing from protective patch because system may lose strength
• **Nasal spray:** to tilt head back; not to swallow or inhale during administration; after smoking is stopped, to use spray up to 8 wk then discontinue over 6 wk by tapering

- **Lozenges:** to allow to dissolve; to avoid swallowing
- **Inhalation:** to use by inhaler for 20 min

NIFEdipine (Rx)

(nye-fed′i-peen)

Adalat CC, Afeditab CR, Apo-Nifed ♣, Apo-Nifed PA ♣, Nifediac CC, Nifedical XL, Procardia, Procardia XL

Func. class.: Calcium channel blocker, antianginal, antihypertensive

Chem. class.: Dihydropyridine

Do not confuse:
NIFEdipine/niCARdipine/niMODipine

ACTION:
Inhibits calcium ion influx across cell membrane during cardiac depolarization; relaxes coronary vascular smooth muscle; dilates coronary arteries; increases myocardial oxygen delivery in patients with vasospastic angina; dilates peripheral arteries

USES:
Chronic stable angina pectoris, variant angina, hypertension

Unlabeled uses: Migraines, migraine prophylaxis; preterm labor, chronic/acute hypertension (pediatrics), diabetic nephropathy, proteinuria, hiccups

CONTRAINDICATIONS:
Hypersensitivity to this product or dihydropyridine; cardiogenic shock

Precautions: Pregnancy (C), breastfeeding, children, hypotension, sick sinus syndrome, 2nd-/3rd-degree heart block, hypotension <90 mm Hg systolic, hepatic injury, renal disease, acute MI, aortic stenosis, GERD, heart failure

DOSAGE AND ROUTES

- **Adult: PO** Immediate release 10 mg tid, increase in 10-mg increments q7-14days, max 180 mg/24 hr or single dose of 30 mg; **SUS REL** 30-60 mg/day, may increase q7-14days, max 90 mg/day

Hypertension

- **Adult: PO EXT REL** 30-60 mg daily, titrate upward as needed, max 90 mg/day (Adalat CC), 120 mg/day (Procardia XL)
- **Child/adolescent (unlabeled): PO EXT REL** 0.25-0.5 mg/kg/day, max 3 mg/kg/day

Acute hypertensive episodes in pediatric patients (unlabeled)

- **Adolescent/child/infant: PO** 0.2-0.5 mg/kg/dose up to 10 mg (total dose)

Migraine prophylaxis (unlabeled)

- **Adult: PO** 30-180 mg/day

Preterm labor (unlabeled)

- **Pregnant female: PO** Immediate release (Procardia, Adalat) 30-mg loading dose then 10-20 mg q4-6hr; use in monitored settings

Hiccups (unlabeled)

- **Adult:** 10-20 mg tid

Available forms: Caps 10, 20 mg; ext rel tabs (CC, XL) 30, 60, 90 mg

Administer:

- Do not break, crush, or chew ext rel tabs, do not use immediate release caps within 7 days of MI, coronary syndrome
- Without regard to meals; avoid grapefruit juice
- Protect caps from direct light, keep in dry area, do not freeze

SIDE EFFECTS

CNS: *Headache,* fatigue, drowsiness, *dizziness,* anxiety, depression, weakness, insomnia, light-headedness, paresthesia, tinnitus, blurred vision, nervousness, tremor, *flushing*

CV: **Dysrhythmias**, edema, hypotension, palpitations, tachycardia

GI: Nausea, vomiting, diarrhea, gastric upset, constipation, increased LFTs, dry mouth, flatulence, gingival hyperplasia

GU: *Nocturia, polyuria*

HEMA: Bruising, bleeding, petechiae

INTEG: Rash, pruritus, flushing, hair loss, **Stevens-Johnson syndrome, toxic epidermal necrolysis, exfoliative dermatitis**

MISC: Sexual difficulties, cough, fever, chills

N

PHARMACOKINETICS

Metabolized by liver; excreted in urine 60%-80% (metabolites), feces 15%; protein binding 92%-98%

PO: Onset 30 min-1 hr, peak 0.5-2 hr, duration 6-8 hr, half-life 2-5 hr, well absorbed

PO-ER: Duration 24 hr

INTERACTIONS

⚠ Contraindicated with strong CYP3A4 inducers

Increase: level of digoxin, phenytoin, cycloSPORINE, prazosin, carBAMazepine

⚠ **Increase:** NIFEdipine, toxicity—cimetidine, ranitidine

Increase: effects of β-blockers, antihypertensives

Decrease: antihypertensive effect—NSAIDs

Decrease: effects of quiNIDine

Decrease: NIFEdipine level—smoking

Drug/Herb

Increase: effect—ginkgo biloba, ginseng, hawthorn

Decrease: effect—ephedra, melatonin, St. John's wort, yohimbe

Drug/Food

Increase: NIFEdipine level—grapefruit juice

Drug/Lab Test

Increase: CPK, LDH, AST

Positive: ANA, direct Coombs' test

NURSING CONSIDERATIONS

Assess:

- **Anginal pain:** location, intensity, duration, character, alleviating, aggravating factors
- Cardiac status: B/P, pulse, respiration, ECG at baseline and periodically, in those taking antihypertensives, β-blockers, monitor B/P often
- Potassium, renal, hepatic studies periodically during treatment
- For bruising, petechiae, bleeding
- **GI obstruction:** Ext rel products have been associated with rare reports of obstruction in those with strictures and no known GI disease
- **Serious skin disorders:** rash that starts suddenly, fever, cutaneous lesions that may have pustules present; discontinue product

Evaluate:

- Therapeutic response: decreased anginal pain, B/P, activity tolerance

Teach patient/family:

- To avoid hazardous activities until stabilized on product, dizziness is no longer a problem
- To limit caffeine consumption; to avoid alcohol products
- To avoid OTC products unless directed by prescriber without regard to meals, Adelat CC should be taken on empty stomach
- That empty tab shells may appear in stools and is not significant
- **Hypertension:** to comply with all areas of medical regimen: diet, exercise, stress reduction, product therapy
- To change position slowly because orthostatic hypotension is common
- ⚠ To notify prescriber of dyspnea, edema of extremities, nausea, vomiting, severe ataxia, severe rash; changes in pattern, frequency, severity of angina
- To increase fluid intake to prevent constipation
- To check for gingival hyperplasia and report promptly
- Not to discontinue abruptly; to gradually taper

TREATMENT OF OVERDOSE:

Defibrillation, atropine for AV block, vasopressor for hypotension

nilotinib (Rx)

(nye-loe'ti-nib)

Tasigna

Func. class.: Antineoplastic—miscellaneous

Chem. class.: Protein-tyrosine kinase inhibitor

ACTION:
Inhibits BCR-ABL tyrosine kinase created in patients with chronic myeloid leukemia (CML)

USES:
Chronic phase/accelerated phase Philadelphia-chromosome–positive CML that is resistant or intolerant to imatinib

CONTRAINDICATIONS:
Pregnancy (D), breastfeeding, hypersensitivity

Black Box Warning: Hypokalemia, hypomagnesemia, QT prolongation

Precautions: Children, females, geriatric patients, active infections, anemia, cardiac disease, bone marrow suppression, cholestasis, diabetes, gelatin hypersensitivity, infertility, galactose-free diet, lactase deficiency, neutropenia, pancreatitis, thrombocytopenia

Black Box Warning: Hepatic disease

DOSAGE AND ROUTES

- **Adult: PO** 400 mg q12hr, continue until disease progression or unacceptable toxicity

Adjustment after discontinuation of a strong CYP3A4 inducer

- **Adult: PO** reduce to 400 mg/bid

For those taking a strong CYP3A4 inhibitor

- **Adult: PO** reduce dose to 300 mg/day

QT prolongation

- **QTcF >480 msec:** withhold dose

Myelosuppression

- **ANC 1 × 109/L or platelets <50 × 109/L:** withhold dose

Hepatic dose

- **Adult: PO** (Child-Pugh A/B/C) newly diagnosed CML 200 mg bid, then escalation to 300 mg bid initially

Available forms: Caps 150, 200 mg

Administer:

- Do not break, crush, or chew caps; if whole capsule cannot be swallowed, disperse capsule contents in 1 tsp. applesauce
- Without regard to meals; separate doses by 12 hr; make-up dose should not be taken if dose is missed
- Store at 59° F-86° F (15° C- 30° C)

SIDE EFFECTS

CNS: Headache, dizziness, fatigue, fever, flushing, paresthesia

CV: **QT prolongation**, palpitations, **torsades de pointes**, AV block

GI: *Nausea,* **hepatotoxicity, vomiting, dyspepsia,** *anorexia, abdominal pain,* constipation, **pancreatitis,** diarrhea, xerostomia

HEMA: **Neutropenia, thrombocytopenia, anemia, pancytopenia**

INTEG: *Rash,* alopecia, erythema

META: Hyperamylasemia, hyperbilirubinemia, hyperglycemia, hyperkalemia, hypocalcemia, hyponatremia, hypomagnesemia

MISC: Diaphoresis, anxiety

MS: Arthralgia, myalgia, back or bone pain, muscle cramps

RESP: Cough, dyspnea

SYST: **Bleeding, tumor lysis syndrome**

PHARMACOKINETICS

Protein binding 98%, metabolized by CYP3A4, plasma levels 3 hr, elimination half-life 17 hr

INTERACTIONS

- Product interactions are numerous
- Do not use with phenothiazines, pimozide, ziprasidone

⚠ **Increase:** QT prolongation—class IA/III antidysrhythmics, some phenothiazines, β agonists, local anesthetics, tricyclics, haloperidol, chloroquine, droperidol, pentamidine; CYP3A4 inhibitors (amiodarone, clarithromycin, erythromycin, telithromycin, troleandomycin), arsenic trioxide, levomethadyl; CYP3A4 substrates (methadone, pimozide, QUEtiapine, quiNIDine, risperiDONE, ziprasidone)

⚠ **Increase:** hepatotoxicity—acetaminophen

Increase: concs—ketoconazole, itraconazole, erythromycin, clarithromycin

Increase: plasma concs of simvastatin, calcium channel blockers

Increase: plasma conc of warfarin; avoid use with warfarin, use low-molecular-weight anticoagulants instead

Decrease: concs—dexamethasone, phenytoin, carBAMazepine, rifampin, PHENobarbital

Drug/Herb

Decrease: conc—St. John's wort

Drug/Food

Increase: plasma concs—grapefruit juice

NURSING CONSIDERATIONS

Assess:

- **Tumor lysis syndrome:** maintain hydration, correct uric acid before use with this product

> **Black Box Warning: QT prolongation** can occur; monitor ECG, left ventricular ejection fraction (LVEF) at baseline periodically; hypertension, assess for chest pain, palpitations, dyspnea

> **Black Box Warning: Hepatotoxicity:** monitor LFTs before treatment and monthly; if liver transaminases $>5 \times$ IULN, withhold until transaminase levels return to $<2.5 \times$ IULN

- **Myelosuppression:** CBC, differential, platelet count; for bleeding: epistaxis, rectal, gingival, upper GI, genital and wound bleeding; tumor-related hemorrhage may occur rapidly
- ANC and platelets; if ANC $<1 \times 10^9$/L and/or platelets $<50 \times 10^9$/L, stop until ANC $>1.5 \times 10^9$/L and platelets $>75 \times 10^9$/L
- **Electrolytes:** calcium, potassium, magnesium, sodium; lipase, phosphate; hypokalemia, hypomagnesemia should be corrected before use

Evaluate:

- Therapeutic response: decrease in progression of disease

Teach patient/family:

- To report adverse reactions immediately: SOB, bleeding
- About reason for treatment, expected results
- That many adverse reactions may occur
- To avoid persons with known upper respiratory tract infections; immunosuppression is common
- To watch for signs, symptoms of low potassium or magnesium

nisoldipine (Rx)

(nye-sole′dih-peen)

Sular

Func. class.: Calcium channel blocker, antihypertensive

Chem. class.: Dihydropyridine

Do not confuse:
nisoldipine/NIFEdipine/niMODipine

ACTION: Inhibits calcium ion influx across the cell membrane, thereby resulting in the dilation of peripheral arteries

USES: Essential hypertension, alone or in combination with other antihypertensives

Unlabeled uses: Variant (Prinzmetal's) angina, stable angina pectoris

CONTRAINDICATIONS: Hypersensitivity to this product or dihydropyridines; sick sinus syndrome; 2nd-/3rd-degree heart block; aortic stenosis

Precautions: Pregnancy (C), breastfeeding, children, geriatric patients, CHF, hypotension <90 mm Hg systolic, hepatic injury, renal disease, acute MI, unstable angina, CAD, cardiogenic shock

DOSAGE AND ROUTES

Hypertension

• **Adult: PO** 17 mg/day initially, may increase by 8.5 mg/wk, usual dose 17-34 mg/day, max 34 mg/day

• **Geriatric/hepatic dose: PO** 8.5 mg/day, increase based on patient response

Variant (Prinzmetal's) angina/ stable angina pectoris (unlabeled)

• **Adult: PO** 17-34 mg/day, max 34 mg/day

Hepatic dose

• **Adult: PO** 8.5 mg/day

Available forms: Ext rel tabs 8.5, 17, 20, 25.5, 30, 34, 40 mg

Administer:

PO route

• Swallow whole; do not break, crush, or chew

• Once daily as whole tablet; avoid high-fat foods, grapefruit juice

SIDE EFFECTS

CNS: Headache, fatigue, drowsiness, dizziness, anxiety, depression, nervousness, insomnia, lightheadedness, paresthesia, tinnitus, psychosis, somnolence, ataxia, confusion, malaise, migraine, flushing

CV: Dysrhythmia, edema, CHF, hypotension, palpitations, MI, pulmonary edema, tachycardia, syncope, AV block, angina, chest pain, ECG abnormalities

GI: Nausea, vomiting, diarrhea, gastric upset, constipation, increased LFTs, dry mouth, dyspepsia, dysphagia, flatulence

GU: Nocturia, hematuria, dysuria

HEMA: Anemia, leukopenia, petechiae

INTEG: Rash, pruritus

MISC: Sexual difficulties, cough, nasal congestion, SOB, wheezing, epistaxis, dyspnea, gingival hyperplasia, chills, fever, gout, sweating

PHARMACOKINETICS

Metabolized by liver, excreted in urine, peak 6-12 hr, protein binding 99%, half-life 7-12 hr

INTERACTIONS

Increase: effects of β-blockers, antihypertensives, digoxin

Increase: nisoldipine level—CYP3A4 inhibitors, cimetidine, ranitidine, azole antifungals

Decrease: nisoldipine effect—CYP3A4 inducers, hydantoins

Drug/Herb

Increase: B/P—ephedra, melatonin

Decrease: B/P—hawthorn

Decrease: nisoldipine effect—St. John's wort, ginseng, ginkgo biloba

Drug/Food

Increase: nisoldipine level—high-fat foods

Increase: hypotensive effect—grapefruit juice

NURSING CONSIDERATIONS

Assess:

• Cardiac status: B/P, pulse, respiration, ECG before treatment and periodically

• **CHF:** weight gain, jugular venous distention, edema, crackles, I&O ratios

• **Angina:** frequency, severity of attacks, if angina worsens, report immediately

Evaluate:

• Therapeutic response: decreased B/P

Teach patient/family:

• To avoid hazardous activities until stabilized on product, dizziness is no longer a problem

• To report nausea, dizziness, swelling, SOB, palpitations, severe headache

• To avoid OTC products unless directed by prescriber; to avoid grapefruit juice

• About the importance of complying with all areas of the medical regimen: diet, exercise, stress reduction, product therapy

• To rise slowly to prevent orthostatic hypotension

• If dose is missed, to take as soon as remembered; not to double dose

• How to perform B/P monitoring at home

TREATMENT OF OVERDOSE:

Defibrillation, atropine for AV block, vasopressor for hypotension

N

nitazoxanide (Rx)

(nye-taz-ox'a-nide)

Alinia

Func. class.: Antiprotozoal

ACTION: Interferes with DNA/RNA synthesis in protozoa

USES: Diarrhea caused by *Cryptosporidium parvum* or *Giardia lamblia*

CONTRAINDICATIONS: Hypersensitivity

Precautions: Pregnancy (B), breastfeeding, children <1 yr or >11 yr, renal/hepatic disease, diabetes mellitus (contains sucrose), HIV, immunocompromised patients

DOSAGE AND ROUTES

- **Adult: PO** 500 mg q12hr × 3 days
- **Child 4-11 yr: PO** 10 ml (200 mg) q12hr × 3 days
- **Child 12-47 mo: PO** 5 ml (100 mg) q12hr × 3 days

Available forms: Powder for oral susp 100 mg/5 ml; tab 500 mg

Administer:

PO route

- With food
- Shake oral suspension before giving, discard after 1 wk

SIDE EFFECTS

CNS: *Dizziness, fever, headache*

CV: Hypotension

GI: *Nausea,* anorexia, flatulence, increased appetite, enlarged salivary glands, abdominal pain, diarrhea, vomiting

HEMA: Anemia, leukopenia, neutropenia

INTEG: Pruritus, sweating

MISC: Increased creatinine, pale yellow eye discoloration, rhinitis, discolored urine, infection, malaise

PHARMACOKINETICS

Peak 1-4 hr; excreted in urine, bile, feces; hydrolyzed to active metabolite, which undergoes conjugation; metabolite protein binding >99%

INTERACTIONS

- Competes for binding sites: other highly protein-bound products; phenytoin, salicylates

Drug/Lab Test

Increase: creatinine, GPT

NURSING CONSIDERATIONS

Assess:

- Signs of infection
- Bowel pattern before, during treatment

Evaluate:

- Therapeutic response: C&S negative for organism, decreased diarrhea

Teach patient/family:

- To take with food; shake susp well before each dose, discard susp after 7 days

nitrofurantoin (Rx)

(nye-troe-fyoor'an-toyn)

Furadantin, Macrobid, Macrodantin, Novo-Furantoin ✦

Func. class.: Urinary tract antiinfective

Chem. class.: Synthetic nitrofuran derivative

ACTION: Inhibits bacterial acetyl-CoA inteference with carbohydrate metabolism

USES: Urinary tract infections caused by *Escherichia coli, Klebsiella, Pseudomonas, Proteus vulgaris, Proteus morganii, Serratia, Citrobacter, Staphylococcus aureus, Staphylococcus epidermidis, Enterococcus, Salmonella, Shigella*

CONTRAINDICATIONS:
Infants <1 mo, hypersensitivity, anuria, severe renal disease CCr <60 ml/min, at term pregnancy (38-42 wk), labor, delivery, cholestatic jaundice due to nitrofurantoin therapy

Precautions: Pregnancy (B), breastfeeding, geriatric patients, G6PD deficiency, GI disease, diabetes

DOSAGE AND ROUTES

Active infections

- **Adult: PO** 50-100 mg qid after meals
- **Child: PO** 5-7 mg/kg/day in 4 divided doses

Chronic suppression

- **Adult: PO** 50-100 mg q PM
- **Child: PO** 1-2 mg/kg/day in PM or 0.5-1 mg/kg q12hr if dose not well tolerated

Available forms: Caps 25, 50, 100 mg; susp 25 mg/5 ml; macrocrystal caps (Macrodantin) 25, 50, 100 mg; Macrobid cap 100 mg (25 macrocrystals, 75 monohydrate)

Administer:

PO route

- Give with meals
- Do not break, crush, chew, or open tabs, caps, store in original container
- Two daily doses if urine output is high or if patient diabetic
- Use calibrated device to measure liquid product; may mix water, fruit juice; rinse mouth after liquid product; staining of teeth may occur

SIDE EFFECTS

CNS: *Dizziness, headache,* drowsiness, peripheral neuropathy, chills, confusion, vertigo

CV: **Bundle branch block**, chest pain

GI: *Nausea, vomiting, abdominal pain, diarrhea,* **cholestatic jaundice**, loss of appetite, **pseudomembranous colitis, hepatitis, pancreatitis**

HEMA: **Anemia, agranulocytosis, hemolytic anemia, leukopenia, thrombocytopenia**

INTEG: Pruritus, rash, urticaria, angioedema, alopecia, tooth staining, **exfoliative dermatitis, Stevens-Johnson syndrome**

MS: Arthralgia, myalgia, numbness, peripheral neuropathy

RESP: Cough, dyspnea, **pneumonitis, pulmonary fibrosis or infiltrate**

SYST: **Superinfection**, SLE-like syndrome

PHARMACOKINETICS

PO: Half-life 20-60 min; crosses blood-brain barrier, placenta; enters breast milk; excreted as inactive metabolites in liver, unchanged in urine; protein binding 60%-90%

INTERACTIONS

Increase: antagonistic effect—norfloxacin

Increase: levels of nitrofurantoin—probenecid

Decrease: absorption of magnesium trisilicate antacid

Drug/Lab Test

Increase: BUN, alk phos, bilirubin, creatinine, blood glucose

NURSING CONSIDERATIONS

Assess:

- Blood count during chronic therapy, LFTs, pulmonary function tests
- **Urinary tract infection:** burning, pain on urination; fever; cloudy, foul-smelling urine; I&O ratio: C&S before treatment, after completion; serum creatinine, BUN
- **Pseudomembraneous colitis: diarrhea with mucus, abdominal pain, fever, fatigue, anorexia; may be treated with vancomycin or metroNIDAZOLE**
- CNS symptoms: insomnia, vertigo, headache, drowsiness, seizures

⚠ **Hepatotoxicity: yellowing of skin or eyes, dark urine, clay-colored stools; monitor AST, ALT**

⚠ **Pulmonary fibrosis, pneumonitis: dyspnea, tachypnea, persistent cough**

⚠ **Serious skin disorders: fever, flushing, rash, urticaria, pruritus**

- **Peripheral neuropathy:** paresthesias (more common in diabetes mellitus, electrolyte imbalances, vit B deficiency, debilitated patients)

Evaluate:

- Therapeutic response: decreased dysuria, fever; negative C&S

Teach patient/family:

- To notify prescriber of continued symptoms of UTI, fever, myalgias, arthralgias, numbness or tingling of extremities
- To take with food or milk; to avoid alcohol
- To protect susp from freezing; shake well before taking
- That product may cause drowsiness; to seek aid with walking, other activities; not to drive or operate machinery while taking medication
- That diabetics should monitor blood glucose levels
- That product may turn urine rust-yellow to brown

⚠ **Pseudomembranous colitis:** fever; diarrhea with mucus, pus, or blood; report immediately

nitrofurazone topical

See Appendix B

nitroglycerin (Rx)

(nye-troe-gli′ser-in)

extended release caps (Rx)

Nitro-Time

topical ointment (Rx)

Nitro-Bid

rectal ointment

Rectiv

SL (Rx)

Nitrostat

translingual spray (Rx)

Nitrolingual

transdermal (Rx)

Minitran, Nitro-Dur

Func. class.: Coronary vasodilator, antianginal

Chem. class.: Nitrate

Do not confuse:
Nitro-Bid/Nicobid

ACTION: Decreases preload and afterload, which are responsible for decreasing left ventricular end-diastolic pressure, systemic vascular resistance; dilates coronary arteries, improves blood flow through coronary vasculature, dilates arterial and venous beds systemically

USES: Chronic stable angina pectoris, prophylaxis of angina pain, CHF, acute MI, controlled hypotension for surgical procedures, anal fissures

Unlabeled uses: Pulmonary hypertension, hemorrhoids, retained placenta

CONTRAINDICATIONS: Hypersensitivity to this product or nitrites; severe anemia, increased intracranial pressure, cerebral hemorrhage, closed-angle glaucoma, cardiac tamponade, cardiomyopathy, constrictive pericarditis

Precautions: Pregnancy (C), breastfeeding, children, postural hypotension, severe renal/hepatic disease, acute MI, abrupt discontinuation, hyperthyroidism

DOSAGE AND ROUTES

• **Adult: SL** Dissolve tab under tongue when pain begins; may repeat q5min until relief occurs; take ≤3 tabs/15 min; use 1 tab prophylactically 5-10 min before activities; **SUS CAP** q6-12hr on empty stomach; **TOP** 1-2 in q8hr, increase to 4 in q4hr as needed; **IV** 5 mcg/min then increase by 5 mcg/min q3-5min; if no response after 20 mcg/min, increase by 10-20 mcg/min until desired response; **TRANS PATCH** apply a pad daily to a site free of hair; remove patch at bedtime to provide 10-12 hr nitrate-free interval to avoid tolerance

• **Child: IV** Initially 0.25-0.5 mcg/kg/min, titrate to patient response, usual dose 1-3 mcg/kg/min transmucosal

Anal fissures (Rectiv)

• **Adult: Rectal** Apply 1 inch of 0.4% ointment q12hr × 3 wk

Available forms: Translingual aero 0.4 mg/metered spray; sus rel tabs 2.6, 6.5, 9 mg; SL tabs 0.3, 0.4, 0.6 mg; topical oint 2%; trans syst 0.1, 0.2, 0.3, 0.4, 0.6, 0.8 mg/hr; inj sol 25 mg/250 ml, 50 mg/250 ml, 50 mg/500 ml, 100 mg/250 ml, 200 mg/500 ml; rectal ointment 0.4% (Rectiv)

Administer:

• **Topical ointment** should be measured on papers supplied; use paper to spread on nonhairy area of chest, abdomen, thigh skin; thin layer spread over 2-3 inches; do not rub

PO route

• Swallow sus rel products whole; do not break, crush, or chew

• With 8 oz water on empty stomach (oral tablet) 1 hr before or 2 hr after meals

• **SL:** should be dissolved under tongue, not swallowed

• **Aerosol** sprayed under tongue **(nitrolingual)**, not inhaled; prime before 1st-time use or if product has not been used in >6 wk; press valve head with forefinger

Transdermal route

• Apply new TD patch daily; remove after 12-14 hr to prevent tolerance

Rectal route

• Cover finger with plastic wrap, disposable glove, or finger cot; lay finger alongside 1-inch dosing line on carton; squeeze tube until equal to 1-inch dosing line; insert covered finger gently into anal canal no further than 1st finger joint and apply to sides; wash hands thoroughly; if too painful, apply directly to outside of anus

Continuous IV INF route

• Diluted in D_5, D_5W, 0.9% NaCl for inf to 200-400 mcg/ml, depending on patient's fluid status; common dilution 50 mg/250 ml, use controlled inf device; use glass inf bottles, non–polyvinyl-chloride inf tubing; titrate to patient response; do not use filters

Y-site compatibilities: Acyclovir, alfentanil, amikacin, aminocaproic acid, aminophylline, amiodarone, amphotericin B cholesteryl, amphotericin B lipid complex, amphotericin B liposome, anidulafungin, argatroban, ascorbic acid, atenolol, atracurium, atropine, azaTHIOprine, aztreonam, benztropine, bivalirudin, bleomycin, bumetanide, buprenorphine, butorphanol, calcium chloride/gluconate, CARBOplatin, caspofungin, cefamandole, ceFAZolin, cefmetazole, cefonicid, cefoperazone, cefotaxime, cefoTEtan, cefOXitin, cefTAZidime, ceftizoxime, cefTRIAXone, cefuroxime, cephalothin, cephapirin, chloramphenicol, chlorproMAZINE, cimetidine, cisatracurium, CISplatin, clindamycin, cloNIDine, cyanocobalamin, cyclophosphamide, cycloSPORINE, cytarabine, DACTINomycin, dexamethasone, digoxin, diltiazem, diphenhydrAMINE, DOBUTamine, DOCEtaxel, DOPamine, doxacurium, DOXOrubicin, doxycycline, drotrecogin alfa, enalaprilat, ePHEDrine, EPINEPHrine, epirubicin, epoetin alfa, eptifiba-

N

tide, ertapenem, erythromycin, esmolol, etoposide, famotidine, fenoldopam, fentaNYL, fluconazole, fludarabine, fluorouracil, folic acid, ganciclovir, gatifloxacin, gemcitabine, gemtuzumab, gentamicin, glycopyrrolate, granisetron, heparin, hydrocortisone, HYDROmorphone, hydrOXYzine, IDArubicin, ifosfamide, imipenem-cilastatin, indomethacin, insulin (regular), irinotecan, isoproterenol, ketorolac, labetalol, lidocaine, linezolid, LORazepam, magnesium sulfate, mannitol, mechlorethamine, meperidine, metaraminol, methicillin, methotrexate, methoxamine, methyldopate, methylPREDNISolone, metoclopramide, metroNIDAZOLE, mezlocillin, micafungin, miconazole, midazolam, milrinone, minocycline, mitoXANtrone, morphine, moxalactam, mycophenolate, nafcillin, nalbuphine, naloxone, nesiritide, netilmicin, niCARdipine, nitroprusside, norepinephrine, octreotide, ondansetron, oxacillin, oxaliplatin, oxytocin, PACLitaxel, palonosetron, pamidronate, pancuronium, pantoprazole, papaverine, PEMEtrexed, penicillin G potassium/sodium, pentamidine, pentazocine, PENTobarbital, PHENobarbital, phentolamine, phenylephrine, phytonadione, piperacillin, piperacillin-tazobactam, polymyxin B, potassium chloride, procainamide, prochlorperazine, promethazine, propofol, propranolol, protamine, pyridoxine, quiNIDine, quinupristin-dalfopristin, ranitidine, remifentanil, ritodrine, rocuronium, sodium bicarbonate, succinylcholine, SUFentanil, tacrolimus, teniposide, theophylline, thiamine, thiopental, thiotepa, ticarcillin, ticarcillin-clavulanate, tigecycline, tirofiban, tobramycin, tolazoline, trimetaphan, urokinase, vancomycin, vasopressin, vecuronium, verapamil, vinCRIStine, vinorelbine, voriconazole, warfarin, zoledronic acid

SIDE EFFECTS

CNS: *Headache, flushing, dizziness*
CV: *Postural hypotension,* tachycardia, collapse, syncope, palpitations
GI: Nausea, vomiting
INTEG: Pallor, sweating, rash

PHARMACOKINETICS

Metabolized by liver, excreted in urine, half-life 1-4 min
SUS REL: Onset 20-45 min, duration 3-8 hr
SL: Onset 1-3 min, duration 30 min
TRANSDERMAL: Onset 30 min-1 hr, duration 12-24 hr
AEROSOL: Onset 2 min, duration 30-60 min
TOPICAL OINT: Onset 30-60 min, duration 2-12 hr
IV: Onset 1-2 min, duration 3-5 min

INTERACTIONS

• Severe hypotension, CV collapse: alcohol
Increase: effects of β-blockers, diuretics, antihypertensives, calcium channel blockers
Increase: fatal hypotension—sildenafil, tadalafil, vardenafil; do not use together
Increase: nitrate level—aspirin
Decrease: heparin—IV nitroglycerin
Drug/Lab Test
Increase: urine catecholamine, urine VMA
False increase: cholesterol

NURSING CONSIDERATIONS

Assess:
• **Pain:** duration, time started, activity being performed, character
• Orthostatic B/P, pulse before and after administration
• Tolerance if taken over long period
• Headache, lightheadedness, decreased B/P; may indicate a need for decreased dosage
Evaluate:
• Therapeutic response: decrease, prevention of anginal pain
Teach patient/family:
• To place buccal tab between lip and gum above incisors or between cheek and gum

• To keep tabs in original container; to replace q6mo because effectiveness is lost; to keep away from heat, moisture, light
• That if 3 SL tabs in 15 min do not relieve pain, to seek immediate medical attention
• To avoid alcohol
• That product may cause headache; that tolerance usually develops; to use nonopioid analgesic
• That product may be taken before stressful activity: exercise, sexual activity
• That SL may sting when product comes in contact with mucous membranes
• To avoid hazardous activities if dizziness occurs
• To comply with complete medical regimen
• To make position changes slowly to prevent fainting
⚠ Never to use erectile dysfunction products (sildenafil, tadalafil, vardenafil); may cause severe hypotension, death

⚠ **HIGH ALERT**

nitroprusside (Rx)

(nye-troe-pruss′ide)

Nitropress

Func. class.: Antihypertensive, vasodilator

ACTION: Directly relaxes arteriolar, venous smooth muscle, thereby resulting in reduction in cardiac preload and afterload

USES: Hypertensive crisis/urgency/induction; to decrease bleeding by creating hypotension during surgery; acute CHF

Unlabeled uses: Postoperative hypertension, mitral regurgitation

CONTRAINDICATIONS: Hypersensitivity, hypertension (compensatory) due to aortic coarctation or AV shunting, acute CHF associated with reduced peripheral vascular resistance, AV shunt, Leber's disease, toxic amblyopia

Black Box Warning: Cyanide toxicity

Precautions: Anemia, increased intracranial pressure, pregnancy (C), breastfeeding, children, geriatric patients, hypovolemia, electrolyte imbalances, renal/hepatic disease, hypothyroidism

Black Box Warning: Hypotension

DOSAGE AND ROUTES

• **Adult and child: IV INF** 0.25-10 mcg/kg/min; max 10 mcg/kg/min

Renal dose

• **Adult: IV INF** CCr <60 ml/min, maintain doses <3 mcg/kg/min to reduce thiocyanate accumulation

Available forms: Inj 50 mg/2 ml

Administer:

• Antidote is sodium thiosulfate

Continuous IV INF route

• Depending on B/P reading q15min
• Reconstitute 50 mg/2-3 ml of D_5W, further dilute in 250, 500, or 1000 ml of D_5W to 200, 100, 50 mcg/ml, respectively; use inf pump only; wrap bottle with aluminum foil to protect from light; observe for color change in inf; discard if highly discolored (blue, green, dark red); titrate to patient response, protect from light

Y-site compatibilities: Alfentanil, alprostadil, amikacin, aminocaproic acid, aminophylline, amphotericin B lipid compex, amphotericin B liposome, anidulafungin, argatroban, atenolol, atropine, aztreonam, benztropine, bivalirudin, bleomycin, bumetanide, buprenorphine, butorphanol, calcium chloride/gluconate, CARBOplatin, cefamandole, ceFAZolin, cefmetazole, cefonicid, cefoperazone, cefotaxime, cefoTEtan, cefOXitin, cefTAZidime, ceftizoxime, cefTRIAXone, cefuroxime, cephalothin, chloramphenicol, cimetidine, CISplatin, clindamycin, cyanocobalamin, cyclophosphamide, cycloSPORINE, cytarabine, DACTINomycin, DAPTOmycin, dexamethasone, digoxin, diltiazem,

DOCEtaxel, DOPamine, doxacurium, DOXOrubicin, doxycycline, enalaprilat, ePHEDrine, EPINEPHrine, epirubicin, epoetin alfa, eptifibatide, ertapenem, esmolol, etoposide, famotidine, fenoldopam, fentaNYL, fluconazole, fludarabine, fluorouracil, folic acid, furosemide, ganciclovir, gatifloxacin, gemcitabine, gemtuzumab, gentamicin, glycopyrrolate, granisetron, heparin, hydrocortisone, HYDROmorphone, IDArubicin, ifosfamide, inamrinone, indomethacin, insulin (regular), isoproterenol, ketorolac, labetalol, lidocaine, linezolid, LORazepam, magnesium sulfate, mannitol, mechlorethamine, meperidine, metaraminol, methicillin, methoxamine, methyldopate, methylPREDNISolone, metoclopramide, metoprolol, metroNIDAZOLE, mezlocillin, micafungin, miconazole, midazolam, milrinone, minocycline, morphine, moxalactam, multiple vitamins injection, nafcillin, nalbuphine, naloxone, nesiritide, netilmicin, niCARdipine, nitroglycerin, norepinephrine, octreotide, ondansetron, oxacillin, oxaliplatin, oxytocin, PACLitaxel, palonosetron, pamidronate, pancuronium, pantoprazole, penicillin G potassium/sodium, pentamidine, PENTobarbital, PHENobarbital, phentolamine, phenylephrine, phytonadione, piperacillin, piperacillin-tazobactam, polymyxin B, potassium chloride/phosphates, procainamide, propofol, propranolol, protamine, pyridoxine, ranitidine, ritodrine, rocuronium, sodium acetate/bicarbonate, succinylcholine, SUFentanil, tacrolimus, teniposide, theophylline, thiamine, ticarcillin, ticarcillin-clavulanate, tigecycline, tirofiban, tobramycin, tolazoline, trimetaphan, urokinase, vancomycin, vasopressin, vecuronium, verapamil, vinCRIStine, zoledronic acid

SIDE EFFECTS

CNS: *Dizziness, headache,* agitation, twitching, decreased reflexes, *restlessness*

CV: *Bradycardia,* ECG changes, tachycardia, *hypotension*

GI: Nausea, vomiting, abdominal pain

INTEG: Pain, irritation at inj site, sweating

MISC: Cyanide, thiocyanate toxicity, flushing, hypothyroidism

PHARMACOKINETICS

IV: Onset 1-2 min, duration 1-10 min, half-life 2 min; metabolized in liver, excreted in urine

INTERACTIONS

Increase: severe hypotension—ganglionic blockers, volatile liquid anesthetics, halothane, enflurane, circulatory depressants

Drug/Herb

Increase: antihypertensive effect—hawthorn

NURSING CONSIDERATIONS

Assess:

- Electrolytes: potassium, sodium, chloride, CO_2, CBC, serum glucose, serum methemoglobin if pulmonary O_2 levels are decreased; use IV 1-2 mg/kg methylene blue given over several min for methemoglobinemia, ABGs
- Renal studies: catecholamines, BUN, creatinine
- Hepatic studies: AST, ALT, alk phos

Black Box Warning: **Hypotension:** B/P by direct means if possible; check ECG continuously; pulse, jugular venous distention; PCWP; rebound hypertension may occur after nitroprusside is discontinued, give only with emergency equipment nearby, rapid decrease in B/P may occur

- Weight daily, I&O

⚠ **Thiocyanate, lactate, cyanide toxicity:** obtain levels daily if inf >3 mcg/kg/min; thiocyanate level should be ≤1 mmol/L; thiocyanate toxicity includes confusion, weakness, seizures, hyperreflexia, psychosis, tinnitus, coma

- Nausea, vomiting, diarrhea

• Edema in feet, legs daily; skin turgor, dryness of mucous membranes for hydration status
• Crackles, dyspnea, orthopnea q30min
• For decrease in bicarbonate, Pco_2 blood pH, acidosis

Evaluate:
• Therapeutic response: decreased B/P, decreasing symptoms of cardiogenic shock or cardiac pump failure

Teach patient/family:
• To report headache, dizziness, loss of hearing, blurred vision, dyspnea, faintness, pain at IV site
• About the reason for giving product and expected results

nizatidine (OTC, Rx)

(ni-za'ti-deen)

Axid, Axid AR

Func. class.: H_2-receptor antagonist
Chem. class.: Substituted thiazole

ACTION: Blocks H_2-receptors, thereby reducing gastric acid output

USES: Benign gastric and duodenal ulceration, prevention of duodenal ulcer recurrence, symptomatic relief of gastroesophageal reflux, heartburn prevention

CONTRAINDICATIONS: Hypersensitivity

Precautions: Pregnancy (B), breastfeeding, renal/hepatic impairment (reduce dose in renal impairment)

DOSAGE AND ROUTES

Gastric and duodenal ulcer
• **Adult: PO** 300 mg at night or 150 mg bid for 4-8 wk; maintenance 150 mg at night

Prophylaxis of duodenal ulcer
• **Adult: PO** 150 mg/day at bedtime

Gastroesophageal reflux
• **Adult and child ≥12 yr: PO** 150 mg bid × ≤12 wk, max 300 mg/day

Heartburn prevention
• **Adult: PO** 75 mg before eating bid

Renal dose
• **Adult: PO** CCr 20-50 ml/min, give 150 mg every other day; CCr <20 ml/min, give 150 mg q72hr

Available forms: Caps 150, 300 mg; tabs 75 mg

Administer:
• With meals for prolonged product effect; antacids 1 hr before or 1 hr after product; at bedtime if taken daily

SIDE EFFECTS

CNS: Headache, somnolence, confusion, abnormal dreams, dizziness
CV: **Cardiac dysrhythmias, cardiac arrest**
ENDO: Gynecomastia
GI: Elevated hepatic enzymes, **hepatitis,** jaundice, nausea
HEMA: **Thrombocytopenia, agranulocytosis, aplastic anemia**
INTEG: Pruritus, sweating, urticaria, **exfoliative dermatitis**
METAB: Hyperuricemia
MS: Myalgia
RESP: **Bronchospasm, laryngeal edema, pneumonia**

PHARMACOKINETICS

Partially metabolized by liver, excreted by kidneys, plasma half-life 1-2.8 hr, 70% absorbed orally, small amount (0.1% of plasma conc) enters breast milk, 35% bound to plasma proteins

INTERACTIONS

Increase: **GI obstruction risk—NIFEdipine (ext rel tabs)**
Increase: effect of—mefloquine
Decrease: effect of—ketoconazole, itraconazole, atazanavir, cefditoren, cefpodoxime, delavirdine, gefitinib, raltegravir

Drug/Lab Test
Increase: ALT, AST, serum creatinine
False negative: allergy skin tests

N

NURSING CONSIDERATIONS

Assess:

- **GI pain:** epigastric, abdominal, character, alleviating factors, hematemesis, occult blood in stool, heartburn, GERD

⚠ **Agranulocytosis:** CBC with differential if patient receiving long-term therapy

Evaluate:

- Decreased GI pain, heartburn, GERD; resolution of gastric, duodenal ulcers

Teach patient/family:

- That gynecomastia, impotence may occur, are reversible
- To avoid driving or other hazardous activities until stabilized on product; that dizziness may occur
- To avoid black pepper, caffeine, alcohol, harsh spices, extremes in temperature of food
- To avoid OTC preparations: aspirin, cough, cold preparations

TREATMENT OF OVERDOSE:

Symptomatic and supportive therapy is recommended; activated charcoal, emesis, or lavage may reduce absorption

⚠ *HIGH ALERT*

norepinephrine (Rx)

(nor-ep-i-nef′rin)

Levophed

Func. class.: Adrenergic

Chem. class.: Catecholamine

Do not confuse:
norepinephrine/EPINEPHrine

ACTION: Causes increased contractility and heart rate by acting on β-receptors in heart; also acts on α-receptors, thereby causing vasoconstriction in blood vessels; B/P is elevated, coronary blood flow improves, and cardiac output increases

USES: Acute hypotension, shock

CONTRAINDICATIONS: Hypersensitivity to this product or cyclopropane/halothane anesthesia, sulfites; ventricular fibrillation, tachydysrhythmias, pheochromocytoma, hypotension, hypovolemia

Precautions: Pregnancy (C), breastfeeding, geriatric patients, arterial embolism, peripheral vascular disease, hypertension, hyperthyroidism, cardiac disease

Black Box Warning: Extravasation

DOSAGE AND ROUTES

- **Adult: IV INF** 0.5-1 mcg/min titrated to B/P; maintenance 2-4 mcg/min; max 30 mcg/min
- **Child: IV INF** 0.1-0.2 mcg/kg/min titrated to B/P; max 2 mcg/kg/min

Available forms: Inj 1 mg/ml

Administer:

- Plasma expanders for hypovolemia

Continuous IV INF route

- Dilute with 500-1000 ml D_5W or D_5/0.9% NaCl; average dilution 4 mg/1000 ml diluent (4 mcg base/ml); give as inf 2-3 ml/min; titrate to response
- Store reconstituted sol in refrigerator ≤24 hr, protect from light, store unopened product at room temperature, do not use discolored sol

Y-site compatibilities: Alfentanil, amikacin, amiodarone, anidulafungin, argatroban, ascorbic acid, atenolol, atracurium, atropine, aztreonam, benztropine, bivalirudin, bleomycin, bumetanide, buprenorphine, butorphanol, calcium chloride/gluconate, CARBOplatin, caspofungin, cefamandole, ceFAZolin, cefmetazole, cefonicid, cefoperazone, cefotaxime, cefoTEtan, cefOXitin, cefTAZidime, ceftizoxime, ceftobiprole, cefTRIAXone, cefuroxime, cephalothin, chloramphenicol, chlorproMAZINE, cimetidine, cisatracurium, CISplatin, clindamycin, cloNIDine, cyanocobalamin, cyclophosphamide, cycloSPORINE, cytarabine, DAPTOmycin, dexamethasone, digoxin, diltiazem, diphenhydrAMINE, DOBUTamine, DOCEtaxel, DOPamine, doripenem, doxycycline, enalaprilat, ePHEDrine, EPINEPHrine, epirubicin, epoetin alfa, ertapenem, erythromycin, esmolol, etoposide, famotidine, fenoldopam, fentaNYL, fluconazole, fludarabine, gatifloxacin, gemcitabine, gentamicin, glycopyrrolate, granisetron, heparin, hydrocortisone, HYDROmorphone, hydrOXYzine, IDArubicin, ifosfamide, imipenem-cilastatin, irinotecan, isoproterenol, ketorolac, labetalol, lidocaine, linezolid, LORazepam, magnesium sulfate, mannitol, mechlorethamine, meperidine, meropenem, metaraminol, methicillin, methotrexate, methoxamine, methyldopate, methylPREDNISolone, metoclopramide, metoprolol, metroNIDAZOLE, mezlocillin, micafungin, miconazole, midazolam, milrinone, minocycline, mitoXANtrone, morphine, moxalactam, multiple vitamins injection, mycophenolate, nafcillin, nalbuphine, naloxone, netilmicin, niCARdipine, nitroglycerin, nitroprusside, octreotide, ondansetron, oxacillin, oxaliplatin, oxytocin, PACLitaxel, palonosetron, pamidronate, pancuronium, papaverine, PEMEtrexed, penicillin G potassium/sodium, pentamidine, pentazocine, phenylephrine, phytonadione, piperacillin, piperacillin-tazobactam, polymyxin B, potassium chloride, procainamide, prochlorperazine, promethazine, propofol, propranolol, protamine, pyridoxine, quiNIDine, ranitidine, remifentanil, ritodrine, succinylcholine, SUFentanil, tacrolimus, teniposide, theophylline, thiamine, thiotepa, ticarcillin, ticarcillin-clavulanate, tigecycline, tirofiban, tobramycin, tolazoline, trimetaphan, urokinase, vancomycin, vasopressin, vecuronium, verapamil, vinCRIStine, vinorelbine, vitamin B complex with C, voriconazole, zoledronic acid

SIDE EFFECTS

CNS: *Headache,* anxiety, dizziness, insomnia, restlessness, tremor, **cerebral hemorrhage**

CV: *Palpitations, tachycardia, hypertension, ectopic beats, angina*

GI: *Nausea, vomiting*

GU: Decreased urine output

INTEG: **Necrosis, tissue sloughing with extravasation, gangrene**

RESP: Dyspnea

SYST: **Anaphylaxis**

PHARMACOKINETICS

IV: Onset 1-2 min; metabolized in liver; excreted in urine (inactive metabolites); crosses placenta

INTERACTIONS

Increase: dysrhythmias—general anesthetics

- **Incompatible with alkaline solutions: sodium, bicarbonate**
- **Severe hypertension: guanethidine**

⚠ **Do not use within 2 wk of MAOIs, antihistamines, ergots, methyldopa, oxytocics, tricyclics, guanethidine because hypertensive crisis may result**

Increase: B/P—oxytocics

Increase: pressor effect—tricyclics, MAOIs

Decrease: norepinephrine action—α-blockers

NURSING CONSIDERATIONS

Assess:

- I&O ratio; notify prescriber if output <30 ml/hr
- B/P, pulse q2-3min after parenteral route, ECG during administration continuously; if B/P increases, product is decreased, CVP or PWP during inf if possible
- Paresthesias and coldness of extremities; peripheral blood flow may decrease

Black Box Warning: Extravasation: inj site: tissue sloughing

- **Sulfite sensitivity, which may be life-threatening**

Evaluate:

- Therapeutic response: increased B/P with stabilization, adequate tissue perfusion

Teach patient/family:
- About the reason for product administration; to report dyspnea, dizziness, chest pain

TREATMENT OF OVERDOSE:
Administer fluids, electrolyte replacement

norethindrone (Rx)
(nor-eth-in′drone)

Aygestin, Camila, Errin ♣, Heather, Jolivette, Micronor, Nora-BE, Nor-QD, Ortho Micronor

Func. class.: Progestogen

ACTION: Inhibits the secretion of pituitary gonadotropins, which prevents follicular maturation and ovulation; stimulates growth of mammary tissue; antineoplastic action against endometrial cancer

USES: Uterine bleeding (abnormal), amenorrhea, endometriosis, contraception

CONTRAINDICATIONS: Pregnancy (X), breast cancer, hypersensitivity, thromboembolic disorders, reproductive cancer, genital bleeding (abnormal, undiagnosed), liver tumors, hepatic disease

Precautions: Breastfeeding, hypertension, asthma, blood dyscrasias, CHF, diabetes mellitus, depression, migraine headache, seizure disorders, bone/gallbladder/renal/hepatic disease, family history of breast or reproductive tract cancer, smoking, HIV

DOSAGE AND ROUTES

Amenorrhea, abnormal uterine bleeding (Aygestin)
- **Adult: PO** 2.5-10 mg/day on days 5-25 of menstrual cycle

Endometriosis (Aygestin)
- **Adult: PO** 5 mg/day × 2 wk then increased by 2.5 mg/day × 2 wk up to 15 mg/day, may continue for 6-9 mo

Contraception
- **Adult: PO** 0.35 mg on 1st day of menses then 0.35 mg/day

Available forms: Tabs (Aygestin) 5 mg; tabs 0.35 mg

Administer:
- Titrated dose; use lowest effective dose
- One dose in AM; do not interrupt between pill packs; give at roughly same time of day
- Without regard to meals
- Store in dark area

SIDE EFFECTS

CNS: *Dizziness, headache,* migraines, depression, fatigue

CV: Hypotension, thrombophlebitis, edema, thromboembolism, CVA, stroke, PE, MI

EENT: Diplopia

GI: *Nausea,* vomiting, anorexia, cramps, increased weight, cholestatic jaundice

GU: Amenorrhea, cervical erosion, breakthrough bleeding, dysmenorrhea, vaginal candidiasis, breast changes, (gynecomastia, testicular atrophy, impotence), endometriosis, spontaneous abortion, *breast tenderness*

INTEG: Rash, urticaria, acne, hirsutism, alopecia, oily skin, seborrhea, purpura, melasma

META: Hyperglycemia

PHARMACOKINETICS
Duration 24 hr, excreted in urine, feces; metabolized in liver

INTERACTIONS

Decrease: progestin effect—barbiturates, carBAMazepine, fosphenytoin, phenytoin, rifampin

Drug/Herb

Decrease: contraception—St. John's wort

Drug/Food

Increase: caffeine level—caffeine

Drug/Lab Test
Increase: LDL
Decrease: GTT, HDL, alk phos

NURSING CONSIDERATIONS

Assess:
- Weight daily: notify prescriber of weekly weight gain >5 lb
- B/P at beginning of treatment and periodically
- I&O ratio; be alert for decreasing urinary output, increasing edema
- Hepatic studies: ALT, AST, bilirubin periodically during long-term therapy
- Edema, hypertension, cardiac symptoms, jaundice, thromboembolism
- Mental status: affect, mood, behavioral changes, depression
- Hypercalcemia
- Breast exam, pap smear

Evaluate:
- Therapeutic response: decreased abnormal uterine bleeding, absence of amenorrhea

Teach patient/family:
- About cushingoid symptoms

⚠ To report breast lumps, vaginal bleeding, amenorrhea, edema, jaundice, dark urine, clay-colored stools, dyspnea, headache, blurred vision, abdominal pain, numbness or stiffness in legs, chest pain; impotence or gynecomastia (men)
- To take at same time of day; not to interrupt between pill packs
- To report suspected pregnancy immediately, to wait ≥3 mo after stopping product to become pregnant, pregnancy (X); to use backup contraception methods for 48 hr if treatment is not begun on the first day of menstruation
- To avoid smoking; CV reactions may occur
- That product does not protect against HIV, STDs
- That product may mask onset of menopause

norfloxacin ophthalmic

See Appendix B

nortriptyline (Rx)

(nor-trip′ti-leen)

Arentyl ♣, Pamelor

Func. class.: Antidepressant, tricyclic
Chem. class.: Dibenzocycloheptene—secondary amine

Do not confuse:
nortriptyline/amitriptyline

ACTION: Blocks reuptake of norepinephrine and serotonin into nerve endings, thereby increasing action of norepinephrine and serotonin in nerve cells

USES: Major depression

Unlabeled uses: Chronic pain management, PMDD, social phobia, neuropathy, panic disorder, enuresis, migraine prophylaxis

CONTRAINDICATIONS: Hypersensitivity to tricyclics, recovery phase of MI, seizure disorders, prostatic hypertrophy

Precautions: Breastfeeding, suicidal patients, severe depression, increased intraocular pressure, closed-angle glaucoma, urinary retention, cardiac/hepatic disease, hyperthyroidism, electroshock therapy, elective surgery, pregnancy (C), carBAMazepine hypersensitivity

Black Box Warning: Children, suicidal ideation

DOSAGE AND ROUTES

- **Adult: PO** 25 mg tid or qid; may increase to 150 mg/day; may give daily dose at bedtime
- **Adolescent: PO** 1-3 mg/kg/day in 3-4 divided doses or daily at bedtime, max 150 mg/day
- **Child 6-12 yr (unlabeled): PO** 1-3 mg/kg/day in 3-4 divided doses, max 150 mg/day
- **Geriatric: PO** 10-25 mg at bedtime, increase by 10-25 mg at weekly intervals to desired dose; usual maintenance 75 mg/day, max 150 mg/day

Available forms: Caps 10, 25, 50, 75 mg; sol 10 mg/5 ml

Administer:

- Store in tight, light-resistant container at room temperature
- Increased fluids, bulk in diet if constipation occurs
- Without regard to meals
- Dosage at bedtime for oversedation during day; may take entire dose at bedtime; geriatric patients may not tolerate once daily dosing
- Gum, hard candy, frequent sips of water for dry mouth
- **Oral solution:** with fruit juice, water, or milk to disguise taste

SIDE EFFECTS

CNS: *Dizziness, drowsiness,* confusion, headache, anxiety, tremors, stimulation, weakness, insomnia, nightmares, EPS (geriatric patients), increased psychiatric symptoms, seizures

CV: *Orthostatic hypotension,* ECG changes, *tachycardia,* hypertension, palpitations, dysrhythmias

EENT: *Blurred vision,* tinnitus, mydriasis, dry eyes

ENDO: SIADH, hyponatremia, hypothyroidism

GI: *Constipation, dry mouth,* nausea, vomiting, paralytic ileus, increased appetite, cramps, epigastric distress, jaundice, hepatitis, stomatitis, weight gain

GU: *Urinary retention,* acute renal failure, sexual dysfunction

HEMA: Agranulocytosis, thrombocytopenia, eosinophilia, leukopenia

INTEG: Rash, urticaria, sweating, pruritus, photosensitivity

SYST: Serotonin syndrome

PHARMACOKINETICS

PO: Steady state 4-19 days; metabolized by liver; excreted by kidneys; crosses placenta; excreted in breast milk; half-life 18-28 hr, protein binding 93%-95%

INTERACTIONS

⚠ **Increase:** QT prolongation—class IA/III antidysrhythmics, some phenothiazines, β agonists, local anesthetics, tricyclics, haloperidol, chloroquine, droperidol, pentamidine; CYP3A4 inhibitors (amiodarone, clarithromycin, erythromycin, telithromycin, troleandomycin), arsenic trioxide, levomethadyl; CYP3A4 substrates (methadone, pimozide, QUEtiapine, quiNIDine, risperiDONE, ziprasidone)

- Heavy smoking: decreased product effect

⚠ Hyperpyretic crisis, seizures, hypertensive episode: MAOI

Increase: effects of direct-acting sympathomimetics (EPINEPHrine), alcohol, barbiturates, benzodiazepines, CNS depressants, products increasing QT interval, other anticholinergics

⚠ **Increase:** serotonin syndrome, neuroleptic malignant syndrome—SSRIs, SNRIs, serotonin receptor agonists, linezolid; methylene blue (IV)

Decrease: effects of guanethidine, cloNIDine, indirect-acting sympathomimetics (ePHEDrine)

Drug/Herb

Increase: CNS effect—kava, valerian

Decrease: nortriptyline level—St. John's wort

Drug/Lab Test

Increase: serum bilirubin, blood glucose, alk phos

Decrease: VMA, 5-HIAA

False increase: urinary catecholamines

NURSING CONSIDERATIONS

Assess:

Black Box Warning: **Suicidal thoughts/behaviors in children/young adults:** not approved for children, monitor for suicidal ideation in depression, adolescents, young adults

- Monitor for glaucoma exacerbation and paralytic ileus
- B/P (lying, standing), pulse q4hr; if systolic B/P drops 20 mm Hg, hold product, notify prescriber; VS q4hr in patients with CV disease

• Blood studies: thyroid function tests, LFTs, serum nortriptyline level/target 50-150 mg/ml if patient is receiving long-term therapy
• Weight weekly; appetite may increase with product
⚠ **PR, QT prolongation:** ECG for flattening of T wave, bundle branch block, AV block, QT prolongation, dysrhythmias in cardiac patients; assess for chest pain, palpitations, dyspnea
• EPS primarily in geriatric patients: rigidity, dystonia, akathisia, preferred tricyclic in geriatric patients
• Mental status changes: mood, sensorium, affect, suicidal tendencies, increase in psychiatric symptoms, depression, panic
• Urinary retention, constipation; constipation is more likely to occur in children
⚠ **Withdrawal symptoms:** headache, nausea, vomiting, muscle pain, weakness; do not usually occur unless product was discontinued abruptly
• Alcohol intake; if alcohol is consumed, hold dose until AM
• **Serotonin syndrome, neuroleptic malignant syndrome:** assess for increased heart rate, shivering, sweating, dilated pupils, tremors, high B/P, hyperthermia, headache, confusion; if these occur, stop product, administer serotonin antagonist if needed **(rare)**
• Assistance with ambulation during beginning therapy because drowsiness/dizziness occurs; safety measures including side rails, primarily for geriatric patients

Evaluate:
• Therapeutic response: decreased depression

Teach patient/family:
• That therapeutic effects may take 2-3 wk; only small quantities may be dispersed
• To use caution when driving, during other activities requiring alertness because of drowsiness, dizziness, blurred vision
• To avoid alcohol ingestion, other CNS depressants; to avoid MAOIs within 14 days
• Not to discontinue medication quickly after long-term use; may cause nausea, headache, malaise
• To wear sunscreen or large hat because photosensitivity occurs
⚠ To immediately report urinary retention, worsening depression, suicidal thoughts/behaviors

TREATMENT OF OVERDOSE:
ECG monitoring; lavage, activated charcoal; administer anticonvulsant

nystatin (Rx)
(nye-stat′in)
Bio-Statin, Pediaderm AF, Nyamyc
Func. class.: Antifungal
Chem. class.: Amphoteric polyene

ACTION: Interferes with fungal DNA replication; binds sterols in fungal cell membrane, which increases permeability, leaking of cell nutrients

USES: *Candida* species causing oral, intestinal infections

CONTRAINDICATIONS: Hypersensitivity
Precautions: Pregnancy (C)

DOSAGE AND ROUTES
Oral infection
• **Adult/adolescent/child: SUSP** 400,000-600,000 units qid; use ½ dose in each side of mouth; swish and swallow; use for at least 48 hr after symptoms resolved
• **Infant: SUSP** 200,000 units qid (100,000 units in each side of mouth)
• **Newborn and premature infant: SUSP** 100,000 units qid
• **Adult/child: TROCHES** 200,000-400,000 units qid × ≤2 wk

GI infection
• **Adult: PO** 500,000-1,000,000 units tid

N

Cutaneous candidiasis

• **Adult/child: Top cream/ointment** apply to affected area bid; **powder** apply to affected area bid-tid

Vaginal candidiasis

• **Adult vag tab:** 100,000 units daily at bedtime × 14 days

Available forms: Tabs 500,000 units; oral caps 500,000, 1,000,000 units, bulk powder

Administer:

• Store at room temperature for oral susp; tabs in tight, light-resistant containers at room temperature

PO route

• Oral susp dose by placing ½ in each cheek, then swallow; do not mix with food

• Topical dose after cleansing area; mouth may be swabbed; very moist lesions best treated with topical powder

SIDE EFFECTS

GI: Nausea, vomiting, anorexia, diarrhea, cramps

INTEG: Rash, urticaria (rare)

PHARMACOKINETICS

PO: Little absorption, excreted in feces

NURSING CONSIDERATIONS

Assess:

• **Allergic reaction:** rash, urticaria, irritated oral mucous membranes; product may have to be discontinued

• Obtain culture, histologic tests to confirm organism

• Predisposing factors: antibiotic therapy, pregnancy, diabetes mellitus, sexual partner infection (vaginal infections)

Evaluate:

• Therapeutic response: culture negative for *Candida*

Teach patient/family:

• That long-term therapy may be needed to clear infection; to complete entire course of medication

• To avoid commercial mouthwashes for mouth infection

• To shake susp before measuring each dose

• To notify prescriber of irritation; product may have to be discontinued

nystatin topical

See Appendix B

nystatin vaginal antifungal

See Appendix B

RARELY USED

ocriplasmin
(ok-ri-plas′min)

Jetrea

Func. class.: Ophthalmic agent

USES: Symptomatic vitreomacular adhesion

DOSAGE AND ROUTES

• **Adult: Intravitreal INJ** 0.125 mg (0.1 ml of diluted solution) by intravitreal injection to the affected eye once as a single dose

octreotide (Rx)
(ok-tree′oh-tide)

SandoSTATIN, SandoSTATIN LAR Depot

Func. class.: Growth hormone, antidiarrheal

Chem. class.: Synthetic octapeptide

ACTION: A potent growth hormone similar to somatostatin

USES: **SandoSTATIN:** acromegaly, improves symptoms of carcinoid tumors, vasoactive intestinal peptide tumors (VIPomas); **LAR Depot:** long-term maintenance of acromegaly, carcinoid tumors, VIPomas

Unlabeled uses: GI fistula, variceal bleeding, diarrheal conditions, pancreatic fistula, irritable bowel syndrome, dumping syndrome, short bowel syndrome, insulinoma, hepatorenal syndrome

CONTRAINDICATIONS: Hypersensitivity

Precautions: Pregnancy (B), breastfeeding, children, geriatric patients, diabetes mellitus, hypothyroidism, renal disease

DOSAGE AND ROUTES

Acromegaly

• **Adult: SUBCUT/IV** (SandoSTATIN) 50-100 mcg bid-tid, adjust q2wk based on growth hormone levels or **IM** (SandoSTATIN LAR) 20 mg q4wk × 3 mo, adjust based on growth hormone levels

VIPomas

• **Adult: SUBCUT/IV** (SandoSTATIN) 200-300 mcg/day in 2-4 doses for 2 wk, or **IM** (SandoSTATIN LAR) 20 mg q4wk × 2 mo, adjust dose

Flushing/diarrhea in carcinoid tumors

• **Adult: SUBCUT/IV** (SandoSTATIN) 100-600 mcg/day in 2-4 doses for 2 wk, titrated to patient response or **IM** (SandoSTATIN LAR) 20 mg q4wk × 2 mo, adjust dose

GI fistula

• **Adult: SUBCUT** (SandoSTATIN) 50-200 mcg q8hr

Antidiarrheal in AIDS patients (unlabeled)

• **Adult: SUBCUT** (SandoSTATIN) 50 mcg q8hr prn, increase to 500 mcg q8hr

Irritable bowel syndrome (unlabeled)

• **Adult: SUBCUT** (SandoSTATIN) 100 mcg in single dose, up to 125 mcg bid

Dumping syndrome (unlabeled)

• **Adult: SUBCUT** (SandoSTATIN) 50-150 mcg/day

Variceal bleeding (unlabeled)

• **Adult: IV** (SandoSTATIN) 25-50 mcg/hr **CONT IV INF** for 18 hr-5 days

Available forms: Inj (SandoSTATIN) 0.05, 0.1, 0.2, 0.5, 1 mg/ml; inj powder for susp (LAR depot) 10, 20, 30 mg/5 ml

Administer:

• Store in refrigerator for unopened amps, vials or at room temperature for 2 wk; protect from light; do not use discolored or cloudy sol

IM route

• Reconstitute with diluent provided; give in gluteal region

SUBCUT route

• Rotate inj site; use hip, thigh, abdomen

• Avoid using medication that is cold;

O

allow to reach room temperature; do not use LAR depot

IV route

- **IV direct:** give over 3 min; during an emergency carcinoid crisis, give rapid bolus
- **Intermittent IV inf:** dilute in 50-200 ml D_5W, 0.9% NaCl; give over 15-30 min

Y-site compatibilities: Acyclovir, alfentanil, allopurinol, amifostine, amikacin, aminocaproic acid, aminophylline, amiodarone, amphotericin B colloidal, amphotericin B lipid complex, amphotericin B liposome, ampicillin, ampicillin-sulbactam, anidulafungin, argatroban, arsenic trioxide, atenolol, atracurium, azithromycin, aztreonam, bivalirudin, bleomycin, bumetanide, buprenorphine, busulfan, butorphanol, calcium chloride/gluconate, capreomycin, CARBOplatin, carmustine, caspofungin, ceFAZolin, cefepime, cefotaxime, cefoTEtan, cefOXitin, cefTAZidime, ceftizoxime, cefTRIAXone, cefuroxime, chloramphenicol, chlorproMAZINE, cimetidine, ciprofloxacin, cisatracurium, CISplatin, clindamycin, cyclophosphamide, cycloSPORINE, cytarabine, dacarbazine, DACTINomycin, DAPTOmycin, DAUNOrubicin, DAUNOrubicin liposome, dexamethasone, digoxin, diltiazem, diphenhydrAMINE, DOBUTamine, DOCEtaxel, dolasetron, DOPamine, DOXOrubicin, DOXOrubicin liposomal, doxycycline, droperidol, enalaprilat, ePHEDrine, EPINEPHrine, epirubicin, eptifibatide, ertapenem, erythromycin, esmolol, etoposide, famotidine, fenoldopam, fentaNYL, fluconazole, fludarabine, fluorouracil, foscarnet, fosphenytoin, furosemide, gallium nitrate, ganciclovir, gatifloxacin, gemcitabine, gentamicin, glycopyrrolate, granisetron, haloperidol, heparin, hydrALAZINE, hydrocortisone, HYDROmorphone, hydrOXYzine, IDArubicin, ifosfamide, imipenem-cilastatin, insulin (regular), irinotecan, isoproterenol, ketorolac, labetalol, lansoprazole, leucovorin, levofloxacin, lidocaine, linezolid, LORazepam, magnesium sulfate, mannitol, mechlorethamine, melphalan, meperidine, meropenem, mesna, methohexital, methotrexate, methyldopate, methylPREDNISolone, metoclopramide, metoprolol, metroNIDAZOLE, midazolam, milrinone, minocycline, mitoMYcin, mitoXANtrone, mivacurium, morphine, moxifloxacin, mycophenolate, nafcillin, nalbuphine, naloxone, nesiritide, niCARdipine, nitroglycerin, nitroprusside, norepinephrine, ondansetron, oxaliplatin, PACLitaxel, palonosetron, pamidronate, pancuronium, PEMEtrexed, pentamidine, pentazocine, PENTobarbital, PHENobarbital, phenylephrine, piperacillin, piperacillin-tazobactam, polymyxin B, potassium acetate/chloride/phosphates, procainamide, prochlorperazine, promethazine, propranolol, quiNIDine, quinupristin-dalfopristin, ranitidine, remifentanil, rocuronium, sodium acetate/bicarbonate/phosphates, streptozocin, succinylcholine, SUFentanil, sulfamethoxazole-trimethoprim, tacrolimus, teniposide, thiopental, thiotepa, ticarcillin, ticarcillin-clavulanate, tigecycline, tirofiban, tobramycin, topotecan, vancomycin, vasopressin, vecuronium, verapamil, vinBLAStine, vinCRIStine, vinorelbine, voriconazole, zidovudine, zoledronic acid

SIDE EFFECTS

CNS: *Headache, dizziness, fatigue, weakness,* depression, anxiety, tremors, seizure, paranoia

CV: *Sinus bradycardia, conduction abnormalities,* dysrhythmias, chest pain, SOB, thrombophlebitis, ischemia, CHF, hypertension, palpitations, QT prolongation, ST- or T-wave changes

ENDO: *Hypo/hyperglycemia, ketosis, hypothyroidism,* galactorrhea, diabetes insipidus

GI: *Diarrhea, nausea, abdominal pain, vomiting, flatulence, distention, constipation,* hepatitis, increased LFTs, GI bleeding, pancreatitis, cholelithiasis, ileus

GU: UTI

HEMA: Hematoma of inj site, bruise
INTEG: Rash, urticaria, pain; inflammation at inj site
MS: *Joint and muscle pain*

PHARMACOKINETICS

Absorbed rapidly, completely; peak ½ hr (subcut/IV), 2-3 wk (IM); half-life 1.7 hr, duration 12 hr, excreted unchanged in urine

INTERACTIONS

⚠ **Increase:** QT prolongation—class IA/III antidysrhythmics, some phenothiazines, β agonists, local anesthetics, tricyclics, haloperidol, chloroquine, droperidol, pentamidine; CYP3A4 inhibitors (amiodarone, clarithromycin, erythromycin, telithromycin, troleandomycin), arsenic trioxide, levomethadyl; CYP3A4 substrates (methadone, pimozide, QUEtiapine, quiNIDine, risperiDONE, ziprasidone)
Decrease: effect of—cycloSPORINE

Drug/Food

Decrease: absorption of dietary fat, vit B_{12} levels

Drug/Lab Test

Increase: glucose
Decrease: T_4, thyroid function tests, vit B_{12}, glucose

NURSING CONSIDERATIONS

Assess:

- Growth hormone antibodies, IGF-1 at 1- to 4-hr intervals for 8-12 hr after dose **(acromegaly)**; 5-HIAA, plasma serotonin; blood glucose, serotonin levels **(carcinoid tumors)**, plasma substance P, plasma vasoactive intestinal peptide **(VIP) (VIPoma)**
- Thyroid function tests: T_3, T_4, T_7, TSH to identify hypothyroidism
- Fecal fat, serum carotene, somatomedin-C q 14 days, glucose; plasma serotonin levels (carcinoid tumors); plasma vasoactive intestinal peptide levels (VIPoma); serum growth hormone, serum IGF-1 baseline and periodically, diabetes to monitor blood glucose
- **Allergic reaction:** rash, itching, fever, nausea, wheezing

⚠ **Cardiac status:** bradycardia, conduction abnormalities, dysrhythmias; monitor ECG for QT prolongation, low voltage, axis shifts, early repolarization, R/S transition, early wave progression

- Gall bladder disease, pancreatitis: monitor closely

Evaluate:

- Therapeutic response: relief of diarrhea in patients with AIDS; improves symptoms of carcinoid or VIP tumors; data is insufficient regarding whether products decrease size/rate of tumor growth; decreasing symptoms of acromegaly

Teach patient/family:

- That regular assessments are required; diabetics to monitor blood glucose
- About SUBCUT inj if patient or other persons will be giving inj

⚠ That product may cause dizziness, drowsiness, weakness; to avoid hazardous activities if these occur; to report abdominal pain immediately

- That pregnancy may occur in acromegaly because fertility may be restored
- That in diabetes, to monitor glucose regularly

O

ofloxacin (Rx)

(o-flox′a-sin)

Func. class.: Antiinfective
Chem. class.: Fluoroquinolone

ACTION: Interferes with conversion of intermediate DNA fragments into high-molecular-weight DNA in bacteria; inhibits DNA gyrase

USES: Treatment of lower respiratory tract infections (pneumonia, bronchitis), genitourinary infections (prostatitis, UTIs) caused by *Escherichia coli, Klebsiella pneumoniae, Chlamydia trachomatis,* skin and skin-structure infections; otitis media, PID

Unlabeled uses: Leprosy, anthrax, epididymitis, meningococcal infection, prophylaxis, mycobacterium avium complex (MAC), plague, proctitis, traveler's diarrhea, typhoid fever, TB

CONTRAINDICATIONS:

QT prolongation, hypersensitivity to quinolones

Precautions: Pregnancy (C), breastfeeding, children, geriatric patients, renal disease, seizure disorders, excessive sunlight, hypokalemia, colitis

Black Box Warning: Tendon pain/rupture, tendinitis, myasthenia gravis

DOSAGE AND ROUTES

Lower respiratory tract infections/ skin and skin-structure infections

- **Adult: PO** 400 mg q12hr × 10 days

Prostatitis from *E. coli*

- **Adult: PO** 300 mg q12hr × 6 wk

Urinary tract infection

- **Adult: PO** 200 mg q12hr × 3-7 days depending on organism 10 days (complicated)

Renal dose

- **Adult: PO** CCr 20-50 ml/min, give q24hr; CCr <20 ml/min, give 50% of dose q24hr

Hepatic dose

- **Adult (Child-Pugh class C): PO** max 400 mg/day

Available forms: Tabs 200, 300, 400 mg

Administer:

PO route

- 2 hr before or 2 hr after antacids, calcium, iron, zinc products, without regard to food, maintain hydration
- Store at room temperature, protect from light

SIDE EFFECTS

CNS: *Dizziness, headache, fatigue, somnolence,* depression, insomnia, lethargy, malaise, seizures, vertigo

CV: QT prolongation, dysrhythmias, chest pain

EENT: Visual disturbances, pharyngitis

GI: *Diarrhea, nausea, vomiting,* anorexia, flatulence, heartburn, dry mouth, increased AST, ALT, abdominal pain, constipation, pseudomembranous colitis, abnormal taste, xerostomia

HEMA: Blood dyscrasias

INTEG: Rash, pruritus, photosensitivity

MS: Tendinitis, tendon rupture, rhabdomyolysis

SYST: Anaphylaxis, Stevens-Johnson syndrome, toxic epidermal necrolysis

PHARMACOKINETICS

PO: Peak 1-2 hr; half-life 4-8 hr; steady state 2 days; excreted in urine as active product, metabolites; 90%-95% bioavailability

INTERACTIONS

Black Box Warning: Increase: tendon rupture/tendinitis—corticosteroids

- May alter blood glucose levels: antidiabetics
- Possible theophylline toxicity: theophylline

⚠ **Increase:** QT prolongation—class IA/III antidysrhythmics, some phenothiazines, β-agonists, local anesthetics, tricyclics, haloperidol, methadone, chloroquine, clarithromycin, droperidol, erythromycin, pentamidine

Increase: CNS stimulation, seizures—NSAIDs

Increase: anticoagulation—warfarin

Decrease: ofloxacin—sevelamer

Decrease: absorption—antacids with aluminum, magnesium, iron products, sucralfate, zinc products; separate by 2 hr

Drug/Lab Test

Increase: INR

NURSING CONSIDERATIONS

Assess:

Black Box Warning: Tendon rupture/ tendinitis: more common in lung, heart, kidney transplants or geriatric patients; assess for pain or inflammation

• Blood studies: BUN, creatinine, AST, ALT, CBC, blood glucose, INR (warfarin use)

• **CNS symptoms:** insomnia, vertigo, headache, agitation, confusion

Black Box Warning: Myasthenia gravis: product may increase weakness, avoid use

• For overgrowth of infection in long-term treatment

• **Allergic reactions:** rash, flushing, urticaria, pruritus

Black Box Warning: Stop product if these occur

Evaluate:

• Therapeutic response: urine culture, absence of symptoms of infection

Teach patient/family:

• That if dizziness or lightheadedness occur, to ambulate, perform activities with assistance

• To complete full course of therapy; take with plenty of fluids

• To avoid iron- or mineral-containing supplements within 2 hr before or after dose; take without regard to meals

• That allergic reactions usually occur after first dose but may occur later; stop product

• To avoid sun exposure; photosensitivity can occur

• To avoid use with other products unless approved by prescriber

ofloxacin ophthalmic

See Appendix B

OLANZapine (Rx)

(oh-lanz′a-peen)

Zyprexa, Zyprexa IntraMuscular, Zyprexa Relprevv, Zyprexa Zydis

Func. class.: Antipsychotic, neuroleptic

Chem. class.: Thienobenzodiazepine

Do not confuse:
OLANZapine/osalazine
Zyprexa/CeleXA/Zyrtec

ACTION: May mediate antipsychotic activity by both DOPamine and serotonin type 2 (5-HT2) antagonists; may antagonize muscarinic receptors, histaminic (H_1)- and α-adrenergic receptors

USES: Schizophrenia, acute manic episodes with bipolar disorder, acute agitation

Unlabeled uses: Acute psychosis

CONTRAINDICATIONS: Hypersensitivity

Precautions: Pregnancy (C), breastfeeding, geriatric patients, hypertension, cardiac/renal/hepatic disease, diabetes, agranulocytosis, abrupt discontinuation, Asian patients, closed-angle glaucoma, coma, leukopenia, QT prolongation, tardive dyskinesia, torsades de pointes, suicidal ideation, stroke history, TIA

Black Box Warning: Increased mortality in elderly patients with dementia-related psychosis, postinjection delirium/sedation syndrome

DOSAGE AND ROUTES

Schizophrenia

• **Adult: PO** 5-10 mg/day initially, may increase dosage by 5 mg at ≥1 wk intervals, max 20 mg/day; **ORALLY DISINTEGRATING** tabs: open blister pack, place tab on tongue, let disintegrate, swallow; **ext rel inj** (Zyprexa Relprevv) **IM** 150-300 mg q2wk or 405 mg q4wk

• **Geriatric: PO** 5 mg, may increase cautiously at 1-wk intervals, max 20 mg/day

- **Adolescent: PO** 2.5 or 5 mg/day, target 10 mg/day, max 20 mg
- **Child 6-12 yr (unlabeled): PO** 2.5 mg daily, may increase to 5 mg/day after 4-7 days

Bipolar mania

- **Adult: PO** 10-15 mg/day, may increase dose after >24 hr by 5 mg, max 20 mg/day
- **Adolescent: PO** 2.5 or 5 mg/day, target 10 mg/day, max 20 mg/day

Agitation associated with schizophrenia, bipolar I mania

- **Adult: IM** (reg rel) 10 mg once
- **Geriatric: IM** (reg rel) 2.5-5 mg once

Severe behavioral disturbances in geriatric patients (unlabeled)

- **Adult: PO** 2.5-5 mg/day; **acute psychosis PO** 5-10 mg every night

Available forms: Tab 2.5, 5, 7.5, 10, 15, 20 mg; **orally disintegrating tabs** 5, 10, 15, 20 mg (Zyprexia Zydis); **powder for inj** 10 mg; **ext rel powder for susp for inj** 210, 300, 405 mg (Zyprexa Relprevv)

Administer:

- Decreased dose in geriatric patients

PO route

- With full glass of water, milk, food to decrease GI upset
- Store in tight, light-resistant container
- **Orally disintegrating tabs:** open blister pack; place tab on tongue until dissolved; swallow; no water needed; do not break, crush, chew

IM route (Zyprexa Intramuscular)

- Dissolve contents of vials with 2.1 ml sterile water for inj (5 mg/ml); use immediately
- Do not use IV or SUBCUT
- Inject slowly, deep into muscle mass

IM route (Zyprexa Relprevv)

Black Box Warning: Available only through restricted distribution program due to postinjection delirium/sedation syndrome; given at a facility with emergency services; continuous observation

- Use deep IM gluteal inj only
- Use only diluent provided in kit; give q2-4wk using 19-G, 1.5-inch needle in kit; for obese patients, use 19-G, 2-inch or larger needle

SIDE EFFECTS

CNS: EPS: (pseudoparkinsonism, akathisia, dystonia, tardive dyskinesia), seizures, headache, neuroleptic malignant syndrome (rare), agitation, nervousness, hostility, *dizziness,* hypertonia, *tremor,* euphoria, confusion, *drowsiness,* fatigue, *abnormal gait, insomnia, fever*

CV: Hypotension, tachycardia, chest pain, heart failure, sudden death (geriatric patients, IM), orthostatic hypotension, peripheral edema

ENDO: Increased prolactin levels, hypo/hyperglycemia

GI: *Dry mouth, nausea, vomiting, appetite, dyspepsia,* anorexia, *constipation,* abdominal pain, *weight gain,* jaundice, hepatitis

GU: Urinary retention, urinary frequency, enuresis, impotence, amenorrhea, gynecomastia, breast engorgement, premenstrual syndrome

HEMA: Neutropenia

INTEG: Rash

MISC: Peripheral edema, accidental injury, hypertonia, hyperlipidemia

MS: *Joint pain,* twitching

RESP: *Cough, pharyngitis;* fatal pneumonia (geriatric patients, IM)

PHARMACOKINETICS

Well absorbed (60%), metabolized by liver, glucuronidation/oxidation by CYP1A2 and CYP2D6; excreted in urine (57%), feces (30%); 93% bound to plasma proteins; half-life 21-54 hr, extended in geriatric patients; clearance decreased in women, increased in smokers, peak PO 6 hr, IM 15-45 min

INTERACTIONS

⚠ **Increase:** serotonin syndrome, neuroleptic malignant syndrome—SSRIs, SNRIs

Increase: sedation—other CNS depressants, alcohol, barbiturate anesthetics,

antihistamines, sedatives/hypnotics, antidepressants
Increase: OLANZapine levels—CYP1A2 inhibitors (fluvoxaMINE)
Increase: hypotension—antihypertensives, alcohol, diazepam
Increase: anticholinergic effects—anticholinergics
Decrease: OLANZapine levels—CYP1A2 inducers: carBAMazepine, omeprazole, rifampin
Decrease: antiparkinson activity—levodopa, bromocriptine, other DOPamine agonists

Drug/Lab Test

Increase: LFTs, prolactin, CPK

NURSING CONSIDERATIONS

Assess:

Black Box Warning: Postinjection delirium/sedation syndrome (Zyprexa Relprevv); monitor continuously for ≥3 hr after injection; patient must be accompanied when leaving: sedation, coma, delirium, EPS, slurred speech, altered gait aggression, dizziness, weakness, hypertension, seizures; before leaving, confirm that patient is alert, oriented, and free of any other symptoms

- Mental status: orientation, mood, behavior, presence of hallucinations and type before initial administration, monthly; EPS, including akathisia (inability to sit still, no pattern to movements), tardive dyskinesia (bizarre movements of jaw, mouth, tongue, extremities), pseudoparkinsonism (rigidity, tremors, pill rolling, shuffling gait)
- I&O ratio; palpate bladder if low urinary output occurs, urinary retention may be cause, especially in geriatric patients
- Bilirubin, CBC, LFTs
- Urinalysis recommended before, during prolonged therapy
- Affect, orientation, LOC, reflexes, gait, coordination, sleep pattern disturbances
- B/P sitting, standing, lying: take pulse, respirations q4hr during initial treatment; establish baseline before starting treatment; report drops of 30 mm Hg; obtain baseline ECG
- Dizziness, faintness, palpitations, tachycardia on rising

⚠ **Geriatric patients for serious reactions:** fatal pneumonia, heart failure, stroke leading to death (IM)

⚠ **Neuroleptic malignant syndrome:** hyperpyrexia, muscle rigidity, increased CPK, altered mental status, for acute dystonia (check chewing, swallowing, eyes, pill rolling)

- Constipation, urinary retention daily; increase bulk, water in diet
- Weight gain, hyperglycemia, metabolic changes in diabetic patients
- Supervised ambulation until patient stabilized on medication; do not involve patient in strenuous exercise program because fainting is possible; patient should not stand still for long periods

Evaluate:

- Therapeutic response: decrease in emotional excitement, hallucinations, delusion, paranoia, reorganization of patterns of thought, speech

Teach patient/family:

Black Box Warning: About postinjection delirium/sedation syndrome: teach about all symptoms

- To use good oral hygiene; frequent rinsing of mouth; sugarless gum, candy, ice chips for dry mouth
- To avoid hazardous activities until product response is determined
- That orthostatic hypotension occurs often; to rise from sitting or lying position gradually
- To avoid hot tubs, hot showers, tub baths because hypotension may occur
- To avoid abrupt withdrawal of this product because EPS may result; product should be withdrawn slowly
- To avoid OTC preparations (cough, hay fever, cold) unless approved by prescriber because serious product interactions may occur; to avoid use with alcohol, CNS depressants because increased drowsiness may occur

O

• That in hot weather, heat stroke may occur; to take extra precautions to stay cool

TREATMENT OF OVERDOSE:

Lavage if orally ingested; provide airway; do not induce vomiting or use EPINEPHrine

olmesartan (Rx)

(ol-meh-sar'tan)

Benicar

Func. class.: Antihypertensive

Chem. class.: Angiotensin II receptor (type AT_1) antagonist

Do not confuse:
Benicar/Mevacor

ACTION: Blocks the vasoconstrictor and aldosterone-secreting effects of angiotensin II; selectively blocks the binding of angiotensin II to the AT_1 receptor found in tissues

USES: Hypertension, alone or in combination with other antihypertensives

CONTRAINDICATIONS: Hypersensitivity, pregnancy (D)

Precautions: Breastfeeding, children, geriatric patients, hepatic disease, CHF, renal artery stenosis, African descent, hyperkalemia

DOSAGE AND ROUTES

• **Adult: PO** single agent 20 mg/day initially in patients who are not volume depleted; may be increased to 40 mg/day if needed after 2 wk

• **Child ≥6 yr/adolescents ≤16 yr weighing ≥35 kg: PO** 20 mg/day; may increase to max 40 mg/day after 2 wk

• **Child ≥6 yr/adolescents ≤16 yr weighing 20-<35 kg: PO** 10 mg daily; may increase to max 20 mg/day after 2 wk

Available forms: Tabs 5, 20, 40 mg

Administer:

• Without regard to meals

SIDE EFFECTS

CNS: *Dizziness,* fatigue, *headache,* insomnia, syncope
CV: Chest pain, peripheral edema, tachycardia, *hypotension*
EENT: Sinusitis, rhinitis, pharyngitis
GI: *Diarrhea,* abdominal pain
META: Hyperkalemia
MS: Arthralgia, pain, rhabdomyolysis
RESP: *Upper respiratory infection,* bronchitis
SYST: Angioedema

PHARMACOKINETICS

Peak 1-2 hr; excreted in urine, feces; half-life 13 hr; protein binding 99%

INTERACTIONS

Increase: antihypertensive effects—other antihypertensives, diuretics
Increase: hyperkalemia—potassium supplements, potassium-sparing diuretics, ACE inhibitors
Increase: effect of lithium, antioxidants
Decrease: antihypertensive effect—NSAIDs, colesevelan

Drug/Herb
Increase: antihypertensive effect—hawthorn
Decrease: antihypertensive effect—ephedra

NURSING CONSIDERATIONS

Assess:

Black Box Warning: Pregnancy; product can cause fetal death when given during pregnancy (D) 2nd/3rd trimesters

• Response, adverse reactions, especially in renal disease; monitor renal function

• **Hypertension:** B/P, pulse q4hr; note rate, rhythm, quality; electrolytes: sodium, potassium, chloride; baselines for renal, hepatic studies before therapy begins

• **Hypotension:** place supine; may occur with hyponatremia or those with volume depletion; more common in those taking a diuretic also

Evaluate:

• Therapeutic response: decreased B/P

Teach patient/family:

• To comply with dosage schedule, even if feeling better

• To notify prescriber of mouth sores, fever, swelling of hands or feet, irregular heartbeat, chest pain

• That excessive perspiration, dehydration, vomiting, diarrhea may lead to fall in B/P; to consult prescriber if these occur; to maintain adequate hydration

• That product may cause dizziness, fainting, lightheadedness; to avoid hazardous activities

• To rise slowly to sitting or standing position to minimize orthostatic hypotension

Black Box Warning: To notify prescriber immediately if pregnant; not to use during breastfeeding

• To avoid all OTC medications unless approved by prescriber

• That blood glucose may increase and antidiabetic product may need dosage change

• To inform all health care providers of medication use

• To use proper technique for obtaining B/P, acceptable parameters

olopatadine nasal agent

See Appendix B

olopatadine ophthalmic

See Appendix B

olsalazine (Rx)

(ohl-sal′ah-zeen)

Dipentum

Func. class.: Antiinflammatory

Chem. class.: Salicylate derivative

Do not confuse:

olsalazine/OLANZapine

ACTION: Bioconverted to 5-aminosalicylic acid, which decreases inflammation

USES: Maintenance of remission of ulcerative colitis in patients intolerant to sulfasalazine

CONTRAINDICATIONS: Hypersensitivity to this product or salicylates

Precautions: Pregnancy (C), breastfeeding, children <14 yr; impaired renal/hepatic function; severe allergy; bronchial asthma

DOSAGE AND ROUTES

• **Adult: PO** 500 mg bid, max 3 g/day

Available forms: Caps 250 mg

Administer:

• Total daily dose evenly spaced to minimize GI intolerance; give with food

• Store in tight, light-resistant container at room temperature

SIDE EFFECTS

CNS: Headache, hallucinations, depression, vertigo, fatigue, dizziness

GI: Nausea, vomiting, abdominal pain, hepatitis, diarrhea, bloating, pancreatitis

HEMA: **Leukopenia, neutropenia, thrombocytopenia, agranulocytosis, anemia**

INTEG: Rash, dermatitis, urticaria

PHARMACOKINETICS

Partially absorbed, peak $1^1/_2$ hr, half-life 30-90 min (rectal), 2-15 hr (PO), excreted in urine as 5-aminosalicylic acid and metabolites, crosses placenta

INTERACTIONS

Increase: azaTHIOprine toxicity—azaTHIOprine

Increase: myelosuppression—mercaptopurine, thioguanine

Increase: PT, INR—warfarin

Drug/Lab Test

Increase: AST, ALT

NURSING CONSIDERATIONS

Assess:

- **Colitis:** bowel pattern, number of stools, consistency, frequency, pain, mucus before treatment and periodically
- ⚠ **Blood dyscrasias:** skin rash, fever, sore throat, bruising, bleeding, fatigue, joint pain (rare); CBC before treatment and periodically
- **Allergic reaction:** rash, dermatitis, urticaria, pruritus, dyspnea, bronchospasm; allergy to salicylates

Evaluate:

- Therapeutic response: absence of fever, mucus in stools, decreased diarrhea, abdominal pain

Teach patient/family:

- To report diarrhea, rash, bleeding, bruising, fever, hallucinations
- That product may cause dizziness; to avoid hazardous activities until reaction is known
- To take even if feeling better; to take as directed; to take missed dose when remembered, but not to double

omacetaxine

(oh′ma-set-ax′een)

Synribo

Func. class.: Antineoplastic-miscellaneous

Chem. class.: Cephalotaxine ester (derived from the Evergreen tree: *Cephalotaxus harringtonia)*

ACTION:
Inhibits protein synthesis by binding to site of large ribosome subunit, it reduces Bcr-Abl and Mcl-1 independent of direct Bcr-Abl binding, inducing apoptosis

USES:
Chronic or accelerated phase chronic myelogenous leukemia (CML) with resistance or/and intolerance to 2 or more tyrosine kinase inhibitors

CONTRAINDICATIONS:
Pregnancy (D), breastfeeding, hypersensitivity

Precautions: Children, bleeding, diabetes mellitus, female patients, geriatric patients, hyperglycemia, hyperosmolar hyperglycemic state (HHS), infection, anemia, neutropenia, thrombocytopenia

DOSAGE AND ROUTES

- **Adult: SUBCUT** 1.25 mg/m^2 bid × 14 days every 28 days

⚠ ***Grade 4 neutropenia (absolute neutrophil count [ANC] <0.5 × 10^9/L) or grade 3 thrombocytopenia (platelet count <50 × 10^9/L):*** Do not start the next cycle until the ANC is ≥1 × 10^9/L and platelets are ≥50 × 10^9/L; when therapy is resumed, reduce the number of dosing days by 2 days/cycle (initial cycles, from 14 to 12 days; maintenance cycles, from 7 to 5 days)

Available forms: Powder for injection 3.5 mg

Administer:

SUBCUT route

- **Reconstitution:** Add 1 ml of 0.9% saline for injection/vial to 3.5 mg/ml, gently swirl until the solution is clear and completely dissolved, visually inspect for particulate matter and discoloration before use, protect from light
- **Administration:** Calculate dose, withdraw into a syringe, and give as a SUBCUT injection; discard unused portion, use ≤12 hr at room temperature, ≤24 hr when refrigerated

SIDE EFFECTS

CNS: Seizures, weakness, agitation, anxiety, chills, fever, confusion, depression, dizziness, fatigue, headache, insomnia, lethargy, night sweats, tremor

CV: Hypotension, peripheral edema, sinus tachycardia
EENT: Blurred vision, change in taste (dysgeusia)
GI: Nausea, vomiting, diarrhea, abdominal pain, stomatitis
HEMA: Neutropenia, leukopenia, thrombocytopenia, anemia, bleeding
INTEG: Alopecia, rash, skin hyperpigmentation, stomatitis, skin ulcer, purpura, petechiae, pruritus
MISC: Fever
MS: Arthralgia, gout, myalgia, back pain
RESP: Dyspnea, cough, rales, sinus congestion
SYST: Infections

PHARMACOKINETICS

Protein binding ≤50%, half-life 6 hr, peak 30 min

INTERACTIONS

Increase: Bleeding risk—anticoagulants, salicylates, NSAIDs, avoid use when platelets are <50,000 cells/mL

NURSING CONSIDERATIONS

Assess:

Black Box Warning: CBC, differential, platelet count before treatment and weekly, during induction/maintenance, then every 2 wk as needed

⚠ **Pregnancy D:** assess for pregnancy before starting treatment

⚠ **Infection:** monitor temp every 4 hr; fever might indicate beginning of infection

- CV status: B/P, edema, flushing
- CNS changes: confusion, paresthesias, dysesthesia, pain, weakness

⚠ **Hypersensitive reactions, anaphylaxis:** hypotension, dyspnea, angioedema, generalized urticaria; discontinue infusion immediately

⚠ **Bone marrow depression/bleeding:** hematuria, guaiac, bruising or petechiae, mucosa or orifices every 8 hr; obtain prescription for viscous lidocaine (Xylocaine); avoid invasive procedures

- Effects of alopecia on body image; discuss feelings about body changes
- **Diabetes mellitus:** avoid in poorly controlled disease, monitor blood glucose levels frequently

Evaluate:

- Therapeutic response: decreased CML, positive hematologic response

Teach patient/family:

- To report signs of infection: fever, sore throat, flulike symptoms
- To report signs of anemia: fatigue, headache, faintness, SOB, irritability
- To report bleeding; to avoid use of razors, commercial mouthwash
- To avoid use of aspirin, ibuprofen
- That hair may be lost during treatment; that a wig or hairpiece might make patient feel better; that new hair may be different in color and texture
- That pain in muscles and joints 2-5 days after infusion is common

⚠ To use barrier contraception during and for several mo after treatment, pregnancy (D); to avoid breastfeeding

- To avoid receiving vaccinations while taking product

omalizumab (Rx)

(oh-mah-lye-zoo′mab)

Xolair

Func. class.: Antiasthmatic
Chem. class.: Monoclonal antibody

ACTION:

Recombinant DNA-derived humanized IgG murine monoclonal antibody that selectively binds to IgE to limit the release of mediators in the allergic response

USES:

Moderate to severe persistent asthma

Unlabeled uses: Seasonal allergic rhinitis, food allergy

CONTRAINDICATIONS:

Hypersensitivity to hamster protein

Black Box Warning: Hypersensitivity to this product

Precautions: Pregnancy (B), breastfeeding, children <12 yr, acute attacks of asthma, lymphoma, nephrotic disease, bronchospasm, neoplastic disease, status asthmatics

DOSAGE AND ROUTES

- **Adult/adolescent/child ≥12 yr: SUBCUT** 150-375 mg × 2-4 wk; divide inj into 2 sites if dose >150 mg; dose is adjusted based on IgE levels, significant changes in body weight

Available forms: Powder for inj, lyophilized 202.5 mg (150 mg/1.2 ml after reconstitution)

Administer:

SUBCUT route

- Reconstitute using 1.4 ml sterile water for inj (150 mg/1.2 ml or 125 mg/ml); gently swirl to dissolve; allow vial to stand and q5min gently swirl for 5-10 sec to dissolve; some vials may take ≥20 min; do not use if contents do not dissolve within 40 min; should be clear or slightly opalescent; use large-bore needle to withdraw medication; replace needle with small-bore needle
- Given q2-4wk; product is viscous; if >150 mg is given, divide into 2 sites; inj may take 5-10 sec to administer

SIDE EFFECTS

CV: Heart failure, cardiomyopathy, hypotension
HEMA: Serious systemic eosinophilia
INTEG: Pruritus, dermatitis, inj site reactions, rash
MISC: Earache, dizziness, fatigue, pain, malignancies, viral infections, anaphylaxis, thrombocytopenia, headache
MS: Arthralgia, fracture, leg, arm pain
RESP: Sinusitis, upper respiratory infections, pharyngitis, pulmonary hypertension, bronchospasm

PHARMACOKINETICS

Slowly absorbed, peak 7-8 days, half-life 26 days, degradation by liver, excretion in bile

INTERACTIONS

- Use cautiously with live virus vaccines

Drug/Lab Test

Increase: IgE

NURSING CONSIDERATIONS

Assess:

- **Asthma:** respiratory rate, rhythm, depth; auscultate lung fields bilaterally; notify prescriber of abnormalities; monitor pulmonary function tests; serum IgE (may increase and continue for 1 yr)
- **Inj site reactions:** inflammation, edema, redness, warmth at site; may occur within 60 min of inj; may decrease with repeated dosing

Black Box Warning: **Anaphylaxis, allergic reactions:** rash, urticaria, inability to breathe, edema of throat; product should be discontinued; have emergency equipment available; observe for 2 hr; reaction can occur ≤24 hr

Evaluate:

- Therapeutic response: ability to breathe more easily

Teach patient/family:

- That improvement will not be immediate
- Not to stop taking or decrease current asthma medications unless instructed by prescriber
- To avoid live virus vaccines while taking this product

Black Box Warning: To report signs of allergic reaction

omeprazole (OTC, Rx)

(oh-mep′ray-zole)

Good Sense Omeprazole, Losec ♣, PriLOSEC, PriLOSEC OTC

Func. class.: Antiulcer, proton pump inhibitor

Chem. class.: Benzimidazole

Do not confuse:

PriLOSEC/Prinivil/PROzac/predniSONE

ACTION:
Suppresses gastric secretion by inhibiting hydrogen/potassium ATPase enzyme system in gastric parietal cells; characterized as gastric acid pump inhibitor because it blocks the final step of acid production

USES:
Gastroesophageal reflux disease (GERD), severe erosive esophagitis, poorly responsive systemic GERD, pathologic hypersecretory conditions (Zollinger-Ellison syndrome, systemic mastocytosis, multiple endocrine adenomas); treatment of active duodenal ulcers with/without antiinfectives for *Helicobacter pylori*

Unlabeled uses: NSAID-induced ulcer prophylaxis, stress gastritis prophylaxis

CONTRAINDICATIONS:
Hypersensitivity

Precautions: Pregnancy (C), breastfeeding, children

DOSAGE AND ROUTES

Active duodenal ulcers

- **Adult: PO** 20 mg/day × 4-8 wk; associated with *H. pylori* 40 mg q AM and clarithromycin 500 mg tid on days 1-14 then 20 mg/day on days 15-28

Severe erosive esophagitis/poorly responsive GERD

- **Adult: PO** (del rel cap/del rel susp) 20 mg/day × 4-8 wk

Pathologic hypersecretory conditions

- **Adult: PO** 60 mg/day; may increase to 120 mg tid; daily doses >80 mg should be divided

Gastric ulcer

- **Adult: PO** 40 mg/day 4-8 wk
- **Geriatric: PO** ≤20 mg/day

Heartburn (OTC)

- **Adult: PO** 1 del rel tab (20 mg)/day before AM meal with glass of water × 14 days

Available forms: Del rel caps 10, 20, 40 mg; del rel tabs 20 mg; granules for oral susp 2.5, 10 mg (del rel)

Administer:

- Swallow caps whole; do not crush or chew; caps may be opened and sprinkled over applesauce
- Before eating, usually in the AM, separate with other medications
- **Oral susp powder:** give on empty stomach ≥1 hr before food; if there is NG or enteral feeding tube, do not feed 3 hr before or 1 hr after giving product: contents of packet should be mixed with 1-2 tbsp water; add 20 ml water for NG tube; for oral, stir well, drink, add more water, and drink

SIDE EFFECTS

CNS: *Headache, dizziness, asthenia*

CV: Chest pain, angina, tachycardia, bradycardia, palpitations, peripheral edema, **heart failure**

EENT: Tinnitus, taste perversion

GI: *Diarrhea, abdominal pain, vomiting, nausea, constipation, flatulence, acid regurgitation,* abdominal swelling, anorexia, irritable colon, esophageal candidiasis, dry mouth, **hepatic failure**

GU: UTI, urinary frequency, increased creatinine, **proteinuria, hematuria,** testicular pain, glycosuria

HEMA: **Pancytopenia, thrombocytopenia, neutropenia, leukocytosis,** anemia

INTEG: *Rash,* dry skin, urticaria, pruritus, alopecia

META: Hypoglycemia, increased hepatic enzymes, weight gain, hypomagnesemia, hyponatremia, vit B12 deficiency

MISC: *Back pain,* fever, fatigue, malaise

RESP: *Upper respiratory infections, cough,* epistaxis, **pneumonia**

SYST: **Angioedema, exfoliative dermatitis, Stevens-Johnson syndrome, toxic epidermal necrolysis**

PHARMACOKINETICS

Bioavailabity 30%-40%; peak ½-3½ hr; half-life ½-1 hr; protein binding 95%; eliminated in urine as metabolites and in feces; in geriatric patients, elimination rate decreased, bioavailability increased; metabolized by CYP2C19 enzyme system

INTERACTIONS

Increase: bleeding—warfarin
Increase: serum levels of diazepam, phenytoin, flurazepam, triazolam, cycloSPORINE, disulfiram, digoxin
Decrease: effect of iron salts, ketoconazole, cyanocobalamin, calcium carbonate, ampicillin, indinavir, gefitinib
Drug/Lab Test
Increase: alk phos, AST, ALT, bilirubin, gastrin

NURSING CONSIDERATIONS

Assess:

- **GI system:** bowel sounds q8hr, abdomen for pain, swelling, anorexia, blood in stools
- **Electrolyte imbalances:** hyponatremia; hypomagnesemia in patients using product (3 mo-1 yr); if hypomagnesemia occurs, use of magnesium supplements may be sufficient; if severe, discontinue product
- **Hepatic enzymes:** AST, ALT, alk phos during treatment; **blood studies:** CBC, differential during treatment, blood dyscrasias may occur; vit B_{12} in long-term treatment

⚠ **Serious skin reactions:** toxic epidermal necrolysis, Stevens-Johnson syndrome, angioedema, exfoliative dermatitis; fever, sore throat, fatigue, thin ulcers, lesions in mouth, lips; discontinue product; some serious skin disorders may be fatal

Evaluate:

- Therapeutic response: absence of epigastric pain, swelling, fullness, bleeding: decreased GERD, esophagitis symptoms

Teach patient/family:

- To report severe diarrhea; black, tarry stools; abdominal cramps/pain; or continuing headache; product may have to be discontinued
- That, if diabetic, hypoglycemia may occur
- To avoid hazardous activities because dizziness may occur
- To avoid alcohol, salicylates, NSAIDs; may cause GI irritation
- To take as directed, even if feeling better; to take missed dose as soon as remembered; not to double; PriLOSEC OTC takes up to 4 days for full effect

ondansetron (Rx)

(on-dan-seh'tron)

Zofran, Zofran ODT, Zuplenz

Func. class.: Antiemetic

Chem. class.: 5-HT_3 receptor antagonist

Do not confuse:
Zofran/Zantac

ACTION:

Prevents nausea, vomiting by blocking serotonin peripherally, centrally, and in the small intestine

USES:

Prevention of nausea, vomiting associated with cancer chemotherapy, radiotherapy; prevention of postoperative nausea, vomiting
Unlabeled uses: Pruritus (rectal use), alcoholism, hyperemesis gravidarum

CONTRAINDICATIONS:

Hypersensitivity; phenylketonuric hypersensitivity (oral disintegrating tab), torsades de pointes
Precautions: Pregnancy (B), breastfeeding, children, geriatric patients, granisetron hypersensitivity, QT prolongation, torsades de pointes

DOSAGE AND ROUTES

Prevention of nausea/vomiting (cancer chemotherapy)

- **Adult and child 4-18 yr: IV** 0.15 mg/kg infused over 15 min given 30 min before start of cancer chemotherapy, max 16 mg/dose; 0.15 mg/kg given 4 hr and 8 hr after 1st dose or 16 mg as single dose; dilute in 50 ml of D_5 or 0.9% NaCl before giving; **RECT** (unlabeled) 16 mg/day 2 hr before chemotherapy; **PO** 8 mg ½ hr before chemotherapy, repeat 4, 8 hr after 1st dose
- **Child ≥4 yr: PO** 4 mg TID, first dose ½ hr before chemotherapy

Prevention of nausea/vomiting (radiotherapy)

- **Adult: PO** 8 mg tid, may repeat q8hr

Prevention of postoperative nausea/vomiting

- **Adult: IV/IM** 4 mg undiluted over >30 sec before induction of anesthesia
- **Child 2-12 yr: IV** 0.1 mg/kg (≤40 kg); **IV** 4 mg (≥40 kg), give over ≥30 sec

Hepatic dose

- **Adult: PO/IM/IV** max dose 8 mg/day

Hyperemesis gravidarum (unlabeled)

- **Adult: PO/IV** 4-8 mg bid-tid

Pruritus (unlabeled)

- **Adult: PO** 8 mg bid

Alcoholism (unlabeled)

- **Adult: PO** 4 mcg/kg bid

Available forms: Inj 2 mg/ml, 32 mg/50 ml (premixed); tabs 4, 8 mg; oral sol 4 mg/5 ml; oral disintegrating tabs 4, 8 mg; oral dissolving film 4, 8 mg

Administer:

PO route

- **Oral disintegrating tab:** do not push through foil; gently remove; immediately place on tongue to dissolve; swallow with saliva
- **Oral dissolving film:** fold pouch along dotted line to expose tear notch; while folded, tear and remove film; place film on tongue until dissolved; swallow after dissolved; to reach desired dose, administer successive films, allowing each to dissolve before using another

IM route

- Visually inspect for particulate or discoloration
- May give 4 mg undiluted IM; inject deeply in large muscle mass; aspirate

Direct IV route

- Check for discoloration or particulate; if particulate is present, shake to dissolve
- After diluting single dose in 50 ml NS or D_5W, 0.45% NaCl or NS; give over 15 min
- Do not use IV 32 mg/dose in chemotherapy nausea/vomiting due to QT prolongation, max 16 mg/dose (adult)
- Store at room temperature for 48 hr after dilution

Y-site compatibilities: Aldesleukin, amifostine, amikacin, aztreonam, bleomycin, CARBOplatin, carmustine, ceFAZolin, cefmetazole, cefotaxime, cefOXitin, cefTAZidime, ceftizoxime, cefuroxime, chlorproMAZINE, cimetidine, cisatracurium, CISplatin, cladribine, clindamycin, cyclophosphamide, cytarabine, dacarbazine, DACTINomycin, DAUNOrubicin, dexamethasone, diphenhydrAMINE, DOPamine, DOXOrubicin, DOXOrubicin liposome, doxycycline, droperidol, etoposide, famotidine, filgrastim, floxuridine, fluconazole, fludarabine, gallium, gentamicin, haloperidol, heparin, hydrocortisone, HYDROmorphone, hydrOXYzine, ifosfamide, imipenem-cilastatin, magnesium sulfate, mannitol, mechlorethamine, melphalan, meperidine, mesna, methotrexate, metoclopramide, miconazole, mitoMYcin, mitoXANtrone, morphine, PACLitaxel, pentostatin, piperacillin/tazobactam, potassium chloride, prochlorperazine, promethazine, ranitidine, remifentanil, streptozocin, teniposide, thiotepa, ticarcillin, ticarcillin-clavulanate, vancomycin, vinBLAStine, vinCRIStine, vinorelbine, zidovudine

O

SIDE EFFECTS

CNS: *Headache, dizziness, drowsiness, fatigue, EPS*

GI: *Diarrhea, constipation,* abdominal pain, dry mouth

MISC: Rash, bronchospasm (rare), *musculoskeletal pain, wound problems, shivering, fever, hypoxia, urinary retention*

PHARMACOKINETICS

IV: Mean elimination half-life 3.5-4.7 hr, plasma protein binding 70%-76%, extensively metabolized in the liver, excreted 45%-60% in urine

INTERACTIONS

Increase: unconsciousness, hypotension—apomorphine, do not use together

Increase: QT prolongation—other products that prolong QT

Decrease: ondansetron effect—rifampin, carBAMazepine, phenytoin
Drug/Lab Test
Increase: LFTs

NURSING CONSIDERATIONS

Assess:
- Absence of nausea, vomiting during chemotherapy
- Hypersensitivity reaction: rash, bronchospasm (rare)
- **EPS:** shuffling gait, tremors, grimacing, rigidity periodically
- **QT prolongation:** monitoring ECG in those with cardiac disease or receiving other products that increase QT

Evaluate:
- Therapeutic response: absence of nausea, vomiting during cancer chemotherapy

Teach patient/family:
- To report diarrhea, constipation, rash, changes in respirations, or discomfort at insertion site
- Headache requiring analgesic is common

orlistat (Rx, OTC)

(or'lih-stat)

Alli, Xenical

Func. class.: Weight-control agent
Chem. class.: Lipase inhibitor

ACTION: Inhibits the absorption of dietary fats

USES: Obesity management

CONTRAINDICATIONS: Hypersensitivity, chronic malabsorption syndrome, cholestasis
Precautions: Pregnancy (X), children, hypothyroidism, other organic causes of obesity, anorexia nervosa, bulimia, nephrolithiasis, GI disease, diabetes, fat-soluble vitamin deficiency, breastfeeding

DOSAGE AND ROUTES

- **Adult: PO** (Alli) 60 mg, (Xenical) 120 mg tid with each main meal containing fat, max 360 mg/day

Available forms: Caps (Alli) 60 mg, (Xenical) 120 mg
Administer:
- For obesity only if patient on weight-reduction program that includes dietary changes, exercise; patient should be on a diet with 30% of calories from fat; omit dose of orlistat if meal contains no fat

SIDE EFFECTS

CNS: *Insomnia,* depression, anxiety, dizziness, headache, fatigue
GI: *Oily spotting, flatus with discharge, fecal urgency, fatty/oily stool, oily evacuation, fecal incontinence, frequent defecation,* nausea, vomiting, abdominal pain, infectious diarrhea, rectal pain, tooth disorder, hypovitaminosis, hepatic failure, hepatitis, pancreatitis
GU: UTI, vaginitis, menstrual irregularity
INTEG: Dry skin, rash
MS: Back pain, arthritis, myalgia, tendinitis
RESP: Influenza, URI, LRI, EENT symptoms

PHARMACOKINETICS

Minimal absorption, peak 8 hr, 99% protein binding, excretion in feces, half-life 1-2 hr

INTERACTIONS

Increase: lipid-lowering effect—pravastatin
Increase: effects of warfarin
Decrease: absorption—fat-soluble vitamins (A, D, E, K), cycloSPORINE

NURSING CONSIDERATIONS

Assess:
- **Weight status:** before starting therapy, obtain testing to rule out physiologic reactions for weight, obtain thyroid testing, BMI, glucose; weight weekly; diabetic patients may need reduction in oral hypoglycemics

• For misuse in certain populations (anorexia nervosa, bulimia)

⚠ **Hepatotoxicity/pancreatitis:** jaundice, weakness, abdominal pain (rare)

Evaluate:

• Therapeutic response: decrease in weight

Teach patient/family:

• That 60-mg cap can be obtained OTC; 60 mg tid is highest OTC dose

• That safety and effectiveness beyond 2 yr have not been determined

• To read patient information sheet; to discuss unpleasant GI side effects

• To take multivitamin containing fat-soluble vitamins 2 hr before or after orlistat; that psyllium taken with each dose or at bedtime may decrease GI symptoms

• To avoid hazardous activities until stabilized on medication; to discuss unpleasant side effects

• To notify prescriber if pregnancy is planned or suspected pregnancy (X); if breastfeeding take proper fat-soluble vitamins

⚠ **Hepatotoxicity/pancreatitis:** yellowing of skin, eyes; dark urine; weakness; abdominal pain

oseltamivir (Rx)

(oss-el-tam′ih-veer)

Tamiflu

Func. class.: Antiviral

Chem. class.: Neuramidase inhibitor

ACTION: Inhibits influenza virus neuraminidase with possible alteration of virus particle aggregation and release

USES: Prevention and treatment of influenza type A or B

Unlabeled uses: Avian flu (H5N1), avian influenzae A (H5N1), swine flu (H1N1), encephalitis

CONTRAINDICATIONS: Hypersensitivity

Precautions: Pregnancy (C), neonates, breastfeeding, infants, children, geriatric patients, renal/hepatic/pulmonary/cardiac disease, psychosis, viral infection

DOSAGE AND ROUTES

Treatment of influenza

• **Adult/child >40 kg: PO** 75 mg bid × 5 days, begin treatment within 2 days of onset of symptoms

• **Child 23-40 kg and ≥1 yr: PO** 60 mg bid

• **Child 15-23 kg and ≥1 yr: PO** 45 mg bid

• **Child ≤15 kg and ≥1 yr: PO** 30 mg bid

• **Neonates ≥14 days, infants: PO** 3 mg/kg/dose BID × 5 days

Prevention of influenza

• **Adult/child ≥13 yr: PO** 75 mg/day × ≥10 days; begin treatment within 2 days of contact, max use 6 wk

Renal dose

• **Adult: PO** CCr 10-30 ml/min, 75 mg/day × 5 days (treatment); 75 mg every other day or 30 mg/day (prophylaxis)

H1N1 influenzae A virus (swine flu) (unlabeled)

• **Adult/adolescent/child >40 kg: PO** 75 mg bid × 5 days

• **Adolescent/child 24-40 kg: PO** 60 mg bid × 5 days

• **Child >1 yr and 15-23 kg: PO** 45 mg bid × 5 days

• **Child >1 yr and ≤15 kg: PO** 30 mg bid × 5 days

Available forms: Caps 30, 45, 75 mg; powder for oral susp 6 mg/ml

Administer:

• Within 2 days of symptoms of influenza; continue for 5 days

• At least 4 hr before bedtime to prevent insomnia

• Without regard to food; give with food for GI upset

• Take with full glass of water

• Store in tight, dry container

• **Oral susp:** 6 mg/ml conc, take care to administer correct dose; loosen powder from side of bottle, add 55 ml, shake well (6 mg/ml), remove child-resistant cap, push bottle adapter into neck of bottle,

O

close tightly with child-resistant cap to ensure sealing, use within 17 days of preparation when refrigerated or within 10 days at room temperature, write expiration date on bottle, shake well before use, use oral syringe provided but only with markings for 30, 45, 60 mg, confirm that dosing instructions are in same units as syringe provided

SIDE EFFECTS

CNS: *Headache, dizziness,* fatigue, *insomnia,* seizures, delirium, self-injury (children)
ENDO: Hyperglycemia
GI: *Nausea, vomiting,* diarrhea, abdominal pain
INTEG: Toxic epidermal necrolysis, Stevens-Johnson syndrome, erythema multiforme
RESP: Cough

PHARMACOKINETICS

Rapidly absorbed, protein binding 40%-45%, converted to oseltamivir carboxylate (active form), active forms half-life 1-3 hr, metabolite 6-10 hr, excreted in urine (99%), protein binding 3%

INTERACTIONS

- Avoid use with H1N1 virus vaccine, intranasal influenzae vaccine

NURSING CONSIDERATIONS

Assess:
- Bowel pattern before, during treatment
- **Influenza:** fever, fatigue, sore throat, headache, muscle soreness, aches

Evaluate:
- Therapeutic response: absence of fever, malaise, cough, dyspnea in infection

Teach patient/family:
- About aspects of product therapy
- To avoid hazardous activities if dizziness occurs
- To take as soon as symptoms appear; to take full course even if feeling better
- To take missed dose as soon as remembered if within 2 hr of next dose
- ⚠ To stop immediately; to report to prescriber skin rash, delirium, psychosis, hallucinations (child)
- That this product should not be substituted for flu shot
- Will not treat the common cold
- To avoid other products unless approved by prescriber

⚠ **HIGH ALERT**

oxaliplatin (Rx)

(ox-al-i′plat-in)

Eloxatin

Func. class.: Antineoplastic
Chem. class.: 3rd-generation platinum analog

ACTION:
Forms crosslinks, thereby inhibiting DNA replication and transcription; not specific to cell cycle

USES:
Metastatic carcinoma of the colon or rectum in combination with 5-FU/leucovorin
Unlabeled uses: Relapsed or refractory non-Hodgkin's lymphoma; advanced ovarian cancer; breast, head/neck, testicular, pancreatic, gastric cancer; mesothelioma

CONTRAINDICATIONS:
Pregnancy (D), breastfeeding, radiation therapy or chemotherapy within 1 mo, thrombocytopenia, smallpox vaccination

Black Box Warning: Hypersensitivity to this product or other platinum products

Precautions: Children, geriatric patients, pneumococcus vaccination, renal disease

DOSAGE AND ROUTES

Dosage protocols may vary

Colorectal cancer
- **Adult: IV INF** *Day 1:* oxaliplatin 85 mg/m^2 in 250-500 ml D_5W and leucovorin 200 mg/m^2 in D_5W; give both over 2 hr at the same time in separate bags us-

ing a Y-line, followed by 5-FU 400 mg/m^2 **IV BOL** over 2-4 min then 5-FU 600 mg/m^2 **IV INF** in 500 ml D_5W as a 22-hr **CONT INF;** *day 2:* leucovorin 200 mg/m^2 **IV INF** over 2 hr, then 5-FU 400 mg/m^2 **IV BOL** over 2-4 min, then 5-FU 600 mg/m^2 **IV INF** in 500 ml D_5W as a 22-hr **CONT INF;** repeat cycle q2wk

Renal dose

• **Adult: IV** CCr <30 ml/min, reduce starting dose to 65 mg/m^2

Advanced ovarian cancer (unlabeled)

• **Adult: IV** 130 mg/m^2 q3wk as a single agent in those previously treated

Advanced breast cancer (unlabeled)

• **Adult: IV** 130 mg/m^2 on day 1 plus 5-fluorouracil (1000 mg/m^2 **CONT IV INF** days 1-4) q3wk

Pancreatic cancer (unlabeled)

• **Adult: IV** 100 mg/m^2 on day 2, with gemcitabine 1000 mg/m^2 on day 1, repeat q2wk; 625 mg/m^2 bid throughout treatment or fluorouracil 200 mg/m^2/day throughout treatment

Gastric cancer (unlabeled)

• **Adult: IV** 130 mg/m^2 over 2 hr on day 1 with epirubic 50 mg/m^2 and capecitabine

Available forms: Powder for inj 50, 100-mg single-use vials (5 mg/ml); solution for inj 50 mg/10 ml, 100 mg/20 ml, 200 mg/40 ml

Administer:

Intermittent IV INF route

• Premedicate with antiemetics including $5HT_3$ blockers, with or without dexamethasone; prehydration not needed

• Do not reconstitute or dilute with sodium chloride or any chloride-containing sol; do not use aluminum equipment during any preparation or administration; will degrade platinum; do not refrigerate unopened powder or sol; do not freeze; protect from light

• Use cytotoxic handling procedures; prepare in biologic cabinet using gown, gloves, mask; do not allow product to come in contact with skin; use soap and water if contact occurs

• EPINEPHrine, antihistamines, corticosteroids for hypersensitivity reaction

• **Lyophilized powder:** reconstitute vial 50 mg/10 ml or 100 mg/20 ml sterile water for inj or D_5W; after reconstitution, sol may be stored for ≤24 hr in refrigerator; after dilution in 250-500 ml D_5W, may store ≤24 hr in refrigerator or 6 hr at room temperature; infuse over 2 hr

• **Aqueous solution:** dilute in 250-500 ml of D_5W; after dilution, may store ≤24 hr refrigerator, 6 hr at room temperature, infuse over 2 hr

Y-site compatibilities: Alfentanil, amifostine, amikacin, aminocaproic acid, amiodarone, amphotericin B colloidal, amphotericin B lipid complex, amphotericin B liposome, ampicillin, ampicillin-sulbactam, anidulafungin, atenolol, atracurium, azithromycin, aztreonam, bivalirudin, bleomycin, bumetanide, buprenorphine, butorphanol, calcium chloride/gluconate, CARBOplatin, caspofungin, ceFAZolin, cefotaxime, cefoTEtan, cefOXitin, cefTAZidime, ceftizoxime, cefTRIAXone, cefuroxime, chloramphenicol, chlorproMAZINE, cimetidine, ciprofloxacin, cisatracurium, CISplatin, clindamycin, cyclophosphamide, cycloSPORINE, cytarabine, dacarbazine, DACTINomycin, DAPTOmycin, DAUNOrubicin, dexamethasone, digoxin, diltiazem, diphenhydrAMINE, DOBUTamine, DOCEtaxel, dolasetron, DOPamine, doxacurium, DOXOrubicin, doxycycline, droperidol, enalaprilat, ePHEDrine, EPINEPHrine, epirubicin, ertapenem, erythromycin, esmolol, etoposide, famotidine, fenoldopam, fentaNYL, fluconazole, fludarabine, foscarnet, fosphenytoin, furosemide, gatifloxacin, gemcitabine, gemtuzumab, gentamicin, glycopyrrolate, granisetron, haloperidol, heparin, hydrALAZINE, hydrocortisone, HYDROmorphone, hydrOXYzine, IDArubicin, ifosfamide, imipenem-cilastatin, inamrinone, insulin (regular), irinotecan, isoproterenol, ketorolac, labetalol, leucovorin, levofloxacin, levorphanol, lidocaine, linezolid, LORazepam, magne-

O

sium sulfate, mannitol, meperidine, meropenem, mesna, metaraminol, methyldopate, methylPREDNISolone, metoclopramide, metoprolol, metroNIDAZOLE, midazolam, milrinone, minocycline, mitoMYcin, mitoXANtrone, mivacurium, morphine, nafcillin, nalbuphine, naloxone, nesiritide, niCARdipine, nitroglycerin, nitroprusside, norepinephrine, octreotide, ondansetron, PACLitaxel, palonosetron, pancuronium, PEMEtrexed, pentamidine, pentazocine, phenylephrine, piperacillin, polymyxin B, potassium chloride/phosphates, procainamide, prochlorperazine, promethazine, propranolol, quiNIDine, quinupristin-dalfopristin, ranitidine, rocuronium, sodium acetate/phosphates, succinylcholine, SUFentanil, sulfamethoxazole-trimethoprim, tacrolimus, teniposide, theophylline, thiotepa, ticarcillin, ticarcillin-clavulanate, tigecycline, tirofiban, tobramycin, tolazoline, topotecan, trimethobenzamide, vancomycin, vasopressin, vecuronium, verapamil, vinBLAStine, vinCRIStine, vinorelbine, voriconazole, zidovudine, zoledronic acid

SIDE EFFECTS

CNS: Peripheral neuropathy, fatigue, headache, dizziness, insomnia
CV: Cardiac abnormalities, thromboembolism
EENT: *Decreased visual acuity, tinnitus, hearing loss*
GI: *Severe nausea, vomiting, diarrhea, weight loss,* stomatitis, anorexia, gastroesophageal reflux, constipation, dyspepsia, mucositis, flatulence
GU: Hematuria, dysuria, creatinine
HEMA: Thrombocytopenia, leukopenia, pancytopenia, neutropenia, anemia, hemolytic uremic syndrome
INTEG: *Alopecia,* rash, flushing, extravasation, redness, swelling, pain at inj site
META: Hypokalemia
RESP: Fibrosis, dyspnea, cough, rhinitis, URI, pharyngitis
SYST: Anaphylaxis, angioedema

PHARMACOKINETICS

Metabolized in liver, excreted in urine; after administration, 15% of platinum in systemic circulation, 85% either in tissues or being eliminated in urine; half-life 390 hr; protein binding >90%

INTERACTIONS

Increase: bleeding risk—NSAIDs, alcohol, anticoagulants, platelet inhibitors, thrombolytics, salicylates
Increase: oxaliplatin toxicity—tannins
Increase: myelosuppression—myelosuppressive agents, radiation
Increase: nephrotoxicity—aminoglycosides, loop diuretics
Decrease: antibody response—live virus vaccines

Drug/Lab Test
Increase: ALT, AST, bilirubin, creatinine
Decrease: potassium, neutrophils, WBC, platelets

NURSING CONSIDERATIONS

Assess:

⚠ **Bone marrow depression:** CBC, differential, platelet count each cycle; withhold product if WBC is <4000 or platelet count is <100,000; notify prescriber of results

• Renal/hepatic studies: BUN, creatinine, serum uric acid, urine CCr before, electrolytes during therapy; dose should not be given if BUN >19 mg/dl; creatinine <1.5 mg/dl; I&O ratio; report fall in urine output of <30 ml/hr; LFTs

Black Box Warning: **Anaphylaxis:** wheezing, tachycardia, facial swelling, fainting; discontinue product, report to prescriber; resuscitation equipment should be nearby

⚠ **Pulmonary fibrosis:** cough, crackles, dyspnea, pulmonary infiltrate; discontinue immediately, death may occur

• Monitor temperature; may indicate beginning infection

• Hepatic studies before each cycle (bilirubin, AST, ALT, LDH) as needed or monthly

• **Bleeding:** hematuria, guaiac, bruising or petechiae, mucosa or orifices; obtain prescription for viscous lidocaine (Xylocaine)
• Effects of alopecia on body image; discuss feelings about body changes
• Edema in feet, joint pain, stomach pain, shaking
• Comprehensive oral hygiene
• All medications PO if possible; avoid IM inj when platelets <100,000/mm^3
• Increase fluid intake to 2-3 L/day to prevent urate deposits, calculi formation; elimination of product
• Blankets, hat, gloves for cold prevention

Evaluate:
• Therapeutic response: decreased tumor size, spread of malignancy

Teach patient/family:
⚠ To report signs of **infection:** increased temperature, sore throat, flulike symptoms
• To report signs of **anemia:** fatigue, headache, faintness, SOB, irritability
• To report **bleeding;** to avoid use of razors, commercial mouthwash
• To avoid aspirin, ibuprofen, NSAIDs, alcohol; may cause GI bleeding
⚠ To report any changes in breathing, coughing
• That hair may be lost during treatment; that a wig or hairpiece may make patient feel better; that new hair may be different in color, texture
• To report numbness, tingling in face or extremities, poor hearing, or joint pain or swelling
• Not to receive vaccines during treatment
⚠ To use contraception during treatment and for 4 mo after; that product may cause infertility, pregnancy (D)
• **Dysesthesias:** to avoid contact with cold (air, ice, liquid)

oxazepam (Rx)

(ox-ay′ze-pam)

Func. class.: Sedative/hypnotic; antianxiety

Chem. class.: Benzodiazepine, short acting

Controlled Substance Schedule IV

ACTION: Potentiates the actions of GABA, especially in the limbic system and the reticular formation

USES: Anxiety, alcohol withdrawal

Unlabeled uses: Insomnia

CONTRAINDICATIONS: Pregnancy (D), breastfeeding, children <6 yr, hypersensitivity to benzodiazepines, closed-angle glaucoma, psychosis

Precautions: Geriatric patients, debilitated, renal/hepatic disease, depression, suicidal ideation, dementia, sleep apnea, seizure disorder, respiratory depression

DOSAGE AND ROUTES

Anxiety
• **Adult: PO** 10-15 mg tid-qid, max 120 mg/day
• **Geriatric: PO** 10 mg daily-bid, max 60 mg/day tid

Alcohol withdrawal
• **Adult: PO** 15-30 mg tid-qid

Available forms: Caps 10, 15, 30 mg

Administer:
• Without regard to food

SIDE EFFECTS

CNS: *Dizziness, drowsiness,* confusion, headache, anxiety, tremors, fatigue, depression, insomnia, hallucinations, paradoxical excitement, transient amnesia

CV: *Orthostatic hypotension,* ECG changes, tachycardia, hypotension

EENT: *Blurred vision,* tinnitus, mydriasis

GI: Nausea, vomiting, anorexia

HEMA: Leukopenia

INTEG: Rash, dermatitis, itching
SYST: Dependence

PHARMACOKINETICS

Peak 2-4 hr; metabolized by liver; excreted by kidneys; half-life 5-15 hr; crosses placenta, breast milk; protein binding 97%

INTERACTIONS

Increase: oxazepam effects—CNS depressants, alcohol, disulfiram, oral contraceptives
Decrease: oxazepam effects—oral contraceptives, phenytoin, theophylline, valproic acid
Decrease: effects of levodopa
Drug/Herb
Increase: CNS depression—kava, melatonin, valerian
Drug/Lab Test
Increase: AST, ALT, serum bilirubin
Decrease: WBC

NURSING CONSIDERATIONS

Assess:
- B/P (lying, standing), pulse; if systolic B/P drops 20 mm Hg, hold product, notify prescriber
- ⚠ Mental status: mood, sensorium, affect, sleeping pattern, drowsiness, dizziness, **suicidal thoughts/behaviors**
- ⚠ **Physical dependency, withdrawal symptoms:** headache, nausea, vomiting, muscle pain, weakness, tremors, seizures (long-term use)
- Assistance with ambulation during beginning therapy because drowsiness, dizziness occurs
- Safety measures, including side rails

Evaluate:
- Therapeutic response: decreased anxiety, restlessness, insomnia

Teach patient/family:
- That product may be taken without regard to food
- That medication not to be used for everyday stress or used >4 mo unless directed by prescriber; not to take more than prescribed dose because product may be habit forming
- To avoid OTC preparations (cough, cold, hay fever) unless approved by prescriber
- To avoid driving, activities that require alertness because drowsiness may occur
- To avoid alcohol, other psychotropic products unless directed by prescriber
- Not to discontinue product abruptly after long-term use
- To rise slowly because fainting may occur, especially among geriatric patients
- That drowsiness may worsen at beginning of treatment
- To notify prescriber if pregnancy is planned or suspected pregnancy (D)

OXcarbazepine (Rx)

(ox′kar-baz′uh-peen)

Trileptal, Oxtellar XR

Func. class.: Anticonvulsant
Chem. class.: CarBAMazepine analog

ACTION:
May inhibit nerve impulses by limiting influx of sodium ions across cell membrane in motor cortex

USES:
Partial seizures
Unlabeled uses: Trigeminal neuralgia, atypical panic disorder, bipolar disorder

CONTRAINDICATIONS:
Hypersensitivity
Precautions: Pregnancy (C), breastfeeding, children <4 yr, hypersensitivity to carBAMazepine, renal disease, fluid restriction, hyponatremia, abrupt discontinuation, suicidal ideation

DOSAGE AND ROUTES

Seizures, adjunctive therapy
- **Adult: PO** 300 mg bid, may be increased by 600 mg/day in divided doses bid at weekly intervals; maintenance 1200 mg/day; ext rel; 600 mg daily × 1

wk, increase weekly in 600 mg/day increments to 1200-2400 mg daily

• **Child 4-16 yr: PO** 8-10 mg/kg/day divided bid; dose determined by weight, increase by 5 mg/kg/day q3days, max doses weight dependent

Conversion to monotherapy for partial seizures

• **Adult: PO** 300 mg bid with reduction in other anticonvulsants; increase OXcarbazepine by 600 mg/day each week over 2-4 wk; withdraw other anticonvulsants over 3-6 wk; max 2400 mg/day

Initiation of monotherapy for partial seizures

• **Adult: PO** 300 mg bid, increase by 300 mg/day q3days to 1200 mg in divided doses bid, max 2400 mg/day

Renal dose

• **Adult: PO** CCr <30 ml/min, 150 mg bid, increase slowly

Bipolar disorder/trigeminal neuralgia (unlabeled)

• **Adult: PO** 300 mg bid, may increase by ≤600 mg/day

Available forms: Film-coated tabs 150, 300, 600 mg; oral susp 300 mg/5 ml; ext rel tab 150, 300, 600 mg

Administer:

PO route

• Without regard to meals
• **Oral susp:** shake well, use calibrated oral syringe provided, use or discard within 7 days of opening
• **Ext rel:** do not crush, break, or chew
• Store at room temperature

SIDE EFFECTS

CNS: *Headache, dizziness, confusion, fatigue,* feeling abnormal, ataxia, abnormal gait, tremors, anxiety, agitation, worsening of seizures, suicidal thoughts/behaviors

CV: *Hypotension,* chest pain, edema, bradycardia, syncope

EENT: *Blurred vision, diplopia, nystagmus,* rhinitis, sinusitis

ENDO: Hypothyroidism, hot flashes

GI: *Nausea, constipation, diarrhea,* anorexia, vomiting, abdominal pain, gastritis

GU: Urinary frequency, hematuria, menses change

INTEG: Purpura, rash, acne

META: Hyponatremia

RESP: Flu-like symptoms

SYST: Angioedema, anaphylaxis, Stevens-Johnson syndrome, toxic epidermal necrolysis, drug reaction with eosinophilia and systemic symptoms (DRESS)

PHARMACOKINETICS

PO: Onset unknown; peak 4-6 hr; metabolized by liver to active metabolite; terminal half-life 7-9 hr metabolite; inhibits P450 CYP2C19, induces CYP3A4/5, 95% renal extraction

INTERACTIONS

⚠ Contraindicated: MAOIs, ranolazine, nisoldipine

Increase: CNS depression—alcohol

Decrease: effects—felodipine, oral contraceptive, carBAMazepine

Decrease: OXcarbazepine levels—carBAMazepine, PHENobarbital, phenytoin, valproic acid, verapamil

Drug/Herb

Increase: anticonvulsant effect—ginkgo

Decrease: anticonvulsant effect—ginseng, santonica

Drug/Lab Test

Decrease: sodium

NURSING CONSIDERATIONS

Assess:

• Description of seizures: frequency, duration, aura
• Hyponatremia: headache, nausea, confusion, usually within the first 3 mo of treatment, but may occur ≤1 yr, if this product is being used with other products that decrease sodium, monitor sodium levels
• Electrolyte: sodium; T_4; phenytoin (when given together)

O

⚠ **Serious reactions:** angioedema, anaphylaxis, Stevens-Johnson syndrome

- CNS/mental status: mood, sensorium, affect, behavioral changes, confusion, suicidal thoughts/behaviors; if mental status changes, notify prescriber
- Eye problems: need for ophthalmic exams before, during, after treatment (slit lamp, funduscopy, tonometry)
- **Pregnancy:** lack of seizure control due to MHD, a metabolite of OXcarbazepine, monitor seizure control
- May monitor target serum level 12-30 mcg/ml to identify compliance/toxicity

Evaluate:

- Therapeutic response: decreased seizure activity

Teach patient/family:

- To avoid driving, other activities that require alertness
- To take twice a day at same intervals
- Not to discontinue medication quickly after long-term use, seizures may increase
- To inform prescriber if hypersensitive to carBAMazepine; multisystem hypersensitivity may occur; to report fever, other allergic symptoms
- To avoid use of alcohol while taking product
- To use alternative contraception if using hormonal method; to report if pregnancy is planned or suspected (C)
- To report skin rashes immediately, serious skin reactions can occur
- To report suicidal thoughts/behavior immediately

TREATMENT OF OVERDOSE:

Activated charcoal; give 0.9% NaCl (hypotensive state), atropine (bradycardia); use benzodiazepines, barbiturates for seizures

oxybutynin (Rx, OTC)

(ox-i-byoo'ti-nin)

Ditropan ♣, Ditropan XL, Gelnique, Oxytrol ♣, Oxytrol Transdermal, Uromax ♣

Func. class.: Anticholinergic

Chem. class.: Synthetic tertiary amine

Do not confuse:
Ditropan/diazepam

ACTION: Relaxes smooth muscles in urinary tract by inhibiting acetylcholine at postganglionic sites

USES: Antispasmodic for neurogenic bladder, overactive bladder in females (OTC)

CONTRAINDICATIONS: Hypersensitivity, GI obstruction, urinary retention, glaucoma, severe colitis, myasthenia gravis, unstable CV disease

Precautions: Pregnancy (B), breastfeeding, children <12 yr, geriatric patients, suspected glaucoma, cardiac disease, dementia

DOSAGE AND ROUTES

- **Adult: PO** 5 mg bid-tid, max 5 mg qid; **EXT REL** 5-10 mg/day, may increase by 5 mg, max 30 mg/day; **TD** apply 1 patch to abdomen, hip, buttock 2×/wk (q3-4 days); **GEL** apply contents of 1 packet to abdomen, upper arms, shoulders, thighs daily
- **Geriatric: PO** 2.5-5 mg bid-tid, increase by 2.5 mg q several days
- **Child >6 yr: PO** 5 mg bid, max 5 mg tid; **EXT REL** 5 mg/day, max 20 mg/day
- **Child 1-5 yr: PO** 0.2 mg/kg/dose bid-tid

Available forms: Syr 5 mg/5 ml; tabs 5 mg; ext rel tabs 5, 10, 15 mg; TD 3.9 mg/day; top gel 10% (Gelnique)

Administer:

PO route

- Do not crush, break, or chew ext rel tabs
- Without regard to meals

Topical route

- Wash hands; apply to clean, dry, intact skin on abdomen, upper arms/shoulders, thighs; avoid navel, rotate sites
- Squeeze contents into palm of hand or directly on site, rub gently
- Do not bathe, exercise, swim for 1 hr after application
- Allow to dry before putting on clothing
- Do not be near flame, fire, or smoke until gel has dried
- Delivers 100 mg

Transdermal route

- Apply to clean, dry, intact skin on abdomen, hip, buttock; use firm pressure; not affected by showering/bathing; rotate sites
- Delivers 3.9 mg/day

SIDE EFFECTS

CNS: *Anxiety, restlessness, dizziness, somnolence, insomnia, nervousness,* seizures, headache, *drowsiness,* confusion

CV: *Palpitations, sinus tachycardia,* hypertension, peripheral edema, **QT prolongation**

EENT: *Blurred vision, dry eyes,* increased intraocular tension, *dry mouth,* throat

GI: *Nausea, vomiting, anorexia,* abdominal pain, *constipation, dyspepsia,* diarrhea, taste perversion, GERD

GU: Dysuria, impotence, *urinary retention, hesitancy*

MISC: Hyperthermia, anaphylaxis, angioedema

PHARMACOKINETICS

Onset ½-1 hr, peak 3-6 hr, duration 6-10 hr; metabolized by liver, excreted in urine; terminal half-life 2-3 hr

INTERACTIONS

- Altered pharmacokinetic parameters: CYP3A4 inhibitors

Increase: CNS depression—benzodiazepines, sedatives, hypnotics, opioids

Increase: levels of atenolol, digoxin, nitrofurantoin

Increase: anticholinergic effects—antihistamines, amantidine, other anticholinergics

Increase or decrease: levels of phenothiazines

Decrease: levels of acetaminophen, haloperidol, levodopa

Decrease: effects of oxybutynin—CYP3A4 inducers

NURSING CONSIDERATIONS

Assess:

- **Urinary patterns:** distention, nocturia, frequency, urgency, incontinence, I&O ratios; cystometry to diagnose dysfunction, urinary tract infections should be treated

⚠ **Allergic reactions:** rash, urticaria; if these occur, product should be discontinued; angioedema

⚠ **QT prolongation:** ECG for QT prolongation, ejection fraction; assess for chest pain, palpitations, dyspnea

- CNS effects: confusion, anxiety; anticholinergic effects in geriatric patients

Evaluate:

- Urinary status: dysuria, frequency, nocturia, incontinence

Teach patient/family:

- To avoid hazardous activities because dizziness, blurred vision may occur
- To avoid OTC medications with alcohol, other CNS depressants
- To avoid hot weather, strenuous activity because product decreases perspiration
- About the correct application of each product form
- **Transdermal:** change patch 2×/wk; do not use same site within 7 days; dispose of and use container not accessible to pets/children
- To open patch immediately before using

- Do not use during MRI, remove
- **Topical gel:** rotate sites
- Apply to clean, dry skin on abdomen, upper arm/shoulders/thighs
- Gel is flammable

⚠ HIGH ALERT

oxyCODONE (Rx)

(ox-i-koe′done)

ETH-Oxydose, Oxecta, Oxyfast, Oxy-CONTIN, Roxicodone, Supeudol 🍁

oxyCODONE/aspirin (Rx)

Endodan, Percodan

oxyCODONE/acetaminophen (Rx)

Endocet, Magnacet, Percocet, Primalev, Roxicet, Tylox, Xolox

oxyCODONE/ibuprofen (Rx)

Func. class.: Opiate analgesic
Chem. class.: Semisynthetic derivative

Controlled Substance Schedule II

Do not confuse:
Percodan/Decadron
Roxicet/Roxanol
Tylox/Xanax/Trimox/Wymox
OxyCODONE/HYDROcodone
OxyCODONE/OxyCONTIN
Roxicodone/roxanol

ACTION: Inhibits ascending pain pathways in CNS, increases pain threshold, alters pain perception

USES: Moderate to severe pain
Unlabeled uses: Postherpetic neuralgic (cont rel)

CONTRAINDICATIONS: Hypersensitivity, addiction (opiate), asthma, ileus

Black Box Warning: Respiratory depression

Precautions: Pregnancy (B), breastfeeding, child <18 yr, addictive personality, increased intracranial pressure, MI (acute), severe heart disease, renal/hepatic disease, bowel impaction

Black Box Warning: Opioid-naive patients, substance abuse, accidental exposure, potential for overdose/poisoning, status asthmaticus

DOSAGE AND ROUTES

- **Adult: PO** 10-30 mg q4hr (5-15 mg q4-6hr for opiate-naive patients) Conc sol is extremely concentrated; do not use interchangeably; **CONT REL** 10 mg q12hr for opiate-naive patients

Available forms: OxyCODONE: cont rel tabs (OxyCONTIN) 10, 15, 20, 30, 40, 80, 160 mg; immediate rel tabs 5, 7.5, 10, 15, 20, 30 mg; immediate rel caps 5 mg; oral sol 5 mg/5 ml, 20 mg/ml; **oxyCODONE with acetaminophen:** tabs 2.5 mg/300 mg, 2.5 mg/325 mg, 5 mg/300 mg, 5 mg/325 mg, 7.5 mg/325 mg, 7.5 mg/500 mg, 10 mg/325 mg, 10 mg/400 mg, 10 mg/650 mg; caps 5 mg/500 mg; oral sol 5 mg/325 mg/5 ml; **oxyCODONE with aspirin:** 4.88/325 mg; **oxyCODONE with ibuprofen:** 5 mg/400 mg

Administer:

- Store in light-resistant area at room temperature
- OxyCODONE should be titrated from the initial recommended dosage to the dosage required to relieve pain
- There is no maximum dosage of oxyCODONE; however, careful titration is required until tolerance develops to some of the side effects (drowsiness, respiratory depression)

Oral solid formulations

Immediate-release tablets route

- May be administered with food or milk to minimize GI irritation
- **Oxecta brand tablets:** swallow whole; do not crush or dissolve; owing to nature of this formulation, do not pre-

soak, lick, or otherwise wet tablet before administering dose; administer 1 tablet at a time; allow patient to swallow each tablet separately with sufficient liquid to ensure prompt and complete transit through the esophagus; do not use this brand for administration via nasogastric, gastric, or other feeding tubes because it can obstruct feeding tubes

Oral solid formulations

Controlled-release tablets route (OxyCONTIN):

- Administer whole; do not crush, chew, or break in half; taking chewed, broken, or crushed controlled-release tablets could lead to the rapid release and absorption of a potentially toxic dose of oxyCODONE
- **OxyCONTIN brand tablets:** owing to hydro-gelling nature of the 2010 reformulation, do not presoak, lick, or otherwise wet tablet before administering dose; administer 1 tablet at a time; allow patient to swallow each tablet separately with sufficient liquid to ensure prompt and complete transit through the esophagus
- **OxyCODONE controlled-release (OxyCONTIN)** 60-mg and 80-mg tablets are for use only in opioid-tolerant patients
- May be administered without food

Oral liquid formulations

Oral concentrate solution route

- **OxyFAST** is a highly concentrated sol (20 mg oxyCODONE/ml), and care should be taken in dispensing and administering this medication; for ease of administration, the sol may be added to 30 ml of a liquid or semisolid food; if the medication is placed in liquid or food, the patient needs to consume it immediately; do not store diluted oxyCODONE for future use

SIDE EFFECTS

CNS: *Drowsiness, dizziness, confusion, headache, sedation, euphoria,* fatigue, abnormal dreams/thoughts, hallucinations

CV: Palpitations, bradycardia, change in B/P

EENT: Tinnitus, blurred vision, miosis, diplopia

GI: *Nausea, vomiting, anorexia, constipation, cramps,* gastritis, dyspepsia, biliary spasms

GU: Increased urinary output, dysuria, urinary retention

INTEG: *Rash,* urticaria, bruising, flushing, diaphoresis, pruritus

RESP: **Respiratory depression**

PHARMACOKINETICS

PO: Onset 15-30 min, peak 1 hr, duration reg rel 2-6 hr, cont rel 12 hr, metabolized by liver, excreted in urine, crosses placenta, excreted in breast milk, half-life 3-5 hr, protein binding 45%

INTERACTIONS

Increase: effects with other CNS depressants—alcohol, opioids, sedative/hypnotics, antipsychotics, skeletal muscle relaxants

Increase: oxyCODONE level: CYP3A4 inhibitors

Increase: toxicity—cimetidine, MAOIs

Drug/Herb

Increase: sedative effect—kava, St. John's wort, valerian

Drug/Lab Test

Increase: amylase, lipase

O

NURSING CONSIDERATIONS

Assess:

- **Pain:** intensity, location, type, characteristics; need for pain medication by pain/sedation scoring; physical dependence
- I&O ratio; check for decreasing output; may indicate urinary retention
- **CNS changes:** dizziness, drowsiness, hallucinations, euphoria, LOC, pupil reaction
- **Allergic reactions:** rash, urticaria

Black Box Warning: **Respiratory dysfunction:** respiratory depression, character, rate, rhythm; notify prescriber if respirations are <10/min; B/P, pulse

- **Bowel status:** constipation; stimulant laxative may be needed with fluids, fiber

Black Box Warning: **Substance abuse:** assess for substance abuse in patient/family/friends before prescribing: monitor for abuse

Black Box Warning: **Accidental exposure:** dispose of properly away from pets, children

- Assistance with ambulation
- Safety measures: nightlight, call bell within easy reach

Evaluate:

- Therapeutic response: decrease in pain without dependence

Teach patient/family:

- To report any symptoms of CNS changes, allergic reactions
- That physical dependency may result from extended use
- That withdrawal symptoms may occur after long-term use: nausea, vomiting, cramps, fever, faintness, anorexia
- To avoid CNS depressants, alcohol
- To avoid driving, operating machinery if drowsiness occurs

TREATMENT OF OVERDOSE:

Naloxone (Narcan) 0.2-0.8 mg IV, O_2, IV fluids, vasopressors, caution with patients physically dependent on opioids

oxymetazoline nasal agent

See Appendix B

oxymetazoline ophthalmic

See Appendix B

HIGH ALERT

oxymorphone (Rx)

(ox-i-mor′fone)

Opana, Opana ER

Func. class.: Opiate analgesic
Chem. class.: Semisynthetic phenanthrene derivative

Controlled Substance Schedule II

Do not confuse:
oxymorphones/oxyCODONE

ACTION: Inhibits ascending pain pathways in CNS, increases pain threshold, alters pain perception

USES: Moderate to severe pain

CONTRAINDICATIONS: Hypersensitivity, addiction (opiate), asthma, hepatic disease, ileus, intrathecal use, surgery

Black Box Warning: Respiratory depression

Precautions: Pregnancy (B) (short-term), breastfeeding, children <18 yr, addictive personality, increased intracranial pressure, MI (acute), severe heart disease, respiratory depression, renal/hepatic disease, bowel impaction

Black Box Warning: Alcoholism, opioid-naive patients, substance abuse

DOSAGE AND ROUTES

- **Adult: IM/SUBCUT** 1-1.5 mg q4-6hr prn; **IV** 0.5 mg q4-6hr prn; *opiate naive* **PO** (immediate release only) 5-20 mg q4-6hr prn; *opiate naive* **PO-ER** 5 mg q12hr in those requiring around-the-clock dosing

Labor analgesia

- **Adult: IM** 0.5-1 mg

Available forms: Inj 1, 1.5 mg/ml; ER tab 5, 10, 20, 40 mg; ER tab, crush resistant 5, 10, 20, 30, 40 mg; tabs 5, 10 mg

Administer:

- 1 hr before or 2 hr after food (PO)
- With antiemetic for nausea, vomiting
- Do not break, crush, chew ER product
- When pain is beginning to return; determine interval by response
- Store in light-resistant area at room temperature

CONTROLLED REL

- **Opiate naive:** start with lowest dose, titrate upward 5-10 mg q12hr q3-7days to therapeutic response
- When converting from immediate rel to ext rel, give 1/2 daily dose of ext rel product q12hr

SUBCUT route

- Rotate inj sites
- Do not use if respirations are <12/min

IV route

- Give undiluted over 2-3 min

Syringe compatibilities: Glycopyrrolate, hydrOXYzine, ranitidine

SIDE EFFECTS

CNS: *Drowsiness, dizziness, confusion, headache,* hallucinations, increased intracranial pressure, *sedation,* seizures, *euphoria (geriatric patients)*
CV: Palpitations, bradycardia, change in B/P, hypotension
EENT: Tinnitus, blurred vision, miosis, diplopia
GI: *Nausea, vomiting, anorexia, constipation, cramps*
GU: Dysuria, urinary retention
INTEG: *Rash,* urticaria, bruising, flushing, diaphoresis, pruritus
RESP: Respiratory depression

PHARMACOKINETICS

Metabolized by liver, excreted in urine, crosses placenta, half life: PO: 7-9 hr, ext rel: 9-11 hr
PO: Peak 1 hr (fasting)
SUBCUT/IM: Onset 10-15 min, peak 1 1/2 hr, duration 3-6 hr
IV: Onset 5-10 min, peak 15-30 min, duration 3-6 hr

INTERACTIONS

⚠ **Increase:** effects with other CNS depressants—alcohol, opiates, sedative/hypnotics, antipsychotics, skeletal muscle relaxants

⚠ **Increase:** unpredictable effects/reactions—MAOIs

Drug/Herb

Increase: sedative effect—kava, St. John's wort, valerian

Drug/Lab Test

Increase: amylase

NURSING CONSIDERATIONS

Assess:

- **Pain:** location, intensity, type, other characteristics before and 1 hr after (IM) IV 30 min; need for pain medication, physical dependence, give 25%-50% until pain reduction of 50% on pain rating scale, repeat dose may be given at time of peak if previous dose does not control pain and respiratory depression has not occurred; give short-acting opioids for breakthrough pain if patient receiving controlled rel product
- I&O ratio for decreasing output; may indicate urinary retention
- **Bowel status:** constipation; may need stimulative laxative, increased fluids, fiber
- **CNS changes:** dizziness, drowsiness, hallucinations, euphoria, LOC, pupil reaction
- **Allergic reactions:** rash, urticaria

Black Box Warning: **Respiratory dysfunction:** respiratory depression, character, rate, rhythm; notify prescriber if respirations are <10/min

Black Box Warning: **Accidental exposure:** dispose of properly, away from children/pets

Black Box Warning: **Overdose/poisoning:** avoid alcohol ingestion, do not crush, chew, snort, or inject tabs, high abuse potential

Black Box Warning: Opioid-naive patients: ext rel tabs are not to be used immediately postop (12-24 hr after surgery) in these patients

Evaluate:

• Therapeutic response: decrease in pain

Teach patient/family:

• To report any symptoms of CNS changes, allergic reactions
• That physical dependency may result from extended use
• That withdrawal symptoms may occur: nausea, vomiting, cramps, fever, faintness, anorexia
• Not to drive or operate machinery if drowsiness occurs

⚠ Not to use other CNS depressants, alcohol

• To make position changes slowly to prevent orthostatic hypotension

TREATMENT OF OVERDOSE:

Naloxone (Narcan) 0.2-0.8 mg IV, (caution with patients physically dependent on opioids) O_2, IV fluids, vasopressors

⚠ HIGH ALERT

oxytocin (Rx)

(ox-i-toe′sin)

Pitocin

Func. class.: Hormone

Chem. class.: Oxytocic, uterine-active agent

ACTION: Acts directly on myofibrils, thereby producing uterine contraction; stimulates milk ejection by the breast; vasoactive antidiuretic effect

USES: Stimulation, induction of labor; missed or incomplete abortion; postpartum bleeding

CONTRAINDICATIONS: Hypersensitivity, serum toxemia, cephalopelvic disproportion, fetal distress, hypertonic uterus, prolapsed umbilical cord, active genital herpes

Precautions: Cervical/uterine surgery, uterine sepsis, primipara >35 yr, 1st/2nd stage of labor

Black Box Warning: Elective induction of labor

DOSAGE AND ROUTES

Postpartum hemorrhage

• **Adult: IV** 10-40 units in 1000 ml nonhydrating diluent infused at 20-40 mU/min
• **Adult: IM** 3-10 units after delivery of placenta

Contraction stress test (CST)

• **Adult: IV** 0.5 mU/min, increase q20min until 3 contractions within 10 min

Stimulation of labor

• **Adult: IV** 0.5-2 mU/min, increase by 1-2 mU q15-60min until contractions occur then decrease dose

Incomplete abortion

• **Adult: IV INF** 10 units/500 ml D_5W or 0.9% NaCl at 10-20 mU/min, max 30 units/12 hr

Available forms: Inj 10 units/ml

Administer:

IV route

• Use inf pump

Labor induction

• After diluting 10 units/1000 ml of 0.9% NS or D_5 NS run at 1-2 mU/min at 15- to 30-min intervals to begin normal labor; dilute 10-40 mU/min; titrate to control postpartum bleeding; dilute 10 units/500 ml sol; run 10 units-20 mU/ml; administer by only 1 route at a time; use inf pump; rotate inf to provide mixing; do not shake

Control of postpartum bleeding

• Dilute 10-40 units/1000 ml of sol; run at 10-20 mU/min; adjust rate as needed
• With crash cart available on unit (magnesium sulfate at bedside)

Incomplete, inevitable, elective abortion

• Dilute 10 units/500 ml compatible IV sol

Y-site compatibilities: Heparin, hydrocortisone, insulin (regular), meperidine, morphine, potassium chloride, vit B/C, warfarin

SIDE EFFECTS

CNS: **Seizures, tetanic contractions**
CV: Hypo/hypertension, dysrhythmias, increased pulse, bradycardia, tachycardia, PVC
FETUS: Dysrhythmias, jaundice, hypoxia, **intracranial hemorrhage**
GI: Anorexia, nausea, vomiting, constipation
GU: **Abruptio placentae, decreased uterine blood flow**
HEMA: Increased hyperbilirubinemia
INTEG: Rash
RESP: **Asphyxia**
SYST: Water intoxication of mother

PHARMACOKINETICS

IM: Onset 3-7 min, duration 1 hr, half-life 12-17 min
IV: Onset 1 min, duration 30 min, half-life 12-17 min

INTERACTIONS

• Hypertension: vasopressors

Drug/Herb

• Hypertension: ephedra

NURSING CONSIDERATIONS

Assess:

• I&O ratio
• B/P, pulse; watch for changes that may indicate hemorrhage
• Respiratory rate, rhythm, depth; notify prescriber of abnormalities
• Length, intensity, duration of contraction; notify prescriber of contractions lasting >1 min or absence of contractions; turn patient on her side; discontinue oxytocin
• **FHTs, fetal distress; watch for acceleration, deceleration; notify prescriber if problems occur; fetal presentation, pelvic dimensions; turn patient on left side if FHT change in rate, give O_2**
⚠ **Water intoxication; confusion, anuria, drowsiness, headache**

Evaluate:

• Therapeutic response: stimulation of labor, control of postpartum bleeding

Teach patient/family:

• To report increased blood loss, abdominal cramps, fever, foul-smelling lochia
• That contractions will be similar to menstrual cramps, gradually increasing in intensity

Black Box Warning: Elective induction of labor: use only for induction when medically necessary

O

PACLitaxel (Rx)

(pa-kli-tax′el)

PACLitaxel nanoparticle albumin-bound (Rx)

Abraxane

Func. class.: Antineoplastic—miscellaneous

Chem. class.: Taxane

Do not confuse:
PACLitaxel/PARoxetine/Paxil

ACTION: Inhibits reorganization of microtubule network needed for interphase and mitotic cellular functions; causes abnormal bundles of microtubules during cell cycle and multiple esters of microtubules during mitosis

USES: **Taxol:** metastatic carcinoma of the ovary, breast; AIDS-related Kaposi's sarcoma (2nd-line), non–small-cell lung cancer (1st-line), adjuvant treatment for node-positive breast cancer

Unlabeled uses: Advanced head, neck, small-cell lung cancer; non-Hodgkin's lymphoma, adenocarcinoma of the upper GI tract, hormone-refractory prostate cancer, bladder cancer

CONTRAINDICATIONS: Pregnancy (D); hypersensitivity to PACLitaxel or other products with polyoxyethylated castor oil, albumin

Black Box Warning: Neutropenia of <1500/mm^3

Precautions: Breastfeeding, children, females, geriatric patients, cardiovascular/hepatic/renal disease, CNS disorder, bone marrow suppression, dental disease/work, extravasation, herpes, infection, infertility, jaundice, ocular exposure, radiation therapy, thrombocytopenia, vaccination

Black Box Warning: Taxane hypersensitivity, requires a specialized care setting, requires an experienced clinician

DOSAGE AND ROUTES

PACLitaxel

Ovarian carcinoma

• **Adult: IV INF** 135 mg/m^2 given over 24 hr q3wk then CISplatin 75 mg/m^2; or 175 mg/m^2 over 3 hr q3wk; or 175 mg/m^2 over 3 hr

Advanced ovarian carcinoma

• **Adult: IV/INF** 175 mg/m^2 with CISplatin 75 mg/m^2 using a 3-hr regimen q3wk

Breast carcinoma

• **Adult: IV INF** 175 mg/m^2 over 3 hr q3wk × 4 courses

AIDS-related Kaposi's sarcoma

• **Adult: IV INF** 135 mg/m^2 over 3 hr q3wk or 100 mg/m^2 over 3 hr q2wk

1st-line non–small-cell lung cancer

• **Adult: IV INF** 135 mg/m^2/24 hr inf with CISplatin 75 mg/m^2 × 3 wk

Hepatic dose

• **Adult:** for 135 mg/m^2 24-hr IV INF AST/ALT 2-10 × ULN, total bilirubin ≤1.5 mg/dl: 100 mg/m^2; AST/ALT <10 × ULN, total bilirubin 1.6-7.5 mg/dl: 50 mg/m^2; AST/ALT ≥10 × ULN or total bilirubin >7.5 mg/dl; avoid use; for 175/m^2 3-hr IV INF AST/ALT <10 × ULN, total bilirubin 1.26-2 × ULN: 135 mg/m^2; AST/ALT <10 × ULN, total bilirubin 2.01-5 × ULN: 90 mg/m^2; AST/ALT ≥10 × ULN or total bilirubin >5 × ULN: avoid use

Bladder cancer (unlabeled)

• **Adult: IV** 225 mg/m^2 over 3 hr with CARBOplatin

Head/neck cancer (unlabeled)

• **Adult: IV** 40-45 mg/m^2 with CARBOplatin over 24 hr or 175-300 mg/m^2 over 3 hr

Stem cell transplant/bone marrow ablation (unlabeled)

• **Adult: IV** 250-775 mg/m^2 over 24 hr in combination with other chemotherapy

PACLitaxel protein-bound particles

• **Adult: IV** 260 mg/m^2 q3wk

Available forms: Inj 6 mg/ml, 30 mg/5-ml vial, 100 mg/16.7-ml vial, 150 mg/25-ml vial, 300 mg/50-ml vial; powder for inj, lyophilized 100 mg in single-use vials (Abraxane)

Administer:

- CBC, differential, platelet count before treatment and weekly; withhold product if WBC is $<1500/mm^3$ or platelet count is $<100{,}000/mm^3$; notify prescriber
- If CISplatin is given, use after taxane
- Confirmation that dexamethasone was given 12 hr and 6 hr before inf begins
- Store prepared sol up to 27 hr in refrigerator

Continuous IV INF route

- After premedicating with dexamethasone 20 mg PO 12 hr and 6 hr before PACLitaxel, diphenhydrAMINE 50 mg IV ½-1 hr before PACLitaxel and cimetidine 300 mg or ranitidine 50 mg IV ½-1 hr before PACLitaxel
- For extravasation if given by regular IV, not port

PACLitaxel

- After diluting in 0.9% NaCl, D_5, D_5 and 0.9% NaCl, D_5LR (0.3-1.2 mg/ml), chemo dispensing pin or similiar devices with spikes should not be used in vials of Taxol; use in-line filter ≤0.22 micron; may be given as 3-hr or 24-hr inf
- Using only glass bottles, polypropylene, polyolefin bags, and administration sets; do not use PVC inf bags or sets

Y-site compatibilities: Acyclovir, amikacin, aminophylline, ampicillin/sulbactam, bleomycin, butorphanol, calcium chloride, CARBOplatin, cefepime, cefoTEtan, cefTAZidime, cefTRIAXone, cimetidine, CISplatin, cladribine, cyclophosphamide, cytarabine, dacarbazine, dexamethasone, diphenhydrAMINE, DOXOrubicin, droperidol, etoposide, famotidine, floxuridine, fluconazole, fluorouracil, furosemide, ganciclovir, gentamicin, granisetron, haloperidol, heparin, hydrocortisone, HYDROmorphone, ifosfamide, LORazepam, magnesium sulfate, mannitol, meperidine, mesna, methotrexate, metoclopramide, morphine, nalbuphine, ondansetron, pentostatin, potassium chloride, prochlorperazine, propofol, ranitidine, sodium bicarbonate, thiotepa, vancomycin, vinBLAStine, vinCRIStine, zidovudine

Abraxane

Intermittent IV INF route

- Reconstitute vial by injecting 20 ml of 0.9% NaCl; slowly inject 20 ml of 0.9% NaCl over at least 1 min to direct sol flow on wall of vial; do not inject 0.9% NaCl directly onto lyophilized cake (foaming will occur); allow vial to sit for at least 5 min to ensure proper wetting of lyophilized cake; gently swirl or invert vial slowly for ≥2 min until completely dissolved
- Calculate dose by dosing vol/ml = total dose (mg) ÷ 5 (mg/ml)

SIDE EFFECTS

CNS: *Peripheral neuropathy*

CV: Bradycardia, *hypotension,* abnormal ECG, **supraventricular tachycardia (SVT)**

GI: *Nausea, vomiting, diarrhea, mucositis, stomatitis, increased bilirubin, alk phos, AST*

HEMA: **Neutropenia, leukopenia, thrombocytopenia,** anemia, bleeding, infections

INTEG: *Alopecia,* **tissue necrosis, generalized urticaria,** *flushing*

MS: *Arthralgia, myalgia*

RESP: **Pulmonary embolism,** dyspnea

SYST: *Hypersensitivity reactions,* **anaphylaxis, Stevens-Johnson syndrome, toxic epidermal necrolysis, angioedema**

PHARMACOKINETICS

89%-98% of product serum protein bound, metabolized in liver, excreted in bile and urine; terminal half-life 5.3-17.4 hr

INTERACTIONS

Increase: myelosuppression—other antineoplastics, radiation

Increase: DOXOrubicin levels—DOXOrubicin

⚠ **Increase: toxicity, decrease metabolism—ketoconazole; avoid concurrent use**

Increase: bleeding risk—NSAIDs, anticoagulants

Decrease: PACLitaxel metabolism—verapamil, diazepam, cycloSPORINE, teniposide, etoposide, quiNIDine, dexamethasone, vinCRIStine, testosterone
Decrease: PACLitaxel levels—CYP2C8, CYP2C9 inducers
Decrease: immune response—live virus vaccines

Drug/Lab Test
Increase: AST/ALT
Decrease: neutrophils, platelets, WBC, Hgb

NURSING CONSIDERATIONS

Assess:

Black Box Warning: Requires a specialized care setting such as a hospital or facility with management of complications; should be used by a clinician experienced in cytotoxic agents

- **Cardiovascular status:** ECG continuously in CV conditions; monitor for hypotension, sinus bradycardia/tachycardia
- **Peripheral neuropathy:** paresthesias, numbness; during inf, use ice packs on extremities to lessen continued neuropathy; may use acupuncture for some relief; use of ice on extremities when infusing
- **Arthralgia, myalgia:** may begin 2-3 days after inf and continue for 4-5 days; may use analgesics
- **Nausea, vomiting:** premedicate with antiemetics; nausea and vomiting occur often
- Hepatic studies before, during therapy (bilirubin, AST, ALT, LDH) prn or monthly, check for jaundiced skin and sclera, dark urine, clay-colored stool, itchy skin, abdominal pain, fever, diarrhea
- VS during 1st hr of inf, check IV site for signs of infiltration

⚠ **Hypersensitivity reactions, anaphylaxis:** hypotension, dyspnea, angioedema, generalized urticaria; discontinue inf immediately; keep emergency equipment available, monitor continuously during first 30-60 min, then periodically

- **Flush:** for mild to moderate flush, may continue diphenhydrAMINE for ≤48 hr
- Effects of alopecia on body image; discuss feelings about body changes

Evaluate:

- Therapeutic response: decreased tumor size, spread of malignancy

Teach patient/family:

- To report signs of infection: fever, sore throat, flulike symptoms
- To report signs of anemia: fatigue, headache, faintness, SOB, irritability
- To report bleeding; to avoid use of razors, commercial mouthwash; to use soft-bristle toothbrush; to use viscous xylocaine or compounded formula for stomatitis
- To avoid use of aspirin, ibuprofen
- To avoid crowds, persons with known infections
- That hair may be lost during treatment; that a wig or hairpiece may make patient feel better; that new hair may be different in color, texture
- That pain in muscles and joints 2-5 days after inf is common
- To notify prescriber if pregnancy is planned or suspected, pregnancy (D); do not breastfeed
- To avoid receiving vaccinations while taking product

paliperidone (Rx)

(pal-ee-per′i-done)

Invega, Invega Sustenna

Func. class.: Antipsychotic

Chem. class.: Benzisoxazole derivative

Do not confuse:
Invega/Iveegam
paliperidone/risperiDONE

ACTION: Mediated through both DOPamine type 2 (D_2) and serotonin type 2 ($5\text{-}HT_2$) antagonism

USES: Schizophrenia, schizoaffective disorder
Unlabeled uses: Agitation

CONTRAINDICATIONS:

Breastfeeding, geriatric patients, seizure disorders, AV block, QT prolongation, torsades de pointes; hypersensitivity to this product, risperidone

Precautions: Pregnancy (C), children, renal/hepatic disease, obesity, Parkinson's disease, suicidal ideation, diabetes mellitus, hematological disease

Black Box Warning: Dementia-related psychosis (mortality)

DOSAGE AND ROUTES

- **Adult: PO** 6 mg/day, max 12 mg/day; **IM** 234 mg on day 1 then 156 mg 1 wk later; after 2nd dose, give 117 mg each mo; range 39-234 mg, dosage change may be needed when used with CYP3A4 inducers
- **Child/adolescent ≥12 yr and ≥51 kg: PO** 3 mg daily, may increase if needed by 3 mg/day in intervals of >5 days, up to max 12 mg/day; <51 kg max 6 mg/day

Renal dose

- **Adult: PO** CCr 50-79 ml/min, 3 mg/day, max 6 mg/day; **EXT REL/IM** 156 mg on day 1, 117 mg 1 wk later, then 78 mg each mo; CCr 10-49 ml/min, 1.5 mg/day, max 3 mg/day; IM not recommended

Available forms: Ext rel tabs 1.5, 3, 6, 9 mg; ext rel susp for inj 39 mg/0.25 ml, 78 mg/0.5 ml, 117 mg/0.75 ml, 156 mg/ml, 234 mg/1.5 ml

Administer:

- Avoid use with CNS depressants

PO route

- Do not break, crush, or chew ext rel tabs; use plenty of water
- Without regard for food
- Reduced dose for geriatric patients

IM route

- Use for IM only; do not use IV or subcut; inj kits contain prefilled syringe and 2 safety needles; for single use only; shake for 10 sec; **deltoid inj:** ≥90 kg, use 1.5-inch, 22-G needle; <90 kg, use 1-inch, 23-G needle; alternate injections between deltoid muscles; **gluteal inj:** use 1.5-inch, 22-G needle; while holding syringe upright, twist rubber tip clockwise to remove; peel safety needle pouch halfway open; grasp needle sheath using plastic peel pouch; attach needle to Luer connection in clockwise motion; pull needle sheath away using straight pull; bring syringe with attached needle upright to de-aerate, de-aerate, inject; after inj, use finger, thumb, or flat surface to activate needle protection system until click heard; use deltoid × 2 doses

SIDE EFFECTS

CNS: *EPS, pseudoparkinsonism, akathisia, dystonia, tardive dyskinesia; drowsiness, insomnia, agitation, anxiety, headache,* **seizures, neuroleptic malignant syndrome,** dizziness

CV: Orthostatic hypotension, **tachycardia; heart failure, QT prolongation, heart block, dysrhythmias**

EENT: Blurred vision, cough

ENDO: Hyperinsulinemia, weight gain, hyperglycemia, dyslipidemia, diabetes mellitus

GI: *Nausea,* vomiting, *anorexia, constipation,* weight gain in adolescents, xerostomia

GU: Priapism, menstrual irregularities

HEMA: **Agranulocytosis**

PHARMACOKINETICS

Peak 24 hr; elimination half-life 23 hr; excreted 80% urine, 11% feces, protein binding >74%

INTERACTIONS

Increase: sedation—other CNS depressants, alcohol, sedative/hypnotics, opiates

Increase: EPS—other antipsychotics

⚠ **Increase: QT prolongation—class IA, III antidysrhythmics, azole antifungals, tricyclics (high doses), some phenothiazines, β-blockers, chloroquine, pimozide, droperidol, some antipsychotics, abarelix, alfuzosin, amoxapine, apomorphine, dasatinib, dolasetron, flecainide, halogenated anesthetics**

Increase: neurotoxicity—lithium

P

Increase: serotonin syndrome, neuroleptic malignant syndrome—SSRIs, SNRIs
Decrease: effect of paliperidone—carBAMazepine, other CYP3A4 inducers
Decrease: levodopa effect—levodopa
Drug/Lab Test
Increase: prolactin levels

NURSING CONSIDERATIONS

Assess:

Black Box Warning: Mental status: mood, behavior, confusion, orientation, suicidal thoughts/behaviors; dementia, especially in geriatric patients before initial administration

QT prolongation: ECG for QT prolongation, ejection fraction; chest pain, palpitations, dyspnea

- Swallowing of PO medication; check for hoarding, giving of medication to others
- Affect, orientation, LOC, reflexes, gait, coordination, sleep pattern disturbances
- B/P (standing, lying), pulse, respirations; q4hr during initial treatment; establish baseline before starting treatment; report drops of 30 mm Hg; watch for ECG changes
- **Hyperprolactinemia:** sexual dysfunction, decreased menstruation, breast pain
- Dizziness, faintness, palpitations, tachycardia on rising
- **EPS:** akathisia, tardive dyskinesia (bizarre movements of jaw, mouth, tongue, extremities), pseudoparkinsonism (rigidity, tremors, pill rolling, shuffling gait)

Serotonin syndrome, neuroleptic malignant syndrome: hyperthermia, increased CPK, altered mental status, muscle rigidity, fever, seizures, discontinue

- Constipation, urinary retention daily; if these occur, increase bulk and water in diet; monitor for weight gain, especially among adolescents
- Supervised ambulation until patient is stabilized on medication; do not involve patient in strenuous exercise program because fainting is possible; patient should not stand still for a long time
- Increased fluids to prevent constipation
- Sips of water, candy, gum for dry mouth

Evaluate:

- Therapeutic response: decrease in emotional excitement, hallucinations, delusions, paranoia; reorganization of patterns of thought, speech

Teach patient/family:

- That orthostatic hypotension may occur; to rise gradually from sitting or lying position
- To avoid hot tubs, hot showers, tub baths because hypotension may occur
- To avoid abrupt withdrawal of this product because EPS may result; that product should be withdrawn slowly
- To avoid OTC preparations (cough, hay fever, cold) unless approved by prescriber because serious product interactions may occur; to avoid alcohol because increased drowsiness may occur
- To avoid hazardous activities if drowsy or dizzy
- About compliance with product regimen; that nonabsorbable tab shell is expelled in stool
- To report impaired vision, tremors, muscle twitching
- That heat stroke may occur in hot weather; to take extra precautions to stay cool
- To use contraception; to inform prescriber if pregnancy is planned or suspected; do not breastfeed

Black Box Warning: To notify prescriber of suicidal thoughts/behaviors, other changes in behavior

TREATMENT OF OVERDOSE:

Lavage if orally ingested; provide airway; *do not induce vomiting*

palonosetron (Rx)

(pa-lone-o′se-tron)

Aloxi

Func. class.: Antiemetic

Chem. class.: 5-HT_3 receptor antagonist

ACTION: Prevents nausea, vomiting by blocking serotonin peripherally, centrally, and in the small intestine at the 5-HT_3 receptor

USES: Prevention of nausea, vomiting associated with cancer chemotherapy, postoperative nausea/vomiting

CONTRAINDICATIONS: Hypersensitivity

Precautions: Pregnancy (B), breastfeeding, children, geriatric patients, hypokalemia, hypomagnesium, patients taking diuretics

DOSAGE AND ROUTES

• **Adult: PO** 0.5 mg as single dose 1 hr before chemotherapy; **IV** 0.25 mg as single dose over 30 sec ½ hr before chemotherapy, max 25 mg **IV** over q7days

Postoperative nausea/vomiting prophylaxis for ≤24 hr after surgery

• **Adult: IV** 0.075 mg given over 10 sec immediately before induction

Available forms: Inj 0.25/5 ml; caps 0.5 mg

Administer:

PO route

• Give 1 hr before chemotherapy
• May be taken without regard to food
• Store at room temperature
• Do not mix with other products; flush IV line before and after administration with 0.9% NaCl

Direct IV route

• **Chemotherapy nausea/vomiting:** give as single dose over 30 sec

• **Postoperative nausea/vomiting:** give over 10 sec immediately before anesthesia induction

Syringe compatibilities: Dexamethasone

Y-site compatibilities: Alemtuzumab, alfentanil, amifostine, amikacin, aminocaproic acid, aminophylline, amiodarone, amphotericin B liposome, ampicillin, ampicillin/sulbactam, atracurium, atropine, azithromycin, aztreonam, bivalirudin, bleomycin, bumetanide, buprenorphine, busulfan, butorphanol, calcium acetate/chloride/gluconate, CARBOplatin, carmustine, caspofungin, ceFAZolin, cefepime, cefotaxime, cefoTEtan, cefOXitin, cefTAZidime, ceftizoxime, cefTRIAXone, cefuroxime, chloramphenicol, chlorproMAZINE, cimetidine, ciprofloxacin, cisatracurium, CISplatin, clindamycin, cyclophosphamide, cycloSPORINE, cytarabine, dacarbazine, DACTINomycin, dantrolene, DAPTOmycin, DAUNOrubicin, dexamethasone, dexmedetomidine, dexrazoxane, digoxin, diltiazem, diphenhydrAMINE, DOBUTamine, DOCEtaxel, DOPamine, doxacurium, DOXOrubicin hydrochloride, droperidol, enalaprilat, ePHEDrine, EPINEPHrine, epirubicin, eptifibatide, erythromycin, esmolol, etoposide, etoposide phosphate, famotidine, fenoldopam, fentaNYL, fluconazole, fludarabine, fluorouracil, foscarnet, fosphenytoin, furosemide, gemcitabine, gentamicin, glycopyrrolate, haloperidol, heparin, hydrALAZINE, hydrocortisone, HYDROmorphone, IDArubicin, ifosfamide, inamrinone, insulin, irinotecan, isoproterenol, ketorolac, labetalol, leucovorin, levofloxacin, lidocaine, linezolid, LORazepam, magnesium sulfate, mannitol, mechlorethamine, melphalan, meperidine, meropenem, mesna, metaraminol, methotrexate, methyldopate, metoclopramide, metoprolol, metroNIDAZOLE, midazolam, milrinone, mitoMYcin, mitoXANtrone, mivacurium, morphine, nalbuphine, naloxone, neostigmine, nesiritide, niCARdipine, nitroglycerin, nitroprusside, norepinephrine, octreotide, oxaliplatin, oxytocin, PACLi-

P

taxel, pamidronate, pancuronium, pentazocine, PHENobarbital, phentolamine, phenylephrine, piperacillin/tazobactam, potassium acetate/chloride/phosphates, procainamide, prochlorperazine, promethazine, propranolol, quinupristin/dalfopristin, ranitidine, remifentanil, rocuronium, sodium acetate/bicarbonate/phosphates, streptozocin, succinylcholine, SUFentanil, tacrolimus, teniposide, theophylline, thiotepa, ticarcillin/clavulanate, tigecycline, tirofiban, tobramycin, topotecan, trimethobenzamide, trimethoprim/sulfamethoxazole, vancomycin, vasopressin, vecuronium, verapamil, vinBLAStine, vinCRIStine, vinorelbine, zidovudine

SIDE EFFECTS

CNS: *Headache, dizziness, drowsiness, fatigue, insomnia*
GI: *Diarrhea, constipation,* abdominal pain
MISC: Weakness, hyperkalemia, anxiety, rash, bronchospasm (rare), arthralgia, *fever, urinary retention*

PHARMACOKINETICS

62% protein bound; metabolized by liver; unchanged product and metabolites excreted by kidney; terminal elimination half-life 40 hr

INTERACTIONS

- Possible QT prolongation: class 1A antidysrhythmics (disopyramide, procainamide, quiNIDine), class III antidysrhythmics (amiodarone, dofetilide, ibutilide, sotalol), chloroquine, clarithromycin, droperidol, erythromycin, haloperidol, levomethadyl, methadone, pentamidine, some phenothiazines, diuretics (except potassium sparing)

NURSING CONSIDERATIONS

Assess:
- For agents that cause QT prolongation, even if manufacturer has removed QT prolongation from warnings
- Absence of nausea, vomiting during chemotherapy
- **Hypersensitivity reaction:** rash, bronchospasm (rare)
- Cardiac disease: check ECG before use
- Hyperkalemia: monitor potassium

Evaluate:
- Therapeutic response: absence of nausea, vomiting during cancer chemotherapy, postoperatively

Teach patient/family:
- To report diarrhea, constipation, rash, changes in respirations, or discomfort at insertion site
- To avoid alcohol, barbiturates
- Use other antiemetics if nausea occurs

pamidronate (Rx)

(pam-i-drone′ate)

Aredia

Func. class.: Bone-resorption inhibitor, electrolyte modifier
Chem. class.: Bisphosphonate

Do not confuse:
Aredia/Adriamycin

ACTION: Inhibits bone resorption, apparently without inhibiting bone formation and mineralization; adsorbs calcium phosphate crystals in bone and may directly block the dissolution of hydroxyapatite crystals of bone

USES: Moderate to severe Paget's disease, hypercalcemia, osteolytic bone metastases in breast cancer, patients with multiple myeloma
Unlabeled uses: Postmenopausal osteoporosis and prevention, osteoporosis prophylaxis, ankylosing spondylitis, osteogenesis imperfecta, hyperparathyroidism

CONTRAINDICATIONS: Pregnancy (D), hypersensitivity to bisphosphonates
Precautions: Children, nursing mothers, renal dysfunction, poor dentition

DOSAGE AND ROUTES

Hypercalcemia of malignancy

• **Adult: IV INF** 60-90 mg as single dose for moderate hypercalcemia; 90 mg for severe hypercalcemia over 2-24 hr; dose should be diluted in 1000 ml 0.45% NaCl, 0.9% NaCl, or D_5W; wait 7 days before 2nd course

Osteolytic lesions

• **Adult: IV** 90 mg/500 ml of D_5W, 0.45% NaCl, or 0.9% NaCl given over 4 hr each mo (multiple myeloma) or over 2 hr q3-4wk (breast carcinoma)

Paget's disease

• **Adult: IV INF** 30 mg/day given over 4 hr × 3 days

Severe osteogenesis imperfecta (unlabeled)

• **Child: IV** 1.5-3 mg/kg/cycle, cycle dose is divided in 3, administered via slow IV over 4 hr/day × 3 days

Hypercalcemia (hyperparathyroidism) (unlabeled)

• **Adult: IV** 15-60 mg as a single dose

Corticosteroid-induced osteoporosis (unlabeled)

• **Adult: IV** 30 mg q3mo × 1 yr

Ankylosing spondylitis (unlabeled)

• **Adult: IV INF** 60 mg over 4 hr; 6-hr inf for 1st dose

Osteoporosis prophylaxis in Crohn's disease (unlabeled)

• **Adult: IV INF** 30 mg over 1 hr q3mo × 1 yr

Available forms: Powder for inj 30, 90 mg/vial; inj 3, 6, 9 mg/ml

Administer:

IV route

• After reconstituting by adding 10 ml sterile water for inj to each vial (30 mg/10 ml or 90 mg/10 ml, depending on vial used); add to 1000 ml of sterile 0.45%, 0.9% NaCl, D_5W, run over 2-24 hr **(hypercalcemia);** dilute reconstituted sol in 500 ml of 0.9% NaCl, 0.45% NaCl, or D_5W, give over 4 hr **(multiple myeloma, Paget's disease);** dilute reconstituted sol in 250 ml of 0.9% NaCl, 0.45% NaCl, or D_5W; give over 2 hr **(osteolytic bone metastases of breast cancer)**

• Do not mix with calcium-containing inf sol such as Ringer's sol

• Store inf sol up to 24 hr at room temperature

• Reconstituted sol with sterile water may be refrigerated for ≤24 hr

Y-site compatibilities: Acyclovir, alfentanil, allopurinol, amifostine, amikacin, aminocaproic acid, aminophylline, amphotericin B lipid complex, amphotericin B liposome, ampicillin, anidulafungin, atenolol, atracurium, azithromycin, aztreonam, bivalirudin, bleomycin, bumetanide, buprenorphine, butorphanol, CARBOplatin, carmustine, ceFAZolin, cefepime, cefoperazone, cefotaxime, cefoTEtan, cefOXitin, cefTAZidime, ceftizoxime, cefTRIAXone, cefuroxime, chloramphenicol, chlorproMAZINE, cimetidine, ciprofloxacin, cisatracurium, CISplatin, clindamycin, cyclophosphamide, cycloSPORINE, cytarabine, dacarbazine, DAPTOmycin, dexamethasone, dexmedetomidine, dexrazoxane, digoxin, diltiazem, diphenhydrAMINE, DOBUTamine, DOCEtaxel, dolasetron, DOPamine, doxacurium, DOXOrubicin, doxycycline, droperidol, enalaprilat, ePHEDrine, EPINEPHrine, epirubicin, ertapenem, erythromycin, esmolol, etoposide, famotidine, fenoldopam, fentaNYL, fluconazole, fludarabine, fluorouracil, foscarnet, fosphenytoin, furosemide, gallium, ganciclovir, gatifloxacin, gemcitabine, gentamicin, glycopyrrolate, granisetron, haloperidol, heparin, hetastarch 6%, hydrALAZINE, hydrocortisone, HYDROmorphone, hydrOXYzine, ifosfamide, imipenem-cilastatin, inamrinone, insulin (regular), isoproterenol, ketorolac, labetalol, levofloxacin, levorphanol, lidocaine, linezolid, LORazepam, magnesium sulfate, mannitol, mechlorethamine, melphalan, meperidine, meropenem, mesna, metaraminol, methotrexate, methyldopate, methylPREDNISolone, metoclopramide, metoprolol, metroNIDAZOLE, midazolam, milrinone, minocycline, mi-

P

toXANtrone, mivacurium, morphine, mycophenolate, nafcillin, nalbuphine, naloxone, nesiritide, niCARdipine, nitroglycerin, nitroprusside, norepinephrine, octreotide, ondansetron, oxytocin, PACLitaxel, palonosetron, pancuronium, PEMEtrexed, pentamidine, pentazocine, PENTobarbital, PHENobarbital, phenylephrine, piperacillin, polymyxin B, potassium chloride/phosphates, procainamide, prochlorperazine, promethazine, propranolol, quiNIDine, quinupristin-dalfopristin, ranitidine, remifentanil, rocuronium, sodium acetate/bicarbonate/phosphates, succinylcholine, SUFentanil, sulfamethoxazole-trimethoprim, teniposide, theophylline, thiopental, thiotepa, ticarcillin, ticarcillin-clavulanate, tigecycline, tirofiban, tobramycin, tolazoline, topotecan, trimethobenzamide, vancomycin, vasopressin, vecuronium, verapamil, vinBLAStine, vinCRIStine, vinorelbine, voriconazole, zidovudine

SIDE EFFECTS

CNS: *Fever,* fatigue
CV: Hypertension, atrial fibrillation
EENT: Ocular pain, inflammation, vision impairment
GI: Abdominal pain, anorexia, constipation, nausea, vomiting, dyspepsia
GU: Renal failure
HEMA: Thrombocytopenia, anemia, leukopenia
INTEG: Redness, swelling, induration, pain on palpation at site of catheter insertion
META: *Hypokalemia, hypomagnesemia, hypophosphatemia, hypocalcemia,* hypothyroidism
MS: Severe bone pain, myalgia, osteonecrosis of the jaw
RESP: Coughing, dyspnea, upper respiratory tract infection
SYST: Angioedema, anaphylaxis

PHARMACOKINETICS

Rapidly cleared from circulation and taken up mainly by bones, primarily in areas of high bone turnover; eliminated primarily by kidneys; half-life 21-35 hr, terminal half-life in bone is 300 days

INTERACTIONS

Increase: hypokalemia—loop diuretics
Increase: nephrotoxicity—aminoglycosides, NSAIDs, vancomycin, radiopaque contrast agents, cycloSPORINE, tacrolimus
Increase: effect of entecavir
Decrease: pamidronate effect—calcium, vit D

Drug/Lab Test
Increase: creatinine
Decrease: potassium, magnesium, phosphate, calcium, WBC, platelets

NURSING CONSIDERATIONS

Assess:
- **Hypocalcemia:** nausea, vomiting, constipation, thirst, dysrhythmias, hypocalcemia, paresthesia, twitching, laryngospasm, Chvostek's sign, Trousseau's sign; **hypercalcemia:** thirst, nausea, vomiting, dysrhythmias
- Dehydration/hypovolemia: should be corrected during treatment of hypercalcemia, before therapy, maintain adequate urine output
- Monitor WBCs, platelets, electrolytes, creatinine, BUN, Hgb/Hct before beginning treatment
- **Dental health:** cover with antiinfectives for dental extractions
- Temperature may be elevated during the first 3 days after a dose; risk of fever increases as dose increases
- Renal disease: max 90-mg single dose, longer infusions >2 hr may increase risk for renal toxicity
- Bone pain; use analgesics
- I&O, check for fluid overload edema, crackles, increased B/P; BUN, creatinine

Evaluate:
- Therapeutic response: decreased calcium levels

Teach patient/family
- To notify prescriber if pregnancy is planned or suspected, pregnancy (D)

• To report hypercalcemic relapse: nausea, vomiting, bone pain, thirst; unusual muscle twitching, muscle spasms; severe diarrhea, constipation
• To continue with dietary recommendations, including calcium and vit D
• To obtain analgesic from provider for bone pain
• That, if nausea, vomiting occur, small, frequent meals may help
• To report ocular symptoms to prescriber: blurred vision, edema, inflammation

pancrelipase (Rx)

(pan-kre-li′pase)

Creon, DMH ♣, Pancrease ♣, Pancreaze, Pancrecarb MS, Pertzye, Ultrase MT, Ultresa, Viokase, Zenpep

Func. class.: Digestant

Chem. class.: Pancreatic enzyme—bovine/porcine

ACTION:
Pancreatic enzyme needed for the breakdown of substances released from the pancreas

USES:
Exocrine pancreatic secretion insufficiency, cystic fibrosis (digestive aid), steatorrhea, pancreatic enzyme deficiency

CONTRAINDICATIONS:
Allergy to pork

Precautions: Pregnancy (B), ileus, pancreatitis, Crohn's disease, diabetes mellitus

DOSAGE AND ROUTES

Many products listed above are not interchangeable

• **Adult/adolescent/child ≥4 yr (del rel caps: Creon Caps, Zenpap Caps, Pancreaze Caps): PO** 500 lipase units/kg/meal, titrate based on patient response, max 2500 lipase units/kg/meal
• **Child 1-4 yr: PO** 1000 lipase units/kg/meal, titrate based on patient response, max 2500 lipase units/kg/meal

Available forms: Tabs (Viokase) 10, 204; cap, del rel 4, 8, 154 (Pancrecarb MS), 12, 18, 20 (Ultrase MT), Ultrase; cap 3000, 4200, 5000, 6000, 8000, 10,500, 12,000, 15,000, 16,000, 16,800, 24,000, 25,000 units

Administer:
• After antacid or cimetidine; decreased pH inactivates product
• Low-fat diet for GI symptoms
• Have patient sit up during administration; give with meals
• Do not crush, chew del rel products, caps
• Viokase is not interchangeable with other products
• Store in tight container at room temperature

SIDE EFFECTS

ENDO: Hypo/hyperglycemia
GI: Anorexia, nausea, vomiting, diarrhea, cramping, bloating
GU: Hyperuricuria, hyperuricemia

INTERACTIONS

Decrease: absorption—cimetidine, antacids, oral iron
Decrease: effect of acarbose, miglitol

NURSING CONSIDERATIONS

Assess:
• Appropriate height, weight development before and periodically; may be delayed
• I&O ratio; watch for increasing urinary output
• Fecal fat, nitrogen, PT during treatment
• **Diabetes mellitus:** for polyuria, polydipsia, polyphagia; monitor glucose level more frequently
• Pork sensitivity; cross-sensitivity may occur
• Adequate hydration

P

Evaluate:

- Therapeutic response: improved digestion of carbohydrates, protein, fat; absence of steatorrhea

Teach patient/family:

- To notify prescriber of allergic reactions, abdominal pain, cramping, or blood in urine
- To always take with food; not to crush, chew del rel product, caps
- To store at room temperature, away from moisture

HIGH ALERT

pancuronium (Rx)

(pan-kyoo-roe′nee-um)

Func. class.: Neuromuscular blocker (nondepolarizing)

Chem. class.: Synthetic curariform

ACTION: Inhibits transmission of nerve impulses by binding with cholinergic receptor sites, antagonizing action of acetylcholine

USES: Facilitation of endotracheal intubation, skeletal muscle relaxation during mechanical ventilation, surgery, or general anesthesia

CONTRAINDICATIONS: Hypersensitivity to bromide ion

Precautions: Pregnancy (C), breastfeeding, children <2 yr, neuromuscular/cardiac/renal/hepatic disease, electrolyte imbalances, dehydration, previous anaphylactic reactions (other neuromuscular blockers)

Black Box Warning: Respiratory insufficiency

DOSAGE AND ROUTES

- **Adult/child/infant >1 mo: IV** 0.04-0.1 mg/kg initially or 0.05 mg/kg after initial dose of succinylcholine; maintenance 0.01 mg/kg 60-100 min after initial dose then 0.01 mg/kg q25-60min as needed; for obese patients, use ideal body weight
- **Neonate <1 mo: IV** test dose 0.02 mg/kg then 0.03 mg/kg/dose initially, repeat 2× as needed at 5-10 min intervals; maintenance 0.03-0.09 mg/kg/dose q30min-4 hr as needed

Available forms: Inj 1, 2 mg/ml

Administer:

Direct IV route

- May be given undiluted over 1-2 min (1 mg/ml [10-ml vial], 2 mg/ml [2-, 5-ml vial])

Intermittent IV INF route

- Add 100 mg of product to 250 ml D_5W, NS, LR (0.4 mg/ml)
- Store in refrigerator; do not store in plastic; use only fresh sol
- Reassurance if communication is difficult during recovery from neuromuscular blockade
- Frequent (q2hr) instillation of artificial tears, covering of eyes to prevent drying of cornea

Additive compatibilities: Verapamil, ciprofloxacin

Y-site compatibilities: Aminophylline, ceFAZolin, cefuroxime, cimetidine, DOBUTamine, DOPamine, EPINEPHrine, esmolol, fenoldopam, fentaNYL, fluconazole, gentamicin, heparin, hydrocortisone, isoproterenol, levofloxacin, LORazepam, midazolam, morphine, nitroglycerin, ranitidine, trimethoprim-sulfamethoxazole, vancomycin

SIDE EFFECTS

CV: Bradycardia; tachycardia; increased, decreased B/P; ventricular extrasystoles, edema, hypertension

EENT: Increased secretions

INTEG: Rash, flushing, pruritus, urticaria, sweating, salivation

MS: Weakness to prolonged skeletal muscle relaxation

RESP: Prolonged apnea, bronchospasm, cyanosis, respiratory depression, dyspnea

SYST: Anaphylaxis

PHARMACOKINETICS

IV: Onset 3-5 min, dose dependent, peak 3-5 min; metabolized (small amounts), excreted in urine (unchanged), crosses placenta

INTERACTIONS

• Dysrhythmias: theophylline

Increase: neuromuscular blockade—aminoglycosides, clindamycin, enflurane, isoflurane, lincomycin, lithium, local anesthetics, opioid analgesics, polymyxin antiinfectives, quiNIDine, thiazides

Drug/Lab Test

Decrease: cholinesterase

NURSING CONSIDERATIONS

Assess:

• **Respiratory recovery:** decreased paralysis of face, diaphragm, leg, arm, rest of body; allow to recover fully before neurologic assessment

• Electrolyte imbalances (K, Mg); may lead to increased action of product

• VS (B/P, pulse, respirations, airway) until fully recovered; rate, depth, pattern of respirations, strength of hand grip

• I&O ratio; check for urinary retention, frequency, hesitancy

⚠ **Allergic reactions, anaphylaxis:** rash, fever, respiratory distress, pruritus; product should be discontinued

Evaluate:

• Therapeutic response: paralysis of jaw, eyelid, head, neck, rest of body

TREATMENT OF OVERDOSE:

Neostigmine, atropine, monitor VS; may require mechanical ventilation

panitumumab (Rx)

(pan-i-tue′moo-mab)

Vectibix

Func. class.: Antineoplastic—miscellaneous

Chem. class.: Multikinase inhibitor, signal transduction inhibitor

ACTION:
Decreases growth and survival of cancer cells by competitive inhibition of EGF receptor

USES:
EGFR expressing metastatic colorectal cancer; not beneficial with KRAS mutations in codon 12 or 13

CONTRAINDICATIONS:
Hypersensitivity

Precautions: Pregnancy (C), breastfeeding, children, hepatic disease, acute bronchospasm, diarrhea, hamster protein allergy, hypomagnesemia, hypotension, pulmonary fibrosis, sepsis, KRAS mutations, soft tissue toxicities

Black Box Warning: Exfoliative dermatitis, infusion-related reactions

DOSAGE AND ROUTES

• **Adult: IV INF** 6 mg/kg over 60 min every 2 wk; doses >1000 mg over 90 min

Available forms: Sol for inj 20 mg/ml (100 mg/5 ml, 400 mg/20 ml)

Administer:

Intermittent IV INF route

• Give in hospital or clinic setting with full resuscitation equipment

• Only as IV inf using controlled IV inf pump; do not give IV push or bolus; use low-protein binding 0.2- or 0.22-micron in-line filter; flush line with 0.9% NaCl before and after administration

• Give over 60 min through a peripheral line or indwelling catheter; infuse doses of >1000 mg over 90 min

• Dilute in 100 ml of 0.9% NaCl; dilute doses >1000 mg in 150 ml of 0.9%

NaCl; mix by inverting; do not exceed 10 mg/ml; use within 6 hr if stored at room temperature; can be stored between 2° C and 8° C for up to 24 hr

- **Dosage adjustment for inf/dermatologic reaction**
- Grade 1 or 2: reduce inf by 50%; Grade 3 or 4: terminate, permanently discontinue depending on severity/resistance
- Store unopened vials in refrigerator; do not shake; protect from direct sunlight; do not freeze

SIDE EFFECTS

CNS: Fatigue
CV: Peripheral edema
EENT: Ocular irritation, ocular toxicity
GI: *Nausea, diarrhea, vomiting,* anorexia, mouth ulceration, abdominal pain, constipation
HEMA: Thrombophlebitis
INTEG: *Rash,* pruritus, exfoliative dermatitis, skin fissure, angioedema, severe/fatal INF reactions
META: Hypocalcemia, hypomagnesemia, antibody formation
RESP: Bronchospasm, cough, dyspnea, hypoxia, pulmonary fibrosis/embolism, pneumonitis, wheezing, interstitial lung disease

PHARMACOKINETICS

Bioavailability 38%-49%; elimination half-life 7.5 days; peak 3 hr; high-fat meal decreases bioavailability; plasma protein binding 99.5%; metabolized in liver; oxidative metabolism by CYP3A4, glucuronidation by UGT1A9; 77% excreted in feces

INTERACTIONS

- Do not use in combination with other antineoplastics

NURSING CONSIDERATIONS

Assess:

Black Box Warning: **Serious skin disorders:** fever, sore throat, fatigue, then lesions in mouth, lips; withhold product, notify prescriber

- Serum electrolytes periodically (calcium, magnesium)
- **Infection:** increased temperature

Black Box Warning: **Inf reactions:** bronchospasm, fever, chills, hypotension; may require discontinuation, have emergency equipment available

- **Ocular toxicity:** ocular irritation, hyperemia
- **Pulmonary fibrosis:** dyspnea, cough, wheezing; may require discontinuation

Evaluate:

- Therapeutic response: decrease in colon carcinoma progression

Teach patient/family:

⚠ To report adverse reactions immediately: difficulty breathing, mouth sores, skin rash, ocular toxicity

- About reason for treatment, expected results, adverse reactions
- To use contraception while taking product, for 6 mo after treatment; not to breastfeed for ≥2 mo after stopping treatment
- To avoid the sun, use sunscreen while taking product

pantoprazole (Rx)

(pan-toe-pray′zole)

Panto ♣, Pantoloc ♣, Protonix, Prontonix IV

Func. class.: Proton pump inhibitor
Chem. class.: Benzimidazole

ACTION: Suppresses gastric secretion by inhibiting hydrogen/potassium ATPase enzyme system in gastric parietal cell; characterized as gastric acid pump inhibitor because it blocks the final step of acid production

USES: Gastroesophageal reflux disease (GERD), severe erosive esophagitis; maintenance of long-term pathologic hypersecretory conditions, including Zollinger-Ellison syndrome

Unlabeled uses: Duodenal/gastric ulcer, NSAID ulcer prophylaxis, *Helicobacterpylori*-associated ulcer, dyspepsia

CONTRAINDICATIONS:
Hypersensitivity to this product or benzimidazole

Precautions: Pregnancy (C), breastfeeding, children, proton pump hypersensitivity

DOSAGE AND ROUTES

GERD

• **Adult: PO** 40 mg/day × 8 wk, may repeat course

Erosive esophagitis

• **Adult: IV** 40 mg/day × 7-10 day; **PO** 40 mg/day × 8 wk; may repeat **PO** course

Pathologic hypersecretory conditions

• **Adult: PO** 40 mg bid; **IV** 80 mg q12hr, max 240 mg/day

Duodenal ulcer/gastric ulcer/ NSAID ulcer prophylaxis (unlabeled)

• **Adult: PO** 40 mg/day

***H. pylori*–associated ulcers (unlabeled)**

• **Adult: PO** 40 mg bid; may be used with other products

Available forms: Del rel tabs 20, 40 mg; powder for inj 40 mg/vial; del rel granules for susp 40 mg

Administer:

PO route

• Swallow del rel tabs whole; do not break, crush, or chew; take del rel tabs at same time of day

• May take with/without food

• **Suspension:** give in apple juice 30 min before a meal or sprinkled on 1 tbsp of applesauce

• **NG tube:** empty contents of packet of granules into barrel of a 60-ml catheter tip syringe (plunger removed) connected to ≥16F NG tube; add 10 ml apple juice and tap or shake barrel of syringe to empty into the tube; add another 10 ml of apple juice; rinse with additional apple juice until syringe is clear

IV route

• Use of Protonix IV vials with spiked IV system adaptors is not recommended

• Visually inspect for particulate matter and discoloration before use

• Give as an IV inf over 15 min either through a dedicated line or a Y-site; a 2-min slow-injection regimen is also approved; do not give fast IV push

• When using a Y-site, immediately stop use if a precipitation or discoloration occurs

• **Reconstitution of vial**: use 40-mg vial/10 ml NS; do not freeze

• **Two-minute slow IV inf injection:** dilute one or two 40-mg vials with 10 ml NS per vial to 4 mg/ml; store ≤24 hr at room temperature before use; infuse slowly over ≥2 min; do not give with other IV fluids or medications; flush line with D_5W, NS, or LR before and after each dose

• **Fifteen-minute IV inf:** dilute each 40-mg dose with 10 ml NS; the reconstituted vial should be further admixed with 100 ml (for one vial) or 80 ml (for 2 vials) of D_5W, NS, or LR (to 0.4 mg/ml or 0.8 mg/ml, respectively); store ≤6 at room temperature before further dilution; the admixed solution (0.4 mg/ml or 0.8 mg/ml) may be stored at room temperature and must be used within 24 hr from the time of initial reconstitution; infuse over 15 min at 7 ml/min; do not administer with other IV fluids or medications; flush IV line with D_5W, NS, or LR before and after each dose

Y-site compatibilities: Acyclovir, allopurinol, amifostine, amikacin, aminocaproic acid, aminophylline, amoxicillin-clavulanate, amphotericin B liposome, ampicillin, ampicillin-sulbactam, anidulafungin, azithromycin, bleomycin, bumetanide, calcium gluconate, CARBOplatin, carmustine, ceFAZolin, cefOXitin, cefTAZidime, ceftizoxime, cefTRIAXone, cefuroxime, clindamycin, cyclophosphamide, cycloSPORINE, cytarabine, dex-

P

trose 3.3% in sodium chloride 0.3%, digoxin, dimenhyDRINATE, DOCEtaxel, DOPamine, doripenem, doxycycline, enalaprilat, EPINEPHrine, ertapenem, fluorouracil, foscarnet, fosphenytoin, furosemide, ganciclovir, gentamicin, granisetron, heparin, hydrocortisone HYDROmorphone, imipenem-cilastatin, inamrinone, insulin (regular), irinotecan, isoproterenol, magnesium, mannitol, mesna, methohexital, methyldopate, metoclopramide, nafcillin, nitroglycerin, nitroprusside, ofloxacin, oxytocin, PACLitaxel, pentazocine, PENTobarbital, phenylephrine, piperacillin-tazobactam, potassium chloride, procainamide, rifampin, sodium bicarbonate, succinylcholine, SUFentanil, sulfamethoxazole-trimethoprim, teniposide, theophylline, thiopental, ticarcillin, ticarcillin-clavulanate, tigecycline, tirofiban, tobramycin, traMADol, vasopressin, zidovudine

SIDE EFFECTS

CNS: *Headache,* insomnia, asthenia, fatigue, malaise, insomnia, somnolence
GI: *Diarrhea, abdominal pain,* flatulence, pancreatitis, weight changes
INTEG: *Rash*
META: Hyperglycemia, weight gain/loss, hyponatremia, hypomagnesemia
MS: Rhabdomyolysis, myalgia
RESP: Pneumonia
SYST: Stevens-Johnson syndrome, toxic epidermal necrolysis, anaphylaxis, angioedema

PHARMACOKINETICS

Peak 2.4 hr, duration >24 hr, half-life 1.5 hr, protein binding 97%, eliminated in urine as metabolites and in feces; in geriatric patients, elimination rate decreased; some Asian patients (15%-20%) may be poor metabolizers

INTERACTIONS

Increase: pantoprazole serum levels—diazepam, phenytoin, flurazepam, triazolam, clarithromycin
Increase: bleeding—warfarin
Decrease: absorption of these products—sucralfate, calcium carbonate, vit B_{12}, ketoconazole, itraconazole, atazanavir, ampicillin, iron salts
Decrease: clopidogrel effect
Drug/Herb
Decrease: effect of pantoprazole—St. John's wort

NURSING CONSIDERATIONS

Assess:
- **GI system:** bowel sounds q8hr; abdomen for pain, swelling; anorexia
- **Hepatic studies:** AST, ALT, alk phos during treatment
- For vit B_{12} deficiency in patients receiving long-term therapy

⚠ **Serious skin reactions:** toxic epidermal necrolysis, Stevens-Johnson syndrome, exfoliative dermatitis: fever, sore throat, fatigue, thin ulcers; lesions in the mouth, lips

- **Electrolyte imbalances:** hyponatremia; hypomagnesemia in patients using product 3 mo to 1 year; if hypomagnesemia occurs, use of magnesium supplements may be sufficient; if severe, discontinuation of product may be required

⚠ **Rhabdomyolysis, myalgia:** muscle pain, increased CPK; weakness, swelling of affected muscles

Evaluate:
- Therapeutic response: absence of epigastric pain, swelling, fullness

Teach patient/family:
- To report severe diarrhea; black, tarry stools; abdominal pain; product may have to be discontinued
- That hyperglycemia may occur in diabetic patients
- To avoid alcohol, salicylates, NSAIDs; may cause GI irritation
- To notify prescriber if pregnant or planning to become pregnant; not to breastfeed
- To continue taking even if feeling better

paricalcitol (Rx)

(par-ih-cal′sih-tol)

Zemplar

Func. class.: Vit D analog

Chem. class.: Fat-soluble vitamin

ACTION: Reduces parathyroid hormone (PTH) levels; suppresses PTH levels in patients with chronic renal failure with absence of hypercalcemia/hyperphosphatemia; serum PO_4, calcium, CaXP may increase

USES: Hyperparathyroidism in chronic renal failure

Unlabeled uses: Renal osteodystrophy

CONTRAINDICATIONS: Hypersensitivity, hypercalcemia

Precautions: Pregnancy (C), breastfeeding, children, geriatric patients, CV disease, renal calculi

DOSAGE AND ROUTES

• **Adult: IV BOL** 0.04-0.1 mcg/kg (2.8-7 mcg) no more than every other day during dialysis; may increase by 2-4 mcg q2-4wk until target serum intact PTH (1.5-3× nonuremic upper limit of normal) achieved; **PO** 1 mcg/day or 2 mcg 3×/wk (IPTH ≤500 pg/ml); 2 mcg/day or 4 mcg 3×/wk (IPTH >500 pg/ml)

Available forms: Inj 2, 5 mcg/ml; caps 1, 2, 4 mcg

Administer:

PO route

• Daily or 3×/wk; may give without regard to food

IV route

• By IV bolus only

SIDE EFFECTS

CNS: Lightheadedness

CV: Palpitations

GI: Nausea, vomiting, anorexia, dry mouth

OTHER: Pneumonia, edema, chills, fever, flu, sepsis

PHARMACOKINETICS

Crosses placenta, enters breast milk

INTERACTIONS

Decrease: paricalcitol effect—cholestyramine, colestipol, mineral oil, orlistat, corticosteroids, barbiturates, hydantoins, CYP3A4 enzymes (nevirapine, rifampin, bosentan)

Increase: calcium levels—thiazide diuretics, calcium products, vit D supplements

Increase: effect of cardiac glycosides

Altered: paricalcitol effect—CYP3A4 inhibitors (amiodarone, protease inhibitors, systemic azole antifungals, chloramphenicol, clarithromycin, delavirdine, erythromycin)

NURSING CONSIDERATIONS

Assess:

• **Hypocalcemia:** twitching, dysrhythmias, Chvostek's/Trousseu's signs, paresthesia, laryngospasm, prolonged QTc/ST interval

• Serum calcium, serum intact parathyroid hormone concentrations (iPTH), phosphate 2×/wk during initial therapy; after dose is established, take calcium and phosphorus monthly

Evaluate:

• Decreased hypoparathyroidism with chronic renal disease, normal serum calcium, phosphate, iPTH

Teach patient/family:

• To report weakness, lethargy, headache, anorexia, loss of weight

• To report nausea, vomiting, palpitations

• To adhere to dietary regimen of calcium supplementation/phosphorus restriction

• To avoid excessive use of aluminum compounds, antacids

• Not to breastfeed

• Not to take mineral oil, antacids (magnesium) while taking vit D

PARoxetine (Rx)

(par-ox′e-teen)

Paxil, Paxil CR, Pexeva

Func. class.: Antidepressant, SSRI

Chem. class.: Phenylpiperidine derivative

Do not confuse:
PARoxetine/PACLitaxel
Paxil/PACLitaxel/Taxol

ACTION: Inhibits CNS neuron uptake of serotonin but not of norepinephrine or DOPamine

USES: Major depressive disorder, obsessive-compulsive disorder, panic disorder, generalized anxiety disorder, posttraumatic stress disorder, premenstrual disorders, social anxiety disorder
Unlabeled uses: Premature ejaculation, hot flashes, menopause

CONTRAINDICATIONS: Pregnancy (D), hypersensitivity, MAOI use, alcohol use
Precautions: Breastfeeding, geriatric patients, seizure history; patients with history of mania, renal/hepatic disease

Black Box Warning: Children, suicidal ideation

DOSAGE AND ROUTES

Generalized anxiety disorder
- **Adult: PO** 20 mg/day in AM, range 20-50 mg/day

Posttraumatic stress disorder
- **Adult: PO** 20 mg/day, range 20-60 mg/day

Depression
- **Adult: PO** 20 mg/day in AM; after 4 wk, if no clinical improvement is noted, dose may be increased by 10 mg/day each wk to desired response, max 50 mg/day or **CONT REL** 25 mg/day, may increase by 12.5 mg/day/wk up to 62.5 mg/day
- **Geriatric: PO** 10 mg/day, increase by 10 mg to desired dose, max 40 mg/day

Obsessive-compulsive disorder
- **Adult: PO** 40 mg/day in AM, start with 20 mg/day, increase in 10-mg/day increments, max 60 mg/day

Panic disorder
- **Adult: PO** start with 10 mg/day, increase in 10-mg/day increments to 40 mg/day, max 60 mg/day or **CONT REL** 12.5 mg/day, max 75 mg/day

Premenstrual disorders
- **Adult: CONT REL** 12.5 mg/day in AM

Renal dose
- **Adult: PO** CCr 30-60 ml/min, lower doses may be needed; CCr <30 ml/min, 10 mg/day initially, regular rel, max 40 mg/day; **CONT REL** 12.5 mg/day initially, max 50 mg/day

Hepatic dose
- **Adult: PO** 10 mg/day initially, max 40 mg **regular rel**; 12.5 mg/day initially, max 50 mg/day **(cont rel)**

Menopause symptoms/hot flashes (unlabeled)
- **Adult: PO (CONT REL)** 12.5 mg/day, may increase to 25 mg/day after 1 wk

Premature ejaculation (unlabeled)
- **Adult: PO** 20 mg/day

Available forms: Tabs 10, 20, 30, 40 mg; oral susp 10 mg/5 ml; cont rel tab 12.5, 25, 37.5 mg

Administer:
- Do not substitute Pexeva with Paxil, Paxil CR, or generic PARoxetine
- Store at room temperature; do not freeze
- Increased fluids, bulk in diet for constipation, urinary retention
- With food, milk for GI symptoms
- Crushed if patient is unable to swallow medication whole (regular rel only)
- Gum, hard candy, frequent sips of water for dry mouth
- Avoid use with other CNS depressants
- **Oral susp:** shake, measure with oral syringe or calibrated measuring device
- **Cont rel tab:** do not cut, chew, crush; do not give concurrently with antacids

SIDE EFFECTS

CNS: *Headache,* nervousness, insomnia, *drowsiness, anxiety, tremors, dizziness,* fatigue, *sedation,* abnormal dreams, agitation, apathy, euphoria, hallucinations, delusions, psychosis, **seizures, neuroleptic-malignant-syndrome–like reactions**, restless leg syndrome

CV: Vasodilation, postural hypotension, palpitations, bleeding

EENT: Visual changes

GI: *Nausea, diarrhea, dry mouth,* anorexia, dyspepsia, *constipation,* cramps, vomiting, taste changes, flatulence, decreased appetite

GU: Dysmenorrhea, decreased libido, urinary frequency, UTI, amenorrhea, cystitis, impotence; decreased sperm quality, decreased fertility, *abnormal ejaculation (male)*

INTEG: *Sweating,* rash

MS: Pain, arthritis, myalgia, myopathy, myosthenia

RESP: Infection, pharyngitis, nasal congestion, sinus headache, sinusitis, cough, dyspnea, yawning

SYST: Asthenia, fever, abrupt withdrawal syndrome

PHARMACOKINETICS

PO: Peak 5.2 hr, ext rel peak 6-10 hr; metabolized in liver by CYP2D6 enzyme system, unchanged products and metabolites excreted in feces and urine; half-life 21 hr (reg rel); 15-20 hr (cont rel); protein binding 95%

INTERACTIONS

⚠ **Increase:** serotonin syndrome—SSRIs, SNRIs, atypical psychotics, serotonin-receptor agonists, tricyclics, amphetamines, methylphenidate, traMADol

Decrease: level of digoxin

⚠ Do not use with MAOIs, pimozide, thioridazine; potentially fatal reactions can occur

Increase: bleeding—NSAIDs, thrombolytics, salicylates, platelet inhibitors, anticoagulants

Increase: PARoxetine plasma levels—cimetidine

Increase: agitation—L-tryptophan

Increase: side effects—highly protein-bound products

Increase: theophylline levels—theophylline

Increase: toxicity—CYP2D6 inhibitors (aprepitant, delavirdine, imatinib, nefazodone)

Decrease: PARoxetine levels—PHENObarbital and phenytoin

Drug/Herb

- Avoid use with St. John's wort, kava
- Possible serotonin syndrome: St. John's wort
- Hypertensive crisis: ephedra

NURSING CONSIDERATIONS

Assess:

Black Box Warning: **Depression/OCD/anxiety/panic attacks:** mental status: mood, sensorium, affect, suicidal tendencies (especially in child/young adult), increase in psychiatric symptoms, decreasing obsessive thoughts, compulsive behaviors, restrict amount available

- **Postural hypotension:** B/P (lying/standing), pulse q4hr; if systolic B/P drops 20 mm Hg, hold product, notify prescriber; take vital signs q4hr for patients with CV disease
- Hepatic/renal studies: AST, ALT, bilirubin, creatinine
- Weight weekly; appetite may decrease with product, constipation
- ECG for flattening of T wave, bundle branch or AV block, dysrhythmias in cardiac patients
- EPS, primarily in geriatric patients: rigidity, dystonia, akathisia
- **Renal status:** BUN, creatinine, urinary retention
- **Withdrawal symptoms:** headache, nausea, vomiting, muscle pain, weakness; not usual unless product discontinued abruptly, taper over 1-2 wk
- Alcohol intake; if alcohol is consumed, hold dose until morning

⚠ **Serotonin, neuroleptic malignant syndrome:** hallucinations, coma, headache, agitation, shivering, sweating, tachycardia, diarrhea, tremors, hypertension, hyperthermia, rigidity, delirium, coma, myoclonus, agitation, nausea, vomiting

Evaluate:

- Therapeutic response: decreased depression

Teach patient/family:

- That therapeutic effect may take 1-4 wk
- To use caution when driving, performing other activities requiring alertness because of drowsiness, dizziness, blurred vision
- Not to discontinue medication quickly after long-term use; may cause nausea, headache, malaise (abrupt withdrawal syndrome)

Black Box Warning: That depression may worsen, suicidal thoughts/behaviors; children/adolescents, young adults, to notify prescriber

- To avoid alcohol ingestion, OTC products unless approved by prescriber
- To report bleeding, headache, nausea, anxiety, or if depression continues
- To discuss sexual side effects: impotence, possible male infertility while taking product

TREATMENT OF OVERDOSE:

Gastric lavage, airway; for seizures, give diazepam, symptomatic treatment

pazopanib

(paz-oh'pa-nib)

Votrient

Func. class.: Antineoplastic biologic response modifiers/multikinase angiogenesis inhibitor

Chem. class.: Signal transduction inhibitor (STI)

ACTION:

Targets vascular endothelial growth factor receptors; a multikinase angiogenesis inhibitor

USES:

Advanced renal cell carcinoma; soft-tissue sarcoma patients who have received prior chemotherapy

Unlabeled uses: Breast, ovarian cancer

CONTRAINDICATIONS:

Pregnancy (D), hypothyroidism, QT prolongation, MI, wound dehiscence, hypertension

Precautions: Breastfeeding, children, cardiac/renal/hepatic/dental disease, GI bleeding

Black Box Warning: Hepatic disease

DOSAGE AND ROUTES

- **Adult: PO** 800 mg/day without food (1 hr before, 2 hr after a meal), may decrease to 400 mg/day if not tolerated (renal cell cancer); or adjust in 200-mg increments based on toxicity (soft-tissue sarcoma)

Available forms: Tabs 200 mg

Administer:

- Give on an empty stomach (1 hr before or 2 hr after a meal); separate doses by ~24 hr
- Do not crush tablets owing to the potential for an increased rate of absorption, which can affect systemic exposure; only intact, whole tablets should be used
- If a dose is missed, it should not be taken if it is <12 hr until the next dose
- Store at 77°F (25°C)

SIDE EFFECTS

CNS: Intracranial bleeding, headache

CV: Heart failure, hypertension, hypertensive crisis, chest pain, MI, QT prolongation, torsades de pointes

GI: Nausea, hepatotoxicity, vomiting, dyspepsia, GI hemorrhage, anorexia, abdominal pain, GI perforation, pancreatitis, diarrhea

HEMA: Neutropenia, thrombocytopenia, bleeding

INTEG: Rash, alopecia

MISC: Fatigue, epistaxis, pyrexia, hot sweats, increased weight, flulike symptoms, hypothyroidism, hand–foot syndrome

PHARMACOKINETICS

Protein binding 99%

INTERACTIONS

Increase: QT prolongation—class IA/III antidysrhythmics, some phenothiazines, β-agonists, local anesthetics, tricyclics, haloperidol, chloroquine, droperidol, pentamidine; CYP3A4 inhibitors (amiodarone, clarithromycin, erythromycin, telithromycin, troleandomycin), arsenic trioxide, levomethadyl; CYP3A4 substrates (methadone, pimozide, QUEtiapine, quiNIDine, risperiDONE, ziprasidone)

Increase: pazopanib concentrations—CYP3A4 inhibitors (ketoconazole, itraconazole, erythromycin, clarithromycin)

Increase: plasma concentrations of simvastatin, calcium-channel blockers, ergots

Increase: plasma concentration of warfarin; avoid use with warfarin; use low-molecular-weight anticoagulants instead

Decrease: pazopanib concentrations—CYP3A4 inducers (dexamethasone, phenytoin, carBAMazepine, rifampin, PHENobarbital)

Drug/Food

Increase: pazopanib effect—grapefruit juice; avoid use while taking product

Drug/Herb

Decrease: pazopanib concentration—St. John's wort

NURSING CONSIDERATIONS

Assess:

Black Box Warning: Hepatic disease: fatal hepatotoxicity can occur; obtain LFTs baseline and at least every 2 wk × 2 mo, then monthly

⚠ **Fatal Bleeding:** from GI, respiratory, GU tracts, permanently discontinue in those with severe bleeding

⚠ **Palmar-plantar erythrodysesthesia (hand-foot syndrome):** more common in those previously treated; reddening swelling, numbness, desquamation on palms and soles

⚠ **GI perforation/fistula:** discontinue if this occurs, assess for pain in epigastric area, dyspepsia, flatulence, fever, chills

⚠ **Hypertension/hypertensive crisis:** hypertension usually occurs in the first cycle; in those with preexisting hypertension, do not start treatment until B/P is controlled; monitor B/P every wk × 6 wk, then at start of each cycle or more often if needed, temporarily or permanently discontinue for severe uncontrolled hypertension

Evaluate:

- Therapeutic response: decreased in size, spread of tumor

Teach patient/family:

- To report adverse reactions immediately: bleeding
- About reason for treatment, expected results
- That effect on male fertility is unknown

pegfilgrastim (Rx)

(peg-fill-grass′stim)

Neulasta

Func. class.: Hematopoietic agent

Chem. class.: Granulocyte colony-stimulating factor

P

ACTION: Stimulates proliferation and differentiation of neutrophils

USES: To decrease infection in patients receiving antineoplastics that are myelosuppressive; to increase WBC count in patients with product-induced neutropenia

CONTRAINDICATIONS: Hypersensitivity to proteins of *Escherichia coli,* filgrastim

Precautions: Pregnancy (C), breastfeeding, children <45 kg, adolescents, myeloid malignancies, sickle cell disease, leukocytosis, splenic rupture, ARDS, allergic-type reactions, peripheral blood stem cell (PBSC) mobilization

DOSAGE AND ROUTES

- **Adult: SUBCUT** 6 mg, give 1× per chemotherapy cycle

Available forms: Sol for inj 6 mg/0.6 ml

Administer:

SUBCUT route

- Using single-use vials; after dose is withdrawn, do not reenter vial
- Do not use 6-mg fixed dose in infants, children, or others <45 kg
- Inspect sol for discoloration, particulates; if present, do not use
- Do not administer during the period 14 days before and 24 hr after cytotoxic chemotherapy
- Store in refrigerator; do not freeze; may store at room temperature up to 6 hr; avoid shaking, protect from light

SIDE EFFECTS

CNS: Fever, fatigue, headache, dizziness, insomnia, peripheral edema

GI: *Nausea,* vomiting, diarrhea, mucositis, anorexia, constipation, dyspepsia, abdominal pain, stomatitis, splenic rupture

HEMA: Leukocytosis, granulocytopenia, sickle cell crisis, hemoglobin S disease with crisis

INTEG: Alopecia

MISC: Chest pain, hyperuricemia, anaphylaxis, influenza-like illness, angioedema, antibody formation

MS: Skeletal pain

RESP: Respiratory distress syndrome

PHARMACOKINETICS

Half-life: 15-80 hr; 20-38 hr (children)

INTERACTIONS

- Do not use product concomitantly, 2 wk before, or 24 hr after administration of cytotoxic chemotherapy

Increase: release of neutrophils—lithium

Drug/Lab Test

Increase: uric acid, LDH, alk phos

NURSING CONSIDERATIONS

Assess:

⚠ **Allergic reactions, anaphylaxis:** rash, urticaria; discontinue product, have emergency equipment nearby

⚠ **ARDS:** dyspnea, fever, tachypnea, occasionally confusion; obtain ABGs, chest x-ray; product may need to be discontinued

- **Bone pain;** give mild analgesics
- Blood studies: CBC with differential, platelet count before treatment, 2× weekly; neutrophil counts may be increased for 2 days after therapy
- B/P, respirations, pulse before and during therapy

Evaluate:

- Therapeutic response: absence of infection

Teach patient/family:

- How to perform the technique for self-administration if product to be given at home: dose, side effects, disposal of containers and needles; provide instruction sheet
- To notify prescriber immediately of allergic reaction, trouble breathing, abdominal pain

peginterferon alfa-2a (Rx)

(peg-in-ter-feer′on)

Pegasys

peginterferon alfa-2b (Rx)

PegIntron, SYLATRON

Func. class.: Immunomodulator

ACTION: Stimulates genes to modulate many biologic effects, including the inhibition of viral replication; inhibits ion cell proliferation, immunomodulation; stimulates effector proteins; decreases leukocyte, platelet counts

USES: Chronic hepatitis C infections in adults with compensated liver disease; chronic hepatitis B in adults who are HBe AG positive, HBe AG negative; HCV patients coinfected with HIV; nonresponders or relapsers with chronic hepatitis C, malignant melanoma

Unlabeled uses: Adenovirus, coronavirus, encephalomyocarditis virus, herpes simplex types 1 and 2, hepatitis D, acute hepatitis C, HIV, HPV, polio virus, rhinovirus, varicella-zoster, variola, vesicular stomatitis

CONTRAINDICATIONS: Neonates, infants, sepsis; hypersensitivity to interferons, benzyl alcohol, *Escherichia coli* protein

Precautions: Pregnancy (C), breastfeeding, children <18 yr, geriatric patients, thyroid disorders, myelosuppression, renal/hepatic disease, suicidal/homicidal ideation, preexisting ophthalmologic disorders, pancreatitis, hemodialysis

Black Box Warning: Cardiac disease, depression, autoimmune disease, infection, use with ribavirin

DOSAGE AND ROUTES

Pegasys

- **Adult: SUBCUT** 180 mcg weekly × 48 wk; if poorly tolerated, reduce dose to 135 mcg weekly; in some cases, reduction to 90 mcg may be needed

Peg-Intron (chronic hepatitis C with compensated liver disease)

- **Adult >105 kg: SUBCUT** 1.5 mcg/kg/wk plus ribavirin 600 mg in AM and 800 mg in PM plus a HCV NS3/4A protease inhibitor; **86-105 kg:** 150 mcg/0.5 ml (0.5 ml of 150 mcg vial or Redipen) per wk plus ribavirin 1200 mg/day in 2 divided doses plus a HCV NS3/4A protease inhibitor; **81-85 kg:** 120 mcg/0.5 ml (0.5 ml of 120 mcg vial or Redipen) per wk plus ribavirin 1200 mg/day in 2 divided doses plus a HCV NS3/4A protease inhibitor; **76-80 kg:** 120 mcg/0.5 ml (0.5 ml of 120 mcg vial or Redipen) per wk plus ribavirin 400 mg in AM and 600 mg in PM plus a HCV NS3/4A protease inhibitor; **66-75 kg:** 96 mcg/0.4 ml (0.4 ml of 120 mcg vial or Redipen) per wk plus ribavirin 400 mg in AM and 600 mg in PM plus a HCV NS3/4A protease inhibitor; **61-65 kg:** 96 mcg/0.4 ml (0.4 ml of 120 mcg vial or Redipen) per wk plus ribavirin 800 mg/day in 2 divided doses plus a HCV NS3/4A protease inhibitor; **51-60 kg:** 80 mcg/0.5 ml (0.5 ml of 80 mcg vial or Redipen) per wk plus ribavirin 800 mg/day in 2 divided doses plus a HCV NS3/4A protease inhibitor; **40-50 kg:** 64 mcg/0.4 ml (0.4 ml of 80 mcg vial or Redipen) per wk plus ribavirin 800 mg/day in 2 divided doses plus a HCV NS3/4A protease inhibitor; **<40 kg:** 50 mcg/0.5 ml (0.5 ml of 50 mcg vial or Redipen) per wk plus ribavirin 800 mg/day in 2 divided doses plus a HCV NS3/4A protease inhibitor

Malignant Melanoma (SYLATRON only)

- **Adult: SUBCUT** 6 mcg/kg/wk × 8 wk then 3 mcg/kg/wk × ≤5 yr, premedicate with acetaminophen 500-1000 mg 30 min before first dose, prn for subsequent doses

Available forms: Pegasys: inj 180 mcg/0.5 ml; Pegintron: 50, 80, 120, 150 mcg/0.5 ml; SYLATRON 296, 444, 888 mcg powder for inj

Administer:

- In evening to reduce discomfort, to allow patient to sleep through some side effects
- Continue pediatric dose in those who turn 18 yr

Interferon alfa-2a

- Use prefilled syringes; store in refrigerator

Interferon alfa-2b

SUBCUT/IM route

- Reconstitute with 1 ml of provided diluent/10-, 18-, or 50-million unit vials, swirl; sol for inj vials do not need reconstitution

P

SIDE EFFECTS

CNS: *Headache, insomnia, dizziness,* anxiety, hostility, lability, nervousness, depression, fatigue, poor conc, pyrexia, suicidal ideation, homicidal ideation, relapse of drug addiction, emotional lability, mania, psychosis
CV: Ischemic CV events
ENDO: Hypothyroidism, diabetes
GI: *Abdominal pain, nausea, diarrhea, anorexia, vomiting,* dry mouth, fatal colitis, fatal pancreatitis
HEMA: Thrombocytopenia, neutropenia, anemia, lymphopenia
INTEG: *Alopecia, pruritus, rash,* dermatitis
MISC: Blurred vision, inj site reaction, rigors
MS: *Back pain,* myalgia, arthralgia
RESP: Cough, dyspnea

PHARMACOKINETICS

Half-life 15-80 hr, large variability in other pharmacokinetics

INTERACTIONS

- Use caution when giving with theophylline, myelosuppressive agents

Increase: hepatic damage-NNRTIs, NRTIs, protein inhibitors

Drug/Lab Test

Increase: triglycerides, ALT, neutrophils, platelets
Abnormal: thyroid function test

NURSING CONSIDERATIONS

Assess:

- **Neuropsychiatric symptoms:** severe depression with suicidal ideation; monitor q3wk then 8 wk then q6mo
- B/P, blood glucose, ophthalmic exam, pulmonary function
- ALT, HCV viral load; patients who show no reduction in ALT, HCV unlikely to show benefit of treatment after 6 mo
- Platelet counts, heme conc, ANC, serum creatinine conc, albumin, bilirubin, TSH, T_4, AFP

⚠ **Myelosuppression:** hold dose if neutrophil count is <500 × 10^6/L or if platelets are <50 × 10^9/L

- **Hypersensitivity:** discontinue immediately if hypersensitivity occurs
- **Infection:** vital signs, increased WBCs, fever; product may need to be discontinued

⚠ **Colitis/pancreatitis:** may be fatal; diarrhea, fever, nausea, vomiting, severe abdominal pain; if these occur, product should be discontinued

Evaluate:

- Therapeutic response: decreased chronic hepatitis C signs, symptoms; undetectable viral load

Teach patient/family:

- Provide patient or family member with written, detailed information about product
- Use 2 forms of effective contraception throughout treatment and for 6 mo after treatment (men and women) (combination therapy with ribavirin)
- To avoid driving, other hazardous activity if dizziness, confusion, fatigue, somnolence occur
- To use puncture-resistant container for disposal of needles/syringes if using at home

⚠ To report suicidal/homicidal ideation, visual changes, bleeding/bruising, pulmonary symptoms

pegloticase (Rx)

(peg-loe′ti-kase)

Krystexxa

Func. class.: Antigout agent
Chem. class.: Pegylated, recombinant, mammalian urate oxidase enzyme

ACTION: Lowers plasma uric acid conc by converting uric acid to allantoin, which is readily excreted by the kidneys

USES: Chronic gout in patients experiencing treatment failure

CONTRAINDICATIONS:
Hypersensitivity, G6PD deficiency

Precautions: Pregnancy (C), breastfeeding, children/infants/neonates, African-American patients, heart failure

Black Box Warning: Requires specialized setting, experienced clinician

DOSAGE AND ROUTES

- **Adult: IV INF** 8 mg over 2 hr q2wk

Available forms: Sol for inj 8 mg/ml

Administer:

Intermittent IV INF route

- ***Reconstitute:*** visually inspect for particulate matter, discoloration whenever sol/container permits; use aseptic technique; withdraw 8 mg (1 ml) of product/250 ml 0.9% NaCl or 0.45% NaCl; invert several times to mix, do not shake; discard remaining product in vial
- ***Premedicate:*** with antihistamines and corticosteroids in all patients and acetaminophen if deemed necessary to prevent anaphylaxis, inf site reactions
- ***Inf:*** if refrigerated, allow to come to room temperature; do not warm artificially; give over 120 min; do not give IV push or bolus; use inf by gravity feed, syringe-type pump, or inf pump; given in a specialized setting by those who can manage anaphylaxis or inj site reactions; monitor during and for 1 hr after inf; if reaction occurs, slow or stop inf, may be restarted at a slower rate; do not admix
- Store diluted product in refrigerator or at room temperature for up to 4 hr; refrigerator is preferred; protect from light; do not freeze; use within 4 hr of preparation

SIDE EFFECTS

CNS: Dizziness, fatigue, fever

CV: *Chest pain,* **heart failure,** hypotension

GI: *Nausea,* vomiting, diarrhea, constipation

GU: Nephrolithiasis

HEMA: **Anemia**

INTEG: Ecchymosis, *erythema, pruritus, urticaria*

MS: Back pain, arthralgia, muscle spasm

SYST: **Antibody formation, infection, anaphylaxis, inf-related reactions**

RESP: *Dyspnea,* upper respiratory infection

PHARMACOKINETICS

Remains primarily in intravascular space after administration, elimination half-life 2 wk, mean nadir uric acid conc 24-72 hr

Interactions

Do not use with urate-lowering agents (allopurinol, probenecid, febuxostat, sulfinpyrazone)

NURSING CONSIDERATIONS

Assess:

- **Gout:** pain in big toe, feet, knees, redness, swelling, tenderness lasting a few days to weeks; intake of alcohol, purines, if patient is overweight or taking diuretics
- Obtain uric acid levels at baseline, before administration; 2 consecutive uric acid levels of >6 mg/dl may indicate therapy failure; greater chance of anaphylaxis; infection-related reactions

Black Box Warning: Specialized care setting: use only in facility where emergency equipment is available, anaphylaxis may occur

Black Box Warning: Inf reactions: monitor for reactions for ≥1 hr after use

Evaluate:

- Therapeutic response: decrease uric acid levels; relief of pain, swelling, redness in toes, feet, knees

Teach patient/family:

- About reason for inf, expected results
- To notify prescriber during inf of allergic reactions or redness, swelling, pain at inf site
- That continuing follow-up exams and uric acid levels will be needed

P

⚠ HIGH ALERT

PEMEtrexed (Rx)

(pem-ah-trex′ed)

Alimta

Func. class.: Antineoplastic-antimetabolite

Chem. class.: Folic acid antagonist

ACTION: Inhibits multiple enzymes that reduce folic acid, which is needed for cell replication

USES: Malignant pleural mesothelioma in combination with CISplatin; non–small-cell lung cancer as single agent; nonsquamous, non–small-cell lung cancer (1st-line treatment)

Unlabeled uses: Bladder, breast, colorectal, gastric, head/neck, pancreatic, renal cancers

CONTRAINDICATIONS: Pregnancy (D), hypersensitivity, ANC <1500 cells/mm^3, CCr <45 ml/min, thrombocytopenia (<100,000/mm^3), anemia

Precautions: Breastfeeding, children, renal/hepatic disease

DOSAGE AND ROUTES

• **Adult: IV INF** 500-600 mg/m^2 given over 10 min on day 1 of 21-day cycle with CISplatin 75 mg/m^2 infused over 2 hr beginning 1/2 hr after end of PEMEtrexed inf

Renal dose

• **Adult: IV INF** CCr <45 ml/min, not recommended

Available forms: Inj, single-use vials, 100, 500 mg

Administer:

• Store at 77° F, excursions permitted at 59° F to 86° F, not light sensitive, discard unused portions

• Vit B_{12} and low-dose folic acid as prophylactic measure to treat related hematologic, GI toxicity; 400-1000 mcg/day × 7 days before 1st dose and × 21 days after last dose, vit B_{12} 1 mg IM 1 wk before 1st dose and q 3 cycles (9 wk) thereafter

• Premedicate with corticosteroid (dexamethasone) given PO bid day before, day of, and day after administration of PEMEtrexed

Intermittent IV INF route

• Use cytotoxic handling procedures

• Reconstitute 500-mg vial/20 ml 0.9% NaCl inj (preservative free) = 25 mg/ml, swirl until dissolved, further dilute with 100 ml 0.9% NaCl inj (preservative free), give as IV inf over 10 mg

• Use only 0.9% NaCl inj (preservative free) for reconstitution, dilution

• Do not begin a new cycle unless neutrophils (ANC) are ≥1500 cells/mm^3, platelets are ≥100,000 cells/mm^3, CCr is ≥45 ml/min

• **Platelet nadir <50,000/mm^3 regardless of the ANC:** if necessary, delay until platelet count recovery, reduce PEMEtrexed and CISplatin by 50%; grade 3/4 toxicity occurs after 2 reductions, discontinue both products

• **ANC nadir <500/mm^3 when platelet nadir is ≥50,000/mm^3:** if necessary, delay until ANC recovery, reduce PEMEtrexed and CISplatin by 75%; if grade 3/4 toxicity occurs after 2 reductions, discontinue both products

• **CTC Grade 3/4 nonhematologic toxicity including diarrhea requiring hospitalization and excluding neurotoxicity, mucositis, and grade 3 transaminase elevations:** withhold therapy until pretherapy value or condition, reduce by 75% both products; if grade 3 or 4 toxicity occurs after 2 reductions, discontinue both products

• **CTC grade 3/4 mucositis:** withhold therapy until pretherapy condition, reduce 50% of PEMEtrexed; if grade 3 or 4 mucositis occurs after 2 dosage reductions, discontinue both products

• **CTC grade 2 neurotoxicity:** withhold therapy until pretherapy value or condition, reduce dose of CISplatin by 50%

• **CTC grade 3/4 neurotoxicity:** discontinue both products

Y-site compatibilities: Acyclovir sodium, alfentanil, allopurinol, amifostine, amikacin, aminocaproic acid, aminophylline, amiodarone, amphotericin B lipid complex, amphotericin B liposome, ampicillin, ampicillin-sulbactam, atenolol, atracurium, azithromycin, aztreonam, bivalirudin, bleomycin, bumetanide, buprenorphine, butorphanol, CARBOplatin, carmustine, ceftizoxime, cefTRIAXone, cefuroxime, cimetidine, cisatracurium, CISplatin, clindamycin, cyclophosphamide, cycloSPORINE, cytarabine, DACTINomycin, DAPTOmycin, dexamethasone, digoxin, diltiazem, diphenhydrAMINE, DOCEtaxel, dolasetron, DOPamine, doxacurium, enalaprilat, ePHEDrine, EPINEPHrine, eptifibatide, ertapenem, esmolol, etoposide, famotidine, fenoldopam, fentaNYL, fluconazole, fludarabine, fluorouracil, foscarnet, fosphenytoin, furosemide, ganciclovir, gatifloxacin, glycopyrrolate, granisetron, haloperidol, heparin, hydrocortisone, HYDROmorphone, hydrOXYzine, ifosfamide, imipenem-cilastatin, insulin (regular), isoproterenol, ketorolac, labetalol, leucovorin, levofloxacin, lidocaine, linezolid, LORazepam, magnesium, mannitol, meperidine, meropenem, mesna, methyldopate, methylPREDNISolone, metoclopramide, metoprolol, midazolam, milrinone, mitoMYcin, mivacurium, morphine, moxifloxacin, nafcillin, naloxone, nesiritide, nitroglycerin, norepinephrine, octreotide, oxaliplatin, PACLitaxel, pamidronate, pancuronium, PENTobarbital, PHENobarbital, piperacillin-tazobactam, polymyxin B, potassium chloride/phosphates, procainamide, promethazine, propranolol, ranitidine, remifentanil, rocuronium, sodium acetate/bicarbonate/phosphates, succinylcholine, SUFentanil, sulfamethoxazole-trimethoprim, tacrolimus, theophylline, thiopental, thiotepa, ticarcillin, ticarcillin-clavulanate, tigecycline, tirofiban, trimethobenzamide, vancomycin, vecuronium, verapamil, vinBLAStine, vinCRIStine, vinorelbine, zidovudine, zoledronic acid

SIDE EFFECTS

CNS: *Fatigue, fever, mood alteration, neuropathy*

CV: Thrombosis/embolism, *chest pain,* arrhythmia exacerbation

GI: *Nausea, vomiting, anorexia, diarrhea, ulcerative stomatitis, constipation, dysphagia, dehydration*

GU: Renal failure, creatinine elevation

HEMA: Neutropenia, leukopenia, thrombocytopenia, myelosuppression, anemia

INTEG: *Rash, desquamation*

RESP: *Dyspnea*

SYST: Infection with/without neutropenia, radiation recall reaction, toxic epidermal necrolysis, Stevens-Johnson syndrome, anaphylaxis

PHARMACOKINETICS

Not metabolized; excreted in urine (unchanged 70%-90%); not known if excreted in breast milk; half-life 3.5 hr, 81% protein binding

INTERACTIONS

Increase: bleeding risk—NSAIDs, anticoagulants, platelet inhibitors, salicylates, thrombolytics

Decrease: clearance of PEMEtrexed—nephrotoxic products

P

NURSING CONSIDERATIONS

Assess:

⚠ Previous radiation treatments; radiation recall reactions have occurred (erythema, exfoliative dermatitis, pain, burning)

⚠ **Bone marrow depression**: CBC, differential, platelet count; monitor for nadir, recovery; new cycle should not begin if ANC <1500 cells/mm^3, platelets <100,000 cells/mm^3, CCr <45 ml/min

- Renal studies: BUN, serum uric acid, urine CCr, electrolytes before, during therapy
- I&O ratio; report fall in urine output to <30 ml/hr
- Monitor temperature q4hr; fever may indicate beginning infection; no rectal temperature

⚠ **Neurotoxicity:** CTC grade 2: withhold until resolution to at least pretherapy value/condition, reduce CISplatin by 50%; CTC grade 3-4: immediately discontinue product and CISplatin if given in combination

• **Mucositis:** CTC 3/4: withhold until resolution to at least pretherapy value/condition, reduce dose by 50%; if grade 3/4 occurs after 2 dosage reductions, discontinue product and CISplatin

• **Bleeding:** bleeding time, coagulation time during treatment; bleeding: hematuria, guaiac, bruising or petechiae, mucosa or orifices q8hr

• Buccal cavity q8hr for dryness, sores, ulceration, white patches, oral pain, bleeding, dysphagia

⚠ **Severe allergic reaction/toxic epidermal necrolysis:** rash, urticaria, itching, flushing

• Rinsing of mouth tid-qid with water, club soda; brushing of teeth bid-tid with soft brush or cotton-tipped applicators for stomatitis; use unwaxed dental floss

Evaluate:

• Therapeutic response: decreased spread of malignancy

Teach patient/family:

• To report any complaints, side effects to nurse or prescriber: black, tarry stools, chills, fever, sore throat, bleeding, bruising, cough, SOB, dark or bloody urine

• To avoid foods with citric acid, hot or rough texture if stomatitis is present

• To report stomatitis: any bleeding, white spots, ulcerations in mouth to prescriber; to examine mouth daily; to report symptoms to nurse; to use good oral hygiene

• That contraceptive measures recommended during therapy, for ≤8 wk after cessation of therapy; to discontinue breastfeeding because toxicity to infant may occur

• To avoid alcohol, salicylates, live vaccines

• To avoid use of razors, commercial mouthwash

• To eat foods high in folic acid; to take supplements as prescribed

penciclovir topical

See Appendix B

PENICILLINS

penicillin G benzathine (Rx)

(pen-i-sill'in)

Bicillin L-A

penicillin G (Rx)

Pfizerpen

penicillin G procaine (Rx)

penicillin V (Rx)

Apo-Pen-VK ✱, Penicillin VK

Func. class.: Broad-spectrum antiinfective

Chem. class.: Natural penicillin

ACTION: Interferes with cell-wall replication of susceptible organisms; lysis is mediated by cell-wall autolytic enzymes, results in cell death

USES: Respiratory infections, scarlet fever, erysipelas, otitis media, pneumonia, skin and soft-tissue infections, gonorrhea; effective for gram-positive cocci *(Staphylococcus, Streptococcus pyogenes, S. viridans, S. faecalis, S. bovis, S. pneumoniae)*, gram-negative cocci *(Neisseria gonorrhoeae)*, gram-positive bacilli *(Actinomyces, Bacillus anthracis, Clostridium perfringens, C. tetani, Corynebacterium diphtheriae, Listeria monocytogenes)*, gram-negative bacilli *(Escherichia coli, Proteus mirabilis, Salmonella, Shigella, Enterobacter, Streptobacillus moniliformis)*, spirochetes *(Treponema pallidum)*

CONTRAINDICATIONS: Hypersensitivity to penicillins, corn

Precautions: Pregnancy (B), breastfeeding; hypersensitivity to cephalosporins, carbapenem, sulfites; severe renal disease, GI disease, asthma

DOSAGE AND ROUTES

Penicillin G benzathine

Early syphilis

• **Adult: IM** 2.4 million units in single dose

Congenital syphilis

• **Child <2 yr: IM** 50,000 units/kg in single dose, max 2.4 million units as single inj

Prophylaxis of rheumatic fever, glomerulonephritis

• **Adult: IM** 1.2 million units in single dose

• **Child >27 kg: IM** 900,000-1.2 million units as single dose

• **Child ≤27 kg: IM** 300,000-600,000 units as single dose

Upper respiratory infections (group A streptococcal)

• **Adult: IM** 1.2 million units as single dose

• **Child >27 kg: IM** 900,000-1.2 million units as single dose

• **Child <27 kg: IM** 300,000-600,000 units as single dose

Available forms: Inj 600,000 units/ml

Penicillin G

Pneumococcal/streptococcal infections (serious)

• **Adult: IM/IV** 5-24 million units in divided doses q4-6hr

• **Child <12 yr: IV** 150,000-300,000 units/kg/day in 4-6 divided doses; max 24 million units/day

Renal dose

• CCr <10 ml/min, give full loading dose then 1/2 of loading dose q8-10hr

Available forms: Powder for inj 1, 5, 20 million units/vial; inj 1, 2, 3 million units/50 ml

Penicillin G procaine

Moderate to severe pneumococcal infections

• **Adult/child: IM** 600,000-1 million units as single dose or divided bid doses/day for 10 days to 2 wk

Pneumococcal pneumonia

• **Adult/child >12 yr: IM** 600,000-1 million units/day × 7-10 days

Moderately severe group A streptococcal/staphylococcal pneumonia

• **Adult/adolescent/child ≥60 lbs: IM** 600,000-1 million units/day

• **Adolescent/child <60 lbs: IM** 300,000 units/day

Available forms: Inj 600,000, 1,200,000 units/unit dose

Penicillin V

Pneumococcal/staphylococcal infections

• **Adult/adolescents/child >12 yr: PO** 250-500 mg q6hr

• **Child <12 yr: PO** 25-50 mg/kg/day in divided doses q6-8hr; max 2 g/day

Streptococcal infections

• **Adult/adolescent/child ≥12 yr: PO** 125-250 mg q6-8hr × 10 days

• **Child <12 yr and >27 kg: PO** 500 mg q8 or 12 hr × 10 days

• **Child <12 yr and ≤27 kg: PO** 250 mg q8hr or q12hr or 40 mg/kg/day in 3 divided doses × 10 days

Prevention of recurrence of rheumatic fever/chorea

• **Adult: PO** 125-250 mg bid continuously

Vincent's gingivitis/pharyngitis

• **Adult: PO** 250-500 mg q6-8hr

Renal dose

• Dosage reduction indicated with renal impairment (CCr <50 ml/min) based on clinical response, degree of impairment

Available forms: Tabs 250, 500 mg; powder for oral sol 125, 250 mg/5 ml

Administer:

• Store in dry, tight container; oral susp refrigerated 2 wk

Penicillin G benzathine

• No dilution needed, shake well, deep

IM inj in large muscle mass; avoid intravascular inj; aspirate; do not give IV

Penicillin G

• Penicillin G sodium or potassium can be given IM or IV, vials containing 10 or 20 million units not for IM use

Intermittent IV INF route

• Vials/bulk packages: dilute according to manufacturer's directions

• Frozen bags: thaw at room temperature, do not force thaw, no reconstitution needed

• Final conc (100,000-500,000 units/ml—adults; 50,000 units/ml—neonate/infant)

• Total daily dose divided q4-6hr and given over 1-2 hr (adult), 15 min (infant/neonate)

Penicillin G potassium

Y-site compatibilities: Acyclovir, amiodarone, cyclophosphamide, diltiazem, enalaprilat, esmolol, fluconazole, foscarnet, heparin, HYDROmorphone, labetalol, magnesium sulfate, meperidine, morphine, perphenazine, potassium chloride, tacrolimus, theophylline, verapamil, vit B/C

Penicillin G procaine

• No dilution needed; give deep IM inj; avoid intravascular inj; aspirate; do not give IV

Penicillin V

• Orally on empty stomach for best absorption

• Oral susp: tap bottle to loosen, add 1/2 total amount of water, shake, add remaining water, shake; final conc (125 or 250 mg/ml) store in refrigerator after reconstitution, discard after 14 days

SIDE EFFECTS

CNS: Lethargy, hallucinations, anxiety, depression, twitching, coma, seizures, hyperreflexia

GI: *Nausea, vomiting, diarrhea,* increased AST, ALT, abdominal pain, glossitis, colitis, pseudomembranous colitis

GU: Oliguria, proteinuria, hematuria, *vaginitis, moniliasis,* glomerulonephritis, renal tubular damage

HEMA: Anemia, increased bleeding time, bone marrow depression, granulocytopenia, hemolytic anemia

META: Hypo/hyperkalemia, alkalosis, hypernatremia

MISC: Anaphylaxis, serum sickness, Stevens-Johnson syndrome, *local pain,* tenderness and fever with IM inj

PHARMACOKINETICS

Penicillin G benzathine:

IM: Very slow absorption; time to peak 12-24 hr; duration 21-28 days; excreted in urine, feces, breast milk; crosses placenta

Penicillin G:

IV: Peak immediate

IM: Peak 1/4-1/2 hr

PO: Peak 1 hr, duration 6 hr

Excreted in urine unchanged, excreted in breast milk, crosses placenta, half-life 30-60 min

Penicillin G procaine:

IM: Peak 1-4 hr, duration 15 hr, excreted in urine

Penicillin V:

PO: Peak 30-60 min, half-life 30 min, excreted in urine, breast milk

INTERACTIONS

Increase: penicillin effect—aspirin, probenecid

Increase: effect of heparin, methotrexate

Decrease: effect of oral contraceptives, typhoid vaccine

Decrease: antimicrobial effect of penicillin—tetracyclines

Drug/Lab Test

False positive: urine glucose, urine protein

NURSING CONSIDERATIONS

Assess:

• **Infection:** temperature; characteristics of sputum; wounds; urine; stools before, during, after treatment; C&S before therapy; product may be given as soon as culture is taken

• I&O ratio; report hematuria, oliguria because penicillin in high doses is nephrotoxic; renal tests: urinalysis, protein, blood

⚠ Any patient with compromised renal system because product is excreted slowly with poor renal system function; toxicity may occur rapidly

- Hepatic studies: AST, ALT
- Blood studies: WBC, RBC, Hct, Hgb, bleeding time

⚠ **Pseudomembranous colitis:** diarrhea, mucus, pus; bowel pattern before, during treatment

- Respiratory status: rate, character, wheezing, tightness in chest

⚠ Allergies before initiation of treatment, reaction of each medication; because of prolonged action, allergic reaction may be prolonged and severe; watch for anaphylaxis: rash, dyspnea, pruritus, laryngeal edema; skin eruptions after administration of penicillin to 1 wk after discontinuing product

- EPINEPHrine, suction, tracheostomy set, endotracheal intubation equipment
- Adequate fluid intake (2 L) during diarrhea episodes
- Scratch test to assess allergy after securing order from prescriber; usually done when penicillin is only product of choice

Evaluate:

- Therapeutic response: resolution of infection

Teach patient/family:

- To report sore throat, fever, fatigue; may indicate superinfection; CNS effects: depression, hallucinations, seizures
- To wear or carry emergency ID if allergic to penicillins
- To report diarrhea, with blood, pus, mucous to prevent dehydration
- To shake susp well before each dose; to store in refrigerator for up to 2 wk
- To use all medication prescribed
- To use additional contraception if using any of these products

TREATMENT OF ANAPHYLAXIS: Withdraw product; maintain airway; administer EPINEPHrine, aminophylline, O_2, IV corticosteroids

pentamidine (Rx)

(pen-tam′i-deen)

Nebupent, Pentam 300

Func. class.: Antiprotozoal

Chem. class.: Aromatic diamide derivative

ACTION: Interferes with DNA/RNA synthesis in protozoa

USES: Treatment/prevention of *Pneumocystis jiroveci* infections

Unlabeled uses: *Leishmania/Trypanosoma* infections

CONTRAINDICATIONS: Hypersensitivity

Precautions: Pregnancy (C), breastfeeding, children, blood dyscrasias, cardiac/renal/hepatic disease, diabetes mellitus, hypocalcemia, hypo/hypertension, anemia

DOSAGE AND ROUTES

- **Adult and child ≥4 mo: IV/IM** 4 mg/kg/day × 2-3 wk; **NEB** 300 mg via specific nebulizer given q4wk for prevention

Available forms: Inj, aerosol 300 mg/vial; sol for aerosol 60 mg/vial ✦

Administer:

- Store in refrigerator protected from light

Inhalation route

- Through nebulizer, using Raspirgard II jet nebulizer; mix contents in 6 ml of sterile water; do not use low pressure (<20 psi); flow rate should be 5-7 L/min (40-50 psi) air or O_2 source over 30-45 min until chamber is empty

IM route

- 300 mg diluted in 3 ml sterile water; give deep IM by Z-track; if painful by this route, rotate inj site

Intermittent IV INF route

- Reconstitute 300 mg/3-5 ml of sterile water for inj, D_5W, withdraw dose and further dilute in 50-250 ml D_5W, give over 1-2 hr with patient lying down; check B/P often

Y-site compatibilities: Alfentanil, atracurium, atropine, benztropine, buprenorphine, calcium gluconate, CARBOplatin, caspofungin, chlorproMAZINE, cimetidine, CISplatin, cyclophosphamide, cycloSPORINE, cytarabine, DACTINomycin, diltiazem, gatifloxacin, zidovudine

SIDE EFFECTS

CNS: Disorientation, hallucinations, *dizziness,* confusion, drowsiness
CV: Hypotension, ventricular tachycardia, QT prolongation, dysrhythmias
GI: *Nausea, vomiting, anorexia;* increased AST, ALT; acute pancreatitis, metallic taste
GU: Acute renal failure, increased serum creatinine, renal toxicity, decreased urination
HEMA: Anemia, leukopenia, thrombocytopenia
INTEG: Sterile abscess, pain at inj site, pruritus, urticaria, *rash*
META: *Hyperkalemia,* hypocalcemia, hypoglycemia, hypomagnesemia
MISC: Fatigue, fever, chills, night sweats, anaphylaxis, Stevens-Johnson syndrome
RESP: Cough, SOB, bronchospasm (with aerosol), sore throat

PHARMACOKINETICS

IV: Peak 1 hr
IM: Peak 30 min
Excreted unchanged in urine (66%); half-life 9-12 hr (IM), 6 hr (IV)

INTERACTIONS

• Nephrotoxicity: aminoglycosides, amphotericin B, CISplatin, NSAIDs, vancomycin
⚠ Fatal dysrhythmias: erythromycin IV
Increase: QT prolongation—class IA/III antidysrhythmics, some phenothiazines, β-agonists, local anesthetics, tricyclics, haloperidol, chloroquine, droperidol, pentamidine; CYP3A4 inhibitors (amiodarone, clarithromycin, erythromycin, telithromycin, troleandomycin, arsenic trioxide, levomethadyl); CYP3A4 substrates (methadone, pimozide, QUEtiapine, quiNIDine, risperiDONE, ziprasidone)
Increase: myelosuppression—antineoplastics, radiation

Drug/Lab Test
Decrease: WBC, platelets, Hbg, Hct
Increase: BUN, creatitine

NURSING CONSIDERATIONS

Assess:
• Blood tests, blood glucose, CBC, platelets, calcium, magnesium
• I&O ratio; report hematuria, oliguria
• ECG for cardiac dysrhythmias; patient should be lying down when receiving product; severe hypotension may develop; monitor B/P during administration and until B/P stable
• Hepatic studies: AST, ALT
• Renal studies: urinalysis, BUN, creatinine; nephrotoxicity may occur; any patient with compromised renal system; product is excreted slowly with poor renal system function; toxicity may occur rapidly
• Signs of infection, anemia
• Bowel pattern before, during treatment
• Sterile abscess, pain at inj site
• Respiratory status: rate, character, wheezing, dyspnea
• Dizziness, confusion, hallucination
• Allergies before treatment, reaction of each medication; place allergies on chart in bright red letters; notify all people giving products
• Diabetic patients, hypoglycemia may occur, then hyperglycemia with prolonged therapy

Evaluate:
• Therapeutic response: resolution of AIDS-related PCP

Teach patient/family:
• To report sore throat, fever, fatigue (may indicate superinfection)
• To maintain adequate fluid intake

⚠ HIGH ALERT

pentazocine (Rx)

(pen-taz′oh-seen)

Talwin, Talwin NX

Func. class.: Opiate analgesic, antagonist

Chem. class.: Synthetic benzomorphan

Controlled Substance Schedule IV

ACTION: Inhibits ascending pain pathways in CNS, increases pain threshold, alters pain perception

USES: Moderate to severe pain

CONTRAINDICATIONS: Hypersensitivity to this product or sulfites; addiction (opiate)

Precautions: Pregnancy (C), breastfeeding, children <18 yr, addictive personality, increased intracranial pressure, MI (acute), severe heart disease, respiratory depression, renal/hepatic disease, seizure disorder, head trauma, bowel impaction, geriatric patients

DOSAGE AND ROUTES

- **Adult: IV/IM/SUBCUT** 30 mg q3-4hr prn, max 360 mg/day

Labor

- **Adult: IM** 30 mg as a single dose; **IV** 20 mg q2-3hr when contractions are regular, max 2-3 times

Renal dose

- **Adult:** CCr 10-50 ml/min, reduce dose by 25%; CCr <10 ml/min, reduce dose by 50%

Available forms: Inj 30 mg/ml

Administer:

- Store in light-resistant area at room temperature
- With antiemetic if nausea, vomiting occur
- When pain is beginning to return; determine dosage interval by patient response

IM/SUBCUT route

- Give IM deeply into large muscle mass, rotate sites; SUBCUT may cause necrosis with repeated inj

Direct IV route

- Undiluted or diluted in 5 mg/ml of sterile water for inj; give ≤5 mg over 1 min

Syringe compatibilities: Atropine, benzquinamide, butorphanol, chlorproMAZINE, cimetidine, dimenhyDRINATE, diphenhydrAMINE, droperidol, fentaNYL, HYDROmorphone, hydrOXYzine, meperidine, metoclopramide, morphine, perphenazine, prochlorperazine, promazine, promethazine, ranitidine, scopolamine

Y-site compatibilities: Heparin, hydrocortisone, potassium chloride, vit B/C

SIDE EFFECTS

CNS: *Drowsiness, dizziness, confusion, headache, sedation, euphoria,* hallucinations, dreaming, insomnia, lightheadedness

CV: Palpitations, bradycardia, change in B/P, tachycardia, increased B/P (high doses), hypotension, syncope, flushing

EENT: Tinnitus, blurred vision, miosis, diplopia

GI: *Nausea,* vomiting, anorexia, constipation, *cramps,* dry mouth

GU: Increased urinary output, dysuria, urinary retention

HEMA: **Eosinophilia, decreased WBC**

INTEG: *Rash,* urticaria, bruising, flushing, diaphoresis, pruritus, severe irritation at inj sites, **Stevens-Johnson syndrome**

RESP: **Respiratory depression**

PHARMACOKINETICS

Metabolized by liver, excreted by kidneys, crosses placenta, half-life 2-3 hr, extensive first-pass metabolism with <20% entering circulation

IM/SUBCUT: Onset 15-30 min, peak 1-2 hr, duration 2-4 hr

IV: Onset 2-3 min, duration 4-6 hr

P

INTERACTIONS

⚠ Unpredictable reactions: MAOIs

Increase: effects—CNS depressants; alcohol, sedative/hypnotics, antipsychotics, skeletal muscle relaxants

Decrease: effects—opiates

Drug/Lab Test

Increase: amylase

NURSING CONSIDERATIONS

Assess:

- **Pain:** intensity, duration, location before and 1 hr after SUBCUT/IM dose or 30 min after IV dose
- I&O ratio; check for decreasing output; may indicate urinary retention
- Bowel status: constipation; may need stimulant laxatives/stool softeners
- **Withdrawal symptoms** in opiate-dependent patients
- Abscesses, ulcerations, WBC count
- CNS changes: dizziness, drowsiness, hallucinations, euphoria, LOC, pupil reaction
- Allergic reactions: rash, urticaria
- **Respiratory depression:** character, rate, rhythm; notify prescriber if respirations are <10/min
- Need for pain medication, physical dependence
- Assistance with ambulation
- Safety measures: nightlight, call bell within easy reach

Evaluate:

- Therapeutic response: decrease in pain

Teach patient/family:

- To report any symptoms of CNS changes, allergic reactions
- That physical dependency may result from extended use
- That withdrawal symptoms may occur: nausea, vomiting, cramps, fever, faintness, anorexia
- To avoid CNS depressants, alcohol
- To avoid driving, operating machinery if drowsiness occurs
- To use good oral hygiene, frequent rinsing of mouth to decrease dry mouth; to avoid gum, candy if drowsy

TREATMENT OF OVERDOSE:

Naloxone (Narcan) 0.2-0.8 mg IV, O_2, IV fluids, vasopressors

pentoxifylline (Rx)

(pen-tox′ih-fill-in)

TRENtal

Func. class.: Hemorheologic agent

Chem. class.: Dimethylxanthine derivative

ACTION: Decreases blood viscosity, stimulates prostacyclin formation, increases blood flow by increasing flexibility of RBCs; decreases RBC hyperaggregation; reduces platelet aggregation, decreases fibrinogen conc

USES: Intermittent claudication related to chronic occlusive vascular disease

Unlabeled uses: Behçet's syndrome, Kawasaki disease to reduce coronary artery lesions, diabetic neuropathies, sickle cell anemia

CONTRAINDICATIONS: Hypersensitivity to this product or xanthines, retinal/cerebral hemorrhage

Precautions: Pregnancy (C), breastfeeding, children, angina pectoris, impaired renal function, recent surgery, peptic ulceration, cardiac disease, bleeding disorders

DOSAGE AND ROUTES

- **Adult: PO** 400 mg tid with meals; may decrease to bid if side effects occur; must be taken for ≥8 wk for maximal effect

Behçet's syndrome (unlabeled)

- **Adult: PO** 300 mg bid × 2 wk then 300 mg/day

Acute claudication in sickle cell disease/diabetic neuropathy (unlabeled)

- **Adult: PO** 400 mg tid

Kawasaki disease (unlabeled)
• **Child: PO** 20 mg/kg/day in 3 divided doses with aspirin and IVIG
Available forms: Cont rel tabs 400 mg; ext rel tabs 400 mg
Administer:
• Do not break, crush, or chew ext rel tabs
• With meals to prevent GI upset

SIDE EFFECTS

CNS: *Headache,* anxiety, *tremors,* confusion, *dizziness*
GI: *Dyspepsia, nausea, vomiting*

PHARMACOKINETICS

PO: Peak 2-4 hr, half-life ½-1 hr, degradation in liver, excreted in urine

INTERACTIONS

Increase: bleeding risk—warfarin, abciximab, eptifibatide, tirofiban, ticlopidine, clopidogrel, thrombin inhibitor
Increase: theophylline level—theophylline
Increase: hypotension—antihypertensives, nitrates
Increase: pentoxifylline—cimetidine, ciprofloxacin

NURSING CONSIDERATIONS

Assess:
• B/P, respirations of patient also taking antihypertensives; intermittent claudication at baseline and throughout
• PT, Hgb, Hct in patients at risk for hemorrhage
• Serum creatinine/BUN
Evaluate:
• Therapeutic response: decreased pain, cramping, increased ambulation
Teach patient/family:
• That therapeutic response may take 2-4 wk, 8-12 wk to reach full benefit
• To observe feet for arterial insufficiency
• To wear cotton socks, well-fitted shoes; not to go barefoot
• To watch for bleeding, bruises, petechiae, epistaxis
• To avoid smoking to prevent blood vessel constriction
• That there are many drug, herb interactions

perampanel

(per-am′pa-nel)
Fycompa
Func. class.: Anticonvulsant

ACTION:
A noncompetitive AMPA-selective receptor antagonist; inhibits calcium influx

USES:
Partial onset-seizures with/without secondary generalization

CONTRAINDICATIONS:
Hypersensitivity
Precautions: Pregnancy (C), breastfeeding, children <12 yr, geriatric patients, abrupt discontinuation, depression, liver disease, kidney disease, substance abuse, suicidal ideation, driving or operating machinery
Black Box Warning: Bipolar disorder, psychosis, schizophrenia

DOSAGE AND ROUTES

• **Adult/adolescent/child ≥12 yr receiving enzyme-inducing AEDs: PO** 4 mg/day at bedtime, increase dosage every wk by 2 mg/day to 4-12 mg/day at bedtime
• **Adult/adolescent/child ≥12 yr not receiving AEDs: PO** 2 mg/day at bedtime, increase dosage every wk by 2 mg/day up to 4-8 mg/day at bedtime
Available forms: Tabs 2, 4, 6, 8, 10, 12 mg
Administer:
• Store at room temperature away from light

SIDE EFFECTS

CNS: Drowsiness, confusion, memory impairment, dizziness, fatigue, anxiety, ataxia, agitation, emotional lability, euphoria, hostility, lethargy, memory im-

pairment, paranoia, paresthesias, suicidal ideation, vertigo, depression
CV: Peripheral edema
EENT: Blurred vision, diplopia
GI: Constipation, nausea, vomiting, weight gain
MS: Myalgia, arthralgia
SYST: Infection

PHARMACOKINETICS

Protein binding 95%-96%, half-life 105 hr, antagonist of the AMPA receptor

INTERACTIONS

Increase: perampanel effect—CYP3A4 inducers
Increase: depression—benzodiazepines
Increase: CNS depression—alcohol, sedatives, antihistamines, all other CNS depressants
Decrease: contraception—oral/implant contraceptives with levonorgestrel/estrogen
Decrease: perampanel effect—CYP3A4 inhibitors
Drug/Herb
Decrease: perampanel effect—St. John's wort

NURSING CONSIDERATIONS

Assess:

⚠ **Seizures:** aura, location, duration, activity at onset; institute seizure precautions: padded side rails; move objects that might harm patient

Black Box Warning: **Bipolar disorder, psychosis, schizophrenia:** Hostility, aggression, anger, homicidal ideation can occur

⚠ **Suicidal ideation:** assess for suicidal thoughts/behaviors as early as 1 wk and any time during treatment

- Mental status: mood, sensorium, affect, behavioral changes, suicidal thoughts/behaviors; if mental status changes: notify prescriber

⚠ **Abrupt discontinuation:** can increase seizures

Evaluate:

- Therapeutic response: decreased seizure activity

Teach patient/family:

- To avoid driving, other activities that require alertness because dizziness, drowsiness can occur

⚠ Not to discontinue medication quickly after long-term use; to taper over ≥1 wk because withdrawal-precipitated seizures can occur

⚠ To notify prescriber if pregnancy is planned or suspected; to avoid breastfeeding

- Take once daily at bedtime

perindopril (Rx)

(per-in′doe-pril)

Aceon, Coversye ✱

Func. class.: Antihypertensive
Chem. class.: Angiotensin-converting enzyme inhibitor

ACTION: Selectively suppresses the renin-angiotensin-aldosterone system; inhibits ACE; prevents the conversion of angiotensin I to angiotensin II and the dilation of arterial and venous vessels

USES: Hypertension, MI prophylaxis
Unlabeled uses: Heart failure, MI

CONTRAINDICATIONS: Hypersensitivity, history of angioedema

Black Box Warning: Pregnancy (D)

Precautions: Breastfeeding, renal disease, hyperkalemia, hepatic failure, dehydration, bilateral renal artery stenosis, cough, severe CHF, aortic stenosis, African descent

DOSAGE AND ROUTES

Hypertension

- **Adult: PO** 4 mg/day, may increase or decrease to desired response, range 4-8 mg/day; may give in 2 divided doses or as single dose, max 16 mg/day

Patients on diuretics
• **Adult: PO** 2-4 mg/day in 1-2 divided doses, range 4-8 mg/day
Stable CAD
• **Adult: PO** 4 mg/day × 2 wk then increase as tolerated to 8 mg/day
Renal dose
• **Adult: PO** CCr 16-29 ml/min, 2 mg every other day; CCr 30-59 ml/min, 2 mg/day; CCr ≤15 ml/min 2 mg on dialysis days only
Available forms: Tabs scored 2, 4, 8 mg
Administer:
• As single dose or in 2 divided doses, without regard to meals

SIDE EFFECTS

CNS: *Insomnia, dizziness,* paresthesias, *headache,* fatigue, anxiety, depression
CV: *Hypotension,* chest pain, tachycardia, dysrhythmias, syncope, **cardiac arrest**
EENT: *Tinnitus;* visual changes; sore throat; double vision; dry, burning eyes
GI: Nausea, vomiting, colitis, cramps, diarrhea, constipation, flatulence, dry mouth, loss of taste, **liver failure**
GU: **Proteinuria, renal failure,** increased frequency of polyuria or oliguria
HEMA: **Agranulocytosis, neutropenia, bone marrow suppression**
INTEG: Rash, purpura, alopecia, hyperhidrosis
META: *Hyperkalemia*
RESP: Dyspnea, *dry cough,* crackles
SYST: **Angioedema**

PHARMACOKINETICS

Bioavailability 75%; peak 1 hr parent product, 3-7 hr prodrug; protein binding 68%; metabolized by liver (active metabolite perindoprilat); half-life 0.8-10 hr; excreted in urine

INTERACTIONS

Increase: effects of neuromuscular blocking agents, antihypertensives, lithium
Increase: antihypertensive effect—diuretics, NSAIDs
Increase: hypersensitivity—allopurinol
Increase: severe hypotension—diuretics, other antihypertensives
Increase: hyperkalemia—salt substitutes, potassium-sparing diuretics, potassium supplements, angiotensin II receptor antagonists
Decrease: effects of NSAIDs
Decrease: antihypertensive effect—NSAIDs, salicylates
Drug/Herb
Increase: antihypertensive effect—hawthorn
Decrease: antihypertensive effect—ephedra
Drug/Lab Test
Interference: glucose/insulin tolerance tests

NURSING CONSIDERATIONS

Assess:
• **Hypertension:** B/P, pulse q4hr; note rate, rhythm, quality
• Electrolytes: K, Na, Cl during 1st 2 wk of therapy
• Baselines of renal, hepatic studies before therapy begins, 1 wk into therapy
• Skin turgor, dryness of mucous membranes for hydration status, dry mouth
• **CHF:** edema, dyspnea, wet crackles
• **Angioedema: facial swelling, urticaria; product should be discontinued, may be more common in African Americans**
Evaluate:
• Therapeutic response: decreased B/P
Teach patient/family:
• Not to use OTC (cough, cold, allergy) products unless directed by prescriber; to avoid salt substitutes
• To avoid sunlight or to wear sunscreen for photosensitivity
• To comply with dosage schedule, even if feeling better
• **To notify prescriber of mouth sores, sore throat, fever, swelling of hands or feet, irregular heartbeat, chest pains, signs of angioedema**

P

- That excessive perspiration, dehydration, vomiting, diarrhea may lead to fall in B/P; to consult prescriber if these occur
- That product may cause dizziness, fainting; that lightheadedness may occur during 1st few days of therapy; to avoid hazardous activities
- That product may cause skin rash or impaired perspiration; that angioedema may occur and to discontinue product if it occurs
- Not to discontinue product abruptly
- To rise slowly to sitting or standing position to minimize orthostatic hypotension
- To report persistent, sustained cough
- To avoid potassium supplements, salt substitutes

Black Box Warning: To notify prescriber if pregnancy is planned or suspected; pregnancy (D)

TREATMENT OF OVERDOSE:

Lavage, IV atropine for bradycardia, IV theophylline for bronchospasm, digoxin, O_2, diuretic for cardiac failure

pertuzumab

(per-too′zoo-mab)

Perjeta

Func. class.: Antineoplastic biologic-response modifier

Chem. class.: Monoclonal antibody, antineoplastic

ACTION: Blocks liquid-dependent action of human epidermal growth factor-2 (HER2), inhibiting signal pathways

USES: First-line treatment of (HER2) positive metastatic breast cancer with trastuzumab and DOCEtaxel

CONTRAINDICATIONS:

Black Box Warning: pregnancy (category D)

Precautions: Breastfeeding, children, infants, neonates, cardiac arrhythmias, MI, cardiac disease, heart failure, hypertension, infusion-related reactions

DOSAGE AND ROUTES

- **Adult:** IV 840 mg over 60 min, then after 3 wk 420 mg over 30-60 min every 3 wk; give with trastuzumab 8 mg/kg IV over 90 min, then after 3 wk 6 mg/kg over 30-90 min every 3 wk and DOCEtaxel 75 mg/m^2 IV every 3 wk; dosage may be escalated to 100 mg/m^2

Available forms: Solution for inj 420 mg/14 ml

Administer:

- Visually inspect for particulate matter and discoloration

Dilution and preparation

- Withdraw the calculated dose from the vial and add to 250 ml 0.9% sodium chloride to PVC or non-PVC polyolefin infusion bag; do not dilute with dextrose 5% solution
- Dilute in normal saline only; do not mix or dilute with other drugs or dextrose solutions
- Mix the diluted solution by gentle inversion; do not shake

IV infusion

- Administer the diluted solution immediately
- Do not administer as an IV push or bolus
- Give the first dose of 840 mg over 60 min and subsequent 420 mg doses over 30–60 min
- If the diluted solution is not used immediately, store at 2°-8° C for up to 24 hr

Delayed or missed doses

- If time since previous dose is <6 wk, give 420 mg IV (do not wait for next scheduled dose)
- If time since previous dose is ≥6 wk, give 840 mg IV over 60 min, followed 3 wk later by 420 mg IV over 30–60 min repeated every 3 wk
- If DOCEtaxel is discontinued, this product and trastuzumab may continue

SIDE EFFECTS

CNS: Headache, fever, *peripheral neuropathy*, chills, *fatigue*, asthenia
CV: **Heart failure**
EENT: Lacrimation, stomatitis
GI: *Nausea*, vomiting, diarrhea, dysgeusia
HEMA: Anemia, **neutropenia**
MS: Myalgia
RESP: Upper respiratory infection
SYST: **Anaphylaxis**, infection, antibody formation

PHARMACOKINETICS

Median half-life 18 days

NURSING CONSIDERATIONS

Assess:

HER2 overexpression:

• Testing should be done to identify HER2 overexpression before using this product

Decreased left ventricular ejection fraction (LVEF):

⚠ Can occur and is increased in those with a history of prior anthracycline use or radiotherapy to the chest; evaluate LVEF at baseline and every 3 mo; withhold therapy ×3 wk if LVEF is <40% or LVEF is 40%-45% with a 10% or greater absolute decrease from baseline; resume therapy if the LVEF is recovered to >45% or to 40%-45% with <10% absolute decrease at reassessment; if the LVEF has not improved or has declined further, consider permanently discontinuing pertuzumab and trastuzumab after a risk/benefit assessment

Infusion-related reactions/hypersensitivity:

⚠ Assess anaphylactoid reaction, acute infusion reaction, cytokine-release syndrome 60 min after the first infusion, 30 min after other infusions; monitor for pyrexia, chills, fatigue, headache, asthenia, hypersensitivity, and vomiting; if a significant reaction occurs, slow or interrupt the infusion; permanent discontinuation may be needed in severe reactions

Black Box Warning: Pregnancy: Determine if pregnancy is planned or suspected; patients who become pregnant during therapy should report exposure to the Genentech Adverse Event line at 1-888-835-2555 and enroll in the MOTHER pregnancy registry at 1-800-690-6720

Neutropenia:

⚠ Can occur, but occurs more commonly when trastuzumab is also used

Upper respiratory infection:

• Monitor for dyspnea, shortness of breath, fever

Evaluate:

• Decreased size, spread of tumor

Teach patient/family:

Black Box Warning: Counsel women of childbearing age on the need for contraception during and for 6 mo after therapy; advise patients who suspect pregnancy to contact their health care provider immediately; discontinue breastfeeding

⚠ HIGH ALERT

PHENobarbital (Rx)

(fee-noe-bar′bi-tal)

Func. class.: Anticonvulsant
Chem. class.: Barbiturate

Controlled Substance Schedule IV

Do not confuse:
PHENobarbital/PENTobarbital

ACTION: Decreases impulse transmission; increases seizure threshold at cerebral cortex level

USES: All forms of epilepsy, status epilepticus, febrile seizures in children, sedation, insomnia

Unlabeled uses: Neonatal hyperbilirubinemia, chronic cholestasis, neonatal abstinence syndrome, febrile seizures in children

CONTRAINDICATIONS:

Pregnancy (D), hypersensitivity to barbiturates, porphyria, hepatic/respiratory disease

Precautions: Anemia, renal disease, breastfeeding, geriatric patients

DOSAGE AND ROUTES

Seizures

- **Adult: PO** 1-3 mg/kg/day in divided doses bid-tid or total dose at bedtime
- **Child 5-12 yr: PO** 3-6 mg/kg/day in 1-2 divided doses
- **Child 1-5 yr: PO** 6-8 mg/kg/day in 1-2 divided doses
- **Infant: PO** 5-6 mg/kg/day in 1-2 divided doses
- **Neonate: PO** 3-4 mg/kg/day as single dose

Status epilepticus

- **Adult: IV INF** loading dose 15-18 mg/kg, then 10 mg/kg; run no faster than 50 mg/min; may give up to 30 mg/kg
- **Child: IV INF** 5-10 mg/kg; may repeat q10-15min up to 20 mg/kg; run no faster than 50 mg/min

Sedation

- **Adult: PO/IM** 30-120 mg/day in 2-3 divided doses
- **Child: PO** 6 mg/kg/day in 3 divided doses

Preoperative sedation

- **Adult: IM** 100-200 mg 1-1½ hr before surgery
- **Child: PO/IM/IV** 1-3 mg/kg 1-1½ hr before surgery

Available forms: Caps 15 mg; elix 20 mg/5 ml; tabs 15, 30, 32, 60, 65, 100 mg; inj 30, 60, 65, 130 mg/ml

Administer:

PO route

- Tabs may be crushed and mixed with food or fluids
- **Oral sol:** use undiluted or mixed with water or other fluids; use calibrated measuring device

IM route

- IM inj deep in large muscle mass to prevent tissue sloughing; use <5 ml at each site

IV route

- Reconstitute powder for IV with ≥3 ml sterile water for inj; slow IV after dilution with ≥10 ml sterile water for inj regardless of dose; give ≤60 mg/min; titrate to patient response

Y-site compatibilities: Doxapram, enalaprilat, fentaNYL, fosphenytoin, levofloxacin, linezolid, meropenem, methadone, morphine, propofol, SUFentanil

SIDE EFFECTS

CNS: Paradoxic excitement (geriatric patients), drowsiness, lethargy, hangover headache, flushing, hallucinations, **coma, suicidal ideation**

EENT: Miosis, mydriasis

GI: Nausea, vomiting, diarrhea, constipation

HEMA: **Agranulocytosis, megaloblastic anemia, thrombocytopenia, thrombophlebitis**

INTEG: Rash, urticaria, **Stevens-Johnson syndrome, angioedema**, local pain, swelling, necrosis, scaling eczema

RESP: Respiratory depression

PHARMACOKINETICS

Metabolized by liver; crosses placenta; excreted in urine, breast milk; half-life 53-118 hr

PO: Onset 20-60 min, duration 6-10 hr

IM/SUBCUT: Onset 10-30 min, duration 4-6 hr

IV: Onset 5 min, peak 30 min, duration 4-6 hr

INTERACTIONS

Increase: effects—CNS depressants, alcohol, chloramphenicol, valproic acid, disulfiram, nondepolarizing skeletal muscle relaxants, sulfonamides, MAOIs

Increase: orthostatic hypotension—furosemide

Decrease: effects—theophylline, oral anticoagulants, corticosteroids, metroNIDAZOLE, doxycycline, quiNIDine, estrogens, hormonal contraceptives

Drug/Herb
Increase: CNS depression—chamomile, eucalyptus, hops, kava, valerian
Decrease: barbiturate effect—St. John's wort

NURSING CONSIDERATIONS

Assess:

- Mental status: mood, sensorium, affect, memory (long, short)
- **Blood dyscrasias:** fever, sore throat, bruising, rash, jaundice
- **Seizures:** type, duration, precipitating factors
- Blood studies, CBC, serum creatinine, BUN, LFTs during long-term treatment
- Therapeutic blood level periodically: 15-40 mcg/ml
- Respiratory status: rate, rhythm, depth, respiratory depression; have emergency equipment nearby
- **Dependence:** physical or psychological; monitor amount given to patient; if patient suicidal, may save medication for attempt
- **Pain:** use pain scale if product used for pain; product may increase pain level
- Supervision of ambulation because dizziness, drowsiness may occur

Evaluate:

- Therapeutic response: decreased seizures, increased sedation, adequate sleep

Teach patient/family:

- To use exactly as ordered
- To avoid alcohol, other CNS depressants
- To avoid hazardous activities until patient stabilized on product because drowsiness may occur
- Never to withdraw product abruptly because withdrawal symptoms may occur
- That therapeutic effects (PO) may not be seen for 2-3 wk
- To use additional nonhormonal contraception during treatment; to notify prescriber if pregnancy (D) is planned or suspected
- To report suicidal thoughts/behavior immediately

phentolamine (Rx)

(fen-tole'a-meen)
Func. class.: Antihypertensive
Chem. class.: α-Adrenergic blocker

Do not confuse: phentolamine/phentermine

ACTION:
α-Adrenergic blocker, binds to α-adrenergic receptors, dilating peripheral blood vessels, lowering peripheral resistances, lowering blood pressure

USES:
Hypertension; pheochromocytoma; prevention/treatment of dermal necrosis following extravasation of norepinephrine, DOPamine, EPINEPHrine
Unlabeled uses: Impotence, hypertensive crisis due to MAOIs, sympathomimetic amines, heart failure

CONTRAINDICATIONS:
Hypersensitivity, MI, coronary insufficiency, angina, hypotension
Precautions: Pregnancy (C), breastfeeding, dysrhythmia, peptic ulcer disease

DOSAGE AND ROUTES

Treatment of hypertensive episodes in pheochromocytoma

- **Adult: IM/IV** 5 mg, repeat if necessary
- **Child: IV** 0.05-0.1 mg/kg/dose, repeat if necessary; max 5 mg

Diagnosis of pheochromocytoma

- **Adult: IV** 5 mg
- **Child: IV** 0.05 mg/kg; if negative, repeat with 0.1 mg/kg IV

Treatment of necrosis

- **Adult:** 5-10 mg/10 ml **NS** injected into area of norepinephrine extravasation within 12 hr
- **Child:** 0.1-0.2 mg/kg, max 5 mg

Prevention of necrosis

- **Adult:** 10 mg/L of norepinephrine-containing sol
- **Child: IV** 0.1-0.2 mg/kg, max 5 mg

P

Left ventricular heart failure (unlabeled)
- **Adult: IV** 0.17-0.4 mg/min

Erectile dysfunction (unlabeled)
- **Adult: PO** intracavernosal 0.5-1 mg with 30 mg papaverine, max 3× per wk

Hypertensive emergency due to MAOIs, sympathomimetic amines (unlabeled)
- **Adult: IV BOL** 5-15 mg

Available forms: Inj 5 mg/ml

Administer:
- Gum, frequent rinsing of mouth, or hard candy for dry mouth
- With vasopressor available
- After discontinuing all medication for 24 hr
- Treatment during required bed rest, 1 hr after

IV route
- After diluting 5 mg/1 ml sterile H_2O for inj; give 5 mg or less/min; patient to remain recumbent during administration

Continuous IV INF route
- Dilute 5-10 mg/500 ml D_5W, titrate to patient response
- 10 mg/L may be added to norepinephrine in IV sol for prevention of dermal necrosis

Additive compatibilities: DOBUTamine, verapamil
Syringe compatibilities: MethylPREDNISolone, papaverine
Y-site compatibilities: Amiodarone

SIDE EFFECTS

CNS: *Dizziness,* flushing, weakness, cerebrovascular spasm, paresthesias, CVA
CV: *Hypotension, tachycardia, angina, dysrhythmias,* MI
EENT: Nasal congestion
GI: *Dry mouth, nausea, vomiting, diarrhea, abdominal pain*
INTEG: Pruritus, injection site pain
RESP: Nasal congestion

PHARMACOKINETICS

Metabolized in liver, excreted in urine
IM: Peak 15-20 min, duration 3-4 hr
IV: Peak 2 min, duration 10-15 min

INTERACTIONS

Increase: effects of EPINEPHrine, antihypertensives
Increase: hypotension, sildenafil, tadalafil, vardenafil

NURSING CONSIDERATIONS

Assess:
- Weight daily, I&O
- B/P lying, standing before starting treatment, q4hr after
- Nausea, vomiting, diarrhea, edema in feet, legs daily; skin turgor, dryness of mucous membranes for hydration status, postural hypotension, cardiac system: pulse, ECG

Evaluate:
- Therapeutic response: decreased B/P

Teach patient/family:
- That bed rest is required during treatment, 1 hr after

TREATMENT OF OVERDOSE:

Administer norepinephrine; discontinue product

phenylephrine (Rx)

(fen-ill-ef′rin)

Neo-Synephrine

Func. class.: Adrenergic, directacting
Chem. class.: Substituted phenylethylamine

ACTION: Powerful and selective (α_1) receptor agonist that causes the contraction of blood vessels

USES: Hypotension, paroxysmal supraventricular tachycardia, shock; maintain B/P for spinal anesthesia

CONTRAINDICATIONS: Hypersensitivity, ventricular fibrillation, tachydysrhythmias, pheochromocytoma, closed-angle glaucoma, severe hypertension

Precautions: Pregnancy (C), breastfeeding, geriatric patients, arterial embolism, peripheral vascular disease, hyperthyroidism, bradycardia, myocardial disease, severe arteriosclerosis, partial heart block

Black Box Warning: Cardiac disease, extravasation

DOSAGE AND ROUTES

Hypotension

- **Adult: SUBCUT/IM** 2-5 mg; may repeat q10-15min if needed; do not exceed initial dose; **IV** 0.1-0.5 mg; may repeat q10-15min if needed; do not exceed initial dose
- **Child: IM/SUBCUT** 0.1 mg/kg/dose q1-2hr prn

Supraventricular tachycardia

- **Adult: IV BOL** max 0.5 mg; max single dose 1 mg

Shock

- **Adult: IV INF** 10 mg/500 ml D_5W given 100-180 mcg/min (if 20 gtt/ml inf device) then maintenance of 40-60 mcg/min; use inf device
- **Child: IV BOL** 5-20 mcg/kg/dose q10-15min; **IV INF** 0.1-0.5 mcg/kg/min

Available forms: Inj 1% (10 mg/ml)

Administer:

IV route

- Plasma expanders for hypovolemia
- IV after diluting 1 mg/9 ml sterile water for inj; give dose over 1/2-1 min; may be diluted 10 mg/500 ml of D_5W or NS; titrate to response (normal B/P); check for extravasation; check site for infiltration; use inf pump
- Store reconstituted sol if refrigerated for ≤24 hr
- Discard discolored sol

Additive compatibilities: Chloramphenicol, DOBUTamine, lidocaine, potassium chloride, sodium bicarbonate

Y-site compatibilities: Amiodarone, amrinone, cisatracurium, famotidine, haloperidol, remifentanil, zidovudine

SIDE EFFECTS

CNS: *Headache, anxiety, tremors, insomnia, dizziness*

CV: *Palpitations, tachycardia, hypertension, ectopic beats, angina,* reflex bradycardia, **dysrhythmias**

GI: *Nausea, vomiting*

INTEG: Necrosis, tissue sloughing with extravasation, **gangrene**

SYST: **Anaphylaxis**

PHARMACOKINETICS

IM/SUBCUT: Onset 10-15 min, duration 45-60 min

IV: Onset immediate, duration 20-30 min

INTERACTIONS

- Dysrhythmias: general anesthetics, digoxin

⚠ **Do not use within 2 wk of MAOIs because hypertensive crisis may result**

Increase: in B/P—oxytocics

Increase: pressor effect—tricyclics, β-blockers, H_1 antihistamines

Decrease: phenylephrine action—α-blockers

NURSING CONSIDERATIONS

Assess:

- I&O ratio; notify prescriber if output <30 ml/hr
- ECG during administration continuously; if B/P increases, product is decreased
- B/P and pulse q5min after parenteral route
- CVP or PWP during inf if possible
- Paresthesias and coldness of extremities; peripheral blood flow may decrease

Evaluate:

- Therapeutic response: increased B/P with stabilization

Teach patient/family:

- About the reason for administration
- To report pain at inf site or other adverse reactions immediately

TREATMENT OF OVERDOSE:

Administer α-blocker

phenylephrine nasal agent

See Appendix B

phenylephrine ophthalmic

See Appendix B

phenytoin (Rx)

(fen'i-toh-in)

Dilantin, Dilantin Infatabs, Phenytek

Func. class.: Anticonvulsant; antidysrhythmic (IB)

Chem. class.: Hydantoin

ACTION: Inhibits spread of seizure activity in motor cortex by altering ion transport; increases AV conduction

USES: Generalized tonic-clonic seizures; status epilepticus; nonepileptic seizures associated with Reye's syndrome or after head trauma; complex partial seizures

Unlabeled uses: Migraines, diabetic neuropathy, neuropathic pain, paroxysmal atrial tachycardia, ventricular tachycardia

CONTRAINDICATIONS: Pregnancy (D), hypersensitivity, psychiatric condition, bradycardia, SA and AV block, Stokes-Adams syndrome

Precautions: Geriatric patients, allergies, renal/hepatic disease, petit mal seizures, hypotension, myocardial insufficiency, Asian patients positive for HLA-B1502, hepatic failure, acute intermittent porphyria

Black Box Warning: IV use

DOSAGE AND ROUTES

Seizures

- **Adult: PO** 15-20 mg/kg (ext rel) in 3-4 divided doses given q2hr or 400 mg then 300 mg q2hr × 2 doses; maintenance 4-7 mg/kg/day; **IV** 15-20 mg/kg, max 25-50 mg/min, then 100 mg q6-8hr
- **Child: PO** 5 mg/kg/day in 2-3 divided doses, maintenance 4-8 mg/kg/day in 2-3 divided doses, max 300 mg/day; **IV** 15-20 mg/kg at 1-3 mg/kg/min

Status epilepticus

- **Adult: IV** 15-20 mg/kg, max 25-50 mg/min, may give 100 mg q6-8hr thereafter
- **Child: IV** 15-20 mg/kg, max in divided doses 1-3 mg/kg/min

Ventricular dysrhythmias

- **Adult: PO** loading dose 1 g divided over 24 hr then 500 mg/day × 2 days; **IV** 250 mg over 5 min until dysrhythmias subside or until 1 g is given or 100 mg q15min until dysrhythmias subside or until 1 g is given
- **Child: PO** 3-8 mg/kg or 250 mg/m^2/day as single dose or 2 divided doses; **IV** 3-8 mg/kg over several min or 250 mg/m^2/day as single dose or 2 divided doses

Renal/hepatic dose

- Do not use loading dose if CCr <10 ml/min or hepatic failure

Neuropathic pain/diabetic neuropathy (unlabeled)

- **Adult: PO** 300 mg/day in divided doses

Migraine prophylaxis (unlabeled)

- **Adult: PO** 200-400 mg/day

Available forms: Susp 25 mg/5 ml; chewable tabs 50 mg; inj 50 mg/ml; ext rel caps 100, 200, 300 mg; prompt rel caps 100 mg

Administer:

PO route

- Do not interchange chewable product with caps, not equivalent; only ext rel caps to be used for once-a-day dosing
- **Oral susp:** shake well before each dose given via G tube/NG tube; dilute susp before administration; flush tube with 20 ml water after dose; hold tube feedings 1 hr before and 1 hr after dose

• Allow 7-10 days between dosage changes
• Divided PO doses with or after meals to decrease adverse effects
• 2 hr before or after antacid, enteral feeding

Direct IV route

Black Box Warning: Give undiluted at ≤50 mg/min (adult) 1-3 mg/kg/min; 0.5-1 mg/kg/min (neonates), monitor for CV reactions

Intermittent IV INF route

Black Box Warning: Dilute dose in NS to ≤6.7 mg/ml; complete inf within 1 hr of preparation; use 0.22 or 0.55 micron in-line particulate final filter between IV catheter and tubing; flush IV line or catheter with NS before and after use; give at ≤50 mg/min (adult), 0.5-1 mg/kg/min (child, infant, neonate)

Additive compatibilities: Do not admix
Y-site compatibilities: CISplatin, foscarnet, temocillin

SIDE EFFECTS

CNS: Dizziness, insomnia, paresthesias, depression, suicidal tendencies, aggression, headache, confusion, slurred speech, peripheral neuropathy
CV: Hypotension, **ventricular fibrillation, bradycardia, cardiac arrest**
EENT: Nystagmus, diplopia, blurred vision
ENDO: Diabetes insipidus
GI: Nausea, vomiting, constipation, anorexia, weight loss, **hepatitis**, jaundice, gingival hyperplasia, abdominal pain
GU: **Nephritis**, urine discoloration, sexual dysfunction
HEMA: **Agranulocytosis, leukopenia, aplastic anemia, thrombocytopenia, megaloblastic anemia**
INTEG: Rash, **lupus erythematosus, Stevens-Johnson syndrome**, hirsutism, **toxic epidermal necrolysis**
SYST: Hypocalcemia, **purple glove syndrome (IV), DRESS, anaphylaxis**, exacerbates myasthenia gravis

PHARMACOKINETICS

Metabolized by liver, excreted by kidneys, protein binding 90%-95%, half-life 7-42 hr, dose dependent
PO: Onset 2-24 hr, peak $1^1/_2$-$2^1/_2$ hr, duration 6-12 hr
PO-ER: Onset 2-24 hr, peak 4-12 hr, duration 12-36 hr
IV: Onset 1-2 hr, duration 12-24 hr

INTERACTIONS

• **Do not use with delavirdine, decreased response, resistance**
Increase: phenytoin effect—benzodiazepines, cimetidine, tricyclics, salicylates, valproate, cycloSERINE, diazepam, chloramphenicol, disulfiram, alcohol, amiodarone, sulfonamides, FLUoxetine, gabapentin, H_2 antagonists, azole antifungals, estrogens, succinamides, phenothiazines, methylphenidate, felbamate, traZODone
Decrease: phenytoin effects—alcohol (chronic use), antacids, barbiturates, carBAMazepine, rifampin, calcium (high dose), folic acid

Drug/Food

Enteral tube feeding: may decrease absorption of oral product; do not use enteral feedings 2 hr before or 2 hr after dose

Drug/Lab Test

Increase: glucose, alk phos, GGT
Decrease: dexamethasone, metyrapone test serum, PBI, urinary steroids

NURSING CONSIDERATIONS

Assess:

⚠ **Phenytoin hypersensitivity syndrome** 3-12 wk after start of treatment: rash, temperature, lymphadenopathy; may cause hepatotoxicity, renal failure, rhabdomyolysis

⚠ **Serious skin disorders:** for beginning rash that may lead to Stevens-Johnson syndrome or toxic epidermal necrolysis; phenytoin should not be used again; may occur more often among Asian patients with HLA-B 1502

⚠ **Purple glove syndrome** with IV use

- Phenytoin level: toxic level 30-50 mcg/ml, therapeutic level: 7.5-20 mcg/ml; wait ≥1 wk to draw levels
- **Seizures:** duration, type, intensity, precipitating factors; obtain EEG periodically; monitor therapeutic level
- Blood studies: CBC, platelets q2wk until stabilized, then monthly × 12, then q3mo; discontinue product if neutrophils <1600/mm³; renal function: albumin conc; folic acid levels, LFTs

⚠ Mental status: mood, sensorium, affect, memory (long, short), suicidal thoughts/behaviors

- Monitor EKG, B/P, respiratory function during IV loading dose; verify potency of IV access port before IV inf
- Monitor EEG function and serum levels periodically
- **Blood dyscrasias:** fever, sore throat, bruising, rash, jaundice

Evaluate:

- Therapeutic response; decrease in severity of seizures, ventricular dysrhythmias

Teach patient/family:

- That, if diabetic, blood glucose should be monitored
- That urine may turn pink
- Not to discontinue product abruptly because seizures may occur
- **Oral hygiene:** about the proper brushing of teeth using a soft toothbrush, flossing to prevent gingival hyperplasia; about the need to see dentist frequently
- To avoid hazardous activities until stabilized on product
- To carry emergency ID stating product use
- That heavy use of alcohol may diminish effect of product; to avoid OTC medications
- Not to change brands or forms once stabilized on therapy because brands may vary
- Not to use antacids within 2 hr of product
- To use nonhormonal contraception; to notify prescriber if pregnancy is planned or suspected, pregnancy (D)
- To notify prescriber of unusual bleeding, bruising, petechiae (bleeding), clay-colored stools, abdominal pain, dark urine, yellowing of skin or eyes (hepatotoxicity); slurred speech, headache, drowsiness
- To report suicidal thoughts/behaviors immediately

phosphate/biphosphate (OTC)

Fleet Enema, Phospho-Soda

Func. class.: Laxative, saline

ACTION: Increases water absorption in the small intestine by osmotic action; laxative effect occurs by increased peristalsis and water retention

USES: Constipation, bowel or rectal preparation for surgery, exam

CONTRAINDICATIONS: Hypersensitivity, rectal fissures, abdominal pain, nausea, vomiting, appendicitis, acute surgical abdomen, ulcerated hemorrhoids, sodium-restricted diet, renal failure, hyperphosphatemia, hypocalcemia, hypokalemia, hypernatremia, Addison's disease, CHF, ascites, bowel perforation, megacolon, imperforate anus

Black Box Warning: GI obstruction, renal failure

Precautions: Pregnancy (C)

Black Box Warning: Colitis, geriatric hypovolemia, renal disease

DOSAGE AND ROUTES

- **Adult: PO** 20-30 ml (Phospho-Soda)
- **Child: PO** 5-15 ml (Phospho-Soda)
- **Adult and child >12 yr: RECT** 1 enema (118 ml)
- **Child 2-12 yr: RECT** ½ enema (59 ml)

Available forms: Enema 7 g phosphate/19 g biphosphate/118 ml; oral sol 18 g phosphate/48 g biphosphate/100 ml

Administer:

- Alone for better absorption; do not take within 1-2 hr of other products

SIDE EFFECTS

CV: **Dysrhythmias, cardiac arrest,** hypotension, **widening QRS complex**
GI: *Nausea, cramps,* diarrhea
META: Electrolyte, fluid imbalances

PHARMACOKINETICS

Onset 30 min-3 hr, excreted in feces

NURSING CONSIDERATIONS

Assess:

- **Stools:** color, amount, consistency; bowel pattern, bowel sounds, flatulence, distention, fever, dietary patterns, exercise; cramping, rectal bleeding, nausea, vomiting; if these occur, product should be discontinued
- Blood, urine electrolytes if product used often

Evaluate:

- Therapeutic response: decrease in constipation

Teach patient/family:

- Not to use laxatives for long-term therapy because bowel tone will be lost
- That normal bowel movements do not always occur daily
- Not to use in presence of abdominal pain, nausea, vomiting
- To notify prescriber if constipation unrelieved or if symptoms of electrolyte imbalance occur: muscle cramps, pain, weakness, dizziness, excessive thirst
- To maintain fluid consumption

physostigmine ophthalmic

See Appendix B

phytonadione (Rx)

(fye-toe-na-dye′one)

Mephyton, Vit K

Func. class.: Vit K_1, fat-soluble vitamin

ACTION: Needed for adequate blood clotting (factors II, VII, IX, X)

USES: Vit K malabsorption, hypoprothrombinemia, prevention of hypoprothrombinemia caused by oral anticoagulants, prevention of hemorrhagic disease of the newborn

CONTRAINDICATIONS: Hypersensitivity, severe hepatic disease, last few weeks of pregnancy

Precautions: Pregnancy (C), neonates, hepatic disease

Black Box Warning: IV use

DOSAGE AND ROUTES

Hypoprothrombinemia caused by vit K malabsorption

- **Adult: PO/IM** 2.5-25 mg, may repeat or increase to 50 mg
- **Child: PO** 2.5-5 mg
- **Infant: PO/IM** 2 mg

Prevention of hemorrhagic disease of the newborn

- **Neonate: IM** 0.5-1 mg within 1 hr after birth, repeat after 2-3 wk if required

Hypoprothrombinemia caused by oral anticoagulants

- **Adult and child: PO/SUBCUT/IM** 1-10 mg, may repeat 12-48 hr after **PO** dose or 6-8 hr after **SUBCUT/IM** dose based on INR

Available forms: Tabs 5 mg; inj 10 mg/ml, 1 mg/0.5 ml

Administer:

- Store in tight, light-resistant container (PO)

P

Intermittent IV INF route

- After diluting with ≥10 ml D_5NS; give max 1 mg/min

⚠ IV only when other routes not possible (deaths have occurred)

Y-site compatibilities: Alfentanil, amikacin, aminophylline, ascorbic acid, atracurium, atropine, azaTHIOprine, aztreonam, bumetanide, buprenorphine, butorphanol, calcium chloride/gluconate, ceFAZolin, cefonicid, cefoperazone, cefotaxime, cefoTEtan, cefOXitin, cefTAZidime, ceftizoxime, cefTRIAXone, cefuroxime, chloramphenicol, chlorproMAZINE, cimetidine, clindamycin, cyanocobalamin, cycloSPORINE, dexamethasone, digoxin, diphenhydrAMINE, DOPamine, doxycycline, enalaprilat, ePHEDrine, EPINEPHrine, epoetin alfa, erythromycin, esmolol, famotidine, fentaNYL, fluconazole, folic acid, furosemide, ganciclovir, gentamicin, glycopyrrolate, heparin, hydrocortisone, imipenem/cilastatin, indomethacin, insulin, isoproterenol, ketorolac, labetalol, lidocaine, mannitol, meperidine, metaraminol, methoxamine, methyldopate, metoclopramide, metoprolol, metroNIDAZOLE, midazolam, morphine, multivitamins, nafcillin, nalbuphine, naloxone, nitroglycerin, nitroprusside, norepinephrine, ondansetron, oxacillin, oxytocin, papaverine, penicillin G potassium, pentamidine, pentazocine, PENTobarbital, PHENobarbital, phentolamine, phenylephrine, potassium chloride, procainamide, prochlorperazine, propranolol, pyridoxime, ranitidine, sodium bicarbonate, succinylcholine, SUFentanil, theophylline, thiamine, ticarcillin/clavulanate, tobramycin, tolazoline, trimetaphan, urokinase, vancomycin, vasopressin, verapamil, vitamin B with C

SIDE EFFECTS

CNS: Headache, brain damage (large doses)

GI: Nausea, decreased LFTs

HEMA: Hemolytic anemia, hemoglobinuria, hyperbilirubinemia

INTEG: Rash, urticaria

RESP: Bronchospasm, dyspnea, feeling of chest constriction, respiratory arrest

PHARMACOKINETICS

PO/INJ: Metabolized, crosses placenta

INTERACTIONS

Decrease: action of phytonadione—bile acid sequestrants, sucralfate, antiinfectives, salicylates, mineral oil

Decrease: action of warfarin—large dose of product

NURSING CONSIDERATIONS

Assess:

- **Bleeding:** emesis, stools, urine; pressure on all venipuncture sites; avoid all inj if possible
- PT during treatment (2-sec deviation from control time, bleeding time, clotting time); monitor for bleeding, pulse, and B/P

Evaluate:

- Therapeutic response: prevention of hemorrhagic disease of the newborn, resolution of hypoprothrombinemia

Teach patient/family:

- Not to take other supplements, OTC products, prescription products unless directed by prescriber
- About the necessary foods for associated diet
- To avoid IM inj; to use soft toothbrush; not to floss; to use electric razor until coagulation defect corrected
- To report symptoms of bleeding
- About the importance of frequent lab tests to monitor coagulation factors
- To notify all health care providers of use of this product
- To carry emergency ID describing condition and products used

pilocarpine ophthalmic

See Appendix B

pimecrolimus topical
See Appendix B

pioglitazone (Rx)
(pie-oh-glye'ta-zone)
Actos
Func. class.: Antidiabetic, oral
Chem. class.: Thiazolidinedione

ACTION: Specifically targets insulin resistance; an insulin sensitizer; regulates the transcription of a number of insulin-responsive genes

USES: Type 2 diabetes mellitus

CONTRAINDICATIONS: Breastfeeding, children, hypersensitivity to thiazolidinedione, diabetic ketoacidosis

Black Box Warning: NYHA Class III/IV heart failure

Precautions: Pregnancy (C), geriatric patients, geriatric patients with CV disease, renal/hepatic/thyroid disease, edema, polycystic ovary syndrome, bladder cancer, osteoporosis, pulmonary disease, secondary malignancy

DOSAGE AND ROUTES

Monotherapy

- **Adult: PO** 15 or 30 mg/day, may increase to 45 mg/day; with strong CYP2C8, max 15 mg/day; with NYHA class I/II heart failure, max 15 mg/day

Combination therapy

- **Adult: PO** 15 or 30 mg/day with a sulfonylurea, metFORMIN, or insulin; decrease sulfonylurea dose if hypoglycemia occurs; decrease insulin dose by 10%-25% if hypoglycemia occurs or if plasma glucose is <100 mg/dl, max 45 mg/day

Hepatic dose

- Do not use in active hepatic disease or if ALT >2.5 times ULN

Available forms: Tabs 15, 30, 45 mg

Administer:

- Once a day; without regard to meals
- Tabs crushed and mixed with food or fluids for patients with difficulty swallowing

SIDE EFFECTS

CNS: *Headache*

CV: **MI, heart failure, death (geriatric patients)**

ENDO: Hypo/hyperglycemia

MISC: *Sinusitis, upper respiratory tract infection, pharyngitis,* **hepatotoxicity**, edema, weight gain, anemia, macular edema; **risk of bladder cancer (use >1 yr), peripheral/pulmonary edema**

MS: **Rhabdomyolysis**, fractures (females), myalgia

PHARMACOKINETICS

Maximal reduction in FBS after 12 wk; half-life 3-7 hr, terminal 16-24 hr; protein binding >99%

INTERACTIONS

Decrease: effect of atorvastatin

Decrease: effect of oral contraceptives; use alternative contraceptive method

Decrease: pioglitazone effect—CYP2C8 inducers (ketoconazole, fluconazole, itraconazole, miconazole, voriconazole)

Drug/Herb

Increase: hypoglycemia—garlic, green tea, horse chestnut

Drug/Lab Test

Increase: CPK, LFTs, HDL, cholesterol

Decrease: glucose, Hct/Hgb

NURSING CONSIDERATIONS

Assess:

Black Box Warning: Heart failure: do not use in NYHA Class III/IV; excessive/rapid weight gain >5 lb, dyspnea, edema; may need to be reduced or discontinued

- **Bladder cancer:** avoid use in a history of bladder cancer; use of pioglitazone >1 yr has shown an increase in bladder cancer
- **Hypoglycemic reactions:** sweating, weakness, dizziness, anxiety, tremors,

P

hunger; hyperglycemic reactions soon after meals (rare); may occur more often with insulin or other antidiabetics

• **Hepatic disease:** check LFTs periodically: AST, LDH; do not start treatment in active heart disease or if ALT >2.5× upper limit of normal; if treatment has already begun, follow closely with continuing ALT levels; if ALT increases to >3× upper limit of normal, recheck ALT as soon as possible; if ALT remains >3× upper limit of normal, discontinue
• FBS, glycosylated HbA1c, plasma lipids/lipoproteins, B/P, body weight during treatment
• CBC with differential before and during therapy; more necessary in those with anemia, Hct/Hgb (may be decreased in first few months of treatment)

Evaluate:
• Therapeutic response: decrease in polyuria, polydipsia, polyphagia; clear sensorium; absence of dizziness; stable gait; blood glucose A1c improvement

Teach patient/family:
• To self-monitor using a blood glucose meter
• About the symptoms of hypo/hyperglycemia; what to do about each
• That product must be continued on daily basis; about the consequences of discontinuing product abruptly
• To avoid OTC medications or herbal preparations unless approved by prescriber
• That diabetes is a lifelong illness; that product is not a cure, it only controls symptoms
• To notify prescriber if oral contraceptives are used, effect may be decreased; not to use product if breastfeeding
• To report symptoms of hepatic dysfunction: nausea, vomiting, abdominal pain, fatigue, anorexia, dark urine, jaundice
• To report weight gain, edema
• That lab work, eye exams will be needed periodically

piperacillin/tazobactam (Rx)

(pip′er-ah-sill′in/ta-zoe-bak′tam)

Zosyn

Func. class.: Antiinfective, broad spectrum

Chem. class.: Extended-spectrum penicillin, β-lactamase inhibitor

ACTION: Interferes with cell-wall replication of susceptible organisms; tazobactam is a β-lactamase inhibitor that protects piperacillin from enzymatic degradation

USES: Moderate to severe infections: piperacillin-resistant, β-lactamase–producing strains causing infections in respiratory, skin, urinary tract, bone, gonorrhea, pneumonia; effective for resistant *Staphylococcus aureus,* resistant *Escherichia coli, Bacteroides fragilis, Bacteroides ovatus, Bacteroides thetaiotaomicron, Bacteroides vulgatus, Haemophilus influenzae*

Unlabeled uses: Endocarditis

CONTRAINDICATIONS: Hypersensitivity to penicillins; neonates; carbapenem allergy

Precautions: Pregnancy (B), breastfeeding, renal insufficiency in neonates, hypersensitivity to cephalosporins, seizures, GI disease, electrolyte imbalances, biliary obstruction

DOSAGE AND ROUTES

Nosocomial pneumonia
• **Adult: IV** 4.5 g q6hr or 3.375 g q4hr with an aminoglycoside or antipseudomonal fluoroquinolone × 1-2 wk

Appendicitis/peritonitis
• **Adults/adolescents/child ≥40 kg (88 lb): IV** 3.375 g q6hr × 7-10 days
• **Child ≥9 mo and <40 kg: IV** 100 mg/kg (piperacillin)/kg q8hr × 7-10 days
• **Infant 2 mo to <9 mo: IV** 80 mg (piperacillin) q8hr × 7-10 days

Renal dose

• **Adult: IV** CCr 20-40 ml/min, give 3.375 g q6hr (nosocomial pneumonia); give 2.25 g q6hr (all other indications); CCr <20 ml/min, give 2.25 g q6hr (nosocomial pneumonia); give 2.25 g q8hr (all other indications)

Available forms: Powder for inj 2 g piperacillin/0.25 g tazobactam, 3 g piperacillin/0.375 g tazobactam, 4 g piperacillin/0.5 g tazobactam, 36 g piperacillin/4.5 g tazobactam

Administer:

• Separate aminoglycoside from piperacillin to avoid inactivation

• Product after C&S is complete

Intermittent IV INF route

• Reconstitute each 1 g of product/5 ml 0.9% NaCl for inj or sterile water for inj, dextrose 5%; shake well; further dilute in ≥50 ml compatible IV sol, run as int inf over ≥30 min

ADD-Vantage IV solution: Reconstitution:

• Reconstitute with 0.9% Sodium Chloride or D5W in the appropriate flexible diluent container provided; for 500 mg vials, use at least a 100 ml diluent container and for 750 mg and 1 g vials, use only the 250 ml diluent container

• Remove the protective covers from the top of the vial and vial port; remove vial cap (do not access with a syringe) and vial port cover; screw the vial into the vial port until it will go no further to assure a seal; once vial is sealed to the port, do not remove; to activate the contents of the vial, squeeze the bottom of the diluent container gently to inflate the portion of the container surrounding the end of the drug vial; with the other hand, push the drug vial down into the container telescoping walls of the container and grasp the inner cap of the vial through the walls of the container; pull the inner cap from the drug vial; verify the rubber stopper has been pulled out, allowing the drug and diluent to mix; mix the container contents thoroughly

• *Storage after reconstitution:* the admixture solution may be stored for up to 24 hr at room temperature; do not refrigerate or freeze after reconstitution

• Do not use in series connections with flexible containers

Pre-mixed Galaxy IV solution:

• Thaw frozen containers at room temperature (20-25°C or 68-77°F) or under refrigeration (2-8°C or 36-46°F); do not force thaw by immersion in water baths or by microwaving; check for leaks by squeezing bag firmly

• Do not admix

• Contents of the solution may precipitate in the frozen state and should dissolve with little or no agitation once the solution has reached room temperature

• ***Storage:*** the thawed solution is stable for 24 hr at room temperature or for 14 days under refrigeration; do not refreeze thawed product

• Do not use plastic containers in series connections as this could result in an embolism due to residual air being drawn from the primary container before administration of the fluid from the secondary container is complete

IV infusion:

• Infuse IV over at least 30 min; ambulatory intravenous infusion pumps can be used; the solution is stable for up to 12 hr at room temperature

Y-site compatibilities: Alfentanil, allopurinol, amifostine, amikacin, aminocaproic acid, aminophylline, amphotericin B lipid complex, amphotericin B liposome, anidulafungin, argatroban, atenolol, aztreonam, bivalirudin, bleomycin, bumetanide, buprenorphine, busulfan, butorphanol, calcium acetate/chloride/gluconate, CARBOplatin, carmustine, cefepime, chloramphenicol, cimetidine, clindamycin, cyclophosphamide, cycloSPORINE, cytarabine, DACTINomycin, DAPTOmycin, dexamethasone, dexrazoxane, diazepam, digoxin, diphenhydrAMINE, DOCEtaxel, DOPamine, doxacurium, enalaprilat, ePHEDrine, EPINEPHrine, eptifibatide, erythromycin,

P

esmolol, etoposide, fenoldopam, fentaNYL, floxuridine, fluconazole, fludarabine, fluorouracil, foscarnet, fosphenytoin, furosemide, gallium, granisetron, heparin, hydrocortisone, HYDROmorphone, ifosfamide, isoproterenol, ketorolac, lansoprazole, lepirudin, leucovorin, lidocaine, linezolid, LORazepam, magnesium sulfate, mannitol, mechlorethamine, melphalan, meperidine, mesna, metaraminol, methotrexate, methylPREDNISolone, metoclopramide, metoprolol, metroNIDAZOLE, milrinone, morphine, naloxone, nitroglycerin, nitroprusside, norepinephrine, octreotide, ondansetron, oxytocin, PACLitaxel, palonosetron, pamidronate, pancuronium, PEMEtrexed, PENTobarbital, PHENObarbital, phentolamine, phenylephrine, plicamycin, potassium chloride/phosphates, procainamide, ranitidine, remifentanil, riTUXimab, sargramostim, sodium acetate/bicarbonate/phosphates, uccinylcholine, SUFentanil, sulfamethoxazole-trimethoprim, tacrolimus, telavancin, teniposide, theophylline, thiotepa, tigecycline, tirofiban, trimethobenzamide, vasopressin, vinBLAStine, vinCRIStine, voriconazole, zidovudine, zoledronic acid

SIDE EFFECTS

CNS: Lethargy, hallucinations, anxiety, depression, twitching, insomnia, headache, fever, dizziness, seizures, vertigo
CV: Cardiac toxicity, edema
GI: *Nausea, vomiting, diarrhea;* increased AST, ALT; abdominal pain, glossitis, pseudomembranous colitis, constipation, pancreatitis
GU: Oliguria, proteinuria, hematuria, *vaginitis, moniliasis,* glomerulonephritis, renal failure
HEMA: Anemia, increased bleeding time, bone marrow depression, agranulocytosis, hemolytic anemia
INTEG: Rash, pruritus, exfoliative dermatitis
META: Hypokalemia, hypernatremia
SYST: Serum sickness, anaphylaxis, Stevens-Johnson syndrome

PHARMACOKINETICS

Half-life 0.7-1.2 hr; excreted in urine, bile, breast milk; crosses placenta; 33% bound to plasma proteins
IV: Peak completion of IV

INTERACTIONS

Increase: effect of neuromuscular blockers, oral anticoagulants, methotrexate
Increase: piperacillin concs—aspirin, probenecid
Decrease: antimicrobial effect of piperacillin—tetracyclines, aminoglycosides IV

Drug/Lab Test
Increase: eosinophilia, neutropenia, leukopenia, serum creatinine, PTT, AST, ALT, alk phos, bilirubin, BUN, electrolytes
Decrease: Hct, Hgb, electrolytes
False positive: urine glucose, urine protein, Coombs' test

NURSING CONSIDERATIONS

Assess:

- **Infection:** temperature, stools, urine, sputum, wounds
- I&O ratio; report hematuria, oliguria because penicillin in high doses is nephrotoxic; maintain hydration unless contraindicated
- Hepatic studies: AST, ALT before treatment and periodically thereafter
- Blood studies: WBC, RBC, Hct, Hgb, bleeding time before treatment and periodically thereafter; serum potassium
- Renal studies: urinalysis, protein, blood, BUN, creatinine before treatment and periodically thereafter
- C&S before product therapy; product may be given as soon as culture is taken
- **Pseudomembranous colitis:** diarrhea, bloody stools, fever, abdominal cramps; may occur ≤2 mo after treatment; bowel pattern before and during treatment

- Skin eruptions after administration of penicillin to 1 wk after discontinuing product
- Respiratory status: rate, character, wheezing, tightness in chest
- **Anaphylaxis:** wheezing, laryngeal edema, rash, itching; discontinue product, have emergency equipment nearby
- Adequate intake of fluids (2 L) during diarrhea episodes

Evaluate:

- Therapeutic response: absence of fever, purulent drainage, redness, inflammation; culture shows decreased organisms

Teach patient/family:

- That culture may be taken after completed course of medication
- To report sore throat, fever, fatigue (superinfection); CNS effects (anxiety, depression, hallucinations, seizures); **pseudomembranous colitis:** fever, diarrhea with blood, pus, mucous
- To wear or carry emergency ID if allergic to penicillins
- To notify nurse of diarrhea

TREATMENT OF OVERDOSE: Withdraw product, maintain airway, administer EPINEPHrine, aminophylline, O_2, IV corticosteroids for anaphylaxis

pirbuterol (Rx)

(peer-byoo′ter-ole)

Maxair

Func. class.: Bronchodilator

Chem. class.: β-Adrenergic agonist

ACTION: Causes bronchodilation with little effect on heart rate by action on β-receptors, causing increased cAMP and relaxation of smooth muscle

USES: Reversible bronchospasm (prevention, treatment) including asthma; may be given with theophylline or steroids

CONTRAINDICATIONS: Hypersensitivity to sympathomimetics, tachycardia

Precautions: Pregnancy (C), breastfeeding, cardiac disorders, hyperthyroidism, hypertension, diabetes mellitus, prostatic hypertrophy

DOSAGE AND ROUTES

- **Adult and child >12 yr:** INH 1-2 puffs (0.4 mg) q4-6hr; max 12 INH/day

Available forms: Aerosol delivery 0.2 mg pirbuterol/actuation

Administer:

- Store in light-resistant container; do not expose to temperatures over 86° F (30° C)
- After shaking; exhale; place mouthpiece in mouth; inhale slowly; hold breath; remove; exhale slowly
- Gum, sips of water for dry mouth

SIDE EFFECTS

CNS: Tremors, anxiety, insomnia, headache, dizziness, stimulation, restlessness, hallucinations, drowsiness, irritability

CV: Palpitations, tachycardia, hypertension, angina, hypotension, dysrhythmias

EENT: Dry nose and mouth, irritation of nose, throat

GI: Gastritis, nausea, vomiting, anorexia

MS: Muscle cramps

RESP: **Paradoxical bronchospasm**, dyspnea, coughing

PHARMACOKINETICS

INH: Onset 3 min, peak ½-1 hr, duration 5 hr, terminal half-life 2 hr

INTERACTIONS

⚠ **Hypertensive crisis: MAOIs**

Increase: action of other aerosol bronchodilators

Increase: pirbuterol action—tricyclics, antihistamines, levothyroxine

Decrease: pirbuterol action—β-blockers

Drug/Herb

Increase: action of both—cola nut, guarana, yerba maté

P

Increase: effect—green tea (large amounts), guarana

NURSING CONSIDERATIONS

Assess:

• Respiratory function: vital capacity, forced expiratory volume, ABGs, B/P, lung sounds, pulse, characteristics of sputum

⚠ **Paradoxical bronchospasm, that can occur rapidly, hold product, notify prescriber**

• Fluid intake >2 L/day to liquefy thick secretions

Evaluate:

• Therapeutic response: absence of dyspnea, wheezing over 1 hr

Teach patient/family:

• Not to use OTC medications; extra stimulation may occur

• Use of inhaler; review package insert with patient

• To avoid getting aerosol in eyes

• Actuator is for Maxair autoinhaler; do not use with other inhaler canister

• About all aspects of product; avoid smoking, smoke-filled rooms, persons with respiratory infections

• To keep fluid intake >2 L/day to liquefy thick secretions

TREATMENT OF OVERDOSE:

Administer a β-adrenergic blocker

pitavastatin (Rx)

(pit'a-va-stat'-in)

Livalo

Func. class.: Antilipidemic

Chem. class.: HMG-CoA reductase inhibitor

Do not confuse:

pitavastatin/pravastatin

ACTION: Inhibits HMG-CoA reductase enzyme, which reduces cholesterol synthesis; high doses lead to plaque regression

USES: As an adjunct for primary hypercholesterolemia (types Ia, Ib), dysbetalipoproteinemia, elevated triglyceride levels, prevention of CV disease by reduction of heart risk in those with mildly elevated cholesterol

Unlabeled Uses: Atherosclerosis

CONTRAINDICATIONS:

Pregnancy (X), breastfeeding, hypersensitivity, active hepatic disease, cholestasis

Precautions: Past hepatic disease, alcoholism, severe acute infections, trauma, severe metabolic disorders, electrolyte imbalance, seizures, surgery, organ transplant, endocrine disease, females, hypotension, renal disease

DOSAGE AND ROUTES

• **Adult: PO** 2 mg/day, usual range 1-4, max 4 mg/day

Renal dose

• **Adult: PO** CCr 30-<60 ml/min, 1 mg daily, max 2 mg daily; CCr <30 ml/min on hemodialysis, 1 mg daily, max 2 mg daily; CCr <30 ml/min, not recommended

Atherosclerosis (unlabeled)

• **Adult: PO** 4 mg/day

Available forms: Tabs 1, 2, 4 mg

Administer:

• Total daily dose any time of day without regard to meals

• Store in cool environment in tight container protected from light

SIDE EFFECTS

CNS: Headache

GI: Constipation, diarrhea

INTEG: Rash, pruritus, alopecia

MS: Arthralgia, myalgia, **rhabdomyolysis**

RESP: Pharyngitis

PHARMACOKINETICS

Peak 1 hr; metabolized in liver, excreted in urine, feces; half-life 12 hr; protein binding 99%; concs lower in healthy African Americans

INTERACTIONS

Increase: risk for possible rhabdomyolysis—azole antifungals, cycloSPORINE, erythromycin, niacin, gemfibrozil, clofibrate

Increase: levels of pitavastatin—erythromycin

Drug/Herb

Increase: pitavastatin—red yeast rice

Drug/Lab Test

Increase: bilirubin, alk phos, ALT, AST

Interference: thyroid function tests

NURSING CONSIDERATIONS

Assess:

- **Diet**; obtain diet history including fat, cholesterol in diet
- Cholesterol, triglyceride levels periodically during treatment; check lipid panel 6 wk after changing dose
- Hepatic studies at 12 wk after starting treatment, then q6mo; if AST >3× normal, reduce or discontinue; AST, ALT, LFTs may be increased
- Renal studies in patients with compromised renal system: BUN, I&O ratio, creatinine

⚠ **Rhabdomyolysis:** muscle pain, tenderness; obtain CPK baseline; if markedly increased, product may need to be discontinued

Evaluate:

- Therapeutic response: decrease in cholesterol to desired level after 6 wk

Teach patient/family:

- That blood work will be necessary during treatment
- To report blurred vision, severe GI symptoms, headache, muscle pain, weakness, tenderness
- That previously prescribed regimen will continue: low-cholesterol diet, exercise program, smoking cessation
- Not to take product if pregnant (X) or if pregnancy is planned or suspected (notify prescriber); to avoid breastfeeding

plasma protein fraction (Rx)

Plasmanate

Func. class.: Hematological agent

Chem. class.: Plasma volume expander

ACTION: Exerts similar oncotic pressure as human plasma, expands blood volume

USES: Hypovolemic shock, hypoproteinemia, ARDS, preoperative cardiopulmonary bypass, acute hepatic failure, nephrotic syndrome, cardiogenic shock

CONTRAINDICATIONS: Hypersensitivity to this product or albumin; CHF, severe anemia, renal insufficiency, hyponatremia, cardiopulmonary bypass

Precautions: Pregnancy (C), decreased salt intake, decreased cardiac reserve, lack of albumin deficiency, hepatic disease

DOSAGE AND ROUTES

Hypovolemia

- **Adult: IV INF** 250-500 ml (12.5-25 g protein), max 10 ml/min
- **Child: IV INF** 10-30 ml/kg at max 5-10 ml/min

Hypoproteinemia

- **Adult: IV INF** 1000-1500 ml/day, max 8 ml/min

Available forms: Inj 5%

Administer:

IV route

- IV access at distant site from infection or trauma; no dilution required; use inf pump, use large-gauge needle (≥20-G); discard unused portion; infuse slowly to prevent hypotension
- Within 4 hr of opening, discard partially used vials
- Do not use sol that has been frozen
- Adjust rate to changes in B/P
- Store at room temperature, max 86° F

Additive compatibilities: Carbohydrate and electrolyte sol, whole blood, packed RBCs, chloramphenicol, tetracycline

SIDE EFFECTS

CNS: Fever, chills, headache, paresthesias, flushing
CV: Fluid overload, hypotension, erratic pulse
GI: Nausea, vomiting, increased salivation
INTEG: Rash, urticaria, cyanosis
RESP: Altered respirations, dyspnea, PE

PHARMACOKINETICS

Metabolized as a protein/energy source

INTERACTIONS

Drug/Lab Test
False increase: alk phos

NURSING CONSIDERATIONS

Assess:
- Blood studies: Hct, Hgb, electrolytes, serum protein; if serum protein declines, dyspnea, hypoxemia can result
- B/P (decreased), pulse (erratic), respiration during inf
- I&O ratio; urinary output may decrease
- Allergy: fever, rash, itching, chills, flushing, urticaria, nausea, vomiting, hypotension requires discontinuation of inf; use new lot if therapy reinstituted; premedicate with diphenhydrAMINE
- ⚠ Increased CVP reading: distended neck veins indicate circulatory overload; SOB, anxiety, insomnia, expiratory crackles, frothy blood-tinged cough, cyanosis indicate pulmonary overload
- Adequate hydration before administration

Evaluate:
- Therapeutic response: increased B/P, decreased edema, increased serum albumin

plerixafor (Rx)

(pler-ix′a-fore)

Mozobil

Func. class.: Biologic modifier

Chem. class.: Colony-stimulating factor

ACTION: Competitively inhibits the binding of stromal-derived factors, thereby allowing hematopoietic stem cells to mobilize into peripheral blood

USES: For peripheral blood stem cell (PBSC) mobilization for collection and autologous transplant in patients with non-Hodgkin's lymphoma, multiple myeloma; used with a granulocyte colony-stimulating factor (G-CSF)

CONTRAINDICATIONS: Hypersensitivity, breastfeeding, pregnancy (D)
Precautions: Children, renal disease, thrombocytopenia

DOSAGE AND ROUTES

- **Adult: SUBCUT** 0.24 mg/kg daily about 11 hr before initiation of apheresis, give for up to 4 consecutive days; give filgrastim 10 mcg/kg; **SUBCUT** daily each AM beginning 4 days before the 1st evening dose of plerixafor and on each day of apheresis; give filgrastim before procedure

Available forms: Inj 300 mcg/ml, 480 mcg/1.6 ml, 480 mcg/0.8 ml, 3000 mcg/0.5 ml

Administer:

SUBCUT route
- Each single-use vial contains 24 mg of plerixafor (1.2 ml of 20 mg/ml sol); volume calculated by multiplying 0.012 by actual body weight (kg)
- Give 11 hr before apheresis
- Max 40 mg/day or 27 mg/day in patients with renal disease
- Store at room temperature

SIDE EFFECTS

CNS: Syncope, dizziness, fatigue, headache, insomnia, malaise, paresthesias
GI: *Nausea,* vomiting, diarrhea, abdominal pain, constipation
HEMA: **Thrombocytopenia,** leukocytosis
INTEG: Rash, skin irritation, pruritus, inj site reaction, erythema, urticaria
MS: Musculoskeletal pain
RESP: Dyspnea, hypoxia

PHARMACOKINETICS

SUBCUT: 30-60 min, peak mobilization 6-9 hr, 58% protein binding, 70% excreted via kidneys (parent drug), terminal half-life 3-5 hr

INTERACTIONS

Increase: adverse reactions—do not use this product concomitantly with lithium, may increase leukocytosis

NURSING CONSIDERATIONS

Assess:
- Blood studies: CBC, differential
- B/P, respirations, pulse before and during therapy
- Bone pain; give mild analgesics

Evaluate:
- Therapeutic response: collection of stem cells

Teach patient/family:
- About reason for use and expected results
- **Pregnancy (D): notify prescriber if pregnancy is planned or suspected; do not breastfeed**

posaconazole (Rx)

(poe′sa-kon′a-zole)

Noxafil, Posanol ✱

Func. class.: Antifungal—systemic
Chem. class.: Triazole derivative

ACTION: Inhibits a portion of cell-wall synthesis; alters cell membranes and inhibits several fungal enzymes

USES: Prevention of aspergillus, candida infection, oropharyngeal candidiasis in immunocompromised patients, chemotherapy-induced neutropenia, mucocutaneous candidiasis
Unlabeled uses: Aspergillosis, cellulitis, coccidioidomycosis, endocarditis, endophthalmitis, esophageal candidiasis, febrile neutropenia, fungal keratitis, fusariosis, histoplasmosis, infectious arthritis, myocarditis, osteomyelitis, pericarditis, sinusitis, tracheobronchitis

CONTRAINDICATIONS: Hypersensitivity to this product or other systemic antifungals or azoles; fungal meningitis, onchomycosis or dermatomycosis in cardiac dysfunction; use with ergots, sirolimus, CYP3A4 substrates
Precautions: Pregnancy (C), breastfeeding, children, cardiac/hepatic/renal disease

DOSAGE AND ROUTES

- **Adult, adolescent: PO** 800 mg/day in 2-4 divided doses

Thrush
- **Adult: PO** 100 mg bid × 1 day, then 100 mg/day × 13 days

Available forms: Oral susp 200 mg/5 ml

Administer:

PO route
- **Oral susp:** shake well; use calibrated measuring device; take only with full meal or liquid nutritional supplements such as Ensure; rinse measuring device after each use
- Store in tight container in refrigerator; do not freeze

SIDE EFFECTS

CNS: *Headache, dizziness,* insomnia, fever, rigors, weakness, anxiety
CV: Hypo/hypertension, tachycardia, anemia, **QT prolongation, torsades de pointes**
GI: *Nausea, vomiting, anorexia, diarrhea,* cramps, abdominal pain, flatulence, **GI bleeding, hepatotoxicity**

GU: Gynecomastia, impotence, decreased libido

INTEG: *Pruritus,* fever, *rash,* toxic epidermal necrolysis

MISC: *Edema, fatigue,* malaise, hypokalemia, tinnitus, rhabdomyolysis, hypokalemia

PHARMACOKINETICS

Well absorbed, enhanced by food, protein binding 98%-99%, peak 3-5 hr, half-life 35 hr, metabolized in liver, excreted in feces (77% unchanged)

INTERACTIONS

- Do not use with lovastatin, atorvastatin

⚠ **Increase:** QT prolongation—class IA/III antidysrhythmics, some phenothiazines, β agonists, local anesthetics, tricyclics, haloperidol, chloroquine, droperidol, pentamidine; CYP3A4 inhibitors (amiodarone, clarithromycin, erythromycin, telithromycin, troleandomycin), arsenic trioxide, levomethadyl; CYP3A4 substrates (methadone, pimozide, QUEtiapine, quiNIDine, risperiDONE, ziprasidone)

Increase: tinnitus, hearing loss—quiNIDine

Increase: hepatotoxicity—other hepatotoxic products

Increase: severe hypoglycemia—oral hypoglycemics

Increase: sedation—triazolam, oral midazolam

Increase: levels, toxicity—busPIRone, busulfan, calcium-channel blockers, clarithromycin, cycloSPORINE, diazepam, digoxin, felodipine, HMG-CoA reductase inhibitors, indinavir, isradipine, midazolam, niCARdipine, NIFEdipine, niMODipine, phenytoin, quiNIDine, ritonavir, saquinavir, sirolimus, tacrolimus, vinca alkaloids, warfarin

Decrease: posaconazole level—cimetidine, phenytoin

Decrease: posaconazole action—antacids, H_2-receptor antagonists, rifamycin, didanosine

Drug/Food

- Food increases absorption

NURSING CONSIDERATIONS

Assess:

⚠ **Infection:** type of, may begin treatment before obtaining results; temperature, WBC, sputum at baseline and periodically, break-through infections may occur when used with fosamprenavir

- I&O ratio, electrolytes; correct electrolyte imbalances before starting treatment
- For allergic reaction: rash, photosensitivity, urticaria, dermatitis

⚠ **Rhabdomyolysis:** muscle pain, increased CPK; weakness, swelling of affected muscles; if these occur and if confirmed by CPK, product should be discontinued

⚠ **Hepatotoxicity:** nausea, vomiting, jaundice, clay-colored stools, fatigue; hepatic studies (ALT, AST, bilirubin) if patient receiving long-term therapy

⚠ **QT prolongation:** ECG for QT prolongation, ejection fraction; assess for chest pain, palpitations, dyspnea

Evaluate:

- Therapeutic response: decreased symptoms of fungal infection, negative C&S for infecting organism

Teach patient/family:

- That long-term therapy may be needed to clear infection (1 wk-6 mo, depending on infection)
- To avoid hazardous activities if dizziness occurs
- To take 2 hr before administration of other products that increase gastric pH (antacids, H_2-blockers, omeprazole, sucralfate, anticholinergics); to notify health care provider of all medications taken (many interactions)
- About the importance of compliance with product regimen; to use alternative method of contraception
- To notify prescriber of GI symptoms, signs of hepatic dysfunction (fatigue, nausea, anorexia, vomiting, dark urine, pale stools)
- To take during or within 20 min of eating

potassium acetate
potassium bicarbonate (OTC, Rx)

K Effervescent, Klor-Con EF, K-Vescent

potassium bicarbonate and potassium chloride (OTC, Rx)

Neo-K ♣

potassium bicarbonate and potassium citrate (OTC, Rx)

potassium chloride (OTC, Rx)

Epiklor, Klor-Con, K-Tab, Micro-K, Odan K-20 ♣

potassium gluconate (OTC, Rx)

Equaline Potassium Gluconate, Walgreens Finest Natural Potassium Gluconate

Func. class.: Electrolyte, mineral replacement

Chem. class.: Potassium

ACTION: Needed for the adequate transmission of nerve impulses and cardiac contraction, renal function, intracellular ion maintenance

USES: Prevention and treatment of hypokalemia

CONTRAINDICATIONS: Renal disease (severe), severe hemolytic disease, Addison's disease, hyperkalemia, acute dehydration, extensive tissue breakdown

Precautions: Pregnancy (C), cardiac disease, potassium-sparing diuretic therapy, systemic acidosis

DOSAGE AND ROUTES

Hypokalemia (prevention) (bicarbonate, chloride, gluconate)

- **Adult: PO** 20 mEq/day in 1-2 divided doses
- **Child: PO** 1-2 mEq/kg/day in 1-2 divided doses

Hypokalemia, digoxin toxicity (acetate, chloride)

- **Adult:** serum potassium conc >2.5 mEq/L: **IV** max 10 mEq/1 hr with 24-hr max dose 200 mEq, initial dose of 20-40 mEq has been recommended; **PO** 40-100 mEq/day in 2-4 divided doses
- **Child: IV** 0.25-0.5 mEq/kg/dose at 0.25-0.5 mEq/kg/hr; **PO** 2-5 mEq/day in divided doses

Available forms: Tabs for sol 6.5, 25 mEq; ext rel caps 8, 10 mEq; powder for sol 3.3, 5, 6.7, 10, 13.3 mEq/5 ml; tabs 2, 4, 5, 13.4 mEq; ext rel tabs 6.7, 8, 10 mEq; elix 6.7 mEq/5 ml; oral sol 2.375 mEq/5 ml; inj for prep of IV 1.5, 2, 2.4, 3, 3.2, 4.4, 4.7 mEq/ml

Administer:

PO route

- Do not break, crush, or chew ext rel tabs, caps, or enteric products
- With or after meals; dissolve effervescent tabs, powder in 8 oz cold water or juice; do not give IM, SUBCUT
- Caps with full glass of liquid

IV route

- Through large-bore needle to decrease vein inflammation; check for extravasation; in large vein, avoid scalp vein in child (IV)

Potassium acetate

Additive compatibilities: Metoclopramide

Y-site compatibilities: Ciprofloxacin

Potassium chloride

- **Potassium chloride:** must be diluted; concentrated potassium injections fatal

Continuous IV INF route

- Conc max 80 mcg/L for peripheral line; 120 mEq/L for central line
- Dehydrated patients should receive 1 L of potassium-free hydrating solution then

P

infuse 10 mEq/hr; in severe hypokalemia, rate may be 40 mEq/hr

Additive compatibilities: Aminophylline, amiodarone, atracurium, calcium gluconate, cefepime, cephalothin, cephapirin, chloramphenicol, cimetidine, ciprofloxacin, cisatracurium, clindamycin, cloxacillin, corticotropin, cytarabine, dimenhyDRINATE, DOPamine, DOXOrubicin liposome, enalaprilat, erythromycin, floxacillin, fluconazole, fosphenytoin, furosemide, heparin, hydrocortisone, isoproterenol, lidocaine, metaraminol, methicillin, methyldopate, metoclopramide, mitoXANtrone, nafcillin, netilmicin, norepinephrine, oxacillin, penicillin G potassium, phenylephrine, piperacillin, ranitidine, sodium bicarbonate, thiopental, vancomycin, verapamil, vit B/C

Y-site compatibilities: Acyclovir, aldesleukin, allopurinol, amifostine, aminophylline, amiodarone, ampicillin, amrinone, atropine, aztreonam, betamethasone, calcium gluconate, cephalothin, cephapirin, chlordiazePOXIDE, chlorproMAZINE, ciprofloxacin, cladribine, cyanocobalamin, dexamethasone, digoxin, diltiazem, diphenhydrAMINE, DOBUTamine, DOPamine, droperidol, edrophonium, enalaprilat, EPINEPHrine, esmolol, estrogens, ethacrynate, famotidine, fentaNYL, filgrastim, fludarabine, fluorouracil, furosemide, gallium, granisetron, heparin, hydrALAZINE, IDArubicin, indomethacin, insulin (regular), isoproterenol, kanamycin, labetalol, lidocaine, LORazepam, magnesium sulfate, melphalan, meperidine, methicillin, methoxamine, methylergonovine, midazolam, minocycline, morphine, neostigmine, norepinephrine, ondansetron, oxacillin, oxytocin, PACLitaxel, penicillin G potassium, pentazocine, phytonadione, piperacillin/tazobactam, prednisoLONE, procainamide, prochlorperazine, propofol, propranolol, pyridostigmine, remifentanil, sargramostim, scopolamine, sodium bicarbonate, succinylcholine, tacrolimus, teniposide, theophylline, thiotepa, trimethaphan, trimethoenzamide, vinorelbine, warfarin, zidovudine

SIDE EFFECTS

CNS: Confusion
CV: Bradycardia, cardiac depression, dysrhythmias, arrest; peaking T waves, lowered R, depressed RST, prolonged P-R interval, widened QRS complex
GI: *Nausea, vomiting, cramps,* pain, *diarrhea,* ulceration of small bowel
GU: Oliguria
INTEG: Cold extremities, rash

PHARMACOKINETICS

PO: Excreted by kidneys and in feces; onset of action ≈30 min
IV: Immediate onset of action

INTERACTIONS

Increase: hyperkalemia—potassium phosphate IV; products containing calcium or magnesium; potassium-sparing diuretic or other potassium products; ACE inhibitors

NURSING CONSIDERATIONS

Assess:

- **Hyperkalemia:** indicates toxicity; fatigue, muscle weakness, confusion, dyspnea, palpitation; ECG for peaking T waves, lowered R, depressed RST, prolonged P-R interval, widening QRS complex, hyperkalemia; product should be reduced or discontinued, administer sodium bicarbonate (metabolic acidosis)
- Potassium level during treatment (3.5-5 mg/dl is normal level)
- Determine hydration status, I&O ratio; watch for decreased urinary output; notify prescriber immediately
- Cardiac status: rate, rhythm, CVP, PWP, PAWP if being monitored directly

Evaluate:

- Therapeutic response: absence of fatigue, muscle weakness; decreased thirst, urinary output; cardiac changes

Teach patient/family:

- To add potassium-rich foods to diet: bananas, orange juice, avocados, whole grains, broccoli, carrots, prunes, cocoa after product is discontinued

- To avoid OTC products: antacids, salt substitutes, analgesics, vitamin preparations unless specifically directed by prescriber; to avoid licorice in large amounts because it may cause hypokalemia, sodium retention
- To report hyperkalemia symptoms (lethargy, confusion, diarrhea, nausea, vomiting, fainting, decreased output) or continued hypokalemia symptoms (fatigue, weakness, polyuria, polydipsia, cardiac changes)
- To dissolve powder or tablet completely in ≥120 ml water or juice
- About the importance of regular follow-up visits
- That potassium levels will need to be monitored periodically

potassium iodide (Rx)

Lugol's, SSKI, Strong Iodine

Func. class.: Thyroid hormone antagonist

Chem. class.: Iodine product

ACTION: Inhibits secretion of thyroid hormone, fosters colloid accumulation in thyroid follicles, decreases vascularity of gland

USES: Preparation for thyroidectomy, thyrotoxic crisis, neonatal thyrotoxicosis, radiation protectant, thyroid storm

Unlabeled uses: Erythema multiforme, erythema nodosum leprosum (ENL), sporotrichosis, thyroid involution induction

CONTRAINDICATIONS: Pregnancy (D), pulmonary edema, pulmonary TB, bronchitis, hypersensitivity to iodine

Precautions: Breastfeeding, children

DOSAGE AND ROUTES

Hyperthyroidism/thyrotoxicosis

- **Adult and child: PO** (SSKI) 250 mg tid × 10-14 days preoperatively

Preparation for thyroidectomy

- **Adult and child: PO** 3-5 gtt strong iodine sol tid or 1-5 drops SSKI in water tid after meals for 10 days before surgery

Radiation exposure (radioactive iodine)

- **Adult: PO** 130 mg/day (distribution by government/public health officials or OTC purchase)
- **Child ≥3 yr: PO** 65 mg daily
- **Child/infant >1 mo-3 yr: PO** 32 mg/day
- **Neonate: PO** 16 mg/day

Available forms: Oral sol (Lugol's solution) iodine 5%/potassium iodide 10%; oral sol (SSKI) 1 g/ml

Administer:

- Products are not interchangeable
- Strong iodine solution after diluting with water or juice to improve taste
- Through straw to prevent tooth discoloration
- With meals to decrease GI upset
- At same time each day to maintain product level
- At lowest dose that relieves symptoms; discontinue before RAIU

SIDE EFFECTS

CNS: Headache, confusion, paresthesias

EENT: Metallic taste, stomatitis, salivation, periorbital edema, sore teeth and gums, cold symptoms

ENDO: Hypothyroidism, hyperthyroid adenoma

GI: *Nausea, diarrhea, vomiting,* small-bowel lesions, upper gastric pain, metallic taste

INTEG: Rash, urticaria, **angioneurotic edema,** acne, mucosal hemorrhage, fever

MS: Myalgia, arthralgia, weakness

PHARMACOKINETICS

PO: Onset 24-48 hr, peak 10-15 days after continuous therapy, uptake by thyroid gland or excreted in urine, crosses placenta

INTERACTIONS

• Hypothyroidism: lithium, other antithyroid agents

Increased: hyperkalemia—angiotensin II receptor antagonist, ACE inhibitors, potassium salts, potassium-sparing diuretics

Drug/Lab Test

Interference: urinary 17-OHCS

NURSING CONSIDERATIONS

Assess:

• Pulse, B/P, temperature; serum potassium

• I&O ratio; check for edema: puffy hands, feet, periorbit; indicate hypothyroidism

• Weight daily; same clothing, scale, time of day

• T_3, T_4, which is increased; serum TSH, which is decreased; free thyroxine index, which is increased if dosage is too low; discontinue product 3-4 wk before RAIU

⚠ Overdose: peripheral edema, heat intolerance, diaphoresis, palpitations, dysrhythmias, severe tachycardia, fever, delirium, CNS irritability

• **Hypersensitivity:** rash; enlarged cervical lymph nodes may indicate product should be discontinued

• **Hypoprothrombinemia:** bleeding, petechiae, ecchymosis

• Clinical response: after 3 wk should include increased weight, pulse; decreased T_4

• Fluids to 3-4 L/day unless contraindicated

Evaluate:

• Therapeutic response: weight gain; decreased pulse, T_4, size of thyroid gland

Teach patient/family:

• To abstain from breastfeeding after delivery

• To keep graph of weight, pulse, mood

• To avoid OTC products that contain iodine

• That seafood, other iodine products may be restricted

• Not to discontinue product abruptly; that thyroid crisis may occur as part of stress response

• That response may take several mo if thyroid is large

• To discontinue product, notify prescriber of fever, rash, metallic taste, swelling of throat; burning of mouth, throat; sore gums, teeth; severe GI distress, enlargement of thyroid, cold symptoms

pramipexole (Rx)

(pra-mi-pex′ol)

Mirapex, Mirapex ER

Func. class.: Antiparkinson agent

Chem. class.: DOPamine-receptor agonist, non-ergot

ACTION:

Selective agonist for D_2 receptors (presynaptic/postsynaptic sites); binding at D_3 receptor contributes to antiparkinson effects

USES:

Idiopathic Parkinson's disease, restless leg syndrome

CONTRAINDICATIONS:

Hypersensitivity

Precautions: Pregnancy (C), cardiac/renal disease, MI with dysrhythmias, affective disorders, psychosis, preexisting dyskinesias, history of falling asleep during daily activities

DOSAGE AND ROUTES

Parkinson's disease

• **Adult: PO** 0.125 mg tid; increase gradually by 0.125 mg/dose at 5- to 7-day intervals until total daily dose of 4.5 mg/day reached; ER 0.375 mg daily initially then up to 0.75 mg/day; may increase by 0.75 mg/day no more than q5-7days as needed, max 4.5 mg/day

Restless leg syndrome

• **Adult: PO** 0.125 mg 2-3 hr before bedtime, increase gradually, max 0.5 mg/day

Renal dose

• **Adult: PO** CCr 35-59 ml/min, 0.125 mg bid, may increase q5-7days to 1.5 mg bid if required; CCr 15-34 ml/min, 0.125 mg/day, increase q5-7days to 1.5 mg/day

Available forms: Tabs 0.125, 0.25, 0.5, 1, 1.5 mg; ER tab 0.375, 0.75, 1.5, 3.0, 4.5 mg

Administer:

- Adjust dosage to patient response, titrate slowly, taper when discontinuing
- With meals to minimize GI symptoms
- Do not crush, chew, or break ext rel product

SIDE EFFECTS

CNS: *Agitation, insomnia,* psychosis, hallucinations, depression, dizziness, headache, confusion, amnesia, dream disorder, asthenia, dyskinesia, hypersomnolence, sudden sleep onset, impulse control disorders

CV: *Orthostatic hypotension,* edema, syncope, tachycardia, increased B/P, heart rate, heart failure

EENT: Blurred vision, retinal/vision deterioration

ENDO: Antidiuretic hormone secretion (SIADH)

GI: *Nausea, anorexia,* constipation, dysphagia, dry mouth

GU: Impotence, urinary frequency

HEMA: Hemolytic anemia, leukopenia, agranulocytosis

INTEG: Pruritus

PHARMACOKINETICS

Minimally metabolized, peak 2 hr, half-life 8 hr, 8.5-12 hr in geriatric patients

INTERACTIONS

Increase: pramipexole levels—levodopa, cimetidine, ranitidine, diltiazem, triamterene, verapamil, quiNIDine

Decrease: pramipexole levels—DOPamine antagonists, phenothiazines, metoclopramide, butyrophenones

NURSING CONSIDERATIONS

Assess:

- **Parkinson's disease:** involuntary movements: bradykinesia, tremors, staggering gait, muscle rigidity, drooling
- B/P, ECG, respiration during initial treatment; hypo/hypertension should be reported
- Mental status: affect, mood, behavioral changes, depression; complete suicide assessment, worsening of symptoms of restless leg syndrome, impulse control disorders

⚠ **Sleep attacks:** may fall asleep during activities without warning; may need to discontinue medication

- Assistance with ambulation during beginning therapy
- Testing for diabetes mellitus, acromegaly if patient receiving long-term therapy

Evaluate:

- Therapeutic response: movement disorder improves

Teach patient/family:

- That therapeutic effects may take several weeks to a few months
- To change positions slowly to prevent orthostatic hypotension
- To use product exactly as prescribed; if product is discontinued abruptly, parkinsonian crisis may occur; to avoid alcohol, OTC sleeping products
- To notify prescriber if pregnancy is planned or suspected
- To notify prescriber of impulse control disorders: shopping

P

pramlintide (Rx)

(pram′lin-tide)

Symlin

Func. class.: Antidiabetic

Chem. class.: Synthetic human amylin analog

ACTION: Modulates and slows stomach emptying, prevents postprandial rise in plasma glucagon, decreases appetite, leads to decreased caloric intake and weight loss

USES: As an adjunct prandial to insulin therapy for uncontrolled type 1 or type 2 diabetes

CONTRAINDICATIONS:

Hypersensitivity to this product or cresol; gastroparesis

Black Box Warning: Hypoglycemia

Precautions: Pregnancy (C), breastfeeding

DOSAGE AND ROUTES

Type 1 diabetes

• **Adult: SUBCUT** 15 mcg before each meal (≥30 g carbohydrate), titrate up in 15-mcg increments to target dose of 60 mcg/dose; each dose titration should occur after no nausea for 3 days

Type 2 diabetes

• **Adult: SUBCUT** 60 mcg before each meal (≥30 g CHO), titrate up to 120 mcg **SUBCUT** with each meal after no nausea for 3-7 days

Available forms: PEN 60, 120 (1000 mcg/ml solution for j-injection)

Administer:

• Store at room temperature for ≤30 days; keep away from heat and sunlight; refrigerate all other supply

• Premeal insulin should be decreased by 50% when starting and adjusted to therapeutic dose to prevent hypoglycemia

SUBCUT route

• Rotate injection sites, allow solution to warm to room temperature before use

• Take immediately before mealtime or if 30 g of carbohydrates will be consumed

• Do not use if a meal is skipped

• Do not use if discolored; do not give in arm; absorption is variable

SIDE EFFECTS

CNS: *Headache,* fatigue, dizziness, confusion

EENT: Blurred vision

GI: *Nausea, vomiting, anorexia,* abdominal pain

INTEG: Inj site reactions, diaphoresis

META: Hypoglycemia

MS: Arthralgia

RESP: *Cough,* pharyngitis

SYST: *Systemic allergy*

PHARMACOKINETICS

Bioavailability 30%-40%, not extensively bound to blood cells or albumin, 40% bound in plasma, half-life 48 min, metabolized by kidneys, peak 20 min, duration 3 hr

INTERACTIONS

• Do not use with erythromycin, metoclopramide

Increase: effect of acetaminophen

Increase: pramlintide action—antimuscarinics, α-glucosidase inhibitors, diphenoxylate, loperamide, octreotide, opiate agonist, tricyclics

Increase: hypoglycemia—ACE inhibitors, disopyramide, anabolic steroids, androgens, fibric acid derivatives, alcohol, corticosteroids, insulin

Increase: hyperglycemia—phenothiazines

Decrease: hypoglycemia—niacin, dextrothyroxine, thiazide diuretics, triamterene, estrogens, progestins, oral contraceptives, MAOIs

NURSING CONSIDERATIONS

Assess:

• Fasting blood glucose, 2 hr post-prandiol (80-150 mg/dl, normal fasting level; 70-130 mg/dl, normal 2 hr level); A1c may also be drawn to identify treatment effectiveness; also monitor weight, appetite

• **Hypoglycemic reaction** sweating; weakness; dizziness; chills; confusion; headache; nausea; rapid, weak pulse; fatigue; tachycardia; memory lapses; slurred speech; staggering gait; anxiety; tremors; hunger

• **Hyperglycemia:** acetone breath; polyuria; fatigue; polydipsia; flushed, dry skin; lethargy

Evaluate:

• Therapeutic response: decrease in polyuria, polydipsia, polyphagia; clear sensorium; absence of dizziness; stable gait; improving blood glucose, A1c

Teach patient/family:

- That product does not cure diabetes but rather controls symptoms
- To carry emergency ID as diabetic
- To recognize hypoglycemia reaction: headache, fatigue, weakness, fast pulse
- About the dosage, route, mixing instructions, diet restrictions, disease process
- To carry a glucose source (candy or lump sugar, glucose tabs) to treat hypoglycemia
- About the symptoms of ketoacidosis: nausea; thirst; polyuria; dry mouth; decreased B/P; dry, flushed skin; acetone breath; drowsiness; Kussmaul respirations
- That a plan is necessary for diet, exercise; that all food on diet should be eaten; that exercise routine should not vary
- About blood glucose testing; how to determine glucose level
- To avoid OTC products, alcohol unless directed by prescriber
- Not to operate machinery or drive until effect is known
- About how to use pen

TREATMENT OF OVERDOSE:

Glucose 25 g IV or 50 ml dextrose 50% sol or 1 mg glucagon SUBCUT

pramoxine topical

See Appendix B

prasugrel (Rx)

(pra′soo-grel)

Effient

Func. class.: Platelet aggregation inhibitor

Chem. class.: ADP receptor antagonist

ACTION: Inhibits ADP-induced platelet aggregation

USES: Reducing the risk of stroke, MI, vascular death, peripheral arterial disease in high-risk patients

CONTRAINDICATIONS: Hypersensitivity, stroke, TIA

Black Box Warning: Active bleeding

Precautions: Pregnancy (B), breastfeeding, children, geriatric patients, hepatic disease, increased bleeding risk, neutropenia, agranulocytosis, renal disease, surgery, trauma, thrombotic thrombocytopenic purpura, Asian patients, weight <60 kg, CABG, abrupt discontinuation

DOSAGE AND ROUTES

- **Adult/geriatric <75 yr and ≥60 kg:** **PO** 60-mg loading dose then 10 mg daily with aspirin (75-325 mg/day)
- **Adult/geriatric <75 yr and <60 kg:** **PO** 60 mg loading dose then 5 mg daily
- **Geriatric >75 yr:** not recommended

Available forms: Tabs 5, 10 mg

Administer:

- With food to decrease gastric symptoms
- Do not break tablets
- Do not discontinue therapy abruptly

SIDE EFFECTS

CNS: Headache, dizziness

CV: Edema, atrial fibrillation, bradycardia, chest pain, hypo/hypertension

GI: Nausea, vomiting, diarrhea

HEMA: Epistaxis, **leukopenia, thrombocytopenia, neutropenia, anaphylaxis, angioedema, anemia**

INTEG: Rash, hypercholesterolemia

MISC: Fatigue, **intracranial hemorrhage, secondary malignancy, angioedema**

MS: Back pain

PHARMACOKINETICS

Rapidly absorbed; peak 30 min; metabolized by liver (CYP3A4; CYP2B6); excreted in urine, feces; half-life 7 hr

INTERACTIONS

Increase: bleeding risk—anticoagulants, aspirin, NSAIDs, abciximab, eptifibatide, tirofiban, thrombolytics, ticlopidine, SSRIs, treprostinil, rifampin

NURSING CONSIDERATIONS

Assess:

⚠ **Thrombotic/thrombocytic purpura:** fever, thrombocytopenia, neurolytic anemia

- Hepatic studies: AST, ALT, bilirubin, creatinine with long-term therapy
- Blood studies: CBC, differential, Hct, Hgb, PT, cholesterol with long-term therapy

Black Box Warning: Bleeding: may be fatal, decreased B/P in those who have had CABG may be the first indication; bleeding should be controlled while continuing product; may use transfusion; do not use within 1 wk of CABG; may use lower doses in those <60 kg

Evaluate:

- Therapeutic response: absence of stroke, MI

Teach patient/family:

- That blood work will be necessary during treatment
- To report any unusual bruising, bleeding to prescriber; that it may take longer to stop bleeding
- To take with food or just after eating to minimize GI discomfort
- To report diarrhea, skin rashes, subcutaneous bleeding, chills, fever, sore throat
- To tell all health care providers that prasugrel is being used; that product may be held before surgery

pravastatin (Rx)

(pra'va-sta-tin)

Pravachol

Func. class.: Antilipemic

Chem. class.: HMG-CoA reductase enzyme

Do not confuse:

Pravachol/Prevacid/propranolol

ACTION: Inhibits HMG-CoA reductase enzyme, which reduces cholesterol synthesis

USES: As an adjunct for primary hypercholesterolemia (types IIa, IIb, III, IV), to reduce the risk for recurrent MI, atherosclerosis, primary/secondary CV events, stroke, TIAs

CONTRAINDICATIONS: Pregnancy (X), breastfeeding, hypersensitivity, active hepatic disease

Precautions: Past hepatic disease, alcoholism, severe acute infections, trauma, severe metabolic disorders, electrolyte imbalances, renal disease

DOSAGE AND ROUTES

- **Adult: PO** 40 mg/day at bedtime (range 10-80 mg/day); start at 10 mg/day if patient also taking immunosuppressants
- **Adolescent 14-18 yr: PO** 40 mg/day
- **Child 8-13 yr: PO** 20 mg/day
- **Geriatric/renal/hepatic disease: PO** 10 mg/day initially

Renal dose

- **Adult: PO** 10-20 mg daily at bedtime, increase at 4-wk intervals

Available forms: Tabs 10, 20, 40, 80 mg

Administer:

- Without regard to meals, at bedtime
- Give 4 hr after bile acid sequestrants
- Store in cool environment in tight container protected from light

SIDE EFFECTS

CNS: Headache, dizziness, fatigue, confusion
CV: Chest pain
EENT: Lens opacities
GI: Nausea, constipation, diarrhea, flatus, abdominal pain, heartburn, hepatic dysfunction, pancreatitis, hepatitis
GU: Renal failure (myoglobinuria)
INTEG: Rash, pruritus
MS: Muscle cramps, myalgia, myositis, rhabdomyolysis
RESP: Common cold, rhinitis, cough

PHARMACOKINETICS

Peak 1-1½ hr; metabolized by liver; protein binding 50%; excreted in urine 20%, feces 70%, breast milk; crosses placenta; half-life 1.25-2.25 hr

INTERACTIONS

Increase: myopathy, rhabdomyolysis risk—erythromycin, niacin, cycloSPORINE, gemfibrozil, clofibrate, clarithromycin, itraconazole, protease inhibitors
Decrease: bioavailability of pravastatin —bile acid sequestrants
Drug/Herb
Increase: adverse reactions—red yeast rice
Increase: hepatotoxicity—eucalyptus
Decrease: effect—St. John's wort
Drug/Lab Test
Increase: CK, LFTs
Altered: thyroid function tests

NURSING CONSIDERATIONS

Assess:
- Fasting lipid profile: LDL, HDL, triglycerides, cholesterol at baseline, q12wk, then q6mo when stable; obtain diet history
- Hepatic studies: baseline, q12wk, then q6mo for remainder of yr; AST, ALT, LFTs may increase
- Renal studies of patients with compromised renal systems: BUN, I&O ratio, creatinine

⚠ **Rhabdomyolysis:** muscle tenderness, pain; obtain CPK at baseline and if these occur, therapy should be discontinued
Evaluate:
- Therapeutic response: decrease in LDL total cholesterol, triglycerides; increase in HDL

Teach patient/family:
- That blood work will be necessary during treatment

⚠ To report blurred vision, severe GI symptoms, dizziness, headache, muscle pain, weakness, fever
- That regimen will continue: low-cholesterol diet, exercise program

⚠ To report suspected, planned pregnancy; not to use product during pregnancy, pregnancy category (X); not to breastfeed
- **Hepatic disease:** to notify prescriber of lack of appetite, yellow sclera/skin, dark urine, abdominal pain, weakness

prazosin (Rx)
(pray'zoe-sin)
Minipress
Func. class.: Antihypertensive
Chem. class.: α_1-Adrenergic blocker, peripheral

P

ACTION: Blocks α-mediated vasoconstriction of adrenergic receptors, thereby inducing peripheral vasodilation

USES: Hypertension
Unlabeled uses: Benign prostatic hypertrophy to decrease urine outflow obstruction, heart failure, hypertensive urgency, Raynaud's phenomenon, posttraumatic stress disorder (PTSD), scorpion venom poisoning

CONTRAINDICATIONS: Hypersensitivity
Precautions: Pregnancy (C), breastfeeding, children, geriatric patients, prostate cancer, ocular surgery, orthostatic hypotension

DOSAGE AND ROUTES

Hypertension

- **Adult: PO** 1 mg bid or tid increasing to 20 mg/day in divided doses, if required; usual range 6-15 mg/day, max 1 mg initially; max 20 mg/day, some may require 40 mg/day
- **Child (unlabeled): PO** 5 mcg/kg q6hr; max 400 mcg/kg/day or 15 mg/day

Benign prostatic hyperplasia (unlabeled)

- **Adult: PO** 2 mg bid

Raynaud's phenomenon (unlabeled)

- **Adult: PO** 0.5-3 mg bid

CHF (unlabeled)

- **Adult: PO** 1 mg bid-tid, may gradually increase to max 20 mg/day
- **Child: PO** 5 mcg/kg q6hr, may gradually increase to 25 mcg/kg q6hr

Hypertensive urgency (unlabeled)

- **Adult: PO** 10-20 mg, may repeat after 30 min

Available forms: Caps 1, 2, 5 mg

Administer:

- 1st dose at bedtime to avoid fainting
- Without regard to meals
- Store at room temperature

SIDE EFFECTS

CNS: *Dizziness, headache, drowsiness,* anxiety, depression, vertigo, *weakness,* fatigue, syncope

CV: *Palpitations, orthostatic hypotension,* tachycardia, edema, rebound hypertension

EENT: Blurred vision, epistaxis, tinnitus, dry mouth, red sclera

GI: *Nausea,* vomiting, diarrhea, constipation, abdominal pain, pancreatitis

GU: Urinary frequency, incontinence, impotence, priapism; water, sodium retention

PHARMACOKINETICS

Onset 2 hr, peak 2-4 hr, duration 6-12 hr, half-life 2-4 hr; metabolized in liver, excreted via bile, feces (>90%), urine (<10%); protein binding 97%

INTERACTIONS

Increase: hypotensive effects—β-blockers, nitroglycerin, alcohol, phosphodiesterase inhibitors (vardenafil, tadalafil, sildenafil); diuretics, other antihypertensives, MAOIs

Decrease: antihypertensive effect—NSAIDs

Increase: antihypertensive effect—hawthorn

Drug/Lab Test

Increase: urinary norepinephrine, VMA

NURSING CONSIDERATIONS

Assess:

- **Hypertension/CHF:** B/P (sitting, standing) during initial treatment, periodically thereafter; pulse, jugular venous distention
- BUN, uric acid if patient receiving long-term therapy
- Weight daily, I&O; edema in feet, legs daily
- **Benign prostatic hypertrophy (unlabeled):** urinary patterns, frequency, stream, dribbling; flow before, during, and after therapy

Evaluate:

- Therapeutic response: decreased B/P

Teach patient/family:

- That fainting occasionally occurs after 1st dose; to take 1st dose at bedtime; not to drive or operate machinery for 4 hr after 1st dose; that full effect may take 4-6 wk
- To change positions slowly to prevent orthostatic hypotension
- To avoid OTC medications, alcohol unless approved by prescriber
- Not to discontinue abruptly

TREATMENT OF OVERDOSE:

Administer volume expanders or vasopressors, discontinue product, place patient in supine position

prednisoLONE (Rx)

(pred-niss'oh-lone)

Flo-Pred, Millipred, Orapred, Orapred ODT, Prelone, Veripred

Func. class.: Corticosteroid, synthetic

Chem. class.: Glucocorticoid, immediate acting

Do not confuse:
prednisoLONE/predniSONE

ACTION: Decreases inflammation by the suppression of migration of polymorphonuclear leukocytes, fibroblasts; reversal to increase capillary permeability and lysosomal stabilization

USES: Severe inflammation, immunosuppression, neoplasms, asthma

CONTRAINDICATIONS: Hypersensitivity, fungal infections, viral infection, varicella

Precautions: Pregnancy (C), breastfeeding, children, diabetes mellitus, glaucoma, osteoporosis, seizure disorders, ulcerative colitis, CHF, myasthenia gravis, abrupt discontinuation, children, acute MI, GI ulcers, hypertension, hepatitis, psychosis, thromboembolism, peptic ulcer disease, renal disease, Cushing syndrome

DOSAGE AND ROUTES

Primary (Addison's disease)/ secondary adrenocortical insufficiency or for the treatment of congenital adrenal hyperplasia

- **Adults: PO** 5-60 mg **PO** per day as a single dose or divided doses
- **Infants, children, and adolescents: PO** 0.14-2 mg/kg or 4-60 mg/m^2 per day in 3-4 divided doses

Nonsuppurative thyroiditis

- **Adults: PO** 5-60 mg/day as a single dose or in divided doses
- **Infants, children, and adolescents: PO** 0.14-2 mg/kg or 4-60 mg/m^2 **PO** per day given in 3-4 divided doses

Management of symptomatic sarcoidosis; or treatment of hypercalcemia associated with sarcoidosis or with various cancers

- **Adults: PO** 5-60 mg/day as a single dose or divided doses
- **Infants, children, and adolescents: PO** 0.14-2 mg/kg or 4-60 mg/m^2 per day in 3-4 divided doses

Adjunct in rheumatic disorders (ankylosing spondylitis, gout with gouty arthritis, juvenile rheumatoid arthritis (JRA)/juvenile idiopathic arthritis (JIA), post-traumatic osteoarthritis psoriatic arthritis, rheumatoid arthritis) or acute episodes or exacerbation of nonrheumatic inflammation (acute and subacute bursitis, epicondylitis, and acute non-specific tenosynovitis)

- **Adults: PO** 5-60 mg/day as a single dose or in divided doses
- **Infants, children, and adolescents: PO** 0.14-2 mg/kg or 4-60 mg/m^2 per day in 3-4 divided doses

Adjunct in carpal tunnel syndrome (unlabeled)

- **Adults: PO** 20 mg/day × 2 wk, then 10 mg/day for an additional 2 wk relief

For maintenance therapy in selected cases of acute rheumatic carditis, systemic dermatomyositis (polymyositis), systemic lupus erythematosus (SLE); (Unlabeled): temporal arteritis, Churg-Strauss syndrome, mixed connective tissue disease, polyarteritis nodosa, relapsing polychondritis, polymyalgia rheumatica, vasculitis, or Wegener's granulomatosis:

- **Adults: PO** 5-60 mg per day as a single dose or in divided doses
- **Infants, children, and adolescents:** 0.14-2 mg/kg or 4-60 mg/m^2 **PO** per day given in 3-4 divided doses

P

Corticosteroid-responsive respiratory disorders (airway-obstructing hemangioma in infants (unlabeled), aspiration pneumonitis, berylliosis, chronic obstructive pulmonary disease (COPD), laryngotracheobronchitis (croup), Loeffler's syndrome, noncardiogenic pulmonary edema (unlabeled)

- **Adults: PO** 5-60 mg/day as a single dose or in divided doses
- **Infants, children, and adolescents: PO** 0.14-2 mg/kg or 4-60 mg/m² per day given in 3-4 divided doses

Asthma; bronchospasm prophylaxis (unlabeled)

- **Adults and adolescents: PO** 40-80 mg **PO** per day in 1-2 divided doses until the peak expiratory flow (PEF) reaches 70% of predicted or personal best; total course of treatment is 3-10 days
- **Children: PO** 1 mg/kg (up to 60 mg) per day in 2 divided doses until PEF reaches 70% of predicted or personal best; if a patient is given systemic corticosteroids, continue **PO** corticosteroids for a total course of 3-10 days; tapering is not necessary for courses <1 wk

Acute asthma exacerbation on an outpatient basis

- **Adults and adolescents: PO** 40-60 mg **PO** per day as a single dose or in 2 divided doses for 3-10 days
- **Children 5-12 yr: PO** 1-2 mg/kg (up to 60 mg) **PO** per day in 2 divided doses for 3-10 days
- **Infants and children ≤4 yr: PO** 1-2 mg/kg (up to 30 mg) **PO** per day in 2 divided doses for 3-10 days

Long-term prevention of symptoms in severe persistent asthma

- **Adults, adolescents, and children ≥12 yr: PO** 7.5-60 mg once daily in the morning or every other day
- **Infants and children ≤11 yr: PO** 0.25-2 mg/kg **PO** daily given as a single dose each morning or every other day

Hematologic disorders with thrombocytopenia (immune thrombocytopenia/idiopathic thrombocytopenic purpura (ITP), or secondary thrombocytopenia)

- **Adults: PO** 5-60 mg/day as a single dose or in divided doses
- **Children and adolescents (unlabeled): PO** 0.14-2 mg/kg or 4-60 mg/m² per day in 3-4 divided doses

Available forms: Tabs 5 mg; oral dissolving tab 10, 15, 30 mg; oral sol 5 mg/5 ml, 10 mg/5 ml, 15 mg/5 ml, 25 mg/5 ml; syrup 5 mg/5 ml

Administer:

- **Oral sol:** use calibrated measuring device
- **Orally disintegrating tabs:** place on tongue; allow to dissolve, swallow or swallow whole; do not cut, split

SIDE EFFECTS

CNS: *Depression,* headache, mood changes

CV: *Hypertension,* circulatory collapse, thrombophlebitis, embolism, tachycardia

EENT: Fungal infections, increased intraocular pressure, blurred vision

GI: *Diarrhea, nausea, abdominal distention,* GI hemorrhage, increased appetite, pancreatitis

INTEG: Acne, poor wound healing, ecchymosis, petechiae, sweating

MS: Fractures, osteoporosis, weakness, arthralgia, myopathy, tendon rupture

PHARMACOKINETICS

PO: Peak 1-2 hr, duration 2 days

INTERACTIONS

Increase: tendon rupture—quinolones

Increase: side effects—alcohol, salicylates, indomethacin, amphotericin B, digitalis, cycloSPORINE, diuretics

Increase: prednisoLONE action—salicylates, estrogens, indomethacin, oral contraceptives, ketoconazole, macrolide antibiotics

Increase: prednisoLONE effect—CYP3A4 inhibitors

Increase: toxicity—azole antifungals, cycloSPORINE, NSAIDs
Decrease: prednisoLONE action—cholestyramine, colestipol, barbiturates, rifampin, ePHEDrine, phenytoin, theophylline
Decrease: effects of anticoagulants, anticonvulsants, antidiabetics, ambenonium, neostigmine, isoniazid, toxoids, vaccines, anticholinesterases, salicylates, somatrem
Decrease: prednisoLONE effect—CYP3A4 inducers
Drug/Lab Test
Increase: cholesterol, sodium, blood glucose, uric acid, calcium, urine glucose
Decrease: calcium, potassium, T_4, T_3, thyroid ^{131}I uptake test, urine 17-OHCS, 17-KS, PBI
False negative: skin allergy tests

NURSING CONSIDERATIONS

Assess:
- Potassium, blood glucose, urine glucose while patient receiving long-term therapy; hypokalemia, hyperglycemia
- Weight daily; notify prescriber if weekly gain of >5 lb
- B/P q4hr, pulse; notify prescriber if chest pain occurs
- I&O ratio; be alert for decreasing urinary output, increasing edema
- Plasma cortisol levels with long-term therapy; normal level: 138-635 nmol/L SI units when drawn at 8 AM
- **Infection:** increased temperature, WBC, even after withdrawal of medication; product masks infection
- **Potassium depletion:** paresthesias, fatigue, nausea, vomiting, depression, polyuria, dysrhythmias, weakness
- Edema, hypertension, cardiac symptoms
- Mental status: affect, mood, behavioral changes, aggression
- **Adrenal insufficiency:** nausea, vomiting, lethargy, restlessness, confusion, weight loss, hypotension before, during treatment; HPA suppression may be precipitated by abrupt withdrawal

Evaluate:
- Therapeutic response: ease of respirations, decreased inflammation

Teach patient/family:
- That emergency ID as steroid user should be carried
- To notify prescriber if therapeutic response decreases; that dosage adjustment may be needed
- Not to discontinue abruptly; that adrenal crisis can result; to take product exactly as prescribed
- To avoid OTC products: salicylates, cough products with alcohol, cold preparations unless directed by prescriber
- About cushingoid symptoms
- About the symptoms of adrenal insufficiency: nausea, anorexia, fatigue, dizziness, dyspnea, weakness, joint pain

prednisoLONE ophthalmic (Rx)

See Appendix B

predniSONE (Rx)

(pred′ni-sone)

Rayos, Winpred ♣

Func. class.: Corticosteroid
Chem. class.: Intermediate-acting glucocorticoid

Do not confuse:
predniSONE/methylPREDNISolone/prednisoLONE/PriLOSEC

ACTION: Decreases inflammation by increasing capillary permeability and lysosomal stabilization, minimal mineralocorticoid activity

USES: Severe inflammation, neoplasms, multiple sclerosis, collagen disorders, dermatologic disorders, pulmonary fibrosis, asthma
Unlabeled uses: Adjunct for refractory seizures, infantile spasms, acute interstitial nephritis, amyloidosis, autoimmune hepatitis, Behçet's syndrome, Bell's palsy,

P

carpal tunnel syndrome, Churg-Strauss syndrome, dermatomyositis, Duchenne muscular dystrophy, endophthalmitis, Lennox-Gastaut syndrome, lupus nephritis, mixed connective-tissue disease, pericarditis, pneumonia, polyarteritis nodosa, polychondritis, polymyositis, rheumatic carditis, temporal arteritis, TB, Wegener's granulomatosis

CONTRAINDICATIONS:

Fungal infections, hypersensitivity

Precautions: Pregnancy (C), diabetes mellitus, glaucoma, osteoporosis, seizure disorders, ulcerative colitis, CHF, myasthenia gravis, renal disease, esophagitis, peptic ulcer, cataracts, coagulopathy, abrupt discontinuation, children, corticosteroid hypertensitivity, Cushing syndrome, diabetes mellitus, ulcerative colitis, thromboembolism, geriatric patients, acute MI

DOSAGE AND ROUTES

- **Adult: PO** 5-60 mg/day or divided bid-qid
- **Child: PO** 0.05-2 mg/kg/day divided 1-4×/day

Nephrotic syndrome

- **Child: PO** 2 mg/kg/day in divided doses, until urine is protein-free for 3 consecutive days, then 1-1.5 mg/kg/day every other day × 4 wk

Multiple sclerosis

- **Adult: PO** 200 mg/day × 1 wk then 80 mg every other day × 1 mo

Asthma

- **Adult/adolescent: PO** 40-80 mg/day in 1-2 divided doses until PEF is 70% of predicted or best
- **Child: PO** 1 mg/kg (max 60 mg)/day in 2 divided doses until PEF is 70% of predicted or personal best

Available forms: Tabs 1, 2.5, 5, 10, 20, 50 mg; oral sol 5 mg/5 ml; syr 5 mg/5 ml, del rel tab 1, 2, 5 mg

Administer:

- For long-term use, alternate-day therapy recommended to decrease adverse reactions; give in AM to coincide with normal cortisol secretion
- Titrated dose; use lowest effective dose
- With food or milk to decrease GI symptoms
- **Oral sol:** use calibrated measuring device
- **Del rel tab:** swallow whole; do not break, crush, chew; give once a day

SIDE EFFECTS

CNS: Depression, flushing, sweating, headache, mood changes

CV: Hypertension, thrombophlebitis, embolism, tachycardia, fluid retention

EENT: Fungal infections, increased intraocular pressure, blurred vision

GI: Diarrhea, nausea, abdominal distention, GI hemorrhage, increased appetite, pancreatitis

INTEG: Acne, poor wound healing, ecchymosis, petechiae

META: Hyperglycemia

MS: Fractures, osteoporosis, weakness

PHARMACOKINETICS

PO: Well absorbed PO, peak 1-2 hr, duration; del rel peak 6-6$^1/_2$ hr; 1-1$^1/_2$ days, half-life 3$^1/_2$-4 hr, crosses placenta, enters breast milk, metabolized by liver after conversion, excreted in urine

INTERACTIONS

Increase: tendon rupture quinolones

Increase: side effects—alcohol, salicylates, NSAIDs, amphotericin B, digoxin, cycloSPORINE, diuretics

Increase: predniSONE action—salicylates, estrogens, NSAIDs, oral contraceptives, ketoconazole, macrolide antiinfectives

Increase: predniSONE effect—CYP3A4 inhibitors

Decrease: predniSONE effect— CYP3A4 inducers

Decrease: predniSONE action—cholestyramine, colestipol, barbiturates, rifampin, phenytoin, theophylline

Decrease: effects of anticoagulants, anticonvulsants, antidiabetics, ambenonium, neostigmine, isoniazid, toxoids, vaccines, anticholinesterases, salicylates, somatrem

Drug/Herb

Decrease: predniSONE effect—ephedra (ma huang)

Drug/Lab Test

Increase: cholesterol, sodium, blood glucose, uric acid, calcium, urine glucose

Decrease: calcium, potassium, T_4, T_3, thyroid ^{131}I uptake test, urine 17-OHCS, 17-KS, PBI

False negative: skin allergy tests

NURSING CONSIDERATIONS

Assess:

- **Adrenal insufficiency:** nausea, vomiting, anorexia, confusion, hypotension, weight loss before, during treatment; HPA suppression may be precipitated by abrupt withdrawal
- Potassium, blood glucose, urine glucose while patient receiving long-term therapy; hypokalemia and hyperglycemia; plasma cortisol with long-term therapy, normal: 138-635 nmol/L SI units drawn at 8 AM
- Weight daily; notify prescriber of weekly gain of >5 lb
- B/P, pulse; notify prescriber of chest pain; monitor for crackles, dyspnea if edema is present; hypertension, cardiac symptoms
- I&O ratio; be alert for decreasing urinary output, increasing edema
- **Infection:** increased temperature, WBC, even after withdrawal of medication; product masks infection
- Potassium depletion: paresthesias, fatigue, nausea, vomiting, depression, polyuria, dysrhythmias, weakness
- Mental status: affect, mood, behavioral changes, aggression

Evaluate:

- Therapeutic response: ease of respirations, decreased inflammation

Teach patient/family:

- That emergency ID as corticosteroid user should be carried; provide information about product being taken and condition
- To notify prescriber if therapeutic response decreases; that dosage adjustment may be needed
- To avoid vaccinations
- ⚠ Not to discontinue abruptly because adrenal crisis can result
- To avoid OTC products: salicylates, cough products with alcohol, cold preparations unless directed by prescriber
- **Cushingoid symptoms:** moon face, weight gain; symptoms of adrenal insufficiency: nausea, anorexia, fatigue, dizziness, dyspnea, weakness, joint pain
- That product causes immunosuppression; to report any symptoms of infection (fever, sore throat, cough)
- To notify prescriber if pregnancy is planned or suspected; cleft palate, stillbirth, abortion reported

pregabalin (Rx)

(pre-gab′a-lin)

Lyrica

Func. class.: Anticonvulsant

Chem. class.: γ-Aminobutyric acid (GABA) analog

Controlled Substance Schedule V

ACTION: Binds to high-voltage–gated calcium channels in CNS tissues; this may lead to anticonvulsant action similar to the inhibitory neurotransmitter GABA; anxiolytic, analgesics, and antiepileptic properties

USES: Neuropathic pain associated with spinal cord injury/diabetic peripheral neuropathy, partial-onset seizures, postherpetic neuralgia, fibromyalgia

Unlabeled uses: Moderate pain, social anxiety disorder

CONTRAINDICATIONS: Hypersensitivity to this product or gabapentin, abrupt discontinuation

Precautions: Pregnancy (C), breast-feeding, children <12 yr, geriatric patients, renal disease, PR interval prolongation, creatine kinase elevations, CHF (class III, IV), decreased platelets, substance abuse, dependence, glaucoma, myopathy, angioedema history, suicidal behavior

DOSAGE AND ROUTES

Diabetic peripheral neuropathic pain

- **Adult: PO/ORAL SOL** 50 mg tid, may increase to 300 mg/day (max) within 1 wk, adjust in patients with renal disease

Partial-onset seizures

- **Adult: PO/ORAL SOL** 75 mg bid or 50 mg tid; may increase to 600 mg/day (max)

Postherpetic neuralgia

- **Adult: PO/ORAL SOL** 150 mg/day in 2-3 divided doses, may increase to 300 mg/day in 2-3 divided doses; if higher dose is required after 2-4 wk, may increase to 600 mg/day in 2-3 divided doses

Fibromyalgia spinal cord injury, pain

- **Adult: PO/ORAL SOL** 75 mg bid, may increase to 150 mg bid within 1 wk and 225 mg bid after 1 wk

Renal dose

- **Adult: PO** CCr 30-60 ml/min, 75-300 mg/day in 2-3 divided doses; CCr 15-30 ml/min, 25-150 mg/day in 1-2 divided doses; CCr <15 ml/min, 25-75 mg/day as a single dose

Social phobia (unlabeled)

- **Adult: PO** 150-600 mg/day in 3 divided doses

Available forms: Caps 25, 50, 75, 100, 150, 200, 225, 300 mg; oral sol 20 mg/ml

Administer:

- Do not crush or chew caps; caps may be opened and contents put in applesauce or dissolved in juice
- Give without regard to meals
- Gradually withdraw over 7 days; abrupt withdrawal may precipitate seizures
- **Oral sol:** should be written in mg and calculated to mL

SIDE EFFECTS

CNS: Dizziness, fatigue, confusion, euphoria, incoordination, nervousness, neuropathy, tremor, vertigo, somnolence, ataxia, amnesia, abnormal thinking, suicidal ideation, drowsiness

EENT: Dry mouth, blurred vision, nystagmus, amblyopia, sinusitis

GI: Constipation, flatulence, abdominal pain, weight gain, nausea, vomiting, increased appetite

GU: Gynecomastia

HEMA: Ecchymosis, thrombocytopenia

MS: Back pain, rhabdomyolysis, myopathy

OTHER: Pruritus, orgasm/erectile dysfunction, peripheral edema, angioedema

RESP: Dyspnea

PHARMACOKINETICS

Well absorbed, peak 1.5 hr; 90% recovered in urine unchanged; negligible metabolism; not bound to plasma proteins; half-life 6 hr

INTERACTIONS

Increase: weight gain/fluid retention—thiazolidinedione; avoid use if possible

Increase: CNS depression—anxiolytics, sedatives, hypnotics, barbiturates, general anesthetics, opiate agonists, phenothiazines, sedating H_1 blockers, thiazolidinediones, tricyclics, alcohol

Drug/Lab Test

Increase: creatine kinase

Decrease: platelets

NURSING CONSIDERATIONS

Assess:

- **Seizures:** aura, location, duration, activity at onset, use seizure precaution
- **Pain:** location, duration, characteristics if using for diabetic neuropathy

• Renal studies: urinalysis, BUN, urine creatinine q3mo, creatine kinase; if markedly increased, discontinue product

⚠ Mental status: mood, sensorium, affect, behavioral changes, suicidal thoughts/behaviors; if mental status changes, notify prescriber

• **Angioedema/hypersensitivity:** monitor for blisters, hives, rash, dyspnea, wheezing; angioedema; if these occur, discontinue; cross-hypersensitivity with this product and gabapentin may occur

• **Rhabdomyolysis and creatinine kinase elevations:** monitor for muscle pain, tenderness, weakness accompanied by malaise or fever; product should be discontinued

Evaluate:

• Therapeutic response: decreased seizure activity; decrease in neuropathic pain

Teach patient/family:

• To carry emergency ID stating patient's name, products taken, condition, prescriber's name and phone number

• To avoid driving, other activities that require alertness because dizziness, drowsiness may occur

• Not to discontinue medication quickly after long-term use, to taper over ≥1 wk; that withdrawal-precipitated seizures may occur; not to double doses if dose is missed, to take if 2 hr or more before next dose

• To notify prescriber if pregnancy planned or suspected; to avoid breastfeeding

• To report muscle pain, tenderness, weakness when accompanied by fever, malaise, suicidal thoughts/behaviors

• To avoid alcohol, live virus vaccines

TREATMENT OF OVERDOSE:
Lavage, VS, hemodialysis

primaquine (Rx)
(prim′a-kween)

Func. class.: Antimalarial

Chem. class.: Synthetic 8-aminoquinolone

ACTION: Unknown; thought to destroy exoerythrocytic forms by gametocidal action

USES: Malaria caused by *Plasmodium vivax;* in combination with clindamycin for *Pneumocystis jiroveci* pneumonia

CONTRAINDICATIONS: Lupus erythematosus, rheumatoid arthritis; hypersensitivity to this product or idoquinol

Precautions: Pregnancy (C), breastfeeding, methemoglobin reductase deficiency

Black Box Warning: Bone marrow suppression, hemolytic anemia, G6PD deficiency

DOSAGE AND ROUTES

• **Adult: PO** 15-30 mg (base)/day × 2 wk or 45 mg (base)/wk × 8 wk; 26.3-mg tab is 15-mg base

• **Child: PO** 0.5 mg/kg (0.3 mg/base/day) daily × 2 wk

Available forms: Tabs 26.3 mg

Administer:

PO route

• Before or after meals at same time each day to maintain product level; take with food to decrease GI upset

SIDE EFFECTS

CNS: Headache, dizziness

CV: Hypertension, dysrhythmias

EENT: *Blurred vision, difficulty focusing*

GI: *Nausea, vomiting, anorexia,* cramps

HEMA: Agranulocytosis, granulocytopenia, leukopenia, hemolytic anemia, leuko-

P

cytosis, mild anemia, methemoglobinemia
INTEG: Pruritus, skin eruptions, pallor, weakness

PHARMACOKINETICS

PO: Metabolized by liver (metabolites), half-life 3.7-9.6 hr

INTERACTIONS

- Toxicity: quinacrine

Decrease: effect of carBAMazepine, PHENobarbital, phenytoins, rifamycins, nafcillin

Drug/Food

Increase: primaquine effect—food
Decrease: primaquine effect—grapefruit juice

Drug/Lab Test

Increase: WBC
Decrease: WBC, RBC, Hgb

NURSING CONSIDERATIONS

Assess:

- Ophthalmic test if patient receiving long-term treatment or product dosage of >150 mg/day
- Hepatic studies weekly: AST, ALT, bilirubin if patient receiving long-term therapy
- Blood studies: CBC; blood dyscrasias occur
- Allergic reactions: pruritus, rash, urticaria
- Blood dyscrasias: malaise, fever, bruising, bleeding (rare)
- Renal status: dark urine, hematuria, decreased output

⚠ **Hemolytic reaction:** chills, fever, chest pain, cyanosis; product should be discontinued immediately; hemolytic anemia may be severe in patients of Asian, Mediterranean descent

Evaluate:

- Therapeutic response: decreased symptoms of malaria

Teach patient/family:

- To report visual problems, fever, fatigue, dark urine, bruising, bleeding; may indicate blood dyscrasias
- To complete full course of therapy
- That an eye exam will be needed q4-6mo if using product for an extended period of time

primidone (Rx)

(pri′mi-done)

Mysoline, Sertan ✤

Func. class.: Anticonvulsant
Chem. class.: Barbiturate derivative

ACTION: Raises seizure threshold by conversion of product to PHENobarbital, decreases neuron firing

USES: Generalized tonic-clonic (grand mal), complex seizures
Unlabeled uses: Benign familial tremor (essential tremor)

CONTRAINDICATIONS: Pregnancy (D), breastfeeding, hypersensitivity to this product or barbiturates, porphyria
Precautions: Hyperactive children, COPD, renal/hepatic disease, suicidal ideation/behavior, hepatic encephalopathy, sleep apnea

DOSAGE AND ROUTES

- **Adult and child >8 yr: PO** 125-250 mg at bedtime, increase by 125-250 mg/day q3-7days, usual dose 750-1500 mg/day in 3-4 divided doses, max 2 g/day in divided doses
- **Child <8 yr: PO** 50-125 mg at bedtime, increase by 50-125 mg/day q3-7days, usual dose 10-25 mg/kg/day in 3-4 divided doses
- **Neonate: PO** 12-20 mg/kg/day in 2-4 divided doses, start at lower dose and titrate

Benign familial tremor/essential tremor (unlabeled)

- **Adult: PO** 50-62.5 mg, increase as tolerated up to 750 mg/day in 3 divided doses

Renal dose

• **Adult: PO** CCr 10-50 ml/min, increase interval between doses to 8-12 hr; CCr <10 ml/min, increase interval to 12-24 hr

Available forms: Tabs 50, 250 mg; susp 250 mg/5 ml ✦; chew tabs 125 mg ✦

Administer:

PO route

• After shaking liquid susp well
• With food for GI upset
• Tablets crushed and mixed with food or fluid for swallowing difficulties
• Avoid use with CNS depressants

SIDE EFFECTS

CNS: *Stimulation, drowsiness,* irritability, psychosis, ataxia, vertigo, fatigue, emotional disturbances, mood changes, paranoia, suicidal ideation

EENT: Diplopia, nystagmus, edema of eyelids, blurred vision, miosis

GI: *Nausea, vomiting, anorexia,* hepatitis

GU: Impotence

HEMA: Thrombocytopenia, leukopenia, neutropenia, eosinophilia, megaloblastic anemia, decreased serum folate level, lymphadenopathy

INTEG: *Rash,* edema, alopecia, lupuslike syndrome

MS: Osteopenia

RESP: Respiratory depression (dose related)

PHARMACOKINETICS

PO: Peak 4 hr; metabolized in liver; excreted by kidneys, in breast milk; half-life 10-12 hr (primidone)

INTERACTIONS

• Primidone levels decreased by acetaZOLAMIDE, succinimides
• May decrease effect of oral contraceptives, acebutolol, metoprolol, propranolol, tricyclics, phenothiazines, lamoTRIgine, estrogens, non-oral combination contraceptives, progestins

Increase: primidone levels—alcohol, heparin, CNS depressants, isoniazid, nicotinamide, MAOIs

Increase: toxicity—CYP3A4 inhibitors (aprepitant, antiretroviral protease inhibitors, delavirdine, fluconazole, imatinib, voriconazole)

Decrease: primidone effect—CYP3A4 inducers (barbiturates, carBAMazepine, efavirenz, phenytoins, nevirapine)

Drug/Herb

• Avoid use with kava, St. John's wort, valerian

Increase: effect—ginkgo

Decrease: effect—ginseng, santonica

NURSING CONSIDERATIONS

Assess:

• **Seizures:** location, duration, type; folic acid deficiency; fatigue, weakness, neuropathy, depression
• Product level: therapeutic level 5-12 mcg/ml; CBC, LFTs should be obtained q6mo

⚠ Mental status: mood, sensorium, affect, memory (long, short), suicidal thoughts/behaviors

• Respiratory depression, wheezing
• Blood dyscrasias: fever, sore throat, bruising, rash, jaundice

Evaluate:

• Therapeutic response: decreased seizures

Teach patient/family:

• Not to withdraw product quickly because withdrawal symptoms may occur
• To avoid hazardous activities until stabilized on product because drowsiness, dizziness may occur
• To carry emergency ID with condition, medication listed
• To recognize signs of blood dyscrasias; about when to notify prescriber
• To avoid alcohol
• To report rash
• To report suicidal thoughts/behaviors immediately
• **Contraception:** Use additional contraceptive during and for 1 mo after therapy, contraceptive may be reduced

P

probenecid (Rx)

(proe-ben′e-sid)

Func. class.: Uricosuric, antigout agent

Chem. class.: Sulfonamide derivative

ACTION: Inhibits tubular reabsorption of urates, with increased excretion of uric acids

USES: Hyperuricemia in gout, gouty arthritis, adjunct to penicillin treatment

CONTRAINDICATIONS: Hypersensitivity, severe renal/hepatic disease, CCr <50 mg/min, history of uric acid calculus

Precautions: Pregnancy (C), children <2 yr, sulfonamide hypersensitivity, pepticular

DOSAGE AND ROUTES

Adjunct to penicillin

- **Adult/adolescent >15 yr, >50 kg (110 lb): PO** 500 mg qid

Gout/gouty arthritis

- **Adult:** PO 250 mg bid for 1 wk, then 500 mg bid, max 2 g/day; maintenance 500 mg/day × 6 mo

Adjunct in penicillin treatment

- **Adult and adolescent >50 kg: PO** 500 mg qid
- **Child <50 kg: PO** 25 mg/kg, then 40 mg/kg in divided doses qid

Renal dose

- Avoid use if CCr <50 ml/min

Available forms: Tabs 500 mg

Administer:

- After meals or with milk if GI symptoms occur
- Increase fluid intake to 2-3 L/day to prevent urinary calculi

SIDE EFFECTS

CNS: Drowsiness, headache, flushing

CV: Bradycardia

GI: *Gastric irritation, nausea, vomiting, anorexia,* hepatic necrosis

GU: Glycosuria, thirst, frequency, nephrotic syndrome

INTEG: Rash, dermatitis, pruritus, fever

META: *Acidosis, hypokalemia, hyperchloremia,* hyperglycemia

RESP: Apnea, irregular respirations

PHARMACOKINETICS

Peak 2-4 hr, duration 8 hr, half-life 5-8 hr; metabolized by liver; excreted in urine

INTERACTIONS

Increase: effect of acyclovir, barbiturates, allopurinol, benzodiazepines, dyphylline, zidovudine, cephalosporins, penicillins, sulfonamides

Increase: toxicity—sulfa products, dapsone, clofibrate, indomethacin, rifampin, naproxen, methotrexate

Decrease: action of probenecid—salicylates

Drug/Lab Test

Increase: theophylline levels

NURSING CONSIDERATIONS

Assess:

- Uric acid levels (3-7 mg/dl); mobility, joint pain, swelling
- Respiratory rate, rhythm, depth; notify prescriber of abnormalities
- Electrolytes; CO_2 before, during treatment
- Urine pH, output, glucose during beginning treatment

⚠ For CNS symptoms: confusion, twitching, hyperreflexia, stimulation, headache; may indicate overdose

Evaluate:

- Therapeutic response: absence of pain, stiffness in joints

Teach patient/family:

- To avoid OTC preparations (aspirin) unless directed by prescriber; to increase water intake, avoid alcohol, caffeine

procainamide (Rx)

(proe-kane-ah′mide)

Func. class.: Antidysrhythmic (class IA)

Chem. class.: Procaine HCl amide analog

ACTION: Depresses excitability of cardiac muscle to electrical stimulation and slows conduction velocity in atrium, bundle of His, and ventricle; increases refractory period

USES: Life-threatening ventricular dysrhythmias

Unlabeled uses: Atrial fibrillation/flutter, paroxysmal atrial tachycardia, PSVT, Wolff-Parkinson-White (WPW) syndrome

CONTRAINDICATIONS: Hypersensitivity, severe heart block, torsades de pointes

Black Box Warning: Lupus erythematosus

Precautions: Pregnancy (C), breastfeeding, children, renal/hepatic disease, CHF, respiratory depression, cytopenia, dysrhythmia associated with digoxin toxicity, myasthenia gravis, digoxin toxicity

Black Box Warning: Bone marrow failure, cardiac arrhythmias

DOSAGE AND ROUTES

Ventricular tachycardia during CPR

- **Adult: IV** loading dose 20 mg/min; either ventricular tachycardia resolves or patient becomes hypotensive; QRS complex is widened by 50% of original width or total is 17 mg/kg (1.2 g for a 70-kg patient); may give up to 50 mg/min in urgent situations; maintenance: 1-4 mg/min **CONT IV INF; IM** 50 mg/kg/day in divided doses q3-6hr
- **Child: IV** PALS 15 mg/kg over 30-60 min

Renal dose

- **Adult: IV** CCr 35-59 ml/min, give 70% maintenance dose; CCr 15-34 ml/min, give 40%-60% maintenance dose; CCr <15 ml/min, individualize dose

Available forms: Inj 100, 500 mg/ml

Administer:

IM route

- IM inj in deltoid; aspirate to avoid intravascular administration; use only when unable to use IV

Direct IV route

- Dilute each 100 mg/10 ml of 0.9% NaCl; give at max 50 mg/min

Intermittent IV INF route

- Dilute 0.2-1 g/50-500 ml of D_5W (2-4 mg/ml); give over 30-60 min at max 25-50 mg/min; use inf pump

Y-site compatibilities: Alfentanil, amikacin, aminocaproic acid, aminophylline, amiodarone, amphotericin B lipid complex, amphotericin B liposome, anidulafungin, ascorbic acid, atenolol, atracurium, atropine, aztreonam, benztropine, bivalirudin, bleomycin, bumetanide, buprenorphine, butorphanol, calcium chloride/gluconate, caspofungin, ceFAZolin, cefmetazole, cefonicid, cefoperazone, cefotaxime, cefoTEtan, cefOXitin, cefTAZidime, cefTRIAXone, cefuroxime, cephalothin, chlorproMAZINE, cimetidine, cisatracurium, CISplatin, clindamycin, cyanocobalamin, cyclophosphamide, cycloSPORINE, cytarabine, DACTINomycin, DAPTOmycin, dexamethasone, digoxin, diphenhydrAMINE, DOBUTamine, DOCEtaxel, DOPamine, doxacurium, DOXOrubicin, doxycycline, enalaprilat, ePHEDrine, EPINEPHrine, epirubicin, epoetin alfa, eptifibatide, ertapenem, erythromycin, esmolol, etoposide, etoposide phosphate, famotidine, fenoldopam, fentaNYL, fluconazole, fludarabine, fluorouracil, folic acid, furosemide, gatifloxacin, gemcitabine, gentamicin, glycopyrrolate, granisetron, heparin, hydrocortisone, HYDROmorphone, IDArubicin, ifosfamide, indomethacin, insulin (regular), irinotecan, isoproterenol, ketorolac, labetalol, lido-

P

caine, linezolid, LORazepam, magnesium sulfate, mannitol, mechlorethamine, meperidine, metaraminol, methicillin, methotrexate, methoxamine, methyldopate, methylPREDNISolone, metoclopramide, metoprolol, mezlocillin, miconazole, midazolam, mitoXANtrone, morphine, moxalactam, multivitamins, mycophenolate, nafcillin, nalbuphine, naloxone, netilmicin, nitroglycerin, nitroprusside, norepinephrine, octreotide, ondansetron, oxacillin, oxaliplatin, oxytocin, PACLitaxel, palonosetron, pamidronate, pancuronium, pantoprazole, papaverine, PEMEtrexed, penicillin G potassium/sodium, pentamidine, pentazocine, PENTobarbital, PHENobarbital, phenylephrine, phytonadione, piperacillin, piperacillin-tazobactam, polymyxin B, potassium chloride, prochlorperazine, promethazine, propranolol, protamine, pyridoxine, quiNIDine, quinupristin-dalfopristin, ranitidine, remifentanil, ritodrine, rocuronium, sodium bicarbonate, succinylcholine, SUFentanil, tacrolimus, teniposide, theophylline, thiamine, thiotepa, ticarcillin, ticarcillin-clavulanate, tigecycline, tirofiban, tobramycin, tolazoline, trimetaphan, urokinase, vancomycin, vasopressin, vecuronium, verapamil, vinCRIStine, vinorelbine, vitamin B complex/C, voriconazole, zoledronic acid

SIDE EFFECTS

CNS: *Headache, dizziness,* confusion, psychosis, restlessness, irritability, weakness, depression

CV: *Hypotension,* heart block, cardiovascular collapse, arrest, torsades de pointes

GI: Nausea, vomiting, anorexia, diarrhea, hepatomegaly, pain, bitter taste

HEMA: SLE syndrome, agranulocytosis, thrombocytopenia, neutropenia, hemolytic anemia

INTEG: Rash, urticaria, edema, swelling (rare), pruritus, flushing, angioedema

SYST: SLE

PHARMACOKINETICS

Metabolized in liver to active metabolites, excreted unchanged by kidneys (60%), protein binding 15%

IM: Peak 10-60 min, half-life 3 hr

INTERACTIONS

Increase: effects of neuromuscular blockers

Increase: procainamide effects—cimetidine, quiNIDine, trimethoprim, β-blockers, ranitidine

Increase: toxicity—other antidysrhythmics, thioridazine, quinolones

Drug/Lab Test

Increase: ALT, AST, alk phos, LDH, bilirubin

NURSING CONSIDERATIONS

Assess:

> Black Box Warning: **Cardiac dysrhythmias:** ECG continuously if using IV to determine increased PR or QRS segments; discontinue immediately; watch for increased ventricular ectopic beats, maximum need to rebolus

- Therapeutic blood levels, 4-10 mcg/ml or NAPA levels 10-20 mcg/ml

> Black Box Warning: **Bone marrow suppression:** CBC q2wk × 3 mo; leukocyte, neutrophil, platelet counts may be decreased, treatment may need to be discontinued

- I&O ratio; electrolytes (K, Na, Cl), weight weekly, report gain of >2 lb

⚠ **Toxicity:** confusion, drowsiness, nausea, vomiting, tachydysrhythmias, oliguria

- ANA titer; during long-term treatment, watch for lupuslike symptoms
- Cardiac rate, rhythm, character, B/P continuously for fluctuations
- Respiratory status: rate, rhythm, character, lung fields; bilateral crackles may occur in CHF patient; watch for respiratory depression

⚠ CNS effects: dizziness, confusion, psychosis, paresthesias, seizures; product should be discontinued

Evaluate:

- Therapeutic response: decreased dysrhythmias

Teach patient/family:

- That wax matrix may appear in stools
- Not to discontinue without provider's approval

Black Box Warning: To notify prescriber immediately if lupuslike symptoms appear (joint pain, butterfly rash, fever, chills, dyspnea)

Black Box Warning: To notify prescriber of leukopenia (sore mouth, gums, throat) or thrombocytopenia (bleeding, bruising)

- How to take pulse and when to report to prescriber
- To avoid driving, other hazardous activities until product effect is known

TREATMENT OF OVERDOSE:

O_2, artificial ventilation, ECG, administer DOPamine for circulatory depression, diazepam or thiopental for seizures, isoproterenol

procarbazine (Rx)

(proe-kar′ba-zeen)

Matulane

Func. class.: Antineoplastic, alkylating agent

Chem. class.: Hydrazine derivative

ACTION:

Inhibits DNA, RNA, protein synthesis; has multiple sites of action; nonvesicant

USES:

Lymphoma, Hodgkin's disease, cancers resistant to other therapy

Unlabeled uses: Brain, lung malignancies; other lymphomas; multiple myeloma, malignant melanoma, polycythemia vera

CONTRAINDICATIONS:

Pregnancy (D), breastfeeding, hypersensitivity, thrombocytopenia, bone marrow depression

Precautions: Cardiac/renal/hepatic disease, radiation therapy, seizure disorder, anemia, bipolar disorder, Parkinson's disease

Black Box Warning: Requires a specialized care setting and an experienced clinician

DOSAGE AND ROUTES

- **Adult: PO** 2-4 mg/kg/day for 1st wk; maintain dosage of 4-6 mg/kg/day until platelets, WBC fall; after recovery, 1-2 mg/kg/day
- **Child: PO** 50 mg/m²/day for 7 days then 100 mg/m² until desired response, leukopenia, or thrombocytopenia occurs; 50 mg/m²/day maintenance after bone marrow recovery

Available forms: Caps 50 mg

Administer:

- In divided doses and at bedtime to minimize nausea and vomiting
- Nonphenothiazine antiemetic 30-60 min before product and 4-10 hr after treatment to prevent vomiting

SIDE EFFECTS

CNS: Headache, dizziness, insomnia, hallucinations, confusion, **coma,** pain, chills, fever, sweating, paresthesias, **seizures,** peripheral neuropathy

EENT: Retinal hemorrhage, nystagmus, photophobia, diplopia, dry eyes

GI: *Nausea, vomiting,* anorexia, diarrhea, constipation, dry mouth, stomatitis, elevated hepatic enzymes

GU: Azoospermia, cessation of menses

HEMA: **Thrombocytopenia, anemia, leukopenia, myelosuppression, bleeding tendencies,** purpura, petechiae, epistaxis, **hemolysis**

INTEG: *Rash,* pruritus, dermatitis, alopecia, herpes, hyperpigmentation

MS: Arthralgias, myalgias

RESP: Cough, pneumonitis, hemoptysis

SYST: **Secondary malignancy**

P

PHARMACOKINETICS

Half-life 1 hr; concentrates in liver, kidney, skin; metabolized in liver, excreted in urine

INTERACTIONS

⚠ **Increase:** hypotension—meperidine; do not use together

Increase: neuroleptic malignant syndrome, seizures, hyperpyrexia—alcohol, MAOIs, tricyclics, sympathomimetic products, SSRIs, SNRIs

Increase: hypertension—guanethidine, levodopa, methyldopa, reserpine, caffeine

⚠ Life-threatening hypertension: sympathomimetics

Increase: bleeding risk—NSAIDs, anticoagulants, platelet inhibitors, thrombolytics

Increase: CNS depression—barbiturates, antihistamines, opioids, hypotensive agents, phenothiazines

Drug/Food

- Hypertensive crisis: tyramine foods

NURSING CONSIDERATIONS

Assess:

- **Bone marrow suppression:** CBC, differential, platelet count weekly; withhold product if WBC is $<$4000/mm^3 or platelet count is $<$100,000/mm^3; notify prescriber
- **Hepatic/renal disease:** can cause accumulation of drug, increased toxicity; renal studies: BUN; serum uric acid; urine CCr; electrolytes before, during therapy; I&O ratio, report fall in urine output to $<$30 ml/hr; hepatic studies before, during therapy: bilirubin, AST, ALT, alk phos, LDH prn or monthly
- Monitor temperature; fever may indicate beginning infection

Black Box Warning: To be used only in a specialized care setting with emergency equipment

Black Box Warning: To be given only by an experienced clinician knowledgeable in cytotoxic products

- CNS changes: confusion, paresthesias, neuropathies; product should be discontinued

⚠ Tyramine foods in diet; hypertensive crisis can occur

⚠ **Toxicity:** facial flushing, epistaxis, increased PT, thrombocytopenia; product should be discontinued

- **Bleeding:** hematuria, guaiac stools, bruising or petechiae, mucosa or orifices q8hr
- Effects of alopecia on body image; discuss feelings about body changes
- Jaundiced skin, sclera; dark urine, clay-colored stools, itchy skin, abdominal pain, fever, diarrhea
- Buccal cavity for dryness, sores or ulceration, white patches, oral pain, bleeding, dysphagia
- GI symptoms: frequency of stools, cramping
- **Acidosis, signs of dehydration:** rapid respirations, poor skin turgor, decreased urine output, dry skin, restlessness, weakness

Evaluate:

- Therapeutic response: decreasing malignancy

Teach patient/family:

- To report any complaints, side effects to nurse or prescriber: CNS changes, diarrhea, cough, SOB, fever, chills, sore throat, bleeding, bruising, vomiting blood; black, tarry stools
- That hair may be lost during treatment and wig or hairpiece may make patient feel better; that new hair may be different in color, texture
- To avoid sunlight or UV exposure; to wear sunscreen or protective clothing
- To avoid foods with citric acid, hot or rough texture
- To report any bleeding, white spots, ulcerations in mouth to prescriber; to examine mouth daily
- To avoid driving, activities requiring alertness because dizziness may occur
- To use effective contraception; to avoid breastfeeding; that product may cause infertility

• To avoid the ingestion of alcohol, caffeine, tyramine-containing foods; that cold, hay fever, and weight-reducing products may cause serious product interactions; to avoid smoking
• To avoid crowds, persons with infections if granulocytes are low
• To avoid vaccines

prochlorperazine (Rx)

(proe-klor-pair′a-zeen)

Compro

Func. class.: Antiemetic, antipsychotic

Chem. class.: Phenothiazine, piperazine derivative

Do not confuse:
prochlorperazine/chlorproMAZINE

ACTION: Decreases DOPamine neurotransmission by increasing DOPamine turnover through the blockade of the D_2 somatodendritic autoreceptor in the mesolimbic system

USES: Nausea, vomiting, psychotic disorders

Unlabeled uses: Migraine

CONTRAINDICATIONS: Hypersensitivity to phenothiazines, coma; infants, neonates, children <2 yr or <20 lb; surgery

Precautions: Pregnancy (C), breastfeeding, geriatric patients, seizure, encephalopathy, glaucoma, hepatic disease, Parkinson's disease, BPH

Black Box Warning: Increased mortality in elderly patients with dementia-related psychosis

DOSAGE AND ROUTES

Postoperative nausea/vomiting

• **Adult: IM** 5-10 mg 1-2 hr before anesthesia; may repeat after 30 min; **IV** 5-10 mg 15-30 min before anesthesia; **IV INF** 20 mg/L D_5W or **NS** 15-30 min before anesthesia, max 40 mg/day

Severe nausea/vomiting

• **Adult: PO** 5-10 mg tid-qid; **SUS REL** 15 mg/day in AM or 10 mg q12hr; **RECT** 25 mg/bid; **IM** 5-10 mg q3-4hr prn, max 40 mg/day
• **Child 18-39 kg: PO** 2.5 mg tid or 5 mg bid; **IM** 0.132 mg/kg q3-4hr prn, max 15 mg/day
• **Child 14-17 kg: PO/RECT** 2.5 mg bid-tid; **IM** 0.132 mg/kg q3-4hr prn, max 10 mg/day
• **Child 9-13 kg: PO/RECT** 2.5 mg/day-bid; **IM** 0.132 mg/kg q3-4hr prn, max 7.5 mg/day

Antipsychotic

• **Adult and child ≥12 yr: PO** 5-10 mg tid-qid; may increase q2-3days, max 150 mg/day; **IM** 10-20 mg q2-4hr up to 4 doses then 10-20 mg q4-6hr
• **Child 2-12 yr: PO** 2.5 mg bid-tid; **IM** 0.132 mg/kg change to oral ASAP

Antianxiety

• **Adult and child ≥12 yr: PO** 5 mg tid-qid, max 20 mg/day
• **Child 2-12 yr: IM** 0.132 mg/kg change to oral ASAP

Available forms: Tabs 5, 10 mg; supp 25 mg

Administer:

• Avoid other CNS depressants

IM route

• IM inj in large muscle mass; aspirate to avoid IV administration
• Keep patient recumbent for ½ hr

Direct IV route

• No dilution needed; inject directly in a vein ≤5 mg/min; do not give as bolus

Intermittent IV INF route

• May dilute 20 mg/L NaCl and give as inf 15-30 min before anesthesia induction

Syringe compatibilities: Atropine, butorphanol, chlorproMAZINE, cimetidine, diamorphine, diphenhydrAMINE, droperidol, fentaNYL, glycopyrrolate, hydrOXYzine, meperidine, metoclopramide, nalbuphine, pentazocine, perphenazine, promazine, promethazine, ranitidine, scopolamine, SUFentanil

Y-site compatibilities: Amsacrine, calcium gluconate, cisatracurium, CISplatin,

P

cladribine, cyclophosphamide, cytarabine, DOXOrubicin, DOXOrubicin liposome, fluconazole, granisetron, heparin, hydrocortisone, melphalan, methotrexate, ondansetron, PACLitaxel, potassium chloride, propofol, remifentanil, sargramostim, SUFentanil, teniposide, thiotepa, vinorelbine, vit B/C

SIDE EFFECTS

CNS: Neuroleptic malignant syndrome, *extrapyramidal reactions, tardive dyskinesia, euphoria,* depression, *drowsiness,* restlessness, tremor, dizziness, headache
CV: Circulatory failure, tachycardia, hypotension, ECG changes
EENT: Blurred vision
GI: Nausea, vomiting, anorexia, dry mouth, diarrhea, constipation, weight loss, metallic taste, cramps
HEMA: Agranulocytosis
MISC: Impotence
RESP: Respiratory depression

PHARMACOKINETICS

Metabolized by liver; excreted in urine, breast milk; crosses placenta; 91%-99% protein binding
IM: Onset 10-20 min, duration 4-6 hr; children: 12 hr
PO: Onset 30-40 min, duration 3-4 hr
RECT: Onset 60 min, duration 3-4 hr

INTERACTIONS

Increase: anticholinergic action—anticholinergics, antiparkinson products, antidepressants
Increase: CNS depression—CNS depressants
Increase: serotonin syndrome, neuroleptic malignant syndrome—SSRIs, SNRIs
Decrease: prochlorperazine effect—barbiturates, antacids, lithium
Drug/Herb
Increase: CNS depression—chamomile, hops, kava, St. John's wort, valerian
Increase: EPS—kava
Drug/Lab Test
Increase: LFTs, cardiac enzymes, cholesterol, blood glucose, prolactin, bilirubin, PBI, ^{131}I, alk phos, leukocytes, granulocytes, platelets
Decrease: hormones (blood and urine)
False positive: pregnancy tests, urine bilirubin
False negative: urinary steroids, 17-OHCS, pregnancy tests

NURSING CONSIDERATIONS

Assess:
- **EPS:** abnormal movement, tardive dyskinesia, akathisia
- VS, B/P; check patients with cardiac disease more often

⚠ **Neuroleptic malignant syndrome:** seizures, hypo/hypertension, fever, tachycardia, dyspnea, fatigue, muscle stiffness, loss of bladder control; notify prescriber immediately

⚠ CBC, LFTs during course of treatment; blood dyscrasias, hepatotoxicity may occur

- Respiratory status before, during, after administration of emetic; check rate, rhythm, character; respiratory depression can occur rapidly among geriatric or debilitated patients

Evaluate:
- Therapeutic response: absence of nausea, vomiting; reduced anxiety, agitation, excitability

Teach patient/family:
- To avoid hazardous activities, activities requiring alertness because dizziness may occur
- To avoid alcohol
- Not to double or skip doses
- That urine may be pink to reddish brown
- That suppositories may contain coconut/palm oil
- To report dark urine, clay-colored stools, bleeding, bruising, rash, blurred vision
- To avoid sun; wear sunscreen, protective clothing

progesterone (Rx)

(proe-jess'ter-one)

Crinone, Endometrin, First-Progesterone, Prochieve, Prometrium

Func. class.: Progestogen

Chem. class.: Progesterone derivative

ACTION: Inhibits secretion of pituitary gonadotropins, which prevents follicular maturation, ovulation; stimulates growth of mammary tissue; antineoplastic action against endometrial cancer

USES: Contraception, amenorrhea, premenstrual syndrome, abnormal uterine bleeding, endometrial hyperplasia prevention, assisted reproductive technology (ART) gel

Unlabeled uses: Corpus luteum insufficiency, early pregnancy failure, PMS, preterm delivery prophylaxis

CONTRAINDICATIONS: Pregnancy (B), ectopic pregnancy; hypersensitivity to this product, peanuts, or peanut oil; thromboembolic disorders, reproductive cancer, genital bleeding (abnormal, undiagnosed), cerebral hemorrhage, PID, STDs, thrombophlebitis

Black Box Warning: Breast cancer

Precautions: Breastfeeding, hypertension, asthma, blood dyscrasias, CHF, diabetes mellitus, bone disease, depression, migraine headache, seizure disorders, gallbladder/renal/hepatic disease, family history of breast/reproductive tract cancer

Black Box Warning: Cardiac disease, dementia

DOSAGE AND ROUTES

Infertility

- **Adult: VAG** 90 mg/day (micronized gel); 100 mg 2-3 times/day starting day after oocyte retrieval and for ≤10 wk total (insert)

Amenorrhea/functional uterine bleeding

- **Adult: IM** 5-10 mg/day × 6-8 doses

Endometrial hyperplasia prevention

- **Adult: PO** 200 mg/day × 12 days

Assisted reproductive therapy

- **Adult: GEL** 90 mg (8%) vaginally daily for supplementation; 90 mg (8%) vaginally bid for replacement; if pregnancy occurs, continue × 10-12 wk

Corpus luteum insufficiency (unlabeled)

- **Adult: VAG INSERT** 90-100 mg bid-tid starting at oocyte retrieval and continuing up to 10-12 wk gestation

Available forms: Inj 50 mg/ml; vag gel 4%, 8%; caps 100, 200 mg; vag insert 100 mg; vag supp 25, 100, 200, 500 mg; compounding kit 25, 50, 100, 200, 400 mg

Administer:

PO route

- Do not break, crush, or chew caps
- Titrated dose; use lowest effective dose
- In 1 dose in AM
- With food or milk to decrease GI symptoms
- Start progesterone 14 days after estrogen dose if given concomitantly

Vaginal route

- Wait at least 6 hr after any vaginal treatment before using vaginal gel

IM route

- Shake vial, inject deeply into large muscle, aspirate
- Check for particulate matter and discoloration before injecting

SIDE EFFECTS

CNS: *Dizziness, headache,* migraines, depression, *fatigue,* mood swings, dementia, drowsiness

CV: Hypotension, **thrombophlebitis**, edema, **thromboembolism, stroke, pulmonary embolism, MI**

EENT: Diplopia, retinal thrombosis

GI: *Nausea,* vomiting, anorexia, cramps, increased weight, **cholestatic jaundice**, *constipation,* abdominal pain

GU: Amenorrhea, cervical erosion, breakthrough bleeding, dysmenorrhea,

vaginal candidiasis, nocturia, breast changes, *gynecomastia, testicular atrophy, impotence,* endometriosis, spontaneous abortion, breast pain, ectopic pregnancy
INTEG: Rash, urticaria, acne, hirsutism, alopecia, oily skin, seborrhea, purpura, melasma
META: Hyperglycemia
SYST: Angioedema, anaphylaxis

PHARMACOKINETICS

Excreted in urine, feces; metabolized in liver
IM/RECT/VAG: Duration 24 hr

INTERACTIONS

Increase: progesterone effect—CYP3A4 inhibitors (ketoconazole, cimetidine, clarithromycin, danazol, diltiazem, erythromycin, fluconazole, itraconazole, troleandomycin, verapamil, voriconazole)
Decrease: progesterone effect—barbiturates, phenytoin
Drug/Lab Test
Increase: alk phos, nitrogen (urine), pregnanediol, amino acids, factors VII, VIII, IX, X
Decrease: GTT, HDL

NURSING CONSIDERATIONS

Assess:

- **Abnormal uterine bleeding:** vaginal bleeding; obtain pad count, patient menstrual history, breast exam, cervical cytology
- Weight daily; notify prescriber of weekly weight gain of >5 lb
- B/P at beginning of treatment and periodically
- I&O ratio; be alert for decreasing urinary output, increasing edema
- Hepatic studies: ALT, AST, bilirubin periodically during long-term therapy
- Edema, hypertension, cardiac symptoms, jaundice, thromboembolism
- Mental status: affect, mood, behavioral changes, depression
- Hypercalcemia

Evaluate:

- Therapeutic response: decreased abnormal uterine bleeding, absence of amenorrhea

Teach patient/family:

- ⚠ To report breast lumps, vaginal bleeding, edema, jaundice, dark urine, clay-colored stools, dyspnea, headache, blurred vision, abdominal pain, numbness or stiffness in legs, chest pain
- To avoid gel with other vaginal products; if to be used together, to separate by ≥6 hr; for vaginal route, on proper insertion technique
- To report suspected pregnancy
- To monitor blood glucose if diabetic
- To avoid activities requiring mental alertness until effects are realized; can cause dizziness

promethazine (Rx)

(proe-meth'a-zeen)

Phenadoz, Phenergan, Promethagan

Func. class.: Antihistamine, H_1-receptor antagonist, antiemetic
Chem. class.: Phenothiazine derivative

Do not confuse:
Phenergan/Theragran

ACTION: Acts on blood vessels, GI, respiratory system by competing with histamine for H_1-receptor sites; decreases allergic response by blocking histamine

USES: Motion sickness, rhinitis, allergy symptoms, sedation, nausea, preoperative and postoperative sedation
Unlabeled uses: Allergic rhinitis, acute peripheral vestibular nystagmus, hyperemesis gravidarum

CONTRAINDICATIONS: Hypersensitivity, breastfeeding, agranulocytosis, bone marrow suppression, coma, jaundice, Reye's syndrome

Black Box Warning: Infants, neonates, children, intraarterial/SUBCUT administration, extravasation

Precautions: Pregnancy (C), cardiac/renal/hepatic disease, asthma, seizure disorder, prostatic hypertrophy, bladder obstruction, glaucoma, COPD, GI obstruction, ileus, CNS depression, diabetes, sleep apnea, urinary retention

Black Box Warning: IV use

DOSAGE AND ROUTES

Nausea/vomiting

- **Adult: PO/IM/IV/RECT** 12.5-25 mg; q4-6hr prn
- **Child >2 yr: PO/IM/IV/RECT** 0.25-0.5 mg/kg q4-6hr prn

Motion sickness

- **Adult: PO** 25 mg bid, give ½-1 hr before departure then q8-12hr prn
- **Child ≥2 yr: PO/IM/RECT** 12.5-25 mg bid, give ½-1 hr before departure then q8-12hr prn

Sedation

- **Adult: PO/IM** 25-50 mg at bedtime
- **Child ≥2 yr: PO/IM/RECT** 12.5-25 mg at bedtime

Sedation (preoperative/postoperative)

- **Adult: PO/IM/IV** 25-50 mg
- **Child >2 yr: PO/IM/IV** 0.5-1.1 mg/kg

Allergy/rhinitis (unlabeled)

- **Adult: PO** 12.5 mg qid or 25 mg at bedtime
- **Child ≥2 yr: PO** 6.25-12.5 mg tid or 25 mg at bedtime

Hyperemesis gravidarum (unlabeled)

- **Pregnant females: PO/RECT/IM/IV** 12.5-25 mg q4hr

Nystagmus (unlabeled)

- **Adult: PO** 12.5-25 mg q4-6hr for ≤48 hr

Available forms: Tabs 12.5, 25, 50 mg; supp 12.5, 25, 50 mg; inj 25, 50 mg/ml

Administer:

- Avoid use with other CNS depressants

PO route

- With meals for GI symptoms; absorption may slightly decrease
- When used for motion sickness, 30 min-1 hr before travel

IM route

- IM inj deep in large muscle; rotate site

Direct IV route

Black Box Warning: Check for extravasation: burning, pain, swelling at IV site; can cause tissue necrosis

- Do not use if precipitate is present
- Rapid administration may cause transient decrease in B/P
- After diluting each 25-50 mg/9 ml of NaCl for inj; give ≤25 mg/2 min

Syringe compatibilities: Butorphanol, chlorproMAZINE, cimetidine, dihydroergotamine, diphenhyDRAMINE, droperidol, fentaNYL, glycopyrrolate, HYDROmorphone, hydrOXYzine, meperidine, metoclopramide, midazolam, pentazocine, perphenazine, prochlorperazine, promazine, ranitidine, scopolamine

Y-site compatibilities: Alfentanil, amifostine, amikacin, aminocaproic acid, amsacrine, anidulafungin, ascorbic acid, atenolol, atracurium, atropine, aztreonam, benztropine, bivalirudin, bleomycin, bumetanide, buprenorphine, butorphanol, calcium chloride/gluconate, CARBOplatin, caspofungin, chlorproMAZINE, cimetidine, ciprofloxacin, cisatracurium, CISplatin, cladribine, codeine, cyanocobalamin, cyclophosphamide, cycloSPORINE, cytarabine, DACTINomycin, DAPTOmycin, dexmedetomidine, digoxin, diltiazem, diphenhydrAMINE, DOBUTamine, DOCEtaxel, DOPamine, doxacurium, DOXOrubicin, doxycycline, enalaprilat, ePHEDrine, EPINEPHrine, epirubicin, epoetin, eptifibatide, erythromycin, esmolol, etoposide, famotidine, fenoldopam, fentaNYL, filgrastim, fluconazole, fludarabine, gemcitabine, gentamicin, glycopyrrolate, granisetron, HYDROmorphone, hydrOXYzine, IDArubicin, ifosfamide, insulin (regular), irinotecan, isoproterenol, labetalol, levofloxacin, lidocaine, linezolid, LORazepam, magnesium sulfate, mannitol, mechlorethamine, melphalan, meperidine, metaraminol, methoxamine, methyldopate,

P

metoclopramide, metoprolol, metroNIDAZOLE, miconazole, midazolam, milrinone, mitoXANtrone, morphine, mycophenolate, nalbuphine, naloxone, netilmicin, nitroglycerin, norepinephrine, octreotide, ondansetron, oxaliplatin, oxytocin, PACLitaxel, palonosetron, pamidronate, pancuronium, PEMEtrexed, pentamidine, pentazocine, phenylephrine, polymyxin B, procainamide, prochlorperazine, propranolol, protamine, pyridoxine, quiNIDine, quinupristin-dalfopristin, ranitidine, remifentanil, Ringer's, ritodrine, riTUXimab, rocuronium, sargramostim, sodium acetate, succinylcholine, SUFentanil, tacrolimus, teniposide, theophylline, thiamine, thiotepa, tigecycline, tirofiban, TNA, tobramycin, tolazoline, trastuzumab, trimetaphan, vancomycin, vasopressin, vecuronium, verapamil, vinCRIStine, vinorelbine, voriconazole

SIDE EFFECTS

CNS: *Dizziness, drowsiness,* poor coordination, fatigue, anxiety, euphoria, confusion, paresthesia, neuritis, EPS, neuroleptic malignant syndrome
CV: Hypo/hypertension, palpitations, tachycardia
EENT: Blurred vision, dilated pupils, tinnitus, nasal stuffiness; dry nose, throat, mouth; photosensitivity
GI: *Constipation,* dry mouth, nausea, vomiting, anorexia, diarrhea
GU: *Urinary retention,* dysuria, frequency
HEMA: Thrombocytopenia, agranulocytosis, hemolytic anemia
INTEG: Rash, urticaria, photosensitivity
RESP: Increased thick secretions, wheezing, chest tightness; apnea in neonates, infants, young children

PHARMACOKINETICS

Metabolized in liver; excreted by kidneys, GI tract (inactive metabolites)
PO: Onset 20 min, duration 4-12 hr
IV: Onset 3-5 min

INTERACTIONS

Increase: CNS depression—barbiturates, opioids, hypnotics, tricyclics, alcohol
Increase: promethazine effect—MAOIs
Decrease: oral anticoagulants effect—heparin
Drug/Lab Test
False negative: skin allergy test
False positive: urine pregnancy test
Interference: blood grouping (ABO), GTT

NURSING CONSIDERATIONS

Assess:

Black Box Warning: Not to be used in children <2 yr, fatal respiratory depression may occur; use cautiously in children >2 yr, seizures, paradoxical CNS stimulation may occur

- **Antiemetic/motion sickness:** nausea, vomiting before, after dose
- I&O ratio; be alert for urinary retention, frequency, dysuria; product should be discontinued
- ⚠ CBC with differential, LFTs during long-term therapy; blood dyscrasias, jaundice may occur
- Respiratory status: rate, rhythm, increase in bronchial secretions, wheezing, chest tightness
- Cardiac status: palpitations, increased pulse, hypo/hypertension, B/P in those receiving IV doses
- **Neuroleptic malignant syndrome:** fever, confusion, diaphoresis, rigid muscles, elevated CPK, encephalopathy; discontinue product, notify prescriber
- Hard candy, gum, frequent rinsing of mouth for dryness

Evaluate:

- Therapeutic response: absence of running, congested nose; rashes; absence of motion sickness, nausea; sedation

Teach patient/family:

- That product may cause photosensitivity; to avoid prolonged exposure to sunlight
- To notify prescriber of confusion, sedation, hypotension, jaundice, fever

- To avoid driving, other hazardous activity if drowsy
- To avoid concurrent use of alcohol or other CNS depressants
- That product may reduce sweating; that there is a risk of heat stroke
- How to use frequent sips of water, gum to decrease dry mouth

propafenone (Rx)

(pro-paff'e-nown)

Rythmol, Rythmol SR

Func. class.: Antidysrhythmic (class IC)

ACTION: Slows conduction velocity; reduces membrane responsiveness; inhibits automaticity; increases ratio of effective refractory period to action potential duration; β-blocking activity

USES: Sustained ventricular tachycardia, atrial fibrillation (single dose), paroxysmal supraventricular tachycardia (PSVT) prophylaxis, supraventricular dysrhythmias

Unlabeled uses: Wolff-Parkinson-White (WPW) syndrome

CONTRAINDICATIONS: 2nd/3rd-degree AV block, right bundle branch block, cardiogenic shock, hypersensitivity, bradycardia, uncontrolled CHF, sick-sinus syndrome, marked hypotension, bronchospastic disorders, electrolyte imbalance, Brugada syndrome

Precautions: Pregnancy (C), breastfeeding, children, geriatric patients, CHF, hypo/hyperkalemia, nonallergic bronchospasm, renal/hepatic disease, hematologic disorders, myasthenia gravis, COPD

Black Box Warning: Recent MI, cardiac arrhythmias, QT prolongation, torsades de pointes

DOSAGE AND ROUTES

PSVT

- **Adult: PO** 150 mg q8hr; allow 3-4 day interval before increasing dose, max 900 mg/day

Atrial fibrillation

- **Adult: PO** 450 or 600 mg as single dose; SR 225 mg q12hr, may increase to 325 q12hr, max 425 mg q12hr

Available forms: Tabs 150, 225, 300 mg; SR cap 225, 325, 425 mg

Administer:

- Do not break, crush, or chew tabs; swallow whole
- To hospitalized patients because heart monitoring is required
- After hypo/hyperkalemia is corrected
- With dosage adjustment q3-4days
- Without regard to meals

SIDE EFFECTS

CNS: Headache, dizziness, abnormal dreams, syncope, confusion, **seizures**, insomnia, tremor, anxiety, fatigue

CV: **Supraventricular dysrhythmia, ventricular dysrhythmia, bradycardia**, prodysrhythmia, palpitations, AV block, intraventricular conduction delay, AV dissociation, hypotension, chest pain, **asystole**

EENT: Blurred vision, altered taste, tinnitus

GI: *Nausea, vomiting,* constipation, dyspepsia, cholestasis, abnormal hepatic studies, dry mouth, diarrhea, anorexia

HEMA: **Leukopenia, agranulocytosis, granulocytopenia, thrombocytopenia,** anemia, bruising

INTEG: Rash

RESP: Dyspnea

PHARMACOKINETICS

Peak 3-8 hr, half-life 2-10 hr, poor metabolizers 10-32 hr; metabolized in liver; excreted in urine (metabolite)

INTERACTIONS

Increase: propafenone effects—CYP1A2, CYP2D6, CYP3A4 inhibitors (protease inhibitors, quiNINE, PARoxetine, saquinavir, erythromycin, azole antifungals, sertraline, tricyclics)

Increase: QT prolongation—other class IA/IC antidysrhythmics, arsenic trioxide, chloroquine, clarithromycin, droperidol, erythromycin, haloperidol, levomethadyl, methadone, pentamidine, chlorproMAZINE, mesoridazine, thioridazine

Increase: anticoagulation—warfarin

Increase: CNS effects—local anesthetics

Increase: digoxin level—digoxin

Increase: β-blocker effect—propranolol, metoprolol

Increase: cycloSPORINE levels—cycloSPORINE

Decrease: propafenone effect—rifampin, cimetidine, quiNIDine

Drug/Food

Increase: propafenone effect—grapefruit juice

Drug/Herb

Decrease: propafenone effect—St. John's wort

Drug/Lab Test

Increase: CPK

NURSING CONSIDERATIONS

Assess:

- GI status: bowel pattern, number of stools

Black Box Warning: QT/PR prolongation: ECG or Holter monitor before and during therapy

Black Box Warning: CHF: dyspnea, jugular venous distention, crackles, edema in extremities, I&O ratio; check for decreasing output; daily weight

- CBC, ANA titer, LFTs
- Chest x-ray, pulmonary function test during treatment
- Lung fields; bilateral crackles, dyspnea, peripheral edema, weight gain; jugular venous distention may occur in patient with CHF

⚠ **Toxicity:** fine tremors, dizziness, hypotension, drowsiness, abnormal heart rate

Evaluate:

- Therapeutic response: absence of ventricular dysrhythmias; decreasing recurrence of PAF, PSVT

Teach patient/family:

- To avoid hazardous activities until response is known
- To report fever, chills, sore throat, bleeding, SOB, chest pain, palpitations, blurred vision
- To take tab with food, not to use with grapefruit juice or St. John's wort
- To carry emergency ID identifying medication and prescriber
- To avoid abrupt discontinuation of product; to take as prescribed; not to miss, double doses

TREATMENT OF OVERDOSE:

O_2, artificial ventilation, defibrillation ECG; administer DOPamine for circulatory depression, diazepam or thiopental for seizures, isoproterenol

proparacaine ophthalmic

See Appendix B

⚠ HIGH ALERT

propofol (Rx)

(pro′poh-fole)

Diprivan, Fresenius Propoven

Func. class.: General anesthetic

ACTION: Produces dose-dependent CNS depression by activation of GABA receptor

USES: Induction or maintenance of anesthesia as part of balanced anesthetic technique; sedation in mechanically ventilated patients

CONTRAINDICATIONS: Hypersensitivity to this product or soybean oil, egg, benzyl alcohol (some products)

Precautions: Pregnancy (B), breast-feeding, children, geriatric patients, respiratory depression, severe respiratory disorders, cardiac dysrhythmias, labor and delivery, renal disease, hyperlipidemia

DOSAGE AND ROUTES

Anesthesia

• **Adult <55 yr and ASA I/II IV (Diprivan or generic): IV** 40 mg q10sec until induction onset, maintenance 100-200 mcg/kg/min or **IV BOL** 20-50 mg prn, allow 3-5 min between adjustments; Fresenius Propoven 1% **IV** 20-40 q10sec until induction then 3-6 mg/kg/hr

• **Child ≥3 yr or ASA I or II: IV induction:** 2.5-3.5 mg/kg over 20-30 sec when not premedicated or lightly premedicated

• **Child 2 mo-16 yr maintenance: IV** 125-300 mcg/kg/min, lower dose for ASA III or **IV**

ICU sedation

• **Adult: IV** 5 mcg/kg/min over 5 min; may increase by 5-10 mcg/kg/min over 5-10 min until desired response (Diprivan or generic): 0.3-4 mg/kg/hr, max 4 mg/kg/hr (Fresenius Propoven)

Available forms: Inj 10 mg/ml in 20-ml ampule, 50-ml and 100-ml vials

Administer:

IV route

• Shake well before use; dilution is not necessary but, if diluted, use only D_5W to ≥2 mg/ml; give over 3-5 min, titrate to needed level of sedation; use only glass containers when mixing, not stable in plastic; use aseptic technique when transferring from original container

• **Fresenius Propoven 1%:** dilution is not necessary but, if diluted, use D_5W or NS to ≥2 mg/ml (max dilution max 1 part Fresenius Propoven/4 parts D_5W or NS); do not admix; lidocaine can be used to reduce pain at site

• May be given by cont inf; give by inf pump

• Only with resuscitative equipment available; only by qualified persons trained in anesthesia

• Store in light-resistant area at room temperature, use within 6 hr of opening

• If transferred from original container to another container, complete inf within 12 hr (Dipravan), 6 hr (generic propofol)

Y-site compatibilities: Acyclovir, alfentanil, aminophylline, ampicillin, aztreonam, bumetanide, buprenorphine, butorphanol, calcium gluconate, CARBOplatin, ceFAZolin, cefoperazone, cefotaxime, cefoTEtan, cefOXitin, ceftizoxime, cefTRIAXone, cefuroxime, chlorproMAZINE, cimetidine, CISplatin, clindamycin, cyclophosphamide, cycloSPORINE, cytarabine, dexamethasone, diphenhydrAMINE, DOBUTamine, DOPamine, doxycycline, droperidol, enalaprilat, ePHEDrine, EPINEPHrine, esmolol, famotidine, fentaNYL, fluconazole, fluorouracil, furosemide, ganciclovir, glycopyrrolate, granisetron, haloperidol, heparin, hydrocortisone, HYDROmorphone, hydrOXYzine, ifosfamide, imipenem/cilastatin, inamrinone, regular insulin, isoproterenol, ketamine, labetalol, levorphanol, lidocaine, LORazepam, magnesium sulfate, mannitol, meperidine, mezlocillin, miconazole, morphine, nafcillin, nalbuphine, naloxone, nitroglycerin, norepinephrine, ofloxacin, PACLitaxel, PENTobarbital, PHENobarbital, piperacillin, potassium chloride, prochlorperazine, propranolol, ranitidine, scopolamine, sodium bicarbonate, sodium nitroprusside, succinylcholine, SUFentanil, thiopental ticarcillin, ticarcillin/clavulanate, vecuronium, verapamil

Solution compatibilities: if given together via Y-site: D_5W, D_5LR, LR, $D_5/0.45\%$ NaCl, $D_5/0.2\%$ NaCl

SIDE EFFECTS

CNS: Involuntary movement, headache, jerking, fever, dizziness, shivering, tremor, confusion, somnolence, paresthesia, agitation, abnormal dreams, euphoria fatigue, **increased intracranial pressure, impaired cerebral flow, seizures**

CV: *Bradycardia, hypotension,* hypertension, PVC, PAC, tachycardia, abnormal ECG, ST segment depression, asystole, bradydysrhythmias
EENT: Blurred vision, tinnitus, eye pain, strange taste, diplopia
GI: *Nausea, vomiting, abdominal cramping,* dry mouth, swallowing, hypersalivation, pancreatitis
GU: Urine retention, green urine, cloudy urine, oliguria
INTEG: *Flushing, phlebitis, hives, burning/stinging at inj site,* rash, pain of extremities
MS: Myalgia
RESP: Apnea, *cough, hiccups,* dyspnea, hypoventilation, sneezing, wheezing, tachypnea, hypoxia, respiratory acidosis
SYST: Propofol inf syndrome

PHARMACOKINETICS

Onset 15-30 sec, rapid distribution, half-life 1-8 min, terminal half-life 3-12 hr; 70% excreted in urine; metabolized in liver by conjugation to inactive metabolites, 95%-99% protein binding

INTERACTIONS

- Do not use within 10 days of MAOIs

Increase: CNS depression—alcohol, opioids, sedative/hypnotics, antipsychotics, skeletal muscle relaxants, inhalational anesthetics

Drug/Herb

Increase: propofol effect—St. John's wort

NURSING CONSIDERATIONS

Assess:

- Inj site: phlebitis, burning, stinging
- **ECG** for changes: PVC, PAC, ST segment changes; monitor VS
- **Neurologic excitatory symptoms:** movement, tremors, dizziness, LOC, pupil reaction
- Allergic reactions: hives

⚠ **Respiratory dysfunction:** respiratory depression, character, rate, rhythm; notify prescriber if respirations are <10/min

- **Propofol inf syndrome:** rhabdomyolysis, renal failure, hyperkalemia, metabolic acidosis, cardiac dysrhythmias, heart failure usually between 35 and 93 hr after inf begun at >5 mg/kg/hr for >58 hr

Evaluate:

- Therapeutic response: induction of anesthesia

Teach patient/family:

- That product will cause dizziness, drowsiness, sedation; to avoid hazardous activities until drug effect wears off

TREATMENT OF OVERDOSE:

Discontinue product; administer vasopressor agents or anticholinergics, artificial ventilation

propranolol (Rx)

(proe-pran′oh-lole)

Inderal, Inderal LA, InnoPran XL

Func. class.: Antihypertensive, antianginal, antidysrhythmic (class II)

Chem. class.: β-Adrenergic blocker

Do not confuse:
propranolol/Pravachol
Inderal/Toradol/Inderide/Adderall/Imuran

ACTION: Nonselective β-blocker with negative inotropic, chronotropic, dromotropic properties

USES: Chronic stable angina pectoris, hypertension, supraventricular dysrhythmias, migraine prophylaxis, pheochromocytoma, cyanotic spells related to hypertrophic subaortic stenosis, essential tremor, acute MI

Unlabeled uses: Anxiety, prevention of variceal bleeding caused by portal hypertension, akathisia induced by antipsychotics, portal hypertension, sclerodermal renal crisis, unstable angina, infantile capillary hemangioma, lithium-induced tremor

CONTRAINDICATIONS:

Hypersensitivity to this product; cardiogenic shock, AV heart block; bronchospastic disease; sinus bradycardia; bronchospasm; asthma

Precautions: Pregnancy (C), breastfeeding, children, diabetes mellitus, hyperthyroidism, COPD, renal/hepatic disease, myasthenia gravis, peripheral vascular disease, hypotension, cardiac failure, Raynaud's disease, sick sinus syndrome, vasospastic angina, smoking, Wolff-Parkinson-White syndrome, thyrotoxicosis

Black Box Warning: Abrupt discontinuation

DOSAGE AND ROUTES

Dysrhythmias

• **Adult: PO** 10-30 mg tid-qid; **IV BOL** 1-3 mg give 1 mg/min; may repeat after 2 min, may repeat q4hr thereafter

• **Child: PO** 1 mg/kg/day in 2 divided doses; **IV** 0.01-0.1 mg/kg over 5 min

Hypertension

• **Adult: PO** 40 mg bid or 80 mg/day (ext rel) initially; usual dose 120-240 mg/day bid-tid or 120-160 mg/day (ext rel)

• **Child: PO** 0.5-1 mg/kg/day divided q6-12hr

Angina

• **Adult: PO** 10-20 mg bid-qid, increase at 3-7 day intervals up to 160-320 mg/day or 80 mg daily

MI prophylaxis

• **Adult: PO** 180-240 mg/day tid-qid starting 5 days to 2 wk after MI

Pheochromocytoma

• **Adult: PO** 60 mg/day × 3 days preoperatively in divided doses or 30 mg/day in divided doses (inoperable tumor)

Migraine

• **Adult: PO** 80 mg/day (ext rel) or in divided doses; may increase to 160-240 mg/day in divided doses

• **Child >35 kg (unlabeled): PO** 20-40 mg tid

• **Child ≤35 kg (unlabeled): PO** 10-20 mg tid

Essential tremor

• **Adult: PO** 40 mg bid; usual dose 120 mg/day

Acute MI

• **Adult: PO** 180-320 mg/day in 3-4 divided doses; IV 0.1 mg/kg in 3 divided doses at 2-3 min intervals

Anxiety (unlabeled)

• **Adult: PO** 10-80 mg given 1 hr before anxiety-producing event

Scleroderma renal crisis (unlabeled)

• **Adult: PO** 40 mg bid, may increase q3-7days, max 160-480 mg/day

Esophageal varices (portal hypertension) (unlabeled)

• **Adult: PO** 40 mg bid, titrate to heart rate reduction of 25%

Infantile capillary hemangioma (unlabeled)

• **Infant: PO** 2-3 mg/kg/day

Available forms: Ext rel caps 60, 80, 120, 160 mg; tabs 10, 20, 40, 60, 80, 90 mg; inj 1 mg/ml; oral sol 4 mg/ml, 8 mg/ml

Administer:

PO route

• Do not break, crush, chew, or open ext rel cap

• Do not use ext rel cap for essential tremor, MI, cardiac dysrhythmias; do not use InnoPran XL in hypertropic subaortic stenosis, migraine, angina pectoris

• Ext rel caps should be taken daily; InnoPran XL should be taken at bedtime

• May mix oral sol with liquid or semisolid food; rinse container to get entire dose

• With 8 oz water with food; food enhances bioavailability

• Do not give with aluminum-containing antacid; may decrease GI absorption

Direct IV route

• IV undiluted or diluted 10 ml D_5W for inj; give ≤1 mg/min

Intermittent IV INF route

• May be diluted in 50 ml NaCl and run 1 mg over 10-15 min

Y-site compatibilities: Acyclovir, alfentanil, alteplase, amikacin, aminocaproic

P

acid, aminophylline, anidulafungin, ascorbic acid, atracurium, atropine, azaTHIOprine, aztreonam, benztropine, bivalirudin, bleomycin, bumetanide, buprenorphine, butorphanol, calcium chloride/gluconate, CARBOplatin, carmustine, caspofungin, cefamandole, ceFAZolin, cefmetazole, cefonicid, cefoperazone, cefotaxime, cefoTEtan, cefOXitin, cefTAZidime, ceftizoxime, cefTRIAXone, cefuroxime, cephalothin, cephapirin, chloramphenicol, chlorproMAZINE, cimetidine, CISplatin, clindamycin, cyanocobalamin, cyclophosphamide, cycloSPORINE, cytarabine, DACTINomycin, DAPTOmycin, dexamethasone, digoxin, diltiazem, diphenhydrAMINE, DOBUTamine, DOCEtaxel, DOPamine, doxacurium, DOXOrubicin, doxycycline, enalaprilat, ePHEDrine, EPINEPHrine, epirubicin, epoetin alfa, eptifibatide, ertapenem, erythromycin, esmolol, etoposide, etoposide phosphate, famotidine, fenoldopam, fentaNYL, fluconazole, fludarabine, fluorouracil, folic acid, furosemide, ganciclovir, gatifloxacin, gemcitabine, gemtuzumab, gentamicin, glycopyrrolate, granisetron, heparin, hydrocortisone, HYDROmorphone, hydrOXYzine, IDArubicin, ifosfamide, imipenem-cilastatin, inamrinone, irinotecan, isoproterenol, ketorolac, labetalol, levofloxacin, lidocaine, linezolid, LORazepam, magnesium, mannitol, mechlorethamine, meperidine, metaraminol, methicillin, methotrexate, methoxamine, methyldopate, methylPREDNISolone, metoclopramide, metoprolol, metroNIDAZOLE, mezlocillin, miconazole, midazolam, milrinone, minocycline, mitoXANtrone, morphine, moxalactam, multiple vitamins, mycophenolate, nafcillin, nalbuphine, naloxone, nesiritide, netilmicin, nitroglycerin, nitroprusside, norepinephrine, octreotide, ondansetron, oxacillin, oxaliplatin, oxytocin, palonosetron, pamidronate, pancuronium, papaverine, PEMEtrexed, penicillin G potassium/sodium, pentamidine, pentazocine, PENTobarbital, PHENobarbital, phenylephrine, phytonadione, piperacillin, polymyxin B, potassium chloride, procainamide, prochlorperazine, promethazine, propofol, protamine, pyridoxine, quiNIDine, quinupristin-dalfopristin, ranitidine, ritodrine, rocuronium, sodium acetate/bicarbonate, succinylcholine, SUFentanil, tacrolimus, teniposide, theophylline, thiamine, thiotepa, ticarcillin, ticarcillin-clavulanate, tigecycline, tirofiban, tobramycin, tolazoline, trimetaphan, urokinase, vancomycin, vasopressin, vecuronium, verapamil, vinCRIStine, vinorelbine, vitamin B complex/C, voriconazole, zoledronic acid

SIDE EFFECTS

CNS: Depression, hallucinations, *dizziness, fatigue,* lethargy, paresthesias, bizarre dreams, disorientation

CV: Bradycardia, hypotension, CHF, palpitations, AV block, peripheral vascular insufficiency, vasodilation, cold extremities, pulmonary edema, dysrhythmias

EENT: Sore throat, laryngospasm, blurred vision, dry eyes

GI: Nausea, vomiting, diarrhea, colitis, constipation, cramps, dry mouth, hepatomegaly, gastric pain, acute pancreatitis

GU: Impotence, decreased libido, UTIs

HEMA: Agranulocytosis, thrombocytopenia

INTEG: Rash, pruritus, fever, Stevens-Johnson syndrome, toxic epidermal necrolysis

META: Hyperglycemia, hypoglycemia

MISC: Facial swelling, weight change, Raynaud's phenomenon

MS: Joint pain, arthralgia, muscle cramps, pain

RESP: Dyspnea, respiratory dysfunction, *bronchospasm,* cough

PHARMACOKINETICS

Metabolized by liver; crosses placenta, blood-brain barrier; excreted in breast milk; protein binding 90%

PO: Onset 30 min, peak 1-1½ hr, duration 12 hr, half-life 3-8 hr

PO-ER: Peak 6 hr, duration 24 hr, half-life 8-11 hr
IV: Onset 2 min, peak 1 min, duration 2-4 hr

INTERACTIONS

Increase: toxicity—phenothiazines
Increase: propranolol level—propafenone
Increase: effect of calcium channel blockers, neuromuscular blocker
Increase: negative inotropic effects—disopyramide
Increase: β-blocking effect—cimetidine
Increase: hypotension—quiNIDine, haloperidol, prazosin
Decrease: β-blocking effects—barbiturates
Decrease: propranolol levels—smoking

Drug/Herb

- Avoid use with feverfew

Increase: antihypertensive effect—hawthorn
Decrease: antihypertensive effect—ma huang

Drug/Lab Test

Increase: serum potassium, serum uric acid, ALT, AST, alk phos, LDH
Decrease: blood glucose
Interference: glaucoma testing

NURSING CONSIDERATIONS

Assess:

Black Box Warning: **Abrupt withdrawal:** taper over a few weeks, do not discontinue abruptly; dysrhythmias, angina, myocardial ischemia, or MI may occur, taper over at least a few weeks

- B/P, pulse, respirations during beginning therapy; notify prescriber if pulse <50 bpm or systolic B/P <90 mm Hg
- ⚠ ECG continuously if using as antidysrhythmic IV, PCWP (pulmonary capillary wedge pressure), CVP (central venous pressure)
- Hepatic enzymes: AST, ALT, bilirubin; blood glucose (diabetes mellitus)
- Angina pain: duration, time started, activity being performed, character
- Tolerance with long-term use
- Headache, lightheadedness, decreased B/P; may indicate need for decreased dosage; may aggrevate symptoms of arterial insufficiency
- **Fluid overload:** weight daily; report gain of >5 lb
- ⚠ I&O ratio, CCr if kidney damage is diagnosed; fatigue, weight gain, jugular distention, dyspnea, peripheral edema, crackles

Evaluate:

- Therapeutic response: decreased B/P, dysrhythmias

Teach patient/family:

- ⚠ Not to discontinue abruptly; may precipitate life-threatening dysrhythmias, exacerbation of angina, MI; to take product at same time each day, either with or without food consistently; to decrease dosage over 2 wk
- To avoid OTC products unless approved by prescriber; to avoid alcohol
- To avoid hazardous activities if dizzy
- About the importance of compliance with complete medical regimen; to monitor blood glucose, may mask symptoms of hypoglycemia
- To make position changes slowly to prevent fainting
- That sensitivity to cold may occur
- How to take pulse, B/P; to withhold product if <50 bpm or systolic B/P <90 mm Hg

propylhexadrine nasal agent

See Appendix B

P

propylthiouracil (Rx)

(proe-pill-thye-oh-yoor′a-sill)

Propyl-Thyracil ✦

Func. class.: Thyroid hormone antagonist (antithyroid)

Chem. class.: Thioamide

ACTION: Blocks synthesis peripherally of T_3, T_4 (triiodothyronine, thyroxine), inhibits organification of iodine

USES: Preparation for thyroidectomy, thyrotoxic crisis, hyperthyroidism, thyroid storm

CONTRAINDICATIONS: Pregnancy (D), breastfeeding, hypersensitivity, agranulocytosis, hepatitis, jaundice

Precautions: Infants, bone marrow depression, fever

Black Box Warning: Hepatic disease

DOSAGE AND ROUTES

Thyrotoxic crisis

- **Adult and child: PO** 200-400 mg q4hr for 1st 24 hr

Preparation for thyroidectomy

- **Adult: PO** 600-1200 mg/day
- **Child: PO** 10 mg/kg/day in divided doses

Hyperthyroidism

- **Adult: PO** 100 mg tid increasing to 300 mg q8hr if condition is severe; continue to euthyroid state then 100 mg daily-tid
- **Child >6 yr: PO** 50 mg/day divided q8hr, titrate based on TSH/free T_4 levels
- **Neonate (unlabeled): PO** 5-10 mg/kg/day in divided doses q8hr

Available forms: Tabs 50 mg

Administer:

- With meals to decrease GI upset
- At same time each day to maintain product level
- At lowest dose that relieves symptoms

SIDE EFFECTS

CNS: *Drowsiness, headache, vertigo, fever,* paresthesias, neuritis

GI: *Nausea, diarrhea, vomiting,* jaundice, hepatitis, loss of taste, liver failure, death

GU: Nephritis

HEMA: Agranulocytosis, leukopenia, thrombocytopenia, hypothrombinemia, lymphadenopathy, bleeding, vasculitis, periarteritis

INTEG: *Rash, urticaria, pruritus, alopecia, hyperpigmentation,* lupuslike syndrome

MS: Myalgia, arthralgia, nocturnal muscle cramps, osteoporosis

PHARMACOKINETICS

Onset up to 3 wk, peak 6-10 wk, duration 1 wk to 1 mo, half-life 1-2 hr; excreted in urine, bile, breast milk; crosses placenta; conc in thyroid gland

INTERACTIONS

- Bone marrow suppression: radiation, antineoplastics
- Agranulocytosis: phenothiazines

Increase: effects—potassium/sodium iodide, lithium

Decrease: anticoagulant effect—heparin, oral anticoagulants

Drug/Lab Test

Increase: PT, AST, ALT, alk phos

NURSING CONSIDERATIONS

Assess:

- **Hyperthyroidism:** weight loss, nervousness, insomnia, fever, diaphoresis, tremors; **hypothyroidism:** constipation, dry skin, weakness, headache; monitor T_3, T_4, which are increased; serum TSH, which is decreased; free thyroxine index, which is increased if dosage is too low; discontinue product 3-4 wk before RAIU
- Pulse, B/P, temperature
- I&O ratio; check for edema: puffy hands, feet, periorbits; indicates hypothyroidism

• Weight daily; same clothing, scale, time of day

⚠ **Blood dyscrasias:** CBC with differential; leukopenia, thrombocytopenia, agranulocytosis

⚠ **Overdose:** peripheral edema, heat intolerance, diaphoresis, palpitations, dysrhythmias, severe tachycardia, increased temperature, delirium, CNS irritability

⚠ **Hypersensitivity:** rash, enlarged cervical lymph nodes; product may have to be discontinued

• **Hypoprothrombinemia:** bleeding, petechiae, ecchymosis

• Clinical response: after 3 wk should include increased weight, pulse; decreased T_4

• **Bone marrow suppression:** sore throat, fever, fatigue

Black Box Warning: **Hepatotoxicity:** LFTs before, during treatment; jaundice, nausea, vomiting, abdominal pain, anorexia, diarrhea, fatigue

• Fluids to 3-4 L/day unless contraindicated

Evaluate:

• Therapeutic response: weight gain, decreased pulse, decreased T_4, decreased B/P

Teach patient/family:

• To abstain from breastfeeding after delivery

• To take pulse daily

• To report redness, swelling, sore throat, mouth lesions, which indicate blood dyscrasias; to report symptoms of hepatic dysfunction

• To keep graph of weight, pulse, mood

• To avoid OTC products that contain iodine

• That seafood, other iodine products may be restricted

• Not to discontinue product abruptly because thyroid crisis may occur; about stress response

• That response may take several months if thyroid is large

• About the symptoms/signs of overdose: periorbital edema, cold intolerance, mental depression

• About the symptoms of an inadequate dose: tachycardia, diarrhea, fever, irritability

• To take medication as prescribed; not to skip or double dose; that missed doses should be taken when remembered up to 1 hr before next dose

• To carry emergency ID listing condition, medication

protamine (Rx)

(proe′ta-meen)

Func. class.: Heparin antagonist

Chem. class.: Low-molecular-weight protein

ACTION: Binds heparin, thereby making it ineffective

USES: Heparin, LMWH toxicity, hemorrhage

CONTRAINDICATIONS: Hypersensitivity

Precautions: Pregnancy (C), breastfeeding, fish allergy, diabetes, previous exposure to protamine, insulins, heparin rebound or bleeding

DOSAGE AND ROUTES

Heparin overdose

• **Adult and child: IV** 1 mg of protamine/100 units heparin given; administer slowly over 1-3 min; max 50 mg/10 min

Enoxaparin overdose

• **Adult: IV** 1 mg protamine/1 mg enoxaparin

Dalteparin/tinzaparin overdose

• **Adult: IV** 1 mg protamine/100 anti-Xa unit

Available forms: Inj 10 mg/ml

Administer:

• Store at 36° F-46° F (2° C-8° C)

Direct IV route

• After reconstituting 50 mg/5 ml sterile bacteriostatic water for inj; shake, give ≤20 mg over 1-3 min

P

Y-site compatibilities: Alfentanil, amikacin, aminophylline, ascorbic acid, atracurium, atropine, azaTHIOprine, aztreonam, benztropine, bumetanide, buprenorphine, butorphanol, calcium chloride/gluconate, cefTAZidime, chlorproMAZINE, cimetidine, clindamycin, cyanocobalamin, cycloSPORINE, digoxin, diphenhydrAMINE, DOBUTamine, DOPamine, doxycycline, enalaprilat, ePHEDrine, EPINEPHrine, epoetin alfa, erythromycin, esmolol, famotidine, fentaNYL, fluconazole, ganciclovir, gentamicin, glycopyrrolate, hydrOXYzine, imipenem-cilastatin, inamrinone, iohexol, iopamidol, iothalamate, isoproterenol, labetalol, lidocaine, magnesium, mannitol, meperidine, metaraminol, methoxamine, methyldopate, metoclopramide, metoprolol, miconazole, midazolam, minocycline, morphine, multiple vitamins, nalbuphine, naloxone, netilmicin, nitroglycerin, nitroprusside, norepinephrine, ondansetron, oxytocin, papaverine, pentazocine, phenylephrine, polymyxin B, potassium chloride, procainamide, prochlorperazine, promethazine, propranolol, pyridoxine, quiNIDine, ranitidine, Ringer's, ritodrine, sodium bicarbonate, succinylcholine, SUFentanil, theophylline, thiamine, tobramycin, tolazoline, trimetaphan, urokinase, vancomycin, vasopressin, verapamil

SIDE EFFECTS

CNS: Lassitude, flushing
CV: Hypotension, bradycardia, circulatory collapse, capillary leak
GI: Nausea, vomiting, anorexia
HEMA: Bleeding
INTEG: *Rash,* dermatitis, urticaria
RESP: Dyspnea, pulmonary edema, severe respiratory distress, bronchospasm
SYST: Anaphylaxis, angioedema

PHARMACOKINETICS

IV: Onset 5 min, duration 2 hr

NURSING CONSIDERATIONS

Assess:

⚠ **Hypersensitivity:** urticaria, cough, wheezing, have emergency equipment nearby; **allergy to fish;** use with caution; men who have had vasectomies may be more prone to hypersensitivity

- Blood studies (Hct, platelets, occult blood in stools) q3mo
- Coagulation tests (aPTT, ACT) 15 min after dose then again after several hours
- VS, B/P, pulse after 30 min then 3 hr after dose
- Skin rash, urticaria, dermatitis

Evaluate:

- Therapeutic response: reversal of heparin overdose

Teach patient/family:

- Not to take if allergic to fish

pseudoephedrine (OTC, Rx)

(soo-doh-eh-fed′rin)

Elix Sure Cold, Eltor ✤, Nasofed, Sudafed, Sudafed 24 Hour, Sudogest

Func. class.: Adrenergic
Chem. class.: Substituted phenylethylamine

ACTION: Primary activity through α-effects on respiratory mucosal membranes reducing congestion hyperemia, edema; minimal bronchodilation secondary to β-effects

USES: Nasal decongestant, adjunct for otitis media; with antihistamines

CONTRAINDICATIONS: Hypersensitivity to sympathomimetics, closed-angle glaucoma

Precautions: Pregnancy (C), breastfeeding, cardiac disorders, hyperthyroidism, diabetes mellitus, prostatic hypertrophy, hypertension

DOSAGE AND ROUTES

- **Adult and child >12 yr: PO** 60 mg q6hr; **EXT REL** 120 mg q12hr or 240 mg q24hr
- **Geriatric: PO** 30-60 mg q6hr prn
- **Child 6-12 yr: PO** 30 mg q6hr, max 120 mg/day
- **Child 2-6 yr: PO** 15 mg q6hr, max 60 mg/day

Available forms: Ext rel caps 120, 240 mg; oral sol 15 mg, 30 mg/5 ml; drops 7.5 mg/0.8 ml; tabs 30, 60 mg; caps 60 mg; ext rel tabs 120, 240 mg

Administer:

- Avoid taking at or near bedtime; stimulation can occur

SIDE EFFECTS

CNS: *Tremors, anxiety,* stimulation, insomnia, headache, dizziness, hallucinations, **seizures** (geriatric patients)

CV: Palpitations, tachycardia, hypertension, chest pain, **dysrhythmias, CV collapse**

EENT: Dry nose; irritation of nose and throat

GI: *Anorexia, nausea, vomiting,* dry mouth, ischemic colitis

GU: Dysuria

PHARMACOKINETICS

PO: Onset 15-30 min; duration 4-6 hr, 8-12 hr (ext rel); metabolized in liver; excreted in feces and breast milk; terminal half-life 9-16 hr

INTERACTIONS

⚠ Do not use with MAOIs or tricyclics; hypertensive crisis may occur

Increase: effect of this product—urinary alkalizers, adrenergics, β-blockers, phenothiazines, tricyclics

Decrease: effect of this product—urinary acidifiers

NURSING CONSIDERATIONS

Assess:

- Nasal congestion: auscultate lung sounds; check for tenacious bronchial secretions
- B/P, pulse throughout treatment
- CNS side effects in geriatric patients: excitation, seizures, hallucinations

Evaluate:

- Therapeutic response: decreased nasal congestion

Teach patient/family:

- About the reason for product administration
- Ext rel: do not divide, crush, chew, or dissolve
- Do not use within 14 days of MAOIs
- Not to use continuously or to take more than recommended dose because rebound congestion may occur

⚠ To notify prescriber immediately of anxiety; slow or fast heart rate; dyspnea; seizures

- To check with prescriber before using other products because product interactions may occur
- To avoid taking near bedtime because stimulation can occur
- Not to use if stimulation, restlessness, tremors occur
- That use in children may cause excessive agitation

pseudoephedrine nasal agent

See Appendix B

psyllium (OTC, Rx)

(sill'ee-um)

Hydrocil, Leader Fiber Laxative, Metamucil, Natural Fiber, Natural Vegetable Fiber, Reguloid, Wal-Mucil

Func. class.: Bulk laxative

Chem. class.: Psyllium colloid

ACTION: Bulk-forming laxative

USES: Chronic constipation, ulcerative colitis

Unlabeled uses: Diarrhea, diverticulosis, irritable bowel syndrome, hypercholesterolemia

CONTRAINDICATIONS: Hypersensitivity, intestinal obstruction, abdominal pain, nausea, vomiting, fecal impaction

Precautions: Pregnancy (C)

DOSAGE AND ROUTES

- **Adult: PO** 1-2 tsp in 8 oz water bid or tid then 8 oz water or 1 premeasured packet in 8 oz water bid or tid, then 8 oz water
- **Child >6 yr: PO** 1 tsp in 4 oz water at bedtime

Available forms: Chew pieces 1.7, 3.4 g/piece; effervescent powder 3.4, 3.7 g/packet; powder 3.3, 3.4, 3.5, 4.94 g/tsp; wafers 3.4 g/wafer

Administer:

PO route

- Alone for better absorption, separate from other products by 1-2 hr
- In morning or evening (oral dose)
- Immediately after mixing with water or will congeal
- With 8 oz water or juice followed by another 8 oz of fluid

SIDE EFFECTS

GI: *Nausea, vomiting, anorexia, diarrhea,* cramps, intestinal esophageal blockage

PHARMACOKINETICS

Onset 12-72 hr, excreted in feces, not absorbed in GI tract

INTERACTIONS

Decrease: absorption of cardiac glycosides, oral anticoagulants, salicylates

Drug/Herb

Increase: laxative effect—flax, senna

NURSING CONSIDERATIONS

Assess:

- Blood, urine electrolytes if used often
- I&O ratio to identify fluid loss
- **Constipation:** cause of constipation; fluids, bulk, exercise missing; bowel sounds, distention, usual bowel function; color, consistency, amount
- Cramping, rectal bleeding, nausea, vomiting; product should be discontinued

Evaluate:

- Therapeutic response: decrease in constipation, decreased diarrhea with colitis

Teach patient/family:

- To maintain adequate fluid consumption
- That normal bowel movements do not always occur daily
- Not to use in presence of abdominal pain, nausea, vomiting
- To notify prescriber if constipation unrelieved or if symptoms of electrolyte imbalance occur: muscle cramps, pain, weakness, dizziness, excessive thirst

pyridostigmine (Rx)

(peer-id-oh-stig'meen)

Mestinon, Mestinon Timespan, Regonol

Func. class.: Cholinergic; anticholinesterase

Chem. class.: Tertiary amine carbamate

ACTION: Inhibits destruction of acetylcholine, which increases conc at sites

where acetylcholine is released; this facilitates the transmission of impulses across the myoneural junction

USES:
Nondepolarizing muscle relaxant antagonist, myasthenia gravis, pretreatment for nerve gas exposure (military only)

CONTRAINDICATIONS:
Bradycardia; hypotension; obstruction of intestine, renal system; bromide, benzyl alcohol sensitivity; adrenal insufficiency; cholinesterase inhibitor toxicity

Precautions: Pregnancy (C), seizure disorders, bronchial asthma, coronary occlusion, hyperthyroidism, dysrhythmias, peptic ulcer, megacolon, poor GI motility

DOSAGE AND ROUTES

Myasthenia gravis

- **Adult: PO** 600 mg/day in 5-6 divided doses, max 1.5 g/day; **IM/IV** 2 mg or 1/30 of **PO** dose; **SUS REL** 180-540 mg/day or bid at intervals of ≥6 hr
- **Child: PO** 7 mg/kg/day in 5-6 divided doses; **IM/IV** 0.05-0.15 mg/kg/dose

Nondepolarizing neuromuscular blocker antagonist

- **Adult:** 0.6-1.2 mg **IV** atropine then 0.1-0.25 mg/kg/dose
- **Child: IV** 0.1-0.25 mg/kg/dose

Nerve gas exposure prophylaxis (military)

- **Adult: PO** 30 mg q8hr if threat of exposure to Soman gas is anticipated; start several hours before exposure and discontinue upon exposure; after product is discontinued, give antidotes (atropine, pralidoxime)

Available forms: Tabs 60 mg; ext rel tabs 180 mg; syr 60 mg/5 ml; inj 5 mg/ml

Administer:

- Only with atropine sulfate available for cholinergic crisis
- Only after all other cholinergics have been discontinued
- Increased doses for tolerance as ordered
- Larger doses after exercise or fatigue as ordered
- Do not break, crush, or chew sus rel tabs

PO route

- On empty stomach for better absorption

Direct IV route

- Undiluted (5 mg/ml), give through Y-tube or 3-way stopcock, give ≤0.5 mg/min (myasthenia gravis); 5 mg/min (reversal of nondepolarizing neuromuscular blockers)

Syringe compatibilities: Glycopyrrolate

Y-site compatibilities: Heparin, hydrocortisone, potassium chloride, vit B/C

SIDE EFFECTS

CNS: Dizziness, headache, sweating, weakness, **seizures**, incoordination, **paralysis**, drowsiness, LOC

CV: Tachycardia, dysrhythmias, bradycardia, AV block, hypotension, ECG changes, **cardiac arrest**, syncope

EENT: Miosis, blurred vision, lacrimation, visual changes

GI: *Nausea, diarrhea, vomiting, cramps, increased salivary and gastric secretions, peristalsis*

GU: Urinary frequency, incontinence, urgency

INTEG: Rash, urticaria, flushing

RESP: **Respiratory depression, bronchospasm, constriction, laryngospasm, respiratory arrest**

SYST: **Cholinergic crisis**

PHARMACOKINETICS

Metabolized in liver, excreted in urine (unchanged)

PO: Onset 20-30 min, duration 3-6 hr

PO-EXT REL: Onset 30-60 min, duration 6-12 hr

IM/IV/SUBCUT: Onset 2-15 min, duration 2½-4 hr

INTERACTIONS

Increase: action—succinylcholine
Decrease: action—gallamine, metocurine, pancuronium, tubocurarine, atropine
Decrease: pyridostigmine action—aminoglycosides, anesthetics, procainamide, quiNIDine, mecamylamine, polymyxin, magnesium, corticosteroids, antidysrhythmics, quinolones

NURSING CONSIDERATIONS

Assess:

- **Myasthenia gravis:** fatigue, ptosis, diplopia, difficulty swallowing, SOB, hand/gait before, after product; improvement should be seen after 1 hr
- VS, respiration q8hr
- I&O ratio; check for urinary retention or incontinence
- **Toxicity:** bradycardia, hypotension, bronchospasm, headache, dizziness, seizures, respiratory depression; product should be discontinued if toxicity occurs

Evaluate:

- Therapeutic response: increased muscle strength, hand grasp, improved gait, absence of labored breathing (if severe); reversal of nondepolarizing neuromuscular blockers; prevention of nerve gas toxicity

Teach patient/family:

- **Myasthenia gravis:** that product is not a cure, only relieves symptoms
- To wear emergency ID specifying myasthenia gravis, products taken
- To avoid driving, other hazardous activities until effect is known
- To report muscle weakness (cholinergic crisis or underdosage), bradycardia
- Not to drink alcohol
- To take with food to decrease gastric side effects

TREATMENT OF OVERDOSE:

Discontinue product, atropine 1-4 mg IV

pyridoxine (vit B_6) (Rx, OTC)

(peer-i-dox'een)

Equaline Vitamin B6, Neuro-K, Walgreens Finest B-6, Walgreens Gold Seal Vitamin B6

Func. class.: Vit B_6, water soluble

ACTION:
Needed for fat, protein, carbohydrate metabolism; enhances glycogen release from liver and muscle tissue; needed as coenzyme for metabolic transformations of a variety of amino acids

USES:
Vit B_6 deficiency of inborn errors of metabolism, seizures, isoniazid therapy, oral contraceptives, alcoholic polyneuritis
Unlabeled uses: Palmar-plantar erythrodysesthesia syndrome

CONTRAINDICATIONS:
Hypersensitivity
Precautions: Pregnancy (A), breastfeeding, children, Parkinson's disease; patients taking levodopa should avoid supplemental vitamins with >5 mg pyridoxine

DOSAGE AND ROUTES

RDA

- **Adult: PO** (male) 1.7-2 mg; (female) 1.4-1.6 mg
- **Child 9-13 yr: PO** 1 mg/day
- **Child 4-8 yr: PO** 0.6 mg/day
- **Child 1-3 yr: PO** 0.5 mg/day
- **Infant 7-12 mo: PO** 0.3 mg/day

Vit B_6 deficiency

- **Adult: PO** 5-25 mg/day × 3 wk
- **Child: PO** 10 mg until desired response

Pyridoxine deficiency neuritis/ seizure (not drug induced)

- **Adult: PO** *without neuritis* 2.5-10 mg/day after corrected 2-5 mg/day; *with neuritis* 100-200 mg/day × 3wk then 2-5 mg/day

• **Child: PO** *without neuritis* 5-25 mg/day × 3wk then 1.5-2.5 mg/day in a multivitamin; *with neuritis:* 10-50 mg/day × 3wk then 1-2 mg/day
• **Neonate with seizures: IM/IV** 50-100 mg as single dose

Deficiency caused by isoniazid, cycloSERINE, hydrALAZINE, penicillamine
• **Adult: PO** 100-300 mg/day
• **Child: PO** 10-50 mg/day

Prevention of deficiency caused by isoniazid, cycloSERINE, hydrALAZINE, penicillamine
• **Adult: PO** 25-100 mg/day
• **Child: PO** 1-2 mg/kg/day

Palmar-plantar erythrodysesthesia syndrome (unlabeled)
• **Adult: PO** 50-150 mg/day

Available forms: Tabs 10, 25, 50, 100 mg; ext rel tabs 100 mg; inj 100 mg/ml; ext rel caps 150 mg

Administer:

PO route
• Do not break, crush, or chew ext rel tabs/caps

IM route
• Rotate sites; burning or stinging at site may occur
• Z-track to minimize pain

IV route
• Undiluted or added to most IV sol; give ≤50 mg/1 min if undiluted

Syringe compatibilities: Doxapram

SIDE EFFECTS

CNS: Paresthesia, flushing, warmth, lethargy (rare with normal renal function)
INTEG: Pain at inj site

PHARMACOKINETICS

PO/INJ: Half-life 2-3 wk, metabolized in liver, excreted in urine

INTERACTIONS

Decrease: effects of levodopa
Decrease: effects of pyridoxine—oral contraceptives, isoniazid, cycloSERINE, hydrALAZINE, penicillamine, chloramphenicol, immunosuppressants

NURSING CONSIDERATIONS

Assess:
• **Pyridoxine deficiency:** seizures, irritability, cheilitis, conjunctivitis, anemia, confusion, red tongue, weakness, fatigue before and during treatment; monitor pyridoxine levels
• Nutritional status: yeast, liver, legumes, bananas, green vegetables, whole grains
• Blood studies: Hct, Hgb

Evaluate:
• Therapeutic response: absence of nausea, vomiting, anorexia, skin lesions, glossitis, stomatitis, edema, seizures, restlessness, paresthesia

Teach patient/family:
• To avoid vitamin supplements unless directed by prescriber
• To increase meat, bananas, potatoes, lima beans, whole grain cereals in diet
• To take as directed; to continue with follow-up exams, blood work

pyrimethamine (Rx)

(peer-i-meth′a-meen)

Daraprim

Func. class.: Antimalarial, antiprotozoal

Chem. class.: Folic acid antagonist

ACTION: Inhibits folic acid metabolism in parasite, prevents transmission by stopping growth of fertilized gametes

USES: Malaria prophylaxis, *Plasmodium vivax, Pneumocystis jiroveci*

Unlabeled uses: Isosporiasis, pneumocystis pneumonia prophylaxis, toxoplasmic encephalitis prophylaxis

CONTRAINDICATIONS: Hypersensitivity, chloroquine-resistant malaria, megaloblastic anemia caused by folate deficiency

Precautions: Pregnancy (C), breastfeeding, geriatric patients, blood dyscrasias, seizure disorder, G6PD disease, renal/hepatic disease

DOSAGE AND ROUTES

Prophylaxis of malaria

Begin 2 wk before entering endemic area, continue for 6-10 wk after return

- **Adult/child >10 yr: PO** 25 mg/wk
- **Child 4-10 yr: PO** 12.5 mg/wk
- **Child <4 yr: PO** 6.25 mg/wk

Malaria treatment

- **Adult/adolescent/child >10 yr: PO** 25 mg daily × 2 days with a sulfonamide
- **Child 4-10 yr: PO** 25 mg/day × 2 days

Toxoplasmosis

- **Adult: PO** 50-75 mg, then reduce by about 50% for 4-5 wk with 1-4 g sulfadoxine × 1-3 wk, then reduce by 50% for 4-5 wk
- **Child: PO** 1 mg/kg/day in 2 divided doses or 2 mg/kg/day × 3 days then 1 mg/kg/day or divided twice daily × 4 wk, max 25 mg/day

Toxoplasmosis in AIDS patients

- **Adult: PO** 100-200 mg/day × 1-2 days, then 50-100 mg/day × 3-6 wk, then 25-50 mg/day for life (given with clindamycin or sulfADIAZINE)

Isosporiasis (unlabeled)

- **Adult: PO** 75 mg/day with leucovorin 10 mg/day × 14 days

Available forms: Tabs 25 mg; combo tabs 500 mg sulfadoxine/25 mg pyrimethamine

Administer:

PO route

- Leucovorin IM 3-9 mg/day × 3 days if folic acid deficiency occurs
- Before or after meals at same time each day to maintain product level, decrease GI symptoms
- Store in tight, light-resistant container
- **Extemporaneous susp:** tabs may be crushed and mixed with 25 ml distilled water, sucrose-containing solution (1 mg/ml); shake well; stable for 5-7 days at room temperature if mixed with sucrose-containing solutions

SIDE EFFECTS

CNS: Stimulation, irritability, seizures, tremors, ataxia, fatigue, fever

CV: Dysrhythmias

GI: *Nausea, vomiting, cramps, anorexia,* diarrhea, atrophic glossitis, gastritis

HEMA: Thrombocytopenia, leukopenia, pancytopenia, megaloblastic anemia, decreased folic acid, agranulocytosis

INTEG: Skin eruptions, photosensitivity, Stevens-Johnson syndrome

RESP: Respiratory failure

PHARMACOKINETICS

PO: Peak 2 hr, half-life 96 hr, half-life accelerated to 23 hr in AIDS patients, metabolized in liver, highly protein bound, excreted in urine (metabolites)

INTERACTIONS

- Synergistic action: folic acid

Increase: Megaloblastic anemia risk, agranulocytosis, thrombocytopenia—zidovudine

Increase: bone marrow suppression—bone marrow depressants, folate antagonists, radiation therapy

NURSING CONSIDERATIONS

Assess:

- Folic acid level; megaloblastic anemia occurs

⚠ **Blood dyscrasias:** blood studies, CBC, platelets; 2×/wk if dosage is increased

⚠ **Toxicity:** vomiting, anorexia, seizure, blood dyscrasia, glossitis; product should be discontinued immediately

- **Serious skin disorders:** Stevens-Johnson syndrome (swelling of face, lips, throat, fever)

Evaluate:

- Therapeutic response: decreased symptoms of malaria, toxoplasmosis

Teach patient/family:

- To report visual problems, fever, fatigue, bruising, bleeding, sore throat; may indicate **blood dyscrasias**

TREATMENT OF OVERDOSE:

Gastric lavage, short-acting barbiturate, leucovorin, respiratory support if needed

QUEtiapine (Rx)

(kwe-tie′a-peen)

Seroquel, Seroquel XR

Func. class.: Antipsychotic, atypical

Chem. class.: Dibenzothiazepine

ACTION: Functions as an antagonist at multiple neurotransmitter receptors in the brain, including $5HT_{1A}$, $5HT_2$, dopamine D_1, D_2, H_1, and adrenergic α_1, α_2 receptors

USES: Bipolar disorder, bipolar I disorder, depression, mania, schizophrenia

Unlabeled uses: Agitation, dementia, OCD, acute psychosis

CONTRAINDICATIONS: Hypersensitivity, breastfeeding

Precautions: Pregnancy (C), geriatric patients, hepatic/cardiac disease, breast cancer, long-term use, seizures, QT prolongation, brain tumor, hematologic disease, torsades de pointes, cataracts, dehydration, abrupt discontinuation

Black Box Warning: Children, suicidal ideation, increased mortality in elderly patients with dementia-related psychosis

DOSAGE AND ROUTES

Bipolar I disorder

- **Adult: PO** (monotherapy or adjunct to lithium, divalproex), not at increased risk for hypotension 50 mg bid on day 1, 100 mg on day 2 in 2 divided doses as tolerated to 400 mg/day on day 4, range 400-800 mg/day
- **Geriatric, debilitated, or at risk for hypotension: PO** 25 mg bid, titrate upward slowly
- **Child ≥10 yr/adolescents: PO** 25 mg bid day 1, 50 mg, bid day 2, 100 mg bid day 3, 150 mg bid day 4, 200 mg bid beginning day 5 dose adjusted

Schizophrenia

- **Adult: PO** (not at risk for hypotension) 25 mg bid, titrate upward; (XR) 200 mg/day in PM, range 400-800 mg/day

Depressive disorder (inadequate response to antidepressants alone)

- **Adult: PO EXT REL** 50 mg/day in PM on days 1, 2; on day 3, give 150 mg in PM
- **Geriatric, debilitated or at risk for hypotension: PO EXT REL** 50 mg, may increase by 50 mg/day based on response

Available forms: Tabs 25, 50, 100, 200, 300, 400 mg; ext rel tab 50, 150, 200, 300, 400 mg

Administer:

- Reduced dose to geriatric patients
- Anticholinergic agent on order from prescriber for EPS
- Avoid use of CNS depressants
- Store in tight, light-resistant container
- **Immediate release:** without regard to meals
- **Ext rel:** without food or with light meal ≤300 calories; swallow whole; do not split, crush, chew; can switch from immediate release to extended release by giving total daily dose daily

SIDE EFFECTS

CNS: EPS, pseudoparkinsonism, akathisia, dystonia, tardive dyskinesia; *drowsiness,* insomnia, agitation, anxiety, *headache,* **seizures, neuroleptic malignant syndrome,** *dizziness,* dystonia, restless legs

CV: Orthostatic hypotension, **tachycardia, QT prolongation,** CV disease, Parkinson's disease, cardiomyopathy, myocarditis

ENDO: SIADH, hyperglycemia

GI: *Nausea, anorexia, constipation,* abdominal pain, dry mouth

HEMA: **Leukopenia, agranulocytosis**

INTEG: Rash

META: Hyponatremia

MISC: Asthenia, back pain, fever, ear pain

MS: **Rhabdomyolysis**

RESP: Rhinitis

SYST: **Stevens-Johnson syndrome, anaphylaxis**

Q

PHARMACOKINETICS

Extensively metabolized by liver, half-life ≥6 hr, peak 1½ hr, ext rel 6 hr, inhibits P450 CYP3A4 enzyme system, 83% protein binding, excretion: 73% urine, 20% feces

INTERACTIONS

⚠ **Increase:** QT prolongation—class IA/III antidysrhythmics, some phenothiazines, β-agonists, local anesthetics, tricyclics, haloperidol, methadone, chloroquine, clarithromycin, droperidol, erythromycin, pentamidine

Increase: CNS depression—alcohol, opioid analgesics, sedative/hypnotics, antihistamines

Increase: hypotension—alcohol, antihypertensives

Increase: QUEtiapine clearance, decrease QUEtiapine effect—phenytoin, thioridazine, barbiturates, glucocorticoids, carBAMazepine, rifampin

Increase: neurotoxicity—lithium

Increase: QUEtiapine action—fluconazole, itraconazole, ketoconazole (CYP3A4 inhibitors)

Increase: effects of erythromycin

Decrease: QUEtiapine clearance—cimetidine

Decrease: effects of DOPamine agonists, levodopa, LORazepam

Drug/Lab Test

Increase: cholesterol, triglycerides, LFTs, glucose

Decrease: thyroid tests, WBC

NURSING CONSIDERATIONS

Assess:

⚠ CV status: QT prolongation, tachycardia, orthostatic B/P

Black Box Warning: Mental status before initial administration, AIMS assessment; affect, orientation, LOC, reflexes, gait, coordination, sleep pattern disturbances; suicidal thoughts/behaviors (child/young adult); dementia (geriatric patients)

⚠ **Suicide:** restrict amount of product given; usually suicidal thoughts/behaviors occur early during treatment and among children/adolescents/young adults

- Baseline blood glucose, LFTs, neurologic function, ophthalmologic exam, cholesterol profile, weight, monitor glucose often in diabetes mellitus
- B/P standing, lying; pulse, respirations; determine q4hr during initial treatment; establish baseline before starting treatment; report drops of 30 mm Hg; watch for ECG changes
- Dizziness, faintness, palpitations, tachycardia on rising
- **EPS:** including akathisia (inability to sit still, no pattern to movements), tardive dyskinesia (bizarre movements of jaw, mouth, tongue, extremities), pseudoparkinsonism (rigidity, tremors, pill rolling, shuffling gait)

⚠ **Neuroleptic malignant syndrome:** hyperthermia, increased CPK, altered mental status, muscle rigidity, seizures, tachycardia, diaphoresis, hypo/hypertension, fatigue; notify prescriber immediately if symptoms occur

- Constipation, urinary retention daily; if these occur, increase bulk, water in diet
- Supervised ambulation until patient stabilized on medication; do not involve patient in strenuous exercise program because fainting possible; patient should not stand still for long period of time

Evaluate:

- Therapeutic response: decrease in emotional excitement, hallucinations, delusions, paranoia; reorganization of patterns of thought, speech

Teach patient/family:

- Not to become overheated
- Not to use if pregnancy is planned or suspected; not to breastfeed
- To rise slowly to prevent orthostatic hypotension
- To take medication only as prescribed; not to crush, chew ext rel product; not to use with alcohol; to take regular tabs without regard to food, ext rel without food

- That follow-up is necessary, including LFTs, blood glucose, neurologic, cholesterol profile, weight
- If drowsiness occurs, to avoid hazardous activities such as driving; not to stand quickly, may be worse during first few days of dose change
- To avoid use of OTC meds unless directed by prescriber
- To notify prescriber if pregnancy planned, suspected; not to breastfeed
- To notify prescriber immediately of fever, difficulty breathing, fatigue, sore throat, rash, bleeding
- **Suicide:** thoughts/behaviors, primarily among children/adolescents/young adults
- To have eye exam before treatment and q6mo, cataracts may occur

quinapril (Rx)

(kwin′a-pril)

Accupril

Func. class.: Antihypertensive

Chem. class.: Angiotensin-converting enzyme (ACE) inhibitor

Do not confuse:
Accupril/Aciphex

ACTION: Selectively suppresses renin-angiotensin-aldosterone system; inhibits ACE, prevents conversion of angiotensin I to angiotensin II; results in dilation of arterial, venous vessels

USES: Hypertension, alone or in combination with thiazide diuretics; systolic CHF

CONTRAINDICATIONS: Children, hypersensitivity to ACE inhibitors, angioedema

Black Box Warning: Pregnancy (D)

Precautions: Breastfeeding, geriatric patients, impaired renal/hepatic function, dialysis patients, hypovolemia, blood dyscrasias, bilateral renal stenosis, cough, hyperkalemia, aortic stenosis, African descent

DOSAGE AND ROUTES

Hypertension

- **Adult: PO** 10-20 mg/day initially then 20-80 mg/day divided bid or daily (monotherapy); start at 5 mg/day (with diuretics)
- **Geriatric: PO** 10 mg/day, titrate to desired response (monotherapy); start at 2.5 mg/day (with diuretics)

Congestive heart failure

- **Adult: PO** 5 mg bid, may increase weekly until 20-40 mg/day in 2 divided doses

Renal dose

- **Adult: PO** CCr 61-89 ml/min 10 mg daily (hypertension), 5 mg bid (heart failure); CCr 30-60 ml/min, 5 mg/day initially; CCr 10-29 ml/min, 2.5 mg/day initially

Available forms: Tabs 5, 10, 20, 40 mg

Administer:

- Tabs may be crushed if necessary
- Take without regard to food
- Store in airtight container at room temperature

SIDE EFFECTS

CNS: *Headache, dizziness, fatigue,* somnolence, depression, malaise, nervousness, vertigo, syncope

CV: *Hypotension,* postural hypotension, syncope, palpitations, *angina pectoris,* **MI, tachycardia**, vasodilation, chest pain

GI: *Nausea,* diarrhea, constipation, *vomiting,* gastritis, **GI hemorrhage**, dry mouth

GU: Increased BUN, creatinine; decreased libido, impotence

INTEG: **Angioedema**, rash, sweating, photosensitivity, pruritus

META: Hyperkalemia

MISC: Back pain, amblyopia

MS: Myalgia

RESP: *Cough,* pharyngitis, dyspnea

PHARMACOKINETICS

Bioavailability ≥60%, onset <1 hr, peak 1-2 hr, duration 24 hr, protein binding 97%, half-life 2 hr, metabolized by liver (active metabolites quinaprilat), metabolites excreted in urine (60%)/feces (37%)

INTERACTIONS

Increase: hyperkalemia—vasodilators, hydrALAZINE, prazosin, potassium-sparing diuretics, sympathomimetics, potassium supplements, ACE/angiotensin II receptor antagonists
Increase: hypotension—diuretics, other antihypertensives, ganglionic blockers, adrenergic blockers, phenothiazines, nitrates, acute alcohol ingestion
Increase: toxicity of lithium
Decrease: absorption of tetracycline, quinolone antibiotics
Decrease: hypotensive effect of quinapril—NSAIDs

Drug/Lab Test
Increase: potassium, creatinine, BUN, LFTs

NURSING CONSIDERATIONS

Assess:

⚠ **Collagen-vascular disease:** blood studies: neutrophils, decreased platelets; WBC with differential at baseline, periodically, q3mo; if neutrophils <1000/mm³, discontinue treatment

- **Hypertension:** B/P, orthostatic hypotension, syncope, monitor B/P before giving and after 2 hr; African American patients are more resistant to antihypertension effect
- Renal studies: protein, BUN, creatinine; watch for increased levels; may indicate nephrotic syndrome
- Baselines of hepatic studies before therapy, periodically; increased LFTs; uric acid, glucose may be increased
- Potassium levels; hyperkalemia rare
- **CHF:** edema in feet, legs daily; weight daily

⚠ **Allergic reactions:** rash, fever, pruritus, urticaria, swelling of eyes/face/throat/neck, SOB, difficulty breathing; product should be discontinued; angioedema is more common in African American patients

Black Box Warning: For pregnancy (D), 2nd/3rd trimester, if pregnancy is suspected, discontinue use

Evaluate:

- Therapeutic response: decrease in B/P

Teach patient/family:

- Not to discontinue product abruptly
- Not to use OTC products (cough, cold, allergy); not to use salt substitutes containing potassium unless directed by prescriber, avoid high-fat meal at same time as product
- To comply with dosage schedule, even if feeling better
- To rise slowly to sitting or standing position to minimize orthostatic hypotension
- To notify prescriber of mouth sores, sore throat, fever, swelling of hands/feet, irregular heartbeat, chest pain, persistent dry cough
- To report excessive perspiration, dehydration, vomiting, diarrhea; may lead to fall in B/P; maintain adequate hydration
- That product may cause dizziness, fainting, lightheadedness; may occur during first few days of therapy
- That product may cause skin rash, impaired taste perception
- How to take B/P, normal readings for age group

Black Box Warning: **Pregnancy:** to report if pregnancy planned, suspected; pregnancy category (D) 2nd/3rd trimester; do not breastfeed

TREATMENT OF OVERDOSE:

0.9% NaCl IV inf

quiNIDine gluconate (Rx)

(kwin'i-deen)

quiNIDine sulfate (Rx)

Func. class.: Antidysrhythmic (Class IA)
Chem. class.: Quinine dextroisomer

Do not confuse:
quiNIDine/quiNINE

ACTION: Prolongs duration of action potential and effective refractory period, thus decreasing myocardial excitability; anticholinergic properties

USES: Atrial fibrillation, PAT, ventricular tachycardia, atrial flutter, Wolff-Parkinson-White syndrome; PVST, malaria/ IV quiNIDine gluconate

Unlabeled uses: Singultus (hiccups)

CONTRAINDICATIONS: Hypersensitivity, idiosyncratic response, digoxin toxicity, blood dyscrasias, myasthenia gravis, AV block

Precautions: Pregnancy (C), breastfeeding, children, geriatric patients, potassium imbalance, renal/hepatic disease, CHF, respiratory depression, bradycardia, hypotension, syncope

Black Box Warning: Cardiac arrhythmias, MI

DOSAGE AND ROUTES

QuiNIDine gluconate

- **Adult: PO** (ext rel) 324-648 mg q8-12hr; **IM** 600 mg then 400 mg q2hr; **IV** give 16 mg/min

Malaria

- **Adult/adolescents/children/infants: IV Regimen 1:** loading dose 24/kg (15 mg/kg base) over 4 hr, then 12 mg/kg (7.5 mg/kg base) over 4 hr q8hr, starting 8 hr after beginning of loading dose × 7 days in SE Asia, 3 days in Africa or South America, maintenance infs may be replaced with PO if feasible; **Regimen 2:** loading dose IV 10 mg/kg (6.25 base) over 1-2 hr, then 0.02 mg/kg (0.0125/kg/min base) for ≥24 hr

QuiNIDine sulfate

PVST/WPW/atrial fibrillation/flutter

- **Adult: PO** 200-300 mg q6-8hr × 5-8 doses; may increase daily until sinus rhythm restored; max 4 g/day given only after digitalization; maintenance 200-300 mg tid-qid or **EXT REL** 300-600 mg q8-12hr

Hiccups (unlabeled)

- **Adult: PO** 200 mg qid

Available forms: *Gluconate:* ext rel tabs 324 mg; inj gluconate 80 mg/ml; *sulfate:* tabs 200, 300 mg; sus rel tabs 300 mg

Administer:

- AV node blocker (digoxin) before starting quinidine to avoid increased ventricular rate

PO route

- Do not break, crush, chew ext rel products
- With full glass of water on empty stomach; if GI upset occurs, may take with food
- Sus rel forms not interchangeable
- Do not use with grapefruit juice

IM route

- IM inj in deltoid; aspirate to avoid intravascular administration

Intermittent IV INF route

- After diluting 800 mg/50 ml D_5W (16 mg/ml); give ≤1 mg/min, use inf pump; quiNIDine absorbed by PVC tubing, minimize length

Y-site compatibilities: Alfentanil, amikacin, anidulafungin, ascorbic acid, atenolol, atracurium, atropine, benztropine, bleomycin, bumetanide, buprenorphine, butorphanol, calcium gluconate, caspofungin, chlorproMAZINE, cimetidine, CISplatin, cyanocobalamin, cycloSPORINE, DACTINomycin, digoxin, diltiazem, diphenhydrAMINE, DOBUTamine, DOCEtaxel, DOPamine, doxycycline, enalaprilat, ePHEDrine, EPINEPHrine, epoetin alfa, erythromycin, esmolol, etoposide, famotidine, fenoldopam, fentaNYL, fluconazole, fludarabine, gatifloxacin, gemcitabine, gentamicin, glycopyrrolate, granisetron, HYDROmorphone, IDArubicin, imipenem-cilastatin, irinotecan, isoproterenol, labetalol, lidocaine, linezolid, LORazepam, magnesium sulfate, mannitol, mechlorethamine, meperidine, metaraminol, methoxamine, methyldopate, metoclopramide, metoprolol, metroNIDAZOLE, miconazole, milrinone, mitoXANtrone, morphine,

Q

multiple vitamins, mycophenolate, nalbuphine, naloxone, nesiritide, netilmicin, nitroglycerin, norepinephrine, octreotide, ondansetron, oxaliplatin, PACLitaxel, palonosetron, pamidronate, pancuronium, papaverine, pentamidine, pentazocine, phenylephrine, phytonadione, polymyxin B, potassium chloride, procainamide, prochlorperazine, promethazine, propranolol, protamine, pyridoxine, ranitidine, ritodrine, succinylcholine, SUFentanil, tacrolimus, teniposide, theophylline, thiamine, thiotepa, tirofiban, tobramycin, tolazoline, trimetaphan, urokinase, vancomycin, vasopressin, verapamil, vinorelbine, voriconazole, zoledronic acid

SIDE EFFECTS

CNS: *Headache, dizziness,* involuntary movement, confusion, psychosis, restlessness, irritability, syncope, excitement, depression, ataxia

CV: Hypotension, *bradycardia,* PVCs, heart block, CV collapse, arrest, torsades de pointes, widening QRS complex, ventricular tachycardia

EENT: Cinchonism: tinnitus, blurred vision, hearing loss, mydriasis, disturbed color vision

GI: Nausea, vomiting, anorexia, abdominal pain, *diarrhea,* hepatotoxicity

HEMA: Thrombocytopenia, hemolytic anemia, agranulocytosis, hypoprothrombinemia

INTEG: Rash, urticaria, angioedema, swelling, photosensitivity, flushing with severe pruritus

RESP: Dyspnea, respiratory depression

PHARMACOKINETICS

PO: Peak 1-6 hr, duration 6-8 hr, half-life 6-7 hr (prolonged in geriatric patients, cirrhosis, CHF), metabolized in liver, excreted unchanged (10%-50%) by kidneys, protein bound (80%-90%)

INTERACTIONS

- Additive vagolytic effect: anticholinergic blockers

Increase: cardiac depression: other antidysrhythmics, phenothiazines, reserpine

Increase: effects of neuromuscular blockers, digoxin, warfarin, tricyclics, propranolol

Increase: QT prolongation—macrolides, quinolones, tricyclics procainamide, antipsychotics

Increase: quiNIDine effects—cimetidine, sodium bicarbonate, carbonic anhydrase inhibitors, antacids, hydroxide suspensions, amiodarone, verapamil, NIFEdipine, protease inhibitors

Decrease: quiNIDine effects—barbiturates, phenytoin, rifampin, sucralfate, cholinergics

Drug/Herb

Increase: quiNIDine effect—hawthorn

Drug/Food

- Delayed absorption, decreased metabolism: grapefruit juice

Drug/Lab Test

Increase: Hgb

Decrease: platelets

Interference: triamterene therapy interferes with quiNIDine test levels

NURSING CONSIDERATIONS

Assess:

⚠ ECG B/P, pulse continuously during IV, baseline, and periodically (PO) to determine increased PR or QRS segments, QT interval; discontinue product or reduce dose

- Blood levels (therapeutic level 2-7 mcg/ml), CBC, LFTs

⚠ **For cinchonism:** tinnitus, headache, nausea, dizziness, fever, vertigo, tremors; may lead to hearing loss

- **Cardiac toxicity:** asystole, ventricular dysrhythmias, widening QRS, torsades de pointes
- CNS effects: dizziness, confusion, psychosis, paresthesias, seizures; product should be discontinued
- **Hepatotoxicity:** monitor LFTs for first 1-2 mo of treatment

Evaluate:

- Therapeutic response: decreased dysrhythmias

Teach patient/family:

- That if dizziness, drowsiness occurs, to avoid driving or hazardous activities
- To use sunglasses; product may cause sensitivity to light
- To carry emergency ID stating disease, medication use
- How to take pulse; when to notify prescriber
- To avoid all products unless approved by prescriber
- Not to crush, chew ext rel product
- Not to use grapefruit juice with this product
- **QuiNIDine toxicity:** visual changes, nausea, headache, ringing in the ears; report immediately
- To report signs of cinchonism, diarrhea, anorexia, decreased B/P

Q

RABEprazole (Rx)

(rah-bep′rah-zole)

Aciphex, Pariet ♣

Func. class.: Antiulcer, proton pump inhibitor

Chem. class.: Benzimidazole

Do not confuse:
Aciphex/Aricept/Accupril

ACTION: Suppresses gastric secretion by inhibiting hydrogen/potassium ATPase enzyme system in the gastric parietal cells; characterized as a gastric acid pump inhibitor because it blocks the final step of acid production

USES: Gastroesophageal reflux disease (GERD), severe erosive esophagitis, poorly responsive systemic GERD, pathologic hypersecretory conditions (Zollinger-Ellison syndrome, systemic mastocytosis, multiple endocrine adenomas); treatment of active duodenal ulcers with/without antiinfectives for *Helicobacter pylori;* daytime, nighttime heartburn

Unlabeled uses: Gastric ulcer, heartburn, *H. pylori* eradication in children

CONTRAINDICATIONS: Hypersensitivity to this product or proton pump inhibitors (PPIs)

Precautions: Pregnancy (C), breastfeeding, children, Asian patients, diarrhea, geriatric patients, gastric cancer, hepatic/GI disease, IBS, osteoporosis, pseudomembranous colitis, ulcerative colitis, vit B_{12} deficiency

DOSAGE AND ROUTES

Healing of duodenal ulcers

- **Adult: PO** 20 mg/day × ≤4 wk; to be taken after breakfast

Healing of erosive esophagitis or ulcerative GERD

- **Adult: PO** 20 mg/day × 4-8 wk
- **Adolescent and child ≥12 yr: PO** 20 mg/day up to 8 wk

Pathologic hypersecretory conditions

- **Adult: PO** 60 mg/day; may increase to 120 mg in 2 divided doses

Gastric ulcer (unlabeled)

- **Adult: PO** 20 mg/day after AM meal × 3-6 wk

Dyspepsia/heartburn (unlabeled)

- **Adult: PO** 20 mg/day × ≤14 days

Available forms: Del rel tabs 20 mg

Administer:

- **PO:** Do not break, crush, chew del rel tab; after breakfast daily with full glass of water, without regard to food

SIDE EFFECTS

CNS: *Headache, dizziness, asthenia*
CV: Chest pain, angina, tachycardia, bradycardia, palpitations, peripheral edema
EENT: Tinnitus, taste perversion
GI: *Diarrhea, abdominal pain, vomiting, nausea, constipation, flatulence, acid regurgitation,* abdominal swelling, anorexia, irritable colon, esophageal candidiasis, dry mouth, pseudomembranous colitis
GU: UTI, urinary frequency, increased creatinine, proteinuria, hematuria, testicular pain, glycosuria
HEMA: Pancytopenia, thrombocytopenia, neutropenia, leukocytosis, anemia
INTEG: *Rash,* dry skin, urticaria, pruritus, alopecia
META: Hypoglycemia, increased hepatic enzymes, weight gain
MISC: *Back pain,* fever, fatigue, malaise, Stevens-Johnson syndrome
RESP: *Upper respiratory tract infections, cough,* epistaxis, pneumonia

PHARMACOKINETICS

Eliminated in urine as metabolites and in feces, terminal half-life 1-2 hr, metabolized by CYP2C19 enzyme system, protein binding 96.3%

INTERACTIONS

Increase: bleeding risk—warfarin, clopidogrel

Increase: serum levels of RABEprazole—benzodiazepines, phenytoin, clarithromycin, antacids, other proton pump inhibitors, H_2 blockers
Increase: levels of—digoxin, nelfinavir/omeprazole
Decrease: levels of RABEprazole—sucralfate, calcium carbonate, vit B_{12}
Decrease: levels of ketoconazole, itraconazole, iron salts, atazanavir/ritonavir, ampicillin
Drug/Herb
Decrease: RABEprazole—St. John's wort
Drug/Lab Test
Decrease: magnesium

NURSING CONSIDERATIONS

Assess:

- GI system: bowel sounds, abdomen for pain, swelling, anorexia, emesis/stool for occult blood
- **Pseudomembranous colitis:** may occur with most antibiotic therapy; watery diarrhea, abdominal pain, fever
- **Vit B_{12} deficiency/cyanocobalamin/hypomagnesemia:** may occur after 3-12 mo of treatment; use magnesium, vit B_{12}, cyanocobalamin supplement; if severe, discontinuing of product may be needed
- Hepatic studies: AST, ALT, alk phos during treatment; CBC with differential periodically
- ⚠ Serious skin reactions: Stevens-Johnson syndrome
- CBC with differential before, periodically during treatment; blood dyscrasias may occur (rare)
- Obtain susceptibility testing if *H. Pylori* treatment is ineffective, another anti-infective may be needed

Evaluate:

- Therapeutic response: absence of epigastric pain, swelling, fullness; decreased symptoms of GERD after 4-8 wk

Teach patient/family:

- To report severe diarrhea or black, tarry stools; product may have to be discontinued
- That hypoglycemia may occur if patient is diabetic
- To avoid hazardous activities because dizziness, drowsiness may occur
- To avoid alcohol, salicylates, NSAIDs because they may cause GI irritation; to avoid other OTC, herbal products unless approved by prescriber
- To use as directed for length of time prescribed; to take missed dose when remembered; not to double dose
- To notify prescriber if pregnancy planned, suspected

radioactive iodine (sodium iodide) ^{131}I (Rx)

Func. class.: Antithyroid
Chem. class.: Radiopharmaceutical

ACTION:
Converted to protein-bound iodine by thyroid gland for use when needed

USES:
High dose: Thyroid cancer, hyperthyroidism
Low dose: Visualization to determine thyroid cancer, diagnostic aid for thyroid function studies

CONTRAINDICATIONS:
Pregnancy (X), breastfeeding, age <30 yr, recent MI, large nodular goiter, vomiting/diarrhea, acute hyperthyroidism, use of thyroid products

DOSAGE AND ROUTES

Thyroid cancer

- **Adult: PO** 50-150 mCi; may repeat, depending on clinical status

Hyperthyroidism

- **Adult: PO** 4-10 mCi, depending on serum thyroxine level

Available forms: Caps 1-50, 0.8-100 mCi; oral sol 7.05 mCi/ml, 3.5-150 mCi/vial; concentrated oral solution 1000 mCi/ml

R

Administer:

• Only after discontinuing all other antithyroid agents × 5-7 days
• After NPO overnight; food delays action
• During or within 10 days of menstruation
• Do not take antithyroid agents except propranolol, which decreases hyperthyroid symptoms, until total effect of taking ^{131}I has occurred (about 6 wk)

SIDE EFFECTS

CV: Chest pain, tachycardia
EENT: Sore throat, cough
ENDO: Hypothyroidism, hyperthyroid adenoma, transient thyroiditis, goiter
GI: *Nausea, diarrhea, vomiting*
HEMA: Eosinophilia, lymphedema, leukemia, bone marrow depression, leukopenia, anemia, lymph node swelling
INTEG: Alopecia, rash, pruritus
RESP: Cough

PHARMACOKINETICS

PO: Onset 3-6 days; excreted in urine, sweat, feces, breast milk; crosses placenta; excreted in 56 days

INTERACTIONS

Increase: Hypothyroidism—lithium
Decrease: effect of I^{131}—amiodarone
Decrease: uptake—recent intake of stable iodine, thyroid, antithyroid products
Drugj/Lab Test
Decrease: Hgb, T4, TSH

NURSING CONSIDERATIONS

Assess:

• Weight daily with same clothing, scale, time of day
• Blood work, including CBC for blood dyscrasias (leukopenia, thrombocytopenia, agranulocytosis)
• **Overdose:** peripheral edema, heat intolerance, diaphoresis, palpitations, dysrhythmias, severe tachycardia, increased temperature, delirium, CNS irritability
• **Hypersensitivity:** rash, enlarged cervical lymph nodes; product may have to be discontinued
• **Hypoprothrombinemia:** bleeding, petechiae, ecchymosis
• Clinical response: after 3 wk should include increased weight, pulse; decreased T_4
• **Bone marrow depression:** sore throat, fever, fatigue
• Limited contact with patient: 1/2 hr/day for each person
• Adequate rest after treatment
• Fluids to 3-4 L/day for 48 hr to remove agent from body

Evaluate:

• Therapeutic response: weight gain, decreased pulse, decreased T_4, B/P

Teach patient/family:

• To fast overnight before treatment, consume increased fluids for 48 hr after therapy
• To empty bladder often during treatment; to avoid irradiation of gonads
• To report redness, swelling, sore throat, mouth lesions; may indicate blood dyscrasias
• To avoid extended contact with children, spouse for 1 wk
• That bathroom may be used by entire family; to flush toilet several times
• To avoid coughing, expectorating for 24 hr (saliva and vomitus highly radioactive for 6-8 hr)

raloxifene (Rx)

(ral-ox′ih-feen)

Evista

Func. class.: Bone resorption inhibitor
Chem. class.: Hormone modifier, selective estrogen receptor modulator (SERM)

ACTION: Tissue-selective estrogen agonist/antagonist; agonist activity in bone and on lipid metabolism; antagonist activity on breast and uterus; reduces resorption of bone and decreases bone turnover

USES: Prevention, treatment of osteoporosis in postmenopausal women;

breast cancer prophylaxis in postmenopausal women with osteoporosis or in postmenopausal women at high risk for developing the disease

Unlabeled uses: Uterine leiomyomata in postmenopausal women with osteoporosis or in postmenopausal women who are at high risk for developing the disease

CONTRAINDICATIONS:

Pregnancy (X), breastfeeding, hypersensitivity

Black Box Warning: Women with active or history of venous thromboembolic events

Precautions: CV/hepatic disease, cervical/uterine cancer, elevated triglycerides, pulmonary embolism

Black Box Warning: Stroke

DOSAGE AND ROUTES

- **Adult: PO** 60 mg/day, max 60 mg/day

Available forms: Tabs 60 mg

Administer:

- PO: without regard to meals, vit D
- Add calcium supplement if inadequate

SIDE EFFECTS

CNS: Insomnia

CV: Hot flashes, peripheral edema, **thromboembolism, stroke**

EENT: Retinal vein occlusion (rare)

GI: *Nausea,* vomiting, diarrhea, dyspepsia

GU: Vaginitis, leukorrhea, cystitis, *hot flashes,* vaginal bleeding

INTEG: Rash, sweating

META: Weight gain, peripheral edema

MS: Arthralgia, myalgia, *leg cramps,* arthritis

RESP: Sinusitis, pharyngitis, increased cough, pneumonia, laryngitis, bronchitis, **pulmonary embolism,** flulike symptoms

PHARMACOKINETICS

Elimination half-life 28-32 hr; excreted in feces, breast milk; highly bound to plasma proteins

INTERACTIONS

- Administer cautiously with other highly protein-bound products, systemic estrogens

Decrease: action of anticoagulants, dessicated thyroid, levothyroxine, liotrix

Decrease: action of raloxifene—ampicillin, cholestyramine

Drug/Food

Decrease: raloxifene—soy

Drug/Lab Test

Increase: apolipoprotein A-1, hormone-binding globulin

Decrease: total cholesterol, LDL, lipoprotein, apolipoprotein B, serum calcium, albumin, total protein

NURSING CONSIDERATIONS

Assess:

Black Box Warning: History of stroke, TIA, thrombosis, atrial fibrillation, hypertension, smoking; venous thrombosis may occur; avoid prolonged sitting; discontinue 3 days before surgery, other immobilization

- Bone density test at baseline, throughout treatment, bone-specific alk phos, osteocalcin

Evaluate:

- Therapeutic response: prevention, treatment of osteoporosis in postmenopausal women; prevention of breast cancer in postmenopausal women with osteoporosis or in those who are at high risk for developing the disease

Teach patient/family:

Black Box Warning: To discontinue product 72 hr before prolonged bedrest; to avoid staying in one position for long periods

- To take calcium supplements, vit D if intake is inadequate
- To increase exercise using weights
- To stop smoking; to decrease alcohol consumption
- That product does not help to control hot flashes
- ⚠ To report fever, acute migraine, insomnia, emotional distress; urinary tract in-

R

fection, vaginal burning/itching; swelling, warmth, pain in calves

⚠ To notify prescriber if pregnancy planned, suspected, pregnancy (X); to avoid breastfeeding

raltegravir (Rx)

(ral-teg'ra-vir)

Isentress

Func. class.: Antiretroviral

Chem. class.: HIV integrase strand transfer inhibitor (ISTIs)

ACTION: Inhibits catalytic activity of HIV integrase, which is an HIV-encoded enzyme needed for replication

USES: HIV in combination with other antiretrovirals

CONTRAINDICATIONS: Breastfeeding, hypersensitivity

Precautions: Pregnancy (C), children, geriatric patients, hepatic disease, immune reconstitution syndrome, hepatitis, antimicrobial resistance, lactase deficiency

DOSAGE AND ROUTES

- **Adult and adolescent ≥16 yr: PO** 400 mg bid; if using with rifampin, give 800 mg bid; max 800 mg/day with/without food

Available forms: Tabs 400 mg

Administer:

- Do not break, crush, chew tabs
- May give without regard to meals, with 8 oz of water
- Store at room temperature

SIDE EFFECTS

CNS: *Fatigue,* fever, *dizziness, headache,* asthenia, suicidal ideation

CV: MI

GI: *Nausea,* vomiting, diarrhea, abdominal pain, asthenia, gastritis, hepatitis

GU: Oliguria, proteinuria, hematuria, glomerulonephritis, acute renal failure, renal tubular necrosis

HEMA: Anemia, neutropenia

INTEG: Rash, urticaria, pruritus, pain or phlebitis at IV site, unusual sweating, alopecia

META: Hyperamylasia, hyperglycemia

MS: Myopathy, rhabdomyolysis

SYST: Immune reconstitution syndrome

PHARMACOKINETICS

Max absorption 3 hr if taken on an empty stomach; terminal half-life 9 hr; metabolized in the liver by uridine diphosphate glucuronosyltransferase (UGT A1A enzyme system); excreted in feces 51%, urine 32%

INTERACTIONS

Increase: raltegravir effect—proton pump inhibitors, H_2 blockers; UGT1A1 inhibitors (atazanavir)

⚠ **Increase:** rhabdomyolysis, myopathy, elevated CPK—fibric acid derivatives, HMG-CoA reductase inhibitors

Decrease: raltegravir levels—rifampin, efavirenz, tenofovir, tipranavir/ritonavir

Drug/Lab Test

Increase: AST, ALT, GGT, total bilirubin, alk phos, amylase/lipase, CK, serum glucose, total/HDL/LDL cholesterol

Decrease: Hgb, platelets, ANC

NURSING CONSIDERATIONS

Assess:

- **HIV infection:** CD4, T-cell count, plasma HIV RNA, viral load; resistance testing before therapy, at treatment failure

⚠ **Rhabdomyolysis:** Assess for calf pain, increased CPK, product should be discontinued

- Skin eruptions: rash, urticaria, itching
- **Suicidal thoughts/behaviors:** monitor for depression; more common in those with mental illness
- Immune reconstitution syndrome, usually during initial phase of treatment, may give anti-infective before starting

• Monitor total/HDL/LDL cholesterol baseline and periodically; all may be elevated

Evaluate:

• Therapeutic response: improvement in CD4 counts, T-cell counts

Teach patient/family:

• To take as prescribed; if dose missed, to take as soon as remembered up to 1 hr before next dose; not to double dose; not to share with others

• That sexual partners need to be told that patient has HIV; that product does not cure infection, just controls symptoms; does not prevent infecting others

⚠ To report sore throat, fever, fatigue (may indicate superinfection)

• That product must be taken in equal intervals 2×/day to maintain blood levels for duration of therapy

⚠ To notify prescriber immediately of suicidal thoughts/behaviors

• To notify prescriber if pregnancy planned, suspected; to avoid breastfeeding

• To continue with follow-up exams, blood work

ramelteon (Rx)

(rah-mel′tee-on)

Rozerem

Func. class.: Sedative/hypnotic, antianxiety

Chem. class.: Melatonin receptor agonist

ACTION: Binds selectively to melatonin receptors (MT_1, MT_2); thought to be involved in circadian rhythms and in the normal sleep/wake cycle

USES: Insomnia (difficulty with sleep onset)

CONTRAINDICATIONS: Breastfeeding, children, infants, hypersensitivity

Precautions: Pregnancy (C), hepatic disease, alcoholism, COPD, seizure disorder, sleep apnea, suicidal ideation, angioedema, depression, sleep-related behaviors (sleepwalking), schizophrenia, bipolar disorder; alcohol intoxication, hepatic encephalopathy

DOSAGE AND ROUTES

• **Adult: PO** 8 mg within 30 min of bedtime

Hepatic dose

• Do not use with severe hepatic disease; use with caution for mild to moderate hepatic disease

Available forms: Tabs 8 mg

Administer:

• Within 30 min of bedtime for sleeplessness; on empty stomach for fast onset

• Do not break tabs, swallow whole

• Store in tight container in cool environment

SIDE EFFECTS

CNS: Dizziness, somnolence, fatigue, headache, insomnia, depression, complex sleep-related reactions (sleep driving, sleep eating), suicidal thoughts/behaviors

GI: Nausea, diarrhea, dysgeusia, vomiting

MISC: Myalgia, arthralgia, decreased blood cortisol, influenza, upper RI

SYST: Severe allergic reactions, angioedema

PHARMACOKINETICS

Absorbed rapidly; peak 0.75 hr; protein binding 82%; rapid first pass metabolism via liver; 84% excreted in urine, 4% in feces; half-life 2-5 hr

INTERACTIONS

⚠ Possible toxicity: antiretroviral protease inhibitors

⚠ **Increase:** ramelteon effect, toxicity—alcohol; CYP1A2 inhibitors, azole antifungals (ketoconazole, fluconazole), fluvoxaMINE, anxiolytics, sedatives, hypnotics, barbiturates, ciprofloxacin

Decrease: effect of ramelteon—rifampin

Drug/Food

• Prolonged absorption, sleep onset reduced: high-fat/heavy meal

NURSING CONSIDERATIONS

Assess:

⚠ **Severe hypersensitive reactions:** Assess for angioedema (facial swelling), product should be discontinued

- **Sleep characteristics:** type of sleep problem: falling asleep, staying asleep; complex sleep disorders (sleep walking/driving/eating) after taking product
- Mental status: mood, sensorium, affect, memory (long, short term), suicidal ideation
- LFTs: before treatment, periodically

Evaluate:

- Therapeutic response: ability to sleep at night, decreased amount of early morning awakening

Teach patient/family:

- To avoid driving, other activities requiring alertness until product stabilized; drowsiness may continue the next day
- To avoid alcohol ingestion, CNS depressants
- About alternative measures to improve sleep: reading, exercise several hr before bedtime, warm bath, warm milk, TV, self-hypnosis, deep breathing
- To report if pregnancy is planned or suspected (C); to avoid breastfeeding
- To take product immediately before going to bed
- Not to ingest a high-fat/heavy meal before taking product

⚠ To report cessation of menses, galactorrhea (women), decreased libido, infertility, worsening of insomnia, behavioral changes, severe allergic reactions, suicidal thoughts/behaviors, if insomnia gets worse

- Med guide should be given to patient and reviewed

ramipril (Rx)

(ra-mi′pril)

Altace

Func. class.: Antihypertensive

Chem. class.: Angiotensin-converting enzyme inhibitor (ACE)

Do not confuse:

ramipril/enalapril

Altace/alteplase

ACTION: Selectively suppresses renin-angiotensin-aldosterone system; inhibits ACE, prevents conversion of angiotensin I to angiotensin II; results in dilation of arterial, venous vessels

USES: Hypertension, alone or in combination with thiazide diuretics; CHF (post-MI), reduction in risk for MI, stroke, death from CV disorders

Unlabeled uses: Proteinuria due to diabetic nephropathy

CONTRAINDICATIONS: Breastfeeding, children, hypersensitivity to ACE inhibitors, history of ACE-inhibitor–induced angioedema

Black Box Warning: Pregnancy (D) 2nd/3rd trimesters

Precautions: Geriatric patients, impaired renal/hepatic function, dialysis patients, hypovolemia, blood dyscrasias, CHF, renal artery stenosis, cough, African descent, aortic stenosis

DOSAGE AND ROUTES

Hypertension

- **Adult: PO** 2.5 mg/day initially then 2.5-20 mg/day divided bid or daily

CHF post-MI

- **Adult: PO** 1.25-2.5 mg bid; may increase to 5 mg bid

Reduction in risk for MI, stroke, death

- **Adult: PO** 2.5 mg/day × 7 days then 5 mg/day × 21 days, then may increase to 10 mg/day

Renal dose

• **Adult: PO** CCr <40 ml/min, reduce by 50%, titrate upward to max 5 mg/day

Proteinuria due to diabetic nephropathy (unlabeled)

• **Adult: PO** 2.5 mg/day up to 20 mg/day

Available forms: Caps 1.25, 2.5, 5, 10 mg

Administer:

• Without regard to meals
• Caps can be opened, added to food; mixture is stable for 24 hr at room temperature, 48 hr refrigerated
• Store in tight container at ≤86° F (30° C)

SIDE EFFECTS

CNS: *Headache, dizziness,* anxiety, insomnia, paresthesia, *fatigue,* depression, malaise, vertigo, syncope
CV: *Hypotension,* chest pain, palpitations, angina, syncope, dysrhythmia, **heart failure, MI**
EENT: Hearing loss
GI: *Nausea,* constipation, vomiting, dyspepsia, dysphagia, anorexia, diarrhea, abdominal pain, **hepatitis, hepatic failure, pancreatitis, hepatic necrosis**
GU: **Proteinuria,** increased BUN, creatinine, impotence
HEMA: Decreased Hct, Hgb, **eosinophilia, leukopenia, pancytopenia, thrombocytopenia, agranulocytosis (rare)**
INTEG: Rash, sweating, photosensitivity, pruritus
META: *Hyperkalemia*, hyperglycemia
MISC: **Angioedema, toxic epidermal necrolysis, anaphylaxis, Stevens-Johnson syndrome**
MS: Arthralgia, arthritis, myalgia
RESP: *Cough,* dyspnea

PHARMACOKINETICS

Bioavailability >50%-60%, onset 1-2 hr, peak 3-6 hr, duration 24 hr, protein binding 73%, half-life 1-2 hr, 9-18 hr for active metabolite, metabolized by liver (metabolites excreted in urine, feces)

INTERACTIONS

Increase: hyperkalemia—potassium-sparing diuretics, potassium supplement, angiotensin II receptor agonists
Increase: hypotension—diuretics, other antihypertensives, ganglionic blockers, adrenergic blockers, nitrates, acute alcohol ingestion
⚠ **Increase: toxicity—vasodilators, hydrALAZINE, prazosin, potassium-sparing diuretics, sympathomimetics, potassium supplements**
Increase: serum levels of lithium
Decrease: absorption—antacids
Decrease: antihypertensive effect—indomethacin, NSAIDs, salicylates

Drug/Herb

Increase: antihypertensive effect—hawthorn
Decrease: antihypertensive effect—ephedra

Drug/Food

Increase: hyperkalemia—potassium salt substitutes, avoid use

Drug/Lab Test

Increase: LFTs, BUN, creatinine, glucose, potassium
Decrease: RBC, Hgb, platelets

NURSING CONSIDERATIONS

Assess:

⚠ **Collagen-vascular disease (SLE, scleroderma): neutrophils, decreased platelets; WBC with differential at baseline, periodically; if neutrophils <1000/mm³, discontinue treatment**
• **Hypertension:** monitor B/P baseline and regularly, orthostatic hypotension, syncope
• **Renal disease: protein, BUN, creatinine, potassium, sodium at baseline, periodically; increased levels may indicate nephrotic syndrome; renal symptoms: polyuria, oliguria, urinary frequency, dysuria**
• **CHF:** edema in feet, legs daily; weight daily
⚠ **Serious allergic reactions: angioedema, Stevens-Johnson syndrome, rash, fever, pruritus, urticaria; product should**

be discontinued if antihistamines fail to help

• Monitor electrolytes baseline and periodically; potassium may be increased

Evaluate:

• Therapeutic response: decrease in B/P; CHF

Teach patient/family:

• Not to discontinue product abruptly; to comply with dosage schedule, even if feeling better

• Not to use OTC products (cough, cold, allergy) unless directed by prescriber; not to use salt substitutes containing potassium without consulting prescriber

• To rise slowly to sitting or standing position to minimize orthostatic hypotension

⚠ To notify prescriber of mouth sores, sore throat, fever, swelling of hands or feet, irregular heartbeat, chest pain

• To report excessive perspiration, dehydration, vomiting, diarrhea; may lead to fall in B/P; to maintain hydration

• That product may cause dizziness, fainting, lightheadedness; that these may occur during first few days of therapy; to avoid hazardous activities until response is known

• That product may cause skin rash, impaired perspiration

• How to take B/P, normal readings for age group

Black Box Warning: To inform prescriber if pregnancy planned, suspected, pregnancy (D); not to breastfeed

TREATMENT OF OVERDOSE:

0.9% NaCl IV inf, hemodialysis

ranibizumab (Rx)

(ran-ih-biz′oo-mab)

Lucentis

Func. class.: Ophthalmic

Chem. class.: Selective vascular endothelial growth factor antagonist

ACTION: Binds to receptor-binding site of active forms of vascular endothelial growth factor A (VEGF-A) that causes angiogenesis and cell proliferation

USES: Macular degeneration (neovascular) (wet), macular edema after retinal vein occlusion (RVO), diabetic macular edema

CONTRAINDICATIONS: Hypersensitivity, ocular infections

Precautions: Pregnancy (C), breastfeeding, children, retinal detachment, increased intraocular pressure

DOSAGE AND ROUTES

Macular degeneration/macular edema after retinal vein occlusion (RVO)

• **Adult: INTRAVITREAL** 0.5 mg (0.05 ml) monthly

Diabetic macular edema

• **Adults: INTRAVITREAL** 0.3 mg q28day

Available forms: Sol for inj 0.5 mg/0.05 ml

Administer:

• By ophthalmologist via intravitreal injection using adequate anesthesia; use 19-gauge filter

• Store in refrigerator; do not freeze

• Protect from light

SIDE EFFECTS

CNS: Dizziness, headache

EENT: Blepharitis, cataract, conjunctival hemorrhage/hyperemia, detachment of retinal pigment epithelium, dry/irritation/pain in eye, visual impairment, vitreous floaters, ocular infection

GI: Constipation, nausea

MISC: Hypertension, UTI, thromboembolism, nonocular bleeding

RESP: Bronchitis, cough, sinusitis, URI

PHARMACOKINETICS

Elimination half-life 9 days

INTERACTIONS

Increase: severe inflammation—verteporfin photodynamic therapy (PDT)

NURSING CONSIDERATIONS

Assess:

- **Eye changes:** redness; sensitivity to light, vision change; increased intraocular pressure change; report infection to ophthalmologist immediately

Evaluate:

- Therapeutic response: prevention of increasing macular degeneration

Teach patient/family:

- That, if eye becomes red, sensitive to light, painful, or if there is a change in vision, to seek immediate care from ophthalmologist
- About reason for treatment, expected results

ranitidine (Rx, OTC)

(ra-nit′i-deen)

Equaline Heartburn Relief, Nu-Ranit ✤, Top Care Heartburn Relief, Wal-zan, Zantac, Zantac C ✤

ranitidine bismuth citrate

Tritec

Func. class.: H_2-Histamine receptor antagonist

Do not confuse:
ranitidine/amantadine/rimantadine
Zantac/Xanax/Zofran/ZyrTEC

ACTION: Inhibits histamine at H_2-receptor site in parietal cells, which inhibits gastric acid secretion

USES: Duodenal ulcer, Zollinger-Ellison syndrome, gastric ulcers, hypersecretory conditions, gastroesophageal reflux disease, stress ulcers, erosive esophagitis (maintenance), active duodenal ulcers with *Helicobacter pylori* in combination with clarithromycin, systemic mastocytosis, multiple endocrine adenoma syndrome, heartburn

Unlabeled uses: Prevention of aspiration pneumonitis, upper GI bleeding, angioedema, gastritis, urticaria, NSAID-induced ulcer prophylaxis

CONTRAINDICATIONS: Hypersensitivity

Precautions: Pregnancy (B), breastfeeding, child <12 yr, renal/hepatic disease

DOSAGE AND ROUTES

Ranitidine

Duodenal ulcer

- **Adult: PO** 150 mg bid or 300 mg/day after PM meal or at bedtime; maintenance 150 mg at bedtime
- **Infant and child: PO** 2-4 mg/kg bid, max 300 mg/day

Zollinger-Ellison syndrome

- **Adult: PO** 150 mg bid, may increase if needed

Gastric ulcer

- **Adult: PO** 150 mg bid × 6 wk then 150 mg at bedtime
- **Infant and child: PO** 2-4 mg/kg bid, max 300 mg/day

GERD

- **Adult: PO** 150 mg bid

Erosive esophagitis

- **Adult: PO** 150 mg qid for up to 12 wk
- **Child ≥1 mo: PO** 5-10 mg/kg/day in 2-3 divided doses

Renal dose

- **Adult:** CCr <50 ml/min, give 50% of dose or extend dosing interval

NSAID-induced ulcer prophylaxis (unlabeled)

- **Adult: PO** 150 mg bid

Stress gastritis prophylaxis (unlabeled)

- **Adult: IM/INT IV INF** 50 mg q6-8hr

Severe, acute urticaria/angioedema (unlabeled)

- **Adult: INT IV INF** 50 mg with H_1-blocker

Ranitidine bismuth citrate

- **Adult: PO** 400 mg bid × 4 wk with clarithromycin 500 mg tid × 1st 2 wk

Available forms: *Ranitidine:* tabs 75, 150, 300 mg; sol for inj 25 mg/ml; caps

R

150, 300 mg; syr 15 mg/ml; *ranitidine bismuth citrate:* tabs 400 mg

Administer:

PO route

- Antacids 1 hr before or 1 hr after ranitidine
- Without regard to meals; **EFFERdose tab:** dissolve 25 mg/≥5 ml, give after dissolved; swallow whole or dissolve on tongue, do not chew
- Store at room temperature

IM route

- No dilution needed; inject in large muscle mass, aspirate

Direct IV route

- Dilute to max 2.5 mg/ml (50 mg/20 ml) using 0.9% NaCl (nonpreserved) or D_5W, give dose over ≥5 min (max 4 mg/ml)

Intermittent IV INF route

- Dilute to max 0.5 mg/ml with D_5W, NS, give over 15-20 min (5-7 ml/min); premixed ready-to-use bags as 1 mg/ml (50 mg/50 ml), inf over 15-20 min

Continuous 24 hr IV INF route

- **Adult:** dilute 150 mg/250 ml of D_5W or NS; run over 24 hr (6.25 mg/hr or as directed); use inf device; use within 48 hr; *Zollinger-Ellison Syndrome:* dilute in D_5W, NS; max conc 2.5 mg/ml, use inf device

Y-site: Acyclovir, aldesleukin, alemtuzumab, alfentanil, allopurinol, amifostine, amikacin, aminophylline, amphotericin B liposome, amsacrine, anikinra, anidulafungin, ascorbic acid, atracurium, atropine, aztreonam, bivalirudin, bumetanide, buprenorphine, butorphanol, calcium chloride/gluconate, CARBOplatin, ceFAZolin, cefepime, cefonicid, cefoperazone, cefotaxime, cefoTEtan, cefOXitin, cefTAZidime, ceftizoxime, cefTRIAXone, cefuroxime, chloramphenicol, chlorproMAZINE, cimetidine, ciprofloxacin, cisatracurium, CISplatin, clindamycin, cyanocobalamin, cyclophosphamide, cycloSPORINE, cytarabine, DACTINomycin, DAPTOmycin, dexamethasone, dexmedetomidine, digoxin, diltiazem, DOBUTamine, DOCEtaxel, DOPamine, doripenem, doxacurium, doxapram, DOXOrubicin, DOXOrubicin liposome, doxycycline, enalaprilat, ePHEDrine, EPINEPHrine, epirubicin, epoetin alfa, ertapenem, erythromycin, esmolol, etoposide, etoposide phosphate, famotidine, fenoldopam, fentaNYL, filgrastim, fluconazole, fludarabine, fluorouracil, folic acid, foscarnet, furosemide, ganciclovir, gemcitabine, gentamicin, glycopyrrolate, granisetron, heparin, hydrocortisone, HYDROmorphone, IDArubicin, ifosfamide, imipenem/cilastatin, inamrinone, indomethacin, isoproterenol, ketorolac, labetalol, levofloxacin, lidocaine, linezolid, LORazepam, magnesium sulfate, mannitol, mechlorethamine, melphalan, meperidine, metaraminol, methotrexate, methoxamine, methyldopate, methylPREDNISolone, metoclopramide, metoprolol, metroNIDAZOLE, midazolam, milrinone, mitoXANtrone, morphine, multivitamin, nalbuphine, naloxone, nesiritide, niCARdipine, nitroglycerin, nitroprusside, norepinephrine, octreotide, ondansetron, oxacillin, oxaliplatin, oxytocin, PACLitaxel, palonosetron, pancuronium, papaverine, PEMEtrexed, penicillin G, pentamidine, pentazocine, PENTobarbital, PHENobarbital, phentolamine, phenylephrine, phytonadione, piperacillin/tazobactam, potassium chloride, procainamide, prochlorperazine, promethazine, propofol, propranolol, protamine, pyrdoxime, remifentanil, riTUXimab, rocuronium, sargramostim, sodium acetate/bicarbonate, succinylcholine, SUFentanil, tacrolimus, teniposide, theophylline, thiamine, thiopental, thiotepa, ticarcillin/clavulanate, tigecycline, tirofiban, tobramycin, tolazoline, trastuzumab, trimethaphan, urokinase, vancomycin, vecuronium, vinCRIStine, vinorelbine, warfarin, zidovudine, zoledronic acid

SIDE EFFECTS

CNS: Headache, sleeplessness, dizziness, confusion, agitation, depression, hallucination (geriatric patients)

CV: Tachycardia, bradycardia, PVCs
EENT: Blurred vision, increased ocular pressure
GI: Constipation, abdominal pain, diarrhea, nausea, vomiting, **hepatotoxicity**
GU: Impotence, **acute interstitial nephritis (rare)**
INTEG: Urticaria, rash, fever
RESP: **Pneumonia**
SYST: **Anaphylaxis (rare)**

PHARMACOKINETICS

PO: Peak 2-3 hr; duration 8-12 hr; metabolized by liver; excreted in urine (30% unchanged, PO), breast milk; half-life 2-3 hr; protein binding 15%

INTERACTIONS

Increase: effect of pramipexole, procainamide, trospium, triazolam, calcium channel blockers, memantine, saquinavir, adefovir
⚠ **Increase:** GI obstruction risk—NIFEdipine ext rel products
Increase: toxicity—sulfonylureas, procainamide, benzodiazepines, calcium channel blockers
Decrease: absorption of ranitidine—antacids, anticholinergics
Decrease: effects of cephalosporins, iron salts, ketoconazole, itraconazole
Increase: GI obstruction risk—NIFEdipine ext rel products
Drug/Lab Test
Increase: AST, ALT, creatinine
False positive: urine protein (Multistix)

NURSING CONSIDERATIONS

Assess:
- **GI complaints:** nausea, vomiting, diarrhea, cramps, abdominal discomfort, jaundice; report immediately
- I&O ratio, BUN, creatinine, LFTs, serum, stool guaiac before, periodically during therapy

Evaluate:
- Therapeutic response: decreased abdominal pain, heartburn

Teach patient/family:
- To avoid driving, other hazardous activities until stabilized on product
- That product must be continued for prescribed time to be effective
- To notify prescriber if pregnancy planned, suspected; to avoid breastfeeding
- Not to take maximum OTC daily dose for >2 wk
- To take once daily dose before bedtime

ranolazine (Rx)

(ruh-no′luh-zeen)

Ranexa

Func. class: Antianginal

ACTION:
Antianginal, antiischemic; unknown, may work by inhibiting portal fatty-acid oxidation

USES:
Chronic stable angina pectoris; use in patients who have not responded to other treatment options; should be used in combination with other antianginals such as amlodipine, β-blockers, nitrates

Unlabeled uses: Unstable angina

CONTRAINDICATIONS:
Preexisting QT prolongation, hepatic disease (Child-Pugh class A, B, C), hypersensitivity, hypokalemia, renal failure, torsades de pointes, ventricular dysrhythmia, ventricular tachycardia, hepatic cirrhosis

Precautions: Pregnancy (C), breastfeeding, children, geriatric patients, hypotension, renal disease, females at risk for torsades de pointes

DOSAGE AND ROUTES

- **Adult: PO** 500 mg bid, increased to 1000 mg bid based on response; max 1000 mg bid

Available forms: Ext rel tabs 500, 1000 mg

Administer:

- **Ext rel tabs:** do not break, crush, chew tabs; take product as prescribed; do not double or skip dose
- Without regard to meals, bid; do not use with grapefruit juice

SIDE EFFECTS

CNS: *Headache, dizziness,* hallucinations
CV: Palpitations, QT prolongation, orthostatic hypotension
EENT: Tinnitus
GI: Nausea, vomiting, constipation, dry mouth
MISC: Peripheral edema
RESP: Dyspnea

PHARMACOKINETICS

Absorption varied; peak 2-5 hr; half-life 7 hr; extensively metabolized by the liver (CYP3A and less by CYP2D6); excreted in urine (75%), feces (25%); protein binding 62%

INTERACTIONS

Increase: ranolazine action—diltiazem, ketoconazole, macrolide antibiotics, dofetilide, PARoxetine, protease inhibitors, quiNIDine, sotalol, thioridazine, verapamil, ziprasidone
Increase: action of digoxin, simvastatin
Increase: ranolazine absorption, toxicity—antiretroviral protease inhibitors
⚠ **Increase:** QT prolongation and torsades de pointes—class IA/III antidysrythmics, arsenic trioxide, chloroquine, droperidol, haloperidol, levomethadyl, methadone, pentamidine, chlorproMAZINE, mesoridazine, thioridazine, pimozide; CYP3A4 inhibitors (ketoconazole, fluconazole, itraconazole, IV miconazole, voriconazole, diltiazem, verapamil)
Drug/Food

- Do not use with grapefruit, grapefruit juice

NURSING CONSIDERATIONS

Assess:

- **Angina:** characteristics of pain (intensity, location, duration, alleviating/precipitating factors)

⚠ **QT prolongation:** ECG for QT prolongation, ejection fraction; assess for chest pain, palpitations, dyspnea

- Cardiac status: B/P, pulse, respirations
- LFTs, serum creatinine/BUN, magnesium, potassium before treatment, periodically

Evaluate:

- Therapeutic response: decreased anginal pain

Teach patient/family:

- To avoid hazardous activities until stabilized on product, dizziness no longer a problem

⚠ To avoid OTC drugs, grapefruit juice, products prolonging QTc (quiNIDine, dofetilide, sotalol, erythromycin, thioridazine, ziprasidone or protease inhibitors, diltiazem, ketoconazole, macrolide antibiotics, verapamil) unless directed by prescriber; to notify prescriber of palpitations, fainting

- To comply with all areas of medical regimen
- To take as directed; not to skip dose, double doses
- Not to chew or crush; not to use with grapefruit juice
- To notify all health care providers of product use
- To notify prescriber of palpitations, dizziness, edema, dyspnea
- For acute angina, take other products prescribed; this product does not decrease acute attack

rasagiline (Rx)

(ra-sa′ji-leen)

Azilect

Func. class.: Antiparkinson agent

Chem. class.: MAOI, type B

ACTION: Inhibits MAOI type B at recommended doses; may increase DOPamine levels

USES: Idiopathic Parkinson's disease monotherapy or with levodopa

CONTRAINDICATIONS: Breastfeeding; hypersensitivity to this product, MAOIs; pheochromocytoma

Precautions: Pregnancy (C), children, psychiatric disorders, moderate to severe hepatic disorders

DOSAGE AND ROUTES

Monotherapy

- **Adult: PO** 1 mg/day

Adjunctive therapy

- **Adult: PO** 0.5 mg/day, may increase 1 mg/day; change of levodopa dose for adjunct therapy; reduced levodopa dose may be needed

Hepatic dose

- **Adult: PO** 0.5 mg for mild hepatic disease

Concomitant ciprofloxacin, other CYP1A2 inhibitors

- **Adult: PO** 0.5 mg; plasma concs of rasagiline may double

Available forms: Tabs 0.5, 1 mg

Administer:

- ⚠ With meals to prevent nausea; continuing therapy usually reduces or eliminates nausea; do not give with foods/liquids containing large amounts of tyramine
- Reduced dose of carbidopa/levodopa cautiously
- Renal failure: in dialysis, increase dose slowly

SIDE EFFECTS

CNS: Drowsiness, hallucinations, depression, headache, malaise, paresthesia, vertigo, syncope

CV: Angina, hypertensive crisis (ingestion of tyramine products), orthostatic hypotension

GI: *Nausea*, diarrhea, dry mouth, dyspepsia

GU: Impotence, decreased libido

HEMA: Leukopenia

INTEG: Alopecia, skin cancers

MISC: Conjunctivitis, fever, flu syndrome, neck pain, allergic reaction, alopecia

MS: Arthralgia, arthritis, dyskinesia

RESP: Rhinitis

PHARMACOKINETICS

Onset, peak, duration unknown; well absorbed; protein binding >88%-94%; metabolized by CYP1A2 in liver; excreted by kidneys

INTERACTIONS

⚠ Do not give with meperidine, other analgesics because serious reactions (including coma and death) may occur; do not give with sympathomimetics

Increase: levels of rasagiline up to 2-fold—ciprofloxacin, CYP1A2 inhibitors (atazanavir, mexiletine, taurine)

Increase: severe CNS toxicity with antidepressants (tricyclics, SSRIs, SNRIs, mirtazapine, cyclobenzaprine)

⚠ **Increase:** hypertensive crisis—MAOIs

Drug/Herb

- Do not give with St. John's wort, yohimbe

Drug/Food

- Do not give with foods/liquids that have large amounts of tyramine

Drug/Lab Test

Increase: LFTs

Decrease: WBCs

NURSING CONSIDERATIONS

Assess:

- **Parkinson's symptoms:** tremors, ataxia, muscle weakness and rigidity at

R

baseline, periodically; increased dyskinesia, postural hypotension if used in combination with levodopa

• Mental status: hallucinations, confusion, notify prescriber

⚠ **Hypertensive crisis:** severe headache, blurred vision, seizures, chest pain, difficulty thinking, nausea, vomiting, signs of stroke; any unexplained severe headache should be considered to be hypertensive crisis

⚠ **Melanomas;** periodic skin exams by a dermatologist

• Cardiac status: B/P, ECG periodically during beginning treatment

⚠ **Tyramine products:** foods, other medications may lead to hypertensive crisis (tachycardia, bradycardia, chest pain, nausea, vomiting, sweating, dilated pupils)

Evaluate:

• Therapeutic response: improved symptoms in patients with Parkinson's disease

Teach patient/family:

• To change positions slowly to prevent orthostatic hypotension

• To avoid hazardous activities until stabilized on product; that dizziness can occur

• To rinse mouth frequently; to use sugarless gum to alleviate dry mouth

• To take as prescribed; not to miss dose or double doses; to take missed dose as soon as remembered if several hours before next dose

⚠ To prevent hypertensive crisis by avoiding high-tyramine foods (>150 mg)

• To report signs of hypertensive crisis

• To avoid CNS depressants, alcohol

• To notify all providers of product use; to avoid elective surgery, other procedures involving CNS depressants

rasburicase (Rx)

(rass-burr'i-case)

Elitek, Fasturtec ✦

Func. class.: Enzyme

Chem. class.: Recombinant urate-oxidase enzyme

ACTION:
Catalyzes enzymatic oxidation of uric acid into an inactive and soluble metabolite (allantoin)

USES:
To reduce uric acid levels in children with leukemia, lymphoma, solid tumor malignancies who are receiving chemotherapy

CONTRAINDICATIONS:
Hypersensitivity

Black Box Warning: G6PD deficiency (Mediterranean, African descent), hemolytic reactions, methemoglobinemia reactions to product

Precautions: Pregnancy (C), breastfeeding, children <2 yr, anemia

Black Box Warning: Acute bronchospasm, angina, angioedema, atony, African-American and Mediterranean patients, hypotension, urticaria

DOSAGE AND ROUTES

• **Adult/adolescent/infant:** **IV INF** 0.2 mg/kg as single daily dose given as **IV INF** over 1/2 hr × 5 days

Available forms: Powder for inj 1.5, 7.5 mg/vial

Administer:

Intermittent IV INF route

• Reconstitute with diluent provided, add 1 ml of diluent/each vial, swirl, withdraw amount needed, mix with NS to final volume of 50 ml, use within 24 hr, give over 30 min, do not use filter, use different line; if not possible, flush with ≥15 ml of saline before, after use

• Chemotherapy started 4-24 hr after 1st dose

SIDE EFFECTS

CNS: *Headache*, fever
CV: Chest pain, hypotension
GI: *Nausea, vomiting, anorexia, diarrhea, abdominal pain, constipation, dyspepsia, mucositis*
HEMA: **Neutropenia with fever, hemolysis, methemoglobinemia**
INTEG: *Rash*
MISC: Edema
RESP: **Bronchospasm,** wheezing, dyspnea
SYST: **Anaphylaxis, hemolysis, methemoglobinemia, sepsis**

PHARMACOKINETICS

Elimination half-life 16-21 hr

INTERACTIONS

⚠ **Increase:** toxicity—allopurinol

NURSING CONSIDERATIONS

Assess:

- Blood studies: BUN, serum uric acid, urine creatinine clearance, electrolytes, CBC with differential before, during therapy
- Monitor temperature; fever may indicate beginning infection; no rectal temperatures

⚠ **Anaphylaxis:** dyspnea, urticaria, flushing, wheezing, swelling of lips, tongue, throat; have emergency equipment nearby

⚠ **G6PD deficiency, hemolytic reactions, methemoglobinemia;** these patients should not be given this product; screen patients who are at higher risk for these disorders

- GI symptoms: frequency of stools, cramping; if severe diarrhea occurs, fluid, electrolytes may need to be given

Evaluate:

- Therapeutic response: decreased uric acid levels in children when antineoplastics causing high uric acid levels used

Teach patient/family:

- About the reason for therapy, expected results

⚠ To report trouble breathing, jaundice, chest pain

regorafenib

(re′goe-raf′e-nib)

Stivarga

Func. class.: Antineoplastic biologic-response modifier; multikinase inhibitor

Chem. class.: Signal transduction inhibitor (STI)

ACTION: Inhibits tyrosine kinase in patients with colorectal cancer

USES: Metastatic colorectal cancer in those who have received fluoropyrimidine, oxaliplatin, irinotecan-based chemotherapy, an anti-VEGF therapy; and an anti-EGFR therapy if *KRAS* wild type

CONTRAINDICATIONS: Pregnancy (D)

Precautions: Breastfeeding, children, geriatric patients, cardiac/renal/hepatic/dental disease, fistula, GI bleeding or perforation, bone marrow suppression, infection, wound dehiscence, thrombocytopenia, neutropenia, immunosuppression

Black Box Warning: Hepatic disease

DOSAGE AND ROUTES

- **Adult: PO** 160 mg/day with a low-fat breakfast × 21 days of a 28-day cycle, cycles may be repeated

Available forms: Tabs 40 mg

Administer:

- Store at 77°F (25°C)
- Give at the same time each day with a low-fat breakfast that contains less than 30% fat such as 2 slices of white toast with 1 tbsp of low-fat margarine and 1 tbsp of jelly, and 8 oz of skim milk; or 1 cup of cereal, 8 oz of skim milk, 1 slice of toast with jam, apple juice, and 1 cup of coffee or tea
- Swallow tablets whole

R

• If a dose is missed, take as soon as possible that day; do not take 2 doses on the same day

Hand–foot skin reaction

⚠ Reduce to 120 mg (grade 2 palmar–plantar erythrodysesthesia); hold if grade 2 toxicity does not improve in ≤7 days or recurs; hold for ≥7 days in grade 3 toxicity; reduce to 80 mg for recurrent grade 2 toxicity; discontinue if 80 mg is not tolerated

Hypertension

⚠ Hold in grade 2 hypertension

Other severe toxicity (except hepatotoxicity)

⚠ Hold until toxicity resolves in grade 3 or 4 toxicity; consider the risk/benefits of continuing therapy in grade 4 toxicity, reduce dosage to 120 mg; if grade 3 or 4 toxicity recurs, hold until toxicity resolves, then reduce to 80 mg; discontinue in those who do not tolerate 80-mg dose

Hepatic dose

⚠ Baseline mild (Child-Pugh class A) or moderate (Child-Pugh class B): no change; baseline severe hepatic impairment (Child-Pugh class C): use not recommended; AST/ALT elevations during therapy: For grade 3 AST/AST level elevations, hold dose; if therapy is continued, reduce to 120 mg after levels recover; discontinue in those with AST/ALT > 20× ULN; AST/ ALT > 3× ULN and bilirubin >2× ULN; recurrence of AST/ALT >5× ULN despite a reduction to 120 mg

SIDE EFFECTS

CNS: Headache, tremor

CV: Hypertensive crisis, MI

EENT: Blurred vision, conjunctivitis

GI: Hepatotoxicity, GI hemorrhage, diarrhea, GI perforation, xerostomia

HEMA: Neutropenia, thrombocytopenia, bleeding

INTEG: Rash, alopecia

META: Hypokalemia

MISC: Fatigue, decreased weight, hand–foot syndrome, hypothyroidism

PHARMACOKINETICS

Protein binding 99%, metabolized by CYP3A4, UGT1A0, half-life 14-58 hr

INTERACTIONS

Increase: regorafenib concentrations—CYP3A4 inhibitors (ketoconazole, itraconazole, erythromycin, clarithromycin)

Increase: plasma concs of simvastatin, calcium-channel blockers, ergots

Increase: plasma conc of warfarin; avoid use with warfarin; use low-molecular-weight anticoagulants instead

Decrease: regorafenib concs—CYP3A4 inducers (dexamethasone, phenytoin, carBAMazepine, rifampin, PHENobarbital)

Drug/Food

Increase: regorafenib effect—grapefruit juice; avoid use while taking product

Drug/Herb

Decrease: imatinib conc—St. John's wort

NURSING CONSIDERATIONS

Assess:

Black Box Warning: Hepatic disease: fatal hepatotoxicity can occur; obtain LFTs baseline and at least every 2 wk × 2 mo, then monthly

Fatal bleeding:

⚠ From GI, respiratory, GU tracts; permanently discontinue in those with severe bleeding

Palmar–plantar erythrodysesthesia (hand–foot syndrome):

⚠ More common in those previously treated; reddening swelling, numbness, desquamation on palms, soles

GI perforation/fistula:

⚠ Discontinue if this occurs; assess for pain in epigastric area, dyspepsia, flatulence, fever, chills

Hypertension/hypertensive crisis:

⚠ Hypertension usually occurs in the first cycle in those with preexisting hypertension; do not start treatment until B/P is controlled; monitor B/P every wk × 6 wk, then at start of each cycle or more often if needed; temporarily or permanently discontinue for severe uncontrolled hypertension

Evaluate:

• Therapeutic response: decrease in spread or size of tumor

Teach patient/family:

• To report adverse reactions immediately: bleeding, rash

• About reason for treatment, expected results

• That effect on male fertility is unknown

⚠ HIGH ALERT

remifentanil (Rx)

(rem-ih-fin′ta-nill)

Ultiva

Func. class.: Opiate agonist analgesic

Chem. class.: μ-Opioid agonist

Controlled Substance Schedule II

ACTION: Inhibits ascending pain pathways in limbic system, thalamus, midbrain, hypothalamus

USES: In combination with other products for general anesthesia to provide analgesia

CONTRAINDICATIONS: Hypersensitivity

Precautions: Pregnancy (C), breastfeeding, children <12 yr, geriatric patients, increased intracranial pressure, acute MI, severe heart disease, GI/renal/hepatic disease, asthma, respiratory conditions, seizure disorders, bradyarrhythmias

DOSAGE AND ROUTES

• **Adult:** induction **IV** 0.5-1 mcg/kg/min with hypnotic or volatile agent; maintenance with isoflurane (0.4-1.5 MAC) or propofol (100-200 mcg/kg/min); **CONT INF** 0.025-0.2 mcg/kg/min

• **Child 1-12 yr: CONT IV INF** 0.25 mcg/kg/min with isoflurane

• **Full-term neonate and infant up to 2 mo: CONT IV INF** 0.4 mcg/kg/min with nitrous oxide

Available forms: Powder for inj lyophilized 1, 2, 5 mg

Administer:

• Add 1 ml diluent per mg remifentanil

• Shake well; further dilute to a final conc of 20, 25, 50, or 250 mcg/mg

• Store in light-resistant area at room temperature

• Interruption of inf results in rapid reversal (no residual opioid effect within 5-10 min)

Direct IV route

• To be used only during maintenance of general anesthesia; inject into tubing close to venous cannula; give to nonintubated patients over 30-60 sec

Continuous IV INF route

• Use inf device, max 16 hr; do not use same tubing as blood, do not admix

Y-site compatibilities: Acyclovir, alfentanil, amikacin, aminophylline, ampicillin, ampicillin/sulbactam, aztreonam, bumetanide, buprenorphine, butorphanol, calcium gluconate, ceFAZolin, cefepine, cefotaxime, cefoTEtan, cefOXitin, cefTAZidime, ceftizoxime, cefTRIAXone, cefuroxime, cimetidine, ciprofloxacin, cisatracurium, clindamycin, dexamethasone, digoxin, diltiazem, diphenhydrAMINE, DOBUTamine, DOPamine, doxacurium, doxycycline, droperidol, enalaprilat, EPINEPHrine, esmolol, famotidine, fentaNYL, fluconazole, furosemide, ganciclovir, gatifloxacin, gentamicin, haloperidol, heparin, hydrocortisone sodium succinate, HYDROmorphone, hydrOXYzine, imipenem-cilastatin, inamrinone, isoproterenol, ketorolac, lidocaine, LORazepam, magnesium sulfate, mannitol, meperidine, methylPREDNISolone sodium succinate, metoclopramide, metroNIDAZOLE, midazolam, morphine, nalbuphine, nitroglycerin, norepinephrine, ondansetron, phenylephrine, piperacillin, potassium chloride, procainamide, prochlor-

R

perazine, promethazine, ranitidine, SUFentanil, theophylline, thiopental, ticarcillin/clavulanale, tobramycin, vancomycin, zidovudine

Solution compatibilities: D_5, 0.45% NaCl, LR, D_5 LR, 0.9% NaCl

SIDE EFFECTS

CNS: Drowsiness, *dizziness,* confusion, *headache,* sedation, euphoria, delirium, agitation, anxiety

CV: Palpitations, *bradycardia,* change in B/P, facial flushing, syncope, asystole

EENT: Tinnitus, blurred vision, miosis, diplopia

GI: *Nausea, vomiting,* anorexia, constipation, cramps, dry mouth

GU: Urinary retention, dysuria

INTEG: Rash, urticaria, bruising, flushing, diaphoresis, pruritus

MS: Rigidity

RESP: Respiratory depression, apnea

PHARMACOKINETICS

70% protein binding, terminal half-life 3-10 min, excreted in urine; onset: 1-3 min

INTERACTIONS

Increase: respiratory depression, hypotension, profound sedation: alcohol, sedatives, hypnotics, other CNS depressants; antihistamines, phenothiazines, MAOIs

Drug/Herb

Increase: CNS depression—kava

NURSING CONSIDERATIONS

Assess:

- I&O ratio; check for decreasing output; may indicate urinary retention, especially in geriatric patients
- CNS changes; dizziness, drowsiness, hallucinations, euphoria, LOC, pupil reaction
- GI status: nausea, vomiting, anorexia, constipation
- Allergic reactions: rash, urticaria
- **Respiratory dysfunction:** respiratory depression, character, rate, rhythm; notify prescriber if respirations <12/min; CV status; bradycardia, syncope

Evaluate:

- Therapeutic response: maintenance of anesthesia

Teach patient/family:

- To call for assistance when ambulating or smoking; that drowsiness, dizziness may occur
- To make position changes slowly to prevent orthostatic hypotension

repaglinide (Rx)

(re-pag′lih′nide)

Gluconorm ✱, Prandin

Func. class.: Antidiabetic

Chem. class.: Meglitinide

ACTION:

Causes functioning β-cells in pancreas to release insulin, thereby leading to a drop in blood glucose levels; closes ATP-dependent potassium channels in the β-cell membrane; this leads to the opening of calcium channels; increased calcium influx induces insulin secretion

USES:

Type 2 diabetes mellitus

CONTRAINDICATIONS:

Hypersensitivity to meglitinides; diabetic ketoacidosis, type 1 diabetes

Precautions: Pregnancy (C), breastfeeding, children, geriatric patients, thyroid/cardiac disease, severe renal/hepatic disease, severe hypoglycemic reactions

DOSAGE AND ROUTES

- **Adult: PO** 1-2 mg with each meal, max 16 mg/day; adjust at weekly intervals; oral-hypoglycemic–naive patients or patients with A1c <8% should start with 0.5 mg with each meal

Renal/hepatic dose

- **Adult: PO** CCr 20-39 ml/min, 0.5 mg/day; titrate upward cautiously

Available forms: Tabs 0.5, 1, 2 mg

Administer:
- Up to 15-30 min before meals; 2, 3, or 4×/day preprandially
- Skip dose if meal skipped; add dose if meal added
- Store in tight container at room temperature

SIDE EFFECTS

CNS: *Headache, weakness,* paresthesia
CV: Angina
EENT: Tinnitus, sinusitis
ENDO: **Hypoglycemia**
GI: Nausea, vomiting, diarrhea, constipation, dyspepsia, **pancreatitis**
HEMA: **Hemolytic anemia, leukopenia**
INTEG: Rash, allergic reactions
MISC: Chest pain, UTI, allergy
MS: Back pains, arthralgia
RESP: URI, sinusitis, rhinitis, bronchitis

PHARMACOKINETICS

Completely absorbed by GI route; onset 30 min; peak 1 hr; duration <4 hr; half-life 1 hr; metabolized in liver by CYP3A4; excreted in urine, feces (metabolites); crosses placenta; 98% protein bound

INTERACTIONS

⚠ **Do not use with gemfibrozil, isophane insulin (NPH)**
Increase: repaglinide effect—CYP3A4, OATP101, CYP2C9 inhibitors
Increase: in both—levonorgestrel/ethinyl estradiol
Increase: repaglinide metabolism—CYP3A4 inducers: rifampin, barbiturates, carBAMazepine
Increase: repaglinide effect—NSAIDs, salicylates, sulfonamides, chloramphenicol, MAOIs, coumarins, β-blockers, probenecid, gemfibrozil, simvastatin, fenofibrate, deferasirox
Decrease: repaglinide metabolism—CYP3A4 inhibitors: antifungals (ketoconazole, miconazole), erythromycin, macrolides
Decrease: repaglinide action—calcium channel blockers, corticosteroids, oral contraceptives, thiazide diuretics, thyroid preparations, estrogens, phenothiazines, phenytoin, rifampin, isoniazid, PHENObarbital, sympathomimetics

Drug/Herb
Increase: antidiabetic effect—garlic, chromium, horse chestnut

Drug/Food
Decrease: repaglinide level; give before meals
Decrease: repaglinide metabolism—grapefruit juice

Drug/Lab Test
Increase/decrease: glucose

NURSING CONSIDERATIONS

Assess:
⚠ **Hypo/hyperglycemic reaction, which can occur soon after meals: dizziness, weakness, headache, tremors, anxiety, tachycardia, hunger, sweating, abdominal pain, A1c, fasting, postprandial glucose during treatment**

Evaluate:
- Therapeutic response: decrease in polyuria, polydipsia, polyphagia; clear sensorium; absence of dizziness; stable gait; blood glucose, A1c improvement

Teach family/patient:
- About technique for blood glucose monitoring; how to use blood glucose meter
- About the symptoms of hypo/hyperglycemia, what to do about each
- That product must be continued on daily basis; about the consequences of discontinuing product abruptly
- To avoid OTC medications unless ordered by prescriber
- That diabetes is a lifelong illness; that product will not cure disease
- That all food included in diet plan must be eaten to prevent hypoglycemia; that if a meal is omitted, dose should be omitted; to have glucagon emergency kit available; to take repaglinide 15-30 min before meals 2, 3, or 4×/day; to carry emergency ID

TREATMENT OF OVERDOSE:

Glucose 25 g IV via dextrose 50% solution, 50 ml or 1 mg glucagon

retapamulin (topical)

(re-te-pam'you-lin)

Altabax

Func. class.: Topical anti-infective

Chem. class.: Pleuromutilin

ACTION: Antibacterial activity results from inhibition of protein synthesis

USES: For the treatment of impetigo

CONTRAINDICATIONS: Hypersensitivity to this product

Precautions: Children

DOSAGE AND ROUTES

• **Adult/child ≥9 mo: TOP** apply to affected area bid × 5 days

Available forms: Topical ointment 1%

Administer:

Topical route

• Do not use skin products near the eyes, nose, or mouth

• Wash hands before and after use

• **Ointment:** Apply a thin film to the cleansed, affected area and massage gently into affected areas

SIDE EFFECTS

INTEG: Pruritus, irritation, headache, diarrhea, nausea

NURSING CONSIDERATIONS

Assess:

Allergic reaction

• Hypersensitivity, product may need to be discontinued

Infection

• Number of lesions, severity of impetigo

Evaluate:

• Decreased lesions, inf in impetigo

Teach patient/family:

Topical route

• Not to use skin products near the eyes, nose, or mouth

• To wash hands before and after use

• **Ointment:** to apply a thin film to the cleansed, affected area

Rh_o(D) immune globulin standard dose IM (Rx)

BayRho-D, HyperRHO SD, RhoGAM Ultra-Filtered Plus

Rh_o(D) immune globulin microdose IM (Rx)

MICRhoGAM Ultra-Filtered Plus

Rh_o(D) immune globulin IV (Rx)

Rhophylac, WinRho SDF

Func. class.: Immune globulins

ACTION: Suppresses immune response of nonsensitized Rh_o(D or D^u)-negative patients who are exposed to Rh_o(D or D^u)-positive blood

USES: Prevention of isoimmunization in Rh-negative women given Rh-positive blood after abortions, miscarriages, amniocentesis; chronic idiopathic thrombocytopenia purpura (Rhophylac)

CONTRAINDICATIONS: Previous immunization with this product, Rh_o(O)-positive/D^u-positive patient

Black Box Warning: Hemolysis

Precautions: Pregnancy (C)

Black Box Warning: Requires a specialized setting

DOSAGE AND ROUTES

To reduce risk of Rh isoimmunization antepartum/suppression of Rh isoimmunization postpartum following delivery of full-term infant

• **Adult and adolescent ≥16 yr: IM** (BayRho-D [HyperRHO SD] full dose only) 300 mcg (1500 international units) at 28 wk gestation, repeat within 72 hr of delivery of confirmed Rho(D)-positive infant; dose not needed after delivery if delivery within 3 wk of last dose and no fetal maternal hemorrhage of >15 ml of RBC; **IM** (RhoGam only) 300 mcg (1500 international units) at 26-28 wk gestation, repeat within 72 hr even if status of Rho unknown or if 72 hr have passed; **IM/IV** (WinRho SDF only) 300 mcg (1500 international units) at 28 wk gestation; if given earlier during pregnancy, give at 12-wk intervals during pregnancy, 120-mcg (600 international units) dose; IM/IV should be given as soon as possible and preferably within 72 hr of delivery of confirmed Rho(D)-positive infant and even if status unknown; give ≤ 28 days after delivery

Known or suspected massive fetomaternal hemorrhage (>15 ml of fetal RBC or >30 ml of fetal whole blood)

• **Adult and adolescent ≥16 yr: IM** (BayRho-D [HyperRHO SD] full dose only) 300 mcg (1500 international units) per every 15 ml of fetal blood cells or 30 ml of whole blood; multiple syringes may be injected IM at same time in different sites, give within 72 hr of exposure, repeat dose within 72 hr of delivery; **IM** (RhoGAM only) 300 mcg (1500 international units) for every 15 ml of fetal blood cells or 30 ml of whole blood, give total dose within 72 hr of exposure; **IM/IV** (WinRho SDF only) if large fetomaternal hemorrhage suspected, give **IV** 9 mcg (45 international units) or **IM** 12 mcg (60 international units) for every ml of fetal whole blood, give **IV** 600 mcg (3000 international units) q8hr or **IM** 1200 mcg (6000 international units) q12hr until total dose given, total dose should be given within 72 hr of exposure

Threatened abortion at any stage of pregnancy

• **Adult and adolescent ≥16 yr: IM** (BayRho-D [HyperRHO SD] full dose only) 300 mcg (1500 international units) as soon as possible; if given at 13-18 wk gestation, give another 300 mcg (1500 international units) at 26-28 wk gestation; repeat dose within 72 hr of delivery; **IM** (RhoGam only) 300 mcg (1500 international units) as soon as possible and within 72 hr of exposure; **IM/IV** (Rhophylac only) 300 mcg (1500 international units) as soon as possible and within 72 hr; **IM/IV** (WinRho SDF only) 300 mcg (1500 international units) as soon as possible and within 72 hr, repeat dose at 12-wk intervals during pregnancy and give 120 mcg (600 international units) as soon as possible after delivery and within 72 hr

After spontaneous abortion, induced termination of pregnancy, or ectopic pregnancy that occurs at ≤12 wk gestation

• **Adult and adolescent ≥16 yr: IM** (BayRho-D Minidose, HyperRHO Minidose, MICRhoGAM only) 50 mcg (250 international units) as soon as possible, give within 3 hr of spontaneous or surgical removal, if possible within 72 hr

After spontaneous abortion, induced termination of pregnancy, or ruptured tubal pregnancy that occurs at ≥13 wk

• **Adult and adolescent ≥16 yr: IM** (BayRho-D [HyperRHO SD] full dose, RhoGAM only) 300 mcg (1500 international units) as soon as possible and within 72 hr of event

After spontaneous abortion, induced termination of pregnancy, amniocentesis, chorionic villus sampling, abdominal trauma, ruptured tubal pregnancy, or percutaneous umbilical cord sampling at ≤34 wk gestation

• **Adult and adolescent ≥16 yr: IM/IV** (WinRho SDF only) 300 mcg (1500 international units) within 72 hr, repeat at 12-wk intervals during pregnancy, give 120 mcg (600 international units) as soon as possible and preferably within 72 hr of delivery

Available forms: BayRho-D sol for inj 300 mcg/ml (HyperRHO SD sol for inj), MICRhoGAM Ultra Filtered Plus Solution for inj 50 mcg/ml; RhoGam Ultra Filtered Plus Solution for inj 50 mcg; Rhophylac Pre-Filled Syringes Solution for inj 300 mcg/2 ml; WinRho SDF liquid for inj; WinRho powder for inj

R

Administer:
- Store in refrigerator
- BayRho-D (HyperRHO SD), MICRhoGAM, RhoGAM given by IM only; do not give IV
- WinRho SDF and Rhophylac can be given IM or IV
- Inspect for particulate matter; do not use if particulate matter present
- Reconstitute/dilution: no reconstitution or dilution needed for BayRho-D (HyperRHO SD); Rhophylac, MICRhoGAM, RhoGAM, or liquid formulation of WinRho SDF
- WinRho SDF powder for IV use: if giving IV, reconstitute 600 international units or 1500 international units immediately before use with 2.5 ml of sterile diluent; reconstitute 5000 international units with 8.5 ml sterile diluent; add diluent to vial slowly down wall of vial; gently swirl until powder dissolved; do not shake
- WinRho SDF powder for IM use: IV giving IM, reconstitute 600 international units or 1500 international units immediately before use with 1.25 ml of sterile diluent; 5000 international units with 8.5 ml of sterile diluent; add diluent to vial slowly down wall of vial; gently swirl until powder dissolved; do not shake

IM route
- Use aseptic technique, observe for 20 min after administration
- Bring Rhophylac to room temperature before use
- Inject into deltoid muscle of upper arm or anterolateral portion of upper thigh; do not inject into gluteal muscle
- If dose calculated requires multiple vials or syringes, use different sites at same time

IV route
- Use aseptic technique
- WinRho SDF: remove entire contents of vial to obtain calculated dose; if partial vial required for dosage calculation, withdraw entire vial contents to ensure correct calculation; infuse correct calculated dose over 3-5 min; do not infuse with other fluids, products
- Rhophylac: bring to room temperature; infuse by slow IV; observe for 20 min

SIDE EFFECTS

CNS: Lethargy
CV: Hypo/hypertension
INTEG: Irritation at inj site, fever
MISC: Infection, ARDs, anaphylaxis, pulmonary edema, DIC
MS: Myalgia, arthralgia

INTERACTIONS

Decrease: antibody response—live virus vaccines (measles, mumps, rubella)

NURSING CONSIDERATIONS

Assess:
- ⚠ Allergies, reactions to immunizations; previous immunization with product
- ⚠ **Intravascular hemolysis:** back pain, chills, hemoglobinuria, renal insufficiency; usually when WinRho SDF is given in those with immune thrombocytopenia purpura
- Type, crossmatch mother and newborn's cord blood; if mother Rh_0(D)-negative, D^u-negative and newborn Rh_0(D)-positive, product should be given

Evaluate:
- Rh_0(D) sensitivity in transfusion error, prevention of erythroblastosis fetalis for normal vision

Teach patient/family:
- How product works; that product must be given after subsequent deliveries if subsequent babies are Rh-positive
- ⚠ To report immediately: shaking, fever, chills, dark urine, swelling of hands or feet, back pain, SOB (intravascular hemolysis)

riboflavin (vit B_2) (OTC)

(rye'boh-flay-vin)

Func. class.: Vit B_2, water soluble

ACTION: Needed for respiratory reactions by catalyzing proteins

USES:
Vit B_2 deficiency or polyneuritis; cheilosis adjunct with thiamine
Unlabeled uses: Migraine prophylaxis

DOSAGE AND ROUTES

Deficiency
- **Adult: PO** 5-30 mg/day
- **Child ≥12 yr: PO** 3-10 mg/day then 0.6 mg/1000 calories ingested

RDA
- **Adult: PO** (males) 1.3 mg, (females) 1.1 mg

Migraine prophylaxis (unlabeled)
- **Adult: PO** 400 mg/day × 3 mo

Available forms: Tabs 5, 10, 25, 50, 100, 250 mg
Administer:
- With food for better absorption
- Store in airtight, light-resistant container

SIDE EFFECTS
GU: Yellow discoloration of urine
Precautions: Pregnancy (A)

PHARMACOKINETICS
Half-life 65-85 min, 60% protein bound, unused amounts excreted in urine (unchanged)

INTERACTIONS
Increase: riboflavin need—alcohol, probenecid, tricyclics, phenothiazines
Decrease: action of tetracyclines
Drug/Lab Test
- May cause false elevations of urinary catecholamines

NURSING CONSIDERATIONS
Assess:
- Nutritional status: liver, eggs, dairy products, yeast, whole grains, green vegetables

Evaluate:
- Therapeutic response: absence of headache, GI problems, cheilosis, skin lesions, depression; burning, itchy eyes; anemia

Teach patient/family:
- That urine may turn bright yellow
- About the addition of needed foods rich in riboflavin
- To avoid alcohol

rifabutin (Rx)
(riff'a-byoo-ten)
Mycobutin
Func. class.: Antimycobacterial agent
Chem. class.: Rifamycin S derivative

Do not confuse:
rifabutin/rifampin/rifapentine

ACTION:
Inhibits DNA-dependent RNA polymerase in susceptible strains of *Escherichia coli* and *Bacillus subtilis;* mechanism of action against *Mycobacterium avium* unknown

USES:
Prevention of *M. avium* complex (MAC) in patients with advanced HIV infection
Unlabeled uses: *Helicobacter pylori* that has not responded to other treatment

CONTRAINDICATIONS:
Hypersensitivity, active TB, WBC <1000/mm^3 or platelet count <50,000/mm^3
Precautions: Pregnancy (B), breastfeeding, children, hepatic disease, blood dyscrasias

DOSAGE AND ROUTES
- **Adult: PO** 300 mg/day (may take as 150 mg bid); max 600 mg/day

Renal dose
- **Adult: PO** CCr <30 ml/min, reduce by 50%

Available forms: Caps 150 mg
Administer:
- With food if GI upset occurs; better to take on empty stomach 1 hr before or 2 hr after meals; high-fat foods slow absorption; may take in 2 divided doses, may open capsule, mix with applesauce if unable to swallow whole cap
- Antiemetic if vomiting occurs

R

• After C&S completed; monthly to detect resistance

SIDE EFFECTS

CNS: *Headache,* fatigue, anxiety, confusion, insomnia
GI: *Nausea, vomiting, anorexia, diarrhea,* heartburn, hepatitis, discolored saliva, pseudomembranous colitis
GU: *Discolored urine*
HEMA: Hemolytic anemia, eosinophilia, thrombocytopenia, leukopenia
INTEG: *Rash*
MISC: Flulike symptoms, shortness of breath, chest pressure
MS: Asthenia, arthralgia, myalgia

PHARMACOKINETICS

53% absorbed, peak 2-3 hr, duration >24 hr, half-life 45 hr, metabolized in liver (active/inactive metabolites), excreted in urine primarily as metabolites

INTERACTIONS

Increase: levels of rifabutin: ritonavir
Decrease: action of amprenavir, anticoagulants, β-blockers, barbiturates, busPIRone, clofibrate, corticosteroids, cycloSPORINE, dapsone, delavirdine, disopyramide, doxycycline, efavirenz, estrogens, fluconazole, indinavir, ketoconazole, losartan, nelfinavir, nevirapine, oral contraceptives, phenytoin, quiNIDine, saquinavir, sulfonylureas, theophylline, tricyclic antidepressants, zidovudine, zolpidem

Drug/Food

• High-fat diet decreases absorption

Drug/Lab Test

Interference: folate level, vit B_{12}, BSP, gallbladder studies

NURSING CONSIDERATIONS

Assess:

⚠ **Acute TB:** chest x-ray, sputum culture, blood culture, biopsy of lymph nodes, PPD; product should not be given for active TB

• CBC for neutropenia, thrombocytopenia, eosinophilia

⚠ **Pseudomembranous colitis:** diarrhea, abdominal pain/cramping, fever, bloody stools

• Signs of anemia: Hct, Hgb, fatigue

• Hepatic studies weekly: ALT, AST, bilirubin

• Renal status before, each mo: BUN, creatinine, output, specific gravity, urinalysis

• Hepatic status: decreased appetite, jaundice, dark urine, fatigue

Evaluate:

• Therapeutic response: not used for active TB because of risk for development of resistance to rifampin; culture negative

Teach patient/family:

⚠ That patients using oral contraceptives should consider using nonhormonal methods of birth control, may decrease effect; to notify prescriber if pregnancy planned, suspected

• That compliance with dosage schedule, duration necessary

• That scheduled appointments must be kept because relapse may occur

• That urine, feces, saliva, sputum, sweat, tears may be colored red-orange; that soft contact lenses may be permanently stained

⚠ To report flulike symptoms: excessive fatigue, anorexia, vomiting, sore throat; unusual bleeding, yellowish discoloration of skin, eyes; myositis: muscle or bone pain; diarrhea, fever, abdominal cramping, bloody stools

rifampin (Rx)

(rif'am-pin)

Rifadin, Rofact 🍁

Func. class.: Antitubercular
Chem. class.: Rifamycin B derivative

Do not confuse:
rifampin/rifabutin

ACTION:

Inhibits DNA-dependent polymerase, decreases tubercle bacilli replication

USES:
Pulmonary TB, meningococcal carriers (prevention)

Unlabeled uses: Endocarditis, *Haemophilus influenzae type B prophylaxis,* Hansen's disease, *Mycobacterium avium* complex (MAC), orthopedic-device–related infection, pruritus

CONTRAINDICATIONS:
Hypersensitivity to this product, rifamycins; active *Neisseria meningitidis* infection

Precautions: Pregnancy (C), breastfeeding, children <5 yr, hepatic disease, blood dyscrasias

DOSAGE AND ROUTES

Tuberculosis

- **Adult: PO/IV** Max 600 mg/day as single dose 1 hr before meals or 2 hr after meals or 10 mg/kg/day 2-3×/wk
- **Child >5 yr: PO/IV** 10-20 mg/kg/day as single dose 1 hr before meals or 2 hr after meals, max 600 mg/day with other antituberculars
- **6-mo regimen:** 2 mo treatment of isoniazid, rifampin, pyrazinamide, and possibly streptomycin or ethambutol then rifampin and isoniazid 3-4 mo
- **9-mo regimen:** rifampin and isoniazid supplemented with pyrazinamide, streptomycin, or ethambutol

Meningococcal carriers

- **Adult: PO/IV** 600 mg bid × 2 days, max 600 mg/dose
- **Child >5 yr: PO/IV** 10-20 mg/kg × 2 days, max 600 mg/dose
- **Infant 3 mo-1 yr:** 5 mg/kg **PO** bid × 2 days

Prevention of *H. influenzae* type B infection (unlabeled)

- **Adult: PO** 600 mg/day × 4 days
- **Child: PO** 20 mg/kg/day × 4 days

MAC (unlabeled)

- **Adult: PO/IV** 600 mg/day used with ≥3 other active microbials
- **Child: PO/IV** 10-20 mg/kg/day used with ≥3 other active microbials

Endocarditis with prosthetic valves (unlabeled)

- **Adult: PO** 300 mg q8hr with gentamicin and vancomycin
- **Child: PO** 20 mg/kg/day in 2 divided doses with gentamicin and vancomycin, max 900 mg/day

Available forms: Caps 150, 300 mg; powder for inj 600 mg/vial

Administer:

- After C&S completed; monthly to detect resistance
- Do not give IM, SUBCUT

PO route

- On empty stomach, 1 hr before or 2 hr after meals with full glass of water; give with other products for TB
- Antiemetic if vomiting occurs
- Capsules may be opened, mixed with applesauce or jelly

Intermittent IV INF route

- After diluting each 600 mg/10 ml of sterile water for inj (60 mg/ml), swirl, withdraw dose, and dilute in 100 ml or 500 ml of D_5W given as inf over 3 hr; if diluted in 100 ml, give over 1/2 hr; do not admix with other sol or products

Y-site compatibilities: amiodarone, bumetanide, midazolam, pantoprazole, vancomycin

SIDE EFFECTS

CNS: Headache, fatigue, anxiety, drowsiness, confusion

EENT: Visual disturbances

GI: *Nausea, vomiting, anorexia, diarrhea,* **pseudomembranous colitis,** *heartburn,* sore mouth and tongue, **pancreatitis,** increased LFTs

GU: **Hematuria, acute renal failure,** **hemoglobinuria**

HEMA: **Hemolytic anemia, eosinophilia, thrombocytopenia, leukopenia**

INTEG: Rash, pruritus, urticaria

MISC: Flulike symptoms, menstrual disturbances, edema, SOB, **Stevens-Johnson syndrome, toxic epidermal necrolysis, angioedema, anaphylaxis**

MS: Ataxia, weakness

R

PHARMACOKINETICS

PO: Peak 1-4 hr, duration >24 hr, half-life 3 hr, metabolized in liver (active/inactive metabolites), excreted in urine as free product (30% crosses placenta) and in breast milk

INTERACTIONS

⚠ Do not use with protease inhibitors

Increase: hepatotoxicity—isoniazid

Decrease: action of acetaminophen, alcohol, anticoagulants, antidiabetics, β-blockers, barbiturates, benzodiazepines, chloramphenicol, clofibrate, corticosteroids, cycloSPORINE, dapsone, digoxin, doxycycline, haloperidol, hormones, imidazole antifungals, NIFEdipine, oral contraceptives, phenytoin, protease inhibitors, theophylline, verapamil, zidovudine

Increase: LFTs

Decrease: Hgb

Drug/Lab Test

Interference: folate level, vit B_{12}

NURSING CONSIDERATIONS

Assess:

- **Infection:** sputum culture, lung sounds, characteristics of sputum
- Signs of anemia: Hct, Hgb, fatigue
- Hepatic function monthly: ALT, AST, bilirubin, decreased appetite, jaundice, dark urine, fatigue
- Renal status before, each mo: BUN, creatinine, output, specific gravity, urinalysis

⚠ **Serious skin reactions:** fever, sore throat, fatigue, ulcers; lesions in mouth, lips, rash; can be fatal

⚠ **Pseudomembranous colitis:** diarrhea, fever, abdominal pain/cramping, bloody stools; product should be discontinued, prescriber notified

Evaluate:

- Therapeutic response: decreased symptoms of TB, culture negative

Teach patient/family:

- That compliance with dosage schedule, duration necessary
- That scheduled appointments must be kept because relapse may occur
- To avoid alcohol because hepatotoxicity may occur
- That urine, feces, saliva, sputum, sweat, tears may be colored red-orange; that soft contact lenses may be permanently stained

⚠ To report flulike symptoms: excessive fatigue, anorexia, vomiting, sore throat; unusual bleeding; yellowish discoloration of skin, eyes; diarrhea with pus, mucous, blood

- To use nonhormonal form of birth control; to notify prescriber if pregnancy planned, suspected; not to breastfeed

rifapentine (Rx)

(riff'ah-pen-teen)

Priftin

Func. class.: Antitubercular

Chem. class.: Rifamycin derivative

ACTION: Inhibits DNA-dependent polymerase, decreases tubercle bacilli replication

USES: Pulmonary TB; must be used with at least one other antitubercular agent

CONTRAINDICATIONS: Hypersensitivity to rifamycins, porphyria

Precautions: Pregnancy (C), breastfeeding, children <12 yr, geriatric patients, hepatic disease, blood dyscrasias, HIV

DOSAGE AND ROUTES

Intensive phase

- **Adult: PO** 600 mg (four 150-mg tabs) 2×/wk with an interval of 72 hr between doses × 2 mo; must be given with at least 1 other antitubercular agent

Continuation phase

- **Adult: PO** 600 mg weekly × 4 mo in combination with isoniazid or other appropriate antitubercular product

Available forms: Tabs 150 mg

Administer:

- May give with food for GI upset; use other products for TB
- Antiemetic if vomiting occurs
- After C&S completed; monthly to detect resistance

SIDE EFFECTS

CNS: Headache, fatigue, anxiety, dizziness
EENT: Visual disturbances
GI: *Nausea, vomiting, anorexia, diarrhea,* bilirubinemia, hepatitis, increased ALT, AST, *heartburn,* **pancreatitis, pseudomembranous colitis**
GU: **Hematuria,** pyuria, proteinuria, urinary casts, urine discoloration
HEMA: **Thrombocytopenia, leukopenia, neutropenia, lymphopenia,** anemia, **leukocytosis,** purpura, hematoma
INTEG: Rash, pruritus, urticaria, acne
MISC: Increased B/P
MS: Gout, arthrosis

PHARMACOKINETICS

Peak 5-6 hr; half-life 13 hr; metabolized in liver (active/inactive metabolites); excreted in urine, feces, breast milk; protein binding 97%; steady state 10 days; CYP450 3A4, 2C8/9 inducer

INTERACTIONS

⚠ **Do not use with protease inhibitors**
Decrease: action of amitriptyline, anticoagulants, antidiabetics, barbiturates, β-blockers, chloramphenicol, clarithromycin, clofibrate, corticosteroids, cycloSPORINE, dapsone, delavirdine, diazepam, digoxin, diltiazem, disopyramide, doxycycline, fentaNYL, fluconazole, haloperidol, indinavir, itraconazole, ketoconazole, methadone, mexiletine, nelfinavir, NIFEdipine, nortriptyline, oral contraceptives, phenothiazines, phenytoin, progestins, quiNIDine, quiNINE, ritonavir, saquinavir, sildenafil, tacrolimus, theophylline, thyroid preparations, tocainide, verapamil, warfarin, zidovudine
Drug/Food
Increase: absorption with food

Drug/Lab Test
Increase: LFTs, platelets
Decrease: Hgb, WBC
Interference: folate level, vit B_{12}

NURSING CONSIDERATIONS

Assess:

- Baselines of CBC, AST, ALT, bilirubin, platelets
- **Infection:** sputum culture, lung sounds
- Signs of anemia: Hct, Hgb, fatigue
- Hepatic studies monthly: ALT, AST, bilirubin; decreased appetite, jaundice, dark urine, fatigue
- Renal status monthly: BUN, creatinine, output, specific gravity, urinalysis

⚠ **Pseudomembranous colitis:** diarrhea, fever, abdominal pain/cramping, bloody diarrhea; discontinue if present, notify prescriber

Evaluate:

- Therapeutic response: decreased symptoms of TB, culture negative

Teach patient/family:

- That compliance with dosage schedule, duration necessary
- That scheduled appointments must be kept because relapse may occur
- That urine, feces, saliva, sputum, sweat, tears may be colored red-orange; that soft contact lenses, dentures may be permanently stained

⚠ To use alternative method of contraception; that oral contraceptive action may be decreased; to notify prescriber if pregnancy planned, suspected; to avoid breastfeeding

⚠ To report flulike symptoms: excessive fatigue, anorexia, vomiting, sore throat; unusual bleeding, yellowish discoloration of skin, eyes; diarrhea with pus, mucous, blood

R

rifaximin (Rx)

(rif-ax'i-min)

Xifaxan

Func. class.: Antiinfective—miscellaneous

Chem. class.: Analog of rifampin

Do not confuse:
rifaximin/rifampin

ACTION: Binds to bacterial-DNA–dependent RNA polymerase, thereby inhibiting bacterial RNA synthesis

USES: Traveler's diarrhea in those ≥12 yr caused by *E. coli,* hepatic encephalopathy
Unlabeled uses: Crohn's disease, diverticulitis, irritable bowel syndrome

CONTRAINDICATIONS: Hypersensitivity to product, rifamycins; diarrhea with fever, blood in stool
Precautions: Pregnancy (C), breastfeeding, children, geriatric patients

DOSAGE AND ROUTES

Traveler's diarrhea
- **Adult and child ≥12 yr: PO** 200 mg tid × 3 days without regard to meals

Hepatic encephalopathy
- **Adult: PO** 550 mg bid

Crohn's disease (unlabeled)
- **Adult: PO** 200 mg tid × 16 wk

Diverticulitis (unlabeled)
- **Adult: PO** 400 mg bid with mesalamine 800 mg tid × 7 days then 7 days/mo

Irritable bowel syndrome (unlabeled)
- **Adult: PO** 550 mg tid × 14 day

Available forms: Tabs 200, 550 mg
Administer:
- Without regard to food

SIDE EFFECTS

CNS: Abnormal dreams, dizziness, insomnia, *headache,* fatigue, depression
CV: Hypotension, chest pain, peripheral edema, ascites
GI: *Abdominal pain, constipation, defecation urgency, flatulence, nausea, rectal tenesmus,* vomiting, ascites, pseudomembranous colitis
GU: Proteinuria, polyuria, increased urinary frequency
MISC: *Pyrexia,* motion sickness, tinnitus, rash, photosensitivity, exfoliative dermatitis
MS: Arthralgia, muscle pain, myalgia
RESP: Dyspnea, cough, pharyngitis

PHARMACOKINETICS

Low systemic absorption, half-life 1.8-4.5 hr, excreted in feces, peak 1-4 hr

INTERACTIONS

Increase: effect of—afatinib
Drug/Lab Test
Increase: LFTs, potassium
Decrease: blood glucose, sodium

NURSING CONSIDERATIONS

Assess:
- GI symptoms: amount, character of diarrhea; abdominal pain, nausea, vomiting, blood in stool; do not use in those with blood in stool, increased temperature with diarrhea

⚠ Overgrowth of infection, pseudomembranous colitis

Evaluate:
- Therapeutic response: absence of infection

Teach patient/family:

⚠ To discontinue rifaximin, notify prescriber if diarrhea persists >24-48 hr, if diarrhea worsens, or if blood in stools and fever present
- To avoid hazardous activities if dizziness occurs
- To notify prescriber if pregnancy planned, suspected
- That headache, rash, insomnia, abnormal dreams, tinnitus may occur
- To take without regard to food
- To take as directed, consume all of the product prescribed

rilpivirine

Edurant

Func. class.: Antiretroviral

Chem. class.: Nonnucleoside transcriptase inhibitors (NNTIs)

ACTION: Inhibits HIV-1 reverse transcriptase; unlike nucleoside reverse transcriptase inhibitors (NRTIs); it does not compete for binding nor does it require phosphorylation to be active; binds directly to a site on reverse transcriptase causing disruption of the enzyme's active site thereby blocking RNA-dependent and DNA-dependent DNA polymerase activities

USES: HIV in combination with other antiretrovirals

CONTRAINDICATIONS: Hypersensitivity

Precautions: Pregnancy (B), breastfeeding, neonates, infants, children, adolescents <18 yr, immune reconstitution syndrome, antimicrobial resistance, pancreatitis, coinfection hepatitis B or C and HIV, hepatic disease, depression, suicidal ideation, QT prolongation, torsades de pointes, hyperlipidemia, hypertriglyceridemia, hypercholesterolemia, immune reconstitution syndrome

DOSAGE AND ROUTES

Antiretroviral treatment-naive adults (HIV) with other antiretroviral agents

- **Adults:** PO 25 mg/day with a meal

Available forms: Tab 25 mg

Administer:

- Give in combination with other antiretroviral agents; in antiretroviral treatment-naive adults, rilpivirine is utilized as an alternative to efavirenz in NNRTI-based treatment regimens; potential rilpivirine-based treatment regimens combine rilpivirine with either tenofovir plus emtricitabine or lamiVUDine; or abacavir plus emtricitabine or lamiVUDine; or zidovudine plus emtricitabine or lamiVUDine
- Give with a meal
- Store at room temperature away from heat and moisture

SIDE EFFECTS

CNS: Depressed mood, dysphoria, major depression, mood alteration, negative thoughts, suicide attempts, fatigue, headache, dizziness, drowsiness

GI: Nausea, vomiting, abdominal pain, diarrhea, cholecystitis, cholelithiasis, decreased appetite, elevated hepatic enzymes, hyperbilirubinemia, hypercholesterolemia

GU: Glomerulonephritis membranous/glomerulonephritis mesangioproliferative

PHARMACOKINETICS:

Protein binding (99.7%) to albumin; metabolism via oxidation CYP3A; terminal elimination half-life 50 hr, excretion feces (85%), 25% excreted unchanged; urine (6.1%); peak 4-5 hr; increased effect 40% (food), decreased effect 50% (high protein drink)

INTERACTIONS:

Increase: rilpivirine effect—CYP3A4 inhibitors (delavirdine, efavirenz, darunavir, tipranavir, atazanavir, fosamprenavir, indinavir, nelfinavir, aldesleukin IL-2, amiodarone aprepitant, basiliximab, boceprevir, bromocriptine, chloramphenicol, clarithromycin, conivaptan danazol, dalfopristin, dasatinib, diltiazem, dronedarone, erythromycin, ethinyl estradiol, fluconazole, FLUoxetine, fluvoxaMINE, fosaprepitant, imatinib, isoniazid, itraconazole, ketoconazole, lanreotide, lapatinib, miconazole, nefazodone, niCARdipine, octreotide, posaconazole, quiNINE, ranolazine, rifaximin, tamoxifen, telaprevir, telithromycin, troleandomycin, verapamil, voriconazole, zafirlukast)

⚠ **Increase:** QT prolongation—class IA/III antidysrhythmics, some phenothiazines, β-agonists, local anesthetics, tricyclics, chloroquine, droperidol, haloperi-

R

dol, pentamidine; CYP3A4 inhibitors (amiodarone, clarithromycin, erythromycin, telithromycin, troleandomycin, arsenic trioxide, levomethadyl); CYP3A4 substrates (methadone, pimozide, QUEtiapine, quiNIDine, risperiDONE, ziprasidone, lopinavir, saquinavir, fluconazole, posaconazole, dasatinib, dronedarone, lapatinib, octreotide, ranolazine, citalopram, abarelix, alfuzosin, amoxapine, apomorphine, artemether, lumefantrine, asenapine, ofloxacin, ciprofloxacin, cloZAPine, cyclobenzaprine, dolasetron, eribulin, flecainide, gatifloxacin, gemifloxacin, halogenated anesthetics, iloperidone, levofloxacin, maprotiline, mefloquine, moxifloxacin, nilotinib, norfloxacin, OLANZapine, ondansetron, paliperidone, palonosetron, QUEtiapine)

Increase: rilpivirine adverse reactions, fungal infections—fluconazole, voriconazole

Decrease: rilpivirine effect, treatment failure—CYP3A4 inducers (phenytoin, fosphenytoin, barbiturates, OXcarbazepine, carBAMazepine, rifabutin, rifampin, rifapentine, dexamethasone), efavirenz, nevirapine, ritonavir; aminoglutethimide, bexarotene, bosentan, griseofulvin, metyrapone, modafinil, flutamide, nafcillin, pioglitazone, primidone, topiramate); proton pump inhibitors (PPIs)

Decrease: rilpivirine effect, treatment failure—H2 receptor antagonists (cimetidine, famotidine, nizatidine, ranitidine), give 12 hr before or 4 hr after rilpivirine

Decrease: rilpivirine effect—antacids, use >2 hr before or 4 hr after rilpivirine

Drug/Food

Increase: adverse reactions—grapefruit juice

NURSING CONSIDERATIONS

Assess:

- **HIV:** Assess symptoms of HIV including opportunistic infections before and during treatment; some may be life-threatening; monitor plasma HIV RNA, CD4+, CD8 + cell counts, serum β-2 microglobulin, serum ICD+24 antigen levels; treatment failures occur more frequently in those with baseline HIV-1 RNA concs >100,000 copies/ml than in patients with concs <100,000 copies/ml; monitor serum cholesterol, lipid panel
- Antiretroviral drug resistance testing is recommended before initiation of therapy in antiretroviral treatment-naive patients
- For adults and adolescents, initiation of antiretroviral therapy is recommended in any patient with a history of an AIDS-defining infection; with a CD4 ≤500/mm^3; who is pregnant; who has HIV-associated nephropathy; or who is being treated for hepatitis B (HBV) infection

⚠ **Suicidal thoughts/behaviors:** Assess frequently for suicidal ideation; report any increase in depressive symptoms

- Hepatic disease: monitor for elevated hepatic enzymes (>2.5 × ULN); grade 3 and 4 may be higher in patients co-infected with hepatitis B or C

Teach Patient/family:

- That product is not a cure but controls symptoms
- That product must be taken in combination with other prescribed products

riluzole (Rx)

(rill′you-zole)

Rilutek

Func. class.: ALS agent

Chem. class.: Benzathiazole

ACTION: May act by modulating the release of glutamate and inactivating voltage-dependent sodium channels

USES: Amyotropic lateral sclerosis (ALS)

CONTRAINDICATIONS: Hypersensitivity

Precautions: Pregnancy (C), breastfeeding, children, geriatric patients, neutropenia, renal/hepatic disease, cigarette smoking, febrile illness, pneumonia

DOSAGE AND ROUTES

- **Adult: PO** 50 mg q12hr; take 1 hr before or 2 hr after meals

Available forms: Tabs 50 mg

Administer:

- 1 hr before or 2 hr after meals; high-fat meal decreases absorption

SIDE EFFECTS

CNS: Hypertonia, depression, dizziness, insomnia, somnolence, vertigo, paresthesia

CV: Hypertension, tachycardia, phlebitis, palpitation, postural hypertension

GI: Nausea, vomiting, dyspepsia, anorexia, diarrhea, flatulence, stomatitis, dry mouth, increased LFTs, jaundice, abdominal pain

GU: UTI, dysuria

HEMA: Neutropenia

INTEG: Pruritus, eczema, alopecia, exfoliative dermatitis

MS: Arthralgia

RESP: Decreased lung function, rhinitis, increased cough, pneumonia

PHARMACOKINETICS

Well absorbed; extensively metabolized by liver; excretion in urine, feces

INTERACTIONS

Increase: elimination of riluzole—cigarette smoking, rifampin, omeprazole, charcoal-broiled food

Increase: hepatic injury—allopurinol, methyldopa, sulfaSALAzine, leflunomide, methotrexate, tacrine

Increase: LFTs—barbiturates, carBAMazepine

Decrease: elimination of riluzole—caffeine, theophylline, amitriptyline, quinolones

Drug/Food

Decrease: absorption—high-fat meal

NURSING CONSIDERATIONS

Assess:

- Clinical improvement of neurologic function
- Hepatic studies: AST, ALT, bilirubin, GGT at baseline, monthly × 3 mo, then q3mo; monitor LFTs
- Neutropenia <500/mm

Evaluate:

- Therapeutic response: improvement in neurologic status

Teach patient/family:

- To report febrile illness, signs of infection, cardiac/respiratory changes, which may indicate neutropenia
- About the reason for the product, expected results

rimantadine (Rx)

(ri-man′tah-deen)

Flumadine

Func. class.: Synthetic antiviral

Chem. class.: Tricyclic amine

Do not confuse:

rimantadine/amantadine/ranitidine

ACTION: Prevents uncoating of nucleic acid in viral cell, thereby preventing the penetration of the virus to the host; causes the release of dopamine from neurons

USES: Prophylaxis or treatment of influenza type A

CONTRAINDICATIONS: Hypersensitivity to this product, amantadine

Precautions: Pregnancy (C), breastfeeding, children <1 yr, seizure disorders, renal/hepatic disease

DOSAGE AND ROUTES

Influenza type A

Prophylaxis

- **Adult and child >10 yr: PO** 100 mg bid
- **Child 1-10 yr: PO** 5 mg/kg/day, max 150 mg

Treatment

- **Adult: PO** 100 mg bid; start treatment at onset of symptoms; continue for ≥1 wk

R

• **Child ≥10 yr (unlabeled): PO** 200 mg/day either as single dose or in 2 divided doses
• **Child 1-9 yr (unlabeled): PO** 6.6 mg/kg/day in 2 divided doses; max 150 mg/day in 2 divided doses
• **Geriatric: PO** 100 mg/day

Renal/hepatic dose
• **Adult: PO** ≤10 ml/min, 100 mg daily

Available forms: Tabs 100 mg; syr 50 mg/5 ml

Administer:
• Store in tight, dry container
• Give with full glass of water
• Within 48 hr of exposure to influenza; continue for 10 days after contact
• At least 4 hr before bedtime to prevent insomnia
• After meals for better absorption; to decrease GI symptoms
• In divided doses to prevent CNS disturbances: headache, dizziness, fatigue, drowsiness

SIDE EFFECTS

CNS: *Headache, dizziness,* fatigue, depression, hallucinations, tremors, seizures, insomnia, poor conc, asthenia, gait abnormalities, *anxiety,* confusion
CV: Pallor, palpitations, edema
EENT: Tinnitus, taste abnormality, eye pain
GI: *Nausea, vomiting,* constipation, *dry mouth, anorexia, abdominal pain,* diarrhea, dyspepsia
INTEG: Rash

PHARMACOKINETICS

PO: Peak 6 hr, elimination half-life 13-65 hr, plasma protein binding (40%)

INTERACTIONS

Increase: rimantadine conc—cimetidine
Decrease: peak conc of rimantadine—acetaminophen, aspirin; intranasal influenza vaccine (separate by ≥48 hr)

NURSING CONSIDERATIONS

Assess:
⚠ **Seizures;** if seizures occur, product should be discontinued
• Bowel pattern before, during treatment
• CNS effects in geriatric patients, patients with severe renal/hepatic disease
• Skin eruptions, photosensitivity after administration of product
• Respiratory status: rate, character, wheezing, tightness in chest
• Signs of infection

Evaluate:
• Therapeutic response: absence of fever, malaise, cough, dyspnea with infection

Teach patient/family:
• About aspects of product therapy: the need to report dyspnea, dizziness, poor conc, behavioral changes
• To avoid hazardous activities if dizziness occurs

rimexolone ophthalmic

See Appendix B

risedronate (Rx)

(rih-sed′roh-nate)

Actonel, Atelvia

Func. class.: Bone resorption inhibitor
Chem. class.: Bisphosphonate

Do not confuse:
Actonel/Actos

ACTION: Inhibits bone resorption, absorbs calcium phosphate crystal in bone, and may directly block dissolution of hydroxyapatite crystals of bone

USES: Paget's disease; prevention, treatment of osteoporosis in postmenopausal women; glucocorticoid-induced osteoporosis; osteoporosis in men
Unlabeled uses: Osteolytic metastases

CONTRAINDICATIONS:
Hypersensitivity to bisphosphonates, inability to stand or sit upright for ≥30 min, esophageal stricture, achalasia, hypocalcemia

Precautions: Pregnancy (C), breastfeeding, children, renal disease, active upper GI disorders, dental disease, hyperparathyroidism, infection, vit D deficiency, coagulopathy, chemotherapy, asthma

DOSAGE AND ROUTES

Paget's disease

- **Adult: PO** 30 mg/day × 2 mo; patients with Paget's disease should receive calcium and vit D if dietary intake lacking; if relapse occurs, retreatment advised

Treatment/prevention of postmenopausal osteoporosis

- **Adult: PO** 5 mg/day or 35 mg/wk or 75 mg/day × 2 consecutive days 2× monthly or 150 mg/mo

Glucocorticoid osteoporosis

- **Adult: PO** 5 mg/day

Osteoporosis in men

- **Adult: PO** 35 mg/wk

Osteolytic metastases (unlabeled)

- **Adult: PO** 30 mg/day × 6 mo

Renal dose

- **Adult: PO** CCr <30 ml/min, avoid use

Available forms: Tabs 5, 30, 35, 150 mg; tab, gastro-resistant, weekly 35 mg; tab, weekly 35 mg

Administer:

- For 2 mo to be effective for Paget's disease
- With a full glass of water; patient should be in upright position for ½ hr; swallow whole; do not crush, break, chew
- Supplemental calcium and vit D for Paget's disease if instructed by prescriber
- Give daily ≥30 min before meals or give weekly
- Store in cool environment, out of direct sunlight

SIDE EFFECTS

CNS: Dizziness, headache, depression

CV: *Chest pain,* hypertension, atrial fibrillation

GI: *Abdominal pain, diarrhea, nausea,* constipation, esophagitis

MISC: Rash, UTI, pharyngitis, hypocalcemia, hypophosphatemia, increase PTH

MS: Osteonecrosis of the jaw, severe muscle/joint/bone pain, fractures

SYST: Angioedema

PHARMACOKINETICS

Rapidly cleared from circulation, taken up mainly by bones (50%), eliminated primarily through kidneys, absorption decreased by food, terminal half-life 220 hr

INTERACTIONS

Increase: GI irritation—NSAIDs, salicylates

Decrease: absorption of risedronate—aluminum, calcium, iron, magnesium salts, antacids

Drug/Food

Decrease: bioavailability—take ½ hr before food or drinks other than water

Drug/Lab Test

Decrease: calcium

NURSING CONSIDERATIONS

Assess:

- **Paget's disease:** headache, bone pain, increased head circumference
- **Osteoporosis:** in men or postmenopausal women; bone density study before and periodically during treatment
- Phosphate, alk phos, calcium; creatinine, BUN (renal disease)
- **Hypocalcemia:** paresthesia, twitching, laryngospasm, Chvostek's/Trousseau's signs

⚠ **Serious skin reactions:** angioedema

- **Dental health:** assess dental health, provide antiinfectives for dental extraction; cover with antiinfectives before dental extraction

⚠ For atrial fibrillation

R

Evaluate:

- Therapeutic response: increased bone mass, absence of fractures

Teach patient/family:

- To sit upright for ½ hr after dose to prevent irritation
- To comply with diet, vitamin/mineral supplements
- To notify prescriber if pregnancy is suspected or if breastfeeding
- To maintain good oral hygiene
- To notify all health care providers of use
- That musculoskeletal pain may occur within a few days/mo after starting but usually resolves; use acetaminophen
- To exercise regularly; to avoid alcohol, tobacco

risperiDONE (Rx)

(ris-pehr'ih-dohn)

RisperDAL, RisperDAL Consta, RisperDAL M-TAB

Func. class.: Antipsychotic

Chem. class.: Benzisoxazole derivative

Do not confuse:

RisperDAL/reserpine

ACTION: Unknown; may be mediated through both dopamine type 2 (D_2) and serotonin type 2 ($5\text{-}HT_2$) antagonism

USES: Irritability associated with autism, bipolar disorder, mania, schizophrenia

Unlabeled uses: Acute psychosis, agitation, ADHD, dementia, psychotic depression, Tourette's syndrome

CONTRAINDICATIONS: Hypersensitivity, breastfeeding

Precautions: Pregnancy (C), children, geriatric patients, cardiac/renal/hepatic disease, breast cancer, Parkinson's disease, CNS depression, brain tumor, dehydration, diabetes, hematologic disease, seizure disorders, abrupt discontinuation, suicidal ideation, phenylketonuria

Black Box Warning: Increased mortality in elderly patients with dementia-related psychosis

DOSAGE AND ROUTES

- **Adult: PO** 2 mg/day as single dose or in 2 divided doses, adjust dose at intervals of ≥24 hr and 1-2 mg/day as tolerated to 4-8 mg/day; **IM** establish dosing with **PO** before **IM** 25 mg q2wk, may increase to max 50 mg q2wk
- **Adolescent: PO** 0.5 mg/day in AM or PM, adjust dose at intervals of ≥24 hr and 0.5-1 mg/day as tolerated to 3 mg/day
- **Geriatric: PO** 0.5 mg daily-bid, increase by 1 mg/wk; **IM** 25 mg q2wk

Hepatic/renal dose

- **Adult: PO** 0.5 mg bid, increase by 0.5 mg bid, increase to 1.5 mg bid at intervals ≥1 wk

Available forms: Tabs 0.25, 0.5, 1, 2, 3, 4 mg; oral sol 1 mg/ml; orally disintegrating tabs 0.25, 0.5, 1, 2, 3, 4 mg; long-acting inj kit (Risperdal Consta) 12.5, 25, 37.5, 50 mg

Administer:

- Reduced dose in geriatric patients
- Anticholinergic agent on order from prescriber, to be used for EPS
- Avoid use with CNS depressants
- Conventional tabs: give without regard to meals
- **Oral disintegrating tab** (Risperdal M-TAB): do not open blister pack until ready to use; tear 1 of the 4 units apart at perforation; bend corner where indicated; peel back foil; do not push tab through foil; remove from pack and place product on patient's tongue; tab disintegrates in seconds and can be swallowed with/without liquids, do not split or chew
- **Oral sol:** May dilute 3-4 oz of beverage, measure dose using calibrated pipette; not compatible with tea, cola; compatible with water, coffee, orange juice, low-fat milk

IM route

Preparation

• Only suspend in the diluent provided and with the supplied needle; do not substitute any components of the dose pack including the SmartSite Needle-Free Vial Access Device or Needle-Pro device
• Before admixing, remove the dose pack from the refrigerator and allow to come to room temperature for about 30 min before reconstituting
• Remove the plastic colored cap from the vial without removing the gray rubber stopper; wipe top of the gray stopper with an alcohol wipe and allow to dry
• Peel back the blister pouch and remove the Vial Access Device by holding between the white Luer cap and the skirt; do not touch the spike tip at any time
• Place the vial on a hard surface and hold the base; orient the Vial Access Device vertically over the vial so that the spike tip is at the center of the vial's rubber stopper
• With a straight downward push, press the spike tip of the Vial Access Device through the center of the vial's rubber stopper until the device securely snaps onto the vial top; improper placement of the Vial Access Device on the vial could result in leakage of the diluent upon transfer
• Hold the base of the vial and swab the syringe connection point (blue circle) of the Vial Access Device with an alcohol wipe and allow to dry before attaching the syringe
• During syringe assembly steps, avoid over-tightening or syringe component parts might loosen from the syringe body
• To open the syringe, hold syringe only by the white collar and snap off the smooth white cap; do not twist or cut off the white cap; remove white cap together with the rubber tip cap inside
• While holding the white collar of the syringe, insert and press the syringe tip into the blue circle of the Vial Access Device and twist in a clockwise motion to secure the connection of the syringe to the Vial Access Device; hold the skirt of the Vial Access Device during attachment to prevent it from spinning; keep the syringe and the Vial Access Device aligned
• Inject the entire contents of the syringe containing the diluent into the vial
• Holding the plunger down with the thumb, shake the vial vigorously for a minimum of 10 sec; suspension should appear uniform, thick, and milky colored, and all the powder is dispersed in liquid; the microspheres will be visible in liquid, but no dry microspheres remain
• Invert completely and slowly withdraw the contents of the suspension from the vial into the syringe; tear the section of the vial label at the perforation and apply the detached label to the syringe
• While holding the white collar of the syringe, unscrew the syringe from the Vial Access Device, then discard both the vial and the Vial Access Device appropriately
• Select the appropriate color-coded needle provided with the kit; two distinct needles are provided: the needle with the yellow hub and print is for injection into the gluteal muscle (2-inch needle), and the needle with the green hub and print is for deltoid muscles (1-inch needle); they are not interchangeable; do not use the needle intended for gluteal injection for deltoid injection, and vice versa
• Peel the blister pouch of the Needle-Pro safety device open halfway; grasp the transparent needle sheath using the plastic peel pouch; to prevent contamination, do not touch the orange Needle-Pro safety device's Luer connector; while holding the white collar of the syringe, attach the Luer connection of the orange Needle-Pro safety device to the syringe with an easy clockwise twisting motion
• While holding the white collar of the syringe, grasp the transparent needle sheath and seat the needle firmly on the orange Needle-Pro safety device with a push and a clockwise twist; seating the needle will secure the connection between the needle and the orange Needle-Pro safety device

• While holding the white collar of the syringe, pull the transparent needle sheath straight away from the needle; do not twist the sheath because this can loosen the luer connection
• Resuspension is necessary before administration, because settling occurs after reconstitution; resuspend the microspheres in the syringe by shaking vigorously
• Refer to the Instructions for Use section of the product labeling for detailed visual aids that accompany the written instructions

Administration
• Only for IM; do not give IV
• Resuspension is necessary before administration, because settling occurs after reconstitution; resuspend the microspheres in the syringe by shaking vigorously
• Remove any air bubbles by tapping the syringe and slowly depressing the plunger with the needle in an upright position
• Inject the entire contents of the syringe into the upper outer quadrant of the gluteal area or the deltoid muscle; inject immediately after reconstitution to avoid settling; gluteal injections should be alternated between the two buttocks
• After the injection is complete, press the needle into the orange Needle-Pro safety device by gently pressing the orange Needle-Pro safety device against a flat surface with one hand; as the orange Needle-Pro safety device is pressed, the needle firmly engages into the orange Needle-Pro safety device
• Visually confirm full engagement of the needle into the Needle-Pro safety device, then appropriately discard both the used and unused needle provided in the dose pack
• Do not store the vial after reconstitution or the suspension will settle
• Do not combine 2 different dosage strengths of RisperDAL Consta in a single administration
• The dose pack device is for single use only; do not reprocess for subsequent reuse because the integrity of the device may be compromised, leading to a deterioration in performance

Stability after reconstitution
• Once in suspension, the product may remain at room temperature, but it must be used within 6 hr; always resuspend before administration if not used immediately

SIDE EFFECTS

CNS: *EPS, pseudoparkinsonism, akathisia, dystonia, tardive dyskinesia; drowsiness, insomnia, agitation, anxiety, headache,* seizures, neuroleptic malignant syndrome, dizziness, suicidal ideation, head titubation (shaking)
CV: Orthostatic hypotension, tachycardia; heart failure, sudden death (geriatric patients), AV block
EENT: Blurred vision, tinnitus
GI: *Nausea,* vomiting, *anorexia, constipation,* jaundice, weight gain
GU: Hyperprolactinemia, gynecomastia, dysuria
HEMA: Neutropenia, granulocytopenia
MISC: Renal artery occlusion; hyperprolactinemia (child)
MS: Rhabdomyolysis
RESP: Rhinitis; sinusitis, upper respiratory infection, cough

PHARMACOKINETICS

PO: Extensively metabolized by liver to major active metabolite, plasma protein binding 90%, peak 1-2 hr, excreted 90% in urine, terminal half-life 3-24 hr

INTERACTIONS

Increase: seizures—traMADol
⚠ **Increase:** possible death in dementia-related psychosis: furosemide
Increase: sedation—other CNS depressants, alcohol
⚠ **Increase:** serotonin syndrome, neuroleptic malignant syndrome—CYP2D6 inhibitors (SSRIs, SNRIs)
Increase: EPS—other antipsychotics

Increase: risperiDONE excretion—carBAMazepine

⚠ **Increase:** QT prolongation—class IA/III antidysrhythmics, some phenothiazines, β-agonists, local anesthetics, tricyclics, haloperidol, methadone, chloroquine, clarithromycin, droperidol, erythromycin, pentamidine, thioridazine, ziprasidone

Increase: risperiDONE levels: acetylcholinesterase, inhibitors, CYP2D6 inhibitors, SSRIs, valproic acid, verapamil

Decrease: risperiDONE action—CYP2D6 inducers (carBAMazepine, barbiturates, phenytoins, rifampin)

Decrease: levodopa effect—levodopa

Drug/Herb

Decrease: risperiDONE effect—echinacea

Drug/Lab Test

Increase: prolactin levels, blood, glucose, lipids

NURSING CONSIDERATIONS

Assess:

⚠ **Suicidal thoughts/behaviors:** often when depression is lessened; mental status before initial administration

- Bilirubin, CBC, hepatic studies monthly
- Affect, orientation, LOC, reflexes, gait, coordination, sleep pattern disturbances

⚠ **QT prolongation:** B/P standing, lying; pulse, respirations; take these q4hr during initial treatment; establish baseline before starting treatment; report drops of 30 mm Hg; watch for ECG changes

- Dizziness, faintness, palpitations, tachycardia on rising
- **EPS:** akathisia, tardive dyskinesia (bizarre movements of the jaw, mouth, tongue, extremities), pseudoparkinsonism (rigidity, tremors, pill rolling, shuffling gait)

⚠ **Serious reactions in geriatric patient:** fatal pneumonia, heart failure, sudden death, dementia

⚠ **Neuroleptic malignant syndrome:** hyperthermia, increased CPK, altered mental status, muscle rigidity, seizures, change in B/P, fatigue, tachycardia

- Constipation, urinary retention daily; if these occur, increase bulk, water in diet
- Weight gain, hyperglycemia, metabolic changes in diabetes, increased lipids
- Decreased stimuli by dimming lights, avoiding loud noises
- Supervised ambulation until patient stabilized on medication; do not involve patient in strenuous exercise program because fainting is possible; patient should not stand still for a long time
- Increased fluids to prevent constipation
- Store in tight, light-resistant container (PO); unopened vials in refrigerator, protect from light; do not freeze

Evaluate:

- Therapeutic response: decrease in emotional excitement, hallucinations, delusions, paranoia; reorganization of patterns of thought, speech

Teach patient/family:

- That orthostatic hypotension may occur; to rise from sitting or lying position gradually
- To avoid abrupt withdrawal of product because EPS may result; product should be withdrawn slowly
- To avoid OTC preparations (cough, hay fever, cold) unless approved by prescriber; that serious product interactions may occur; to avoid use of alcohol because increased drowsiness may occur
- To avoid hazardous activities if drowsy or dizzy
- To comply with product regimen
- To report impaired vision, tremors, muscle twitching
- That heat stroke may occur in hot weather; to take extra precautions to stay cool; to avoid hot tubs, hot showers, tub baths
- To use contraception; to inform prescriber if pregnancy is planned or suspected; not to breastfeed
- To notify provider of suicidal thoughts/behaviors

TREATMENT OF OVERDOSE:

Lavage if orally ingested; provide airway; *do not induce vomiting*

ritonavir (Rx)

(ri-toe′na-veer)

Norvir

Func. class.: Antiretroviral

Chem. class.: Protease inhibitor

Do not confuse:
ritonavir/Retrovir

ACTION: Inhibits human immunodeficiency virus (HIV-1) protease and prevents maturation of the infectious virus

USES: HIV-1 in combination with at least 2 other antiretrovirals

CONTRAINDICATIONS: Hypersensitivity

Black Box Warning: Coadministration with other drugs

Precautions: Pregnancy (B), breastfeeding, hepatic disease, pancreatitis, diabetes, hemophilia, AV block, hypercholesterolemia, immune reconstitution syndrome, neonates, cardiomyopathy, immune reconstitution syndrome, infants 1-6 mo (overdose)

DOSAGE AND ROUTES

- **Adult and adolescent >16 yr: PO** 600 mg bid; if nausea occurs, begin at ½ dose and gradually increase, max 1200 mg/day
- **Adolescent ≤16 yr and child, infant: PO** 400 mg/m^2 bid up to 1200 mg/day, may start lower and escalate

Available forms: Caps 100 mg; oral sol 80 mg/ml; tab 100 mg

Administer:

Tablet/capsule

- Take with food, swallow whole; do not crush, break, chew
- When switching from cap to tab, more GI symptoms may occur, will lessen over time
- Use dosage titration to minimize side effects
- Store caps in refrigerator
- **Oral sol:** shake well, use calibrated measuring device
- Mix liquid formulation with chocolate milk or liquid nutritional supplement to improve taste
- **Overdose:** infants, children, alcohol 43.2% alcohol, 26.57% propylene glycol oral sol, calculate total amount of alcohol, propylene glycol from all products given

SIDE EFFECTS

CNS: *Paresthesia, headache,* seizures, fever, dizziness, insomnia, asthenia, intracranial bleeding

CV: QT, PR interval prolongation

GI: *Diarrhea,* buccal mucosa ulceration, *abdominal pain, nausea, taste perversion,* dry mouth, *vomiting, anorexia*

INTEG: Rash

MISC: Asthenia, angioedema, anaphylaxis, Stevens-Johnson syndrome, increase lipids, lipodystrophy, toxic epidermal necrolysis

MS: Pain, rhabdomyolysis, myalgia

PHARMACOKINETICS

Well absorbed, 98% protein binding, hepatic metabolism, peak 2-4 hr, terminal half-life 3-5 hr

INTERACTIONS

Black Box Warning: Increase: toxicity—amiodarone, astemizole, azole antifungals, benzodiazepines, buPROPion, CISapride, cloZAPine, desipramine, dihydroergotamine, encainide, ergotamine, flecainide, HMG-CoA reductase inhibitors, interleukins, meperidine, midazolam, pimozide, piroxicam, propafenone, propoxyphene, quiNIDine, ranolazine, saquinavir, terfenadine, triazolam, zolpidem; CYP2D6 inhibitors

Black Box Warning: Increase: QT prolongation—class 1A/III antidysrhythmics, some phenothiazines, β-agonists, local anesthetics, tricyclics, haloperidol, chloroquine, droperidol, pentamidine, CYP3A4 inhibitors (amiodarone, clarithromycin, dasatinib, erythromycin, telithromycin, troleandomycin), arsenic trioxide, levomethadyl, CYP3A4 substrates (methadone, pimozide, QUEtiapine, quiNIDine, risperiDONE, ziprasidone)

Increase: ritonavir levels—fluconazole

Increase: level of both products—clarithromycin, ddI

Increase: levels of—bosentan

Decrease: ritonavir levels—rifamycins, nevirapine, barbiturates, phenytoin

Decrease: levels of anticoagulants, atovaquone, divalproex, ethinyl estradiol, lamotrigine, phenytoin, sulfamethoxazole, theophylline, voriconazole, zidovudine

Drug/Herb

Decrease: ritonavir levels—St. John's wort; avoid concurrent use

• Avoid use with red yeast rice

Drug/Lab Test

Increase: AST, ALT, K, CK, cholesterol, GGT, triglycerides, uric acid

Decrease: Hct, Hgb, RBC, neutrophils, WBC

NURSING CONSIDERATIONS

Assess:

• **HIV:** viral load, CD4 at baseline, throughout therapy; blood glucose, plasma HIV RNA, serum cholesterol/lipid profile; resistance testing before starting therapy and after treatment failure

• **Immune reconstitution syndrome:** may occur with combination therapy; may develop inflammatory response with opportunistic infection (MAC, Graves' disease, Guillain-Barré Syndrome, TB, PCP); may occur during initial treatment or months after

• Signs of infection, anemia

• Hepatic studies: ALT, AST, in those with hepatic disease, monitor q3mo

• Bowel pattern before, during treatment; if severe abdominal pain with bleeding occurs, discontinue product; monitor hydration

• Skin eruptions; rash

⚠ **Rhabdomyolysis:** muscle pain, increased CPK, weakness, swelling of affected muscles; if these occur and if confirmed by CPK, product should be discontinued

⚠ **QT prolongation:** ECG for QT prolongation, ejection fraction; assess for chest pain, palpitations, dyspnea

⚠ **Serious skin disorders:** Stevens-Johnson syndrome, angioedema, anaphylaxis, toxic epidermal necrolysis

Evaluate:

• Therapeutic response: improvement in HIV symptoms; improving viral load, CD4+ T cells

Teach patient/family:

• To take as prescribed; if dose is missed, to take as soon as remembered up to 1 hr before next dose; not to double dose

• That product not a cure for HIV; that opportunistic infections may continue to be acquired

• That redistribution of body fat or accumulation of body fat may occur

• That others may continue to contract HIV from patient

• To avoid OTC, prescription medications, herbs, supplements unless approved by prescriber; not to use St. John's wort because it decreases product's effect

• That regular follow-up exams and blood work will be required

riTUXimab (Rx)

(rih-tuks'ih-mab)

Rituxan

Func. class.: Antineoplastic—miscellaneous; DMARDs

Chem. class.: Murine/human monoclonal antibody

ACTION: Directed against the CD20 antigen that is found on malignant B lymphocytes; CD20 regulates a portion of cell-cycle initiation/differentiation

USES: Non-Hodgkin's lymphoma (CD20 positive, B-cell), bulky disease (tumors >10 cm), rheumatoid arthritis, Wegener's granulomatosis, microscopic polyangitis

Unlabeled uses: Acquired blood factor deficiency, acute lymphocytic leukemia (ALL), Burkitt's lymphoma, chronic lymphocytic leukemia (CLL), hemolytic anemia, human herpesvirus 8, mantle cell lymphoma (MCL), multicentric Castleman's disease, peripheral blood stem cell (PBSC) mobilization, refractory pemphigus vulgaris, relapsing/remitting MS, ITP with dexamethasone, steroid refractory chronic graft-versus-host disease

CONTRAINDICATIONS:

Hypersensitivity, murine proteins

Precautions: Pregnancy (C), breastfeeding, children, geriatric patients, pulmonary/cardiac/renal conditions

Black Box Warning: Exfoliative dermatitis, inf-related reactions, progressive multifocal leukoencephalopathy

DOSAGE AND ROUTES

Relapsed or refractory low-grade or follicular, CD20-positive, B-cell non-Hodgkin's lymphoma (NHL)

- **Adult: IV** 375 mg/m² every wk × 4 doses, may re-treat with 4 more doses of 375 mg/m²/wk

First-line treatment of follicular, CD20-positive, B-cell NHL in combination with chemotherapy

- **Adult: IV** 375 mg/m² on day 1 of each cycle for up to 8 cycles; may be given with cyclophosphamide IV 750 mg/m² on day 1, vinCRIStine IV 1.4 mg/m² (max 2 mg) on day 1, and predniSONE 40 mg/m²/day PO on days 1-5

Single-agent maintenance therapy in patients with follicular, CD20-positive, B-cell NHL (complete or partial response following first-line treatment with riTUXimab in combination with chemotherapy)

- **Adult: IV** 375 mg/m² IV every 8 wk × 12 doses as maintenance therapy starting 8 wk after the completion of induction chemotherapy with 8 doses of riTUXimab with 6-8 cycles of cyclophosphamide, vinCRIStine, and predniSONE 4-6 cycles of cyclophosphamide, DOXOrubicin, vinCRIStine, and predniSONE

Component of the Zevalin (ibritumomab tiuxetan) regimen

- **Adult: IV** As a required component of the ibritumomab regimen; riTUXimab 250 mg/m² given within 4 hr before administration of Yttrium-90 ibritumomab that may occur on day 7, 8, or 9

First-line treatment of diffuse large B-cell, CD20-positive NHL in combination with CHOP or other anthracycline-based chemotherapy regimen

- **Adult 18-59 yr: IV** 375 mg/m² on day 1 of each cycle for up to 8 infs

Single-agent maintenance therapy in patients with low-grade, CD20-positive, B-cell NHL with nonprogressing disease (stable disease or better) following first-line treatment with cyclophosphamide, vinCRIStine, and predniSONE (CVP)

- **Adult: IV** 375 mg/m² weekly × 4 wk repeated every 6 mo × 2 yr (total of 16 doses) as maintenance therapy starting 4 wk after the completion of first-line chemotherapy with 6-8 cycles of cyclophosphamide, vinCRIStine, and predniSONE (CVP)

Available forms: Inj 10 mg/ml (100 mg/10 ml, 500 mg/50 ml)

Administer:

Rheumatoid arthritis:

- Give methylPREDNISolone 100 mg or similar product 30 min before inf to decrease reactions

Intermittent IV INF route

- Hold antihypertensives 12 hr before administration
- After diluting to final conc of 1-4 mg/ml; use 0.9% NaCl, D_5W, gently invert bag to mix; do not mix with other products; give 50 mg/hr initially; if no reaction, increase rate by 50 mg/hr to max 400 mg/hr
- Store vials at 36° F-40° F; protect vials from direct sunlight; inf sol is stable at 36° F-46° F × 24 hr and at room temperature for another 12 hr

Y-site compatibilities: Amcyclovir, amifostine, amikacin, aminophylline, ampicillin, ampicillin/sulbactam, aztreonam, bleomycin, bumetanide, buprenorphine, busulfan, butorphanol, calcium gluconate, CARBOplatin, carmustine, ceFAZolin, cefoperazone, cefotaxime, cefoTEtan, cefOXitin, cefTAZidime, ceftizoxime, cefTRIAXone, cefuroxime, chlorproMAZINE, cimetidine, CISplatin, clindamycin, cyclophosphamide, cytarabine, DACTINomycin, DAUNOrubicin hydrochloride, dexamethasone, dexrazoxane, digoxin, diphenhydrAMINE, DOBUTamine, DOCEtaxel, DOPamine, DOXOrubicin liposome, doxycycline, droperidol, enalaprilat, etoposide phosphate, famotidine, fentaNYL, filgrastim, floxuridine, fluconazole, fludarabine, fluorouracil, ganciclovir, gemcitabine, gentamicin, granisetron, haloperidol, heparin, hydrocortisone, HYDROmorphone, IDArubicin, ifosfamide, imipenem/cilastatin, irinotecan, leucovorin, levorphanol, LORazepam, magnesium sulfate, mannitol, meperidine, mesna, methotrexate, methylPREDNISolone, metoclopramide, metroNIDAZOLE, mitoMYcin, mitoXANtrone, morphine, nalbuphine, netilmicin, PACLitaxel, pentamidine, piperacillin/tazobactam, plicamycin, potassium chloride, prochlorperazine, promethazine, ranitidine, sargramostim, streptozocin, teniposide, theophylline, thiotepa, ticarcillin/clavulanate, tobramycin, trimethoprim/sulfamethoxazole, trimethobenzamide, vinBLAStine, vincCRIStine, vinorelbine, zidovudine

SIDE EFFECTS

CNS: **Life-threatening brain infection (progressive multifocal leukoencephalopathy)**

CV: **Cardiac dysrhythmias, heart failure, hypertension, MI, supraventricular tachycardia,** angina

GI: *Nausea, vomiting, anorexia,* **GI obstruction/perforation**

GU: **Renal failure**

HEMA: **Leukopenia, neutropenia, thrombocytopenia,** anemia

INTEG: *Irritation at site, rash,* **fatal mucocutaneous infections (rare)**

MISC: *Fever,* chills, asthenia, *headache,* **angioedema,** hypotension, myalgia, **bronchospasm, ARDs**

SYST: **Toxic epidermal necrolysis, tumor lysis syndrome, Stevens-Johnson syndrome, exfoliative dermatitis**

PHARMACOKINETICS

Half-life 60-174 hr, binds to CD20 sites or lymphoma cells

INTERACTIONS

Increase: hypotension—antihypertensives, separate by 12 hr

Increase: nephrotoxicity—CISplatin, avoid concurrent use; if used, monitor renal status

Increase: bleeding—anticoagulants

- Avoid with vaccines, toxoids

NURSING CONSIDERATIONS

Assess:

Black Box Warning: Fatal inf reaction: hypoxia, pulmonary infiltrates, ARDS, MI, ventricular fibrillation, cardiogenic shock; most fatal reactions occur with 1st inf; potentially fatal

Black Box Warning: Severe mucocutaneous reactions: Stevens-Johnson syndrome, lichenoid dermatitis, toxic epidermal lysis; occur 1-13 wk after product given, discontinue treatment immediately

Black Box Warning: Tumor lysis syndrome: acute renal failure requiring hemodialysis, hyperkalemia, hypocalcemia, hyperuricemia, hyperphosphatemia; allopurinol and adequate hydration may be needed

Black Box Warning: Multifocal leukoencephalopathy: confusion, dizziness, lethargy, hemiparesis; monitor periodically

- CBC, differential, platelet count weekly; withhold product if WBC is $<3500/mm^3$ or platelet count $<100,000/mm^3$; notify

R

prescriber of results; product should be discontinued

- ECG, serum creatinine/BUN, electrolytes, uric acid
- GI symptoms: frequency of stools, abdominal pain, perforation/obstruction may occur
- **Infection:** fever, increased temperature, flulike symptoms in those with Wegener's granulomatosis and microscopic polyangiitis in those using DMARDs

Evaluate:

- Therapeutic response: prevention of increasing cancer progression

Teach patient/family:

- To avoid use with vaccines, toxoids
- To use contraception during, for up to 12 mo after therapy
- ⚠ To report to prescriber possible infection (cough, fever, chills, sore throat), renal issues (painful urination, back/side pain), bleeding (gums, stools, urine, bruising, emesis, fatigue)
- To avoid OTC products
- To avoid crowds, those with known infections
- To maintain fluid intake

rivaroxaban

Xarelto

Func. class.: Anticoagulant

Chem. class.: Factor Xa inhibitor

ACTION: A novel, oral anticoagulant that selectively and potently inhibits coagulation factor Xa

USES: For deep venous thrombosis (DVT) prophylaxis/treatment, pulmonary embolism (PE), in patients undergoing knee or hip replacement surgery; for stroke prophylaxis and systemic embolism prophylaxis in patients with nonvalvular atrial fibrillation

CONTRAINDICATIONS Severe hypersensitivity

Black Box Warning: Active bleeding

Precautions Pregnancy (C), breastfeeding, neonates, infants, children, adolescents, geriatric patients, moderate or severe hepatic disease (Child-Pugh Class B or C), hepatic disease associated with coagulopathy, creatinine clearance <30 ml/min for use as DVT prophylaxis and <15 ml/min for stroke and systemic embolism prophylaxis in nonvalvular atrial fibrillation, dental procedures, aneurysm, diabetes retinopathy, diverticulitis, endocarditis, GI bleeding, hypertension, obstetric delivery, peptic ulcer disease, stroke, surgery

Black Box Warning: Abrupt discontinuation, epidermal/spinal anesthesia

Patients, especially those with dental disease, should be instructed in proper oral hygiene, including caution in use of regular toothbrushes, dental floss, toothpicks

DOSAGE AND ROUTES

DVT prophylaxis, which may lead to pulmonary embolism (PE) (knee or hip replacement)

- **Adult: PO** 10 mg/day × 12 days after knee replacement surgery or × 35 days after hip replacement; administer the initial dose ≥6-10 hr after surgery once hemostasis has been established

DVT/PE treatment/reduction of risk

- **Adult: PO** 15 mg bid with food × 21 days then 20 mg daily for a total of 6 mo, may continue after 6 mo to reduce risk

Stroke prophylaxis and systemic embolism prophylaxis (with nonvalvular atrial fibrillation)

- Unless pathological bleeding occurs, do not discontinue rivaroxaban in the absence of alternative anticoagulation
- **Adult: PO** 20 mg/day with evening meal (CrCl >50 ml/min)

Converting from warfarin to rivaroxaban

- Discontinue warfarin and start rivaroxaban when INR is <3

Converting from another anticoagulant other than warfarin to rivaroxaban

• Start rivaroxaban 0-2 hr before the next scheduled evening administration of anticoagulant (omit that dose of anticoagulant); for continuous inf of unfractionated heparin, stop the inf and initiate rivaroxaban simultaneously

Converting from rivaroxaban to another anticoagulant with rapid onset (not warfarin)

• Discontinue rivaroxaban and give the first dose of the other anticoagulant (oral or parenteral) at the time that the next dose of rivaroxaban would have been administered

Hepatic Dose

• **Adult: PO** Child-Pugh class B or C: avoid use

Renal Dose

• **Adult: PO** (nonvalvular atrial fibrillation) CCr 15-50 ml/min 15 mg daily; CCr <15 ml/min avoid use; (treatment/prophylaxis of DVT/pulmonary embolism) CCr <30 ml/min avoid use

Available forms: Tabs 10, 15, 20 mg

Administer:

⚠ **For DVT prophylaxis:** give daily without regard to food; give initial dose ≥6-10 hr after surgery when hemostasis has been established

⚠ **For stroke/systemic embolism prophylaxis:** give daily with evening meal

• If dose is not given at correct time, give as soon as possible on the same day

• 15-, 20-mg tabs should be taken with food; for those unable to swallow whole, tabs may be crushed, mixed with applesauce; immediately following administration, instruct to eat; crushed tabs are stable in applesauce for up to 4 hr

• 10-mg tab can be taken without regard to food

Nasogastric (NG) tube or gastric feeding tube:

• Confirm gastric placement of tube

• Crush 15- or 20-mg tab, suspend in 50 ml of water, and administer via NG or gastric feeding tube

• To minimize reduced absorption, avoid administration distal to the stomach

• Enteral feeding should immediately follow administration of a crushed dose

• Crushed tabs are stable in water for up to 4 hr

Missed doses

• Patients receiving 15 mg twice daily should take their missed dose immediately to ensure intake of 30 mg per day; 2 15-mg tabs may be taken at once followed by the regular 15 mg twice daily dose the next day

• For patients receiving once daily dosing, take the missed dose as soon as it is remembered

• Store at room temperature

SIDE EFFECTS

GI: Increased hepatic enzymes, hyperbilirubinemia, jaundice, nausea, cholestasis, cytolytic hepatitis

HEMA: Bleeding, intracranial bleeding, epidural hematoma, GI bleeding, retinal hemorrhage, adrenal bleeding, retroperitoneal hemorrhage, cerebral hemorrhage, subdural hematoma, epidural hematoma, hemiparesis, thrombocytopenia

INTEG: Pruritus, blister, hypersensitivity, anaphylactic reaction, anaphylactic shock

SYST: Stevens-Johnson syndrome

PHARMACOKINETICS

Bioavailability 80%-100%, protein binding (92%-95%) albumin, excreted in urine 66% (36% unchanged, 30% metabolites), 28% in feces (7% unchanged, 21% metabolites), unchanged drug excreted in urine (via active tubular secretion, glomerular filtration); terminal elimination half-life 5-9 hr, peak 2-4 hr; increased effect in hepatic/renal disease, Japanese patients, increased terminal half-life in geriatric patients

R

INTERACTIONS

Increase: rivaroxaban effect, possible bleeding—ketoconazole itraconazole, ritonavir, lopinavir/ritonavir; conivaptan, clarithromycin, erythromycin, salicylates, NSAIDs, other anticoagulants, thrombolytics, platelet inhibitors, niCARdipine, fluconazole

Decrease: rivaroxaban effect—carBAMazepine, phenytoin, rifampin

Increase: rivaroxaban effect in renal impairment—telithromycin, darunavir, mifepristone, nelfinavir, pantoprazole, posaconazole, saquinavir, tamoxifen, lapatinib, azithromycin, diltiazem, verapamil, quiNIDine, ranolazine, dronedarone, amiodarone, felodipine

Drug/Herb:

Decrease: rivaroxaban effect—St. John's wort

Drug/Food:

Increase: rivaroxaban effect in renal disease—grapefruit juice

NURSING CONSIDERATIONS

Assess:

Black Box Warning: **Bleeding:** monitor for bleeding, including bleeding during dental procedures (easy bruising, blood in urine, stools, emesis, sputum, epistaxis); there is no specific antidote

Black Box Warning: **Abrupt discontinuation:** avoid abrupt discontinuation unless an alternative anticoagulant in those with atrial fibrillation; discontinuing puts patients at an increased risk of thrombotic events; if product must be discontinued for reasons other than pathological bleeding, consider administering another anticoagulant

- **Pregnancy/breastfeeding:** pregnancy (C); identify if pregnancy is suspected or planned; pregnancy-related hemorrhage may occur and anticoagulation cannot be monitored with standard laboratory testing; breastfeeding should be discontinued before beginning use of this product

Black Box Warning: **Epidural/spinal anesthesia:** epidural or spinal hematomas that result in long-term or permanent paralysis may occur in patients who have received anticoagulants and are receiving neuraxial anesthesia or undergoing spinal puncture; epidural catheter should not be removed <18 hr after the last dose of rivaroxaban; do not administer the next rivaroxaban dose <6 hr after the catheter removal; delay rivaroxaban administration for 24 hr if traumatic puncture occurs; monitor for neuro changes

- **Hepatic/renal disease:** increase in effect of product in hepatic disease (Child-Pugh Class B or C), hepatic disease with coagulopathy; renal failure/severe renal impairment (creatinine clearance <30 ml/min in DVT prophylaxis and <15 ml/min for stroke or systemic embolism prophylaxis in nonvalvular atrial fibrillation); product should be discontinued in acute renal failure; reduce dose in those with atrial fibrillation and CrCl 15-50 ml/min; monitor renal function periodically (creatinine clearance, BUN)

Evaluate:

- Prevention of DVT, stroke, and systemic embolism

Teach patient/family:

- To report if pregnancy is planned or suspected; not to breastfeed

Black Box Warning: To report bleeding (bruising, blood in urine, stools, sputum, emesis, heavy menstrual flow); use soft toothbrush, electric shaver

- To inform all health care providers of use

Black Box Warning: To avoid abrupt discontinuation without another blood thinner

rivastigmine (Rx)

(riv-as-tig′mine)

Exelon, Exelon Patch

Func. class.: Anti-Alzheimer agent

Chem. class.: Cholinesterase inhibitor

ACTION: Potent, selective inhibitor of brain acetylcholinesterase (AChE) and butyrylcholinesterase (BChE)

USES: Mild to severe Alzheimer's dementia, mild to moderate Parkinson's disease dementia (PDD)

Unlabeled uses: Vascular dementia, dementia with Lewy bodies, Pick's disease

CONTRAINDICATIONS: Hypersensitivity to this product, other carbamates

Precautions: Pregnancy (B), breastfeeding, children, respiratory/cardiac/renal/hepatic disease, seizure disorder, peptic ulcer, urinary obstruction, asthma, increased intracranial pressure, surgery, GI bleeding, jaundice

DOSAGE AND ROUTES

• **Adult: PO** 1.5 mg bid with food; after ≥4 wk, may increase to 3 mg bid; may increase to 4.5 mg bid and thereafter 6 mg bid, max 12 mg/day; **TRANSDERMAL** apply 4.6 mg/24 hr/day, after ≥4 wk may increase to 9.5 mg/24 hr/day; max 13.3 mg/24 hr; for those using 6-12 mg/day PO and switching to transdermal use one 9.5 mg/24 hr daily, for those using <6 mg/day PO and switching to transdermal use one 4.6 mg/24 hr daily

Available forms: Caps 1.5, 3, 4.5, 6 mg; sol 2 mg/ml; transdermal patch 4.6, 9.5, 13.3 mg/24 hr

Administer:

• Oral sol, caps are interchangeable

• With meals; take with morning and evening meal even though absorption may be decreased

• Discontinue treatment for several doses; restart at same or next lower dosage level if adverse reactions cause intolerance

• If treatment is interrupted for more than several days, treatment should be initiated with lowest daily dose and titrated as indicated previously

• **Oral sol:** measure dose with supplied syringe; may be given undiluted or diluted in a small amount of water, fruit juice, or soda; stir diluted sol well; entire amount should be swallowed; stable for 4 hr when mixed

Transdermal route

• Once a day to hairless, clean, dry skin, not in an area that clothing will rub; rotate sites daily; do not apply to same site more than once q14days; remove liner; apply firmly; may be used during bathing, swimming; avoid saunas, excess sunlight; or external heat such as saunas; each 5-cm^2 patch contains 9 mg base, rate of 4.6 mg/24 hr, each 10-cm^2 patch 18 mg base, rate of 9.5 mg/24 hr

SIDE EFFECTS

CV: **QT prolongation, AV block, cardiac arrest**, angina, **MI**, palpitations

CNS: *Tremors, confusion, insomnia,* psychosis, hallucination, depression, dizziness, headache, anxiety, somnolence, fatigue, syncope, EPS, exacerbation of Parkinson's disease

GI: *Nausea, vomiting, anorexia, abdominal distress, flatulence,* diarrhea, constipation, dyspepsia, colitis, eructation, fecal incontinence, **GI bleeding/obstruction**, GERD, gastritis, **pancreatitis**

MISC: Urinary tract infection, asthenia, increased sweating, hypertension, flulike symptoms, weight change

PHARMACOKINETICS

Rapidly and completely absorbed; peak 1 hr, metabolized to decarbamylated metabolite; half-life 1.5 hr; excreted via kidneys (metabolites); clearance lowered in geriatric patients, hepatic disease and increased with nicotine use; 40% protein binding

INTERACTIONS

Increase: synergistic effect—cholinergic agonists, other cholinesterase inhibitors
Increase: metabolism—nicotine
Increase: GI effects—NSAIDs
Decrease: rivastigmine effect—anticholinergics, sedating H_1 blockers, tricyclics, phenothiazines

NURSING CONSIDERATIONS

Assess:

- Hepatic studies: AST, ALT, alk phos, LDH, bilirubin, CBC
- **Severe GI effects:** nausea, vomiting, anorexia, weight loss, diarrhea
- B/P, heart rate, respiration during initial treatment; hypo/hypertension should be reported
- **Cognitive/mental status:** affect, mood, behavioral changes, depression, insomnia; complete suicide assessment
- Assistance with ambulation during beginning therapy; dizziness may occur

Evaluate:

- Therapeutic response: improved mood/cognition

Teach patient/family:

- About the procedure for giving **oral sol**; use instruction sheet provided; how to apply **transdermal** product, to fold in half and throw away, not to get in eyes, to wash hands after application; not to use heating pad, sauna, tanning bed
- To notify prescriber of severe GI effects
- That product may cause dizziness, anorexia, weight loss
- That effect may take weeks or months; not to discontinue abruptly
- To notify prescriber if pregnancy is planned or suspected

rizatriptan (Rx)

(rye-zah-trip′tan)

Maxalt, Maxalt-MLT

Func. class.: Migraine agent
Chem. class.: 5-HT_{1D} receptor agonist, abortive agent-triptan

ACTION: Binds selectively to the vascular 5-$HT_{1B/1D}$ receptor subtype; exerts antimigraine effect; causes vasoconstriction of the cranial arteries

USES: Acute treatment of migraine

CONTRAINDICATIONS: Angina pectoris, history of MI, documented silent ischemia, Prinzmetal's angina, ischemic heart disease, concurrent ergotamine-containing preparations, uncontrolled hypertension, hypersensitivity, basilar or hemiplegic migraine
Precautions: Pregnancy (C), breastfeeding, children, geriatric patients, postmenopausal women, men >40 yr, risk factors for CAD, hypercholesterolemia, obesity, diabetes, impaired renal/hepatic function

DOSAGE AND ROUTES

- **Adult: PO** 5-10 mg single dose, redosing separated by ≥2 hr, max 30 mg/24 hr; use 5 mg for patient receiving propranolol, max 15 mg/24 hr

Available forms: Tabs (Maxalt) 5, 10 mg; orally disintegrating tabs (Maxalt-MLT) 5, 10 mg
Administer:

- **Orally disintegrating tab:** do not open blister until use; peel blister open with dry hands; place tab on patient's tongue, where it will dissolve, and have patient swallow with saliva (contains phenylalanine)

SIDE EFFECTS

CNS: *Dizziness, drowsiness, headache, fatigue,* warm/cold sensations, flushing

CV: **MI, ventricular fibrillation, ventricular tachycardia, coronary artery vasospasm,** palpitations, hypertension, peripheral vascular ischemia, ECG changes
ENDO: Hot flashes, mild increase in growth hormone
GI: *Nausea,* dry mouth, diarrhea, abdominal pain, ischemic colitis
RESP: Chest tightness, pressure, dyspnea

PHARMACOKINETICS

Onset of pain relief 10 min-2 hr; peak 1-1½ hr; duration 14-16 hr; 14% plasma protein binding; metabolized in liver (metabolite); excreted in urine (82%), feces (12%); half-life 2-3 hr

INTERACTIONS

⚠ **Weakness, hyperreflexia, incoordination: SSRIs**
Increase: levels of sibutramine
Increase: rizatriptan action—cimetidine, oral contraceptives, MAOIs, nonselective MAOI (type A and B), isocarboxazide, pargyline, phenelzine, propranolol, tranylcypromine
Increase: vasospastic effects—ergot, ergot derivatives, other 5-HT receptor agonists
Drug/Herb
⚠ **Serotonin syndrome: St. John's wort**

NURSING CONSIDERATIONS

Assess:

- **Migraine symptoms:** visual disturbances, aura, intensity, nausea, vomiting, photophobia
- Stress level, activity, recreation, coping mechanisms
- Neurologic status: LOC, blurring vision, nausea, vomiting, tingling in extremities preceding headache
- **Ingestion of tyramine foods** (pickled products, beer, wine, aged cheese), food additives, preservatives, colorings, artificial sweeteners, chocolate, caffeine, which may precipitate these types of headaches
- Renal status: urine output
- Quiet, calm environment with decreased stimulation: noise, bright light, excessive talking

Evaluate:

- Therapeutic response: decrease in frequency, severity of headache

Teach patient/family:

- **About use of orally disintegrating tab:** instruct patient not to open blister until use, to peel blister open with dry hands, to place tab on tongue, where it will dissolve, and to swallow with saliva (contains phenylalanine)
- To report any side effects to prescriber
- To use alternative contraception while taking product if oral contraceptives are being used
- That product does not prevent or reduce number of migraines; if 1st dose does not relieve pain, do not use more; notify prescriber

roflumilast

Daliresp
Func. class.: Respiratory anti-inflammatory agent
Chem. class.: Phosphodiesterase-4 (PDE4) inhibitor

ACTION: Roflumilast (and the active metabolite roflumilast N-oxide) selectively inhibit phosphodiesterase-4 (PDE4); not a bronchodilator; inhibition of the PDE4 enzyme blocks the hydrolyses and inactivation of cyclic adenosine monophosphate (cAMP), resulting in intracellular cAMP accumulation; decreases inflammatory activity, PDE4 inhibition may affect migration and actions of pro-inflammatory cells (neutrophils, other leukocytes, T-lymphocytes, monocytes, macrophages, fibroblasts)

USES: For the prevention of COPD exacerbations in patients with severe COPD associated with chronic bronchitis and a history of exacerbations

R

CONTRAINDICATIONS: *Moderate to severe hepatic disease* (Child-Pugh B or C)

Precautions: Pregnancy (C), breastfeeding, neonates, infants, children, adolescents, acute bronchospasm, anxiety, insomnia, depression, suicidal ideation or behavior

DOSAGE AND ROUTES

- **Adult: PO** 500 mcg/day

Available forms: Tabs 500 mcg

Administer:

PO route

- Give without regard to meals
- Store at room temperature

SIDE EFFECTS

CNS: Insomnia, anxiety, depression, headache, dizziness, tremors, suicidal ideation

EENT: Rhinitis, sinusitis

GI: *Weight loss, diarrhea, nausea*, anorexia, abdominal pain, dyspepsia, gastritis, vomiting

GU: Urinary tract infection

MS: Back pain, muscle cramps/spasm

SYST: Infections, influenza

PHARMACOKINETICS

80% absolute bioavailability; protein binding 99% (roflumilast); 97% (N-oxide metabolite); low penetration across the blood–brain barrier; extensively metabolized (liver); metabolism by CYP3A4 and CYP1A2 produces (active metabolite N-oxide); half-life parent drug 17 hr, metabolite 30; steady state 4 days (parent drug), 6 days (metabolite); 70% excreted in urine; parent drug peak 1 hr (range, 0.5-2 hr), metabolite peak 8 hr (range, 4-13 hr); contraindicated in moderate to severe hepatic impairment; use with caution in patients with mild hepatic

INTERACTIONS

Increase: roflumilast effect—CYP3A4/CYP1A2 inhibitors (enoxacin, cimetidine, delavirdine, indinavir, isoniazid, itraconazole, dalfopristin, quinupristin, tipranavir)

Increase: roflumilast effect—oral contraceptives (gestodene and ethinyl estradiol)

Decrease: roflumilast effect—CYP3A4 inducers (rifampin, barbiturates, carBAMazepine, phenytoin, erythromycin, ketoconazole, fluvoxaMINE, alcohol, etravirine, ritonavir, bexarotene, rifabutin, OXcarbazepine, nevirapine, modafinil, metyrapone, PHENobarbital, bosentan, dexamethasone)

Altered effect of: fosamprenavir

Drug/Herb

Decrease: roflumilast effect—St. John's wort

NURSING CONSIDERATIONS

Assess:

- Lung sounds and respiratory function baseline and periodically thereafter

⚠ Behavioral changes including mood, depression, suicidal thoughts/behaviors

- Liver function tests baseline and periodically thereafter; if increases in liver function studies occur, product should be discontinued
- Weight; weight loss is common

Evaluate:

- Decreasing exacerbations in COPD

Teach patient/family:

- To take product as directed; not to skip or double doses; to take missed doses as soon as remembered unless almost time for next dose
- Not to use OTC or other products without prescriber approval; not to discontinue other respiratory products unless approved by prescriber
- Not to be used for acute bronchospasm but may be continued during acute asthma attacks

⚠ **Suicidal thoughts/behaviors:** To notify prescriber of worsening depression or suicidal thoughts/behaviors

romiPLOStim (Rx)

(roe-mi-ploe′stim)

Nplate

Func. class.: Hematopoietin

Chem. class.: Thrombopoietin receptor agonist

ACTION: Thrombopoietin-like fusion protein produced by DNA recombinant technology

USES: Chronic idiopathic thrombocytopenic purpura in patients who have had an insufficient response to corticosteroids, immunoglobulins, or splenectomy

CONTRAINDICATIONS: Hypersensitivity to this product or mannitol

Precautions: Pregnancy (C), breastfeeding, children, malignancies, bleeding, bone marrow suppression

DOSAGE AND ROUTES

Thrombocytopenia in chronic idiopathic thrombocytopenic purpura (ITP) with insufficient response to corticosteroids, immunoglobulins, splenectomy

• **Adult: SUBCUT** The initial dosage is 1 mcg/kg SC weekly (based on actual body weight); increase the weekly dosage by 1 mcg/kg until the patient achieves a platelet count ≥50,000/mm^3; do not exceed a maximum weekly dosage of 10 mcg/kg; use the lowest dosage needed to achieve and maintain a platelet count ≥50,000/mm^3; monitor CBC, including platelet counts, weekly until a stable platelet count is achieved with platelets ≥50,000/mm^3 ≥4 wk without dosage adjustment; then monitor the CBC, including platelet counts, monthly; once a stable dosage is achieved, if the platelet count falls to <50,000/mm^3, increase the dosage by 1 mcg/kg/wk; if the platelet count increases to >200,000/mm^3 for 2 consecutive weeks, reduce the dosage by 1 mcg/kg; if the platelet count is >400,000/mm^3, temporarily stop romiPLOStim and continue to monitor the platelet count every wk; once the platelets are <200,000/mm^3, restart, but reduce the previous dosage by 1 mcg/kg/wk; romiPLOStim may be administered concomitantly with other medical ITP therapies; if platelet counts exceed 50,000/mm^3, other medical ITP therapies may be reduced or discontinued; discontinue romiPLOStim if the platelet count does not increase to avoid important bleeding after 4 wk of therapy at max dosage of 10 mcg/kg

Available forms: Inj vials 250, 500 mcg

Administer:

• Refrigerated storage of vials; do not freeze; protect from light; diluted sol is stable refrigerated or at room temperature for 24 hr

SUBCUT route

• Use syringe with 0.01-ml graduations
• Discard any unused portion in vial; do not pool unused portions from vials
• Dilute 250 mcg/0.72 preservative-free sterile water for inj; 500 mcg/1.2 preservative-free sterile water for inj; final conc 500 mcg/ml
• Gently swirl until dissolved; do not shake
• Do not use if discolored or if particulate matter is present
• Inject into outer aspect of upper arm or abdomen except for 2 inches around navel or front aspect of middle thigh; do not use areas that are bruised, scratched, or scarred
• Rotate inj sites

SIDE EFFECTS

CNS: *Dizziness, insomnia, headache,* fatigue

GI: Abdominal pain, dyspepsia, diarrhea

HEMA: **Thromboembolism, thrombosis,** bleeding, myelofibrosis, erythromelalgia

MS: Myalgia

SYST: **Secondary malignancy,** antibody formation

R

PHARMACOKINETICS

Peak 7-50 hr, half-life 1-34 days

INTERACTIONS

- Possible bleeding risk: anticoagulants, NSAIDs, platelet inhibitors, thrombolytics, salicylates

NURSING CONSIDERATIONS

Assess:

- Blood studies: CBC during treatment weekly and for 2 wk after discontinuing
- **Bone marrow suppression:** If cytopenias occur, product should be discontinued; may use a bone marrow biopsy and staining for fibrosis
- **Thromboembolic disease:** do not use to normalize patients; use only in those with thrombocytopenia in idiopathic thrombocytopenic purpura, maintain platelets ≥ 50,000/mm^3

Evaluate:

- Therapeutic response: increase in platelet counts, absence of bleeding

Teach patient/family:

- To report bleeding
- About the reason for product and expected results; to avoid hazardous activities that may cause bleeding
- To report a missed dose to prescriber due to increased risk of bleeding
- That lab tests will be done weekly and dose may be changed; if dose is not changed, lab will be checked monthly; after drug is discontinued, labs will be checked weekly × 2 wk
- To advise prescriber if spleen has been removed, bleeding or clotting problems
- To notify prescriber if pregnancy is planned or suspected, pregnancy (C); if pregnancy occurs, call registry 1-877-675-2831

rOPINIRole (Rx)

(roh-pin′ih-role)

Requip, Requip XL

Func. class.: Antiparkinson agent

Chem. class.: DOPamine-receptor agonist, nonergot

Do not confuse:

rOPINIRole/risperiDONE

ACTION:

Selective agonist for D_2 receptors (presynaptic/postsynaptic sites); binding at D_3 receptor contributes to antiparkinson effects

USES:

Parkinson's disease, restless leg syndrome (RLS)

CONTRAINDICATIONS:

Hypersensitivity

Precautions: Pregnancy (C), dysrhythmias, affective disorder, psychosis, cardiac/renal/hepatic disease

DOSAGE AND ROUTES

Parkinson's disease

- **Adult: PO (regular release)** Initially, 0.25 mg PO tid × 1 wk; gradually titrate at weekly intervals: **Week 2**, 0.5 mg tid; **Week 3**, 0.75 mg tid; **Week 4**, 1 mg tid; **After week 4**, may increase by 1.5 mg/day each week, max 9 mg/day total dosage, and then by 3 mg/day each week, max 24 mg/day
- **PO (ext rel)** initially, 2 mg/day × 1-2 wk, may increase by mg/day at intervals ≥1 wk based upon response; max 24 mg/day; if significant interruption of therapy occurs, retitration may be necessary

Conversion from immediate-release to extended-release tablets

- **Adult: PO** currently taking 0.75-2.25 mg/day imm rel: give 2 mg/day ext rel
- **PO** currently taking 3-4.5 mg/day imm rel: give 4 mg/day ext rel
- **PO** currently taking 6 mg/day imm rel: give 6 mg/day ext rel
- **PO** currently taking 7.5-9 mg/day imm rel: give 8 mg/day ext rel

• **PO** currently taking 12 mg/day imm rel: give 12 mg/day ext rel
• **PO** currently taking 15-18 mg/day imm rel: give 16 mg/day ext rel
• **PO** currently taking 21 mg/day imm rel: give 20 mg/day ext rel
• **PO** currently taking 24 mg/day imm rel: give 24 mg/day ext rel

Restless legs syndrome

• **Adult: PO (reg rel)** Initially, 0.25 mg every day 1-3 hr before bedtime; days 3-7, may increase to 0.5 mg every day; at the beginning of wk 2 (day 8); the dosage may be increased to 1 mg every day × 1 wk; weeks 3-6, dosage may be titrated up by 0.5 mg each wk (from 1.5-3 mg over the 5-wk period) as needed to achieve desired effect; wk 7, may increase dosage to 4 mg/day; dosage is titrated based on clinical response; give all doses 1-3 hr before bedtime

Available forms: Tabs 0.25, 0.5, 1, 2, 3, 4, 5 mg; ext rel tab 2, 4, 8, 12 mg

Administer:

• Product until NPO before surgery
• Adjust dosage to patient response; taper when discontinuing
• With meals to reduce nausea
• **Extended release:** do not chew, crush, or divide
• Testing for diabetes mellitus, acromegaly if patient receiving long-term therapy

SIDE EFFECTS

CNS: *Agitation, insomnia,* psychosis, hallucination, dystonia, depression, dizziness, somnolence, **sleep attacks,** impulse control disorders

CV: *Orthostatic hypotension,* tachycardia, hypo/hypertension, syncope, palpitations

EENT: Blurred vision

GI: *Nausea, vomiting, anorexia, dry mouth,* constipation, dyspepsia, flatulence

GU: Impotence, urinary frequency

HEMA: **Hemolytic anemia, leukopenia, agranulocytosis**

INTEG: Rash, sweating

RESP: Pharyngitis, rhinitis, sinusitis, bronchitis, dyspnea

PHARMACOKINETICS

Peak 1-2 hr, half-life 6 hr, extensively metabolized by liver by P450 CYP1A2 enzyme system, protein binding 40%

INTERACTIONS

Increase: rOPINIRole effect—cimetidine, ciprofloxacin, diltiazem, enoxacin, erythromycin, fluvoxaMINE, mexiletine, norfloxacin, tacrine, digoxin, theophylline, L-dopa

Decrease: rOPINIRole effects—butyrophenones, metoclopramide, phenothiazines, thioxanthenes

NURSING CONSIDERATIONS

Assess:

• **Parkinsonism:** akinesia, tremors, staggering gait, muscle rigidity, drooling
• B/P, respirations during initial treatment; hypo/hypertension should be reported
⚠ **Sleep attacks: drowsiness, falling asleep without warning even during hazardous activities**
• Mental status: affect, mood, behavioral changes, depression; complete suicide assessment; worsening of symptoms in restless leg syndrome

Evaluate:

• Therapeutic response: improvement in movement disorder

Teach patient/family:

• To notify prescriber if pregnancy is planned or suspected, pregnancy (C)
• To take with food to prevent nausea
• To report hallucinations, confusion (usually in geriatric patients)
• That therapeutic effects may take several weeks to a few months
• To change positions slowly to prevent orthostatic hypotension
• To use product exactly as prescribed; if product is discontinued abruptly, parkinsonian crisis may occur
• That drowsiness, sleep attacks may occur; to avoid driving, other hazardous activities until response known
• To avoid alcohol, CNS depressants, cough and cold products

⚠ To notify prescriber if unusual urges occur

rosiglitazone (Rx)

(ros-ih-glit′ah-zone)

Avandia

Func. class.: Antidiabetic, oral

Chem. class.: Thiazolidinedione

ACTION: Improves insulin resistance by hepatic glucose metabolism, insulin receptor kinase activity, insulin receptor phosphorylation

USES: Type 2 diabetes mellitus, alone or in combination with sulfonylureas, metformin, insulin

CONTRAINDICATIONS: Breastfeeding, children, hypersensitivity to thiazolidinediones, diabetic ketoacidosis, jaundice

Black Box Warning: NYHA III, IV acute heart failure, heart failure

Precautions: Pregnancy (C), geriatric patients, thyroid disease, renal/hepatic disease, heart failure, class I, II NYHA

Black Box Warning: MI

DOSAGE AND ROUTES

• **Adult: PO** 4 mg/day or in 2 divided doses, may increase to 8 mg/day or in 2 divided doses after 12 wk; may be added to metformin, sulfonylurea for adult dose

Available forms: Tabs 2, 4, 8 mg

Administer:

• Conversion from other oral hypoglycemic agents if needed; change may be made without gradual dosage change; monitor blood glucose during conversion

• Store in tight container in cool environment

PO route

• Once or in 2 divided doses, without regard to food

• Tabs crushed and mixed with food or fluids for patients with difficulty swallowing

• Only through the REMS Program 1-800-Avandia

SIDE EFFECTS

CNS: Fatigue, *headache*

CV: MI, CHF, death (geriatric patients)

ENDO: Hypo/hyperglycemia

GI: Weight gain, hepatotoxicity, increase total, LDL, HDL cholesterol; decrease free fatty acids, diarrhea

MISC: Accidental injury, URI, sinusitis, anemia, back pain, diarrhea, edema, bone fractures (female), pulmonary/macular/peripheral edema

SYST: Anaphylaxis, Stevens-Johnson syndrome

PHARMACOKINETICS

Maximal reductions in FBS after 6-12 wk; protein binding 99.8%; excreted in urine, feces; elimination half-life 3-4 hr; may be excreted in breast milk

INTERACTIONS

Increase: hypoglycemia—gemfibrozil, fluvoxaMINE, ketoconazole, trimethoprim; monitor glucose

• Avoid concurrent use with insulin, nitrates

• May increase or decrease level: CYP2C5 inducer/inhibitors

Drug/Herb

Increase: antidiabetic effect—garlic, horse chestnut

Drug/Lab Test

Increase: ALT, HDL, LDL, total cholesterol, blood glucose

Decrease: Hgb/Hct

NURSING CONSIDERATIONS

Assess:

Black Box Warning: CHF/MI: dyspnea, crackles, edema, weight gain ≥5 lb, jugular venous distention; may need to change dose or discontinue product, do not use in acute coronary syndrome, NYHA Class 3/IV heart failure

• Hypoglycemic reactions (sweating, weakness, dizziness, anxiety, tremors, hunger), hyperglycemic reactions soon after meals

⚠ Systemic reactions: anaphylaxis, Stevens-Johnson syndrome

⚠ **Hepatotoxicity:** LFTs periodically AST, ALT (if ALT >2.5 × ULN, do not use product)

• Fasting blood sugar, A1c, plasma lipids/lipoproteins, B/P, body weight during treatment

• To use product provider/patient must be enrolled in the Avandia-Rosiglitazone Medicines Access Program

Evaluate:

• Therapeutic response: decrease in polyuria, polydipsia, polyphagia; clear sensorium; absence of dizziness; stable gait; blood glucose, A1c improvement

Teach patient/family:

• To monitor blood glucose; that periodic liver function tests mandatory; to report edema, weight gain

• About the symptoms of hypo/hyperglycemia, what to do about each

• That product must be continued on daily basis; about the consequences of discontinuing the product abruptly

• To avoid OTC medications, herbal preparations, nitrates, or insulin unless approved by prescriber

• That diabetes is lifelong; that product is not a cure, only controls symptoms

• That all food included in diet plan must be eaten to prevent hypoglycemia

• To carry emergency ID and glucagon emergency kit

⚠ To report symptoms of hepatic dysfunction (nausea, vomiting, abdominal pain, fatigue, anorexia, dark urine, jaundice)

• That 2 wk is needed to see reduction in blood glucose level and 2-3 mo needed to see full effect of product

• To notify prescriber if oral contraceptives are used

• Not to use if breastfeeding, may be secreted in breast milk

• That a medication guide should be dispensed with each prescription/refill

rosuvastatin (Rx)

(roe-soo′va-sta-tin)

Crestor

Func. class.: Antilipemic

Chem. class.: HMG-CoA reductase inhibitor

ACTION: Inhibits HMG-CoA reductase enzyme, which reduces cholesterol synthesis

USES: As an adjunct for primary hypercholesterolemia (types IIa, IIb) and mixed dyslipidemia, elevated serum triglycerides, homozygous/heterozygous familial hypercholesterolemia (FH), slowing of atherosclerosis, CV disease prophylaxis, MI, stroke prophylaxis (normal LDL)

CONTRAINDICATIONS: Pregnancy (X), breastfeeding, hypersensitivity, active hepatic disease

Precautions: Children <10 yr, geriatric patients, past hepatic disease, alcoholism, severe acute infections, trauma, hypotension, uncontrolled seizure disorders, severe metabolic disorders, electrolyte imbalances, severe renal impairment, hypothyroidism, Asian patients

DOSAGE AND ROUTES

Hypercholesterolemia

• **Adult: PO** 5-40 mg/day; initial dose 10 mg/day, reanalyze lipid levels at 2-4 wk, adjust dosage accordingly

Homozygous familial hypercholesterolemia

• **Adult: PO** 20 mg/day, max 40 mg; Asian patients 5 mg/day

Dose in patients taking cycloSPORINE/gemfibrozil/lopinavir/ritonavir/atazanavir

• **Adult: PO** 5 mg/day, max 10 mg/day

Asian patients/patients with predisposition for myopathy

• **Adult: PO** 5 mg/day

R

Atherosclerosis slowing
- **Adult: PO** 10 mg/day (for those not taking cycloSPORINE or gemfibrozil)

Heterozygous familial hypercholesterolemia
- **Females ≥1 yr postmenarche and ≥10 yr): PO** 5-20 mg/day individualized

MI/stroke prophylaxis (unlabeled)
- **Adult: PO** 20 mg daily

Renal/hepatic dose
- **Adult: PO** CCr <30 ml/min, 5 mg/day; max 10 mg/day; avoid use with hepatic disease

Available forms: Tabs 5, 10, 20, 40 mg

Administer:
- May be taken at any time of day, with/without food
- Store in cool environment in airtight, light-resistant container

SIDE EFFECTS

CNS: *Headache, dizziness,* insomnia, paresthesia, confusion
GI: *Nausea, constipation, abdominal pain, flatus, diarrhea, dyspepsia, heartburn,* kidney failure, liver dysfunction, vomiting
HEMA: Thrombocytopenia, hemolytic anemia, leukopenia
INTEG: *Rash, pruritus*
MS: *Asthenia, muscle cramps, arthritis, arthralgia, myalgia,* myositis, rhabdomyolysis; leg, shoulder, or localized pain

PHARMACOKINETICS

Peak 3-5 hr, minimal live metabolism (about 10%), 88% protein bound, excreted primarily in feces (90%), crosses placenta, half-life 19 hr, not dialyzable

INTERACTIONS

Increase: hepatotoxicity—alcohol
⚠ **Increase:** myalgia, myositis, rhabdomyolysis—cycloSPORINE, gemfibrozil, niacin, clofibrate, azole antifungals, antiretroviral protease inhibitors, fibric acid derivatives
Increase: bleeding risk—warfarin

Drug/Lab Test
Increase: LFTs

NURSING CONSIDERATIONS

Assess:
- Diet; obtain diet history including fat, cholesterol in diet
- Fasting cholesterol, LDL, HDL, triglycerides at baseline and q4-6wk, then periodically
- Liver function: LFTs at baseline, 12 wk, then q6mo; AST, ALT, LFTs may increase
- Renal function in patients with compromised renal system: BUN, creatinine, I&O ratio

⚠ **Rhabdomyolysis:** muscle pain, tenderness, obtain CPK; if these occur, product may need to be discontinued; for patients with Asian ancestry: increased blood levels, rhabdomyolysis

Evaluate:
- Therapeutic response: decreased LDL, cholesterol, triglycerides, increased HDL, slowing CAD

Teach patient/family:

⚠ To report suspected pregnancy; to use contraception while taking product, pregnancy (X); not to breastfeed

⚠ To report weakness, muscle tenderness, pain, fever, liver injury (jaundice, anorexia, abdominal pain)

- That blood work and follow-up exams will be necessary during treatment
- To report severe GI symptoms, dizziness, headache, muscle pain, weakness
- That previously prescribed regimen will continue: low-cholesterol diet, exercise program, smoking cessation

rufinamide (Rx)

(roo-fin′a-mide)

Banzel

Func. class.: Anticonvulsant
Chem. class.: Triazole derivative

ACTION: May act through action at sodium channels; exact action is unknown

USES:
Lennox-Gastaut syndrome
Unlabeled uses: Partial seizures

CONTRAINDICATIONS:
Hypersensitivity, familial short QT syndrome
Precautions: Pregnancy (C), breastfeeding, children <16 yr, geriatric patients, renal/hepatic disease, depression, dialysis, hazardous activities, suicidal ideation

DOSAGE AND ROUTES
- **Adult: PO** 400-800 mg/day divided bid; increase by 400-800 mg/day q2days to 3200 mg/day
- **Child ≥4 yr: PO** 10 mg/kg/day divided equally bid; increase by 10 mg/kg/day every other day to 45 mg/kg/day or 3200 mg/day, whichever is less

Available forms: Tabs 200, 400 mg; oral susp 40 mg/ml
Administer:
- **PO tabs:** give with food; may give whole, crushed, or halved
- **Oral susp:** shake well before use; use provided adapter and calibrated oral dosing syringe; insert adapter firmly in neck of bottle before use; keep in place for duration of bottle use; insert dosing syringe into adapter; withdraw dose from inverted bottle; replace cap after each use; use within 90 days of opening; give with food

SIDE EFFECTS
CNS: Dizziness, ataxia, drowsiness, fever, seizures, tremors, fatigue, headache, gait disturbance, **suicidal ideation**
EENT: Diplopia, blurred vision, nystagmus
GI: Nausea, hepatitis, vomiting, abdominal pain
HEMA: **Anemia, leukopenia, neutropenia, thrombocytopenia,** lymphadenopathy, DRESS
INTEG: Rash, urticaria
MISC: Edema, hematuria, influenzae, nephrolithiasis, back pain, urinary frequency

PHARMACOKINETICS
Peak 4-6 hr, half-life 6-10 hr, metabolized by liver, excreted by kidneys, 34% protein binding

INTERACTIONS
Increase: rufinamide effect—valproate
Decrease: effect of hormonal contraceptives
Decrease: effect of rufinamide—carBAMazepine, phenytoin, primidone, PHENobarbital
Drug/Lab Test
Increase: LFTs

NURSING CONSIDERATIONS
Assess:
- **Seizures: duration, type, intensity precipitating factors**
- ⚠ **Mental status: mood, sensorium, affect, memory (long, short), increased suicidal thoughts/actions**

Evaluate:
- Therapeutic response: decrease in severity of seizures

Teach patient/family:
- Not to discontinue product abruptly because seizures may occur
- To avoid hazardous activities until stabilized on product
- To carry emergency ID stating product use
- **To notify prescriber if pregnancy is planned or suspected; to use alternative form of contraception because effect of hormonal contraceptives may be decreased**
- To consume adequate fluids
- **To report suicidal thoughts/behavior immediately**
- To report rash/fever to prescriber

R

salicylic acid topical

See Appendix B

salmeterol (Rx)

(sal-met′er-ole)

Serevent ✦, Serevent Diskus

Func. class.: β_2-Adrenergic agonist, bronchodilator

ACTION: Causes bronchodilation by action on β_2 (pulmonary) receptors by increasing levels of cAMP, which relaxes smooth muscle with little effect on heart rate; maintains improvement in FEV from 3 to 12 hr; prevents nocturnal asthma symptoms

USES: Prevention of exercise-induced bronchospasm, COPD, asthma

CONTRAINDICATIONS: Hypersensitivity to sympathomimetics, tachydysrhythmias, severe cardiac disease, monotherapy treatment of asthma

Precautions: Pregnancy (C), breastfeeding, cardiac disorders, hyperthyroidism, diabetes mellitus, hypertension, closed-angle glaucoma, seizures, acute asthma, as a substitute to corticosteroids, QT prolongation

Black Box Warning: Asthma-related death, children <4 yr

DOSAGE AND ROUTES

- **Adult/child ≥4 yr: INH** 50 mcg (1 inhalation as dry powder) q12hr; exercise-induced bronchospasm 50 mcg (1 inhalation) ½-1 hr before exercise

Available forms: Inhalation powder 50 mcg/blister

Administer:

- Gum, sips of water for dry mouth
- Use 1 hr before exercise for exercise-induced bronchospasm prevention
- Store in foil pouch; do not expose to temperature >86° F (30° C); discard 6 wk after removal from foil pouch

SIDE EFFECTS

CNS: *Tremors, anxiety,* insomnia, headache, dizziness, fever

CV: Palpitations, tachycardia, hypo/hypertension, angina, dysrhythmias

EENT: Dry nose, irritation of nose and throat

GI: Heartburn, nausea, vomiting, abdominal pain

MS: Muscle cramps

RESP: Bronchospasm, cough

PHARMACOKINETICS

INH: Onset 30-50 min; peak 4 hr; duration 12 hr; metabolized in liver; excreted in urine, breast milk; crosses placenta, blood-brain barrier; protein binding 94%-98%; terminal half-life 3-5 hr

INTERACTIONS

Increase: CV effect—CYP3A4 inhibitors (itraconazole, ketoconazole, nelfinavir, nefazodone, saquinavir)

Increase: action of aerosol bronchodilators

Increase: action of salmeterol—tricyclics, MAOIs

Decrease: salmeterol action—other β-blockers

NURSING CONSIDERATIONS

Assess:

- **Respiratory function:** vital capacity, forced expiratory volume, ABGs, lung sounds, heart rate and rhythm

⚠ **Paradoxical bronchospasm:** dyspnea, wheezing, chest tightness

Black Box Warning: Children should not use this product as monotherapy for asthma, use only with persistent asthma in those whose symptoms are not well-controlled with a long-term asthma product

Evaluate:

- Therapeutic response: absence of dyspnea, wheezing

Teach patient/family:

- Not to use for exercise-induced bronchospasm; never to exhale into diskus; to hold level, keep mouthpiece dry
- Not to use OTC medications because extra stimulation may occur
- Review package insert with patient
- To avoid getting powder in eyes
- To avoid smoking, smoke-filled rooms, persons with respiratory infections
- Not for treatment of acute exacerbation, a fast-acting β-blocker should be used instead

⚠ To immediately report dyspnea after use if ≥1 canister is used in 2 mo time

TREATMENT OF OVERDOSE:

β_2-Adrenergic blocker

sargramostim (GM-CSF) (Rx)

(sar-gram′oh-stim)

Leukine, GM-CSF

Func. class.: Biologic modifier

Chem. class.: Granulocyte macrophage colony-stimulating factor (GM-CSF)

Do not confuse:

Leukine/leucovorin/Leukeran

ACTION: Stimulates proliferation and differentiation of hematopoietic progenitor cells (granulocytes, macrophages)

USES: Acceleration of myeloid recovery in patients with non–Hodgkin's lymphoma, acute lymphoblastic leukemia, acute myelogenous leukemia, autologous bone marrow transplantation in Hodgkin's disease; bone marrow transplantation failure or engraftment delay, mobilization and transplant of peripheral blood progenitor cells (PBPCs)

Unlabeled uses: Aplastic anemia, Crohn's disease, HIV, ganciclovir- or zidovudine-induced neutropenia, malignant melanoma

CONTRAINDICATIONS: Neonates; hypersensitivity to GM-CSF, benzyl alcohol, yeast products; excessive leukemic myeloid blast in bone marrow, peripheral blood

Precautions: Pregnancy (C), breastfeeding, children; lung/cardiac/renal/hepatic disease; pleural, pericardial effusions, peripheral edema, leukocytosis, mannitol hypersensitivity, hepatic/renal disease

DOSAGE AND ROUTES

Myeloid reconstitution after autologous bone marrow transplantation

- **Adult: IV** 250 mcg/m²/day × 3 wk; give over 2 hr; begin 2-4 hr after bone marrow inf, not <24 hr after last dose of antineoplastics and 12 hr after last dose of radiotherapy, bone marrow transplantation failure, or engraftment delay

Acceleration of myeloid recovery

- **Adult: IV** 250 mcg/m²/day × 14 days; give over 2 hr; may repeat in 7 days; may repeat 500 mcg/m²/day × 14 days after another 7 days if no improvement

Mobilization of PBPCs

- **Adult: IV/SUBCUT** 250 mcg/m²/day during collection of PBPCs

After PBPC transplantation

- **Adult: IV/SUBCUT** 250 mcg/m²/day until ANC >1500 cells/mm³ × 3 days

Aplastic anemia (unlabeled)

- **Adult: SUBCUT** 250-500 mcg/day or 5 mcg/kg/day × 14-90 days, used with erythropoietin or immunosuppressive therapy

Malignant melanoma (unlabeled)

- **Adult: SUBCUT** 125 mcg/m²/day × 14 days; alternate with 14 days off

HIV (unlabeled)

- **Adult: SUBCUT** 250 mcg/day 3×/wk for ≤20 mo

Available forms: Powder for inj lyophilized 250 mcg; sol for inj 500 mcg/ml

Administer:

- Store in refrigerator; do not freeze

SUBCUT route

- No further dilution of reconstituted sol

S

is needed; take care not to inject intradermally

Intermittent IV INF route

- After reconstituting with 1 ml sterile water for inj without preservative; do not reenter vial; discard unused portion; direct reconstitution sol at side of vial; rotate contents; do not shake
- Dilute in 0.9% NaCl inj to prepare IV inf; if final conc is <10 mcg/ml, add human albumin to make final conc of 0.1% to NaCl before adding sargramostim to prevent adsorption; for a final conc of 0.1% albumin, add 1 mg human albumin/1 ml 0.9% NaCl inj run over 2 hr **(bone marrow transplant or failure of graft)**; over 4 hr **(chemotherapy for AML)**; over 24 hr as cont inf **(PBPCs)**; give within 6 hr after reconstitution

Y-site compatibilities: Amikacin, aminophylline, aztreonam, bleomycin, butorphanol, calcium gluconate, CARBOplatin, carmustine, ceFAZolin, cefepime, cefotaxime, cefoTEtan, ceftizoxime, cefTRIAXone, cefuroxime, cimetidine, CISplatin, clindamycin, cyclophosphamide, cycloSPORINE, cytarabine, dacarbazine, DACTINomycin, dexamethasone, diphenhydrAMINE, DOPamine, DOXOrubicin, doxycycline, droperidol, etoposide, famotidine, fentaNYL, floxuridine, fluconazole, fluorouracil, furosemide, gentamicin, granisetron, heparin, IDArubicin, ifosfamide, immune globulin, magnesium sulfate, mannitol, mechlorethamine, meperidine, mesna, methotrexate, metoclopramide, metroNIDAZOLE, minocycline, mitoXANtrone, netilmicin, pentostatin, piperacillin/tazobactam, potassium chloride, prochlorperazine, promethazine, ranitidine, teniposide, ticarcillin, ticarcillin-clavulanate, trimethoprim-sulfamethoxazole, vinBLAStine, vinCRIStine, zidovudine

SIDE EFFECTS

CNS: Fever, malaise, CNS disorder, weakness, chills, dizziness, syncope, headache

CV: Transient supraventricular tachycardia, peripheral edema, pericardial effusion, hypotension, tachycardia

GI: Nausea, vomiting, diarrhea, anorexia, GI hemorrhage, stomatitis, liver damage, hyperbilirubinemia

GU: Urinary tract disorder, abnormal kidney function

HEMA: Blood dyscrasias, hemorrhage

INTEG: Alopecia, rash, peripheral edema

MS: Bone pain, myalgia

RESP: Dyspnea

PHARMACOKINETICS

Half-life elimination: IV 60 min, SUBCUT 2-3 hr; detected within 5 min after administration, peak 2 hr

INTERACTIONS

- Do not use product concomitantly with antineoplastics or radiation; separate by ≥24 hr

Increase: myeloproliferation—lithium, corticosteroids

Drug/Lab Test

Increase: bilirubin, BUN, creatinine, eosinophils, LFTs, leukocytes

NURSING CONSIDERATIONS

Assess:

⚠ Blood studies: CBC, differential count before treatment, 2 × weekly; leukocytosis may occur (WBC >50,000 cells/mm^3, ANC >20,000 cells/mm^3), platelets; if ANC >20,000/mm^3 or 10,000/mm^3 after nadir has occurred or platelets >500,000/mm^3, reduce dose by 1/2 or discontinue; if blast cells occur, discontinue

- Renal, hepatic studies before treatment: BUN, creatinine, urinalysis; AST, ALT, alk phos; 2 × weekly monitoring is needed in renal/hepatic disease
- **Hypersensitivity**, rashes, local inj site reactions; usually transient
- Body weight, hydration status; increased fluid retention in cardiac disease; pulmonary function
- Constitutional symptoms: asthenia, chills, fever, headache, malaise

• Myalgia, arthralgia in legs, feet; use analgesics, antipyretics
• **Gasping syndrome in neonates:** due to benzyl alcohol hypersensitivity

Evaluate:
• Therapeutic response: WBC and differential recovery

Teach patient/family:
• To report dyspnea to health care provider
• Review all aspects of product use

saxagliptin (Rx)
(sax-a-glip′tin)
Onglyza
Func. class.: Antidiabetic, oral
Chem. class.: Dipeptidyl-peptidase-4 inhibitor (DPP-4 inhibitor)

Do not confuse:
saxagliptin/sitaGLIPtin

ACTION: Slows the inactivation of incretin hormones; improves glucose homeostasis, improves glucose-dependent insulin synthesis, lowers glucagon secretions, and slows gastric emptying time

USES: In adults, type 2 diabetes mellitus as monotherapy or in combination with other antidiabetic agents

CONTRAINDICATIONS: Hypersensitivity, angioedema
Precautions: Pregnancy (B), geriatric patients, GI obstruction, surgery, thyroid/renal/hepatic disease, trauma, diabetic ketoacidosis (DKA), type 1 diabetes

DOSAGE AND ROUTES
• **Adult: PO** 2.5-5 mg; may use with other antidiabetic agents other than insulin; if used with insulin, a lower dose may be needed; max 2.5 mg with strong 3A415 inhibitors

Renal dose
• **Adult: PO** CCr ≤50 ml/min, 2.5 mg daily

Available forms: Tabs 2.5, 5 mg

Administer:
PO route
• May be taken with/without food
• Conversion from other antidiabetic agents; change may be made with gradual dosage change
• Store in tight containers at room temperature

SIDE EFFECTS
CNS: *Headache*
ENDO: Hypoglycemia (renal impairment)
GI: *Nausea, vomiting,* abdominal pain, pancreatitis
INTEG: Urticaria, angioedema, anaphylaxis
MISC: Lymphopenia, peripheral edema

PHARMACOKINETICS
Rapidly absorbed, excreted by the kidneys (unchanged 24%), terminal half-life 2.5 hr, 3.1 hr metabolite, peak 2 hr, duration 24 hr

INTERACTIONS
Increase: hypoglycemia—androgens, insulins, β-blockers, cimetidine, corticosteroids, salicylates, MAOIs, fibric acid derivatives, FLUoxetine, insulin, sulfonylureas, ACE inhibitors; CYP3A4/5 inhibitors

Drug/Herb
Increase: antidiabetic effect—garlic, horse chestnut

Drug/Lab Test
Decrease: lymphocytes, glucose

NURSING CONSIDERATIONS
Assess:
• **Hypoglycemic reactions** (sweating, weakness, dizziness, anxiety, tremors, hunger); monitor blood glucose, HbA1c
renal studies: BUN, creatinine during treatment

Evaluate:
• Therapeutic response: decrease in polyuria, polydipsia, polyphagia; clear sensorium; absence of dizziness; stable gait, blood glucose at normal level

Teach patient/family:
- To perform regular self-monitoring of blood glucose using blood-glucose meter
- About the symptoms of hypo/hyperglycemia; what to do about each
- That product must be continued on daily basis; about consequences of discontinuing product abruptly
- To avoid OTC medications, alcohol, digoxin, exenatide, insulins, nateglinide, repaglinide, and other products that lower blood glucose unless approved by prescriber
- That diabetes is lifelong; that this product is not a cure, only controls symptoms
- That all food included in diet plan must be eaten to prevent hypo/hyperglycemia
- To carry emergency ID
- To take product without regard to food
- To notify prescriber when surgery, trauma, stress occurs because dose may need to be adjusted

scopolamine (Rx)

(skoe-pol′a-meen)

Maldemar, Scopace, Transderm Scop

Func. class.: Cholinergic blocker
Chem. class.: Belladonna alkaloid

ACTION: Inhibits acetylcholine at receptor sites in autonomic nervous system, which controls secretions, free acids in stomach; blocks central muscarinic receptors, which decreases involuntary movements

USES: Preoperatively to produce amnesia, sedation and to decrease secretions; motion sickness, parkinsonian symptoms

CONTRAINDICATIONS: Hypersensitivity, closed-angle glaucoma, myasthenia gravis, GI/GU obstruction, hypersensitivity to belladonna, barbiturates

Precautions: Pregnancy (C), breastfeeding, children, geriatric patients, prostatic hypertrophy, CHF, hypertension, dysrhythmia, gastric ulcer, renal/hepatic disease, hiatal hernia, GERD, ulcerative colitis, hyperthyroidism

DOSAGE AND ROUTES

Motion sickness
- **Adult: TD** 1 patch 4 hr before travel and q3days

Parkinsonian symptoms
- **Adult: PO** 0.4-0.8 mg q8hr

Preoperatively
- **Adult: IM/IV/SUBCUT** 0.32-0.65 mg; **TD** apply 1 patch PM before surgery or 1 hr before c-section

Nausea and vomiting
- **Adult: SUBCUT** 0.6-1 mg
- **Child: SUBCUT** 0.006 mg/kg; max 0.3 mg/dose

Available forms: Inj 0.4 mg/ml; TD patch 72 hr (1.5 mg)

Administer:
- Parenteral dose with patient recumbent to prevent postural hypotension
- Parenteral dose slowly; keep patient in bed for ≥1 hr after dose
- With or after meals for GI upset; may give with fluids other than water
- At bedtime to avoid daytime drowsiness in patients with parkinsonism
- With analgesic to avoid behavioral changes when given preoperatively
- Store at room temperature in light-resistant container

Direct IV route
- Dilute with equal sterile water; give over 2-3 min

Syringe compatibilities: Atropine, benzquinamide, butorphanol, chlorproMAZINE, cimetidine, diamorphine, dimenhyDRINATE, diphenhydrAMINE, droperidol, fentaNYL, glycopyrrolate, HYDROmorphone, hydrOXYzine, meperidine, metoclopramide, midazolam, morphine, nalbuphine, oxyCODONE, pentazocine, PENTobarbital, perphenazine, prochlorperazine, promazine, promethazine, ranitidine, SUFentanil, thiopental

Y-site compatibilities: FentaNYL, heparin, hydrocortisone, HYDROmorphone, methadone, morphine, potassium chloride, propofol, SUFentanil, vit B/C

SIDE EFFECTS

CNS: Confusion, anxiety, restlessness, irritability, delusions, hallucinations, headache, sedation, depression, incoherence, dizziness, excitement, delirium, flushing, weakness, fatigue, loss of memory
CV: Palpitations, tachycardia, postural hypotension, paradoxical bradycardia
EENT: Blurred vision, photophobia, dilated pupils, difficulty swallowing, mydriasis, cycloplegia
GI: *Dryness of mouth, constipation,* nausea, vomiting, abdominal distress, **paralytic ileus**
GU: Urinary hesitancy, retention
INTEG: Urticaria, dry skin
MISC: Suppression of breastfeeding, nasal congestion, decreased sweating

PHARMACOKINETICS

Excreted in urine, bile, feces (unchanged), half-life 8 hr
SUBCUT/IM: Peak 30-60 min, duration 7 hr
IV: Peak 10-15 min, duration 2 hr

INTERACTIONS

Increase: anticholinergic effect—alcohol, opioids, antihistamines, phenothiazines, tricyclics

NURSING CONSIDERATIONS

Assess:
- VS periodically
- I&O ratio; retention commonly causes decreased urinary output
- **Parkinsonism, EPS:** shuffling gait, muscle rigidity, involuntary movements; affect, mood, CNS depression, worsening of mental symptoms during early therapy
- Urinary hesitancy, retention; palpate bladder if retention occurs
- Constipation; increase fluids, bulk, exercise
- Tolerance during long-term therapy; dose may have to be increased or changed
- Hard candy, frequent drinks, sugarless gum to relieve dry mouth

Evaluate:
- Therapeutic response: decreased secretions

Teach patient/family:
- Not to discontinue product abruptly; to taper off over 1 wk
- To avoid driving, other hazardous activities because drowsiness may occur
- To avoid OTC medication: cough, cold preparations with alcohol, antihistamines unless directed by prescriber
- **Transdermal route:** to apply with clean, dry hands; to wash, dry hands before and after applying to surface behind ear; to press patch firmly
- To avoid hazardous activities, activities requiring alertness because dizziness may occur
- To change patch q72hr
- To apply at least 4 hr before traveling
- If blurred vision, severe dizziness, drowsiness occurs, to discontinue use, use another type of antiemetic, or to rotate patch to other ear
- To read label of all OTC medications; if any scopolamine is found in product, avoid use
- To keep out of children's reach

scopolamine ophthalmic

See Appendix B

selegiline (Rx)

(se-le'ji-leen)

Eldepryl, Emsam, Zelapar
Func. class.: Antiparkinson agent
Chem. class.: MAOI, type B

Do not confuse:
Eldepryl/enalapril

ACTION: Increased dopaminergic activity by inhibition of MAO type B activity; not fully understood

USES:
Adjunct management of Parkinson's disease for patients being treated with levodopa/carbidopa who had poor response to therapy; depression (transdermal)

Unlabeled uses: Alzheimer's disease, depression

CONTRAINDICATIONS:
Children/adolescents (suicide/hypertensive crisis), hypersensitivity, breastfeeding

Precautions: Pregnancy (C)

DOSAGE AND ROUTES

- **Adult: PO** 10 mg/day given with levodopa/carbidopa in divided doses, 5 mg at breakfast and lunch; after 2-3 days, begin to reduce dose of levodopa/carbidopa 10%-30%; **ORAL DISINTEGRATING** 1.25 mg (1 tab) × 6 wk or more initially then 2.5 mg (2 tabs) dissolved on tongue daily before breakfast; max 2.5 mg/day; **TRANSDERMAL** 6 mg/24 hr initially, increase by 3 mg/24 hr at ≥2 wk, up to 12 mg/24 hr if needed

Alzheimer's disease (unlabeled)

- **Adult: PO** 5 mg bid AM, PM

Available forms: Tabs 5 mg; caps 5 mg; oral disintegrating tabs 1.25 mg; transdermal 6 mg/24 hr (20 mg/20 cm^2), 9 mg/24 hr (30 mg/30 cm^2), 12 mg/24 hr (40 mg/40 cm^2)

Administer:

PO route

⚠ Do not use in children due to risk for hypertensive crisis

- Product until NPO before surgery
- Adjust dosage to response
- With meals; limit protein taken with product
- Dosing bid in AM and afternoon; avoid PM or bedtime dosing
- At doses of <10 mg/day because of risks associated with nonselective inhibition of MAO
- **Oral disintegrating tab:** peel back foil; remove tab; do not push through foil; place tab on tongue; allow to dissolve; swallow with saliva

Transdermal route

- Apply to dry, intact skin on upper torso, upper thigh, or outer surface of upper arm q12hr

SIDE EFFECTS

CNS: Increased tremors, chorea, restlessness, blepharospasm, increased bradykinesia, grimacing, tardive dyskinesia, dystonic symptoms, involuntary movements, increased apraxia, hallucinations, *dizziness,* mood changes, nightmares, delusions, lethargy, apathy, overstimulation, sleep disturbances, headache, migraine, numbness, muscle cramps, confusion, anxiety, tiredness, vertigo, personality change, back/leg pain, suicide in child/adolescent, suicidal ideation in adults

CV: Orthostatic hypotension, hypo/hypertension, dysrhythmia, palpitations, angina pectoris, tachycardia, edema, sinus bradycardia, syncope, hypertensive crisis (children)

EENT: Diplopia, dry mouth, blurred vision, tinnitus

GI: Nausea, vomiting, constipation, weight loss, anorexia, diarrhea, heartburn, rectal bleeding, poor appetite, dysphagia, xerostomia

GU: Slow urination, nocturia, prostatic hypertrophy, urinary hesitation, retention, frequency, sexual dysfunction

INTEG: Increased sweating, alopecia, hematoma, rash, photosensitivity, facial hair

RESP: Asthma, SOB

PHARMACOKINETICS

Absorption (tab) 40-90 min, (oral disintegrating tab) 10-15 min; peak ½-2 hr; rapidly metabolized (active metabolites: *N*-desmethyldeprenyl, amphetamine, methamphetamine); metabolites excreted in urine; half-life 10 hr, orally disintegrating tab 1.3 hr; protein binding up to 85%

INTERACTIONS

⚠ Fatal interaction: opioids (especially meperidine); do not administer together

⚠ Serotonin syndrome (confusion, seizures, fever, hypertension, agitation); death—FLUoxetine, PARoxetine, sertraline, fluvoxaMINE (discontinue 5 wk before selegiline treatment); do not use together

⚠ Fatal interaction: do not use with tricyclics

Increase: side effects of levodopa/carbidopa

Increase: unusual behavior, psychosis—dextromethorphan

Increase: hypotension—antihypertensives

Drug/Lab Test

Decrease: VMA

False positive: urine ketones, urine glucose

False negative: urine glucose (glucose oxidase)

False increase: uric acid, urine protein

NURSING CONSIDERATIONS

Assess:

- **Parkinson's symptoms:** decreased rigidity, unsteady gait, weakness, tremors
- Cardiac status: tachycardia/bradycardia; B/P, respiration throughout treatment
- Mental status: affect, mood, behavioral changes, depression; perform suicide assessment on all patients, suicidal ideation may occur

⚠ Opioids; if patient has received, do not administer selegiline; fatal reactions have occurred

- Assistance with ambulation during beginning therapy

Evaluate:

- Therapeutic response: decrease in akathisia, improved mood

Teach patient/family:

- To change positions slowly to prevent orthostatic hypotension
- **Hypertensive crisis:** nausea, vomiting, sweating, agitation, change in mental status, headache, chest pain; to notify prescriber immediately
- **Serotonin syndrome:** to report twitching, sweating, shivering, diarrhea to prescriber immediately
- To use product exactly as prescribed; if discontinued abruptly, parkinsonian crisis may occur
- **Pregnancy:** to report if pregnancy is planned or suspected, pregnancy (C); to avoid breastfeeding
- To use during the day to prevent insomnia
- To avoid heating pads, hot tubs when using transdermal products
- To avoid hazardous activities until response is known
- To avoid foods high in tyramine: cheese, pickled products, wine, beer, large amounts of caffeine

⚠ Not to exceed recommended dose of 10 mg (PO) because this might precipitate hypertensive crisis; to report severe headache, other unusual symptoms

TREATMENT OF OVERDOSE:

IV fluids for hypertension, IV dilute pressure agent for B/P titration

selenium topical

See Appendix B

sertraline (Rx)

(ser′tra-leen)

Zoloft

Func. class.: Antidepressant

Chem. class.: SSRI

Do not confuse:

Zoloft/Zocor

ACTION: Inhibits serotonin reuptake in CNS; increases action of serotonin; does not affect dopamine, norepinephrine

USES: Major depressive disorder, obsessive-compulsive disorder (OCD), posttraumatic stress disorder (PTSD), panic disorder, social anxiety disorder, premenstrual dysphoric disorder (PMDD)

Unlabeled uses: Premature ejaculation, pruritus with cholestatic liver disease, hot flashes during menopause, breast cancer patients taking tamoxifen; men with prostate cancer secondary to androgen-deprivation therapy, separation anxiety disorder

CONTRAINDICATIONS:

Hypersensitivity to this product or SSRIs

Precautions: Pregnancy (C), breastfeeding, geriatric patients, renal/hepatic disease, epilepsy, recent MI, latex sensitivity (dropper of oral conc)

Black Box Warning: Children, suicidal ideation

DOSAGE AND ROUTES

Major depression/OCD

- **Adult/geriatric patient/adolescent (unlabeled): PO** 25-50 mg/day; may increase to max of 200 mg/day; do not change dose at intervals of <1 wk; administer daily in AM or PM
- **Child 6-12 yr (unlabeled): PO** 25 mg/day, increase by 25-50 mg/wk

Premenstrual dysphoric disorder

- **Adult: PO** 50-150 mg nightly

Premature ejaculation (unlabeled)

- **Adult: PO** 50 mg/day

Pruritus (unlabeled)

- **Adult: PO** 50-100 mg/day

Hot flashes (unlabeled)

- **Adult: PO** 50 mg/day × 4 wk; men with prostate cancer (androgen deprivation), 50-100 mg/day

Hepatic dose

- **Adult: PO** use lower dose or less frequent dosing intervals

Available forms: Tabs 25, 50, 100 mg; oral sol 20 mg/ml

Administer:

- Store at room temperature; do not freeze
- Increased fluids, bulk in diet for constipation, urinary retention
- With food, milk for GI symptoms
- Crushed if patient is unable to swallow medication whole
- Sugarless gum, hard candy, frequent sips of water for dry mouth
- **Oral conc:** dilute before use with 4 oz (1/2 cup) of water, orange juice, ginger ale, or lemon/lime soda; do not mix with other liquids
- Avoid use with other CNS depressants
- Dropper contains latex

SIDE EFFECTS

CNS: *Insomnia, agitation, somnolence, dizziness, headache, tremors, fatigue,* paresthesia, twitching, confusion, ataxia, gait abnormality (geriatric patients), seizures, neuroleptic malignant-syndrome–like reaction, serotonin syndrome, suicidal ideation, anxiety, drowsiness

CV: Palpitations, chest pain

EENT: Vision abnormalities, yawning, tinnitus, intraocular pressure

ENDO: SIADH (geriatric patients); diabetes mellitus

GI: *Diarrhea, nausea, constipation, anorexia, dry mouth,* dyspepsia, *vomiting, flatulence,* weight gain/loss, hepatitis

GU: *Male sexual dysfunction,* micturition disorder

INTEG: Increased sweating, rash, hot flashes

MISC: Hyponatremia, neonatal abstinence syndrome

PHARMACOKINETICS

PO: Peak 4.5-8.4 hr; steady state 1 wk; plasma protein binding 98%; elimination half-life 26 hr; extensively metabolized; metabolite excreted in urine, bile; weak inhibitor of CYP3A4, moderate inhibitor of CYP2D6

INTERACTIONS

- Altered lithium levels: lithium
- Disulfiram reaction: disulfiram and oral conc due to alcohol content

⚠ Fatal reactions: MAOIs, pimozide

Increase: sertraline levels—cimetidine, warfarin, other highly protein-bound products

Increase: effects of antidepressants (tricyclics), diazepam, TOLBUTamide, warfarin, benzodiazepines, SUMAtriptan, phenytoin, cloZAPine
Increase: bleeding risk—anticoagulants, NSAIDs, thrombolytic, platelet inhibitors, salicylates
Increase: serotonin syndrome, neuroleptic malignant syndrome—SSRIs, SNRIs, serotonin-receptor agonists, tricyclics, sibutramine, traZODone, busPIRone, linezolid, traMADol

Drug/Herb

Increase: of SSRI, serotonin syndrome—St. John's wort, SAM-e, tryptophan; do not use together
Increase: CNS effect—kava kava, valerian

Drug/Lab Test

Increase: AST, ALT
False positive: urine screen for benzodiazepines

NURSING CONSIDERATIONS

Assess:

Black Box Warning: Mental status: mood, sensorium, affect, suicidal tendencies (child/young adult), increase in psychiatric symptoms, depression, panic attacks, OCD, PTSD, social anxiety disorder

⚠ **Serotonin syndrome** (hyperthermia, hypertension, rigidity, delirium, coma, myoclonus) or neuroleptic malignant-like syndrome (muscle cramps, fever, unstable B/P, agitation, tremors, mental changes)

⚠ **Bleeding** (platelet serotonin depletion): GI bleeding, ecchymoses, epistaxis, hematomas, petechiae, hemorrhage

- LFTs, thyroid function tests, growth rate, weight at baseline and periodically
- B/P (lying/standing), pulse q4hr; if systolic B/P drops 20 mm Hg, hold product, notify prescriber; VS q4hr in patients with CV disease
- Weight weekly; appetite may decrease with product
- Urinary retention, constipation, especially in geriatric patients
- Alcohol consumption; hold dose until morning

Evaluate:

- Therapeutic response: significant improvement in depression, OCD

Teach patient/family:

- That therapeutic effect may take ≥1 wk
- To use caution when driving, performing other activities requiring alertness because drowsiness, dizziness, blurred vision may occur
- Not to discontinue medication quickly after long-term use; may cause nausea, headache, malaise
- To avoid alcohol
- May be taken without regard to food
- **Serotonin syndrome:** agitation, nausea, vomiting, diarrhea, twitching, sweating, shivering; report to prescriber immediately
- To notify prescriber if pregnant or if planning to become pregnant, pregnancy (C) or breastfeeding

Black Box Warning: That suicidal thoughts/behaviors may occur in children/adolescents

sildenafil (Rx)

(sil-den′a-fill)

Revatio, Viagra

Func. class.: Erectile agent, antihypertensive, peripheral vasodilator
Chem. class.: Phosphodiesterase type-5 inhibitor

Do not confuse:
Viagra/Allegra

ACTION: Enhances the effect of nitric oxide (NO) by inhibiting phosphodiesterase type 5 (PDE5), which is necessary for degrading cGMP in the corpus cavernosum

USES: Treatment of erectile dysfunction, improvement in exercise ability, pulmonary hypertension

S

Unlabeled uses: Sexual dysfunction (women); lower urinary tract symptoms and erectile dysfunction (with alfuzosin); altitude sickness, Raynaud's disease

CONTRAINDICATIONS:
Hypersensitivity to this product or nitrates

Precautions: Pregnancy (B), anatomical penile deformities, sickle cell anemia, leukemia, multiple myeloma, retinitis pigmentosa, bleeding disorders, active peptic ulceration, CV/renal/hepatic disease, multiproduct antihypertensive regimens, geriatric patients

DOSAGE AND ROUTES

Erectile dysfunction (Viagra only)

- **Adult male <65 yr: PO** 50 mg 1 hr before sexual activity; may be increased to 100 mg or decreased to 25 mg; max 1×/day
- **Adult ≥65 yr (male): PO** 25 mg as needed about 1 hr before sexual activity

Renal/hepatic dose

- **Adult: PO** (Child-Pugh A, B) 25 mg, take 1 hr before sexual activity; max 1×/day; CCr <30 ml/min, 25 mg starting dose

Pulmonary hypertension (Revatio only)

- **Adult: PO** 20 mg tid; take 4-6 hr apart; **IV BOL** 10 mg tid

Pulmonary hypertension induced by altitude sickness (unlabeled)

- **Adult: PO** 40 mg 6-8 hr after arriving at 14,272 ft, then 40 mg tid × 6 days

Anorgasmy in antidepressant therapy/sexual dysfunction in women (unlabeled)

- **Adult: PO** 50 mg 60-90 min before sexual activity

Available forms: Tabs 20, 25, 50, 100 mg; sol for inj 10 mg/12.5 ml

Administer:

- **Erectile dysfunction:** give approximately 1 hr before sexual activity; do not use more than 1×/day; give on empty stomach for better absorption
- **Pulmonary hypertension:** give 3×/day, 4-6 hr apart

SIDE EFFECTS

CNS: *Headache, flushing, dizziness,* transient global amnesia, seizures

CV: MI, sudden death, CV collapse, TIAs, ventricular dysrhythmias, CV hemorrhage

MISC.: *Dyspepsia, nasal congestion, UTI, abnormal vision, diarrhea, rash,* nonarteritic ischemic optic neuropathy, hearing loss, priapism, sickle cell crisis

PHARMACOKINETICS

Rapidly absorbed; bioavailability 40%; metabolized by P45 CYP3A4, 2C9 in the liver (active metabolites); terminal half-life 4 hr; peak 15-30 min; reduced absorption with high-fat meal; excreted in feces, urine

INTERACTIONS

⚠ Do not use with nitrates; fatal fall in B/P

Increase: sildenafil levels—cimetidine, erythromycin, ketoconazole, itraconazole, antiretroviral protease inhibitors, tacrolimus

Decrease: sildenafil levels—CYP-450 inducers, rifampin, barbiturates, bosentan, carBAMazepine, dexamethasone, phenytoin, nevirapine, rifabutin, troglitazone; antacids

Decrease: B/P—α-blockers, alcohol, amlodipine, angiotensin II receptor blockers

Drug/Food

Increase: product effect—grapefruit

Decrease: absorption—high-fat meal

NURSING CONSIDERATIONS

Assess:

- **Phosphodiesterase type 5 inhibitors with lopinavir/ritonavir (Kaletra):** assess for hypotension, visual changes, prolonged erection, syncope; give only 25 mg q48hr and monitor for adverse reactions
- Erectile dysfunction before use

⚠ Any severe loss of vision while taking this or any similar products; products should not be used

⚠ Use of organic nitrates that should not be used with this product

• **MI, sudden death, CV collapse:** those with an MI within 6 mo, resting hypotension <90/50, resting hypertension >170/100, fluid depletion should use this product cautiously; may occur right after sexual activity to days afterward
• **Sickle cell crisis (vasoocclusive crisis):** when used for pulmonary hypertension, may require hospitalization
• Cardiac status, hemodynamic parameters, exercise tolerance in pulmonary hypertension: B/P, pulse

Evaluate:
• Therapeutic response: decreasing pulmonary hypertension/improved exercise tolerance; ability to perform sexually (male)

Teach patient/family:
• That product does not protect against sexually transmitted diseases, including HIV
• That product absorption is reduced with a high-fat meal
• That product should not be used with nitrates in any form
• That tabs may be split
⚠ To notify prescriber immediately and to stop taking product if vision/hearing loss occurs or erection lasts >4 hr
• That erections lasting >4 hr may occur; report to prescriber immediately
• Do not use more than 100 mg in 24 hr

silodosin (Rx)

(si-lo′do-seen)

Rapaflo

Func. class.: Selective α_1-adrenergic blocker, BPH agent

Chem. class.: Sulfamoylphenethylamine derivative

ACTION: Binds preferentially to α_{1A}-adrenoceptor subtype located mainly in the prostate

USES: Symptoms of benign prostatic hyperplasia (BPH)

CONTRAINDICATIONS: Hypersensitivity, renal failure, hepatic disease

Precautions: Pregnancy (B), breastfeeding, children, females, geriatric patients, renal/hepatic disease, hypotension, ocular surgery, orthostatic hypotension, prostate cancer, syncope

DOSAGE AND ROUTES

• **Adult: PO** 8 mg/day with meal; max 8 mg/day

Renal dose

• **Adult: PO** CCr 30-49 ml/min, 4 mg/day; CCr <30 ml/min, not recommended

Available forms: Cap 4, 8 mg

Administer:
• Give with meal at same time each day
• Cap may be opened and sprinkled in applesauce
• Store at room temperature; protect from light and moisture

SIDE EFFECTS

CNS: *Dizziness, headache,* asthenia, insomnia, syncope
CV: Orthostatic hypotension
EENT: Nasal congestion, rhinorrhea, sinusitis
GI: Diarrhea, abdominal pain, jaundice
GU: Abnormal ejaculation, priapism, urinary incontinence
HEMA: Purpura

PHARMACOKINETICS

Decreased absorption with high-fat/high-calorie meal, half-life of metabolite 24 hr, metabolized in liver, excreted via urine, extensively protein bound (97%)

INTERACTIONS

Increase: silodosin effect—CYP3A4 inhibitors (clarithromycin, itraconazole, ritonavir, antiretroviral protease inhibitors, aprepitant, chloramphenicol, conivaptan, dalfopristin, danazol, delavirdine, efavirenz, fosaprepitant, fluconazole, fluvoxaMINE, imatinib, isoniazid, mifepristone, nefazodone, tamoxifen, telithromycin, troleandomycin, voriconazole, zileuton, zafirlukast)

S

Drug/Food
Increase: silodosin effect—grapefruit juice
Drug Lab/Test
Increase: LFTs

NURSING CONSIDERATIONS

Assess:
- **Prostatic hyperplasia:** change in urinary patterns at baseline and throughout treatment, testing for prostate cancer, before administration is recommended
- BUN, uric acid, urodynamic studies (urinary flow rates, residual volume)
- I&O ratios, weight daily; edema, report weight gain or edema
- B/P, monitor for orthostatic hypotension

Evaluate:
- Therapeutic response: decreased symptoms of BPH

Teach patient/family
- Not to drive, operate machinery until effect known
- Not to use with grapefruit juice
- To take with same meal each day

silver nitrate 1% ophthalmic
See Appendix B

silver nitrate 1% sulfacetamide sodium ophthalmic
See Appendix B

silver sulfADIAZINE topical
See Appendix B

simethicone (OTC, Rx)
(si-meth'i-kone)
Barriere ✦, Equaline Extra Strength Gas Relief, Gas Relief ✦, Gas-Relief, Gas-X, Good Sense Ultra Strength Gas Relief, Mylanta Gas Relief, Mylanta Gas, Mylicon, Ovol ✦, Phazyme, Top Care Gas Relief Extra Strength, Walgreens Gas Relief
Func. class.: Antiflatulent

Do not confuse:
Mylicon/Mylanta Gas

ACTION: Disperses/prevents mucus gas pockets in GI system, lowers surface tension of gas bubbles

USES: Flatulence
Unlabeled uses: Dyspepsia

CONTRAINDICATIONS: Hypersensitivity, GI obstruction/perforation
Precautions: Pregnancy (C), abdominal pain, fistula, hiatal hernia

DOSAGE AND ROUTES
- **Adult and child >12 yr: PO** 40-125 mg after meals and at bedtime prn, max 500 mg/day
- **Child 2-12 yr: PO** 40-50 mg after meals and at bedtime prn, max 240 mg/day
- **Child <2 yr: PO** 20 mg qid prn

Available forms: Chew tabs 40, 150, 166 mg; tabs 60, 80, 95, 125 mg; drops 20 mg/0.3 ml, 95 mg/1.425 ml; caps 95, 180 mg; soft gel caps 125, 180 mg; oral dissolving film 62.5 mg
Administer:
- After meals, at bedtime; shake susp well before giving; chew tabs should be chewed

SIDE EFFECTS
GI: Belching, rectal flatus, diarrhea

NURSING CONSIDERATIONS

Assess:

- Reason for excess gas production, decreased bowel sounds, recent surgery, other GI conditions

Evaluate:

- Therapeutic response: reduction of abdominal gas, discomfort

Teach patient/family:

- That tabs must be chewed
- To shake susp well before pouring

simvastatin (Rx)

(sim-va-sta'tin)

Zocor

Func. class.: Antilipemic

Chem. class.: HMG-CoA reductase inhibitor

Do not confuse:

Zocor/Cozaar/Zoloft

ACTION: Inhibits HMG-CoA reductase enzyme, which reduces cholesterol synthesis

USES: As an adjunct for primary hypercholesterolemia (types IIa, IIb), isolated hypertriglyceridemia (Frederickson type IV), and type III hyperlipoproteinemia, CAD, heterozygous familial hypercholesterolemia; MI/stroke prophylaxis

CONTRAINDICATIONS: Pregnancy (X), breastfeeding, hypersensitivity, active hepatic disease

Precautions: Past hepatic disease, alcoholism, severe acute infections, trauma, severe metabolic disorders, electrolyte imbalances, Chinese patients

DOSAGE AND ROUTES

- **Adult: PO** 20-40 mg/day in PM initially; usual range 5-40 mg/day in PM, max 40 mg/day for most patients, max 80 mg/day for patients taking 80 mg/day chronically without myopathy; dosage adjustments may be made in ≥4-wk intervals; those taking verapamil and amiodarone max 20 mg/day; max <80 mg for Chinese patients taking lipid-modifying niacin doses
- **Child/adolescent ≥10 yr including girls ≥1 yr postmenarche: PO** 10 mg in PM, range 10-40 mg/day

With diltiazem/verapamil/dronedarone

- **Adult: PO** 5-10 mg in PM, max 10 mg/day

With amiodarone, amoLODIPine, ranolazine

- **Adult: PO** 5-20 mg daily in evening, max 20 mg/day

Heterozygous familial hypercholesterolemia

- **Adolescent 10-17 yr: PO** 10 mg/day, max 40 mg/day

Available forms: Tabs 5, 10, 20, 40, 80 mg

Administer:

- Total daily dose in evening
- Store in cool environment in tight container protected from light

SIDE EFFECTS

CNS: Headache, cognitive impairment

GI: Nausea, constipation, diarrhea, dyspepsia, flatus, abdominal pain, **liver dysfunction, pancreatitis**, hyperglycemia

INTEG: Rash, pruritus

MS: Muscle cramps, myalgia, **myositis, rhabdomyolysis**, myopathy

RESP: Upper respiratory tract infection

PHARMACOKINETICS

Metabolized in liver (active metabolites); >98% protein bound; excreted primarily in bile, feces (60%), kidneys (15%); peak 1-2 hr; half-life 3 hr

INTERACTIONS

⚠ **Do not use with cycloSPORINE, gemfibrozil**

⚠ **Increase: effects of warfarin**

⚠ **Increase: rhabdomyolysis, myalgia; do not use concurrently—niacin, erythromycin, clofibrate, clarithromycin, ketoconazole, itraconazole, protease inhibitors, macrolide antibiotics, danazol,**

delavirdine, nefazodone, verapamil, diltiazem, amiodarone, azole antifungals, telithromycin, voriconazole
Increase: serum level of digoxin
Drug/Herb
Increase: effect—red yeast, rice
Decrease: effect—St. John's wort
Drug/Lab Test
Increase: CK, LFTs

NURSING CONSIDERATIONS

Assess:

- Diet history: fat consumption; baseline and lipid profile: LDL, HDL, TG, cholesterol
- Hepatic studies at baseline, after 4-6 wk, periodically thereafter; AST, ALT, may increase

⚠ **Rhabdomyolysis:** muscle tenderness, increased CPK levels (10× above upper normal limit); therapy should be discontinued, more likely in those receiving >80 mg/day, first year of treatment, those ≥65 yr, females

- Renal studies in patients with compromised renal systems: BUN, I&O ratio, creatinine

Evaluate:

- Therapeutic response: decrease in LDL, total cholesterol, triglycerides; increase in HDL; slowing CAD

Teach patient/family:

- That blood work will be necessary during treatment
- To report severe GI symptoms, headache
- That previously prescribed regimen will continue: low-cholesterol diet, exercise program
- To notify prescriber if pregnancy is suspected or planned; pregnancy (X); not to breastfeed

sirolimus (Rx)

(seer-oh-lie′mus)

Rapamune

Func. class.: Immunosuppressant
Chem. class.: Macrolide

ACTION: Produces immunosuppression by inhibiting T-lymphocyte activation and proliferation

USES: Organ transplants to prevent rejection; recommended use is with cycloSPORINE and corticosteroids

CONTRAINDICATIONS: Breastfeeding, hypersensitivity to this product, components of product
Precautions: Pregnancy (C), children <13 yr, severe cardiac/renal/hepatic disease; diabetes mellitus, hyperkalemia, hyperuricemia, hypertension, interstitial lung disease, hyperlipidemia

Black Box Warning: Lymphomas, infection, other malignancies

DOSAGE AND ROUTES

- **Adult/adolescent ≥40 kg: PO** 2 mg/day with 6 mg loading dose
- **Child >13 yr weighing <40 kg (88 lb): PO** 1 mg/m²/day, 3 mg/m² loading dose

Hepatic dose

- **Adult/child ≥13 yr/<40 kg: PO** reduce by 33% for maintenance dose (mild to moderate hepatic impairment); reduce by 50% for maintenance dose (severe hepatic impairment)

Available forms: Oral sol 1 mg/ml; tabs 0.5 mg, 1 mg, 2 mg
Administer:

- Prophylaxis for *Pneumocystis jiroveci* pneumonia for 1 yr after transplantation; prophylaxis for CMV is recommended for 90 days after transplantation in those at increased risk for CMV
- All medications PO if possible; avoid IM inj; bleeding may occur
- For 3 days before transplant surgery; patients should be placed in protective

isolation; give at same time of day; give 4 hr after cycloSPORINE oral sol or caps; do not give with grapefruit juice

• Use amber oral dose syringe and withdraw amount of oral sol needed from bottle; empty correct dose into plastic/glass container holding 60 ml of water/orange juice; stir vigorously and have patient drink at once; refill container with additional 120 ml water/orange juice; stir vigorously and have patient drink at once; if using a pouch, squeeze entire contents into container; follow above directions

• Store protected from light, refrigerate; stable for 30 days after opening (sol)

• Do not crush, chew; store tabs at room temperature

SIDE EFFECTS

CNS: *Tremors, headache, insomnia, paresthesia,* chills, fever

CV: Hypertension, *atrial fibrillation, CHF, hypotension, palpitation, tachycardia,* peripheral edema, **thrombosis**

EENT: Blurred vision, photophobia

GI: Nausea, vomiting, diarrhea, constipation, **hepatotoxicity**

GU: UTIs, **albuminuria, hematuria, proteinuria, renal failure, nephrotic syndrome,** increased creatinine

HEMA: **Anemia, leukopenia, thrombocytopenia, purpura, pancytopenia**

INTEG: *Rash, acne,* photosensitivity

META: Hyperglycemia, increased creatinine, edema, hypercholesterolemia, *hyperlipemia,* hypophosphatemia, weight gain, hypo/hyperkalemia, hyperuricemia, hypomagnesemia, hypertriglyceridemia

MS: Arthralgia

RESP: **Pleural effusion, atelectasis,** *dyspnea,* pneumonitis, pulmonary embolism/fibrosis

SYST: **Lymphoma, exfoliative dermatitis**

PHARMACOKINETICS

Rapidly absorbed; peak 1 hr single dose, 2 hr multiple dosing; protein binding 92%; extensively metabolized by CYP3A4 enzyme system, half-life 57-63 hr

INTERACTIONS

⚠ **Increase:** angioedema—ACE inhibitors, angiotensin-II–receptor antagonists, cephalosporins, iodine-containing radiopaque contrast media, neuromuscular blockers, NSAIDs, penicillins, salicylates, thrombolytics

Increase: blood levels—antifungals, calcium channel blockers, cimetidine, danazol, erythromycin, cycloSPORINE, metoclopramide, bromocriptine, HIV-protease inhibitors

Decrease: blood levels—carBAMazepine, PHENobarbital, phenytoin, rifamycin, rifapentine

Decrease: effect of vaccines

Drug/Herb

• St. John's wort: may decrease the effect of sirolimus

Drug/Food

• Alters bioavailability; use consistently with/without food; do not use with grapefruit juice

Drug/Lab Test

Increase: LFTs, alk phos, lipids, triglycerides, total cholesterol, BUN, creatinine, LDH, phosphate

Decrease: platelets, sodium

Increase or decrease: magnesium, glucose, calcium

NURSING CONSIDERATIONS

Assess:

• Blood levels in patients who may have altered metabolism, trough level ≥15 ng/ml are associated with increased adverse reactions; monitor trough concs in all patients

• Creatinine/BUN, CBC, serum potassium

• **Lipid profile:** cholesterol, triglycerides; lipid-lowering agent may be needed

Black Box Warning: Infection and development of lymphoma; only those experienced in immunosuppressant therapy and organ transplantation should use this product; use only in renal transplant

• **High risk:** those with Baniff grade 3 acute rejection or vascular rejection before cycloSPORINE withdrawal, dialysis dependent, creatinine >4.5 mg/dl, African descent, re-transplants, multiorgan transplant, high panel of reactive antibodies

• **Pulmonary fibrosis, pulmonary effusion, pneumonitis:** dyspnea, cough, hypoxia; some fatal cases have occurred

• **Wound dehiscence and anastomotic disruption:** wound, vascular, airway, ureteral, biliary, inhibition of growth factors; do not combine with corticosteroids

• **Anaphylaxis, angioedema, exfoliative dermatitis:** more common when given with ACE inhibitors; do not use if a hypersensitivity reaction occurs

⚠ **Bone marrow suppression:** Hgb, WBC, platelets during treatment each mo; if leukocytes <3000/mm³ or platelets <100,000/mm³, product should be discontinued or reduced; decreased hemoglobin level

⚠ **Hepatotoxicity:** alk phos, AST, ALT, amylase, bilirubin, dark urine, jaundice, itching, light-colored stools; product should be discontinued

Evaluate:

• Therapeutic response: absence of graft rejection; immunosuppression with autoimmune disorders

Teach patient/family:

⚠ To report fever, rash, severe diarrhea, chills, sore throat, fatigue; serious infections may occur; clay-colored stools, cramping **(hepatotoxicity)**

• To avoid crowds, persons with known infections to reduce risk for infection

• To use contraception before, during, for 12 wk after product discontinued; to avoid breastfeeding

• To use sunscreen, protective clothing to prevent burns, skin cancer

• Not to use with grapefruit juice

• To avoid vaccines

• That lifelong use will be required to prevent rejection

• That continuing follow-up exams and blood work will be required

• Not to get on skin

• How to use product

• To take with or without regard to food, at same time, consistently

• Take 4 hr after cycloSPORINE

sitaGLIPtin (Rx)

(sit-a-glip′tin)

Januvia

Func. class.: Antidiabetic, oral

Chem. class.: Dipeptidyl-peptidase-4 inhibitor (DPP-4 inhibitor)

ACTION: Slows the inactivation of incretin hormones; improves glucose homeostasis, improves glucose-dependent insulin secretion, lowers glucagon secretions, and slows gastric emptying time

USES: Type 2 diabetes mellitus as monotherapy or in combination with other antidiabetic agents

CONTRAINDICATIONS: Angioedema, diabetic ketoacidosis (DKA)

Precautions: Pregnancy (B), geriatric patients, GI obstruction, surgery, thyroid/renal/hepatic disease, trauma, breastfeeding, pancreatitis, hypercortisolism, hyperglycemia, hyperthyroidism, hypogylcemia, ileus, pituitary insufficiency, surgery, type 1 diabetes mellitus, diabetic ketoacidosis, adrenal insufficiency, burns

DOSAGE AND ROUTES

• **Adult: PO** 100 mg/day; may use with antidiabetic agents other than insulin

Renal dose

• **Adult: PO** CCr 30-50 ml/min, 50 mg daily; CCr <30 ml/min, 25 mg daily

Available forms: Tabs 25, 50, 100 mg

Administer:

• May be taken with/without food

• Do not split, crush, chew; swallow whole

• Conversion from other antidiabetic agents; change may be made with gradual dosage change

• Store in tight container at room temperature

SIDE EFFECTS

CNS: *Headache*
ENDO: Hypoglycemia
GI: *Nausea, vomiting,* abdominal pain, diarrhea, pancreatitis, constipation
GU: Acute renal failure
MISC: *Peripheral edema*
SYST: Anaphylaxis, Stevens-Johnson syndrome, angioedema

PHARMACOKINETICS

Rapidly absorbed, excreted by the kidneys (unchanged 79%), terminal half-life 12.4 hr, peak 1-4 hr

INTERACTIONS

Increase: sitaGLIPtin level—cimetidine, disopyramide
Increase: levels of digoxin
Increase: hypoglycemia—androgens, insulins, β-blockers, cimetidine, corticosteroids, salicylates, MAOIs, fibric acid derivatives, FLUoxetine, sulfonylureas
Decrease: antidiabetic effect—thiazide diuretics, ACE inhibitors, protease inhibitors, sympathomimetics, ARIPiprazole, cloZAPine, OLANZapine, QUEtiapine, risperiDONE, ziprasidone, phenytoin, fosphenytoin, phenothiazines, estrogens, progestins, oral contraceptives
Drug/Herb
Increase: antidiabetic effect—garlic, green tea, horse chestnut
Drug/Lab Test
Increase: creatinine, LFTs

NURSING CONSIDERATIONS

Assess:
- **Hypoglycemic reactions:** sweating, weakness, dizziness, anxiety, tremors, hunger; hyperglycemic reactions soon after meals

⚠ **Serious skin reactions:** swelling of face, mouth, lips, dyspnea, wheezing

⚠ **Pancreatitis:** severe abdominal pain, nausea, vomiting; discontinue product

- **Renal studies:** BUN, creatinine during treatment, especially in geriatric patients or those with renal disease
- Glycosylated hemoglobin A1c; monitor blood glucose (BG) as needed

Evaluate:
- Therapeutic response: decrease in polyuria, polydipsia, polyphagia; clear sensorium; absence of dizziness; stable gait, blood glucose, A1c improvement

Teach patient/family:
- To perform regular self-monitoring of blood glucose using blood-glucose meter
- About the symptoms of hypo/hyperglycemia; what to do about each; to carry emergency ID
- To notify prescriber if pregnancy is planned, suspected
- That product must be continued on daily basis; about consequences of discontinuing product abruptly; to continue health regimen (diet, exercise)
- To avoid OTC medications, alcohol, digoxin, exenatide, insulins, nateglinide, repaglinide, and other products that lower blood glucose unless approved by prescriber
- That diabetes is a lifelong illness; that product is not a cure, only controls symptoms
- That all food included in diet plan must be eaten to prevent hypo/hyperglycemia
- To immediately notify prescriber of hypersensitivity reactions (rash, swelling of face, trouble breathing)

sodium bicarbonate (Rx, OTC)

Baking soda, Neut, Sellymin ✦
Func. class.: Alkalinizer
Chem. class.: $NaHCO_3$

ACTION: Orally neutralizes gastric acid, which forms water, NaCl, CO_2; increases plasma bicarbonate, which buffers H^+ ion conc; reverses acidosis IV

USES: Acidosis (metabolic), cardiac arrest, alkalinization (systemic/urinary) antacid, salicylate poisoning

S

Unlabeled uses: Contrast media nephrotoxicity prevention

CONTRAINDICATIONS:
Metabolic/respiratory alkalosis, hypochloremia, hypocalcemia

Precautions: Pregnancy (C), children, CHF, cirrhosis, toxemia, renal disease, hypertension, hypokalemia, breastfeeding, hypernatremia, Bartter's syndrome, Cushing syndrome, hyperaldosteronism

DOSAGE AND ROUTES

Acidosis, metabolic (not associated with cardiac arrest)

- **Adult and child: IV INF** 2-5 mEq/kg over 4-8 hr depending on CO_2, pH, ABGs

Cardiac arrest

- **Adult and child: IV BOL** 1 mEq/kg of 7.5% or 8.4% sol, then 0.5 mEq/kg q10min, then doses based on ABGs
- **Infant: IV** 1 mEq/kg over several min (use only the 0.5 mEq/ml [4.2%] sol for inj)

Alkalinization of urine

- **Adult: PO** 325 mg to 2 g qid or 48 mEq (4 g) then 12-24 mEq q4hr
- **Child: PO** 84-840 mg/kg/day (1-10 mEq/kg) in divided doses q4-6hr

Antacid

- **Adult: PO** 300 mg to 2 g chewed, taken with water daily-qid

Available forms: Tabs 300, 325, 600, 650 mg; inj 4.2%, 5%, 7.5%, 8.4%

Administer:

PO route

- Chew antacid tablets and drink 8 oz water
- Do not take antacid with milk because milk-alkali syndrome may result

Direct IV route

- Use for cardiac emergencies, not used often in cardiac arrest
- Use ampules or prefilled syringes only; give by rapid bolus dose; flush with NS before, after use

Continuous IV INF route

- Diluted in an equal amount of compatible sol given 2-5 mEq/kg over 4-8 hr, max 50 mEq/hr; slower rate in children
- Extravasation with IV administration (tissue sloughing, ulceration, necrosis)

Y-site compatibilities: Acyclovir, amifostine, asparaginase, aztreonam, bivalirudin, bumetanide, ceFAZolin, cefepime, cefTAZidime, ceftizoxime, cefTRIAXone, chloramphenicol, cimetidine, cladribine, clindamycin, cyclophosphamide, cycloSPORINE, cytarabine, DAPTOmycin, DAUNOrubicin, dexamethasone sodium phosphated exmedetomidine, digoxin, DOCEtaxel, DOXOrubicin, enalaprilat, ertapenem, erythromycin, esmolol, etoposide, etoposide phosphate, famotidine, fentaNYL, filgrastim, fluconazole, fludarabine, furosemide, gallium nitrate, gemcitabine, gentamicin, granisetron, heparin, hydrocortisone sodium succinate, ifosfamide, indomethacin, insulin, ketorolac, labetalol, levofloxacin, lidocaine, linezolid, LORazepam, magnesium sulfate, melphalan, mesna, meperidine, methylPREDNISolone sodium succinate, metoclopramide, metoprolol, metroNIDAZOLE, milrinone, morphine, nafcillin, nitroglycerin, nitroprusside, PACLitaxel, palonosetron, pantoprazole, PEMEtrexed, penicillin G potassium, phenylephrine, phytonadione, piperacillin/tazobactam, potassium chloride, procainamide, propranolol, propofol, protamine, ranitidine, remifentanil, tacrolimus, teniposide, thiotepa, ticarcillin/clavulanate, tirofiban, tobramycin, tolazoline, vasopressin, vit B complex with C, voriconazole

SIDE EFFECTS

CNS: Irritability, headache, confusion, stimulation, tremors, *twitching, hyperreflexia,* tetany, weakness, seizures of alkalosis

CV: Irregular pulse, cardiac arrest, water retention, edema, weight gain

GI: Flatulence, *belching, distention*

META: *Metabolic alkalosis*
MS: Muscular twitching, tetany, irritability

PHARMACOKINETICS

PO: Onset rapid, duration 10 min
IV: Onset 15 min, duration 1-2 hr, excreted in urine

INTERACTIONS

Increase: effects—amphetamines, mecamylamine, quiNINE, quiNIDine, pseudoephedrine, flecainide, anorexiants, sympathomimetics
Increase: sodium and decrease potassium—corticosteroids
Decrease: effects—lithium, chlorproPAMIDE, barbiturates, salicylates, benzodiazepines, ketoconazole, corticosteroids
Drug/Lab Test
Increase: sodium, lactate
Decrease: potassium

NURSING CONSIDERATIONS

Assess:

- Respiratory and pulse rate, rhythm, depth, lung sounds; notify prescriber of abnormalities
- **Fluid balance** (I&O, weight daily, edema); notify prescriber of fluid overload; assess for edema, crackles, shortness of breath
- Electrolytes, blood pH, PO_2, HCO_3^-, during treatment; ABGs frequently during emergencies
- Weight daily with initial therapy
- **Alkalosis:** irritability, confusion, twitching, hyperreflexia stimulation, slow respirations, cyanosis, irregular pulse
- **Milk-alkali syndrome:** confusion, headache, nausea, vomiting, anorexia, urinary stones, hypercalcemia
- For GI perforation secondary to carbon dioxide in GI tract; may lead to perforation if ulcer is severe enough

Evaluate:

- Therapeutic response: ABGs, electrolytes, blood pH, HCO_3^- WNL

Teach patient/family:

- Not to take antacid with milk because milk-alkali syndrome may result; not to use antacid for >2 wk
- ⚠ To notify prescriber if indigestion accompanied by chest pain; trouble breathing; diarrhea; dark, tarry stools; vomit that looks like coffee grounds; swelling of feet/ankles
- About sodium-restricted diet; to avoid use of baking soda for indigestion

sodium polystyrene sulfonate (Rx)

(po-lee-stye′reen)

Kayexalate, Kionex, SPS

Func. class.: Potassium-removing resin

Chem. class.: Cation exchange resin

ACTION:

Removes potassium by exchanging sodium for potassium in body, primarily in large intestine

USES:

Hyperkalemia in conjunction with other measures

CONTRAINDICATIONS:

Hypersensitivity to saccharin or parabens that may be in some products, GI obstruction, neonate (reduced gut motility)

Precautions: Pregnancy (C), geriatric patients, renal failure, CHF, severe edema, severe hypertension, sodium restriction, constipation, GI bleeding, hypocalcemia

DOSAGE AND ROUTES

- **Adult: PO** 15 g daily-qid; **RECT** enema 30-50 g q1-2hr initially prn then q6hr prn
- **Child (unlabeled): PO** 1 g/kg q6hr prn; **RECT** 1 g/kg q2-6hr prn

Available forms: Powder for susp 453.6 g, 454 g; oral susp 15 g/60 ml

S

Administer:

PO route

- **Powdered resin:** each dose of the powdered resin is usually given orally as a suspension in water or in a syrup; usually, the amount of fluid ranges from 20 to 100 ml, depending on the dosage, or 3-4 ml/g of resin; suspensions should be freshly prepared and not stored for >24 hr; the suspension may be also be introduced into the stomach via a tube, or the powdered resin may be mixed with the patient's food; the powder should not be mixed with foods or liquids that contain a large amount of potassium (bananas or orange juice)

Rectal route

- Precede retention enema with a cleansing enema
- Instruct patient to lie down on left side with lower leg extended and the upper leg flexed for support, or place the patient in the knee–chest position; gently insert a soft, large (28 Fr) rubber tube into the rectum for a distance of about 20 cm; the tip should be well into the sigmoid colon; tape the tube in place; suspend the sodium polystyrene sulfonate powdered resin in 100 ml of an aqueous vehicle (water or sorbitol) that has been warmed to body temperature and introduce through the tube by gravity; the particles should be kept suspended by stirring the suspension during administration; alternatively, 120-180 ml of a commercially available suspension may be administered as a retention enema after the suspension has been warmed to body temperature; following administration, flush the tube with 50-100 ml of fluid and clamp the tube and leave in place; the suspension should be retained in the colon for at least 30-60 min or for several hours, if possible
- After several hours have passed, administer a cleansing enema using a non–sodium-containing solution at body temperature; up to 2 quarts of fluid may be necessary; drain fluid through a Y-tube connection; observe the drainage if sorbitol was used

SIDE EFFECTS

GI: Constipation, anorexia, nausea, vomiting, diarrhea (sorbitol), fecal impaction, gastric irritation

META: Hypocalcemia, hypokalemia, hypomagnesemia, sodium retention

INTERACTIONS

Increase: hypokalemia—loop diuretics, cardiac glycosides

Increase: metabolic alkalosis—magnesium/calcium antacids

⚠ **Increase:** colonic necrosis—sorbitol; do not use concurrently

Decrease: effect of—lithium, thyroid hormones

NURSING CONSIDERATIONS

Assess:

- **Hyperkalemia:** confusion, dyspnea, weakness, dysrhythmias; ECG for spiked T waves, depressed ST segments, prolonged QT and widening QRS complex
- Bowel function daily; note consistency of stools, times/day
- **Hypotension:** confusion, irritability, muscular pain, weakness
- **Electrolytes:** serum potassium, calcium, magnesium, sodium; acid–base balance
- I&O ratio, weight daily; crackles, dyspnea, jugular venous distension, edema
- **Digoxin toxicity** (nausea, vomiting, blurred vision, anorexia, dysrhythmias) in those receiving digoxin

Evaluate:

- Therapeutic response: potassium level 3.5-5 mg/dl

Teach patient/family:

- About reason for medication and expected results
- To follow a low potassium diet, provide sample diet
- To avoid laxatives, antacids, electrolyte-based products unless approved by prescriber

solifenacin (Rx)

(sol-i-fen′a-sin)

VESIcare

Func. class.: Urinary antispasmodic, anticholinergic

Chem. class.: Anti-muscarinic

ACTION: Relaxes smooth muscles in urinary tract by inhibiting acetylcholine at postganglionic sites

USES: Overactive bladder (urinary frequency, urgency, incontinence)

CONTRAINDICATIONS: Hypersensitivity, uncontrolled closed-angle glaucoma, urinary retention, gastric retention

Precautions: Pregnancy (C), breastfeeding, children, geriatric patients, renal/hepatic disease, controlled closed-angle glaucoma, bladder outflow obstruction, GI obstruction, decreased GI motility, history of QT prolongation

DOSAGE AND ROUTES

- **Adult: PO** 5 mg/day, max 10 mg/day

Renal/hepatic dose

- **Adult: PO** (Child-Pugh B) max 5 mg/day; CCr ≤30 ml/min, 5 mg/day

Available forms: Tabs 5, 10 mg

Administer:

PO route

- Without regard to meals
- Swallow product whole with water, liquid

SIDE EFFECTS

CNS: Anxiety, paresthesia, fatigue, *dizziness,* headache, confusion, delirium, depression, drowsiness

CV: Chest pain, hypertension, QTc prolongation, peripheral edema, palpitations, sinus tachycardia

EENT: *Vision abnormalities, xerophthalmia,* nasal dryness

GI: *Nausea, vomiting, anorexia,* abdominal pain, *constipation, dry mouth,* dyspepsia

GU: Dysuria, urinary retention, frequency, UTI

INTEG: Rash, pruritus, angioedema, exfoliative dermatitis, erythema multiforme, anaphylaxis

MISC: Hyperthermia

RESP: Bronchitis, cough, pharyngitis, upper respiratory tract infection

PHARMACOKINETICS

90% absorbed; 98% protein bound; extensively metabolized by CYP3A4; excreted in urine 69% (metabolites), feces 22%; terminal half-life 45-68 hr

INTERACTIONS

Increase: QT prolongation—Class IA, III antidysrhythmias

Increase: CNS depression—sedatives, hypnotics, benzodiazepines, opioids

Increase: effects—CYP3A4 inhibitors (ketoconazole, clarithromycin, diclofenac, doxycycline, erythromycin, isoniazid, nefazodone, propofol, protease inhibitors, verapamil), max dose 5 mg

Decrease: effects—CYP3A4 inducers (carBAMazepine, nevirapine, PHENobarbital, phenytoin)

Drug/Herb

Decrease: effects—St. John's wort

Drug/Food

Increase: effect—grapefruit juice

Drug/Lab Test

Increase: LFTs

NURSING CONSIDERATIONS

Assess:

- **Urinary patterns:** distention, nocturia, frequency, urgency, incontinence
- **Allergic reactions:** rash
- **Angioedema:** of the face, lips, tongue, larynx

⚠ Cardiac patients: monitor ECG for QTc prolongation; avoid products that increase QT prolongation

Evaluate:

- Decreasing dysuria, frequency, nocturia, incontinence

Teach patient/family:

- To avoid hazardous activities because dizziness may occur
- That constipation, blurred vision may occur; to notify prescriber if abdominal pain with constipation occurs
- To take without regard to food
- To swallow tab whole; do not split, crush, chew
- To call prescriber if severe abdominal pain or constipation lasts for ≥3 days
- That heat prostration may occur if used in hot environment, sweating is decreased
- About **anticholinergic effects:** blurred vision, constipation, urinary retention, hyperthermia

somatropin (Rx)

(soe-ma-troe′pin)

Genotropin, Humatrope, Norditropin, Norditropin Flexpro, Nutropin, Nutropin AQ, Omnitrope, Saizen, Serostim, Tev-Tropin, Zorbtive

Func. class.: Pituitary hormone

Chem. class.: Growth hormone

Do not confuse:
somatropin/SUMAtriptan/somatrem

ACTION: Stimulates growth; somatropin is similar to natural growth hormone; both preparations were developed with the use of recombinant DNA

USES: Pituitary growth hormone deficiency (hypopituitary dwarfism), children with human growth hormone deficiency/growth failure, AIDS wasting syndrome, cachexia, adults with somatropin deficiency syndrome (SDS), short stature in Noonan syndrome, SHOX deficiency, Turner's syndrome, Prader-Willi syndrome

CONTRAINDICATIONS: Hypersensitivity to benzyl alcohol, creosol; closed epiphyses, intracranial lesions, acute respiratory failure, Prader-Willi syndrome with obesity, trauma

Precautions: Pregnancy (C), breastfeeding, newborn, geriatric patients, diabetes mellitus, hypothyroidism, prolonged treatment in adults, scoliosis, sleep apnea, chemotherapy, respiratory disease, glycerin hypersensitivity (with formulations that contain these products)

DOSAGE AND ROUTES

Genotropin

- **Child: SUBCUT** 0.16-0.24 mg/kg/wk divided into 6 or 7 daily inj; give in abdomen, thigh, buttocks
- **Adult: SUBCUT** 0.04-0.08 mg/kg/wk divided into 6-7 daily doses

Humatrope

- **Adult: IM** 0.006 international units/kg/day, max 0.0125 units/kg/day
- **Child: SUBCUT/IM** 0.18 mg/kg divided into equal doses either on 3 alternate days or 6×/wk, max weekly dose 0.3 mg/kg

Nutropin/Nutropin AQ (growth hormone deficiency)

- **Child: SUBCUT** 0.3 mg/kg/wk

Serostim

- **Adult: SUBCUT** at bedtime >55 kg, 6 mg; 45-55 kg, 5 mg; 35-45 kg, 4 mg

Norditropin

- **Child: SUBCUT** 0.024-0.034 mg/kg 6-7×/wk

Accretropin

- **Child: SUBCUT** 0.18-0.3 mg/kg/wk divided into 6 or 7 equal daily inj

Replacement of GH in GH deficiency

- **Adult: SUBCUT** (Saizen) 0.005 mg/kg/day; may increase after 4 wk to max 0.01 mg/kg/day

Available forms: Powder for inj (lyophilized) 1.5 mg (4 international units/ml), 4 mg (12 international units/vial), 5 mg (13 international units/vial), 5 mg (15 international units/vial) rDNA origin, 5.8 mg (15 international units/ml), 6 mg

(18 international units/ml), 8 mg (24 international units/vial), 10 mg (26 international units/vial); inj 10 mg (30 international units/vial), 5 mg/1.5 ml, 10 mg/1.5 ml, 15 mg/1.5 ml

Administer:

- Give IM or subcut; do not use IV
- Discontinue therapy if final height is achieved or epiphyseal fusion occurs
- Visually inspect parenteral products for particulate matter and discoloration

Reconstitution and storage

Accretropin:

- No reconstitution needed; before use, swirl the contents of the vial with a gentle rotary motion, do not shake; the solution should be clear
- Store unopened vials refrigerated, do not freeze, protect from light; once opened, the vial may be stored refrigerated ≤14 days

Genotropin:

- This product is supplied as a powder, filled in a two-chamber cartridge with the active substance in the front chamber and the diluent in the rear chamber; the product is available in a 5-mg cartridge (green tip) and a 12-mg cartridge (purple tip); the 5- and 12-mg cartridges may be used with the Genotropin Pen or the Genotropin Mixer; genotropin is also available, in various doses ranging from 0.2 mg to 2 mg, in single use, auto-mix devices called Genotropin Miniquicks
- **Cartridges:** store cartridges refrigerated before reconstitution, do not freeze; protect from light; a reconstitution device supplied by the manufacturer is used to mix the powder and the diluent; after the powder and diluent are mixed, gently tip the cartridge upside down a few times until powder is completely dissolved; *do not shake;* this can cause denaturation of the protein; if the solution is cloudy, do not use; following reconstitution, the 5-mg cartridge contains a 5 mg/ml solution and the 12 mg cartridge contains a 12 mg/ml solution; the 5-mg and 12-mg cartridges contain overfill; the cartridges contain diluent with preservative (m-cresol) and may be stored refrigerated ≤28 days after reconstitution; do not use the 5-mg and 12-mg cartridges in patients with m-cresol hypersensitivity
- **Genotropin Minquicks:** after dispensing but before reconstitution, store at ≤77° ≤3 mo; a reconstitution device supplied by the manufacturer is used to mix the powder and diluent; 10 different strengths are available that each deliver a fixed volume of 0.25 ml; this product contains a diluent with no preservative, refrigerate after reconstitution and use within 24 hr; use the reconstituted solution only once and discard any remaining solution

Humatrope:

- Before reconstitution, store refrigerated
- **Vials:** reconstitute each 5-mg vial with 1.5-5 ml of the supplied diluent (contains m-cresol as a preservative) or bacteriostatic water for injection (contains benzoyl alcohol as a preservative); sterile water for injection may be used for patients with a hypersensitivity to m-cresol and benzoyl alcohol; direct the liquid against the glass vial wall; swirl vial with a gentle rotary motion until contents are dissolved completely; do not shake; if the solution is cloudy, do not use; small, colorless particles may be present after refrigeration, this is not unusual for solutions containing proteins; vials reconstituted with the diluent or bacteriostatic water are stable for 14 days when stored refrigerated; for vials reconstituted with sterile water, use the vial only once and discard the unused portion; if not used immediately, refrigerate and use within 24 hr; avoid freezing reconstituted solutions
- **Cartridges:** reconstitute cartridges using *only* the supplied diluent syringe; the cartridges are designed for use only with the Humatrope injection device; once reconstituted, the cartridges are stable for up to 28 days when stored refrigerated; store the injection device without the needle attached; avoid freezing reconstituted solutions

S

Norditropin:

• Do not use reconstituted solution if it is cloudy or contains particulate matter

• Before use, store refrigerated

• Reconstitution of the cartridges is not required; the cartridge is intended for use only with the NordiPen injector; a prefilled, disposable pen, NordiFlex Pen injector, is also available; each cartridge size (5 mg, 10 mg, or 15 mg per 1.5-ml cartridge) has a color-coded corresponding pen, which is graduated to deliver an appropriate dose based on the solution's conc; NordiPen and NordiFlex Pen allow administration of a minimum 0.25-mg dose to a maximum 4.5-mg dose, depending on cartridge concs; follow directions provided in NordiPen injector instruction booklet

• After a cartridge has been inserted into the NordiPen injector or once a NordiFlex pen is in use, the pen should be stored refrigerated and used within 4 wk; alternatively, the 5-mg and 10-mg cartridges may be stored in the pen at room temperature, no higher than 77°, for up to 3 wk; NovoFine needles are recommended for administration; wipe the stopper of the pen cartridge with rubbing alcohol

Nutropin:

• Before reconstitution, store refrigerated

• Reconstitute each 5-mg vial with 1-5 ml bacteriostatic water for injection (benzyl alcohol preserved) and each 10-mg vial with 1-10 ml of bacteriostatic water for injection (benzyl alcohol preserved); if using for newborns, reconstitute with sterile water for injection; direct the liquid against the glass vial wall; swirl vial with a gentle rotary motion until contents are dissolved completely; do not shake; if the solution is cloudy after reconstitution or refrigeration, do not use; small, colorless particles may be present after refrigeration, this is not unusual for solutions containing proteins

• Solutions reconstituted with bacteriostatic water for injection are stable for 14 days refrigerated

• Solutions reconstituted with sterile water for injection should be used immediately and only once; discard any unused portions; avoid freezing reconstituted solutions

Nutropin AQ:

• Does not require reconstitution; solution should be clear; small, colorless particles may be present after refrigeration, this is not unusual for solutions containing proteins; allow vial or pen cartridge to come to room temperature, and gently swirl; if solution is cloudy, do not use

• **Vials:** before inserting needle, wipe the vial septum with rubbing alcohol or antiseptic solution to prevent contamination by microorganisms that may be introduced by repeated needle insertions; administer using sterile disposable syringes and needles; use syringes with small enough volume that the prescribed dose may be drawn from the vial with reasonable accuracy

• **Pen cartridge:** two strengths are available, 10 mg and 20 mg; intended for use only with Nutropin AQ Pen; each pen and cartridge are color coded to ensure accurate placement of the 10-mg or 20-mg cartridge into the appropriate pen; do not use the 20-mg cartridge in the pen intended for the 10-mg cartridge, and vice versa; wipe septum of pen cartridge with rubbing alcohol or antiseptic solution to prevent contamination by microorganisms that may be introduced by repeated needle insertions; administer using sterile, disposable needles; follow the directions provided in the Nutropin AQ Pen Instructions for Use; the Nutropin AQ 10 pen allows administration of a minimum 0.1-mg dose to a maximum 4-mg dose, in 0.1-mg increments; the Nutropin AQ 20 pen allows administration of a minimum 0.2-mg dose to a maximum 8-mg dose, in 0.2-mg increments

• **Prefilled device:** a prefilled multidose, dial-a-dose device is available in 3 strengths; administer using disposable needles; follow the directions provided in the Nutropin AQ NuSpin Instructions for Use; the Nutropin AQ Nuspin 5 allows administration of a minimum dose of 0.05 mg to a maximum dose of 1.75 mg, in increments of 0.05 mg; the Nutropin AQ Nuspin 10 allows administration of a minimum dose of 0.1 mg to a maximum dose of 3.5 mg, in increments of 0.1 mg; the Nutropin AQ Nuspin 20 allows administration of a minimum dose of 0.2 mg to a maximum dose of 7 mg, in increments of 0.2 mg

• After initial use, vials, cartridges, and prefilled devices are stable for 28 days refrigerated; avoid freezing; vials, cartridges, and prefilled devices are light sensitive; protect from light

Omnitrope:

• Before reconstitution, store vials refrigerated; store in the carton; Omnitrope is sensitive to light

• **Vials:** reconstitute the vial with diluent provided using a sterile disposable syringe; swirl the vial gently, but do not shake; if the solution is cloudy after reconstitution, the contents must not be injected; after reconstitution, the 1.5-mg vial may be refrigerated ≤24 hr; the 1.5-mg vial does not contain a preservative and should only be used once; discard any remaining solution; the 5.8-mg vial diluent contains benzoyl alcohol as a preservative; after reconstitution, the contents must be used within 3 wk; after the first injection, store the 5.8-mg vial in the carton, to protect from light, in the refrigerator; avoid freezing

• **Omnitrope Pen 5 cartridge:** each 5-mg cartridge must be inserted into the Omnitrope Pen 5 delivery system; follow the directions provided in the Omnitrope Instructions for Use; the cartridge contains benzoyl alcohol as a preservative; after the first use, store refrigerated ≤28 days; protect from light, avoid freezing

• **Omnitrope Pen 10 cartridge:** each 10-mg cartridge must be inserted into the Omnitrope Pen 10 delivery system; follow the directions provided in the Omnitrope Instructions for Use; after the first use, store refrigerated ≤28 days; protect from light, avoid freezing

Saizen:

• Before reconstitution, store at room temperature

• **Vials:** reconstitute each 5-mg vial with 1-3 ml bacteriostatic water for injection; reconstitute each 8.8-mg vial with 2-3 ml bacteriostatic water for injection (benzoyl alcohol preserved); in patients with hypersensitivity to benzyl alcohol, the vials may be mixed with sterile water for injection; direct the liquid against the glass vial wall; swirl vial with a gentle rotary motion until contents are dissolved completely; do not shake; the solution should be clear; if it is cloudy immediately after reconstitution or refrigeration, do not use; small colorless particles may be present after refrigeration, this is not unusual for solutions containing proteins; after reconstitution, store vials mixed with bacteriostatic water for injection refrigerated and use within 14 days; for vials mixed with sterile water for injection, the solution should be used immediately, and any unused portion should be discarded; avoid freezing

• **Cartridges:** available in 4 mg and 8.8 mg click.easy cartridges for use in a compatible injection device; a reconstitution device supplied by the manufacturer is used to mix the Saizen with accompanying diluent containing metacresol; cartridges reconstituted with the diluent containing metascresol are stable under refrigeration for ≤21 days; avoid freezing

Serostim:

• Before reconstitution, store vials and diluent at room temperature (15-30° C; 59–86° F)

• **Vials:** reconstitute the 5-mg or 6-mg vials with 0.5-1 ml of supplied diluent (sterile water for injection); reconstitute the 4-mg vial with 0.5-1 ml of bacteriostatic water for injection (benzoyl alcohol preserved) and the 8.8-mg vial with 1-2 ml of bacteriostatic water for injection (benzoyl alcohol preserved); direct the liquid against the glass vial wall; swirl vial with a gentle rotary motion until contents are dissolved completely; do not shake; the solution should be clear; if it is cloudy immediately after reconstitution or refrigeration, do not use; small colorless particles may be present after refrigeration, this is not unusual for solutions containing proteins; if reconstituted with sterile water for injection, use within 24 hr; if reconstituted with bacteriostatic water for injection (benzoyl alcohol preserved), the solution is stable for up to 14 days under refrigeration (2-8° C or 36-46° F); avoid freezing

• **Cartridges:** available in 8.8-mg click. easy cartridges for use in a compatible injection device; a reconstitution device is supplied by the manufacturer and is used to mix the Serostim with accompanying diluent containing metacresol; after reconstitution, cartridges are stable under refrigeration for ≤21 days; avoid freezing

Serostim LQ:

• Before use, store refrigerated

• Available in 6-mg single-use cartridges that do not require reconstitution; administer using sterile disposable syringes and needles

• Bring to room temperature before use; discard single-use cartridge after use, even if some drug remains; discard cartridges after the expiration date stated on the product; do not freeze; protect from light

Tev-Tropin:

• Before reconstitution, store refrigerated

• Reconstitute each 5-mg vial with 1-5 ml bacteriostatic 0.9% sodium chloride (benzoyl alcohol preserved) for injection; direct the liquid against the glass vial wall; swirl vial with a gentle rotary motion until contents are dissolved completely; do not shake; the solution should be clear; if it is cloudy immediately after reconstitution, do not inject; small, colorless particles may be present after refrigeration, this is not unusual for solutions containing proteins; when administering to newborns, reconstitute with sterile normal saline for injection that is unpreserved

• Solution reconstituted with bacteriostatic 0.9% sodium chloride is stable for 14 days when stored refrigerated; solution reconstituted with sterile normal saline should be used only once; discard any remaining solution; avoid freezing

Valtropin:

• Before dispensing, store vials and diluent refrigerated; after dispensing to patients, may be stored at or below 77° F for up to 3 mo

• Reconstitute each 5-mg vial with the entire contents of the accompanying diluent, which contains metacresol as a preservative; if patients are allergic to metacresol, sterile water for injection may be used; direct the liquid against the glass vial wall; swirl vial with a gentle rotary motion until contents are dissolved completely; do not shake; the solution should be clear, if it is cloudy or contains particulate matter immediately after reconstitution or after refrigeration, do not inject; the final conc of the reconstituted solution is 3.33 mg/ml

• After reconstitution with the provided diluent, solutions may be stored refrigerated for up to 14 days; after reconstitution with sterile water for injection, use only one dose of Valtropin per vial and discard the unused portion if not used immediately

Zorbtive:

• Unreconstituted vials of drug and diluent may be stored at room temperature until expiration date

• Reconstitute each vial of 4 mg, 5 mg, or 6 mg with 0.5-1 ml sterile water for injection, USP; reconstitute each 8.8 mg with 1-2 ml bacteriostatic water for injection (0.9% benzyl alcohol preserved); in newborns or patients with a benzoyl alcohol hypersensitivity, sterile water for injection may be used; review manufacturer's labeling for expected concs; direct the liquid against the glass vial wall; swirl vial with a gentle rotary motion until contents are dissolved completely; do not shake; the solution should be clear, if it is cloudy after reconstitution or refrigeration, do not use; small colorless particles may be present after refrigeration, this is not unusual for solutions containing proteins

- After reconstitution with sterile water for injection, use the solution immediately and discard any unused portion; when using bacteriostatic water for injection, reconstituted solutions are stable for up to 14 days refrigerated; avoid freezing vials of drug or diluent or reconstituted vials

IM route

- Inject deeply into a large muscle; aspirate before injection; rotate injection sites daily

SUBCUT route

- Volumes >1 ml of reconstituted solution is not recommended; do not inject intradermally
- Allow refrigerated solutions to come to room temperature before injection
- Subcutaneous injections may be given in the thigh, buttocks, or abdomen; rotate injection sites daily

SIDE EFFECTS

CNS: Headache, growth of intracranial tumor, fever, aggressive behavior
ENDO: Hyperglycemia, ketosis, hypothyroidism, thyroid hormone replacement may be needed
GI: Nausea, vomiting, pancreatitis
GU: *Hypercalciuria*
INTEG: Rash, urticaria, pain; inflammation at inj site, hematoma
MS: Tissue swelling, joint and muscle pain
SYST: Antibodies to growth hormone

PHARMACOKINETICS

Half-life 15-60 min, duration 7 days, metabolized in liver

INTERACTIONS

Increase: epiphyseal closure—androgens, thyroid hormones
Decrease: growth—glucocorticosteroids
Decrease: insulin, antidiabetic effect—dosage adjustment may be needed

Drug/Lab Test

Increase: glucose, urine glucose
Decrease: glucose thyroid hormones

NURSING CONSIDERATIONS

Assess:

- Signs/symptoms of diabetes
- Growth hormone antibodies if patient fails to respond to therapy
- Thyroid function tests: T_3, T_4, T_7, TSH to identify hypothyroidism
- **Allergic reaction:** rash, itching, fever, nausea, wheezing
- **Hypercalciuria:** urinary stones; groin, flank pain; nausea, vomiting, urinary frequency, hematuria, chills
- Growth rate, bone age of child at intervals during treatment
- **Respiratory infection:** in those with Prader-Willi syndrome, may have sleep apnea, upper airway obstruction; discontinue if obstruction occurs
- Rapid growth: assess for slipped capital femoral epiphysis may also occur in endocrine disorders
- Monitor ophthalmologic status baseline and periodically; intracranial hypertension may occur

Evaluate:

- Therapeutic response: growth in children

Teach patient/family:

- That treatment may continue for years; that regular assessments are required
- To maintain a growth record; to report knee/hip pain or limping
- That treatment is very expensive
- Identify creosol or benzyl-alcohol hypersensitivity before use

S

sotalol (Rx)

(sot'ah-lahl)

Betapace, Betapace AF, Sorine

Func. class.: Antidysrhythmic group III

Chem. class.: Nonselective β-blocker

ACTION: Blockade of β_1- and β_2-receptors leads to antidysrhythmic effect, prolongs action potential in myocardial fibers without affecting conduction, prolongs QT interval, no effect on QRS duration

USES: Life-threatening ventricular dysrhythmias; Betapace AF: to maintain sinus rhythm with symptomatic atrial fibrillation/flutter

Unlabeled uses: Atrial fibrillation prophylaxis, cardiac surgery, PSVT, Wolff-Parkinson-White (WPW) syndrome

CONTRAINDICATIONS: Hypersensitivity to β-blockers, cardiogenic shock, heart block (2nd/3rd degree), sinus bradycardia, CHF, bronchial asthma, CCr <40 ml/min

Black Box Warning: Congenital or acquired long QT syndrome, hypokalemia

Precautions: Pregnancy (B), breastfeeding, major surgery, diabetes mellitus, renal/thyroid disease, COPD, well-compensated heart failure, CAD, nonallergic bronchospasm, electrolyte disturbances, bradycardia, peripheral vascular disease

Black Box Warning: Cardiac dysrhythmias, torsades de pointes, ventricular dysrhythmias, ventricular fibrillation

DOSAGE AND ROUTES

- **Adult: PO** initial 80 mg bid, may increase to 240-320 mg/day
- **Child >2 yr with normal renal function (unlabeled): PO** 30 mg/m² tid, adjust dose gradually after ≥36 hr to max 60 mg/m² tid

Renal dose

- **Adult: PO** CCr 30-60 ml/min, give q24hr; CCr 10-29 ml/min, give q36-48hr; CCr <10 ml/min, individualize dose

Life-threatening ventricular dysrhythmias

- **Adult (CCr 40-60 ml/min): IV** 75 mg over 5 hr daily, monitor QTc at end of each inf during initiation and titration; 80 mg PO = 75 mg IV; 120 mg PO = 112.5 mg IV; 160 mg PO = 150 mg IV

Betapace AF

- **Adult: PO** initial 80 mg bid, titrate upward to 120 mg bid during initial hospitalization, monitor QTc interval for 2-4 hr after each dose

Renal dose (Betapace AF)

- **Adult: PO** CCr >60 ml/min, give q12hr; CCr 40-60 ml/min, give q24hr; CCr <40 ml/min, do not use

Available forms: Tabs 80, 120, 160, 240 mg; (Betapace AF) 80, 120, 160 mg; inj 150 mg/10 ml (15 mg/ml)

Administer:

PO route

- Before, at bedtime; tab may be crushed or swallowed whole; give 1 hr before or 2 hr after meals
- Reduced dosage in renal dysfunction
- Betapace and Betapace AF are not interchangeable
- Do not give within 2 hr of antacids
- Store in dry area at room temperature; do not freeze

IV route

- Dilute to vol of either 120 ml or 300 ml with D_5W, LR
- **75 mg dose:** withdraw 6 ml sotalol inj (90 mg), add 114 ml dilute to make 120 ml (0.75 mg/ml); or withdraw 6 ml sotalol inj (90 mg), add 294 ml dilute to make 300 ml (0.3 mg/ml)
- **112.5 mg dose:** withdraw 9 ml sotalol inj (135 ml), add 111 ml dilute to make 120 ml (1.125 mg/ml); or withdraw 9 ml sotalol (135 mg), add 291 ml dilute to make 300 ml (0.45 mg/ml)
- **150 mg dose:** withdraw 12 ml of sotalol (180 mg), add 108 ml to make 120 ml (1.5 mg/ml); or withdraw 12 ml so-

talol (180 mg), add 288 ml to make 300 ml (0.6 mg/ml)
• Use inf pump and infuse 100 or 250 ml over 5 hr at a constant rate

SIDE EFFECTS

CNS: Dizziness, mental changes, drowsiness, fatigue, headache, catatonia, depression, anxiety, nightmares, paresthesia, lethargy, insomnia, decreased conc
CV: Prodysrhythmia, prolonged QT, orthostatic hypotension, bradycardia, CHF, chest pain, ventricular dysrhythmias, AV block, peripheral vascular insufficiency, palpitations, torsades de pointes; life-threatening ventricular dysrhythmias (Betapace AF)
EENT: Tinnitus, visual changes, sore throat, double vision; dry, burning eyes
GI: Nausea, vomiting, diarrhea, dry mouth, flatulence, constipation, anorexia, indigestion
GU: Impotence, dysuria, ejaculatory failure, urinary retention
HEMA: Agranulocytosis, thrombocytopenic purpura (rare), thrombocytopenia, leukopenia
INTEG: Rash, alopecia, urticaria, pruritus, fever, diaphoresis
MISC: Facial swelling, decreased exercise tolerance, weight change, Raynaud's disease
MS: Joint pain, arthralgia, muscle cramps, pain
RESP: Bronchospasm, dyspnea, wheezing, nasal stuffiness, pharyngitis

PHARMACOKINETICS

PO: Onset 1-2 hr, peak 2-4 hr, duration 8-12 hr, half-life 12 hr, excreted unchanged in urine, crosses placenta, excreted in breast milk, protein binding 0%

INTERACTIONS

Black Box Warning: Increase: QT prolongation—class IA/III antidysrhythmics, some phenothiazines, β agonists, local anesthetics, tricyclics, haloperidol, chloroquine, droperidol, pentamidine; CYP3A4 inhibitors (amiodarone, clarithromycin, erythromycin, telithromycin, troleandomycin), arsenic trioxide, levomethadyl; CYP3A4 substrates (methadone, pimozide, QUEtiapine, quiNIDine, risperiDONE, ziprasidone)

Increase: hypoglycemia effect—insulin
Increase: effects of lidocaine
Increase: hypotension—diuretics, other antihypertensives, nitroglycerin
Decrease: β-blocker effects—sympathomimetics
Decrease: bronchodilating effects of theophylline, β_2-agonists
Decrease: hypoglycemic effects of sulfonylureas

Drug/Lab Test
False increase: urinary catecholamines
Interference: glucose, insulin tolerance tests

Drug/Herb
• Do not use with hawthorn

NURSING CONSIDERATIONS

Assess:
• I&O, weight daily; edema in feet, legs daily
• B/P, pulse q4hr; note rate, rhythm, quality
• Potassium, magnesium levels

Black Box Warning: Requires a specialized care setting: for a minimum of at least 3 days on maintenance dose with continuous ECG monitoring, creatinine clearance; calculate before dosing

Black Box Warning: Cardiogenic shock, acute pulmonary edema: do not use, effect can further depress cardiac output

⚠ **QT syndrome:** apical/radial pulse before administration: notify prescriber of any significant changes; monitor ECG continuously (Betapace AF); use QT interval to determine patient eligibility; baseline QT must be ≤450 msec, if ≥500 msec, frequency or dosage must be decreased or drug discontinued
• Baselines of renal studies before therapy begins
• **Abrupt discontinuation:** do not discontinue abruptly, taper over 1-2 wk

S

- Dose should be adjusted slowly, with at least 3 days between changes; monitor ECG for QT interval
- Monitor electrolytes (hypokalemia, hypomagnesia) may increase dysrhythmias

Evaluate:

- Therapeutic response: absence of life-threatening dysrhythmias

Teach patient/family:

- Not to discontinue product abruptly; to taper over 2 wk or may precipitate angina; to take exactly as prescribed
- Not to use antacids or OTC products containing α-adrenergic stimulants (nasal decongestants, OTC cold preparations) unless directed by prescriber
- To report bradycardia, dizziness, confusion, depression, fever
- To take pulse at home; advise patient when to notify prescriber
- To avoid alcohol, smoking, sodium intake
- To carry emergency ID to identify product being taken, allergies
- To avoid hazardous activities if dizziness present
- To report symptoms of CHF including difficulty breathing, especially on exertion or when lying down; night cough, swelling of extremities
- To wear support hose to minimize effects of orthostatic hypotension
- To monitor blood glucose if diabetic
- That hospitalization will be required for ≥3 days

TREATMENT OF OVERDOSE:

Lavage, IV atropine for bradycardia, IV theophylline for bronchospasm, digoxin, O_2, diuretic for cardiac failure; hemodialysis is useful for removal; administer vasopressor (norepinephrine) for hypotension, isoproterenol for heart block

spironolactone (Rx)

(speer′on-oh-lak′tone)

Aldactone, Novo-Spiroton ✱

Func. class.: Potassium-sparing diuretic

Chem. class.: Aldosterone antagonist

Do not confuse:

Aldactone/Aldactazide

ACTION:

Competes with aldosterone at receptor sites in distal tubule, thereby resulting in the excretion of sodium chloride and water and the retention of potassium and phosphate

USES:

Edema of CHF, hypertension, diuretic-induced hypokalemia, primary hyperaldosteronism (diagnosis, short-term treatment, long-term treatment), edema of nephrotic syndrome, cirrhosis of liver with ascites

Unlabeled uses: CHF, hirsutism in women, bronchopulmonary dysplasia (BPD), PMS, polycystic ovary syndrome, acne vulgaris, premenstrual syndrome

CONTRAINDICATIONS:

Hypersensitivity, anuria, severe renal disease, hyperkalemia

Precautions: Breastfeeding, dehydration, hepatic disease, renal impairment, electrolyte imbalances, metabolic acidosis, gynecomastia, pregnancy (C)

Black Box Warning: Secondary malignancy

DOSAGE AND ROUTES

Edema/hypertension

- **Adult: PO** 25-200 mg/day in 1-2 divided doses

CHF

- **Adult: PO** 12.5-25 mg/day; max 50 mg/day

Edema

- **Child: PO** 1.5-3.3 mg/kg/day as single dose or in divided doses

Hypertension

- **Child (unlabeled): PO** 1.5-3.3 mg/kg/day in divided doses

Hypokalemia

- **Adult: PO** 25-100 mg/day; if **PO**, potassium supplements must not be used

Primary hyperaldosteronism diagnosis

- **Adult: PO** 400 mg/day × 4 days or 4 wk depending on test, then 100-400 mg/day maintenance

Edema (nephrotic syndrome, CHF, hepatic disease)

- **Adult: PO** 100 mg/day given as single dose or in divided doses, titrate to response
- **Child: PO** 1.5-3.3 mg/kg/day or 60 mg/m^2/day given daily or in 2-4 divided doses

Renal dose

- **Adult: PO** CCr 10-50 ml/min; give dose q12-24hr; CCr <10 ml/min, avoid use

Polycystic ovary syndrome/hirsutism in women (unlabeled)

- **Adult: PO** 50-200 mg in 1-2 divided doses

Available forms: Tabs 25, 50, 100 mg

Administer:

- In AM to avoid interference with sleep
- With food; if nausea occurs, absorption may be decreased slightly
- Effect may take 2 wk

SIDE EFFECTS

CNS: *Headache,* confusion, drowsiness, lethargy, ataxia

ELECT: Hyperchloremic metabolic acidosis, hyperkalemia, hyponatremia

ENDO: Impotence, gynecomastia, irregular menses, amenorrhea, postmenopausal bleeding, hirsutism, deepening voice, breast pain

GI: *Diarrhea,* cramps, bleeding, gastritis, *vomiting,* anorexia, nausea, hepatocellular toxicity

HEMA: Agranulocytosis

INTEG: *Rash, pruritus,* urticaria

PHARMACOKINETICS

Onset 24-48 hr, peak 48-72 hr, metabolized in liver, excreted in urine, crosses placenta, protein binding >90%, terminal half-life 10-35 hr

INTERACTIONS

Increase: action of antihypertensives, digoxin, lithium

Increase: hyperchloremic acidosis in cirrhosis—cholestyramine

Increase: hyperkalemia—potassium-sparing diuretics, potassium products, ACE inhibitors, salt substitutes

Decrease: effect of anticoagulants, monitor INR/PT

Decrease: effect of spironolactone—ASA, NSAIDs

Drug/Food

Increase: hyperkalemia—potassium rich foods, potassium salt substitutes

Drug/Herb

Increase: hypotension—hawthorn, horse chestnut

Decrease: antihypertensive effect—ephedra

Increase: severe photosensitivity—St. John's wort

Increase: BUN, potassium

Decrease: sodium, magnesium

Drug/Lab Test

Interference: 17-OHCS, 17-KS, radioimmunoassay, digoxin assay

NURSING CONSIDERATIONS

Assess:

- **Hypokalemia:** polyuria, polydipsia; dysrhythmias, inluding a U wave on ECG
- **Hyperkalemia:** weakness, fatigue, dyspnea, dysrhythmias, confusion, fatigue
- Electrolytes: sodium, chloride, potassium, BUN, serum creatinine, ABGs, CBC
- Weight, I&O daily to determine fluid loss; effect of product may be decreased if used daily; ECG periodically with long-term therapy
- Signs of metabolic acidosis: drowsiness, restlessness

S

• Confusion, especially in geriatric patients; take safety precautions if needed
• **Hydration:** skin turgor, thirst, dry mucous membranes

Evaluate:
• Therapeutic response: improvement in edema of feet, legs, sacral area daily if medication is being used in CHF

Teach patient/family:
• To avoid foods with high potassium content: oranges, bananas, salt substitutes, dried apricots, dates; to avoid potassium salt substitutes
• That drowsiness, ataxia, mental confusion may occur; to observe caution when driving
• To notify prescriber of cramps, diarrhea, lethargy, thirst, headache, skin rash, menstrual abnormalities, deepening voice, breast enlargement
• To take in AM, to prevent sleeplessness
• To avoid hazardous activities until reaction is known
• To notify prescriber if pregnancy is planned or suspected, pregnancy (C); not to breastfeed

TREATMENT OF OVERDOSE:
Lavage if taken orally; monitor electrolytes, administer IV fluids, monitor hydration, renal, CV status

stavudine d4t (Rx)
(sta′vyoo-deen)
Zerit
Func. class.: Antiretroviral
Chem. class.: Nucleoside reverse transcriptase inhibitor (NRTI)

ACTION: Prevents replication of HIV by the inhibition of the enzyme reverse transcriptase; causes DNA chain termination

USES: Treatment of HIV-1 in combination with other antiretrovirals

CONTRAINDICATIONS: Hypersensitivity to this product or zidovudine; didanosine, zalcitabine; severe peripheral neuropathy

Black Box Warning: Lactic acidosis

Precautions: Breastfeeding, advanced HIV infection, bone marrow suppression, renal disease, peripheral neuropathy, osteoporosis, obesity

Black Box Warning: Pregnancy (C), hepatic disease, pancreatitis

DOSAGE AND ROUTES
• **Adult >60 kg: PO** 40 mg q12hr
• **Adult <60 kg: PO** 30 mg q12hr
• **Child <30 kg: PO** 1 mg/kg q12hr
• **Child ≥30 kg, ≤60 kg: PO** 30 mg q12hr
• **Child >60 kg: PO** 40 mg q12hr

Renal dose
• **Adult: PO** CCr 26-50 ml/min, reduce by 50%, give q12hr; CCr 10-25 ml/min, reduce by 50%, give q24hr

Available forms: Caps 15, 20, 30, 40 mg; powder for oral sol 1 mg/ml

Administer:
• With/without meals; absorption does not appear to be lowered when taken with food
• Use after hemodialysis
• Every 12 hr around the clock
• Shake suspension well before using

SIDE EFFECTS
CNS: *Peripheral neuropathy,* insomnia, anxiety, depression, dizziness, confusion, *headache,* chills/fever, malaise, neuropathy
CV: Chest pain, vasodilation, hypertension
EENT: Conjunctivitis, abnormal vision
GI: Hepatotoxicity, *diarrhea, nausea, vomiting,* anorexia, dyspepsia, constipation, stomatitis, pancreatitis
HEMA: Bone marrow suppression, leukopenia, macrocytosis
INTEG: *Rash,* sweating, pruritus, benign neoplasms
MISC: Lactic acidosis, asthenia, lipodystrophy

MS: Myalgia, arthralgia
RESP: Dyspnea, pneumonia, asthma

PHARMACOKINETICS

Excreted in urine, breast milk; peak 1 hr; half-life: elimination 1-1.6 hr

INTERACTIONS

Increase: myelosuppression—other myelosuppressants
Increase: peripheral neuropathy—lithium, dapsone, chloramphenicol didanosine, ethambutol, hydrALAZINE, phenytoin, vinCRIStine, zalcitabine
Increase: stavudine levels—probenecid
Decrease: stavudine effect—methadone, zidovudine

NURSING CONSIDERATIONS

Assess:

Black Box Warning: **Lactic acidosis and severe hepatomegaly with steatosis:** death may result

⚠ **Pancreatitis:** severe upper abdominal pain, radiating to back, nausea, vomiting usually when used in combination with didanosine

- Blood studies: WBC, differential, RBC, Hct, Hgb, platelets, serum amylase, lipase
- Renal tests: urinalysis, protein, blood, serum creatinine
- C&S before product therapy; product may be given as soon as culture taken
- Bowel pattern before, during treatment
- Weakness, tremors, confusion, dizziness; product may have to be decreased, discontinued
- Viral load, CD4 counts, plasma HIV RNA at baseline and throughout treatment
- **Peripheral neuropathy:** tingling, pain in extremities; discontinue product

Evaluate:

- Therapeutic response: decreased symptoms of HIV

Teach patient/family:

- **About the signs of peripheral neuropathy:** burning, weakness, pain, prickling feeling in extremities
- That product should not be given with antineoplastics
- That product is not a cure for AIDS but will control symptoms
- To notify prescriber if sore throat, swollen lymph nodes, malaise, fever occur; that other products may be needed to prevent other infections
- That, even with use of product, patient may pass AIDS virus to others
- That follow-up visits are necessary; that serious toxicity may occur; that blood counts must be done q2wk

⚠ That serious product interactions may occur if other medications are ingested; to see prescriber before taking chloramphenicol, dapsone, CISplatin, didanosine, ethambutol, lithium, antifungals, antineoplastics

Black Box Warning: To notify prescriber if pregnancy is planned or suspected, fatal lactic acidosis may occur, pregnancy (C), avoid breastfeeding

- That product may cause fainting or dizziness

⚠ HIGH ALERT

succinylcholine (Rx)

(suk-sin-ill-koe′leen)

Anectine, Quelicin

Func. class.: Neuromuscular blocker (depolarizing, ultra short)

ACTION: Inhibits transmission of nerve impulses by binding with cholinergic receptor sites, thus antagonizing action of acetylcholine; causes release of histamine

USES: Facilitation of endotracheal intubation, skeletal muscle relaxation during orthopedic manipulations

S

CONTRAINDICATIONS:

Hypersensitivity, malignant hyperthermia, trauma

Precautions: Pregnancy (C), breastfeeding, geriatric or debilitated patients, cardiac disease, severe burns, fractures (fasciculations may increase damage) electrolyte imbalances, dehydration, neuromuscular/respiratory/cardiac/renal/hepatic disease, collagen diseases, glaucoma, eye surgery

Black Box Warning: Children <2 yr, hyperkalemia, myopathy, rhabdomyolysis

DOSAGE AND ROUTES

- **Adult:** **IV** 0.3-1.1 mg/kg, max 150 mg, maintenance 0.04-0.07 mg/kg q5-10min as needed; **CONT IV INF** dilute to conc of 1-2 mg/ml in D_5W or NS 10-100 mcg/kg/min
- **Child:** **IV** initially 1-2 mg/kg; **CONT IV INF** not recommended

Available forms: Inj 20, 50, 100 mg/ml; powder for inj 100, 500 mg/vial, 1 g/vial

Administer:

- Give IV or IM; only experienced clinicians familiar with the use of neuromuscular blocking drugs should administer or supervise the use of this product
- Visually inspect parenteral products for particulate matter and discoloration before use
- Monitor heart rate and mechanical ventilator status during use
- Store in refrigerator, powder at room temperature; close tightly

IM route

- Recommended for infants and other patients in whom a suitable vein is not accessible
- Inject into a large muscle, preferably high into the deltoid muscle; aspirate before injection

Rapid IV injection route

- Owing to tachyphylaxis and prolonged apnea, this method is not recommended for prolonged procedures; rapid IV injection of succinylcholine can result in profound bradycardia or asystole in pediatric patients; as with adults, the risk increases with repeated doses; pretreatment with atropine may be needed
- No dilution of injection solution is necessary
- Inject rapidly IV over 10-30 sec

Continuous IV inf route

- Not recommended for infants and children owing to risk of malignant hyperthermia
- This route is preferred for long surgical procedures owing to possible tachyphylaxis and prolonged apnea associated with administration of repeated fractional doses
- Dilute succinylcholine to a conc of 1-2 mg/ml with D_5W, D_5NS, NS, or 1/6 M sodium lactate injection; one gram of the powder for injection or 20 ml of a 50-mg/ml solution may be added to 1 L or 500 ml of diluent to give solutions containing 1 or 2 mg/ml, respectively; alternatively, 500 mg of the powder for injection or 10 ml of a 50 mg/ml solution may be added to 500 ml or 250 ml of diluent to give solutions containing 1 or 2 mg/ml, respectively
- Infuse IV at a rate of 2.5 mg/minute (range = 0.5-10 mg/min); adjust rate based on patient's response and requirements

Additive compatibilities: Amikacin, cephapirin, isoproterenol, meperidine, methyldopate, morphine, norepinephrine, scopolamine

Syringe compatibilities: Heparin

Y-site compatibilities: Etomidate, heparin, potassium chloride, propofol, vit B/C

SIDE EFFECTS

CV: Bradycardia, tachycardia; increased, decreased B/P; sinus arrest, dysrhythmias, edema

EENT: Increased secretions, intraocular pressure

HEMA: Myoglobulinemia

INTEG: Rash, flushing, pruritus, urticaria

MS: Weakness, muscle pain, fasciculations, prolonged relaxation, myalgia, rhabdomyolysis

RESP: Prolonged apnea, bronchospasm, cyanosis, respiratory depression, wheezing, dyspnea
SYST: Anaphylaxis, angioedema

PHARMACOKINETICS

Hydrolyzed in blood, excreted in urine (active/inactive metabolites)
IM: Onset 2-3 min, duration 10-30 min
IV: Onset 1 min, peak 2-3 min, duration 6-10 min

INTERACTIONS

Increase: dysrhythmias: theophylline
Increase: neuromuscular blockade—aminoglycosides, β-blockers, cardiac glycosides, clindamycin, lincomycin, procainamide, quiNIDine, local anesthetics, polymyxin antibiotics, lithium, opioids, thiazides, enflurane, isoflurane, magnesium salts, oxytocin
Drug/Herb
- Blocks succinylcholine: melatonin

NURSING CONSIDERATIONS

Assess:
- Electrolyte imbalances (potassium, magnesium); may lead to increased action of product
- VS (B/P, pulse, respirations, airway) until fully recovered; rate, depth, pattern of respirations, strength of hand grip
- I&O ratio; check for urinary retention, frequency, hesitancy
- **Recovery:** decreased paralysis of face, diaphragm, leg, arm, rest of body
- **Allergic reactions:** rash, fever, respiratory distress, pruritus; product should be discontinued
- **Myopathy, rhabdomyolysis:** in pediatric patients (rare)
- Reassurance if communication is difficult during recovery from neuromuscular blockade; postoperative stiffness is normal, soon subsides

Evaluate:
- Therapeutic response: paralysis of jaw, eyelid, head, neck, rest of body

TREATMENT OF OVERDOSE:

Neostigmine, atropine, monitor VS; may require mechanical ventilation

sucralfate (Rx)

(soo-kral′fate)

Carafate, Sulcrate ♣

Func. class.: Protectant, antiulcer
Chem. class.: Aluminum hydroxide, sulfated sucrose

Do not confuse:
Carafate/Cafergot

ACTION: Forms a complex that adheres to ulcer site, adsorbs pepsin

USES: Duodenal ulcer, oral mucositis, stomatitis after radiation of head and neck
Unlabeled uses: Gastric/aphthous ulcers, gastroesophageal reflux, NSAID-induced ulcer prophylaxis, proctitis, stomatitis, stress gastritis prophylaxis, *C. difficile*

CONTRAINDICATIONS: Hypersensitivity
Precautions: Pregnancy (B), breastfeeding, children, renal failure; hypoglycemia (diabetics)

DOSAGE AND ROUTES

Duodenal ulcers
- **Adult: PO** 1 g qid 1 hr before meals, at bedtime
- **Child: PO** 40-80 mg/kg/day divided

Aphthous ulcer/stomatitis (unlabeled)
- **Adult: PO** 5-10 ml (500 mg-1 g) swished in mouth for several min; spit or swallow qid

Gastric ulcer/NSAID-induced ulcer prophylaxis/esophagitis/GERD (unlabeled)
- **Adult: PO** 1 g qid, 1 hr before meals and at bedtime

Available forms: Tabs 1 g; oral susp 1 g/10 ml

S

Administer:

PO route

• Do not crush or chew tabs; tabs may be broken or dissolved in water

• Do not take antacids 30 min before or after sucralfate

• On an empty stomach 1 hr before meals or other medications and at bedtime

• Store at room temperature

SIDE EFFECTS

CNS: Drowsiness, dizziness

ENDO: Hyperglycemia (diabetes mellitus)

GI: *Dry mouth, constipation,* nausea, gastric pain, vomiting, bezoar (for critically ill patients)

INTEG: Urticaria, rash, pruritus

PHARMACOKINETICS

PO: Duration up to 6 hr

INTERACTIONS

Decrease: action of tetracyclines, phenytoin, fat-soluble vitamins, digoxin, ketoconazole, theophylline

Decrease: absorption of fluoroquinolones

Decrease: absorption of sucralfate—antacids, cimetidine, ranitidine

NURSING CONSIDERATIONS

Assess:

• **GI symptoms:** abdominal pain, blood in stools

• **Hypoglycemia:** may occur in patients with diabetes mellitus; monitor blood glucose carefully

Evaluate:

• Therapeutic response: absence of pain, GI complaints

Teach patient/family:

• To take on empty stomach

• To take full course of therapy; not to use for >8 wk; to avoid smoking

• To avoid antacids within ½ hr of product

• To increase fluids, bulk, exercise to lessen constipation

sulfamethoxazole-trimethoprim (Rx)

(trye-meth′oh-prim–sul-fa-meth-ox′a-zole)

Bacter-Aid DS, Bactrim DS, Novo-Trimel ✦, Nu-Cotrimox ✦, Septra, Septra DS, Sultrex, SMZ/TMP

Func. class.: Antiinfective

Chem. class.: Sulfonamide—miscellaneous

ACTION:

Sulfamethoxazole (SMZ) interferes with the bacterial biosynthesis of proteins by competitive antagonism of PABA when adequate levels are maintained; trimethoprim (TMP) blocks the synthesis of tetrahydrofolic acid; the combination blocks 2 consecutive steps in the bacterial synthesis of essential nucleic acids and protein

USES:

UTI, otitis media, acute and chronic prostatitis, shigellosis, chancroid, traveler's diarrhea, *Enterobacter* sp., *Escherichia coli, Haemophilus influenzae (beta-lactamase negative), Haemophilus influenzae (beta-lactamase positive), Klebsiella* sp., *Morganella morganii, Pneumocystis carinii, Pneumocystis jiroveci, Proteus mirabilis, Proteus* sp., *Proteus vulgaris, Shigella flexneri, Shigella sonnei, Streptococcus pneumonia;* **may also be effective for** *Acinetobacter baumannii, Actinomadura madurae, Actinomadura pelletierii, Bordetella pertussis, Burkholderia pseudomallei, Cyclospora cayetanensis, Haemophilus ducreyi, Isospora belli, Klebsiella granulomatis, Legionella micdadei, Legionella pneumophila, Listeria monocytogenes, Moraxella catarrhalis, Neisseria gonorrhoeae, Nocardia asteroides, Nocardia brasiliensis, Nocardia otitidiscaviarum, Pediculus capitis, Plasmodium falciparum, Providencia* sp.,

Salmonella sp., *Serratia* sp., *Shigella* sp., *Staphylococcus aureus (MRSA)*, *Staphylococcus aureus (MSSA)*, *Staphylococcus epidermidis*, *Stenotrophomonas maltophilia*, *Streptococcus pyogenes* (group A beta-hemolytic streptococci), *Streptomyces somaliensis*, *Toxoplasma gondii*, *Vibrio cholerae*, *Viridans streptococci*, *Yersinia enterocolitica*

CONTRAINDICATIONS:

Breastfeeding, infants <2 mo; hypersensitivity to trimethoprim or sulfonamides; pregnancy at term, megaloblastic anemia, CCr <15 ml/min

Precautions: Pregnancy (C), geriatric patients, infants, renal disease, G6PD deficiency, impaired hepatic/renal function, possible folate deficiency, severe allergy, bronchial asthma, UV exposure, porphyria, hyperkalemia, hypothyroidism

DOSAGE AND ROUTES

Based on TMP content

UTI

- **Adult: PO** 160 mg TMP q12hr × 10-14 days
- **Child: PO** 8 mg/kg TMP/day in 2 divided doses q12hr (treatment): 2 mg/kg/day (prophylaxis)

Otitis media

- **Child: PO** 8 mg/kg TMP/day in 2 divided doses q12hr × 10 days

Chronic bronchitis

- **Adult: PO** 160 mg TMP q12hr × 10-14 days

***Pneumocystis jiroveci* pneumonitis**

- **Adult and child: PO** 15-20 mg/kg TMP daily in 4 divided doses q6hr × 14-21 days; **IV** 15-20 mg/kg/day (based on TMP) in 3-4 divided doses for ≤14 days
- Dosage reduction necessary in moderate to severe renal impairment (CCr <30 ml/min)

Available forms: Tabs 80 mg trimethoprim/400 mg sulfamethoxazole, 160 mg trimethoprim/800 mg sulfamethoxazole; susp 200 mg-40 mg/5 ml, 800 mg-160 mg/20 ml; IV 16 mg/80 mg/ml

Administer:

PO route

- Medication after C&S; repeat C&S after full course of medication
- With resuscitative equipment, EPINEPHrine available; severe allergic reactions may occur
- Without regard to meals
- With full glass of water to maintain adequate hydration; increase fluids to 2 L/day to decrease crystallization in kidneys
- Store in tight, light-resistant container at room temperature

Intermittent IV INF route

- After diluting 5 ml of product/125 ml D_5W, run over 1-1½ hr, if using Septra ADD-Vantage vials dilute each 10 ml vial in ADD-Vantage diluent containers containing 250 ml of D5W, infuse over 60-90 min, change site q48-72hr

Y-site compatibilities: Acyclovir, aldesleukin, allopurinol, amifostine, amphotericin B cholesteryl, atracurium, aztreonam, cefepime, cyclophosphamide, diltiazem, DOXOrubicin liposome, enalaprilat, esmolol, filgrastim, fludarabine, gallium, granisetron, HYDROmorphone, labetalol, LORazepam, magnesium sulfate, melphalan, meperidine, morphine, pancuronium, perphenazine, piperacillin/tazobactam, remifentanil, sargramostim, tacrolimus, teniposide, thiotepa, vecuronium, zidovudine

SIDE EFFECTS

CNS: Headache, insomnia, hallucinations, depression, vertigo, fatigue, anxiety, **seizures, product fever**, chills, **aseptic meningitis**

CV: Allergic myocarditis

EENT: Tinnitus

GI: *Nausea, vomiting, abdominal pain*, stomatitis, **hepatitis**, glossitis, pancreatitis, diarrhea, **enterocolitis**, anorexia, **pseudomembranous colitis**

GU: Renal failure, toxic nephrosis; increased BUN, creatinine; crystalluria

HEMA: Leukopenia, neutropenia, thrombocytopenia, agranulocytosis, hemolytic

anemia, hypoprothrombinemia, Henoch-Schönlein purpura, methemoglobinemia, eosinophilia I

INTEG: Rash, dermatitis, urticaria, Stevens-Johnson syndrome, erythema, photosensitivity, pain, inflammation at inj site, toxic epidermal necrolysis, erythema multiforme

RESP: Cough, SOB

SYST: Anaphylaxis, SLE

PHARMACOKINETICS

PO: Rapidly absorbed; peak 1-4 hr; half-life 8-13 hr; excreted in urine (metabolites and unchanged), breast milk; crosses placenta; 68% bound to plasma proteins; TMP achieves high levels in prostatic tissue and fluid

INTERACTIONS

Increase: thrombocytopenia—thiazide diuretics

Increase: potassium levels—potassium-sparing diuretics, potassium supplements

Increase: hypoglycemic response—sulfonylurea agents

Increase: anticoagulant effects—oral anticoagulants

Increase: levels of dofetilide

Increase: crystalluria—methenamine

Increase: bone marrow depressant effects—methotrexate

Decrease: hepatic clearance of phenytoin, CYP2C9, CYP3A4 inducers

Decrease: response—cycloSPORINE

Drug/Lab Test

Increase: creatinine, bilirubin

Decrease: Hgb, platelets

NURSING CONSIDERATIONS

Assess:

- I&O ratio; note color, character, pH of urine if product administered for UTI; output should be 800 ml less than intake; if urine is highly acidic, alkalization may be needed
- Renal studies: BUN, creatinine, urinalysis with long-term therapy
- Type of infection; obtain C&S before starting therapy
- Blood dyscrasias, skin rash, fever, sore throat, bruising, bleeding, fatigue, joint pain
- **Allergic reaction:** rash, dermatitis, urticaria, pruritus, dyspnea, bronchospasm; product should be discontinued at 1st sign of rash; AIDS patients more susceptible

Evaluate:

- Therapeutic response: absence of pain, fever; C&S negative

Teach patient/family:

- To take each oral dose with full glass of water to prevent crystalluria; to drink 8-10 glasses of water/day; to take product on an empty stomach 1 hr before meals, 2 hr after meals
- To complete full course of treatment to prevent superinfection
- To avoid sunlight; to use sunscreen to prevent burns
- To avoid OTC medications (aspirin, vit C) unless directed by prescriber
- To use alternative contraceptive measures; that decreased effectiveness of oral contraceptives may occur
- To notify prescriber if skin rash, sore throat, fever, mouth sores, unusual bruising, bleeding occur; to notify prescriber of CNS effects: anxiety, depression, hallucinations, seizures

sulfaSALAzine (Rx)

(sul-fa-sal′a-zeen)

Azulfidine, Azulfidine EN-tabs, Salazopyrin ✦

Func. class.: GI antiinflammatory, antirheumatic (DMARD)

Chem. class.: Sulfonamide

Do not confuse:

sulfaSALAzine/sulfiSOXAZOLE

ACTION: Prodrug to deliver sulfapyridine and 5-aminosalicylic acid to colon; antiinflammatory in connective tissue also

USES:
Ulcerative colitis; RA; juvenile RA (Azulfidine EN-tabs)
Unlabeled uses: Crohn's disease

CONTRAINDICATIONS:
Pregnancy at term, children <2 yr; hypersensitivity to sulfonamides or salicylates; intestinal, urinary obstruction; porphyria
Precautions: Pregnancy (B), breastfeeding, impaired renal/hepatic function, severe allergy, bronchial asthma, megaloblastic anemia

DOSAGE AND ROUTES

Bowel disease
- **Adult: PO** 3-4 g/day in divided doses; maintenance 2 g/day in divided doses q6hr
- **Child ≥6 yr: PO** 40-60 mg/kg/day in 4-6 divided doses then 30 mg/kg/day in 4 doses, max 2 g/day

Rheumatoid arthritis
- **Adult: PO** 0.5-1 g/day then increase daily dose by 500 mg/wk to 2 g/day in 2-3 divided doses

Juvenile rheumatoid arthritis
- **Child ≥6 yr: PO** 30-50 mg/kg/24 hr in 2 divided doses

Renal dose
- Modify dose based on renal impairment, response

Crohn's disease (unlabeled)
- **Adult: PO** 1 g/15 kg, max 5 g/day

Available forms: Tabs 500 mg; oral susp 250 mg/5 ml; del rel tabs 500 mg
Administer:
- Do not break, crush, chew del rel tabs
- With full glass of water to maintain adequate hydration; increase fluids to 2 L/day to decrease crystallization in kidneys
- Total daily dose in evenly spaced doses and after meals to help minimize GI intolerance
- Store in tight, light-resistant container at room temperature

SIDE EFFECTS
CNS: Headache, confusion, insomnia, hallucinations, depression, vertigo, fatigue, anxiety, seizures, product fever, chills
CV: Allergic myocarditis
GI: *Nausea, vomiting, abdominal pain,* stomatitis, hepatitis, glossitis, pancreatitis, diarrhea
GU: Renal failure, toxic nephrosis, increased BUN, creatinine, crystalluria
HEMA: Leukopenia, neutropenia, thrombocytopenia, agranulocytosis, hemolytic anemia
INTEG: Rash, dermatitis, urticaria, Stevens-Johnson syndrome, erythema, photosensitivity
SYST: Anaphylaxis

PHARMACOKINETICS
PO: Partially absorbed; peak 1½-6 hr; duration 6-12 hr; half-life 6 hr; excreted in urine as sulfaSALAzine (15%), sulfapyridine (60%), 5-aminosalicylic acid, metabolites (20%-33%); excreted in breast milk; crosses placenta

INTERACTIONS
Increase: leukopenia risk—thiopurines (azaTHIOprine, mercaptopurine)
Increase: hypoglycemic response—oral hypoglycemics
Increase: anticoagulant effects—oral anticoagulants
Decrease: effect of cycloSPORINE, digoxin, folic acid
Decrease: renal excretion of methotrexate
Drug/Food
Decrease: iron/folic acid absorption
Drug/Lab Test
False positive: urinary glucose test

NURSING CONSIDERATIONS
Assess:
- Renal studies: BUN, creatinine, urinalysis (long-term therapy)

⚠ **Blood dyscrasias:** skin rash, fever, sore throat, bruising, bleeding, fatigue, joint pain; monitor CBC before therapy and q3mo

S

⚠ **Allergic reaction:** rash, dermatitis, urticaria, pruritus, dyspnea, bronchospasm

- **Ulcerative colitis, proctitis, other inflammatory bowel disease:** character, amount consistency of stools; abdominal pain, cramping, blood, mucous
- **Rheumatoid arthritis:** assess mobility, joint swelling, pain, ability to complete activities of daily living

Evaluate:

- Therapeutic response: absence of fever, mucus in stools, pain in joints

Teach patient/family:

- To take each oral dose with full glass of water to prevent crystalluria
- That contact lenses, urine, skin may be yellow-orange
- To avoid sunlight or use sunscreen to prevent burns
- To notify prescriber of skin rash, sore throat, fever, mouth sores, unusual bruising, bleeding

SUMAtriptan (Rx)

(soo-ma-trip′tan)

ALSUMA Auto-injector, Imitrex, Sumavel DosePro, Zecuity

Func. class.: Antimigraine agent

Chem. class.: 5-HT_1B/D receptor agonist, abortive agent, triptan

Do not confuse:
SUMAtriptan/somatropin

ACTION: Binds selectively to the vascular 5-HT_1B/D receptor subtype; exerts antimigraine effect; causes vasoconstriction in cranial arteries

USES: Acute treatment of migraine with/without aura and cluster headache

CONTRAINDICATIONS: Angina pectoris, history of MI, documented silent ischemia, Prinzmetal's angina, ischemic heart disease, IV use, concurrent ergotamine-containing preparations, uncontrolled hypertension, hypersensitivity, basilar or hemiplegic migraine

Precautions: Pregnancy (C), breastfeeding, children <18 yr, geriatric patients, postmenopausal women, men >40 yr, risk factors for CAD, hypercholesterolemia, obesity, diabetes, impaired renal/hepatic function, overuse

DOSAGE AND ROUTES

- **Adult: SUBCUT** ≤6 mg; may repeat in 1 hr; max 12 mg/24 hr; **PO** 25 mg with fluids, if no relief in 2 hr, give another dose, max 200 mg/day; **NASAL** single dose of 5, 10, or 20 mg in 1 nostril, may repeat in 2 hr, max 40 mg/24 hr; 1 puff each nostril q2hr; TD 1 patch (6.5 mg/4 hr) after application push activation button

Hepatic dose

- **Adult: PO** 25 mg; if no response after 2 hr, give ≤50 mg

Available forms: Inj 4, 6 mg/0.5 ml; tabs 25, 50, 100 mg; nasal spray 5 mg/100 mcl-U; dose spray device 20 mg/100 mcl-U; transdermal patch 6.5 mg/4 hr

Administer:

PO route

- Swallow tabs whole; do not break, crush, or chew
- Take tabs with fluids as soon as symptoms appear; may take a 2nd dose >4 hr; max 200 mg/24 hr

SUBCUT route

SUBCUT only just below the skin; avoid IM or IV administration; use only for actual migraine attack

- Give 1st dose supervised by medical staff to patients with coronary artery disease or those at risk for CAD

Nasal route

- May give as 2 sprays of 5 mg in 1 nostril or 1 spray in each nostril (10 mg)

Transdermal route

- Do not cut; apply to upper arm or thigh of dry intact skin; do not use over scars, tattoos, cuts, scratches, burns, abrasions
- Apply another patch if headache is not

relieved ≥ 2 hr after 1st patch; push activation button within 15 min of applying or patch will not work; do not bathe, shower, swim; may be taped with medical tape if needed
- Do not use with MRI
- Remove slowly; cleanse with soap and water; may cause redness
- Dispose of after folding in half

SIDE EFFECTS

CNS: *Tingling, hot sensation, burning, feeling of pressure, tightness, numbness, dizziness, sedation,* headache, anxiety, fatigue, cold sensation
CV: *Flushing,* **MI**, hypo/hypertension
EENT: Throat, mouth, nasal discomfort; vision changes
GI: Abdominal discomfort
INTEG: Inj site reaction, sweating
MS: *Weakness, neck stiffness,* myalgia
RESP: Chest tightness, pressure

PHARMACOKINETICS

Onset of pain relief 10 min-2 hr, peak 10-20 min; 10%-20% plasma protein binding; metabolized in liver (metabolite); excreted in urine, feces; nasal spray half-life 2 hr

INTERACTIONS

Increase: vasospastic effects: ergot, ergot derivatives
Increase: serotonin syndrome—SSRIs, SNRIs, serotonin-receptor agonists, sibutramine
Increase: SUMAtriptan effect—MAOIs
Drug/Herb
- Serotonin syndrome: SAM-e, St. John's wort

NURSING CONSIDERATIONS

Assess:
- **Migraine:** type of pain, aura; alleviating, aggravating factors; sensitivity to light, noise

⚠ **Serotonin syndrome:** delirium, coma, agitation, diaphoresis, hypertension, fever, tremors; may resemble neuroleptic malignant syndrome in patients taking SSRIs, SNRIs
- B/P; signs, symptoms of coronary vasospasms, ECG
- Tingling, hot sensation, burning, feeling of pressure, numbness, flushing, inj site reaction
- Stress level, activity, recreation, coping mechanisms
- Neurologic status: LOC, blurring vision, nausea, vomiting, tingling in extremities preceding headache
- Ingestion of tyramine foods (pickled products, beer, wine, aged cheese), food additives, preservatives, colorings, artificial sweeteners, chocolate, caffeine, which may precipitate these types of headaches
- Renal function, urinary output
- Quiet, calm environment with decreased stimuli: noise, bright light, excessive talking

Evaluate:
- Therapeutic response: decrease in frequency, severity of migraine

Teach patient/family:
- To report chest pain, tightness; sudden, severe abdominal pain; swelling of eyelids, face, lips; skin rash to prescriber immediately
- To notify prescriber if pregnancy is planned or suspected; to use contraception while taking product
- **Nasal spray:** to use 1 spray in 1 nostril; may repeat if headache returns; not to repeat if pain continues after 1st dose
- To have a dark, quiet environment
- To avoid hazardous activities if dizziness, drowsiness occur
- To avoid alcohol; may increase headache
- To use SUBCUT inj technique, nasal route if prescribed
- That product does not reduce number of migraines; to be used for acute migraine; to use as symptoms occur

SUNItinib (Rx)

(soo-nit′-in-ib)

Sutent

Func. class.: Antineoplastic—miscellaneous

Chem. class.: Protein-tyrosine kinase inhibitor

ACTION: Inhibits multiple receptor tyrosine kinases (RTKs); some are responsible for tumor growth

USES: Gastrointesitnal stromal tumors (GIST) after disease progression or intolerance to imatinib; advanced renal carcinoma, pancreatic neuroendocrine tumors (pNET) in patients with unresectable locally advanced/metastatic disease

CONTRAINDICATIONS: Pregnancy (D), breastfeeding, hypersensitivity

Precautions: Children, geriatric patients, active infections, QT prolongation, torsades de pointes, stroke, heart failure

Black Box Warning: Hepatic disease

DOSAGE AND ROUTES

Gastrointestinal stromal tumors (GIST)/renal cell cancer

• **Adult: PO** 50 mg/day × 4 wk then 2 wk off; may increase or decrease dose by 12.5 mg; if administered with CYP3A4 inducers, give 87.5 mg/day; if given with CYP3A4 inhibitors, give 37.5 mg/day

Pancreatic neuroendocrine (pNET)

• **Adult: PO** 37.5 mg daily continuously, increase or decrease by 12.5 mg based on tolerance, avoid potent CYP3A4 inhibitors/inducers; if used with CYP3A4 inhibitors, decrease SUNItinib dose to minimum of 25 mg/day; if used with CYP3A4 inducers, increase SUNItinib to max 62.5 mg/day

Available forms: Caps 12.5, 25, 50 mg

Administer:

• With meal and large glass of water to decrease GI symptoms

• Store at 25° C (77° F)

SIDE EFFECTS

CNS: CNS hemorrhage, headache, dizziness, insomnia, seizures, fatigue

CV: Hypertension, left ventricular dysfunction, QT prolongation, cardiotoxicity, torsades de pointes, thrombotic microangiopathy, cardiac arrest, thromboembolism

ENDO: Hypo/hyperthyroidism

GI: *Nausea,* hepatotoxicity, vomiting, dyspepsia, *anorexia, abdominal pain,* altered taste, *constipation,* stomatitis, mucositis, pancreatitis, diarrhea, GI bleeding/perforation

GU: Nephrotic syndrome

HEMA: Neutropenia, thrombocytopenia, hemolytic anemia, leukopenia

INTEG: *Rash, yellow skin discoloration,* depigmentation of hair or skin, alopecia, necrotizing fasciitis, pyoderma gangrenosum

MS: Pain, arthralgia, myalgia, myopathy, rhabdomyolysis

RESP: Cough, dyspnea, pulmonary embolism

SYST: Bleeding, electrolyte abnormalities, hand-foot syndrome, serious infection

PHARMACOKINETICS

Protein binding 95%; metabolized by CYP3A4; excreted in feces, small amount in urine; peak plasma levels 6-12 hr; terminal half-life 40-60 hr (SUNItinib); active metabolite 80-110 hr

INTERACTIONS

⚠ **Increase:** microangiopathic hemolytic anemia—bevacizumab; avoid concurrent use

⚠ **Increase:** QT prolongation—class IA/III antidysrhythmics, some phenothiazines, β agonists, local anesthetics, tricyclics, haloperidol, chloroquine, droperidol, pentamidine; CYP3A4 inhibitors

(amiodarone, clarithromycin, erythromycin, telithromycin, troleandomycin), arsenic trioxide, levomethadyl; CYP3A4 substrates (methadone, pimozide, QUEtiapine, quiNIDine, risperiDONE, ziprasidone)
Increase: hepatotoxicity—acetaminophen
Increase: plasma concs of simvastatin, calcium channel blockers; warfarin; avoid use with warfarin, use low-molecular-weight anticoagulants instead
Decrease: SUNItinib concs—dexamethasone, phenytoin, carBAMazepine, rifampin, PHENobarbital

Drug/Herb
Decrease: SUNItinib conc—St. John's wort

Drug/Food
Increase: plasma concs—grapefruit juice

NURSING CONSIDERATIONS

Assess:

⚠ ANC and platelets; if ANC $<1 \times 10^9$/L and/or platelets $<50 \times 10^9$/L, stop until ANC $>1.5 \times 10^9$/L and platelets $>75 \times 10^9$/L; if ANC $<0.5 \times 10^9$/L and/or platelets $<10 \times 10^9$/L, reduce dosage by 200 mg; if cytopenia continues, reduce dosage by another 100 mg; if cytopenia continues for 4 wk, stop product until ANC $\geq 1 \times 10^9$/L

⚠ **CV status:** hypertension, QT prolongation can occur; monitor left ventricular ejection fraction (LVEF), (MUGA) at baseline, periodically; ECG

⚠ **Renal toxicity:** if bilirubin $>3 \times$ IULN, withhold SUNItinib until bilirubin levels return to $<1.5 \times$ IULN; electrolytes

⚠ **Hepatotoxicity:** monitor LFTs before treatment, monthly; if liver transaminases $>5 \times$ IULN, withhold SUNItinib until transaminase levels return to $<2.5 \times$ IULN

- **CHF:** adrenal insufficiency in those experiencing trauma
- Bleeding: epistaxis; rectal, gingival, upper GI, genital, wound bleeding; tumor-related hemorrhage may occur rapidly
- Nutritious diet with iron, vitamin supplement, low fiber, few dairy products

Evaluate:

- Therapeutic response: decrease in size of tumor

Teach patient/family:

- To report adverse reactions immediately: SOB, bleeding
- About reason for treatment, expected result
- That many adverse reactions may occur: high B/P, bleeding, mouth swelling, taste change, skin discoloration, depigmentation of hair/skin
- To avoid persons with known upper respiratory infections; that immunosuppression is common
- To avoid grapefruit juice

⚠ To report if pregnancy is planned or suspected, pregnancy (D)

suprofen ophthalmic

See Appendix B

S

tacrolimus (Rx) (PO, IV)

(tak-row'lim-us)

Prograf

tacrolimus (topical) (Rx)

Protopic

Func. class.: Immunosuppressant

Chem. class.: Macrolide

ACTION: Produces immunosuppression by inhibiting T-lymphocytes

USES: Organ transplants to prevent rejection; **topical:** atopic dermatitis

Unlabeled uses: Severe recalcitrant psoriasis, contact dermatitis, GVHD prophylaxis/disease, pancreas/heart/kidney/liver/lung/small bowel transplant rejection, uveitis, ulcerative colitis, nephrotic syndrome, lichen sclerosus

CONTRAINDICATIONS: Children <2 yr (topical); hypersensitivity to this product or to some kinds of castor oil (IV); long-term use (topical)

Precautions: Pregnancy (C), breastfeeding, severe renal/hepatic disease; diabetes mellitus, hyperkalemia, hyperuricemia, hypertension, acute bronchospasm, African American patients, heart failure, seizures, QT prolongation

Black Box Warning: Children <12, lymphomas, infection, neoplastic disease, neonates, infants, requires a specialized setting, requires an experienced clinician

DOSAGE AND ROUTES

Kidney transplant rejection prophylaxis

- **Adult: IV** 0.03-0.05 mg/kg/day as **CONT INF,** give no sooner than 6 hr after transplantation

Liver transplant rejection prophylaxis

- **Adult: PO** 0.10-0.15 mg/kg/day in 2 divided doses q12hr; give no sooner than 6 hr after transplantation; **IV** 0.03-0.05 mg/kg/day as **CONT INF;** give no sooner than 6 hr after transplantation

Heart transplant rejection prophylaxis

- **Adult: PO** 0.075 mg/kg/day in 2 divided doses q12hr; give no sooner than 6 hr after transplantation; **IV** 0.01 mg/kg/day as **CONT INF;** give no sooner than 6 hr after transplantation

Atopic dermatitis

- **Adult: TOP** use 0.03% or 0.1% ointment; apply bid × 7 days after clearing of signs
- **Child ≥ 2-15 yr: TOP** 0.03% ointment; apply bid × 7 days after clearing of signs

Graft-versus-host disease (Orphan drug)

- **Adult and adolescent: IV** 0.1 mg/kg/day in 2 divided doses given with other immunosuppressants or **PO** 0.3 mg/kg/day in 2 divided doses
- **Child: CONT IV INF** 0.1 mg/kg/day

Graft-versus-host prophylaxis (unlabeled)

- **Adult: CONT IV INF** 0.03 mg/kg/day starting 1-2 days before bone marrow transplant; **PO** 0.12 mg/kg/day in 2 divided doses
- **Adolescent and child: PO** 0.12 mg/kg/day in 2 divided doses

Heart transplant rejection (unlabeled)

- **Adult: IV** 0.05 mg/kg/day or **PO** 0.2-0.3 mg/kg/day in 2 divided doses; adjust to maintain whole blood conc 7-15 ng/ml

Lung transplant rejection (unlabeled)

- **Adult: PO** 0.15 mg/kg/day, maintain 12-hr trough, whole blood conc 1-1.5 ng/ml

Small bowel transplant rejection (unlabeled)

- **Adult: IV** 0.1-0.15 mg/kg/day then **PO** 0.3 mg/kg/day in divided doses

Contact dermatitis (unlabeled)

- **Adult: TOP** 0.1% ointment, apply bid × 8 wk

Available forms: Inj 5 mg/ml; caps 0.5, 1, 5 mg; ointment 0.03%, 0.1%

Administer:

PO route

- Give on empty stomach; food decreases absorption
- For several days before transplant surgery, patients should be placed in protective isolation

Topical route

- Apply thin layers to affected skin only; rub in gently
- Do not use occlusive dressings
- Use on small area of skin
- Topical ointment has risk of developing cancer; use only when other options have failed

Continuous IV INF route

- After diluting in 0.9% NaCl or D_5W to 0.004 to 0.02 mg/ml as continuous inf over 24 hr

Y-site compatibilities: Alemtuzumab, alfentanil, amifostine, amikacin, aminophylline, amiodarone, amphotericin B colloidal, amphotericin B liposome, anidulafungin, argatroban, atracurium, aztreonam, benztropine, bivalirudin, bleomycin, bumetanide, buprenorphine, busulfan, butorphanol, calcium acetate/chloride/gluconate, CARBOplatin, carmustine, caspofungin, ceFAZolin, cefoperazone, cefotaxime, cefoTEtan, cefOXitin, cefTAZidime, ceftizoxime, cefTRIAXone, cefuroxime, chloramphenicol, chlorproMAZINE, cimetidine, ciprofloxacin, cisatracurium, CISplatin, clindamycin, cyclophosphamide, cycloSPORINE, cytarabine, DACTINomycin, DAPTOmycin, dexamethasone, dexmedetomidine, dexrazoxane, digoxin, diltiazem, diphenhydrAMINE, DOBUTamine, DOCEtaxel, dolasetron, DOPamine, doripenem, doxacurium, DOXOrubicin hydrochloride, doxycycline, droperidol, enalaprilat, ePHEDrine, EPINEPHrine, epirubicin, ertapenem, erythromycin, esmolol, etoposide, etoposide phosphate, famotidine, fenoldopam, fentaNYL, fluconazole, fludarabine, foscarnet, fosphenytoin, gemcitabine, gentamicin, glycopyrrolate, granisetron, haloperidol, heparin, hydrALAZINE, hydrocortisone, HYDROmorphone, IDArubicin, ifosfamide, imipenem/cilastatin, inamrinone, insulin, isoproterenol, ketorolac, labetalol, leucovorin, levofloxacin, levorphanol, lidocaine, linezolid, LORazepam, magnesium sulfate, mannitol, mechlorethamine, meperidine, meropenem, mesna, metaraminol, methotrexate, methyldopate, methylPREDNISolone, metoclopramide, metoprolol, metroNIDAZOLE, micafungin, midazolam, milrinone, mitoMYcin, mitoXANtrone, mivacurium, morphine, multivitamins, nafcillin, nalbuphine, naloxone, nesiritide, niCARdipine, nitroglycerin, nitroprusside, norepinephrine, octreotide, ondansetron, oxacillin, oxaliplatin, oxytocin, PACLitaxel, palonosetron, pancuronium, PEMEtrexed, penicillin G, pentamidine, pentazocine, perphenazine, phentolamine, phenylephrine, piperacillin/tazobactam, potassium chloride/phosphates, procainamide, prochlorperazine, promethazine, propranolol, quinapristin/dalfopristin, ranitidine, remifentanil, rocuronium, sodium acetate/bicarbonate/phosphates, streptozocin, succinylcholine, SUFentanil, teniposide, theophylline, thiotepa, ticarcillin/clavulanate, tigecycline, tirofiban, tobramycin, tolazoline, trimethobenzamide, vancomycin, vasopressin, vecuronium, verapamil, vinCRIStine, vinorelbine, voriconazole, zidovudine, zoledronic acid

SIDE EFFECTS

CNS: *Tremors, headache,* insomnia, paresthesia, chills, fever, **seizures, posterior reversible encephalopathy syndrome, BK-virus–associated nephropathy**

CV: Hypertension, myocardial hypertrophy, **prolonged QTc,** cardiomyopathy

EENT: Blurred vision, photophobia

GI: Nausea, vomiting, diarrhea, constipation, **GI bleeding**

GU: UTIs, **albuminuria, hematuria, proteinuria, renal failure, hemolytic uremic syndrome**

T

HEMA: Anemia, leukocytosis, thrombocytopenia, purpura
INTEG: Rash, flushing, itching, alopecia
META: Hirsutism, hyperglycemia, hyperuricemia, hypo/hyperkalemia, hypomagnesemia
MS: Back pain, muscle spasms
RESP: Pleural effusion, atelectasis, dyspnea, interstitial lung disease
SYST: Anaphylaxis, infection, malignancy

PHARMACOKINETICS

PO: Extensively metabolized, half-life 10 hr, 75% protein binding

INTERACTIONS

⚠ **Increase:** QT prolongation—class IA/III antidysrhythmics, some phenothiazines, β agonists, local anesthetics, tricyclics, haloperidol, chloroquine, droperidol, pentamidine; CYP3A4 inhibitors (amiodarone, clarithromycin, erythromycin, telithromycin, troleandomycin), arsenic trioxide, levomethadyl; CYP3A4 substrates (methadone, pimozide, QUEtiapine, quiNIDine, risperiDONE, ziprasidone); do not use together
⚠ **Increase:** toxicity—aminoglycosides, CISplatin, cycloSPORINE
Increase: blood levels—antifungals, calcium channel blockers, cimetidine, danazol, mycophenolate, mofetil
Decrease: blood levels—carBAMazepine, PHENobarbital, phenytoin, rifamycin
Decrease: effect of live-virus vaccines

Drug/Herb
Decrease: immunosuppression—astragalus, echinacea, melatonin
Decrease: effect—ginseng, St. John's wort

Drug/Food
Increase: effect—grapefruit juice
Decreased absorption: food

Drug/Lab Test
Increase: glucose, BUN, creatinine
Increase or decrease: LFTs, potassium
Decrease: magnesium, Hgb, platelets

NURSING CONSIDERATIONS

Assess:
- Blood studies: Hgb, WBC, platelets during treatment monthly; if leukocytes <3000/mm^3 or platelets <100,000/mm^3, product should be discontinued or reduced; decreased hemoglobin level may indicate bone marrow suppression
- Hepatic studies: alk phos, AST, ALT, amylase, bilirubin; for hepatotoxicity: dark urine, jaundice, itching, light-colored stools; product should be discontinued
- Serum creatinine/BUN, serum electrolytes, lipid profile, serum tacrolimus conc

⚠ **Anaphylaxis:** rash, pruritus, wheezing, laryngeal edema; stop inf, initiate emergency procedures
- **QT prolongation:** ECG, ejection fraction; assess for chest pain, palpitations, dyspnea

Evaluate:
- Therapeutic response: absence of graft rejection; immunosuppression in patients with autoimmune disorders

Teach patient/family:

PO route
- To report fever, rash, severe diarrhea, chills, sore throat, fatigue; that serious infections may occur; to report clay-colored stools, cramping **(hepatotoxicity), nephrotoxicity,** signs of diabetes mellitus
- To avoid crowds, persons with known infections to reduce risk for infection; to avoid eating raw shellfish
- To avoid exposure to natural or artificial sunlight
- Not to breastfeed while taking product
- That repeated lab tests will be needed during treatment
- To avoid vaccines
- Not to use with alcohol, grapefruit
- To report if pregnancy is planned or suspected

Black Box Warning: To report symptoms of lymphoma

tadalafil (Rx)

(tah-dal'a-fil)

Adcirca, Cialis

Func. class.: Impotence agent

Chem. class.: Phosphodiesterase type 5 inhibitor

ACTION: Inhibits phosphodiesterase type 5 (PDE5); enhances erectile function by increasing the amount of cGMP, which causes smooth muscle relaxation and increased blood flow into the corpus cavernosum; improves erectile function for up to 36 hr

USES: Treatment of erectile dysfunction, benign prostatic hyperplasia (BPH) with or without erectile dysfunction pulmonary arterial hypertension (PAH) (Adcirca only)

Unlabeled uses: Sexual dysfunction in males receiving antidepressants

CONTRAINDICATIONS: Newborns, children, women, hypersensitivity, patients taking organic nitrates either regularly and/or intermittently, patients taking any α-adrenergic antagonist other than 0.4 mg once-daily tamsulosin

Precautions: Pregnancy (B) although not indicated for females, anatomic penile deformities, sickle cell anemia, leukemia, multiple myeloma, CV/renal/hepatic disease, bleeding disorders, active peptic ulcer, prolonged erection

DOSAGE AND ROUTES

Erectile dysfunction

- **Adult: PO** (Cialis) 10 mg taken before sexual activity; dose may be reduced to 5 mg or increased to max 20 mg; usual max dosing frequency is 1×/day; once-daily dosing 2.5 mg/day at same time each day

BPH

- **Adult: PO** 5 mg qd at the same time every day

Renal dose

- **Adult: PO** CCr 51-80 ml/min: no adjustment for erectile dysfunction, 20 mg/day initially for pulmonary hypertension; CCr 31-50 ml/min, 5 mg/day, max 10 mg q48hr; CCr <30 ml/min, max 5 mg q72hr

Hepatic dose

- **Adult: PO** (Child-Pugh A, B) max 10 mg/day or 20 mg/day, (pulmonary hypertension) max 40 mg/day; (Child-Pugh C) not recommended

Concomitant medications

- Ketoconazole, itraconazole, ritonavir, max 10 mg q72hr

Pulmonary hypertension

- **Adult: PO (Adcirca only)** 40 mg daily
- **Adult taking ritonavir: PO** 20 mg daily initially then increase to 40 mg daily as tolerated

Male sexual dysfunction (from antidepressants) (unlabeled)

- **Adult: PO** (Cialis) 10-20 mg before sexual activity

Available forms: Tabs 2.5, 5, 10, 20 mg; PO tab (Adcirca) 20 mg

Administer:

- Product should not be used with nitrates in any form
- **Sexual dysfunction:** give before sexual activity; do not use more than 1×/ day
- **Pulmonary hypertension:** give Adcirca with/without meals

SIDE EFFECTS

CNS: *Headache, flushing, dizziness,* seizures, transient global amnesia

CV: Hypotension, QT prolongation

INTEG: Stevens-Johnson syndrome, exfoliative dermatitis, urticaria

MISC: Back pain/myalgia, *dyspepsia, nasal congestion, UTI,* blurred vision, changes in color vision, *diarrhea,* pruritus, priapism, nonarteritic ischemic optic neuropathy (NAION), hearing loss

PHARMACOKINETICS

Rapidly absorbed; metabolized by liver by CYP3A4; terminal half-life 17.5 hr; peak ½-6 hr; excreted primarily as metabolites in feces, urine; excreted 61% in feces, 36% in urine; 94% protein bound; rate and extent of absorption not influenced by food

INTERACTIONS

⚠ Do not use with nitrates because of unsafe drop in B/P, which could result in MI or stroke

Increase: tadalafil levels—itraconazole, ketoconazole, ritonavir (although not studied, may also include other HIV protease inhibitors)

Decrease: B/P—alcohol, α-blockers, amLODIPine, angiotensin II receptor blockers, enalapril

Decrease: effects of tadalafil—bosentan, antacids

Drug/Food

Increase: tadalafil effect—grapefruit

NURSING CONSIDERATIONS

Assess:

• **Cialis:** underlying cause of erectile dysfunction before treatment; use of organic nitrates that should not be used with this product; any severe loss of vision while taking this or any similar products

• **Adcirca:** hemodynamic parameters at baseline and periodically

Evaluate:

• Therapeutic response: ability to engage in sexual intercourse, improvement in exercise ability in pulmonary hypertension

Teach patient/family:

• To take 1 hr before sexual activity

• Not to drink large amounts of alcohol

• That product does not protect against sexually transmitted diseases, including HIV

• That product has no effect in the absence of sexual stimulation; to seek medical help if erection lasts >4 hr

• To tell physician about all medicines, vitamins, herbs being taken, especially ritonavir, indinavir, ketoconozole, itraconazole, erythromycin, nitrates, α-blockers; that tadalafil is contraindicated for use with α-blockers except 0.4 mg/day tamsulosin

⚠ To notify prescriber immediately and to stop taking product if vision, hearing loss occur or if erection lasts >4 hr or if chest pain occurs

tamoxifen (Rx)

(ta-mox′i-fen)

Apo-Tamox ♣, Nolvadex ♣, Tamofen ♣

Func. class.: Antineoplastic

Chem. class.: Antiestrogen hormone

ACTION: Inhibits cell division by binding to cytoplasmic estrogen receptors; resembles normal cell complex but inhibits DNA synthesis and estrogen response of target tissue

USES: Advanced breast carcinoma not responsive to other therapy in estrogen-receptor–positive patients (usually postmenopausal), prevention of breast cancer, after breast surgery/radiation for ductal carcinoma in situ

Unlabeled uses: Mastalgia, to reduce pain/size of gynecomastia, ovulation stimulation, malignant carcinoid tumor, carcinoid syndrome, metastatic melanoma, desmoid tumors, McCune-Albright syndrome (female pediatric patients), osteoporosis, bipolar I disorder, infertility, precocious puberty, gynecomastia, mastalgia

CONTRAINDICATIONS: Pregnancy (D), breastfeeding, hypersensitivity

Black Box Warning: Thromboembolic disease

Precautions: Women of childbearing age, leukopenia, thrombocytopenia, cataracts

Black Box Warning: Endometrial cancer, stroke

DOSAGE AND ROUTES

Breast cancer

- **Adult: PO** 20-40 mg/day for 5 yr; doses >20 mg/day, divide AM/PM

High risk for breast cancer

- **Adult: PO** 20 mg/day × 5 yr

DCIS

- **Adult: PO** 20 mg/day × 5 yr

McCune-Albright syndrome/ precocious puberty (unlabeled)

- **Child 2-10 yr (girls): PO** 20 mg/day for ≤1 yr

Bipolar I disorder (unlabeled)

- **Adult: PO** 40 mg bid

Stimulation of ovulation with infertility (unlabeled)

- **Adult: PO** 20-80 mg/day × 5 days

Mastalgia (unlabeled)

- **Adult (female): PO** 10-20 mg/day × 3-6 mo

Mastalgia/gynecomastia in men with prostate cancer (unlabeled)

- **Adult (male): PO** 20 mg/day for ≤1 yr

Available forms: Tabs 10, 20 mg

Administer:

- Do not break, crush, or chew tabs
- Antacid before oral agent; give product after evening meal, before bedtime; give with food or fluids for GI symptoms
- Antiemetic 30-60 min before product to prevent vomiting
- Store in light-resistant container at room temperature

SIDE EFFECTS

CNS: *Hot flashes, headache, lightheadedness,* depression, mood changes

CV: Chest pain, **stroke**, fluid retention, flushing

EENT: Ocular lesions, retinopathy, cataracts, corneal opacity, blurred vision (high doses)

GI: *Nausea, vomiting,* altered taste (anorexia)

GU: Vaginal bleeding, pruritus vulvae, **uterine malignancies**, *altered menses, amenorrhea*

HEMA: **Thrombocytopenia, leukopenia,** DVT

INTEG: *Rash,* alopecia

META: Hypercalcemia

RESP: **Pulmonary embolism**

PHARMACOKINETICS

PO: Peak 4-7 hr, half-life 7 days (1 wk terminal), metabolized in liver, excreted primarily in feces

INTERACTIONS

⚠ **Increase:** **risk for death from breast cancer—PARoxetine**

Increase: bleeding—anticoagulants

Increase: tamoxifen levels—bromocriptine

Increase: thromboembolic events—cytotoxics

⚠ **Increase:** **toxicity—CYP3A4 inhibitors (aprepitant, antiretroviral protease inhibitors, clarithromycin, danazol, delavirdine, diltiazem, erythromycin, fluconazole, FLUoxetine, fluvoxaMINE, imatinib, ketoconazole, mibefradil, nefazodone, telithromycin, voriconazole)**

Decrease: tamoxifen levels—aminoglutethimide, rifamycin

Decrease: letrozole levels—letrozole

Decrease: tamoxifen effect—CYP3A4 inducers (barbiturates, bosentan, carBAMazepine, efavirenz, phenytoins, nevirapine, rifabutin, rifampin)

Decrease: tamoxifen effects—CYP2D6 inhibitors (antidepressants)

Drug/Herb

- Avoid use with St. John's wort, dong qui, black cohosh

Drug/Lab Test

Increase: serum calcium, T_4, AST, ALT, cholesterol, triglycerides

NURSING CONSIDERATIONS

Assess:

- **CBC, differential, platelet count weekly; withhold product if WBC count is <3500 or platelet count is <100,000; notify prescriber; breast exam, mammogram, pregnancy test, bone mineral density, LFTs, serum calcium, serum lipid profile, periodic eye exams (cataracts, retinopathy)**

Black Box Warning: Bleeding q8hr: hematuria, guaiac, bruising, petechiae, mucosa, or orifices

• Effects of alopecia on body image; discuss feelings about body changes

⚠ **Uterine malignancies, symptoms of stroke, pulmonary embolism** that may occur in women with ductal carcinoma in situ (DCIS) and women at high risk for breast cancer

⚠ **Severe allergic reactions:** rash, pruritus, urticaria, purpuric skin lesions, itching, flushing

• Bone pain; may give analgesics; pain usually transient

Evaluate:

• Therapeutic response: decreased tumor size, spread of malignancy

Teach patient/family:

• To use nonhormonal contraception during and for 2 mo after discontinuing treatment, pregnancy (D)

• To notify prescriber of stroke: blurred vision, headache, weakness on one side of the body; PE: chest pain, fainting, sweating, difficulty breathing

• To report any complaints, side effects to prescriber; that use may be 5 yr

• To increase fluids to 2 L/day unless contraindicated

• To wear sunscreen, protective clothing, sunglasses

• That vaginal bleeding, pruritus, hot flashes are reversible after discontinuing treatment

• To immediately report decreased visual acuity, which may be irreversible; about need for routine eye exams; that care providers should be told about tamoxifen therapy

• To report vaginal bleeding immediately

• **Tumor flare**—increase in size of tumor, increased bone pain—may occur and will subside rapidly; may take analgesics for pain

• That premenopausal women must use mechanical birth control because ovulation may be induced

• That hair may be lost during treatment; that a wig or hairpiece may make patient feel better; that new hair may be different in color, texture

tamsulosin (Rx)

(tam-sue-lo′sen)

Flomax

Func. class.: Selective α_1-peripheral adrenergic blocker, BPH agent

Chem. class.: Sulfamoylphenethylamine derivative

Do not confuse:
Flomax/Fosamax/Volmax

ACTION: Binds preferentially to α_{1A}-adrenoceptor subtype, which is located mainly in the prostate

USES: Symptoms of benign prostatic hyperplasia (BPH)

CONTRAINDICATIONS: Hypersensitivity

Precautions: Pregnancy (B), breastfeeding, children, hepatic disease, CAD, severe renal disease, prostate cancer; cataract surgery (floppy iris syndrome)

DOSAGE AND ROUTES

• **Adult: PO** 0.4 mg/day increasing to 0.8 mg/day if required after 2-4 wk

Available forms: Caps 0.4 mg

Administer:

• Without regard to food

• Swallow caps whole; do not break, crush, or chew

• Give 1/2 hr after same meal each day

• If treatment is interrupted for several days, restart at lowest dose (0.4 mg/day)

• Store in tight container in cool environment

SIDE EFFECTS

CNS: *Dizziness, headache,* asthenia, insomnia

CV: Chest pain, orthostatic hypotension

EENT: Amblyopia, floppy iris syndrome

GI: Nausea, diarrhea, dysgeusia

GU: Decreased libido, abnormal ejaculation, priapism
INTEG: Rash, pruritus, urticaria
MS: Back pain
RESP: Rhinitis, pharyngitis, cough
SYST: Angioedema

PHARMACOKINETICS

Peak 4-5 hr, duration 9-15 hr, half-life 9-13 hr, metabolized in liver, excreted via urine, extensively protein bound (98%)

INTERACTIONS

Increase: B/P—prazosin, terazosin, doxazosin, α-blockers, vardenafil
Increase: toxicity—cimetidine

NURSING CONSIDERATIONS

Assess:

- **Prostatic hyperplasia:** change in urinary patterns at baseline and throughout treatment; I&O ratios, weight daily; edema; report weight gain or edema
- **Orthostatic hypotension:** monitor B/P, standing, sitting

Evaluate:

- Therapeutic response: decreased symptoms of benign prostatic hyperplasia

Teach patient/family:

- Not to drive or operate machinery for 4 hr after 1st dose or after dosage increase
- To continue to take even if feeling better
- To advise providers of all products, herbs taken
- To make position changes slowly because orthostatic hypotension may occur
- To take ½ hr before same meal each day
- To teach about priapism (rare)
- Not to crush, break, chew

tapentadol (Rx)

(ta-pen′ta-dol)

Nucynta, Nucynta ER

Func. class.: Analgesic, misc.

Chem. class.: μ-Opioid receptor agonist

Controlled Substance Schedule II

ACTION:

Centrally acting synthetic analgesic; μ-opioid agonist activity is thought to result in analgesia; inhibits norepinephrine uptake

USES:

Moderate to severe pain, diabetic peripheral neuropathy

CONTRAINDICATIONS:

Hypersensitivity, asthma, ileus, respiratory depression

Black Box Warning: Respiratory depression

Precautions: Pregnancy (C), breastfeeding, children <18 yr, increased intracranial pressure, MI (acute), severe heart disease, respiratory depression, renal/hepatic disease, GI obstruction, ulcerative colitis, sleep apnea, seizure disorder

Black Box Warning: Accidental exposure, avoid ethanol, substance abuse

DOSAGE AND ROUTES

- **Adult: PO** 50-100 mg q4-6hr, may give 2nd dose ≥1 hr after 1st dose, max 700 mg on day 1, max 600 mg/day thereafter; ext rel 50 mg q12hr (opioid-naive), titrate to 100-250 mg q12hr, max 250 mg q12hr

Available forms: Tabs 50, 75, 100 mg; tabs ext rel 50, 100, 150, 200, 250 mg

Administer:

- With antiemetic if nausea, vomiting occur
- When pain is beginning to return; determine dosage interval by response

T

• Do not crush, chew, break, or use with alcohol, ext rel product
• Preferred analgesic in those with altered cytochrome P450 or mild hepatic, mild to moderate renal disease
• Store in light-resistant area at room temperature

SIDE EFFECTS

CNS: *Drowsiness, dizziness, confusion, headache, euphoria,* hallucinations, restlessness, syncope, anxiety, flushing, psychological dependence, insomnia, lethargy, tremors, seizures
CV: Palpitations, bradycardia, hypo/hypertension, orthostatic hypotension, sinus tachycardia
GI: *Nausea, vomiting, anorexia, constipation, cramps,* gastritis, dyspepsia, biliary spasms
GU: Urinary retention/frequency
INTEG: *Rash,* urticaria, diaphoresis, pruritus
RESP: Respiratory depression, cough
SYST: Anaphylaxis, infection, serotonin syndrome

PHARMACOKINETICS

Bioavailability 32%, extensively metabolized by liver, excreted in urine 99%, terminal half-life 4 hr, protein binding 20%

INTERACTIONS

Increase: effects with other CNS depressants—alcohol, opioids, sedative/hypnotics, antipsychotics, skeletal muscle relaxants
Increase: toxicity—MAOIs
Increase: serotonin syndrome—SSRIs, SNRIs, serotonin-receptor agonists, tricyclics
Drug/Herb
Increase: sedative effect—kava, St. John's wort, valerian

NURSING CONSIDERATIONS

Assess:
• **Pain:** intensity, location, type, characteristics; need for pain medication by pain/sedation scoring; physical dependence
• I&O ratio; check for decreasing output; may indicate urinary retention
• CNS changes: dizziness, drowsiness, hallucinations, euphoria, LOC, pupil reaction
• **Serotonin syndrome:** increased heart rate, shivering, sweating, dilated pupils, tremors, high B/P, hyperthermia, headache, confusion; if these occur, stop product, administer serotonin antagonist if needed
• Allergic reactions: rash, urticaria, anaphylaxis

Black Box Warning: **Respiratory dysfunction:** respiratory depression, character, rate, rhythm; notify prescriber if respirations are <10/min; B/P, pulse

• Assistance with ambulation
• Safety measures: nightlight, call bell within easy reach

Evaluate:
• Therapeutic response: decrease in pain

Teach patient/family:
• To report any symptoms of CNS changes, allergic reactions, seizures, serotonin syndrome
• That physical dependency may result from extended use
• That withdrawal symptoms may occur: nausea, vomiting, cramps, fever, faintness, anorexia
• To avoid CNS depressants, alcohol
• To avoid driving, operating machinery if drowsiness, dizziness occur
• To change positions slowly to decrease orthostatic hypotension
• To notify prescriber if pregnancy planned, suspected or if breastfeeding

tbo-filgrastim

(fil-gras′tim)

Neutrophil

Func. class.: Biologic modifier

Chem. class.: Short-acting granulocyte colony-stimulating factor (G-CSF)

ACTION: Stimulates proliferation and differentiation of neutrophils by binding to G-CSF receptors

USES: Chemotherapy-induced neutropenia prophylaxis to reduce the duration of severe neutropenia in nonmyeloid malignancies

CONTRAINDICATIONS: Hypersensitivity

Precautions: pregnancy (C), breastfeeding, children, sickle cell disease, chemotherapy, neoplastic disease

DOSAGE AND ROUTES

- **Adult: SUBCUT** 5 mcg/kg/day

Available forms: Solution for injection 300 mcg/0.5 ml, 480 mcg/0.8 ml

Administer:

SUBCUT route

- Visually inspect for particulate matter and discoloration before use; the solution should be clear; do not shake
- Injection sites include the front of the middle thighs, the upper outer areas of the buttocks, the upper back portion of the upper arms, or the abdomen except for the 2 inches around the navel; vary sites daily
- After removing the needle shield, expel unneeded volume based on the needed dose; discard unused portions
- Inject SUBCUT, avoiding any area that is tender, red, bruised, or hard or that has stretch marks or scars; push the plunger as far as it will go, because injection of the entire contents is necessary to activate the needle guard
- Remove needle from skin while keeping plunger depressed; slowly let the plunger go to allow the empty syringe to move inside the device and guard the needle

SIDE EFFECTS

CNS: Fever, headache

GI: Nausea, vomiting, diarrhea, anorexia, abdominal pain

HEMA: **Excessive leukocytosis**

INTEG: Alopecia

RESP: **Acute respiratory distress syndrome (ARDS)**

SYST: **Anaphylaxis, angioedema, sickle-cell crisis, splenic rupture**

PHARMACOKINETICS

Half-life 3.2-3.8 hr

INTERACTIONS

Increase: adverse reactions—do not use this product concomitantly with antineoplastics, lithium

NURSING CONSIDERATIONS

Assess:

- Blood studies: CBC, platelet count before treatment and twice weekly; neutrophil counts may be increased for 2 days after therapy
- B/P, respirations, pulse before and during therapy
- Bone pain; give mild analgesics

Evaluate:

- Absence of severe neutropenia in cancer chemotherapy

Teach patient/family:

- About the technique for self-administration: dose, side effects, disposal of containers and needles; provide instruction sheet

T

telaprevir

Incivek

Func. class.: Antivirals, antihepatitis agent

Chem. class.: NS3/4A protease inhibitor

ACTION: Prevents hepatitis C viral (HCV) replication by blocking the proteolytic activity of HCV NS3/4A serine protease; hepatitis C virus NS3/4A serine protease is an enzyme responsible for the conversion of HCV encoded polyproteins to mature/functioning viral proteins; these proteins (NS4A, NS4B, NS5A, NS5B) are needed for viral replication

USES: Chronic hepatitis C

CONTRAINDICATIONS: Pregnancy (X) in combination, male partners of women who are pregnant

Precautions: Breastfeeding, neonates, infants, children, adolescents <18 yr of age, anemia, neutropenia, thrombocytopenia, HIV, hepatitis B, decompensated hepatic disease, liver or other organ transplants

Black Box Warning: Serious rash

DOSAGE AND ROUTES

Chronic hepatitis C infection (genotype 1) in compensated liver disease without cirrhosis who are previously untreated or have relapsed after treatment with interferon and ribavirin therapy

- **Adults:** PO 1125 mg bid with peginterferon alfa and ribavirin; duration is determined by patient's HCV RNA level at treatment wk 4, 12; if the HCV RNA is undetectable at wk 4, 12, give three-drug regimen for 12 wk then give an additional 12 wk of only peginterferon alfa and ribavirin (24 wk total); if the HCV RNA is detectable but ≤1000 international units/ml at wk 4 or 12, give three-drug regimen × 12 wk then an additional 36 wk of only peginterferon alfa and ribavirin (48 wk total)

Without cirrhosis who are previously partial or null responders to interferon and ribavirin therapy/or those with cirrhosis:

- **Adults:** PO 1125 mg bid with peginterferon alfa and ribavirin; give three-drug regimen × 12 wk then another 36 wk (48 wk total) of only peginterferon alfa and ribavirin

Available forms: Tab 375 mg

Administer:

- Only use in combination with peginterferon alfa and ribavirin; never give as monotherapy

⚠ Discontinue in all patients with hepatitis C virus RNA concs ≥1000 IU/ml at wk 4 or 12 or a confirmed detectable HCV RNA conc at wk 24

⚠ Any contraindication to peginterferon alfa or ribavirin also applies to this product; see ribavirin or peginterferon alfa monographs for additional information regarding contraindications and warning associated with these products

- For relief of mild to moderate rashes, give oral antihistamines and/or topical corticosteroids; treatment with systemic steroids is not recommended
- Give with food, not low-fat
- Store tabs at room temperature

SIDE EFFECTS

CNS: Fatigue

GI: Hemorrhoids, anorectal discomfort, pruritus ani, rectal burning, nausea, diarrhea, vomiting, dysgeusia, hyperbilirubinemia

HEMA: Anemia, decreased Hgb, lymphopenia, leukopenia, neutropenia, thrombocytopenia

INTEG: Drug reaction with eosinophilia, and systemic symptoms (DRESS), Stevens-Johnson Syndrome (SJS), rash, pruritus, severe rashes (bullous rash, vesicular rash, and skin ulcerations)

META: Increased uric acid

PHARMACOKINETICS

59%-76% plasma protein binding, primarily to α 1-acid glycoprotein and albumin; metabolized extensively in liver by hydrolysis, oxidation, reduction by CYP3A4; excreted in feces (82%), exhaled (9%), excreted in urine (1%); elimination half life 4-4.7 hr; steady state half life 9-11 hr, peak 4-5 hr; food increases effect; steady state is reduced by 15% in mild hepatic disease, 46% in moderate hepatic disease (Child-Pugh class B)

INTERACTIONS

⚠ **Increase:** life-threatening reactions of each product—alfuzosin, ergots (dihydroergotamine, ergotamine, ergonovine, methylergonovine), cisapride, pimozide, lovastatin, simvastatin, ezetimibe, niacin with simvastatin and boceprevir; triazolam, oral midazolam; sildenafil, tadalafil (pulmonary arterial hypertension); do not use concurrently

Increase: effect and adverse reactions of each product—phosphodiesterase type 5 (PDE5) inhibitors (for erectile dysfunction), atorvastin, lidocaine, atorvastatin, amiodarone, bepridil, flecainide, propafenone, quiNIDine, digoxin, ketoconazole, itraconazole, posaconazole, voriconazole, desipramine, traZODone, erythromycin, clarithromycin, telithromycin, rifabutin, dexamethasone, felodipine, niCARdipine, NIFEdipine, budesonide, bosentan, warfarin, astemizole, boceprevir, bupivacaine, busPIRone, cevimeline, chloroquine, cilostazol, cinacalcet, clomiPRAMINE, clonazePAM, clopidogrel, cloZAPine, cyclobenzaprine, dapsone, dextromethorphan, diazepam, diclofenac, disopyramide, disulfiram, dolasetron, donepezil, colchicine, cycloSPORINE, tacrolimus, sirolimus, buprenorphine, salmeterol, vardenafil, ALPRAZolam, citalopram, amLODIPine, diltiazem, nisoldipine, verapamil, systemic corticosteroids, acetaminophen, alfentanil, aliskiren, almotriptan, alosetron, ARIPiprazole, dutasteride, ebastine, eplerenone, erlotinib, omeprazole, estazolam, eszopiclone, ethosuximide, exemestane, finasteride, flunitrazepam, flurazepam, galantamine, gefitinib, granisetron, halofantrine, haloperidol, HYDROcodone, ifosfamide, imipramine, indiplon, isradipine, ivermectin, ixabepilone, losartan, meloxicam, mirtazapine, montelukast, nateglinide, oxybutynin, oxyCODONE, palonosetron, paricalcitol, prasugrel, praziquantel, quazepam, QUEtiapine, quinacrine, ramelteon, repaglinide, ropivacaine, selegiline, sibutramine, sitaxsentan, solifenacin, SUFentanil, SUNItinib, theophylline, aminophylline, tiaGABine, tinidazole, tolterodine, traMADol, venlafaxine, amitriptyline, carvedilol, DAUNOrubicin, loratadine, desloratadine, lansoprazole, dexlansoprazole, DOCEtaxel, DOXOrubicin, droperidol, eletriptan, etoposide, fentaNYL, fexofenadine, glyburide, irinotecan, loperamide, maraviroc, mefloquine, mitoMYcin, morphine, nortriptyline, ondansetron, PACLitaxel, plicamycin, risperiDONE, sertraline, silodosin, teniposide, terfenadine, testosterone, tolvaptan, vinBLAStine, vinCRIStine and others; use cautiously, may need to reduce dose

Increase: hyperkalemia—drospirenone

Decrease: estrogen levels—ethinyl estradiol

Decrease: telaprevir effect—CYP3A4 inhibitors (phenytoin, carBAMazepine, PHENobarbital, rifampin)

Decrease: effect of—methadone

Possible treatment failure: efavirenz, ritonavir, atazanavir, lopinavir with ritonavir

Drug/Herb

⚠ Do not use with St. John's wort

NURSING CONSIDERATIONS

Assess:

⚠ **Pregnancy:** pregnancy (X) combination therapy; obtain a pregnancy test before, monthly during, and for 6 mo after treatment is completed; those who are not willing to practice strict contraception should not receive treatment with these products; report any cases of prenatal ribavirin exposure to the Ribavirin Pregnancy registry at (800) 593-2214

⚠ **Anemia:** monitor hemoglobin before, at treatment wk 4, 8, 12, and as needed; if Hgb is <10 g/dL, decrease ribavirin dosage; if Hgb is <8.5 g/dL, discontinuation of therapy is recommended; telaprevir dosage should not be altered based on adverse reactions; anemia may be managed through ribavirin dose modifications; never alter the dose of telaprevir; if anemia persists despite a reduction in ribavirin dose, consider discontinuing telaprevir; if management of anemia requires permanent discontinuation of ribavirin, treatment with telaprevir must also be permanently discontinued; once telaprevir has been discontinued, it must not be restarted; monitor CBC with differential at treatment wk 4, 8, 12, and at other treatment points

Black Box Warning: **Serious rash:** Toxic epidermal necrolysis, Stevens-Johnson syndrome: eosinophilia, fever, mucosal skin erosion, mucosal ulceration, target lesions; if serious skin reaction occurs, immediately discontinue all components of the three-drug regimen and refer the patient for urgent medical care

• Thyroid function tests, LFTs, serum bilirubin/creatinine, BUN, bilirubin, serum electrolytes, serum uric acid, baseline and periodically during treatment

Evaluate:

• Decreasing hepatitis C viral infection

Teach patient/family:

⚠ To use two forms of effective contraception (intrauterine devices and barrier methods) during treatment and for 6 mo after treatment; pregnancy (X), not to breastfeed

⚠ That if a serious skin reaction occurs, to seek urgent medical care; all components of the triple-drug regimen must be discontinued immediately

telavancin (Rx)

(tel-a-van′sin)

Vibativ

Func. class.: Antiinfective, miscellaneous

Chem. class.: Lipoglycopeptide

ACTION: Inhibits bacterial cell wall synthesis, disrupts cell membrane integrity, blocks glycopeptides

USES: Skin/skin-structure infections caused by *Enterococcus faecalis, E. faecium, Staphylococcus aureus* (MSRA), *S. aureus* (MSSA), *S. epidermidis, S. haemolyticus, Streptococcus agalactiae* (group B), *S. dysgalactiae, S. pyogenes* (group A β-tremolytic), *S. anginosus, S. intermedius, S. constellates,* nosocomial pneumonia caused by susceptible gram-positive bacteria

Unlabeled uses: Bacteremia

CONTRAINDICATIONS: Hypersensitivity

Precautions: Breastfeeding, children, geriatric patients, renal disease, antimicrobial resistance, diabetes mellitus, diarrhea, GI disease, heart failure, hypertension, pseudomembranous colitis, QT prolongation, vancomycin hypersensitivity

Black Box Warning: Pregnancy (C)

DOSAGE AND ROUTES

Skin/skin-structure infections

• **Adult: IV INF** 10 mg/kg over 60 min q24hr × 7-14 days

Nosocomial pneumonia

• **Adult: IV INF** 10 mg/kg q24hr × 7-21 days

Renal dose

• **Adult: IV** CCr 30-50 ml/min 7.5 mg/kg q24hr; CCr 10-29 ml/min 10 mg/kg q48hr

Available forms: Lyophilized powder for inj 250, 750 mg

Administer:

- Use only for susceptible organisms to prevent drug-resistant bacteria
- Antihistamine if red-man syndrome occurs: decreased B/P; flushing of neck, face
- Avoid IM, subcut use

Intermittent IV INF route

- After reconstitution with 15 ml D_5W sterile water for inj; 0.9% NaCl (15 mg/ml) 250-mg vial; add 45 ml to 750-mg vial (15 mg/ml) for dose of 150-800 mg; further dilute with 100-250 ml of compatible sol; for dose <150 mg or >800 mg, further dilute to conc of 0.6-8 mg/ml with compatible sol; give over 60 min; reconstituted or diluted sol is stable for 4 hr room temperature, 7 hr refrigerated; **avoid rapid IV; may cause Red man's syndrome**

Y-site compatibility: Amphotericin B lipid complex (Abelcet), ampicillin-sulbactam, azithromycin, calcium gluconate, caspofungin, cefepime, cefTAZidime, cefTRIAXone, ciprofloxacin, dexamethasone, diltiazem, DOBUTamine, DOPamine, doripenem, doxycycline, ertapenem, famotidine, fluconazole, gentamicin, hydrocortisone, labetalol, magnesium sulfate, mannitol, meropenem, metoclopramide, milrinone, norepinephrine, ondansetron, pantoprazole, phenylephrine, piperacillin-tazobactam, potassium chloride/phosphates, ranitidine, sodium bicarbonate, sodium phosphates, tigecycline, tobramycin, vasopressin

SIDE EFFECTS

CNS: Anxiety, chills, flushing, headache, insomnia, dizziness
CV: **QT prolongation**, irregular heartbeat
EENT: Hearing loss
GI: **Nausea**, vomiting, **pseudomembranous colitis**, abdominal pain, constipation, diarrhea, metallic/soapy taste
GU: **Nephrotoxicity**, *increased BUN, creatinine,* **renal failure**, foamy urine
HEMA: **Leukopenia, eosinophilia, anemia, thrombocytopenia**
INTEG: Chills, fever, rash, thrombophlebitis at inj site; urticaria, pruritus, necrosis (Red man's syndrome)
SYST: **Anaphylaxis, superinfection**

PHARMACOKINETICS

Onset rapid, half-life 8-9 hr, excreted in urine (76%), protein binding 90%, hepatic metabolism

INTERACTIONS

⚠ **Increase: otoxicity or nephrotoxicity—aminoglycosides, cephalosporins, colistin, polymyxin, bacitracin, CISplatin, amphotericin B, nondepolarizing muscle relaxants, cidofovir, tacrolimus, IV pentamine acyclovir, adefovir, cycloSPORINE, foscarnet, ganciclovir, pamidronate, streptozocin, zoledronic acid, NSAIDs, salicylates, ACE inhibitors**

⚠ **Increase: QT prolongation—class IA, III antidysrhythmics; some phenothiazines; chloroquine, clarithromycin, droperidol, dronedarone, erythromycin, haloperidol, levomethadye, methadone, pimozide, ziprasidone**

Drug/Lab Test

False increase: INR, PT, PTT

NURSING CONSIDERATIONS

Assess:

- **Infection:** WBC, urine, stools, sputum, characteristics of wound throughout treatment C&S
- **I&O ratio; report hematuria, oliguria; nephrotoxicity may occur**

Black Box Warning: Obtain a pregnancy test before use; if a woman has taken this product during pregnancy, the national registry should be notified at 866-658-4228

- Auditory function during, after treatment; hearing loss; ringing, roaring in ears; product should be discontinued
- B/P during administration; sudden drop may indicate Red man syndrome; also flushing, pruritus, rash
- Respiratory status: rate, character, wheezing, tightness in chest

• Allergies before treatment, reaction to each medication
• EPINEPHrine, suction, tracheostomy set, endotracheal intubation equipment on unit; anaphylaxis may occur
• Adequate intake of fluids (2 L/day) to prevent nephrotoxicity

Evaluate:
• Therapeutic response: negative culture

Teach patient/family:
• About all aspects of product therapy; that culture may be taken after completed course of medication
• To report sore throat, fever, fatigue; could indicate superinfection; diarrhea **(pseudomembranous colitis)**; hearing loss
• To use contraception while taking this product; not to breastfeed; to notify prescriber if pregnancy is planned or suspected

telbivudine (Rx)

(tel-bi′vyoo-deen)

Sebivo ✦, Tyzeka

Func. class.: Antiretroviral

Chem. class.: Nucleoside reverse transcriptase inhibitor (NRTI)

ACTION: Inhibits replication of HBV DNA polymerase, which inhibits HBV replication

USES: Treatment of chronic hepatitis B

CONTRAINDICATIONS: Hypersensitivity, breastfeeding

Precautions: Pregnancy (B), children, severe renal disease, anemia, organ transplant, dialysis, HIV, obesity, alcoholism; Hispanic or African descendents (safety not established)

Black Box Warning: Impaired hepatic function, lactic acidosis

DOSAGE AND ROUTES

• **Adult and adolescent >16 yr: PO** 600 mg/day; max 600 mg/day

Renal dose
• **Adult: PO** CCr 30-49 ml/min, 600 mg tab q48hr or 400 mg oral sol daily; CCr <30 ml/min (not requiring dialysis), 600 mg tab q72hr or 200 mg oral sol daily

Available forms: Tabs 600 mg

Administer:
• With/without food with a full glass of water
• Store at room temperature

SIDE EFFECTS

CNS: *Fever, headache, malaise,* weakness, *dizziness, insomnia*
EENT: Taste change, hearing loss, photophobia
GI: *Nausea, vomiting, diarrhea, anorexia,* abdominal pain, hepatomegaly
INTEG: *Rash*
MISC: Lactic acidosis
MS: Myalgia, arthralgia, muscle cramps
RESP: Cough

PHARMACOKINETICS

Excreted by kidneys (unchanged), steady state 5-7 days, protein binding 3.3%, terminal half-life 40-49 hr, peak 1-4 hr

INTERACTIONS

• Altered telbivudine levels: any agent altering renal function
• Do not use with pegylated interferon α-2a

Increase: myopathy risk possible—HMG-CoA reductase inhibitors, fibric acid derivatives, penicillamine, zidovudine, ZDV, cycloSPORINE, erythromycin, niacin, azole antifungals, corticosteroids, hydrochloroquine

NURSING CONSIDERATIONS

Assess:

Black Box Warning: LFTs, hepatitis B serology, creatine kinase, periodically, monitor HBV DNA after 24 wk; if viral suppression incomplete (≥300 copies/ml), start alternate therapy; monitor HBV DNA q6mo

Black Box Warning: Lactic acidosis: obtain baseline liver function tests; if elevated, discontinue treatment; discontinue even if liver function tests normal but lactic acidosis, hepatomegaly present; may be fatal

Evaluate:

• Therapeutic response: decreasing hepatitis B serology

Teach patient/family:

• That GI complaints and insomnia may resolve after 3-4 wk of treatment

• That product does not cure hepatitis B and does not stop its spread to others

• That follow-up visits must be continued

• That serious product interactions may occur if OTC products are ingested; to check with prescriber before taking

• That product may cause dizziness; to avoid hazardous activities until response is known

• To report symptoms of cough, difficulty sleeping, excessive headache, muscle pain/weakness

telmisartan (Rx)

(tel-mih-sar′tan)

Micardis

Func. class.: Antihypertensive

Chem. class.: Angiotensin II receptor (Type AT_1) antagonist

ACTION: Blocks the vasoconstricting and aldosterone-secreting effects of angiotensin II; selectively blocks the binding of angiotensin II to the AT_1 receptor found in tissues

USES: Hypertension, alone or in combination; stroke, MI prophylaxis (>55 yr) in patients unable to take ACE inhibitors

Unlabeled uses: Heart failure, proteinuria in diabetic nephropathy

CONTRAINDICATIONS: Hypersensitivity

Black Box Warning: Pregnancy (D) 2nd/3rd trimesters

Precautions: Pregnancy (C) 1st trimester, breastfeeding, children, geriatric patients; hypersensitivity to ACE inhibitors; renal/hepatic disease, renal artery stenosis, dialysis, CHF, hyperkalemia, hypotension, hypovolemia, African descent

DOSAGE AND ROUTES

• **Adult: PO** 40 mg/day; range 20-80 mg/day

Stroke, MI prophylaxis

• **Adult >55 yr: PO** 80 mg/day

Available forms: Tabs 20, 40, 80 mg

Administer:

• Without regard to meals

• Increased dose to African American patients or consider alternative agent; B/P response may be reduced

• Do not remove from blister pack until ready to use

SIDE EFFECTS

CNS: Dizziness, insomnia, *anxiety,* headache, fatigue, syncope

GI: Diarrhea, dyspepsia, *anorexia, vomiting*

META: Hyperkalemia

MS: Myalgia, pain

RESP: *Cough, upper respiratory infection,* sinusitis, pharyngitis

SYST: **Angioedema**

PHARMACOKINETICS

Onset of antihypertensive activity 3 hr, peak 0.5-1 hr, extensively metabolized, terminal half-life 24 hr, protein binding 99.5%, excreted in feces >97%, B/P response is less in African American patients

INTERACTIONS

Increase: digoxin peak/trough concs—digoxin

Increase: antihypertensive action—diuretics, other antihypertensives, NSAIDs

Increase: hyperkalemia—potassium-sparing diuretics, potassium salt substitutes, ACE inhibitors

Decrease: antihypertensive effect—NSAIDs, salicylates

Drug/Lab Test

Increase: LFTs

NURSING CONSIDERATIONS

Assess:

- B/P, pulse standing, lying; note rate, rhythm, quality; if severe hypotension occurs, place in supine position and give IV NS
- Baselines of renal, hepatic, electrolytes studies before therapy begins
- **CHF:** edema in feet, legs daily; jugular venous distention; dyspnea, crackles

Evaluate:

- Therapeutic response: decreased B/P

Teach patient/family:

- To comply with dosage schedule, even if feeling better; to take at same time of day; that therapeutic effect may take 2-4 wk
- To notify prescriber of mouth sores, fever, swelling of hands or feet, irregular heartbeat, chest pain, decreased urine output
- Not to stop abruptly
- That excessive perspiration, dehydration, vomiting, diarrhea may lead to fall in blood pressure; to consult prescriber if these occur
- That product may cause dizziness, fainting, lightheadedness; to avoid hazardous activities until response is known

Black Box Warning: To notify prescriber if pregnancy is planned or suspected, pregnancy (D) 2nd/3rd trimester

- To notify prescriber of all prescriptions, OTC products, and supplements taken; to rise slowly from sitting to prevent drop in B/P
- **Overdose:** dizziness, bradycardia, or tachycardia

temazepam (Rx)

(te-maz′e-pam)

Restoril

Func. class.: Sedative/hypnotic

Chem. class.: Benzodiazepine, short to intermediate acting

Controlled Substance Schedule IV (USA), Schedule F (Canada)

Do not confuse:
temazepam/flurazepam

ACTION: Produces CNS depression at limbic, thalamic, hypothalamic levels of the CNS; may be mediated by neurotransmitter γ-aminobutyric acid (GABA); results are sedation, hypnosis, skeletal muscle relaxation, anticonvulsant activity, anxiolytic action

USES: Insomnia, short-term treatment (generally 7-10 days)

CONTRAINDICATIONS: Pregnancy (X), breastfeeding, hypersensitivity to benzodiazepines

Precautions: Children <15 yr, geriatric patients, anemia, renal/hepatic disease, suicidal individuals, drug abuse, psychosis, acute closed-angle glaucoma, seizure disorders, angioedema, sleep-related behaviors (sleepwalking), intermittent porphyria, COPD, dementia, myasthenia gravis

DOSAGE AND ROUTES

- **Adult: PO** 7.5 to 30 mg at bedtime
- **Geriatric: PO** 7.5 mg at bedtime

Available forms: Caps 7.5, 15, 22.5, 30 mg

Administer:

- ½-1 hr before bedtime for sleeplessness
- On empty stomach for fast onset; may be taken with food if GI symptoms occur
- Avoid use with CNS depressants; serious CNS depression may result

• Store in tight container in cool environment

SIDE EFFECTS

CNS: *Lethargy, drowsiness, daytime sedation,* dizziness, confusion, lightheadedness, headache, anxiety, irritability, complex sleep-related reactions (sleep driving, sleep eating), fatigue

CV: Chest pain, pulse changes, hypotension

EENT: Blurred vision

GI: Nausea, vomiting, diarrhea, heartburn, abdominal pain, constipation, anorexia

SYST: **Severe allergic reactions**

PHARMACOKINETICS

Onset 30 min, peak 1-2 hr, duration 6-8 hr, half-life 10-20 hr, metabolized by liver, excreted by kidneys, crosses placenta, excreted in breast milk, 98% protein binding

INTERACTIONS

Increase: effects of cimetidine, disulfiram, oral contraceptives

Increase: action of both products—alcohol, CNS depressants

Increase: effect of temazepam—probenecid

Decrease: effect of antacids, theophylline, rifampin

Drug/Herb

Increase: CNS depression—hops, kava, valerian, chamomile, skullcap

Drug/Food

Decrease: temazepam effect—caffeine

Drug/Lab Test

Increase: ALT, AST

NURSING CONSIDERATIONS

Assess:

• Mental status: mood, sensorium, affect, memory (long, short), orientation

• Type of sleep problem: falling asleep, staying asleep, baseline, periodically

• **Dependency:** restrict amount given to patient, assess for physical/psychological dependency; high-level risk for abuse

• Assistance with ambulation after receiving dose

Evaluate:

• Therapeutic response: ability to sleep at night, decreased early morning awakening if taking product for insomnia

Teach patient/family:

• To avoid driving, other activities requiring alertness until stabilized

• To avoid alcohol ingestion, other CNS depressants

• That effects may take 2 nights for benefits to be noticed

• To limit to 7-10 days of continuous use

• About alternative measures to improve sleep: reading, exercise several hours before bedtime, warm bath, warm milk, TV, self-hypnosis, deep breathing

• Not to discontinue abruptly, withdraw gradually

• That complex sleep-related behaviors may occur: sleep driving/eating

• That hangover, memory impairment are common in geriatric patients but less common than with barbiturates

⚠ To notify prescriber if pregnancy is planned or suspected, pregnancy (X); to use contraception while taking this product

TREATMENT OF OVERDOSE:

Lavage, activated charcoal; monitor electrolytes, VS

temozolomide (Rx)

(tem-oh-zole′oh-mide)

Temodar

Func. class.: Antineoplastic-alkylating agent

Chem. class.: Imidazotetrazine derivative

ACTION: Prodrug that undergoes conversion to MTIC; MTIC action prevents DNA transcription

USES: Anaplastic astrocytoma with relapse, glioblastoma multiforme, malignant glioma

Unlabeled uses: Metastatic melanoma

CONTRAINDICATIONS:

Pregnancy (D), breastfeeding; hypersensitivity to this product, carbazine, or gelatin

Precautions: Geriatric patients, radiation therapy, renal/hepatic disease, bone marrow suppression, infection, myelosuppression

DOSAGE AND ROUTES

Anaplastic astrocytoma

- **Adult: PO** adjust dose based on nadir neutrophil and platelet counts 150 mg/m^2/day × 5 days during a 28-day cycle

Glioblastoma multiforme

- **Adult: PO/IV** 75 mg/m^2/day × 42 days with focal radiotherapy then maintenance of 6 cycles

Refractory anaplastic astrocytoma

- **Adult: IV** 150 mg/m^2/day over 90 min on days 1-5 q28days, may increase to 200 mg/m^2/day on days 1-5 q28days if hematologic parameters permit

Available forms: Caps 5, 20, 100, 140, 180, 250 mg; powder for inj 100 mg

Administer:

PO route

- Do not break, crush, chew, open caps
- Antiemetic 30-60 min before product to prevent vomiting
- Caps 1 at a time with 8 oz of water at same time of day
- Fluids IV or PO before chemotherapy to hydrate patient
- If caps accidentally damaged, do not allow contact with skin or inhale
- Use proper procedures for handling/disposing of chemotherapy products
- Give on empty stomach at bedtime to prevent nausea/vomiting
- Store in light-resistant container in a dry area

IV route

- Bring vial to room temperature; discard if cloudy
- Inject 41 ml sterile water for inj into vial (2.5 mg/ml)
- Gently swirl; do not shake

Intermittent IV INF route

- Withdraw up to 40 ml from each vial to make total dose; transfer to empty 250-ml PVC inf bag; flush before and after inf
- Run over 90 min
- Use reconstituted sol within 14 hr, including inf time
- Do not admix

SIDE EFFECTS

CNS: Seizures, *hemiparesis, dizziness, poor coordination, amnesia, insomnia, paresthesia, somnolence, paresis, ataxia, anxiety, dysphagia, depression, confusion*

GI: *Nausea, anorexia, vomiting,* abdominal pain, constipation

GU: Urinary incontinence, UTI, frequency

HEMA: Thrombocytopenia, leukopenia, anemia, myelosuppression, neutropenia

INTEG: *Rash, pruritus*

MISC: Headache, fatigue, asthenia, fever, edema, back pain, weight increase, diplopia

RESP: URI, pharyngitis, sinusitis, coughing

SYST: Anaphylaxis, secondary malignancy

PHARMACOKINETICS

Absorption complete, rapid; crosses blood-brain barrier; excreted in urine, feces; half-life 1.8 hr; peak 1 hr

INTERACTIONS

- Do not use within 24 hr of sargramostim, filgrastim, G-CSF

Increase: myelosuppression—radiation, other antineoplastics

Increase: bleeding risk—NSAIDs, anticoagulants, platelet inhibitors, thrombolytics

Decrease: antibody reaction—live virus vaccines, toxoids

Decrease: action of digoxin

Drug/Food

Decrease: drug absorption

Drug/Lab Test

Decrease: Hgb, platelets, WBC, neutrophils

NURSING CONSIDERATIONS

Assess:

- Tumor response during treatment
- CBC on day 22 (21 days after 1st dose), CBC weekly until recovery if ANC is $<1.5 \times 10^9$/L and platelets $<100 \times 10^9$/L; do not administer to patients who do not tolerate 100 mg/m²; myelosuppression usually occurs late during the treatment cycle
- Seizures throughout treatment; mental status
- Monitor temperature; may indicate beginning infection
- Hepatic studies before, during therapy (bilirubin, AST, ALT, LDH), as needed or monthly
- Bleeding: hematuria, guaiac, bruising, petechiae, mucosa or orifices

Evaluate:

- Therapeutic response: decreased tumor size, spread of malignancy

Teach patient/family:

- To report signs of infection: fever, sore throat, flulike symptoms
- To report signs of anemia: fatigue, headache, faintness, SOB, irritability
- To report bleeding; to avoid use of razors, commercial mouthwash
- To notify prescriber if pregnancy is planned or suspected, pregnancy (D); not to breastfeed

temsirolimus (Rx)

(tem-sir-oh′li-mus)

Torisel

Func. class.: Biologic response modifier

Chem. class.: Kinase inhibitor, mTOR antagonist

ACTION: Inhibits mammalian target of rapamycin (mTOR), a protein kinase

USES: Renal cell carcinoma

Unlabeled uses: Mantle cell lymphoma, relapsed/refractory high-grade malignant gliomas, neuroblastoma, rhabdomyosarcoma in children

CONTRAINDICATIONS: Pregnancy (D), breastfeeding; hypersensitivity to this product or to sirolimus; polysorbate 80

Precautions: Children <13 yr, females, severe pulmonary/renal/hepatic disease (bilirubin >1-1.5×ULN or AST >ULN but bilirubin ≤ULN), diabetes mellitus, hyperkalemia, hyperuricemia, hypertension, bone marrow suppression, hypertriglyceridemia/hyperlipidemia, surgery, brain tumors

DOSAGE AND ROUTES

- **Adult: IV** 25 mg over 30-60 min weekly; treat until disease progression or severe toxicity occurs

Hepatic dose

- **Adult: IV** (mild impairment) bilirubin >1-1.5×ULN or AST >ULN but bilirubin ≤ULN; reduce to 15 mg/wk; (moderate or severe impairment, do not use)

Available forms: 25 mg/ml sol for inj kit

Administer:

- Using in-line filter ≤5 microns and inf pump
- Premedicate with 25-50 mg diphenhydrAMINE IV 30 min before dose; if reaction occurs, stop for ½-1 hr, may resume at slower rate
- Over 30-60 min, complete inf within 6 hr
- Dilute product with 1.8 ml provided diluent, result is 3 ml (10 mg/ml); invert to mix well; withdraw required amount and inject rapidly into 250 ml of NS; do not use PVC inf bags/sets
- Protect from light during preparation; use only glass

SIDE EFFECTS

CNS: *Headache,* seizures

CV: Hypertension, thrombophlebitis

ENDO: Hypertriglyceridemia, hyperlipidemia, hyperglycemia

GI: Nausea, vomiting, diarrhea, constipation, bowel perforation
GU: UTIs, albuminuria, hematuria, proteinuria, renal failure, mucositis
HEMA: Anemia, leukopenia, thrombocytopenia
INTEG: *Rash,* pruritus
META: Metabolic acidosis, hyperglycemia, hyperlipidemia
RESP: Interstitial lung disease
SYST: Lymphoma

PHARMACOKINETICS

Rapidly absorbed, peak 0.5-2 hr, extensively metabolized via liver by P450 3A4, eliminated via feces

INTERACTIONS

Increase: blood levels—CYP3A4 inhibitors, antifungals, calcium channel blockers, cimetidine, clarithromycin, danazol, erythromycin, cycloSPORINE, metoclopramide, bromocriptine, HIV-protease inhibitors, benzodiazepines, HMG-CoA reductase inhibitors
Increase: toxicity—SUNItinib
Decrease: blood levels—CYP3A4 inducers, carBAMazepine, dexamethasone, PHENobarbital, phenytoin, rifamycin, rifapentine
Decrease: effect of vaccines—avoid with vaccines

Drug/Herb

Decrease: sirolimus effect—St. John's wort
Increase: effect—ginseng, maitake, mistletoe
Decrease: immunosuppression—astragalus, echinacea, melatonin

Drug/Food

- Alters bioavailability; use consistently with/without food; do not use with grapefruit juice

NURSING CONSIDERATIONS

Assess:

- Cardiac status: B/P, heart rate
- Interstitial lung disease
- Hypersensitivity reactions: anaphylaxis
- Lipid profile: cholesterol, triglycerides; lipid-lowering agent may be needed; blood glucose
- ⚠ Infection and development of lymphoma
- ⚠ Blood studies: Hgb, WBC, platelets during treatment monthly
- Renal studies: BUN, creatinine, phosphate potassium; proteinuria, hematuria, albuminemia may indicate renal failure
- **Hepatic disease:** monitor liver function tests at baseline and periodically

Evaluate:

- Therapeutic response: decreased time of progression of renal cell carcinoma

Teach patient/family:

- To report fever, rash, severe diarrhea, chills, sore throat, fatigue; that serious infections may occur; to report clay-colored stools, cramping **(hepatotoxicity),** excessive thirst, urinary frequency, new or worsening breathing problems, blood in stool, abdominal pain
- To avoid crowds, persons with known infections to reduce risk of infection
- ⚠ To use contraception before, during, for 12 wk after product discontinued; to avoid breastfeeding; that men should also use reliable contraception during and for 12 wk after cessation of product, pregnancy (D)

⚠ HIGH ALERT

tenecteplase (TNK-tPA) (Rx)

(ten-ek′ta-place)

TNKase

Func. class.: Thrombolytic
Chem. class.: Tissue plasminogen activator

ACTION: Activates conversion of plasminogen to plasmin (fibrinolysin): plasmin breaks down clots (fibrin), fibrinogen, factors V, VII; occlusion of venous access lines

USES: Acute myocardial infarction, coronary artery thrombosis

CONTRAINDICATIONS:

Hypersensitivity, arteriovenous malformation, aneurysm, active bleeding, intracranial/intraspinal surgery or trauma within 2 mo, CNS neoplasms, severe hypertension, severe renal/hepatic disease, history of CVA, increased ICP/stroke

Precautions: Pregnancy (C), breastfeeding, children, geriatric patients, arterial emboli from left side of heart, hypocoagulation, subacute bacterial endocarditis, rheumatic valvular disease, cerebral embolism/thrombosis/hemorrhage, intraarterial diagnostic procedure or surgery (10 days), recent major surgery, dysrhythmias, hypertension

DOSAGE AND ROUTES

Total dose, max 50 mg based on patient's weight

- **Adult <60 kg: IV BOL** 30 mg, give over 5 sec
- **Adult ≥60-<70 kg: IV BOL** 35 mg, give over 5 sec
- **Adult ≥70-<80 kg: IV BOL** 40 mg, give over 5 sec
- **Adult ≥80-<90 kg: IV BOL** 45 mg, give over 5 sec
- **Adult ≥90 kg: IV BOL** 50 mg, give over 5 sec, max 50 mg total dose

Available forms: Powder for inj, lyophilized 50 mg

Administer:

Intermittent IV INF route

- As soon as thrombi identified; not useful for thrombi >1 wk old
- Cryoprecipitate or fresh frozen plasma if bleeding occurs
- Heparin after fibrinogen level >100 mg/dl; heparin inf to increase PTT to 1.5-2× baseline for 3-7 days; IV heparin with loading dose is recommended
- Aseptically withdraw 10 ml of sterile water for inj from diluent vial; use red cannula syringe-filling device; inject all contents of syringe into product vial; direct into powder, swirl, withdraw correct dose; discard any unused sol; stand shield with dose vertically on flat surface and passively recap red cannula; remove entire shield assembly by twisting counter-clockwise; give by IV BOL
- IV therapy: use upper-extremity vessel that is accessible to manual compression
- If product not used immediately, refrigerate; use within 8 hr; not compatible with dextrose; flush dextrose-containing lines with saline before and after administration

SIDE EFFECTS

CV: Dysrhythmias, hypotension, pulmonary edema, **pulmonary embolism, cardiogenic shock, cardiac arrest, heart failure, myocardial reinfarction, myocardial rupture, tamponade, pericarditis, pericardial effusion, thrombosis, CVA**

HEMA: Decreased Hct, **bleeding**

INTEG: Rash, urticaria, phlebitis at IV inf site, itching, flushing

SYST: GI, GU, intracranial, retroperitoneal bleeding, surface bleeding, anaphylaxis

PHARMACOKINETICS

IV: Onset immediate, half-life 20-24 min, metabolized by liver

INTERACTIONS

Increase: bleeding—aspirin, indomethacin, phenylbutazone, anticoagulants, antithrombolytics, glycoprotein IIb/IIIa inhibitors, dipyridamole, clopidogrel, ticlopidine, NSAIDs, cefamandole, cefoperazone, cefoTEtan, SSRIs, SNRIs

Drug/Herb

Increase: risk of bleeding—feverfew, garlic, ginger, ginkgo, green tea, horse chestnut

Drug/Lab Test

Increase: INR, PT, PTT

NURSING CONSIDERATIONS

Assess:

- **Allergy:** fever, rash, itching, chills; mild reaction may be treated with antihistamines
- **Cholesterol embolism, blue-toe syndrome, renal failure, MI, cerebral/spinal cord/bowel/retinal infarction, hypertension; can be fatal**

T

⚠ **Bleeding** during 1st hr of treatment; hematuria, hematemesis, bleeding from mucous membranes, epistaxis, ecchymosis; may require tranfusion (rare), continue to assess for bleeding for 24 hr

- Blood studies (Hct, platelets, PTT, PT, TT, aPTT) before starting therapy; PT or aPTT must be <2× control before starting therapy; PTT or PT q3-4hr during treatment
- Hypersensitive reactions: fever, rash, dyspnea; product should be discontinued
- VS, B/P, pulse, respirations, neurologic signs, temperature at least q4hr; temperature >104° F (40° C) indicates internal bleeding; systolic pressure increase >25 mm Hg should be reported to prescriber

⚠ Neurologic changes that may indicate intracranial bleeding

⚠ Retroperitoneal bleeding: back pain, leg weakness, diminished pulses

- Bed rest during entire course of treatment
- Avoidance of venous or arterial puncture, inj, rectal temperature, any invasive treatment
- Treatment of fever with acetaminophen or aspirin
- Pressure for 30 sec to minor bleeding sites; inform prescriber if this does not attain hemostasis; apply pressure dressing

Evaluate:

- Therapeutic response: resolution of MI

Teach patient/family:

- About proper dental care to avoid bleeding
- To notify prescriber immediately of sudden, severe headache
- To notify prescriber of bleeding; hypersensitivity; fast, slow, or uneven heart rate; feeling of faintness; blood in urine, stools; nose bleeds

tenofovir (Rx)

(ten-oh-foh′veer)

Viread

Func. class.: Antiretroviral

Chem. class.: Nucleoside reverse transcriptase inhibitor (NRTI)

ACTION: Inhibits replication of HIV virus by competing with the natural substrate and then incorporating into cellular DNA by viral reverse transcriptase, thereby terminating cellular DNA chain

USES: HIV-1 infection with at least 2 other antiretrovirals, hepatitis B

CONTRAINDICATIONS: Hypersensitivity

Black Box Warning: Lactic acidosis

Precautions: Pregnancy (B), breastfeeding, children, geriatric patients, renal disease, CCr <60 ml/min, osteoporosis, immune reconstitution syndrome

Black Box Warning: Hepatic disease, hepatitis

DOSAGE AND ROUTES

- **Adult: PO** 300 mg/day with meal; if used with didanosine, give tenofovir 2 hr before or 1 hr after didanosine
- **Child ≥2 yr: PO** 8 mg/kg daily approximate; ≥35 kg 300 mg daily; 28-34 kg 250 mg daily; 22-27 kg 200 mg daily; 17-21 kg 150 mg daily

Renal dose

- **Adult: PO** CCr 30-49 ml/min, 300 mg q48hr; CCr 10-29 ml/min, 300 mg q72-96hr; CCr <10 ml/min, not recommended

Available forms: Tabs 150, 200, 250, 300 mg; oral powder 40 mg/scoop

Administer:

- Without regard to food
- 2 hr before or 1 hr after didanosine (if used)
- Store at 25° C (77° F)
- **Oral powder:** use scoop provided, mix powder into 2-4 oz (1/4-1/2 cup) of

applesauce, yogurt, do not mix with liquid, product is bitter, use immediately after mixing, clean scoop

SIDE EFFECTS

CNS: *Headache, asthenia*
GI: *Nausea, vomiting, diarrhea,* anorexia, *flatulence, abdominal pain,* pancreatitis
GU: Renal failure, renal tubular acidosis/necrosis, Fanconi syndrome
HEMA: Neutropenia, osteopenia
INTEG: *Rash,* angioedema
META: Lactic acidosis, hypokalemia, hypophosphatemia
MS: Myopathy, rhabdomyolysis
SYST: Lipodystrophy

PHARMACOKINETICS

Rapidly absorbed, distributed to extravascular space, excreted unchanged in urine 70%-80%, terminal half-life 17 hr, peak 1-2 hr

INTERACTIONS

Increase: tenofovir level—cidofovir, acyclovir, valacyclovir, ganciclovir, valganciclovir
Increase: level of didanosine when given with tenofovir
Increase: tenofovir level—any product that decreases renal function

NURSING CONSIDERATIONS

Assess:
- Viral load, CD4+ T-cell count, plasma HIV RNA, serum creatinine/BUN/phosphate
- Resistance testing at start of therapy and at treatment failure
- Hepatic studies: AST, ALT, bilirubin; amylase, lipase, triglycerides periodically during treatment
- Bone, renal toxicity: if bone abnormalities are suspected, obtain tests; serum phosphorus, creatinine
- **Lactic acidosis, severe hepatomegaly with steatosis, Fanconi syndrome:** obtain baseline liver function tests; if elevated, discontinue treatment; discontinue even if liver function tests normal but lactic acidosis, hepatomegaly present; may be fatal

Evaluate:
- Therapeutic response: decrease in signs, symptoms of HIV

Teach patient/family:
- To take without regard to food
- That GI complaints resolve after 3-4 wk of treatment
- Not to breastfeed while taking this product
- That product must be taken daily even if patient feels better
- That follow-up visits must be continued because serious toxicity may occur; that blood counts must be done q2wk
- That product will control symptoms but is not a cure for HIV; that patient is still infectious, may pass HIV virus on to others
- That other products may be necessary to prevent other infections
- That changes in body fat distribution may occur
- To notify prescriber of symptoms of lactic acidosis

terazosin (Rx)

(ter-ay′zoe-sin)
Func. class.: Antihypertensive
Chem. class.: α-Adrenergic blocker

ACTION: Decreases total vascular resistance, which is responsible for a decrease in B/P; this occurs by the blockade of α_1-adrenoreceptors

USES: Hypertension, as a single agent or in combination with diuretics or β-blockers; BPH

CONTRAINDICATIONS: Hypersensitivity
Precautions: Pregnancy (C), breastfeeding, children, prostate cancer, syncope

T

DOSAGE AND ROUTES

Hypertension

• **Adult: PO** 1 mg at bedtime, may increase dose slowly to desired response; max 20 mg/day divided q12hr

Benign prostatic hyperplasia

• **Adult: PO** 1 mg at bedtime, gradually increase up to 5-10 mg; max 20 mg divided q12hr

Available forms: Caps 1, 2, 5, 10 mg

Administer:

• Give dose at bedtime; patient should not operate machinery because fainting may occur
• If treatment is interrupted for several days, restart with initial dose
• Without regard to food; feeding tube: place cap in 60 ml of warm tap water; stir until liquid spills from ruptured shell (5 min); stir until cap dissolves; draw solution into oral syringe; give through feeding tube; flush with water
• Store at room temperature

SIDE EFFECTS

CNS: *Dizziness, headache, drowsiness,* anxiety, depression, vertigo, weakness, fatigue, syncope

CV: *Palpitations, orthostatic hypotension,* tachycardia, *edema,* rebound hypertension

EENT: Blurred vision, epistaxis, tinnitus, dry mouth, red sclera, nasal congestion, sinusitis

GI: *Nausea,* vomiting, diarrhea, constipation, abdominal pain

GU: Urinary frequency, incontinence, impotence, priapism

RESP: Dyspnea, cough, pharyngitis, nasal congestion

PHARMACOKINETICS

Half-life 9-12 hr; protein binding 90%-94%; metabolized in liver; excreted in urine, feces, peak 2-3 hr, onset 15 min, duration 24 hr

INTERACTIONS

Increase: hypotensive effects—β-blockers, nitroglycerin, verapamil, other antihypertensives, alcohol, phosphodiesterase (PDE5) inhibitors (vardenafil, tadalafil, sildenafil)

Decrease: hypotensive effects—estrogens, NSAIDs, sympathomimetics, salicylates

Drug/Herb

Increase: antihypertensive effect—hawthorn

Decrease: antihypertensive effect—ephedra

Drug/Lab Test

Decrease: Hgb, WBC, platelets, albumin

NURSING CONSIDERATIONS

Assess:

• **BPH:** urinary patterns (hesitancy, frequency, change in stream, dribbling, dysuria, urgency)
• **Hypertension:** crackles, dyspnea, orthopnea q30min; orthostatic B/P, pulse, jugular venous distention q4hr; weight daily, I&O
• BUN, uric acid if patient receiving long-term therapy

Evaluate:

• Therapeutic response: decreased B/P, edema in feet, legs; decreased symptoms of BPH

Teach patient/family:

• That fainting occasionally occurs after 1st dose; not to drive or operate machinery for 4 hr after 1st dose or after an increase in dose; to take 1st dose at bedtime
• To rise slowly from sitting or lying position
• Not to discontinue abruptly; not to drink alcohol
• **Hypertension:** to continue with regimen including diet, exercise

terbinafine oral (Rx)

(ter-bin'a-feen)

Lamisil

Func. class.: Antifungal

Chem. class.: Synthetic allylamine derivative

Do not confuse:
terbinafine/terbutaline
LamISIL/LaMICtal

ACTION: Interferes with cell-membrane permeability of fungi such as *Trichophyton rubrum, Trichophyton mentagrophytes, Trichophyton tonsurans, Epidermophyton floccosum, Microsporum canis, Microsporum audouinii, Microsporum gypseum, Candida;* broad-spectrum antifungal

USES: (Oral) onychomycosis of toenail or fingernail due to dermatophytes, tinea capititis/corporis/cruris/pedis/versicolor

Unlabeled uses: Cutaneous candidiasis, tinea versicolor

CONTRAINDICATIONS: Hypersensitivity

Precautions: Pregnancy (B), breastfeeding, children, chronic/active hepatic disease, renal disease GFR ≤50 ml/min, immunosuppression

DOSAGE AND ROUTES

- **Adult: PO** 250 mg/day × 6 wk (fingernail); × 12 wk (toenail)

Available forms: Tabs 250 mg; oral granules 125, 187.5 mg

Administer:

- **PO:** without regard to food
- Store at <25° C (77° F); protect from light
- **Granules:** take with food; sprinkle packet contents on pudding or nonacidic soft food; swallow without chewing; do not use fruit-based foods

SIDE EFFECTS

CNS: Depression

EENT: Tinnitus, hearing impairment

GI: Diarrhea, dyspepsia, vertigo, abdominal pain, nausea, hepatitis

HEMA: **Neutropenia**

INTEG: Rash, pruritus, urticaria, **Stevens-Johnson syndrome**, photosensitivity

MISC: Headache, hepatic enzyme changes, taste, visual/olfactory disturbance

PHARMACOKINETICS

Peak 1-2 hr, >99% protein binding, half-life 36 hr

INTERACTIONS

Increase: levels of dextromethorphan

Increase: terbinafine clearance—rifampin

Increase: clearance of cycloSPORINE

Decrease: terbinafine clearance—cimetidine

Decrease: metabolism of—CYP2D6 (antidysrthymics IC, III, amoxapine, atomoxetine, cloZAPine

Drug/Herb

- Side effects: cola nut, guarana, yerba maté, tea (black, green), coffee

Drug/Lab Test

Increase: LFTs

NURSING CONSIDERATIONS

Assess:

- Hepatic studies (ALT, AST) before beginning treatment; do not use in presence of hepatic disease
- CBC in treatment >6 wk
- Continuing infection: increased size, number of lesions

Evaluate:

- Therapeutic response: decrease in size, number of lesions

Teach patient/family:

- That treatment may take 10 wk (toenail), 4 wk (fingernail)
- **To notify prescriber of nausea, vomiting, fatigue, jaundice, dark urine, clay-colored stool, RUQ pain; may indicate hepatic dysfunction**

terbinafine topical

See Appendix B

terbutaline (Rx)

(ter-byoo'ta-leen)

Bricanyl ✦

Func. class.: Selective β_2-agonist; bronchodilator

Chem. class.: Catecholamine

Do not confuse:
terbutaline/TOLBUTamide/terbinafine

ACTION:
Relaxes bronchial smooth muscle by direct action on β_2-adrenergic receptors through the accumulation of cAMP at β-adrenergic receptor sites; bronchodilation, diuresis, CNS, cardiac stimulation occur; relaxes uterine smooth muscle

USES:
Bronchospasm

Unlabeled uses: Premature labor, nonresponsive status asthmaticus in children (IV)

CONTRAINDICATIONS:
Hypersensitivity to sympathomimetics, closed-angle glaucoma, tachydysrhythmias

Precautions: Pregnancy (B), breastfeeding, geriatric patients, cardiac disorders, hyperthyroidism, diabetes mellitus, prostatic hypertension, hypertension, seizure disorder

Black Box Warning: Labor

DOSAGE AND ROUTES

Bronchospasm

- **Adult and child >12 yr: PO** 2.5-5 mg q8hr; **SUBCUT** 0.25 mg q15-30min, max 0.5 mg in 4 hr
- **Adolescent ≤15 yr and child ≥12 yr: PO** 2.5 mg tid, max 7.5 mg/day
- **Child 6-11 yr (unlabeled): PO** 0.05 mg/kg q8hr, may increase slowly

Renal dose

- **Adult: PO** CCr 10-50 ml/min, 50% of dose; CCr <10 ml/min, avoid use

Tocolytic (preterm labor) (unlabeled)

- **Adult: SUBCUT** 0.25 mg q20min to 6 hr, hold if pulse >120 bpm

Available forms: Tabs 2.5, 5 mg; inj 1 mg/ml

Administer:

- With food; may be crushed
- 2 hr before bedtime to avoid sleeplessness

IV route (unlabeled)

- Only used if subcut is ineffective
- IV after diluting each 5 mg/1 L D_5W for inf
- IV, run 5 mcg/min; may increase 5 mcg q10min, titrate to response; after ½-1 hr, taper dose by 5 mcg; switch to PO as soon as possible
- Store at room temperature; do not use discolored sol

Y-site compatibilities: Insulin (regular)

SIDE EFFECTS

CNS: Tremors, anxiety, insomnia, headache, dizziness, stimulation

CV: Palpitations, tachycardia, hypertension, dysrhythmias, cardiac arrest, QT prolongation

GI: Nausea, vomiting

META: Hypokalemia, hyperglycemia

RESP: Paradoxical bronchospasm, dyspnea

PHARMACOKINETICS

PO: Onset ½ hr, peak 1-2 hr, duration 4-8 hr, half-life 3.4 hr

SUBCUT: Onset 6-15 min, peak ½-1 hr, duration 1½-4 hr, half-life 5.7 hr

INTERACTIONS

- Hypertensive crisis: MAOIs

Increase: effects of both products—other sympathomimetics

Decrease: action—β-blockers; do not use together

Drug/Herb
Increase: effect—green tea (large amounts), guarana

NURSING CONSIDERATIONS

Assess:

- Respiratory function: vital capacity, forced expiratory volume, ABGs, B/P, pulse, respiratory pattern, lung sounds, sputum before and after treatment
- Tolerance in patients receiving long-term therapy; dose may have to be changed; monitor for rebound bronchospasm

⚠ **Paradoxical bronchospasm:** dyspnea, wheezing; keep emergency equipment nearby

- **Labor:** maternal heart rate, B/P, contraction, fetal heart rate; can inhibit uterine contractions, labor; monitor for hypoglycemia
- Increase in fluids of >2 L/day

Evaluate:

- Therapeutic response: absence of dyspnea, wheezing

Teach patient/family:

- Not to use OTC medications because extra stimulation may occur
- About all aspects of product; to avoid smoking, smoke-filled rooms, persons with respiratory infections
- To increase fluids by >2 L/day; to allow 15 min between inhalation of product and inhaled product containing steroid
- To take on time; if missed, not to make up after 1 hr; to wait until next dose

terconazole vaginal antifungal

See Appendix B

teriflunomide

(ter′i-floo′noe-mide)

Aubagio

Func. class.: Multiple sclerosis agent
Chem. class.: Pyrimidine synthesis inhibitor

ACTION: Antiproliferative effects including peripheral T- and B-lymphocytes, might reduce inflammatory demyelination

USES: Reduction of the frequency of relapses or remitting MS

CONTRAINDICATIONS: Hypersensitivity

Black Box Warning: Pregnancy (X)

Precautions: Breastfeeding, alcoholism, diabetes mellitus, eosinophilic pneumonia, hepatitis, jaundice, male-mediated teratogenicity, pneumonitis, pulmonary disease/fibrosis, sarcoidosis, TB, vaccination

Black Box Warning: Hepatic disease

DOSAGE AND ROUTES

- **Adult: PO** 7 or 14 mg/day

Available forms: Tabs 7, 14 mg

Administer:

PO route

- May be taken without regard to food

SIDE EFFECTS

CNS: Anxiety, headache
CV: Palpitations, hypertension, MI
EENT: Blurred vision, conjunctivitis, sinusitis
GI: Nausea, vomiting, diarrhea, cystitis
HEMA: **Leukopenia, lymphopenia, neutropenia**
INTEG: Acne vulgaris, alopecia, pruritus
META: Weight loss
MISC: Infection, cystitis

T

PHARMACOKINETICS

Protein binding >99%, median half-life 18-19 days, peak 1-4 hr

INTERACTIONS

• Do not use with leflunomide, live virus vaccines

Increase: teriflunomide effect—cycloSPORINE, eltrombopag, gefitinib

Increase: hepatotoxicity—methotrexate, HMG-CoA reductase inhibitors

Increase: hematologic toxicity—zidovudine

Increase: effect of—oral contraceptives, repaglinide, pioglitazone, rosiglitazone, PACLitaxel, naproxen, topotecan, bosentan, furosemide

Decrease: effect of—warfarin, alosetron, DULoxetine, theophylline, tiZANidine, quiNINE, tamoxifen, bendamustine, rasagiline, rOPINIRole, selegiline, propafenone, mexiletine, lidocaine, anagrelide, cloZAPine, cinacalcet, caffeine, monitor closely

Decrease: effect of teriflunomide—cholestyramine, activated charcoal

NURSING CONSIDERATIONS

Assess:

• CNS symptoms: anxiety, confusion, vertigo

• GI status: diarrhea, vomiting, abdominal pain

• Cardiac status: tachycardia, palpitations, vasodilation, chest pain

Evaluate:

• Therapeutic response: decreased symptoms of MS

Teach patient/family:

• That blurred vision can occur

• To notify prescriber if pregnancy is planned or suspected

• Not to change dosing or stop taking without advice of prescriber

teriparatide (Rx)

(tah-ree-par′ah-tide)

Forteo

Func. class.: Parathyroid hormone (rDNA)

Chem. class.: Teriparatide

ACTION: Contains human recombinant parathyroid hormone to stimulate new bone growth

USES: Postmenopausal women with osteoporosis, men with primary or hypogonadal osteoporosis who are at high risk for fracture, glucocorticoid-induced osteoporosis

Unlabeled uses: Hypoparathyroidism

CONTRAINDICATIONS: Hypersensitivity, increased baseline risk for osteosarcoma (Paget's disease, open epiphyses; previous bone radiation), bone metastases, history of skeletal malignancies, other metabolic bone diseases, preexisting hypercalcemia

Precautions: Pregnancy (C), breastfeeding, children, urolithiasis, hypotension, use >2 yr, cardiac disease

Black Box Warning: Secondary malignancy

DOSAGE AND ROUTES

• **Adult: SUBCUT** 20 mcg/day up to 2 yr

Available forms: Prefilled pen delivery device (delivers 20 mcg/day)

Administer:

SUBCUT route

• Give by SUBCUT using disposable pen only; inject in thigh or abdomen; lightly pinch fold of skin; insert needle; release skin; inject at 90-degree angle over 5 sec; rotate inj sites

• Have patient sit or lie down; orthostatic hypotension may occur

• Protect from freezing, light; refrigerate pen
• Store refrigerated; do not freeze; may be used for 28 days after first inj

SIDE EFFECTS

CNS: Dizziness, headache, insomnia, depression, vertigo
CV: Hypertension, angina, syncope
GI: Nausea, diarrhea, dyspepsia, vomiting, constipation
INTEG: Rash, sweating
MISC: Pain, asthenia, hyperuricemia
MS: Arthralgia, leg cramps, back/leg pain, weakness, osteosarcoma (rare)
RESP: Rhinitis, cough, pharyngitis, pneumonia, dyspnea

PHARMACOKINETICS

SUBCUT: Extensively and rapidly absorbed, metabolized by liver, excreted by kidneys, terminal half-life 1 hr, onset rapid, peak 1/2 hr, duration 3 hr

INTERACTIONS

Increase: digoxin toxicity: digoxin
Drug/Lab Test
Increase: calcium, uric acid, urinary calcium
Decrease: phosphorous, magnesium

NURSING CONSIDERATIONS

Assess:

Black Box Warning: Secondary malignancy: osteosarcoma, dependent on length of treatment; those at higher risk for osteosarcoma should not use this product

• Uric acid, magnesium, creatinine, BUN, urine pH, vit D, phosphate for normal serum levels; serum calcium may be transiently increased after dosing (max at 4-6 hr after dose)
• Bone pain, headache, fatigue, changes in LOC, leg cramps
• **Signs of persistent hypercalcemia:** nausea, vomiting, constipation, lethargy, muscle weakness
• Nutritional status: diet for sources of vit D (milk, some seafood), calcium (dairy products, dark green vegetables), phosphates (dairy products)

Evaluate:
• Therapeutic response: increased bone mineral density

Teach patient/family:
• About the symptoms of hypercalcemia
• About foods rich in calcium, vit D
• How to use delivery device, dispose of needles; not to share pen with others; to use at same time of day
• To sit or lie down if dizziness or fast heartbeat occurs after 1st few doses
• To rotate administration sites
• To store pen in refrigerator; pen may be used for 28 days

tesamorelin
Egrifta
Func. class.: Pituitary hormone, growth hormone modifiers

ACTION: Binds to growth hormone releasing factor receptors on the pituitary somatotroph cells; binding stimulates the production, release of endogenous growth hormone (GH)

USES: Treatment of excess abdominal fat in HIV-infected patients with lipodystrophy

CONTRAINDICATIONS: Hypersensitivity to this product or mannitol, neoplastic disease, pregnancy X, *disruption of the hypothalamic-pituitary axis* (hypothalamic-pituitary-adrenal (HPA) suppression) due to hypophysectomy, hypopituitarism, pituitary tumor/surgery, radiation therapy of the head or head trauma, IV/IM administration
Precautions: Breastfeeding, CABG, diabetes, diabetic retinopathy, edema, geriatric patients, children, infants, adolescents

T

DOSAGE AND ROUTES

• **Adult:** SUBCUT 2 mg/day

Available forms: Powder for injection 1 mg

Administer:

Subcut route:

• Visually inspect parenteral products for particulate matter and discoloration before use

• To reconstitute, inject 2.1 ml sterile water for inj into the 2-mg vial; use syringe with needle already attached; to avoid foaming, push plunger in slowly with needle on a slight angle so sterile water goes down the inside wall of vial; with needle and syringe attached to vial, keep vial upright and gently roll vial for 30 sec until mixed; do not shake; withdraw 2.1 ml of the reconstituted sol

• To administer, take syringe out of vial, place needle cap on its side against a clean flat surface, do not touch needle, hold syringe and slide needle into cap, push cap all the way or until it snaps shut; do not touch cap until it covers needle completely; remove needle, insert a 0.5 inch 27-G safety injection needle onto syringe; use immediately; throw away any unused product or used sterile water for inj, sol should be clear, do not use if discolored, cloudy, or has particles; slight foaming is acceptable; inject subcut into abdomen; avoid scar tissues, bruises, or navel; rotate inj sites in abdomen; slowly push plunger down until all sol has been injected

• After removing needle from skin, flip back needle shield until it snaps, covering the injection needle completely; keep pressing until you hear a click, which means the injection needle is protected

• Use a piece of sterile gauze to rub the inj site clean; if bleeding, apply pressure to site with gauze for 30 sec; if bleeding continues, apply a bandage to site

• Properly dispose of used syringe, needles, vial, and sterile water for injection bottle in a sharps container

SIDE EFFECTS

CNS: Depression, peripheral neuropathy paresthesias, hypoesthesia, spasms, flushing, night sweats, insomnia, headache

CV: Chest pain, palpitations, hypertension, edema, peripheral edema

GI: Nausea, vomiting, upper abdominal pain, dyspepsia, diarrhea

INTEG: Pruritus, urticaria, rash, flushing, injection site reactions

MS: Arthralgia, joint swelling, stiffness, myalgias, carpal tunnel syndrome

RESP: Upper respiratory tract infection

SYST: Secondary malignancy

PHARMACOKINETICS

Half-life 26 min in healthy patients, 38 min in those with HIV infection; peak 0.15 hr

INTERACTIONS

Decrease: effect of—simvastatin, ritonavir, cortisone, predniSONE

NURSING CONSIDERATIONS

Assess:

• **Lipodystrophy:** Assess for sunken cheeks, thinning arms and legs, fat accumulation in the abdomen, jaws, and back of neck; after treatment these should lessen

• Monitor glycosylated hemoglobin A1c (HbA1c), serum IGF-1 concs, ophthalmologic exam

Evaluate:

• Decreasing lipodystrophy in HIV patients

testosterone cypionate (Rx)
Depo-Testosterone
testosterone enanthate (Rx)
Delatestryl
testosterone gel (Rx)
AndroGel, FORTESTA, Testim
testosterone pellets (Rx)
Testopel
testosterone transdermal (Rx)
Androderm
testosterone buccal (Rx)
Striant
testosterone topical solution
Axiron

Func. class.: Androgenic anabolic steroid

Chem. class.: Halogenated testosterone derivative

Controlled Substance Schedule III

ACTION: Increases weight by building body tissue; increases potassium, phosphorus, chloride, nitrogen levels, bone development

USES: Female breast cancer, hypogonadism, eunuchoidism, male climacteric, oligospermia, impotence, vulvar dystrophies, low testosterone levels, delayed male puberty (inj)

Unlabeled uses: Weight loss in AIDS patients, andropause, anemia, cryptorchidism, lichen sclerosis, microphallus, transsexualism

CONTRAINDICATIONS: Pregnancy (X), breastfeeding, severe cardiac/renal/hepatic disease, hypersensitivity, genital bleeding (rare), male breast/prostate cancer

Precautions: Diabetes mellitus, CV disease, MI, urinary tract disorders, prostate cancer, hypercalcemia

Black Box Warning: Children, accidental exposure

DOSAGE AND ROUTES

Replacement

- **Adult: IM (enanthate or cypionate)** 50-400 mg q2-4wk; topical sol (Axiron) 60 mg (2 pump actuations); apply each AM
- **Adult (male) and child: SUBCUT (pellets)** 150-450 mg (2-6 pellets) inserted q3-6mo
- **Adult: TRANSDERMAL (Testoderm, Androderm)** 4-6 mg applied q24hr; **GEL (AndroGel)** 1% 5 mg applied q24hr, once daily; 1.62% 40.5 mg (2 pump actuations) q AM; topical sol **(Axiron)** 30 mg/actuation **BUCCAL** 1 buccal system (30 mg) to the gum region q12hr before meals/PM

Breast cancer

- **Adult: IM** 200-400 mg q2-4wk (cypionate or enanthate)

Delayed male puberty

- **Child >12 yr: IM** ≤100 mg/mo for ≤6 mo

Available forms: Enanthate: inj 200 mg/ml; **cypionate:** inj 100, 200 mg/ml; pellets 75 mg; **transdermal** 2, 4, mg/24 hr; **gel** 1%, 1.62%, 10 mg/actuation; **buccal system** 30 mg; **topical sol** 30 mg/actuation

Administer:

- Titrated dose; use lowest effective dose
- IM inj deep into upper outer quadrant of gluteal muscle
- **Transdermal patches:** Testoderm to skin of scrotum; Androderm to skin of back, upper arms, thighs, abdomen; area must be dry shaved; may be reapplied after bathing, swimming

T

• **Gel:** products not interchangeable; dosage and administration for AndroGel 1% differs from that of AndroGel 1.62%; apply daily to clean, dry area on shoulders, upper arms, or abdomen; women, children should not touch treated skin

Buccal system route

• Do not chew or swallow buccal system

• Rotate sites; place above incisor tooth on either side of mouth

• Open packet; place rounded side of surface against gum and hold firmly in place with finger over lip for 30 sec; if product falls off, replace with new system; discard in trash can away from children or pets

Topical solution route

• Using the provided applicator, apply the solution to clean, dry, intact skin of the axilla, preferably at the same time each morning; do not apply to any other part of the body; allow the solution to dry completely before dressing; if an antiperspirant or deodorant is used, apply at least 2 min before applying the solution; the pump must be primed before the first use by fully depressing the pump mechanism 3×; discard any solution that is released during the priming; to dispense the solution, position the nozzle over the applicator cup and carefully depress the pump once fully; the cup should be filled with no more than 1 pump actuation (30 mg); with the applicator upright, place it up into the axilla and wipe steadily down and up into the axilla; do not use fingers or hand to rub the solution; if multiple applications are necessary for the required dose, alternate application between the left and right axilla; when repeat application to the same axilla is necessary, allow the solution to dry completely before the next application; after use, rinse the applicator under running water and pat dry with tissue; wash hands with soap and water

• Following application, allow the site to dry before putting on clothing

• Direct contact of the medicated skin with the skin of another person can result in the transfer of residual testosterone and absorption by the other person; to reduce accidental transfer, the patient should cover the application site(s) with clothing (e.g., a T-shirt) after the solution has dried; the application site should be washed with soap and water before any skin-to-skin contact regardless of the length of time since application; in the case of direct contact, the other person should wash the area of contact with soap and water as soon as possible

• Patients should be advised that the topical solution is flammable; therefore, fire, flame, and smoking should be avoided during use

• Advise patients to avoid swimming or washing the application site until 2 hr following application of solution

SIDE EFFECTS

CNS: Dizziness, headache, fatigue, tremors, paresthesias, flushing, sweating, anxiety, lability, insomnia, carpal tunnel syndrome

CV: Increased B/P

EENT: Conjunctival edema, nasal congestion

ENDO: Abnormal glucose tolerance test

GI: Nausea, vomiting, constipation, weight gain, cholestatic jaundice

GU: Hematuria, amenorrhea, vaginitis, decreased libido, decreased breast size, clitoral hypertrophy, testicular atrophy, gynecomastia, large prostate

HEMA: Polycythemia

INTEG: Rash, acneiform lesions, oily hair/skin, flushing, sweating, acne vulgaris, alopecia, hirsutism

MS: Cramps, spasms

PHARMACOKINETICS

PO: Metabolized in liver; excreted in urine, breast milk; crosses placenta

INTERACTIONS

• Edema: ACTH, adrenal steroids, buPROPion

Increase: effects of oxyphenbutazone

Increase: PT—anticoagulants

Decrease: glucose levels may alter need for oral antidiabetics, insulin

Drug/Lab Test

Increase: serum cholesterol, blood glucose, urine glucose

Decrease: serum calcium, serum potassium, T_4, T_3, thyroid ^{131}I uptake test, urine 17-OHCS, 17-KS, PBI

NURSING CONSIDERATIONS

Assess:

- Weight daily; notify prescriber if weekly weight gain >5 lb
- B/P q4hr, Hgb/HCT
- I&O ratio; be alert for decreasing urinary output, increasing edema
- Growth rate, bone age in children; growth rate may be uneven (linear/bone growth) with extended use
- Electrolytes: potassium, sodium, chlorine, calcium; cholesterol
- Hepatic studies: ALT, AST, bilirubin
- Edema, hypertension, cardiac symptoms, jaundice
- Mental status: affect, mood, behavioral changes, aggression
- **Signs of masculinization in female:** increased libido, deepening of voice, decreased breast tissue, enlarged clitoris, menstrual irregularities; male: gynecomastia, impotence, testicular atrophy
- **Hypercalcemia:** lethargy, polyuria, polydipsia, nausea, vomiting, constipation; product may have to be decreased
- **Hypoglycemia** in diabetic patients; oral antidiabetic action is increased
- Diet with increased calories, protein; decrease sodium if edema occurs

Evaluate:

- Therapeutic response: 4-6 wk with osteoporosis

Teach patient/family:

- That product must be combined with complete health plan: diet, rest, exercise
- To notify prescriber if therapeutic response decreases, if edema occurs
- About changes in sex characteristics: priapism, gynecomastia, increased libido
- That women should report menstrual irregularities, voice changes, acne, facial hair growth; if pregnancy is planned or suspected
- That 1-3-mo course is necessary for response with breast cancer
- About the proper application of patches

tetracaine ophthalmic

See Appendix B

tetracaine topical

See Appendix B

tetracycline (Rx)

(tet-ra-sye′kleen)

Apo-Tetra ✦, Nu-Tetra ✦

Func. class.: Broad-spectrum antiinfective

Chem. class.: Tetracycline

ACTION: Inhibits protein synthesis and phosphorylation in microorganisms; bacteriostatic

USES: Syphilis, *Chlamydia trachomatis,* gonorrhea, lymphogranuloma venereum; uncommon gram-positive, gram-negative organisms; rickettsial infections

CONTRAINDICATIONS: Pregnancy (D), breastfeeding, children <8 yr, hypersensitivity to tetracyclines

Precautions: Renal/hepatic disease, UV exposure

DOSAGE AND ROUTES

Susceptible gram-positive/gram-negative infections

- **Adult: PO** 250-500 mg q6hr
- **Child >8 yr: PO** 25-50 mg/kg/day in divided doses q6hr

Chlamydia trachomatis

• **Adult: PO** 500 mg qid × 7 days

Syphilis

• **Adult and adolescent: PO** 500 mg qid × 2 wk; if syphilis duration >1 yr, must treat 28 days

Brucellosis

• **Adult: PO** 500 mg q6hr × 3 wk with **IM** 1 g streptomycin bid × 1st wk then daily × 2nd wk

Urethral, endocervical, rectal infections *(C. trachomatis)*

• **Adult: PO** 500 mg qid × 7 days

Acne

• **Adult and adolescent: PO** 250 mg q6hr then 125-500 mg/day or every other day

Renal dose

• **Adult: PO** CCr 51-90 ml/min, give dose q8-12hr; CCr 10-50 ml/min, give dose q12-24hr; CCr <10 ml/min, give dose q24hr

Available forms: Caps 250, 500 mg

Administer:

• After C&S obtained
• 2 hr before or after iron products; 1 hr after antacid products
• Should be given on empty stomach (1 hr before or 2 hr after meals)
• Store in tight, light-resistant container at room temperature

SIDE EFFECTS

CNS: Fever, headache, paresthesia

CV: Pericarditis

EENT: Dysphagia, glossitis, decreased calcification, discoloration of deciduous teeth, oral candidiasis, oral ulcers

GI: *Nausea,* abdominal pain, *vomiting, diarrhea,* anorexia, enterocolitis, hepatotoxicity, flatulence, abdominal cramps, epigastric burning, stomatitis, hepatitis, pseudomembranous colitis

GU: *Increased BUN,* azotemia, acute renal failure

HEMA: Eosinophilia, neutropenia, thrombocytopenia, leukocytosis, hemolytic anemia

INTEG: *Rash, urticaria, photosensitivity, increased pigmentation,* exfoliative dermatitis, pruritus, angioedema, Stevens-Johnson syndrome

MISC: Increased intracranial pressure, candidiasis

PHARMACOKINETICS

PO: Peak 2-3 hr; duration 6 hr; half-life 6-12 hr; excreted in urine, breast milk; crosses placenta; 65% protein bound

INTERACTIONS

⚠ Fatal nephrotoxicity: methoxyflurane

Increase: effect of warfarin, digoxin

Decrease: effect of tetracycline—antacids, sodium bicarbonate, alkali products, iron, cimetidine

Decrease: effect of penicillins

Drug/Herb

• Photosensitivity: dong quai

Drug/Food

Decrease: tetracycline effect—dairy products

Drug/Lab Test

Increase: BUN, LFTs

NURSING CONSIDERATIONS

Assess:

⚠ **Pseudomembranous colitis:** diarrhea, abdominal pain, fever, fatigue, anorexia; possible anemia, elevated WBC count, low serum albumin; stop product; usually either vancomycin or IV metroNIDAZOLE is given

• Signs of anemia: Hct, Hgb, fatigue
• I&O ratio
• Blood studies: PT, CBC, AST, ALT, BUN, creatinine if on prolonged therapy
• **Allergic reactions:** rash, itching, pruritus
• **Serious skin reactions:** angioedema, Stevens-Johnson syndrome, exfoliative dermatitis; report immediately after stopping product
• Nausea, vomiting, diarrhea; administer antiemetic, antacids as ordered

⚠ **Superinfection:** fever, malaise, redness, pain, swelling, drainage, perineal itching, diarrhea, changes in cough or sputum if on prolonged therapy

Evaluate:

• Therapeutic response: absence of lesions, negative C&S, resolution of infection, prevention of malaria

Teach patient/family:

• To avoid sun exposure; that sunscreen does not seem to decrease photosensitivity

• That all prescribed medication must be taken to prevent superinfection

• To avoid milk products, antacids or to separate by 2 hr; to take with full glass of water; to take 1 hr before bedtime to prevent esophageal ulceration

• That tooth discoloration may occur

• To notify prescriber immediately of diarrhea with pus, mucous, fever, abdominal pain

⚠ To notify prescriber if pregnancy is planned or suspected, pregnancy (D)

• Not to use outdated products, Fanconi syndrome (nephrotoxicity) may occur

tetrahydrozoline nasal agent

See Appendix B

tetrahydrozoline ophthalmic

See Appendix B

theophylline (Rx)

(thee-off'i-lin)

Phyllocontin ✦, Theochron, Theo-24, Uniphyl

Func. class.: Bronchodilator, spasmolytic

Chem. class.: Methylxanthine

ACTION: Exact mechanism unknown, relaxes smooth muscle of respiratory system by blocking phosphodiesterase, which increases cAMP; increased cAMP alters intracellular calcium ion movements; produces bronchodilation, increased pulmonary blood flow, relaxation of respiratory tract

USES: Bronchial asthma, bronchospasm associated with chronic bronchitis, emphysema, apnea during infancy for respiratory/myocardial stimulation

Unlabeled uses: Methotrexate toxicity, sleep apnea, status asthmaticus

CONTRAINDICATIONS: Hypersensitivity to xanthines, tachydysrhythmias

Precautions: Pregnancy (C), breastfeeding, children, geriatric patients, CHF, cor pulmonale, hepatic disease, diabetes mellitus, hyperthyroidism, hypertension, seizure disorder, irritation of the rectum/lower colon, alcoholism, active peptic ulcer disease

DOSAGE AND ROUTES

Reversible airway obstruction (bronchospasm prophylaxis) associated with asthma or chronic obstructive pulmonary disease (COPD, emphysema, chronic bronchitis)

Calculate initial mg/kg dose based on ideal body weight as theophylline distributes poorly into body fat

Acute exacerbations of the symptoms and reversible airflow obstruction associated with asthma and other chronic lung diseases (COPD) along with inhaled β-2 selective agonists and systemic corticosteroids

• **Adult <60 yr/adolescent ≥16 yr: IV CONT INF** expressed in theophylline; initially, 0.4 mg/kg/hr; max 900 mg/day (otherwise healthy nonsmokers); patients who smoke may require an increased dose; give 0.2 mg/kg/hr (≥400 mg/day) in CHF, cor pulmonale, liver dysfunction, sepsis with multiorgan failure, shock; reduced doses may be needed in patients receiving other drugs that decrease theophylline clearance; adjust dosage based on serum concs

• **Adult ≥60 yr: IV CONT INF** initially, 0.3 mg/kg/hour; give 0.2 mg/kg/hr in CHF, cor pulmonale, liver dysfunction, sepsis with multiorgan failure, shock; reduced doses may be needed in patients receiving other drugs that decrease theophylline clearance; adjust dosage based serum concs; max 400 mg/day unless serum conc and patient condition require a higher dose

• **Child ≥12 yr/adolescent ≥15 yr: IV CONT INF** initially, 0.5 mg/kg/hr (healthy nonsmokers (up to 900 mg/day); 0.7 mg/kg/hr in smokers; 0.2 mg/kg/hr (up to 400 mg/day) in cardiac decompensation, cor pulmonale, liver dysfunction, sepsis with multiorgan failure, or shock); adjust dosage based on serum concs

• **Child 9-11 yr: IV CONT INF** initially, 0.7 mg/kg/hr; reduced doses may be needed in patients receiving other drugs that decrease theophylline clearance; give 0.2 mg/kg/hr (up to 400 mg/day) in cardiac decompensation, cor pulmonale, liver dysfunction, sepsis with multiorgan failure, or shock; adjust dosage based on serum concs

• **Child 1-8 yr: IV CONT INF** initially, 0.8 mg/kg/hr; reduced doses may be needed in patients receiving other drugs that decrease theophylline clearance; 0.2 mg/kg/hr (up to 400 mg/day) in cardiac decompensation, cor pulmonale, liver dysfunction, sepsis with multiorgan failure, or shock; adjust dosage based on serum concs

• **Infant 6-52 wk: IV CONT INF** calculate initial dosage using the following equation: (0.008 × age in wk) + 0.21 = theophylline dosage in mg/kg/hr IV; reduced doses may be needed in patients with risk factors for reduced theophylline clearance (cimetidine therapy, cardiac or liver dysfunction, renal impairment in infant <3 mo); adjust dosage based on serum concs

For maintenance therapy of asthma or COPD

• **Adult/adolescent/child >45 kg: PO** (expressed in theophylline and reg release products) PO initially, 300 mg/day in divided doses q6-8hr; after 3 days, if tolerated, increase dose to 400 mg/day in divided doses q6-8hr; after 3 more days, if tolerated, increase dose to 600 mg/day PO in divided doses q6-8hr; adjust dose to maintain therapeutic range; doses of 400-1600 mg/day may be needed, max 400 mg/day in patients with risk factors for decreased theophylline clearance (elderly patients >60 yr); NIH recommends initial dose for asthma of 10 mg/kg/day (max: 300 mg/day) titrated to serum conc of 5-15 mcg/ml (usual max: 800 mg/day in ≥12 yr and 16 mg/kg/day in children 1-11 yr)

• **Child/adolescent ≤15 yr old and ≤45 kg: PO** initially, 12-14 mg/kg (max 300 mg) per day in divided doses q4-6hr; after 3 days, if tolerated, may increase to 16 mg/kg (max 400 mg) per day in divided doses q4-6hr; after 3 more days, if tolerated and if needed, increase dose to 20 mg/kg (max 600 mg) per day in divided doses q4-6hr; adjust dose to maintain therapeutic range (doses of 10-36 mg/kg/day in child 1-9 yr may be needed); max 16 mg/kg (400 mg) per day in patients with risk factors for decreased theophylline clearance or who cannot receive recommended serum conc monitoring; NIH recommends initial dose for asthma of 10 mg/kg/day (max 300 mg/day) titrated to serum conc of 5-15 mcg/ml (usual max 800 mg/day in patients ≥12 yr and 16 mg/kg/day in children 1-11 yr)

• **Neonate/infant ≤52 wk: PO** calculate initial dose using the following equation: [(0.2 × age in wks) + 5] × (kg body weight) = theophylline dosage in mg/day; in infants ≤26 wk, divide dose and administer q8hr and in infant >26 wk, divide dose and give q6hr; adjust dose to maintain a steady-state serum conc of 5-10 mcg/ml in neonates and 10-15 mcg/ml in infants; NIH recommends initial dose for asthma of 10 mg/kg/day titrated to theophylline serum conc of 5-15 mcg/ml, max in mg/kg/day = [(0.2 × age in weeks) + 5]

• Extended-release products are to be used only for chronic disease management; do not use in treatment of acute symptoms of asthma and reversible bronchospasm

• **Adult/adolescents/child ≥12 yr and >45 kg: (Theo-24 Cap) PO** initially, 300-400 mg/day q24hr; evening dosing is not recommended; after 3 days, if tolerated, increase to 400-600 mg/day q24hr; adjust dose to maintain therapeutic range; doses of 400-1600 mg/day may be needed; max 400 mg/day in patients with decreased theophylline clearance, (elderly >60 yr), and in those who cannot receive recommended serum conc monitoring

• **Child >12 yr/adolescent <16 yr and <45 kg: (Theo-24 Cap) PO** initially, 12-14 mg/kg (max 300 mg) per day q24hr; evening dosing is not recommended; after 3 days, if tolerated, may increase to 16 mg/kg (max 400 mg) per day; after 3 more days, if tolerated and if needed, increase to 20 mg/kg (max 600 mg) per day; adjust dose to maintain therapeutic range; max 16 mg/kg/day up to 400 mg/day in patients with decreased theophylline clearance

Treatment of methotrexate toxicity (unlabeled)

• **Adult/adolescents/child ≥3 yr: IV** 2.5 mg/kg over 45-60 min

Available forms: Inj 250 mg/10 ml, 500 mg/20 ml, 100 mg/100 ml in 0.45% NaCl, 200 mg/100 ml in 0.45% NaCl; rect supp 250, 500 mg; tabs 100, 200 mg; ext rel tab 100, 200, 300, 400, 450, 600 mg; ext rel cap 100, 200, 300, 400; elixir 80 mg/15 ml

Administer:

• Avoid IM inj; pain and tissue damage may occur

PO route

• Avoid giving with food

Continuous IV INF route

• Only clear sol; flush IV line before dose

• May be **diluted** for IV inf in 100-200 ml in D_5W, $D_{10}W$, $D_{20}W$, 0.9% NaCl, 0.45% NaCl, LR

• **Give** loading dose over 1/2 hr; max rate of inf 25 mg/min, use inf pump; after loading dose, give by cont inf

• Store diluted sol for 24 hr if refrigerated

Syringe compatibilities: Heparin, metoclopramide, PENTobarbital, thiopental

Y-site compatibilities: Allopurinol, amifostine, amphotericin B, amrinone, aztreonam, cefTAZidime, cholesteryl sulfate complex, cimetidine, cladribine, DOXOrubicin liposome, enalaprilat, esmolol, famotidine, filgrastim, fluconazole, fludarabine, foscarnet, gallium, granisetron, heparin sodium with hydrocortisone sodium succinate, labetalol, melphalan, meropenem, netilmicin, PACLitaxel, pancuronium, piperacillin/tazobactam, potassium chloride, propofol, ranitidine, remifentanil, sargramostim, tacrolimus, teniposide, thiotepa, tolazoline, vecuronium

SIDE EFFECTS

CNS: Anxiety, restlessness, insomnia, *dizziness,* **seizures,** headache, lightheadedness, muscle twitching, tremors

CV: *Palpitations, sinus tachycardia,* hypotension, flushing, **dysrhythmias,** edema

GI: *Nausea, vomiting,* diarrhea, dyspepsia, anal irritation (suppositories), epigastric pain, reflux, anorexia

GU: Urinary frequency, SIADH

INTEG: Flushing, urticaria

MISC: Hyperglycemia

RESP: Tachypnea, increased respiratory rate

PHARMACOKINETICS

Metabolized by liver (caffeine); excreted in urine; crosses placenta; appears in breast milk; half-life 6.5-10.5 hr; half-life increased in geriatric patients, hepatic disease, CHF, neonates, premature infants; protein binding 40%, crosses into CSF

PO: Onset 1/4 hr, peak 1-2 hr, duration 6-8 hr, well absorbed

PO-ER: Onset unknown, peak 4-7 hr, duration 8-12 hr, well absorbed slowly
IV: Onset rapid, duration 6-8 hr
RECT: Onset erratic, peak 1-2 hr, duration 6-8 hr, supp absorbed erratically, sol absorbed quickly

INTERACTIONS

• Dose-dependent reversal of neuromuscular blockade
• Dysrhythmias: halothane
• May increase or decrease theophylline levels: carBAMazepine, loop diuretics, isoniazid

Increase: action of theophylline, toxicity—cimetidine, nonselective β-blockers, erythromycin, clarithromycin, oral contraceptives, corticosteroids, interferons, fluoroquinolones, disulfiram, mexiletine, fluvoxaMINE, high doses of allopurinol, influenza vaccines, interferon, benzodiazepines
Increase: adverse reactions—tetracyclines
Increase: elimination—smoking
Decrease: effects of lithium
Decrease: effect of theophylline—nicotine products, adrenergics, barbiturates, phenytoin, ketoconazole, rifampin

Drug/Herb
Increase: effects—cola tree, guarana, yerba maté, tea (black, green), horsetail, ginseng, Siberian ginseng
Decrease: effects—St. John's wort

Drug/Food
Increase: effect—xanthines
Increase: elimination by low-carbohydrate, high-protein diet; charcoal-broiled beef
Decrease: elimination by high-carbohydrate, low-protein diet

Drug/Lab Test
Increase: plasma-free fatty acids

NURSING CONSIDERATIONS

Assess:
• Theophylline blood levels (therapeutic level is 10-20 mcg/ml); toxicity may occur with small increase above 20 mcg/ml, especially in geriatric patients; whether theophylline was given recently (24 hr)
• Monitor I&O; diuresis occurs; dehydration may occur in geriatric patients or children
• Liver function tests: periodically, lower doses may be required in those with moderate to severe hepatic disease
• Respiratory rate, rhythm, depth; auscultate lung fields bilaterally; notify prescriber of abnormalities, monitor pulmonary function tests
• Allergic reactions: rash, urticaria; if these occur, product should be discontinued

Evaluate:
• Therapeutic response: decreased dyspnea, respiratory stimulation for infants, clear lung fields bilaterally

Teach patient/family:
• To take doses as prescribed; not to skip dose; not to double dose
• To check OTC medications, current prescription medications for ePHEDrine; will increase CNS stimulation; not to drink alcohol or caffeine products (tea, coffee, chocolate, colas); to avoid large amounts of charcoal-grilled beef
• To avoid hazardous activities; dizziness may occur
• If GI upset occurs, to take product with 8 oz water; to avoid food because absorption may be decreased
⚠ To notify prescriber immediately about **toxicity:** insomnia, anxiety, nausea, vomiting, rapid pulse, seizures, flushing, headache, diarrhea
• To notify prescriber about changes in smoking habits because a change in dose may be required; to avoid smoking because it decreases drug blood levels and terminal half-life
• To increase fluids to 2 L/day to decrease secretion viscosity

thiamine (vit B_1) (PO-OTC; IV, IM-Rx)

Vitamin B-1

Func. class.: Vit B_1

Chem. class.: Water soluble

Do not confuse:
thiamine/Tenormin

ACTION: Needed for pyruvate metabolism, carbohydrate metabolism

USES: Vit B_1 deficiency or polyneuritis, cheilosis adjunct with thiamine beriberi, Wernicke-Korsakoff syndrome, pellagra, metabolic disorders, alcoholism

CONTRAINDICATIONS: Hypersensitivity

Precautions: Pregnancy (A)

DOSAGE AND ROUTES

RDA

- **Adult: PO** (Males) 1.2-1.5 mg; (females) 1.1 mg; (pregnancy) 1.4 mg; (breastfeeding) 1.4 mg
- **Child 9-13 yr: PO** 0.9 mg
- **Child 4-8 yr: PO** 0.6 mg
- **Child 1-3 yr: PO** 0.5 mg
- **Infant 7 mo-1 yr: PO** 0.3 mg
- **Neonate and infant ≤6 mo: PO** 0.2 mg

Beriberi

- **Adult: PO** 5-30 mg daily or in 3 divided doses × 1 mo; **IM/IV** 5-30 mg daily or in 3 divided doses then convert to **PO**
- **Infant/child: PO** 10-50 mg daily × 2 wk then 5-10 mg daily × 1 mo; **IV/IM** 10-25 mg/day × 2 wk then 5-10 mg daily × 1 mo

Available forms: Tabs 50, 100, 250, 500 mg; inj 100 mg/ml; enteric-coated tabs 20 mg

Administer:

IM route

- By IM inj; rotate sites if pain and inflammation occur; do not mix with alkaline sol; Z-track to minimize pain

Direct IV route

- Undiluted at 100 mg/ml over 5 min

Continuous IV INF route

- Diluted in compatible IV sol

Y-site compatibilities: Famotidine

SIDE EFFECTS

CNS: Weakness, restlessness

CV: Collapse, pulmonary edema, hypotension

EENT: Tightness of throat

GI: Hemorrhage, *nausea, diarrhea*

INTEG: Angioneurotic edema, cyanosis, sweating, warmth

SYST: Anaphylaxis

PHARMACOKINETICS

PO/INJ: Unused amounts excreted in urine (unchanged)

NURSING CONSIDERATIONS

Assess:

- **Anaphylaxis (IV only):** swelling of face, eyes, lips, throat, wheezing
- **Thiamine deficiency:** anorexia, weakness/pain, depression, confusion, blurred vision, tachycardia
- Nutritional status: yeast, beef, liver, whole or enriched grains, legumes
- Application of cold to help decrease pain

Evaluate:

- Therapeutic response: absence of nausea, vomiting, anorexia, insomnia, tachycardia, paresthesias, depression, muscle weakness

Teach patient/family:

- About the necessary foods to be included in diet: yeast, beef, liver, legumes, whole grains

thioridazine (Rx)

(thye-or-rid'a-zeen)

Func. class.: Antipsychotic (typical)

Chem. class.: Phenothiazine piperidine

Do not confuse:
thioridazine/thiothixene/Thorazine

T

ACTION: Depresses cerebral cortex, hypothalamus, limbic system, which control activity, aggression; blocks neurotransmission produced by dopamine at synapse; exhibits strong α-adrenergic and anticholinergic blocking action; mechanism for antipsychotic effects is unclear

USES: Psychotic disorders, schizophrenia, behavioral problems in children, anxiety, major depressive disorders, organic brain syndrome

Unlabeled uses: Behavioral symptoms associated with dementia in geriatric patients

CONTRAINDICATIONS: Children <2 yr, hypersensitivity, coma, CNS depression

Black Box Warning: QT prolongation, cardiac dysrhythmias

Precautions: Pregnancy (C), breastfeeding, seizure disorders, hypertension, hepatic/pulmonary disease, renal failure, BPH, glaucoma, phenothiazine hypersensitivity, suicidal ideation, smoking, Reye's syndrome, Parkinson's disease

Black Box Warning: Cardiac disease, dementia, AV block, bundle branch block, torsades de pointes

DOSAGE AND ROUTES

Psychosis

- **Adult: PO** 25-100 mg tid, max 800 mg/day; dose gradually increased to desired response then reduced to minimum maintenance

Depression/behavioral problems/organic brain syndrome

- **Adult: PO** 25 mg tid, range from 10 mg bid-qid to 50 mg tid-qid, max 800 mg/day for short period
- **Geriatric: PO** 10-25 mg daily-tid, increase 4-7 days by 10-25 mg to desired dose, max 800 mg/day for short period
- **Child 2-12 yr: PO** 0.5-3 mg/kg/day in divided doses, max 3 mg/kg/day

Available forms: Tabs 10, 25, 50, 100 mg

Administer:

- Antiparkinsonian agent on order from prescriber for EPS
- Avoid use with CNS depressants
- Antacids separated by 2 hr or more
- Store in tight, light-resistant container; avoid contact with skin

SIDE EFFECTS

CNS: *EPS: pseudoparkinsonism, akathisia, dystonia, tardive dyskinesia,* seizures, *headache,* confusion, neuroleptic malignant syndrome, dizziness, drowsiness

CV: Orthostatic hypotension, cardiac arrest, ECG changes, tachycardia, QT prolongation, torsades de pointes

EENT: Blurred vision, glaucoma, dry eyes

GI: *Dry mouth, nausea, vomiting, anorexia, constipation,* diarrhea, jaundice, weight gain

GU: Urinary retention, urinary frequency, enuresis, impotence, amenorrhea, gynecomastia, ejaculation dysfunction, priapism

HEMA: Anemia, leukopenia, leukocytosis, agranulocytosis

INTEG: *Rash,* photosensitivity, dermatitis

RESP: Laryngospasm, dyspnea, respiratory depression

PHARMACOKINETICS

PO: Onset erratic; peak 2-4 hr; metabolized by liver; excreted in urine, breast milk; crosses placenta; half-life 26-36 hr; 91%-99% protein binding

INTERACTIONS

⚠ **Increase:** QT prolongation—class IA/III antidysrhythmics, some phenothiazines, β agonists, local anesthetics, tricyclics, haloperidol, chloroquine, droperidol, pentamidine; CYP3A4 inhibitors (amiodarone, clarithromycin, erythromycin, telithromycin, troleandomycin), arsenic trioxide, levomethadyl; CYP3A4 substrates (methadone, pimozide, QUEtiapine, quiNIDine, risperiDONE, ziprasidone), CYP2D6 inhibitors

• Oversedation: other CNS depressants, alcohol, barbiturate anesthetics

Increase: anticholinergic effects—anticholinergics

Decrease: antiparkinson's agent effects

Decrease: thioridazine effect—lithium, barbiturates

Decrease: antihypertensive effect—centrally acting antihypertensives

Decrease: absorption—aluminum hydroxide, magnesium hydroxide antacids

Drug/Lab Test

Increase: LFTs, prolactin, bilirubin, alk phos

Decrease: Hct/Hgb, platelets, granulocytes, leukocytes, neutrocytes, eosinophils

NURSING CONSIDERATIONS

Assess:

• Mental status before 1st dose
• I&O ratio; palpate bladder if low urinary output occurs, urinary retention may be the cause
• Bilirubin, CBC, LFTs monthly
• Urinalysis recommended before and during prolonged therapy
• Affect, orientation, LOC, reflexes, gait, coordination, sleep pattern disturbances
• B/P standing, lying; include pulse and respirations q4hr during initial treatment; establish baseline before starting treatment; report drops of 30 mm Hg
• Dizziness, faintness, palpitations, tachycardia on rising
• **EPS** including akathisia (inability to sit still, no pattern to movements), tardive dyskinesia (bizarre movements of jaw, mouth, tongue, extremities), pseudoparkinsonism (rigidity, tremors, pill rolling, shuffling gait)

⚠ **Neuroleptic malignancy syndrome:** altered mental status, muscle rigidity, increased CPK, hyperthermia, dyspnea, fatigue

⚠ **QT prolongation:** ECG, ejection fraction; assess for chest pain, palpitations, dyspnea

• Constipation, urinary retention daily; increase bulk, water in diet
• Supervised ambulation if needed until stabilized on medication; do not involve patient in strenuous exercise program because fainting is possible; patient should not stand still for long periods
• Sips of water, candy, gum for dry mouth

Evaluate:

• Therapeutic response: decrease in emotional excitement, hallucinations, delusions, paranoia; reorganization of patterns of thought, speech

Teach patient/family:

• That orthostatic hypotension occurs frequently; to rise from sitting or lying position gradually; to avoid hazardous activities until stabilized on medication
• To avoid hot tubs, hot showers, tub baths because hypotension may occur
• To avoid abrupt withdrawal of thioridazine because EPS may result; that product should be withdrawn slowly
• To avoid OTC preparations (cough, hay fever, cold) unless approved by prescriber; that serious product interactions may occur; to avoid use with alcohol because increased drowsiness may occur
• To use sunscreen to prevent burns
• About compliance with product regimen
• About the necessity for meticulous oral hygiene because oral candidiasis may occur
• To report sore throat, malaise, fever, bleeding, mouth sores; if these occur, CBC should be drawn and product discontinued; that product may cause vision impairment, should be reported to prescriber
• That in hot weather, heat stroke may occur; to take extra precautions to stay cool
• That product may cause discoloration of urine

TREATMENT OF OVERDOSE:

Lavage if orally ingested, provide an airway; do not induce vomiting; CV monitoring, continuous ECG

T

thyroid USP (desiccated) (Rx)

(thye′roid)

Armour Thyroid, Bio-Throid, Nature Thyroid, NP Thyroid

Func. class.: Thyroid hormone

Chem. class.: Active thyroid hormone in natural state and ratio

ACTION: Increases metabolic rates, increases cardiac output, O_2 consumption, body temperature, blood volume; growth, development at cellular level

USES: Hypothyroidism, cretinism (juvenile hypothyroidism), myxedema

CONTRAINDICATIONS: Adrenal insufficiency, MI, thyrotoxicosis, porcine protein hypersensitivity

Black Box Warning: Obesity treatment

Precautions: Pregnancy (A), breastfeeding, geriatric patients, angina pectoris, hypertension, ischemia, cardiac disease

DOSAGE AND ROUTES

Hypothyroidism

- **Adult: PO** 30 mg/day, increased by 15 mg/mo until desired response; maintenance dose 60-120 mg/day
- **Geriatric: PO** 7.5-15 mg/day, increase dose q6-8wk until desired response

Cretinism/juvenile hypothyroidism

- **Child: PO** 15 mg/day, then 30 mg/day after 2 wk, then 60 mg/day after another 2 wk; maintenance dose 60-180 mg/day

Myxedema

- **Adult: PO** 15 mg/day, double dose q2wk, maintenance 60-180 mg/day

Available forms: Tabs 16, 32, 60, 65, 98, 130, 195, 260, 325 mg; enteric-coated tabs 32, 65, 130 mg; sugarcoated tabs 32, 65, 130, 195 mg; caps 65, 130, 195, 325 mg

Administer:

- In AM if possible as single dose to decrease sleeplessness; separate iron, calcium products by 4 hr
- At same time each day to maintain product level
- Only for hormone imbalances; not to be used for obesity, male infertility, menstrual disorders, lethargy
- Lowest dose that relieves symptoms
- Removal of medication 4 wk before RAIU test

SIDE EFFECTS

CNS: *Insomnia, tremors,* headache, thyroid storm

CV: *Tachycardia, palpitations, angina, dysrhythmias,* hypertension, cardiac arrest

GI: Nausea, diarrhea, increased or decreased appetite, cramps

MISC: Menstrual irregularities, weight loss, sweating, heat intolerance, fever

PHARMACOKINETICS

PO: Peak 12-48 hr, half-life 6-7 days

INTERACTIONS

Increase: effects of anticoagulants, sympathomimetics, tricyclics, catecholamines

Decrease: thyroid absorption—bile acid sequestrants, aluminum, magnesium, calcium

Decrease: thyroid effects—estrogens

Drug/Herb

Decrease: thyroid effect—soy

Drug/Lab Test

Increase: CPK, LDH, AST, PBI, blood glucose

Decrease: thyroid function tests

NURSING CONSIDERATIONS

Assess:

Black Box Warning: **Obesity treatment:** use can lead to serious or life-threatening toxicity

- B/P, pulse before each dose
- I&O ratio

• Weight daily in same clothing, using same scale, at same time of day
• Height, growth rate of child
• T_3, T_4, which are decreased; radioimmunoassay of TSH, which is increased; radio uptake, which is decreased if dosage is too low
• PT may require decreased anticoagulant; check for bleeding, bruising
• **Hyperthyroidism:** increased nervousness, excitability, irritability; may indicate too high of a dose of medication, usually after 1-3 wk of treatment
• **Hypothyroidism:** lethargy, cold intolerance, weight gain, constipation, muscle cramps; may indicate too low of a dose
• Cardiac status: angina, palpitation, chest pain, change in VS

Evaluate:
• Therapeutic response: absence of depression; increased weight loss, diuresis, pulse, appetite; absence of constipation, peripheral edema, cold intolerance; pale, cool, dry skin; brittle nails, alopecia, coarse hair, menorrhagia, night blindness, paresthesias, syncope, stupor, coma, rosy cheeks

Teach patient/family:
• That hair loss will occur in child, is temporary
• To report excitability, irritability, anxiety; indicates overdose
• Not to switch brands unless directed by prescriber
• That strong odor is normal
• That hypothyroid child will show almost immediate behavior/personality change
• That treatment product is not to be taken to reduce weight
• To avoid OTC preparations with iodine; to read labels
• To separate iron, calcium products by 4 hr
• To avoid iodine food, iodized salt, soybeans, tofu, turnips, some seafood, some bread

tiaGABine (Rx)
(tie-ah-ga′been)
Gabitril
Func. class.: Anticonvulsant

Do not confuse:
tiaGABine/tiZANidine

ACTION: Inhibits reuptake and metabolism of GABA, may increase seizure threshold; structurally similar to GABA; tiaGABine binding sites in neocortex, hippocampus

USES: Adjunct treatment of partial seizures in adults and children ≥12 yr

CONTRAINDICATIONS: Hypersensitivity

Precautions: Pregnancy (C), breastfeeding, children <12 yr, geriatric patients, renal/hepatic disease, suicidal thoughts/behaviors, status epilepticus, mania, bipolar disorder, abrupt discontinuation, depression

DOSAGE AND ROUTES
When not given with a CYP3A4 enzyme, effect of tiaGABine is doubled; lower doses are indicated
• **Adult (those receiving an enzyme-inducing antiepileptic product): PO** 4 mg/day in divided doses, may increase by 4-8 mg/wk until desired response, max 56 mg/day
• **Child 12-18 yr: PO** 4 mg/day, may increase by 4 mg at beginning of wk 2; may increase by 4-8 mg/wk until desired response; max 32 mg/day

Hepatic dose
• **Adult: PO** reduce dose or increase dosing interval

Available forms: Tabs 2, 4, 12, 16 mg

Administer:
• Store at room temperature, away from heat and light
• With food

SIDE EFFECTS

CNS: *Dizziness, anxiety,* somnolence, ataxia, confusion, *asthenia,* unsteady gait, depression, suicidal ideation, seizures, tremors, hostility, EEG changes
CV: Vasodilation, tachycardia, hypertension
ENDO: Goiter, hypothyroidism
GI: Nausea, vomiting, diarrhea, increased appetite
INTEG: Pruritus, rash, Stevens-Johnson syndrome, alopecia, hyperhidrosis
MS: Myalgia
RESP: Pharyngitis, coughing

PHARMACOKINETICS

Absorption >95%; peak 45 min; protein binding 96%; metabolized in the liver via CYP3A4; half-life 7-9 hr without enzyme inducers, 2-5 hr with enzyme inducers

INTERACTIONS

- Lower doses may be needed when used with valproate

Increase: CNS depression—CNS depressants, alcohol
Decrease: tiaGABine effect—sevelamer
Decrease: effect—carBAMazepine, PHENobarbital, phenytoin, primidone
Drug/Food
Decrease: rate of absorption—high-fat meal

NURSING CONSIDERATIONS

Assess:

- Renal studies: urinalysis, BUN, urine creatinine q3mo
- Hepatic studies: ALT, AST, bilirubin
- **Seizures:** location, duration, presence of aura; assess for weakness
- Withdraw gradually to prevent seizures
- May cause status epilepticus and unexplained death

⚠ Mental status: mood, sensorium, affect, behavioral changes, suicidal thoughts/behaviors; if mental status changes, notify prescriber, hypomania may be present before suicide attempt

- Assistance with ambulation during early part of treatment; dizziness occurs
- Seizure precautions: padded side rails; move objects that may harm patient

Evaluate:

- Therapeutic response: decreased seizure activity; document on patient's chart

Teach patient/family:

- To carry emergency ID stating patient's name, products taken, condition, prescriber's name and phone number
- To avoid driving, other activities that require alertness
- Not to discontinue medication quickly after long-term use
- To take with food
- To notify prescriber if pregnancy is planned or suspected; avoid breastfeeding
- To report suicidal thoughts, behaviors immediately

TREATMENT OF OVERDOSE:

Lavage, VS

ticagrelor

Brilinta
Func. class.: Platelet inhibitor
Chem. class.: ADP receptor antagonist

ACTION: Reversibly bind to the platelet receptor, preventing platelet activation

USES: Arterial thromboembolism prophylaxis in acute coronary syndrome (ACS) (unstable angina, acute MI), including in patients undergoing percutaneous coronary intervention (PCI)

CONTRAINDICATIONS: Hypersensitivity, severe hepatic disease

Black Box Warning: Bleeding, intracranial bleeding

Precautions: Pregnancy (C), breastfeeding, infants, neonates, children, GI bleeding, hepatic disease, abrupt discontinuation

Black Box Warning: Coronary artery bypass graft surgery (CABG), surgery

⚠ Nurse Alert

DOSAGE AND ROUTES

• **Adult: PO** loading dose 180 mg with aspirin (usually 325 mg PO); then, give 90 mg bid with aspirin 75-100 mg/day, do not give maintenance doses of aspirin >100 mg/day

Available forms: Tab 90 mg

Administer:

PO route

• May be taken without regard to food
• Discontinue 5-7 days before surgery
• May be crushed (90 mg tab) and mixed with purified water, 100 mg (PO) or 50 ml (NG); ensure entire dose is given by flushing mortar, syringe, NG tube with 2 additional 50 ml of water
• Store at room temperature, in original container in dry place

SIDE EFFECTS

CNS: Headache, dizziness, fatigue
CV: Hypertension, hypotension, chest pain, atrial fibrillation, bradyarrhythmias, syncope, ventricular pauses
GI: Nausea, diarrhea
HEMA: Serious, fatal bleeding
MISC: Back pain, hyperuricemia, gynecomastia
RESP: Dyspnea, cough

PHARMACOKINETICS

Absolute bioavailability 36%, protein binding (>99%), metabolism by CYP3A4, weak P-glycoprotein substrates and inhibitors, elimination for product and metabolite are hepatic and biliary, 84% excreted in feces, 26% in urine, half-life is 7 hr for ticagrelor, 9 hr for metabolite, maximum inhibition of platelet aggregation (IPA) effect 2 hr, maintained ≥8 hr, peak 1.5 hr product, 2.5 hr metabolite

INTERACTIONS

Increase: bleeding risk—CYP3A4 inhibitors (ketoconazole, itraconazole, voriconazole, clarithromycin, telithromycin, nefazodone, ritonavir, lopinavir, ritonavir, saquinavir, nelfinavir, indinavir, atazanavir, delavirdine, isoniazid, dalfopristin, quinupristin, tipranavir
Decrease: ticagrelor action—CYP3A4 inducers (rifampin, dexamethasone, phenytoin, carBAMazepine, PHENobarbital)
Increase: effect of—simvastatin lovastatin
Increase: bleeding risk—NSAIDs, anticoagulants, platelet inhibitors
Increase or decrease: digoxin

Drug/Lab Test

Increase: serum creatinine

NURSING CONSIDERATIONS

Assess:

• **Thromboembolism:** Monitor CBC with differential with platelet count baseline and periodically during treatment

Black Box Warning: Bleeding: Assess for bleeding that may occur when aspirin is combined with this product; some bleeding can be fatal

• **Abrupt discontinuation:** do not discontinue abruptly, may increase risk for MI, stent thrombosis, death

Evaluate:

• Prevention of thromboembolism

Teach patient/family:

• To take only as prescribed; not to skip or double doses; if a dose is missed, to take next dose at scheduled time

Black Box Warning: To notify prescriber of chills, fever, bruising, bleeding

• Not to use any prescription, OTC products, herbs without approval of prescriber; products with aspirin, NSAIDs may cause bleeding
• To notify all health care providers of product use
• That product can be taken without regard to meals
• That it may take longer for bleeding to stop

ticarcillin/clavulanate (Rx)

Timentin

Func. class.: Broad-spectrum antiinfective

Chem. class.: Extended-spectrum penicillin, β-lactamase inhibitor

ACTION: Interferes with cell-wall replication of susceptible organisms; osmotically unstable cell wall swells, bursts from osmotic pressure; clavulanate inhibits β-lactamase and protects against enzymatic degradation of ticarcillin

USES: Respiratory, soft-tissue, and urinary tract infections; bacterial septicemia; effective for gram-positive cocci *(Staphylococcus aureus, Streptococcus faecalis, Streptococcus pneumoniae)*, gram-negative cocci *(Neisseria gonorrhoeae)*, gram-positive bacilli *(Clostridium perfringens, Clostridium tetani)*, gram-negative bacilli *(Bacteroides, Fusobacterium nucleatum, Escherichia coli, Proteus mirabilis, Salmonella, Morganella morganii, Proteus rettgeri, Enterobacter, Pseudomonas aeruginosa, Serratia)*; and *Peptococcus, Peptostreptococcus*, and *Eubacterium*

CONTRAINDICATIONS: Neonates, hypersensitivity to penicillins

Precautions: Pregnancy (B), hypersensitivity to cephalosporins, renal disease

DOSAGE AND ROUTES

Systemic/urinary tract infections, moderate/severe infections

- **Adult ≥60 kg: IV INF** 3.1 g q4-6hr
- **Adult <60 kg: IV INF** 200-300 mg/kg/day q4-6hr
- **Child >60 kg: IV INF** 3.1 g q4-6hr
- **Child <60 kg: IV INF** 200-300 mg/kg/day q4hr
- **Full-term neonates/infants <3 mo (unlabeled): IV** 50 mg/kg q4hr for severe infections; **IV** 50 mg/kg q6hr for mild to moderate infections

Mild to moderate infections

- **Child ≥60 kg: IV INF** 3.1 g q6hr
- **Child <60 kg: IV INF** 200 mg/kg/day q6hr

Renal dose

- **Adult: IV INF** loading dose 3.1 g; CCr 60 ml/min, 3.1 g q4hr; CCr 30-60 ml/min, 2 g q4hr; CCr 10-30 ml/min, 2 g q8hr; CCr <10 ml/min, 2 g q12hr; CCr <10 ml/min with hepatic dysfunction, 2 g q24hr

Available forms: Inj 3 g ticarcillin, 0.1 g clavulanate; IV inf 3 g ticarcillin, 0.1 g clavulanate; powder for inj 3 g ticarcillin, 0.1 g clavulanate

Administer:

- Product after C&S, give ≥q1hr before bactericidal anti-infectives, change IV site q48hr

Intermittent IV INF route

- After diluting ≤3.1 g/13 ml of sterile water or NaCl (200 mg/ml), shake; may further dilute in ≥50-100 ml NS, D_5W, or LR sol and run over ½ hr
- Store reconstituted sol 12-24 hr at room temperature, or 3-7 days refrigerated

Y-site compatibilities: Allopurinol, amifostine, amikacin, anidulafungin, atropine, aztreonam, bivalirudin, bumetanide, ceFAZolin, cefepime, cefotaxime, cefOXitin, cefTAZidime, ceftizoxime, cefTRIAXone, cefuroxime, chloramphenicol, cimetidine, clindamycin, cyclophosphamide, cycloSPORINE, dexamethasone, dexmedetomidine, digoxin, diltiazem, diphenhydrAMINE, DOCEtaxel, DOPamine, DOXOrubicin liposome, doxycycline, enalaprilat, EPINEPHrine, esmolol, etoposide phosphate, famotidine, fenoldopam, filgrastim, fluconazole, furosemide, gemcitabine, gentamicin, granisetron, heparin, hydrocortisone, HYDROmorphone, imipenem/cilastatin, insulin, isoproterenol, labetalol, levofloxacin, lidocaine, linezolid, LORazepam, melphalan, meperidine, methylPREDNISolone, metoclopramide, metoprolol, metroNIDAZOLE, milrinone, morphine, nitroglycerin, nitroprusside, norepinephrine, ondansetron, palonosetron, pantoprazole, PEMEtrexed, penicillin G potassium, perphenazine, phenylephrine, procainamide, propofol, propranolol, ranitidine, remifentanil, sargramostim, sodium bicarbonate, tacrolimus, teniposide, theophylline, thiotepa, tirofiban, tobramycin, vasopressin, verapamil, vinorelbine, voriconazole

SIDE EFFECTS

CNS: Anxiety, **seizures**, confusion, drowsiness

GI: *Nausea, vomiting, diarrhea;* increased AST, ALT; abdominal pain, glossitis, colitis, **pseudomembranous colitis, hepatotoxicity**

HEMA: Anemia, increased bleeding time, **bone marrow depression, granulocytopenia**

INTEG: Rash, urticaria, **toxic epidermal necrolysis,** pain at injection site

META: Hypokalemia, hypernatremia

SYST: **Anaphylaxis, Stevens-Johnson syndrome,** overgrowth of organisms

PHARMACOKINETICS

IV: Peak 30-45 min, duration 4 hr, half-life 64-68 min, excreted in urine

INTERACTIONS

⚠ **Increase: bleeding—anticoagulants**

Increase: methotrexate level—methotrexate, uric acid, bilirubicin, BUN, creatining, alk, phos, LDH

Increase: ticarcillin concs—probenecid, sulfipyrazone

Decrease: antimicrobial effect of ticarcillin—tetracyclines, aminoglycosides IV, chloramphenicol, macrolides, sulfonamides

Decrease: effect—**oral contraceptives,** erythromycin

Drug/Lab Test

False positive: urine glucose, urine protein, Coombs' test

Increase: LFTs, sodium, eosinophils, INR bleeding time

Decrease: Hgb, potassium, platelets, WBC, granulocytes

NURSING CONSIDERATIONS

Assess:

- Infection: WBC, wound, temperature, sputum, urine, baseline and periodically

⚠ **Pseudomembranous colitis: diarrhea, abdominal pain, fever, fatigue, anorexia; possible anemia, elevated WBC count, low serum albumin; stop product; usually either vancomycin or IV metroNIDAZOLE is given**

⚠ **Serious skin reactions: Stevens-Johnson syndrome, toxic epidermal necrolysis; anaphylaxis: wheezing, rash, laryngeal edema; have emergency equipment nearby**

- Hepatic studies: AST, ALT
- Blood studies: WBC, RBC, Hct, Hgb, bleeding time, platelets, baseline and periodically
- Renal studies: BUN, creatinine, sodium, potassium
- Bowel pattern before, during treatment
- Skin eruptions after administration of penicillin to 1 wk after discontinuing product
- EPINEPHrine, suction, tracheostomy set, endotracheal intubation equipment

Evaluate:

- Therapeutic response: resolution of infection

Teach patient/family:

- To report persistent diarrhea with blood, pus, mucous, or fever
- That culture may be taken after completed course of medication
- To report sore throat, fever, fatigue (may indicate superinfection); CNS effects (anxiety, depression, hallucinations, seizures)
- To wear or carry emergency ID if allergic to penicillins
- **To use alternative birth control method instead of hormonal**

TREATMENT OF OVERDOSE:

Withdraw product, maintain airway, administer EPINEPHrine, O_2, IV corticosteroids for anaphylaxis

ticlopidine (Rx)

(tye-cloe′pi-deen)

Func. class.: Platelet aggregation inhibitor

Chem. class.: Thienopyridine compound

ACTION:

Irreversible inhibition of platelet aggregation through antagonism of ADP

USES:

Reducing the risk for stroke in high-risk patients

Unlabeled uses: Intermittent claudication, chronic arterial occlusion, subarachnoid hemorrhage, uremic patients with AV shunts/fistulas, open heart surgery, coronary artery bypass grafts, primary glomerulonephritis, sickle cell disease, diabetic retinopathy

CONTRAINDICATIONS:

Hypersensitivity, severe hepatic disease, active bleeding, coagulopathy

Black Box Warning: Agranulocytosis, neutropenia, thrombocytopenia, thrombotic thrombocytopenic purpura (TTP)

Precautions: Pregnancy (B), breastfeeding, children, geriatric patients, past hepatic disease, renal disease, increased bleeding risk, peptic ulcer disease, surgery

Black Box Warning: Anemia, hematologic disease

DOSAGE AND ROUTES

- **Adult: PO** 250 mg bid with food

Available forms: Tabs 250 mg

Administer:

- With food to decrease gastric symptoms
- Discontinue when absolute neutrophil count falls during treatment to <1200/mm^3 or platelets <80,000/mm^3; may use methylPREDNISolone IV 20 mg to provide normal bleeding time in 2 hr

SIDE EFFECTS

CNS: Dizziness, headache, weakness

EENT: Tinnitus, epistaxis

GI: Nausea, vomiting, *diarrhea,* GI discomfort, cholestatic jaundice, hepatitis, increased cholesterol, LDL, VLDL, triglycerides

GU: Hematuria

HEMA: Bleeding (epistaxis, hematuria, conjunctival hemorrhage, GI bleeding), agranulocytosis, neutropenia, thrombocytopenia, thrombotic thombocytopenic purpura

INTEG: *Rash,* pruritus

META: Hypercholesterolemia, hypertriglyceridemia

PHARMACOKINETICS

Peak 1-3 hr; metabolized by liver; excreted in urine, feces; half-life increases with repeated dosing, initially 12-36 hr; antiplatelet effect 2-5 days; 98% protein binding

INTERACTIONS

Increase: levels of CYP2C19, CYP2DC substrates, phenytoin, fosphenytoin, ambrisentan, theophylline

Increase: bleeding tendencies—anticoagulants, salicylates, thrombolytics, NSAIDs, abciximab, eptifibatide, tirofiban, thrombin inhibitors, SSRIs, aspirin

Increase: effects of ticlopidine—cimetidine

Decrease: plasma levels of ticlopidine—antacids

Decrease: plasma levels of digoxin, cycloSPORINE

Drug/Herb

Increase: bleeding risk—ginger, ginkgo, garlic, feverfew, horse chestnut, green tea

NURSING CONSIDERATIONS

Assess:

- Hepatic studies: AST, ALT, bilirubin, creatinine with long-term therapy

Black Box Warning: Blood dyscrasias, bone marrow depression, do not use in those with a history of these conditions; blood studies: CBC; CBC q2wk × 3 mo, Hct, Hgb, PT with long-term therapy

⚠ Bleeding time at baseline and throughout treatment; levels may be 2-5× normal limit

Evaluate:

• Therapeutic response: absence of stroke

Teach patient/family:

• That blood work will be necessary during treatment

• To report any unusual bleeding to prescriber

• To report side effects such as diarrhea, skin rashes, subcut bleeding, signs of cholestasis (jaundiced skin and sclera, dark urine, light-colored stools)

• That product should be discontinued 10-14 days before surgery; not to double a missed dose

• That there are many product and herbal interactions to avoid all OTC products unless approved by prescriber

tigecycline (Rx)

(tye-ge-sye′kleen)

Tygacil

Func. class.: Broad-spectrum antiinfective

Chem. class.: Glycylcyclines

ACTION: Inhibits protein synthesis and phosphorylation in microorganisms; bacteriostatic structurally similar to the tetracyclines

USES: Complicated skin/skin-structure infections *(Escherichia coli, Enterococcus faecalis* [vancomycin-susceptible only] *Staphylococcus aureus, Streptococcus agalactiae, S. anginosus* group, *S. pyogenes, Bacteroides fragilis;* complicated intraabdominal infections *[Citrobacter freundii] Enterobacter cloacae, E. coli, Klebsiella oxytoca, K. pneumoniae, E. faecalis* [vancomycin-susceptible only], *S. aureus* [methicillin-susceptible only], *S. anginosus* group, *B. fragilis, Bacteroides thetaiotaomicron, B. uniformis, B. vulgatus, Clostridium perfringens, Peptostreptococcus micros);* community-acquired pneumonia

CONTRAINDICATIONS: Pregnancy (D), breastfeeding, children <18 yr, hypersensitivity to tigecycline

Precautions: Renal/hepatic disease, hypersensitivity to tetracyclines, ventilator-associated/hospital-acquired pneumonias

DOSAGE AND ROUTES

• **Adult: IV** 100 mg then 50 mg q12hr, **IV INF** given over 30-60 min q12hr; given for 5-14 days, depending on infection

Hepatic dose

• **Adult: IV** (Child-Pugh C) 100 mg then 25 mg q12hr

Available forms: Powder for inj, lyophilized 50 mg

Administer:

• Tigecycline allergy test before using, obtain C&S, do not begin treatment before results or if susceptible organism is strongly suspected

Intermittent IV INF route

• Reconstitute each vial with 5.3 ml of 0.9% NaCl or D_5 (10 mg/ml); swirl to dissolve; immediately withdraw 5 ml of reconstituted sol and add to 100-ml IV bag for inf (1 mg/ml); may be yellow or orange; if not, sol should be discarded; do not give if particulate matter is present, use a dedicated IV line or Y-site, flush with NS before and after use, give over 1/2 hr

• Store in tight, light-resistant container at room temperature, diluted sol at room temperature for up to 24 hr, 6 hr in vial, and remaining time in IV bag, ≤48 hr refrigerated

Y-site compatibilities: Acyclovir, alfentanil, allopurinol, amifostine, amikacin, aminocaproic acid, aminophylline, am-

T

photericin B liposome, ampicillin, ampicillin/sulbactam, argatroban, azithromycin, aztreonam, bivalirudin, bumetanide, buprenorphine, butorphanol, calcium chloride/gluconate, CARBOplatin, carmustine, caspofungin, ceFAZolin, cefepime, cefotaxime, cefoTEtan, cefOXitin, cefTAZidime, ceftizoxime, cefTRIAXone, cefuroxime, cimetidine, ciprofloxacin, cisatracurium, CISplatin, clindamycin, cyclophosphamide, cycloSPORINE, cytarabine, dacarbazine, DACTINomycin, DAPTOmycin, DAUNOrubicin hydrochloride, dexamethasone, dexmedetomidine, dexrazoxane, digoxin, diltiazem, diphenhydrAMINE, DOBUTamine, DOCEtaxel, dolasetron, DOPamine, doripenem, DOXOrubicin hydrochloride, DOXOrubicin liposome, droperidol, enalaprilat, EPINEPHrine, eptifibatide, ertapenem, erythromycin, esmolol, etoposide, etoposide phosphate, famotidine, fenoldopam, fentaNYL, fluconazole, fludarabine, fluorouracil, foscarnet, fosphenytoin, furosemide, ganciclovir, gemcitabine, gentamicin, glycopyrrolate, granisetron, haloperidol, heparin, hydrocortisone, HYDROmorphone, ifosfamide, imipenem/cilastatin, insulin, irinotecan, isoproterenol, ketorolac, labetalol, lansoprazole, lepirudin, leucovorin, levofloxacin, lidocaine, linezolid, LORazepam, magnesium sulfate, mannitol, mechlorethamine, melphalan, meperidine, meropenem, mesna, methohexital, methotrexate, methyldopa, metoclopramide, metoprolol, metroNIDAZOLE, midazolam, milrinone, mitoMYcin, mitoXANtrone, morphine, moxifloxacin, mycophenolate, nafcillin, nalbuphine, naloxone, nesiritide, nitroglycerin, nitroprusside, norepinephrine, octreotide, ondansetron, oxaliplatin, oxytocin, PACLitaxel, palonosetron, pamidronate, pancuronium, pantoprazole, PEMEtrexed, pemtamidine, pentazocin, PENTobarbital, PHENobarbital, phenylephrine, piperacillin/tazobactam, potassium acetate/chloride/phosphate, procainamide, prochlorperazine, promethazine, propofol, propranolol, ranitidine, remifentanil, rocuronium, sodium acetate/bicarbonate/phosphate, streptozocin, succinylcholine, SUFentanil, tacrolimus, teniposide, theophylline, thiopental, thiotepa, ticarcillin/clavulanate, tirofiban, tobramycin, topotecan, trimethoprim/sulfamethoxazole, vancomycin, vasopressin, vecuronium, vinBLAStine, vinCRIStine, vinorelbine, zidovudine, zoledronic acid

SIDE EFFECTS

CNS: Headache, dizziness, insomnia
CV: Hypo/hypertension, phlebitis
EENT: Tooth discoloration
GI: *Nausea, vomiting, diarrhea,* anorexia, constipation, dyspepsia, abdominal pain, hepatotoxicity, hepatic failure, pseudomembranous colitis
HEMA: Anemia, leukocytosis, thrombocytopenia
INTEG: *Rash,* pruritus, sweating, photosensitivity
META: Increased ALT, AST, BUN, lactic acid, alk phos, amylase; hyperglycemia, hypokalemia, hypoproteinemia, bilirubinemia
MISC: Back pain, fever, abnormal healing, abdominal pain, abscess, asthenia, infection, pain, peripheral edema, local reactions
RESP: Cough, dyspnea
SYST: Anaphylaxis

PHARMACOKINETICS

Not extensively metabolized, 22% of unchanged product excreted in urine, terminal half-life 42 hr, primarily biliary excreted, protein binding 71%-89%

INTERACTIONS

Increase: effect of warfarin
Decrease: effect of oral contraceptives

Drug/Lab Test

Increase: amylase, LFTs, alk phos, BUN, creatinine, LDH, WBC, INR, PTT, PT
Decrease: potassium, calcium, sodium, Hgb/Hct, platelets

NURSING CONSIDERATIONS

Assess:

⚠ **Pseudomembranous colitis:** diarrhea, abdominal pain, fever, fatigue, anorexia; possible anemia, elevated WBC level, low serum albumin; stop product; usually either vancomycin or IV metroNIDAZOLE is given

- Signs of anemia: Hct, Hgb, fatigue
- Blood studies: PT, CBC, AST, ALT, BUN creatinine

⚠ **Allergic reactions:** rash, itching, pruritus, angioedema

- Serious allergic skin reactions: Stevens-Johnson anaphylaxis
- Nausea, vomiting, diarrhea; administer antiemetic, antacids as ordered
- **Toxicity:** pseudotumor cerebri, photosensitivity, antianabolic actions (azotemia, BUN, hypophosphatemia, metabolic acidosis) tigecycline is structurally similar to tetracycline; pancreatitis, hyperamylasemia; (may be fatal) if these occur, discontinue, improvement usually occurs after product is discontinued
- **Overgrowth of infection:** fever, malaise, redness, pain, swelling, drainage, perineal itching, diarrhea, changes in cough or sputum

Evaluate:

- Therapeutic response: decreased temperature, absence of lesions, negative C&S

Teach patient/family:

- To avoid sun exposure; sunscreen does not seem to decrease photosensitivity

⚠ To avoid pregnancy while taking this product; fetal harm may occur; to avoid breastfeeding

- To report infection, increase in temperature; to report burning, pain at inj site

timolol (Rx)

(tye′moe-lole)

Apo-Timol ✱, Novo-Timol ✱

Func. class.: Antihypertensive

Chem. class.: Nonselective β-blocker

ACTION: Competitively blocks stimulation of β-adrenergic receptor within vascular smooth muscle (decreases rate of SA node discharge, increases recovery time); slows conduction of AV node and decreases heart rate, which decreases O_2 consumption in myocardium; also decreases renin-aldosterone-angiotensin system; at high doses, inhibits β_2-receptors in bronchial system

USES: Mild to moderate hypertension, migraine prophylaxis, to decrease mortality after MI

Unlabeled uses: Tremors, angina pectoris

CONTRAINDICATIONS: Hypersensitivity to β-blockers, cardiogenic shock, heart block (2nd/3rd degree), sinus bradycardia, CHF, cardiac failure, severe COPD, asthma

Precautions: Pregnancy (C), breastfeeding, major surgery, diabetes mellitus, COPD, well-compensated heart failure, nonallergic bronchospasm, peripheral vascular disease, thyroid/renal/hepatic disease

Black Box Warning: Abrupt discontinuation

DOSAGE AND ROUTES

Hypertension

- **Adult: PO** 10 mg bid or 20 mg/day, may increase by 10 mg q7days, max 60 mg/day
- **Geriatric patients: PO** initiate dose cautiously

Myocardial infarction

- **Adult: PO** 10 mg bid beginning 1-4 wk after MI for ≥2 yr

Migraine headache prevention

• **Adult: PO** 10 mg bid or 20 mg/day; may increase to 30 mg/day, 20 mg in AM, 10 mg in PM; discontinue if not effective after 8 wk

Available forms: Tabs 5, 10, 20 mg

Administer:

• PO before or immediately after meals, at bedtime; tab may be crushed or swallowed whole

• Reduced dosage in renal dysfunction

• Store at room temperature; do not freeze

SIDE EFFECTS

CNS: *Insomnia, dizziness,* hallucinations, anxiety, fatigue, depression, headache

CV: Hypotension, bradycardia, CHF, edema, chest pain, claudication, angina, AV block, ventricular dysrhythmias

EENT: *Visual changes;* sore throat; *double vision;* dry, burning eyes

GI: *Nausea,* vomiting, ischemic colitis, diarrhea, *abdominal pain,* mesenteric arterial thrombosis, flatulence, constipation

GU: Impotence, urinary frequency

HEMA: Agranulocytosis, thrombocytopenia, purpura

INTEG: Rash, alopecia, pruritus, fever

META: Hypoglycemia

MUSC: *Joint pain, muscle pain*

RESP: Bronchospasm, *dyspnea,* cough, crackles, nasal stuffiness

PHARMACOKINETICS

Peak 1-2 hr; half-life 4 hr; metabolized by liver; excreted in urine, breast milk; protein binding $<10\%$

INTERACTIONS

Increase: hypotension, bradycardia—hydrALAZINE, methyldopa, prazosin, anticholinergics, alcohol, reserpine, nitrates

Increase: effects of β-blockers, calcium channel blockers

Decrease: antihypertensive effects—NSAIDs, sympathomimetics, thyroid, salicylates

Decrease: hypoglycemic effects—insulin, sulfonylureas

Decrease: bronchodilation—theophyllines

Drug/Lab Test

Increase: renal, hepatic studies, uric acid

Interference: glucose, insulin tolerance test

NURSING CONSIDERATIONS

Assess:

Black Box Warning: Abrupt discontinuation: may result in myocardial ischemia, MI, severe hypotension, ventricular dysrhythmias in those with preexisting cardiovascular disease

• **Headaches:** location, severity, duration, frequency at baseline and throughout treatment

• I&O, weight daily

• B/P during initial treatment, periodically thereafter, pulse q4hr; note rate, rhythm, quality

• Apical/radial pulse before administration; notify prescriber of any significant changes

• Baselines of renal, hepatic studies before therapy begins

• Edema in feet, legs daily

Evaluate:

• Therapeutic response: decreased B/P after 1-2 wk

Teach patient/family:

• To take before or immediately after meals

Black Box Warning: Not to discontinue product abruptly; to taper over 2 wk; may cause precipitate angina

• Not to use OTC products containing α-adrenergic stimulants (nasal decongestants, cold preparations) unless directed by prescriber

• To report bradycardia, dizziness, confusion, depression, fever, sore throat, SOB to prescriber

• Product masks hypoglycemia; monitor blood sugar

- To take pulse at home; advise when to notify prescriber
- To avoid alcohol, smoking, sodium intake
- To comply with weight control, dietary adjustments, modified exercise program
- To carry emergency ID to identify product, allergies
- To avoid hazardous activities if dizziness is present
- To report symptoms of **CHF:** difficulty breathing, especially on exertion or when lying down; night cough; swelling of extremities
- To take medication at bedtime; to wear support hose to minimize effect of orthostatic hypotension

TREATMENT OF OVERDOSE:
Lavage, IV atropine for bradycardia, IV theophylline for bronchospasm, digoxin, O_2, diuretic for cardiac failure, hemodialysis; administer vasopressor (norepinephrine)

timolol (ophthalmic)
(tie-moe′lol)

Betimol, Istalol, Timoptic, Timoptic-XE

Func. class.: Antiglaucoma

Chem. class.: β-Blocker

ACTION:
Can decrease aqueous humor and increase outflows

USES:
Treatment of chronic open-angle glaucoma and ocular hypertension

CONTRAINDICATIONS:
Hypersensitivity, AV block, heart failure, bradycardia, sick sinus syndrome, asthma

Precautions: Abrupt discontinuation, pregnancy, breastfeeding, children, COPD, depression, diabetes mellitus, myasthenia gravis, hyperthyroidism, pulmonary disease, angle-closure glaucoma

DOSAGE AND ROUTES
- **Adult:** instill 1 drop in each affected eye bid (0.25% solution) initially; if no response, 1 drop in each affected eye bid (0.5% solution) or 1 drop of gel in each affected eye daily

Available forms: Ophthalmic solution 0.25, 0.5%; ophthalmic gel 0.25%, 0.5%

Administer:
- For ophthalmic use only
- Do not touch the tip of the dropper to the eye, fingertips, or other surface to prevent contamination
- Wash hands before and after use
- Tilt head back slightly and pull the lower eyelid down with the index finger to form a pouch; squeeze the prescribed number of drops into the pouch; close eyes to spread drops; to avoid excessive systemic absorption, apply finger pressure on the lacrimal sac for 1-2 min after use
- If >1 topical ophthalmic drug product is being used, the drugs should be administered at least 5 min apart
- Administer other topically applied ophthalmic medications at least 10 min before timolol gel-forming solution
- To avoid contamination or the spread of infection, do not use dropper for more than one person
- Some products contain the preservative benzalkonium chloride, which can be absorbed by soft contact lenses; remove contact lenses before administration of the solution; lenses may be reinserted 15 min after administration
- Decreased intraocular pressure can take several weeks, monitor IOP after a month

SIDE EFFECTS
CNS: *Insomnia,* headache, *dizziness,* anxiety, depression, headache, nightmares, *fatigue*

CV: Palpitations, heart failure, hypotension

EENT: Eye stinging/burning, tearing, photophobia, visual disturbances

T

GI: Nausea, dry mouth
PULM: Bronchospasm

PHARMACOKINETICS

Onset 30 min, peak 1-2 hr, duration 12-24 hr

INTERACTIONS

Increase: β-Blocking effect—oral β-blockers
Increase: Intraocular pressure reduction—topical miotics, dipivefrin, EPINEPHrine, carbonic anhydrase inhibitors; this may be beneficial
Increase: B/P, severe—when abruptly stopping cloNIDine
Increase: Depression of AV nodal conduction, bradycardia, or hypotension—adenosine, cardiac glycosides, disopyramide, other antiarrhythmics, class 1C antiarrhythmic drugs (flecainide, propafenone, moricizine, encainide quiNIDine, calcium-channel blockers, or drugs that significantly depress AV nodal conduction)
Increase: AV block nodal conduction, induce AV block—high doses of procainamide
Increase: Antihypertensive effect—other antihypertensives

NURSING CONSIDERATIONS

Assess:

⚠ **Systemic absorption**: When used in the eye, systemic absorption is common with the same adverse reactions and interactions

• Glaucoma: monitor intraocular pressure

Evaluate:

• Decreasing intraocular pressure

Teach patient/family:

• That product is for ophthalmic use only
• Not to touch the tip of the dropper to the eye, fingertips, or other surface to prevent contamination
• To wash hands before and after use
• To tilt the head back slightly and pull the lower eyelid down with the index finger to form a pouch; squeeze the prescribed number of drops into the pouch; close eyes to spread drops
• To apply finger pressure on the lacrimal sac for 1-2 min following use to prevent excessive systemic absorption
• To administer drugs at least 5 min apart if more than one topical ophthalmic drug product is being used
• To administer other topically applied ophthalmic medications at least 10 min before timolol gel-forming solution
• To not use dropper for more than one person to avoid contamination or the spread of infection
• That some products contain the preservative benzalkonium chloride, which may be absorbed by soft contact lenses; to remove contact lenses before administration of the solution; that lenses may be reinserted 15 min after administration

tinidazole (Rx)

(tye-ni′da-zole)

Tindamax

Func. class.: Antiprotozoal
Chem. class.: Nitroimidazole derivative

ACTION: Interferes with DNA/RNA synthesis in protozoa

USES: Amebiasis, giardiasis, trichomoniasis
Unlabeled uses: *Bacteroides* sp., *Clostridium* sp., *Eubacterium* sp., *Fusobacterium* sp., *Peptococcus* sp., *Peptostreptococcus* sp., gingivitis, urethritis, *Veillonella* sp.

CONTRAINDICATIONS: Pregnancy, breastfeeding; hypersensitivity to this product or nitroimidazole derivative
Precautions: Children, geriatric patients, hepatic disease, CNS depression, blood dyscrasias, candidiasis, seizures, viral infection, alcoholism, pregnancy (C)

Black Box Warning: Secondary malignancy

DOSAGE AND ROUTES

Intestinal amebiasis/amebic involvement of the liver

- **Adult: PO** 2 g daily × 3 days
- **Child ≥3 yr/adolescent: PO** 50 mg/kg/day × 3 days, max 2 g/day

Giardiasis

- **Adult: PO** 2 g as a single dose
- **Child ≥3 yr: PO** 50 mg/kg as a single dose, max 2 g

Trichomoniasis

- **Adult: PO** 2 g as a single dose

Bacterial vaginosis

- **Adult (nonpregnant woman): PO** 2 g/day × 2 days with food or 1 g/day × 5 days with food

Available forms: Tabs 250, 500 mg

Administer:

- Tabs can be crushed and mixed with artificial cherry syrup for children
- With food to increase plasma concs, minimize epigastric distress and other GI effects

SIDE EFFECTS

CNS: *Dizziness, headache,* **seizures,** *peripheral neuropathy,* malaise, fatigue

GI: *Nausea, vomiting,* anorexia, increased AST/ALT, constipation, abdominal pain, indigestion, altered taste

HEMA: **Leukopenia,** neutropenia

INTEG: Pruritus, urticaria, *rash,* oral candidiasis

SYST: **Angioedema,** cramping

PHARMACOKINETICS

Peak 1½ hr; metabolized extensively in liver; excreted unchanged (20%-25%) in urine, (12%) feces; half-life 12-14 hr; crosses blood-brain barrier

INTERACTIONS

- Do not use within 2 wk of disulfiram

Increase: tinidazole action—CYP3A4 inhibitors (cimetidine, ketoconazole): increased action of tinidazole

Increase: action of anticoagulants, cycloSPORINE, tacrolimus, fluorouracil, hydantoins, lithium

Decrease: tinidazole action—CYP3A4 inducers (PHENobarbital, rifampin, phenytoin); cholestyramine, oxytetracycline: decreased action of tinidazole

Drug/Herb

Increase or decrease: tinidazole level—St. John's wort

Drug/Lab Test

Increase: triglycerides, LDH, AST/ALT, glucose

Decrease: WBCs

NURSING CONSIDERATIONS

Assess:

- **Giardiasis:** obtain 3 stool samples several days apart beginning q3-4wk after treatment
- **Amebic liver abscess:** monitor CBC, ESR, amebic gel diffusion test, ultrasound; also total and differential leukocyte count

Black Box Warning: Secondary malignancy: avoid unnecessary use

- Signs of infection, anemia
- Bowel pattern before, during treatment

Evaluate:

- Therapeutic response: decrease in infection as evidenced by negative culture

Teach patient/family:

- To take with food to increase plasma concs, minimize epigastric distress and other GI effects; not to use alcoholic beverages during or for 3 days after treatment
- **Trichomoniasis:** both partners should be treated at the same time
- To avoid alcohol; may cause disulfiram reaction
- To avoid doing hazardous activities until reaction is known
- That product causes taste
- Not to use OTC, Rx, or herbal products unless approved by prescriber

T

HIGH ALERT

tinzaparin (Rx)

(tin-zay-par'in)

Innohep

Func. class.: Anticoagulant

Chem. class.: Unfractionated porcine heparin

ACTION: Inhibits the effect of factor Xa

USES: Treatment of DVT, PE after abdominal, knee, or hip surgery or after knee or hip replacement, cerebral thromboembolism, thrombosis prophylaxis

Unlabeled uses: Antiphospholipid antibody syndrome, arterial thromboembolism prophylaxis, cerebral thromboembolism, DVT prophylaxis, PE prophylaxis, thrombosis prophylaxis

CONTRAINDICATIONS: Hypersensitivity to this product, heparin, pork or benzyl alcohol, sulfites; hemophilia, leukemia with bleeding, peptic ulcer disease, thrombocytopenic purpura, heparin-induced thrombocytopenia

Precautions: Pregnancy (B), breastfeeding, children, geriatric patients, alcoholism, severe renal/hepatic disease, blood dyscrasias; severe, uncontrolled hypertension; subacute bacterial endocarditis, acute nephritis; geriatric patients >70 yr (renal disease with DVT/PE)

Black Box Warning: Spinal/epidural anesthesia, lumbar puncture

DOSAGE AND ROUTES

Treatment of DVT

- **Adult: SUBCUT** 175 anti-Xa international units/kg/day ≥6 days and until adequate anticoagulation with warfarin (therapeutic INR ≥2 for 2 consecutive days)
- **Adolescent/child 10-16 yr: SUBCUT** 275 anti-Xa international units/kg daily; adjust dose to maintain anti-Xa 0.5-1 international units/ml drawn 4 hr after inj
- **Child 5-10 yr: SUBCUT** 200 anti-Xa international units/kg daily; adjust dose to maintain anti-Xa of 0.5-1 international units drawn 4 hr after injection

Prophylaxis of DVT in orthopedic procedures (unlabeled)

- **Adult:** 75 anti-Xa units/kg/day started 12-24 hr after surgery for up to 14 days

Prophylaxis of DVT/ thromboembolism/PE (unlabeled)

- **Adult:** 3500 anti-Xa units (50 anti-Xa units/kg) daily beginning 1-2 hr before surgery and continued for 7-10 days

Available forms: Inj 20,000 international units/1 ml

Administer:

- Only after screening patient for bleeding disorders
- For 6 days, until warfarin has been given to result in adequate coagulation
- SUBCUT only; do not give IM/IV
- To recumbent patient; give SUBCUT; rotate inj sites (left/right anterolateral, left/right posterolateral abdominal wall)
- Insert whole length of needle into skin fold held with thumb and forefinger

Only this product when ordered; not interchangeable with heparin (unfractionated) or LMWHs

- At same time each day to maintain steady blood levels
- Do not massage area or aspirate when giving SUBCUT inj
- Do not mix with other products or inf fluids
- Alternate inj sites
- For excessive bruising at inj site, may use ice before SUBCUT inj
- Store at 77° F (25° C); do not freeze

SIDE EFFECTS

CNS: Fever, confusion, dizziness, insomnia

CV: Angina, dysrhythmias, peripheral edema, tachycardia, hypo/hypertension

GI: Nausea, constipation, flatulence, dyspepsia, **hepatitis**
GU: UTI, hematuria, urinary retention, dysuria
HEMA: Hemorrhage, **anemia, thrombocytopenia**, bleeding
INTEG: Ecchymosis, inj site reaction
MISC: Headache, chest/back pain, hypersensitivity
SYST: **Stevens-Johnson syndrome**

PHARMACOKINETICS

Onset 2-3 hr, max antithrombin activity (4-5 hr), elimination half-life 3-4 hr

INTERACTIONS

Increase: action of tinzaparin—oral anticoagulants, salicylates, thrombolytics, NSAIDs, platelet inhibitors, ticlopidine, clopidogrel
Drug/Herb
Increase: bleeding risk—feverfew, garlic, ginger, ginkgo, horse chestnut green tea

NURSING CONSIDERATIONS

Assess:

Black Box Warning: Spinal/epidermal anesthesia, lumbar puncture: monitor neurological impairment; if impairment occurs, urgent treatment needed

- Blood studies (Hct, platelets, occult blood in stools), anti-Xa; thrombocytopenia may occur
- **Bleeding:** gums, petechiae, ecchymosis, black tarry stools, hematuria, decreased Hct, Hgb
- Hypersensitivity (fever, urticaria, chills); report to prescriber
- Inj site reactions; do not use IM inj

Evaluate:

- Therapeutic response: resolution of DVT

Teach patient/family:

- To report any signs of bleeding: gums, under skin, urine, stools

TREATMENT OF OVERDOSE:

Protamine 1 mg/100 anti-Xa international units of tinzaparin

tioconazole vaginal antifungal

See Appendix B

tiotropium (Rx)

(ty-oh′tro-pee-um)

Spiriva HandiHaler

Func. class.: Anticholinergic, bronchodilator

Chem. class.: Synthetic quaternary ammonium compound

Do not confuse:
Spiriva/Inspra

ACTION: Inhibits interaction of acetylcholine at receptor sites on the bronchial smooth muscle, thereby resulting in decreased cGMP and bronchodilation

USES: COPD; for the long-term treatment and once-daily maintenance of bronchospasm associated with COPD, including chronic bronchitis and emphysema

CONTRAINDICATIONS: Hypersensitivity to this product, atropine, or its derivatives
Precautions: Pregnancy (C), breastfeeding, children, geriatric patients, closed-angle glaucoma, prostatic hypertrophy, bladder neck obstruction, renal disease

DOSAGE AND ROUTES

- **Adult: INH** content of 1 cap/day (18 mcg) using HandiHaler inhalation device

Available forms: Powder for INH 18 mcg in blister packs containing 6 caps with inhaler; 30 caps with inhaler
Administer:
Inhalation route

- **Caps are for INH only; do not swallow**
- Immediately before administration, peel back foil until cap is visible and to "stop" line; remove cap from blister cav-

T

ity; open dust cap of HandiHaler by pulling upward, then open mouthpiece; place cap in center chamber; firmly close mouthpiece until it clicks, leaving dust cap open
• When finished taking dose, remove used capsule and dispose of it; close mouthpiece and dust cap; store
• Rinse mouth after use

SIDE EFFECTS

CNS: Depression, paresthesia
CV: Chest pain, increased heart rate
EENT: Dry mouth, blurred vision, glaucoma
GI: Vomiting, abdominal pain, constipation, dyspepsia
GU: Urinary difficulty, urinary retention, UTI
INTEG: Rash, angioedema
MISC: Candidiasis, flulike syndrome, herpes zoster, infections, angina pectoris
MS: Arthritis myalgic leg/skeletal pain
RESP: *Cough, sinusitis, upper respiratory tract infection,* epistaxis, pharyngitis

PHARMACOKINETICS

Half-life 5-6 days in animals, does not cross blood-brain barrier, very little metabolized in the liver, excreted in urine, 72% protein binding

INTERACTIONS

• Anticholinergics: avoid use with other anticholinergics
Drug/Lab Test
Increase: cholesterol, glucose

NURSING CONSIDERATIONS

Assess:
• **Respiratory status:** dyspnea, rate, breath sounds before and during treatment; pulmonary function tests at baseline and periodically; upper respiratory infections, cough, sinusitis
• Tolerance over long-term therapy; dose may have to be increased or changed
• Patient's ability to use HandiHaler

Evaluate:
• Therapeutic response: ability to breathe easier

Teach patient/family:
• Signs of closed-angle glaucoma (eye pain, blurred vision, visual halos)
• That product is used for long-term maintenance, not for immediate relief of breathing problems; that effect takes 20 min, lasts 24 hr
• To avoid getting the powder in the eyes; may cause blurred vision and pupil dilation
• To hold HandiHaler with mouthpiece upward; to press button in once, completely, and release; this allows for medication to be released
• To breathe out completely; not to breathe into mouthpiece at any time
• To raise device to mouth and close lips tightly around mouthpiece
• With head upright, to breathe in slowly and deeply, but allow the cap to vibrate; to breathe until the lungs fill; to hold breath and remove mouthpiece; to resume normal breathing
• To rinse mouth after use; to use hard candy or regular oral hygiene to reduce dry mouth
• To report immediately blurred vision, eye pain, halos
• To keep caps in sealed blisters before use; to store at room temperature

tipranavir (Rx)

(ti-pran′a-veer)

Aptivus

Func. class.: Antiretroviral
Chem. class.: Protease inhibitor

ACTION: Inhibits human immunodeficiency virus (HIV) protease, thereby preventing the maturation of the virus

USES: HIV in combination with other antiretrovirals

CONTRAINDICATIONS: Hypersensitivity

Black Box Warning: Hepatic disease (Child-Pugh B, C)

Precautions: Pregnancy (C), breastfeeding, children, renal disease, history of renal stones, sulfa allergy, hemophilia, diabetes mellitus, pancreatitis, alcoholism, immune reconstitution syndrome, surgery, trauma, infection

Black Box Warning: Intracranial bleeding, hepatitis

DOSAGE AND ROUTES

Reduce dose with mild to moderate hepatic impairment and ketoconazole coadministration

- **Adult: PO** 500 mg coadministered with ritonavir 200 mg bid with food
- **Adolescent and child ≥2 yr: PO** 14 mg/kg given with ritonavir 6 mg/kg bid or 375 mg/m^2 given with ritonavir 150 mg/m^2 bid, max 500 mg with ritonavir 200 mg bid

Available forms: Caps 250 mg; oral sol 100 mg/ml

Administer:

- Swallow cap whole; do not break, crush, chew; store caps in refrigerator before use; after opening, store at room temperature; use within 60 days
- After meals
- In equal intervals around the clock to maintain blood levels
- Give oral sol using calibrated dosing syringe or 5-ml oral syringe provided; store oral sol at room temperature; use within 60 days after opening bottle

SIDE EFFECTS

CNS: *Headache, insomnia,* dizziness, somnolence, fatigue, *fever,* **intracranial bleeding**

GI: *Diarrhea, abdominal pain, nausea, vomiting,* anorexia, dry mouth, **hepatitis B or C, fatalities when given with ritonavir, pancreatitis**

GU: Nephrolithiasis

INTEG: *Rash,* urticaria, lipodystrophy, **serious rash**

MS: Pain

OTHER: Asthenia, **insulin-resistant hyperglycemia,** *hyperlipidemia,* **ketoacidosis**

PHARMACOKINETICS

Terminal half-life 6 hr, peak 3 hr, plasma protein binding 99.9%, steady state 7-10 days, metabolism CYP3A4, 80% fecal excretion

INTERACTIONS

Life-threatening dysrhythmias: amiodarone, astemizole, cisapride, ergots, flecainide, midazolam, pimozide, propafenone, quiNIDine, rifabutin, rifampin, terfenadine, triazolam

Increase: myopathy, rhabdomyolysis—HMG-CoA reductase inhibitors (lovastatin, simvastatin)

Increase: tipranavir levels—ketoconazole, delavirdine, itraconazole

Increase: levels of both products—clarithromycin, zidovudine

Increase: levels of tipranavir—oral contraception

Decrease: tipranavir levels—rifamycins, fluconazole, nevirapine, efavirenz

Drug/Herb

Decrease: tipranavir levels—St. John's wort; avoid concurrent use

Drug/Food

Decrease: tipranavir absorption—grapefruit juice; high-fat, high-protein foods

Drug/Lab Test

Increase: AST/ALT, cholesterol, blood glucose, amylase, lipase, triglycerides

NURSING CONSIDERATIONS

Assess:

- Signs of infection, anemia; presence of other sexually transmitted diseases

Black Box Warning: Hepatic studies: ALT, AST; total bilirubin, amylase; all may be elevated, discontinue in those with hepatic insufficiency or hepatitis or AST/ALT 10 × upper limit or AST/ALT 5-10 × ULN and total bilirubin 2.5 × ULN

- **HIV:** Viral load, CD4, plasma HIV RNA, serum cholesterol profile, serum triglycerides during treatment

• Bowel pattern before, during treatment; if severe abdominal pain with bleeding occurs, product should be discontinued; monitor hydration
• **Serious rash:** if serious rash occurs, product should be discontinued
• Immune reconstitution syndrome: has been reported with combination antiretroviral therapy, patients may develop pain (MAC, CMV, PcP, TB) and autoimmune disease months after treatment
• Intracranial bleeding: more common in those with trauma, surgery, or those taking antiplatelets or anticoagulants
• Cushingoid symptoms: buffalo hump, facial/peripheral wasting, breast enlargement, central obesity

Evaluate:
• Therapeutic response: improving CD4 counts, viral load

Teach patient/family:
• To take as prescribed; if dose is missed, to take as soon as remembered up to 1 hr before next dose; not to double dose
• That product must be taken in equal intervals around the clock to maintain blood levels for duration of therapy
⚠ That hyperglycemia may occur; to watch for increased thirst, weight loss, hunger, dry, itchy skin; to notify prescriber
• That product does not cure AIDS, only controls symptoms; not to donate blood

⚠ **HIGH ALERT**

tirofiban (Rx)

(tie-roh-fee′ban)

Aggrastat

Func. class.: Antiplatelet
Chem. class.: Glycoprotein IIb/IIIa inhibitor

ACTION: Antagonist of platelet glycoprotein (GP) IIb/IIIa receptor that prevents binding of fibrinogen and von Willebrand's factor, which inhibits platelet aggregation

USES: Acute coronary syndrome in combination with heparin

CONTRAINDICATIONS: Hypersensitivity, active internal bleeding, stroke, major surgery, severe trauma within 30 days, intracranial neoplasm, aneurysm, hemorrhage, acute pericarditis, platelets $<100{,}000/mm^3$, history of thrombocytopenia, coagulopathy, systolic B/P >180 mm Hg or diastolic B/P >110 mm Hg

Precautions: Pregnancy (B), breastfeeding, children, geriatric patients, renal disease, bleeding tendencies, hypertension, platelets $<150{,}000/mm^3$

DOSAGE AND ROUTES

• **Adult: IV** 0.4 mcg/kg/min × 30 min then 0.1 mcg/kg/min for 12-24 hr after angioplasty or atherectomy

Renal dose
• **Adult: IV** CCr <30 ml/min, 0.2 mcg/kg/min × 30 min then 0.05 mcg/kg/min during angiography and for 12-24 hr after angioplasty

Available forms: Inj 50 ml vials; inj premixed bag 50 mcg/ml in 100, 250 ml

Administer:

Intermittent IV INF route
• Dilute inj: withdraw and discard 100 ml from 500-ml bag of sterile 0.9% NaCl or D_5W and replace this vol with 100 ml of tirofiban inj from 2 vials
• Tirofiban inj for sol is premixed in containers of 500-ml 0.9% NaCl (50 mg/ml), infuse over 30 min
• Minimize other arterial/venous punctures; IM inj, catheter use, intubation, to reduce bleeding risk

Y-site compatibilities: Acyclovir, alfentanil, allopurinol, amifostine, amikacin, aminocaproic acid, aminophylline, amiodarone, ampicillin, ampicillin/sulbactam, anidulafungin, argatroban, arsenic trioxide, atracurium, atropine, azithromycin, aztreonam, bivalirudin, bleomycin, bumetanide, buprenorphine, butorphanol, calcium chloride/gluconate, capreomycin, CARBOplatin, carmustine, caspofungin, ceFAZolin, cefepime, cefotaxime, cefoTEtan, cefOXitin, cefTAZidime, ceftizoxime, cefTRIAXone, cefuroxime, chloramphenicol, chlorproMAZINE, cimetidine, ciprofloxacin, cisatracurium, CISplatin, clindamycin, cyclophosphamide, cycloSPORINE, cytarabine, DACTINomycin, DAPTOmycin, dexamethasone, dexmedetomidine, dexrazoxane, digoxin, diltiazem, diphenhydrAMINE, DOBUTamine, DOCEtaxel, dolasetron, DOPamine, doxacurium, DOXOrubicin, DOXOrubicin liposome, doxycycline, droperidol, enalaprilat, ePHEDrine, EPINEPHrine, epirubicin, eptifibatide, ertapenem, erythromycin, esmolol, etoposide, etoposide phosphate, famotidine, fenoldopam, fentaNYL, fluconazole, fludarabine, fluorouracil, foscarnet, fosphenytoin, furosemide, ganciclovir, gemcitabine, gentamicin, glycopyrrolate, granisetron, haloperidol, heparin, hydrALAZINE, hydrocortisone, HYDROmorphone, IDArubicin, ifosfamide, imipenem/cilastatin, insulin, irinotecan, isoproterenol, ketorolac, labetalol, leucovorin, lidocaine, linezolid, LORazepam, magnesium sulfate, mannitol, mechlorethamine, melphalan, meperidine, meropenem, mesna, methylhexital, methotrexate, methyldopate, methylPREDNISolone, metoclopramide, metoprolol, metroNIDAZOLE, midazolam, milrinone, mitoXANtrone, morphine, mycophenolate, nafcillin, nalbuphine, naloxone, nesiritide, niCARdipine, nitroglycerin, nitroprusside, norepinephrine, octreotide, ondansetron, oxaliplatin, oxytocin, PACLitaxel, palonosetron, pamidronate, pancuronium, pantoprazole, PEMEtrexed, PENTobarbital, PHENobarbital, phentolamine, phenylephrine, piperacillin/tazobactam, potassium acetate, potassium chloride/phosphates, procainamide, prochlorperazine, promethazine, propranolol, quinupristin/dalfopristin, ranitidine, remifentanil, rocuronium, sodium acetate/bicarbonate, streptozocin, succinylcholine, SUFentanil, tacrolimus, teniposide, theophylline, thiopental, thiotepa, ticarcillin/clavulanate, tigecycline, tobramycin, topotecan, vancomycin, vasopressin, vecuronium, verapamil, vinBLAStine, vinCRIStine, vinorelbine, voriconazole, zidovudine, zoledronic acid

SIDE EFFECTS

CNS: Dizziness, headache
CV: Bradycardia, hypotension
GI: Nausea, vomiting
HEMA: **Bleeding, thrombocytopenia**
INTEG: *Rash*
MISC: Dissection, edema, pain in legs/pelvis, sweating
SYST: **Anaphylaxis**

PHARMOCOKINETICS

Half-life 2 hr; excretion via urine, feces; plasma clearance 20%-25% lower in geriatric patients with CAD; renal insufficiency decreases plasma clearance

INTERACTIONS

Increase: bleeding—aspirin, heparin, NSAIDs, abciximab, eptifibatide, clopidogrel, ticlopidine, dipyridamole, cefamandole, cefoTEtan, cefoperazone, valproic acid, heparins, thrombin inhibitors, SSRIs, SNRIs

NURSING CONSIDERATIONS

Assess:

⚠ Bleeding: platelet counts, Hct, Hgb before treatment, within 6 hr of loading dose, and at least daily thereafter; watch for bleeding from puncture sites, catheters, or in stools, urine; discontinue if platelets <100,000/mm³

Evaluate:

- Therapeutic response: treatment of acute coronary syndrome

Teach patient/family:

- That it is necessary to quit smoking to prevent excessive vasoconstriction
- About signs, symptoms of bleeding and low platelets
- That there are many product and herbal interactions

T

tiZANidine (Rx)

(ti-za′nih-deen)

Zanaflex

Func. class.: Skeletal muscle relaxant, α_2-adrenergic agonist

Chem. class.: Imidazoline

Do not confuse:
tiZANidine/tiaGABine

ACTION:
Increases presynaptic inhibition of motor neurons and reduces spasticity by α_2-adrenergic agonism

USES:
Acute/intermittent management of increased muscle tone associated with spasticity, symptoms of MS

Unlabeled uses: Tension headache, low back pain, trigeminal neuralgia

CONTRAINDICATIONS:
Hypersensitivity

Precautions: Pregnancy (C), breastfeeding, children, geriatric patients, hypotension, renal/hepatic disease

DOSAGE AND ROUTES

- **Adult: PO** 8 mg q6-8hr, max 36 mg/24 hr

Renal dose

- **Adult: PO** CCr <25 ml/min, start with lower dose

Available forms: Tabs 2, 4 mg; caps 2, 4, 6 mg

Administer

- Consistently either with/without food; food may affect absorption
- Titrate doses carefully
- Avoid use with other CNS depressants

SIDE EFFECTS

CNS: Somnolence, dizziness, speech disorder, dyskinesia, nervousness, hallucination, psychosis

CV: Hypotension, bradycardia

GI: Dry mouth, vomiting, increased ALT, abnormal LFTs, constipation

OTHER: Blurred vision, urinary frequency, pharyngitis, rhinitis, tremors, rash, muscle weakness

PHARMACOKINETICS

Completely absorbed, widely distributed, peak 1-2 hr, duration 3-6 hr, half-life 2.5 hr, protein binding 30%, metabolized by liver; excreted in urine, feces

INTERACTIONS

Increase: CNS depression—alcohol, other CNS depressants

Increase: tiZANidine levels—other CYP1A2 inhibitors (acyclovir, amiodarone, famotidine, mexiletine, enoxacin, norfloxacin, propafenone, tacrine, verapamil, zileuton, oral contraceptives ciprofloxacin), fluvoxaMINE; avoid concurrent use

Increase: hypotension—antihypertensives

Increase: effect of rasagiline

Drug/Herb

Increase: CNS depression—kava, St. John's wort

Drug/Lab Test

Increase: alk phos, AST, ALT, serum glucose

NURSING CONSIDERATIONS

Assess:

- **Muscle spasticity** at baseline and throughout treatment
- **Hypotension;** gradual dosage increase should lessen hypotensive effects; have patient rise slowly from supine to upright; watch those patients receiving antihypertensives for increased effects
- Increased sedation, dizziness, hallucinations, psychosis; product may need to be discontinued
- Vision by ophthalmic exam; corneal opacities may occur
- Hepatic studies: 1, 3, 6 mo during treatment and periodically thereafter

Evaluate:

- Therapeutic response: decreased muscle spasticity

Teach patient/family:

• To rise slowly from lying or sitting to upright position to prevent orthostatic hypotension

• To ask for assistance if dizziness, sedation occur; to avoid drinking alcohol; to avoid operating machinery, driving until effects known

• To discontinue gradually

tobramycin (Rx)

(toe-bra-mye′sin)

TOBI, TOBI Podhaler

Func. class.: Antiinfective

Chem. class.: Aminoglycoside

ACTION: Interferes with protein synthesis in bacterial cell by binding to ribosomal subunits, thereby causing inaccurate peptide sequences to form in protein chain causing bacterial death

USES: Severe systemic infections of CNS, respiratory, GI, urinary tract, bone, skin, soft tissues; cystic fibrosis (nebulizer) for *acinetobacter calcoaceticus, citrobacter* sp., *enterobacter aerogenes, enterobacter* sp., *enterococcus* sp., *escherichia coli, haemophilus aegyptius, haemophilus influenzae (beta-lactamase negative), haemophilus influenzae (beta-lactamase positive), klebsiella pneumoniae, klebsiella* sp., *moraxella lacunata, morganella morganii, neisseria* sp., *proteus mirabilis, proteus vulgaris, providencia* sp., *pseudomonas aeruginosa, serratia* sp., *staphylococcus aureus (MSSA), staphylococcus epidermidis, staphylococcus* sp., *streptococcus pneumoniae, streptococcus* sp.; may also be used for the following: *acinetobacter* sp., *aeromonas* sp., *bacillus anthracis, salmonella* sp., *shigella* sp.

Unlabeled uses: Endocarditis, febrile neutropenia

CONTRAINDICATIONS: Hypersensitivity to aminoglycosides

Black Box Warning: Pregnancy (D), severe renal disease

Precautions: Breastfeeding, geriatric patients, neonates, mild renal disease, myasthenia gravis, Parkinson's disease

Black Box Warning: Hearing deficits, neuromuscular disease

DOSAGE AND ROUTES

• **Adult: IM/IV** 3 mg/kg/day in divided doses q8hr; may give up to 6 mg/kg/day in divided doses q8-12hr; once-daily dosing (pulse dosing) (unlabeled) **IV** 5-7 mg/kg, dosing intervals determined using nomogram, based on random levels drawn 8-12 hr after 1st dose

• **Child: IM/IV** 6-7.5 mg/kg/day in 3-4 equal divided doses

• **Child ≥6 yr: NEB** 300 mg bid in repeating cycles of 28 days on/28 days off of product; give **INH** over 10-15 min using a handheld PARI LC PLUS reusable nebulizer with DeVilbiss Pulmo-Aid compressor

• **Neonate <1 wk: IM/IV** ≤4 mg/kg/day divided q12hr

Cystic fibrosis with *Pseudomonas aeruginosa*

• **Adult/adolescent/child ≥6 yr: IV** 2.5-3.3 mg/kg q8hr, **NEB** 300 mg via inhalation bid × 28 days, then 28 days after; **ORAL INH** (TOBI Podhaler) 112 mg bid ×28 day, then 28 days after; may repeat

Renal dose

Conventional dosing:

• Multiply the serum creatinine (mg/100 ml) by 6 to determine the dosing; to decrease the dose divide the standard dose by the serum creatinine (mg/100 ml) to determine the lower recommended dose

Interval adjustment of extended-interval dosing of 5 or 7 mg/kg (unlabeled):

• Adjust doses based on serum concentrations and organism MIC; CCr 40-59 ml/min: 5 or 7 mg/kg **IV** q36hr; CCr 20-39 ml/min: 5 or 7 mg/kg **IV** q48hr; CCr <20 ml/min: 5 or 7 mg/kg **IV** once, then follow serial levels to determine time of next dose (serum concentration <1 mcg/ml)

Dose adjustment of extended dosing of 5 mg/kg (unlabeled)

• Adjust doses based on serum concentrations and organism MIC: CCr >80 ml/min: no dosage adjustment is needed: CCr 60-79 ml/min: 4 mg/kg **IV** q24hr; CCr 50 ml/min: 3.5 mg/kg **IV** q24hr; CrCl 40 ml/min: 2.5 mg/kg **IV** q24hr; CrCl <30 ml/min: use traditional dosing

Available forms: Inj 10, 40 mg/ml; powder for inj 1.2 g; neb sol 300 mg/5 ml; powder for inh 28 mg

Administer:

• After obtaining specimen for C&S; begin treatment before results

• Product in evenly spaced doses to maintain blood level; separate aminoglycosides and penicillins by ≥1 hr

• Use only on susceptible organisms to prevent development of product-resistant bacteria

IM route

• IM inj in large muscle mass; rotate inj sites, aspirate

• Draw peak 1 hr after dose, trough right before next dose; absorption erratic

Inhalation route: (TOBI Podhaler)

• Use with Podhaler device; do not swallow caps; use device for 7 days then discard

• Keep caps in blister pack until ready to use; administer other inhaled products or chest physiotherapy before

• While holding base of Podhaler device, unscrew lid, stand upright, unscrew mouthpiece; while holding body, tear blister card in half lengthwise along precut lines, peel back foil, place cap in chamber at top of device, reattach mouthpiece and tighten; with mouthpiece pointed down, press blue button down with thumb, release, exhale completely, place mouth over mouthpiece, close lips, inhale with single breath, hold 5 sec, exhale normally away from device; after a few normal breaths, repeat, unscrew mouthpiece, and remove cap; cap should be empty; repeat process 3 more times (total 4 caps); after use, reattach mouthpiece and wipe with clean, dry cloth

Intermittent IV INF route

• Visually inspect sol; do not use if discolored or particulate is present

• **Vantage vials** are for IV only and only for exactly 60 or 80 mg

• Diluted in 50-100 ml 0.9% NaCl D_5W ($D_{10}W$, Ringer's, LR); infuse over 20-60 min; volume for pediatric patients needs and should be sufficient to allow for 20-60 min inf

Y-site compatibilities: Acyclovir, aldesleukin, alfentanil, alprostadil, amifostine, aminophylline, amiodarone, amsacrine, anidulafungin, ascorbic acid, atracurium, atropine, aztreonam, bivalirudin, bretylium, bumetanide, buprenorphine, butorphanol, calcium chloride/gluconate, CARBOplatin, caspofungin, chloramphenicol, cimetidine, ciprofloxacin, cisatracurium, CISplatin, clindamycin, cyanocobalamin, cyclophosphamide, cycloSPORINE, cytarabine, DACTINomycin, DAPTOmycin, dexmedetomidine, digoxin, diltiazem, diphenhydrAMINE, DOBUTamine, DOCEtaxel, DOPamine, doripenem, doxacurium, DOXOrubicin hydrochloride, DOXOrubicin liposome, doxycycline, enalaprilat, ePHEDrine, EPINEPHrine, epirubicin, epoetin alfa, ertapenem, esmolol, etoposide, etoposide phosphate, famotidine, fenoldopam, fentaNYL, filgrastim, fluconazole, fludarabine, fluorouracil, foscarnet, furosemide, gemcitabine, gentamicin, glycopyrrolate, granisetron, HYDROmorphone, ifosfamide, imipenem/cilastatin, isoproterenol, ketorolac, labetalol, levofloxacin, lidocaine, linezolid, LORazepam, magnesium sulfate, mannitol, mechlorethamine, melphalan, meperidine, metaraminol, methicillin, methotrexate, methoxamine, methyldopate, methylPREDNISolone, metoclopramide, metoprolol, metroNIDAZOLE, miconazole, midazolam, milrinone, minocycline, mitoXANtrone, morphine, moxalactam, multiple vitamins, nafcillin, nalbuphine, naloxone, niCARdipine, nitroglycerin, nitroprusside, norepinephrine, octreotide, ondansetron, oxaliplatin, oxytocin, PA-

CLitaxel, palonosetron, pantoprazole, papaverine, penicillin G, pentazocine, perphenazine, PHENobarbital, phentolamine, phenylephrine, phytonadione, potassium chloride, procainamide, prochlorperazine, promethazine, propranolol, protamine, pyridoxime, quinupristin/dalfopristin, ranitidine, remifentanil, riTUXimab, rocuronium, sodium acetate/bicarbonate, succinylcholine, SUFentanil, tacrolimus, teniposide, theophylline, thiamine, thiotepa, ticarcillin/clavulanate, tigecycline, tirofiban, tolazoline, trastuzumab, trimethaphan, urokinase, vancomycin, vasopressin, vecuronium, verapamil, vinCRIStine, vinorelbine, voriconazole, zidovudine

SIDE EFFECTS

CNS: Confusion, depression, numbness, tremors, seizures, muscle twitching, neurotoxicity, dizziness, vertigo
CV: Hypo/hypertension, palpitation
EENT: Ototoxicity, deafness, visual disturbances, tinnitus
GI: *Nausea, vomiting, anorexia;* increased ALT, AST, bilirubin, hepatomegaly, hepatic necrosis, splenomegaly
GU: Oliguria, hematuria, renal damage, azotemia, renal failure, nephrotoxicity
HEMA: Agranulocytosis, thrombocytopenia, leukopenia, eosinophilia, anemia
INTEG: *Rash,* burning, urticaria, dermatitis, alopecia

PHARMACOKINETICS

Plasma half-life 2-3 hr, prolonged in neonates; not metabolized; excreted unchanged in urine; crosses placental barrier; poor penetration into CSF
IM: Onset rapid, peak 1 hr, duration 8 hr
IV: Onset immediate, peak 30 min, duration 8 hr

INTERACTIONS

Increase: ototoxicity, neurotoxicity, nephrotoxicity—other aminoglycosides, amphotericin B, polymyxin, vancomycin, ethacrynic acid, furosemide, mannitol, methoxyflurane, CISplatin, cephalosporins, bacitracin, acyclovir, penicillins, cidofovir

Drug/Lab Test
Increase: eosinophils, BUN, creatinine, AST, ALT, LDH, alk phos, glucose
Decrease: potassium, calcium, sodium, magnesium, WBC, granulocytes, platelets

NURSING CONSIDERATIONS

Assess:
- Weight before treatment; dosage is usually based on ideal body weight but may be calculated on actual body weight
- I&O ratio, urinalysis daily for proteinuria, cells, casts; report sudden change in urine output
- VS during inf; watch for hypotension, change in pulse
- IV site for thrombophlebitis, including pain, redness, swelling q30min; change site if needed; apply warm compresses to discontinued site
- Serum aminoglycoside conc; serum peak drawn at 30-60 min after IV inf or 60 min after IM inj, trough drawn just before next dose, peak 4-10 mcg/ml, trough 0.5-2 mcg/ml

Black Box Warning: Renal impairment: CCr, BUN, serum creatinine; lower dosage should be given in renal impairment (CCr $<$80 ml/min); monitor electrolytes: potassium, sodium, chloride, magnesium monthly if patient receiving long-term therapy

Black Box Warning: Deafness by audiometric testing; ringing, roaring in ears; vertigo; assess hearing before, during, after treatment

- **Overgrowth of infection:** fever, malaise, redness, pain, swelling, perineal itching, diarrhea, stomatitis, change in cough, sputum
- **Vestibular dysfunction:** nausea, vomiting, dizziness, headache; product should be discontinued if severe
- Adequate fluids of 2-3 L/day unless contraindicated to prevent irritation of tubules

T

• Supervised ambulation, other safety measures with vestibular dysfunction

Evaluate:

• Therapeutic response: absence of fever, draining wounds, negative C&S after treatment

Teach patient/family:

• To report headache, dizziness, symptoms of overgrowth of infection, renal impairment

Black Box Warning: To report loss of hearing; ringing, roaring in ears; feeling of fullness in head

Black Box Warning: To notify prescriber if pregnancy is planned or suspected; pregnancy (D)

Nebulizer

• To use other therapies first, then tobramycin

TREATMENT OF OVERDOSE:

Hemodialysis; monitor serum levels of product

tobramycin ophthalmic

See Appendix B

tocilizumab (Rx)

(toe′si-liz′oo-mab)

Actemra

Func. class.: DMARDs (disease-modifying antirheumatoid drugs)/tumor necrosis factor (TNF) modifier

ACTION: Interleukin-6 (IL-6) receptor inhibiting monoclonal antibody

USES: Rheumatoid arthritis, active systemic juvenile idiopathic arthritis

CONTRAINDICATIONS: Hypersensitivity

Precautions: Breastfeeding, pregnancy (C); risk for GI perforation, active hepatic disease, severe neutropenia/thrombocytopenia, demyelinating disorders

Black Box Warning: Invasive fungal infection, active TB

DOSAGE AND ROUTES

• **Adult: IV** 4 mg/kg over 1 hr q4wk, may increase to 8 mg/kg q4wk based on clinical response, max dose 800 mg/inf; do not initiate if ANC $>$2000, platelets $<$100,000

Juvenile idiopathic arthritis

• **Child $\geq$2 yr/adolescent $\geq$30 kg: IV** 8 mg/kg over 1 hr q2wk

• **Child $\geq$2 yr/adolescent $<$30 kg: IV** 10 mg/kg over 1 hr

Available forms: Sol for inj 80 mg/4 ml, 200 mg/10 ml, 400 mg/20 ml

Administer:

Intermittent IV INF route

• Visually inspect for particulate matter, discoloration before administration whenever sol and container permit; colorless to pale yellow liquid

• From 100-ml inf bag or bottle, withdraw vol of 0.9% sodium chloride inj equal to vol of tocilizumab sol required for patient's dose

• Slowly add tocilizumab from each vial into inf bag or bottle; gently invert bag to avoid foaming; fully diluted sols are compatible with polypropylene, polyethylene, polyvinyl chloride inf bags and polypropylene, polyethylene, glass inf bottles

• Fully diluted sol for inf may be stored refrigerated or at room temperature for $\leq$24 hr and should be protected from light; do not use unused product remaining in vials; no preservatives

• Allow the fully diluted sol to reach room temperature before inf

• Give over 60 minutes with inf set; do not administer as IV push or bolus

• Do not infuse concomitantly in same IV line with other drugs

SIDE EFFECTS

CNS: Headache, dizziness

CV: Hypertension

GI: Perforation, abdominal pain, gastritis, mouth ulcerations

HEMA: Neutropenia, thrombocytopenia

INTEG: Rash, inf reactions
RESP: Upper respiratory infections, nasopharyngitis, bronchitis
SYST: **Serious infections, anaphylaxis,** inf-related reactions, antitocilizumab antibody formation

INTERACTIONS

Decrease: product level—cycloSPORINE, theophylline warfarin
Decreased: level of CYP3A4 substrates (hormonal contraceptives, omeprazole, atorvastatin, simvastatin)
- Do not give with live virus vaccines
- Avoid use with TNF modifiers, DMARDs, immunosuppressives due to increased risk of infection

PHARMACOKINETICS

Half-life approx 6 days with single dose, approx 11 days with multiple (steady-state) doses

NURSING CONSIDERATIONS

Assess:
- **Rheumatoid arthritis:** ROM, pain, stiffness at baseline q1-2wk
- Blood studies: CBC with differential, LFTs, platelet count, serum lipid profile at baseline and periodically
- **Infection** before treatment and periodically; obtain TB screening before beginning treatment, invasive fungal infections; discontinue if infection occurs during administration; may use antituberculosis therapy before tocilizumab in past history of latent or active TB when adequate course of treatment cannot be confirmed and those with a negative TB with risk factors for infections

Evaluate:
- Therapeutic response: ability to move more easily with less pain

Teach patient/family:
- That this treatment must continue unless safety or effectiveness is an issue
- About reason for use and expected results
- To avoid live vaccines, bring immunizations up to date before treatment

Black Box Warning: To report signs, symptoms of infection, including TB and hepatitis B
- To notify prescriber if pregnancy or suspected pregnancy (C); not to use if breastfeeding

tofacitinib

(toe′fa-sye′ti-nib)

Xeljanz

Func. class.: Antirheumatic agent (disease modifying), immunomodulator/biologic DMARD
Chem. class.: Janus kinase inhibitor

ACTION: Affects the signaling pathway of Janus kinase

USES: Rheumatoid arthritis (moderately to severely active) in those who have taken methotrexate with inadequate response or intolerance

CONTRAINDICATIONS: Hypersensitivity
Precautions: Pregnancy (C), breastfeeding, neonates, infants, children, geriatric patients, neoplastic disease, ulcerative colitis, neutropenia, peptic ulcer disease
Active infections, risk of lymphomas/leukemias, TB, posttransplant lymphoproliferative disorder (PTLD), kidney disease, diabetes mellitus, HIV, hypercholesterolemia

Black Box Warning: Infection, secondary malignancy

DOSAGE AND ROUTES

- **Adult: PO** 5 mg bid if receiving potent CYP3A4 inhibitor, then 5 mg/day if receiving moderate CYP3A4 inhibitor and 2C19 inhibitors

Available forms: Tabs 5 mg
Administer:
PO route
- Without regard to food

T

SIDE EFFECTS

CNS: Headache, paresthesias, insomnia, fatigue
CV: Hypertension
GI: Abdominal pain, nausea, liver damage, dyspepsia, vomiting, diarrhea, gastritis, GI perforation, steatosis
HEMA: Anemia, lymphocytosis, lymphopenia, neutropenia
INTEG: Rash, pruritus
MISC: Increased cancer risk, risk of infection (TB, invasive fungal infections, other opportunistic infections), may be fatal, posttransplant lymphoproliferative disorder (PTLD)

PHARMACOKINETICS

Bioavailability 70%, protein binding 40% (albumin), metabolism mediated by CYP3A4, half-life 3 hr, peak 0.5-1 hr

INTERACTIONS

• Do not use with TNF modifiers, vaccines, potent immunosuppressants, other biologic DMARDS
Increase: tofacinib effect—CYP3A4 inhibitors (amprenavir, boceprevir, delavirdine, ketoconazole, indinavir, itraconazole, dalfopristin/quinupristin, ritonavir, tipranavir, fluconazole, isoniazid, miconazole)
Decrease: tofacinib effect—CYP3A4 inducers (rifampin, rifapentine, rifabutin, primidone, phenytoin, PHENobarbital, nevirapine, nafcillin, modafinil, griseofulvin, etravirine, efavirenz, barbiturates, bexarotene, bosentan, carBAMazepine, enzalutamide, dexamethasone)
Drug/Lab Test
Increase: LFTs, cholesterol
Decrease: neutrophils, lymphocytes, Hct, Hgb

NURSING CONSIDERATIONS

Assess:
• Monitor lipid profile, Hct/Hgb WBC, LFTs
• **RA:** Pain, stiffness, ROM, swelling of joints before, during treatment

Black Box Warning: **Active infection, including localized infection:** evaluate and test patients for latent or active TB before use; treat with antimycobacterials before use of product; this product increases the risk of serious including fatal infections (pulmonary or extrapulmonary TB; invasive fungal infections; and bacterial, viral, and opportunistic infections); during and after use, monitor for infection including TB in those who tested negative for latent TB before use; if a serious infection develops, interrupt receipt until the infection is controlled

Black Box Warning: **Secondary malignancy:** lymphoma and other malignancies have been noted with product use

⚠ **Epstein–Barr virus–associated posttransplant lymphoproliferative disorder (PTLD):** in kidney transplant patients when used with this product and immunosupressives
• **Liver disease:** not recommended in severe liver disease, impairment; dose modification is needed with moderate liver impairment, monitor LFTs
⚠ **GI perforation:** assess in those with diverticulitis, peptic ulcer disease, or ulcerative colitis
⚠ **Immunosuppression:** obtain neutrophil and lymphocyte counts before use; do not start the product in lymphocyte count <500 cells/mm^3 or ANC <1000 cells/mm^3; for ANC >1000 cells/mm^3, monitor neutrophil counts after 4-8 wk and every 3 mo thereafter; lymphocyte count >500 cells/mm^3, monitor lymphocyte counts every 3 mo
⚠ **Anemia:** determine Hgb, do not start in Hgb <9 g/dl; in Hgb ≥ 9 g/dl, monitor Hgb after 4-8 wk and every 3 mo thereafter
⚠ **Pregnancy (C)/breastfeeding:** use during pregnancy only if the potential benefit justifies the potential risk to the fetus; if pregnancy occurs, enrollment in the pregnancy registry is encouraged by calling 1-877-311-8972; discontinue product or breastfeeding; serious adverse reactions can occur in nursing infants

Black Box Warning: Neoplastic disease (lymphomas/leukemias)

Evaluate:

- Therapeutic response: decreased inflammation, pain in joints, decreased joint destruction

Teach patient/family:

- Not to have vaccines while taking this product
- Not to take any live virus vaccines during treatment
- To report signs of infection, allergic reaction

⚠ Pregnancy (C): teach patient to report if pregnancy is planned or suspected; do not breastfeed

tolcapone (Rx)

(toll′cah′pone)

Tasmar

Func. class.: Antiparkinson agent

Chem. class.: COMT inhibitor

ACTION: Inhibits COMT; used as adjunct to levodopa/carbidopa therapy

USES: Parkinson's disease

CONTRAINDICATIONS: Hypersensitivity, rhabdomyolysis

Black Box Warning: Hepatic disease

Precautions: Pregnancy (C), breastfeeding, cardiac/renal disease, hypertension, asthma, history of rhabdomyolysis

DOSAGE AND ROUTES

- **Adult: PO** 100-200 mg tid with levodopa/carbidopa therapy; max 600 mg/day, discontinue if no benefit after 3 wk

Available forms: Tabs 100, 200 mg

Administer:

- Only to be used if levodopa/carbidopa does not provide satisfactory results
- Give without regard to food

SIDE EFFECTS

CNS: Dystonia, dyskinesia, dreaming, *fatigue, headache, confusion,* psychosis, hallucination, dizziness, sleep disorders

CV: *Orthostatic hypotension,* chest pain, hypotension

EENT: Cataract, eye inflammation

GI: *Nausea, vomiting, anorexia, abdominal distress,* diarrhea, constipation, **fatal hepatic failure,** increased LFTs

GU: UTI, urine discoloration, uterine tumor, micturition disorder, hematuria

HEMA: Hemolytic anemia, leukopenia, agranulocytosis

INTEG: Sweating, alopecia

MS: Rhabdomyolysis

PHARMACOKINETICS

Rapidly absorbed, peak 2 hr, protein binding 99%, extensively metabolized, half-life 2-3 hr, excreted in urine (60%)/feces (40%)

INTERACTIONS

Increase: CNS depression—CNS depressants

- May influence pharmacokinetics of α-methyldopa, DOBUTamine, apomorphine, isoproterenol
- Inhibition of normal catecholamine metabolism: MAOIs, MAO-B inhibitor may be used

NURSING CONSIDERATIONS

Assess:

Black Box Warning: **Hepatic disease:** AST, ALT, alk phos, LDH, bilirubin, CBC; monitor ALT, AST q2wk × 1 yr, then q4wk × 6 mo, then q8wk thereafter; if LFTs elevated, product should not be used

- Involuntary movements of parkinsonism: akinesia, tremors, staggering gait, muscle rigidity, drooling
- B/P, respiration during initial treatment; hypo/hypertension should be reported
- Mental status: affect, mood, behavioral changes

Evaluate:

- Therapeutic response: decrease in akathisia, increased mood

Teach patient/family:
- To change positions slowly to prevent orthostatic hypotension
- That urine, sweat may change color
- That food taken within 1 hr before meals or 2 hr after meals decreases action of product by 20%; may be taken without regard to food
- To notify prescriber if pregnancy is planned or suspected, pregnancy (C)
- That CNS changes may occur, hallucinations, involuntary movement
- To avoid hazardous activities until reaction is known; dizziness may occur
- That nausea and diarrhea are common

Black Box Warning: To report signs of hepatic injury: clay-colored stools, jaundice, fatigue, appetite loss, lethargy, fatigue, itching, right upper abdominal pain

- To report nausea, vomiting, anorexia; that nausea may occur at beginning of treatment

tolnaftate topical

See Appendix B

tolterodine (Rx)

(toll-tehr′oh-deen)

Detrol, Detrol LA

Func. class.: Overactive bladder product

Chem. class.: Muscarinic receptor antagonist

ACTION: Relaxes smooth muscles in urinary tract by inhibiting acetylcholine at postganglionic sites

USES: Overactive bladder (urinary frequency, urgency), urinary incontinence

CONTRAINDICATIONS: Hypersensitivity, uncontrolled closed-angle glaucoma, urinary retention, gastric retention

Precautions: Pregnancy (C), breastfeeding, children, renal/hepatic disease, controlled closed-angle glaucoma, bladder obstruction, QT prolongation, decreased GI motility

DOSAGE AND ROUTES

- **Adult and geriatric: PO** 2 mg bid; may decrease to 1 mg bid; **EXT REL** 4 mg/day, may decrease to 2 mg/day if needed, max 4 mg/day

Hepatic/renal dose
- **Adult: PO** 1 mg bid (50% dose) or **EXT REL** 2 mg/day; CCr ≤30 ml/min, reduce by 50%

Available forms: Tabs 1, 2 mg; ext rel caps 2, 4 mg

Administer:
- Whole; take with liquids; do not crush, chew, or break ext rel product; without regard to meals

SIDE EFFECTS

CNS: Anxiety, paresthesia, fatigue, *dizziness, headache;* increasing dementia, memory impairment

CV: Chest pain, hypertension, QT prolongation

EENT: Vision abnormalities, xerophthalmia

GI: *Nausea, vomiting, anorexia,* abdominal pain, constipation, dry mouth, dyspepsia

GU: Dysuria, urinary retention, frequency, UTI

INTEG: Rash, pruritus

RESP: Bronchitis, cough, pharyngitis, upper respiratory tract infection

SYST: Angioedema, Stevens-Johnson syndrome

PHARMACOKINETICS

Rapidly absorbed; highly protein bound; extensively metabolized by CYP2D6; a portion of the population may be poor metabolizers; excreted in urine, feces

INTERACTIONS

- Do not use in those with known hypersensitivity-festerodine

⚠ **Increase:** QT prolongation—class IA/III antidysrhythmics, some phenothiazines, β-agonists, local anesthetics, tricyclics, haloperidol, methadone, chloroquine, clarithromycin, droperidol, erythromycin, pentamidine

Increase: action of tolterodine—antiretroviral protease inhibitors, macrolide antiinfectives, azole antifungals

Increase: anticholinergic effect—antimuscarinics

Increase: urinary frequency—diuretics

Drug/Food

- Food increases bioavailability of tolterodine

Drug/Lab

Increase: LFTs, bilirubin

NURSING CONSIDERATIONS

Assess:

- **Urinary patterns:** distention, nocturia, frequency, urgency, incontinence

⚠ **Serious skin disorders:** angioedema, Stevens-Johnson syndrome; allergic reactions: rash; if this occurs, product should be discontinued

⚠ **QT prolongation:** ECG, ejection fraction; assess for chest pain, palpitations, dyspnea

Black Box Warning: Fatal hepatic injury: increased LFTs, bilirubin during first 18 mo of therapy in those with autosomal dominant polycystic kidney disease; assess for fatigue, anorexia, right upper abdominal pain, dark urine, jaundice; if these occur, discontinue product and do not restart if cause is liver injury

Evaluate:

- Decreasing dysuria, frequency, nocturia, incontinence

Teach patient/family:

- To avoid hazardous activities; dizziness may occur
- Not to drink liquids before bedtime
- About the importance of bladder maintenance
- Not to breastfeed
- To report signs of infection, skin effects

tolvaptan (Rx)

(tole-vap′tan)

Samsca

Func. class.: Antihypertensive

Chem. class.: Vasopressin receptor antagonist, V2

ACTION:

Arginine vasopressin (AVP) antagonist with affinity for V2 receptors; level of circulating AVP in circulating blood is critical for the regulation of water and the electrolyte balance, and it is usually elevated with euvolemic/hypervolemic hyponatremia

USES:

Hypervolemic/euvolemic hyponatremia with heart failure, cirrhosis, SIADH

CONTRAINDICATIONS:

Hypersensitivity, hypovolemia, anuria

Precautions: Pregnancy (C), breastfeeding, children, dehydration, geriatric, hyperkalemia, autosomal dominant PKD

Black Box Warning: Alcoholism, malnutrition, hepatic disease

DOSAGE AND ROUTES

- **Adult: PO** 15 mg daily; after 24 hr, may increase to 30 mg daily; max 60 mg/day for ≤30 days

Available forms: Tab 15, 30 mg

Administer:

- PO with/without food
- Avoid fluid restriction for first 24 hr
- Initiate in hospital setting

SIDE EFFECTS

CNS: Fever, dizziness

CV: **Ventricular fibrillation, DIC, stroke, thrombosis**

GI: *Nausea,* vomiting, *constipation,* colitis, hepatic injury

GU: Polyuria

HEMA: **Bleeding**

META: *Dehydration, hyperglycemia,* hyperkalemia, hypernatremia

MS: Rhabdomyolysis
RESP: Respiratory depression, pulmonary embolism

PHARMACOKINETICS

Peak 2-4 hr, protein binding 99%, metabolized by CYP3A4, terminal half-life 12 hr

INTERACTIONS

Increase: concs of tolvaptan—CYP3A4 inhibitors (efavirenz, fosamprenavir, quiNINE); P-gp inhibitors (cycloSPORINE, azithromycin, mefloquine, palperidone, propafenone, quiNIDine, testosterone)
Decrease: conc of tolvaptan—CYP3A4 inducers (carBAMazepine, dexamethasone, etravirine, flutamide, griseofulvin, metyrapsone, modafinil, nafacillin, nevirapine, OXcarbazepine, phenytoin, rifampin, rifabutin, rifapendine, topiramate)
Drug/Herb
Decreased: tolvaptan effect: CYP3A4 inducer (St. John's wort)
Drug/Food
- Grapefruit juice: do not use together

NURSING CONSIDERATIONS

Assess:
- Renal, hepatic function
- Frequent sodium vol status; overly rapid correction of sodium conc (12 mEq/L per 24 hr), may result in osmotic demyelination syndrome, may occur in alcoholism, severe malnutrition, advanced liver disease, syndrome of inappropriate antidiuretic hormone
- CV status: ventricular fibrillation, hypertension; monitor B/P, pulse
- Monitor electrolytes (sodium, potassium)

Evaluate:
- Therapeutic response: correction of serum sodium levels

Teach patient/family:
- To avoid pregnancy, breastfeeding while taking this product
- About administration procedure and expected results
- To report difficulty swallowing, speaking, seizures, dizziness, drowsiness; embolism may be the cause
- To drink fluid in response to thirst
- Not to use grapefruit juice

topiramate (Rx)

(toh-pire′ah-mate)

Topamax, Topamax Sprinkle, Topiragen

Func. class.: Anticonvulsant—miscellaneous

Chem. class.: Monosaccharide derivative

ACTION: May prevent seizure spread as opposed to an elevation of seizure threshold, increases GABA activity

USES: Partial seizures in adults and children 2-16 yr old; tonic-clonic seizures; seizures with Lennox-Gastaut syndrome; migraine prophylaxis
Unlabeled uses: Infantile spasms, bipolar disorder, alcohol dependence, absence seizures, neuropathic pain, cluster headaches, mania, bulimia nervosa

CONTRAINDICATIONS: Hypersensitivity, metabolic acidosis, pregnancy (D)
Precautions: Breastfeeding, children, renal/hepatic disease, acute myopia, secondary closed-angle glaucoma, behavioral disorders, COPD, dialysis, encephalopathy, status asthmaticus, status epilepticus, surgery, paresthesias, maculopathy, nephrolithiasis

DOSAGE AND ROUTES

Adjunctive therapy for seizures
- **Adult/adolescent/child ≥10 yr: PO** 25-50 mg/day initially, titrate by 25-50 mg/wk, up to 200-400 mg/day in 2 divided doses
- **Child 2-9 yr: PO** week 1: 25 mg in PM, then 25 mg bid if tolerated (week 2), then increase by 25-50 mg/day each

week as tolerated over 5 to 7-wk titration period, maintenance given in 2 divided doses; <11 kg, minimum 150 mg/day, max 250 mg/day; 12-22 kg, minimum 200 mg/day, max 300 mg/day; 23-31 kg, minimum 200 mg/day, max 350 mg/day; 32-38 kg, minimum 250 mg/day, max 350 mg/day; >38 kg, minimum 250 mg/day, max 400 mg/day

Migraine prophylaxis

• **Adult: PO** 25 mg/day initially, increase by 25 mg/day/wk up to 100 mg/day in 2 divided doses

Renal dose

• **Adult: PO** CCr <70 ml/min, give 1/2 dose

Atonic/atypical absence/myoclonic seizures (unlabeled)

• **Adult/adolescent >16 yr: PO** 50 mg/day, titrate slowly by 50 mg/wk to 100-300 mg tid

• **Child 2-16 yr: PO** 0.5-1 mg/kg, max 25 mg, initially daily × 7 days then increase by 0.5-1 mg/kg/day weekly up to 3-6 mg/kg/day in divided doses

Refractory infantile spasms (unlabeled)

• **Child: PO** 25 mg/day, may increase by 25 mg q2-3days until spasms controlled, max 24 mg/kg/day

Alcoholism (unlabeled)

• **Adult: PO** 25 mg/day, titrated to max 300 mg/day in divided doses

Neuropathic pain (unlabeled)

• **Adult: PO** 12.5-25 mg daily or bid × 4 wk then double dose q4wk to max 100-200 mg/day in divided doses

Bipolar disorder (unlabeled)

• **Adult: PO** 25 mg/day then increase by 25-mg increments to 200 mg/day

Available forms: Tabs 25, 50, 100, 200 mg; sprinkle caps 15, 25 mg

Administer:

• Swallow tabs whole; do not break, crush, or chew tabs; very bitter
• May take without regard to meals
• Sprinkle cap can be given whole or opened and sprinkled on soft food; do not chew, drink water after sprinkle
• Store at room temperature away from heat, light

SIDE EFFECTS

CNS: *Dizziness, fatigue*, cognitive disorders, insomnia, *anxiety*, depression, paresthesia, *memory loss, tremors*, motor retardation, **suicidal ideation**, poor balance, ataxia

CV: Flushing, chest pain

EENT: Diplopia, *vision abnormality*

GI: Diarrhea, *anorexia, nausea, dyspepsia*, abdominal pain, constipation, dry mouth, **pancreatitis**

GU: Breast pain, dysmenorrhea, menstrual disorder

INTEG: Rash, alopecia

MISC: Weight loss, leukopenia, metabolic acidosis, increased body temperature; **unexplained death (epilepsy)**

RESP: Upper respiratory tract infection, pharyngitis, sinusitis

PHARMACOKINETICS

Well absorbed, peak 2 hr, terminal half-life 19-25 hr, excreted in urine (55%-97% unchanged), crosses placenta, excreted in breast milk, protein binding (9%-17%), steady state 4 days

INTERACTIONS

Increase: renal stones—carbonic anhydrase inhibitors

Increase: effect of amitriptyline

Increase: CNS depression—alcohol, CNS depressants

Increase: topiramate levels—metformin, hydrochlorothiazide, lamoTRIgine

Decrease: levels of oral contraceptives, estrogen, digoxin, valproic acid, lithium, risperiDONE

Decrease: topiramate levels—phenytoin, carBAMazepine, valproic acid, probenecid

NURSING CONSIDERATIONS

Assess:

• **Seizures:** location, type, duration, aura
• **Bipolar disorder:** mood, behavior
• Renal studies: urinalysis, BUN, urine creatinine, electrolytes q3mo; symptoms of renal colic

• Hepatic studies: ALT, AST, bilirubin if patient receiving long-term treatment
• CBC during long-term therapy (anemia); serum bicarbonate (metabolic acidosis)
• Migraines: pain location, duration; alleviating factors
⚠ Mental status: mood, sensorium, affect, behavioral changes, **suicidal thoughts/behaviors;** if mental status changes, notify prescriber
• Body weight, evidence of cognitive disorder
• Assistance with ambulation during early part of treatment; dizziness occurs
• Seizure precautions: padded side rails, move objects that may harm patient

Evaluate:
• Therapeutic response: decreased seizure activity

Teach patient/family:
• To carry emergency ID stating patient's name, products taken, condition, prescriber's name, and phone number
• To avoid driving, other activities that require alertness
• Not to discontinue medication quickly after long-term use
• To notify prescriber immediately of blurred vision, periorbital pain
• To maintain adequate fluid intake
• About administration procedure and expected results
• To use nonhormonal contraceptive; that effect of oral contraceptives is decreased, pregnancy (D)
• To drink plenty of fluids to prevent kidney stones
• May need to increase amount of food consumed; weight loss may occur

⚠ HIGH ALERT

topotecan (Rx)

(toh-poh-tee′kan)

Hycamtin

Func. class.: Antineoplastic, natural; topoisomerase inhibitor

Chem. class.: Camptothecin analog

ACTION: Antitumor product with topoisomerase-I–inhibitory activity; topoisomerase I relieves torsional strain in DNA by causing single-strand breaks; also causes double-strand DNA damage

USES: Metastatic ovarian cancer after failure of traditional chemotherapy; relapsed small-cell lung cancer; cervical cancer

Unlabeled uses: Non–small-cell lung cancer (NSCLC), rhabdomyosarcoma

CONTRAINDICATIONS: Pregnancy (D), breastfeeding, hypersensitivity, severe bone marrow depression

Black Box Warning: Neutropenia

Precautions: Children, renal disease, gelatin hypersensitivity

DOSAGE AND ROUTES

Metastatic carcinoma of the ovary after failure of first or subsequent chemotherapy; SCLC-sensitive disease after failure of first-line therapy

• **Adult: IV INF** 1.5 mg/m^2 over 30 min daily × 5 days starting on day 1 of 21-day course × 4 courses; may be reduced to 0.25 mg/m^2 for subsequent courses if severe neutropenia occurs; relapsed small-cell lung cancer (SCLC) in those with a prior complete or partial response, ≥45 days from end of first-line treatment. **PO** 2.3 mg/m^2/day on days 1-5 of 21-day course

Renal dose

• **Adult: IV** CCr 20-39 ml/min, 0.75 mg/m²/day × 5 days starting on day 1 of 21-day course

Available forms: Lyophilized powder for inj 4 mg; caps 0.25, 1 mg

Administer:

• Store caps in refrigerator; IV INF unopened at room temperature; protect both from light

PO route

• Do not break, crush, chew, or open caps; protect from light
• Take without regard to food

Intermittent IV inf route

• Visually inspect for particulate matter and discoloration before use
• Reconstitute each 4-mg vial with 4 ml sterile water for injection; use immediately; no preservative
• Withdraw the appropriate volume of the reconstituted solution; dilute further; dilute in 0.9% NaCl or D5W before administration
• The reconstituted solution is yellow or yellow-green
• Topotecan injection diluted for inf is stable at room temperature with normal light for 24 hr
• Infuse over 30 min

SIDE EFFECTS

CNS: Arthralgia, *asthenia, headache,* myalgia, *pain,* weakness
GI: *Abdominal pain, constipation,* diarrhea, obstruction, *nausea,* stomatitis, *vomiting;* increased ALT, AST; anorexia
HEMA: **Neutropenia, leukopenia, thrombocytopenia, anemia, sepsis**
INTEG: *Total alopecia*
RESP: Dyspnea, cough, **interstitial lung disease**

PHARMACOKINETICS

Rapidly and completely absorbed, excreted in urine and feces as metabolites, half-life 2.8 hr, 7%-35% bound to plasma proteins

INTERACTIONS

• Avoid use with P-glycoprotein, breast cancer resistance protein inhibitors (amiodarone, clarithromycin, diltiazem, erythromycin, indinavir), quiNIDine, testosterone, verapamil, tamoxifen, itraconazole, mefloquine, RU-486, niCARdipine, vaccines, toxoids

Increase: myelosuppression when used with CISplatin

Increase: bleeding risk—NSAIDs, anticoagulants, thrombolytics, platelet inhibitors

Drug/Food

• Avoid use with grapefruit juice

NURSING CONSIDERATIONS

Assess:

• Hepatic studies: AST, ALT, alk phos, which may be elevated; creatinine, BUN

Black Box Warning: CBC, differential, platelet count weekly; withhold product if WBC is <3500/mm³ or platelet count is <100,000/mm³; notify prescriber of results; product should be discontinued

• Buccal cavity for dryness, sores or ulcerations, white patches, oral pain, bleeding, dysphagia
• **Interstitial lung disease (ILD): fever, cough, dyspnea, hypoxia, may be fatal**
• Increased fluid intake to 2-3 L/day to prevent dehydration unless contraindicated
• Rinsing of mouth tid-qid with water, club soda; brushing of teeth bid-tid with soft brush or cotton-tipped applicator for stomatitis; use unwaxed dental floss

Evaluate:

• Therapeutic response: decreased tumor size, spread of malignancy

Teach patient/family:

• That total alopecia may occur; that hair grows back but is different in color and texture
• To avoid foods with citric acid or hot or rough texture if stomatitis is present; to drink adequate fluids
• To report stomatitis and any bleeding, white spots, ulcerations in mouth; to examine mouth daily; to report symptoms

Black Box Warning: To report signs of anemia: fatigue, headache, faintness, SOB, irritability

- To use effective contraception during treatment and for ≤6 mo after, pregnancy (D); to avoid breastfeeding
- To avoid OTC products without approval of prescriber
- To avoid driving or other activities requiring alertness
- To avoid vaccines, toxoids

toremifene (Rx)

(tor-em′ih-feen)

Fareston

Func. class.: Antineoplastic

Chem. class.: Antiestrogen hormone

ACTION: Inhibits cell division by binding to cytoplasmic estrogen receptors; resembles normal cell complex but inhibits DNA synthesis and estrogen response of target tissue

USES: Advanced breast carcinoma not responsive to other therapy in estrogen-receptor–positive patients (usually postmenopausal)

Unlabeled uses: Prostate cancer prophylaxis

CONTRAINDICATIONS: Pregnancy (D), hypersensitivity, history of thromboembolism

Black Box Warning: QT prolongation

Precautions: Breastfeeding, children, leukopenia, thrombocytopenia, cataracts, hypercalcemia, hepatic disease, endometrial hyperplasia

DOSAGE AND ROUTES

- **Adult: PO** 60 mg/day

Available forms: Tabs 60 mg

Administer:

- Antacid before oral agent; give product after evening meal, before bedtime
- Antiemetic 30-60 min before product to prevent vomiting
- Store in light-resistant container at room temperature

SIDE EFFECTS

CNS: *Hot flashes, headache, lightheadedness,* depression

CV: CHF, MI, PE, chest pain, angina

EENT: Ocular lesions, retinopathy, corneal opacity, blurred vision (high doses)

GI: *Nausea, vomiting,* altered taste (anorexia)

GU: Vaginal bleeding, pruritus vulvae

HEMA: Thrombocytopenia, leukopenia, thrombosis

INTEG: Rash, alopecia, *sweating*

META: Hypercalcemia

RESP: Pulmonary embolism

PHARMACOKINETICS

Peak 3 hr, excreted primarily in feces, 99.5% protein binding, terminal half-life 5-6 days, metabolized by liver

INTERACTIONS

- May increase effect of warfarin

⚠ **Increase:** QT prolongation—class IA/III antidysrhythmics, some phenothiazines, β agonists, local anesthetics, tricyclics, haloperidol, chloroquine, droperidol, pentamidine; CYP3A4 inhibitors (amiodarone, clarithromycin, erythromycin, telithromycin, troleandomycin), arsenic trioxide, levomethadyl; CYP3A4 substrates (methadone, pimozide, QUEtiapine, quiNIDine, risperiDONE, ziprasidone)

⚠ **Increase:** toxicity—CYP3A4 inhibitors (aprepitant, antiretroviral protease inhibitors, clarithromycin, danazol, delavirdine, diltiazem, erythromycin, fluconazole, FLUoxetine, fluvoxaMINE, imatinib, ketoconazole, mibefradil, nefazodone, telithromycin, voriconazole)

Decrease: toremifene effect—CYP3A4 inducers (barbiturates, bosentan, carBAMazepine, efavirenz, phenytoins, nevirapine, rifabutin, rifampin)

Drug/Herb

- Avoid use with St. John's wort

Drug/Lab Test

Increase: serum calcium

NURSING CONSIDERATIONS

Assess:

- CBC, differential, platelet count weekly; withhold product if WBC is <3500/mm³ or platelet count is <100,000/mm³; notify prescriber; LFTs, serum calcium
- **Bleeding:** hematuria, guaiac, bruising, petechiae, mucosa or orifices q8hr
- Effects of alopecia on body image; discuss feelings about body changes

Severe allergic reactions: rash, pruritus, urticaria, purpuric skin lesions, itching, flushing

Black Box Warning: QT prolongation: ECG, ejection fraction; assess for chest pain, palpitations, dyspnea

- Increase fluid intake to 2-3 L/day to prevent dehydration

Evaluate:

- Therapeutic response: decreased tumor size, spread of malignancy

Teach patient/family:

- That vaginal bleeding, pruritus, hot flashes are reversible after discontinuing treatment
- To report immediately decreased visual acuity, which may be irreversible; about need for routine eye exams; that health care providers should be told about tamoxifen therapy
- To report vaginal bleeding immediately
- **That tumor flare**—increase in size of tumor, increased bone pain—may occur and will subside rapidly; to take analgesics for pain
- That premenopausal women must use mechanical birth control because ovulation may be induced, pregnancy (D)
- That hair may be lost during treatment; that a wig or hairpiece may make patient feel better; that new hair may be different in color, texture

torsemide (Rx)

(tor'suh-mide)

Demadex

Func. class.: Loop diuretic

Chem. class.: Sulfonamide derivative

ACTION: Acts on loop of Henle by inhibiting absorption of chloride, sodium, water

USES: Treatment of hypertension and edema with CHF, ascites

CONTRAINDICATIONS: Infants, hypersensitivity to sulfonamides, anuria

Precautions: Pregnancy (B), breastfeeding, diabetes mellitus, dehydration, severe renal disease, electrolyte depletion, hypovolemia, syncope, ventricular dysrhythmias

DOSAGE AND ROUTES

CHF

- **Adult: PO/IV** 10-20 mg/day, may increase as needed, max 200 mg/day

Edema in chronic renal failure

- **Adult: PO/IV** 20 mg/day, may increase to 200 mg/day

Hepatic cirrhosis in combination with aldosterone antagonist/potassium-sparing diuretic

- **Adult: PO/IV** 5-10 mg/day, may increase as needed, max 40 mg/day

Hypertension

- **Adult: PO** 5 mg/day, may increase to 10 mg/day

Available forms: Tabs 5, 10, 20, 100 mg; inj 10 mg/ml

Administer:

PO route

- In AM to avoid interference with sleep if using product as diuretic
- With food or milk if nausea occurs; absorption may be decreased slightly

IV route

- Visually inspect for particulate matter and discoloration before use

• Give either slowly as an IV bolus or as a continuous IV inf
• Ampules are single-use containers; discard any unused portion after opening

IV bolus
• No dilution is necessary if given as a slow IV injection; give over 2 min
• If administered through an IV line, flush the line with 0.9% sodium chloride for injection before and after administration

Continuous IV inf
• If administered through an IV line, flush the line with 0.9% sodium chloride for injection before and after use
• Stable for 24 hr at room temperature

SIDE EFFECTS

CNS: *Headache, dizziness,* asthenia, insomnia, nervousness
CV: Orthostatic hypotension, chest pain, ECG changes, circulatory collapse, ventricular tachycardia, edema
EENT: *Loss of hearing,* ear pain, tinnitus, blurred vision
ELECT: *Hypokalemia, hypochloremic alkalosis, hyponatremia,* metabolic alkalosis
ENDO: *Hyperglycemia, hyperuricemia*
GI: *Nausea,* diarrhea, dyspepsia, cramps, constipation
GU: *Polyuria,* renal failure, glycosuria
INTEG: *Rash,* photosensitivity, pruritus
MS: Cramps, stiffness
RESP: Rhinitis, cough increase

PHARMACOKINETICS

PO: Rapidly absorbed; duration 6 hr; breast milk; crosses placenta; half-life 3.5 hr; protein binding 97%-99%, cleared through hepatic metabolism

INTERACTIONS

Increase: toxicity—lithium, nondepolarizing skeletal muscle relaxants, digoxin
Increase: action of antihypertensives, oral anticoagulants, nitrates
Increase: ototoxicity—aminoglycosides, CISplatin, vancomycin
Decrease: antihypertensive effect of torsemide—indomethacin, carBAMazepine, PHENobarbital, phenytoin, rifampin, NSAIDs

Drug/Herb
• Severe photosensitivity: St. John's wort

Drug/Lab Test
Increase: BUN, creatinine, uric acid, blood glucose, cholesterol
Decrease: potassium, magnesium, chloride sodium

NURSING CONSIDERATIONS

Assess:
• Hearing when giving high doses
• Weight, I&O daily to determine fluid loss; effect of product may be decreased if used daily
• B/P lying, standing; postural hypotension may occur
• Electrolytes: potassium, sodium, chlorine; include BUN, blood glucose, CBC, serum creatinine, blood pH, ABGs, uric acid, calcium, magnesium
• Glucose in urine of diabetic patients
• **Signs and symptoms of metabolic alkalosis:** drowsiness, restlessness
• **Signs and symptoms of hypokalemia:** postural hypotension, malaise, fatigue, tachycardia, leg cramps, weakness
• Rashes, temperature elevation daily
• Confusion, especially in geriatric patients; take safety precautions if needed

Evaluate:
• Therapeutic response: improvement in edema of feet, legs, sacral area daily if medication is being used with CHF

Teach patient/family:
• To rise slowly from lying, sitting position
• To recognize adverse reactions: muscle cramps, weakness, nausea, dizziness, tinnitus
• To take with food or milk for GI symptoms; to limit alcohol use
• To take early during the day to prevent nocturia
• To use sunscreen, protective clothing to prevent sunburn

TREATMENT OF OVERDOSE:
Lavage if taken orally; monitor electrolytes; administer dextrose in saline; monitor hydration, CV, renal status

traMADol (Rx)
(tram′a-dole)

Conzip, Ryzolt, Ultram, Ultram ER, Zytram ♣

Func. class.: Analgesic—miscellaneous

Do not confuse:
traMADol/Toradol

ACTION:
Binds to μ-opioid receptors, inhibits reuptake of norepinephrine, serotonin

USES:
Management of moderate to severe pain, chronic pain

Unlabeled uses: Restless leg syndrome (RLS), postoperative shivering, arthralgia/myalgia, bone/dental/neuropathic pain, headache, osteoarthritis

CONTRAINDICATIONS:
Hypersensitivity, acute intoxication with any CNS depressant, alcohol, asthma, respiratory depression

Precautions: Pregnancy (C), breastfeeding, children, geriatric patients, seizure disorder, renal/hepatic disease, respiratory depression, head trauma, increased intracranial pressure, acute abdominal condition, drug abuse, depression, suicidal ideation

DOSAGE AND ROUTES

Mild to moderate pain
- **Adult: PO** 25 mg daily, titrate by 25 mg ≥3 days to 100 mg/day (25 mg qid), then may increase by 50 mg ≥3 days to 200 mg (50 mg qid), then 50-100 mg q4-6hr, max 400 mg/day, use caution in geriatric patients
- **Geriatric >75 years: PO** <300 mg/day in divided doses

Moderate to severe chronic pain
- **Adult: PO-ER** (Ultram ER) 100 mg daily, titrate upward q5days in 100-mg increments, max 300 mg/day; (Ryzolt) 100 mg, titrate upward q2-3days in 100-mg increments, max 300 mg/day; products are not interchangeable

Renal dose
- **Adult: PO** CCr <30 ml/min, give q12hr, max 200 mg/day; do not use ext rel tab

Hepatic dose
- **Adult (Child-Pugh C): PO** 50 mg q12hr; do not use ext rel tab

Restless leg syndrome (RLS) (unlabeled)
- **Adult: PO** 50-150 mg/day × 15-24 mo

Available forms: Tabs 50 mg; ext rel tab 100, 200, 300 mg; ext rel caps 100, 200, 300 mg

Administer:
- **Ext rel products** (Ryzolt/Utram ER) are not interchangeable
- Do not break, crush, or chew ext rel product
- With antiemetic for nausea, vomiting
- When pain is beginning to return; determine dosage interval by patient response
- With or without food; ER: always give with food, or always give on empty stomach
- Store in cool environment; protected from sunlight

SIDE EFFECTS

CNS: Dizziness, CNS stimulation, somnolence, headache, anxiety, confusion, euphoria, **seizures,** hallucinations, sedation, **neuroleptic-malignant-syndrome–like reactions**

CV: Vasodilation, orthostatic hypotension, tachycardia, hypertension, abnormal ECG

EENT: Visual disturbances

GI: Nausea, constipation, vomiting, dry mouth, diarrhea, abdominal pain, anorexia, flatulence, *GI bleeding*

GU: Urinary retention/frequency, menopausal symptoms, dysuria, menstrual disorder

T

INTEG: Pruritus, rash, urticaria, vesicles, flushing
SYST: Anaphylaxis, Stevens-Johnson syndrome, toxic epidermal necrolysis, serotonin syndrome

PHARMACOKINETICS

Rapidly and almost completely absorbed, steady state 2 days, peak 1.5 hr, duration 6 hr, terminal half-life 7.9-8.8 hr, may cross blood-brain barrier, extensively metabolized, 30% excreted in urine as unchanged product, protein binding 20%

INTERACTIONS

- Inhibition of norepinephrine and serotonin reuptake: MAOIs; use together with caution

Increase: CNS depression—alcohol, sedatives, hypnotics, opiates
⚠ **Increase:** serotonin syndrome—SSRIs, SNRIs, serotonin-receptor agonists
Increase: traMADol levels—CYP3A4 inhibitors (aprepitant, antiretroviral protease inhibitors, clarithromycin, danazol, delavirdine, diltiazem, erythromycin, fluconazole, FLUoxetine, fluvoxaMINE, imatinib, ketoconazole, mibefradil, nefazodone, telithromycin, voriconazole)
Decrease: traMADol effects—CYP3A4 inducers (barbiturates, bosentan, carBAMazepine, efavirenz, phenytoins, nevirapine, rifabutin, rifampin)
Decrease: levels of traMADol—carBAMazepine
Drug/Herb
- Avoid use with St. John's wort

Increase: CNS depression—chamomile, hops, kava, skullcap, valerian
Drug/Lab Test
Increase: creatinine, hepatic enzymes
Decrease: Hgb

NURSING CONSIDERATIONS

Assess:
- **Pain:** location, type, character, give before pain becomes extreme
- **Respiratory depression:** withhold if respirations <12/min
- I&O ratio: check for decreasing output; may indicate urinary retention
- Need for product; dependency
- Bowel pattern; for constipation, increase fluids, bulk in diet
- CNS changes: dizziness, drowsiness, hallucinations, euphoria, LOC, pupil reaction
- **Hypersensitivity:** usually after beginning treatment
- Increased side effects in renal/hepatic disease

⚠ **Serotonin syndrome, neuroleptic malignant syndrome:** increased heart rate, shivering, sweating, dilated pupils, tremors, high B/P, hyperthermia, headache, confusion; if these occur, stop product, administer serotonin antagonist if needed
- Assistance with ambulation
- Safety measures: side rails, night-light, call bell within easy reach

Evaluate:
- Therapeutic response: decrease in pain

Teach patient/family:
- To report any symptoms of CNS changes, allergic reactions, serotonin syndrome, seizures
- That drowsiness, dizziness, and confusion may occur; to avoid hazardous activities
- To make position changes slowly because orthostatic hypotension may occur
- To avoid OTC medications, herbs, supplements, CNS depressants, and alcohol unless approved by prescriber
- Not to discontinue abruptly, taper

trandolapril (Rx)

(tran-doe′la-prill)

Mavik

Func. class.: Antihypertensive
Chem. class.: Angiotension-converting enzyme inhibitor

ACTION: Selectively suppresses renin-angiotensin-aldosterone system; inhibits ACE; prevents conversion of angiotensin I to angiotensin II, dilates arterial and venous vessels, lowers B/P

USES:
Hypertension, heart failure, left ventricular dysfunction post MI

CONTRAINDICATIONS:
Breastfeeding, hypersensitivity, history of angioedema

Black Box Warning: Pregnancy (D) 2nd/3rd trimester

Precautions: Geriatric patients, hyperkalemia, hepatic disease, bilateral renal stenosis, after kidney transplant, aorta/mitral valve stenosis, cirrhosis, severe renal disease, untreated CHF, autoimmune disease, cough, pregnancy (C) 1st trimester, African descent

DOSAGE AND ROUTES

Hypertension

• **Adult: PO** 1 mg/day; 2 mg/day in African-Americans; make dosage adjustment ≥1 wk; max 8 mg/day

Heart failure, left ventricular dysfunction post MI

• **Adult: PO** 1 mg/day, titrate upward to 4 mg/day if tolerated, 2-4 yr

Renal/hepatic dose

• **Adult: PO** CCr <30 ml/min or hepatic disease, 0.5 mg/day

Available forms: Tabs 1, 2, 4 mg

Administer:

• Discontinue diuretic 2-3 days before starting this product; if not possible, decrease initial dose to 0.5 mg
• Make dosage changes ≥1 wk
• Give without regard to food
• Space antacids by 2 hr after dose

SIDE EFFECTS

CNS: *Dizziness, syncope,* paresthesias, headache, fatigue, drowsiness, depression, sleep disturbances, anxiety, syncope
CV: *Hypotension,* **MI**, palpitations, angina, TIAs, **stroke**, *bradycardia,* dysrhythmias
GI: Nausea, vomiting, cramps, diarrhea, constipation, **pancreatitis**, *dyspepsia*
GU: **Proteinuria, renal failure**
HEMA: **Agranulocytosis, neutropenia, leukopenia, anemia**
INTEG: Rash, purpura, pruritus, **angioedema**
MISC: Hyperkalemia, hyponatremia, impotence, *myalgia,* **angioedema,** muscle cramps, *asthenia,* hypocalcemia, gout
RESP: Dyspnea, *cough*

PHARMACOKINETICS

Peak 4-10 hr, duration 24 hr, half-life 6-10 hr, metabolized by liver (active metabolite trandolaprilat), excreted in urine (33%), feces (66%), protein binding 65%-94%

INTERACTIONS

Increase: effects—phenothiazines, diuretics
Increase: severe hypotension—diuretics, other antihypertensives, NSAIDs
Increase: potassium levels—salt substitutes, potassium-sparing diuretics, potassium supplements, ACE inhibitors
Increase: effects of ergots, neuromuscular blocking agents, antihypertensives, hypoglycemics, barbiturates, reserpine, levodopa, lithium
Decrease: effects of trandolapril—antacids, NSAIDs, salicylates

Drug/Lab Test

Increase: potassium, LFTs, BUN, creatinine
Decrease: sodium, WBC

NURSING CONSIDERATIONS

Assess:

• **Hepatotoxicity (rare):** increased LFTs, jaundice, fulfillment hepatic necrosis; if jaundice occurs, discontinue product

• **Angioedema** of the face, edema of the extremities, mucus membranes; may need to discontinue
• **Hyperkalemia:** monitor electrolytes, check potassium
• B/P, pulse q4hr; note rate, rhythm, quality
• Baselines of renal, hepatic studies before therapy begins
• Edema in feet, legs daily

• **Symptoms of CHF:** edema, dyspnea, wet crackles

Evaluate:

• Therapeutic response: decreased B/P

Teach patient/family:

• Not to use OTC (cough, cold, or allergy) products unless directed by prescriber, to separate antacids by 2 hr; to avoid salt substitutes

• To comply with dosage schedule even if feeling better

• To notify prescriber of mouth sores, sore throat, fever, swelling of hands/feet, irregular heartbeat, chest pain, signs of angioedema

• That excessive perspiration, dehydration, vomiting, diarrhea may lead to fall in B/P; to consult prescriber if these occur

• That product may cause dizziness, fainting; lightheadedness may occur during 1st few days of therapy

• That product may cause skin rash or impaired perspiration

• Not to discontinue product abruptly, taper

• To rise slowly to sitting or standing position to minimize orthostatic hypotension

Black Box Warning: To notify prescriber if pregnancy is planned or suspected; pregnancy (D)

HIGH ALERT

trastuzumab (Rx)

(tras-tuz'uh-mab)

Herceptin

Func. class.: Antineoplastic—miscellaneous

Chem. class.: Humanized monoclonal antibody

ACTION: DNA-derived monoclonal antibody selectively binds to extracellular portion of human epidermal growth factor receptor 2; it inhibits the proliferation of cancer cells

USES: Breast cancer; metastatic with overexpression of HER2, early breast cancer (adjuvant, neoadjuvant), gastric cancer; previously untreated HER2 overexpressing metastatic gastric or gastroesophageal junction adenocarcinoma with CISplatin, 5-fluorouracil or capecitabine

CONTRAINDICATIONS: Pregnancy (D); hypersensitivity to this product, Chinese hamster ovary cell protein

Precautions: Breastfeeding, children, geriatric patients, pulmonary disease, anemia, leukopenia

Black Box Warning: Cardiac disease, respiratory distress syndrome, respiratory insufficiency, inf-related reactions, cardiomyopathy

DOSAGE AND ROUTES

Breast cancer

• Several regimens may be used

• **Adult: IV** 4 mg/kg given over 90 min then maintenance 2 mg/kg given over 30 min; do not give as IV push or bol; may be given in combination with other antineoplastics

Gastric cancer

• **Adult: IV** 8 mg/kg over 90 min on day 1 then 6 mg/kg over 30-90 min q21days from day 22, give with CISplatin 80 mg/m^2 on day 1 plus 5-fluorouracil 800 mg/m^2 **CONT INF** on days 1-5 or capecitabine 1000 mg/m^2 bid on days 1-14, repeat cycle q3wk

Available forms: Lyophilized powder 440 mg

Administer:

• Acetaminophen as ordered to alleviate fever and headache

Intermittent IV INF route

• After reconstituting vial with 20 ml bacteriostatic water for inj, 1.1% benzyl alcohol preserved (supplied) to yield 21 mg/ml; mark date on vial 28 days from reconstitution date; if patient is allergic to benzyl alcohol, reconstitute with sterile water for inj; use immediately; infuse over 90 min; q3wk give 8 mg/kg loading

dose over 90 min; subsequent 6 mg/kg dose may be given over 30-60 min
• Do not mix or dilute with other products or dextrose sol

SIDE EFFECTS

CNS: *Dizziness, numbness, paresthesias,* depression, *insomnia,* neuropathy, peripheral neuritis
CV: Tachycardia, CHF
GI: Nausea, vomiting, *anorexia, diarrhea,* abdominal pain, hepatotoxicity, dysgeusia
HEMA: *Anemia,* leukopenia
INTEG: Rash, acne, herpes simplex
META: Edema, peripheral edema
MISC: *Flulike symptoms; fever, headache, chills*
MS: Arthralgia, *bone pain*
RESP: *Cough, dyspnea, pharyngitis, rhinitis,* sinusitis, pneumonia, pulmonary edema/fibrosis, acute respiratory distress syndrome (ARDS)
SYST: Anaphylaxis, angioedema

PHARMACOKINETICS

Half-life 1-32 days

INTERACTIONS

Increase: bleeding risk—warfarin
Increase: cardiomyopathy—anthracyclines, cyclophosphamide; avoid use
Decrease: immune response—vaccines, toxoids

NURSING CONSIDERATIONS

Assess:
• CBC, HER2 overexpression

Black Box Warning: CHF, other cardiac symptoms: dyspnea, coughing; gallop; obtain full cardiac workup including ECG, echo, multigated acquisition scan

• Symptoms of infection; may be masked by product
• CNS reaction: LOC, mental status, dizziness, confusion

⚠ **Hypersensitive reactions, anaphylaxis**

Black Box Warning: Inf reactions that may be fatal: fever, chills, nausea, vomiting, pain, headache, dizziness, hypotension; discontinue product

• **Pulmonary toxicity:** dyspnea, interstitial pneumonitis, pulmonary hypertension, ARDs; can occur after inf reaction, those with lung disease may have more severe toxicity

Perform/provide:
• Increased fluid intake to 2-3 L/day

Evaluate:
• Therapeutic response: decrease in size of tumors

Teach patient/family:
• To take acetaminophen for fever
• To avoid hazardous tasks because confusion, dizziness may occur
• To report signs of infection: sore throat, fever, diarrhea, vomiting
• That emotional lability is common; to notify prescriber if severe or incapacitating
• To use contraception while taking this product; pregnancy (D); to avoid breastfeeding

Black Box Warning: To report pain at inf site

travoprost ophthalmic

See Appendix B

traZODone (Rx)

(tray′zoe-done)

Oleptro

Func. class.: Antidepressant—miscellaneous
Chem. class.: Triazolopyridine

Do not confuse:
traZODone/traMADol

ACTION: Selectively inhibits serotonin uptake by brain; potentiates behavorial changes

USES: Depression

T

Unlabeled uses: Alcoholism, anxiety, panic disorder, insomnia

CONTRAINDICATIONS:

Hypersensitivity to tricyclics

Precautions: Pregnancy (C), suicidal patients, severe depression, increased intraocular pressure, closed-angle glaucoma, urinary retention, cardiac/hepatic disease, hyperthyroidism, electroshock therapy, elective surgery, bleeding, abrupt discontinuation, bipolar disorder, breastfeeding, dehydration, hyponatremic, hypovolemia, recovery phase of MI, seizure disorders, prostatic hypertrophy, family history of long QT

Black Box Warning: Suicidal ideation in children/adolescents

DOSAGE AND ROUTES

Depression

- **Adult: PO** 150 mg/day in divided doses, may increase by 50 mg/day q3-4days, max 400 mg/day (outpatient), 600 mg/day (inpatients); **EXT REL** 150 mg in PM, may increase gradually by 75 mg/day q3days, max 375 mg/day
- **Child 6-18 yr (unlabeled): PO** 1.5-2 mg/kg/day in divided doses, may increase q3-4days up to 6 mg/kg/day or 400 mg/day, whichever is less
- **Geriatric: PO** 25-50 mg at bedtime, increase by 25-50 mg q3-7days to desired dose, usually 75-150 mg/day

Alcoholism (unlabeled)

- **Adult: PO** 50-100 mg/day

Panic disorder (unlabeled)

- **Adult: PO** 150 mg in divided doses, may increase by 50 mg/day q3-4days

Insomnia (unlabeled)

- **Adult: PO** 50 mg at bedtime

Available forms: Tabs 50, 100, 150, 300 mg; ext rel tabs 150, 300 mg

Administer:

- Increased fluids, bulk in diet if constipation occurs, especially in geriatric patients
- With food, milk for GI symptoms
- Dosage at bedtime for oversedation during day; may take entire dose at bedtime; geriatric patients may not tolerate daily dosing
- Avoid use of CNS depressants
- Do not crush, break, chew ext rel product
- Store in tight, light-resistant container at room temperature

SIDE EFFECTS

CNS: *Dizziness, drowsiness,* confusion, headache, anxiety, tremors, stimulation, weakness, insomnia, nightmares, EPS (geriatric patients), increase in psychiatric symptoms, **suicide in children/adolescents**

CV: *Orthostatic hypotension, ECG changes, tachycardia,* **hypertension**, palpitations

EENT: *Blurred vision,* tinnitus, mydriasis

GI: *Diarrhea, dry mouth,* nausea, vomiting, **paralytic ileus**, increased appetite, cramps, epigastric distress, jaundice, **hepatitis**, stomatitis, constipation

GU: *Urinary retention,* **acute renal failure**, priapism

HEMA: Agranulocytosis, thrombocytopenia, eosinophilia, leukopenia

INTEG: Rash, urticaria, sweating, pruritus, photosensitivity

PHARMACOKINETICS

Peak 1 hr without food, 2 hr with food; metabolized by liver (CYP3A4); excreted by kidneys, in feces; half-life 4.4-7.5 hr

INTERACTIONS

⚠ **Hyperpyretic crisis, seizures, hypertensive episode: MAOIs; do not use within 14 days of traZODone**

Increase: toxicity, serotonin syndrome—FLUoxetine, nefazodone, other SSRIs, SNRIs, linezolid; methylene blue (IV)

Increase: effects of direct-acting sympathomimetics (EPINEPHrine), alcohol, barbiturates, benzodiazepines, CNS depressants, digoxin, phenytoin, carBAMazepine

Increase: effects of traZODone—CYP3A4, 2D6 inhibitors (phenothiazines, protease inhibitors, azole antifungals)

Increase or decrease: effects of warfarin
Decrease: effects of guanethidine, cloNIDine, indirect-acting sympathomimetics (ePHEDrine)

Drug/Herb
Increase: serotonin syndrome—SAM-e, St. John's wort
Increase: CNS depression—hops, kava, lavender, valerian

Drug/Lab Test
Increase: LFTs
Decrease: Hgb

NURSING CONSIDERATIONS

Assess:

• B/P lying, standing; pulse q4hr; if systolic B/P drops 20 mm Hg, hold product, notify prescriber; take vital signs q4hr in patients with CV disease
• Blood studies: CBC, leukocytes, differential
• Hepatic studies: AST, ALT, bilirubin
• Weight weekly; appetite may increase with product
• ECG for flattening of T wave, bundle branch block, AV block, dysrhythmias in cardiac patients
• EPS, primarily in geriatric patients: rigidity, dystonia, akathisia

Black Box Warning: Mental status changes: mood, sensorium, affect, suicidal tendencies, increase in psychiatric symptoms, depression, panic; observe for suicidal behaviors in children/adolescents, not approved for children

• Urinary retention, constipation; constipation most likely in children
• **Withdrawal symptoms:** headache, nausea, vomiting, muscle pain, weakness; not usual unless product discontinued abruptly
• Alcohol consumption; hold dose until morning

⚠ **Serotonin syndrome, neuroleptic malignant syndrome:** increased heart rate, shivering, sweating, dilated pupils, tremors, high B/P, hyperthermia, headache, confusion; if these occur, stop product, administer serotonin antagonist if needed

• Assistance with ambulation during beginning therapy for drowsiness, dizziness

Evaluate:

• Therapeutic response: decreased depression

Teach patient/family:

• That therapeutic effects may take 2-3 wk; to take product before bedtime; not to crush, chew ext rel product
• To use caution when driving, performing other activities requiring alertness because of drowsiness, dizziness, blurred vision
• To avoid alcohol ingestion
• Not to discontinue medication quickly after long-term use; may cause nausea, headache, malaise
• To report urinary retention, priapism >4 hr immediately
• To wear sunscreen or large hat because photosensitivity occurs

Black Box Warning: That suicidal thoughts/behaviors may occur (adolescents/children)

• To notify prescriber if pregnancy is planned or suspected, pregnancy (C), avoid breastfeeding
• To rise slowly to prevent dizziness

TREATMENT OF OVERDOSE:

ECG monitoring; lavage, activated charcoal; administer anticonvulsant, atropine for bradycardia

treprostinil (Rx)

(treh-prah'stin-ill)

Remodulin, Tyvaso

Func. class.: Antihypertensive, vasodilators

Chem. class.: Tricyclic benzidine prostacyclin analog

ACTION: Direct vasodilation of pulmonary, systemic arterial vascular beds; inhibition of platelet aggregation

USES: Pulmonary arterial hypertension (PAH) NYHA class II through IV

Unlabeled uses: Pulmonary arterial hypertension in children/adolescents, pediatric patients transitioning from epoprostenol to treprostinil, claudication

CONTRAINDICATIONS:
Hypersensitivity to this product, other prostacyclin analogs

Precautions: Pregnancy (B), breastfeeding, children, geriatric patients, past renal/hepatic disease, thromboembolic disease, abrupt discontinuation, IV administration

DOSAGE AND ROUTES

Pulmonary arterial hypertension (WHO Group 1)

- **Adult: SUBCUT INF** 1.25 ng/kg/min by **CONT INF**, may reduce to 0.625 ng/kg/min if not tolerated; may increase by 1.25 ng/kg/min weekly for first 4 wk then 2.5 ng/kg/min/wk for remainder of inf; **ORAL INH** 3 breaths via Tyvaso inh system qid

Hepatic dose

- **Adult: SUBCUT INF 0.625 ng/kg ideal body weight/min; increase cautiously**

Available forms: Inj 1, 2.5, 5, 10 mg/ml; neb sol 1.74 mg/2.9 ml

Administer:

- Sudden decreased doses, abrupt withdrawal may worsen pulmonary arterial hypertension symptoms

SUBCUT route

- By continuous inf
- No dilution required

Oral INH route

- Avoid skin, eyes; do not take orally; use Tyvaso Inhalation System only
- Patient should have backup Optineb-ir device to avoid interruptions
- Follow instructions for use, cleaning
- Do not mix with other medications in Optineb-ir device
- Twist off cap; squeeze total contents into medicine cup; vol is sufficient for 4 treatments

Continuous IV INF route

- By surgically placed CV catheter using ambulatory inf pump
- IV pump, product, patient education can be obtained from Priority Healthcare in United States
- Must be diluted with sterile water for inj or 0.9% NaCl
- Conc should be calculated using this formula: diluted conc = [dose (ng/kg/min) × weight (kg) × 0.00006] / inf rate (ml/hr)]

SIDE EFFECTS

CNS: Dizziness, headache, syncope
CV: Vasodilation, *hypotension, edema,* right ventricular heart failure
GI: Nausea, *diarrhea*
INTEG: *Rash,* pruritus
OTHER: Jaw pain, cough, throat irritation
SYST: Inf site reactions, pain; increased risk for infection

PHARMACOKINETICS

Metabolized by liver; excreted in urine, feces; terminal half-life 2-4 hr; 90% protein binding

INTERACTIONS

- Excessive hypotension: diuretics, antihypertensives, vasodilators, MAOIs, β-blockers, calcium channel blockers

Increase: bleeding tendencies—anticoagulants, aspirin, NSAIDs, thrombin inhibitors, SSRIs

NURSING CONSIDERATIONS

Assess:

- **Hypertension:** monitor B/P, baseline and periodically
- Hepatic studies: AST, ALT, bilirubin, creatinine with long-term therapy

⚠ **Blood studies: CBC; CBC q2wk × 3 mo, Hct, Hgb, PT with long-term therapy, ABGs**

⚠ **Bleeding time at baseline, throughout treatment; levels may be 2-5× normal limit**

Evaluate:

- Avoid abrupt discontinuation
- Therapeutic response: decreased pulmonary arterial hypertension (PAH)

Teach patient/family:

- That blood work will be necessary during treatment; that treatment may last for years
- To report side effects such as diarrhea, skin rashes
- That therapy will be needed for prolonged periods of time, sometimes years
- To prevent infection, aseptic technique must be used for preparation, administration of treprostinil
- That there are many product, herbal interactions
- About signs, symptoms of bleeding; blood in urine, stools
- How to use inhaled solution; how to care for equipment

tretinoin (vit A acid, retinoic acid) (Rx)

(tret'i-noyn)

Avita, Renova, Retin-A, Retin-A Micro, Stieva-A ✤

Func. class.: Vit A acid, acne product; antineoplastic (miscellaneous)

Chem. class.: Tretinoin derivative

ACTION: (Topical) Decreases cohesiveness of follicular epithelium, decreases microcomedone formation; (PO) induces maturation of acute promyelocytic leukemia, exact action is unknown

USES: (Topical) Acne vulgaris (grades 1-3); (PO) acute promyelocytic leukemia, facial wrinkles, photoaging

Unlabeled uses: Acne rosacea, actinic keratosis, ichthyosis, Kaposi's sarcoma, keloids, keratosis follicularis, melasma

CONTRAINDICATIONS: Hypersensitivity to retinoids or sensitivity to parabens

Black Box Warning: Pregnancy (D) (PO)

Precautions: Pregnancy (C) (topical), breastfeeding, eczema, sunburn, sun exposure

Black Box Warning: Rapid-evolving leukocytosis, respiratory compromise, acute promyelocytic leukemia differentiation syndrome

DOSAGE AND ROUTES

- **Adult and child: TOP** cleanse area, apply 0.025%-0.1% cream or 0.05% liquid gel at bedtime, cover lightly

Promyelocytic leukemia

- **Adult: PO** 45 mg/m²/day given as 2 evenly divided doses until remission, discontinue treatment 30 days after remission or 90 days of treatment, whichever is first

Available forms: Cream 0.01%, 0.02%, 0.025%, 0.05%, 0.1%; gel 0.01%, 0.025%, 0.04%, 0.05%, 0.1%; liquid 0.05%; caps 10 mg

Administer:

Topical route

- Once daily before bedtime; cover area lightly using gauze; use gloves to apply
- Store at room temperature
- Handwashing after application

SIDE EFFECTS

PO

CNS: *Headache, fever, sweating,* fatigue

CV: Cardiac dysrhythmias, pericardial effusion

GI: *Nausea, vomiting,* **hemorrhage,** *abdominal pain, diarrhea, constipation, dyspepsia, distention, hepatitis*

Topical

INTEG: Rash, stinging, warmth, redness, erythema, blistering, crusting, peeling, contact dermatitis, hypo/hyperpigmentation, dry skin, pruritus, scaly skin, retinoic acid syndrome (RAS)

META: Hypercholesterolemia, hypertriglyceridemia

RESP: Pneumonia, upper respiratory tract disease

PHARMACOKINETICS

PO: Terminal half-life 0.5-2 hr

TOPICAL: Poor systemic absorption

INTERACTIONS

- Use with caution: medicated, abrasive soaps; cleansers that have a drying effect; products with high conc of alcohol astringents (topical)

Increase: peeling—medication containing agents such as sulfur, benzoyl peroxide, resorcinol, salicylic acid (topical)

Increase: plasma concs of tretinoin—ketoconazole (PO)

⚠ **Increase:** ICP, risk of pseudotumor cerebri—tetracyclines; do not use together

Increase: photosensitivity—retinoids, quinolones, phenothiazines, sulfonamides, sulfonylureas, thiazide diuretics

Increase: thrombotic complications—aninocaproic acid, aprotinin, tranexamic acid

Drug/Lab Test

Increase: AST, ALT

NURSING CONSIDERATIONS

Assess:

Topical route

- Area of body involved, what helps or aggravates condition; cysts, dryness, itching; lesions may worsen at beginning of treatment

PO route

- Hepatic function, coagulation, hematologic parameters; also cholesterol, triglycerides

Evaluate:

- Therapeutic response: decrease in size, number of lesions

Teach patient/family:

Topical route

- To avoid application on normal skin; to avoid getting cream in eyes, nose, other mucous membranes; not to use product on areas with cuts, scrapes
- To use cream/gel by applying a thin layer to affected skin; to rub gently; to use liquid; to apply with fingertip or cotton swab
- To avoid sunlight, sunlamps; to use protective clothing, sunscreen
- That treatment may cause warmth, stinging; that dryness, peeling will occur
- That cosmetics may be used over product; not to use shaving lotions
- That rash may occur during first 1-3 wk of therapy
- That product does not cure condition, only relieves symptoms
- That therapeutic results may be seen in 2-3 wk but may not be optimal until after 6 wk

PO route

Black Box Warning: To notify prescriber if pregnancy is planned or suspected; pregnancy (D) PO

tretinoin topical

See Appendix B

triamcinolone (Rx)

(trye-am-sin′oh-lone)

Aristospan, Kenalog-10, Kenalog-40, Tac-3, Triesence

Func. class.: Corticosteroid, synthetic

Chem. class.: Glucocorticoid, intermediate acting

See ophthalmic in Appendix B

ACTION: Decreases inflammation by suppression of migration of polymorphonuclear leukocytes, fibroblasts; reversal of increased capillary permeability and lysosomal stabilization

USES: Severe inflammation, immunosuppression, neoplasms, asthma (steroid dependent); collagen, respiratory, dermatologic/rheumatic disorders

CONTRAINDICATIONS: Hypersensitivity, neonatal prematurity; epidural/intrathecal administration (Triamcinolone acetonide injections [Kenalog]), systemic fungal infections

Precautions: Pregnancy (C), breastfeeding, diabetes mellitus, glaucoma, osteoporosis, seizure disorders, ulcerative colitis, CHF, myasthenia gravis, renal disease, esophagitis, peptic ulcer, acne, cataracts, coagulopathy, head trauma, children <2 yr, psychosis, idiopathic thrombocytopenia, acute glomerulonephritis, amebiasis, fungal infections, nonasthmatic bronchial disease, AIDS, TB, adrenal insufficiency, acute bronchospasm, acne rosacea, Cushing syndrome, acute MI, thromboembolism

DOSAGE AND ROUTES

• **Adult: IM** (acetonide) 40 mg/wk; (acetonide) 5-48 mg into neoplasms; (acetonide) 2-40 mg into joint or soft tissue; (hexacetonide) 0.5 mg/in^2 of affected intralesional skin; (hexacetonide) 2-20 mg into joint or soft tissue

Severe/incapacitating allergic conditions such as asthma

• **Adult: IM** (Trivaris) 60 mg, titrate, usual range 40-80 mg

• **Child: IM** (Trivaris) 0.11-1.6 mg/kg/day (3.2-48 mg/m^2/day) in 3-4 divided doses

Available forms: Inj 3, 10, 40 mg/ml acetonide; inj 20, 5 mg/ml hexacetonide

Administer:

IM route

• After shaking susp (parenteral)
• Titrated dose; use lowest effective dose
• IM inj deep in large muscle mass; rotate sites; avoid deltoid; use 21-G needle
• Avoid SUBCUT administration, may damage tissue

SIDE EFFECTS

CNS: *Depression,* headache, mood changes

CV: *Hypertension,* **circulatory collapse, embolism,** tachycardia, edema

EENT: Fungal infections, increased intraocular pressure, blurred vision

GI: *Diarrhea, nausea, abdominal distention,* **GI hemorrhage,** *increased appetite,* **pancreatitis**

HEMA: **Thrombocytopenia**

INTEG: Acne, poor wound healing, ecchymosis, petechiae

MS: Fractures, osteoporosis, weakness

PHARMACOKINETICS

PO/IM: Peak 1-2 hr, half-life 2-5 hr

INTERACTIONS

Increase: side effects—alcohol, salicylates, indomethacin, amphotericin B, digoxin, cycloSPORINE, diuretics, quinolones

Increase: action of triamcinolone—salicylates, estrogens, indomethacin, oral contraceptives, ketoconazole, macrolide antiinfectives, carBAMazepine

Decrease: action of triamcinolone—cholestyramine, colestipol, barbiturates, rifampin, ePHEDrine, phenytoin, theophylline

Decrease: effects of anticoagulants, anticonvulsants, antidiabetics, ambenonium, neostigmine, isoniazid, toxoids, vaccines, anticholinesterases, salicylates, somatrem

Drug/Herb

• Hypokalemia: aloe, cascara, senna

Drug/Lab Test

Increase: cholesterol, sodium, blood glucose, uric acid, calcium, urine glucose

Decrease: Calcium, potassium, T_4, T_3, thyroid ^{131}I uptake test, urine 17-OHCS, 17-KS, PBI

False negative: skin allergy tests

NURSING CONSIDERATIONS

Assess:

• Potassium, blood glucose, urine glucose while patient receiving long-term therapy; hypokalemia and hyperglycemia
• Weight daily; notify prescriber if weekly gain of >5 lb
• B/P, pulse; notify prescriber if chest pain occurs
• I&O ratio; be alert for decreasing urinary output, increasing edema
• Plasma cortisol levels during long-term therapy (normal level: 138-635 nmol/L SI units when drawn at 8 AM)
• **Infection:** increased temperature, WBC even after withdrawal of medication; product masks infection

T

• Potassium depletion: paresthesias, fatigue, nausea, vomiting, depression, polyuria, dysrhythmias, weakness
• Edema, hypertension, cardiac symptoms
• Mental status: affect, mood, behavioral changes, aggression
• Assistance with ambulation for patient with bone-tissue disease to prevent fractures

Evaluate:
• Therapeutic response: ease of respirations, decreased inflammation

Teach patient/family:
• That emergency ID as corticosteroid user should be carried; not to discontinue abruptly, taper dose
• To notify prescriber if therapeutic response decreases; that dosage adjustment may be needed
• To avoid OTC products: salicylates, alcohol in cough products, cold preparations unless directed by prescriber; to avoid live vaccines
• About cushingoid symptoms
• About the symptoms of adrenal insufficiency: nausea, anorexia, fatigue, dizziness, dyspnea, weakness, joint pain

triamcinolone nasal agent
See Appendix B

triamcinolone ophthalmic
See Appendix B

triamcinolone (topical)
(try-am-sin′oh-lone)
Kenalog, Triderm
Func. class.: Corticosteroid, topical

ACTION: Crosses cell membrane to attach to receptors to decrease inflammation, itching; inhibits multiple inflammatory cytokines

USES: Inflammation/itching in corticosteroid-responsive dermatoses on the skin or inflammation in the mouth

CONTRAINDICATIONS: Hypersensitivity, use on face, ear canal, infections

Precautions: Pregnancy (C), breastfeeding, children

DOSAGE AND ROUTES
• Apply to the affected areas bid-qid

Available forms: Aerosol 0.2 mg; paste (dental) 0.1%; lotion, cream, ointment 0.025%; ointment 0.05%; lotion, cream, ointment 0.1%; ointment, cream 0.5%

Administer:
Topical route
• May be used with occlusive dressings
• **Cream/Ointment/Lotion:** apply sparingly in a thin film and rub gently into the cleansed, slightly moist affected area; may use gloves to apply cream/ointment/lotion
• **Paste:** apply without rubbing, press into lesion until film develops
• **Spray:** spray a small amount of preparation onto the lesion

SIDE EFFECTS
ENDO: HPA axis suppression, Cushing syndrome
INTEG: Burning, folliculitis, pruritus, dermatitis, hypopigmentation
META: Hyperglycemia; glycosuria

PHARMACOKINETICS
Absorption varies

INTERACTIONS
Increase: blood glucose

NURSING CONSIDERATIONS
Assess:
• Skin reactions: burning pruritus, folliculitis, mouth lesions

Evaluate:
• Decreasing itching, inflammation on the skin, decreasing mouth lesions

Teach patient/family:
- How to use each product

triamcinolone (topical-oral) (Rx, OTC)
(trye-am-sin′oh-lone)

Kenalog in Orabase, Oralone Dental

Func. class.: Topical anesthetic

Chem. class.: Synthetic fluorinated adrenal corticosteroid

ACTION: Binds with steroid receptors, decreases inflammation

USES: Oral pain

CONTRAINDICATIONS: Hypersensitivity, application to large areas; presence of fungal, viral, or bacterial infections of mouth or throat

Precautions: Pregnancy (C), children <6 yr, sepsis, denuded skin, geriatric patients

DOSAGE AND ROUTES
- **Adult: TOP** Press 1/4 inch into affected area until film appears, repeat bid-tid

Available forms: Paste 0.1%

Administer:
- After cleansing oral cavity after meals

SIDE EFFECTS
INTEG: Rash, irritation, sensitization

NURSING CONSIDERATIONS
Assess:
- Allergy: rash, irritation, reddening, swelling
- Infection: if affected area is infected, do not apply

Evaluate:
- Therapeutic response: absence of pain in affected area

Teach patient/family:
- To report rash, irritation, redness, swelling
- How to apply paste

triazolam (Rx)
(trye-ay′zoe-lam)

Apo-Triazo ✤, Halcion

Func. class.: Sedative-hypnotic, antianxiety

Chem. class.: Benzodiazepine, short acting

Controlled Substance Schedule IV (USA), Targeted (CDSA IV) (Canada)

Do not confuse:
Halcion/Haldol/halcinonide

ACTION: Produces CNS depression at limbic, thalamic, hypothalamic levels of CNS; may be mediated by neurotransmitter γ-aminobutyric acid (GABA); results are sedation, hypnosis, skeletal muscle relaxation, anticonvulsant activity, anxiolytic action

USES: Insomnia, sedative/hypnotic

CONTRAINDICATIONS: Pregnancy (X), breastfeeding, hypersensitivity to benzodiazepines

Precautions: Children <15 yr, geriatric patients, anemia, renal/hepatic disease, suicidal individuals, drug abuse, psychosis, acute closed-angle glaucoma, seizure disorders, angioedema, respiratory disease, depression, sleep-related behaviors (sleep walking), intermittent porphyria, myasthenia gravis, Parkinson's disease

DOSAGE AND ROUTES
- **Adult: PO** 0.125-0.5 mg at bedtime, max 0.5 mg/day
- **Geriatric: PO** 0.0625-0.125 mg at bedtime, max 0.25 mg/day

Available forms: Tabs 0.125, 0.25 mg

Administer:
- After trying conservative measures for insomnia
- 1/2 hr before bedtime for sleeplessness
- On empty stomach for fast onset; may be taken with food if GI symptoms occur

T

• Avoid use with CNS depressants; serious CNS depression may result

SIDE EFFECTS

CNS: *Headache, lethargy, drowsiness, daytime sedation,* dizziness, confusion, lightheadedness, anxiety, irritability, amnesia, poor coordination, complex sleep-related reactions: sleep driving, sleep eating

CV: Chest pain, pulse changes, ECG changes

GI: Nausea, vomiting, diarrhea, heartburn, abdominal pain, constipation, hepatic injury

SYST: Severe allergic reactions

PHARMACOKINETICS

Onset 15-30 min, duration 6-8 hr, metabolized by liver, excreted by kidneys (inactive metabolites), crosses placenta, excreted in breast milk, half-life 1.5-5.5 hr

INTERACTIONS

• Smoking may decrease hypnotic effect

Increase: triazolam levels—CYP3A4 inhibitors, protease inhibitors

⚠ **Increase:** effects of cimetidine, disulfiram, erythromycin, clarithromycin, probenecid, isoniazid, oral contraceptives; do not use concurrently

Increase: action of both products—alcohol, CNS depressants

Decrease: effect of antacids, theophylline, rifampin, smoking

Drug/Herb

Increase: CNS depression—chamomile, hops, kava, lavender, valerian

Drug/Food

• Grapefruit may increase action, avoid concurrent use

Drug/Lab Test

Increase: ALT, AST, serum bilirubin

Decrease: RAI uptake

False increase: urinary 17-OHCS

NURSING CONSIDERATIONS

Assess:

• Blood studies: Hct, Hgb, RBC if blood dyscrasias suspected (rare)

• Hepatic studies: AST, ALT, bilirubin if hepatic damage has occurred

• Mental status: mood, sensorium, affect, memory (long, short term), insomnia, withdrawal symptoms, excessive sedation, impaired coordination

• Blood dyscrasias: fever, sore throat, bruising, rash, jaundice, epistaxis (rare)

• Type of sleep problem: falling asleep, staying asleep

• Assistance with ambulation after receiving dose

Evaluate:

• Therapeutic response: ability to sleep at night, decreased amount of early morning awakening if taking product for insomnia

Teach patient/family:

• To use reliable contraception; pregnancy (X)

• That dependence is possible after long-term use

• To avoid driving, other activities requiring alertness until product is stabilized

• To avoid alcohol ingestion

• That effects may take 2 nights for benefits to be noticed; that product is for short-term use only; to use for 7-10 continuous nights

• About alternative measures to improve sleep: reading, exercise several hours before bedtime, warm bath, warm milk, TV, self-hypnosis, deep breathing

• That complex sleep-related behaviors (sleep eating/driving) may occur

• That hangover common in geriatric patients but less common than with barbiturates; that rebound insomnia may occur for 1-2 nights after discontinuing product; to discontinue by decreasing dose by 50% q2 nights until 0.125 mg for 2 nights, then stop

TREATMENT OF OVERDOSE:

Lavage, activated charcoal; monitor electrolytes, VS

trifluridine ophthalmic

See Appendix B

trimethobenzamide (Rx)

(trye-meth-oh-ben′za-mide)

Tigan

Func. class.: Antiemetic, anticholinergic

Chem. class.: Ethanolamine derivative

ACTION: Acts centrally by blocking chemoreceptor trigger zone, which in turn acts on vomiting center

USES: Nausea, vomiting

CONTRAINDICATIONS: Children (parenterally), hypersensitivity to opioids, shock

Precautions: Pregnancy (C), children, geriatric patients, cardiac dysrhythmias, acute febrile illness, encephalitis, gastroenteritis, dehydration, electrolyte imbalances, Reye's syndrome

DOSAGE AND ROUTES

Nausea/vomiting

- **Adult: IM** 200 mg 3-4×/day; **PO** 300 mg 3-4×/day

Postoperative

- **Adult: IM** 200 mg followed by 2nd dose 1 hr later

Renal dose

- **Adult: IM** CCr 15-30 ml/min, give 50% of dose

Available forms: Caps 300 mg; inj 100 mg/ml

Administer:

PO route

- Capsules may be swallowed whole, chewed, allowed to dissolve

IM route

- Inj in large muscle mass; aspirate to avoid IV administration; inj not to be used in children or infants

Syringe compatibilities: Butorphanol, glycopyrrolate, HYDROmorphone, midazolam, nalbuphine

SIDE EFFECTS

CNS: *Drowsiness,* headache, dizziness, confusion, disorientation, **coma, seizures,** depression, *vertigo,* EPS

CV: Hypo/hypertension, palpitation, **cardiac dysrhythmias**

EENT: Dry mouth, blurred vision, photosensitivity

GI: Nausea, diarrhea, vomiting, difficulty swallowing

INTEG: Rash, urticaria, fever, chills, flushing, hyperpyrexia

PHARMACOKINETICS

Metabolized by liver, excreted by kidneys

PO: Onset 20-40 min, duration 3-4 hr

IM: Onset 15-35 min, duration 2-3 hr

INTERACTIONS

Increase: effect—CNS depressants, alcohol

NURSING CONSIDERATIONS

Assess:

- Nausea, vomiting before, after treatment
- VS, B/P; check patients with cardiac disease more often
- **Signs of toxicity of other products or masking of symptoms of disease: brain tumor, intestinal obstruction**
- Observe for drowsiness, dizziness

Evaluate:

- Therapeutic response: decreased nausea, vomiting

Teach patient/family:

- To avoid hazardous activities, activities requiring alertness because dizziness may occur; to request assistance with ambulation
- To avoid alcohol, other depressants
- To keep out of children's reach

tropicamide ophthalmic

See Appendix B

trospium (Rx)

(trose'pee-um)

Sanctura, Sanctura XR

Func. class.: Anticholinergic, urinary antispasmodic

Chem. class.: Muscarinic receptor antagonist

ACTION: Relaxes smooth muscles in bladder by inhibiting acetylcholine effect on muscarinic receptors

USES: Overactive bladder (urinary frequency, urgency)

CONTRAINDICATIONS: Hypersensitivity, uncontrolled closed-angle glaucoma, urinary retention, gastric retention, myasthenia gravis

Precautions: Pregnancy (C), breastfeeding, children, geriatric patients, renal/hepatic disease, controlled closed-angle glaucoma, ulcerative colitis, intestinal atony, bladder outflow obstruction

DOSAGE AND ROUTES

- **Adult <75 yr: PO** 20 mg bid 1 hr before meals or on empty stomach; **ER** 60 mg in AM
- **Geriatric ≥75 yr: PO** titrate down to 20 mg/day based on response and tolerance

Renal dose

- **Adult: PO** CCr <30 ml/min, 20 mg/day at bedtime, ext rel product not recommended

Available forms: Tabs 20 mg; caps ext rel 60 mg

Administer:

- 1 hr before meals or on empty stomach (reg rel); in AM (ext rel) ≥1 hr before meal

SIDE EFFECTS

CNS: Fatigue, dizziness, headache
CV: Tachycardia
EENT: Dry eyes, vision abnormalities
GI: Flatulence, abdominal pain, *constipation, dry mouth,* dyspepsia
GU: Urinary retention, UTI
INTEG: Dry skin, **angioedema**
MISC: Heat stroke, fever

PHARMACOKINETICS

Rapidly absorbed (10%); peak 5-6 hr; protein bound (50%-85%); extensively metabolized; excreted in urine, feces; excreted in urine by active tubular secretion; half-life 20 hr

INTERACTIONS

Increase: drowsiness—CNS depressants, alcohol
Increase or decrease: trospium effect—products excreted by active renal secretion (aMILoride, digoxin, morphine), metformin, quiNIDine, procainamide, ranitidine, tenofovir, triamterene, vancomycin

Drug/Food

Decrease: absorption—high-fat meal

NURSING CONSIDERATIONS

Assess:

- **Urinary patterns:** distention, nocturia, frequency, urgency, incontinence, voiding patterns

Evaluate:

- Therapeutic response: correction of urinary status: absence of dysuria, frequency, nocturia, incontinence

Teach patient/family:

- To avoid hazardous activities because dizziness may occur
- That alcohol may increase drowsiness
- About anticholinergic effects that may occur
- That overheating may occur with strenuous exercise
- To avoid all other products unless approved by prescriber

undecylenic acid topical
See Appendix B

unoprostone ophthalmic
See Appendix B

ustekinumab (Rx)
(us′te-kin′ue-mab)

Stelara

Func. class.: Antipsoriatic agent

ACTION: Interleukin (IL)-12, IL-23 Antagonist

USES: Plaque psoriasis

CONTRAINDICATIONS: Hypersensitivity, sepsis, active infections

Precautions: Pregnancy (B), breastfeeding, children ≤18 yr, geriatric patients, surgery, TB, diabetes mellitus, immunosuppression

DOSAGE AND ROUTES

- **Adult ≥100 kg: SUBCUT** 90 mg, repeat in 4 wk, then 90 mg q12wk starting wk 16
- **Adult ≤100 kg: SUBCUT** 45 mg, repeat in 4 wk, then 45 mg q12wk starting wk 16

Available forms: Solutions for inj 45 mg/0.5 mc

Administer:

- Visually inspect for particulate matter or discoloration; solution should be slightly yellow and may contain a few small translucent or white particles; do not use if discolored, cloudy, or if foreign particulate matter is present; do not shake
- Use at 27-G, 0.5 inch needle
- May be administered subcut into upper arm, abdomen, or thigh; rotate inj sites

SIDE EFFECTS

CNS: Headache, leukoencephalopathy

HEMA: Bleeding

INTEG: *Inj site reaction,* pruritus, skin irritation, erythema

SYST: Serious infections, malignancies

PHARMACOKINETICS

Maximum serum conc: 13.5 days after a single 45 mg subcut dose, 7 days after a single 90 mg subcut dose; half-life 14.9-45.6 days

INTERACTIONS

- Do not give concurrently with vaccines; immunizations should be brought up to date before treatment
- Avoid use with immunosuppressives

NURSING CONSIDERATIONS

Assess:

- For inj site pain, swelling

Evaluate:

- Therapeutic response: decreased plaque psoriasis

Teach patient/family:

- That product must be continued for prescribed time to be effective
- Not to receive live vaccinations during treatment
- To notify prescriber of possible infection (upper respiratory or other)

U

valACYclovir (Rx)

(val-a-sye′kloh-vir)

Valtrex

Func. class.: Antiviral

Chem. class.: Synthetic purine nucleoside analog

Do not confuse:
valACYclovir/valGANciclovir
Valtrex/Valcyte

ACTION:
Interferes with DNA synthesis by conversion to acyclovir, thereby causing decreased viral replication, time of lesional healing

USES:
Treatment or suppression of herpes zoster (shingles), genital herpes, herpes labialis (cold sores), varicella, varicella-zoster

Unlabeled uses: CMV with advanced HIV, posttransplant patients, Bell's palsy, herpes simplex virus prophylaxis, acute retinal necrosis (ARN), encephalitis

CONTRAINDICATIONS:
Hypersensitivity to this product or acyclovir, valGANciclovir

Precautions: Pregnancy (B), breastfeeding, geriatric patients, hepatic/renal disease, electrolyte imbalance, dehydration, penciclovir, famciclovir, ganciclovir, hypersensitivity, varicella

DOSAGE AND ROUTES

Herpes zoster (shingles)

- **Adult: PO** 1 g tid × 1 wk

Genital herpes (suppressive, initial)

- **Adult: PO** 1 g bid × 10 days initially

Genital herpes (recurrent episodes)

- **Adult: PO** 500 mg bid × 3 days

Genital herpes (suppressive therapy)

- **Adult: PO** 1 g/day with normal immune function; 500 mg/day for those with ≤9 recurrences/yr; 500 mg bid for HIV-infected patients with CD4 count ≥100

Reduction of transmission

- **Adult: PO** 500 mg/day for source partner

Herpes labialis

- **Adult: PO** 2 g bid × 1 day at 1st sign of lesions

Varicella (chickenpox) in immunocompetent patients

- **Adolescent and child ≥2 yr: PO** 20 mg/kg/dose tid × 5 days, max 3 g/day; start at 1st sign, preferably within 24 hr of rash

Renal dose

- **Adult: PO** CCr 30-49 ml/min, 1 g q12hr (for regimens 1 g q8hr); 1 g q12hr × 1 day (herpes labialis); CCr 10-29 ml/min, 1 g q24hr (genital herpes/herpes zoster); 500 mg q24hr (recurrent genital herpes); CCr <10 ml/min, 500 mg q24hr (genital herpes/herpes zoster), 500 mg q24hr (recurrent genital herpes)

Available forms: Tabs 500 mg, 1 g

Administer:

- As soon as possible (herpes labialis, genital herpes); within 24 hr of rash (varicella)
- Within 72 hr of outbreak (herpes zoster)
- Without regard to food
- Caps may be made into susp by pharmacy
- Store at room temperature; protect from light, moisture

SIDE EFFECTS

CNS: Tremors, lethargy, *dizziness, headache,* weakness, depression

ENDO: *Dysmenorrhea*

GI: *Nausea,* vomiting, diarrhea, abdominal pain, constipation, *increased AST*

HEMA: Thrombocytopenic purpura, hemolytic uremic syndrome

INTEG: *Rash*

PHARMACOKINETICS

Onset unknown; terminal half-life 2½-3½ hr; converted to acyclovir that crosses placenta, enters breast milk; excreted in urine primarily as acyclovir; protein binding 13.5%-17.9%

INTERACTIONS

Increase: blood levels of valACYclovir—cimetidine, probenecid; only significant with renal disease

Drug/Lab Test

Increase: LFTs, creatinine

Decrease: WBC, platelets

NURSING CONSIDERATIONS

Assess:

- **Infection;** characteristics of lesions; therapy should be started at 1st sign or symptom of herpes; most effective within 72 hr of outbreak

⚠ Thrombocytopenic purpura, hemolytic uremic syndrome; may be fatal

- C&S before product therapy; product may be taken as soon as culture is taken; repeat C&S after treatment; determine presence of other sexually transmitted diseases
- Bowel pattern before, during treatment
- Skin eruptions: rash
- Allergies before treatment, reaction of each medication

Evaluate:

- Therapeutic response: absence of itching, painful lesions; crusting and healed lesions

Teach patient/family:

- To take as prescribed; if dose is missed, to take as soon as remembered up to 2 hr before next dose; not to double dose
- That product may be taken orally before infection occurs; that product should be taken when itching or pain occurs, usually before eruptions
- That partners need to be told that patient has herpes because they can become infected; that condoms must be worn to prevent reinfections
- That product does not cure infection, just controls symptoms; that product does not prevent infection of others

valGANciclovir (Rx)

(val-gan-sy′kloh-veer)

Valcyte

Func. class.: Antiviral

Chem. class.: Synthetic nucleoside

Do not confuse:

valGANciclovir/valACYclovir

Valcyte/Valtrex

ACTION: Metabolized to ganciclovir; inhibits replication of human cytomegalovirus in vivo and in vitro by selective inhibition of viral DNA synthesis

USES: Cytomegalovirus (CMV) retinitis in immunocompromised persons, including those with AIDS, after indirect ophthalmoscopy confirms diagnosis; prevention of CMV with transplantation; prevention of CMV in at-risk patient going through transplant (kidney, heart, pancreas)

Unlabeled uses: Colitis, Epstein-Barr virus, esophagitis, herpes simplex type 1, 2; human herpesvirus 6, 8; multicentric Castleman's disease, varicella-zoster virus

CONTRAINDICATIONS: Breastfeeding, hypersensitivity to ganciclovir, valACYclovir; absolute neutrophil count $<500/mm^3$; platelet count $<25,000/mm^3$; hemodialysis; liver transplantation

Precautions: Pregnancy (C), children, geriatric patients, renal function impairment; hypersensitivity to acyclovir, penciclovir, famciclovir

Black Box Warning: Preexisting cytopenias, secondary malignancy, infertility, anemia

V

DOSAGE AND ROUTES

Treatment of CMV

- **Adult and adolescent (unlabeled): PO** induction 900 mg bid × 21 days with food; maintenance 900 mg/day with food

Transplant (CMV prophylaxis)

• **Adult/adolescent >16 yr: PO** 900 mg/day with food starting 10 days before transplantation until day 100 after transplantation

• **Infant ≥4 mo/child/adolescent ≤16 yr: PO** give within 10 days of heart/kidney transplant; calculate dose as 7 × BSA × CCr, give as single daily dose

Renal dose

• **Adult: PO** CCr ≥60 ml/min, same as above; CCr 40-59 ml/min, 450 mg bid for 21 days then 450 mg/day; CCr 25-39 ml/min, 450 mg/day then 450 mg q2days; CCr 10-24 ml/min, 450 mg q2days then 450 mg 2×/week

Available form: Tabs 450 mg, powder for oral sol 50 mg/ml

Administer:

PO tab

• With food for better absorption; avoid getting product on skin; do not break

Oral sol

• Measure 9 ml purified water in graduated cylinder, shake bottle to loosen powder, add 1/2 liquid, shake well, add remaining water, shake; remove child-resistant cap and push bottle adapter into neck of bottle, close with cap, give using dispenser provided

• Store liquid in refrigerator; do not freeze; throw away any unused after 49 days

SIDE EFFECTS

CNS: *Fever,* chills, coma, *confusion,* abnormal thoughts, dizziness, bizarre dreams, *headache, insomnia,* psychosis, tremors, somnolence, *paresthesia, weakness,* seizures

EENT: Retinal detachment with CMV retinitis

GI: Abnormal LFTs, *nausea, vomiting, anorexia, diarrhea, abdominal pain,* hemorrhage

GU: Hematuria, increased creatinine, BUN

HEMA: Granulocytopenia, thrombocytopenia, irreversible neutropenia, anemia, eosinophilia

INTEG: *Rash,* alopecia, *pruritus,* urticaria, pain at site, phlebitis, Stevens-Johnson syndrome

MISC: Local and systemic infections, sepsis

PHARMACOKINETICS

Metabolized to ganciclovir, which has a half-life of 3-4 1/2 hr; excreted by kidneys (unchanged); crosses blood-brain barrier, CSF

INTERACTIONS

Increase: Severe granulocytopenia: immunosuppressants, zidovudine, antineoplastics, radiation; do not use together

Increase: toxicity—dapsone, pentamidine, flucytosine, vinCRIStine, vinBLAStine, adriamycin, DOXOrubicin, amphotericin B, trimethoprim-sulfamethoxazole combinations or other nucleoside analogs, cycloSPORINE

Increase: effect of both drugs—mycophenolate

Increase: seizures—imipenem-cilastatin

Increase: effect of didanosine; monitor for adverse effects, toxicity

Decrease: renal clearance of valGANciclovir—probenecid

Drug/Food

Increase: absorption, high-fat meal

Drug/Lab Test

Increase: creatinine

Decrease: RBC/WBC, Hct/Hgb

NURSING CONSIDERATIONS

Assess:

• CMV retinitis by ophthalmoscopy before beginning treatment, q2wk; culture for CMV retinitis

Black Box Warning: Leukopenia/neutropenia/thrombocytopenia: WBCs, platelets q2days during 2×/day dosing then q1wk; leukopenia with daily WBC count in patients with prior leukopenia with other nucleoside analogs or for whom leukopenia counts are <1000 cells/mm^3 at start of treatment

• Serum creatinine or CCr ≥q2wk

Evaluate:

- Therapeutic response: decreased symptoms of CMV

Teach patient/family:

- That product does not cure condition; that regular ophthalmologic and blood tests necessary
- That major toxicities may necessitate discontinuing product
- To use contraception during treatment; that infertility may occur; that men should use barrier contraception for 90 days after treatment
- To take with food
- ⚠ Blood dyscrasias: bruising, bleeding, petechiae; seizures, dizziness; to avoid hazardous activities
- To use sunscreen to prevent burns

valproate (Rx)

(val′proh-ate)

Depacon

valproic acid (Rx)

(val′proh-ik)

Depakene, Stavzor

divalproex sodium (Rx)

(dye-val′proh-ex)

Depakote, Depakote ER, Epival ♣

Func. class.: Anticonvulsant, vascular headache suppressant

Chem. class.: Carboxylic acid derivative

ACTION: Increases levels of γ-aminobutyric acid (GABA) in the brain, which decreases seizure activity

USES: Simple (petit mal), complex (petit mal), absence, mixed, seizures; manic episodes associated with bipolar disorder, prophylaxis of migraine, adjunct for schizophrenia, tardive dyskinesia, aggression in children with ADHD, organic brain syndrome, mania, migraines; tonic-clonic (grand mal), myoclonic seizures

Unlabeled uses: Rectal for seizures (valproic acid)

CONTRAINDICATIONS: Hypersensitivity, urea cycle disorders

Black Box Warning: Pregnancy (D), hepatic disease, pancreatitis

Precautions: Breastfeeding, geriatric patients

Black Box Warning: Children <2 yr

DOSAGE AND ROUTES

Epilepsy

- **Adult and child: PO** 10-15 mg/kg/day in 2-3 divided doses, may increase by 5-10 mg/kg/day weekly, max 60 mg/kg/day in 2-3 divided doses; **IV** ≤ 20 mg/min over 1 hr

Status epilepticus refractory to diazepam IV (unlabeled)

- **Adult: RECT** 400-600 mg PR as enema or wax-based suppository (not commercially available)
- **Child: RECT** 20 mg/kg/dose

Mania (divalproex sodium)

- **Adult: PO** 750 mg/day in divided doses, max 60 mg/kg/day or 3000 mg/day

Mania (valproic acid: Stavzor)

- **Adult: DEL REL CAP** 750 mg/day in divided doses

Migraine (divalproex sodium)

- **Adult: PO** 250 mg bid, may increase to 1000 mg/day if needed or 500 mg (Depakote ER) daily × 7 days, then 1000 mg/day

Available forms: *Valproate:* inj 100 mg/ml; *valproic acid:* caps 250 mg; oral solution 250 mg/5 ml; del rel cap (Stavzor) 125, 250, 500 mg; *divalproex:* gastro-resistant tabs 125, 250, 500 mg; ext rel tabs 250, 500 mg cap gastro-resistant; sprinkle cap 125 mg

Administer:

PO route

- Swallow tabs or caps whole; do not break, crush, or chew ext rel tabs
- Sprinkle cap contents on food

V

• Oral solution alone; do not dilute with carbonated beverage; do not give oral solution to patients with sodium restrictions
• Give with food or milk to decrease GI symptoms

IV route
• Dilute dose with ≥50 ml D_5W, NS, LR
• Run over 60 min (20 mg/min)

SIDE EFFECTS

CNS: *Sedation, drowsiness,* dizziness, headache, incoordination, depression, hallucinations, behavioral changes, tremors, aggression, weakness, **coma, suicidal ideation, hypothermia**
CV: Hypo/hypertension, chest pain, palpitations, peripheral edema
EENT: Visual disturbances, taste perversion
GI: *Nausea, vomiting, constipation, diarrhea, dyspepsia,* anorexia, cramps, **hepatic failure, pancreatitis, toxic hepatitis,** stomatitis, weight gain, dry mouth
GU: Enuresis, irregular menses
HEMA: **Thrombocytopenia, leukopenia, lymphocytosis,** increased PT, bruising, epistaxis, **pancytopenia**
INTEG: *Rash,* alopecia, photosensitivity, dry skin
META: Hyperammonemia, SIADH
RESP: Dyspnea

PHARMACOKINETICS

Metabolized by liver; excreted by kidneys, in breast milk; crosses placenta; half-life 6-16 hr; 90% protein binding
PO: Peak 4 hr (regular rel); 4-17 hr (ext rel)

INTERACTIONS

Increase: valproic acid toxicity level—erythromycin, felbamate, salicylates, NSAIDs, rifampin
Increase: CNS depression—alcohol, opioids, barbiturates, antihistamines, MAOIs, sedative/hypnotics, tricyclics
Increase: action of, possible toxicity phenytoin, carBAMazepine, ethosuximide, barbiturates, zidovudine, LORazepam, rufinamide, lamoTRIgine
Increase: bleeding—warfarin
Decrease: seizure threshold—tricyclics

Drug/Lab Test
False positive: ketones, urine
Interference: thyroid function tests
Increase: LFTs, bleeding time, ammonia
Decrease: sodium

NURSING CONSIDERATIONS

Assess:
• **Seizure disorder:** location, aura, activity, duration; seizure precautions should be in place
⚠ **Mental status: bipolar disorder: mood, activity, sleeping/eating, behavior; suicidal thoughts/behaviors**
• **Migraines:** frequency, intensity, alleviating factors
• Blood studies: Hct, Hgb, RBC, serum folate, PT/PTT, serum ammonia, platelets, vit D if patient receiving long-term therapy

Black Box Warning: Hepatic studies: AST, ALT, bilirubin; hepatic failure has occurred

• **Hyperammonemic encephalopathy: can be fatal in those with urea cycle disorders (UCD); lethargy, confusion, coma, CV, respiratory changes; discontinue**
• Blood levels: therapeutic level 50-125 mcg/ml, during seizures

Black Box Warning: **Pancreatitis;** may be fatal; report immediately nausea, vomiting, anorexia, abdominal pain

Evaluate:
• Therapeutic response: decreased seizures

Teach patient/family:
• That physical dependency may result from extended use
• To avoid driving, other activities that require alertness
• To drink plenty of fluids
• **Not to discontinue medication quickly after long-term use because seizures may result**

- To report visual disturbances, rash, diarrhea, abdominal pain, light-colored stools, jaundice, protracted vomiting, weakness to prescriber
- To use contraception while taking this product; pregnancy (D)
- **Overdose symptoms: heart block, coma**
- **To report immediately suicidal thoughts/behaviors**

valsartan (Rx)

(val′sahr-tan)

Diovan

Func. class.: Antihypertensive

Chem. class.: Angiotensin II receptor antagonist (Type AT_1)

Do not confuse:
Diovan/Dioval

ACTION: Blocks the vasoconstrictor and aldosterone-secreting effects of angiotensin II; selectively blocks the binding of angiotensin II to the AT_1 receptor found in tissues

USES: Hypertension, alone or in combination in patients >6 yr, CHF, post MI with left ventricular dysfunction/failure in stable patients

CONTRAINDICATIONS: Hypersensitivity, severe hepatic disease, bilateral renal artery stenosis

Black Box Warning: Pregnancy (D) 2nd/3rd trimester

Precautions: Breastfeeding, children, geriatric patients, hypersensitivity to ACE inhibitors; CHF, hypertrophic cardiomyopathy aortic/mitral valve stenosis, CAD, angioedema, renal/hepatic disease, hyperkalemia, hypovolemia, African descent

DOSAGE AND ROUTES

Hypertension

- **Adult: PO** 80 or 160 mg/day alone or in combination with other antihypertensives, may increase to 320 mg/day
- **Geriatric: PO** adjust on clinical response; may start with lower dose
- **Child and adolescent 6-16 yr: PO** 1.3 mg/kg/day, max 40 mg/day

CHF

- **Adult: PO** 40 mg bid, up to 160 mg bid

Post MI

- **Adult: PO** 20 mg bid as early as 12 hr post MI, may be titrated within 7 days to 40 mg bid, then titrate to maintenance of 160 mg bid

Available forms: Tabs 40, 80, 160, 320 mg

Administer:

- Without regard to meals

SIDE EFFECTS

CNS: *Dizziness, insomnia,* drowsiness, vertigo, headache, fatigue
CV: Angina pectoris, 2nd-degree AV block, **cerebrovascular accident,** hypotension, **MI,** ***dysrhythmias***
EENT: Conjunctivitis
GI: *Diarrhea,* abdominal pain, nausea, **hepatotoxicity**
GU: Impotence, **nephrotoxicity, renal failure**
HEMA: *Anemia,* neutropenia
META: Hyperkalemia
MISC: Vasculitis, **angioedema**
MS: Cramps, myalgia, pain, stiffness
RESP: *Cough*

PHARMACOKINETICS

Onset up to 2 hr; peak 2-4 hr; duration 24 hr; extensively metabolized; protein binding 95%; half-life 6 hr; excreted in feces, urine, breast milk

INTERACTIONS

- **Do not use with aliskiren**

Increase: effects of lithium, antidiabetics
Increase: hyperkalemia—potassium-sparing diuretics, potassium supplements, ACE inhibitors, cycloSPORINE
Increase: Valsartan level—rifampin, ritonavir, gemfibrozil, telithromycin
Decrease: antihypertensive effects—NSAIDs, salicylates

V

Drug/Herb
Increase: antihypertensive effect—hawthorn
Decrease: antihypertensive effect—ephedra, ma huang
Drug/Food
Increase: hyperkalemia—salt substitutes with potasssium

NURSING CONSIDERATIONS

Assess:

- B/P, pulse q4hr lying, sitting, standing; note rate, rhythm, quality periodically
- Blood studies; BUN, creatinine, LFTs, potassium total/direct bilirubin before treatment
- **Angioedema:** facial swelling; SOB; edema in feet, legs daily
- Skin turgor, dryness of mucous membranes for hydration status; correct volume depletion before initiating therapy

Evaluate:

- Therapeutic response: decreased B/P

Teach patient/family:

- To comply with dosage schedule, even if feeling better; that, if dose is missed, to take it as soon as possible unless it is within 1 hr of next dose
- To notify prescriber of fever, swelling of hands or feet, irregular heartbeat, chest pain, dizziness, persistent cough
- That excessive perspiration, dehydration, diarrhea may lead to fall in blood pressure; to consult prescriber if these occur; to maintain hydration
- That product may cause dizziness, fainting, lightheadedness; to rise slowly to sitting or standing position to minimize orthostatic hypotension; to take B/P readings
- To avoid potassium supplements and foods, salt substitutes

Black Box Warning: Not to take product if pregnant (D) or breastfeeding

- **Overdose symptoms:** bradycardia or tachycardia, circulatory collapse

vancomycin (Rx)

(van-koe-mye'sin)

Vancocin

Func. class.: Antiinfective—miscellaneous
Chem. class.: Tricyclic glycopeptide

ACTION: Inhibits bacterial cell wall synthesis, blocks glycopeptides

USES: *Actinomyces* sp., *bacillus* sp., *clostridium difficile, clostridium* sp., *enterococcus faecalis, enterococcus faecium, enterococcus* sp., *lactobacillus* sp., *listeria monocytogenes, staphylococcus aureus (MRSA), staphylococcus aureus (MSSA), staphylococcus epidermidis, staphylococcus* sp., *streptococcus agalactiae (group B streptococci), streptococcus bovis, streptococcus pneumoniae, streptococcus pyogenes (group A beta-hemolytic streptococci), viridans streptococci;* may be effective against *corynebacterium jeikeium, corynebacterium* sp., pseudomembranous colitis, staphylococcal enterocolitis, endocarditis prophylaxis for dental procedures, bacteremia, joint/bone infections, osteomyelitis, pneumonia, septicemia
Unlabeled uses: Bacterial infection prophylaxis, brain abscess, endocarditis prophylaxis, meningitis, orthopedic device related infection, peritonitis, surgical infection prophylaxis, vancomycin desensitization, ventriculitis

CONTRAINDICATIONS: Hypersensitivity to this product or corn
Precautions: Pregnancy (B) PO, (C) IV; breastfeeding, neonates, geriatric patients, renal disease, hearing loss

DOSAGE AND ROUTES

Serious staphylococcal infections

- **Adult:** IV 500 mg q6-8hr or 1 g q12hr
- **Child:** IV 40-60 mg/kg/day divided q6-8hr

• **Neonate: IV** 15 mg/kg initially, then 10 mg/kg q8-24hr

Pseudomembranous/ staphylococcal enterocolitis

• **Adult: PO** 125 mg qid × 10-14 days

• **Child: PO (unlabeled)** 40 mg/kg/day divided q6hr × 7-10 days, max 2 g/day

Endocarditis prophylaxis

• **Adult: IV** 2 g divided

• **Child: IV** 20 mg/kg over 1 hr given 1 hr before procedure

Renal dose

• **Adult: IV** 15-20 mg/kg loading dose in seriously ill; individualize all other doses

Available forms: Cap 125, 250 mg; powder for inj 500 mg, 1, 5, 10 g; Dextrose sol for inj 500 mg/100 ml, 750 mg/150 ml, 1 g/200 ml

Administer:

• Use only for susceptible organisms to prevent product-resistant bacteria

• Antihistamine if red-man syndrome occurs: decreased B/P, flushing of neck, face; stop or slow inf

• Dose based on serum conc

• **PO:** without regard to food, swallow whole

• Store at room temperature for ≤2 wk after reconstitution

Intermittent IV INF route

• After reconstitution with 10 ml sterile water for inj (500 mg/10 ml); further dilution is needed for IV, 500 mg/100 ml 0.9% NaCl, D_5W given as intermittent inf over 1 hr; decrease rate of inf if red-man syndrome occurs

Continuous IV INF route (unlabeled)

• May infuse 1-2 g in volume to give over 24 hr if intermittent IV route cannot be used

Y-site compatibilities: Acetylcysteine, acyclovir, alatrofloxacin, aldesleukin, alemtuzumab, alfentanil, allopurinol, alprostadil, amifostine, amikacin, amino acids injection, aminocaproic acid, amiodarone, amoxicillin-clavulanate, amsacrine, anidulafungin, argatroban, ascorbic acid injection, atenolol, atracurium, atropine, azithromycin, benztropine, bleomycin, bretylium, bumetanide, buprenorphine, butorphanol, calcium chloride/gluconate, CARBOplatin, carmustine, caspofungin, cefpirome, chlorproMAZINE, cimetidine, ciprofloxacin, cisatracurium, CISplatin, clarithromycin, clindamycin, codeine, cyanocobalamin, cyclophosphamide, cycloSPORINE, cytarabine, DACTINomycin, DAUNOrubicin liposome, dexamethasone, dexmedetomidine, dexrazoxane, digoxin, diltiazem, diphenhydrAMINE, DOBUTamine, DOCEtaxel, dolasetron, DOPamine, doripenem, doxacurium, doxapram, DOXOrubicin, DOXOrubicin liposomal, doxycycline, enalaprilat, ePHEDrine, EPINEPHrine, epirubicin, eptifibatide, ertapenem, erythromycin, esmolol, etoposide, etoposide phosphate, famotidine, fenoldopam, fentaNYL, filgrastim, fluconazole, fludarabine, folic acid (as sodium salt), gallium, gemcitabine, gentamicin, glycopyrrolate, granisetron, HYDROmorphone, hydrOXYzine, ifosfamide, insulin, regular, irinotecan, isoproterenol, isosorbide, ketamine, labetalol, lactated Ringer's injection, lepirudin, levofloxacin, lidocaine, linezolid, LORazepam, magnesium sulfate, mannitol, mechlorethamine, melphalan, meperidine, meropenem, metaraminol, methyldopate, metoclopramide, metoprolol, metroNIDAZOLE, midazolam, milrinone, minocycline, mitoXANtrone, morphine, morphine, multiple vitamins injection, mycophenolate, nalbuphine, naloxone, nesiritide, netilmicin, niCARdipine, nitroglycerin, nitroprusside, norepinephrine, octreotide, ofloxacin, ondansetron, oxacillin, oxaliplatin, oxytocin, PACLitaxel (solvent/surfactant), palonosetron, pamidronate, pancuronium, papaverine, PEMEtrexed, penicillin G potassium/sodium, pentamidine, pentazocine, PENTobarbital, perphenazine, PHENobarbital, phentolamine, phenylephrine, phytonadione, piritramide, polymyxin B, potassium acetate/chloride, procainamide, prochlorperazine, promethazine, propranolol, protamine, pyridoxine, quiNIDine, ranitidine, remifentanil, rifampin, Ringer's injection, riTUXimab,

V

sodium acetate/bicarbonate/citrate, succinylcholine, SUFentanil, tacrolimus, teniposide, thiamine, thiotepa, tigecycline, tirofiban, TNA (3-in-1), tobramycin, tolazoline, TPN (2-in-1), trastuzumab, urapidil, vasopressin, vecuronium, verapamil, vinBLAStine, vinCRIStine, vinorelbine, voriconazole, zidovudine, zoledronic acid

SIDE EFFECTS

CNS: Headache
CV: Cardiac arrest, vascular collapse (rare), hypotension, peripheral edema
EENT: *Ototoxicity, permanent deafness,* tinnitus, nystagmus
GI: Nausea, pseudomembranous colitis
GU: Nephrotoxicity, *increased BUN, creatinine, albumin,* fatal uremia
HEMA: Leukopenia, eosinophilia, neutropenia
INTEG: Chills, fever, rash, thrombophlebitis at inj site, urticaria, pruritus, necrosis (red-man syndrome), skin/subcutaneous tissue disorders
MS: Back pain
RESP: Wheezing, dyspnea
SYST: Anaphylaxis, superinfection

PHARMACOKINETICS

PO: Absorption poor
IV: Onset rapid, peak 1 hr, half-life 4-8 hr, excreted in urine (active form)

INTERACTIONS

Increase: ototoxicity or nephrotoxicity—aminoglycosides, cephalosporins, colistin, polymyxin, bacitracin, CISplatin, amphotericin B, methotrexate, NSAIDs, cycloSPORINE, acyclovir, adefovir, capreomycin, foscarnet, ganciclovir, pamidronate, IV pentamidine, streptozocin, tacrolimus, zoledronic acid
• Do not use with cholestyramine, colestipol, cidofovir
Increase: lactic acidosis, metFORMIN
Increase: neuromuscular effects—nondepolarizing muscle relaxants

Drug/Lab Test

Increase: BUN/creatinine, eosinophils
Decrease: WBC

NURSING CONSIDERATIONS

Assess:
• **Infection:** WBC, urine, stools, sputum, characteristics of wound throughout treatment
• I&O ratio; report hematuria, oliguria; nephrotoxicity may occur, BUN, creatinine
• Serum levels: peak 1 hr after 1-hr inf 25-40 mg/L, trough before next dose 5-10 mg/L, especially in renal disease
• C&S
• Auditory function during, after treatment; hearing loss, ringing, roaring in ears; product should be discontinued
• B/P during administration; sudden drop may indicate red-man syndrome
• Skin eruptions
• **Red man syndrome:** flushing of neck, face, upper body, arms, back, may lead to anaphylaxis; slow IV inf to >1 hr
• EPINEPHrine, suction, tracheostomy set, endotracheal intubation equipment on unit; anaphylaxis may occur
• Adequate intake of fluids (2 L/day) to prevent nephrotoxicity

Evaluate:
• Therapeutic response: absence of fever, sore throat; negative culture

Teach patient/family:
• About all aspects of product therapy; about the need to complete entire course of medication to ensure organism death (7-10 days); that culture may be taken after completed course of medication
• To report sore throat, fever, fatigue; could indicate superinfection
• That product must be taken in equal intervals around the clock to maintain blood levels

vardenafil (Rx)

(var-den′a-fil)

Levitra, Staxyn

Func. class.: Erectile dysfunction agent

Chem. class.: Phosphodiesterase type 5 inhibitor

ACTION:
Inhibits phosphodiesterase type 5 (PDE5), enhances erectile function by increasing the amount of cGMP, which in turn causes smooth muscle relaxation and increased blood flow into the corpus cavernosum

USES:
Treatment of erectile dysfunction

CONTRAINDICATIONS:
Hypersensitivity, coadministration of α-blockers or nitrates, renal failure, congenital or acquired QT prolongation

Precautions: Pregnancy (B); not indicated for women, children, or newborns; hepatic impairment, retinitis pigmentosa, anatomical penile deformities, sickle cell anemia, leukemia, multiple myeloma, bleeding disorders, active peptic ulceration, CV/renal disease

DOSAGE AND ROUTES

- **Adult: PO** 10 mg taken 1 hr before sexual activity; dose may be reduced to 5 mg or increased to max 20 mg; max dosing frequency once daily; orally disintegrating tab 10 mg 60 min before sexual activity; do not use with potent CYP3A4 inhibitors
- **Geriatric >65 yr: PO** 5 mg initially, titrated as needed/tolerated

Hepatic dose

- **Adult: PO** (Child-Pugh B) 5 mg, max 10 mg/day

Concomitant medications

- Ritonavir, max 2.5 mg q72hr; for indinavir, ketoconazole 400 mg/day and itraconazole 400 mg/day, max 2.5 mg/day; for ketoconazole 200 mg/day, itraconazole 200 mg/day and erythromycin max 5 mg/day

Available forms: Tabs 2.5, 5, 10, 20 mg; orally disintegrating tab 10 mg

Administer:

- Approximately 1 hr before sexual activity; do not use more than once daily; orally disintegrating tabs are not interchangeable with film-coated tabs
- Without regard to food; avoid taking with high-fat meal
- **Oral disintegrating tab:** place on tongue, allow to dissolve, do not use water

SIDE EFFECTS

CNS: *Headache, flushing, dizziness, insomnia,* **seizures,** transient global amnesia

CV: Hypertension, **MI, CV collapse,** chest pain

EENT: Conjunctivitis, tinnitus, photophobia, diminished vision, glaucoma, hearing loss

GU: Abnormal ejaculation, priapism

MISC: Rash, GERD, GGTP increased, **NAION (nonarteritic ischemic optic neuropathy),** dyspepsia

MS: Myalgia, arthralgia, neck pain

RESP: Rhinitis, sinusitis, dyspnea, pharyngitis, epistaxis

PHARMACOKINETICS

Rapidly absorbed, bioavailability 15%, protein binding 95%, metabolized by liver, terminal half-life 4-5 hr, onset 20 min, peak ½-1½ hr, duration <5 hr, reduced absorption with high-fat meal, primarily excreted in feces (91%-95%)

INTERACTIONS

⚠ Do not use with nitrates because of unsafe decrease in B/P, which could result in MI or stroke

⚠ Serious dysrhythmias: class IA/III antiarrhythmics, clarithromycin, droperidol, procainamide, quiNIDine, quinolones; do not use concurrently

Increase: hypotension—α-blockers, protease inhibitors, metoprolol, NIFEdip-

ine, alcohol, amLODIPine, angiotensin II receptor blockers; do not use concurrently
Increase: vardenafil levels—erythromycin, azole antifungals (ketoconazole, itraconazole), cimetidine, antiretroviral protease inhibitors
Drug/Food
Decrease: absorption—high-fat meal
Drug/Lab Test
Increase: CK

NURSING CONSIDERATIONS

Assess:
- Erectile dysfunction and cause before treatment
- ⚠ Any severe loss of vision while taking this or similar products; products should not be used
- ⚠ Use of organic nitrates that should not be used with this product

Teach patient/family:
- That product does not protect against STDs, including HIV
- That product absorption is reduced with high-fat meal
- That product should not be used with nitrates in any form; to inform physician of all medications being taken
- That product has no effect in the absence of sexual stimulation; that patient should seek immediate medical attention if erections last >4 hr
- ⚠ To notify prescriber immediately and stop taking product if vision loss occurs

varenicline (Rx)

(var-e-ni′kleen)
Champix ✦, Chantix
Func. class.: Smoking cessation agent
Chem. class.: Nicotine receptor agonist

ACTION: Partial agonist for nicotine receptors; partially activates receptors to help curb cravings; occupies receptors to prevent nicotine binding

USES: Smoking deterrent

CONTRAINDICATIONS: Hypersensitivity, eating disorders
Precautions: Pregnancy (C), breastfeeding, children <18 yr, geriatric patients, renal disease, recent MI, angioedema

Black Box Warning: Bipolar disorder, depression, schizophrenia, suicidal ideation

DOSAGE AND ROUTES

Smoking cessation
- **Adult: PO** therapy should begin 1 wk before smoking stop date (i.e., take product plus tobacco for 7 days); titrate for 1 wk; days 1 through 3, 0.5 mg/day; days 4 through 7, 0.5 mg bid; day 8 through end of treatment, 1 mg bid; treatment is for 12 wk and may be repeated for another 12 wk

Renal dose
- **Adult: PO** CCr ≤50 ml/min, titrate to max 0.5 mg bid

Available forms: Tabs 0.5, 1 mg; Chantix continuing month PAK; Chantix starting month PAK
Administer:
- Do not break, crush, or chew tabs
- Increased fluids, bulk in diet if constipation occurs
- After eating with a full glass of water
- Sugarless gum, hard candy, or frequent sips of water for dry mouth

SIDE EFFECTS

CNS: Headache, agitation, dizziness, insomnia, abnormal dreams, fatigue, malaise, behavior changes, depression, suicidal ideation, amnesia, hallucinations, hostility, mania, psychosis, tremors
CV: Dysrhythmias, hypo/hypertension, palpitations, tachycardia, angina, MI
EENT: *Blurred vision,* tinnitus
GI: *Nausea, vomiting,* anorexia, *dry mouth,* increased/decreased appetite, *constipation,* flatulence, GERD
GU: Erectile dysfunction, urinary frequency, menstrual irregularities

INTEG: Rash, pruritus, angioedema, Stevens-Johnson syndrome
MISC: Weight loss or gain
RESP: Dyspnea, rhinorrhea

PHARMACOKINETICS

Elimination half-life 24 hr; metabolism minimal; 93% excreted unchanged in urine; steady state 4 days

NURSING CONSIDERATIONS

Assess:

- **Smoking history:** motivation for smoking cessation, years used, amount each day
- Renal function in geriatric patients; cardiac status in cardiac disease
- Smoking cessation after 12 wk; if progress not made, product may be used for additional 12 wk

Black Box Warning: Neuropsychiatric symptoms: mood, sensorium, affect; behavior changes, agitation, depression, suicidal ideation; suicide has occurred; possible worsening of depression, schizophrenia, bipolar disorder

Evaluate:

- Therapeutic response: smoking cessation

Teach patient/family:

- That treatment for smoking cessation lasts 12 wk and that another 12 wk may be required
- To use caution when driving, performing other activities requiring alertness; blurred vision may occur
- To set a date to quit smoking and to initiate treatment 1 wk before that date
- How to titrate product
- Not to use with nicotine patches unless directed by prescriber; may increase B/P
- To notify prescriber if pregnancy planned or suspected
- About common side effects to be expected

Black Box Warning: To notify prescriber immediately of change in thought/behavior (suicidal ideation, hostility, depression), stop product

vasopressin (Rx)

(vay-soe-press′in)

Pitressin, Pressyn ✤

Func. class.: Pituitary hormone
Chem. class.: Lysine vasopressin

ACTION: Promotes the reabsorption of water via action on the renal tubular epithelium; causes vasoconstriction

USES: Diabetes insipidus (nonnephrogenic/nonpsychogenic), abdominal distention postoperatively, bleeding esophageal varices

CONTRAINDICATIONS: Hypersensitivity, chronic nephritis

Precautions: Pregnancy (C), breastfeeding, CAD, asthma, vascular/renal disease, migraines, seizures

DOSAGE AND ROUTES

Diabetes insipidus

- **Adult: IM/SUBCUT** 5-10 units bid-qid as needed; **CONT IV INF** 0.0005 units/kg/hr (0.05 milliunit/kg/hr), double dose q30min as needed
- **Child: IM/SUBCUT** 2.5-10 units bid-qid as needed; **IM/SUBCUT** 1.25-2.5 units q2-3days (Pitressin Tannate) for chronic therapy

Abdominal distention

- **Adult: IM** 5 units then q3-4hr, increasing to 10 units if needed (aqueous)

Available forms: sol for inj 20 units/ml; 10 units/0.5 ml

SIDE EFFECTS

CNS: Drowsiness, headache, lethargy, flushing, vertigo
CV: Increased B/P, dysrhythmias, cardiac arrest, shock, chest pain, MI
EENT: Nasal irritation, congestion, rhinitis
GI: Nausea, heartburn, cramps, vomiting, flatus
GU: Vulval pain, uterine cramping

V

MISC: Tremors, sweating, vertigo, urticaria, bronchial constriction

PHARMACOKINETICS

Nasal: Onset 1 hr; duration 3-8 hr; half-life 15 min; metabolized in liver, kidneys; excreted in urine

INTERACTIONS

Increase: antidiuretic effect—tricyclics, carBAMazepine, chloropromide, fludrocortisone, clofibrate, urea

Decrease: antidiuretic effect—lithium, demeclocycline

NURSING CONSIDERATIONS

Assess:

- Pulse, B/P when giving product IV or IM
- I&O ratio, weight daily; fluid/electrolyte balance; check for edema in extremities; if water retention is severe, diuretic may be prescribed
- **Water intoxication:** lethargy, behavioral changes, disorientation, neuromuscular excitability
- Small doses may precipitate coronary adverse effects; keep emergency equipment nearby

Evaluate:

- Therapeutic response: absence of severe thirst, decreased urine output, osmolality

Teach patient/family:

- To measure and record I&O
- To avoid alcohol, all OTC medications unless approved by prescriber

vemurafenib

Zelboraf

Func. class.: Biologic response modifiers; signal transduction inhibitors (STIs)

ACTION:

Inhibitor of some mutated forms of BRAF serine threonine kinase, thereby blocking cellular proliferation in melanoma cells with the mutation; inhibits other kinases including CRAF, ARAF, wild-type BRAF, SRMA, ACK1, MAP4H5, and FGR; potent adenosine triphosphate-competitice inhibitor of RAFs, with a modest preference for mutant BRAF and CRAF compared with wild-type BRAF

USES:

Unresectable or metastatic malignant melanoma with V600E mutation of the BRAF gene

Precautions: Pregnancy (D), breastfeeding, children, infants, neonates, hepatic disease, QT prolongation, secondary malignancy, torsades de pointes, hypokalemia, hypomagnesia, sunlight exposure

DOSAGE AND ROUTES

- **Adult: PO** 960 mg (4 tabs) bid about q12hr

Dose adjustments for toxicity due to symptomatic adverse reactions or QTc prolongation

Grade 1 or tolerable grade 2 adverse events: no dosage change; **intolerable grade 2 or grade 3 adverse events (1st episode):** interrupt treatment until toxicity resolves to grade ≤1, when resuming, reduce dosage to 720 mg (3 tabs) bid; **intolerable grade 2 or grade 3 adverse events (2nd episode):** interrupt treatment until toxicity resolves to grade ≤1, when resuming, reduce dose to 480 mg (2 tabs) bid; **intolerable grade 2 or grade 3 adverse events (3rd episode)** discontinue treatment permanently; **grade 4 adverse events (1st episode):** discontinue permanently or interrupt until toxicity resolves to grade ≤1, when resuming, reduce dose to 480 mg (2 tabs) bid; **grade 4 adverse events (2nd episode):** discontinue permanently

Available forms: Tabs 240 mg

Administer:

- Continue until disease progresses or unacceptable toxicity occurs
- Missed doses can be taken up to 4 hr before the next dose is due; take about 12 hr apart, take without regard to meals

• Swallow whole with a full glass of water; do not crush or chew
• Store at room temperature in original container

SIDE EFFECTS

CNS: Fatigue, fever, asthenia, headache, dizziness, peripheral neuropathy, muscle paralysis
CV: QT prolongation, atrial fibrillation, peripheral edema, hypotension
EENT: Uveitis, blurred vision, iritis, photophobia
GI: Nausea, diarrhea, vomiting, constipation, dysgeusia, decreased appetite, weight loss
INTEG: Alopecia, pruritus, hyperkeratosis, maculopapular rash, actinic keratosis, xerosis/dry skin, papular rash, palmar-plantar erythrodysesthesia (hand and foot syndrome), photosensitivity
MS: Arthralgia, myalgias, extremity pain, musculoskeletal pain, back pain, arthritis
RESP: Cough
SYST: Secondary malignancy, anaphylaxis

PHARMACOKINETICS

>99% protein binding (albumin and alpha-1 acid glycoprotein) an inhibitor of CYP1A2, 2A6, 2C9, 2C19, 2D6, 3A4/5 CYP1A2 inhibitor, a weak CYP2D6 inhibitor, and a CYP3A4 inducer; elimination 94% in feces, 1% in urine, half-life 57 hr

INTERACTIONS

Increase: vemurafenib effect—CYP3A4/CYP1A2 inhibitors (enoxacin, cimetidine, delavirdine, indinavir, isoniazid, itraconazole, dalfopristin, quinupristin, tipranavir)
⚠ **Increase:** QT prolongation, torsades de pointes—arsenic trioxide, certain phenothiazines (chlorproMAZINE, mesoridazine, thioridazine), grepafloxacin, levomethadyl, pentamidine, probucol, sparfloxacin, troleandomycin, class IA antiarrhythmics (disopyramide, procainamide, quiNIDine), class III antiarrhythmics (amiodarone, dofetilide, ibutilide, sotalol), clarithromycin, ziprasidone, pimozide, haloperidol, halofantrine, quiNIDine, chloroquine, dronedarone, droperidol, erythromycin, methadone, posaconazole, propafenone, saquinavir, abarelix, amoxapine, apomorphine, asenapine, β-agonists, ofloxacin, eribulin, ezogabine, flecainide, gatifloxacin, gemifloxacin, halogenated anesthetics, iloperidone, levofloxacin, local anesthetics, magnesium sulfate, potassium sulfate, sodium, maprotiline, moxifloxacin, nilotinib, norfloxacin, ciprofloxacin, OLANZapine, paliperidone, some phenothiazines (fluPHENAZine, perphenazine, prochlorperazine, trifluoperazine), telavancin, tetrabenazine, tricyclic antidepressants, venlafaxine, vorinostat, citalopram, alfuzosin, cloZAPine, cyclobenzaprine, dolasetron, palonosetron, QUEtiapine, rilpivirine, SUNItinib, tacrolimus, vardenafil, indacaterol, dasatinib, fluconazole, lapatinib, lopinavir/ritonavir, mefloquine, octreotide, ondansetron, ranolazine, risperiDONE, telithromycin, vemurafenib
Decrease: vemurafenib effect—CYP3A4 inducers (rifampin, barbiturates, carBAMazepine, phenytoin, erythromycin, ketoconazole, fluvoxaMINE, alcohol, etravirine, ritonavir, bexarotene, rifabutin, OXcarbazepine, nevirapine, modafinil, metyrapone)
Drug/Lab Test:
Increase: serum creatinine, LFTs, alkaline phosphatase, bilirubin

NURSING CONSIDERATIONS

Assess:
• **Hepatic disease:** Liver function test (LFT) abnormalities, altered bilirubin levels, may occur; monitor LFTs and bilirubin levels before treatment, then monthly; more frequent testing is needed in those with grade 2 or greater toxicities; laboratory alterations should be managed with dose reduction, treatment interruption, or discontinuation
⚠ **QT prolongation**: has been reported with the use of this product; therefore,

V

should be avoided in patients with QT prolongation; monitor ECG and electrolytes in those with congestive heart failure, bradycardia, electrolyte imbalance (hypokalemia, hypomagnesemia), or those who are taking concomitant medications known to prolong the QT interval; treatment interruption, dosage adjustment, treatment discontinuation may be needed in those who develop QT prolongation

⚠ **Pregnancy/breastfeeding:** Identify if pregnancy is planned or suspected, pregnancy (D); avoid breastfeeding

- Serum electrolytes

Evaluate:

- Decreased spread of malignancy

Teach patient/family:

- That missed doses can be taken up to 4 hr before the next dose is due to maintain the twice-daily regimen

⚠ Teach patient to use reliable contraception; both women and men of childbearing age should use adequate contraceptive methods during therapy and for at least 90 days after completing treatment, pregnancy (D)

venlafaxine (Rx)

(ven-la-fax′een)

Effexor ✦, Effexor XR

Func. class.: Antidepressant—SNRI

Chem. class.: SNRI

ACTION: Potent inhibitor of neuronal serotonin and norepinephrine uptake, weak inhibitor of dopamine; no muscarinic, histaminergic, or α-adrenergic receptors in vitro

USES: Prevention/treatment of major depression; depression at the end of life; long-term treatment of general anxiety disorder, panic disorder, social anxiety disorder (Effexor XR only)

Unlabeled uses: Hot flashes, premenstrual dysphoric disorder (PMDD), headache, neuropathic pain, fibromyalgia, diabetic neuropathy

CONTRAINDICATIONS: Hypersensitivity

Precautions: Pregnancy (C), breastfeeding, geriatric patients, mania, hypertension, seizure disorder, recent MI, cardiac/renal/hepatic disease, eosinophilic pneumonia, desvenlafaxine hypersensitivity, bipolar disorder, interstitial lung disease

Black Box Warning: Children, suicidal ideation

DOSAGE AND ROUTES

Depression

- **Adult: PO** 75 mg/day in 2-3 divided doses; taken with food, may be increased to 150 mg/day; if needed, may be further increased to 225 mg/day; increments of 75 mg/day at intervals of ≥4 days; some hospitalized patients may require up to 375 mg/day in 3 divided doses; **EXT REL** 37.5-75 mg PO daily, max 225 mg/day; give XR daily

Anxiety disorders

- **Adult: PO** 75 mg/day or 37.5 mg/day × 4-7 days initially, max 225 mg/day

Renal dose

- **Adult: PO** CCr 10-70 ml/min, reduce dose by 25%-50%; CCr <10 ml/min, reduce dose by 50%

Hepatic dose

- **Adult: PO** Moderate impairment, 50% of dose

Hot flashes (unlabeled)

- **Adult (male, prostate cancer): PO** 12.5 mg bid × 4 wk; females 37.5-75 mg/day

Neuropathic pain, diabetic neuropathy, headache, fibromyalgia (unlabeled)

- **Adult: PO** 37.5-75 mg/day, max 75 mg bid (regular rel) or 150 mg/day (ext rel)

Premenstrual dysphoric disorder (PMDD) (unlabeled)

- **Adult female: PO** 50-200 mg/day, start at 50 mg/day for 1st cycle, titrate upward

Available forms: Tabs scored 25, 37.5, 50, 75, 100 mg; ext rel cap (Effexor XR) 37.5, 75, 150 mg; ext rel tab 37.5, 75, 150, 225 mg

Administer:

- With food, milk for GI symptoms; do not crush, chew caps; caps can be opened and contents sprinkled on applesauce, given with full glass of water
- Sugarless gum, hard candy, frequent sips of water for dry mouth
- Avoid use with CNS depressants
- Give in small amounts because of suicide potential, especially at beginning of therapy
- Store in tight container at room temperature; do not freeze

SIDE EFFECTS

CNS: Emotional lability, *dizziness, weakness,* apathy, ataxia, headache, tremors, hypertonia, euphoria, hallucinations, hostility, insomnia, anxiety, **suicidal ideation in children/adolescents, seizures, neuroleptic-malignant-syndrome–like reaction**

CV: Angina pectoris, hypertension, **tachycardia, change in QTc interval,** increased cholesterol, extrasystoles, syncope

EENT: *Abnormal vision,* taste, *ear pain*

GI: *Dysphagia, eructation, nausea,* anorexia, dry mouth, colitis, gastritis, gingivitis, *constipation,* stomatitis, stomach and mouth ulceration

GU: *Anorgasmia,* abnormal ejaculation, urinary frequency, decreased libido, impotence, menstrual changes, *impaired urination*

HEMA: Thrombocytopenia, leukocytosis, leukopenia, **abnormal bleeding**

INTEG: Ecchymosis, brittle nails, dry skin, photosensitivity, sweating, **Stevens-Johnson syndrome; angioedema (ext rel)**

META: *Peripheral edema, weight loss,* edema, glycosuria, hyperlipemia, hypokalemia

MS: Arthritis, bone pain, tenosynovitis, arthralgia

RESP: *Bronchitis, dyspnea,* cough

SYST: **Neonatal abstinence syndrome**

PHARMACOKINETICS

Well absorbed; extensively metabolized in liver by CYP2D6 to active metabolite; 87% of product recovered in urine; 27% protein binding; half-life 5, 11 hr (active metabolite), respectively

INTERACTIONS

⚠ **Hyperthermia, rigidity, rapid fluctuations of vital signs, mental status changes, neuroleptic malignant syndrome: MAOIs**

Increase: bleeding risk—salicylates, NSAIDs, platelet inhibitors, anticoagulants

Increase: venlafaxine effect—cimetidine

Increase: CNS depression—alcohol, opioids, antihistamines, sedative/hypnotics

Increase: levels of cloZAPine, desipramine, haloperidol, warfarin

Increase: serotonin syndrome—sibutramine, SUMAtriptan, traZODone, traMADol, SSRIs, serotonin receptor agonist, linezolid, methylene blue, tryptophan

Decrease: effect of indinavir

Decrease: venlafaxine effect—cyproheptadine

Drug/Herb

- Serotonin syndrome: St. John's wort, tryptophan

Increase: CNS depression—kava, valerian

Drug/Lab Test

Increase: alk phos, bilirubin, AST, ALT, BUN, creatinine, serum cholesterol, CPK, LDH

False positive: amphetamines, phencyclidine

NURSING CONSIDERATIONS

Assess:

Black Box Warning: Mental status: mood, sensorium, affect, increase in psychiatric symptoms; depression, panic; assess for suicidal ideation in children/adolescents

- B/P lying, standing; pulse q4hr; if systolic B/P drops 20 mm Hg, hold product, notify prescriber; take VS q4hr in patients with CV disease

⚠ Bleeding: GI, ecchymosis, epistaxis, hematomas, petechiae, hemorrhage
- Blood studies: CBC, differential, leukocytes, cardiac enzymes if patient is receiving long-term therapy
- Hepatic studies: AST, ALT, bilirubin
- Weight weekly; weight loss or gain; appetite may increase; peripheral edema may occur; monitor cholesterol
- **Withdrawal symptoms:** flulike symptoms, headache, nervousness, agitation, nausea, vomiting, muscle pain, weakness; not usual unless product is discontinued abruptly

⚠ Serotonin syndrome, neuroleptic malignant syndrome: increased heart rate, shivering, sweating, dilated pupils, tremors, high B/P, hyperthermia, headache, confusion; if these occur, stop product, administer serotonin antagonist if needed; usually worse if given with linezolid, methylene blue, tryptophan
- Assistance with ambulation during beginning therapy because drowsiness, dizziness occur

Evaluate:
- Therapeutic response: decreased depression, anxiety; increased well-being

Teach patient/family:
- To notify prescriber of rash, hives, allergic reactions, bleeding
- To use with caution when driving, performing other activities requiring alertness because of drowsiness, dizziness, blurred vision

> **Black Box Warning:** That worsening of symptoms, suicidal thoughts/behaviors may occur in children/young adults

- To avoid alcohol ingestion
- Not to discontinue medication abruptly after long-term use; may cause nausea, headache, malaise
- To wear sunscreen or large hat because photosensitivity occurs
- To avoid pregnancy or breastfeeding while taking this product; birth defects have occurred when used in the 3rd trimester
- To monitor B/P with hypertension

TREATMENT OF OVERDOSE:

ECG monitoring; lavage, activated charcoal; administer anticonvulsant; may require whole-bowel irrigation for ext rel product

verapamil (Rx)

(ver-ap′a-mill)

Apo-Verap ✦, Calan, Calan SR, Covera-HS ✦, Isoptin SR, Nu-Verap ✦, Verelan, Verelan PM

Func. class.: Calcium channel blocker; antihypertensive; antianginal, antidysrhythmic (class IV)

Chem. class.: Diphenylalkylamine

ACTION: Inhibits calcium ion influx across cell membrane during cardiac depolarization; produces relaxation of coronary vascular smooth muscle; dilates coronary arteries; decreases SA/AV node conduction; dilates peripheral arteries

USES: Chronic stable, vasospastic, unstable angina; dysrhythmias, hypertension, supraventricular tachycardia, atrial flutter or fibrillation

Unlabeled uses: Prevention of migraine headaches, claudication, mania

CONTRAINDICATIONS: Sick sinus syndrome, 2nd-/3rd-degree heart block, hypotension <90 mm Hg systolic, cardiogenic shock, severe CHF, Lown-Ganong-Levine syndrome, Wolff-Parkinson-White syndrome

Precautions: Pregnancy (C), breastfeeding, children, geriatric patients, CHF, hypotension, hepatic injury, renal disease, concomitant β-blocker therapy

DOSAGE AND ROUTES

Angina
- **Adult: PO** 80-120 mg tid, increase weekly, max 480 mg/day

Dysrhythmias

• **Adult: PO** 240-480 mg/day in 3-4 divided doses in digitalized patients

• **Adult: IV BOL** 5-10 mg (0.075-0.15 mg/kg) over 2 min, may repeat 10 mg (0.15 mg/kg) ½ hr after 1st dose

• **Child 1-15 yr: IV BOL** 0.1-0.3 mg/kg over ≥2 min, repeat in 30 min, max 5 mg in single dose

• **Child 0-1 yr: IV BOL** 0.1-0.2 mg/kg over ≥2 min, may repeat after 30 min

Hypertension

• **Adult: PO** 80 mg tid, may titrate upward; **EXT REL** 120-240 mg/day as single dose, may increase to 240-480 mg/day

Hepatic disease/geriatric patients/compromised ventricular function

• **Adult: PO** 40 mg tid initially, increase as tolerated

Claudication due to PVD (unlabeled)

• **Adult: PO** 120-480 mg/day in divided doses

Mania (unlabeled)

• **Adult: PO** 160-320 mg/day in divided doses, may be given with lithium

Migraine prophylaxis (unlabeled)

• **Adult: PO** 80 mg tid

Available forms: Tabs 40, 80, 120 mg; ext rel tabs 120, 180, 240 mg; inj 2.5 mg/ml in ampules, syringes, vials; ext rel caps 100, 200, 240, 300 mg

Administer:

PO route

• **Reg rel:** Give without regard to food

Ext rel route

• Do not crush or chew ext rel products; caps may be opened and contents sprinkled on food; do not dissolve chew cap contents

• Before meals, at bedtime; give ext rel product with food

Direct IV route

• Undiluted through Y-tube or 3-way stopcock of compatible sol; give over 2 min or 3 min for geriatric patients, with continuous ECG and B/P monitoring, discard unused solution

• Do not use IV with IV β-blockers, may cause AV nodal blockade

Y-site compatibilities: Alfentanil, amikacin, argatroban, ascorbic acid, atracurium, atropine, aztreonam, bivalirudin, bumetanide, buprenorphine, butorphanol, calcium chloride/gluconate, CARBOplatin, caspofungin, ceFAZolin, cefonicid, cefotaxime, cefoTEtan, cefOXitin, ceftizoxime, cefTRIAXone, cefuroxime, chlorproMAZINE, cimetidine, ciprofloxacin, clindamycin, cyanocobalamin, cyclophosphamide, cycloSPORINE, cytarabine, DACTINomycin, DAPTOmycin, dexamethasone, dexmedetomidine, digoxin, diltiazem, diphenhydrAMINE, DOBUTamine, DOCEtaxel, DOPamine, doxacurium, DOXOrubicin hydrochloride, doxycycline, enalaprilat, ePHEDrine, EPINEPHrine, epirubicin, epoetin alfa, eptifibatide, erythromycin, esmolol, etoposide, etoposide phosphate, famotidine, fenoldopam, fentaNYL, fluconazole, fludarabine, gemcitabine, gentamicin, glycopyrrolate, granisetron, heparin, hydrALAZINE, hydrocortisone, HYDROmorphone, ifosfamide, imipenem/cilastatin, inamrinone, insulin, isoproterenol, ketorolac, labetalol, levofloxacin, lidocaine, linezolid, LORazepam, magnesium sulfate, mannitol, mechlorethamine, meperidine, metaraminol, methotrexate, methoxamine, methyldopate, methylPREDNISolone, metoclopramide, metoprolol, metroNIDAZOLE, miconazole, midazolam, milrinone, mitoXANtrone, morphine, multivitamins, nalbuphine, naloxone, nesiritide, nitroglycerin, nitroprusside, norepinephrine, octreotide, ondansetron, oxaliplatin, oxytocin, PACLitaxel, palonosetron, papaverine, PEMEtrexed, penicillin G, pentamidine, pentazocine, phentolamine, phenylephrine, phytonadione, piperacillin/tazobactam, potassium chloride, procainamide, prochlorperazine, promethazine, propranolol, protamine, pyridoxime, quinupristin/dalfopristin, ranitidine, rocuronium, sodium acetate, succinylcholine, SUFentanil, tacrolimus, teniposide, theophylline, thiamine, ticarcillin/clavulanate, tirofiban, tobramycin, tolazoline, trimethaphan, uroki-

V

nase, vancomycin, vasopressin, vecuronium, vinCRIStine, vinorelbine, voriconazole

SIDE EFFECTS

CNS: *Headache, drowsiness,* dizziness, anxiety, depression, weakness, insomnia, confusion, lightheadedness, asthenia, fatigue
CV: *Edema,* **CHF**, bradycardia, hypotension, palpitations, AV block, **dysrhythmias**
GI: *Nausea,* diarrhea, gastric upset, *constipation,* increased LFTs
GU: Impotence, gynecomastia, nocturia, polyuria
HEMA: Bruising, petechiae, bleeding
INTEG: Rash, bruising
MISC: Gingival hyperplasia
SYST: **Stevens-Johnson syndrome**

PHARMACOKINETICS

Metabolized by liver, excreted in urine (70% as metabolites)
PO: Onset variable; peak 3-4 hr; duration 17-24 hr; half-life (biphasic) 4 min, 3-7 hr (terminal)
IV: Onset 3 min, peak 3-5 min, duration 10-20 min

INTERACTIONS

Increase: hypotension—prazosin, quiNIDine, fentaNYL, other antihypertensives, nitrates
Increase: effects of verapamil—β-blockers, cimetidine, clarithromycin, erythromycin, monitor for CV effects
Increase: levels of digoxin, theophylline, cycloSPORINE, carBAMazepine, nondepolarizing muscle relaxants
Decrease: effects of lithium
Decrease: antihypertensive effects—NSAIDs
Drug/Food
Increase: hypotensive effects—grapefruit juice
Drug/Herb
Increase: verapamil effect—ginseng, ginkgo
Increase: hypertension ephedra (ma huang)
Decrease: verapamil effect—St. John's wort
Drug/Lab Test
Increase: AST, ALT, alk phos, BUN, creatinine, serum cholesterol

NURSING CONSIDERATIONS

Assess:

• **Cardiac status:** B/P, pulse, respiration, ECG intervals (PR, QRS, QT); notify prescriber if pulse <50 bpm, systolic B/P <90 mm Hg

⚠ **CHF: I&O ratios, weight daily; crackles, weight gain, dyspnea, jugular venous distention**

• Renal, hepatic studies during long-term treatment, serum potassium periodically

Evaluate:

• Therapeutic response: decreased anginal pain, decreased B/P, dysrhythmias

Teach patient/family:

• To increase fluids, fiber to counteract constipation
• How to take pulse, B/P before taking product; to keep record or graph
• To avoid hazardous activities until stabilized on product, dizziness no longer a problem
• To limit caffeine consumption; to avoid alcohol products
• To avoid OTC or grapefruit products unless directed by prescriber
• To comply with all areas of medical regimen: diet, exercise, stress reduction, product therapy
• To change positions slowly to prevent syncope
• Not to discontinue abruptly; chest pain may occur

⚠ **To report chest pain, palpitations, irregular heart beats, swelling of extremities, skin irritation, rash, tremors, weakness**

• To notify prescriber if pregnancy is planned; may breastfeed (American Academy of Pediatrics)
• That cap (Covera-HS) 🍁 may appear in stool

TREATMENT OF OVERDOSE: Defibrillation, atropine for AV block, vasopressor for hypotension, IV calcium

vidarabine ophthalmic

See Appendix B

vigabatrin (Rx)

(vye-ga'ba-trin)

Sabril

Func. class.: Anticonvulsant

Chem. class.: GABA transaminase inhibitor

ACTION: May inhibit reuptake and metabolism of GABA; may increase seizure threshold; structurally similar to GABA

USES: Adjunct treatment of partial seizures in adults and children ≥12 yr, infantile spasm

CONTRAINDICATIONS: Hypersensitivity to this product

Precautions: Pregnancy (C), breastfeeding, children <2 yr, geriatric patients, renal/hepatic disease, suicidal thoughts/behaviors, abrupt discontinuation

Black Box Warning: Visual disturbance

DOSAGE AND ROUTES

Partial seizures

- **Adult: PO** 500 mg bid, titrate in 500-mg increments at weekly intervals up to 1.5 g bid

Infantile spasm

- **Infant >1 mo, child ≤2 yr: PO** 50 mg/kg/day in 2 divided doses, titrate in 25 to 50-mg/kg/day increments q3days, max 150 mg/kg/day

Renal dose

- **Adult: PO** CCr 50-80 ml/min, reduce dose by 25%; CCr 30-50 ml/min, reduce dose by 50%; CCr 10-30 ml/min, reduce dose by 75%

Available forms: Tabs 500 mg; powder for oral solution 500 mg

Administer:

PO route (tab)

- Give without regard to meals

PO route (oral sol)

- Reconstitute immediately before using
- Empty contents of appropriate number of packets into clean cup
- For each packet, dissolve 10 ml water, conc 50 mg/ml; do not use other liquids
- Stir until dissolved, sol should be clear
- Use calibrated oral syringe to measure correct dosage
- Discard any unused sol
- Store at room temperature

SIDE EFFECTS

CNS: Headache, memory impairment, *dizziness,* irritability, lethargy, **malignant hyperthermia**, insomnia, **suicidal ideation**

CV: Edema

EENT: **Visual impairment**

GI: Nausea, vomiting, diarrhea, increased appetite, abdominal pain, GI bleeding, hemorrhoids, weight gain, constipation

GU: Impotence, dysmenorrhea

HEMA: Anemia

INTEG: Pruritus, rash

RESP: Coughing, **respiratory depression, pulmonary embolism**

PHARMACOKINETICS

Absorption >95%, no protein binding, widely distributed, not metabolized, excretion in urine 80% parent drug, excretion slowed in renal disease, peak 2 hr, half-life 7.5 hr

INTERACTIONS

Increase: CNS depression—CNS depressants

Increase: Serious ophthalmic effects (glaucoma, retinopathy): azaTHIOprine, chloroquine, corticosteroids, deferoxamine, ethambutol, hydroxychloroquine, interferons, loxapine, mecasermin, rh-IGF-1, pentostatin, phenothiazine, phosphodiesterase inhibitors, tamoxifen, thiothixene; avoid concurrent use

V

Drug/Lab Test
Decrease: ALT/AST

NURSING CONSIDERATIONS

Assess:

Visual impairment: Prescribers must be registered with the SHARE program due to risk of permanent vision loss; if no clinical response in 2-4 wk of pediatric patients or 3 mo in adults discontinue, provide vision assessment before and after ≤ 4 wk at least q3mo, and 3-6 mo after stopping product

- Renal studies: urinalysis, BUN, urine creatinine q3mo in those with renal disease
- Hepatic studies: ALT, AST, bilirubin
- Description of seizures: location, duration, presence of aura
- Mental status: mood, sensorium, affect, behavioral changes; if mental status changes, notify prescriber
- Assistance with ambulation during early part of treatment; dizziness occurs
- Seizure precautions: padded side rails; move objects that may harm patient

Evaluate:

- Therapeutic response: decreased seizure activity; document on patient's chart

Teach patient/family:

- To carry emergency ID stating patient's name, products taken, condition, prescriber's name and phone number
- To avoid driving, other activities that require alertness
- Not to discontinue medication quickly after long-term use
- To notify prescriber if pregnancy is planned or suspected
- To report suicidal thoughts/behaviors immediately
- To avoid alcohol; drowsiness, dizziness may occur

vilazodone (Rx)

(vil-az′oh-done)

Viibryd

Func. class.: Antidepressant, miscellaneous

ACTION: Novel antidepressant unrelated to other antidepressants, enhances serotonergic action by a dual mechanism

USES: Major depression

CONTRAINDICATIONS: Concomitant use of MOA inhibitors or within 14 days after discontinuing MOA inhibitor or within 14 days after discontinuing vilazodone

Precautions: Pregnancy (C), labor, infants, geriatric patients, abrupt discontinuation, bipolar disorder, bleeding, operating machinery, ECT, hepatic disease, hyponatremia, hypovolemia, substance abuse, history of seizures, serotonin syndrome, neuroleptic malignant syndrome; use with serotonin precursors (e.g., tryptophan) or serotonergic drugs; suicidal ideation, worsening depression or behavior

Black Box Warning: Children, suicidal ideation

DOSAGE AND ROUTES

- **Adult: PO** 10 mg × 7 days, then 20 mg × 7 days, then 40 mg/day; if taking potent CYP3A4 inhibitor, max 20 mg/day

Available forms: Tabs 10, 20, 40 mg

Administer:

- With food to increase absorption
- Do not use within 2 wk of MAOIs
- Store at room temperature, away from moisture, heat

SIDE EFFECTS

CNS: Restlessness, dizziness, drowsiness, fatigue, mania, insomnia, migraine, neuroleptic-malignant–like syndrome, paresthesias, seizures, suicidal ideation, tremors, night sweats, dream disorders

CV: Palpitations, ventricular extrasystole
EENT: Cataracts, blurred vision
GI: *Nausea,* vomiting, flatulence, *diarrhea, xerostomia,* altered taste, gastroenteritis, increased appetite
GU: Decreased libido, ejaculation disorder, increased frequency of urination, sexual dysfunction
HEMA: Bleeding, decreased platelets
INTEG: Sweating
MS: Arthralgia
SYST: Neonatal abstinence syndrome, withdrawal, **serotonin syndrome**

INTERACTIONS

⚠ **Do not use within 2 wk of MAO inhibitors**

⚠ **Increased: serotonin syndrome—SSRIs, SNRIs, serotonin receptor agonists, selegiline, busPIRone, dextromethorphan, ergots, fenfluramine, dexfluramine, lithium, meperidine, fentaNYL, methylphenidate, dexmethylphenidate, metoclopramide, mirtazapine, nefazodone, pentazocine, phenothiazines, haloperidol, loxapine, thiothixene, molindone, amphetamines**

Increased: bleeding—anticoagulants, thrombolytics, platelet inhibitors, salicylates, NSAIDs

Increased: vilazodone levels—CYP3A4 inhibitors (ketoconazole, erythromycin, efavirenz, dronedarone, clarithromycin and others)

Decrease: vilazodone effect—CYP3A4 inducers

Drug/Food

- Avoid use with grapefruit juice

Drug/Herb

Increased: serotonin syndrome—St. John's wort

Drug/Lab Test

Decrease: sodium

PHARMACOKINETICS

Protein binding 96%-99%, metabolized by liver by CYP3A4 (major) and CYP2C19 and CYP2D (minor) and non-CYP pathways, peak 4-5 hr, half-life 25 hr

NURSING CONSIDERATIONS

Assess:

Black Box Warning: Mental status: orientation, mood behavior initially and periodically; initiate suicide precautions if indicated; history of seizures, mania

- Renal/hepatic status: hyponatremia
- **Abrupt discontinuation: do not discontinue abruptly, taper, monitor for symptoms of withdrawal; if intolerable, resume previous dose and decrease more slowly**
- **Serotonin syndrome/neuroleptic malignant syndrome: nausea, vomiting, sedation, sweating, facial flushing, high B/P; discontinue product**

Evaluate:

- Therapeutic response: remission of depressive symptoms

Teach patient/family:

- To take as directed; not to double dose
- To avoid abrupt discontinuation unless approved by prescriber
- Not to drive or operate machinery until effects are known
- Not to use other products unless approved by prescribed

⚠ **To contact prescriber regarding the following: allergic reactions; personality changes (aggression, anxiety, anger, hostility); extreme sleepiness or drowsiness; feeling confused, nervous, restless, or clumsy; numbness, tingling, or burning pain in hands, arms, legs, or feet; tremors; unusual behavior or thoughts about hurting oneself**

- To notify prescriber if pregnancy is planned or suspected; to avoid breastfeeding

⚠ ***HIGH ALERT***

vinBLAStine (VLB) (Rx)

(vin-blast′een)

Func. class.: Antineoplastic

Chem. class.: Vinca alkaloid

Do not confuse:

vinBLAStine/vinCRIStine/vinorelbine

ACTION: Inhibits mitotic activity, arrests cell cycle at metaphase; inhibits RNA synthesis, blocks cellular use of glutamic acid needed for purine synthesis; vesicant

USES: Breast, testicular cancer, lymphomas, neuroblastoma; Hodgkin's/non-Hodgkin's lymphoma; mycosis fungoides, histiocytosis, Kaposi's sarcoma, Langerhans cell histiocytosis

Unlabeled uses: Lung, bladder, prostate cancer; desmoid tumor, malignant melanoma

CONTRAINDICATIONS: Pregnancy (D), breastfeeding, infants, hypersensitivity, leukopenia, granulocytopenia, bone marrow suppression, infection

Black Box Warning: Intrathecal use

Precautions: Renal/hepatic disease, tumor lysis syndrome

Black Box Warning: Extravasation

DOSAGE AND ROUTES

Breast cancer

- **Adult: IV** 4.5 mg/m^2 on day 1 of every 21 days in combination with DOXOrubicin and thiotepa

Hodgkins disease

- **Adult: IV** 6 mg/m^2 on days 1 and 15 of every 28 days with DOXOrubicin, bleomycin, dacarbazine (ABVD regimen)
- **Child: IV** 2.5-6 mg/m^2/day once q1-2wk × 3-6 wk, max weekly dose 12.5 mg/m^2

Available forms: Inj, powder 10 mg for 10 ml IV; sol for inj 1 mg/ml

Administer:

- Antiemetic 30-60 min before product and prn to prevent vomiting

Black Box Warning: Hyaluronidase 150 units/ml in 1 ml NaCl, warm compress for extravasation for vesicant activity treatment

Black Box Warning: Do not administer intrathecally; fatal

IV inj route

- After diluting 10 mg/10 ml NaCl; give through Y-tube or 3-way stopcock or directly over 1 min

Intermittent IV INF route

- Further dilute in 50-100 ml of NS, infuse over 15-30 min

Additive compatibilities: Bleomycin

Syringe compatibilities: Bleomycin, CISplatin, cyclophosphamide, droperidol, fluorouracil, leucovorin, methotrexate, metoclopramide, mitoMYcin, vinCRIStine

Y-site compatibilities: Allopurinol, amifostine, amphotericin B cholesteryl, aztreonam, bleomycin, CISplatin, cyclophosphamide, DOXOrubicin, DOXOrubicin liposome, droperidol, filgrastim, fludarabine, fluorouracil, granisetron, heparin, leucovorin, melphalan, methotrexate, metoclopramide, mitoMYcin, ondansetron, PACLitaxel, piperacillin/tazobactam, sargramostim, teniposide, thiotepa, vinCRIStine, vinorelbine

SIDE EFFECTS

CNS: Paresthesias, peripheral neuropathy, depression, headache, seizures, malaise

CV: Tachycardia, orthostatic hypo/hypertension

GI: *Nausea, vomiting,* ileus, *anorexia, stomatitis, constipation,* abdominal pain, GI/rectal bleeding, hepatotoxicity, pharyngitis

GU: Urinary retention, renal failure, hyperuricemia

HEMA: Thrombocytopenia, leukopenia, myelosuppression, agranulocytosis, granulocytosis, aplastic anemia, neutropenia, pancytopenia

INTEG: *Rash, alopecia,* photosensitivity, extravasation, tissue necrosis

META: SIADH

RESP: Fibrosis, pulmonary infiltrate, bronchospasm

SYST: Tumor lysis syndrome (TLS)

PHARMACOKINETICS

Half-life (triphasic) <5 min, 50-155 min, 23-85 hr; metabolized in liver; excreted in urine, feces; crosses blood-brain barrier

INTERACTIONS

Increase: Synergism: bleomycin
Increase: Bronchospasm: mitoMYcin
• Do not use with radiation
Increase: bleeding risk—NSAIDs, anticoagulants, thrombolytics, antiplatelets
Increase: toxicity, bone marrow suppression—antineoplastics
Increase: action of methotrexate
Increase: adverse reactions—live virus vaccines
⚠ **Increase:** toxicity—CYP3A4 inhibitors (aprepitant, antiretroviral protease inhibitors, clarithromycin, danazol, delavirdine, diltiazem, erythromycin, fluconazole, FLUoxetine, fluvoxaMINE, imatinib, ketoconazole, mibefradil, nefazodone, telithromycin, voriconazole)
Decrease: vinBLAStine effect—CYP3A4 inducers (barbiturates, bosentan, carBAMazepine, efavirenz, phenytoins, nevirapine, rifabutin, rifampin)

Drug/Herb
• Avoid use with St. John's wort

Drug/Lab Test
Increase: uric acid, bilirubin
Decrease: Hgb, platelets, WBC

NURSING CONSIDERATIONS

Assess:
⚠ CBC, differential, platelet count weekly; withhold product if WBC is <2000/mm³ or platelet count is <75,000/mm³; notify prescriber; RBC, Hct, Hgb may be decreased, nadir occurs on days 4-10 and continues for another 1-2 wk
• **Tumor lysis syndrome:** monitor for hyperkalemia, hyperphosphatemia, hyperuricemia; usually occurs with leukemia, lymphoma; alkalinization of the urine, allopurinol should be used to prevent urate nephrophy; monitor electrolytes and renal function (BUN, uric acid, urine CCR)
• Hepatitis: transient hepatitis may occur with continuous IV
• Bleeding: hematuria, guaiac, bruising, petechiae, mucosa or orifices
• **Bronchospasm:** can be life-threatening; usually occurs when giving mitoMYcin
• Effects of alopecia on body image; discuss feelings about body changes
• Sensitivity of feet/hands, which precedes neuropathy
• Jaundiced skin, sclera; dark urine, clay-colored stools, itchy skin, abdominal pain, fever, diarrhea
• Buccal cavity q8hr for dryness, sores, ulcerations, white patches, oral pain, bleeding, dysphagia

Black Box Warning: Local irritation, pain, burning, discoloration at inj site, extravasation

• Symptoms indicating severe allergic reaction: rash, pruritus, urticaria, purpuric skin lesions, itching, flushing
• Increased fluid intake to 2-3 L/day to prevent urate deposits, calculi formation
• Brushing of teeth bid-tid with soft brush or cotton-tipped applicator for stomatitis; use unwaxed dental floss

Evaluate:
• Therapeutic response: decreased tumor size, spread of malignancy

Teach patient/family:
• To report any complaints or side effects to nurse or prescriber
• To report any changes in breathing or coughing; to avoid exposure to persons with infection
• That hair may be lost during treatment; that a wig or hairpiece may make patient feel better; that new hair may be different in color, texture
• To report change in gait or numbness in extremities; may indicate neuropathy
• To avoid foods with citric acid, hot or rough texture
• To report any bleeding, white spots, ulcerations in mouth to prescriber; to examine mouth daily
• To wear sunscreen, protective clothing, sunglasses
• To avoid receiving vaccinations

⚠ **Pregnancy:** pregnancy (D); to avoid breastfeeding; that product may cause male infertility

- To notify prescriber if pregnancy is planned or suspected; to avoid persons with known infections
- **Infection:** to report sore throat, flulike symptoms

⚠ ***HIGH ALERT***

vinCRIStine (VCR) (Rx)

(vin-kris′teen)

Vincasar PFS

Func. class.: Antineoplastic—miscellaneous

Chem. class.: Vinca alkaloid

Do not confuse:
vinCRIStine/vinBLAStine/vinorelbine

ACTION: Inhibits mitotic activity, arrests cell cycle at metaphase; inhibits RNA synthesis, blocks cellular use of glutamic acid needed for purine synthesis; vesicant

USES: Lymphomas, neuroblastoma, Hodgkin's disease, acute lymphoblastic and other leukemias, rhabdomyosarcoma, Wilms' tumor, non-Hodgkin's lymphoma, malignant glioma, soft-tissue sarcoma

Unlabeled uses: Lung, breast, colorectal, head/neck, osteogenic sarcomas; small-cell lung cancer, trophoblastic disease

CONTRAINDICATIONS: Pregnancy (D), breastfeeding, infants, hypersensitivity, radiation therapy

Black Box Warning: Intrathecal use

Precautions: Renal/hepatic disease, hypertension, neuromuscular disease

Black Box Warning: Extravasation

DOSAGE AND ROUTES

- **Adult: IV** 0.4-1.4 mg/m^2/wk, max 2 mg
- **Child: IV** 1-2 mg/m^2/wk, max 2 mg

Available forms: Inj 1 mg/ml; powder for inj 5 mg/vial

Administer:

- Antiemetic 30-60 min before product and prn
- Antispasmodic for GI symptoms

Black Box Warning: Do not give intrathecally; fatal

IV route

- After diluting with diluent provided or 1 mg/10 ml sterile water or NaCl; give through Y-tube or 3-way stopcock or directly over 1 min; do not use 5-mg vial for single doses

Black Box Warning: Hyaluronidase 150 units/ml in 1 ml NaCl; apply warm compress for extravasation

Additive compatibilities: Bleomycin, cytarabine, fluorouracil, methotrexate

Syringe compatibilities: Bleomycin, CISplatin, cyclophosphamide, doxapram, DOXOrubicin, droperidol, fluorouracil, heparin, leucovorin, methotrexate, metoclopramide, mitomycin, vinBLAStine

Y-site compatibilities: Allopurinol, amifostine, amphotericin B cholesteryl, aztreonam, bleomycin, CISplatin, cladribine, cyclophosphamide, DOXOrubicin, DOXOrubicin liposome, droperidol, filgrastim, fludarabine, fluorouracil, granisetron, heparin, leucovorin, melphalan, methotrexate, metoclopramide, mitoMYcin, ondansetron, PACLitaxel, piperacillin/tazobactam, sargramostim, teniposide, thiotepa, vinBLAStine, vinorelbine

SIDE EFFECTS

CNS: *Decreased reflexes, numbness, weakness, motor difficulties,* CNS depression, cranial nerve paralysis, seizures, peripheral neuropathy

CV: Orthostatic hypotension

EENT: *Diplopia*

GI: *Nausea, vomiting, anorexia, stomatitis, constipation,* **paralytic ileus,** *abdominal pain,* **hepatotoxicity**
GU: **Renal tubular obstruction**
HEMA: **Thrombocytopenia, leukopenia, myelosuppression, anemia**
INTEG: *Alopecia,* extravasation
SYST: **Tumor lysis syndrome (TLS)**

PHARMACOKINETICS

Half-life (triphasic) <5 min, 50-155 min, 23-85 hr; metabolized in liver; excreted in bile, feces; crosses placental, blood-brain barrier

INTERACTIONS

Decrease: immune response—vaccines, toxoids
Decrease: digoxin level—digoxin
Decrease: vinCRIStine effect—CYP3A4 inducers (barbiturates, bosentan, carBAMazepine, efavirenz, phenytoins, nevirapine, rifabutin, rifampin)
Increase: Neurotoxicity: peripheral nervous system products
• Do not use with radiation
• **Acute pulmonary reactions: mitoMYcin c**
⚠ **Increase: toxicity—CYP3A4 inhibitors (aprepitant, antiretroviral protease inhibitors, clarithromycin, danazol, delavirdine, diltiazem, erythromycin, fluconazole, FLUoxetine, fluvoxaMINE, imatinib, ketoconazole, mibefradil, nefazodone, telithromycin, voriconazole)**

Drug/Herb
• Avoid use with St. John's wort

Drug/Lab Test
Increase: uric acid
Decrease: Hgb, WBC, platelets, sodium

NURSING CONSIDERATIONS

Assess:
• CBC, differential, platelet count before each dose; withhold product if WBC is <4000/mm^3 or platelet count is <75,000/mm^3; notify prescriber; RBC, Hct, Hgb; may be decreased
• **Bronchospasm:** more common with mitoMYcin
• Hepatic studies before, during therapy (bilirubin, AST, ALT, LDH) as needed or monthly
• Sensitivity of feet/hands, which precedes neuropathy
• **Tumor lysis syndrome: hyperkalemia, hyperphosphatemia, hyperuricemia, hypocalcemia; more common in leukemia, lymphoma; use alkalinization of urine with allopurinol, monitor electrolytes, renal function (BUN, urine, CCR, uric acid)**

Black Box Warning: Extravasation: pain, swelling, poor blood return; if extravasation occurs, local inj of hyaluronidase and moderate heat to area may help disperse product

Intrathecal administration
• **Bleeding:** hematuria, guaiac, bruising, petechiae, mucosa or orifices q8hr
• Effects of alopecia on body image; discuss feelings about body changes
• Buccal cavity q8hr for dryness, sores, ulcerations, white patches, oral pain, bleeding, dysphagia
• Symptoms indicating severe allergic reaction: rash, pruritus, urticaria, purpuric skin lesions, itching, flushing
• Brushing of teeth bid-tid with soft brush or cotton-tipped applicator for stomatitis; use unwaxed dental floss

Evaluate:
• Therapeutic response: decreased tumor size, spread of malignancy

Teach patient/family:
• To report change in gait or numbness in extremities; may indicate neuropathy
• To report any bleeding, white spots, or ulcerations in mouth to prescriber; to examine mouth daily
• To increase bulk, fluids, exercise to prevent constipation
• **Infection: to report sore throat, fever, flulike symptoms; avoid persons with known infection**
• To avoid vaccinations
• That hair may be lost; that hair will grow back but with different texture, color
• **Pregnancy (D): to notify prescriber if pregnancy is planned or suspected**

⚠ To use effective contraception during and for 2 mo after therapy, pregnancy (D), to avoid breastfeeding

⚠ **HIGH ALERT**

vinorelbine (Rx)

(vi-nor'el-bine)

Navelbine

Func. class.: Antineoplastic—miscellaneous

Chem. class.: Semisynthetic vinca alkaloid

ACTION:
Inhibits mitotic spindle activity, arrests cell cycle at metaphase; inhibits RNA synthesis, blocks cellular use of glutamic acid needed for purine synthesis; vesicant

USES:
Unresectable advanced non–small-cell lung cancer (NSCLC) stage IV; may be used alone or in combination with CISplatin for stage III or IV NSCLC

Unlabeled uses: Hodgkin's disease, breast/ovarian/head/neck cancer, desmoid tumor

CONTRAINDICATIONS:
Pregnancy (D), breastfeeding, infants, hypersensitivity, granulocyte count <1000 cells/mm^3 pretreatment

Black Box Warning: Severe neutropenia, intrathecal administration

Precautions: Children, geriatric patients, renal/hepatic/pulmonary/neurologic disease, anemia, bone marrow suppression

Black Box Warning: Extravasation

DOSAGE AND ROUTES

- **Adult:** IV 30 mg/m^2/wk
- **ANC 1000-1499:** Give 50% of dose; <1000, hold dose; <1000 × 3 wk, discontinue

Hepatic dose

- **Adult:** IV total bilirubin 2.1-3 mg/dl 15 mg/m^2/wk; total bilirubin ≥3 mg/dl 7.5 mg/m^2/day

Available forms: Inj 10 mg/ml

Administer:

Black Box Warning: Do not give intrathecally; fatal

- Antiemetic 30-60 min before product and prn to prevent vomiting

Intermittent IV INF route

- Dilute to 0.5-2 mg/ml with 0.9% NaCl, 0.45% NaCl, D_5W, D_5/0.45% NaCl, LR, Ringer's sol, give over 6-10 min into Y-site or central line, flush line

Black Box Warning: Hyaluronidase 150 units/ml in 1 ml NaCl, warm compress for extravasation for vesicant activity treatment

Continuous IV INF route

- 40 mg/m^2 q3wk after IV bol of 8 mg/m^2; may be given in combination with DOXOrubicin, fluorouracil, CISplatin

Y-site compatibilities: Amikacin, aztreonam, bleomycin, bumetanide, buprenorphine, butorphanol, calcium gluconate, CARBOplatin, carmustine, cefotaxime, cefTAZidime, ceftizoxime, chlorproMAZINE, cimetidine, CISplatin, clindamycin, cyclophosphamide, cytarabine, dacarbazine, DACTINomycin, DAUNOrubicin, dexamethasone, diphenhydrAMINE, DOXOrubicin, DOXOrubicin liposome, doxycycline, droperidol, enalaprilat, etoposide, famotidine, filgrastim, floxuridine, fluconazole, fludarabine, gallium, gentamicin, granisetron, haloperidol, heparin, hydrocortisone, HYDROmorphone, hydrOXYzine, IDArubicin, ifosfamide, imipenem-cilastatin, LORazepam, mannitol, mechlorethamine, melphalan, meperidine, mesna, methotrexate, metoclopramide, metroNIDAZOLE, minocycline, mitoXANtrone, morphine, nalbuphine, netilmicin, ondansetron, plicamycin, streptozocin, teniposide, ticarcillin, ticarcillin-clavulanate, tobramycin, vancomycin, vinBLAStine, vinCRIStine, zidovudine

SIDE EFFECTS

CNS: Paresthesias, peripheral neuropathy, depression, headache, **seizures,** weakness, jaw pain, asthenia
CV: Chest pain
GI: *Nausea, vomiting,* ileus, *anorexia, stomatitis,* constipation, abdominal pain, *diarrhea,* **hepatotoxicity, GI obstruction/perforation**
HEMA: **Neutropenia, anemia, thrombocytopenia, granulocytopenia**
INTEG: *Rash, alopecia,* photosensitivity, inj site reaction, necrosis
META: SIADH
MS: Myalgia
RESP: SOB, **dyspnea, pulmonary edema, acute bronchospasm, acute respiratory distress syndrome (ARDS)**

PHARMACOKINETICS

Half-life 27-43 hr; peak 1-2 hr; highly bound to platelets, lymphocytes; metabolized in liver; excreted in feces; small amount unchanged in kidneys

INTERACTIONS

Increase: bleeding risk—NSAIDs, anticoagulants

⚠ **Increase: toxicity—CYP3A4 inhibitors (aprepitant, antiretroviral protease inhibitors, clarithromycin, danazol, delavirdine, diltiazem, erythromycin, fluconazole, FLUoxetine, fluvoxaMINE, imatinib, ketoconazole, mibefradil, nefazodone, telithromycin, voriconazole)**

Decrease: vinorelbine effect—CYP3A4 inducers (barbiturates, bosentan, carBAMazepine, efavirenz, phenytoins, nevirapine, rifabutin, rifampin)

Drug/Herb
- Avoid use with St. John's wort

Drug/Lab Test
Increase: LFTs, bilirubin
Decrease: Hgb, WBC, platelets

NURSING CONSIDERATIONS

Assess:
- B/P (baseline, q15min) during administration

Black Box Warning: CBC, differential, platelet count before each dose; withhold product if WBC is <4000/mm^3 or platelet count is <75,000/mm^3; notify prescriber of results; recovery will take 3 wk

- **Bronchospasm: more common with mitoMYcin; also dyspnea, wheezing; may be treated with oxygen, bronchodilators, corticosteroids, especially if there is underlying pulmonary disease**
- Respiratory status: dyspnea, crackles, unproductive cough, chest pain, tachypnea
- Renal studies: BUN, serum uric acid, urine CCr before, during therapy; I&O ratio; report fall in urine output to <30 ml/hr; decreased hyperuricemia
- **Infection,** cold, fever, sore throat; notify prescriber if these occur; effects of alopecia on body image
- **Bleeding:** hematuria, guaiac, bruising, petechiae, mucosa or orifices, no rectal temperatures; avoid IM inj; apply pressure to venipuncture sites
- Hepatic function tests: AST, ALT, bilirubin, LDH

⚠ **Severe allergic reactions: rash, pruritus, urticaria, itching, flushing, bronchospasm, hypotension; EPINEPHrine and crash cart should be nearby**

- **Neurologic status: numbness, pain, tingling, loss of Achilles reflex, weakness, palsies**
- Brushing of teeth bid-tid with soft brush or cotton-tipped applicator for stomatitis; unwaxed dental floss

Evaluate:
- Therapeutic response: decreased tumor size, spread of malignancy

Teach patient/family:
- **To report change in gait or numbness in extremities, continuing constipation; may indicate neurotoxicity**
- To examine mouth daily for bleeding, white spots, ulcerations; to notify prescriber
- **Infection: report sore throat, fever, flu-like symptoms**
- To avoid crowds, people with infections, vaccinations, OTC products

Pregnancy: notify prescriber if pregnancy is planned or suspected; to use effective contraception during treatment and for ≥2 mo after product is discontinued, pregnancy (D); to avoid breastfeeding

• That hair may be lost; that hair will grow back but with different texture, color

vismodegib

(vis′moe-deg′ib)

Erivedge

Func. class.: Antineoplastic biologic response modifier

Chem. class.: Signal transduction inhibitor (STI)

ACTION: A hedgehog (Hh) signaling pathway inhibitor

USES: Patients who have metastatic basal cell carcinoma, locally advanced, that has recurred after surgery and who are not candidates for surgery/radiation

CONTRAINDICATIONS: Hypersensitivity, breastfeeding

Black Box Warning: Intrauterine fetal death, male-mediated teratogenicity, pregnancy (D)

Precautions: Children, blood donation

DOSAGE AND ROUTES

• **Adult: PO** 150 mg/day

Available forms: Cap 150 mg

Administer:

• Give without regard to food

• Swallow whole, do not open or crush caps

• If a dose is missed, do not take additional dose, take at usual time

• Store at 77°F (25°C)

SIDE EFFECTS

GI: Nausea, vomiting, dysgeusia
GU: Amenorrhea, azotemia
INTEG: Alopecia
META: Hyponatremia
MISC: Fatigue, decreased weight
MS: Arthralgia

PHARMACOKINETICS

Protein binding >99%, elimination half-life 4 days

INTERACTIONS

Increase: effect of each product—Pgp inhibitors (amiodarone, clarithromycin, cycloSPORINE, diltiazem, erythromycin, indinavir, itraconazole, ketoconazole, nelfinavir, niCARdipine, propafenone, quiNIDine, ritonavir, saquinavir, tacrolimus, tamoxifen, verapamil); CYP2C19 substrates (amitriptyline, clomiPRAMINE, imipramine, citalopram, diazepam, phenytoin, PHENobarbital, lansoprazole, omeprazole, pantoprazole, RABEprazole, esomeprazole, clopidogrel, proguanil, propranolol, carisoprodol, chloramphenicol, cyclophosphamide, indomethacin, nelfinavir, nilutamide, progesterone, teniposide, warfarin)

NURSING CONSIDERATIONS

Assess:

Black Box Warning: Pregnancy (D): verify pregnancy status of all women within 7 days before starting therapy; effective contraception is needed during and for 7 mo after treatment; men receiving this product should use condoms with spermicide (even after vasectomy) during sexual intercourse with female partners and for 2 mo after the last dose; report exposure during pregnancy to the Genentech Adverse Event Line

Evaluate:

• Therapeutic response: decreased spread of tumor

Teach patient/family:

Black Box Warning: Pregnancy (D): teach patient to notify their provider immediately if pregnancy is suspected (or in a female partner for male patients); effective contraception is needed during and for 7 mo after treatment; men receiving this product should use condoms with spermicide (even after vasectomy) during sexual intercourse with female partners and for 2 mo after the last dose; if product is used during pregnancy or if the patient becomes pregnant during use, the woman (or female partner for male patients) should be apprised of the potential hazard to the fetus; encourage exposed women (either directly or through seminal fluid) to participate in the ERIVEDGE pregnancy pharmacovigilance program

• About reason for treatment, expected results

vitamin A (Rx, PO-OTC, Rx-IM)

Aquasol A, Vitamin A

Func. class.: Vitamin, fat soluble

Chem. class.: Retinol

ACTION: Needed for normal bone, tooth development; visual dark adaptation; skin disease; mucosa tissue repair; assists with production of adrenal steroids, cholesterol, RNA

USES: Vit A deficiency

CONTRAINDICATIONS: Pregnancy (X) (IM), hypersensitivity to vit A, malabsorption syndrome, hypervitaminosis A, IV administration

Precautions: Pregnancy (C) (PO), breastfeeding, impaired renal function, children, hepatic disease, infants, alcoholism, hepatitis

DOSAGE AND ROUTES

• **Adult and child >8 yr: PO** 100,000-500,000 international units/day × 3 days then 50,000/day × 2 wk; dose based on severity of deficiency; maintenance 10,000-20,000 international units for 2 mo

• **Child 1-8 yr: IM** 5000-15,000 international units/day × 10 days

• **Infant <1 yr: IM** 5000-15,000 international units × 10 days

Maintenance

• **Child 4-8 yr: IM** 15,000 international units/day × 2 mo

• **Child <4 yr: IM** 10,000 international units/day × 2 mo

Available forms: Caps 10,000, 25,000, 50,000 international units; drops 5000 international units; inj 50,000 international units/ml; tabs 10,000, 25,000, 50,000 international units

Administer:

PO route

• With food (PO) for better absorption

• Do not administer IV because of risk of anaphylactic shock; IM only

• Oral preparations not indicated for vit A deficiency in those with malabsorption syndrome

• Store in tight, light-resistant container

IM route

• Give deep in large muscle mass; do not use deltoid muscle for administration of >1 ml

SIDE EFFECTS

CNS: Headache, **increased intracranial pressure, intracranial hypertension**, lethargy, malaise

EENT: Gingivitis, papilledema, exophthalmos, inflammation of tongue and lips

GI: Nausea, vomiting, anorexia, abdominal pain, **jaundice**

INTEG: Drying of skin, pruritus, increased pigmentation, night sweats, alopecia

META: Hypomenorrhea, hypercalcemia

MS: Arthralgia, retarded growth, hard areas on bone

V

PHARMACOKINETICS

Stored in liver, kidneys, fat; excreted (metabolites) in urine, feces

INTERACTIONS

Increase: levels of vit A—corticosteroids, oral contraceptives
Decrease: absorption of vit A—mineral oil, cholestyramine, colestipol
Drug/Lab Test
False increase: bilirubin, serum cholesterol

NURSING CONSIDERATIONS

Assess:

- Nutritional status: yellow and dark green vegetables, yellow/orange fruits, vit-A–fortified foods, liver, egg yolks
- **Vit A deficiency:** decreased growth; night blindness; dry, brittle nails; hair loss; urinary stones; increased infection; hyperkeratosis of skin; drying of cornea

Evaluate:

- Therapeutic response: increased growth rate, weight; absence of dry skin and mucous membranes, night blindness

Teach patient/family:

- That, if dose is missed, it should be omitted
- That ophthalmic exams may be required periodically throughout therapy
- Not to use mineral oil while taking this product
- To notify prescriber of nausea, vomiting, lip cracking, loss of hair, headache
- Not to take more than prescribed amount

TREATMENT OF OVERDOSE:

Discontinue product

vitamin E (OTC)

Aquasol E
Func. class.: Vit E
Chem. class.: Fat soluble

ACTION: Needed for digestion and metabolism of polyunsaturated fats; decreases platelet aggregation, blood clot formation; promotes normal growth and development of muscle tissue, prostaglandin synthesis

USES: Vit E deficiency, impaired fat absorption, hemolytic anemia in premature neonates, prevention of retrolental fibroplasia, sickle cell anemia, supplement for malabsorption syndrome

CONTRAINDICATIONS: IV use in infants
Precautions: Pregnancy (A), anemia, breastfeeding, hypoprothrombinemia

DOSAGE AND ROUTES

Deficiency

- **Adult: PO** 60-75 international units/day
- **Child: PO** 1 international units/kg (malabsorption)

Prevention of deficiency

- **Adult: PO** 30 international units/day; **TOP** apply to affected areas
- **Infant: PO** 5 international units/day

Available forms: Caps 100, 200, 400, 500, 600, 1000 international units; tabs 100, 200, 400 international units; drops 50 mg/ml; chew tabs 400 units; ointment; cream; lotion; oil

Administer:

PO route

- Administer with or after meals
- Chew chewable tabs well
- Sol may be dropped in mouth or mixed with food
- Store in tight, light-resistant container

Topical route

- To moisturize dry skin

SIDE EFFECTS

CNS: Headache, fatigue
CV: Increased risk for thrombophlebitis
EENT: Blurred vision
GI: Nausea, cramps, diarrhea
GU: Gonadal dysfunction
INTEG: Sterile abscess, contact dermatitis

META: Altered metabolism of hormones (thyroid, pituitary, adrenal), altered immunity
MS: Weakness

PHARMACOKINETICS

PO: Metabolized in liver, excreted in bile

INTERACTIONS

Increase: action of oral anticoagulants
Decrease: absorption—cholestyramine, colestipol, mineral oil, sucralfate

NURSING CONSIDERATIONS

Assess:
• Nutritional status: wheat germ; dark green, leafy vegetables; nuts; eggs; liver; vegetable oils; dairy products; cereals
Evaluate:
• Therapeutic response: absence of hemolytic anemia, adequate vit E levels, improvement in skin lesions, decreased edema
Teach patient/family:
• About the necessary foods for diet
• To omit dose if missed
• To avoid vitamin supplements unless directed by prescriber

voriconazole (Rx)

(vohr-i-kahn′a-zol)

Vfend

Func. class.: Antifungal, systemic
Chem. class.: Triazole derivative

Do not confuse:
Vfend/Venofer

ACTION: Inhibits fungal CYP 450-mediation demethylation; needed for biosynthesis; causes leakage from cell membrane

USES: Invasive aspergillosis, serious fungal infections (*Candida* sp., *Scedosporium apiospermum, Fusarium* sp., *Monosporium, Apiospermum*)
Unlabeled uses: *Acremonium sp., Blastomyces dermatitidis, Coccidioides immitis, Cryptococcus neoformans,* febrile neutropenia, fungal keratitis, *Histoplasma capsulatum,* oropharyngeal candidiasis, *Rhodotorula sp., Scedosporium sp.,* cutaneous aspergillosis, candidemia (premature neonates), fungal infections in children ≥12 yr

CONTRAINDICATIONS: Pregnancy (D), breastfeeding, children, hypersensitivity, severe bone marrow depression, severe hepatic disease
Precautions: Renal disease (IV); patients of Asian/African descent; cardiomyopathy, cholestasis, chemotherapy, lactase deficiency, visual disturbances, renal failure, pancreatitis, QT prolongation, hypokalemia; ventricular dysrhythmias, torsades de pointes

DOSAGE AND ROUTES

Esophageal candidiasis
• **Adult/geriatric/child ≥12 yr and ≥40 kg: PO** 200 mg q12hr; **<40 kg**, 100 mg q12hr
• **Adult/geriatric/child ≥12 yr: IV INF** Loading dose 6 mg/kg q12hr × 2 dose then 3-4 mg/kg q12hr; may switch to oral dosing
Candidemia of the skin, kidney, bladder wall, abdomen (nonneutropenic patients)
• **Adult/child ≥12 yr: IV** loading dose 6 mg/kg q12hr × 24 hr then 3-4 mg/kg q12hr × ≥14 days and ≥7 days after resolution of symptoms; **PO** after loading dose **>40 kg** 200 mg q12hr × ≥14 days and ≥7 days after resolution of symptoms; **<40 kg** 100 mg q12hr × ≥14 days and ≥7 days after resolution of symptoms
Invasive aspergillosis
• **Adults/adolescents: IV** 6 mg/kg q12hr (loading dose) then 4 mg/kg q12hr, may reduce to 3 mg/kg q12hr if intolerable
• **Child ≥12 yr: IV** 6 mg/kg q12hr, then 4 mg/kg q12hr

V

CNS blastomycosis/blastomycosis meningitis (unlabeled)

• **Adult: PO** 200-400 mg bid × at least 12 mo and until resolution of CSF abnormalities

Renal dose

• **Adult: PO** CCr <50 ml/min, use orally only

Hepatic dose

• **Adult: PO** 6 mg/kg q12hr × 2 doses then 2 mg/kg q12hr or 100 mg q12hr if >40 kg; 50 mg q12hr if <40 kg

Available forms: Tabs 50, 200 mg; powder for inj, lyophilized 200 mg voriconazole, powder for oral susp 45 g (40 mg/ml after reconstitution)

Administer:

PO route

• Oral susp: tap bottle; add 46 ml of water to bottle; shake well; remove cap; push bottle adaptor into neck of bottle; replace cap; write expiration date (14 days); shake well before each use; administer using only oral dispenser supplied, 1 hr before or after meals; tabs and susp may be interchanged

• Store at room temperature (powder, tabs)

Intermittent IV INF route

• Product only after C&S confirms organism, product needed to treat condition; make sure product used in life-threatening infections

• Reconstitute powder with 19 ml water for inj to 10 mg/ml; shake until dissolved; infuse over 1-2 hr at conc of ≤5 mg/ml; do not admix with other products, 4.2% sodium bicarbonate inf

Y-site compatibilities: Acyclovir, alfentanil, allopurinol, amifostine, amikacin, aminocaproic acid, aminophylline, amiodarone, amphotericin B liposome, ampicillin, ampicillin/sulbactam, anidulafungin, azithromycin, aztreonam, bivalirudin, bleomycin, bumetanide, buprenorphine, butorphanol, calcium acetate/chloride/gluconate, CARBOplatin, carmustine, caspofungin, ceFAZolin, cefotaxime, cefoTEtan, cefOXitin, cefTAZidime, ceftizoxime, cefTRIAXone, chloramphenicol, chlorproMAZINE, cimetidine, ciprofloxacin, cisatracurium, CISplatin, clindamycin, cyclophosphamide, cytarabine, dacarbazine, DACTINomycin, DAPTOmycin, DAUNOrubicin, dexamethasone, dexmedetomidine, dexrazoxane, digoxin, diltiazem, diphenhydrAMINE, DOBUTamine, DOCEtaxel, dolasetron, DOPamine, doripenem, doxacurium, doxycycline, droperidol, enalaprilat, ePHEDrine, EPINEPHrine, epirubicin, ertapenem, erythromycin, esmolol, etoposide, etoposide phosphate, famotidine, fenoldopam, fentaNYL, fluconazole, fludarabine, fluorouracil, foscarnet, fosphenytoin, furosemide, ganciclovir, gemcitabine, gentamicin, glycopyrrolate, granisetron, haloperidol, heparin, hydrALAZINE, hydrocortisone, ifosfamide, imipenem/cilastatin, inamrinone, insulin, irinotecan, isoproterenol, ketorolac, labetalol, leucovorin, levofloxacin, lidocaine, linezolid, LORazepam, magnesium sulfate, mannitol, mechlorethamine, melphalan, meperidine, meropenem, mesna, metaraminol, methohexital, methotrexate, methyldopate, methylPREDNISolone, metoclopramide, metoprolol, metroNIDAZOLE, midazolam, milrinone, mitoMYcin, morphine, nafcillin, nalbuphine, naloxone, niCARdipine, nitroglycerin, norepinephrine, octreotide, ondansetron, oxaliplatin, oxytocin, PACLitaxel, pamidronate, pancuronium, pentamidine, pentazocine, PENTobarbital, PHENobarbital, phentolamine, phenylephrine, piperacillin/tazobactam, potassium chloride/phosphates, procainamide, promethazine, propranolol, quinupristin/dalfopristin, remifentanil, rocuronium, sodium acetate/bicarbonate/phosphates, streptozocin, succinylcholine, SUFentanil, tacrolimus, teniposide, theophylline, thiotepa, ticarcillin/clavulanate, tirofiban, tobramycin, topotecan, trimethobenzamide, trimethoprim/sulfamethoxazole, vancomycin, vasopressin, vecuronium, verapamil, vinBLAStine, vinCRIStine, vinorelbine, zidovudine

SIDE EFFECTS

CNS: *Headache,* paresthesias, peripheral neuropathy, ***hallucinations,*** psychosis, EPS, depression, Guillain-Barré syndrome, insomnia, suicidal ideation, dizziness, fever

CV: **Tachycardia,** hypo/hypertension, vasodilation, **atrial arrhythmias, atrial fibrillation, AV block, bradycardia, CHF, MI, QT prolongation, torsades de pointes,** peripheral edema

EENT: Blurred vision, eye hemorrhage, **visual disturbances**

GI: *Nausea, vomiting, anorexia, diarrhea,* cramps, **hemorrhagic gastroenteritis, acute hepatic failure, hepatitis, intestinal perforation, pancreatitis**

GU: *Hypokalemia,* azotemia, **renal tubular necrosis, permanent renal impairment, anuria, oliguria**

HEMA: Anemia, **eosinophilia,** hypomagnesemia, **thrombocytopenia, leukopenia, pancytopenia**

INTEG: *Burning, irritation,* pain, necrosis at inj site with extravasation, dermatitis, *rash,* photosensitivity

MISC: Respiratory disorder

SYST: **Stevens-Johnson syndrome, toxic epidermal necrolysis, sepsis;** melanoma (photosensitivity reactions)

PHARMACOKINETICS

By CYP3A4, CYP2C9 enzymes; max serum conc 1-2 hr after dosing; eliminated via hepatic metabolism; protein binding 58%; elimination half-life 6 hr (dose dependent)

INTERACTIONS

Increase: effects of benzodiazepines, calcium channel blockers, cycloSPORINE, ergots, HMG-CoA reductase inhibitors, pimozide, quiNIDine, prednisoLONE, sirolimus, sulfonylureas, tacrolimus, vinca alkaloids, warfarin, rifabutin, proton pump inhibitors, NNRTIs, protease inhibitors, phenytoin

Increase: nephrotoxicity—other nephrotoxic antibiotics (aminoglycosides, CISplatin, vancomycin, cycloSPORINE, polymyxin B)

Increase: hypokalemia—corticosteroids, digoxin, skeletal muscle relaxants, thiazides

⚠ **Increase: QT prolongation—class IA/III antidysrhythmics, some phenothiazines, β agonists, local anesthetics, tricyclics, haloperidol, chloroquine, droperidol, pentamidine; CYP3A4 inhibitors (amiodarone, clarithromycin, erythromycin, telithromycin, troleandomycin), arsenic trioxide, levomethadyl; CYP3A4 substrates (methadone, pimozide, QUEtiapine, quiNIDine, risperiDONE, ziprasidone)**

Drug/Herb

- Do not use with St. John's wort

Drug/Food

- Avoid use with high-fat meals, take 1 hr before or after meal

Drug/Lab Test

Increase: AST/ALT, alk phos, creatinine, bilirubin

Decrease: Hgb/Hct, platelets, WBC

NURSING CONSIDERATIONS

Assess:

- VS q15-30min during first inf; note changes in pulse, B/P
- I&O ratio; watch for decreasing urinary output, change in specific gravity; discontinue product to prevent permanent damage to renal tubules
- Blood studies: CBC, potassium, sodium, calcium, magnesium, q2wk; BUN, creatinine weekly
- Weight weekly; if weight increases >2 lb/wk, edema is present; renal damage should be considered

⚠ **Renal toxicity: increasing BUN, serum creatinine; if BUN is >40 mg/dl or if serum creatinine >3 mg/dl, product may be discontinued or dosage reduced**

⚠ **Hepatotoxicity: increasing AST, ALT, alk phos, bilirubin, baseline and periodically**

- **Allergic reaction:** dermatitis, rash; product should be discontinued, antihistamines (mild reaction) or epinephrine (severe reaction) administered

V

• **Hypokalemia:** anorexia, drowsiness, weakness, decreased reflexes, dizziness, increased urinary output, increased thirst, paresthesias
• **Ototoxicity:** tinnitus (ringing, roaring in ears), vertigo, loss of hearing (rare); visual disturbance
• **QT prolongation:** ECG, ejection fraction; assess for chest pain, palpitations, dyspnea

Evaluate:

• Therapeutic response: decreased fever, malaise, rash, negative C&S for infecting organism

Teach patient/family:

• That long-term therapy may be needed to clear infection (2 wk-3 mo, depending on type of infection)
• To notify prescriber of bleeding, bruising, soft-tissue swelling, dark urine, persistent nausea or diarrhea, headache, rash, yellow skin/eyes
• Take 1 hr before or after meal (PO)
• Do not drive at night because of vision changes
• Avoid strong, direct sunlight
• Women of childbearing age should use effective contraceptive, pregnancy (D)

⚠ HIGH ALERT

warfarin (Rx)
(war'far-in)
Coumadin, Jantoven
Func. class.: Anticoagulant
Chem. class.: Coumarin derivative

Do not confuse:
Coumadin/Cardura/Compazine

ACTION:
Interferes with blood clotting by indirect means; depresses hepatic synthesis of vit-K–dependent coagulation factors (II, VII, IX, X)

USES:
Antiphospholipid antibody syndrome, arterial thromboembolism prophylaxis, DVT, MI prophylaxis, after MI, stroke prophylaxis, thrombosis prophylaxis

Unlabeled uses: Angina, unstable angina

CONTRAINDICATIONS:
Pregnancy (X), breastfeeding, hypersensitivity, hemophilia, leukemia with bleeding, peptic ulcer disease, thrombocytopenic purpura, hepatic disease (severe), malignant hypertension, subacute bacterial endocarditis, acute nephritis, blood dyscrasias, eclampsia, preeclampsia, hemorrhagic tendencies; surgery of CNS, eye; traumatric surgery with large open surface, bleeding tendencies of GI/GU/respiratory tract, stroke, aneurysms, pericardial effusion, spinal puncture, major regional/lumbar block anesthesia

Black Box Warning: Bleeding

Precautions: Geriatric patients, alcoholism, CHF, debilitated patients, trauma, indwelling catheters, severe hypertension, active infections, protein C deficiency, polycythemia vera, vasculitis, severe diabetes, Asian patients (CYP2C9), protein C, S deficiency

DOSAGE AND ROUTES
- **Adult: PO/IV** 2.5-10 mg/day × 2-4 days then titrated to INR/PT
- **Adolescent/child/infant: PO/IV** 0.2 mg/kg/day × 2 days titrated to INR

Available forms: Tabs 1, 2, 2.5, 3, 4, 5, 6, 7.5, 10 mg; inj 5 mg powder for inj

Administer:

PO route
- At same time each day to maintain steady blood levels without regard to food; food decreases rate but not extent of absorption; do not change brands
- Tabs whole or crushed
- Avoiding all IM inj that may cause bleeding
- Store in tight container

Direct IV route
- Reconstitute with 2.7 ml sterile water for inj (2 mg/ml); do not use sol that is discolored or that has particulates
- Give over 1-2 min into peripheral vein

Y-site compatibilities: Amikacin, ascorbic acid, bivalirudin, ceFAZolin, cefTRIAXone, DOPamine, EPINEPHrine, heparin, lidocaine, metaraminol, morphine, nitroglycerin, oxytocin, potassium chloride, ranitidine

SIDE EFFECTS
CNS: *Fever,* dizziness, fatigue, headache, lethargy
CV: Angina, chest pain, edema, hypotension, syncope
GI: *Diarrhea,* nausea, vomiting, anorexia, stomatitis, cramps, **hepatitis**, cholestatic jaundice
GU: **Hematuria**
HEMA: **Hemorrhage, agranulocytosis, leukopenia, eosinophilia**, anemia, ecchymosis, petechiae
INTEG: *Rash,* dermatitis, urticaria, alopecia, pruritus
MISC: Epistaxis, hemoptysis, mouth ulcers, taste disturbances, priapism, dyspnea
MS: Bone fractures
SYST: **Anaphylaxis**, coma, cholesterol, microembolism, **exfoliative dermatitis, purple toe syndrome**

W

PHARMACOKINETICS

PO: Onset 12-24 hr, peak $1^1/_2$-4 days, duration 3-5 days, effective half-life 20-60 hr; metabolized in liver; excreted in urine, feces (active/inactive metabolites); crosses placenta, 99% bound to plasma proteins

INTERACTIONS

Increase: warfarin action—allopurinol, amiodarone, chloral hydrate, chloramphenicol, cimetidine, clofibrate, cotrimoxazole, COX-2 selective inhibitors, dextrothyroxine, diflunisal, disulfiram, erythromycin, ethacrynic acids, furosemide, glucagon, heparin, HMG-CoA reductase inhibitors, indomethacin, isoniazid, mefenamic acid, metroNIDAZOLE, mifepristone, NSAIDs, oxyphenbutazones, penicillins, phenylbutazone, quiNIDine, quinolone antiinfectives, RU-486, salicylates, sulfinpyrazone, sulfonamides, sulindac, SSRIs, steroids, thrombolytics, thyroid, tricyclics

Increase: toxicity—oral sulfonylureas, phenytoin

Decrease: warfarin action—aprepitant, azaTHIOprine, barbiturates, bile acid sequestrants, bosentan, carBAMazepine, dicloxacillin, estrogens, ethchlorvynol, factor IX/VIIa, griseofulvin, nafcillin, oral contraceptives, phenytoin, rifampin, sucralfate, sulfaSALAzine, thyroid, vit K, vit K foods

Drug/Herb

Increase: risk for bleeding—angelica, anise, basil, chamomile, chondroitin, dong quai, evening primrose, feverfew, garlic, ginger, ginkgo, ginseng, horse chestnut, kava, licorice, melatonin, red yeast rice, saw palmetto

Decrease: anticoagulant effect—coenzyme Q10, St. John's wort

Drug/Lab Test

Increase: T_3 uptake, LFTs

Decrease: uric acid

NURSING CONSIDERATIONS

Assess:

Black Box Warning: Blood studies (Hct, PT, platelets, occult blood in stools) q3mo; INR: in hospital daily after 2nd or 3rd dose; when in therapeutic range for 2 consecutive days, monitor 2-3× wk for 1-2 wk then less frequently, depending on stability of INR results; *Outpatient:* monitor every few days until stable dose then periodically thereafter, depending on stability of INR results, usually at least monthly

Black Box Warning: Bleeding gums, petechiae, ecchymosis, black tarry stools, hematuria; fatal hemorrhage can occur

⚠ Fever, skin rash, urticaria

Evaluate:

- Therapeutic response: decrease in deep venous thrombosis

Teach patient/family:

- To avoid OTC preparations that may cause serious product interactions unless directed by prescriber
- To carry emergency ID identifying product taken
- About the importance of compliance
- To report any signs of bleeding: gums, under skin, urine, stools; to use soft-bristle toothbrush to avoid bleeding gums; to use electric razor
- To avoid hazardous activities (e.g., football, hockey, skiing), dangerous work
- About the importance of avoiding unusual changes in vitamin intake, diet, or lifestyle
- To inform all health care providers of anticoagulant intake

TREATMENT OF OVERDOSE:

Administer vit K

xylometazoline nasal agent

See Appendix B

zafirlukast (Rx)

(za-feer'loo-cast)

Accolate

Func. class.: Bronchodilator

Chem. class.: Leukotriene receptor antagonist

ACTION: Antagonizes the contractile action of leukotrienes (LTD_4, LTE_4) in airway smooth muscle; inhibits bronchoconstriction caused by antigens

USES: Prophylaxis and chronic treatment of asthma in adults/children >5 yr

Unlabeled uses: Allergic rhinitis

CONTRAINDICATIONS: Hypersensitivity, hepatic encephalopathy

Precautions: Pregnancy (B), breastfeeding, children, geriatric patients, hepatic disease, Churg-Strauss syndrome, acute bronchospasm

DOSAGE AND ROUTES

- **Adult and child ≥12 yr: PO** 20 mg bid
- **Child 5-11 yr: PO** 10 mg bid

Available forms: Tabs 10, 20 mg

Administer:

- 1 hr before or 2 hr after meals; absorption may be decreased if given with food

SIDE EFFECTS

CNS: Headache, dizziness, suicidal ideation, insomnia, fever

GI: Nausea, diarrhea, abdominal pain, vomiting, dyspepsia, hepatic failure, hepatitis

HEMA: Agranulocytosis

OTHER: Infections, pain, asthenia, myalgia, fever, increased ALT, urticaria, rash, angioedema

PHARMACOKINETICS

Rapidly absorbed, peak 3 hr, 99% protein binding (albumin), extensively metabolized, inhibits CYP2C9 and 3A4 enzyme systems, excreted in feces, clearance reduced in geriatric patients, hepatic impairment, half-life 10 hr

INTERACTIONS

Increase: plasma levels of zafirlukast—aspirin

Increase: PT—warfarin

Decrease: plasma levels of zafirlukast—erythromycin, theophylline

Drug/Food

Decrease: bioavailability

NURSING CONSIDERATIONS

Assess:

⚠ **Adult patients carefully for symptoms of Churg-Strauss syndrome** (rare), including eosinophilia, vasculitic rash, worsening pulmonary symptoms, cardiac complications, neuropathy; may be caused by reducing oral corticosteroids

- Respiratory rate, rhythm, depth; auscultate lung fields bilaterally; notify prescriber of abnormalities; not to be used for acute bronchospasm in acute asthma
- **Hepatic/renal/pancreatic/visual function:** monitor liver function tests

Evaluate:

- Therapeutic response: ability to breathe more easily

Teach patient/family:

- Avoid driving at night (blurred vision); direct sunlight (photosensitivity), taking 1 hr before or after a meal
- To check OTC medications, current prescription medications that may increase stimulation
- To avoid hazardous activities because dizziness may occur
- That, if GI upset occurs, to take product with 8 oz water; to avoid taking with food if possible because absorption may be decreased
- To notify prescriber of nausea, vomiting, diarrhea, abdominal pain, fatigue, jaundice, anorexia, flulike symptoms (hepatic dysfunction)
- Not to use for acute asthma episodes
- Not to take if breastfeeding
- To take even if symptom free

Z

zaleplon (Rx)

(zal'eh-plon)

Sonata

Func. class.: Hypnotic, nonbarbiturate
Chem. class.: Pyrazolopyrimidine

Controlled Substance Schedule IV

ACTION:
Binds selectively to omega-1 receptor of the $GABA_A$ receptor complex; results are sedation, hypnosis, skeletal muscle relaxation, anticonvulsant activity, anxiolytic action

USES:
Insomnia (short-term treatment)

CONTRAINDICATIONS:
Hypersensitivity, severe hepatic disease

Precautions: Pregnancy (C), breastfeeding, children <15 yr, geriatric patients, respiratory/renal/hepatic disease, psychosis, angioedema, depression, sleep-related behaviors (sleep walking)

DOSAGE AND ROUTES

- **Adult: PO** 10 mg at bedtime; may increase dose to 20 mg at bedtime if needed; 5 mg may be used in low-weight persons
- **Geriatric: PO** 5 mg at bedtime; may increase if needed

Available forms: Caps 5, 10 mg

Administer:

- Immediately before bedtime for sleeplessness
- On empty stomach for fast onset
- Store in tight container in cool environment

SIDE EFFECTS

CNS: *Lethargy, drowsiness, daytime sedation,* dizziness, confusion, anxiety, amnesia, depersonalization, hallucinations, hyperesthesia, paresthesia, somnolence, tremors, vertigo, complex sleep related reactions: sleep driving, sleep eating

CV: Chest pain, peripheral edema

EENT: Vision change, ear/eye pain, hyperacusis, parosmia

GI: Nausea, abdominal pain, constipation, anorexia, colitis, dyspepsia, dry mouth

MISC: Asthenia, fever, headache, myalgia, dysmenorrhea

MS: Myalgia, back pain, arthritis

RESP: Bronchitis

SYST: Severe allergic reactions

PHARMACOKINETICS

Rapid onset, metabolized by liver extensively, excreted by kidneys (inactive metabolites), half-life 1 hr, onset, peak 1 hr, duration 3-4 hr

INTERACTIONS

Increase: effect of zaleplon—cimetidine

Decrease: zaleplon bioavailability—CYP3A4 inducers

Drug/Food

- Prolonged absorption, sleep onset reduced: high-fat/heavy meal

NURSING CONSIDERATIONS

Assess:

- Mental status: mood, sensorium, affect, memory (long, short term), excessive sedation, impaired coordination
- **Sleep disorder:** type of sleep problem: falling asleep, staying asleep; monitor for complex sleep disorders

Evaluate:

- Therapeutic response: ability to sleep at night, decreased amount of early morning awakening

Teach patient/family:

- To avoid driving or other activities requiring alertness until product is stabilized
- To avoid alcohol ingestion
- That product may cause memory problems, dependence (if used for longer periods of time), changes in behavior/thinking, complex sleep-related behaviors (sleep eating/driving)
- That product is for short-term use only

• To take immediately before going to bed
• Not to ingest a high-fat/heavy meal before taking

zanamivir (Rx)

(zan'ah-mih-veer)

Relenza

Func. class.: Antiviral
Chem. class.: Neuramidase inhibitor

ACTION: Inhibits neuramidase enzyme needed for influenza virus replication

USES: Treatment of influenza types A and B for patients who have been symptomatic for ≤2 days
Unlabeled uses: Prophylaxis against influenza; swine flu (H1N1)

CONTRAINDICATIONS: Hypersensitivity
Precautions: Pregnancy (C), breastfeeding, children <7 yr, geriatric patients, respiratory disease, angioedema, milk protein hypersensitivity, Reye's syndrome

DOSAGE AND ROUTES

• **Adult/child >7 yr: INH** 2 inhalations (two 5-mg blisters) q12hr × 5 days; on the 1st day, 2 doses should be taken with at least 2 hr between doses
H1N1 (swine flu) (unlabeled)
• **Adult/adolescent/child ≥7 yr: INH** 2 bid × 5 days
Available forms: Blisters of powder for inhalation: 5 mg
Administer:
• Within 2 days of symptoms of influenza; continue for 5 days
• Give patient Patient's Instructions for Use, review all points before using delivery system
• Do not use as nebulized sol or in mechanical ventilation
• Store in tight, dry container

SIDE EFFECTS

CNS: *Headache, dizziness,* **seizures**, fatigue; self-injury, delirium (child)
EENT: Ear, nose, throat infections
GI: *Nausea, vomiting,* diarrhea
RESP: Nasal symptoms, cough, sinusitis, bronchitis, **bronchospasm**
SYST: **Angioedema**

PHARMACOKINETICS

Half-life 2½-5 hr, not metabolized, excreted in urine unchanged

INTERACTIONS

• May decrease intranasal influenzae vaccine; separate by ≥48 hr; do not restart antiviral products for ≥2 wk

NURSING CONSIDERATIONS

Assess:
• Skin eruptions, photosensitivity after administration of product
• Respiratory status: rate, character, wheezing, tightness in chest
• Allergies before initiation of treatment, reaction of each medication
• Signs of infection
Evaluate:
• Therapeutic response: absence of fever, malaise, cough, dyspnea with infection
Teach patient/family:
• That product does not reduce transmission risk of influenza to others
• That patients with asthma or COPD should carry a fast-acting inhaled bronchodilator because bronchospasm may occur; to use scheduled inhaled bronchodilators before using product
• To avoid hazardous activities if dizziness occurs

Z

zidovudine (Rx)

(zye-doe′-vue-deen)

Novo-AZT ✱, Retrovir

Func. class.: Antiretroviral

Chem. class.: Nucleoside reverse transcriptase inhibitor (NRTI)

Do not confuse:
Retrovir/ritonavir

ACTION: Inhibits replication of HIV-1 virus by incorporating into cellular DNA by viral reverse transcriptase, thereby terminating the cellular DNA chain

USES: Used in combination with at least 2 other antiretrovirals for HIV-1 infection

Unlabeled uses: Epstein-Barr virus, hepatitis B, human T-lymphotropic virus type I (HTLV-I), thrombocytopenia

CONTRAINDICATIONS: Hypersensitivity

Precautions: Pregnancy (C), breastfeeding, children, granulocyte count <1000/mm^3 or Hgb <9.5 g/dl, severe renal disease, obesity

Black Box Warning: Impaired hepatic function, anemia, lactic acidosis, myopathy, neutropenia

DOSAGE AND ROUTES

HIV infections with other antiretrovirals

- **Adult: PO (tabs, caps, syrup)** 600 mg/day in divided doses, either 200 mg tid or 300 mg bid in combination with other antiretrovirals; **IV** 1 mg/kg q4hr, initiate **PO** as soon as possible up to 1000 mg
- **Child 25 kg to <30 kg: PO (tabs/caps)** 500 mg/day divided bid
- **Child 19 kg to <25 kg: PO (tabs/caps)** 400 mg/day divided bid
- **Infant/child ≥4 wk and 13 to <19 kg: PO (tabs/caps)** 300 mg/day divided bid
- **Infant/child ≥4 wk and 7 to <13 kg: PO (tabs/caps)** 200 mg/day divided bid
- **Infant/child ≥4 wk and 5 to <7 kg: PO (tabs/caps)** 150 mg/day divided bid
- **Adolescent/child ≥30 kg: PO (tabs/caps/syrup)** 300 mg bid or 200 mg tid
- **Infant ≥4 wk, child/adolescent ≥9 to <30 kg: PO (syrup)** 18 mg/kg/day divided bid or tid
- **Infant ≥4 wk and 4 to <9 kg: PO (syrup)** 24 mg/kg/day divided bid or tid
- **Neonate (unlabeled): PO (syrup)** 2 mg/kg q6hr
- **Premature neonate (unlabeled): PO (syrup)** 2 mg/kg q12hr, increase to 2 mg/kg q8hr at 2 wk for neonates ≥30 wk gestation or at 4 wk for neonates <30 wk gestation

Prevention of maternal–fetal HIV transmission

- **Neonates ≥34 wk: PO** 2 mg/kg/dose q6hr × 6 wk beginning 8-12 hr after birth; **IV** 1.5 mg/kg/dose over 30 min q6hr until able to take **PO**
- **Maternal (>14 wk gestation): PO** 100 mg 5×/day until start of labor then during labor/delivery **IV** 2 mg/kg over 1 hr followed by **IV INF** 1 mg/kg/hr until umbilical cord clamped

Prevention of HIV after needlestick

- **Adult: PO** 200 mg tid plus lamiVUDine 150 mg bid plus a protease inhibitor for high-risk exposure; begin within 2 hr of exposure

Available forms: Caps 100; tabs 300 mg; inj 10 mg/ml; oral syr 50 mg/5 ml

Administer:

- By mouth; capsules should be swallowed whole
- Trimethoprim-sulfamethoxazole, pyrimethamine, or acyclovir as ordered to prevent opportunistic infections; if these products are given, watch for neurotoxicity
- Store in cool environment; protect from light

Intermittent IV INF route

- After diluting each 1 mg/0.25 ml or more D_5W to ≤4 mg/ml; give over 1 hr
- Protect unopened product from light; use diluted sol within 24 hr room temperature, 48 hr refrigerated

Y-site compatibilities: Acyclovir, allopurinol, amifostine, amikacin, amphotericin B, amphotericin B cholesteryl, aztreonam, cefepime, cefTAZidime, cefTRIAXone, cimetidine, cisatracurium, clindamycin, dexamethasone, DOBUTamine, DOPamine, DOXOrubicin liposome, erythromycin, filgrastim, fluconazole, fludarabine, gentamicin, granisetron, heparin, imipenem-cilastatin, LORazepam, melphalan, metoclopramide, morphine, nafcillin, ondansetron, oxacillin, PACLitaxel, pentamidine, phenylephrine, piperacillin, piperacillin-tazobactam, potassium chloride, ranitidine, remifentanil, sargramostim, teniposide, thiotepa, tobramycin, trimethoprim-sulfamethoxazole, trimetrexate, vancomycin, vinorelbine

SIDE EFFECTS

CNS: *Fever, headache, malaise,* diaphoresis, *dizziness, insomnia,* paresthesia, somnolence, chills, tremors, twitching, anxiety, confusion, depression, lability, vertigo, loss of mental acuity, **seizures,** malaise

EENT: Taste change, hearing loss, photophobia

GI: *Nausea, vomiting, diarrhea, anorexia,* cramps, *dyspepsia, constipation,* dysphagia, *flatulence,* rectal bleeding, mouth ulcer, abdominal pain, **hepatomegaly**

GU: Dysuria, polyuria, urinary frequency, hesitancy

HEMA: **Granulocytopenia, anemia**

INTEG: *Rash,* acne, pruritus, urticaria

MS: Myalgia, arthralgia, muscle spasm

RESP: Dyspnea

SYST: **Lactic acidosis**

PHARMACOKINETICS

PO: Rapidly absorbed from GI tract, peak $^1/_2$-$1^1/_2$ hr, metabolized in liver (inactive metabolites), excreted by kidneys, protein binding 38%, terminal half-life $^1/_2$-3 hr

INTERACTIONS

Increase: bone marrow depression—antineoplastics, radiation, ganciclovir, valganciclovir, trimethoprim-sulfamethoxazole

Increase: zidovudine level—methadone, atovaquone, fluconazole, probenecid, trimethoprim, valproic acid; may need to reduce zidovudine dose

Decrease: zidovudine levels—interferons, NRTIs, DOXOrubicin, ribavirin, stavudine; avoid concurrent use

Drug/Lab Test

Decrease: platelets

Increase: LFTs

NURSING CONSIDERATIONS

Assess:

• Blood dyscrasias (anemia, granulocytopenia): bruising, fatigue, bleeding, poor healing

Black Box Warning: Blood counts q2wk; watch for decreasing granulocytes, Hgb; if low, therapy may have to be discontinued and restarted after hematologic recovery; blood transfusions may be required; viral load, CD4 counts, LFTs, plasma HIV RNA, serum creatinine/BUN at baseline and throughout treatment

Black Box Warning: Lactic acidosis, severe hepatomegaly with steatosis: Obtain baseline liver function tests, if elevated, discontinue treatment; discontinue even if liver function tests are normal but lactic acidosis, hepatomegaly are present; may be fatal

Evaluate:

• Therapeutic response: decreased viral load, increased CD4 counts, decreased symptoms of HIV

Teach patient/family:

• That GI complaints and insomnia resolve after 3-4 wk of treatment

• That product not cure for AIDS but will control symptoms

• To notify prescriber of sore throat, swollen lymph nodes, malaise, fever because other infections may occur

Z

• That patient is still infective, may pass AIDS virus on to others
• That follow-up visits must be continued because serious toxicity may occur; that blood counts must be done q2wk; that blood transfusions may be needed for severe anemia
• That product must be taken bid or tid
• That serious product interactions may occur if OTC products are ingested; to check with prescriber before taking aspirin, acetaminophen, indomethacin
• That other products may be necessary to prevent other infections
• That product may cause fainting or dizziness

zinc (Rx, otc)

Galzin

Func. class.: Trace element; nutritional supplement

ACTION: Needed for adequate healing, bone and joint development (23% zinc)

USES: Prevention of zinc deficiency, adjunct to vit A therapy

Unlabeled uses: Wound healing

Precautions: Pregnancy (C) parenteral; breastfeeding, neonates, hypocupremia, neonatal prematurity, renal disease

DOSAGE AND ROUTES

Dietary supplement (elemental zinc)

• **Adult/adolescent/pregnant female: PO** 11-13 mg/day
• **Adult/lactating female: PO** 12-14 mg/day × 12 mo
• **Adult/adolescent male ≥14 yr: PO** 11 mg/day
• **Adult female ≥19 yr: PO** 8 mg/day
• **Adolescent female ≥14 yr: PO** 9 mg/day
• **Child 9-13 yr: PO** 8 mg/day
• **Child 4-8 yr: PO** 5 mg/day
• **Child 1-3 yr: PO** 3 mg/day
• **Infant 7-12 mo: PO** 3 mg/day
• **Infant birth to 6 mo: PO** 2 mg/day (adequate intake)

Nutritional supplement (IV)

• **Adult: IV** 2.5-4 mg/day; may increase by 2 mg/day if needed
• **Child 1-5 yr: IV** 50 mcg/kg/day

Wound healing

• **Adult: PO** 50 mg tid until healed (elemental zinc)

Available forms: Tabs 66, 110 mg; caps 220 mg; inj 1 mg, 5 mg/ml

Administer:

• With meals to decrease gastric upset; avoid dairy products

SIDE EFFECTS

GI: Nausea, vomiting, cramps, heartburn, ulcer formation

OVERDOSE: Diarrhea, rash, dehydration, restlessness

INTERACTIONS

Decrease: absorption of fluoroquinolones—tetracyclines

Drug/Food

Decrease: absorption of PO zinc—dairy products, caffeine

NURSING CONSIDERATIONS

Assess:

• **Zinc deficiency:** poor wound healing, absence of taste, smell, slowing growth
• Alkaline phosphatase, HDL monthly in long term therapy
• Zinc levels during treatment

Evaluate:

• Therapeutic response: absence of zinc deficiency

Teach patient/family:

• That element must be taken for 2-3 mo to be effective
• To immediately report nausea, diarrhea, rash, severe vomiting, restlessness, abdominal pain, tarry stools

ziprasidone (Rx)

(zi-praz'ih-dohn)

Geodon, Zeldox ♣

Func. class.: Antipsychotic/neuroleptic

Chem. class.: Benzisoxazole derivative

ACTION: Unknown; may be mediated through both dopamine type 2 (D_2) and serotonin type 2 ($5\text{-}HT_2$) antagonism

USES: Schizophrenia, acute agitation, acute psychosis, bipolar disorder, mania, psychotic depression

Unlabeled uses: Tourette's syndrome, agitation

CONTRAINDICATIONS: Breastfeeding, hypersensitivity, acute MI, heart failure, QT prolongation

Precautions: Pregnancy (C), children, geriatric patients, cardiac/renal/hepatic disease, breast cancer, diabetes, seizure disorders, AV block, CNS depression, abrupt discontinuation, agranulocytosis, ambient temperature increase, suicidal ideation, torsades de pointes, strenuous exercise

Black Box Warning: Increased mortality in geriatric patients with dementia-related psychosis

DOSAGE AND ROUTES

Schizophrenia

- **Adult: PO** 20 mg bid with food, adjust dosage every 2 days upward to max of 80 mg bid; **IM** 10-20 mg; may give 10 mg q2hr; doses of 20 mg may be given q4hr; max 40 mg/day (acute episodes); switch to PO as soon as possible

Bipolar disorder

- **Adult: PO** 40 mg bid with food; on day 2 increase to 60 or 80 mg bid, then adjust to response; maintenance as adjunct to lithium/valproate 40-80 mg bid

Tourette's syndrome (unlabeled)

- **Adolescent and child ≥7 yr: PO** 5 mg/day, may increase in divided doses to 20 mg bid

Available forms: Caps 20, 40, 60, 80 mg; inj 20 mg/ml single-dose vials

Administer:

PO route

- Take cap whole and with food
- Reduced dose in geriatric patients
- Anticholinergic agent on order from prescriber to be used for EPS
- Store in tight, light-resistant container

IM route

- Add 1.2 ml sterile water for inj to vial; shake vigorously until dissolved; do not admix; give only IM; give deeply in large muscle; do not mix with other in same syringe

SIDE EFFECTS

CNS: *EPS, pseudoparkinsonism, akathisia, dystonia, tardive dyskinesia; drowsiness, insomnia, agitation, anxiety, headache,* **seizures, neuroleptic malignant syndrome,** dizziness, tremors, facial droop

CV: Orthostatic hypotension, **tachycardia, prolonged QT/QTc,** hypertension, **sudden death, heart failure (geriatric patients), torsades de pointes**

EENT: Blurred vision, diplopia

ENDO: Metabolic changes, hyperprolactinemia (rare)

GI: *Nausea,* vomiting, *anorexia, constipation,* jaundice, weight gain, diarrhea, dry mouth, abdominal pain

GU: Enuresis, urinary incontinence, gynecomastia, impotence, priapism

MS: Decreased bone density

RESP: Rhinitis, dyspnea, infection, cough

PHARMACOKINETICS

IM: Peak 60 min

PO: Extensively metabolized by liver to major active metabolite, plasma protein binding 99%, peak 6-8 hr, terminal half-life 7 hr

Z

INTERACTIONS

⚠ **Increase:** QT prolongation—class IA/III antidysrhythmics, some phenothiazines, β-agonists, local anesthetics, tricyclics, haloperidol, methadone, chloroquine, clarithromycin, droperidol, erythromycin, pentamidine, moxifloxacin

Increase: sedation—other CNS depressants, alcohol

Increase: EPS, possible neurotoxicity—other antipsychotics, lithium

Increase: ziprasidone excretion—carBAMazepine, barbiturates, phenytoin, rifampin

Increase: ziprasidone level—ketoconazole

Increase: hypotension—antihypertensives

Increase: serotonin syndrome, neuroleptic malignant syndrome—SSRIs, SNRIs

NURSING CONSIDERATIONS

Assess:

- Mental status before initial administration, AIMS assessment

> **Black Box Warning:** Geriatric patient with dementia closely; heart failure, sudden death have occurred

- Bilirubin, CBC, LFTs, fasting blood glucose, cholesterol profile; potassium, magnesium when taken with loop/thiazide diuretics monthly
- Urinalysis before, during prolonged therapy
- B/P standing and lying; also pulse, respirations; take these q4hr during initial treatment; establish baseline before starting treatment; report drops of 30 mm Hg; watch for ECG changes; QT prolongation may occur
- Dizziness, faintness, palpitations, tachycardia on rising; metabolic changes, weight gain
- **EPS,** including akathisia (inability to sit still, no pattern to movements), tardive dyskinesia (bizarre movements of the jaw, mouth, tongue, extremities), pseudoparkinsonism (rigidity, tremors, pill rolling, shuffling gait)

⚠ **Neuroleptic malignant syndrome, serotonin syndrome:** hyperthermia, increased CPK, altered mental status, muscle rigidity

- Constipation, urinary retention daily; if these occur, increase bulk and water in diet
- Supervised ambulation until patient is stabilized on medication; do not involve patient in strenuous exercise program because fainting is possible; patient should not stand still for a long time
- Increased fluids to prevent constipation

Evaluate:

- Therapeutic response: decrease in emotional excitement, hallucinations, delusions, paranoia; reorganization of patterns of thought, speech

Teach patient/family:

- That orthostatic hypotension may occur; to rise from sitting or lying position gradually
- To avoid hot tubs, hot showers, tub baths because hypotension may occur
- To avoid abrupt withdrawal of product because EPS may result; that product should be withdrawn slowly
- To avoid OTC preparations (cough, hay fever, cold) unless approved by prescriber because serious product interactions may occur; to avoid use with alcohol because increased drowsiness may occur
- To avoid hazardous activities if drowsy or dizzy
- To report impaired vision, tremors, muscle twitching
- In hot weather, that heat stroke may occur; to take extra precautions to stay cool

TREATMENT OF OVERDOSE:

Lavage if orally ingested; provide airway; *do not induce vomiting*

ziv-aflibercept

(ziv-a-flih′ber-sept)

Zaltrap

Func. class.: Antineoplastic biologic response modifier

Chem. class.: Signal transduction inhibitor (STI)

ACTION: An angiogenesis inhibitor, a fusion protein that binds to vascular endothelial growth factors (VEGF-A, VEGF-B) and placental growth factor 1 and 2

USES: Metastatic colorectal cancer that is resistant or has progressed after an Oxaliplatin-containing regimen in combination with 5-fluorouracil, leucovorin, irinotecan (FOLFIRI)

CONTRAINDICATIONS: Hypersensitivity

Precautions: infertility, male-mediated teratogenicity, encephalopathy, hypertension, dental work, breastfeeding, children, neonates, geriatric patients, infection, neutropenia, pregnancy (C)

Black Box Warning: Bleeding, GI bleeding/perforation, intracranial bleeding, surgery

DOSAGE AND ROUTES

• **Adult: IV** 4 mg/kg over 1 hr on day 1 every 2 wk in combination with the FOLFIRI regimen (irinotecan 180 mg/m^2 over 90 min on day 1 with dl-racemic leucovorin 400 mg/m^2 over 2 hr) (infused at the same time, in same Y-line; then on day 1 by 5-fluorouracil 400 mg/m^2 as a bolus, then 2400 mg/m^2 as a 46-hr cont IV inf)

Available forms: Solution for injection 100 mg/4 ml; 200 mg/8 ml

Administer:

• Give before FOLFIRI chemotherapy; visually inspect for particulate matter and discoloration before use

Dilution and preparation

• Withdraw the calculated dose; add to 0.9% sodium chloride or dextrose 5% solution to 0.6-8 mg/mL; use polyvinyl chloride (PVC) infusion bags containing bis (2-ethylhexyl) phthalate (DHEP) or polyolefin infusion bags; do not re-enter the vial after first puncture; discard any unused portion; do not mix or combine with other drugs in the same infusion bag; the diluted solution may be stored refrigerated for ≤4 hr; discard any unused portion in the infusion bag

IV infusion

• Give diluted solution over 1 hr using a 0.2 micron polyethersulfone filter; do not use nylon or polyvinylidene fluoride (PVDF) filters; do not give IV push or bolus; do not mix or combine with other drugs in the same IV line; give using an infusion set made of one of the following: PVC containing DEHP, DEHP-free PVC containing trioctyl-trimellitate (TOTM), polypropylene, polyethylene-lined PVC, or polyurethane

Dosage adjustments for recurrent or severe hypertension

⚠ Hold until B/P is controlled and then permanently reduce dose to 2 mg/kg; discontinue in hypertensive crisis or hypertensive encephalopathy

Dosage adjustments for proteinuria (2 g/24 hr)

⚠ Hold until proteinuria is <2 g/24 hr; if proteinuria recurs, hold therapy until proteinuria is <2 g/24 hr, then reduce to 2 mg/kg; discontinue in nephrotic syndrome or thrombotic microangiopathy

SIDE EFFECTS

CNS: Intracranial bleeding, headache, dizziness

CV: Hypertensive crisis, hypertension, stroke

GI: Nausea, hepatotoxicity, dyspepsia, GI hemorrhage, abdominal pain, GI perforation

GU: Proteinuria, hematuria

HEMA: Neutropenia, leukopenia

Z

INTEG: Rash, pruritus, alopecia, hypersensitivity
MISC: Fatigue, epistaxis, night sweats, decreased weight, flulike symptoms, infection
RESP: Dyspnea, pulmonary embolism

PHARMACOKINETICS

Elimination half-life 6 days

NURSING CONSIDERATIONS

Assess:

⚠ **Severe bleeding:** GI bleeding, intracranial bleeding, and pulmonary hemorrhage/hemoptysis may be fatal; monitor patients for signs and symptoms of bleeding

⚠ **GI perforation:** some cases are fatal; monitor patients for signs and symptoms of GI perforation; discontinue therapy if GI perforation develops

- **Poor wound healing**: hold ≥4 wk before elective surgery; after major surgery, do not restart for ≥4 wk and until the surgical wound is entirely healed

⚠ **Severe hypertension/hypertensive crisis:** usually occurs within the first 2 cycles (grade 3 or 4 hypertension); monitor B/P every 2 wk or more often if needed; treatment with antihypertensives may be needed; product may need to be discontinued

⚠ **Severe proteinuria/nephrotic syndrome/thrombotic microangiopathy (TMA):** monitor urine protein by dipstick analysis and urinary protein–to–creatinine ratio (UPCR); obtain a 24-hr urine collection for a UPCR >1; for proteinuria of 2 g/24 hr, temporarily hold doses until proteinuria is <2 g/24 hr; if proteinuria recurs, hold doses until proteinuria is <2 g/24 hr, then permanently reduce; discontinue in nephrotic syndrome or TMA

⚠ **Febrile neutropenia, neutropenic infection/sepsis:** monitor CBC with differential at baseline and before each cycle; hold FOLFIRI until the neutrophil count is $\geq 1.5 \times 10^9$/L

⚠ **Geriatric toxicity:** assess for diarrhea, dizziness, asthenia, weight loss, and dehydration that can indicate toxicity

⚠ **Pregnancy C/Breastfeeding:** highly effective contraception during and up to 3 mo after the last dose is needed for all patients of reproductive potential because infertility and male-mediated teratogenicity can occur; infertility can occur but is reversible within 18 wk after stopping product: do not breastfeed

Evaluate:

- Therapeutic response: decrease in spread of size of tumor

Teach patient/family:

Pregnancy (C)/breastfeeding:

⚠ That highly effective contraception should be used during and up to 3 mo after the last dose in all patients of reproductive potential, infertility and male-mediated teratogenicity can occur; infertility is reversible within 18 wk after stopping product; do not breastfeed

- About reason for treatment, expected results

⚠ Notify prescriber immediately of bleeding, severe abdominal pain, poor wound healing

zoledronic acid (Rx)

(zoh′leh-drah′nick ass′id)

Reclast, Zometa

Func. class.: Bone-resorption inhibitor

Chem. class.: Bisphosphonate

ACTION: Potent inhibitor of osteoclastic bone resorption; inhibits osteoclastic activity, skeletal calcium release caused by stimulating factors released by tumors; reduction of abnormal bone resorption is responsible for therapeutic effect with hypercalcemia; may directly block dissolution of hydroxyapatite bone crystals

USES: Moderate to severe hypercalcemia associated with malignancy; multiple myeloma; bone metastases from solid tumors (used with antineoplastics); active Paget's disease; osteoporosis, glucocorticoid-induced osteoporosis, osteoporosis prophylaxis in postmenopausal women

CONTRAINDICATIONS:

Pregnancy (D), breastfeeding; hypersensitivity to this product or bisphosphonates; hypocalcemia

Precautions: Children, geriatric patients, renal dysfunction, asthmatic patients, acute bronchospasm, anemia, chemotherapy, coagulopathy, dehydration, dental disease, diabetes mellitus, renal disease, electrolyte imbalance, hypertension, hypovolemia, infection, multiple myeloma, phosphate hypersensitivity

DOSAGE AND ROUTES

Hypercalcemia of malignancy

- **Adult: IV INF** 4 mg, given as single inf over ≥15 min; may re-treat with 4 mg if serum calcium does not return to normal within 1 wk

Multiple myeloma/metastatic bone lesions

- **Adult: IV INF** 4 mg, give over 15 min q3-4wk

Osteoporosis

- **Adult: IV INF** 5 mg over ≥15 min q12mo

Active Paget's disease

- **Adult: IV INF** 5 mg over ≥15 min

Osteoporosis prophylaxis (Reclast), postmenopausal women

- **Adult: IV INF** 5 mg every other year

Osteoporosis prophylaxis (Reclast) when taking systemic glucocorticoids

- **Adult: IV** 5 mg every yr

Early breast cancer (unlabeled)

- **Adult: IV** 4 mg q6mo with goserelin 3.6 mg SUBCUT monthly and tamoxifen 20 mg daily or anastrozole 1 mg daily for 3 yr

Available forms: (Zometa) sol for inj 4 mg/5 ml; (Reclast) inj 5 mg/100 ml

Administer:

- Saline hydration must be performed before administration; urine output should be 2 L/day during treatment; do not overhydrate patient

IV route

Zometa

- Administer after reconstituting by adding 5 ml of sterile water for inj to each vial then add to ≥100 ml of sterile 0.9% NaCl, D_5W; run over ≥15 min
- Administer in separate IV line from all other products

Reclast

- No further dilution required
- Infuse over ≥15 min at constant rate; max 5 mg

SIDE EFFECTS

CNS: Dizziness, headache, anxiety, confusion, insomnia, agitation

CV: Hypotension, leg edema, **atrial fibrillation**, chest pain

GI: Abdominal pain, anorexia, constipation, nausea, diarrhea, vomiting, taste change

GU: UTI, possible reduced renal function, **renal damage**

META: Anemia, hypokalemia, hypomagnesemia, hypophosphatemia, hypocalcemia, increased serum creatinine

MISC: *Fever, chills, flulike symptoms*

MS: Severe bone pain, *arthralgias, myalgias,* **osteonecrosis of jaw**

PHARMACOKINETICS

Rapidly cleared from circulation, taken up mainly by bones, not metabolized, eliminated primarily by kidneys, approximately 50% eliminated in urine within 24 hr, max effect 7 days; terminal half-life 167 hr, protein binding 22%

INTERACTIONS

- Hypomagnesemia, hypokalemia: digoxin

Decrease: effect of zoledronic acid—calcium, vit D

Decrease: serum calcium, aminoglycosides, loop diuretics

Z

• Do not mix with calcium-containing inf sol such as lactated Ringer's sol

Increase: nephrotoxicity—aminoglycosides, NSAIDs, radiopaque contrast agents

Drug/Lab Test

Increase: creatinine

Decrease: calcium, phosphorous, magnesium, potassium, Hct/Hgb, RBC, platelets, WBC

NURSING CONSIDERATIONS

Assess:

• Renal tests, calcium, phosphate, magnesium, potassium; creatinine, BUN; if creatinine elevated, hold treatment

• **Hypocalcemia:** paresthesia, twitching, laryngospasm; Chvostek's/Trousseau's signs

• Dental status; cover with antiinfectives for dental extraction

• Atrial fibrillation

• Sol reconstituted with sterile water may be stored under refrigeration for up to 24 hr

• Acetaminophen before and for 72 hr after to decrease pain

Evaluate:

• Therapeutic response: decreased calcium levels, increased bone density

Teach patient/family:

• To report hypercalcemic relapse: nausea, vomiting, bone pain, thirst

• To continue with dietary recommendations, including calcium and vit D; to take a multiple vitamin daily as well as 500 mg of calcium, 400 international units vit D with multiple myeloma

• If nausea or vomiting occurs, to eat small, frequent meals, to use lozenges or chewing gum

• If bone pain occurs, to notify prescriber to obtain analgesics

• To avoid use during pregnancy (D)

• To continue good oral hygiene

ZOLMitriptan (Rx)

(zole-mih-trip′tan)

Zomig, Zomig-ZMT

Func. class.: Migraine agent, abortive

Chem. class.: 5-HT_{1B}/5HT_{1D} receptor agonist (triptan)

ACTION: Binds selectively to the vascular 5-HT_{1B}/5HT_{1D} receptor subtype, exerts antimigraine effect; causes vasoconstriction in cranial arteries

USES: Acute treatment of migraine with/without aura

CONTRAINDICATIONS: Angina pectoris, history of MI, documented silent ischemia, ischemic heart disease, uncontrolled hypertension, hypersensitivity, basilar or hemiplegic migraine, risk of CV events

Precautions: Pregnancy (C), breastfeeding, children, postmenopausal women, men >40 yr, geriatric patients, risk factors for CAD, hypercholesterolemia, obesity, diabetes, impaired renal/hepatic function

DOSAGE AND ROUTES

• **Adult: PO** start at ≤2.5 mg (tab may be broken), may repeat after 2 hr, max 10 mg/24 hr; **NASAL** 1 spray in 1 nostril at onset of migraine, repeat in 2 hr if no relief

Available forms: Tabs 2.5, 5 mg; orally disintegrating tabs 2.5, 5 mg; nasal spray 5 mg

Administer:

• Take with fluids as soon as symptoms of migraine occur

• **Orally disintegrating tab:** do not crush or chew; allow to dissolve on tongue

SIDE EFFECTS

CNS: *Tingling, hot sensation, burning, feeling of pressure, tightness, numbness, dizziness, sedation*

CV: Palpitations, chest pain
GI: Abdominal discomfort, nausea, dry mouth, dyspepsia, dysphagia
MISC: Odd taste (spray)
MS: *Weakness, neck stiffness,* myalgia
RESP: Chest tightness, pressure

PHARMACOKINETICS

Duration 2-3½ hr; 25% plasma protein binding; half-life 3-3½ hr; metabolized in liver (metabolite); excreted in urine (60-80%), feces (20-40%)

INTERACTIONS

⚠ Extended vasospastic effects: ergot, ergot derivatives
⚠ Do not use within 2 wk of MAOIs
⚠ Weakness, hyperreflexia, incoordination: SSRIs (FLUoxetine, fluvoxaMINE, PARoxetine, sertraline)
Increase: half-life of ZOLMitriptan—cimetidine, oral contraceptives
Increase: ZOLMitriptan levels—sibutramine
Drug/Herb
• Serotonin syndrome: SAM-e, St. John's wort
Drug/Lab Test
Increase: alk phos

NURSING CONSIDERATIONS

Assess:
• Tingling, hot sensation, burning, feeling of pressure, numbness, flushing
• Neurologic status: LOC, blurring vision, nausea, vomiting, tingling in extremities preceding headache
• Ingestion of tyramine foods (pickled products, beer, wine, aged cheese), food additives, preservatives, colorings, artificial sweeteners, chocolate, caffeine, which may precipitate these types of headaches
• **Serotonin syndrome** if also taking SSRI
Evaluate:
• Therapeutic response: decrease in frequency, severity of headache
Teach patient/family:
• To report any side effects to prescriber
• To use contraception while taking product
• That product does not prevent or reduce number of migraines
• To report chest pain, rash, swelling of face
• Not to double doses; if second dose is needed, wait at least 2 hr

zolpidem (Rx)
(zole′pih-dem)
Ambien, Ambien CR, Edluar, Zolpimist
Func. class.: Sedative/hypnotic
Chem. class.: Imidazopyridine

Controlled Substance Schedule IV

ACTION: Produces CNS depression at limbic, thalamic, hypothalamic levels of CNS; may be mediated by neurotransmitter γ-aminobutyric acid (GABA); results are sedation, hypnosis, skeletal muscle relaxation, anticonvulsant activity, anxiolytic action

USES: Insomnia, short-term treatment; insomnia with difficulty of sleep onset/maintenance (ext rel)
Unlabeled uses: Head trauma

CONTRAINDICATIONS: Hypersensitivity to benzodiazepines
Precautions: Pregnancy (C), breastfeeding, children <18 yr, geriatric patients, anemia, renal/hepatic disease, suicidal individuals, drug abuse, psychosis, seizure disorders, angioedema, depression, respiratory disease, sleep apnea, sleep-related behaviors (sleepwalking), myasthenia gravis, pulmonary disease, next morning impairments; females (lower dose needed)

DOSAGE AND ROUTES

• **Adult: PO** 5 mg (women), 5-10 mg (men) at bedtime × 7-10 days only; total max dose 10 mg; **EXT REL** 6.25

Z

(women), 6.25-12.5 mg (men) immediately before bedtime, may be useful for ≤24 wk in people 18-64 yr with primary insomnia, max 12.5 mg/day; **oral spray** (Zolpimist) 5 mg (women), 5-10 mg (men) immediately before bedtime, max 10 mg/day; **SL** (Edluar) 5 mg (women), 5-10 mg (men) just before bedtime

• **Geriatric: PO** 5 mg at bedtime; **EXT REL** 6.25, 12.5 mg; SL: 5 mg at bedtime

Available forms: Tabs 5, 10 mg; ext rel tabs 6.25, 12.5 mg; SL: 1.75, 3.5, 5, 10 mg; oral spray 5 mg/spray

Administer:

PO route

• Do not break, crush, or chew ext rel
• Take with full glass of water
• ½-1 hr before bedtime (PO); right before retiring (ext rel)
• On empty stomach for fast onset; may be taken with food if GI symptoms occur
• Store in tight container in cool environment

Spray route

• Prime before first use or if pump is not used for ≥14 days
• Do not use spray with or after a meal

Sublingual route

• Separate blister pack at perforation; peel paper and push product through; place product under tongue; allow to dissolve before swallowing; do not take with water

SIDE EFFECTS

CNS: Headache, lethargy, drowsiness, daytime sedation, dizziness, confusion, lightheadedness, anxiety, irritability, amnesia, poor coordination, complex sleep-related reactions (sleep driving, sleep eating), depression, somnolence, suicidal ideation, abnormal thinking/behavioral changes

CV: Chest pain, palpitations

GI: Nausea, vomiting, diarrhea, heartburn, abdominal pain, constipation

HEMA: Leukopenia, granulocytopenia (rare)

MISC: Myalgia

SYST: Severe allergic reactions, angioedema, anaphylaxis

PHARMACOKINETICS

PO: Onset up to 1.5 hr, metabolized by liver, excreted by kidneys (inactive metabolites), crosses placenta, excreted in breast milk, half-life 2-3 hr

INTERACTIONS

Increase: action of both products—alcohol, CNS depressants

Increase or decrease: zolpidem levels—CYP3A4 inhibitors/inducers

Decrease: zolpidem effect—rifamycins

NURSING CONSIDERATIONS

Assess:

⚠ Mental status: mood, sensorium, affect, memory (long, short term), excessive sedation, impaired coordination, **suicidal thoughts/behaviors**

• Blood dyscrasias: fever, sore throat, bruising, rash, jaundice, epistaxis (rare)
• Type of sleep problem: falling asleep, staying asleep

Evaluate:

• Therapeutic response: ability to sleep at night, decreased amount of early morning awakening if taking product for insomnia

Teach patient/family:

• That dependence is possible after long-term use

⚠ That complex sleep-related behaviors may occur (sleep driving/eating)

• To avoid driving or other activities requiring alertness until dosage is stabilized
• To avoid alcohol ingestion
• That effects may take 2 nights for benefits to be noticed; next morning impairment may occur
• Not to use during pregnancy, breastfeeding
• That hangover is common in geriatric patients but less common than with barbiturates; that rebound insomnia may occur for 1-2 nights after discontinuing product; not to discontinue abruptly; to taper
• Not to crush, chew, break ext rel tabs
• To prime spray pump before using

TREATMENT OF OVERDOSE:
Lavage, activated charcoal; monitor electrolytes, VS

zonisamide (Rx)
(zone-is'a-mide)

Zonegran

Func. class.: Anticonvulsant

Chem. class.: Sulfonamides

ACTION:
May act through action at sodium and calcium channels, but exact action is unknown; serotonergic action

USES:
Epilepsy, adjunctive therapy for partial seizures

Unlabeled uses: Bipolar disorder (mania)

CONTRAINDICATIONS:
Hypersensitivity to this product or sulfonamides

Precautions: Pregnancy (C), breastfeeding, children <16 yr, geriatric patients, allergies, renal/hepatic disease; psychiatric condition, hepatic failure

DOSAGE AND ROUTES

- **Adult and child >16 yr:** 100 mg/day, may increase after 2 wk to 200 mg/day, may increase q2wk, max dose 600 mg/day

Available forms: Caps 25, 50, 100 mg

Administer:

- Without regard to food; swallow whole

SIDE EFFECTS

CNS: Dizziness, insomnia, paresthesias, depression, fatigue, headache, confusion, somnolence, agitation, irritability, speech disturbance, suicidal ideation, seizures, status epilepticus

EENT: Diplopia, verbal difficulty, speech abnormalities, taste perversion, amblyopia, pharyngitis, rhinitis, tinnitus, nystagmus

GI: Nausea, constipation, anorexia, weight loss, diarrhea, dyspepsia, dry mouth, abdominal pain

GU: Kidney stones

HEMA: Aplastic anemia, granulocytopenia (rare); ecchymosis

INTEG: Rash, pruritus

MISC: Flulike symptoms

SYST: Stevens-Johnson syndrome, metabolic acidosis

PHARMACOKINETICS

Peak 2-6 hr, half-life in RBCs 105 hr, metabolized by liver, excreted by kidneys, protein binding 40%

INTERACTIONS

Decrease: half-life of zonisamide—carBAMazepine, phenytoin, PHENobarbital

Altered product levels: CYP3A4 inhibitors/inducers

Increase: CNS depression—alcohol

Drug/Herb

Increase: effect of this product—St. John's wort

Drug/Food

- Do not use with grapefruit

Drug/Lab Test

Increase: BUN, creatinine

NURSING CONSIDERATIONS

Assess:

- **Seizures:** duration, type, intensity, precipitating factors
- Renal function: albumin conc, BUN, urinalysis, creatinine, serum bicarbonate at baseline and periodically

⚠ Mental status: mood, sensorium, affect, memory (long, short term), **suicidal thoughts/behaviors**

- Stevens-Johnson syndrome, aplastic anemia, fulminant hepatic necrosis; may cause death; monitor for rashes and hypersensitive reactions
- Rash, hypersensitivity reactions
- Obtain bicarbonate before treatment/periodically; metabolic acidosis may occur in children

Z

Evaluate:

- Therapeutic response: decrease in severity of seizures

Teach patient/family:

- Not to discontinue product abruptly because seizures may occur
- To avoid hazardous activities until stabilized on product
- To carry emergency ID stating product use
- To notify prescriber of rash immediately; to notify prescriber of back pain, abdominal pain, blood in urine; to increase fluid intake to reduce risk of kidney stones
- To notify prescriber if pregnancy planned, suspected
- To avoid grapefruit
- To notify prescriber of sore throat, fever, easy bruising
- To report suicidal thoughts, behaviors immediately

Appendix A

Selected new drugs

ado-trastuzumab

(a′doe-tras-tooz′ue-mab)

Kadcyla

Func. class: Antineoplastic-biologic response modifier

Chem. class: Signal transduction inhibitors (STIs), humanized anti-HER2 antibody

ACTION: Humanized anti-HER2 monoclonal antibody that is linked to DM1, a small molecule microtubular inhibitor; once the antibody is bound to the HER2 receptor, the complex is internalized and the DM1 is released to bind with tubulin to lead to apoptosis

USES: Breast cancer; metastatic with overexpression of HER2, who previously received trastuzumab and a taxane, separately or in combination

CONTRAINDICATIONS: Hypersensitivity to this product, Chinese hamster ovary cell protein, pregnancy (D)

Precautions: Breastfeeding, children, pulmonary disease, acute bronchospasm, anticoagulant, Asian patients, asthma, COPD, extravasation, fever, hepatitis, human anti-human antibody, hypotension, neuropathy, hepatotoxicity; cardiomyopathy/CHF; interstitial lung disease/pneumonitis; extravasation; thrombocytopenia

DOSAGE AND ROUTES

- **Adult:** IV 3.6 mg/kg over 30-90 min q3wk; give first infusion over 90 min; if tolerated, give over 30 min

Dosage adjustments for toxicities:

Hepatotoxicity:

- AST/ALT >5 to $\leq 20\times$ ULN: withhold, resume at a reduced dose when AST/ALT is $\leq 5\times$ ULN; first dose reduction: reduce the dose to 3 mg/kg; second dose reduction: reduce the dose to 2.4 mg/kg; requirement for further dose reduction: discontinue
- AST/ ALT $>20\times$ ULN: discontinue
- Total bilirubin >3 to $\leq 10\times$ ULN: withhold, resume treatment at a reduced dose when total bilirubin recovers to ≤ 1.5; first dose reduction: reduce the dose to 3 mg/kg; second dose reduction: reduce the dose to 2.4 mg/kg; requirement for further dose reduction: discontinue treatment
- Total bilirubin $>10\times$ ULN: permanently discontinue; permanently discontinue treatment in patients with AST/ALT $>3\times$ ULN and total bilirubin $>2\times$ ULN; permanently discontinue treatment in patients diagnosed with nodular regenerative hyperplasia (NRH)
- Left ventricular ejection fraction *(LVEF); LVEF 40%-45% and decrease is <10% points from baseline:* continue treatment; repeat LVEF assessment within 3 wk; *LVEF 40%-45% and decrease is ≥10% points from baseline:* withhold; repeat LVEF assessment within 3 wk; if LVEF remains ≥10% points from baseline, discontinue; *LVEF <40%:* withhold; repeat LVEF assessment within 3 wk; if LVEF remains <40%, discontinue; *symptomatic congestive heart failure (CHF)*: discontinue

Thrombocytopenia:

- Platelet count 25,000/mm^3 to <50,000/mm^3: withhold; resume treatment at same dose when platelet count recovers to ≥75,000/mm^3

• Platelet count <25,000/mm^3: withhold; resume treatment at a reduced dose when platelet count recovers to ≥75,000/mm^3: first dose reduction: reduce the dose to 3 mg/kg; second dose reduction: reduce the dose to 2.4 mg/kg; requirement for further dose reduction: discontinue

Pulmonary toxicity:

• Permanently discontinue in patients diagnosed with interstitial lung disease or pneumonitis

Peripheral neuropathy:

• Withhold in patients experiencing grade 3 or 4 peripheral neuropathy; resume treatment upon resolution to ≤grade 2

Available forms: Lyophilized powder 100, 160 mg

Administer:

IV route

• Visually inspect for particulate matter and discoloration before use

• Give as (IV) inf with a 0.22 micron in-line filter; do not administer as an IV push or bolus

• Use cytotoxic handling procedures; do not mix with, or administer as an inf with, other IV products

Reconstitution:

• Slowly inject 5 ml of sterile water for injection into each 100-mg vial, or 8 ml of sterile water for injection into each 160-mg vial to yield a single-use solution of 20 mg/ml

• Direct the stream of sterile water toward the wall of the vial and not directly at the cake or powder

• Gently swirl the vial to aid in dissolution; do not shake

• After reconstitution, withdraw desired amount from the vial and dilute immediately in 250 ml of 0.9% sodium chloride; do not use dextrose 5% solution; gently invert the bag to mix the solution in order to avoid foaming

• The reconstituted single-use product does not contain a preservative; use the diluted solution immediately or store at 2° C-8° C (36-46° F) for up to 24 hr after reconstitution; discard any unused drug after 24 hr; do not freeze

IV inf

• Closely monitor for possible subcutaneous infiltration during drug administration

• **First infusion**: give over 90 min; the inf rate should be slowed or interrupted if the patient develops an infusion-related reaction; patients should be observed for at least 90 min following the initial dose for fever, chills, or other inf-related reactions; permanently discontinue for life-threatening inf-related reactions

• **Subsequent infusions**: administer over 30 min if prior infs were well tolerated; the inf rate should be slowed or interrupted if the patient develops an inf-related reaction; patients should be observed for at least 30 min after the inf; permanently discontinue for life-threatening inf-related reactions

SIDE EFFECTS

CNS: Dizziness, insomnia, neuropathy, chills, fatigue, fever, flushing, headache, neuropathy

CV: Heart failure, tachycardia

EENT: Blurred vision, conjunctivitis, stomatitis

GI: Diarrhea, nausea, vomiting, abdominal pain, constipation, dyspepsia, hepatotoxicity

HEMA: Anemia, bleeding, thrombocytopenia

INTEG: Rash

MS: Arthralgia

RESP: Cough, dyspnea, bronchospasm, pneumonitis, interstitial lung disease

SYST: Anaphylaxis

Other: Elevated LFTs, hand-foot syndrome

PHARMACOKINETICS

93% protein binding, metabolized in the liver

INTERACTIONS

Increase: bleeding risk—warfarin; avoid use with CYP3A4 inhibitors with

strong CYP3A4 inhibitors; platelet inhibitors; anticoagulants
- Do not give with other IV products; do not give with 5% dextrose

NURSING CONSIDERATIONS

Assess:
- Liver function tests; pregnancy test CBC, HER2 overexpression

Black Box Warning: CHF, other cardiac symptoms: dyspnea, coughing; gallop; obtain full cardiac workup including ECG, echo, MUGA

- Symptoms of infection; may be masked by product
- CNS reaction: LOC, mental status, dizziness, confusion

Black Box Warning: Hypersensitive reactions, anaphylaxis

- Inf reactions that may be fatal: fever, chills, nausea, vomiting, pain, headache, dizziness, hypotension; discontinue product
- Pulmonary toxicity: dyspnea, interstitial pneumonitis, pulmonary hypertension, ARDs; can occur after inf reaction; those with lung disease may have more severe toxicity

Evaluate:
- Therapeutic response: decrease in size of tumors

Black Box Warning: Hepatic disease: may be fatal

Teach patient/family:
- To take acetaminophen for fever
- To avoid hazardous tasks because confusion, dizziness may occur
- To report signs of infection: sore throat, fever, diarrhea, vomiting; bleeding; decreased heart function/SOB with exertion; neuropathy; liver toxicity
- That emotional liability is common; to notify prescriber if severe or incapacitating
- To use contraception while taking this product and for additional 6 mo after discontinuing this drug; pregnancy (D); to avoid breastfeeding
- To report pain at inf site

afatinib
(a-fat′i-nib)

Gilotrif

Func. class.: Antineoplastic biologic response modifiers

Chem. class.: Signal transduction inhibitors (STIs), epidermal growth factor receptor, tyrosine kinase inhibitor

Do not confuse:
Afatinib/afinitor/axitinib

ACTION:
Selective inhibitor of EGFR (ErbB1), HER2 (ErbB2), and HER4 (ErbB4); irreversible, covalent binding of intracellular tyrosine kinase inhibiting (and causing regression) tumor growth by decreasing EGFR signal transduction, cell cycle arrest, and inhibition of angiogenesis

USES:
Treatment of non-small cell lung cancer whose tumors have epidermal growth factor receptor Exon 19 deletions or 21 substitution mutations

CONTRAINDICATIONS:
Pregnancy (D), hypersensitivity

Precautions: Contact lens, dehydration, inflammation, keratitis, ocular disease, pneumonitis, pulmonary disease, renal disease, respiratory distress syndrome, serious rash, skin disease, diarrhea, hepatic disease, ocular disease

DOSAGE AND ROUTES
- **Adult: PO** 40 mg daily

Dose adjustments for toxicities:
- **Hepatotoxicity:** Hold therapy in patients who develop worsening liver function; when toxicity resolves to grade 1 or less, resume therapy at a reduced dose (10 mg/day less than the dose causing hepatotoxicity); permanently discontinue for severe drug-induced hepatic impairment or if hepatotoxicity persists at a dose of 20 mg/day

• **Grade 2 or higher renal toxicity (CCr >1.5 × ULN):** Hold until the toxicity resolves to grade 1 or less (CCr <1.5 × ULN), then resume at a reduced dose (10 mg/day less than the dose causing nephrotoxicity); permanently discontinue if nephrotoxicity persists at a dose of 20 mg/day

• **Grade 2 diarrhea, lasting 2 or more consecutive days while taking anti-diarrheal medication, or any grade 3 diarrhea:** Hold until the toxicity resolves to grade 1 or less, then resume therapy at a reduced dose (10 mg/day less than the dose causing toxicity); anti-diarrheal therapy should continue until no loose bowel movement for 12 hr; permanently discontinue if toxicity persists at a dose of 20 mg/day

• **Grade 2 cutaneous reactions that last more than 7 days or are intolerable:** Hold until the toxicity resolves to grade 1 or less, then resume at a reduced dose (10 mg/day less than the dose causing toxicity); permanently discontinue for life-threatening bullous, blistering, or exfoliative skin lesions, or if toxicity persists at a dose of 20 mg/day

• **Confirmed interstitial lung disease, persistent ulcerative keratitis, symptomatic left ventricular dysfunction, or any severe/intolerable adverse reaction occurring at a dose of 20 mg/day:** Permanently discontinue

• **Any grade 3 or higher adverse event:** Hold until the toxicity resolves to grade 1 or less, then resume therapy at a reduced dose (10 mg/day less than the dose causing toxicity); permanently discontinue if toxicity persists at a dose of 20 mg/day

• **Permanently discontinue for any severe or intolerable adverse event that occurs at a dose of 20 mg/day.**

Dosage Guidance in Patients on P-glycoprotein (P-gp) inhibitors/ inducers:

• P-gp inhibitors: If use of a P-gp inhibitor is required, reduce the initial daily afatinib dosage to 30 mg/day if not tolerated; resume original dose after discontinuation of the P-gp inhibitor as tolerated

• P-gp inducers: If use of a P-gp inducer is required, increase the initial daily afatinib dosage to 50 mg/day as tolerated; resume original dose 2-3 days after discontinuation of the P-gp inducer

Available forms: Tabs 20, 30, 40 mg

Administer:

PO route

• Give on empty stomach 1 hr before, 2 hr after food; give at same time of day

• Do not take a missed dose if within 12 hr of the next dose

SIDE EFFECTS

CNS: Fatigue, fever

CV: Heart failure

EENT: Blurred vision, conjunctivitis

GI: Diarrhea, nausea, vomiting, stomatitis, decreased appetite

HEMA: Anemia, neutropenia, leukopenia, epistaxis

INTEG: Rash, pruritus, acne vulgaris, photosensitivity, nail bed infections

MS: Arthralgia

RESP: Cough, dyspnea, acute respiratory distress syndrome, interstitial lung disease, pneumonitis

Other: Elevated LFTs, infection, hand and foot syndrome, dehydration, renal failure, cystitis, hypokalemia

PHARMACOKINETICS

95% protein binding, half-life 37 hr, primarily excreted as unchanged drug in feces; peak 2-5 hr after dose; time to steady state is approx. 8 days

INTERACTIONS

Increase: afatinib—P-gp inhibitors effect

Drug/Herb

Increase: afatinib concentration—St. John's wort

Drug/Food Test

Increase: afatinib effect—grapefruit juice; avoid use while taking product

NURSING CONSIDERATIONS

Assess:

Black Box Warning: Myelosuppression: anemia, neutropenia; obtain a weekly × 1 mo, then monthly as needed; LFTs every mo × 3 mo, then as clinically indicated; hepatic failure may occur

Evaluate: Therapeutic response: decrease in progression of disease

Teach patient/family:

- To report adverse reactions immediately; bleeding
- About reason for treatment, expected results
- To use effective contraception during treatment and up to 30 days after discontinuing treatment
- To treat skin rash with topicals and oral antibiotics; use loperamide for diarrhea; if any side effect is severe or persistent, contact prescriber
- That complicated dosing changes may occur based on toxicity or drug-drug interaction
- To take on empty stomach 1 hr before or 2 hr after meal; not to take missed dose within 12 hr of next scheduled dose

canagliflozin

(kan′a-gli-floe′zin)

Invokana

Func. class.: Oral antidiabetic

Chem class.: SGLT 2 inhibitor

ACTION: Blocks reabsorption by the kidney, increases glucose excretion, lower blood glucose concentrations

USES: Type 2 diabetes mellitus, with diet and exercise

CONTRAINDICATIONS: Dialysis, renal failure, hypersensitivity, breastfeeding, diabetic ketoacidosis

Precautions: Pregnancy (C), children, renal/hepatic disease, hypothyroidism, hyperglycemia, hypotension, pituitary insufficiency, type 1 diabetes mellitus, malnutrition, fever, dehydration, adrenal insufficiency, geriatrics

DOSAGE AND ROUTES

Renal Dose

- **Adult PO** eGFR 45-59 ml/min/1.73 m^2, max 100 mg/day; <45 ml/min/1.73 m^2

Available forms: Tabs 100, 300 mg

Administer:

PO route

- Once/day with first meal of the day

SIDE EFFECTS

CNS: Dizziness, fatigue

GI: Abdominal pain, pancreatitis, constipation, nausea

GU: Cystisis, candidiasis, urinary frequency, polyclipsia, polyuria

INTEG: Photosensitivity, rash, pruritus

META: Hypercholesterolemia, lipidemia, hypoglycemia, hyperkalemia, hypermagnesemia, hyperphosphatemia

MISC: Bone fractures

PHARMACOKINETICS

99% protein binding, metabolized by UGT1A9, UGT2B4, excreted 33% in urine

INTERACTIONS

- Do not use gatifloxacin

Increase: hypoglycemia-sulfonylureas, insulin, MAOIs, salycilates, fibric acid derivatives, bile acid sequestrants, ACE inhibitors, angiotensin II receptor antagonists, beta blockers

Increase or decrease: glycemic control-androgens, lithium, bortezomib, quinolones

Decrease: effect hyperglycemia—digestive enzymes, intestinal absorbents, thiazide diuretics, loop diuretics, corticosteroids, estrogen, progestins, oral contraceptives, sympathomimetics, isoniazid, phenothiazines; protease inhibitors, atypical antipsychotics, carbonic anhydrase inhibitors, cycloSPORINE, tacrolimus, baclofen

NURSING CONSIDERATIONS

Assess:

• Hypoglycemia (weakness, hunger, dizziness, tremors, anxiety, tachycardia, sweating), hyperglycemia; even though product does not cause hypoglycemia, if patient is on sulfonylureas or insulin, hypoglycemia may be additive; if hypoglycemia occurs, treat with dextrose, or, if severe, with IV glucagon
• For stress, surgery, or other trauma that may require a change in dose
• A1c q3mo, monitor serum glucose; 1 hr PP throughout treatment; serum cholesterol, serum creatinine/ BUN, serum electrolytes

Evaluate:

• Therapeutic response: improved signs/ symptoms of diabetes mellitus (decreased polyuria, polydipsia, polyphagia); clear sensorium, absence of dizziness, stable gait

Teach patient/family:

• The symptoms of hypo/hyperglycemia, what to do about each
• That medication must be taken as prescribed; explain consequences of discontinuing medication abruptly; that insulin may need to be used for stress, including trauma, fever, surgery
• To avoid OTC medications and herbal supplements unless approved by health care provider
• That diabetes is a lifelong illness; that the diet and exercise regimen must be followed; that this product is not a cure
• To carry emergency ID and glucose source; to avoid sugar, because sugar is blocked by acarbose
• That blood glucose monitoring is required to assess product effect
• That GI side effects may occur

crofelemer

(kroe-fel′e-mer)

Fulyzaq

Func. class: Antidiarrheal
Chem. class: Red sap of Croton lechleri plant

ACTION: Blocks chloride channel and high volume water loss in diarrhea

USES: Non-infectious diarrhea in those with HIV/AIDS using antiretrovirals

CONTRAINDICATIONS: Hypersensitivity

Precautions: Pregnancy (C), breastfeeding, black patients, children/adolescents, GI disease, infection, malabsorption syndrome, pancreatitis

DOSAGE AND ROUTES

• **Adult: PO** 125 mg bid

Available forms: Delayed rel tabs 125 mg

Administer:

• Do not break, crush, or chew
• Without regard to meals

SIDE EFFECTS

CNS: Dizziness, depression
GI: Nausea, constipation, abdominal pain, anorexia, flatulence
INTEG: Acne vulgaris, contact dermatitis
MISC: Arthralgia, cough, increased urinary frequency

INTERACTIONS

Increase: serious constipation, bowel obstruction—alosetron (IBS)
Increase: constipation—may occur with antimuscarinics, opiate agonists

NURSING CONSIDERATIONS

Assess:

• Stools: volume, color, characteristic, frequency; bowel pattern before protein rebound constipation

- Electrolytes (Potassium, Sodium, Chlorine), hydration status
- Monitor effect in black patients; may be less effective

Evaluate: Therapeutic response: decreased diarrhea

Teach patient/family:

- To avoid OTC products unless directed by prescriber
- If drowsiness occurs, not to operate machinery

dabrafenib

(da-braf′e-nib)

Tafinlar

Func. class.: Antineoplastic

Chem. class.: Signal transduction inhibitor, kinase inhibitor

ACTION: Inhibits kinase, inhibitor against mutated forms of BRAF kinases in melanoma cells

USES: Unresectable or metastatic BRAD V600E-mutated malignant melanoma

CONTRAINDICATIONS: Pregnancy (D), hypersensitivity

Precautions: Breastfeeding, children, infection, dehydration, diabetes mellitus, fever, G6PD deficiency, hemolytic anemia, hyperglycemia, hypotension, infertility, iritis, renal failure, secondary malignancy

DOSAGE AND ROUTES

- **Adult: PO** 150 mg q12hr until disease progression; avoid strong CYP3A4/CYP2C8 inhibitors or inducers

Available forms: Caps 50, 75 mg

Administer:

PO route

- Swallow whole
- If dose is missed, take within 6 hr of missed dose; if >6 hr have passed, skip dose
- Space doses q12hr
- Take at least 1 hr before or 2 hr after a meal

SIDE EFFECTS

CNS: Headache, fever

GI: Pancreatitis

INTEG: Rash, alopecia

MISC: Arthralgia, myalgia, back pain

OTHER: Hyperglycemia, hypophosphatemia, hyponatremia, secondary malignancy, hand/foot syndrome

PHARMACOKINETICS

Protein binding 99.7%, half-life 8 hr (dabrafenib), 10 hr, 21-22 hr metabolites, excreted 71% (feces), 23% (urine)

INTERACTIONS

Altered: dabrafenib concentrations—CYP3A4 inhibitors (ketoconazole, itraconazole, erythromycin, clarithromycin)

Decrease: dabrafenib concentrations—CYP3A4 inducers (dexamethasome, phenytoin, carBAMazepine, rifampin, PHENobarbital), antacids, proton-pump inhibitors

Drug/Herb

Decrease: dabrafenib concentrations St. John's Wort

Drug/Food Test

Increase: dabrafenib effect—grapefruit juice; avoid use while taking product

NURSING CONSIDERATIONS

Assess:

- Toxicity: fever, grade 2, 3

Evaluate:

- Therapeutic response: decrease in melanoma progression

Teach patient/family:

- To report adverse reactions immediately
- About reason for treatment, expected results
- To use effective contraception during treatment and up to 30 days after discontinuing treatment, pregnancy (D)

dapagliflozin

(dap′a-gli-floe′zin)

Farxiga

Func. class.: Oral antidiabetic

Chem. class.: SGLT 2 inhibitor

ACTION: Blocks reabsorption of glucose by the kidney, increases glucose excretion, lower blood glucose concentrations

USES: Type 2 diabetes mellitus, with diet and exercise

CONTRAINDICATIONS: Dialysis, renal failure, hypersensitivity, breastfeeding, diabetic ketoacidosis

Precautions: Pregnancy (C), children, renal/hepatic disease, hypothyroidism, hyperglycemia, hypotension, bladder cancer, hypercholesterolemia, pituitary insufficiency, type 1 diabetes mellitus, malnutrition, fever, dehydration, adrenal insufficiency, geriatrics, genital fungal infections, hypoglycemia

DOSAGE AND ROUTES

- **Adult: PO** 5 mg in AM; may increase to 10 mg q day if needed

Renal Dose

- **Adult: PO** eGFR $\geq$60 ml/min/1.73 m^2 no change, eGFR $<$60 ml/min, do not use

Available forms: Tabs 5, 10 mg

Administer:

PO route

- Once/day in AM without regard to food

SIDE EFFECTS

CNS: Dizziness, fatigue

GI: Abdominal pain, pancreatitis, constipation, nausea

GU: Cystitis, candidiasis, urinary frequency, polydipsia, polyuria, increased serum creatinine; renal impairment/failure; infections

INTEG: Photosensitivity, rash, pruritus

META: Hypercholesterolemia, lipidemia, hypoglycemia, hyperkalemia, hypomagnesemia, hypo/hyperphosphatemia

MISC: Bone fractures, hypotension; dehydration; orthostatic hypotension; hypersensitivity; new bladder cancer; increased hematocrit

PHARMACOKINETICS

91% protein binding, primary excretion in urine, half-life 12.9 hr; primarily metabolized by O-glucuronidation by UGT1A9; minor CYP3A4; Cmax is less than 2 hr

INTERACTIONS

- Do not use gatifloxacin

Increase: hypoglycemia—sulfonylureas, insulin, MAOIs, salicylates, fibric acid derivatives, bile acid sequestrates, ACE inhibitors, angiotensin II receptor antagonists, beta blockers

Increase or decrease: glycemic control—androgens, lithium, bortezomib, quinolones

Decrease: effect hyperglycemia—digestive enzymes, intestinal absorbents, thiazide diuretics, loop diuretics, corticosteroids, estrogen, progestins, oral contraceptives, sympathomimetics, isoniazid, phenothiazines, protease inhibitors, atypical antipsychotics, carbonic anhydrase inhibitors, cycloSPORINE, tacrolimus, baclofen

NURSING CONSIDERATIONS

Assess:

- Hypoglycemia (weakness, hunger, dizziness, tremors, anxiety, tachycardia, sweating), hyperglycemia; even though product does not cause hyperglycemia, if patient is on sulfonylureas or insulin, hypoglycemia may be additive; if hypoglycemia occurs, treat with dextrose, or, if severe, with IV glucagon; HbA1c; lipid panel; renal function tests; volume status; blood pressure

Evaluate:

- Therapeutic response: improved signs/symptoms of diabetes mellitus (decreased polyuria, polydipsia, polyphagia); clear sensorium, absence of dizziness, stable gait; HbA1c WNL

Teach patient/family:
- The symptoms of hypo/hyperglycemia, what to do about each
- That medication must be taken as prescribed; explain consequences of discontinuing medication abruptly; that insulin may need to be used for stress, including trauma, fever, surgery
- To avoid OTC medications and herbal supplements unless approved by health care provider
- That diabetes is a lifelong illness; that the diet and exercise regimen must be followed; that this product is not a cure
- To carry emergency ID and glucose source
- That blood glucose monitoring is required to assess product effect
- That GI side effects may occur
- That there is a risk of renal impairment, dehydration, and new bladder cancer

dolutegravir

(dole-oo-teg'ra-vir)

Tivicay

Func. class: Antiretroviral

Chem class: HIV integrase strand transfer inhibitor (ISTIs)

ACTION: Inhibits catalytic activity of HIV integrase, which is an HIV-encoded enzyme needed for replication

USES: HIV in combination with other retrovirals

CONTRAINDICATIONS: Breastfeeding, hypersensitivity

Precautions: Pregnancy (C), children, geriatric patients, hepatic disease, immune reconstitution syndrome, hepatitis, antimicrobial resistance, lactase deficiency

DOSAGE AND ROUTES

- **Adult and child ≥12 yr and ≥40 kg (treatment naïve or treatment experienced but integrase strand transfer inhibitor naïve): PO** 50 mg daily; if given with efavires, fosanprenavir/ritonavir, tipranavir/ritonavir, or rifampin, give 50 mg bid

Available forms: Tabs 50 mg

Administer:
- May give without regard to meals, with 8 oz of water
- Store at room temperature
- Give 2 hr before or 6 hr after cation-containing antacids or laxatives, sucralfate, oral iron, oral calcium, or buffered products

SIDE EFFECTS

CNS: Fatigue, fever, dizziness, headache, asthenia, suicidal ideation

CV: MI

GI: Nausea, vomiting, diarrhea, abdominal pain, asthenia, gastritis, hepatitis

INTEG: Rash, pruritus, urticaria

META: Hyperglycemia

SYST: Immune reconstitution syndrome

PHARMACOKINETICS

Peak 2-3 hr, steady state 5 days, terminal half-life 14 hr, 98% protein binding, metabolized in the liver, excreted in feces 53%, urine 31%

INTERACTIONS

Decreased: effect of dolutegravir—antacids, laxatives/ sucralfate, oral iron, oral calcium, buffered products

Decreased: levels—rifampin efavirenz, tenofovir, tipranavir/ritonavir

Drug/Herb
- Avoid concurrent use with St. John's wort

NURSING CONSIDERATIONS

Assess:
- HIV infection: CD4, T-cell count, plasma HIV RNA, viral load; resistance testing before treatment, at treatment failure

• Drug resistance testing before use in treatment naïve patients
• Immune reconstitution syndrome, usually during initial phase of treatment; may be anti-infective before starting
• Monitor total HDL/LDL cholesterol baseline and periodically; all may be elected

Evaluate:
• Therapeutic response: improvement in cell counts, T-cell counts

Teach patient/family:
• To take as prescribed; if dose missed to take as soon as remembered up to 1 hr before next dose; not to double dose; not to share with others
• That sexual partners need to be told that patient has HIV; that product does not cure infection, just controls symptoms, does not prevent infecting others
• To report sore throat, fever, fatigue (may indicate superinfection)
• To notify prescriber if pregnancy is planned or suspected; to avoid breastfeeding and to continue follow up exams and work

eslicarbazepine
(es′lye-kar-bay′ze-peen)

Aptiom

Func. class: Anticonvulsant, misc
Chem. class: Voltage-gated sodium channel (VGSC) blocker

ACTION: Exact mechanism unknown; a voltage-gated sodium-channel blocker inhibits repetitive neuronal firing

USES: Partial seizures, adjunctive treatment

CONTRAINDICATIONS: Hypersensitivity to this product or OXcarbazepine

Precautions: Breastfeeding, abrupt discontinuation, depression, driving/operating machinery, ethanol intoxication, hepatic disease, renal disease, hyponatremia, suicidal ideation, pregnancy (C)

DOSAGE AND ROUTES
• **Adult: PO** 400 mg daily; after 1 wk, increase to 800 mg daily, max 1200 mg daily

Renal dose
• **Adult: PO** CCr <50 ml/min 200 mg daily; after 2 wk, increase to 400 mg daily, max 600 mg daily

Available forms: Tab 200, 400, 600, 800 mg

Administer:
• May be taken without regard to food
• May be crushed or whole

SIDE EFFECTS
CNS: Drowsiness, dizziness, amnesia, depression, insomnia, lethargy, memory impairment, confusion, fatigue, headache, speech disturbance, suicidal thoughts/behaviors, tremors
CV: Hypertension, peripheral edema
EENT: Blurred vision, nystagmus, diplopia
GI: Nausea, constipation, diarrhea, hypercholesterolemia/hypertriglyceridemia, vomiting, hepatotoxicity
GU: Cystitis
INTEG: Rash, Stevens-Johnson syndrome, toxic epidermal necrolysis, anaphylaxis, angioedema
META: Hyponatremia
RESP: Cough

PHARMACOKINETICS
Peak 1-4 hr; metabolized by liver; moderate CYP2C19 inhibitor; weak/moderate CYP3A4 inducer; steady state 4-5 days; excreted in urine, feces; half-life 13-20 hr; protein binding <40%

INTERACTIONS
Decrease: effects of bedaquiline, boceprevir, bosutinib, cabozantinib, cobicistat, elvitegravir, emtricitabine, crizotinib, cycloSPORINE, dronedarone, erlotinib, fosamprenavir, galantamine, gefitinib, HYDROcodone, maraviroc, oxy-

CODONE, paliperidone, perm panel, pimozide, praziquantel, QUEtiapine, ranolazine, rilpivirine
Decrease: eslicarbazepine effect CYP1A2, CYP2C19 substrates
Decrease: effect of CYP3A inducers

NURSING CONSIDERATIONS

Assess:

• Seizures: character, location, duration, intensity, frequency, presence of aura
• Hepatic studies: ALT, AST, bilirubin; sodium

Black Box Warning: Mental status: mood, sensorium, affect, behavioral changes, suicidal thoughts/behaviors; if mental status changes, notify prescriber

• Eye problems: need for ophthalmic examinations before, during, after treatment (slit lamp, fundoscopy, tonometry)
• Allergic reaction: purpura, red, raised rash; if these occur, product should be discontinued

Evaluate:

• Therapeutic response: decreased seizure activity; document on patients chart

Teach patient/family:

• To carry emergency ID stating patient's name, products taken, condition, prescriber's name, and phone number
• To avoid driving, other activities that require alertness usually for the first 3 days of treatment
• Not to discontinue medication quickly after long term use
• To notify if pregnancy is planned or suspected, pregnancy (C); to avoid breastfeeding
• Not to abruptly discontinue drug
• To report signs of decreased renal function, dizziness, increased cholesterol, ocular toxicity, suicide risk, skin rashes
• To take with or without food; that tablet can be crushed

ibrutinib

(eye-broo′ti-nib)
Imbruvica
Func. class.: Antineoplastic-biologic response modifier
Chem. class.: Signal Transduction inhibitor (STIs)

ACTION: Irreversible inhibitor of Bruton's tyrosine kinases in B cells responsible for tumor growth

USES: Recurrent mantle cell lymphoma who have received at least 1 prior treatment

CONTRAINDICATIONS: Pregnancy (D), breastfeeding, hypersensitivity
Precautions: Children, geriatric patients, active infections, anticoagulant therapy, bleeding, hepatic/renal disease, neutropenia, surgery

DOSAGE AND ROUTES

• **Adult: PO** 560 mg (4×140 mg caps) q day
• **Dosage adjustment for ≥grade 3 non-hematologic, ≥grade 3 neutropenia with infection or fever, or grade 4 hematologic toxicities:** Interrupt therapy; resume upon recovery to grade 1 or baseline as indicated below:
• **First occurrence:** Resume dosing at original dose (daily dose = 560 mg/day)
• **Second occurrence:** Reduce dose by 1 capsule (daily dose = 420 mg/day)
• **Third occurrence:** Reduce dose by 2 capsules (daily dose = 280 mg/day)
• **Fourth occurrence:** Discontinue

Available forms: Caps 140 mg

Administer:

• At same time of day with water, if dose is missed, take as soon as possible on same day, do not double
• Do not open, break, chew cap

SIDE EFFECTS:

CNS: Fatigue, fever
CV: Hypertension, atrial fibrillation, peripheral edema
EENT: Sinusitis
GI: Nausea, vomiting, dyspepsia, anorexia, abdominal pain, constipation, stomatitis, diarrhea, GI bleeding
GU: Increased serum creatinine, UTI
HEMA: Neutropenia, thrombocytopenia, anemia, bleeding, epistaxis, transient lymphocytosis
INTEG: Rash, skin infections
MS: Pain, arthralgia, muscle cramps
RESP: Cough, dyspnea
SYST: Secondary malignancy, infection

PHARMACOKINETICS

Protein binding 97.3%; metabolized by CYP3A4/CYP2D6; primarily excreted in feces, small amount in urine; peak 1-2 hr, terminal half-life 4-8 hr

INTERACTIONS

Increase: ibrutinib effect—avoid use with moderate or strong CYP3A4 inhibitors
Decrease: ibrutinib effect—avoid use with moderate or strong CYP3A4 inducers
Drug/Herb
Decrease: SUNItinib concentration—St. John's wort
Drug/Food Test
Increase: plasma concentrations grapefruit juice

NURSING CONSIDERATIONS

Assess:
- **Bleeding:** bruising, grade 3 or higher bleeding events may occur; CBC
- Hepatic/renal function; signs & symptoms infections

Evaluate:
- Therapeutic response: decrease in progression of disease

Teach patient/family:
- To report adverse reactions immediately: SOB bleeding
- About reason for treatment, expected result
- That many adverse reactions may occur: high B/P, bleeding, mouth swelling
- To avoid persons with known upper respiratory infections; that immunosuppression is common
- To avoid grapefruit juice or medications, herbs; there are many interactions
- To report if pregnancy is planned or suspected, pregnancy (D)
- To report bleeding, severe infections, renal toxicity (maintain hydration), development of second malignancies, diarrhea (contact physician if it persists)
- To take with water (avoid food because it increases drug levels) at same time each day; do not open, break, or chew

levomilnacipran

(lee′voe-mil-na′si-pran)

Fetzima

Func. class: Antidepressant
Chem. class: Serotonin

ACTION: May potentiate serotonergic, adrenergic activity in the CNS; is a potent inhibitor of adrenal serotonin and norepinephrine reuptake

USES: Major depressive disorder in adults

CONTRAINDICATIONS: Alcohol intoxication, alcoholism, closed angle glaucoma, hepatic disease, hepatitis, jaundice, hypersensitivity

Precautions: Pregnancy (C), breastfeeding, geriatric patients, mania, hypertension, renal/cardiac disease, seizures, increased intraocular pressure, anorexia, bleeding, dehydration, diabetes, hypotension, hypovolemia, orthostatic hypotension, abrupt product withdrawal

Black Box Warning: Children, suicidal ideation

DOSAGE AND ROUTES

- **Adult: PO** 20 mg/day ×2 days, then 40 mg/day; may increase in increments of 40 mg at intervals of at least 2 days, max 120 mg/day; max 80 mg/day (strong CYP3A4 inhibitors therapy)

Renal Dose

- **Adult: PO** CCr 30-59 ml/min max 80 mg/day; 15-29 ml/min, max 40 mg/day; <15 ml/min; avoid use

Available forms: Ext rel caps 20, 40, 80, 120 mg

Administer:

- Swallow cap whole; do not break, crush, or chew; do not sprinkle on food or mix with liquid
- Without regard to food
- Give at the same time each day

SIDE EFFECTS

CNS: Dizziness, agitation, hallucinations, seizures, drowsiness, mania, migraine, paresthesias, serotonin syndrome, suicidal ideation, syncope

CV: Hypertension, palpitations, dysrhythmia, sinus tachycardia

EENT: Teeth grinding, blurred vision

GI: Constipation, diarrhea, nausea, vomiting, anorexia, dry mouth, abdominal pain

GU: Urinary retention

INTEG: Stevens-Johnson syndrome

SYST: Serotonin syndrome, Stevens Johnson syndrome

PHARMACOKINETICS

Peak 6-8 hr metabolized (CYP2D6) in the liver; 22% protein binding, half-life 12 hr

INTERACTIONS

- Do not use with linezolid or methylene blue IV
- Increase levomilnacipran effect—CYP34A Inhibitors

Black Box Warning: Hyperthermia, rigidity, rapid fluctuations of vital signs, mental status changes, neuroleptic malignant syndrome-MAOIs coadministration contraindicated within 14 days of MAOI

Increase: Serotonin syndrome, SSRIs, serotonin receptor agonists, SNRIs

Increase: bleeding risk—anticoagulants, antiplatelets, salicylates, NSAIDs

Drug/Herb

- Serotonin syndrome: St. John's wort

Increase: CNS depression—kava valerian

NURSING CONSIDERATIONS

Assess:

Black Box Warning: Depression: mood, sensorium, affect, suicidal tendencies, increase in psychiatric symptoms; depression, panic, monitor, children weekly face to face during first 4 wk, or dosage change, then every other week for the next 4 wk, then at 12 wk

- B/P lying, standing; pulse q4hr; if systolic B/P drops 20 mm Hg, hold product, notify prescriber; take VS q4hr in patients with CV disease
- **Hepatic studies:** AST ALT, bilirubin
- **Withdrawal symptoms:** headache, nausea, vomiting, muscle pain, weakness; not common unless product is discontinued abruptly

Black Box Warning: Serotonin syndrome: nausea/vomiting, dizziness, facial flush, shivering, sweating

Evaluate:

- Therapeutic response: decreased depression

Teach patient/family:

- About signs and symptoms of bleeding (GI bleeding, nosebleed, ecchymosis, bruising)
- To use with caution when driving and forming other activities requiring alertness because of drowsiness and blurred vision
- To avoid alcohol ingestion, MAOIs, other CNS depressants
- Not to discontinue medication quickly after long term use; may cause headache, malaise; taper

Black Box Warning: That clinical worsening and suicidal risk may occur

• To notify prescriber if pregnancy is planned or suspected or if breastfeeding
• Improvement may occur in 4-8 wk or up to 12 wk (geriatric patients)

macitentan

(ma′si-ten′tan)

Opsumit

Func. class.: Antihypertensive
Chem. class.: Vasodilator/endothelin receptor antagonist

ACTION:
Prevents the binding of ET-1, to ETA and ETB receptors on human pulmonary arterial smooth muscle

USES:
Pulmonary arterial hypertension; WHO Group 1 to delay disease progression

CONTRAINDICATIONS:
Pregnancy (X), breastfeeding, hypersensitivity

Precautions: Anemia, hepatic disease, pulmonary edema

DOSAGE AND ROUTES

• **Adult: PO** 10 mg/day

Available forms: Tabs 10 mg

Administer:

• Do not break, crush, chew tabs, without regard to food; if a dose is missed, then take when remember to (do not take more than one tab per day)
• Do not discontinue abruptly

SIDE EFFECTS:

CNS: Headache
GI: Hepatotoxicity
GU: Decreased sperm counts
HEMA: Anemia
RESP: Pharyngitis, pulmonary edema, bronchitis

PHARMACOKINETICS

Rapidly absorbed, peak 2 hr, protein binding 99%, metabolized by CYP3A4, CYP2C19, terminal half-life 15 hr, effective half-life 9 hr

INTERACTIONS

Increase: Macitentan effect CYP3A4 inhibitors

Drug/Herb

• Need for macitentan dosage change: St. John's wort, ephedra (ma huang)

NURSING CONSIDERATIONS

Assess:

• **Pulmonary status:** improvement in breathing, ability to exercise; pulmonary edema that may indicate veno-occlusive disease
• **Blood studies:** CBC with differential; Hct, Hgb may be decreased
• **Liver function tests:** AST, ALT, bilirubin

Black Box Warning: Assess pregnancy status before giving this product; pregnancy (X)

Evaluate:

• Therapeutic response: decrease in B/P, decreased shortness of breath

Teach patient/family:

• The importance of complying with dosage schedule, even if feeling better
• To notify if pregnancy is planned or suspected (if pregnant, product will need to be discontinued, pregnancy test done monthly); to use 2 contraception methods while taking this product
• Not to use OTC products including herbs and supplements, unless approved by prescriber
• To report to prescriber immediately: dizziness, faintness, chest pain, palpitations, uneven or rapid heart rate, headache, edema, weight gain
• Do not split, crush, or chew tabs; if a dose is missed, take as soon as remembered (do not take more than one tab per day); there are many drug interactions
• The signs and symptoms of hepatotoxicity

mipomersen

(mye′poe-mer-sen)

Kynamro

Func. class: Antilipemics

Chem. class: Antisense oligonucleotide

ACTION: Inhibits synthesis of the principal apolipoprotein of LDL, VLDL, binds to messenger ribonucleic acid (mRNA)

USES: Decreasing LDL, total cholesterol, apolipoprotein B, non-high density lipoprotein cholesterol to reduce LDL, total cholesterol, apolipoprotein B, non-high density lipoprotein cholesterol (homozygous familial hypercholesterolemia)

CONTRAINDICATIONS:

Black Box Warning: Hepatic disease

Precautions: Pregnancy (B), breastfeeding, dialysis, alcohol ingestion, geriatrics, proteinuria, renal disease, low density lipoprotein apheresis

DOSAGE AND ROUTES

Homozygous familial hypercholesterolemia

- **Adult: Subcut** 200 mg weekly

Renal/hepatic dose

- **Adult: SUBCUT** do not use in severe renal, hepatic disease

Available forms: Solution for injection 200 mg/ml

Administer:

- Give on the same day weekly; if a dose is missed, give $\geq$3 days from the next weekly dose
- Monitor ALT, AST, alkaline phosphatase, and total bilirubin before start of therapy; monitor lipid levels $\geq$q3-4mo for the first year
- Monitor LDL-C level after 6 mo
- Dose adjustments for elevated transaminases during treatment
- ALT or AST $\geq 3\times$ and $<5\times$ ULN: confirm elevation with a repeat test within 1 wk; if confirmed, withhold product; obtain other tests if not already obtained (total bilirubin, alkaline phosphatase, INR) to identify the probable cause; if resuming product after transaminases resolve to $<3\times$ ULN, consider monitor liver-related tests more frequently
- ALT or AST $\geq 5\times$ ULN: withhold, obtain additional liver-related tests if not already obtained (total bilirubin, alkaline phosphatase, INR), and identify the probable cause; if resuming product after transaminases resolve to $<3\times$ ULN, monitor liver-related tests more frequently

SIDE EFFECTS

CNS: Fatigue, headache, fever
CV: Hypertension, palpitations
GI: Abdominal pain, vomiting
GU: Glomerulonephritis, proteinuria
MS: Musculoskeletal pain
SYST: Angioedema

PHARMACOKINETICS

Half-life 1-2 mo protein binding $>$90%

INTERACTIONS

Increase: hepatotoxicity risk—acetaminophen, methotrexate, tetracyclines, tamoxifen

NURSING CONSIDERATIONS

Assess:

- Determine if the LDL-C reduction achieved is sufficient to warrant the potential risk of liver toxicity
- Assess geriatric patients: increased risk for hypertension, peripheral edema, hepatic steatosis
- Hypercholesterolemia: diet history: fat content, lipid levels (triglycerides, LDL, HDL, cholesterol); LFTs are at baseline, periodically during treatment

Evaluate:

- Therapeutic response: decreasing LDL, total cholesterol, apolipoprotein B, non-high density lipoprotein cholesterol

Teach patient/family:
- That compliance is needed
- That risk factors should be decreased: high fat diet, smoking, alcohol consumption, absence of exercise
- To notify prescriber if pregnancy is suspected, planned, or if breastfeeding
- To notify prescriber of dietary/herbal supplements

obinutuzmab

(oh'bi-nue-tooz'ue-mab)

Gazyva

Func. class.: Antineoplastic; biologic response modifier

ACTION: A recombinant, human monoclonal antibody that binds to the gastric B-lymphocyte-associated antibody; action is indirect, possible through T-cell-mediated anti-tumor responses

USES: Chronic lymphocytic leukemia

CONTRAINDICATIONS: Hypersensitivity

Precautions: Pregnancy (C), breastfeeding, cardiac disease, children, human antichimeric antibody (HACA), human antimurine antibody (HAMA), infection, inf-related reactions, neutropenia, pulmonary disease, thrombocytopenia, tumor lysis syndrome, vaccination

Black Box Warning: Hepatitis, progressive multifocal leukoencephalopathy

DOSAGE AND ROUTES

- **Adult: IV cycle 1** 100 mg over 4 hr (day 1); then 900 mg (0 mg/hr, increased by 50 mg/hr q30min, to max 400 mg/hr) (day 2); then 1000 mg (100 mg/hr, increased by 100 mg/hr q30min to max 400 mg/hr (day 8, day 115); **cycle 2-6**; 1000 mg (100 mg/hr increased by 100 mg/hr q30min to max 400 mg/hr (day 1 repeat q28days)

Available forms: Sol for inj 1000 mg/40 ml

Administer:

IV intermittent inf route
- Due to the risk of hypotension, consider withholding antihypertensive medications for 12 hr before, during, and for the 1st hr after use until blood pressure is stable
- Give antimicrobial prophylaxis to neutropenic patients throughout treatment; consider antiviral and antifungal prophylaxis as needed
- **Premedication for cycle 1, days 1 and 2:** acetaminophen 650-1000 mg, and diphenhydrAMINE 50 mg at least 30 min before inf, dexamethasone 20 mg IV or methylPREDNISolone 80 mg IV at least 1 hr before inf
- **Premedication for cycle 1, days 8 and 15 and cycles 2-6, day 1:** acetaminophen 650-1000 mg at least 30 min before inf; those with any inf-related reaction with the previous inf should also receive diphenhydrAMINE 50 mg at least 30 min before the inf; if the patient had a grade 3 inf-related reaction with the previous dose or has a lymphocyte count >25 ×109/L, additionally administer dexamethasone 20 mg IV or methylPREDNISolone 80 mg IV at least 1 hr before inf
- Use in a facility to adequately monitor and treat inf reactions
- Visually inspect parenteral products for particulate matter and discoloration before use
- Prepare all doses in 0.9% NaCl; do not admix; use a final conc of 0.4-4 mg/mL; give as an IV inf only

Reconstitution: Cycle 1, days 1 and 2:
- Withdraw 4 mL (100 mg) from the vial and dilute into 100 mL 0.9% NaCL use on day 1; mix by gentle inversion; do not shake, use immediately
- Withdraw the remaining 36 mL (900 mg) and dilute into 250 mL 0.9% NaCl for use on day 2, mix by gentle inversion; do not shake

Cycle 1, days 8 and 15; cycles 2-6:
- Withdraw 40 mL (1000 mg) from the vial and dilute into 250 mL 0.9% NaCl; mix by gentle inversion; do not shake

- Store following reconstitution: store at 2° C-8° C (36° F-46° F) for up to 24 hr; do not freeze; allow to come to room temperature before administration; use a dedicated line
- ***Day 1 (100 mg dose):*** dive at an initial rate of 25 mg/hr over 4 hr; do not increase the inf rate
- ***Day 2 (900 mg dose)***: give at an initial rate of 50 mg/hr ×30 min; if no hypersensitivity or infusion-related events occur, increase the infusion rate by 50 mg/hr q30min to a max rate of 400 mg/hr; if a grade 1-2 inf-related reaction occurs, the inf should be temporarily interrupted or the rate reduced; the inf may be resumed or continued at a reduced rate upon improvement of the patient's symptoms; if the reaction does not recur, the rate may be increased as appropriate for the current cycle and dose; if a grade 3 hypersensitivity or inf-related event develops, the inf should be temporarily interrupted; upon improvement of the patient's symptoms, the inf can be resumed at half the rate being used at the time that the reaction occurred; if the reaction does not recur, the rate may be increased as appropriate for the current cycle and dose; if a grade 4 hypersensitivity or inf-related event develops, discontinue the inf and do not resume
- ***Subsequent infs (1000 mg dose):*** give at an initial rate of 100 mg/hr for 30 min; if no hypersensitivity or inf-related events occur, increase the inf rate by 100 mg/hr q30min to a max rate of 400 mg/hr; if a grade 1-2 inf-related reaction occurs, the inf should be temporarily interrupted or the rate reduced; the inf may be resumed or continued at a reduced rate upon improvement of the patient's symptoms; if the reaction does not recur, the rate may be increased as appropriate for the current cycle and dose; if a grade 3 hypersensitivity or inf-related event develops, the infusion should be temporarily interrupted; upon improvement of the patient's symptoms, the inf can be resumed at half the rate being used at the time that the reaction occurred; if the reaction does not recur, the rate may be increased as appropriate for the current cycle and dose; if a grade 4 hypersensitivity or inf-related event develops, discontinue the inf and do not resume

SIDE EFFECTS

CNS: Headache, fever, chills, flushing
CV: Cardiac arrest, MI, sinus tachycardia, hypertension
GI: Constipation, decreased appetite, diarrhea, hepatitis/hepatic failure nausea, vomiting
HEMA: Neutropenia, thrombocytopenia, lymphopenia, leukopenia
META: Lower potassium/sodium/calcium, aluminum, higher potassium/uric acid
RESP: Wheezing, dyspnea
SYST: Tumor lysis syndrome

PHARMACOKINETICS

Terminal half-life 28.4 days

NURSING CONSIDERATIONS

Assess:

Black Box Warning: Hepatitis B: reactivation of HBV in those who are HBsAg positive, HBsAg negative, and core antibody anti-HBc positive; may result in fulminant hepatitis, hepatic failure, or death

Black Box Warning: Progressive multifocal leukoencephalopathy (PML): notify prescriber any new, worsening neurological signs/symptoms (ataxia, visual changes, confusion)

- **Tumor lysis syndrome:** can occur within 24 hr of 1st infusion; those with high tumor burden or lymphocyte count >25 ×10 to the 9th/l are at increased risk; monitor serum creatinine, potassium, calcium, uric acid, phosphate closely
- **Severe/life-threatening inf reactions:** 2/3 have a reaction to 1st dose; consider withholding antihypertensives for 12 hr before, during, and after 1st hr of inf

Evaluate:

- Decreased disease progression

Teach patient/family:

- About the reason for treatment and expected results

pomalidomide

(pom–a-lid′oh-mide)

Pomalyst

Func. class.: Antineoplastic, biologic response modifier, hormone

Chem. class.: TNF modifier

ACTION: Inhibits growth of tumor cells and induces apoptosis, can be used in those resistant to lenalidomide

USES: Multiple myeloma in those who have received ≥2 treatments including lenalidomide and bortezomib and disease has progressed within 60 days of completion of the treatment

CONTRAINDICATIONS: Breastfeeding, hypersensitivity

Black Box Warning: Pregnancy (X)

Precautions: Children geriatric patients, accidental exposure, bone marrow suppression, uterine bleeding, dental disease, fungal/viral infections, smoking

Black Box Warning: Thrombocytopenia

DOSAGE AND ROUTES

- **Adult: PO** 4 mg on days 1-21

Hepatic/renal dose:

- Adult: PO bilirubin >2 mg/dl and AST/ALT >3× ULN or CCr >3 mg/dl, do not use

Available forms: Tabs 1, 2, 3, 4 mg

Administer:

- With or without dexamethasone, may use dexamethasone 40 mg on days 1, 8, 15, 22 of each cycle, if >75 yr, decrease dexamethasone to 20 mg/dose

SIDE EFFECTS

CNS: Dizziness, fatigue, fever, headache, peripheral neuropathy

CV: Chest pain

GI: Constipation, diarrhea, nausea/vomiting

HEMA: Leukopenia, neutropenia, thrombocytopenia

META: Hypokalemia

MS: Arthralgia, back pain

RESP: Cough, dyspnea, pulmonary embolism, epistaxis

SYST: Secondary malignancy

PHARMACOKINETICS

12%-44% protein binding

INTERACTIONS

Increase: pomalidomide effect—CYP3A4 inhibitors (amprenavir, boceprevir, delavirdine, ketoconazole, indinavir, itraconazole, dalfopristin/quinupristin, ritonavir, tipranavir, fluconazole, isoniazid, miconazole); P-gb inhibitors; avoid using together

Decrease: pomalidomide effect—CYP3A4 inducers (rifampin, rifapentine, rifabutin, primidone, phenytoin, PHENObarbital, nevirapine, nafcillin, modafinil, griseofulvin, etravirine, efavirenz, barbiturates, bexarotene, bosentan, carBAMazepine, enzalutamide, dexamethasone); avoid using together

Drug/Herb

- Decreased pomalidomide effect—St. John's wort, avoid using together

Drug/Smoking

Decrease: pomalidomide effect—cigarettes

NURSING CONSIDERATIONS

Assess:

Black Box Warning: Blood studies: Hct, Hgb, thrombolytic disease may occur

Black Box Warning: For pregnancy before treatment, pregnancy (X)

Evaluate:

- Therapeutic response: decreased growth of tumor cells

Team patient/family:

• To avoid driving or hazardous activity during beginning of treatment

Black Box Warning: Advise patient to use adequate contraception, pregnancy (X)

riociguat

(rye′oh-sig′ue-at)

Adempas

Func. class.: Antihypertensive/vasodilator

Chem. class.: Guanylate cyclase stimulator

ACTION: Guanylate cyclase stimulator also a vasoconstrictor

USES: WHO group I pulmonary arterial hypertension and WHO group IV persistent/recurrent chronic thromboembolic pulmonary hypertension

CONTRAINDICATIONS: Breastfeeding hypersensitivity

Black Box Warning: Pregnancy (X)

Precautions: Hypotension, hypovolemia, pulmonary edema, tobacco smoking

DOSAGE AND ROUTES

• **Adult: PO** 1 mg tid; may start at 0.5 mg tid if needed; if systolic B/P remains >95 mmHg increased by 0.5 mg tid; increase no sooner than 2 wk, max 2.5 mg tid

Renal/hepatic dose

• **Adult:** Not recommended in CCr <15 ml/min or Child-Pugh C

Available forms: Tabs 0.5, 1, 1.5, 2, 2.5 mg

Administer:

• tid without regard to food
• Do not discontinue abruptly
• If dose is missed, take at next scheduled dose; if stopped for ≥3 days, begin with initial dose

SIDE EFFECTS

CNS: Dizziness
CV: Hypotension, peripheral edema, palpitations
EENT: Sinusitis, rhinitis
GI: Constipation, gastritis, gastroesophageal reflux, nausea, vomiting
GU: Decreased sperm counts
HEMA: Anemia, bleeding, nosebleeds
RESP: Pulmonary edema

PHARMACOKINETICS

95% protein binding, peak 1.5 hr, metabolized by P-gb, CYP1A1, CYP3A4, CYP3A, CYP2C8, CYP2J2, terminal half-life 15 hr, half-life 12 hr

INTERACTIONS

• Do not use with nitrates, nitric oxide donors, phosphodiesterase-5 inhibitors
• Decreased riociguat effect: tobacco smoking

Increase: riociguat—CYP3A4 inhibitors (amprenavir, aprepitant, atazanavir, clarithromycin, conivaptan, cycloSPORINE, dalfopristin, danazol, darunavir, erythromycin, estradiol, imatinib, itraconazole, ketoconazole, nefazodone, nelfinavir, propoxyphene, quinupristin, ritonavir, RU-486, saquinavir, tamoxifen, telithromycin, troleandomycin, zafirlukast)

Drug/Herb

• Need for dosage change: St. John's wort, ephedra (ma huang)

NURSING CONSIDERATIONS

Assess:

• Pulmonary status: improvement in breathing, ability to exercise; pulmonary edema may indicate venoocclusive disease

Black Box Warning: Assess pregnancy status before giving this product; pregnancy (X)

Evaluate:

• Therapeutic response: decrease in B/P; decreased shortness of breath

Teach patient/family:
- The importance of complying with dosage schedule even if feeling better

Black Box Warning: To notify if pregnancy is planned or suspected (if pregnant, product will need to be discontinued, pregnancy test done monthly); to use 2 contraception methods while taking this product

- Not to use OTC products including herbs, supplements unless approved by prescriber

simeprevir
(sim-e′pre-vir)

Olysio

Func. class.: Antiviral, anti-hepatitis agent

ACTION: Prevents formation of mature viral proteins by inhibiting hepatitis C viral (HCV) replication by blocking proteolytic activity of the hepatitis C virus NS3/4A protease

USES: Chronic C infection, hepatitis, (HCV, genotype 1) in adults with compensated liver disease

CONTRAINDICATIONS: Hypersensitivity, male-mediated teratogenicity, pregnancy (X) in combination

Precautions: Pregnancy (C), (alone), breastfeeding, Asian patients, liver transplant, UV exposure, hepatic disease, serious rash, children

DOSAGE AND ROUTES
- **Adult: PO** 150 mg daily with food with peginterferon alfa and ribavirin ×12 wk

Available forms: Cap 150 mg

Administer:
- By mouth with food; swallow whole
- Do not use monotherapy; if peginterferon alfa or ribavirin are discontinued, discontinue simeprevir and do not restart
- Obtain NS3 Q8OK polymorphism test before starting treatment
- Treatment naïve/prior relapse: after initial 3 drug treatment, give another 12 wk with peginterferon alfa and ribavirin × another 12 wk; if HCV RNA concs ≥25 interstitial units/ml, discontinue all 3 drugs
- Prior non responders (partial/null): after initial 3 drug treatments, give an additional 36 wk of peginterferon alfa and ribavirin; monitor HCV RNA at wk 4, 12, 24; if ≥25 interstitial units/ml, discontinue products

SIDE EFFECTS
EENT: Blurred vision, conjunctivitis
GI: Hyperbilirubinemia, nausea
INTEG: Exfoliative dermatitis, pruritus, photosensitivity, vasculitis
MISC: Rash, dyspnea

PHARMACOKINETICS
99.9% protein binding, excreted 91% feces half-life 40 hrs, metabolized primarily by CYP3A4, also a mild inhibitor of intestinal CYP3A4 and CYP1A2

INTERACTIONS
Increase: effects of CYP3A4 inhibitors
Increase: simeprevir effects—bromocriptine, chloramphenicol, cimetidine, dalfopristin/quinupristin, danazol, FLUoxetine, isoniazid, lanreotide, octreotide, zafirlukast, erythromycin

NURSING CONSIDERATIONS
Assess:
- Allergic reaction: rash, pruritus, exfoliative dermatitis
- Asian patients: levels may be 3.4 fold higher
- Pregnancy: if planned or suspected; if pregnant call the Pregnancy Registry 800-258-4263, pregnancy (X) in combination
- Liver function tests; serum HCV RNA concentrations at wk 4, 12, and 24 and at end of treatment; pregnancy tests

Evaluate:
- Therapeutic response: decreased symptoms of chronic hepatitis C

Teach patient/family:
• That optimal duration of treatment is unknown; that product is not a cure; that transmission may still occur
• To avoid use with other medications unless approved by prescriber
• To use protective clothing, sunscreen; photosensitivity may occur and can be severe
• Not to stop abruptly unless directed; worsening of hepatitis may occur
• To notify prescriber if pregnancy is planned or suspected; to avoid breastfeeding; use 2 reliable forms of contraception; do not try to conceive for at least 6 mo after discontinuation of this drug combination
• Teach patient to use cautiously in sulfonamide allergy; do not take the drug as single agent; must use as combination therapy

sofosbuvir
(soe-fos′bue-vir)
Sovaldi
Func. class.: Antiviral, anti hepatitis agent
Chem. class.: Adenosine monophosphate analog

ACTION: Inhibits hepatitis C virus RNA polymerase by incorporating the polymerase into the viral RNA; also acts as a chain terminator

USES: Chronic hepatitis C (genotypes 1, 2, 3, 4) with compensated liver disease

CONTRAINDICATIONS: Hypersensitivity, pregnancy (X) in combination; male-mediated teratogenicity
Precautions: Pregnancy (B), breastfeeding, children, hepatic/renal disease

DOSAGE AND ROUTES
Genotype 1, 4
• **Adult: PO** 400 mg daily with peginterferon alfa and ribavirin ×12 wk; may consider use for genotype 1 with on ribavirin ×24 wk
Genotype 2
• **Adult: PO** 400 mg daily with ribavirin ×12 wk
Genotype 3
• **Adult: PO** 400 mg daily with ribavirin ×24 wk
Available forms: Tabs 400 mg
Administer:
• By mouth without regard to food
• Do not use as mono therapy

SIDE EFFECTS
CNS: Headache, chills, weakness, fatigue, fever, insomnia
GI: Diarrhea, hyperbilirubinemia
MISC: Rash, pruritus, neutropenia, anemia, myalgia

PHARMACOKINETICS
PO: Peak ½-2 hr, excreted by kidneys 80%, 61%-65% protein binding; half-life 0.4-27 hr

INTERACTIONS
Decrease: sofosbuvir-P-glycoprotein (P-gp) inducers (carBAMazepine, PHENobarbital, phenytoin, rifampin); OXcarbazepine, rifabutin, rifapentine, tipranavir; avoid concurrent use

NURSING CONSIDERATIONS
Assess:
• Severe renal disease/GFR <30 ml/min/1.73 m^2: monitor BUN, creatinine
• Geriatric patients more carefully; may develop renal, cardiac symptoms more rapidly
• Pregnancy: if planned or suspected; if pregnant call the Pregnancy Registry 800-258-4263; obtain pregnancy test before starting treatment
Evaluate:
• Therapeutic response: decreased symptoms of chronic hepatitis C

Teach patient/family:

- That optimal duration of treatment is unknown; that product is not a cure; that transmission may still occur
- To avoid use with other medications unless approved by prescriber
- Not to stop abruptly unless directed; worsening of hepatitis may occur
- To notify prescriber if pregnancy is planned or suspected; use 2 forms of reliable contraception; avoid breastfeeding

teduglutide

(te′due-gloo′tide)

Gattex

Func. class.: Functional GI disorder agent

Chem. class.: Recombinant glucagon-like peptide-2 analog

ACTION: Increases intestinal, portal blood flow, inhibits gastric acid secretion, decreases gastric motility

USES: Short bowel syndrome, dependent on parenteral support

Precautions: Pregnancy (C), breastfeeding, diarrhea, pancreatitis, renal disease, electrolyte imbalances, GI obstruction, heart failure, neoplastic disease, biliary tract disease, cardiac disease, gastric cancer, GI disease

DOSAGE AND ROUTES

- **Adult: SUBCUT** 0.05 mg/kg daily

Available forms: Powder for injection 5 mg

Administer:

Subcut route

- ***Reconstitution:*** Slowly inject the 0.5 ml of preservative-free sterile water for injection provided in the prefilled syringe into the vial; allow to stand for 30 sec and gently roll the vial between your palms for 15 sec; do not shake, allow to stand 2 min; if undissolved powder is present, roll the vial again until all material is dissolved; if the product remains undissolved after the second attempt, do not use; discard unused portion; use within 3 hr after reconstitution
- ***Subcut injection:*** Calculate dose, withdraw into a syringe, and give; do not use IV, IM; if a dose is missed, it should be given as soon as possible on the same day; two doses should not be given on the same day; alternate sites

SIDE EFFECTS

CNS: Headache, fatigue

GI: Nausea, abdominal pain, cholecystitis, cholestasis, GI obstruction, pancreatitis, vomiting

PHARMACOKINETICS

Half-life 2 hr; 1.3 hr (short bowel syndrome)

INTERACTIONS

Increase: absorption of—benzodiazepine, carBAMazepine, cycloSPORINE, digoxin, disopyramide, ethosuximide, flecainide, levothyroxine, lithium, phenytoin, procainamide, quiNIDine, sirolimus, tacrolimus, theophylline, valproic acid, warfarin

NURSING CONSIDERATIONS

Assess:

- GI symptoms: nausea, abdominal pain
- Monitor alk phos, amylase, bilirubin, serum electrolytes, lipase baseline, q6mo

Evaluate:

- Therapeutic response: increased absorption of nutrients

Teach patient/family:

- To notify prescriber of GI symptoms

trametinib

(tra-me′ti-nib)

MeKinist

Func. class.: Antineoplastic biologic response modifiers

Chem. class.: Signal transduction inhibitors (STIs), tyrosine kinase inhibitor

ACTION: Inhibits MEK-1, MEK-2 tyrosine kinase created in patients with malignant melanoma

USES: Unresectable or metastatic BRAD V600E or BRAF V600K mutated malignant melanoma

CONTRAINDICATIONS: Pregnancy (D), hypersensitivity

Precautions: Breastfeeding, children, diarrhea, geriatric patients, hepatic disease, bone marrow suppression, infection, thrombocytopenia, neutropenia, immunosuppression

DOSAGE AND ROUTES

• **Adult: PO** 2 mg daily; may be used in combination with dacarbazine or PACLitaxel in those with BRAF V600E

Management of treatment-related toxicity

Cutaneous toxicity:

• *Grade 2 rash:* reduce dose by 0.5 mg (e.g., 2 mg/day to 1.5 mg/day) or discontinue in patients who are receiving trametinib 1 mg/day; in patients with an intolerable grade 2 rash that does not improve within 3 wk of a dosage reduction, withhold trametinib for up to 3 wk; if the rash is improved within 3 wk, resume therapy at a lower dose (reduce the previous dose by 0.5 mg); discontinue in patients who are receiving trametinib 1 mg/day; if the rash does not improve within 3 wk, permanently discontinue

• *Grade 3 or 4 rash:* withhold for up to 3 wk; if the rash is improved within 3 wk, resume at a lower dose (reduce the previous dose by 0.5 mg); discontinue in patients who are receiving trametinib 1 mg/day; if the rash does not improve within 3 wk, permanently discontinue therapy

Cardiac toxicity:

• Asymptomatic cardiac toxicity and an absolute decrease in LVEF of ≥10% from baseline and is below institutional lower limits of normal (LLN) from pretreatment value: withhold for up to 4 wk; if the LVEF is improved within 4 wk, resume at a lower dose (reduce the previous dose by 0.5 mg); discontinue in patients who are receiving 1 mg/day; if the LVEF does not improve to normal within 4 wk, permanently discontinue

• Symptomatic CHF or an absolute decrease in LVEF of >20% from baseline and is below institutional LLN: permanently discontinue

Ocular toxicity:

• *Grade 2 or 3 retinal pigment epithelial detachment (RPED):* withhold for up to 3 wk; if the RPED improves to grade 1 or less within 3 wk, resume at a lower dose (reduce the previous dose by 0.5 mg); discontinue therapy in patients who are receiving 1 mg/day; if the RPED does not improve to at least grade 1 within 3 wk, permanently discontinue

• Retinal vein occlusion: permanently discontinue therapy

Pulmonary toxicity

• Interstitial lung disease/pneumonitis: permanently discontinue

Other toxicity:

• *Grade 3 toxicity:* withhold for up to 3 wk: if the toxicity improves to grade 1 or less within 3 wk, resume at a lower dose (reduce the previous dose by 0.5 mg); discontinue in patients who are receiving 1 mg/day; if the toxicity does not improve to at least grade 1 within 3 wk, permanently discontinue

• *Grade 4 toxicity:* permanently discontinue

Available forms: Tabs 0.5, 2 mg

Administer:

- Give 1 hr before or 2 hr after a meal
- If dose is missed, take within 12 hr of missed dose; if >12 hr have passed, skip dose
- Follow cytotoxic handling procedures

SIDE EFFECTS

CNS: Dizziness
CV: Heart failure, hypertension, cardiomyopathy
EENT: Ocular hemorrhage, retinal detachment, blurred vision
GI: Diarrhea, nausea, vomiting, abdominal pain, stomatitis
GU: Hematuria
INTEG: Rash, pruritus, acne, folliculitis, erythema
MS: Rhabdomyolysis
OTHER: Elevated LFTs, Hand-foot syndrome
RESP: Cough, dyspnea, pleural effusion, pneumonitis, edema

PHARMACOKINETICS

Protein binding 97.4%, peak 1.5 hr, excreted in feces

INTERACTIONS

- None known

NURSING CONSIDERATIONS

Assess:

- LFTs every mo ×3 mo, then as clinically indicated

Evaluate:

- Therapeutic response: decrease in progression of disease

Teach patient/family:

- To report adverse reactions immediately, bleeding
- About reason for treatment, expected results
- To use effective contraception during treatment and up to 30 days after discontinuing treatment

vortioxetine

(vor'tye-ox'e-teen)

Brintellix

Func. class.: Antidepressant

Chem. class.: Serotonin modulator

ACTION:
Reuptake inhibition at the serotonin transporter and agonist, or antagonist effects at serotonin receptors

USES:
Major depressive disorder in adults

CONTRAINDICATIONS:
Hypersensitivity, MAOI therapy

Precautions: Pregnancy (C), breastfeeding, seizure disorder, hypersensitivity, bipolar disorder, hyponatremia, hypovolemia

DOSAGE AND ROUTES

- **Adult: PO** 10 mg/day, may start with 5 mg/day initially, increase to 20 mg/day as tolerated, max 20 mg/day; poor metabolizers of CYP2D6 max 10 mg/day

Available forms: Tabs 5, 10, 20 mg
Administer: Without regard to food

SIDE EFFECTS

CNS: Flushing, mania, serotonin syndrome, vertigo, dizziness, suicidal attempts
GI: Nausea, diarrhea, dyspnea, constipation, vomiting, flatulence
GU: Impotence
INTEG: Pruritus
SYST: Serotonin syndrome, neonatal abstinence syndrome

PHARMACOKINETICS

Protein binding 98%, excreted in urine (59%), feces (26%)

INTERACTIONS

Increase: effect of tricyclics; use cautiously
Increase: serotonin syndrome—serotonin receptor agonists, SSRIs, traMADol,

lithium, MAOIs, traZODone, SNRIs, (venlafaxine, DULoxetine)
Increase: bleeding risk—NSAIDs, salicylates, thrombolytics, anticoagulants, antiplatelets
Increase: CNS effects—barbiturates, sedative/hypnotics, other CNS depressants
Decrease: vortioxetine levels—carBAMazepine
Drug/Herb:
Increase: serotonin syndrome, St. John's wort

NURSING CONSIDERATIONS

Assess:

> **Black Box Warning:** Mental status: mood, sensorium, affect, suicidal tendencies, increase in psychiatric symptoms, depression, panic

- **Serotonin syndrome:** increased heart rate, sweating, dilated pupils, tremors, twitching, hyperthermia, agitation
- Alcohol consumption; if alcohol is consumed, hold dose until AM
- **Sexual dysfunction:** impotence

Evaluate:

- Therapeutic response: decreased depression
- Children, suicidal ideation

Teach patient/family:

- That therapeutic effect may take several wk
- To use caution when driving, performing other activities that require alertness because of drowsiness, dizziness, blurred vision; to report signs, symptoms or bleeding
- To avoid alcohol, other CNS depressants

> **Black Box Warning:** That suicidal ideas, behaviors may occur in children or young adults

- To notify prescriber if pregnant, planning to become pregnant, or breastfeeding

> **Black Box Warning:** About the effects of serotonin syndrome: nausea/vomiting, tremors; if symptoms occur, to discontinue immediately, notify prescriber

APPENDIX A

Appendix B

Ophthalmic, otic, nasal, and topical products

OPHTHALMIC PRODUCTS

ANESTHETICS

lidocaine (Rx)
Akten

proparacaine (Rx)
(proe-par′a-kane)
Alcaine, Diocaine ♣, Parcaine

tetracaine (Rx)
(tet′ra-kane)
Tetcaine, TetraVisc

ANTIHISTAMINES

alcaftadine (Rx)
Lastacaft

emedastine (Rx)
(ee-med′ah-steen)
Emadine

olopatadine (Rx)
(oh-loh-pat′ah-deen)
Pataday, Patanol

ANTIINFECTIVES

azithromycin (Rx)
(ay-zi-thro-my′sin)
AzaSite

besifloxacin (Rx)
(be′si-flox′a-sin)
Besivance

ganciclovir (Rx)
(gan-sye′kloe-vir)
Zirgan

levofloxacin (Rx)
(lee-voh-floks′a-sin)

natamycin (Rx)
(nat-a-mye′sin)
Natacyn

ofloxacin (Rx)
(oh-flox′a-sin)
Ocuflox

silver nitrate 1% (Rx)
silver nitrate

sulfacetamide sodium (Rx)
(sul-fa-seet′a-mide)
Bleph-10

tobramycin (Rx)
(toe-bra-mye′sin)
Tobrasol, Tobrex

trifluridine (Rx)
(trye-floor′i-deen)
Viroptic

β-ADRENERGIC BLOCKERS

betaxolol (Rx)
(beh-tax′oh-lole)
Betoptic S

levobunolol (Rx)
(lee-voe-byoo′no-lole)
Betagen

metipranolol (Rx)
(met-ee-pran′oh-lole)
OptiPranolol

CARBONIC ANHYDRASE INHIBITORS

brinzolamide (Rx)
(brin-zoh′la-mide)
Azopt

CHOLINERGICS (Direct-acting)

acetylcholine Rx)
(ah-see-til-koe′leen)
Miochol-E

carbachol (Rx)
(kar′ba-kole)
Isopto Carbachol

pilocarpine (Rx)
(pye-loe-kar′peen)
Isopto Carpine, Pilopine HS

CHOLINESTERASE INHIBITORS

echothiophate (Rx)
(ek-oh-thye′eh-fate)
Phospholine Iodide

CORTICOSTEROIDS

fluocinolone (Rx)
(floo-oh-sin′oh-lone)
Retisert

fluorometholone (Rx)
(flure-oh-meth′oh-lone)
Flarex, FML, Liquifilm, FML Forte, FML S.O.P.

loteprednol (Rx)
(loe-tee-pred′nole)
Alrex, Lotemax

prednisoLONE (Rx)
(pred-niss′oh-lone)
Econopred, Omnipred, Pred Mild, Pred-Forte

rimexolone (Rx)
(ri-mex′a-lone)
Vexol

MYDRIATICS

atropine (Rx)
(a′troe-peen)
Atropine Care, Isopto Atropine

cyclopentolate (Rx)
(sye-kloe-pen′toe-late)
Cyclogyl, Cylate

homatropine (Rx)
(home-a′troe-peen)
Isopto Homatropine

phenylephrine (OTC)
(fen-ill-ef′rin)
Neofrin, Mydfrin

scopolamine (Rx)
(skoe-pol′a-meen)
Isopto Hyoscine

tropicamide (Rx)
(troe-pik′a-mide)
Mydral, Mydriacyl

NONSTEROIDAL ANTIINFLAMMATORIES

diclofenac (Rx)
(dye-kloe′fen-ak)

flurbiprofen (Rx)
(flure-bih-proh′fen)
Ocufen

nepafenac (Rx)
(ne-pa-fen′ak)
Ilevro, Nevanac

SYMPATHOMIMETICS

apraclonidine (Rx)
(a-pra-klon′i-deen)
Iopidine

OPHTHALMIC DECONGESTANTS/ VASOCONSTRICTORS

Iodoxamide
(loe-dox′a-mide)
Alomide

oxymetazoline (Rx)
(ox-i-meth′oh-leen)
Visine L.R.

tetrahydrozoline (OTC)
(tet-ra-hye-dro′zoe-leen)
Opticlear, Visine Original

MISCELLANEOUS OPHTHALMICS

travoprost (Rx)
(tra′voe-prost)
Travatan Z

unoprostone (Rx)
(un-oh-proe′stone)
Rescula

β-Adrenergic blockers

ACTION: Reduces production of aqueous humor by unknown mechanism

USES: Ocular hypertension, chronic open-angle glaucoma

Anesthetics

ACTION: Decreases ion permeability by stabilizing neuronal membrane

USES: Cataract extraction, tonometry, gonioscopy, removal of foreign objects, corneal suture removal, glaucoma surgery (ophthalmic); pruritus, sunburn, toothache, sore throat, cold sores, oral pain, rectal pain and irritation, control of gagging (topical)

Antiinfectives

ACTION: Inhibits folic acid synthesis by preventing PABA use, which is necessary for bacterial growth

Uses: Conjunctivitis, superficial eye infections, corneal ulcers, prophylaxis against infection after removal of foreign matter from the eye

Antiinflammatories

ACTION: Decreases inflammation, resulting in decreased pain, photophobia, hyperemia, cellular infiltration

USES: Inflammation of eye, eyelids, conjunctiva, cornea; uveitis, iridocyclitis, allergic conditions, burns, foreign bodies, postoperatively in cataract

Carbonic anhydrase inhibitor

ACTION: Converted to EPINEPHrine, which decreases aqueous production and increases outflow

Uses: Open-angle glaucoma, ocular hypertension

Direct-acting miotic

ACTION: Acts directly on cholinergic receptor sites; induces miosis, spasm of accommodation, fall in intraocular pressure, caused by stimulation of ciliary, pupillary sphincter muscles, which leads to pulling away of iris from filtration angle, resulting in increased outflow of aqueous humor

Uses: Primary glaucoma, early stages of wide-angle glaucoma (less useful in advanced stages), chronic open-angle glaucoma, acute closed-angle glaucoma before emergency surgery; also neutralizes mydriatics used during eye exam; may be used alternately with mydriatics to break adhesions between iris and lens

CONTRAINDICATIONS: Hypersensitivity

Precautions: Pregnancy, breastfeeding, children, aphakia, hypersensitivity to carbonic anhydrase inhibitors, sulfonamides, thiazide diuretics, ocular inhibitors, hepatic/renal insufficiency

SIDE EFFECTS

CNS: Headache
CV: Hypertension, tachycardia, dysrhythmias
EENT: Burning, stinging
GI: Bitter taste

NURSING CONSIDERATIONS

Assess:
- Ophth exams and intraocular pressure readings
- Blood counts; hepatic, renal function tests and serum electrolytes during long-term treatment

Perform/provide:
- Storage at room temperature away from light

Evaluate:
- Positive therapeutic response
- Absence of increased intraocular pressure

Teach patient/family:
- How to instill drops
- That product may cause burning, itching, blurring, dryness of eye area

NASAL AGENTS

NASAL ANTIHISTAMINES
NASAL DECONGESTANTS

azelastine (Rx)
(ay-zell′ah-steen)
Astelin

desoxyephedrine (OTC)
(des-oxy-e-fed′rin)
Vicks Vapor Inhaler

EPINEPHrine (OTC)
(ep-i-neff′rin)
Adrenalin

oxymetazoline (OTC)
(ox-i-met-az′oh-leen)
12-Hour Nasal, Afrin 12-Hour Original, Afrin 12-Hour Original Pump Mist, Afrin Severe Congestion with Menthol, Afrin Sinus with Vapornase, Afrin No-Drip 12-Hour, Afrin No-Drip 12-Hour Extra Moisturizing, Dristan, Mucinex Full Force, Mucinex Moisture Smart, Neo-Synephrine 12-Hour, Vicks Sinex

phenylephrine (OTC)
(fen-ill-eff′rin)
4-Way Fast Acting Nasal Spray, Neo-Synephrine

tetrahydrozoline (OTC)
(tet-ra-hye-dro′zoe-leen)
Tyzine

NASAL CORTICOSTEROIDS

budesonide (Rx)
(byoo-des′oh-nide)
Rhinocort Aqua

fluticasone (Rx)
(floo-tic′a-son)
Flonase, Veramyst

triamcinolone (Rx)
(trye-am-sin′oh-lone)
Nasacort AQ

NONSTEROIDAL ANTIINFLAMMATORY

ketorolac (Rx)
(kee′toe-role′ak)
Sprix

ACTION: Produces vasoconstriction (rapid, long acting) of arterioles, thereby decreasing fluid exudation, mucosal engorgement by stimulation of α-adrenergic receptors in vascular smooth muscle

USES: Nasal congestion

CONTRAINDICATIONS:

Hypersensitivity to sympathomimetic amines

Precautions: Pregnancy (C), children <6 yr, geriatric, diabetes, CV disease, hypertension, hyperthyroidism, increased intracranial pressure, prostatic hypertrophy, glaucoma

DOSAGE AND ROUTES

Desoxyephedrine

• **Adult and child >6 yr: 1-2 INH** in each nostril q2hr or less

EPINEPHrine

• **Adult and child >6 yr:** Apply with swab, drops, spray prn

Oxymetazoline

• **Adult and child >6 yr: INSTILL** 2-3 gtt or sprays to each nostril bid

• **Child 2-6 yr: INSTILL** 2-3 gtt or sprays 0.025 sol bid, max 3 days

Phenylephrine

• **Adult and child >12 yr:** 2-3 drops/spray (0.25-0.5) in each nostril q3-4hr or less; or 2-3 drops/spray (1%) in each nostril q4hr or less

• **Child 6-12 yr:** 2-3 drops/spray (0.25%) in each nostril q3-4hr

• **Infant >6 mo:** 1-2 drops (0.16%) in each nostril q3hr

Tetrahydrozoline

• **Adult and child >6 yr:** 2-4 drops (0.1%) q3-4hr prn or 3-4 sprays in each nostril q4hr prn

• **Child 2-6 yr:** 2-3 drops (0.05%) in each nostril q4-6hr prn

Available forms: Nasal sol

Administer:

• Having patient tilt head back, squeeze bulb to create a vacuum, and draw correct amount of sol into dropper; insert 2 gtt of sol into nostril; repeat in other nostril

• Store in light-resistant container; do not expose to high temperature or let sol come into contact with aluminum

• For <4 consecutive days

• Environmental humidification to decrease nasal congestion, dryness

SIDE EFFECTS

CNS: Anxiety, restlessness, tremors, weakness, insomnia, dizziness, fever, headache

EENT: Irritation, burning, sneezing, stinging, dryness, rebound congestion

GI: Nausea, vomiting, anorexia

INTEG: Contact dermatitis

NURSING CONSIDERATIONS

Assess:

• For redness, swelling, pain in nasal passages before and during treatment

• For systemic absorption; hypertension, tachycardia; notify prescriber; systemic absorption occurs at high doses or after prolonged use

Evaluate:

• Therapeutic response: decreased nasal congestion

Teach patient/family:

• That stinging may occur for several applications; drying of mucosa may be decreased by environmental humidification

• To notify prescriber if irregular pulse, insomnia, dizziness, or tremors occur

• Proper administration to avoid systemic absorption

• To rinse dropper with very hot water to prevent contamination

TOPICAL GLUCOCORTICOIDS

betamethasone (augmented) (Rx)
(bay-ta-meth′a-sone)
Diprolene

desonide (Rx)
(dess′oh-nide)
LoKara, Verdeso Foam

fluocinolone (Rx)
(floo-oh-sin′oh-lone)
Derma-Smoothe/F, Synalar

flurandrenolide (Rx)
(flure-an-dren′oh-lide)
Cordran, Cordran SP, Drenison 1/4 ✱, Drenison Tape ✱

halcinonide (Rx)
(hal-sin′oh-nide)
Halog

ACTION: Antipruritic, antiinflammatory

USES: Psoriasis, eczema, contact dermatitis, pruritus; usually reserved for severe dermatoses that have not responded to less potent formulation

CONTRAINDICATIONS: Hypersensitivity, viral infections, fungal infections
Precautions: Pregnancy (C)

DOSAGE AND ROUTES

- **Adult and child:** Apply to affected area

Administer:
- Only to affected areas; do not get in eyes
- Leaving site uncovered or lightly covered; occlusive dressing is not recommended—systemic absorption may occur
- Use only on dermatoses; do not use on weeping, denuded, or infected area
- Cleansing before application of product
- Continuing treatment for a few days after area has cleared
- Store at room temperature

SIDE EFFECTS

INTEG: *Acne, atrophy, epidermal thinning, purpura, striae*

NURSING CONSIDERATIONS

Assess:
- Temp; if fever develops, product should be discontinued
- For systemic absorption, increased temp, inflammation, irritation

Evaluate:
- Therapeutic response: absence of severe itching, patches on skin, flaking

Teach patient/family:
- To avoid sunlight on affected area, burns may occur
- To limit treatment to 14 days

TOPICAL ANTIFUNGALS

nystatin (OTC)
(nye-stat′in)
Nyamyc, Nystop, Pediaderm, Pedia-dri

selenium (OTC)
(see-leen′ee-um)
Dandrex, Selsun, Selsun Blue

terbinafine (OTC)
(ter-bin′a-feen)
Lamisil

tolnaftate (OTC)
(tole-naf′tate)
Absorbine Athlete's Foot Cream, Fungi-Guard, Tinactin

undecylenic acid (OTC)
(un-deh-sih-len′ik)

ACTION: Interferes with fungal cell membrane permeability

USES:
Tinea cruris, tinea pedis, diaper rash, minor skin irritations; amphotericin B is used for candida infections

CONTRAINDICATIONS:
Hypersensitivity

Precautions: Pregnancy (B), breastfeeding, children

DOSAGE AND ROUTES

- Massage into affected area, surrounding area daily or bid, continue for 7-14 days, not to exceed 4 wk

Administer:

Topical route

- To affected area, surrounding area; do not cover with occlusive dressings
- Store below 86° F (30° C)

SIDE EFFECTS

INTEG: Burning, stinging, dryness, itching, local irritation

NURSING CONSIDERATIONS

Assess:

- Skin for fungal infections; peeling, dryness, itching before and throughout treatment
- For continuing infection; increased size, number of lesions

Evaluate:

- Therapeutic response: decrease in size, number of lesions

Teach patient/family:

- To apply with glove to prevent further infection; not to cover with occlusive dressings
- That long-term therapy may be needed to clear infection (2 wk-6 mo depending on organism); compliance is needed even after feeling better
- Proper hygiene; hand-washing technique, nail care, use of concomitant top agents if prescribed
- To avoid use of OTC creams, ointments, lotions unless directed by prescriber
- To use medical asepsis (hand washing) before, after each application; to change socks and shoes once a day during treatment of tinea pedis
- To report to health care prescriber if infection persists or recurs; if blisters, burning, oozing, swelling occur
- To avoid alcohol because nausea, vomiting, hypertension may occur
- To use sunscreen or avoid direct sunlight to prevent photosensitivity
- To notify health care provider of sore throat, fever, skin rash, which may indicate overgrowth of organisms

TOPICAL ANTIINFECTIVES

bacitracin (OTC)
(bass-i-tray′sin)

mafenide (Rx)
(ma′fe-nide)
Sulfamylon

salicylic acid (Rx)
(sal′i-sil′ik)

silver sulfADIAZINE (Rx)
(sul-fa-dye′a-zeen)
Flamazine ✱, Silvadene, SSD, SSD AF, Thermazene

tretinoin (Rx)
(treh′tih-noyn)
Atralin, Avita, Retin-A, Renova

ACTION:
Interferes with bacterial protein synthesis

USES:
Skin infections, minor burns, wounds, skin grafts, primary pyodermas, otitis externa

CONTRAINDICATIONS:
Hypersensitivity, large areas, burns, ulcerations

Precautions: Pregnancy (C), breastfeeding, impaired renal function, external ear or perforated eardrum

Administer:

- Enough medication to cover lesions completely

- After cleansing with soap, water before each application; dry well
- To less than 20% of body surface area when patient has impaired renal function

SIDE EFFECTS

INTEG: Rash, urticaria, scaling, redness

NURSING CONSIDERATIONS

Assess:

- Allergic reaction: burning, stinging, swelling, redness
- For signs of nephrotoxicity or ototoxicity

Perform/provide:

- Storage at room temperature in dry place

Evaluate:

- Therapeutic response: decrease in size, number of lesions

TOPICAL ANTIVIRALS

acyclovir (Rx)
(ay-sye′kloe-ver)
Zovirax

penciclovir (Rx)
(pen-sye′kloe-ver)
Denavir

ACTION: Interferes with viral DNA replication

USES: Simple mucocutaneous herpes simplex, in immunocompromised clients with initial herpes genitalis

CONTRAINDICATIONS: Hypersensitivity

Precautions: Pregnancy (C), breastfeeding

Administer:

- Using finger cot or rubber glove to prevent further infection
- Enough medication to cover lesions completely
- After cleansing with soap, water before each application; dry well

SIDE EFFECTS

INTEG: Rash, urticaria, stinging, burning, pruritus, vulvitis

NURSING CONSIDERATIONS

Assess:

- Allergic reaction: burning, stinging, swelling, redness, rash, vulvitis, pruritus

Perform/provide:

- Storage at room temperature in dry place

Evaluate:

- Therapeutic response: decrease in size, number of lesions

Teach patient/family:

- Not to use in eyes or when there is no evidence of infection
- To apply with glove to prevent further infection
- To avoid use of OTC creams, ointments, lotions unless directed by prescriber
- To use medical asepsis (hand washing) before, after each application and avoid contact with eyes
- To adhere strictly to prescribed regimen to maximize successful treatment outcome
- To begin taking product when symptoms arise

TOPICAL ANESTHETICS

benzocaine (OTC)
(ben'zoe-kane)
Americaine Anesthetic, Anbesol Maximum Strength, Baby Anbesol, Biozene, Boil-Ease, Comfort Caine, Ora-Jel

dibucaine (OTC)
(dye'byoo-kane)
Nupercainal

lidocaine (Rx, OTC)
(lye'doe-kane)
LMX, Solarcaine, Xylocaine, Xylocaine 10% oral, Xylocaine Viscous, Zilactin-L

pramoxine (OTC)
(pra-mox'een)
Pramox, Prax, Sarna

tetracaine (Rx, OTC)
(tet'ra-cane)
Pontocaine, Viractin

ACTION: Inhibits conduction of nerve impulses from sensory nerves

USES: Oral irritation, sore throat, toothache, cold sore, canker sore, sunburn, minor cuts, insect bites, pain, itching

CONTRAINDICATIONS: Hypersensitivity, infants <1 yr, application to large areas

Precautions: Pregnancy (C), children <6 yr, sepsis, denuded skin

DOSAGE AND ROUTES

- **Adult and child: TOP** apply qid as needed; **RECT** insert tid and after each BM

SIDE EFFECTS

INTEG: Rash, irritation, sensitization

NURSING CONSIDERATIONS

Assess:
- Pain: location, duration, characteristics before and after administration
- For infection: redness, drainage, inflammation; this product should not be used until infection is treated

Perform/provide:
- Storage in tight, light-resistant container; do not freeze, puncture, or incinerate aerosol container

Evaluate:
- Therapeutic response: decreased redness, swelling, pain

Teach patient/family:
- To avoid contact with eyes
- Not to use for prolonged periods: use for <1 wk; if condition remains, prescriber should be contacted

TOPICAL MISCELLANEOUS

docosanol (OTC)
(doe-koe'san-ole)
Abreva

pimecrolimus (Rx)
(pim-eh-croh'lim-us)
Elidel

ACTION: Docosanol unknown; pimecrolimus may bind with macrophilin and inhibit calcium-dependent phosphatase

USES: Docosanol applied to fever blisters to promote more rapid healing; pimecrolimus used to treat mild to moderate atopic dermatitis in nonimmunocompromised patients ≥2 yr who are unresponsive to other treatment

CONTRAINDICATIONS: Hypersensitivity

Precautions: Pregnancy (C), breastfeeding, dermal infections

DOSAGE AND ROUTES

Docosanol

• **Adult: TOP** Rub into blisters 5×/day until healing occurs

Pimecrolimus

• **Adult and child ≥2 yr: TOP** Apply thin layer 2×/day and rub in, use as long as needed

Administer:

• To skin, rub in gently

SIDE EFFECTS

Pimecrolimus

INTEG: Burning

NURSING CONSIDERATIONS

Assess:

• Skin condition (color, pain, inflammation) before and after administration

• For signs and symptoms of skin infections (redness, draining lesions); if present, avoid use of product (pimecrolimus)

Evaluate:

• Therapeutic response: decreased inflammation, redness

Teach patient/family:

• To avoid contact between medication and eyes

• To discontinue use of product when condition clears

VAGINAL ANTIFUNGALS

butoconazole (OTC)
(byoo-toh-kone′ah-zole)
Gynazol-1

miconazole (OTC)
(mye-kon′a-zole)
Monistat, Monistat 3, Monistat 7, Monistat Dual Pak

terconazole (OTC)
(ter-kone′ah-zole)
Terazol 7, Terazol 3, Zazole

tioconazole (OTC)
(tye-oh-kone′ah-zole)
Gyne-Trosyd ♣, Vagistat-1

ACTION: Interferes with fungal DNA replication; binds sterols in fungal cell membranes, which increases permeability, leaking of nutrients

USES: Vaginal, vulval, vulvovaginal candidiasis (moniliasis)

CONTRAINDICATIONS: Hypersensitivity

Precautions: Pregnancy, breastfeeding, children <2 yr

DOSAGE AND ROUTES

Butoconazole

• **Adult: VAG** 5 g (1 applicator) at bedtime × 3-6 days

Miconazole

• **Adult:** 200 mg supp at bedtime × 3 days or 100 mg supp × 1 wk

Terconazole

• **Adult:** VAG 5 g (1 applicator) at bedtime

Tioconazole

• **Adult:** 1 applicator at bedtime

Administer:

Topical route

• One full applicator every night high into the vagina

• Store at room temperature in dry place

SIDE EFFECTS

GU: Vulvovaginal burning, itching, pelvic cramps
INTEG: Rash, urticaria, stinging, burning
MISC: Headache, body pain

NURSING CONSIDERATIONS

Assess:
- For allergic reaction: burning, stinging, itching, discharge, soreness

Evaluate:
- Therapeutic outcome: decrease in itching or white discharge (vaginal)

Teach patient/family:
- About asepsis (hand washing) before, after each application
- To apply with applicator only; to avoid use of any other vaginal product unless directed by prescriber; sanitary napkin may prevent soiling of undergarments
- To abstain from sexual intercourse until treatment is completed; reinfection and irritation may occur
- To notify prescriber if symptoms persist

OTIC ANTIINFECTIVES

ciprofloxacin (Rx)

Cetraxel

ACTION: Inhibits protein synthesis in susceptible microorganisms

USES: Ear infection (external), short-term use

CONTRAINDICATIONS: Hypersensitivity, perforated eardrum
Precautions: Pregnancy (C)
Administer:
- After removing impacted cerumen by irrigation
- After cleaning stopper with alcohol
- After restraining child if necessary
- After warming sol to body temp

SIDE EFFECTS

EENT: Itching, irritation in ear
INTEG: Rash, urticaria

NURSING CONSIDERATIONS

Assess:
- For redness, swelling, fever, pain in ear, which indicates superinfection

Evaluate:
- Therapeutic response: decreased ear pain

Teach patient/family:
- The correct method of instillation using aseptic technique, including not touching dropper to ear
- That dizziness may occur after instillation

Appendix C Vaccines and toxoids

GENERIC NAME	TRADE NAME	USES	DOSAGE AND ROUTES	CONTRAINDICATIONS
advenovirus vaccine		Prevention of advenovirus types 4/7	Adult ≤50 yr/adolescent ≥17 yr in the military PO 1 tab of each 4/7 or a single dose	Pregnancy
avian influenza A (H5N1) virus vaccine		Prophylaxis	Adult: IM 1 ml (90 mcg) 2 doses, 28 days apart	IV
anthrax vaccine	BioThrax	Pre-/postexposure prophylaxis	**Preexposure** Adult: SUBCUT 0.5 ml at 0, 2, 4 wk, then 0.5 ml at 6, 12, 18 mo **Postexposure** Adult: SUBCUT 0.5 ml 0, 2, 4 wk, with antibiotics	Hypersensitivity
BCG vaccine	TICE BCG	TB exposure	Adult and child >1 mo: 0.2-0.3 ml Child <1 mo: Reduce dose by 50% using 2 ml of sterile water after reconstituting	Hypersensitivity, hypogamma-globulinemia, positive TB test, burns
cholera vaccine	No trade name	Immunization for cholera outside the United States	Adult and child >10 yr: IM/SUBCUT 2× of 0.5 ml, 7-30 days before traveling to cholera areas Booster is used q6mo 0.5 ml prn	Hypersensitivity, acute febrile illness
diphtheria and tetanus toxoids, adsorbed	No trade name	Induces antitoxins to provide immunity to diphtheria and tetanus	Adult and child ≥7 yr: IM (adult strength) 0.5 ml q4-8wk × 2 doses, then 3rd dose 6-12 mo after 2nd dose, booster IM 0.5 ml q10yr Child 1-6 yr: IM (pediatric strength) 0.5 ml q4wk × 2 doses, booster 6-12 mo after 2nd dose Infant 6 wk-1 yr: IM (pediatric strength) 0.5 ml q4wk × 3 doses, booster 6-12 mo after 3rd dose	Hypersensitivity to mercury, thimerosal; immunocompromised patients; radiation; corticosteroids; acute illness
diphtheria and tetanus toxoids and whole-cell pertussis vaccine (DPT, DTP)	DTwP, Tr-Immunol	Prevention of diphtheria, tetanus, pertussis	Doses vary Check product information	Hypersensitivity, active infection, poliomyelitis outbreak, immunosuppression, febrile illness

Continued

Appendix c **Vaccines and toxoids—cont'd**

GENERIC NAME	TRADE NAME	USES	DOSAGE AND ROUTES	CONTRAINDICATIONS
diphtheria and tetanus toxoids and acellular pertussis vaccine	Adacel, Boostrix, Daptacel, Infanrix, Tripedia			
diphtheria, tetanus, pertussis, haemophilus, polio IPV	Pentacel	Immunity to diphtheria, tetanus, pertussis, haemophilus, polio IPV	Infant >6 wk and child ≤5 yr: IM 0.5 ml at 2, 4, 6, and 15-18 mo	Hypersensitivity, polio outbreak, acute infection, immunosuppression
diphtheria, tetanus, pertussis, polio vaccine IPV	Kinrix	Immunity to diphtheria, tetanus, pertussis, polio vaccine IPV	Child: IM 0.5 ml	Hypersensitivity, polio outbreak, acute infection, immunosuppression
H1N1 influenza A (swine flu) virus vaccine	Influenza A (H1N1)	Immunity to H1N1	Adult <50 yr, adolescent, child ≥2 yr: Intranasal 1 dose (roughly 0.1 ml) into each nostril; child 2-9 repeat dose ≥4 wk later Adult, adolescent, child ≥3 yr: IM 0.5 ml as a single dose; child 3-9 yr repeat dose ≥4 wk later (Sanofi) (CSL); child 4-9 yr repeat dose ≥4 wk later (Novartis); infants ≥6 mo-child <36 mo: IM 0.25 ml, repeat in 4 wk (Sanofi) Adult: IM 0.5 ml as a single dose (GSK)	
haemophilus b conjugate vaccine, diphtheria CRM_{197} protein conjugate (HbOC)	HibTITER	Polysaccharide immunization of children 2-6 yr against *H. influenzae* b, conjugate	**HibTITER (IM only)** Child: IM 0.5 ml Child 2-6 mo: 0.5 ml q2mo × 3 inj	Hypersensitivity, febrile illness, active infection
haemophilus b conjugate vaccine, meningococcal protein conjugate (PRP-OMP)	PedvaxHIB	Immunization of child 2, 4, 6 mo	Child 7-11 mo: Previously unvaccinated 0.5 ml q2mo inj Child 12-14 mo: Previously unvaccinated 0.5 ml × 1 inj **PedvaxHIB (IM only)** Child 2-14 mo: 0.5 ml × 2 inj at 2, 4 mo of age (6 mo dose not needed), then booster at 12-18 mo against invasive disease Child ≥15 mo: Previously unvaccinated 0.5 ml inj	

hepatitis A vaccine, inactivated	Havrix, VAQTA	Active immunization against hepatitis A virus	Adult: IM 1440 EL units (Havrix) or 50 units (VAQTA) as a single dose; booster dose is the same given at 6, 12 mo Child 2-18 yr: IM 720 EL units (Havrix) or 25 units (VAQTA) as a single dose, booster dose is the same given at 6, 12 mo	Hypersensitivity
hepatitis B vaccine, recombinant	Engerix-B, Recombivax HB	Immunization against all subtypes of hepatitis B virus	Varies widely	Hypersensitivity to this vaccine or yeast
herpes zoster virus vaccine	Zostavax	Prevention of herpes zoster	Adult ≥60 yr: SUBCUT 0.65 ml	<60 yr, child, infant, AIDS, IM/IV, leukemia, lymphoma, pregnancy
human papillomavirus recombinant vaccine, quadrivalent	Gardasil	Prevention of HPV types 6, 11, 16, 18, cervical cancer, genital warts, precancerous dysplasic lesions, anal cancer/anal intraepithelial neoplasia	Adult up to 26 yr and child >9 yr to 26 yr: IM give as 3 separate doses; 1st dose as elected; 2nd dose 2 mo after 1st dose; 3rd dose 6 mo after 1st dose	Child <9 yr, pregnancy, breastfeeding, geriatric, active disease, hypersensitivity
influenza virus vaccine	Afluria, FluMist, Fluogen, Flu-Shield, Fluviral*, Fluvirin, Fluzone, influenza virus vaccine, trivalent	Prevention of Russian, Chilean, Philippine influenza	Adult and child >12 yr: IM 0.5 ml in 1 dose Adult 18-64 yr: ID 0.1 ml as a single dose Child 3-12 yr: IM 0.5 ml, repeat in 1 mo (split) unless 1978-1985 vaccine was given; also given nasal Child 6 mo to 3 yr: IM 0.25 ml, repeat in 1 mo (split) unless 1978-1985 vaccine was given; also given nasal child ≤2 yr	Hypersensitivity, active infection, chicken egg allergy, Guillain-Barré syndrome, active neurologic disorders
Japanese encephalitis virus vaccine, inactivated	JE-VAX	Active immunity against Japanese encephalitis (JE)	Adult and child ≥3 yr: SUBCUT 1 ml, days 0, 7, 30; booster SUBCUT 1 ml 2 yr after last dose Child 1-3 yr: SUBCUT 0.5 ml, days 0, 7, 30; booster SUBCUT 0.5 ml 2 yr after last dose	Hypersensitivity to murine, thimerosal; allergic reactions to previous dose

Continued

Appendix c Vaccines and toxoids—cont'd

GENERIC NAME	TRADE NAME	USES	DOSAGE AND ROUTES	CONTRAINDICATIONS
Lyme disease vaccine (recombinant OspA)	LYMErix	Immunization against Lyme disease	Adult and adolescent 15-70 yr: IM 30 mcg in deltoid, repeat at 1, 12 mo after first dose	Hypersensitivity, antibiotic refractory Lyme arthritis
measles and rubella virus vaccine, live attenuated	M-R-Vax II	Immunity to measles and rubella by antibody production	Adult and child ≥15 mo: SUBCUT 0.5 ml (1000 units)	Hypersensitivity, immunocompromised patients, active untreated TB, cancer, blood dyscrasias, radiation, corticosteroids, pregnancy; allergic reactions to neomycin, eggs
measles, mumps, and rubella vaccine, live	M-M-R-II	Prevention of measles, mumps, rubella	Adult: SUBCUT 1 vial; 2 vials separated by 1 mo, in person born after 1957 Child >15 mo and adult: SUBCUT 0.5 ml	Hypersensitivity, blood dyscrasias, anemia, active infection, immunosuppression; egg, chicken allergy; pregnancy, febrile illness, neomycin allergy, neoplasms
measles, mumps, rubella, varicella	ProQuad	Immunity to measles, mumps, rubella, varicella	Child: SUBCUT 0.5 ml	Hypersensitivity to eggs, neomycin, cancer, radiation, corticosteroids, blood dyscrasias, active untreated TB
measles virus vaccine, live attenuated	Attenuvax	Immunity to measles by antibody production	Adult and child ≥15 mo: SUBCUT 0.5 ml (1000 units), 1 dose 15 mo, 2nd dose age 4-6 or 11, or 12	Hypersensitivity to eggs, neomycin; cancer, radiation, corticosteroids, pregnancy, immunocompromised patients, blood dyscrasias, active untreated TB
meningococcal diphtheria toxoid conjugate vaccine	Menomune, Menactra, Menveo	Prophylaxis to meningococcal meningitis/diphtheria	All doses are different; check product information	Latex hypersensitivity
meningococcal polysaccharide vaccine	Menomune-A/C/Y/W-135, Menactra	Prophylaxis to meningococcal meningitis	Adult and child >2 yr: SUBCUT 0.5 ml	Hypersensitivity to thimerosal, pregnancy, acute illness
mumps virus vaccine, live	Mumpsvax	Active immunity to mumps	Adult and child ≥1 yr: SUBCUT 0.5 ml (20,000 units)	Hypersensitivity to eggs, neomycin; cancer, radiation, corticosteroids, pregnancy, immunocompromised patients, blood dyscrasias, active untreated TB

plague vaccine	No trade name	Active immunity to *Yersinia pestis* plague	Adult: IM 1 ml, then 0.2 ml in 4-12 wk, then 0.2 ml 5-6 mo after 2nd dose; booster 0.1-0.2 ml q6mo when in plague area	Hypersensitivity to phenol, sulfites, formaldehyde, beef, soy, casein; pregnancy, coagulation disorders
pneumococcal 7-valent conjugate vaccine	Prevnar	Immunity against *Streptococcus pneumoniae*	Child: IM 0.5 ml × 3 doses (7-11 mo); × 2 doses (12-23 mo); × 1 dose >2-9 yr	Hypersensitivity to diphtheria toxoid or this product
pneumococcal vaccine, polyvalent	Pneumovax 23, Pnu-Imune 23	Pneumococcal immunization	Adult and child >2 yr: IM/SUBCUT 0.5 ml	Hypersensitivity, Hodgkin's disease, ARDS
poliovirus vaccine, live, oral, trivalent (TOPV) poliovirus vaccine (IPV)	Orimune, IPOL	Prevention of polio	Adult and child >2 yr: PO 0.5 ml, given q8wk × 2 doses, then 0.5 ml ½-1 yr after dose 2 Infant: PO 0.5 ml at 2, 4, 18 mo; booster at 4-6 yr; may also be given: IPV at 2, 4 mo, then TOPV at 12-18 mo, booster at 4-6 yr	Hypersensitivity, active infection, allergy to neomycin/streptomycin, immunosuppression, vomiting, diarrhea
rabies vaccine, adsorbed	No trade name	Active immunity to rabies	**Preexposure** Adult and child: IM 1 ml day 0, 7, 21, or 28 days (total 3 doses); booster IM 1 ml prn q2-5yr **Postexposure** Adult and child not vaccinated: IM 20 international units/kg of human rabies immune globulin (HRIG), give 5 total doses of 1-ml inj of rabies vaccine on days 0, 3, 7, 14, 28	Severe hypersensitivity to previous inj of vaccine, thimerosol
rabies vaccine, human diploid cell (HDCV)	Imovax Rabies, Imovax Rabies I.D.	Active immunity to rabies	**Preexposure** Adult and child: IM 1 ml day 0, 7, 21, or 28 (total 4 doses) **Postexposure** Adult and child: IM 1 ml on day 0, 3, 7, 14, 28 (total 5 doses)	No contraindications

*Canada only.

Continued

Appendix c Vaccines and toxoids—cont'd

GENERIC NAME	TRADE NAME	USES	DOSAGE AND ROUTES	CONTRAINDICATIONS
rotovirus	RotaTeq, Rotarix	Prevents rotovirus	Infant: PO 3 doses given between 6 and 32 wk of age; 1st dose between 6-12 wk of age; 2nd and 3rd doses q4-10wk	Hypersensitivity to this product or latex, immunocompromised, blood products given within 6 wk, lymphatic disorders
rubella and mumps virus vaccine, live	Biavax II	Immunity to rubella and mumps by antibody production	Adult and child $\geq$1 yr: SUBCUT 0.5 ml	Hypersensitivity to eggs, neomycin; cancer, radiation, corticosteroids, pregnancy, immunocompromised patients, blood dyscrasias, active untreated TB
rubella virus vaccine, live attenuated (RA 27/3)	Meruvax II	Immunity to rubella by antibody production	Adult and child $\geq$1 yr: SUBCUT 0.5 ml (1000 units)	Hypersensitivity to eggs, neomycin; cancer, radiation, corticosteroids
smallpox vaccine	ACAM 2000, Dry Vax	Prevention of smallpox	See package insert	No contraindications
tetanus toxoid, adsorbed/tetanus toxoid	No trade name	Tetanus toxoid: Used for prophylactic treatment of wounds	Adult and child: IM 0.5 ml q4-6wk $\times$ 2 doses, then 0.5 ml 1 yr after dose 2 (adsorbed); SUBCUT/IM 0.5 ml q4-8wk $\times$ 3 doses, then 0.5 ml $^1/_2$-1 yr after dose 3, booster dose 0.5 ml q10yr	Hypersensitivity, active infection, poliomyelitis outbreak, immunosuppression
typhoid vaccine, parenteral typhoid vaccine, oral	No trade name Vivotif Berna Vaccine	Active immunity to typhoid fever	Adult: PO 1 cap 1 hr before meals $\times$ 4 doses, booster q5yr Adult and child >10 yr: SUBCUT 0.5 ml, repeat in 4 wk, booster q3yr Child 6 mo-10 yr: SUBCUT 0.25 ml, repeat in 4 wk, booster q3yr	Parenteral: Systemic or allergic reaction, acute respiratory or other acute infection, intensive physical exercise in high temperatures Oral: Hypersensitivity, acute febrile illness, suppressive or antibiotic products
typhoid Vi polysaccharide vaccine	Typhim Vi	Active immunity to typhoid fever	Adult and child $\geq$2 yr: IM 0.5 ml as a single dose, reimmunize q2yr 0.5 ml IM, if needed	Hypersensitivity, chronic typhoid carriers
varicella virus vaccine	Varivax	Prevention of varicella-zoster (chickenpox)	Adult and child $\geq$13 yr: SUBCUT 0.5 ml, 2nd dose SUBCUT 0.5 ml 4-8 wk later	Hypersensitivity to neomycin; blood dyscrasias, immunosuppression, active untreated TB, acute illness, pregnancy, diseases of lymphatic system

Nurse Alert

yellow fever vaccine	YF-Vax	Active immunity to yellow fever	Adult and child ≥9 mo: SUBCUT 0.5 ml deeply, booster q10yr Child 6-9 mo: same as above if exposed	Hypersensitivity to egg or chicken embryo protein, pregnancy, child <6 mo, immunodeficiency
zoster vaccine, live	Zostavax	Herpes zoster prevention	Reconstitute immediately after removing from freezer; give SUBCUT as a single dose; inject total amount of single-dose vial	Immunosuppression; neomycin, gelatin allergy; children, TB, pregnancy (C)

Recent FDA drug approvals

Generic	Trade	Use
droxidopa	Northera	For the treatment of neurogenic orthostatic hypotension
elosulfase alfa	Vimizim	For the treatment of Mucopolysaccharidosis type IVA
metreleptin	Myalept	For the treatment of congenital or acquired generalized lipodystrophy
tasimelteon	Hetlioz	For the treatment of non-24-hour sleep-wake disorder in the totally blind

Index

Drug names and/or monographs indexed with SEE Evolve are found on your Evolve Resources at: http://elsevier.com/nursingdrugupdates/Skidmore/NDR.

Entries can be identified as follows: DISEASES/DISORDERS, *DRUG CATEGORIES,* generic names, Trade Names.

Entries can be identified as follows: DISEASES/DISORDERS, *DRUG CATEGORIES*, generic names, Trade Names.

Entries can be identified as follows: DISEASES/DISORDERS, *DRUG CATEGORIES,* generic names, Trade Names.

Entries can be identified as follows: DISEASES/DISORDERS, *DRUG CATEGORIES*, generic names, Trade Names.

Entries can be identified as follows: DISEASES/DISORDERS, *DRUG CATEGORIES,* generic names, Trade Names.

Entries can be identified as follows: DISEASES/DISORDERS, *DRUG CATEGORIES*, generic names, Trade Names.

Entries can be identified as follows: DISEASES/DISORDERS, *DRUG CATEGORIES*, generic names, Trade Names.

Entries can be identified as follows: DISEASES/DISORDERS, *DRUG CATEGORIES,* generic names, Trade Names.

Entries can be identified as follows: DISEASES/DISORDERS, *DRUG CATEGORIES*, generic names, Trade Names.

Entries can be identified as follows: DISEASES/DISORDERS, *DRUG CATEGORIES,* generic names, Trade Names.

Formulas

Surface area rule:

$$\text{Child dose} = \frac{\text{Surface area } (m^2)}{1.73\ m^2} \times \text{Adult dose}$$

Calculating strength of a solution:

Solution Strength: *Desired Solution*:

$$\frac{x}{100} = \frac{\text{Amount of drug desired}}{\text{Amount of finished solution}}$$

Calculating flow rate for IV:

$$\text{Rate of flow} = \frac{\text{Amount of fluid} \times \text{Administration set calibration}}{\text{Running time}}$$

$$\frac{x}{1} = \frac{(\text{ml})\ (\text{gtt/min})}{\text{min}}$$

Calculation of medication dosages:

Formula method:

$$\frac{\text{Amount ordered}}{\text{Amount on hand}} \times \text{Vehicle} = \text{Number of tablets, capsules, or amount of liquid}$$

Vehicle is the drug form or amount of liquid containing the dosage. Amounts used in calculation by formula must be in same system.

Ratio–proportion method:

1 tablet:tablet in mg on hand::x tablet order in mg

Know or have::Want to know or order

Multiply means and extremes, divide both sides by known amount to get x. Amounts used in equation must be in same system.

Dimensional analysis method:

$$\text{Order in mg} \times \frac{\text{1 tablet or capsule}}{\text{What 1 tablet or capsule is in mg}} = \text{Tablets or capsules to be given}$$

If amounts are in different systems:

$$\text{Order in mg} \times \frac{\text{1 tablet or capsule}}{\text{What 1 tablet or capsule is in g}} \times \frac{1}{1000\text{ mg}} = \text{Tablets or capsules to be given}$$

Temperature conversion:

$F = C \times 9/5 + 32$

$C = 5/9\ (F - 32)$

Nomogram for calculation of body surface area

Place a straight edge from the patient's height in the left column to the patient's weight in the right column. The point of intersection on the body surface area column indicates the body surface area (BSA). (Reproduced in Behrman RE, Kliegman RM, Jenson HB: *Nelson textbook of pediatrics,* ed 18, Philadelphia, 2011, WB Saunders; Nomogram modified from data of E. Boyd by CD West.)

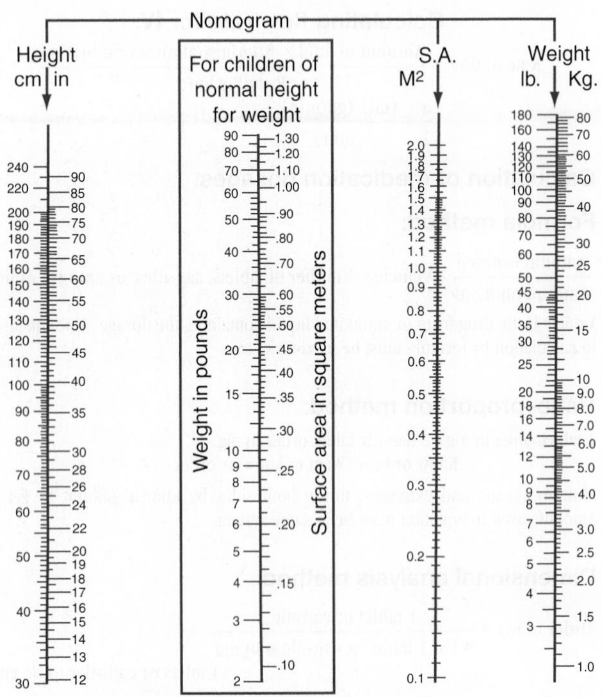

Alternative (Mosteller's formula):

$$\text{Surface area } (m^2) = \sqrt{\frac{\text{Height (cm)} \times \text{Weight (kg)}}{3600}}$$

Antitoxins and antivenins

GENERIC NAME	TRADE NAME	USES	DOSAGE AND ROUTES	CONTRAINDICATIONS
Black widow spider antivenin (*Lactrodectus mactans*)	No trade name	Black widow spider bite	Adult and child: IM 2.5 ml, 2nd dose may be given if severe; give in anterolateral thigh, obtain test for sensitivity before inj	Hypersensitivity to this product or horse serum
Crotalidae antivenom, polyvalent	No trade name	Rattlesnake bite	Adult and child: IV 20-150 ml depending on seriousness of bite, may give additional doses based on response	Hypersensitivity
Diphtheria antitoxin, equine	No trade name	Diphtheria	Adult and child: IM/slow IV 20,000-120,000 units, may give additional doses after 24 hr	Hypersensitivity
Micrurus fulvius antivenin	No trade name	East/Texas coral snake bite	Adult and child: IV 30-50 ml, give through running IV line of normal saline, give 1st 1-2 ml over 4-5 min, watch for allergic reaction	Hypersensitivity
Scorpion antivenin (centruroides sculpturatus equine)	Anascorp	Scorpion stings	Adult and child: IV 3 vial/50 ml NS given over 10 min, may give other doses 1 vial/50 ml NS over 10 min q30-60min	N/A

Abbreviations

ABG arterial blood gas
ADA American Diabetes Association
ADH antidiuretic hormone
ALT alanine aminotransferase
ANA antinuclear antibody
APLA antiphospholipid antibody syndrome
APTT activated partial thromboplastin time
ASA acetylsalicylic acid, aspirin
AST aspartate aminotransferase (SGOT)
AV atrioventricular
bid twice a day
BPH benign prostatic hypertrophy
BPM beats per minute
BUN blood urea nitrogen
CAD coronary artery disease
CBC complete blood cell count
CCr creatinine clearance
CHF congestive heart failure
CNS central nervous system
CONT continuous
COPD chronic obstructive pulmonary disease
CPAP continuous positive airway pressure
CPK creatine phosphokinase
CPS carbamoyl phosphate synthetase
C&S culture and sensitivity
CSF cerebrospinal fluid
CTCL cutaneous T-cell lymphoma
CV cardiovascular
CVA cerebrovascular accident
CVP central venous pressure
D&C dilatation and curettage
DIC diffuse intravascular coagulation
DIR INF direct infusion
D_5W 5% glucose in distilled water
DVT deep vein thrombosis
ECG electrocardiogram (EKG)
EDTA ethylenediamine tetraacetic acid
EEG electroencephalogram
EPS extrapyramidal symptom
ESR erythrocyte sedimentation rate
EXT REL extended release
FBS fasting blood sugar
FHT fetal heart tones
FSH follicle-stimulating hormone
GABA γ-aminobutyric acid
GPC giant papillary conjunctivitis
gr grain
GT glucose tolerance test
GU genitourinary
GVHD graft-versus-host disease
H_2 histamine$_2$
hCG human chorionic gonadotropin
Hct hematocrit
HDCV human diploid cell rabies vaccine
Hgb hemoglobin
H & H hematocrit and hemoglobin
5-HIAA 5-hydroxyindoleacetic acid
HIV human immunodeficiency virus (AIDS)
HR heart rate
IBD inflammatory bowel disease
IC intracardiac
ICP intracranial pressure
ID intradermal
IgG immunoglobulin G
IM intramuscular
INF infusion
INH inhalation
inj injection
I&O intake and output
INT intermittent
IPPB intermittent positive-pressure breathing
IT intrathecal
ITP idiopathic thrombocytopenic purpura
IUD intrauterine device
IV intravenous
IVP intravenous pyelogram
K potassium
LDH lactic dehydrogenase
LE lupus erythematosus
LFT liver function test
LH luteinizing hormone
LOC level of consciousness
LR lactated Ringer's solution
LT leukotriene
m minim
m^2 square meter
MAC monitored anesthesia care
MAOI monoamine oxidase inhibitor
mcg microgram
mEq milliequivalent
mg milligram
MI myocardial infarction
ml milliliter